D1389481

Instrumentation Reference Book

Instrumentation Reference Book

Edited by
B E Noltingk
BSc, PhD, CPhys, FInstP, CEng, FIEE
With specialist contributors

Butterworths
London · Boston · Singapore
Sydney · Toronto · Wellington

First published 1988
Reprinted 1990

Butterworth International Edition, 1990
ISBN 0-408-06426-9

British Library Cataloguing in Publication Data

Instrumentation reference book.
1. Measuring instruments
I. Noltingk, B.E.
620'.0044 QC100.5

ISBN 0-408-01562-4

Library of Congress Cataloging-in-Publication Data

Instrumentation reference book / edited by B.E. Noltingk with specialist contributors.
p. cm.
Bibliography: p.
Includes index.
ISBN 0-408-01562-4 :
1. Physical instruments—Handbooks, manuals, etc. 2. Engineering instruments—Handbooks, manuals, etc. 3. Scientific apparatus and instruments industry—Great Britain—Directories. I. Noltingk, B.E.
QC53.I574 1988
530'.7—dc19 87-26002
CIP

Photoset by Mid-County Press,
2a Merivale Road, London SW15 2NW

Printed and bound in Great Britain by
Hartnolls Ltd, Bodmin

Preface

Instrumentation is not a clearly defined subject, having what might be called a 'fuzzy frontier' with many other subjects. Look for books about it, and in most libraries you are liable to find them widely separated along the shelves, classified under several different headings. Instrumentation is barely recognized as a science or technology in its own right. That raises some difficulties for writers in the field and indeed for would-be readers. We hope that what we are offering here will prove to have helped with clarification.

A reference book should of course be there for people to refer to for the information they need. The spectrum is wide: students, instrument engineers, instrument users and potential users who just want to explore possibilities. And the information needed in real life is a mixture of technical and commercial matters. So while the major part of the *Instrumentation Reference Book* is a technical introduction to many facets of the subject, there is also a commercial part where manufacturers etc. are listed. Instrumentation is evolving, perhaps even faster than most technologies, emphasizing the importance of relevant research; we have tried to recognize that by facilitating contact with universities and other places spear-heading development.

One need for information is to ascertain where more information can be gained. We have catered for this in several ways, notably with references at the ends of chapters to more specialized books and, more unusually, with thumb-nail reviews of some general books in Part 6.

Many agents have come together to produce the *Instrumentation Reference Book* and to whom thanks are due: those who have written, those who have drawn, and those who have painstakingly checked facts. I should especially thank Caroline Mallinder and Elizabeth Alderton who produced order out of chaos in the compilation of long lists of names and addresses. Thanks should also go elsewhere in the Butterworth hierarchy for the original germ of the idea that this could be a good addition to their family of reference books. In a familiar tradition, I thank my wife for her tolerance and patience about time-consuming activities such as telephoning, typing and travelling – or at the least for limiting her natural intolerance and impatience of my excessive indulgence in them!

B.E.N.
Dorking
1988

Contents

List of Contributors

Part 1

C S Bahra, BSc, MSc, CEng, MIMechE, was formerly Development Manager at Transducer Systems Ltd

G Fowles was formerly a Senior Development Engineer with the Severn-Trent Water Authority after some time as a Section Leader in the Instrumentation Group of the Water Research Centre

E H Higham, MA, CEng, FIEE, MIMechE, MIERE, MInstMC, is Visiting Fellow at the Department of Fluid Engineering and Instrumentation at Cranfield Institute of Technology after a long career with Foxboro Great Britain Ltd

W M Jones, BSc, DPhil, FInstP, is a Reader in the Physics Department at the University College of Wales

B E Noltingk, BSc, PhD, CEng, FIEE, FInstP, is now a Consultant after some time as Head of the Instrumentation Section at the Central Electricity Research Laboratories

D J Pacey, BSc, FInstP, was, until recently, a Senior Lecturer in the Physics Department at Brunel University

W L Snowsill, BSc, was formerly a Research Officer in the Control and Instrumentation Branch of the Central Electricity Research Laboratories

P H Sydenham, ME, PhD, FInstMC, FIIC, AMIAust, is Head of and Professor at the School of Electronic Engineering in the South Australian Institute of Technology

K Walters, MSc, PhD, is a Professor in the Department of Mathematics at the University College of Wales

Part 2

Sir Claud Hagart-Alexander Bt., BA, MInstMC, DL, formerly worked in instrumentation with ICI Ltd. He is now a director of Instrumentation Systems Ltd

The chapters on Chemical Analysis have been contributed by a team from the Central Electricity Research Laboratories. **W G Cummings,** BSc, CChem, FRSC, MInstE, MInstMC and **A C Smith,** BSc, CChem, FRSC, MInstP, are both former Heads of the Analytical Chemistry Section; **C K Laird,** BSc, PhD, CChem, MRSC and **D B Meadowcroft,** BSc, PhD, CPhys, FInstP, FICorrST work in the Chemistry Branch and **K Torrance,** BSc, PhD, is in the Materials Branch

Part 3

D Aliaga Kelly, BSc, CPhys, MInstP, MAmPhysSoc, MSRP, FSAS, is now retired after working for many years as Chief Physicist with Nuclear Enterprises Ltd

J Kuehn, FInstAcoust, is Managing Director of Bruel & Kjaer (UK) Ltd

M L Sanderson, BSc, PhD, is Director of the Centre for Fluid Instrumentation at Cranfield Institute of Technology

A W S Tarrant, BSc, PhD, CPhys, FInstP, FCIBSE, is Director of the Engineering Optics Research Group at the University of Surrey

The chapter on Non-destructive Testing has been contributed by a team from the Scottish School of Non-destructive Testing, namely
W McEwan, BSc, CEng, MIMechE, FWeldInst (Director)
G Muir, BSc, MSc, MIM, MInstNDT, MWeldInst, CEng, FIQA
G Burns, BSc, PhD, AMIEE (now with the Glasgow College of Technology)
R Service, MSc, FInstNDT, MWeldInst, MIM, MICP, CEng, FIQA
R Cumming, BSc, FIQA
A McNab, BSc, PhD (now with the University of Strathclyde)

Part 4

A Danielsson, CEng, FIMechE, FInstMC, is with Wimpey Engineering Ltd. He was a member of the BS working party developing the Code of Practice for Instrumentation in Process Control Systems: Installation–Design

C I Daykin, MA, is Director of Research and Development at Automatic Systems Laboratories Ltd

J G Giles, TEng, has been with Ludlam Sysco Ltd for a number of years

E H Higham, MA, CEng, FIEE, MIMechE, MIERE, MInstMC, is Visiting Fellow at the Department of Fluid Engineering and Instrumentation at Cranfield Institute of Technology after a long career with Foxboro Great Britain Ltd

E G Kingham, CEng, FIEE, was formerly at the Central Electricity Research Laboratories

B E Noltingk, BSc, PhD, CEng, FIEE, CPhys, FInstP, is now a Consultant after some time as Head of the Instrumentation Section at the Central Electricity Research Laboratories

M L Sanderson, BSc, PhD, is Director of the Centre for Fluid Instrumentation at Cranfield Institute of Technology

L C Towle, BSc, CEng, MIMechE, MIEE, MInstMC, is a Director of the MTL Instruments Group Ltd

Part 5

J Barron, BA, MA(Cantab), is a Lecturer at the University of Cambridge

D R Heath, BSc, PhD, is with Rank Xerox Ltd

C Kindell, is with AMP of Great Britain Ltd

T Kingham, is with AMP of Great Britain Ltd

F F Mazda, DFH, MPhil, CEng. MIEE, MBIM, is with Rank Xerox Ltd

J Riley, is with AMP of Great Britain Ltd

M G Say, MSc, PhD, CEng, ACGI, DIC, FIEE, FRSE, is Professor Emeritus of Electrical Engineering at Heriot-Watt University

K R Sturley, BSc, PhD, FIEE, FIEEE, is a Telecommunications Consultant

L W Turner, CEng, FIEE, FRTS, is a Consultant Engineer

Introduction

1 Techniques and applications

Much instrumentation work can be categorized in two ways, by *techniques* or by *applications*. In the former, one scientific field, such as radioactivity or ultrasonics, is considered and all the ways in which it can be used to make useful measurements are surveyed. In the latter, it is the application, the measurement of a particular quantity, that forms the primary division. In a single chapter different techniques are covered. Under flowmetering, for instance, methods using tracers, ultrasonics or pressure measurement are included. This book is mainly applications oriented, but in a few cases, notably pneumatics and the employment of nuclear technology, the technique has been the primary unifying theme.

2 Accuracy

A supremely important question in all matters of instrumentation is the accuracy with which a measurement is made. This arises so universally that it deserves consideration at the outset, though the accuracy achievable in different situations is a recurring theme in many chapters. An attitude of robust scepticism is healthy in instrument engineers: they should tend to distrust what they are told about equipment. Equally they should hesitate to accept their own reasoning about systems they have assembled. They should demand evidence and preferably proof. Moreover, they should give their minds to the question of what accuracy is really needed. They may have to argue with a client, whose tendency may well be to ask for an arbitrarily high accuracy, regardless of effort and cost. Once a figure is agreed there must be constant alertness whether any factors have crept in that will increase errors.

Accuracy is important but complex. A first analysis distinguishes 'systematic' and 'random' errors in an instrument. The concept is based on the idea of making a series of measurements of the same quantity under the same conditions. There will generally be differences in the results, but even their average may differ from the 'true value'. This is an abstract concept, but it is often replaceable by the 'conventional true value' that has been measured by an instrument or technique known to have a very high accuracy. The difference can be equated roughly with systematic error and can be eliminated by calibrating. Calibration is referred to in several later chapters. Calibration is concerned with the chain linking a local measurement to fundamental standards and with the 'traceability' of this chain.

The phrase *random errors* gives a hint that probability comes into the assessment of the results of measurement, but in a deterministic world where does the randomness come from? Some variations in readings, though clearly observed, are difficult to explain, but most random errors can be treated statistically without knowing their cause. In most cases it is assumed that the probability of error is such that errors in individual measurements have a normal distribution about the mean, which is zero if there is no systematic error. This implies that errors should be quoted, not as indicating that it is impossible that a true value lies outside a certain range, though that is, in fact, sometimes done, but that a certain probability can be attached to its whereabouts. The probability grows steadily larger as the range where it might be grows wider.

When we consider a measurement chain with several links, the two approaches give increasingly different figures. For if we think of possibilities/impossibilities then we must allow that the errors in each link can be extreme and in the same direction, calling for a simple addition when calculating the possible total error. On the other hand, this is *improbable* so the 'chain error' that corresponds to a given probability, e_c, is appreciably smaller. In fact, statistically,

$$e_c = \sqrt{e_1^2 + e_2^2 + \ldots}$$

where e_1, e_2, etc. are the errors in the different links, each corresponding to the same probability as e_c.

The causes of random errors can often be appreciated by thinking of 'influence quantities'. Most devices intended to measure some physical quantity make some response to other quantities: they come under their *influence*. Even in the simple case of a tape measure being used to measure an unknown length, the tape will itself expand with temperature, so giving a false reading unless the influence is allowed for. It is desirable for any instrument to be insensitive to influence quantities and for the user to be aware of them. Whether they are realized or not, their effects can be reduced by calibrating under conditions as close as possible to those holding for the live measurement. Influence quantities may be complicated. Thus it is not only temperature—often important—that can affect an instrument, but also rate of change of temperature and temperature differences between various items in apparatus.

One particular factor that could be thought of as an influence quantity is the direction in which the quantity to be measured is changing. Many instruments give slightly different readings according to whether, as it changes, the particular value of interest is approached from above or below. This phenomenon is called 'hysteresis'.

'Non-linearity errors' arise if it is assumed that an instrument's output is exactly proportional to some quantity and there is a discrepancy from this law. The error can be specified as the maximum departure of the true input/output curve from the idealized straight line approximating it: it may be noted that this does not cover changes in 'incremental gain', the term used for the local slope of the input/output curve. Special cases of the accuracy of conversion from digital to analogue signals and vice versa are discussed in sections 4.3.1 and 4.4.5 of Part 4. Calibration at sufficient intermediate points in the range of an instrument can cover systematic non-linearity and the advent of sophisticated data-processing facilities, which allow easy correction, has reduced the problems it introduces.

Special terms used in the discussion above are defined in BS 5233 along with numerous others.

The general approach to errors that we have outlined follows a statistical approach to a static situation. In communications theory, working frequencies and time available are emphasized, and this approach may gain importance in instrument technology. Sensors often feed to electronic amplifiers and the electronic noise in the amplifier may be an ultimate source of error. This will mean that more accurate results can be expected when a longer time is available in which to make the measurement.

There is, of course, some danger of making a highly accurate measurement of the wrong thing! This could happen from faulty system analysis throwing up the wrong quantity as relevant to process control. More directly significant for instrumentation, the operation of measurement could disturb the quantity measured. This can happen in most fields: a flow-meter can obstruct flow and reduce the velocity to be measured, an over-large temperature sensor can cool the material studied or a low-impedance voltmeter can reduce the potential it is monitoring. Part of the instrument engineer's task is to foresee and avoid errors resulting from the effect his instrument has on the system it is being used to study.

3 Environment

Instruments must always be selected in full consciousness of the environment in which they will be used. On plant there may be extremes of temperature, vibration, dust much more severe than are found in the laboratory. There are two sorts of ill effects: false readings from exceptional values of influence quantities and irreversible breakdown of equipment. Sometimes manufacturers specify limits to working conditions, sometimes these have to be estimated from a general judgement of robustness.

The reliability of equipment—the likelihood of its continuing to work satisfactorily over long periods—is always a key feature. We go into it more deeply in Part 4 and it must be taken into account when choosing apparatus for all applications.

4 Units

The introductory chapters to some books have discussed the theme of what system of units is used there. Fortunately the question is becoming obsolete because SI units are adopted nearly everywhere, and certainly in this book. In a few areas, where other units still have some usage, we have listed the relations for the benefit of those who are still more at home with the older expressions.

5 References

British Standards Institution, *Glossary of terms used in Metrology*, BS 5233 (1975)

Dietrich, D. F., *Uncertainty, Calibration and Probability: the Statistics of Scientific and Industrial Measurement*, Adam Hilger, London (1973)

Topping, J., *Errors of Observation and their Treatment*, Chapman and Hall, London (1972)

Part 1

Mechanical Measurements

1 Measurement of flow

G. FOWLES

1.1 Introduction

Flow measurement is a technique used in any process requiring the transport of a material from one point to another (for example, bulk supply of oil from a road tanker to a garage holding tank). It can be used for quantifying a charge for material supplied or maintaining and controlling a specific rate of flow. In many processes, plant efficiency will depend on being able to measure and control flow accurately.

However it is applied, a flow-measurement system will need to be compatible with the process or material being measured whilst being capable of producing the desired accuracy and repeatability.

It is often said that, 'The ideal flowmeter should be non-intrusive, inexpensive, have absolute accuracy, infinite repeatability and run forever without maintenance.' Unfortunately, such a device does not yet exist, although some manufacturers may claim that it does. Over recent years, however, many improvements have been made to established systems, and new products utilizing novel techniques are continually being introduced onto the market. The 'ideal flowmeter', may not in fact be so far away and now more than ever potential users must be fully aware of the systems at their disposal.

1.2 Basic principles of flow measurement

Before a full understanding of the operational techniques of existing or new measurement systems can be gained, it is necessary to develop a knowledge of the basic theory of flow measurement and the derivation of flow formulae.

Flow can be measured as either a volumetric quantity or an instantaneous velocity (this is normally translated into a flow rate). If Figure 1.1 is examined the interdependence of these measurements can be established.

$$\text{flow rate} = \text{velocity} \times \text{area} = \frac{\text{m}}{\text{s}} \cdot \text{m}^2 = \frac{\text{m}^3}{\text{s}}$$

$$\text{quantity} = \text{flow rate} \times \text{time} = \frac{\text{m}^3}{\text{s}} \cdot \text{s} = \text{m}^3$$

If, as above, flow rate is recorded for a period of time, the quantity is equal to the area under the curve (shaded area). This can be established automatically by many instruments and the process is called integration. It may be carried out by an instrument's integrator either electrically or mechanically.

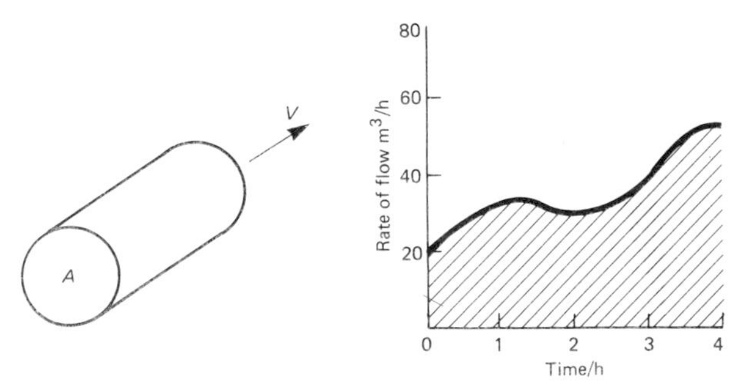

Figure 1.1 Flow–time graph.

1.2.1 Streamlined and turbulent flow

Streamlined flow in a liquid is a phenomenon best described by example. Reynolds did a considerable amount of work on this subject and Figure 1.2 illustrates the principle of streamlined flow (also called laminar flow).

A thin filament of coloured liquid is introduced into a quantity of water flowing through a smooth glass tube. The paths of all fluid particles will be parallel to the tube walls and therefore the coloured liquid travels in a straight line, almost as if it were a tube within a tube. However, this state is velocity- and viscosity-dependent and as velocity is increased, a point is

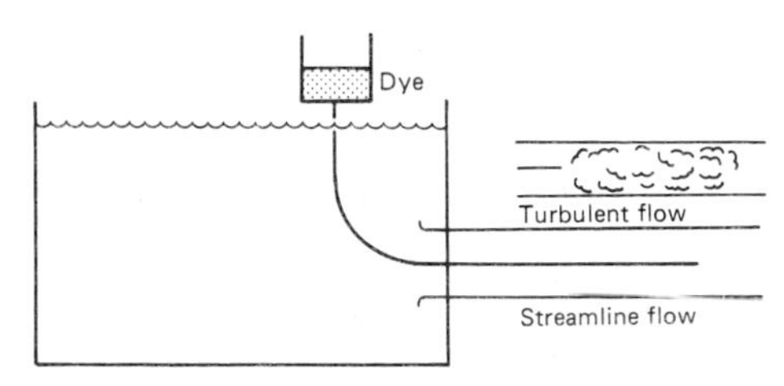

Figure 1.2 Reynolds's experiment.

reached (critical velocity) when the coloured liquid will appear to disperse and mix with the carrier liquid. At this point the motion of the particles of fluid are not all parallel to the tube walls but have a transverse velocity also. This form of flow pattern is called turbulent flow. Summarizing therefore, for velocities below the critical velocity, flow is said to be streamlined or laminar and for velocities above the critical value flow is said to be turbulent—this situation is most common in practice.

Reynolds formulated his data in a dimensionless form

$$Re = \frac{D.v.\rho}{\mu} \tag{1.1}$$

where Re is Reynolds number, D is the diameter of the throat of the installation, v is velocity, ρ is density of fluid, and μ is absolute viscosity. Flow of fluid in pipes is expected to be laminar if the Reynolds number is less than 2000 and turbulent if it is greater than 4000. Between these values is the critical zone. If systems have the same Reynolds number and are geometrically similar they are said to have dynamic similarity.

1.2.1.1 Flow profile

The velocity across the diameter of a pipe varies and the distribution is termed the velocity profile of the system. For laminar flow the profile is parabolic in nature, the velocity at the centre of the pipe being twice the mean velocity. For turbulent flow, after a sufficient straight pipe run the flow profile becomes fully developed where velocity at the centre of the pipe is only about 1.2 times the mean velocity and is the preferred situation to allow accurate flow measurement.

1.2.1.2 Energy of a fluid in motion

Before establishing the use of Reynolds number in universal flow formulae, it is worthwhile considering the forms in which energy will be represented in a fluid in motion. The basic types of energy associated with a moving fluid are:

(a) Potential energy or potential head.
(b) Kinetic energy.
(c) Pressure energy.
(d) Heat energy.

1.2.1.3 Potential energy

This is the energy the fluid has by virtue of its position or height above some fixed level. For example 1 m^3 of liquid of density ρ_1 kg/m^3 will have a mass of ρ_1 kg and would require a force of 9.81 ρ_1 N to support it at a point where the gravitational constant g is 9.81 m/s. Therefore if it is at a height of z metres above a reference plane it would have 9.81 $\rho_1 z$ joules of energy by virtue of its height.

1.2.1.4 Kinetic energy

This is the energy a fluid has, by virtue of its motion. 1 m^3 of fluid of density ρ_1 kg/m^3 with a velocity V_1 m/s would have a kinetic energy of $\frac{1}{2}\rho_1 V_1^2$ joules.

1.2.1.5 Pressure energy

This is the energy a fluid has by virtue of its pressure. For example a fluid having a volume v_1 m^3 and a pressure of p_1 N/m^2 would have a pressure energy of $p_1 v_1$ joules.

1.2.1.6 Internal energy

The fluid will also have energy by virtue of its temperature (i.e. heat energy). If there is resistance to flow in the form of friction, other forms of internal energy will be converted into heat energy.

1.2.1.7 Total energy

The total energy E of a fluid is given by the equation

$$\begin{aligned}\text{total energy } (E) = {} & \text{potential energy} \\ & + \text{kinetic energy} \\ & + \text{pressure energy} \\ & + \text{internal energy} \\ E = {} & \text{P.E.} + \text{K.E.} + \text{PR.E.} + \text{I.E.}\end{aligned} \tag{1.2}$$

1.2.2 Viscosity

The frictional resistance that exists in a flowing fluid is called viscosity and is discussed in more detail in the next chapter. Briefly, the particles of fluid actually in contact with the walls of the channel are at rest, while those at the centre of the channel move at maximum velocity. Thus, the layers of fluid near the centre which are moving at maximum velocity, will be slowed down by the slower moving layers and the slower moving layers will be speeded up by the faster moving layers.

Dynamic viscosity of a fluid is expressed in units of Ns/m^2. Thus a fluid has a dynamic viscosity of 1 Ns/m^2 if a force of 1 N is required to move a plane of 1 m^2 in area at a speed of 1 m/s parallel to a fixed plane, the moving plane being 1 m away from the fixed plane and the space between the planes being completely filled with the fluid. This is illustrated diagrammatically in Figure 1.3.

Thus for parallel flow lines

$$\text{dynamic viscosity } \mu = \frac{\text{force}(F)}{\text{area}(A) \times \text{velocity}(v)} \tag{1.3}$$

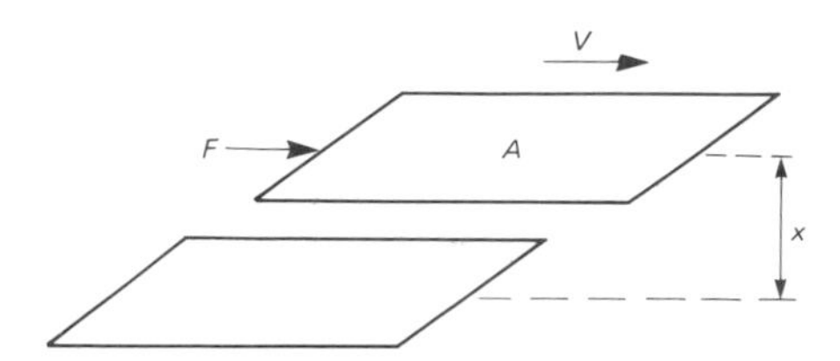

Figure 1.3 Determination of dynamic viscosity.

or, if a velocity gradient exists,

$$\mu = \frac{F}{A \, \mathrm{d}v/\mathrm{d}x} \tag{1.4}$$

It is sometimes useful to know the ratio of the dynamic viscosity of a fluid to its density at the same temperature. This is called the kinematic viscosity of the fluid.

$$\text{kinematic viscosity at } T\,°\text{C} = \frac{\text{dynamic viscosity at } T\,°\text{C}}{\text{density at } T\,°\text{C}} \tag{1.5}$$

For liquids the viscosity decreases with increase of temperature at constant pressure whilst for gases viscosity will increase with increasing temperature, at a constant pressure.

It is viscosity that is responsible for the damping out or suppression of flow disturbances caused by bends and valves in a pipe; the energy that existed in the swirling liquid is changed into heat energy.

1.2.3 Bernoulli's theorem

All formulae for the derivation of fluid flow in a closed pipe are based on Bernoulli's theorem. This states that in a steady flow, without friction, the sum of potential energy, kinetic energy and pressure energy is a constant along any streamline.

If we now consider a closed pipe or channel (Figure 1.4) in which two sections are examined (section 1 and section 2); due to a restriction, orifice or hydraulic gradient there is a pressure or head loss in the transition from section 1 to section 2. If 1 kg of fluid enters at section 1 and there is no accumulation of fluid between section 1 and section 2 then 1 kg of fluid must leave at section 2.

The energy of the fluid at section 1

= potential energy + kinetic energy + pressure energy + internal energy

$$= 1 \,.\, Z_1 \,.\, g + \tfrac{1}{2} \,.\, 1 \,.\, V_1^2 + p_1 \,.\, v_1 + I_1 \tag{1.6}$$

The energy of the fluid at section 2

$$= 1 \,.\, Z_2 \,.\, g + \tfrac{1}{2} \,.\, 1 \,.\, V_2^2 + p_2 \,.\, v_2 + I_2 \tag{1.7}$$

and since energy cannot leave the channel nor be created or destroyed,

total energy at section 1 = total energy at section 2

$$Z_1 \,.\, g + \frac{V_1^2}{2} + p_1 \,.\, v_1 + I_1 = Z_2 \,.\, g + \frac{V_2^2}{2} + p_2 \,.\, v_2 + I_2 \tag{1.8}$$

Now, if the temperature of the fluid remains the same the internal energy remains the same and

$$I_1 = I_2 \tag{1.9}$$

and equation (1.8) reduces to

$$Z_1 \,.\, g + \frac{V_1^2}{2} + p_1 \,.\, v_1 = Z_2 \,.\, g + \frac{V_2^2}{2} + p_2 \,.\, v_2 \tag{1.10}$$

This equation applies to liquids and ideal gases.

Now consider liquids only. These can be regarded as being incompressible and their density and specific volume will remain constant along the channel and

$$v_1 = v_2 = \frac{1}{\rho_1} = \frac{1}{\rho_2} = \frac{1}{\rho} \tag{1.11}$$

and equation (1.10) may be rewritten as,

$$Z_1 \,.\, g + \frac{V_1^2}{2} + \frac{p_1}{\rho} = Z_2 \,.\, g + \frac{V_2^2}{2} + \frac{p_2}{\rho} \tag{1.12}$$

Dividing by g, this becomes,

$$Z_1 + \frac{V_1^2}{2g} + \frac{p_1}{\rho \,.\, g} = Z_2 + \frac{V_2^2}{2g} + \frac{p_2}{\rho \,.\, g} \tag{1.13}$$

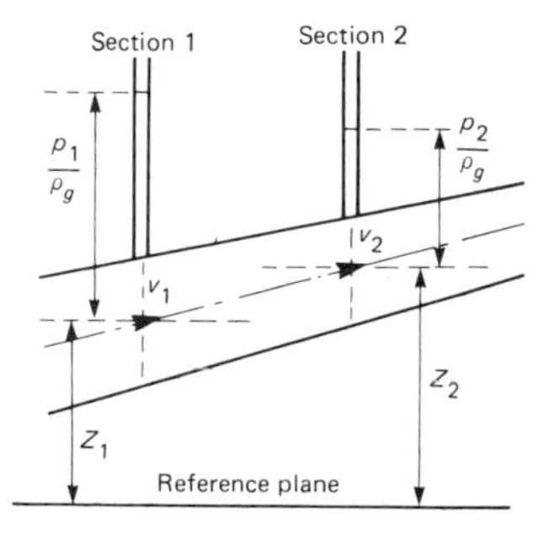

Parameter	At section 1	At section 2	Units
Area	A_1	A_2	m^2
Velocity	V_1	V_2	m/s
Pressure	p_1	p_2	N/m^2
Density	ρ_1	ρ_2	Kg/m^3
Specific volume of 1Kg	$v_1 = 1/\rho_1$	$v_2 = 1/\rho_2$	m^3
Height of centre of gravity above reference plane	Z_1	Z_2	m
Internal energy per Kg	I_1	I_2	J

Figure 1.4 Hydraulic conditions for pipe flow.

Referring back to Figure 1.4 it is obvious that there is a height differential between the upstream and downstream vertical connections representing sections 1 and 2 of the fluid. Considering first the conditions at the upstream tapping, the fluid will rise in the tube to a height $p_1/\rho . g$ above the tapping or $p_1/\rho . g + Z_1$ above the horizontal level taken as the reference plane. Similarly the fluid will rise to a height $p_2/\rho . g$ or $p_2/\rho . g + Z_2$ in the vertical tube at the downstream tapping.

The differential head will be given by

$$h = \left(\frac{p_1}{\rho . g} + Z_1\right) - \left(\frac{p_2}{\rho . g} + Z_2\right) \qquad (1.14)$$

but from equation (1.13) we have

$$\left(\frac{p_1}{\rho . g} + Z_1\right) + \frac{V_1^2}{2g} = \left(\frac{p_2}{\rho . g} + Z_2\right) + \frac{V_2^2}{2g}$$

or $$\left(\frac{p_1}{\rho . g} + Z_1\right) - \left(\frac{p_2}{\rho . g} + Z_2\right) = \frac{V_2^2}{2g} - \frac{V_1^2}{2g}$$

Therefore

$$h = \frac{V_2^2}{2g} - \frac{V_1^2}{2g} \qquad (1.15)$$

and

$$V_2^2 - V_1^2 = 2gh \qquad (1.16)$$

Now the volume of liquid flowing along the channel per second will be given by Q m^3 where,

$$Q = A_1 . V_1 = A_2 . V_2$$

or $$V_1 = \frac{A_2 . V_2}{A_1}$$

Now substituting this value in equation (1.16):

$$V_2^2 - V_2^2 \frac{A_2^2}{A_1^2} = 2gh$$

or $$V_2^2(1 - A_2^2/A_1^2) = 2gh \qquad (1.17)$$

dividing by $(1 - A_2^2/A_1^2)$ equation (1.17) becomes

$$V_2^2 = \frac{2gh}{1 - A_2^2/A_1^2} \qquad (1.18)$$

and taking the square root of both sides

$$V_2 = \frac{\sqrt{2gh}}{\sqrt{(1 - A_2^2/A_1^2)}} \qquad (1.19)$$

Now A_2/A_1 is the ratio (area of section 2)/(area of section 1) and is often represented by the symbol m. Therefore

$$\left(1 - \frac{A_2^2}{A_1^2}\right) = 1 - m^2$$

and

$$\frac{1}{\sqrt{[1 - (A_2^2/A_1^2)]}} \text{ may be written as } \frac{1}{\sqrt{(1 - m^2)}}$$

This is termed the velocity of approach factor often represented by E. Equation (1.19) may be written

$$V_2 = E\sqrt{2gh} \qquad (1.20)$$

and $$Q = A_2 . V_2 = A_2 . E\sqrt{2gh} \text{ m}^3/\text{s} \qquad (1.21)$$

Mass of liquid flowing per second $= W = \rho . Q = A_2 . \rho . E\sqrt{2gh}$ kg also since $\Delta p = hp$,

$$Q = A_2 . E\sqrt{\frac{2g\,\Delta p}{\rho}} \text{ m}^3/\text{s} \qquad (1.22)$$

$$W = A_2 . E\sqrt{2g\rho . \Delta p} \text{ kg/s} \qquad (1.23)$$

1.2.4 Practical realization of equations

The foregoing equations apply only to streamlined flow. To determine actual flow it is necessary to take into account various other parameters. In practice flow is rarely streamlined, but is turbulent. However, the velocities of particles across the stream will be entirely random and will not affect the rate of flow very considerably.

In developing the equations, effects of viscosity have also been neglected. In an actual fluid the loss of head between sections will be greater than that which would take place in a fluid free from viscosity.

In order to correct for these and other effects another factor is introduced into the equations for flow. This factor is the discharge coefficient C and is given by the equation

Discharge coefficient

$$C = \frac{\text{actual mass rate of flow}}{\text{theoretical mass rate of flow}}$$

or if the conditions of temperature, density, etc., are the same at both sections it may be written in terms of volume.

$$C = \frac{\text{actual volume flowing}}{\text{theoretical volume flowing}}$$

It is possible to determine C experimentally by actual tests. It is a function of pipe size, type of pressure tappings and Reynolds number.

Equation (1.22) is modified and becomes

$$Q = C . A_2 . E\sqrt{\frac{2g . \Delta p}{\rho}} \qquad (1.24)$$

This is true for flow systems where Reynolds number is

above a certain value (20 000 or above for orifice plates). For lower Reynolds numbers and for very small or rough pipes the basic coefficient is multiplied by a correction factor Z whose value depends on the area ratio, the Reynolds number and the size and roughness of the pipe. Values for both C and Z are listed with other relevant data in BS 1042 Part 1 1964.

We can use differential pressure to measure flow.

Consider a practical example of a device having the following dimensions:

Internal diameter of upstream pipe	D mm
Orifice or throat diameter	d mm
Pressure differential produced	h mm water gauge
Density of fluid at upstream tapping	ρ kg/m^2
Absolute pressure at upstream tapping	p bar

Then introducing the discharge coefficient C, the correction factor and the numerical constant, the equation for quantity rate of flow Q m^3/h becomes

$$Q=0.012\,52C\,.\,Z\,.\,E\,.\,d^2\sqrt{\frac{h}{\rho}} \tag{1.25}$$

and the weight or mass rate of the flow W kg/h is given by

$$W=0.012\,52C\,.\,Z\,.\,E\,.\,d^2\sqrt{h\rho} \tag{1.26}$$

1.2.5 Modification of flow equations to apply to gases

Gases differ from liquids in that they are compressible. If the gas under consideration can be regarded as an ideal gas (most gases are ideal when well away from their critical temperatures and pressures) then the gas obeys several very important gas laws. These laws will now be stated.

1.2.5.1 Dry gases

(a) Boyle's law This states that the volume of any given mass of gas will be inversely proportional to its absolute pressure provided temperature remains constant. Thus, if a certain mass of gas occupies a volume v_0 at an absolute pressure p_0 and a volume v_1 at an absolute pressure p then

$$p_0\,.\,v_0=p\,.\,v_1$$

$$\text{or}\quad v_1=v_0\,.\,p_0/p \tag{1.27}$$

(b) Charles's law This states that if the volume of a given mass of gas occupies a volume v_1 at a temperature T_0 kelvin, then its volume v at T kelvin is given by

$$v_1/T_0=v/T \qquad\text{or}\qquad v=v_1\,.\,T/T_0 \tag{1.28}$$

(c) The ideal gas law In the general case p, v and T change. Suppose a mass of gas at pressure p_0 and temperature T_0 kelvin has a volume v_0 and the mass of gas at pressure p and temperature T has a volume v, and that the change from the first set of conditions to the second set of conditions takes place in two stages.

(a) Change the pressure from p_0 to p at a constant temperature. Let the new volume be v_1. From Boyle's law:

$$p_0\,.\,v_0=p\,.\,v_1 \qquad\text{or}\qquad v_1=v_0\,.\,p_0/p$$

(b) Change the temperature from T_0 to T at constant pressure. From Charles's law:

$$v_1/T_0=v/T$$

Hence, equating the two values of v_1

$$v_0\,.\,p_0/p=v\,.\,T_0/T$$

$$p_0\,.\,v_0/T_0=pv/T=\text{constant} \tag{1.29}$$

If the quantity of gas considered is 1 mole, i.e. the quantity of gas that contains as many molecules as there are atoms in 0.012 kg of carbon-12, this constant is represented by R, the gas constant, and equation (1.29) becomes:

$$pv=R_0\,.\,T$$

where $R_0=8.314$ J/Mol K and p is in N/m^2 and v is in m^3.

Adiabatic expansion When a gas is flowing through a primary element the change in pressure takes place too rapidly for the gas to absorb heat from its surroundings. When it expands owing to the reduction in pressure it does work, so that if it does not receive energy it must use its own heat energy, and its temperature will fall. Thus the expansion that takes place owing to the fall in pressure does not obey Boyle's law, which applies only to an expansion at constant temperature. Instead it obeys the law for adiabatic expansion of a gas:

$$p_1\,.\,v_1^\gamma=p_2\,.\,v_2^\gamma \quad\text{or}\quad p\,.\,v^\gamma=\text{constant} \tag{1.30}$$

where γ is the ratio of the specific heats of the gas

$$\gamma=\frac{\text{specific heat of a gas at constant pressure}}{\text{specific heat of a gas at constant volume}}$$

and has a value of 1.40 for dry air and other diatomic gases, 1.66 for monatomic gases such as helium and about 1.33 for triatomic gases such as carbon dioxide.

If a fluid that is being metered cannot be regarded as being incompressible, another factor is introduced into the flow equations, to correct for the change in volume due to the expansion of the fluid while passing through the restriction. This factor is called the expansibility factor ε and has a value of unity for incompressible fluids. For ideal compressible fluids expanding without any change of state the value can be calculated from the equation

$$\varepsilon = \sqrt{\left(\frac{\gamma r^{2/\gamma}}{\gamma-1}\frac{1-m^2}{1-m^2 r^{2/\gamma}}\frac{1-r^{(\gamma-1)/\gamma}}{1-r}\right)}$$

where r is the ratio of the absolute pressures at the upstream and downstream tappings (i.e. $r = p_1/p_2$) and γ is the ratio of the specific heat of the fluid at constant pressure to that at constant volume. This is detailed in BS 1042 Part 1 1964.

In order that working fluid flow equations can be applied to both liquids and gases the factor ε is introduced and the equations become:

$$Q = 0.012\,52 CZ\varepsilon Ed^2\sqrt{h/\rho}\ \text{m}^3/\text{h} \quad (1.32)$$

$$W = 0.012\,52 CZ\varepsilon Ed^2\sqrt{h\rho}\ \text{kg/h} \quad (1.33)$$

$$\varepsilon = 1 \text{ for liquids}$$

1.2.5.2 Critical flow of compressible fluids

It can be shown theoretically for flow through a convergent tube such as a nozzle that the value of r at the throat cannot be less than a critical value r_c. When the pressure at the throat is equal to this critical fraction of the upstream pressure, the rate of flow is a maximum and cannot be further increased except by raising the upstream pressure. The critical pressure ratio is given by the equation

$$2r_c^{(1-\gamma)/\gamma} + (\gamma-1)m^2 . r_c^{2/\gamma} = \gamma - 1 \quad (1.34)$$

The value of r is about 0.5 but it increases slightly with increase of m and with decrease of specific heat ratio. Values of r are tabulated in BS 1042 Part 1 1964.

The basic equation for critical flow is obtained by substituting $(1-r_c)\rho$ for Δp in equation (1.23) substituting r_c for r in equation (1.31) and the equation becomes

$$W = 1.252U . d^2\sqrt{\rho p}\ \text{kg/h} \quad (1.35)$$

where

$$U = C\sqrt{(\gamma/2)r_c . (\gamma-1)/\gamma} \quad (1.36)$$

The volume rate of flow (in m^3/h) is obtained by dividing the weight ratio of flow by the density (in kg/m^3) of the fluid at the reference conditions.

1.2.5.3 Departure from gas laws

At room temperature and at absolute pressures less than 10 bar most common gases except carbon dioxide behave sufficiently like an ideal gas that the error in flow calculations brought about by departure from the ideal gas laws is less than 1 per cent. In order to correct for departure from the ideal gas laws a deviation coefficient K (given in BS 1042 Part 1 1964) is used in the calculation of densities of gases where the departure is significant. For ideal gases $K = 1$.

1.2.5.4 Wet gases

The above modification applies to dry gases. In practice many gases are wet, being a mixture of gas and water vapour. Partial pressure due to saturated water vapour does not obey Boyle's law.

Gas humidity is discussed in Chapter 6 of Part 2. If the temperature and absolute pressure at the upstream tapping and the state of humidity of the gas are known, a correction factor can be worked out and applied to obtain the actual mass of gas flowing.

Gas density is given by the equation

$$\rho = 6.196\left[\delta\frac{(p-p_v)}{k.T} + \frac{0.622p_v}{T}\right]\text{kg/m}^3 \quad (1.37)$$

where δ is specific gravity of dry gas relative to air, T is temperature in kelvin, p is pressure in mbar at the upstream tapping, p_v is partial pressure in mbar of the water vapour, k is the gas law deviation at temperature T, and ρ is gas density. For dry gas pv is zero and the equation becomes

$$\rho = 6.196\frac{\delta p}{kT}\ \text{kg/m}^3 \quad (1.38)$$

1.3 Fluid flow in closed pipes

1.3.1 Differential-pressure devices

By far the most common technique for measurement of fluid flow in pipes is that which utilizes the pressure drop caused by a constriction in the pipeline. As already shown in the derivation of Bernoulli's equation in the previous section, a constriction will cause an increase in fluid velocity in the area of that constriction, which in turn will result in a corresponding pressure drop across the constriction. This differential pressure (d.p.) is a function of the flow velocity and density of the fluid and is shown to be a square root relationship, see equation (1.24).

A flowmeter in this category would normally comprise a primary element to develop a differential pressure and a secondary element to measure it. The secondary element is effectively a pressure transducer, and operational techniques are discussed in Chapter 9, so no further coverage will be given here. However there are various types of primary element and these deserve further consideration. The main types of interest are: orifice plate, venturi, nozzle, Dall, rotameter, gate meter, Gilflo element, and target meter.

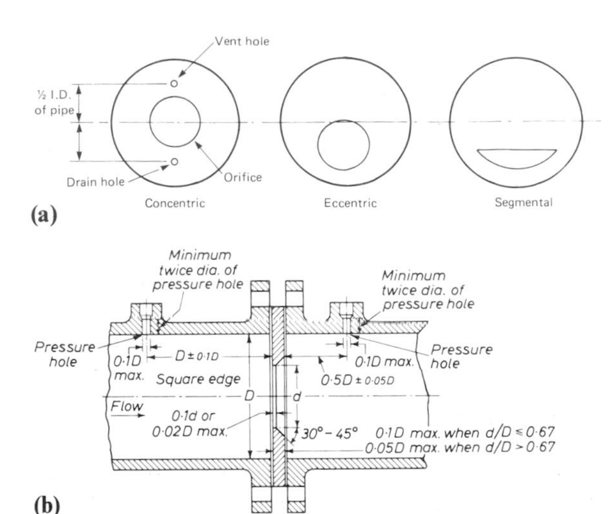

Figure 1.5 (a) Orifice plate types. (b) Concentric orifice plate with D and $D/2$ tappings mounted between flange plates. Courtesy, British Standards Institution.

1.3.1.1 Orifice plate

In its simplest form this comprises a thin steel plate with a circular orifice of known dimensions located centrally in the plate. This is termed a concentric orifice plate, see Figure 1.5(a). The plate would normally be clamped between adjacent flange fittings in a pipeline, a vent hole and drain hole being provided to prevent solids building up and gas pockets developing in the system, see Figure 1.5(b).

The differential pressure is measured by suitably located pressure tappings on the pipeline on either side of the orifice plate. These may be located in various positions depending on the application (e.g. corner, D and $D/2$, or flange tappings), and reference should be made to BS 1042 Part 1 1964 for correct application. Flow rate is determined from equation (1.24).

This type of orifice plate is inadequate to cope with difficult conditions experienced in metering dirty or viscous fluids and gives a poor disposal rate of condensate in flowing steam and vapours. Modified designs are utilized to overcome these problems in the form of segmental or eccentric orifice plates as shown in Figure 1.5(a).

The segmental orifice provides a method for measuring the flow of liquids with solids in suspension. It takes the form of a plate which covers the upper cross-section of the pipe leaving the lower portion open for the passage of solids to prevent their build-up.

The eccentric orifice is used on installations where condensed liquids are present in gas-flow measurement or where undissolved gases are present in the measurement of liquid flow. It is also useful where pipeline drainage is required.

To sum up the orifice plate:

Advantages
1 Inherently simple in operation
2 No moving parts
3 Long-term reliability
4 Inexpensive

Disadvantages
1 Square root relationship
2 Poor turn-down ratio
3 Critical installation requirements
4 High irrecoverable pressure loss

1.3.1.2 Venturi tube

The basic construction of the classical venturi tube is shown in Figure 1.6. It comprises a cylindrical inlet section followed by a convergent entrance into a cylindrical throat and a divergent outlet section. A complete specification may be found by reference to BS 1042 Part 1 1964 and relevant details are repeated here:

(a) Diameter of throat. The diameter d of the throat shall be not less than $0.224D$ and not greater than $0.742D$, where D is the entrance diameter.
(b) Length of throat. The throat shall have a length of $1.0d$.
(c) Cylindrical entrance section. This section shall have an internal diameter D and a length of not less than $1.0d$.
(d) Conical section. This shall have a taper of $10\frac{1}{2}°$. Its length is therefore $2.70(D-d)$ within $\pm 0.24(D-d)$.
(e) Divergent outlet section. The outlet section shall have an inclined angle of not less than 5° and not greater than 15°. Its length shall be such that the exit diameter is not less than $1.5d$.

In operation the fluid passes through the convergent entrance, increasing velocity as it does so, resulting in a differential pressure between the inlet and throat. This differential pressure is monitored in the same way

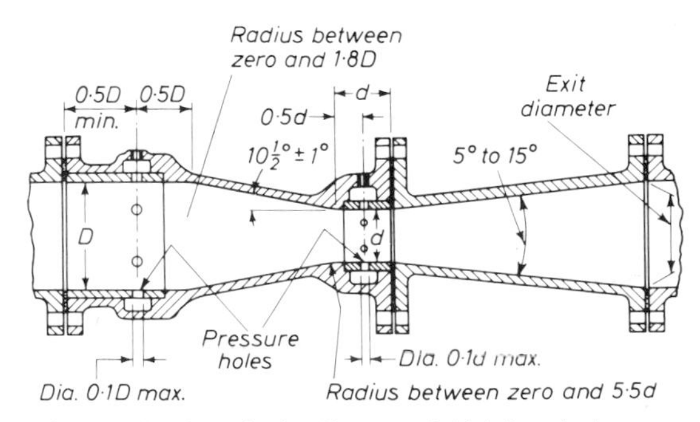

Figure 1.6 Venturi tube. Courtesy, British Standards Institution.

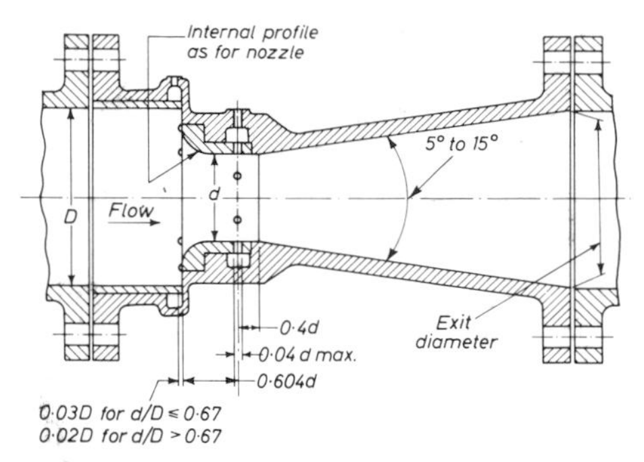

Figure 1.7 Venturi nozzle. Courtesy, British Standards Institution.

as for the orifice plate, the relationship between flow rate and differential being as defined in equation (1.24).

Location of pressure tappings The upstream pressure tapping is located in the cylindrical entrance section of the tube $0.5D$ upstream of the convergent section and the downstream pressure tapping is located in the throat at a distance $0.5D$ downstream of the convergent section. Pressure tappings should be sized so as to avoid accidental blockage.

Generally the tappings are not in the form of a single hole but several equally spaced holes connected together in the form of an annular ring sometimes called a piezometer ring. This has the advantage of giving a true mean value of pressure at the measuring section.

Application The venturi is used for applications where there is a high solids content or where high pressure recovery is desirable. The venturi is inherently a low head loss device and can result in an appreciable saving of energy.

To sum up the venturi tube:

Advantages
1 Simple in operation
2 Low head loss
3 Tolerance of high solids content
4 Long-term reliability
5 No moving parts

Disadvantages
1 Expensive
2 Square root pressure–velocity relationship
3 Poor turn-down ratio
4 Critical installation requirements

1.3.1.3 Nozzles

The venturi effect is also utilized in other more inexpensive forms, the most common being the venturi nozzle.

Venturi nozzle This is in effect a shortened venturi tube. The entrance cone is much shorter and has a curved profile. The inlet pressure tap is located at the mouth of the inlet cone and the low pressure tap in the plane of minimum section as shown in Figure 1.7. This reduction in size is taken a stage further in the flow nozzle.

Flow nozzle Overall length is again reduced greatly. The entrance cone is bell-shaped and there is no exit cone. This is illustrated in Figure 1.8. The flow nozzle is not suitable for viscous liquids but for other applications it is considerably cheaper than standard venturi tube. Also, due to the smooth entrance cone there is less resistance to fluid flow through the nozzle and a lower value of m may be used for a given rate of flow. Its main area of use therefore is in high-velocity mains where it will produce a substantially smaller pressure drop than an orifice plate of similar m number.

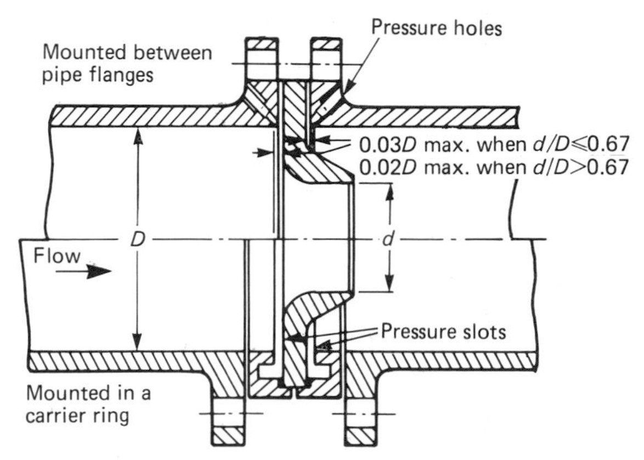

Figure 1.8 Flow nozzle. Courtesy, British Standards Institution.

1.3.1.4 Dall tube

This is another variation of the venturi tube and gives a higher differential pressure but a lower head loss than the conventional venturi tube. Figure 1.9 shows a cross-section of a typical Dall flow tube. It consists of a short straight inlet section, a convergent entrance section, a narrow throat annulus and a short divergent recovery cone. The whole device is about 2 pipe-diameters long.

A shortened version of the Dall tube, the Dall orifice or insert is also available; it is only 0.3 pipe-diameter long. All the essential Dall tube features are retained in a truncated format as shown in Figure 1.10.

Pressure loss All of the differential pressure devices discussed so far cause an irrecoverable pressure loss of varying degree. In operation it is advantageous to keep this loss as low as possible and this will often be a major factor in the selection criteria of a primary

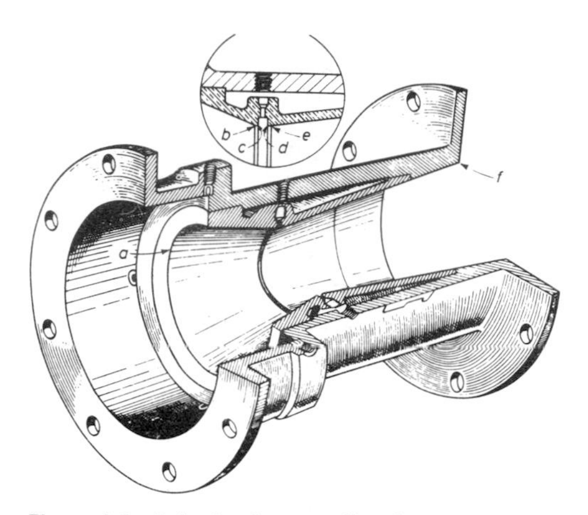

Figure 1.9 Dall tube. Courtesy, Kent Instruments Ltd.

Throat pressure connection

d_1 d_2

Upstream pressure connection

Figure 1.10 Dall insert. Courtesy, British Standards Institution.

element. The pressure loss curves for nozzles, orifices and venturi tubes are given in Figure 1.11.

Installation requirements As already indicated installation requirements for differential-pressure devices are quite critical. It is advisable to install primary elements as far downstream as possible from flow disturbances, such as bends, valves and reducers. These requirements are tabulated in considerable detail in BS 1042 Part 1 1964 and are reproduced in part in Appendix 1.1.

1.3.1.5 Variable-orifice meters

So far the devices discussed have relied on a constriction in the flowstream causing a differential pressure varying with flow rate. Another category of differential-pressure device relies on maintaining a nominally constant differential pressure by allowing effective area to increase with flow. The principal defices to be considered are: rotameter, gate meter and Gilflo.

Rotameter This is shown schematically in Figure 1.12(a). In a tapered tube the upward stream of fluid supports the float where the force on its mass due to gravity is balanced against the flow force determined by the annular area between the float and the tube and the velocity of the stream. The float's position in the tube is measured by a graduated scale and its position is taken as an indication of flow rate.

Many refinements are possible, including the use of magnetic coupling between the float and external devices to translate vertical movement into horizontal and develop either electrical transmission or alarm actuation. Tube materials can be either metal or glass depending on application. Figure 1.12(b) shows an exploded view of a typical rotameter.

Gate meter In this type of meter the area of the orifice may be varied by lowering a gate either manually or by an automatically controlled electric motor. The gate is moved so as to maintain a constant pressure drop across the orifice. The pressure drop is measured by pressure tappings located upstream and downstream of the gate as shown in Figure 1.13(a). The position of the gate is indicated by a scale. As the rate of flow through the orifice increases, the area of the orifice is increased. If all other factors in equation (1.21) except

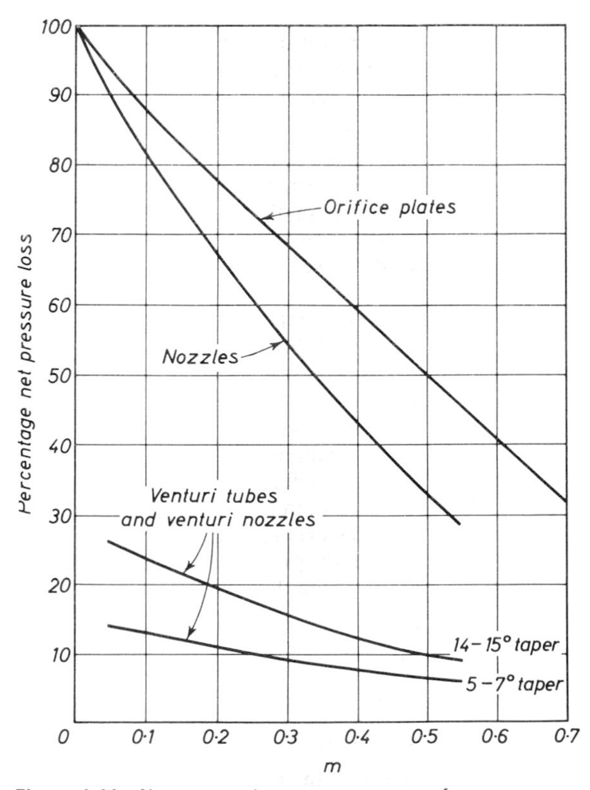

Figure 1.11 Net pressure loss as a percentage of pressure difference. Courtesy, British Standards Institution.

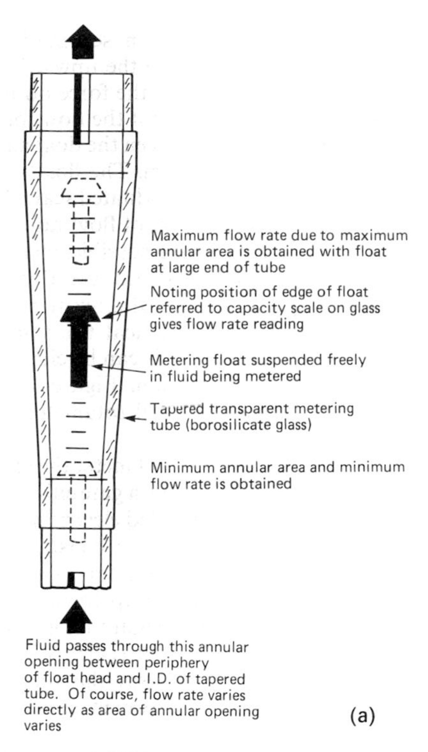

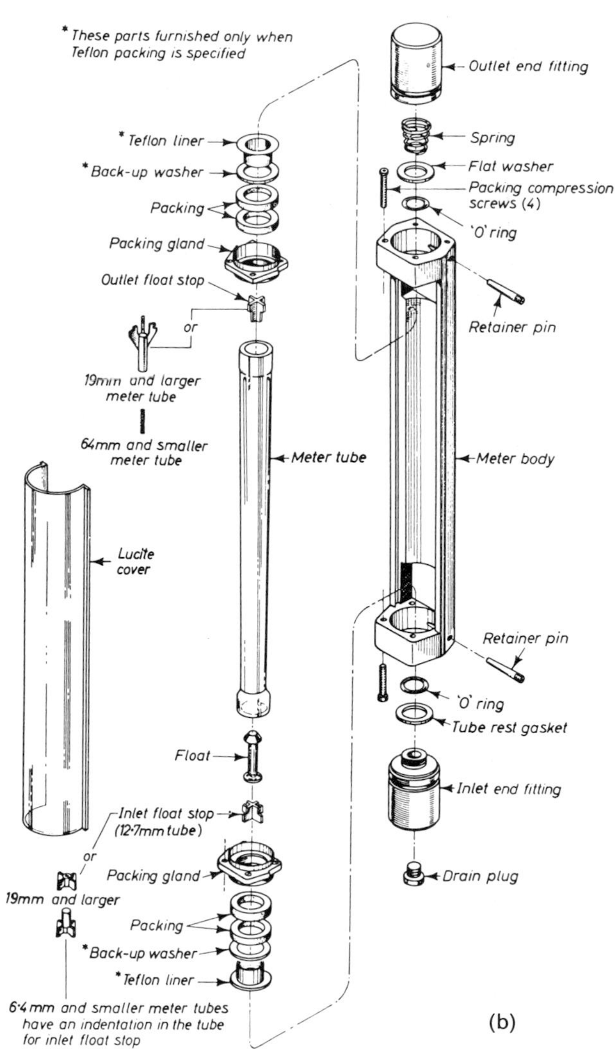

Figure 1.12 (a) Rotameter—principle of operation. Courtesy, Fischer & Porter Ltd. (b) Rotameter—exploded view. Courtesy, Fischer & Porter Ltd.

area A_2 are kept constant the flow through the orifice will depend upon the product $A_2 . E$ or $A_2/\sqrt{[1-(A_2/A_1)^2]}$. As A_2 increases, $(A_2/A_1)^2$ increases and $[1-(A_2/A_1)^2]$ decreases and therefore $1/\sqrt{[1-(A_2/A_1)^2]}$ increases.

The relationship between A_2 and flow is not linear. If the vertical movement of the gate is to be directly proportional to the rate of flow, the width of the opening A_2 must decrease towards the top as shown in Figure 1.13(a).

The flow through the meter can be made to depend directly upon the area of the orifice A_2 if instead of the normal static pressure being measured at the upstream tapping the impact pressure is measured. In order to do this the upstream tap is made in the form of a tube with its open end facing directly into the flow as shown in Figure 1.13(b). It is in effect a pitot tube (see section on point-velocity measurement).

The differential pressure is given by equation (1.15), where h is the amount the pressure at the upstream tap is greater than that at the downstream tap:

$$h=\frac{V_2^2}{2g}-\frac{V_1^2}{2g} \tag{1.39}$$

Now, at the impact port, $V_2=0$

therefore $h_1=V_1^2/2g$

where h_1 is the amount the impact pressure is greater than the normal upstream static pressure. Thus the difference between impact pressure and the pressure measured at the downstream tap will be h_2 where

$$h_2=h+h_1$$

$$=\frac{V_2^2}{2g}-\frac{V_1^2}{2g}+\frac{V_1^2}{2g}=\frac{V_2^2}{2g} \tag{1.40}$$

Therefore the velocity V_2 through the section A_2 is given by $V_2=\sqrt{(2g . h_2)}$. The normal flow equations for the type of installation shown in Figure 1.13(b) will be the same for other orifices but the velocity of approach factor is 1 and flow is directly proportional to A_2. The opening of the gate may therefore be made

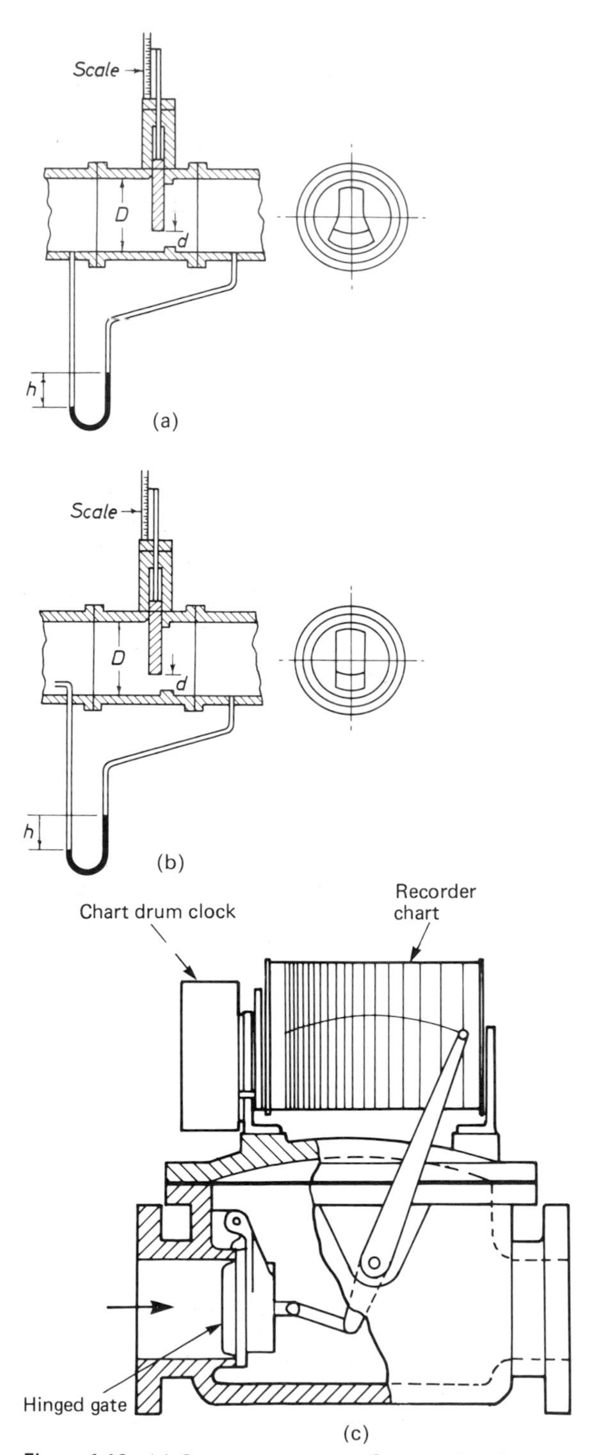

Figure 1.13 (a) Gate-type area meter. Courtesy, American Society of Mechanical Engineers. (b) Gate-type area meter corrected for velocity of approach. Courtesy, American Society of Mechanical Engineers. (c) Weight-controlled hinged-gate meter.

rectangular and the vertical movement will be directly proportional to flow.

The hinged gate meter is another version of this type of device. Here a weighted gate is placed in the flowstream, its deflection being proportional to flow. A mechanical linkage between the gate and a recorder head provides flow indication. It is primarily used for applications in water mains where the user is interested in step changes rather than absolute flow accuracy. The essential features of this device are shown in Figure 1.13(c).

The 'Gilflo' primary sensor The Gilflo metering principle was developed to overcome the limitations of the square law fixed orifice plate in the mid 1960s. Its construction is in two forms; the Gilflo 'A', Figure 1.14(a) sizes 10 to 40 mm has an orifice mounted to a strong linear bellows fixed at one end and with a shaped cone positioned concentrically in it. Under flow conditions the orifice moves axially along the cone creating a variable annulus across which the

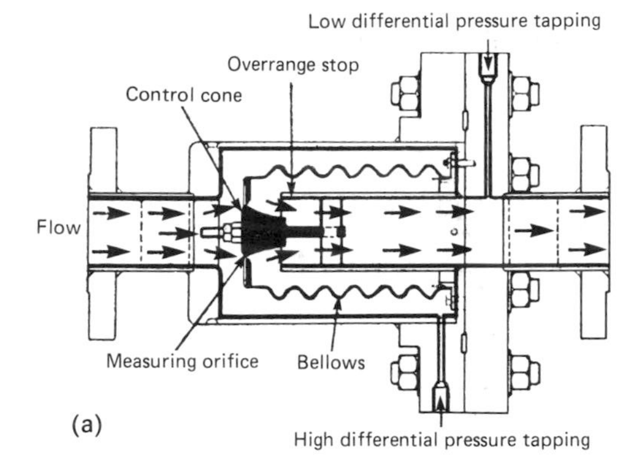

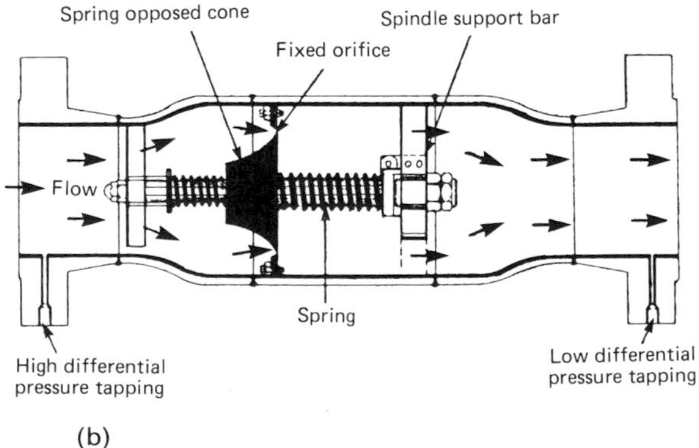

Figure 1.14 (a) The essentials of Gilflo A. As flow increases the measuring orifice moves along the control cone against the spring bellows. Courtesy, Gervase Instruments Ltd. (b) Gilflo B extends the principle to higher flows. Now the orifice is fixed and the control cone moves against the spring. Courtesy, Gervase Instruments Ltd.

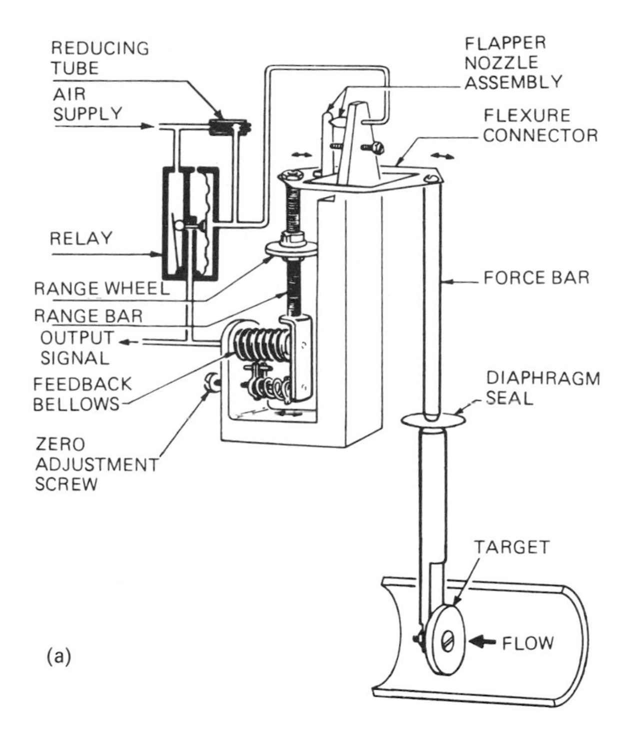

Figure 1.15 (a) Target flow transmitter working principle. Courtesy, Foxboro-Yoxall Ltd. (b) The 13 series pneumatic target flow transmitter. Courtesy, The Foxoboro Company.

differential pressure varies. Such is the relationship of change that the differential pressure is directly proportional to flowrate enabling a rangeability of up to 100:1.

The Gilflo 'B', Figure 1.14(b) sizes 40 to 300 mm standard, has a fixed orifice with a shaped cone moving axially against the resistance of a spring, again producing a linear differential pressure and a range of up to 100:1.

The Gilflo 'A' has a water equivalent range of 0–5 to 0–350 litres/minute and the Gilflo 'B' 0–100 to 0–17 500 litres/minute.

The main application for Gilflo-based systems is on saturated and superheated steam, with pressures up to 200 bar and temperatures up to +500 °C.

1.3.1.6 Target flowmeter

Although not strictly a differential-pressure device it is generally categorized under that general heading. The primary and secondary element form an integral unit and differential pressure tappings are not required. It is particularly suited for measuring the flow of high-viscosity liquids; hot asphalt, tars, oils and slurries at pressures up to 100 bar and Reynolds numbers as low as 2000. Figure 1.15 shows the meter and working principles.

The liquid impinging on the target will be brought to rest so that pressure increases by $V^2/2g$ in terms of head of liquid so that the force F on the target will be

$$F = K\gamma' V_1^2 A_t/2 \text{ N} \qquad (1.41)$$

where γ' is the mass per unit volume in kg/m^3. The area of the target is A_t measured in m^3, K is a constant, and V_1 is the velocity in m/s of the liquid through the annular ring between target and pipe.

If the pipe diameter is D m, and the target diameter d m, then area A of the annular space equals $\pi(D^2-d^2)/4$ m^2.

Therefore volume flow rate is

$$Q = A\,.\,V_1 = \frac{\pi(D^2-d^2)}{4}\sqrt{\frac{8F}{K\gamma'\pi d^2}}$$

$$= \frac{C(D^2-d^2)}{d^2}\sqrt{\frac{F}{\gamma'}} \text{ m}^3/\text{s} \qquad (1.42)$$

where C is a new constant including the numerical factors. Mass flow rate is

$$W = Q\gamma' = \frac{C(D^2-d^2)}{d}\sqrt{F\gamma'} \text{ kg/s} \qquad (1.43)$$

The force F is balanced through the force bar, by the air pressure in the bellows so that a 0.2 to 1.0 bar signal proportional to the square root of flow is obtained.

Flow ranges available vary from 0–52.7 to 0–123 litres/minute for the 19 mm size at temperatures up to 400 °C, to from 0–682 to 0–2273 litres/minute for the 100 mm size at temperatures up to 260 °F. Meters are also available for gas flow.

The overall accuracy of the meter is ±0.5 per cent with repeatability of ±0.1 per cent.

1.3.2 Rotating mechanical meters for liquids

Rotating mechanical flowmeters derive a signal from a moving rotor which is rotated at a speed proportional to the fluid flow velocity. Most of these meters are velocity-measuring devices except for positive-displacement meters which are quantity or volumetric in operation. The principal types to be discussed are: positive-displacement, rotating vane, angled propeller meter, bypass meter, helix meter, and turbine meter.

1.3.2.1 Positive-displacement

Positive-displacement meters are widely used on applications where high accuracy and good repeatability are required. Accuracy is not affected by pulsating flow and accurate measurement is possible at higher liquid viscosities than with many other flowmeters. Positive-displacement meters are frequently used in oil and water undertakings for accounting purposes.

The principle of the measurement is that as the liquid flows through the meter, it moves a measuring element which seals off the measuring chamber into a series of measuring compartments which are successively filled and emptied. Thus, for each complete cycle of the measuring element a fixed quantity of liquid is permitted to pass from the inlet to the outlet of the meter. The seal between the measuring element and the measuring chamber is provided by a film of the measured liquid. The number of cycles of the measuring element is indicated by means of a pointer moving over a dial driven from the measuring element by suitable gearing.

The extent of error, defined as the difference between the indicated quantity and the true quantity and expressed as a percentage of the true quantity, is dependent on many factors, among them being:

(a) The amount of clearance between the rotor and the measuring chamber through which liquid can pass unmetered.
(b) The amount of torque required to drive the register. The greater the torque the greater the pressure drop across the measuring element which in turn determines the leakage rate past the rotor.
(c) The viscosity of the liquid to be measured. Increase in viscosity will also result in increased pressure drop across the measuring element, but this is compensated for by the reduction in flow through the rotor clearances for a given pressure drop.

The accuracy of measurement attained with a positive-displacement meter varies very considerably from one design to another, with the nature and condition of the liquid measured, and with the rate of flow. Great care should be taken to choose the correct meter for an application.

The most common forms of positive-displacement meters are: rotary piston, reciprocating piston, nutating disc, fluted spiral rotor, sliding vane, rotating vane, and oval gear.

Rotary piston The rotary-piston flowmeter is most common in the water industry where it is used for metering domestic supplies. It consists of a cylindrical working chamber which houses a hollow cylindrical piston of equal length. The central hub of the piston is guided in a circular motion by two short inner cylinders. The piston and cylinder are alternately filled and emptied by the fluid passing through the meter. A slot in the sidewall of the piston is removed so that a partition extending inward from the bore of the working chamber can be inserted. This has the effect of restricting the movement of the piston to a sliding motion along the partition. The rotary movement of the piston is transmitted via a permanent-magnet coupling from the drive shaft to a mechanical register. The basic design and principle of operation of this meter is shown diagrammatically in Figure 1.16.

Reciprocating piston A reciprocating meter can be either of single- or multi-piston type, this being dependent on the application. This type of meter exhibits a wide turn-down ratio (e.g. 300:1), with extreme accuracy of ± 0.1 per cent and can be used for a wide range of liquids. Figure 1.17 illustrates the operating principle of this type of meter.

Suppose the piston is at the bottom of its stroke. The valve is so arranged that inlet liquid is admitted below the piston, causing it to travel upwards and the liquid above the piston to be discharged to the outlet pipe. When the piston has reached the limit of its travel, the top of the cylinder is cut off from the outlet side, and opened to the inlet liquid supply. At the same time the bottom of the cylinder is opened to the outlet side but cut off from the inlet liquid. The pressure of the incoming liquid will therefore drive the piston downwards, discharging the liquid from below the piston to the outlet pipe. The process is then repeated.

As the piston reciprocates, a ratchet attached to the piston rod provides an actuating force for an incremental counter, each count representing a predetermined quantity of liquid.

Nutating-disc type This type of meter is similar in principle to the rotary-piston type. In this case, however, the gear train is driven not by a rotating piston but by a movable disc mounted on a concentric sphere. The basic construction is shown in Figure 1.18.

The liquid enters the left side of the meter, alternately above and below the disc, forcing it to rock (nutate) in a circular path without rotating about its own axis. The disc is contained in a spherical working chamber and is restricted from rotating about its own

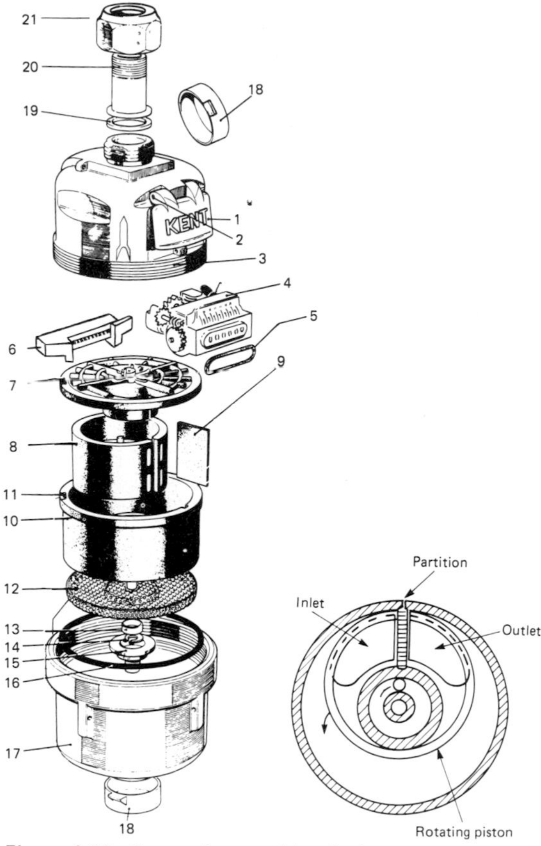

Figure 1.16 Rotary-piston positive-displacement meter. Courtesy, Kent Meters Ltd. 1. Lid. 2. Hinge pin. 3. Counter housing complete with lid and hinge pin. 4. Counter with worm reduction gear and washer. 5. Counter washer. 6. Ramp assembly. 7. Top plate assembly comprising top plate only; driving spindle; driving dog; dog retaining clip. 8. Piston. 9. Shutter. 10. Working chamber only. 11. Locating pin. 12. Strainer—plastic. Strainer—copper. 13. Strainer cap. 14. Circlip. 15. Non-return valve. 16. O ring. 17. Chamber housing. 18. Protective caps for end threads.

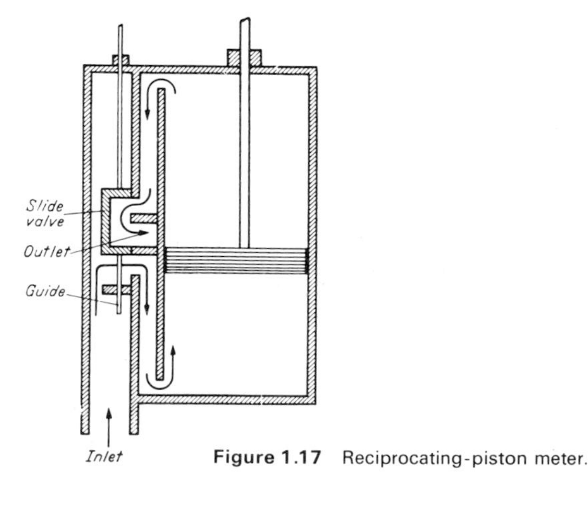

Figure 1.17 Reciprocating-piston meter.

axis by a radial partition that extends vertically across the chamber. The disc is slotted to fit over this partition. The spindle protruding from the sphere traces a circular path and is used to drive a geared register.

This type of meter can be used for a wide variety of liquids—disc and body materials being chosen to suit.

Fluted-spiral-rotor type (*rotating-impeller type*) The principle of this type of meter is shown in Figure 1.19. The meter consists of two fluted rotors supported in sleeve-type bearings and mounted so as to rotate rather like gears in a liquid tight case. The clearance between the rotors and measuring chambers is kept to a minimum. The shape of the rotors is designed so that a uniform uninterrupted rotation is produced by the liquid. The impellers in turn rotate the index of a counter which shows the total measured quantity.

This type of meter is used mainly for measuring crude and refined petroleum products covering a range of flows up to 3000 m^3/h at pressures up to 80 bar.

Sliding-vane type The principle of this type is illustrated in Figure 1.20. It consists of an accurately machined body containing a rotor revolving on ball bearings. The rotor has four evenly spaced slots, forming guides for four vanes. The vanes are in contact with a fixed cam. The four cam-followers follow the contour of the cam, causing the vanes to move radially. This ensures that during transition through the measuring chamber the vanes are in contact with the chamber wall.

The liquid impact on the blades causes the rotor to revolve allowing a quantity of liquid to be discharged.

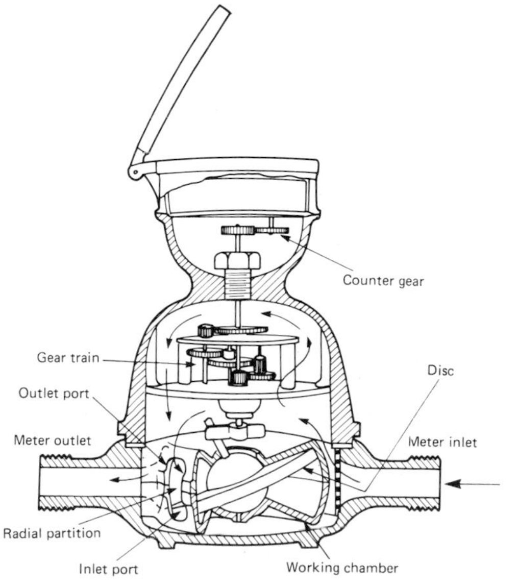

Figure 1.18 Nutating-disc meter.

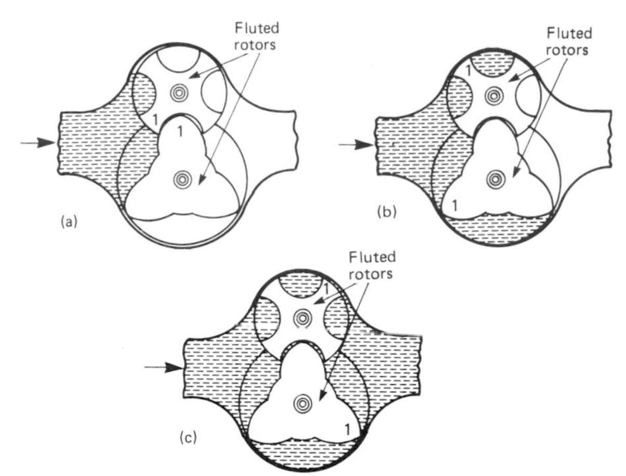

Figure 1.19 Fluted spiral rotor type of meter.

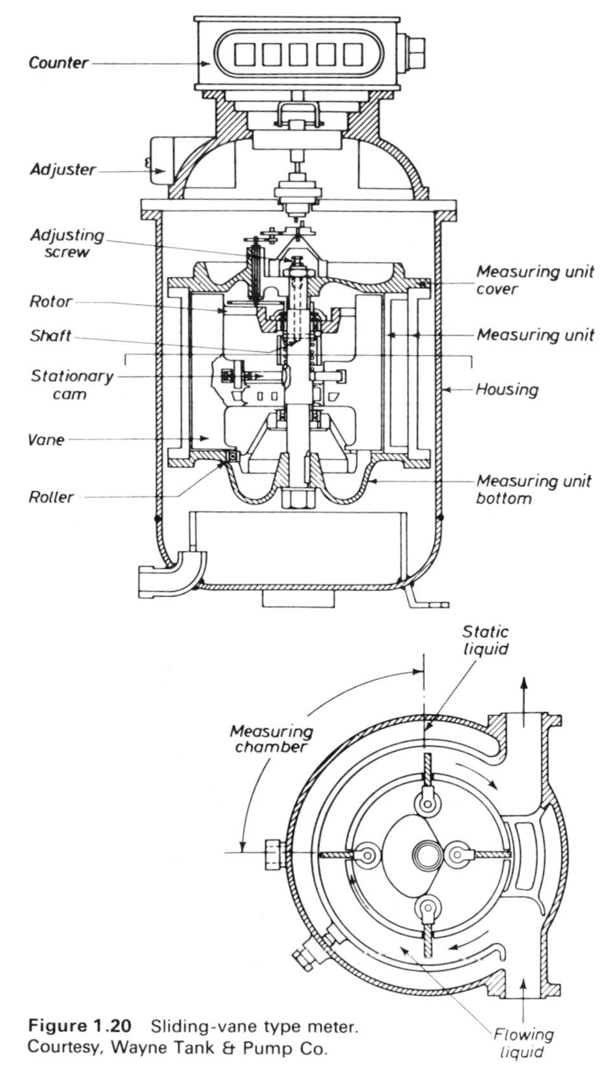

Figure 1.20 Sliding-vane type meter. Courtesy, Wayne Tank & Pump Co.

The number of revolutions of the rotor is a measure of the volume of liquid passed through the meter.

Rotating-vane type This meter is similar in principle to the sliding-vane meter but the measuring chambers are formed by four half-moon-shaped vanes spaced equidistant on the rotor circumference. As the rotor is revolved, the vanes turn to form sealed chambers between the rotor and the meter body. Accuracy of ± 0.1 per cent is possible down to 20 per cent of the rated capacity of the meter.

Oval-gear type This type of meter consists of two intermeshing oval gearwheels which are rotated by the fluid passing through it. This means that for each revolution of the pair of wheels a specific quantity of liquid is carried through the meter. This is shown diagrammatically in Figure 1.21. The number of revolutions is a precise measurement of the quantity of liquid passed. A spindle extended from one of the gears can be used to determine the number of revolutions and convert them to engineering units by suitable gearing.

Oval-gear meters are available in a wide range of materials, in sizes from 10 to 400 mm and suitable for pressures up to 60 bar and flows up to 1200 m^3/h. Accuracy of ± 0.25 per cent of rate of flow can be achieved.

1.3.2.2 Rotating vane

This type of meter operates on the principle that the incoming liquid is directed to impinge tangentially on the periphery of a free spinning rotor. The rotation is monitored by magnetic, or photo-electric pick-up, the frequency of the output being proportional to flow rate, or alternatively by a mechanical register connected through gearing to the rotor assembly as shown in Figure 1.22.

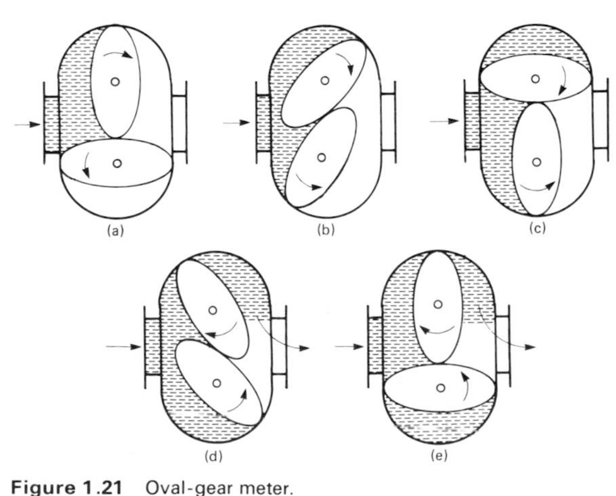

Figure 1.21 Oval-gear meter.

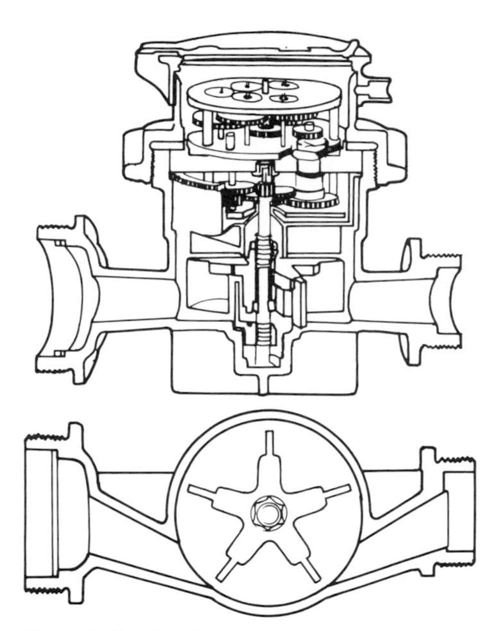

Figure 1.22 Rotating-vane type meter.

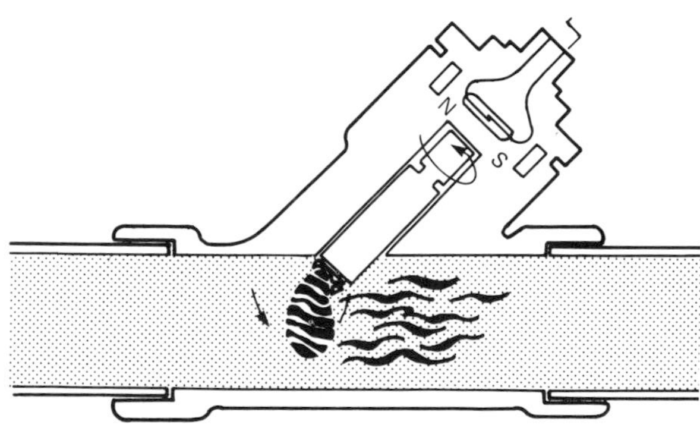

Figure 1.23 Angled-propeller meter.

Accuracy is dependent on calibration and turn-down ratios up to 20:1 can be achieved. This device is particularly suited to low flow rates.

1.3.2.3 Angled-propeller meter

The propeller flowmeter comprises of a Y-type body with all components apart from the propeller being out of the liquid stream. The construction of this type of meter is shown in Figure 1.23. The propeller has three blades and is designed to give maximum clearance in the measuring chamber, thereby allowing maximum tolerance of suspended particles. The propeller body is angled at 45° to the main flowstream and liquid passing through the meter rotates it at a speed proportional to flow rate. As the propeller goes through each revolution, encapsulated magnets generate pulses through a pick-up device, the number of pulses being proportional to flow rate.

1.3.2.4 Bypass meter

In this type of meter (also known as a shunt meter) a proportion of the liquid is diverted from the main flowstream by an orifice plate into a bypass configuration. The liquid is concentrated through nozzles to impinge on the rotors of a small turbine located in the bypass, the rotation of the turbine being proportional to flow rate.

This type of device can give moderate accuracy over a 5:1 turn-down ratio and is suitable for liquids, gases and steam.

1.3.2.5 Helix meter

In this type of meter the measuring element takes the form of a helical vane mounted centrally in the measuring chamber with its axis along the direction of flow as shown in Figure 1.24. The vane consists of a hollow cylinder with accurately formed wings. Owing to the effect of the buoyancy of the liquid on the cylinder, friction between its spindle and the sleeve bearings is small. The water is directed evenly onto the vanes by means of guides.

Transmission of the rotation from the undergear to the meter register is by means of ceramic magnetic coupling.

The body of the meter is cast iron and the mechanism and body cover is of thermoplastic injection moulding. The meter causes only small head loss in operation and is suited for use in water-distribution mains. It is available in sizes from 40 mm up to 300 mm, respective maximum flow rates being 24 m^3/h and 1540 m^3/h, with accuracy of ± 2 per cent over 20:1 turn-down ratio.

1.3.2.6 Turbine meter

This type of meter consists of a practically friction-free rotor pivoted along the axis of the meter tube and designed in such a way that rate of rotation of the rotor is proportional to the rate of flow of fluid through the meter. This rotational speed is sensed by means of an electric pick-off coil fitted to the outside of the meter housing as shown in Figure 1.25(a).

The only moving component in the meter is the rotor and the only component subject to wear is the rotor bearing assembly. However, with careful choice of materials (e.g. tungsten carbide for bearings) the meter should be capable of operating for up to 5 years without failure.

In the Kent turbine flowmeter range the rotor is designed so that the pressure distribution of the

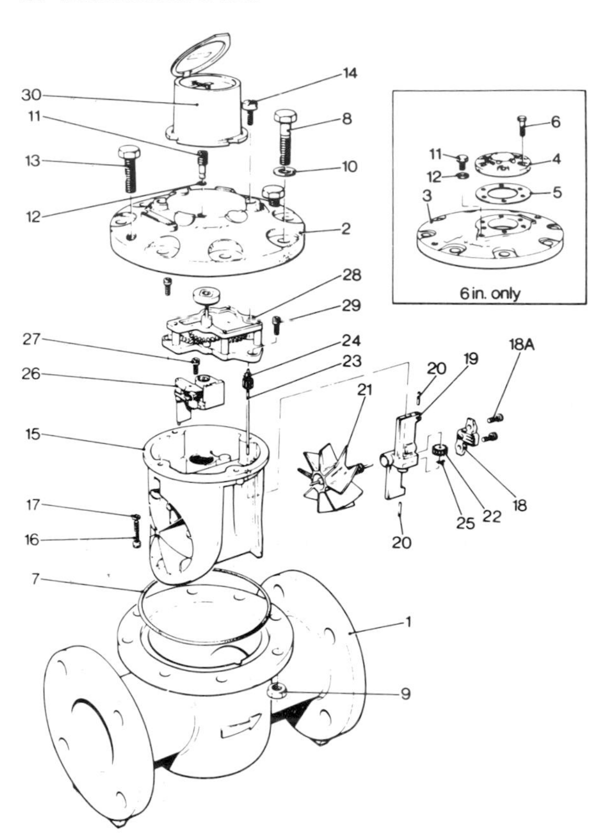

Figure 1.24 Helix meter, exploded view. 1. Body. 2. Top cover with regulator plug and regulator sealing ring. 3. Top cover plate. 4. Joint plate. 5. Joint plate gasket. 6. Joint plate screws. 7. Top cover sealing ring. 8. Body bolt. 9. Body bolt unit. 10. Body bolt washer. 11. Regulator plug. 12. Regulator plug sealing ring. 13. Joint breaking screw. 14. Counter box screw. 15. Measuring element. 16. Element securing screw. 17. Element securing screw washer. 18. Back bearing cap assembly. 19. Back vane support. 20. Tubular dowel pin. 21. Vane. 22. Worm wheel. 23. Vertical worm shaft. 24. First pinion. 25. Drive clip. 26. Regulator assembly. 27. Regulator assembly screw. 28. Undergear. 29. Undergear securing screw. 30. Register.

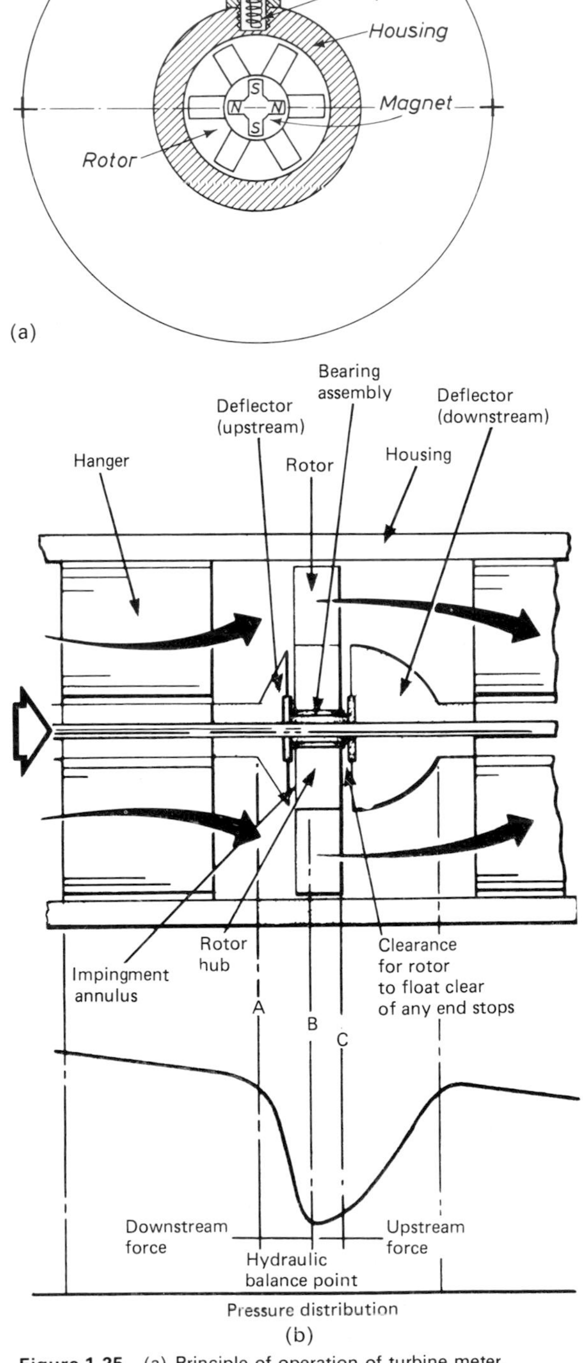

Figure 1.25 (a) Principle of operation of turbine meter. (b) Pressure distribution through turbine meter.

process liquid helps to suspend the rotor in an 'axial' floating position, thereby eliminating end-thrust and wear, improving repeatability and extending the linear flow range. This is illustrated in Figure 1.25(b).

As the liquid flows through the meter, there is a small gradual pressure loss up to point A caused by the rotor hangers and housing. At this point the area through which flow can take place reduces and velocity increases, resulting in a pressure minimum at point B. By the time the liquid reaches the downstream edge of the rotor (C), the flow pattern has re-established itself and a small pressure recovery occurs which causes the rotor to move hard upstream in opposition to the downstream forces. To counteract

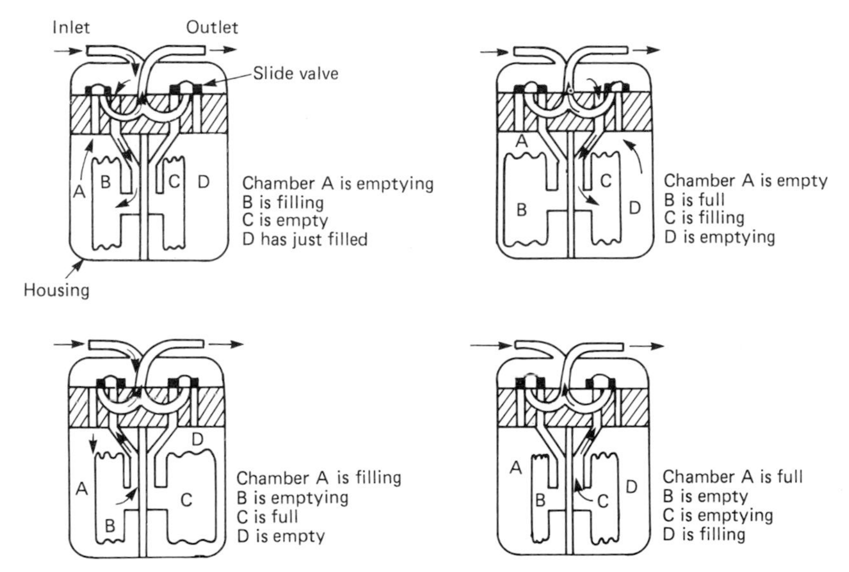

Figure 1.26 Diaphragm meter—stages of operation.

this upstream force the rotor hub is designed to be slightly larger in diameter than the outside diameter of the deflector cone to provide an additional downstream force. A hydraulic balance point is reached with the rotor floating completely clear of any end stops.

The turbine meter is available in a range of sizes up to 500 mm with linearity better than ± 0.25 per cent and repeatability better than ± 0.02 per cent and can be bi-directional in operation. To ensure optimum operation of the meter it is necessary to provide a straight pipe section of 10 pipe-diameters upstream and 5 pipe-diameters downstream of the meter. The addition of flow straighteners is sometimes necessary.

1.3.3 Rotating mechanical meters for gases

The principal types to be discussed are positive displacement, deflecting vane, rotating vane, and turbine.

1.3.3.1 Positive displacement

Three main types of meter come under this heading. They are diaphragm meter, wet gas meter (liquid sealed drum), and rotary displacement meter.

Diaphragm meter (bellows type) This type of meter has remained fundamentally the same for over 100 years and is probably the most common kind of meter in existence. It is used in the UK for metering the supply of gas to domestic and commercial users.

The meter comprises a metal case having an upper and a lower section. The lower section consists of four chambers, two of which are enclosed by flexible diaphragms that expand and contract as they are charged and discharged with the gas being metered. Figure 1.26 illustrates the meter at four stages of its operating cycle.

Mechanical readout is obtained by linking the diaphragms to suitable gearing since each cycle of the diaphragms discharges a known quantity of gas. This type of meter is of necessity highly accurate and trouble-free and the performance is governed by the regulations of the Department of Trade and Industry.

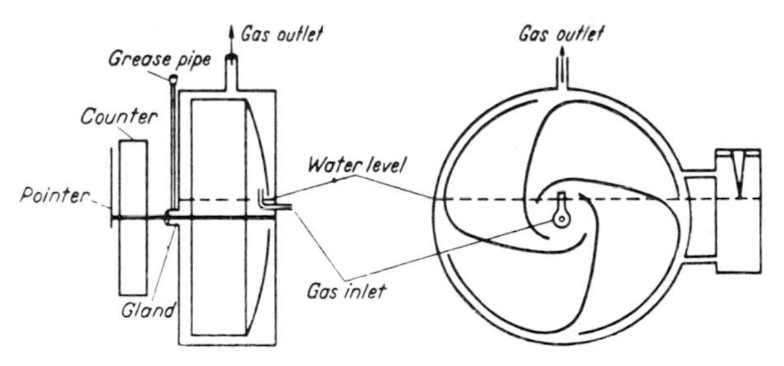

Figure 1.27 Liquid sealed drum type gas meter.

Liquid sealed drum This type of meter differs from the bellows type of meter in that the sealing medium for the measuring chambers is not solid but is water or some other suitable liquid.

The instrument is shown in section in Figure 1.27. It consists of an outer chamber of tinned brass plate or Staybrite steel sheeting containing a rotary portion. This rotating part consists of shaped partitions forming four measuring chambers made of light-gauge tinplate or Staybrite steel, balanced about a centre spindle so that it can rotate freely. Gas enters by the gas inlet near the centre and leaves by the outlet pipe at the top of the outer casing. The measuring chambers are sealed off by water or other suitable liquid which fills the outer chamber to just above the centre line. The level of the water is so arranged that when one chamber becomes unsealed to the outlet side, the partition between it and the next chamber seals it off from the inlet side. Thus, each measuring chamber will, during the course of a rotation, deliver a definite volume of gas from the inlet side to the outlet side of the instrument. The actual volume delivered will depend upon the size of the chamber and the level of the water in the instrument. The level of the water is therefore critical and is maintained at the correct value by means of a hook type of level indicator in a side chamber which is connected to the main chamber of the instrument. If the level becomes very low, the measuring chambers will become unsealed and gas can pass freely through the instrument without being measured; while if the level is too high, the volume delivered at each rotation will be too small and water may pass back down the inlet pipe. The correct calibration is obtained by adjusting the water level.

When a partition reaches a position where a small sealed chamber is formed connected to the inlet side there is a greater pressure on the inlet side than on the outlet side. There will therefore be a force that moves the partition in an anticlockwise direction, and so increases the volume of the chamber. This movement continues until the chamber is sealed off from the inlet pipe but opened up to the outlet side, while at the same time the chamber has become open to the inlet gas but sealed off from the outlet side. This produces continuous rotation. The rotation operates a counter which indicates complete rotations and fractions of rotation, and can be calibrated in actual volume units. The spindle between the rotor and the counter is usually made of brass and passes through a grease-packed gland. The friction of this gland, together with the friction in the counter gearing, will determine the pressure drop across the meter, which is found to be almost independent of the speed of rotation. This friction must be kept as low as possible, for if there is a large pressure difference between inlet and outlet sides of the meter, the level of the water in the measuring chambers will be forced down, causing errors in the volume delivered; and at low rates of flow the meter will rotate in a jerky manner.

It is very difficult to produce partitions of such a shape that the meter delivers accurate amounts for fractions of a rotation; consequently the meter is only approximately correct when fractions of a rotation are involved.

The mass of gas delivered will depend upon the temperature and pressure of the gas passing through the meter. The volume of gas is measured at the inlet pressure of the meter, so if the temperature and the density of the gas at s.t.p. are known it is not difficult to calculate the mass of gas measured. The gas will of course be saturated with water vapour and this must be taken into account in finding the partial pressure of the gas.

Rotating-impeller type This type of meter is similar in principle to the rotating-impeller type meter for liquids and could be described as a two-toothed gear pump. It is shown schematically in Figure 1.28. Although the meter is usually manufactured almost entirely from cast iron, other materials may be used if desired. The meter basically consists of two impellers housed in a casing and supported on rolling element bearings. A clearance of a few thousandths of an inch between the impellers and the casing prevents wear, with the result that the calibration of the meter remains constant throughout its life. The leakage rate is only a small fraction of 1 per cent and this is compensated for in the gearing counter ratio. Each lobe of the impellers has a scraper tip machined onto its periphery to prevent deposits forming in the measuring chamber. The impellers are timed relative to each other by gears fitted to one or both ends of the impeller shafts.

The impellers are caused to rotate by the decrease in pressure which is created at the meter outlet following the use of gas by the consumer. Each time an impeller passes through the vertical position a pocket of gas is momentarily trapped between the impeller and the casing. Four pockets of gas are therefore trapped and expelled during each complete revolution of the index shaft. The rotation of the impellers is transmitted to the meter counter by suitable gearing so that the counter reads directly in cubic metres or cubic feet. As the meter records the quantity of gas passing through it at the conditions prevailing at the inlet it is necessary to correct the volume indicated by the meter index for various factors. These are normally pressure,

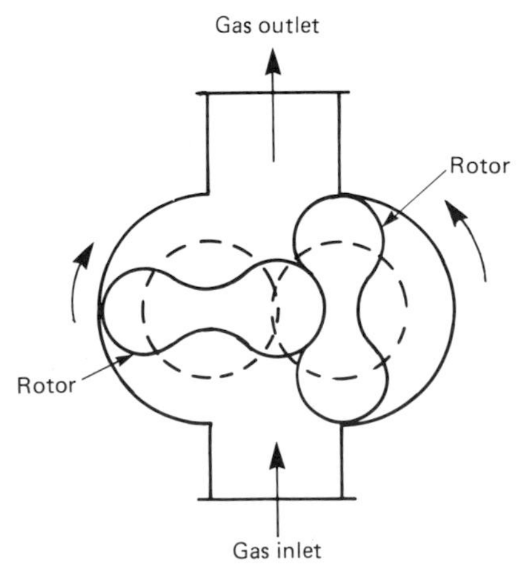

Figure 1.28 Rotary displacement meter.

temperature and compressibility. Corrections can be carried out manually if the conditions within the meter are constant. Alternatively the correction can be made continuously and automatically by small mechanical or electronic computers if conditions within the meter vary continuously and by relatively large amounts. Meters can also drive, through external gearing, various types of pressure- or temperature-recording devices as required.

Meters of this type are usually available in pressures up to 60 bar and will measure flow rates from approximately 12 m^3/h up to 10 000 m^3/h. Within these flow rates the meters will have a guaranteed accuracy of ± 1.0 per cent, over a range of from 5 to 100 per cent of maximum capacity. The pressure drop across the meter at maximum capacity is always less than 50 mm wg. These capacities and the pressure loss information are for meters operating at low pressure; the values would be subject to the effects of gas density at high pressure.

1.3.3.2 Deflecting-vane type: velometers

The principle of this type of instrument is similar to that of the same instrument for liquids. The construction however has to be different, for the density of a gas is usually considerably less than that of a liquid. As the force per unit area acting on the vane depends upon the rate of change of momentum and momentum is mass multiplied by velocity, the force will depend upon the density and upon the velocity of the impinging gas. The velocity of gas flow in a main is usually very much greater (6 to 10 times) than that of liquid flow but this is not sufficient to compensate for the greatly reduced density. (Density of dry air at 0 °C and 760 mm is 0.0013 g/ml while density of water is 1 g/ml.)

The vane must therefore be considerably larger when used for gases or be considerably reduced in weight. The restoring force must also be made small if an appreciable deflection is to be obtained.

The simple velometer consists of a light vane which travels in a shaped channel. Gas flowing through the channel deflects the vane according to the velocity and density of the gas, the shape of the channel and the restoring torque of the hairspring attached to the pivot of the vane.

The velometer is usually attached to a 'duct jet' which consists of two tubes placed so that the open end of one faces upstream while the open end of the other points downstream. The velometer then measures the rate of flow through the pair of tubes and as this depends upon the lengths and size of connecting pipes and the resistance and location of the pressure holes, each assembly needs individual calibration.

The main disadvantages of this simple velometer are the effects of hot or corrosive gases on the vane and channel. This disadvantage may be overcome by measuring the flow of air through the velometer produced by a differential air pressure equal to that produced by the 'duct jet'. In this way the hot gases do not pass through the instrument and so it is not damaged.

1.3.3.3 Rotating-vane type

Anemometers As in the case of the deflecting-vane type the force available from gases to produce the rotation of a vane is considerably less than that available in the measurement of liquids. The vanes must therefore be made light or have a large surface area. The rotor as a whole must be accurately balanced and the bearings must be as friction-free as possible and may be in the form of a multi-cap or multiple-fan blade design, the speed of rotation being proportional to air speed.

Rotary gas meter The rotary meter is a development of the air meter type of anemometer and is shown in Figure 1.29. It consists of three main assemblies: the body, the measuring element and the multi-point index driven through the intergearing. The lower casing (1) has integral in-line flanges (2) and is completed by the bonnet (3) with index glass (4) and bezel (5).

The measuring element is made up of an internal tubular body (6), which directs the flow of gas through a series of circular ports (7) onto a vaned anemometer (8). The anemometer is carried by a pivot (9) which runs in a sapphire–agate bearing assembly (10), the upper end being steadied by a bronze bush (11).

The multi-pointer index (12) is driven by an integear (13) supported between index plates (14). The index assembly is positioned by pillars (15) which are secured to the top flange of the internal tubular body.

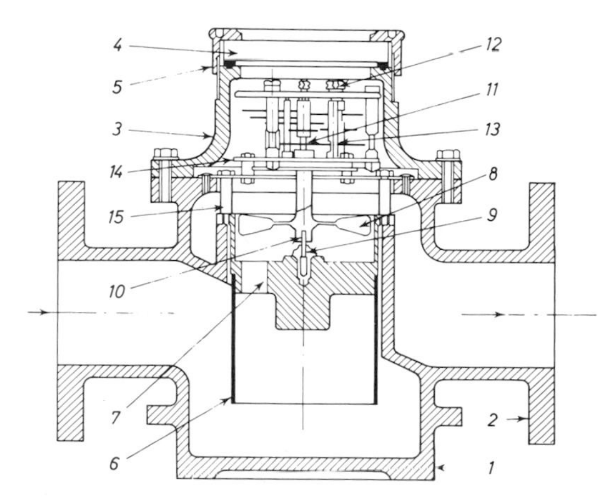

Figure 1.29 Diagrammatic section of a rotary gas meter. Courtesy, Parkinson & Cowan Compteurs.

The meter casing is made of cast iron whilst the anemometer is made from aluminium. The larger sizes have a separate internal tubular body made from cast iron, with a brass or mild steel skirt which forms part of the overall measuring element.

Its area of application is in the measurement of gas flow in industrial and commercial installations at pressures up to 1.5 bar and flows up to 200 m^3/h giving accuracy of ± 2 per cent over a flow range of 10:1.

1.3.3.4 Turbine meter

The gas turbine meter operates on the same principle as the liquid turbine meter previously described although the design is somewhat different since the densities of gases are much lower than those of liquids—high gas velocities are required to turn the rotor blades.

1.3.4 Electronic flowmeters

Either the principle of operation of flowmeters in this category is electronically based or the primary sensing is by means of an electronic device. Most of the flowmeters discussed in this section have undergone considerable development in the last 5 years and the techniques outlined are a growth area in flowmetering applications. They include electromagnetic flowmeters, ultrasonic flowmeters, oscillatory flowmeters, and cross-correlation techniques.

1.3.4.1 Electromagnetic flowmeters

The principle of operation of this type of flowmeter is based on Faraday's law of electromagnetic induction which states that if an electric conductor moves in a magnetic field, an electromotive force (e.m.f.) is induced whose amplitude is dependent on the force of the magnetic field, the velocity of the movement and the length of the conductor such that,

$$E \propto BlV \tag{1.44}$$

where E is e.m.f., B is magnetic field density, l is length of conductor, and V is rate at which the conductor is cutting the magnetic field. The direction of the e.m.f. with respect to the movement and the magnetic field is given by Fleming's right-hand generator rule.

If the conductor now takes the form of a conductive liquid an e.m.f. is generated in accordance with Faraday's law. It is useful at this time to refer to BS 5792 1980 which states: 'If the magnetic field is perpendicular to an electrically insulating tube through which a conductive liquid is flowing, a maximum potential difference may be measured between two electrodes positioned on the wall of the tube such that the diameter joining the electrodes is orthogonal to the magnetic field. The potential difference is proportional to the magnetic field strength, the axial velocity and the distance between the electrodes'. Hence the axial velocity and rate of flow can be determined. This principle is illustrated in Figure 1.30(a).

Figure 1.30(b) shows the basic construction of an electromagnetic flowmeter. It consists of a primary device, which contains the pipe through which the liquid passes, the measurement electrodes and the magnetic field coils and a secondary device, which provides the field-coil excitation and amplifies the output of the primary device and converts it to a form suitable for display, transmission and totalization.

The flow tube, which is effectively a pipe section, is lined with some suitable insulating material (dependent on liquid type) to prevent short circuiting of the electrodes which are normally button-type mounted flush with the liner. The field coils wound around the outside of the flow tube are usually epoxy-resin-encapsulated to prevent damage by damp or liquid submersion.

Field-coil excitation To develop a suitable magnetic field across the pipeline it is necessary to drive the field coil with some form of electrical excitation. It is not possible to use pure d.c. excitation due to the resulting polarization effect on electrodes and subsequent electrochemical action, so some form of a.c. excitation is employed. The most common techniques are: sinusoidal and non-sinusoidal (square wave, pulsed d.c., or trapezoidal).

Sinusoidal a.c. excitation Most early electromagnetic flowmeters used standard 50 Hz mains voltage as an excitation source for the field coils and in fact most systems in use today operate on this principle. The signal voltage will also be a.c. and is normally capacitively coupled to the secondary electronics to avoid any d.c. interfering potentials. This type of system has several disadvantages. Due to a.c. excitation the transformer effect produces interfering voltages. These are caused by stray pick-up by the signal cables from the varying magnetic field. It has a high power consumption and suffers from zero drift caused by the above interfering voltages and electrode contamination. This necessitates manual zero control adjustment.

These problems have now been largely overcome by the use of non-sinusoidal excitation.

Non-sinusoidal excitation Here it is possible to arrange that rate of change of flux density $dB/dt = 0$ for part of the excitation cycle and therefore there is no transformer action during this period. The flow signal is sampled during these periods and is effectively free from induced error voltages.

Square-wave, pulsed and trapezoidal excitations

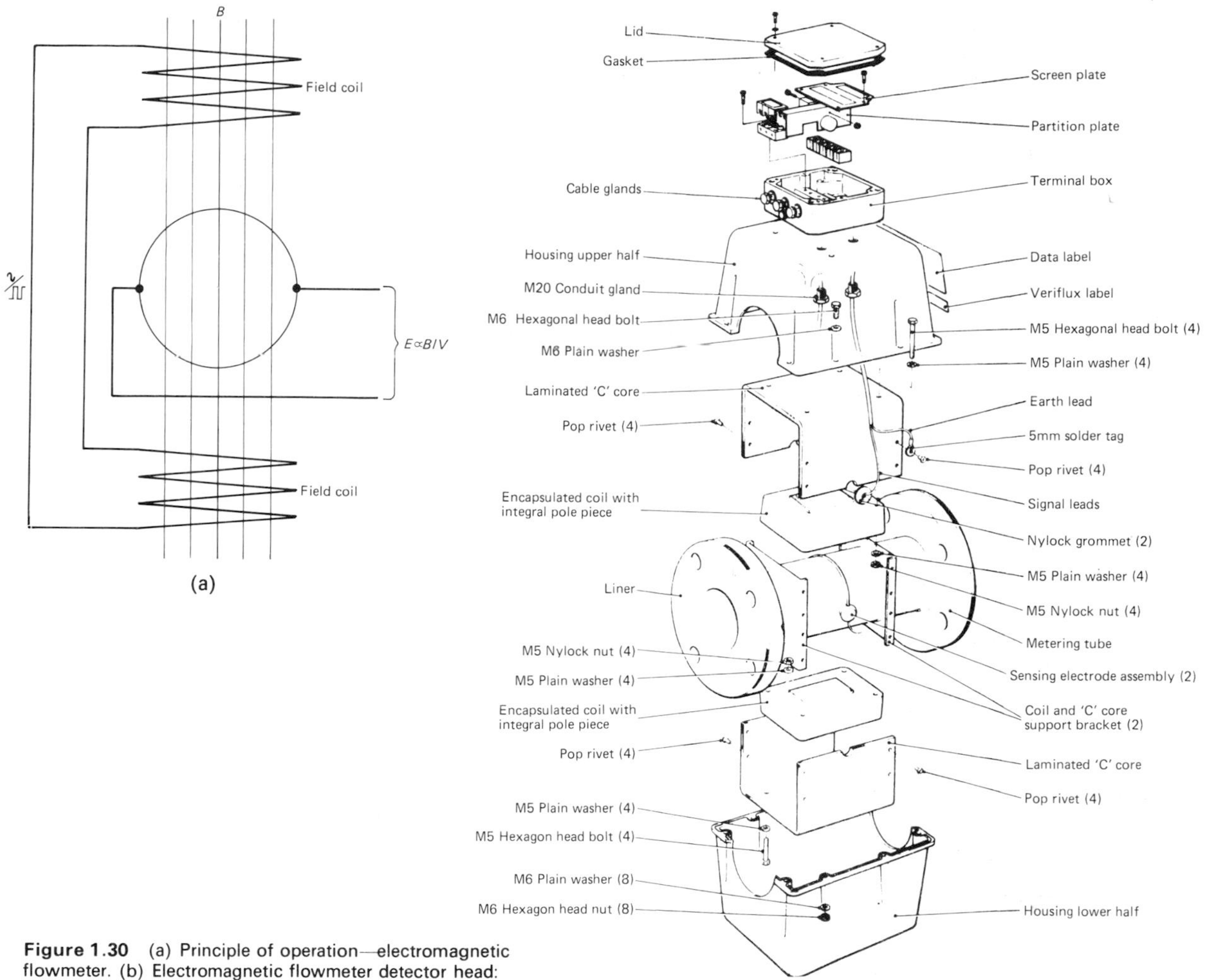

Figure 1.30 (a) Principle of operation—electromagnetic flowmeter. (b) Electromagnetic flowmeter detector head: exploded view.

have all been employed initially at frequencies around 50 Hz but most manufacturers have now opted for low-frequency systems (2–7 Hz) offering the benefits of minimum power consumption (i.e. only 20 per cent of the power used by a comparative 50 Hz system), automatic compensation for interfering voltages, automatic zero adjustment and tolerance of light build-up of material on electrode surfaces.

An example of this type of technique is illustrated in Figure 1.31 where square-wave excitation is used. The d.c. supply to the coils is switched on and off at approximately 2.6 Hz with polarity reversal every cycle. Figure 1.31(a) shows the ideal current waveform for pulsed d.c. excitation but, because of the inductance of the coils, this waveform cannot be entirely achieved. The solution as shown in Figure 1.31(b) is to power the field coils from a constant-current source giving a near square-wave excitation. The signal produced at the measuring electrodes is shown in Figure 1.31(c). The signal is sampled at five points during each measurement cycle as shown, microprocessor techniques being utilized to evaluate and separate the true flow signal from the combined flow and zero signals as shown in the equation in Figure 1.31(c).

Area of application Electromagnetic flowmeters are suitable for measuring a wide variety of liquids such as dirty liquids, pastes, acids, slurries and alkalis; accuracy is largely unaffected by changes in temperature, pressure, viscosity, density or conductivity. Although in the case of the latter conductivities must be greater than 1 microhm/cm.

Installation The primary element can be mounted in any attitude in the pipework although care should be taken to ensure that when the flowmeter is mounted

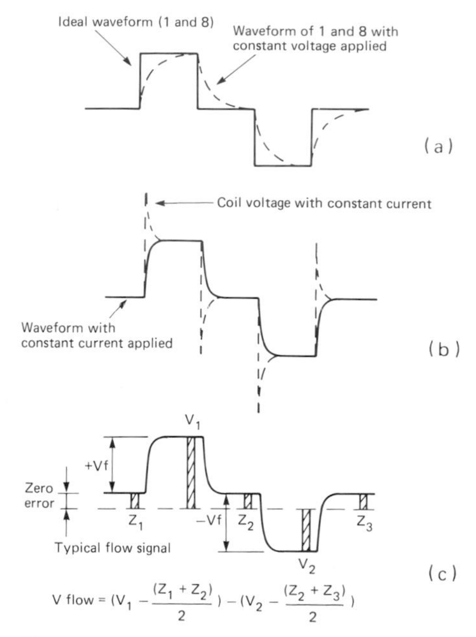

Figure 1.31 Electromagnetic flowmeter—pulsed d.c. excitation. Courtesy, Flowmetering Instruments Ltd.

horizontally, the axis of the electrodes be in the horizontal plane.

Where build-up of deposits on the electrodes is a recurring problem there exist two alternatives for consideration:

(a) Ultrasonic cleaning of electrodes.
(b) Utilize capacitive electrodes which do not come into contact with the flowstream and therefore insulating coatings have no effect.

It should be noted that on insulated pipelines earthing rings will normally be required to ensure that the flowmeter body is at the same potential as that of the flowing liquid to prevent circulating current and interfering voltages occurring.

The accuracy of the flowmeter can be affected by flow profile and the user should allow at least 10 straight pipe-diameters upstream and 5 straight pipe-diameters downstream of the primary element to ensure optimum conditions. Also to ensure system accuracy it is essential that the primary element should remain filled with the liquid being metered at all times. Entrained gases will cause similar inaccuracy.

For further information on installation requirements the reader is referred to the relevant sections of BS 5792 1980.

Flowmeters are available in sizes from 32 mm to 1200 mm nominal bore to handle flow velocities from 0–0.5 m/s to 0–10 m/s with accuracy of ± 1 per cent over a 10:1 turn-down ratio.

1.3.4.2 Ultrasonic flowmeters

Ultrasonic flowmeters measure the velocity of a flowing medium by monitoring interaction between the flowstream and an ultrasonic sound wave transmitted into or through it. Many techniques exist, the two most commonly applied being Doppler and transmissive (time of flight). These will now be dealt with separately.

Doppler flowmeters These make use of the well-known Doppler effect which states that the frequency of sound changes if its source or reflector moves relative to the listener or monitor. The magnitude of the frequency change is an indication of the speed of the sound source or sound reflector.

In practice the Doppler flowmeter comprises a housing in which two piezo-electric crystals are potted, one being a transmitter and the other a receiver, the whole assembly being located on the pipe wall as shown in Figure 1.32. The transmitter transmits ultrasonic waves of frequency F_1 at an angle θ to the flowstream. If the flowstream contains particles, entrained gas or other discontinuities, some of the transmitted energy will be reflected back to the receiver. If the fluid is travelling at velocity V, the frequency of the reflected sound as monitored by the receiver can be shown to be F_2 such that

$$F_2 = F_1 \pm 2V.\cos\theta.\frac{F_1}{C}$$

where C is the velocity of sound in the fluid. Rearranging:

$$V = \frac{C(F_2 - F_1)}{2.F_1.\cos\theta}$$

which shows that velocity is proportional to the frequency change.

The Doppler meter is normally used as an inexpensive clamp-on flowmeter, the only operational constraints being that the flowstream must contain discontinuities of some kind (the device will not monitor clear liquids) and the pipeline must be acoustically transmissive.

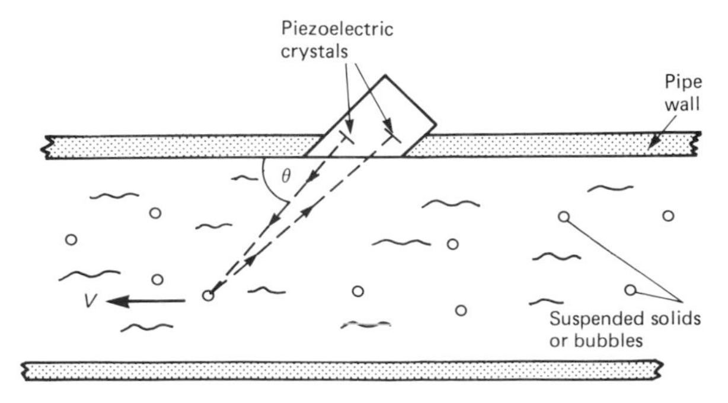

Figure 1.32 Principle of operation—Doppler meter.

Accuracy and repeatability of the Doppler meter are somewhat suspect and difficult to quantify since its operation is dependent on flow profile, particle size and suspended solids concentration. However, under ideal conditions and given the facility to calibrate *in situ* accuracies of ± 5 per cent should be attainable. This type of flowmeter is most suitable for use as a flow switch or for flow indication where absolute accuracy is not required.

Transmissive flowmeters Transmissive devices differ from Doppler flowmeters in that they rely on transmission of an ultrasonic pulse through the flowstream and therefore do not depend on discontinuities or entrained particles in the flowstream for operation.

The principle of operation is based on the transmission of an ultrasonic sound wave between two points, first in the direction of flow and then opposing flow. In each case the time of flight of the sound wave between the two points will have been modified by the velocity of the flowing medium and the difference between the flight times can be shown to be directly proportional to flow velocity.

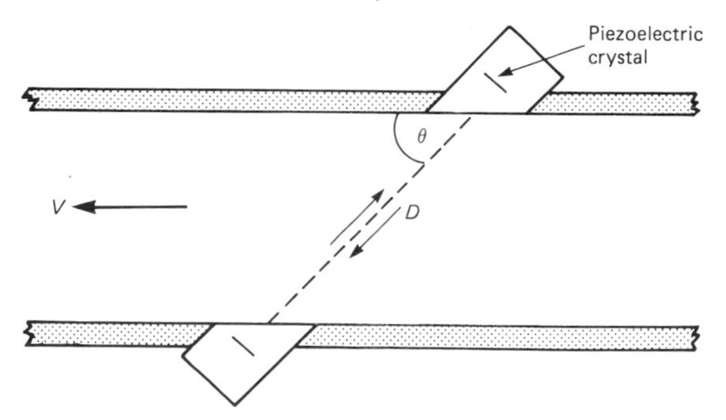

Figure 1.33 Principle of operation—time-of-flight ultrasonic flowmeter.

In practice the sound waves are not generated in the direction of flow but at an angle across it as shown in Figure 1.33. Pulse transit times downstream T_1 and upstream T_2 along a path length D can be expressed as: $T_1 = D/(C+V)$ and $T_2 = D/(C-V)$, where C is the velocity of sound in the fluid and V is the fluid velocity. Now

$$T = T_1 - T_2 = 2DV/(C^2 - V^2) \tag{1.44}$$

Since V^2 is very small compared to C^2 it can be ignored. It is convenient to develop the expression in relation to frequency and remove the dependency on the velocity of sound (C). Since $F_1 = 1/T_1$ and $F_2 = 1/T_2$ and average fluid velocity $\bar{V} = V/\cos\theta$ equation (1.44) is developed to:

$$F_1 - F_2 = (2\bar{V}\cos\theta)/D$$

The frequency difference is calculated by an electronic converter which gives an analogue output proportional to average fluid velocity. A practical realization of this technique operates in the following manner.

A voltage-controlled oscillator generates electronic pulses from which two consecutive pulses are selected. The first of these is used to operate a piezo-electric ceramic crystal transducer which projects an ultrasonic beam across the liquid flowing in a pipe. This ultrasonic pulse is then received on the other side of the pipe, where it is converted back to an electronic pulse. The latter is then received by the 'first-arrival' electronics, comparing its arrival time with the second pulse received directly. If the two pulses are received at the same time, the period of time between them equates to the time taken for the first pulse to travel to its transducer and be converted to ultra-sound, to travel across the flowstream, to be reconverted back to an electronic pulse and travel back to the first-arrival position.

Should the second pulse arrive before the first one, then the time between pulses is too short. Then the first-arrival electronics will step down the voltage to the voltage-controlled oscillator (VCO), reducing the resulting frequency. The electronics will continue to reduce voltage to the VCO in steps, until the first and second pulses are received at the first-arrival electronics at the same time. At this point, the periodic time of the frequency will be the same as the ultrasonic flight time, plus the electronic delay time.

If, now, a similar electronic circuit is used to project an ultrasonic pulse in the opposite direction to that shown, another frequency will be obtained which, when subtracted from the first, will give a direct measure of the velocity of the fluid in the pipe, since the electronic delays will cancel out.

In practice, the piezo-electric ceramic transducers used act as both transmitters and receivers of the ultrasonic signals and thus only one is required on each side of the pipe.

Typically the flowmeter will consist of a flowtube containing a pair of externally mounted, ultrasonic transducers and a separate electronic converter/transmitter as shown in Figure 1.34(a). Transducers may be wetted or non-wetted and consist of a piezo-electric crystal sized to give the desired frequency (typically 1–5 MHz for liquids and 0.2–0.5 MHz for gases). Figure 1.34(b) shows a typical transducer assembly.

Due to the fact that the flowmeter measures velocity across the centre of the pipe it is susceptible to flow profile effects and care should be taken to ensure sufficient length of straight pipe upstream and downstream of the flowtube to minimize such effects. To overcome this problem, some manufacturers use multiple-beam techniques where several chordal velocities are measured and the average computed. However, it is still good practice to allow for approximately 10 upstream and 5 downstream

(a)

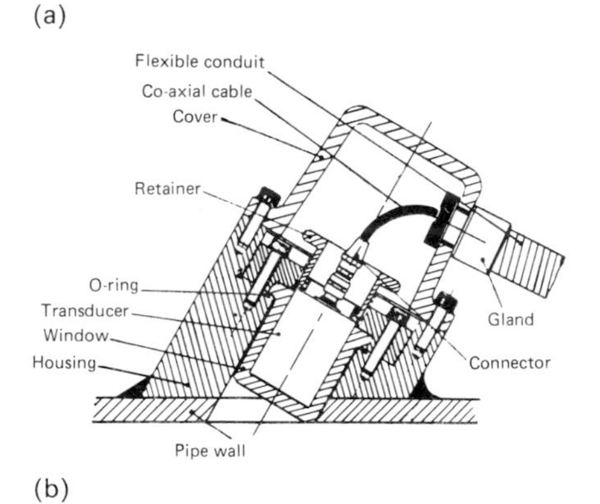

(b)

Figure 1.34 (a) Ultrasonic flowmeter. Courtesy, Bestobell Sparling Ltd. (b) Transducer assembly.

diameters of straight pipe. Also since this type of flowmeter relies on transmission through the flowing medium, fluids with a high solids or gas-bubble content cannot be metered.

This type of flowmeter can be obtained for use on liquids or gases for pipe sizes from 75 mm nominal bore up to 1500 mm or more for special applications and it is bi-directional in operation. Accuracy of better than ± 1 per cent of flow rate can be achieved over a flow range of 0.2 to 12 metres per second.

This technique has also been successfully applied to open channel and river flow and is also now readily available as a clamp-on flowmeter for closed pipes, but accuracy is dependent on knowledge of each installation and *in situ* calibration is desirable.

1.3.4.3 Oscillatory flowmeters

The operating principle of flowmeters in this category is based on the fact that if an obstruction of known geometry is placed in the flowstream the fluid will start to oscillate in predictable manner. The degree of oscillation is related to fluid flow rate. The two main types of flowmeter in this category are: vortex-shedding flowmeter and swirl flowmeter.

The vortex flowmeter This type of flowmeter operates on the principle that if a bluff (i.e. non-streamlined) body is placed in a flowstream vortices will be detached or shed from the body. The principle is illustrated in Figure 1.35.

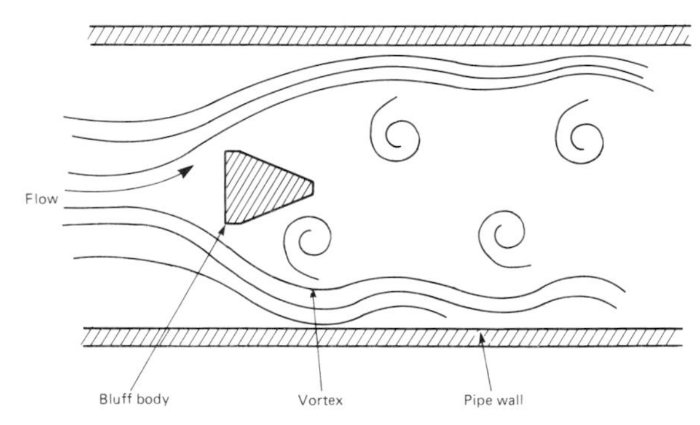

Figure 1.35 Vortex shedding.

The vortices are shed alternately to each side of the bluff body, the rate of shedding being directly proportional to flow velocity. If this body is fitted centrally into a pipeline the vortex-shedding frequency is a measure of the flow rate.

Any bluff body can be used to generate vortices in a flowstream, but for these vortices to be regular and well defined requires careful design. Essentially, the body must be non-streamlined, symmetrical and capable of generating vortices for a wide Reynolds number range. The most commonly adopted bluff body designs are shown in Figure 1.36.

These designs all attempt to enhance the vortex-shedding effect to ensure regularity or simplify the detection technique. If the design (d) is considered it will be noted that a second non-streamlined body is placed just downstream of the vortex-shedding body. Its effect is to reinforce and stabilize the shedding. The width of the bluff body is determined by pipe size and a rule-of-thumb guide is that the ratio of body width to pipe diameter should not be less than 0.2.

Sensing methods Once the bluff-body type has been selected we must adopt a technique to detect the vortices. Various methods exist, the more popular techniques being as follows:

(a) Ultrasonic. Where the vortices pass through an ultrasonic beam and cause refraction of this beam resulting in modulation of the beam amplitude.

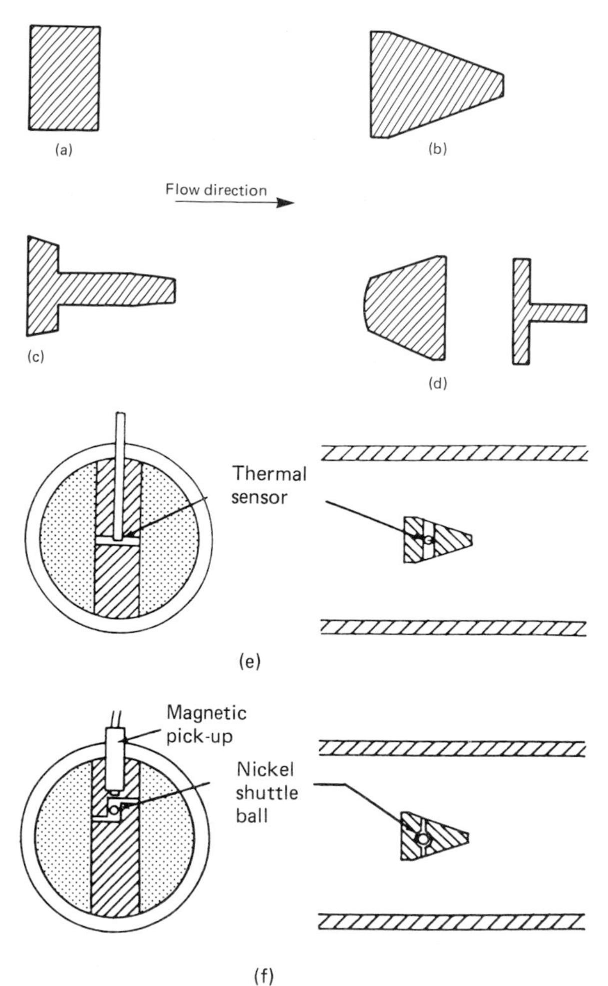

Figure 1.36 (a)–(d) Bluff body shapes. (e) Thermal sensor, Courtesy, Neptune Measurement Ltd. (f) Shuttle ball sensor. Courtesy, Neptune Measurement Ltd.

(b) Thermal (Figure 1.36(e)). Where a thermistor-type sensor is located in a through passage across the bluff body and behind its face. The heated thermistor will sense alternating vortices due to the cooling effect caused by their passage and an electrical pulse output is obtained.
(c) Oscillating disc. Sensing ports on both sides of the flow element cause a small disc to oscillate. A variable-reluctance pick-up detects the disc's oscillation. This type is particularly suited to steam or wet-gas flow.
(d) Capacitance. Metal diaphragms are welded on opposite sides of the bluff body, the small gaps between the diaphragms and the body being filled with oil. Interconnecting ports allow transfer of oil between the two sides. An electrode is placed close to each plate and the oil used as a dielectric. The vortices alternately deform the diaphragm plates causing a capacitance change between the diaphragm and electrode. The frequency of changes in capacitance is equal to the shedding frequency.
(e) Strain. Here the bluff body is designed such that the alternating pressures associated with vortex shedding are applied to a cantilevered section to the rear of the body. The alternating vortices create a cyclic strain on the rear of the body which is monitored by an internal strain gauge.
(f) Shuttle ball (Figure 1.36(f)). The shuttle technique uses the alternating pressures caused by vortex shedding to drive a magnetic shuttle up and down the axis of a flow element. The motion of the shuttle is detected by a magnetic pick-up.

The output derived from the primary sensor is a low-frequency signal dependent on flow; this is then applied to conditioning electronics to provide either analogue or digital output for display and transmission. The calibration factor (pulses per m^3) for the vortex meter is determined by the dimensions and geometry of the bluff body and will not change.

Installation parameters for vortex flowmeters are quite critical. Pipe flange gaskets upstream and at the transmitter should not protrude into the flow and to ensure a uniform velocity profile there should be 20 diameters of straight pipe upstream and 5 diameters downstream. Flow-straighteners can be used to reduce this requirement if necessary.

The vortex flowmeter has wide-ranging applications in both gas and liquid measurement providing the Reynolds number lies between 2×10^3 and 1×10^5 for gases and 4×10^3 and 1.4×10^5 for liquids. The output of the meter is independent of the density, temperature and pressure of the flowing fluid and represents the flow rate to better than ± 1 per cent of full scale giving turn-down ratios in excess of 20:1.

The swirlmeter Another meter that depends on the oscillatory nature of fluids is the swirlmeter shown in Figure 1.37. A swirl is imparted to the body of flowing fluid by the curved inlet blades which give a tangential component to the fluid flow. Initially the axis of the fluid rotation is the centre line of the meter, but a change in the direction of the rotational axis (precession) takes place when the rotating liquid enters the enlargement, causing the region of highest velocity to rotate about the meter axis. This produces an oscillation or precession, the frequency of which is proportional to the volumetric flow rate. The sensor, which is a bead thermistor heated by a constant-current source, converts the instantaneous velocity changes into a proportional electrical pulse output. The number of pulses generated is directly proportional to the volumetric flow.

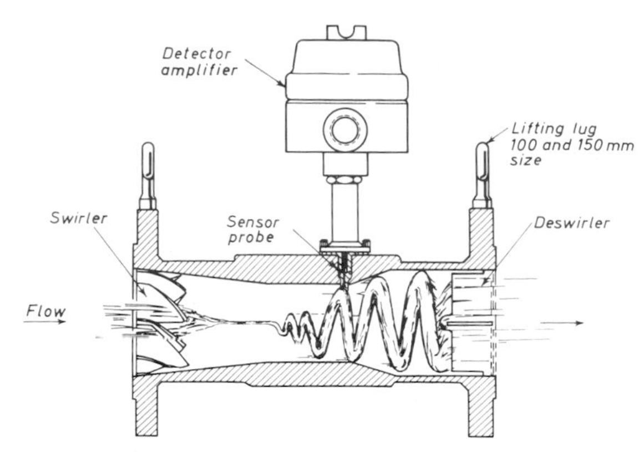

Figure 1.37 Cutaway view of the Swirlmeter. Courtesy, Fischer & Porter Ltd.

The operating range of the swirlmeter depends upon the specific application, but typical for liquids are 3.5 to 4.0 litres per minute for the 25 mm size to 1700 to 13 000 litres per minute for the 300 mm size. Typical gas flow ranges are 3 to 35 m^3/h for the 25 mm size to 300 to 9000 m^3/h for the 300 mm size. Accuracy of ± 1 per cent of rate is possible with repeatability of ± 0.25 per cent of rate.

1.3.4.4 Cross-correlation

In most flowing fluids there exist naturally occurring random fluctuations such as density, turbulence and temperature which can be detected by suitably located transducers. If two such transducers are installed in a pipeline separated by a distance L as shown in Figure 1.38, the upstream transducer will pick up a random fluctuation t seconds before the downstream transducer and the distance between the transducers divided by the transit time t will yield flow velocity. In practice the random fluctuations will not be stable and are compared in a cross-correlator which has a peak response at transit time T_p and correlation velocity $V = L/T_p$ metres per second.

This is effectively a non-intrusive measurement and could in principle be developed to measure flow of most fluids. Very few commercial cross-correlation systems are in use for flow measurement because of the slow response time of such systems. However with the use of microprocessor techniques processing speed has been increased significantly and several manufacturers are now producing commercial systems for industrial use. Techniques for effecting the cross-correlation operation are discussed in Part 4.

1.3.5 Mass flowmeters

The measurement of mass flow rate can have certain advantages over volume flow rate, i.e. pressure, temperature and specific gravity do not have to be considered. The main interfering parameter to be avoided is that of two-phase flow where gas/liquid, gas/solid or liquid/solid mixtures are flowing together in the same pipe. The two phases may be travelling at different velocities and even in different directions. This problem is beyond the scope of this book but the user should be aware of the problem and ensure where possible that the flow is as near homogeneous as possible (by pipe-sizing or meter-positioning) or that the two phases are separately metered.

Methods of measurement can be categorized under two main headings: true mass-flow measurement in which the measured parameter is directly related to mass flow rate, and inferential mass-flow measurement in which volume flow rate and fluid density are measured and combined to give mass flow rate. Since volume flow rate and density measurement are discussed elsewhere only true mass-flow measurement will be dealt with here.

1.3.5.1 True mass-flow measurement methods

Fluid-momentum methods (a) Angular momentum. This type of device consists of two turbines on separate axial shafts in the meter body. The upstream turbine is rotated at constant speed and imparts a swirling motion to the fluid passing through it. On reaching the downstream turbine, the swirling fluid attempts to impart motion onto it; however, this turbine is constrained from rotating by a calibrated spring. The meter is designed such that on leaving the downstream turbine all angular velocity will have been removed from the fluid and the torque produced on it is proportional to mass flow.

This type of device can be used for both gases and liquids with accuracies of ± 1 per cent.

(b) Gyroscopic/Coriolis mass flowmeter. Mass flowmeters in this category use the measurement of torque developed when subjecting the fluid stream to a Coriolis acceleration,* as a measure of mass flow rate.

* On a rotating surface there is an inertial force acting on a body at right angles to its direction of motion in addition to the ordinary effects of motion of the body. This force is known as a Coriolis force.

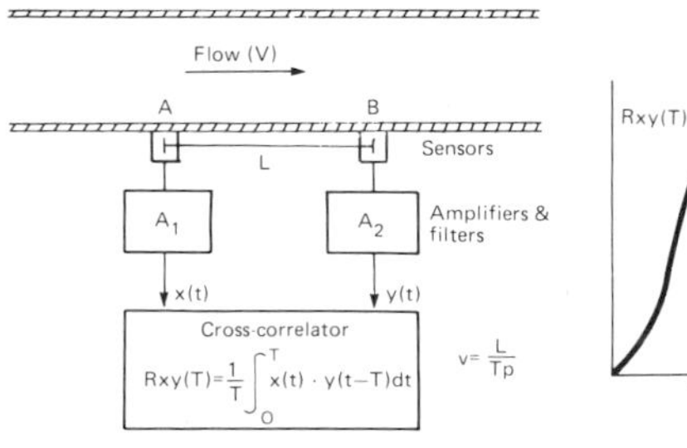

Figure 1.38 Cross correlation meter.

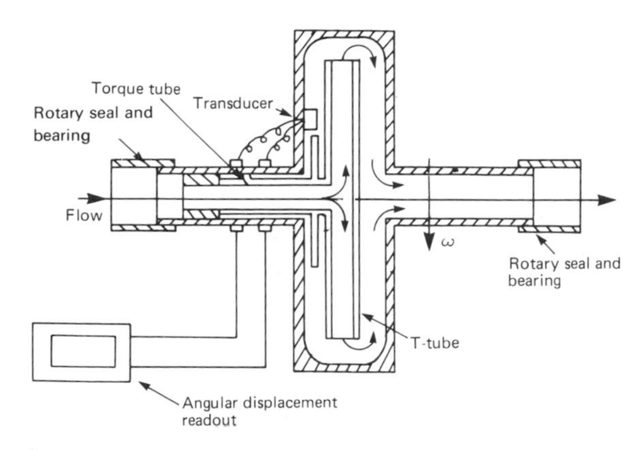

Figure 1.39 Early form of Coriolis mass flowmeter.

An early application of this technique is illustrated in Figure 1.39.

The fluid enters a T-shaped tube, flow being equally divided down each side of the T, and then recombines into a main flowstream at the outlet from the meter. The whole assembly is rotated at constant speed causing an angular displacement of the T-tube which is attached to the meter casing through a torque tube. The torque produced is proportional to mass flow rate.

This design suffered from various problems mainly due to poor sealing of rotating joints or inadequate speed control. However recent developments have overcome these problems as shown in Figure 1.40.

The mass flowmeter consists of a U-tube and a T-shaped leaf spring as opposite legs of a tuning fork. An electromagnet is used to excite the tuning fork, thereby subjecting each particle within the pipe to a Coriolis-type acceleration. The resulting forces cause an angular deflection in the U-tube inversely proportional to the stiffness of the pipe and proportional to the mass flow rate. This movement is picked up by optical transducers mounted on opposite sides of the U-tube, the output being a pulse that is width-modulated proportional to mass flow rate. An oscillator/counter digitizes the pulse width and provides an output suitable for display purposes.

This system can be used to measure the flow of liquids or gases and accuracies better than ± 0.5 per cent of full scale are possible.

Pressure-differential methods In its classical form the meter consists of four matched orifice plates installed in a Wheatstone bridge arrangement. A pump is used to transfer fluid at a known rate from one branch of the bridge into another to create a reference flow. The resultant differential pressure measured across the bridge is proportional to mass flow rate.

Thermal mass flowmeter This version of a mass flowmeter consists of a flowtube, an upstream and downstream temperature sensor and a heat source as illustrated in Figure 1.41. The temperature sensors are effectively active arms of a Wheatstone bridge. They are mounted equidistant from the constant-temperature heat source such that for no flow conditions, heat received by each sensor is the same and the bridge remains in balance. However, with increasing flow, the downstream sensor receives progressively more heat than the upstream sensor causing an imbalance to occur in the bridge circuit. The temperature difference is proportional to mass flow rate and an electrical output representing this is developed by the bridge circuit.

This type of mass flowmeter is most commonly applied to the measurement of gas flows within the ranges 2.5×10^{-10} to 5×10^{-3} kg/s and accuracy of ± 1 per cent of full scale is attainable.

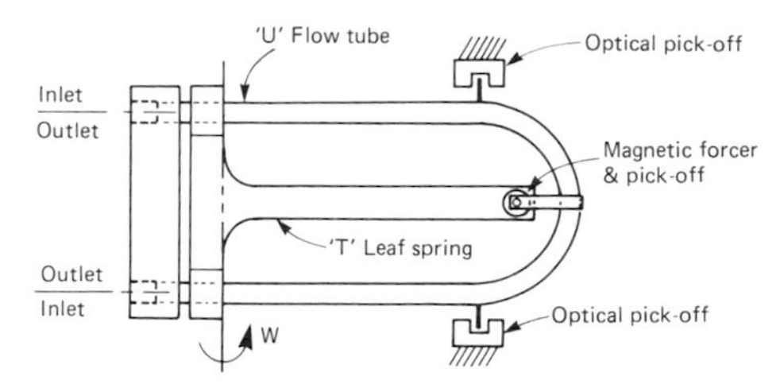

Figure 1.40 Gyroscopic/Coriolis mass flowmeter.

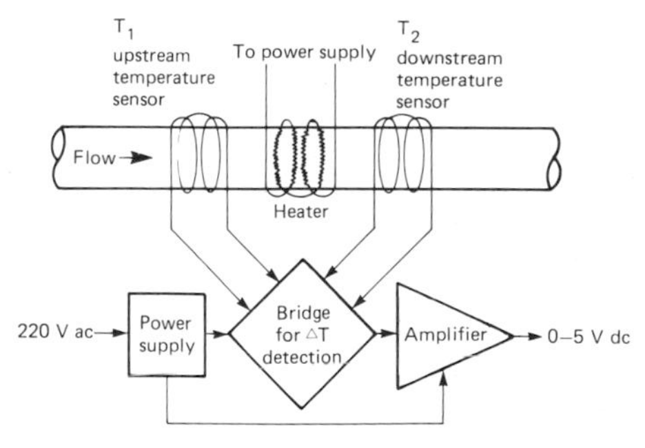

Figure 1.41 Thermal mass flowmeter. Courtesy, Emerson Electric Co.

1.4 Flow in open channels

Flow measurement in open channels is a requirement normally associated with the water industry. Flow in rivers, sewers (part-filled pipes) and regular-shaped channels may be measured by the following methods:

(a) Head/area method. Where a structure is built into the flowstream to develop a unique head/flow relationship as in,
   (i) The weir which is merely a dam over which liquid is allowed to flow, the depth of liquid

over the sill of the weir being a measure of the rate of flow.

(ii) The hydraulic flume, an example being the venturi flume in which the channel is given the same form in the horizontal plane as a section of a venturi tube while the bottom of the channel is given a gentle slope up the throat.

(b) Velocity/area method. Where measurement of both variables, i.e. head and velocity, is combined with the known geometry of a structure to determine flow.

(c) Dilution gauging.

1.4.1 Head/area method

1.4.1.1 Weirs

Weirs may have a variety of forms and are classified according to the shape of the notch or opening.

The simplest is the rectangular notch, or in certain cases the square notch.

The V or triangular notch is a V-shaped notch with the apex downwards. It is used to measure rates of flow which may become very small. Owing to the shape of the notch the head is greater at small rates of flow with this type than it would be for the rectangular notch.

Notches of other forms, which may be trapezoidal or parabolic, are designed so that they have a constant discharge coefficient, or a head that is directly proportional to the rate of flow.

The velocity of the liquid increases as it passes over the weir because the centre of gravity of the liquid falls. Liquid that was originally at the level of the surface above the weir can be regarded as having fallen to the level of the centre of pressure of the issuing stream. The head of liquid producing the flow is therefore equal to the vertical distance from the centre of pressure of the issuing stream to the level of the surface of the liquid upstream.

If the height of the centre of pressure above the sill can be regarded as being a constant fraction of the height of the surface of the liquid above the sill of the weir, then the height of the surface above the sill will give a measure of the differential pressure producing the flow. If single particles are considered, some will have fallen a distance greater than the average but this is compensated for by the fact that others have fallen a smaller distance.

The term 'head of a weir' is usually taken to mean the same as the depth of the weir, and is measured by the height of the liquid above the level of the sill of the weir just upstream of where it begins to curve over the weir, and is denoted by H and usually expressed in metres.

Rectangular notch Consider the flow over the weir in exactly the same way as the flow through other primary differential-pressure elements. If the cross-section of the stream approaching the weir is large in comparison with the area of the stream over the weir, then the velocity V_1 at section 1 upstream can be neglected in comparison with the velocity V_2 over the weir, and in equation (1.16) $V_1=0$ and the equation becomes:

$$V_2^2=2gh \qquad \text{or} \qquad V_2=\sqrt{(2gh)}$$

The quantity of liquid flowing over the weir will be given by:

$$Q=A_2V_2$$

But the area of the stream is BH, where H is the depth over the weir and B the breadth of the weir, and h is a definite fraction of H.

By calculus it can be shown that for a rectangular notch

$$Q=\tfrac{2}{3}BH\sqrt{(2gH)} \qquad (1.45)$$

$$=\tfrac{2}{3}B\sqrt{(2gH^3)}\ \text{m}^3/\text{s} \qquad (1.46)$$

The actual flow over the weir is less than that given by equation (1.45) for the following reasons:

(a) The area of the stream is not BH but something less, for the stream contracts at both the top and bottom as it flows over the weir as shown in Figure 1.42 making the effective depth at the weir less than H.

(b) Owing to friction between the liquid and the sides of the channel, the velocity at the sides of the channel will be less than that at the middle. This effect may be reduced by making the notch narrower than the width of the stream as shown in Figure 1.43. This, however, produces side-contraction of the stream. Therefore $B_1=B$ should be at least equal to $4H$ when the side-contraction is equal to $0.1H$ on both sides, so that the effective width becomes $B-0.2H$.

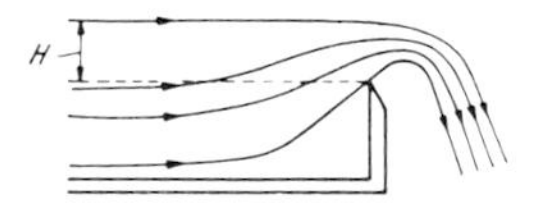

Figure 1.42 Rectangular notch, showing top and bottom of contraction.

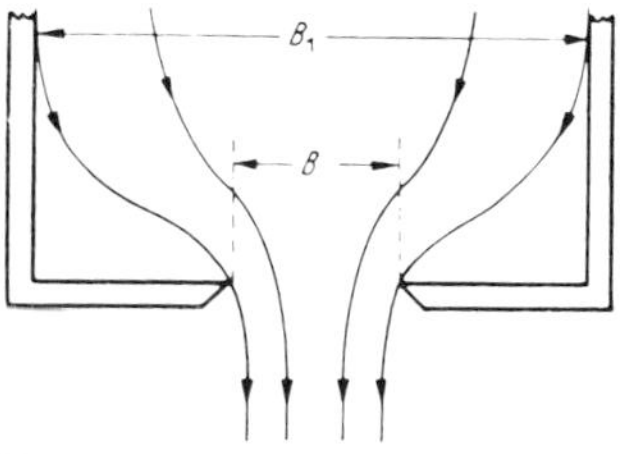

Figure 1.43 Rectangular notch, showing side-contraction.

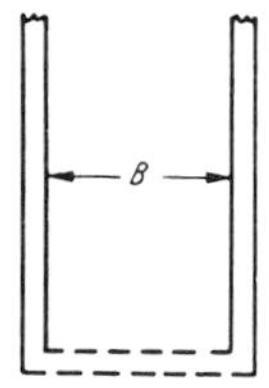

Figure 1.44 Rectangular notch, showing side plates.

When it is required to suppress side-contraction and make the measurement more reliable, plates may be fitted as shown in Figure 1.44 so as to make the stream move parallel to the plates as it approaches the weir.

To allow for the difference between the actual rate of flow and the theoretical rate of flow, the discharge coefficient C, defined as before, is introduced and equation (1.46) becomes:

$$Q = \tfrac{2}{3} CB\sqrt{(2gH^3)}\ \mathrm{m^3/s} \tag{1.47}$$

The value of C will vary with H and will be influenced by the following factors, which must remain constant in any installation if its accuracy is to be maintained: (a) the relative sharpness of the upstream edge of the weir crest, (b) the width of the weir sill. Both these factors influence the bottom-contraction and influence C, so the weir sill should be inspected from time to time to see that it is free from damage.

In developing the above equations it was assumed that the velocity of the liquid upstream of the weir could be neglected. As the rate of flow increases, this is no longer possible and a velocity of approach factor must be introduced. This will influence the value of C, and as the velocity of approach increases it will cause the observed head to become less than the true or total head so that a correcting factor must be introduced.

Triangular notch If the angle of the triangular notch is θ as shown in Figure 1.45, $B = 2H\tan(\theta/2)$. The position of the centre of pressure of the issuing stream will now be at a different height above the bottom of the notch from what it was for the rectangular notch. It can be shown by calculus that the numerical factor involved in the equation is now $\frac{4}{15}$. Substituting this factor and the new value of A_2 in equation (1.47):

$$Q = \frac{4}{15} CB\sqrt{(2gH^3)}\ \mathrm{m^3/s}$$

$$= \frac{4}{15} C2H \tan\frac{\theta}{2} \sqrt{(2gH^3)}$$

$$= \frac{8}{15} C \tan\frac{\theta}{2} \sqrt{(2gH^5)} \tag{1.48}$$

Experiments have shown that θ should have a value between 35° and 120° for satisfactory operation of this type of installation.

While the cross-section of the stream from a triangular weir remains geometrically similar for all values of H, the value of C is influenced by H. The variation of C is from 0.57 to 0.64, and takes into account the contraction of the stream.

If the velocity of approach is not negligible the value of H must be suitably corrected as in the case of the rectangular weir.

Installation and operation of weirs

(a) Upstream of a weir there should be a wide, deep, and straight channel of uniform cross-section, long enough to ensure that the velocity distribution in the stream is uniform. This approach channel may be made shorter if baffle plates are placed across it at the inlet end to break up currents in the stream.
(b) Where debris is likely to be brought down by the stream, a screen should be placed across the approach channel to prevent the debris reaching the weir. This screen should be cleaned as often as necessary.
(c) The upstream edge of the notch should be maintained square or sharp-edged according to the type of installation.
(d) The weir crest should be level from end to end.
(e) The channel end wall on which the notch plate is mounted should be cut away so that the stream may fall freely and not adhere to the wall. To ensure this happens a vent may be arranged in the side wall of the channel so that the space under the falling water is open to the atmosphere.
(f) Neither the bed, nor the sides of the channel downstream from the weir should be nearer the weir than 150 mm and the water level downstream should be at least 75 mm below the weir sill.
(g) The head H may be measured by measuring the height of the level of the stream above the level of the weir sill, sufficiently far back from the weir to ensure the surface is unaffected by the flow. This measurement is usually made at a distance of at least $6H$ upstream of the weir. It may be made by any appropriate method for liquids as described in the section on level measurement: for example, the hook gauge, float-operated mechanisms, air purge systems or ultrasonic techniques. It is often more convenient to measure the level of the liquid in a 'stilling well' alongside the channel at the appropriate distance above the notch. This well is

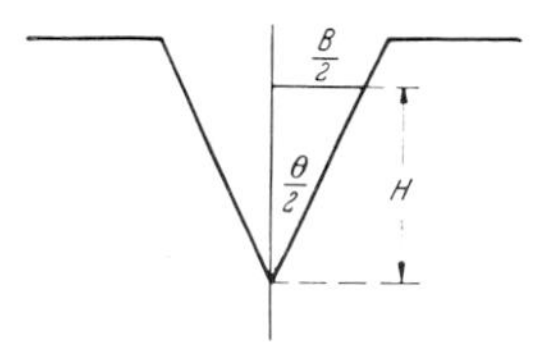

Figure 1.45 Triangular notch (V-notch).

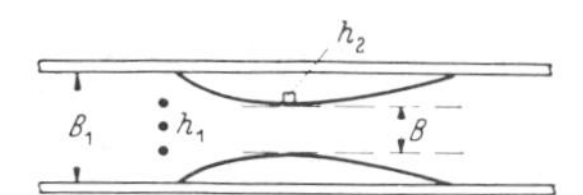

Figure 1.46 Hydraulic flume (venturi type).

connected to the weir chamber by a small pipe or opening near the bottom. Liquid will rise in the well to the same height as in the weir chamber and will be practically undisturbed by currents in the stream.

1.4.1.2 Hydraulic flumes

Where the rate of fall of a stream is so slight that there is very little head available for operating a measuring device or where the stream carries a large quantity of silt or debris, a flume is often much more satisfactory than a weir. Several flumes have been designed, but the only one we shall consider here is the venturi flume. This may have more than one form, but where it is flat-bottomed and of the form shown in Figure 1.46 the volume rate of flow is given by the equation

$$Q = CBh_2 \sqrt{\frac{2g(h_1 - h_2)}{1-(Bh_2/B_1h_1)^2}} \text{ m}^3/\text{s} \qquad (1.49)$$

where B_1 is width of channel, B is width of the throat, h_1 is depth of water measured immediately upstream of the entrance to the converging section, and h_2 is minimum depth of water in the throat. C is discharge coefficient whose value will depend upon the particular outline of the channel and the pattern of the flow. Tests on a model of the flume may be used to determine the coefficient provided that the flow in the model and in the full-sized flume are dynamically similar.

The depths of water h_1 and h_2 are measured as in the case of the weir by measuring the level in wells at the side of the main channel. These wells are connected to the channel by small pipes opening into the channel near or at the bottom.

As in the case of the closed venturi tube a certain minimum uninterrupted length of channel is required before the venturi is reached, in order that the stream may be free from waves and vortices.

By carefully designing the flume, it is possible to simplify the actual instrument required to indicate the flow. If the channel is designed in such a manner that the depth in the exit channel at all rates of flow is less than a certain percentage of the depth in the entrance channel, the flume will function as a free-discharge outlet. Under these conditions, the upstream depth is independent of the downstream conditions, and the depth of water in the throat will maintain itself at a certain critical value, at which the energy of the water is at the minimum whatever the rate of flow. When this is so, the quantity of water flowing through the channel is a function of the upstream depth h_1 only, and may be expressed by the equation:

$$Q = kh_1^{3/2}$$

where k is a constant for a particular installation and can be determined.

It is now necessary to measure h_1 only, and this may be done by means of a float in a well, connected to the upstream portion of the channel. This float operates an indicated recording and integrating instrument.

The channel is usually constructed of concrete, the surface on the inside of the channel being made smooth to reduce the friction between water and channel. Flumes of this kind are used largely for measuring flow of water or sewerage and may be made in a very large variety of sizes to measure anything from the flow of a small stream to that of a large river.

1.4.2 Velocity/area methods

In these methods volume flow rate is determined by measurement of the two variables concerned (mean velocity and head), since the rate of flow is given by the equation

$$Q = V \,.\, A \text{ m}^3/\text{s}$$

where area A is proportional to head or level.

The head/level measurement can be made by many of the conventional level devices described in Chapter 5 and will not therefore be dealt with here. Three general techniques are used for velocity measurement, these being turbine current meter, electromagnetic, and ultrasonic. The techniques have already been discussed in the section on closed pipe flow and application only will be described here.

1.4.2.1 Turbine current meter

In a current-meter gauging, the meter is used to give point velocity. The meter is sited in a predetermined cross-section in the flowstream and the velocity obtained. Since the meter only measures point velocity it is necessary to sample throughout the cross-section to obtain mean velocity.

The velocities that can be measured in this way range from 0.03 to 3.0 m/s for a turbine meter with a propeller of 50 mm diameter. The disadvantage of a current-meter gauging is that it is a point and not a continuous measurement of discharge.

1.4.2.2 Electromagnetic method

In this technique Faraday's law of electromagnetic induction is utilized in the same way as for closed-pipe flow measurement (Section 1.3.4.1). That is, $E \propto BlV$, where E is e.m.f. generated, B is magnetic field strength,

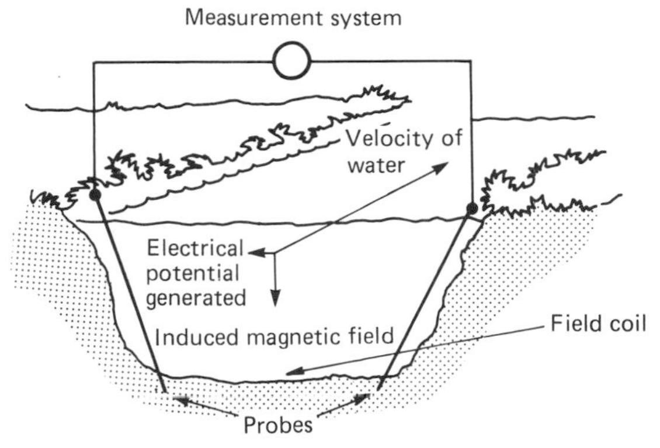

Figure 1.47 Principle of electromagnetic gauge. Courtesy, Plessey Electronic Systems Ltd.

l is width of river or channel in metres, and V is average velocity of the flowstream.

This equation only applies if the bed of the channel is insulated, similar to the requirement for pipe flowmeters. In practice it is costly to insulate a riverbed and where this cannot be done, riverbed conductivity has to be measured to compensate for the resultant signal attenuation.

In an operational system a large coil buried under the channel is used to produce a vertical magnetic field. The flow of water through the magnetic field causes an e.m.f. to be set up between the banks of the river. This potential is sensed by a pick-up electrode at each bank. This is shown diagrammatically in Figure 1.47.

1.4.2.3 Ultrasonic method

As for closed-pipe flow two techniques are available, single-path and multi-path, both relying on time-of-flight techniques as described in Section 1.3.4.2. Transducers capable of transmitting and receiving acoustic pulses are staggered along either bank of the river or channel. In practice the acoustic path is approximately 60° to the direction of flow, but angles between 30° and 60° could be utilized. The smaller the angle the longer the acoustic path. Path lengths up to 400 metres can be achieved.

1.4.3 Dilution gauging

This technique is covered in detail in the section on flow calibration but basically the principle involves injecting a tracer element such as brine, salt or radioactive solution and estimating the degree of dilution caused by the flowing liquid.

1.5 Point velocity measurement

It is often desirable in flow studies and survey work to be able to measure the velocity of liquids at points within the flow pattern inside both pipes and open channels to determine either mean velocity or flow profile. The following techniques are most common: laser Doppler anemometer, hot-wire anemometer, pitot tube, insertion electromagnetic, insertion turbine, propeller-type current meter, insertion vortex, and Doppler velocity probe.

1.5.1 Laser Doppler anemometer

This uses the Doppler shift of light scattered by moving particles in the flowstream to determine particle velocity and hence fluid flow velocity. It can bc uscd for both gas and liquid flow studies and is used in both research and industrial applications.

Laser Doppler is a non-contact technique and is particularly suited to velocity studies in systems that would not allow the installation of a more conventional system; for example, around propellers and in turbines.

1.5.2 Hot-wire anemometer

The hot-wire anemometer is widely used for flow studies in both gas and liquid systems. Its principle of operation is that a small electrically heated element is placed within the flowstream; the wire sensor is typically 5 μm diamerer and approximately 5 mm long. As flow velocity increases it tends to cool the heated element. This change in temperature causes a change in resistance of the element proportional to flow velocity.

1.5.3 Pitot tube

The pitot tube is a device for measuring the total pressure in a flowstream (i.e. impact/velocity pressure and static pressure) and the principle of operation is as follows.

If a tube is placed with its open end facing into the flowstream (Figure 1.48) then the fluid impinging on the open end will be brought to rest and its kinetic energy converted into pressure energy. The pressure build-up in the tube will be greater than that in the free

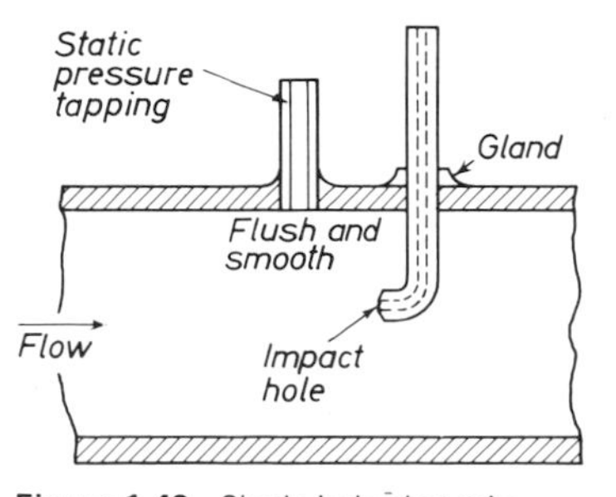

Figure 1.48 Single hole pitot tube.

stream by an amount termed the 'impact pressure'. If the static pressure is also measured, the differential pressure between that measured by the pitot tube and the static pressure will be a measure of the impact pressure and therefore the velocity of the stream. In equation (1.15) h the pressure differential or impact pressure developed is given by $h=(V_2^2/2g)-(V_1^2/2g)$ where $V_2=0$. Therefore, $h=-V_1^2/2g$, i.e. the pressure increases by $V_1^2/2g$. The negative sign indicates that it is an increase in pressure and not a decrease.

Increase in head:

$$h=V_1^2/2g \quad \text{or} \quad V_1^2=2gh \quad \text{i.e.} \quad V_1=\sqrt{(2gh)} \tag{1.50}$$

However, since this is an intrusive device not all of the flowstream will be brought to rest on the impact post; some will be deflected round it. A coefficient C is introduced to compensate for this and equation (1.50) becomes:

$$V_1=C\sqrt{(2gh)} \tag{1.51}$$

If the pitot tube is to be used as a permanent device for measuring the flow in a pipeline the relationship between the velocity at the point of its location to the mean velocity must be determined. This is achieved by traversing the pipe and sampling velocity at several points in the pipe, thereby determining flow profile and mean velocity.

For more permanent types of pitot-tube installation an Annubar may be used as shown in Figure 1.49. The pressure holes are located in such a way that they measure the representative dynamic pressure of equal annuli. The dynamic pressure obtained at the four holes facing into the stream is then averaged by means of the 'interpolating' inner tube (Figure 1.49(b)), which is connected to the high-pressure side of the manometer.

The low-pressure side of the manometer is connected to the downstream element which measures the static pressure less the suction pressure. In this way a differential pressure representing the mean velocity along the tube is obtained enabling the flow to be obtained with an accuracy of ± 1 per cent of actual flow.

1.5.4 Electromagnetic velocity probe

This type of device is basically an inside-out version of the electromagnetic pipeline flowmeter discussed earlier, the operating principle being the same. The velocity probe consists of either a cylindrical or an ellipsoidal sensor shape which houses the field coil and two diametrically opposed pick-up electrodes.

The field coil develops an electromagnetic field in the region of the sensor and the electrodes pick up a voltage generated which is proportional to the point velocity. The probe system can be used for either open-channel or closed-pipe flow of conducting liquids.

1.5.5 Insertion turbine

The operating principle for this device is the same as for a full-bore pipeline flowmeter.

It is used normally for pipe-flow velocity measurement in liquids and consists of a small turbine

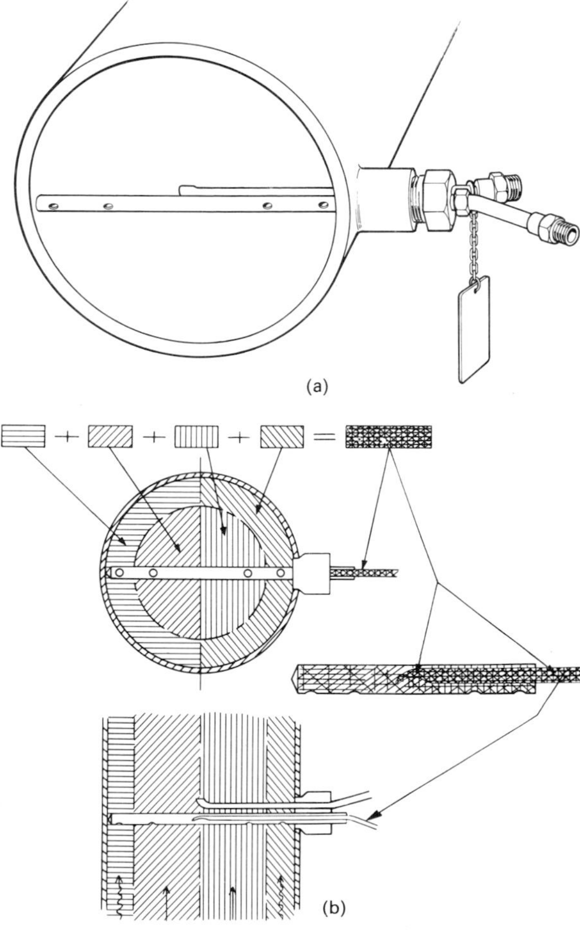

Figure 1.49 The Annubar. Courtesy, A.E.D. International Ltd.

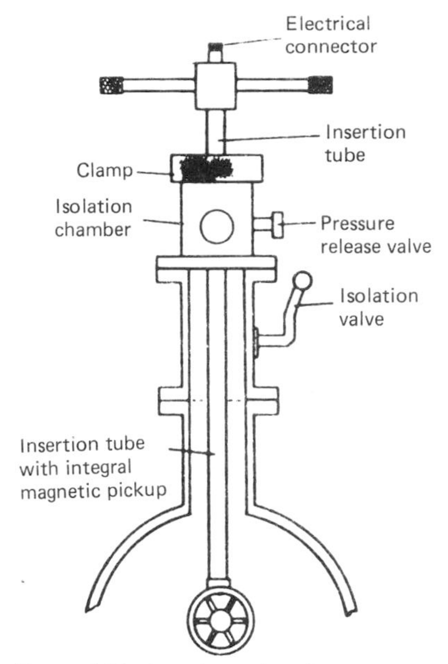

Figure 1.50 Insertion turbine flowmeter.

housed in a protective rotor cage as shown in Figure 1.50. In normal application the turbine meter is inserted through a gate valve assembly on the pipeline; hence it can be installed under pressure and can be precisely located for carrying out a flow traverse. Also, given suitable conditions it can be used as a permanent flowmetering device in the same way as the pitot tube. The velocity of the turbine is proportional to liquid velocity but a correction factor is introduced to compensate for errors caused by blockage in the flowstream caused by the turbine assembly.

1.5.6 Propeller-type current meter

Similar to the turbine in operation, this type of velocity probe typically consists of a five-bladed PVC rotor (Figure 1.51) mounted in a shrouded frame. This device is most commonly used for river or stream gauging and has the ability to measure flow velocities as low as 2.5 cm/s.

1.5.7 Insertion vortex

Operating on the same principle as the full-bore vortex meter previously described, the insertion-vortex meter consists of a short length of stainless-steel tube surrounding a centrally situated bluff body. Fluid flow through the tube causes vortex shedding. The device is normally inserted into a main pipeline via a flanged T-piece and is suitable for pipelines of 200 mm bore and above. It is capable of measuring flow velocities from 0.1 m/s up to 20 m/s for liquids and from 1 m/s to 40 m/s for gases.

1.5.8 Ultrasonic Doppler velocity probe

This device again is more commonly used for open-channel velocity measurement and consists of a streamlined housing for the Doppler meter already described.

1.6 Flowmeter calibration methods

There are various methods available for the calibration of flowmeters and the requirement can be split into two distinct categories: *in situ* and laboratory. Calibration of liquid flowmeters is generally somewhat more straightforward than that of gas flowmeters since liquids can be stored in open vessels and water can often be utilized as the calibrating liquid.

Figure 1.51 Propeller-type current meter. Courtesy, Nixon Instrumentation Ltd.

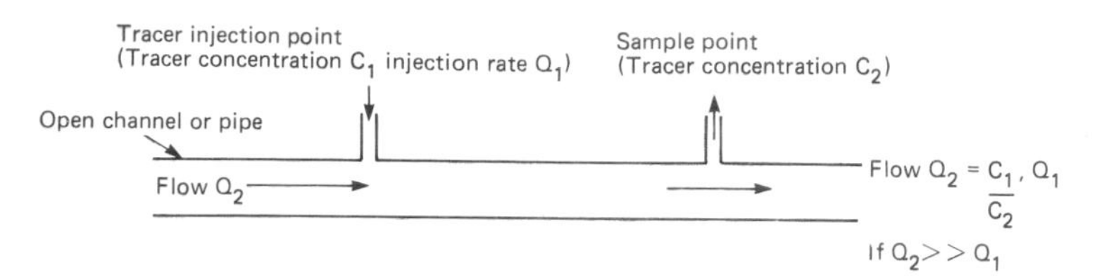

Figure 1.52 Dilution gauging by tracer injection.

1.6.1 Flowmeter calibration methods for liquids

The main principles used for liquid flowmeter calibration are *in situ*: insertion-point velocity and dilution gauging/tracer method; laboratory: master meter, volumetric, gravimetric, and pipe prover.

1.6.1.1 In-situ calibration methods

Insertion-point velocity One of the simpler methods of *in situ* flowmeter calibration utilizes point-velocity measuring devices (see Section 1.5) where the calibration device chosen is positioned in the flowstream adjacent to the flowmeter being calibrated and such that mean flow velocity can be measured. In difficult situations a flow traverse can be carried out to determine flow profile and mean flow velocity.

Dilution gauging/tracer method This technique can be applied to closed-pipe and open-channel flowmeter calibration. A suitable tracer (chemical or radioactive) is injected at an accurately measured constant rate and samples are taken from the flowstream at a point downstream of the injection point where complete mixing of the injected tracer will have taken place. By measuring the tracer concentration in the samples the tracer dilution can be established and from this dilution and the injection rate the volumetric flow can be calculated. This principle is illustrated in Figure 1.52. Alternatively a pulse of tracer material may be added to the flowstream and the time taken for the tracer to travel a known distance and reach a maximum concentration is a measure of the flow velocity.

1.6.1.2 Laboratory calibration methods

Master meter For this technique a meter of known accuracy is used as a calibration standard. The meter to be calibrated and the master meter are connected in series and are therefore subject to the same flow regime. It must be borne in mind that to ensure consistent accurate calibration the master meter itself must be subject to periodic recalibration.

Volumetric method In this technique, flow of liquid through the meter being calibrated is diverted into a tank of known volume. When full this known volume can be compared with the integrated quantity registered by the flowmeter being calibrated.

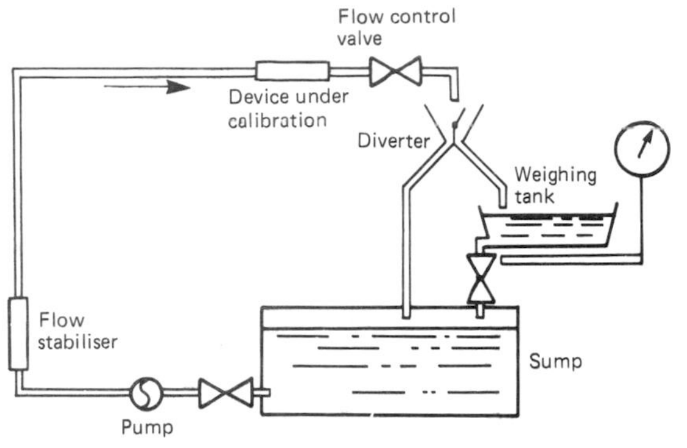

Figure 1.53 Flowmeter calibration by weighing. Courtesy, British Standards Institution.

Gravimetric method Where the flow of liquid through the meter being calibrated is diverted into a vessel that can be weighed either continuously or after a pre-determined time, the weight of the liquid is compared with the registered reading of the flowmeter being calibrated, (see Figure 1.53).

Pipe prover This device sometimes known as a 'meter prover' consists of a U-shaped length of pipe and a piston or elastic sphere. The flowmeter to be calibrated is installed on the inlet to the prover and the sphere is forced to travel the length of the pipe by the flowing liquid. Switches are inserted near both ends of the pipe and operate when the sphere passes them. The swept volume of the pipe between the two switches is determined by initial calibration and this known volume is compared with that registered by the flowmeter during calibration. A typical pipe-prover loop is shown in Figure 1.54.

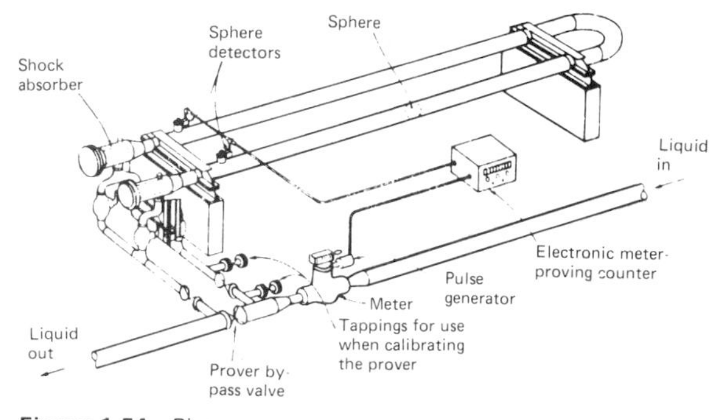

Figure 1.54 Pipe prover.

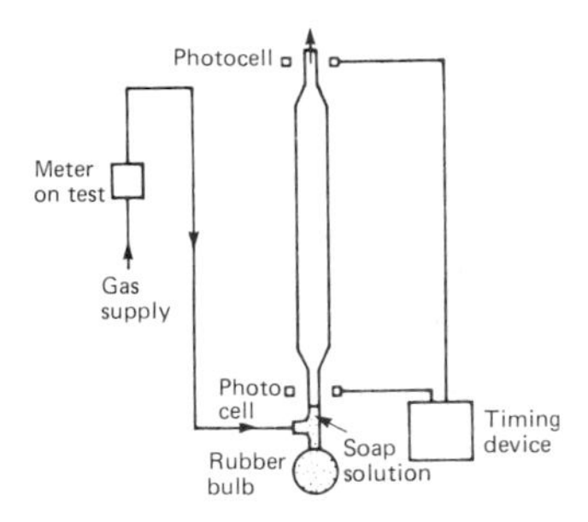

Figure 1.55 Gas flowmeter calibration—soap-film burette.

1.6.2 Flowmeter calibration methods for gases

Methods suitable for gas flowmeter calibration are *in situ* as for liquids; and laboratory: soap-film burette, water-displacement method, and gravimetric.

1.6.2.1 Laboratory calibration methods

Soap-film burette This method is used to calibrate measurement systems with gas flows in the range of 10^{-7} to 10^{-4} m^3/s. Gas flow from the meter on test is passed through a burette mounted in the vertical plane. As the gas enters the burette a soap film is formed across the tube and travels up it at the same velocity as the gas. By measuring the time of transit of the soap film between graduations of the burette it is possible to determine flow rate. A typical calibration system is illustrated in Figure 1.55.

Water-displacement method In this method a cylinder closed at one end is inverted over a water bath as shown in Figure 1.56. As the cylinder is lowered into the bath a trapped volume of gas is developed. This gas can escape via a pipe connected to the cylinder out through the flowmeter being calibrated. The time of the fall of the cylinder combined with the knowledge of the volume/length relationship leads to a determination of the amount of gas displaced which can be compared with that measured by the flowmeter under calibration.

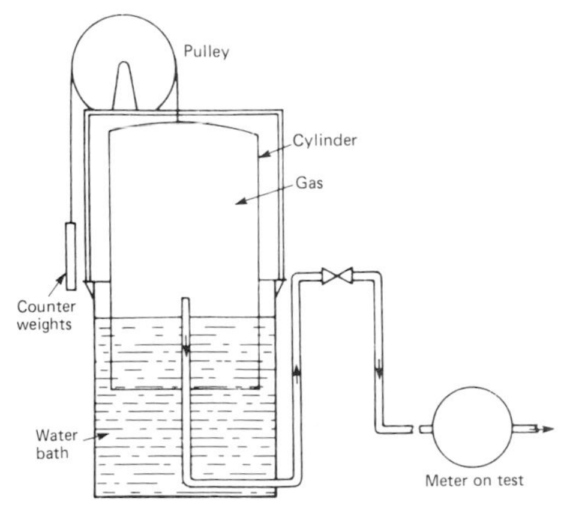

Figure 1.56 Water displacement method (bell prover).

Gravimetric method Here gas is diverted via the meter under test into a gas-collecting vessel over a measured period of time. By weighing the collecting vessel before diversion and again after diversion the difference will be due to the enclosed gas, and flow can be determined. This flow can then be compared with that measured by the flowmeter.

It should be noted that the cost of developing laboratory flow calibration systems as outlined can be quite prohibitive and it may be somewhat more cost-effective to have systems calibrated by the various national standards laboratories (such as NEL and SIRA) or by manufacturers, rather than committing capital to what may be an infrequently used system.

1.7 References

BS 1042, *Methods for the Measurement of Fluid Flow in Pipes, Part 1 Orifice Plates, Nozzles & Venturi Tubes, Part 2a Pitot Tubes*, (1964)
BS 3680, *Methods of Measurement of Liquid Flow in Open Channels*, (1969–1983)
BS 5781, *Specification for Measurement & Calibration Systems*, (1979)
BS 5792, *Specification for Electromagnetic Flowmeters*, (1980)
BS 6199, *Measurement of Liquid Flow in Closed Conduits Using Weighting and Volumetric Methods*, (1981)
Cheremisinoff, N. P., *Applied Fluid Flow Measurement*, Dekker, (1979)
Durrani, T. S. and Greated, C. A., *Laser Systems in Flow Measurement*, Plenum, (1977)
Haywood, A. T. J., *Flowmeters—A Basic Guide and Sourcebook for Users*, Macmillan, (1979)
Henderson, F. M., *Open Channel Flow*, Macmillan, (1966)
Holland, F. A., *Fluid Flow for Chemical Engineers*, Arnold, (1973)
International Organization for Standardization, ISO 3354, (1975) *Measurement of Clean Water Flow in Closed Conduits (Velocity Area Method Using Current Meters)*
Linford, A., *Flow Measurement and Meters*, E. & F.N. Spon
Miller, R. W., *Flow Measurement Engineering Handbook*, McGraw-Hill, (1982)
Shercliff, J. A., *The Theory of Electromagnetic Flow Measurement*, Cambridge University Press, (1962)
Watrasiewisy, B. M. and Rudd, M. J., *Laser Doppler Measurements*, Butterworth, (1975)

Appendix 1.1 Minimum lengths of straight pipeline upstream of device*

	Minimum number of pipe diameters for Cases A to F listed below								
	(a) Minimum length of straight pipe immediately upstream of device								*(b) Minimum length between first upstream fitting and next upstream fitting*
Diameter ratio d/D less than:	*0.22*	*0.32*	*0.45*	*0.55*	*0.63*	*0.7*	*0.77*	*0.84*	
Area ratio m less than:	*0.05*†	*0.1*	*0.2*	*0.3*	*0.4*	*0.5*	*0.6*	*0.7*	
Fittings producing symmetrical disturbances									
Case A. Reducer (reducing not more than 0.5*D* over a length of 3*D*) Enlarger (enlarging not more than 2*D* over a length of 1.5*D*) Any pressure difference device having an area ratio *m* not less than 0.3	16	16	18	20	23	26	29	33	13
Case B. Gate valve fully open (for ¾ closed see Case H)	12	12	12	13	16	20	27	38	10
Case C. Globe valve fully open (for ¾ closed see Case J)	18	18	20	23	27	32	40	49	16
Case D. Reducer (any reduction including from a large space)	25	25	25	25	25	26	29	33	13
Fittings producing asymmetrical disturbances in one plane									
Case E. Single bend up to 90°, elbow, Y-junction, T-junction (flow in either but not both branches)	10	10	13	16	22	29	41	56	15
Case F. Two or more bends in the same plane, single bend of more than 90°, swan	14	15	18	22	28	36	46	57	18
Fittings producing asymmetrical disturbances and swirling motion									
Case G‡. Two or more bends, elbows, loops or Y-junctions in different planes, T-junction with flow in both branches	34	35	38	44	52	63	76	89	32
Case H‡. Gate valve up to ¾ closed§ (for fully open see Case B)	40	40	40	41	46	52	60	70	26
Case J‡. Globe valve up to ¾ closed§ (for fully open see Case C)	12	14	19	26	36	60	80	100	30
Other fittings									
Case K. All other fittings (provided there is no swirling motion)	100	100	100	100	100	100	100	100	50

* See Subclauses 47*b* and 47*c*.
† For area ratios less than 0.015, or diameter ratios less than 0.125 see Subclause 47*b*.
‡ If swirling motion is eliminated by a flow straightener (Appendix F) installed downstream of these fittings they may be treated as Class F, B and C respectively.
§ The valve is regarded as three quarters closed when the area of the opening is one quarter of that when fully open.

2 Measurement of viscosity

K. WALTERS and W. M. JONES

2.1 Introduction

In the *Principia* published in 1687, Sir Isaac Newton postulated that 'the resistance which arises from the lack of slipperiness of the parts of the liquid, other things being equal, is proportional to the velocity with which parts of the liquid are separated from one another' (see Figure 2.1). This 'lack of slipperiness' is what we now call *viscosity*. The motion in Figure 2.1 is referred to as steady simple shear flow and if τ is the relevant shear stress producing the motion and γ is the velocity gradient ($\gamma = U/d$), we have

$$\tau = \eta\gamma \qquad (2.1)$$

η is sometimes called the coefficient of viscosity, but it is now more commonly referred to simply as the viscosity. An instrument designed to measure viscosity is called a *viscometer*. A viscometer is a special type of *rheometer* (defined as an instrument for measuring rheological properties) which is limited to the measurement of viscosity.

The SI units of viscosity are the pascal second = 1 Nsm^{-2} (= 1 kg m^{-1} s^{-1}) and Nsm^{-2}. The c.g.s. unit is the poise (= 0.1 kg m^{-1} s^{-1}) or the poiseuille (= 1 Nsm^{-2}). The units of kinematic viscosity v (= η/ρ, where ρ is the density) are m^2 s^{-1}. The c.g.s. unit is the stokes (St) and 1 cSt = 10^{-6} m^2 s^{-1}.

For simple liquids like water, the viscosity can depend on the pressure and temperature, but not on the velocity gradient (i.e. shear rate). If such materials satisfy certain further formal requirements (e.g. that they are inelastic), they are referred to as *Newtonian* viscous fluids. Most viscometers were originally designed to study these simple Newtonian fluids. It is now common knowledge, however, that most fluid-like materials one meets in practice have a much more complex behaviour and this is characterized by the adjective 'non-Newtonian'. The most common expression of non-Newtonian behaviour is that the viscosity is now dependent on the shear rate γ and it is usual to refer to the *apparent viscosity* $\eta(\gamma)$ of such fluids, where, for the motion of Figure 2.1,

$$\tau = \eta(\gamma)\gamma \qquad (2.2)$$

In the next section, we shall argue that the concept of viscosity is intimately related to the flow field under investigation (e.g. whether it is steady simple shear flow or not) and in many cases it is more appropriate and convenient to define an *extensional viscosity* η_ε corresponding to a steady uniaxial extensional flow. Now, although there is a simple relation between the (extensional) viscosity η_ε and the (shear) viscosity η in the case of Newtonian liquids (in fact, $\eta_\varepsilon = 3\eta$ for Newtonian liquids) such is not the case in general for non-Newtonian liquids, and this has been one of the motivations behind the emergence of a number of *extensional viscometers* in the last decade (see Section 2.5).

Most fluids of industrial importance can be classified as non-Newtonian: liquid detergents, multigrade oils, paints, printing inks and molten plastics are obvious examples (see, for example, Walters (1980)), and no chapter on 'the measurement of viscosity' would be complete without a full discussion of the application of viscometry to these complex fluids. This will necessitate an initial discussion of such important concepts as yield stress and thixotropy (which are intimately related to the concept of viscosity) and this is undertaken in the next section.

2.2 Newtonian and non-Newtonian behaviour

We have already indicated that for Newtonian liquids, there is a linear relation between shear stress τ and shear rate γ. For most non-Newtonian materials, the *shear-thinning* behaviour shown schematically in Figure 2.2 pertains. Such behaviour can be represented by the viscosity/shear-rate rheogram of Figure 2.3,

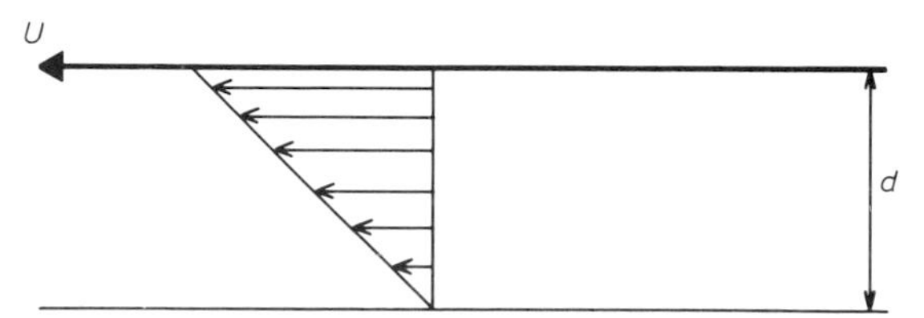

Figure 2.1 Newton's postulate.

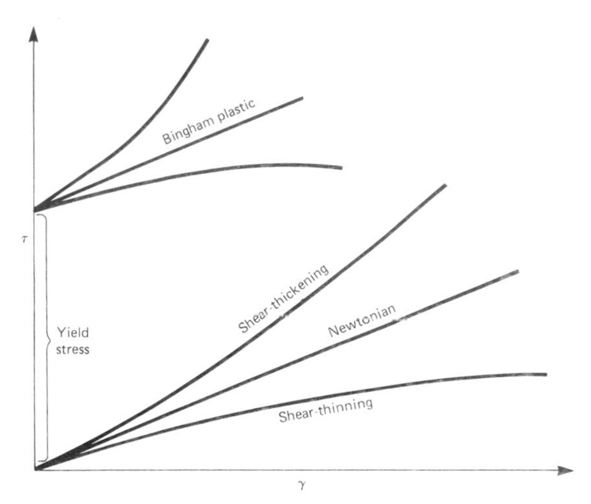

Figure 2.2 Representative (τ, γ) rheograms.

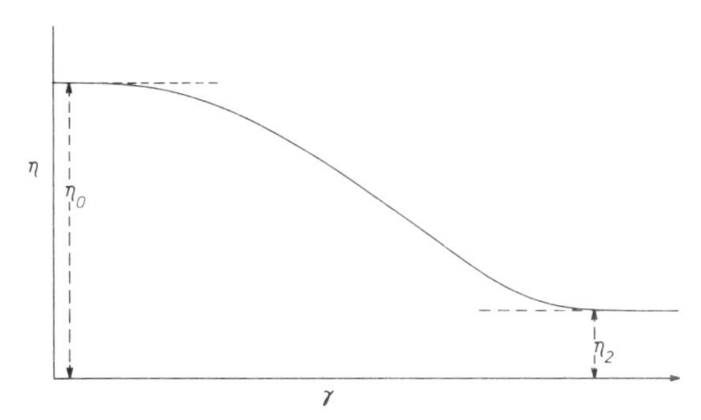

Figure 2.3 Schematic diagram of typical shear-thinning behaviour.

where we see that the viscosity falls from a 'zero-shear' value η_0 to a lower (second-Newtonian) value η_2. The term 'pseudo-plasticity' was once used extensively to describe such behaviour but this terminology is now less popular. In the lubrication literature, shear thinning is often referred to as 'temporary viscosity loss'.

Some non-Newtonian fluids, cornflour suspensions for example, show the opposite type of behaviour in which the viscosity increases with shear rate (Figure 2.2). This is called 'shear thickening'. In old-fashioned texts, the term 'dilatancy' was often used to describe such behaviour.

For many materials over a limited shear-rate range a logarithmic plot of τ against γ is linear, so that

$$\tau = K\gamma^n \tag{2.3}$$

When $n>1$, these so-called 'power-law fluids' are shear-thickening and when $n<1$, they are shear-thinning.

An important class of materials will not flow until a critical stress, called the 'yield stress', is exceeded. These 'plastic' materials can exhibit various kinds of behaviour above the yield stress as shown in Figure 2.2. If the rheogram above the yield stress is a straight line, we have what is commonly referred to as a Bingham plastic material.

In addition to the various possibilities shown in Figure 2.2, there are also important 'time-dependent' effects exhibited by some materials; these can be grouped under the headings 'thixotropy' and 'anti-thixotropy'. The shearing of some materials at a *constant* rate can result in a substantial lowering of the viscosity with time, with a gradual return to the initial viscosity when the shearing is stopped. This is called thixotropy. Paints are the most obvious examples of thixotropic materials. As the name suggests, anti-thixotropy involves an *increase* in viscosity with time at a constant rate-of-shear.

Clearly, the measurement of the shear viscosity within an industrial context is a non-trivial task and requires an initial appreciation of material behaviour. Is the material Newtonian or non-Newtonian? Is thixotropy important? And so on.

As if to complicate an already complex situation, it has recently been realized that many industrial processes involve more extensional deformation than shear flow, and this has been the motivation behind the search for extensional viscometers, which are constructed to estimate a material's resistance to a stretching motion of the sort shown schematically in Figure 2.4. In this case, it is again necessary to define an appropriate stress T and rate of strain κ, and to define the extensional viscosity η_ε by

$$T = \eta_\varepsilon \kappa \tag{2.4}$$

For a Newtonian liquid, η_ε is a constant ($\equiv 3\eta$). For some non-Newtonian liquids, the extensional viscosity can take very high values indeed, and it is this exceptional resistance to stretching in some materials, together with the practical importance of extensional flow, which makes the study of extensional viscosity so important. The reader is referred to the book *Elongational Flows* by Petrie (1979) for a detailed treatise on the subject. The recent text by Dealy (1982) on polymer-melt rheometry is also recommended in this context.

A detailed assessment of the importance of non-Newtonian effects is given in the text *Rheometry: Industrial Applications* (Walters, 1980) which contains a general discussion of basic principles in addition to

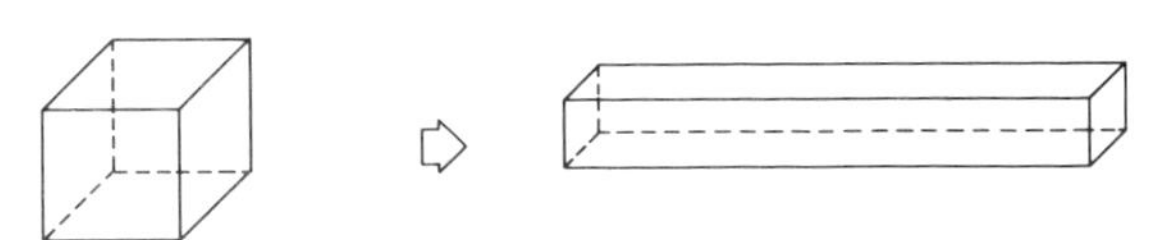

Figure 2.4 Uniaxial extensional deformation.

an in-depth study of various industrial applications.

The popular book on viscometry by Van Wazer *et al.* (1963) and that of Wilkinson (1960) on non-Newtonian flow are now out of date in some limited respects, but they have stood the test of time remarkably well and are recommended to readers, provided the dates of publication of the books are appreciated. More modern treatments, developed from different but complementary viewpoints, are given in the books by Lodge (1974), Walters (1975) and Whorlow (1980). The text by Dealy (1982) already referred to is limited to polymer-melt rheometry, but much of the book is of general interest to those concerned with the measurement of viscosity.

2.3 Measurement of the shear viscosity

It is clearly impracticable to construct viscometers with the infinite planar geometry associated with Newton's postulate (Figure 2.1), especially in the case of mobile liquid systems, and this has led to the search for convenient geometries and flows which have the same basic steady simple shear flow structure. This problem has now been resolved and a number of the so-called 'viscometric flows' have been used as the basis for viscometer design. (The basic mathematics is non-trivial and may be found in the texts by Coleman *et al.* (1966), Lodge (1974) and Walters (1975).) Most popular have been (i) capillary (or Poiseuille) flow, (ii) circular Couette flow, and (iii) cone-and-plate flow. For convenience, we shall briefly describe each of these flows and give the simple operating formulae for Newtonian liquids, referring the reader to detailed texts for the extensions to non-Newtonian liquids. We also include in Section 2.3.4 a discussion of the parallel-plate rheometer, which approximates closely the flow associated with Newton's postulate.

2.3.1 Capillary viscometer

Consider a long capillary with a circular cross-section of radius a. Fluid is forced through the capillary by the application of an axial pressure drop. This pressure drop P is measured over a length L of the capillary, far enough away from both entrance and exit for the flow to be regarded as 'fully-developed' steady simple shear flow. The volume rate of flow Q through the capillary is measured for each pressure gradient P/L and the viscosity η for a Newtonian liquid can then be determined from the so-called Hagen–Poiseuille law:

$$Q=\frac{\pi P a^4}{8\eta L} \tag{2.5}$$

The non-trivial extensions to (2.5) when the fluid is non-Newtonian may be found in Walters (1975), Whorlow (1980) and Coleman *et al.* (1966). For example, in the case of the power-law fluid (2.3), the formula is given by

$$Q=\frac{\pi n a^3}{(3n+1)}\left(\frac{aP}{2KL}\right)^{1/n} \tag{2.6}$$

One of the major advantages of the capillary viscometer is that relatively high shear-rates can be attained.

Often, it is not possible to determine the pressure gradient over a restricted section of the capillary and it is then necessary, especially in the case of non-Newtonian liquids, to study carefully the pressure losses in the entry and exit regions before the results can be interpreted correctly (see, for example, Dealy (1982) and Whorlow (1980)). Other possible sources of error include viscous heating and flow instabilities. These and other potential problems are discussed in detail by Dealy (1982), Walters (1975) and Whorlow (1980).

The so-called 'kinetic-energy correction' is important when it is not possible to limit the pressure drop measurement to the steady simple shear flow region and when this is taken over the complete length L of the capillary. For a *Newtonian* fluid, the kinetic energy correction is given (approximately) by

$$P=P_0-\frac{1.1\rho Q^2}{\pi^2 a^4} \tag{2.7}$$

where P is the pressure drop required in (2.5), P_0 is the measured pressure drop and ρ is the density of the fluid.

Since a gas is highly compressible, it is more convenient to measure the *mass* rate of flow $\dot{m}$. Equation (2.5) has then to be replaced by (see, for example, Massey (1968))

$$\eta=\frac{\pi a^4 \bar{p} M P}{8\dot{m} R T L} \tag{2.8}$$

where $\bar{p}$ is the mean pressure in the pipe, M is the molecular weight of the gas, R is the gas constant per mole and T is the Kelvin temperature. The kinetic-energy correction (2.7) is still valid and must be borne in mind, but in the case of a gas, this correction is usually very small. A 'slip correction' is also potentially important in the case of gases, but only at low pressures.

In commercial capillary viscometers for non-gaseous materials, the liquids usually flow through the capillaries under gravity. A good example is the Ostwald viscometer (Figure 2.5). In this b, c and d are fixed marks and there are reservoirs at D and E. The amount of liquid must be such that at equilibrium one meniscus is at d. To operate, the liquid is sucked or blown so that the other meniscus is now a few millimetres above b. The time t for the level to fall from

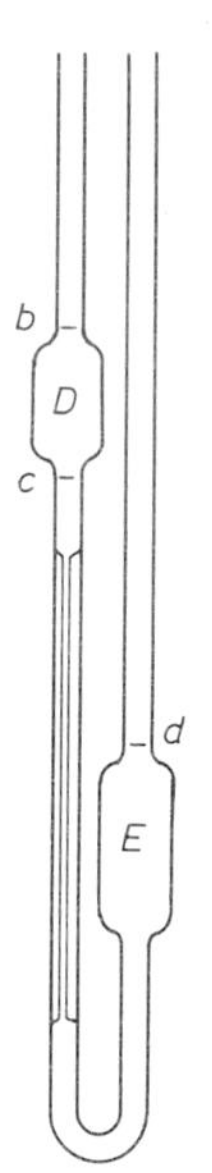

Figure 2.5 Schematic diagram of an Ostwald viscometer.

b to c is measured. The operating formula is of the form

$$\nu = At - B/t \tag{2.9}$$

where ν is the kinematic viscosity ($\equiv \eta/\rho$). The second term on the right-hand side of equation (2.9) is a correction factor for end effects. For any particular viscometer, A and B are given as calibration constants. Viscometers with pipes of different radii are supplied according to British Standards specifications and a 'recommended procedure' is also given in B.S. Publication 188: 1957.

Relying on gravity flow alone limits the range of measurable stress to between 1 and 15 Nm^{-2}. The upper limit can be increased to 50 Nm^{-2} by applying a known steady pressure of inert gas over the left-hand side of the U-tube during operation.

2.3.2 Couette viscometer

The most popular rotational viscometer is the Couette concentric-cylinder viscometer. Fluid is placed in the annulus between two concentric cylinders (regarded as infinite in the interpretation of data) which are in relative rotation about their common axis. It is usual for the outer cylinder to rotate and for the torque required to keep the inner cylinder stationary to be measured, but there are variants, as in the Brookfield viscometer, for example, where a cylindrical bob (or sometimes a disc) is rotated in an expanse of test liquid and the torque on this same bob is recorded, see Section 2.4.

If the outer cylinder of radius r_0 rotates with angular velocity Ω_0 and the inner cylinder of radius r_1 is stationary, the torque C per unit length of cylinder on the inner cylinder for a Newtonian liquid is given by

$$C = \frac{4\pi\Omega_0 r_1^2 r_0^2 \eta}{(r_0^2 - r_1^2)} \tag{2.10}$$

so that measurement of C at each rotational speed Ω_0 can be used to determine the viscosity η. The extensions to (2.10) when the fluid is non-Newtonian are again non-trivial (unless the annular gap is very small) but the relevant analysis is contained in many texts (see, for example Walters (1975) and Whorlow (1980)). With reference to possible sources of error, end effects are obvious candidates as are flow instabilities, misalignment of axes and viscous heating. A detailed discussion of possible sources of error is to be found in Dealy (1982), Walters (1975) and Whorlow (1980).

2.3.3 Cone-and-plate viscometer*

Consider the cone-and-plate arrangement shown schematically in Figure 2.6. The cone rotates with angular velocity Ω_0 and the torque C required to keep the plate stationary is measured. The gap angle θ_0 is usually very small (<4°) and, in the interpretation of results, edge effects are neglected. It is then easy to show that for a Newtonian liquid, the operating formula is

$$C = \frac{2\pi a^3 \Omega_0}{3\theta_0}\eta \tag{2.11}$$

where a is the radius of the cone.

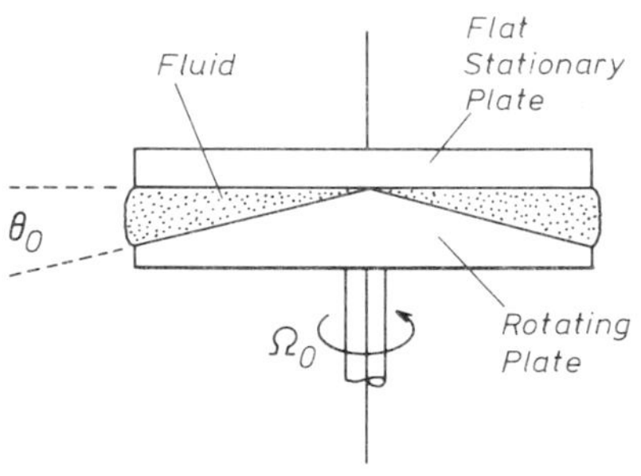

Figure 2.6 Basic cone-and-plate geometry.

In contrast to the capillary-flow and Couette-flow situations, the operating formula for non-Newtonian fluids is very similar to (2.11) and is in fact given by

$$C = \tfrac{2}{3}\pi a^3 \gamma \eta(\gamma) \tag{2.12}$$

where the shear rate γ is given by

$$\gamma = \Omega_0/\theta_0 \tag{2.13}$$

which is (approximately) constant throughout the test

* The torsional-flow rheometer in which the test fluid is contained between parallel plates is similar in operation to the cone-and-plate rheometer, but the data interpretation is less straightforward, except in the Newtonian case (see, for example, Walters (1975)).

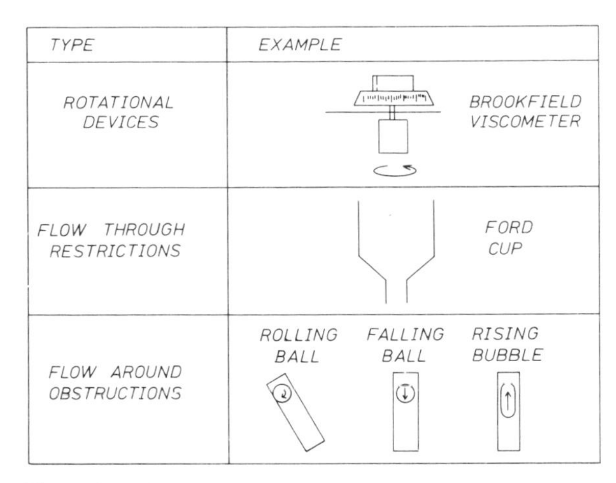

Figure 2.7 Classes of industrial viscometers.

fluid, provided θ_0 is small ($<4^\circ$ say). This is an important factor in explaining the popularity of cone-and-plate flow in non-Newtonian viscometry. Indeed, it is apparent from (2.12) and (2.13) that measurements of the torque C as a function of rotational speed Ω_0 immediately yield apparent viscosity/shear-rate data.

Sources of error in the cone-and-plate viscometer have been discussed in detail by Walters (1975) and Whorlow (1980). Measurements on all fluids are limited to modest shear rates ($<100\ \mathrm{s}^{-1}$) and this upper bound is significantly lower for some fluids like polymer melts.

Time-dependent effects such as thixotropy are notoriously difficult to study in a systematic way and the constant shear rate in the gap of a cone-and-plate viscometer at least removes one of the complicating factors. The cone-and-plate geometry is therefore recommended for the study of time-dependent effects.

For the rotational viscometer designs discussed thus far, the shear rate is fixed and the corresponding stress is measured. For plastic materials with a yield stress this may not be the most convenient procedure and the last decade has seen the emergence of constant-stress devices, in which the shear *stress* is controlled and the resulting motion (i.e. shear rate) recorded. The Deer rheometer is the best known of the constant-stress devices and at least three versions of such an instrument are now commercially available. The cone-and-plate geometry is basic in current instruments.

2.3.4 Parallel-plate viscometer

In the parallel-plate rheometer, the test fluid is contained between two parallel plates mounted vertically; one plate is free to move in the vertical direction, so that the flow is of the plane-Couette type and approximates that associated with Newton's postulate.

A mass M is attached to the moving plate (of area A) and this produces a displacement x of the plate in a time t. If the plates are separated by a distance h, the relevant shear stress τ is given by

$$\tau = Mg/A \tag{2.14}$$

and the shear rate γ by

$$\gamma = x/th \tag{2.15}$$

so that the viscosity η is determined from

$$\eta = Mgth/xA \tag{2.16}$$

Clearly, this technique is only applicable to 'stiff' systems, i.e. liquids of very high viscosity.

2.4 Shop-floor viscometers

A number of *ad hoc* industrial viscometers are very popular at the shop-floor level of industrial practice and these usually provide a very simple and convenient method of determining the viscosity of *Newtonian* liquids. The emphasis on the 'Newtonian' is important since their application to non-Newtonian systems is far less straightforward (see, for example, Walters and Barnes (1980)). Three broad types of industrial viscometer can be identified (see Figure 2.7). The first type comprises simple rotational devices such as the Brookfield viscometer, which can be adapted in favourable circumstances to provide the apparent viscosity of non-Newtonian systems (see, for example, Williams (1979)). The instrument is shown schematically in Figure 2.8. The pointer and the dial rotate together. When the disc is immersed, the test fluid exerts a torque on the disc. This twists the spring and the pointer is displaced relative to the dial. For Newtonian liquids (and for non-Newtonian liquids in

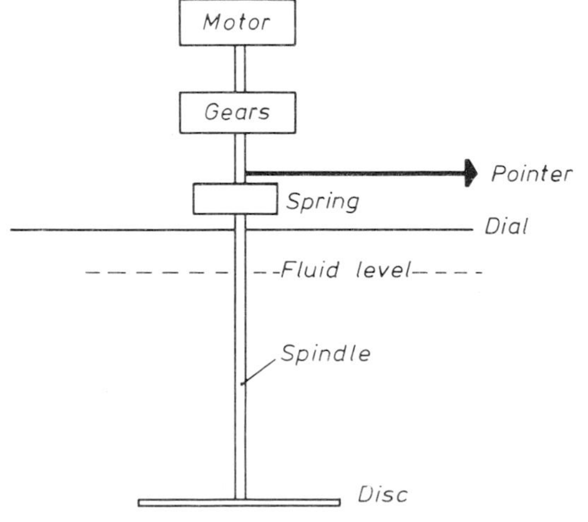

Figure 2.8 Schematic diagram of the Brookfield viscometer.

favourable circumstances) the pointer displacement can be directly related to the viscosity of the test sample.

The second type of industrial viscometer involves what we might loosely call 'flow through constrictions' and is typified by the Ford-cup arrangement. The idea of measuring the viscosity of a liquid by timing its efflux through a hole at the bottom of a cup is very attractive. It is simple to operate, inexpensive, and the apparatus can be made very robust. Historically, the cup device was probably one of the first forms of viscometer ever used and today there are over 50 versions of the so-called flow cups.

Often, results from flow cups are simply expressed as 'time in seconds' (e.g. 'Redwood seconds') but for Newtonian liquids these can be converted to kinematic viscosity ν through an (approximate) formula of the form

$$\nu = At - B/t \tag{2.17}$$

where A and B are constants which depend on the cup geometry (see, for example, Walters and Barnes (1980)).

The second term on the right-hand side of (2.17) is essentially a kinetic-energy correction. For Newtonian liquids, A and B can be determined by carrying out standard experiments on liquids of known kinematic viscosity.

A major disadvantage of the standard Ford cup so far as *non*-Newtonian liquids is concerned is that only one time can be taken, i.e. the time taken for the cup to empty. Such a measurement leads to a single (averaged) viscosity for a complicated deformation regime and this is difficult to interpret consistently for rheologically complex fluids. Indeed, liquids with different rheologies as regards shear viscosity, extensional viscosity and elasticity may behave in an identical fashion in a Ford-cup experiment (Walters and Barnes, 1980) so that shop-floor instruments of this sort should be used with extreme caution when dealing with *non*-Newtonian liquids. The same applies to the 'flow-around-obstacle' viscometers of Figure 2.7. Typical examples of this type of viscometer are the Glen Creston falling-ball viscometer and the Hoeppler rolling-ball instrument (see, for example, Van Wazer *et al.* (1963) and Cheng (1979). Rising-bubble techniques may also be included in this category.

2.5 Measurement of the extensional viscosity

Many industrial processes, especially in the polymer industries, involve a high extensional-flow component and there is an acknowledged need for *extensional viscometers*. The construction of such devices is, however, fraught with difficulties. For example, it is difficult to generate an extensional deformation over a sufficient deformation-rate range. Indeed, many of the most popular and sophisticated devices for work on polymer melts (such as those constructed at B.A.S.F. in Germany) cannot reach the steady state required to determine the extensional viscosity η_ε defined in (2.4). Therefore, they are, as yet, of unproven utility. A full discussion of the subject of extensional viscometry within the context of polymer-melt rheology is provided by Dealy (1982).

In the case of more mobile liquid systems, it is difficult to generate flows which approximate to steady uniaxial extension and the most that can reasonably be hoped for is that instruments will provide an *estimate* of a fluid's resistance to stretching flows (see, for example, Chapter 1 of Walters (1980)). With this important proviso, the Ferguson Spin-Line Rheometer is a commercially-available instrument which can be used on mobile liquids to provide extensional viscosity information.

2.6 Measurement of viscosity under extremes of temperature and pressure

Any of the techniques discussed above can be adapted to study the effect of temperature and pressure on viscosity, provided the apparatus can accommodate the extremes prevailing.

It is important to emphasize that viscosity is very sensitive to temperature. For example, the viscosity of water changes by 3 per cent per kelvin. It is therefore essential to control the temperature and to measure it accurately.

Pressure is also an important variable in some studies. In the case of lubricating oils, for example, high pressures are experienced during use and it is necessary to know the pressure-dependence of viscosity for these fluids.

At temperatures near absolute zero measurements have been concerned with the viscosity of liquid helium. Recently, special techniques have been developed. Webeler and Allen (1972) measured the attenuation of torsional vibrations initiated by a cylindrical quartz crystal. Vibrating-wire viscometers have also been used. The resonant frequency and damping of the oscillations of a wire vibrating transversely in a fluid depend on the viscosity and density of the fluid. References to this and other work are given in Bennemann and Ketterson (1976).

To study the effect of high pressure, Abbot *et al.* (1981) and Dandridge and Jackson (1981) have observed the rate of fall of a sphere in lubricants exposed to high pressures ($\sim$3 GPa). Galvin, Hutton and Jones (1981) used a capillary viscometer at high pressure to study liquids over a range of temperatures (0 to 150 °C) and shear rates (0 to 3×10^5 sec^{-1}) with pressures up to 0.2 GPa. Kamal and Nyun (1980) have also adapted a capillary viscometer for high-pressure work.

2.7 On-line measurements

It is frequently necessary to monitor the viscosity of a fluid 'on line' in a number of applications, particularly when the constitution or temperature of the fluid is likely to change. Of the viscometers described in this chapter, the capillary viscometer and the concentric-cylinder viscometer are those most conveniently adapted for such a purpose. For the former, for example, the capillary can be installed directly in series with the flow and the pressure difference recorded using suitably placed transducers and recorders. The corresponding flow rate can be obtained from a metering pump.

Care must be taken with the on-line concentric-cylinder apparatus as the interpretation of data from the resulting *helical* flow is not easy.

Other on-line methods involve obstacles in the flow channel; for example, a float in a conical tube will arrive at an equilibrium position in the tube depending on the rate of flow and the kinematic viscosity of the fluid. The parallel-plate viscometer has also been adapted for on-line measurement. These and other on-line techniques are considered in detail in *The Instrument Manual* (1975).

2.8 Accuracy and range

The ultimate absolute accuracy obtained in any one instrument cannot be categorically stated in a general way. For example, using the Ostwald viscometer, reproducible measurements of time can be made to 0.3 per cent. But to achieve this absolutely, the viscometer and the fluid must be scrupulously clean and the precise amount of fluid must be used. The temperature within the viscometer must also be uniform and be known to within 0.1 °C. Obviously, this can be achieved, but an operator might well settle for 1 to 2 per cent accuracy and find this satisfactory for his purpose with less restriction on temperature measurement and thermostating. Similar arguments apply to the Couette viscometer but here, even with precise research instruments, an accuracy of 1 per cent requires very careful experimentation.

The range of viscosities and rates of shear attainable in any type of viscometer depend on the dimensions of the viscometer, e.g. the radius of the capillary in the capillary viscometer and the gap in a Couette viscometer.

By way of illustration, we conclude with a table of values claimed by manufacturers of instruments within each type, but we emphasize that no one instrument will achieve the entire range quoted.

Table 2.1

Viscometer type	*Lowest viscosity (poise)*	*Highest viscosity (poise)*	*Shear-rate range* (s^{-1})
Capillary	2×10^{-3}	10^{3}	1 to 1.5×10^{4}
Couette	5×10^{-3}	4×10^{7}	10^{-2} to 10^{4}
Conc-and-plate	10^{-3}	10^{10}	10^{-4} to 10^{3}
Brookfield type	10^{-2}	5×10^{5}	10^{-3} to 5×10^{6}
Falling-ball, rolling-ball	10^{-4}	10^{4}	indeterminate

2.9 References

Abbott, L. H., Newhall, D. H., Zibberstein, V. A. and Dill, J. F., *A.S.L.E. Trans.*, **24**, 125, (1981)

Bennemann, K. H. and Ketterson, J. B. (eds), *The Physics of Liquid and Solid Helium*, Wiley, (Part 1, 1976; Part 2, 1978)

Cheng, D. C.-H., 'A comparison of 14 commercial viscometers and a home-made instrument', Warren Spring Laboratory LR 282 (MH), (1979)

Coleman, B. D., Markovitz, H. and Noll, W., *Viscometric Flows of Non-Newtonian Fluids*, Springer-Verlag, (1966)

Dandridge, A. and Jackson, D. A., *J. Phys. D*, **14**, 829, (1981)

Dealy, J. M., *Rheometers for Molten Plastics*, Van Nostrand, (1982)

Galvin, G. D., Hutton, J. F. and Jones, B. J., *Non-Newtonian Fluid Mechanics*, **8**, 11, (1981)

The Instrument Manual, United Trade Press, p. 62, (5th edn, 1975)

Kamal, M. R. and Nyun, H., *Polymer Eng. and Science*, **20**, 109, (1980)

Lodge, A. S., *Body Tensor Fields in Continuum Mechanics*, Academic Press, (1974)

Massey, R. S., *Mechanics of Fluids*, Van Nostrand, (1968)

Petrie, C. J. S., *Elongational Flows*, Pitman, (1979)

Van Wazer, J. R., Lyon, J. W., Kim, K. Y. and Colwell, R. E., *Viscosity and Flow Measurement*, Wiley-Interscience, (1963)

Walters, K., *Rheometry*, Chapman & Hall, (1975)

Walters, K. (ed.), *Rheometry: Industrial Applications*, Wiley, (1980)

Walters, K. and Barnes, H. A., *Proc. 8th Int. Cong. on Rheology, Naples, Italy*, p. 45, Plenum Press, (1980)

Webeler, R. W. H. and Allen, G., *Phys. Rev.*, **A5**, 1820, (1972)

Whorlow, R. W., *Rheological Techniques*, Wiley, (1980)

Wilkinson, W. L., *Non-Newtonian Fluids*, Pergamon Press, (1960)

Williams, R. W., *Rheol. Acta*, **18**, 345, (1979)

3 Measurement of length

P. H. SYDENHAM

3.1 Introduction

Length is probably the most measured physical parameter. This measurand (the term used here for the parameter that is to be measured) is known under many alternative names—displacement, movement, motion.

Length is often the intermediate stage of systems used to measure other parameters. For example, a common method of measuring fluid pressure is to use the force of the pressure to elongate a metal element, a length sensor then being used to give an electrical output related to pressure.

Older methods were largely mechanical giving readout suited to an observer's eyes. The possibility of using electrical and radiation techniques to give electronic outputs is now much wider. Pneumatic techniques are also quite widely used and these are discussed in Part 4.

Length can now be measured through over thirty decadic orders. Figure 3.1 is a chart of some common methods and their ranges of use. In most cases only two to three decades can be covered with a specific geometrical scaling of a sensor's configuration.

This chapter introduces the reader to the commonly used methods that are used in the micrometre to sub-kilometre range.

For further reading, it may be noted that most instrumentation books contain one chapter, or more, on length measurement of the modern forms, examples being Mansfield (1973), Norton (1969), Oliver (1971) and Sydenham (1984, 1983). Mechanical methodology is more generally reported in the earlier literature on the subjects of mechanical measurements, tool-room gauging and optical tooling. Some such books are Batson and Hyde (1931), Hume (1970), Kissam (1962), Rolt (1929) and Sharp (1970). In this aspect of length measurement the value of the older books should not be overlooked for they provide basic understanding of technique that is still relevant today in proper application of modern electronic methods.

For the microdisplacement range see Garratt (1979), Sydenham (1969; 1972) and Woltring (1975); for larger ranges see Sydenham (1968) and Sydenham (1971). Neubert (1975) has written an excellent analysis of transducers, including those used for length measurement.

3.2 The nature of length

Efficient and faithful measurement requires an appreciation of the nature of the measurand and of the pitfalls that can arise for that particular physical system domain.

Length, as a measured parameter, is generally so self-evident that very little is ever written about it at the philosophical level. Measurement of length is apparently simple to conceptualize and it appears easy to devise methods for converting the measurand into an appropriate signal.

It is an experimental finding that space can be described in terms of three length parameters. Three coordinate numbers suffice to describe the position of a point in space regardless of the kind of coordinate framework used to define that point's coordinates. Where some of the mechanical degrees of freedom are constrained the number of measurement coordinates can be reduced in order to define position. Thus to measure position along a defined straight line only requires one length-sensing system channel; to plot position in a defined plane requires two sensors.

Length measurements fall into two kinds, those requiring determination of the absolute value in terms of the defined international standard and those that determine a change in length of a gauge length interval (relative length). In the latter case the measurement problem is usually greatly simplified for there is no need to determine the gauge interval length to high accuracy. Thus, to measure the length of a structure in absolute terms is a different kind of problem from measuring strains induced in the structure.

Descriptive terminology is needed to simplify general description of the measuring range of a length sensor and a classification into microdisplacement, industrial, surveying, navigation and celestial is included in Figure 3.1.

The actual range of a length sensor is not necessarily that of the size of the task. For example, to measure strain over a long test interval may make use of a long-

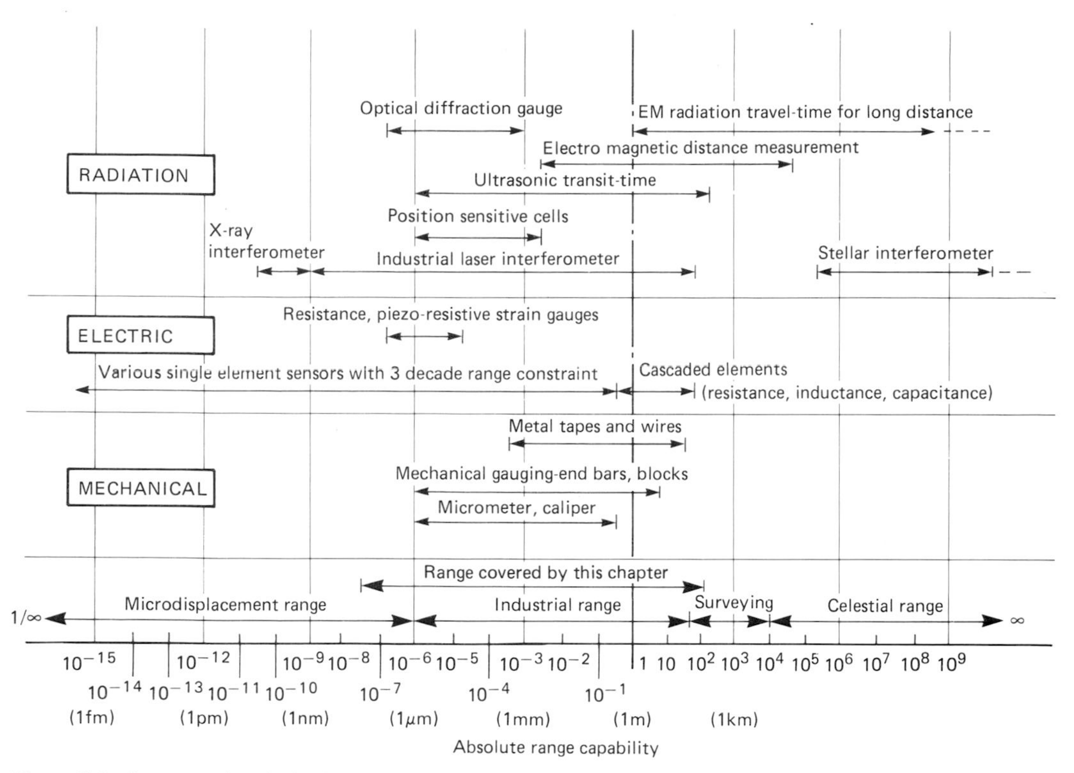

Figure 3.1 Ranges and methods of measuring length.

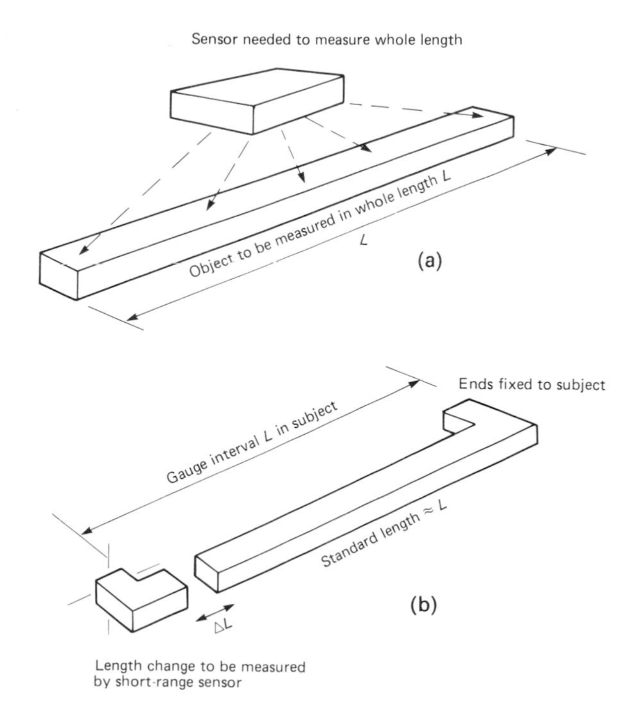

Figure 3.2 Absolute (a) and relative (b) length-measurement situations.

range, fixed-length, standard structure which is compared with the object of interest using a short-range sensor to detect the small differences that occur (see Figure 3.2(b)). Absolute whole length measurement, Figure 3.2(a), requires a sensor of longer range.

It is often possible to measure a large length by adding together successive intervals. This is obvious when using a single ruler to span a length greater than itself. The same concept is often applied using multiple electronic sensors placed in line or by stepping a fixed interval along the whole distance counting the number of coarse intervals and subdividing the last partial interval by some other sensor that has finer sensing detail.

Some mention is appropriate on the choice of use of non-contact or contacting methods. In the former (see Figure 3.3), the length measurement is made by a method that does not mechanically contact the subject. An example is the use of an optical interferometer to monitor position of a machine slide. It does not impose significant force on the slide and, as such, does not alter the measurand value by its presence.

Contacting methods must be used with some caution lest they alter the measurement value due to the mechanical forces imposed by their presence.

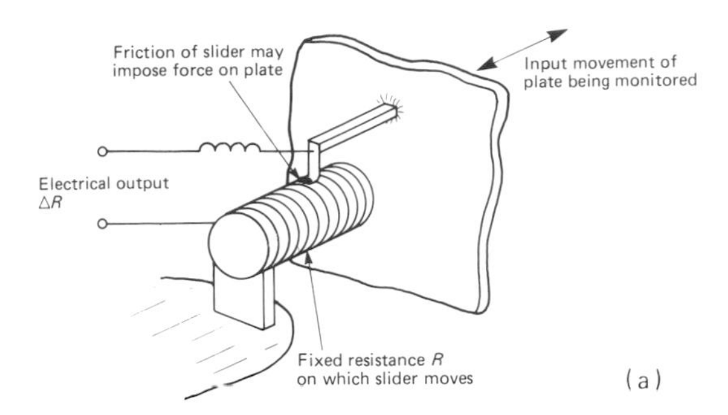

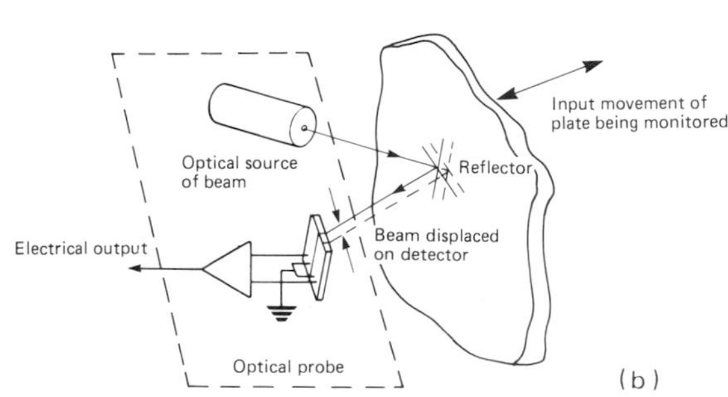

Figure 3.3 Examples of contacting and non-contacting length measurements. (a) Contacting, using variable resistance. (b) Non-contacting, using an optical probe.

3.3 Derived measurements

3.3.1 Derived from length measurement alone

Length (m) comes into other measurement parameters, including relative length change (m/m), area (m^2), volume (m^3), angle (m/m), velocity ($m\,s^{-1}$) and acceleration ($m\,s^{-2}$). To measure position, several coordinate systems can be adopted. Figure 3.4 shows those commonly used. In each instance the general position of a point P will need three measurement numbers, each being measured by separate sensing channels.

The cartesian (or rectangular) system shown in Figure 3.4(a), is that most adopted for ranges less than a few tens of metres. Beyond that absolute size it becomes very difficult to establish an adequately stable and calibratable framework. Errors can arise from lack of right angles between axes, from errors of length sensing along an axis and from the imperfection of projection out from an axis to the point.

The polar system of Figure 3.4(b) avoids the need for an all-encompassing framework, replacing that problem with the practical need for a reference base from which two angles and a length are determined. Errors arise here in definition of the two angles and in the length measurement which, now, is not restricted to a slide-way. Practical angle measurement reaches practical and cost barriers at around one arc second of discrimination. This method is well suited to such applications as radar tracking of aircraft or plotting of location under the sea.

The above two systems of coordinate framework are those mostly adopted. A third alternative which is less used, has in principle, the least error sources. This is the triangular system shown as Figure 3.4(c). In this method three lengths are measured from a triangle formed of three fixed lengths. Errors arise only in the three length measurements with respect to the base triangle and in their definition in space. Where two or more points in space are to be monitored, then their relative position can be obtained accurately even if the base triangle moves in space. The major practical problem in adopting this method is that the three length measurements each require tracking arrangements to keep them following the point. The accuracy of pointing, however, is only subject to easily tolerated cosine forms of error which allow relatively poor following ability to give quite reasonable values.

The three alternatives can also be combined to provide other arrangements but in each case there will always be the need to measure three variables (as combinations of at least one length with length and/or angle) to define point position in a general manner. Where a translational freedom is constrained the need to measure reduces to a simpler arrangement needing less sensing channels.

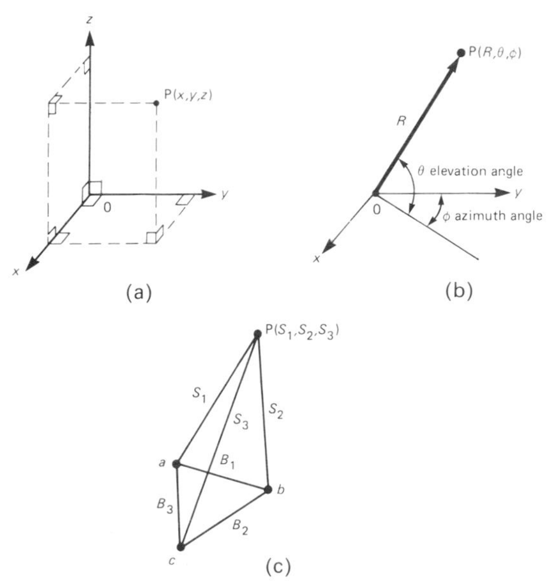

Figure 3.4 Coordinate systems that can be used to locate position in space. (a) Cartesian, rectangular, frame for three lengths. (b) Two polar directions and a length. (c) Triangulated lengths from a base triangle.

3.4 Standards and calibration of length

With very little exception length measurements are now standardized according to SI measurement unit definitions, length being one of the seven base units. It is defined in terms of the unit called the metre.

Until early 1982 the metre was defined in terms of a given number of wavelengths of krypton-86 radiation. Over the 1970 decade, however, it was becoming clear that there were improved methods available that would enable definition with reduced uncertainty.

The first was to make use, in the manner already adopted, of the radiation available from a suitable wavelength-stabilized laser source for this is easier to produce and is more reproducible than krypton radiation. At first sight this might be an obvious choice to adopt but a quite different approach was also available, that which was recommended in 1982.

The numerical value of the speed of light c ($c = 299\,792\,458$ m s^{-1}) is the result of numerical standards chosen for the standards of time and of length. Thus the speed of light, as a numerical value, is not a fundamental constant.

Time standards (parts in 10^{14} uncertainty) are more reproducible in terms of uncertainty than length (parts in 10^8 uncertainty) so if the speed of light is defined as a fixed number then, in principle, the time standard will serve as the length standard provided suitable apparatus exists to make the conversion from time to length via the constant c.

Suitable equipment and experimental procedures have now been proven as workable. By choosing a convenient value for c that suited measurement needs (that given above) it was, in 1982, agreed by the signatories of the committee responsible for standardization of the metre that the new definition should be 'The metre is the length of the path travelled by light in vacuum during the fraction (1/299 792 458) of a second.'

This new definition takes reproducibility of length definition from parts in 10^8 to parts in 10^{10}. In common industrial practice, few people require the full capability of the standard but adequate margin is needed to provide for loss of uncertainty each time the standards are transferred to a more suitable apparatus.

To establish the base standard takes many months of careful experimental work using very expensive apparatus. This method is not applicable to the industrial workplace due to reasons of cost, time and

Figure 3.5 Laser interferometer being used to measure length in an industrial application. Courtesy, Hewlett-Packard Inc.

apparatus complexity. The next level of uncertainty down is, however, relatively easily provided in the form of the industrial laser interferometer that has been on the market for several years. Figure 3.5 is such an equipment in an industrial application. The nature of the laser system shown is such that it has been given approval by the National Bureau of Standards (NBS) for use without traceable calibration for, provided the lengths concerned are not too small, it can give an uncertainty of around 1 part in 10^8 which is adequate for most industrial measurements.

Optical interferometer systems operate with wavelengths of the order of 600 nm. Subdivision of the wavelength becomes difficult at around 1/1000 of the wavelength making calibration of sub-micrometre displacements very much less certain than parts in 10^8. In practice, lengths of a micrometre cannot be calibrated to much better than 1 per cent of the range.

Laser interferometers are easy to use and very precise, but they too are often not applicable due to high cost and unsuitability of equipment. A less expensive calibration method that sits below the interferometer in the traceable chain uses the mechanical slip and gauge block family. These are specially treated and finished pieces of steel whose lengths between defined surfaces are measured using a more certain method, such as an interferometer. The length values are recorded on calibration certificates. In turn the blocks are used as standards to calibrate micrometers, go/no-go gauges and electronic transducers. Mechanical gauges can provide of the order of 1 in 10^6 uncertainties.

For lengths over a few metres, solid mechanical bars are less suitable as standard lengths due to handling reasons. Flexible tapes are used which are calibrated against the laser interferometer in standards facilities such as that shown in Figure 3.6. Tapes are relatively cheap and easy to use in the field compared with the laser interferometer. They can be calibrated to the order of a part in 10^6.

For industrial use little difficulty will be experienced in obtaining calibration of a length measuring device. Probably the most serious problem to be faced is that good calibration requires considerable time: the standard under calibration must be observed for a time in order to ensure that it does have the long-term stability needed to hold the calibration.

3.5 Practice of length measurement for industrial use

3.5.1 General remarks

A large proportion of industrial range measurements can be performed quite adequately using simple mechanical gauging and measuring instruments. If, however, the requirement is for automatic measurement such as is needed in automatic inspection, or in closed-loop control, then the manual methods must be replaced by transducer forms of length sensor.

Figure 3.6 Tape-calibration facility at the National Measurement Laboratory New South Wales, Australia. Courtesy, CSIRO.

In many applications the speed of response needed is far greater than the traditional mechanical methods can yield. Numerically controlled mills, for instance, could not function without the use of electronic sensors that transduce the various axial dimensions into control signals.

Initially, that is in the 1950s, the cost of electronic sensors greatly exceeded that of the traditional mechanical measuring tools and their servicing required a new breed of technician. Most of these earlier shortcomings are now removed and today the use of electronic sensing can be more productive than the use of manually-read micrometers and scales because of the reduced cost of the electronic part of the sensing system and the need for more automatic data processing. There can be little doubt that solely mechanical instruments will gradually become less attractive in many uses.

3.5.2 Mechanical length-measuring equipment

Measurement of length from a micrometre to fractional metres can be performed inexpensively using an appropriate mechanical instrument. These group into the familiar range of internal and external micrometers, sliding-jaw calipers and dial gauges. Accuracy obtained with these depends much upon the quality of manufacture.

Modern improvements, which have been incorporated into the more expensive units, include addition of electronic transduction to give direct digital readout of the value making them easier to read and suitable for automatic recording.

Another improvement, for the larger throat micrometers, has been the use of carbon-fibre for throat construction. This has enabled the frame to be lighter for a given size allowing larger units and increased precision.

For the very best accuracy and precision work, use is made of measuring machines incorporating manually or automatically read, optical scales. Figure 3.7 shows the modern form of such measuring machines. This is capable of a guaranteed accuracy of around 1 μm in length measurements up to its full range capability of 300 mm. Larger machines are also made covering the range of around 4 m: these machines have also been adapted to provide electronic readout, see Section 3.6.

Where measurement of complex shapes is important, the use of measuring machines can be quite tedious; more speedy, direct methods can be used when the accuracy needed is not to the highest limits. In this aspect of tool-room measurement the optical profile projector may be applicable.

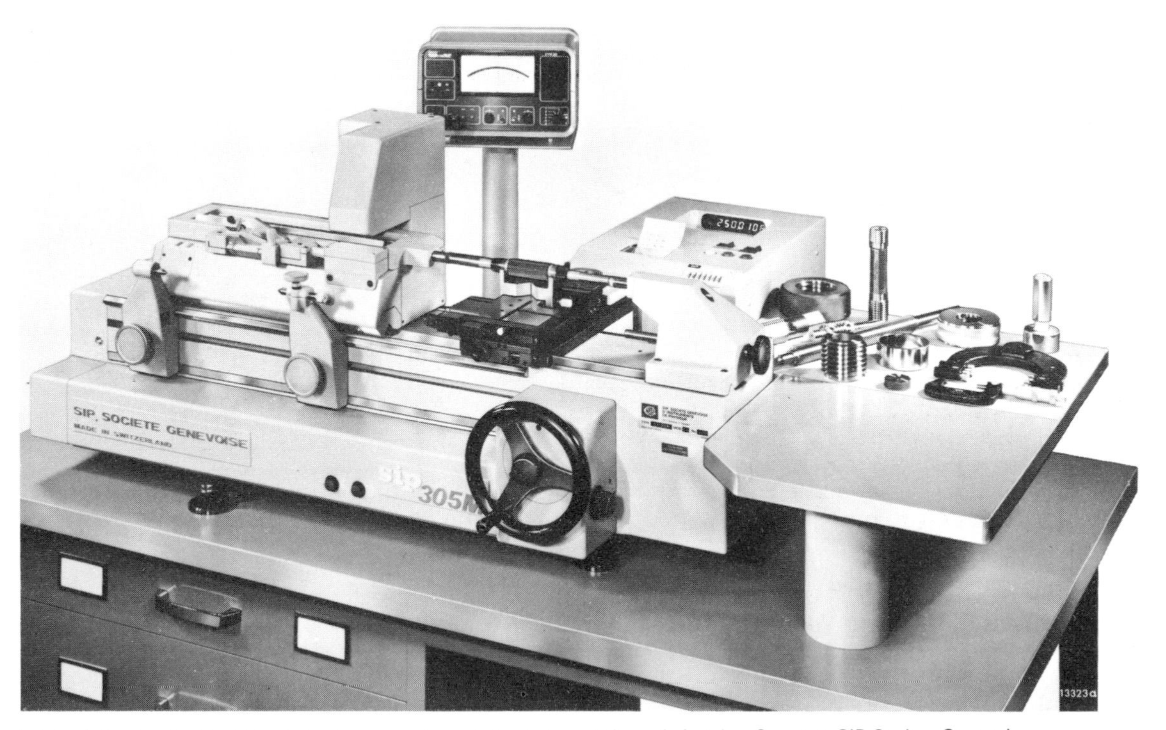

Figure 3.7 Automatically-read length-measuring machine incorporating ruled scales. Courtesy, SIP Society Genevoise d'Instruments de Physique.

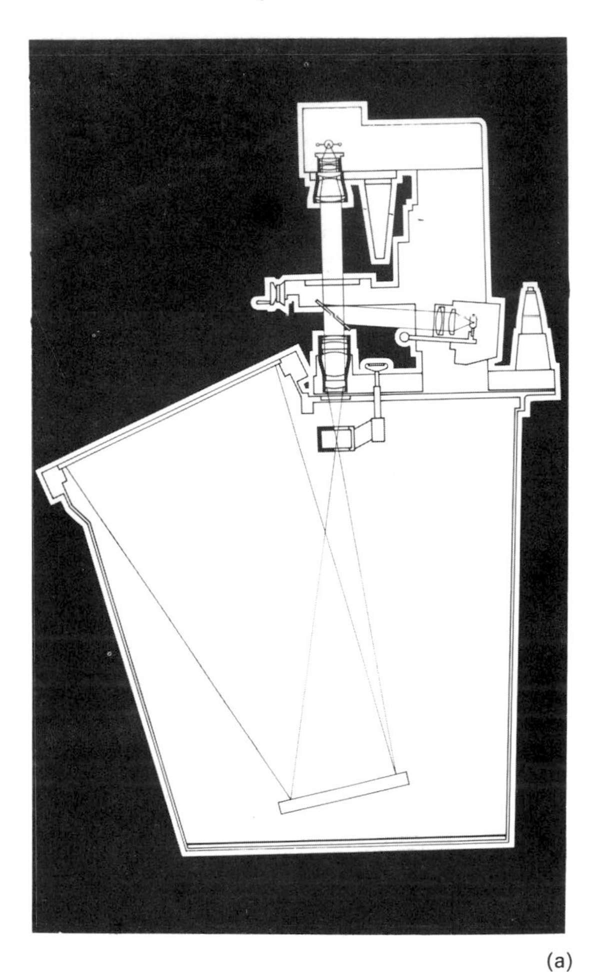

(a)

(b)

(c)

Figure 3.8 Measurement of geometry and lengths using the optical projector. (a) Optical system schematic. (b) Projected and enlarged image of profile of a component. (c) Oblique episcopic lighting to show surface detail.
Courtesy, Henri Hauser Ltd.

In these (see Figure 3.8(a)), the outline, Figure 3.8(b), of the article to be measured is projected on to a large screen. This can then be compared with a profile template placed around the image. It is also possible to project an image of the surface of an article, Figure 3.8(c), on to the viewing screen. The two forms of lighting can also be used in combination.

3.5.3 Electronic length measurement

Any physical principle that relates a length to a physical variable has the potential to be used for converting length to another equivalent signal. The most used output signal is the electronic form. Thus, most length sensors use a transduction relationship that converts length into an electrical entity either by a direct path or via one or more indirect conversion stages.

Most methods for smaller ranges make use of electromechanical structures, in which electrical resistance, inductance or capacitance vary, or make use of time and spatial properties of radiation. Basic cells of such units are often combined to form larger range devices having similar discrimination and dynamic range to those given by the best mechanical measuring machines.

For best results differential methods are utilized where practicable for this reduces the inherent errors (no transducers are perfect!) of the various systems by providing an in-built mechanism that compensates for some deficiencies of the transducer principle adopted. For example, to measure displacement it is possible to use two electrical plates forming a capacitor. As their separation varies, the capacitance alters to give a corresponding change in electrical signal. To use only one plate-pair makes the system directly susceptible to variations in the dielectric constant of the material between the plates; small changes in the moisture content in an air-gap can give rise to considerable error. By placing two plate-pairs in a differential connection, the effect of the air moisture can largely be cancelled out.

3.5.3.1 Electrical resistance

In essence some mechanical arrangement is made in which the electrical resistance between two ends of an interval is made to vary as the interval changes length.

Methods divide into two groups—those in which the resistance of the whole sensor structure remains constant, length being taken off as the changing position of a contact point, and those in which the bulk properties of the structure are made to change as the whole structure changes length.

In the first category is the slide-wire in which a single wire is used. A coiled system can provide a larger resistance gradient which is generally more suited to signal levels and impedance of practical electronic circuitry.

Figure 3.9(a) is a general schematic of sliding-contact length sensors. A standard voltage V_s is applied across the whole length of the resistance unit. The output voltage V_{out} will be related to the length l being measured as follows.

$$V_{out} = \frac{V_s}{L} \cdot l$$

Given that V_s and L are constant, V_{out} gives a direct measure of length l. The resistance unit can be supplied, V_s, with either d.c. or a.c. voltage. Errors can arise in the transduction process due to non-uniform heating effects causing resistance and length L change to the unit but probably the most important point is that the readout circuit must not load the resistance, for in that case the output-to-length relationship does not hold in a linear manner as shown above.

Sliding-contact sensors are generally inexpensive but can suffer from granularity as the contact moves from wire to wire, in wound forms, and from noise caused by the mechanical contact of the wiper moving on the surface of the wire. Wear can also be a problem as can the finite force imposed by the need to maintain the wiper in adequate contact. These practical reasons often rule them out as serious contenders for an application. Their use is, however, very simple to understand and apply. The gradient of the resistance with position along the unit can be made to vary according to logarithmic, sine, cosine and other progressions. The concept can be formed in either a linear or a rotary form. Discrimination clearly depends upon the granularity of the wire diameter in the wound types; one manufacturer has reduced this by sliding a contact along (rather than across) the wound wire as a continuous motion.

Resistance units can cover the range from around a millimetre to a metre with discrimination of the order of up to 1/1000 of the length. Non-linearity errors are of the same order.

The frequency response of such units depends more upon the mechanical mass to be moved during dynamic changes because the electrical part can be made to have low inductance and capacitance, these being the two electrical elements that decide the storage of electrical energy and hence slowness of electrical response.

Signal-to-noise performance can be quite reasonable but not as good as can be obtained with alternative inductive and capacitive methods; the impedance of a resistance unit set to give fine discrimination generally is required to be high with subsequent inherent resistance noise generation. These units are variously described in most general instrumentation texts.

The alternative method, Figure 3.9(b), makes use of strain of the bulk properties of the resistance, the most used method being the resistance strain gauge. As

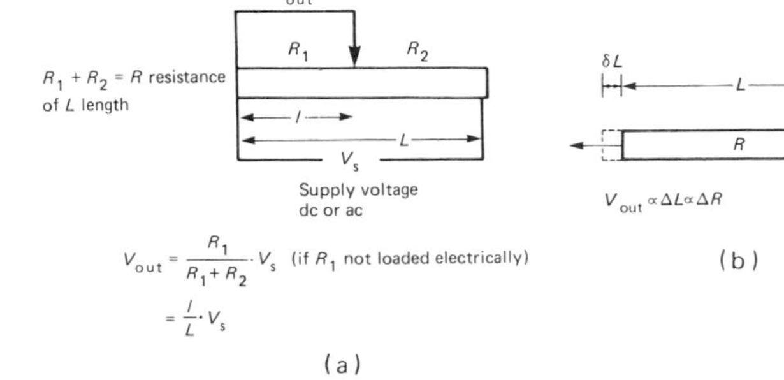

Figure 3.9 Electrical-resistance length sensors. (a) Sliding contact along fixed resistance unit. (b) Resistance change due to change of bulk properties of a resistance element induced by strain of the element.

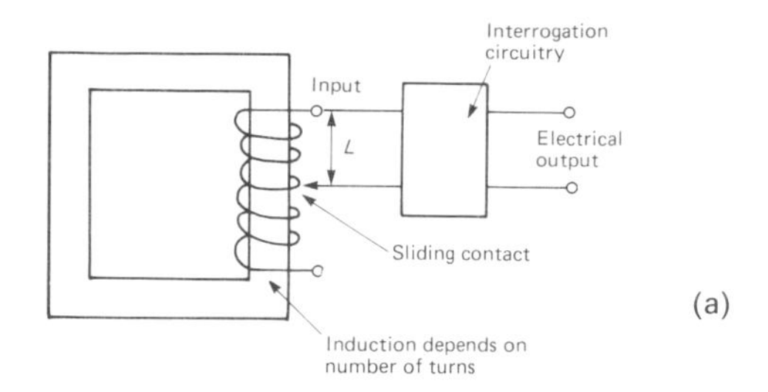

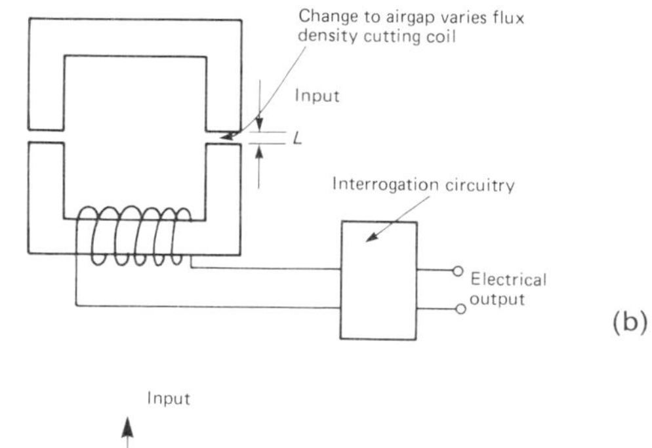

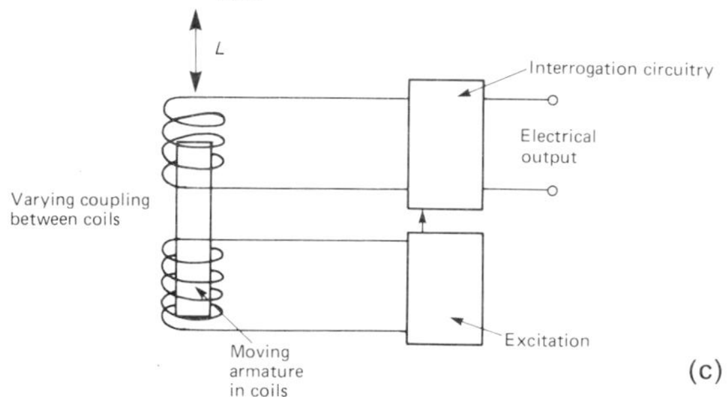

Figure 3.10 Electrical-inductance forms of sensor. (a) Turns variation with length change. (b) Reluctance variation with length change. (c) Mutual inductance change with length change.

strain gauges are the subject of Chapter 4 they will not be discussed further at this stage.

Al alternative bulk resistance method that sometimes has application is to use a material, such as carbon in disc form, using the input length change to alter the force of surface contact between the discs. This alters the pile resistance. The method requires considerable force from the input and, therefore, has restricted application. It does, however, have high electric current-carrying capability and can often be used to drive directly quite powerful control circuits without the need for electronic amplification. The bulk properties method can only transduce small relative length changes of an interval. Practical reasons generally restrict its use to gauge intervals of a few millimetres and to strains of that interval of around 1 per cent.

3.5.3.2 Electrical magnetic inductive processes

In general, the two main groups that use electrical inductive processes are those that vary the inductance value by geometry change, and those that generate a signal by the law of electromagnetic induction.

An electrical inductance circuit component is formed by current-carrying wire(s) producing a magnetic field which tends to impede current change in dynamic current situations. The use of a magnetic circuit enhances the effect; it is not absolutely essential, but is generally found in inductive sensors. Change of the magnetic field distribution of an inductor changes its inductance. Length sensors make use of this principle by using a length change of the mechanical structure of an inductance to vary the inductance. This can be achieved by varying the turns, changing the magnetic circuit reluctance or by inducing effects by mutual inductance. Various forms of electric circuit are then applied to convert the inductance change to an electronic output signal.

Figure 3.10 shows these three options in their primitive form. In some applications such simple arrangements may suffice but the addition of balanced, differential, arrangements and use of phase-sensitive detection, where applicable, is often very cost-effective for the performance and stability are greatly improved. Now described, in more detail, are examples of the mainly used forms of Figures 3.10(b) and 3.10(c).

Figure 3.11 shows a single-coil proximity detector that is placed close to a suitable, high magnetic permeability plate attached to or part of the subject. The sensor would be mounted around 2 mm from the plate. As the plate moves relative to the unit the reluctance of the iron circuit, formed by the unit, the plate and the air-gap, varies as the air-gap changes. When the unit has a permanent magnet included in the magnetic circuit then movement will generate a voltage without need for separate electronic excitation. It will not, however, produce a distance measurement when the system is stationary unless excited by a continuous a.c. carrier signal.

Where possible two similar variable-reluctance units are preferred, mounted one each side of the moving object, and connected into a bridge configuration giving common-mode rejection of

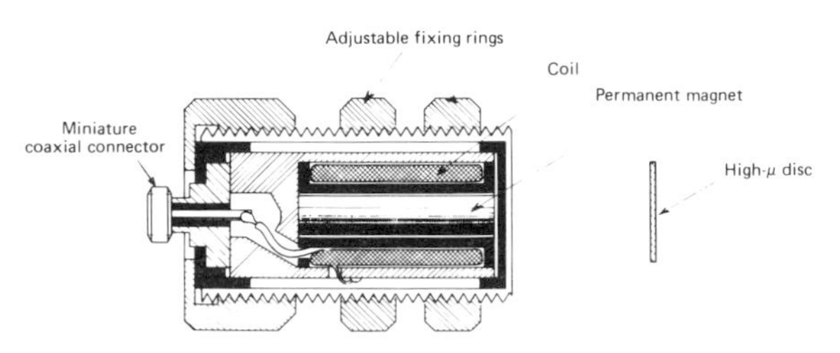

Figure 3.11 Magnetic-reluctance proximity sensor. Courtesy, Bruel & Kjaer Ltd.

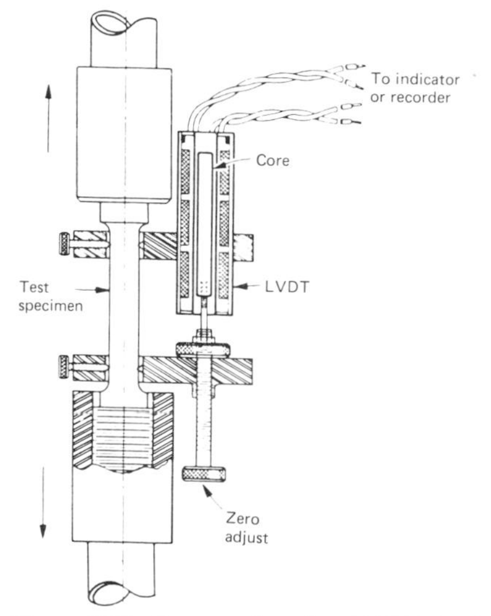

Figure 3.12 Cross-section of an LVDT inductive length sensor used to measure length change of a tensile test specimen. Courtesy, Schaevitz Engineering.

unwanted induced noise pick-up. These arrangements are but two of many possible forms that have been applied. Variable-reluctance methods are characterized by their relatively short range, poor linearity over longer ranges and the possible need to move a considerable mass in making the measurement with consequent restricted dynamic performance.

Mutual-inductance methods also exist in very many forms including the equivalent of Figure 3.11. Probably the most used is the linear variable-differential transformer LVDT. Figure 3.12 shows a cross-section through a typical unit mounted for monitoring length change of a tensile test specimen. A magnetic material core, normally a ferrite rod, moves inside three coils placed end to end. The centre coil is fed from an a.c. excitation supply thus inducing voltages into the outer two coils. (It can also be wound over the other two outer coils.) The two generated voltages will be equal when the core is positioned symmetrically. The voltage rises in one coil relative to the other when the core is off-centre. Difference between the two voltages is, therefore, related to the position of the core, and the relation can be made linear. Without circuitry to detect in which direction the core has moved from the null position, the output will be an a.c. signal of the excitation frequency which changes in amplitude with position and having direction in the signal as its phase.

Practical use generally requires a d.c. output signal (actually a signal having frequency components in it that are present in the measurand's movement) with direction information as signal polarity. This is easily achieved, at marginal additional expense, by the use of phase-sensitive detection (also known as lock-in detection or carrier demodulation). Figure 3.13(a) shows a block diagram of the subsystem elements that form a complete LVDT length-measuring system. Figure 3.13(b) shows the output relationship with position of the core. Modern units now often supply the phase-sensitive detection circuits inside the case of the sensor; these are known as d.c. LVDT units. Considerable detail of the operation of these and variations on the theme are available in Herceg (1976). Detail of phase-sensitive detection is in Part 4 and Sydenham (1982b) where further references will be found in a chapter on signal detection by D. Munroe.

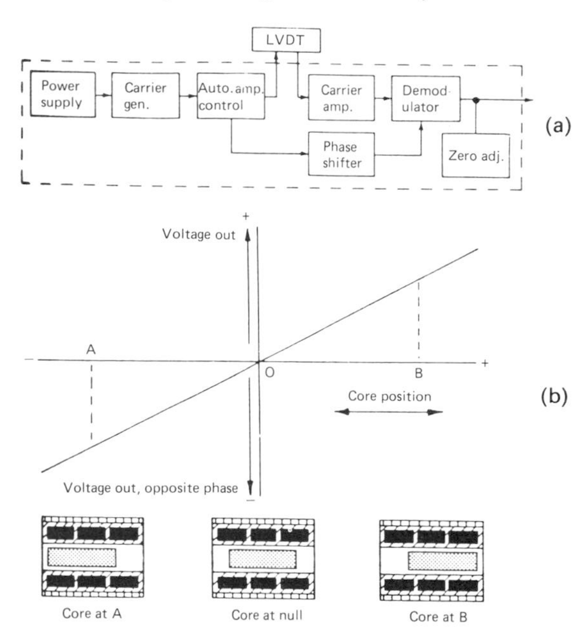

Figure 3.13 Phase-sensitive detection system used with practical LVDTs. (a) Block diagram of electronic system. (b) Input–output characteristics after phase-sensitive detection. Courtesy, Schaevitz Engineering.

A simpler non-transformer form of the LVDT arrangement can be used in which the need for a separate central excitation coil is avoided. With reference to the LVDT unit shown in Figure 3.12 the two outer coils only would be used, their inner ends being joined to form a centre-tapped linearly-wound inductor. This split inductor is then placed in an a.c. bridge to which phase-sensitive detection is applied in a similar manner to recover a polarized d.c. output signal.

Inductive sensors of the mutual-inductance form are manufactured in a vast range of sizes providing for length detection from as small as atomic diameters (sub-nanometre) out to a maximum range of around ± 250 mm. They are extremely robust, quite linear, very reliable and extremely sensitive if well designed. By mounting the core on the measurand object and the body on the reference frame it is also possible to arrange for non-interacting, non-contact measurement. Frequency response depends upon the carrier frequency used to modulate the coils; general practice sets that at around 50 kHz as the upper limit but this is not a strong constraint on design. Attention may also need to be paid to mechanical resonances of the transducer.

The second class of magnetically inductive length sensors, the magneto-electric kind, are those in which part of the electromagnetic circuit moves to generate a voltage caused by flux-cutting. For these the relevant basic expression is $e = -(N\,\mathrm{d}\phi/\mathrm{d}t)$ where e is the voltage generated as N turns are cut by flux ϕ in time t. These are strictly velocity, not displacement sensors, for the output is proportional to velocity. Integration of the signal (provided it has sufficient magnitude to be detected, for at low speeds the signal induced may be very small) will yield displacement. This form of sensor is covered in more detail in Chapter 6.

Magneto-electric sensors are prone to stray magnetic field pick-up for that can introduce flux-cutting that appears to be signal. Good shielding is essential to reduce effects of directly detected fields and also for the more subtle induced eddy current fields that are produced in adjacent non-magnetic materials if they are electrically conducting. Magnetic shielding is a highly developed engineering process that requires special materials and heat treatments to obtain good shielding. Manufacturers are able to provide units incorporating shielding and to advise users in this aspect. It is not simply a matter of placing the unit in a thick, magnetic material case! Herceg (1976) is a starting point on effective use of LVDT units and other inductive sensors.

When larger range is required than can be conveniently accommodated by the single unit (desired discrimination will restrict range to a given length of basic cell) it is possible to add inductive sensor cells, end-to-end, using digital logic to count the cells along the interval with analogue interrogation of the cell at the end of the measurand interval. This hybrid approach was extensively developed to provide for the metre distances required in numerically controlled machine tools. A form of long inductive sensor that has stood the test of time is the flat, 'printed', winding that is traversed by a sensing short, flat coil. Each cell of this continuous grid comprises a coil pair overlapped by the sense coil that forms a flat profile LVDT form of sensor.

Angles can be measured electro-magnetically using devices called Synchros. These inherently include means for transmitting the information to a distance and are therefore described under Telemetry in Part 4.

3.5.3.3 Electrical-capacitance sensors

Electrical capacitance stores electrical energy in the electric field form; electrical inductors store energy in the magnetic field form. Electromagnetic theory relates the two forms of field and thus most concepts applied to magnetic-inductance sensing are applicable to electrical-capacitance structures.

It is, therefore, also possible to sense length change by using the input length to alter the structure of a capacitance assembly or to cause shape change to a solid material thereby directly generating charge. The electrical capacitance of a structure formed by two electrically conducting plates is given by

$$C = \frac{\varepsilon A}{l}$$

where C is the capacitance, ε the dielectric constant of the material between the plates in the area A where they overlap at a distance of separation l.

Thus a length sensor can be formed by varying the value of l which gives an inverse relationship between length input and capacitance change or by varying one length of the two plates that combine to provide the overlap area A; this latter alternative can give a direct relationship. It is also sometimes practical to vary ε by inserting a moving piece of different dielectric into the capacitance structure.

Simpler forms make use of a single capacitance structure, but for these, variation in ε can introduce error due to humidity and pressure changes of air, the most commonly used dielectric. Differential systems are more satisfactory. Figure 3.14 shows the basic

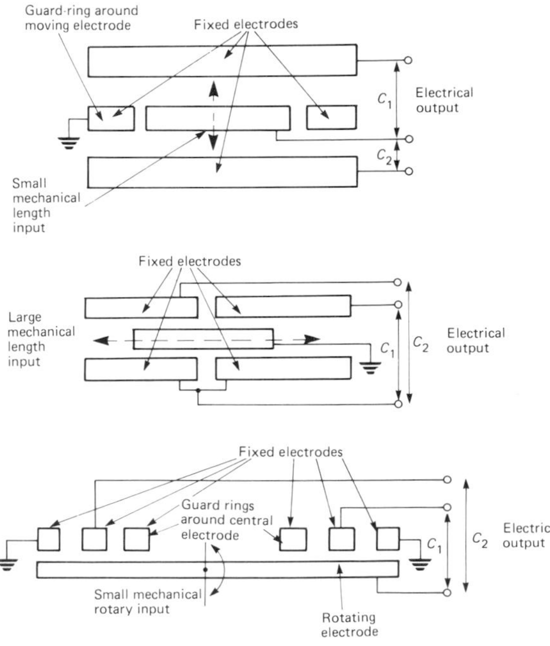

Figure 3.14 Some differential capacitance length-sensing structures.

configuration of differential capacitance sensors. They can be formed from flat plates or from cylindrical sections, whichever is convenient. The cylindrical form of the second design has been used in a highly accurate alternative to the LVDT. The guard rings shown are not needed if some non-linearity can be accepted.

Capacitance systems are characterized by high output impedance, need for relatively high excitation voltages, accuracy of plate manufacture and small plate clearance dimensions.

Potential sensitivity of the alternatives (inductive and capacitive) is virtually the same in practice: each having its own particular signal-to-noise problems. For low-sensitivity use, capacitance devices are more easily manufactured than the inductive alternatives.

Noise occurs in capacitance systems from charge pick-up produced by stray voltage potentials in the high impedance of the capacitance assembly. It is reduced by appropriate shielding and use of earthed guard plates that surround the active plates to collect and dump the unwanted charge. Capacitance structures lend themselves more for original equipment design rather than as ready-made sensor units applied after the basic plant is designed. This is because the layout of the working plant to be sensed can often provide directly one or more plates of the sensor as a non-contacting arrangement. For example, to monitor position of a pendulum in a tilt sensor it is straightforward to use the pendulum bob as the central plate that moves inside two plates added one to each side.

In general, therefore, it will be found that commercial sensor makers offer more inductive systems than capacitive alternatives, whereas in scientific and research work the tendency is to use capacitance systems as they are easier to implement at the prototype stage. Commercial suppliers also wish to offer a product that is self-contained and ready to apply and, therefore, a unit that can be verified before delivery.

At extreme limits of discrimination (sub-nanometres) capacitance sensing can be shown to be superior to inductive arrangements if properly designed. As with inductive systems they also need a.c. excitation to obtain a length change response for slowly moving events.

Forces exerted by the magnetic and electric fields of the two alternatives can be designed to be virtually non-existent. Use in sensitive force balance systems allows the same plates to be used to apply a d.c. voltage to the central plate of a differential system. The plate can thus be forced back to the central null position at the same time as a higher frequency excitation signal is applied for position detection.

Although the plate-separation capacitance method fundamentally provides an inverse relationship to length change the signal can be made directly proportional by placing the sensing capacitance in the feedback path of an operational amplifier.

3.5.4 Use of electromagnetic and acoustic radiation

Radiation ranging from the relatively long radio wavelengths to the short-wavelength X-rays in the electromagnetic (EM) radiation spectrum, and from audio to megahertz frequencies in the acoustic spectrum, has been used in various ways to measure length.

In the industrial range the main methods adopted have been those based upon optical and near-optical radiation, microwave EM and acoustic radiation. These are now discussed in turn.

3.5.4.1 Position-sensitive photocells

An optical beam, here to be interpreted as ranging in wavelength from infrared ($\approx 10\ \mu m$) to the short visible ($\approx 0.4\ \mu m$), can be used in two basically different ways to measure length. The beam can be used to sense movements occurring transverse to it or longitudinally with it.

Various position-sensitive optical detectors have been devised to sense transverse motion. Their range is relatively small, being of the order of only millimetres. They are simple to devise and apply and can yield discrimination and stability of the order of a micrometre.

Figure 3.15 outlines the features of the structure of the three basic kinds that have been designed. Consider that of Figure 3.15(a). A beam having uniform radiation intensity across its cross-section falls upon two equal characteristic photocells. When the beam straddles the two cells, thereby providing equal illumination of each, the differentially connected output will be zero. At that point any common-mode noise signals will be largely cancelled out by the system. Such systems have good null stability. As the beam moves to one side of the null the differential output rises proportionally until all of the beam's illumination falls on one cell alone. Direction of movement is established by the polarity of the output signal. Once the beam has become fully placed on one cell the output is saturated and remains at its maximum. These cells can be manufactured from one silicon slice by sawing or diffusing a non-conducting barrier in the top junction layer or can be made from separate cells placed side by side. Four cells, placed in two perpendicular directions in a plane, can be used to sense two axes of motion.

Linearity of the output of these cells depends upon their terminating conditions. Working range can be seen to be equal to twice the beam width which should not exceed the width of the half detector size. Sensitivity depends upon the level of beam illumination so it is important to have constant beam intensity to obtain good results.

In some applications the light-beam cross-section

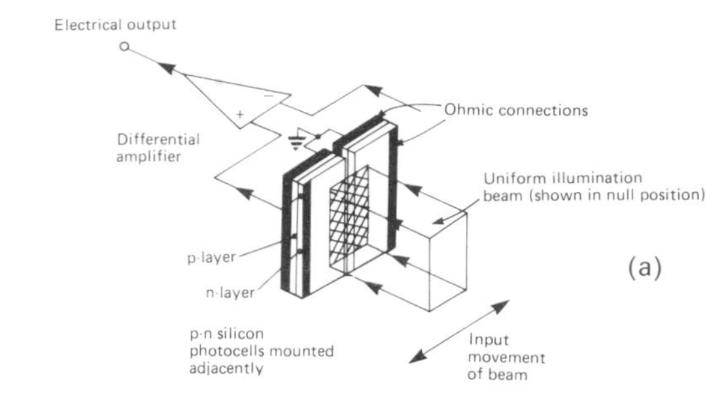

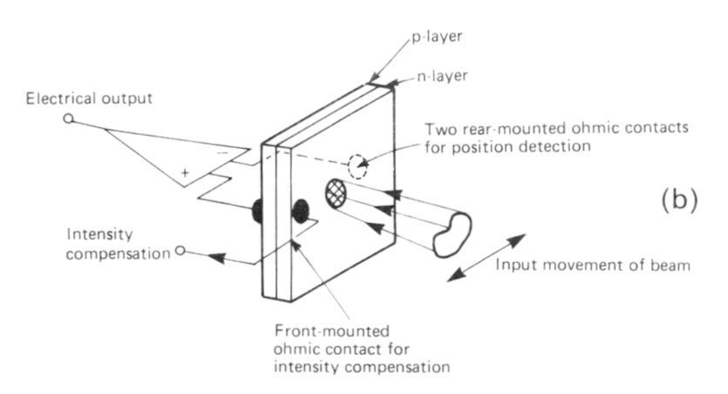

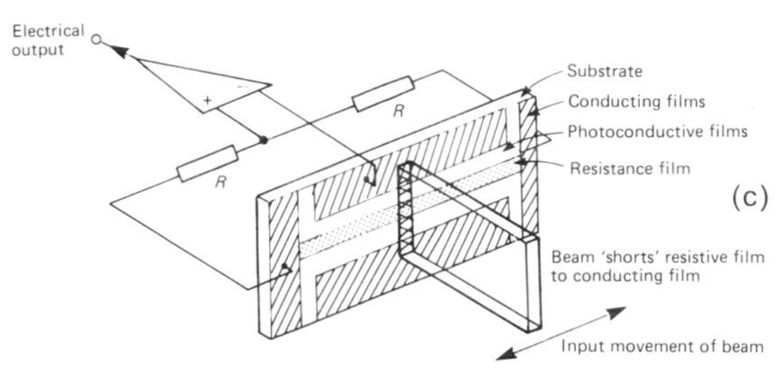

Figure 3.15 Optical position-sensitive detectors. (a) Split cell. (b) Lateral effect cell. (c) Photopotentiometer.

may vary with distance to the cell. In such cases the so-called Wallmark or lateral effect cell, Figure 3.15(b), may be more appropriate. In this case the two contacts are not rectifying as in the Figure 3.15(a) case but are, instead, ohmic.

It has been shown that the voltage produced between the two ohmic contacts is related to the position of the centroid of the beam's energy and to the intensity of the whole beam. Addition of the rectifying contact on the other side of the cell enables correction to be made for intensity changes making this form of cell able to track the movements of a spot of radiation that changes both in intensity and in size. Here also the beam must be smaller than the full working region of the cell surface. Detection limits are similar to those of the split cell form.

This cell has enjoyed a resurgence of interest in its design for the original logarithmic voltage-to-position characteristic can quite easily be arranged to be effectively linear by driving the cell into the appropriate impedance amplifier. It, too, is able to sense the motion of a beam in two axes simultaneously by the use of two additional contacts placed at right angles to those shown.

Optical position-sensitive photocells, such as these, have found extensive use in conjunction with laser sources of radiation in order to align floors, ceilings and pipes and to make precise mechanical measurement of geometry deviations in mechanical structures.

The third form of optical position detector is the photopotentiometer shown in Figure 3.15(c). This form, although invented several years before micro-electronic methods (it uses thick-film methods of manufacture) has also found new interest due to its printable form. The input beam of light falls across the junction of the conducting and resistive films causing, in effect, a wiper contact action in a Wheatstone bridge circuit. The contact is frictionless and virtually stepless. The range of these units is larger than for the position-sensitive photocells but they are rather specialized; few are offered on the market. Their response is somewhat slow (10 ms) compared with the cells detailed above (which have microsecond full scale times) due to the time response of the photoconductive materials used. The light beam can be arranged to move by the use of a moving linear shutter, a moving light source or a rotating mirror.

Moiré fringe position sensing methods make use of mechanical shuttering produced by ruled lines on a scale. These produce varying intensity signals, at a reference position location, that are in a fixed phase relationship, see Figure 3.16. These signals are interrogated to give coarse, whole line cycle counts with cycle division ranging from simple four-times digital division up to around 1 part in 100 of a cycle by the use of analogue subdivision methods. Moiré fringe scales are able to provide large range (by butting sections)

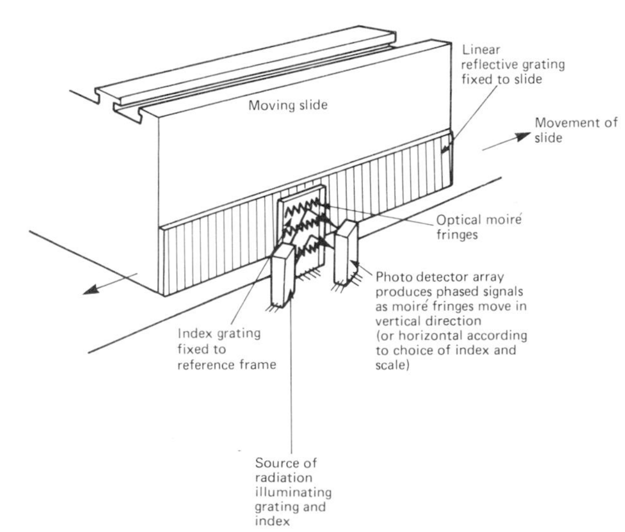

Figure 3.16 Moiré fringes used to measure length.

and a discrimination level at the subdivision chosen which can be as small as 1 μm. Accuracy is limited by the placement of the lines on the scale, by the mounting of the scale and by the temperature effects that alter its length. The method is suitable for both linear and rotary forms of sensing.

Although there was much developmental activity in moiré methods in the 1950 to 1965 period their design has stabilized now that the practicalities have been realized.

Moiré methods have also found use for strain investigation, a subject discussed in Chapter 4.

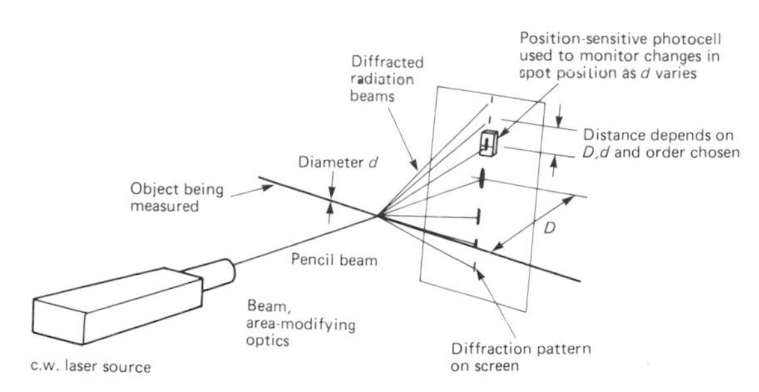

Figure 3.17 Use of diffraction to gauge fine diameters.

The easily procured coherent properties of continuous wave laser radiation has provided a simple, yet highly effective method for monitoring the diameter of small objects such as wire as it is being drawn. The radiation, being coherent, diffracts at the intersection with the wire producing a number of diffraction beams at points past the wire. Figure 3.17 shows the method. As the wire size reduces the spots diverge giving this method improved precision as the object size becomes smaller.

The position of the spots can be sensed with a position-sensitive photocell as described above. Where digital output of beam movement is needed it is also possible to use a linear array of photodetectors each operating its own level detector or being scanned, sequentially, into a single output channel. Linear arrays with over 200 elements are now commonplace.

Optical lenses and mirrors can be used to alter the beam movement geometry in order to scale the subject's movement amplitude to suit that of an optical position detector.

3.5.4.2 Interferometry and transit-time methods

Where long length (metres and above) measurements are needed it is possible to use a suitable beam of radiation sensed in its longitudinal direction. Several different methods are available.

If the beam has time-coherent properties (either superimposed or inherently in the carrier) it will then be possible to use interference methods to detect phase differences between a reference part of the beam and that sent to the subject to be reflected back. Laser, microwave sources and coherently generated acoustic radiation each can be used in this interference mode. The shorter the wavelength of the radiation the smaller the potential discrimination that can be obtained. Thus with optical interferometers it is practical to detect length differences as small as a nanometre but only around a few millimetres when microwave radiation is used.

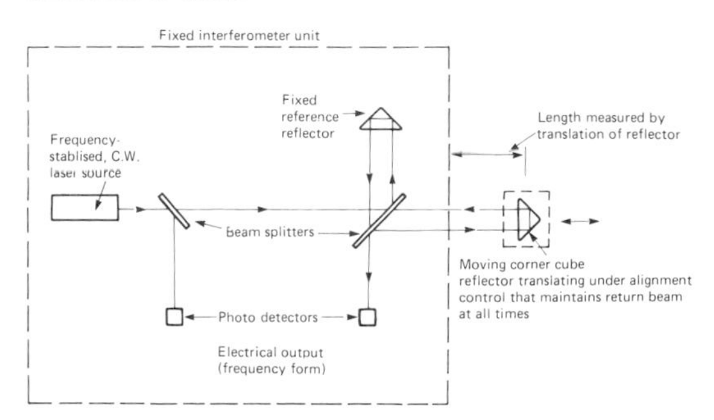

Figure 3.18 Basic layout of the frequency output form of laser length-measuring interferometer.

Figure 3.18 shows the basic layout of the optical elements needed to form a laser-based interferometer. That shown incorporates frequency stabilization and Zeeman splitting of the radiation features that give the system a highly stable and accurate measurement capability and a frequency, rather than amplitude form, of output representing length. A commercial unit is shown in Figure 3.5. Corner cubes are used instead of the flat mirrors originally used in the classical Michelson interferometer. They make adjustment very straightforward for the angle of the cube to the incoming radiation is not critical. Usually one corner cube is held fixed to provide a reference arm. The other corner cube must be translated from a datum position to give length readings for the method is inherently incremental. Allowing both corner cubes to move enables the system to be configured to also measure small angles to extreme precision.

The laser interferometer is, without doubt, the superior length-measuring instrument for general-purpose industrial work but its cost, line-of-sight beam movement restriction, incremental nature and need for path condition control (for the best work) does eliminate it from universal use. Its dynamic measuring range covers from micrometres to tens of metres, a range not provided by any other electronic length sensor.

Interferometry requires a reflector that gives adequate energy return without wavefront distortion. At optical wavelengths the reflecting surface must be of optical quality. Where very fine, micrometre, discrimination is not required the use of microwave radiation allows interferometry systems that can operate directly on to the normal machined surface of components being machined. Acoustic methods can yield satisfactory results when the accuracy needed is only of the order of 1 part in 1000.

Radiation methods can also make use of the time of flight of the radiation. For light this is around 300 mm in a nanosecond and for acoustic vibration from 300 mm to 6 m in a millisecond, depending upon the medium. In 'time-of-flight' methods the radiation, which can here be incoherent, is modulated at a convenient frequency or simply pulsed. The radiation returning from the surface of interest is detected and the lapsed time to go and return is used to calculate the distance between the source and the target. These methods do not have the same discrimination potential that is offered by interferometry but can provide in certain applications (for EDM systems used over kilometre ranges) uncertainty of the order of a few parts in 10^6. By the use of more than one modulation frequency it is possible to provide absolute ranging by this method, a feature that is clearly required for long distance measurements. The need for a controlled movement path over which the reflector must traverse the whole distance, as is required in incremental interferometers, is unworkable in surveying operations.

The interference concept shown in Figure 3.18 for one-dimensional movement can be extended to three-dimensional measurement application in the holograph method. Holography provides a photographic image, in a plate known as a hologram, that captures the three-dimensional geometric detail of an object. The object of interest is flooded with coherent radiation of which some reflects from the surfaces to be optically combined with reference radiation taken directly from the source. As the two are coherent radiations their wavefronts combine to form a flat two-dimensional interference pattern. The hologram bears little pictorial resemblance to the original object and has a most unexpected property. When the hologram is illuminated by coherent light the object can be seen, by looking through the illuminated hologram, as an apparent three-dimensional object.

This basic procedure can be used in several forms to provide highly discriminating measurements of the shape of objects.

A first method places a similar object to that for which a hologram has been made in the image space of that of the hologram. This, in effect, superimposes the standard object over the real object. Differences between the two can then be decided by eye. This is not a very precise method but does suit some inspection needs.

In another method for using holography a second hologram is formed on the same plate as the first was exposed on. The combined pair is developed as a single plate. When viewed, as explained above, this will reproduce an apparent object on which are superimposed fringes that represent shape differences between the two units. Each fringe width, as a guide, represents detailed differences of the order of the wavelength of the radiation. This form of holography is, therefore, a very powerful method for detecting small differences. It has been used, for example, to detect imperfections in car tyres (by slightly altering the internal pressure) and to investigate shape changes in gas cylinders. It is very suitable for non-destructive testing but is expensive and somewhat slow in its use.

Fast moving objects can also be gauged using optical holography in the so-called time-lapse pulse holography method. Two holograms are exposed on top of each other on an undeveloped plate as mentioned above but in this situation they are formed by the same object which presents itself periodically at known times, for example a turbine blade rotating inside an aircraft engine. The laser source is pulsed as the object passes using synchronized electronic circuitry.

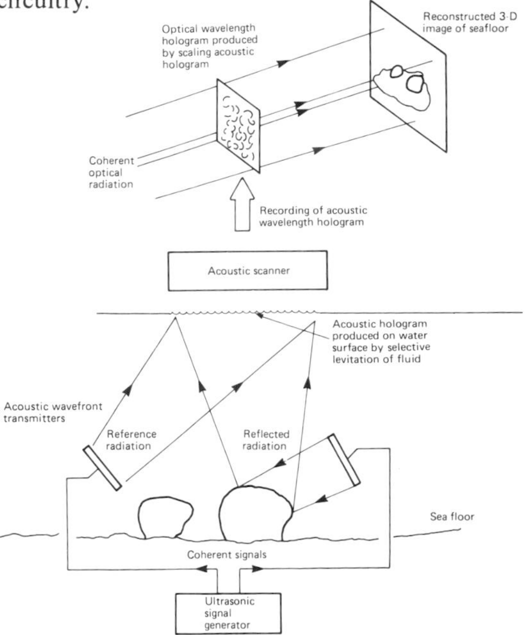

Figure 3.19 Holography applied to sea-floor mapping. Two radiations are used to convert the image size for reasons of convenience.

Holography is suitable for use with any form of coherent radiation; optical, microwave and acoustic systems have been reported. It is also possible to mix the radiations used at various stages in order to produce, and view, the hologram with different absolute size scales. For example, a sea-floor sand-profile mapping system, Figure 3.19, uses acoustic radiation inside the sea space to obtain an acoustic interference hologram which is then viewed by optical radiation for reasons of convenience.

The most serious disadvantages of holography are the cost of the apparatus, slowness to produce an output and difficulties in obtaining numerical measurements from the recorded information.

3.5.5 Miscellaneous methods

The above descriptions have shown that even for a few restricted classes of sensor there are many principles that can be used to produce transduced length signals. A comprehensive coverage would require several volumes on this parameter alone. This short sub-section is included to emphasize the availability of many more methods that may be appropriate in given circumstances. Many of the unusual methods are less likely to be marketed for potential sales would not justify quantity manufacture. In applications of the aerospace industry, in original equipment needs of science and in industrial testing, in development and in isolated applications they may be the most viable methods to adopt. It should not be construed that lack of commercial interest implies that a method is necessarily unworkable. Here are a few of these less commonly used methods.

Magneto-resistive sensing elements are those in which their electrical resistance varies with the level of ambient magnetic field. These can be used as linear, or as proximity, sensors by moving the sensor relative to a field which is usually provided by a permanent magnet.

Thickness of a layer being deposited in a deposition chamber can be measured by several means. One way is to monitor the mass build-up on a test sample placed in the chamber alongside that being coated. Another method is to directly monitor the change in optical transmission during deposition.

Statistical calculation on the signal strength of an ionizing radiation source can be used to determine distance from the source.

Pressure formed, or liquid displaced, can be used to drive a pressure- or volume-sensitive device in order to measure movement of the driving element. This method has been used in the measurement of volumetric earth strain as it can provide very sensitive detection due to the cube-law relationship existing between volume change input and length output.

The following chapter deals specifically with strain measurements. Chapter 6 on vibration, and Chapter 5, on level measurement, each include descriptions of length sensors.

3.6 Automatic gauging systems

Tool-room and factory gauging has its roots in the use of manually-read measuring machines and tools such as those shown in Figures 3.7 and 3.8. These required, in their non-electronic forms, high levels of skill in their use and are very time-consuming. In general, however, they can yield the best results given that the best machines are used.

The advent of electronic sensing of length and angle has gradually introduced a transformation in the measurement practices in the tool-room and on the factory floor. The cost of providing, using manual procedures, the very many inspection measurements needed in many modern production processes has often proven to be uneconomic and far too slow to suit automatic production plants. As an example piston manufacture for a car engine can require over twenty length parameters to be measured for each piston produced on the final grinding machine. The time available to make the measurements is of the order of fractions of minutes. It has, in such instances become cost-effective to design and install automatic measuring machines that generate the extensive data needed.

Automatic measuring systems are characterized by their ability to deliver electronic signals representing one or many length dimensions. In simple applications the operator places the component in a preset, multi-probe system. In totally automated plant use is often made of pick-and-place robots to load the inspection machines.

Automatic inspection systems began their development in the 1950s when they were required to complement the then emerging numerically controlled metal-working machine tools. They made use of similar measuring sensors as did the tools but differed from the metal working machine in several ways.

Where inspection machines are hand-operated the operator can work best when the system effectively presents no significant inertial forces to the input probe as it is moved. This can be achieved by a design that minimizes the moving masses or by the use of closed-loop sensor control that effectively reduces the sluggish feel due to the inertia of the moving mass present. For small-size systems (those around a metre in capacity) multi-point inspection needs can be met economically by the use of short-range length sensors. These come into contact with the surfaces to be measured as the component is placed in the test set-up. Values are recorded, stored and analysed. The component may need to be rotated to give total coverage of the surfaces of interest. Figure 3.20 shows such an apparatus being used to automatically inspect several length dimensions of a gearbox shaft.

When the size of the object to be inspected is large the use of multiple probes can be too expensive and a single probe may be used to check given locations as a serial operation. Manual methods of point-to-point movement have, in some applications, given way to automatic, surface contour-following, probes and to the use of robot arms that are preprogrammed to move as required; see Figure 3.21.

The reliability of electronics, its cost-effectiveness and its capability to be rapidly structured into circuits that can suit new data-processing, recording and display situations has made transducer measurement of length parameters a strong competitor for the traditional manually-operated measuring machines. In most cases, however, the quality of length measurements made with automatic transducer methods still rests largely upon the mechanical structures of the sensing systems and upon the user's appreciation of the effects of temperature, operation of transducer and presentation to the subject, all of which can generate sources of error.

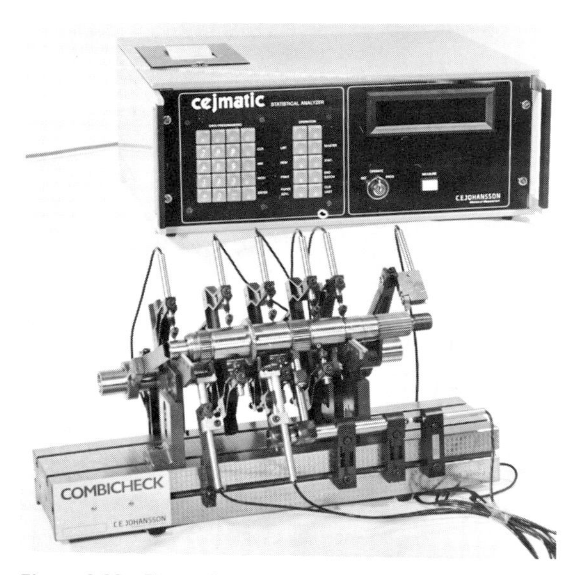

Figure 3.20 Electronic gauge heads being used in a versatile test apparatus set up to inspect several length parameters of a gearbox shaft. Courtesy, C.E. Johansson.

3.7 References

Batson, R. G. and Hyde, J. H. *Mechanical Testing; Vol. 1. Testing of Materials of Construction*, Chapman and Hall, London, (1931)

Garratt, J. D. 'Survey of displacement transducers below 50 mm', *J. Phys. E: Sci. Instrum.*, **12**, 563–574, (1979)

Herceg, E. E. *Handbook of Measurement and Control*, Schaevitz Engineering, Pennsauken, (1976)

Hume, K. J. *Engineering Metrology*, Macdonald, London, (1970)

Kissam, P. *Optical Tooling for Precise Manufacture and Alignment*, McGraw-Hill, New York, (1962)

Mansfield, P. H. *Electrical Transducers for Industrial Measurement*, Butterworths, London, (1973)

Neubert, K. K. P. *Instrument Transducers*, Clarendon, Oxford, 2nd ed. (1975)

Norton, H. N. *Handbook of Transducers for Electronic Measuring Systems*, Prentice-Hall, Englewood Cliffs, New Jersey, (1969)

Oliver, F. J. *Practical Instrumentation Transducers*, Pitman, London, (1971)

Rolt, R. H. *Gauges and Fine Measurements*, 2 vols, Macmillan, London, (1929)

Sharp, K. W. B. *Practical Engineering Metrology*, Pitman, London, (1970)

Sydenham, P. H. 'Linear and angular transducers for positional control in the decametre range', *Proc. IEE*, **115**, 7, 1056–1066, (1968)

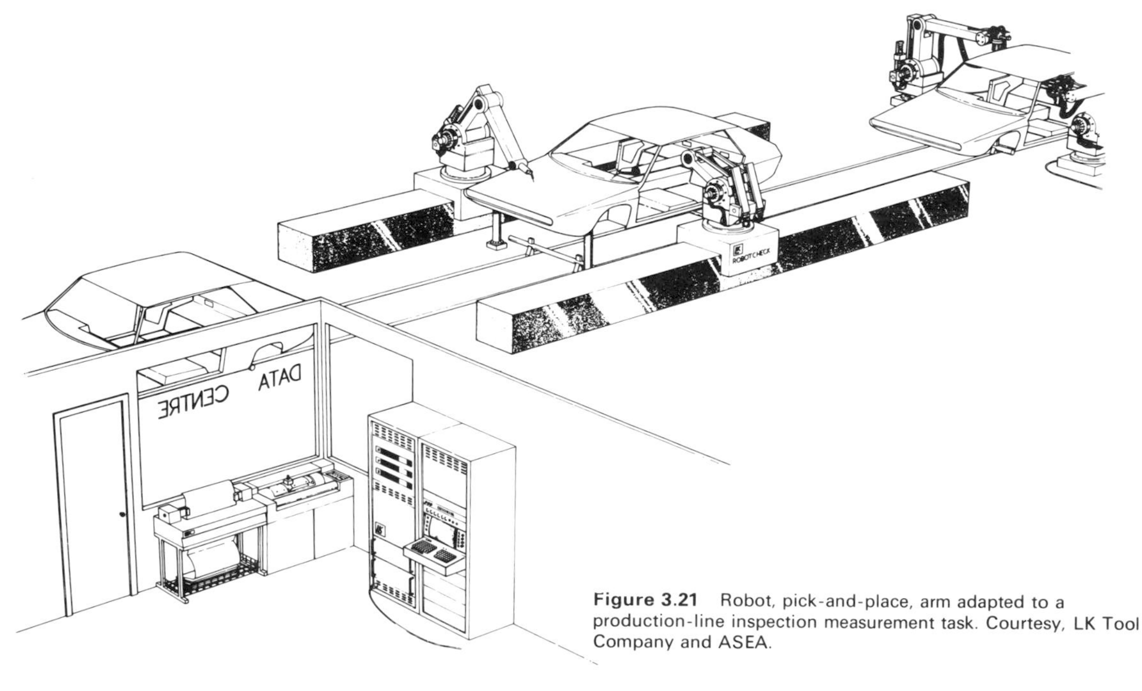

Figure 3.21 Robot, pick-and-place, arm adapted to a production-line inspection measurement task. Courtesy, LK Tool Company and ASEA.

Sydenham, P. H. 'Position sensitive photo cells and their application to static and dynamic dimensional metrology', *Optica Acta*, **16**, 3, 377–389, (1969)

Sydenham, P. H. 'Review of geophysical strain measurement', *Bull., N.Z., Soc. Earthquake Engng.*, **4**, 1, 2–14, (1971)

Sydenham, P. H. 'Microdisplacement transducers', *J. Phys., E. Sci., Instrum.*, **5**, 721–733, (1972)

Sydenham, P. H. *Transducers in Measurement and Control*, Adam Hilger, Bristol; or ISA, Research Triangle, (1984).

Sydenham, P. H. 'The literature of instrument science and technology', *J. Phys. E.: Sci. Instrum.*, **15**, 487–491, (1982a)

Sydenham, P. H. *Handbook of Fundamentals of Measurement Systems Vol. 1 Theoretical Fundamentals*, Wiley, Chichester, (1982b)

Sydenham, P. H. *Handbook of Measurement Science Vol. 2 Fundamentals of Practice*, Wiley, Chichester, (1983)

Woltring, H. J. 'Single and dual-axis lateral photodetectors of rectangular shape', *IEEE TRANS ED*, 581–590, (1975)

4 Measurement of strain

B. E. NOLTINGK

4.1 Strain

A particular case of length measurement is the determination of strains, i.e. the small changes in the dimensions of solid bodies as they are subjected to forces. The emphasis on such measurements comes from the importance of knowing whether a structure is strong enough for its purpose or whether it may fail in use.

The interrelation between stress (the force per unit area) and strain (the fractional change in dimension) is a complex one, involving in general three dimensions, particularly if the material concerned is not isotropic, i.e. does not have the same properties in all directions. A simple stress/strain concept is of a uniform bar stretched length-wise, for which Young's modulus of elasticity is defined as the ratio stress:strain, i.e. the force per unit of cross-sectional area divided by the fractional change in length

$$E=\frac{F}{A}\div\frac{\Delta l}{l}$$

The longitudinal extension is accompanied by a transverse contraction. The ratio of the two fractions (transverse contraction)/(longitudinal extension) is called Poisson's ratio, denoted by μ, and is commonly about 0.3. While we have talked of increases in length, called positive strain, similar behaviour occurs in compression, which is accompanied by a transverse expansion.

Another concept is that of shear. Consider the block PQRS largely constrained in a holder, Figure 4.1. If this is subjected to a force F as shown, it will distort, PQ moving to P′Q′. The 'shear strain' is the ratio PP′/PT, i.e. the angle PTP′ or $\Delta\theta$ (which equals angle QUQ′) and the *modulus of rigidity* is defined as (shear stress)/(shear strain) or

$$C=\frac{F}{A}\div\Delta\theta$$

when A is the area PQ × depth of block. In practical situations, shear strain is often accompanied by bending, the magnitude of which is governed by Young's modulus.

There is some general concern with stress and strain at all points in a solid body, but there is a particular interest in measuring strains on surfaces. It is only there that conventional strain gauges can readily be used and wide experience has been built up in interpreting results gained with them. At a surface, the strain normal to it can be calculated because the stress is zero (apart from the exceptional case of a high fluid pressure being applied), but we still do not usually know the direction or magnitude of strains in the plane of the surface so that for a complete analysis three strains must be measured in different directions.

4.2 Bonded resistance strain gauges

In the early 1940s, the bonded resistance strain gauge was introduced and it has dominated the field of strain measurement ever since. Its principle can be seen from Figure 4.2. A resistor R is bonded to an insulator I, which in turn is fixed to the substrate S whose strain is to be measured. (The word 'substrate' is not used universally with this meaning; it is adopted here for its convenience and brevity.) When S is strained, the change in length is communicated to R if the bonding is adequate; it can be shown that the strain will be transmitted accurately even through a mechanically compliant insulator provided there is sufficient overlap, i.e. if I is larger than R by several times the thickness of either of them. Strains of interest are commonly very small; for elastic behaviour, where concern is usually concentrated, strains do not exceed about 10^{-3}. Many metals break if they are stretched by a few per cent and changes in length of a few parts in a million are sometimes of interest but when these are used to produce even small changes in the resistance of R we can take advantage of the precision with which

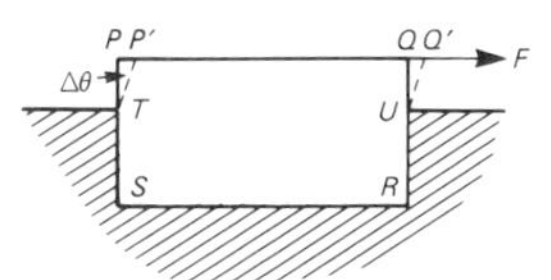

Figure 4.1 Shear strain.

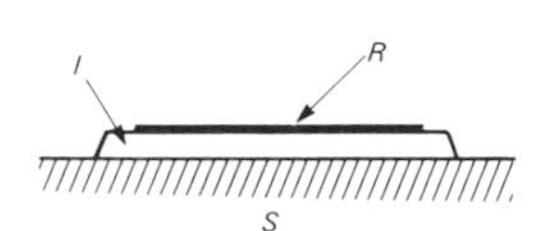

Figure 4.2 Principle of resistance strain gauge.

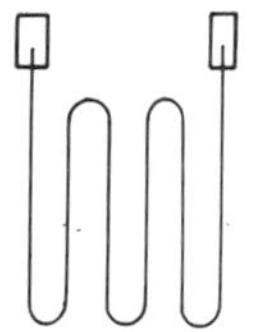

Figure 4.3 Layout of wire gauge.

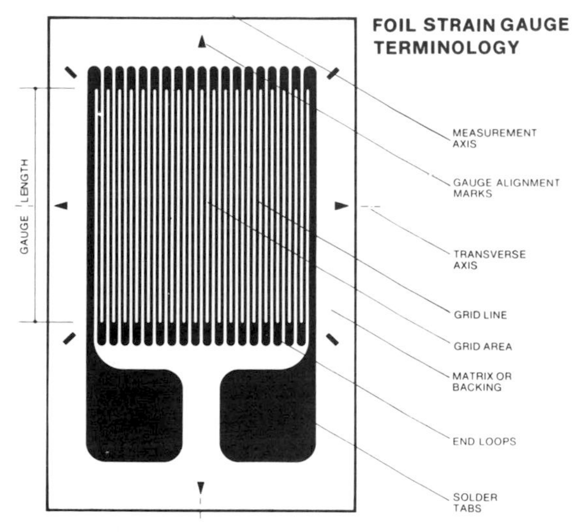

Figure 4.4 Shape of foil gauge. Courtesy, Micro-Measurements Division, Measurements Group Inc.

resistance can be measured in order to get a precise figure for strain.

The resistance of a conductor of length l, cross-sectional area A and resistivity ρ is

$$R=\frac{\rho l}{A}$$

When a strain $\Delta l/l$ is imparted, it causes a fractional change of resistance.

$$\frac{\Delta R}{R}=\left(1+2\mu+\frac{l}{\rho}\frac{\Delta\rho}{\Delta l}\right)\frac{\Delta l}{l}$$

since there will be a Poisson contraction in A and there may also be a change in resistivity. The ratio $(\Delta R/R)/(\Delta l/l)$ is called the gauge factor of a strain gauge. If there were no change in resistivity, it would be $1+2\mu$ or about 1.6 for most metals, whereas it is found to be 2 or more, demonstrating that the sensitivity of strain gauges is increased by a change in ρ.

Nickel alloys are commonly used as the strain-sensitive conductor, notably Nichrome (nickel–chromium) and Constantan (copper–nickel). As well as paper, epoxy resins and polyimide films are used for the backing insulator.

Strain gauges are available commercially as precision tools; units supplied in one batch have closely similar characteristics, notably a quoted gauge factor.

4.2.1 Wire gauges

It is easier to measure resistances accurately when their values are not too low and this will also help to avoid complications in allowing for the effect of lead resistance in series with the active gauge element. On the other hand gauges should not be too big, in order to measure strain effectively 'at a point'; this calls for dimensions of the order of a centimetre, or perhaps only a few millimetres where very localized strains are to be studied. Both considerations point to the need for very fine wire and diameters of 15–30 micrometres are used. The effective length is increased by having several elements side by side as shown in Figure 4.3. Larger tags are attached at the ends of the strain-sensitive wire for connecting leads to.

4.2.2 Foil gauges

An alternative to using wire is to produce the conductor from a foil—typically 4 micrometres thick—by etching. Figure 4.4 illustrates a typical shape. Foil gauges have the advantage that their flatness makes adhesion easier and improves heat dissipation (see p. 71) as well as allowing a wider choice of shape and having the tags for the leads integral with the strain-sensitive conductor, and are in fact more widely used now than wire gauges.

4.2.3 Semiconductor gauges

Another version of strain gauge employs semi-conductor material, commonly silicon. Because the resistivity is higher, the sensitive element can be shorter, wider and simpler: Figure 4.5. The great advantage of semiconductor strain gauges is that their resistivity can be very sensitive to strain, allowing them to have gauge factors many times (typically 50) greater than those of simple metals, but they tend to have higher temperature sensitivity and are less linear. They can be made integral with structural components and are used in this way for pressure measurement (see Chapter 9).

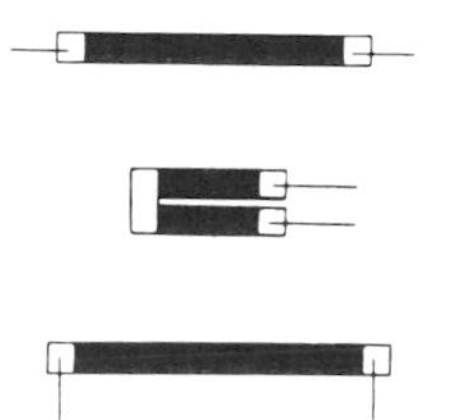

Figure 4.5 Examples of semi-conductor gauges. Courtesy, Kulite Semiconductor Products Inc.

4.2.4 Rosettes

We pointed out earlier that a full analysis of strain involves measurements in more than one direction. In fact, three measurements are required on a surface because strain can be represented as an ellipse, for which the magnitudes and directions of the axes must be established. The directions chosen for strain measurements are commonly either at 120° or at 45° and 90° to each other.

If we are dealing with large structures, it may be expected that strain will only vary gradually across a surface and three closely spaced individual gauges can be thought of as referring to the same point. When there is little room to spare, it is desirable to have the three gauges constructed integrally, which anyhow simplifies installation. Such a unit is called a *rosette*. The three units may be either close together in one plane or actually stacked on top of each other (Figure 4.6).

Figure 4.6 Rosette of gauges. Courtesy, Micro-Measurements Division, Measurements Group Inc.

4.2.5 Residual stress measurement

The state of the surface at the time when a strain gauge is bonded to it has of course to be taken as the strain zero relative to which subsequent changes are measured. The gauge essentially measures increments of strain with increments of load. For many purposes of calculating stresses and predicting life, this is the most important thing to do.

However, during fabrication, and before a gauge can be attached, some stresses can be locked up in certain parts and it may be desirable to know these. This cannot be done with any accuracy non-destructively but if we deliberately remove some material the observed strain changes in neighbouring material can tell us what forces were previously applied through the now absent material. One technique is to strain-gauge a small area of interest, noting the changes in the gauge readings as that area is freed by trepanning. An alternative procedure is to drill a simple hole inside an array of strain gauges that remain attached to the main surface; changes in the strain they show can again indicate what the residual stress was. An array for this purpose is shown in Figure 4.7.

4.3 Gauge characteristics

We have discussed the *gauge factor* at some length; that is what enables resistance to be used at all to measure strain. Other features of strain gauges are important for successful instrumentation. Information about the characteristics of particular gauges is available from manufacturers.

4.3.1 Range

The materials that strain gauges are made from cannot be expected to stretch by more than a few per cent at most and still retain their properties in linear relationships; generally non-linearity is introduced before permanent damage occurs. Metals vary in the strain range over which they can be used; semi-conductors have an appreciably shorter range. Although their limited range is an obvious theoretical restriction on the use of strain gauges, they can in fact cover most of the common field of interest for metals and other hard structural materials. Strain gauges are not generally suitable for use on rubber.

Figure 4.7 Array of gauges for measuring residual stress. Courtesy, Micro-Measurements Division, Measurements Group Inc.

4.3.2 Cross-sensitivity

We have so far described the action of a strain gauge in terms of strain in the direction of the length of its conductor: this is the strain it is intended to measure. But, as explained on p. 66, some strain is generally present in the substrate also in a direction at right angles to this, and gauges are liable to respond in some degree to this. For one thing, part of the conducting path may be in that direction; for another the variation of resistivity with strain is a complex phenomenon. The cross-sensitivity of a gauge is seldom more than a few per cent of its direct sensitivity and for foil gauges can be very small, but it should be taken into account for the most accurate work.

4.3.3 Temperature sensitivity

The resistance of a strain gauge, as of most things, varies with temperature. The magnitude of the effect may be comparable with the variations from the strain to be measured, and a lot of strain gauge technology has been devoted to ensuring that results are not falsified in this way.

Several effects must be taken account of. Not only does the resistance of an unstrained conductor vary with temperature but the expansion coefficients of the gauge material and of the substrate it is bonded to mean that temperature changes cause dimensional changes apart from those, resulting from stress, that it is desired to measure.

It is possible to eliminate these errors by compensation. Gauge resistance is commonly measured in a bridge circuit (see p. 71) and if one of the adjacent bridge arms consists of a similar strain gauge (called a dummy) mounted on similar but unstressed material whose temperature follows that of the surface being strained, then thermal, but not strain effects will cancel and be eliminated from the output.

Self-temperature compensated gauges are made in which the conductor material is heat treated to make its resistivity change with temperature in such a way as to balance out the resistance change from thermal expansion. Because the expansion coefficient of the substrate has an important effect, these gauges are specified for use on a particular material. The commonly matched materials are ferritic steel (coefficient 11×10^{-6} K^{-1}), austenitic steel (16×10^{-6} K^{-1}) and aluminium (23×10^{-6} K^{-1}).

4.3.4 Response times

In practice, there are few fields of study where strain gauges do not respond quickly enough to follow the strain that has been imposed. An ultimate limit to usefulness is set by the finite time taken for stress waves to travel through the substrate, which means that different parts of a strain gauge could be measuring different phases of a high-frequency stress cycle. But with stress-wave velocities (in metals) of the order 5000 m/s, a 10 mm gauge can be thought of as giving a point measurement at frequencies up to 10–20 kHz. Of course it is necessary that the measuring circuits used should be able to handle high-frequency signals.

It must be noted that strain gauges essentially measure the change in strain from the moment when they are fixed on. They do not give absolute readings.

Very slowly varying strains present particular measurement problems. If a strain gauge is to be used over periods of months or years without an opportunity to check back to its zero reading, then errors will be introduced if the zero has drifted. Several factors can contribute to this: creep in the cement or the conductor, corrosion or other structural changes. Drift performance depends on the quality of the installation; provided that it has been carried out to high standards, gauges used at room temperature should have their zero constant to a strain of about 10^{-6} over months. At high temperatures it is a different matter; gauges using ceramic bonding can be used with difficulty up to 500/600 °C but high-temperature operation is a specialized matter.

4.4 Installation

Sometimes strain gauges are incorporated in some measuring device from the design stage. More often they are used for a stress survey of a pre-existing structure. In either case it is most important to pay very close attention to correct mounting of the gauges and other details of installation. The whole operation depends on a small unit adhering securely to a surface, generally of metal. Very small changes in electrical resistance have then to be measured, necessitating close control of any possible resistances that may be either in series or parallel to that of interest—i.e. from leads or leakage.

Of course it is important to ensure that a gauge is mounted at the correct site—often best identified by tape. It may be noted that gauges can be mounted on cylindrical surfaces with quite small radii, but any double curvature makes fixing very difficult. We have already referred to the use of another gauge for temperature compensation. The introduction of any such 'dummies' must be thought out; it is possible that an active gauge can be used for compensation, so doubling the signal, if a place can be identified where the strain will be equal but opposite to that at the primary site, e.g. the opposite side of a bending beam.

The surface where the gauge is to be fixed must be thoroughly cleaned—probably best by abrasion followed by chemical degreasing. Cements commonly used are cellulose nitrate up to 100 °C, epoxy up to

200 °C and ceramic above that where special techniques must be used. Gauge manufacturers may specify a particular cement for use with their product.

After the gauge is fixed down, its leads should be fastened in position and connected (by soldering or spot-welding) to the gauge. It is most important for leads to be mounted securely to withstand the vibration they may be subject to; in practice there are more failures of leads than in strain gauges themselves.

Unless the installation is in a friendly environment, it must then be protected by covering with wax, rubber, or some such material. The chief purpose of this is to exclude moisture. Moisture could cause corrosion and, also serious, an electrical leakage conductance. It must be remembered that 10^8 ohms introduced in parallel with a 350-ohm gauge appears as a 3 in a million reduction in the latter; such a paralleling can be caused between leads or by an earth leakage depending on the circuit configuration and gives a false indication of strain of 2×10^{-6}.

The various stages of installation are illustrated in Figure 4.8.

Figure 4.8 Stages of installing gauges. Courtesy, HBM.

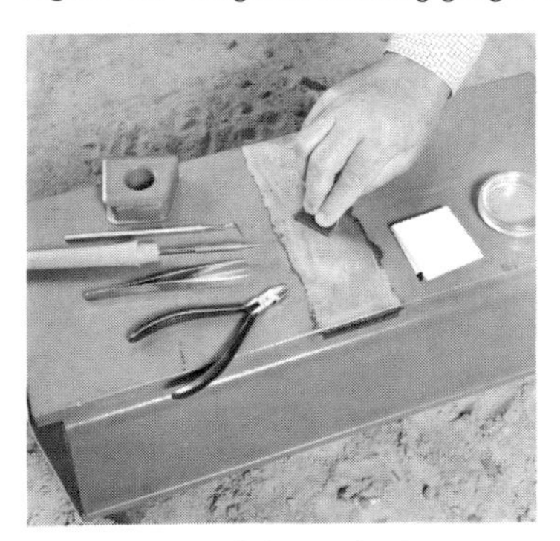

1. Roughening of the application area

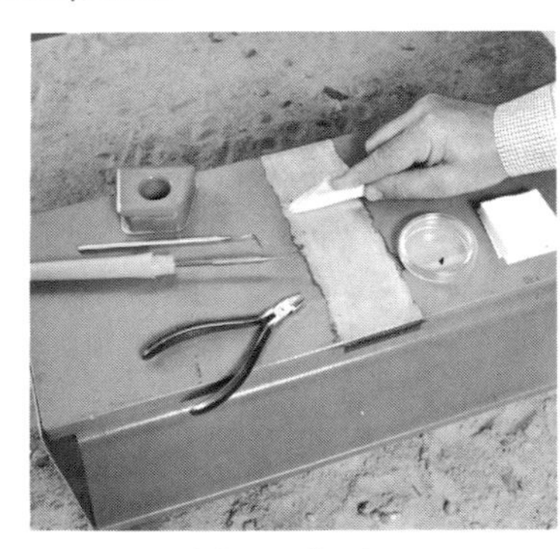

2. Cleaning of the application area

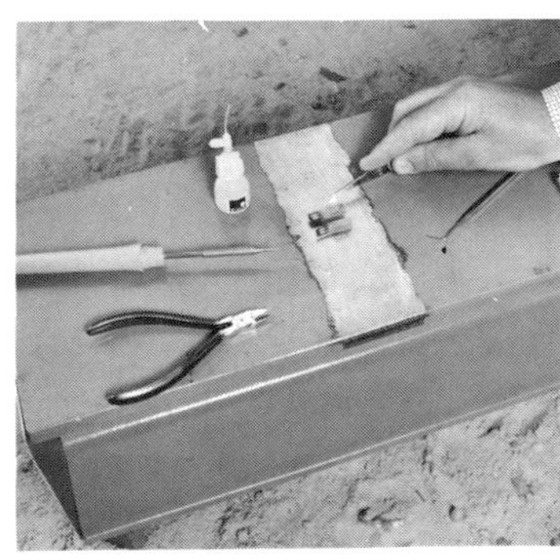

3. Positioning of the strain guage with tape

4. Applying the adhesive

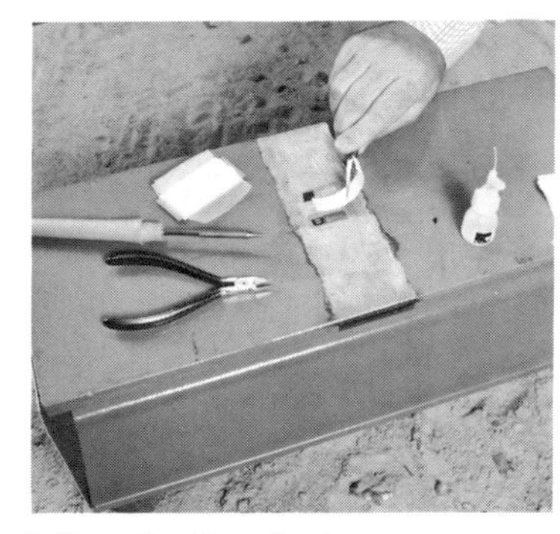

5. Spreading the adhesive

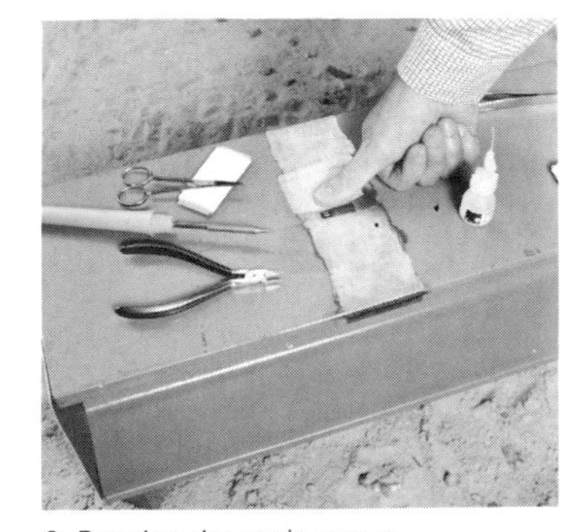

6. Pressing the strain guage

7. Soldering the strain gauge ribbons

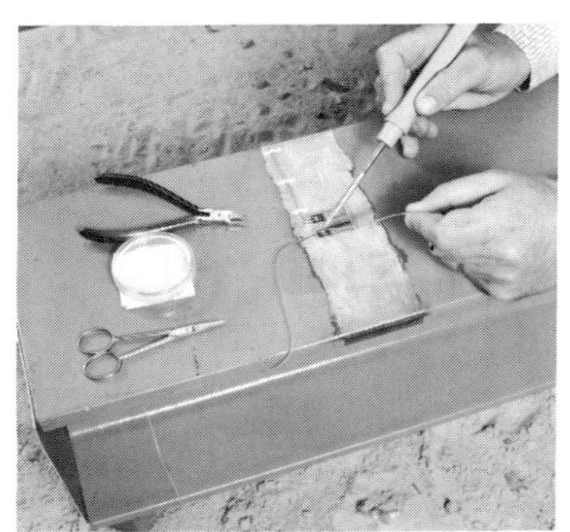

8. Soldering and fixing the cables

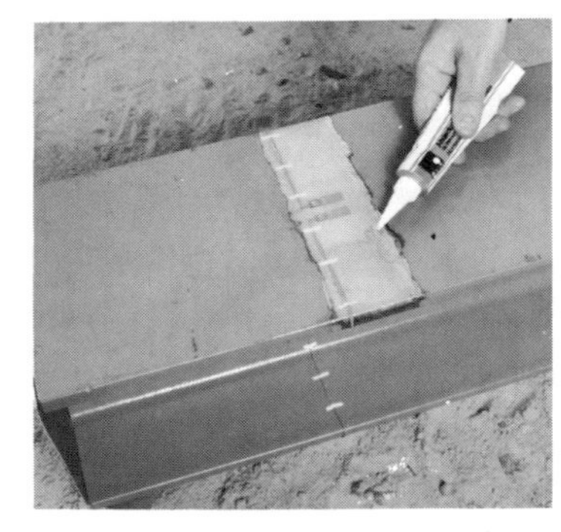

9. Covering the measurement point

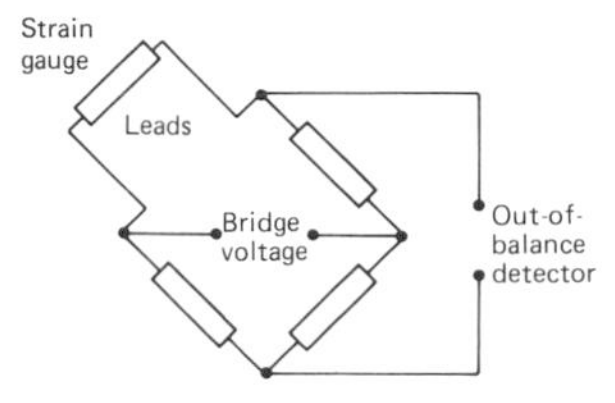

Figure 4.9 Simple bridge circuit (quarter-bridge).

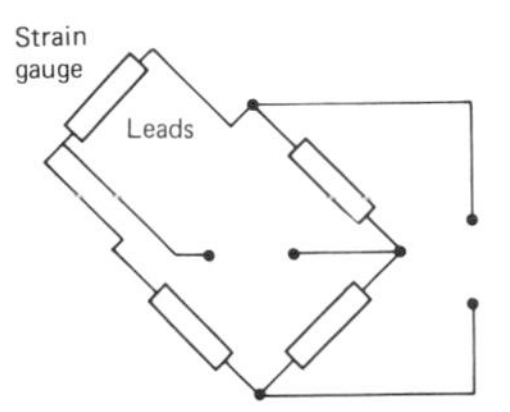

Figure 4.10 Quarter-bridge circuit with three long leads.

4.5 Circuits for strain gauges

For measurement of its small resistance changes, a strain gauge is generally connected in a Wheatstone bridge. This may be energized with d.c., but a.c.—at frequencies of the order of kilohertz—is commoner; a.c. has the advantage of avoiding errors from the thermocouple potentials that can arise in the leads when the junctions of dissimilar metals are at different temperatures.

Gauges are often mounted some distance from their associated measuring equipment and care must be taken that the long leads involved do not introduce errors. In a simple gauge configuration (Figure 4.9), the two leads will be directly in series with the live gauge, and any changes in their resistance, for instance from temperature, will be indistinguishable from strain. This can be overcome by having three leads to the gauge (Figure 4.10); one lead is now in series with each of two adjacent arms, so giving compensation (for equal-ratio arms) provided changes in one lead are reproduced in the other. The third lead, going to the power source, is not critical. These are called 'quarter-bridge' arrangements.

A 'half-bridge' set-up is sometimes used (Figure 4.11). This is when two strain gauges are both used in the measurement as explained on p. 69.

The third possibility is a 'full-bridge' (Figure 4.12) when all four arms consist of gauges at the measurement site, the four long leads being those connecting the power source and out-of-balance detector. Their resistances are not critical so it is not necessary even to ensure that changes are equal. As with most bridge circuits, the power source and the detector can be interchanged. We have called the power source 'bridge voltage', implying that the supply is at constant potential; it can alternatively come as a constant current and this has some advantages for linearity.

Bridges can be balanced, to take up component tolerance, by fine adjustment of series or parallel elements in the arms. Instead, the zero can be set within the amplifier that commonly forms part of the detector. Changing a high resistance across a strain gauge can be used to simulate a known strain and so calibrate all the circuit side of the measuring system. It is possible to have measurements made in terms of the adjustment needed to re-balance a bridge after a strain has occurred but more often the magnitude of the out-of-balance signal is used as an indication.

The larger the voltage or current applied to a strain gauge bridge, the higher will its sensitivity be. The practical limit is set by self-heating in the gauge. If a gauge is appreciably hotter than its substrate, temperature errors are introduced. Compensation from a similar effect in another gauge cannot be relied on because the cooling is unlikely to be identical in the two cases. Self-heating varies a lot with the details of an installation but, with metal substrates, can generally be ignored below 1 milliwatt per square millimetre of gauge area.

We have described the basic circuitry as it concerns a single strain gauge. Tests are often made involving large numbers of gauges. For these, there is available elaborate equipment that allows a multiplicity of gauges to be scanned in quick succession, or simultaneous recordings made of a number of high-speed dynamic strain measurements.

4.6 Vibrating wire strain gauge

Although bonded-resistance strain gauges are the type that has much the widest application, one or two other principles are made use of in certain situations. One of these is that of the vibrating-wire strain gauge.

If a wire is put under tension, then its natural

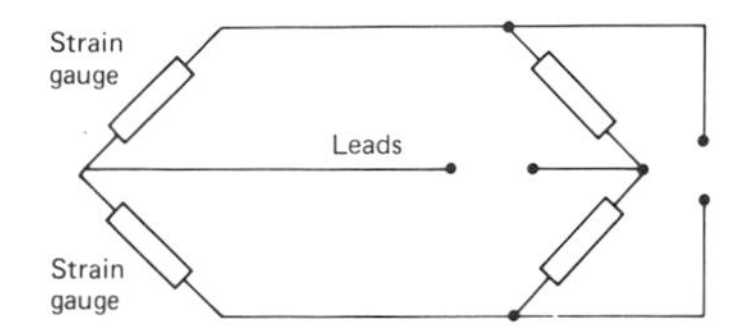

Figure 4.11 Half-bridge circuit.

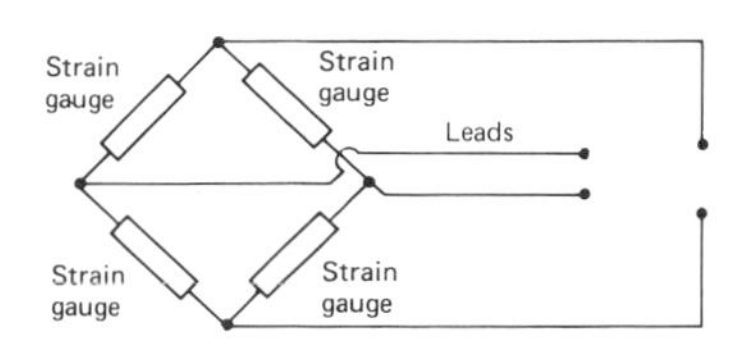

Figure 4.12 Full-bridge circuit.

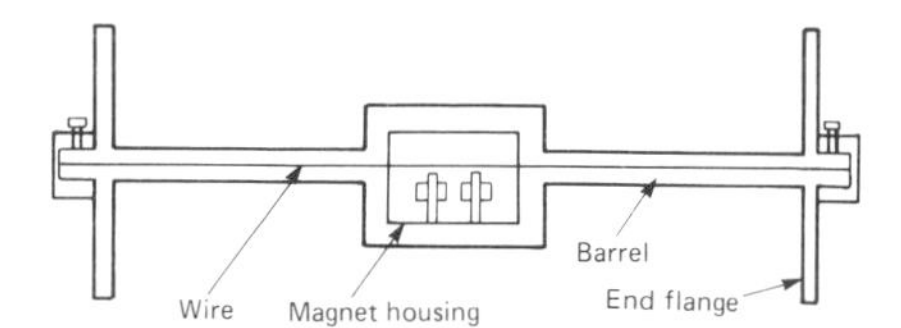

Figure 4.13 Vibrating-wire strain gauge. Courtesy, Strainstall Ltd.

frequency of vibration (in its first mode) is

$$f = \frac{1}{2l}\sqrt{\frac{T}{m}}$$

where l is its length, T its tension and m its mass per unit length.

The fixing points of the wire can be bonded to material whose strain changes are to be measured. As the latter stretches, l changes and, more significantly, T changes following Hooke's law. Strain can then be determined by monitoring natural frequency, easily done if the wire is magnetic and a solenoid is placed nearby to attract it. Wire diameter is typically 0.25 mm, operating frequency 1 kHz. The commonest circuit arrangement is to excite the wire with a single current pulse, measuring the frequency of the damped vibrations that ensue. A sketch of a typical device is given in Figure 4.13.

With the number of items that go to make up a vibrating-wire gauge, it is considerably larger (typically 100 mm long) than a bonded-resistance gauge. Because of the large force needed to stretch it, thought must be given to the gauge's mounting.

In fact, the largest application of such gauges has been to the measurement of *internal* strains in concrete, where both these factors are attended to. By embedding the gauge in the concrete when the concrete is cast, with electrical leads coming out to some accessible point, good bonding to the end-points is ensured. A large gauge-length is desirable in order to average the properties of the material, which is very inhomogeneous on a scale up to a few centimetres. By choosing appropriate dimensions for the components of a vibrating-wire strain gauge, it is possible to make its effective elastic modulus the same as that of the concrete it is embedded in; the stress distribution in the bulk material will not then be changed by the presence of the gauge. A strain range up to 0.5 per cent can be covered.

It has been found that vibrating-wire strain gauges can be very stable; the yielding that might be expected in the vibrating wire can be eliminated by pre-straining before assembly. With careful installation and use at room temperature, the drift over months can correspond to strains of less than 10^{-6}.

4.7 Capacitive strain gauges

It is possible to design a device to be fixed to a substrate so that when the latter is strained the electrical *capacitance* (rather than the *resistance*) of the former is changed. Figure 4.14 is a diagram showing the principles of such a gauge. When the feet are moved nearer together, the arched strips change curvature and the gap between the capacitor plates P changes. The greater complexity makes these devices several times more expensive than simple bonded-resistance strain gauges, and they are seldom used except when their unique characteristic of low drift at high temperature (up to 600 °C) is important. They are most commonly fixed to metal structures by spot-welding. Although the capacitance is only about a picofarad, it can be measured accurately using appropriate bridge circuits; because both plates are live to earth, the effects of cable capacitance can be largely eliminated.

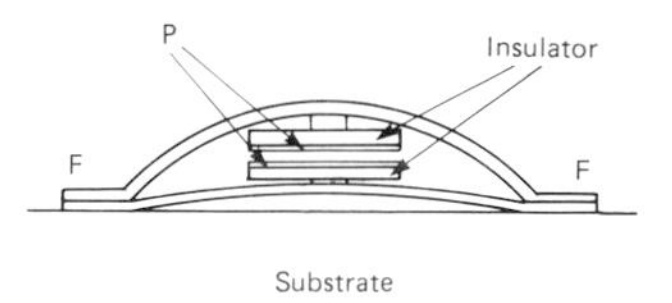

Figure 4.14 Capacitive strain gauge. Courtesy, G. V. Planar Ltd.

4.8 Surveys of whole surfaces

A strain gauge only gives information about what is happening in a small region. Sometimes it is desirable to take an overview of a large area.

4.8.1 Brittle lacquer

One way of surveying a large area is to use brittle lacquer, though the technique is much less accurate than using strain gauges and does not work at all unless the strains are large. It has particular value for deciding where to put strain gauges for more accurate measurements.

A layer of special lacquer is carefully and consistently applied to the surface to be studied. The lacquer is dried under controlled conditions, whereupon it becomes brittle and therefore cracks if and when the part is strained above a certain threshold. Moreover, the higher the surface strain, the closer together are the cracks. It is best to coat a calibration bar in the same way and at the same time as the test part. Bending the bar then gives a range of known strains along it and the crack pattern observed at different points on the live surface can be compared with that seen on the calibration bar.

In this way, the critical points where there are the highest stresses on the surface of a structure can be

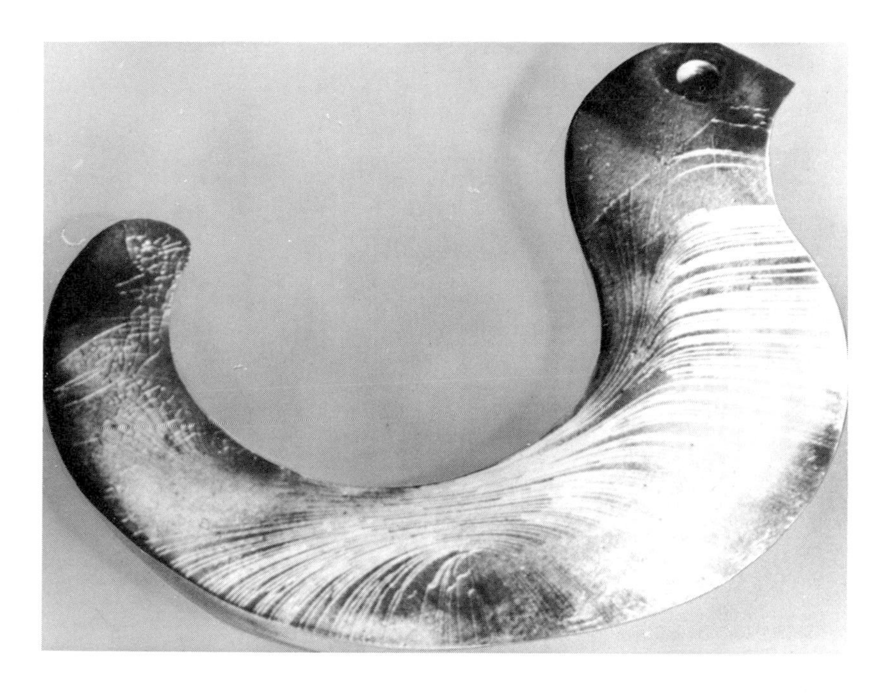

Figure 4.15 Brittle lacquer used on a specimen. Courtesy, Photolastic Division, Measurements Group Inc.

quickly recognized and the strain levels identified within about ± 25 per cent (of those levels) over a range of strains from 0.05 per cent to 0.2 per cent. Because the cracks form perpendicularly to the maximum principal stress, the technique has the considerable additional advantage of showing the directions of principal stresses all over a structure. Figure 4.15 shows the sort of crack pattern that can be observed.

4.8.2 Patterns on surfaces

Large strains can be determined simply by inscribing a pattern on a surface and noting how it changes. Figure 4.16 shows the changes of shape observed in one such investigation.

The sensitivity to smaller strains can be increased by using moiré fringes. When a fine grating is seen through another grating that is comparable but not identical, dark and brighter regions—called 'fringes'—will alternate. The dark regions correspond to where spaces in one grating block light that has come through gaps in the other grating; bright regions arise when the gaps are superposed. The small difference between the gratings can be one of orientation, or of separation of their elements. For instance, if one grating has 1000 lines per centimetre and its neighbour 1001 lines, there will be a dark fringe every centimetre, with bright ones in between them. Now suppose a 1000-line grating is etched on a surface and another placed just above it; a strain of 10^{-3} in the surface will change the former into a 1001-line grating and mean that fringes (seen in this case in light reflected off the

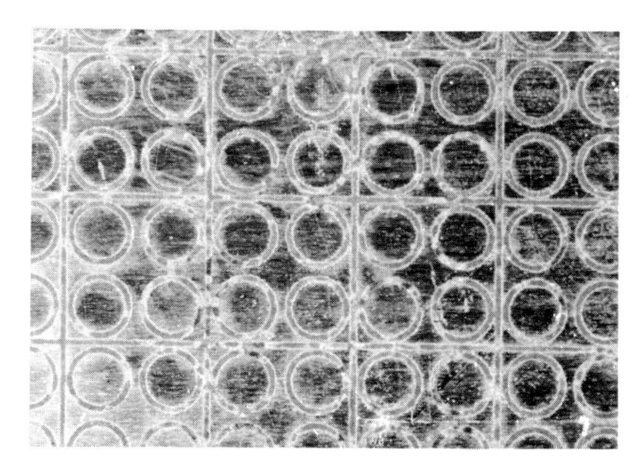

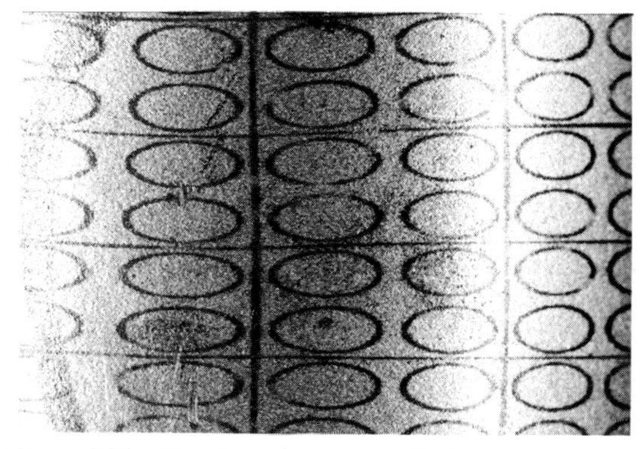

Figure 4.16 Distortion of a pattern under large strain. Courtesy, South Australian Institute of Technology.

Figure 4.17 Polariscope in use. Courtesy, Sharples Stress Engineers Ltd.

surface) appear every centimetre. The larger the strain, the more closely spaced the fringes.

Fringes appearing parallel to the original grating lines will be a measure of the direct strain of one grating relative to the other. Fringes appearing perpendicular to the original grating lines will indicate rotation of one grating relative to the other. In general the fringes will be at an angle to the grating lines, therefore one pair of gratings will give a direct strain and a shear strain component. Two pairs, i.e. a grid or mesh, will give two direct strains and the associated shear strain and will thus permit a complete surface strain determination to be made, i.e. principal strains and principal strain directions.

4.9 Photo-elasticity

For many years, the phenomenon of photo-elasticity has been employed in experimental studies of stress distribution in bodies of various shapes. The effect made use of is the birefringence that occurs in some materials and its dependence on stress.

Light is in general made up of components polarized in different directions; if, in a material the velocities of these components are different, the material is said to be birefringent. The birefringence is increased (often from zero in stress-free material) by stress, commonly being proportional to the difference in stress in the two directions of polarization. The effect was originally discovered in glass; synthetic materials, notably epoxy resins, are much more commonly used now.

In practice, a model of the structure to be examined is made out of photo-elastic material. This is placed in a rig or *polariscope* such as the one shown in Figure 4.17, and loaded in a way that corresponds to the load imposed on the original. By re-combining components of the light ray with polarizations parallel to the two principal stresses, fringes are produced and their position and number give information about strain in the model. The first-order fringe occurs when there is a phase difference of 360° between the components; the nth order when there is a $360n°$ difference.

As discussed for strain gauges, both the direction and the magnitude of the principal stresses come into stress analysis, and this complicates the situation. To find directions, a polariscope system as indicated in Figure 4.18 can be used; if the axes of polarizer and analyser are at right angles and parallel to the principal stresses, there will be no interference, just a black spot

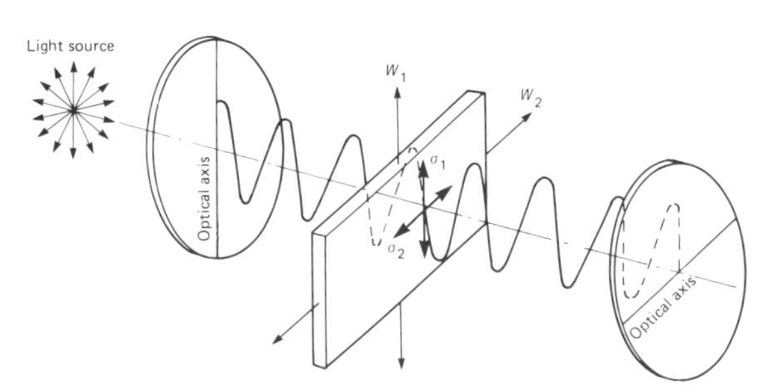

Figure 4.18 Use of a polariscope to determine principal stress directions. Courtesy, Sharples Stress Engineers Ltd.

whatever the load may be. So the 'isoclinics'—loci of points having the same direction of principal stresses—can be established.

For example, if a two-dimensional model of a loaded notched beam were examined in such a crossed polariscope two types of fringes would be observed: a black fringe, the isoclinic, joining all points where the principal stress directions were parallel to the axes of polarization, and coloured fringes, the 'isochromatics', contours of equal principal stress difference. The first-order isochromatic would pass through all points of the model where the stress had a particular value of $P-Q$, where P and Q are the two principal stresses. Similarly the nth-order isochromatic would pass through all points where the stress had n times that value.

By using simple tensile calibration strips it is possible to determine the value which corresponds to each fringe order. Since the stress normal to the unloaded boundaries of the model is zero, i.e. $Q=0$, it is a relatively simple matter to determine the stress all along the unloaded boundaries. Determination of the stresses in the interior of the model is also possible but requires complex stress separation.

Normally a monochromatic light source is used, but white light has the advantage that the first-order fringe can be distinguished from higher orders.

Birefringence all along the ray path through the model will contribute to the phase difference between the two optical components. The effect measured is therefore an integral of stress along that path. This means that fringe patterns in photo-elasticity are most easily interpreted with thin—or effectively two-dimensional—models. There is a technique for studies in three dimensions. This is to load the model in an oven at a temperature high enough to 'anneal' out the birefringence. Subsequent slow cooling means that when the load is removed at room temperature bi-refringence is locked into the unloaded model, which can then be carefully sliced; examination of each slice in a polariscope will show the original stresses in that particular region.

4.10 References

Holister, G. S. *Experimental Stress Analysis, Principles and Methods*, Cambridge University Press (1967)

Kuske, A. and Robertson, G. *Photoelastic Stress Analysis*, John Wiley and Sons (1974)

Theocaris, P. S. *Moiré Fringes in Strain Analysis*, Pergamon Press (1969)

Window, A. L. and Holister, G. S. (eds) *Strain Gauge Technology*, Allied Science Publishers (1982)

5 Measurement of level and volume

P. H. SYDENHAM

5.1 Introduction

Many industrial and scientific processes require knowledge of the quantity of content of tanks and other containers. In many instances it is not possible, or practical, to directly view the interior. Parameters of interest are generally level of the contents, volume, or simply the presence of substances. Sensors may be needed, Figure 5.1, to detect high or low level states for alarm or control use, to provide proportional readout of the level with respect to a chosen datum, for automatic control purposes or for manually read records.

Simple installations may be able to adopt inexpensive manual methods such as a dip-stick. Automatic control applications, however, will require control signals for operation of a process actuator or alarm. For these cases there are available several options of output signal which include indirect electric contacts and electronic proportional outputs, and direct flow and pneumatic valving.

As well as the obvious liquid substances, such as water, oil, petroleum and milk, level sensors have been successfully applied to measurement of solids such as flour, mineral ores, food grain and even potatoes and coal. Two-phase systems are also often measured, examples being determination of liquid and froth levels in beer making and for mineral slurries.

Due to the extensive need for this basic process parameter there are available very many commercial equipments; their installation details are important. A useful tutorial introduction to level measurement is

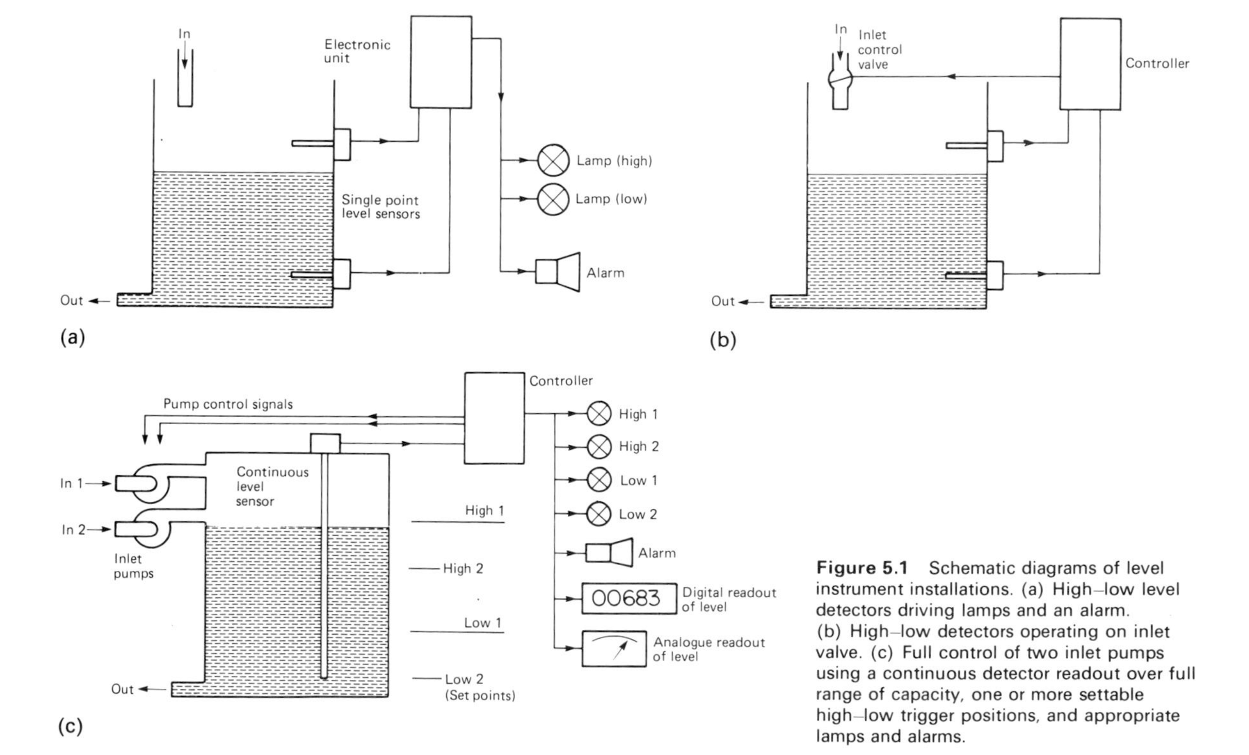

Figure 5.1 Schematic diagrams of level instrument installations. (a) High–low level detectors driving lamps and an alarm. (b) High–low detectors operating on inlet valve. (c) Full control of two inlet pumps using a continuous detector readout over full range of capacity, one or more settable high–low trigger positions, and appropriate lamps and alarms.

available in Lazenby (1980). Norton (1969), O'Higgins (1966) and Miller (1975) also make valuable contributions.

5.2 Practice of level measurement

5.2.1 Installation

Suppliers of level-sensing systems generally provide good design support for installation, enabling the prospective user to appreciate the practical problems that arise and offering wide variation in the sensor-packaging and systems arrangements.

Positioning of the sensor head, or heads, should be chosen with due regard for the problems of turbulence caused by contents flowing in and out. Positioning in order to control errors is also important. For example, when a stilling tube is placed external to the container, its contents may be at a different temperature to those in the container. The complete sensor may need to be fully removable without imposing the need to empty the container.

It is often necessary to incorporate followers, such as shown in Figure 5.2, in the sensing system to constrain the unwanted degrees of freedom of such components as floats.

Corrosion effects, caused by the contents, on the components of the sensing arrangements must also be carefully considered. High temperatures, corrosive materials and abrasion in granular-material measurement can progressively alter the characteristics of the system by producing undue friction, changing the mass of floats and simply reducing the system to an unworkable state.

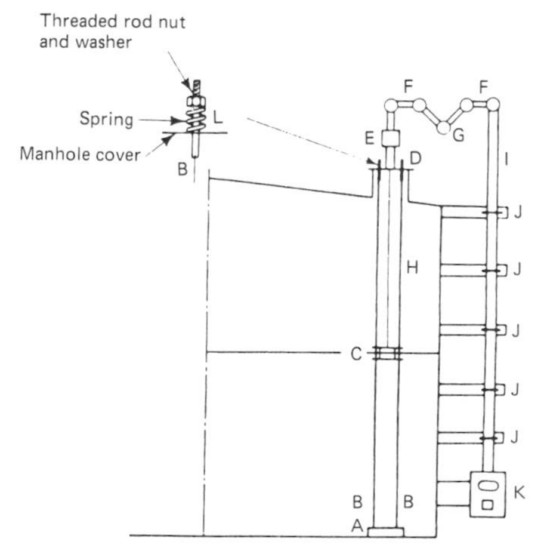

Figure 5.2 Installation of float-type level indicator with guide wires on a fixed-roof tank. A, guide wire anchor. Wires may be anchored to a heavy weight or the bottom of the tank. B, float guide wires. C, float having sliding guides which move freely on the guide wires. D, manhole sufficiently large for float and anchor weight to pass through. E, flexible joint, F, pulley housings, G, vapour seal (if required). H, float tape. I, tape conduit. J, sliding guides. K, gauge head. L, guide wire tension adjustment.

Where the requirement is to operate simple high, or low, level alarms or switch controls the level sensor should preferably have in-built back-lash which provides a toggle action (hysteresis) so that the sensing contacts do not dither at the control point. Some systems incorporate this in their mechanical design (an example being given later in Figure 5.5), others in the electronic circuits.

Gauges based on the use of a radioactive source will need installation and operation within nucleonic device standards. These gauges use very small strength sources and pose an absolutely minimal hazard in themselves. They are described in Volume 3.

If installed without due consideration, the installation itself may introduce a hazard due to the nature of the mechanical components of the system causing blockages in the system.

5.2.2 Sources of error

A first group of errors are those associated with problems of defining the distributed contents by use of a single measurement parameter made at one point in the extended surface of the whole. As a general guide, definition of the surface, which is used to decide the contents, can be made to around 0.5 mm in industrial circumstances. Methods that sense the surface can be in error due to surface tension effects that introduce hysteresis. Where the quantity of a particular substance is of concern, any build-up of sediment and other unwanted residues introduces error.

Granular materials will not generally flow to form a flat surface as do liquids. The angle of repose of the material and the form of input and output porting will give rise to error in volume calculations if the actual geometry is not allowed for in use of a single point sensor.

Turbulence occurring at the sensor, caused by material flow in the container or by vibrations acting on the container, may also be a source of error. It is common practice to mount the sensor in some form of integrating chamber that smooths out transient dynamic variations. A common method is the use of a stilling pipe or well that is allowed to fill to the same level as the contents via small holes—see Figures 5.3 and 5.4. The time rate responses of such still tubes, however, become important in fast moving systems as they can introduce phase-shift and amplitude attenuation.

Changes of the mass of floats, due to sediment build-up or corrosion, will alter the depth of immersion of float sensors. A systematic error also exists due to the actual depth to which the float sinks to provide its necessary buoyancy force. This varies with material density which often varies with temperature.

A second class of errors arise due to temperature and, to a lesser extent, pressure changes to the

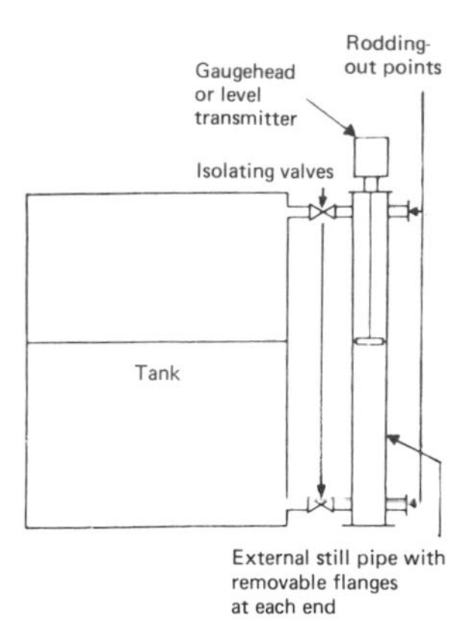

Figure 5.3 Integrating chamber used to average transient variation. Internal still-pipe.

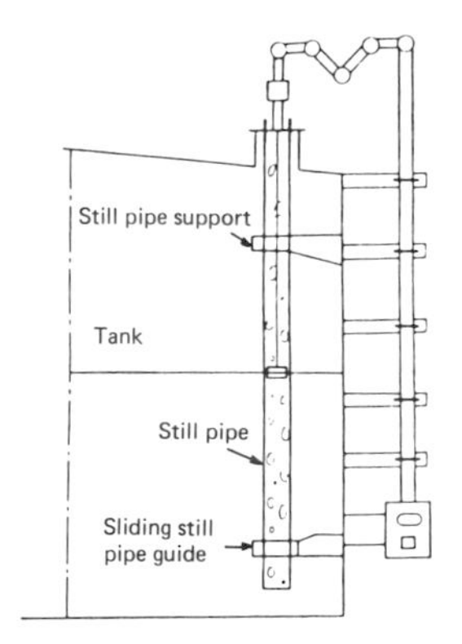

Figure 5.4 Integrating chamber used to average transient motion. External still-pipe.

contents. Where the required measurement is a determination of the volume or mass of the contents, use is made of level as an indirect step toward that need. All materials change volume wich changing temperature. It may therefore, be necessary to provide temperature measurements so that level outputs can be corrected.

For some forms of level sensor external still tubes should be situated to retain the same temperature as that of the tank because localized heating, or cooling, can cause the contents of the still tube to have a different density from that existing in the tank. Methods that are based upon use of buoyancy chambers, that produce a measurement force rather than following the surface, will produce force outputs that vary with temperature due to altered buoyancy upthrust as the density of the fluid changes.

Floats are generally made from waterproofed cork, stainless steel, copper and plastic materials. The material used may need to be corrosion-resistant. Where the contents are particularly corrosive, or otherwise inhospitable to the components of the sensing systems, it is preferable to reduce, to the absolute miñimum, the number of subsystem parts that are actually immersed in the contents.

Considerable use is made of magnetic coupling between the guided float and the follower. Figure 5.5 shows one such arrangement.

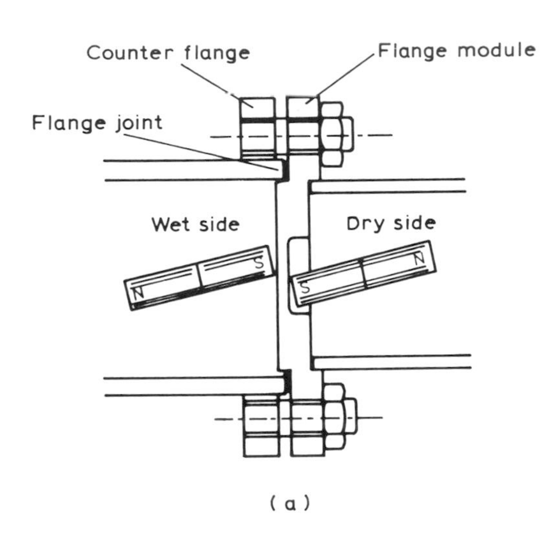

Figure 5.5 Repulsion of like magnetic poles providing non-immersed sensing and toggle action to a float sensor. One commercial form. Courtesy, BESTA Ltd.

Nucleonic gauging offers the distinct advantage, see Figure 5.6, that no part of the level-detecting system need be inside the container. It is discussed further in Volume 3.

Finally, on general choice of level-sensing system, Table 5.1 from Lazenby (1980) provides guidelines for selecting appropriate methods.

Table 5.1 Some guidelines for selecting a suitable level sensor. Adapted from Lazenby (1980).

Is remote control or indication desirable?	A YES excludes mechanical float gauges, sight glasses, dip-sticks and other devices designed specifically for local control.
With level indicators, is time spent in taking a reading important?	A YES excludes dipsticks, sight glasses, mechanical float-gauges and certain balance systems.
Can the sensor contact the material being measured?	A NO eliminates all but ultrasonics, radiation, radar, and optical and load cells.
Must weight be measured, rather than height?	A YES means the choice is limited to load cells; but in uniform-sided tanks, other devices such as capacitance meters can be calibrated for weight, particularly if a liquid is being measured.
Are there objections to mechanical moving parts?	A YES gives one a choice of sight glasses—and capacitance, ultrasonic, radiation, conductivity, radar, load-cell, optical, thermistor, and bubbler devices.
Is the application in a liquid?	A YES eliminates vibrators and certain paddle types.
Is the application for a powdered or granular material?	A YES eliminates dipsticks, sight glasses, floats, thermistors, conductivity devices, pressure (except pressure switches) instruments, bubblers, displacers, sight glasses.
Do level indicators have to be accurate to about 2%?	A YES eliminates thermistors, vibrators, paddles, optical devices, suspended tilting switches, and conductivity instruments, as those types only provide control. All other types can be considered, but the poorest accuracies probably come from float gauges and radiation instruments.
Do level indicators need to have an accuracy that is a lot better than 1%?	A YES reduces the list to dipsticks, some of the displacers, and balance devices.

5.3 Calibration of level-measuring systems

Contents that are traded for money, such as petrol and alcohol, must be measured to standards set by the relevant Weights and Measures authority. Official approval of the measuring system and its procedures of use and calibration is required. In such cases the intrinsic value of the materials will decide the accuracy of such measurements and this often means that the system and calibrations must comply to very strict codes and be of the highest accuracy possible.

The use of the indirect process of determining volumetric or mass contents, based upon a level measurement, means that a conversion coefficient, or chart of coefficients, must be prepared so that the level measurements can be converted into the required measurement form.

Calibration tables for a large fabricated tank are most easily prepared using the original engineering construction drawings. This, however, is not an accurate or reliable method.

A more accurate, and traceable, method (known as 'strapping') is to actually survey the container's dimensions after it is built. This can be a time-consuming task. The values can be used to calculate the volume corresponding to different levels. From this is compiled a conversion chart for manual use or a 'look-up table' for computer use.

By far the most accurate method, however, is a direct volumetric calibration of the container by which a suitable fluid (usually water) is pumped in and out, to provide two readings, the tank passing it through an accurate flow-metering station. Whilst this is in process level data are recorded, enabling the conversion factors to be provided for each level measurement value. These results will often require correction for temperature as has been already discussed.

Highly accurate level measurement requires continuous monitoring of the various error sources described earlier so that on-going corrections can be made. A continuous maintenance programme is needed to clean floats and electrodes and to remove unwanted sediment.

In many instances the use of hand dipping is seen as the on-going calibration check of the level measurement level. For this, rods or tapes are used to observe the point where the contents wet the surface along the mechanical member. Obviously this cannot be used for dry substances; for those the rod or tape is lowered until the end rests on the surface.

In each case it is essential to initially establish a permanent measurement datum, either as the bottom of the tank where the rod strikes or as a fiducial mark at the top. This mark needs to be related to the transducer system's readout and to the original calibration.

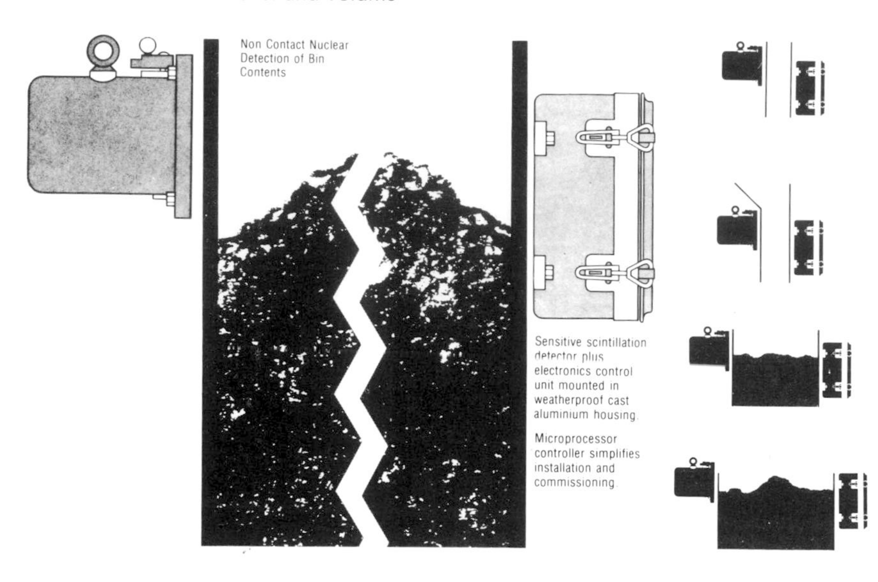

Figure 5.6 Nucleonic gauging system needs mounting only on the outside of existing pipes or containers. Courtesy, Mineral Control Instrumentation.

5.4 Methods providing full-range level measurement

Methods used to measure or control level in a container group into those that can provide continuous readout over the whole range of level experienced and those that sense a small level change using this to operate alarms and switches, or to control the motion of a full-range follower mechanism that, in turn, drives a secondary output transducer or displacement.

Attention is first given to the full range methods that have found wide acceptance.

5.4.1 Sight gauges

A device for reading the level of contents within a closed container, such as a steam boiler, is the simple externally mounted sight-glass. This generally consists of a tube of toughened glass connected through unions and valves into the tank wall. The diameter of the tube must be large enough not to cause 'climb' of the contents due to capillary action; clearly then the level will follow that of the contents. Figure 5.7(a) gives the basic structure of such a device. Where the contents are under pressure it will be necessary to use safety devices to control pressure release in case of tube breakage. Valves are usually incorporated to allow the whole fitting to be removed without having to depressurize the container. Figure 5.7(b) shows a configuration incorporating these features.

A modern development of the above sight gauge concept is the magnetic level indicator shown in Figure 5.8. As the float, containing a bar magnet, follows the liquid surface in the still tube the individual, magnetically actuated, flaps rotate to expose a differently coloured surface that is coated with luminous paint. The float is chosen to suit the specific gravity of the fluid.

The magnetic action also operates individual switches for control and alarm purposes. Discrimination is to around 5 mm. Magnets must not be operated beyond their Curie point at which temperature they lose their desired properties. As a guide these systems can measure liquids under pressures up to 300 bars at temperatures up to 400 °C.

In some circumstances it may be possible to view the surface of the liquid from above but not from the side. In this case a hook gauge, Figure 5.9, can be used to enable the observer to detect when the end of the dipstick rod just breaks the surface.

5.4.2 Float-driven instruments

The magnetic indicator described above is one of a class of level indicators that use a float to follow the liquid surface. Where a float is used to drive a mechanical linkage that operates a remotely located readout device of the linkage motion, there is need to ensure that the linkage geometry does not alter the force loading imposed on the float, for this will alter its immersion depth and introduce error. Frictional forces exerted by the linkage can also introduce error.

Compensation for changes in linkage weight, as a float moves, is achieved by using such mechanisms as counterbalance masses and springs. Figure 5.10 shows the construction of a sophisticated form that uses a pre-wound 'Neg'ator' (also called 'Tensator') spring torque motor that has its torque characteristic tailored to vary as more tape is to be supported during windout.

The production costs of precision mechanical

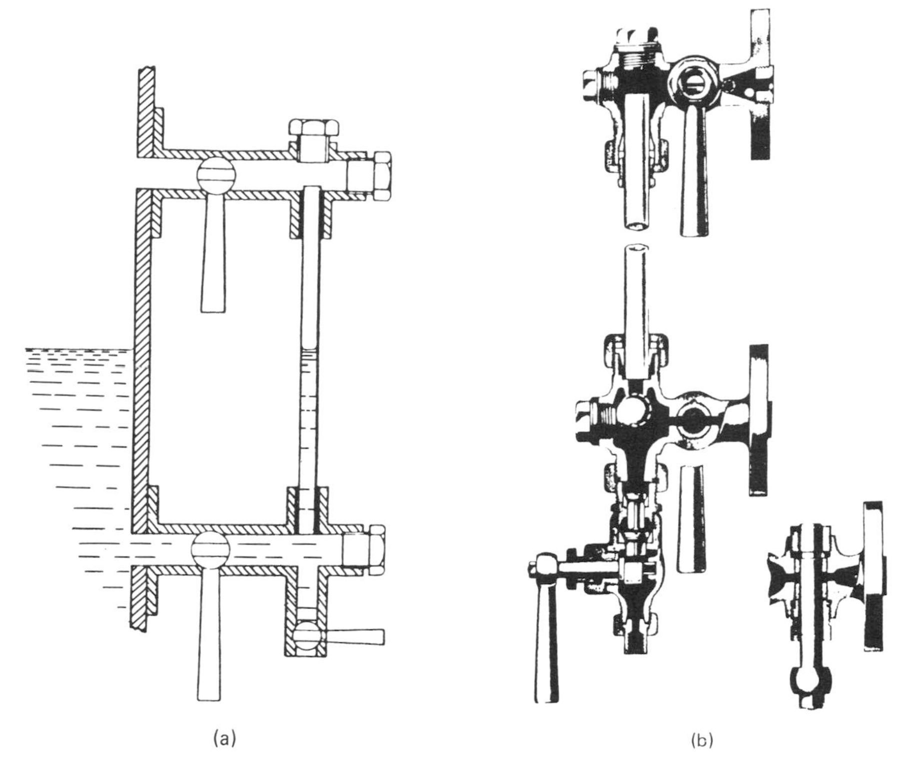

Figure 5.7 Sight-glass level indicator. (a) Basic schematic. (b) Sight-glass with automatic cut-off. Courtesy, Hopkinsons Ltd.

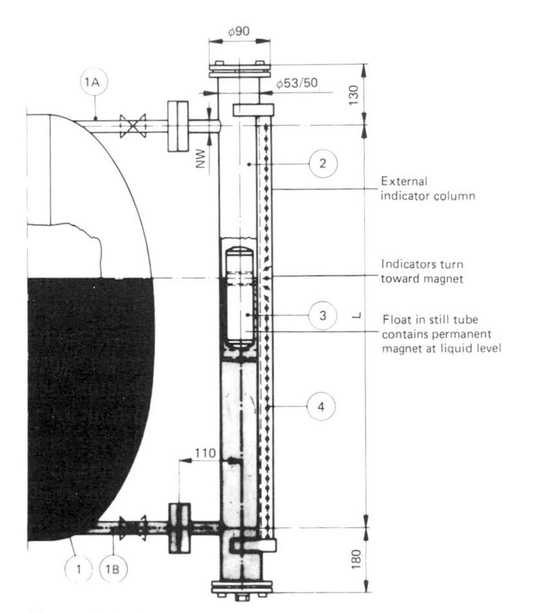

Figure 5.8 Schematic of magnetic level indicator installation. Courtesy, Weka-Besta Ltd.

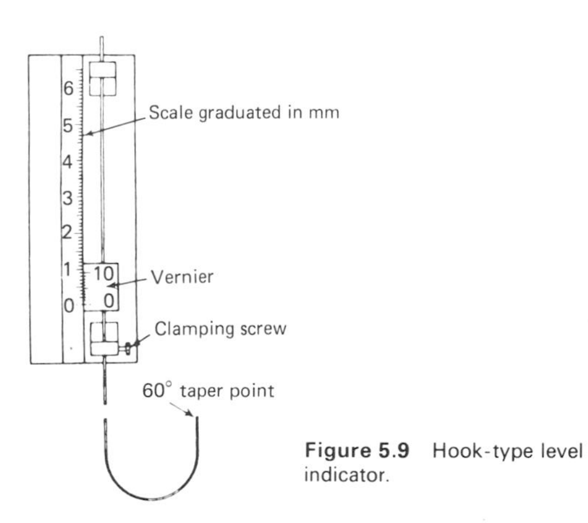

Figure 5.9 Hook-type level indicator.

systems can make them less attractive than electronic equivalents but such systems do have the advantages that no electrical power supply is needed and that the system is easily understood by a wider range of plant operators.

5.4.3 Capacitance probes

The electrical capacitance C between two adjacent electrically conducting surfaces of area A, separated by

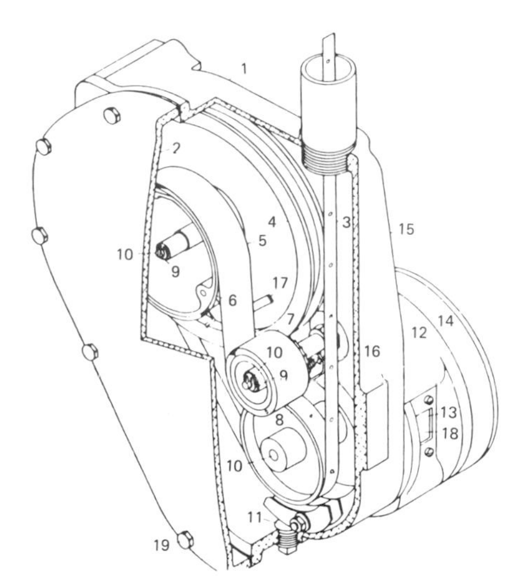

Figure 5.10 Spring torque motor compensated float-type transfer system. 1, Precision cast main housing. 2. Side cover. 3, Perforated steel tape type 316 stainless. 4, Moulded thermosetting phenolic tape drum. 5, Broad Neg'ator Motor, stainless steel. 6, Power drum. 7, Storage drum. 8, Precision made sprocket. 9, P.T.F.E. bearings, 10, Type 316 stainless steel shafts. 11, Drain plug, 12, Digital counter housing. 13, Reading window. 14, Stainless steel band covers adjustment slots. 15, Operation checker handle (out of view). 16, Operation checker internal assembly. 17, Neg'ator motor guide, stainless steel. 18, Counter assembly in the chamber beyond tank pressure and vapours. 19, Cap screws drilled for sealing. Courtesy, Whessoe Ltd.

distance d, is given by

$$C=\varepsilon\frac{A}{d}$$

The constant of proportionality ε is the dielectric constant of the material between the plates. An electrode is suspended in the container, electrically insulated from it. Presence of liquid or granular material around the electrode alters the capacitance between the electrode and the walls. The capacitance is sensed by electronic circuitry. Figure 5.11 is a cut-away view of one form.

The electrode is tailored to the situation; forms include rigid metal rods, flexible cables, and shielded tubes. Capacitance sensors rely on uniform contact being maintained between the contents and a long thin electrode. Where they are used for level sensing of granular materials, such as wheat, the material has a tendency to pile non-uniformly around the electrode, producing what is known as 'rat-holing'. Placing the electrode at an angle to the vertical helps reduce this as it alters the angle of repose of the material helping it to follow the stem more consistently. As the method provides continuous readout of level over its full electrode length, circuitry can also be used to provide multiple on-off setpoints for alarms and control functions. The same principle is used for single point sensing, in which case a simpler electrode and circuitry can be used. Electrical potential and power are usually low enough to eliminate hazards.

Weighing of the contents

The volume of a container's contents can of course be inferred from weight measurements; these are discussed in Chapter 8.

5.4.4 Upthrust buoyancy

A long vertical tubular float will exert an upward force proportional to the depth of immersion in the fluid. These are also sometimes referred to as 'displacers'. The float does not rise to follow the surface but is used instead to exert a force that is usually converted into a torque by a radius arm with a counteracting torque shaft. Force-balance can also be used to determine the upthrust force. Figure 5.12 is an assembly view of one design.

Upthrust depends upon the specific gravity of the fluid so instruments employing it must be calibrated for a stated density. Density varies with temperature. For the best accuracy correction is needed; some reduction in the actual error magnitude, however, occurs due to the float becoming a little larger in volume as its temperature increases.

5.4.5 Pressure sensing

Providing the contents behave as a liquid that flows to equalize pressures at a given depth (some granular materials may not fulfil this requirement) then pressure acting upon a given area at the bottom of a tank is proportional only to the density of the fluid and the head of pressure. In most cases density can be assumed to be uniform, thereby allowing a pressure sensor, placed on the bottom, to be used as a measure of tank level. Pressure gauges are described in Chapter 9.

Lying in the same class are purge methods. The pressure needed to discharge gas or liquid from a nozzle placed at the bottom of the tank depends upon the head of liquid above and its density. Bubblers, as these are called, are simple to arrange and give readout as a pressure-gauge reading that can be read directly or transduced into a more suitable form of signal.

Obviously bubblers do not work in granular materials. The addition of small quantities of liquid or gas must not significantly affect the contents of the tank.

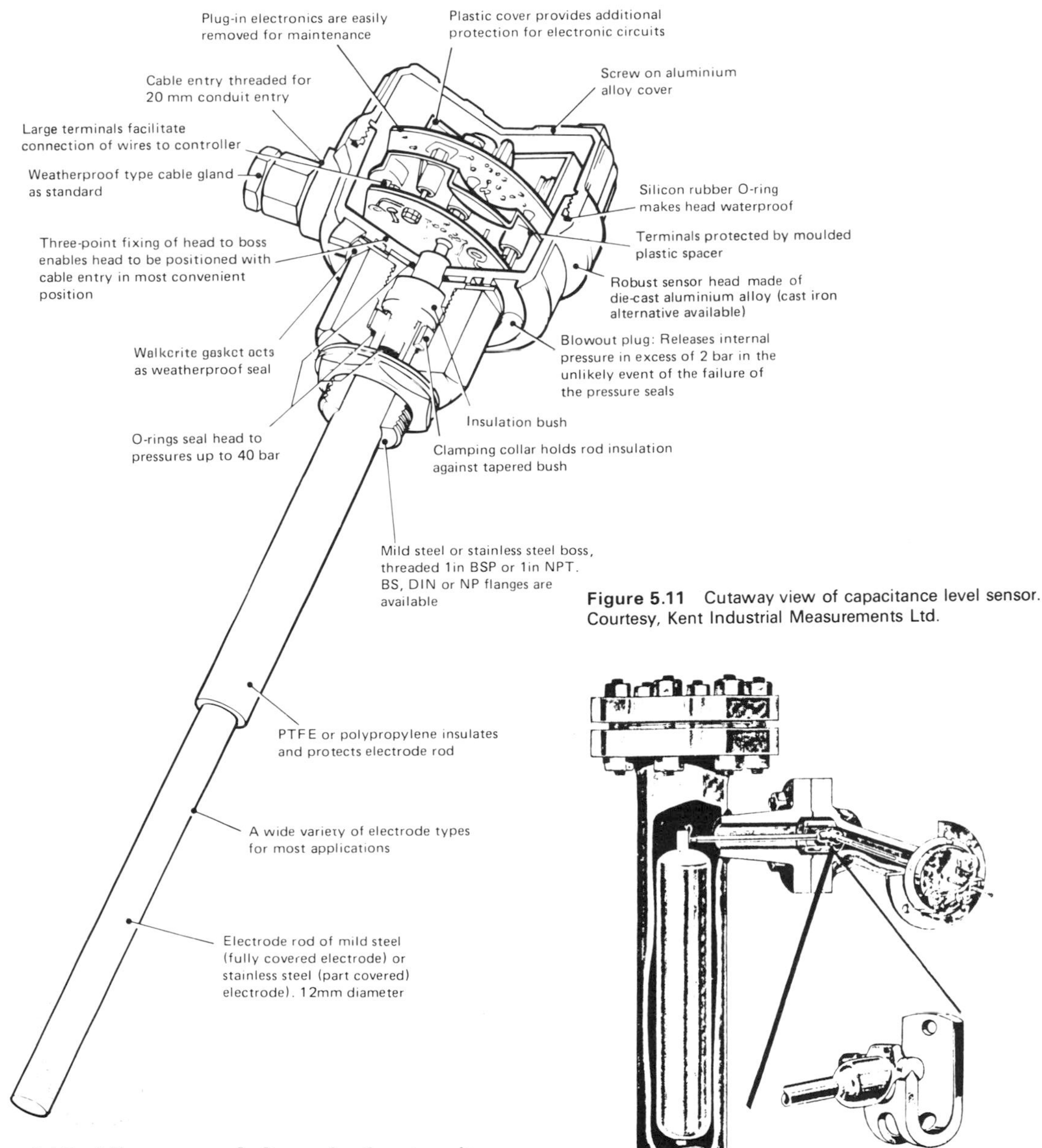

Figure 5.11 Cutaway view of capacitance level sensor. Courtesy, Kent Industrial Measurements Ltd.

Figure 5.12 Assembly view of Fisher torque tube unit. Courtesy, GEC Elliot Control Valves Ltd.

5.4.6 Microwave and ultrasonic, time-transit, methods

A source of coherent radiation, such as ultra-sound or microwaves, can be used to measure the distance from the surface to the top of the tank or the depth of the contents. In essence a pulse of radiation is transmitted down to the surface where some proportion is bounced back by the reflecting interface formed at the surface. The same concept can be used with the waves being sent upward through the material to be reflected downward from the surface. With relatively sophisticated electronic circuitry it is possible to measure the flight time and, given that the velocity of the waves is known, the distance may then be calculated.

Many variations exist upon this basic theme. The choice of radiation, use from above or from below and of frequency depend much upon the material and the accuracy sought.

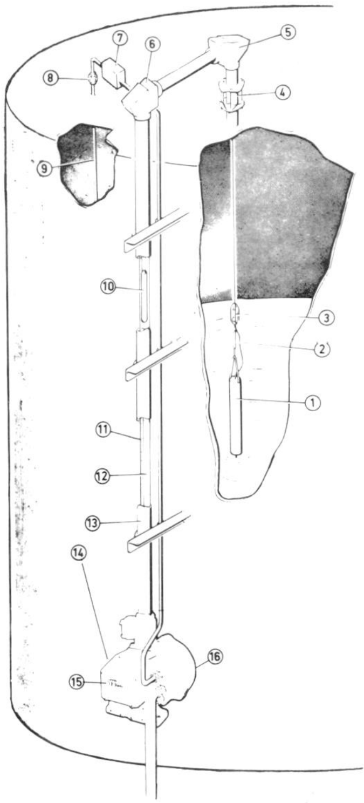

Figure 5.13 Schematic of self-balancing level gauge using RF surface sensing. Courtesy, GEC-Elliot Process Instruments Ltd. 1. Sensing element. 2. Tape insulator. 3. Tape counter-weight. 4. Flexible coupling used on fixed roof only. 5. Pulley box over tank. 6. Pulley box over tankside unit. 7. Temperature cable junction box. 8. Temperature bulb mounting kit. 9. Averaging resistance thermometer bulb. 10. Cable counter-weight. 11. Stainless steel perforated measuring tape. 12. Radio frequency cable. 13. 65 mm dia. standpipe. 14. Servo-electronic box. 15. Level indication. 16. Tape retrieval housing.

Although pulses are sent the repetition rate is fast enough for the output to appear continuous. The method can be made suitable for use in hazardous regions.

5.4.7 Force or position balance

In these methods a short-range sensor, such as a float resting in the surface or a surface sensor of electronic nature, is used to provide automatic control signals that take in, or let out, cable or wire so that the sensor is held at the same position relative to the surface. Figure 5.13 gives the arrangement of one such system that makes use of a radio-frequency surface sensor to detect the surface.

Self-balancing level sensors offer extreme ranges and variable forces exerted by changing mechanical linkage geometry are made negligible. Very high accuracies can be provided, the method being virtually an automated tape measure.

5.5 Methods providing short-range detection

In many applications there is only a need to sense the presence or absence of contents in order to operate on–off switches or alarms. In some continuous-reading systems a short proportional range is needed to control the driven measuring member. This section addresses short-range detectors of level.

5.5.1 Magnetic

Movement of a permanent magnet floating in the surface of the liquid can be sensed by using the magnet to operate a switch contact set. Figure 5.5 shows a system actuated by a rising radius arm and incorporating a toggle snap action. An alternative arrangement uses the rising magnet to close a magnetic reed switch contact, as shown in Figure 5.14, either as a coaxial arrangement, or as a proximity sensor.

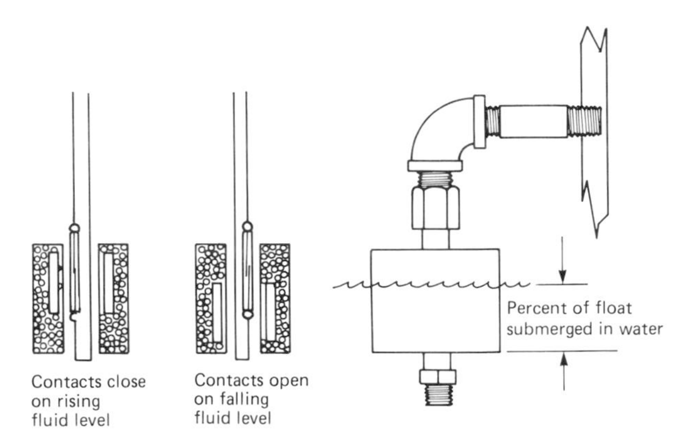

(a)

(b)

Figure 5.14 Magnetic level switch using magnetic reed contact set. (a) Coaxial. (b) Proximity.

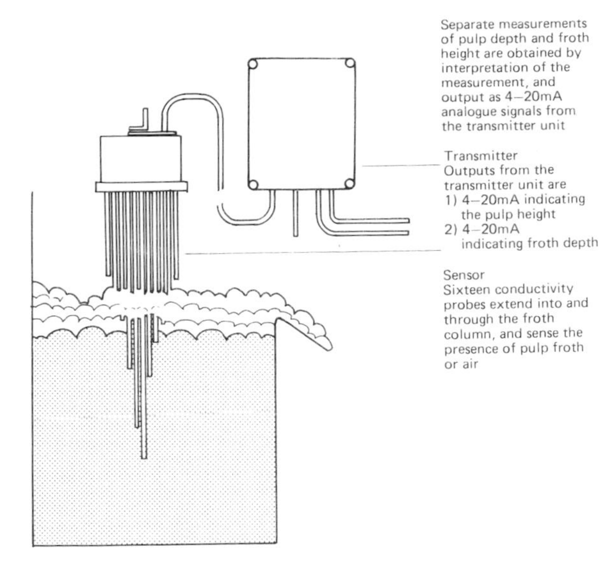

Figure 5.15 Multiple-conductivity probe sensing pulp and froth layers over set increments. Courtesy, Mineral Control Instrumentation.

5.5.2 Electrical conductivity

Liquids such as sewage, sea water and town supply water, which contain dissolved salts, have conductivities higher than pure water. The conductivity of most liquids is much higher than that of air so an electrical circuit, depending on current flow, can discriminate between air and liquid and so detect the interface. Figure 5.15 is a multiple-probe system used to distinguish the various layers in a pulp–froth–air system, as is found in mineral processing.

Conductivity probes are used for digital monitoring of the level of boiler water. Conductivity can also be used to provide continuous range measurement, for as the liquid rises up an electrode the resistivity between the electrode and a reference surface changes in a proportional manner.

5.5.3 Infrared

When fluid wets an optical surface the reflectance of that surface changes considerably enabling detection of liquid when it rises to cover the optical component.

The optical arrangement that is commonly used is a prism, arranged as shown in Figure 5.16. Infrared

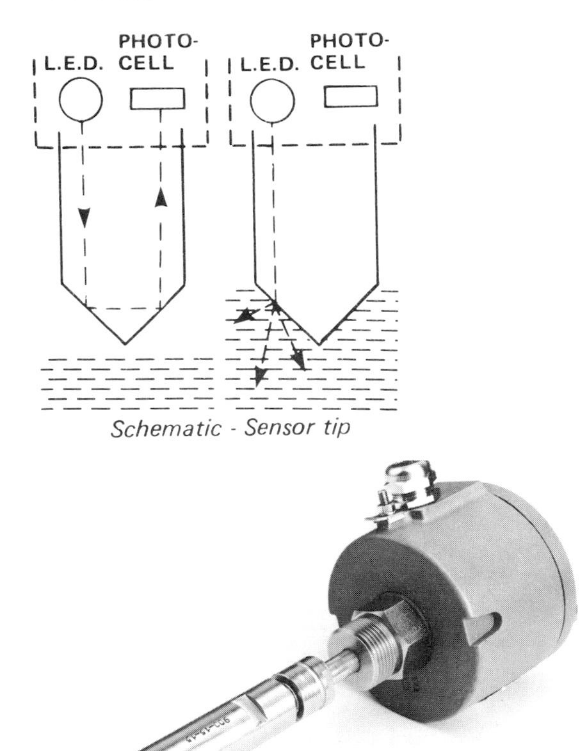

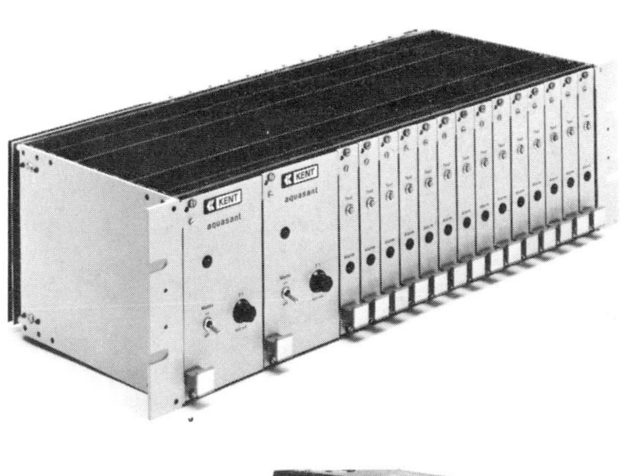

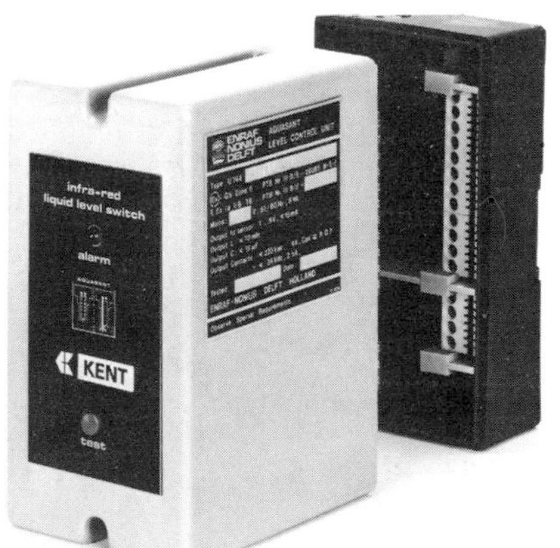

Figure 5.16 Infrared discrete-position level sensor. Principle and physical equipment. Courtesy, Kent Industrial Measurements Ltd.

radiation is used, this being easily produced with light-emitting diodes and readily detected. When the prism outer surface is wetted the majority of the radiation passes through the quartz-glass prism into the liquid, dropping the signal level given out by the photocell.

This method does not require the installation of electrical connections into the tank and, therefore, lends itself to use where intrinsically safe working is needed. Typical discrimination is around 1 mm.

5.5.4 Radio frequency

This form of surface sensor is used in the system shown in Figure 5.13. The tank gauge unit contains an RF (radio-frequency) oscillator tuned to around 160 MHz. Its signal, modulated at 50 Hz, is transmitted to the sensing probe located on the end of the cable line. The probe is a tuned antenna set to be resonant at the carrier frequency. When the tip is brought close to the liquid its resonant frequency is altered. Demodulation at the probe produces a 50 Hz signal that is fed back along the cable as a voltage level depending upon the relationship between oscillation frequency and the resonance of the antenna. This is compared with a reference voltage to produce an error signal that is used to drive the cable to place the probe at the present null position with respect to the surface.

5.5.5 Miscellaneous methods

The following are some of the other principles that have been used to sense the presence of a liquid. The turning moment exerted to rotate a turning paddle will vary when material begins to cover the paddle. The resonant frequency of a vibrating tuning fork will change as it becomes immersed. The electrical resistance of a heated thermistor will vary depending upon its surroundings.

5.6 References

Lazenby, B. 'Level monitoring and control', *Chemical Engineering*, **87**, 1, 88–96, (1980)

Miller, J. T. (ed.) *The Instrument Manual*, United Trade Press, (5th edn, 1975)

Norton, H. N. *Handbook of Transducers for Electronic Measuring Systems*, Prentice-Hall, Englewood Cliffs, New Jersey, (1969)

O'Higgins, P. J. *Basic Instrumentation—Industrial Measurement*, McGraw-Hill, New York, (1966)

6 Vibration

P. H. SYDENHAM

6.1 Introduction

6.1.1 Physical considerations

Vibration is the name given to measurands describing oscillatory motion. Several different measurable parameters may be of interest—relative position, velocity, acceleration, jerk (the derivative of acceleration) and dynamic force being those most generally desired.

For each parameter it may be the instantaneous value, the average value or some other descriptor that is needed. Accuracy of the order of 1 part in 100 is generally all that is called for.

Vibration, in the general sense, occurs as periodic oscillation, as random motion or as transient motion, the latter more normally being referred to as shock when the transient is large in amplitude and brief in duration.

Vibration can occur in linear or rotational forms of motion, the two being termed respectively translational or torsional vibrations. In many ways the basic understanding of each is similar because a rotational system also concerns displacements. Translational forms are outlined in the following description; there will usually exist an equivalent rotational system for all arrangements described.

In vibration measurement it is important to decide whether or not a physically attached mechanical sensor can be used, corresponding to a contacting or non-contacting technique.

Adequate measurement of vibration measurands can be a most complex problem. The requirement is to determine features of motion of a point, or an extended object, in space relative to a reference framework, see Figure 6.1.

A point in space has three degrees of freedom. It can translate in one or more of three directions when referred to the cartesian coordinate system. Rotation of a point has no meaning in this case. Thus to monitor free motion of a point object requires three measurement sensing channels.

If the object of interest has significant physical size it must be treated as an extended object in which the rotations about each of the three axes, described above, provide a further three degrees of freedom. Thus to monitor the free motion of a realistic object may need up to six sensors, one for each degree of freedom.

In practice some degrees of freedom may be nominally constrained (but are they really?) possibly eliminating the need for some of the six sensors. Practical installation should always contain a test that evaluates the degree of actual constraint because sensors will often produce some level of output for the directions of vibration they are not primarily measuring. This is called their cross-axis coupling factor, transverse response or some such terminology.

In many installations the resultant of the motion vector may lie in a constant fixed direction with time. In such cases, in principle, only one sensor will be required provided it can be mounted to sense in that direction. If not, as is often the case, more than one unit will be required, the collective signals then being combined to produce the single resultant.

The potential frequency spectrum of vibration parameters extends, as shown in Figure 6.2, from very slow motions through frequencies experienced in machine tools and similar mechanical structures to the supersonic megahertz frequencies of ultra-sound. It is not possible to cover this range with general-purpose sensors. Each application will need careful

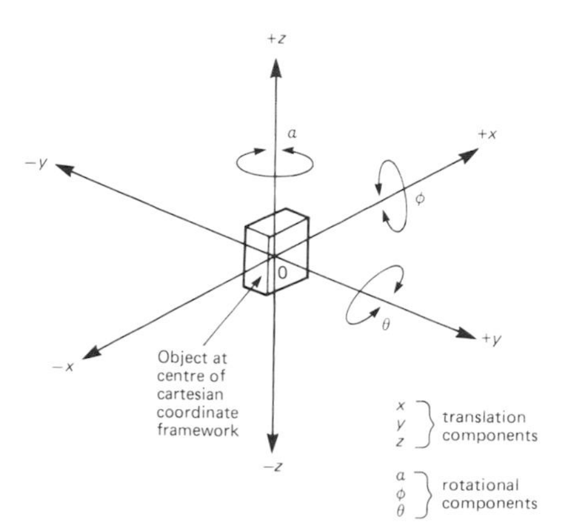

Figure 6.1 Possible motions of an extended object in space relative to a cartesian framework.

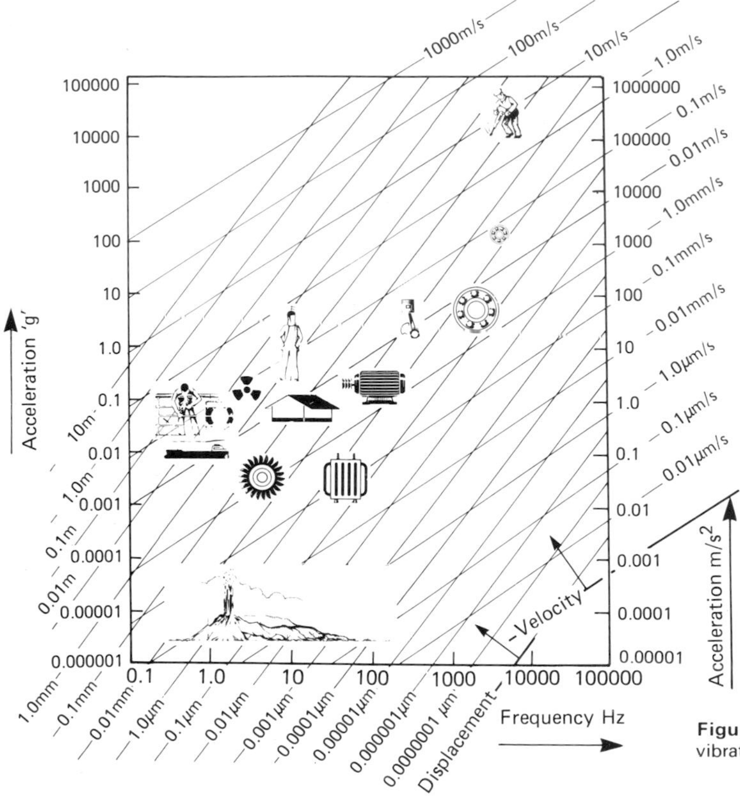

Figure 6.2 Frequency spectrum and magnitude of vibration parameters. Courtesy, Brüel & Kjaer.

consideration of its many parameters to decide which kind of sensor should be applied in order to make the best measurement.

A complicating factor in vibration measurement can be the distributed nature of mechanical systems. This leads to complex patterns of vibration, demanding care in the positioning of sensors.

Mechanical systems, including human forms, given as an example in Figure 6.3, comprise mass, spring compliance (or stiffness) and damping components. In the simplest case, where only one degree of freedom exists, linear behaviour of this combination can be well described using linear mathematical theory to model the time behaviour as the result of force excitation or some initial position displacement.

Vibration can be measured by direct comparison of instantaneous dimensional parameters relative to some adequately fixed datum point in space. The fixed point can be on an 'independent' measurement framework (fixed reference method) or can be a part that remains stationary because of its high inertia (seismic system).

In general a second-order linear system output response q_o is related to an input function q_i by the

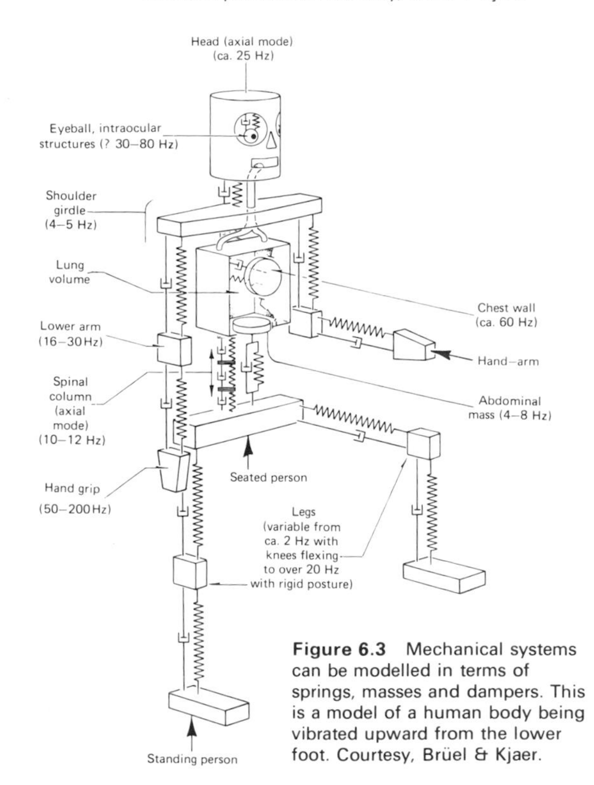

Figure 6.3 Mechanical systems can be modelled in terms of springs, masses and dampers. This is a model of a human body being vibrated upward from the lower foot. Courtesy, Brüel & Kjaer.

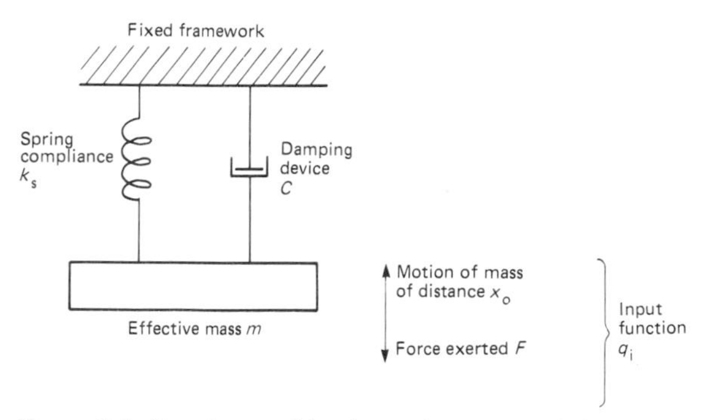

Figure 6.4 One-degree-of-freedom spring, mass and damper system model.

differential equation

$$\frac{a_2\,\mathrm{d}^2q_o}{\mathrm{d}t^2}+\frac{a_1\,\mathrm{d}q_o}{\mathrm{d}t}+a_0q_o=q_i$$

(spring–mass–damper system) (input driving function)

For the specific mechanical system of interest here, given in Figure 6.4, this becomes

$$\frac{m\,\mathrm{d}^2x_0}{\mathrm{d}t^2}+\frac{c\,\mathrm{d}x_0}{\mathrm{d}t}-k_sx_0=q_i$$

where m is the effective mass (which may need to include part of the mass of the spring element or be composed entirely of it), c is the viscous damping factor and k_s the spring compliance (expressed here as length change per unit of force applied).

Where the damping effect is negligible, the system will have a frequency at which it will naturally vibrate if excited by a pulse input. This natural frequency ω_n is given by

$$\omega_n=\sqrt{\frac{k_s}{m}}$$

Presence of damping will alter this value, but as the damping rises the system is less able to provide continuous oscillation.

The static sensitivity is given by the spring constant, either as k_s the spring compliance or as its reciprocal, that is, expressed as force per unit extension.

The influence of damping is easily described by a dimensionless number, called the damping ratio, which is given by

$$\xi=\frac{c}{2\sqrt{k_s\cdot m}}$$

It is usually quoted in a form that relates its magnitude with respect to that at $\xi=1$.

These three important parameters are features of the spring–mass–damper system and are independent of the input driving function.

Such systems have been extensively analysed when excited by the commonly met input forcing functions (step, impulse, ramp, sinusoid). A more general theory for handling any input function other than these is also available. In practice the step, impulse, and continuous sinusoidal responses are used in analyses, as they are reasonably easy to apply in theory and in practical use.

As the damping factor ξ increases the response to a transient step force input (applied to the mass) can vary from sinusoidal oscillation at one extreme (underdamped) to a very sluggish climb to the final value (overdamped). These responses are plotted in Figure 6.5. In the case of continuous sinusoidal force input the system frequency response varies as shown in Figure 6.6. Note the resonance build-up at ω_n which is limited by the degree of damping existing. Thus the damping of the system to be measured or of the sensor, if it is of the seismic kind, can be of importance as a modifier of likely system responses. As damping increases the system response takes on the form of the lower first-

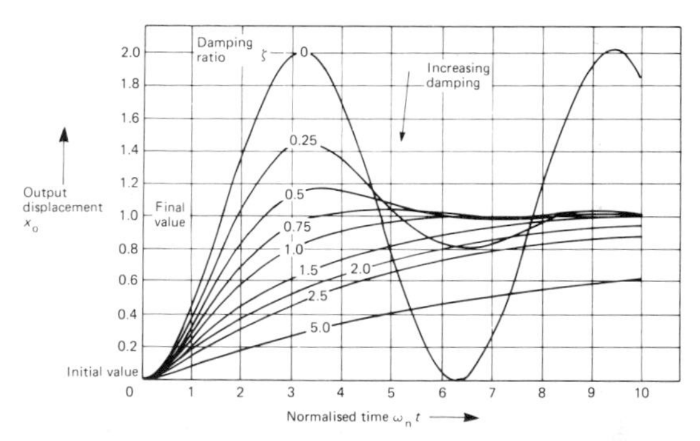

Figure 6.5 Displacement responses of second-order system to input step of force.

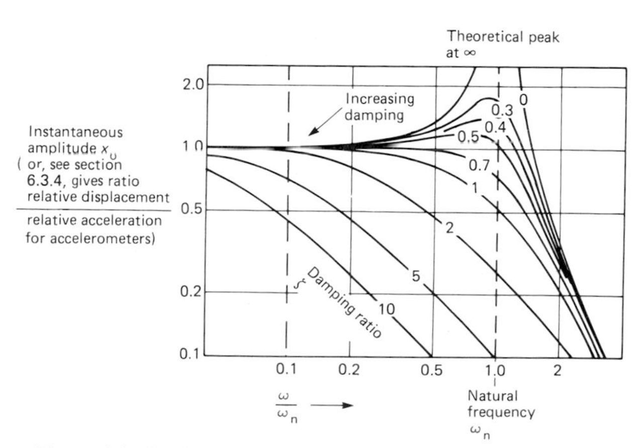

Figure 6.6 Displacement responses of second-order system to continuous sinusoidal force input. The same curves relative displacements of a seismic mass to the acceleration of the mass. See section 6.3.4.

order, exponential response, system and it cannot oscillate.

Useful introductions to this aspect of vibrations are to be found in Oliver (1971), the dynamic behaviour of systems being expounded in more depth in Crandall (1959), Harris and Crede (1961), Sydenham (1983), Trampe-Broch (1980), and Wallace (1970).

The above discussion, given with respect to vibration of the measurand, is also the basis of understanding the operation of seismic vibration sensors, as will be seen later.

It is a property of second-order systems, therefore, to have a natural frequency of vibration. This is the frequency at which they vibrate when given impulse energy that is not overridden by continuous forced vibrations. Thus a sensing system that is second-order and not damped will (due to noise energy inputs) produce outputs at its natural frequency that are not correlated with frequencies occurring in the system of interest. Use of seismic vibration sensors must, therefore, recognize these limitations.

In practice it is also often more convenient to sense a measurand by an indirect means and obtain the desired unit by mathematical processing. For example, accelerometers are conveniently used to obtain forces (from force = mass × acceleration) and hence stresses and strains. Acceleration signals can be twice integrated with respect to time to yield displacement. Sensors that operate as velocity transducers can yield displacement by single integration.

Integration is generally preferred to differentiation as the former averages random noise to a smaller value compared to the signal whereas the latter, in reverse, can deteriorate the signal-to-noise ratio. Mathematical signal manipulation is common practice in vibration measurement as a means to derive other related variables.

6.1.2 Practical problems of installation

With vibration measurement it is all too easy to produce incorrect data. This section addresses several important installation conditions that should be carefully studied for each new application.

6.1.2.1 Cross-coupling

Transducers may exhibit cross-axis coupling. Wise practice, where possible, includes a test that vibrates the sensor in a direction perpendicular to the direction of normal use. Rotational sensitivity may also be important. These tests can be avoided each time they are used if the sensors are precalibrated for this source of error and, of course, are still within calibration. Sensors that have no such parameter quoted should be regarded as potential sources of error until proven otherwise.

6.1.2.2 Coupling compliance

The compliance of the bond made between the sensor and the surface it is mounted on must be adequately stiff. If not, the surface and the sensor form a system that can vibrate in unpredictable ways. As an example an insufficiently stiff mounting can give results that produce much lower frequency components than truly exist. In extreme cases the sensor can be shaken free as it builds up the unexpectedly low resonance frequency of the joint to dangerous amplitude levels. As a guide the joint should be at least ten times stiffer than the sensor so that the resonant frequency of the joint is well above that of the sensor.

6.1.2.3 Cables and pre-amplifiers

Certain types of sensor, notably the piezo-electric kind, are sensitive to spurious variation in capacitance and charge. Sources of such charges are the tribo-electric effect of vibrating cables (special kinds are used, the design of which allows for movement of the cable), varying relative humidity that alters electric field leakage (this becomes important when designing long-term installations) and pre-amplifier input condition variations.

6.1.2.4 Influence errors

Ideally the sensor should operate in a perfect environment wherein sources of external error, called influence parameters, do not occur. In vibration sensing possible influence error sources include temperature variation of the sensor, possible magnetic field fluctuations (especially at radio frequency) and existing background acoustic noise vibrations. Each of these might induce erroneous signal.

A good test for influence parameters is to fully connect the system, observing the output when the measurand of interest is known to be at zero level. Where practical the important error inputs can be systematically varied to see the sensor response. Many a vibration measurement has finally been seen to be worthless because some form of influence error turned out to be larger than the true signal from the measurand. Vibrations apparently occurring at electric mains frequency (50 or 60 Hz) and harmonics thereof are most suspect. Measurement of mechanical vibration at these frequencies is particularly difficult because of the need to separate true signal from influence error noise.

6.1.2.5 Subject loading by the sensor

Vibration sensors contain mass. As this mass is made smaller the sensitivity usually falls. Addition of mass to a vibrating system can load the mass of that system, causing shifts in frequency. For this reason manu-

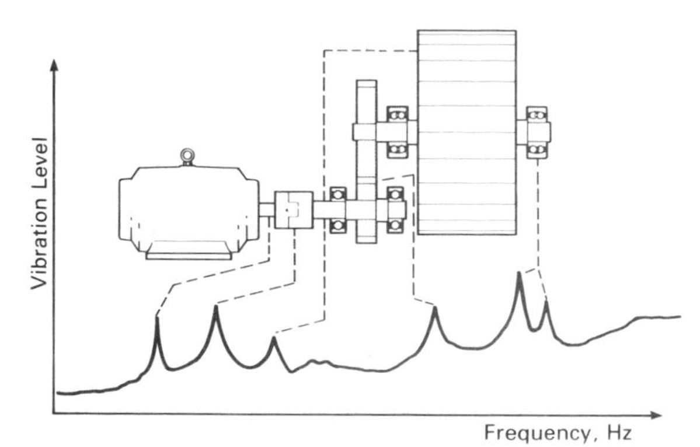

Figure 6.7 In machine health monitoring the normal vibration levels of parts of the installation are recorded to provide a normal signature. Variations of this indicate changes in mechanical conditioning. Courtesy, Brüel & Kjaer.

facturers offer a wide range of attached type sensors. Provided the mass added is, say, 5 per cent or less of the mass of interest, then the results will be reasonable. Cables can also reduce mechanical compliance, reducing the system amplitude. Where a system is particularly sensitive to loading, the use of non-contact, fixed-reference, methods may be the only way to make a satisfactory measurement.

6.1.2.6 Time to reach equilibrium

When damping of a structure is small, the time taken for a resonance to build up to its peak value is large. When using forced vibration to seek such a resonance, it is therefore important not to sweep the excitation input frequency too rapidly.

6.1.3 Areas of application

When searching for information about a measurement technique it is usually helpful to have an appreciation of the allied fields which use the same equipment. Vibration, of course, will be of interest in very many applications but a small number can be singled out as the main areas to which commercial marketing forces have been directed.

6.1.3.1 Machine health monitoring

A significant field of interest is that of machine health, or condition, monitoring; failures can often be avoided by 'listening' to the sounds and vibrations made by the system. An example is shown in Figure 6.7. Vibration and other forms of sensor are applied to the operating system, first whilst running in early life, and then at periodic intervals during life. If the frequency/amplitude data (the so-called signature) has changed then this can provide diagnostic information suggesting which component is beginning to fail. It can then be conveniently replaced before a major, untimely, breakdown occurs. Introduction to this aspect is to be found in Bently Nevada (1982) and Wells (1981).

6.2 Amplitude calibration

Static amplitude (displacement) is easily calibrated using a standardized micrometer, displacement sensor or optical interferometry. Dynamic calibrations may be made either by comparison, using a technique of known accuracy and frequency response or by using a calibrated vibration generator.

6.2.1 Accelerometer calibration

Figure 6.8 shows outlines of three methods for the calibration of accelerometers and other vibration-

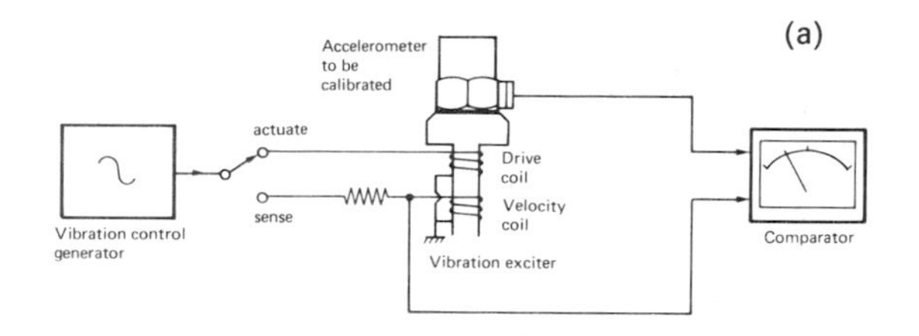

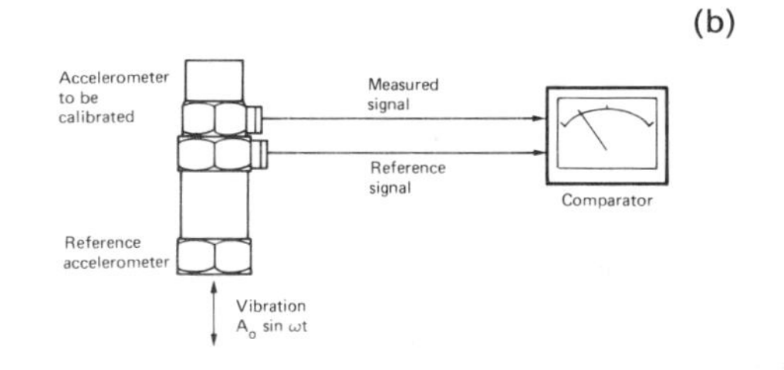

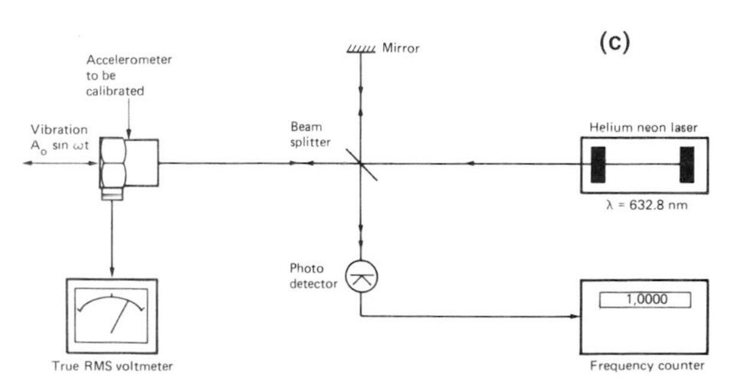

Figure 6.8 Two alternatives for calibrating accelerometers. (a) Calibrated vibration exciter shaking accelerometer at calibrated levels—the reciprocity method. (b) Back-to-back, calibration of a calibrated accelerometer against one to be calibrated—comparison method. (c) Absolute measurement using optical interferometry.

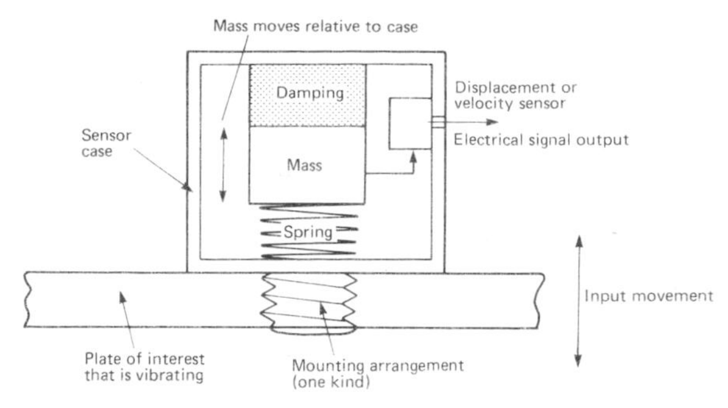

Figure 6.9 Schematic layout of open-loop, seismic-form, vibration sensor.

measuring sensors. Calibration is normally performed at 500 rad s^{-1}.

Other methods that can be used are to subject the accelerometer to accelerations produced by the earth's force. Simple pseudo-static rotation of an accelerometer in the vertical plane will produce accelerations in the 0 to ± 1 g range (g is used here for the earth's acceleration). Larger values can be obtained by whirling the accelerometer on the extremity of a rotating arm of a calibrating centrifuge, or it can be mounted on the end of a hanging pendulum.

6.2.2 Shock calibration

Short-duration acceleration, as produced by impact, requires different approaches to calibration. Accelerations can exceed 10 000 g and last for only a few milliseconds.

A commonly used method is to produce a calibrated shock by allowing a steel ball to free-fall on to an anvil on which is mounted the sensor. This method provides an absolute calibration but, as with all of the above described methods, has uncertainties associated with the practical method. In this case one source of error is caused by the difficulty of releasing a ball to begin its downward path without imparting some velocity at time zero.

6.2.3 Force calibration

Static forces can be calibrated by applying 'dead-weights' to the force sensor, the 'weights' being calibrated masses. (See Chapter 7.)

Dynamic forces arising in vibration can more easily be determined using the relationship force = mass × acceleration. A shaking table is used to produce known accelerations on a known mass. In this way the forces exerted on the accelerometer can be determined along with the corresponding output voltage or current needed to produce transducer sensitivity constants.

Space does not permit greater explanation but there are several detailed accounts of vibration sensor calibration available in the literature—Endevco (1980), Harris and Crede (1961), Herceg (1972), Norton (1969), Oliver (1971) and Trampe-Broch (1980). National and international standards are extensively listed in Brüel and Kjaer (1981).

6.3 Sensor practice

6.3.1 Mass–spring seismic sensors

Whereas the fixed reference methods do have some relevance in the practical measurement of vibration the need for a convenient datum is very often not able to be met. In the majority of vibration measurements use is made of the mass–spring, seismic, sensor system.

Given the correct spring–mass damping combination a seismic system attached to a vibrating surface can yield displacement, velocity or acceleration data. Unfortunately the conflicting needs of the three do not enable one single design to be used for all three cases. However, it is often possible to derive one variable from another by mathematical operations on the data.

Two forms of seismic sensor exist. The first, called open-loop, makes use of the unmodified response of the mass moving relative to the case to operate either a displacement or a velocity sensing transducer. The second form closes the loop (and is, therefore, referred to as a closed-loop or servo seismic sensor) using the output signal to produce an internal force that retains the mass in the same relative position with respect to the case, the magnitude of the force being a measure of the vibration parameter.

6.3.1.1 Open-loop sensors

The fundamental arrangement of the open-loop seismic sensor form is as given in Figure 6.9. Actual construction can vary widely depending upon how the spring force and damping are provided and upon the form of the sensor used.

The spring element can be produced as a distinct mechanical element, Figure 6.10 is an example made

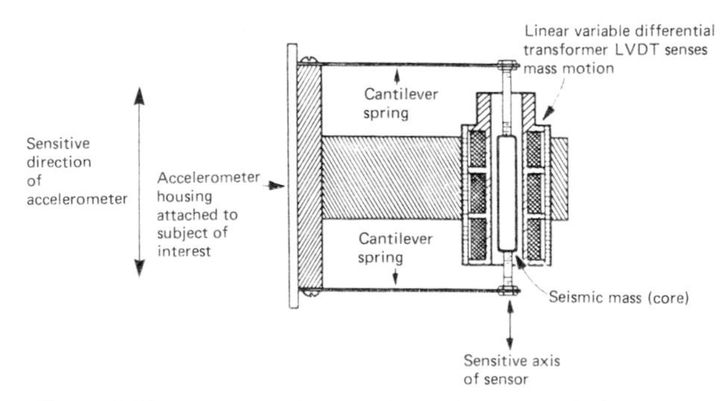

Figure 6.10 Diagrammatic view of a spring–mass seismic sensor that uses parallel flexure-strip spring suspension and inductive sensor of mass displacement. Courtesy, Schaevitz Engineering.

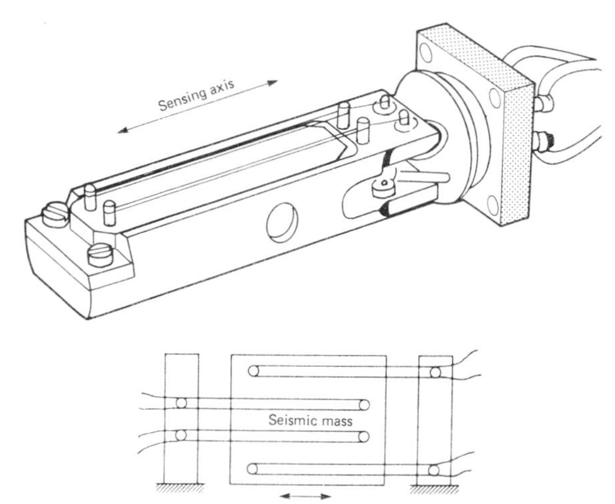

Figure 6.11 Displacement sensing of mass motion in accelerometers can be achieved by many methods. This unit uses unbonded strain gauges. Courtesy, Statham Instruments Inc.

with flexure strips; alternatively perforated membranes, helical coils, torsional strips and the like can be used. Otherwise, the compliance of the mass itself may be the spring, for example in the piezo-electric crystal which also acts as the sensing element.

Important design parameters of a spring are the compliance, amplitude range, fatigue life, constancy of rate with time, temperature and other influence effects and the suitability to be packaged to produce a suitable sensor unit. Except for the highest natural frequency sensors the masses used can be regarded as completely rigid compared with the spring element.

Rotary forms of the linear arrangement, shown in Figure 6.9, are also available.

Sensing methods that have been used include electrical-resistance sliding potentiometer, variable inductance (see for instance Figure 6.10), variable reluctance, variable capacitance, electrical metallic strain gauges (bonded and unbonded as in Figure 6.11) and semiconductor strain gauges, piezo-electric crystal and magnetostrictive elements, position-sensitive optical detectors and electromagneto principles (that provide direct velocity sensing).

Sensors are often encapsulated. The encapsulation takes many forms ranging from miniature units of total weight around 1 g through to 0.5 kg units where the sensing mass must be physically large. As simultaneous measurements in two or three directions are often required, seismic sensors are also made that consist of two or three units, as shown in Figure 6.12, mounted in different directions.

Compensation for temperature is needed in many designs. This is either performed in the electronic circuitry or by incorporating some form of thermo-mechanical device into the spring–mass–sensor layout.

6.3.1.2 Servo accelerometers

The performance of open-loop seismic sensors can be improved with respect to their sensitivity, accuracy and output signal amplitude by forming the design into a closed-loop form.

Figure 6.13 gives the schematic diagram of one kind of closed-loop system which is based upon a moving-coil actuator and capacitance displacement sensor. The mass upon which the acceleration is to be exerted is able to rotate on the end of a freely supported arm. This is attached to the electrical coil placed in the permanent magnetic field supplied by the magnet assembly. Acceleration applied to the mass attempts to rotate the arm causing displacement. This unbalances the capacitive displacement sensor monitoring the relative position of the mass. Displacement signals

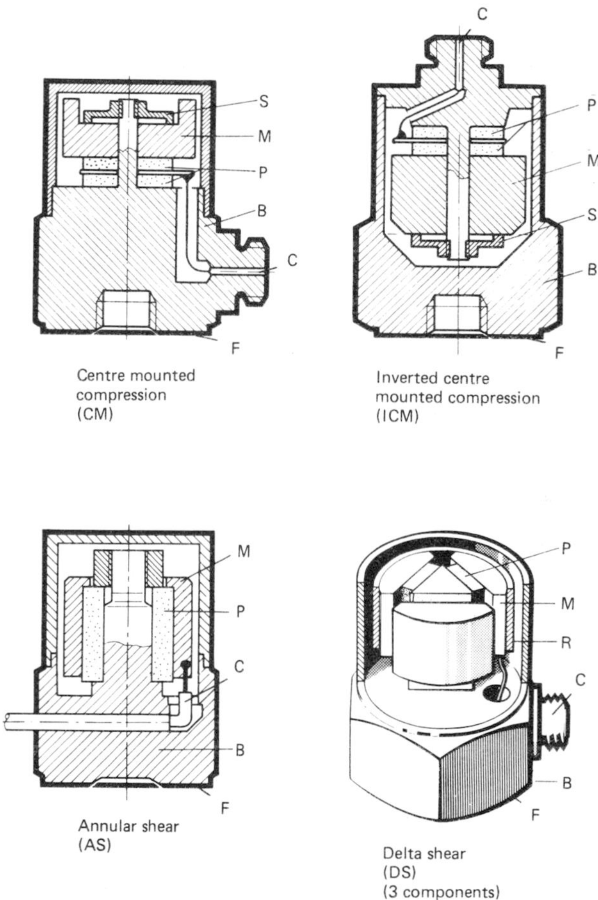

Figure 6.12 Examples of single- and three-axis accelerometers based on the piezo-electric sensor. Courtesy, Brüel & Kjaer.

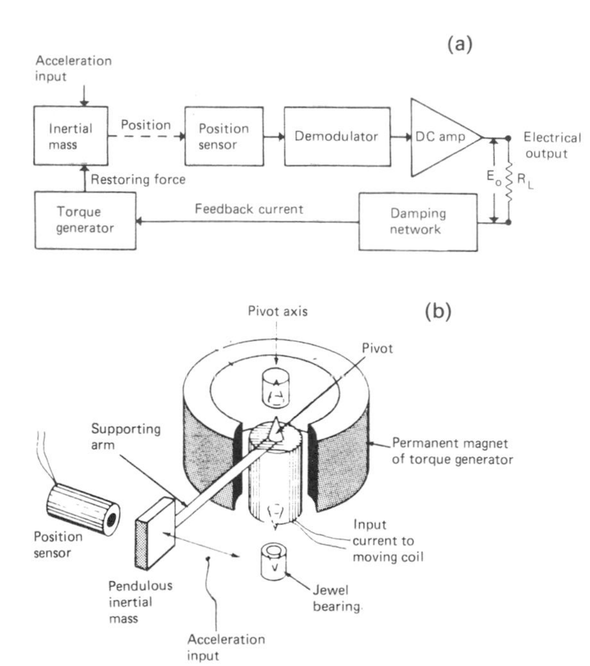

Figure 6.13 Basic component layout of one form of closed-loop accelerometer. Courtesy, Schaevitz Engineering.

produce an input to the difference-sensing amplifier. The amplifier drives a corresponding electric current into the coil causing the arm to rotate back to the null displacement position. Provided the loop response is rapid enough the mass will be retained in a nearly constant place relative to the displacement sensor. Acceleration variations are, thereby, converted to variations in coil current. In this way the displacement sensor is used in the, preferred, null-balance mode wherein error of linearity and temperature shift are largely avoided. Only the more easily achieved proportionality between coil current and force is important. Servo instruments are further described in Jones (1982), Herceg (1972) and Norton (1969).

6.3.2 Displacement measurement

Where a fixed reference method is applicable it is possible to employ any suitable displacement sensor. Spurious phase shift in the system may be a concern but in some applications phase is unimportant. The reader is directed to Chapter 3 for an introduction to displacement devices.

It is sometimes more convenient to integrate the signal from a velocity transducer mounted to make use of a fixed reference.

Where a fixed reference method is inconvenient one of several forms of seismic sensor system can be employed as follows.

The second-order equations of motion, given in Section 6.1.1 for a mass moving relative to a fixed reference frame (the mode for studying the movement of vibrating objects), can be reworked to provide response curves relating the displacement amplitude of the seismic mass to the amplitude of its case. This is the seismic sensor output. Figure 6.14 is the family of response curves showing the effects of operating frequency and degree of damping. Given that the case is moving in sympathy with the surface of interest it can be seen, from the curves, that for input vibration frequencies well above the natural frequency of the seismic sensor the measured output displacements will be a true indication (within a percent or so) of the movements of the surface. This form of seismic displacement sensor is also often called a vibrometer. It is possible to lower the frequency of operation by using a damping factor with a nominal value of 0.5. This, however, does introduce more rapidly changing phase shift error with frequency which may be important. The lowest frequency of use above which the response remains virtually flat is seen to be where the various damping factor curves approach the horizontal line equal to unity ratio, to within the allowable signal tolerance.

Given that the chosen damping remains constant and that a system does follow the second-order response it is also possible to provide electronic frequency compensation that can further lower the useful frequency of operation by a small amount.

Thus to directly measure displacement amplitudes with a seismic sensor it must have a natural frequency set to be below the lowest frequency of interest in the subject's vibration spectrum. In this mode the seismic

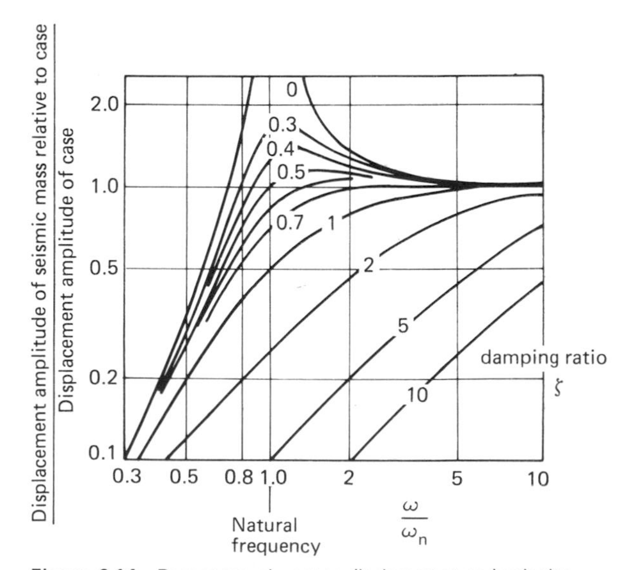

Figure 6.14 Responses relevant to displacement and velocity sensing with seismic sensors.

mass virtually remains stationary in space acting as a fixed reference point. It is also clear that the seismic sensor cannot measure very low frequencies for it is not possible to construct an economical system having a low enough resonant frequency.

The curves are theoretical perfections and would apparently indicate that the seismic sensor, in this case, will have flat response out to infinite frequency. This is not the case in practice for as the frequency of vibration rises the seismic sensor structure begins to resonate at other frequencies caused by such mechanisms as the spring vibrating in modes other than the fundamental of the system.

Given that the low frequency range of accelerometers can extend down to less than 1 Hz (see Section 6.3.4) it may often be more practical to twice integrate an accelerometer signal, in order to derive displacement amplitude, rather than to make use of a direct-reading displacement design of seismic sensor.

6.3.3 Velocity measurement

The prime method used to generate a direct velocity signal makes use of the law of electromagnetic induction. This gives the electrical voltage e generated as N turns of an electric coil cut magnetic flux ϕ over time t as

$$e = -N\frac{d\phi}{dt}$$

Velocity sensors are self-generating, producing a voltage output that is proportional to the velocity at which a set of turns moves through a constant and uniform magnetic field.

Many forms of this kind of sensor exist. The commonly used, moving-coil, arrangement comprises a cylindrical coil vibrating inside a magnetic field that is produced by a permanent electromagnet. A commonly seen arrangement, Figure 6.15(a), is that typified by the reversible-role moving-coil, loudspeaker movement. For this form the output voltage V_{out} is given by

$$V_{out} = -Blv \,.\, 10^{-9}$$

where B is the flux density in tesla, l the effective length of conductor contributing in total flux-cutting and v the instantaneous velocity, expressed in m s^{-1}, of the coil relative to the magnet. Given that the design ensures that the field is uniform in the path of the fixed conductor length the sensor can provide very linear output.

The sensors were adopted in early seismology studies because of their inherently high output at relatively low velocities. The coil impedance will generally be low, enabling signals with good signal-to-noise ratios to be generated along with reduction of error caused by variations in lead length and type. They are, however, large with resultant mass and rigidity problems. They tend to have relatively low resonant frequencies (tens of hertz), which restricts use to the lower frequencies. Output tends to be small at the higher frequencies. It will be apparent that these sensors cannot produce signals at zero velocity because no relative movement occurs to generate flux-cutting.

A second variation of the self-generating velocity

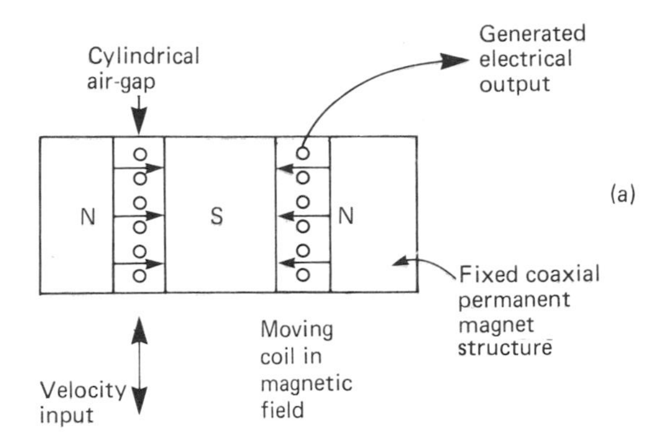

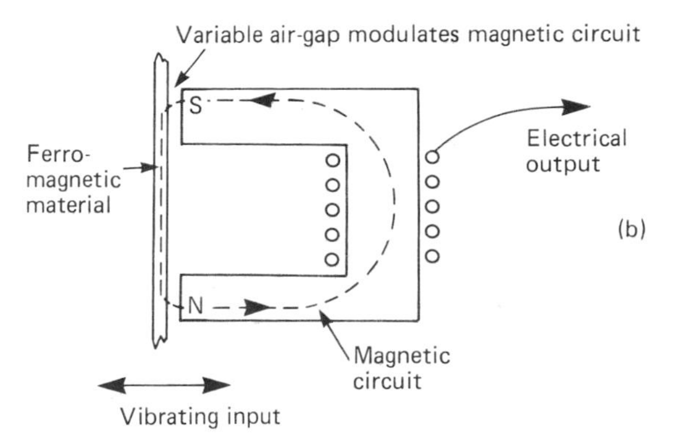

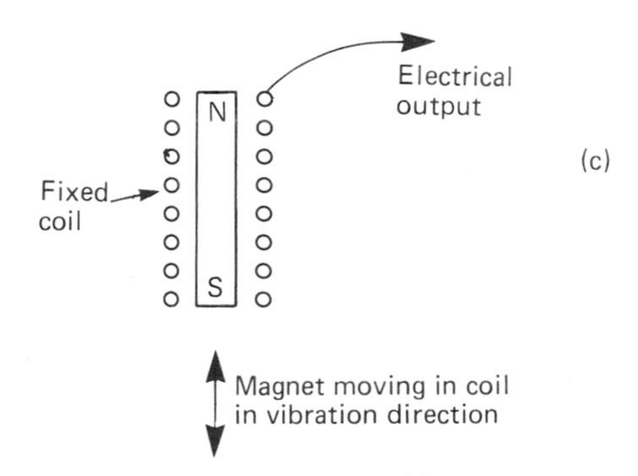

Figure 6.15 Forms of velocity sensor. (a) Moving coil. (b) Variable reluctance. (c) Moving permanent magnet.

sensor, the variable-reluctance method, uses a series magnetic circuit containing a permanent magnet to provide permanent magnetic bias. A part of this circuit, the armature, is made so that the effective air-gap is varied by the motion to be monitored. Around the magnetic circuit is placed a pick-up coil. As the armature moves the resulting flux variation cuts the coil generating a signal that can be tailored by appropriate design to be linear with vibration amplitude. This form of design has the advantage that the armature can readily be made as part of the structure to be monitored, as shown in Figure 6.15(b). This version is not particularly sensitive, for the air-gap must be at least as large as the vibration amplitude. As an example a unit of around 12 mm diameter, when used with a high magnetic permeability moving disc set at 2 mm distance, will produce an output of around 150 mV/m s^{-1}.

A third method uses a permanent magnet as the mass supported on springs. One example is shown in Figure 6.15(c). Vibration causes the magnet to move relatively to the fixed coil thereby generating a velocity signal. This form can produce high outputs, one make having a sensitivity of around 5 V/m s^{-1}.

Where a fixed reference cannot be used this form of sensor, instead of a displacement sensor, can be built into the seismic sensor arrangement. In such cases the vibrating seismic sensor will then directly produce velocity signals. These will follow the general responses given in Figure 6.14. From those curves it can be seen that there is a reasonably flat response above the natural frequency which is inherently quite low.

6.3.4 Acceleration measurement

The fixed-reference method of measuring acceleration is rarely used, most determinations being made with the seismic form of sensor. For the seismic sensor system the mass and the spring compliance are fixed. Consideration of the $F=m.a$ law and spring compliance shows that displacement of the mass relative to the sensor case is proportional to the acceleration of the case. This means that the curves, plotted in Figure 6.6 (for sinusoidal input of force to a second-order system), are also applicable as output response curves of accelerometers using displacement sensing. In this use the vertical axis is interpreted as the relative displacement of the mass from the case for a given acceleration of the case.

The curves show that a seismic sensor will provide a constant sensitivity output representing sensor acceleration from very low frequencies to near the natural frequency of the spring–mass arrangement used. Again the damping ratio can be optimized at around 0.5–0.6 and electronic compensation added (if needed) to raise the upper limit a little further than the resonance point.

At first sight it might, therefore, appear that a single, general-purpose design could be made having a very high resonant frequency. This, however, is not the case for the deflection of the spring (which is a major factor deciding the system output sensitivity) is proportional to $1/\omega_n^2$. In practice this means that as the upper useful frequency limit is extended the sensor sensitivity falls off. Electronic amplification allows low signal output to be used but with additional cost to the total measuring system.

At the low-frequency end of the accelerometer response the transducers become ineffective because the accelerations produce too small a displacement to be observed against the background noise level.

6.3.4.1 *Typical sensors*

As a guide to the ranges of capability available, one major manufacturer's catalogue offers accelerometers with sensitivities ranging from a small 30 μV/ms^{-2} through to 1 V/ms^{-2} with corresponding sensor weights of 3 g and 500 g and useful frequency ranges of 1–60 000 Hz and 0.2–1000 Hz. Sensors have been constructed for even higher frequencies but these must be regarded as special designs. A selection are shown in Figure 6.16.

The many constraints placed upon the various performance parameters of a particular seismic sensor can be shown on a single chart such as Figure 6.17, Harris and Crede (1961).

As the accelerometer spring is often required to be stiff compared with that of the seismic displacement sensor it will not always need to make use of coiling, a device for decreasing the inherent spring constant of a material. Accelerometer springs may occur as stamped rigid plates, as flat cusped spring washers, or as a sufficiently compliant clamping bolt. In the case of piezo-sensitive material use is often made of the compliance of the material.

Figure 6.16 A range of accelerometers is required to cover the full needs of vibration measurement. Courtesy, Inspek Supplies, New South Wales.

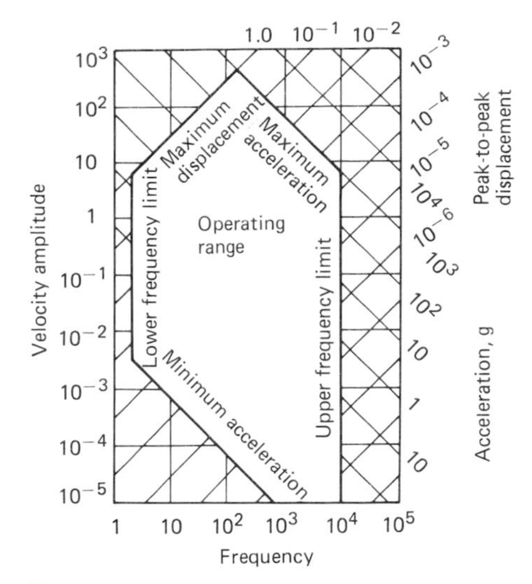

Figure 6.17 Useful linear operating range of an individual, seismic vibration sensor can be characterized with this form of chart. Courtesy, McGraw-Hill.

6.3.4.2 *Response to complex waveforms*

The response curves given relate to seismic sensors excited by sinusoidal signals. To predict the behaviour of a certain sensor, such as an accelerometer, when used to measure other continuous or discrete waveforms it is first necessary to break down the waveform into its Fourier components. The response, in terms of amplitude and phase, to each of these is then added to arrive at the resultant response. It has been stated above that damping can be added to extend the useful bandwidth of a seismic sensor. However, where this is done it generally increases the phase shift variation with frequency. A signal comprising many frequencies will, therefore, produce an output that depends largely on the damping and natural frequency values of the sensor. A number of responses are plotted, such as that in Figure 6.18, in Harris and Crede (1961) to which the reader is referred. Generally the damping value for best all-round results is that near the critical value.

6.3.4.3 *The piezo-electric sensor*

Numerous sensing methods have been devised to measure the motion of the mass in a seismic sensor. We discuss here the most commonly used method: others are described in Endevco (1980), Harris and Crede (1961), Herceg (1972), Norton (1969) and Oliver (1971).

Force applied to certain crystalline substances, such as quartz, produces between two surfaces of a suitably shaped crystal an electric charge that is proportional to the force. This charge is contained in the internal electrical capacitance formed by the high-dielectric material and two deposited conducting surfaces. The descriptive mathematical relation for this effect is

$$q = a \,.\, F \,.\, k_s$$

where q is the electrical charge generated by force F (in newtons) applied across the faces of a piezo-electric device having a mechanical compliance of spring rate k_s (m N^{-1}) and a more complex material constant a (of dimensions C m^{-1}).

The constant a depends on many factors including the geometry of the crystal, position of electrodes and material used. Typical materials now used (natural quartz is less sensitive and, therefore, less applicable) include barium titanate with controlled impurities, lead zirconate, lead niobate and many that are trade secrets. The material is made from loose powder that, after shaping, is fired at very high temperature. Whilst cooling, the blocks are subjected to an electric field that polarizes the substance.

The sensitivity of these so-called 'PZT materials' is temperature-dependent through the charge sensitivity and the capacitance value, both of which alter with temperature. These changes do not follow simple linear laws. Such materials have a critical temperature, called the 'Curie point'. They must never be taken above it. The Curie point varies from 120 °C for the

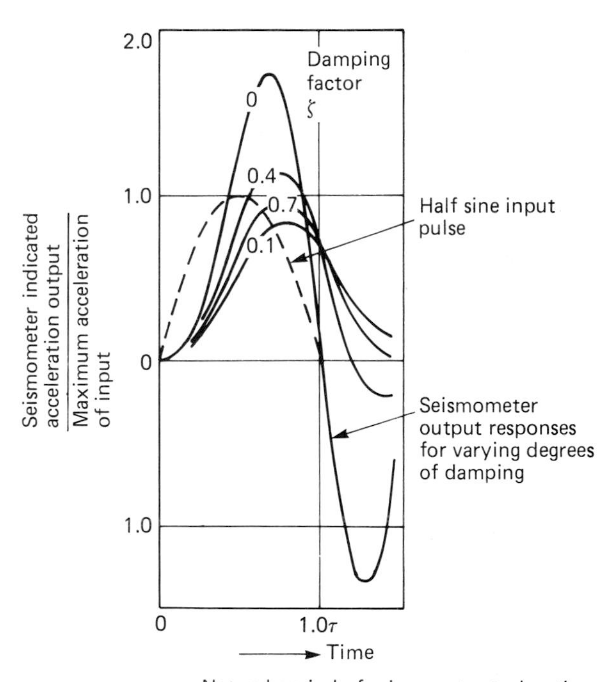

Figure 6.18 Example of response, at various damping factor levels, of a seismic accelerometer to a complex forcing input—a half-sine wave of similar period to that of the natural resonance period of the sensor.

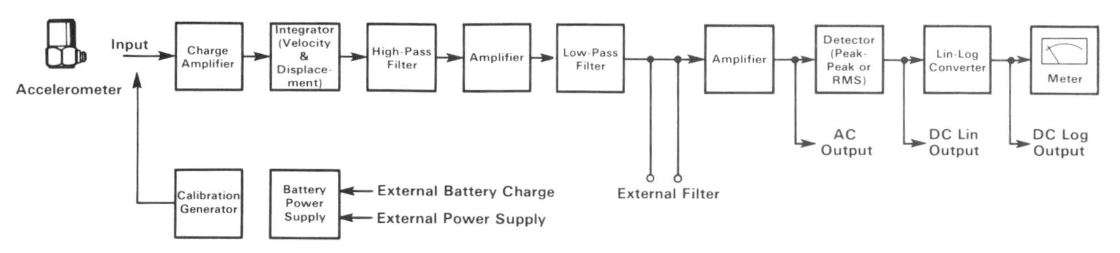

Figure 6.19 Block diagram of vibration measuring system showing functions that may be required. Courtesy, Brüel & Kjaer.

simpler barium titanate forms ranging up to values close to 600 °C. For the interested reader more explanation is to be found in Brüel and Kjaer (1976), Endevco (1980), Harris and Crede (1961), Klaasen (1978) and Trampe-Broch (1980) and in the detailed information provided by the makers of PZT materials.

To read the charge of a PZT sensor an electronic amplifier that converts charge magnitude to a voltage equivalent is used. The nature of the system provides no true d.c. response.

In practice the PZT sensors used to measure acceleration can be operated down to around 0.1 Hz, dependent on the amplitude to be measured. With natural resonant frequency that can be made relatively very high (up to 100 000 Hz in some designs), PZT sensors provide a useful frequency range that can cover most vibration needs. The system response, however, relies not only on the sensor but upon the cables and the pre-amplifier used with the PZT unit.

The PZT material can be used in pure compression, shear or bending, to produce the charge. Figure 6.12 gives some examples of commercially available PZT accelerometers. The sensor design is amenable to the combination of three units giving the three translation components of vibration.

PZT material itself contributes only of the order of 0.03 of critical damping. If no additional damping is added, PZT transducers must not be used too close to their resonant frequency. Mounting arrangements within the case will also add some additional damping. Some designs make use of an additional spring element, some use an additional spring to precompress the PZT element so that it remains biased in compression under all working amplitudes: this makes for more linear operation.

Typical sensor sensitivities range from 0.003 pC/m s^{-2} up to 1000 pC/m s^{-2}, implying that the following pre-amplifier units will also need to vary considerably.

6.3.4.4 Amplifiers for piezo-electric sensors

An amplifier for reading out the state of the PZT sensor is one that has very high input impedance, an adequate frequency response and low output impedance. Adjustment of gain and filtering action

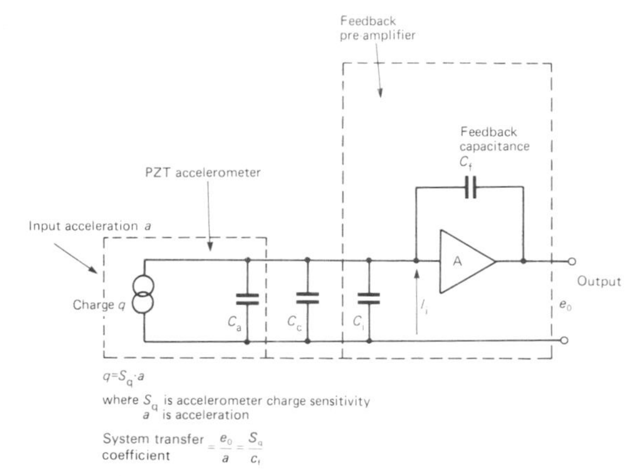

Figure 6.20 Equivalent circuit for piezo-electric sensor when interrogated, as a charge-generating device, by an operational amplifier technique.

and integration to yield velocity and displacement are usually also needed to provide easy use for a variety of applications. Figure 6.19 is a typical system incorporating most features that might be needed.

The amplifier could be designed to see the sensor either as a voltage source or as a charge source. The latter is preferred for, using modern electronic-feedback operational amplifier techniques, the effect of cable, sensor and amplifier capacitances can be made negligible (which, in the voltage-reading method, is not the case). Cable length is, therefore, of no consequence. This is justified as follows.

Figure 6.20 is the relevant equivalent circuit for a PZT accelerometer that is connected to an operational amplifier (the pre-amplifier) via a cable. It includes the dominant capacitances that occur.

It can be shown, see Trampe-Broch (1980) for example, that the use of feedback in this way and a very high amplifier gain A gives

$$e_0 = \frac{S_q a}{C_f}$$

This shows that the user need only define the sensor charge sensitivity S_q and the feedback capacitance C_f in order to be able to relate output voltage from the pre-amplifier to the acceleration of the sensor.

6.3.5 Measurement of shock

Shock is a sudden short impulse of applied force that generates very large acceleration (100 000 g can arise) and is not recurrent. It can be regarded as a once-only occurrence of a vibration waveform, although sometimes it is used to describe a short burst of oscillation.

Understanding the behaviour of a given vibration sensor requires Fourier analysis of its response to a truncated waveshape. The mathematics becomes more complex. Theoretical study does lead to the generalization that as the waveform becomes more like a single pulse of high magnitude and very short duration the frequency band of the sensor must be widened if the delivered output is to be a satisfactory replica of the actual input vibration parameter. Fidelity increases as the period of the natural frequency of the sensor becomes shorter than the pulse length. An idea of the variation of responses with natural frequency and damping is available in graphs given in Harris and Crede (1961). An example is that in Figure 6.18.

The very large forces exerted on the transducer require a design that recognizes the need to withstand large transient forces without altering mechanical strains in the sensor system.

Well-designed shock sensors can accurately measure single half-sinewave pulses as short as 5 μs. Some amount of ringing in the output is usually tolerated in order to provide measurement of very short duration shocks.

6.4 Literature

There exist many general books on the kinds of transducers that are in use. An IMEKO bibliography, Sydenham (1983), is a useful entry point to the literature of measurement technology.

Of the many general instrument texts that are available, very few actually address the subject of vibration as a distinct chapter. Where included, relevant material will be found under such headings as velocity and acceleration measurement, accelerometers, position sensing and piezo-electric systems. Texts containing a chapter-length introductory discussion include Herceg (1972), Norton (1969) and Oliver (1971).

There are, as would be expected, some (but only a few) works entirely devoted to vibration and related measurands. The following will be of value to readers who require more than the restricted introduction that a chapter such as this provides—Brüel and Kjaer (1975, 1982), Endevco (1980), Harris and Crede (1961), Trampe-Broch (1980) and Wallace (1970).

The various trade houses that manufacture vibration-measuring and -testing equipment also often provide extensive literature and other forms of training aids to assist the uncertain user.

6.5 References

Bently Nevada, *Bently Book One* (Application notes on vibration and machines), Bently Nevada, Minden, USA, (1982)

Brüel & Kjaer, *Vibration Testing Systems*, Brüel & Kjaer, Naerum, Denmark, (1975)

Brüel & Kjaer, *Piezoelectric Accelerometer and Vibration Preamplifier Handbook*, Brüel & Kjaer, Naerum, Denmark, (1976)

Brüel & Kjaer, *Acoustics, Vibration & Shock, Luminance and Contrast, National and International Standards and Recommendations*, Brüel & Kjaer, Naerum, Denmark, (1981)

Brüel & Kjaer, *Measuring Vibration—an Elementary Introduction*, Brüel & Kjaer, Naerum, Denmark, (1982)

Crandall, S. H. *Random Vibration*, Wiley, New York, (1959)

Endevco, *Shock and Vibration Measurement Technology*, Endevco Dynamic Instrument Division, San Juan Capistrano, USA, (1980)

Harris, C. M. and Crede, C. E. *Shock and Vibration Handbook Vol. 1, Basic Theory and Measurements*, McGraw-Hill, New York, (1961, reprinted in 1976)

Herceg, E. E. *Handbook of Measurement and Control*, HB-72, Schaevitz Engineering, Pennsauken, USA, (1972, revised 1976)

Jones, B. E. 'Feedback in instruments and its applications' in *Instrument Science and Technology*, Jones, B. E. (ed.), Adam Hilger, Bristol, (1982)

Klaasen, K. B. 'Piezoelectric accelerometers' in *Modern Electronic Measuring Systems*, Regtien, P. P. L., (ed.), Delft University Press, Delft, (1978)

Norton, H. N. *Handbook of Transducers for Electronic Measuring Systems*, Prentice-Hall, Englewood Cliffs, New Jersey, (1969)

Oliver, F. J. *Practical Instrumentation Transducers*, Pitman, London, (1971)

Sydenham, P. H. *Handbook of Measurement Science—Vol. 2. Fundamentals of Practice*, Wiley, Chichester, (1983)

Trampe-Broch, J. *Mechanical Vibration and Shock Measurements*, Brüel & Kjaer, Naerum, Denmark, (1980)

Wallace, R. H. *Understanding and Measuring Vibrations*, Wykeham Publications, London, (1970)

Wells, P. 'Machine condition monitoring', Proceedings of structured course, Chisholm Institute of Technology, Victoria, Australia, (1981)

7 Measurement of force

C. S. BAHRA

7.1 Basic concepts

If a body is released, it will start to fall with an acceleration due to gravity or acceleration of free fall of its location. We denote by g the resultant acceleration due to attraction of the earth upon the body and the component of acceleration due to rotation of the earth about its axis. The value of g varies with location and height and this variation is about 0.5 per cent between the equator and the poles. The approximate value of g is 9.81 m/s^2. A knowledge of the precise value of g is necessary to determine gravitational forces acting on known masses at rest, relative to the surface of the earth in order to establish practical standards of force. Practical standards of dead-weight calibration of force-measuring systems or devices are based on this observation.

It is necessary to make a clear distinction between the units of weight-measuring (mass-measuring) and force-measuring systems. The weight-measuring systems are calibrated in kilograms while the force-measuring systems are in newtons. Mass, force and weight are defined as follows:

Mass. The mass of a body is defined as the quantity of matter in that body and it remains unchanged when taken to any location. The unit of mass is the kilogram (kg).

Force. Force is that which produces or tends to produce a change of velocity in a body at rest or in motion. Force has magnitude, direction and a point of application. It is related to the mass of a body through Newton's second law of motion which gives: force = mass × acceleration.

Unit of force. In the International System of units, the unit of force is the newton (N) and it is that force which when applied to a mass of one kilogram, gives it an acceleration of one metre per second per second (m/s^2).

Weight. Weight F of a body of mass m at rest relative to the surface of the earth is defined as the force exerted on it by gravity: $F = mg$, where g is the acceleration due to gravity.

The main purpose of this chapter is to review the most commonly used force measurement methods and to discuss briefly the principles employed in their design, limitations and use. It is not intended to give a too detailed description of mathematical and physical concepts, but enough information to allow an interested reader to read further.

7.2 Force measurement methods

Force measurement methods may be divided into two categories, direct comparison and indirect comparison. In a direct comparison method, an unknown force is directly compared with a gravitational force acting on a known mass. A simple analytical balance is an example of this method. An indirect comparison method involves the use of calibrated masses or transducers and a summary of indirect comparison methods is given below:

(a) Lever-balance methods.
(b) Force-balance method.
(c) Hydraulic pressure measurement.
(d) Acceleration measurement.
(e) Elastic elements.

Note that the lever-balance methods include examples of both direct and indirect comparisons, but to maintain continuity of information, they are described under one heading.

7.3 Lever-balance methods

7.3.1 Equal-lever balance

A simple analytical balance is an example of an equal-lever balance which consists of a 'rigid' beam pivoted on a knife-edge, as shown in Figure 7.1. An unknown force F_1 is compared directly with a known force F_2. When the beam is in equilibrium the sum of moments about the pivot is zero.

$$F_1 a - F_2 a = 0$$
$$\therefore \quad F_1 = F_2$$

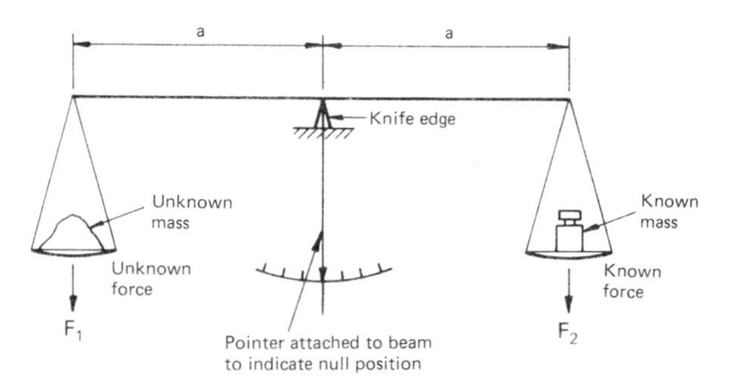

Figure 7.1 Equal-lever balance.

This type of balance is mainly used for weighing chemicals. It gives direct reading, and can weigh up to about 1000 kg with high accuracy if required.

7.3.2 Unequal-lever balance

Figure 7.2 shows a typical arrangement of an unequal-lever balance which can be used for measuring large masses or forces. The balance is obtained by sliding a known mass along the lever. At equilibrium, we have

$$F_1 a = F_2 b$$

$$F_1 = F_2 b/a$$

$$\therefore \quad F_1 \propto b$$

The right-hand side of the beam can therefore be used as a measure of force.

This type of balance is extensively used in materials-testing machines and weight measurement. The balance is normally bulky and heavy, but can be made very accurate.

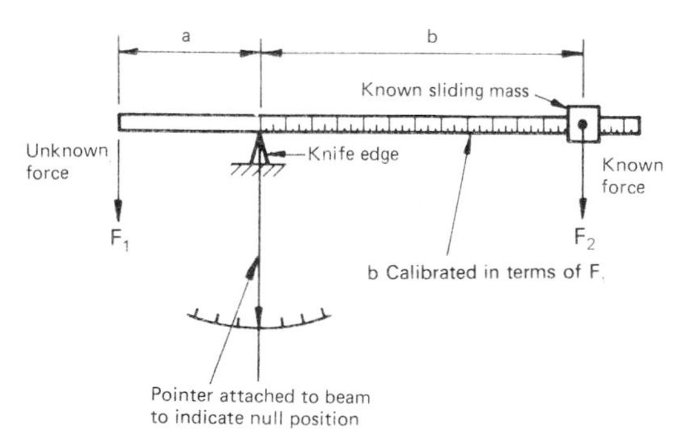

Figure 7.2 Unequal-lever balance.

7.3.3 Compound lever balance

Figure 7.3 shows a compound lever balance which is used for measuring very large masses or forces. Using a number of ratio levers, the applied force is reduced to a level which is just sufficient to actuate a spring within the indicator dial head. The balance is calibrated in units of mass.

7.4 Force-balance methods

Figure 7.4 shows an electronic force-balance type force-measuring system. The displacement caused by the applied force is sensed by a displacement transducer. The displacement transducer output is fed to a servo amplifier to give an output current which flows through a restoring coil and exerts a force on the permanent magnet. This force adjusts itself always to balance the applied force. In equilibrium, the current flowing through the force coil is directly proportional

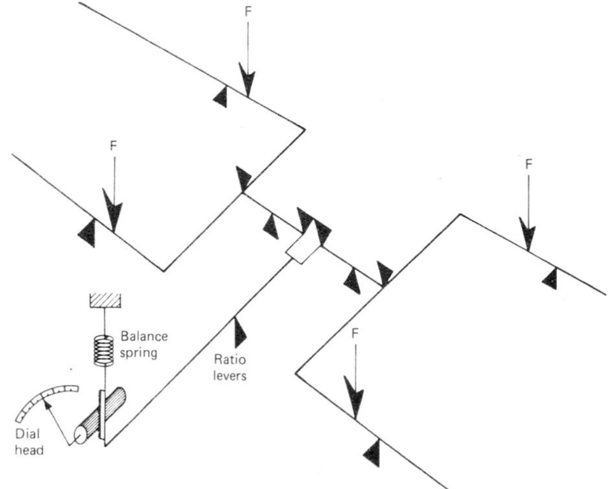

Figure 7.3 Compound-lever balance.

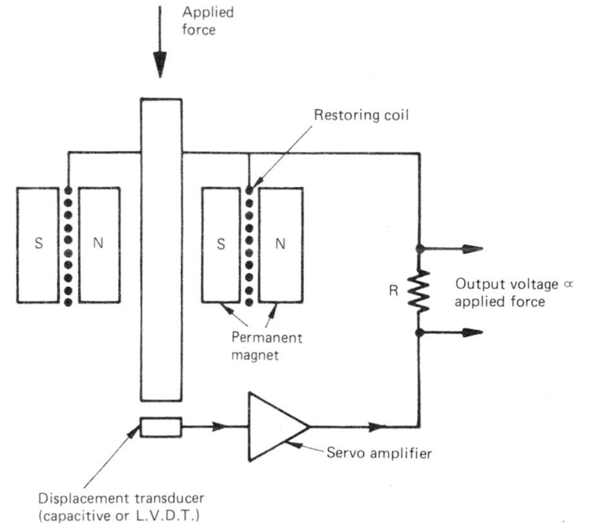

Figure 7.4 Force-balance system.

to the applied force. The same current flows through the resistor R and the voltage drop across this resistor is a measure of the applied force. Being a feedback system, such a device must be thoroughly checked for stability.

The force-balance system gives high stability, high accuracy, negligible displacement and is suitable for both static and dynamic force measurement. The range of this type of instrument is from 0.1 N to about 1 kN. It is normally bulky and heavy and tends to be expensive.

7.5 Hydraulic pressure measurement

The change in pressure due to the applied force may be used for force measurement. Figure 7.5 shows a general arrangement of a hydraulic load cell. An oil-filled chamber is connected to a pressure gauge and is sealed by a diaphragm. The applied force produces a pressure increase in the confined oil and is indicated on the pressure gauge calibrated to give direct reading of force. If an electrical output is required, an electrical pressure transducer may be used in place of the pressure gauge.

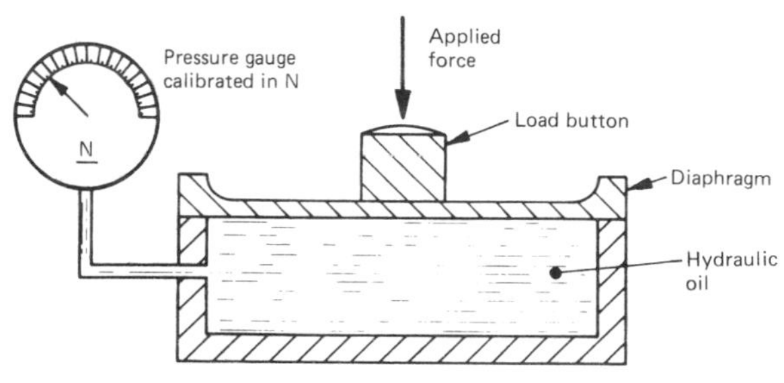

Figure 7.5 Force measurement using hydraulic load cell.

Hydraulic load cells are stiff, with virtually no operational movement, and they can give local or remote indication. They are available in force ranges up to 5 MN with system accuracy of the order of 0.25 to 1.0 per cent.

7.6 Acceleration measurement

As mentioned earlier, force is a product of mass and acceleration. If the acceleration $\ddot{x}$ of a body of known mass m is known, then the force Fx causing this acceleration can be found from the relationship:

$$Fx = m\ddot{x}$$

The acceleration is measured by using a calibrated accelerometer as shown in Figure 7.6. In practice, this method may be used for measuring dynamic forces associated with vibrating masses, and is discussed further in Chapter 6.

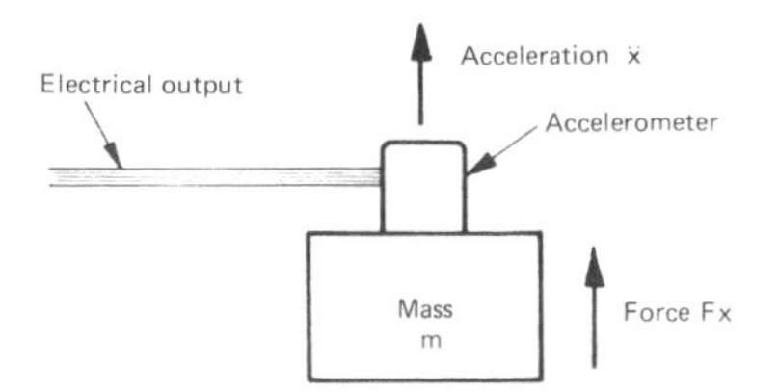

Figure 7.6 Force measurement using accelerometer.

7.7 Elastic elements

A measuring system basically consists of three elements; a transducer, a signal conditioner and a display or recorder. In this section, we will discuss various types of transducers, based on small displacements of elastic elements. In general, a transducer is defined as a device which changes information from one form to another. For the purpose of this discussion, a force transducer is defined as a device in which the magnitude of the applied force is converted into an electrical output, proportional to the applied force. The transducers are divided into two classes: active and passive. A passive transducer requires an external excitation voltage whereas an active transducer does not require an electrical input.

In general, a transducer consists of two parts: a primary elastic element which converts the applied force into a displacement and a secondary sensing element which converts the displacement into an electrical output. The elastic behaviour of the elastic element is governed by Hooke's law which states that the relationship between the applied force and the displacement is linear, provided the elastic limit of the material is not exceeded. The displacement may be sensed by various transducing techniques; some of them are examined in this section.

7.7.1 Spring balances

The extension of a spring may be used as a measure of the applied force and this technique is employed in the design of a spring balance as shown in Figure 7.7. This type of balance is a relatively low cost, low accuracy device and can be used for static force measurement.

7.7.2 Proving rings

A proving ring is a high-grade steel ring-shaped element with integral loading bosses as shown in Figure 7.8. Under the action of a diametral force, the ring tends to distort. The amount of distortion is directly proportional to the applied force. For low accuracy requirements, the distortion is measured using a dial gauge or a micrometer whereas for high accuracy applications, a displacement transducer such

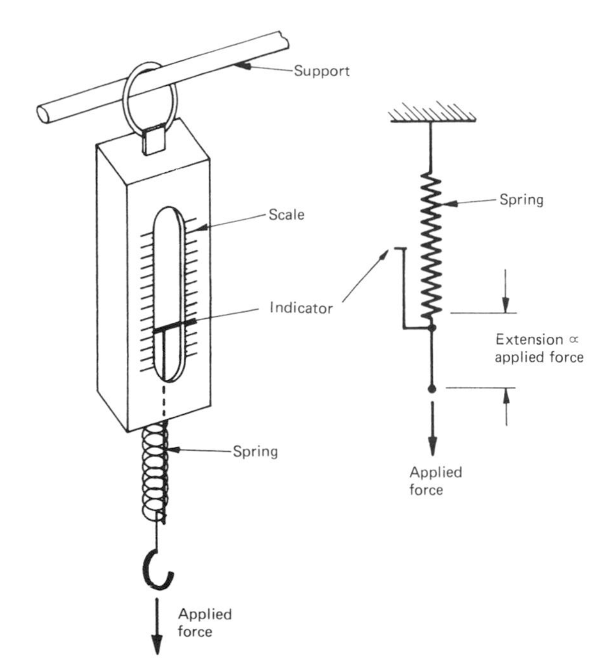

Figure 7.7 Spring balance.

as a linear variable differential transformer may be used. See Chapter 3.

Proving rings are high precision devices which are extensively used to calibrate materials-testing machines. They may be used both in tension and compression, with a compressive force range of the order of 2 kN to 2000 kN with accuracy from 0.2 to 0.5 per cent.

7.7.3 Piezo-electric transducers

A typical arrangement of a piezo-electric transducer is shown in Figure 7.9. When the transducer is subjected to the applied force, a proportional electrical charge appears on the faces of the piezo-electric element. The charge is also a function of force direction. The piezo-electric transducer differs from a conventional (passive) transducer in two respects. First, it is an active system, and secondly, the deflection at rated load is no more than a few thousandths of a millimetre, whereas the corresponding deflection for the conventional system may amount to several tenths of a millimetre.

This type of transducer has high stiffness, resulting in a very high resonant frequency. Because charge can leak away through imperfect insulation, it is unsuitable for measuring steady forces. It is mainly used in vibration studies and is discussed further in Chapter 6. It has small size, rugged construction, is sensitive to temperature changes and is capable of measuring compressive forces from a few kilonewtons to about 1 meganewton with accuracy from 0.5 to 1.5 per cent.

7.7.4 Strain-gauge load cells

7.7.4.1 Design

Bonded strain-gauge load cells are devices producing an electrical output which changes in magnitude when a force or weight is applied, and which may be displayed on a readout instrument or used in a control device. The heart of the load cell is the bonded-foil strain gauge which is an extremely sensitive device, whose electrical resistance changes in direct proportion to the applied force. See Chapter 4.

A load cell comprises an elastic element, normally machined from a single billet of high tensile steel alloy, precipitation hardening stainless steel, beryllium copper or other suitable material, heat-treated to optimize thermal and mechanical properties. The element may take many forms, such as hollow or solid column, cantilever, diaphragm, shear member, or ring. The design of the element is dependent on the load range, type of loading and operational requirements. The gauges are bonded on to the element to measure the strains generated and are usually connected into a four-arm Wheatstone bridge configuration. On larger elements, to get a true average of the strains, often 8, 16

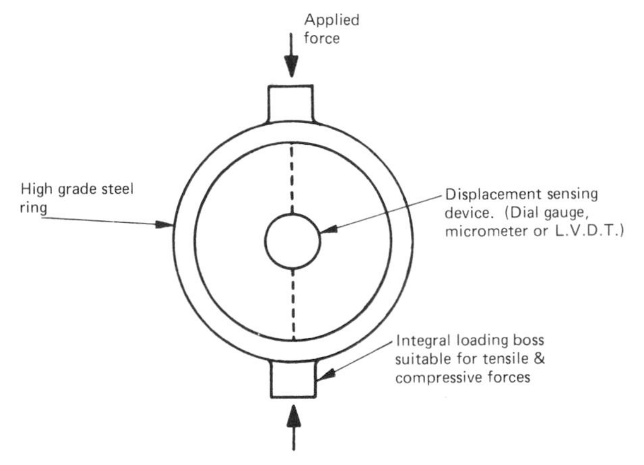

Figure 7.8 Proving ring fitted with displacement-sensing device.

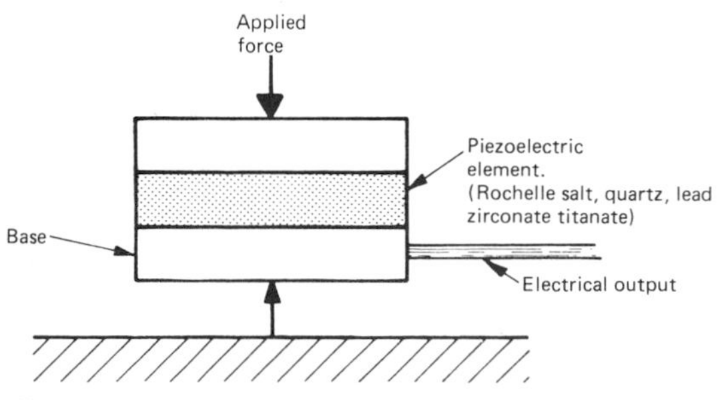

Figure 7.9 Piezo-electric force transducer.

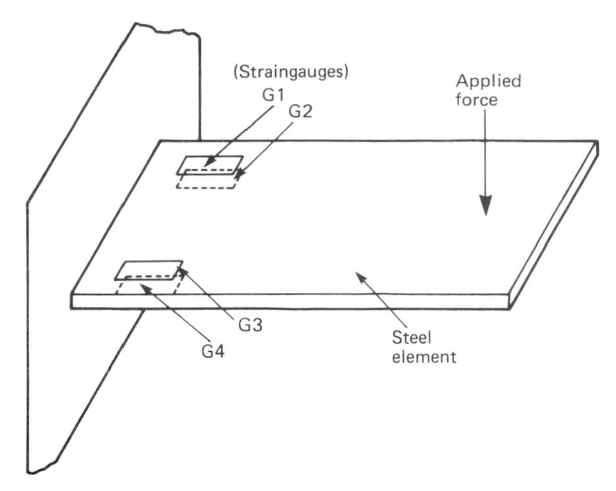

Figure 7.10 Cantilever load cell.

or even 32 gauges are used. To illustrate the working principle, a cantilever load cell is shown in Figure 7.10. Figure 7.11 shows a bridge circuit diagram that includes compensation resistors for zero balance and changes of zero and sensitivity with temperature. To achieve high performance and stability and to minimize glue line thickness, the gauges are often installed on flat-sided elements.

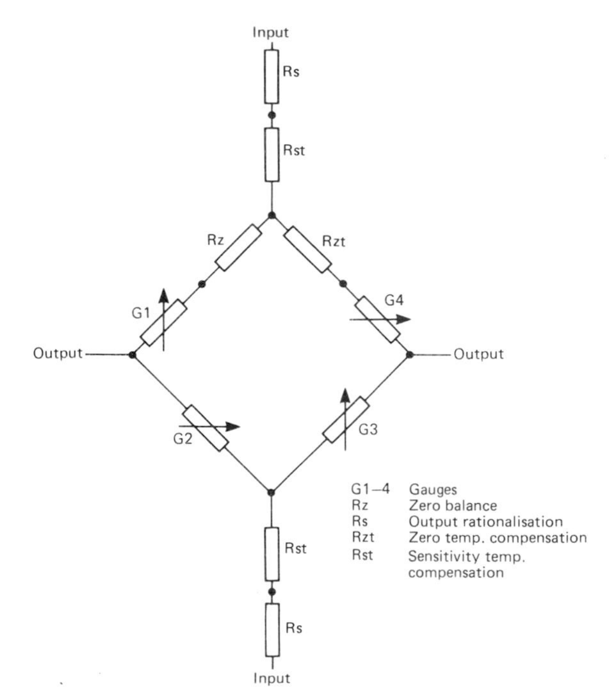

Figure 7.11 Load cell bridge circuit with compensation resistors.

The complete assembly is housed within a protective case with sealing sufficient to exclude the external environment, but capable of allowing the deformation of the element to occur when the force is applied. In some cases, restraining diaphragms minimize the effect of side-loading.

After assembly, the elements are subjected to a long series of thermal and load cycling to ensure that remaining 'locked-up' stresses in the element and bonding are relieved, so that the units will give excellent long-term zero stability.

Figure 7.12 shows some commercially available compression load cells that have been successfully used for monitoring the tension in the mooring legs of a North Sea platform.

7.7.4.2 Selection and installation

There are five basic types of cell on the market: compression, tension, universal (both compression and tension), bending and shear. The main factors influencing the selection of cell type are:

(a) The ease and convenience (and hence the cost) of incorporating a cell into the weigher structure.
(b) Whether the required rated load and accuracy can be obtained in the type of cell.

Other considerations include low profile, overload capacity, resistance to side-loads, environmental protection and a wide operating temperature range.

To retain its performance, a cell should be correctly installed into the weigher structure. This means the structure of the weigher, such as vessel, bin, hopper or platform, is the governing factor in the arrangement of the load cells. The supporting structure is also to be considered since it will carry the full weight of the vessel and contents. Difficulties caused by mis-application leading to poor performance and unreliability fall into three main headings:

(a) A non-axial load is applied.
(b) Side-loads are affecting the weight reading.
(c) Free-axial movement of the load is restricted.

Figure 7.13 shows how normal, non-axial and side-loading affects a column stress member. Under normal loading conditions (A) the active strain gauges go into equal compression; however, under non-axial (B) or side-loading (C) conditions, asymmetrical compression results, causing readout errors.

Examples of correct and incorrect fitments are shown in Figure 7.14. The support bracket D is cantilevered out too far and is liable to bend under load. The bracket is applying a load to the side of the vessel, which itself exaggerates this effect as the vessel is not strong enough to support it. The beam also deflects under load, rotating the load cell away from the vertical. The correct example E shows how the errors can be overcome.

Figure 7.12 Compression load cells.

In weighing installations it is important that there is unimpeded vertical movement of the weigh vessel. Obviously this is not possible where there are pipe fittings or stay rods on the vessel, but the vertical stiffness must be kept within allowable limits. One of the most satisfactory ways of reducing the spring rates is to fit flexible couplings in the pipework, preferably in a horizontal mode, and after (for example) the discharge valve so that they are not subject to varying stiffness due to varying pressure (see F and G in Figure 7.14). Where possible, entry pipes should be free of contact with the vessel (refer to H and I).

7.7.4.3 Applications

Load cells have many applications including weight and force measurement, weigh platforms, process control systems, monorail weighing, beltweighers, aircraft, freight and baggage weighing and conversion of a mechanical scale to an electronic scale. Over the past few years, the industrial weighing field has been dominated by load cells because electrical outputs are ideal for local and remote indication and to interface with microprocessors and minicomputers.

Key features of load cells are:

(a) Load range 5 N to 40 MN.

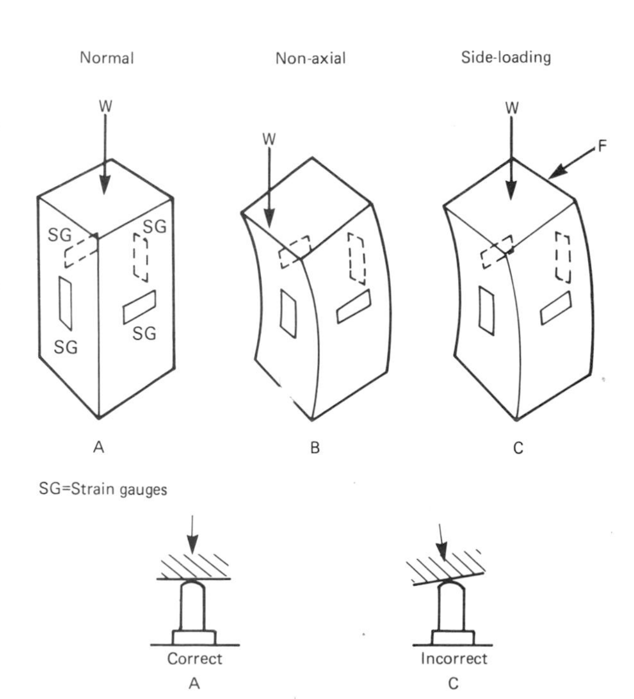

Figure 7.13 Effects of normal, non-axial and side-loading.

(b) Accuracy 0.01 to 1.0 per cent.
(c) Rugged and compact construction.
(d) No moving parts and negligible deflection under load.
(e) Hermetically sealed and thermal compensation.
(f) High resistance to side-loads and withstand overloads.

7.7.4.4 *Calibration*

Calibration is a process that involves obtaining and recording the load cell output while a direct known input is applied in a well-defined environment. The load cell output is directly compared against a primary or secondary standard of force. A primary standard of force includes dead-weight machines with force range up to about 500 kN; higher forces are achieved with machines having hydraulic or mechanical amplification.

A secondary standard of force involves the use of high precision load cells and proving rings with a calibration standard directly traceable to the National Standard at the National Physical Laboratory in Teddington, Middlesex, or the equivalent standards in other countries. The choice of the standards to be used for a particular calibration depends on the range and the location of the device to be calibrated.

The foregoing has indicated some force-measurement methods. Others are many and varied and no attempt has been made to cover all types. To simplify the selection of a method for a particular application, the main parameters of the methods discussed are summarized in Table 7.1.

7.8 Further developments

Advancing technology, improvements in manufacturing techniques and new materials have permitted increased accuracy and improved design of bonded strain-gauge load cells since their introduction about 30 years ago. Now the microprocessor is available, and therefore further design improvements in these devices are expected.

New transducing techniques are being constantly researched; a number of them have been well studied or are being considered, including gyroscopic force transducers, fibre optics, microwave cavity resonator and thin-film transducing techniques. The thin-film techniques are well documented and therefore are briefly discussed.

Pressure transducers based on vacuum-deposited thin-film gauges are commercially available and attempts are being made to apply these techniques to load cells. The advantages of these techniques are as follows:

(a) Very small gauge and high bridge resistance.
(b) Intimate contact between the element and gauge. No hysteresis or creep of a glue line.
(c) Wide temperature range (−200 °C to +200 °C).

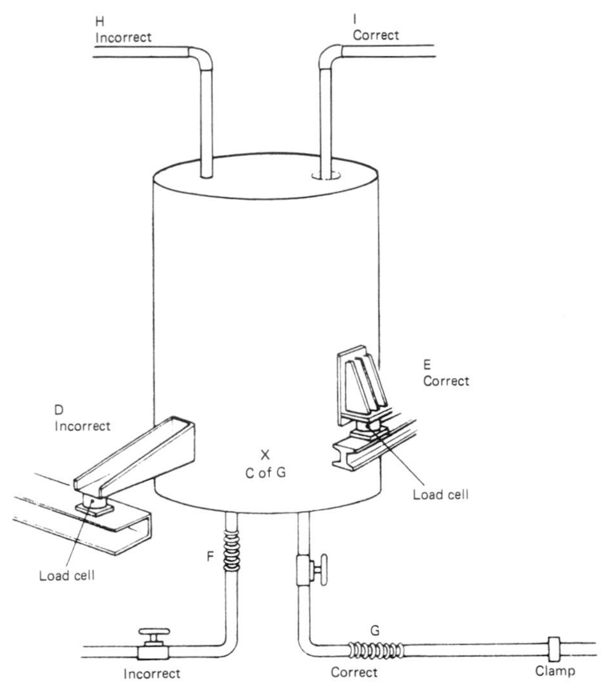

Figure 7.14 Examples of correct and incorrect fitments.

Table 7.1 Summary of main parameters of force-measuring methods

Method	*Type of loading*	*Force range, N (approx.)*	*Accuracy % (approx.)*	*Size*
Lever balance	Static	0.001 to 150 k	Very high	Bulky and heavy
Force-balance	Static/dynamic	0.1 to 1 k	Very high	Bulky and heavy
Hydraulic load cell	Static/dynamic	5 k to 5 M	0.25 to 1.0	Compact and stiff
Spring balance	Static	0.1 to 10 k	Low	Large and heavy
Proving ring	Static	2 k to 2 M	0.2 to 0.5	Compact
Piezo-electric transducer	Dynamic	5 k to 1 M	0.5 to 1.5	Small
Strain-gauge load cell	Static/dynamic	5 to 40 M	0.01 to 1.0	Compact and stiff

(d) Excellent long-term stability of the bridge.
(e) Suitability for mass production.

The techniques are capital-intensive and are generally suitable for low force ranges.

7.9 References

Adams, L. F. *Engineering Measurements and Instrumentation*, The English Universities Press, (1975)

Cerni, R. H. and Foster, L. E. *Instrumentation for Engineering Measurement*, John Wiley and Sons, (1962)

Mansfield, P. H. *Electrical Transducers for Industrial Measurement*, Butterworth, (1973)

Neubert, H. K. P. *Instrument Transducers*, Clarendon Press, (2nd edition, 1975)

WEIGHTECH 79, Proceedings of the Conference on Weighing and Force Measurement; Hotel Metropole, Brighton, England 24–26 September 1979

8 Measurement of density

E. H. HIGHAM

8.1 General

The measurement (and control) of liquid density is critical to a great number of industrial processes. But although density in itself can be of interest, it is usually more important as a way of inferring composition, or concentration of chemicals in solution, or of solids in suspension. Because the density of gases is very small, the instruments for that measurement have to be very sensitive to small changes. They will be dealt with separately at the end of the chapter.

In considering the measurement and control of density or relative density* of liquids, the units used in the two factors should be borne in mind. Density is defined as the mass per unit volume of a liquid and is expressed in such units as kg/m^3, g/l or g/ml.

Relative density, on the other hand, is the ratio of the mass of a volume of liquid to the mass of an equal volume of water at 4 °C (or some other specified temperature), the relative density of water being taken as 1.0. Both density and relative density are temperature-dependent and, for high precision, the temperature at which a measurement is made will have to be known, so that any necessary compensation can be introduced.

The majority of industrial liquid-density instruments are based on the measurement of: weight, buoyancy or hydrostatic head, but measuring systems based on resonant elements or radiation techniques are also used.

8.2 Measurement of density using weight

The actual weighing of a sample of known volume is perhaps the simplest practical application of this principle. Various methods for continuous weighing have been devised, but the most successful involves the use of a horizontal U-shaped tube with flexible couplings at a pivot point.

One example of this type of instrument is the Fisher Controls Company Mark V Gravitrol Density Meter shown in Figure 8.1. In it, the process fluid passes via flexible connectors into the tube loop which is supported towards the curved end on a link associated with the force-balance measuring system. In the pneumatic version of the instrument, the link is attached towards one end of the weighbeam which itself is supported on cross flexure pivots and carries an adjustable counterbalance weight on the opposite side. Also attached to the weighbeam is a dash-pot to reduce the effect of vibration induced by the flow of the process fluid or by the environment in which the instrument is located.

In operation, the counterbalance weight is positioned to achieve balance with the tube loop filled with fluid having a density within the desired working range and the span adjustment is set to its mid position. Balance is achieved when the force applied by the feedback bellows via the pivot and span-adjustment mechanism to the weighbeam causes it to take up a position in which the feedback loop comprising the flapper nozzle and pneumatic relay generates a pressure which is both applied to the feedback bellows and used as the output signal. A subsequent increase in the density of the process fluid causes a minute clockwise rotation of the weighbeam with the result that the flapper is brought closer to the nozzle and so

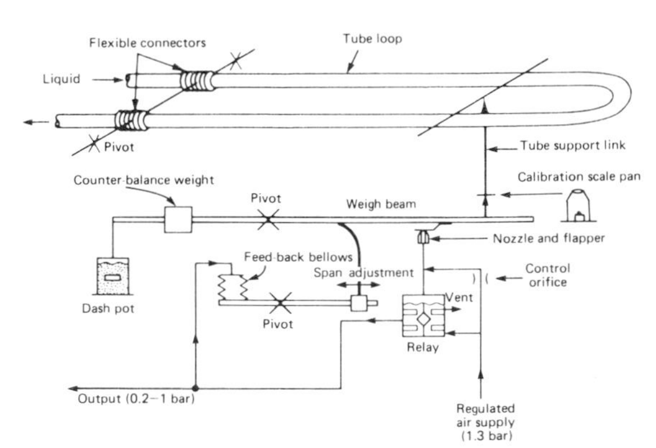

Figure 8.1 Gravitrol density meter. Courtesy, Fischer Controls Ltd.

* The term 'specific gravity' is often used for relative density. However it is not included in the S1 System of Units and BS350 points out that it is commonly used when the reference substance is water

increases the back pressure. This change is amplified by the relay and applied to the feedback bellows which in turn applies an increased force via the span-adjustment system until balance is restored.

An electronic force-balance system is also available which serves the same function as the pneumatic force-balance system just described. The basic calibration constants for each instrument are determined at the factory in terms of the weight equivalent to a density of 1.0 kg/dm^3. To adjust the instrument for any particular application, the tube loop is first emptied. Then weights corresponding to the lower range value are added to the calibration scale-pan and the counter-balance weight is adjusted to achieve balance.

Further weights are then added, representing the required span and the setting of the span adjustment is varied until balance is restored. The two procedures are repeated until the required precision is achieved. The pneumatic output, typically 20–100 kPa, then measures the change in density of the flowing fluid. It can be adjusted to operate for spans between 0.02 and 0.5 kg/dm^3 and for fluids having densities up to 1.6 kg/dm^3. The instrument is of course suitable for measurement on 'clean' liquids as well as slurries or fluids with entrained solid matter. In the former case a minimum flow velocity of 1.1 m/s is recommended and in the latter case at least 2.2 m/s to avoid deposition of the entrained solids.

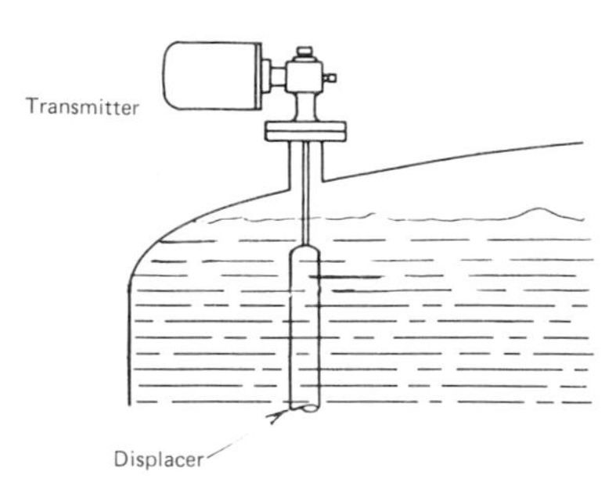

Figure 8.2 Buoyancy transducer and transmitter with tank.

8.3 Measurement of density using buoyancy

Buoyancy transmitters operate on the basis of Archimedes' principle: that a body immersed in a liquid is buoyed upward by a force equal to the weight of the liquid displaced. The cross-sectional area of a buoyancy transmitter displacer is constant over its working length, so that the buoyant force is proportional to the liquid density, see Figure 8.2.

With the arrangement of the force-balance mechanism shown in Figure 8.3, the force on the transmitter force bar must always be in the downward direction. Thus, the displacer element must always be heavier than the liquid it displaces. Displacers are available in a wide selection of lengths and diameters to satisfy a variety of process requirements.

Buoyancy transmitters are available for mounting either on the side of a vessel or for top entry and can be installed on vessels with special linings such as glass, vessels in which a lower connection is not possible. They are also suitable for density measurements in enclosed vessels where either the pressure or level may fluctuate and they avoid the need for equalizing legs or connections for secondary compensating instrumentation, such as repeaters. These transmitters are also suitable for applications involving high temperatures.

Turbulence is sometimes a problem for buoyancy transmitters. When this occurs, the most simple (and often the least expensive) solution is the installation of a stilling well or guide rings. Another alternative is to use a cage-mounted buoyancy transmitter, as shown in Figure 8.4. With this configuration, the measurement is outside the vessel and therefore isolated from the turbulence.

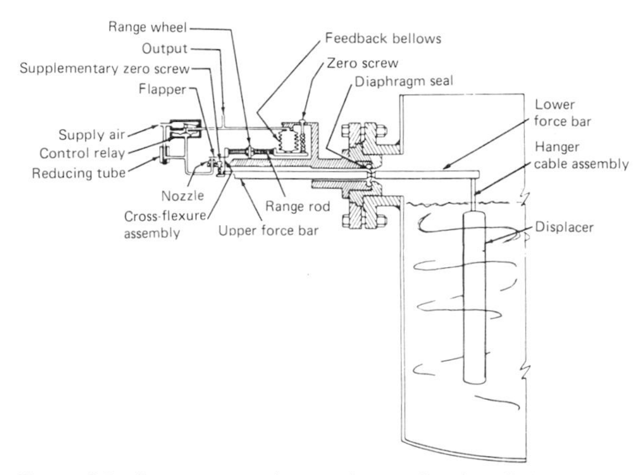

Figure 8.3 Buoyancy transducer and transmitter installation. Courtesy, The Foxboro Company.

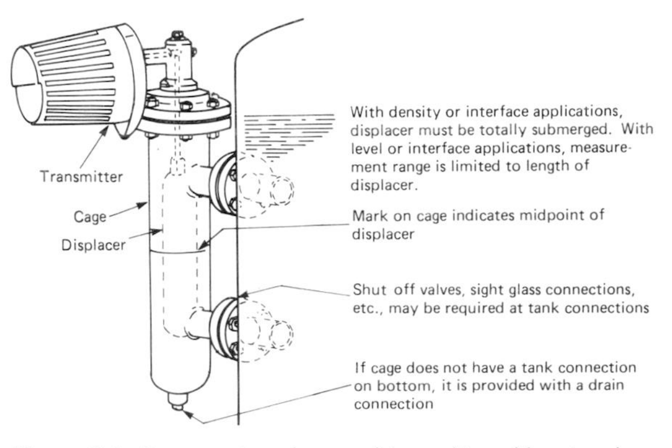

Figure 8.4 Buoyancy transducer and transmitter with external cage. Courtesy, The Foxboro Company.

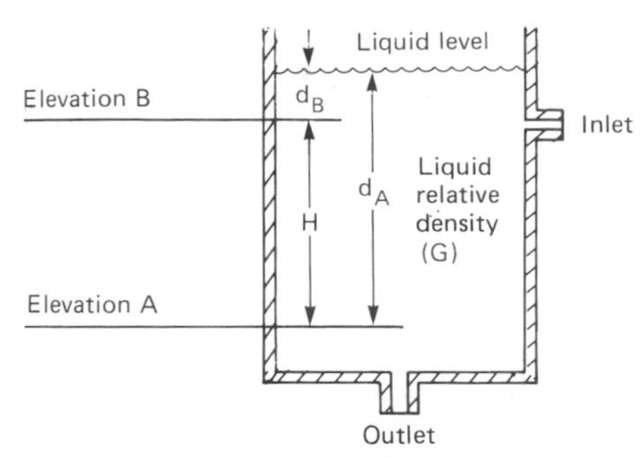

Figure 8.5 Density measurement—hydrostatic head.

8.4 Measurement of density using hydrostatic head

The hydrostatic-head method, which continuously measures the pressure variations by a fixed height of liquid, has proved to be suitable for many industrial processes. Briefly, the principle, illustrated in Figure 8.5, is as follows.

The difference in pressure between any two elevations (A and B) below the surface is equal to the difference in liquid head pressure between these elevations. This is true regardless of variation in liquid level above elevation B. This difference in elevation is represented by dimension H. Dimension H must be multiplied by the relative density of the liquid to obtain the difference in head. This is usually measured in terms of millimetres of water.

To measure the change in head resulting from a change in relative density from G_1 to G_2, it is necessary only to multiply H by the difference between G_1 and G_2. Thus

$$P = H(G_2 - G_1)$$

and if both H and P are measured in millimetres then the change in density is

$$(G_2 - G_1) = P/H$$

It is common practice to measure only the span of actual density changes. Therefore, the instrument 'zero' is 'suppressed' to the minimum head pressure to be encountered; this allows the entire instrument measurement span to be devoted to the differential caused by density changes. For example, if G_1 is 0.6 and H is 3 metres, then the zero suppression value should be 1.8 metres of water. The two principal relationships which must be considered in selecting a measuring device are:

$$\text{span} = H(G_2 - G_1)$$
$$\text{zero suppression value} = H \,.\, G_1$$

For a given instrument span, a low density span requires a greater H dimension (a deeper tank). For a given density span, a low span measuring device permits a shallower tank. Figure 8.6 shows values of H plotted against gravity spans.

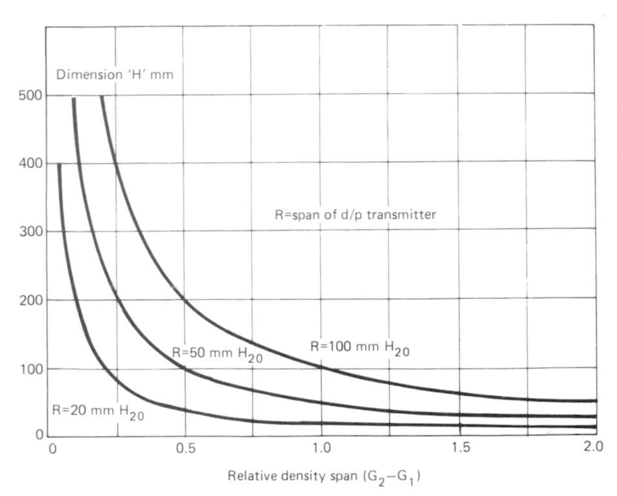

Figure 8.6 Relative density span versus H for various spans.

8.4.1 General differential-pressure (d/p) transmitter methods

There is a variety of system arrangements for determining density from measurements of hydrostatic head using differential-pressure (d/p) transmitters. Although flange-mounted d/p transmitters are often preferred, pipe-connected transmitters can be used on liquids where crystallization or precipitation in stagnant pockets will not occur.

These d/p transmitter methods are usually superior to those in which the d/p transmitter operates in conjunction with bubble tubes and can be used in either pressure or vacuum systems. However, all require the dimensions of the process vessel to provide a sufficient change of head to satisfy the minimum span of the transmitter.

8.4.2 D/p transmitter with overflow tank

Constant-level, overflow tanks permit the simplest instrumentation, as shown in Figure 8.7. Only one d/p transmitter is required. With H as the height of liquid above the transmitter, the equations are still:

$$\text{span} = H(G_2 - G_1)$$
$$\text{zero suppression value} = H \,.\, G_1$$

8.4.3 D/p transmitter with a wet leg

Applications with level or static pressure variations require compensation. There are three basic arrangements for density measurement under these conditions. First, when a seal fluid can be chosen that is always denser than the process fluid and will not mix

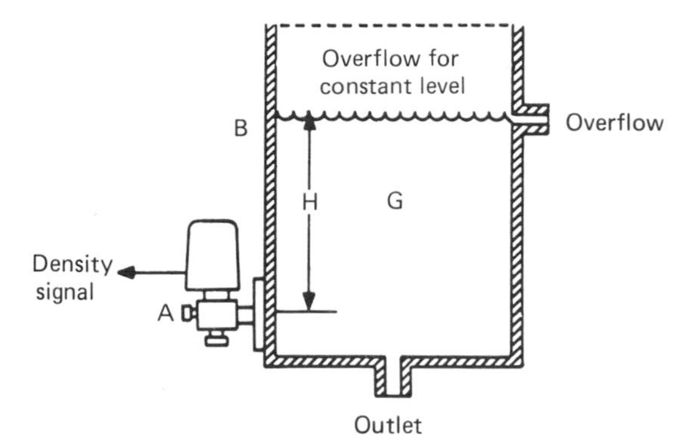

Figure 8.7 Density measurement with constant head.

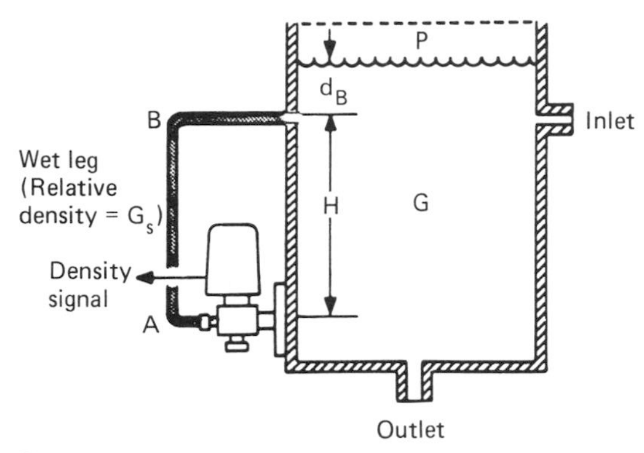

Figure 8.8 Density measurement with wet leg.

with it, the method shown in Figure 8.8 is adequate. This method is used extensively on hydrocarbons with water in the wet leg. For a wet-leg fluid of specific gravity G_S, an elevated zero transmitter must be used. The equations become:

$$\text{span} = H(G_2 - G_1)$$
$$\text{zero elevation value} = H(G_S - G_1)$$

When there is no suitable wet-leg seal fluid, but the process liquid will tolerate a liquid purge, the method shown in Figure 8.9 can be used. To ensure that the process liquid does not enter the purged wet leg, piping to the process vessel should include an appropriate barrier, either gooseneck or trap, as shown in Figure 8.10. Elevation or suppression of the transmitter will depend on the difference in specific gravity of the seal and process liquids. Here, the equations are:

$$\text{span} = H(G_2 - G_1)$$
$$\text{zero suppression value} = H(G_1 - G_S), \quad \text{when } G_1 > G_S$$
$$\text{zero elevation value} = H(G_S - G_1), \quad \text{when } G_S > G_1$$

Ideally, the purge liquid has a specific gravity equal to G, which eliminates the need for either suppression or elevation.

8.4.4 D/p transmitter with a pressure repeater

When purge liquid cannot be tolerated, there are ways to provide a 'mechanical seal' for the low-pressure leg, or for both legs if needed. Figure 8.11 shows the use of a pressure repeater for the upper connection. In one form, this instrument reproduces any pressure existing at the B connection from full vacuum to about 250 Pa positive pressure. In another form this instrument will reproduce any pressure from 7 kPa to 700 kPa. The repeater transmits the total pressure at elevation B to the low-pressure side of the d/p transmitter. In this way, the pressure at elevation B is subtracted from the pressure at elevation A.

The lower transmitter, therefore, measures density (or $H \,.\, G$, where G is the specific gravity of the liquid). The equations for the lower transmitter are:

$$\text{span} = H(G_2 - G_1)$$
$$\text{zero suppression value} = H \,.\, G_1$$

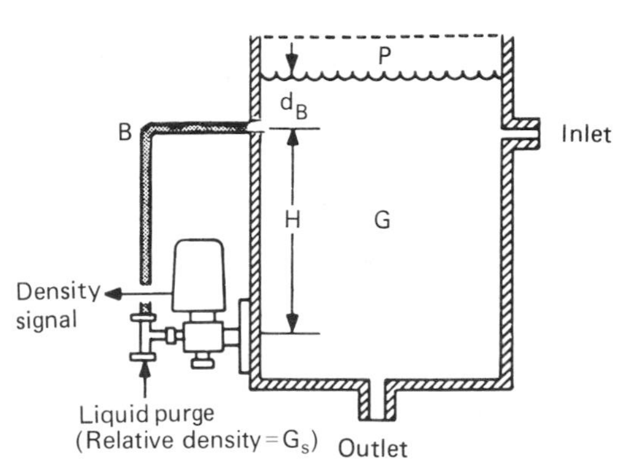

Figure 8.9 Density measurement with purge liquid.

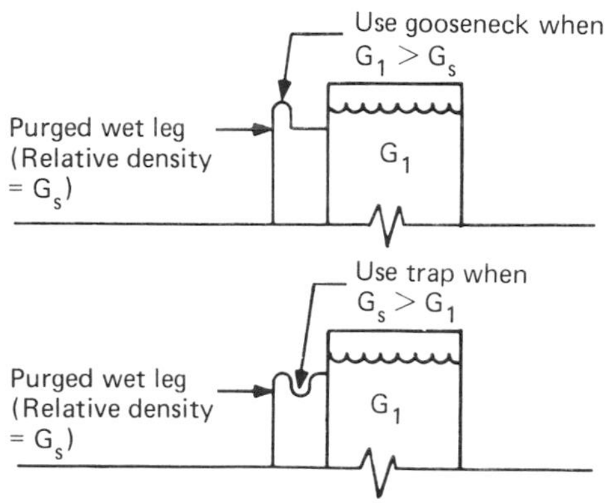

Figure 8.10 Purge system with gooseneck and trap.

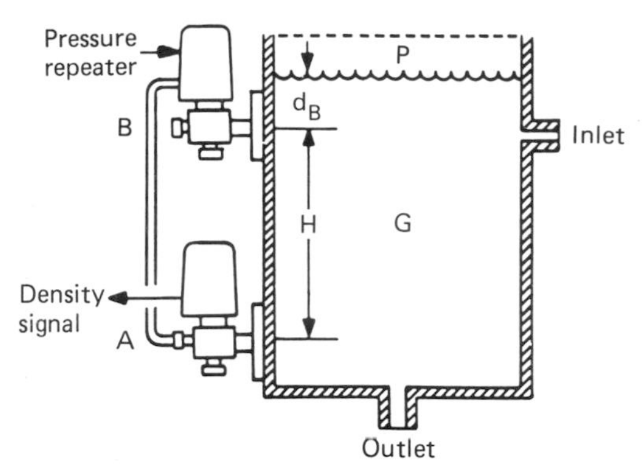

Figure 8.11 Density measurement with pressure repeater.

The equation for the upper repeater is:

$$\text{output (maximum)} = (d_B \text{ max})/(G_2) + P \text{ max}$$

where d_B is the distance from elevation B to the liquid surface, and P is the static pressure on the tank, if any.

Special consideration must be given when the repeater method is used for vacuum applications, where the total pressure on the repeater is less than atmospheric. In some instances density measurement is still possible. Vacuum application necessitates biasing of the repeater signal or providing a vacuum source for the repeater relay. In this case, there are restrictions on allowable gravity spans and tank depths.

8.4.5 D/p transmitter with flanged or extended diaphragm

Standard flanged and extended diaphragm transmitter applications are illustrated in Figure 8.12(a) and (b), respectively. An extended diaphragm transmitter may be desirable in order to place the capsule flush with or inside the inner wall of the tank. With this instrument, pockets in front of the capsule where build-up may occur are eliminated.

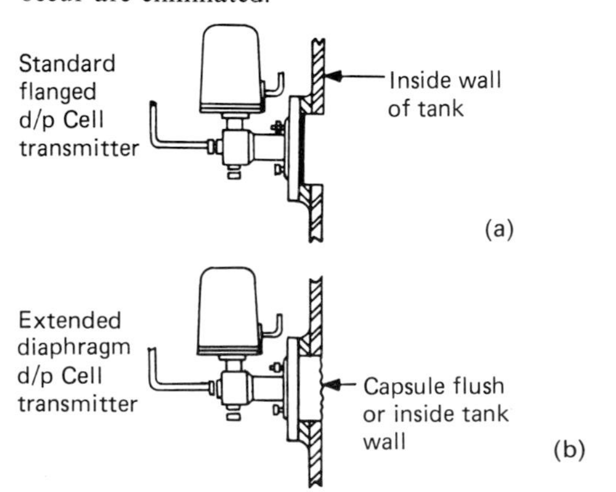

Figure 8.12 D/p cell with flanged or extended diaphragm.

8.4.6 D/p transmitter with pressure seals

If the process conditions are such that the process fluid must not be carried from the process vessel to the d/p transmitter then a transmitter fitted with pressure seals can be used as shown in Figure 8.13. Apart from the additional cost, the pressure seals reduce the sensitivity of the measurement and any mismatch in the two capillary systems can cause further errors. However, the system can be used for either open or closed vessels.

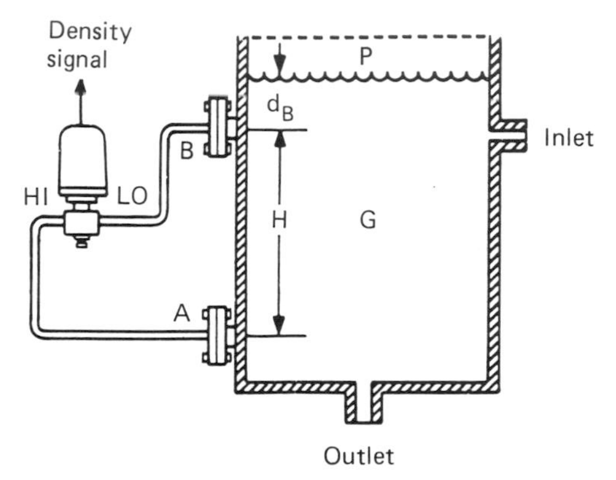

Figure 8.13 D/p cell with pressure seals.

8.4.7 D/p transmitter with bubble tubes

This very simple system, illustrated in Figure 8.14, involves two open-ended tubes, terminated with 'V'

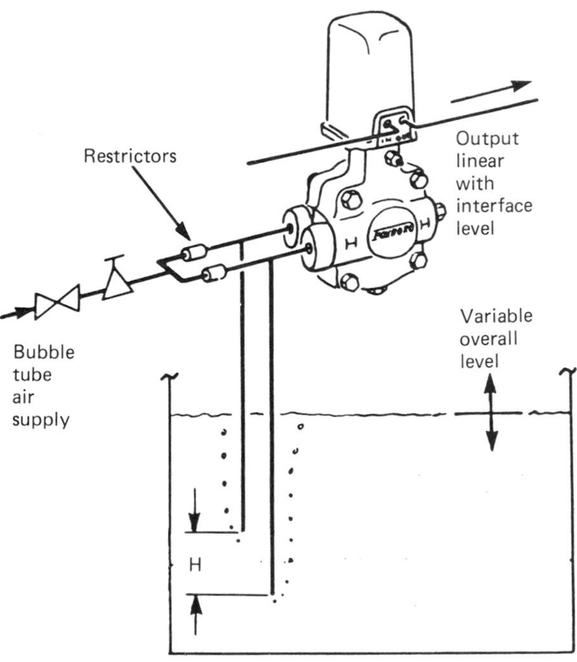

Figure 8.14 D/p cell with bubble tubes.

notches. These are immersed in the liquid with the 'V' notches separated by a known fixed vertical distance H and purged with a low but steady flow of air (or inert gas) at a suitable pressure.

A d/p transmitter connected between these tubes with the higher-pressure side associated with the lower 'V' notch, measures the difference P in hydrostatic pressure at the two points. This is equal to the density × the vertical distance between the two 'V' notches:

$$\text{density} = P/H$$

Although this method is very simple and effective, it is unsuitable for closed vessels or for liquids that may crystallize or involve precipitation which might block the bubble tubes and so give rise to erroneous results.

8.4.8 Other process considerations

Agitation in a process tank where density measurement is made must be sufficient to ensure uniformity of the liquid. But the velocity of fluid at the points where head pressure is measured must be sufficiently low to avoid a significant measurement error. Locations of side-mounted transmitters should be sufficiently high above the bottom of the tank to avoid errors due to them becoming submerged in the sediment that tends to collect there.

8.5 Measurement of density using radiation

Density measurements by this method are based on the principle that absorption of gamma radiation increases with increasing specific gravity of the material measured. These are discussed in Volume 3.

The principal instrumentation includes: a constant gamma source, a detector, and an indicating or recording instrument. Variations in radiation passing through a fixed volume of flowing process liquid are converted into a proportional electrical signal by the detector.

8.6 Measurement of density using resonant elements

Several density-measuring instruments are based on the measurement of the resonant frequency of an oscillating system such as a tube filled with the fluid under test or a cylinder completely immersed in the medium. Examples of each are described in the succeeding sections.

8.6.1 Liquid density measurement

The Solartron Liquid Density Transducer Type 7830 is shown in Figure 8.15. The sensing element comprises a single smooth bore tube through which flows the fluid to be measured. The tube is fixed at each end into heavy nodal masses which are isolated from the outer case by bellows and ligaments. Located along the tube are the electromagnetic drive and pick-up coil assemblies. In operation, the amplifier maintains the tube oscillating at its natural frequency.

Since the natural frequency of oscillation of the tube is a function of the mass per unit length, it must also be a function of the density of the flowing fluid. It also follows that the tube should be fabricated from material having a low and stable coefficient of expansion. If for reasons of corrosion or wear this is not possible, it is important that the temperature is

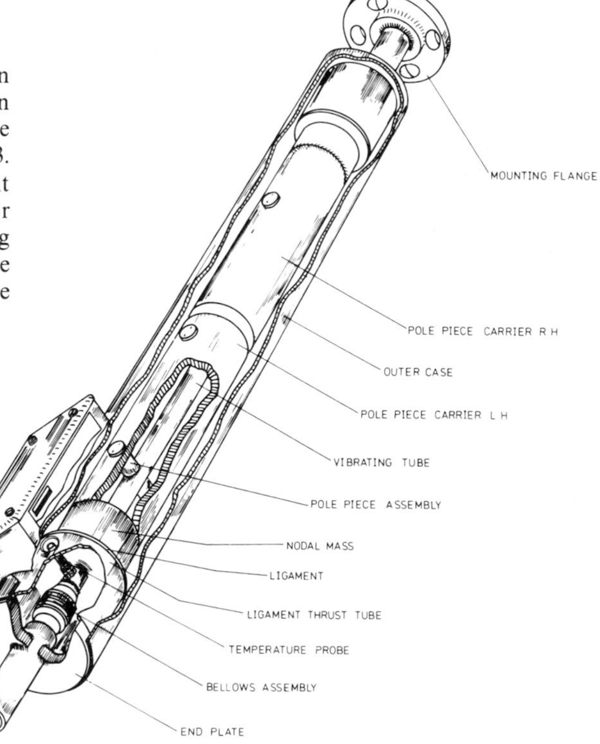

Figure 8.15 Solartron liquid density transducer. Courtesy, Schlumberger Electronics (UK) Ltd.

measured and a suitable correction applied to the density value determined from the resonant frequency.

Typically, the tube vibrates at about 1.3 kHz (when filled with water) and with an amplitude of about 0.025 mm. Densities up to 3000 kg/m^3 can be measured with an accuracy of 0.2 kg/m^3 and a repeatability of 0.02 kg/m^3. This contrasts with accuracies of only about 1 per cent of span that can be achieved with other methods, unless extreme care is taken.

The response is continuous throughout its operating range with no adjustments of span or zero. Recalibration is effected by adjustment of the constants in the associated readout or signal conditioning circuits. The density–frequency relation is given by

$$d = K_0\left(\frac{T^2}{T_0^2} - 1\right)$$

where P is the density of the measured fluid, K_0 is constant for the transducer, T_0 is the time period of oscillation under vacuum conditions, and T is the time period of oscillation under operating conditions.

It is noteworthy that, although the relation between density and the period of the oscillation strictly obeys a square law, it is linear within 2 per cent for a change in density of 20 per cent. For narrower spans the error is proportionally smaller.

8.6.2 Gas density measurements

The relationship between temperature, pressure and volume of a gas is given by

$$PV = nZR_0T$$

where P is the absolute pressure, V is the volume, n is the number of moles, Z is the compressibility factor, R_0 is the Universal gas constant and T is the absolute temperature. Use of the mole in this equation eliminates the need for determining individual gas constants, and the relationship between the mass of a gas m, its molecular weight Mw and number of moles is given by

$$n = m/Mw$$

When the compressibility factor Z is 1.0 the gas is called *ideal* or *perfect*. When the specific heat is assumed to be only temperature dependent the gas is referred to as *ideal*. If the *ideal* relative density RD of a gas is defined as the ratio of molecular weight of the gas to that of air, then

$$RD = \frac{Mw_{gas}}{Mw_{air}}$$

whereas the *real* relative density is defined as the ratio of the density of the gas to that of air, which is

$$RD = \frac{\rho_{gas}}{\rho_{air}}$$

for a particular temperature and pressure.

The above equation can be rearranged as a density equation thus

$$\rho = \frac{m}{V} = \frac{SGMw_{air}P}{ZR_0T}$$

Most relative density measuring instruments operate at pressures and temperatures close to ambient conditions and hence measure *real* relative density rather than the *ideal* relative density which is based on molecular weights and does not take into account the small effects of compressibility. Hence

$$\underset{\text{(real)}}{RD} = \left(\frac{\rho_{gas}}{\rho_{air}}\right)_{TP}$$

where T and P are close to ambient conditions.

Substituting the equation leads to P

$$\underset{\text{(ideal)}}{RD} = \left(\frac{ZT}{P_{gas}}\right) \times \left(\frac{P}{ZT_{air}}\right)_{TP} \times \underset{\text{(real)}}{RD}$$

For most practical applications this leads to

$$RD = \left(\frac{Z_{gas}}{Z_{air}}\right)_{TP} \underset{\text{(real)}}{RD}$$

Thus, the signal from the density transducer provides an indication of the molecular weight or specific gravity of the sampled gas.

The measurement can be applied to almost any gas provided that it is clean, dry and non-corrosive. The accuracy is typically 0.1 per cent of reading and the repeatability 0.02 per cent.

To measure the lower densities of gases, a more sensitive sensing element than that described for measurements on liquids is required. The Solartron Gas Density Transducer Type 7810 shown in Figure 8.16 achieves this by using a thin-walled cylinder resonated in the hoop or radial mode. The maximum amplitude of vibration occurs at the middle of the cylinder with nodes at each end and it is therefore clamped at one end with a free node-forming ring at the other end.

The cylinder is immersed in the gas whose density is to be measured, and it is thus not stressed due to the pressure of the gas. Gas in contact with the cylinder is brought into oscillation and effectively increases the mass of the vibrating system, thereby reducing its resonant frequency.

Oscillation is maintained electromagnetically by positioning drive and pick-up coils inside the cylinder

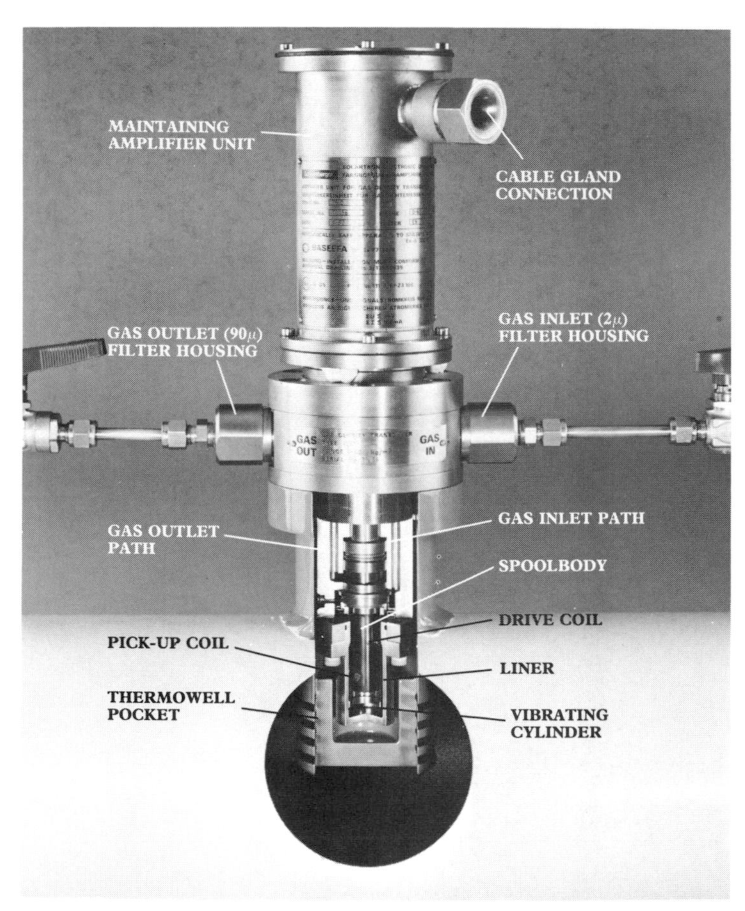

Figure 8.16 Solartron gas density transducer type 7810. Courtesy, Schlumberger Electronics (UK) Ltd.

and connecting them to a maintaining amplifier. The coils are mounted at right angles to each other to minimize stray coupling and phased so that the induced signal is proportional to the velocity, thereby reducing the effect of viscous damping.

A low temperature coefficient is obtained by constructing the cylinder from material having a low temperature coefficient of expansion. The cylinder wall thickness varies from 0.05 to 0.15 mm according to the required density range, the corresponding density ranges varying from 0 to 60 kg/m^3 and 40 to 400 kg/m^3.

The relation between the time period of oscillation of the transducer and gas density d is given by

$$d = 2d_0 \frac{(\tau - \tau_0)}{\tau_0} \left[1 + \frac{K}{2} \left(\frac{\tau - \tau_0}{\tau_0} \right) \right]$$

where τ is the measured time period of oscillation, τ_0 is the time period of oscillation under vacuum conditions, and d_0 and K are the calibration constants for each transducer.

An alternative method for measuring gas density involves a cylindrical test cell in which a hollow spinner is rotated at constant speed. This develops a differential pressure between the centre and ends of the spinner which is directly proportional to the density of the gas and can be measured by any standard differential pressure measuring device. A calibration constant converts the differential pressure to density for the actual conditions of temperature and pressure in the cell.

A sample flow of gas through the cell is induced by connecting it across a small restriction inserted in the main line to create an adequate pressure drop. The restriction could be determined from the square root of the differential pressure across the orifice plate multiplied by the differential pressure developed in the density cell. However, it is important to ensure that the flow of the gas through the density cell is not a significant proportion of the total flow. It is also important to apply a correction if there is any difference between the temperature and pressure of the gas in the density transducer and that in the main stream.

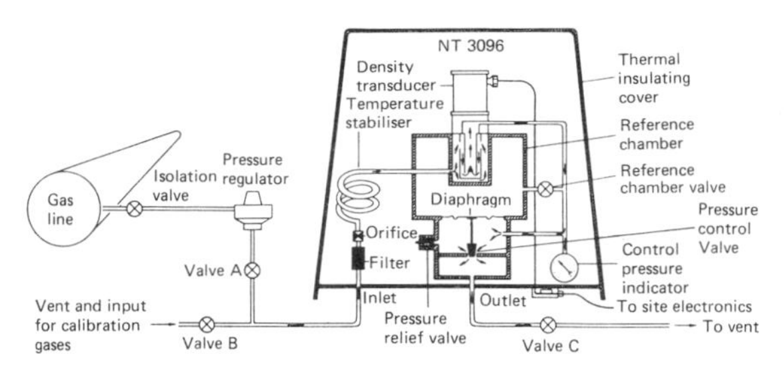

Figure 8.17 Relative density measurement using resonant element. Courtesy, Schlumberger Electronics (UK) Ltd.

8.6.3 Relative density of gases

The Solartron Specific Gravity Transducer Type 3096, shown in Figure 8.17 utilizes the density sensor described in the previous section to measure relative density of gases. In it, the sample of gas and the reference gas are stabilized at the same temperature by coils within thermal insulation. The reference chamber is a constant volume containing a fixed quantity of gas, any variation in temperature is compensated by a change in pressure which is transmitted to the sample gas by a flexible diaphragm. Having achieved pressure and temperature equalization by using a reference gas, a direct relationship between density and relative density can be realized.

9 Measurement of pressure

E. H. HIGHAM

9.1 What is pressure?

When a fluid is in contact with a boundary it produces a force at right angles to that boundary. The force per unit area is called the pressure. In the past, the distinction between mass and force has been blurred because we live in an environment in which every object is subjected to gravity and is accelerated towards the centre of the earth unless restrained. As explained in Chapter 8, the confusion is avoided in the SI system of units (Système International d'Unites) where the unit of force is the newton and the unit of area is a square metre so that pressure, being force per unit area, is measured in newtons per square metre and the unit, known as the pascal, is independent of the acceleration due to gravity.

The relation between the pascal and other units used for pressure measurements is shown in Table 9.1.

There are three categories of pressure measurements, namely absolute pressure, gauge pressure and differential pressure. The absolute pressure is the difference between the pressure at a particular point in a fluid and the absolute zero of pressure, i.e. a complete vacuum. A barometer is one example of an absolute pressure gauge because the height of the column of mercury measures the difference between the atmospheric pressure and the 'zero' pressure of the Torricellian vacuum that exists above the mercury column.

When the pressure-measuring device measures the difference between the unknown pressure and local atmospheric pressure the measurement is known as gauge pressure.

When the pressure-measuring device measures the difference between two unknown pressures, neither of which is atmospheric pressure, then the measurement is known as the differential pressure.

A mercury manometer is used in Figure 9.1 to illustrate these three measurements.

9.2 Pressure measurement

There are three basic methods for pressure measurement. The simplest method involves balancing the unknown pressure against the pressure produced by a column of liquid of known density. The second method involves allowing the unknown pressure to act on a known area and measuring the resultant force either directly or indirectly. The third method involves allowing the unknown pressure to act on an elastic member (of known area) and measuring the resultant stress or strain. Examples of these methods are described in the following sections.

9.2.1 Pressure measurements by balancing a column of liquid of known density

The simplest form of instrument for this type of measurement is the U-tube.

Consider a simple U-tube containing a liquid of density ρ as shown in Figure 9.2. The points A and B are at the same horizontal level and the liquid at C stands at a height h mm above B.

Then the pressure at A
= the pressure at B
= atmospheric pressure + pressure due to column of liquid BC
= atmospheric pressure $+h\rho$

If the liquid is water the unit of measure is mmH_2O and if the liquid is mercury then the unit of measure is mmHg. The corresponding SI unit is the pascal and

$$1\ mmH_2O = 9.806\,65\ Pa$$
$$1\ mmHg = 133.322\ Pa$$

For a system such as this it must be assumed that the density of the fluid in the left-hand leg of the manometer (Figure 9.2) is negligible compared with

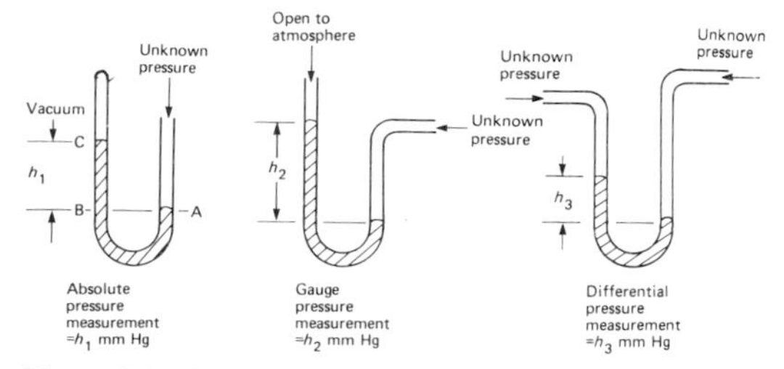

Figure 9.1 Comparison of types of pressure measurements.

Table 9.1 Pressure measurements

	Pascal *Pa*	*Bar* *bar*	*Millibar* *mbar*	*Standard atmosphere* *atm*	*Kilogram force per square cm* *kgf/cm²*	*Pound force per square inch* *lbf/in²*	*Torr*	*Milli-metre of water* *mmH₂O*	*Milli-metre of mercury* *mmHg*	*Inch of water* *inH₂O*	*Inch of mercury* *inHg*
Pa	1	10^{-5}	10^{-2}	$9.869\,23 \times 10^{-6}$	$1.019\,72 \times 10^{-5}$	$1.450\,38 \times 10^{-4}$	$7.500\,62 \times 10^{-3}$	$1.019\,72 \times 10^{-1}$	$7.500\,62 \times 10^{-3}$	$4.014\,63 \times 10^{-3}$	$2.953\,00 \times 10^{-4}$
bar	10^5	1	10^3	$9.869\,23 \times 10^{-1}$	1.019 72	14.5038	$7.500\,62 \times 10^{-2}$	$1.019\,72 \times 10^{-2}$	$7.500\,62 \times 10^{-2}$	$4.014\,63 \times 10^{-8}$	29.5300
mbar	10^2	10^{-3}	1	$9.869\,23 \times 10^{-4}$	$1.019\,72 \times 10^{-3}$	$1.450\,38 \times 10^{-2}$	$7.500\,62 \times 10^{-1}$	$1.019\,72 \times 10$	$7.500\,62 \times 10^{-5}$	$4.014\,62 \times 10^{-1}$	$2.953\,00 \times 10^{-2}$
atm	$1.013\,25 \times 10^5$	1.013 25	$1.013\,25 \times 10^3$	1	1.033 23	$1.469\,59 \times 10$	$7.600\,00 \times 10^2$	$1.033\,23 \times 10^5$	760	$4.067\,83 \times 10^2$	29.9213
kgf/cm²	98 066.5	0.980 665	980.665	0.967 841	1	14.2233	735.559×10^2	10^4	$7.355\,59 \times 10^2$	$3.937\,00 \times 10^2$	28.9590
lbf/in²	6894.76	0.068 947 6	68.9476	$6.804\,60 \times 10^{-2}$	$7.030\,70 \times 10^{-2}$	1	51.7149	$7.030\,69 \times 10^2$	51.7149	27.6798	2.036 02
torr	133.322	$1.333\,22 \times 10^{-3}$	1.333 22	$1.315\,79 \times 10^{-3}$	$1.359\,51 \times 10^{-3}$	$1.933\,68 \times 10^{-2}$	1	13.5951	1	53.5240	$3.937\,01 \times 10^{-2}$
mmH₂O	9.806 65	$9.806\,65 \times 10^{-5}$	$9.806\,65 \times 10^2$	$9.678\,41 \times 10^{-5}$	10^{-4}	$1.422\,33 \times 10^{-3}$	$7.355\,59 \times 10^{-2}$	1	$7.355\,59 \times 10^{-2}$	$3.937\,01 \times 10^{-2}$	$2.895\,90 \times 10^{-3}$
mmHg	133.322	$1.333\,22 \times 10^{-3}$	1.333 22	$1.315\,79 \times 10^{-3}$	$1.359\,51 \times 10^{-3}$	$1.933\,68 \times 10^{-2}$	1	13.5951	1	53.5240	$3.937\,01 \times 10^{-2}$
inH₂O	249.089	$2.490\,89 \times 10^{-3}$	2.490 89	$2.458\,31 \times 10^{-3}$	2.54×10^{-3}	$3.612\,72 \times 10^{-2}$	1.868 32	25.4	1.868 32	1	$7.355\,59 \times 10^{-2}$
inHg	3386.39	$3.386\,39 \times 10^{-2}$	33.8639	$3.342\,11 \times 10^{-2}$	$3.453\,16 \times 10^{-2}$	0.491 154	25.4000	$3.453\,16 \times 10^{-2}$	25.4000	13.5951	1

NB. Extracts from British Standards are reproduced by permission of the British Standards Institution, 2 Park Street, London W1A 2BS from whom complete copies can be obtained.

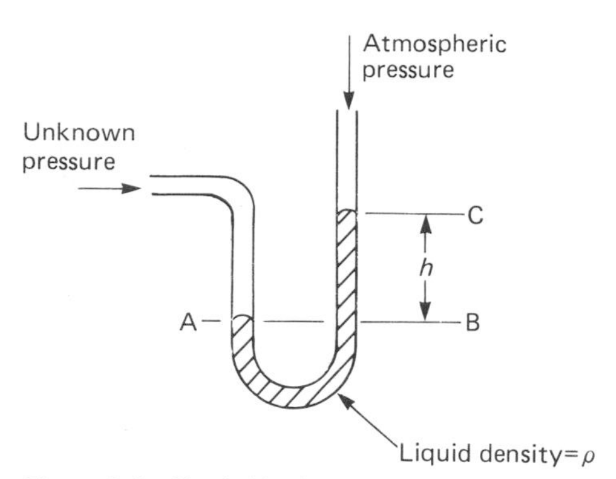

Figure 9.2 Simple U-tube manometer.

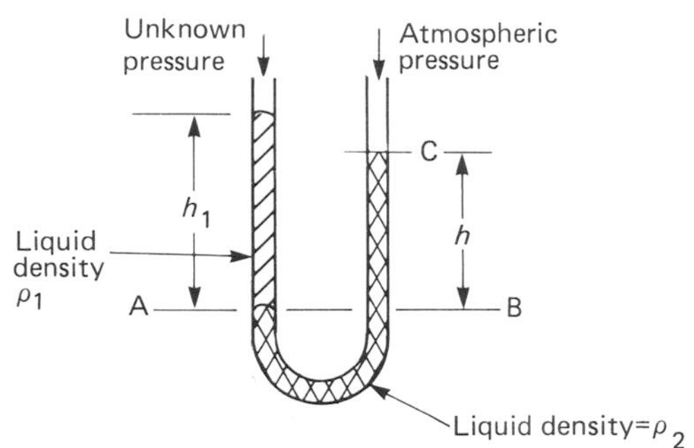

Figure 9.3 Manometer with wet leg connection.

the manometer liquid. If this is not so then a correction must be applied to allow for the pressure due to the fluid in the gauge and connecting pipes. Referring to Figure 9.3, we have

Pressure at A = pressure at B

$$P\text{(gauge pressure)} = \rho_1 h_1 + \text{atmospheric pressure}$$
$$= \rho_2 h + \text{atmospheric pressure}$$

or

$$P = \rho_2 h - \rho_1 h_1$$

(Gauge pressure because the atmospheric pressure is superimposed on each manometer leg measurement.)

If the manometer limbs have different diameters as in the case for a well-type manometer, shown in Figure 9.4, then the rise in one leg does not equal the fall in the other. If the well has a cross-sectional area A and the tube has an area a, then the loss of liquid in one unit must equal the gain of liquid in the other. Hence $h_m A = h_2 a$ so that $h_2 = h_m A/a$.

For a simple U-tube measurement the applied pressure $P = (h_2 + h_m)\rho$. If the left-hand leg of the manometer becomes a wet leg with fluid density then

$$P + (h_1 + h_2)\rho_2 = (h_2 + h_m)\rho_1$$

so that

$$P = (h_2 + h_m)\rho_1 - (h_1 + h_2)\rho_2$$

If both manometer legs are wet then

$$P + (h_1 + h_2)\rho_2 = (h_2 + h_m)\rho_1 + (h_1 - h_m)\rho_2$$
$$P = (h_2 + h_m)\rho_1 + (h_1 - h_m)\rho_2 - (h_1 + h_2)\rho_2$$
$$= h_2\rho_1 + h_m\rho_1 + h_1\rho_2 - h_m\rho_2 - h_1\rho_2 - h_2\rho_2$$
$$= \rho_1(h_2 + h_m) - \rho_2(h_m + h_2)$$
$$= (h_2 + h_m)(\rho_1 - \rho_2)$$
$$= h_m(A/a + 1)(\rho_1 - \rho_2)$$

Effect of temperature The effect of variations in temperature have been neglected so far but for accurate work the effect of temperature on the densities of the fluids in the manometer must be taken into account and the effect of temperature on the scale should not be overlooked. For most applications it is sufficient to consider the effect of temperature only on the manometer liquid, in which case the density ρ at any temperature T can be taken to be:

$$\rho = \frac{\rho_0}{1 + \beta(T - T_0)}$$

where ρ_0 is the density at base conditions, β is the coefficient of cubic expansion, T_0 is the base temperature, and T is the actual temperature.

9.2.2 Pressure measurements by allowing the unknown pressure to act on a known area and measuring the resultant force

9.2.2.1 Dead-weight testers

The simplest technique for determining a pressure by measuring the force that is generated when it acts on a known area is illustrated by the dead-weight tester, but

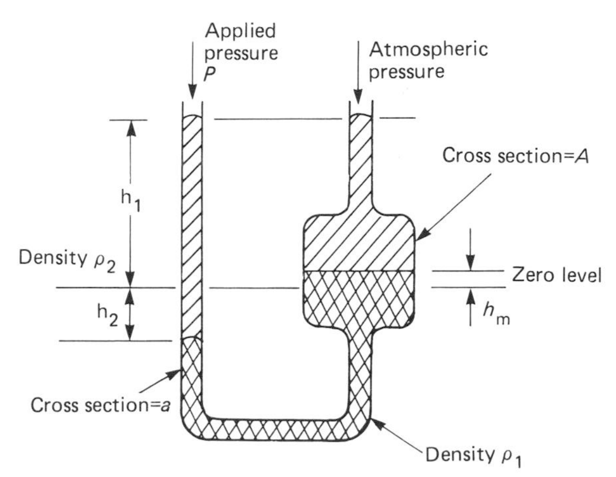

Figure 9.4 Manometer with limbs of different diameters.

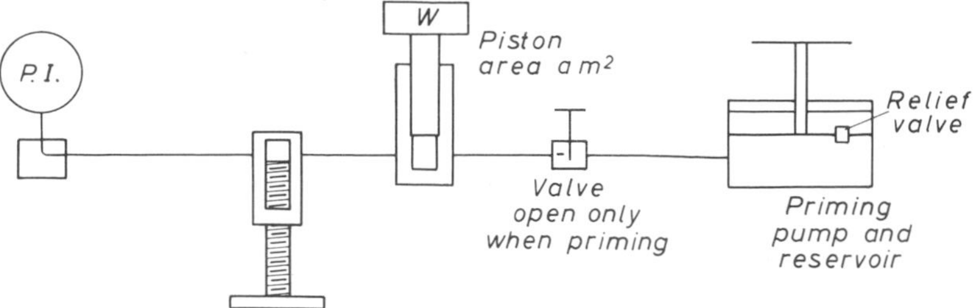

Figure 9.5 Basic system of dead weight tester.

this system is used for calibrating instruments rather than measuring unknown pressures.

The basic system is shown diagrammatically in Figure 9.5. It comprises a priming pump and reservoir, an isolating valve, the piston carrying the weight, a screw press and the gauge under test. In operation, the screw press is set to its zero position, weights representing the desired pressure are applied to the piston and the priming pump is operated to pressure the system. The valve is then shut and the screw press is adjusted until the pressure in the system is sufficient to raise the piston off its stops. If the frictional forces on the piston are neglected then the pressure acting on it is p newtons per square metre and if its area is a square metres, then the resultant force is pa N. This will support a weight $W = pa$ N.

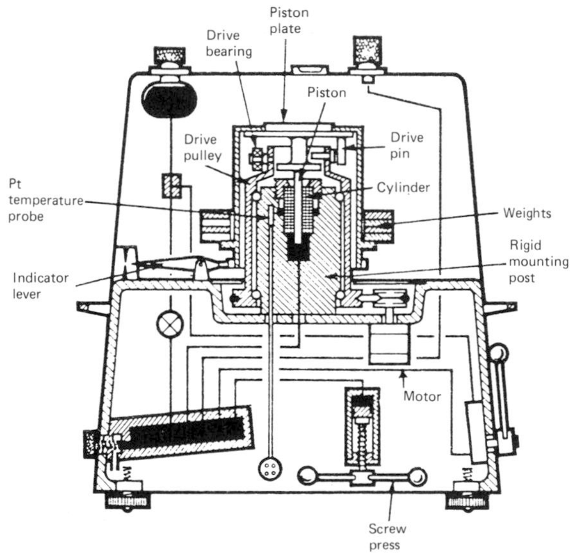

Figure 9.6 Arrangement of a precision dead weight tester. Courtesy, Desgranges and Hout.

The accuracy depends on the precision with which the piston and its associated cylinder are manufactured and on eliminating the effect of friction by rotating the piston whilst the reading is taken.

The Desgranges and Huot range of primary pressure standards is a very refined version of the dead-weight testers. Figure 9.6 shows a sectional drawing of an oil-operated standard. For this degree of precision it is important to ensure that the piston area and gravitational forces are constant so that the basic relation between the mass applied to the piston and the measured pressure is maintained. The instrument therefore includes levelling screws and bubble indicators.

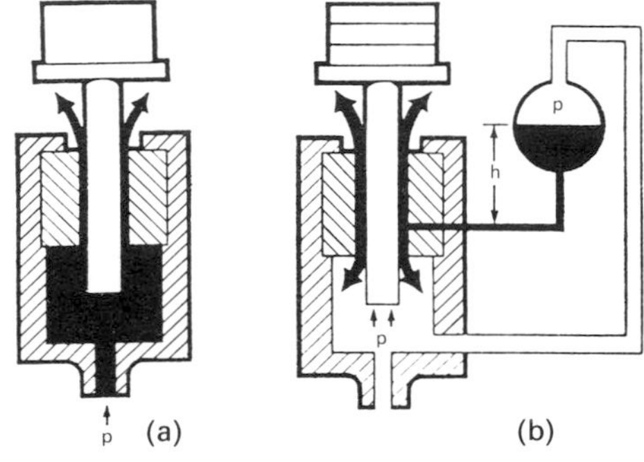

Figure 9.7 Lubrication of the piston. (a) Oil-operated system. (b) Gas-operated system. Courtesy, Desgranges and Hout.

Side stresses on the piston are avoided by loading the principal weights on a bell so that their centre of gravity is well below that of the piston. Only the fractional weights are placed directly on the piston plate and the larger of these are designed to stack precisely on the centre line.

The mobility of the piston in the cylinder assembly determines the sensitivity of the instrument and this requires an annulus that is lubricated by liquid even when gas pressures are being measured. Figure 9.7(a) and (b) show how this is achieved.

The system for liquids is conventional but for gases lubricant in the reservoir passes into the annulus between the piston and cylinder. The gas pressure is applied both to the piston and to the reservoir so that there is always a small hydraulic head to cause lubricant to flow into the assembly.

Rotation of the piston in the cylinder sets up radial forces in the lubricating fluid which tend to keep the piston centred, but the speed of rotation should be constant and the drive itself should not impart vibration or spurious forces. This is achieved by arranging the motor to drive the cylinder pulley via an oval drive pulley which is therefore alternatively accelerating and decelerating. The final drive is via the bearing on to the pin secured to the piston plate. In this way, once the piston is in motion, it rotates freely until it has lost sufficient momentum for the drive bearing to impart a small impulse which accelerates the piston. This ensures that it is rotating freely for at least 90 per cent of the time.

The piston and cylinder are machined from tungsten carbide to a tolerance of 0.1 micrometre so that the typical clearance between them is 0.5 micrometre. A balance indicator which tracks a soft iron band set in the bell shows the position of the piston and allows fluid head corrections for the most precise measurements.

The principal weights are fabricated in stainless steel and supplied in sets up to 50 kg according to the model chosen. The mass of the bell (typically 0.8 kg) and the piston plate assembly (typically 0.2 kg) must be added to the applied mass.

A complete set of piston and cylinder assemblies allows measurements to be made in the ranges from 0.1 to 50 bar to 2.0 to 1000 bar, the uncertainty of measurement being $\pm 5 \times 10^{-4}$ or less for the 'N' class instruments and $\pm 1 \times 10^{-4}$ or less for the 'S' class instruments.

Another type of dead-weight tester is the pneumatic dead weight tester. This is a self-regulating primary pressure standard. An accurate pressure is generated by establishing equilibrium between the air pressure on the underside of a spherical ball on top of which are loaded weights of known mass. The arrangement is shown in Figure 9.8.

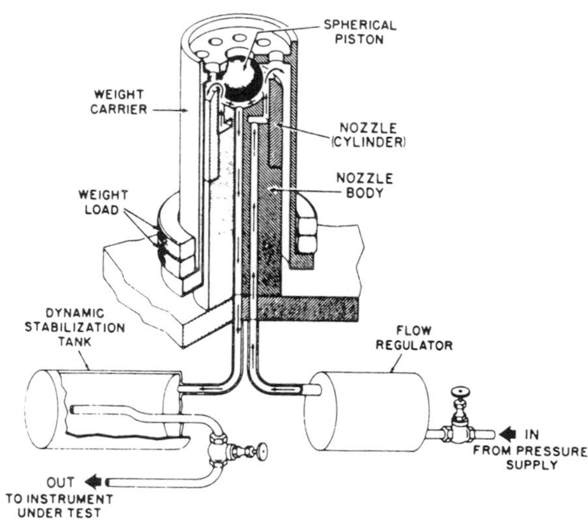

Figure 9.8 Self-regulating primary pressure standard. Courtesy, Amatek Inc.

The precision spherical ball is floated within a tapered stainless-steel nozzle. A regulated flow of air is introduced below the ball and the tapered nozzle so that it is lifted towards the annulus. Equilibrium is established as soon as the ball floats and the vented flow equals the fixed flow and the pressure is proportional to the weight loaded on the ball. In operation the ball is centred by the dynamic film of air which eliminates contact between the ball and nozzle.

When the weights are changed, the position of the ball also changes and in so doing it affects the air flow. The input regulator responds to this change by adjusting the flow of air so that pressure under the sphere and hence the output pressure is regulated at the new value.

9.2.3 Pressure measurement by allowing the known pressure to act on a flexible member and measuring the resultant motion

The great majority of pressure gauges utilize a Bourdon tube, stacked diaphragms, or a bellows to sense the pressure. The applied pressure causes a change in the shape of the sensor that is used to move a pointer with respect to a scale.

9.2.3.1 Bourdon tubes

The simplest form of Bourdon tube comprises a tube of oval cross-section bent into a circle. One end is sealed and attached via an adjustable connecting link to the lower end of a pivoted quadrant. The upper part of the quadrant is the toothed segment which engages in the teeth of the central pinion which carries the pointer that moves with respect to a fixed scale. Backlash between the quadrant and pinion is minimized by a delicate hair-spring. The other end of the tube is open so that the pressure to be measured can be applied via the block to which it is fixed and which also carries the pressure connection and provides the datum for measurement of the deflection.

If the internal pressure exceeds the external pressure the shape of the tube changes from oval towards circular with the result that it becomes straighter. The movement of the free end drives the pointer mechanism so that the pointer moves with respect to the scale. If the internal pressure is less than the external pressure, the free end of the tube moves towards the block, causing the pointer to move in the opposite direction.

The tube may be made from a variety of elastic materials depending on the nature of the fluid whose pressure is to be measured (phosphor bronze, beryllium copper and stainless steel are used most widely but for applications involving particularly corrosive fluids alloys, such as K-Monel, are used). The thickness of the tube and the material from which it is

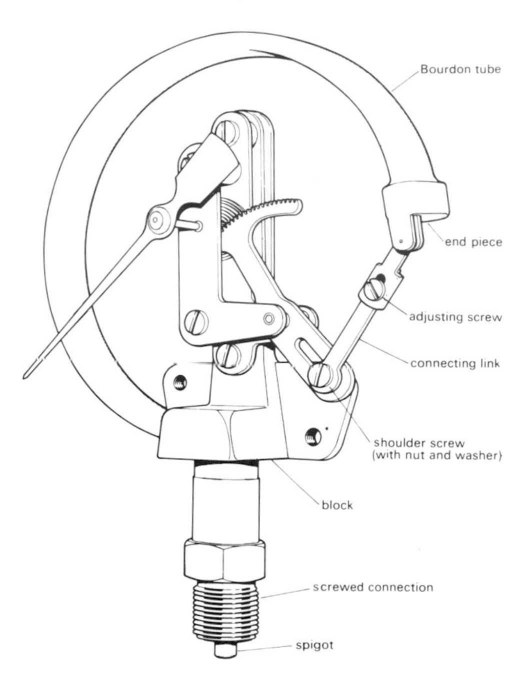

Figure 9.9 Mechanism of Bourdon tube gauge. Courtesy, Budenberg Gauge Co. Ltd.

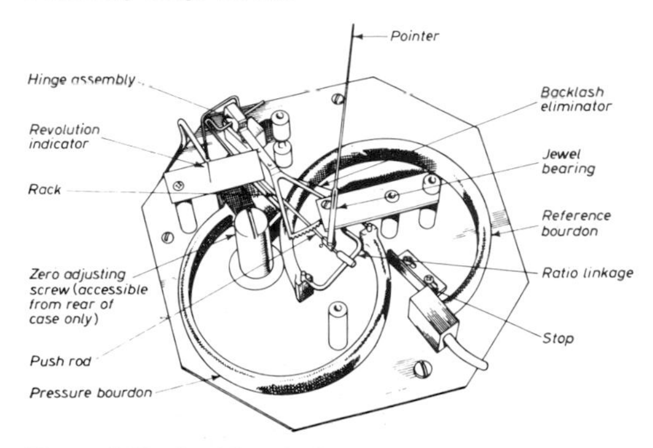

Figure 9.10 Precision absolute pressure gauge. Courtesy, Wallace & Tiernan Ltd.

to be fabricated are selected according to the pressure range, but the actual dimensions of the tube determine the force available to drive the pointer mechanism. The construction of a typical gauge is shown in Figure 9.9.

The performance of pressure gauges of this type varies widely, not only as a result of their basic design and materials of construction, but also because of the conditions under which they are used. The principal sources of error are hysteresis in the Bourdon tube, changes in its sensitivity due to changes of temperature, frictional effects, and backlash in the pointer mechanism. A typical accuracy is ± 2 per cent of span.

Much higher precision can be achieved by attention to detail and one example is illustrated in Figure 9.10 which shows a gauge for measuring absolute pressure. It includes two Bourdon tubes, one being completely evacuated and sealed to provide the reference whilst the unknown pressure is applied to the other Bourdon tube. The free ends of the Bourdon tubes are connected by a ratio linkage which through a push rod transmits the difference in the movement of the free ends to a rack assembly which in turn rotates the pinion and pointer. Jewel bearings are used to minimize friction and backlash is eliminated by maintaining a uniform tension for all positions of the rack and pinion through the use of a nylon thread to connect a spring on the rack with a grooved pulley on the pinion shaft.

The Bourdon tubes are made of Ni-Span C which has a very low coefficient of expansion and good resistance to corrosion. As both Bourdon tubes are subjected to the same atmospheric pressure, the instrument maintains its accuracy for barometric pressure changes of ± 130 mmHg. The dial diameter is 216 mm and the full range of the instrument is covered by two revolutions of the pointer giving an effective scale length of 1.36 m. The sensitivity is 0.0125 per cent and the accuracy 0.1 per cent of full scale. The ambient temperature effect is less than 0.01 per cent of full scale per kelvin.

9.2.3.2 Spiral and helical Bourdon tubes

The amount of the movement of the free end of a Bourdon tube varies inversely as the wall thickness and is dependent on the cross-sectional shape. It also varies directly with the angle subtended by the arc through which the tube is formed. By using a helix or spiral to increase the effective angular length of the tube, the movement of the free end is similarly increased and the need for further magnification is reduced. Examples of these constructions are shown in Figures 9.11 and 9.12. They avoid the necessity for the toothed quadrant with the consequent reduction of backlash and frictional errors. In general, the spiral configuration is used for low pressures and the helical form for high pressures.

Figure 9.11 Helical Bourdon tube. Courtesy, The Foxoboro Company.

Figure 9.12 Spiral Bourdon tube. Courtesy, The Foxboro Company.

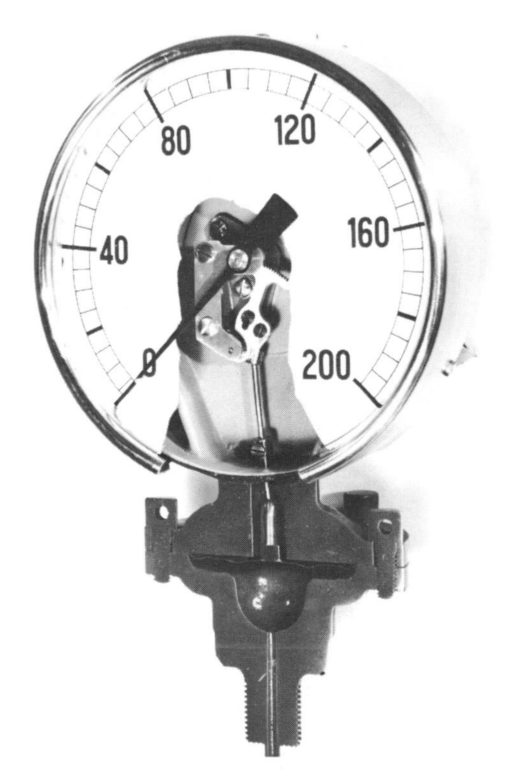

Figure 9.13 Schaffer pressure gauge. Courtesy, Budenberg Gauge Co. Ltd.

9.2.3.3 Diaphragm pressure elements

There are two basic categories of diaphragm elements, namely stiff metallic diaphragms and slack diaphragms associated with drive plates.

The simplest form of diaphragm gauge is the Schaffer gauge shown in Figure 9.13. It consists of a heat-treated stainless-steel corrugated diaphragm about 65 mm in diameter and held between two flanges. The unknown pressure is applied to the underside of the diaphragm and the resultant movement of the centre of the diaphragm is transmitted through a linkage to drive the pointer as in the Bourdon gauge. The upper flange is shaped to provide protection against the application of over-range pressures.

In the Schaffer gauge it is the elastic properties of the metallic diaphragm which govern the range and accuracy of the measurement. An aneroid barometer (Figure 9.14) also uses a corrugated diaphragm but it is supplemented by a spring. The element consists of a flat circular capsule having a corrugated lid and base and is evacuated before being sealed. It is prevented from collapse by a spring which is anchored to a bridge and attached to the top centre of the capsule. Also attached at this point is a lever which acts through a bell crank and lever mechanism to rotate the pointer. When the atmospheric pressure increases the capsule contracts so that the pointer is caused to rotate in one direction. Conversely when the atmospheric pressure falls the capsule expands and the pointer is driven in the opposite direction.

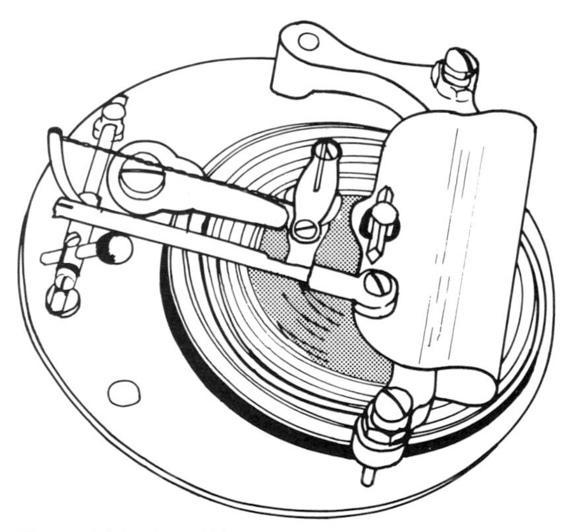

Figure 9.14 Aneroid barometer.

A further example of an instrument employing stiff diaphragms augmented by springs is shown in Figures 9.15 and 9.16. This instrument has largely superseded the bell-type mercury pressure manometer previously widely used for measuring differential associated with orifice-plate flowmeters, partly because of the increased cost, but more particularly because of the health hazards associated with mercury.

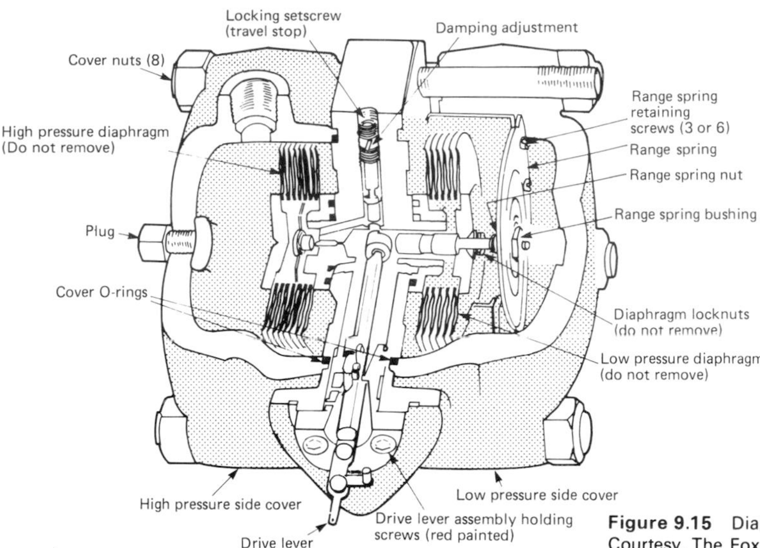

Figure 9.15 Diaphragm type differential pressure transmitter. Courtesy, The Foxoboro Company.

The diaphragm elements (7) and (2) are made up from pairs of corrugated diaphragms with a spacing ring stitch-welded at the central hole. These assemblies are then stitch-welded at their circumference to form a stack. This configuration ensures that when excess pressure is applied to the stack the individual corrugations nest together whilst the stack spacing rings come together to form a metal-to-metal stop.

The diaphragm stacks (7) and (2) are mounted on the central body together with the range spring (3) and drive unit (4). Pressure-tight covers (8) form the high- and low-pressure chambers. The diaphragm stacks (2) and (7) are interconnected via the damping valve (1) and fitted internally with a liquid which remains fluid under normal ambient conditions.

An increase in pressure in the high-pressure chamber compresses the diaphragm stack (7) and in so doing displaces fluid via the damping valve (1) into stack (2) causing it to expand until the force exerted by the range spring balances the initial change in pressure. The deflection of the range spring is transmitted to the inner end of the drive unit, which being pivoted at a sealed flexure (5) transfers the motion to the outer end of the drive shaft (4) where it can be used to operate a pen arm.

A bimetallic temperature-compensator (6) is mounted inside the stack (7) and adjusts the volume of that stack to compensate for the change in volume of the fill liquid resulting from a change of temperature. The instrument is suitable for operating at pressures up to 140 bar and spans between 50 and 500 mbar can be provided by selecting suitable combinations of the range springs which are fabricated from Ni-Span C to make them substantially insensitive to changes of temperature.

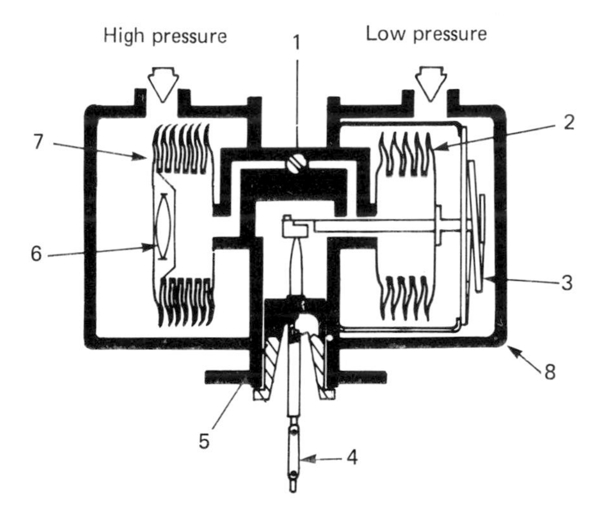

Figure 9.16 Components of the differential pressure transmitter. Courtesy, The Foxboro Company.

Bellows elements With the development of the hydraulic method for forming bellows, many of the pressure-sensing capsules previously fabricated from corrugated diaphragms have been replaced by bellows which are available in a variety of materials. The spring rate or modulus of compression of a bellows varies directly as the modulus of elasticity of the material from which it is formed and proportionally to the third power of the wall thickness. It is also inversely proportional to the number of convolutions and to the square of the outside diameter of the bellows.

The combined effect of variations in the elastic properties of the materials of construction and manufacturing tolerance results in appreciable variations in

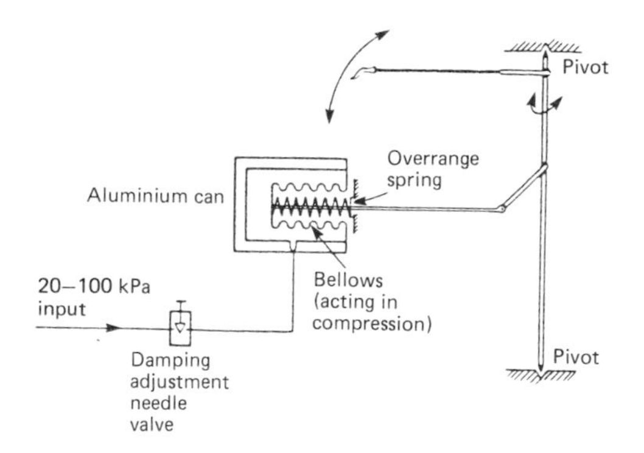

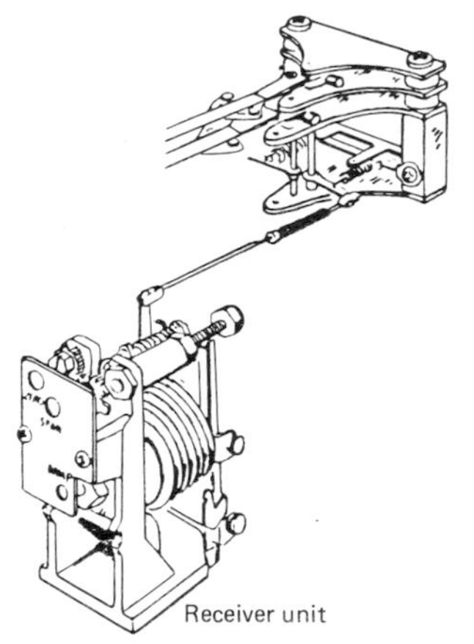

Figure 9.17 Pneumatic receiver using a bellows. Courtesy, The Foxboro Company.

the bellows spring rate, not only from one batch to another but also within a batch. For some applications this may not be particularly significant but, when it is, the effect can be reduced by incorporating a powerful spring into the assembly.

Figure 9.17 shows a pneumatic receiver, i.e. a unit specifically designed for measurements in the range 20 to 100 kPa which is one of the standard ranges for transmission in pneumatic systems.

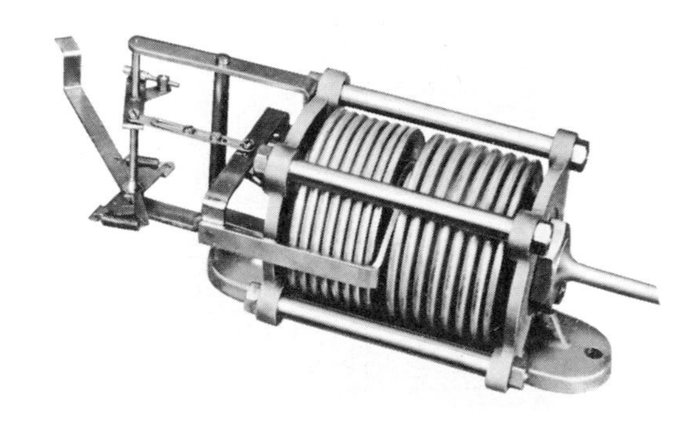

Figure 9.18 Bellows assembly for an absolute pressure gauge. Courtesy, Foxboro-Yoxall Ltd.

Figure 9.18 shows a bellows assembly for the measurement of absolute pressure. It comprises two carefully matched stainless-steel bellows, one of which is evacuated to a pressure of less than 0.05 mmHg and sealed. The unknown pressure is applied to the other bellows. The two bellows are mounted within a frame and connected together via a yoke which transmits the bellows motion via a link to a pointer or the pen arm of a recorder.

9.2.3.5 Slack-diaphragm pressure elements with drive plates

These are a further example of pressure sensors in which the force produced by the pressure acting on a fixed area is opposed by a spring. The most common application is the measurement of furnace draught where a span between 100 Pa and 2.5 kPa above atmospheric may be required. A typical instrument of this type is shown in Figure 9.19. To respond to such

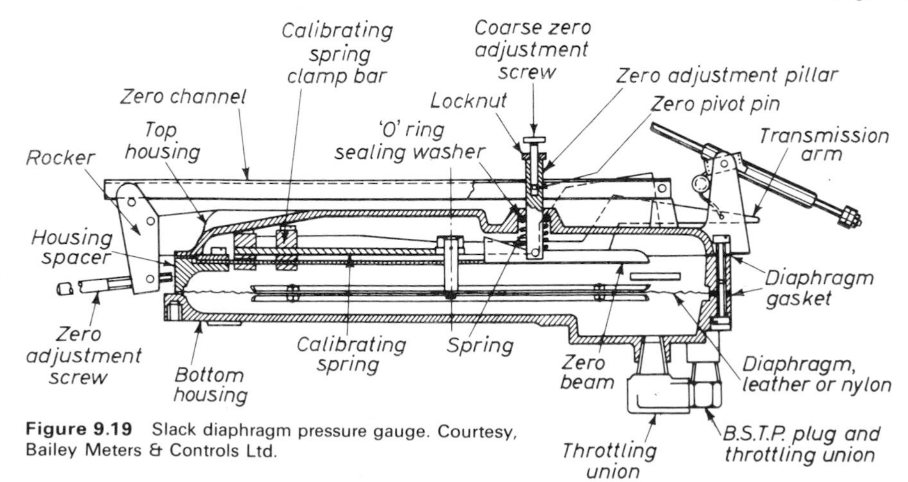

Figure 9.19 Slack diaphragm pressure gauge. Courtesy, Bailey Meters & Controls Ltd.

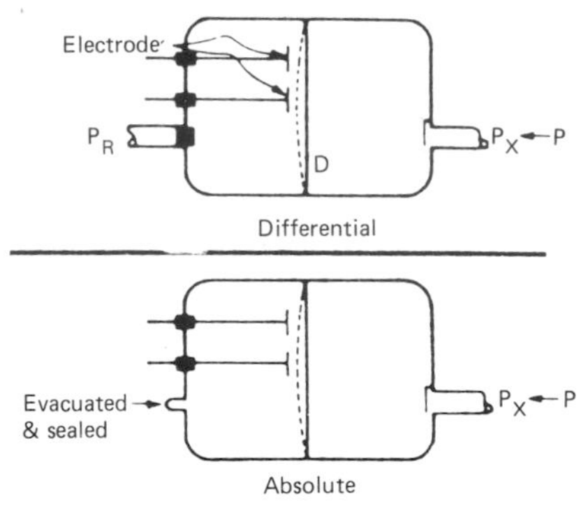

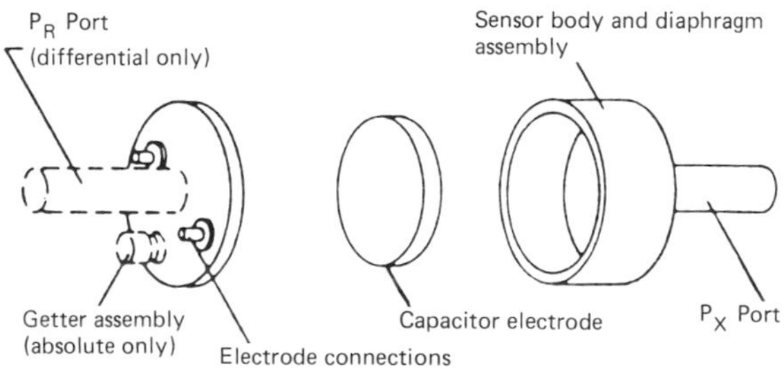

Figure 9.20 Capacitance manometer sensor. Courtesy, MKS Instruments Inc.

small pressures a large-area diaphragm is required. It is made from very thin non-porous material and supported on both sides over a large portion of its area by drive plates. Travel of the diaphragm is limited by the drive plates coming to rest against stops to provide overload protection. Motion of the diaphragm is produced by the difference in pressure across it and is opposed by a flat beryllium copper spring. The span of the instrument is adjusted by varying the effective length of the spring, and hence its rate.

For measuring gauge pressure, the movement of the diaphragm is transmitted to the pointer directly by a suitable mechanism. For measuring differential pressure, both sides of the chamber containing the diaphragm are sealed and the motion is transmitted through the seal by a magnetic coupling.

9.2.3.6 *Capacitance manometers*

The application of electronic techniques to measure the deflection of a diaphragm and hence to infer pressure has resulted in major improvements in both sensitivity and resolution as well as providing means for compensating for nonlinear effects. One of the devices in which these techniques have been applied is the capacitance manometer shown diagrammatically in Figure 9.20.

For such a sensor it is important that the diaphragm and sensor body are capable of withstanding a wide range of process fluids including those which are highly corrosive. It is also important for them to have thermal coefficients which closely match those of the electrode assembly and screening material. 'Inconel' is a suitable material for the body and diaphragm, whilst 'Fosterite' with either nickel or palladium is used for the electrode assembly. With these materials pressures as low as 10^{-3} Pa can be measured reliably.

The tensioned metal diaphragm is welded into the sensor body and the electrode assembly is located in the body at the correct position with respect to the diaphragm. If the sensor is to be used for absolute pressure measurements, the sensor-body assembly is completed by welding in place a cover which carries the two electrode connections and the getter assembly. If on the other hand the sensor is to be used for differential pressure measurements, then provision is made for connecting the reference pressure.

The hysteresis error for such a sensor is normally less than 0.01 per cent of the reading; for sensors having spans greater than 100 Pa the error is almost immeasurable. The non-linearity is the largest source of error in the system apart from temperature effects and is usually in the order of 0.05 per cent of reading and is minimized by selective adjustments in the associated electronic circuits.

Errors due to ambient temperature changes affect both the zero and span. Selection of the optimum materials of construction results in a zero error of approximately 0.02 per cent of span per kelvin and a span error of approximately 0.06 per cent of span per kelvin. The span error can be reduced to 0.0005 per cent by including a temperature sensor in the body of the pressure sensor and developing a corresponding correction in the measuring circuits. The zero error can be reduced to 0.002 per cent by including a nulling circuit.

9.2.3.7 *Quartz electrostatic pressure sensors*

There are two principal types of quartz sensors used for pressure measurement. The first are those in which the applied force causes an electrostatic charge to be developed across the crystal which is then measured by a charge amplifier and the resultant signal used to provide an indication of the applied force.

The second category (Section 9.2.3.9) involves the use of the quartz in some form of resonator whose frequency is modified as a result of the applied force.

The Kistler type 601 and 707 series shown in Figure 9.21 are an example of quartz electrostatic sensor. The assemblies utilize the transverse piezo-electric effect illustrated in Figure 9.22. The application of a force F in the direction of one of the neutral axes Y sets up an electrostatic charge on the surfaces of the polar axis x at right angles to it. The magnitude of this charge depends on the dimensions of the quartz crystal and by selecting a suitable shape it is possible to secure a high charge yield combined with good linearity and low

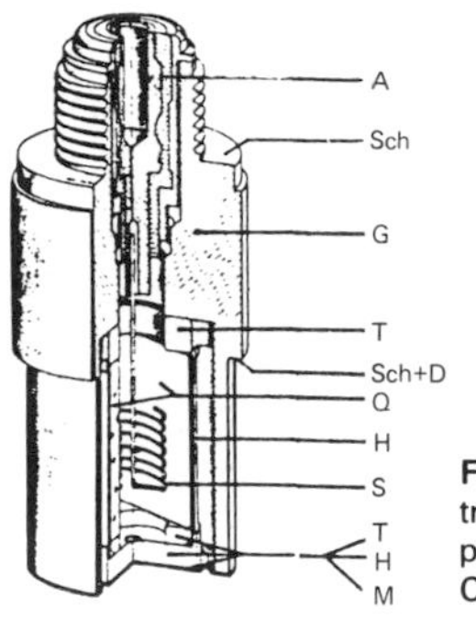

Figure 9.21 Pressure transducer using transverse piezo-electric effect of quartz. Courtesy, Kistler Instruments Ltd.

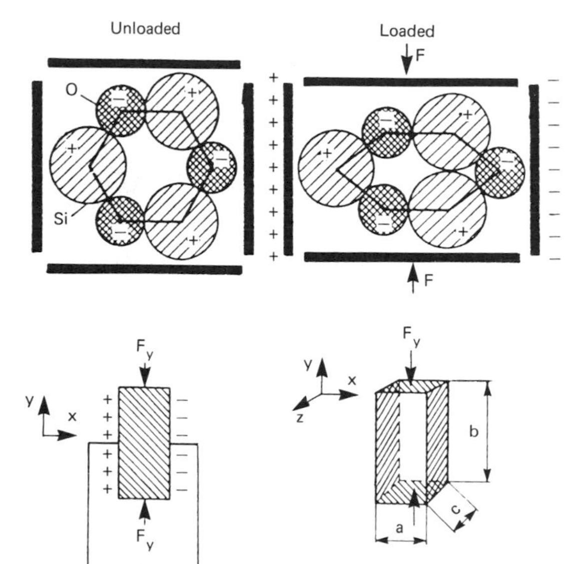

Figure 9.22 Principle of transverse piezo-electric effect. Courtesy, Kistler Instruments Ltd.

temperature sensitivity. Similarly, the principle of the longitudinal piezo-electric effect is illustrated in Figure 9.23.

A typical transducer is assembled from three quartz stacks Q (Figure 9.21) joined rigidly to the holder G by means of a preload sleeve H and temperature compensator T. The pressure to be measured acts on the diaphragm M where it is converted into the force which is applied to the three quartz stacks. The contact faces of the quartz are coated with silver and a central noble metal coil S conducts charge to the connector A.

The outer faces of the quartz are connected to the housing. With this configuration linearities of between 0.2 and 0.3 per cent are achieved for spans up to 25 MPa and the sensors have a uniform response up to about 30 kHz with a peak of about 100 kHz. Because there must be a finite leakage resistance across the

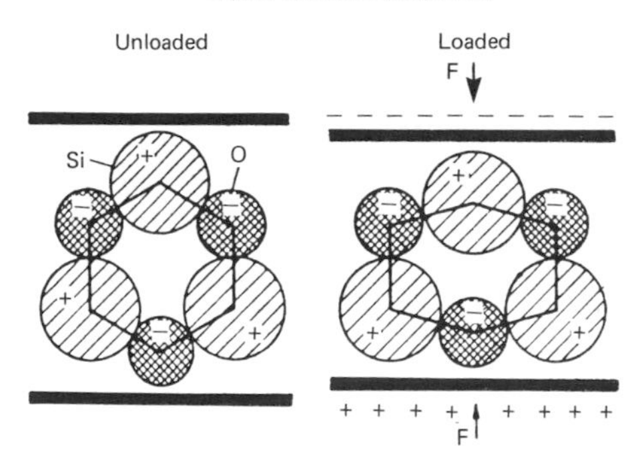

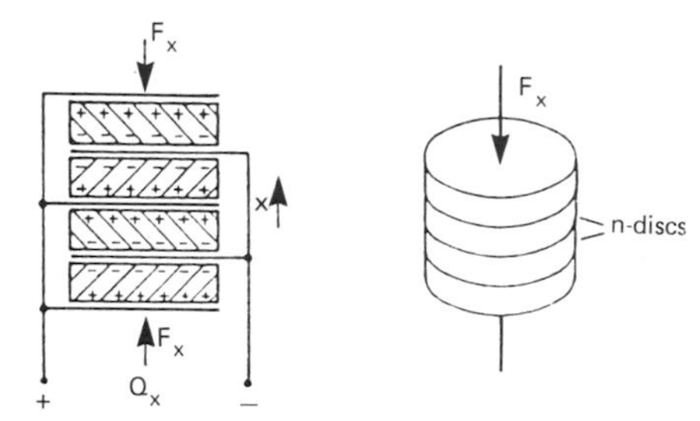

Figure 9.23 Principle of longitudinal piezo-electric effect. Courtesy, Kistler Instruments Ltd.

sensor, such devices cannot be used for static measurements. The low frequency limit is of the order of 1 Hz, depending on the sensitivity. The type of charge amplifier associated with these sensors is shown in Figure 9.24. It comprises a high-gain operational amplifier with MOSFET input stage to ensure that the input impedance is very high, and capacitor feedback to ensure that the charge generated on the quartz transducer is virtually completely compensated. It can be shown that the output voltage from the amplifier is $-Q/C_g$ where Q is the charge generated by the quartz sensor and C_g is the feedback capacitance. Thus the system is essentially insensitive to the influence of the input cable impedance.

Sensors such as these are characterized by their high stability, wide dynamic range, good temperature stability, good linearity and low hysteresis. They are available in a very wide variety of configurations for pressure ranges from 200 kPa to 100 MPa.

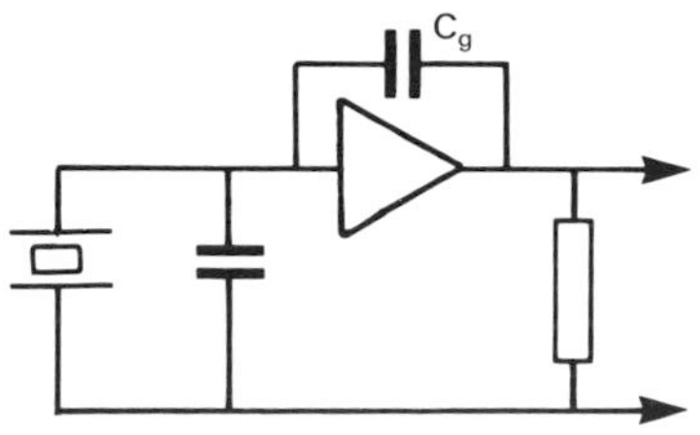

Figure 9.24 Charge amplifier associated with piezo-electric effect sensor.

Figure 9.25 Piezo-resistive pressure transducer. Courtesy, Kistler Instruments Ltd.

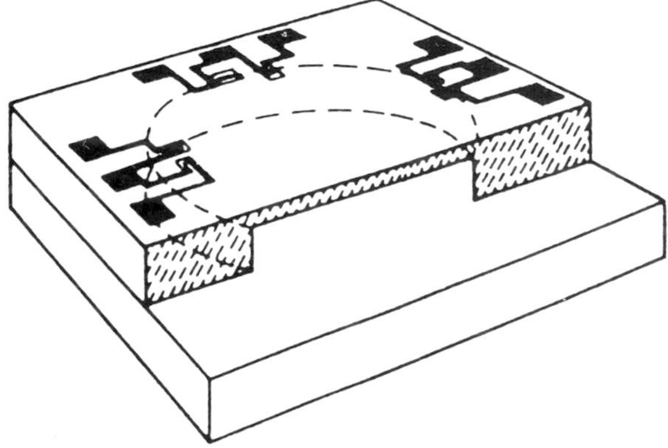

Figure 9.26 Schematic drawing of pressure sensing element. Courtesy, Kistler Instruments Ltd.

9.2.3.8 *Piezo resistive pressure sensors*

For many metals and some other solid materials, the resistivity changes substantially when subjected to mechanical stress. Strain gauges, as described in Chapter 4, involve this phenomenon, but the particular characteristics of silicon allow construction of a thin diaphragm that can be deflected by an applied pressure and can have resistors diffused into it to provide a means for sensing the deflection. An example of this is the Kistler 4000 series as shown in Figure 9.25 for which the pressure-sensing element is shown in Figure 9.26.

Because the stress varies across the diaphragm, four pairs of resistors are diffused into a wafer of n-type silicon, each pair having one resistor with its principal component radial and one with its principal component circumferential. As described later, this provides means for compensating the temperature-sensitivity of the silicon. Mechanically they form part of the diaphragm but they are isolated electrically by the p–n junction so that they function as strain gauges. The diaphragm is formed by cutting a cylindrical recess on the rear surface of the wafer using initially ultrasonic or high-speed diamond machining and finally chemical etching. This unit is then bonded to a similar unprocessed chip so that a homogeneous element is produced. If it is desired to measure absolute pressures the bonding is effected under a vacuum. Otherwise the cavity behind the diaphragm is connected via a hole in the base chip and a bonded tube to the atmospheric or reference pressure. The schematic arrangements of two such transducers are shown in Figure 9.27.

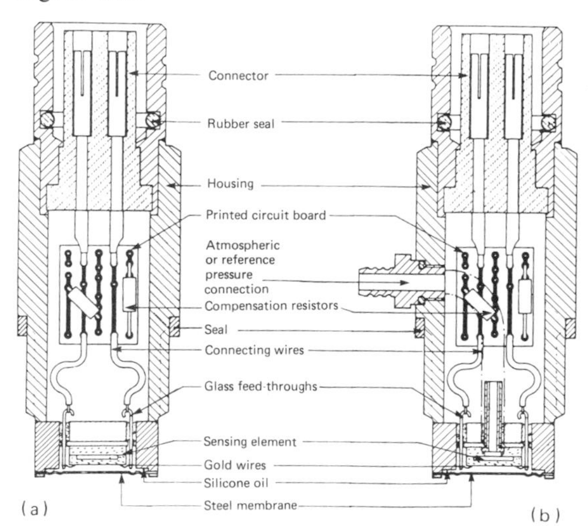

Figure 9.27 Cross section of piezo-resistive pressure transducer. (a) For absolute pressure. (b) For gauge pressure. Courtesy, Kistler Instruments Ltd.

The mechanical strength of silicon depends largely on the state of the surface and in general this imposes an upper limit of about 100 MPa on the pressures that can be measured safely by the sensors. The lower limit is about 100 kPa and is determined by the minimum thickness to which the diaphragm can be manufactured reliably.

Both the gauge factor G and resistance R of the diffused resistors are sensitive to changes of temperature and the sensors need to be associated with some form of compensating circuits. In some instances this is provided by discrete resistors closely associated with the gauge itself. Others utilize hybrid circuits, part of which may be on the chip itself. Individual compensation is always required for the zero offset, the measurement span and temperature stabilization of the zero. Further improvement in the performance can be achieved by compensating for the nonlinearity and the effect of temperature on the span.

9.2.3.9 *Quartz resonant pressure sensors*

A second method of utilizing the piezo-electric effect is to observe the change in resonant frequency of the

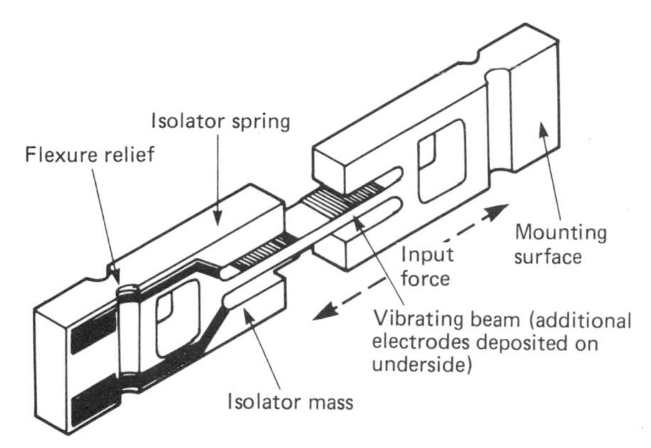

Figure 9.28 Resonant piezo-electric force sensor. Courtesy, Paroscientific Inc.

piezo-electric sensor resulting from application of the force developed by the unknown pressure acting on a flexible diaphragm of fixed area. One such sensor utilizing this principle to measure absolute pressures is shown diagrammatically in Figure 9.28.

It depends for its operation on a fixed beam oscillating in its first flexural mode and mounted in an isolation system that effectively decouples it from the structures to which it is attached. The entire sensor is fabricated from a single piece of quartz to minimize energy loss to the joints and the cut is orientated to minimize the effect of temperature changes. Four electrodes are vacuum-deposited on the beam and, with diagonally opposite pairs of electrodes connected to an oscillator, the beam is maintained in oscillation at its resonant frequency. The response to the imposed electric field is illustrated in Figure 9.29. The Q of the sensor may be as high as 40 000 so that not more than a few milliwatts are needed to sustain the oscillation. Under tension, the resonant frequency increases and under compression it is reduced.

Figure 9.30 shows the embodiment of the sensor in an absolute pressure transducer. The pressure to be measured is applied to a bellows which converts it into an upward force which acts on the force arm causing it to pivot about the centrally located cross-flexures. As a result, a compressive force is applied to the quartz sensor causing its resonant frequency to be reduced. Because the quartz element is essentially rigid the entire mechanical structure is constrained to minute deflections and hysteretic effects are virtually eliminated. If the bellows, together with its associated flexures and levers, is enclosed in a structure which is evacuated, this not only provides the zero pressure reference, it also eliminates the mass loading and damping effects on the quartz resonator (directly increasing the Q) and further minimizes the effect of ambient temperature changes. Balance weights are included in the system of levers so that the centre of gravity of the mechanism coincides with the centre of the flexures. This minimizes errors in response due to vibration or acceleration forces.

The relation between the applied pressure P and the oscillation frequency is given by

$$P = A\left(1 - \frac{T_0}{T}\right) - B\left(1 - \frac{T_0}{T}\right)^2$$

where T_0 is the periodic time when the applied pressure is zero, T is the periodic time when the applied pressure is P, and A and B are calibration constants for the transducer.

The nominal frequency of the sensor at zero pressure is 40 kHz and this changes to about 36 kHz at full-scale pressure. A convenient method of measuring the output is to allow the signal to gate the signal from a 10 MHz clock. Thus by timing 1000 cycles of the resonant sensor the applied pressure can be determined with a resolution of 0.003 per cent of full scale in approximately 25 milliseconds.

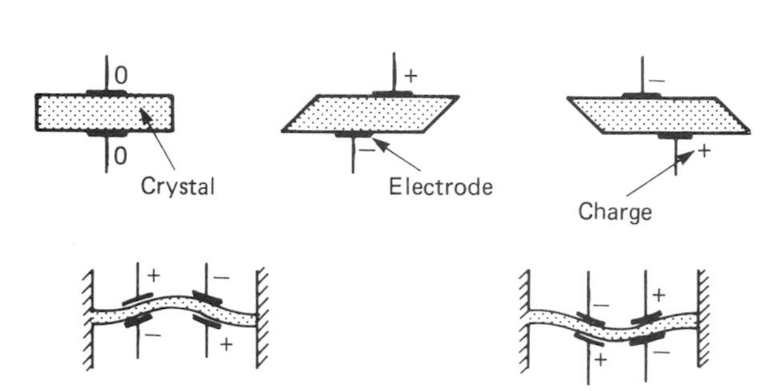

Figure 9.29 Oscillator mode for piezo-electric force sensor. Courtesy, Paroscientific Inc.

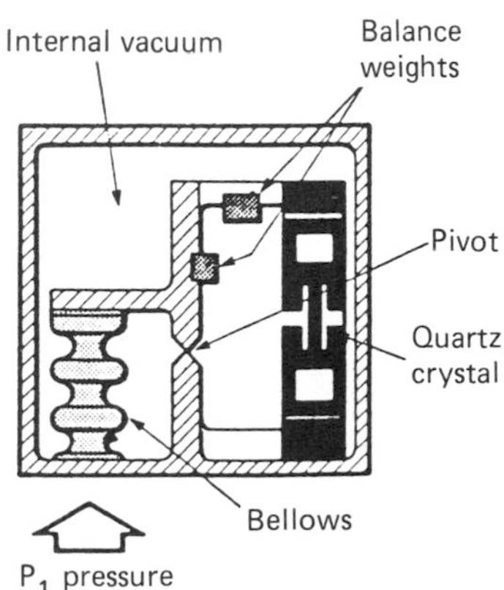

Figure 9.30 Configuration of force sensor in absolute pressure transducer. Courtesy, Paroscientific Inc.

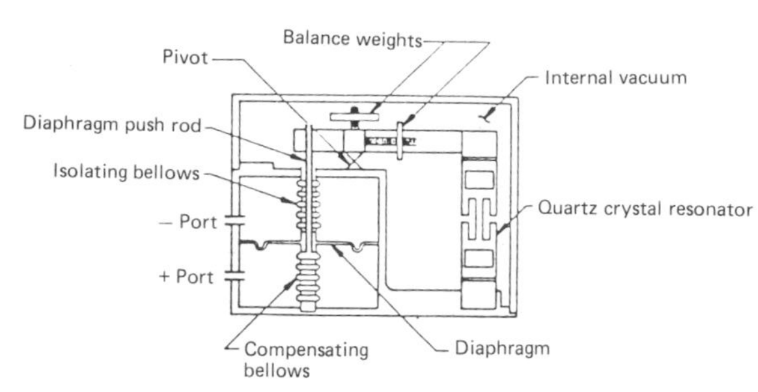

Figure 9.31 Configuration of force sensor in differential pressure transducer. Courtesy, Paroscientific Inc.

The sensor can be adapted to measure differential pressures by mounting a pair of matched bellows in opposition to act on the suspension arm as shown in Figure 9.31. However, additional errors arise from sensitivity to the static pressure due to imperfections in the bellows and their matching. This has led to the development of a transducer similar in many respects to the established designs of differential pressure transmitters.

Typical specifications:

For absolute pressure measurements with spans between 0 to 100 kPa and 0 to 6 MPa or differential pressure measurements with spans between 0 to 40 kPa and 0 to 275 kPa.

Repeatability	0.005% of span
Hysteresis	0.005% of span
Temperature coefficient of zero	0.007% of span/kelvin
Temperature coefficient of reading	0.0045% of reading/kelvin
Operating temperature range	−55 to 110 °C
Power requirement	6 milliwatts
Mass	0.17 kg
Approximate size	22 × 40 × 40 mm

A third group of pressure sensors is based on strain-gauge technology (see Chapter 4), in which the resistance-type strain sensors are connected in a Wheatstone bridge network. To achieve the required long-term stability and freedom from hysteresis, the strain sensors must have a molecular bond to the deflecting member which in addition must also provide the necessary electrical isolation over the operating temperature range of the transducer.

This can be achieved by first sputtering the electrical isolation layer on the stainless-steel sensor beam or diaphragm and then sputtering the thin-film strain-gauge sensors on top of this. An example of this type of sensor is the CEC Instrumentation 4201 series shown in Figure 9.32.

The pressure inlet adaptor is fabricated from precipitation-hardened stainless steel and has a deep recess between the mounting thread and diaphragm chamber to isolate the force-summing diaphragm from the mounting and other environmental stresses.

The transducer is modular in construction to allow the use of alternative diaphragm configurations and materials. For most applications the diaphragm is stainless steel and the thickness is selected according to the required measurement range. For some applications, enhanced corrosion resistance is required, in which case Inconel 625 or other similar alloys may be used as the diaphragm material, but to retain the same margin of safety a thicker member is usually required and this in turn reduces the sensitivity.

The sensor is a sputtered thin-film strain gauge in which the strain-gauge pattern is bonded into the structure of the sensor assembly on a molecular basis and the sensor assembly itself is welded into the remaining structure of the transducer. The stainless-steel header which contains the electrical feed-through to the temperature-compensation compartment is also welded into the structure of the transducer.

This welding, in conjunction with the ceramic firing technique used for the electrical feed-through con-

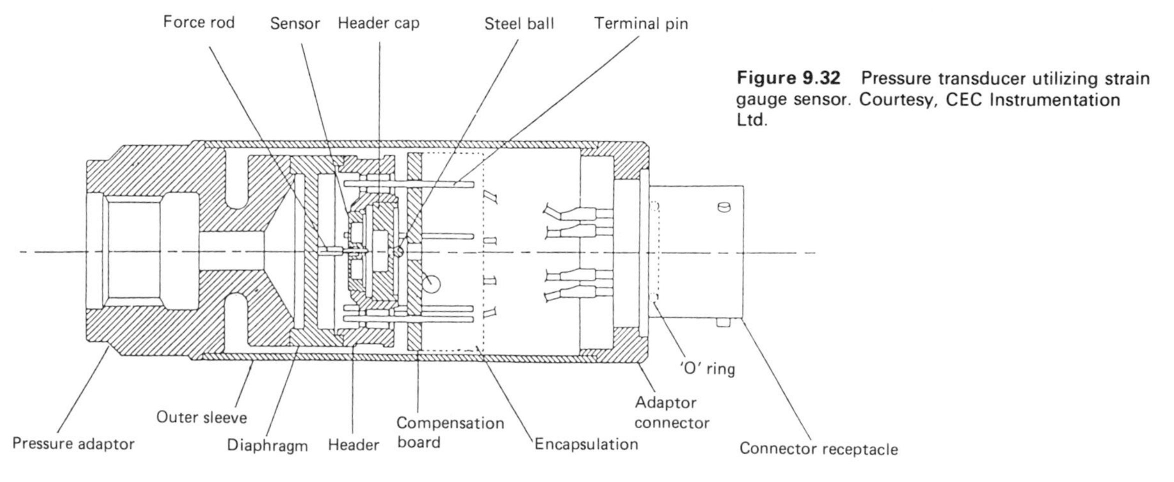

Figure 9.32 Pressure transducer utilizing strain gauge sensor. Courtesy, CEC Instrumentation Ltd.

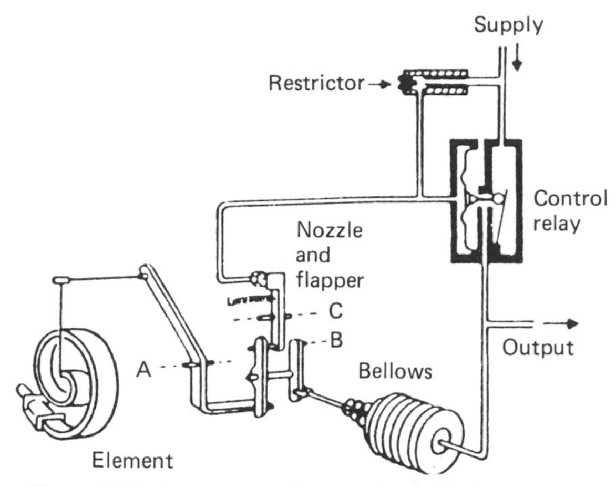

Figure 9.33 Arrangement of pneumatic motion-balance transmitter. Courtesy, The Foxboro Company.

nections, provides secondary containment security of 50 MPa for absolute gauges and those with a sealed reference chamber.

Sensors of this type are available with ranges from 0 to 100 kPa up to 0 to 60 MPa with maximum non-linearity and hysteresis of 0.25 to 0.15 per cent respectively and a repeatability of 0.05 per cent of span. The maximum temperature effect is 0.15 per cent of span per kelvin.

9.3 Pressure transmitters

In the process industries, it is often necessary to transmit the measurement signal from a sensor over a substantial distance so that it can be used to implement a control function or can be combined with other measurement signals in a more complex scheme.

The initial development of such transmission systems was required for the petroleum and petrochemical industries where pneumatic control schemes were used most widely, because they could be installed in plants where explosive or hazardous conditions could arise and the diaphragm actuator provided a powerful and fast-acting device for driving the final operator. It followed that the first transmission systems to be evolved were pneumatic and were based on the standardized signal range 20 to 100 kPa.

The first transmitter utilized a motion-balance system, i.e. one in which the primary element produces a movement proportional to the measured quantity, such as a Bourdon tube, in which movement of the free end is proportional to the applied pressure. However, these transmitters were rather sensitive to vibration and have, in general, been superseded by force-balance systems. But pneumatic transmission itself is unsuitable when the distance involved exceeds a few hundred metres, because of the time delay and response lag which occur.

Consequently, an equivalent electronic system has been evolved. In this, a current in the range 4 to 20 mA d.c. and proportional to the span of the measured quantity is generated by the sensor and transmitted over a two-wire system. The advantage of this system is that there is virtually no delay or response lag, and the transmitted signal is not affected by changes in the characteristic of the transmission line. Also there is sufficient power below the live zero (i.e. 4 mA) to operate the sensing device. Such systems have the additional advantage that they are more easily configurated in complex control schemes than the corresponding pneumatic transmitters. Telemetry and pneumatic systems are discussed further in Chapters 5 and 7 of Part 4.

9.3.1 Pneumatic motion-balance pressure transmitters

Figure 9.33 shows the arrangement of a typical pneumatic motion-balance transmitter in which the sensor is a spiral Bourdon tube. Changes in the measured variable, which could be pressure, or temperature in the case of a filled thermal system, cause the free end of the Bourdon tube to move. This movement is transmitted via a linkage to the lever that pivots about the axis A. The free end of this lever bears on a second lever that is pivoted at its centre so that the movement is transmitted to a third lever that is free to pivot about the axis C. The initial movement is thus transferred to the flapper of the flapper/nozzle system. If as a result, the gap between the flapper and nozzle is increased, the nozzle back-pressure falls and this in turn causes the output pressure from the control relay to fall. As this pressure is applied to the bellows the change causes the lever pivoted about the axis B to retract so that the lever pivoted about the axis C moves the flapper towards the nozzle. This causes the nozzle back-pressure to rise until equilibrium is established. For each value of the measurement there is a definite flapper/nozzle relationship and therefore a definite output signal.

9.3.2 Pneumatic force-balance pressure transmitters

There are many designs of pneumatic force-balance transmitters, but in the Foxboro Company design the same force-balance mechanism is used in all the pressure and differential pressure transmitters. It is shown in Figure 9.34 and its basic function is to convert a force applied to its input point into a proportional pneumatic signal for transmission, such as 20 to 100 kPa.

The force to be measured may be generated by a Bourdon tube, a bellows or a diaphragm assembly and applied to the free end of the force bar. This is pivoted

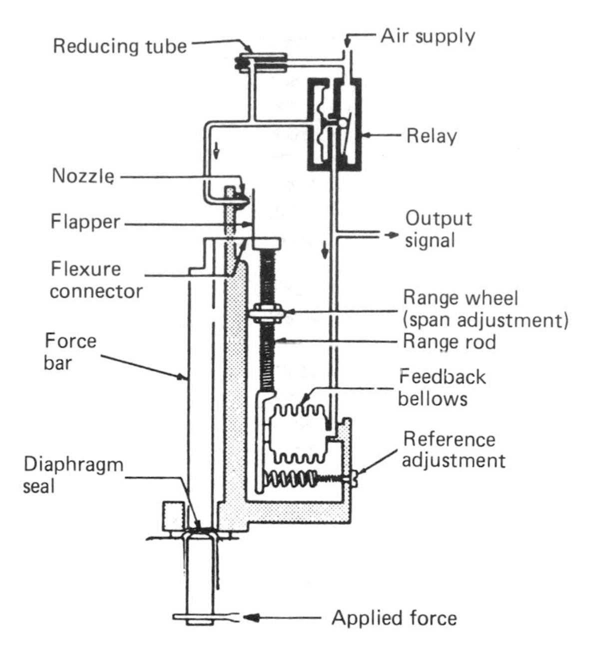

Figure 9.34 Arrangement of pneumatic force-balance transmitter. Courtesy, The Foxboro Company.

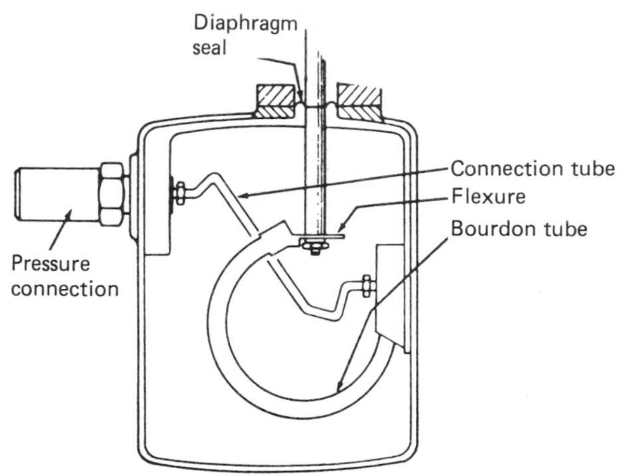

Figure 9.35 Bourdon tube primary element arranged for operation in conjunction with a force-balance mechanism. Courtesy, The Foxboro Company.

at the diaphragm seal, which in some instruments also provides the interface between process fluid and the force-balance mechanism, so that an initial displacement arising from the applied force appears amplified at the top of the force bar where it is transmitted via the flexure connector to the top of the range rod. If the applied force causes movement to the right, the flapper uncovers the nozzle with the result that the nozzle back-pressure falls. This change is magnified by the 'relay' whose output is applied to the feedback bellows thereby producing a force which balances the force applied initially. The output signal is taken from the 'relay' and by varying the setting of the range wheel the sensitivity or span can be adjusted through a range of about 10 to 1. By varying the primary element pressures from about 1.3 kPa to 85 MPa and differential pressures from 1 kPa to 14 MPa may be measured.

Figures 9.35–9.38 show some of the alternative primary elements which can be used in conjunction with this force-balance mechanism to measure gauge differential and absolute (high and low) pressures.

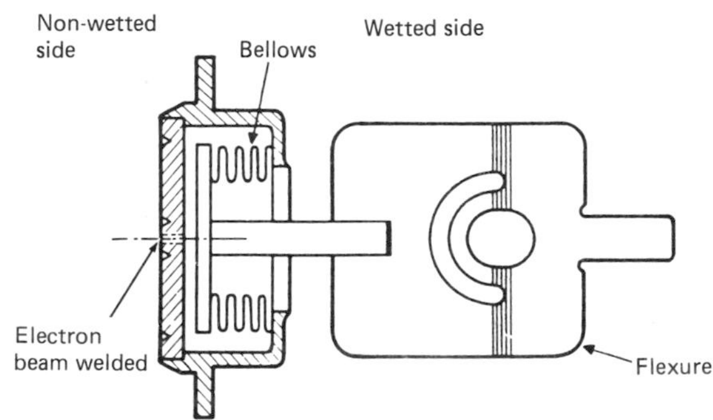

Figure 9.36 Bellows type primary element for absolute pressure measurements. Courtesy, The Foxboro Company.

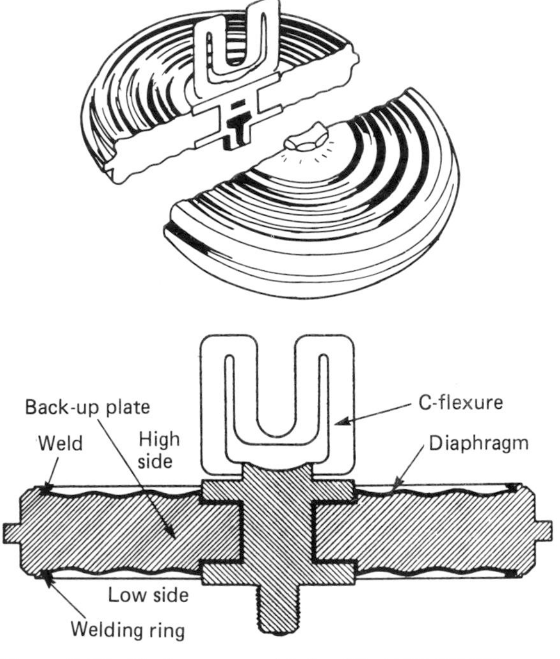

Figure 9.37 Diaphragm assembly for differential pressure measurements. Courtesy, The Foxboro Company.

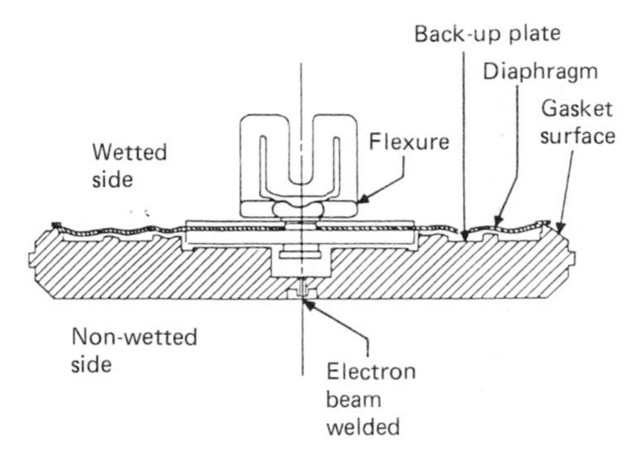

Figure 9.38 Diaphragm assembly for low absolute pressure measurements. Courtesy, The Foxboro Company.

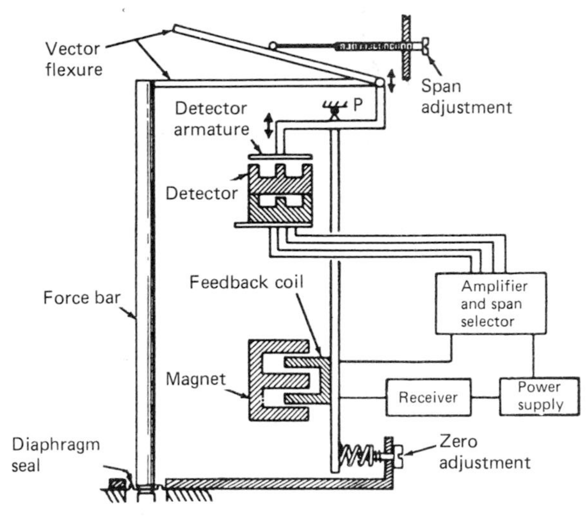

Figure 9.39 Arrangement of electronic force-balance mechanism. Courtesy, The Foxboro Company.

9.3.3 Electronic force-balance pressure transmitters

As with the pneumatic force-balance transmitters, there are many designs of electronic force-balance transmitters, but the Foxboro series serves to illustrate the principal mechanism. Again the mechanism shown in Figure 9.39 is used for a series of transmitters. Its basic function is to measure the force derived by a primary element as a result of applying the pressure to be measured.

The force is applied via a flexure to the lower end of the force bar which pivots on the diaphragm seal that in some instruments also serves to isolate the process fluid from the measuring system. The corresponding force at the top of the force bar is transmitted to the vector assembly, the effective angle of which can be varied by the span adjustment. A secondary lever system, attached at the apex of the vector assembly, is pivoted at the point P and carries both the detector armature and the coil of the feedback motor.

In operation, a change in the force applied to the force bar is applied to the vector assembly, causing a slight movement at the apex. This is transferred to the detector armature, a ferrite disc which moves with respect to the detector coils and so modifies the coupling between the two coils. These coils are connected in an oscillator circuit so that a slight movement of the armature with respect to the coils causes a substantial change in the oscillator output which after rectification is applied to the feedback motor in such a manner that the initial disturbing force is balanced.

The current flowing through the feedback coil is used as the output signal from the transmitter. Fine adjustments of the span are effected by varying the effective angle of the vector flexures. But larger changes are made by changing the effective gain of the oscillator/amplifier circuit shown in Figure 9.40.

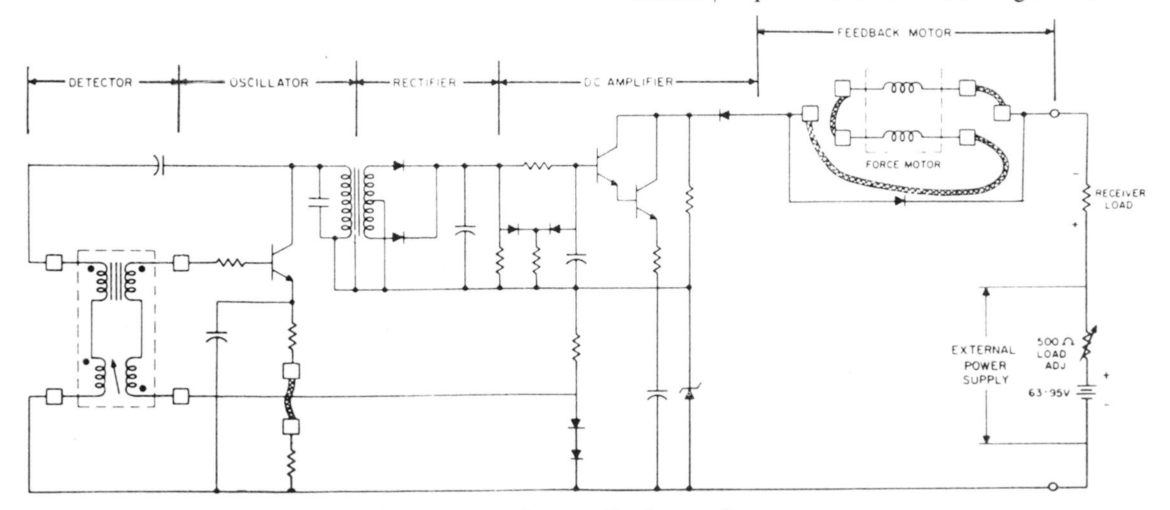

Figure 9.40 Circuit for electronic force-balance system. Courtesy, The Foxboro Company.

As with the pneumatic force-balance transmitters there is a wide range of both gauge and absolute pressure sensors that can be operated in conjunction with the force-balance mechanism as well as differential pressure sensors as shown in Figures 9.41–9.44. These provide absolute and gauge pressure measurements for ranges from 1.3 kPa to 85 MPa and differential pressures from 1 kPa to 14 MPa with an accuracy of 0.5 per cent of span.

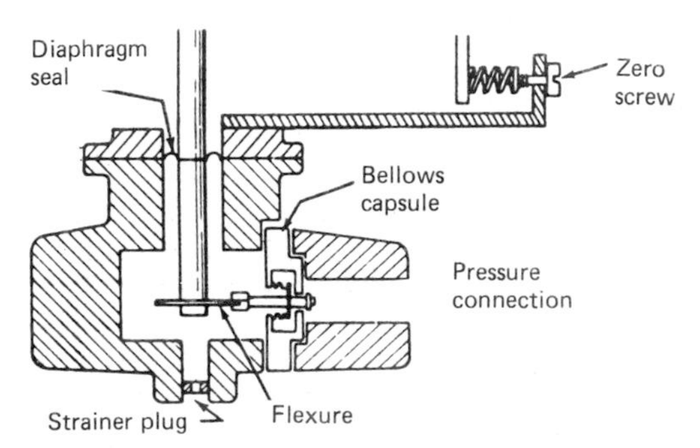

Figure 9.41 Bellows type primary element for gauge pressure measurements. Courtesy, The Foxboro Company.

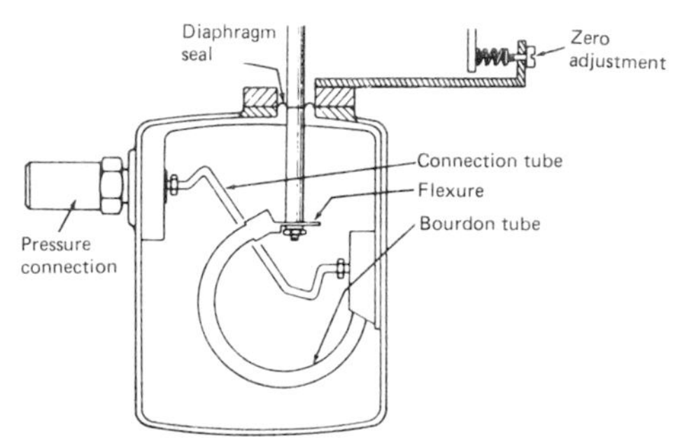

Figure 9.42 Bourdon type primary element. Courtesy, The Foxboro Company.

9.3.4 Force-measuring pressure transmitters

In addition to the force-balance pressure transmitters previously described, there are now transmitters which measure pressure by measuring the deflection of an elastic member resulting from the applied pressure. One of these is the Foxboro 820 series of transmitters, in which the force is applied to a pre-stressed wire located in the field of a permanent magnet. The wire is an integral part of an oscillator circuit which causes the wire to oscillate at its resonant (or natural) frequency. For an ideal system, the resonant frequency is a function of the length, the square root of tension and the mass of the wire.

The associated electronic circuits include the oscillator as well as the components to convert oscillator frequency into a standard transmission signal such as 4 to 20 mA d.c. As shown in Figure 9.45 the oscillator signal passes via a pulse-shaper to two frequency-converters arranged in cascade, each of which produces an output proportional to the product of the applied frequency and its input voltage so that the output of the second converter is proportional to the square of the frequency and therefore to the tension in the wire. The voltage is therefore directly proportional to the force produced by the primary element which in turn is proportional to the measured pressure.

The configuration of the resonant-wire system for primary elements such as a helical Bourdon tube, a gauge pressure, a differential pressure and absolute pressure is shown in Figures 9.46–9.49 respectively. Vibrating wires are also used as strain gauges, as discussed in Chapter 4.

A second category of pressure transmitters involves the measurement of the deflection of a sensing diaphragm which is arranged as the movable electrode between the fixed plates of a differential capacitor. An

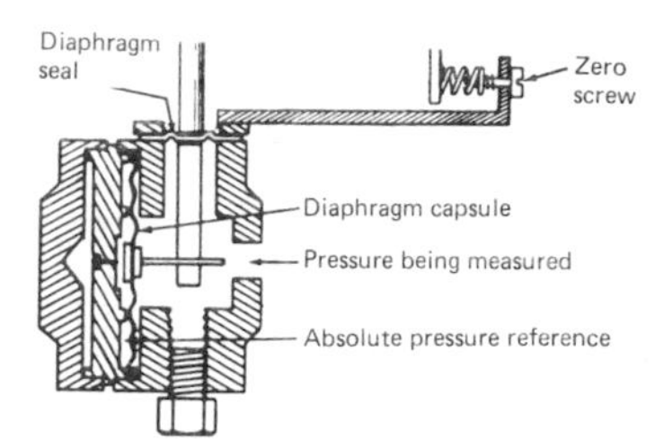

Figure 9.43 Diaphragm assembly for low absolute pressure measurements. Courtesy, The Foxboro Company.

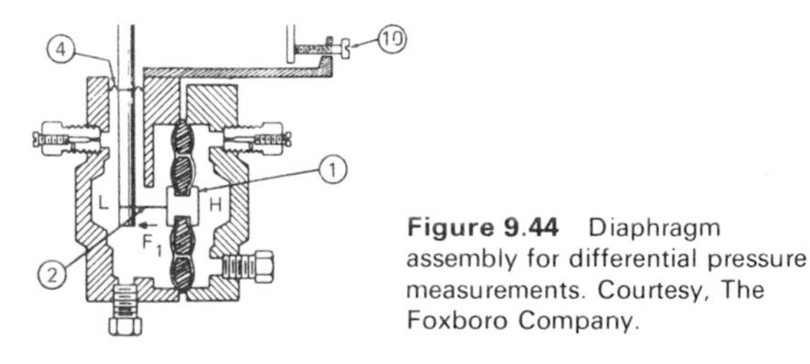

Figure 9.44 Diaphragm assembly for differential pressure measurements. Courtesy, The Foxboro Company.

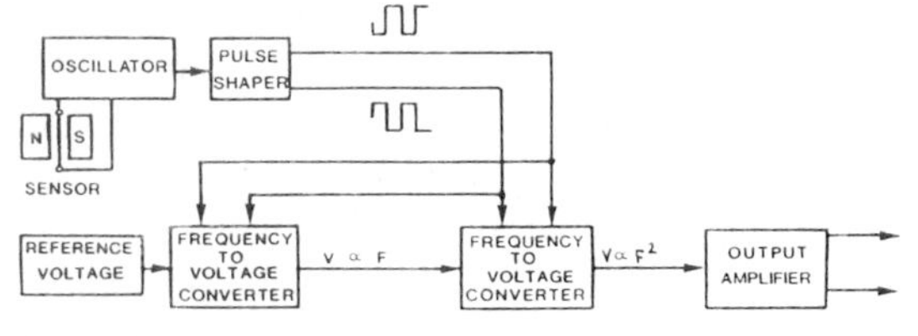

Figure 9.45 Functional diagram of electronic circuit for resonant wire pressure transmitter. Courtesy, The Foxboro Company.

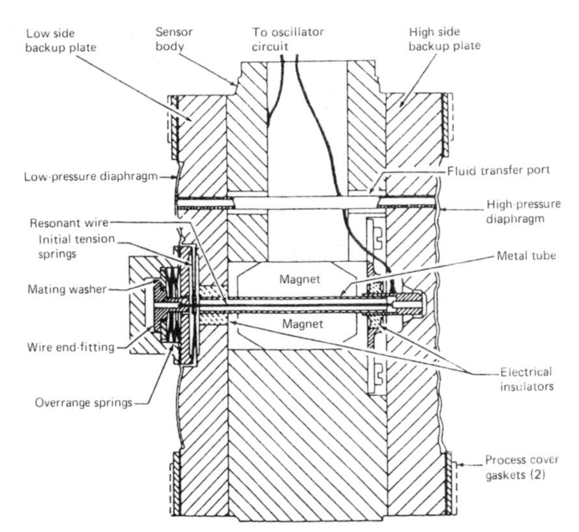

Figure 9.46 Arrangement of diaphragm assembly for differential pressure measurements. Courtesy, The Foxboro Company.

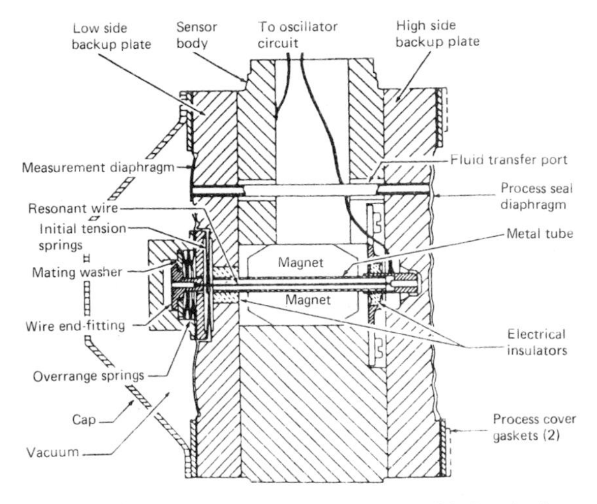

Figure 9.47 Arrangement of diaphragm assembly for absolute pressure measurements. Courtesy, The Foxboro Company.

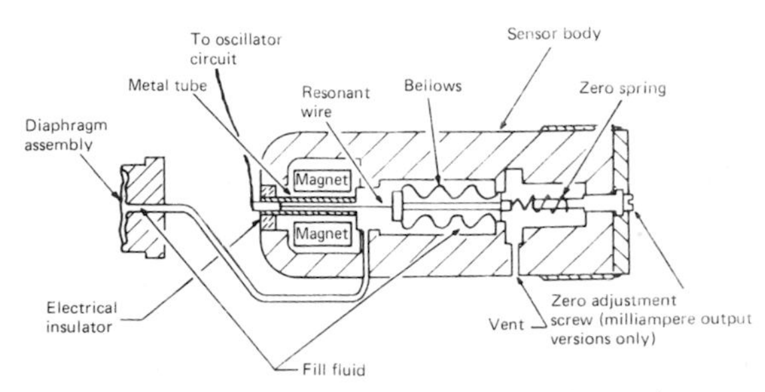

Figure 9.48 Arrangement of gauge pressure element for resonant wire sensor. Courtesy, The Foxboro Company.

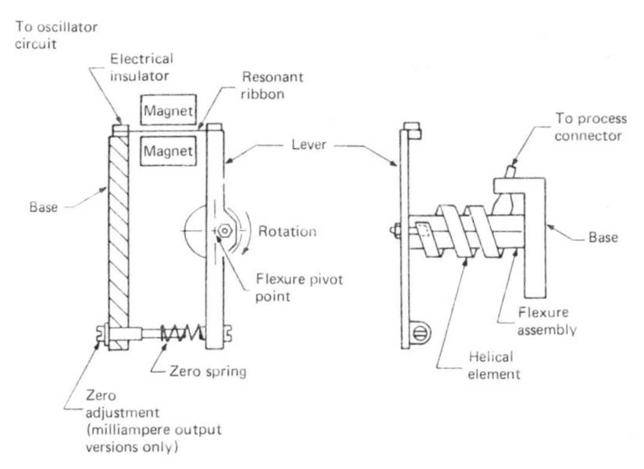

Figure 9.49 Arrangement of helical Bourdon tube for resonant wire sensor. Courtesy, The Foxboro Company.

example of this is the Siemens Teleperm K Transmitter. The arrangement of the measuring cell for differential pressures and absolute pressures is shown in Figure 9.50.

It is a flat cylindrical unit sealed at both ends by flexible corrugated diaphragms which provide the interface between the process fluid and the sensor. Under overload conditions, these seat on matching corrugations machined in the housing. The sensor comprises a hollow ceramic chamber divided into two by the sensing diaphragm. The interiors of the chambers are filled with liquid and sealed by a ring diaphragm. The interior walls of both chambers are metallized and form the fixed plates of the differential capacitor whilst the sensing diaphragm forms the movable plates of the capacitor.

When the measuring cell is subjected to a differential pressure, the sensing diaphragm is displaced slightly,

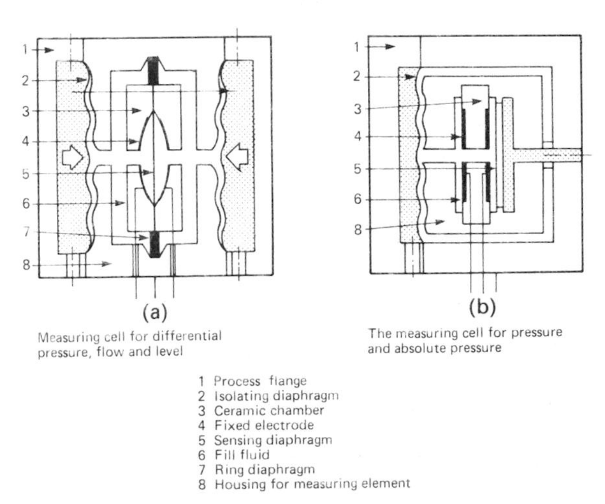

Figure 9.50 Arrangement of capacitance type differential and absolute pressure sensors. Courtesy, Siemens Ltd.

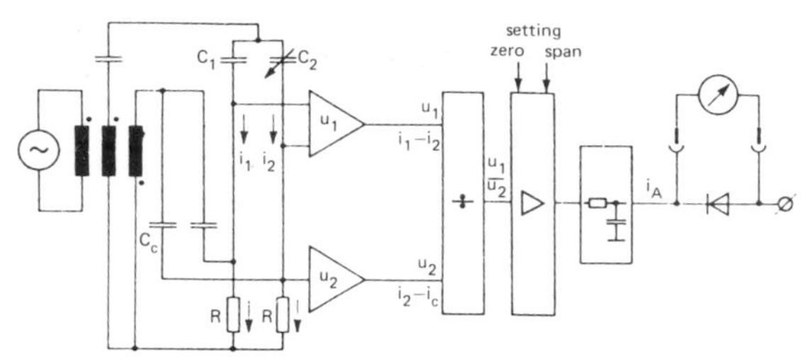

Figure 9.51 Functional diagram of electronic circuit for gauge and absolute pressure capacitance sensor. Courtesy, Siemens Ltd.

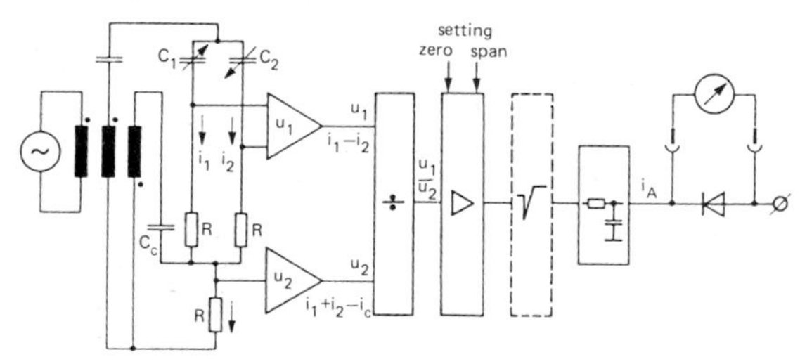

Figure 9.52 Functional diagram of electronic circuit for differential pressure capacitance sensor. Courtesy, Siemens Ltd.

causing a change in capacitance which is converted by the associated measuring circuit into a standard transmission signal such as 4 to 20 mA d.c.

For measuring absolute pressures, one side of the measuring cell is evacuated to provide the reference pressure and for measuring gauge pressures, one side is vented to atmosphere.

Under these conditions the stiffness of the isolating diaphragm determines the range. For high pressures the diaphragm is augmented by a spring.

Figure 9.51 shows the basic circuit for absolute pressure and gauge pressure transmitters whilst Figure 9.52 shows the corresponding details for the differential pressure transmitter which includes a square root extraction circuit.

In both models the sensing diaphragm acts as the moving electrode in the capacitor detector. The effective values of the capacitors are as follows

$$C_1 = \frac{A\varepsilon}{d_0} + C_S$$

$$C_2 = \frac{A\varepsilon}{d_0 + \Delta d} + C_S$$

where A is effective electrode area, ε is permittivity of the dielectric fluid, d_0 is effective distance between fixed electrodes and sensing electrode, Δd is displacement of sensing electrode, and C_S is stray capacitance. From this it follows that

$$\frac{C_1 - C_2}{C_2 - C_S} = \frac{\Delta d}{d_0}$$

which is the same as the deflection constant for the sensing diaphragm so that Δd is proportional to the differential pressure which can therefore be measured by an all-bridge network. To compensate for the effect of stray capacitances a capacitor C is included in the bridge circuit but supplied with a voltage in anti-phase to that applied to C_1 and C_2.

If the impedances of the capacitors C_1, C_2 and C_c are high compared with their associated resistors then the currents flowing through them are proportional to the capacitances, so that the output from amplifier U_1 is proportional to $(i_1 - i_2)$ and from amplifier U_2 $(i_2 - i_C)$. When these two signals are applied to a dividing stage, the resultant signal is proportional to the displacement of the sensing electrode and hence to the applied pressure. For the differential pressure transmitter, both C_1 and C_2 are variable but it can be shown that

$$\frac{\Delta d}{d_0} = \frac{C_1 - C_2}{C_1 + C_2 - 2C_S}$$

Applying the same conditions as before, this leads to

$$\frac{\Delta d}{d_0} = \frac{i_1 - i_2}{i_1 + i_2 - i_C}$$

As $(i_1 - i_2)$ and $(i_1 + i_2 - i_C)$ are proportional to the input signals of amplifier U_1 and U_2 it follows that the output from the dividing stage is proportional to the applied differential pressure.

Most differential pressure transmitters are used in conjunction with orifice plates to measure flow. If therefore the output from the dividing stage is applied to a square root extracting circuit then its output becomes proportional to flow rate.

9.4 References

An Introduction to Process Control, Pub. 105B, The Foxboro Company, (1986)

Hewson, *J. E., Process Instrumentation Manifolds: their Selection and Use.* Instrument Society of America, (1981)

Lyons, J. L., *The Designer's Handbook of Pressure-Sensing Devices.* Van Nostrand Reinhold, (1980)

Neubert, H. K. P., *Instrument Transducers, an Introduction to their Performance and Design.* Clarendon Press, 1975)

10 Measurement of vacuum

D. J. PACEY

10.1 Introduction

10.1.1 Systems of measurement

The term *vacuum* refers to the range of pressures below atmospheric, and the measurement of vacuum is thus the measurement of such pressures. Pressure is defined as force divided by area, and the SI unit is the newton/metre^2 (Nm^{-2}) or pascal (Pa). Pressure may also be stated in terms of the height of a column of a suitable liquid, such as mercury or water, that the pressure will support. The relation between pressure units currently in use is shown in Table 10.1.

In engineering, it has long been customary to take atmospheric pressure as the reference, and to express pressures below this as 'pounds per square inch of vacuum', or 'inches of vacuum' when using a specified liquid. The continual changes in atmospheric pressure, however, will lead to inaccuracy unless they are allowed for. It is preferable to use zero pressure as the reference, and to measure pressures above this. Pressures expressed in this way are called *absolute pressures*.

10.1.2 Methods of measurement

Since pressure is defined to be force/area, its measurement involves directly or indirectly the measurement of the force exerted upon a known area. A gauge which does this is called an *absolute gauge*, and allows the pressure to be obtained from a *reading* and known physical quantities associated with the gauge, such as areas, lengths, sometimes temperatures, elastic constants, etc. The pressure when obtained is independent of the composition of the gas or vapour which is present.

Many technological applications of vacuum use the long free paths, or low molecular incidence rates that vacuum makes available. These require pressures that are only a very small fraction of atmospheric, where the force exerted by the gas is too small to be measured, making absolute gauges unusable. In such cases non-absolute gauges are used which measure pressure indirectly by measuring a pressure-dependent physical property of the gas, such as thermal conductivity, ionizability, or viscosity. These gauges always require calibration against an absolute gauge, for each gas that is to be measured. Commercial gauges are usually calibrated by the manufacturer using dry air, and will give true readings only when dry air is present. In practice it is difficult to be certain of the composition of the gases in vacuum apparatus, thereby causing errors. This problem is overcome in the following way. When a gauge using variation of thermal conductivity indicates a pressure of 10^{-1} Pa, this would be recorded as an *equivalent dry air pressure of* 10^{-1} *Pa*. This means that the thermal conductivity of the unknown gases present in the vacuum apparatus has the same value as that of air at 10^{-1} Pa, and not that the pressure is 10^{-1} Pa.

10.1.3 Choice of non-absolute gauges

Since the gauge referred to above measures thermal conductivity, it is particularly useful for use on vacuum apparatus used for making vacuum flasks, or in which low-temperature experiments are carried out, and in which thermal conductivity plays an important part. Similarly an ionization gauge would be suitable in the case of apparatus used for making radio valves and cathode ray tubes in which the ionizability of the gases is important. In general, it is desirable to match as far

Table 10.1 Relation between pressure units

	N/m^2 (Pa)	*torr*	*mb*	*atm*
N/m^2 (Pa)	1	7.50×10^{-3}	10^{-2}	9.87×10^{-6}
torr	133.3	1	1.333	1.316×10^{-3}
mb	100	0.750	1	9.87×10^{-4}
atm	1.013×10^5	760	1.013×10^3	1

as possible, the physical processes in the gauge with those in the vacuum apparatus.

10.1.4 Accuracy of measurement

Having chosen a suitable gauge, it is necessary to ensure that the pressure in the gauge head is the same as that in the vacuum apparatus. Firstly the gauge head is connected at a point as close as possible to the point where the pressure is to be measured and by the shortest and widest tube available. Secondly, sufficient time must be allowed for pressure equilibrium to be obtained. This is particularly important when the pressure is below 10^{-1} Pa, and when ionization gauges, which interact strongly with the vacuum apparatus, are used. Times of several minutes are often required. When non-absolute gauges are used, even under ideal conditions, the accuracy is rarely better than ± 20 per cent, and in a carelessly operated ionization gauge worse than ± 50 per cent. Representative values for the mid-range accuracy of various gauges are given in Table 10.2 at the end of this chapter along with other useful information.

10.2 Absolute gauges

10.2.1 Mechanical gauges

These gauges measure the pressure of gases and vapours by making use of the mechanical deformation of tubes or diaphragms when exposed to a pressure difference. If one side of the sensitive element is exposed to a good vacuum, the gauge is absolute.

10.2.1.1 The Bourdon tube gauge

A conventional gauge of this type can be used to measure pressure down to 100 Pa if carefully made. Its construction is described in Chapter 9 on pressure measurement.

10.2.1.2 The quartz spiral gauge

This gauge measures differential pressures over a range of 100 Pa from any chosen reference pressure. It is suitable for use with corrosive gases or vapours.

Construction The sensitive element is a helix of 0.5 mm diameter quartz tubing usually 20 mm in diameter, 30 mm long to which the vacuum is applied internally. The helix coils and uncoils in response to pressure changes, the motion being measured by observing the movement of a light spot reflected from a small mirror attached to its lower end. The whole assembly is mounted in a clear glass or quartz enclosure, which can be brought to any desired reference pressure. If this is zero, the gauge indicates absolute pressure.

10.2.1.3 Diaphragm gauge

This gauge measures pressures of gases and vapours down to 10 Pa. Its construction is described in Chapter 9.

10.2.2 Liquid manometers

These gauges measure the pressure of gases and vapours from atmospheric to about 1 Pa by balancing the force exerted by the gas or vapour against the weight of a column of liquid, usually mercury, water or oil. These devices provide the simplest possible means of pressure measurement.

Construction The construction of various forms of liquid manometer is described in Chapter 9.

Operation For measuring relative pressures, the open manometer shown in Figure 10.1(a) may be used. In this case the difference h in levels may be taken to express the *vacuum* directly in inches of water, or by use of the formula $p = h\rho g$, where ρ is the density of the liquid, and g is the acceleration due to gravity, the

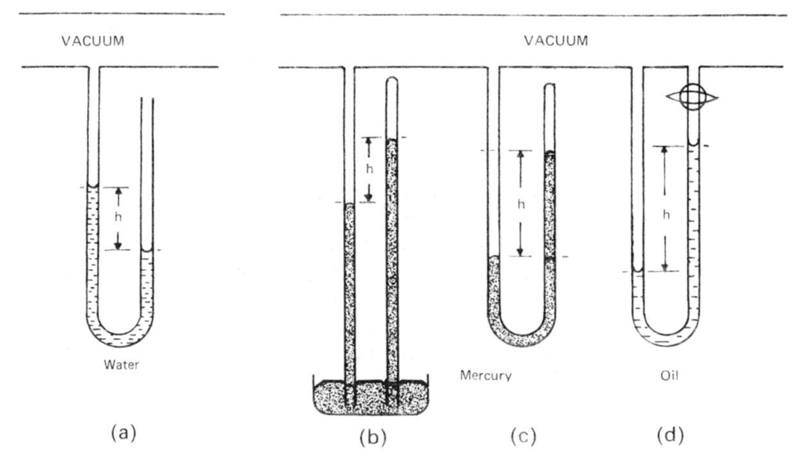

Figure 10.1 Liquid manometers.

vacuum may be expressed in SI units. The measurement of absolute pressures requires a vacuum reference which may be obtained in several ways as shown in Figure 10.1(b), (c), (d). A barometer tube is used in diagram (b), immersed in the same liquid pool, in this case usually mercury, as the manometer tube. A more compact form is provided by the closed manometer, shown in (c). This again uses mercury, and the space in the closed limb is evacuated. A useful version of the closed manometer which can be used with oil is shown in (d), where the tap may be opened when the apparatus is at zero pressure, and closed immediately before taking measurements. When oil is used in vacuum measurement, difficulty will be experienced with the liberation of dissolved gases, which must be removed slowly by slow reduction of the pressure.

10.2.3 The McLeod gauge (1878)

Function This gauge measures the pressure of gases only, from 5×10^{-4} Pa to atmospheric by measuring the force exerted by a sample of the gas of known volume after a known degree of compression.

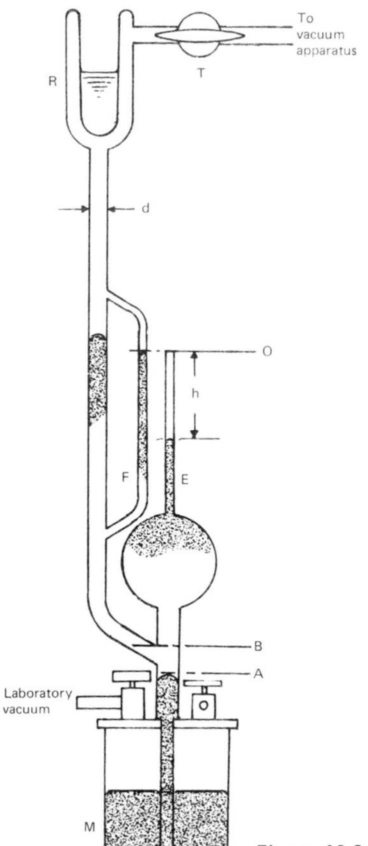

Figure 10.2 The McLeod gauge.

Construction This is of glass as shown in Figure 10.2 and uses mercury as the manometric liquid. The measuring capillary E and the reference capillary F are cut from the same length of tube to equalize capillary effects. A trap R refrigerated with liquid nitrogen or solid carbon dioxide excludes vapours, and prevents the escape of mercury vapour from the gauge. The tap T maintains the vacuum in the gauge when it is transferred to another system.

Operation The mercury normally stands at A, allowing the bulb and measuring capillary to attain the pressure p in the vacuum apparatus; at pressures below 10^{-2} Pa several minutes are required for this. To take a reading, the mercury is raised by slowly admitting air to the mercury reservoir M. When the mercury passes B, a sample of gas of volume V is isolated, and the pressure indicated will be that in the gauge at this instant. The mercury in the reference capillary is then brought to O, the level of the top of the measuring capillary and the length h of the enclosed gas column, of area a, is measured. The mercury is then returned to A, by reducing the pressure in the reservoir M, thus preparing the gauge for a further measurement.

Calculation of the pressure Applying Boyle's law to the isolated sample of gas, we have

original pressure × original volume
= final pressure × final volume

or

$$pV = (h\rho g + p)ah$$

where ρ is the density of mercury and g is the acceleration due to gravity.

Thus

$$p = \frac{ah^2}{V - ah}\rho g$$

When measuring low pressures, the final volume ah is very much less than the original volume v.

Hence

$$P = \frac{a}{V} . \rho g . h^2$$

Showing that $p \propto h^2$, giving a square law scale.

The value of a is found by weighing a pellet of mercury of known length inserted in the measuring capillary, and V is determined by weighing the quantity of distilled water that fills the bulb and measuring capillary from the level A, to its closed end. Both of these measurements are carried out by the manufacturer, who then calculates values of h corresponding to a series of known pressures, which can then be read directly on the scale.

The Ishii effect The use of a refrigerated trap with this gauge, which is necessary in almost every instance, leads to a serious underestimation of the pressure, which is greater for gases with large molecules. First noted by Gaede in 1915, and thoroughly investigated by Ishii and Nakayama in 1962 the effect arises from the movement of mercury vapour from the pool at A, which passes up the connecting tube, and then condenses in the refrigerated trap. Gas molecules encountered by the mercury vapour stream are carried along with it, and are removed from the bulb, producing a lowering of the pressure there. The effect is greater for large molecules since they provide large targets for the mercury vapour stream. The error may be reduced by reducing the mercury vapour flow, by cooling the mercury pool at A artificially, or by reducing the diameter d of the connecting tube. In the latter case the response time of the gauge to pressure changes will be lengthened.

Approximate errors for a gauge in which $d = 1$ cm, and the mercury temperature is 300 K are -4 per cent for helium, -25 per cent for nitrogen, and -40 per cent for xenon.

10.3 Non-absolute gauges

10.3.1 Thermal conductivity gauges

Function These gauges measure the pressure of gases and vapours from 1000 Pa to 10^{-1} Pa by making use of the changes in *thermal conductivity* which take place over this range. Separate calibration against an absolute gauge is required for each gas. Since the sensitive element used is an electrically heated wire, these gauges are known as *hot-wire gauges*.

10.3.1.1 Thermocouple gauge (Voege 1906)

Construction An electrically heated wire operating at a temperature of about 320 K is mounted inside a glass or metal envelope connected to the vacuum apparatus. A thermocouple attached to the centre of the wire enables its variation in temperature due to pressure changes to be observed. For simplicity the construction shown in Figure 10.3 may be used. Four parallel lead-through wires pass through one end of the envelope, and two noble metal thermocouple wires are fixed across diagonal pairs. The wires are welded at their intersection so that one dissimilar pair forms the hot wire, and the other, the thermocouple.

Operation A stabilized electrical supply S is connected to the hot wire, and adjusted to bring it to the required temperature. The resistance R connected in series with the millivoltmeter M is used to set the zero of the instrument. A rise in the pressure increases the heat loss from the wire, causing a fall of temperature. This results in a reduction of thermocouple output registered by M, which is scaled to read pressure of dry air.

10.3.1.2 The Pirani gauge (Pirani 1906)

Construction An electrically heated platinum or tungsten wire, operating at a temperature of 320 K is mounted along the axis of a glass or metal tube, which is connected to the vacuum apparatus. Changes in the pressure cause temperature changes in the wire, which are followed by using the corresponding changes in its electrical resistance. The wire temperature may also be affected by variations of room temperature and these are compensated by use of an identical dummy gauge head sealed off at a low pressure. The gauge and dummy heads are shown in Figures 10.4 and 10.5. For reasons of economy, the dummy head is often replaced by a bobbin of wire having the same resistance and temperature coefficient of resistance, mounted close to the gauge head.

Operation The gauge and dummy heads form adjacent arms of a Wheatstone bridge circuit as shown in Figure 10.6. This arrangement also compensates for the effects of temperature changes on the resistance of the leads connecting the gauge head to the control unit, thereby allowing remote indications of pressure. Resistances Y and Z form the other arms of the bridge,

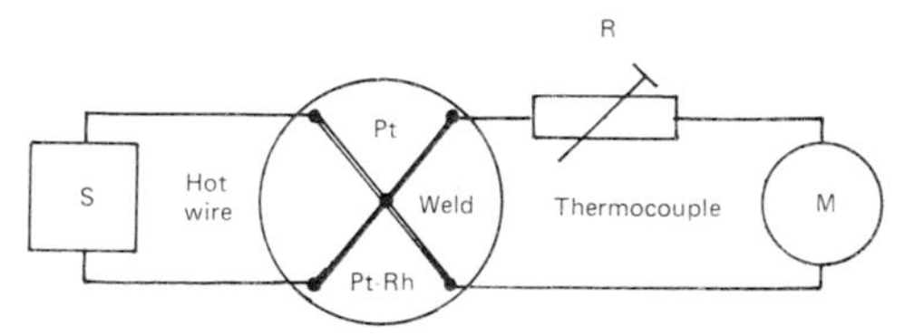

Figure 10.3 The thermocouple gauge.

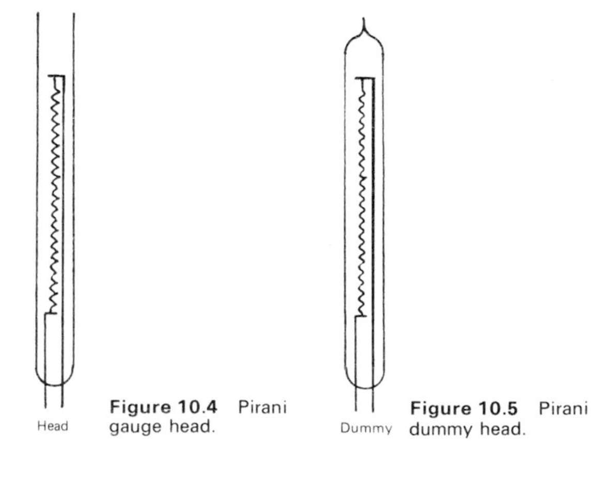

Figure 10.4 Pirani gauge head.

Figure 10.5 Pirani dummy head.

Z being variable for use in zero setting. Power is obtained from a stabilized supply S, and the meter M measures the out-of-balance current in the bridge.

The wire is brought to its operating temperature, and the bridge is balanced giving zero current through M, when the pressure is low. A rise in pressure causes an increase of heat loss and a fall in temperature of the wire. If the power input is kept constant, the wire temperature falls, causing a fall of resistance which produces a current through M, which is calibrated to read pressure of dry air. Alternatively, the fall in wire temperature may be opposed by increasing the input voltage so that the wire temperature remains constant. The input voltage then depends on the pressure, and the meter measuring the voltage can be scaled to read pressure. The constant balance of the bridge is maintained by a simple electronic circuit. This arrangement is effective in extending the high pressure sensitivity of the gauge to 10^4 Pa or higher, since the wire temperature is maintained at this end of the pressure range.

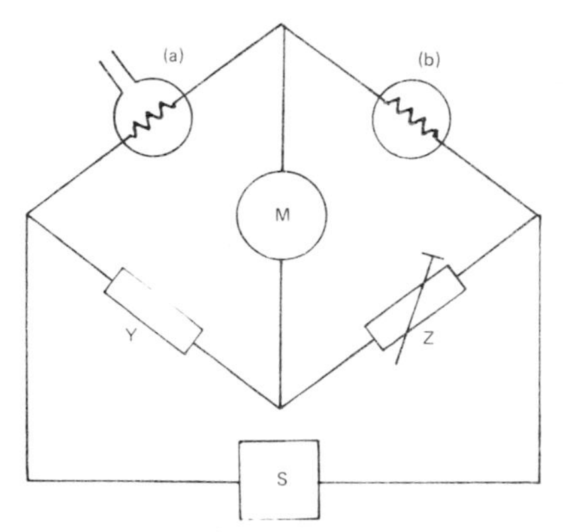

Figure 10.6 Pirani gauge circuit.

10.3.1.3 The thermistor gauge

This gauge closely resembles the Pirani gauge in its construction except that a small bead of semiconducting material takes the place of the metal wire as the sensitive element. The thermistor bead is made of a mixture of metallic oxides, about 0.2 mm in diameter. It is mounted on two platinum wires 0.02 mm in diameter, so that a current may be passed through it. Since the semiconductor has a much greater temperature coefficient of resistance than a metal, a greater sensitivity is obtained. Furthermore, on account of its small size it requires less power, allowing the use of batteries, and it responds more rapidly to sudden changes of pressure.

10.3.2 Ionization gauges

These gauges measure the pressure of gases and vapours over the range 10^3 Pa to 10^{-8} Pa by making use of the current carried by ions formed in the gas by the impact of electrons. In the cold-cathode gauges, the electrons are released from the cathode by the impact of ions, whilst in the hot-cathode gauges the electrons are emitted by a heated filament.

10.3.2.1 The discharge-tube gauge

Construction This is the simplest of the cold-cathode ionization gauges and operates over the range from 10^3 Pa to 10^{-1} Pa. The gauge head shown in Figure 10.7 consists of a glass tube about 15 cm long and 1 cm in diameter, connected to the vacuum apparatus. A flat or cylindrical metal electrode attached to a glass/metal seal is mounted at each end. Aluminium is preferable as it does not readily disintegrate to form metal films on the gauge walls during use. A stable power supply with an output of 2.0 kV at 2.0 mA is connected across the electrodes, in series with a resistor R of about 2 MΩ to limit the current, and a 1 mA meter M scaled to read the pressure.

Operation When the gauge is operating, several distinct luminous glows appear in the tube, the colours of which depend on the gases present. These glows are called the positive column P and negative glow N and result from ionization of the gas. The process is illustrated in Figure 10.8, where a positive ion striking the cathode C releases an electron. The electron is accelerated towards the anode, and after travelling some distance encounters a gas molecule and produces the ionization which forms the negative glow. Ions

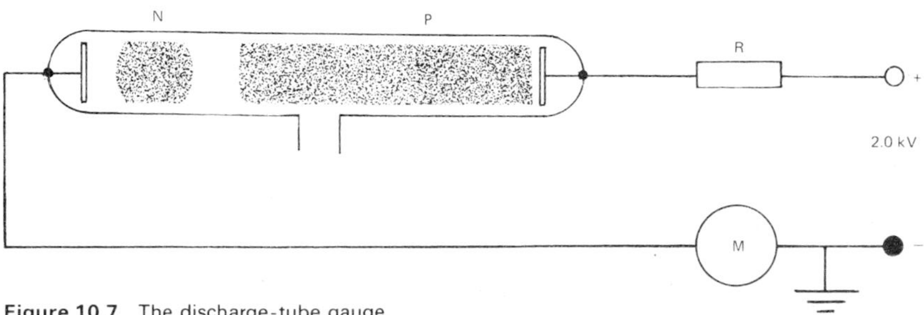

Figure 10.7 The discharge-tube gauge.

from the negative glow are attracted to the cathode where further electrons are emitted. This process, though continuous when established, requires some initial ions to start it. These may be formed by traces of radioactivity in the environment, though some delay may be experienced after switching on. The operating pressure range is determined by the electron path lengths. Above 10^3 Pa the motion of the electrons is impeded by the large number of gas molecules, whilst below about 10^{-1} Pa the electrons travel the whole length of the tube without meeting a gas molecule, and ionization ceases.

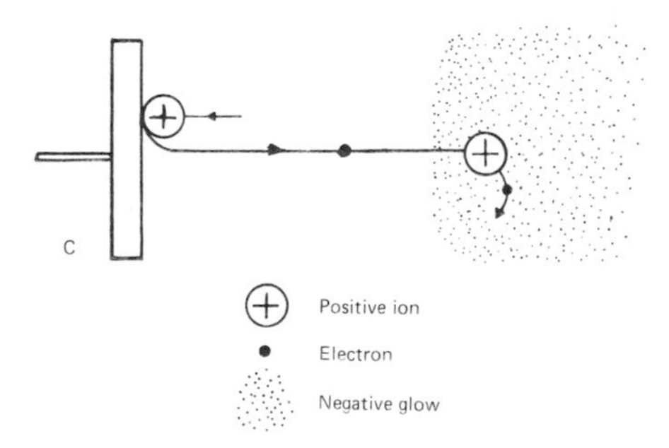

Figure 10.8 The production of carriers in the glow discharge.

10.3.2.2 The Penning ionization gauge (Penning 1937)

This cold-cathode gauge is sensitive, simple and robust, and therefore finds wide industrial application. It measures the pressure of gases and vapours over the range from 1 Pa to 10^{-5} Pa. This is shown in Figure 10.9.

Construction A glass or metal envelope connected to the vacuum apparatus houses two parallel cathode plates C of non-magnetic material separated by about 2 cm, and connected together electrically. Midway between these, and parallel to them is a wire anode ring A. Attached to the outside of the envelope is a permanent-magnet assembly which produces a transverse magnetic flux density B of about 0.03 T. This greatly increases the electron path length and enables a glow discharge to be maintained at low pressures. An alternative construction, of greater sensitivity, due to Klemperer (1947) is shown in Figure 10.10. This uses a non-magnetic cylindrical cathode C about 30 mm in diameter and 50 mm long, which may form the gauge envelope, along the axis of which is a stiff wire anode A, 1 mm in diameter. An axial magnetic flux density B of about 0.03 T is provided by a cylindrical permanent magnet or solenoid.

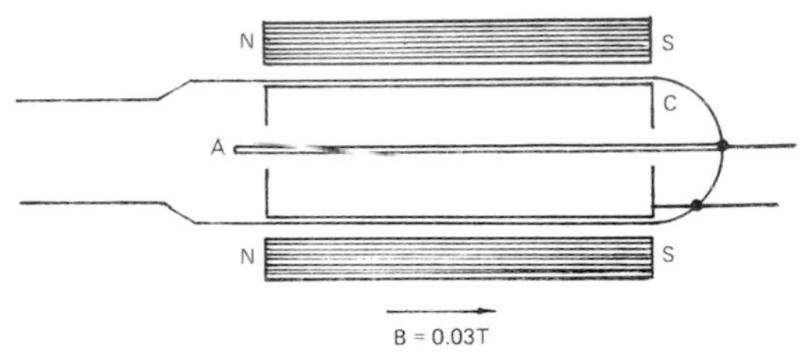

Figure 10.10 Cylindrical form of Penning gauge.

Operation A stable 2.0 kV power supply capable of supplying 2 mA is connected in series with a 1 mA meter M scaled to read pressure. The 2 MΩ ballast resistor R limits the current to 1 mA at the upper end of the pressure range. Electron paths in the gauge are shown in Figure 10.11. The electrode assembly, shown in section, is divided for purposes of explanation into two regions. On the right-hand side the magnetic field is imagined to be absent, and an electron from the cathode oscillates through the plane of the anode ring several times before collection, thereby increasing the electron path length. On the left-hand side, the presence of the magnetic flux causes a helical motion around the oscillatory path causing a still greater increase. The combined effect of these two processes is to bring about electron paths many metres in length, confined within a small volume. All gas discharges are subject to abrupt changes in form when the pressure varies. These mode changes lead to a sudden change in gauge current of 5 to 10 per cent as the pressure rises, which is not reversed until the pressure is reduced below the level at which it occurred. The effect, shown in Figure 10.12, is known as hysteresis, and causes ambiguity since a given pressure p is associated with two slightly different currents i_1 and i_2. By careful

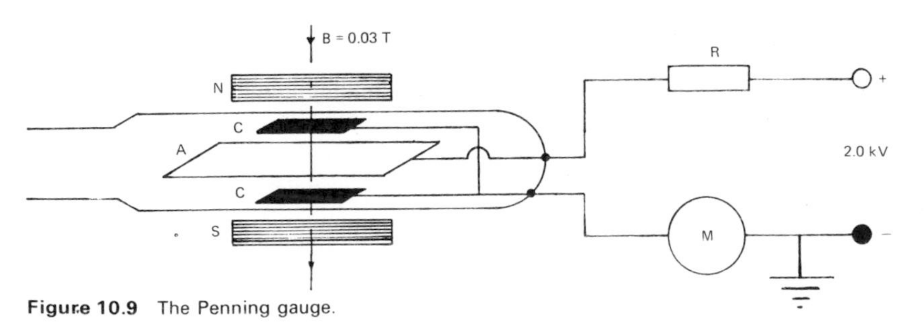

Figure 10.9 The Penning gauge.

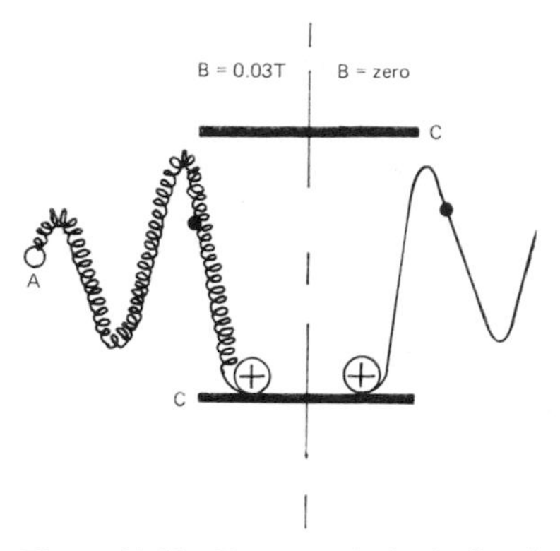

Figure 10.11 Electron paths in the Penning gauge.

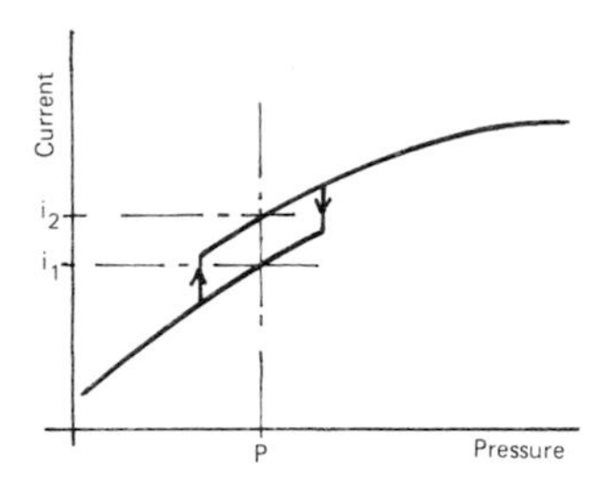

Figure 10.12 Hysteresis in the Penning gauge.

design, the effect may be minimized, and made to appear outside the operating range of the gauge.

10.3.2.3 The hot-cathode ionization gauge (Buckley 1916)

This is the most sensitive available gauge, and has the unique property that its reading is directly proportional to pressure over the range from 100 Pa to 10^{-8} Pa.

Construction The gauge head shown in Figure 10.13 is a special triode valve, usually with a hard glass envelope. Stainless steel may also be used, or a nude form of gauge in which the gauge electrodes are inserted directly into the vacuum vessel. The filament F, of heavy-gauge tungsten, operates at about 2000 K, and may be readily damaged by accidental inrushes of air. A filament of iridium coated with thorium oxide operating at a lower temperature is almost indestructible. Around the filament is a molybdenum or tungsten grid G also heavily constructed, and outside the grid is a cylindrical ion-collector C of nickel. Since the ion current received by this electrode is very small, special care is taken with its insulation.

Operation The gauge head is furnished with stable electrical supplies as shown in Figure 10.14. The filament is heated to produce electrons which are attracted to the grid where a fraction of them is immediately collected. The remainder oscillate several times through the grid wires before collection, forming ions by collision with the gas molecules. The electron current i_- is measured by M_1 and is usually between 0.1 mA and 5.0 mA. Ions formed between the grid and ion-collector constitute an ion current i_+ shown by M_2. Electrons are prevented from reaching the ion-collector by the application of a negative bias of 20 V. Ions formed between the filament and the grid are attracted by the filament where their impact etches its surface and shortens its life. This is particularly the case when the gauge is operated at pressures above 1 Pa, and if active gases such as oxygen are present.

Outgassing Since the gauge is highly sensitive, the gas molecules covering the electrodes and envelope must be removed by heating them to the highest safe temperature. For the filament, the required temperature can be obtained by increasing the filament

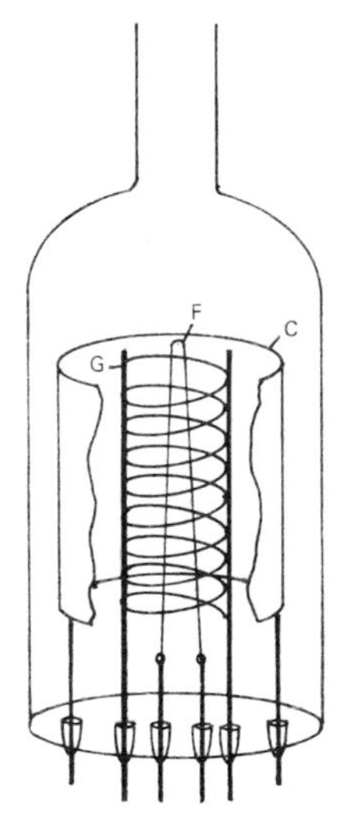

Figure 10.13 The hot-cathode ionization gauge.

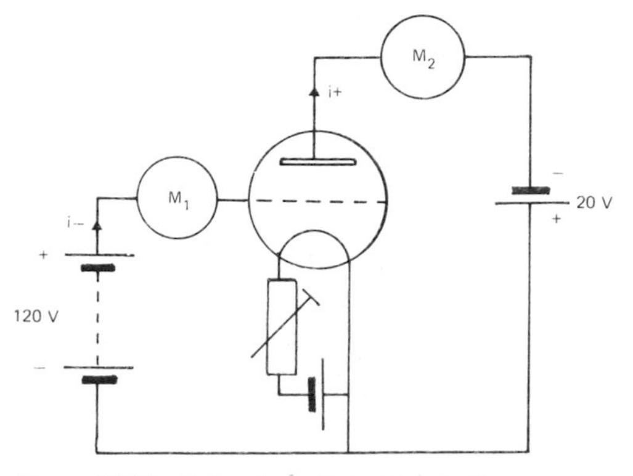

Figure 10.14 Hot-cathode ionization gauge.

current, whilst the grid and ion-collector may be heated by electron bombardment, using the filament as the source. Commercial gauge control units make provision for this treatment. The envelope is heated in an oven, or by means of a hot-air gun.

Pumping During operation, the gauge removes gas molecules from the vacuum apparatus, and thus behaves as a pump. Two processes are involved: at the filament one takes place in which molecules of active gases combine to form stable solid compounds; in the other, at the ion-collector, the positive ions embed themselves beneath its surface. The speed of pumping can be reduced by lowering the filament temperature and by reducing the rate of collection of ions.

Relationship between ion current and pressure If the pressure is p, the ion current i_+, and the electron current i_-, it is found for a given gas, say nitrogen, that

$$i_+ \propto p i_- \tag{10.1}$$

Therefore

$$i_+ = K p i_- \tag{10.2}$$

where K is a constant called the *gauge factor* for nitrogen. The SI unit is Pa^{-1}, and its value for an average gauge is $0.1\ \mathrm{Pa}^{-1}$. The value for a particular gauge is given by the maker. For other gases

$$i_+ = C K p i_- \tag{10.3}$$

where C is the relative sensitivity, with respect to nitrogen. Its approximate value for various gases is as follows:

Gas	He	H_2	N_2	Air	Ar	Xe	Organic vapours
C	0.16	0.25	1.00	1.02	1.10	3.50	>4.0

Equation (10.3) shows that for a given gas and value of i_-

$$i_+ \propto p$$

This valuable property of the gauge is obtained by stabilizing i_- by means of an electronic servo system which controls the filament temperature, and can switch the filament off if the gauge pressure becomes excessive.

10.3.2.4 *The Bayard–Alpert ionization gauge (1950)*

The soft X-ray effect The conventional ionization gauge described in Section 10.3.2.3 is not able to measure pressures below 10^{-6} Pa, due to the presence of a spurious current in the ion-collector circuit produced by processes occurring in the gauge, and independent of the presence of gas. This current is produced by soft X-rays generated when electrons stike the grid. The phenomenon is called the *soft X-ray effect*. The wavelength λ of this radiation is given by

$$\lambda = 1200/V\ \mathrm{nm} \tag{10.4}$$

where V is the grid potential. If this is 120 volts then $\lambda = 10$ nm. Radiation of this wavelength cannot escape through the gauge envelope, but when absorbed by the ion-collector causes the emission of photoelectrons which are collected by the grid. In the collector circuit the loss of an electron cannot be distinguished from the gain of a positive ion, so that the process results in a steady spurious ion current superimposed on the true ion current. The spurious current is about 10^{-10} A and is of the same order of magnitude as true ion current at 10^{-6} Pa. It is therefore difficult to measure pressures below this value with this type of gauge. A modified design due to Bayard and Alpert shown in Figure 10.15 enables the area of the ion-collector, and hence the spurious current, to be reduced by a factor of 10^3, thereby extending the range to 10^{-9} Pa.

Construction The filament F is mounted outside a cylindrical grid G, having a fine wire collector C of 0.1 mm diameter tungsten mounted along its axis. The electrodes are mounted in a glass envelope which has a transparent conducting coating W on its inner wall to stabilize the surface potential.

Operation The gauge is operated in the same manner as a conventional ionization gauge. Electrons from the filament oscillate through the grid volume and form ions there. These ions are pushed by the positive potential of the grid towards the ion-collector, which forms their only means of escape. Thus despite its small area, the electrode is an extremely good collector of ions.

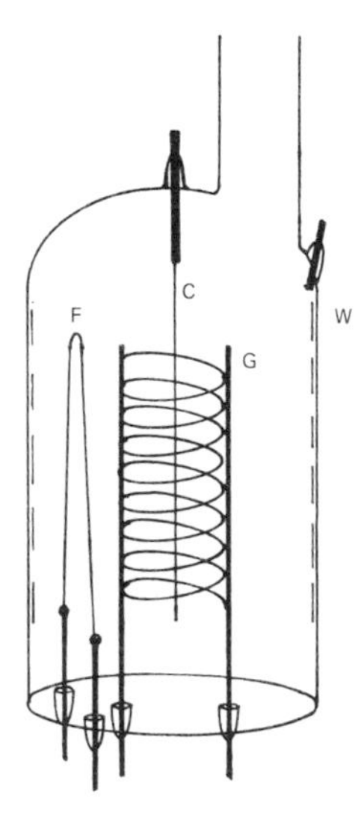

Figure 10.15 The Bayard–Alpert ionization gauge.

Table 10.2 Properties of gauges

Gauge	*Pressure range (Pa)*	*Accuracy ±%*	*Cost**	*Principal advantages*	*Principal limitations*
Bourdon tube	10^5–10^2	10	A	Simple. Robust.	Poor accuracy below 100 Pa.
Quartz spiral	10^5–10	10	B	Reads differential pressures.	Rather fragile.
Diaphragm	10^5–10	5	B	Good general-purpose gauge.	Zero setting varies.
Liquid manometers	10^5–10^2	5–10	A	Simple. Direct reading.	Vapour may contaminate vacuum.
McLeod	10^5–5×10^{-4}	5–10	C	Wide pressure range. Used for calibration.	Intermittent. Measures GAS pressures only.
Thermocouple	10^3–10^{-1}	20	B	Simple. Robust. Inexpensive.	Response not instantaneous.
Pirani	10^3–10^{-2}	10	C	Robust.	Zero variation due to filament contamination.
Thermistor	10^3–10^{-2}	10	C	Fast response. Low current consumption.	—
Discharge tube	10^3–1	20	B	Very simple. Robust.	Limited pressure range.
Penning	1–10^{-5}	10–20	C	Sensitive. Simple. Robust.	Large pumping effect. Hysteresis.
Hot-cathode ion	10^2–10^{-6}	10–30	D	Sensitive. Linear scale. Instantaneous response.	Filament easily damaged. Needs skilful use.
Bayard–Alpert	1–10^{-8}	10–30	D	As above, with better low-pressure performance.	As above.

* Scale of costs (£): A 0–50; B 50–200; C 200–400; D 400–600.

10.4 References

Carpenter, L. G., *Vacuum Technology*, Hilger, (1970)
Leck, J. H., *Pressure Measurement in Vacuum Systems*, Institute of Physics, (1964)
Pirani, M. and Yarwood, J., *Principles of Vacuum Engineering*, Chapman & Hall, (1961)
Ward, L. and Bunn, J. P., *Introduction to the Theory and Practice of High Vacuum Technology*, Butterworth, (1967)

11 Particle sizing

W. L. SNOWSILL

11.1 Introduction

The size of particles is an extremely important factor in their behaviour. To name but a few examples, it affects their chemical reactivity, their optical properties, their performance in a gas stream, and the electrical charge they can acquire. The methods used for assessing size are often based on one or more of these effects.

Particulate technology is a complex subject, and the major factor in this complexity is the variety of the physical and chemical properties of the particles. What appears to the naked eye as a simple grey powder can be a fascinating variety of shapes, colours and sizes when viewed under a microscope. Particles can be solid or hollow, or filled with gas. The surface structure, porosity, specific gravity etc. can have a profound effect on their behaviour. Their ability to absorb moisture or to react with other chemicals in the environment or with each other can make handling very difficult as well as actually affecting the size of the particles. The size analyst has to combat the problem of particles adhering to each other because of chemical reactions, mechanical bonding or electrostatic charging, and the problem increases as the size decreases. At the same time he must be aware that the forces applied to keep them separate may be enough with friable particles to break them.

Sampling is a crucial factor when measurements are made on particles. The essential points are:

(a) To be of any value at all, the sample must be representative of the source.
(b) Steps must be taken to avoid the sample changing its character before or during analysis.
(c) Particulate material when poured, vibrated or moved in any way tends to segregate itself. The coarser particles tend to flow down the outside of heaps, rise to the top of any vibrating regime, and be thrown to the outside when leaving a belt feeder. These factors need to be given careful consideration especially when attempting to subdivide samples.

11.2 Characterization of particles

Most particles are not regularly shaped so that it is not possible to describe the size uniquely. To overcome this problem, the standard procedure is to use the diameter of equivalent spheres. However, an irregularly shaped particle can have an almost limitless number of different equivalent spheres depending on the particular parameter chosen for equivalence.

For example, the diameter of a sphere with an equivalent volume would be different from that with an equivalent surface area.

Consider a cubic particle with edge of length x. The diameter of an equivalent volume sphere would be $x(6/\pi)^{1/3}$, i.e. $1.24x$. The diameter of an equivalent surface area sphere would be $x(6/\pi)^{1/2}$, i.e. $1.38x$. The chosen equivalent is usually related to the method of analysis. It is sensible to select the method of analysis to suit the purpose of the measurement but in some cases this is complicated by practical and economic considerations.

Sometimes, the equivalent diameter is not particularly relevant to the process, whereas the actual measurement made is relevant. In such cases, the size is sometimes quoted in terms of the parameter measured. A good example of this is terminal velocity (see Section 11.3). If, for example, information is required to assess the aerodynamic effect of a gas stream on particles, terminal velocity is more relevant than particle size. Even if the particles are spherical, conversion can be complicated by the possible variations in particle density. The term 'vel' is sometimes used to denote particle size. A 1 vel particle has a free falling speed of $10\,\mathrm{mm\,s^{-1}}$ in still dry air at s.t.p.

It is important when equivalent diameters are quoted that the basis (equivalent mass, volume, surface area, projected area, etc.) is clearly stated.

11.2.1 Statistical mean diameters

Microscopic examination of an irregularly shaped particle suggests other methods of assessing the mean

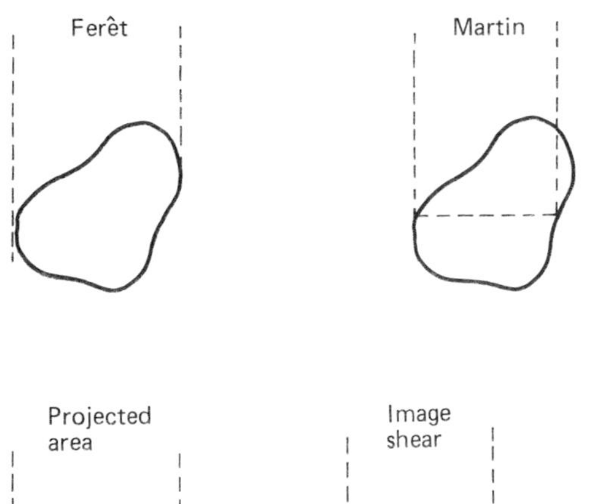

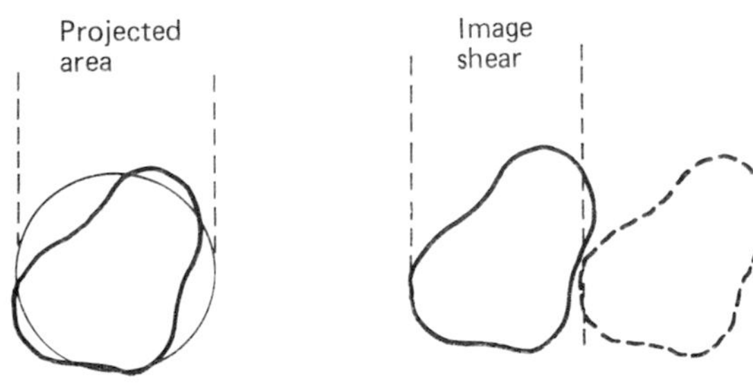

Figure 11.1 Statistical diameters.

diameter. Consider a large number of identical particles, 'truly' randomly orientated on a microscope slide. The mean of a given measurement made on each of the particles but in the same direction (relative to the microscope) would yield a statistical mean diameter. The following is a series of statistical mean diameters that have been proposed (see Figure 11.1):

(a) Ferêt's diameter: the mean of the overall width of a particle measured in all directions.
(b) Martin's diameter: the mean of the length of a chord bisecting the projected area of the particle measured in all directions.
(c) Projected area diameter: the mean of the diameters of circles having the same area as the particle viewed in all directions.
(d) Image shear diameter: the mean of the distances that the image of a particle needs to be moved so that it does not overlap the original outline of the particle, measured in all directions.

In microscopy, because particles tend to lie in a stable position on the slide, measurements as above of a group of particles would not be 'truly' randomly orientated. In these circumstances, the above diameters are 'two-dimensional statistical mean diameters'.

11.3 Terminal velocity

The terminal velocity of a particle is that velocity resulting from the action of accelerating and drag forces. Most commonly it is the free falling speed of a particle in still air under the action of gravity. The relationship between the terminal velocity of a particle and its diameter depends on whether the flow local to the particle is laminar or turbulent. In laminar flow, the particle falls more quickly than in turbulent flow where particles tend to align themselves for maximum drag and the drag is increased by eddies in the wake of the particles.

The general equation for the drag force F on a particle is:

$$F = Kd^n V^n \eta^{2-n} \rho_0^{n-1}$$

where K is the drag coefficient depending on shape, surface, etc., d is the particle dimension (diameter of a sphere), V is relative velocity, η is fluid viscosity, ρ_0 is fluid density, and n varies from 1 for laminar flow to 2 for turbulent flow.

For some regularly shaped particles K can be calculated. For example for a sphere in laminar flow, $K = 3\pi$.

Hence from the above, we find for laminar flow spheres

$$F = 3\pi d V \eta$$

and this is known as Stokes's law.

By equating drag force and gravitational force we can show

$$3\pi d V_T \eta = \frac{\pi d^3}{6}(\rho - \rho_0)g$$

where ρ is particle density, and V_T is terminal velocity. Thus,

$$V_T = (\rho - \rho_0)\frac{gd^2}{18\eta}$$

If the terminal velocity of irregularly shaped particles is measured together with ρ and η, the value obtained for d is the Stokes diameter. Sometimes the term 'aerodynamic' diameter is used, denoting an equivalent Stokes sphere with unit density. Stokes diameters measured with spheres are found to be accurate (errors <2 per cent) if the Reynolds number

$$Re = \frac{\rho_0 V d}{\eta}$$

is less than 0.2. At higher values of Re, the calculated diameters are too small. As Re increases, n increases progressively to 2 for $Re > 1000$ when the motion is fully turbulent and according to Newton the value of K for spheres reduces to $\pi/16$.

For very small particles where the size approaches the mean free path of the fluid molecules (~ 0.1 μm for dry air at s.t.p.) the drag force is less than that predicted by Stokes. Cunningham devised a correction for Stokes's equation:

$$F=3\pi dV\eta\frac{1}{1+(b\lambda/d)}$$

where λ is the mean free path and b depends on the fluid (e.g. for air at s.t.p. dry, $b\simeq 1.7$).

11.4 Optical effects caused by particles

When light passes through a suspension of particles, some is absorbed, some scattered and a proportion is unaffected, the relative proportions depending on the particle size, the wavelength of the light and the refractive indices of the media. The molecules of the fluid also scatter light.

Some optical size-analysis methods infer size from measurements of the transmitted, i.e. unaffected light, others measure the scattered light. Some operate on suspensions of particles, others on individual particles.

The theory of light-scattering by particles is complicated. Rayleigh's treatment which applies only to particles whose diameter $d \ll \lambda$ (the wavelength) shows that the intensity of scattered light is proportional to d^6/λ^4. It also shows that the scattering intensity varies with the observation angle and this also depends on d. As size d approaches λ however, the more rigorous treatment of Mie indicates that the scattering intensity becomes proportional to d^2, i.e. particle cross-sectional area, but that the effective area is different from the geometrical area by a factor K, known as the scattering coefficient or efficiency factor, which incorporates d, λ and the refractive index. Where $d \ll \lambda$ the two theories are similar. In the region around $d=\lambda$, however, K oscillates (typically between about 1.5 and 5) tending towards a mean of 2. Beyond about $d=5\lambda$, the value of K becomes virtually 2, i.e. the effective cross-sectional area of a particle is twice the geometrical area. As d/λ increases the preferred scattering angle reduces and becomes more distinct and forward scattering predominates (diffraction).

If the light is not monochromatic, the oscillation of K is smoothed to a mean of about 2.

The ratio of the intensity of the transmitted light I_T to the incident light I_0 is given by the Lambert–Beer law

$$\frac{I_T}{I_0}=\exp\left(-K\frac{a}{A}\right)$$

where a is the total projected area of the particles in the light beam, A is the area of the beam and again K is the scattering coefficient. This is often simplified to

$$\text{optical density } D=\log_{10} I_0/I_T = 0.4343K(a/A)$$

The scattering coefficient is sometimes called the 'particle extinction coefficient'. This should not be confused with extinction coefficient ξ. If the transmission intensity of a beam of light changes from I_0 to I_t in a path length L

$$I_T/I_0=\xi L$$

where K is contained within ξ.

Extinction $=\ln I_0/I_t$ is the Napierian equivalent of optical density.

Although the value of K has been shown to be virtually 2, the scattering angle for larger particles ($\sim 30\ \mu$m) is small and about half the light is forward-scattered. It follows that depending on the observation distance and the size of the sensor, much of the forward-scattered light could be received and the effective value of K in the above expression could be as low as 1. It will be apparent that the effect of a distribution of particles on light transmission is not a simple function of the projected area.

Bearing in mind the above limitations on K, it is possible to estimate the transmitted light intensity through a distribution of particles by summing the area concentrations within size bands. In each band of mean diameter d, the effective area a/A is $1.5KcL/\rho d$, where c is the mass concentration, L is the optical path length, and ρ is the particle density.

11.5 Particle shape

Although we can attribute to a particle an equivalent diameter, for example a Ferêt diameter d_F, this does not uniquely define the particle. Two particles with the same Ferêt diameter can have a very different shape. We can say that the volume of an equivalent sphere is

$$\frac{\pi}{6}(d_F)^3$$

but we must recognize that the actual volume V is probably very different. Heywood has proposed the use of shape coefficients. We can assign a coefficient $\alpha_{V,F}$ to a particle such that

$$V=\alpha_{V,F}(d_F)^3$$

Thus, if we use another method of size analysis which in fact measures particle volume V, knowing d_F we can calculate $\alpha_{V,F}$. Similarly, by measuring particle surface area S, we can assign a coefficient $\alpha_{S,F}$ so that

$$S=\alpha_{S,F}(d_F)^2$$

$\alpha_{V,F}$ is called the volume shape coefficient (based on Ferêt diameter) and $\alpha_{S,F}$ is called the surface shape coefficient (based on Ferêt diameter).

Clearly, there are other shape coefficients, and they can be associated with other diameters.

The ratio $\alpha_S/\alpha_V=\alpha_{S,V}$ is called the surface volume shape coefficient.

The subject is covered by BS 4359 (1970) Pt. III which includes definitions and tables of the various coefficients for a number of regular shapes: cubes, ellipsoids, tetrahedra, etc., and a number of commonly occurring particles. The coefficients α_S, α_V and $\alpha_{S,V}$ together provide a very good indication of particle shape, in a quantified form.

11.6 Methods for characterizing a group of particles

We have already established a number of alternative 'diameters' to be used to characterize particles. There are also several ways of characterizing groups of particles. They are all assessments of the quantities of particles within 'diameter' bands, but the quantities can be numbers of particles, mass of particles, volume, surface area, etc. As with particle equivalent diameters, it is important that the basis of the analysis is made clear.

There are also several methods for expressing the results of a size analysis. Perhaps the most obvious is tabulation and a contrived example of this is given in Table 11.1 which shows the masses of particles

Table 11.1 Alternative methods of tabulating the same size analysis

(a)

Size band (μm)	*% mass in band*
0–5	0.1
5–10	2.4
10–15	7.5
15–20	50.0
20–25	27.0
25–30	12.5
>30	0.5

(b)

Stated size	*% less than stated size*
5	0.1
10	2.5
15	10.0
20	60.0
25	87.0
30	99.5

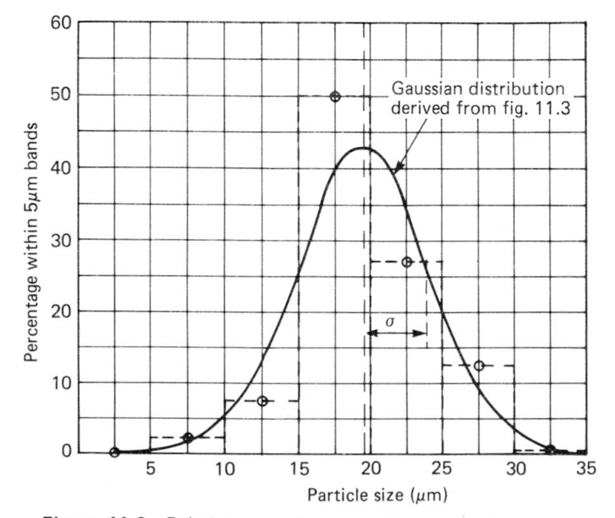

Figure 11.2 Relative percentage mass–frequency plot.

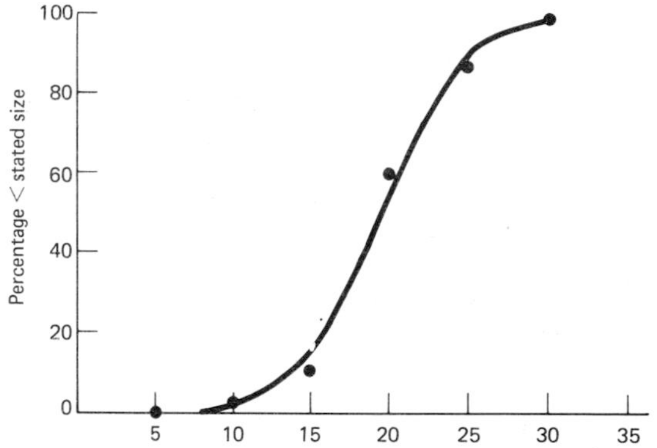

Figure 11.3 Cumulative percentage mass–frequency plot using linear scales.

contained within 5 μm size fractions from 0 to 40 μm. The main disadvantage is that it requires considerable experience to recognize what could be important differences between samples. Such differences are much more readily apparent if the results are plotted graphically. One method is to plot the quantity obtained, be it mass, volume, surface area or number of particles in each size fraction against size, both on a linear scale. This is called a relative frequency plot and Table 11.1 has been transferred in this way to Figure 11.2.

Students with a little understanding of statistics will be tempted to compare this with a normal or Gaussian distribution (also shown). In practice, Gaussian distributions are not very common with powder samples but this simple example is useful to illustrate a principle.

11.6.1 Gaussian or normal distributions

The equation for a Gaussian distribution curve is

$$y = \frac{1}{\sigma\sqrt{2\pi}} \exp - \left[\frac{(x-\bar{x})^2}{2\sigma^2}\right]$$

where $\int y\,dx$, the area under the curve, represents the total quantity of sample (again number, mass, volume, etc.) and is made equal to 1. The symbol $\bar{x}$ represents the arithmetic mean of the distribution and σ the standard deviation of the distribution is a measure of the spread. These two parameters uniquely define a Gaussian distribution. It can be shown that 68.26 per cent of the total area under the curve is contained between the boundaries $x = \bar{x} \pm \sigma$. In this case we have plotted the values of $y\,\delta x$ for equal 5 μm increments. We could just as easily have drawn a histogram. At this point it should be stated that if any one of the distributions should turn out to be 'normal' or Gaussian, then none of the other plots, i.e. number, volume, surface area distributions, will be Gaussian.

An advantage of the above presentation is that small

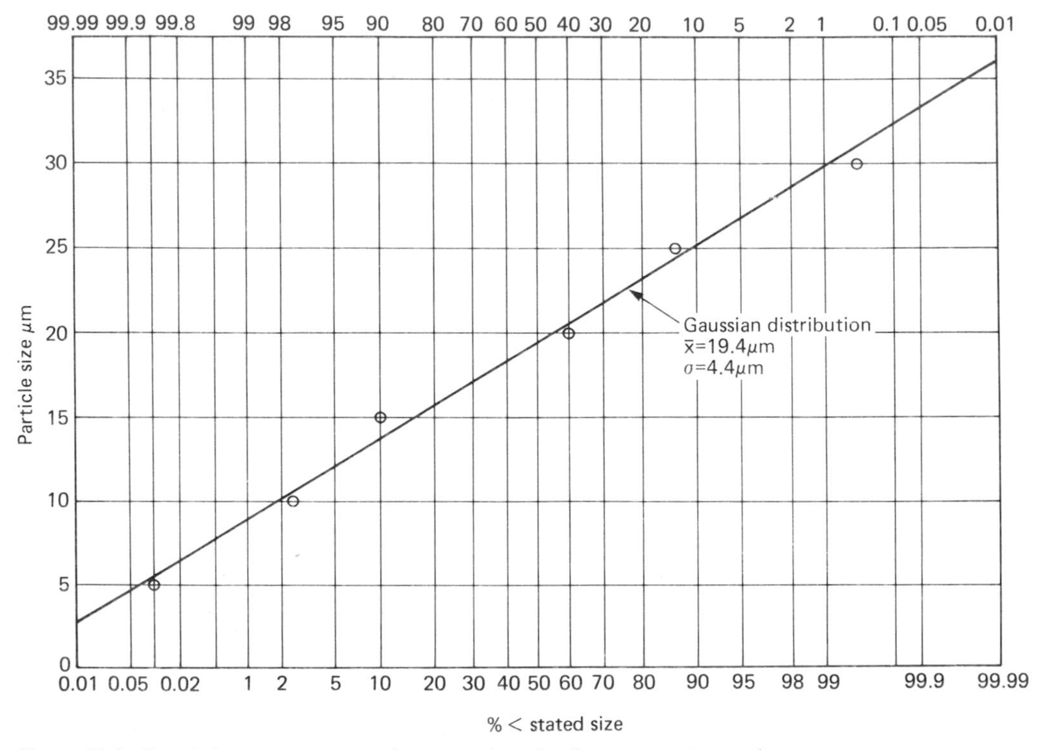

Figure 11.4 Cumulative percentage mass-frequency plot using linear × percentage scales.

differences between samples would be readily apparent. However, it would be useful to be able to measure easily the values of x and σ and this is not the case with the above. Two alternatives are possible. One is to plot a cumulative percentage frequency diagram, again on linear axes, as in Figure 11.3. In this case one plots the percentage less (or greater) than given sizes. Alternatively, one can plot the same information on linear-probability paper where one axis, the percentage frequency axis, is designed so that a Gaussian distribution will give a straight line, as in Figure 11.4. In a non-exact science such as size analysis, the latter has distinct advantages, but in either case the arithmetic mean $\bar{x}$ is the value of x at the 50 per cent point and the value of σ can be deduced as follows. Since 68.26 per cent of a normal distribution is contained between the values $x=\bar{x}+\sigma$ and $x=\bar{x}-\sigma$, it follows that

$$\sigma = x_{84\%} - \bar{x} = \bar{x} - x_{16\%}$$
$$= \tfrac{1}{2}(x_{84\%} - x_{16\%})$$

because $x_{84\%} - x_{16\%}$ covers the range of 68 per cent of the total quantity.

The closeness of fit to a Gaussian distribution is much more obvious in Figure 11.4 than in Figures 11.2 and 11.3. With probability paper, small differences or errors at either extreme produce an exaggerated effect on the shape of the line. This paper can still be used when the distribution is not 'normal' but in this case, the line will not be straight and standard deviation is no longer meaningful. If the distribution is not 'normal' the 50 per cent size is not the arithmetic mean but is termed the median size. The arithmetic mean needs to be calculated from

$$\bar{x} = \sum (\text{percentage in size fraction} \times \text{mean of size fraction})/100$$

and the basis on which it is calculated (mass, surface area, volume or particle number) has to be stated. Each will give a different mean and median value.

11.6.2 Log-normal distributions

It is unusual for powders to occur as Gaussian distributions. A plot as in Figure 11.2 would typically be skewed towards the smaller particle sizes. Experience has shown, however, that powder distributions often tend to be log-normal. Thus a percentage frequency plot with a logarithmic axis for the particle size reproduces a close approximation to a symmetrical curve and a cumulative percentage plot on log-probability paper often approximates to a straight line, Figure 11.5.

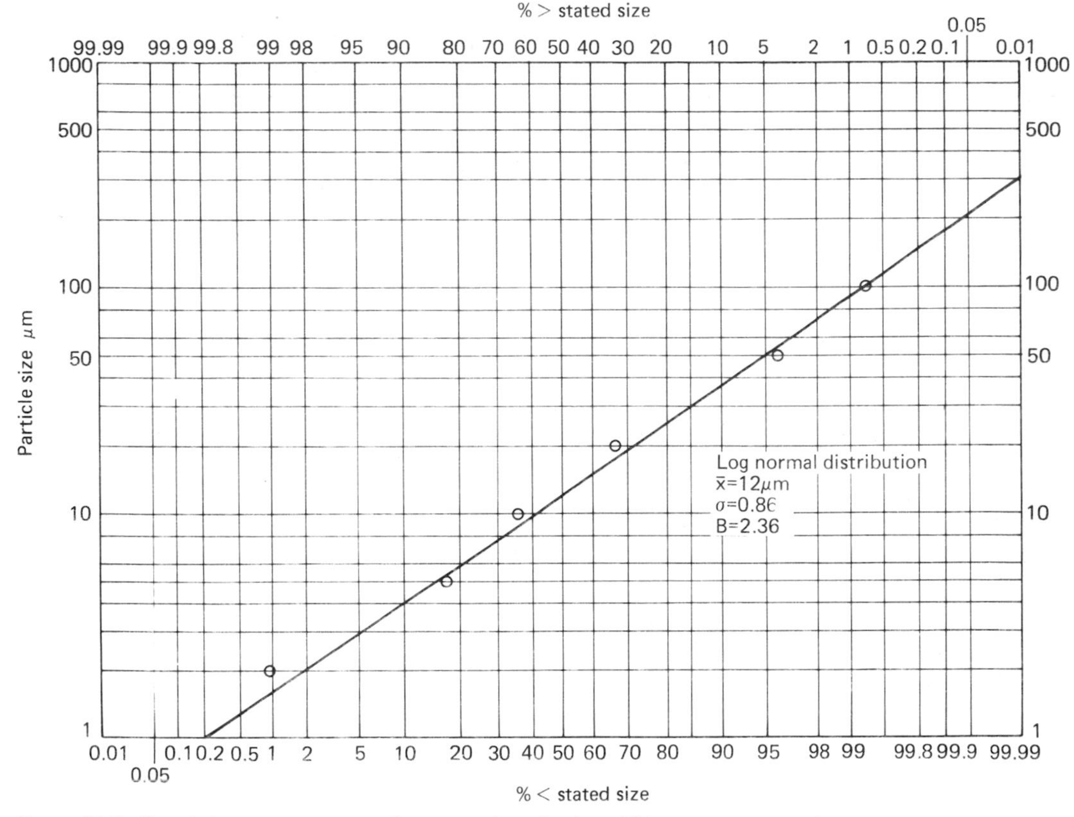

Figure 11.5 Cumulative percentage mass–frequency plot using base-10 log × percentage scales.

In a true log-normal distribution, the equation becomes

$$y=\frac{1}{\sigma\sqrt{2\pi}}\exp\left[-\frac{(\ln x-\overline{\ln x})^2}{2\sigma^2}\right]$$

where now it is $\int y\mathrm{d}(\ln x)$, the area under the curve using a log axis which represents the total quantity, and σ now refers to the log distribution and is not the same as before. The expression $\overline{\ln x}$ is the arithmetic mean of the logarithms of the size so that $\bar{x}$ is now the geometric mean of the distribution. On a cumulative percentage diagram, $\bar{x}$, the geometric mean particle size is the 50 per cent size and σ is found from

$$\sigma=\ln x_{84}-\ln\bar{x}=\ln\bar{x}-\ln x_{16}$$

$$=\ln\frac{x_{84}}{\bar{x}} \qquad =\ln\frac{\bar{x}}{x_{16}}$$

$$=\tfrac{1}{2}\ln\frac{x_{84}}{x_{16}}$$

If x is plotted on base-10 logarithm × probability paper,

$$\sigma=\tfrac{1}{2}\ln 10\log_{10}\frac{x_{84}}{x_{16}}$$

$$=1.15\log_{10}\frac{x_{84}}{x_{16}}$$

Again $\bar{x}$ and σ define the distribution.

Sometimes σ is replaced by $\ln B$ to show that it is the standard deviation of a log-normal distribution

$$B=\sqrt{(x_{84}/x_{16})}$$

Again, if the cumulative percentage plot is not truly linear the derivation of the standard deviation is not truly meaningful and the 50 per cent size is then the median size. However, in practice such curves are commonly used for comparing size analyses and it is sometimes useful for mathematical treatment to draw an approximate straight line.

A feature of a log normal distribution is that if one method of treatment, for example a mass/particle size analysis, demonstrates log-normal properties, then all the other methods will also be log-normal. Clearly the values of $\bar{x}$ will be different. Log-probability diagrams are particularly useful when the range of particle sizes is large.

11.6.3 Rosin–Rammler distributions

Some distributions are extremely skewed, for example, ground coal. Rosin and Rammler have developed a further method for obtaining a straight-line cumulative percentage frequency graph. If the percentage over size x is R, it has been found that

$$\log\log(100/R) = K + n\log x$$

where K is a constant and n a characteristic for the material.

The Rosin–Rammler distribution is included for completeness but its use is not generally recommended.

Sometimes when a distribution covers a wide range of sizes, more than one analysis method has to be used. It is not unusual for a discontinuity to occur in the graphs at the change-over point and this can be due to shape or density effects (see shape factor, Section 11.5).

11.7 Analysis methods that measure size directly

11.7.1 Sieving

Sieving is the cheapest, most popular and probably the most easily understood method of size analysis. It also covers a very wide range of sizes, it being possible to buy sieves (screens) ranging in mesh size from 5 μm up to several centimetres. However, sieving of fine materials requires special techniques and the British Standard 410:(1962) indicates a lower limit of 45 μm. Sieves are made in a variety of materials from non-metallic (e.g. polyester) to stainless steel. The common method of construction is woven wire or fabric but the smallest mesh sizes are electroformed holes in plates. The British Standard gives minimum tolerances on mesh size and wire spacing, etc. American, German and I.S.O. standards are also applicable with very similar criteria. The British Standard nomenclature is based on the number of wires in the mesh per inch. Thus B.S. sieve number 200 has 200 wires per inch and with a specified nominal wire diameter of 52 μm has a nominal aperture size of 75 μm square. In principle all particles less than 75 μm diameter in a sample of spherical particles will pass through a B.S. number 200 sieve. The sample is placed in the uppermost of a stack of sieves covering the range of diameters of interest arranged in ascending order of size from the bottom. The powder is totally enclosed by means of a sealed base and lid. The stack is agitated until the particles have found their appropriate level and the mass in each sieve noted. The tolerance on mesh size introduces a measure of uncertainty on the band widths, and clearly irregularly shaped particles with one or more dimensions larger than the nominal size could still pass through. It is customary therefore to quote particle size when sieving in terms of 'sieve' diameters.

Sieving is by no means as straightforward as it may first appear. For example it is relatively easy for large particles to block the apertures to the passage of small particles (blinding) and there is a statistical uncertainty whether a small particle will find its way to a vacant hole. Both of these are very dependent on the quantity of material placed on the sieve and as a general rule this quantity should be kept small. It is not possible to give an arbitrary figure for the quantity since it will depend on the material and its size distribution, particle shape and surface structure, its adhesive qualities and to some extent the sieve itself. The same comments apply to sieving time. Optimum times and quantities can only be found by experiment, assessing the variation in size grading produced by both factors. Generally a reduction in the quantity is more advantageous than an increase in sieving time and it is normally possible to obtain repeatable results with less than about 10 minutes sieving. The analyst is cautioned that some friable materials break up under the action of sieving.

A number of manufacturers produce sieves and sieving systems using various methods of mechanical agitation, some of which include a rotary action. The objective, apart from reducing the tedium, is to increase the probability of particles finding vacant holes by causing them to jump off the mesh and to return in a different position. Figure 11.6 is an example of one that uses vibration. The vibration can be adjusted in amplitude, and it can be pulsed.

Figure 11.6 Fritsch Analysette sieve shaker.

A feature of any vibrating mechanism is that parts of it can resonate and this is particularly relevant to sieving where it is possible for the sieve surface to contain systems of nodes and anti-nodes so that parts of the surface are virtually stationary. This can be controlled to some extent by adjustment of the amplitude, but although the top sieve surface may be visible, through a transparent lid, the lower surfaces are not and they will have different nodal patterns. One solution to this dilemma is to introduce into each sieve a small number (5 to 10) of 10 mm diameter agate spheres. These are light enough not to damage the sieves or, except in very friable materials, the particles, but they break up the nodal patterns and therefore increase the effective area of the sieve.

A useful feature of dry sieving is that it can be used to obtain closely sized samples for experimental purposes in reasonable quantities.

Although most sieving is performed in the dry state, some difficult materials and certainly much finer sieves can be used in conjunction with a liquid, usually water, in which the particles are not soluble. The lid and base of the sieve stack are replaced by fitments adapted for the introduction and drainage of the liquid with a pump if necessary.

Sieving systems are now commercially available which introduce either air or liquid movement alternately up and down through the sieves to prevent blinding and to assist the particles through the sieves.

Results are usually quoted in terms of percentage of total mass in each size range, the material being carefully removed from the sieve with the aid of a fine soft brush. In wet sieving, the material is washed out, filtered, dried at 105 °C and weighed. Many powders are hygroscopic so that the precaution of drying and keeping in a dessicator until cool and therefore ready for weighing is a good general principle. Convection currents from a hot sample will cause significant errors.

11.7.2 Microscope counting

11.7.2.1 Basic methods

With modern microscopes, the analyst can enjoy the benefits of a wide range of magnifications, a large optical field, stereoscopic vision and zoom facilities, together with back and top illumination. These and calibrated field stop graticules have considerably eased the strain of microscope counting as a method of size analysis but it is still one of the most tedious. It has the advantage, however, that, as well as being able to size particles, the microscope offers the possibility of minute examination of their shape, surface structure, colour, etc. It is often possible to identify probable sources of particles.

The optical microscope can be used to examine particles down to sizes approaching the wavelength of light. At these very small sizes however, interference and diffraction effects cause significant errors, and below the wavelength of light the particles are not resolvable. Microscope counting is covered by British Standard 3406.

Smaller particles, down to 0.001 μm diameter can be examined using the electron microscope.

The two major disadvantages of microscopy are: the restricted depth of focus which means that examination of a sample with a wide size range involves continual re-focusing and the real possibility of missing 'out-of-focus' particles during a scan; it depends more than most other methods on good representativeness.

Several techniques are available for the preparation of slides, the all important factor being that the sample is fully representative. The particles also need to be well separated from each other. A common method is to place a small fraction of the sample onto the slide with one drop of an organic fluid such as methanol or propanol which disperses the particles. Subsequent evaporation leaves the particles suitably positioned. The fluid obviously must not react with the particles, but it must have the ability to 'wet' them. Agitation with a soft brush can help. If the particles have a sticky coating on them, it may be necessary to remove this first by washing. One technique is to agitate, perhaps in water, to allow ample time for all the particles to settle and then to pour off the fluid carefully, repeating as necessary. Obviously the particles must not be soluble in the fluid. Sometimes the material is first agitated in a fluid and then one drop of the particle-laden fluid transferred to the slide. Representativeness can only be tested by the repeatability of results from a number of

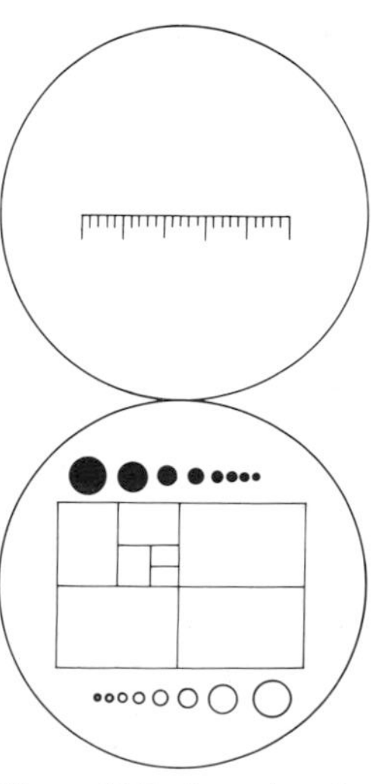

Figure 11.7 Examples of eyepiece graticules.

samples. Techniques have been devised for transferring samples from suspension onto films within the suspension. It is sometimes possible to collect samples directly onto sticky slides coated with grease, gelatine or even rubber solution.

Earlier methods of microscope counting involved the use of an optical micrometer by which a cross-hair could be aligned with each side of each particle in turn and the difference measured. As can be imagined this was slow and tedious and now the most commonly used methods involve calibrated field graticules. The graticules are engraved with a scale, Figure 11.7, nominally divided for example into 20 μm, 100 μm and 1 mm steps. Calibration is dependent on the magnification and this is usually finely adjustable by moving the objective relative to the eyepiece slightly, or by adjusting the zoom if available. Calibration is effected by comparing the field graticule with a stage graticule, similarly and accurately engraved. When set, the stage graticule is replaced by the sample slide.

The slide is scanned in lateral strips, each strip an order or so wider than the largest particles, the objective being to cover the whole of the slide area containing particles. Typically one edge of a chosen reference particle will be aligned with a major graticule line using the longitudinal stage adjustment. The slide will then be traversed laterally along that line and all particles to the right of that line will be counted, and measured using the eyepiece scale. The slide will then be traversed longitudinally to the right until the original particle is in the same relative position but for example five major graticule lines further over and the counting process is repeated for particles within the strip formed by the two lines. The process involves selecting new reference particles as necessary. To avoid duplication, if a particle lies on one of the strip edge-lines, it is counted as if it were in the strip to the right. Particles are allocated to size bands suitably chosen to give say 10 points on the distribution curve. The tedium is relieved if operators work in pairs, one observing, one recording, alternately.

Some graticules have been designed containing systems of opaque and open circles, the sizes arranged in various orders of progression. This can assist the classification of particles by comparison into size bands, each bounded by one of the circles.

When sizing irregularly shaped particles, microscope counting introduces a bias because the particles tend to lie in their most stable orientation. By making measurement of a distribution of randomly orientated particles on a slide along a fixed direction, one obtains a two-dimensional statistical mean diameter.

The method outlined above using a line graticule measures the mean two-dimensional Ferêt diameter, the direction being fixed parallel with the long edge of the slide. It is equally possible to measure the mean two-dimensional Martin's diameter, projected area or image shear diameter.

To obtain three-dimensional mean diameters, it is necessary to take special steps to collect and fix the particles on the slide in a truly random fashion.

11.7.2.2 Semi-automatic microscope size analysers

Semi-automatic methods are those where the actual counting is still done by the analyst but the task, especially the recording, is simplified or speeded up.

The Watson image-shearing eyepiece is a device that replaces the normal eyepiece of a microscope and produces a double image, the separation of which can be adjusted using a calibrated knob along a fixed direction, which again can be preset. The image spacing can be calibrated using a stage graticule. In the Watson eyepiece the images are coloured red and green to distinguish them. The technique in this case is to divide the slide into a number of equal areas, to set the shear spacing at one of a range of values and to count the number of particles in each area with image-shear diameters less than or equal to each of the range.

Some methods have been developed for use with photomicrographs, particularly with the electron microscope. Usually an enlargement of the print, or a projection of it onto a screen, is analysed using comparison aids. The Zeiss–Endter analyser uses a calibrated iris to produce a variable-diameter light spot which can be adjusted to suit each particle. The adjustment is coupled electrically via a multiple switch to eight counters so that pressing a foot switch automatically allocates the particle to one of eight preset ranges of size. A hole is punched in the area of each particle counted to avoid duplication.

11.7.2.3 Automatic microscope size analysers

Several systems have been developed for the automatic scanning of either the microscope field or of photomicrographs. In one type, the system is similar to a television scanning system and indeed the field appears on a television screen. Changes in intensity during the scan can be converted to particle size and the information analysed in a computer to yield details of size in terms of a number of the statistical diameters. It can also calculate shape factors. One system, the Quantimet, now uses digital techniques, dividing the field into 650 000 units.

11.7.3 Direct optical methods

Malvern Instruments Ltd. use a laser and forward-scattering to analyse particles in suspension. The parallel light from the laser passes through a lens, producing an intense spot at the focus. Light falling on any particle is diffracted and is brought to a focus in a system of Fraunhofer diffraction rings centred on the axis and in the focal plane of the lens. The diameters correspond to the preferred scattering angles which are a function of the diameters of the particles. The intensity of each ring is proportional to the total cross-sectional area of the particles of that size. The variation of intensity therefore reflects the dize distribution. Irregularly shaped particles produce blurred rings. A multi-ringed sensor located in the focal plane of the lens passes information to a computer which calculates the volume distribution. It will also assess the type of distribution, whether normal, log-normal or Rosin–Rammler.

The position of the particles relative to the axis of the lens does not affect the diffraction pattern so that movement at any velocity is of no consequence. The method therefore works 'on-line' and has been used to analyse oil fuel from a spray nozzle. The claimed range is from 1 to more than 500 μm.

There are distinct advantages in conducting a size analysis 'on-line'. Apart from obtaining the results generally more quickly, particles as they occur in a process are often agglomerated, i.e. mechanically bound together to form much larger groups which exhibit markedly different behavioural patterns. This is important for example in pollution studies. Most laboratory techniques have to disperse agglomerates produced in sampling and storage. This can be avoided with an on-line process.

11.8 Analysis methods that measure terminal velocity

As already discussed, the terminal velocity of a particle is related to its size and represents a useful method of analysis particularly if the area of interest is aerodynamic. Methods can be characterized broadly into: sedimentation, elutriation and impaction.

11.8.1 Sedimentation

A group of particles, settling for example under the influence of gravity, segregates according to the terminal velocities of the particles. This phenomenon can be used in three ways to grade the particles.

(a) The particles and the settling medium are first thoroughly mixed and changes in characteristics of the settling medium with time and depth are then measured.
(b) The particles and settling medium are mixed as in (a) and measurements are then made on the sediment collecting at the base of the fluid column.
(c) The particles are introduced at the top of the fluid column and their arrival at the base of the column is monitored.

Group (a) is sometimes termed incremental, i.e. increments of the sedimenting fluid are analysed.

Group (b) is sometimes termed cumulative referring to the cumulative effect on the bottom of the column.

Group (c) is also cumulative, but it is sometimes distinguished by the term 'two-layer', i.e. at the initiation of the experiment there are two separate fluids, the upper one thin compared with the lower and containing all the particles, the lower one clear.

11.8.1.1 Incremental methods

Consider at time $t=0$ a homogeneous distribution of particles containing some special ones with terminal velocity V. Ignoring the minute acceleration period, after a time t_1 all the special particles will have fallen a distance $h=Vt_1$. The concentration of those special ones below the depth h will have remained unchanged except on the bottom. Above the depth h, however, all the special particles will have gone. The same argument applies to any sized particle except that the values of h and V are obviously different.

It follows that a measurement of the concentration of all the particles at depth h and time t is a measurement of the concentration of those particles with a terminal velocity less than V, and we have therefore a method of measuring the cumulative distribution.

The following methods use this general principle.

Andreasen's pipette This consists of a relatively large ($\sim$550 ml) glass container with a pipette fused into its ground-glass stopper. A 1 per cent concentration of the sample, suitably dispersed in a chosen liquid, is poured into the container to a set level exactly 200 mm above the lower tip of the pipette. Means are provided to facilitate the withdrawal of precise 10 ml aliquots from the container via the pipette.

After repeated inversions of the container to give thorough mixing, the particles are allowed to sediment. At preselected times after this, e.g. 1 minute, 2 minutes, 4 minutes, etc., 10 ml samples are withdrawn and their solids content weighed. Corrections are applied for the change in depth as samples are removed. Samples are removed slowly over a 20-second period centred around the selected times to minimize errors caused by the disturbance at the pipette tip. The results yield a cumulative mass/terminal velocity distribution which can be converted to mass/size etc., as already discussed. With suitable choice of liquid, the method can be used for particles ranging in size from about 1–100 μm. The conditions should be controlled to be Stokesian, i.e. laminar flow, and of course the terminal velocity is appropriate to the fluid conditions, i.e. it depends on the liquid density and viscosity which are also dependent on temperature.

Density-measuring methods Several techniques involve the measurement of the density of the fluid/particle mixture at different times or at different depths at the same time. Initially, after mixing, the density is uniform and depends on the fluid, the particle densities and the mass concentration of particles. After a period of sedimentation, the density varies with depth.

One method of measuring the density uses a hydrometer. This is complicated by allowance having to be made for the change in the overall height of the fluid caused by the immersion of the hydrometer tube. Also, any intruding object causes a discontinuity in the settling medium, some particles settling on the upper surface of the hydrometer bulb, none settling into the volume immediately below the bulb. Motion around the bulb is not vertical. These problems tend to outweigh the basic simplicity of the method.

A neat method of hydrometry which overcomes the overall height problem uses a number of individual hydrometers, commonly called divers. Each consists of a small body which is totally immersed in the fluid, but with its density individually adjusted so that after the preselected time, the divers will be distributed vertically within the sedimenting column, each at the depth appropriate to its density. Accurate measurement of the positions yields the required information. The main problem in this case is being able to see the divers.

A specific-gravity balance can be used. The change with time in the buoyancy of a ball suspended from one arm of a balance at a fixed depth again gives the information required.

Photosedimentation In a photosedimentometer, the sedimentation is observed by a lamp and photocell system (see Section 11.3). The observation distance is small and for particles greater in size than about 15 μm, the value of K, the scattering coefficient, progressively reduces from 2 to 1. We know that

$$\text{optical density } D = 0.4343Ka/A$$

where a/A is the area concentration. With no particles present in the liquid, let the values of D and K be D_0 and K_0. With all particles present, thoroughly mixed, let the corresponding values be D_1 and K_1. At time t and depth h, $h = Vt$, where V is the upper limit of the terminal velocity of the particles present. If the corresponding values of D and K are D_V and K_V, the fractional surface area of particles with terminal velocity less than V is given by

$$\left(\frac{D_V}{K_V} - \frac{D_0}{K_0}\right) \bigg/ \left(\frac{D_1}{K_1} - \frac{D_0}{K_0}\right)$$

We thus have a method of measuring cumulative surface area terminal velocity distribution.

Proprietary sedimentometers are available which measure D at a fixed height, or scan the whole settlement zone. It is usual to assume $K_V = K_1 = K_0$ and to compensate the result appropriately from supplied tables or graphs. Most photosedimentometers use narrow-beam optics in an attempt to restrict the light to maintain the value of K as 2. The WASP (wide-angle scanning photosedimentometer) has the photocell close to the fluid so that most of the diffracted light is also received and the value of K is nearer 1. The 200 mm settling column is scanned automatically at a fixed rate and the optical density continuously recorded giving a graph which can then be evaluated as a cumulative mass size or cumulative surface area size distribution.

X-ray sedimentation is similar to photosedimentation except that X-rays replace light and the intensity of transmission is dependent on the mass of the particles rather than the surface area. Again

$$I_T = I_0 \exp(-Kc)$$

where c is the mass concentration of the particles and K is a constant. The x-ray density is

$$D = \log_{10} \frac{I_0}{I_T}$$

11.8.1.2 Cumulative methods

Sedimentation balance Consider at time $t = 0$ a homogeneous suspension of particles contained in a column which includes in its base a balance pan.

Let W_1 be the mass of particles with terminal velocity greater than V_1. If h is the height of the column, at time $t_1 = h/V_1$ all those particles will have arrived on the balance pan. However the mass M_1 on the pan will also include a fraction of smaller particles that started part-way down. It can be shown that

$$M_1 = W_1 + t\frac{\mathrm{d}M}{\mathrm{d}t}$$

and measurements of M and t can be used to evaluate W. British Standard 3406 part 2 suggests that values of M should be observed at times t following a geometrical progression, for example 1, 2, 4, 8, etc. seconds. Then $t/\mathrm{d}t$ is constant, in this case 2. It follows that comparing the nth and the $(n-1)$th terms in the time progression,

$$W_n = M_n - 2(M_n - M_{n-1})$$

The final value of M is assumed to be equal to the initial mass introduced.

An alternative method, useful if M is continuously recorded, is to construct tangents as in Figure 11.8. Then W is the intercept on the M axis. Unfortunately, because of the inaccuracy of drawing tangents, the method is not very precise, especially if the overall time is protracted, with a wide size distribution. The

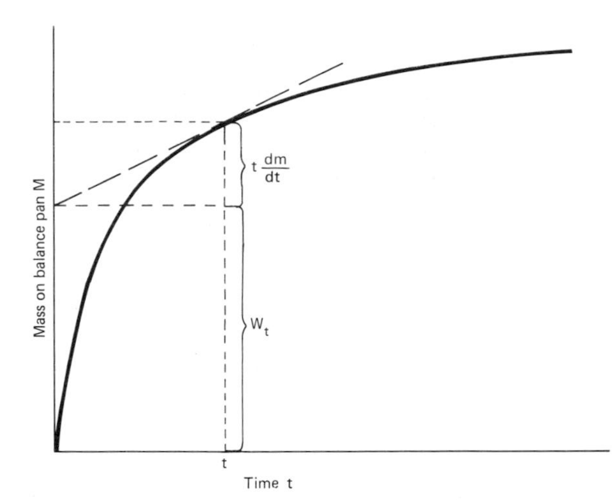

Figure 11.8 Sedimentation balance—plot of mass against time.

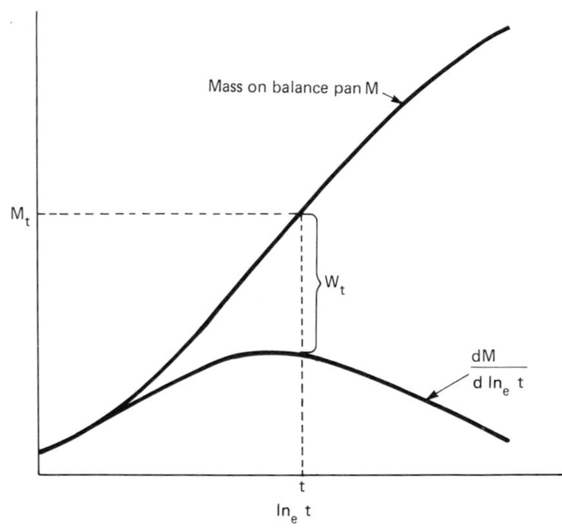

Figure 11.9 Sedimentation balance—plots of M and $dM/d \ln t$ against $\ln t$.

method can be improved by replotting M against $\ln t$ instead of t.

Since

$$\frac{dM}{dt} = \frac{1}{t}\frac{dM}{d(\ln t)}$$

the above expression can be rewritten

$$M = W + \frac{dM}{d(\ln t)}$$

A plot of M against t on logarithmic paper, Figure 11.9, enables tangents to be drawn with greater precision making it possible to compute $dM/d(\ln t)$, the gradient at time t. From a further plot of $dM/d(\ln t)$ against $\ln t$ on the same graph, W can be derived by difference. The method relies on none of the initial material being lost to the sides of the column or round the edges of the pan and that the initial quantity beneath the pan is insignificant. These factors do lead to errors.

Several commercial liquid sedimentation balances are available, notably Sartorious, Shimadzu and Bostock. They have means for introducing the homogeneous suspension into the column above the balance pan. In some, the fluid beneath the pan initially contains no dust. The pan is counterbalanced incrementally by small amounts to minimize pan movement although some is bound to occur and this causes pumping, i.e. some of the finer particles transfer from above to below the pan. All the columns are contained in thermostatic jackets, constancy of temperature being important to the constancy of fluid viscosity and density.

Sedimentation columns A sedimentation column works on the same principle as a sedimentation balance but instead of weighing the sediment continuously the sediment is removed at preset times and weighed externally enabling a higher quality balance to be used.

The ICI sedimentation column, Figure 11.10, tapers at the bottom to a narrow tube a few centimetres long. It is fitted with taps top and bottom. A side branch in the narrow section connects to a clear fluid reservoir.

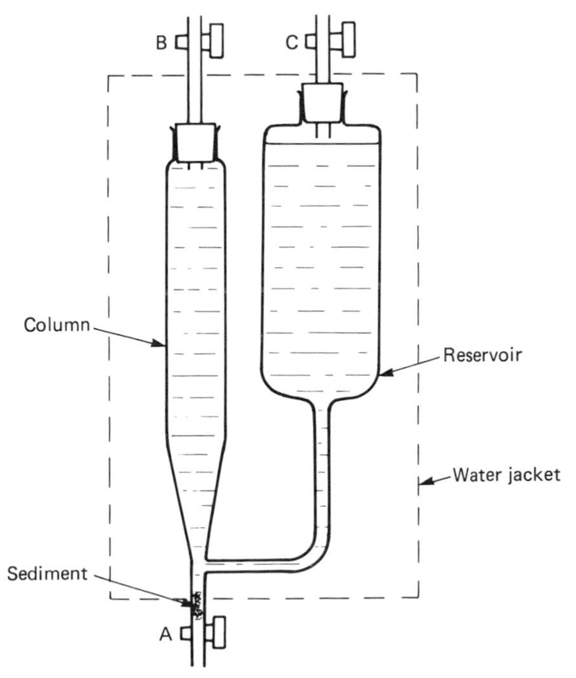

Figure 11.10 ICI sedimentation column.

The dust sample is introduced into the main column and mixed using a rising current of compressed air through tap A, with tap B also open and tap C closed. At time $t=0$, taps A and B are closed and C is opened. Particles sediment into the narrow tube below the side branch. At preset times, tap A is opened allowing clear fluid to wash the sediment into a centrifuge tube or beaker. Negligible sedimenting fluid is lost. The sediment is filtered, dried and weighed.

The BCURA sedimentation column also uses an air supply for mixing, Figure 11.11. In this case, however, the lower tap and narrow tube are replaced by a length of 1 mm capillary tubing. With the top tap closed, surface tension in the capillary supports the column of fluid. At the prescribed times, a container of clear fluid is brought up around the open capillary, breaking the surface tension, and the sediment passes out into the container. The container is then removed. In principle, no sedimenting fluid is lost, but usually a small initial loss occurs until a partial vacuum forms at the top of the column.

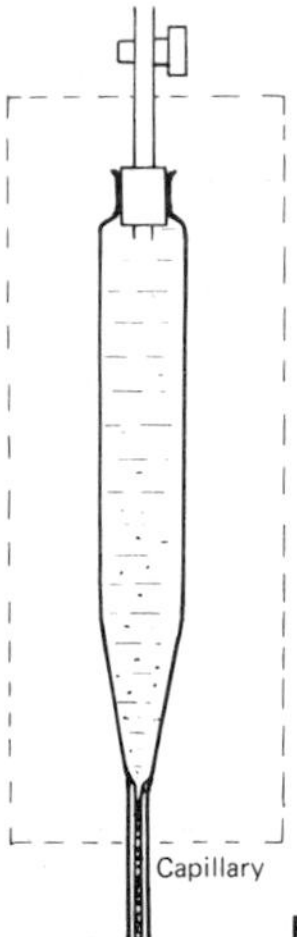

Figure 11.11 BCURA sedimentation column.

The above systems have the advantage of cheapness, but they are subject to errors because the tapered walls affect sedimentation and particles adhere to the tapered walls.

Manometric methods have been used to assess the sedimentation. A manometer fused into the side of the column near the base and filled with the clear fluid will register the change in mean pressure of the column with time and this of course depends on the mass of material still in suspension. Pressure differences are very small.

Beta particle back-scattering has been used to measure the mass of material at the base of a column. The intensity is proportional to the atomic number and the thickness, i.e. the weight of sediment, provided the thickness does not build up to saturation level.

Decanting If a homogeneous fluid/sample mixture is allowed to settle for time t_1 seconds and the fluid down to a depth h is then removed using a pipette, particles removed would all have a terminal velocity less than h/t_1. In the decanting method, this process is repeated several times, replacing the removed fluid with clear fluid and re-mixing until the supernatant fluid is clear. The process is then repeated but with a shorter time, t_2. The removed fluids are analysed for dust content, each containing the fraction of the total mass of material of terminal velocity between h/t_n and h/t_{n-1}. The accuracy depends on the precision of h, t, the rate of removal of fluid, and the number of repeated decantations at each value of t; so it is not high.

Two-layer methods If the upper layer is significantly thinner than the lower layer at time $t=0$, then after a time $t=t_1$, the only material to have reached the base of the column will be those particles with terminal velocity greater than h/t_1, where h is the height of the column, and a measurement of the weight of those particles gives the cumulative distribution directly. Liquid two-layer methods are not common because of the difficulties of arranging the starting condition. The Granulometer uses a shutter system to maintain separation until $t=0$ and a manometer to measure the pressure change as particles sediment.

The Sharples Micromerograph uses an air column approximately 2 m tall. The sample is injected using a puff of compressed nitrogen between two concentric cones to disperse the particles and break up the agglomerates. An electronic balance coupled to a pen recorder monitors the arrival of particles at the base of the column. The column is jacketed to maintain thermal stability. Errors are experienced due to the loss of some fine material to the walls of the column. These can be reduced to some extent by anti-static treatment. The time scale of the recorder is adjustable, fast for the initial phase, slow for the later phase of the sedimentation. With friable particles care has to be exercised at the injection point.

Centrifugal methods For small particles, gravitational systems are very slow-acting. There is also a lower limit to the size of particle that can be measured because of the effects of Brownian motion and convection. While it is possible to use gravitational systems for particles as small as 1 μm in water and about 0.5 μm in air, in practice the lower limit is usually taken to be about 5 μm. These problems can be reduced by the use of centrifugal enhancement of the settling velocity.

The theory for centrifugal settling from a homogeneous fluid mixture is complicated in particular because the terminal velocity varies with distance from the centre of rotation. Approximations have to be made which limit the usefulness of the techniques.

The theory for two-layer systems is exact and therefore these are more attractive. Unfortunately, a phenomenon known as 'streaming' can occur, in which particles tend to agglomerate and accelerate in a bunch instead of individually, behaving as a large particle; this renders the results useless. Streaming has been prevented by using extremely low concentrations of particles in the starting layer. A technique using a third, intermediate layer of immiscible fluid has also been used successfully.

Theories always depend on the applicability of Stokes's law to the particle motion and this imposes a restriction on the maximum particle size, depending on the speed of rotation. Further problems exist for larger particles with respect to the acceleration and decleration of the centrifuge.

In spite of the many problems, the techniques have advantages for very small particles, as small as 0.01 μm and systems are available paralleling many of the gravitational techniques. The most promising appear to be the three-layer methods and the use of optical detection devices.

11.8.2 Elutriation

A group of particles suspended in a fluid which is moving upwards at a velocity V will undergo separation, those particles with a terminal velocity less than V travelling upwards, the others settling downwards. This process is called elutriation. The fluid is usually water or air depending on the particle sizes. Strictly, an elutriator is a classifier rather than a particle sizer. It divides the group of particles into those above and those below a given cut size. In the perfect system, the fluid would clarify except for the few particles with terminal velocity exactly V, and the settled particles could be removed and would contain no undersized particles. The system, however, is not perfect. Proper Stokesian conditions do not exist because practical considerations make the tubes too short compared with their diameters. Also, the velocity at the cylinder walls is considerably less than at the centre, causing a circulation of some particles up the centre and down the walls. The cut point therefore is not sharp. A multiple elutriator consists of a number of elutriators connected in series but with increasing diameters. Each section therefore collects progressively smaller particles.

Both water and air elutriators are commercially available. In use, the sample is introduced into the smallest section and the fluid velocity set to a preset value. When the fluid has cleared in all the sections, the system is switched off and the sediment in each weighed. The last section is usually terminated by a filter. In some designs, the separating section is short and followed by a conical reducing zone to hasten removal of the undersize particles to the next section.

The elutriators described so far are counter-flow elutriators. Acceleration (i.e. gravity) and drag are in opposite directions. Elutriators have been designed for transverse flow. If the sample of particles is introduced slowly into the upper level of a horizontal laminar stream of fluid, two-layer sedimentation takes place but the sediment is separated horizontally, different-sized particles following different trajectories. The particles are collected on judiciously placed pre-weighed sticky plates.

11.8.2.1 Centrifugal elutriation

Elutriation is also used in the centrifugal field, usually with air as the fluid. In principle, air with entrained particles travels inwards against the centrifugal force. Particles with terminal velocities sufficiently low also pass inwards; the rest pass outwards.

The natural flow of air in a spinning system is radially outwards. In order therefore to obtain a counterflow, the air must be driven. In some systems a separate pump is used. In others, the air is introduced at the centre, passes radially outwards under centrifugal force, turns 180° to travel radially inwards for a short distance, of the order of half a radius, and finally through another 180° to pass to the circumference. Elutriation takes place in the inwards-flowing section. In this case, no pump is necessary because the net airflow is outwards.

Adjustment of either rotation speed or air velocity will affect the cut size. Air velocity is usually set by a throttling mechanism.

A variety of centrifuge systems is available, their design and size making them particularly suitable for different size ranges. Some for example are very sensitive in the range 2–12 μm, while others are better at larger sizes. Some are capable of a continuous and large throughput which makes them especially suitable for producing quantities of closely-sized particles. The devices are then being used for their original purpose which is classification rather than size grading.

A cyclone is a centrifugal device normally used for extracting dust from carrier gases. It consists of a conically shaped vessel. The dusty gas is drawn tangentially into the base of the cone, takes a helical route towards the apex where the gas turns sharply back along the axis and is withdrawn axially through the base. The device is a classifier in which only dust with terminal velocity less than a given value can pass through the vortex formed and out with the gas. The particle cut-off diameter is calculable for given conditions. Systems have been designed using a set of cyclones in series with increasing gas velocities for size analysis.

The Cyclosizer Analyser uses liquid instead of gas and has five equal-sized cyclones with different inlet and outlet diameters to obtain the velocity variation. The cyclones are arranged, apex uppermost. Thus the coarse particles retained in a given section fall back many times for reclassification, thereby obtaining good separation. In this case the range is 8–50 μm.

11.8.3 Impaction

When a fluid containing a suspension of particles is made to turn a corner, particles with terminal velocity in excess of a value determined by the fluid velocity and the geometry of the bend are deposited or impacted. A cascade impactor consists of a series of orifices each accurately positioned above a collector plate. The orifices can be round holes or slots. The holes in successive stages are reduced in size to increase the impaction velocity. The particles pass through the holes and are either deposited on the adjacent plate or pass on to the next stage. There are typically between six and ten stages covering an aerodynamic size range from about 0.4 μm to 15 μm.

The Andersen cascade impactor is designed to work 'on-line' incorporating a nozzle for isokinetic sampling from a gas stream. A precyclone removes particles $>15\,\mu$m and a filter catches those $<0.4\,\mu$m. The Sierra, designed for room or atmospheric air measurement, covers a range 0.05 μm to 10 μm. The collected particles are removed and weighed. California Measurements Inc. markets an instrument with piezoelectric crystal mass monitors at each of ten stages giving immediate automatic readout.

Impaction surfaces frequently require the aid of an adhesive to prevent re-entrainment from one stage to the next.

11.9 Analysis methods that infer size from some other property

The methods discussed so far either measure size directly or measure a fluid-dynamic effect dependent on the particle terminal velocity which, although dependent on size, is also affected by density, shape and surface structure. The following methods do not measure size directly nor are they dependent on terminal velocity.

11.9.1 Coulter counter

The Coulter counter uses the principle that the electrical resistance of a conducting liquid is increased by the addition of an insulating material. Particles are assessed individually. To obtain adequate sensitivity the volume of liquid measured must be similar to the volume of the particle.

These criteria are achieved by containing the electrolyte in two chambers separated by a narrow channel containing an orifice, the dimensions of which are accurately known. An electric current from a constant-current source passes through the orifice from one chamber to the other. The voltage across the orifice is therefore directly proportional to the resistance of the orifice.

The sample, suitably dispersed, is placed in one of the chambers. An accurately controlled volume of the well-agitated electrolyte is then passed through the orifice. The concentration of the sample (of the order of 0.1 per cent) is such that particles pass through individually. Each particle causes a voltage pulse and a pulse-height analyser increments one of a set of counters, each representing a size maximum.

The theory of the Coulter counter is complicated, particularly for randomly shaped particles but it has been shown that, to a first approximation, the pulse height is directly proportional to particle volume, errors being less than 6 per cent when the particle size is less than 40 per cent of the orifice diameter. This size limitation also represents a reasonable practical limitation to avoid blockage of the orifice. Although the resistivity of the particles should affect the result, in practice, surface-film effects make this insignificant. The method also works with conducting particles.

The lower limit on particle size is set by the electronic noise in the circuit and in practical terms is usually taken to be about 4 per cent of the orifice diameter. Orifices are available ranging in size from 10 μm up to 1 mm giving a particle-size range (using different orifices) from 0.4 μm to 400 μm. Samples containing a wide range of sizes need to be wet-sieved to remove those larger than 40 per cent orifice size. With small sizes, as always, care has to be exercised to avoid contamination. Bubbles can cause false signals. Aqueous or organic electrolytes can be used. It is usual to calibrate the instrument using a standard sample of latex polymer particles. The technique permits several runs on each sample/electrolyte mix and it is easy to test the effect of changes of concentration, which should be negligible if dilution is adequate.

Models are available with 16-channel resolution and the output can be in tabular form giving particle frequency volume or particle cumulative volume or can be in the form of an automatic plot.

11.9.2 Hiac automatic particle sizer

The Hiac analyser can be considered to be the optical equivalent of the Coulter. In this, the cylindrical orifice is replaced by a two-dimensional funnel-shaped orifice which guides the particle stream through a very narrow light beam located at its throat. A sensor measures the obscuration caused by each particle as it passes through. The responses are proportional to the particle cross-sectional areas and are sorted by a pulse-height analyser. A range of orifices is available to suit particle sizes from 2 μm up to 9 mm, each covering a size range in the ratio 30:1. Although the measurement on a given particle is along one axis, an irregularly shaped particle will not tend to be orientated in any particular way so that statistically the area measured will be a mean cross-sectional area. The optical method has an advantage over the conductivity method in that it can operate in any liquid or gas provided it is translucent over the very short optical path length (typically 2–3 mm of fluid). Scattering-coefficient problems are reduced by calibration using standard samples. The instrument has been used 'on-line' to measure the contamination of hydraulic fluid. The number of particles in a fixed volume is found by counting while timing at a constant flow rate.

11.9.3 Climet

The Climet method involves measuring the light scattered from individual particles which are directed accurately through one focus of an elliptical mirror. Light is focused onto the particles and it is claimed that about 90 per cent of the scattered light is detected by a photomultiplier at the other focus of the ellipse. Direct light is masked. The response is pulse-height analysed.

11.9.4 Adsorption methods

In some processes, a knowledge of the surface area of particles is more important than the actual size. Optical techniques may not be appropriate since these tend to give the 'smoothed' surface area rather than the true surface area—including roughness and pores. Some gases adsorb onto the surface of substances, i.e. they form a layer which may be only one molecule thick on the surface. This is not to be confused with absorption in which the substance is porous and the gases penetrate the substance. The subject is too large for a complete treatment here. The principle is that the particles are first 'cleaned' of all previous adsorbates and then treated with a specific adsorbate, e.g. nitrogen, which is then measured by difference from the adsorbate source. If the layer is monomolecular, the surface area is readily calculated; corrections have to be applied because this is generally not the case. Results are given in terms of surface area/volume or surface area/mass.

11.10 References

For a more comprehensive treatment of this whole complex subject, *Particle Size Measurement* by T. Allen, and *Particle Size Analysis in Industrial Hygiene* by L. Silverman, C. E. Billings and M. W. First are recommended.

Part 2

Measurement of Temperature and Chemical Composition

1 Temperature measurement

C. HAGART-ALEXANDER

1.1 Temperature and heat

1.1.1 Application considerations

Temperature is one of the most frequently used process measurements. Almost all chemical processes and reactions are temperature-dependent. Not infrequently in chemical plant, temperature is the only indication of the progress of the process. Where the temperature is critical to the reaction, a considerable loss of product may result from incorrect temperatures. In some cases, loss of control of temperature can result in catastrophic plant failure with the attendant damage and possibly loss of life.

Another area where accurate temperature measurement is essential is in the metallurgical industries. In the case of many metal alloys, typically steel and aluminium alloys, the temperature of the heat treatment the metal receives during manufacture is a crucial factor in establishing the mechanical properties of the finished product.

There are many other areas of industry where temperature measurement is essential. Such applications include steam raising and electricity generation, plastics manufacture and moulding, milk and dairy products and many other areas of the food industries.

Then, of course, where most of us are most aware of temperature is in the heating and air-conditioning systems which make so much difference to people's personal comfort.

Instruments for the measurement of temperature, as with so many other instruments, are available in a wide range of configurations. Everyone must be familiar with the ubiquitous liquid-in-glass thermometer. There is then a range of dial thermometers with the dial attached directly to the temperature measuring element, i.e. local reading thermometers. Remote reading instruments are also available where the measuring system operates the dial directly through a length of metal capillary tubing. The distance between the sensing 'bulb' and the dial, or readout, of these instruments is limited to about thirty metres. Where the temperature readout is required at a long distance from the location of the sensing element there are two main options; either an electrical measuring technique such as a thermocouple or resistance thermometer can be used or where the distances between the plant measurement locations and the control room are very long it is usually better to use temperature transmitters. These instruments use the same types of probes as other temperature measuring instruments. The transmitting mechanism is normally attached directly to the probe. It may also have a local readout facility as well as its transmitting function which is to convert the measurement effect into a pneumatic or electrical signal suitable for transmission over long distances' (See Part 4 on transmitters).

Temperature measurement effects are also used directly for simple control functions such as switching on and off an electric heater or the direct operation of a valve, i.e. thermostats.

1.1.2 Definitions

For the understanding of temperature measurement it is essential to have an appreciation of the concepts of temperature and other heat-related phenomena.

1.1.2.1 Temperature

The first recorded temperature measurement was carried out by Galileo at the end of the sixteenth century. His thermometer depended on the expansion of air. Some form of scale was attached to his apparatus for he mentions 'degrees of heat' in his records.

As with any other measurement, it is necessary to have agreed and standardized units of measurement. In the case of temperature the internationally recognized units are the kelvin and the degree Celsius. The definitions of these units are set out in Section 1.2.

One must differentiate between heat and temperature. The effect of temperature is the state of agitation, both oscillation and rotation of molecules in a medium. The higher the temperature of a body the greater the vibrational energy of its molecules and the greater its potential to transfer this molecular kinetic energy to another body. Temperature is the potential to cause heat to move from a point of higher temperature to one of lower temperature. The rate of heat transfer is a function of that temperature difference.

1.1.2.2 Heat

Heat is thermal energy. The quantity of heat in a body is proportional to the temperature of that body, i.e. it is its heat capacity multiplied by its absolute temperature.

Heat is energy and as such is measured in units of energy. Heat is measured in joules. (Before the international agreements on the SI system of units heat was measured in calories. One calorie was approximately 4.2 joules.)

1.1.2.3 Specific heat capacity

Different materials absorb different amounts of heat to produce the same temperature rise. The specific heat capacity, or more usually the specific heat, of a substance is the amount of heat which, when absorbed by 1 kg of that substance, will raise its temperature by one kelvin

$$\text{Specific heat capacity} = \text{J kg}^{-1}\ \text{K}^{-1}$$

1.1.2.4 Thermal conductivity

The rate at which heat is conducted through a body depends upon the material of the body. Heat travels very quickly along a bar of copper, for instance, but more slowly through iron. In the case of non-metals, ceramics or organic substances, the thermal conduction occurs more slowly still. The heat conductivity is not only a function of the substance but also the form of the substance. Plastic foam is used for heat insulation because the gas bubbles in the foam impede the conduction of heat. Thermal conductivity is measured in terms of:

$$\frac{\text{energy} \times \text{length}}{\text{area} \times \text{time} \times \text{temperature difference}}$$

$$\text{thermal conductivity} = \frac{\text{J.m}}{\text{m}^2\text{.s.K}} = \text{J.m}^{-1}\text{.s}^{-1}\text{.K}^{-1}$$

1.1.2.5 Latent heat

When a substance changes state from solid to liquid or from liquid to vapour it absorbs heat without change of temperature. If a quantity of ice is heated at a constant rate its temperature will rise steadily until it reaches a temperature of 0 °C; at this stage the ice will continue to absorb heat with no change of temperature until it has all melted to water. Now as the heat continues to flow into the water the temperature will continue to rise but at a different rate from before due to the different specific heat of water compared to ice. When the water reaches 100 °C the temperature rise will again level off as the water boils, changing state from water to steam. Once all the water has boiled to steam the temperature will rise again but now at yet another rate dependent on the specific heat of steam. This is illustrated in Figure 1.1.

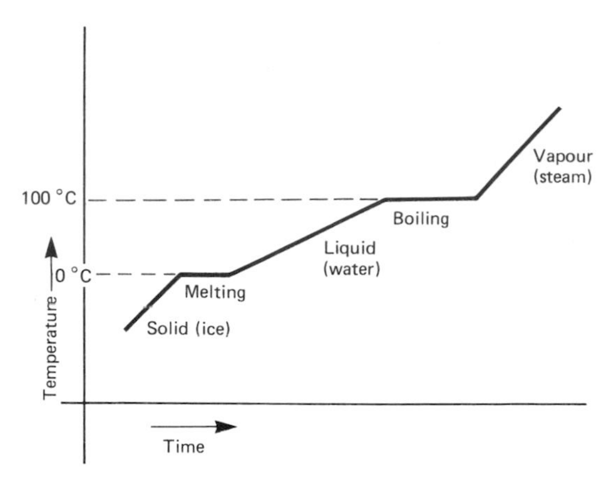

Figure 1.1 Increase of temperature during change of state of a mass of water under conditions of constant energy input.

The amount of heat required to convert a kilogram of a substance from solid state to liquid state is the 'latent heat of fusion'.

Likewise the 'latent heat of evaporation' is the amount of heat required to convert a kilogram of liquid to vapour.

This levelling of temperature rise during change of state accounts for the constant freezing temperatures and constant boiling temperatures of pure materials. The units of measurement of latent heat are joules per kilogram:

$$\text{latent heat} = \text{J.kg}^{-1}$$

1.1.2.6 Thermal expansion

Expansion of solids When a solid is heated, it increases in volume. It increases in length, breadth and thickness. The increase in length of any side of a solid will depend upon the original length l_0, the rise in temperature t, and the coefficient of linear expansion α.

The coefficient of linear expansion may be defined as the increase in length per unit length when the temperature is raised 1 K. Thus, if the temperature of a rod of length l_0, is raised from 0 °C to t °C, then the new length, l_t, will be given by:

$$l_t = l_0 + l_0 . \alpha t = l_0(1 + \alpha t) \qquad (1.1)$$

The value of the coefficient of expansion varies from substance to substance and the coefficients of linear expansion of some common materials are given in Table 1.1.

Table 1.1 Coefficients of linear expansion of solids
Extracted from *Tables of Physical and Chemical Constants* by Kaye and Laby (Longmans). The values given are per kelvin and, except where some temperature is specified, for a range about 20 degrees

Substance	α *(ppm)*
Aluminium	25.5
Copper	16.7
Gold	13.9
Iron (cast)	10.2
Lead	29.1
Nickel	12.8
Platinum	8.9
Silver	18.8
Tin	21.4
Brass (typical)	18.9
Constantan (Eureka), 60 Cu, 40 Ni	17.0
Duralumin	22.6
Nickel steel,	
10% Ni	13.0
30% Ni	12.0
36% Ni (Invar)	−0.3 to +2.5
40% Ni	6.0
Steel	10.5 to 11.6
Phosphor bronze, 97.6 Cu, 2 Sn, 0.2 P	16.8
Solder, 2 Pb, 1 Sn	25
Cement and concrete	10
Glass (soda)	8.5
Glass (Pyrex)	3
Silica (fused) −80° to 0 °C	0.22
Silica (fused) 0° to 100 °C	0.50

The increase in area with temperature, i.e. the coefficient of superficial expansion is approximately twice the coefficient of linear expansion. The coefficient of cubic expansion is almost three times the coefficient of linear expansion.

In engineering practice it is necessary, especially in large structures, to make allowance for thermal expansion. For instance bridges are built with expansion joints. Many instruments are designed with temperature compensation to accommodate thermal expansion. Thermal expansion can be made use of for temperature measurement, as is dealt with in Section 1.3.

If great accuracy is required when measuring lengths with a scale made of metal, allowance should be made for the increase in length of the scale when its temperature is greater than that at which it was calibrated. Owing to the expansion of the scale, a length which was originally l_1 at the temperature t_1, at which the scale was calibrated, will have increased to l_2 where

$$l_2 = l_1[1 + \alpha(t_2 - t_1)] \tag{1.2}$$

Here t_2 is the temperature at which the measurement is made, and α the coefficient of expansion of the metal of the scale.

A 1 mm division on the scale will therefore now measure

$$1 + \alpha(t_2 - t_1)\ \text{mm} \tag{1.3}$$

An actual length l_2 mm will therefore measure

$$\frac{l_2}{1 + \alpha(t_2 - t_1)} \tag{1.4}$$

The length will therefore appear to be smaller than it actually is. To make this error negligibly small, secondary standards of length are made of Invar, a nickel steel alloy whose linear coefficient of expansion is nearly zero.

Expansion of liquids and gases In dealing with the expansion of liquids and gases it is necessary to consider the volume expansion, or cubical expansion. Both liquids and gases have to be held by a container, which will also expand, so that the apparent expansion of the liquid or gas will be less than the true or absolute expansion. The true coefficient of expansion of a liquid is equal to the coefficient of apparent expansion plus the coefficient of cubical expansion of the containing vessel. Usually the expansion of a gas is so much greater than that of the containing vessel that the expansion of the vessel may be neglected in comparison with that of the gas.

The coefficient of expansion of a liquid may be defined in two ways. First, there is the zero coefficient of expansion, which is the increase in volume per degree rise in temperature, divided by the volume at 0 °C, so that volume V_t at temperature t is given by:

$$V_t = V_0(1 + \beta t) \tag{1.5}$$

where V_0 is the volume at 0 °C and β is the coefficient of cubical expansion.

There is also the mean coefficient of expansion between two temperatures. This is the ratio of the increase in volume per degree rise of temperature, to the original volume. That is,

$$\beta = \frac{V_{t_2} - V_{t_1}}{V_{t_1}(t_2 - t_1)} \tag{1.6}$$

where V_{t_1} is the volume at temperature t_1, and V_{t_2} is the volume at temperature t_2.

This definition is useful in the case of liquids that do not expand uniformly, e.g. water.

1.1.2.7 Radiation

There are three ways in which heat may be transferred: conduction, convection, and radiation. Conduction is, as already covered, the direct transfer of heat through matter. Convection is the indirect transfer of heat by the thermally induced circulation of a liquid or gas; in 'forced convection', the circulation is increased by a fan or pump. Radiation is the direct transfer of heat (or other form of energy) across space. Thermal radiation is electromagnetic radiation and comes within the infrared, visible and ultraviolet regions of the electromagnetic spectrum. The demarcation between these three classes of radiation is rather indefinite but as a guide the wavelength bands are shown in Table 1.2.

Table 1.2 Wavelengths of thermal radiation

Radiation	*Wavelength* (μm)
Infrared	100–0.8
Visible light	0.8–0.4
Ultraviolet	0.4–0.01

So far as the effective transfer of heat is concerned the wavelength band is limited to about 10 μm in the infrared and to 0.1 μm in the ultraviolet. All the radiation in this band behaves in the same way as light. The radiation travels in straight lines, may be reflected or refracted and the amount of radiant energy falling on a unit area of a detector is inversely proportional to the square of the distance between the detector and the radiating source.

1.2 Temperature scales

To measure and compare temperatures it is necessary to have agreed scales of temperature. These temperature scales are defined in terms of physical phenomena which occur at constant temperatures. The temperatures of these phenomena are known as 'fixed points'.

1.2.1 Celsius temperature scale

The Celsius temperature scale is defined by international agreement in terms of two fixed points, the ice point and the steam point. The temperature of the ice point is defined as zero degrees Celsius and the steam point as one hundred degrees Celsius.

The ice point is the temperature at which ice and water exist together at a pressure of 1.0132×10^5 $N.m^{-2}$ (originally one standard atmosphere=760 mm of mercury). The ice should be prepared from distilled water in the form of fine shavings and mixed with ice-cold distilled water.

The steam point is the temperature of distilled water boiling at a pressure of 1.0132×10^5 $N.m^{-2}$. The temperature at which water boils is very dependent on pressure. At a pressure p, $N.m^{-2}$ the boiling point of water t_p in degrees Celsius is given by

$$t_p = 100 + 2.795 \times 10^{-4}(p - 1.013 \times 10^{-5}) - 1.334 \times 10^{-9}(p - 1.013 \times 10^5)^2 \quad (1.7)$$

The temperature interval of 100 °C between the ice point and the steam point is called the fundamental interval.

1.2.2 Kelvin, absolute or thermodynamic temperature scale

The earlier scales of temperature depended upon the change with temperature of some property, such as size, of a substance. Such scales depended upon the nature of the substance selected. About the middle of the nineteenth century, Lord Kelvin defined a scale of temperature in terms of the mechanical work which may be obtained from a reversible heat engine working between two temperatures, and which, therefore, does not depend upon the properties of a particular substance. Kelvin divided the interval between the ice and steam points into 100 divisions so that one kelvin represents the same temperature interval as one Celsius degree. (The unit of the Kelvin or thermodynamic temperature scale is the 'kelvin'.) The definition of the kelvin is the fraction 1/273.16 of the thermodynamic temperature of the triple point of water. This definition was adopted by the thirteenth meeting of the General Conference for Weights and Measures in 1967 (13th CGPM, 1967).

It has also been established that an ideal gas obeys the gas law $PV = RT$, where T is the temperature on the absolute or kelvin scale and where P is the pressure of the gas, V is the volume occupied and R is the universal gas constant. Thus, the behaviour of an ideal gas forms a basis of temperature measurement on the absolute scale. Unfortunately the ideal gas does not exist, but the so-called permanent gases, such as hydrogen, nitrogen, oxygen and helium, obey the law very closely, provided the pressure is not too great. For other gases and for the permanent gases at greater pressures, a known correction may be applied to allow for the departure of the behaviour of the gas from that of an ideal gas. By observing the change of pressure of a given mass of gas at constant volume, or the change of volume of the gas at constant pressure, it is possible to measure temperatures on the absolute scale.

The constant-volume gas thermometer is simpler in form, and is easier to use, than the constant-pressure gas thermometer. It is, therefore, the form of gas thermometer which is most frequently used. Nitrogen has been found to be the most suitable gas to use for temperature measurement between 500 and 1500 °C, while at temperatures below 500 °C hydrogen is used. For very low temperatures, helium at low pressure is used.

The relationship between the kelvin and Celsius scales is such that zero degrees Celsius is equal to 273.15 K

$$t = T - 273.15 \quad (1.8)$$

where t represents the temperature in degrees Celsius and T is the temperature kelvin.

It should be noted that temperatures on the Celsius scale are referred to in terms of degrees Celsius, °C, temperatures on the absolute scale are in kelvins, K, no degree sign being used. For instance the steam point is written in Celsius, 100 °C, but on the Kelvin scale 373.15 K.

1.2.3 International Practical Temperature Scale of 1968 (IPTS-68)

The gas thermometer, which is the final standard of reference, is, unfortunately, rather complex and cumbersome, and entirely unsuitable for industrial use. Temperature measuring instruments capable of a very high degree of repeatability are available. Use of these instruments enables temperatures to be reproduced to a very high degree of accuracy, although the actual value of the temperature on the thermodynamic scale is not known with the same degree of accuracy. In order to take advantage of the fact that temperature scales may be reproduced to a much higher degree of accuracy than they can be defined, an International Practical Temperature Scale was adopted in 1929 and revised in 1948. The latest revision of the scale was in 1968 (IPTS-68) and this is the scale used in this book. The 1948 scale is still used in many places in industry. The differences between temperatures on the two scales are small, frequently within the accuracy of commercial instruments. Table 1.3 shows the deviation of the 1948 scale from the 1968 revision.

Table 1.3 Deviation of IPTS-68 from IPTS-48

t_{68} (°C)	$t_{68} - t_{48}$
−200	0.022
−150	−0.013
0	0.000
50	0.010
100	0.000
200	0.043
400	0.076
600	0.150
1000	1.24

The International Practical Temperature Scale is based on a number of defining fixed points each of which has been subject to reliable gas thermometer or radiation thermometer observations and these are linked by interpolation using instruments which have the highest degree of reproducibility. In this way the International Practical Temperature Scale is conveniently and accurately reproducible and provides means for identifying any temperature within much narrower limits than is possible on the thermodynamic scale.

The defining fixed points are established by realizing specified equilibrium states between phases of pure substances. These equilibrium states and the values assigned to them are given in Table 1.4.

Table 1.4 Defining fixed points of the IPTS-68[1]

Equilibrium state	*Assigned value of International Practical temperature* T_{68}	t_{68}
Triple point of equilibrium hydrogen	13.81 K	−259.34 °C
Boiling point of equilibrium hydrogen at pressure of 33 330.6 kN . m^{-2}	17.042 K	−256.108 °C
Boiling point of equilibrium hydrogen	20.28 K	−252.87 °C
Boiling point of neon	27.102 K	−246.048 °C
Triple point of oxygen	54.361 K	−218.789 °C
Boiling point of oxygen	90.188 K	−182.962 °C
Triple point of water[3]	273.16 K	0.01 °C
Boiling point of water[2][3]	373.15 K	100 °C
Freezing point of zinc	692.73 K	419.58 °C
Freezing point of silver	1235.08 K	961.93 °C
Freezing point of gold	1337.58 K	1064.43 °C

(1) Except for the triple points and one equilibrium hydrogen point (17.042 K) the assigned values of temperature are for equilibrium states at a pressure $p_0 = 1$ standard atmosphere (101.325 kN . m^{-2}).

In the realization of the fixed points small departures from the assigned temperatures will occur as a result of the differing immersion depths of thermometers or the failure to realize the required pressure exactly. If due allowance is made for these small temperature differences, they will not affect the accuracy of realization of the Scale.

(2) The equilibrium state between the solid and liquid phases of tin (freezing point of tin has the assigned value of $t_{68} = 231.9681$ °C and may be used as an alternative to the boiling point of water.

(3) The water used should have the isotopic composition of ocean water.

The scale distinguishes between the International Practical Kelvin Temperature with the symbol T_{68} and the International Practical Celsius Temperature with the symbol t_{68} the relationship between T_{68} and t_{68} is

$$t_{68} = T_{68} - 273.15 \text{ K} \quad (1.9)$$

The size of the degree is the same on both scales, being 1/273.16 of the temperature interval between absolute zero and the triple point of water (0.01 °C). Thus, the interval between the ice point 0 °C and the boiling point of water 100 °C is still 100 Celsius degrees. Temperatures are expressed in kelvins below 273.15 K (0 °C) and degrees Celsius above 0 °C.

Temperatures between and above the fixed points given in Table 1.4 can be interpolated as follows.

From 13.81 K to 630.74 °C the standard instrument is the platinum resistance thermometer. The thermometer resistor must be strain-free, annealed pure platinum having a resistance ratio $W(T_{68})$ defined by

$$W(T_{68})=\frac{R(T_{68})}{R(273.15\ \mathrm{K})} \qquad (1.10)$$

where R is the resistance, which must not be less than 1.392 50 ohms at $T_{68}=373.15$ K, i.e. the resistance ratio

$$\frac{R(100\ ^\circ\mathrm{C})}{R(0\ ^\circ\mathrm{C})}$$

is greater than the ratio 1.3920 of the 1948 scale, i.e. the platinum must be purer.

Below 0 °C the resistance–temperature relationship of the thermometer is found from a reference function and specified deviation equations. From 0 °C to 630.74 °C two polynomial equations provide the resistance temperature relationship. This will be discussed further in the section on resistance thermometers.

From 630.74 °C to 1064.43 °C the standard instrument is the platinum 10 per cent rhodium/platinum thermocouple, the electromotive force–temperature relationship of which is represented by a quadratic equation and is discussed in the appropriate section.

Above 1337.58 K (1064.43 °C) the scale is defined by Planck's law of radiation with 1337.58 K as the reference temperature and the constant c_2 has a value 0.014 388 metre kelvin. This will be discussed in the section on radiation thermometers.

In addition to the defining fixed points the temperatures corresponding to secondary points are given. These points, particularly the melting or freezing points of metals, form convenient workshop calibration points for temperature measuring devices (see Table 1.5).

Table 1.5 Secondary reference points (IPTS-68)

Substance	*Equilibrium state*	*Temperature (K)*
Normal hydrogen	TP	13.956
Normal hydrogen	BP	20.397
Neon	TP	24.555
Nitrogen	TP	63.148
Nitrogen	BP	77.342
Carbon dioxide	Sublimation point	194.674
Mercury	FP	234.288
Water	Ice point	273.15
Phenoxy benzine	TP	300.02
Benzoic acid	TP	395.52
Indium	FP	429.784
Bismuth	FP	544.592
Cadmium	FP	594.258
Lead	FP	600.652
Mercury	BP	629.81
Sulphur	BP	717.824
Copper/aluminium eutectic	FP	821.38
Antimony	FP	903.89
Aluminium	FP	933.52
Copper	FP	1357.6
Nickel	FP	1728
Cobalt	FP	1767
Palladium	FP	1827
Platinum	FP	2045
Rhodium	FP	2236
Iridium	FP	2720
Tungsten	FP	3660

TP: triple point; FP: freezing point; BP: boiling point.

Table 1.6 Comparison of temperature scales

	K	°C	°F	°R
Absolute zero	0	−273.15	−523.67	0
Boiling point O_2	90.19	−182.96	−361.33	162.34
Zero Fahrenheit	255.37	−17.78	0	459.67
Ice point	273.15	0	32	491.67
Steam point	373.15	100	212	671.67
Freezing point of silver	1235.08	961.93	1763.47	2223.14

1.2.4 Fahrenheit and Rankine scales

These two temperature scales are now obsolescent in Britain and the United States, but as a great deal of engineering data, steam tables etc. have been published using the Fahrenheit and Rankine temperature a short note for reference purposes is relevant.

Fahrenheit This scale was proposed in 1714. Its original fixed points were the lowest temperature obtainable with ice and water which was taken as zero. Human blood heat was made 96 degrees (98.4 on the modern scale). On this scale the ice point is at 32 °F and the steam point at 212 °F. There does not appear to be any formal definition of the scale.

To convert from the Fahrenheit to Celsius scale, if t is the temperature in Celsius and f the temperature in Fahrenheit

$$t=\tfrac{5}{9}(f-32) \qquad (1.11)$$

Rankine The Rankine scale is the thermodynamic temperature corresponding to Fahrenheit. Zero in Rankine is of course the same as zero kelvin. On the

Rankine scale the ice point is at 491.67 °R. Zero Fahrenheit is 459.67 °R. To convert temperature from Fahrenheit to Rankine, where R is the Rankine temperature

$$R = f + 459.67 \tag{1.12}$$

Table 1.6 illustrates the relationship between the four temperature scales. It has to be emphasized again that the Fahrenheit and Rankine scales are essentially obsolete.

1.2.5 Realization of temperature measurement

Techniques for temperature measurement are very varied. Almost any temperature-dependent effect may be used for temperature measurement. Sections 1.3–1.6 describe the main techniques for temperature measurement used in industry. However, in laboratories or under special industrial conditions, a wider range of instruments is available. In Table 1.7 is a summary of the more usually used measuring instruments in the range quoted. All measuring instruments require to be calibrated against standards. In the case of temperature the standards are the defining fixed points on the IPTS-68. These fixed points are not particularly easy to achieve in workshop conditions. Although the secondary points are intended as workshop standards it is more usual, in most instrument workshops, to calibrate against high grade instruments whose calibration is traceable to the IPTS-68 fixed points.

1.3 Measurement techniques – direct effects

Instruments for measuring temperature described in this section are classified according to the nature of the change in the measurement probe produced by the change of temperature. They have been divided into four classes: liquid expansion, gas expansion, change of state, and solid expansion.

1.3.1 Liquid-in-glass thermometers

The glass thermometer must be the most familiar of all thermometers. Apart from its industrial and laboratory use it finds application in both domestic and medical fields.

1.3.1.1 Mercury-filled glass thermometer

The coefficient of cubical expansion of mercury is about eight times greater than that of glass. If, therefore, a glass container holding mercury is heated, the mercury will expand more than the container. At a high temperature, the mercury will occupy a greater fraction of the volume of the container than at a low temperature. If, then, the container is made in the form of a bulb with a capillary tube attached, it can be so arranged that the surface of the mercury is in the capillary tube, its position along the tube will change with temperature and the assembly used to indicate temperature. This is the principle of the mercury-in-glass thermometer.

The thermometer, therefore, consists simply of a

Table 1.7 Temperature measurement techniques

Range (K)	*Technique*	*Application*	*Resolution* (K)
0.01–1.5	Magnetic susceptance of paramagnetic salt	Laboratory	0.001
0.1–50	Velocity of sound in acoustic cavity	Laboratory standard	0.0001
0.2–2	Vapour pressure	Laboratory standard	0.001
1.5–100	Germanium resistance thermometer	Laboratory standard	0.0001
1.5–100	Carbon resistance thermometer	Laboratory	0.001
1.5–1400	Gas thermometer	Laboratory	0.002
		Industrial	1.0
210–430	Silicon P–N junction	Laboratory	0.1
		Industrial	—
4–500	Thermistor	Laboratory	0.001
		Industrial	0.1
11–550	Quartz crystal oscillator	Laboratory	0.001
		Industrial	—
15–1000	Platinum resistance thermometer	Standard	0.000 01
		Industrial	0.1
20–2700	Thermocouple	General-purpose	1.0
30–3000	Sound velocity in metal rod	Laboratory	1%
130–950	Liquid-in-glass	General-purpose	0.1
130–700	Bimetal	Industrial	1–2
270–5000	Total radiation thermometer	Industrial	10
270–5000	Spectrally selective radiation thermometer	Industrial	2

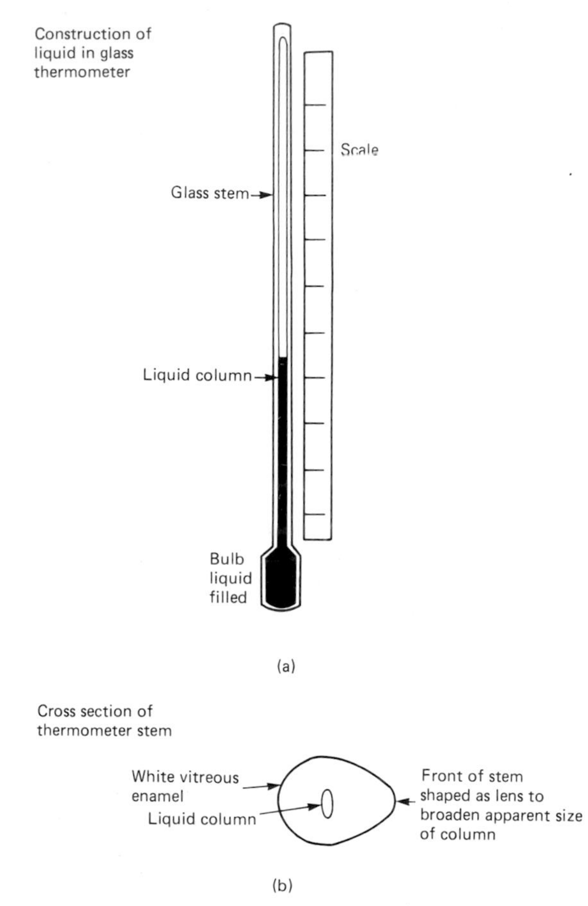

Figure 1.2 Mercury-in-glass thermometer: (a) thermometer and scale, (b) cross-section of thermometer stem.

stem of suitable glass tubing having a very small, but uniform, bore. At the bottom of this stem there is a thin-walled glass bulb. The bulb may be cylindrical or spherical in shape, and has a capacity very many times larger than that of the bore of the stem. The bulb and bore are completely filled with mercury, and the open end of the bore sealed off either at a high temperature, or under vacuum, so that no air is included in the system. The thermometer is then calibrated by comparing it with a standard thermometer in a bath of liquid whose temperature is carefully controlled.

When the standard thermometer and the thermometer to be calibrated have reached equilibrium with the bath at a definite temperature, the point on the glass of the thermometer opposite the top of the mercury meniscus is marked. The process is repeated for several temperatures. The intervals between these marks are then divided off by a dividing machine. In the case of industrial thermometers, the points obtained by calibration are transferred to a brass plate, which is then fixed with the tube into a suitable protecting case to complete the instrument.

The stem of the thermometer is usually shaped in such a way that it acts as a lens, magnifying the width of the mercury column. The mercury is usually viewed against a background of glass which has been enamelled white. Figure 1.2 shows the typical arrangement for a liquid-in-glass thermometer.

Mercury-in-glass thermometers are available in three grades. The limits of error of grades A and B are specified by the National Physical Laboratory, and are given in the tables in the British Standards Code No. 1041 on 'Temperature measurement'. Grade C is a commercial grade of thermometer, and no limits of accuracy are specified. Thermometers of this grade are, of course, cheaper than those of the other grades and their price varies, to a certain degree, according to their accuracy. They should be compared from time to time during use with thermometers of known accuracy. See section on errors due to ageing.

Whenever possible, thermometers should be calibrated, standardized, and used, immersed up to the reading (i.e. totally immersed) as this avoids errors that are due to the fact that the emergent column of mercury and the glass stem are at a lower temperature than that of the bulb, and are therefore not expanded by the same amount. Errors introduced in this way should be allowed for if accurate readings are required, particularly at high temperatures. Some thermometers are, however, calibrated for 'partial immersion', and should be used immersed to the specified depth.

When reading a thermometer an observer should keep his eye on the same level as the top of the mercury column. In this way errors due to parallax will be avoided.

Figure 1.3 shows the effect of observing the thermometer reading from the wrong position. When viewed from (a) the reading is too high. Taken from (b) the reading is correct, but from (c) it is too low.

A mercury-in-glass thermometer has a fairly large thermal capacity (i.e. it requires quite an appreciable amount of heat to change its temperature by one degree), and glass is not a very good conductor of heat. This type of thermometer will, therefore, have a definite thermal lag. In other words, it will require a definite time to reach the temperature of its surroundings. This time should be allowed for before any reading is taken. If there is any doubt as to whether the thermometer has reached equilibrium with a bath of liquid having a constant temperature, then readings should be taken at short intervals of time. When the reading remains constant the thermometer must be in equilibrium with the bath. If the temperature is varying rapidly the thermometer may never indicate the temperature accurately, particularly if the tested medium is a gas.

Glass thermometers used in industry are usually

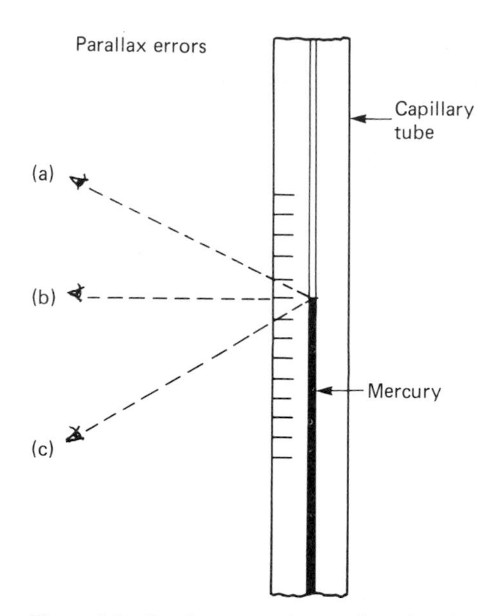

Figure 1.3 Parallax errors when reading glass thermometer.

protected by metal sheaths. These sheaths may conduct heat to or from the neighbourhood of the thermometer bulb and cause the thermometer to read either high or low according to the actual conditions prevailing. A thermometer should, therefore, be calibrated, whenever possible, under the conditions in which it will be used, if accurate temperature readings are required. If, however, the main requirement is that the temperature indication be consistent for the same plant temperature, then an error introduced is not so important, so long as the conditions remain the same, and the error is constant.

Errors due to ageing It is often assumed that provided a mercury-in-glass thermometer is in good condition it will always give an accurate reading. This is not always so, particularly with cheap thermometers. A large error may be introduced by changes in the size of the bulb due to ageing. When glass is heated to a high temperature, as it is when a thermometer is made, it does not, on cooling, contract to its original volume immediately. Thus, for a long time after it has been made the bulb continues to contract very slowly so that the original zero mark is too low on the stem, and the thermometer reads high. This error continues to increase over a long period, and depends upon the type of glass used in the manufacture of the thermometer. In order to reduce to a minimum the error due to this cause, during manufacture thermometers are annealed by baking for several days at a temperature above that which they will be required to measure, and then cooled slowly over a period of several days.

Another error due to the same cause is the depression of the zero when a thermometer is cooled rapidly from a high temperature. When cooled, the glass of the thermometer bulb does not contract immediately to its original size so that the reading on the thermometer at low temperature is too low, but returns to normal after a period of time. This period depends upon the nature of the glass from which the bulb is made.

High temperature thermometers Mercury normally boils at 357 °C at atmospheric pressure. In order to extend the range of a mercury-in-glass thermometer beyond this temperature, the top end of the thermometer bore is enlarged into a bulb having a capacity of about 20 times that of the bore of the stem. This bulb, together with the bore above the mercury, is then filled with nitrogen or carbon dioxide at a sufficiently high pressure to prevent the mercury boiling at the highest temperature at which the thermometer will be used. In order to extend the range to 550 °C, a pressure of about 20 bar is required. In spite of the existence of this gas at high pressure, there is a tendency for mercury to vaporize from the top of the column at high temperatures and to condense on the cooler portions of the stem in the form of minute globules which will not join up again with the main bulk of the mercury. It is, therefore, inadvisable to expose a thermometer to high temperatures for prolonged periods.

At high temperatures the correction for the temperature of the emergent stem becomes particularly important, and the thermometer should, if possible, be immersed to the top of the mercury column. Where this is not possible, the thermometer should be immersed as far as conditions permit, and a correction made to the observed reading, for the emergent column. To do this, the average temperature of the emergent column should be found by means of a short thermometer placed in several positions near to the stem. The emergent column correction may then be found from the formula:

$$\text{correction} = 0.0016(t_1 - t_2)n \text{ on Celsius scale}$$

where t_1 is the temperature of the thermometer bulb, t_2 is the average temperature of the emergent column, and n is the number of degrees exposed. The numerical constant is the coefficient of apparent expansion of mercury in glass.

1.3.1.2 Use of liquids other than mercury

In certain industrial uses, particularly in industries where the escape of mercury from a broken bulb might cause considerable damage to the products, other liquids are used to fill the thermometer. These liquids are also used where the temperature range of the mercury-in-glass thermometer is not suitable. Table 1.8 lists them together with their range of usefulness.

Table 1.8 Liquids used in glass thermometers (expansion type)

Liquid	*Temperature range* (°C)
Mercury	−35 to +510
Alcohol	−80 to +70
Toluene	−80 to +100
Pentane	−200 to +30
Creosote	−5 to +200

1.3.1.3 Mercury-in-glass electric contact thermometer

A mercury-in-glass thermometer can form the basis of a simple on/off temperature controller which will control the temperature of an enclosure at any value between 40 °C and 350 °C.

Mercury is a good electrical conductor. By introducing into the bore of a thermometer two platinum contact wires, one fixed at the lower end of the scale and the other either fixed or adjustable from the top of the stem, it is possible to arrange for an electrical circuit to be completed when a predetermined temperature is reached. The current through the circuit is limited to about 25 mA. This current is used to operate an electronic control circuit. Contact thermometers find applications in laboratories for the temperature control of water baths, fluidized beds and incubators. With careful design, temperature control to 0.1 K can be attained.

Formerly, fixed temperature contact thermometers were used for the temperature control of quartz crystal oscillator ovens, but now this duty is more usually performed by thermistors or semiconductor sensors which can achieve better temperature control by an order of magnitude.

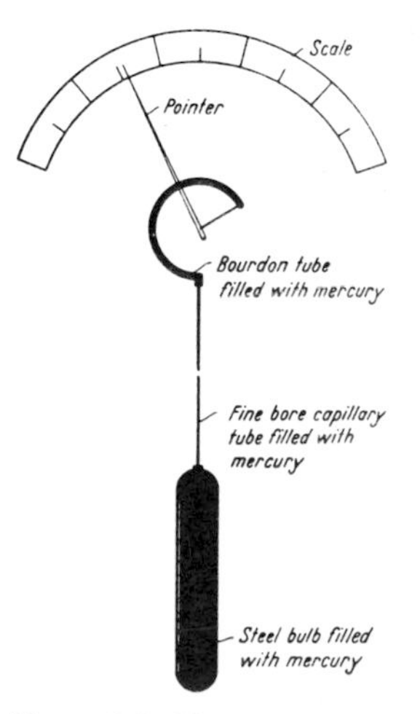

Figure 1.4 Mercury-in-steel thermometer.

1.3.2 Liquid-filled dial thermometers

1.3.2.1 Mercury-in-steel thermometer

Two distinct disadvantages restrict the usefulness of liquid-in-glass thermometers in industry – glass is very fragile, and the position of the thermometer for accurate temperature measurement is not always the best position for reading the scale of the thermometer.

These difficulties are overcome in the mercury-in-steel thermometer shown in Figure 1.4. This type of thermometer works on exactly the same principle as the liquid-in-glass thermometer. The glass bulb is, however, replaced by a steel bulb and the glass capillary tube by one of stainless steel. As the liquid in the system is now no longer visible, a Bourdon tube is used to measure the change in its volume. The Bourdon tube, the bulb and the capillary tube are completely filled with mercury, usually at a high pressure. When suitably designed, the capillary tube may be of considerable length so that the indicator operated by the Bourdon tube may be some distance away from the bulb. In this case the instrument is described as being a 'distant reading' or 'transmitting' type.

When the temperature rises, the mercury in the bulb expands more than the bulb so that some mercury is driven through the capillary tube into the Bourdon tube. As the temperature continues to rise, increasing amounts of mercury will be driven into the Bourdon tube, causing it to uncurl. One end of the Bourdon tube is fixed, while the motion of the other end is communicated to the pointer or pen arm. As there is a large force available the Bourdon tube may be made robust and will give good pointer control and reliable readings.

The Bourdon tube may have a variety of forms, and the method of transmitting the motion to the pointer also varies. Figure 1.5 shows one form of Bourdon tube in which the motion of the free end is transmitted to the pointer by means of a segment and pinion. The free end

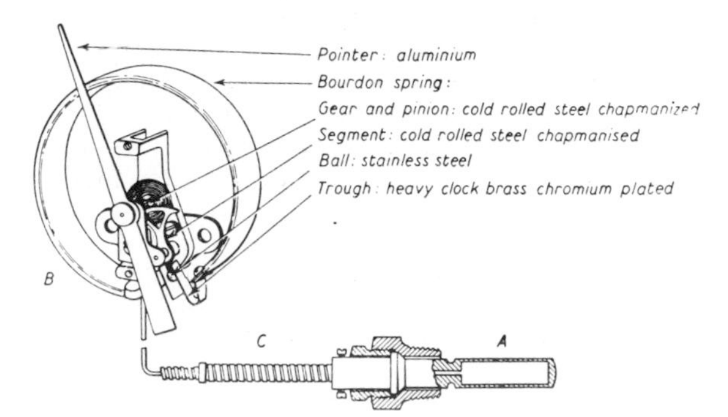

Figure 1.5 Construction of mercury-in-steel thermometer (courtesy The Foxboro Company).

Figure 1.6 Multi-turn Bourdon tube.

of the tube forms a trough in which a stainless steel ball at the end of the segment is free to move. The ball is held against the side of the trough by the tension in the hair-spring. By using this form of construction lost motion and angularity error are avoided, and friction reduced to a minimum. Ambient-temperature compensation may be obtained by using a bimetallic strip, or by using twin Bourdon tubes in the manner described under the heading of capillary compensation.

Figure 1.6 shows a Bourdon tube having a different form, and a different method of transmitting the motion to the pen arm. This Bourdon tube is made of steel tube having an almost flat section. A continuous strip of the tubing is wound into two coils of several turns. The coils are arranged one behind the other so that the free end of each is at the centre while the outer turn of the coils is common to both, as can be seen in the illustration. One end of the continuous tube – the inner end of the back coil – is fixed and leads to the capillary tube; while the other end – the inner end of the front coil – is closed, and is attached to the pointer through a small bimetallic coil which forms a continuation of the Bourdon tube. This bimetallic coil compensates for changes brought about in the elastic properties of the Bourdon tube and in the volume of the mercury within the Bourdon tube due to ambient temperature changes.

This particular formation of the tube causes the pointer to rotate truly about its axis without the help of bearings, but bearings are provided to keep the pointer steady in the presence of vibration. The friction at the bearings will, therefore, be very small as there is little load on them. As the end of the Bourdon tube rotates the pointer directly, there will be no backlash.

Thermometer bulbs The thermometer bulb may have a large variety of forms depending upon the use to which it is put. If the average temperature of a large enclosure is required, the bulb may take the form of a considerable length of tube of small diameter either arranged as a U or wound into a helix. This form of bulb is very useful when the temperature of a gas is being measured, for it presents a large surface area to the gas and is therefore more responsive than the forms having a smaller surface area for the same cubic capacity.

In the more usual form, the bulb is cylindrical in shape and has a robust wall: the size of the cylinder depends upon many factors, such as the filling medium and the temperature range of the instrument, but in all cases, the ratio of surface area to volume is kept at a maximum to reduce the time lag in the response of the thermometer.

The flange for attaching the bulb to the vessel in which it is placed also has a variety of forms depending upon whether the junction has to be gas-tight or not, and upon many other factors. Figure 1.7 shows some forms of bulbs.

The capillary tube and its compensation for ambient temperature The capillary tube used in the mercury-in-steel thermometer is usually made from stainless steel, as mercury will amalgamage with other metals. Changes of temperature affect the capillary and the mercury it contains, and hence the thermometer reading; but if the capillary has a very small capacity, the error owing to changes in the ambient temperature will be negligible.

Where a capillary tube of an appreciable length is used, it is necessary to compensate for the effects brought about by changes in the temperature in the

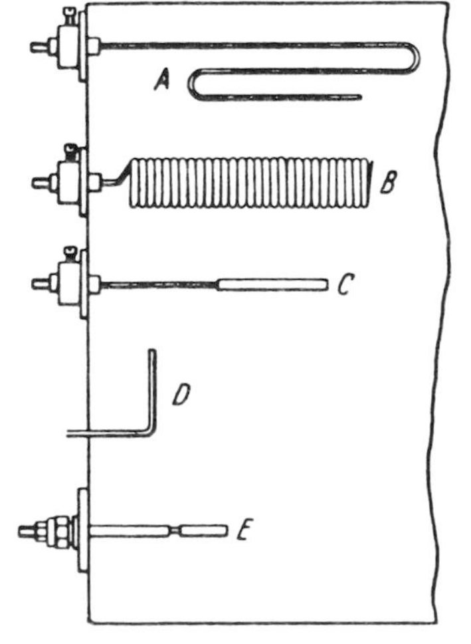

Figure 1.7 Forms for bulbs for mercury-in-steel thermometers.

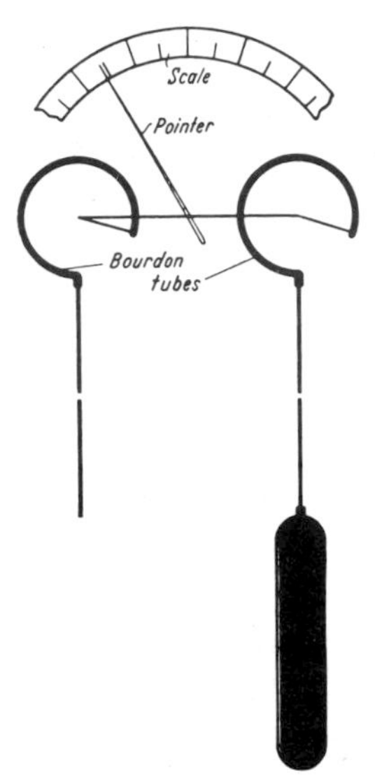

Figure 1.8 Ambient temperature compensation of mercury-in-steel thermometer.

neighbourhood of the tube. This may be done in a number of ways. Figure 1.8 illustrates a method which compensates not only for the changes of temperature of the capillary tube, but also for the changes of temperature within the instrument case. In order to achieve complete temperature compensation two thermal systems are used, which are identical in every respect except that one has a bulb and the other has not. The capillary tubes run alongside each other, and the Bourdon tubes are in close proximity within the same case. If the pointer is arranged to indicate the difference in movement between the free ends of the two Bourdon tubes, then it will be indicating an effect which is due to the temperature change in the bulb only. If compensation for case temperature only is required, then the capillary tube is omitted in the compensating system, but in this case the length of capillary tube used in the uncompensated system should not exceed about 8 metres.

Another method of compensating for temperature changes in the capillary tube is to use a tube of comparatively large bore and to insert into the bore a wire made of Invar, or other alloy with a very low coefficient of expansion. Mercury has a coefficient of cubical expansion about six times greater than that of stainless steel. If the expansion of the Invar wire may be regarded as being negligibly small, and the wire is arranged to fill five-sixths of the volume of the capillary bore, then the increase in the volume of the mercury which fills the remaining one-sixth of the bore will exactly compensate for the increase in volume of the containing capillary tube. This method requires the dimensions both of the bore of the capillary tube and of the diameter of the wire insert to be accurate to within very narrow limits for accurate compensation. The insert may not necessarily be continuous, but may take the form of short rods, in which case it is, however, difficult to eliminate all trapped gases.

Compensation for changes in the temperature of the capillary tube may also be achieved by introducing compensating chambers, of the form shown in Figure 1.9, at intervals along the length of the capillary tube. These chambers operate on exactly the same principle as the Invar-wire-insert type of capillary tube, but the proportion of the chamber occupied by the Invar is now arranged to compensate for the relative increase in volume of the mercury within the chamber and in the intervening length of capillary tube.

1.3.2.2 Other filling liquids

Admirable though mercury may be for thermometers, in certain circumstances it has its limitations, particularly at the lower end of the temperature scale. It is also very expensive to weld mercury systems in stainless steel. For these and other reasons, other liquids are used in place of mercury. Details of the liquids used in liquid-in-metal thermometers, with their usual temperature ranges, are given in Table 1.9. Comparison with Table 1.8 shows that liquids are used for different temperature ranges in glass and metal

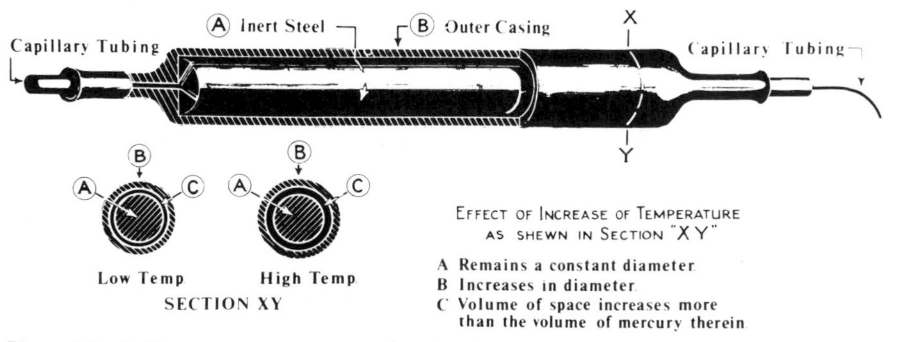

Figure 1.9 Ambient temperature compensation chamber.

Table 1.9 Liquids used in metal thermometers (expansion type)

Liquid	*Temperature range* (°C)
Mercury	−39 to +650
Xylene	−40 to +400
Alcohol	−46 to +150
Ether	+20 to +90
Other organic liquids	−87 to +260

thermometers. In general, in metal thermometers, liquids can be used up to higher temperatures than in glass thermometers as they can be filled to higher pressures.

When liquids other than mercury are used, the bulb and capillary tube need no longer be made of steel. The material of the bulb may, therefore, be chosen from a wide range of metals and alloys, and is selected to give the maximum resistance to any corrosive action which may be present where the bulb is to be used.

The capillary tube, too, may be made from a variety of materials, although copper and bronze are the most common. When capillary tubes are made from materials other than stainless steel, it may be necessary to protect them from corrosion or mechanical damage. This may be done by covering the tube with thermal insulation material – formerly asbestos was used – and winding the whole in a heavy spiral of bronze. In cases where a bronze outer casing is likely to be damaged either by acid fumes or mechanically, it may be replaced by a stainless steel spiral which results in a much stronger but slightly less flexible construction. For use in damp places, or where the tube is liable to be attacked by acid fumes, the capillary and bronze spiral may be protected by a covering of moulded rubber, polyvinyl chloride, or rubber-covered woven-fabric hose. For use on chemical plants such as sulphuric acid plants both the capillary tube and the bulb are protected by a covering of lead.

The construction of the liquid-in-metal thermometer is the same as that of the mercury-in-steel thermometer, and compensation for changes in ambient temperature may be achieved in the same ways.

Further facts about liquid-in-metal thermometers will be found in Table 1.11 on p. 17 showing the comparison of the various forms of non-electrical dial thermometers.

In installations where liquid-filled instruments with very long capillaries are used, care must be taken to see that there is not a significant height difference between the bulb location and that of the instrument. If there is a large height difference, the pressure due to the column of liquid in the capillary will be added to (or subtracted from) the pressure due to the expansion of the liquid in the bulb resulting in a standing error in the temperature reading. This problem is at its worst with mercury-filled instruments. Instruments with double capillary ambient temperature compensation, Figure 1.7, are, of course, also compensated for static head errors.

1.3.3 Gas-filled instruments

The volume occupied by a given mass of gas at a fixed pressure is a function of both the molecular weight of the gas and its temperature. In the case of the 'permanent gases' provided the temperature is significantly above zero kelvin the behaviour of a gas is represented by the equation

$$pv = RT \tag{1.13}$$

where p is pressure in $N.m^{-2}$, v is volume in m^3, T is the temperature in K and R is the gas constant with a value of $8.314\ J.mol^{-1}.K^{-1}$.

If, therefore, a certain volume of inert gas is enclosed in a bulb, capillary and Bourdon tube, and most of the gas is in the bulb, then the pressure as indicated by the Bourdon tube, may be calibrated in terms of the temperature of the bulb. This is the principle of the gas-filled thermometer.

Since the pressure of a gas maintained at constant volume increases by 1/273 of its pressure at 0 °C for every degree rise in temperature, the scale will be linear provided the increase in volume of the Bourdon tube, as it uncurls, can be neglected in comparison with the total volume of gas.

An advantage of the gas-filled thermometer is that the gas in the bulb has a lower thermal capacity than a similar quantity of liquid, so that the response of the thermometer to temperature changes will be more rapid than that for a liquid-filled system with a bulb of the same size and shape.

The coefficient of cubical expansion of a gas is many times larger than that of a liquid or solid (air, 0.003 7; mercury, 0.000 18; stainless steel, 0.000 03). It would therefore appear at first sight, that the bulb for a gas-filled system would be smaller than that for a liquid-filled system. The bulb must, however, have a cubical capacity many times larger than that of the capillary tube and Bourdon tube, if the effects of ambient temperature changes upon the system are to be negligible.

It is extremely difficult to get accurate ambient temperature compensation in any other way. The change in dimensions of the capillary tube due to a temperature change is negligible in comparison with the expansion of the gas. Introducing an Invar wire into the capillary bore would not be a solution to the problem, because the wire would occupy such a large proportion of the bore that extremely small variations in the dimensions of the bore or wire would be serious.

Placing an exactly similar capillary and Bourdon

tube alongside that of the measuring system, and measuring the difference in the change of the two Bourdons does not give accurate compensation. This can be seen from the following example. Suppose the capillary and Bourdon tube have a capacity equal to 1/100 part of the capacity of the whole system. Let the ambient temperature rise by 10 °C, while the bulb remains at the same temperature. In the compensating system the pressure will increase by 10/273 of the pressure at 0 °C. In the measuring system this temporary increase in pressure in the capillary tube will soon be reduced by gas flowing into the bulb from the capillary tube until the pressures in the bulb and capillary are the same. Thus, the increase in pressure in the measuring system will only be about one-hundredth of the pressure increase in the compensating system.

More accurate compensation would be obtained by having a compensating system which also included a bulb maintained at ambient temperature, but this again would not give the completely accurate compensation.

Further facts about gas expansion thermometers will be found in Table 1.11, in which certain forms of dial thermometers are compared.

1.3.4 Vapour pressure thermometers

Suppose a container is partially filled with liquid and the space above the liquid is completely evacuated. The molecules of liquid will be in motion, and will be moving in an entirely random manner. From time to time, molecules having a vertical component of velocity will reach the surface of the liquid. If this vertical component of velocity is great enough, a molecule will be able to leave the liquid in spite of the fact that other molecules are attracting it back into the liquid. Thus, after a time, a number of liquid molecules will occupy the space above the liquid. These molecules, too, will be in a state of random motion, and from time to time molecules will leave the vapour and pass back into the liquid.

At first, the rate at which molecules are returning to the liquid will be less than the rate at which they leave, and the vapour above the liquid is said to be 'unsaturated'. Eventually, however, as the number of molecules in the vapour state increases, the rate at which the molecules leave the liquid will be exactly equal to the rate at which they return, and the quantity – and, therefore, the pressure – of the vapour in the space will remain constant. When this is so, the vapour is said to be 'saturated'.

If the temperature of the container is raised, the velocity of the molecules of liquid will be increased. The number of molecules now having sufficient energy to leave the liquid will be increased, and a greater number will leave the liquid. The rate at which molecules leave the liquid will now be greater than the rate at which they return so that the vapour pressure in the space will increase. It will continue to increase until the saturated vapour pressure for the new temperature is reached, when the molecules of vapour will again be returning to the liquid at the same rate as that at which liquid molecules leave.

If, instead of having a fixed top, the container has a movable piston, then, if the volume of the space is increased without the temperature changing, the vapour pressure will temporarily fall as the same number of vapour molecules are occupying a larger space. Now, however, the rate at which the molecules return to the liquid will be reduced, and will not be as great as the rate at which they leave the liquid. The pressure of the vapour will therefore increase, until the rate at which molecules return to the liquid again balances the rate at which they leave. The vapour pressure returns, therefore, to the 'saturated' vapour pressure for the particular temperature.

Table 1.10 Liquids used in vapour pressure thermometers

Liquid	*Critical temperature* (°C)	*Boiling point* (°C)	*Typical ranges available* (°C)
Argon	−122	−185.7	Used for measuring very low temperatures down to −253 °C in connection with the liquefaction of gases
Methyl chloride	143	−23.7	0 to 50
Sulphur dioxide	157	−10	30 to 120
Butane (n)	154	−0.6	20 to 80
Methyl bromide		4.6	30 to 85
Ethyl chloride	187	12.2	30 to 100
Diethyl ether	194	34.5	60 to 160
Ethyl alcohol	243	78.5	30 to 180
Water	375	100	120 to 220
Toluene	321	110.5	150 to 250

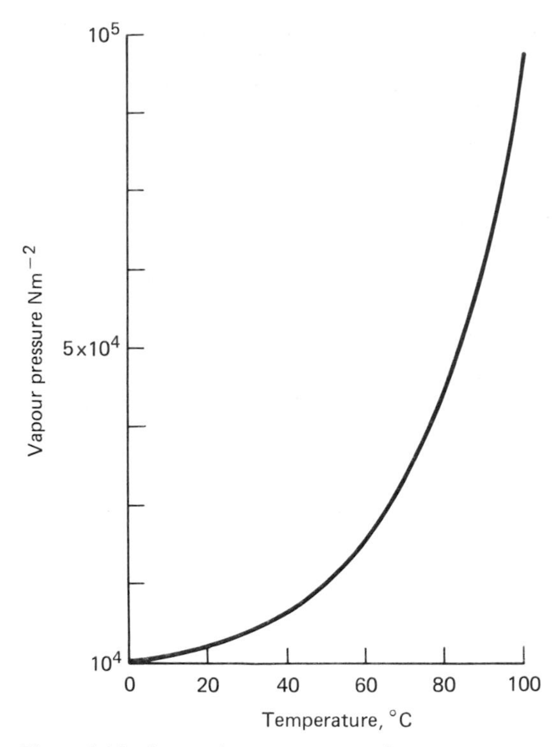

Figure 1.10 Saturated vapour pressure of water.

In the same way, if the volume of the space is reduced, molecules will leave the vapour at a greater rate than that at which they leave the liquid, so that the vapour pressure, which was temporarily increased, will fall until it is again the saturated vapour pressure of the liquid at the particular temperature.

Thus, provided there is always liquid and vapour present, the saturated vapour pressure of the liquid depends only upon its temperature, and is independent of the size of the container.

If a thermometer system similar to that described for gas expansion thermometers is arranged so that the system contains both liquid and vapour and the interface between liquid and vapour is in the bulb, that is, at the temperature whose value is required, then the vapour pressure as measured by the Bourdon tube will give an indication of the temperature. This indication will be completely independent of the volume of the bulb, the capillary and the Bourdon tube and therefore independent of expansion due to ambient temperature changes.

The saturated vapour pressure of a liquid is not linear with temperature. Figure 1.10 shows the temperature—vapour pressure relationship for a typical liquid. The form of the vapour pressure graphs for other volatile liquids is of a similar form. It will be seen that pressure versus temperature is non-linear. A thermometer based on vapour pressure will have a scale on which the size of the divisions increases with increasing temperature.

The realization of a vapour instrument is essentially the same as a gas-filled instrument except that in the latter the whole instrument is filled with a permanent gas while in the former the bulb is filled partly with liquid and partly with gas. This arrangement is shown diagrammatically in Figure 1.11(a).

Many liquids are used for vapour-pressure-actuated thermometers. The liquid is chosen so as to give the required temperature range, and so that the usual operating temperature comes within the widely spaced graduations of the instrument. In some forms of the instrument, a system of levers is arranged to give a linear portion to the scale over a limited portion of its range. By suitable choice of filling liquid, a wide variety of ranges is available, but the range for any particular filling liquid is limited. The choice of material for bulb construction is also very wide. Metals – such as copper, steel, Monel metal, tantalum – may be used. Table 1.10 shows a number of liquids commonly used for vapour-pressure thermometers together with their useful operating ranges.

In the instrument, shown diagrammatically in Figure 1.11(a) a quantity of liquid partially fills the bulb. The surface of the liquid in the bulb should be at the temperature which is being measured. The method by which the vapour pressure developed in the bulb is transmitted to the Bourdon tube will depend upon whether the temperature of the capillary tube and Bourdon tube is above or below that of the bulb.

If the ambient temperature of the capillary and Bourdon tube is above that of the bulb, then they will be full of vapour, which will transmit the vapour pressure, as shown in Figure 1.11(a). When the ambient temperature increases, it will cause the vapour in the capillary and Bourdon tube to increase in pressure temporarily, but this will cause vapour in the bulb to condense until the pressure is restored to the saturated vapour pressure of liquid at the temperature of the bulb.

Vapour pressure instruments are not usually satisfactory when the temperature being measured at the bulb is near the ambient temperature of the capillary and the Bourdon tube. In particular, significant measurement delays occur as the measured temperature crosses the ambient temperature. These delays are caused by the liquid distilling into or out of the gauge and capillary, Figure 1.11(b).

If there is a significant level difference between the bulb and the gauge, an error will be produced when liquid distils into the capillary due to the pressure head from the column of liquid.

When rapid temperature changes of the bulb occur passing through ambient temperature the movement of the instrument pointer may be quite erratic due to

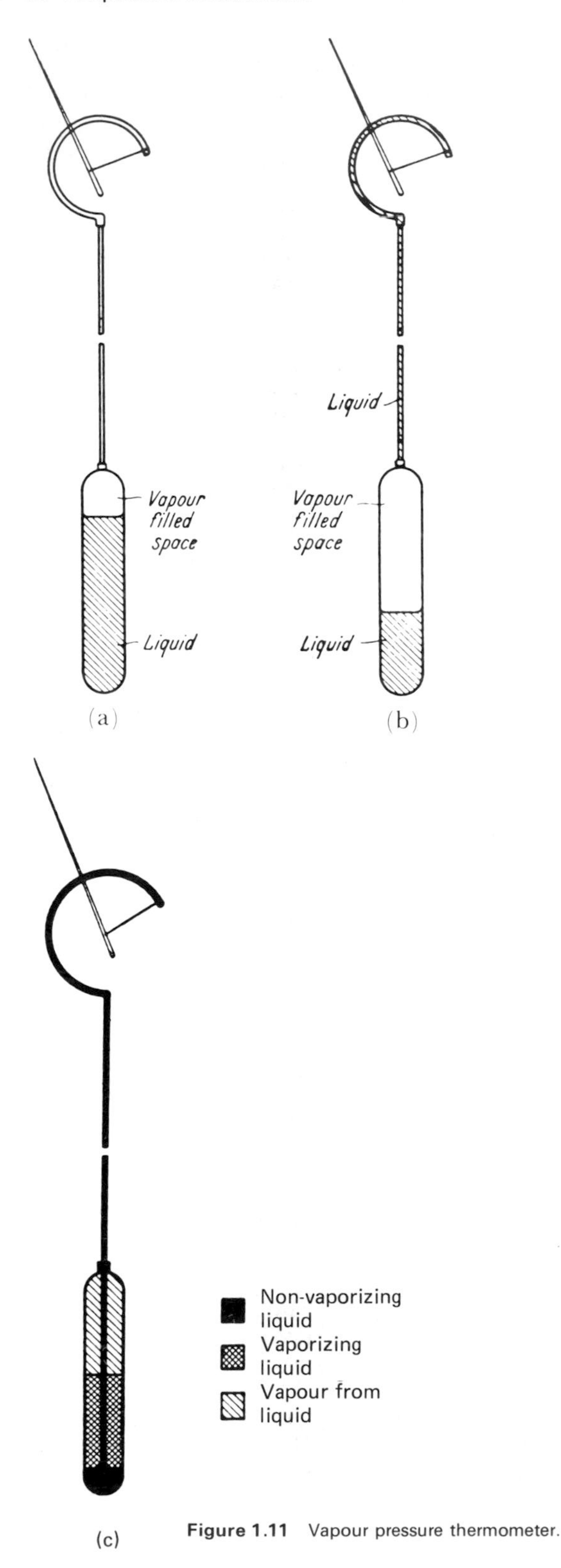

Figure 1.11 Vapour pressure thermometer.

the formation of bubbles in the capillary.

In order to overcome the defects brought about by distillation of the liquid into, and out of, the capillary and Bourdon tubes, these tubes may be completely filled with a non-vapourizing liquid which serves to transmit the pressure of the saturated vapour from the bulb to the measuring system. To prevent the non-vaporizing liquid from draining out of the capillary tube, it is extended well down into the bulb, as shown in Figure 1.11(c), and the bulb contains a small quantity of the non-vaporizing fluid. The non-vaporizing fluid will still tend to leave the capillary tube unless the bulb is kept upright.

Vapour pressure thermometers are very widely used because they are less expensive than liquid- and gas-filled instruments. They also have an advantage in that the bulb can be smaller than for the other types.

The range of an instrument using a particular liquid is limited by the fact that the maximum temperature for which it can be used must be well below the critical temperature for that liquid. The range is further limited by the non-linear nature of the scale.

In Table 1.11 the three types of fluid-filled thermometers are compared.

1.3.5 Solid expansion

Thermal expansion of solids, usually metals, forms the basis of a wide range of inexpensive indicating and control devices. These devices are not particularly accurate, typically errors of as much as ± 5 K or more may be expected, but due to their low cost they find wide application especially in consumer equipment. As indicated earlier in this section this technique is also used to provide temperature compensation in many instruments.

The temperature sensitive elements using solid expansion fall into two groups: rod sensing probes and bimetal strips.

There are so many applications that only one or two examples will be given to illustrate the techniques.

1.3.5.1 Rod sensing probes

The widest application of this technique is for immersion thermostats for use in hot water temperature control. Figure 1.12 shows diagrammatically the operation of an immersion thermostat. The microswitch is operated by the thermal expansion of the brass tube. The reference length is provided by a rod of low thermal expansion such as Invar. These thermostats, though not particularly accurate and having a switching differential of several kelvin, provide a very rugged and reliable control system for a non-critical application such as domestic hot water

Table 1.11 Comparison of three types of dial thermometers

	Liquid-in-metal	*Gas expansion (constant volume)*	*Vapour pressure*
Scale	Evenly divided.	Evenly divided.	Not evenly divided. Divisions increase in size as the temperature increases. Filling liquid chosen to give reasonably uniform scale in the neighbourhood of the operating temperatures.
Range	Wide range is possible with a single filling liquid, particularly with mercury. By choice of suitable filling liquid, temperatures may be measured between −200 °C and 570 °C, but not with a single instrument.	Usually has a range of at least 50 °C between −130 °C and 540 °C. Can be used for a lower temperature than mercury in steel.	Limited for a particular filling liquid, but with the choice of a suitable liquid almost any temperature between −50 °C and 320 °C may be measured. Instrument is not usually suitable for measuring temperatures near ambient temperatures owing to the lag introduced when bulb temperature crosses ambient temperature.
Power available to operate the indicator.	Ample power is available so that the Bourdon tube may be made robust and arranged to give good pointer control.	Power available is very much less than that from liquid expansion.	Power available is very much less than that from liquid expansion.
Effect of difference in level of bulb and Bourdon tube.	When the system is filled with a liquid at high pressure, errors due to difference of level between bulb and indicator will be small. If the difference in level is very large, a correction may be made.	No head error, as the pressure due to difference in level is negligible in comparison with the total pressure in the system.	Head error is not negligible, as the pressure in the system is not large. Error may be corrected over a limited range of temperature if the ratio pressure to deflection of the pointer can be considered constant over that range. In this case the error is corrected by resetting the pointer.
Effect of changes in barometric pressure.	Negligible.	May produce a large error. Error due to using the instrument at a different altitude from that at which it was calibrated may be corrected by adjusting the zero. Day to day variations in barometric pressure may be corrected for in the same way.	Error may be large, but may be corrected by resetting the pointer as for head error. Day to day errors due to variation in barometric pressure may be corrected by zero adjustment.
Capillary error.	Compensation for change in ambient temperature obtained as described in text (page 12).	Difficult to eliminate (see page 13).	No capillary error.
Changes in temperature at the indicator.	Compensation obtained by means of a bimetallic strip.	Compensation obtained by means of bimetallic strip.	Errors due to changes in the elasticity of the Bourdon tube are compensated for by means of a bimetallic strip.
Accuracy.	$\pm\frac{1}{2}\%$ of range to 320 °C $\pm 1\%$ of range above 320 °C.	$\pm 1\%$ of differential range of the instrument if the temperature of the capillary and Bourdon tube does not vary too much.	$\pm 1\%$ of differential range even with wide temperature variation of the capillary and Bourdon tube.

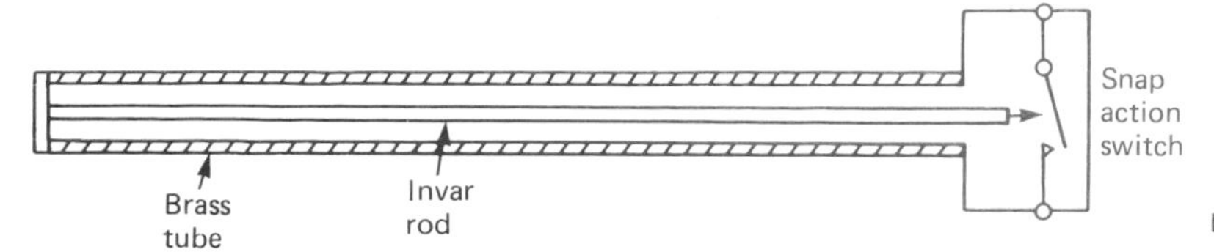

Figure 1.12 Rod thermostat.

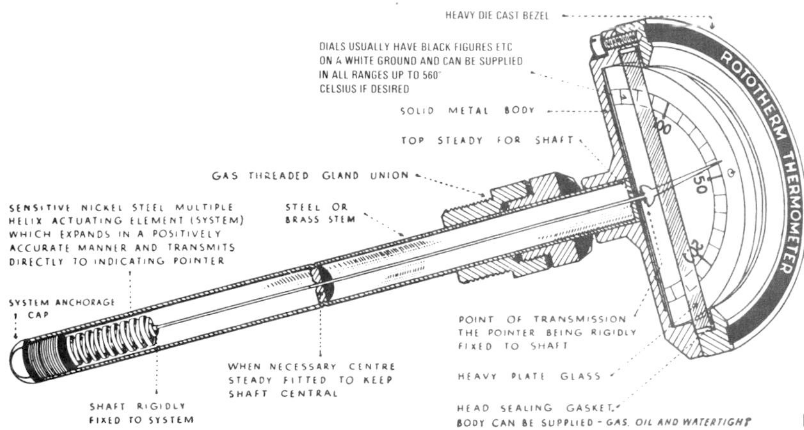

Figure 1.13 Dial thermometer.

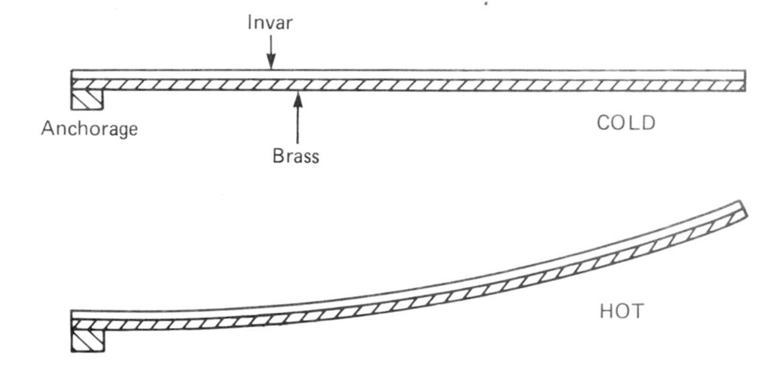

Figure 1.14 Action of bimetal strip.

control. Life spans of fifty years continuous operation are fairly typical.

Figure 1.13 shows another rod application. In this case to achieve greater sensitivity the expanding component is coiled.

1.3.5.2 Bimetal strip thermometer

Bimetal strips are fabricated from two strips of different metals with different coefficients of thermal expansion bonded together to form, in the simplest case, a cantilever. Typical metals are brass and Invar. Figure 1.14 illustrates this principle. As the temperature rises the brass side of the strip expands more than the Invar side, resulting in the strip curling, in this case upwards.

In this 'straight' form a bimetal strip can form part of a micro-switch mechanism thus forming a temperature-sensitive switch or thermostat.

To construct a thermometer the bimetal element is coiled into a spiral or helix. Figure 1.15 shows a typical coiled thermometer element.

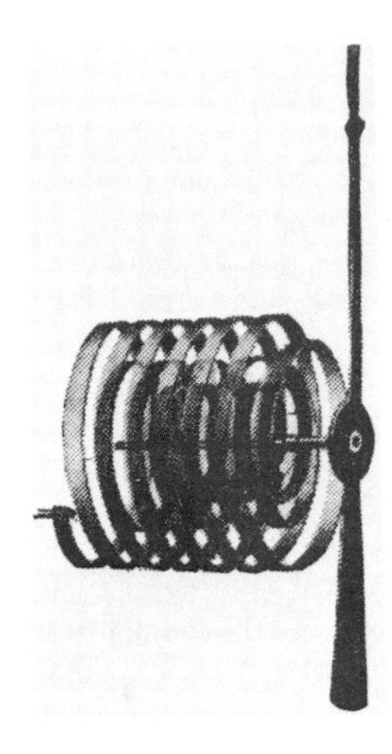

Figure 1.15 Helical bimetal strip.

A long bimetal strip, consisting of an Invar strip welded to a higher expansion nickel–molybdenum alloy wound around without a break into several compensated helices, arranged coaxially one within the other forms the temperature-sensitive element of an instrument which may be designed to measure temperature. This method of winding the strip enables a length, sufficient to produce an appreciable movement of the free end, to be concentrated within a small space. It also makes it possible to keep the thermal capacity of the element and its stem at a low value, so the instrument will respond rapidly to small temperature changes.

The helices in the winding are so compensated that any tendency towards lateral displacement of the spindle in one helix is counteracted by an opposite tendency on the part of one or more of the other helices. Thus, the spindle of the instrument is fully floating, retaining its position at the centre of the scale without the help of bearings. The instrument is, therefore, not injured by mechanical shocks which would damage jewelled bearings.

This particular design also results in the angular rotation of the spindle being proportional to the change in temperature for a considerable temperature range. The instrument has a linear temperature scale, and can be made to register temperatures up to 300 °C to within ±1 per cent of the scale range.

Due to its robust construction, this instrument is used on many industrial plants, and a slightly modified form is used in many homes and offices to indicate room temperature. It can be made for a large variety of temperature ranges and is used in many places where the more fragile mercury-in-glass thermometer was formerly used.

1.4 Measurement techniques – electrical

1.4.1 Resistance thermometers

All metals are electrical conductors which at all but very low temperatures offer resistance to the passage of electric current. The electrical resistance exhibited by a conductor is measured in ohms. The proportional relationship of electrical current and potential difference is given by Ohm's law:

$$R = E/I \tag{1.14}$$

where R is resistance in ohms, E is potential difference in volts, and I is current in amperes. Different metals show widely different resistivities. The resistance of a conductor is proportional to its length and inversely proportional to its cross-sectional area, i.e.

$$R = \rho \frac{L}{A} \tag{1.15}$$

or

$$\rho = R \frac{A}{L} \tag{1.16}$$

where R is resistance of the conductor, ρ is resistivity of the material, L is length of the conductor, and A is cross-sectional area of the conductor. The units of resistivity are ohms . metre.

The resistivity of a conductor is temperature-dependent. The temperature coefficient of resistivity is positive for metals, that is, the resistance increases with temperature, and for semiconductors the temperature coefficient is negative. As a general guide at normal ambient temperatures the coefficient of resistivity of most elemental metals lies in the region of 0.35 per cent to 0.7 per cent per kelvin.

Table 1.12 Resistivities of different metals

Metal	*Resistivity at 293 K microhms . metre*	*Temperature coefficient of resistivity* (K^{-1})
Aluminium	282.4	0.0039
Brass (yellow)	700	0.002
Constantan	4900	10^{-5}
Copper (annealed)	172.4	0.003 93
Gold	244	0.0034
Iron (99.98%)	1000	0.005
Mercury	9578	0.000 87
Nichrome	10 000	0.0004
Nickel	780	0.0066
Platinum (99.85%)	11 060	0.003 927
Silver	159	0.0038
Tungsten	560	0.0045

1 Resistivities of metals are dependent on the purity or exact composition of alloys. Some of the above figures represent average values.
2 Temperature coefficients of resistivity vary slightly with temperature. The above values are for 20 °C.

Table 1.12 shows the resistivity and temperature coefficients for a number of common metals: both elements and alloys.

The metals most used for resistance measurement are platinum, nickel and copper. These metals have the advantage that they can be manufactured to a high degree of purity and consequently they can be made with very high reproducibility of resistance characteristics. Copper has the disadvantage of a low resistivity resulting in inconveniently large sensing elements and has the further disadvantage of poor resistance to corrosion resulting in instability of electrical characteristics. The main area of application of copper for resistance thermometers is in electronic instrumentation where it is in a controlled environment and where an essentially linear temperature characteristic is required.

1.4.1.1 Platinum resistance thermometers

Platinum is the standard material used in the resistance thermometer which defines the International Practical Temperature Scale, not because it has a particularly high coefficient of resistivity, but because of its stability in use. In fact, a high coefficient is not, in general, necessary for a resistance thermometer material as resistance values can be determined with a high degree of accuracy using suitable equipment and taking adequate precautions.

Platinum, having the highest possible coefficient of resistivity, is considered the best material for the construction of thermometers. A high value of this coefficient is an indication that the platinum is of high purity. The presence of impurities in resistance thermometer material is undesirable, as diffusion, segregation and evaporation may occur in service, resulting in a lack of stability of the thermometer. The temperature coefficient of resistivity is also sensitive to internal strains so that it is essential that the platinum should be annealed at a temperature higher than the maximum temperature of service. The combination of purity and adequate annealing is shown by a high value of the ratio of the resistances at the steam and ice points. To comply with the requirements of the International Practical Temperature Scale of 1968 this ratio must exceed 1.392 50.

It is essential that the platinum element is mounted in such a way that it is not subject to stress in service.

Platinum is used for resistance thermometry in industry for temperatures up to 800 °C. It does not oxidize, but must be protected from contamination. The commonest cause of contamination of platinum resistance thermometers is contact with silica, or silica-bearing refractories, in a reducing atmosphere. In the presence of a reducing atmosphere, silica is reduced to silicon which alloys with platinum making it brittle. Platinum resistance thermometers may be used for temperatures down to about 20 K.

For measuring temperatures between 1 K and 40 K doped germanium sensors are usually used, while carbon resistors are used between 0.1 K and 20 K. Above 20 K platinum has a greater temperature coefficient of resistivity and has a greater stability. Between 0.35 K and 40 K a new resistance thermometer material (0.5 atomic % iron–rhodium) is also used.

Calibration of resistance thermometers To conform with IPTS-68 the resistance of the thermometer at temperatures below 0 °C is measured at a number of defining points and the calibration is obtained by difference from a reference function W which is defined and tabulated in the scale. The differences from the function ΔW are expressed by polynomials, the coefficients of which are obtained from calibration at fixed points for each of the ranges 13.81–20.28 K, 20.28–54.361 K, 54.361–90.188 K and 90.188–273.15 K. The last mentioned range was formerly defined by the Callendar–Van Dusen equation but the difference from the reference function given by the equation

$$\Delta W = A + B\left[\frac{t_{68}}{100\,^{\circ}\mathrm{C}} - 1\right]\frac{t_{68}}{100} \qquad (1.17)$$

is now used, where t_{68} is the temperature in °C and the constants A and B are determined by measurements of W at 100 °C and −182.962 °C (90.188 K).

For the range 0 °C to 630.74 °C the Callendar equation is still used but a correction term is added so that the calibration procedure is to measure the resistance of the thermometer at 0 °C (obtained by way of the triple point of water), the boiling point of water (100 °C) and the freezing point of zinc 419.58 °C on the 1968 scale (formerly 419.505 °C on the 1948 scale). The Callendar equation is then used to determine the intermediate value of t':

$$t' = \frac{1}{\alpha}(W(t') - 1) + \delta\left(\frac{t'}{100\,^{\circ}\mathrm{C}}\right)\left(\frac{t'}{100\,^{\circ}\mathrm{C}} - 1\right) \qquad (1.18)$$

The procedure is then to correct t' by an amount which varies with temperature but is the same for all thermometers which meet the specification of the scale:

$$t_{68} = t' + 0.45\left(\frac{t'}{100\,^{\circ}\mathrm{C}}\right)\left(\frac{t'}{100\,^{\circ}\mathrm{C}} - 1\right) \times\left(\frac{t'}{419.58\,^{\circ}\mathrm{C}} - 1\right)\left(\frac{t'}{630.74\,^{\circ}\mathrm{C}} - 1\right)\,^{\circ}\mathrm{C} \qquad (1.19)$$

The value of α for a given specimen of platinum is the same on the 1948 and 1968 scales but the value of δ changes because of the change in the assigned zinc point; for example, a δ coefficient of 1.492 on the old scale becomes 1.497 on the new.

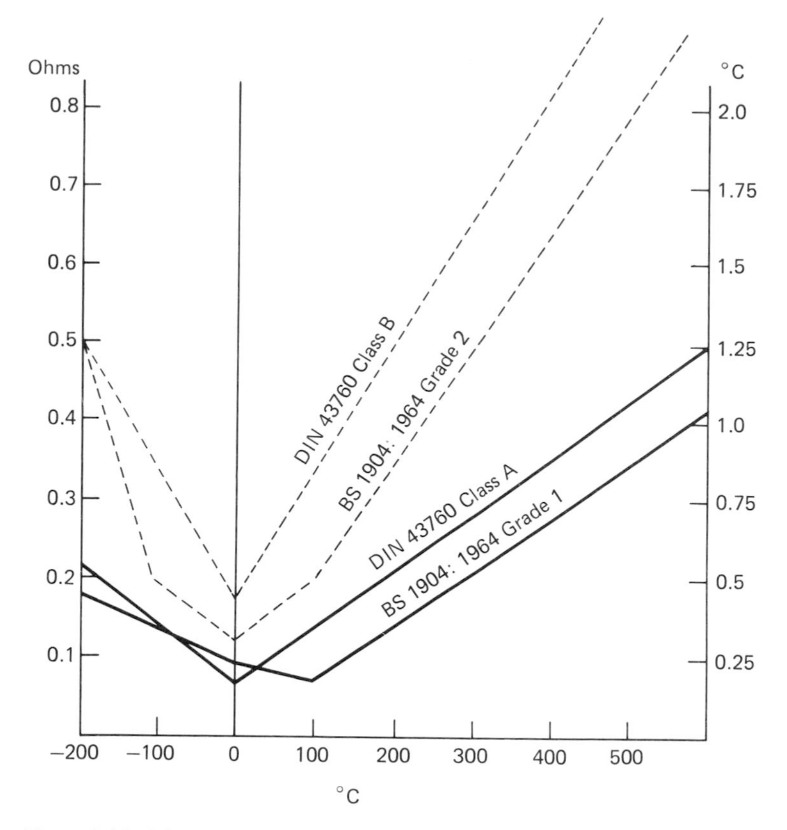

Figure 1.16 BS and DIN specifications for 100 ohm platinum resistance thermometers (courtesy Kent Industrial Instruments Ltd.)

Industrial resistance thermometers Industrial platinum resistance thermometers manufactured in Britain and Europe in general conform to BS 1904, 1964 or DIN 43 760 (1980). BS 1904 is currently being revised to conform with DIN 43 760 (1980). Figure 1.16 shows the tolerances in ohms and kelvin permitted by these two standards. Thermometers are usually produced to Grade 2 or Class B for general industrial use but they are available to Grade 1 or Class A. The normal resistance value is 100 ohms and the fundamental interval (the increase of resistance between 0 °C and 100 °C) is 38.5 ohms.

The calibration of a resistance thermometer is based on IPTS-68 and is usually carried out by comparison with a standard resistance thermometer. Platinum resistance sensors may be designed for any range within the limits of 15 K and 800 °C and may be capable of withstanding pressures up to 600 bar and vibration up to 60 g's, or more, at frequencies up to 2000 Hz. The size of the sensitive element may be as small as 2 mm diameter by 8 mm long, in the case of the miniature fast response elements, to 6 mm diameter by 50 mm long in the more rugged types. A wide range of sensor designs is available, the form used depending upon the duty and the speed of response required. Some typical forms of construction are illustrated in Figure 1.17. Figure 1.17(a) shows a high temperature form in which the spiral platinum coil is bonded at one edge of each turn with high temperature glass inside cylindrical holes in a ceramic rod. In the high accuracy type, used mainly for laboratory work, the coil is not secured at each turn but is left free to ensure a completely strain-free mounting, Figure 1.17(b). Where a robust form, suitable for use in aircraft and missiles or any severe vibration condition is required, the ceramic is in solid rod form and the bifilar wound platinum coil is sealed to the rod by a glass coating as shown in Figure 1.17(c). Where the sensor is intended for use for measuring surface temperatures, the form shown in Figure 1.17(d) is used. In all forms, the ceramic formers are virtually silica-free and the resistance element is sealed in with high temperature glass to form an impervious sheath which is unaffected by most gases and hydrocarbons. The external leads, which are silver or platinum of a diameter much larger than the wire of the resistance element, are welded to the fine

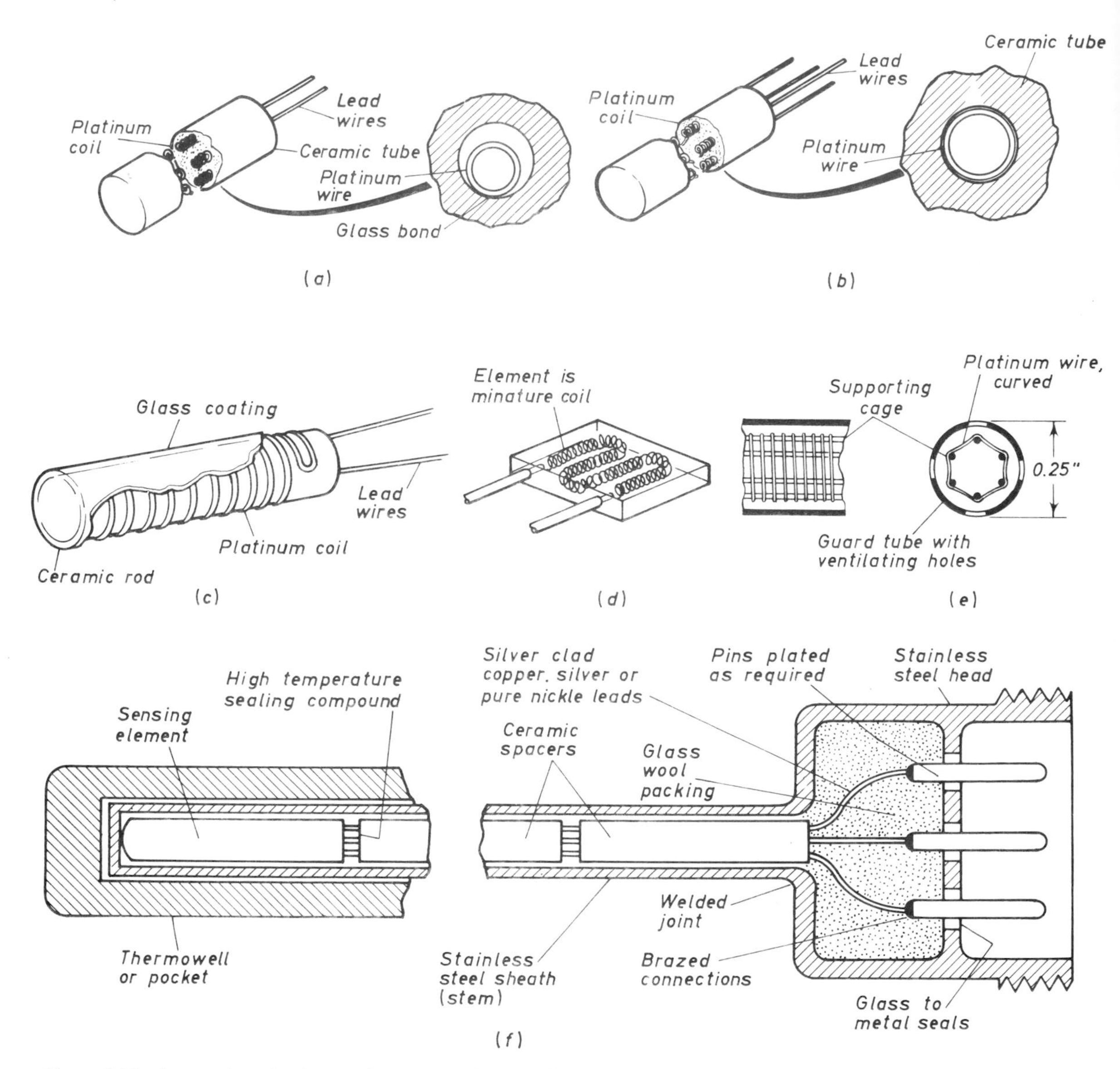

Figure 1.17 Construction of resistance thermometers (courtesy Rosemount Engineering).

platinum wire wholly inside the glass seal.

The inductance and capacitance of elements are made as low as possible in order to allow their use with a.c. measuring instruments. Typically the elements shown will have self-inductance of 2 μH per 100 Ω and the element self-capacitance will not exceed 5 pF. The current passed through a resistance thermometer to measure the resistance must be limited to minimize errors by self-heating of the resistance element. Typical maximum acceptable current is 10 mA for a 100 ohm thermometer.

The rate of response of a resistance thermometer is a function of its construction and encapsulation. A heavy industrial type may have a response of one or two minutes when plunged into water while a naked type, like that shown in Figure 1.17(e), will be only a few milliseconds under the same conditions. Figure 1.17(f) shows the cross-section of a resistance thermometer encapsulated in a metal tube. Figure 1.18 shows a range of typical industrial resistance thermometers.

A more recent development of resistance ther-

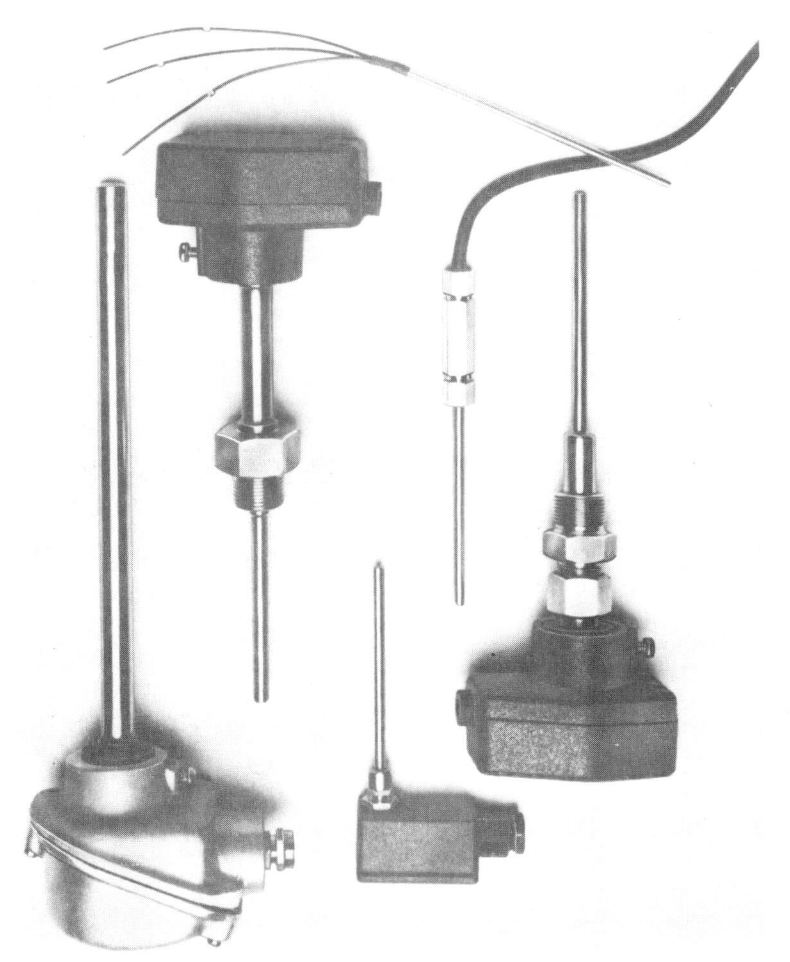

Figure 1.18 Typical industrial resistance thermometers (courtesy Kent Industrial Instruments Ltd.).

mometers has been the replacement of the wire-wound element of the conventional resistance thermometer by a metallized film track laid down on a glass or ceramic substrate. These thermometer elements are made by similar techniques to those used for making hybrid integrated electronic circuits. After the laying down of the metallized film the film is trimmed by a laser to achieve the required parameters.

1.4.1.2 Nickel resistance thermometers

Nickel forms an inexpensive alternative to platinum for resistance thermometers. The usable range is restricted to $-200\,^\circ C$ to $+350\,^\circ C$. But the temperature coefficient of resistivity of nickel is 50 per cent higher than that of platinum which is an advantage in some instruments. Nickel resistance thermometers find wide use in water-heating and air-conditioning systems.

As mentioned above, the current through a resistance thermometer sensor must be kept low enough to limit self-heating. However in some applications such as flow meters, anemometers and psychrometers, the self-heating effect is used, the final temperature of the sensor being a function of the flow rate of the process fluid or air. See also Chapter 1 of Part 1.

1.4.1.3 Resistance thermometer connections

When resistance thermometers are located at some distance from the measuring instrument the electrical resistance of the connecting cables will introduce errors of reading. This reading error will, of course, vary as the temperature of the cables changes. However, this error can be compensated by the use of extra conductors. Since, normally, the change of resistance of a resistance thermometer is measured in a Wheatstone bridge circuit or a modified Wheatstone

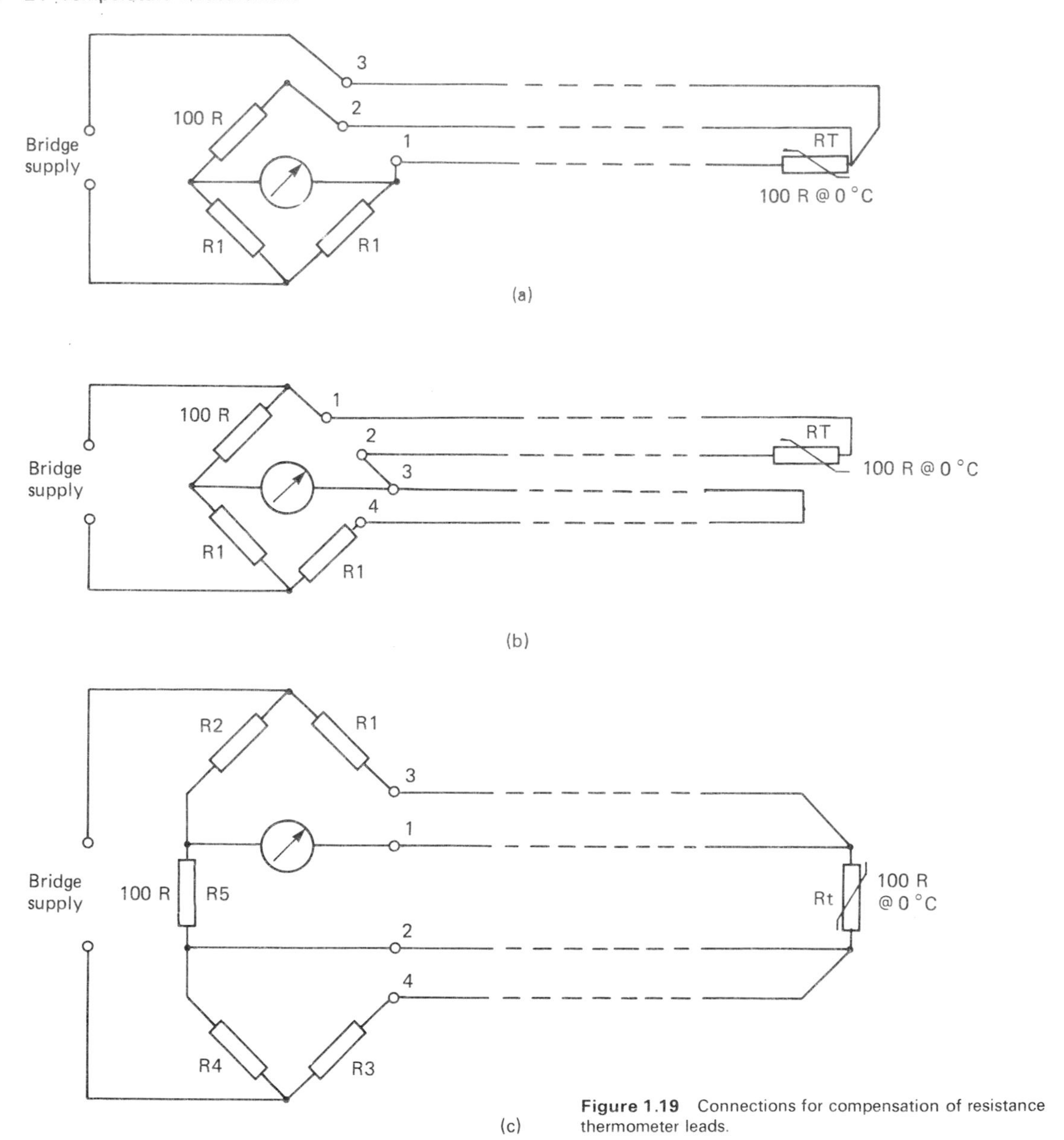

Figure 1.19 Connections for compensation of resistance thermometer leads.

bridge, the compensating conductors can be connected in the opposite side of the bridge. In this way bridge unbalance is only a function of the change of resistance of the thermometer element. Figure 1.19(a) shows three-wire compensation. The resistance of wire 1 is added to that of the resistance thermometer but is balanced by wire 2 in the reference side of the bridge. Wire 3 supplies the power to the bridge. In Figure 1.19(b) four-wire compensation is shown. The resistance of wires 1 and 2 which connect to the resistance thermometer are compensated by the resistance of wires 3 and 4 which are connected together at the resistance thermometer and are again in the opposite arm of the bridge. A Kelvin double bridge is illustrated in Figure 1.19(c). Resistors R1 and R3 set up a constant current through the resistance thermometer. Resistors R2 and R4 set up a constant current in the reference resistor R5 such that the voltage V_R is equal to the voltage V_t across the resistance thermometer when it is at 0 °C. At any other temperature $V_t = I_t R_t$ and the meter will indicate the difference between V_t and V_R which will be proportional to the temperature. The indicator must have a very high resistance so that the current in conductors 1 and 2 is essentially zero. Refer to Part 3.

1.4.2 Thermistors

1.4.2.1 Negative temperature coefficient thermistors

An alternative to platinum or nickel for resistance thermometer sensing elements is a semiconductor composed of mixed metal oxides. The composition of these materials depends on the particular properties required. Combinations of two or more of the following oxides are used: cobalt, copper, iron, magnesium, manganese, nickel, tin, titanium, vanadium and zinc. Devices made of these materials are called thermistors. They consist of a piece of the semiconductor to which two connecting wires are attached at opposite sides or ends. Thermistors have a negative temperature coefficient; that is, as the temperature rises the electrical resistance of the device falls. This variation of resistance with temperature is much higher than in the case of metals. Typical resistance values are 10 kilohms at 0 °C and 200 ohms at 100 °C. This very high sensitivity allows measurement or control to a very high resolution of temperature differences. The accuracy is not as good as for a metallic resistance thermometer owing to the difficulty in controlling the composition of the thermistor material during manufacture. The resolution differs across the usable span of the devices due to their non-linearity. With the right choice of device characteristics it is nevertheless possible to control a temperature to within very close limits: 0.001 degree Celsius temperature change is detectable.

The total range that can be measured with thermistors is from -100 °C to $+300$ °C. However, the span cannot be covered by one thermistor type – four or five types are needed.

The physical construction of thermistors covers a wide range. The smallest are encapsulated in glass or epoxy beads of 1–2.5 mm diameter, bigger ones come as discs 5–25 mm diameter or rods 1–6 mm diameter and up to 50 mm length. The bigger devices are able to pass quite high currents and so operate control equipment directly without need of amplifiers. Thermistors are also available in metal encapsulations like those used for platinum resistance thermometers.

The big disadvantage of thermistors is that their characteristics are non-linear. The temperature coefficient of resistivity α at any temperature within the range of a sensor is given by:

$$\alpha = -B/T^2 \qquad (1.20)$$

where B is the characteristic temperature constant for that thermistor and T is temperature in kelvin. The units of α are ohms . K^{-1}.

Most thermistors have a specified resistance at 20 °C or 25 °C. To determine the resistance at any other temperature equation (1.21) is used:

$$R_2 = R_1 \exp\left(\frac{B}{t_2} - \frac{B}{t_1}\right) \qquad (1.21)$$

where R_1 is resistance of thermistor at temperature t_1(°C) and R_2 is resistance of thermistor at temperature t_2(°C).

Thermistors are available described as curve-matched. These devices are manufactured to fine tolerances and are interchangeable with an error of less than ± 0.2 per cent. However, they are expensive and are only available in a limited range of formats.

In general most thermistors are manufactured with tolerances of 10 to 20 per cent. Instrumentation for use with these devices must have provision for trimming out the error. Thermistors do not have the stability of platinum resistance thermometers. Their characteristics tend to drift with time. Drifts of up to 0.1 °C or more can be expected from some types over a period of some months.

1.4.2.2 Positive temperature coefficient thermistors

Positive temperature coefficient (PTC) thermistors are manufactured from compounds of barium, lead and strontium titanates. PTC thermistors are primarily designed for the protection of wound equipment such as transformers and motors. The characteristics of these devices have the general shape shown in Figure 1.20. The resistance of PTC thermistors is low and

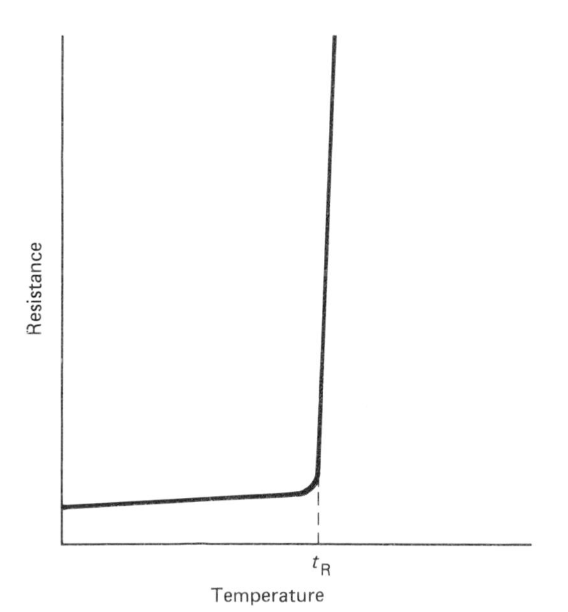

Figure 1.20 Resistance temperature characteristic for PTC thermistor.

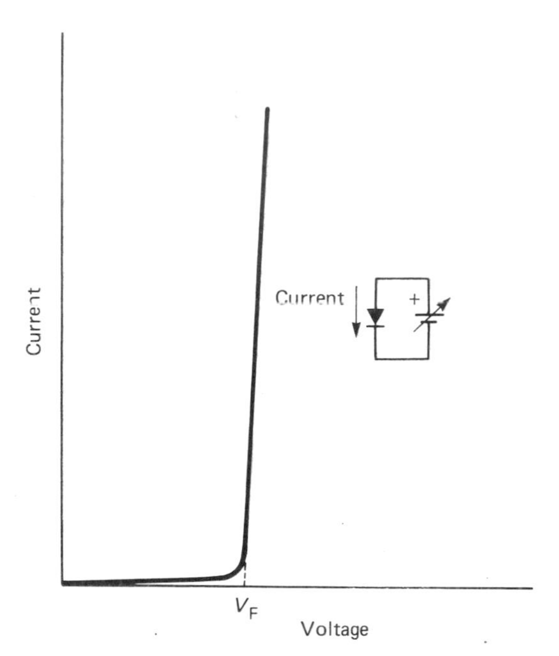

Figure 1.21 Forward bias characteristic of silicon diode.

relatively constant with temperature at low temperature. At temperature T_R the increase of resistance with temperature becomes very rapid. We refer to T_R as the reference or switching temperature.

In use PTC thermistors are embedded in the windings of the equipment to be protected. They are connected in series with the coil of the equipment contactor or protection relay. If the temperature of the windings exceeds temperature T_R the current becomes so small that power is effectively disconnected from the equipment.

1.4.3 Semiconductor temperature measurement

1.4.3.1 Silicon junction diode

Recently silicon semiconductors have been entering the field of temperature measurement. Figure 1.21 shows the forward bias characteristic of a silicon diode. At voltages below V_f, the forward conduction voltage, virtually no current flows. Above V_f the diode passes current. The voltage V_f represents the energy required by current carriers to cross the junction space charge. The value of V_f varies between diode types. It is typically 700 millivolts at 20 °C. The voltage V_f has a temperature coefficient which is essentially the same for all silicon devices of -2 mV per degree Celsius. The forward voltage against temperature characteristic is very nearly linear over the temperature range of -50 °C to $+150$ °C. This voltage change with temperature is substantial and as the characteristic is linear it makes a very useful measurement or control signal. There are two principal disadvantages to silicon diodes as control elements. The negative coefficient Figure 1.22 is not fail-safe. If the control loop is controlling a heater, breakage of the diode wires would be read by the controller as low temperature and full power would be applied to the heaters. The second disadvantage is the rather limited temperature range. Also if a silicon diode is heated above about 200 °C it is completely destroyed, effectively becoming a short circuit.

1.4.3.2 Temperature-sensing integrated circuits

The temperature characteristic of a silicon junction can be improved if the measuring diode is incorporated in an integrated circuit containing an amplifier. Devices are available either to provide an output current proportional to temperature or an output voltage proportional to temperature. Figure 1.23(a) shows the basis of such a device. Figure 1.23(b) shows the circuit of the Analog Devices temperature sensor type AD 590. The operating range of this device is -55 °C to $+150$ °C. The temperature is sensed by the emitter–base junctions of two transistors. If two identical transistors are operated at a constant ratio r of collector current densities then the difference in V_t in their base emitter voltages is given by equation (1.22):

$$V_t = \frac{KT}{q} . \ln r \qquad (1.22)$$

where K is Boltzmann's constant ($1.380\,66 \times 10^{-23}$ J.K^{-1}), q is the electron charge ($1.602\,19 \times 10^{-19}$ coulomb) and T is temperature in kelvins. It can be seen that V_t is directly proportional to temperature in kelvins. The voltage is converted to a temperature-

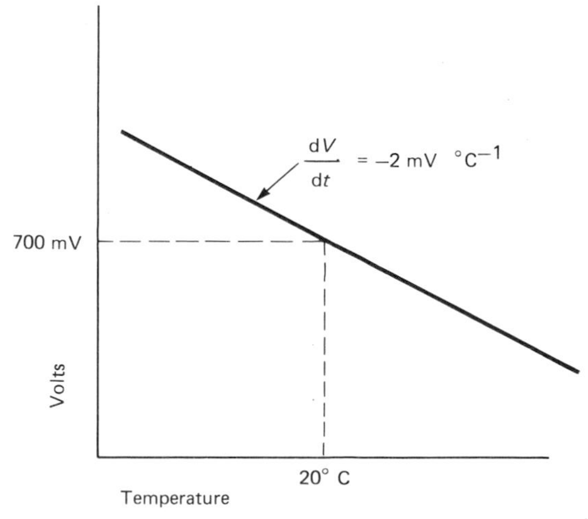

Figure 1.22 Temperature characteristic of silicon diode.

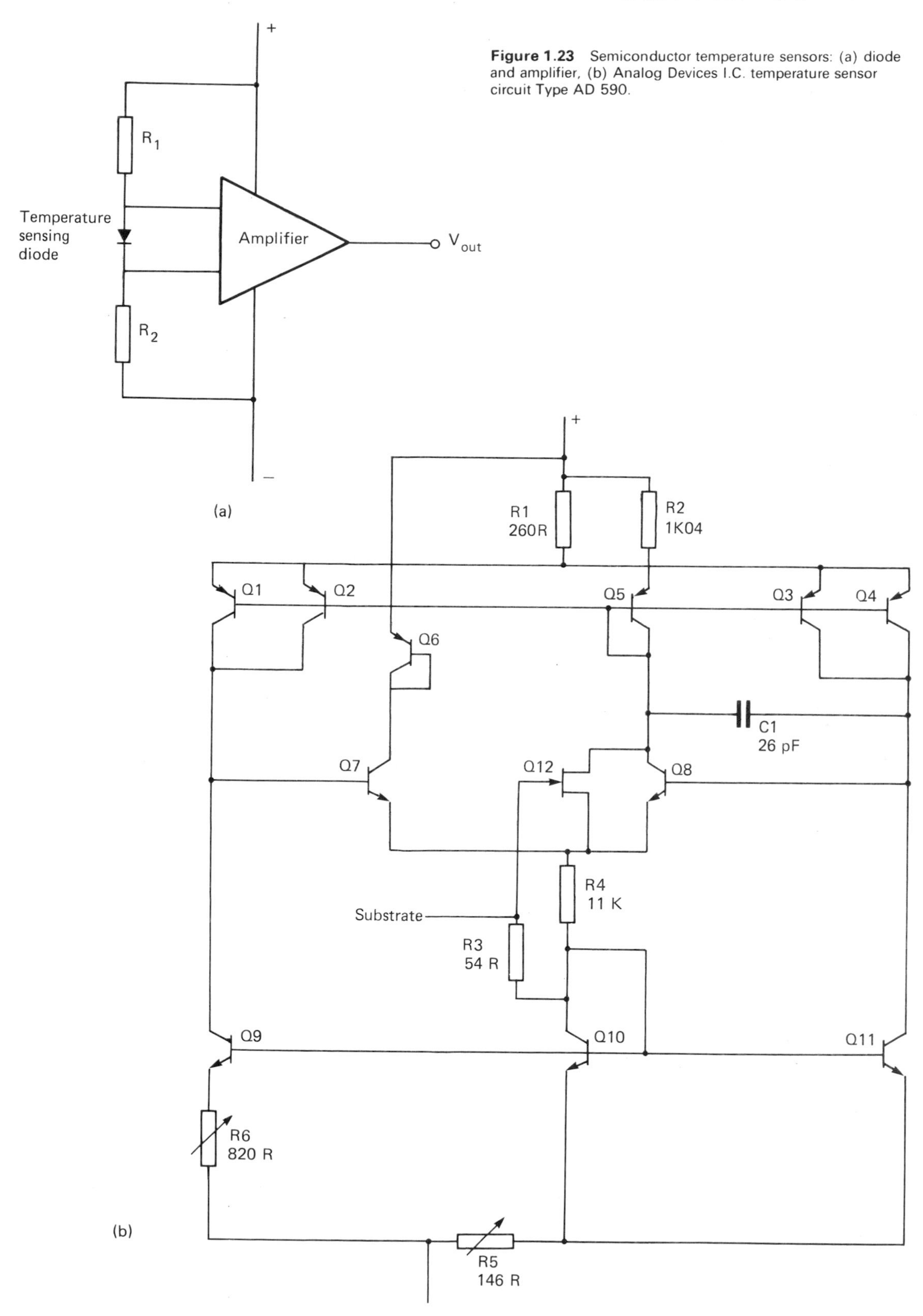

Figure 1.23 Semiconductor temperature sensors: (a) diode and amplifier, (b) Analog Devices I.C. temperature sensor circuit Type AD 590.

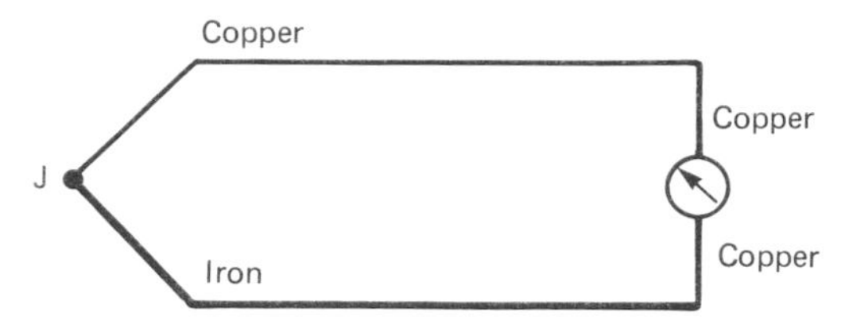

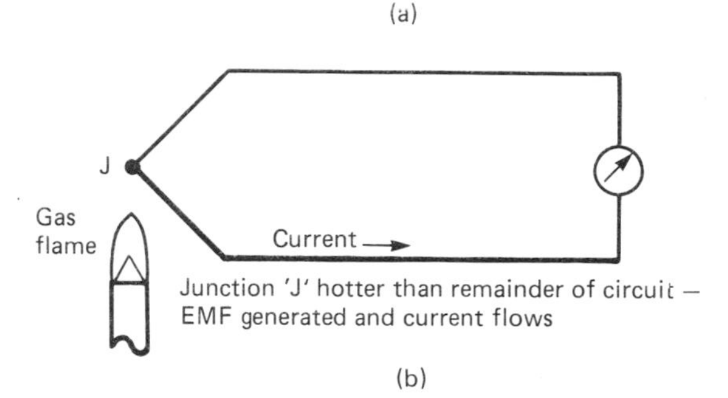

Figure 1.24 Basic thermocouple circuit.

dependent current I_t by low temperature coefficient thin film resistors R5 and R6. These resistors are laser-trimmed to give the required tolerance at 25 °C. Transistors Q_8 and Q_{11} provide the temperature-dependent voltage V_t. The remaining transistors provide the amplification to give the output current of one microampere per kelvin. The transistor Q_{10} supplies the bias and substrate leakage currents for the circuit. The device is packaged in a transistor can or ceramic capsule or it can be supplied as the naked chip for encapsulation into other equipment.

1.5 Measurement techniques – thermocouples

1.5.1 Thermoelectric effects

If an electrical circuit consists of entirely metallic conductors and all parts of the circuit are at the same temperature there will be no electromotive force in the circuit and therefore no current flows. However, if the circuit consists of more than one metal and if junctions between two metals are at different temperatures then there will be an e.m.f. in the circuit and a current will flow. Figure 1.24 illustrates this effect. The e.m.f. generated is called a thermoelectric e.m.f. and the heated junction is a thermocouple.

1.5.1.1 Seebeck effect

In 1821 Seebeck discovered that if a closed circuit is formed of two metals, and the two junctions of the metals are at different temperatures, an electric current will flow round the circuit. Suppose a circuit is formed by twisting or soldering together at their ends, as shown in Figure 1.25, wires of two different metals such as iron and copper. If one junction remains at room temperature, while the other is heated to a higher temperature, a current is produced, which flows from copper to iron at the hot junction, and from iron to copper at the cold one.

Seebeck arranged a series of 35 metals in order of their thermoelectric properties. In a circuit made up of any two of the metals, the current flows across the hot junction from the earlier to the later metal of the series. A portion of his list is as follows: Bi—Ni—Co—Pd—Pt—U—Cu—Mn—Ti—Hg—Pb—Sn—Cr—Mo—Rh—Ir—Au—Zn—W—Cd—Fe—As—Sb—Te.

1.5.1.2 Peltier effect

In 1834 Peltier discovered that when a current flows across the junction of two metals heat is absorbed at the junction when the current flows in one direction and liberated if the current is reversed. Heat is absorbed when a current flows across an iron–copper junction from copper to iron, and liberated when the current flows from iron to copper. This heating effect should not be confused with the Joule heating effect, which being proportional to I^2R, depends only upon the size of the current and the resistance of the conductor and does not change to a cooling effect when the current is reversed. The amount of heat liberated, or absorbed, is proportional to the quantity of electricity which crosses the junction, and the amount liberated, or absorbed, when unit current passes for a unit time is called the Peltier coefficient.

As heat is liberated when a current does work in overcoming the e.m.f. at a junction, and is absorbed when the e.m.f. itself does work, the existence of the Peltier effect would lead one to believe that the junction of the metals is the seat of the e.m.f. produced in the Seebeck effect. It would appear that an e.m.f. exists across the junction of dissimilar metals, its direction being from copper to iron in the couple considered. The e.m.f. is a function of the conduction

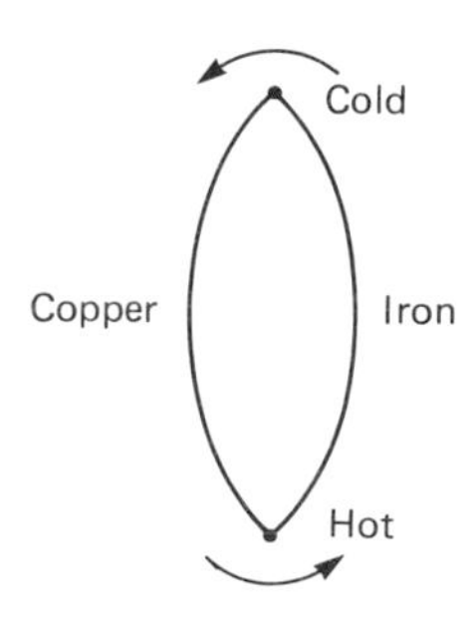

Figure 1.25 Simple thermocouple.

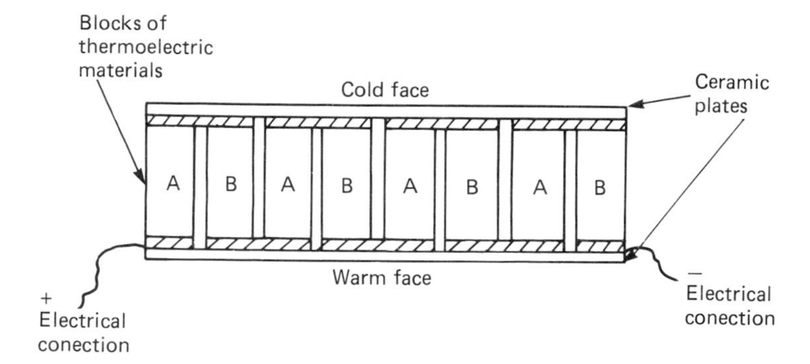

Figure 1.26 Peltier cooler.

electron energies of the materials making up the junction. In the case of metals the energy difference is small and therefore the e.m.f. is small. In the case of semiconductors the electron energy difference may be much greater, resulting in a higher e.m.f. at the junction. The size of this e.m.f. depends not only on the materials making up the junction but also upon the temperature of the junction. When both junctions are at the same temperature, the e.m.f. at one junction is equal and opposite to that at the second junction, so that the resultant e.m.f. in the circuit is zero. If, however, one junction is heated, the e.m.f. across the hot junction is greater than that across the cold junction, and there will be a resultant e.m.f. in the circuit which is responsible for the current:

$$\text{e.m.f. in the circuit} = P_2 - P_1$$

where P_1 is the Peltier e.m.f. at temperature T_1, and P_2 is the Peltier e.m.f. at temperature T_2 where $T_2 > T_1$. Peltier cooling is used in instrumentation where a small component is required to be cooled under precise control. Figure 1.26 shows diagrammatically the construction of such a cooler. The conductors and junctions have a big cross-section to minimize IR heating. The warmer face is clamped to a suitable heat sink while the cold face has the component to be cooled mounted in contact with it. Typical size for such a unit is of the order of 5–25 mm. The conductors in Peltier coolers may be either metals or semiconductors; in the latter case they are called Frigistors.

1.5.1.3 Thomson effect

Reasoning on the basis of the reversible heat engine, Professor William Thomson (later Lord Kelvin) pointed out that if the reversible Peltier effect was the only source of e.m.f., it would follow that if one junction was maintained at a temperature T_1, and the temperature of the other raised to T_2, the available e.m.f. should be proportional to $(T_2 - T_1)$. It may be easily shown that this is not true. If the copper–iron thermocouple, already described, is used, it will be found that on heating one junction while the other is maintained at room temperature, the e.m.f. in the circuit increases at first, then diminishes, and passing through zero, actually becomes reversed. Thomson, therefore, concluded that in addition to the Peltier effects at the junctions there were reversible thermal effects produced when a current flows along an unequally heated conductor. In 1856, by a laborious series of experiments, he found that when a current of electricity flows along a copper wire whose temperature varies from point to point, heat is liberated at any point P when the current at P flows in the direction of the flow of heat at P, that is when the current is flowing from a hot place to a cold place, while heat is absorbed at P when the current flows in the opposite direction. In iron, on the other hand, the heat is absorbed at P when the current flows ih the direction of the flow of heat at P, while heat is liberated when the current flows in the opposite direction from the flow of heat.

1.5.1.4 Thermoelectric diagram

It will be seen that the Seebeck effect is a combination of the Peltier and Thomson effects and will vary according to the difference of temperature between the two junctions, and with the metals chosen for the couple. The e.m.f. produced by any couple with the junctions at any two temperatures may be obtained from a thermoelectric diagram suggested by Professor Tait in 1871. On this diagram the thermoelectric line for any metal is a line such that the ordinate represents the thermoelectric power (defined as the rate of change of e.m.f. acting round a couple with the change of temperature of one junction) of that metal with a standard metal at a temperature represented by the abscissa. Lead is chosen as the standard metal as it does not show any measurable Thomson effect. The ordinate is taken as positive when, for a small difference of temperature, the current flows from lead to the metal at the hot junction. If lines a and b (Figure 1.27) represent the thermoelectric lines for two metals A and B then the e.m.f. round the circuit formed by the two metals, when the temperature of the cold junction

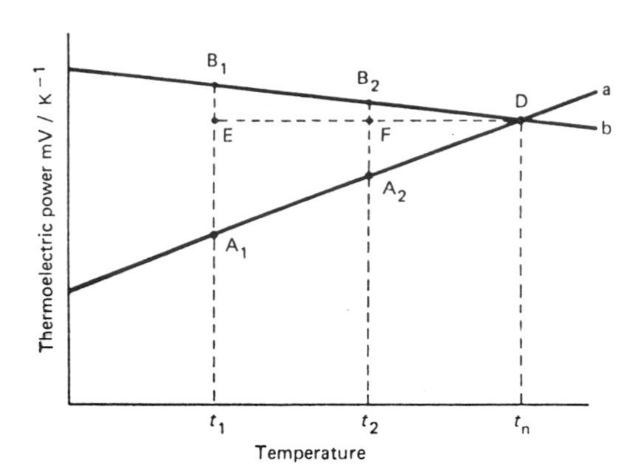

Figure 1.27 Thermoelectric diagram of two metals.

is t_1 and that of the hot junction is t_2, will be the difference in the areas of triangles A_1B_1D and A_2B_2D. Now the area of the triangle is

$$A_1B_1D=\tfrac{1}{2}(A_1B_1\times ED) \tag{1.23}$$

and area

$$A_2B_2D=\tfrac{1}{2}(A_2B_2\times FD) \tag{1.24}$$

$$\text{The e.m.f.}=\tfrac{1}{2}(A_1B_1\times ED)-\tfrac{1}{2}(A_2B_2\times FD) \tag{1.25}$$

Since triangles A_1B_1D and A_2B_2D are similar triangles the sides A_1B_1 and A_2B_2 are proportional to ED and FD respectively.

Therefore: $\text{e.m.f.}\propto ED^2-FD^2$

But: $ED=t_n-t_1$ and $FD=t_n-t_2$

So: $\text{e.m.f.}\propto (t_n-t_1)^2-(t_n-t_2)^2$

$$\propto (t_1-t_2)\left(\frac{t_1+t_2}{2}-t_n\right)$$

$$\text{Or}\quad \text{e.m.f.}=K(t_1-t_2)\left(\frac{t_1+t_2}{2}-t_n\right) \tag{1.26}$$

where K is a constant which together with t_n must be obtained experimentally for any pair of metals. The temperarure t_n is called the neutral temperature. Equation (1.26) shows that the e.m.f. in any couple is proportional to the difference of temperature of the junctions and also to the difference between the neutral temperature and the average temperature of the junctions. The e.m.f. is zero either if the two junctions are at the same temperature or if the average of the temperature of the two junctions is equal to the neutral temperature. Figure 1.28 shows the graph of the e.m.f. of a zinc–iron thermocouple with temperature.

1.5.1.5 Thermoelectric inversion

This reversal of the thermoelectric e.m.f. is 'thermoelectric inversion'.

Figure 1.29 shows the thermoelectric lines for several common materials. It will be seen that the lines for iron and copper cross at a temperature of 275 °C. If the temperature of the cold junction of iron and copper is below 270 °C and the temperature of the other junction is raised, the thermoelectric e.m.f. of the circuit (represented by a trapezium) will increase until the temperature of the hot junction reaches 275 °C (when the e.m.f. is represented by a triangle). Further increase in the temperature of the hot junction will result in a decrease in the thermoelectric e.m.f. (the e.m.f. represented by the second triangle will be in the opposite sense). When the average temperature of the two junctions is 275 °C, or what comes to the same thing, the sum of the two temperatures is 550 °C, the areas of the two triangles will be equal and there will be no thermoelectric e.m.f.: 275 °C is the 'neutral temperature' for the copper–iron couple. With circuits of other materials, the neutral point will occur at different temperatures. Further increase in the temperature of the hot junction will produce a thermoelectric e.m.f. in the opposite direction: from iron to copper at the hot junction, which will again increase with increasing temperature of the hot junction as was seen with zinc and iron in Figure 1.28.

In choosing two materials to form a thermocouple to measure a certain range of temperature, it is very important to choose two which have thermoelectric lines which do not cross within the temperature range, that is, the neutral temperature must not fall within the range of temperature to be measured. If the neutral temperature is within the temperature range, there is some ambiguity about the temperature indicated by a certain value of the thermoelectric e.m.f., for there will be two values of the temperature of the hot junction for

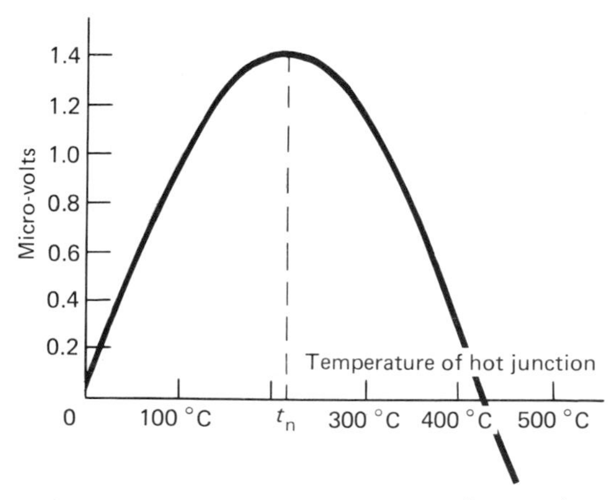

Figure 1.28 Temperature/emf curve for zinc/iron couple.

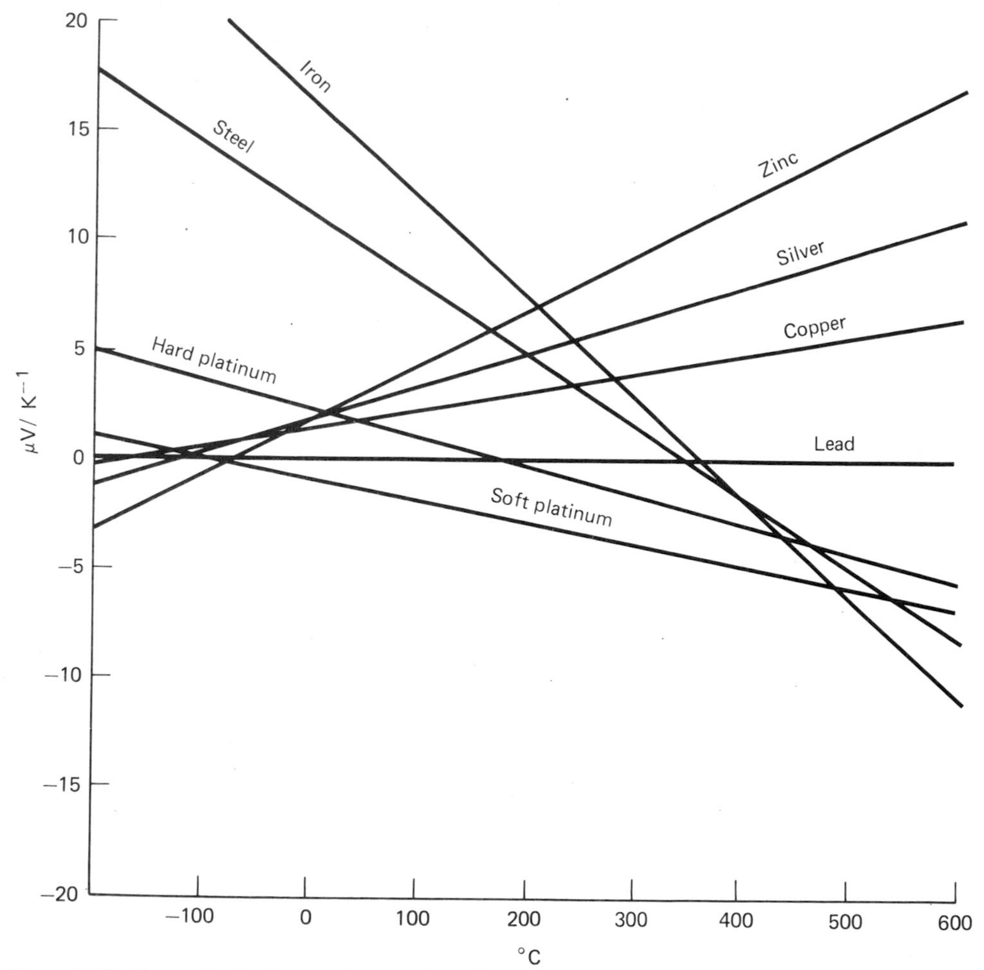

Figure 1.29 Thermoelectric diagrams for several metals.

which the thermoelectric e.m.f. will be the same. For this reason tungsten–molybdenum thermocouples must not be used at temperatures below 1250 °C.

1.5.1.6 Addition of thermoelectric e.m.f.s

In measuring the e.m.f. in any circuit due to thermoelectric effects, it is usually necessary to insert some piece of apparatus, such as a millivoltmeter, somewhere in the circuit, and since this generally involves the presence of junctions other than the two original junctions, it is important to formulate the laws according to which the e.m.f.s produced by additional junctions may be dealt with. These laws, discovered originally by experiment, have now been established theoretically.

Law of intermediate metals In a thermoelectric circuit composed of two metals A and B with junctions at temperatures t_1 and t_2 the e.m.f. is not altered if one or both the junctions are opened and one or more other metals are interposed between metals A and B, provided that all the junctions by which the single junction at temperature t_1 may be replaced are kept at t_1, and all those by which the junction at temperature t_2 may be replaced are kept at t_2.

This law has a very important bearing on the application of thermocouples to temperature measurement, for it means that, provided all the apparatus for measuring the thermoelectric e.m.f., connected in the circuit at the cold junction, is kept at the same temperature, the presence of any number of junctions of different metals will not affect the total e.m.f. in the circuit. It also means that if another metal is introduced into the hot junction for calibration purposes it does not affect the thermoelectric e.m.f., provided it is all at the temperature of the hot junction.

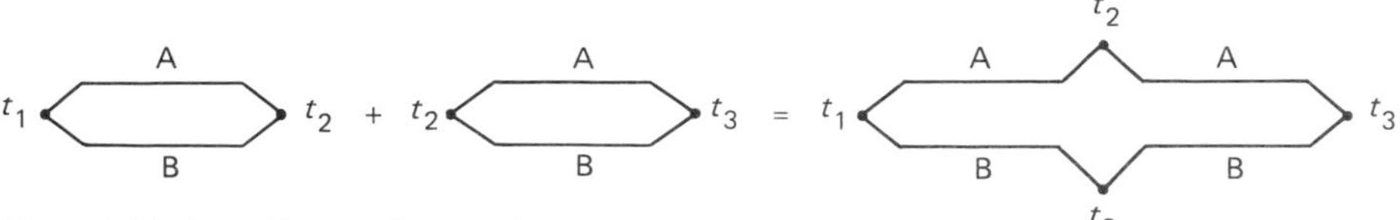

Figure 1.30 Law of intermediate metals.

Law of intermediate temperatures The e.m.f. E_{1-3} of a thermocouple with junctions at temperatures t_1 and t_3 is the sum of the e.m.f.s of two couples of the same metals, one with junctions at temperatures t_1 and t_2 (e.m.f. $= E_{1-2}$), and the other with junctions at t_2 and t_3 (e.m.f. $= E_{2-3}$), see Figure 1.30:

$$E_{1-2}+E_{2-3}=E_{1-3} \tag{1.27}$$

This law is the basis upon which thermocouple measuring instruments can be manufactured.

1.5.1.7 *Cold junction compensation*

It is not normally practical in industrial applications to have thermocouple cold junctions maintained at 0 °C, but with the cold junctions at ambient temperature cold junction compensation is required. To achieve cold junction compensation consider a thermocouple with its hot junction at t °C and its cold junction at ambient, its e.m.f. being E_{a-t}. The instrument must indicate an e.m.f. equivalent to having the cold junction at 0 °C, i.e. an e.m.f. of E_{0-t}. This requires that an e.m.f. must be added at E_{a-t} to provide the required signal:

$$E_{0-t}=E_{a-t}+E_{0-a} \tag{1.28}$$

The voltage E_{0-a} is called the cold junction compensation voltage.

This cold junction compensation e.m.f. can be provided automatically by the use of a temperature-sensitive element such as a resistance thermometer, thermistor or semiconductor sensor in the thermocouple circuit. Figure 1.31 shows such a circuit. In this circuit R_1, R_2 and R_3 are temperature-stable resistors and R_t is a resistance thermometer. The bridge is balanced when all components are at 0 °C and the e.m.f. appearing between points A and B is zero. As the temperature changes from 0 °C an e.m.f., which is the unbalance voltage of the bridge, exists across AB. This voltage is scaled by setting R_4 such that the e.m.f. AB is equal to E_{0-a} in equation (1.28).

Mechanical cold junction compensation An alternative cold junction compensation technique is used when a simple non-electronic thermometer is required. In this technique the thermocouple is connected directly to the terminals of a moving-coil galvanometer. A bimetal strip is connected mechanically to the mechanical zero adjustment of the instrument in such a way that the instrument zero is offset to indicate the ambient temperature. The e.m.f. E_{a-t} is then sufficient to move the pointer upscale to indicate the true temperature of the thermocouple.

1.5.1.8 *Thermocouple circuit considerations*

Galvanometer instruments A thermocouple circuit is like any other electrical circuit. There are one or more sources of e.m.f., which can be batteries, a generator or in this case the hot and cold junctions. There is a load, the indicator and there are electrical conductors, which have resistance, to connect the circuit together. The current in this circuit is, as always, governed by Ohm's law:

$$I=E/R \tag{1.29}$$

where I is the current, E is the e.m.f. and R is the total circuit resistance.

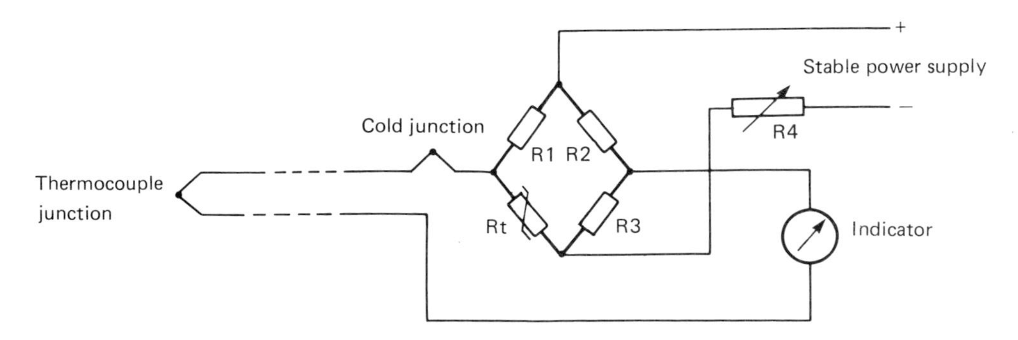

Figure 1.31 Bridge circuit to provide cold junction compensation.

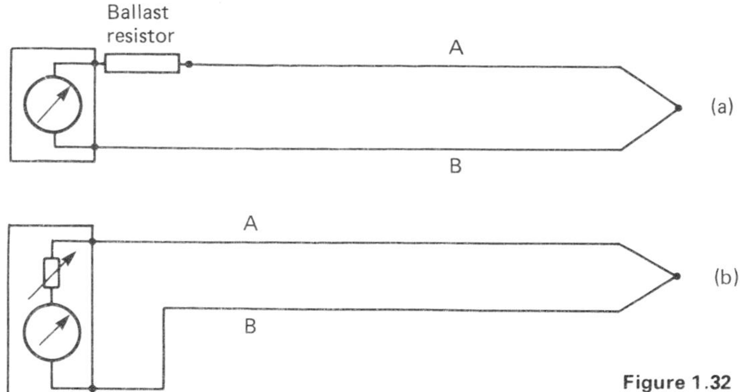

Figure 1.32 Use of ballast resistor: (a) external to instrument, (b) adjustable ballast mounted inside instrument.

In a practical thermocouple thermometer the resistance consists of the sum of the resistances of the thermocouple, the compensating cable (see Section 1.5.3.9) and the indicating instrument. Galvanometer-type thermocouple indicators with mechanical cold junction compensation as described in the previous section, are designed either to be used with an external circuit of stated resistance (this resistance value is usually marked on the dial) or they have an internal adjustable resistor. In the latter case the resistance of the external circuit must not exceed a stated maximum value and the adjustable resistor is adjusted to give the specified total circuit value. Where no internal resistance adjustment is provided the instrument must be used together with an external ballast resistor, see Figure 1.32(a). This resistor must be mounted as near as possible to the indicating instrument to ensure its being at the same temperature as the cold junction compensating mechanism. The usual practice when installing one of these instruments is to wind the ballast resistor with constantan wire on a small bobbin. The length of constantan wire is chosen to make up the required total resistance. On some instruments the bobbin is made integral with one of the indicator terminals. Figure 1.32(b) shows the arrangement with the ballast resistor integral with the indicating instrument.

Potentiometric instruments One way in which to circumvent the critical external resistor is to use a potentiometric indicating device. In a potentiometric device the thermocouple e.m.f. is opposed by an equal and opposite potential from the potentiometer; there is then no current in the circuit and therefore the circuit resistance value is irrelevant.

Potentiometric thermocouple indicators used to be quite common but are now not met so often. However, if the thermocouple indicator is, as it frequently is, a strip chart recorder, it is almost certain to be a potentiometric instrument. Figure 1.33(a) shows the potentiometric arrangement diagrammatically.

Electronic instruments In modern electronic instruments for thermocouple indication, whether they be analogue or digital devices, the input circuit 'seen' by the thermocouple is a high impedance amplifier. Again there is negligible current in the thermocouple circuit and as the resistance of the thermocouple circuit is of the order of 100 ohms while the amplifier input is likely to be a megohm or more the effect of the external circuit resistance is negligible. Electronic instruments allow their designer much more versatility for cold junction compensation. Instead of the bridge circuit of Figure 1.31 it is possible to arrange the cold junction correction after the input amplifier. This has the advantage that the voltage levels being worked with may be of the order of several volts amplitude instead of a few millivolts, making it easier to get a higher degree of accuracy for compensation. Figure 1.33(b) shows a block diagram of such an arrangement. Thermocouple input circuits are available as encapsulated electronic modules. These modules contain input amplifier and cold junction compensation. Since the cold junction consists of the input connections of the module, the connections and the cold junction sensor can be accurately maintained at the same temperature by encapsulation, giving very accurate compensation. These modules can be very versatile. Many are available for use with any of the normal thermocouples. The cold junction compensation is set to the thermocouple in use by connecting a specified value resistor across two terminals of the module. These modules are very convenient to use: they mount like other components directly on to an instrument circuit board. Typical module size is a 25 mm cube. Where the thermocouple instrument is based on a microcomputer the cold junction compensation can be done by software, the microcomputer being pro-

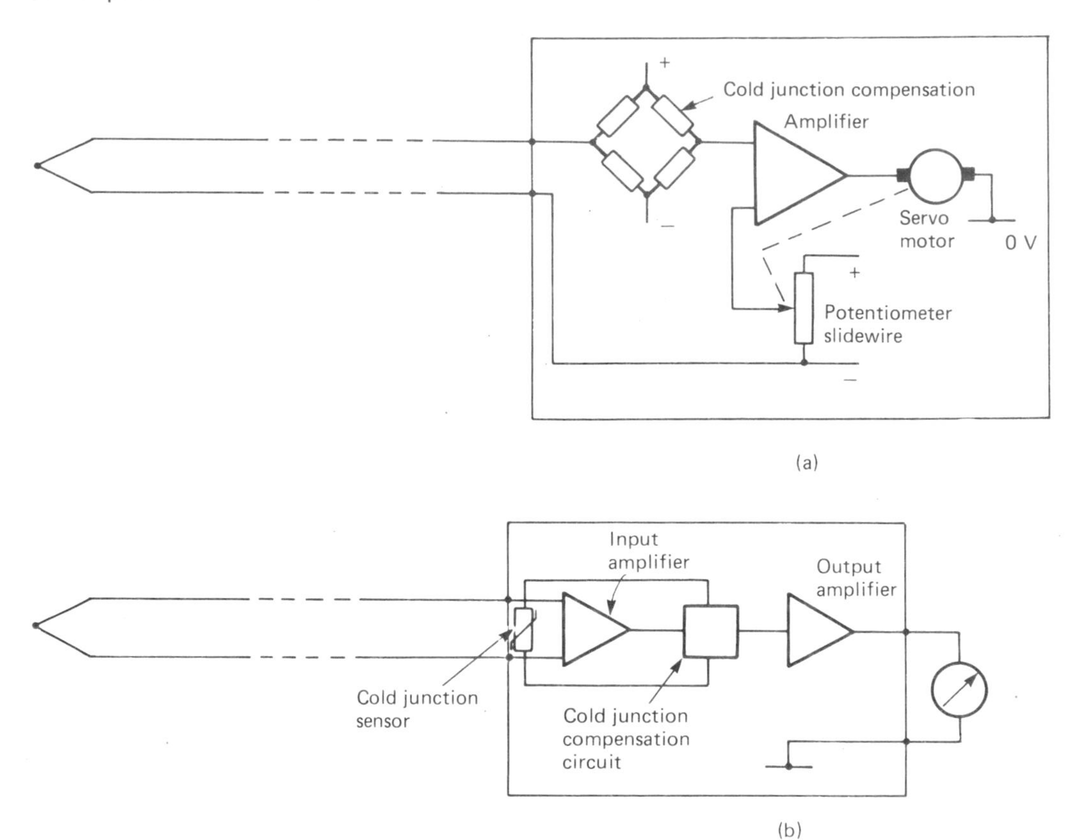

Figure 1.33 Cold junction compensation: (a) in conjunction with potentiometric indicating instrument, (b) alternative arrangement for cold junction compensation.

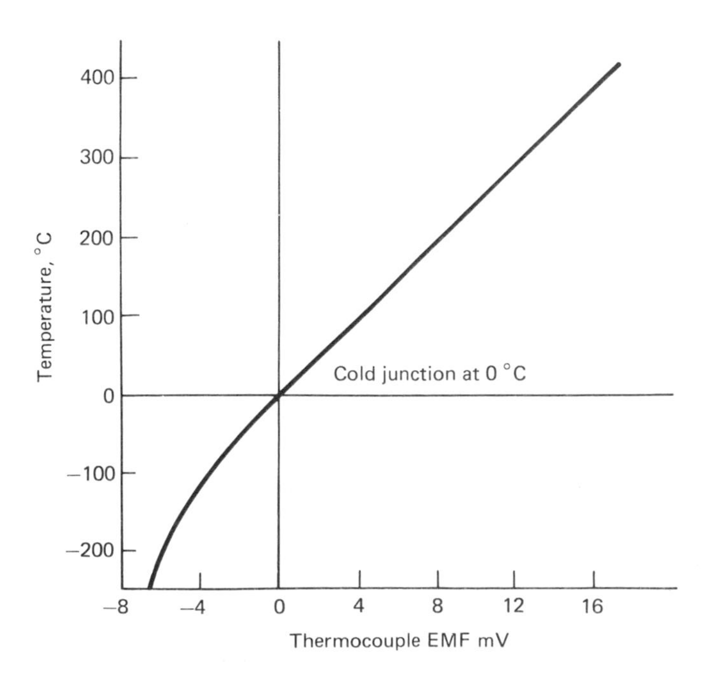

Figure 1.34 Type K thermocouple characteristic.

grammed to add the compensation value to the thermocouple output. In all electronic equipment for thermocouple signal processing the location of the sensor for cold junction temperature sensing is critical. It must be very close to the cold junction terminals and preferably in physical contact with them.

1.5.2 Thermocouple materials

Broadly, thermocouple materials divide into two arbitrary groups based upon cost of the materials, namely, base metal thermocouples and precious metal thermocouples.

1.5.2.1 Base metal thermocouples

The most commonly used industrial thermocouples are identified for convenience by type letters. The main types, together with the relevant British Standard specification and permitted tolerance on accuracy, are shown in Table 1.13. Also shown are their output e.m.f.s with the cold junction at 0 °C. These figures are given to indicate the relative sensitivities of the various couples. Full tables of voltages against hot junction temperatures are published in BS 4937. The standard also supplies the equations governing the thermocouple e.m.f.s for convenience for computer programming purposes. These equations are essentially square law; however, provided a thermocouple is used at temperatures remote from the neutral temperature its characteristic is very nearly linear. Figure 1.34 shows a plot of the characteristic for type K thermocouple. It can be seen that for temperatures in the range −50 °C to 400 °C the characteristic is approximately linear. The commonly used base metal thermocouples are types E, J, K, and T. Of these J and K are probably the most usual ones. They have a high e.m.f. output and type K is reasonably resistant to corrosion. Type T has a slight advantage, where the temperature measurement points are very remote from the instrumentation, that as one conductor is copper the overall resistance of the circuit can be lower than for other types. Table 1.14 shows the accuracy limits and recommended temperature measurement ranges as specified by the United States Standard ANSI MC 96 1 (1975).

1.5.2.2 Precious metal thermocouples

Thermocouples types B, R, and S clearly carry a considerable cost penalty and normally are only used when essential for their temperature range or their relatively high resistance to chemical attack. Their temperature top limit is 1500 °C for continuous use or 1650 °C for intermittent, spot reading, applications. This compares with 1100 °C continuous and 1300 °C intermittent for type K.

Errors in type R and S thermocouple readouts result from strain, contamination and rhodium drift.

The effect of strain is to reduce the e.m.f. resulting in low readings. The effect of strain may be removed by annealing the thermocouple. Installations should be designed to minimize strain on the thermocouple wires.

Contamination is by far the most common cause of thermocouple error and often results in ultimate mechanical failure of the wires. Elements such as Si, P, Pb, Zn, and Sn combine with platinum to form low melting point eutectics and cause rapid embrittlement and mechanical failure of the thermocouple wires. Elements such as Ni, Fe, Co, Cr, and Mn affect the e.m.f. output of the thermocouple to a greater or lesser degree, but contamination by these elements does not result in wire breakage and can only be detected by regular checking of the accuracy of the thermocouple. Contamination can be avoided by careful handling of the thermocouple materials before use and by the use of efficient refractory sheathing. Care should be taken to prevent dirt, grease, oil or soft solder coming into contact with the thermocouple wires before use. If the atmosphere surrounding the thermocouple sheath contains any metal vapour, the sheath must be impervious to such vapours.

Rhodium drift occurs if a rhodium–platinum limb is maintained in air for long periods close to its upper temperature limit. Rhodium oxide will form and volatilize, and some of this oxide can settle on, and react with, the platinum limb causing a fall in e.m.f. output. This is a comparatively slow process and is therefore only of significance in installations where the maximum stability and repeatability are required. Type B thermocouples are less susceptible to rhodium drift than types R or S, but type B has a lower e.m.f. than R and S and is subject to higher errors.

To overcome these disadvantages Pallador I which has a thermal e.m.f. comparable with iron/constantan, and Pallador II which provides a noble metal alternative to nickel–chromium/nickel–aluminium have been introduced. The positive limb of Pallador I is 10 per cent iridium–platinum and the negative limb is 40 per cent palladium–gold, while the positive limb of Pallador II is $12\frac{1}{2}$ per cent platinum–palladium and the negative 46 per cent palladium–gold. The maximum operating range for Pallador I thermocouples is 1000 °C and the maximum operating temperature for Pallador II is 1200 °C when protected by a 10 per cent rhodium–platinum sheath. (Pallador is the registered trade mark of Johnson Matthey Metals Ltd.)

The corresponding base metal thermocouple wires may be used as the compensating lead and an accuracy

Table 1.13 Thermocouples to British Standards

Type	*Conductors (positive conductor first)*	*BS 4937 Part No.*	*BS 1041, Part 4: 1966 Tolerance on temperature*	*Output for indicated temperature Cold junction at 0 °C*	*Service temperature. Max intermittant service in brackets*
B	Platinum: 30% Rhodium Platinum: 6% Rhodium	Part 7: 1974	0 to 1100 °C ± 3 °C 1100 to 1550 °C ± 4 °C	1.241 mV at 500 °C	*0 to 1 1500 °C (1700 °C)* Better life expectancy at high temperature than types R & S
E	Nickel: Chromium/Constantan (Chromel/Constantan) (Chromel/Advance)	Part 6: 1974	0 to 400 °C ± 3 °C	6.317 mV at 100 °C	*−200 to 850 °C (1100 °C)* Resistant to oxidizing atmospheres
J	Iron/Constantan	Part 3: 1973	0 to 300 °C ± 3 °C 300 to 850 °C ± 1%	5.268 mV at 100 °C	*−200 to 850 °C (1100 °C)* Low cost, suitable for general use
K	Nickel:Chromium/Nickel:Aluminium (Chromel/Alumel), (C/A), (T1/T2)	Part 4: 1973	0 to 400 °C ± 3 °C 400 to 1100 °C ± 0.75%	4.095 mV at 100 °C	*−200 to 1100 °C (1300 °C)* Good general purpose, best in oxidizing atmosphere
R	Platinum:13% Rhodium/Platinum	Part 2: 1973	0 to 1100 ± 1 °C 1100 to 1400 °C ± 2 °C	4.471 mV at 500 °C	*0 to 1500 °C (1700 °C)* High temperature
S	Platinum:10% Rhodium/Platinum	Part 1: 1973	1400 °C ± 3 °C	4.234 mV at 500 °C	corrosion resistant
T	Copper/Constantan; (Copper/Advance) (Cu/Con)	Part 5: 1974	0 to 100 °C ± 1 °C 100 to 400 °C ± 1%	4.277 mV at 100 °C	*−250 °C to 400 °C (500 °C)* High resistance to corrosion by water
	Rhodium:Iridium/Rhodium	Composition and accuracy to be agreed with manufacturer		Typically 6.4 mV at 1200 °C	0 to 2000 °C (2100 °C)
	Tungsten:Rhenium 5%/Tungsten: Rhenium 26%	Accuracy to be agreed with manufacturer		8.890 mV at 500 °C	0 to 2300 °C (2600 °C)
	Tungsten/Molybdenum	Composition and accuracy to be agreed with manufacturer		—	1250 to 2600 °C

Table 1.14 Thermocouples to American Standard ANSI MC 96 (1975)

Type	*Temperature range* (°C)	*Standard quality (whichever is greater)*	*Special quality (whichever is greater)*
B	800 to 1700	±0.5%	–
E	0 to 900	±1.7 °C or ±0.5%	±1.0 °C or ±0.4%
	–200 to 0	±1.7 °C or ±1%	–
J	0 to 750	±2.2 °C or ±0.75%	±1.1 °C or ±0.4%
K	0 to 1.200	±2.2 °C or ±0.75%	±1.1 °C or ±0.4%
	–200 to 0	±2.2 °C or ±2%	–
R, S	0 to 1450	±1.5 °C or ±2.5%	±0.6 °C or ±0.1%
T	0 to 350	±1.0 °C or 0.75%	±0.5 °C or ±0.4%
	–22 to 0	±1.0 °C or 1.5%	–

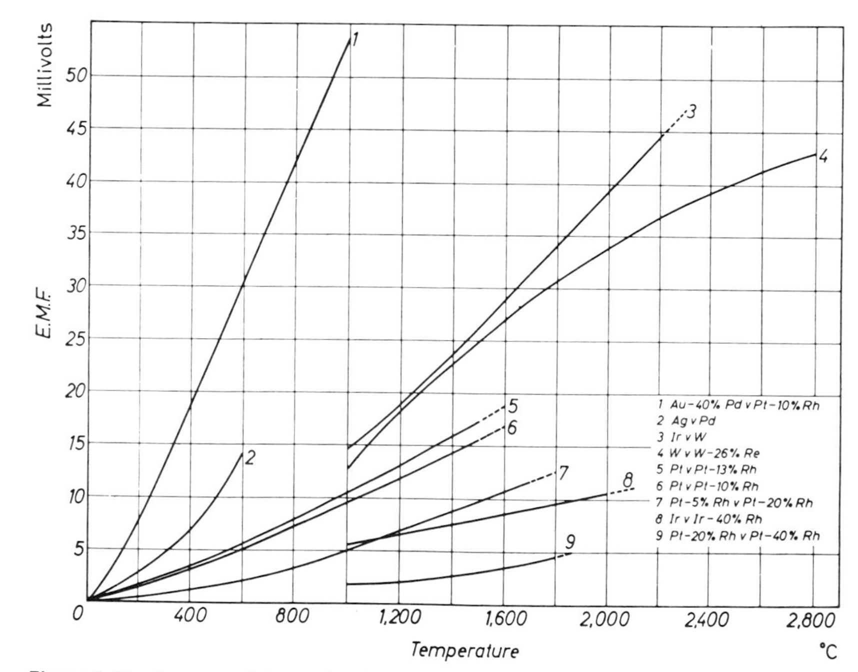

Figure 1.35 Summary of thermoelectric properties of precious metal thermocouples. Broken lines indicate areas for intermittent service.

of ±1 per cent will be attained on an instrument calibrated on the base metal characteristics. When the instrument is calibrated on the Pallador temperature–e.m.f. relationship an accuracy of ±2 K over the whole operating range is attainable.

Noble metal thermocouples may also be used for measuring cryogenic temperatures. Iron–gold/nickel–chromium or iron–gold/silver (normal silver with 0.37 atomic per cent gold) may be used for temperatures from 1 K to above 300 K.

Noble metal thermocouples are often used in the 'metal-clad' form with magnesia or alumina powder as the insulant. This form of construction is described in Section 1.5.3.2.

The following sheath materials are used: nickel, stainless steel, inconel in 1.6 and 3.2 mm sizes, and 5 per cent rhodium-plated and 10 per cent rhodium–platinum both in 1.0 mm sizes. For high temperature work other special thermocouples have been developed, tungsten 5 per cent rhenium/tungsten 20

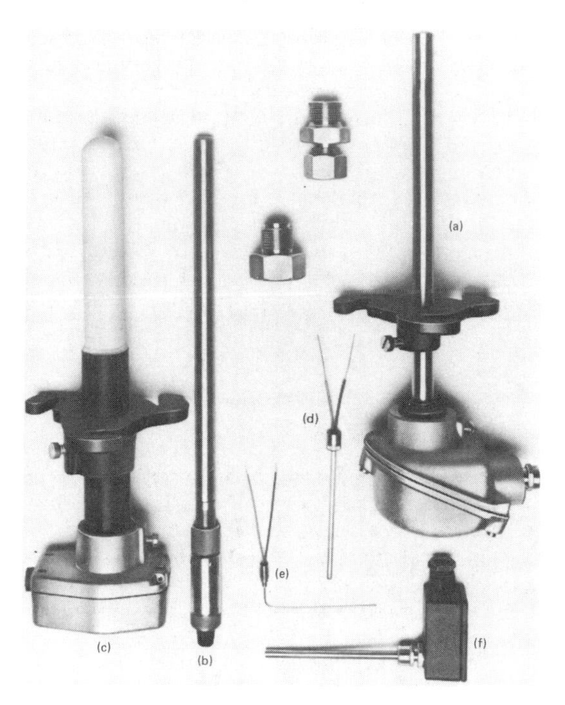

Figure 1.36 Examples of industrial thermocouple probes (courtesy Kent Industrial Measurements Ltd.).

per cent rhenium for use in hydrogen, vacuum and inert gas atmospheres up to 2320 °C and tungsten/molybdenum and tungsten/iridium for temperatures up to 2100 °C.

There is quite a wide range of precious metal thermocouples available. Types B, R and S are specified in BS 4937. These three are based only on platinum and rhodium. Gold, iridium, other 'platinum metals' and silver are also not uncommonly used. Figure 1.35 shows the characteristics of some of the options available.

1.5.3 Thermocouple construction

Thermocouples, like resistance thermometers and other temperature sensors, are available in a wide range of mechanical constructions.

1.5.3.1 Plain wire thermocouples

For use in protected environments, such as for laboratory use or inside otherwise enclosed equipment, plain wire thermocouples can be used. They are also used in plant where the fastest possible response is required. However, they suffer from the obvious disadvantage that they are both fragile and liable to chemical attack. The wires are available insulated with PVC or glass fibre sleeving, or for use with higher temperatures the wires can be insulated with refractory ceramic beads or sleeves.

1.5.3.2 Sheathed thermocouples

Thermocouples for use in plant situation, where robust construction is required or where they need to be interchangeable with other types of temperature measurement equipment, are available sheathed in steel or stainless steel designed for direct insertion into process vessels or for use in a thermometer pocket. Figure 1.36(a) and (b) show typical insertion probes. Where thermocouples are to be immersed in very corrosive process fluids or into very high temperature locations they are available constructed in ceramic sheaths as in Figure 1.36(c). Sheathed thermocouples, especially the ceramic ones, suffer from a slow response time, typically a minute or more. However, the locations where they are essential for their mechanical properties are usually in heavy plant where temperatures do not normally move fast in any case.

1.5.3.3 Mineral-insulated thermocouples

Probably the most versatile format for thermocouples is the mineral-insulated (MI) construction. In this form the thermocouples are made from mineral-insulated cable similar in concept to the MI cable used for electrical wiring applications. It differs however in that the conductors are of thermocouple wire and the sheath is usually stainless steel. The insulation, however, is similar, being in the form of finely powdered and densely compacted ceramic, usually aluminium oxide or magnesium oxide. Figure 1.36 shows MI thermocouples at (d), (e) and (f).

They are available in diameters from 1 millimetre up to 6 millimetres and can be supplied in any length required. The junction can be either insulated (a) or welded (b) to the tip of the sheath as shown in Figure 1.37. The latter arrangement has the advantage of very quick response. For some applications the junction being connected to the plant earth via the sheath tip can be unacceptable so in such cases insulated thermocouples must be used. The principal advantages are their quick response and mechanical flexibility, being able to be bent into almost any shape. Care must be taken if re-using MI thermocouples, for though they can be straightened or re-bent to a new shape this cannot be done too often. Either the wires break or the insulation gets displaced and the thermocouple becomes short-circuited. As shown in Figures 1.36 and

1.38, MI thermocouples can be supplied fitted with a variety of terminations. A further useful advantage of MI thermocouples is that the cable can be bought in rolls together with suitable terminations and the thermocouples can be made up to the required specifications on site. Also in situations where robust cabling is required, MI thermocouple cable can be used in lieu of compensating cable (see Section 1.5.3.9).

1.5.3.4 Surface contact thermocouples

Thermocouples for the measurement of the surface temperature of objects such as pipes or other components or plant items are available. On pipes a surface measurement makes a simple but not very accurate non-invasive temperature measurement. For higher temperatures or more rugged applications thermocouples are available embedded in a metal plate designed to be clamped or welded to the component to be measured. For lower temperature applications, below about 200 °C, or for use in protected environments, self-adhesive contact surface thermocouples are supplied. In these probes the thermocouple is embedded in a small plastic pad coated on one face with a suitable contact adhesive.

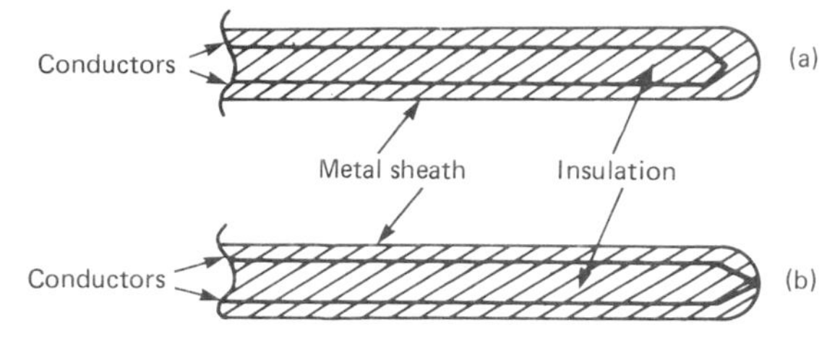

Figure 1.37 Mineral insulated thermocouples: (a) insulated junction, (b) junction welded to sheath.

1.5.3.5 Hot metal thermocouples

Where it is necessary to make spot measurements of the temperature of hot metal billets, very simple test prods are available which consist of a two-pronged 'fork'. The two prongs are made of the two thermocouple metals with sharpened points. When both prongs are in contact with the hot metal two junctions are formed, metal A to the billet and the billet to metal B. If the billet is large and enough time is allowed for the tips of the prongs to reach the temperature of the

Crimp on threaded seal type
Stainless steel 8 mm x 1.0 – 6g ISO metric externally threaded.
Sealing: epoxy resin
Max. operating temperature: 105° C

7.75 mm 9.5 mm
Nominal length A
18 mm
PVC insulated -flexible stranded tails 7/0.2 mm

Pot seal type
Stainless steel screw-on pot
Sealing: plastic sealing compound
Max. operating temperature: 135° C

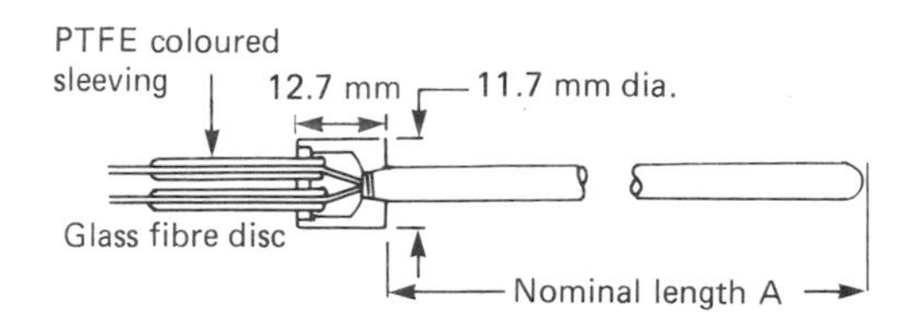

Connector head type
Die cast aluminium alloy connector head
Cable entry gland: 6 mm nylon, 10 mm nylon or 6/10 mm brass
Max. operating temperature: 105 ° C
See photograph on page 12.

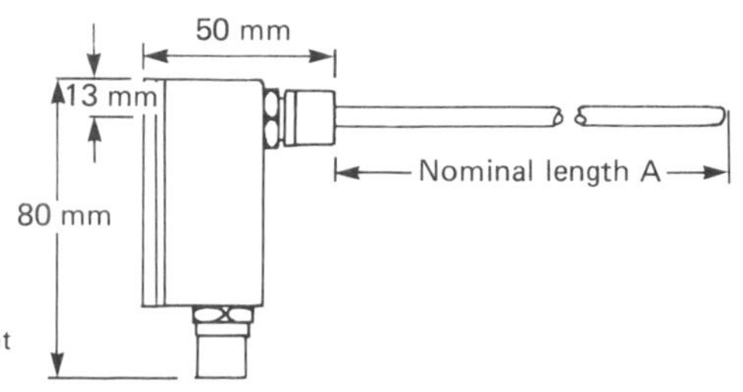

Figure 1.38 MI thermocouple terminations (courtesy Kent Industrial Measurements Ltd.).

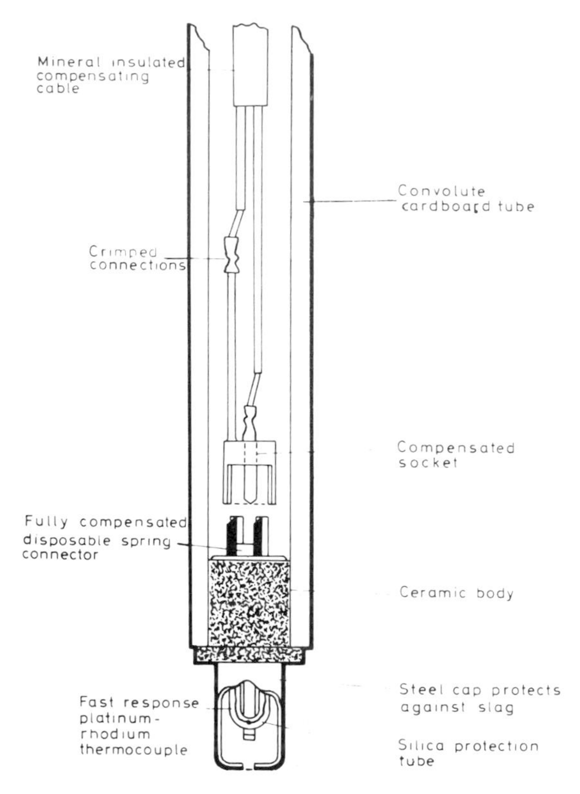

Figure 1.39 Liquid metal thermocouple.

billet then both junctions will be at the same temperature and the error thermal e.m.f.s cancel. This makes a simple, quick and very inexpensive way of measuring hot metal temperatures. The points of the prongs are screwed to the main assembly and are expendable. As soon as they lose their sharpness or begin to get corroded they can be changed.

1.5.3.6 Liquid metal thermocouples

When measuring the temperature of liquid metals such as steel it is desirable to use an expendable probe. The cost of a fully protected probe would be very high and the response time slow. For checking the temperature of liquid steel a dipstick probe can be used. The probe itself is robust and constructed with a socket of thermocouple material in the end. A disposable platinum–rhodium/platinum thermocouple itself lasts in the molten metal for a few seconds, long enough to take a temperature measurement. Figure 1.39 shows this arrangement.

1.5.3.7 Thermopiles

Where a very small temperature rise is to be measured many thermocouples may be connected in series. All the hot junctions are on the object whose temperature is to be measured and all the cold junctions are kept at a constant and known temperature. Where a quick temperature response is required these thermocouples can be of very thin wire of about 25 μm diameter. A speed of response of the order of 10 milliseconds can be achieved. Typical applications of thermopiles are to be found in infrared radiation measurement. This subject is dealt with in Section 1.6.

1.5.3.8 Portable thermocouple instruments

With the development, over the last decade of micro-electronic equipment, portable electrical thermometers have become very popular. They are available with either analogue or digital readouts. The analogue instruments are about the size of an analogue multi-meter, the digital instruments are about the size of a pocket calculator. While most of these instruments use type K thermocouples they are available for use with other thermocouple materials. There are also portable thermometers available using resistance thermometer or thermistor sensors. However, the thermocouple instruments are on the whole the most popular. The more sophisticated instruments have the option to use more than one type of thermocouple: a switch on the instrument sets it for the type in use. They are also available with a switched option to read out in Celsius or Fahrenheit. A range of hand-held probes are supplied for use with these instruments. Figure 1.40 shows some of the options available. The spring-loaded thermocouples are for surface contact measurements, hypodermic probes are supplied for such applications as temperature measurements in food such as meat where it may be an advantage to know the internal temperature of the material. Figure 1.41 shows typical analogue and digital readout instruments.

1.5.3.9 Thermocouple compensating cable

Ideally a thermocouple connects back to the reading instrument with cables made of the same metals as the thermocouple. This does however have two disadvantages in industrial conditions. Firstly many thermocouple metals have high electrical resistance. This means that on long runs, which on a big plant may be up to 1000 metres or more, heavy gauge conductors must be used. This is not only expensive but also makes the cables difficult to handle. Secondly in the case of precious metal thermocouples, types B, R and S for instance, the cost would be very high indeed. To overcome these problems compensating cables are used, see Figure 1.42. These cables are made of base metal and are of lower resistivity than the thermocouple material. The alloys used have thermoelectric properties that essentially match the thermocouples

General purpose thermocouple

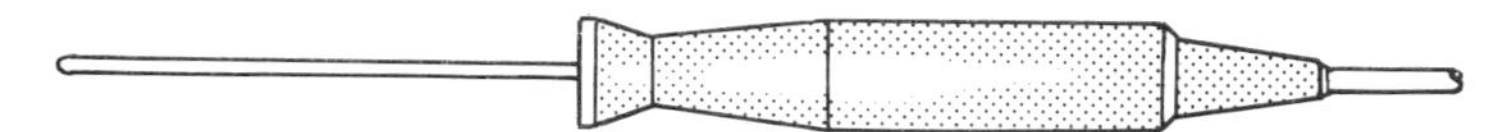

Spring loaded thermocouple
for surface temperature measurement

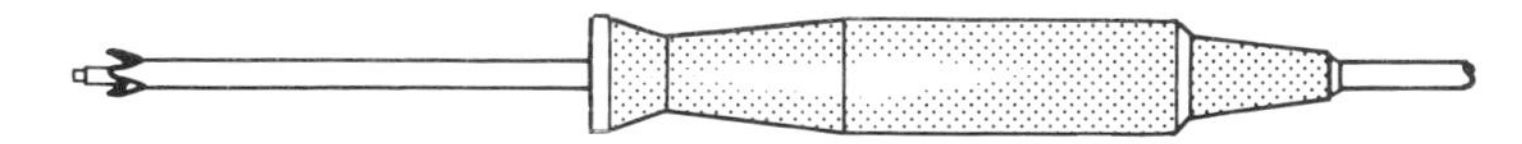

Hypodermic thermocouple
for internal temperature measurement of soft plastic, etc.

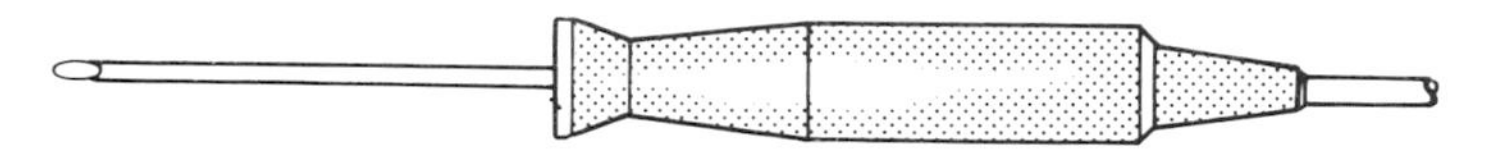

Figure 1.40 Hand-held thermocouple probes.

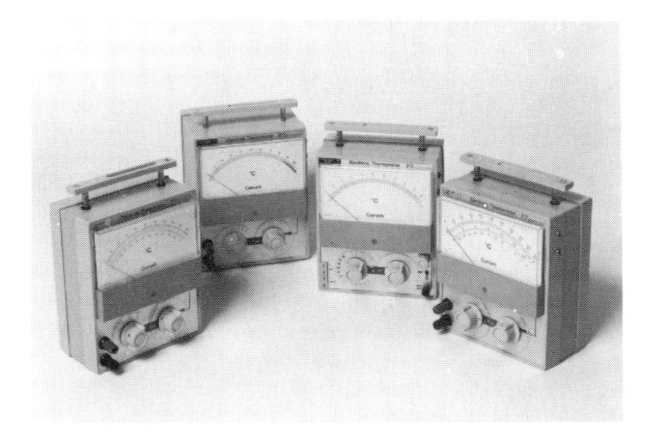

Figure 1.41 Portable analogue and digital thermocouple instruments (courtesy Comark Electronics Ltd.).

Figure 1.42 Thermocouple compensating cable.

themselves over a limited ambient temperature range.
Examples of compensating cables are:

Type	*Composition*	*Thermo-couples compensated*	*Temperature limitations*
U	Copper/copper-nickel	R and S	0–50 °C
Vx	Copper/Constantan	K	0–80 °C

Other base metal thermocouples such as types J and T comprise relatively inexpensive and low resistance metals. They are therefore normally installed using cables consisting of the same metals as the thermocouples themselves.

1.5.3.10 Accuracy consideration

The very extensive use of thermocouples stems from their great versatility combined with their low cost. However, as seen in Tables 1.13 and 1.14, thermocouples have a fairly wide permitted tolerance. This is due to the fact that most metals used for thermocouples are alloys and it is not possible to manufacture alloys to the same reproducibility as pure metals. It must be said that, in general, manufacturers do manufacture their thermocouples to better tolerance than BS 4937 demands. But, where the highest accuracy is required, it is essential to calibrate thermocouples on installation and to recalibrate them at regular intervals to monitor any deterioration due to corrosion or diffusion of foreign elements into the hot junction.

Where high accuracy is required it is necessary to calibrate first the thermocouple readout instrument and then the thermocouple itself in conjunction with the instrument.

The calibration of instruments can be done with a precision millivolt source which injects a signal equivalent to the temperature difference between the ambient or cold junction temperature and a temperature in the region in which the thermocouple is to be used.

To calibrate or check thermocouples the hot junction must be kept at an accurately known temperature. This can be done by inserting it into a heated isothermal block. An isothermal block is a block of metal, large compared with the thermocouple being measured and made of copper or aluminium. The block has provision for heating it and in some cases cooling. It is well insulated from the environment and is provided with suitable holes for inserting various sizes of thermocouple. Where not so high precision is required the thermocouple can be immersed in a heated fluidized sand bath. This consists of an open vessel fitted with a porous bottom (usually made of sintered metal). Heated air is forced up through the bottom. The vessel is filled with carefully graded sand. With the air coming up through it the sand behaves like a liquid. It takes up the temperature of the air. The sand is a good heat transfer medium. The apparatus makes a most convenient way of calibrating temperature probes. Where maximum accuracy is essential the thermocouple should be calibrated against one of the IPTS-68 secondary reference points. Table 1.5 shows some of the points.

In carrying out these calibrations the whole installation needs to be calibrated: thermocouple readout instrument together with compensating cable. In cases where very high accuracy is required, compensating cable should not be used, the conductors should be thermocouple metal for the full length of the installation.

There is now on the market some very versatile equipment for thermocouple calibration. Figure 1.43 shows a microprocessor-controlled calibrator. Its facilities include thermocouple simulation for types E, J, K, R, S and T, thermocouple output measurement with cold junction compensation and resistance thermometer simulation. Tests can be static or dynamic using ramp functions. Resolution is to 0.1 K and accuracy is ± 0.01 per cent of reading.

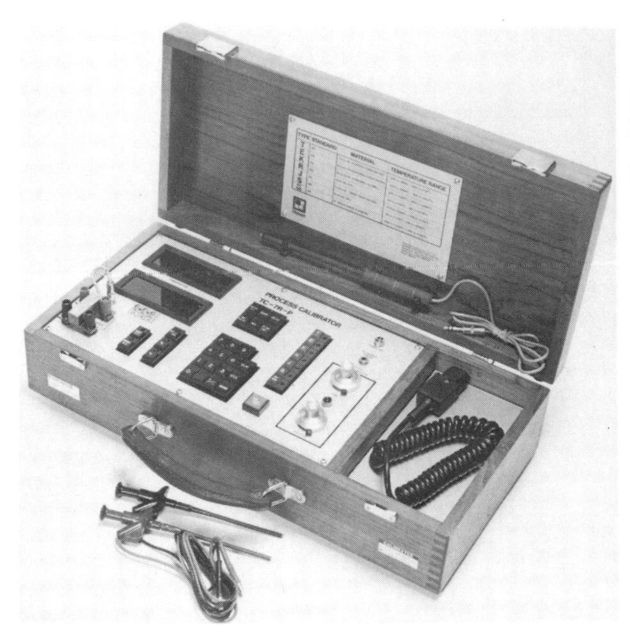

Figure 1.43 Thermocouple calibration equipment (courtesy Haven Automation Ltd.).

As with any other type of temperature measurement the location of the thermocouple junctions is critical. This is just as important for the cold junction as for the hot junction. It must be remembered that there may well be a temperature gradient over quite short distances in an instrument and unless the cold junction temperature sensor is in close thermal contact with the cold junction itself a reading error of several degrees Celsius may result. This problem is at its worst with mains electricity powered measuring instruments where there is a certain amount of heat liberated by the power unit.

The point to remember is that it is not usually adequate to measure the air temperature in the vicinity of the cold junctions. The sensor should be in good thermal contact with them.

An obvious point, but one which surprisingly often causes trouble, is the mismatch between the thermocouple and the measuring instrument. The obvious mismatch is using the wrong type of thermocouple or compensating cable.

In the case of galvanometric instruments inaccuracies occur if sufficient care has not been taken in the winding of the make-up resistor or if the thermocouple has been changed and the new external circuit resistance not checked. Careless location or make-up of the ballast resistor so that one of the cold junction terminals is too remote from the cold junction compensating element causes variable errors of several degrees as the ambient temperature changes. Where the ballast resistor required is of a low value, 10 ohms or so, the best arrangement may well be to use a coil of compensating cable of the right resistance.

1.6 Measurement techniques – radiation thermometers

1.6.1 Introduction

As was mentioned in Section 1.1, thermal energy may be transferred from one body to another by radiation as well as by conduction. The amount of thermal energy or heat leaving a body by radiation and the wavelength of that radiation are functions of the temperature of the body.

This dependence on temperature of the characteristics of radiation is used as the basis of temperature measurement by radiation thermometers.

1.6.1.1 Black body radiation

An ideal black body is one that at all temperatures will absorb all radiation falling on it without reflecting any whatever in the direction of incidence. The absorptive power of the surface, being the proportion of incident radiation absorbed, will be unity. Most surfaces do not absorb all incident radiation but reflect a portion of it. That is, they have an absorptive power of less than unity.

A black body is also a perfect radiator. It will radiate more radiation than a body with an absorptive power of less than unity. The emissive power is called the 'emissivity' of a surface. The emissivity is the ratio of the radiation emitted at a given temperature compared to the radiation from a perfectly black body at the same temperature.

The total emissivity of a body is the emissive power over the whole band of thermal radiation wavelengths and is represented by ε_t. When only a small band of wavelengths is considered the term 'spectral emissivity' is used and a subscript is added defining the wavelength band, e.g., $\varepsilon_{1.5}$ indicates the emissivity at 1.5 μm wavelength.

The emissivity of surfaces is not usually the same over all wavelengths of the spectrum. In general the emissivity of metals is greater at shorter wavelengths and the emissivity of oxides and refractory materials is greater at longer wavelengths. Some materials may have a very low emissivity at a particular wavelength band and higher emissivities at shorter and longer wavelength. For instance, glass has an emissivity of almost zero at 0.65 μm.

Realization of a black body radiator A black body radiator is achieved in practice by an enclosure, A in Figure 1.44, having a relatively small orifice B from which black body radiation is emitted. The inside walls of the enclosure must be at a uniform temperature. To show that the orifice B behaves as a black body consider the ray of radiation C entering the chamber through B. The ray will suffer many reflections on the inside walls of the enclosure before it emerges at B. Provided the walls of the chamber are not perfectly reflecting the total energy of the radiation will have been absorbed by the many reflections before the ray can emerge. The orifice is then totally absorbing all radiation that enters it. It is a black body.

To show that the orifice must also radiate as a black body first consider a body in a radiant flux at any single wavelength. If that body did not radiate energy at that wavelength as fast as it absorbed it, it would rapidly get warmer than its environment. In practice a body will be at thermal equilibrium with its surroundings so it must be radiating energy as it receives it.

Therefore the emissivity ε of a body must equal its absorbance α. The orifice B which is a black body absorber must also be a black body radiator.

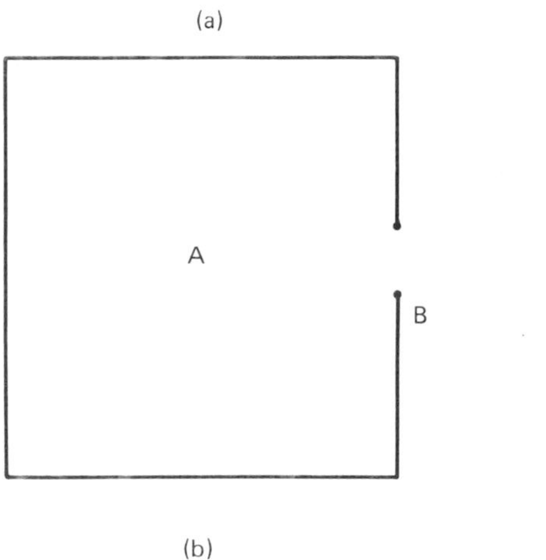

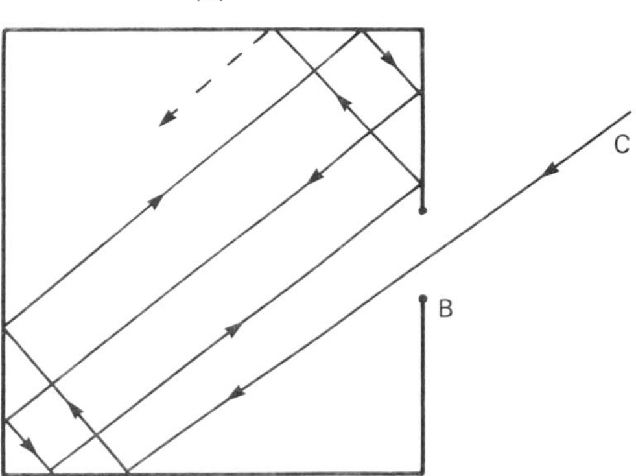

Figure 1.44 (a) Black body radiator, (b) absorption of ray of radiation by black body radiator.

In practice a sighting hole in a furnace will radiate as a black body if the furnace and its contents are in thermal equilibrium and provided it does not contain a gas or flame which absorbs or radiates preferentially in any wavelength band. However, the radiation from the sighting hole will only be black body radiation provided everything in the furnace is at the same temperature. When all objects in the furnace are at the same temperature all lines of demarcation between them will disappear. If a cold object is introduced to the furnace it will be absorbing more energy than it is radiating, the rest of the furnace will be losing more radiation than it receives. Under these conditions the radiation will no longer be black body radiation but will be dependent upon the emissivity of the furnace walls.

Prevost's theory of exchanges Two bodies A and B in a perfectly heat-insulated space will both be radiating and both be absorbing radiation. If A is hotter than B it will radiate more energy than B. Therefore B will receive more energy than it radiates and consequently its temperature will rise. By contrast body A will lose more energy by radiation than it receives so its temperature will fall. This process will continue until both bodies reach the same temperature. At that stage the heat exchanged from A to B will be equal to that exchanged from B to A.

A thermometer placed in a vessel to measure gas temperature in that vessel will, if the vessel walls are cooler than the gas, indicate a temperature lower than the gas temperature because it will radiate more heat to the vessel walls than it receives from them.

Black body radiation: Stefan–Boltzmann law The total power of radiant flux of all wavelengths R emitted into the frontal hemisphere by a unit area of a perfectly black body is proportional to the fourth power of the temperature Kelvin:

$$R = \sigma T^4 \tag{1.30}$$

where σ is the Stefan–Boltzmann constant, having an accepted value of $5.670\,32 \times 10^{-8}$ W . m^{-2} . K^{-4}, and T is the temperature Kelvin.

This law is very important, as most total radiation thermometers are based upon it. If a receiving element at a temperature T_1 is arranged so that radiation from a source at a temperature T_2 falls upon it, then it will receive heat at the rate of σT_2^4, and emit it at a rate of σT_1^4. It will, therefore, gain heat at the rate of $\sigma(T_2^4 - T_1^4)$. If the temperature of the receiver is small in comparison with that of the source, then T_1^4 may be neglected in comparison with T_2^4, and the radiant energy gained will be proportional to the fourth power of the temperature Kelvin of the radiator.

1.6.1.2 The distribution of energy in the spectrum: Wien's laws

When a body is heated it appears to change colour. This is because the total energy and distribution of radiant energy between the different wavelengths, is changing as the temperature rises. When the temperature is about 500 °C the body is just visibly red. As the temperature rises, the body becomes dull red at 700 °C, cherry red at 900 °C, orange at 1100 °C, and finally white hot at temperatures above 1400 °C. The body appears white hot because it radiates all colours in the visible spectrum.

It is found that the wavelength of the radiation of the maximum intensity gets shorter as the temperature rises. This is expressed in Wien's displacement law:

$$\lambda_m T = \text{constant}$$
$$= 2898\ \mu\text{m} . \text{K} \tag{1.31}$$

where λ_m is the wavelength corresponding to the radiation of maximum intensity, and T is the temperature Kelvin. The actual value of the spectral radiance at the wavelength λ_m is given by Wien's second law:

$$L_{\lambda_m} = \text{constant} \times T^5 \tag{1.32}$$

where L_{λ_m} is the maximum value of the spectral radiance at any wavelength, i.e. the value of the radiance at λ_m, and T is the temperature Kelvin. The constant does not have the same value as the constant in equation (1.31). It is important to realize that it is only the maximum radiance at one particular wavelength which is proportional to T^5, the total radiance for all wavelengths is given by the Stefan–Boltzmann law, i.e. it is proportional to T^4.

Wien deduced that the spectral concentration of radiance, that is, the radiation emitted per unit solid angle per unit area of a small aperture in a uniform temperature enclosure in a direction normal to the area in the range of wavelengths between λ and $\lambda + \delta\lambda$ is $L_\lambda . \delta\lambda$ where

$$L_\lambda = \frac{C_1}{\lambda^5 . e^{C_2/\lambda T}} \tag{1.33}$$

where T is the temperature Kelvin and C_1 and C_2 are constants. This formula is more convenient to use and applies with less than 1 per cent deviation from the more refined Planck's radiation law used to define IPTS-68 provided $\lambda T < 3 \times 10^3$ m . K.

In 1900 Planck obtained from theoretical considerations based on his quantum theory, the expression

$$L_\lambda = \frac{C_1}{\lambda^5(e^{C_2/\lambda T} - 1)} \tag{1.34}$$

where the symbols have the same meaning, and $C_2 = 0.014\,388$ m . K.

These laws also enable the correction to be calculated for the presence of an absorbing medium

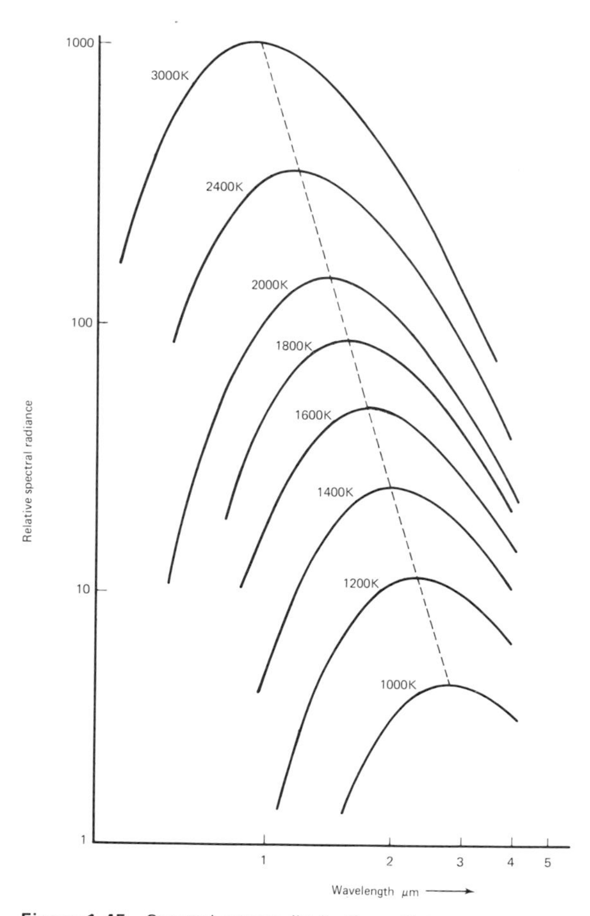

Figure 1.45 Spectral energy distribution with temperature.

such as glass in the optical pyrometer, and also the correction required for changes in the spectral emissive power of the radiating surface.

The variation of spectral radiance with wavelength and temperature of a black body source is given by Figure 1.45.

1.6.2 Radiation thermometer types

Since the energy radiated by an object is a function of its absolute temperature this is a suitable property for the non-contact and non-intrusive measurement of temperature. Instruments for temperature measurement by radiation are called radiation thermometers. The terms pyrometer or radiation pyrometer were formerly used.

There are four principal techniques for the measurement of temperature by the radiation from a hot body: total radiation, pyroelectric, photoelectric, and optical.

Instruments using the first three of these techniques are normally constructed in the same general physical form. Figure 1.46 shows the general format of one of these instruments. It consists of a cylindrical metal body made of aluminium alloy, brass or plastic. One end of the body carries a lens, which depending on the wavelength range required, consists of germanium, zinc sulphide, quartz, glass or sapphire. The opposite end carries the electrical terminations for connecting the sensing head to its signal conditioning module. A typical size of such a sensing head is 250 mm long by 60 mm diameter. A diagrammatic sketch of the construction of the instrument is shown in Figure 1.47. Infrared energy from a target area on the object whose temperature is to be measured is focused by the lens onto the surface of the detector. This energy is converted to an electrical signal which may be amplified by a head amplifier on the circuit board. Power is supplied to the instrument and the output transmitted down a cable which is connected to terminals in the termination box. In instruments working in the near-infrared region where the lens is transparent to visible light a telescope can be provided, built into the instrument, so that it can be focused and aligned by looking through the lens.

A primary advantage of radiation thermometers, especially when used to measure high temperatures, is that the instrument measuring head can be mounted remote from the hot zone in an area cool enough not to exceed the working temperature of the semiconductor electronics, typically about 50–75 °C. However, where the instrument has to be near the hot region, such as attached to the wall of a furnace, or where it needs to be of rugged construction, it can be housed in an air- or water-cooled housing. Such a housing is shown in Figure 1.48.

The function of the lens as indicated above is to concentrate the radiation from the source onto the surface of the sensor. This also has the great advantage that the instrument reading is substantially independent of the distance from the source, provided

Figure 1.46 General-purpose radiation thermometer (courtesy Land Infrared Ltd.).

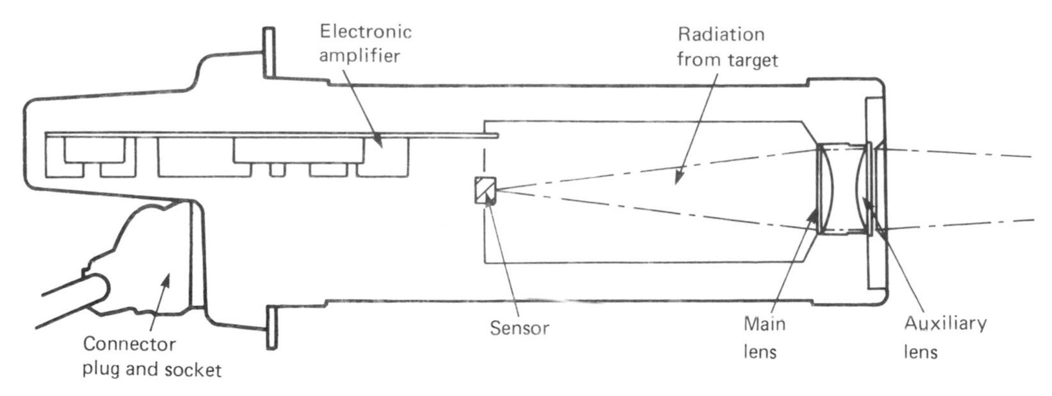

Figure 1.47 Diagram of radiation thermometer.

the source is large enough for its image to fully fill the area of the sensor. The lens material depends on the wavelength to be passed. This will normally be a function of the temperature range for which the instrument is specified. For lower temperatures the lens material will be chosen to give a wide wavelength passband. For higher temperatures a narrower passband may be acceptable. Of course the higher the temperature to be measured the shorter the wavelength that needs to be passed by the lens. Table 1.15 shows the wavelength passband of some lens materials.

Table 1.15 Wavelengths transmitted by lens materials

Lens material	*Passband* (μm)
Pyrex	0.3–2.7
Fused silica	0.3–3.8
Calcium fluoride	0.1–10
Arsenic trisulphide	0.7–12
Germanium	2–12
Zinc selenide	0.5–15

To achieve a wider wavelength range the focusing can be achieved with a concave mirror. Figure 1.49 shows diagrammatically the general arrangement of a reflection instrument.

A special application of mirror focusing for radiation thermometry is in the temperature measurement of stars and other astronomic bodies. The thermopile, or more usually a semiconductor detector, is cooled with liquid nitrogen or helium to increase its sensitivity to very small amounts of radiation. It is located at the focus of a reflecting astronomical telescope. The telescope is directed to the body whose temperature is to be measured so that its image is focused on the detector. The whole assembly forms a very sensitive radiation thermometer with the ability to detect temperatures down to a few tens of kelvins.

1.6.2.1 Total radiation thermometer

In this type of instrument, the radiation emitted by the body whose temperature is required, is focused on a suitable thermal-type receiving element. This receiving element may have a variety of forms. It may be a resistance element, which is usually in the form of a very thin strip of blackened platinum, or a thermocouple or thermopile. The change in temperature of the receiving element is then measured as has already been described.

In a typical radiation thermopile a number of thermocouples made of very fine strips are connected

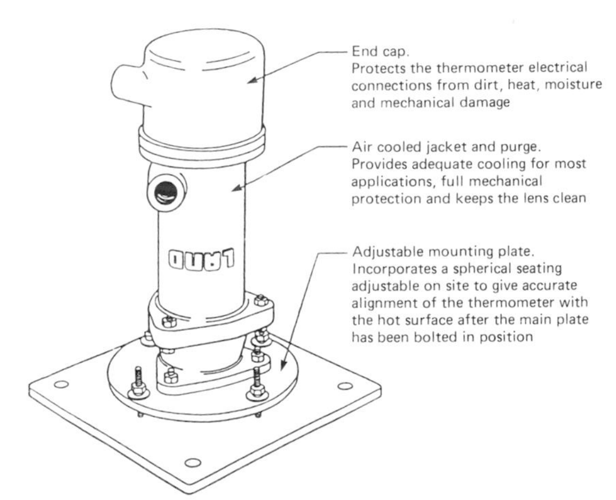

Figure 1.48 Air-cooled housing for radiation thermometer (courtesy Land Infrared Ltd.).

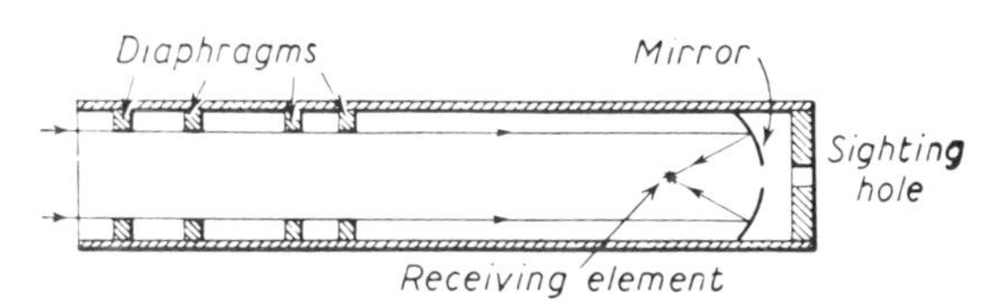

Figure 1.49 Mirror-focused radiation thermometer (courtesy Land Infrared Ltd.).

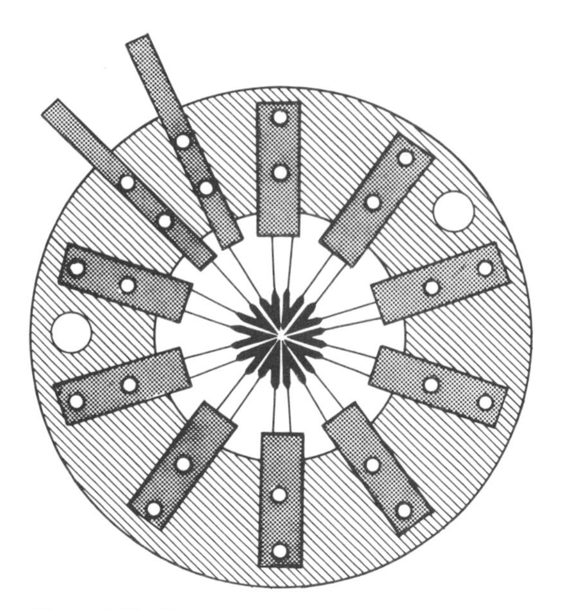

Figure 1.50 Thermopile for use in total radiation pyrometer.

in series and arranged side by side, or radially as in the spokes of a wheel, so that all the hot junctions, which are blackened to increase the energy-absorbing ability, fall within a very small target area. The thermoelectric characteristics of the thermopiles are very stable because the hot junctions are rarely above a few hundred degrees Celsius, and the thermocouples are not exposed to the contaminating atmosphere of the furnace. Stability and the fact that it produces a measurable e.m.f. are the main advantages of the thermopile as a detector. In addition, thermopiles have the same response to incoming radiant energy regardless of wavelength within the range 0.3–20 μm. The main disadvantage of the thermopile is its comparatively slow speed of response which depends upon the mass of the thermocouple elements, and the rate at which heat is transferred from the hot to the cold junctions. Increase in this rate of response can only be attained by sacrificing temperature difference with a resultant loss of output. A typical industrial thermopile of the form shown in Figure 1.50 responds to 98 per cent of a step change in incoming radiation in 2 seconds. Special thermopiles which respond within half a second are obtainable but they have a reduced e.m.f. otuput.

In order to compensate for the change in the thermopile output resulting from changes in the cold junction temperature an ambient temperature sensor is mounted by the cold junctions. Alternative thermal detectors to thermopiles are also used. Thermistors and pyroelectric detectors are currently in use. The advantage of thermistors is that they can be very small and so have a quick speed of response. Their main disadvantage is their non-linearity, though this is not so great a disadvantage as with a direct measurement of temperature because provision has to be made to linearize the radiated energy signal anyway.

Correction for emissivity When the temperature of a hot object in the open is being measured, due regard must be given to the correction required for the difference between the emissivity of the surface of the object and that of a perfectly black body.

The total radiant flux emitted by the source will be given by

$$R = \varepsilon A T^4 \qquad (1.35)$$

where ε is the total emissivity of the body, A is the area from which radiation is received, σ is the Stefan–Boltzmann constant, and T the actual temperature of the body.

This flux will be equal to that emitted by a perfectly black body at a temperature T_a, the apparent temperature of the body:

$$R = \sigma A T_a^4 \qquad (1.36)$$

Equating the value of R in equations (1.35) and (1.36):

$$\varepsilon \sigma A T^4 = \sigma A T_a^4$$

$$T^4 = \frac{T_a^4}{\varepsilon}$$

$$T = \frac{T_a}{\sqrt[4]{\varepsilon}} \qquad (1.37)$$

The actual correction to be applied to the apparent temperature is given in Figure 1.51. Table 1.16 shows the emissivity of some metals at different temperatures.

The radiation from a hot object can be made to approximate much more closely to black body radiation by placing a concave reflector on the surface. If the reflectivity of the reflecting surface is r, then it can be shown that the intensity of the radiation which would pass out through a small hole in the reflector is given by

$$R = \frac{\varepsilon}{1 - r(1 - \varepsilon)} \sigma T^4 \qquad (1.38)$$

where R is the radiation intensity through the hole, ε is the emissivity of the surface, σ is the Stefan–Boltzmann constant, and T the temperature in kelvin. With a gold-plated hemisphere, the effective emissivity of a surface of emissivity 0.6 is increased by this method to a value of 0.97.

Surface radiation thermometer A surface radiation thermometer manufactured by Land Infrared Ltd. uses the above principle, see Figure 1.52. This instrument uses a thermopile sited on a small hole in a gold-

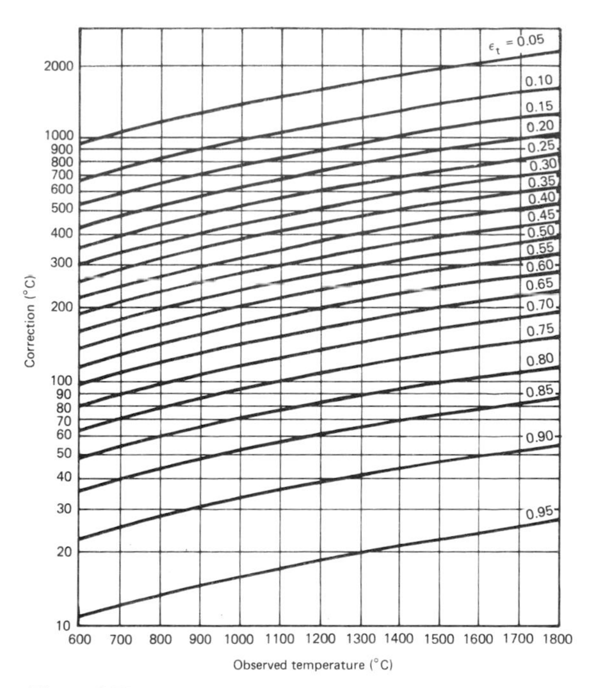

Figure 1.51 Emissivity corrections to the readings of a total radiation thermometer.

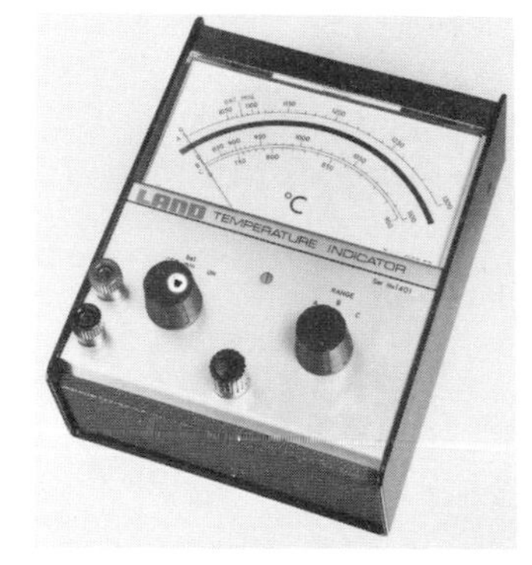

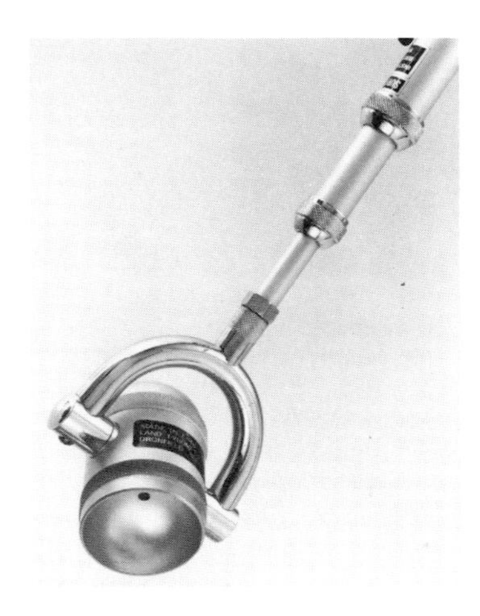

Figure 1.52(a)

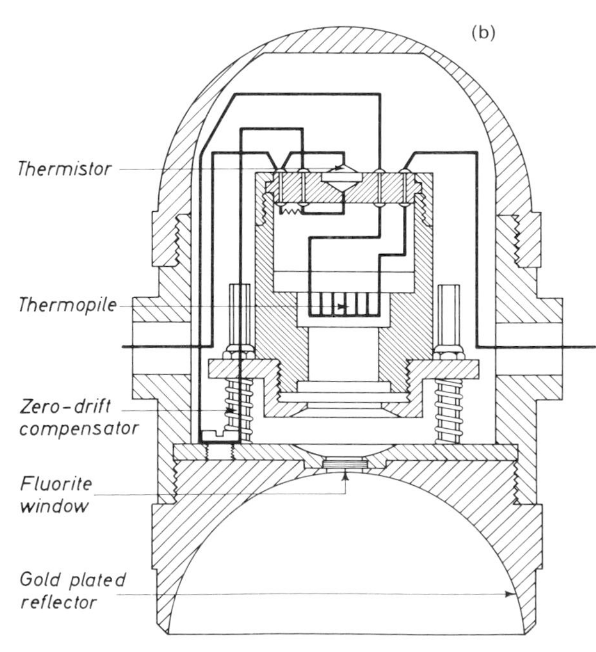

Figure 1.52 (a) Surface radiation thermometer (courtesy Land Infrared Ltd.). (b) cross-section diagram of Land surface radiation thermometer.

plated hemisphere mounted on the end of a telescopic arm.

Gold is chosen for the reflecting surface because it is the best reflector of infrared radiation known, and is not easily tarnished. The hole in the reflector is closed by a fluorite window which admits a wide range of radiation to the thermopile but excludes dirt and draughts. This pyrometer will give accurate surface temperature readings for most surfaces, other than bright or lightly oxidized metals, without any significant error due to surface emissivity changes. The standard instrument covers a temperature range of from 100 to 1300 °C on three scales. A special low temperature version is available for the range 0 to 200 °C. The indicator gives a reading in 5 to 6 seconds, and the pyrometer should not be left on the hot surface for more than this length of time, particularly at high temperatures. The thermistor bridge provides compensation for changes in the sensitivity of the thermopile at high temperatures, but if the head is too hot to touch it is in danger of damage to soldered joints, insulation etc.

Table 1.16 Total emissivity of miscellaneous materials

Total emissivity of unoxidized metals

Material	*25 °C*	*100 °C*	*500 °C*	*1000 °C*	*1500 °C*	*2000 °C*
Aluminium	0.022	0.028	0.060	—	—	—
Bismuth	0.048	0.061	—	—	—	—
Carbon	0.081	0.081	0.079	—	—	—
Chromium	—	0.08	—	—	—	—
Cobalt	—	—	0.13	0.23	—	—
Columbium	—	—	—	—	0.19	0.24
Copper	—	0.02	—	(Liquid 0.15)	—	—
Gold	—	0.02	0.03	—	—	—
Iron	—	0.05	—	—	—	—
Lead	—	0.05	—	—	—	—
Mercury	0.10	0.12	—	—	—	—
Molybdenum	—	—	—	0.13	0.19	0.24
Nickel	0.045	0.06	0.12	0.19	—	—
Platinum	0.037	0.047	0.096	0.152	0.191	—
Silver	—	0.02	0.035	—	—	—
Tantalum	—	—	—	—	0.21	0.26
Tin	0.043	0.05	—	—	—	—
Tungsten	0.024	0.032	0.071	0.15	0.23	0.28
Zinc	(0.05 at 300 °C)					
Brass	0.035	0.035	—	—	—	—
Cast Iron	—	0.21	—	(Liquid 0.29)		—
Steel	—	0.08	—	(Liquid 0.28)		—

Total emissivity ε_t of miscellaneous materials

Material	*Temp.* (°C)	ε_t	*Material*	*Temp.* (°C)	ε_t
Aluminium (oxidized)	200	0.11	Lead (oxidized)	200	0.63
	600	0.19	Monel (oxidized)	200	0.43
Brass (oxidized)	200	0.61		600	0.43
	600	0.59	Nickel (oxidized)	200	0.37
Calorized copper	100	0.26		1200	0.85
	500	0.26	Silica brick	1000	0.80
Calorized copper (oxidized)	200	0.18		1100	0.85
	600	0.19	Steel (oxidized)	25	0.80
Calorized steel (oxidized)	200	0.52		200	0.79
	600	0.57		600	0.79
Cast iron (strongly oxidized)	40	0.95	Steel plate (rough)	40	0.94
	250	0.95		400	0.97
Cast iron (oxidized)	200	0.64	Wrought iron (dull	25	0.94
	600	0.78	oxidized)	350	0.94
Copper (oxidized)	200	0.60	20Ni—25Cr—55Fe	200	0.90
	1000	0.60	(oxidized)	500	0.97
Fire Brick	1000	0.75	60Ni—12Cr—28Fe	270	0.89
Gold enamel	100	0.37	(oxidized)	560	0.82
	100	0.74		100	0.87
Iron (oxidized)	500	0.84	80Ni—20Cr (oxidized)	600	0.87
	1200	0.89		1300	0.89
Iron (rusted)	25	0.65			

Source: 'Temperature, its measurement & control' in *Science & Industry*, American Institute of Physics, Reinhold Publishing Co. (1941).

The instrument may be used to measure the mean emissivity of a surface for all wavelengths up to about 10 μm. This value can be used for the correction of total radiation thermometer readings. A black hemispherical insert is provided with the instrument which can be clipped into the hemispherical reflector to cover the gold. If two measurements are made, one with the gold covered and the other with the gold exposed, the emissivity can readily be deduced from the two measurements. A graph provided with the instrument enables the emissivity to be derived easily from the two readings, while a second graph gives an indication of the error involved in the temperature measurement of the hot body.

Calibration of total radiation thermometers A total radiation thermometer may be calibrated by sighting it through a hole into a black body enclosure of known temperature. A special spherical furnace was developed by the British Iron and Steel Research Association for this purpose. The furnace consisted of a sphere 0.3 m in diameter consisting of a diffusely reflecting material. For temperatures up to 1300 °C stainless steel, 80Ni 20Cr alloy, or nickel may be used. For temperatures up to 1600 °C silicon carbide is necessary, and for temperatures up to 3000 °C graphite may be used provided it is filled with argon to prevent oxidation. The spherical core is uniformly wound with a suitable electrical heating element, completely enclosed in a box containing thermal insulation. For calibration of radiation thermometers up to 1150 °C a hole of 65 mm diameter is required in the cavity, but above this temperature a 45 mm hole is sufficient.

Where the larger hole is used a correction for the emissivity of the cavity may be required for very accurate work. Two sheathed thermocouples are usually placed in the furnace, one near the back and the other just above the sighting hole. Comparison of the two measured temperatures indicates when the cavity is at a uniform temperature.

Calibration may be carried out by comparing the thermometer and thermocouple temperature, or the test thermometer may be compared with a standard radiation thermometer when both are sighted on to the radiating source which may or may not be a true black body.

Cylindrical furnaces may also be used with a thermocouple fitted in the sealed end of the cylinder, which is cut on the inside to form a series of 45° pyramids.

A choice of three aperture sizes is available at the open end. For temperatures up to 1100 °C the furnace is made of stainless steel but for higher temperatures refractory materials are used. For further details see *The Calibration of Thermometers* (HMSO 1971). Figure 1.53 shows typical black body furnaces.

Furnace temperature by radiation thermometer Conditions in a furnace which might otherwise be considered as perfectly black body conditions may be upset by the presence of flame, smoke or furnace gases. In these conditions, a total radiation thermometer generally indicates a temperature between that of the furnace atmosphere and the temperature which would be indicated if such an atmosphere were not present.

A thick luminous flame may shield the object almost completely. Non-luminous flames radiate and absorb energy only in certain wavelength bands, principally because of the presence of carbon dioxide and water vapour. The error due to the presence of these gases can be reduced by using a lens of Pyrex which does not transmit some of these wavelengths, so that the instrument is less affected by variations in quantity of these gases. Where appreciable flame, smoke and gas are present it is advisable to use a closed-ended sighting tube, or provide a purged sighting path by means of a blast of clean dry air.

Errors in temperature measurement can also occur owing to absorption of radiation in the cold atmosphere between a furnace and the thermometer. To ensure that the error from this source does not exceed 1 per cent of the measured temperature, even on hot damp days, the distance between thermometer lens and furnace should not exceed 1.5 m if a glass lens is used, 1 m if the lens is silica, and 0.6 m if it is of fluorite.

1.6.2.2 Pyroelectric techniques

Pyroelectric detectors for thermal radiation are a comparatively recent introduction. Pyroelectric materials, mainly ceramics, are materials whose

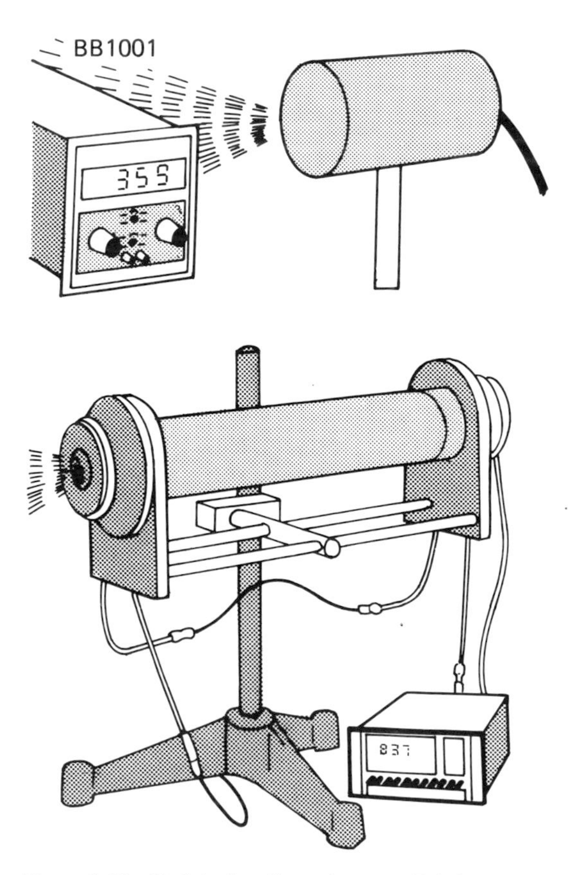

Figure 1.53 Black body radiators (courtesy Polarisers Technical Products).

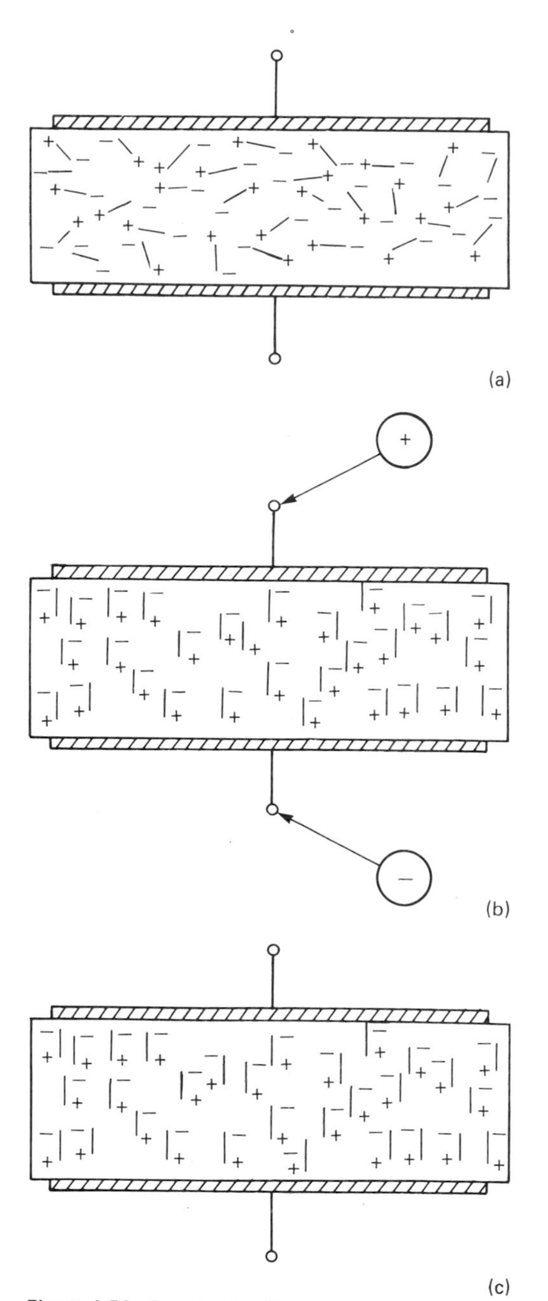

Figure 1.54 Pyroelectric effect.

molecules have a permanent electric dipole due to the location of the electrons in the molecules. Normally these molecules lie in a random orientation throughout the bulk of the material so that there is no net electrification. Also at ambient temperatures the orientations of the molecules are essentially fixed. If the temperature is raised above some level characteristic to the particular material, the molecules are free to rotate. This temperature is called the Curie temperature by analogy with the magnetic Curie temperature.

If a piece of pyroelectric ceramic is placed between two electrodes at ambient temperature the molecular dipoles are fixed in a random orientation, Figure 1.54(a). If it is then heated above its Curie temperature and an electrical potential applied to the electrodes, thus generating an electric field in the ceramic, the molecules will all align themselves parallel to the field, Figure 1.54(b). On cooling the ceramic back to ambient temperature and then removing the applied potential the molecules remain aligned. Figure 1.54(c). The amount of the polarization of the ceramic and therefore the magnitude of the resulting external electric field is a constant Σ which is a function of the material. If the field due to the applied voltage was E and the polarization P then

$$P=\Sigma E \tag{1.39}$$

If the temperature of the polarized pyroelectric ceramic is raised the molecular dipoles, which are anyway oscillating about their parallel orientation, will oscillate through a greater angle. Figure 1.55 shows one molecular dipole of length x and charge $\pm q$. Its electric moment is qx. If then the dipole oscillates through an average angle of $\pm\theta$ the effective length will be z where

$$z=x\cos\theta \tag{1.40}$$

The angle θ will increase with increasing temperature, thus reducing the electric moment of all the molecular dipoles. The electric moment or polarization of the whole piece of pyroelectric ceramic is of course the sum of all the molecular dipoles. Thus as the temperature rises the polarization of the whole piece of material gets less.

The Curie point is the temperature at which the oscillatory energy of the molecular dipoles is such that they can rotate freely into any position allowing them to return to their random orientation.

As stated above the electric moment M of the whole slice of ceramic is the sum of all the molecular dipole

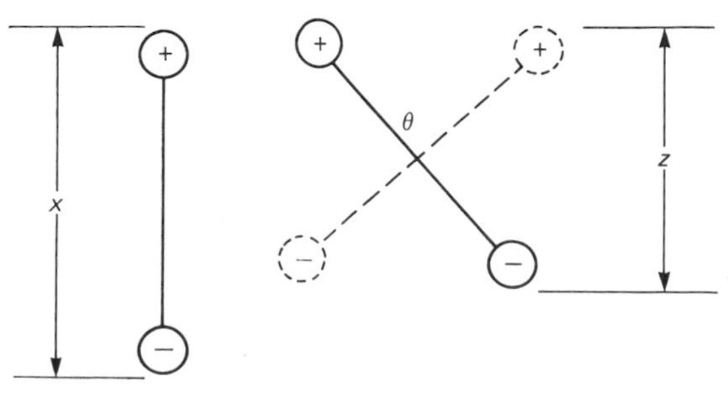

Figure 1.55 Mechanism of pyroelectric effect.

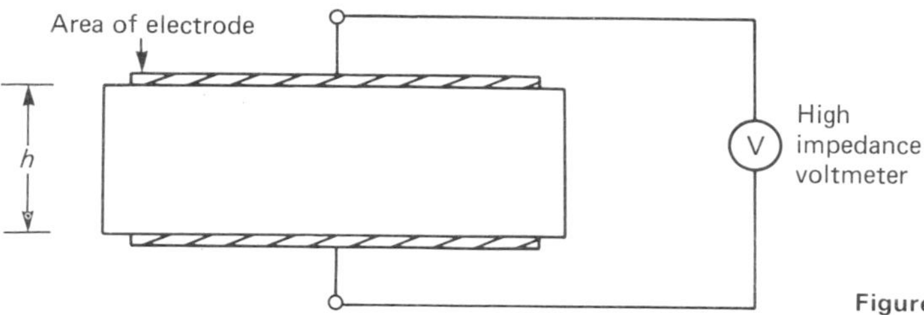

Figure 1.56 Pyroelectric detector.

moments:

$$M = PAh \quad (1.41)$$

where P is the dipole moment per unit volume, h is the thickness of the slice and A is the electrode area; see Figure 1.56.

If the electric charge at the two surfaces of the slice of pyroelectric ceramic is Q_s this has a dipole moment of $Q_s . h$, so that

$$Q_s = PA \quad (1.42)$$

If the temperature of the material rises the polarization is reduced and therefore Q_s becomes less. But if the electrodes are connected by an external circuit to an electrometer or other high impedance detector Q_s is normally neutralized by a charge Q on the electrodes. A reduction of Q_s therefore results in an excess charge on the electrodes and therefore a voltage V is detected.

$$V = Q/C \quad (1.43)$$

where C is the electrical capacitance of the device, for a temperature change of δT the change of charge δQ is given by

$$\delta Q = \Omega . A . \delta T \quad (1.44)$$

where Ω is the pyroelectric coefficient of the material. Therefore the voltage change will be

$$\delta V = \delta Q/C = \Omega A\, \delta T/C \quad (1.45)$$

where C is the electrical capacitance between the electrodes. The pyroelectric coefficient Ω is a function of temperature reducing with a non-linear characteristic to zero at the Curie temperature.

When used as a detector in a radiation thermometer, radiation absorbed at the surface of the pyroelectric slice causes the temperature of the detector to rise to a new higher level. At the start the charge on the electrodes will have leaked away through the external electrical circuit so there will have been zero voltage between the electrodes. As the slice heats up a voltage is detected between the two electrodes. When the device reaches its new temperature, losing heat to its environment at the same rate as it is receiving heat by radiation, the generation of excess charge on the electrodes ceases, the charge slowly leaks away through the electrical circuit and the detected voltage returns to zero. The device detects the change of incident radiation. To detect a constant flux of radiation, i.e. to measure a constant temperature, it is necessary to 'chop' the incident radiation with a rotating or oscillating shutter.

The physical construction of a pyroelectric radiation thermometer is essentially identical to a total radiation instrument except for the location of the radiation-chopping shutter just in front of the detector. Figure 1.57(a) shows the location and Figure 1.57(b) a typical profile of the optical chopper in a pyroelectric radiation thermometer. Figure 1.57(c) shows the graph against time of the chopped radiation together with the resulting electrical signal.

1.6.2.3 Optical (disappearing filament) thermometer

Optical radiation thermometers provide a simple and accurate means for measuring temperatures in the range 600 °C to 3000 °C. Since their operation requires the eye and judgement of an operator they are not suitable for recording or control purposes. However, they provide an effective way of making spot measurements and for calibration of total radiation thermometers.

In construction an optical radiation thermometer is similar to a telescope. However, a tungsten filament lamp is placed at the focus of the objective lens. Figure 1.58 shows the optical arrangement of an optical radiation thermometer. To use the instrument the point where the temperature is required to be known is viewed through the instrument. The current through the lamp filament is adjusted so that the filament disappears in the image. Figure 1.59 shows how the filament looks in the eyepiece against the background of the object, furnace or whatever is to have its temperature measured. At (a) the current through the filament is too high and it looks bright against the light from the furnace, at (c) the current is too low while at (b) the filament is at the same temperature as the background. The temperature of the filament is known from its electrical resistance. Temperature readout is achieved either by a meter measuring the current through the filament or by temperature calibrations on the control resistor regulating the current through the lamp. The filter in the eyepiece shown in Figure 1.58

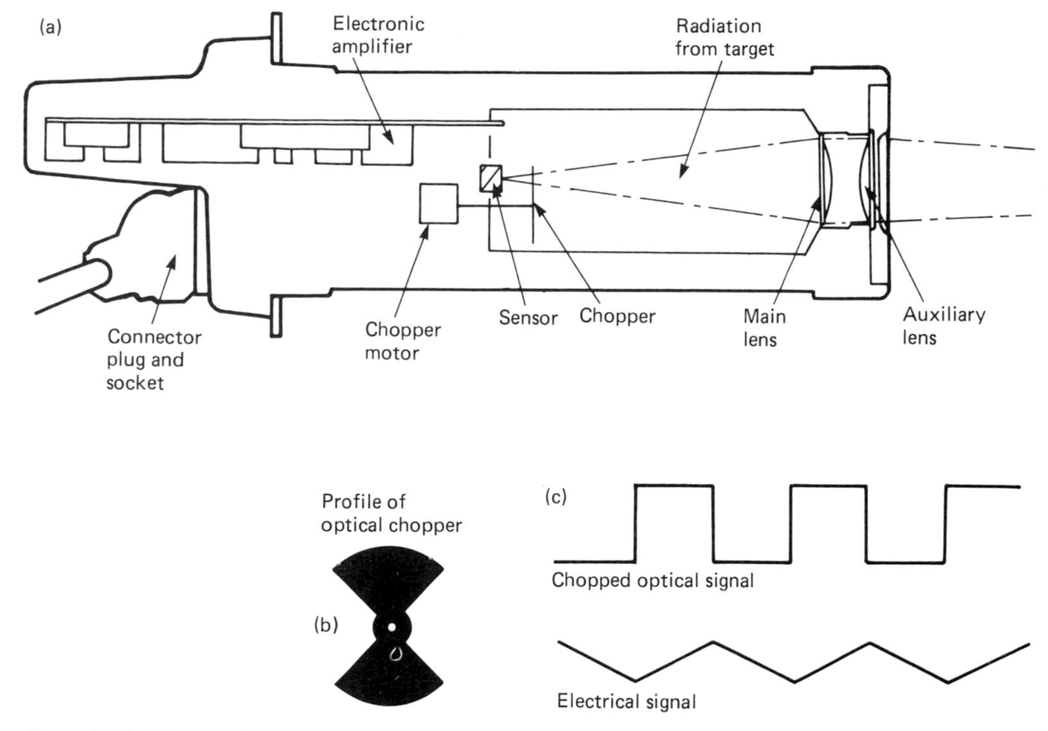

Figure 1.57 Diagram of pyroelectric radiation thermometer.

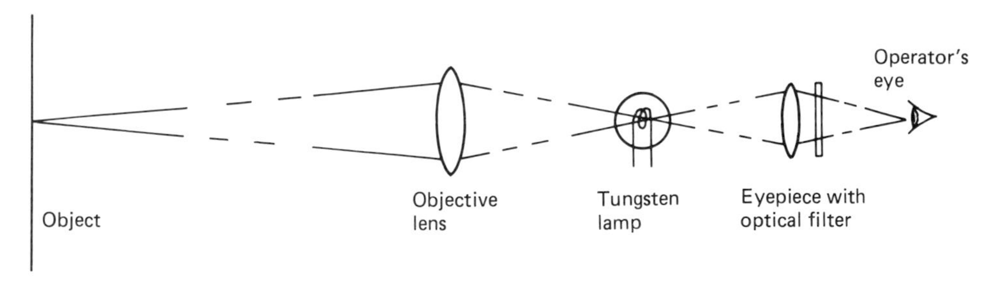

Figure 1.58 Optical system of disappearing filament thermometer.

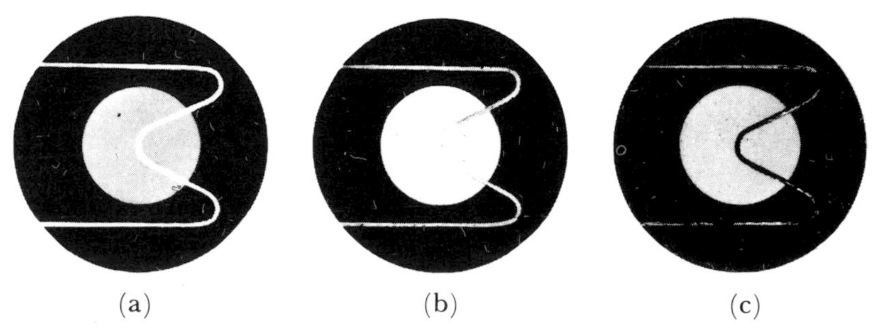

Figure 1.59 Appearance of image in optical thermometer.

passes light at a wavelength around 0.65 μm.

Lamps for optical thermometers are not normally operated at temperatures much in excess of 1500 °C. To extend the range of the instrument beyond this temperature a neutral filter of known transmission factor can be placed in the light path before the lamp. The measurement accuracy of an optical thermometer is typically ±5 K between 800 °C and 1300 °C and ±10 K between 1300 °C and 2000 °C.

Corrections for non-black-body conditions Like the total radiation thermometer, the optical thermometer is affected by the emissivity of the radiation source and by any absorption of radiation which may occur between the radiation source and the instrument.

The spectral emissivity of bright metal surfaces at 0.65 μm is greater than the total emissivity *e*, representing the average emissivity over all wavelengths. The correction required for the departure from black body conditions is therefore less than in the case of total radiation thermometers.

Due to the fact that a given change of temperature produces a much larger change in radiant energy at 0.65 μm than produced in the average of radiant energy over all wavelengths, the readings of an optical radiation thermometer require smaller corrections than for a total radiation instrument.

The relationship between the apparent temperature T_a and the true temperature T is given by equation (1.46) which is based on Wien's law

$$\frac{1}{T}-\frac{1}{T_a}=\frac{\lambda \log_{10}\varepsilon_\lambda}{6245} \qquad (1.46)$$

where λ is the wavelength in micrometers (usually 0.65 μm) and ε_λ is the spectral emissivity at wavelength λ.

1.6.2.4 Photoelectric radiation thermometers

The reading obtained with an optical thermometer shows a lower temperature error than a total radiation thermometer. This is because the emissivity error for a given temperature and a known emissivity is proportional to the wavelength of the radiation used to make the measurement. For instance in the case of oxidized steel at 1000 °C with an emissivity of 0.8 a total radiation thermometer will have an error in excess of 50 degrees while the optical thermometer reading will be within 20 degrees. However the optical thermometer has two major drawbacks. First it is only suitable for spot measurements and requires a skilled operator to use it. Secondly it is not capable of a quick response and is totally unsuitable for control purposes.

Photoelectric radiation thermometers are ideally suited to the short wavelength application. Structurally they are essentially identical to a total radiation thermometer except that the thermal sensor is replaced by a photodiode.

A photodiode is a semiconductor diode, which may be either a silicon or germanium junction diode constructed so that the incident radiation can reach the junction region of the semiconductor. In the case of germanium the diode will be a plain P–N junction, in the case of silicon it may be either a P–N or P–I–N junction. In service the diodes are operated with a voltage applied in the reverse, i.e. non-conduction, direction. Under these conditions the current carriers, i.e. electrons, in the semiconductor do not have sufficient energy to cross the energy gap of the junction. However, under conditions of incident radiation some electrons will gain enough energy to cross the junction. They will acquire this energy by collision with photons. The energy of photons is inversely proportional to the wavelength. The longest wavelength of photons that will, on impact, give an electron enough energy to cross the junction dictates the long wave end of the spectral response of the device. The short wavelength end of the response band is limited by the transparency of the semiconductor material. The choice of germanium or silicon photodiodes is dictated by the temperature and therefore the wavelength to be measured. Silicon has a response of about 1.1 μm to 0.4 μm. The useful passband of germanium lies berween 2.5 μm and 1.0 μm. The exact passband of photodiodes varies somewhat from type to type depending on the manufacturing process used, but the above figures are typical. Normally the range of wavelengths used is reduced to a narrower passband than that detected by the semiconductor sensor. For instance, for general applications above 600 °C a narrow passband centred on 0.9 μm is usually used. Wherever possible silicon is to be preferred as it will tolerate higher ambient temperatures than germanium and in general it has the higher speed of response. Small P–I–N photodiodes can have a frequency response up to several hundred megahertz while P–N devices more usually have a response of several kilohertz.

Like all other semiconductor devices the electrical output of photodiodes is temperature-dependent. It is therefore necessary to construct these radiation thermometers with thermistors or resistance thermometers in close proximity to the photodiode to provide ambient temperature compensation.

1.6.2.5 Choice of spectral wavelength for specific applications

It might seem at first sight that apart from optical radiation thermometers the obvious choice should be to use a total radiation thermometer so as to capture as much as possible of the radiant emission from the target to achieve the maximum output signal. However as already mentioned above, except at the

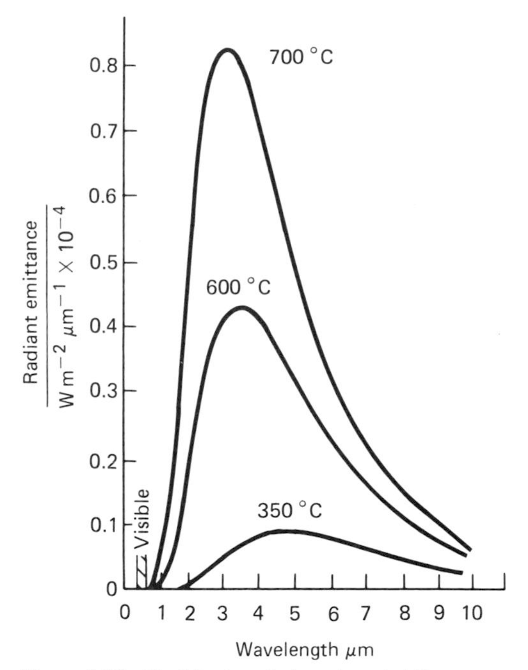

Figure 1.60 Black body radiation characteristics.

lowest temperature ranges, there are several reasons for using narrower wavelength bands for measurement.

Effect of radiant emission against wavelength One reason relates to the rate at which the radiant emission increases with temperature. An inspection of Figure 1.60 will show that the radiant emission at 2 μm increases far more rapidly with temperature than it does at say 6 μm. The rate of change of radiant emission with temperature is always greater at shorter wavelengths. It is clear that the greater this rate of change the more precise the temperature measurement and the tighter the temperature control. On the other hand this cannot be carried to extremes because at a given short wavelength there is a lower limit to the temperature that can be measured. For example, the eye becomes useless below about 600 °C. For these reasons alone we can understand the general rule that the spectral range of the appropriate infrared thermometer shifts to longer wavelengths as the process temperature decreases.

Emittance, reflectance and transmittance Another important reason for the use of different spectral regions relates to the specific emission characteristics of particular target materials. The curves of Figure 1.60 show the emission characteristics of the ideal emitter or black body. No material can emit more strongly than a black body at a given temperature. As discussed previously, however (p. 43), many materials can and do emit less than a black body at the same temperature in various portions of the spectrum. The ratio of the radiant emittance at wavelength λ of a material to that of a black body at the same temperature is called spectral emittance (ε_λ). The value of ε_λ for the substance can range between 0 and 1, and may vary with wavelength. The emittance of a substance depends on its detailed interaction with radiation. A stream of radiation incident on the surface of a substance can suffer one of three fates. A portion may be reflected. Another portion may be transmitted through the substance. The remainder will be absorbed and degraded to heat. The sum of the fraction reflected r, the fraction transmitted t and the fraction absorbed a will be equal to the total amount incident on the substance. Furthermore, the emittance ε of a substance is identical to the absorptance a and we can write

$$\varepsilon = a = 1 - t - r \tag{1.47}$$

For the black body the transmittance and relectance are zero and the emittance is unity. For any opaque substance the transmittance is zero and

$$\varepsilon = 1 - r \tag{1.48}$$

An example of this case is oxidized steel in the visible and near-infrared where the transmittance is 0, the reflectance is 0.20 and the emittance is 0.80. A good example of a material whose emittance characteristics change radically with wavelength is glass. Figure 1.61 shows the overall transmission of soda-lime glass. The reflectance of the glass is about 0.03 or less through most of the spectral region shown. At wavelengths below about 2.6 μm the glass is very highly transparent and the emittance is essentially zero. Beyond 2.6 μm the glass becomes increasingly opaque. From this it is seen that beyond 4 μm glass is completely opaque and the emittance is above 0.98.

This example of glass clearly illustrates how the detailed characteristics of the material can dictate the choice of the spectral region of measurement. For example, consider the problem of measuring and controlling the temperature of a glass sheet during manufacture at a point where its temperature is 900 °C. The rule that suggests a short wavelength infrared thermometer, because of the high temperature, obviously fails. To use the region around 1 μm would be useless because the emittance is close to 0. Furthermore, since the glass is highly transparent the radiation thermometer will 'see through' the glass and can give false indications because of a hot wall behind the glass. One can recognize that glass can be used as an effective 'window' with a short wavelength radiation thermometer. By employing the spectral region between 3 and 4 μm the internal temperature of the glass can be effectively measured and controlled. By operating at 5 μm or more the surface temperature of the glass is

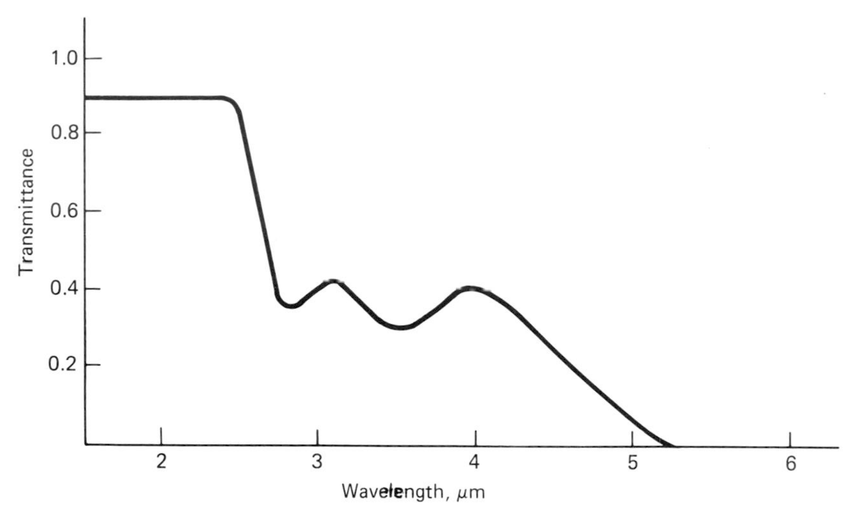

Figure 1.61 Transmittance of one millimetre of soda-lime glass.

measured. Each of these cases represents a practical application of infrared thermometry.

Atmospheric transmission A third important consideration affecting the choice of spectral region is that of the transmission of the atmosphere between the target substance and the radiation thermometer. The normal atmosphere always contains a small but definite amount of carbon dioxide and a variable amount of water vapour. Carbon dioxide strongly absorbs radiation between 4.2 and 4.4 µm and the water vapour absorbs strongly between 5.6 and 8.0 µm and also somewhat in the region 2.6 to 2.9 µm; see Figure 1.62. It is obvious that these spectral regions should be avoided, particularly in the region of the water bands. If this is not done the temperature calibration will vary with path length and also humidity. If the air temperature is comparable to or higher than the target temperature the improperly designed infrared thermometer could provide temperature measurements strongly influenced by air temperatures.

1.6.2.6 Signal conditioning for radiation thermometers

Although the output of a radiation thermometer can be used directly in a voltage or current measuring instrument this is unsatisfactory for two prime reasons. First the energy radiated by a hot body is a function of the fourth power of absolute temperature resulting in a very non-linear scale. Secondly the radiation detectors are themselves sensitive to ambient temperature. This requires either that the radiation thermometer be maintained at a constant temperature or alternatively an ambient temperature sensor is mounted beside the radiation sensor to provide a signal for temperature correction.

To compensate for these two deficiencies in the signal suitable electronic circuits must be used to provide linearization of the signal and to provide automatic temperature correction. It is also necessary to provide correction for the emissivity of the target. Typically the instrument itself carries a small 'head amplifier' to bring the signal up to a suitable level for transmission to the readout instrument. This head amplifier also provides the required ambient tempera-

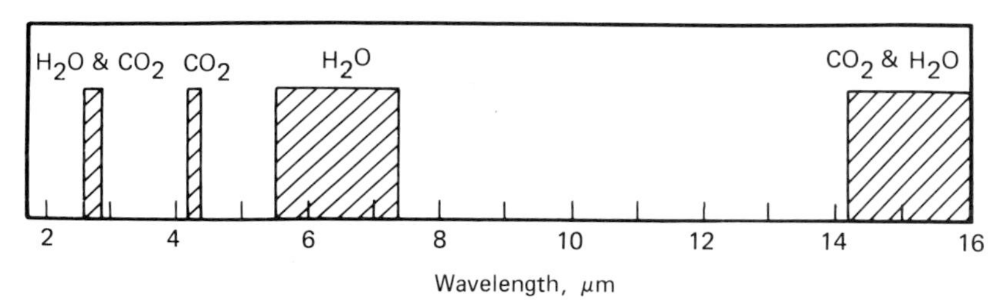

Figure 1.62 Atmospheric absorption of infrared radiation.

Figure 1.63 Radiation thermometer transmitter (courtesy Sirius Instruments Ltd.).

ture compensation circuits. The linearization and compensation for emissivity are provided at the readout module.

Some modern instruments provide the whole signal conditioning circuitry in the main instrument itself. Figure 1.63 shows such an instrument. In this equipment the output is a 4 to 20 milliamp signal linear with temperature and compensated for ambient temperature.

With the growing use of microprocessors in instrumentation several manufacturers are introducing instruments where the linearization and compensation are performed by a microcomputer.

1.6.2.7 Radiation thermometer applications

Infrared thermometers are currently used in a wide range of laboratory and industrial temperature control applications. A few low temperature examples include extrusion, lamination and drying of plastics, paper and rubber, curing of resins, adhesives and paints, and cold rolling and forming of metals.

Some high temperature examples include forming, tempering and annealing of glass, smelting, casting, rolling, forging and heat treating of metals, and calcining and firing of ceramics and cement.

In short, the infrared thermometer can be used in almost any application in the range 0 to 3600 °C where its unique capabilities can turn a seemingly impossible measurement and control problem into a practical working process. Many processes now controlled manually can be converted into continuous, automated systems.

1.7 Temperature measurement considerations

1.7.1 Readout

1.7.1.1 Local readout

If temperature requires to be measured at a particular point on say a chemical plant, what considerations govern the choice of instrument? The obvious first choice to most people is a liquid-in-glass thermometer. However this requires that one must be able to get close enough to read the instrument accurately. A better solution is a dial thermometer. The type of instrument chosen will of course depend upon the accuracy and repeatability required. In general and especially on bigger plants local temperature measurement is for general surveillance purposes only, the measurement is probably not essential but is provided as a cross-check on the control instruments to provide operator confidence. An inexpensive bimetal thermometer is probably adequate. If greater accuracy is required then a capillary-type thermometer (see Sections 1.3.2–1.3.4) with short capillary can be used, or where high accuracy is necessary an electrical technique may be specified. In the case of furnaces a portable radiation instrument may be the best choice.

Of course, on small plant not controlled from a separate control room all measurements will probably be local measurements. It is mainly in this situation that the higher accuracy local readout is required.

1.7.1.2 Remote reading thermometers

The first question to ask in the selection of remote reading instruments is: what is the distance between the measurement point and the readout location? If that distance is less than say 100 metres, capillary instruments may well be the best solution. However if the distance is near the top limit vapour pressure instruments will probably be ruled out. They may also not be usable if there is likely to be big ambient temperature variation at the readout point or along the length of the capillary.

The next question is: what is the height difference between the thermometer bulb and the readout position? Long vertical runs using liquid-in-metal thermometers can cause measurement offsets due to the liquid head in the vertical capillary adding to (or subtracting from) the pressure at the instrument Bourdon tube. In the case of height differences greater than say 10 metres liquid thermometers are likely to be unsuitable. This then reduces the choice to gas-filled instruments. A further consideration when specifying instrumentation on a new plant is that it is convenient from itinerary considerations to use as many instruments of the same type as possible. The choice of instrument is then dictated by the most stringent requirement.

On large installations where many different types of instrument are being installed and especially where pneumatic instrumentation is used, capillary instruments can run into an unexpected psychological hazard. Not infrequently a hard-pressed instrument technician has, on finding he has a too long capillary, been known to cut a length out of the capillary and rejoint the ends with a compression coupling. The result is of course disaster to the thermometer. Where on installation the capillary tube is found to be significantly too long it must be coiled neatly in some suitable place. The choice of that place may depend on the type of instrument. In gas-filled instruments the location of the spare coil is irrelevant but especially with vapour pressure instruments it wants to be in a position where it will receive the minimum of ambient temperature excursions to avoid introduction of measurement errors.

For installations with long distances between the point of measurement and the control room it is almost essential to use an electrical measurement technique. For long runs resistance thermometers are to be preferred to thermocouples for two principal reasons. First the copper cables used for connecting resistance bulbs to their readout equipment are very much less expensive than thermocouple wire or compensating cable. Secondly the resistance thermometer signal is at a higher level and lower impedance than most thermocouple signals and is therefore less liable to electrical interference.

An added advantage of electrical measurements is that, whether the readout is local or remote, the control engineer is given wider options as to the kinds of readout available to him. Not only does he have a choice of analogue or digital readout but he can also have a wider range of analogue readouts, since they are not limited to a rotary dial.

1.7.1.3 Temperature transmitters

On big installations or where a wide variety of different measurements are being made with a wide range of instrumentation it is more usual to transfer the signal from the measurement point to the control area by means of temperature transmitters. This has the great advantage of allowing standardization of the readout equipment. Also in the case of electrical transmission, by say a 4–20 milliamp signal, the measurement is much less liable to degradation from electrical interference. Also the use of temperature transmitters allows the choice of measurement technique to be unencumbered by considerations of length of run to the readout location.

The choice of electrical or pneumatic transmission is usually dictated by overall plant policy rather than the needs of the particular measurement, in this case temperature. However, where the requirement is for electrical temperature measurement for accuracy or other considerations the transmission will also need to be electrical. See Part 4 (Telemetry).

1.7.1.4 Computer-compatible measurements

With the increasing use of computer control of plants there is a requirement for measurements to be compatible. The tendency here is to use thermocouples, resistance thermometer, or where the accuracy does not need to be so high, thermistors as the measuring techniques. The analogue signal is either transmitted to an interface unit at the control room or to interface units local to the measurement. The latter usually provides for less degradation of the signal.

As most industrial temperature measurements do not require an accuracy much in excess of 0.5 per cent it is usually adequate for the interface unit to work at eight-bit precision. Higher precision would normally only be required in very special circumstances.

1.7.1.5 Temperature controllers

While thermometers, in their widest sense of temperature measurement equipment are used for readout purposes, probably the majority of temperature measurements in industrial applications are for control purposes. There are therefore many forms of dedicated temperature controllers on the market. As briefly described in Section 1.3.5.1, the simplest of these is a thermostat.

Thermostats A thermostat is a device in which the control function, usually electrical contacts but sometimes some other control function such as a valve, is directly controlled by the measurement action. The instrument described in Section 1.3.5.1 uses solid expansion to operate electrical contacts, but any of the other expansion techniques may be used. In automotive applications the thermostat in an engine cooling system is a simple valve directly operated either by vapour pressure or change of state, e.g. the change of volume of wax when it melts.

Thermostats, however, are very imprecise controllers. In the first place their switching differential (the difference in temperature between switch-off and switch-on) is usually several kelvin. Secondly the only adjustment is setpoint.

Contact dial thermometers A first improvement on a thermostat is the use of a contact dial thermometer. The dial of this instrument carries a second pointer, the position of which can be set by the operator. When the indicating pointer reaches the setpoint pointer they make electrical contact with one another. The current that then flows between the pointers operates an electrical relay which controls the load. In this case the switching differential can be very small, typically a fraction of a kelvin.

Proportional temperature controllers Dedicated one-, two- or three-term temperature controllers are available in either pneumatic or electronic options. The use of such controllers is mainly confined to small plants where there is a cost advantage in avoiding the use of transmitters.

In the case of pneumatic controllers the input measurement will be liquid, vapour pressure or gas expansion. The Bourdon tube or bellows used to measure the pressure in the capillary system operates directly on the controller mechanism.

However in recent years there has been an enormous increase in the number of electronic temperature controllers. The input to these instruments is from either a thermocouple or a resistance thermometer. The functions available in these controllers vary from on/off control to full three-term proportional, integral and derivative operation. Some of the more sophisticated electronic controllers use an internal microprocessor to provide the control functions. Some units are available with the facility to control several temperature control loops. Of course the use of an internal microprocessor can make direct computer compatibility a simple matter.

1.7.2 Sensor location considerations

To obtain accurate temperature measurement careful consideration must be given to the siting of temperature sensing probes. Frequently in industrial applications temperature measuring equipment does not live up to the expectations of the plant design engineer. The measurement error is not infrequently ten or even twenty times the error tolerance quoted by the instrument manufacturer.

Large measurement errors in service may be due to the wrong choice of instrument but more frequently the error is due to incorrect location of the measurement points. Unfortunately the location of temperature sensors is dictated by the mechanical design of the plant rather than by measurement criteria.

1.7.2.1 Immersion probes

To minimize errors in the measurement of the temperature of process fluids, whether liquid or gas, it is preferable to insert the sensor so that it is directly immersed in the fluid. The probe may be directly dipped into liquid in an open vessel, inserted through the wall of the vessel, or inserted into a pipe.

Measurement of liquid in vessels Temperature measurement of liquid in a plant vessel may illustrate the dilemma of the control engineer when faced with mechanical problems. Consider Figure 1.64 which represents a vessel filled with liquid and stirred by a double anchor agitator. The ideal place to measure the temperature would be somewhere near the centre of

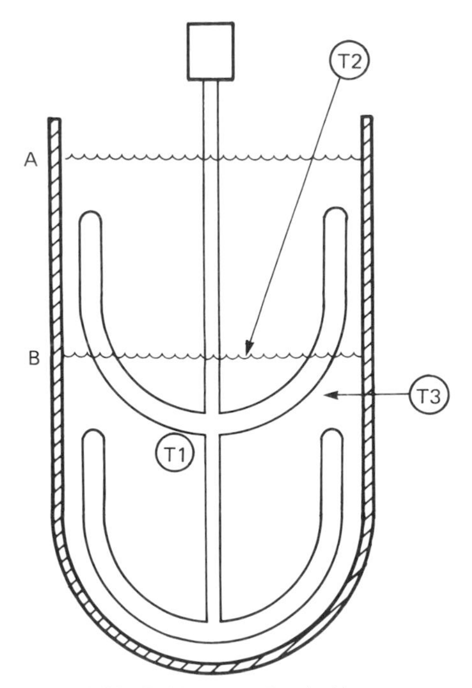

Figure 1.64 Problems associated with temperature measurement in a stirred vessel.

the mass at say T1. The best arrangement would seem to be a dip probe T2. But even though the design level of the liquid is at A in operation the liquid level may fall as low as B leaving probe T2 dry. The only remaining possibility is T3. This is not a very good approach to T1 and is subject to error due to conduction of heat from or to the vessel wall.

An approach that can be used if the temperature measurement is critical is to mount a complete temperature measuring package onto the shaft of the agitator. Wires are then brought up the shaft out of the vessel from whence the temperature signal can be taken off with slip rings, inductively coupled or radio telemetered to a suitable receiver. This is of course only possible where the temperature of the process is within the operating range of the electronics in the measurement package. The use of slip rings is not very satisfactory as they add unreliability, but in the absence of slip rings the package must also carry its own power supply in the form of batteries.

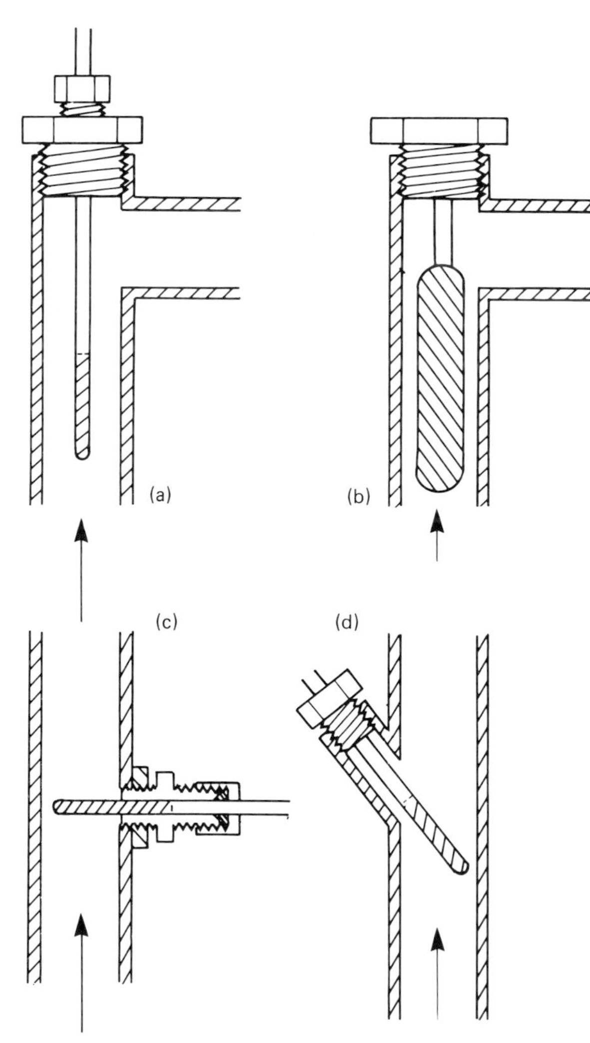

Figure 1.65 Problems associated with location of thermometer probe in pipe: (a) preferred arrangement, (b) probe obstructing pipe, (c) sensitive area of probe not fully immersed, (d) alternative preferred arrangement, sensitive portion of probe shaded.

Probes in pipes or ducts There is frequently a requirement to measure the temperature of a fluid flowing in a pipe. This is usually straightforward but there are still points to watch out for. Figure 1.65 shows three possible configurations for insertion into a pipe. The most satisfactory arrangement is to insert the thermometer probe into the pipe at a bend or elbow. Figure 1.65(a) shows this arrangement. Points to note are:

(a) To ensure that the probe is inserted far enough for the sensitive length to be wholly immersed and far enough into the fluid to minimize thermal conduction from the sealing coupling to the sensor.
(b) To insert the probe into the direction of flow as indicated. The reasons for this are to keep the sensor ahead of the turbulence at the bend which could cause an error due to local heating and to remove the effects of cavitation that could occur at the tip of a trailing probe. Figure 1.65(b) shows the problem that can arise in small pipes where the probe can cause serious obstruction to the flow.

Where it is not possible to put the thermometer at a bend in the pipe we can insert it radially provided the pipe is big enough. Great care should be taken to ensure complete immersion of the sensitive portion of the probe. Figure 1.65(c) illustrates this problem. A better solution is diagonal insertion as shown at (d). Again the probe should point into the direction of flow.

When measuring temperature in large pipes or ducts it must be remembered that the temperature profile across the pipe may not be constant. This is especially true for large flue stacks and air-conditioning ducts. The centre liquid or gas is usually hotter (or colder in refrigerated systems) than that at the duct wall. In horizontal ducts carrying slow-moving air or gas the gas at the top of the duct will be significantly hotter than that at the bottom of the duct. In these circumstances careful consideration must be given as to how a representative measurement can be obtained; it may well be necessary to make several measurements across the duct and average the readings.

1.7.2.2 Radiation errors

Gas temperature measurements present extra problems compared with temperature measurements in liquids. The difficulties arise from two sources. First the relatively low thermal conductivity and specific

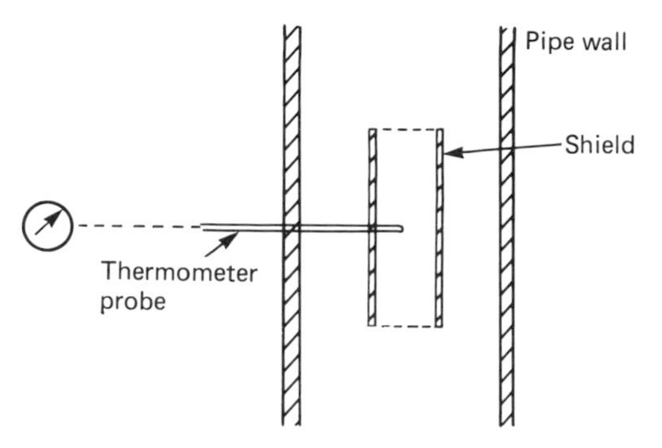

Figure 1.66 Radiation shield for gas temperature measurement.

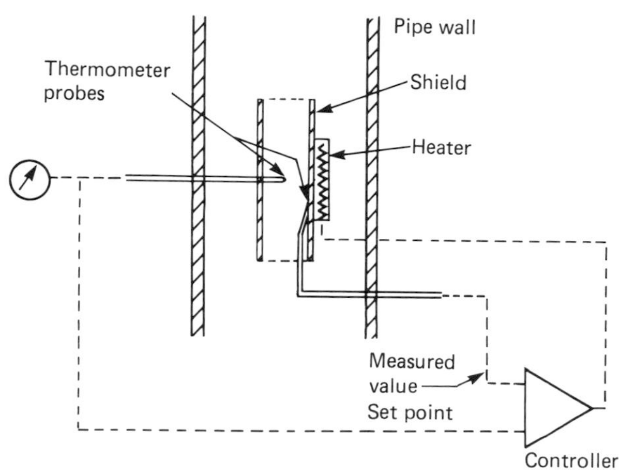

Figure 1.67 Active radiation shield.

heat of gases result in a poor heat transfer from the gas to the sensing element. This results in a slow response to temperature changes. Secondly since most gases are transparent at least to a substantial part of the thermal radiation spectrum significant measurement errors are lilely to occur, as mentioned in Section 1.6. Consider a thermometer bulb inserted into a pipe containing a gas stream. The walls of the pipe or duct are likely to be at a different temperature to the gas, probably but not necessarily cooler. This means that while the thermometer is being warmed by receiving heat by contact with the gas it is also losing heat by radiation to the pipe wall and if the wall is cooler than the gas the thermometer will lose more heat than it receives and will therefore register a lower temperature than the true gas temperature. Likewise if the pipe wall is hotter than the gas then the thermometer reading will be too high. This error can be reduced by surrounding the sensitive part of the thermometer probe with a cylindrical shield with its axis parallel to the pipe axis. This shield will reach a temperature intermediate between that of the pipe wall and that of the gas, Figure 1.66. Where more precise measurements are required an active shield may be employed. In this case a second thermometer is attached to the shield which is also provided with a small heater. This heater's output is controlled via a controller so that the two thermometers, the one in the gas and the one on the shield, always indicate identical temperatures. In this state the thermometer will be receiving exactly the same amount of radiation from the shield as it radiates back to the shield. Figure 1.67 shows this arrangement.

1.7.2.3 Thermometer pockets, thermowells

The direct immersion of temperature sensing probes into process fluid, while being the optimum way to get an accurate measurement, has its disadvantages. First it has disadvantages from the maintenance point of view: normally the sensing probe cannot be removed while the plant is on stream. Secondly in the case of corrosive process streams special corrosion-resistant materials may need to be used. Standard temperature gauges are normally only available in a limited range of materials, typically brass, steel, stainless steel or ceramic, so a sheath or thermometer pocket or thermowell can be used to protect the temperature sensing probe.

The use of a thermometer pocket does degrade the measurement accuracy of the instrumentation.

Figure 1.68 shows a thermometer pocket mounted in the wall of a steam-jacketed process vessel. The thermometer probe receives heat from the wall of the pocket by conduction where it touches it and by radiation at other places. The inner wall of the pocket receives heat from the process fluid and by conduction in this case from the steam jacket of the vessel. In the case of a short pocket the heat conducted along the

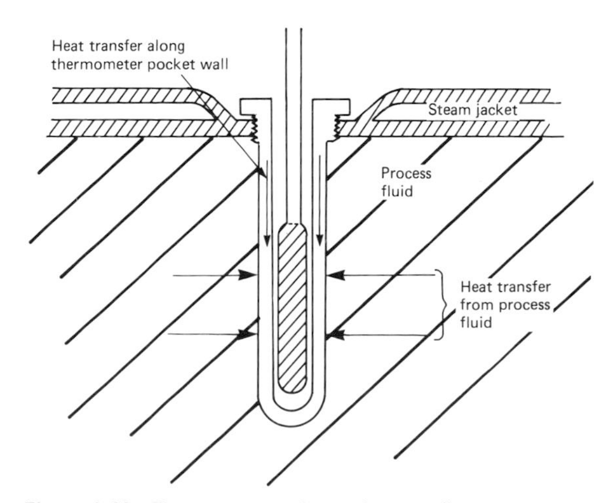

Figure 1.68 Thermometer pocket or thermowell.

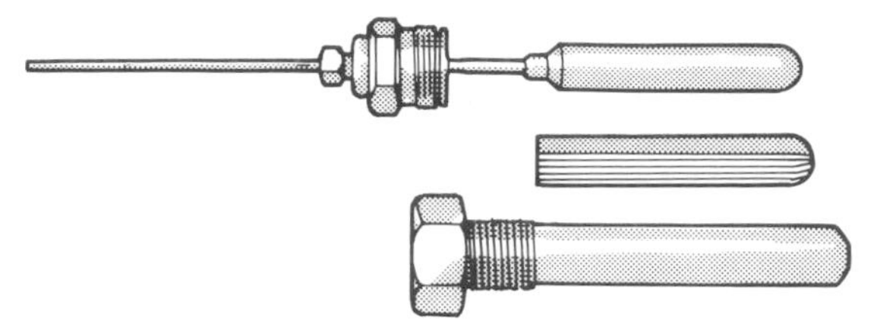

Figure 1.69 Taylor thermospeed separable well system (courtesy Taylor Instruments).

pocket can cause a significant measurement error, causing too high a reading. In the situation where the outer jacket of the vessel is used for cooling the vessel, for example, a cooling water jacket, the heat flow will be away from the sensing probe and consequently the error will be a low measurement. This conduction error is only significant where the thermometer pocket is short or where the pocket is inserted into a gas stream. To minimize the error the length of the pocket should be at least three times the length of the sensitive area of the probe.

The use of a thermowell or pocket will also slow down the speed of response of an instrument to temperature changes. A directly immersed thermometer probe will typically reach thermal equilibrium within 30 to 90 seconds. However the same probe in a thermometer pocket may take several minutes to reach equilibrium. This delay to the instrument response can be improved in those cases where the pocket is mounted vertically pointed downwards or in any position where the closed end is generally lower than the mouth, by filling it with a heat-transfer liquid. This liquid is usually a silicone oil.

An alternative method for improving the rate of heat transfer between the pocket and the bulb is illustrated in Figure 1.69. A very thin corrugated aluminium or bronze sleeve is inserted between the bulb and pocket on one side. This forces the bulb over to the other side, ensuring metal-to-metal contact on this side, while on the other side, the sleeve itself, being made of aluminium which has a high thermal conductivity, provides a reasonable path for the heat. In addition the bulb should be placed well down in the pocket to reduce the possibility of errors due to heat conducted by the pocket to the outside with consequent reduction of the temperature at the bulb.

The errors associated with thermal conduction along the thermometer pocket are of course more critical in the case of gas temperature measurement, as the thermal transfer from gas to thermometer is not nearly as good as it is from liquid.

1.7.2.4 Effect of process fluid flow rate

Two sources of error in temperature measurement are clearly identified.

Frictional heating Where the process fluid flows past a probe at high velocity there is, especially in the case of gases, a frictional heating effect. The magnitude of the effect is not easily evaluated but it is advisable if possible to site the probe at a location where the fluid velocity is low.

Conductive cooling Resistance thermometers and thermistors depend for their operation on an electric current flowing through them. This current causes a small heating effect in the sensor. When such a sensor is used for liquid temperature measurement the relatively high specific heat of most liquids ensures that this heat is removed and the sensor temperature is that of the liquid. However, in gas measurement the amount of heat removed is a function of the gas velocity and thus a variable source of error can arise dependent on flow rate. In a well designed instrument this error should be very small but it is a potential source of error to be borne in mind.

Cavitation Liquid flowing past a thermometer probe at high speed is liable to cause cavitation at the downstream side of the probe. Apart from any heating effect of the high flow rate the cavitation will generate noise and cause vibration of the probe. This vibration is likely in due course to cause deterioration or premature catastrophic failure of the probe.

1.7.2.5 Surface temperature measurement

Where the temperature of a surface is to be measured this can be done either with a temperature probe cemented or clamped to the surface or where a spot measurement is to be made a sensor can be pressed against the surface. In the former arrangement, which is likely to be a permanent installation, the surface in the region of the sensor itself can be protected from heat loss by lagging with thermally insulating material. Provided heat losses are minimized the measurement error can be kept small. Errors can be further reduced where the sensor is clamped to the surface by coating the surface and the sensor with heat-conducting grease. This grease is normally a silicone grease heavily loaded with finely ground alumina. A grease loaded with beryllium oxide has better heat transfer properties. However, since beryllium oxide is very toxic this grease must be handled with the greatest of care.

Where spot measurements are to be made, using for instance a hand-held probe, it is difficult to get accurate readings. The normal practice is to use a probe mounted on a spring so that it can take up any reasonable angle to press flat against the surface to be measured. The mass of the probe tip is kept as small as possible, usually by using a thermocouple or thermistor, to keep the thermal mass of the probe to a minimum. Again accuracy can be improved somewhat by using thermally conducting grease. Figure 1.40 shows a typical hand-held probe.

1.7.3 Miscellaneous measurement techniques

Temperature measurement may be the primary measurement required for the control of a plant. There are, however, many cases where temperature measurement is a tool to get an indication of the conditions in a plant. For instance in distillation columns it is more convenient and quicker to judge the compositions of the offtake by temperature measurement than to install on-line analysers and as a further bonus the cost of temperature measurement is very significantly less than the cost of analysers.

The reverse situation also exists where it is not possible to gain access for a thermometer to the region where the temperature requires to be known. In this instance some indirect measurement technique must be resorted to. One case of indirect measurement that has already been dealt with at some length is the case of radiation thermometers.

1.7.3.1 Pyrometric cones

At certain definite conditions of purity and pressure, substances change their state at fixed temperatures. This fact forms a useful basis for fixing temperatures, and is the basis of the scales of temperature.

For example, the melting points of metals give a useful method of determining the electromotive force of a thermocouple at certain fixed points on the International Practical Temperature Scale as has been described.

In a similar way, the melting points of mixtures of certain minerals are used extensively in the ceramic industry to determine the temperature of kilns. These minerals, being similar in nature to the ceramic ware, behave in a manner which indicates what the behaviour of the pottery under similar conditions is likely to be. The mixtures, which consist of silicate minerals such as kaolin or china clay (aluminium silicate), talc (magnesium silicate), felspar (sodium aluminium silicate), quartz (silica), together with other minerals such as calcium carbonate, are made up in the form of cones known as Seger cones. By varying the composition of the cones, a range of temperature between 600 °C and 2000 °C may be covered in convenient steps.

A series of cones is placed in the kiln. Those of lower melting point will melt, but eventually a cone is found which will just bend over. This cone indicates the temperature of the kiln. This can be confirmed by the fact that the cone of next higher melting point does not melt.

Since the material of the cone is not a very good conductor of heat, a definite time is required for the cone to become fluid, so that the actual temperature at which the cone will bend will depend to a certain extent upon the rate of heating. In order to obtain the maximum accuracy, which is of the order of ± 10 °C, the cones must, therefore, be heated at a controlled rate.

1.7.3.2 Temperature-sensitive pigments

In many equipment applications it is necessary to ensure that certain components do not exceed a specified temperature range. A typical case is the electronics industry where it is essential that semiconductor components remain within their rather limited operating range, typically -5 °C to 85 °C or for equipment to military specification -40 °C to 125 °C. These components are too small to fix all but the finest thermocouples to them. To deal with this situation temperature-sensitive paints can be used. These paints contain pigments which change colour at known temperatures with an accuracy of ± 1 °C. The pigments are available either having a reversible or a non-reversible colour change, the latter being the more usually used. In the case above, a semiconductor component in an electronic machine can have two spots of paint put on its case having colour changes at say 0 °C and 110 °C. On subsequent inspection, perhaps after equipment failure, it can be seen at once whether that component has been outwith its temperature tolerance.

As an alternative to paint these pigments are available on small self-adhesive labels. In either case they are available for temperatures within the range of 0 °C to about 350 °C in steps of about 5 degrees.

1.7.3.3 Liquid crystals

A number of liquids, mainly organic, when not flowing tend to form an ordered structure with for instance all the molecules lying parallel to one another. This structure is maintained against the thermal agitation by weak intermolecular bonding such as hydrogen bonding. These bonds hold the structure until the weak bonds between the molecules get broken as will occur when the liquid begins to flow. The structure can also be changed by electric fields, magnetic fields or temperature. Different compounds respond to different stimuli. Most poeple will be familiar with the liquid crystal displays on digital watches and pocket calculators. These displays use compounds sensitive to electric fields.

However, in this section we are interested in those liquid crystalline compounds that respond primarily to temperature. The compounds involved are a group of compounds derived from or with molecular structures similar to cholesterol. They are therefore

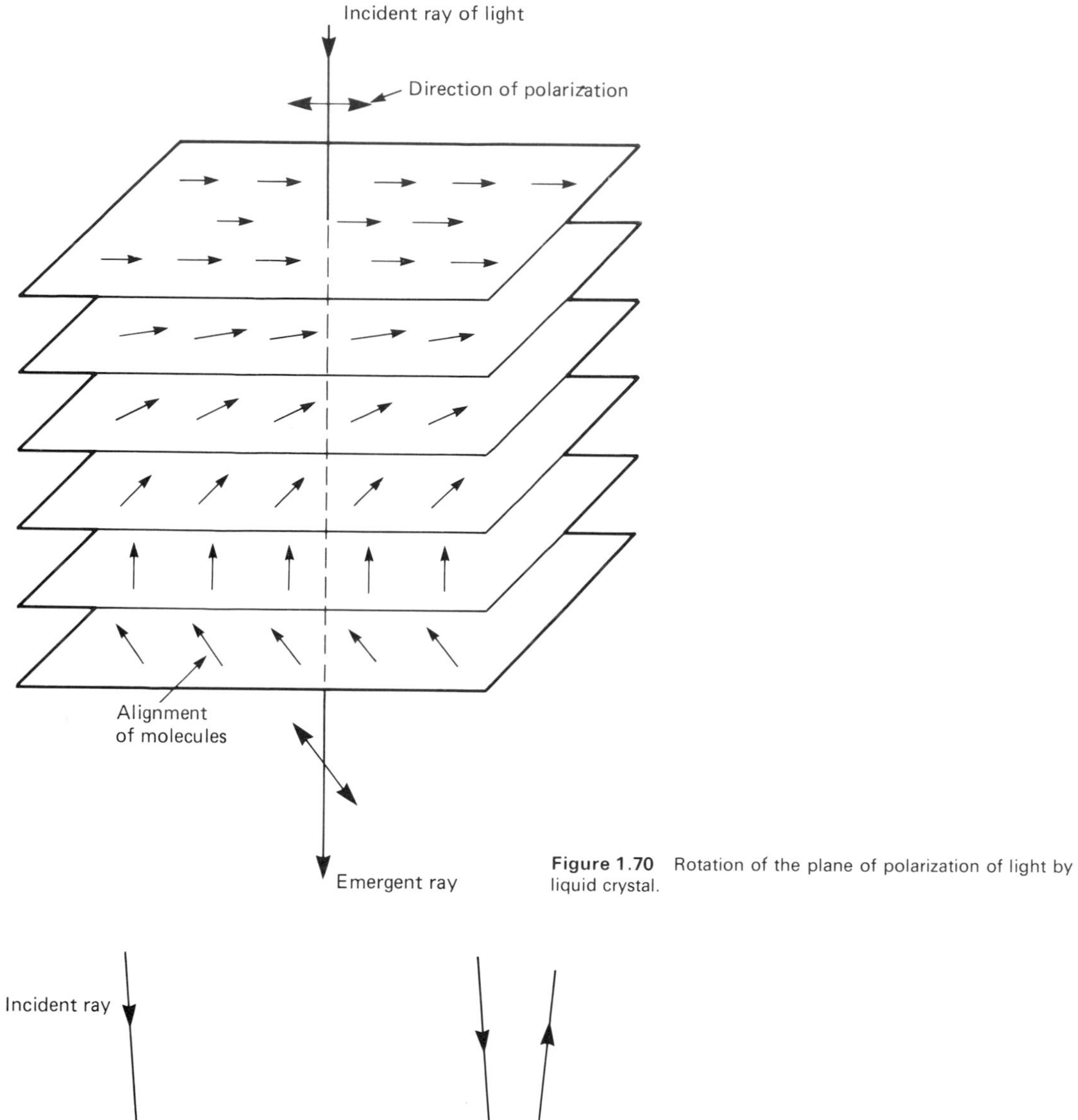

Figure 1.70 Rotation of the plane of polarization of light by liquid crystal.

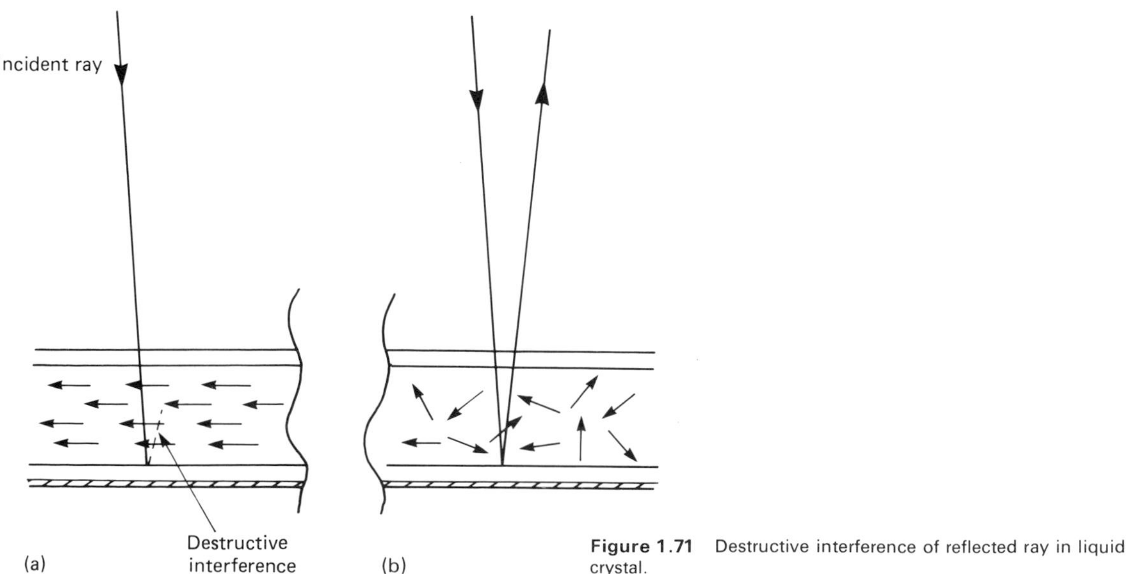

Figure 1.71 Destructive interference of reflected ray in liquid crystal.

called cholesteric compounds. Cholesteric liquids are extremely optically active as a consequence of their forming a helical structure. The molecules have a largely flat form and as a result lie in a laminar arrangement. However, the molecules have side groups which prevent them lying on top of one another in perfect register. The orientation of one layer of molecules lies twisted by a small angle compared to the layer below. This helical structure rotates the plane of polarization of light passing through the liquid in a direction perpendicular to the layers of molecules. Figure 1.70 illustrates this effect diagrammatically. The optical effect is very pronounced, the rotation of polarization being of the order of 1000° per millimetre of path length. The laminar structure can be enhanced by confining the cholesteric liquid between two parallel sheets of suitable plastic. The choice of polymer for this plastic is based on two prime requirements, first it is required to be transparent to light and secondly it should be slightly chemically active so that the liquid crystal molecules adjacent to the surface of the polymer are chemically bonded to it with their axes having the required orientation.

When used for temperature measurement the liquid crystal is confined between two sheets of transparent plastic a few tens of micrometres apart. The outer surface of one plastic layer is coated with a reflective layer, see Figure 1.71. In (a) a light ray enters the sandwich and travels to the bottom face where it is reflected back. Since the liquid crystal is in its ordered form it is optically active. The reflected ray interferes destructively with the incident ray and the sandwich looks opaque. In (b), however, the liquid crystal is above the temperature at which the ordered structure breaks up. The material is no longer optically active and the light ray is reflected back in the normal way – the material looks transparent.

The temperature at which the ordered structure breaks up is a function of the exact molecular structure. Using polarized light a noticeable change in reflected light occurs for a temperature change of 0.001 K. In white light the effect occurs within a temperature range of 0.1 K. Both the appearance of the effect and the exact temperature at which it occurs can be affected by addition of dyes or other materials.

1.7.3.4 Thermal imaging

In Section 1.6 the measurement of temperature by infrared and visual radiation was discussed in some detail. This technique can be extended to measure surface temperature profiles of objects. This is known as thermal imaging. The object to be examined is scanned as for television but at a slower rate and in the infrared region instead of the optical part of the spectrum. The signal so obtained is displayed on a visual display unit. This then builds up an image of the object as 'seen' by the infrared radiation from its surface. As well as producing a 'picture' of the object, the temperature of the surface is indicated by the colour of the image, producing a temperature map of the surface. Surface temperatures can be so imaged to cover a wide range from sub-ambient to very high temperatures. The technique has a very high resolution of temperature of the order of a small fraction of a Kelvin. Applications are to be found in such diverse fields as medicine and geological survey from space.

The technique is dealt with in very much greater detail in Part 3, Chapter 2.

1.7.3.5 Turbine blade temperatures

In the development and design of gas turbines there is a requirement to measure the temperature and

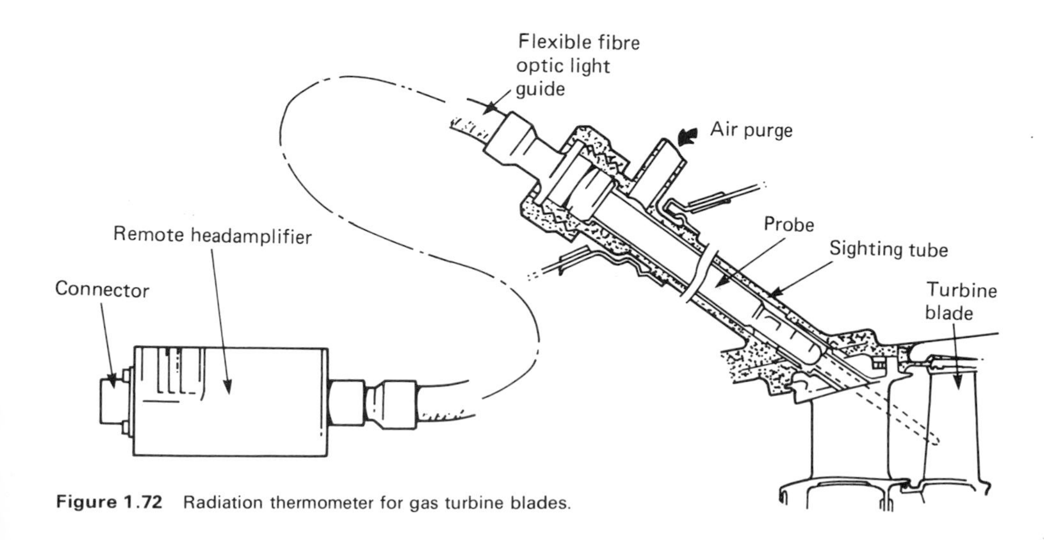

Figure 1.72 Radiation thermometer for gas turbine blades.

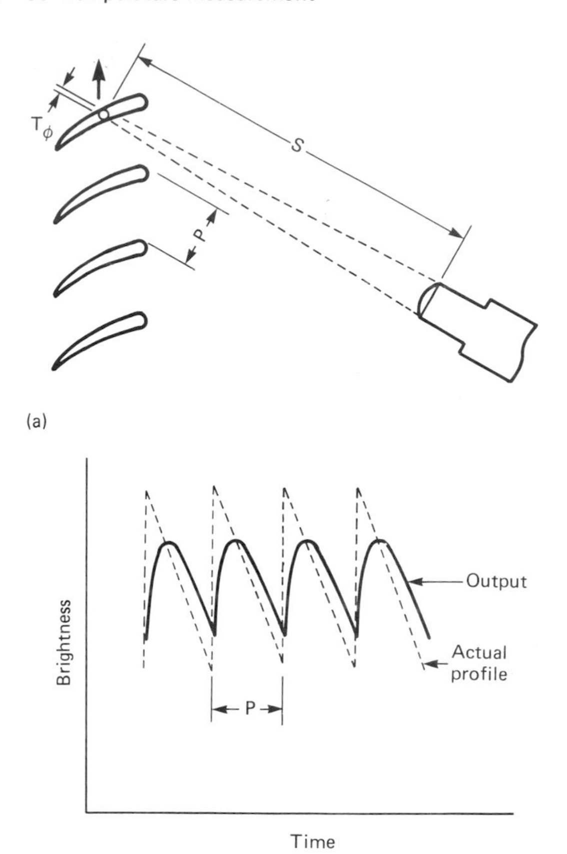

Figure 1.73 Measurement of the temperature profile of gas turbine blades: (a) geometry of focusing of thermometer, (b) temperature profile as 'seen' by radiation thermometer and electrical output.

temperature profile of the turbine rotor blades. This presents some problems as the turbine may be running at speeds of the order of 25 000 revolutions per minute. The rotor may consist of say 50 blades so that the time available to measure each blade temperature profile as it passes a point will be about a microsecond. A technique has been developed by Land Infrared Ltd. to carry out this measurement using fibre optic radiation thermometers. In this arrangement a small optical probe is inserted through the turbine wall and focused onto the rotor blades. The probe is connected by a fibre optic cable to a detector head amplifier unit nearby. Figure 1.72 shows a schematic diagram of focusing a measurement head. By designing the probe so that it focuses on a very small target area it is possible to 'read' a turbine blade temperature profile as it passes the target spot. Figure 1.73 shows the installation arrangement schematically at (A) and at (B) shows the theoretical and actual signal from the radiation thermometer. The degradation between the theoretical and actual signal is a function of the speed of response of the detector and the frequency bandwidth of the electronics. The theoretical signal consists of a sawtooth waveform. The peak represents the moment when the next blade enters the target area. The hottest part of the blade is its leading edge, the temperature falling towards the trailing edge. The signal falls until the next blade enters the field. The output from the thermometer can be displayed, after signal conditioning, on an oscilloscope or can be analysed by computer.

1.8 References

ASTM, *Manual on Use of Thermo-couples in Temperature Measurement*, ASTM Special Technical Publication 470B, (1981)

Billing, B. F. and Quinn, T. J. (eds), *Temperature Measurement 1975*, Adam Hilger, (1975)

Eckert, E. R. G. and Goldstein, R. J. (eds), *Measurements in Heat Transfer*, McGraw-Hill, (1976)

HMSO, *The Calibration of Thermometers*, (1971)

Kinzie, P. A. *Thermo-couple Temperature Measurement*, Wiley, (1973)

Quinn, T. J., *Temperature*, Academic Press, (1983)

2 Chemical analysis – introduction

W. G. CUMMINGS

2.1 Introduction to chemical analysis

Fifty years ago analytical chemistry depended almost entirely on measurements made gravimetrically and by titrimetry and students were taught that the essential steps in the process were sampling, elimination of interfering substances, the actual measurement of the species of concern and finally the interpretation of results. Each step required care, and, often, substances were analysed completely so that the components could be checked to total to within an acceptable reach of 100 per cent.

Classical analytical methods are still used from time to time, generally for calibrating instruments, but during the last thirty years the analytical chemistry scene has changed considerably. Spectroscopy and other physical methods of analysis are now widely used and a comprehensive range of chemical measuring instruments has been developed for specific techniques of analysis. This has meant that chemical analysis is now carried out as a cooperative effort by a team of experts, each having extensive knowledge of his own specialist technique, e.g. infrared absorption, emission spectrography, electrochemistry, gas chromatography, while also having considerable knowledge of the capabilities of the methods used by other members of the team.

Thus the analytical chemist has become more than just a chemist measuring the chemical composition of a substance; he is now a problem solver with two more steps in the analytical process – one at the beginning, 'definition of the problem', and another at the end, 'solution to the problem'. This means that the analytical chemist may measure things other than narrowly defined chemical composition – he may decide, for example, that pH measurements are better than analysis of the final product for controlling a process or that information on the valency states of compounds on the surface of a metal is more important than determining its composition.

Many elegant techniques have now become available for the analytical chemist's armoury with beautifully constructed electronic instruments, many complete with microprocessors or in-built computers. However, the analytical chemist should beware of becoming obsessed solely with the instruments that have revolutionized analytical chemistry and remember that the purpose of his work is to solve problems. He must have an open and critical mind so as to be able to evaluate the analytical instruments available – it is not unknown for instrument manufacturers in their enthusiasm for a new idea to emphasize every advantage of a technique without mentioning major disadvantages. It should also be remembered that, although modern analytical instrumentation can provide essential information quickly, misleading information can equally easily be obtained by inexperienced or careless operators and chemical measuring instruments must be checked and recalibrated at regular intervals.

Choosing the correct analytical technique or instrument can be difficult because several considerations have to be taken into account. First of all one must ensure that the required range of concentrations can be covered with an accuracy and precision that is acceptable for the required purpose. Then one must assess the frequency with which a determination must be made in order to set the time required for an analysis to be made or the speed of response of an instrument. This is particularly important if control of an on-going process depends on results of an analysis but is of less importance when the quality of finished products is being determined where ease of handling large numbers of samples may be paramount. Many requirements are conflicting and decisions have to be made on speed versus accuracy, cost versus speed, cost versus accuracy, and correct decisions can only be made with a wide knowledge of analytical chemistry and of the advantages and limitations of the many available analytical techniques. An important consideration is the application of the analytical instrument. This can be in a laboratory, in a rudimentary laboratory or room in a chemical plant area or working automatically on-stream. It is obvious that automatic on-stream instrumentation will be much more complex and expensive than simple laboratory instruments because the former must withstand the hostile environment of chemical plant and be capable of coping with temperature changes and plant variables without loss of accuracy. Such instruments have to be constructed to work for long continuous periods without exhibiting untoward drift or being adversely affected by the materials in the plant stream being monitored. Laboratory instruments on the other hand can be much more simple. Here the essential is a robust, easy-to-use instrument for a unique determination. Temperature compensation can be made by manual adjustment of controls at the time of making a determination and the instrument span can be set by use of standards each time the instrument is used. Thus there is no problem with drift. Laboratory instruments in general-purpose laboratories, however, can be as complex and costly as on-stream instruments but with different requirements. Here flexibility to carry out several determinations on a wide variety of samples is of prime importance but again temperature compensation and span adjustment can be carried out manually each time a determination is made. More expensive instruments use microprocessors to do such things automatically and these are becoming common in modern laboratories. Finally, although the cost of an analytical instrument depends on its complexity and degree of automation, there are other costs which should not be forgotten. Instrument maintenance charges can be appreciable and there is also the cost of running an instrument. The latter can range from almost nothing in the case of visible and ultraviolet spectrometers to several thousand pounds a year for argon supplies to inductively coupled plasma spectrometers. Many automatic analytical instruments require the preparation of reagent solutions and this, too, can involve an appreciable man-power requirement – also something which should be costed.

More detailed analysis of the factors affecting the costing of analytical chemistry techniques and instrumentation is beyond the scope of this chapter, but other chapters in this reference book give details and comparisons of analytical instrumentation for many applications. It is arranged with large chapters on electrochemical and spectrochemical techniques and further chapters on the applications of gas analysis and humidity measuring instruments. For completeness, the remainder of this chapter contains brief descriptions of chromatography, thermal analysis and polarography.

2.2 Chromatography

2.2.1 General chromatography

Around 1900 M.S. Tswett used the adsorbing power of solids to separate plant pigments and coined the term chromatography for the method. It was then not used for twenty years; then the method was rediscovered and used for the separation of carotenes, highly unsaturated hydrocarbons to which various animal and plant substances (e.g. butter and carrots) owe their colour.

Chromatography is thus a separating procedure with the actual measurement of the separated substance made by another method, such as ultraviolet absorption or thermal conductivity, but as it is such a powerful analytical tool it will be dealt with here as an analytical method.

All chromatographic techniques depend on the differing distributions of individual compounds in a mixture between two immiscible phases as one phase (the mobile phase) passes through or over the other (the stationary phase). In practice the mixture of compounds is added to one end of a discrete amount of stationary phase (a tubeful) and the mobile phase is then introduced at the same end and allowed to pass along the stationary phase. The mixture of compounds is eluted, the compound appearing first at the other end of the stationary phase being that which has the smallest distribution into the stationary phase. As the separated compounds appear at the end of the stationary phase they are detected either by means of unique detectors or by general-purpose detectors which sense the compound only as an impurity in the mobile phase.

The apparatus used varies according to the nature of the two phases. In gas chromatography the mobile phase is a gas with the stationary phase either a solid or a liquid. This is described in detail in Chapter 5. Liquid chromatography covers all techniques using liquid as a mobile phase – these are column chromatography (liquid/liquid or liquid/solid), paper chromatography and thin layer chromatography.

2.2.2 Paper chromatography and thin layer chromatography

In paper chromatography the separation is carried out on paper, formerly on ordinary filter papers but more recently on papers specially manufactured for the purpose. These are made free from metallic impurities and have reproducible thickness, porosity and arrangement of cellulose fibres.

The paper used (which must not have been dried) contains adsorbed water and so paper chromatography can be regarded as an absorption process. However, the characteristics of the paper can be changed by applying specific liquids to it. Silicone oils, paraffin oil, petroleum jelly and rubber latex can be used to give a paper with non-polar liquid phases. Specially treated papers are also available, such as those containing ion exchange resins. Papers for paper chromatography can also be made of glass fibres or nylon as well as cellulose.

In thin layer chromatography, instead of using paper, a thin layer of an adsorbing substance such as

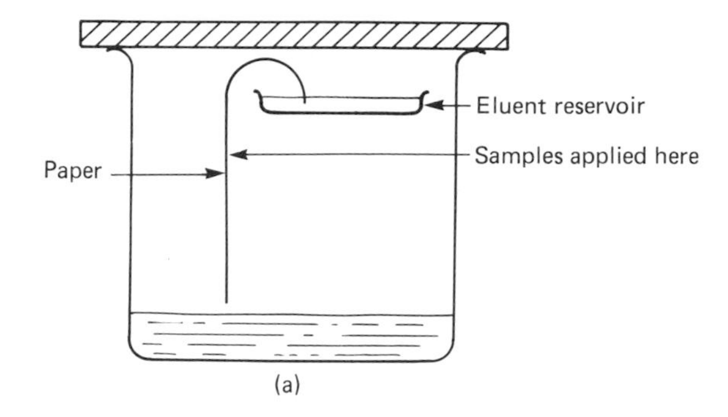

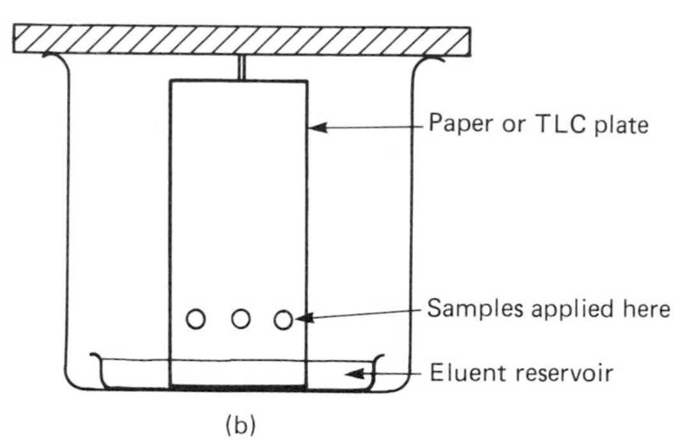

Figure 2.1 Apparatus for paper or thin-layer chromatography: (a) descending eluent used with paper chromatography, (b) ascending eluent used with paper chromatography or TLC.

silica gel is coated onto a glass or plastic plate. A very small volume of sample (~30 μl) is transferred onto one end of the plate which is then placed in a closed tank dipping into a solvent, the mobile phase. As the mobile phase moves along the plate the components of the sample are separated into a series of spots at different distances from the sample starting position. Figure 2.1 shows alternative arrangements. The location of the spots can be identified by their colour, or if colourless by spraying the plate with a reagent that produces a visible colour (or UV-detectable absorbance) with the compounds of interest. The position of the spots identifies the compound, the intensity of the colour, the concentration.

To establish a method for a particular mixture of compounds one has to select suitable adsorbents, solvents or mixtures of solvents, and a sensitive and selective reagent for detecting the separated compounds. There are many textbooks which discuss this in detail and give applications of the technique.

The apparatus used for measuring the separated substances in both paper and thin layer chromatography is quite straightforward laboratory-type equipment, for example, visible/ultraviolet spectrometers to determine the colour density or the UV absorbance of the spots.

Thin layer chromatography is generally found to be more sensitive than paper chromatography, development of the chromatogram is faster and it is possible to use a wider range of mobile phases and reagents to detect the position of the spots. Uses include the determination of phenols, carcinogenic polynuclear aromatic hydrocarbons, non-ionic detergents, oils, pesticides, amino acids and chlorophylls.

2.2.2.1 High performance liquid chromatography

Although liquid chromatography in columns was used by Tswett at the beginning of this century, an improved, quantitative version of the technique, high performance liquid chromatography (HPLC), has been fully developed and used only recently. By using precision instruments determination of trace organic and inorganic materials at concentrations of 10^{-6} to 10^{-12} g are possible. There are also several advantages of HPLC over other chromatographic techniques. HPLC is more rapid and gives better separations than classical liquid chromatography. It also gives better reproducibility, resolution and accuracy than thin layer chromatography, although the latter is generally the more sensitive technique. A large variety of separation methods is available with HPLC: liquid/liquid; liquid/solid; ion exchange and exclusion chromatography; but, again, the sensitivity obtainable is less than with gas chromatography.

Classical column liquid chromatography, in which the mobile liquid passed by gravity through the column of stationary phase, was used up to about 1946–50. In these methods a glass column was packed with a stationary phase such as silica gel and the sample added at the top of the column. Solvent, the mobile phase, was then added at the top of the column and this flowed through under the force of gravity until the sample components were either separated in the column or were sequentially eluted from it. In the latter case components were identified by refractive index or absorption spectroscopy. This type of elution procedure is slow (taking several hours) and the identification of the components of the sample is difficult and time-consuming.

Modern high performance liquid chromatography equipment has considerably better performance and is available from many chemical measuring instrument manufacturers. The main parts of a general-purpose HPLC apparatus are as shown in Figure 2.2.

The system consists of a reservoir and degassing system, a gradient device, a pump, a pulse dampener, a pre-column, a separating column and a detector.

Reservoir and degassing system The capacity of the reservoir is determined by the analysis being carried out, generally 1 litre is suitable. If oxygen is soluble in the solvent being used, it may need to be degassed. This can be done by distilling the solvent, heating it with stirring, or by applying a reduced pressure.

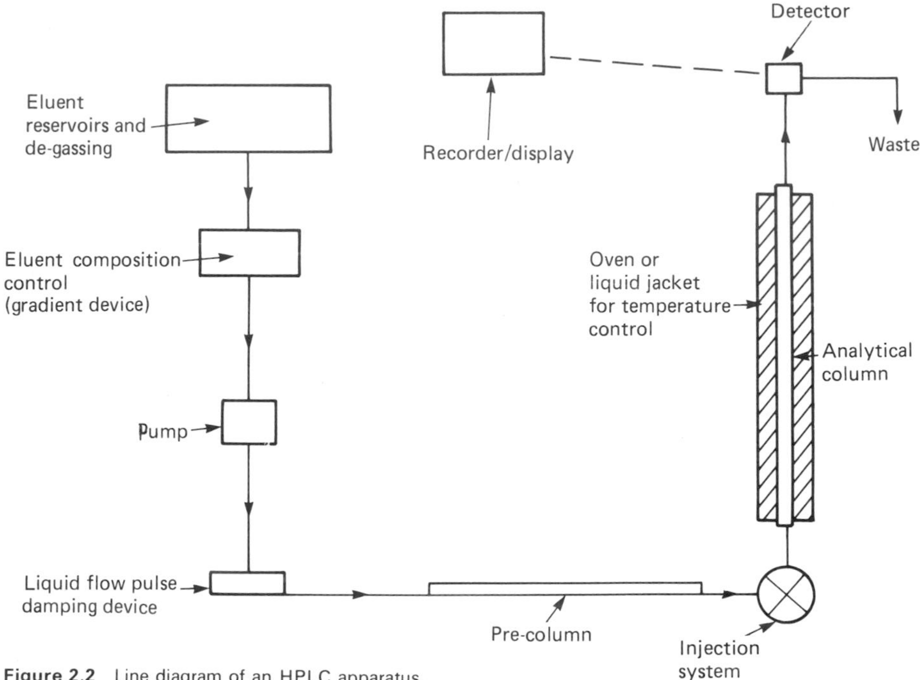

Figure 2.2 Line diagram of an HPLC apparatus.

Gradient devices If one wishes to change the composition of the mobile phase during the separation this can be done by allowing another solvent to flow by gravity into a stirred mixing vessel that contains the initial solvent and feeds the pump. This change of solvent mix is known as generating a solvent gradient.

A better way is to pump the solvents separately into a mixing tube; the desired gradient (composition) can be obtained by programming the pumps. This is elegant but expensive.

Pumps Suitable pumps deliver about 10 ml of solvent per minute at pressures up to 70 bar. These can be pressurized reservoirs, reciprocating pumps, motor-driven syringes or pneumatically operated syringes. It is essential to arrange for pulseless liquid flow and pulse damping may be required. This can be done by using small-bore tubes of small volume or by using sophisticated constant pressure control equipment.

Pre-column The solvent (the mobile phase) must be presaturated with the stationary liquid phase in the pre-column so that the stationary phase is not stripped off the analytical column.

Sample introduction Samples can be injected onto the analytical column by injection by syringe through a septum or by means of a sample loop. Injection via a septum can be difficult because of the very high pressures in the column – an alternative is stop-flow injection, where the solvent flow is stopped, the sample injected and then solvent flow and pressure restored. However, this can cause problems from the packing in the column shifting its position.

Analytical columns Very smooth internal walls are necessary for efficient analytical columns and very thick-walled glass tubing or stainless steel are the preferred materials. Connections between injection ports, columns and detectors should be of very low volume and inside diameters of components should be of similar size. Tubing of 2–3 mm internal diameter is most often used and temperature control is sometimes necessary. This can be done by water-jacketing or by containing the columns within air-ovens.

Stationary phases A very wide variety of materials can be used as solid stationary phases for HPLC – a summary of materials to use has been compiled (R. E. Majors, *Am. Lab.*, **4**(5), 27, May 1972). Particle sizes must be small: e.g. 35–50 μm and 25–35 μm.

There are various methods of packing the stationary phase into the column. Materials such as ion exchange resins, which swell when they come into contact with a solvent, must be packed wet as a slurry. Other materials are packed dry with the column being vibrated to achieve close packing. Packed columns should be evaluated before use for efficiency (a theoretical plate height of about 0.1 mm), for permeability (pressure required), and speed. (Theoretical plate height is a measure of the separating efficiency of a column analogous to the number of separating plates in a liquid distillation column.)

Guidance on column packing materials can be obtained from manufacturers such as Pechiney-St Gobain, Waters Associates, E.M. Laboratories, Reeve Angel, Dupont and Separations Group.

Mobile phase The mobile phase must have the correct 'polarity' for the desired separation, low viscosity, high purity and stability, and compatibility with the detection system. It must also dissolve the sample and wet the stationary phase.

Detectors Commercially available detectors used in HPLC are fluorimetric, conductiometric, heat of absorption detector, Christiansen effect detector, moving wire detector, ultraviolet absorption detector and the refractive index detector. The last two are the most popular.

Ultraviolet detection requires a UV-absorbing sample and a non-UV-absorbing mobile phase. Temperature regulation is not usually required.

Differential refractometers are available for HPLC but refractive index measurements are temperature-sensitive and good temperature control is essential if high sensitivity is required. The main advantage of the refractive index detector is wide applicability.

HPLC has been applied successfully to analysis of petroleum and oil products, steroids, pesticides, analgesics, alkaloids, inorganic substances, nucleotides, flavours, pharmaceuticals and environmental pollutants.

2.3 Polarography and anodic stripping voltammetry

2.3.1 Polarography

Polarography is an electrochemical technique and a specific polarographic sensor for the on-stream determination of oxygen in gas streams is described in Chapter 5. However, there are also many laboratory polarographic instruments; these are described briefly here together with the related technique of anodic stripping voltammetry.

2.3.1.1 Direct current polarography

In polarography an electrical cell is formed with two electrodes immersed in the solution to be analysed. In the most simple version of the technique (d.c. polarography) the anode is a pool of mercury in the bottom of the cell (although it is often preferable to use a large capacity calomel electrode in its place) and the cathode consists of a reservoir of mercury connected to a fine glass capillary with its tip below the surface of the solution. This arrangement allows successive fine drops of mercury to fall through the solution to the anode at the rate of one drop of mercury every 3 or 4 seconds. Figure 2.3 shows the arrangement in practice. The voltage applied across the two electrodes is slowly increased at a constant rate and the current flowing is measured and recorded. Figure 2.4 shows the step type of record obtained, the oscillations in the magnitude of the current are due to the changing surface area of the mercury drop during the drop life.

The solutions to be analysed must contain an 'inert' electrolyte to reduce the electrical resistance of the solution and allow diffusion to be the major transport mechanism. These electrolytes can be acids, alkalis, or citrate, tartrate and acetate buffers, as appropriate. The cells are designed so that oxygen can be removed from the solution by means of a stream of nitrogen, for otherwise the step given by oxygen would interfere with other determinations. The voltage range can run from +0.2 to −2.2 volts with respect to the calomel electrode. At the positive end the mercury electrode itself oxidizes; at the negative end the 'inert' electrolyte is reduced.

The potential at which reduction occurs in a given base electrolyte, conventionally the half-wave potential, is characteristic of the reducible species under consideration and the polarogram (the record obtained during polarography) thus shows the reducible species present in the solution. The magnitude of the diffusion current is a linear function of the concentration of the ion in solution. Thus, in Figure 2.4, $E_{1/2}$ is characteristic of cadmium in a hydrochloric acid electrolyte and I_d is a measure of the amount of cadmium. The limit of detection for d.c. polarography is about 1 ppm.

2.3.1.2 Sampled d.c. polarography

One disadvantage of the simple polarographic technique is that the magnitude of diffusion current has to be measured on a chart showing current oscillations (Figure 2.4). As these are caused by the changing surface area of the mercury drop during its lifetime an improvement can be made by using sampled d.c. polarography in which the current is measured only during the last milliseconds of the drop life. To do this the mercury drop time must be mechanically controlled. The resulting polarogram has the same shape as the d.c. polarogram, but is a smooth curve without large oscillations.

2.3.1.3 Single-sweep cathode ray polarography

Another modification to d.c. polarography is single-

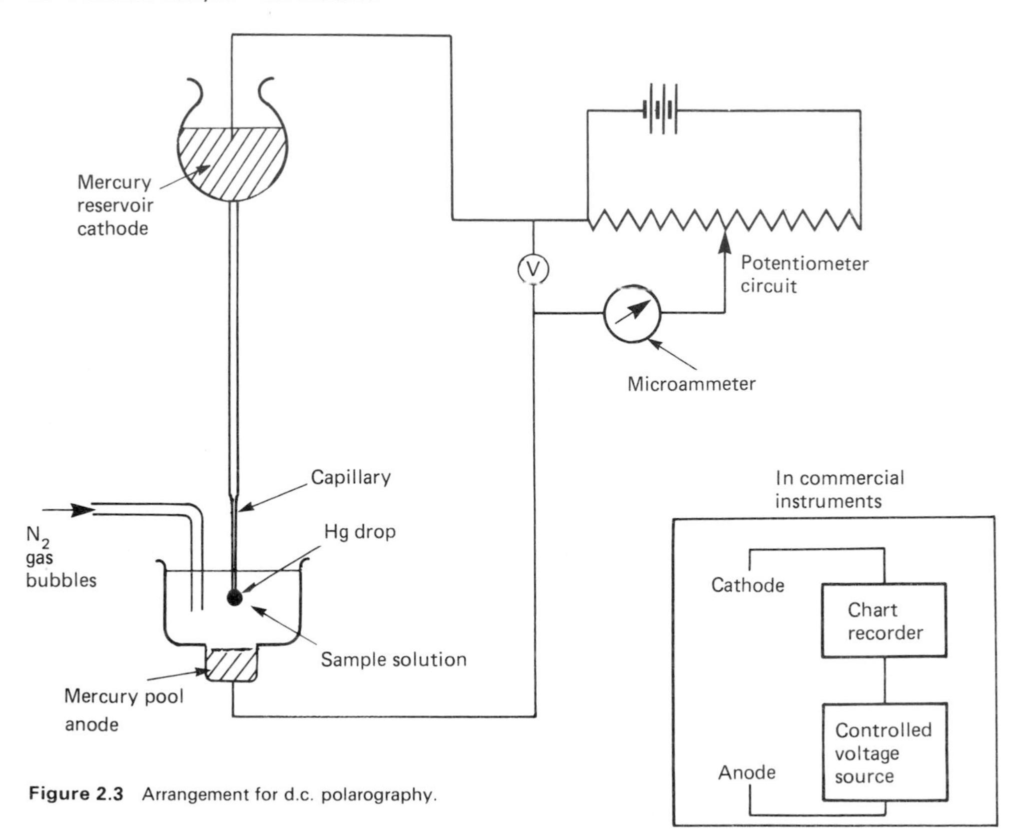

Figure 2.3 Arrangement for d.c. polarography.

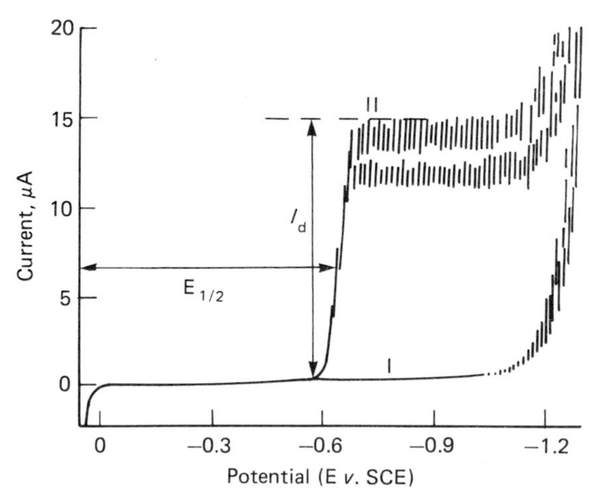

I : DC Polarogram 1*M* HCl
II : DC Polarogram of 5.0×10^{-4} *M* Cd (ii) in 1*M* HCl
I_d : Diffusion current
$E_{1/2}$: Half-wave potential

Figure 2.4 Polarograms of cadmium in hydrochloric acid (reprinted by courtesy of EG & G Princeton Applied Research and EG & G Instruments Ltd.).

sweep cathode ray polarography. Here an increasing d.c. potential is applied across the cell but only once in the life of every mercury drop. Drop times of about 7 seconds are used; the drop is allowed to grow undisturbed for 5 seconds at a preselected fixed potential, and a voltage sweep of 0.3 volts per second is applied to the drop during the last 2 seconds of its life. The sharp decrease in current when the drop falls is noted by the instrument and the sweep circuits are then automatically triggered back to zero. After the next 5 seconds drop growing time another voltage sweep is initiated, is terminated by the drop fall, and so on. The use of a long persistence cathode ray tube enables the rapid current changes to be followed easily with the trace remaining visible until the next sweep. Permanent records can be made by photography.

A characteristic of this technique is the peaked wave (Figure 2.5(a)) obtained compared with classical d.c. polarography. This peak is not a polarographic maximum, but is due to the very fast voltage sweep past the deposition potential causing the solution near the drop surface to be completely stripped of its reducible species. The current therefore falls and eventually flattens out at the diffusion current level. The peak height is proportional to concentration in the same way as the diffusion current level but

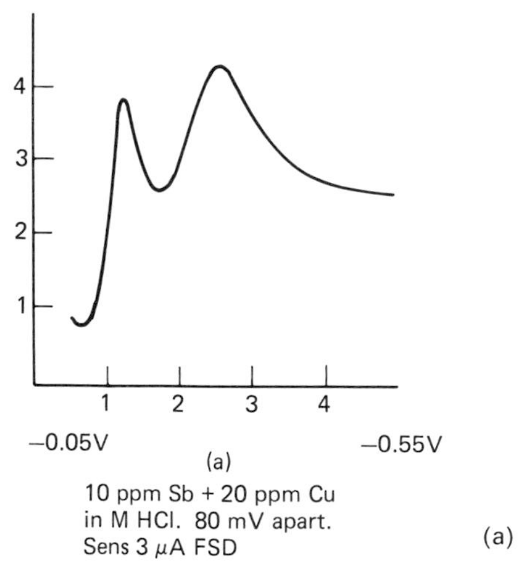

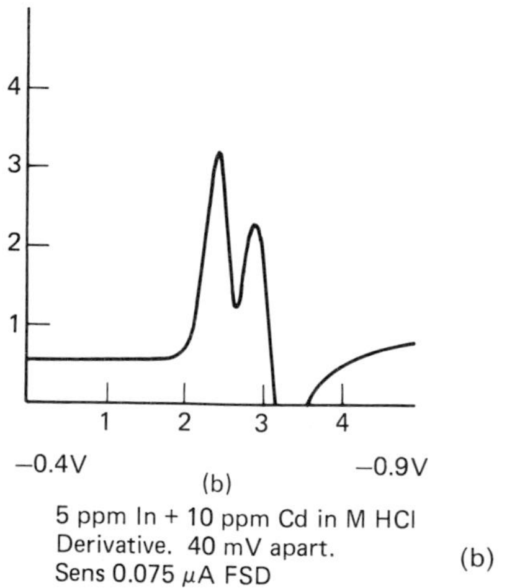

Figure 2.5 Single-sweep cathode ray polarograms (courtesy R. C. Rooney) (a) direct; (b) derivative.

sensitivity is increased. Resolution between species is enhanced by the peaked waveform and even this can be improved by the use of a derivative circuit, see Figure 2.5(b). Also, because of the absence of drop growth oscillations, more electronic amplification can be used. This results in the sensitivity of the method being at least ten times that of conventional d.c. polarography.

2.3.1.4 *Pulse polarography*

The main disadvantage of conventional d.c. polarography is that the residual current, due mainly to the capacitance effect continually charging and discharging at the mercury drop surface, is large compared with the magnitude of the diffusion current when attempting to determine cations at concentrations of 10^{-5} mol l^{-1} or below. Electronic methods have again been used to overcome this difficulty and the most important techniques are pulse and differential pulse polarography.

In normal pulse polarography the dropping mercury electrode is held at the initial potential to within about 60 milliseconds of the end of the drop life. The potential is then altered in a stepwise manner to a new value and held there for the remainder of the drop life. During the last 20 milliseconds of this the current is measured and plotted against the applied potential. Each new drop has the potential increased to enable the whole range of voltage to be scanned. The change in current that occurs when the voltage is stepped comes from the current passed to charge the double-layer capacitance of the electrode to the new potential. This decays very rapidly to zero. There is also a Faradaic current which is observed if the potential is stepped to a value at which an oxidation or reduction reaction occurs. This decays more slowly and is the current that is measured. This technique gives detection limits from 2 to 10 times better than d.c. polarography, Figure 2.6, but it is still not as sensitive as differential pulse polarography.

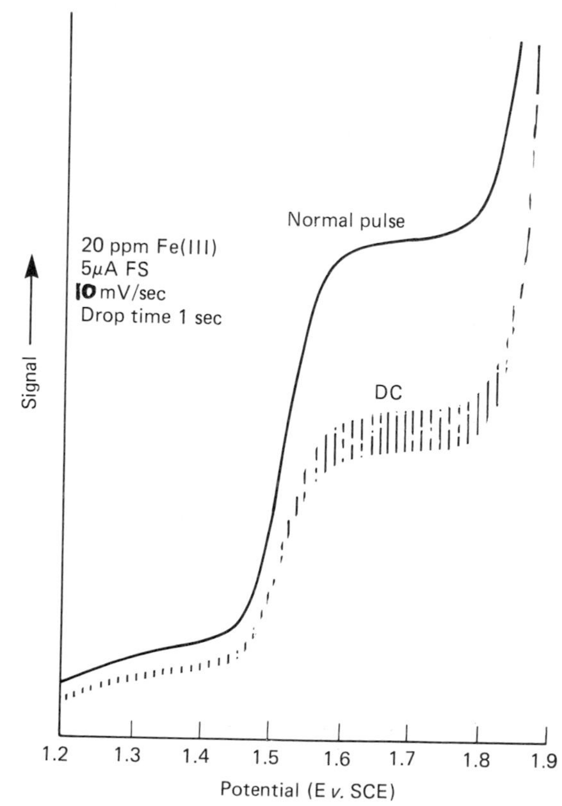

Figure 2.6 Normal pulse and d.c. polarograms for iron in ammonium tartrate buffer, pH 9 (reprinted by courtesy of EG & G Princeton Applied Research and EG & G Instruments Ltd.).

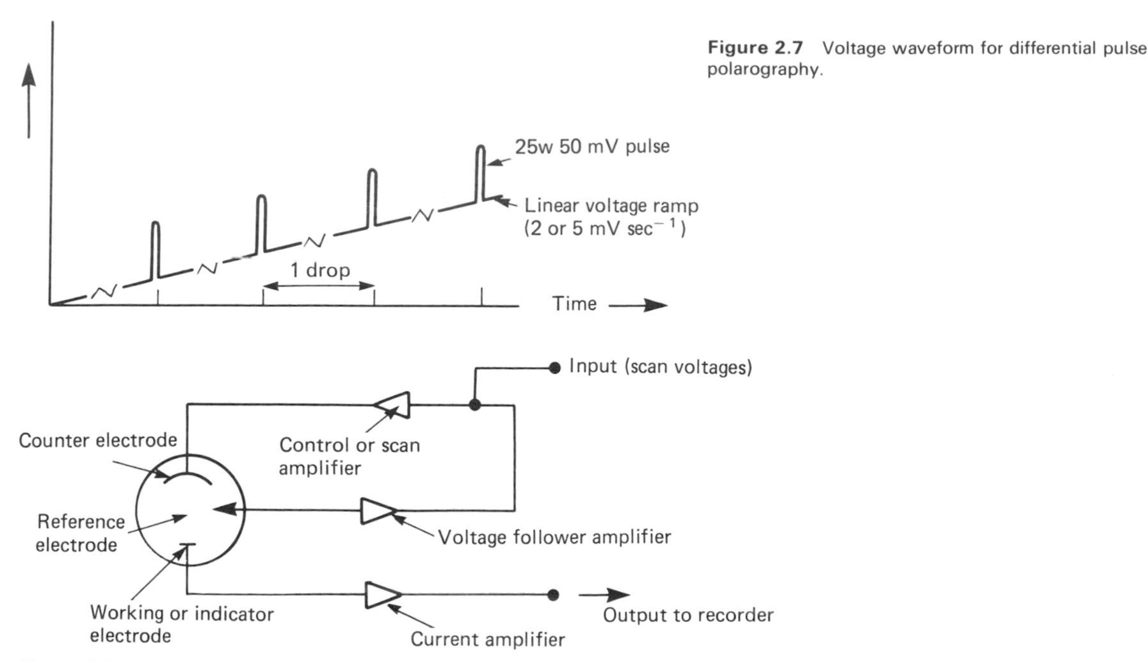

Figure 2.7 Voltage waveform for differential pulse polarography.

Figure 2.8 Practical arrangement for differential pulse polarography.

2.3.1.5 Differential pulse polarography

The most important of modern polarographic techniques is that of differential pulse polarography. Here a 25 or 50 mV amplitude pulse is superimposed at fixed time intervals on the normal linear increasing voltage of 2 or 5 $mV\,s^{-1}$ with the mercury drop being dislodged mechanically and so arranged that the pulse occurs once during the lifetime of each drop, Figure 2.7. The current is measured over a period of about 0.02 seconds just before the pulse is applied and during 0.02 seconds towards the end of the drop life. The difference between the two measurements is recorded as a function of the applied d.c. potential. In practice, a three-electrode potentiostatic arrangement is used, Figure 2.8. The polarograms obtained in this way are peak shaped (Figure 2.9), there is increased resolution between any two species undergoing reduction, and a great increase in sensitivity which is mainly a function of the reduction in measured capacitance current. There is a linear relationship between peak height and the concentration of the species being determined and limits of detection can be as low as 10^{-8} $mol\,l^{-1}$. The sensitivity of the technique can be varied by varying the pulse height; the peak height increases with increased pulse height but the resolution between peaks suffers (Figure 2.10). A comparison of the sensitivities of d.c., sampled d.c., normal pulse and differential pulse polarography is shown in Figure 2.11.

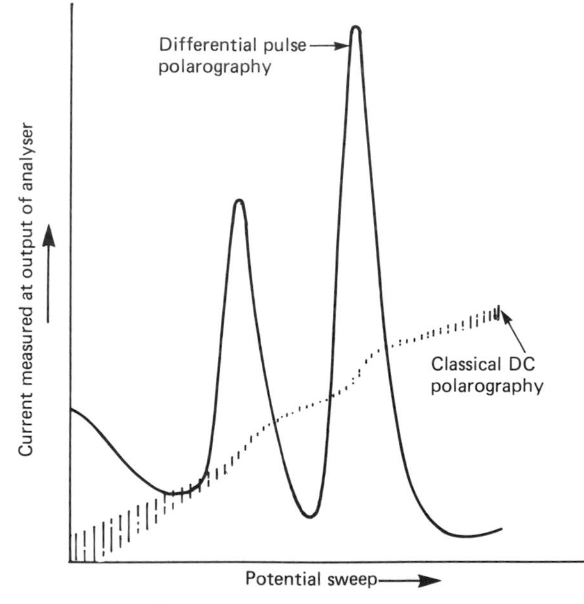

Figure 2.9 Differential pulse polarogram.

2.3.1.6 Applications of polarography

Polarographic methods can be used for analysing a wide range of materials. In metallurgy Cu, Sn, Pb, Fe, Ni, Zn, Co, Sb and Bi can be determined in light and zinc-based alloys, copper alloys and aluminium bronze; the control of effluents is often carried out

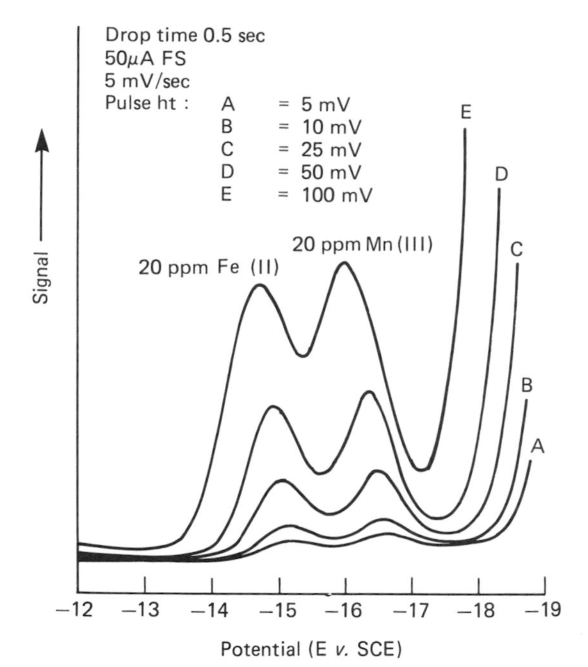

Figure 2.10 Effect of pulse height on peak height and resolution (reprinted by courtesy of EG & G Princeton Applied Research and EG & G Instruments Ltd.).

using polarographic methods. Cyanide concentrations down to ~0.1 ppm can be determined and sludges and sewage samples as well as fresh and sea waters can be analysed. Trace and toxic elements can be determined polarographically in foodstuffs and animal feed, in soils and in pharmaceutical products. In the latter, some compounds are themselves polarographically reducible or oxidizable, for example, ascorbic acid, riboflavin, drugs such as phenobarbitone and ephedrine and substances such as saccharine. Body fluids, plastics and explosives can also be analysed by polarographic techniques.

2.3.2 Anodic stripping voltammetry

Anodic stripping voltammetry is really a reversed polarographic method. Metals that are able to form amalgams with mercury, e.g. Pb, Cu, Cd and Zn can be cathodically plated onto a mercury drop using essentially the same instrumentation as for polarography and then the amalgamated metal is stripped off again by changing the potential on the mercury drop linearly with time in an anodic direction. By recording the current as a function of potential, peaks are observed corresponding to the specific species present in the test solution; the heights of the peaks are proportional to concentration.

In practice, it is not very convenient to use a mercury drop as cathode and several other types of electrode have been used including a rotating ring-disc electrode. The most often used, especially for water and environmental analysis is a wax-treated mercury-coated graphite rod. This, together with a silver/silver chloride reference electrode and a platinum counter electrode are immersed in the test solution (Figure 2.12) and the plating out and metal stripping carried out. Figure 2.13 illustrates the plating and stripping steps and Figure 2.14 shows a typical recording of the peak heights of Cd, In, Pb, Cu and Bi. As with polarography, various electronic modifications have been made to the basic technique and the stripping step has also been carried out with a.c. or pulsed voltages superimposed on the linear variation of d.c. voltage. Details of these systems can be found in reviews of the subject. Equipment for this technique is available at reasonable cost and units can be obtained for simultaneous plating of up to 12 samples with sequential recording of the stripping stages.

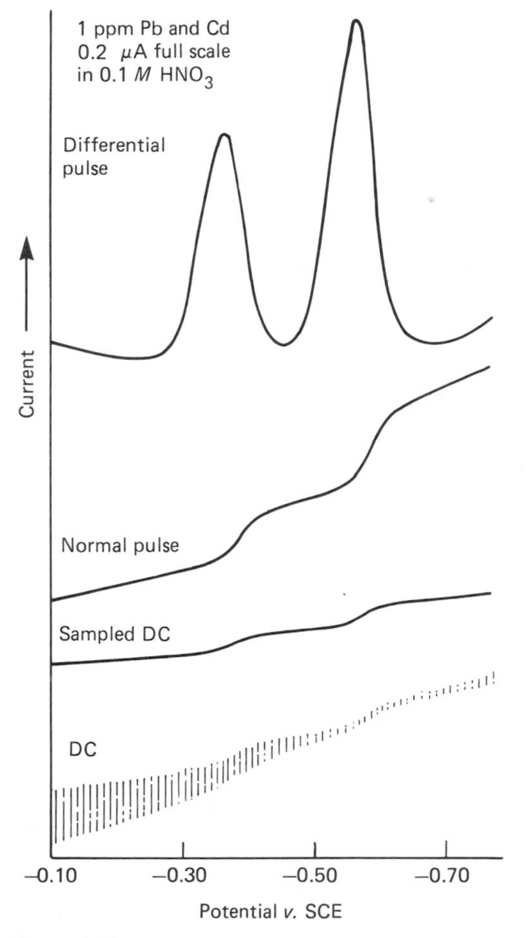

Figure 2.11 Comparison of polarographic modes (reprinted by courtesy of EG & G Princeton Applied Research and EG & G Instruments Ltd.).

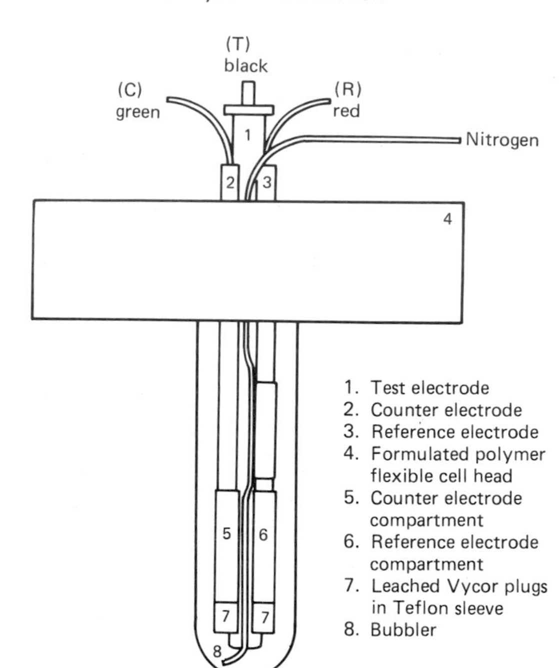

Figure 2.12 Cell arrangement for anodic stripping voltammetry (courtesy International Laboratory).

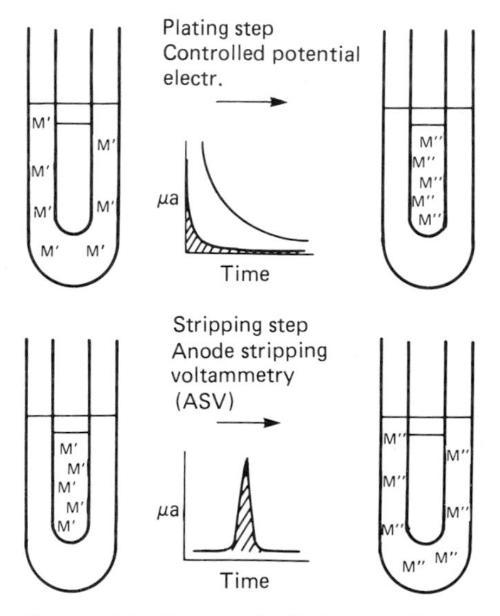

Figure 2.13 Plating and stripping steps (courtesy International Laboratory).

With anodic stripping voltammetry small samples (mg) can be used or very low concentrations of species determined because the plating step can be used as a concentration step. Plating times from 5 to 30 minutes are common depending on the required speed and accuracy of the analysis. Figure 2.14 was obtained using a 30-minute plating time. Good precision and accuracy can be obtained in concentration ranges as low as 0.1 to 10 μg per litre and this, combined with the fact that small samples can be used, means that the technique is most attractive for trace-metal characterization in the analysis of air, water, food, soil and biological samples.

2.4 Thermal analysis

No work on instrumental methods of determining chemical composition would be complete without mention of thermal analysis. This is the name applied to techniques where a sample is heated or cooled while some physical property of the sample is recorded as a function of temperature. The main purpose in making such measurements is most often not to evaluate the variation of the physical property itself but to use the thermal analysis record to study both the physical and chemical changes occurring in the sample on heating.

There are three main divisions of the technique depending on the type of parameter recorded on the thermal analysis curve. This can be (a) the absolute value of the measured property such as sample weight, (b) the difference between some property of the sample and that of a standard material, e.g. their temperature difference (these are differential measurements), and (c) the rate at which the property is changing with temperature or time, e.g. the weight loss, these are derivative measurements.

A convention has grown up for thermal analysis nomenclature and recommendations of the International Confederation for Thermal Analysis are that the term 'thermogravimetry' (TG) be used for measuring sample weight, 'derivative thermogravimetry' (DTG) for rate of weight loss, and 'differential thermal analysis' (DTA) for measuring the temperature difference between sample and standard. There are also many other terms relating to specific

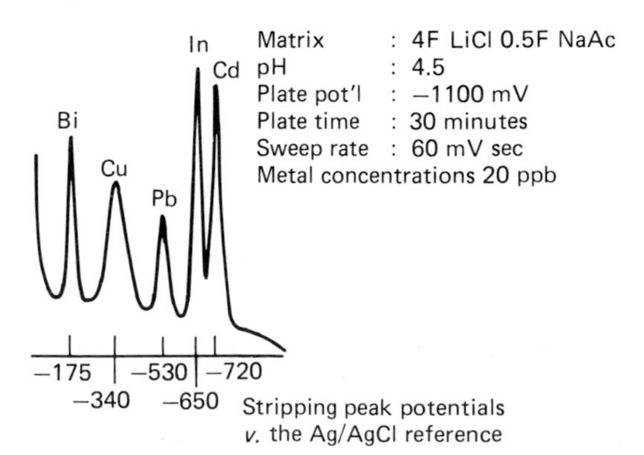

Figure 2.14 Stripping peak potentials (courtesy International Laboratory).

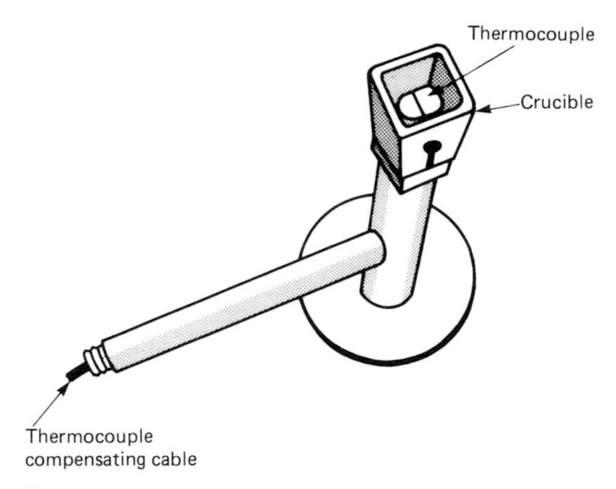

Figure 2.15 Cup for thermal analysis of cast iron.

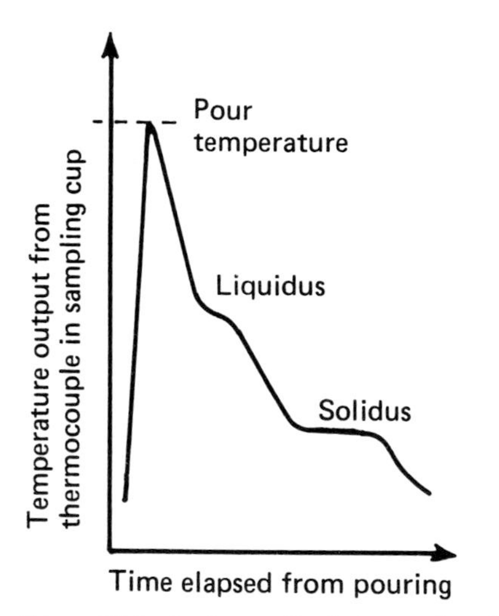

Figure 2.16 Cooling profile during cooling of liquid cast iron.

heat measurement, magnetic susceptibility, evolved gases, etc.

During the past 20 years a wide choice of commercially available equipment has become available and thermal analysis is now widely used as a tool in research and product control.

One particular application is to the composition of cast iron in terms of its carbon, silicon and phosphorus content, which can be calculated from the temperatures at which it freezes. As it is an alloy the freezing occurs at two temperatures, the liquidus and the solidus temperatures. At both temperatures the change of state of the metal releases latent heat. The temperatures at which the liquidus and solidus occur can be measured by the use of equipment made by Kent Industrial Measurements Ltd. To make the measurement a sample of liquid iron is poured into a special cup made from resin-bonded sand into which a small type K thermocouple is mounted, Figure 2.15. As the iron cools and passes through its two changes of state its temperature is monitored by the thermocouple. The graph showing the cooling against time, Figure 2.16, has two plateaux, one at the liquidus and one at the solidus. To complete the analysis the signal from the thermocouple is processed by a microcomputer which calculates and prints out the required analysis.

Figures 2.17–2.22 show other applications of thermogravimetry and derivative thermogravimetry to commercial samples and are largely self-explanatory.

In commercial thermal analysis instruments the sample is heated at a uniform rate while its temperature and one or more of its physical properties are measured and recorded. A typical arrangement is shown in Figure 2.22(a). The measuring unit has a holder to fix the position of the sample in the furnace, a means of controlling the atmosphere around the sample, a thermocouple for measuring the sample

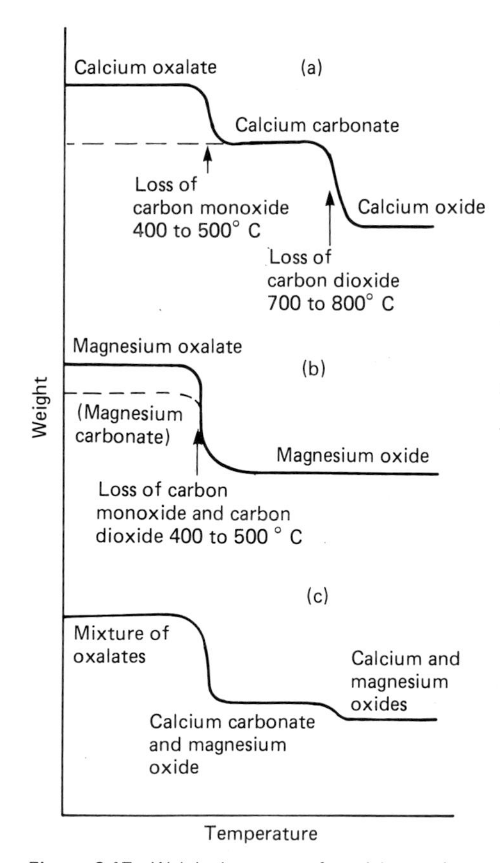

Figure 2.17 Weight–loss curves for calcium and magnesium oxalates and a precipitated mixture. (Reproduced by permission from *Thermal Analysis* by T. Daniels, published Kogan Page, Ltd.)

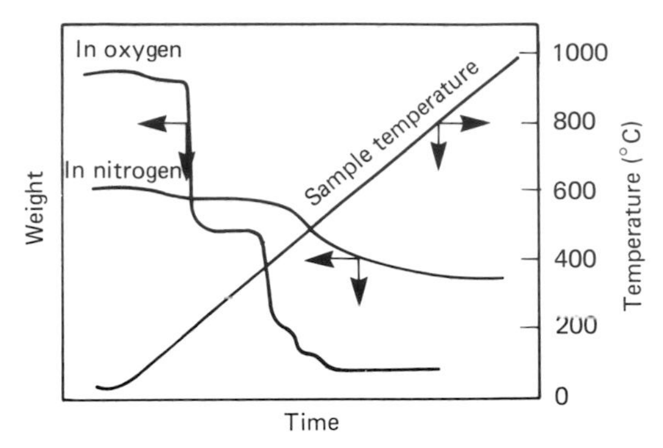

(a) TG CURVES FOR A COAL SAMPLE IN OXYGEN AND NITROGEN (FISHER TG SYSTEM)

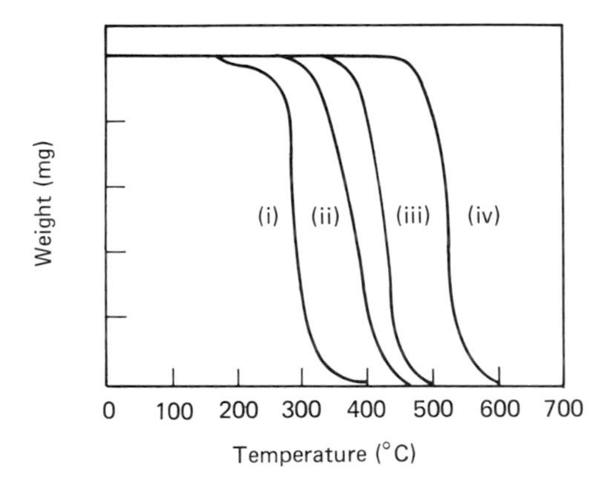

(b) TG CURVES FOR (i) POLYHEXAFLUOROPROPYLENE, (ii) POLYPROPYLENE, (iii) POLYETHYLENE, AND (iv) POLYTETRAFLUOROETHYLENE (Du Pont TG SYSTEM)

Figure 2.18 Thermal and thermo-oxidative stability of organic materials. (Reproduced by permission from *Thermal Analysis* by T. Daniels, published Kogan Page, Ltd.)

temperature and the sensor for the property to be measured, e.g. a balance for measuring weight. The design of the property sensor has to be such that it will function accurately over a wide temperature range and it is most important to ensure that the atmosphere around the sample remains fixed, be it an inert gas, a reactive gas or a vacuum.

The temperature control unit consists of a furnace and a programming unit, the function of which is to alter the sample temperature (not the furnace temperature) in a predetermined manner. The recording unit receives signals from the property sensor and the sample thermocouple, amplifies them and displays them as a thermal analysis curve. Figure 2.22(b) shows arrangements for differential instruments where the sample material and a reference material are placed in identical environments with sensors to measure the difference in one of their properties. The differential signal is amplified and recorded as in the basic system. In derivative instruments (Figure 2.22(c)) a derivative generator, such as an electro-optical device or an electronic unit is incorporated to compute the derivative of an input signal. Generally, both the derivative signal and the signal from the property being measured are recorded on the thermal analysis curve. It is, of course, possible to combine both modifications, thereby recording the derivative of a differential signal.

Most measuring units are designed specifically for a particular thermal analysis technique but furnaces, programmers, amplifiers and recorders are common to all types of instrument. Instrument manufacturers therefore generally construct a basic control unit containing programming and recording facilities to which can be connected modules designed for specific thermal analysis techniques.

Detailed description of the design of thermal

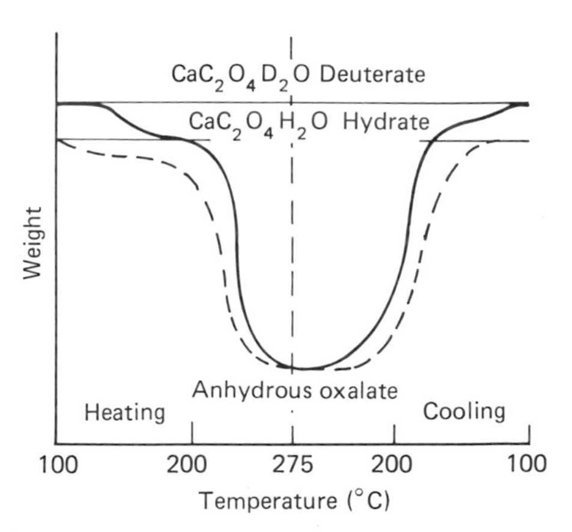

(a) TG PLOTS FOR CALCIUM OXALATE HYDRATE AND DEUTERATE ON HEATING AND COOLING IN A VAPOUR ATMOSPHERE

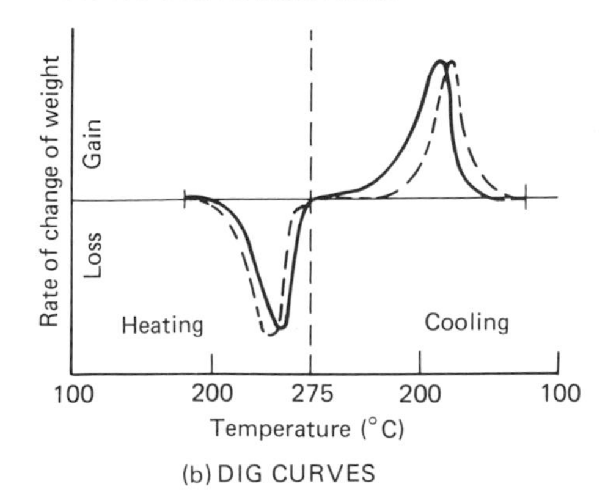

(b) DIG CURVES

Figure 2.19 The use of vapour atmospheres in TG. (Reproduced by permission from *Thermal Analysis* by T. Daniels, published Kogan Page, Ltd.)

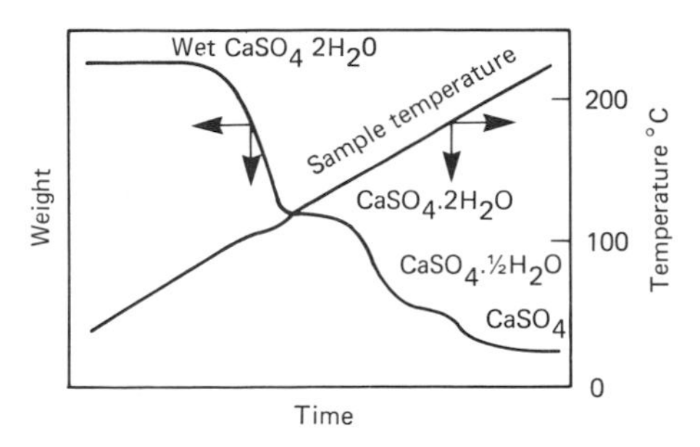

(a) EVALUATION OF THE WATER CONTENT OF GYPSUM

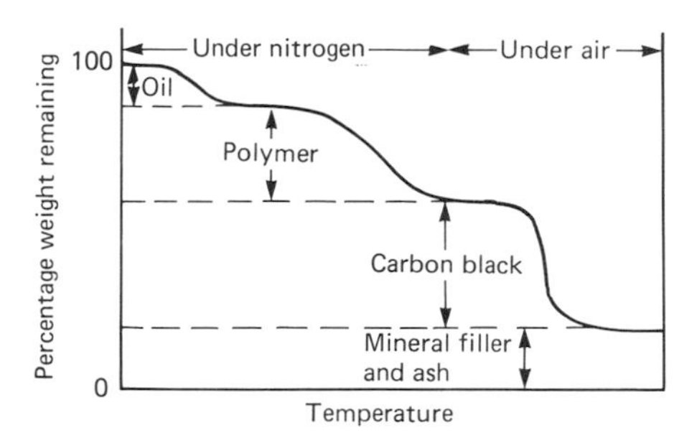

(b) ANALYSIS OF A GUM ELASTOMER (AFTER MAURER,11)

Figure 2.20 Analysis of commercial materials by TG. (Reproduced by permission from *Thermal Analysis* by T. Daniels, published Kogan Page, Ltd.)

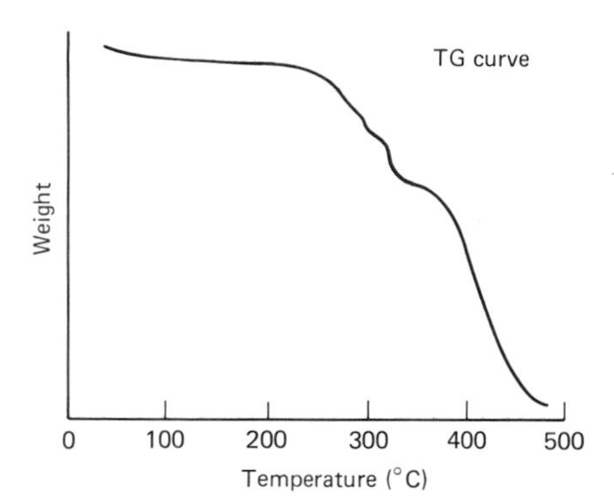

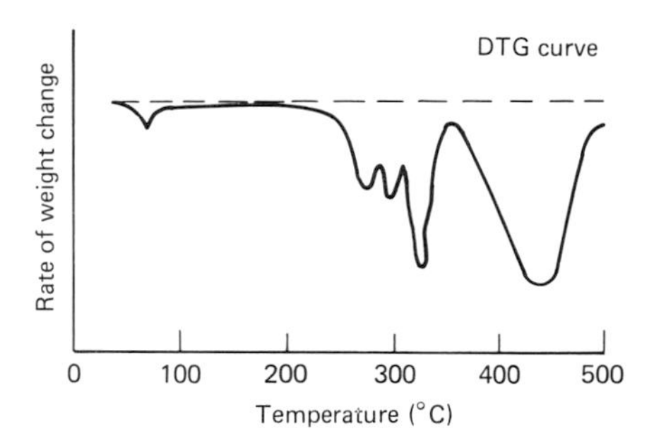

Figure 2.21 Dehydration and reduction of $\alpha Fe_2O_3 . H_2O$ on heating in hydrogen. (Reproduced by permission from *Thermal Analysis* by T. Daniels, published Kogan Page, Ltd.)

analysis instruments, their applications and the precautions necessary to ensure good results are beyond the scope of this volume, but there are several well written books on the topic.

2.5 Further reading

Bristow, P. A., *Liquid Chromatography in Practice*, Lab. Data Florida, (ISBN 0 9504933 1 1)

Daniels, T., *Thermal Analysis*, Kogan Page, (1973)

Fried, B. and Sherma, J., *Thin Layer Chromatography: Techniques and Applications*, Marcel Dekker, New York, (1982)

Heyrovsky, J. and Zuman, P., *Practical Polarography*, Academic Press, (1968)

Kirkland, J. J. (ed.), *Modern Practice of Liquid Chromatography*, Wiley Interscience, New York, (1971)

Meites, L., *Polarographic Techniques* (2nd edn), Interscience, (1965)

Perry, S. G., Amos, R. and Brewer, P. I., *Practical Liquid Chromatography*, Plenum, New York, (1972)

Snyder, L. R. and Kirkland, J. J., *Introduction to Modern Liquid Chromatography*, Wiley Interscience, New York, (1974)

Touchstone, J. C. and Rogers, D. (eds), *Thin Layer Chromatography Quantitative, Environmental and Clinical Applications*, Wiley, New York, (1980)

Wendland, W. W., *Thermal Methods of Analysis*, Interscience, (1964)

Wiedemann, H. G. (ed.), *Thermal Analysis* Vols 1–3, Birkhäuser Verlag, Basle and Stuttgart, (1972)

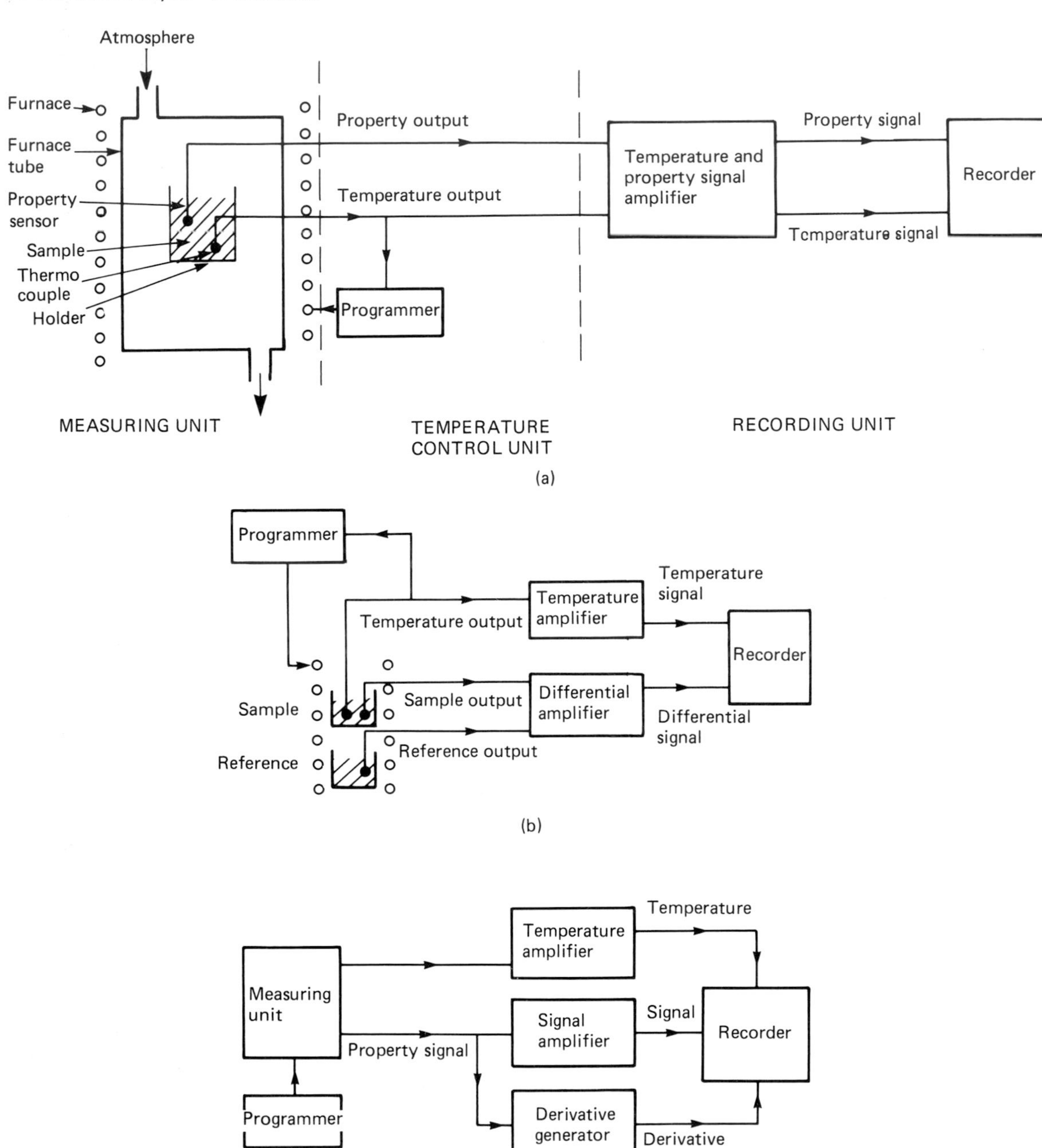

Figure 2.22 Construction of thermal analysis instruments. (Reproduced by permission from *Thermal Analysis* by T. Daniels, published Kogan Page, Ltd.): (a) basic thermal analysis system, (b) differential instrument, (c) derivative instrument.

3 Chemical analysis – spectroscopy

A. C. SMITH

The analysis of substances by spectroscopic techniques is a rather specialized field and cannot be covered in full depth in a book such as this. However, some fifteen techniques will be covered – giving the basic principles for each, descriptions of commercial instruments and where possible, their use as on-line analysers.

Details of other techniques may be found in modern physics textbooks and greater detail of those techniques which are described may be found in literature provided by instrument manufacturers such as Pye Unicam, Perkin-Elmer, Hilgers, Applied Research Laboratories. There are also many textbooks devoted to single techniques.

Some aspects of measurements across the electromagnetic spectrum are dealt with in Part 3, Chapter 2.

3.1 Absorption and reflection techniques

3.1.1 Infrared

Measurement of the absorption of infrared radiation enables the quantity of many gases in a complex gas mixture to be measured in an industrial environment. Sometimes this is done without restricting the infrared frequencies used (Dispersive). Sometimes only a narrow frequency band is used (Non-dispersive).

3.1.1.1 Non-dispersive infrared analysers

Carbon monoxide, carbon dioxide, nitrous oxide, sulphur dioxide, methane and other hydrocarbons and vapours of water, acetone, ethyl alcohol, benzene and others may be measured in this way. (Oxygen, hydrogen, nitrogen, chlorine, argon and helium do not absorb infrared radiation and are therefore ignored.) An instrument to do this is illustrated in Figure 3.1(a). Two beams of infrared radiation of equal energy are interrupted by a rotating shutter which allows the beams to pass intermittently but simultaneously through an analysis cell assembly and a parallel reference cell, and hence into a Luft-pattern detector.

The detector consists of two sealed absorption

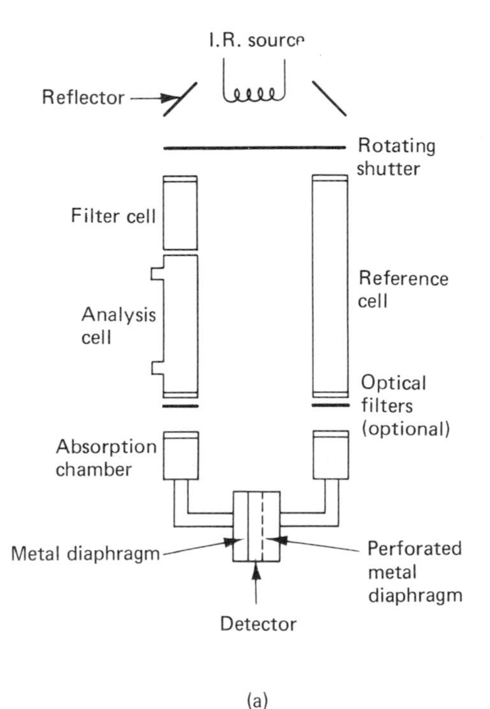

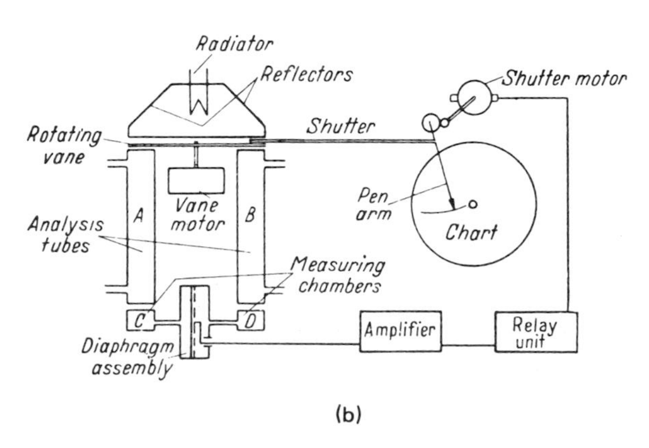

Figure 3.1 (a) Luft-type infrared gas analyser (courtesy Grubb Parsons), (b) infrared gas analyser of the concentration recorder.

chambers separated by a thin metal diaphragm. This diaphragm, with an adjacent perforated metal plate, forms an electrical capacitor. The two chambers are filled with the gas to be detected so that the energy characteristic of the gas to be measured is selectively absorbed.

The reference cell is filled with a non-absorbing gas. If the analysis cell is also filled with a non-absorbing gas equal energy enters both sides of the detector. When the sample is passed through the analysis cell, the component to be measured absorbs some of the energy to which the detector is sensitized, resulting in an imbalance of energy causing the detector diaphragm to be deflected and thus changing the capacitance. This change is measured electrically and a corresponding reading is obtained on the meter.

Any other gas also present in the sample will not affect the result unless it has absorption bands which overlap those of the gas being determined. In this event, filter tubes containing the interfering gas or gases can be included in one or both optical paths, so that the radiations emerging from these tubes will contain wavelengths which can be absorbed by the gas to be detected but will contain very little radiation capable of being absorbed by the interfering gases in the sample, since such radiations have already been removed.

The length of absorption tube to be used depends upon the gas being estimated and the concentration range to be covered. The energy absorbed by a column of gas l cm long and containing a concentration c of absorbing component is approximately $Elkc$, where E is the incident energy and k is an absorption constant, provided that kcl is small compared with unity. Thus at low concentrations it is advantageous to use long absorption paths provided kcl remains small and the relationship between energy absorbed and the measured concentration remains reasonably linear. At higher concentrations the energy absorbed is $E[1-\exp(-kcl)]$, and the relationship between energy absorbed and concentration departs greatly from linearity when absorption exceeds 25 per cent. When the absorption reaches this value it is, therefore, necessary to reduce the length of the absorption cell, and the product $c \times l$ should be kept approximately constant.

The most convenient method of calibrating the instrument is to pass mixtures of the pure gas and air of known composition through the measuring cell and note the output for each concentration of measured gas. For day-to-day checking a simple internal calibrating device is fitted, and it is only necessary to adjust the sensitivity control until a standard deflection is obtained.

The instrument is usually run from a.c. mains through a constant voltage transformer. Where utmost stability is required an a.c. voltage stabilizer may be used, as the constant voltage transformer converts frequency variations to voltage changes. Generally, the instrument is insensitive to temperature changes, although the gas sensitivity depends on the temperature and pressure of the sample gas in the absorption tube, since it is the number of absorbing molecules in the optical path which determines the meter deflection. For instruments sensitive to water vapour the detecting condenser has a temperature coefficient of sensitivity of 3 per cent per kelvin and it is therefore necessary to maintain the detector at a constant temperature.

The approximate maximum sensitivity to certain gases is given in Table 3.1.

Table 3.1 Sensitivity of non-dispersive infrared analyser

Gas	*Minimum concentration for full-scale deflection, (Vol. %)*	*Gas*	*Minimum concentration for full-scale deflection, (Vol. %)*
CO	0.05	NO_2	0.1
CO_2	0.01	SO_2	0.02
H_2O	0.1	HCN	0.1
CH_4	0.05	Acetone	0.25
C_2H_4	0.1	Benzene	0.25
N_2O	0.01		

Errors due to zero changes may be avoided by the use of a null method of measurement illustrated in Figure 3.1(b). The out-of-balance signal from the detector is amplified, rectified by a phase-sensitive rectifier, and applied to a servo system which moves a shutter to cut off as much energy from the radiation on the reference side as has been absorbed from the analysis side, and so restore balance. The shutter is linked to the pen arm which indicates the gas concentration.

On-line infrared absorption meter using two wavelengths In order to overcome the limitations of other infrared analysers and provide a rugged reliable drift-free analyser for continuous operation on a chemical plant, ICI Mond Division developed an analyser based on the comparison of the radiation absorbed at an absorption band with that at a nearby wavelength. By use of this comparison method many of the sources of error such as the effect of variation in the source intensity, change in the detector sensitivity or fouling of the measurement cell windows are greatly reduced.

The absorption at the measurement wavelength (λ_m) is compared with the nearby reference wavelength (λ_r) at which the measured component does not absorb. The two measurements are made alternately using a single absorption path and the same source and detecting system.

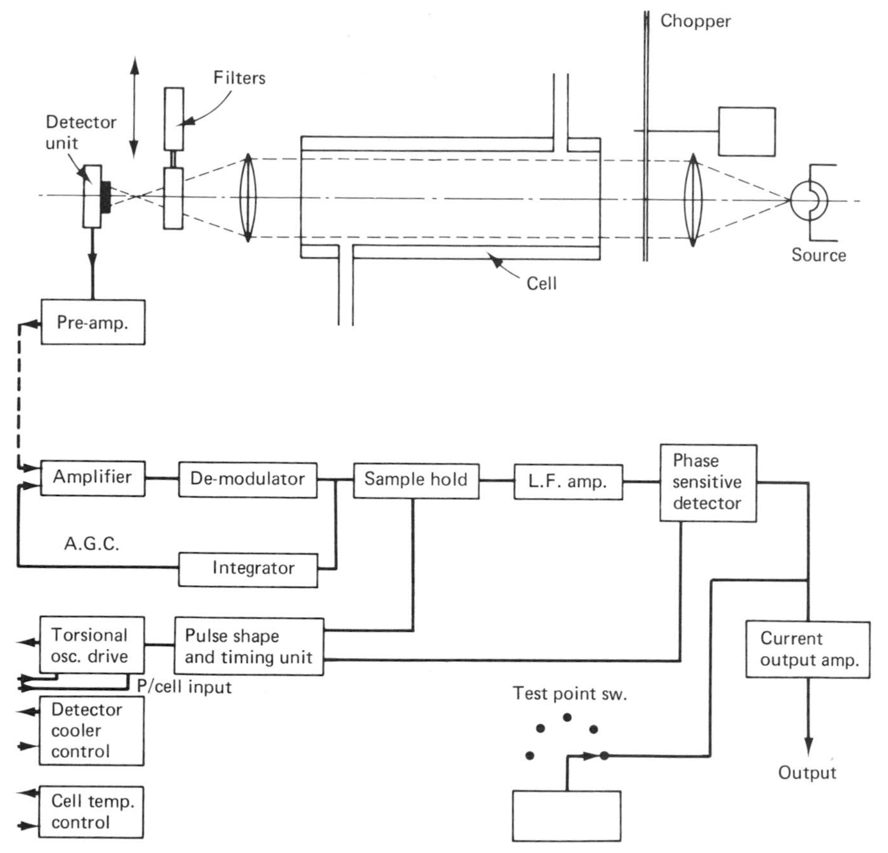

Figure 3.2 Dual wavelength comparison method (courtesy Feedback Instruments Ltd.).

The principle of the system is illustrated in Figure 3.2. The equipment consists of two units, the optical unit and the electronics unit, which are connected by a multicore cable. The source unit contains a sealed infrared source which consists of a coated platinum coil at the focus of a calcium fluoride collimating lens. A chopper motor with sealed bearings rotates a chopper disc which modulates the energy beam at 600 Hz. The source operates at low voltage, and at a temperature well below the melting point of platinum. It is sealed in a nitrogen atmosphere. Energy from the source passes through the absorption cell to the detector unit. A calcium fluoride lens focuses the energy onto an indium antimonide detector. This is mounted on a Peltier cooler in a sealed unit. The temperature is detected by a thermistor inside the sealed module. A pre-amplifier mounted in the detector unit amplifies the signal to a suitable level for transmission to the electronics unit. Between the lens and the detector module two interference filters, selected for the measurement and reference wavelengths, are interposed alternately in the beam, at about 6 Hz, so that the detector receives chopped energy at a level corresponding alternately to the measurement and reference transmission levels. Its output is a 600 Hz carrier modulated at 6 Hz.

The two filters are mounted on a counterbalanced arm, attached to a stainless steel torsion band. An iron shoe at the opposite end of the arm moves in and out of the gap in an electromagnet. It also cuts two light beams which illuminate two silicon phototransistors. The light is provided by two aircraft-type signal lamps which are under-run to ensure very long life. A drive circuit in the electronics unit causes the system to oscillate at its own natural frequency. One of the photocells provides positive feedback to maintain the oscillation, and the other provides negative feedback to control the amplitude. There are no lubricated parts in the detector unit, and the whole can be hermetically sealed if desired.

The absorption cell is a thick-walled tube with heavy flanges. Standard construction is in mild steel, nickel-plated, but type 316 stainless steel construction is available where required. The windows are of calcium fluoride, sealed with Viton O-rings and retaining rings. A heater wire is wound on the cell, and the sample gas passes through a tube in thermal contact along the length of the cell before entering it at the end. Provision is made for rodding out tubes and entries in case of blockage. A thermistor embedded in the cell wall detects the cell temperature which is controlled by a circuit in the electronics unit. The cell is thermally insulated and sealed inside a plastics bellows. The enclosed space is coupled to the purge system. The two end units each have a sealing window so there is a double seal between the cell and the interior of the detector and source units. Since the source is inside a further sealed module, there is minimal danger of the hot source being exposed to leakage from the sample cell. The gaps between the three units are normally sealed with neoprene gaskets and the whole device is sufficiently well sealed to maintain a positive purge pressure of at least 2 cm water gauge with a purge gas consumption of 8.3 cm^3/s. For use with highly flammable sample gases, the sealing gaskets at either end of the absorption cell may be replaced by vented gaskets. In this case a relatively large purge flow may be maintained around the cell, escaping to atmosphere across the windows. Thus, any leak at the windows can be flushed out.

To facilitate servicing on site the source, detector, torsional vibrator, lamps, pre-amplifier and source voltage control are all removable without the use of a soldering iron. Since the single-beam system is tolerant to window obscuration and the internal walls of the absorption cell are not polished, cell cleaning will not be required frequently, and in many cases adequate cleaning may be achieved *in situ* by passing solvent or detergent through the measuring cell. There is no need to switch the instrument off while doing this. If it becomes necessary the cell can be very quickly removed and disassembled.

The electronics unit contains the power supplies together with signal processing circuits, temperature control circuits, output and function check meter operating controls and signal lamps. The housing is of cast-aluminium alloy, designed for flush panel mounting. The circuitry is mostly on five plug-in printed circuit boards. The indicating meter, controls and signal lamps are accessible through a window in the door. The unit is semi-sealed, and a purge flow may be connected if sealed glands are used at the cable entry. The signal processing circuits are contained on two printed circuit boards. Output from the pre-amplifier is applied to a gain-controlled amplifier which produces an output signal of 3 V peak-to-peak mean. Thus the mean value of $I_r + I_m$ is maintained constant. The signal is demodulated and smoothed to obtain the 6 Hz envelope waveform. A sample-and-hold circuit samples the signal level near the end of each half-cycle of the envelope, and this produces a square wave whose amplitude is related to $I_r - I_m$. Since $I_r + I_m$ is held constant, the amplitude is actually proportional to $(I_r - I_m)/(I_r + I_m)$ which is the required function to give a linearized output in terms of sample concentration. This signal is amplified and passed to a phase-sensitive detector, consisting of a pair of gating transistors which select the positive and negative half-cycles and route them to the inverting and non-inverting inputs of a differential amplifier. The output of this amplifier provides the 0–5 V output signal.

The synchronizing signals for the sample/hold and phase-sensitive detector circuits are derived from the torsional oscillator drive circuit via appropriate time delays. The instrument span is governed by selection of feedback resistors in the low frequency amplifier, and a fine trim is achieved by adjusting the signal level at the gain-controlled amplifier. This is a preset adjustment – no operator adjustment of span is considered necessary or desirable. A front panel zero adjustment is provided. This adds an electrical offset signal at the phase-sensitive detector. The system is normally optically balanced (i.e. $I_r = I_m$) at some specified concentration of the measured variable (usually zero).

The current output and alarm circuits are located on a separate printed circuit board. The voltage output is applied to an operational amplifier with selected feedback and offset signals to produce 0–10 mA, 5–20 mA or 10–50 mA output. The required output is obtained by soldered selector links. The output current is unaffected by load resistances up to 1 kΩ at 50 mA, or 5 kΩ at 10 mA.

A front panel alarm-setting potentiometer provides a preset signal which is compared with the analyser output voltage in a differential amplifier. The output of this opens a relay if the analyser output exceeds a preset value, which may be either a low or high analyser output as required. The alarm condition is indicated by two signal lamps on the panel and the system can be arranged to operate external alarms, or shut-down circuits.

The power to the cell heater and the detector cooler is controlled from a bridge circuit containing thermistors which detect the temperatures of the absorption cell and detector.

The indicating meter on the front panel has a calibrated output scale, and is used in conjunction with a selector switch to monitor key points in the circuit, in particular the degree of obscuration in the measuring cell. By choosing the appropriate absorption bands the analyser may be made suitable for a wide range of gases or liquids. For gases, it may be used for CO_2, CO, SO_2, CH_4, C_2H_6, C_2H_4, C_6H_6, C_2H_2, NH_3, N_2O, NO, NO_2, $COCl_2$, H_2O, with ranges of 0–300 ppm and 0–100 per cent.

It may also be used for measuring water in ketones, hydrocarbons, organic acids, alcohols, glucols and oils. The accuracy is ± 1 per cent and the response time for 90 per cent change is 3 s.

The instrument is marketed by Anatek Ltd. as the

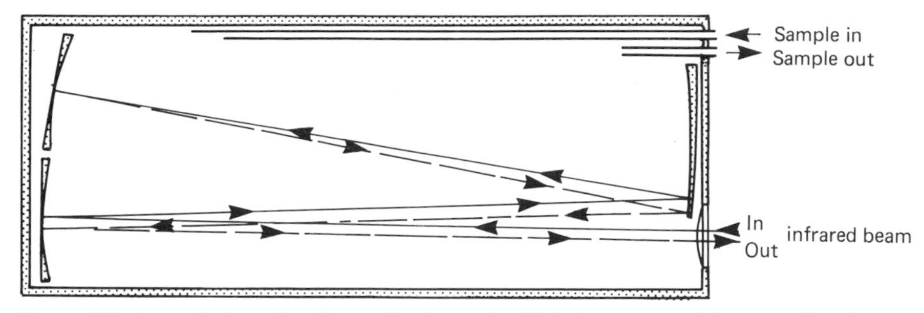

Figure 3.3 Internal view of multiple reflections of variable long path cell (courtesy The Foxboro Company).

PSA 401 process stream analyser.

Another instrument based on the same principle is the Miran II Infra Red process analyser – the chief difference being the sample cell used for gas and liquid streams. These cells are either long path gas cells or multiple internal reflection cells. The gas cells which are normally manufactured in stainless steel have a variable path length (see Figure 3.3). Energy passes through the sample gas and reflects one or more times off the mirrors in the cell before striking the detector. The path length can be adjusted between 0.75 and 20.25 metres by suitable adjustment of the mirrors. These gas cells are used to analyse the presence of low concentrations of components in gases or for those gases requiring a long path length to enhance sensitivity at a weak analytical wavelength.

In a multiple internal reflection cell, the infrared beam is directed along or around an optical crystal through which the beam passes (Figure 3.4). As the beam is reflected on the sample crystal interface, it slightly penetrates the liquid. These penetrations form a path whose length is dependent on the number of reflections. The energy is absorbed at the analytical wavelength proportionally to concentration just as in other types of cells. The crystal used is made of KRS (a composite of thallium bromide and iodide). Ordinary transmission cells have limited applicability for high concentrations, viscous or aqueous streams. In many cases, the infrared beam is grossly attenuated or the sample cannot be pumped through such cells. Multiple internal reflection overcomes these problems.

The applications to which this instrument has been put include (a) for gases: the determination of phosgene in methane and plastic production, methane and carbon dioxide in synthetic and natural gases in the range 1 ppm to 100 per cent, (b) for liquids: water in acetone distillation, petroleum waste treatments, urea in fertilizer production and isocyanates in urethane and plastic production in the range 50 ppm to 50 per cent and (c) for solids: the percentage weight of film coatings such as inks and polymers and film thickness for nylon and polythene (up to 0.025 mm).

3.1.1.2 Dispersive infrared analysis

The previous section was devoted to analysis using only one absorption frequency. However, all organic compounds give rise to a spectrum in the infrared in which there are many absorption frequencies giving a complete fingerprint of that compound. Dispersive infrared can be used, amongst other things, to identify a substance, for the determination of molecular structure for reaction kinetic studies, and for studies of hydrogen bonding.

In Figure 3.5 is shown a simplified layout of a typical double-beam spectrophotometer. A source provides radiation over the whole infrared spectrum, the monochromator disperses the light and then selects a narrow frequency range, the energy of which is measured by a detector – the latter transforms the energy received into an electrical signal which is then amplified and registered by a recorder or stored in a computer for further processing. The light path and ultimate focusing on the detector is determined by precision manufactured mirrors.

Light from the radiation source S is reflected by mirrors M_1 and M_2 to give identical sample and reference beams. Each of these focuses upon vertical entrance slits S_1 and S_2, the sample and reference cells being positioned in the two beams near their foci. Transmitted light is then directed by a mirror M_3 on to a rotating sector mirror (or oscillating plane mirror) M_4. The latter serves first to reflect the sample beam towards the monochromator entrance slit S_3 and then as it rotates (or oscillates), to block the sample beam and allow the reference beam to pass on to the entrance slit. A collimating mirror M_5 reflects parallel light to a prism P, through which it passes only to be reflected

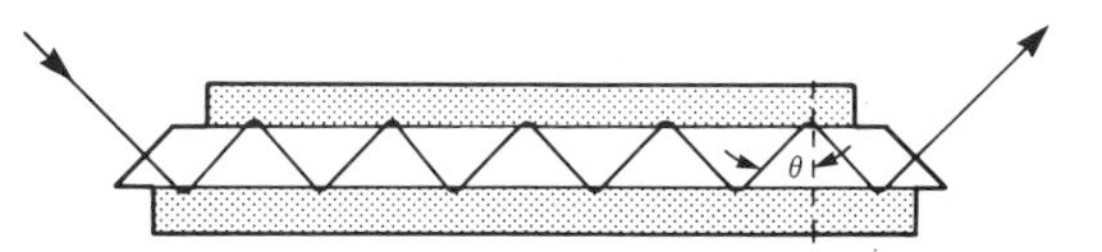

Figure 3.4 Principle of MIR sampling technique (courtesy Foxboro Analytical).

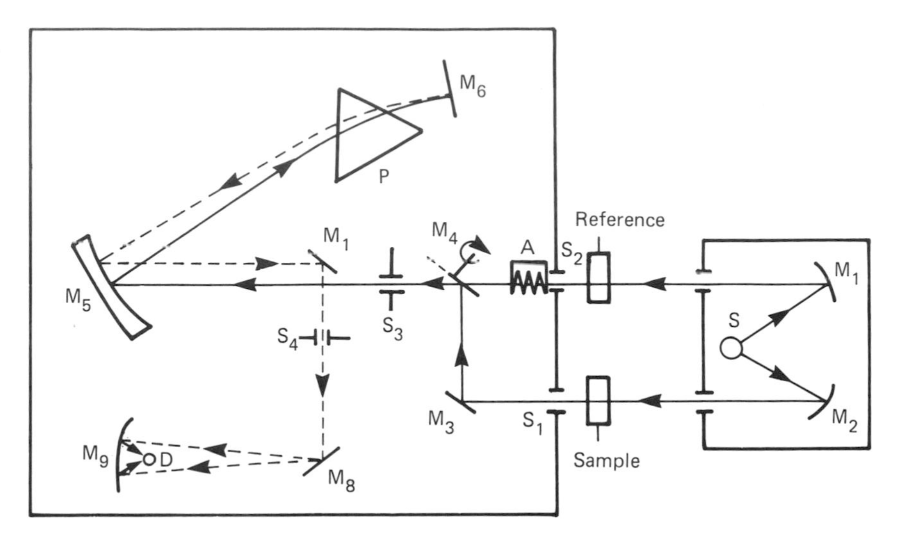

Figure 3.5 Simplified spectrophotometer.

back again through the prism by a rotatable plane mirror M_6. The prism disperses the light beam into its spectrum. A narrow range of this dispersed light becomes focused on a plane mirror M_7 which reflects it out through the exit slit. A further plane mirror M_8, reflects the light to a condenser M_9 which focuses it sharply on the detector D. When the energy of the light transmitted by both sample and reference cells is equal, no signal is produced by the detector. Absorption of radiation by the sample results in an inequality of the two transmitted beams falling on the detector and a pulsating electrical signal is produced. This is amplified and used to move an attenuator A across the reference beam, cutting down the transmitted light until an energy balance between the two beams is restored. The amount of reference beam reduction necessary to balance the beam energies is a direct measure of the absorption by the sample.

The design and function of the major instrument components now described have a significant influence on its versatility and operational accuracy.

Source IR radiation is produced by electrically heating a Nernst filament (a high resistance, brittle element composed chiefly of the powdered sintered oxides of zirconium, thorium and cerium held together by a binding material) or a Globar (SiC) rod. At a temperature in the range 1100–1800 °C depending on the filament material, the incandescent filament emits radiation of the desired intensity over the wavelength range 0.4–40 μm.

Monochromator The slit width, and optical properties of the components are of paramount importance. The wavelength range covered by different prisms is shown in Table 3.2. Gratings allow better resolution than is obtainable with prisms.

Detector This is usually a bolometer or thermocouple. Some manufacturers use a Golay pneumatic detector which is a gas-filled chamber that undergoes a pressure rise when heated by radiant energy. One wall of the chamber functions as a mirror and reflects a light beam directed at it onto a photocell – the output of the photocell bearing a direct relation to the gas chamber expansion.

The infrared spectra of liquids and gases may be obtained by direct study of undiluted specimens. Solids, however, are usually studied after dispersion in one of a number of possible media. These involve reduction of the solid to very small particles which are then diluted in a mill, pressed into an alkali halide disc at 1500–3300 bar or spread as pure solid on a cell plate surface.

Table 3.2 Prism frequency ranges

Prism material	*Glass*	*Quartz*	*CaF_2*	*LiF*	*NaCl*	*KBr(CsBr)*	*CsI*
Useful frequency range (cm^{-1})	above 3500	above 2860	5000–1300	5000–1700	5000–650	1100–285	1000–200
Wavelength range (μm)	below 2.86	below 3.5	2.0–7.7	2.0–5.9	2–15.4	9–35	10–50

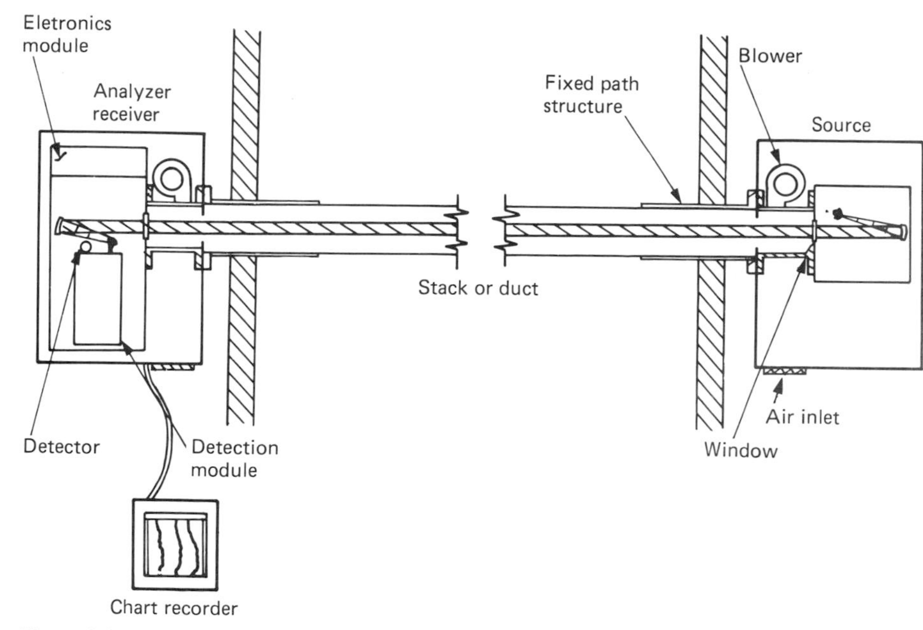

Figure 3.6 EDC flue gas analyser system (courtesy Environmental Data Corp.).

The interpretation of the spectra – particularly of mixtures of compounds – is a complex problem and readers should consult textbooks on infrared analysis.

3.1.2 Absorption in UV, visible and IR

One instrument that uses absorption in the UV, visible and IR is the Environmental Data Corporation stack-gas monitoring system. It is designed to measure from one to five component gases simultaneously. Depending on requirements, the components may include CO_2, NO, CO, SO_2, H_2O, NH_3, hydrocarbons and opacity or any other gases with selected spectral absorption bands in the UV, visible or IR. The basis of the system is shown in Figure 3.6. It consists of a light source, receiver, mounting hardware and recorder. Each gas monitoring channel is similar in basic operation and calibration. The instrumentation can be mounted on a stack, duct or other gas stream. A polychromatic beam of light, from a source in an enclosure on one side, is collimated and then passed through the gas to an analyser on the opposite side. Signals proportional to the gas concentrations are transmitted from analyser to recorder.

Most gases absorb energy in only certain spectral regions. Their spectra are often quite complex with interspersed absorbing and non-absorbing regions. The analyser section of the instrument isolates the wavelengths characteristic of the gases of interest and measures their individual intensities. Both the intensity at a specific wavelength where the gas uniquely absorbs (A) and the intensity at a nearby region where the gas is non-absorbing (B) are alternately measured with a single detector 40 times per second. Any light level change, whether due to source variation, darkening of the window, scattering by particulates, water drops or aerosols in the gas stream affects both A and B leaving the ratio unchanged. This ratio gives a reading that is free of interferences, instrumental drift etc. Most gases obey approximately Beer's law:

$$B = A\,e^{-\alpha cl}$$

or

$$\ln(B/A) = -\alpha cl$$

or

$$c = \frac{\ln(A/B)}{\alpha l}$$

where α is absorption coefficient (known), l is path length (fixed), and c is sample concentration (unknown).

The system response is almost instantaneous and is averaged by damping circuits to typically one second.

The stack gas is separated from the source and analyser enclosures by means of optical surfaces, such as mirrors or windows. These windows are kept clean by an air curtain system. Self-contained blowers continually renew the air curtains, preventing the gases from contacting the windows directly (see Figure 3.7).

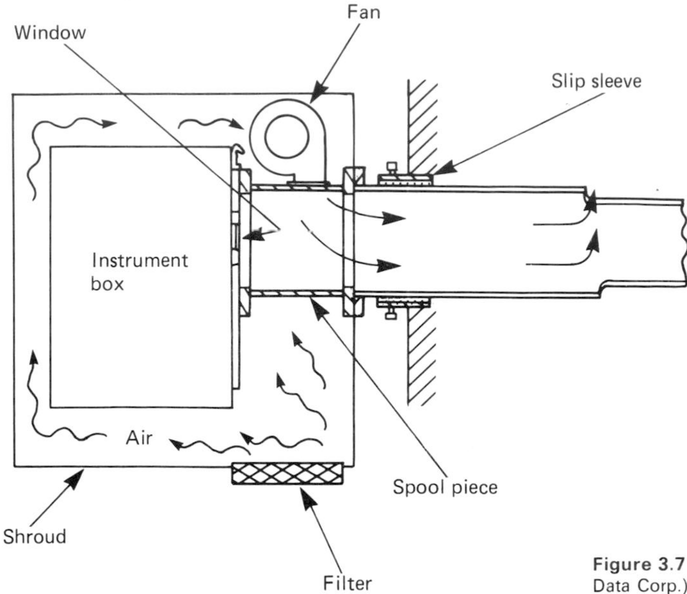

Figure 3.7 EDC flue gas analyser (courtesy Environmental Data Corp.).

The flow volume and pressure of the purge air is designed for each application to allow a well defined shear by the flue gas. Thus a known and fixed path length is provided.

When measuring opacity the instrument measures the reduction in transmission in the visible portion of the spectrum.

Typical ranges covered by the instrument are:

NO	0–25 ppm to 0–5000 ppm
CO	0–500 ppm to 0–3000 ppm
CO_2	0–15%
SO_2	0–25 ppm to 0–10 000 ppm
C–H	0–25 ppm to 0–6000 ppm
H_2O	0–1000 ppm to 0–80%
NH_3	0–100 ppm

3.1.3 Absorption in the visible and ultraviolet

Two instruments are worthy of note here. The first is the Barringer remote sensing correlation spectrometer designed for the quantitative measurement of gases such as nitrogen oxides or sulphur dioxide in an optical path between the instrument and a suitable source of visible and ultraviolet radiant energy. The sensor is designed for maximum versatility in the remote measurement of gas clouds in the atmosphere using the day sky or ground-reflected solar illumination as the light source. It may also be used with artificial sources such as quartz-iodine or high pressure Xe lamps. Very simply the sensor contains two telescopes to collect light from a distant source, a two-grating spectrometer for dispersion of the incoming light, a disc-shaped exit mask or correlator and an electronics system (see Figure 3.8). The slit arrays are designed to correlate sequentially in a positive and negative sense with absorption bands of the target gas by rotation of the disc in the exit plane. The light modulations are detected by photomultiplier tubes and processed in the electronics to produce a voltage output which is proportional to the optical depth (expressed in ppm meters) of the gas under observation. The system automatically compensates for changes in average source light intensity in each channel. The basic principle of this method rests on comparison of energy in selected proportions of the electromagnetic spectrum where absorption by the target gas occurs in accordance with the Beer–Lambert law of absorption.

Typically, this instrument covers the range 1–1000 ppm m or 100–10 000 ppm m, this unit being the product of the length of the optical path through the gas and the average concentration (by volume) over that length.

The second instrument which covers absorption in the visible in liquids, is the Brinkmann Probe Colorimeter. This instrument is basically a standard colorimeter consisting of a tungsten light source, the output from which passes through one of a series of interchangeable filters covering the wavelength range 420–880 nm, then through a light pipe at the end of which is a probe cell. This cell has a reflecting mirror at one end and so the optical path length is twice the length of the cell. The light then returns to the instrument via a second light pipe to a photomultiplier, the output of which is amplified and fed to a recorder in the usual way. This instrument is ideal for

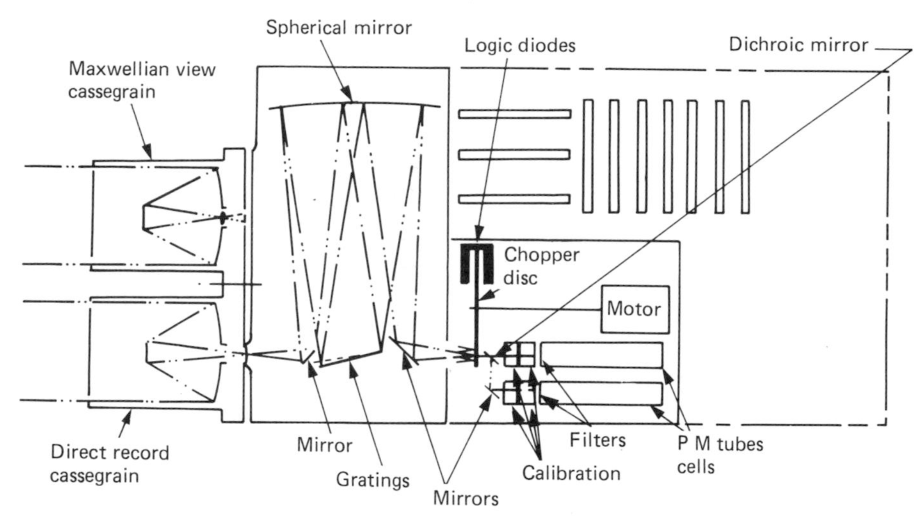

Figure 3.8 Barringer remote sensing correlation spectrometer.

measuring turbidity in liquids and has the advantage that very small volumes of liquid (down to 0.5 ml) may be examined. Its other uses include general quality control, chemical analyses, pollution control and food processing. Most of these applications make use of the fact that different elements will form coloured solutions with reagents. The absorption of these coloured solutions is then proportional to the concentration of that particular element.

3.1.4 Measurements based on reflected radiation

Just as measurements of moisture, or other components, may be made by comparison at two wavelengths of transmitted infrared radiation, the method will work equally well by measuring the attenuation when infrared is reflected or back-scattered. The principle is illustrated in Figure 3.9.

For water measurement of paper or granulated material on a conveyor belt, the intensity of the reflected beam at the moisture absorption wavelength of 1.93 μm may be compared with the intensity at a reference wavelength of 1.7 μm. The beams are produced by interposing appropriate filters contained in a rotating disc in front of a lamp producing appropriate radiation. The radiation is then focused onto the measured material, and the reflected beam focused onto a lead sulphide photoelectric cell. By measuring the ratio of the intensity of radiation at two wavelengths, the effects of source variation, detector sensitivity and drift in the electronic circuitry are minimized. Furthermore, calibration has shown that for a number of materials the results are substantially independent of the packing density.

However, if the measured material is a strong absorber of radiation a powerful source of radiation such as a water-cooled quartz halogen lamp may be necessary.

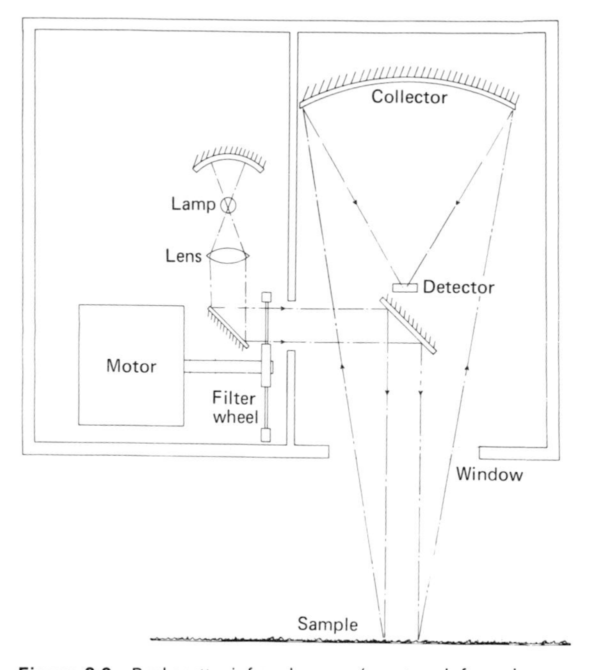

Figure 3.9 Backscatter infrared gauge (courtesy Infra-red Engineering Ltd.).

With this type of instrument on-line measurement of the moisture content of sands, clay, dust or flake, refractory mixtures, paper, textiles, feeding stuffs and a wide range of other materials may be undertaken with an accuracy of ± 1 per cent of instrument full scale.

3.1.5 Chemiluminescence

When some chemical reactions take place, energy may be released as light. This phenomenon is known as chemiluminescence. There are many instruments which make use of this effect for the determination of the concentration of oxides of nitrogen and for ozone. The principles are described in Chapter 5.

3.2 Atomic techniques – emission, absorption and fluorescence

3.2.1 Atomic emission spectroscopy

This is one of the oldest of techniques employed for trace analysis. Because of its relative simplicity, sensitivity and ability to provide qualitative information quickly, it has been widely used in both industrial and academic analytical problems. It can be used for the analysis of metals, powders and liquids and is used extensively in the steel and non-ferrous alloy industries, and the advent of inductively coupled plasma sources for producing spectra has made the technique invaluable for the analysis of some 70 elements in solution – down to concentrations of 1 ppb and less.

The basic principles of the technique are as follows.

Each atom consists of a nucleus around which revolve a set of electrons. Normally these electrons follow orbits immediately adjacent to the nucleus. If energy is imparted to the atom by means of a flame or an electric arc or spark, then it undergoes excitation and its electrons move into orbits further removed from the nucleus. The greater the energy, the further from the nucleus are the orbits into which the electrons are moved. When sufficient energy is imparted to the electron, it may be torn from the atom and the atom becomes a positively charged ion. Atoms will not remain in this excited state, especially when removed from the source of energy and they return to their original states with electrons falling to lower orbits. This electron transition is accompanied by a quantum of light energy. The size of this pulse of light energy and its wavelength depend on the positions of the orbits involved in the transition.

The energy emitted is

$$E = h\nu$$

where h is Planck's constant, and ν is the frequency of the radiation. Or

$$E = hc/\lambda$$

where c is the velocity of light and λ the wavelength. Hence the greater the light energy quantum, the shorter is the wavelength of the light emitted.

Only the outer, valence electrons participate in the emission of spectral lines. The number of valence electrons in an atom differs for chemical elements. Thus the alkali elements, sodium, lithium, potassium etc. contain only one electron in their outer shell and these elements have simple spectra. Such elements as manganese and iron have five or six valence electrons and their spectra are very complex. Generally speaking, the structure of an atom is closely bound up with its optical spectrum. Thus if a mixture of atoms (as found in a sample) are excited by applying energy, then quantities of light are emitted at various wavelengths, depending on the elements present. The intensity of light corresponding to one element bears a relationship to the concentration of that element in the sample.

In order to sort out the light emitted, use is made of a spectroscope. In Figures 3.10–3.12 are shown the layout of a medium quartz spectroscope, a Littrow spectrograph and a spectroscope using a diffraction grating. This last employs the principle, due to Rowland, of having the grating on a concave surface. There are many other configurations. In all cases, each instrument contains three main components, a slit, a dispersive device such as a prism or diffraction grating to separate radiation according to wavelength, and a suitable optical system to produce the spectrum lines which are monochromatic images of the slit. These images may be recorded on a photographic plate, or by suitable positioning of exit slits, mirrors and photomultiplier tubes, the light intensity may be recorded electronically.

3.2.1.1 *Dispersive devices*

Prisms Prisms are usually made of glass or quartz and their dispersive ability is based on the variation of the index of refraction with wavelength. As the incident light beam enters the transparent material, it bends towards the normal according to Snell's law:

$$n_1 \sin i = n_2 \sin r$$

where n_1 is the refractive index of air, n_2 is the refractive index of the prism material, i is angle of incidence, and r is angle of refraction. Shorter wavelengths are deviated

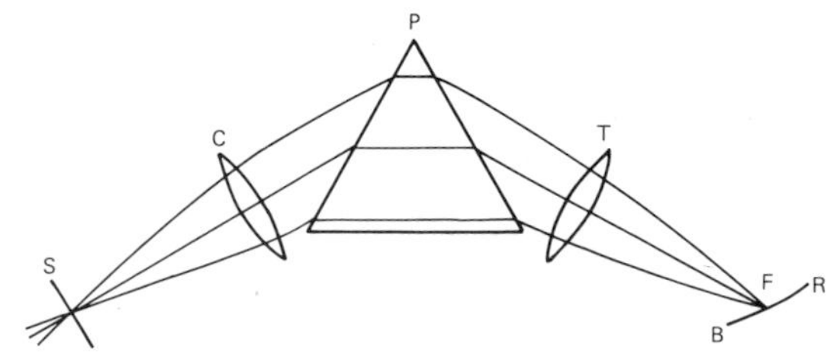

Figure 3.10 Optical system of a simple spectroscope. S, silt; C, collimator lens; P, prism; T, telescope lens; F, curve along which the various parts of the spectrum are in focus; B, blue or short wavelength part; R, red or long wavelength part.

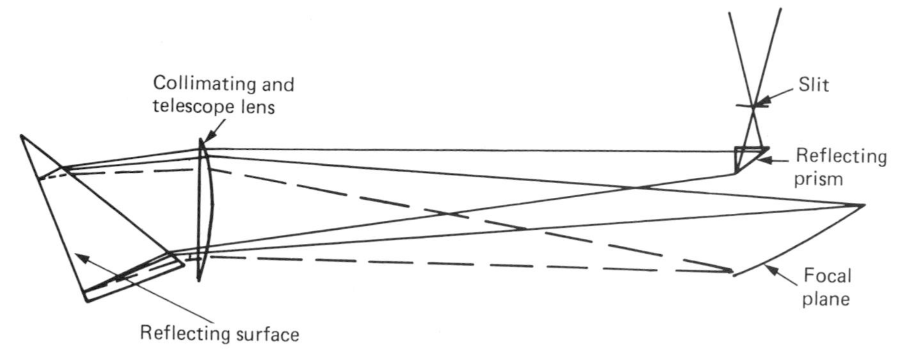

Figure 3.11 Diagram of the optical system of a Littrow spectrograph. The lens has been reversed to reduce scattered light.

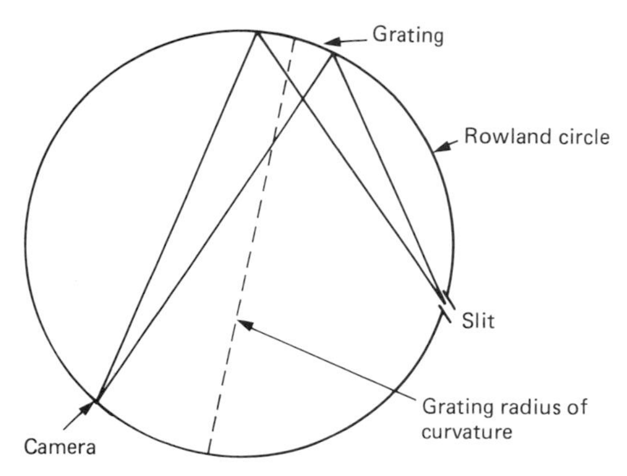

Figure 3.12 Elements of Rowland circle.

more than longer ones. The resulting dispersion is greater for the UV than for IR wavelengths.

Gratings Gratings may be considered as a large number of parallel, close, equidistant slits or diffracting lines. The equation $n\lambda = 2d \sin \theta$ shows the dependence of θ upon the wavelength of the incident light, where n is an integer, λ is the wavelength of incident light, d is the distance between the lines, and θ is the angle between the diffracted beam and the normal incident beam.

Modern gratings offer the spectroscopist uniform dispersion and coverage of a wide spectral range. Today, nearly all manufacturers have turned almost exclusively to grating instruments.

3.2.1.2 Vacuum spectrographs

Many elements, particularly the non-metallic ones, have their most persistent lines in the spectral region 150–220 nm. Light of these wavelengths is absorbed by air, and instruments are manufactured in which the optical paths are evacuated to overcome this problem.

3.2.1.3 Excitation – spectroscopic sources

Many factors are considered in the choice of a source. Sample form, necessary sensitivity and the elements which must be determined are the most critical. The main sources used are (a) a d.c. arc, (b) a high voltage condensed spark, (c) an arc triggered by a high voltage spark, (d) flames, (e) plasma jets and (f) inductively coupled plasmas. A recent form of excitation consists of evaporating a non-conducting sample by means of a laser and exciting the vapour with a high voltage spark.

3.2.1.4 Standards

In order to achieve a quantitative estimation of the impurity concentrations, some form of standard sample of known purity must be analysed under exactly the same conditions as the unknown samples and the intensity of the spectral lines compared. Thus a spectrochemical laboratory may have many thousands of standards covering the whole range of materials likely to require analysis.

3.2.1.5 Applications

There are very few on-line instruments employing atomic emission techniques, but mention should be made of a continuous sodium monitor for boiler/feed water. The water is nebulized into a flame, the sodium emission is isolated by means of a monochromator and the intensity measured by means of a photomultiplier and associated electronics. Standard solutions are automatically fed into the instrument from time to time to check the calibration.

In both the steel and non-ferrous alloy industries, large grating spectroscopes are used to control the composition of the melts before they are finally poured. A complete analysis for some 30–40 elements can be made within 2 minutes of a small sample being taken. Suitable additions are then made to the melt to satisfy the required composition specification. In these cases the output from the instrument is fed to a computer which is programmed to produce actual elemental concentrations and also the necessary amounts required to be added to known weights of melts in the furnaces for them to be of the correct composition. Analysis of water samples or samples in solution can be carried out using an inductively coupled plasma direct reading spectrometer. Some 60 elements can be determined in each sample every two minutes. The source is ionized argon pumped inductively from an r.f. generator into which the sample is nebulized. Temperatures of about 8500 °C are achieved. Many instruments of this type are now manufactured and have been of great value to the water industry and to environmental chemists generally – in particular, those instruments manufactured by A.R.L., Philips and Javell Ash. Limits of detection are of the order of 1 ppb (parts per 10^9) with an accuracy of about 10 per cent.

3.2.2 Atomic absorption spectroscopy

In emission spectroscopy, as we have already seen, the sample is excited, the emitted radiation dispersed and the intensities of the selected lines in the emission spectrum measured. If self-absorption and induced emission are neglected, then the integrated intensity of emission of a line is given by

$$\int I_v \, dv = CN_j F$$

where N_j is the number of atoms in the higher-energy level involved in the transition responsible for the line, F is the oscillation strength of the line, and C is a constant dependent upon the dispersing and detecting systems. Assuming that the atoms are in thermal equilibrium at temperature T, then the number of atoms in the excited state, of excitation energy E_j is given by

$$N_j = N_0 \frac{P_j}{P_0} \exp(-E_j/KT)$$

where N_0 is the number of atoms in the ground state, P_j and P_0 are statistical weights of the excited and ground states respectively, and K is Boltzmann's constant. For a spectral term having a total quantum number J_1, P is equal to $2J_1+1$. From the above equations, it can be seen that the emitted intensity depends on T and E_j. Examples of the variation of N_j/N_0 with temperature are given in Table 3.3.

In nearly all cases, the number of atoms in the lowest excited state is very small compared with the number of atoms in the ground state and the ratio only becomes appreciable at high temperatures. The strongest resonance lines of most elements have wavelengths less than 600 nm and as temperatures in the flames used are normally less than 3000 °K, the value of N_j will be negligible compared with N_0.

In absorption, consider a parallel beam of radiation of intensity I_0, frequency v incident on an atomic vapour of thickness l cm, then if I_v is the intensity of the transmitted radiation, and K_v is the absorption coefficient of the vapour at frequency v, then

$$I_v = I_0 \exp(-E_v l)$$

From classical dispersion theory

$$\int K_v dv = \frac{\pi e^2}{mc} N_v f$$

where m and e are the electronic mass and charge respectively, c is the velocity of light, N_v the number of atoms/cm^3 capable of absorbing radiation of frequency v, and f the oscillator strength (the average number of electrons per atom capable of being excited by the incident radiation). Thus, for a transition initiated from the ground state, where N_v is for all practical purposes equal to N_0 (the total number of atoms/cm^3), the integrated absorption is proportional to the concentration of free atoms in the absorbing medium. The theoretical sensitivity is therefore increased because all the atoms present will take part in the absorption whereas in the emission techniques only a very small number are excited and are used for detection.

In practice, the flame, into which the solution is nebulized, is treated as if it were the cell of absorbing solution in conventional spectrophotometry. The absorbance in the flame of light of a resonant wave-

Table 3.3 Values of N_j/N_0 for various resonance lines

Resonance line	*Transition*	P_j/P_0	N_j/N_0			
			T = 2000 K	*T = 3000 K*	*T = 4000 K*	*T = 5000 K*
Cs 852.1 nm	$^2S_{1/2}-^2P_{3/2}$	2	4.4×10^{-4}	7.24×10^{-3}	2.98×10^{-2}	6.82×10^{-2}
K 766.5 nm	$^2S_{1/2}-^2P_{3/2}$	2	2.57×10^{-4}	4.67×10^{-3}	1.65×10^{-2}	3.66×10^{-2}
Na 589.0 nm	$^2S_{1/2}-^2P_{3/2}$	2	9.86×10^{-6}	5.88×10^{-4}	4.44×10^{-3}	1.51×10^{-2}
Ca 422.7 nm	$^1S_0-^1P_1$	3	1.21×10^{-7}	3.69×10^{-5}	6.03×10^{-4}	3.33×10^{-3}
Zn 213.8 nm	$^1S_0-^1P_1$	3	7.29×10^{-15}	5.58×10^{-10}	1.48×10^{-7}	4.32×10^{-6}

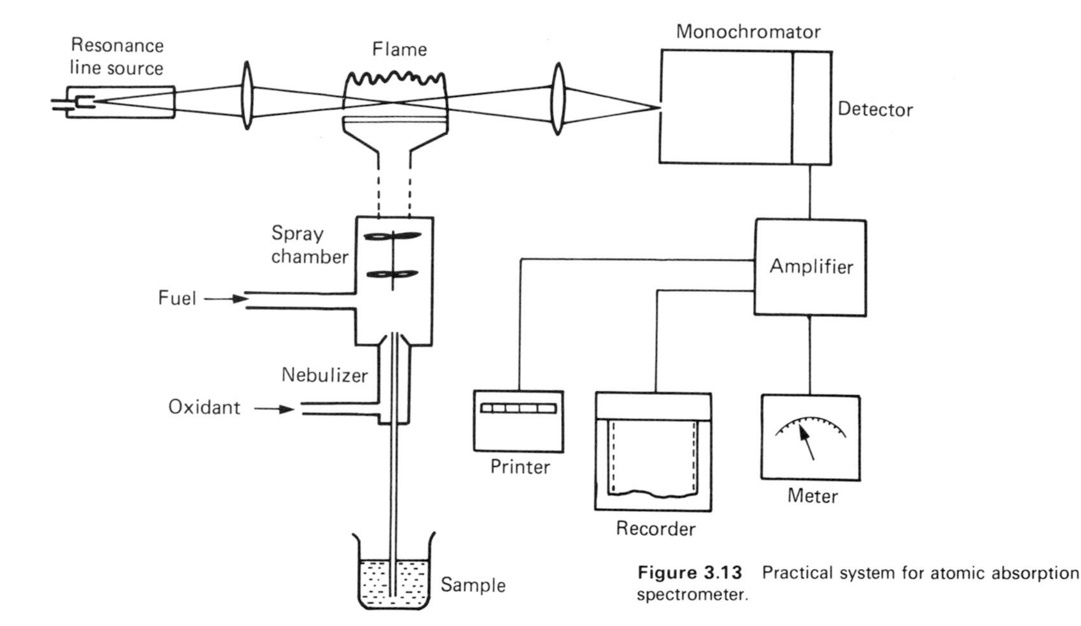

Figure 3.13 Practical system for atomic absorption spectrometer.

length of a particular element, is a direct measure of the concentration of atoms of that element in solution being nebulized into the flame. A practical system for an atomic absorption spectrometer is shown in Figure 3.13.

When only small volumes of sample are available the flame may be replaced by a graphite tube or rod furnace. Small volumes (10 μl) are placed on the graphite and the latter is heated resistively in stages to about 3000 °C and the absorption of a resonant wavelength measured as a pulse. The sensitivity of this technique is such that very low concentrations of some elements may be determined (~0.001 ppm). The limit of detection using a flame varies from element to element from less than 1 ppm up to about 50 ppm. The technique has found wide use in analysis of solutions in virtually every industry – from 'pure' water analysis to the analysis of plating solutions, from soil extracts to effluent from a steel works.

There are many manufacturers of atomic absorption spectrophotometers and the modern instruments are very highly automated. The resonant line source is usually a high intensity hollow cathode lamp and up to ten of these may be contained in a turret so that each is used in turn. The flames are usually air–propane, air–acetylene or nitrous oxide–acetylene – the hotter flames being necessary to atomize the more refractory elements. The output from the monochromator and detector is usually handled by a microprocessor, so that once the instrument has been calibrated, results are automatically printed out as concentrations. Another instrument based on atomic absorption is the mercury vapour detector. A mercury vapour lamp is the resonant source and the detector is tuned to the mercury line at 253.6 nm. Air to be sampled is passed through a tube located between source and detector and the absorption is a measure of the mercury vapour in the air. There are many instruments manufactured for this purpose and all are very sensitive with limits of detection of around 0.1 ppm by volume.

3.2.3 Atomic fluorescence spectroscopy

This is a technique closely allied to atomic absorption. To initiate atomic fluorescence, neutral atoms in a flame cell are excited as in atomic absorption, i.e. by absorption of a characteristic radiation. Fluorescence occurs when these atoms are de-activated by the emission of radiation at the same or a different wavelength. The fluorescent wavelength is characteristic of the atoms in question and its intensity is proportional to the atomic concentration. In practice, initiation is achieved with a high intensity source and the fluorescent signal emitted by the atomic vapour is examined at right angles by passing it into a radiation detection system. Very briefly the basic equation relating the intensity of a fluorescent signal to atomic concentration is

$$F = 2.303\phi I_0 e_A lcp$$

where F is the intensity of fluorescent radiation, ϕ the quantum efficiency (which factor has to be used to account for energy losses by processes other than a fluorescence), I_0 is the intensity of the excitation radiation, e_A the atomic absorptivity at the wavelength of irradiation, l the flame path length, c the concentration of the neutral atom absorbing species, and p a proportionality factor relating to the fraction of the total fluorescence observed by the detector. Thus, $F = K\phi I_0 c$ for a particular set of instrumental conditions, and c is proportional to F and F will increase if the intensity of the irradiating source is increased.

There are four types of atomic fluorescence.

Resonance fluorescence This is the most intense type of fluorescence and most widely used in practice. It occurs when the fluorescent and excitation wavelengths are the same, that is the atom is excited from the ground state to the first excited state and then emits fluorescent energy on de-activation to the ground state.

Direct line fluorescence Here, the valence electron is excited to an energy level above the first excited state. It is then de-activated to a lower energy level (not the ground state) and fluorescent energy is emitted. The wavelength of fluorescence is longer than the excitation wavelength, e.g. the initiation of thallium fluorescence at 535 nm by a thallium emission at 377.6 nm.

Stepwise fluorescence This entails excitation of the atom to a high energy level. The atom is then de-activated to the first excited state. There, it emits resonance radiation on returning to the ground state, e.g. the emission of sodium fluorescence at 589 nm, following excitation at 330.3 nm.

Sensitized fluorescence This occurs when the atom in question is excited by collision with an excited atom of another species and normal resonance fluorescence follows. Thallium will fluoresce at 377.6 nm and 535 nm following a collision of neutral thallium atoms with mercury atoms excited at 253.7 nm.

An instrument used to determine trace amounts of elements in solution by atomic fluorescence very simply consists of (a) an excitation source which can be a high intensity hollow cathode lamp, a microwave-excited electrodeless discharge tube, some spectral discharge lamps or more recently, a tunable dye laser, (b) a flame cell or a graphite rod as in atomic absorption, and (c) a detection system to measure the fluorescence at right angles to the line between source and flame. The detection system is usually a simple monochromator or narrow band filter followed by a photomultiplier tube, amplifier and recording device. Limits of detection are achieved which are much lower than those obtained by atomic absorption because it is easier to measure small signals against a zero background than to measure small differences in large signals as is done in atomic absorption. Detection limits as low as 0.0001 ppm are quoted in the literature.

3.3 X-ray spectroscopy

3.3.1 X-ray fluorescence spectroscopy

Many books have been written about this technique and only a brief outline is given here.

The technique is analogous to atomic emission spectroscopy in that characteristic x-radiation arises from energy transferences involved in the rearrangement of orbital electrons of the target element following ejection of one or more electrons in the excitation process. The electronic transitions involved are between orbits nearer to the nucleus (see Figure 3.14).

Thus if an atom is excited by an electron beam or a beam of x-rays, electronic transitions take place and characteristic x-radiation is emitted for that atom. If, after collimation, these x-rays fall on to a crystal lattice – which is a regular periodic arrangement of atoms – a diffracted beam will only result in certain directions, depending on the wavelength of the x-rays λ, the angle of incidence θ, and atomic spacing within the crystal d. Bragg's law for the diffraction of x-rays states that $n\lambda = 2d \sin \theta$. Thus the K_α, K_β, L_α, L_β, M_α etc. x-radiations will be diffracted at different angles. These fluorescent radiations are then collimated and detected by a variety of detectors. The intensity of these radiations is a measure of the concentration of that particular atom. Thus if a sample containing many elements is subjected to x-radiation, fluorescent radiation for all the elements present will be spread out into a

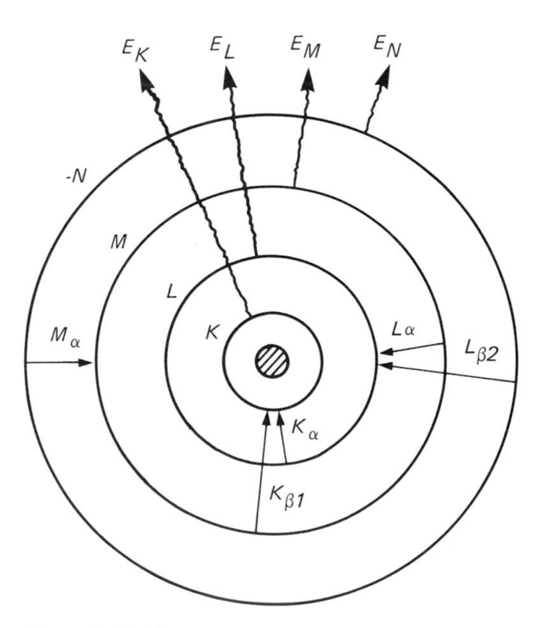

Figure 3.14 Transitions giving x-radiation. $(E)_{K_\alpha} = E_K - E_L$; $(E)_{K_{\beta 1}} = E_K - E_M$; $(E)_{L_\alpha} = E_L - E_M$; $(E)_{L_{\beta 2}} = E_L - E_N$; $(E)_{M_\alpha} = E_M - E_N$.

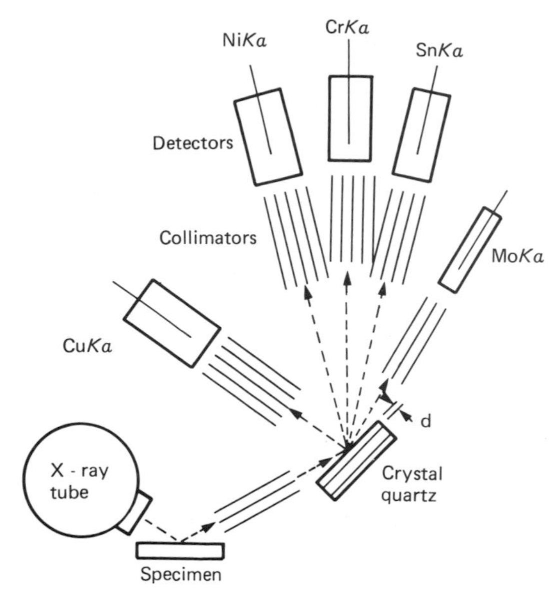

Figure 3.15 Multi-channel spectrometer having 5 collimator – detector channels arranged to receive 5 different analyte lines, each from a different crystallographic plane (*hkil*) from the same quartz crystal.

spectrum, depending on the elements present and the crystal being used (see Figure 3.15).

All modern x-ray fluorescence spectrometers use this layout. The source of x-rays is usually an x-ray tube, the anode of which is chromium, tungsten or rhodium. All types of sample can be analysed, ranging from metals, through powders to solutions. The collimator systems are based on series of parallel plates. As their purpose is to limit the divergence of the x-ray beam and provide acceptable angular resolution, the distance between the plates must be such that the divergence embraces the width of the diffraction profile of the crystal. In general, this entails a spacing between plates of 200–500 μm.

Most modern instruments can accommodate six analysing crystals, any one of which can be automatically placed in the fluorescent x-ray beam. A list of the types of crystal used is shown in Table 3.4. The detectors are either gas flow proportional counters or scintillation counters. (See Volume 3, Nuclear Instrumentation Technology.) The instruments are microprocessor-controlled and this varies the output of the x-ray source, chooses the correct crystals and controls the samples going into the instrument. A small computer analyses the output from the detectors and (having calibrated the instrument for a particular sample type) calculates the concentration of the elements being analysed – allowing for matrix and inter-element effects. Instruments of this type, made by Philips, Siemens, and ARL are widely used in the metallurgical industry as the technique – although capable of low limits of detection – is very accurate for major constituents in a sample, such as copper in brass. Analysis of atmospheric particulate pollution is carried out using x-ray fluorescence. The sample is filtered on to a paper and the paper and deposit analysed.

A portable instrument, which uses a radioactive isotope as a source is used to monitor particular elements (depending on settings) in an ore sample before processing. This instrument is now marketed by Nuclear Enterprises. See Part 3 (Measurements employing nuclear techniques).

Electron probe microanalysis is a technique which is based on the same principle as x-ray fluorescence, electrons being the exciting source, but by using electronic lenses the electron beam can be focused onto a very small area of a sample and so analysis of areas as

Table 3.4 Analysing crystals

Crystal	*Reflection plane*	*2d spacing* (Å) (1 Å = 0.1 nm)	*Lowest atomic number detectable*	
			K series	L series
Topaz	(303)	2.712	V (23)	Ce (58)
Lithium fluoride	(220)	2.848	V (23)	Ce (58)
Lithium fluoride	(200)	4.028	K (19)	In (49)
Sodium chloride	(200)	5.639	S (16)	Ru (44)
Quartz	$(10\bar{1}1)$	6.686	P (15)	Zr (40)
Quartz	$(10\bar{1}0)$	8.50	Si (14)	Rb (37)
Penta erythritol	(002)	8.742	Al (13)	Rb (37)
Ethylenediamine tartrate	(020)	8.808	Al (13)	Br (35)
Ammonium dihydrogen phosphate	(110)	10.65	Mg (12)	As (23)
Gypsum	(020)	15.19	Na (11)	Cu (29)
Mica	(002)	19.8	F (9)	Fe (26)
Potassium hydrogen phthalate	$(10\bar{1}1)$	26.4	O (8)	V (23)
Lead stearate		100	B (5)	Ca (20)

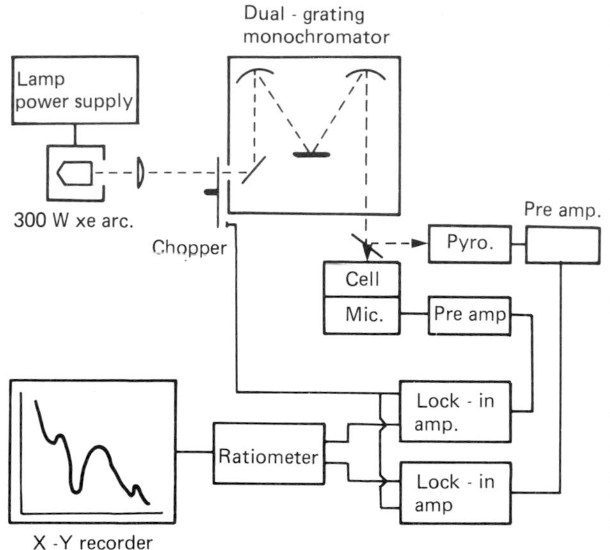

Figure 3.16 Photo-acoustic spectrometer layout.

small as 0.1 μm diameter can be carried out. The technique can be used for looking at grain boundaries in metallurgical specimens and plotting elemental maps in suspected heterogeneous alloys. Again this is a technique which is very specialized.

A further allied technique is photoelectron spectroscopy (PES) or Electron Spectroscopy for Chemical Analysis (ESCA). In Figure 3.14, showing the transitions within an atom to produce x-rays, it is seen that some electrons are ejected from the various shells in the atom. The energy of these electrons is characteristic of that atom and so by producing an energy spectrum of electrons ejected from a sample when the latter is subjected to x-ray or intense UV radiation, the presence of different elements and their concentrations can be determined. It should be pointed out that this technique is essentially a surface technique and will only analyse a few monolayers of sample. Instruments are manufactured by Vacuum Generators.

3.3.2 X-ray diffraction

This is a technique which is invaluable for the identification of crystal structure. In the section on x-ray fluorescence it was seen that crystals diffract x-rays according to Bragg's law:

$$n\lambda = 2d \sin \theta$$

Thus if a small crystal of an unidentified sample is placed in an x-ray beam, the x-rays will be diffracted equally on both sides of the sample to produce an x-ray pattern on a film placed behind the sample. The position of the lines on the film (i.e. their distance from the central beam) is a function of the crystal lattice structure and by reference to standard x-ray diffraction data, the crystals in the sample are identified. Again this is a specialized technique and beyond the scope of this book.

Manufacturers of x-ray fluorescence spectrometers also make x-ray diffraction spectrometers. Typical uses for an instrument are the identification of different types of asbestos, and corrosion deposit studies.

3.4 Photo-acoustic spectroscopy

An instrument marketed by EDT Research makes use of this technique to study both liquid and solid samples. Figures 3.16 and 3.17 give schematic diagrams of the instrument and cell. Radiation from an air-cooled high pressure xenon arc source, fitted with an integral parabolic mirror, is focused onto a variable speed rotating light chopper mounted at the entrance slit of a high radiance monochromator. The monochromator has two gratings to enable optical acoustic spectra to be obtained in the UV, visible and near-infrared. The scanning of the monochromator is completely automatic over the spectral range covered and a range of scan rates can be selected. The exit and entrance slits provide variable band passes of width 2–16 nm in the UV and 8–64 nm in the IR. A reflective beam-splitter passes a fraction of the dispersed radiation to a pyroelectric detector to provide source compensation and a reference signal. Source radiation is then focused onto the specially designed opto-acoustic cell and sample-holder assembly. The sample cell contains a sensitive microphone and pre-amplifier. Special cells are used for different applications. Absorption of the radiation by the molecular species in the sample occurs and is converted to kinetic energy. The sample temperature fluctuates and causes a variation in the pressure of the gas surrounding the sample. This pressure variation is monitored by the microphone. The amplitude of the microphone signal is recorded as a function of the wavelength of the incident radiation to give an absorption spectrum of

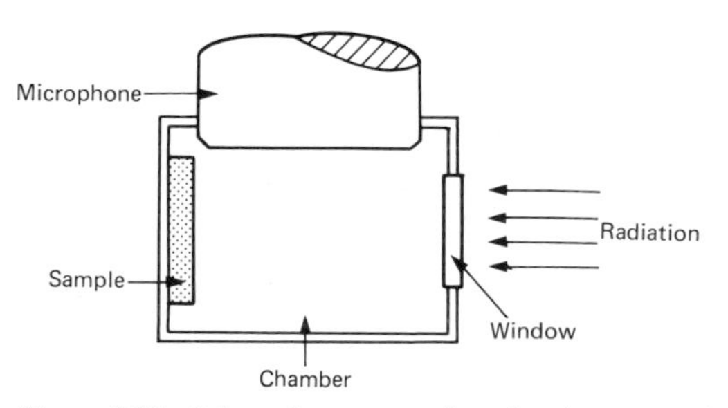

Figure 3.17 Schematic representation of a photo-acoustic cell employed for the examination of solid materials.

the sample. Typical applications include the identification of foodstuffs, blood and bloodstains, paints and inks, papers and fabrics, and pharmaceutical materials.

3.5 Microwave spectroscopy

The portion of the electromagnetic spectrum extending approximately from 1 mm (300 000 MHz) to 30 cm (1000 MHz) is called the microwave region. Spectroscopic applications of microwaves consist almost exclusively of absorption work in gaseous samples. With some exceptions, the various types of spectra are distinguished by their energy origins. As mentioned earlier, in the visible and UV regions the transitions between electronic energy states are directly measurable as characteristics of elements, and vibrational and rotational energies of molecules are observed only as perturbation effects. In the infrared region the vibrational spectra are observed directly as characteristic of functional groups with rotational energies observed as perturbation effects. In the microwave region transitions between rotational energies of molecules are observed directly as characteristic of absorbing molecules as a whole with nuclear effects as first-order perturbations. In the radio frequency (r.f.) region, the nuclear effects are directly observable. (Especially important today is the observation in the microwave region of paramagnetic resonance absorption (PMR) and also nuclear magnetic resonance. Both these techniques will be discussed briefly in a later section.) As in any other type of absorption spectroscopy, the instrument required consists of a source of radiation, a sample cell and detector. Unlike optical spectrometers, the microwave spectrometer is a completely electronic instrument requiring no dispersive components because the source is monochromatic and any frequency can be chosen and measured with very high precision. The most common type of source is the Klystron, a specially designed high-vacuum electron tube. The output is monochromatic under any given set of conditions and different types are available to cover various parts of the microwave spectrum. The sample cell is usually a waveguide and the detector could be a silicon crystal, although bolometers and other heat-type detectors are sometimes used. In addition to the three basic components a complete spectrometer includes provision for modulation of the absorption spectrum, an a.c. amplifier for the detector output, a final indicator consisting of a CRT or strip recorder, a sweep generator to vary synchronously the source frequency, a gas sample handling system and necessary power supplies.

Since the lines in a microwave spectrum are usually completely resolved, it is only necessary to compare these measured frequencies against tables of the frequencies observed for known substances in order to identify molecules. Quantitative analysis is somewhat more complex, but is based on the fact that the integrated intensity and the product of the peak height and half-width of a microwave absorption line can be directly related to the concentration of molecules per unit volume. The technique is used extensively in isotopic analysis.

3.5.1 Electron paramagnetic resonance (EPR)

This is really a special part of microwave spectroscopy because it usually involves the absorption of microwave radiation by paramagnetic substances in a magnetic field. A typical layout of a spectrometer is given in Figure 3.18. The electromagnet has a homo-

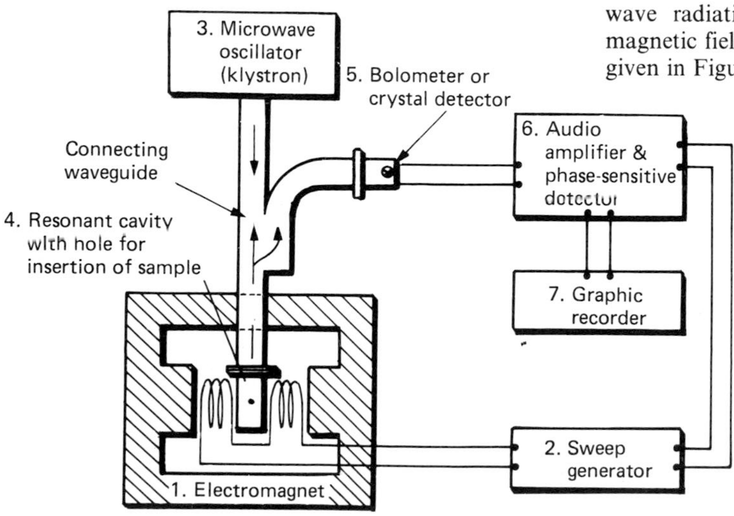

Figure 3.18 Block diagram of electron paramagnetic resonance spectrometer.

geneous gap field H which can be swept continuously from near zero to over 50 microtesla. The sweep generator produces small modulations of the main field H at the centre of the air-gap. The sample cavity resonates at the Klystron frequency.

The electron, like the proton, is a charged particle; it spins and therefore has a magnetic field. It spins much faster than a proton and so has a much stronger magnetic field. Because of this and being lighter than a proton, it precesses much more rapidly in a magnetic field. Thus when microwaves travel down a waveguide and produce a rotating magnetic field at any fixed point, it can serve to flip over electron magnets in matter, just as a rotating field in a coil flips protons. If a sample is placed on the sidewall of the waveguide and the microwave radiation, applied to the external magnetic field, causes the electrons to precess, then when the precession rate reaches a resonance value and the electrons flip, they extract energy from the microwaves and the reading on the recorder dips accordingly.

If the electron has not only a magnetic moment along its own spin axis but also one associated with its circulation in an atomic orbit, the electron will possess a total magnetic moment equal to the vector sum of the magnetic moments. The ratio of the total magnetic moment to the spin value is a constant for a given atom in a given environment and is called the gyromagnetic ratio or spectroscopic splitting factor for that particular electron. The fact that these ratios differ for various atoms and environments and that local magnetic fields depend on the structure of the matter permit spectral separation and EPR spectroscopy. Not all atoms and molecules are susceptible to this technique; in substances in which electrons are paired, magnetism is neutralized. But for unpaired electrons, electronic resonance occurs. This effect is observed in unfilled conduction bands, transition element ions, free radicals, impurities in semiconductors and, as might be expected, applications in the biological field are fruitful. The most common use is the paramagnetic oxygen analyser.

3.5.2 Nuclear magnetic resonance spectroscopy

When atomic nuclei – the hydrogen proton is the simplest – are placed in a constant magnetic field of high intensity and subjected to a radio frequency alternating field, a transfer of energy takes place between the high frequency field and the nucleus to produce a phenomenon known as 'nuclear magnetic resonance'.

If a system of nuclei in a magnetic field is exposed to radiation of frequency ν such that the energy of a quantum of radiation $h\nu$ is exactly equal to the energy difference between two adjacent nuclear energy levels, then energy transitions may occur in which the nuclei may 'flip' back and forth from one orientation to another. A quantum of energy is equally likely to tip a nucleus in either direction, so that there is a net absorption of energy from the radiation only when the number of nuclei in one energy level exceeds the number in another. Under these conditions a nuclear magnetic resonance spectrum is observed. Applications of this technique include such problems as locating hydrogen atoms in solids, measuring bond lengths, crystal imperfections, and determination of crystalline and amorphous fractions in polymers.

3.6 Neutron activation

Gamma ray spectroscopy is the technique by which the intensities of various gamma energies emanating from a radioactive source are measured. See Volume 3 (Measurements employing nuclear techniques). It can be used for qualitative identification of the components of radionuclide mixtures and for quantitative determination of their relative abundance. Such a situation arises in neutron activation analysis. This is a technique of chemical analysis for extremely minute traces down to ppb (parts per 10^9) of chemical elements in a sample. It employs a beam of neutrons for activation of isotopes which can then be identified, with counters, by the radioactive characteristics of the new nuclear species. This technique has been applied for the trace analysis of many elements in a variety of materials, from coal ash to catalysts, halides in phosphors, and trace impurities in many metals.

3.7 Mass spectrometers

The mass spectrometer is capable of carrying out quick and accurate analysis of a wide variety of solids, liquids and gases and has a wide range of application in process monitoring and laboratory research. When combined with the gas chromatograph it provides an extremely powerful tool for identifying and quantifying substances which may be present in extremely small quantities.

While the optical spectrometer resolves a beam of light into components according to their wavelengths, a mass spectrometer resolves a beam of positive ions into components according to their mass/charge ratio, or if all carry single elementary charges, according to their masses. As with the optical spectrometer the mass spectrometer may be used to identify substances and to measure the quantity present.

The original mass spectrometer was devised by F. W. Aston about 1919 to measure the mass of individual positive ions. The accuracy of the instrument enabled the different masses of what appeared to be chemically identical atoms to be

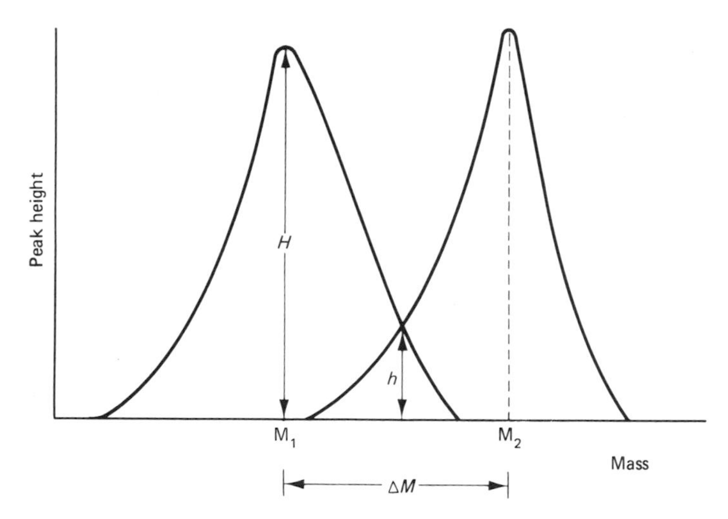

Figure 3.19 Peak separation for a mass spectrometer.

measured, resulting in the discovery of isotopes. Considerable development has taken place over the years, resulting in very versatile instruments having very high resolving power and sensitivity.

The resolving power of a mass spectrometer is a measure of its ability to separate ions having a very small difference in mass. If two ions of masses M_1 and M_2 differing in mass by ΔM give adjacent peaks in their spectrum as shown in Figure 3.19 and the height of peak is H above the baseline, then on the 10 per cent valley definition the peaks are said to be resolved if the height of the valley h is less than or equal to 10 per cent of the peak H, i.e.

$$(h/H) \leqslant 10 \text{ per cent}$$

The resolution is then $M_1/\Delta M$, e.g. if the peaks representing two masses 100.000 and 100.005 are separated by a 10 per cent valley, the resolution of the instrument is 100.000/0.005, i.e. 20 000. Instruments with a resolution of 150 000 are readily available. The sensitivity on the other hand is a measure of the smallest detectable quantity of the substance being identified. An example of the extreme sensitivity of modern instruments is that at a resolution of 1000, 3 ng/s of a compound, relative molecular mass 300, will give a spectrum with a signal-to-noise ratio of 10:1 for a peak having an intensity of 5 per cent of the base peak when a mass range of 10:1 is scanned in 3 s.

The mass spectrometer has a very wide range of use in process monitoring and laboratory research. It is used in refineries for trace element survey, analysis of lubricating oils and identifying and quantifying the substances in mixtures of organic compounds. Its use in detecting and measuring the concentration of pollutants in air, water and solids is rapidly increasing, also its use in biochemical analysis in medicine and other fields, particularly the analysis of drugs in biological extracts.

By means of a double-beam instrument an unknown sample may be compared with a standard so that the unknown components are readily identified and the concentration measured. By suitable modifications an instrument can be made to provide an energy analysis of electrons released from the surface of a sample by x-radiation, or ultraviolet light.

3.7.1 Principle of the classical instrument

There are many different types of mass spectrometers but the ones described here are the most commonly used.

In all types the pressure is reduced to about 10^{-5} N/m^2 in order to reduce collisions between particles in the system. The spectrometer consists of an inlet system by which the sample is introduced into the region in which ions of the sample are produced. The separation of ions according to their mass-to-charge ratio may be achieved by magnetic or electric fields or by a combination of both. The differences between the various types of mass spectrometer lie in the manner in which the separation is achieved. In the instrument illustrated in Figure 3.20 the ions are accelerated by an electrical potential through accelerating and defining slits into the electrostatic analyser, where ions having energies within a restricted band are brought to a focus at the monitor slit which intercepts a portion of the ion beam. They then enter the electromagnetic analyser which gives direction and mass focusing. This double focusing results in ions of all masses being focused simultaneously along a given plane. The ions can be recorded photographically on a plate over a period of time to give a very high sensitivity and reduction of the effects of ion-beam fluctuation.

Alternatively, the accelerating or deflecting field may be arranged so that ions of a given mass are focused on a detector which may consist of a plate or, if initial amplification of the charge is required, onto an electron multiplier or scintillation detector. By arranging the deflecting field to change in a predetermined manner, the instrument may be arranged to scan a range of masses and so record the abundance of ions of each particular mass. Such a record is known as a 'mass spectrum' and mathematical analysis of this mass spectrum enables the composition of the sample to be determined. Mass spectra obtained under constant conditions of ionization depend upon the structure of the molecules from which the ions originate. Each substance has its own characteristic mass spectrum, and the mass spectrum of a mixture may therefore be analysed in terms of the spectra of the pure components, and the percentage of the different substances in the mixture calculated.

Analysis of the mass spectrum of a mixture may involve the solution of a large number of simultaneous equations, which can be accomplished using a microprocessor or a small computer.

3.7.2 Inlet systems

The mode of introduction of the sample into the ion source is dependent upon the nature of the sample and in particular its volatility.

The simplest system designed to introduce reference compounds into the ion source includes a 35 cm^3 reservoir into which the compound is injected through a septum. Flow into the ion source is through a molecular leak and a shut-off valve is provided. Facilities for pumping out the system and obtaining temperatures up to 100 °C are provided.

Relatively volatile gases and liquids may be introduced by a probe attached to a small reservoir into which the sample is injected and from which it flows to the ion source at a controlled rate. The temperature of the system may be controlled between ambient and 150 °C.

For less volatile substances an all-glass heated system may be used. Glass is used for the system so that catalytic decomposition of the sample is reduced to a minimum. The system can be operated at temperatures up to 350 °C and incorporates its own controlled heating and temperature-monitoring facilities. It includes both large and small reservoirs to enable a wide range of quantities of liquid or solid samples to be introduced.

To introduce less volatile and solid samples into the ion chamber a probe may be used. The sample is loaded onto the tip of the probe, which is inserted into the ion source through a two-stage vacuum lock.

The probe may be heated or cooled independently of the ion chamber as required from −50 to +350 °C. The temperature is measured by a platinum resistance thermometer, forming part of the temperature control system, which enables the temperature to be set from the instrument control panel.

Effluents from a gas chromatograph column usually flow at about 50 cm^3/min and consist mainly of carrier gas. In order to reduce the flow, the gas is passed through a molecular separator designed to remove as much as possible of the carrier gas but permitting the significant components to pass into the mass spectrometer.

3.7.3 Ion sources

In the system shown the ions are produced by a spark passed between electrodes formed from the sample by applying a controlled pulsed r.f. voltage. Positive ions representative of the sample are produced in the discharge and are accelerated through a simple ion gun. This beam is defined by resolving slits before it passes into the analyser section.

Other methods may be employed in order to produce ions of the sample which are impelled towards the exit slit by a small positive potential in the ion chamber. These methods involve increasing the energy of the sample by some form of radiation. Organic compounds require photons of energy up to 13 eV to produce ionization so that a high energy beam of short wavelength radiation is sufficient. Where energies greater than 11 eV are required window materials become a problem so that the photon source has to emit radiation directly into the ion source. A helium discharge at 21.21 eV provides a convenient source of photons capable of ionizing all organic compounds.

Electrons emitted by a heated filament and accelerated by about 70 eV and directed across the ion chamber may also be used to ionize many substances. While 70 eV produces the maximum ion yield, any voltage down to the ionization voltage of the compound studied may be used.

The electric field production near a sharp point or edge at a high potential will have a high potential gradient and may be used to produce ions. Ions can also be formed by the collision of an ion and a molecule. This method can produce stable but unusual ions, e.g.

$$CH_4^+ + CH_4 \rightarrow CH_5^+ + CH_3$$

and is most efficient at pressures of about 10^{-2} N/m^2.

It is most important to realize that the process of producing ions from molecules will in many cases split up the original molecule into a whole range of ions of simpler structure and the peak of maximum height in the spectrum does not necessarily represent the ion of

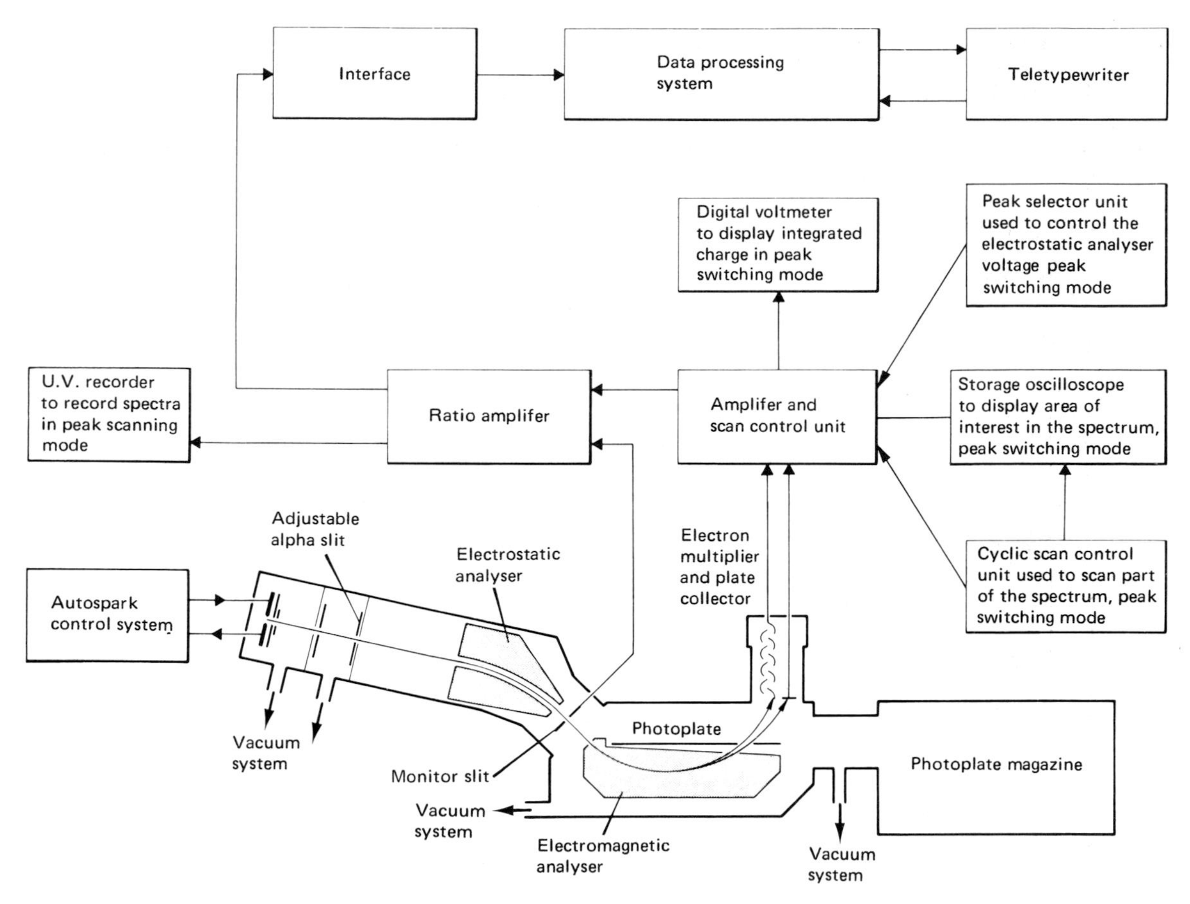

Figure 3.20 Schematic diagram of the complete system of a spark source mass spectrometer (courtesy Kratos Ltd.).

the original molecule. For example the mass spectrum of *m*-xylene $C_6H_4(CH_3)_2$ may contain 22 peaks of different m/e values, and the peak of maximum height represents a m/e ratio of 91 while the ions having the next highest peak have a m/e ratio of 106.

3.7.4 Separation of the ions

The mass spectrometer shown in Figure 3.20 employs the Mattauch-Herzog geometry but other forms of geometry achieve a similar result.

The positive ions representative of the sample produced in the ion source are accelerated by a controlled electrostatic field in a simple gun, the spread of the ions being controlled by the resolving slits. If an ion of mass m and charge e can be regarded as starting from rest, then its velocity v after falling through a potential V volts will be represented by the equation

$$\tfrac{1}{2}mv^2 = eV$$

The ion beam then passes through the electrostatic analyser where it passes between two smooth curved plates which are at different potentials, such that an electrostatic field B exists between them which is at right angles to the path of the ions. The centrifugal force on the ions will therefore be given by

$$mv^2/r = eB$$

Combining the equations we see that the radius of curvature r of the path will be given by

$$r = mv^2/eB = 2eV/eB = 2V/B$$

Thus, the curvature of the path of all ions will be dependent upon the accelerating and deflecting fields only and independent of the mass/charge ratio. Therefore, if the field B is kept constant the electrostatic analyser focuses the ions at the monitor slit in accordance with their translational energies. The monitor slit can be arranged to intercept a given portion of the beam. The energy-focused ion beam is then passed through the electromagnetic analyser where a magnetic field at right angles to the electrostatic field is applied (i.e. at right angles to the plane of the diagram). Moving electric charges constitute an electric current so that if each carries a charge e, and moves with a velocity v, at right angles to a uniform magnetic field H, each particle will be subject to a force F where $F=Hev$ in a direction given by Fleming's left-hand rule, i.e. in a direction mutually at right angles to the magnetic field and the direction of the stream. Thus the ions will move in a curved path radius r such that

$$mv^2/r=Hev$$

or

$$r=mv^2/Hev=mv/He$$

but

$$mv^2=2eV \quad \text{or} \quad v=\sqrt{(2eV/m)}$$

$$\therefore \; r=(m/eH)\sqrt{(2eV/m)}$$

or

$$r^2=(m^2/e^2H^2)(2eV/m)$$
$$=(2V/H^2)(e/m)$$

or

$$m/e=(H^2r^2)/2V$$

At constant values of the electrostatic and electromagnetic fields all ions of the same m/e ratio will have the same radius of curvature. Thus, after separation in the electromagnetic analyser, ions having a single charge will be brought to a focus along definite lines on the photographic plate according to their mass, starting with the lowest mass on the left-hand edge of the plate and increasing to the highest mass on the right.

The ions will therefore give rise to narrow bands on the photographic plate and the density of these bands will be a measure of the number of ions falling on the band. The sensitivity range of the plate is limited, and it is necessary to make several exposures for increasing periods of time to record ions which have a large ratio of abundance. By using long exposure, ions which are present in very low abundances may be accurately measured. The intensity of the photographic lines after development of the plate may be compared with a microphotometer similar to that used with optical spectrometers.

As all ions are recorded simultaneously, ion beam fluctuations affect all lines equally and the photographic plate also integrates the ions over the whole of the exposure.

The instantaneous monitor current may be measured and used to control the sparking at the electrodes at optimum by adjusting the gap between the electrodes.

The integrated monitor current is a guide to the exposure, and the range of masses falling on the photographic plate may be controlled by adjustment of the value of the electrostatic and magnetic fields.

The plate collector and the electron multiplier detection systems enable quantitative analysis to be carried out with greater speed and precision than with the photographic plate detector. For high sensitivity the ions may be caused to fall on the first dynode of the electron multiplier and the final current further amplified, and recorded on the ultraviolet sensitive strip recorder. The logarithmic ratio of the monitor and collector signals is used in recording spectra in order to minimize the errors due to variations in the ion beam.

In the peak switching mode the operator can select the peaks of interest and display them on an oscilloscope and examine them with greater precision. Increasing the resolving power of the instrument will enable what may initially appear to be a single peak to be split up into its components representing ions differing in mass by a small amount.

Provision is made for changing the amplification in logarithmic steps so that a wide range of abundances may be measured. Where a rapid qualitative and semi-quantitative analysis is required for a very wide range of masses, consecutive masses are swept across the multiplier collector by allowing the magnet current to decay from a preset value at a preset rate while the accelerating voltage is kept constant. Values of ion current from the individual ion species received at the detector are amplified and instantaneously compared with a fraction of the total ion current at the monitor by means of two logarithmic amplifiers which feed into a summing amplifier. This gives a signal proportional to the relative ion concentrations, which can be recorded on the ultraviolet sensitive strip recorder and has the form shown in Figure 3.21.

Where large amounts of data are generated the output from the ratio detector of the electrical detection system can be fed through a suitable interface into a data acquisition and processing system. If necessary this system can be programmed to print out details of the elements present in the sample and an indication of their concentration.

3.7.5 Other methods of separation of ions

3.7.5.1 Time-of-flight mass spectrometer

This type of instrument is shown schematically in Figure 3.22. It has a relatively low resolution but a very fast response time.

In this instrument, the ions are accelerated through a potential V, thus acquiring a velocity v given by:

$$\tfrac{1}{2}mv^2=eV \quad \text{or} \quad v=[2V(e/m)]^{1/2}$$

If the ions then pass through a field-free (drift) region of length d, to the detector the time of transit t will be d/v. That is,

$$t=d/[2V(e/m)]^{1/2}=[(e/m)2d^2V]^{-1/2}$$

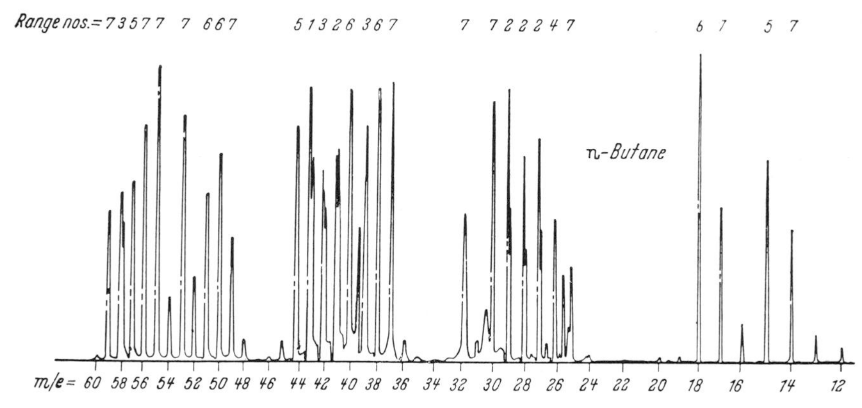

Figure 3.21 Ultraviolet-sensitive strip recording.

Thus, the ions will arrive at the detector after times proportional to $(m/e)^{1/2}$. The time intervals between the arrival of ions of different mass at the detector is usually very short and the mass spectrum is most conveniently displayed on a cathode ray tube. The time-of-flight mass spectrometer occupies a unique place in mass spectrometry as it provides a simple rapid measurement of the abundance of various isotopes or elements comprising a sample. In practice, 10 000 to 100 000 spectra can be scanned per second. With the aid of suitable electronic circuitry it is possible to monitor reaction rates and to investigate reaction profiles of only 100 μs duration. Longer length drift tubes have also contributed to improved mass resolution. It is also possible to scan from 0 to 900 atomic mass units in 1.5 seconds and also, to prevent multiplier saturation when very large ion peaks are present near smaller peaks, appropriate 'gating' peaking can be applied to the multiplier. Thus, it is possible to suppress mass 40 without interfering with the recording of mass 39 or 41. This has extended the practical range of sensitivity in identifying gas chromatograph effluent by orders of magnitude.

3.7.5.2 Quadrupole mass spectrometer

This type of instrument is particularly suited to vacuum system monitoring and to a wide range of gas analysis. Although it has a relatively modest resolving power (about 16 000 maximum) it has the advantages of compactness, robustness and relatively low cost.

Ions, produced by bombarding the sample with electrons from a filament assembly, are extracted electrostatically from the ionizer and focused by electrostatic lenses into the quadrupole mass filtering system. The latter consists of two pairs of metal rods, precisely aligned and housed in a chamber at a pressure of 2.6×10^{-4} N/m^2. One pair is connected to a source of d.c. voltage, while the other is supplied by a radio frequency voltage. Combination of the d.c. and r.f. voltages creates a hyperbolic potential distribution. The applied voltages increase uniformly from zero to a given maximum and then drop to zero again – a voltage sweep which is then repeated. Most ions entering the quadrupole field will undergo an oscillating trajectory of increasing amplitude so that they will eventually be collected on one of the electrodes. However, at any given time, ions of one

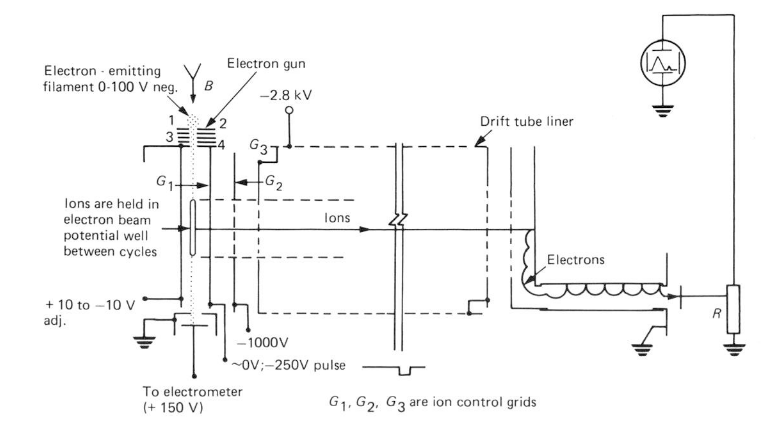

Figure 3.22 Time-of-flight spectrometer.

specific mass/charge ratio are deflected as much to one electrode as to another and are passed by the filter.

As the voltages are swept from zero to their maximum values, the entire mass range is scanned. After passing through the mass filter, the ions impinge on an electron multiplier and a signal proportional to the collected ion current can be displayed on an oscilloscope or recorder. As the voltages increase, the position of the mass peaks is linearly related to mass, making the spectrum easy to interpret. The instrument covers mass ranges up to about 400 amu. Modern instruments are able to detect partial pressures in the 10^{-13} torr range. They are equipped with variable mass scanning sweeps so that rapidly changing concentrations of gases can be monitored on a continuing basis. There are many other types of ion separators, for details on these, the reader should consult textbooks devoted to mass spectroscopy. Among these types are multiple magnet systems, the cycloidal mass spectrometer, cyclotron resonance types and r.f. mass filters.

3.8 Bibliography

Bertin, Eugene P., *Principles and Practice of X-ray Spectrographic Analysis*, Plenum Press, (1970)

Ebdon, L., *An Introduction to Atomic Absorption Spectroscopy – A Self Teaching Approach*, Heyden, (1982)

Jenkins, Ron, Gould, R. W. and Gedcke, Dale, *Quantitative X-ray Spectrometry*, Marcel Dekker, (1981)

Price, W. J., *Spectrochemical Analysis by Atomic Absorption*, Heyden, (1979)

Royal Society of Chemistry, Annual Reports on Analytical Atomic Spectroscopy

Slavin, W., *Atomic Absorption Spectroscopy* (2nd edn), Wiley, (1978)

Tertian, R. and Claisse, F., *Principles of Quantitative X-ray Fluorescence Analysis*, Heyden, (1982)

Welvy, E. L. (ed.), *Modern Fluorescence Spectroscopy*, Plenum Press, (1981)

White, Frederick A., *Mass Spectrometry in Science and Technology*, Wiley, (1968)

4 Chemical analysis – electrochemical techniques

W. G. CUMMINGS and K. TORRANCE

4.1 Acids and alkalis

In order to appreciate electrochemical techniques of chemical analysis it is necessary to have an understanding of how substances dissociate to form ions.

All acids dissociate when added to water to produce hydrogen ions in the solution, e.g. nitric acid:

$$HNO_3 \rightleftharpoons H^+ + NO_3^-$$

The extent to which dissociation takes place varies from acid to acid, and increases with increasing dilution until, in very dilute solutions, almost all the acid is dissociated.

According to the ionic theory, the characteristic properties of acids are attributed to the hydrogen ions (H^+) which they produce in solution. Strong acids (nitric, sulphuric, hydrochloric) are those that produce a large concentration of hydrogen ions when added to water. As a result the solutions are excellent conductors of electricity. Weak acids like carbonic acid (H_2CO_3) and acetic acid (CH_3COOH) when dissolved in water produce small concentrations of hydrogen ions and their solutions are poor conductors of electricity.

The strength of a weak acid is indicated by its dissociation constant K which is defined as

$$K = \frac{[A^-][H^+]}{[HA]}$$

where $[A^-]$ is the molar concentration of the acidic ions, $[H^+]$ is the concentration of hydrogen ions, and $[HA]$ is the concentration of undissociated acid.

The dissociation constant K varies with temperature but, at a given temperature, if a little more acid is added to the solution, a portion of it dissociates immediately to restore the relative amount of ions and undissociated acid to the original value.

Similarly the typical properties of alkalis in solution are attributed to hydroxyl ions (OH^-). Strong alkalis such as sodium hydroxide (NaOH) produce large concentrations of hydroxyl ions when added to water but weak alkalis such as ammonium hydroxide (NH_4OH) are only slightly ionized in water and produce much smaller concentrations of hydroxyl ions.

As with weak acids, the strength of a weak base is indicated by its dissociation constant

$$K = \frac{[B^+][OH^-]}{[BOH]}$$

where $[B^+]$ is the concentration of alkaline ions, $[OH^-]$ is the concentration of hydroxyl ions, and $[BOH]$ is the concentration of undissociated alkali.

Strong electrolytes have no dissociation constant; the expression for strong acids $[A^-][H^+]/[HA]$, and the corresponding expression for alkalis, vary considerably with change in concentration. With strong acids and alkalis the apparent degree of ionization can be taken as a measure of the strength of the acid or base.

So far it has been assumed that the effective concentrations or active masses could be expressed by the stoichiometric concentrations but, according to modern thermodynamics, this is not strictly true. For a binary electrolyte $AB \rightleftharpoons A^+ + B^-$ the correct equilibrium equation is:

$$K_a = \frac{a_{A^+} \times a_{B^-}}{a_{AB}}$$

where a_{A^+}, a_{B^-} and a_{AB} represent the activities of A^+, B^- and AB and K_a is the thermodynamic dissociation constant. The thermodynamic quantity 'activity' is related to concentration by a factor called the activity coefficient, i.e. activity = concentration × activity coefficient.

Using this concept, the thermodynamic activity coefficient is

$$K_a = \frac{[A^+][B^-]}{[AB]} \times \frac{f_{A^+} \times f_{B^-}}{f_{AB}}$$

where f refers to the activity coefficients and the square brackets to the molar concentrations.

The activity coefficients of un-ionized molecules do not differ much from unity and so for weak electrolytes in which the ionic concentration, and therefore the ionic strength is low, the error introduced by neglecting the difference between the actual values of the activity coefficients of the ions, f_{A^+} and f_{B^-}, and unity is small (less than 5 per cent). Hence for weak electrolytes, the constants obtained by using the simpler equation $K=[A^+][B^-]/[AB]$ are sufficiently precise for the purposes of calculation in quantitative analysis. Strong electrolytes are assumed to be completely dissociated and no correction for activity coefficients needs to be made for dilute solutions.

However, the concept of activity is important in potentiometric techniques of analysis (described later). The activity coefficient varies with concentration and for ions it varies with the charge and is the same for all *dilute* solutions having the same ionic strength. The activity coefficient depends upon the total ionic strength of a solution (a measure of the electrical field existing in the solution) and for ion-selective work it is often necessary to be able to calculate this. The ionic strength I is given by

$$I=0.5\sum C_i Z_i^2$$

where C_i is the ionic concentration in moles per litre of solution at Z_i is the charge of the ion concerned. Thus the ionic strength of 0.1 M nitric acid solution (HNO_3) containing 0.2 M barium nitrate $[Ba(NO_3)_2]$ is given by

$$\begin{aligned}&0.5[0.1 \text{ (for } H^+) + 0.1 \text{ (for } NO_3^-)\\ &\qquad +0.2\times 2^2 \text{ (for } Ba^{++})\\ &\qquad +0.4\times 1 \text{ (for } NO_3^-)]\\ &=0.5[1.4]=0.7\end{aligned}$$

4.2 Ionization of water

As even the purest water possesses a small but definite electrical conductivity, water itself must ionize to a very slight extent into hydrogen and hydroxyl ions:

$$H_2O \rightleftharpoons H^+ + OH^-$$

This means that at any given temperature

$$\frac{a_{H^+}\times a_{OH^-}}{a_{H_2O}}=\frac{[H^+].[OH^-]}{[H_2O]}\times\frac{f_{H^+}.f_{OH^-}}{f_{H_2O}}=K$$

where a_x, $[X]$ and f_x refer to the activity, concentration and activity coefficient of the species X, and K is a constant.

As water is only slightly ionized, the ionic concentrations are small and the activity coefficients of the ions can therefore be regarded as unity. The activity coefficient of the un-ionized molecule H_2O may also be taken as unity and the above expression therefore reduces to

$$\frac{[H^+]\times[OH^-]}{[H_2O]}=K$$

In pure water too, because there is only very slight dissociation into ions, the concentration of the undissociated water $[H_2O]$ may also be considered constant and the equation becomes $[H^+]\times[OH^-]=K_w$. The constant K_w is known as the ionic product of water.

Strictly speaking, the assumptions that the activity coefficient of water is constant and that the activity coefficients of the ions are unity are only correct for pure water and for very dilute solutions where the ionic strength is less than 0.01. In more concentrated solutions the ionic product for water will not be constant but, as activity coefficients are generally difficult to determine, it is common usage to use K_w.

The ionic product of water, K_w, varies with temperature and is given by the equation

$$\begin{aligned}\log_{10} K_w &= 14.00-0.0331(t-25)\\ &\quad +0.000\,17(t-25)^2\end{aligned}$$

where t is the temperature in °C.

Conductivity measurements show that, at 25 °C, the concentration of hydrogen ions in water is 1×10^{-7} mol litre^{-1}. The concentration of hydroxyl ions equals that of the hydrogen ions therefore $K_w=[H^+]\times[OH^-]=10^{-14}$. If the product of $[H^+]$ and $[OH^-]$ in aqueous solution momentarily exceeds this value, the excess ions will immediately recombine to form water. Similarly if the product of the two ionic concentrations is momentarily less than 10^{-14}, more water molecules will dissociate until the equilibrium value is obtained. Since the concentrations of hydrogen and hydroxyl ions are equal in pure water it is an exactly neutral solution. In aqueous solutions where the hydrogen ion concentration is greater than 10^{-7}, the solution is acid; if the hydrogen ion concentration is less than 10^{-7} the solution is alkaline.

4.3 Electrical conductivity

4.3.1 Electrical conduction in liquids

As early as 1833, Faraday realized that there are two classes of substances which conduct electricity. In the first class are the metals and alloys, and certain non-metals such as graphite, which conduct electricity without undergoing any chemical change. The flow of the current is due to the motion of electrons within the conductor, and the conduction is described as metallic, or electronic.

In the second class are salts, acids and bases which, when fused or dissolved in water, conduct electricity owing to the fact that particles, known as ions, carrying positive or negative electric charges move in opposite directions through the liquid. It is this motion of electrically charged particles which constitutes the current. Liquids which conduct electricity in this manner are known as electrolytes.

4.3.2 Conductivity of solutions

The passage of current through an electrolyte generally obeys Ohm's law and the current-carrying ability of any portion of electrolyte is termed its conductance and has the units of reciprocal resistance ($1/\Omega$), siemens (S). The specific current-carrying ability of an electrolyte is called its conductivity and consequently has the units of $S\ m^{-1}$.

The conductivity of electrolytes varies greatly with their concentration because dilution (a) increases the proportion of the dissolved electrolyte which forms ions in solution but (b) tends to reduce the number of these ions per unit of volume. In order to measure the first effect alone another term, molar conductivity, Λ, is defined,

$$\Lambda\ (S\ m^2/mol) = \kappa/c,$$

where κ is the conductivity and c is the concentration in $mol\ m^{-3}$. Although these are the basic SI units most work is reported using volume units of cm^3 since the litre is a convenient volume for laboratory use and Λ is usually in units of $S\ cm^2/mol$.

Table 4.1 Limiting ionic conductivities at 25 °C

Cation	λ° *S cm²/mol*	*Anion*	λ° *S cm²/mol*
H^+	349.8	OH^-	199.1
Li^+	38.7	F^-	55.4
Na^+	50.1	Cl^-	76.4
K^+	73.5	Br^-	78.1
NH_4^+	73.6	I^-	76.8
$(CH_3)_2NH_2^+$	51.9	NO_3^-	71.5
$\frac{1}{2}Mg^{2+}$	53.1	ClO_4^-	64.6
$\frac{1}{2}Ca^{2+}$	59.5	Acetate	40.9
$\frac{1}{2}Cu^{2+}$	53.6	$\frac{1}{2}SO_4^{2-}$	80.0
$\frac{1}{2}Zn^{2+}$	52.8	$\frac{1}{2}CO_3^{2-}$	69.3

At infinite dilution the ions of an electrolyte are so widely separated by solvent molecules that they are completely independent and the molar conductivity is equal to the sum of the ionic conductivities, λ°, of the cation and anion, i.e.

$$\Lambda_\infty = \lambda_-^\circ + \lambda_+^\circ$$

The values of λ° are the values for unit charge, referred to as equivalent ionic conductivities at infinite dilution. The general case is

$$\Lambda_\infty = z_+ n_+ \lambda_+^\circ + z_- n_- \lambda^\circ$$

where z is the charge on the ion and n the number of these ions produced by dissociation of one molecule of the salt, e.g.

$$\lambda_\infty(LaCl_3) = 3 \times 1 \times \lambda_{La}^\circ + 1 \times 3 \times \lambda_{Cl}^\circ$$

Since, for example, the ionic conductivity of the chloride ion is the same in all chloride salts then the molar conductivity at infinite dilution of any chloride salt can be calculated if the corresponding value for the cation is known. Values of ionic conductivities at infinite dilution at 25 °C are given in Table 4.1.

Providing the concentration of a fully dissociated salt is less than about 10^{-4} mol/l then the conductivity κ at 25 °C can be calculated from

$$\kappa\ (S\ cm^{-1}) = zn(\lambda_+^\circ + \lambda_-^\circ)c\ 10^{-3}$$

or

$$\kappa\ (\mu S\ cm^{-1}) = zn(\lambda_+^\circ + \lambda_-^\circ)c\ 10^{3}$$

where c is the concentration in mol/l.

Values of limiting ionic conductivities in aqueous solution are highly temperature-dependent and in some cases the value increases five- or sixfold over the temperature range 0–100 °C (see Table 4.2). These changes are considered to be due mainly to changes in the viscosity of water and the effect this has on the mobility and hydration of the ions.

Table 4.2 Ionic conductivities between 0 and 100 °C (S cm²/mol)

Ion	*0°*	*5°*	*15°*	*18°*	*25°*	*35°*	*45°*	*55°*	*100°*
H^+	225	250.1	300.6	315	349.8	397.0	441.4	483.1	630
OH^-	105	—	165.9	175.8	199.1	233.0	267.2	301.4	450
Li^+	19.4	22.7	30.2	32.8	38.7	48.0	58.0	68.7	115
Na^+	26.5	30.3	39.7	42.8	50.1	61.5	73.7	86.8	145
K^+	40.7	46.7	59.6	63.9	73.5	88.2	103.4	119.2	195
Cl^-	41.0	47.5	61.4	66.0	76.4	92.2	108.9	126.4	212
Br^-	42.6	49.2	63.1	68.0	78.1	94.0	110.6	127.8	—
I^-	41.4	48.5	62.1	66.5	76.8	92.3	108.6	125.4	—
NO_3^-	40.0	—	—	62.3	71.5	85.4	—	—	195
ClO_4^-	36.9	—	—	58.8	67.3	—	—	—	185
Acetate	20.1	—	—	35	40.9	—	—	—	—
$\frac{1}{2}Mg^{2+}$	28.9	—	—	44.9	53.0	—	—	—	165
$\frac{1}{2}Ca^{2+}$	31.2	—	46.9	50.7	59.5	73.2	88.2	—	180
$\frac{1}{2}SO_4^-$	41	—	—	68.4	80.0	—	—	—	260

4.3.3 Practical measurement of electrical conductivity

From the foregoing, it can be seen that measurement of electrical conductivity enables concentration to be determined.

4.3.3.1 Alternating current cells with contact electrodes

Conductivity cells provide the means of conducting a small, usually alternating, current through a precise volume of liquid whose conductivity we wish to know. At its simplest, this process involves the measurement of the resistance between two electrodes of fixed shape and constant distance apart. The relationship between the specific conductivity κ of the solution and the resistance R across the electrodes includes a cell constant 'a' such that

$$\kappa = a/R$$

If we express the conductivity in units of S cm^{-1} then the cell constant has the dimension of cm^{-1}. In order to simplify the electrical circuits of the measuring instruments it is customary to maintain the resistance of conductivity cells between the limits of 10 and 100 000 Ω. The conductivity of aqueous solutions varies from pure water with a conductivity of about 5 μS/m to those of concentrated electrolytes with conductivities as high as 1000 S/m. In order to keep within these resistance limits it is necessary, therefore, to have cells with a range of cell constants from 0.01 to 100 cm^{-1}. A working guide to the most appropriate value of cell constant for any given range of conductivity is shown in Table 4.3.

Table 4.3 Guide to cell constant for known conductivity range

Conductivity range, μS cm^{-1}	*Cell constant cm^{-1}*
0.05 to 20	0.01
1 to 200	0.1
10 to 2000	1
100 to 20 000	10
100 to 200 000	50

In order to measure the conductivity accurately it is necessary to know the cell constant accurately. It is usual to determine the cell constant by preferably (a) measuring the conductance when the cell is filled with a solution whose conductivity is accurately known or, failing that, (b) comparing the measured conductance with that obtained from a cell of known cell constant when both cells contain the same solution at the same temperature.

The only solutions whose conductivities are known with sufficient accuracy to be used for reference purposes are aqueous solutions of potassium chloride. This salt should be of the highest purity, at least analytical reagent grade and dried thoroughly in an oven at 120 °C before preparing solutions by dissolving in de-ionized water whose conductivity is less than 2 μS/cm at room temperature. The most accurate reference solutions are prepared by weight and the two most useful solutions are given in Table 4.4.

Table 4.4 Standard solutions for cell calibration

*Solution g KCl/1000 g solution**	*κ at 18 °C S m^{-1}*	*κ at 25 °C S m^{-1}*
(A) 7.4191	1.1163	1.2852
(B) 0.7453	0.12201	0.14083

* All values are 'mass in vacuo'.

For many purposes a simpler procedure can be followed. This involves weighing only the potassium chloride and preparing solutions by volume at 20 °C, these details are given in Table 4.5.

Table 4.5 Standard solutions (volumetric) for cell calibration

Solution	*κ at 18 °C*	*κ at 25 °C*
(A′) 7.4365 g KCl/l at 20 °C	1.1167 S m^{-1}	1.2856 S m^{-1}
(B′) 0.7440 g KCl/l at 20 °C	0.1221 S m^{-1}	0.1409 S m^{-1}
(C′) 100 ml of solution B′ made up to 1 litre at 20 °C	—	146.93 μS cm^{-1}*

* For the highest accuracy the conductivity of the dilution water should be added to this value.

Calibration of conductivity cells by these solutions requires considerable care if accurate values of cell constants are to be determined. The importance of temperature control cannot be over-emphasized since the conductivity of the potassium chloride solution will change by over 2 per cent per kelvin. Alternatively the cell constant can be determined by the comparison technique with identical, rather than standard conditions in both the 'known' and 'unknown' cell. Equally important as the effect of temperature, is that of polarization in these cells where the electrodes contact the solution and conduct a significant current.

The extent of polarization depends on a number of factors, the most important of which are the nature of the electrode surface and the frequency of the a.c. signal applied to the cell. The restrictions that polarization errors, arising from electrode material, impose on the choice of cell means that cells with

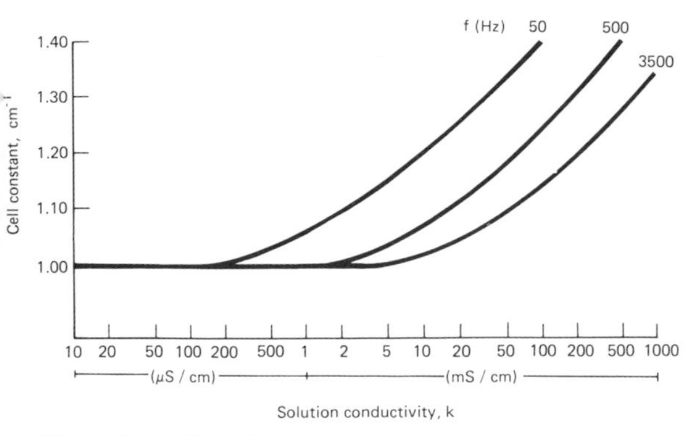

Figure 4.1 Effect of frequency on the useful range of a cell with titanium carbide coated stainless steel electrodes (courtesy F. Oehme, Polymetron).

bright metal electrodes are best suited for measurements of low conductivities where the proportion of the total resistance due to polarization is very small. Treated or coated electrodes are suitable for low ($\sim 0.05\ \mu S\ cm^{-1}$) to intermediate ($\sim 0.1\ S\ m^{-1}$) conductivities provided that the frequency of the a.c. voltage is in the range normally found in commercial instruments (50–1000 Hz).

Polarization in all the cells we have been discussing can be reduced by increasing the frequency of the applied voltage. This can best be appreciated by considering Figure 4.1 in which the apparent cell constant over a range of conductivities is plotted against three values of a.c. frequency. The true value of the cell constant was 1 cm^{-1} and it can be seen that the highest frequency, 3.5 kHz, gave the true value for the cell constant over the widest concentration range. Unfortunately increase of frequency can introduce capacitative errors into the measurement, particularly from the signal cable and in many applications the choice of operating frequency is a compromise. Although variable frequency conductivity meters are available as laboratory instruments (e.g. Philips Model PW 9509, High Performance Conductivity Meter) such a facility is not usually found on industrial instruments. In this case it is necessary to consider the range of conductivities to be measured, together with the chemical and physical nature of the solutions to be measured before specifying the operating frequency. All determinations of cell constant should be carried out at this frequency.

Cell construction The materials used in cell construction must be unaffected by the electrolyte and the insulation between the electrodes must be of a high quality and not absorb anything from the process liquid.

A wide range of materials are at present available covering a wide range of pressures, temperatures and process fluids. The body may be made of glass, epoxy resins, plastics such as PTFE, pure or reinforced, PVC, Perspex, or any other material suitable for the application, but it must not be deformed in use by temperature or pressure, otherwise the cell constant will change.

The electrodes may be parallel flat plates or rings of metal or graphite cast in the tube forming the body, or in the form of a central rod with a concentric tubular body.

One common form of rod-and-tube conductivity cell consists of a satinized stainless steel rod-electrode surrounded by a cylindrical stainless steel electrode, having holes to permit the sample to flow freely through the cell. This is surrounded by an intermediate cylinder also provided with holes, and two O-rings which together with the tapered inner end form a pressure-tight seal onto the outer body when the inner cell is withdrawn for cleaning, so that the measured solution can continue to flow and the cell be replaced without interruption of the process. The outer body is screwed into the line through which the measured solution flows. Figure 4.2(a) shows the inserted cell as it is when in use, and (b) the withdrawn measuring element with the intermediate sleeve forming a seal on the outer body. The cell may be used at 110 °C up to 7 bar pressure.

Many manufacturers offer a type of flow-through conductivity cell with annular graphite electrodes, one

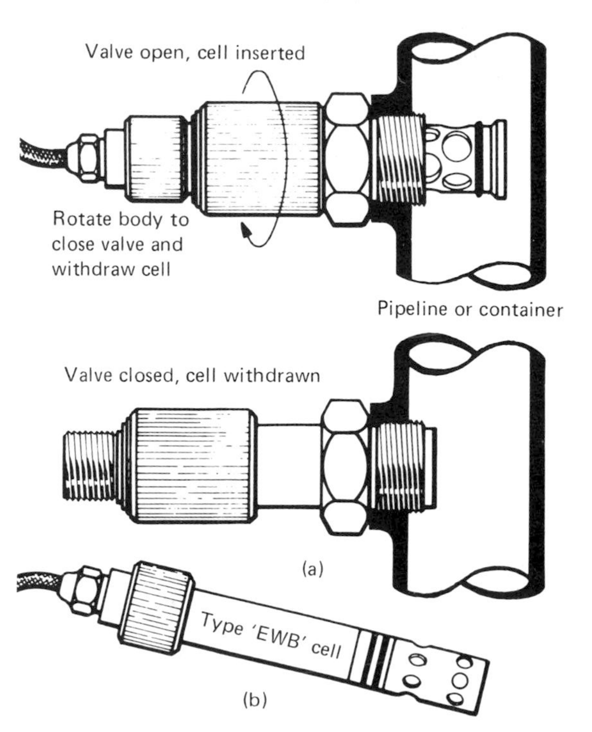

Figure 4.2 Retractable conductivity cell (courtesy Kent Industrial Measurements Ltd., Analytical Instruments).

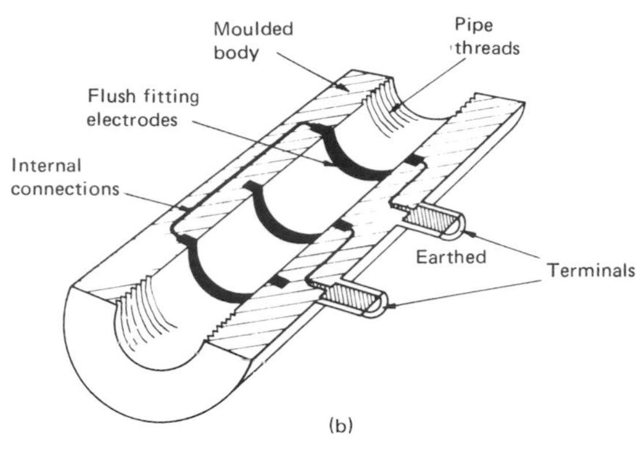

Figure 4.3 Flow-through cell (courtesy Kent Industrial Measurements Ltd., Analytical Instruments).

form of which is shown in Figure 4.3. It consists of three annular rings of impervious carbon composition material equally spaced within the bore of an epoxy resin moulded body. Conduction through the solution within the cell takes place between the central electrode and the two outer rings, which are connected to the earthed terminal of the measuring instrument; thus electrical conduction is confined entirely within the cell, where it is uninfluenced by the presence of adjoining metal parts in the pipe system. This pattern of cell, having a simple flow path, is ideally suited to the exacting requirements of dialysate concentration monitoring in the artificial kidney machine. Screw-in patterns of this cell are also generally available.

The use of an impervious carbon composition material for the electrodes substantially eliminates polarization error and provides conducting surfaces that do not require replatinization or special maintenance, other than periodic, but simple and infrequent cleaning by means of a bottle brush. Typical operating temperature and pressure limits for this type of cell are 100 °C and 7 bar.

Measuring cells should be installed in positions where they are adequately protected from mechanical shock by passing traffic, dampness and extremes of temperature. Where a flow-line cell is connected directly in the electrolyte pipe, suitable support should be given to the pipes to ensure that the cell is under no mechanical strain, and that the pipe threads in a rigid system are straight and true. Dip pattern cells should be installed so that moving parts in a tank, e.g. agitators, are well clear of the cells.

Where measuring cells are installed in pipework, it is essential that they are positioned in a rising section of the system to ensure that each cell is always full of electrolyte, and that pockets of air are not trapped. Alternatively, they may be installed in the bottom member of a U formed in horizontal pipework. In this case, screw-in cells should be in the top or side of the pipe so that sediment cannot settle in them.

Cleaning and maintenance of cells Periodic inspection and cleaning of conductivity cells is essential to ensure that the electrode surfaces are free from contamination which would otherwise alter the electrode area and effective cell constant. The frequency of such procedures is mainly dependent on the nature of the samples but the design of the cells and the accuracy required for the measurement will also have to be taken into consideration. All new cells should be thoroughly cleaned before installation and these cleaning procedures depend on the design of the cell and the electrode material.

Platinized electrodes Cleaning of these electrodes constitutes a major drawback in their application because no form of mechanical cleaning should be attempted. A suitable cleaning solution consists of a stirred mixture of 1 part by volume isopropyl alcohol, 1 part of ethyl ether and 1 part hydrochloric acid (50%). Alternatively, the sensitivity of the electrodes can frequently be restored by immersion in a 10–15% solution of hydrochloric or nitric acid for about two minutes. The electrodes should be thoroughly rinsed with water before being returned to service.

Annular graphitic electrodes Cleaning should be carried out with a 50% solution of water/detergent using a bottle brush. After thorough brushing with this solution, the cell bore should be rinsed several times in distilled water and then viewed. Looking through the bore towards a source of illumination, the surface

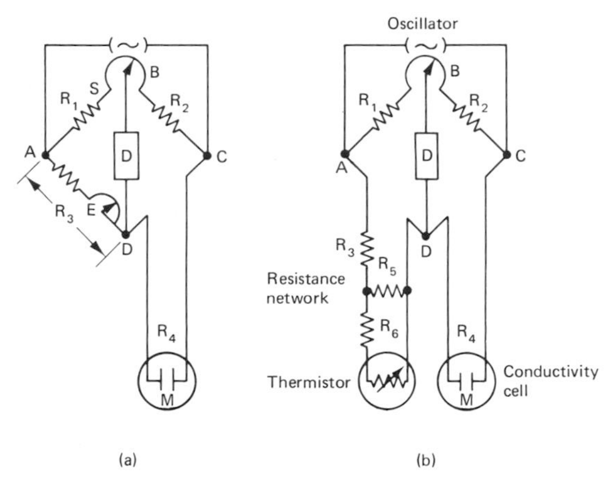

Figure 4.4 Measurement of conductance using Wheatstone bridge: (a) simple circuit, (b) thermistor temperature-corrected circuit.

should be evenly wetted with no dry patches where the water has peeled away. If dry patches appear rapidly, indicating that a thin film of grease is present, the surface is not clean.

Stainless steel and Monel A feature of many stainless steel cells is the frosted appearance of the electrodes which is essential to reduce polarization. It is most important that this frosting is not polished away by the regular use of abrasive cleaners. This type of cell may be cleaned with a 50% water detergent solution and a bottle brush.

In the case of screw-in cells the outer electrode may be removed to facilitate cleaning, but on no account should the central electrode be disturbed, as this will impair the accuracy of the electrical constant of the cell. In cases where metal cells have become contaminated with adherent particulate matter, such as traces of magnetite or other metal oxides, ultrasonic cleaning in the detergent solution has been shown to be effective.

In all cleaning processes care should be taken to keep the external electrical contact, cable entries and plugs dry.

Instruments for conventional a.c. measurement The conductance of a cell may be measured (a) by Wheatstone bridge methods or (b) by direct measurement of the current through the cell when a fixed voltage is applied.

Wheatstone bridge methods The actual conductance of the cell is usually measured by means of a self-balancing Wheatstone bridge of the form shown in Figure 4.4 and described in detail in Part 3.

Direct measurement of cell conductance The conductance of a cell may be measured directly by the method indicated in Figure 4.5. The current is directly proportional to the conductance so the output from the current amplifier is applied to the indicator and recorder. Temperature compensation is achieved by connecting a manual temperature compensator in the amplifier circuit, or a resistance bulb may be used to achieve automatic compensation.

Multiple-electrode cells From the foregoing discussion on errors introduced by polarization together with the importance of constancy of electrode area, it can be appreciated that two-electrode conductivity cells have their limitations. In circumstances where accurate measurements of conductivity are

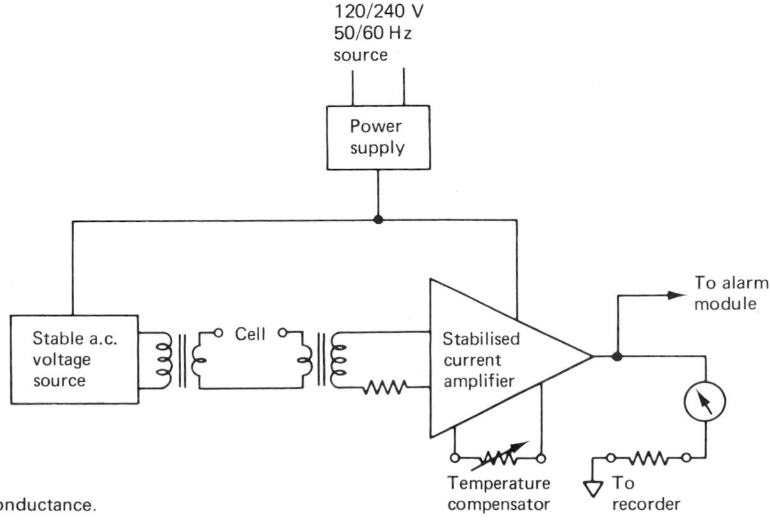

Figure 4.5 Direct measurement of cell conductance.

required in solutions of moderate or high conductivity or in solutions which can readily contaminate the electrode surfaces, multiple-electrode cells should be considered.

In its simplest form, a multiple-electrode cell has four electrodes in contact with the solution. An outer pair operate similarly to those in a conventional two-electrode cell and an a.c. current is passed through the solution via these electrodes. The voltage drop across a segment of the solution is measured potentiometrically at a second or inner pair of the electrodes, and this drop will be proportional to the resistivity or inversely proportional to the conductivity of the solution. Four-electrode cells can be operated in either the constant-current or constant-voltage mode but the latter is the more popular and will be described further. In this form of measurement the voltage at the inner electrode pair is maintained at a constant value by varying the current passed through the solution via the outer electrodes. The current flowing in the cell will be directly proportional to the conductivity, and can be measured as indicated in Figure 4.6.

The circuit shown in the figure is considerably simplified and there are multiple-electrode cells available from a number of manufacturers which contain additional electrodes whose function is to minimize stray current losses in the cell, particularly for solutions flowing through earthed metal pipework.

Since there is imperceptible current flowing through the voltage sensing electrodes, cells of this type are free from the restrictions imposed by polarization. Therefore multiple-electrode cells can be standardized with any of the potassium chloride solutions given in Tables 4.4 and 4.5. The precaution previously stated about constancy of temperature during any determination of cell constant must still be observed.

Multiple-electrode cells are available with cell constants from 0.1 to 10 cm^{-1} and can therefore be used over a wide range of solution conductivities. However, their most valuable applications are when contamination or polarization is a problem.

Temperature compensation The conductivity of a solution is affected considerably by change of temperature, and each solution has its own characteristic conductivity–temperature curve. Figure 4.7 shows how different these characteristics can be. When it is required to measure composition rather than absolute conductivity it is therefore essential to use a temperature compensator to match the solution.

Manual compensators consist of a variable and a fixed resistor in series. The temperature scale showing the position of the contact on the variable resistance is calibrated so that the resistance of the combined resistors changes by the same percentage of the value of conductivity of the solution at 25 °C as does the solution. The scale becomes crowded at the upper end, thus limiting the span of the compensator to about 70 °C.

Aqueous solutions containing very low ($\mu g\,l^{-1}$) concentrations of electrolytes must have more elaborate compensation to allow for the non-linear conductivity–temperature characteristic of pure water. This type of compensation system is applied in all conductivity transmitters (either with two-electrode or multiple-electrode cells) designed for accurate operation in the range up to 0.5 $\mu S\,cm^{-1}$.

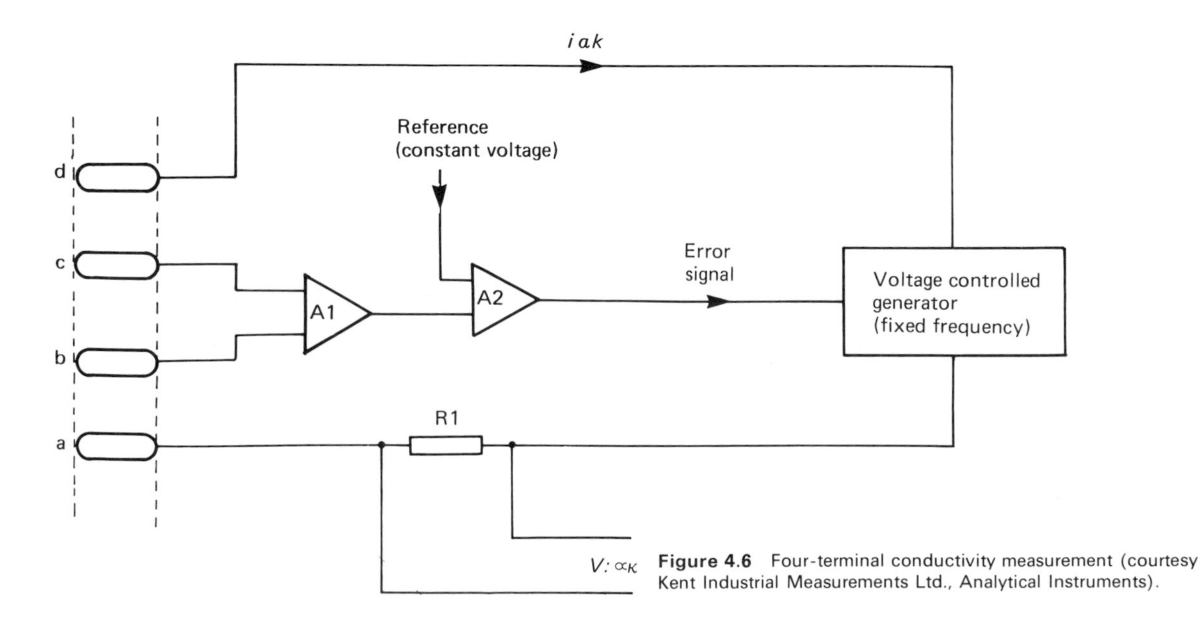

Figure 4.6 Four-terminal conductivity measurement (courtesy Kent Industrial Measurements Ltd., Analytical Instruments).

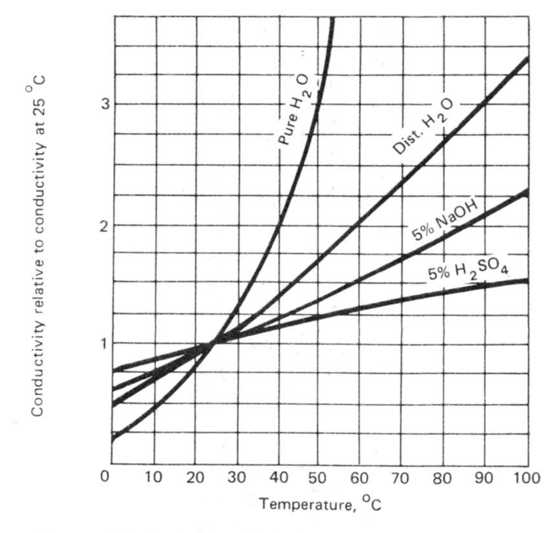

Figure 4.7 Variation of solution conductivity with temperature.

4.3.3.2 Electrodeless method of measuring conductivity

The principle of the method is to measure the resistance of a closed loop of solution by the extent to which the loop couples two transformer coils. The liquid to be measured is enclosed in a non-conducting pipe, or a pipe lined with a non-conducting material. Three forms of measuring units are available, as shown in Figure 4.8. As the method is most successful with full scale resistances of 10–1000 Ω, relatively large bore pipe may be used, reducing the possible errors due to solid deposition or film formation.

Figure 4.8(a) shows the form used for immersion in a large volume of solution. For measurements on a solution flowing through a pipe the arrangement shown in Figure 4.8(b) is used. If the liquid contains suspended solids or fibres, wide bore non-conducting pipe fitted with metallic end-pieces connected together with a length of wire to complete the circuit may sometimes be used (Figure 4.8(c)).

The principle of the measuring system is shown in Figure 4.9. Figure 4.9(a) shows the simple circuit which consists of two transformers. The first has its primary winding, the input toroid, connected to an oscillator operating at 3 or 18 kHz and as its secondary the closed loop of solution. The closed loop of solution forms the primary of the second transformer and its secondary is the output toroid. With constant input voltage the output of the system is proportional to the conductivity of the solution. The receiver is a high impedance voltage measuring circuit which amplifies and rectifies the output and displays it on a large indicator.

In order to eliminate effects of source voltage and changes in the amplifier characteristics a null balance system may be provided as shown in Figure 4.9(b). An additional winding is provided on each toroid and the position of the contact is adjusted on the main slide-wire to restore the system to the original balanced state by means of the balancing motor operated by the amplified out-of-balance signal in the usual way.

The electrodeless measurement of conductivity has obvious advantages in applications where the solution is particularly corrosive or has a tendency to foul or mechanically abrade the electrodes. Typical of these applications are measurements in oleum, hot concentrated sodium hydroxide and slurries. In addition, this technique is ideal for applications in concentrated electrolytes (not necessarily aggressive) such as estuarine or sea waters where polarization errors

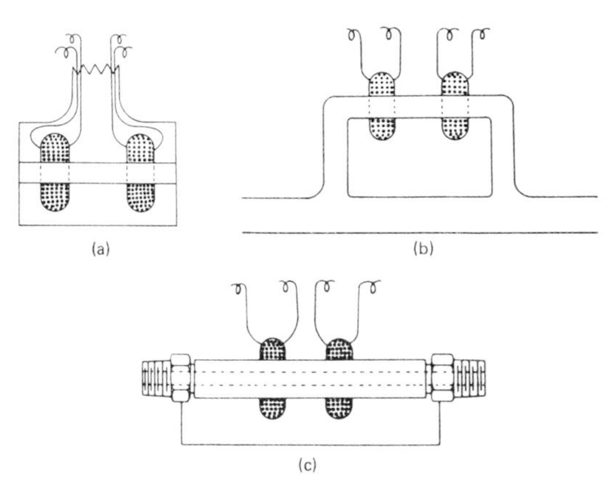

Figure 4.8 Electrodeless conductivity cells.

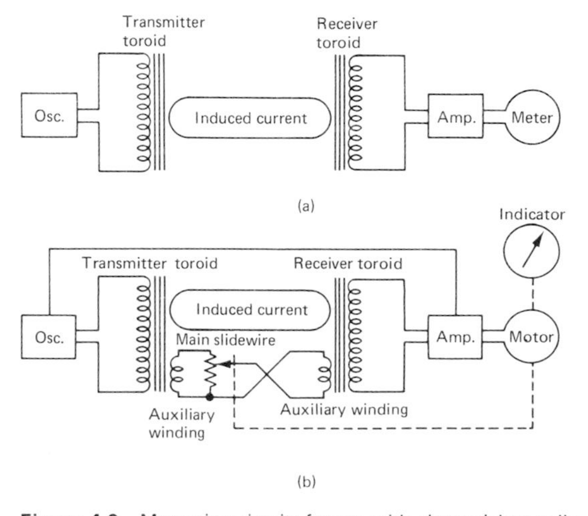

Figure 4.9 Measuring circuits for use with electrodeless cells (courtesy Beckman Instruments Inc.): (a) direct reading, (b) balanced bridge.

would be considerable in a conventional cell. Temperature compensation is normally incorporated.

4.3.4 Applications of conductivity measurement

The measurement of electrical conductivity is the simplest and probably the most sensitive method of providing a non-specific indication of the dissolved solids, or more correctly the ionic, content of a solution. If the number of ionic species in solution are few then it may be possible to use conductivity as a measure of the concentration of a particular component. Undoubtedly the robust nature of conductivity measurements has led to its use in circumstances where its non-specific response gives rise to errors in interpretation of concentration. Consequently, any successful instrumental application of conductivity as a concentration sensor has to ensure that the species of interest is the dominating ion or the only ion (together with its counter-ion of opposite charge) whose concentration is changing. With these restrictions it can be appreciated that determinations of concentrations by conductivity measurements are often supported by additional analyses or preceded by a physical or chemical separation of the species of interest.

4.3.4.1 Conductivity and water purity

Water of the highest purity is increasingly being used for industrial purposes, for example, the manufacture of electronic components and the preparation of drugs. Other examples of large-scale uses include process steam and feedwater for high pressure boilers. In all these cases conductivity provides the most reliable measurement of water purity in circumstances where contamination from non-conducting impurities is considered to be absent. The conductivity of pure water is highly temperature-dependent due to the increase in the dissociation of water molecules into hydrogen and hydroxyl ions of water, K_w, with temperature. The extent of this can be seen in Table 4.6.

The conductivity of pure water can be calculated at any temperature provided values of λ°_{OH}, λ°_H, K_w, the dissociation constant of water, and the density of water d are known at the appropriate temperature.

$$\kappa(\mu S\ cm^{-1}) = (\lambda^\circ_H + \lambda^\circ_{OH})d \,.\, \sqrt{K_w} \,.\, 10^3$$

In the application under consideration here (i.e. the use of pure water) the exact nature of the ionic species giving rise to a conductivity greater than that of pure water are of no interest but it is useful to note how little impurity is required to raise the conductivity. For example, at 25 °C only about 10 $\mu g\ l^{-1}$ of sodium (as sodium chloride) are required to increase the conductivity to twice that of pure water.

Table 4.6 Pure water, conductivity from 0 to 100 °C

Temperature (°C)	*Conductivity* ($\mu S\ cm^{-1}$)	*Resistivity* (°C)
0	0.0116	86.0
5	0.0167	60.0
10	0.0231	43.3
15	0.0314	31.9
20	0.0418	23.9
25	0.0548	18.2
30	0.0714	14.0
35	0.0903	11.1
40	0.1133	8.82
45	0.1407	7.11
50	0.1733	5.77
60	0.252	3.97
70	0.346	2.89
80	0.467	2.14
90	0.603	1.66
100	0.788	1.27

4.3.4.2 Condensate analyser

The purity of the water used in the steam–water circuit of power stations is particularly important for the prevention of corrosion. An essential component of such a circuit is the condenser wherein the steam from the turbines is condensed before returning to the boiler. On one side of the condenser tubes is the highly pure steam and water from the turbines and on the other is cooling water chosen for its abundance (e.g. river water or estuarine water) rather than its chemical purity. Any leakage of this cooling water through the condenser tubes leads to the ingress of unwanted impurities into the boiler and therefore must be immediately detected. Direct measurement of conductivity would detect significant ingress of say sodium chloride from estuarine water but it would not be capable of detecting small leakages since the conductivity of the condensate would be dominated by the alkaline additives carried over in the steam from the boiler. A better method of detection of leakage is to pass the condensate through a cation exchange column in the H^+-form, then measuring the conductivity. Using this procedure, all cations in the condensate are exchanged for hydrogen ions and the solution leaving the column will be weakly acidic if any salts have entered through the condenser. Otherwise, the effluent from the column will ideally be pure water since the cations of the alkaline boiler water additives (NH_4OH, $NaOH$) will be exchanged and recombine as,

$$H^+ + OH^- \rightleftharpoons H_2O$$

A secondary advantage of such a system is the enhancement of the conductivity due to replacement of cations by hydrogen ions which gives about a fivefold enhancement in ionic conductance. This is particularly important with very low leak rates.

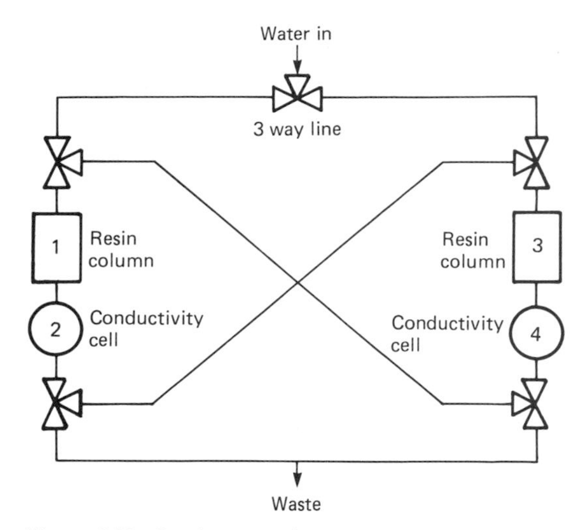

Figure 4.10 Condensate analyser.

A schematic diagram of an instrument based on the above principles is given in Figure 4.10. The incoming sample flows at about 400 ml min^{-1} through a H^+-form cation exchange column (1), 500 mm deep and 50 mm in diameter, and then to a flow-through conductivity cell (2). The effluent from the cell flows to waste via an identical column/cell system (3 and 4) which is held in reserve. Since there will be no exchange on this second column it will not be depleted and the constant flow of water or weak acid keeps it in constant readiness for instant replacement of column (1) when the latter becomes exhausted. The measured conductivity can be recorded and displayed and, where necessary, alarms set for notification of specific salt ingress levels. In the case of power stations using estuarine water for cooling the condensers the condensate analyser can be used to give a working guide to the salt going forward to the boiler (see Table 4.7).

Table 4.7 Relationship between conductivity and salt fed to the boiler

Conductivity at 25 °C (μS cm^{-1})	*Chloride in condensate* (ppm)	*Salt going forward to boiler* (g NaCl/Tonne)
0.137	0.01	0.0165
0.604	0.05	0.0824
1.200	0.10	0.1649
1.802	0.15	0.2473
2.396	0.20	0.3298
6.003	0.50	0.8265

4.3.4.3 Conductivity ratio monitors

These instruments measure the conductivities at two points in a process system continuously and compare the ratio of the measurements with a preset ratio. When the measured ratio reaches the preset value, a signal from the monitor can either operate an alarm or initiate an action sequence or both.

One application of this type of dual conductivity measurement is to control the regeneration frequency of cation exchange units (usually in the H^+-form) in water treatment plants. The conductivity at the outlet of such a unit will be higher than at the inlet since cations entering the ion exchange bed will be replaced by the much more conductive hydrogen ion ($\lambda^{\circ}_{H}=350$, $\lambda^{\circ}_{Na}=50$). For example, an inlet stream containing 10^{-4} mol l^{-1} of sodium chloride will have ratios of 3.5, 3.3 and 2.3 for 100, 90 and 50 per cent exchange respectively. A value corresponding to the acceptable extent of exchange can then be set on the instrument. Reverse osmosis plants use ratio monitors to measure the efficiency of their operation and these are usually calibrated in percentage rejection or passage.

This type of operational control is most effective when the chemical constituents of the inlet stream do not vary greatly, otherwise the ratio will be subject to errors from unconsidered ionic conductivities.

4.3.4.4 Ion chromatography

Although conductivity measurements have a non-specific response they can, when combined with a separation technique, provide extremely sensitive and versatile detectors of chemical concentration. The best example of this is in ion chromatography which in recent years has been shown to be an invaluable instrumental technique for the identification and measurement of the concentration of ions, particularly at low levels, in aqueous solution.

The general principles of chromatography are outlined in Chapter 2. In an ion chromatograph, a small volume of sample is injected into a carrier or eluent electrolyte stream. The eluent together with the sample is carried forward under high pressure (5–50 bar) to an ion exchange column where chromatographic separation of either the cations (+ve) or anions (−ve), depending on the nature of the exchanger, takes place. The ion exchange material in these chromatographic separator columns is fundamentally the same as conventional ion exchange resins but the exchange sites are limited to the surface of very fine resin beads. This form of exchanger has been shown to have the characteristics required for rapid separation and elution of the ionic components in the order expected from the general rules of ion exchange (e.g. Cl^- before Br^- before SO_4^{2-}). At this stage the conductivity can be monitored and the elution peaks corresponding to the separated ionic components measured as increases superimposed on the relatively high background conductivity of the eluent. This is the procedure used in the ion chromatograph manu-

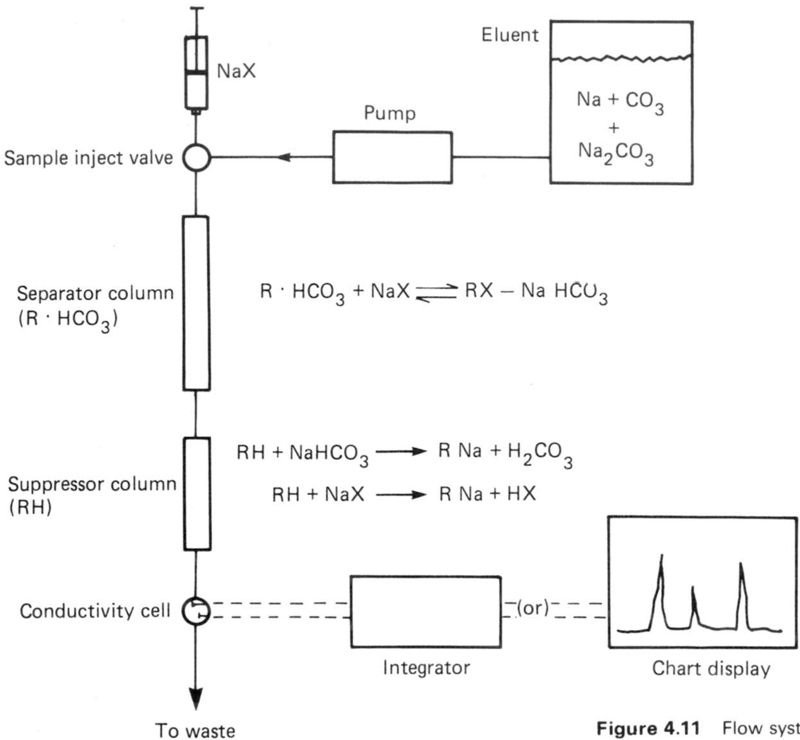

Figure 4.11 Flow system for anion chromatography.

factured by Wescan Instruments Inc. In another instrument manufactured by The Dionex Corporation, the eluent from the separator column passes through a second ion exchange column where the ions of opposite charge to those which have been separated chromatographically are all converted to a common form. This second column, termed a 'suppressor column', reduces the background conductivity of the eluent and thus ensures that conductivity changes due to the sample constitute a significant portion of the total measured conductivity. With a system such as this, the retention time identifies the elution peak and the area under the peak is a measure of the concentration of the ionic species giving rise to it. In many cases peak heights rather than areas can be used as the indicators of concentration, thus simplifying the measurement since an integrator is not required. For most purposes this is adequate since sharp elution peaks are obtained by keeping mixing to a minimum by use of very narrow bore transmission tubing combined with a conductivity cell whose volume is of the order of 6 μl. In cells of this size polarization resistance can be considerable due to the proximity of the electrodes.

A schematic outline of the main features of a typical system for the determination of anions is given in Figure 4.11.

In this particular example the eluent consisting of a mixture of 2.4×10^{-3} mol l^{-1} sodium carbonate and 3×10^{-3} mol l^{-1} sodium bicarbonate has a conductivity of about 700 μS cm^{-1}. The separator column consists of a strong base anion exchanger (R.HCO_3) mainly in the bicarbonate form and the suppressor column is a strong acid cation exchanger in the H^+-form (R.H). After the eluent has passed through the cation exchange it will be weakly acid carbonic acid (H_2CO_3) having a conductivity level of about 25 μS cm^{-1} and with this much reduced base conductivity level it is possible to detect quantitatively the small changes due to the acids (H.X) from the sample anions.

4.3.4.5 Sulphur dioxide monitor

A technique used to measure the concentration of sulphur dioxide in air in the parts per hundred million (pphm) range is based on the measurement of the change in the conductivity of a reagent before and after it has absorbed sulphur dioxide. The principle of the measurement is to absorb the sulphur dioxide in hydrogen peroxide solution, thus forming sulphuric acid which increases the electric conductivity of the absorbing reagent.

Continuous measurements can be made by passing air upwards through an absorption column down which the hydrogen peroxide absorbing solution is

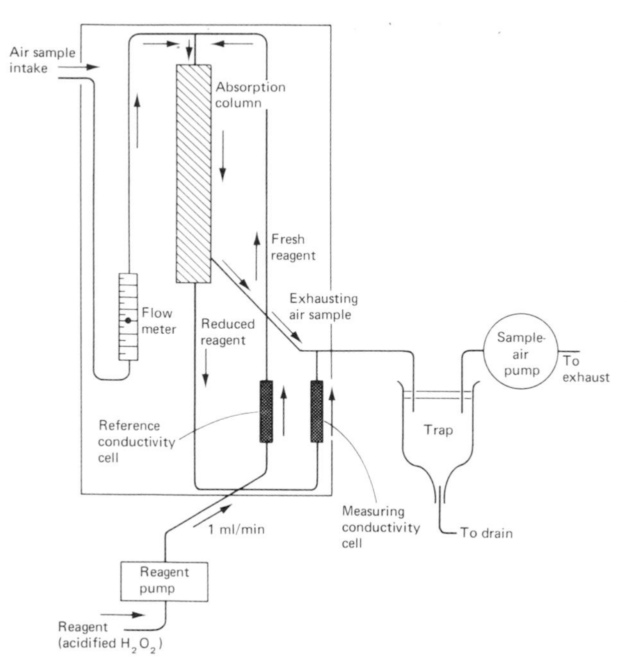

Figure 4.12 Continuous sulphur dioxide monitor.

flowing. Provided flow rates of air and hydrogen peroxide reagent are maintained constant, the sulphur dioxide concentration is proportional to the measured conductivity of the hydrogen peroxide reagent. Figure 4.12 is a diagram of suitable apparatus.

4.3.4.6 Salt-in-crude-oil monitor

A rapid continuous measurement of the salt in crude oil before and after desalting is based on the measurement of the conductivity of a solution to which a known quantity of crude oil has been added. The sample of crude oil is continuously circulated through a loop in the measurement section of the 'salt-in-crude monitor'. When the test cycle is initiated, solvent (xylene) is introduced from a metering cylinder into the analyser cell. A sample is then automatically diverted from the sample circulating loop into a metering cylinder calibrated to deliver a fixed quantity of crude oil into the analysis cell. A solution containing 63% n-butanol, 37% methanol, 0.25% water is then metered into the analysis cell from another calibrated cylinder.

The cell contents are thoroughly mixed by a magnetic stirrer, then the measuring circuit is energized and an a.c. potential is applied between two electrodes immersed in the liquid. The resulting a.c. current is displayed on a milliammeter in the electrical control assembly, and a proportional d.c. millivolt signal is transmitted from the meter to a suitable recorder.

At the end of the measuring period, a solenoid valve is opened automatically to drain the contents of the measuring cell to waste. The minimum cycle time is about 10 minutes.

Provision is made to introduce a standard sample at will to check the calibration of the instrument. Salt concentrations between 1 and 200 kg salt per 1000 m^3 crude oil can be measured with an accuracy of $\pm 5\%$ and a repeatability of 3% of the quantity being measured.

4.4 The concept of pH

4.4.1 General theory

Ionic concentrations were discussed in Section 4.2. The range of hydrogen ion concentrations met in practice is very wide; also when dealing with small concentrations it is inconvenient to specify hydrogen or hydroxyl concentrations. A method proposed by S. P. L. Sörenson in 1909 is now used universally – this is the concept of a hydrogen ion exponent or pH defined as:

$$\mathrm{pH} = -\log_{10}[\mathrm{H}^+] = \log_{10}\frac{1}{[\mathrm{H}^+]}$$

Thus pH is the logarithm to base 10 of the reciprocal of the hydrogen ion concentration. The advantage of this nomenclature is that all values of acidity and alkalinity between those of solutions molar with respect to hydrogen and hydroxyl ions can be expressed by a series of positive numbers between 0 and 14. Thus a neutral solution with $[\mathrm{H}^+] = 10^{-7}$ has a pH of 7. If the pH is less than 7 the solution is acid, if greater than 7, the solution is alkaline.

It must be realized that pH measuring devices measure the effective concentration, or activity, of the hydrogen ions and not the actual concentration. In very dilute solutions of electrolyte the activity and concentration are identical. As the concentration of electrolyte in solution increases above 0.1 mol/litre however, the measured value of pH becomes a less reliable measure of the concentration of hydrogen ions. In addition, as the concentration of a solution increases the degree of dissociation of the electrolyte decreases.

A dilute solution of sulphuric acid is completely dissociated and the assumption that $\mathrm{pH} = -\log 2(H_2SO_4)$ is justified. (The 2 occurs because each molecule of acid provides two hydrogen ions.) Anhydrous sulphuric acid is only slightly dissociated, the degree of dissociation rising as the pure acid is diluted.

A maximum hydrogen ion concentration occurs in the neighbourhood of 92% H_2SO_4, but, at this concentration, the difference between actual hydrogen ion concentration and the activity of the hydrogen ions is large, and the measured pH minimum of about -1.4

occurs at a much lower sulphuric acid content.

A more reliable indication of the ionic behaviour of a solution will be obtained if we define pH in terms of the hydrogen ion activity aH^+ so that

$$pH = \log_{10}(1/aH^+) = -\log_{10} aH^+$$

where aH is related to the hydrogen ion concentration cH^+ by the equation

$$aH^+ = fH^+cH^+$$

where fH^+ is the activity coefficient, see Section 4.1. The pH values of common acids, bases and salts are given in Table 4.8.

Table 4.8 pH values of common acids, bases and salts

Compound	*Molarity*	*pH*
Acid benzoic	(Saturated)	2.8
Acid boric	0.1	5.3
Acid citric	0.1	2.1
Acid citric	0.01	2.6
Acid hydrochloric	0.1	1.1
Acid oxalic	0.1	1.3
Acid salicylic	(Saturated)	2.4
Acid succinic	0.1	2.7
Acid tartaric	0.1	2.0
Ammonia, aqueous	0.1	11.3
Ammonium alum	0.05	4.6
Ammonium chloride	0.1	4.6
Ammonium oxalate	0.1	6.4
Ammonium phosphate, primary	0.1	4.0
Ammonium phosphate, secondary	0.1	7.9
Ammonium sulphate	0.1	5.5
Borax	0.1	9.2
Calcium hydroxide	(Saturated)	12.4
Potassium acetate	0.1	9.7
Potassium alum	0.1	4.2
Potassium bicarbonate	0.1	8.2
Potassium carbonate	0.1	11.5
Potassium dihydrogen citrate	0.1	3.7
Potassium dihydrogen citrate	0.02	3.8
Potassium hydrogen oxalate	0.1	2.7
Potassium phosphate, primary	0.1	4.5
Sodium acetate	0.1	8.9
Sodium benzoate	0.1	8.0
Sodium bicarbonate	0.1	8.3
Sodium bisulphate	0.1	1.4
Sodium carbonate	0.1	11.5
Sodium carbonate	0.01	11.0
Sodium hydroxide	0.1	12.9
Sodium phosphate, primary	0.1	4.5
Sodium phosphate, secondary	0.1	9.2
Sodium phosphate, tertiary	0.01	11.7
Sulphamic acid	0.01	2.1

4.4.2 Practical specification of a pH scale

As the value of pH defined as $-\log_{10}$ (hydrogen ion activity) is extremely difficult to measure it is necessary to ensure that, when different workers state a pH value they mean the same thing. An operational definition of pH has been adopted in British Standard 1647:1961. The e.m.f. E_X of the cell

Pt H_2/soln. X/conc. KCl soln./ref. electrode

is measured and likewise the e.m.f. E_S of the cell

Pt H_2/soln. S/conc. KCl soln./ref. electrode

both cells being at the same temperature throughout and the reference electrodes and bridge solutions being identical in the two cells.

The pH of the solution X denoted by pH(X) is then related to the pH of the solution S denoted by pH(S) by the definition:

$$pH(X) - pH(S) = (E_X - E_S)/(2.3026\, RT/F)$$

where R is the gas constant, T is temperature in kelvins, and F is the Faraday constant. Thus defined, pH is a pure number.

To a good approximation, the hydrogen electrodes in both cells may be replaced by other hydrogen-responsive electrodes, e.g. glass or quinhydrone. The two bridge solutions may be of any molarity not less than 3.5 mol/kg provided they are the same.

4.4.3 pH standards

The difference between the pH of two solutions having been defined as above, the definition of pH can be completed by assigning at each temperature a value of pH to one or more chosen solutions designated as standards. In BS 1647 the chosen primary standard is a solution of pure potassium hydrogen phthalate having a concentration of 0.05 mol/litre.

This solution is defined as having a pH value of 4.000 at 15 °C and the following values at other temperatures between 0 and 95 °C:

Between 0 and 55 °C

$$pH = 4.000 + 1/2[(t-15)^2/100]$$

Between 55 and 95 °C

$$pH = 4.000 + 1/2[(t-15)^2/100] - (t-55)/500$$

Other standard buffer solutions are given on p. 120.

The e.m.f. E_X is measured and likewise the e.m.f. E_1 and E_2 of similar cells with solution X replaced by standard solutions S_1 and S_2, so that E_1 and E_2 are on either side of and as near as possible to E_X. The pH of the solution X is then obtained by assuming linearity between pH and E, i.e.

$$(pHX - pH\,S_1)/(pH\,S_2 - pH\,S_1) = (E_X - E_1)/(E_2 - E_1)$$

4.4.4 Neutralization

When acid and base solutions are mixed, they combine to form a salt and water, e.g.

hydrochloric acid		sodium hydroxide		sodium chloride		water HOH
	+		=		+	
H^+Cl^-	+	Na^+OH^-	=	Na^+Cl^-	+	(largely
(dissociated)		(dissociated)		(dissociated)		undissociated)

Thus, if equal volume of equally dilute solutions of strong acid and strong alkali are mixed, they yield neither an excess of H^+ ions nor of OH^- ions and the resultant solution is said to be neutral. The pH value of such a solution will be 7.

4.4.5 Hydrolysis

Equivalent amounts of acid and base when mixed will produce a neutral solution only when the acids and bases used are strong electrolytes. When a weak acid or base is used, hydrolysis occurs. When a salt such as sodium acetate, formed by a weak acid and a strong base, is present in water, the solution is slightly alkaline because some of the H^+ ions from the water are combined with acetic radicals in the relatively undissociated acetic acid, leaving an excess of OH^- ions, thus:

sodium acetate	+ water→	acetic acid	+ sodium hydroxide
Na^+Ac^-	+ HOH→	HAc +	Na^+OH^-
(dissociated)	(largely undissociated)		(dissociated)

The pH value of the solution will therefore be greater than 7. Experiment shows it to be 8.87 in 0.1 mol/litre solution at room temperature.

Similarly, ammonium chloride (NH_4Cl), the salt of a weak base and a strong acid, hydrolyses to form the relatively undissociated ammonium hydroxide (NH_4OH), leaving an excess of H^+ ions. The pH value of the solution will therefore be less than 7. Experiment shows it to be 5.13 at ordinary temperatures in a solution having a concentration of 0.1 mol/litre.

A neutralization process therefore does not always produce an exactly neutral solution when one mole of acid reacts with one mole of base.

4.4.6 Common ion effect

All organic acids and the majority of inorganic acids are weak electrolytes and are only partially dissociated when dissolved in water. Acetic acid, for example, ionizes only slightly in solution, a process represented by the equation

$$HAc \rightleftharpoons H^+ + Ac^-$$

Its dissociation constant at 25 °C is only 1.8×10^{-5}, i.e.

$$([H^+][Ac^-])/[HAc] = 1.8 \times 10^{-5} \text{ mol/litre}$$

or

$$[H^+][Ac^-] = 1.8 \times 10^{-5} [HAc]$$

Therefore in a solution of acetic acid of moderate concentration, the bulk of the acid molecules will be undissociated, and the proportion present as acetic ions and hydrogen ions is small. If one of the salts of acetic acid, such as sodium (NaAc) is added to the acetic acid solution, the ionization of the acetic acid will be diminished. Salts are, with very few exceptions, largely ionized in solution, and consequently when sodium acetate is added to the solution of acetic acid the concentration of acetic ions is increased. If the above equation is to continue to hold, the reaction $H^+ + Ac^- \rightarrow HAc$ must take place, and the concentration of hydrogen ions is reduced and will become extremely small.

Most of the acetic ions from the acid will have recombined; consequently the concentration of un-ionized acid will be practically equal to the total concentration of the acid. In addition, the concentration of acetic ions in the equilibrium mixture due to the acid will be negligibly small, and the concentration of acetic ions will, therefore, be practically equal to that from the salt. The pH value of the solution may, therefore, be regulated by the strength of the acid and the ratio [salt]/[acid] over a wide range of values.

Just as the ionization of a weak acid is diminished by the addition of a salt of the acid, so the ionization of a weak base will be diminished by the addition of a salt of the base, e.g. addition of ammonium chloride to a solution of ammonium hydroxide. The concentration of hydroxyl ions in the mixture will be given by a similar relationship to that obtained for hydrogen ions in the mixture of acid and salt, i.e.

$$[OH^-] = K[\text{alkali}]/[\text{salt}]$$

4.4.7 Buffer solutions

Solutions of a weak acid and a salt of the acid such as acetic acid mixed with sodium acetate and solutions of a weak base and one of its salts, such as ammonium hydroxide mixed with ammonium chloride (as explained above in Section 4.4.6) undergo relatively little change of pH on the further addition of acid or alkali and the pH is almost unaltered on dilution. Such solutions are called buffer solutions; they find many applications in quantitative chemical analysis. For example, many precipitations are made in certain ranges of pH values and buffer solutions of different values are used for standardizing pH measuring equipment.

Buffer solutions with known pH values over a wide range can be prepared by varying the proportions of the constituents in a buffer solution; the value of the pH is given by

$$\text{pH} = \log_{10}\left(\frac{1}{K}\right) + \log_{10}\frac{[\text{salt}]}{[\text{acid}]}$$

The weak acids commonly used in buffer solutions include phosphoric, boric, acetic, phthalic, succinic and citric acids with the acid partially neutralized by alkali or the salt of the acid used directly. Their preparation requires the use of pure reagents and careful measurement and weighing but it is more important to achieve correct proportions of acid to salt than correct concentration. An error of 10 per cent in the volume of water present may be ignored in work correct to 0.02 pH units.

National Bureau of Standards (USA) standard buffer solutions have good characteristics and for pH 4, pH 7 and pH 9.2 are available commercially, as pre-weighed tablets, sachets of powder or in solution form. Those unobtainable commercially are simple to prepare provided analytical grade reagents are used dissolved in water with a specific conductance not exceeding 2 μS/cm.

4.5 Electrode potentials

4.5.1 General theory

When a metallic electrode is placed in a solution, a redistribution of electrical charges tends to take place. Positive ions of the metal enter the solution leaving the electrode negatively charged, and the solution will acquire a positive charge. If the solution already contains ions of the metal, there is a tendency for ions to be deposited on the electrode, giving it a positive charge. The electrode eventually reaches an equilibrium potential with respect to the solution, the magnitude and sign of the potential depending upon the concentration of metallic ions in the solution and the nature of the metal. Zinc has such a strong tendency to form ions that the metal forms ions in all solutions of its salts, so that it is always negatively charged relative to the solution. On the other hand, with copper, the ions have such a tendency to give up their charge that the metal becomes positively charged even when placed in the most dilute solution of copper salt.

This difference between the properties of zinc and copper is largely responsible for the e.m.f. of a Daniell cell (Figure 4.13). When the poles are connected by a wire, sudden differences of potential are possible (a) at the junction of the wires with the poles, (b) at the junction of the zinc with the zinc sulphate, (c) at the junction of the zinc sulphate with the copper sulphate, (d) at the junction of the copper with the copper sulphate. The e.m.f. of the cell will be the algebraic sum of these potential differences.

In the measurement of the electrode potential of a metal, a voltaic cell similar in principle to the Daniell cell is used. It can be represented by the scheme

Metal 1	Solution containing ions of metal 1	Solution containing ions of metal 2	Metal 2

Under ordinary conditions, when all the cell is at the same temperature, the thermoelectric e.m.f. at the junctions of wires and electrodes will vanish.

The potential difference which arises at the junction of the solutions, known as the 'liquid junction potential', or 'diffusion potential', is due to the difference in rate of diffusion across the junction of the liquids of the cations and anions. If the cations have a greater rate of diffusion than the anions then the solution into which the cations are diffusing will acquire a positive charge, and the solution which the cations are leaving will acquire a negative charge. Therefore there is a potential gradient across the boundary. If the anions have the greater velocity, the direction of the potential gradient will be reversed. The potential difference at the junction of the two liquids may be reduced to a negligible value either by having present in the two solutions relatively large and equal concentrations of an electrolyte, such as potassium nitrate, which produces ions which diffuse with approximately equal velocities or by inserting between

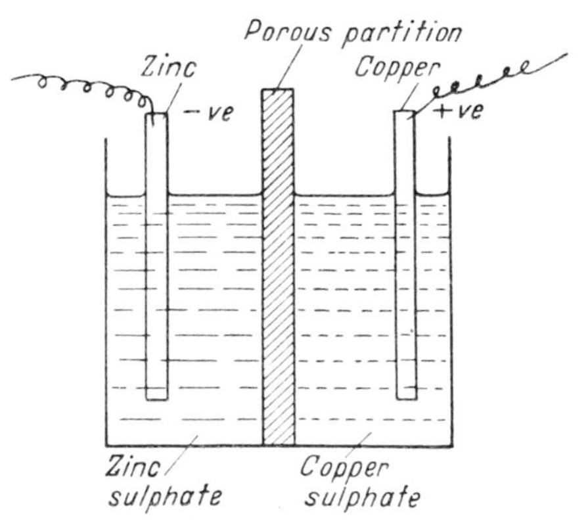

Figure 4.13 Daniell cell.

the two solutions a 'salt bridge' consisting of a saturated solution of potassium chloride or of ammonium or potassium nitrate. These salts produce ions whose diffusion rates are approximately equal.

When salt bridges are used in pH work, the liquid junction potentials are reduced to less than 1 mV unless strong acids or alkalis are involved. If an excess of neutral salt is added to the acid or alkali, the liquid junction potential will be reduced. Thus the error involved is rarely measurable on industrial instruments.

All measurements of the e.m.f. of cells give the potential of one electrode with respect to another. In the Daniell cell, all that can be said is that the copper electrode is 1 volt positive with respect to the zinc electrode. It is not possible to measure the potential of a single electrode as it is impossible to make a second contact with the solution without introducing a second metal–solution interface. Practical measurement always yields a difference between two individual electrode potentials.

In order to assign particular values to the various electrode potentials an arbitrary zero is adopted; all electrode potentials are measured relative to that of a standard hydrogen electrode (potential taken as zero at all temperatures). By convention, the half cell reaction is written as a reduction and the potential designated positive if the reduction proceeds spontaneously with respect to the standard hydrogen electrode, otherwise the potential is negative.

The standard hydrogen electrode consists of a platinum electrode coated with platinum black, half immersed in a solution of hydrogen ions at unit activity (1.228 M HCl at 20 °C) and half in pure hydrogen gas at one atmosphere pressure. In practice, however, it is neither easy nor convenient to set up a hydrogen electrode, so subsidiary reference electrodes are used, the potential of which relative to the standard hydrogen electrode has previously been accurately determined. Practical considerations limit the choice to electrodes consisting of a metal in contact with a solution which is saturated with a sparingly soluble salt of the metal and which also contains an additional salt with a common anion. Examples of these are the silver/silver chloride electrode ($Ag/AgCl_{(s)}KCl$) and the mercury/mercurous chloride electrode ($Hg/Hg_2Cl_{2s}KCl$) known as the calomel electrode. In each case the potential of the reference electrode is governed by the activity of the anion in the solution, which can be shown to be constant at a given temperature.

4.5.2 Variation of electrode potential with ion activity (the Nernst equation)

The most common measurement of electrode potential is in the measurement of pH, i.e. hydrogen ion activity, and selective ion activity, p(ion). The circuit involved is as shown in Figure 4.14.

The measured potential is the algebraic sum of the potentials developed within the system, i.e.

$$E = E_{\text{Int.ref.}} + E_s + E_j - E_{\text{Ext.ref.}}$$

where $E_{\text{Int.ref.}}$ is the e.m.f. generated at the internal reference inside the measuring electrode, E_s is the e.m.f. generated at the selective membrane, E_j is the e.m.f. generated at the liquid junction, and $E_{\text{Ext.ref.}}$ is the e.m.f. generated at the external reference electrode.

At a fixed temperature, with the reference electrode potentials constant and the liquid junction potentials zero the equation reduces to

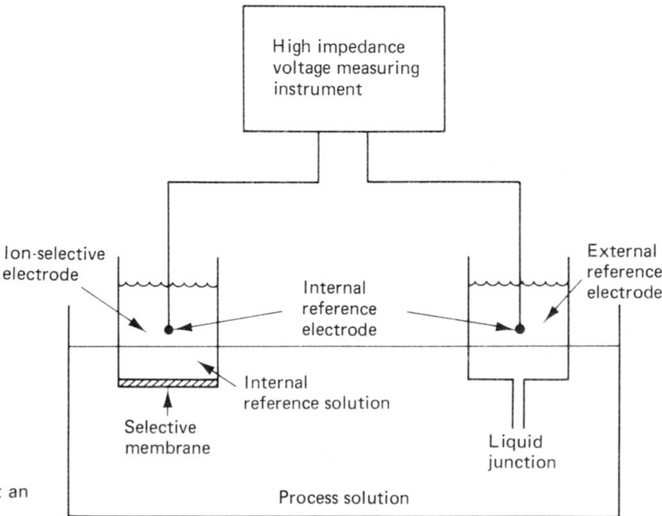

Figure 4.14 Method of measuring potential developed at an ion-selective membrane.

$$E = E' + E_s$$

where E' is a constant.

The electrode potential generated is related to the activities of the reactants and products that are involved in the electrode reactions.

For a general half cell reaction

$$\text{oxidized form} + n \text{ electrons} \rightarrow \text{reduced form}$$

or

$$aA + bB + \ldots + ne^- \rightarrow xX + yY + \ldots$$

the electrode potential generated can be expressed by the Nernst equation

$$E = E_0 + \frac{RT}{nF} \ln \frac{\text{OXID}}{\text{RED}} \text{ volts}$$

or

$$E = E_0 + 2.303 \frac{RT}{nF} \log_{10} \frac{[A]^a . [B]^b}{[X]^x . [Y]^y} \text{ volts}$$

where R is the molar gas constant (8.314 joule . mol^{-1} K^{-1}), T is absolute temperature in kelvins, F is the Faraday constant (96 487 coulomb . mol^{-1}), and n is the number of electrons participating in the reaction according to the equation defining the half cell reaction. The value of the term $2.303RT/nF$ is dependent upon the variables n and T and reduces to $0.059/n$ volts at 25 °C and $0.058/n$ volts at 20 °C.

An ion-selective electrode (say selective to sodium ions) is usually constructed so that the ion activity of the internal reference solution inside the electrode is constant and the Nernst equation reduces at constant temperature to

$$E = E_0 + \frac{RT}{nF} \ln a$$

where E_0 includes all the constants and a is the activity of the sodium ion. As sodium is a positive ion with one charge

$$E = E_0 + 59.16 \log_{10}(a) \text{ mV at } 25\,°\text{C}$$

This equation shows that a tenfold increase in ion activity will increase the electrode potential by 59.16 mV.

If the ion being measured is doubly charged the equation becomes

$$E = E_0 + \frac{59.16}{2} \log_{10}(a) \text{ mV at } 25\,°\text{C}$$

The applicability of these equations assume that the ion-selective electrode is sensitive uniquely to one ion. In most cases in practice, the electrode will respond to other ions as well but at a lower sensitivity. The equation for electrode potential thus becomes:

$$E = E_0 + 59.16 \log_{10}(a_1 + K_2 a_2 + \ldots) \text{ mV}$$

where $K_2 a_2$ etc. represents the ratio of the sensitivity of the electrode of the ion 2 to that of ion 1. The literature on ion-selective electrodes provided by manufacturers usually gives a list of interfering ions and their sensitivity ratios.

4.6 Ion-selective electrodes

Whereas, formerly, ion-selective electrodes were used almost exclusively for measuring hydrogen ion activity (pH), many electrodes have now been developed to respond to a wide range of selected ions. These electrodes are classified into five groups according to the type of membrane used.

4.6.1 Glass electrodes

The glass electrode (Figure 4.15(a)) used for pH measurement is designed to be selective to hydrogen ions but, by choosing the composition of the glass membrane, glass electrodes selective to sodium, potassium, ammonium, silver and other univalent cations can be made.

4.6.2 Solid state electrodes

In these electrodes the membrane consists of a single crystal or a compacted disc of the active material. In Figure 4.15(b) the membrane isolates the reference solution from the solution being measured. In Figure 4.15(c) the membrane is sealed with a metal backing with a solid metal connection. A solid state electrode selective to fluoride ions employs a membrane of lanthanum fluoride (LaF_3). One which is selective to sulphide ions has a membrane of silver sulphide. There are also electrodes available for measurement of Cl^-, Br^-, I^-, Ag^+, Cu^{2+}, Pb^{2+}, Cd^{2+}, and CN^- ions.

4.6.3 Heterogeneous membrane electrodes

These are similar to the solid state electrodes but differ in having the active material dispersed in an inert matrix. Electrodes in this class are available for Cl^-, Br^-, I^-, S^{2-}, and Ag^+ ions.

4.6.4 Liquid ion exchange electrodes

In this type of electrode (Figure 4.15(d)) the internal reference solution and the measured solution are separated by a porous layer containing an organic liquid of low water solubility. Dissolved in the organic phase are large molecules in which the ions of interest are incorporated. The most important of these electrodes is the calcium electrode, but other electrodes in this class are available for the determination of Cl^-, ClO_4^-, NO_3^-, Cu^{2+}, Pb^{2+}, and BF_4^- ions. The liquid

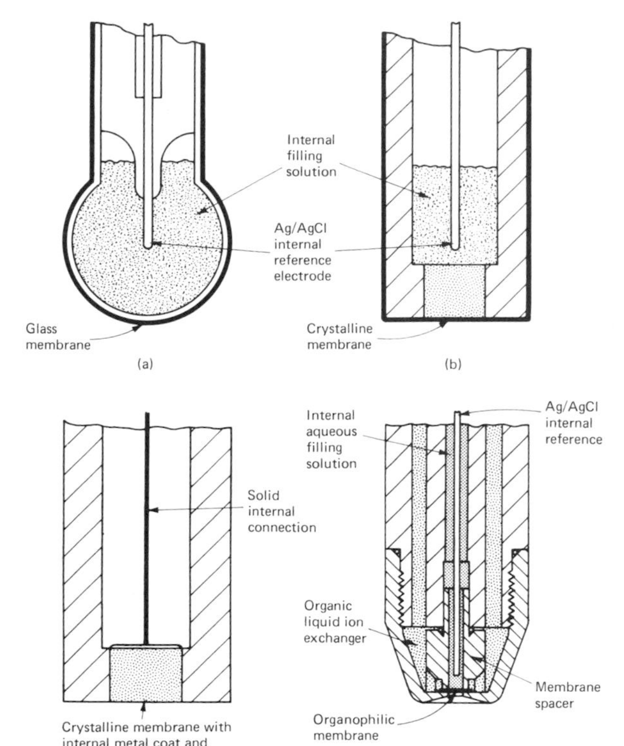

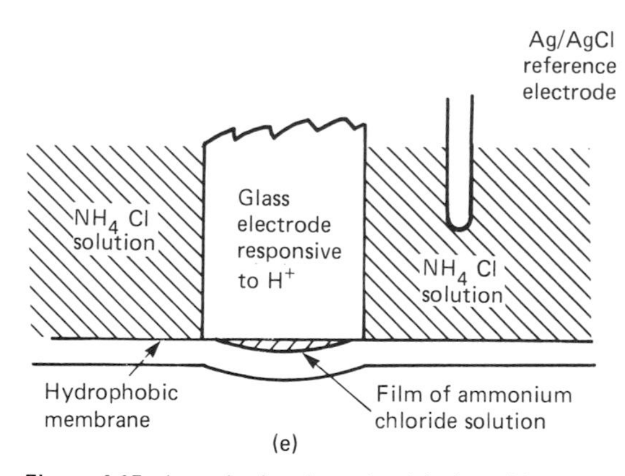

Figure 4.15 Ion-selective electrodes: (a) glass, (b) crystalline membrane with internal reference electrode, (c) crystalline membrane with solid connection, (d) liquid ion exchange, (e) gas sensing membrane (courtesy Orion Research Inc.).

ion exchange electrodes have more restricting chemical and physical limitations than the glass or solid state electrodes but they may be used to measure ions which cannot yet be measured with a solid state electrode.

4.6.5 Gas-sensing membrane electrodes

These electrodes are not true membrane electrodes as no current passes across the membrane. They are complete electrochemical cells, monitored by an ion-selective electrode as the internal chemistry is changed by the ion being determined passing from the sample solution across the membrane to the inside of the cell.

An example is an ammonia electrode (Figure 4.15(e)). The sensing surface of a flat-ended glass pH electrode is pressed tightly against a hydrophobic polymer membrane which is acting as a seal for the end of a tube containing ammonium chloride solution. A silver/silver chloride electrode is immersed in the bulk solution. The membrane permits the diffusion of free ammonia (NH_3), but not ions, between the sample solution and the film of ammonium chloride solution. The introduction of free ammonia changes the pH of the internal ammonium chloride solution which is sensed by the internal glass pH electrode.

4.6.6 Redox electrodes

In elementary chemistry a substance is said to be oxidized when oxygen is combined with it and said to be reduced when oxygen is removed from it. The definition of oxidation and reduction may, however, be extended. Certain elements, e.g. iron and tin, can exist as salts in more than one form. Iron, for example can be combined with sulphuric acid in the form of ferrous iron, valency 2, or ferric iron, valency 3.

Consider the reaction:

ferrous sulphate	+	chlorine	=	ferric chloride	+	ferric sulphate
$6FeSO_4$	+	$3Cl_2$	=	$2FeCl_3$	+	$2Fe_2(SO_4)_3$

The ferrous sulphate is oxidized to ferric sulphate; chlorine is the oxidizing agent. In terms of the ionic theory, the equation may be written

$$6Fe^{2+} + 3Cl_2 = 6Fe^{3+} + 6Cl^-$$

i.e. each ferrous ion loses an electron and so gains one positive charge. When a ferrous salt is oxidized to a ferric salt each mole of ferrous ions gains one mole (1 faraday) of positive charges or loses one mole of negative charges, the negative charge so lost being taken up by the oxidizing agent (chlorine). Oxidation, therefore, involves the loss of electrons; reduction, the gain of electrons. Thus the oxidation of a ferrous ion to ferric ion can be represented by the equation

$$Fe^{2+} - e = Fe^{3+}$$

When a suitable electrode, such as an inert metal which is not attacked by the solution and which will not catalyse side reactions, is immersed in a solution containing both ferrous and ferric ions, or some other substance in the reduced and oxidized state, the electrode acquires a potential which will depend upon the tendency of the ions in the solution to pass from a higher or lower state of oxidation. If the ions in solution tend to become oxidized (i.e. the solution has reducing properties) the ions tend to give up electrons to the electrode which will become negatively charged relative to the solution. If, on the other hand, the ions in solution tend to become reduced (i.e. the solution has oxidizing properties), then the ions will tend to take up electrons from the electrode and the electrode will become positively charged relative to the solution. The sign and magnitude of the electrode potential, therefore, gives a measure of the oxidizing or reducing power of the solution, and the potential is called the oxidation–reduction or redox potential of the solution, E_h. The potential E_h may be expressed mathematically by the relationship

$$E_h = E_0 + (RT/nF)\log_{10}(a_o/a_r)$$

where a_o is the activity of the oxidized ion and a_r is the activity of the reduced ion.

To measure the oxidation potential it is necessary to use a reference electrode to complete the electrical circuit. A calomel electrode is often used for this (see Section 4.7 below).

The measuring electrode is usually either platinum or gold, but other types are used for special measurements: as, for example, the hydrogen electrode for use as a primary standard and the quinhydrone electrode for determining the pH of hydrofluoric acid solutions. However, the latter two electrodes do not find much application in industrial analytical chemistry.

4.7 Potentiometry and specific ion measurement

4.7.1 Reference electrodes

All electrode potential measurements are made relative to a reference electrode and the e.m.f. generated at this second contact with the solution being tested must be constant. It should also be independent of temperature changes (or vary in a known manner), be independent of the pH of the solution and remain stable over long periods.

Standard hydrogen electrodes are inconvenient (p. 130) and in practice three types of reference are commonly used.

Silver/silver chloride electrode This consists of a silver wire or plate, coated with silver chloride, in contact with a salt bridge of potassium chloride saturated with silver chloride. The concentration of the potassium chloride may vary from one type of electrode to another but concentrations of 1.00 or 4.00 mol per litre or a saturated solution are quite common. This saturated type of electrode has a potential of −0.199 V relative to a hydrogen electrode. It has a variety of physical forms which are discussed below.

Mercury/mercurous chloride or calomel electrode The metal used is mercury which has a high resistance to corrosion and being fluid at ambient temperature cannot be subject to strain. The mercury is in contact with either mercurous chloride or in some electrodes with mercurous chloride and potassium chloride paste. Contact with the measured solution is through a salt bridge of potassium chloride whose concentration may be 3.8 mol per litre or some other concentration appropriate to the application. Contact with the mercury is usually made by means of a platinum wire which may be amalgamated. The calomel, saturated potassium chloride, electrode has a potential relative to the hydrogen electrode of −0.244 V.

Where the use of potassium salt is precluded by the condition of use, it may be replaced by sodium sulphate, the bridge solution having a concentration of 1 mol per litre.

Whatever the type of the reference electrode, contact must be made between the salt bridge and the measured solution. Two common methods are through a ceramic plug whose shape and porosity govern the rate at which the salt bridge solution diffuses out and the process solution diffuses into and contaminates the bridge solution. If the plug is arranged to have a small cross-sectional area relative to its length, the rate of diffusion is very small (say less than 0.02 cm^3/day) and the electrode can be considered to be sealed and is used until it becomes unserviceable. It is then replaced by a similar electrode.

Where the application warrants it a high rate of diffusion from the electrode has to be tolerated (say 1 or 2 cm^3/day), so the relative dimensions and porosity of the plug are changed, or it is replaced by a glass sleeve which permits relatively fast flow of salt bridge solution, thus reducing the rate and degree of fouling of the junction. In these circumstances, the electrode is refilled on a routine basis, or a continuous supply of bridge solution is arranged into the electrode at the appropriate pressure for the application.

A wide range of electrodes are illustrated in Figures 4.16–4.19. The choice of the appropriate reference electrode for the application is vital, and consideration must be given to the pressure, temperature and nature of the process stream. The accuracy of the measurement and the frequency of maintenance depends upon the correct choice of electrode. The e.m.f. of the reference electrode will only remain constant provided satisfactory contact is made by the salt bridge, so the junction must not become plugged by suspended solids, viscous liquids, or reaction products of the

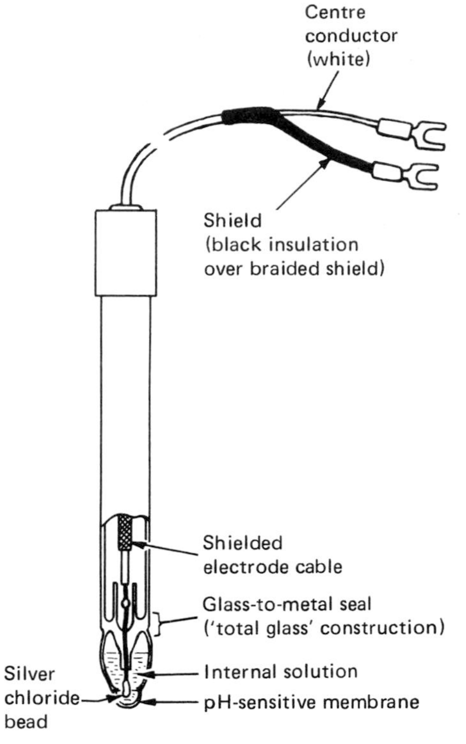

Figure 4.16 pH measuring electrode (courtesy The Foxboro Company).

process stream. Where this is a danger, the faster flow type of plug must be used. Many routine measurements can, however, be made with the non-flowing electrode, thus avoiding the necessity of refilling, or arranging a pressurized continuous supply. Flowing types of junctions are usually required where an accuracy of ± 0.02 pH units (± 1 or 2 mV) is required, where frequent or large temperature or composition changes occur, or where the process fluid is such that it is prone to foul the junction.

The temperature of operation will influence the choice of concentration of the filling solutions. Potassium chloride solution having a concentration of 4 mol per litre saturates and starts to precipitate solids at about 19 °C, and will freeze at -4 °C, while if the concentration is reduced to 1 mol per litre the solution will freeze at -2 °C without becoming saturated. Thus, no precipitation will take place in the solution of lower concentration. Although not damaging, precipitated potassium chloride and associated silver chloride will tend to clog reference junctions and tubes, decreasing electrolyte flow rate, and increasing the risk of spurious potentials. For these reasons, flowing reference electrodes are not recommended for low temperature applications unless provision is made to prevent freezing or precipitation in the electrode and any associated hardware.

When materials such as sulphides, alkali phosphates or carbonates, which will react with silver, are present in the process stream, either non-flowing electrodes, or electrodes containing potassium chloride at 1 mol per litre should be used. The diffusion rate of silver can be neglected in the non-flowing type, and the solubility of silver chloride in potassium chloride at a concentration of 1 mol per litre is only 1 or 2 per cent of that in a solution at 4 mol per litre.

High temperatures with wide fluctuations are best handled by potassium chloride solution at 1 mol per litre.

4.7.2 Measurement of pH

Glass electrode Almost all pH measurements are best made with a glass electrode (the earliest of the ion-selective electrodes), the e.m.f. being measured relative to a reference electrode. The glass electrode can be made to cover practically the whole of the pH scale and is unaffected by most chemicals except hydrofluoric acid. It can also be used in the presence of oxidizing or reducing agents without loss of measuring accuracy.

The electrode consists of a thin membrane of

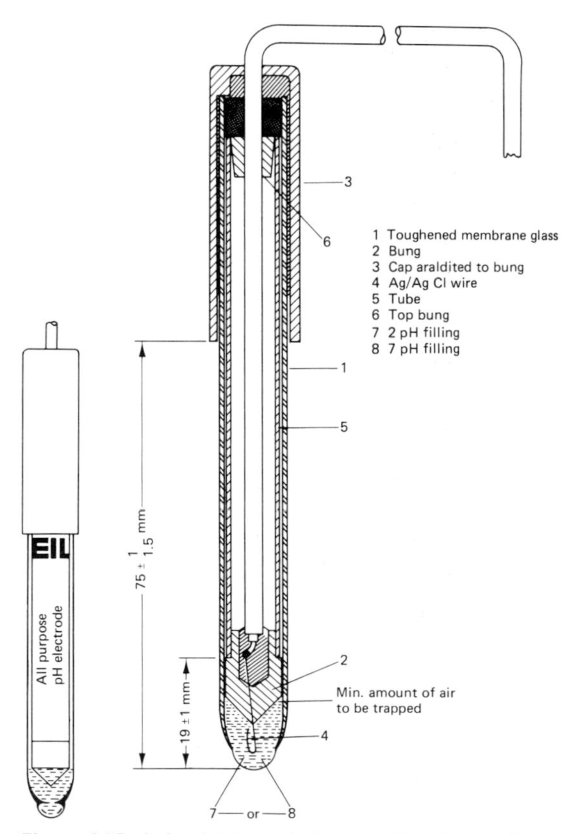

Figure 4.17 Industrial electrode (courtesy Kent Industrial Measurements Ltd., Analytical Instruments).

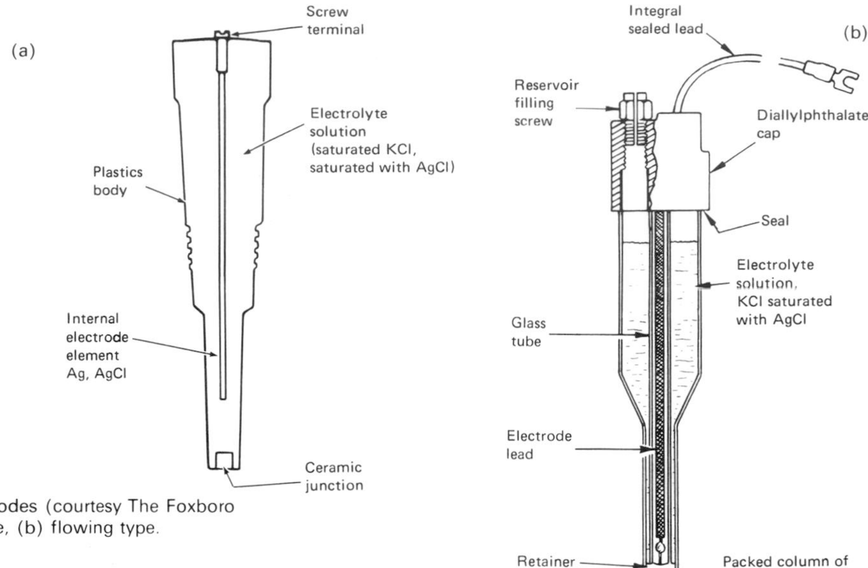

Figure 4.18 Reference electrodes (courtesy The Foxboro Company): (a) sealed electrode, (b) flowing type.

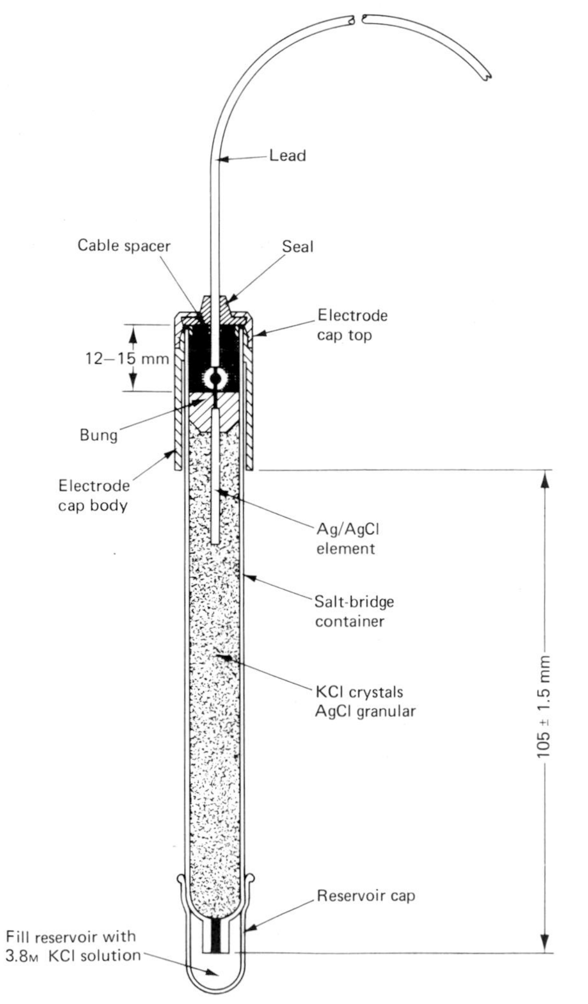

sodium-ion-selective glass sealed onto the end of a glass tube that has no ion-selective properties. The tube contains an internal reference solution in which is immersed the internal reference electrode and this is connected by a screened lead to the pH meter. The internal reference electrode is almost always a silver/silver chloride electrode although, recently, Thalamid electrodes* have sometimes been used. The internal reference solution contains chloride ions to which the internal silver/silver chloride reference electrode responds and hydrogen ions to which the electrode as a whole responds. The ion to which the glass electrode responds, hydrogen in the case of pH electrodes, is determined by the composition of the glass membrane.

* The Thalamid electrode is a metal in contact with a saturated solution of the metallic chloride. Thallium is present as a 40 per cent amalgam and the surface is covered with solid thallous chloride. The electrode is immersed in saturated potassium chloride solution. Oxygen access is restricted to prevent the amalgam being attacked. The advantage of the Thalamid electrode is that there is scarcely any time lag in resuming its electrode potential after a temperature change.

Figure 4.19 Sealed silver/silver chloride electrode (courtesy Kent Industrial Measurements Ltd., Analytical Instruments).

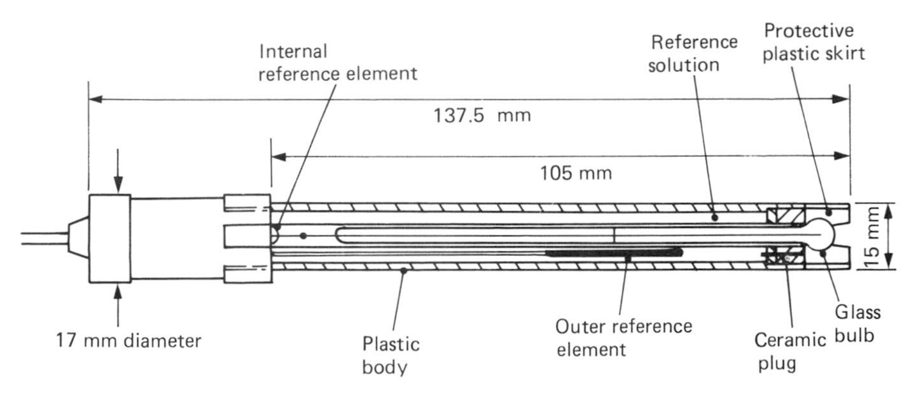

Figure 4.20 Combined reference electrode and glass electrode for pH measurement (courtesy Kent Industrial Measurements Ltd., Analytical Instruments).

A glass pH electrode can be represented as

reference electrode	test solution a_{H^+}	glass membrane	internal reference solution $a'_{H^+}+a'_{Cl}$	AgCl	Ag

Glass electrodes for pH measurement are of three main types (a) general-purpose for wide ranges of pH over wide ranges of temperature, (b) low temperature electrodes (less than 10 °C) which are low resistance electrodes and are generally unsuitable for use above pH 9 to 10, (c) high pH and/or high temperature electrodes (greater than 12 pH units). Glass electrodes are manufactured in many forms some of which are shown in Figures 4.16 and 4.20. Spherical membranes are common but hemispherical or conical membranes are available to give increased robustness where extensive handling is likely to occur. Electrodes with flat membranes can be made for special purposes such as measurement of the pH of skin or leather and micro-electrodes are available but at great expense. Combination glass and reference electrodes (Figure 4.20) can be obtained and some electrodes can be steam-sterilized.

New electrodes supplied dry should be conditioned before use as the manufacturer recommends or by leaving them overnight in 0.1 mol/litre^{-1} hydrochloric acid. Electrodes are best not allowed to dry out and they should be stored in distilled or demineralized water at temperatures close to those at which they are to be used. The best treatment for pH electrodes for high pH ranges is probably to condition them and store them in borax buffer solution.

Electrical circuits for use with glass electrodes For measurement of pH the e.m.f. in millivolts generated by the glass electrode compared with that of the reference electrode has to be converted to a pH scale, that is, one showing an increase of one unit for a decrease in e.m.f. of approximately 60 mV. The pH scale requires the use of two controls – the calibration control and the slope control. The latter may not always be identified as such on the pH meter as it often acts in the same way as the temperature compensation control. The slope and temperature compensation controls adjust the number of millivolts equivalent to one pH unit. The calibration control relates the measured e.m.f. to a fixed point on the pH scale.

A typical pH measuring system (glass electrode and reference electrode immersed in a solution) may have a resistance of several hundred megohms. To obtain an accurate measurement of the e.m.f. developed at the measuring electrode, the electrical measuring circuit must have a high input impedance and the insulation resistance of the electrical leads from the electrodes to the measuring circuit must be extremely high ($\sim 10^5$ MΩ – a 'Megger' test is useless). The latter is best achieved by keeping the electrode leads as short as possible and using the best moisture-resistant insulating materials available (e.g. polythene or silicone rubber).

The usual method of measurement is to convert the developed e.m.f. into a proportional current by means of a suitable amplifying system. The essential requirements of such a system have been met completely by modern electronic circuits and one system uses an amplifier with a very high negative feedback ratio. This means that the greater part of the input potential is balanced by a potential produced by passing the meter current through an accurately known resistor, as shown in Figure 4.21. If the p.d. V_0, developed across the feedback resistance is a very large fraction of the measured potential V_1, then the input voltage V is a very small fraction of V_1, and

$$I_0=(V_1-V)/R, \text{ approaches } V_1/R$$

With modern integrated circuit techniques it is possible to obtain an amplifier with a very high input impedance and very high gain, so that little or no current is drawn from the electrodes.

Such a system is employed in the pH-to-current converter shown in Figure 4.22 which employs zener

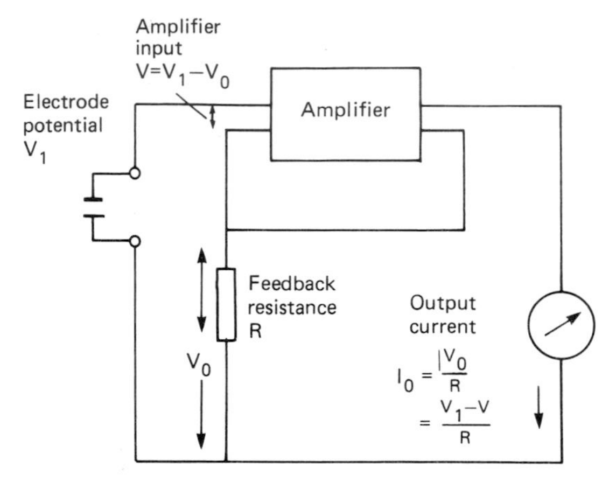

Figure 4.21 Principle of d.c. amplifier with negative feedback (courtesy Kent Industrial Measurements Ltd.).

diode stabilized supplies and feedback networks designed to give a high gain, high input impedance diode bridge amplifier.

The d.c. imbalance signal resulting from the pH signal, asymmetry correcting potential and the feedback voltage, changes the output of a capacity balance diode bridge. This output feeds a transistor amplifier which supplies feedback and output proportional to the bridge error signal. Zener diode stabilized and potentiometer circuits are used to provide continuous adjustment of span, elevation, and asymmetry potential over the entire operating range of the instrument.

The input impedance of the instrument is about $1 \times 10^{12}\,\Omega$ and the current taken from the electrodes less than 0.5×10^{-12} A.

The principle of another system which achieves a similar result is shown in Figure 4.23. It uses a matched

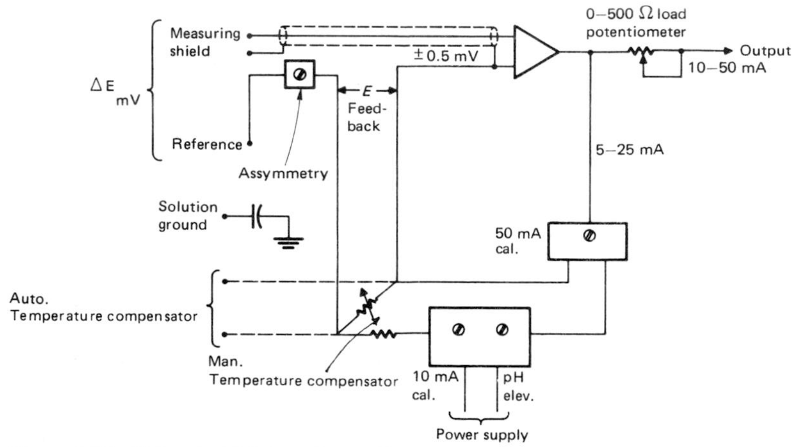

Figure 4.22 High gain, high impedance pH-to-current converter (courtesy The Foxboro Company).

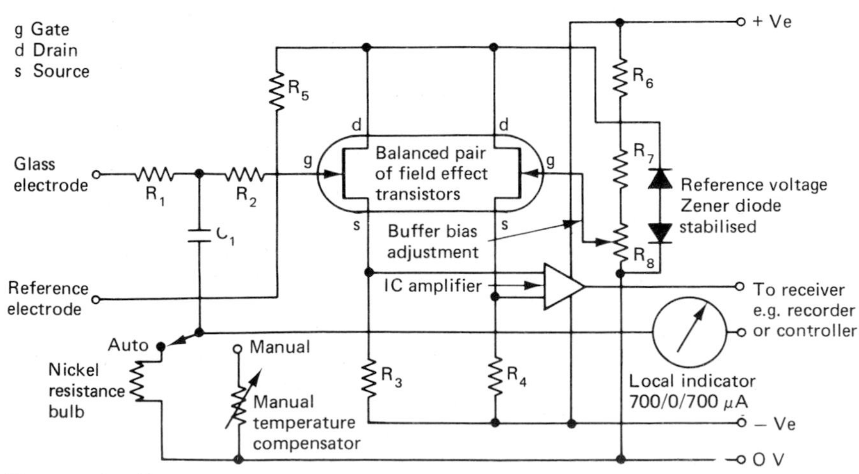

Figure 4.23 pH measuring circuit using field effect transistors.

pair of field effect transitors (FETs) housed in a single can. Here the e.m.f. produced by the measuring electrode is fed to the gate of one of the pair. The potential which is applied to one side of the high gain operational amplifier will be governed by the current which flows through the transistor and its corresponding resistance R_3. The potential applied to the gate of the second FET is set by the buffer bias adjustment which is fed from a zener stabilized potential supply. The potential developed across the second resistance R_4 which is equal in resistance to R_3 will be controlled by the current through the second of the pair of matched FETs. Thus the output of the operational amplifier will be controlled by the difference in the potentials applied to the gates of the FETs, that is, to the difference between the potential developed on the measuring electrode and the highly stable potential set up in the instrument. Thus, the current flowing through the local and remote indicators will be a measure of the change of potential of the measuring electrode.

If the e.m.f. given by the glass electrode is plotted against pH for different temperatures it will be seen that there is a particular value of the pH at which the e.m.f. is independent of temperature. This point is known as the 'iso-potential point'.

If the iso-potential point is arranged to be the locus of the slope of the measuring instrument, the pH measuring circuit can be modified to include a temperature sensor arranged to change the negative feed-back so that the circuit compensates for the change in slope of the e.m.f./pH relationship. It is important to realize that the temperature compensation only corrects for the change in the electrode response due to temperature change and the iso-potential control setting therefore enables pH electrodes calibrated at one temperature to be used at another. The iso-potential control does *not* compensate for the actual change in pH of a solution with temperature. Thus if pH is being measured to establish the composition of a solution one must carry out the measurements at constant temperature.

A few commercial pH meters have a variable iso-potential control so that they can be used with several different combinations of electrodes but it is more generally the case that pH meters have fixed iso-potential control settings and can only be used with certain combinations of pH and reference electrodes. It is strongly recommended that, with fixed iso-potential control settings, both the glass and reference electrodes be obtained from the manufacturer of the pH meter. Temperature compensation circuits generally work only on the pH and direct activity ranges of a pH meter and not on the millivolt, expanded millivolt and relative millivolt ranges.

Modern pH meters with analogue displays are scaled 0 to 14 pH units with the smallest division on the scale equivalent to 0.1 unit giving the possibility of estimating 0.02 pH units by interpolation. The millivolt scale is generally 0 to 1400 mV with a polarity switch, or -700 to $+700$ mV without one. The smallest division is 10 mV, allowing estimation to 2 mV. Many analogue meters have a facility of expanding the scale so that the precision of the reading can be increased up to 10 times. Digital outputs are also available with the most sensitive ones reading to 0.001 pH unit (unlikely to be meaningful in practice) or 0.1 mV. Instruments incorporating microprocessors are also now available – these can calculate the concentration of substances from pH measurements and give readout in concentration units. Blank and volume corrections can be applied automatically.

Precision and accuracy Measurements reproducible to 0.05 pH units are possible in well buffered solutions in the pH range 3 to 10. For routine measurements it is rarely possible to obtain a reproducibility of better than ± 0.01 pH units.

In poorly buffered solutions reproducibility may be no better than ± 0.1 pH unit and accuracy may be lost by the absorption of carbon dioxide or by the presence of suspensions, sols and gels. However, measured pH values can often be used as control parameters even when their absolute accuracies are in doubt.

Sodium ion error Glass electrodes for pH measurement are selective for hydrogen ions, not uniquely responsive to them, and so will also respond to sodium and other ions especially at alkaline pH values (more than about 11). This effect causes the pH value to be underestimated. Sodium ions produce the greatest error, lithium ions about a half, potassium ions about a fifth and other ions less than a tenth of the error due to sodium ions. One can either standardize the electrode in an alkaline buffer solution containing a suitable concentration of the appropriate salt or, better, use the special lithium and caesium glass electrodes developed for use in solutions of high alkalinity. These are less prone to interference. For a given glass electrode at a stated measuring temperature the magnitude of the error can be found from tables provided by electrode manufacturers. An example is shown in Figure 4.24.

Temperature errors The calibration slope and standard potential of ion-selective electrodes (including glass pH electrodes) are affected by temperature. If the pH is read directly off the pH scale, some form of temperature correction will be available, but often only for the calibration slope and not for the standard potential. If measurements are made at a temperature different from that at which the electrode was calibrated there will be an error. This will be small if the meter has an iso-potential setting. For the most accurate work the sample and buffer solutions should

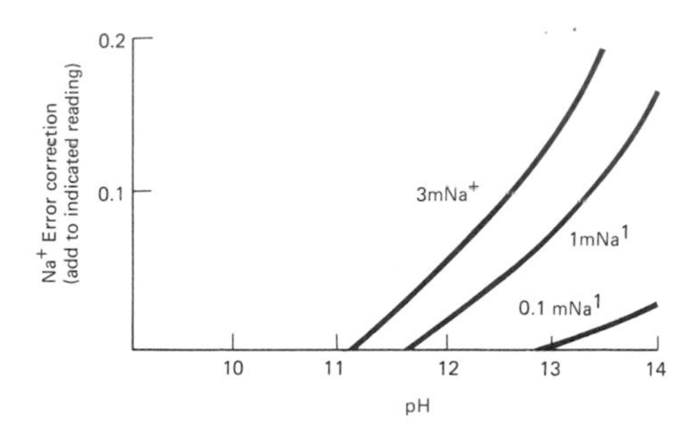

Figure 4.24 Relationship of pH and Na ion error (courtesy Kent Industrial Measurements Ltd., Analytical Instruments).

be at the same temperature, even if iso-potential correction is possible.

Stirring factor In well buffered solutions it may not be necessary to stir when making pH measurements. However, it is essential in poorly buffered solutions.

The hydrogen electrode The hydrogen electrode, consisting in practice of a platinum plate or wire coated with platinum block (a finely divided form of the metal) can measure hydrogen ion activity when hydrogen is passed over the electrode. However this electrode is neither easy nor convenient to use in practice and is now never used in industrial laboratories nor on plant.

The antimony electrode The antimony electrode is simply a piece of pure antimony rod (~12 mm diameter, 140 mm long), housed in a protective plastic body resistant to acid attack, see Figure 4.25. The protruding antimony rod when immersed in a solution containing dissolved oxygen becomes coated with antimony trioxide Sb_2O_3 and the equilibria governing the electrode potential are:

$$Sb \rightarrow Sb^{3+} + 3e^-$$

$$Sb_2O_3 + 6H^+ \rightarrow 2Sb^{3+} + 3H_2O, \quad K = \frac{[Sb^{3+}]}{[H^+]^3}$$

However, there are many possible side reactions depending on the pH and the oxidizing conditions; salt effects are large. There is therefore difficulty in calibrating with buffer solutions; stirring temperature and the amount of oxygen present all have rather large effects. A reproducibility of about 0.1 pH unit is the best that is normally attained, the response is close to Nernstian over the pH range 2 to 7 and the response time can be as short as 3 minutes but is often about 30 minutes.

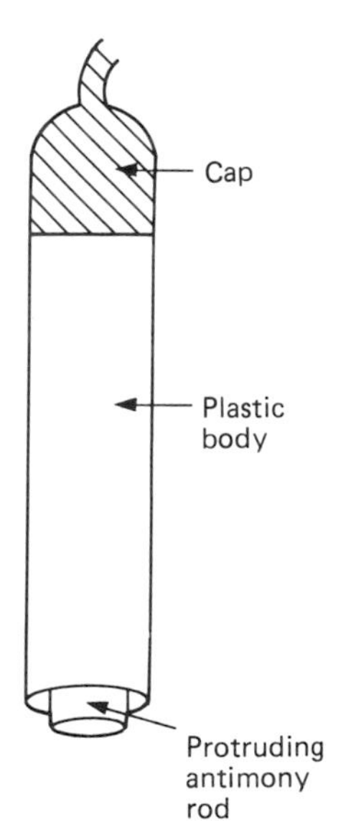

Figure 4.25 Antimony electrode.

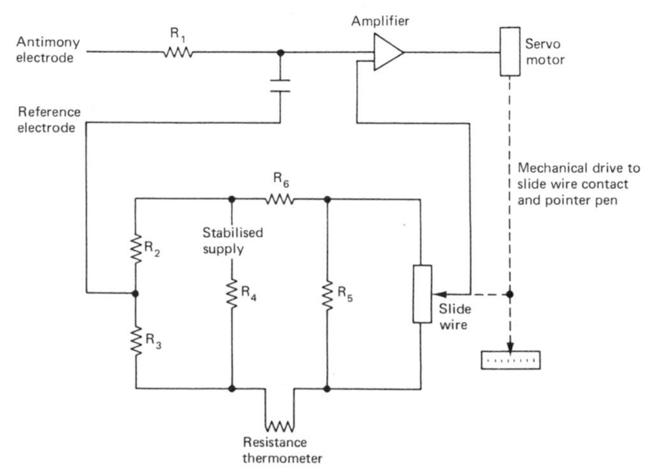

Figure 4.26 Low impedance measuring circuit for use with antimony electrodes.

The outstanding advantage of the antimony electrode is its ruggedness and for this reason it has been used for determining the pH of soils. Also, of course, it is indispensable for solutions containing hydrofluoric acid which attack glass. If the electrode becomes coated during use its performance can be restored by grinding and polishing the active surface and then reforming the oxide film by immersion in oxygenated water before using in deoxygenated solutions.

However, there is much more uncertainty to every aspect of behaviour of the antimony electrode than with the glass electrode and even the fragile glass electrodes of years ago with their limited alkaline range displaced the antimony electrode when accurate pH measurements were required. Modern glass electrodes are excellent in respect of robustness and range and antimony electrodes are not much used apart from the specialized applications already mentioned. In these, the resistance of the measuring system is low, so a simple low impedance electrical circuit can be used with them, for example a voltmeter or a potentiometric type of system as described in Volume **0**. Figure 4.26 shows the principle of such a system. Any difference between the electrode e.m.f. and that produced across the potentiometer will be amplified and applied to the servo-motor which moves the slide-wire contact to restore balance.

Industrial pH systems with glass electrodes Two types of electrode systems are in common use: the continuous-flow type of assembly, and the immersion, or dip-type of assembly.

Continuous-flow type of assembly The physical form of the assembly may vary a little from one manufacturer to another but Figure 4.27 illustrates a typical assembly designed with reliability and easy maintenance in mind. Constructed in rigid PVC throughout, it operates at pressure up to 2 bar and temperatures up to 60 °C. For higher temperatures and pressures the assembly may be made from EN 58J stainless steel, flanged and designed for straight-through flow when pressures up to 3 bar at temperatures up to 100 °C can be tolerated. It accommodates the standard measuring electrode, usually of toughened glass.

A reservoir for potassium chloride (or other electrolyte) forms a permanent part of the electrode holder. A replaceable reference element fits into the top of the reservoir, and is held in place by an easily detachable clamp nut. A microceramic plug at the lower end of the reservoir ensures slow electrolyte leakage (up to six months continuous operation without attention is usually obtained). The ceramic junction is housed in a screw-fitting plug, and is easily replaceable.

The close grouping of electrodes makes possible a small flow cell, and hence a fast pH response at low flow rates. An oil-filled reservoir built into the electrode holder houses a replaceable nickel wire resistance element, which serves as a temperature compensator. (This is an optional fitment.)

The flow through the cell creates some degree of turbulence and thus minimizes electrode coating and sedimentation.

The integral junction box is completely weatherproof and easily detachable. Electrode cables and the output cable are taken via individual watertight compression fittings into the base of the junction box. A desiccator is included to absorb moisture which may be trapped when the cover is removed and replaced.

Two turns of the lower clamp nut allow the entire electrode unit to be detached from the flow cell and hence from the process fluid. The electrodes can be immersed easily in buffer solution.

Immersion type Basically this assembly is similar to the flow type except that the flow cell is replaced by a protecting guard which protects the electrode but allows a free flow of solution to the electrodes. Also the upper cap is replaced by a similarly moulded tube which supports the electrode assembly, but brings the terminal box well above the electrode assembly so that the terminals are clear of the liquid surface when the assembly is in the measured solution. Immersion depths up to 3 m are available.

Electrode assemblies should be designed so that the electrodes can be kept wet when not in use. It is often possible to arrange for the easy removal of the assembly from the process vessel so that it can be immersed in a bucket filled with process liquid, water or buffer solution during shut-down.

The design of the assembly is often modified, to suit the use. For example, in measuring the pH of pulp in a paper beater the electrodes and resistance bulb are mounted side by side in a straight line and then inclined downstream at about 45° from the vertical so that they present no pockets to collect pulp and are self-cleaning.

When the assembly is immersed in a tank, care must be taken in the siting to ensure the instrument is measuring the properties of a representative sample; adequate mixing of the process material is essential. Sometimes it is more convenient to circulate the contents of a tank through a flow type of assembly and then return the liquid to the tank.

The main cause of trouble in electrode assemblies is the fouling of the electrodes. In order to reduce this, two forms of self-cleaning are available and the choice of method is dependent on the application. Where the main cause of trouble is deposits on the glass electrode and mechanical cleaning is required, this may be achieved by the cleaning attachment shown on a dip

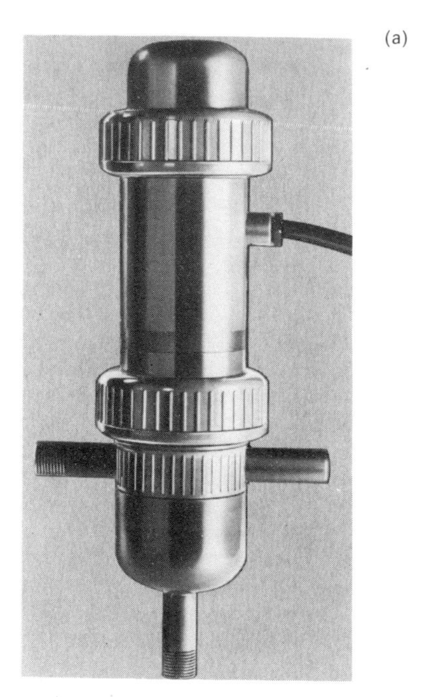

(a)

(b)

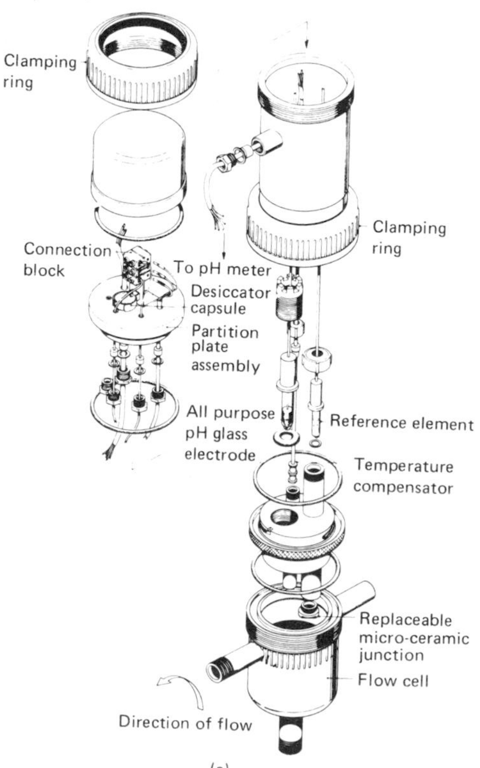

(c)

Figure 4.27 Flow-type of electrode system (courtesy Kent Industrial Measurements Ltd., Analytical Instruments): (a) external view, (b) upper section detaches for easy buffering, (c) exploded view showing the components.

system in Figure 4.28. The pneumatically driven rubber membrane wipes the electrode, providing a simple reliable cleaning action. It is driven by compressed air at preset intervals from a controller which incorporates a programmed timer mechanism that governs the frequency of the wiping action. The cleaning attachment is constructed entirely of polypropylene and 316 stainless steel, except for the rubber wiper which may be replaced by a polypropylene brush type should this be more suitable.

Alternatively an ultrasonic generator operating at 25 kHz, can be fitted to the electrode assembly, this greatly increasing the periods between necessary electrode cleaning.

4.7.3 Measurement of redox potential

When both the oxidized and reduced forms of a substance are soluble in water the old-fashioned metal redox electrode is useful – an equilibrium being set up between the two forms of the substance and the electrons in the metal electrode immersed in the solution. Again a reference electrode, generally calomel, has to be used and determinations can be made either by using the redox electrode as an indicator during titrations or by direct potentiometric determination. Arrangements are similar to those for a pH electrode. Redox electrodes, too, can be immersed directly in a liquid product stream when monitoring on plant. The high impedance e.m.f. measuring circuits as used for pH electrode systems are completely satisfactory but, as metal redox electrodes are low resistance systems, low impedance e.m.f. measuring

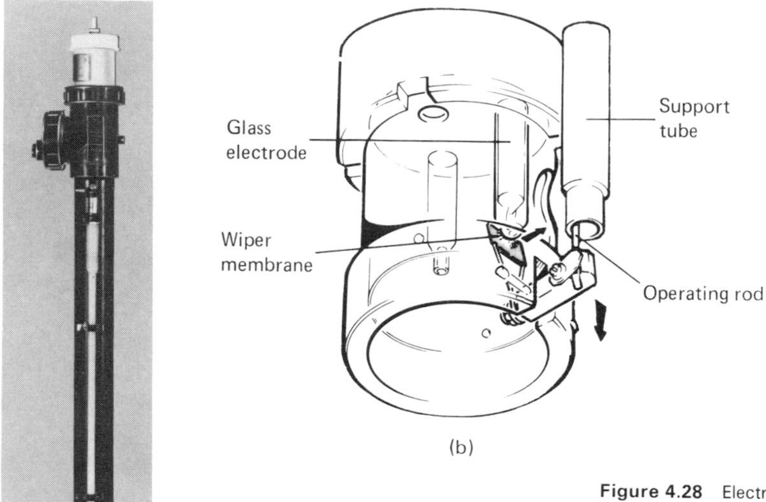

Figure 4.28 Electrode cleaning (courtesy Kent Industrial Measurements Ltd., Analytical Instruments): (a) assembly, (b) detail of cleaning attachment.

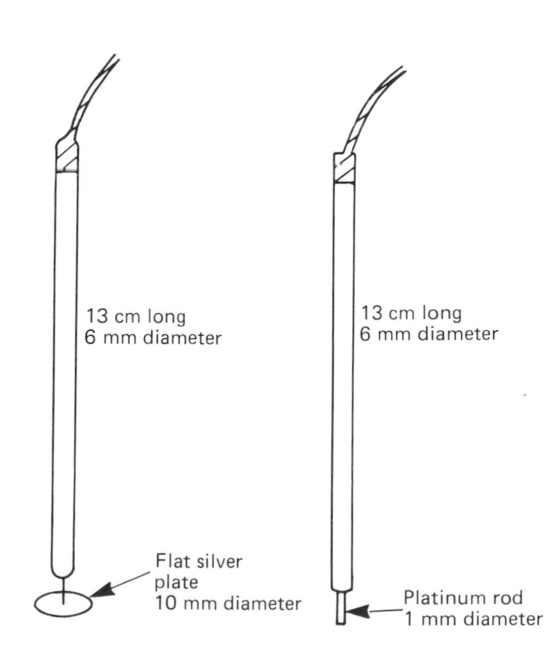

Figure 4.29 Examples of metal redox electrodes.

circuits may also be used as for the antimony pH electrode. (The latter is also a metal redox electrode.)

Apart from the antimony electrode, platinum, silver and gold electrodes (Figure 4.29) are available commercially and simple electrodes for use with separate reference and combination electrodes can be obtained both for laboratory and industrial use.

Analytical chemistry applications of redox electrodes include determination of arsenic, cyanides, hydrogen peroxide, hypochlorite or chlorine, ferrous iron, halides, stannous tin, and zinc. The silver electrode is widely used for halide determination. Platinum electrodes are suitable for most other determinations with the exception of when cyanide is being oxidized with hypochlorite (for example, in neutralizing the toxic cyanide effluent from metal plating baths). In this case a gold electrode is preferable.

4.7.4 Determination of ions by ion-selective electrodes

General considerations The measurement of the concentration or the activity of an ion in solution by means of an ion-selective electrode is as simple and rapid as making a pH measurement (the earliest of ion-selective electrodes). In principle it is necessary only to immerse the ion-selective and reference electrodes in the sample, read off the generated e.m.f. by means of a suitable measuring circuit and obtain the result from a calibration curve relating e.m.f. and concentration of the substance being determined. The difference from pH determinations is that most ion-selective electrode applications require the addition of a reagent to buffer or adjust the ionic strength of the sample before measurement of the potential. Thus, unlike measurement of pH and redox potentials, ion-selective electrodes cannot be immersed directly in a plant stream of liquid product and a sampling arrangement has to be used. However this can usually be done quite simply.

pH and pIon meters High impedance e.m.f. measuring circuits must be used with most ion-selective electrodes and are basically the same as used for measuring pH with a glass electrode. The pH meters measure e.m.f. in millivolts and are also scaled in pH units. Provided the calibration control on the pH meter (which relates the measured e.m.f. to a fixed point on the pH scale) has a wide enough range of adjustment, the pH scale can be used for any univalent positive ion, for example, measurements with a sodium-selective electrode can be read on the meter as a pNa scale (or $-\log C_{Na}$). Measurements with electrodes responding to divalent or negative ions cannot be related directly to the pH scale. However, manufacturers generally make some modifications to pH meters to simplify measurements with ion-selective electrodes and the modified meters are called 'pIon meters'. Scales are provided, analogous to the pH scale, for ions of various valencies and/or a scale that can be calibrated to read directly in terms of concentration or valency. Meters manufactured as pIon meters generally also have pH and millivolt scales. To date pIon scales only cover ions with charges of ± 1 and ± 2 because no ion-selective electrodes for determining ions of high charge are yet available commercially. Direct activity scales read in relative units only and so must be calibrated before use in the preferred measurement units.

As with pH meters, pIon meters can be obtained with analogue and digital displays, with integral microprocessors, with recorder and printer outputs and with automatic standardization. Temperature compensation can be incorporated but although ion-selective and reference electrode combinations have iso-potential points, the facility of being able to set the iso-potential control has so far been restricted to pH measurement. On dual pH/pIon meters the iso-potential control (if it exists) should be switched out on the pIon and activity scales if one wishes to make a slope correction when working with an ion-selective electrode at constant temperature.

For the best accuracy and precision pIon meters should be chosen that can discriminate 0.1 mV for direct potentiometry; 1 mV discrimination is sufficient when using ion-selective electrodes as indicators for titrimetric methods.

Practical arrangements For accurate potentiometry the temperature of the solution being analysed and the electrode assembly should be controlled and ideally all analyses should be carried out at the same temperature, e.g. by using a thermostatically controlled water bath. Solutions must also be stirred, otherwise the e.m.f. developed by the electrode may not be representative of the bulk of the solution. A wide range of stirring speeds is possible but too slow a speed may give long response times and too high a speed may generate heat in the solution. Precautions must also be taken to minimize contamination.

Taking all these items into account, best results in the laboratory can be obtained by mounting the electrodes in a flow cell through which the test solution is being pumped, see Figure 4.30. This is a mandatory arrangement for on-stream instruments and in the laboratory in cases where the ion concentration being determined is close to the limit of detection of the electrode.

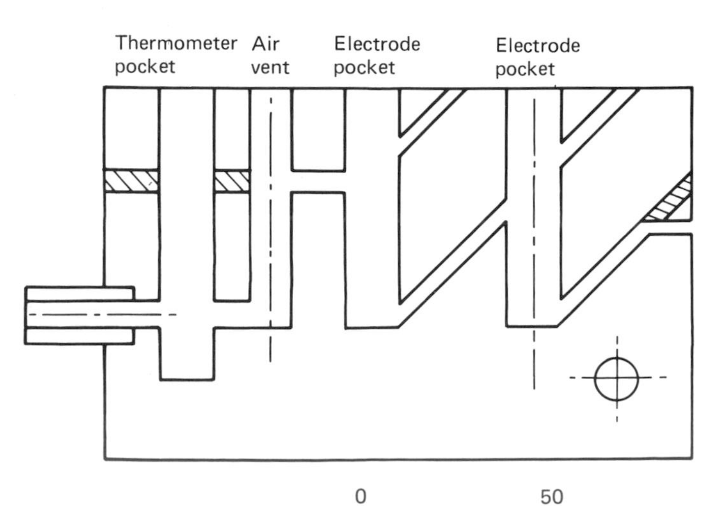

Figure 4.30 Flow cell for ion-selective electrodes.

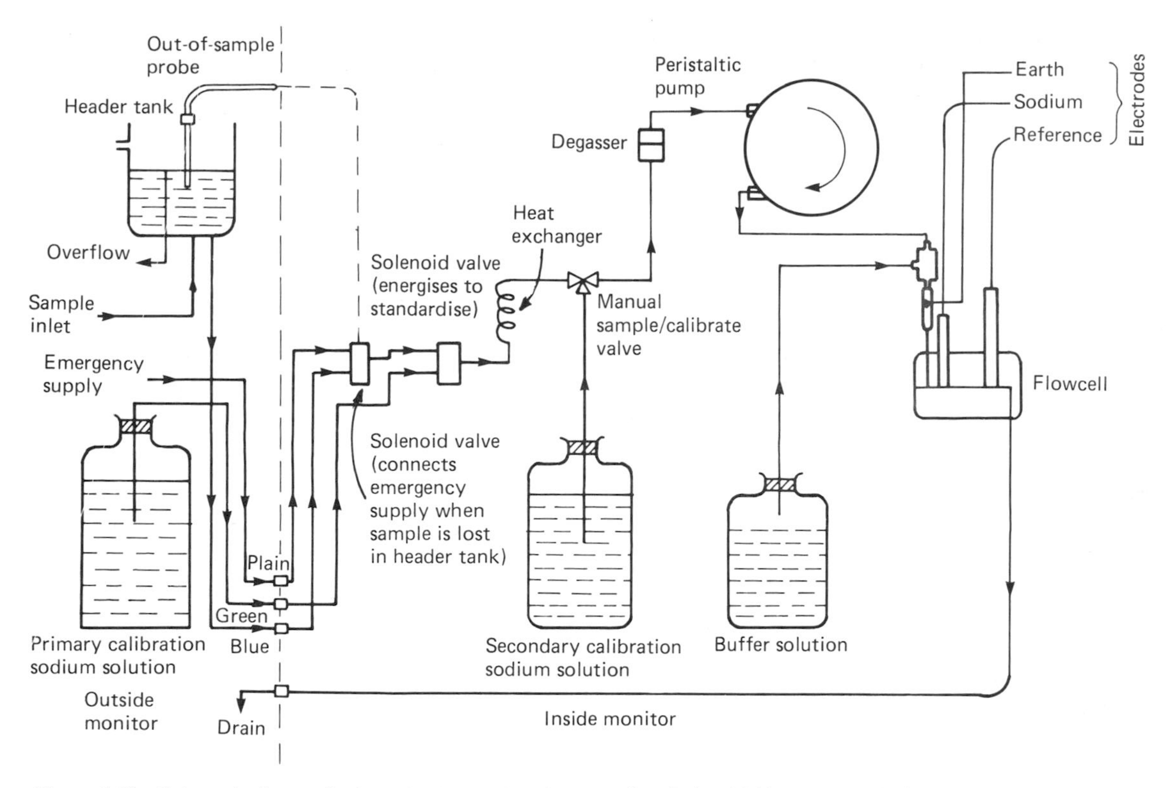

Figure 4.31 Schematic diagram for ion-selective monitor (courtesy Kent Industrial Measurements Ltd., Analytical Instruments).

Flow cells should be constructed of a material that will not contaminate a sample with the ion being determined; the flow rates of the solution must be high enough to provide 'stirring' but low enough that sample volumes are kept low. There should be good displacement of a previous sample by an incoming one, solution from the reference electrode should not reach the measuring electrode and, when liquid flow through a flow cell stops, the cell must retain liquid around the electrodes to prevent them drying out. Finally a flow cell should be water-jacketed so that its temperature can be controlled. Suitable flow cells can be machined out of Perspex and are available commercially.

Pumps used must be capable of pumping at least two channels simultaneously at different rates, the larger volume for the sample and the lesser for the reagent solution. Peristaltic pumps are the most frequently used. It follows that all interconnecting tubing and other components in contact with the sample must be inert with respect to the ion being determined.

As direct potentiometric determination of ions by ion-selective electrodes requires more frequent calibration than the more stable pH systems, industrially developed ion-selective electrode systems often incorporate automatic recalibration. This makes them more expensive than pH measuring systems. A typical scheme for an ion-selective monitor (in this case for sodium) is shown in Figures 4.31 and 4.32.

Sample water flows to the constant head unit and is then pumped anaerobically at a constant rate into the flow cell where it is equilibrated with ammonia gas obtained by pumping a stream of air through ammonia solution. (Instead of ammonia gas a liquid amine could be used and this would then be the buffer liquid delivered by the second channel of the pump.) The sample then flows through the flow cell to contact the ion-selective and reference electrodes and then to a drain.

Automatic chemical standardization takes place at preset intervals (in this case once every 24 hours) with provision for manual initiation of the sequence at any time. The standardization sequence commences by activating a valve to stop the sample flow and to allow a sodium ion solution of known strength (the standard sodium solution) to be pumped into the flow cell. When the electrodes have stabilized in the new solution, the amplifier output is compared with a

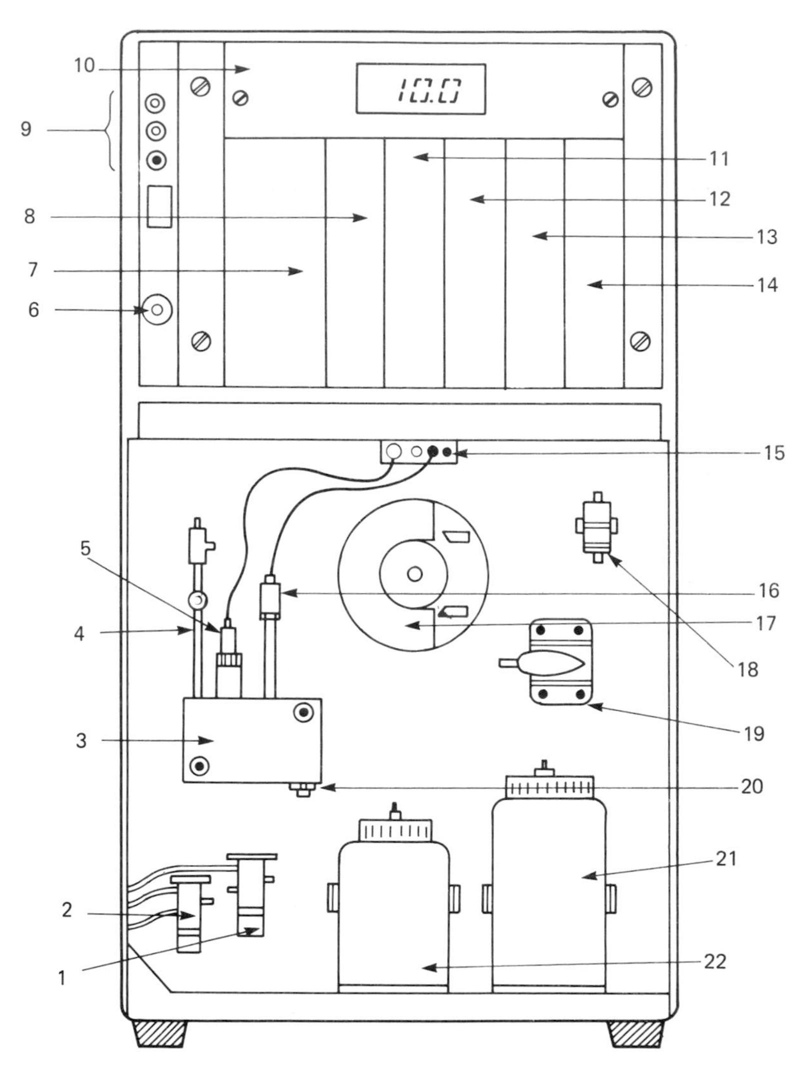

Figure 4.32 Diagrammatic arrangement of components for an ion-selective monitor (courtesy Kent Industrial Measurements Ltd., Analytical Instruments). 1, solenoid valve (energizes during standardization to connect primary standard solution); 2, solenoid valve (energizes to admit emergency sample supply when sample is lost in the header tank); 3, flow cell; 4, earthing tube; 5, sodium electrode; 6, SUPPLY ON lamp (illuminates when power is connected to the monitor); 7, 8020 100 amplifier; 8, 8033 200 current output module; 9, SERVICE lamp (red) and ON-LINE lamp (green) with push-button (optional feature); 10, digital display module (linear motor readout optional); 11, 8060 300 compensation module; 12, 8021 400 alarm and temperature control module; 13, 8020 500 power supply; 14, 8020 600 function module; 15, electrodes connection point (junction box); 16, refillable calomel reference electrode; 17, peristaltic pump; 18, gas debubbler; 19, manual SAMPLE/CALIBRATE valve; 20, flow cell drain; 21, secondary standard solution container (1 litre) (heat exchanger located behind the panel at this point); 22, buffer solution container (500 ml)

preset standard value in the auto-compensation unit and any error causes a servo-potentiometer to be driven so as to adjust the output signal to the required value. The monitor is then returned to measurement of the sample. The standardization period lasts 30 minutes, a warning lamp shows that standardization is taking place and any alarm and control contacts are disabled. It is also possible to check the stability of the amplifier and, by a manual introduction of a second sodium standard, to check and adjust the scale length.

Conditioning and storage of electrodes The manufacturer's instructions regarding storage and pretreatment of electrodes should be followed closely. The general rules are that (a) glass electrodes should not be allowed to dry out because reconditioning may not be successful, (b) solid state electrodes can be stored in deionized water, for long periods, dry-covered with protective caps and generally ready for use after rinsing with water, (c) gas-sensing membranes and liquid ion-exchange electrodes must never be allowed to dry out, (d) reference electrodes are as important as the measuring electrodes and must be treated exactly as specified by the manufacturer. The element must not be allowed to dry out, as would happen if there were insufficient solution in the reservoir.

Ion-selective electrodes available and application areas There is a very wide range of electrodes available. Not only are there many specific ion monitors but several manufacturers now market standardized modular assemblies which only need different electrodes, different buffer solutions and minor electrical adjustments for the monitors to cope with many ion determinations.

Table 4.9 Available ion-selective electrodes

Solid-state membrane electrodes	*Glass membrane electrodes*	*Liquid ion exchange membrane electrodes*	*Gas-sensing electrodes*
Fluoride	pH	Calcium	Ammonia
Chloride	Sodium	Calcium + magnesium	Carbon dioxide
Bromide	Potassium	(i.e. water hardness)	Sulphur dioxide
Iodide			Nitrous oxide
Thiocyanate			Hydrogen sulphide
Sulphide			Hydrogen fluoride
Silver		Barium	
Copper		Nitrate	
Lead		Potassium	
Cadmium			
Cyanide			
Redox			
pH (antimony)			

Table 4.9 shows the ion-selective electrodes available for the more common direct potentiometric determination of ions.

Ion-selective electrodes, as their name implies, are selective rather than specific for a particular ion. A potassium electrode responds to some sodium ion activity as well as to potassium, and this can be expressed as:

$$E_{\text{measured}} = \text{constant} \pm S \log (a_{\text{potassium}^+} + K\, a_{\text{Na}^+})$$

where K is the selectivity coefficient of this electrode to sodium and $0 < K < 1$.

Thus the fraction K of the total sodium activity will behave as though it were potassium. The *smaller* the value of K, the more selective that electrode is to potassium, i.e. the *better* it is. To identify a particular selectivity coefficient the data are best written in the form:

$$K_{\text{potassium}^+/\text{sodium}^+} = 2.6 \times 10^{-3}$$

This shows that the selectivity of potassium over sodium for the potassium electrode is about 385:1, i.e. $1/(2.6 \times 10^{-3})$. It is important to note that selectivity coefficients are not constant, but vary with the concentration of both primary and interferent ions and the coefficients are, therefore, often quoted for a particular ion concentration. They should be regarded as a guide to the effectiveness of an electrode in a particular measurement and not for use in precise calculations, particularly as quoted selectivity coefficients vary by a factor of 10 or more. For accurate work the analyst should determine the coefficient for himself for his own type of solution.

Direct potentiometric determination of ions by means of ion-selective electrodes has many applications. Examples are determination of pH, sodium and chloride in feedwater, condensate and boiler water in power stations; cyanide, fluoride, sulphide and chloride in effluents, rivers and lakes; fluoride, calcium and chloride in drinking water and sea water; bromide, calcium, chloride, fluoride, iodide, potassium and sodium in biological samples; calcium, chloride, fluoride and nitrate in soils; sulphur dioxide in wines and beer; chloride and calcium in milk; sulphide and sulphur dioxide in the paper-making industry; fluoride, calcium, chloride, nitrate and sulphur dioxide in foodstuffs, pH in water and effluents, papers, textiles, leather and foodstuffs, and calcium, chloride, fluoride and potassium in pharmaceuticals.

4.8 Common electrochemical analysers

4.8.1 Residual chlorine analyser

When two dissimilar metal electrodes are immersed in an electrolyte, and connected together, current will flow due to the build-up of electrons on the more electropositive electrode. The current will soon stop, however, owing to the fact that the cell will become polarized.

If, however, a suitable depolarizing agent is added, a current will continue to flow, the magnitude of which will depend upon the concentration and nature of the ions producing the depolarization. Thus, by choice of suitable materials for the electrodes and arranging for the addition of the depolarizing agent which is in fact the substance whose concentration is to be measured, amperometric analysers may be made to measure the concentration of a variety of chemicals. In some instruments a potential difference may be applied to the electrodes, when the current is again a linear function of the concentration of the depolarizing agent.

The sensitivity of the analyser is sometimes increased by using buffered water as the electrolyte so

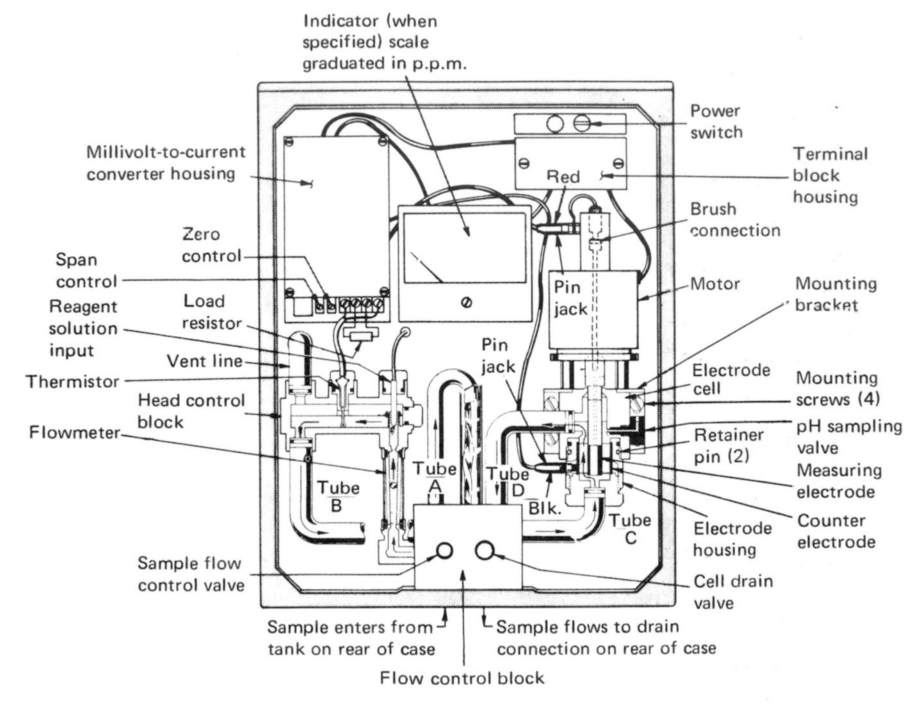

Figure 4.33 Residual chlorine analyser (courtesy Fischer & Porter).

that the cell operates at a definite pH. Amperometric instruments are inherently linear in response, but special steps have to be taken in order to make them specific to the substance whose concentration is to be measured, because other substances may act as depolarizing agents and so interfere with the measurement. When the interfering substances are known steps may be taken to remove them.

Where the instrument is intended to measure pollutants in air or gas, the gas to be tested is either bubbled through a suitable cell or arranged to impinge upon the surface of the liquid in the cell. In these cases interfering gases can be removed by chemical or molecular filters in the sampling system.

This form of instrument may be used to detect halogens, such as chlorine, in air and instruments with ranges from 0–0.5 to 0–20 ppm are available measuring with an accuracy of $\pm 2\%$ and a sensitivity of 0.01 ppm. By altering the electrolyte the instrument may be changed to measure the corresponding acid vapours, i.e. HCl, HBr and HF. One type of instrument for measuring chlorine in water is shown in Figure 4.33.

The sample stream is filtered in the tank on the back of the housing, and then enters the analyser unit through the sample flow control valve and up the metering tube into the head control block where reagent (buffer solution to maintain constant pH) is added by means of a positive displacement feed pump.

Buffered sample flows down tube B, through the flow control block and up tube C to the bottom of the electrode cell assembly. Sample flow rate is adjusted to approximately 150 millilitres per minute. Flow rate is not critical since the relative velocity between the measuring electrode and the sample is established by rotating the electrode at high speed.

In the electrode cell assembly, the sample passes up through the annular space between the concentrically mounted outer (copper) reference electrode and the inner (gold) measuring electrode and out through tube D to the drain. The space between the electrodes contains plastic pellets which are continuously agitated by the swirling of the water in the cell. The pellets keep the electrode surfaces clear of any material which might tend to adhere. The measuring electrode is coupled to a motor which operates at 1550 rev/min. The electrical signal from the measuring electrode is picked up by a spring-loaded brush on top of the motor and the circuit is completed through a thermistor for temperature compensation, precision resistors and the instationary copper electrode.

The composition of the electrodes is such that the polarization of the measuring electrode prevents current flow in the absence of a strong oxidizing agent. The presence of the smallest trace of strong oxidizer, such as chlorine (hypochlorous acid), will permit a current to flow by oxidizing the polarizing layer. The amplitude of the self-generated depolarization current is proportional to the concentration of the strong oxidizing agent. The generated current is passed through a precision resistor and the millivoltage across the resistor is then measured by the indicating or recording potentiometer. This instrument is calibrated to read in terms of the type (free or total) of residual chlorine measured. When measuring total residual chlorine, potassium iodide is added to the buffer. This reacts with the free and combined chlorine to liberate iodine in an amount equal to the total chlorine. The iodine depolarizes the cell in the same manner as hypochlorous acid, and a current directly proportional to the total residual chlorine is generated.

4.8.2 Polarographic process oxygen analyser

An instrument using the amperometric (polarographic) method of measurement is an oxygen analyser used for continuous process measurement of oxygen in flue gas, inert gas monitoring and other applications.

The key to the instrument is the rugged sensor shown in Figure 4.34. The sensor contains a silver anode and a gold cathode that are protected from the sample by a thin membrane of PTFE. An aqueous KCl solution is retained in the sensor by the membrane and forms the electrolyte in the cell (Figure 4.35).

Oxygen diffuses through the PTFE membrane and reacts with the cathode according to the equation:

$$4e^- + O_2 + 2H_2O \rightarrow 4OH^-$$

The corresponding anodic reaction is

$$Ag + Cl^- \rightarrow AgCl + e^-$$

For the reaction to continue, however, an external potential (0.7 volts) must be applied between cathode and anode. Oxygen will then continue to be reduced at the cathode, causing the flow of a current, the magnitude of which is proportional to the partial pressure of oxygen in the sample gas.

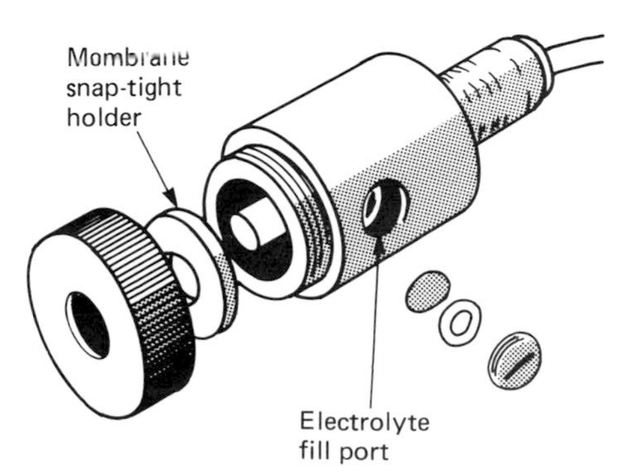

Figure 4.34 Process oxygen analyser (courtesy Beckman Instruments Inc.).

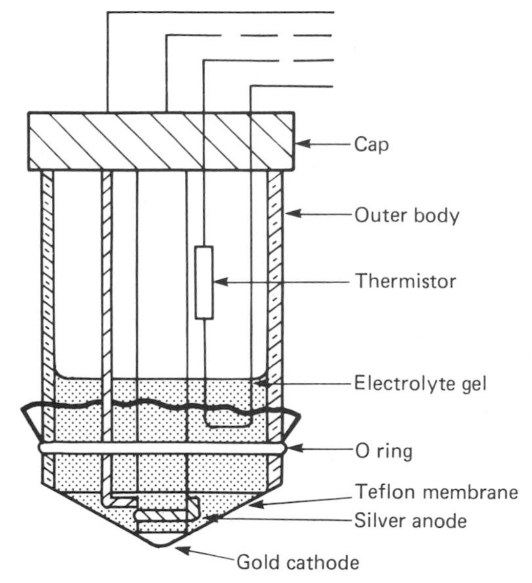

Figure 4.35 Diagram of polarographic oxygen sensor (courtesy Institute of Measurement and Control).

The only materials in contact with the process are PVC and PTFE and the membrane is recessed so that it does not suffer mechanical damage. The cell needs to be recharged with a new supply of electrolyte at 3- or 6-monthly intervals depending on the operating conditions and the membrane can be replaced easily should it be damaged.

The cell current is amplified by a solid state amplifier which gives a voltage output which can be displayed on an indicator or recorded. The instrument has a range selection switch giving ranges of 0–1, 0–5, 0–10 or 0–25 per cent oxygen and a calibration adjustment. The calibration is checked by using a reference gas, or air when the instrument should read 20.9 per cent oxygen on the 0–25 per cent scale. The instrument has an accuracy of ±1 per cent of scale range at the calibration temperature but an error of ±3 per cent of the readily will occur for a 16 °C departure in operating temperature.

When in use the sensor may be housed in an in-line type housing or in a dip-type of assembly, usually made of PVC suitable for pressures up to 3.5 bar.

4.8.3 High temperature ceramic sensor oxygen probes

Just as an electrical potential can be developed at a glass membrane which is a function of the ratio of the hydrogen concentrations on either side, a pure zirconia tube maintained at high temperature will develop a potential between its surfaces which is a function of the partial pressure of oxygen which is in

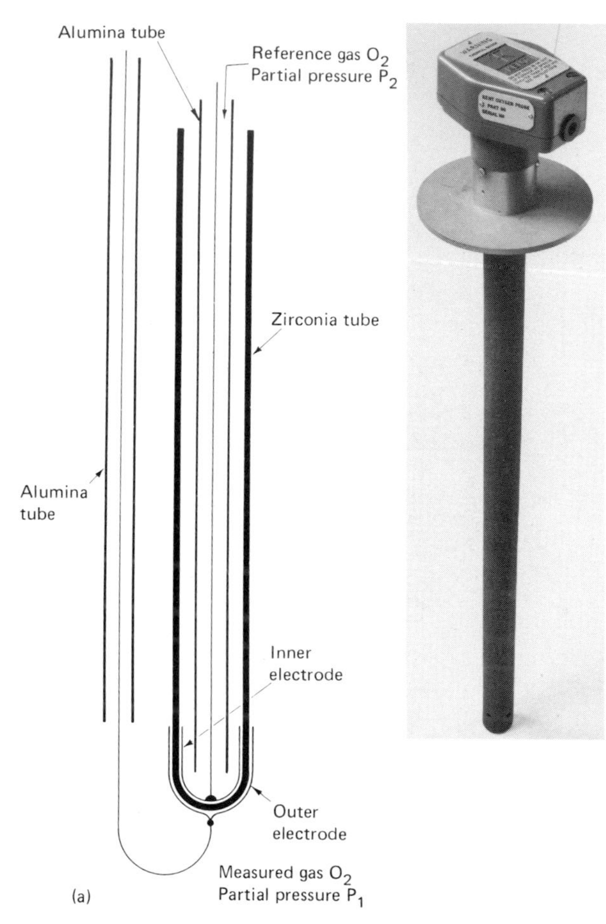

Figure 4.36 Oxygen probe (courtesy Kent Instruments).

contact with its surfaces. This is the principle involved in the oxygen meter shown in Figure 4.36.

The potential developed is given by the Nernst equation:

$$E_s = (RT/4F)\{\ln[\text{internal partial pressure of } O_2^{4-} \text{ ions}] / [\text{external partial pressure of } O_2^{4-} \text{ ions}]\}$$

Thus, if the potential difference between the surfaces is measured by platinum electrodes in contact with the two surfaces a measure may be made of the ratio of the partial pressure of the oxygen inside and outside the probe. If dry instrument air (20.9 per cent oxygen) is fed into the inside of the probe, the partial pressure of oxygen inside the tube may be regarded as constant, so that the electrical potential measured in a similar manner to that adopted in pH measurement will be a measure of the concentration of the oxygen in the atmosphere around the measuring probe. Thus by positioning the probe in a stack or flue where the temperature is above 600 °C a direct measurement of the oxygen present may be made. (In another manufacturer's instrument the probe is maintained at a temperature of 850 °C by a temperature-controlled heating element.) The instrument illustrated can operate from 600 to 1200 °C, the reading being corrected for temperature, which is measured by a thermocouple. The probe is protected by a silicon carbide sheath. The zirconia used is stabilized with calcium.

Standard instruments have ranges of oxygen con centration of 20.9–0.1 per cent, 1000–1 ppm, 10^{-5}–10^{-25} partial pressure and can measure oxygen with an accuracy of better than ± 10 per cent of the reading.

As temperatures in excess of 600 °C must be used some of the oxygen in the sample will react with any combustible gas present, e.g. carbon monoxide and hydrocarbons. Thus the measurement will be lower than the correct value but will still afford a rapid means of following changes in the oxygen content of a flue gas caused by changes in combustion conditions.

4.8.4 Fuel cell oxygen-measuring instruments

Galvanic or fuel cells differ from polarographic cells and the high temperature ceramic sensors in that they are power devices in their own right, that is, they require no external source of power to drive them. One manufacturer's version is shown in Figure 4.37.

A lead anode is made in that geometric form that maximizes the amount of metal available for reaction with a convex disc as the cathode. Perforations in the cathode facilitate continued wetting of the upper surface with electrolyte and ensure minimum internal resistance during the oxygen sensing reaction. The surfaces of the cathode are plated with gold and then covered with a PTFE membrane. Both electrodes are immersed in aqueous potassium hydroxide electrolyte. Diffusion of oxygen through the membrane enables

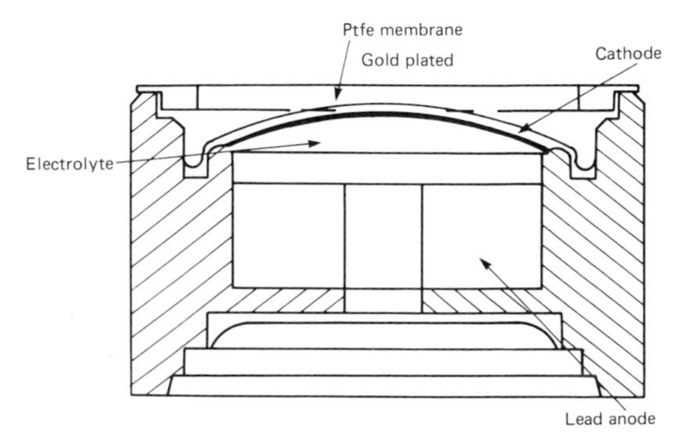

Figure 4.37 Diagrammatic micro-fuel cell oxygen sensor (courtesy Analysis Automation).

the following reactions to take place:

Cathode $4e^- + O_2 + 2H_2O \rightarrow 4OH^-$

Anode $Pb + 2OH^- \rightarrow PbO + H_2O + 2e$

Overall
cell reaction $2Pb + O_2 \rightarrow PbO$

The electrical output of the cell can be related to the partial pressure of oxygen on the gas side of the membrane in a manner analogous to that described for membrane-covered polarographic cells. In this instance, however, because there is no applied potential and no resultant hydrolysis of the electrolyte, absence of oxygen in the sample corresponds to zero electrical output from the cell. There is a linear response to partial pressure of oxygen and a single point calibration, e.g. on air, is sufficient for most purposes.

The main limitation of this type of oxygen sensor is the rate of diffusion of oxygen across the membrane; this determines the speed of response and, at low oxygen partial pressure, this may become unacceptably slow. However, to overcome this, one type of fuel cell oxygen sensor has a completely exposed cathode, i.e. not covered with a PTFE membrane.

In common with all membrane cells, the response of the micro-fuel cell is independent of sample flow rate but the cell has a positive temperature-dependence. This is accommodated by incorporating negative temperature coefficient thermistors in the measuring circuit. These fuel cells have sufficient electrical output to drive readout meters without amplification. However where dual or multi-range facilities are required some amplification may be necessary.

4.8.5 Hersch cell for oxygen measurement

This galvanic cell differs from fuel cells in that a third electrode is added to the cell and a potential applied to provide anodic protection to the anode. In one manufacturer's cell (Figure 4.38) the cathode is silver and the anode cadmium. The third electrode is platinum. The anodic protection limits the cadmium current to a few microamperes and extends the life of the cadmium. However, this arrangement gives an electrical output from the cell which is non-linear with oxygen partial pressure and it is necessary for the signal to be passed through a 'shaping' circuit for the readout to be given in concentration units. Calibration is carried out by generating a predetermined concentration of oxygen in a sample by electrolysis, and

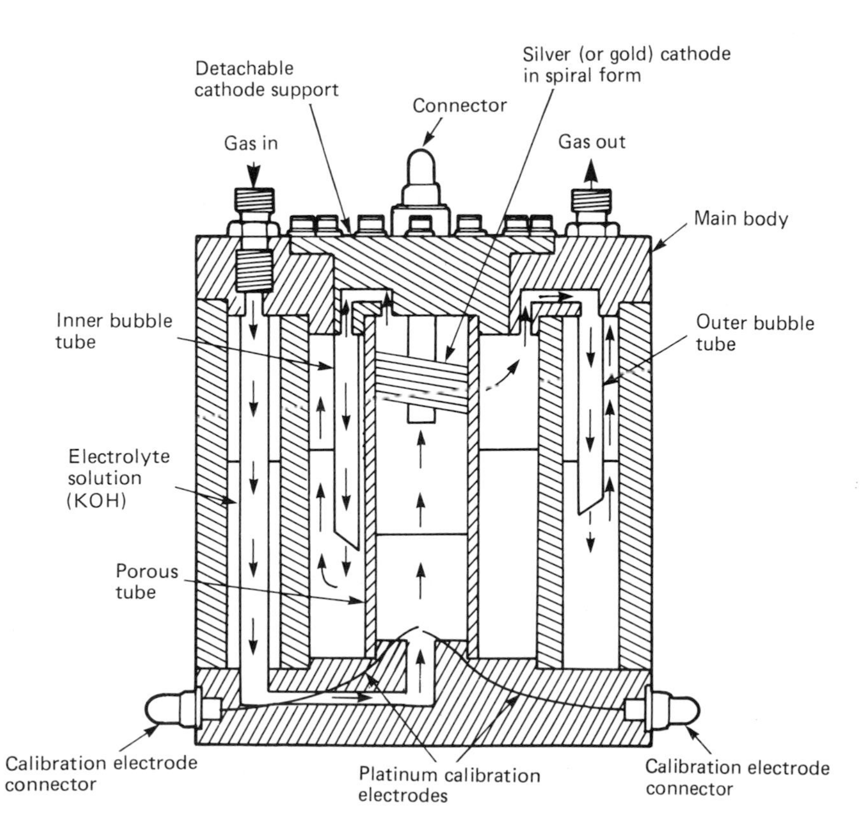

Figure 4.38 Cross-section of Hersch cell (courtesy Anacon (Instruments) Ltd.).

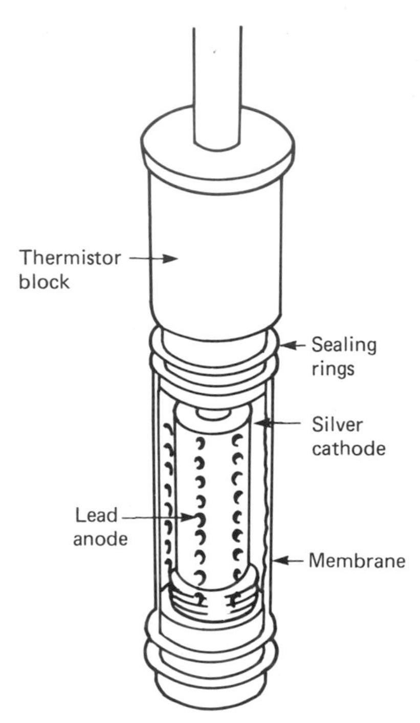

Figure 4.39 Diagram of Mackereth oxygen sensor assemblies (courtesy Kent Industrial Measurements Ltd., Analytical Instruments).

electrodes for this are incorporated in the cell. When dry gas samples are being used they must be humidified to prevent the water-based electrolyte in the cell from drying out.

4.8.6 Sensor for oxygen dissolved in water

Electrochemical sensors with membranes for oxygen determination can be applied to measuring oxygen dissolved in water; both polarographic and galvanic sensors can be used.

A most popular type of sensor is the galvanic Mackereth electrode. The cathode is a perforated silver cylinder surrounding a lead anode with an aqueous electrolyte of potassium bicarbonate (Figure 4.39). The electrolyte is confined by a silicone rubber membrane which is permeable to oxygen but not to water and interfering ions.

The oxygen which diffuses through the membrane is reduced at the cathode to give a current proportional to the oxygen partial pressure. Equations for the reactions have already been given (p. 139).

Accurate temperature control is essential (6 per cent error per degree) and thermistor- or resistance-thermometer-controlled compensation circuits are generally used. Working ranges can be from a few μg O_2/litre of water up to 200 per cent oxygen saturation. The lead anode is sacrificial and electrodes therefore

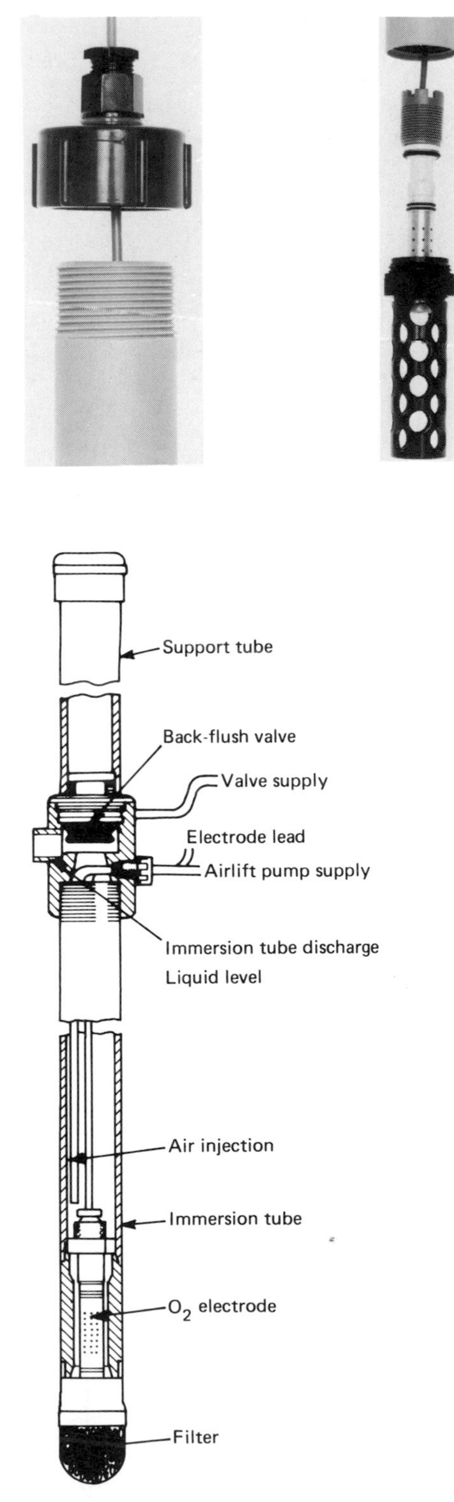

Figure 4.40 Varieties of Mackereth oxygen sensor assemblies (courtesy Kent Industrial Measurements Ltd., Analytical Instruments).

have to be refurbished according to the actual design and the total amount of oxygen that has diffused into the cell. Cells are calibrated using water containing known amounts of oxygen. Indicating meters or recorders can be connected and manufacturers offer both portable instruments and equipment for permanent installation with timing devices, water pumps, etc. There are also several variations on the basic design of electrodes to cope with oxygen determination in water plant, rivers, lakes, sewage tanks, etc. (see Figure 4.40). One of those shown includes a patented assembly incorporating water sampling by air lift – air reversal gives calibration check and filter clean.

4.8.7 Coulometric measurement of moisture in gases and liquids

Moisture from gases (or vaporized from liquids) can be absorbed by a layer of desiccant, generally phosphoric anhydride (P_2O_5), in contact with two platinum or rhodium electrodes. A d.c. voltage is applied to electrolyse the moisture, the current produced being directly proportional to the mass of moisture absorbed (Faraday's law of electrolysis). The response of such an instrument obviously depends on the flow rate of gas which is set and controlled accurately at a predetermined rate so that the current measuring meter can be calibrated in vppm moisture. Details are given in Chapter 6.

4.9 Further reading

Bailey, P. L., *Analysis with Ion-selective Electrodes*, Heyden, (1976)

Bates, R. G., *The Determination of pH* (2nd edn), Wiley Interscience, (1973)

Durst, R. A. (ed.), *Ion Selective Electrodes*, National Bureau of Standards Special Publication 314, Dept. of Commerce, Washington DC, (1969)

Eisenman, G., *Glass Electrodes for Hydrogen and Other Cations*, Edward Arnold, London/Marcel Dekker, New York, (1967)

Freiser, H. (ed.), *Ion-selective Electrodes in Analytical Chemistry* Vol. I, Plenum Press, New York, (1978)

Ives, G. J. and Janz, D. J. G., *Reference Electrodes, Theory and Practice*, Wiley Interscience, (1961)

Midgley, D. and Torrance, K., *Potentiometric Water Analysis*, Wiley Interscience, (1978)

Perrin, D. D. and Dempsey, B., *Buffers for pH and Metal Ion Control*, Chapman and Hall, London, (1974)

Sawyer, D. T. and Roberts, J. L. Jr., *Experimental Electrochemistry for Chemists*, Wiley Interscience, (1974)

5 Chemical analysis – gas analysis

C. K. LAIRD

5.1 Introduction

The ability to analyse one or more components of a gas mixture depends on the availability of suitable detectors which are responsive to the components of interest in the mixture and which can be applied over the required concentration range. Gas detectors are now available which exploit a wide variety of physical and chemical properties of the gases detected, and the devices resulting from the application of these detection mechanisms show a corresponding variety in their selectivity and range of response. In a limited number of applications it may be possible to analyse a gas mixture merely by exposure of the sample to a detector which is specific to the species of interest, and thus obtain a direct measure of its concentration. However, in the majority of cases no sufficiently selective detector is available and the gas sample requires some pretreatment, e.g. drying or removal of interfering components, to make it suitable for the proposed detector. In these cases a gas analysis system must be used.

A block diagram of the components of a typical gas analyser is given in Figure 5.1. The sample is taken into the instrument either as a continuous stream or in discrete aliquots and is adjusted as necessary in the sampling unit to the temperature, pressure and flow-rate requirements of the remainder of the system. Any treatment of the sample, for example separation of the sample into its components, removal of interfering components or reaction with an auxiliary gas is carried out and the sample is passed to the detector. The signal from the detector is amplified if necessary and processed to display or record the concentration of the components of interest in the sample.

In many gas analysers the time lag between sampling and analysis is reduced to a minimum by taking a continuous stream of sample at a relatively high flow rate, and arranging for only a small proportion to enter the analyser, the remainder being bypassed to waste. Provision is also normally made to check the zero by passing a sample, free of the species to be analysed, to the detector, and the instrument may also include facilities for calibration by means of a

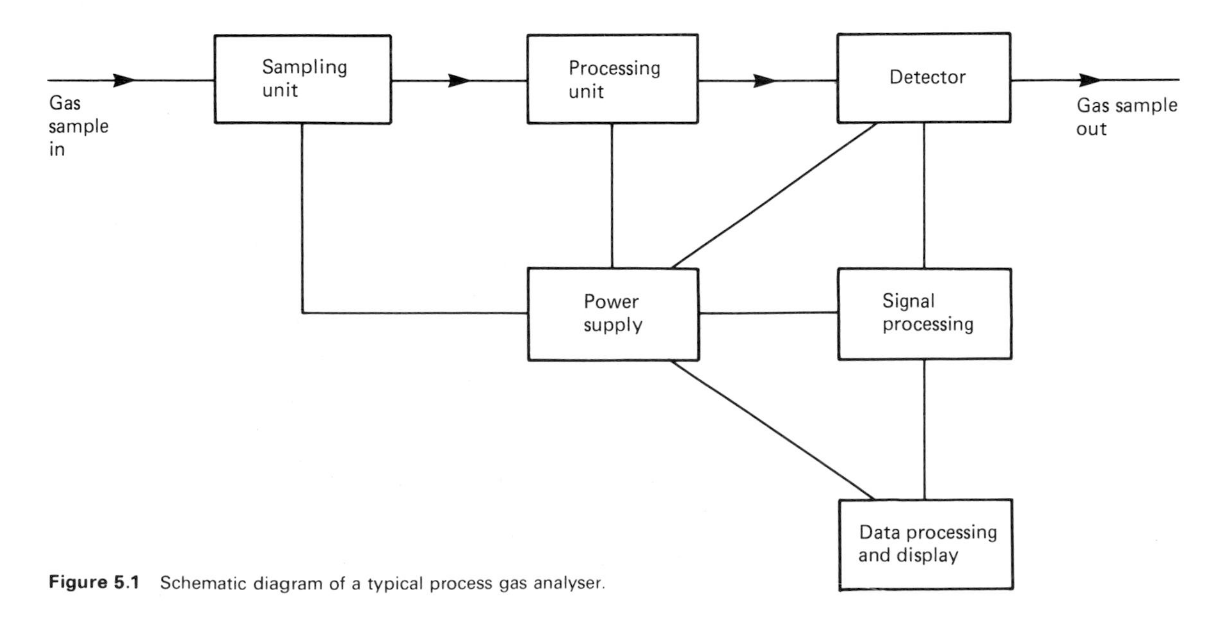

Figure 5.1 Schematic diagram of a typical process gas analyser.

'span' switch which feeds a sample of known concentration to the analyser.

For certain applications there may be a choice between the use of a highly selective detector, with relatively little pretreatment of the sample, or use of a detector which responds to a wider range of chemical species, the sample being separated into its components before it reaches the detector. In the special case of gas chromatography the sample is separated on the basis of the different times taken by each component to pass through a tube or column packed with adsorbent. The outlet gas stream may then be passed through a single detector, or through more than one detector in series or switched between detectors to analyse several components of the original sample mixture. By choice of columns, operating conditions and detectors, a gas-chromatographic analysis system may be built up individually tailored to analyse several different preselected components in a single aliquot taken from a gas sample. Because of its importance in process analysis, gas chromatography is given particularly detailed treatment.

In addition to the analysis techniques described in this chapter, a number of spectroscopic methods are given under that heading in Chapter 3, while some electrochemical methods are outlined in Chapter 4.

5.2 Separation of gaseous mixtures

Although detectors have been developed which are specific to particular gases or groups of gases, for example flammable gases or total hydrocarbons there is often a need to separate the sample into its components, or to remove interfering species, before the sample is passed to the detector. A non-specific detector, such as a katharometer, may also be used to measure one component of a gas mixture by measuring the change in detector response which occurs when the component of interest is removed from the gas mixture.

Methods for separating gaseous mixtures may be grouped under three main headings.

Chemical reaction A simple example of chemical separation is the use of desiccants to remove water from a gas stream. The percentage of carbon dioxide in blast furnace gas may be determined by measuring the thermal conductivity of the gas before and after selective removal of the carbon dioxide by passing the gas through soda-lime. Similarly the percentage of ammonia gas, in a mixture of nitrogen, hydrogen and ammonia may be measured by absorbing the ammonia in dilute sulphuric acid or a suitable solid absorbent.

Physical methods The most powerful physical technique for separation of gases is mass spectrometry, described in Chapter 3 – though only minute quantities can be handled in that way. Gases may also be separated by diffusion, for example, hydrogen may be removed from a gas stream by allowing it to diffuse through a heated tube of gold– or silver–palladium alloy.

Physico-chemical methods: chromatography Gas chromatography is one of the most powerful techniques for separation of mixtures of gases or (in their vapour phase) volatile liquids. It is relatively simple and widely applicable. Mixtures of permanent gases, such as oxygen, nitrogen, hydrogen, carbon monoxide, and carbon dioxide can easily be separated, and when applied to liquids, mixtures such as benzene and cyclohexane can be separated even though their boiling points differ by only 0.6 K. Separation of such mixtures by other techniques such as fractional distillation would be extremely difficult.

5.2.1 Gas chromatography

Chromatography is a physical or physico-chemical technique for the separation of mixtures into their components on the basis of their molecular distribution between two immiscible phases. One phase is normally stationary and is in a finely divided state to provide a large surface area relative to volume. The second phase is mobile and transports the components of the mixture over the stationary phase. The various types of chromatography are classified according to the particular mobile and stationary phases employed in each (see Chapter 2). In gas chromatography the mobile phase is a gas, known as the carrier gas, and the stationary phase is either a granular solid (gas–solid chromatography) or a granular solid coated with a thin film of non-volatile liquid (gas–liquid chromatography). In gas–solid chromatography the separation is effected on the basis of the different adsorption characteristics of the components of the mixture on the solid phase, while in gas–liquid chromatography the separation mechanism involves the distribution of the components of the mixture between the gas and stationary liquid phases. Because the components of the mixture are transported in the gaseous phase, gas chromatography is limited to separation of mixtures whose components have significant vapour pressures, and this normally means gaseous mixtures or mixtures of liquids with boiling points below approximately 450 K.

The apparatus for gas chromatography, known as the gas chromatograph, consists of a tube or column to contain the stationary phase, and itself contained in an environment whose temperature can be held at a constant known value or heated and cooled at controlled rates. The column may be uniformly packed with the granular stationary phase (packed column chromatography) and this is normally used in process instruments. However, it has been found that columns of the highest separating performance are obtained if the column is in the form of a capillary tube, with the solid or liquid stationary phase coated on its inner walls (capillary chromatography). The carrier-gas mobile phase is passed continuously through the column at a constant controlled and known rate. A facility for introduction of known volumes of the mixture to be separated into the carrier-gas stream is provided in the carrier-gas line upstream of the column, and a suitable detector, responsive to changes in the composition of the gas passing through it, is connected to the downstream end of the column.

To analyse a sample, an aliquot of suitable known volume is introduced into the carrier-gas stream, and the output of the detector is continuously monitored. Due to their interaction with the stationary phase, the components of the sample pass through the column at different rates. The processes affecting the separation are complex, but in general, in gas–solid chromatography the component which is least strongly adsorbed is eluted first, while in gas–liquid chromatography the dominant process is the solubility of the components in the liquid stationary phase. Thus the separation achieved depends on the nature of the sample and stationary phase, on the length and temperature of the column, and on the flow rate of the carrier gas, and these conditions must be optimized for a particular analysis.

The composition of the gas passing through the detector alternates between pure carrier gas, and mixtures of the carrier gas with each of the components of the sample. The output record of the detector, known as the chromatogram, is a series of deflections or peaks, spaced in time and each related to a component of the mixture analysed.

A typical chromatogram of a mixture containing five components, is shown in Figure 5.2. The first 'peak' (A) at the beginning of the chromatogram is a pressure wave or unresolved peak caused by momentary changes in carrier-gas flow and pressure during the injection of the sample. The recording of the chromatogram provides a visual record of the analysis, but for qualitative analysis each peak must be identified on the basis of the time each component takes to pass through the column by use of single pure compounds or mixtures of known composition. For quantitative analysis the apparatus must be calibrated by use of standard gas mixtures or solutions to relate the detector response to the concentration of the determinand in the initial mixture.

A significant advantage of gas chromatography is that several components of a sample may be analysed essentially simultaneously in a single aliquot extracted from a process stream. However, sampling is on a regular discrete basis rather than continuous, so that the chromatograph gives a series of spot analyses of a sample stream, at times corresponding to the time of sample injection into the instrument. Before a new sample can be analysed, it is necessary to be certain that all the components of the previous sample have

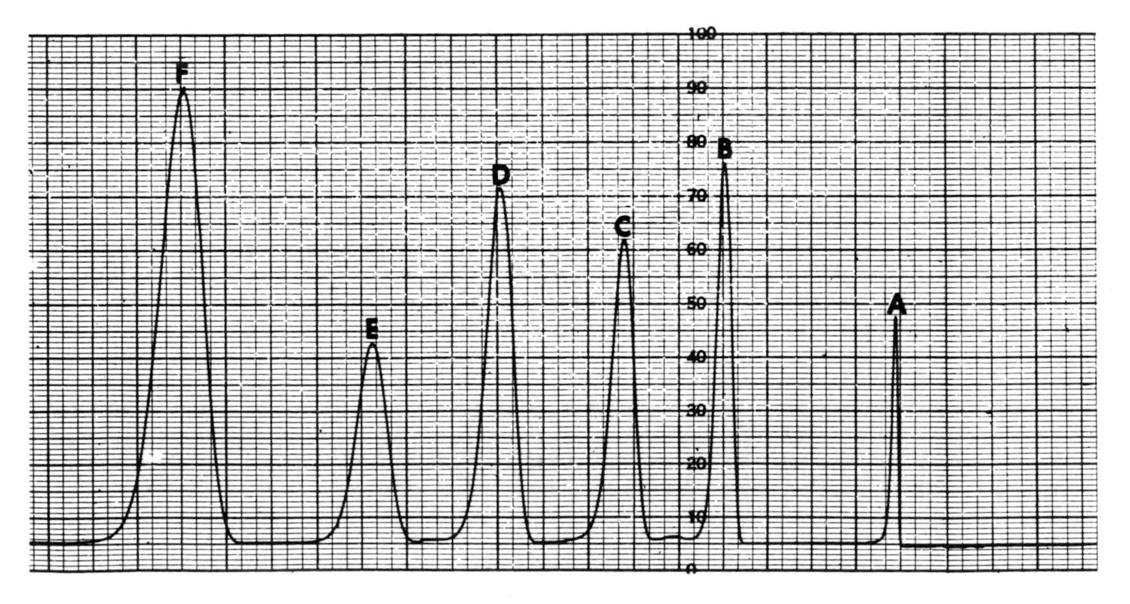

Figure 5.2 Chromatogram of a sample containing five components.

been eluted from the column. It is therefore advantageous to arrange the analytical conditions so that the sample is eluted as quickly as possible, consistent with adequate resolution of the peaks of interest.

5.3 Detectors

5.3.1 Thermal conductivity detector (TCD)

The thermal conductivity detector is among the most commonly used gas detection devices. It measures the change in thermal conductivity of a gas mixture, caused by changes in the concentration of the species it is desired to detect.

All matter is made up of molecules which are in constant rapid motion. Heat is the energy possessed by a body by virtue of the motion of the molecules of which it is composed. Raising the temperature of the body increases the energy of the molecules by increasing the velocity of the molecular motion.

In solids the molecules do not alter their position relative to one another but vibrate about a mean position, while in a liquid the molecules vibrate about mean positions, but may also move from one part of the liquid to another. In a gas the molecular motion is almost entirely translational: the molecules move from one part of the gas to another, only impeded by frequent intermolecular collisions and collisions with the walls of the vessel. The collisions with the walls produce the pressure of the gas on the walls. In a so-called 'perfect gas' the molecules are regarded as being perfectly elastic so no energy is dissipated by the intermolecular collisions.

Consideration of the properties of a gas which follow as a consequence of the motion of its molecules is the basis of the kinetic theory. Using this theory Maxwell gave a theoretical verification of laws which had previously been established experimentally. These included Avogadro's law, Dalton's law of partial pressures and Graham's law of diffusion.

Since heat is the energy of motion of the gas molecules, transfer of heat, or thermal conductivity, can also be treated by the kinetic theory. It can be shown that the thermal conductivity K of component S is given by

$$K_S = \tfrac{1}{2}\rho\tilde{v}\lambda C_v$$

where ρ is the gas density, $\tilde{v}$ is the mean molecular velocity, λ is the mean free path, and C_v is the specific heat at constant volume. Thus thermal conductivity depends on molecular size, mass and temperature.

The quantity $\tilde{v}\lambda$ is the diffusion coefficient D of the gas, and the thermal conductivity can be written

$$K_S = \tfrac{1}{2}D\rho C_v$$

According to this treatment, the thermal conductivity of the gas is independent of pressure. This is found to be true over a wide range of pressures, provided that the pressure does not become so high that the gas may no longer be regarded as being a perfect gas. At very low pressures, the conductivity of the gas is proportional to its pressure, and this is the basis of the operation of the Knudsen hot-wire manometer or Pirani gauge (see Chapter 10 of Part 1).

It can be shown that the conductivity K_T of a pure gas at absolute temperature T, varies with temperature according to the equation

$$K_T = K_0\left[b + \frac{273}{b} + T\right]\left[\frac{T}{273}\right]^{3/2}$$

where K_0 is the thermal conductivity at 0 °C and b is a constant.

The relative thermal conductivities of some gases, relative to air as 1.00, are given in Table 5.1.

Table 5.1 Relative thermal conductivities of some common gases

Gas	*Conductivity*
Air	1.00
Oxygen	1.01
Nitrogen	1.00
Hydrogen	4.66
Chlorine	0.32
Carbon monoxide	0.96
Carbon dioxide	0.59
Sulphur dioxide	0.32
Water vapour	1.30
Helium	4.34

It can be shown that the conductivity of a binary mixture of gases is given by

$$K = \frac{K_1}{1 + A\left(\dfrac{1-x_1}{x_1}\right)} + \frac{K_2}{1 + B\left(\dfrac{x_1}{1-x_1}\right)}$$

where A and B are constants known as the Wasiljewa constants, K_1 and K_2 are the conductivities of the pure gases, and x_1 is the molar fraction of component 1.

In gas analysis, conductivities of pure gases are of limited value, and it is much more important to know how the conductivity of a mixture varies with the proportion of the constituent gases. However, as shown above, the relationship between the conductivity of a mixture of gases and the proportion of the constituents is complicated. When collisions occur between molecules of different gases the mathematics of the collisions are no longer simple, and the relationship between conductivity and the proportions of the constituents depends upon the molecular and physical constants of the gases, and on the intermolecular forces during a collision. In practice thermal conductivity instruments are therefore calibrated by establishing the required composition–conductivity curves experimentally.

Several forms of gas sensor based on thermal conductivity have been developed. The majority use the hot-wire method of measuring changes in conductivity, with the hot-wire sensors arranged in a Wheatstone bridge circuit.

5.3.1.1 Katharometer

A wire, heated electrically and maintained at constant temperature, is fixed along the axis of a cylindrical hole bored in a metal block which is also maintained at a constant temperature. The cylindrical hole is filled with the gas under test. The temperature of the wire reaches an equilibrium value when the rate of loss of heat by conduction, convection and radiation is equal to the rate of production of heat by the current in the wire. In practice, conduction through the gas is the most important source of heat loss. End-cooling, convection, radiation and thermal diffusion effects, though measurable, account for so small a part (less than 1 per cent each) of the total loss that they can satisfactorily be taken care of in the calibration. Most instruments are designed to operate with the wire mounted vertically, to minimize losses by convection. Convective losses also increase with the pressure of the gas, so the pressure should be controlled for accurate conductivity measurements in dense gases. The heat loss from the wire depends on the flow rate of gas in the sensor. In some instruments errors due to changes in gas flow are minimized because the gas does not flow through the cell but enters by diffusion, but otherwise the gas flow rate must be carefully controlled.

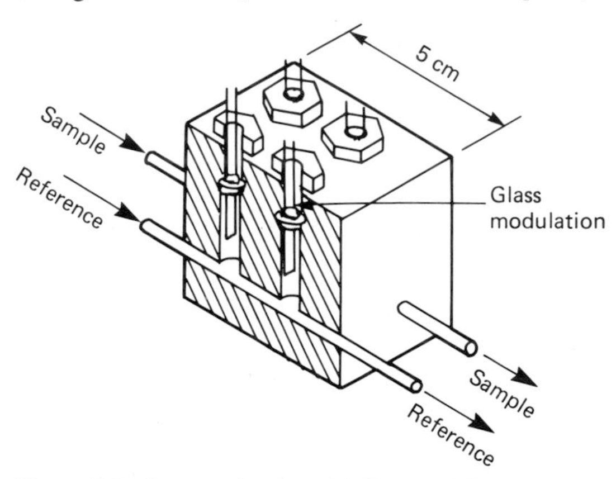

Figure 5.3 Cutaway drawing of 4-filament diffusion katharometer cell.

The resistance of the wire depends on its temperature; thus by measuring the resistance of the wire, its temperature may be found, and the wire is effectively used as a resistance thermometer. The electrical energy supplied to the wire to maintain the excess temperature is a measure of the total heat loss by conduction, convection and radiation. To measure the effects due to changes in the conductivity of the gas only, the resistance of the hot wire in a cell containing the gas to be tested is compared with the resistance of an exactly similar wire in a similar cell containing a standard gas. This differential arrangement also lessens the effects of changes in the heating current and the ambient temperature conditions. In order to increase the sensitivity two measuring and two reference cells are often used, and this arrangement is usually referred to as a 'katharometer'.

In the katharometer four filaments with precisely matched thermal and electrical characteristics are mounted in a massive metal block, drilled to form cells and gas paths. A cutaway drawing of a 4-filament cell is shown in Figure 5.3. Depending on the specific purpose, the filaments may be made of tungsten, tungsten–rhenium alloy, platinum or other alloys. For measurements in highly reactive gases gold-sheathed tungsten filaments may be used. The filaments are connected in a Wheatstone bridge circuit, which may be supplied from either a regulated-voltage or regulated-current power supply. The circuit for a constant-voltage detector is shown in Figure 5.4. The

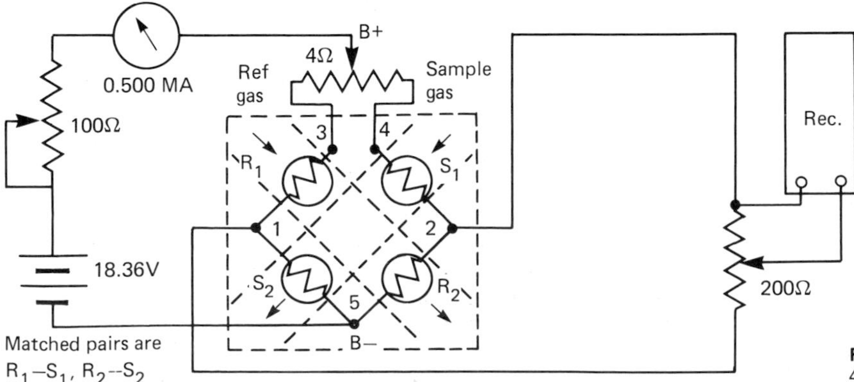

Figure 5.4 Circuit for 4-filament katharometer cell.

detector is balanced with the same gas in the reference and sample cells. If a gas of different thermal conductivity enters the sample cell, the rate of loss of heat from the sample filaments is altered, so changing their temperature and hence resistance. The change in resistance unbalances the bridge and the out-of-balance voltage is recorded as a measure of the change in gas concentration. The katharometer can be calibrated by any binary gas mixture, or for a gas mixture which may be regarded as binary, e.g. carbon dioxide in air.

A theory of the operation of the katharometer bridge follows. This is simplified but is insufficiently rigid for calibrations to be calculated. Small variations in the behaviour of individual filaments also mean that each bridge must be calibrated using mixtures of the gas the instrument is to measure.

Assume that the four arms of the bridge (Figure 5.4) have the same initial resistance R_1 when the bridge current is flowing and the same gas mixture is in the reference and sample cells. Let R_0 be resistance of filament at ambient temperature, R_1 working resistance (i.e. resistance when a current I flows), I current through one filament (i.e. half bridge current), and T wire temperature above ambient.

Then, at equilibrium, energy input is equal to heat loss

$$I^2R_1 = K_1T \quad (5.1)$$

where K_1 is a constant proportional to the thermal conductivity of the gas as most of the heat loss is by conduction through the gas. A simple expression for the working resistance is

$$R_1 = R_0(1+\alpha T) \quad (5.2)$$

where α is the temperature coefficient of resistance of the filament material. Then, from equations (5.1) and (5.2):

$$I^2R_1R_0\alpha = K_1(R_1 - R_0) \quad (5.3)$$

Then

$$R_1 = \frac{K_1R_0}{(K_1 - R_0I^2\alpha)}$$

$$= R_0 + \frac{K_1R_0}{(K_1 - R_0I^2\alpha)} - R_0$$

$$= R_0 + \frac{K_1R_0 - K_1R_0 + I^2R_0^2}{(K_1 - I^2R_0\alpha)}$$

$$= R_0 + \frac{I^2R_0}{(K_1 - I^2R_0\alpha)} \quad (5.4)$$

From equation (5.3), if $R_1 - R_0$ is small compared with R_1, then K_1 must be large compared with $I^2R_0\alpha$ and the term $I^2R_0\alpha$ can be ignored. Then

$$R_1 = R_0 + (I^2R_0^2\alpha/K_1) \quad (5.5)$$

If the two measurement filaments have a total resistance of R_1 and the reference filaments of R_2, the output voltage of the bridge E is given by

$$E = I(R_1 - R_2) \quad (5.6)$$

Combining equations (5.5) and (5.6):

$$E = I^3R_0^2\alpha[(1/K_1) - (1/K_2)] \quad (5.7)$$

where K_1 and K_2 are proportional to the conductivities of the gases in each pair of cells.

Equation (5.7) shows that the output is proportional to the cube of the bridge current but in practice the index is usually between $I^{2.5}$ and I^3. For accurate quantitative readings the bridge current must be kept constant.

This equation also shows that the output is proportional to the difference between the reciprocals of the thermal conductivities of the gases in each pair of cells. This is usually correct for small differences in thermal conductivity but does not hold for large differences.

These conditions show that the katharometer has maximum sensitivity when it is used to measure the concentration of binary or pseudo-binary gas mixtures whose components have widely different thermal conductivities and when the bridge current is as high as possible. The maximum bridge current is limited by the need to avoid overheating and distortion of the filaments, and bridge currents can be highest when a gas of high thermal conductivity is in the cell. When the katharometer is used as the detector in gas chromatography, hydrogen or helium, which have higher thermal conductivities than other common gases, is often used as the carrier gas, and automatic circuits may be fitted to reduce the current to the bridge to prevent overheating.

For maximum sensitivity, especially when it is necessary to operate the detector at low temperatures, the hot-wire filaments may be replaced by thermistors. A thermistor is a thermally sensitive resistor having a high negative coefficient of resistance, see Chapter 1. In the same manner as with hot wires, the resistance of the conductor is changed (in this case lowered) by the passage of current. Thermistor katharometers usually have one sensing and one reference element, the other resistors in the Wheatstone bridge being external resistors.

Except in the case of thermally unstable substances the katharometer is non-destructive, and it responds universally to all substances. The sensitivity is less than that of the ionization detectors, but is adequate for many applications. The detector is basically simple, and responds linearly to concentration changes over a wide range. It is used in gas chromatography and in a variety of custom-designed process analysers.

5.3.2 Flame ionization detector (FID)

An extensive group of gas detectors is based on devices in which changes in ionization current inside a chamber are measured. The ionization process occurs when a particle of high energy collides with a target particle which is thus ionized. The collision produces positive ions and secondary electrons which may be moved towards electrodes by application of an electric field, giving a measurable current, known as the ionization current, in the external circuit.

The FID utilizes the fact that, while a hydrogen–oxygen flame contains relatively few ions (10^7 ions cm^{-3}), it does contain highly energetic atoms. When trace amounts of organic compounds are added to the flame the number of ions increases (to approximately 10^{11} ions cm^{-3}) and a measurable ionization current is produced. It is assumed that the main reaction in the flame is

$$CH + O \rightarrow CHO + e$$

However, the FID gives a small response to substances that do not contain hydrogen, such as CCl_4 and CS_2. Hence it is probable that the reaction above is preceded by hydrogenation to form CH_4 or CH_3 in the reducing part of the flame. In addition to the ionization reactions, recombination also occurs, and the response of the FID is determined by the net overall ionization reaction process.

A schematic diagram of an FID is shown in Figure 5.5 and a cross-sectional view of a typical detector is shown in Figure 5.6. The sample gas, or effluent from a gas-chromatographic column, is fed into a hydrogen–air flame. The jet itself serves as one electrode and a second electrode is placed above the flame. A potential is applied across these electrodes. When sample molecules enter the flame, ionization occurs yielding a current which, after suitable amplification, may be displayed on a strip chart recorder.

The FID is a mass-sensitive, rather than concentration-sensitive, detector. This means that it does not respond to the concentration of a component entering it, but rather produces a signal which is proportional to the amount of organic material entering it per unit time. The ion current is effectively proportional to the number of carbon atoms present in the flame, and the sensitivity of the detector may be expressed as the mass of carbon passing through the flame per second required to give a detectable signal. A typical figure is 10^{-11} g C/sec.

The FID is sensitive to practically all organic substances, but is insensitive to inorganic gases and

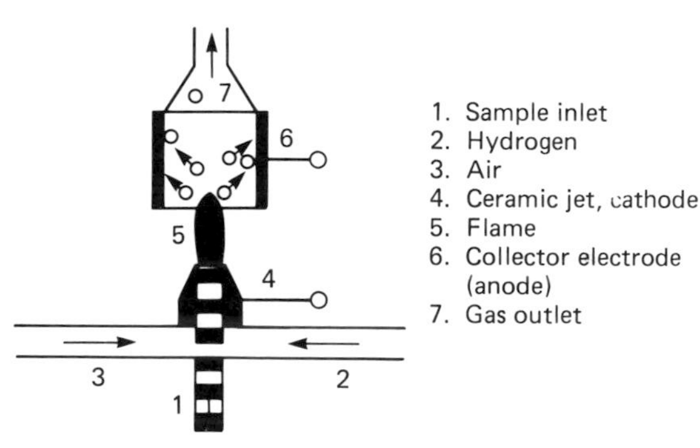

Figure 5.5 Flame ionization detector – schematic.

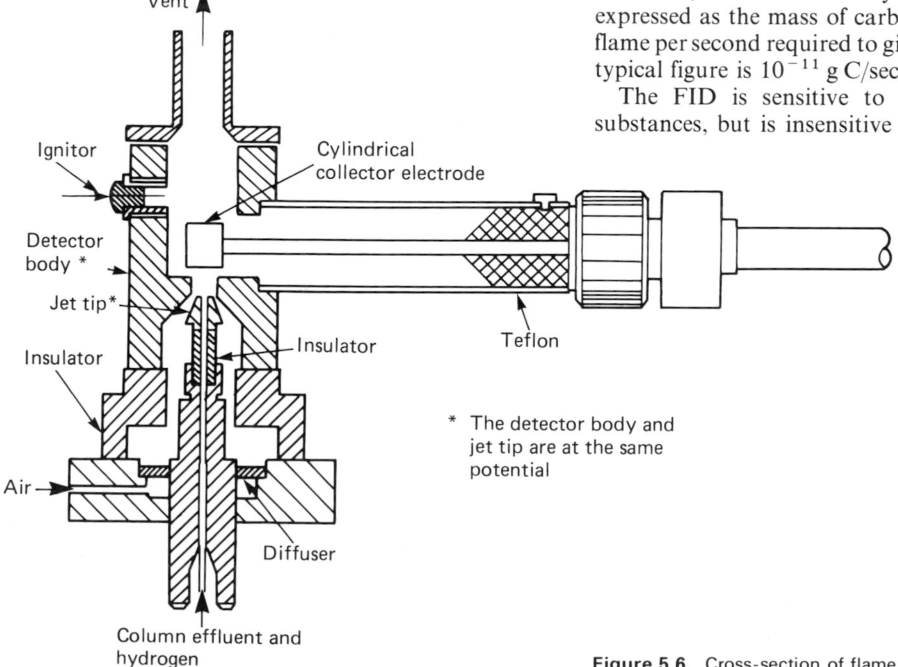

Figure 5.6 Cross-section of flame ionization detector.

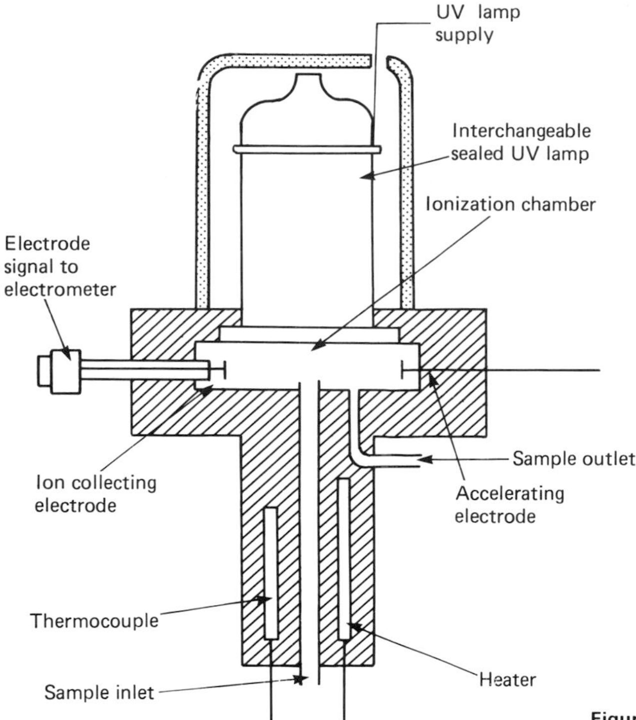

Figure 5.7 Photo-ionization detector.

water. It has a high sensitivity, good stability, wide range of linear response and low effective volume. It is widely used as a gas-chromatographic detector, and in total hydrocarbon analysers.

5.3.3 Photo-ionization detector (PID)

The photo-ionization detector (Figure 5.7) has some similarities to the flame ionization detector, and like the FID, it responds to a wide range of organic and also to some inorganic molecules. An interchangeable sealed lamp produces monochromatic radiation in the UV region. Molecules having ionization potentials less than the energy of the radiation may be ionized on passing through the beam. In practice, molecules with ionization potentials just above the photon energy of the incident beam may also be ionized, due to a proportion being in excited vibrational states. The ions formed are driven to a collector electrode by an electric field and the ion current is measured by an electrometer amplifier.

The flame in the FID is a high energy ionization source and produces highly fragmented ions from the molecules detected. The UV lamp in the PID is of lower quantum energy leading to the predominant formation of molecular ions. The response of the PID is therefore determined mainly by the ionization potential of the molecule, rather than the number of carbon atoms it contains. In addition the ionization energy in the PID may be selected by choice of the wavelength of the UV source, and the detector may be made selective in its response. The selectivity obtainable by use of three different UV lamps is shown in Figure 5.8. The ionization potentials of N_2, He, CH_3CN, CO and CO_2 are above the energy of all the lamps, and the PID does not respond to these gases.

The PID is highly sensitive, typically to picogram levels of organic compounds, and has a wide linear range. It may be used for direct measurements in gas streams or as a gas-chromatographic detector. When used as a detector in gas chromatography any of the commonly used carrier gases is suitable. Some gases, such as CO_2, absorb UV radiation and their presence may reduce the sensitivity of the detector.

5.3.4 Helium ionization detector

Monatomic gases, such as helium or argon, can be raised to excited atomic states by collision with energetic electrons emitted from a β-source. The metastable atomic states are themselves highly energetic and lose their energy by collision with other atomic or molecular species. If the helium contains a small concentration of a gas whose ionization potential is less than the excitation of the metastable helium atoms, ions will be formed in the collision, so

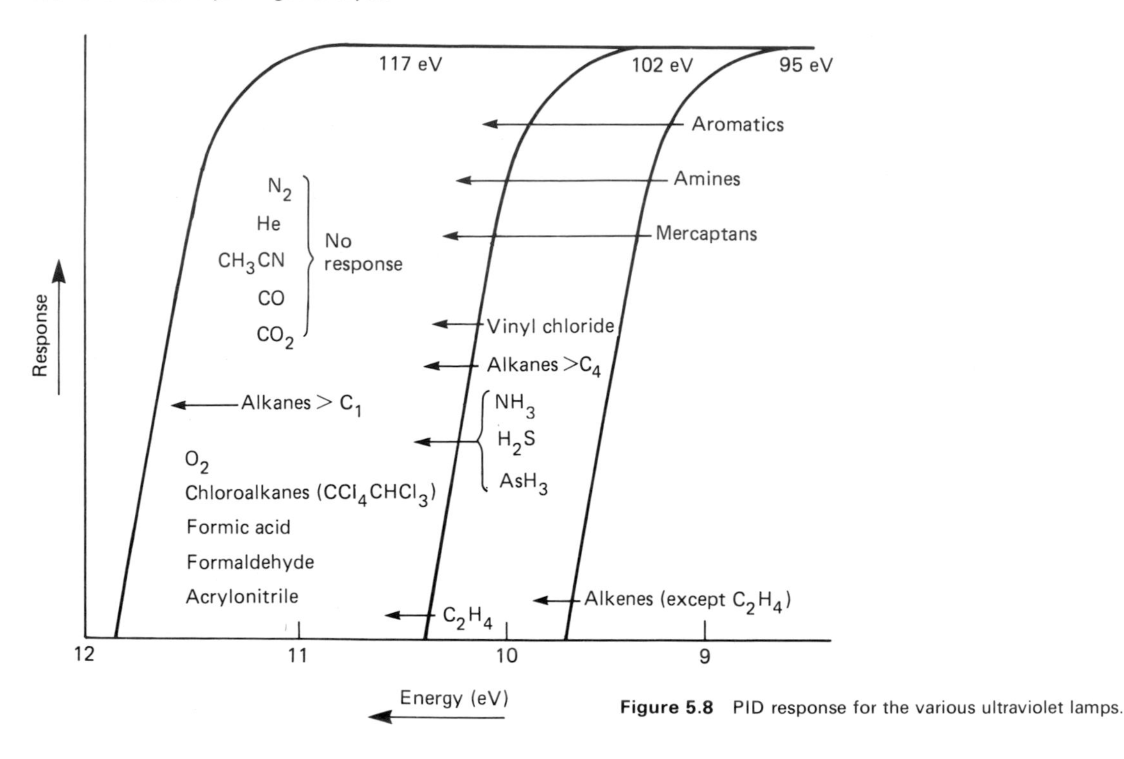

Figure 5.8 PID response for the various ultraviolet lamps.

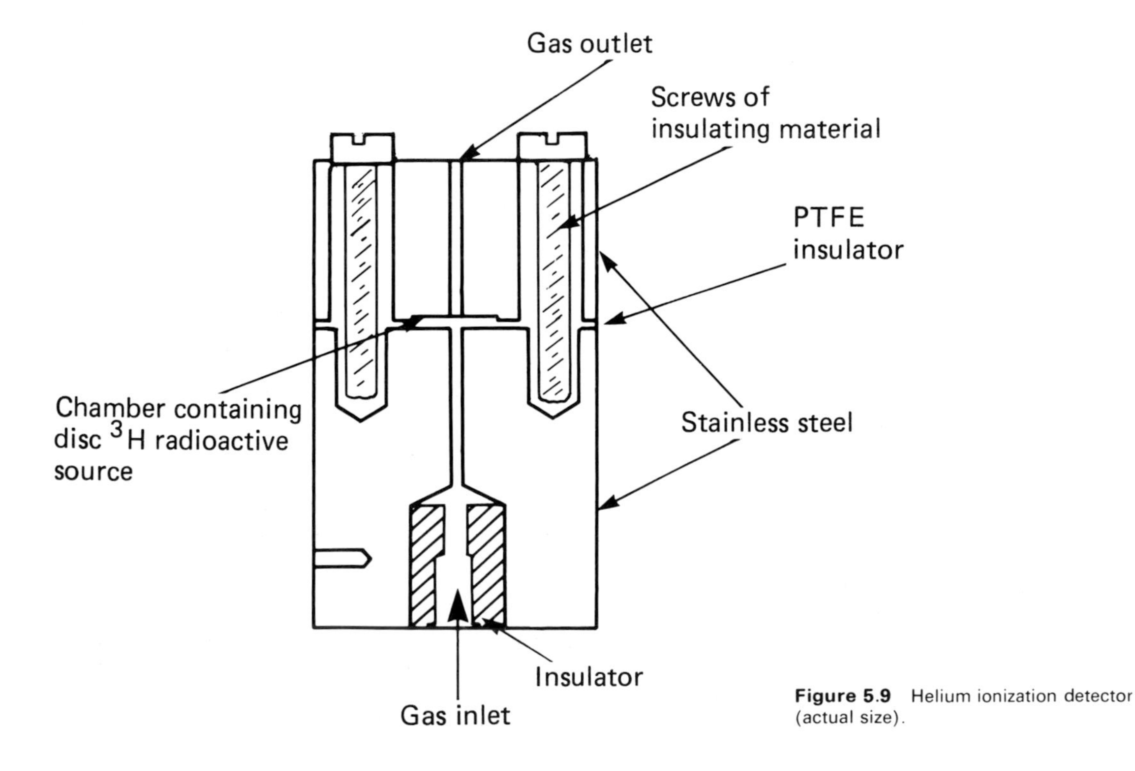

Figure 5.9 Helium ionization detector (actual size).

A — Inlet for carrier gas and anode
B — Diffuser – made of 100 mesh brass gauze
C — Source of ionizing radiation
D — Gas outlet and cathode

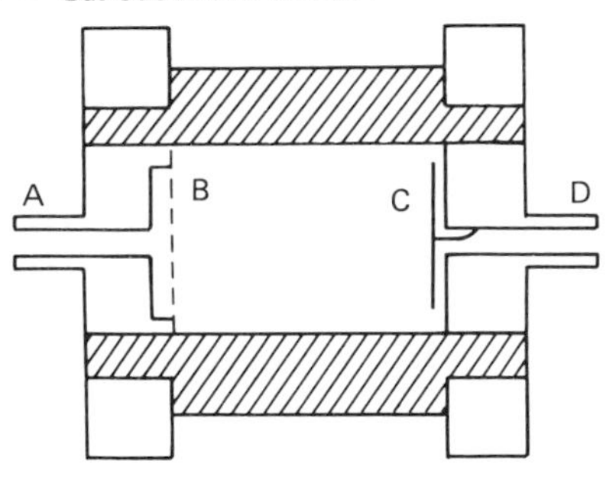

Brass

PTFE

Figure 5.10 Electron capture detector.

increasing the current-carrying capacity of the gas. This is the basis of the helium ionization detector.

The main reactions taking place can be represented as

$$He + e \rightarrow He^* + e$$

$$He^* + M \rightarrow M^+ + He + e$$

where M is the gas molecule forming ions. However, other collisions can occur, for example between metastable and ground-state helium atoms, or between metastable atoms, which may also result in ion formation.

The helium ionization detector (Figure 5.9) typically consists of a cylindrical chamber, approximately 1 cm in diameter and a few millimetres long, containing a β-emitting radioactive source. The ends of the cylinder are separated by an insulator and form electrodes. The detector is used as part of a gas-chromatographic system, with helium as the carrier gas.

It can be shown that the ionization mechanism described above depends on the number of atoms formed in metastable states. It can also be shown that the probability of formation of metastable states depends on the primary electron energy and on the intensity of the applied electric field. The reaction exhibits the highest cross-section for electrons with an energy of about 20 eV, and a field strength of 500 V/cm torr. Tritium (^3H) sources of 10–10 GBq or ^{63}Ni β-sources of 400–800 MBq activity are usually used, but the free path of the β-particles is very short, and the performance of the detector is strongly dependent on its geometry.

The helium ionization detector is used in gas chromatography, when its ability to measure trace levels of permanent gases is useful. However, the carrier gas supply must be rigorously purified.

5.3.5 Electron capture detector

The electron capture detector (Figure 5.10) consists of a cell containing a β-emitting radioactive source, purged with an inert gas. Electrons emitted by the radioactive source are slowed to thermal velocities by collision with the gas molecules, and are eventually collected by a suitable electrode, giving rise to a standing current in the cell. If a gas with greater electron affinity is introduced to the cell, some of the electrons are 'captured' forming negative ions, and the current in the cell is reduced. This effect is the basis of the electron capture detector. The reduction in current is due both to the difference in mobility between electrons and negative ions, and to differences in the rates of recombination of the ionic species and electrons.

The radioactive source may be tritium or ^{63}Ni, with ^{63}Ni usually being preferred since it allows the detector to be operated at higher temperatures, thus lessening the effects of contamination. A potential is applied between the electrodes which is just great enough to collect the free electrons. Originally, the detector was operated under d.c. conditions, potentials up to 5 volts being used, but under some conditions space charge effects produced anomalous results. Present detectors use a pulsed supply, typically 25 to 50 volts, 1 microsecond pulses at intervals of 5 to 500 microseconds. Either the pulse interval is selected and the change in detector current monitored, or a feedback system maintains a constant current and the pulse interval is monitored.

The electron capture detector is extremely sensitive to electronegative species, particularly halogenated compounds and oxygen. To obtain maximum sensitivity for a given compound, the choice of carrier gas, pulse interval or detector current and detector temperature must be optimized.

The electron capture detector is most often used in gas chromatography, with argon, argon–methane mixture or nitrogen as carrier gas, but it is also used in leak or tracer detectors. The extreme sensitivity of the ECD to halogenated compounds is useful, but high purity carrier gas and high stability columns are required to prevent contamination. Under optimum conditions, 1 part in 10^{12} of halogenated compounds such as Freons, can be determined.

5.3.6 Flame photometric detector (FPD)

Most organic and other volatile compounds containing sulphur or phosphorus produce chemiluminescent species when burned in a hydrogen-rich flame. In a flame photometric detector (Figure 5.11) the sample gas passes into a fuel-rich H_2/O_2 or H_2/air mixture which produces simple molecular species and excites them to higher electronic states. These excited species subsequently return to their ground states and emit characteristic molecular band spectra. This emission is monitored by a photomultiplier tube through a suitable filter, thus making the detector selective to either sulphur or phosphorus. It may also be sensitive to other elements, including halogens and nitrogen.

The FPD is most commonly used as a detector for sulphur-containing species. In this application, the response is based on the formation of excited S_2 molecules, S_2^*, and their subsequent chemiluminescent emission. The original sulphur-containing molecules are decomposed in the hot inner zone of the flame, and sulphur atoms are formed which combine to form S_2^* in the cooler outer cone of the flame. The exact mechanism of the reaction is uncertain, but it is believed that the excitation energy for the $S_2 \rightarrow S_2^*$ transition may come from the formation of molecular hydrogen or water in the flame, according to the reactions

$$H + H + S_2 \rightarrow S_2^* + H_2 \text{ (4.5 eV)}$$

$$H + OH + S_2 \rightarrow S_2^* + H_2O \text{ (5.1 eV)}$$

As the excited S_2 molecule reverts to the ground state it emits a series of bands in the range 300–450 nm, with the most intense bands at 384.0 and 394.1 nm. The 384.0 nm emission is monitored by the photomultiplier tube.

The FPD is highly selective and sensitive, but the response is not linearly proportional to the mass-flow rate of the sulphur compound. Instead, the relationship is given by:

$$I_{S_2} = I_0[S]^n$$

where I_{S_2} is the observed intensity of the emission (photomultiplier tube output), [S] is the mass-flow rate of sulphur atoms (effectively the concentration of the sulphur compound) and n is a constant, found to be between 1.5 and 2, depending on flame conditions. Commercial analysers employing the FPD often incorporate a linearizing circuit to give an output which is directly proportional to sulphur mass-flow. The detector response is limited to two or three orders of magnitude.

The FPD is highly selective, sensitive (10^{-11} g) and relatively simple, but has an extremely non-linear response. It is used in gas chromatography and in sulphur analysers.

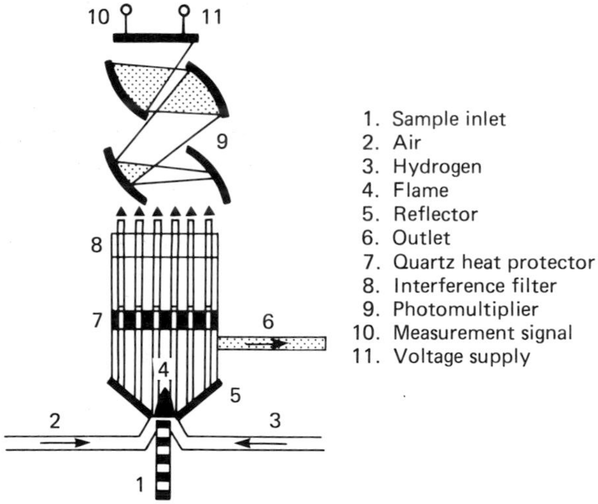

1. Sample inlet
2. Air
3. Hydrogen
4. Flame
5. Reflector
6. Outlet
7. Quartz heat protector
8. Interference filter
9. Photomultiplier
10. Measurement signal
11. Voltage supply

Figure 5.11 Flame photometric detector.

5.3.7 Ultrasonic detector

The velocity of sound in a gas is inversely proportional to the square root of its molecular weight. By measuring the speed of sound in a binary gas mixture, its composition can be deduced, and this technique is the basis of the ultrasonic detector (Figure 5.12). A

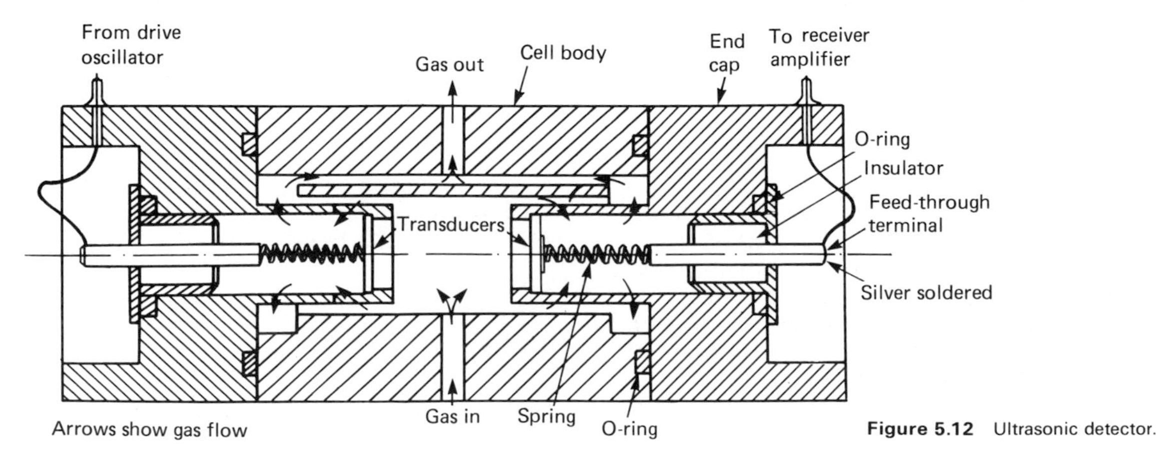

Figure 5.12 Ultrasonic detector.

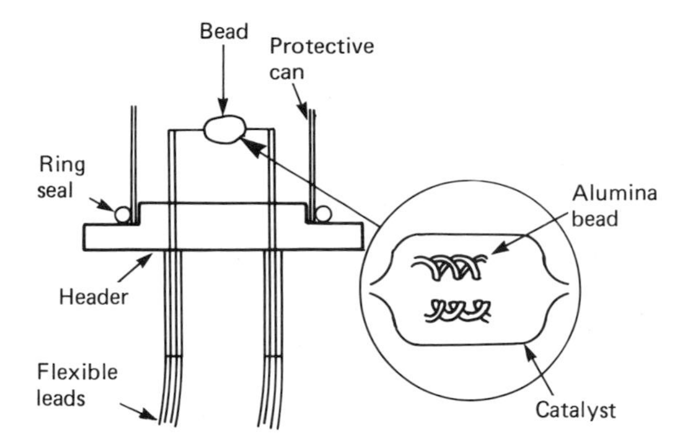

Figure 5.13 Catalytic gas-sensing element.

quartz crystal transducer located at one end of the sample cell sound tube acts as the emitter and an identical crystal located at the other end of the sound tube acts as the receiver. To obtain efficient transfer of sound energy between the gas and the transducers, the detector must be operated at above atmospheric pressure, and the gas in the cell is typically regulated to 1 to 7 bar gauge depending on the gas. The phase shift of the sound signal traversing the cell between the emitter and receiver is compared to a reference signal to determine the change in speed of sound in the detector.

The detector is most often used in gas chromatography. It has a universal response, and the output signal is proportional to the difference in molecular weight between the gaseous species forming the binary mixture. When used as a gas-chromatographic detector it has good sensitivity (10^{-9}–10^{-10} g) and a wide linear dynamic range (10^6), and allows a wide choice of carrier gas. However, precise temperature control is required, and the electronic circuitry is complex. It may be a useful alternative where flames cannot be used, or where a katharometer would not respond to all components in a mixture.

5.3.8 Catalytic detector (pellistor)

Catalytic gas detectors operate by measuring the heat output resulting from the catalytic oxidation of flammable gas molecules to carbon dioxide and water vapour at a solid surface. By use of a catalyst, the temperature at which the oxidation takes place is much reduced compared with gas phase oxidation. The catalyst may be incorporated into a solid state sensor containing an electrical heater and temperature-sensing device. A stream of sample gas is fed over the sensor, and flammable gases in the sample are continuously oxidized, releasing heat and raising the temperature of the sensor. Temperature variations in the sensor are monitored to give a continuous record of the flammable-gas concentration in the sample.

The most suitable metals for promoting the oxidation of molecules containing C—H bonds, such as methane and other organic species, are those in Group 8 of the Periodic Table, particularly platinum and palladium. The temperature sensor is usually a platinum resistance thermometer, wound in a coil and also used as the electrical heater for the sensor. The resistance is measured by connecting the sensor as one arm of a Wheatstone bridge and measuring the out-of-balance voltage across the bridge.

The construction of a typical catalytic sensing element is shown in Figure 5.13. A coil of 50 μm platinum wire is mounted on two wire supports which also act as electrical connections. The coil is embedded in porous ceramic material, usually alumina, to form a bead about 1 mm long. The catalyst material is impregnated on the outside of the bead. This type of catalytic sensor is often called a 'pellistor'. The choice of catalyst, and of the treatment of the outside of the bead, for example by inclusion of a diffusion layer, influences the overall sensitivity of the sensor, and the relative sensitivity to different gases. The sensitivity and selectivity are also influenced by the choice of catalyst and by the temperature at which the sensor is operated. Palladium and its oxides are the most widely used catalysts; they have the advantage that they are much more active than platinum, enabling the sensor to be operated at the lowest possible temperature. The sensor is mounted in a protective open-topped can as shown, so that the gas flow to the sensor is largely diffusion-controlled.

The Wheatstone bridge network commonly used with a catalytic sensor is shown in Figure 5.14. The sensing element forms one arm of the bridge, and the second arm is occupied by a compensator element.

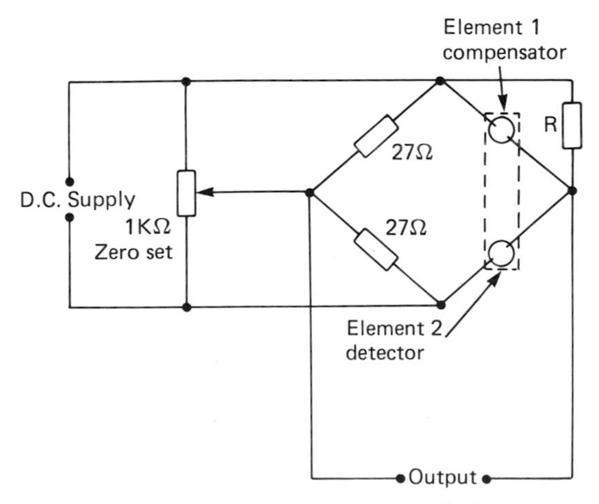

Figure 5.14 Wheatstone bridge network used with catalytic detector.

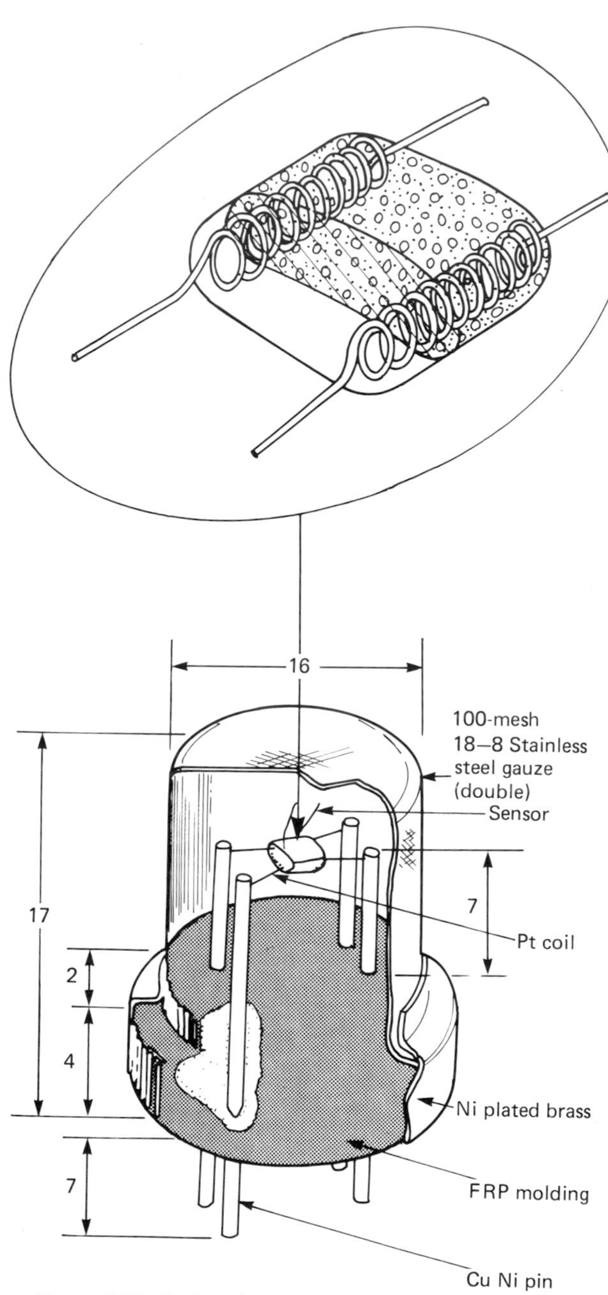

Figure 5.15 Semiconductor sensor.

This is a ceramic bead element, identical in construction to the sensor, but without the catalytic coating. The sensor and compensator are mounted close together in a suitable housing so that both are exposed to the same sample gas. The pellistor or catalytic sensor is the basis of the majority of portable flammable-gas detectors.

5.3.9 Semiconductor detector

The electrical conductivity of many metal oxide semiconductors, particularly those of the transition and heavy metals, such as tin, zinc and nickel, is changed when a gas molecule is adsorbed on the semiconductor surface. Adsorption involves the formation of bonds between the gas molecule and the semiconductor, by transfer of electrical charge. This charge transfer changes the electronic structure of the semiconductor, changing its conductivity. The conductivity changes are related to the number of gas molecules adsorbed on the surface, and hence to the concentration of the adsorbed species in the surrounding atmosphere.

A typical semiconductor detector is shown in Figure 5.15. The semiconducting material is formed as a bead, about 2–3 mm in diameter, between two small coils of platinum wire. One of the coils is used as a heater, to raise the temperature of the bead so that the gas molecules it is desired to detect are reversibly adsorbed on the surface, and the resistance of the bead is measured by measuring the resistance between the two coils. The bead is mounted in a stainless-steel gauze enclosure (Figure 5.15) to ensure that molecules diffuse to the semiconductor surface, thus ensuring that the device is as free as possible from the effects of changes in the flow rate of the sample gas.

Semiconductor detectors are mainly used as low-cost devices for detection of flammable gases. A

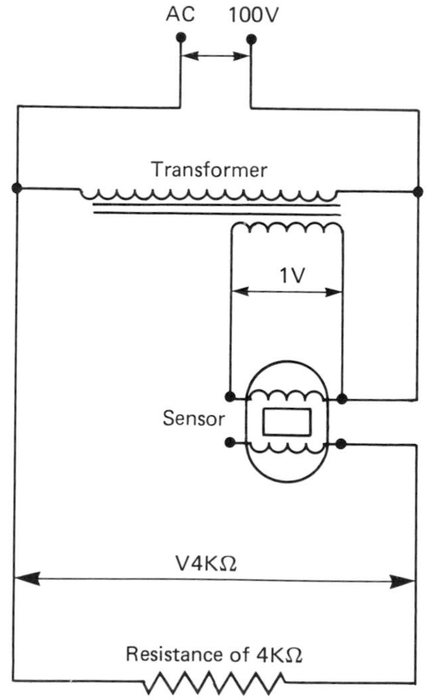

Figure 5.16 Measuring circuit for semiconductor sensor.

Table 5.2 Properties and applications of gas detectors

Detector	*Applicability*	*Selectivity*	*Carrier or bulk gas*	*Lower limit of detection (grams)*	*Linear range*	*Typical applications*
Thermal conductivity	Universal	non-selective	He, H_2	10^{-6}–10^{-7}	10^4	Analysis of binary or pseudo-binary mixtures; gas chromatography
Flame ionization	Organic compounds	non-selective	N_2	10^{-11}	10^6	Gas chromatography; hydrocarbon analysers
Photo-ionization	Organic compounds except low molecular weight hydrocarbons	limited	N_2	10^{-11}–10^{-12}	10^7	Gas chromatography
Helium ionization	Trace levels of permanent gases	non-selective	He	10^{-11}	10^4	Gas chromatography
Electron capture	Halogenated and oxygenated compounds	response is highly compound-dependent	Ar, N_2, $N_2 + 10\%\ CH_4$	10^{-12}–10^{-13}	10^3	Gas chromatography, tracer gas detectors, explosive detectors
Flame photometric	Sulphur and phosphorus compounds	selective to compounds of S or P	N_2, He	10^{-11}	5×10^2 (S) 10^3 (P)	Gas chromatography, sulphur analysers
Ultrasonic detector	Universal	non-selective, mainly low molecular weight	H_2, He, Ar, N_2, CO_2	10^{-9}–10^{-10}	10^6	Gas chromatography
Catalytic (pellistor)	Flammable gases	selective to flammable gases	Air	*		Flammable gas detectors
Semiconductor	Flammable gases, other gases	limited	Air	*		Low-cost flammable gas detectors

* The performance of these detectors depends on the individual design and application.

suitable power-supply and measuring circuit is shown in Figure 5.16. The main defect of the devices at present is their lack of selectivity.

5.3.10 Properties and applications of gas detectors

The properties and applications of the most commonly used gas detectors are summarized in Table 5.2.

5.4 Process chromatography

On-line or process gas chromatographs are instruments which incorporate facilities to carry out automatically the analytical procedure for chromatographic separation, detection and measurement of predetermined constituents of gaseous mixtures. Samples are taken from process streams and are presented, in a controlled manner and under known conditions, to the gas chromatograph. Successive analyses may be made, on a regular timed basis, on aliquots of sample taken from a single stream, or by use of suitable stream-switching valves, a single process chromatograph may carry out automatic sequential analyses on process streams originating from several different parts of the plant.

The main components of a typical process chromatograph system are shown in Figure 5.17. These components are: a supply of carrier gas to transport the sample through the column and detector, a valve for introduction of known quantities of sample, a chromatographic column to separate the sample into its components, a detector and associated amplifier to sense and measure the components of the sample in the carrier-gas stream, a programmer to actuate the operations required during the analytical sequence and to control the apparatus, and a display or data-processing device to record the results of the analyses.

5.4.1 Sampling system

The sampling system must present a homogeneous and representative sample of the gas or liquid to be analysed, to the gas chromatograph. In process chromatography a continuous stream of the sample is taken, usually by means of a fast bypass loop, and treated as necessary for example by drying, filtering or adjustment of temperature or pressure. Discrete volumes of the treated sample stream are periodically injected into the carrier gas stream of the chromatograph by means of a gas (or liquid) sampling valve. The chromatograph is normally supplied with sample from the point or points to be monitored by use of permanently installed sampling lines. However, where

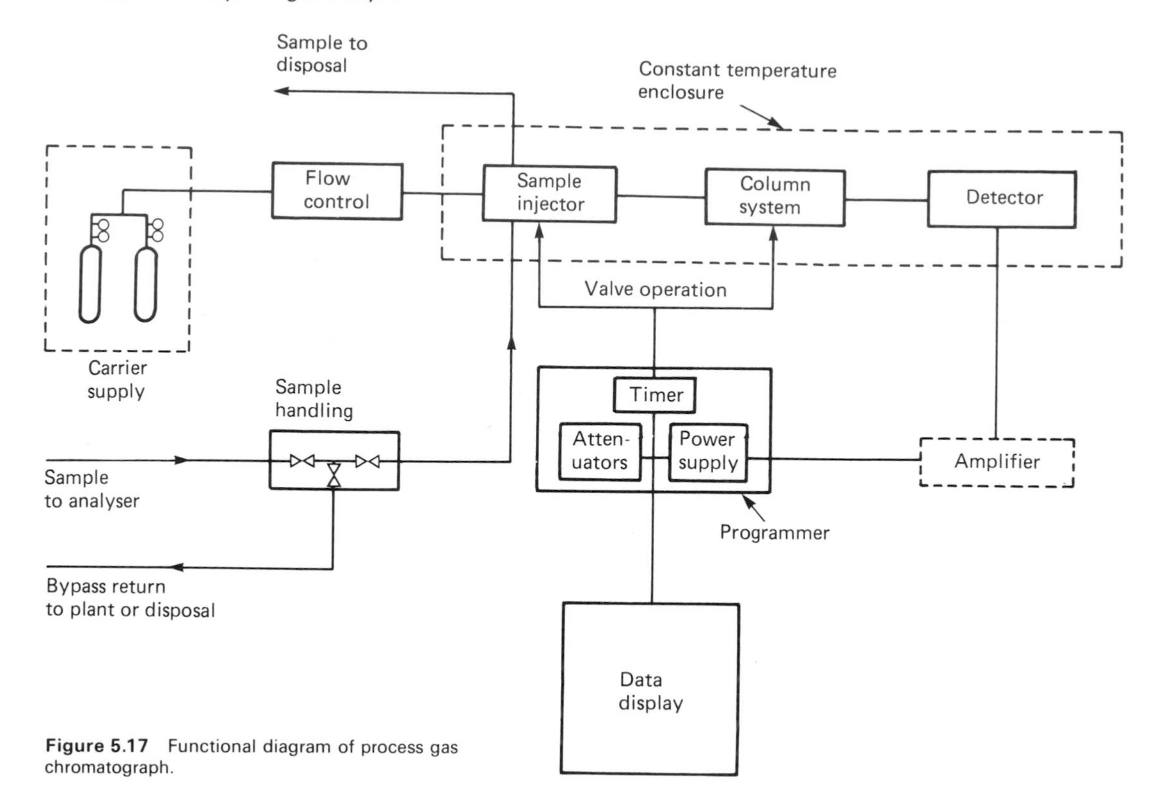

Figure 5.17 Functional diagram of process gas chromatograph.

the frequency of analysis does not justify the installation of special lines, samples may be collected in suitable containers for subsequent analysis. Gas samples may be collected under pressure in metal (usually stainless steel) cylinders or at atmospheric pressure in gas pipettes, gas sampling syringes or plastic bags. For analysis of gases at very low concentrations such as the determination of pollutants in ambient air, the pre-column or adsorption tube concentration technique is often used. The sample is drawn or allowed to diffuse through a tube containing a granular solid packing to selectively adsorb the components of interest. The tube is subsequently connected across the sample loop ports of the gas sampling valve on the chromatograph and heated to desorb the compounds to be analysed into the carrier-gas stream.

It is essential that the sample size should be constant for each analysis, and that it is introduced into the carrier gas stream rapidly as a well-defined slug. The sample should also be allowed to flow continuously through the sampling system to minimize transportation lag. Chromatographic sampling or injection valves are specially designed changeover valves which enable a fixed volume, defined by a length of tubing (the sample loop) to be connected in either one of two gas streams with only momentary interruption of either stream. The design and operation of a typical sampling valve is shown in Figure 5.18. The inlet and outlet tubes terminate in metal (usually stainless steel) blocks with accurately machined and polished flat faces. A slider of soft plastic material, with channels or holes machined to form gas paths, is held against the polished faces and moved between definite positions to fill the loop or inject the sample. The main difference between 'gas' and 'liquid' sampling valves is in the size of sample loop. In the 'gas' sampling valve the loop is formed externally, and typically has a volume in the range 0.1–10 ml. For liquid sampling the volumes required are smaller and the loop is formed in the internal channels of the valve and may have a volume as small as 1 μl. In process chromatography, sampling valves are normally fitted with electric or pneumatic actuators so that they may be operated automatically by the programmer at predetermined times during the analytical sequence.

When it is required to change between columns or detectors during an analysis, similar types of valves are required. The number of ports, and the arrangement of the internal channels, may be tailored for the individual application. Figure 5.19 shows an arrangement where a single valve is used for sample injection and backflushing in a chromatograph with two analytical columns in series. The sample is injected

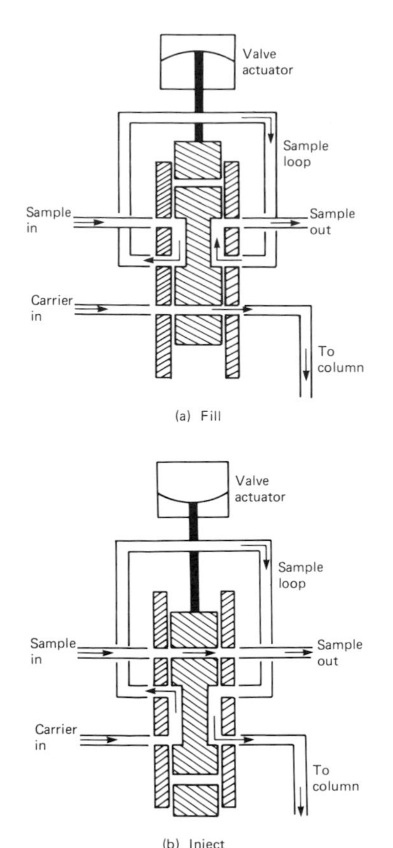

Figure 5.18 Gas-sampling valve (schematic).

onto column 1, which is chosen so that the components of interest are eluted first, and pass to column 2. At a predetermined time, the valve is switched to refill the sample loop and to reverse the flow of carrier gas to column 1, while the forward flow is maintained in column 2 to effect the final separation of the components of the sample. By this means components of no interest, such as high-boiling compounds or solvents, can be 'backflushed' to waste before they reach column 2 or the detector, thus preserving the performance of the columns and speeding the analytical procedure.

The gas sample must arrive at the sampling valve at or only slightly above atmospheric pressure, at a flow rate typically in the range 10–50 ml min^{-1}, and be free from dust, oil or abrasive particles. The sampling system may also require filters, pressure or flow controllers, pumps and shut-off valves for control and processing of the sample stream. All the components of the system must be compatible with the chemical species to be sampled, and must be capable of withstanding the range of pressures and temperatures expected.

Many applications require analysis of two or more process streams with one analyser. In these instances a sample line from each stream is piped to the analyser and sample lines are sequentially switched through solenoid valves to the sampling valve. When multi-stream analysis is involved, inter-sample contamination must be prevented. Contamination of samples can occur through valve leakage and inadequate flushing of common lines. To ensure adequate flushing, the capacity of common lines is kept to a minimum and the stream selection valves are timed so that while the sample from one stream is being analysed, the sample from the next stream is flowing through all common lines.

Prevention of inter-sample contamination from valve leakage is accomplished by locating valves with respect to pressure drops such that any leakage will flow to vent rather than intermix in common lines. A typical flow arrangement for gas supplies to a chromatograph for multi-stream application is shown in Figure 5.20. This is designed to ensure that the sample and other supplies are delivered at the correct flow rate

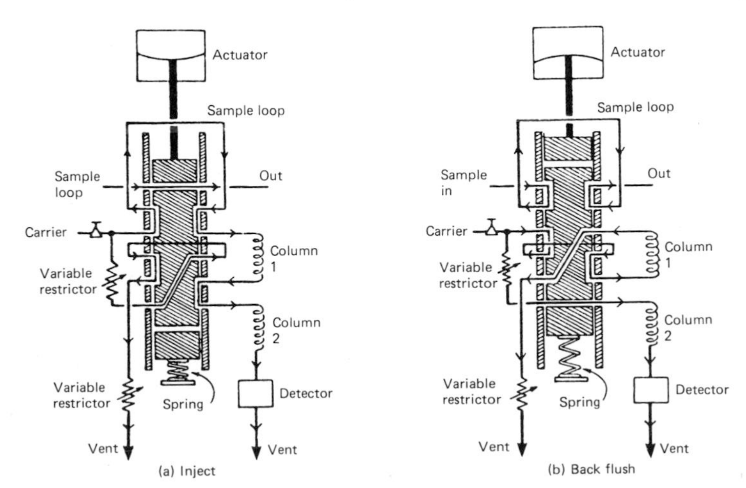

Figure 5.19 Schematic diagram of sample and backflush valve.

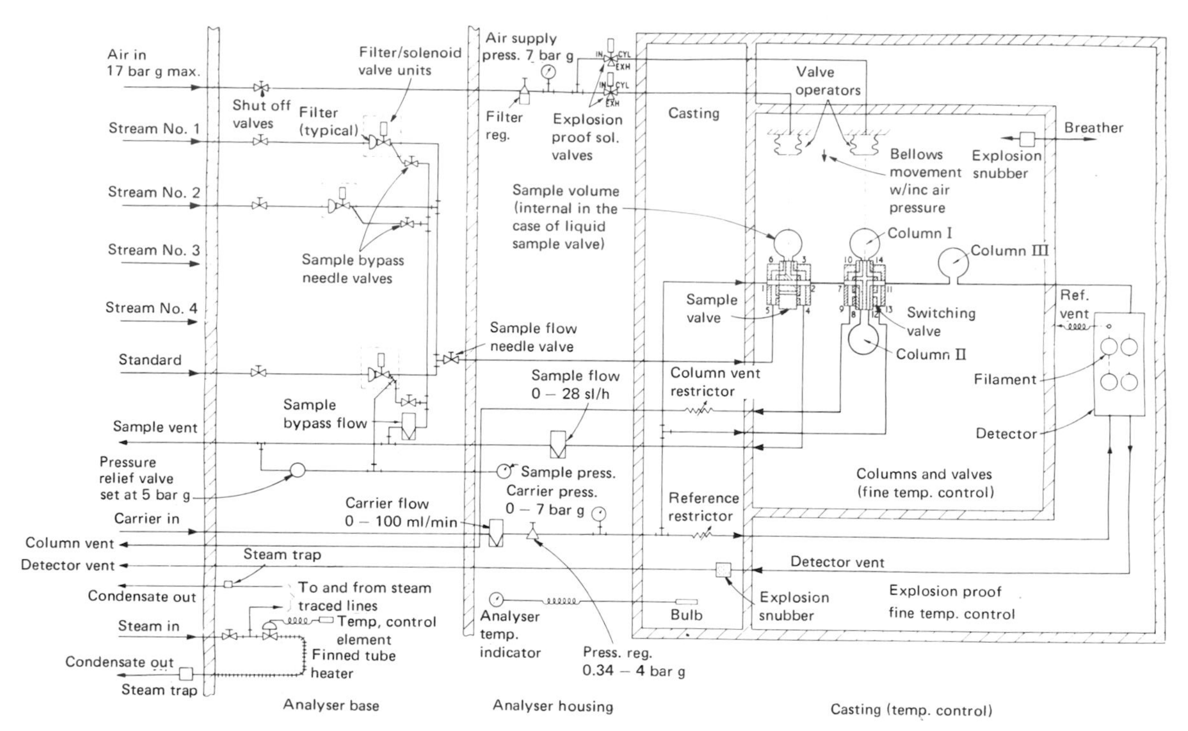

Figure 5.20 Flow diagram of multi-stream chromatograph with thermal conductivity detector (courtesy The Foxboro Company).

and pressure. A pressure-relief valve is fitted to protect the sampling valve from excessive pressure, and shut-off valves are fitted on all services except bottled gas lines.

In some applications additional conditioning of the sample is required. Typical of these would be trace-heating of sample lines to maintain a sample in a gaseous state, vaporization to change a liquid to a gas and elimination of stream contaminants by mechanical or chemical means.

5.4.2 Carrier gas

The carrier gas transports the components of the sample over the stationary phase in the chromatographic column. The carrier gas must not react with the sample, and for maximum efficiency when using long columns, it is advantageous to use a gas of low viscosity. However, the most important criterion in choosing a carrier gas is often the need to ensure compatibility with the particular detector in use.

The primary factors determining the choice of carrier gas are the effect of the gas on component resolution and detector sensitivity. The carrier gas and type of detector are chosen so that the eluted components generate large signals. For this reason, helium is generally used with thermal conductivity cells because of its high thermal conductivity. Hydrogen has a higher thermal conductivity and is less expensive than helium, but because of precautions necessary when using hydrogen, helium is preferred where suitable.

Specific properties of a particular carrier gas are exploited in other types of detectors, for example helium in the helium ionization detector. In special instances a carrier gas other than that normally associated with a particular detector may be used for other reasons. For example, to measure hydrogen in trace quantities using a thermal conductivity detector, it is necessary to use a carrier gas other than helium because both helium and hydrogen have high and similar thermal conductivities. Accordingly, argon or nitrogen is used because either has a much lower thermal conductivity than hydrogen, resulting in a larger difference in thermal conductivity and greater output.

The flow rate of carrier gas affects both the retention time of a compound in the column, and the shape of the chromatographic peak and hence the amplitude of the detector signal. It is therefore essential for the flow rate to be readily adjustable to constant known values. The gas is usually supplied from bottles, with pressure reducing valves to reduce the pressure to a level compatible with the flow control equipment, and sufficient to give the required flow rate through the column and detector.

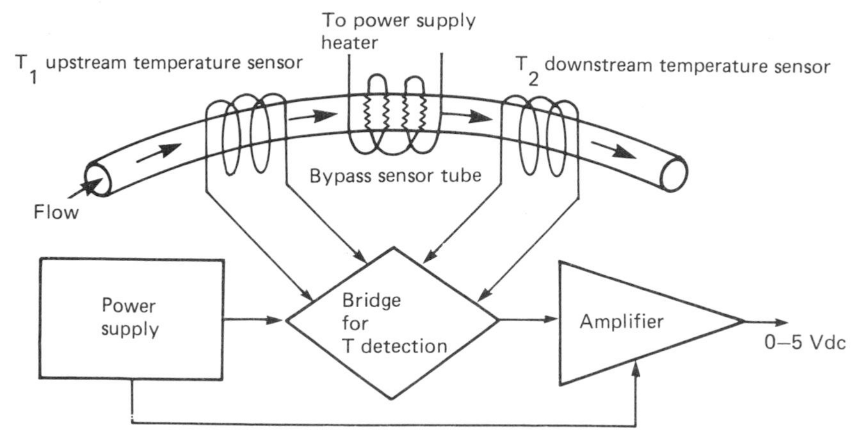

Figure 5.21 Principle of operation of electronic mass-flow controller (courtesy Brooks).

The flow rate of carrier gas may be measured and controlled either mechanically or electronically. Mechanical controllers are either precision pressure regulators which maintain a constant pressure upstream of the column and detector, or differential pressure regulators which maintain a constant pressure drop across a variable restriction. The principle of operation of one type of electronic flow controller is shown in Figure 5.21. A proportion of the gas stream is diverted via a narrow tube, fitted with an electric heating coil as shown. Sensor coils of resistance wire are wound on the tube upstream and downstream of the heating coil. Heat at a constant rate is supplied to the heating coil. Gas passing through the tube is heated by the coil, and some heat is transferred to the downstream sensor. The sensor coils are connected in a Wheatstone bridge circuit. The out-of-balance signal from the bridge, caused by the difference in temperatures and hence resistance of the upstream and downstream coils, depends on the mass-flow rate of gas through the tube, and on the specific heat of the gas. (See also Part 1, p. 31.) The signal, suitably amplified, can be used to give a direct readout of the flow rate of gas through the tube, and can be used to control the flow by feeding the signal to open or close a solenoid valve in the main gas line downstream of the sensing device.

In cases where the carrier gas flow rate is controlled mechanically, a rotameter is provided to indicate the flow rate. However, the best indication of correct flow is often the analysis record itself, as the retention times of known components of the sample should remain constant from one injection to the next.

5.4.3 Chromatographic column

The separating columns used in process chromatographs are typically 1–2 m lengths of stainless steel tubing, 3–6 mm outer diameter, wound into a helix for convenient housing, and packed with a solid absorbent. Separation of permanent gases is normally carried out on columns packed with molecular sieve. These are synthetic zeolites, available in a range of effective pore diameters. Porous polymeric materials have been developed which are capable of separating a wide range of organic and inorganic molecules, and use of these proprietary materials gives much more predictable column performance than when liquid-coated solids are used. In addition the polymeric materials are thermally stable and do not suffer from 'bleed' or loss of the liquid stationary phase at high temperatures which can give rise to detector noise or drift in the baseline of the chromatogram.

One or more columns packed with these materials can normally be tailored to the needs of most process analyses. However, in certain cases it may be necessary to include valves to switch between columns or detectors during the analysis, or to divert the carrier gas to waste to prevent a certain component, for example a solvent present in high concentration, from reaching the detector. These switching operations are referred to as backflushing, or heart-cutting if the unwanted peak occurs in the middle of the chromatogram.

5.4.4 Controlled temperature enclosures

Many components of the gas chromatograph, including the injection valve, columns and detectors require to be kept at constant temperatures or in environments whose temperature can be altered at known rates, and separate temperature-controlled zones are usually provided in the instrument.

Two general methods are used to distribute heat to maintain the temperature-sensitive components at constant temperatures (±0.1 K or better) and to minimize temperature gradients. One uses an air bath, and the other metal-to-metal contact (or heat sink). The former depends on circulation of heated air and the latter upon thermal contact of the temperature-sensitive elements with heated metal.

An air bath has inherently fast warm-up and comparatively high temperature gradients and offers the advantage of ready accessibility to all components within the temperature-controlled compartment. The air bath is most suitable for temperature programming and is the usual method for control of the temperature of the chromatographic column.

Metal-to-metal contact has a slower warm-up but relatively low temperature gradients. It has the disadvantage of being a greater explosion hazard, and may require the analyser to be mounted in an explosion-proof housing resulting in more limited accessibility and more difficult servicing. The detectors are often mounted in heated metal blocks for control of temperature.

The choice of the method of heating and temperature control may depend on the location where the instrument is to be used. Instruments are available with different degrees of protection against fire or explosion hazard. For operation in particularly hazardous environments, for example where there may be flammable gases, instruments are available where the operation, including temperature control, valve switching, and detector operation is entirely pneumatic, with the oven being heated by steam.

5.4.5 Detectors

A gas-chromatographic detector should have a fast response, linear output over a wide range of concentration, be reproducible and have high detection sensitivity. In addition the output from the detector must be zero when pure carrier gas from the chromatographic column is passing through the detector.

In process chromatography, the most commonly used detectors are the thermal conductivity and flame ionization types. Both have all the desirable characteristics listed above, and one or other is suitable for most commonly analysed compounds: the thermal conductivity detector is suitable for permanent gas analysis and also responds universally to other compounds while the flame ionization detector responds to almost all organic compounds. In addition these detectors can be ruggedly constructed for process use and can be used with a wide range of carrier gases. Most other detectors have disadvantages in comparison with these two, for example, fragility, non-linear response or a requirement for ultra-pure carrier-gas supplies, and although widely used in laboratory chromatographs, their application to process instruments is restricted.

The helium ionization detector may be used for permanent gas analyses at trace levels where the katharometer is insufficiently sensitive, and the ultrasonic detector may be a useful alternative in applications where a flame cannot be used or where a katharometer cannot be used for all components in a mixture. The selective sensitivity of the electron capture detector to halogenated molecules may also find occasional application. A comprehensive list of gas-detecting devices, indicating which are suitable for use in gas chromatography, is given in Table 5.2 (p. 157).

5.4.6 Programmers

Analysis of a sample by gas chromatography requires the execution of a series of operations on or by the instrument at predetermined times after the analytical sequence is initiated by injection of the sample. Other instrumental parameters must also be continuously monitored and controlled. Process gas chromatographs incorporate devices to enable the analytical sequence to be carried out automatically, and the devices necessary to automate a particular instrument are usually assembled into a single module, known as the programmer or controller.

At the most basic level the programmer may consist of mechanical or electromechanical timers, typically of the cam-timer variety, to operate relays or switches at the appropriate time to select the sample stream to be analysed, operate the injection valve and start the data recording process, combined with a facility to correct the output of the chromatograph for baseline drift. Some chromatographs with built-in microprocessors may incorporate the programmer as part of the central control and data acquisition facility, or the programmer itself may contain a microprocessor and be capable of controlling and monitoring many more of the instrumental parameters as well as acting as a data-logger to record the output of the chromatograph. Computer-type microprocessor-based integrators are available for laboratory use, and in many cases these have facilities to enable them to be used as programmers for the automation of laboratory gas chromatographs.

When the process chromatograph is operated in the automatic mode, all the time-sequenced operations are under programmer control. These will typically include operations to control the gas chromatograph and sampling system, such as sample stream selection, sample injection, column or detector switching, automatic zero and attenuation adjustment, and back-flushing. The programmer will also carry out at least some initial processing of the output data, by, for example, peak selection. It is also necessary for a process instrument to incorporate safety devices to prevent damage to itself or to the surroundings in the event of a malfunction, and also to give an indication of faults which may lead to unreliable results. Functions which may be assigned to the programmer include: fault detection and identification, alarm generation and automatic shutdown of the equipment when a fault is detected.

In addition to the automatic mode of operation the programmer must allow the equipment to be operated manually for start-up, maintenance and calibration.

5.4.7 Data-processing systems

The output from a gas chromatograph detector is usually an electrical signal, and the simplest method of data presentation is the chromatogram of the sample, obtained by direct recording of the detector output on a potentiometric recorder. However, the complexity of the chromatograms of typical mixtures analysed by chromatography means that this simple form of presentation is unsuitable for direct interpretation or display, and further processing is required. The data-processing system of a process chromatograph must be able to identify the peaks in the chromatogram corresponding to components of interest in the sample, and it must measure a suitable parameter of each peak which can be related to the concentration of that component of the sample. In addition the system should give a clear indication of faults in the equipment.

Identification of the peaks in the chromatogram is made on the basis of retention time. Provided that instrumental parameters, particularly column temperature and carrier-gas flow rate remain constant, the retention time is characteristic of a given compound on a particular column. Small changes in operating conditions may change the retention times, so the data-processing system must identify retention times in a suitable 'window' as belonging to a particular peak. In addition retention times may show a long-term drift due to column ageing; and the data-processing system may be required to compensate for this.

Relation of the output signal to the concentration of the component of interest may be made on the basis either of the height of the peak or the area under it. In both cases a calibration curve must be prepared beforehand by analysis of standard mixtures, and in the more sophisticated systems, this information can be stored and the necessary calculations carried out to give a printed output of the concentrations of the components of interest for each analysis. Automatic updating of the calibration may also be possible. The simplest data-processing systems relate peak height to concentration, but it is usually better to measure peak areas, particularly for complex chromatograms, as this gives some automatic compensation for changes in peak shape caused by adventitious changes in operating conditions. In this case the data-processing system must incorporate an integrator.

5.4.7.1 Display of chromatographic data

A refinement of the basic record of the complete chromatogram of the sample is to select and display only peaks corresponding to species of interest, each species being assigned to a separate recorder channel so that successive analyses enable changes in the concentration of each species to be seen. The peaks may be displayed directly or in bar form as shown in Figure 5.22.

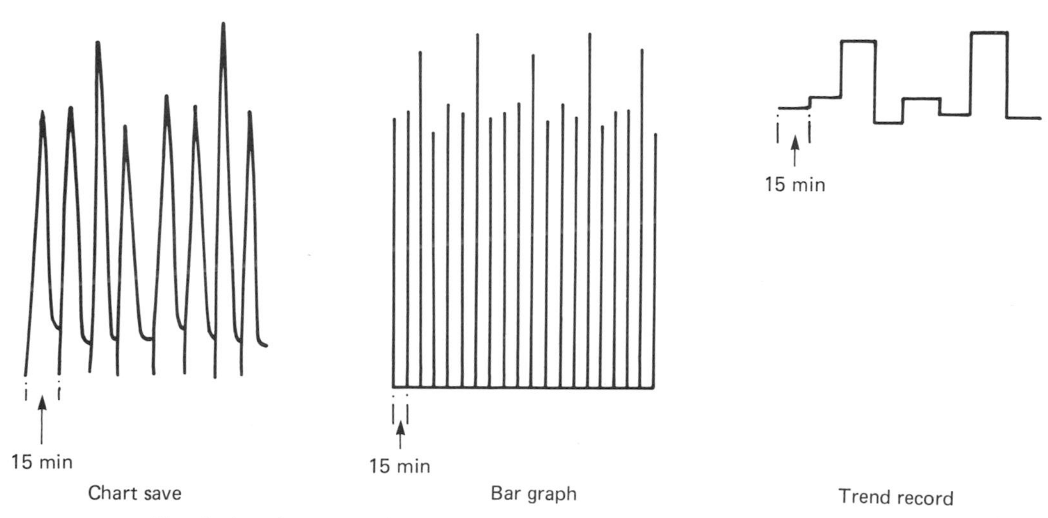

Figure 5.22 Methods of display of chromatographic data.

For trend recording a peak selector accepts the output from the process chromatograph, detects the peak height for each selected measured component and stores the data. The peak heights are transferred to a memory unit which holds the value of the height for each peak until it is updated by a further analysis. The output of this unit may be displayed as a chart record of the change in concentration of each measured species. An example of this type of output is shown in Figure 5.22.

5.4.7.2 *Gas-chromatographic integrators*

A variety of gas-chromatographic integrators are available to provide a measure of the areas under the peaks in a chromatogram. The area is obtained by summation of a number of individual measurements of the detector output during a peak, and the number reported by the integrator is typically the peak area expressed in millivolt-seconds. Integrators differ in the method of processing the individual readings of detector output, and in the facilities available in the instrument for further processing of the peak area values. In all instruments the analogue output signal from the gas chromatograph is first converted to digital form. In simpler integrators an upward change in the baseline level, or in the rate of baseline drift, is taken as the signal to begin the summation process which continues until the baseline level, or a defined rate of baseline drift, is regained. As the instrument has to be aware of the baseline change before it can begin integration, a proportion, usually negligibly small, of each peak is inevitably lost, the amount depending on the settings of the slope sensitivity and noise-rejection controls. This difficulty is obviated in the so-called 'computing' integrators by storing the digitized detector readings in a memory so that a complete peak, or series of merged peaks, can be stored and integrated retrospectively. Baseline assignment can then also be made retrospectively. In the most sophisticated models the memory is large enough to store data corresponding to a complete chromatogram. Use is also made of the memory to provide facilities for automatic computation of calibration curves and the integrator may then provide a printed output record giving the concentrations of each component of interest.

5.4.8 Operation of a typical process chromatograph

As an example of process chromatography, the operation of a single-stream instrument designed for high-speed on-line measurement of the concentration of a single component, or group of components, is described. The chromatograph is shown schematically in Figure 5.23, and consists of an analyser, a processor and a power unit.

The analyser unit contains those parts of the system required for sample handling and separation and detection of the components. There is a single column

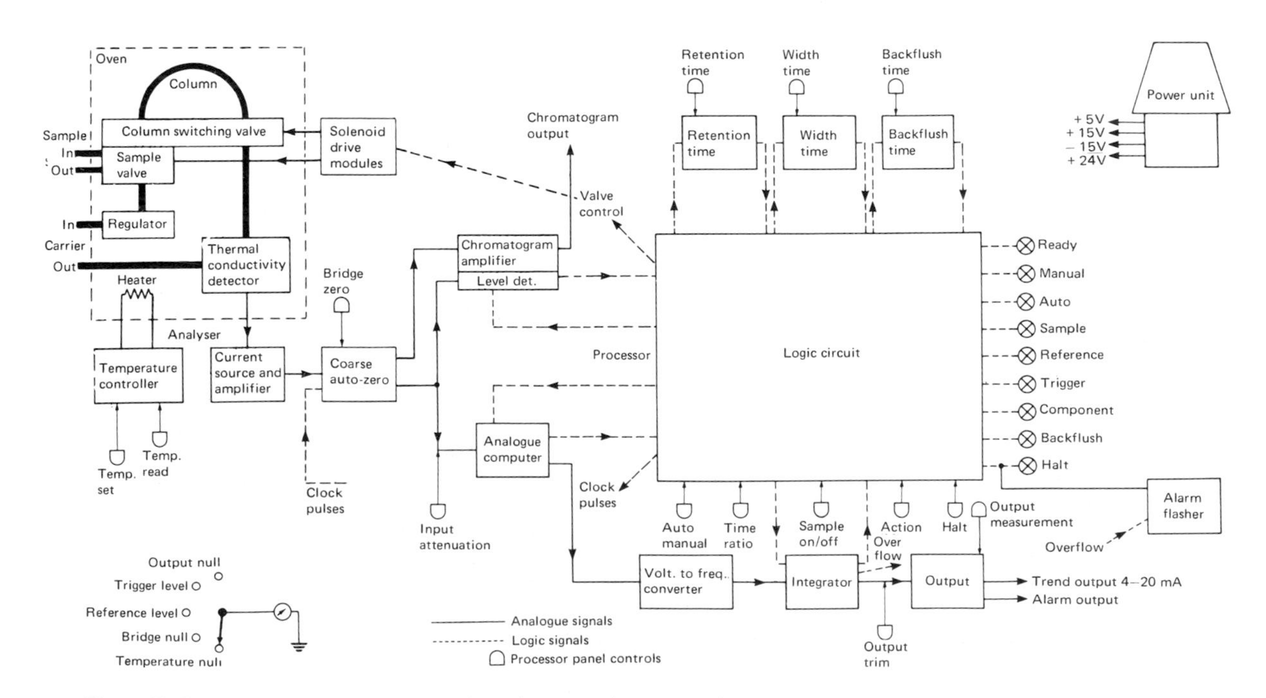

Figure 5.23 Schematic diagram of single-channel process chromatograph.

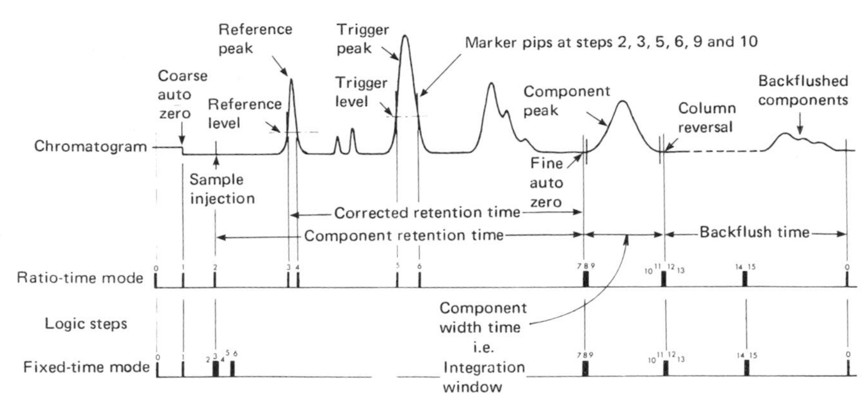

Figure 5.24 Chromatogram, showing logic and switching steps.

and thermal conductivity detector housed in a temperature-controlled zone at the top of the unit, with the associated electronics beneath. The packing and length of the small-bore column are chosen to suit the application, and the carrier-gas regulator is designed for high stability under low-flow conditions.

The small-volume thermal conductivity type detector uses thermistor elements to produce the output signals with high speed and stability. The electronic circuit modules mounted in the lower half of the main case control the oven temperature, power the detector and amplify its output, and provide power pulses to operate the valve solenoids.

The processor contains the electronic circuits which control the sequential operation of the total system. It times the operation of the simple injection and column switching valves, selects and integrates a chromatographic peak, and updates the trend output signal.

The power unit provides the low voltage regulated supplies for the analyser and the processor and may be mounted up to 3 metres from the processor.

A typical chromatogram of a sample analysed by the instrument is shown in Figure 5.24, annotated to show the various switching and logic steps during the analysis.

The operation of the chromatograph can be either on a fixed-time or ratio-time basis. In fixed-time operation the sample injection is the start of the time cycle. At preset times the 'integration window' is opened and closed, to coincide with the start and finish of the emergence of the components from the column. While the window is open the detector signal is integrated to give a measure of the concentration of the component. Other operations such as column switching and automatic zeroing are similarly timed from the sample injection. For fixed-time operation to be reliable, pressure and flow rate of carrier gas, and temperature and quantity of stationary phase in the column must be closely controlled.

Many of the problems associated with fixed-time operation may be avoided by use of ratio-time operation. In this mode of operation the retention time of components is measured from an early reference peak (corrected retention time, see Figure 5.24) instead of from the time of sample injection. The ratio of two corrected retention times (retention ratio) is less affected by changes in the critical column parameters. The corrected retention time for an early trigger peak is used to predict the time of emergence of the component of interest, that is, the integration window.

For the system to be able to operate in the ratio mode, it is necessary to have two specific peaks in the chromatogram in advance of the peak of the component of interest.

Reference peak The reference peak is due to the first component eluted from the column, with a very low retention time (such as air), and is used as the start point for the ratio timing. If a suitable component is not consistently present in the process sample, one can be injected into the column at the same time as the sample, by using the second loop of the sample valve.

Trigger peak The trigger peak must appear on the chromatogram between the reference and component peaks. It must be self-evident by virtue of size, and it must be consistent in height and width. As with the reference peak it can be from a component of the process sample, or injected separately. Alternatively it can be a negative peak derived by using a doped carrier gas. The logic circuits measure the time between reference and trigger peaks and use this, together with the preset ratio value, to compute the time for the start of the integration window. Similarly the trigger peak width is used to define the width of the window. At the start of integration the value of the signal level is stored. The integrator then measures the area under the component peak for the period of the window opening. At this point the signal level is again measured and compared with the stored start value to determine whether any baseline shift has occurred. The integration is corrected for any baseline shifts.

The final value of the integration is stored and used to give an output signal which represents the concentration of the component. As this signal is updated after each analysis the output shows the trend of the concentration.

After the completion of integration, the column is backflushed in order to remove any later components, the duration of the backflushing being ratioed from the analysis time. Alternatively, for those applications requiring a measurement such as 'total heavies', the peak of the total backflushed components can be integrated.

There are some applications where the ratio-time mode cannot be used. Typically, the measurement of a very early component, such as hydrogen, precludes the existence of earlier reference and trigger peaks. Operation of the various functions is then programmed using the fixed-time mode. Selection of the required mode is made using a switch on the processor.

Manual operation This mode of operation, selected by the 'auto/manual' switch on the front panel, provides a single analysis which is followed by column backflushing and the normal 'halt' condition. Single analyses are initiated by operation of the 'action' push-button, provided that the previous analysis has been completed. This mode of operation is used during initial programming or servicing.

5.5 Special gas analysers

5.5.1 Paramagnetic oxygen analysers

Many process analysers for oxygen make use of the fact that oxygen, alone among common gases, is paramagnetic.

5.5.1.1 Basic principles

The strength of a magnet is expressed as its magnetic moment. When a material, such as a piece of soft iron, is placed in a magnetic field, it becomes magnetized by induction and the magnetic moment of the material divided by its volume is known as the intensity of magnetization. The ratio of the intensity of magnetization to the intensity of the magnetic field is called the volume susceptibility k of the material. All materials show some magnetic effect when placed in a magnetic field, but apart from elements such as iron, nickel and cobalt and alloys such as steel, all known as ferromagnetics, the effect is very small, and intense magnetic fields are required to make it measurable.

Substances which are magnetized in the opposite direction to that of the applied field (so that k is negative) are called diamagnetics. Most substances are diamagnetic and the value of the susceptibility is usually very small. The most strongly diamagnetic substance is bismuth.

The magnetic properties of a substance can be related to its electronic structure. In the oxygen molecule, two of the electrons in the outer shell are unpaired. Because of this the magnetic moment of the molecule is not neutralized as is the commoner case, and the permanent magnetic moment is the origin of oxygen's paramagnetic properties.

A ferro- or paramagnetic substance when placed in a magnetic field in a vacuum or less strongly paramagnetic medium tries to move from the weaker to the stronger parts of the field. A diamagnetic material, in a magnetic field in a vacuum or medium of algebraically greater susceptibility tries – although the effect is very small – to move from the stronger to the weaker parts of the field. Thus when a rod of ferromagnetic or paramagnetic substance is suspended between the poles of a magnet it will set with its length along the direction of the magnetic field. A rod of bismuth, on the other hand, placed between the poles of a powerful electromagnet will set at right angles to the field.

It has been shown experimentally that for paramagnetic substances the susceptibility is independent of the strength of the magnetizing field but decreases with increase of temperature acording to the Curie–Weiss law:

$$\text{atomic susceptibility} = \frac{\text{relative atomic mass}}{\text{density}} \times \text{volume susceptibility} = C/(T-\theta)$$

where T is the absolute temperature and C and θ are constants.

The susceptibilities of ferromagnetic materials vary with the strength of the applied field, and above a certain temperature (called the Curie temperature and characteristic of the individual material) ferromagnetics lose their ability to retain a permanent magnetic field and show paramagnetic behaviour. The Curie temperature of iron is 1000 K.

The susceptibility of diamagnetic substances is almost independent of the magnetizing field and the temperature.

The paramagnetic properties of oxygen are exploited in process analysers in two main ways; the so-called 'magnetic wind' or thermal magnetic instruments, and magnetodynamic instruments.

5.5.1.2 Magnetic wind instruments

The magnetic wind analyser, originally introduced by Hartmann and Braun, depends on the fact that oxygen, as a paramagnetic substance, tends to move

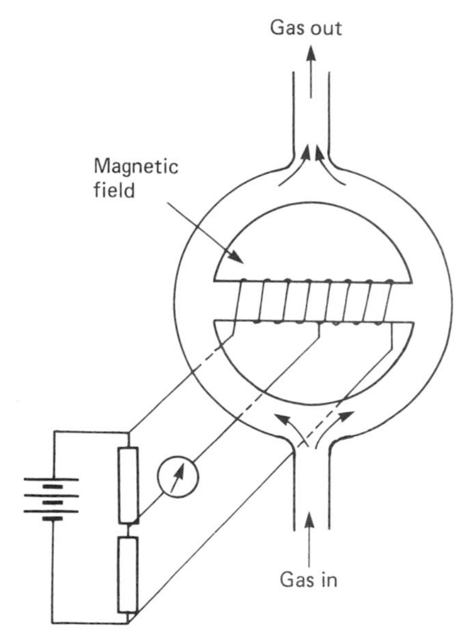

Figure 5.25 Magnetic wind oxygen analyser (courtesy Taylor Analytics).

from the weaker to the stronger part of a magnetic field, and that the paramagnetism of oxygen decreases as the temperature is raised.

$$\frac{\text{volume susceptibility}}{\text{density}}=\frac{C}{(T-\theta)} \quad \text{(Curie–Weiss law)}$$

$$\text{i.e. volume susceptibility}=\frac{C}{(T-\theta)}\times\text{density}$$

But for a gas, the density is proportional to $1/T$ where T is the absolute temperature. Thus

$$\text{volume susceptibility}=\frac{C}{(T^2-\theta T)}$$

The principle of the magnetic wind instrument is shown in Figure 5.25. The measuring cell consists of a circular annulus with a horizontal bypass tube on the outside of which are wound two identical platinum heating coils. These two coils form two arms of a Wheatstone bridge circuit, the bridge being completed by two external resistances. The coils are heated by means of the bridge current, supplied by a d.c. source of about 12 V. The winding on the left is placed between the poles of a very powerful magnet. When a gas sample containing oxygen enters the cell, the oxygen tends to flow into the bypass tube. Here it is heated so that its magnetic susceptibility is reduced. The heated gas is pushed along the cross-tube by other cold gas entering at the left. This gas flow cools the filaments, the left coil more than the right, and so changes their resistance, as in the flow controller mentioned in Section 5.4.2. The change in resistance unbalances the Wheatstone bridge and the out-of-balance e.m.f. is measured to give a signal which is proportional to the oxygen content of the gas.

This type of oxygen analyser is simple and reasonably robust, but it is subject to a number of errors. The instrument is temperature-sensitive: an increase in temperature causes a decrease in the out-of-balance e.m.f. of about 1 per cent per kelvin. This can be automatically compensated by a resistance thermometer placed in the gas stream near the cell. The calibration depends on the pressure of the gas in the cell.

Another error arises from the fact that the analyser basically depends on the thermal conductivity of the gas passing through the cross-tube. Any change in the composition of the gas mixed with the oxygen changes the thermal balance and so gives an error signal. This is known as the carrier-gas effect.

To a first approximation the out-of-balance e.m.f. is given by

$$e=kC_o$$

where e is the e.m.f., C_o is the oxygen concentration and k is a factor which varies with the composition of the carrier gas, and depends on the ratio of the volumetric specific heat to the viscosity of the carrier gas. For a binary mixture of oxygen with one other gas, k is a constant, and the out-of-balance e.m.f. is directly proportional to the oxygen concentration. For ternary or more complex mixtures, the value of k is constant only if the composition of the carrier gas remains constant.

Values of k for a number of common gases are given in Table 5.3 for an e.m.f. measured in volts and oxygen concentration measured in volume per cent. The value of k for a mixture can be calculated by summing the partial products:

$$k=(C_A k_A+C_B k_B)/100$$

where C_A and C_B are the percentage concentrations of

Table 5.3 *k* values for common gases

Gas	*k*	*Gas*	*k*
Ammonia	2.21	Nitrogen	1.00
Argon	0.59	Nitric oxide	0.94
Carbon dioxide	1.54	Nitrous oxide	1.53
Carbon monoxide	1.01	Oxygen	0.87
Chlorine	1.52	Sulphur dioxide	1.96
Helium	0.59	Water vapour	1.14
Hydrogen	1.11		

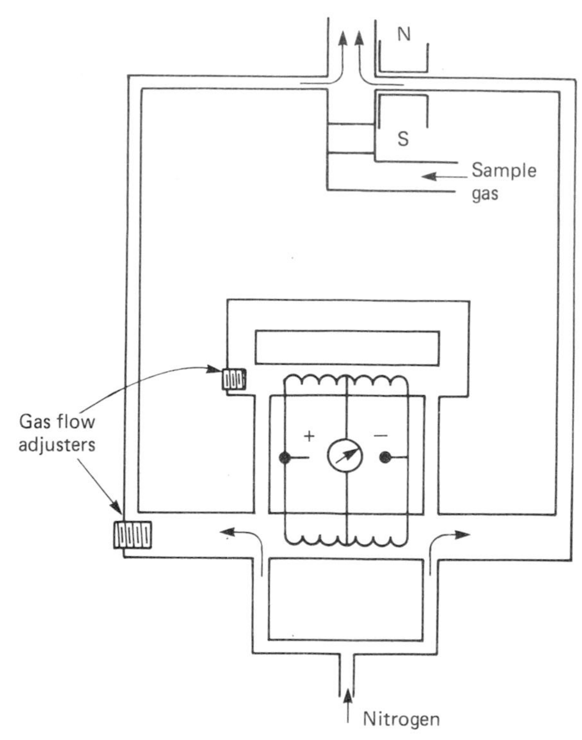

Figure 5.26 Quincke oxygen analyser (courtesy Taylor Analytics).

components A and B and k_A and k_B are the corresponding values of k.

Convective flow or misalignment of the sensor may also change the thermal balance and cause errors. In the case of flammable gases, errors may be caused if they can burn at the temperature in the cross-tube. This type of analyser is therefore usually considered to be unsuitable for oxygen measurements in hydrocarbon vapours.

5.5.1.3 Quincke analyser

The Quincke analyser is shown in Figure 5.26. A continuous stream of nitrogen enters the cell and is divided into two streams which flow over the arms of filaments of a Wheatstone bridge circuit. The flows are adjusted to balance the bridge to give zero output. One of the nitrogen streams passes the poles of a strong magnet while the other stream passes through a similar volume but without the magnetic field.

The sample gas enters the cell as shown and is mixed with the nitrogen streams immediately downstream of the magnetic field. Oxygen in the sample gas tends to be drawn into the magnetic field, causing a pressure difference in the arms of the cell and changing the flow-pattern of the nitrogen over the arms of the Wheatstone bridge. The out-of-balance e.m.f. is proportional to the oxygen concentration of the sample gas.

Because the sample gas does not come into contact with the heated filaments, the Quincke cell does not suffer from the majority of the errors present in magnetic wind instruments, but it does require a separate supply of nitrogen.

5.5.1.4 Magnetodynamic instruments

Magnetic wind instruments are susceptible to hydrocarbon vapours and to any change in the carrier gas producing a change in its thermal conductivity. These difficulties led to the development by Pauling of a measuring cell based on Faraday's work on determination of magnetic susceptibility by measuring the force acting on a diamagnetic body in a non-uniform magnetic field.

5.5.1.5 Magnetodynamic oxygen analyser

In the Pauling cell, two spheres of glass or quartz, filled with nitrogen, which is diamagnetic, are mounted at the ends of a bar to form a dumb-bell. The dumb-bell is mounted horizontally on a vertical torsion suspension, and is placed between the specially-shaped poles of a powerful permanent magnet. The gas to be measured surrounds the dumb-bell. If oxygen is present it is drawn into the field and so displaces the spheres of the dumb-bell which are repelled from the strongest parts of the field, so rotating the suspension until the torque produced is equal to the deflecting couple on the spheres, see Figure 5.27. If the oxygen content of the gas in the cell changes, there will be a change in the force acting on the spheres, which will take up a new

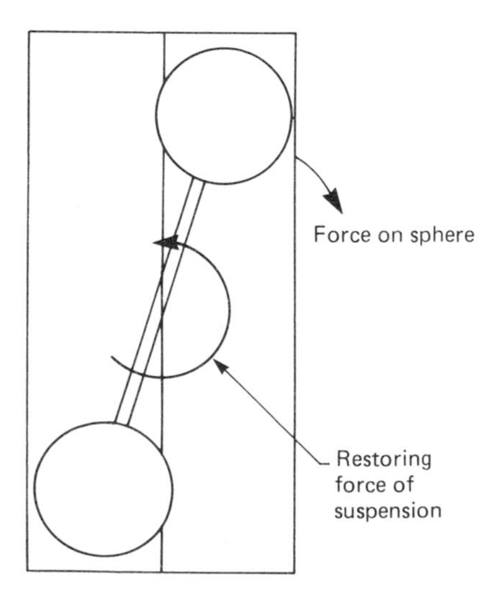

Figure 5.27 Magnetodynamic oxygen measuring cell (courtesy Taylor Analytics).

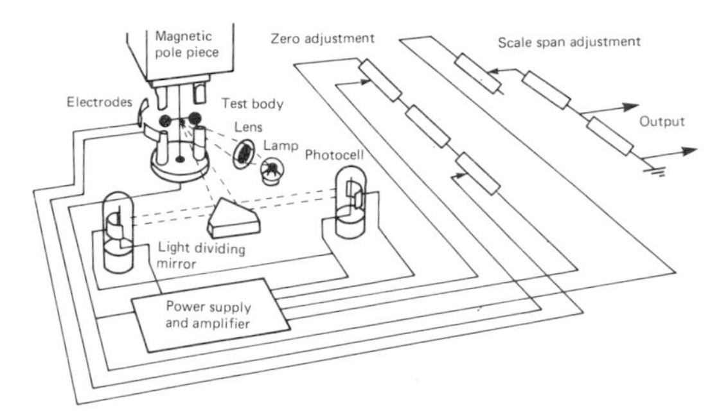

Figure 5.28 Bendix oxygen analyser.

position. The magnitude of the force on the dumb-bell may be measured in a number of ways, but a small mirror is commonly attached to the middle of the arm, and the deflection measured by focusing a beam of light on the mirror. The deflection may either be measured directly, or a force balance system may be used whereby the deflection of the dumb-bell is detected but an opposing force is applied to restore it to the null position.

Two different designs of oxygen analyser, based on the magnetodynamic principle, are shown in Figures 5.28 and 5.29. In the Bendix instrument the suspension is a quartz fibre and the restoring force is produced electrostatically by the electrodes adjacent to the dumb-bell. One electrode is held above ground potential and the other below ground potential by the amplifier controlled from the matched photocells upon which the light from the mirror falls. In the Servomex instrument (Figure 5.29) the suspension is platinum, and the restoring force is produced electrically in a single turn of platinum wire connected to the rest of the electronics through the platinum suspension. Electromagnetic feedback is used to maintain the dumb-bell in the zero position, and the current required to do this is a measure of the oxygen content of the gas.

The deflecting couple applied to the dumb-bell by the magnetic field depends on the magnetic susceptibility of the surrounding gas. The magnetic susceptibilities of all common gases at 20 °C are very small (nitrogen, -0.54×10^{-8}; hydrogen, -2.49×10^{-8}; carbon dixode, -0.59×10^{-8}) compared to that of oxygen ($+133.6 \times 10^{-8}$) and the susceptibility of the gas will depend almost entirely on the concentration of oxygen. This type of analyser is not influenced by the thermal conductivity of the gas, and is unaffected by hydrocarbons. However, the susceptibility of oxygen varies considerably with temperature. This may be overcome by maintaining the instrument at a constant temperature above ambient, or the temperature of the measuring cell may be detected and the appropriate temperature correction applied electronically. The reading also depends on the pressure of gas in the cell.

This type of analyser is suitable for measuring the oxygen content of hydrocarbon gases, but paramagnetic gases interfere and must be removed. The most important of these is nitric oxide (susceptibility $+59.3 \times 10^{-8}$), but nitrogen peroxide and chlorine dioxide are also paramagnetic. If the concentration of these gases in the sample is reasonably constant, the instrument may be zeroed on a gas sample washed in acid chromous chloride, and the oxygen measured in the usual way.

5.5.2 Ozone analyser

Continuous analysers for ozone are based on the chemiluminescent flameless reaction of ozone with

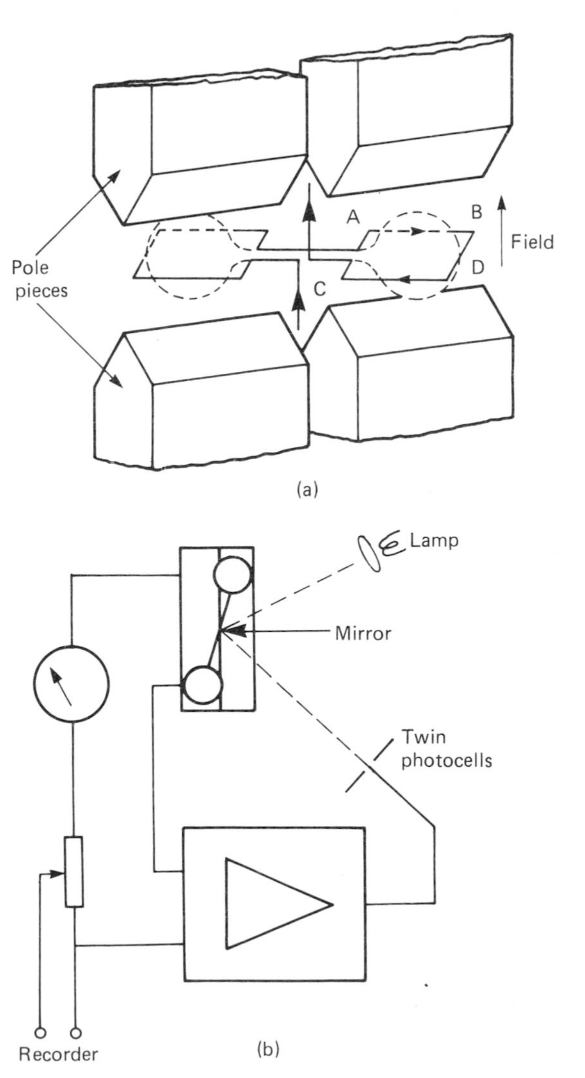

Figure 5.29 Servomex oxygen analyser (courtesy Taylor Analytics): (a) measuring cell, (b) electronic circuit.

ethylene. The light emission from the reaction, centred at 430 nm, is measured by a photomultiplier and the resulting amplified signal is a measure of the concentration of ozone in the sample stream. The flow diagram, and functional block diagram of a typical portable ozone analyser are given in Figure 5.30. The chemiluminescent light emission from the reaction chamber is a direct function of the ambient temperature, and therefore the temperature is regulated to 50 °C. The photomultiplier is contained in a thermoelectrically cooled housing maintained at 25 °C to ensure that short- and long-term drift is minimized. The instrument is capable of measuring ozone levels in the range 0.1 to 1000 ppb.

5.5.3 Oxides of nitrogen analyser

Analysers for oxides of nitrogen – NO, NO_x (total oxides of nitrogen), NO_2 – are based on the chemiluminescent reaction of nitric oxide (NO) and ozone to produce nitrogen dioxide (NO_2). About 10 per cent of the NO_2 is produced in an electronically excited state, and undergoes a transition to the ground state,

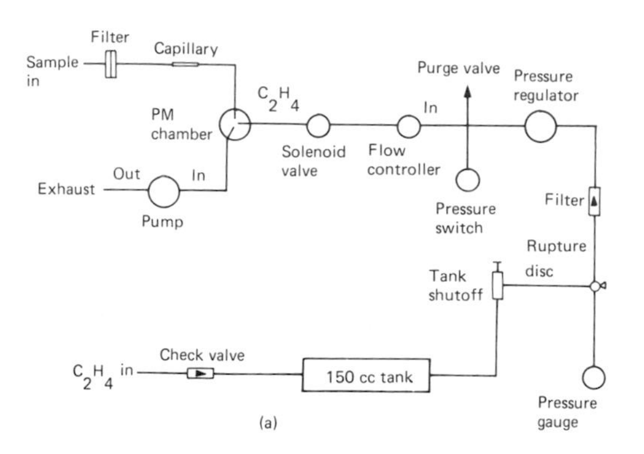

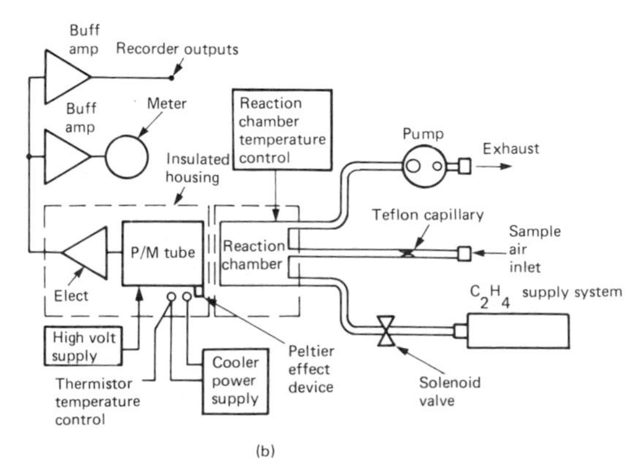

Figure 5.30 Ozone analyser (courtesy Columbia Scientific Industries Corp.): (a) flow diagram, (b) functional block diagram.

emitting light in the wavelength range 590–2600 nm:

$$NO + O_3 \rightarrow NO_2^* + O_2$$

$$NO_2^* \rightarrow NO_2 + h\nu$$

The intensity of the light emission is proportional to the mass-flow rate of NO through the reaction chamber and is measured by a photomultiplier tube.

Analysis of total oxides of nitrogen (NO_x) in the sample is achieved by passing the gases through a stainless steel tube at 600–800 °C. Under these conditions, most nitrogen compounds (but not N_2O) are converted to NO which is then measured as above. Nitrogen dioxide (NO_2) may be measured directly by passing the air sample over a molybdenum catalyst to reduce it to NO, which is again measured as above, or the NO_2 concentration may be obtained by automatic electronic subtraction of the NO concentration from the NO_x value.

The flow system of a nitrogen oxides analyser is shown in Figure 5.31. Ozone is generated from ambient air by the action of UV light

$$3O_2 \xrightarrow{h\nu} 2O_3$$

and a controlled flow rate of ozonized air is passed to the reaction chamber for reaction with NO in the air sample, which is passed through the chamber at a controlled flow of $1\ l\ min^{-1}$. By selection of a switch to operate the appropriate solenoid valves, a span gas may be directed to the reaction chamber, or a zero calibration may be carried out by shutting off the flow of ozonized air to the reactor. The three-way solenoid valve downstream of the converter is switched to permit NO analysis when bypassing the converter, and NO_x analysis when the sample is passed through the converter. The analyser can measure ozone in air in the range 5 ppb to 25 ppm, with a precision of $\pm 1\%$.

5.5.4 Summary of special gas analysers

The operating principles of analysers for the most commonly measured gases are given in Table 5.4.

5.6 Calibration of gas analysers

None of the commonly-used gas detectors is absolute; that is, they are devices where the output signal from the detector for the gas mixture under test is compared with that for mixtures of the bulk gas containing known concentrations of the determinand. The use of standard gas mixtures is analogous to the use of standard solutions in solution chemistry, but their preparation and handling present some peculiar problems. As in solution chemistry, the calibration gas

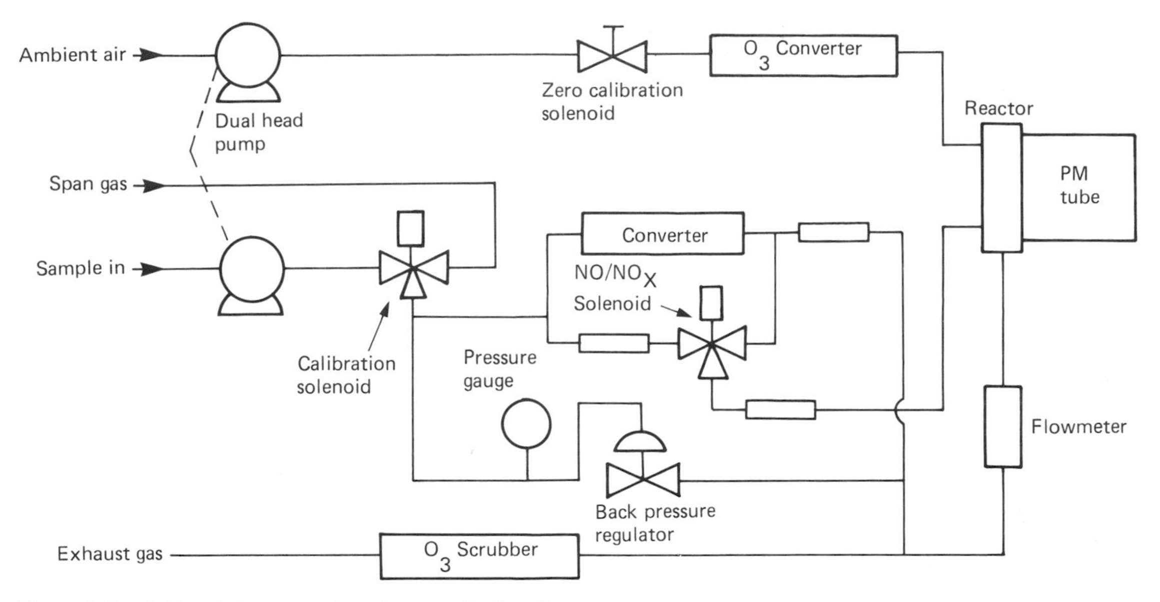

Figure 5.31 Oxides of nitrogen analyser (courtesy Beckman).

Table 5.4 Measurement principles of special gas analysers

Gas	*Measurement principle*
Oxygen	Paramagnetism
	Electrochemical sensor
	Fuel cell
Ozone	Chemiluminescence
	Electrochemical sensor
Nitrogen oxides	Chemiluminescence
Carbon dioxide	Infrared spectrometry
Carbon monoxide	Infrared spectrometry
	Electrochemical sensor
Sulphur oxides	Flame photometry
Hydrocarbons	Flame ionization detector
	Infrared spectrometry
	Catalytic detector
Flammable gases	Catalytic detector
	Semiconductor detector
Hydrogen sulphide	Semiconductor detector
	Flame photometry
	Electrochemical sensor

mixtures should reflect, as closely as possible, the composition of the samples it is desired to measure. Ideally a number of standard mixtures, whose concentration covers the range of samples to be measured, should be used to establish the response curve of the instrument or detector. However, for routine calibration where the response curve has previously been established or is well known, it is usual to calibrate gas analysers by use of a 'zero' gas mixture which is free of the determinand and establishes the zero of the instrument, and one or more 'span' gases containing concentrations of the determinand close to those it is desired to measure.

The accuracy to which a gas mixture can be prepared depends on the number and nature of the components, and on their concentrations. For gas mixtures prepared under pressure in cylinders, it is useful to specify two parameters, the filling and analytical tolerances. The filling tolerance describes the closeness of the final mixture to its original specification, and depends mainly on the concentrations of the components. Thus, while it may be possible to fill a cylinder with a component gas at the 50 per cent level to a tolerance of ± 2.5 per cent or ± 5 per cent of the component (that is the cylinder would contain between 47.5 and 52.5 per cent of the component), at the 10 vpm level the tolerance would typically be ± 5 vpm or ± 50 per cent of the component, and the cylinder would contain between 5 and 15 vpm of the component. The analytical tolerance is the accuracy with which the final mixture can be described, and depends on the nature of the mixture and the analytical techniques employed. Accuracies achievable are typically in the range from ± 2 per cent of component or ± 0.2 vpm at the 10 vpm level to ± 1 per cent of component or ± 0.5 per cent at the 50 per cent level. However, these figures are strongly dependent on the actual gases involved, and the techniques available to analyse them.

Gas mixtures may be prepared by either static or dynamic methods. In the static method known quantities of the constituent gases are admitted to a suitable vessel and allowed to mix, while in the dynamic method streams of the gases, each flowing at a known rate, are mixed to provide a continuous stream of the sample mixture. Cylinders containing supplies of the standard mixtures prepared under pressure are usually most convenient for fixed instruments such as process gas chromatographs, while portable instruments are often calibrated by mixtures prepared dynamically. Where mixtures containing low concentrations of the constituents are needed, adsorptive effects may make the static method inapplicable, while the dynamic method becomes more complex for mixtures containing large numbers of constituents.

Before any gas mixture is prepared, its properties must be known, particularly if there is any possibility of reaction between the components, over the range of pressures and concentrations expected during the preparation.

5.6.1 Static methods

Static gas mixtures may be prepared either gravimetrically or by measurement of pressure. Since the weight of gas is usually small relative to the weight of the cylinder required to contain it, gravimetric procedures require balances which have both high capacity and high sensitivity, and the buoyancy effect of the air displaced by the cylinder may be significant. Measurement of pressure is often a more readily applicable technique.

After preparation gas mixtures must be adequately mixed to ensure homogeneity, usually by prolonged continuous rolling of the cylinder. Once mixed, they should remain homogeneous over long periods of time. Any concentration changes are likely to be due to adsorption on the cylinder walls. This is most likely to happen with mixtures containing vapours near their critical pressures, and use of such mixtures should be avoided if possible.

5.6.2 Dynamic methods

5.6.2.1 Gas flow mixing

Gas mixtures of known concentration may be prepared by mixing streams of two or more components, each of which is flowing at a known rate. The concentration of one gas in the others may be varied by adjustment of the relative flow rates, but the range of concentration available is limited by the range of flows which can be measured with sufficient accuracy. Electronic mass-flow controllers are a convenient method of flow measurement and control.

5.6.2.2 Diffusion-tube and permeation-tube calibrators

Standard gas mixtures may be prepared by allowing the compound or compounds of interest to diffuse through a narrow orifice or to permeate through a membrane, into a stream of the base gas which is flowing over the calibration source at a controlled and known rate.

Typical designs of diffusion and permeation tubes are shown in Figure 5.32. In both cases there is a reservoir of the sample, either a volatile liquid or a liquefied gas under pressure, to provide an essentially constant pressure, the saturation vapour pressure, upstream of the diffusion tube or permeation membrane. After an initial induction period it is found that, provided the tube is kept at constant temperature, the permeation or diffusion rate is constant as long as there is liquid in the reservoir. The tube can then be calibrated gravimetrically to find the diffusion or permeation rate of the sample. The concentration of the sample in the gas stream is then given by

$$C = RK/F$$

where C is the exit gas concentration, R is the diffusion or permeation rate, K is the reciprocal density of the sample vapour, and F is the gas flow rate over the calibration device. The diffusion or permeation rate depends on the temperature of the tube, and on the molecular weight and vapour pressure of the sample. Additionally, the diffusion rate depends on the length

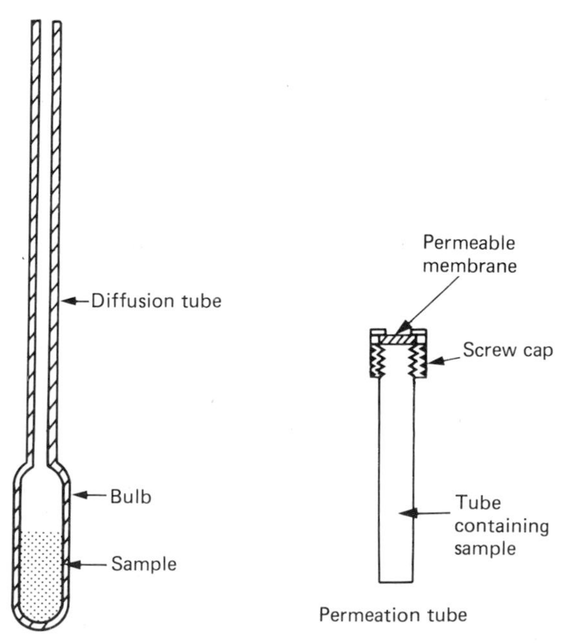

Figure 5.32 Cross-sectional diagrams of diffusion and permeation tube calibration sources.

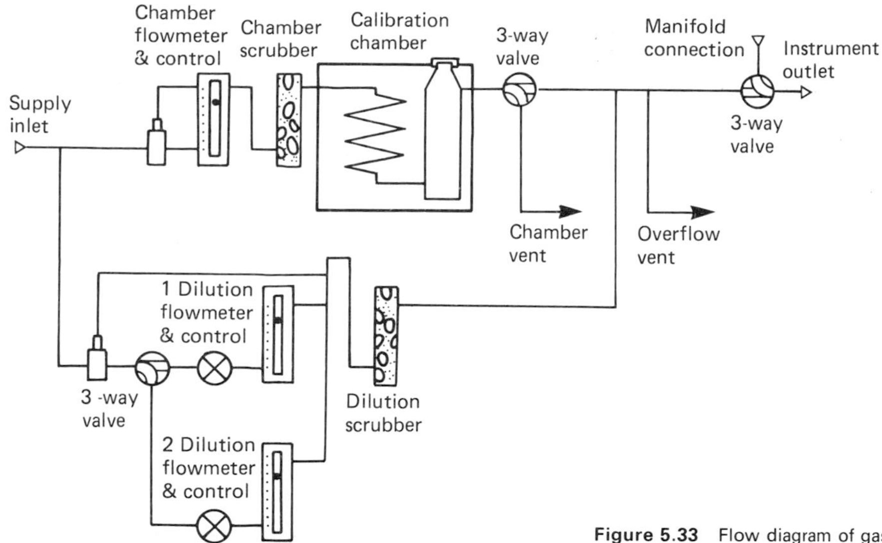

Figure 5.33 Flow diagram of gas calibrator.

and inner diameter of the capillary tube and the permeation rate depends on the nature, area and thickness of the permeation membrane. Data are available for a large number of organic and inorganic vapours to allow tubes to be designed with the required diffusion or permeation rate, and the exact rate for each tube is then established empirically.

The temperature-dependence of diffusion or permeation means that the tubes must be carefully thermostated for accurate calibrations. The empirical equation for the temperature-dependence of permeation rate is:

$$\log \frac{R_2}{R_1} = 2950\left(\frac{1}{T_1} - \frac{1}{T_2}\right)$$

where R_1 is permeation rate at T_1 K and R_2 is permeation rate at T_2 K. The permeation rate changes by approximately 10 per cent for every 1 K change in temperature. Thus the temperature of the permeation tube must be controlled to within 0.1 K or better if 1 per cent accuracy in the permeation rate, and thus the concentration that is being developed, is to be achieved.

The flow diagram of a typical calibrator for use with diffusion or permeation tubes is shown in Figure 5.33. The gas supply is scrubbed before passing through a thermostated coil and over the calibration source or sources in the calibration chamber. Secondary streams of purified gas may be added to the effluent gas stream to adjust the final concentration to the range required.

The diffusion or permeation technique is especially useful for generating standard mixtures at low concentrations, for example of organic compounds in air for calibration of environmental monitors, air pollution monitors etc., and the calibrator can be made portable for field use. The range of compounds which can be used is limited by their saturation vapour pressure; if this is too low, the diffusion or permeation rates, and hence the concentrations available, are very small, while compounds with high saturation vapour pressures present problems in construction and filling of the calibration tubes.

5.6.2.3 *Exponential dilution*

In the exponential dilution technique a volume of gas contained in a vessel, in which there is perfect and instantaneous mixing, is diluted by passing a stream of a second gas through the vessel at a constant flow rate. It can be shown that, under these conditions, the concentration of any gaseous species in the vessel, and hence the instantaneous concentration in the effluent stream of diluent gas, decays according to the law·

$$C = C_0 \exp\left(-\frac{Ut}{V}\right)$$

where C is the concentration of the diluted species at time t, C_0 is the initial concentration, U is the flow rate of diluent gas, and V is the volume of the vessel.

The vessel may either be filled with the gaseous species to be analysed, in which case the concentration decays from an initial value of 100 per cent, or the vessel may be filled with the diluent gas and a known volume of the gas of interest may be injected into the diluent gas just upstream of the dilution vessel at the start of the experiment. In either case the concentration of the species of interest in the effluent gas stream may be calculated at any time after the start of the dilution.

The exponential dilution vessel is typically a spherical or cylindrical glass vessel of 250–500 ml capacity, fitted with inlet and outlet tubes, and a septum cap or gas sampling valve for introduction of the gas to be diluted. The vessel must be fitted with a stirrer, usually magnetically driven, and baffles to ensure that mixing is as rapid and homogeneous as possible. The diluent gas flows through the vessel at a constant known flow rate, usually in the range 20–30 ml min^{-1}. For a vessel of the dimensions suggested above, this gives a tenfold dilution in approximately 30 minutes.

The exponential dilution technique is a valuable calibration method especially suitable for use at very low concentrations. It is also valuable for studying or verifying the response of a detector over a range of concentrations. However it should be noted that strict adherence to a known exponential law for the decay of concentrations in the vessel depends on the attainment of theoretically perfect experimental conditions which cannot be achieved in practice.

Changes in the flow rate of the diluent gas or in the temperature or pressure of the gas in the dilution vessel and imperfect or non-instantaneous mixing in the vessel lead to unpredictable deviations from the exponential decay law. Deviations also occur if the determinand is lost from the system by adsorption on the walls of the vessel. Since the technique involves extrapolation from the known initial concentration of the determinand in the diluting gas, any deviations are likely to become more important at the later stages of the dilution. If possible it is therefore advisable to restrict the range of the dilution to two or three orders of magnitude change in concentration. Where the gas to be diluted is introduced to the dilution vessel by injection with a valve or syringe, the accuracy and precision of the entire calibration curve resulting from the dilution is limited by the accuracy and precision of the initial injection.

5.7 Further reading

Cooper, C. J. and De Rose, A. J., 'The analysis of gases by chromatography', *Pergamon Series in Analytical Chemistry*, Vol. 7, Pergamon, (1983)

Cullis, C. F. and Firth, J. G. (eds), *Detection and Measurement of Hazardous Gases*, Heinemann, (1981)

Grob, R. L. (ed.), *Modern Practice of Gas Chromatography*, Wiley, (1977)

Jeffery, P. F. and Kipping, P. J., *Gas Analysis by Gas Chromatography*, International Series of Monographs in *Analytical Chemistry*, Vol. 17, Pergamon, (1972)

Sevcik, J., *Detectors in Gas Chromatography*, *Journal of Chromatography Library*, Vol. 4, Elsevier, (1976)

Also review artices in *Analytical Chemistry*, and manufacturers' literature

6 Chemical analysis – moisture measurement

D. B. MEADOWCROFT

6.1 Introduction

The measurement and control of the moisture content of gases, liquids and solids is an integral part of many industries. Numerous techniques exist, none being universally applicable, and the instrument technologist must be able to choose the appropriate measurement technique for his application. It is particularly important to measure moisture because of its presence in the atmosphere, but it is awkward because it is a condensable vapour which will combine with many substances by either physical adsorption or chemical reaction. Moisture measurement may be needed to ensure the level remains below a prescribed value or within a specified band, and the range of concentrations involved can be from less than one part per million to percentage values.

A few examples will illustrate the range of applications:

Gases In gas-cooled nuclear reactors the moisture level of the coolant has to be within a prescribed band (e.g. 250–500 volume parts per million) or below a certain value (e.g. 10 vppm) depending on the type of reactor. Rapid detection of small increases due to leaks from the steam generators is also essential. Moisture must be excluded from semiconductor device manufacture, and glove boxes are fitted with moisture meters to give an alarm at, say, 40 vppm. Environmental control systems need moisture measurement in order to control the humidity, and even tumble driers can be fitted with sensors to automatically end the clothes drying cycle.

Liquids The requirement is usually to ensure the water contamination level is low enough. Examples are the prevention of corrosion in machinery, breakdown of transformer oil and loss of efficiency of refrigerants or solvents.

Solids Specified moisture levels are often necessary for commercial reasons. Products sold by weight (e.g. coal, ore, tobacco, textiles) can most profitably have moisture contents just below the maximum acceptable limit. Some textiles and papers must be dried to standard storage conditions to prevent deterioration caused by excessive wetness and to avoid the waste of overdrying as the moisture would be picked up again during storage. Finally, many granulated foods must have a defined moisture content.

The purpose of this chapter is to introduce the reader to the major measurement techniques which are available. The three states, gas, liquid and solid, will be treated separately. In addition, many commercial instruments measure some parameter which changes reproducibly with moisture concentration and these instruments must be regularly calibrated by the user. The chapter therefore ends with a discussion of the major calibration techniques which the average user must be willing to employ when using such instruments.

First it is necessary to clarify a further aspect of moisture measurement which can confuse the newcomer, which is to define the large number of units which are used, particularly for gases, and show how they are interrelated.

6.2 Definitions

6.2.1 Gases

Although water vapour is not an ideal gas, for most hygrometry purposes, and to gain an understanding of the units involved, it is sufficient to assume water vapour does behave ideally. The basic unit of moisture in a gas against which other units can readily be referred is *vapour pressure*, and Dalton's Law of Partial Pressures can be assumed to hold if the saturated vapour pressure is not exceeded.

In environmental applications the unit often used is *relative humidity* which is the ratio in per cent of the actual vapour pressure in a gas to the saturation vapour pressure of water at that temperature. It is therefore temperature-dependent but is independent of the pressure of the carrier gas.

For chemical measurements the concentration of moisture is usually required. The *volume concentration* is given by the vapour pressure of moisture divided by the total pressure, often multiplied by 10^6 to give

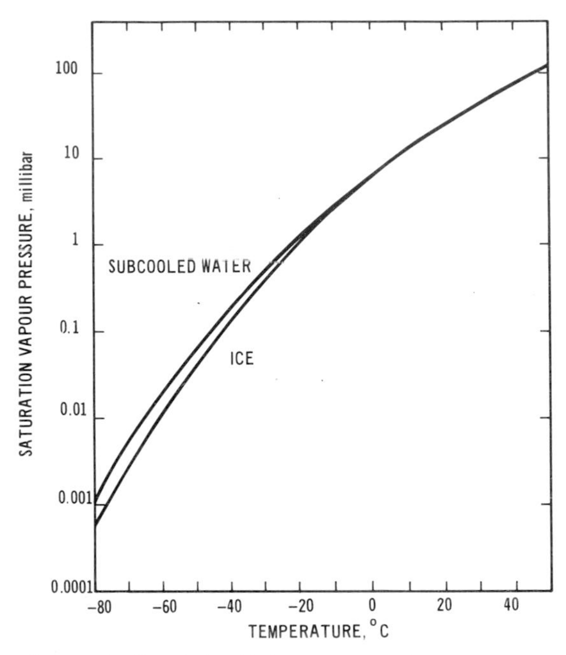

Figure 6.1 The relationship between saturation vapour pressure and dew point and frost point temperatures.

volume parts per million (vppm). The concentration by *weight* in wppm is given by the volume concentration multiplied by the molecular weight of water and divided by that of the carrier gas. Meteorologists often call the weight concentration the 'mixing ratio' and express it in g/kg.

When the prime aim is to avoid condensation the appropriate unit is the *dew point* which is the temperature at which the vapour pressure of the moisture would become saturated with respect to a plane surface. Similarly the *frost point* refers to the formation of ice. The relationship between dew and frost points and saturated vapour pressure is derived from thermodynamic and experimental work and is shown in Figure 6.1. It should be noted that below 0 °C the dew point and frost point differ. It is possible for supercooled water to exist below 0 °C, which can give some ambiguity, but this is unlikely very much below 0 °C (certainly not below −40 °C). In addition it can be seen that the saturated vapour pressure increases by an order of magnitude every 15–20 degrees so that in the range −80 °C to 50 °C dew point there is a vapour pressure change of five orders of magnitude. Table 6.1 lists the vapour pressure for dew or frost point between −90 °C and +50 °C.

Table 6.2 gives the interrelationships between these various units for some typical values.

Table 6.1 The relationship between dew/frost point and vapour pressure (μbar which is equivalent to vppm at 1 bar total pressure)

Frost point (°C)	*Saturated vapour pressure* (μbar)	*Frost point* (°C)	*Saturated vapour pressure* (μbar)	*Dew point* (°C)	*Saturated vapour pressure* (μbar)
−90	0.10	−40	128	0	6110
−80	0.55	−36	200	4	8120
−75	1.22	−32	308	8	10 700
−70	2.62	−28	467	12	14 000
−65	5.41	−24	700	16	19 200
−60	10.8	−20	1030	20	23 400
−56	18.4	−16	1510	25	31 700
−52	30.7	−12	2170	30	41 800
−48	50.2	−8	3100	40	73 000
−44	81.0	−4	4370	50	120 000

Table 6.2 Some examples of the relationships between the various units for moisture in gases

Dew/frost point (°C)	*Vapour pressure* (μbar *or* vppm at 1 bar)	*RH at 20 °C ambient* (%)	*Mixing ratio in air* (g/kg)
−70	2.5	0.01	1.5×10^{-3}
−45	72	0.3	0.045
−20	1030	4.4	0.64
0	6110	26	3.8
10	12 300	53	7.6
20	23 400	100	14.5

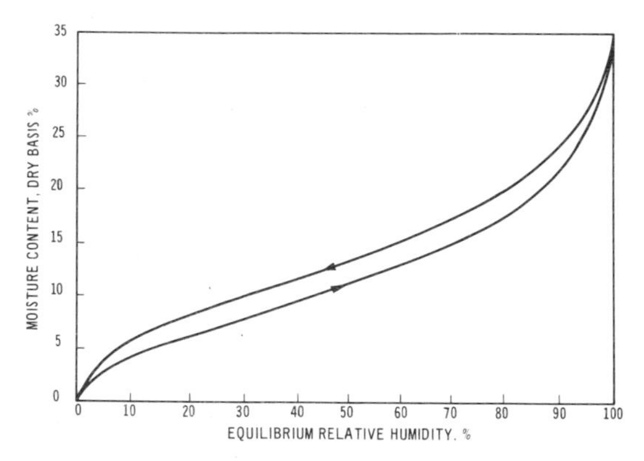

Figure 6.2 The relationship between the moisture content of a substance and the equilibrium relative humidity of the surrounding gas, for the example of wool.

6.2.2 Liquids and solids

Generally measurements are made in terms of *concentration*, either as a percentage of the total wet weight of the sample (e.g. in the ceramics industry for clay) or of the dry weight (e.g. in the textile industry where the moisture concentration is called 'regain'). In addition if a liquid or solid is in equilibrium with the gas surrounding it, the *equilibrium relative humidity* of the gas can be related to the moisture content of the solid or liquid by experimentally derived isotherms (e.g. Figure 6.2), or by Henry's law for appropriate non-saturated liquids. For liquids which obey Henry's law the partial vapour pressure of the moisture P is related to the concentration of water dissolved in the liquid by $W=KP$ where K is Henry's law constant. K can be derived from the known saturation values of the particular liquid, i.e. $K=W_s/P_s$ where W_s and P_s are respectively saturation concentration and saturation vapour pressure at a given temperature.

6.3 Measurement techniques

Techniques which allow automatic operation have the important advantage that they can be used for process control. We therefore concentrate our attention here on such techniques. Again, those available for gases, liquids and solids will be discussed separately.

6.3.1 Gases

There is a huge choice of techniques for the measurement of moisture in gases, reflecting the large number of ways in which its presence is manifested. The techniques range from measuring the extension of hair in simple wall-mounted room monitors to sophisticated electronic instruments. To some extent the choice of technique depends on the property required – dew point, concentration or relative humidity. Only the major techniques are discussed here. More extensive treatments are given in the bibliography.

6.3.1.1 Dew point instruments

The determination of the temperature at which moisture condenses on a plane mirror can be readily estimated (Figure 6.3) using a small mirror whose temperature can be controlled by a built-in heater and thermoelectric cooler. The temperature is measured by a thermocouple or platinum resistance thermometer just behind the mirror surface and the onset of dew is detected by the change of reflectivity measured by a lamp and photocell. A feedback circuit between the cell output and the heater/cooler circuit enables the dew point temperature to be followed automatically. Systematic errors can be very small and such intruments are used as secondary standards, yet with little loss of sophistication they can be priced competitively for laboratory and plant use. Mirror contamination can be a problem in dirty gases and in some instruments the mirror is periodically heated to reduce the effect of contamination. Condensable carrier gases which condense at similar temperatures to the moisture invalidate the technique. It is an ideal method if the dew point itself is required, but if another unit is to be derived from it accurate temperature measurements are essential because of the rapid change in vapour pressure with dew point temperature (see Section 6.2.1).

6.3.1.2 Coulometric instruments

The gas is passed at a constant rate through a sampling tube in which the moisture is absorbed onto a film of partially hydrated phosphoric anhydride (P_2O_5) coated on two platinum electrodes (Figure 6.4). A d.c. voltage is applied across the electrodes to decompose the water, the charge produced by the electrolysis being directly proportional to the mass of water absorbed (Faraday's law). Thus the current depends

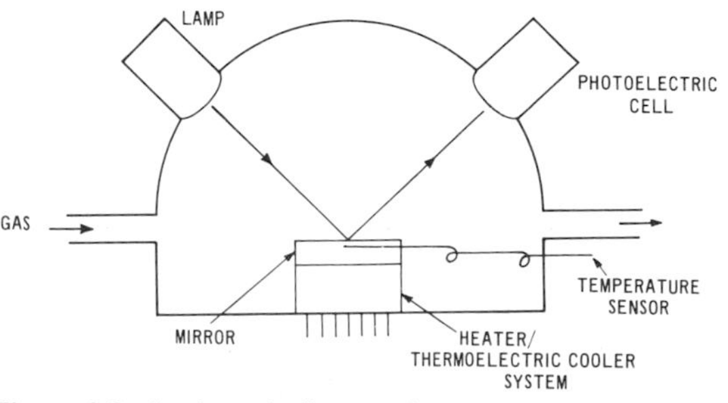

Figure 6.3 A schematic diagram of a sensor of a dew point mirror instrument.

on the flow rate, which must be set and controlled accurately at a predetermined rate (usually 100 ml min^{-1}) so that the current meter can be calibrated directly in ppm. Several points are worth making:

(a) The maximum moisture concentration measurable by this technique is in the range 1000–3000 vppm but care must be taken to ensure surges of moisture level do not wash off the P_2O_5.
(b) There is generally a zero leakage current equivalent to a few ppm. To allow for this error, when necessary, the current should be measured at two flow rates and the difference normalized to the flow for 100 ml min^{-1}.
(c) Platinum electrodes are not suitable for use in gases containing significant amounts of hydrogen. The platinum can catalyse the recombination of the electrolysed oxygen and this water is also electrolysed giving inaccurate measurements. Gold or rhodium elements reduce this effect.
(d) In the absence of recombination and gas leaks the response of a coulometric instrument can be regarded as absolute for many purposes.
(e) Cells which work at pressure can be obtained. This can increase the sensitivity at low moisture levels as it is possible to use a flow rate of 100 ml min^{-1} at the measuring pressure, which does not increase the velocity of gas along the element and hence does not impair the absorption efficiency of the P_2O_5.

6.3.1.3 *Infrared instruments*

Water vapour absorbs in the 1–2 μm infrared range, and infrared analysers (see Chapter 3) can be successfully used as moisture meters. For concentrations in the vppm range the path length has to be very long and high sample flow rates of several litres per minute can be necessary to reduce the consequent slow response time. Both single-beam instruments, in which the zero baseline is determined by measuring the absorption at a nearby non-absorbing wavelength, and double-beam instruments, in which a sealed parallel cell is used as reference, can be used. Single-beam instruments are less affected by deposits on the cell windows and give better calibration stability in polluted gases.

6.3.1.4 *Electrical sensor instruments*

There are many substances whose electrical impedance changes with the surrounding moisture level. If this absorption process is sufficiently reproducible on a thin film the impedance, measured at either an audio frequency or a radio frequency, can be calibrated in terms of moisture concentration or relative humidity. Materials used in commercial instruments include polymers, tantalum oxide, silicon oxide, chromium oxide, aluminium oxide, lithium chloride mixed with plastic, and carbon-loaded plastics which change length and hence resistance with moisture level. Many such instruments are available commercially, particularly using an anodized aluminium oxide layer which has a very narrow columnar pore structure (Figure 6.5), but ageing and other deterioration processes can occur so that regular calibration is essential. A major advantage of such sensors is that as no imposed gas flow is necessary they can simply be placed in the gas to be measured – for example an environmental chamber. In addition they can be used at high pressure, they have a wide response range (typically 50 °C to −80 °C dew point for a single

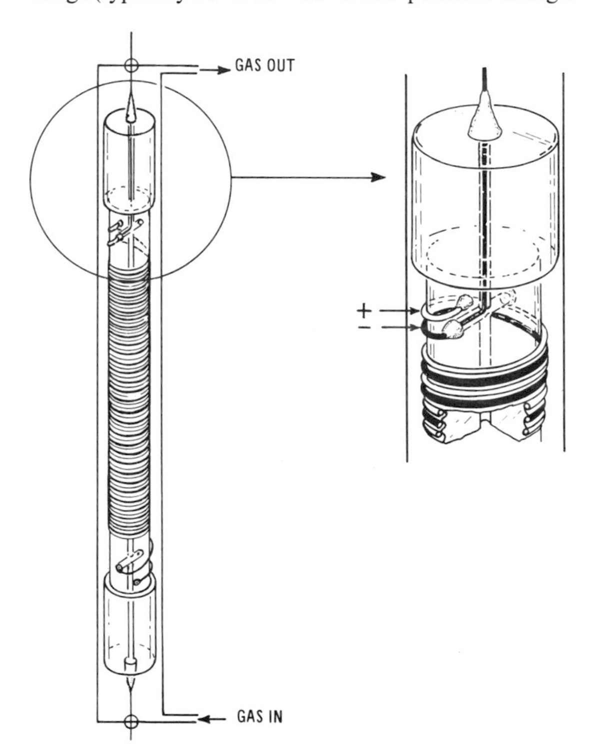

Figure 6.4 A schematic diagram of a sensor of a coulometric instrument.

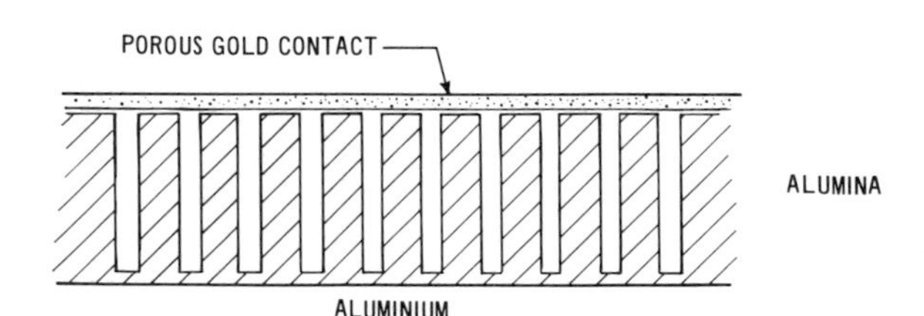

Figure 6.5 An idealized representation of the pore structure of anodized alumina. The pores are typically less than 20 nm in diameter and more than 100 μm deep. A porous gold layer is deposited on the alumina for electrical contact when used as a hygrometer sensor.

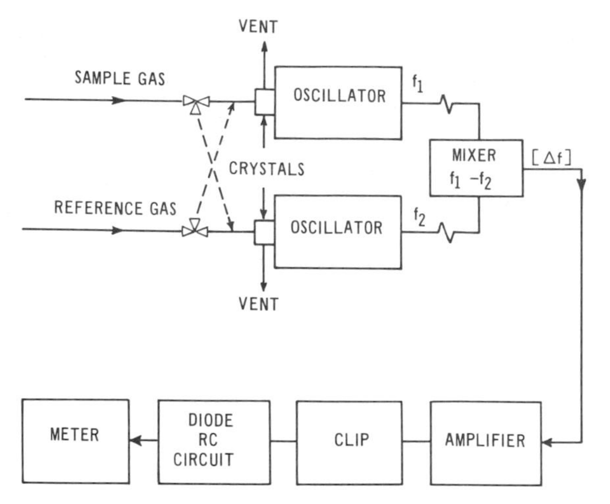

Figure 6.6 A block diagram of the arrangement of a piezoelectric humidity instrument (courtesy Du Pont Instruments (UK) Ltd.).

aluminium oxide sensor), have a rapid response, and are generally not expensive. These advantages often outweigh any problems of drift and stability, and the requirement for regular calibration, but they must be used with care.

6.3.1.5 Quartz crystal oscillator instrument

The oscillation frequency of a quartz crystal coated with hygroscopic material is a very sensitive detector of the weight of absorbed water because very small changes in frequency can be measured. In practice as shown in Figure 6.6, two quartz crystal oscillators are used, and the wet and a dry gas are passed across them alternately, usually for 30 seconds at a time. The frequency of crystal oscillation is about $9 . 10^6$ Hz, and that of the crystal exposed to the wet gas will be lowered and that of the crystal exposed to the dry gas will rise. The resultant audio frequency difference is extracted, amplified and converted to voltage to give a meter response whose maximum value on each 30-second cycle is a measure of the moisture level. The range of applicable concentrations is 1–3000 vppm and at lower levels the fact that the value after a certain time is measured rather than an equilibrium value means that the instrument can have a more rapid response than alternative methods (sample lines, however, often determine response time). Because the crystals see the sample gas for equal times contamination of the two crystals should be similar, and the frequency difference little affected, resulting in stability. However, regular calibration is still necessary and the complexity of the instrument makes it expensive.

6.3.1.6 Automatic psychrometers

The measurement of the temperature difference between a dry thermometer bulb and one surrounded by a wet muslin bag fed by a wick is the classical meteorological humidity measurement. This is called psychrometry, and automated instruments are available. The rate of evaporation depends on the gas flow as well as on the relative humidity, but generally a flow rate greater than $3\,m\,s^{-1}$ gives a constant temperature depression. It is most useful at high relative humidities with accurate temperature measurements.

6.3.2 Liquids

6.3.2.1 Karl Fischer titration

The Karl Fischer reagent contains iodine, sulphur dioxide and pyridine (C_5H_5N) in methanol; the iodine reacts quantitatively with water as follows:

$$[3C_5H_5N + I_2 + SO_2] + H_2O \rightarrow 2C_5H_5NHI + C_5H_5NSO_3$$

$$C_5H_5NSO_3 + CH_3OH \rightarrow C_5H_5NHSO_4CH_3$$

If a sample containing water is titrated with this reagent the end-point at which all the H_2O has been reacted is indicated by a brown colour showing the presence of free iodine. This is the basic standard technique, and is incorporated into many commercial instruments, with varying levels of automation. In process instruments the end-point is determined electrometrically by amperometric, potentiometric, or coulometric methods (see Chapter 3). In the amperometric method two platinum electrodes are polarized and when free iodine appears they are depolarized and the resultant current is measured to define the end-point. Potentiometrically, the potential of an indicator electrode is monitored against a calomel electrode and the end-point is characterized by a sudden change in potential. Coulometrically, iodine is generated by a constant electrolysing current from a modified reagent and the time taken to reach the end-point gives the mass of water in the sample. This last technique lends itself to automatic operation, with samples injected sequentially or, in one instrument, the moisture in a sample flow is measured continuously by mixing with standardized reagent, and the electrolysis current is a measure of the mass flow of water.

6.3.2.2 Infrared instruments

The same comments apply as for gases (Section 6.3.1.3), but sample cell lengths are usually shorter, in the range 1–100 mm. It is an attractive method for on-line analysis but care must be taken that other components in the liquid do not interfere with the

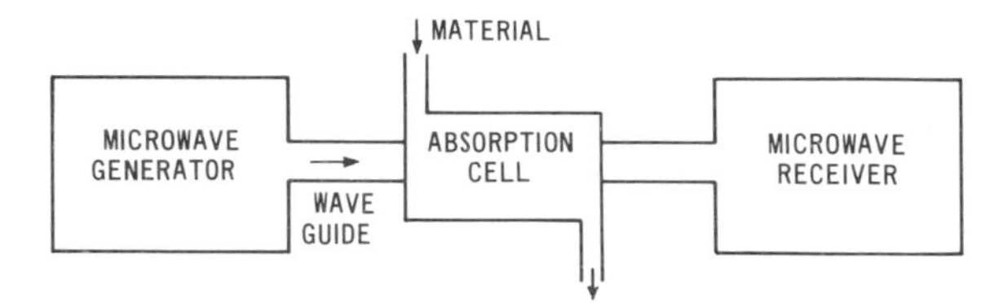

Figure 6.7 The basic concept for measuring moisture by microwave absorption.

measurement. Single-beam instruments are most often used.

6.3.2.3 Vapour pressure methods

As discussed in Section 6.2.2 the equilibrium relative humidity above a liquid can be used to determine the moisture content in the liquid. Either the relative humidity in a closed volume above the liquid can be measured, or a sensor which responds to the moisture vapour pressure in the liquid can be immersed in the liquid. The aluminium oxide sensor (Section 6.3.1.4) can be used, either above the liquid because it does not require a gas flow rate, or within the liquid because though the aluminium oxide pores will adsorb water molecules they will not adsorb the liquid molecules. These techniques are not appropriate if suspended free water is present in the liquid.

One manufacturer has developed a system in which the sensor is a moisture-permeable plastic tube which is immersed in the liquid. A fixed quantity of initially dry gas is circulated through the tube and the moisture in the gas is measured by an optical dew point meter. When equilibrium is reached the dew point measured equals that of the moisture in the liquid.

6.3.2.4 Microwave instruments

The water molecule has a dipole moment with rotational vibration frequencies which give absorption in the microwave S-band (2.6–3.95 GHz) and X-band (8.2–12.4 GHz) suitable for moisture measurement (Figure 6.7). The S-band needs path lengths four times longer than the X-band for a given attenuation and therefore the microwave band as well as cell dimensions can be chosen to give a suitable attenuation. Electronic developments are causing increased interest in this technique.

6.3.2.5 Turbidity/nephelometer

Undissolved water must be detected in aviation fuel during transfer. After thorough mixing the fuel is divided into two flows – one is heated to dissolve all the water before it passes into a reference cell; the other passes directly into the working cell. Light beams split from a single source pass through the cells and suspended water droplets in the cell scatter the light and a differential output is obtained from the matched photoelectric detectors on the two cells. 0 to 40 ppm moisture can be detected at fuel temperatures of −30 to 40 °C.

6.3.3 Solids

The range of solids in which moisture must be measured commercially is wide and many techniques are limited to specific materials and industries. In this book just some of the major methods are discussed.

6.3.3.1 Equilibrium relative humidity

The moisture level of the air immediately above a solid can be used to measure its moisture content. Electrical probes as discussed in Section 6.3.1.4 are generally used, and if appropriate can be placed above a moving conveyor. If a material is being dried its temperature is related to its equilibrium relative humidity and a temperature measurement can be used to assess the extent of drying.

6.3.3.2 Electrical impedance

Moisture can produce a marked increase in the electrical conductivity of a material and, because of water's high dielectric constant, capacitance measurements can also be valuable. Electrical resistance measurements of moisture in timber and plaster are generally made using a pair of sharp pointed probes (Figure 6.8(a)) which are pushed into the material, the meter on the instrument being calibrated directly in percentage moisture. For on-line measurements of granular materials electrodes can be rollers, plates (Figure 6.8(b)), or skids but uniform density is essential. A difficulty with this and other on-line methods which

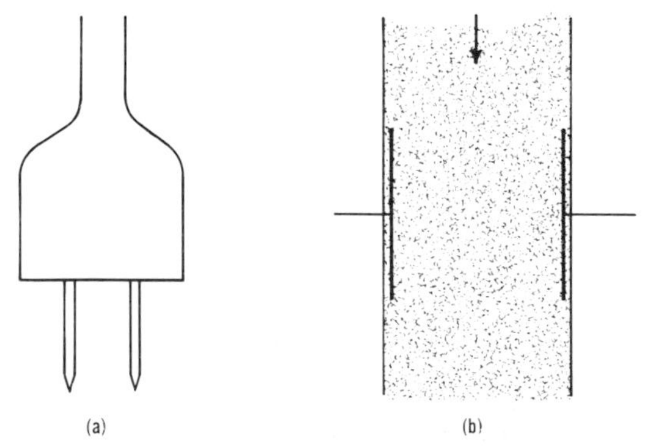

Figure 6.8 Two techniques for electrical measurements of moisture in solids: (a) pointed probes for insertion in wood, plaster, etc. to measure resistance, (b) capacitance plates to measure moisture in flowing powder or granules.

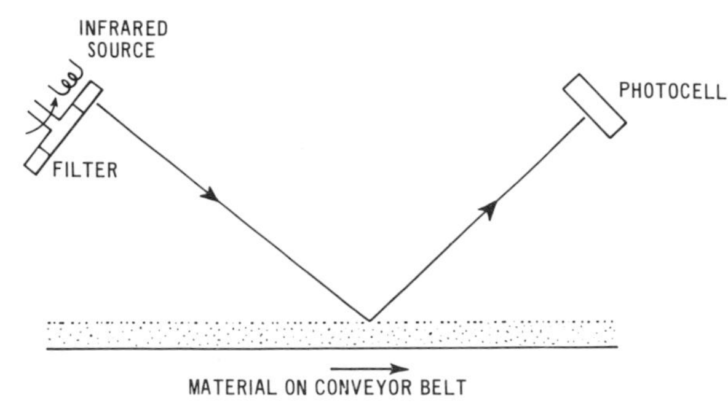

Figure 6.9 The principle of infrared reflectance used to measure moisture in a solid on a conveyor.

require contact between the sensor and the material is that hard materials will cause rapid erosion of the sensor.

6.3.3.3 *Microwave instruments*

Most comments appropriate to liquids also apply to solids, but, as above, constant packing density is necessary. For sheet materials, such as paper or cloth, measurement is simple, the sheet passing through a slot in the waveguide. For granular materials uniform material density is achieved by design of the flow path; alternatively extruders or compactors can be useful.

6.3.3.4 *Infrared instruments*

The basic difference from measurements in gases and liquids is that, for solids, reflectance methods (see Figure 6.9) are usually used rather than transmission methods. Single-beam operation is used with a rotating absorption and reference frequency filter to give regular zero readings. The calibration of a reflectance method can be substantially independent of the packing density as it measures only the surface concentration. For material on a conveyor belt a plough is often used in front of the sensing position to ensure a measurement more typical of the bulk. The method is not suitable for poorly reflecting materials, e.g. carbon and some metal powders.

6.3.3.5 *Neutron moderation*

Hydrogen nuclei slow down ('moderate') fast neutrons and therefore if a fast neutron source is placed over a moist material with a slow neutron detector adjacent, the detector output can be used to indicate the moisture concentration. The concentration of any other hydrogen atoms in the material and its packing density must be known. This technique is described in Chapter 4 of Part 3. Nuclear magnetic resonance can also be used to detect hydrogen nuclei as a means of measuring moisture content.

6.4 Calibration

It will be seen from the above sections that many moisture measurement techniques are not absolute and must be calibrated, generally at very regular intervals. It must first be emphasized that the absolute accuracy of moisture measurement, particularly in gases, is not usually high. Though it is possible to calibrate moisture detectors for liquids or solids to 0.1 to 1.0 per cent, such accuracies are the exception rather than the rule for gases. Figure 6.10 shows the accuracies of some of the techniques discussed in this chapter compared with the absolute gravimetric standard of the US National Bureau of Standards.

6.4.1 Gases

First of all the difficulties of making accurate moisture measurements must be stressed. This is particularly so at low levels, say less than 100 vppm because as all materials absorb moisture to some extent sample lines must come to equilibrium as well as the detector. At low moisture levels this can take hours particularly at low flow rates. A rapid-flow bypass line can be valuable. Patience is mandatory and if possible the outputs of the instruments should be recorded to establish when stable conditions are achieved. Many plastics are permeable to moisture and must never be used. At high moisture levels copper, Teflon, Viton, glass or quartz can be satisfactorily used but at low levels stainless steel is essential. Finally, at high moisture levels it must be remembered that the sample lines and detectors must be at least 10 kelvins hotter than the dew point of the gas.

There are two basic calibration methods which can, with advantage, be combined. Either a sample gas is passed through a reference hygrometer and the

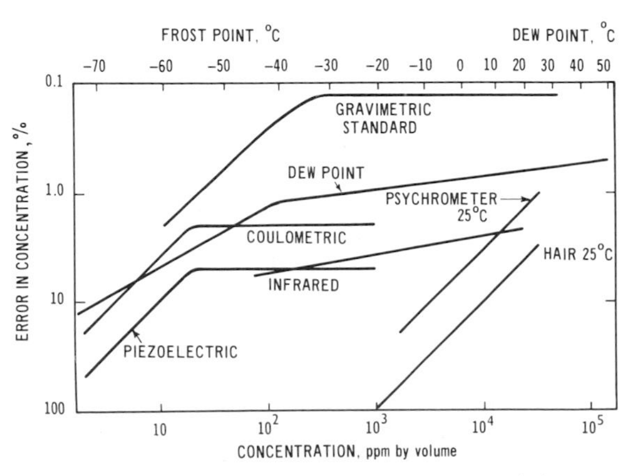

Figure 6.10 The accuracy of some of the major techniques for measuring moisture in gases, after Wexler (1970).

instrument under test, or a gas of known humidity is generated and passed through the instrument under test. Obviously it is ideal to double-check the calibration by using a known humidity and a reference hygrometer.

The most suitable reference hygrometer is the dew point meter, which can be readily obtained with certified calibration traceable to a standard instrument. For many applications less sophisticated dew point instruments would be adequate and coulometric analysers are possible for low moisture levels. At high levels gravimetric methods can be used but they are slow and tedious and difficult to make accurate.

There are a range of possible humidity sources, some of which are available commercially, and the choice depends on the facilities available and the application:

(a) A plastic tube, permeable to moisture, held in a thermostatically controlled water bath, will give a constant humidity for a given flow rate. Some manufacturers sell such tubes precalibrated for use as humidity sources but obviously the method is not absolute and the permeation characteristics of the tubes may change with time.
(b) Gas cylinders can be purchased with a pre-determined moisture level which does not significantly drift because of the internal surface treatment of the cylinder. However to prevent condensation in the cylinder the maximum moisture level is limited to about 500 vppm even with a cylinder pressure of only 10 bar. They are most suitable for spot checks of instruments on site.
(c) If an inert gas containing a known concentration of hydrogen is passed through a bed of copper oxide heated to $\sim 350\,^{\circ}C$ the hydrogen is converted to water vapour. This method relies on the measurement and stability of the hydrogen content which is better than for moisture. The generated humidity is also independent of flow rate.
(d) Water can be continuously injected into a gas stream using either an electrically driven syringe pump or a peristaltic pump. The injection point should be heated to ensure rapid evaporation. The method can be used very successfully, syringes in particular allowing a very wide range to be covered.
(e) If a single humidity level can be generated, a range can be obtained using a flow mixing system, but to achieve sufficient accuracy mass flow meters will probably be necessary.

6.4.2 Liquids

The basic absolute method is that of the Karl Fischer titration which was described in Section 6.3.2.1.

6.4.3 Solids

There are several methods which allow the absolute moisture level of a solid to be determined, but, for all of them, samples of the specific substance being measured by the process technique must be used. The most common technique is of course to weigh a sample, dry it, and then weigh again. Drying temperature and time depend on the material; if necessary the temperature must be limited to avoid decomposition, loss of volatile components or absorption of gases from the atmosphere. Balances can be obtained with a built-in heater, which give a direct reading of moisture content for a fixed initial sample weight. Other favoured techniques include measuring the water vapour given off by absorbing it in a desiccant to avoid the effects of volatiles; the Karl Fischer method again; or mixing the substance with calcium carbide in a closed bomb and measuring the pressure of acetylene produced. The method must be carefully chosen to suit the substance and process technique being used. Finally it is worth noting that rather than an absolute calibration, calibration directly in terms of the desired quality of the substance in the manufacturing process may be the most appropriate.

6.5 Bibliography

Mitchell, J. and Smith, D:, *Aquametry. Part 1, A Treatise on Methods for the Determination of Water*, Chemical Analysis Series No 5, Wiley, New York, (1977)
Mitchell, J. and Smith, D., *Aquametry. Part 2, The Karl Fischer Reagent*, Wiley, New York, (1980)
Verdin, A., *Gas Analysis Instrumentation*, Macmillan, London, (1973)
Wexler, A., 'Electric hygrometers', National Bureau of Standards Circular No 586, (1957)
Wexler, A. (ed.), *Humidity and Moisture* (3 volumes), papers presented at a conference, Reinhold, New York, (1965)
Wexler, A., 'Measurement of humidity in the free atmosphere near the surface of the Earth', *Meteorological Monographs*, **11**, 262–82, (1970)

Part 3

Electrical and Radiation Measurements

1 Electrical measurements

M. L. SANDERSON

1.1 Units and standards of electrical measurement

1.1.1 SI electrical units

The *ampere* (*A*) is the SI base unit. (Goldman and Bell, 1982; Bailey, 1982.) The 9th General Conference of Weights and Measures (CGPM) in 1948 adopted the definition of the ampere as that constant current which, if maintained in two straight parallel conductors of infinite length, of negligible circular cross section, and placed 1 m apart in vacuum, would produce between these conductors a force equal to 2×10^{-7} newton per metre of length. The force/unit length, F/l, between two such conductors separated by a distance d when each is carrying a current I A is given by:

$$\frac{F}{l}=\frac{\mu_0 I^2}{2\pi d}$$

where μ_0 is the permeability of free space. Thus inherent in this definition of the ampere is the value of μ_0 as exactly $4\pi\times10^{-7}$ N/A^2.

The derived SI electrical units are defined as follows.

The *volt* (*V*), the unit of potential difference and electromotive force, is the potential difference between two points of a conducting wire carrying a constant current of 1 A, when the power dissipated between these points is equal to 1 W.

The *ohm* (Ω), the unit of electrical resistance, is the electric resistance between two points of a conductor when a constant potential difference of 1 V, applied to these points, produces in the conductor a current of 1 A, the conductor not being the seat of any electromotive force.

The *coulomb* (*C*), the unit of quantity of electricity, is the quantity of electricity carried in 1 s by a current of 1 A.

The *farad* (*F*), the unit of capacitance, is the capacitance of a capacitor between the plates of which there appears a potential difference of 1 V when it is charged by a quantity of electricity of 1 C.

The *henry* (*H*), the unit of electric inductance, is the inductance of a closed circuit in which an electromotive force of 1 V is produced when the electric current varies uniformly at the rate of 1 A/s.

The *weber* (*Wb*), the unit of magnetic flux, is the flux which, linking a circuit of one turn would produce in it an electromotive force of 1 V if it were reduced to zero at a uniform rate in 1 s.

The *tesla* (*T*) is a flux density of 1 Wb/m^2.

1.1.2 Realization of the SI base unit

The definition of the SI ampere does not provide a suitable 'recipe' for its physical realization. The realization of the ampere has thus traditionally been undertaken by means of the Ayrton–Jones current balance (Vigoureux, 1965, 1971).

The force, F_x, in a given direction between two electrical circuits carrying the same current I is given by

$$F_x=I^2.\frac{dM}{dx}$$

where M is the mutual inductance between them.

In the current balance the force between current-carrying coils is weighed against standard masses. The principle of the balance is shown in Figure 1.1. The balance has two suspended coils and two pairs of fixed coils through which the same current flows. If the

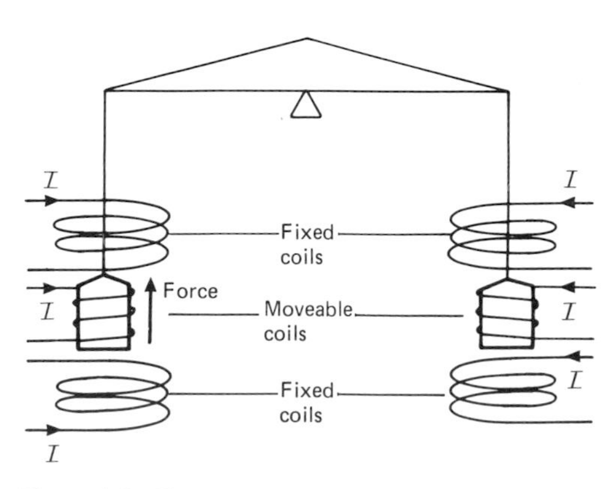

Figure 1.1 The current balance.

upper and lower coils of the fixed pair carry current in the same direction then the suspended coil experiences no force. If, however, the currents in the coils of the fixed pair are in opposite directions then the suspended coil experiences an upward or downward force. The force, F_x, is counterbalanced by the weight of a known mass, m, and thus

$$mg = I^2 . \frac{dM}{dx}$$

where g is the acceleration due to gravity.

I can be determined absolutely, i.e. in terms of the mechanical base units of mass, length and time, if dM/dx is known. dM/dx can be calculated from dimensional measurements made on the suspended and fixed coils. Changes in current direction and averaging the masses required to restore the balance condition enable the effects of interactive forces between opposite sides of the balance and of external magnetic fields to be eliminated.

Typically the accuracy of realization of the ampere using the current balance has a probable error of several parts in 10^6. One of the major causes of this inaccuracy is the relative magnitude of the force generated by the coils when compared with the mass of the suspended coils. Alternative techniques for the absolute determination of the ampere have been suggested. These include the use of the proton gyromagnetic ratio, γ_p, in conjunction with weak and strong magnetic field measurements (Dix and Bailey, 1975; Vigoureux, 1971) and the measurement of the force on a coil in a magnetic field together with the measurement of the potential induced when the coil moves in the same magnetic field (Kibble *et al.*, 1983).

1.1.3 National primary standards

Because the accuracy of realization of the SI ampere by the current balance is significantly poorer than the precision of intercomparison of standard cells and resistors, and also because of the difficulty of storing the realized value of the ampere, most National Standards Laboratories use standard cell banks and resistors as their maintained primary standards. Intercomparison of these national standards is regularly made through the International Bureau of Weights and Measures (BIPM) at Sèvres in France. Figure 1.2 (taken from Dix and Bailey, 1975) shows the UK primary standards which are maintained by the National Physical Laboratory (NPL). This figure also shows the relationships of the primary standards to the absolute reference standards; to the national low-frequency a.c. standards; and to the primary standards of other countries. Table 1.1 lists the UK national d.c. and low-frequency standards apparatus. Radiofrequency and microwave standards at NPL are listed in Table 1.2 (Steele *et al.*, 1975). Submillimetre wave measurements and standards are given by Stone *et al.* (1975).

These electrical standards are similar to standards held by other national laboratories; for example, the National Bureau of Standards (NBS) in the USA and Physikalisch-Technische Bundesanstalt (PTB) in West Germany and others elsewhere.

1.1.3.1 Standard cells

The UK primary standard of voltage is provided by a bank of some thirty Weston saturated mercury cadmium cells, the construction of a single cell being shown in Figure 1.3. The electrodes of the cell are mercury and an amalgam of cadmium and mercury. The electrolyte of cadmium sulphate is kept in a saturated condition over its operating temperature range by the presence of cadmium sulphate crystals. The pH of the electrolyte has a considerable effect on the stability of the emf of the cell and has an optimal value of 1.4 ± 0.2 (Froelich, 1974). The mercurous sulphate paste over the anode acts as a depolarizer. For details concerning the construction, maintenance and characteristics of such cells and their use the reader is directed to the NBS monograph listed in the references.

The nominal value of the emf generated by the saturated Weston cell is 1.01865 V at 20°C. Cells constructed from the same materials at the same time will have emfs differing by only a few μV. Cells produced at different times will have emfs differing by between 10 and 20 μV. The stability of such cells can be of order a few parts in 10^7 per year. They can be intercompared by back-to-back measurements to 1 part in 10^8. The internal resistance of the cell is approximately 750 Ω.

The variation of the cell emf with temperature can be described by the equation

$$\begin{aligned} V_T = V_{20} &- 4.06 \times 10^{-5}(T-20) \\ &- 9.07 \times 10^{-7}(T-20)^2 + 6.6 \times 10^{-9}(T-20)^3 \\ &- 1.5 \times 10^{-10}(T-20)^4 \end{aligned}$$

where V_T is the emf of the cell at a temperature T°C and V_{20} is its emf at 20°C. For small temperature variations about 20°C the cell has a temperature coefficient of $-40.6\ \mu$V/K.

To produce a source of emf with high stability it is necessary to maintain the cells in a thermostatically controlled enclosure. At NPL the standard cell enclosure contains up to 54 cells housed in groups of nine in separate copper containers in an air enclosure which has a temperature stability of better than 1 mK/h and a maximum temperature difference between any two points in the enclosure of less than

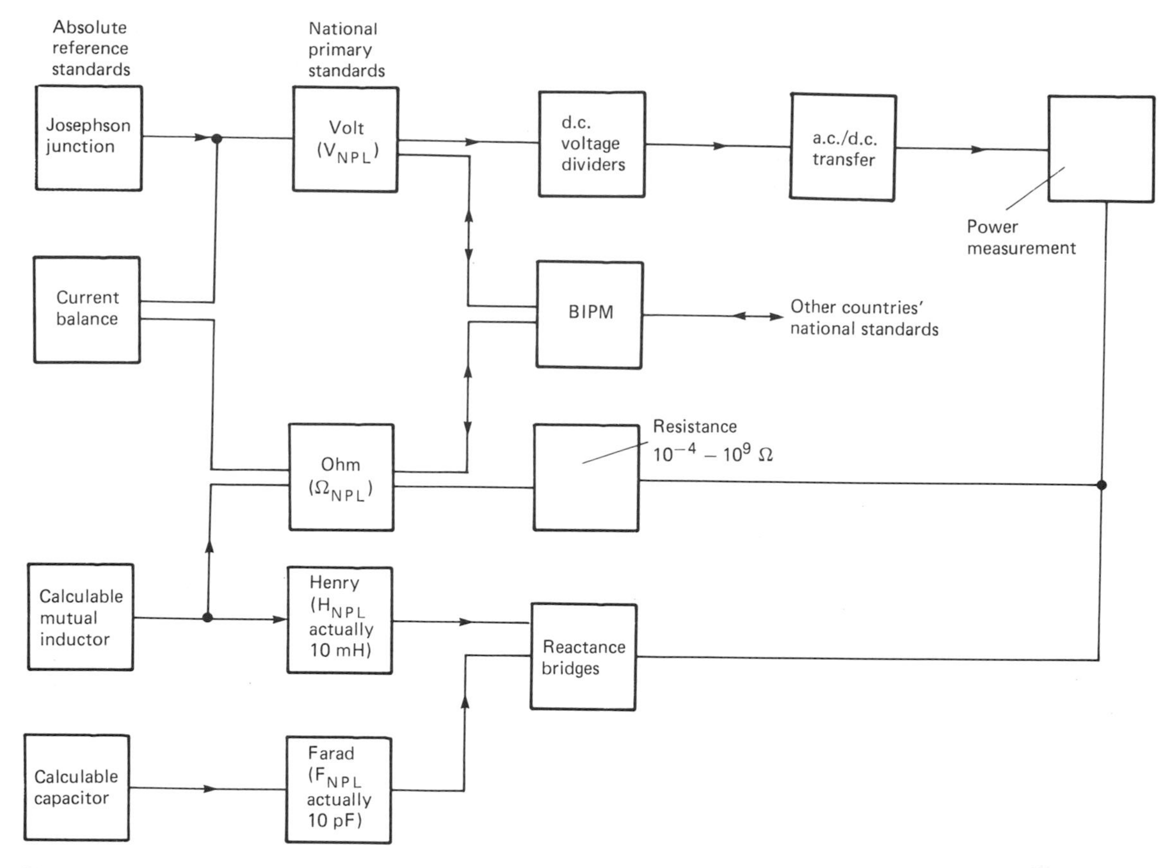

Figure 1.2 UK National Standards (from Dix and Bailey, 1975).

5 μK. Measurement of the emfs of the cells is effected under computer control.

1.1.3.2 Monitoring the absolute value of national voltage standards by means of the Josephson effect

Although intercomparison of standard cells can be undertaken to a high degree of precision, and such intercomparisons demonstrate that standard cells can be produced which show a high degree of stability with respect to each other, such measurements do not guarantee the absolute value of such cells. The Josephson effect (Josephson, 1962), is now used widely as a means of monitoring the absolute value of national standards of voltage maintained by standard cell banks. The effect allows the voltage maintained by standard cells to be related to a frequency, f, and the Josephson constant $2e/h$; e is the charge on the electron and h is Planck's constant.

The Josephson junction effect shown in Figure 1.4 predicts that if a very thin insulating junction between two superconductors is irradiated with rf energy of frequency f then the voltage–current relationship will exhibit distinct steps, as shown in Figure 1.4(b). The magnitude of one voltage step is given by

$$\Delta V=\frac{h}{2e}\cdot f$$

Thus voltage is related by the Josephson effect to the frequency of the rf radiation and hence to the base unit time. A value of 483 594.0 GHz/V has been ascribed to the Josephson constant, $2e/h$, with an uncertainty of ± 5 parts in 10^7.

The insulating junction can be produced in several ways, one of the simplest being to produce a dry solder joint between two conductors. For an irradiation frequency of 10 GHz the voltage steps are approximately 20 μV. By using the potential difference between a number of steps is is possible to produce a useable voltage of a few millivolts. Figure 1.5 shows a system employing another application of the Josephson junction known as a superconducting

Table 1.1 UK d.c. and low-frequency standards (from Dix and Bailey, 1975)

Absolute reference standards	*National primary standards*	*Other national standards apparatus*		
		Diesselhorst potentiometer (1 in 10^8)	Volt ratio box Voltage dividers	Power measurement
Josephson-junction system (1 in 10^7)	Standard cells (3 in 10^8)	Cell comparator (1 in 10^8)	D.C./A.C. thermal transfer	Electrostatic wattmeter
			Electrostatic voltmeter Inductive dividers	Dynamometer wattmeters Calibrated loads
	Standard 1 Ω resistors (1 in 10^7)	Wheatstone bridge (1 in 10^8)	High-current bridge	Electronic sources and amplifiers
		Current comparator (1 in 10^8)	High-resistance bridge	Rotary generators
		Build-up resistors (2 in 10^8)	Potentiometer Current-comparator potentiometer	Reference-measurement transformers
		Standard resistors	Standard resistors 10^{-4}–10^9 Ω	Transformer-measurement systems
Campbell mutual inductor 10 mH (1 in 10^6)		Inductance bridge Standard inductors 1 μH to 10 H (2 in 10^5)		Magnetic measurement Permeameters Vibrating-coil magnetometer
Current balance		Phase-angle standards for *L*, *C* and *R*		Magnetic-susceptibility balance
				Epstein-square magnetic-loss system
		Capacitance bridge		Local-loss tester
Calculate capacitor 0.4 pF (2 in 10^7)	Standard capacitors 10 pF (2 in 10^7)	Standard capacitors 10 pF to 1 nF (5 in 10^7)	Standard capacitors 10 nF to 1 μF	Magnetic-tape calibration

quantum interferometric detector (squid) as the detector in a superconducting potentiometer. This technique enables the comparison of the Josephson junction emf with the emf of a standard cell to be made with an accuracy of 1 part in 10^7. Further details of the techniques involved can be found in Dix and Bailey (1975).

1.1.3.3 Standard resistors

The desirable characteristics of standard resistors are that they should be stable with age, have a low temperature coefficient of resistance and be constructed of a material which exhibits only small thermoelectric emf effects with dissimilar materials. The UK national primary standard of resistance consists of a group of standard 1-Ω resistors wound from Ohmal, an alloy with 85 per cent copper, 11 per cent manganese and 4 per cent nickel, and freely supported by combs on a former. The resistors are immersed in oil. With such resistors it is possible to obtain a stability of 1 part in 10^7/yr.

1.1.3.4 Absolute determination of the ohm

The ohm can be determined absolutely by means of the Campbell mutual inductor, whose mutual inductance can be determined from geometric measurements made on the coils forming the inductor (Rayner, 1967). When such mutual inductors are used in Campbell's bridge as shown in Figure 1.6 the balance conditions are

$$R.r+\omega^2.M_1.M_2=0$$

and

$$M_1.R_s=L.r$$

where L is the loop inductance and R is its resistance.

Thus the first equation can be used to determine the product $R.r$ in terms of the SI base units of length and

Table 1.2 UK rf and microwave standards (from Steele *et al.*, 1975)

Quantity	*Method*	*Frequency (GHz)*	*Level*	*Uncertainty (95% confidence)*
Power in 14 mm coaxial line	Twin calorimeter	0–8.5	10–100 mW	0.2–0.5%
Power in 7 mm coaxial line	Twin calorimeter	0–18	10–100 mW	Under development
Power in WG16 (WR90)	Microcalorimeter	9.0, 10.0, 12.4	10–100 mW	0.2%
Power in WG18 (WR62)	Microcalorimeter	13.5, 15.0, 17.5	10–100 mW	0.2%
Power in WG22 (WR28)	Microcalorimeter	35	10–100 mW	0.5%
Power in WG26 (WR12)	Twin calorimeter	70	10 mW	0.8%
Attenuation	w.b.c.o. piston	0.0306	0–120 dB	0.002 dB
Attenuation in 14 mm coaxial line	w.b.c.o. piston	0–8.5	0–80 dB	0.001 dB/10 dB
Attenuation in WG11A (WR229), WG15 (WR112), WG16 (WR90), WG18 (WR62), WG22 (WR28), WG26 (WR12)	Modulated sub carrier		0–100 dB	From 0.002 dB at low values up to 0.02 dB at 100 dB, for v.s.w.r. < 1.05
Impedance				
Lumped conductance	rf bridge	1×10^{-3}	10 μS–1 S	0.1%
Lumped capacitance	rf bridge	1×10^{-3}	1 pF–10 μF	0.1%
Coaxial conductance	Woods bridge	5×10^{-3}–30×10^{-3}	0–40 mS	0.1% + 0.001 mS
		30×10^{-3}–200×10^{-3}		0.2% + 0.001 mS
Coaxial capacitance	Woods bridge	5×10^{-3}	0–40 pF	0.1% + 0.001 pF
(these refer to major components only)				
			K	
Noise temperature				
in 14 mm coaxial line	Thermal	1–2	10^4	About 1.5 K transfer standards calibrate to 110 K
in WG10 (WR284)	Thermal	2.75, 3.0, 3.5	10^4	
in WG14 (WR137)	Thermal	6.0, 7.0, 8.0	10^4	
in WG16 (WR90)	Thermal	9.0, 10.0, 11.2	10^4	
in WG18 (WR62)	Thermal	13.5, 15.0	10^4	
in WG22 (WR28)	Thermal	35	10^4	
in WG11A (WR229)	Cryogenic		77	0.15 K: transfer standards calibrate to 0.6 K
in WG15 (WR112)	Cryogenic		77	

time. The ratio of the two resistances R and r can be found using a bridge technique, and thus r can be determined absolutely. This absolute determination has a probable error of 2 parts in 10^6.

An alternative method for the absolute determination of the ohm employs the Thompson–Lampard calculable capacitor (Thompson and Lampard, 1956). This capacitor has a value which can be determined from a knowledge of the velocity of light and a single length measurement.

Consider a cylindrical electrode structure having the symmetrical cross section shown in Figure 1.7(a) in which neighbouring electrodes are separated only by small gaps; Thompson and Lampard showed that the cross capacitances per unit length C_1 and C_2 are related by

$$\exp\left(-\frac{\pi C_1}{\varepsilon_0}\right)+\exp\left(-\frac{\pi C_2}{\varepsilon_0}\right)=1$$

Because of symmetry $C_1=C_2$ and the cross capacitance per metre, C, is given by

$$C=\frac{\varepsilon_0 \log_e 2}{\pi} \quad \text{F/m}$$

Since the velocity of light, c, is given by

$$c^2=\frac{1}{\varepsilon_0\mu_0}$$

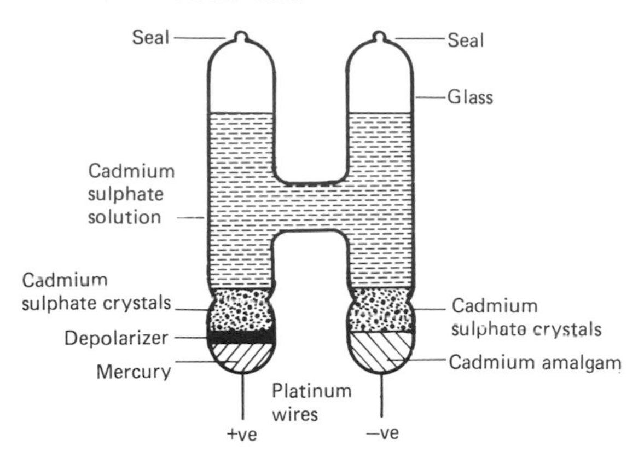

Figure 1.3 Weston standard cell.

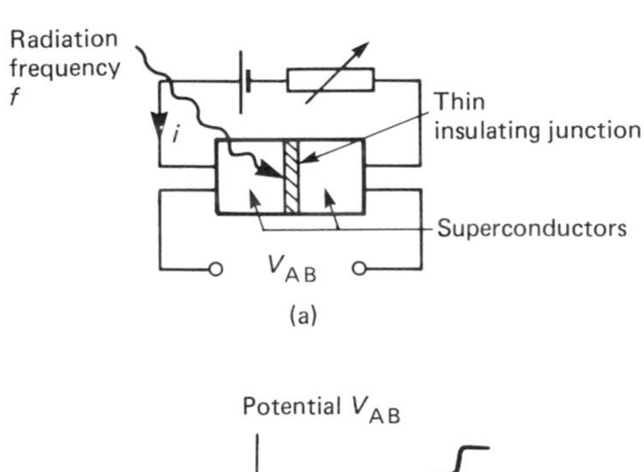

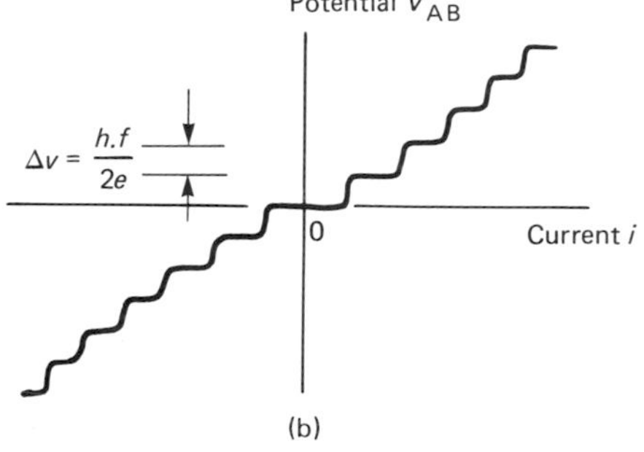

Figure 1.4 (a) Josephson junction effect; (b) voltage/current characteristic of Josephson junction.

and the value of μ_0 is, by definition, $4\pi \times 10^{-7}$, then if the velocity of light is known the capacitance per metre of the capacitor can be determined. C has a value of 1.953 548 5 pF/m.

By inserting a movable guard electrode as shown in Figure 1.7(b), the position of which can be determined by means of an optical interference technique, it is possible to generate changes in capacitance which can be determined absolutely. The change in capacitance obtained can be compared with the capacitance of a standard 10-pF capacitor and hence by means of the chain shown in Figure 1.7(c) used to determine the

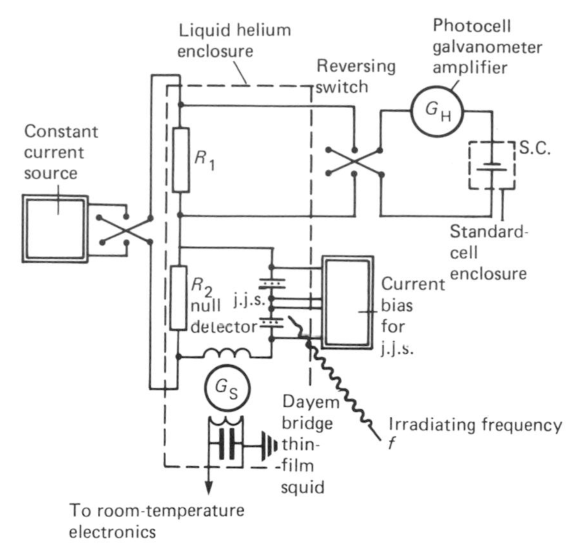

Figure 1.5 Voltage comparison system using Josephson junction (from Dix and Bailey, 1975).

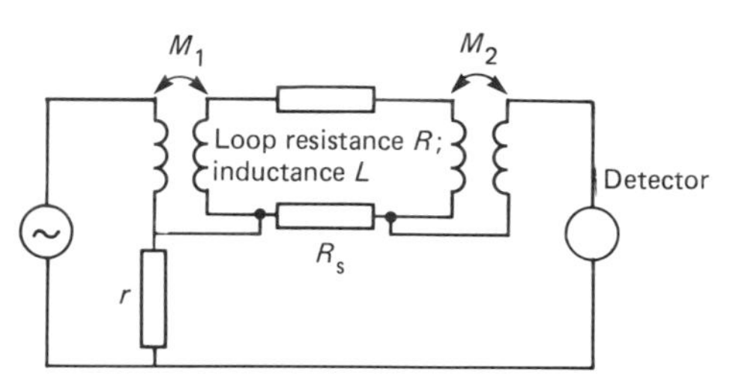

Figure 1.6 Campbell bridge.

absolute value of the ohm. The accuracy of this determination is typically 1 part in 10^7.

1.2 Measurement of d.c. and a.c. current and voltage using indicating instruments

The most commonly used instruments for providing an analogue indication of direct or alternating current or voltage are the permanent magnet-moving coil, moving iron and dynamometer instruments. Other indicating instruments include thermocouple and electrostatic instruments, the latter based on the attraction between two charged plates. This section provides a description of the basic principles of operation of such instruments. Further details can be found in Golding and Widdis (1963), Harris (1966), Gregory (1973) and Tagg (1974). The accuracy specification and the assessment of influence factors upon direct-acting indicating electrical measuring

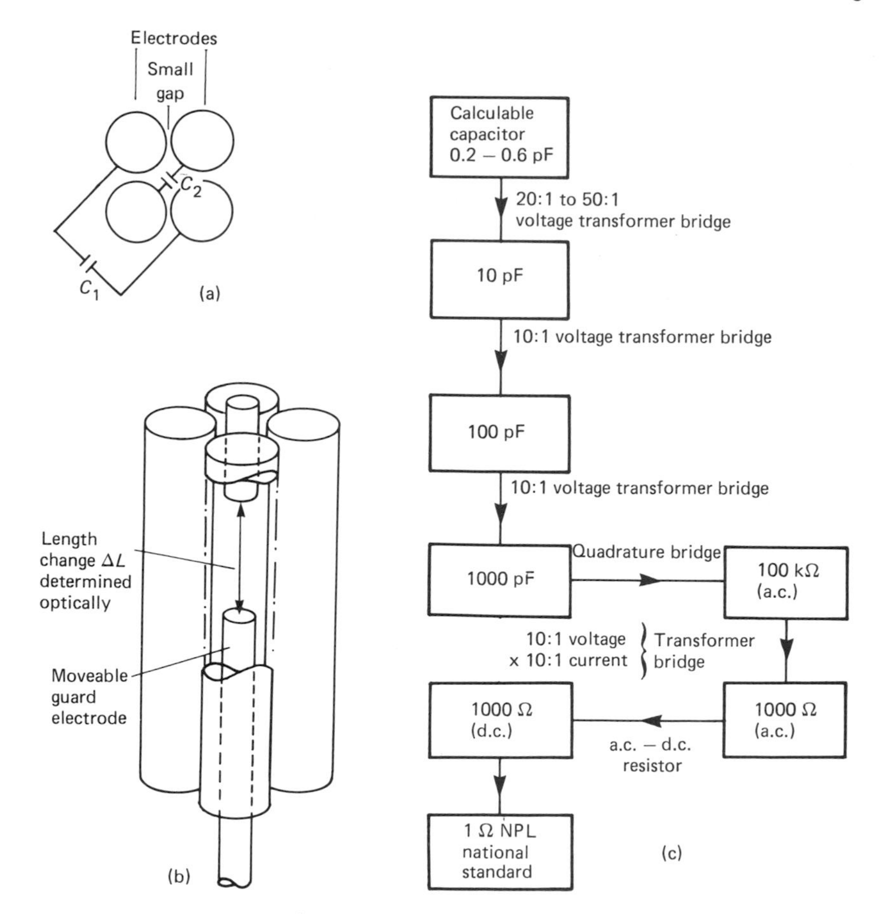

Figure 1.7 (a) Cross-section of Thompson–Lampard capacitor; (b) variable capacitor; (c) comparison chain for Thompson–Lampard capacitor.

instruments and their accessories are set out in BSI 89: 1977 (British Standards Institution, 1977). This is equivalent to IEC 51:1973.

1.2.1 Permanent magnet–moving coil instruments

Permanent magnet–moving coil instruments are based upon the principle of the D'Arsonval moving-coil galvanometer, the movement of which is also used in light spot galvanometers, pen and ultra-violet recorders. A typical construction for a moving-coil instrument is shown in Figure 1.8(a). The current to be measured is passed through a rectangular coil wound on an insulated former, which may be of copper or aluminium, to provide eddy-current damping. The coil is free to move in the gap between the soft iron pole pieces and core of a permanent magnet employing a high-coercivity material such as Columax, Alcomax or Alnico. The torque produced by the interaction of the current and the magnetic field is opposed by control springs which are generally flat or helical phosphor-bronze springs. These also provide the means by which the current is supplied to the coil. The bearings for the movement are provided by synthetic sapphire jewels and silver-steel or stainless-steel pivots. Alternative means of support can be provided by a taut band suspension, as shown in Figure 1.8(b). This has the advantage of removing the friction effects of the jewel and pivot but is more susceptible to damage by shock loading. The pointer is usually a knife-edge one, and for high-accuracy work it is used in conjunction with a mirror to reduce parallax errors.

The torque, T_g, generated by the interaction of the current, i, and the magnetic field of flux density B is given by

$$T_g = N . B . h . b . i$$

where h and b are the dimensions of the coil having N turns. This is opposed by the restoring torque, T_r, produced by the spring

$$T_r = k\theta$$

where k is the spring constant.

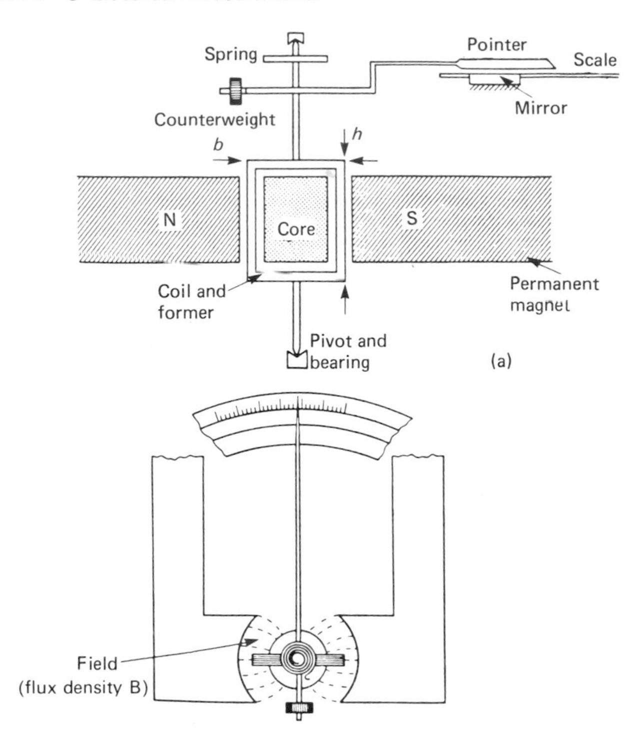

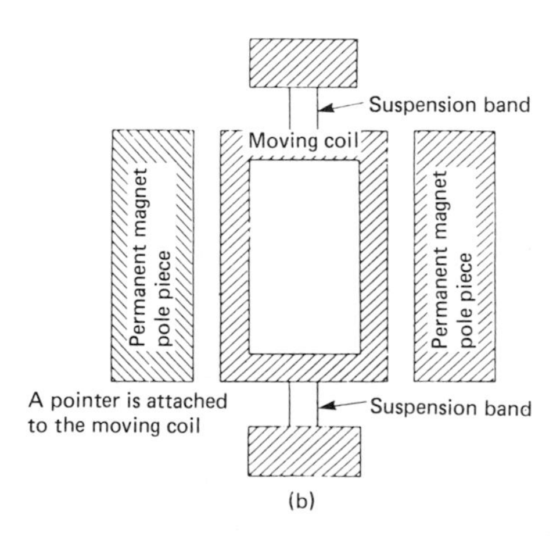

Figure 1.8 (a) Elements of a permanent magnet-moving coil instrument; (b) taut band suspension system.

Under static conditions these two torques are equal and opposite, and thus

$$\theta = \frac{N.B.h.b.i}{k} = S.i$$

where S is the sensitivity of the instrument.

Under dynamic conditions the generated torque, T_g, is opposed by inertial, damping and spring-restoring forces, and thus

$$T_g = J\frac{d^2\theta}{dt^2} + D.\frac{d\theta}{dt} + k\theta$$

where J is the inertia of the moving system, D is its damping constant and k is the spring constant.

Damping can be provided by an air damper or by eddy-current damping from the shorted turn of the former or from the coil and external circuit. For eddy-current damping

$$D = h^2b^2B^2\left(\frac{N^2}{R} + \frac{1}{R_f}\right)$$

where R represents the resistance of the coil circuit and R_f represents the resistance of the coil making up the former.

The instrument thus has a second-order transfer function given by

$$G(s) = \frac{\theta(s)}{i(s)} = \frac{(k/J).S}{s^2 + (D/J).s + (k/J)}$$

Comparing this transfer function with the standard second-order transfer function

$$G(s) = \frac{K\omega_n^2}{s^2 + 2\xi\omega_n s + \omega_n^2}$$

The natural frequency of the instrument is given by $\omega_n = \sqrt{(k/J)}$, and its damping factor $\xi = D/2\sqrt{(Jk)}$.

If $D^2 > 4\,kJ$ then $\xi > 1$ and the system is overdamped. The response to a step input of current magnitude I at $t = 0$ is given by

$$\theta(t) = S.I\left\{1 - \frac{\xi + \sqrt{(\xi^2 - 1)}}{2\sqrt{(\xi^2 - 1)}}e^{[-\xi + \sqrt{(\xi - 1)}]\omega_n t} + \frac{\xi - \sqrt{(\xi^2 - 1)}}{2\sqrt{(\xi^2 - 1)}}e^{[-\xi - \sqrt{(\xi^2 - 1)}]\omega_n t}\right\}$$

If $D^2 = 4\,kJ$ then $\xi = 1$ and the system is critically damped. The response to the step input is given by

$$\theta(t) = S.I[1 - (1 + \omega_n t)\,e^{-\omega_n t}]$$

If $D^2 < 4\,kJ$ then $\xi < 1$ and the system is underdamped. The response to the step input is given by

$$\theta(t) = S.I.\left\{1 - \frac{e^{-\xi\omega_n t}}{\sqrt{(1 - \xi^2)}}.\sin\left[\sqrt{(1 - \xi^2)}\omega_n t + \phi\right]\right\}$$

and $\phi = \cos^{-1}\xi$.

These step responses are shown in Figure 1.9.

1.2.1.1 Range extension

The current required to provide full-scale deflection (FSD) in a moving-coil instrument is typically in the

Figure 1.9 Second-order system responses. (a) Overdamped; (b) critically damped; (c) underdamped.

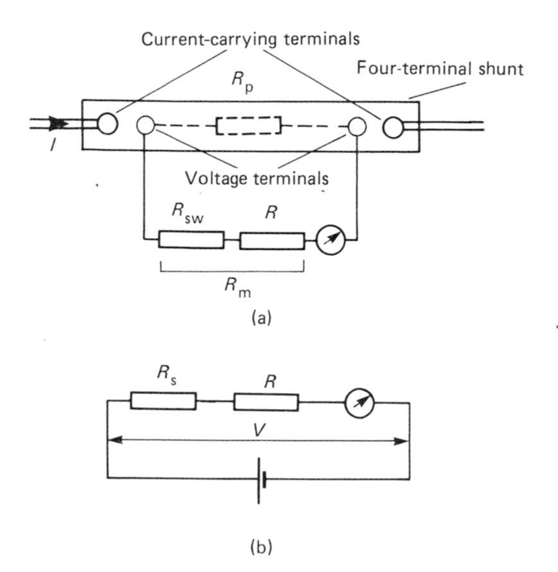

Figure 1.10 (a) Ammeter using a moving-coil instrument; (b) voltmeter using a moving-coil instrument.

range of 10 μA to 20 mA. D.C. current measurement outside this range is provided by means of resistive shunts, as shown in Figure 1.10(a). The sensitivity of the shunted ammeter, S_A, is given by

$$S_A = \frac{\theta}{I} = \frac{R_p}{R_p + R_m} \cdot S$$

where R_p is the resistance of the shunt, R_m is the resistance of the coil and swamping resistance and S is the sensitivity of the unshunted movement.

High-current ammeters usually employ a movement requiring 15 mA for FSD. The shunts are usually four-terminal devices made of manganin. The voltage drop across the instrument is 0.075 V and thus the power dissipated in the shunt is approximately 0.075 I W. Table 1.3 gives the power dissipation in the shunt for various current ratings.

For use as a d.c. voltmeter the sensitivity, S_v, is given by

$$S_v = \frac{\theta}{V} = \frac{S}{R_s + R}$$

where R_s is the series resistance, R is the resistance of the coil and S is the sensitivity of the movement (Figure 1.10(b)).

The value of the series resistance depends on the sensitivity of the moving coil. For a movement with a FSD of 10 mA it is 100 Ω/V. If FSD requires only 10 μA then the resistance has a value of 100 000 Ω/V. Thus for a voltmeter to have a high input impedance the instrument movement must have a low current for FSD.

1.2.1.2 Characteristics of permanent magnet–moving coil instruments

Permanent magnet–moving coil instruments have a stable calibration, low power consumption, a high torque to weight ratio, and can provide a long uniform scale. They can have accuracies of up to 0.1 per cent of FSD. With the use of shunts or series resistors they can cover a wide range of current and voltage. The errors due to hysteresis effects are small and they are generally unaffected by stray magnetic fields. It is possible to adjust the damping in such instruments to any required value. The major errors are likely to be caused by friction in the bearings and changes in the resistance of the coil with temperature. Copper wire which is used for the coil has a temperature coefficient of +0.4%/K. When used as a voltmeter this temperature variation is usually swamped by the series resistance. When used as an ammeter with manganin shunts it is necessary to swamp the coil resistance with a larger resistor, usually manganin, as shown in Figure 1.10(a). This has the effect of more closely matching the temperature coefficient of the coil/swamp resistance combination to that of the shunt, thus effecting a constant current division between the

Table 1.3 Power dissipated in shunt for various current ratings

Current (A)	*Power dissipated (W)*
1	0.075
2	0.150
5	0.375
10	0.75
20	1.50
50	3.75
100	7.50
200	15.00
500	37.50
1000	75.00

instrument and the shunt over a given temperature range.

1.2.1.3 A.C. voltage and current measurement using moving-coil instruments

The direction of the torque generated in a moving-coil instrument is dependent on the instantaneous direction of the current through the coil. Thus an alternating current will produce no steady-state deflection.

Moving-coil instruments are provided with an a.c. response by the use of a full-wave bridge rectifier, as shown in Figure 1.11. The bridge rectifier converts the

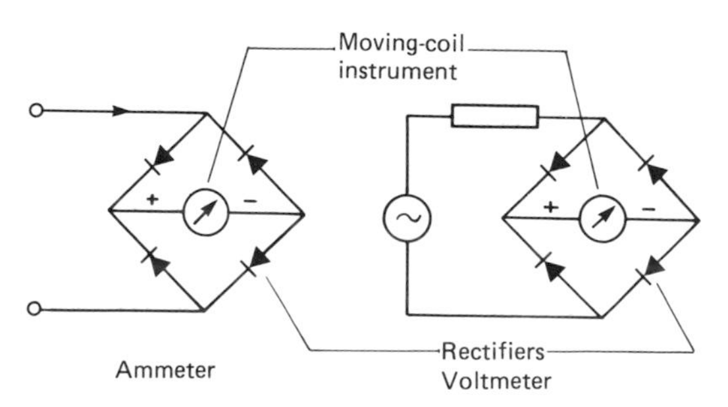

Figure 1.11 A.C. current and voltage measurement using a rectifier–moving-coil instrument.

a.c. signal into a unidirectional signal through the moving-coil instrument which then responds to the average d.c. current through it. Such instruments measure the mean absolute value of the waveform and are calibrated to indicate the rms value of the wave on the assumption that it is a sinusoid.

For a periodic current waveform $I(t)$ through the instrument the mean absolute value, I_{mab}, is given by

$$I_{mab} = \frac{1}{T}\int_0^T |I(t)| \, . \, dt$$

and its rms value is given by

$$I_{rms} = \sqrt{\left[\frac{1}{T}\int_0^T I^2(t) \, . \, dt\right]}$$

where T is the period of the wave.

The Form Factor (FF) for the current waveform is defined as

$$FF = \frac{I_{rms}}{I_{mab}} = \frac{\sqrt{\left[(1/T)\int_0^T I^2(t).dt\right]}}{(1/T)\int_0^T |I(t)|.dt}$$

For a sinusoid $I(t) = \hat{I} \, . \sin \omega t$ the rms value is $\hat{I}/\sqrt{2}$ and its mean absolute value is $2\hat{I}/\pi$. The Form Factor for a sinusoid is thus 1.11. Rectifier instruments indicate $1.11 \, . \, I_{mab}$. For waveforms which are not sinusoidal rectifier instruments will provide an indication which will have an error of

$$\left(\frac{1.11 - FF}{FF}\right) \times 100\%$$

Figure 1.12 shows several waveforms with their Form Factors and the errors of indication which occur if they are measured with mean absolute value measuring—rms scaled instruments. This Form Factor error also occurs in the measurement of a.c. current and voltage using digital voltmeters which employ rectification for the conversion from a.c. to d.c.

As a current-measuring device the diodes should be selected for their current-carrying capability. The non-linear characteristics of the diodes make range extension using shunts impractical and therefore it is necessary to use rectifier instruments with current transformers (see section 1.2.3). The forward diode drop places a lower limit on the voltage which can be measured accurately and gives such instruments a typical minimum FSD of 10 V. When used as a voltmeter the variation of the diode forward drop with temperature can provide the instrument with a sensitivity to ambient temperature. It is possible to design such instrumentes to provide an accuracy of 1 per cent of FSD from 50 Hz to 10 kHz.

1.2.1.4 Multimeters

These are multi-range devices using a permanent magnet–moving coil instrument. They enable the measurement of d.c. and a.c. current and voltage, and resistance. One of the most common instruments of this type is the AVO Model 8 Mark 6 (Thorn–EMI). Table 1.4 gives the specification for this instrument and Figure 1.13 shows the circuit diagram. The basic movement has a full-scale deflection of 50 μA and therefore this gives the instrument a sensitivity of 20 000 Ω/V on its d.c. voltage ranges. The three ranges of resistance operate by measuring the current passing through the resistance on applying a d.c. voltage supplied from internal batteries. A zero control on these ranges, used with the instrument probes shorted together, enables compensation for changes in the emf of the internal batteries to be made.

1.2.1.5 Electronic multimeters

By using the electronic input in Figure 1.14(a) it is possible to achieve a high input impedance irrespective of the voltage range. This is used as shown in Figure 1.14(b) to measure current, resistance and a.c. quantities. For current measurement the maximum input voltage can be made to be the same on all ranges. Resistance measurements can be made with

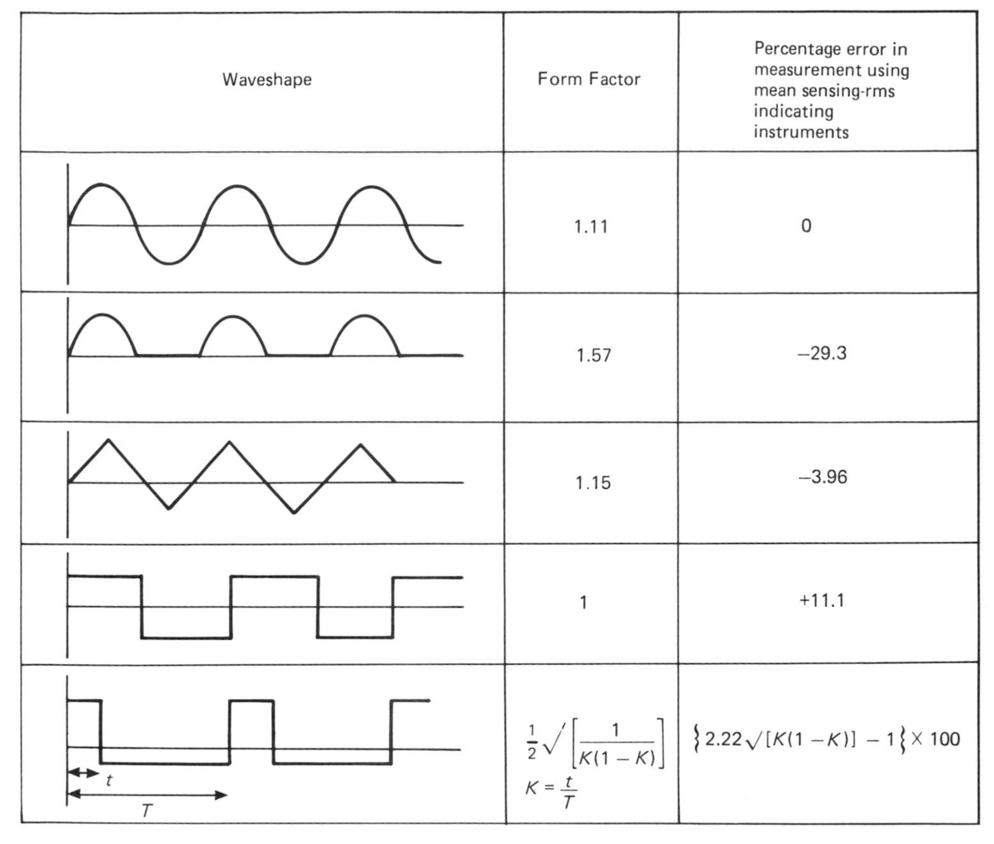

Waveshape	Form Factor	Percentage error in measurement using mean sensing-rms indicating instruments
	1.11	0
	1.57	–29.3
	1.15	–3.96
	1	+11.1
t, T	$\frac{1}{2}\sqrt{\left[\frac{1}{K(1-K)}\right]}$ $K = \frac{t}{T}$	$\{2.22\sqrt{[K(1-K)]} - 1\} \times 100$

Figure 1.12 Waveform Form Factors and errors of indication for rectifier instruments.

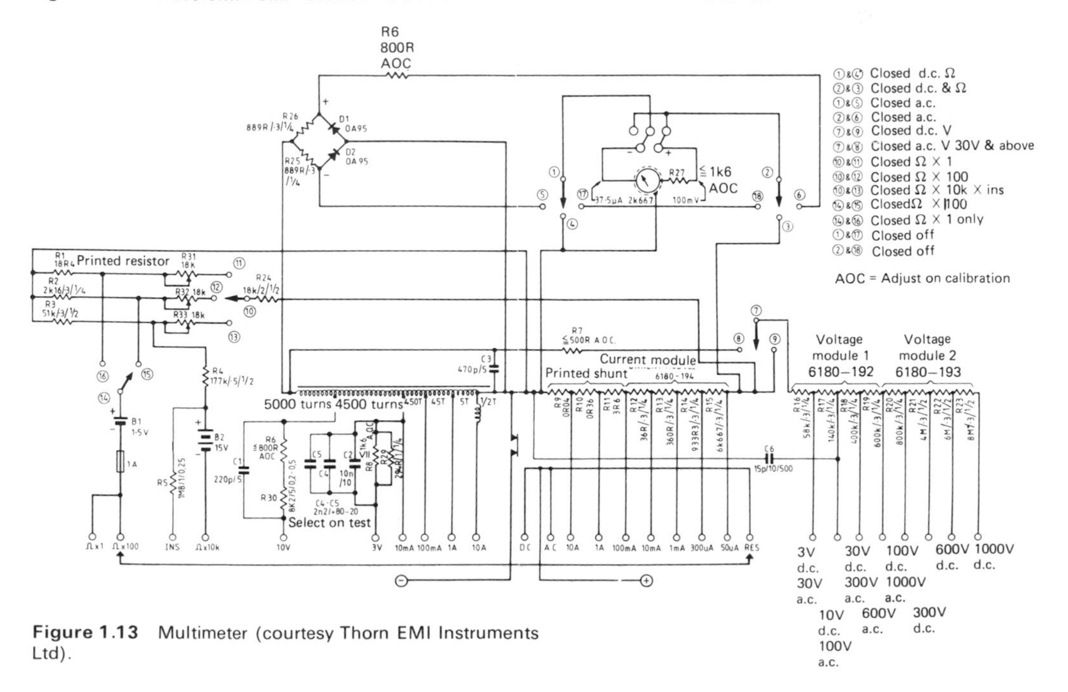

Figure 1.13 Multimeter (courtesy Thorn EMI Instruments Ltd).

Table 1.4 Multimeter specification

D.C. voltage	8 ranges: 100 mV, 3, 10, 30, 100, 300, 600 V, 1 kV
D.C. current	7 ranges: 50 μA, 300 μA, 1, 10, 100 mA, 1 A and 10 A
A.C. voltage	7 ranges: 3, 10, 30, 100, 300, 600 V, 1 kV
A.C. current	4 ranges: 10 mA, 100 mA, 1 A and 10 A
Resistance	3 ranges: ×1:0–2 kΩ ×100:0–200 kΩ ×10 k:0–20 MΩ
Source for resistance measurement	One 15 V type B121 battery (for ×10 k range) One 1.5 V type SP2 single cell (for ×1, ×100 range)
Accuracy	D.C. ±1% fsd A.C. (150 Hz) ±2% fsd Resistance ±3% centre scale
Sensitivity	D.C. 20 000 Ω/V all ranges A.C. 100 Ω/V 3 V range 1 000 Ω/V 10 V range 2 000 Ω/V all other ranges
Overload protection	High speed electromechanical cut-out with a fuse on the two lower resistance ranges
Decibels	−10 to +55 using a.c. voltage scale
Voltage drop at terminals	D.C. 100 mV on 50 μA range, approx. 400 mV on other ranges A.C. less than 450 mV at 10 A
Frequency response a.c. voltage range (up to 300 V)	< ±3% discrepancy between 50 Hz reading and readings taken between 15 Hz and 15 kHz

lower voltage drops across the resistors and with a linear indication. A.C. quantities are measured using rectification and mean or peak sensing. Table 1.5 gives the specification of such an instrument (Hewlett-Packard HP410C General Purpose Multi-Function Voltmeter).

1.2.2 Moving-iron instruments

There are two basic types of moving-iron instrument—the attraction and repulsion types shown in Figure 1.15. In the attraction type a piece of soft iron in the form of a disc is attracted into the coil which is in the form of a flat solenoid. Damping of the instrument is provided by the air-damping chamber. The shape of the disc can be used to control the scale shape. In the repulsion instrument two pieces of iron, either in the

Table 1.5 Electronic multimeter specification

D.C. voltmeter
Voltage ranges: ±15 mV to ±1500 V full scale in 15, 50 sequence (11 ranges)
Accuracy: ±2% of full scale on any range
Input resistance: 100 MΩ ±1% on 500 mV range and above, 10 MΩ ±3% on 150 mV range and below

AC voltmeter
Voltage ranges: 0.5 V to 300 V full scale in 0.5, 1.5, 5 sequence (7 ranges)
Frequency range: 20 Hz to 700 MHz
Accuracy: ±3% of full scale at 400 Hz for sinusoidal voltages from 0.5 V–300 V rms. The a.c. probe responds to the positive peak–above-average value of the applied signal. The meter is calibrated in rms
Frequency response: ±2% from 100 Hz to 50 MHz (400 Hz ref.); 0 to −4% from 50 MHz to 100 MHz; ±10% from 20 Hz to 100 Hz and from 100 MHz to 700 MHz
Input impedance: input capacitance 1.5 pF, input resistance >10 MΩ at low frequencies. At high frequencies, impedance drops off due to dielectric loss
Safety: the probe body is grounded to chassis at all times for safety. All a.c. measurements are referenced to chassis ground

D.C. ammeter
Current ranges: ±1.5 μA to ±150 mA full scale in 1.5, 5 sequence (11 ranges)
Accuracy: ±3% of full scale on any range
Input resistance: decreasing from 9 kΩ on 1.5 μA range to approximately 0.3 Ω on the 150 mA range
Special current ranges: ±1.5, ±5 and ±15 μA may be measured on the 15, 50 and 150 mV ranges using the d.c. voltmeter probe, with ±5% accuracy and 10 MΩ input resistance

Ohmmeter
Resistance range: resistance from 10 Ω to 10 MΩ centre scale (7 ranges)
Accuracy: zero to midscale: ±5% of reading or ±2% of midscale, whichever is greater; ±7% from midscale to scale value of 2; ±8% from scale value of 2 to 3; ±9% from scale value of 3 to 5; ±10% from scale value of 5 to 10
Maximum input: D.C.: 100 V on 15, 50 and 150 mV ranges, 500 V on 0.5 to 15 V ranges, 1600 V on higher ranges. A.C.: 100 times full scale or 450 V p, whichever is less

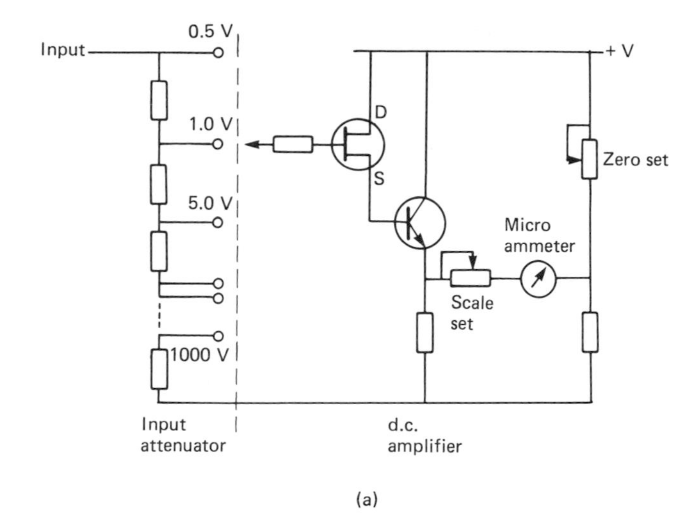

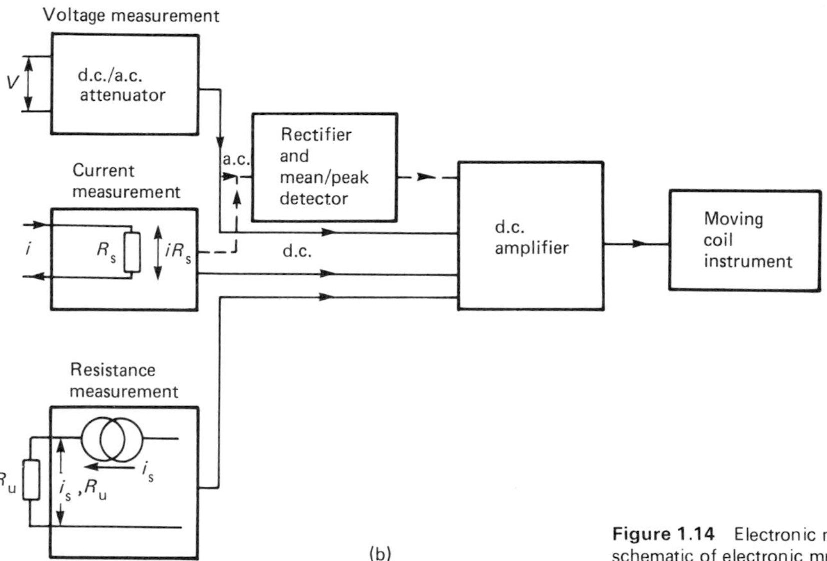

Figure 1.14 Electronic multimeter. (a) Electronic input; (b) schematic of electronic multimeter.

form of rods or vanes, one fixed and the other moveable, are magnetized by the field current to be measured. In both instruments the torque, T_g, generated by the attraction or repulsion is governed by

$$T_g = \frac{1}{2} \cdot \frac{dL}{d\theta} \cdot i^2$$

where L is the inductance of the circuit. The restoring torque, T_r, is produced by a spring:

$$T_r = k\theta$$

and thus

$$\theta = \frac{1}{2} \cdot \frac{1}{k} \cdot \frac{dL}{d\theta} \cdot i^2$$

The deflection of the instrument is proportional to the mean square of the current and thus the instrument provides a steady-state deflection from an a.c. current. The scales of such instruments are usually calibrated in terms of rms values and they tend to be non-linear, being cramped at the lower end.

Friction in the bearings of the instrument causes error. Hysteresis effects in the iron of the instrument give rise to different indications for increasing and decreasing current. Errors can also be caused by the presence of stray magnetic fields. Variation in ambient temperature causes changes in the mechanical dimensions of the instrument, alters the permeability of the iron and changes the resistance of the coil. This last effect is the most important. Used as an ammeter

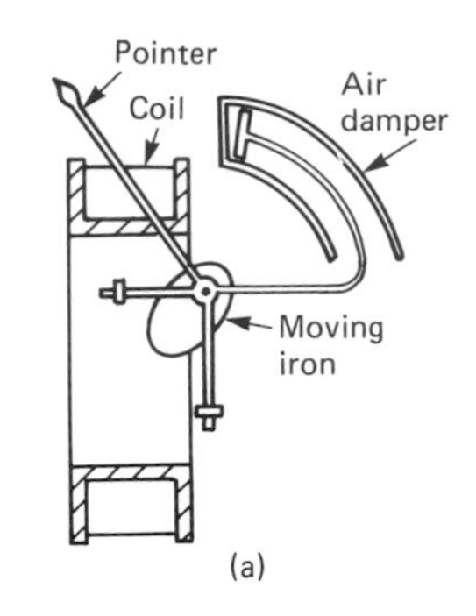

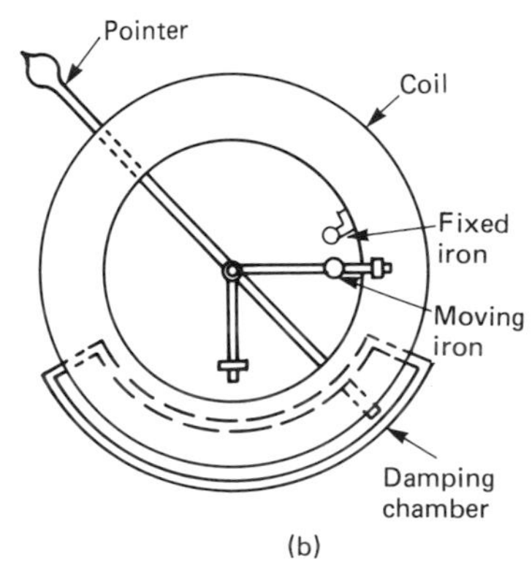

Figure 1.15 Moving-iron instrument. (a) Attraction; (b) repulsion (from Tagg, 1974).

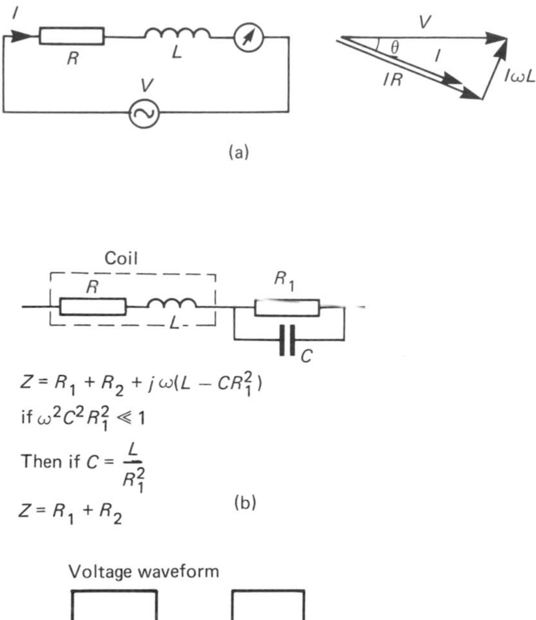

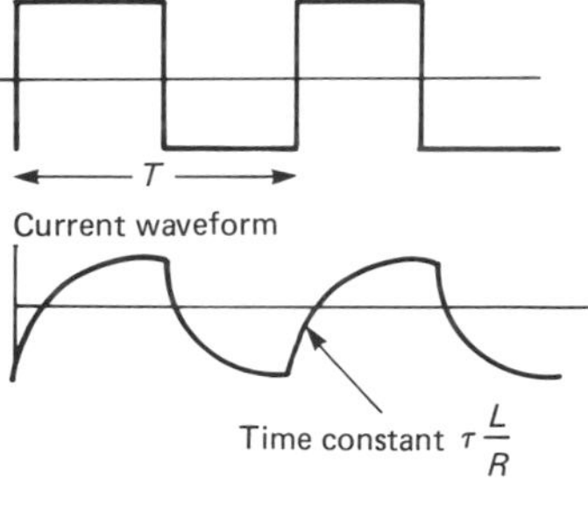

Percentage error on rms reading =

$$\frac{2\tau}{T}\left(e^{-T/2\tau} - e^{-T/\tau}\right) \times 100\%$$

for $T \gg \tau$

(c)

Figure 1.16 (a) Inductance effects in moving-iron voltmeters (b) compensation for effect of inductance; (c) errors in measurement of non-sinusoidal waveforms.

the change in resistance causes no error, but when used as a voltmeter the change in resistance of the copper winding of +0.4%/K causes the sensitivity of the voltmeter to change. This effect is usually reduced by using a resistance in series with the coil wound with a wire having a low temperature coefficient. The inductance of the instrument can also cause changes in its sensitivity with frequency when used as a voltmeter. This is shown in Figure 1.16(a). At a given angular frequency ω, the error of reading of the voltmeter is given by $(\omega^2L^2)/(2R^2)$, where L is its inductance and R its resistance. Figure 1.16(b) shows a compensation method for this error.

Although the moving-iron instrument is a mean square indicating instrument errors can be introduced when measuring the rms value of a non-sinusoidal voltage waveform. These errors are caused by the peak flux in the instrument exceeding the maximum permitted flux and also by attenuation of the harmonic current through the instrument by the time constant of the meter, as shown in Figure 1.16(c).

Moving-iron instruments are capable of providing an accuracy of better than 0.5 per cent of FSD. As ammeters they have typical FSDs in the range of 0.1–30 A without shunts. The minimum FSD when they are used as voltmeters is typically 50 V with a low input impedance of order 50 Ω/V. Their frequenc response is limited by their high inductance and stra capacitance to low frequencies, although instrument are available which will measure at frequencies up t 2500 Hz. Moving-iron instruments have relativel high power requirements and therefore they ar unsuitable for use in high-impedance a.c. circuits.

1.2.3 A.C. range extension using current and voltage transformers

In section 1.2.1.1, extension of the range of permanen magnet–moving coil instruments using current shunt and resistive voltage multipliers was described. Th

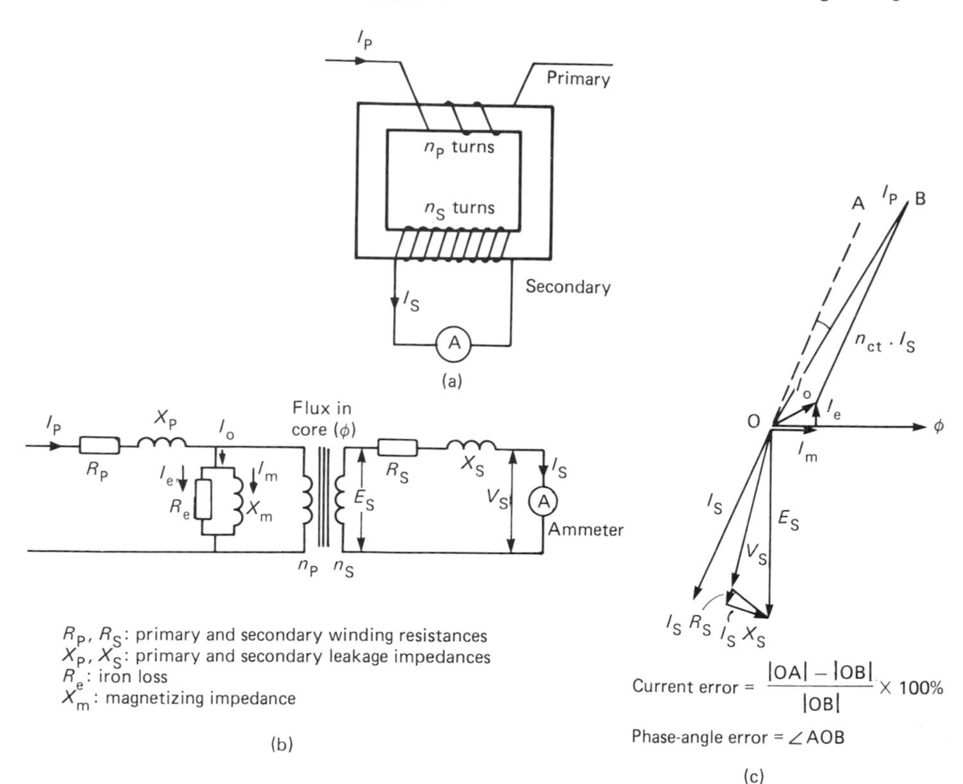

Figure 1.17 (a) Current transformer; (b) equivalent circuit; (c) phasor diagram of current transformer; (d) current ratio and phase-angle errors for current transformer.

same techniques can be applied in a.c. measurements. However, in power measurements with large currents the power dissipated in the shunt becomes significant (see Table 1.3). For high voltage measurements the resistive voltage multiplier provides no isolation for the voltmeter. For these reasons, range extension is generally provided by the use of current and voltage transformers. These enable single range ammeters and voltmeters, typically with FSDs of 5 A and 110 V, respectively, to be used.

The principle of the current transformer (ct) is shown in Figure 1.17(a) and its equivalent circuit is shown in Figure 1.17(b). The load current being measured flows through the primary winding whilst the ammeter acts as a secondary load. The operation of the ct depends upon the balance of the ampere turns (the product of current and turns) produced by the primary and secondary windings. If the transformer is ideal with no magnetizing current or iron loss then

$$\frac{I_P}{I_S} = n_{ct}$$

where n_{ct} is the current transformer turns ratio given by

$$n_{ct}=\frac{n_S}{n_P}$$

The ct is generally constructed with a toroidal core of a high-permeability, low-loss material such as mumetal or strip-wound silicon steel. This construction minimizes the magnetizing current, iron loss and leakage flux, ensuring that the actual primary to secondary current ratio is close to the inverse-turns ratio.

Figure 1.17(c) shows the effect of magnetizing current and iron loss on the relative magnitudes and phases of the primary and secondary currents.

Two errors of cts can be identified in Figure 1.17(c). These are the current or ratio error and the phase angle error or phase displacement. The current or ratio error is defined as

$$\frac{\text{Rated ratio } (I_P/I_S)-\text{actual ratio } (I_P/I_S)}{\text{Actual ratio } (I_P/I_S)}\times 100\%$$

The phase-angle error or phase displacement is the phase angle between the primary and secondary current phasors drawn in such a way (as in Figure 1.17(c)) that for a perfect transformer there is zero phase displacement. When the secondary current leads the primary current the phase displacement is positive.

These errors are expressed with respect to a particular secondary load which is specified by its burden and power factor. The burden is the VA rating of the instrument at full load current. A typical burden may be 15 VA with a power factor of 0.8 lagging. Figure 1.17(d) shows typical current and phase angle errors for a ct as a function of secondary load current. BS 3938:1973 sets limits on ratio and displacement errors for various classes of ct (British Standards Institution, 1973).

The ampere turn balance in the current transformer is destroyed if the secondary circuit is broken. Under these circumstances a high flux density results in the core which will induce a high voltage in the secondary winding. This may break down the insulation in the secondary winding and prove hazardous to the operator. It is therefore important not to open-circuit a current transformer whilst the primary is excited.

Voltage transformers (vts) are used to step down the primary voltage to the standard 110-V secondary voltage. Figure 1.18(a) shows the connection of such a transformer and Figure 1.18(b) shows its equivalent circuit. For an ideal transformer

$$\frac{V_P}{V_S}=n_{vt}$$

where n_{vt} is the voltage transformer turns ratio given by

$$n_{vt}=\frac{n_P}{n_S}$$

Figure 1.18(c) shows the phasor diagram of an actual voltage transformer. The two errors of voltage transformers are the voltage or ratio error and the phase-angle error or phase displacement.

The voltage error is defined to be

$$\frac{\text{Rated voltage ratio } (V_P/V_S) - \text{actual ratio } (V_P/V_S)}{\text{Actual voltage ratio } (V_P/V_S)}\times 100\%$$

The phase displacement is the phase displacement between the primary and secondary voltages as shown in Figure 1.18(c), and is positive if the secondary voltage leads the primary voltage. Figure 1.18(d) shows typical curves for the voltage ratio and phase-angle errors for a vt as a function of secondary voltage. BS 3941:1974 sets out specifications for voltage transformers (British Standards Institution, 1974).

Ratio errors are significant in cts and vts when they are used in current and voltage measurement. Both ratio errors and phase-angle errors are important when cts and vts are used to extend the range of wattmeters (see section 1.4).

1.2.4 Dynamometer instruments

The operation of the dynamometer instrument is shown in Figure 1.19. The instrument has two air- or iron-cored coil systems—one fixed and the other pivoted and free to rotate. The torque, T_g, generated by the interaction of the two currents is given by

$$T_g=\frac{dM}{d\theta}\,.\,i_1\,.\,i_2$$

and the restoring torque produced by the control springs is given by

$$T_r=k\,.\,\theta$$

Thus the deflection, θ, is given by

$$\theta=\frac{1}{k}\cdot\frac{dM}{d\theta}\,.\,i_1\,.\,i_2$$

Now if the same current flows through both coils then the steady-state deflection is proportional to the mean square of the currrent. Alternatively, if swamping resistances are employed the instrument can be used as a voltmeter. The scale of such instruments is usually calibrated in rms quantities and thus is non-linear. Air-cored instruments have no errors due to hysteresis effects, but the absence of an

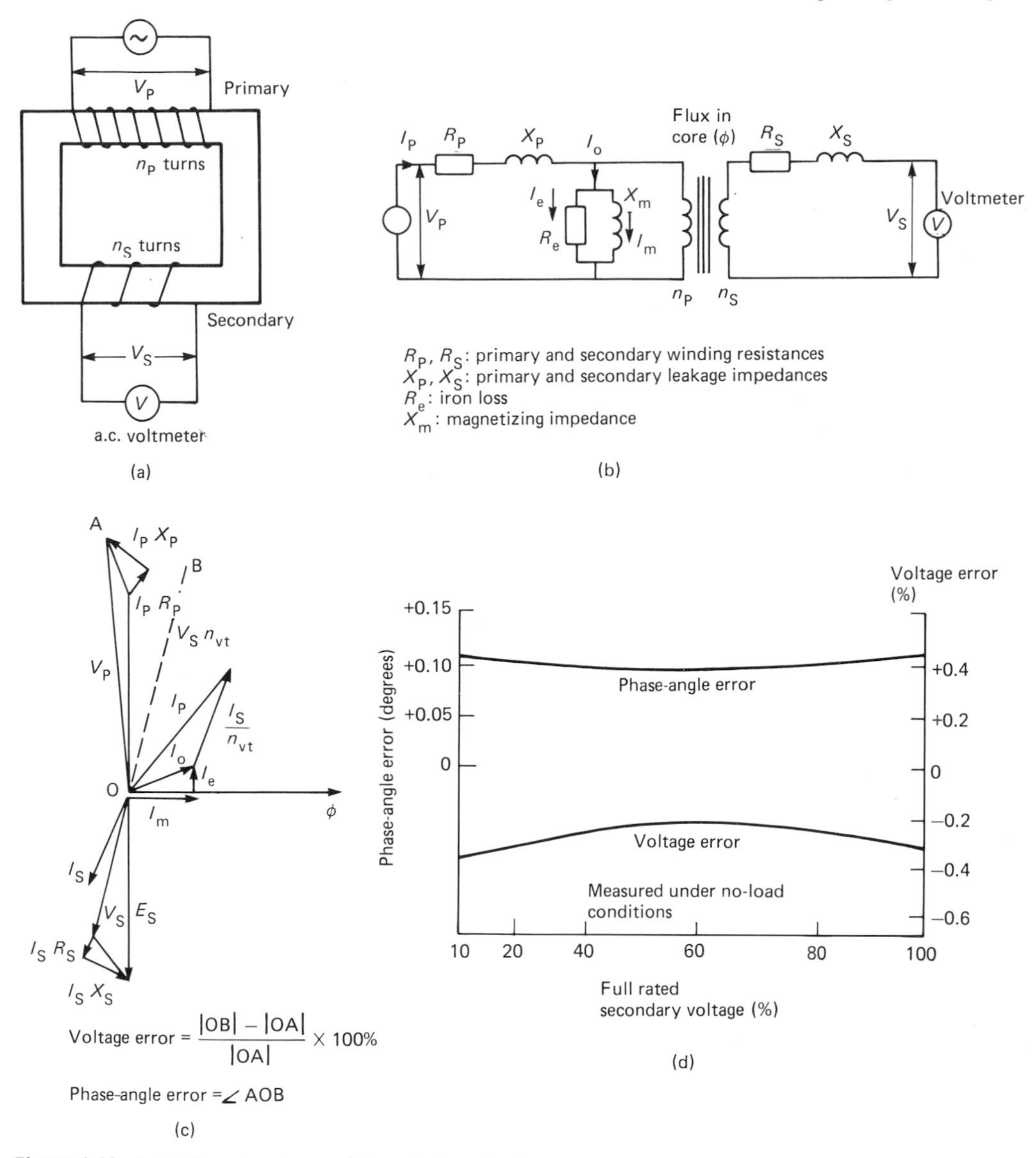

Figure 1.18 (a) Voltage transformer; (b) equivalent circuit; (c) phasor diagram of voltage transformer; (d) voltage and phase-angle errors for voltage transformer.

iron core requires the coils to have a large number of ampere turns to provide the necessary deflecting torque. This results in a high power loss to the circuit to which the instrument is connected. The torque-to-weight ratio is small and therefore friction effects are more serious, and the accuracy of these instruments can be affected by stray magnetic fields. Dynamometer instruments tend to be more expensive than other types of ammeter and voltmeter. The most important use of the dynamometer principle is in the wattmeter (see section 1.4.1).

1.2.5 Thermocouple instruments

Figure 1.20 shows the elements of a thermocouple instrument. These are a heating element which usually consists of a fine wire or a thin-walled tube in an evacuated glass envelope, a thermocouple having its hot junction in thermal contact with the heating element and a permanent magnet–moving coil millivoltmeter. Thermocouple instruments respond to the heating effect of the current passing through the heating element and are thus mean-square sensing devices and provide an indication which is independent of the current waveshape. They are capable of operating over a wide frequency range. At low frequencies (less than 10 Hz) their operation is limited by pointer vibration caused by the thermal response of the wire. At high frequencies (in excess of 10 MHz) their operation is limited by the skin effect altering the resistance of the heating element.

Thermocouple instruments have FSDs typically in the range 2–50 mA and are usually calibrated in rms values. The scale is thus non-linear. They are fragile and have only a limited over-range capability before the heating element is melted by overheating. The frequency range of the instrument as a voltmeter is limited by the ability to produce non-reactive series resistors.

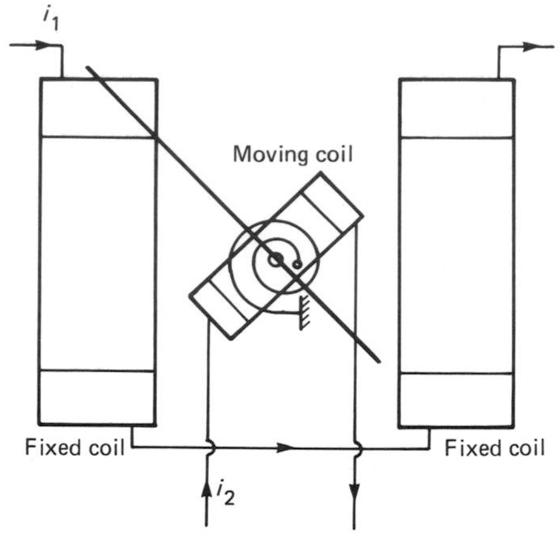

Figure 1.19 Dynamometer instrument.

1.2.6 Electrostatic instruments

Electrostatic instruments which may be used as voltmeters and wattmeters depend for their operation on the forces between two charged bodies. The torque between the fixed and moving vane in Figure 1.21(a) is given by

$$T = \frac{1}{2} \cdot \frac{dC}{d\theta} \cdot V^2$$

where C is the capacitance between the plates.

The usual form of the electrostatic voltmeter is the four-quadrant configuration shown in Figure 1.21(b). There are two possible methods of connection for such a voltmeter. These are the heterostatic and idiostatic connections shown in Figure 1.21(c). Commercial instruments usually employ the idiostatic connection, in which the needle is connected to one pair of quadrants. In this configuration the torque produced is proportional to the mean square value of the voltage. If the instrument is scaled to indicate the rms value then the scale will be non-linear. The torques produced by electrostatic forces are small and multicellular devices of the form shown in Figure 1.21(d) are used to increase the available torque. Multicellular instruments can be used for voltages in the range 100–1000 V. Electrostatic instruments have the advantage of a capacitive high input impedance. They are fragile and expensive, and therefore their use is limited to that of a transfer standard between a.c. and d.c. quantities.

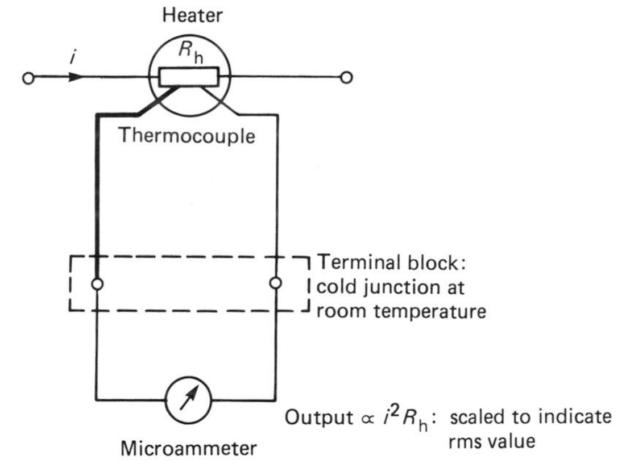

Figure 1.20 Thermocouple instrument.

1.3 Digital voltmeters and digital multimeters

Analogue indicating instruments provide a simple and relatively cheap method of indicating trends and changes in measured quantities. As voltmeters, direct indicating instruments have low input impedance. At best they provide only limited accuracy and this is achieved only with considerable skill on the part of the observer. Their speed of response is also slow. Digital instruments, in contrast, can provide high input impedance, high accuracy and resolution, and a high speed of measurement. They provide an indication to the observer which is free from ambiguity and requires no interpolation.

1.3.1 Analogue-to-digital conversion techniques

Fundamental to both digital voltmeters (DVMs), whose functions are limited to the measurement of d.c. and a.c. voltage, and digital multimeters (DMMs), whose functions may include voltage, current and resistance measurement, is an analogue-to-digital converter (ADC). ADCs are dealt with in detail in Part 4 and also in Owens (1983), Arbel (1980) and Sheingold (1977). In this section consideration is limited to the successive-approximation, dual-ramp and pulse-width techniques.

ADCs take an analogue signal whose amplitude can vary continuously and convert it into a digital form which has a discrete number of levels. The number of levels is fixed by the number of bits employed in the conversion and this sets the resolution of the conversion. For a binary code having N bits there are 2^N levels. Since the digital representation is discrete there is a range of analogue values which all have the same digital representation. Thus there is a quantization uncertainty of $\pm 1/2$ Least Significant Bit (LSB), and this is in addition to any other errors which may occur in the conversion itself. ADCs used in DVMs and DMMs are either sampling ADCs or integrating ADCs, as shown in Figure 1.22. Sampling ADCs provide a digital value equivalent to the voltage at one time instant. Integrating ADCs provide a digital value equivalent to the average value of the input over the period of the measurement. The successive-approximation technique is an example of a sampling ADC. The dual-ramp and pulse-width techniques described below are examples of integrating ADCs. Integrating techniques require a longer time to perform their measurement but have the advantage of providing noise- and line-frequency signal rejection.

1.3.1.1 Successive-approximation ADCs

This technique is an example of a feedback technique which employs a digital-to-analogue converter (DAC) in such a way as to find the digital input for the DAC whose analogue output voltage most closely corresponds to the input voltage which is to be converted. Detailed consideration of DACs is found in Part 4.

Figure 1.23(a) shows an N-bit $R-2R$ ladder network DAC. The output of this device is an analogue voltage given by

$$V_{out} = \frac{V_{ref}}{2^N} \sum_{n=0}^{N-1} a_n 2^n$$

where the a_n take values of either 1 or 0, dependent on the state of the switches, and $-V_{ref}$ is the reference voltage.

The successive-approximation technique shown in Figure 1.23(b) employs a decision-tree approach to the conversion problem. The control circuitry on the first cycle of the conversion sets the most significant bit of the DAC (MSB), bit a_{N-1}, to 1 and all the rest of the bits to 0. The output of the comparator is examined. If it is a 0, implying that the analogue input is greater than the output, then the MSB is maintained at a 1, otherwise it is changed to a 0. The next cycle determines whether the next most significant bit is a 1 or a 0. This process is repeated for each bit of the DAC. The conversion period for the successive-approximation ADC technique is fixed for a given ADC irrespective of the signal level and is equal to $N\tau$, where N is the number of bits and τ is the cycle time for determining a single bit. Integrated circuit successive-approximation logic-generating chips are available to be used in conjunction with standard DACs and comparators to produce medium-speed ADCs. A typical 8-bit ADC will have a conversion time of 10 μs. Successive-approximation ADCs are limited to 16 bits, equivalent to a five-decade conversion.

1.3.1.2 Dual-ramp ADCs

The dual-ramp conversion technique is shown in Figure 1.24 and operates as follows:

The input voltage, V_{in}, is switched to the input of the integrator for a fixed period of time t_1, after which the integrator will have a value of

$$\frac{-V_{in} \cdot t_1}{RC}$$

The reference voltage $-V_{ref}$ is then applied to the integrator and the time is then measured for the output of the integrator to ramp back to zero. Thus

$$\frac{V_{in} \cdot t_1}{RC} = \frac{V_{ref} \cdot t_2}{RC}$$

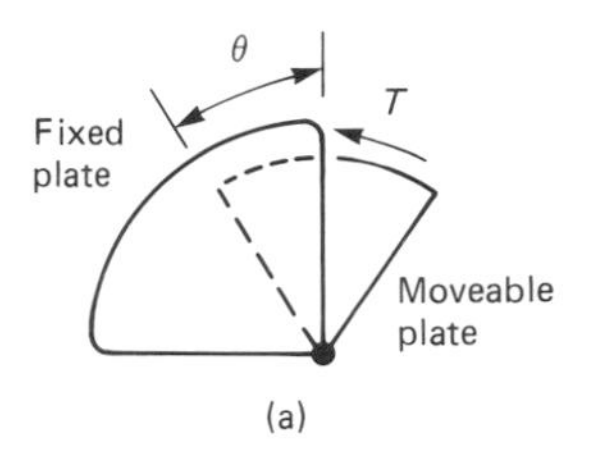

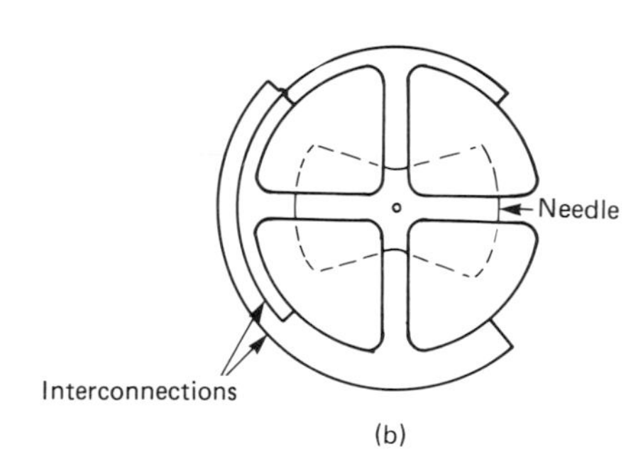

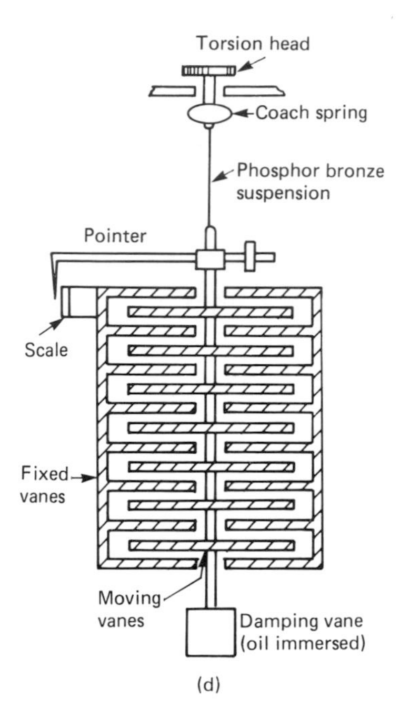

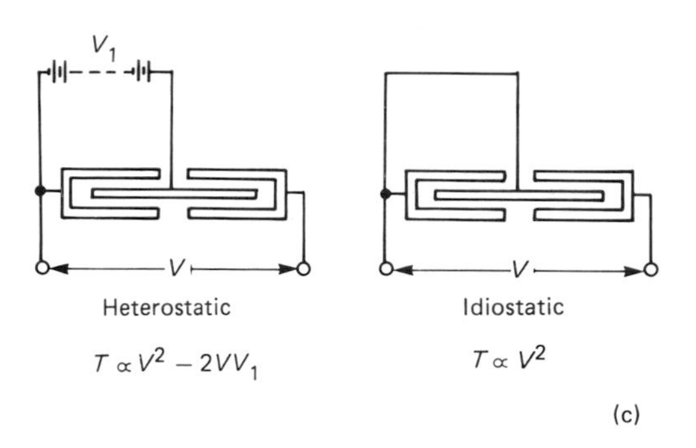

Figure 1.21 (a) Principle of electrostatic voltmeter; (b) four-quadrant electrostatic voltmeter; (c) heterostatic and idiostatic connections; (d) multicellular electrostatic voltmeter.

from which

$$\frac{t_2}{t_1}=\frac{V_{in}}{V_{ref}}$$

If t_1 corresponds to a fixed number of counts, n_1, of a clock having a period τ and t_2 is measured with the same clock, say, n_2 counts, then

$$n_2=\frac{V_{in}}{V_{ref}}\cdot n_1$$

The values of the R and C components of the integrator do not appear in the defining equation of the ADC neither does the frequency of the reference clock. The only variable which appears explicitly in the defining equation is the reference voltage. The effect of the offset voltage on the comparator will be minimized as long as its value remains constant over the cycle and also providing it exhibits no hysteresis. Modifications of the technique employing quad-slope integrators are available which reduce the effects of switch leakage current and offset voltage and bias current in the integrator to second-order effects (Analog Devices, 1984). Errors caused by non-linearity of the integrator limit the conversion using dual-ramp techniques to five decades.

The dual-ramp conversion technique has the advantage of line-frequency signal rejection

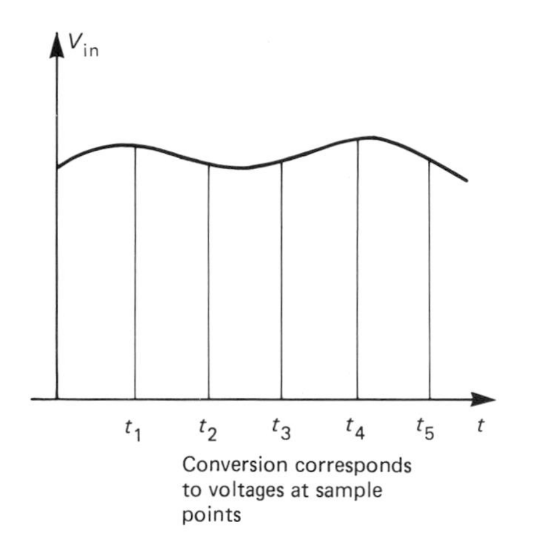

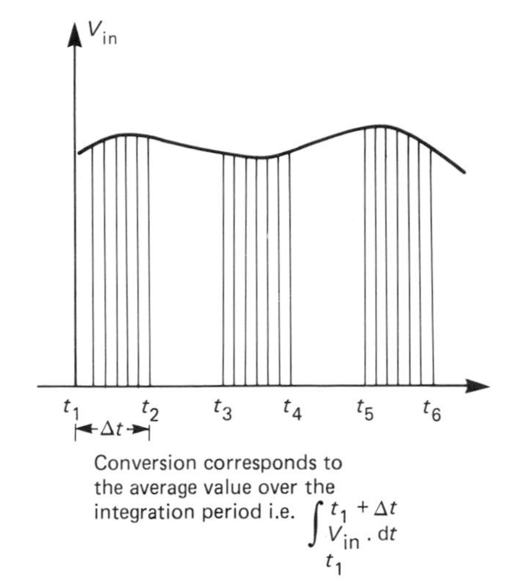

Figure 1.22 Sampling and integrating ADCs.

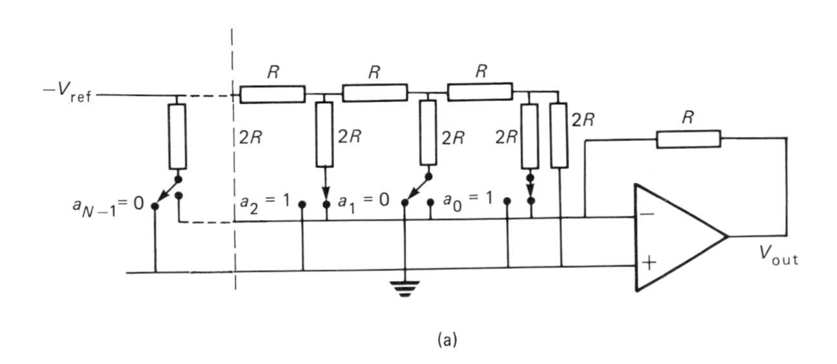

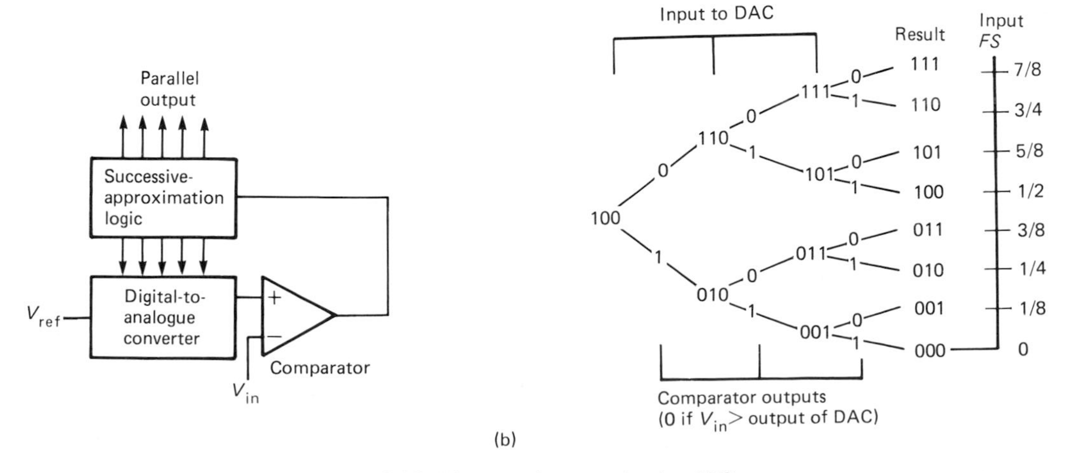

Figure 1.23 (a) $R-2R$ ladder network DAC; (b) successive-approximation ADC.

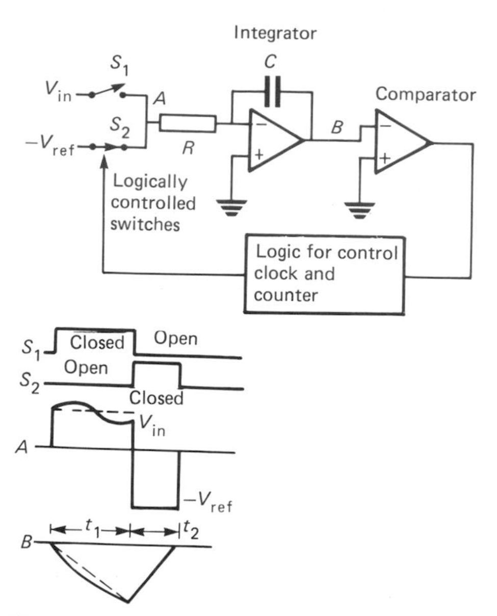

Figure 1.24 Dual-slope ADC.

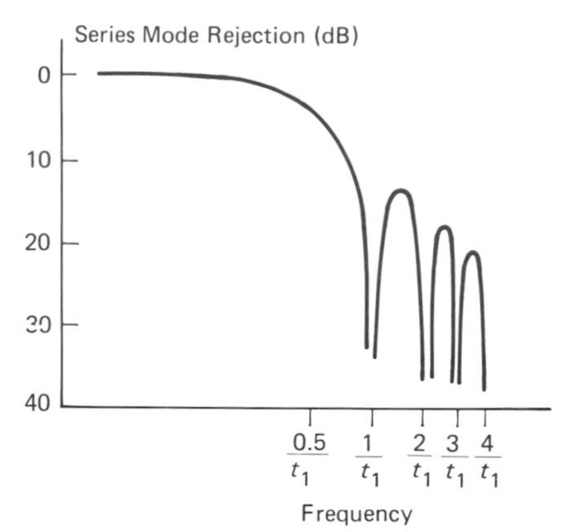

Figure 1.25 Series Mode Rejection for dual-slope ADC.

(Gumbrecht, 1972). If the input is a d.c. input with an a.c. interference signal superimposed upon it,

$$V_{in} = V_{d.c.} + V_{a.c.} \sin(\omega t + \phi)$$

where ϕ represents the phase of the interference signal at the start of the integration, then the value at the output of the integrator, V_{out}, at the end of the period t_1, is given by

$$V_{out} = \frac{-V_{d.c.}t_1}{RC} - \frac{1}{RC}\int_0^{t_1} V_{a.c} \sin(\omega t + \phi)$$

If the period t_1 is made equal to the period of the line frequency then the integral of a line-frequency signal or any harmonic of it over the period will be zero, as shown in Figure 1.24. At any other frequency it is possible to find a value of ϕ such that the interference signal gives rise to no error. It is also possible to find a value ϕ_{max} such that the error is a maximum. It can be shown that the value of ϕ_{max} is given by

$$\tan \phi_{max} = \frac{\sin \omega t_1}{(1 - \cos \omega t_1)}$$

The series or normal mode rejection of the ADC is given as the ratio of the maximum error produced by the sine wave to the peak magnitude of the sine wave. It is normally expressed (in dBs) as

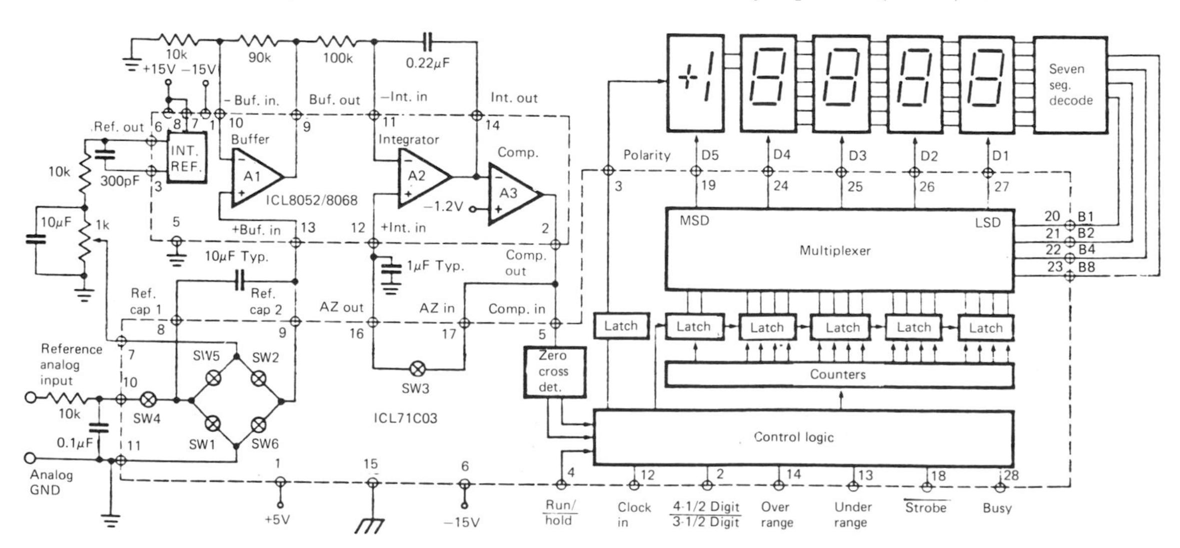

(a)

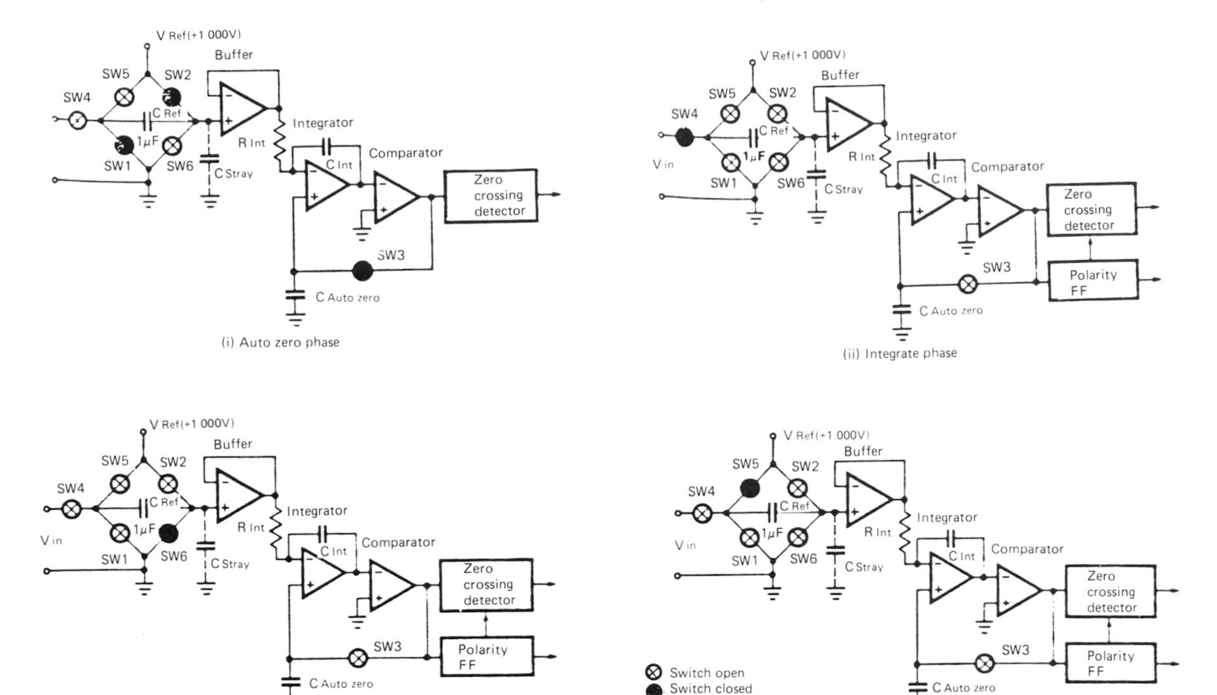

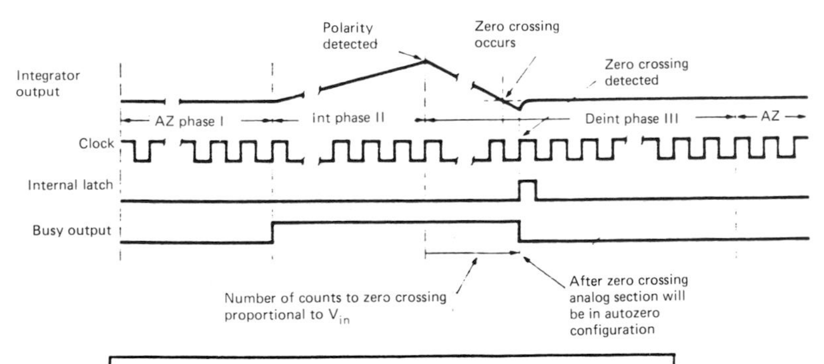

	Counts		
	Phase I	Phase II	Phase III
4-1/2 digit	10 001	10 000	20 001
3-1/2 digit	1 001	1 000	2 001

(c)

Figure 1.26 Dual-slope integrated-circuit chip set. (a) Function block diagram; (b) operating phase of converter (courtesy Intersil Datel (UK) Ltd); (c) timing.

Series Mode Rejection (SMR)

$$= -20\log_{10}\frac{\omega t_1}{\cos\phi_{\max}-\cos(\omega t_1+\phi_{\max})}$$

A plot of the SMR of the dual-slope ADC is shown in Figure 1.25. It can be seen that ideally it provides infinite SMR for any frequency given by n/t_1, $n=1, 2, 3, \ldots$. Practically, the amount of rejection such an ADC can provide is limited because of non-linear effects, due to the fact that the period t_1 can only be defined to a finite accuracy and that the frequency of the signal to be rejected may drift. However, such a technique can easily provide 40 dB of line-frequency rejection.

Figure 1.26 shows a schematic diagram of a commercially available dual-slope integrated-circuit chip set.

1.3.1.3 Pulse-width ADCs

A simple pulse-width ADC is shown in schematic form in Figure 1.27. The ADC employs a voltage-controlled monostable to produce a pulse whose width is proportional to the input voltage. The width of the pulse is then measured by means of a reference clock. Thus the counter has within it at the end of the conversion period a binary number which corresponds to the analogue input. The accuracy of the technique depends on the linearity and stability of the voltage to pulse-width converter and the stability of the reference clock. High-speed conversion requires the use of a high-frequency clock. By summing the counts over longer periods of time the effect of line frequency and noise signals can be integrated out.

A modified pulse-width technique for use in precision voltmeters is shown in Figure 1.28 (Pitman, 1978; Pearce, 1983). Precision pulses generated by chopping +ve and −ve reference voltages are fed into the input of an integrator which is being forced to ramp up and down by a square wave. The ramp waveform applied to the two comparators generates two pulse trains which are used to gate the reference voltages. In the absence of an input voltage, feedback ensures that the width of the +ve and −ve pulses will be equal. The outputs of the comparators are fed to an up–down counter. For the duration of the +ve pulse the counter counts up and during the −ve pulses it counts down. Thus ideally with no input the count at the end of the integration period will be zero.

If an input is applied to the integrator the width of the +ve and −ve pulse widths are adjusted by the

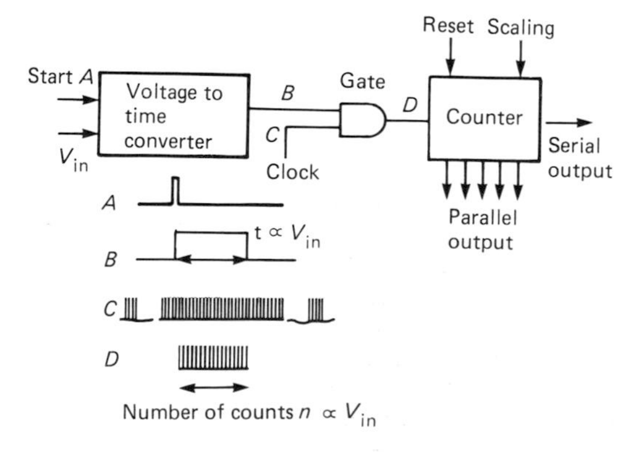

Figure 1.27 Pulse-width ADC.

feedback mechanism, as shown in Figure 1.28. If the period of the square wave is approximately 312 μs and the clock runs at approximately 13 MHz then it is possible to provide a reading with a resolution of 1 part in 4000 over a single period. Figure 1.29 shows the variation of the pulse widths for a time-varying input. By extending the integration period to 20 ms the resolution becomes 1 part in 260 000 and significant rejection of 50-Hz line frequency is achieved. The method allows trading to occur between resolution and speed of measurement.

1.3.1.4 Voltage references in ADCs

In all ADCs a comparison is made with some reference voltage. Therefore in order to have accurate conversion it is necessary to have an accurate and stable voltage source. Most digital voltmeters use zener diodes to generate their reference voltages, although high-precision devices often have facilities to employ standard cells to provide the reference voltage. Two types of zener devices are commonly used. The compensated zener diode is a combination of a zener junction and a forward-biased junction in close proximity to the zener junction so that the temperature coefficient can be set to a few ppm by choosing the correct zener current. Active zener diodes have a temperature controller built into the silicon chip around the zener diode. Table 1.6 (taken from Spreadbury, 1981) compares the characteristics of these two types of zener devices with bandgap devices and the Weston standard cell.

1.3.2 Elements in DVMs and DMMs

The ADC is the central element of a DVM or DMM. The ADC is, however, a limited input range device operating usually on unipolar d.c. signals. Figure 1.30 shows the elements of a complete DVM or DMM.

1.3.2.1 D.C. input stage and guarding

The d.c. input stage provides high input impedance together with attenuation/amplification and polarity sensing of the signal to ensure that the voltage applied to the ADC is of the correct magnitude and polarity.

DVMs and DMMs are often used to measure small d.c. or a.c. signals superimposed on much larger common-mode signals. For example, in measuring the output signal from a d.c. Wheatstone bridge, as shown in Figure 1.31(a), the common-mode voltage is half the bridge supply. If a transducer is situated some distance away from its associated DVM the common-mode signal may be generated by line-frequency ground currents as shown in Figure 1.31(b)

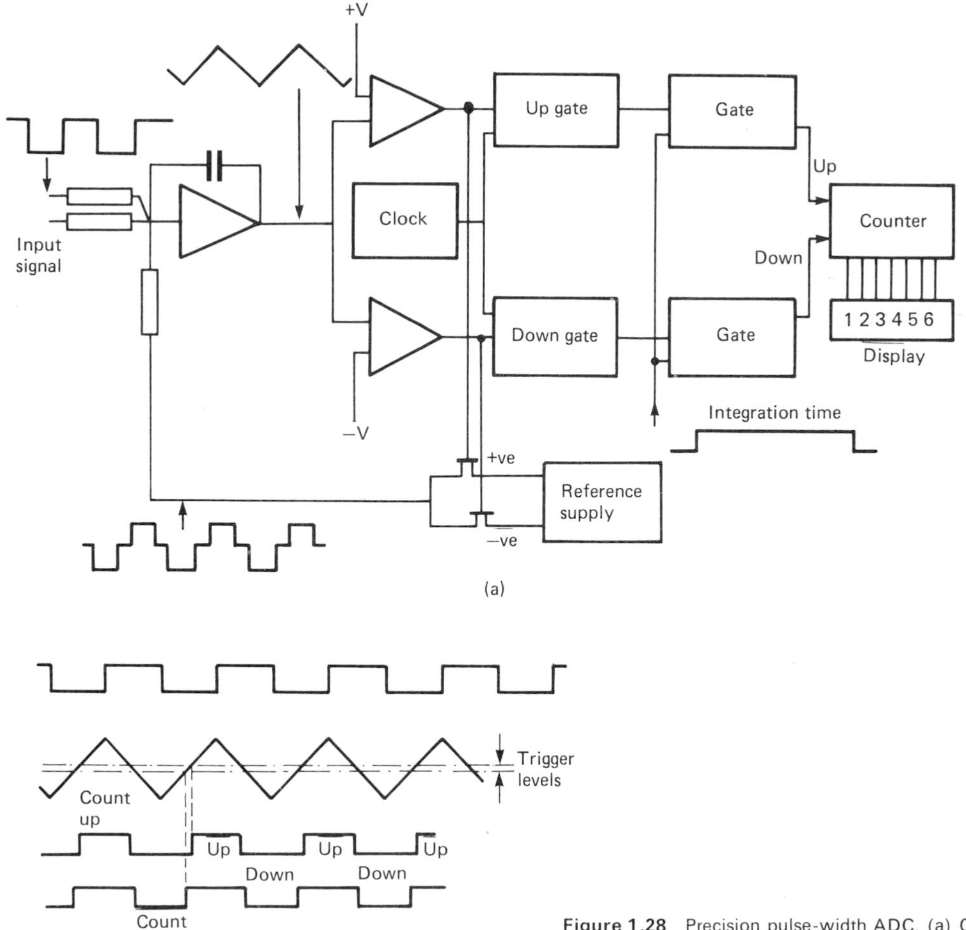

Figure 1.28 Precision pulse-width ADC. (a) Circuit; (b) timing (courtesy Solartron Instruments Ltd).

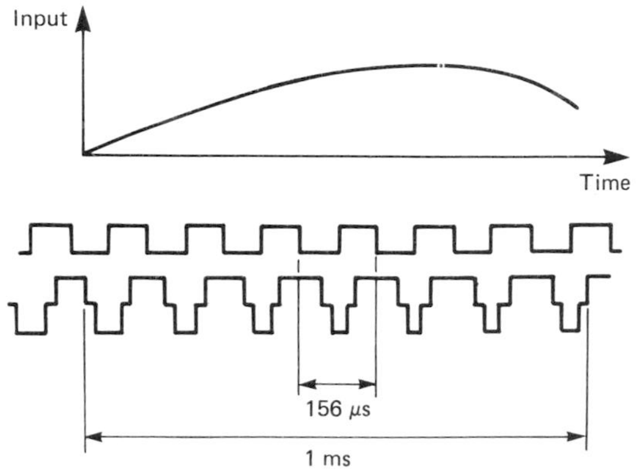

Figure 1.29 Effect of time-varying input on pulse-width ADC (courtesy Solartron Instruments Ltd).

and thus the potential to be measured may be superimposed on an a.c. line-frequency common-mode signal. Figure 1.31(c) shows the equivalent circuit for the measurement circuit and the input of the DVM or DMM. R_A and R_B represent the high and low side resistances of the measurement circuit, R_{in} the input resistance of the DVM or DMM and R_i and C_i the leakage impedance between the low terminal of the instrument and power ground. The leakage impedance between the high terminal and the instrument ground can be neglected because the high side is usually a single wire whereas the low side often consists of a large metal plate or plane. The divider consisting of R_B and R_i and C_i converts common-mode signals to input signals. Typically R_i is $10^9\ \Omega$ and C_i may be as high as 2.5 nF. For specification purposes

Table 1.6 Reference voltage sources

	Weston cell	*Compensated Zener*	*Active Zener*	*Bandgap device*
Stable level, *V*	1.018	6.4	7	1
Temperature coefficient parts in 10^6 per deg C	−40	1	0.2	30
Internal resistance	500 Ω	15 Ω at 7.5 mA	½ Ω at 1 mA	½ Ω
	(in all cases, with op. amp. can be reduced to 0.001 Ω)			
Ageing, parts in 10^6 per year	0.1 to 3	2 to 10	20	100
Noise, μV rms	.0.1	1	7	6

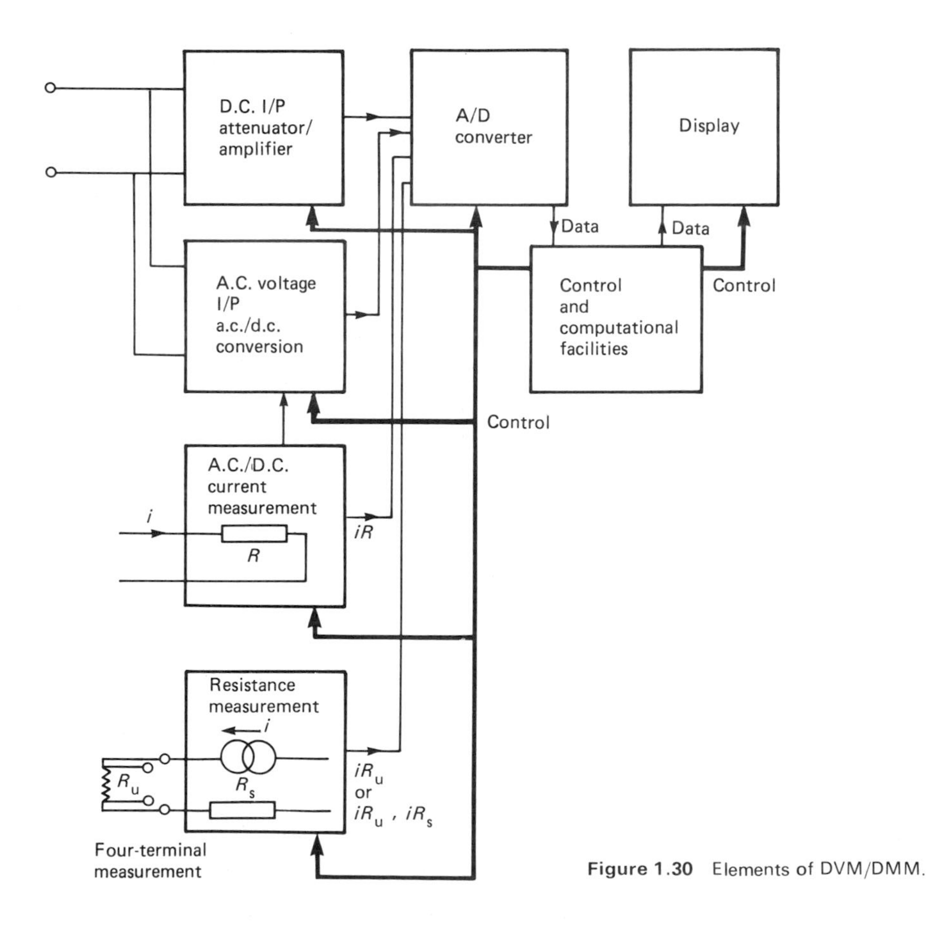

Figure 1.30 Elements of DVM/DMM.

R_A is taken as zero and R_B is taken as 1 kΩ. Thus at d.c. the common-mode rejection is −120 dB and at 50 Hz it is −62 dB.

The common-mode rejection can be improved by the addition of an input guard. This is shown in Figure 1.31(d) and can be considered as the addition of a metal box around the input circuit. This metal box is insulated both from the input low and the power ground. It is available as a terminal of the input of the instrument. If the guard is connected to the low of the measurement circuit then the effect of current flow between the low terminal and guard is eliminated since they are at the same potential. The potential dividing action now occurs between the residual leakage

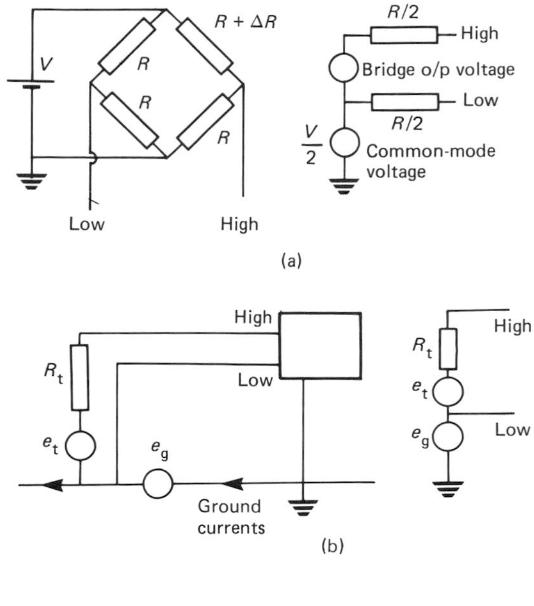

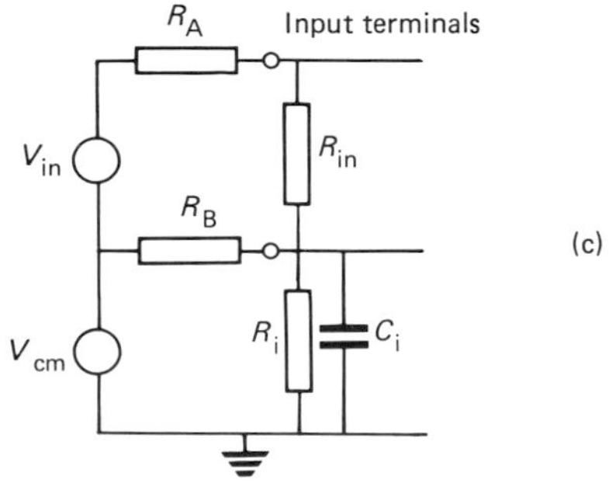

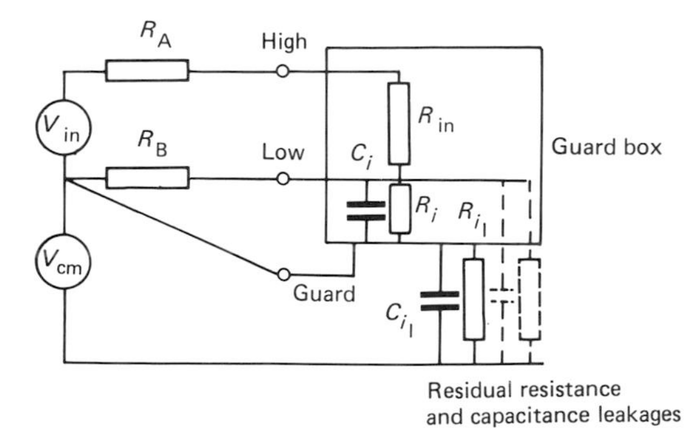

Figure 1.31 (a) Common-mode signals in bridge measurements; (b) ground current-generated common-mode signals; (c) input equivalent circuit; (d) input guarding.

impedance between low and power ground in the presence of the guard. The value of these leakage impedances are of order 10^{11} Ω and 2.5 pF. The d.c. common-mode rejection has now been increased to −160 dB and the 50 Hz common-mode rejection to −122 dB. Thus a d.c. common-mode signal of 100 V will produce an input voltage of 1 μV and a 20-V, 50 Hz common-mode signal will produce an input of less than 20 μV.

In situations where there is no common-mode signal the guard should be connected to the signal low, otherwise unwanted signals may be picked up from the guard.

1.3.2.2 a.c./d.c. conversion

Two techniques are commonly used in a.c. voltage and current measurement using digital instruments. Low-cost DVMs and DMMs employ a mean absolute value measurement–rms indicating technique similar to that employed in a.c. current and voltage measurement using a permanent magnet–moving coil instrument. By the use of operational techniques as shown in Figure 1.32 the effect of the forward diode drop can be reduced and thus precision rectification can be achieved. However, because the instrument is then not rms sensing but relies on the waveform being sinusoidal for correct indication this technique suffers from the Form Factor errors shown in section 1.2.1.3.

True rms measurement can be obtained either by use of analogue electronic multipliers and square-root extractors, as shown in Figure 1.33(a), or by the use of thermal converters, as shown in Figure 1.33(b). High-precision instruments employ vacuum thermocouples to effect an a.c./d.c. transfer. Brodie (1984) describes an a.c. voltmeter using such a technique which provides a measurement accuracy of 160 ppm for any signal level from 100 mV to 125 V in a frequency band from 40 Hz to 20 kHz. This voltmeter is capable of measuring over a range from 12.5 mV to 600 V in a frequency band from 10 Hz to 1 MHz with reduced accuracy.

In true rms sensing instruments the manufacturer often specifies the maximum permissible crest factor for the instrument. The crest factor is the ratio of the peak value of the periodic signal to its rms value. Typically the maximum permissible crest factor is 5.

1.3.2.3 Resistance and current measurement

Resistance measurement is provided by passing a known current through the resistor and measuring the voltage drop across it. Four-terminal methods, as shown in Figure 1.30, enable the effect of lead resistance to be reduced. High-precision DMMs employ ratiometric methods in which the same current is passed through both the unknown resistance and a standard resistance, and the unknown resistance is computed from the ratio of the voltages developed across the two resistances and the value of the standard resistor. a.c. and d.c. current measurements

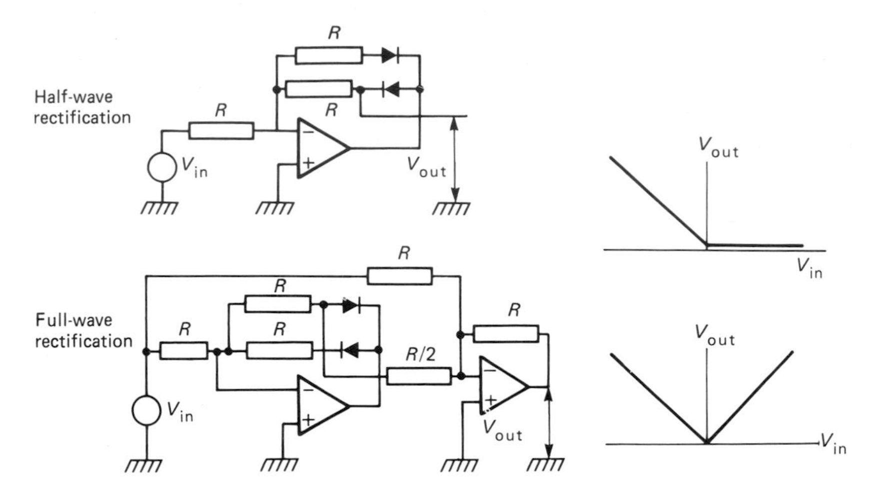

Figure 1.32 A.C. signal precision rectification.

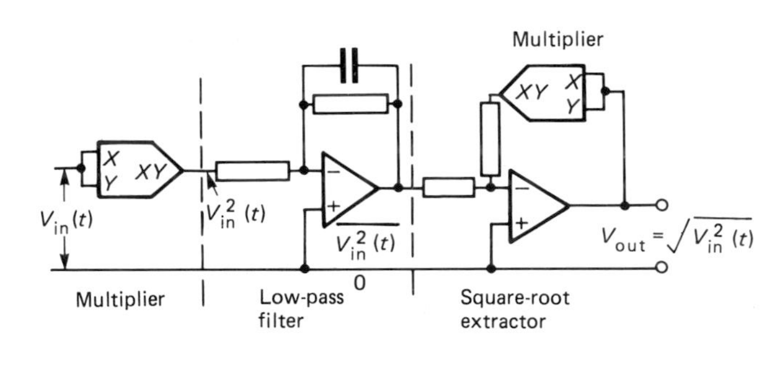

(a)

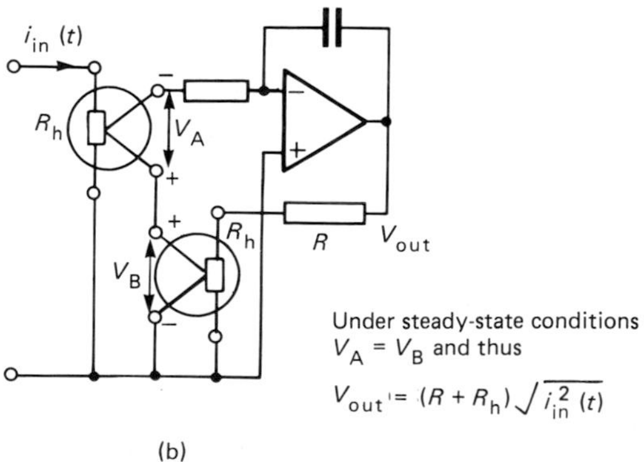

(b)

Figure 1.33 (a) RMS evaluation by analogue multiplication; (b) a.c./d.c. conversion using thermal techniques.

use a shunt across which a voltage is developed. This voltage is then measured by the ADC.

1.3.2.4 Control and post-measurement computational facilities

The control element in DVMs and DMMs is increasingly provided by a microprocessor. The use of the microprocessor also enables the digital instrument to provide the user with a large range of post-measurement storage and computational facilities. These may include:

(1) The collection and storage of a set of readings with a given time interval between readings.
(2) The application of scaling and offset calculations to the readings to provide an output of the form

$y = mx + c$, where x is the reading and m and c are constants input by the operator. This enables the measured value to be output in engineering units.

(3) Testing readings to ascertain whether they are within preset limits. In this mode the instrument may either display hi–lo-pass or may count the number of readings in each category.
(4) The calculation and display of the percentage deviation from a given set point input by the operator.
(5) Calculation of the ratio of the measured value to some value input by the operator.
(6) Storing the maximum and minimum value of the measured variable.
(7) Generating statistical data from a given set of measurements to provide the sample average, standard deviation, variance or rms value.
(8) Digital filtering of the measured variable to provide a continuous average, an average over n readings or a walking window average over n readings.

1.3.2.5 Output

The visual display of DVMs and DMMs is commonly provided by light-emitting diodes (LEDs) or liquid crystal displays (LCDs). The relative merits of each of these displays is considered in the section on displays in Volume 4. If the results are to be communicated to further digital systems the output may be provided as either a parallel binary or binary-coded decimal (BCD) output. Many DVMs and DMMs are fitted with the standard IEEE-488 or RS232 parallel or serial interfaces which allow data and control to pass between the instrument and a host control computer. The characteristics of IEEE-488 and RS232 interfaces are considered in Part 4.

1.3.3 DVM and DMM specifications

DVMs and DMMs cover a wide range of instruments from hand-held, battery-operated multimeters, through panel meters and bench instruments to standards laboratory instruments. These digital instruments are specified primarily by their resolution, accuracy and speed of reading. The resolution of the instrument, which may be higher than its accuracy, corresponds to the quantity indicated by a change in the least significant digit of the display. Typically, digital instruments have displays which are between $3\frac{1}{2}$ and $8\frac{1}{2}$ digits. The half digit indicates that the most significant digit can only take the value 1 or 0. Thus a $3\frac{1}{2}$-digit instrument has a resolution of 1 part in 2000 and an $8\frac{1}{2}$-digit one has a resolution of 1 part in 2×10^8. The accuracy of the instrument is specified as $\pm(x$ per cent of reading $(R) + y$ per cent of scale $(S) + n$ digits).

Table 1.7 gives condensed specifications for comparison of a hand-held $3\frac{1}{2}$-digit DMM, a $5\frac{1}{2}$-digit intelligent multimeter and an $8\frac{1}{2}$-digit standards laboratory DVM. The accuracies quoted in Table 1.7 are only for guidance, and for complete specifications the reader should consult the specification provided by the manufacturer.

1.4 Power measurement

For a two-terminal passive network if the instantaneous voltage across the network is $v(t)$ and the instantaneous current through it is $i(t)$ then the instantaneous power, $p(t)$, taken or returned to the source is given by

$$p(t) = v(t) \, . \, i(t)$$

For a linear network, if $v(t)$ is sinusoidal, i.e.

$$v(t) = \hat{V} \, . \sin \omega t$$

then $i(t)$ must be of the form

$$i(t) = \hat{I} \sin(\omega t + \phi)$$

and the instantaneous power, $p(t)$, is given by

$$p(t) = v(t) \, . \, i(t) = \hat{V}\hat{I} \sin \omega t \sin(\omega t + \phi)$$

The average power dissipated by the network is given by

$$P = \frac{1}{T} \int_0^T p(t) \, . \, \mathrm{d}t$$

where T is the period of the waveform and thus

$$P = \frac{\omega}{2\pi} \int_0^{2\pi/\omega} \hat{V}\hat{I} \sin \omega t \sin(\omega t + \phi) \, . \, \mathrm{d}t$$

Therefore P is given by

$$P = \frac{\hat{V}\hat{I}}{2} \, . \cos \phi$$

The rms voltage, V, is given by

$$V = \frac{\hat{V}}{\sqrt{2}}$$

and the rms current, I, is given by

$$I = \frac{\hat{I}}{\sqrt{2}}$$

Thus the average power dissipated by the network is given by

$$P = VI \cos \phi$$

($\cos \phi$ is known as the power factor).

Table 1.7 Comparison of digital voltmeter specifications

	3½-digit multimeter (Fluke 8026B)	*5½-digit intelligent multimeter (Thurlby 1905A)*	*8½-digit Standards Laboratory DVM (Solartron 7081)*
D.C. voltage ranges	199.9 mV–1000 V	210.000 mV–1100.00 V	0.1 V–1000 V
Typical accuracy	±(0.1 %R + 1 digit)	±(0.015 %R + 0.0015 %S + 2 digits)	Short-term stability ±(1.2 ppm R + 0.3 ppm S)
Input impedance	10 MΩ on all ranges	>1 GΩ on lowest two ranges 10 MΩ on remainder	>10 GΩ on 3 lowest ranges 10 MΩ on remainder
A.C. voltage ranges	199.9 mV–750 rms	210.00 mV–750 V rms	0.1 V–1000 V rms
Type	True rms sensing crest factor 3:1	Mean sensing/rms calibrated for sinusoid	True rms sensing crest factor 5:1 short-term stability
Typical accuracy	±(1 %R + 3 digits)	±(2 %R + 10 digits)	±(0.05 %R + 0.03 %S)
Frequency range	45 Hz–10 kHz	45 Hz–20 kHz	10 Hz–100 kHz
Input impedance	10 MΩ‖100 pF	10 MΩ‖47 pF	1 MΩ‖150 pF
D.C. current ranges	1.999 mA–1.999 A	210.000 μA–2100.00 mA	
Typical accuracy	±(0.75 %R + 1 digit)	±(0.1 %R + 0.0015 %S + 2 digits)	
Voltage burden	0.3 V max. on all ranges except 1.999 A range. Max. burden on 1.999 A range 0.9 V	0.25 V max. on all ranges except 2100 mA range. Max. burden on 2100 mA range 0.75 V	Not applicable
A.C. current ranges	1.999 mA–1.999 A	210.00 μA–2100.0 mA	
Type	True rms sensing crest factor 3:1	Mean sensing/rms calibrated for sinusoid	
Typical accuracy	±(1.5 %R + 2 digits)	±(0.3 %R + 5 digits)	Not applicable
Frequency range	45 Hz–1 kHz	45 Hz–500 Hz	
Voltage burden	0.3 V max. on all ranges except 1.999 A range. Max. burden on 1.999 A range 0.9 V	0.25 V max. on all ranges except 2100.0 mA range. Max. burden on 2100.0 mA range 0.75 V	
Resistance ranges	199.9 Ω–19.99 MΩ	210.000 Ω–21.000 MQ	0.1 kΩ–1000 MΩ
Typical accuracy	±(0.1 %R + 1 digit)	±(0.04 %R + 0.0015 %S + 2 digits)	Short-term stability (2 ppm R + 0.4 ppm S)
Current employed	Max. current 0.35 mA on 199.9 Ω range	Max. current 1 mA on 210.000 Ω range	Max. current 1 mA on 0.1, 1 and 10 kΩ ranges
Speed of reading		3 per second	100 per second to 1 per 51.2 s
Common-mode rejection ratio	>100 dB at d.c., 50 Hz, and 60 Hz with 1 kΩ unbalance for d.c. ranges >60 dB for 50 and 60 Hz with 1 kΩ unbalance on a.c. ranges	>120 dB at d.c. or 50 Hz	Effective CMR [CMR + SMR] with 1 kΩ unbalance 5½–8½ digit >140 dB at 50(60) Hz >120 dB at 400 Hz for a.c. measurement >40 dB at 50(60) Hz
Series mode rejection	>60 dB at 50 Hz or 60 Hz	>60 dB at 50 Hz	5½–8½ digits >70 dB at 50(60) or 400 Hz

continued

Table 1.7 *Continued*

Additional notes	Battery operated with LCD display Also provides conductance measurement and continuity testing	LED display Intelligent functions include: scaling and offsetting, percentage deviation, low–hi-pass, max–min, filtering, averaging and data logging RS232 and IEEE-488 interfaces True rms option available	LED display Intelligent functions include: ratio, scaling and offsetting, digital filtering, statistics, limits, time—real or elapsed, history file with 1500 numeric readings or 500 readings with time and channel mode RS232 and IEEE-488 interfaces

1.4.1 The three-voltmeter method of power measurement

By using a non-inductive resistor and measuring the three voltages shown in Figure 1.34(a) it is possible to measure the power dissipation in the load without using a wattmeter. Figure 1.34(b) shows the phasor diagram for both leading and lagging power factors.

From the phasor diagram by simple trigonometry,

$$V_A^2 = V_B^2 + V_C^2 + 2V_B V_C \cos\phi$$

and

$$V_B = IR$$

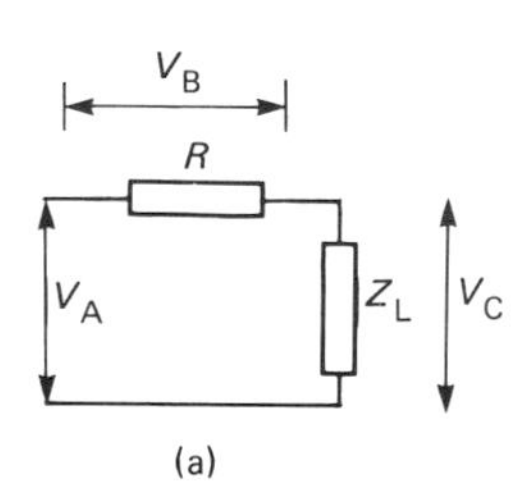

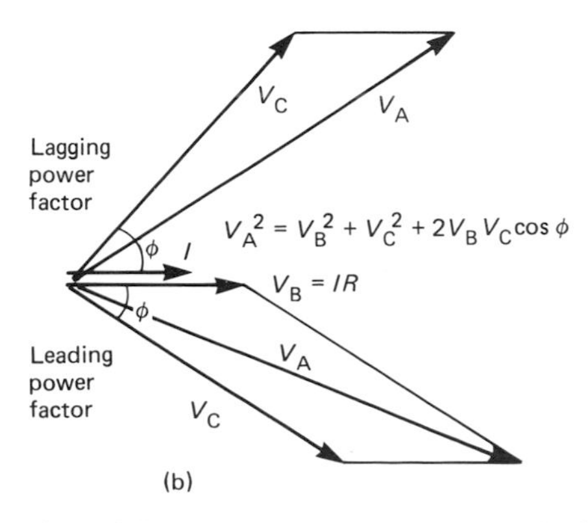

Figure 1.34 (a) Power measurement using the three-voltmeter method; (b) phasor diagram for the three-voltmeter method.

Since the average power dissipated in the load is given by

$$P = V_C I \cos\phi$$

then

$$P = \frac{V_A^2 - V_B^2 - V_C^2}{2R}$$

and the power factor $\cos\phi$ is given by

$$\cos\phi = \frac{V_A^2 - V_B^2 - V_C^2}{2V_B V_C}$$

1.4.2 Direct-indicating analogue wattmeters

Direct-indicating analogue wattmeters employ the dynamometer, induction, electrostatic or thermocouple principles. These are shown in Figures 1.35 and 1.36. Of these, the dynamometer is the most commonly used. In the dynamometer wattmeter shown in Figure 1.35(a) the current into the network is passed through the fixed coils whilst the moving coil carries a current which is proportional to the applied voltage. The series resistance in the voltage coil is non-inductive. The restoring torque is provided by a spring thus the mean deflection of the wattmeter from section 1.2.4 is given by

$$\theta = \frac{1}{k} \cdot \frac{1}{R_s} \cdot \frac{dM}{d\theta} \cdot V \cdot I \cdot \cos\phi$$

The primary errors in dynamometer wattmeters occur as a consequence of magnitude and phase errors in the voltage coil and power loss in the wattmeter itself. Other errors are caused by the capacitance of the voltage coil and eddy currents.

If the resistance and inductance of the voltage coil are R_V and L_V respectively and if R_S is the resistance in series with the voltage coil then the current through the voltage coil at an angular frequency ω has a

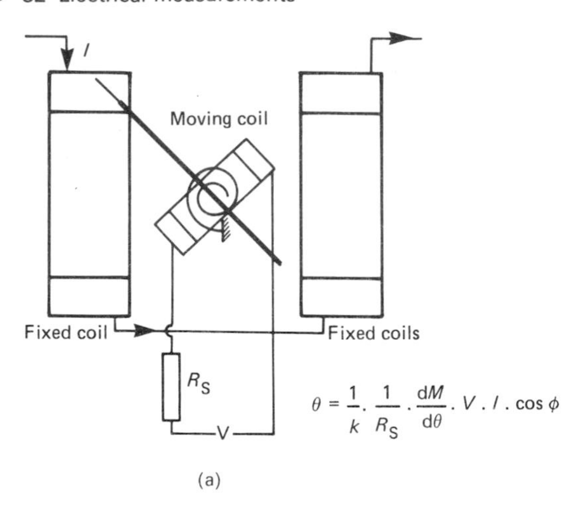

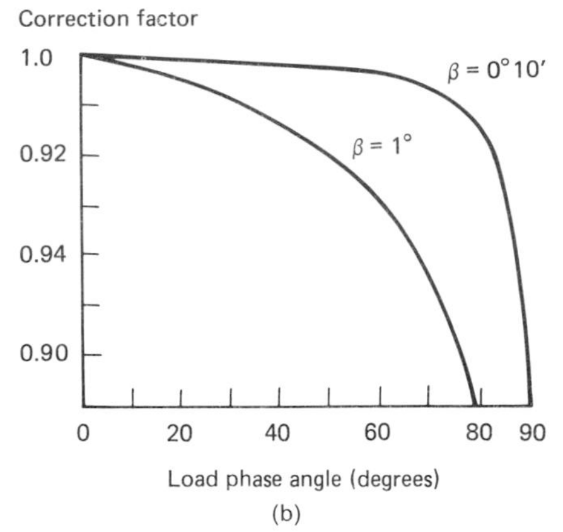

Figure 1.35 (a) Dynamometer wattmeter; (b) Wattmeter correction factors

magnitude given by

$$I_V = \frac{V}{\sqrt{[(R_V+R_S)^2+\omega^2 L_V^2]}}$$

with a phase angle, β, given by

$$\beta = \tan^{-1}\frac{\omega L_V}{(R_V+R_S)} \simeq \frac{\omega L_V}{(R_V+R_S)}$$

Thus altering the frequency alters both the sensitivity and phase angle of the voltage coil.

If the load circuit has a lagging power factor, $\cos\phi$, the wattmeter true indication will be

$$\frac{\cos\phi}{\cos\beta . \cos(\phi-\beta)} \times \text{actual indication}$$

and the error as a percentage of actual indication will be

$$\frac{\sin\beta}{(\cos\phi+\sin\beta)} \times 100\%$$

The wattmeter reads high on lagging power factors. Figure 1.35(b) shows the correction factors for $\beta = 1°$ and $\phi = 0°10'$.

The induction wattmeter in Figure 1.36(a) operates on a principle similar to the shaded pole induction

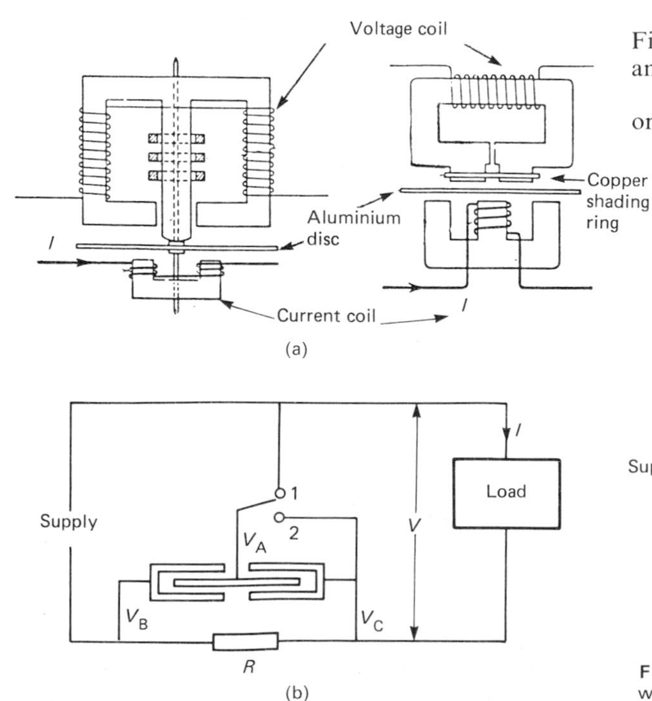

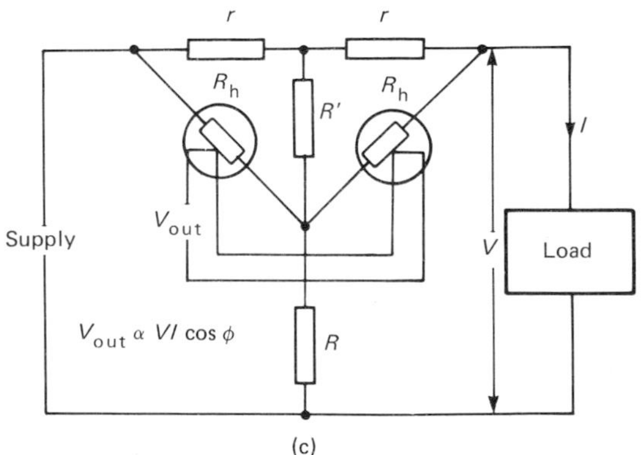

Figure 1.36 (a) Induction wattmeter; (b) electrostatic wattmeter; (c) thermocouple wattmeter

watt–h meter described in section 1.5 in that the torque is generated by the interaction of eddy currents induced in a thin aluminium disc with the imposed magnetic fields. The average torque generated on the disc is proportional to the average power. In the induction wattmeter the generated torque is opposed by a spring and thus it has a scale which can be long and linear.

In the electrostatic wattmeter shown in Figure 1.36(b) with the switch in position 1 the instantaneous torque is given by

$$T \propto (v_A - v_B)^2 - (v_A - v_C)^2$$

and thus

$$T \propto 2R\left(v \, . \, i + \frac{i^2R}{2}\right)$$

where v and i are the instantaneous load voltage and current, respectively.

If this torque is opposed by a spring then the average deflection will be given by

$$\theta \propto 2R\left(VI \cos \phi + \frac{I^2R}{2}\right)$$

i.e. the average power dissipated in the load plus half the power dissipated in R.

With the switch in position 2 the instantaneous torque is given by

$$T \propto (v_A - v_B)^2$$

and the average deflection will be given by

$$\theta \propto R(I^2R)$$

i.e. the power dissipated in R.

Thus from these two measurements the power in the load can be computed.

In the compensated thermal wattmeter employing matched thermocouples as shown in Figure 1.36(c) the value of the resistance R' is chosen such that

$$R' = \frac{(R_h + r) \, . \, R}{r}$$

The output of the wattmeter can then be shown to be given by

$$V_{out} = \frac{k \, . \, r}{(r + R_h)(r + R)} \, . \, V \, . \, I \, . \cos \phi$$

where k is a constant of the thermocouples.

In the compensated thermal wattmeter there are no errors due to the power taken by either the current or voltage circuits.

Dynamometer wattmeters are capable of providing an accuracy of order 0.25 per cent of FSD over a frequency range from d.c. to several kHz. Induction wattmeters are suitable only for use in a.c. circuits and require constant supply voltage and frequency for accurate operation. The electrostatic wattmeter is a standards instrument having no waveform errors and suitable for measurements involving low power factors such as the measurement of iron loss, dielectric loss and the power taken by fluorescent tubes. Thermocouple wattmeters are capable of providing measurements up to 1 MHz with high accuracy.

1.4.3 Connection of wattmeters

There are two methods of connecting a dynamometer wattmeter to the measurement circuit. These are shown in Figures 1.37(a) and (b). In the connection shown in Figure 1.37(a) the voltage coil is connected to the supply side of the current coil. The wattmeter therefore measures the power loss in the load plus the power loss in the current coil. With the wattmeter connected as in Figure 1.37(b) the current coil takes the current for both the load and the voltage coil. This method measures the power loss in the load and in the voltage coil. For small load currents the voltage drop in the current coil will be small, therefore the power loss in this coil will be small and the first method of connection introduces little error. For large load currents the power loss in the voltage coil will be small compared with the power loss in the load and the second method of connection is to be preferred.

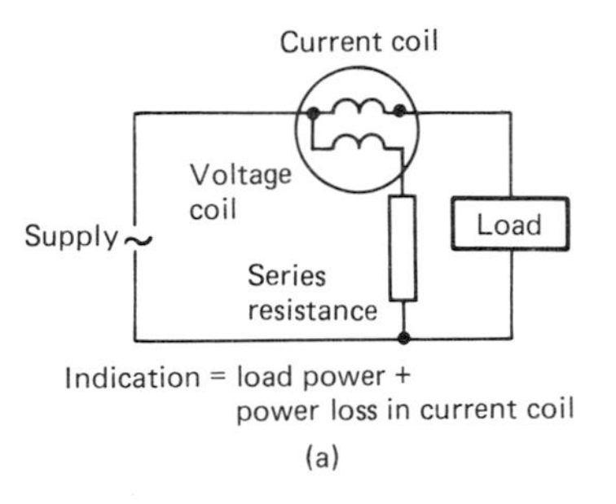

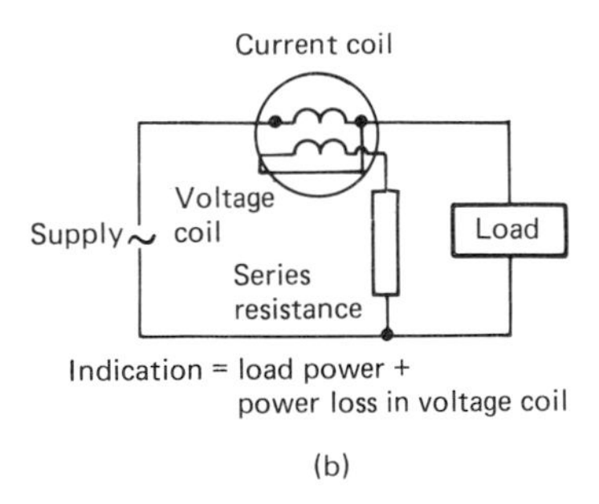

Figure 1.37 Wattmeter connection.

Compensated wattmeters of the type shown in

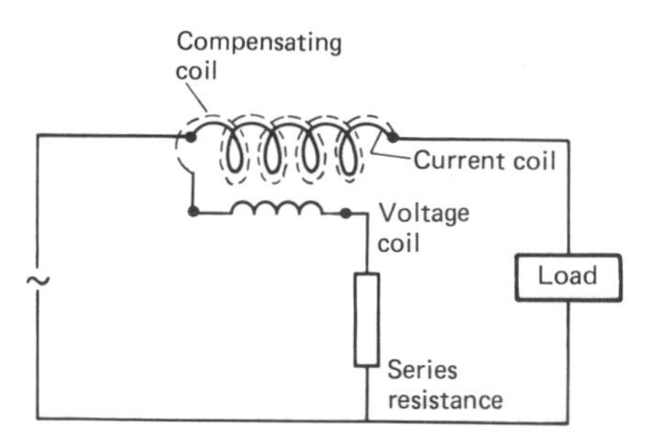

Figure 1.38 Compensated dynamometer wattmeter.

Figure 1.38 employ a compensating coil in series with the voltage windings. This compensating coil is identical to the current coil and tightly wound with it to produce a magnetic field in opposition to the main magnetic field due to the load current. Thus the effect of the voltage coil current is eliminated, and therefore the wattmeter connected in the manner shown in Figure 1.38 shows no error due to the power consumption in the voltage coil.

For electronic wattmeters the power loss in the voltage detection circuit can be made to be very small and thus the second method of connection is to be preferred.

The current and voltage ranges of wattmeters can be extended by means of current and voltage transformers as shown in Figure 1.39. These transformers introduce errors in the measurement as outlined in section 1.2.3.

1.4.4 Three-phase power measurement

For an n conductor system the power supplied can be measured by n wattmeters if they are connected with each wattmeter having its current coil in one of the conductors and its potential coil between the conductor and a single common point. This method of measurement is shown for both star- and delta-

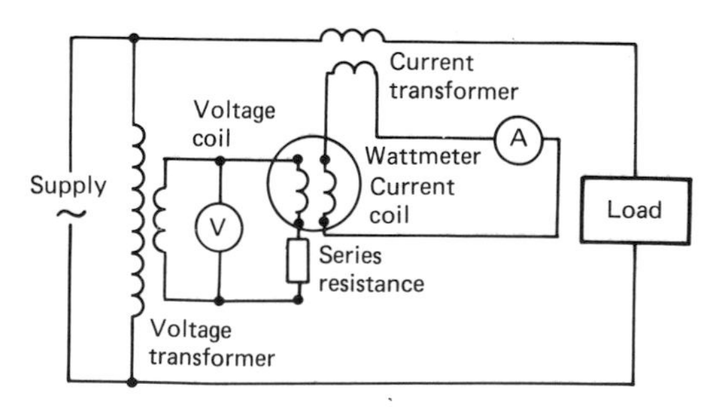

Figure 1.39 Wattmeter used with instrument transformers.

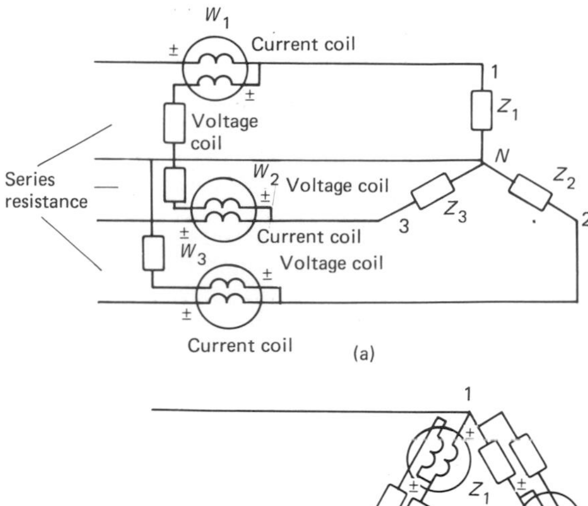

Figure 1.40 (a) Power measurement in a star-connected three-phase load using three wattmeters; (b) power measurement in a delta-connected three-phase load using three wattmeters.

connected three-phase systems in Figures 1.40(a) and (b). The power dissipated in the three-phase system is given by

$$P = W_1 + W_2 + W_3$$

Blondel's theorem states that if the common point for the potential coil is one of the conductors then the number of wattmeters is reduced by one. Thus it is possible to measure the power in a three-phase system using only two wattmeters, irrespective of whether the three-phase system is balanced. This method is shown in Figures 1.41(a) and (b). The phasor diagram for a star-connected balanced load is shown in Figure 1.41(c). The total power dissipated in the three-phase system is given by

$$P = W_1 + W_2$$

i.e. the power dissipated is the algebraic sum of the indications on the wattmeters.

It should be noted that if the voltage applied to the voltage coil of the wattmeter is more than 90 degrees out of phase with the current applied to its current coil then the wattmeter will indicate in the reverse direction. It is necessary under such circumstances to reverse the direction of the voltage winding and to count the power measurement as negative. If the power factor of the load is 0.5 so that I_1 lags 60 degrees

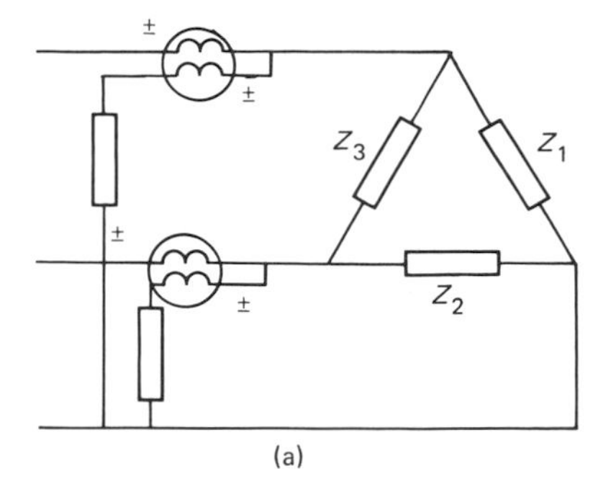

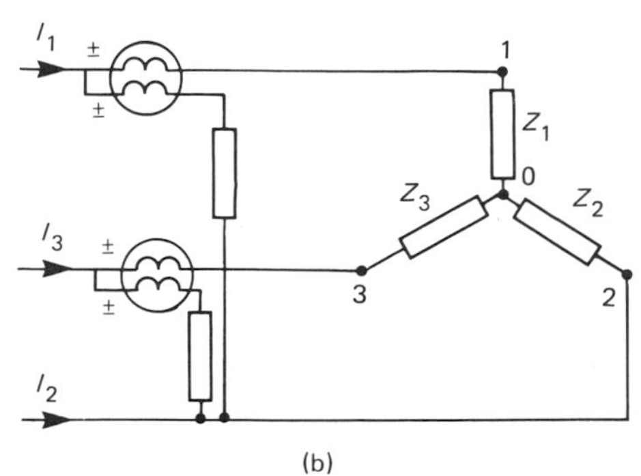

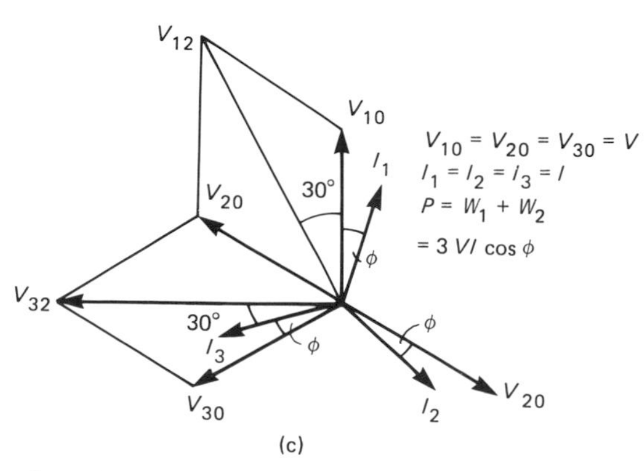

Figure 1.41 (a) Two-wattmeter method of power measurement in a three-phase delta-connected load; (b) two-wattmeter method of power measurement in a three-phase star-connected load; (c) phasor diagram for two-wattmeter method in a balanced star-connected load.

behind V_{10} then the phase angle between V_{12} and I_1 is 90 degrees and wattmeter W_1 should read zero.

It is also possible in the case of a balanced load to obtain the power factor from the indication on the two wattmeters, since

$$W_2 - W_1 = \sqrt{3}\,.\,V.\,I\,.\sin\phi$$

and therefore

$$\tan\phi = \sqrt{3}\,\frac{(W_2 - W_1)}{(W_1 + W_2)}$$

If the three-phase system is balanced then it is possible to use a single wattmeter in the configuration shown in Figure 1.42. With the switch in position 1 the indication on the wattmeter is given by

$$W_1 = \sqrt{3}\,.\,V.\,I\cos(30+\phi)$$

With the switch in position 2 the wattmeter indicates

$$W_2 = \sqrt{3}\,.\,V.\,I\,.\cos(30-\phi)$$

The sum of these two readings is therefore

$$W_1 + W_2 = 3VI\cos\phi = P$$

i.e. the total power dissipated in the system.

1.4.5 Electronic wattmeters

The multiplication and averaging process involved in wattmetric measurement can be undertaken by electronic means as shown in Figure 1.43. Electronic wattmeters fall into two categories, depending on whether the multiplication and averaging is continuous or discrete.

In the continuous method the multiplication can be by means of a four-quadrant multiplier as shown in Figure 1.44(a) (Simeon and McKay, 1981); time-division multiplication as in Figure 1.44(b) (Miljanic *et al*, 1978); or by the use of a Hall-effect multiplier as in Figure 1.44(c) (Bishop and Cohen, 1973).

The sampling wattmeter shown in Figure 1.45 takes simultaneous samples of both the voltage and current waveforms, digitizes these values and provides multiplication and averaging using digital techniques (Dix, 1982; Matouka, 1982).

If the voltage and current waveforms have fundamental and harmonic content with a fundamental period T then the instantaneous power can be written as a Fourier series:

$$p(t) = P + \sum_{k=1}^{\infty} p_k\,.\sin\left(\frac{2\pi kt}{T} + \rho_k\right)$$

where P is the average power.

If the waveforms are uniformly sampled n times over m periods then the time t_j of the jth sample is given by

$$t_j = j\,.\frac{m}{n}\,.\,T$$

and the measured average power W is given by

$$W = \frac{1}{n}\sum_{j=0}^{n-1} p(t_j)$$

The error between the measured and true mean values is given by

$$W - P = \frac{1}{n}\sum_{k=1}^{\infty} p_k \sum_{j=0}^{n-1} \sin\left(\frac{2\pi k\,.\,j\,.\,m)}{n} + \rho_k\right)$$

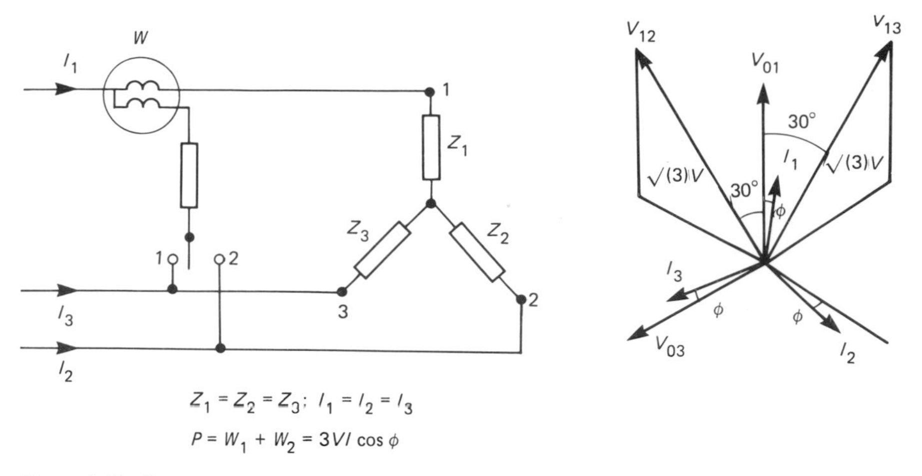

Figure 1.42 One-wattmeter method for balanced three-phase systems.

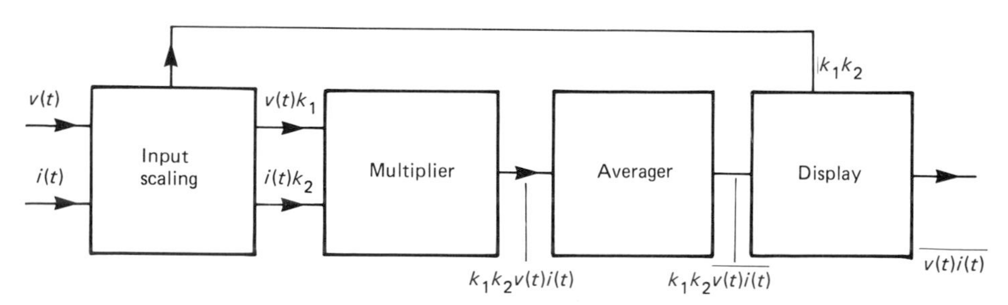

Figure 1.43 Electronic wattmeter.

It can be shown (Clarke and Stockton, 1982; Rathore, 1984), that the error of measurement is given by

$$|W-P| = \left|\sum_{k>0}^{*} p_k \sin(\rho_k)\right| \leqslant \sum_{k>0}^{*} |p_k|$$

where $\sum^*$ indicates summation over those terms where $k.m/n$ is an integer, i.e. those harmonics of the power signal whose frequencies are integer multiples of the sampling frequency.

Matouka (1982) has analysed other sources of error in sampling wattmeters including amplifier, offset, sampled data, amplitude and time quantization, and truncation errors.

Continuous analogue methods employing analogue multipliers are capable of providing measurement of power typically up to 100 kHz. The Hall-effect technique is capable of measurement up to the region of several GHz and can be used in power measurement in a waveguide. Using currently available components with 15-bit A/D converters the sampling wattmeter can achieve a typical uncertainty of 1 part in 10^4 at power frequencies. Table 1.8 gives the characteristics of an electronic wattmeter providing digital display.

1.4.6 High-frequency power measurement

At high frequencies average power measurement provides the best method of measuring signal amplitude because power flow, unlike voltage and current, remains constant along a loss-less transmission line. Power measurements are made by measuring the thermal effects of power or by the use of a square law device such as a diode (Hewlett Packard, 1978; Fantom, 1985).

Static calorimetric techniques employ a thermally insulated load and a means for measuring the rise in temperature caused by the absorbed rf power. Flow calorimeters consist of a load in which absorbing liquid such as water converts the rf power into heat, together with a circulating system and a means for measuring the temperature rise of the circulating liquid. Because of their potentially high accuracy, calorimetric methods are used as reference standards. However, because of the complexity of the measurement systems they are not easily portable.

Commercially available thermal techniques employ either thermistors or thermocouple detectors. Figure

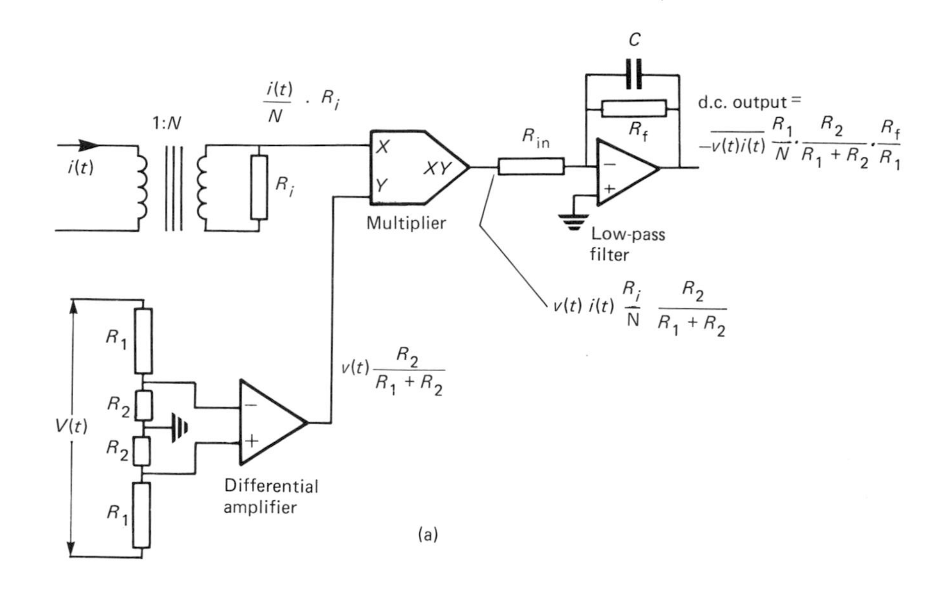

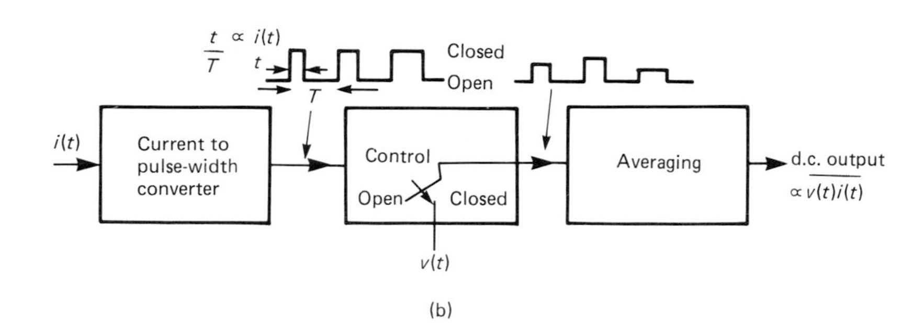

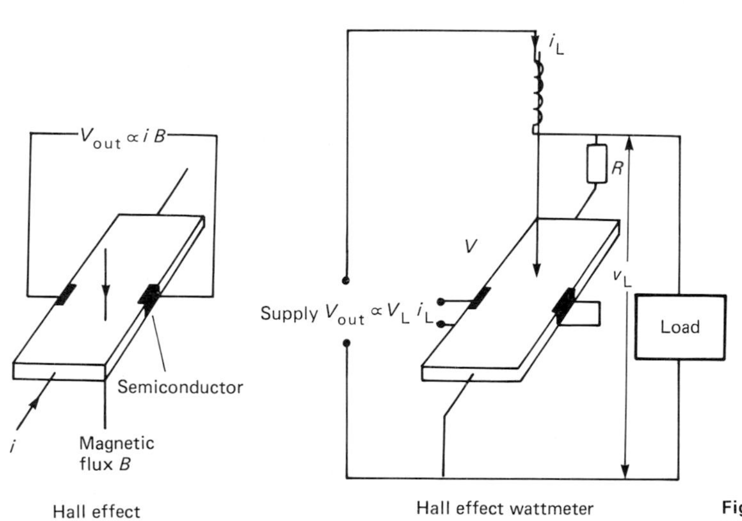

Figure 1.44 (a) Four-quadrant analogue multiplier wattmeter; (b) time-division multiplication wattmeter; (c) Hall-effect wattmeter.

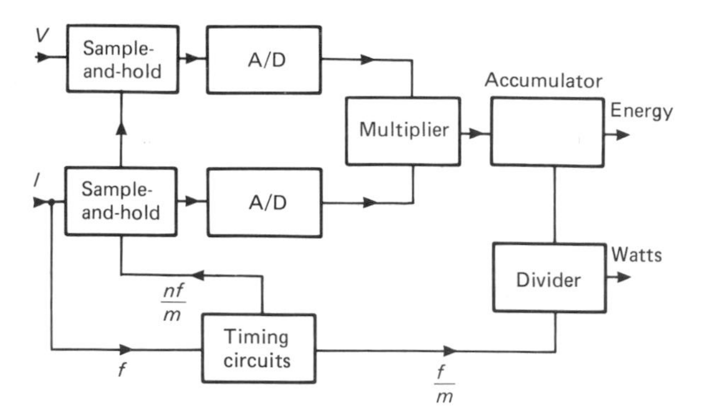

Figure 1.45 Sampling wattmeter (from Dix, 1982).

Table 1.8 Electronic wattmeter specification

Valhalla Scientific Digital Power Analyser
Model 2100 range/resolution table

True rms voltage ranges	*True rms current ranges* 0.2000 A	2.000 A	20.00 A
150.00 V	30.00 W	300.0 W	3000 W
300.0 V	60.00 W	600.0 W	6000 W
600.0 W	120.00 W	1200.0 W	12000 W
	True watts ranges		

Performance specifications

A.C./D.C. CURRENT (true rms)
Crest factor response: 50:1 for minimum rms input, linearly decreasing to 2.5:1 for full-scale rms input
Peak indicator: Illuminates at 2.5 × full scale
Minimum input: 5% of range
Maximum input: 35 A peak, 20 A d.c. or rms; 100 A d.c. or rms for 16 mS without damage
Overrange: 150% of full scale for d.c. up to maximum input

A.C./D.C. VOLTAGE (true rms)
Crest factor response: 50:1 for minimum rms input, linearly decreasing to 2.5:1 for full-scale rms input
Minimum input: 5% of range
Maximum input: 600 V d.c. or rms a.c., 1500 V peak
Maximum common mode: 1500 V peak, neutral to earth
Peak indicator: Illuminates at 2.5 × full scale

WATTS (true power—VI cos ϕ)
Power factor response: Zero to unity leading or lagging
Accuracy: (V-A-W 25°C ± 5°C, 1 year)
D.C. and 40 Hz to 5 kHz: 0.25% of reading ±6 digits
5 Hz to 10 kHz: ±0.5% of reading ±0.5% of range
10 kHz to 20 kHz: ±1% of reading ±1% of range (2 A range only)

GENERAL SPECIFICATIONS
Displays: Dual 4½-digit large high-intensity 7-segment LED
Operating temperature range: 0–50°C
Temperature coefficient: ±0.025% of range per °C from 0°C to 20°C and 30–50°C
Conversion rate: Approximately 600 mS
Power: 115/230 V a.c. ± 10%, 50–60 Hz. 5 W

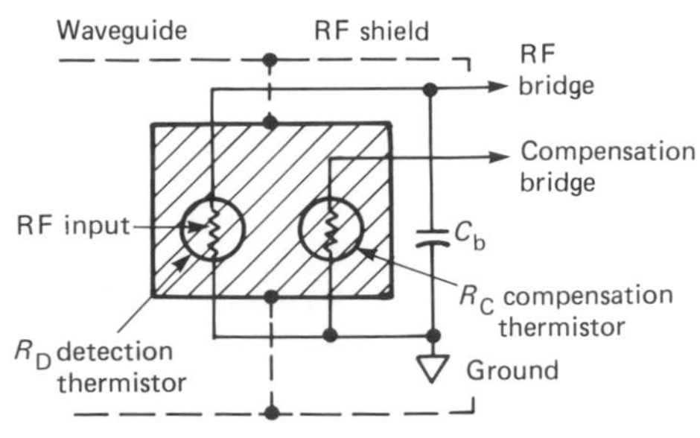

Figure 1.46 Equivalent circuit of a thermistor rf power detector (from Hewlett Packard, 1978).

1.46 shows the equivalent circuit of a thermistor system. The detecting thermistor is in either a coaxial or waveguide mount. The compensating thermistor is in close thermal contact with the detecting thermistor but shielded from the rf power.

Figure 1.47 shows a thermistor power meter employing two self-balancing d.c. bridges. The bridges are kept in balance by adjusting their supply voltages. With no applied rf power V_c is made equal to V_{rf0}, i.e. the value of V_{rf} with no applied rf energy. After this initialization process ambient temperature changes in both bridges track each other.

If rf power is applied to the detecting thermistor then V_{rf} decreases such that

$$P_{rf}=\frac{V_{rf0}^2}{4R}-\frac{V_{rf}^2}{4R}$$

where R is the resistance of the thermistor, and since

$$V_{rf0}=V_c$$

then the rf power can be calculated from

$$P_{rf}=\frac{1}{4R}(V_c-V_{rf})(V_c+V_{rf})$$

The processing electronics performs this computation on the output signals from the two bridges.

1.5 Measurement of electrical energy

The energy supplied to an electrical circuit over a time period T is given by

$$E=\int_0^T p(t)\,.\,dt$$

The most familiar instrument at power frequencies for the measurement of electrical energy is the watt-hour meter used to measure the electrical energy supplied to consumers by electricity supply undertakings. The most commonly used technique is

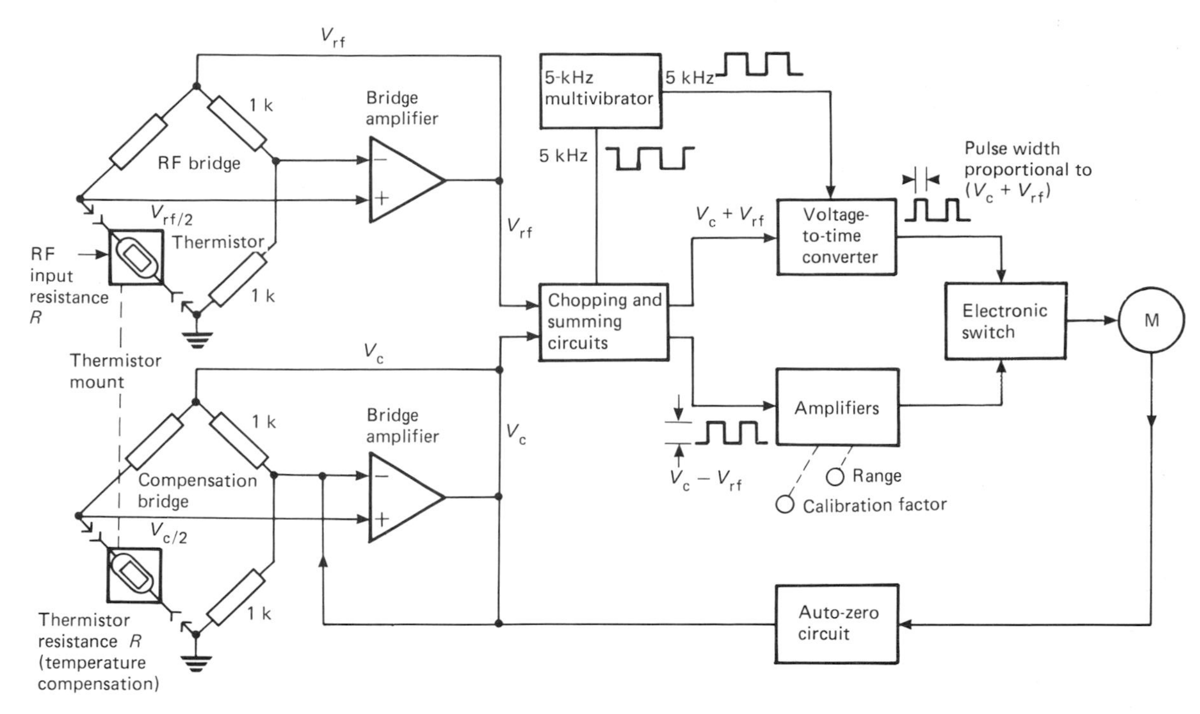

Figure 1.47 Thermistor rf power meter (from Hewlett Packard, 1978).

the shaded pole induction watt-hour meter, shown in schematic form in Figure 1.48(a). This is essentially an induction motor whose output is absorbed by its braking system and dissipated in heat. The rotating element is an aluminium disc and the torque is produced by the interaction of the eddy currents induced in the disc with the imposed magnetic fields. The instantaneous torque is proportional to

$$(\phi_v i_i - \phi_i i_v)$$

where ϕ_v is the flux generated by the voltage coil, ϕ_i is the flux generated by the current coil, i_v is the eddy current generated in the disc by the voltage coil and i_i is the eddy current generated in the disc by the current coil.

The relative phases of these quantities are shown in Figure 1.48(b). The flux generated by the current coil is in phase with the current and the flux generated by the voltage coil is adjusted to be exactly in quadrature with the applied voltage by means of the copper shading ring on the voltage magnet.

The average torque, T_g, can be shown to be proportional to the power

$$T_g \propto VI\cos\phi$$

The opposing torque, T_b, is provided by eddy-current braking and thus is proportional to the speed of rotation of the disc, N, as shown in Figure 1.48(c).

Equating the generated and braking torques,

$$T_b = T_g; \qquad \text{and} \qquad N \propto VI\cos\phi$$

and therefore the speed of rotation of the disc is proportional to the average power and the integral of the number of revolutions of the disc is proportional to the total energy supplied. The disc is connected via a gearing mechanism to a mechanical counter which can be read directly in watt-hours.

1.6 Power-factor measurement

Power-factor measurement is important in industrial power supply since generating bodies penalize users operating on poor power factors because this requires high current-generating capacity but low energy transfer. It is possible to employ the dynamometer principle to provide an indicating instrument for power factor. This is shown in Figure 1.49. The two movable coils are identical in construction but orthogonal in space. The currents in the two coils are equal in magnitude but time displaced by 90 degrees. There is no restoring torque provided in the instrument and the movable coil system aligns itself so that there is no resultant torque. Thus:

$$VI\cos\phi \, . \frac{dM_1}{d\theta} + VI\cos(\phi - 90°)\frac{dM_2}{d\theta} = 0$$

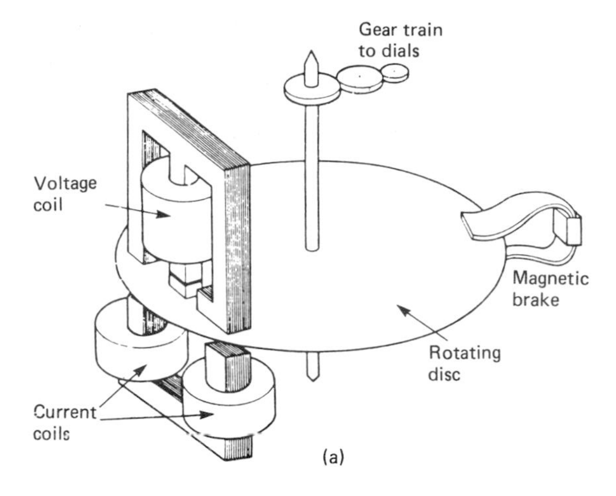

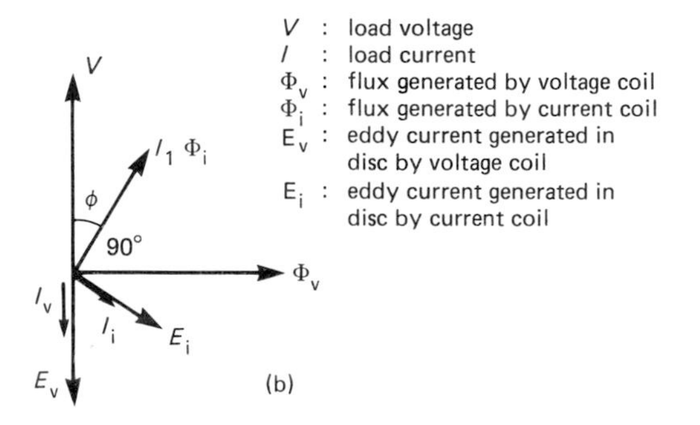

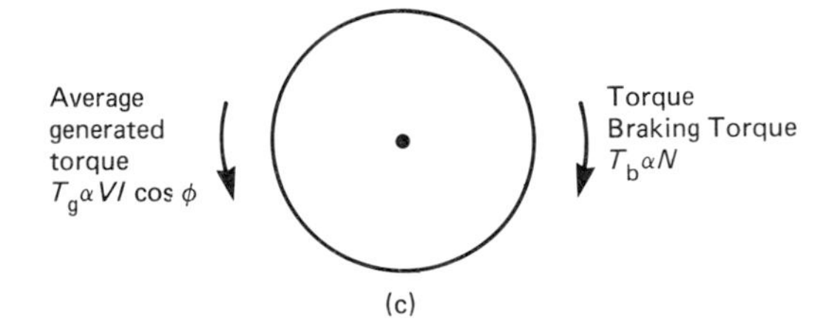

Figure 1.48 (a) Watt-hour meter; (b) phasor diagram of fluxes and eddy currents in watt-hour meter; (c) torque balance in a watt-hour meter.

If the mutual inductance between the current carrying coil and the voltage coil 1 is given by

$$M_1 = k_1 \cos\theta$$

and if the mutual inductance between the current-carrying coil and the voltage coil 2 is given by

$$M_2 = k_1 \sin\theta$$

then the rest position of the power factor instrument occurs when

$$\theta = \phi$$

The dial of the instrument is usually calibrated in terms of the power factor, as shown in Figure 1.49. The method can also be applied to power-factor measurement in balanced three-phase loads. (Golding and Widdis, 1963).

1.7 The measurement of resistance, capacitance and inductance

The most commonly used techniques for the measurement of these quantities are those of bridge measurement. The word 'bridge' refers to the fact that in such measurements two points in the circuit are bridged by a detector which detects either a potential difference or a null between them. Bridges are used extensively by National Standards Laboratories to maintain electrical standards by facilitating the calibration and intercomparison of standards and substandards. They are used to measure the resistance, capacitance and inductance of actual components, and do this by comparison with standards of these quantities. For details of the construction of standard resistors, capacitors and inductors the reader should consult Hague and Foord (1971) and Dix and Bailey (1975). In a large number of transducers non-electrical quantities are converted into corresponding changes in resistance, capacitance or inductance, and this has led to the use of bridges in a wide variety of scientific and industrial measurements.

1.7.1 D.C. bridge measurements

The simplest form of a d.c. four-arm resistance bridge is the Wheatstone bridge, which is suitable for the measurement of resistance typically in the range from

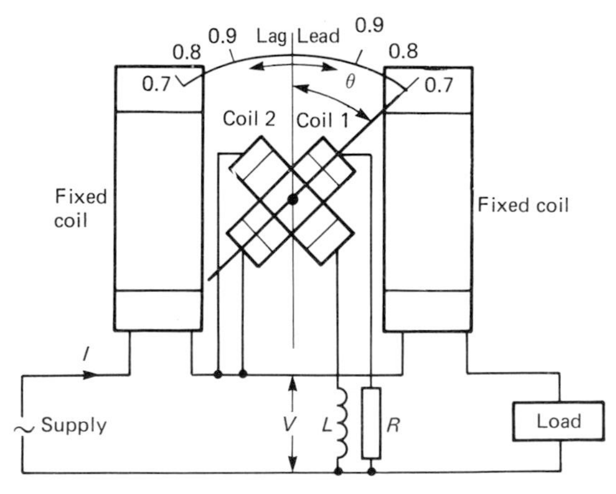

Figure 1.49 Power-factor instrument.

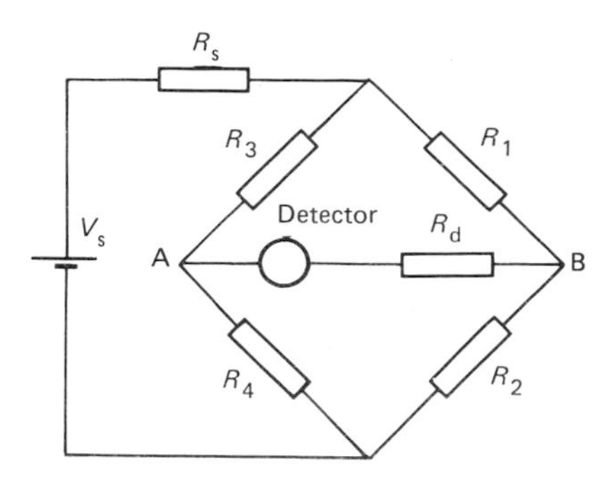

Figure 1.50 Wheatstone bridge.

1 Ω to 10 MΩ and is shown in Figure 1.50. The bridge can be used in either a balanced, i.e. null, mode or a deflection mode. In the balanced mode the resistance to be measured is R_1 and R_3 is a variable standard resistance. R_2 and R_4 set the ratio. The detector, which may be either a galvanometer or an electronic detector, is used to detect a null potential between the points A and B of the bridge. A null occurs when

$$R_1 = \frac{R_2}{R_4} \cdot R_3$$

The bridge is balanced either manually or automatically using the output signal from the detector in a feedback loop to find the null position. The null condition is independent of the source resistance, R_s, of the voltage source supplying the bridge or the sensitivity or input resistance, R_d, of the detector. These, however, determine the precision with which balance condition can be determined. The sensitivity, S, of the bridge can be expressed as

$$S = \frac{\text{Bridge output voltage, } V_{out}, \text{ for a change } \Delta R_1 \text{ in } R_1}{\text{Bridge supply voltage}}$$

Near the balance condition for a given fractional change, δ, in R_1 given by

$$\delta = \frac{\Delta R_1}{R_1}$$

the sensitivity is given by

$$S = \frac{\delta R_d}{\sum_{i=1}^{4} R_i + R_d[2 + (R_2/R_4) + (R_4/R_3)] + R_s[2 + (R_3/R_1) + (R_1/R_3)] + R_d R_s \sum_{i=1}^{4} (1/R_i)}$$

With an electronic detector R_d can be made large and if R_s is small then S is given by

$$S = \frac{\delta}{[2 + (R_3/R_4) + (R_4/R_3)]}$$

which has a maximum value of $\delta/4$ when $(R_3/R_4) = 1$.

The unbalanced mode is shown in Figure 1.51(a) and is often used with strain gauges (Volume 1, Chapter 4). R_1 is the active strain gauge and R_2 is the dummy gauge subject to the same temperature changes as R_1 but no strain. The output from the bridge is given by

$$V_{out} = \frac{V_s}{2}\left\{1 - \frac{1}{[1 + (\delta/2)]}\right\}$$

where

$$\delta = \frac{\Delta R}{R}$$

For $\delta \ll 1$ the output of the bridge is linearly related to the change in resistance, i.e.

$$V_{out} = \frac{V_s}{4} \cdot \delta$$

Self-heating generally limits the bridge supply voltage and hence the output voltage. Amplification of the bridge output voltage has to be undertaken with an amplifier having a high common-mode rejection ratio (CMRR) since the output from the bridge is in general small and the common-mode signal applied to

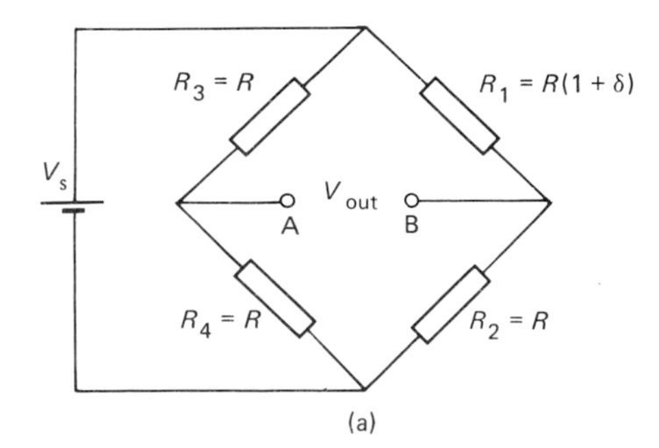

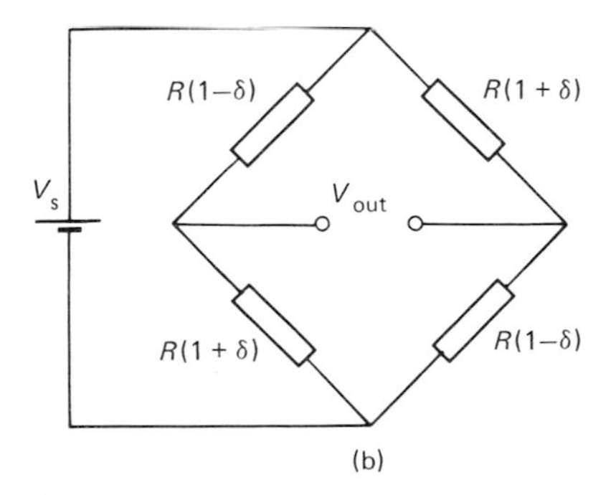

Figure 1.51 (a) Unbalanced Wheatstone bridge; (b) unbalanced Wheatstone bridge with increased sensitivity.

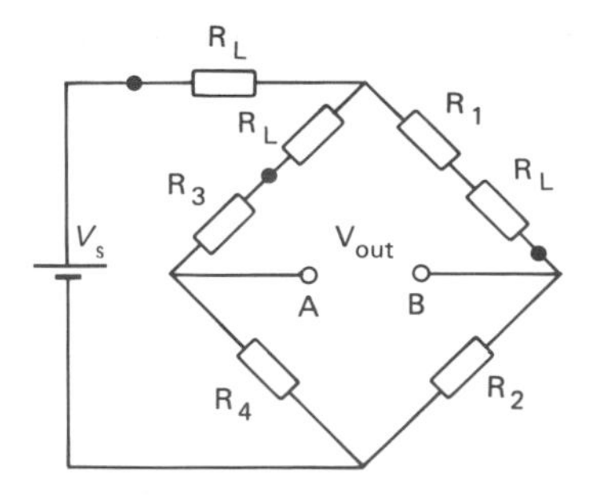

R_1 is unknown resistance
R_L represents lead resistances
● Connections to R_1

$$R_4 = R_2$$

Balance condition:

$$\frac{R_1 + R_L}{R_2} = \frac{R_3 + R_L}{R_4}$$

$$\therefore \quad R_1 = R_3$$

Out-of-balance condition:

$$R_2 = R_3 = R_4 = R$$

$$R_1 = R(1 + \delta)$$

$$V_{out} = \frac{V_s \delta}{4}(1 - \beta)$$

$$\beta = \frac{R_L[R(4+\delta) + R_L]}{R^2(3 + 2\delta) + RR_L(4 + \delta) + R_L^2}$$

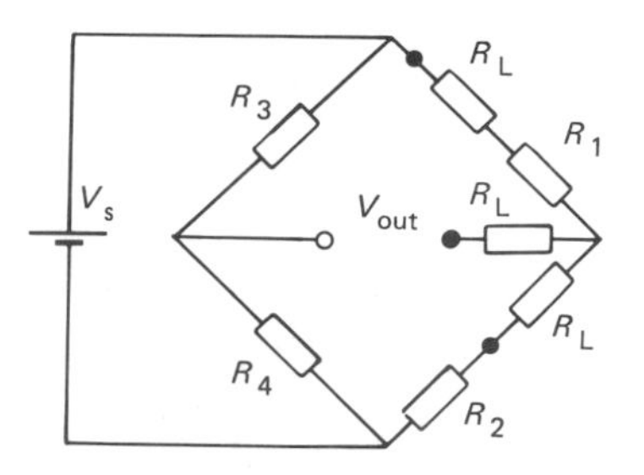

R_1 is unknown resistance
R_L represents lead resistances
● Connections to R_1

$$R_3 = R_4$$

Balance condition:

$$\frac{R_1 + R_L}{R_2 + R_L} = \frac{R_3}{R_4}$$

$$\therefore \quad R_1 = R_2$$

Out-of-balance condition:

$$R_2 = R_3 = R_4 = R$$

$$R_1 = R(1 + \delta)$$

$$V_{out} = \frac{V_s \delta}{4}(1 - \beta')$$

$$\beta' = \frac{R_L}{R_L + R[1 + (\delta/2)]}$$

Figure 1.52 Three-lead measurements using a Wheatstone bridge.

the amplifier is $V_s/2$. Further details of amplifiers suitable for use as bridge detectors can be found in the section on amplifiers in Part 4.

The output from a strain gauge bridge can be increased if four gauges are employed, with two being in tension and two in compression, as shown in Figure 1.51(b). For such a bridge the output is given by

$$V_{out} = V_s \,.\, \delta$$

Strain gauges and platinum resistance thermometers may be situated at a considerable distance from the bridge and the long leads connecting the active element to the bridge will have a resistance which will vary with temperature. Figure 1.52 shows the use of the Wheatstone bridge in three lead resistance measurement where it can be seen that close to balance the effect of the lead resistance and its temperature variation is approximately self-cancelling and that the cancelling effect deteriorates the further the bridge condition departs from balance. Figure 1.53 shows the use of Smith and Muller bridges to eliminate the lead resistance of a four-lead platinum resistance thermometer. (See also Part 2, Chapter 1).

1.7.1.1 Low-resistance measurement

Contact resistance causes errors in the measurement of low resistance, and therefore in order to accurately define a resistance it is necessary to employ the four-terminal technique shown in Figure 1.54. The outer two terminals are used to supply the current to the resistance and the inner two, the potential terminals, determine the precise length of conductor over which the resistance is defined.

Measurement of low resistance is undertaken using the Kelvin double bridge shown in Figure 1.55(a). R_1 is the resistance to be measured and R_2 is a standard resistance of the same order of magnitude as R_1. The link between them which is sometimes referred to as the yoke has resistance r. The current through R_1 and

R_1 is unknown resistance; $R_2 = R_4$; ● Connections to unknown resistance; R_{L1}, R_{L2}, R_{L3}, R_{L4} are lead resistances

Bridge connections for first balance

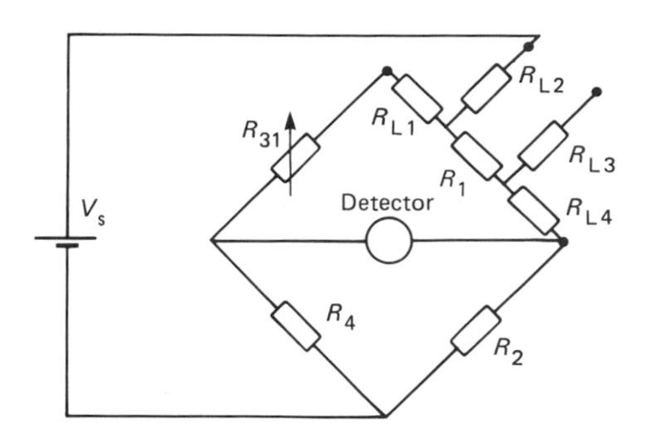

Balance condition:

$$R_1 + R_{L4} = R_{31} + R_{L1}$$

Bridge connections for second balance

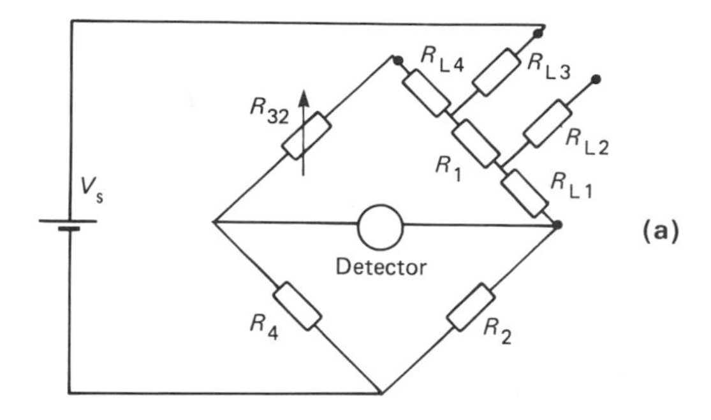

(a)

Balance condition:

$$R_1 + R_{L1} = R_{32} + R_{L4}$$

Thus $$R_1 = \frac{R_{31} + R_{32}}{2}$$

R_1 is unknown resistance; $R_3 = R_4$; R_{L1}, R_{L2}, R_{L3}, R_{L4} are lead resistances

Bridge connections for first balance

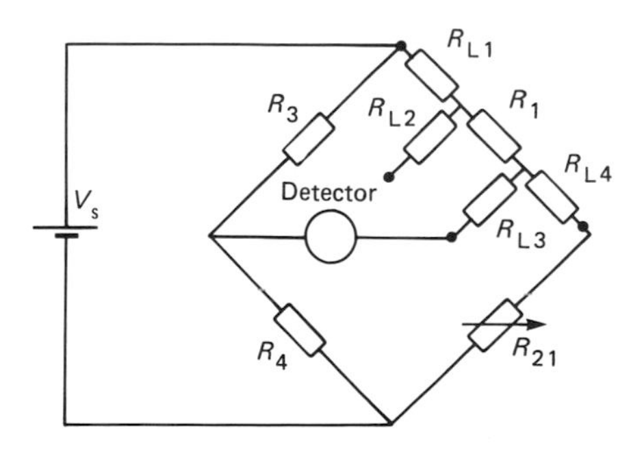

Balance condition:

$$R_1 + R_{L1} = R_{21} + R_{L4}$$

Bridge connections for second balance

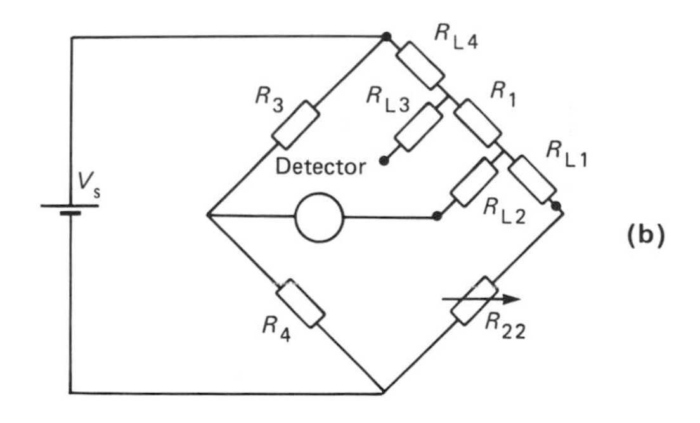

(b)

Balance condition:

$$R_1 + R_{L4} = R_{22} + R_{L1}$$

Thus $$R_1 = \frac{R_{21} + R_{22}}{2}$$

Figure 1.53 (a) Smith bridge for four-lead platinum resistance thermometer measurement; (b) Muller bridge for four-lead platinum resistance thermometer measurement.

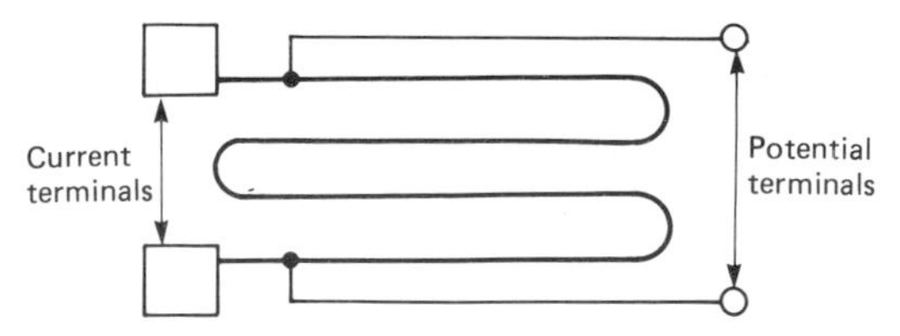

Figure 1.54 A four-terminal resistance.

R_2 is regulated by R. R_3, R_4, r_3, r_4 are four resistances of which either R_3 and r_3 or R_4 and r_4 are variable, and for which

$$\frac{R_3}{R_4}=\frac{r_3}{r_4}$$

The delta star transformation applied to the bridge as shown in Figure 1.55(b) apportions the yoke resistance between the two sides of the bridge. The

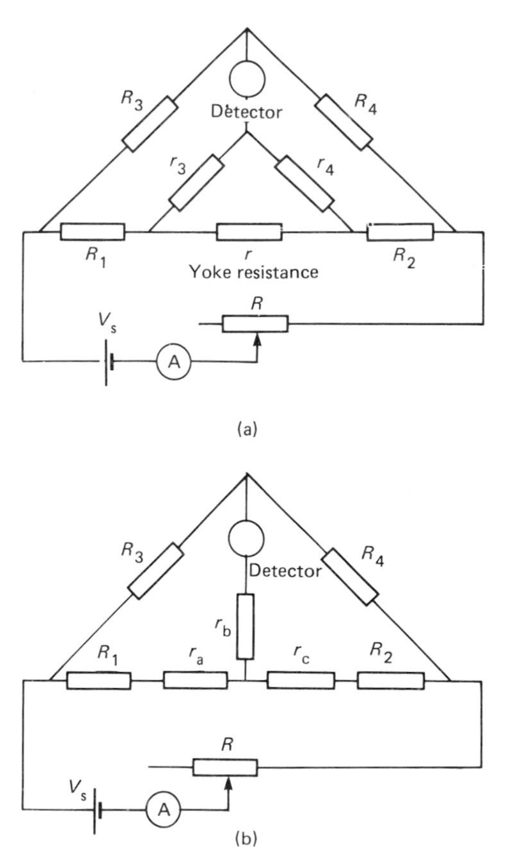

Figure 1.55 (a) Kelvin double bridge; (b) equivalent circuit of Kelvin double bridge.

balance condition is given by

$$\frac{R_1+r_a}{R_2+r_c}=\frac{R_3}{R_4}; \qquad r_a=\frac{r_3 \,.\, r}{(r_3+r_4+r)};$$

$$r_c=\frac{r_4 \,.\, r}{(r_3+r_4+r)}$$

and thus the unknown resistance R_1 is given by

$$R_1=\frac{R_3}{R_4}(R_2+r_c)-r_a$$

$$=\frac{R_3}{R_4} \,.\, R_2+\frac{r_4 \,.\, r}{r_3+r_4+r}\left(\frac{R_3}{R_4}-\frac{r_3}{r_4}\right)$$

The term involving the yoke resistance r can be made small by making r small and also by making

$$\frac{R_3}{R_4}=\frac{r_3}{r_4}$$

The bridge can be used to measure resistances typically from 0.1 $\mu\Omega$ to 1 Ω. For high precision the effect of thermally generated emfs can be eliminated by reversing the current in R_1 and R_2 and rebalancing the bridge. The value of R_1 is then taken as the average of the two measurements.

1.7.1.2 High-resistance measurement

Modified Wheatstone bridges can be used to measure high resistance up to 10^{15} Ω. The problems in such measurements arise from the difficulty of producing stable high-value standard resistors and errors caused by shunt-leakage resistance.

The problem of stable high-resistance values can be overcome by using the bridge with lower value and therefore more stable resistances. This leads to bridges which have larger ratios and hence reduced sensitivity. By operating the bridge with R_4 as the variable element then as $R_1 \rightarrow \infty$, $R_4 \rightarrow 0$.

The shunt leakage is made up of leakage resistance across the leads, the terminals of the bridge and also across the unknown resistor itself. High-value standard resistors are constructed with three terminals. In the bridge arrangement shown in Figure 1.56(a) R_{sh1} shunts R_3 and thus if $R_1 \gg R_3$ this method of connection decreases the effect of the leakage resistance. The only effect of R_{sh2} is to reduce the sensitivity of the balance condition.

Figure 1.56(b) shows a d.c. form of the Wagner earthing arrangement used to eliminate the effect of leakage resistance. The bridge balance then involves balancing the bridge with the detector across BC by adjusting R_6 and then balancing the bridge with the detector across AB by adjusting R_4. The procedure is

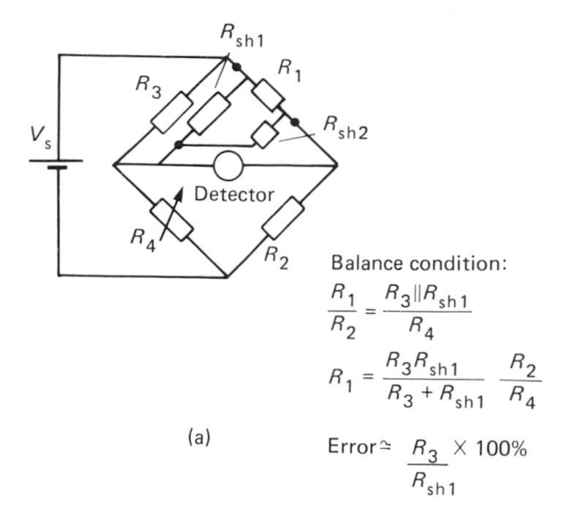

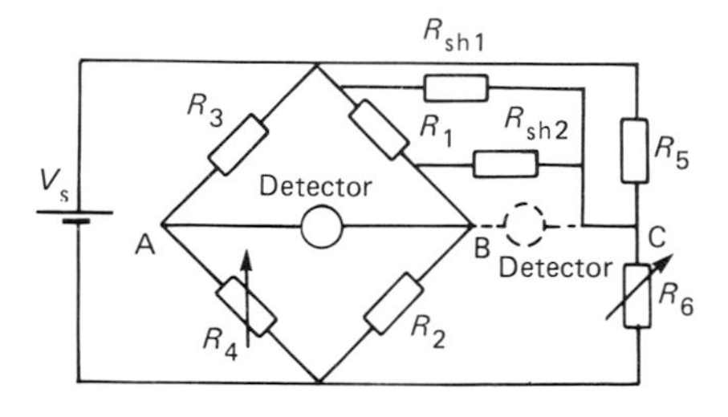

Balance conditions:
With detector across AB

$$\frac{R_1}{R_2} = \frac{R_3}{R_4}$$

With detector across BC

$$\frac{R_1}{R_2} = \frac{R_{sh1} \| R_5}{R_6}$$

(b)

Figure 1.56 (a) Wheatstone bridge for use with three-terminal high resistances; (b) d.c. Wagner earthing arrangement.

then repeated until a balance is achieved under both conditions. The first balance condition ensures that there is no potential drop across R_{sh2} and thus no current flows through it.

1.7.2 A.C. equivalent circuits of resistors, capacitors and inductors

Resistors, capacitors and inductors do not exist as pure components. They are in general made up of combinations of all three impedance elements. For example, a resistor may have both capacitive and inductive parasitic elements. Figure 1.57 shows the complete equivalent circuits for physical realizations of the three components together with simplified equivalent circuits which are commonly used. Further details of these equivalent circuits can be found in Oliver and Cage (1971).

At any one frequency any physical component can be represented by its complex impedance $Z = R \pm jX$ or its admittance $Y = G \mp jB$. Since $Y = 1/Z$ and $Z = 1/Y$ then

$$R = \frac{G}{G^2 + B^2}; \qquad X = \frac{-B}{G^2 + B^2}$$

and

$$G = \frac{R}{R^2 + X^2}; \qquad B = \frac{-X}{R^2 + X^2}$$

These two representations of the component correspond to series and parallel equivalent circuits. If at a given frequency the impedance is $Z = R + jX$ then the equivalent circuit at that frequency in terms of ideal components is a resistor in either series or parallel with an inductor, as shown in Figure 1.58(a). This figure also gives the conversion formulae between the two representations. For components whose impedance at any given frequency is given by $Z = R - jX$ the equivalent circuits are series or parallel combinations of a resistor and a capacitor, as in Figure 1.58(b).

The quality factor, Q, is a measure of the ability of a reactive element to act as a pure storage element. It is defined as

$$Q = \frac{2\pi \times \text{maximum stored energy in the cycle}}{\text{Energy dissipated per cycle}}$$

The dissipation factor, D, is given by

$$D = \frac{1}{Q}$$

The Q and D factors for the series and parallel inductive and capacitive circuits are given in Figure 1.58. From this figure it can be seen that Q is given by $\tan\theta$ and D by $\tan\delta$, where δ is the loss angle. Generally, the quality of an inductance is measured by its Q factor and the quality of a capacitor by its D value or loss angle.

1.7.3 Four-arm a.c. bridge measurements

If the resistive elements of the Wheatstone bridge are replaced by impedances and the d.c. source and detector are replaced by their a.c. equivalents, as shown in Figure 1.59, then if Z_1 is the unknown impedance the balance condition is given by

$$Z_1 = \frac{Z_2 Z_3}{Z_4}; \qquad R_1 + jX_1 = \frac{(R_2 + jX_2)(R_3 + jX_3)}{(R_4 + jX_4)}$$

Ideal component	*Equivalent circuit of physical realization*	*Simplified equivalent circuit*
Resistance R	$R_d = \frac{1}{\omega C_p D}$ R_{dc}: d.c. resistance R_e: eddy-current loss and skin effect changes R_d: dielectric loss in C_p and C_d L: inductance C_p: lumped capacitance C_d: distributed capacitance	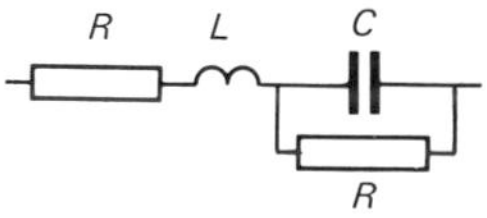 R L C
Capacitance C	C_0: electrostatic capacitance C_a: increase in C_0 caused by interfacial polarization with time constant $R_a C_a$ L: series inductance R_s: series resistance R_l: leakage resistance R_d: dielectric loss	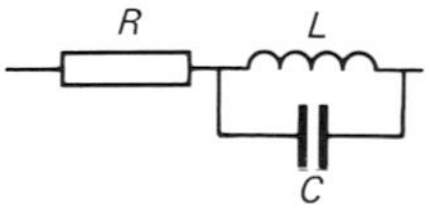 R L C R
Inductance L	L_0: inductance R_c: winding resistance R_e: eddy-current loss R_h: hysteresis loss (iron-cored inductors) C: capacitance R_d: dielectric loss	R L C

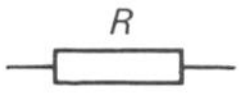

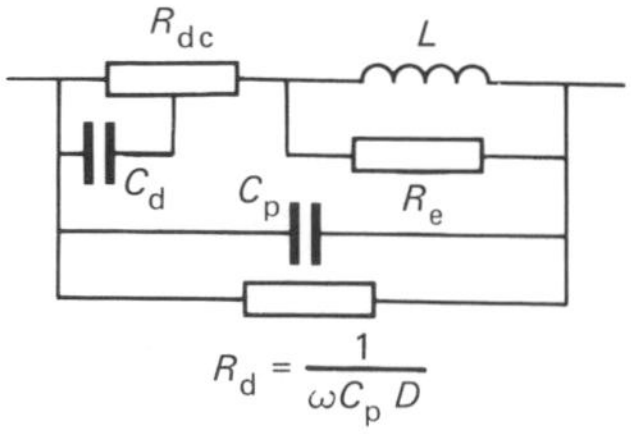

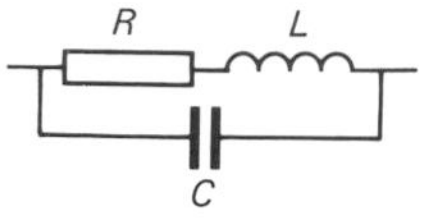

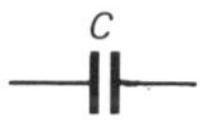

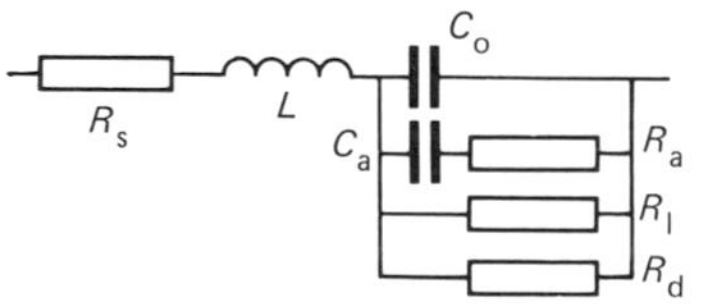

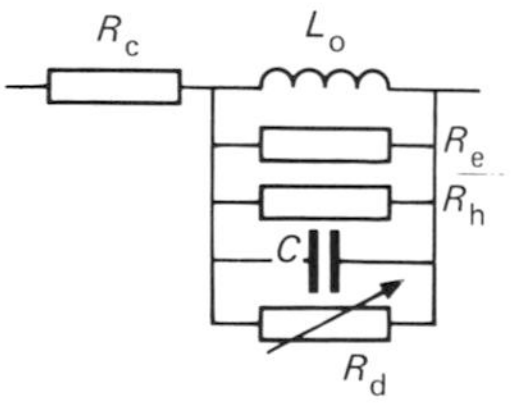

Figure 1.57 Equivalent circuit for physical realizations of resistance, capacitance and inductance.

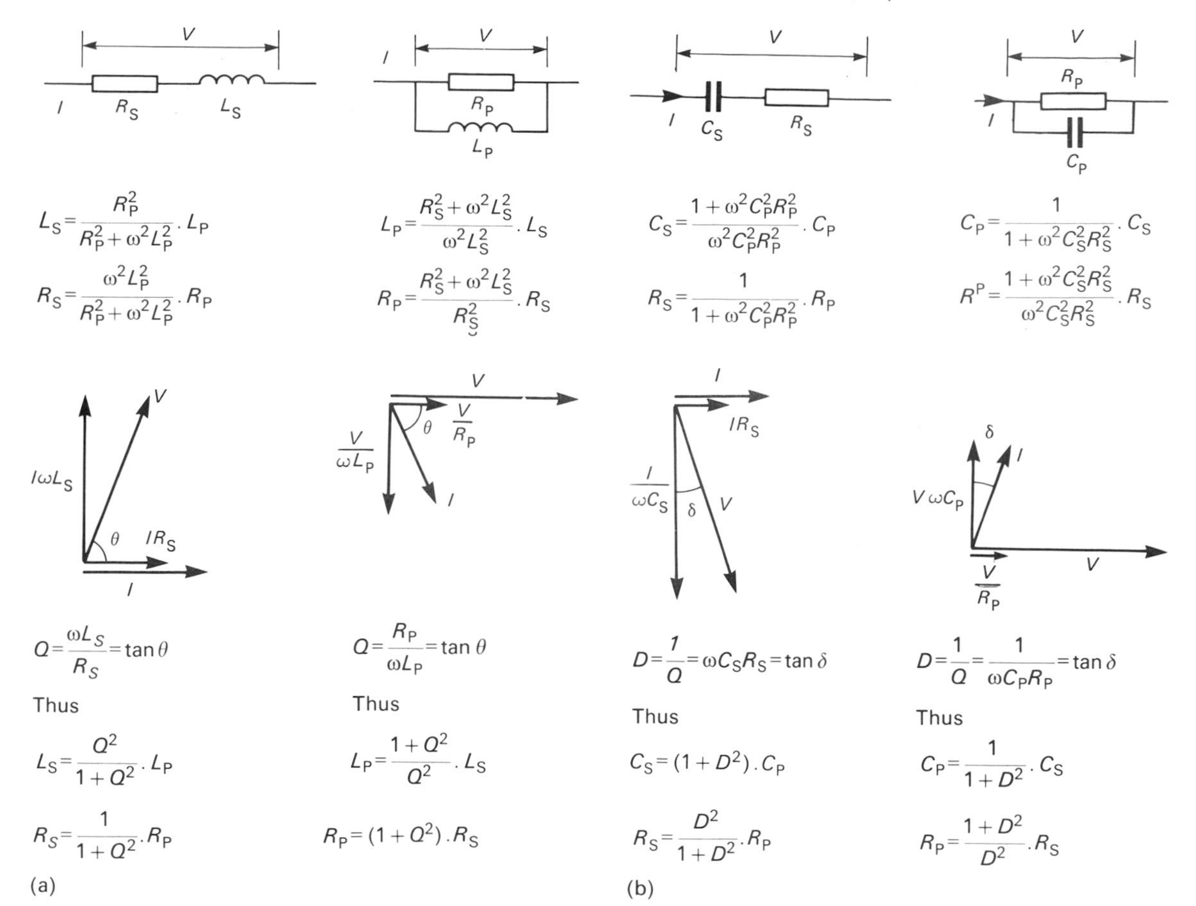

Figure 1.58 (a) Equivalent series/parallel resistor and inductor circuits; (b) equivalent series/parallel resistor and capacitor circuits.

or

$$|Z_1| = \frac{|Z_2||Z_3|}{|Z_4|} \quad \text{and}$$

$$\angle Z_1 = \angle Z_2 + \angle Z_3 - \angle Z_4$$

There are therefore a very large number of possible bridge configurations. The most useful can be classified according to the following scheme due to Ferguson. Since the unknown impedance has only two parameters R_1 and X_1, it is therefore sufficient to adjust only two of the six available parameters on the right-hand side of the balance equation. If the adjustment for each parameter of the unknown impedance is to be independent then the variables should be adjusted in the same branch. Adjusting the parameters R_2, X_2 is the same as adjusting parameters R_3, X_3, and thus four-arm bridges can be classified into one of two types, either ratio bridges or product bridges.

In the ratio bridge the adjustable elements in either Z_2 or Z_3 are adjacent to the unknown impedance and the ratio either Z_3/Z_4 or Z_2/Z_4 must be either real or imaginary but not complex if the two elements in the balance condition are to be independent. In product bridges the balance is achieved by adjusting the elements in Z_4, which is opposite the unknown. For the adjustments to be independent requires $Z_2 . Z_3$ to be real or imaginary but not complex.

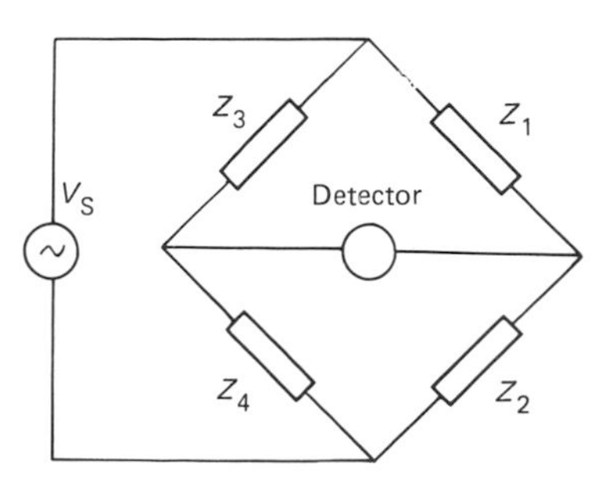

Figure 1.59 A.C. four-arm bridge.

Bridge	*Circuit*	*Balance conditions*	*Notes*
Maxwell	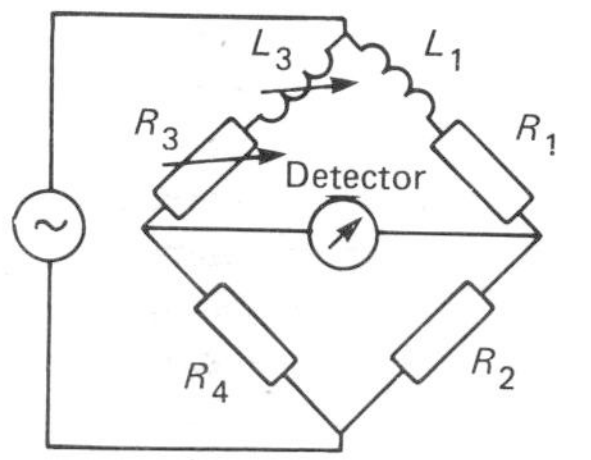	$L_1 = \frac{R_2}{R_4} L_3$ $R_1 = \frac{R_2}{R_4} R_3$	Ratio bridge with inductive and resistive standards for the measurement of the series inductance and resistance of an unknown inductor; balance condition is frequency independent and therefore purity of source is unimportant; a parallel form of the bridge can be used to measure the parallel components of an unknown inductance
Maxwell–Wien	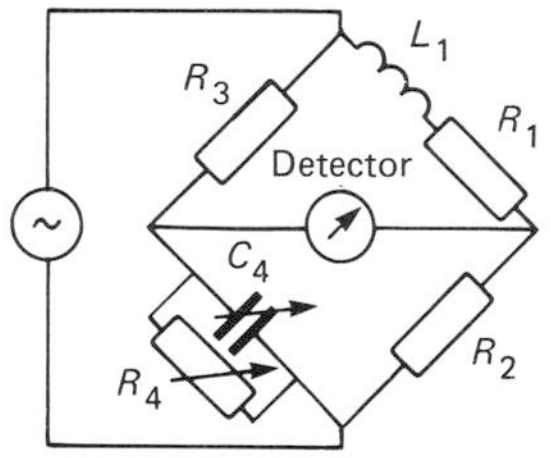	$L_1 = R_2 R_3 C_4$ $R_1 = \frac{R_2 R_3}{R_4}$ $Q_1 = \omega C_4 R_4$	Product bridge employing capacitive and resistive standards for the measurement of the series inductance and resistance of an unknown inductor; widely used for the measurement of inductance; if C_4 and R_4 are variable bridge measures L_1 and R_1; if R_4 and R_2 or R_3 are variable bridge measures L_1 and Q_1
Hay	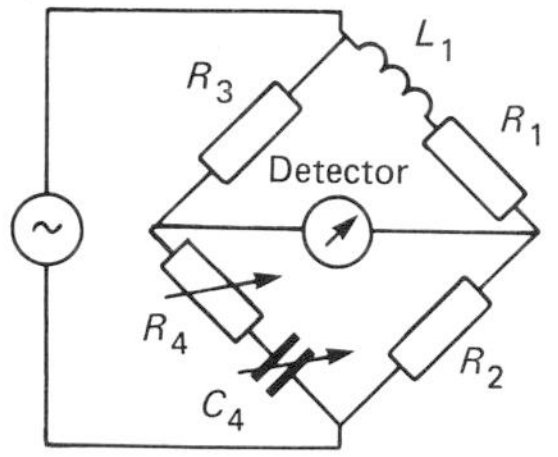	$L_1 = \frac{R_2 R_3 C_4}{1 + \omega^2 C_4^2 R_4^2}$ $R_1 = \frac{R_2 R_3 \omega^2 C_4^2 R_4^2}{(1 + \omega^2 C_4^2 R_4^2)}$ $Q_1 = \frac{1}{\omega C_4 R_4}$	Product bridge employing capacitive and resistive standards for the measurement of the series inductance and resistance of an unknown inductor; suitable for the measurement of a.c. inductance in the presence of d.c. bias current; usedfor the measurement of inductances with high L and Q
Owen	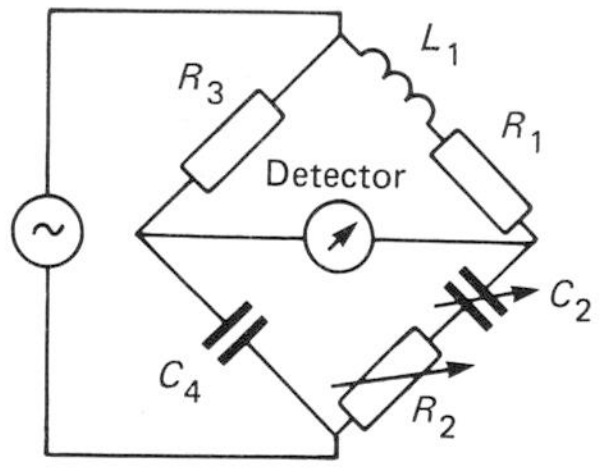	$L_1 = C_4 R_3 . R_2$ $G_1 = \frac{1}{R_1} = \frac{1}{C_4 R_3} . C_2$	Ratio bridge employing capacitive and resistive standards for the measurement of the series inductance and conductance of an unknown inductor; used as a high-precision bridge

Figure 1.60 A.C. four-arm bridges for the measurement of capacitance and inductance.

Bridge	*Circuit*	*Balance conditions*	*Notes*
Series capacitance component bridge	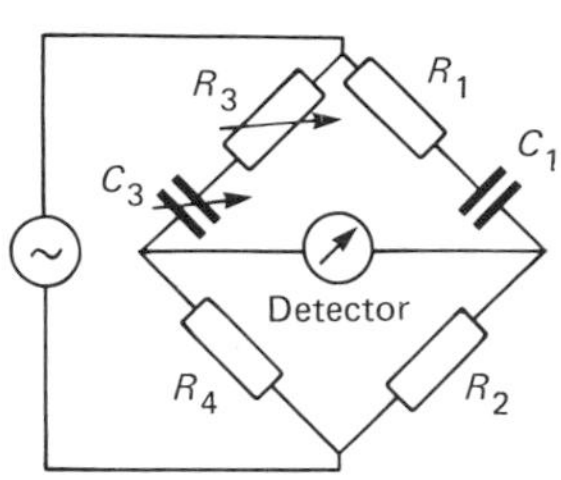	$C_1=\frac{R_4}{R_2}.C_3$ $R_1=\frac{R_2}{R_4}.R_3$ $D_1=\omega C_3R_3$	Ratio bridge employing capacitive and resistive standards for the measurement of the series capacitance and resistance of an unknown capacitor; widely used for the measurement of capacitance; if C_3 and R_3 are variable bridge measures C_1 and R_1; if R_3 and R_4 are variable bridge measures C_1 and D_1
Parallel capacitance component bridge	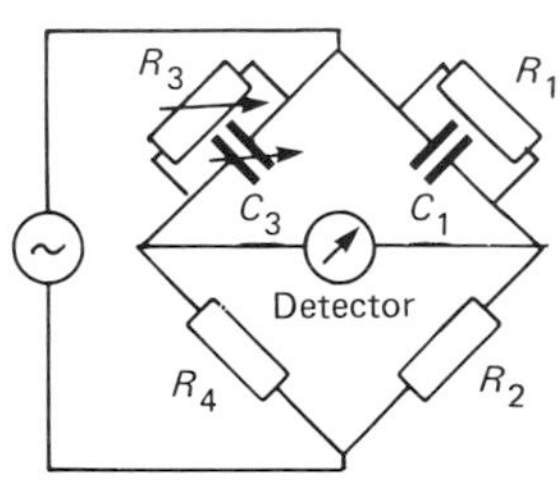	$C_1=\frac{R_4}{R_2}.C_3$ $R_1=\frac{R_4}{R_2}.R_3$ $D_1=\frac{1}{\omega C_3R_3}$	Ratio bridge employing capacitive and resistive standards for the measurement of the parallel capacitance and resistance of an unknown capacitor; used particularly for high D capacitor measurement
Maxwell–Wien	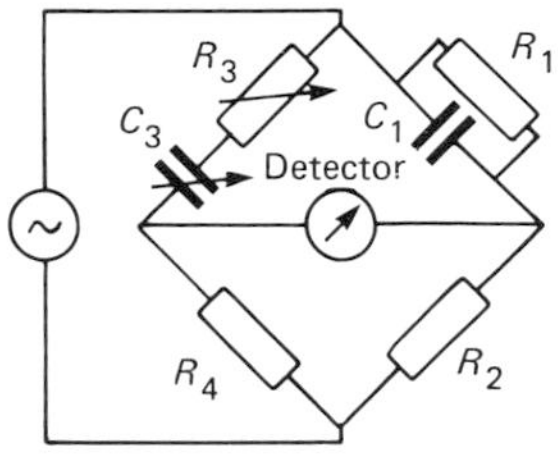	$C_1=\frac{R_4}{R_2}.\frac{C_3}{1+\omega^2C_3^2R_3^2}$ $R_1=\frac{R_2}{R_4}.\frac{1+\omega^2C_3^2R_3^2}{\omega^2C_3^2R_3}$ $D_1=\omega C_3R_3$	Ratio bridge employing capacitive and resistive standards for the measurement of the parallel capacitance and resistance of an unknown capacitor; used as a frequency-dependent circuit in oscillators
Schering	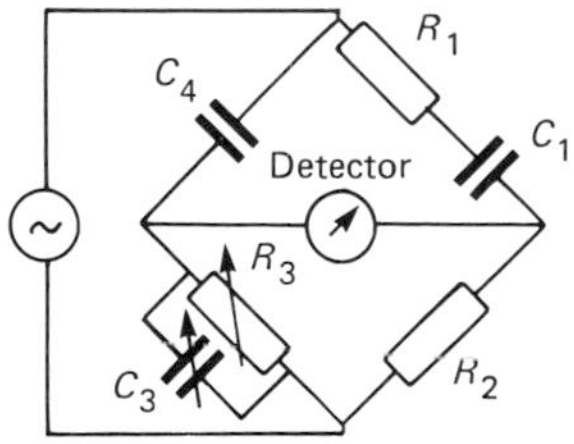	$C_1=\frac{C_4}{R_2}.R_3$ $R_1=\frac{R_2}{C_4}.C_3$ $D_1=\omega C_3R_3$	Product bridge employing capacitive and resistive standards for the measurement of the parallel capacitance and resistance of an unknown capacitor; used for measuring dielectric losses at high voltage and r.f. measurements

Figure 1.60 (cont.)

Figure 1.60 gives examples of a range of commonly used four-arm bridges for the measurement of C and L. For further details concerning the application of such bridges the reader should consult Hague and Foord (1971).

1.7.3.1 Stray impedances in a.c. bridges

Associated with the branches, source and detector of an a.c. bridge there are distributed capacitances to ground. The use of shields around these elements enable the stray capacitances to be defined in terms of their location, magnitude and effect. Figure 1.61(a) shows these capacitances and Figure 1.61(b) shows the equivalent circuit with the stray capacitances transformed to admittances across the branches of the bridge and the source and detector. The stray admittances across the source and detector do not affect the balance condition. The balance condition of the bridge in terms of the admittances of the branches and the admittances of the stray capacitances across them is given by

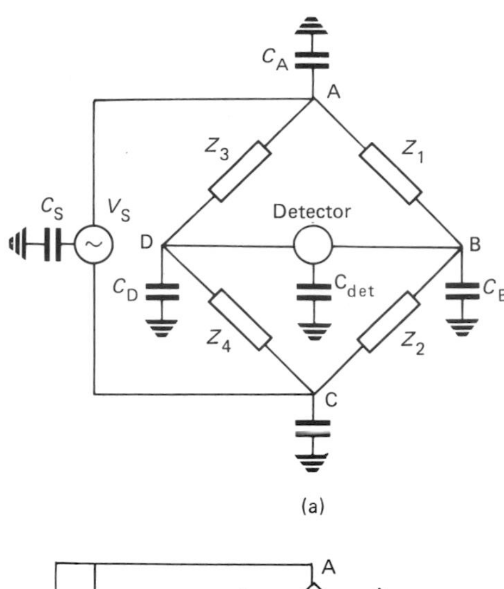

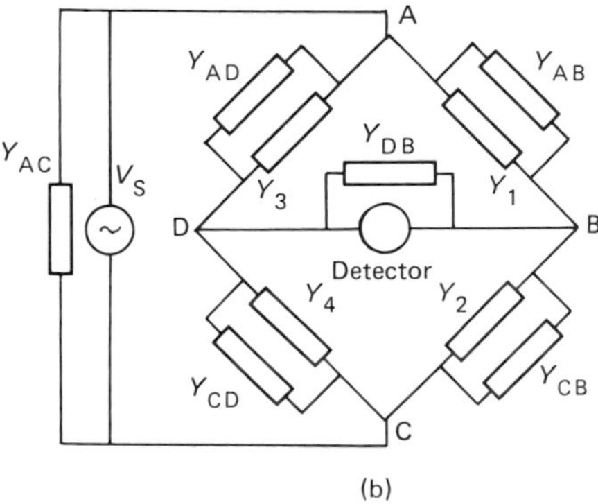

Figure 1.61 (a) Stray capacitances in a four-arm a.c. bridge; (b) equivalent circuit of an a.c. four-arm bridge with stray admittances.

$$(Y_1+Y_{AB})(Y_4+Y_{CD})=(Y_3+Y_{AD})(Y_2+Y_{CB})$$

where, for example

$$Y_{AB}=\frac{Y_A Y_B}{Y_A+Y_B+Y_C+Y_D}=\frac{Y_A Y_B}{\Delta};$$

$$\Delta=Y_A+Y_B+Y_C+Y_D$$

and thus the balance condition is given by

$$(Y_1 Y_4-Y_2 Y_3)$$
$$+\frac{1}{\Delta}(Y_1 Y_C Y_D+Y_4 Y_A Y_B-Y_3 Y_C Y_B-Y_2 Y_A Y_D)=0$$

If the stray capacitances are to have no effect on the balance condition then this must be given by

$$Y_1 Y_4=Y_2 Y_3$$

and the second term of the balance condition must be zero. It can be easily shown that this can be achieved by either

$$\frac{Y_A}{Y_C}=\frac{Y_1}{Y_2}=\frac{Y_3}{Y_4} \quad \text{or} \quad \frac{Y_B}{Y_D}=\frac{Y_1}{Y_3}=\frac{Y_2}{Y_4}$$

Thus the stray impedances to ground have no effect on the balance condition if the admittances at one opposite pair of branch points are in the same ratio as the admittances of the pairs of branches shunted by them.

The Wagner earthing arrangement shown in Figure 1.62 ensures that points D and B of the balanced bridge are at ground potential and thus the effect of stray impedances at these points is eliminated. This is

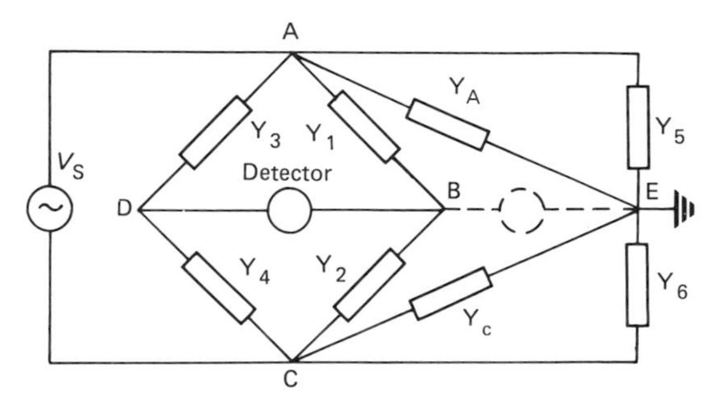

Figure 1.62 Wagner earthing arrangement.

achieved by means of an auxiliary arm of the bridge consisting of the elements Y_5 and Y_6. The bridge is first balanced with the detector between D and B by adjusting Y_3. The detector is moved between B and E and the auxiliary bridge balanced by adjusting Y_5 and Y_6. This ensures that point B is at earth potential. The two balancing processes are repeated until the bridge balances with the detector in both positions. The balance conditions for the main bridge and the auxiliary arm are then given by

$$Y_1 Y_4=Y_2 Y_3 \quad \text{and}$$
$$Y_3(Y_6+Y_C)=Y_4(Y_5+Y_A)$$

1.7.4 Transformer ratio bridges

These bridges, which are also called inductively coupled bridges, largely eliminate the problems associated with stray impedances. They also have the advantage that only a small number of standard resistors and capacitors is needed. Such bridges are therefore commonly used as universal bridges to measure the resistance, capacitance and inductance of components having a wide range of values at frequencies up to 250 MHz.

The element which is common to all transformer

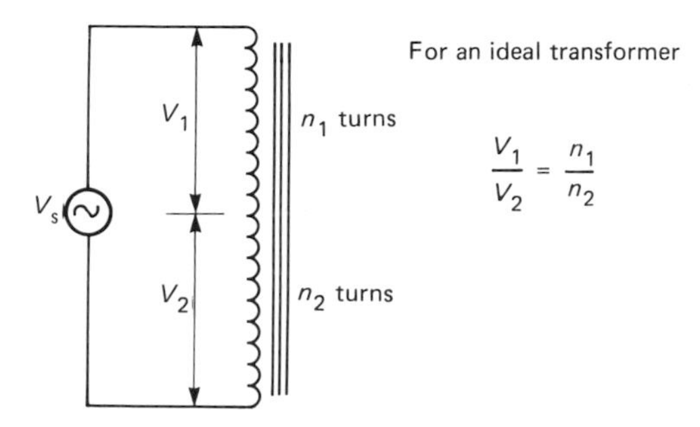

Figure 1.63 Tapped transformer winding.

ratio bridges is the tapped transformer winding, shown in Figure 1.63. If the transformer is ideal then the windings have zero leakage flux, which implies that all the flux from one winding links with the other, and zero winding resistance. The core material on which the ideal transformer is wound has zero eddy-current and hysteresis losses. Under these circumstances the ratio of the voltages V_1 to V_2 is identical to the ratio of the turns n_1 to n_2, and this ratio is independent of the loading applied to either winding of the transformer.

In practice the transformer is wound on a tape-wound toroidal core made from a material such as supermalloy or supermumetal which has low eddy-current and hysteresis loss and also high permeability. The coil is wound as a multistranded rope around the toroid with individual strands in the rope joined in series as shown in Figure 1.64. This configuration minimizes the leakage inductance of the windings. The windings are made of copper with the largest cross-sectional area to minimize their resistance. Figure 1.65 shows an equivalent circuit of such a transformer. L_1 and L_2 are the leakage inductances of the windings; R_1 and R_2 are the winding resistances; M is the mutual inductance between the windings; and R represents hysteresis and eddy-current loss in the core.

The ratio error from the ideal value of n_1/n_2 is given

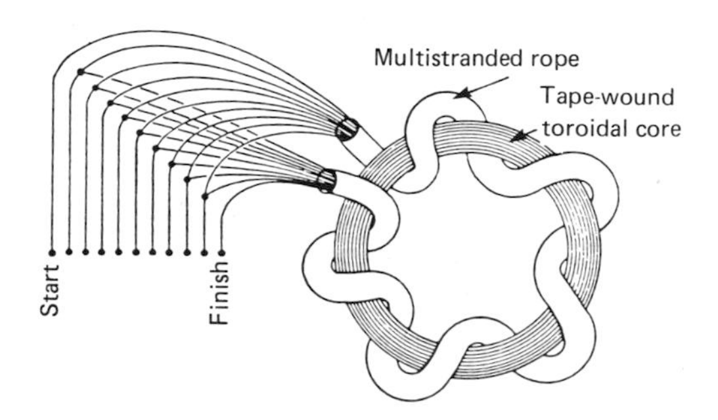

Figure 1.64 Construction of a toroidal tapped transformer.

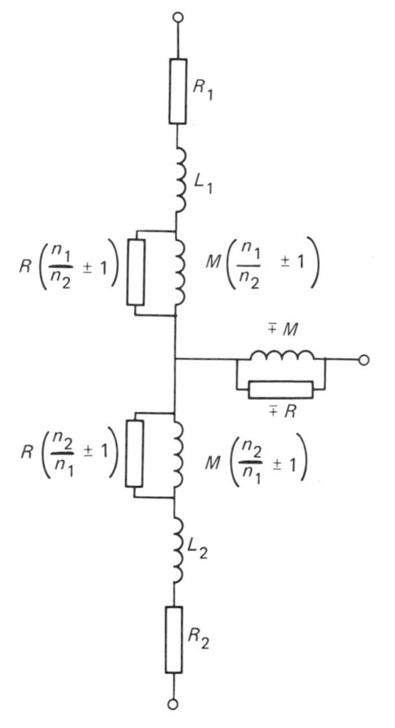

Figure 1.65 Equivalent circuit of a tapped transformer.

approximately by

$$\frac{n_2(R_1+j\omega L_1)-n_1(R_2+j\omega L_2)}{(n_1+n_2)}\cdot\left(\frac{1}{R}+\frac{1}{j\omega M}\right)\times 100\%$$

and this error can be made to be less than 1 part in 10^6.

The effect of loading is also small. An impedance Z applied across the n_2 winding gives a ratio error of

$$\frac{(n_1/n_2)(R_2+j\omega L_2)+(n_2/n_1)(R_1+j\omega L_1)}{[(n_1+n_2)/n_2]\,.\,Z}\times 100\%$$

For an equal bridge, with $n_1=n_2$, this is

$$\frac{R_2+j\omega L_2}{Z}\times 100\%$$

which is approximately the same error as if the transformer consisted of a voltage source with an output impedance given by its leakage inductance and the winding resistance. These can be made to be small and thus the effective output impedance of the transformer is low; therefore the loading effect is small. The input impedance of the winding seen by the a.c. source is determined by the mutual inductance of the windings (which is high) and the loss resistance (which is also high).

Multi-decade ratio transformers as shown in Figure 1.66 use windings either with separate cores for each

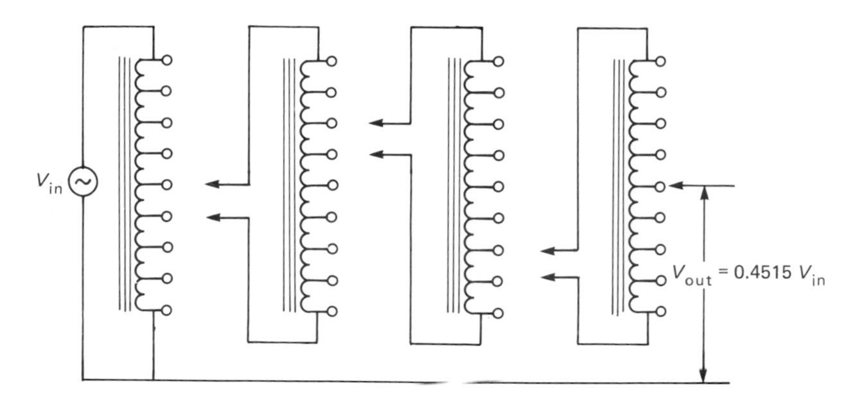

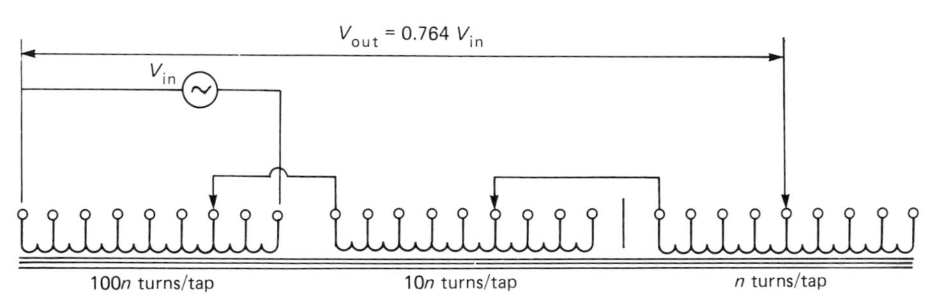

Figure 1.66 Multi-decade ratio transformers.

decade or all wound on the same core. For the multi-core transformer the input for the next decade down the division chain is the output across a single tap of the immediately higher decade. For the windings on a single core the number of decades which can be accommodated is limited by the need to maintain the volts/turn constant over all the decades, and therefore the number of turns per tap at the higher decade becomes large. Generally a compromise is made between the number of cores and the number of decades on a single core.

1.7.4.1 Bridge configurations

There are three basic bridge configurations, as shown in Figure 1.67. In Figure 1.67(a) the detector indicates a null when

$$\frac{Z_1}{Z_2}=\frac{V_1}{V_2}$$

and for practical purposes

$$\frac{V_1}{V_2}=\frac{n_1}{n_2}=n$$

Thus

$$Z_1=nZ_2; \quad |Z_1|=n|Z_2| \quad \text{and} \quad \angle Z_1=\angle Z_2$$

The bridge can therefore be used for comparing like impedances.

The three-winding voltage transformer shown in Figure 1.67(b) has the same balance condition as the bridge in Figure 1.67(a). However, in the three-winding bridge the voltage ratio can be made more nearly equal to the turns ratio. The bridge has the disadvantage that the leakage inductance and winding resistance of each section is in series with Z_1 and Z_2 and therefore the bridge is most suitable for the measurement of high impedances.

Figure 1.67(c) shows a double ratio transformer bridge in which the currents I_1 and I_2 are fed into a second double-wound transformer. The detector senses a null condition when there is zero flux in the core of the second transformer. Under these conditions for an ideal transformer

$$I_1n_1'=I_2n_2'; \qquad \frac{I_1}{I_2}=\frac{n_2'}{n_1'}=\frac{1}{n'}$$

and the second transformer presents zero input impedance. Therefore since

$$\frac{I_1}{I_2}=\frac{Z_2}{Z_1}\cdot\frac{n_1}{n_2}=\frac{Z_2}{Z_1}n$$

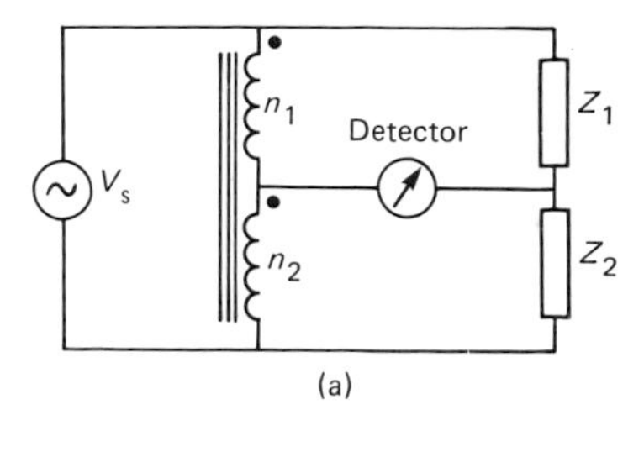

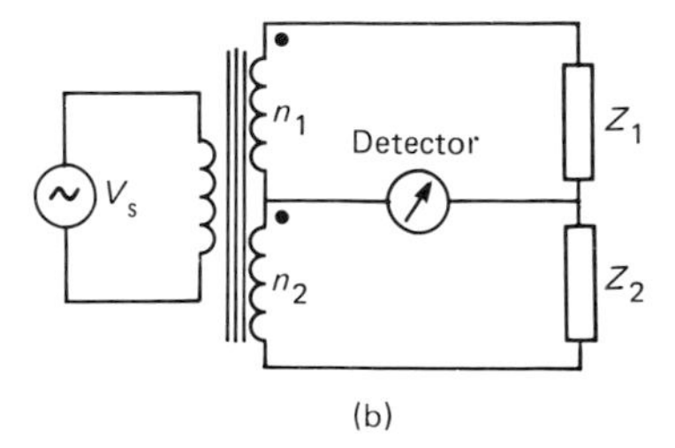

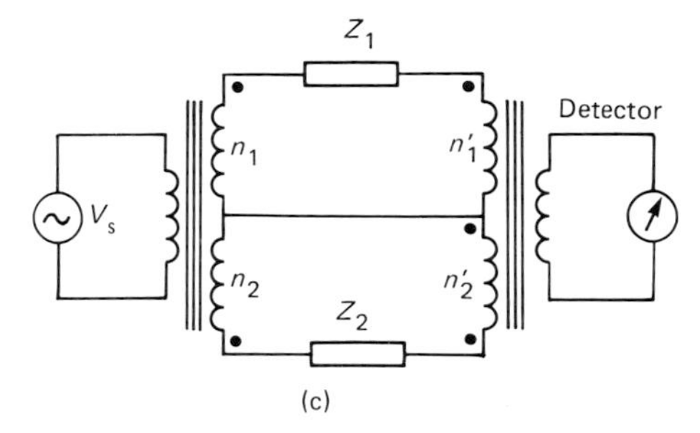

Figure 1.67 (a) Autotransformer ratio bridge; (b) double-wound transformer ratio bridge; (c) double ratio bridge.

then

$$Z_1 = nn'Z_2; \qquad |Z_1| = nn'|Z_2| \qquad \text{and}$$

$$\angle Z_1 = \angle Z_2$$

By using the two ratios this bridge extends the range of measurement which can be covered by a small number of standards.

Figure 1.68 shows a universal bridge for the measurement of R, C and L. In the figure only two decades of the inductive divider which control the voltages applied to the bank of identical fixed capcitors and resistors are shown. The balance condition for the bridge when connected to measure capacitance is given by

$$C_u = \frac{n'_2}{n'_1}\left(\frac{n_2}{10} + \frac{n_4}{100}\right) . C_S$$

and

$$\frac{1}{R_u} = \frac{n'_2}{n'_1}\left(\frac{n_1}{10} + \frac{n_3}{100}\right) \cdot \frac{1}{R_S}$$

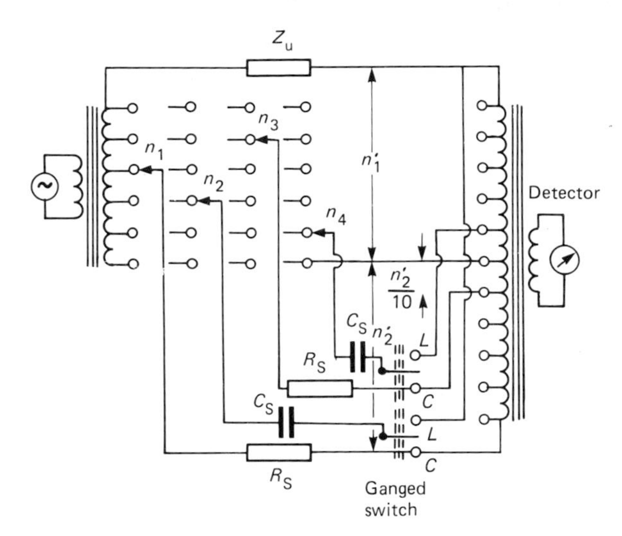

Figure 1.68 Universal bridge.

When measuring inductance the current through the capacitor and inductor are summed into the current transformer and the value of capacitance determined is the value which resonates with the inductance. For an unknown inductance its measured values in terms of its parallel equivalent circuit is given by

$$L_{up} = \frac{1}{\omega^2 C_u}; \qquad \frac{1}{R_{up}} = \frac{1}{R_u}$$

where the values of C_u and R_u are given in the above equations. The value of ω is chosen such that it is a multiple of ten and therefore the values of L_{up} and C_u are reciprocal. The values of L_{up} and R_{up} can be converted to their series equivalent values using the equations in section 1.7.2.

The transformer ratio bridge can also be configured to measure low impedances, high impedances and network and amplifier characteristics. The ampere turn balance used in ratio bridges is also used in current comparators employed in the calibration of current transformers and for intercomparing four-terminal impedances. Details of these applications can be found in Gregory (1973), Hague and Foord (1971) and Oliver and Cage (1971). The current comparator principle can also be extended to enable current comparison to be made at d.c. (Dix and Bailey, 1975).

Transformer ratio bridges are often used with capacitive and inductive displacement transducers because they are immune to errors caused by earth-leakage impedances and since they offer an easily constructed, stable and accurately variable current or voltage ratio (Hugill, 1983; Neubert, 1975).

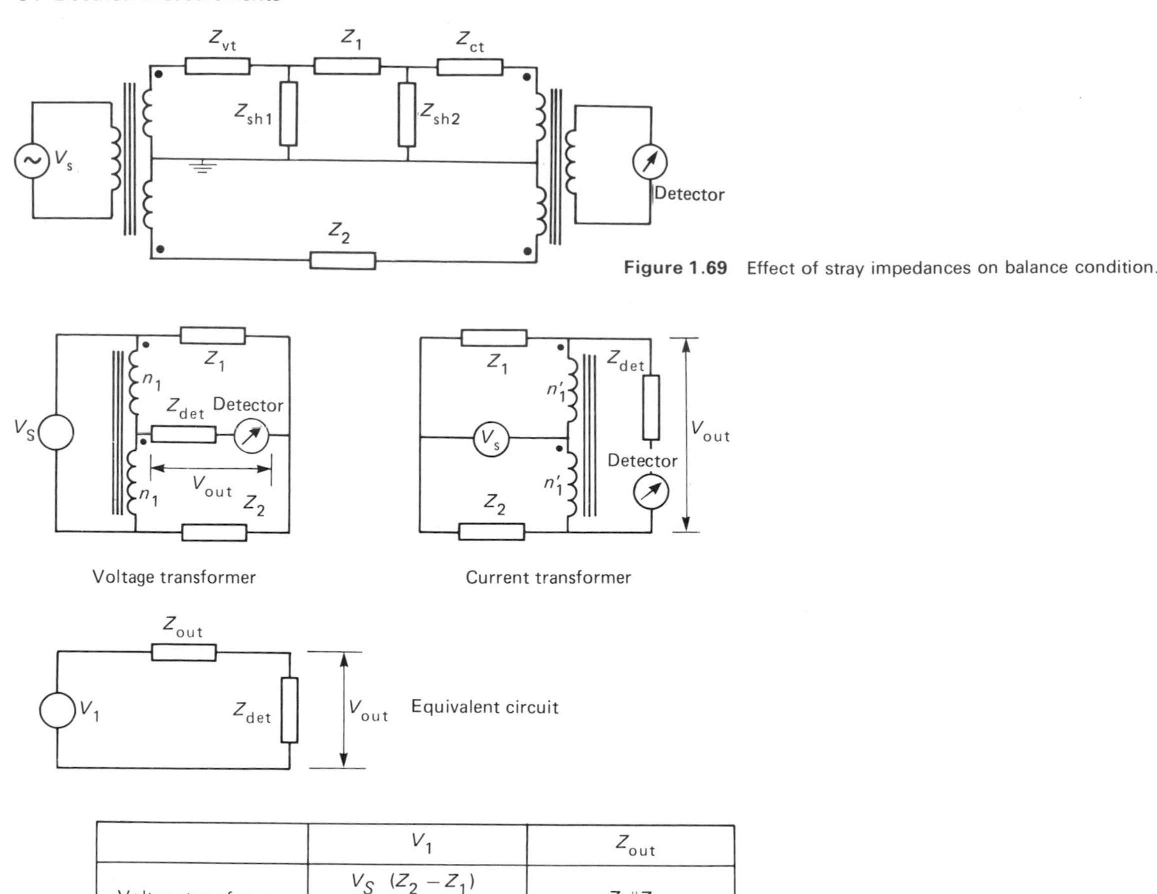

Figure 1.69 Effect of stray impedances on balance condition.

	V_1	Z_{out}
Voltage transformer	$\frac{V_S}{2}\frac{(Z_2 - Z_1)}{(Z_2 + Z_1)}$	$Z_1 \Vert Z_2$
Current transformer	$\frac{2V_S\,(Z_2 - Z_1)}{(Z_2 + Z_1 + Z_1 Z_2/Z_c)}$	$Z_1 \Vert 2Z_c + Z_2 \Vert 2Z_c$

$Z_C = j\omega L_C$: L_C is inductance of ratio arms
$L_c = M_c$ mutual inductance of ratio arms

Figure 1.70 Unbalanced inductively coupled bridge.

1.7.4.2 The effect of stray impedances on the balance condition of inductively coupled bridges

Figure 1.69 shows the unknown impedance with its associated stray impedances Z_{sh1} and Z_{sh2}. The balance condition of the bridge is unaffected by Z_{sh1} since the ratio of V_1 to V_2 is unaffected by shunt loading. At balance the core of the current transformer has zero nett flux. There is no voltage drop across its windings and hence there is no current flow through Z_{sh2}. Z_{sh2} has therefore no effect on the balance condition. Thus the bridge rejects both stray impedances. This enables the bridge to measure components *in situ* whilst still connected to other components in a circuit. In practice if the output impedance of the voltage transformer has a value Z_{vt} and the current transformer has an input impedance of Z_{ct} then the error on the measurement of Z_1 is given approximately by

$$\left(\frac{Z_{vt}}{Z_{sh1}} + \frac{Z_{ct}}{Z_{sh2}}\right) \times 100\%$$

1.7.4.3 The use of inductively coupled bridges in an unbalanced condition

The balance condition in inductively coupled bridges is detected as a null. The sensitivity of the bridge determines the output under unbalance conditions and therefore the precision with which the balance can be found. Figure 1.70 shows the two-winding voltage and current transformers and their equivalent circuits.

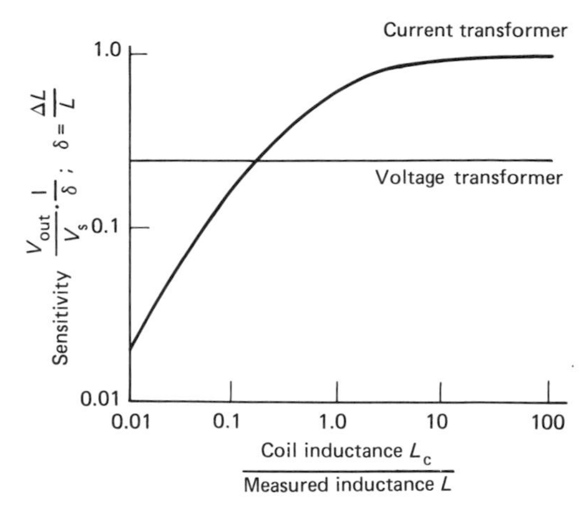

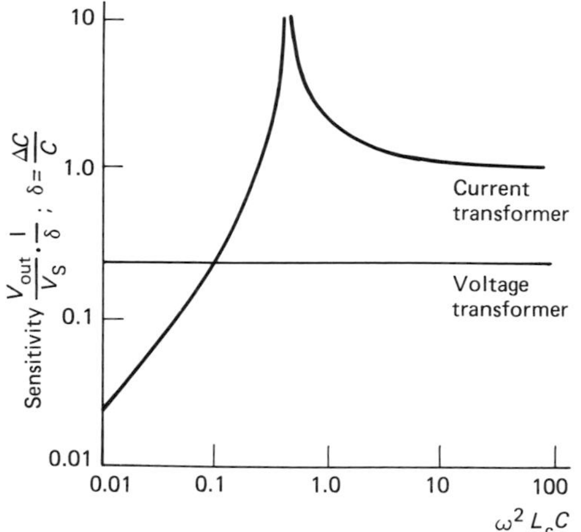

Figure 1.71 Sensitivity of current and voltage transformer bridges.

Figure 1.71 shows the sensitivities of the two bridges when used with capacitive and inductive elements. The capacitors form a resonant circuit with the current transformer and for frequencies below the resonant frequency the sensitivity of the bridge is dependent on both ω, the angular excitation frequency of the bridge, and L_c, the self-inductance of the winding as shown in Figure 1.71. The dependence of the sensitivity on ω and L_c can be reduced at the cost of reduced sensitivity (Neubert, 1975).

1.7.4.4 Autobalancing ratio bridges

By employing feedback as shown in Figure 1.72 the transformer ratio bridge can be made to be self-balancing. The high-gain amplifier ensures that at balance the current from the unknown admittance Y_u is balanced by the current through the feedback resistor. Thus at balance

$$V_1 Y_u n_1' = \frac{V_{out}}{R} \cdot n_2'$$

with

$$V_1 = \hat{V}_1 \sin \omega t$$

$$V_{out} = \hat{V}_{out} \sin(\omega t + \phi)$$

and

$$Y_u = G_u + jB_u$$

$$G_u = \frac{n_2'}{n_1'} \cdot \frac{1}{R} \cdot \frac{V_{out}}{V_1} \cos \phi;$$

$$B_u = \frac{n_2'}{n_1'} \cdot \frac{1}{R} \cdot \frac{V_{out}}{V_1} \cdot \sin \phi$$

The amplifier output and a signal 90 degrees shifted from that output are then passed into two phase-sensitive detectors. These detectors employ reference voltages which enable the resistive and reactive components of the unknown to be displayed.

Windings can be added to the bridge which enable the bridge to measure the difference between a standard and the unknown.

1.7.5 High-frequency impedance measurement

As the frequency of measurement is increased the parasitic elements associated with real components begin to dominate the measurement. Therefore rf bridges employ variable capacitors (typically less than 1000 pF) as the adjustable elements in bridges and fixed resistors whose physical dimensions are small. A bridge which can be constructed using these elements is the Schering bridge, shown in Figure 1.60. Great care has to be taken with shielding and wiring layout in rf bridges to avoid large coupling loops. The impedance range covered by such bridges decreases as the frequency is raised. At microwave frequencies all the wiring is coaxial, discrete components are no longer used and impedance measurements can only be undertaken for impedances close to the characteristic impedance of the system. Further details of high-frequency measurements can be found in Oliver and Cage (1971) and Somlo and Hunter (1985).

The bridged T and parallel T circuits (shown in Figure 1.73 together with their balance conditions) can be used for measurements at rf frequencies. The parallel T measurement technique has the advantage that the balance can be achieved using two grounded variable capacitors.

Resonance methods can also be used for the measurement of components at high frequencies. One of the most important uses of resonance in component measurement is the Q meter, shown in Figure 1.74. In measuring inductance as shown in Figure 1.74(a) the variable capacitor C, which forms a series-resonant circuit with L_{us}, is adjusted until the detector detects resonance at the frequency f. The resonance is detected as a maximum voltage across C. At resonance Q is given by

$$Q=\frac{V_c}{V_{in}}=\frac{V_L}{V_{in}}$$

and L_{us} is given by

$$L_{us}=\frac{1}{4\pi^2 f^2 C}$$

The value of R_{us} is given by

$$R_{us}=\frac{1}{2\pi f C Q}$$

The self-capacitance of an inductor can be determined by measuring the value of C, say C_1, which resonates with it at a frequency f together with value of C, say C_2, which resonates with the inductance at $2f$. Then C_0, the self-capacitance of the coil, is given by

$$C_0=\frac{C_1-4C_2}{3}$$

In Figure 1.74(b) the use of the Q meter to measure the equivalent parallel capacitance and resistance of a capacitor is shown. Using a standard inductor at a frequency f, the capacitor C is adjusted to a value $C_1 C_c$ at which resonance occurs. The unknown capacitor is connected across C and the value of C is adjusted until resonance is found again. If this value is C_2 then the unknown capacitor C_{up} has a value given by

$$C_{up}=C_1-C_2$$

Its dissipation factor, D, is given by

$$D=\frac{Q_1-Q_2}{Q_1 Q_2}\cdot\frac{C_1}{C_1-C_2}$$

where Q_1 and Q_2 are the measured Q values at the two resonances. Its parallel resistance, R_{up}, is given by

$$R_{up}=\frac{Q_1 Q_2}{Q_1-Q_2}\cdot\frac{1}{2\pi f C_1}$$

The elements of the high-frequency equivalent circuit of a resistance in Figure 1.74(c) can also be measured. At a given frequency, f, the capacitor C is adjusted to a value C_1 such that it resonates with L. The resistor is then connected across the capacitor and the value of C adjusted until resonance is re-established. Let this value of C be C_2. If the values of Q at the resonances are Q_1 and Q_2, respectively, then values of the unknown elements are given by

$$R_{up}=\frac{Q_1 Q_2}{(Q_1-Q_2)}\cdot\frac{1}{2\pi f C_1}$$

$$C_{up}=C_1-C_2$$

and

$$L_{up}=\frac{1}{(2\pi f)^2 . C_{up}}$$

1.8 Digital frequency and period/time-interval measurement

These measurements, together with frequency ratio, phase difference, rise and fall time, and duty-factor measurements employ digital counting techniques and are all fundamentally related to the measurement of time.

The SI unit of time is defined as the duration of 9 192 631 770 periods of the radiation corresponding to the transition between the $F=4$, $m_f=0$ and $F=3$, $m_f=0$ hyperfine levels of the ground state of the caesium-133 atom. The unit is realized by means of the caesium-beam atomic clock in which the caesium beam undergoes a resonance absorption corresponding to the required transition from a microwave source. A feedback mechanism maintains the frequency of the microwave source at the resonance frequency. The SI unit can be realized with an uncertainty of between 1 part in 10^{13} and 10^{14}. Secondary standards are provided by rubidium gas cell resonator-controlled oscillators or quartz crystal oscillators. The rubidium oscillator uses an atomic resonance effect to maintain the frequency of a quartz oscillator by means of a frequency-lock loop. It provides a typical short-term stability (averaged over a 100-s period) of five parts in 10^{13} and a long-term stability of one part in 10^{11}/month. Quartz crystal oscillators provide inexpensive secondary standards with a typical short-term stability (averaged over a 1-s period) of five parts in 10^{12} and a long-term stability of better than one part in 10^8/month. Details of time and frequency standards can be found in Hewlett Packard (1974).

Dissemination of time and frequency standards is also undertaken by radio broadcasts. Radio stations transmit waves whose frequencies are known to an uncertainty of a part in 10^{11} or 10^{12}. Time-signal broadcasting on a time scale known as Coordinated Universal Time (UTC) is coordinated by the Bureau International de L'Heure (BIH) in Paris. The BIH annual report details the national authorities responsible for time-signal broadcasts, the accuracies

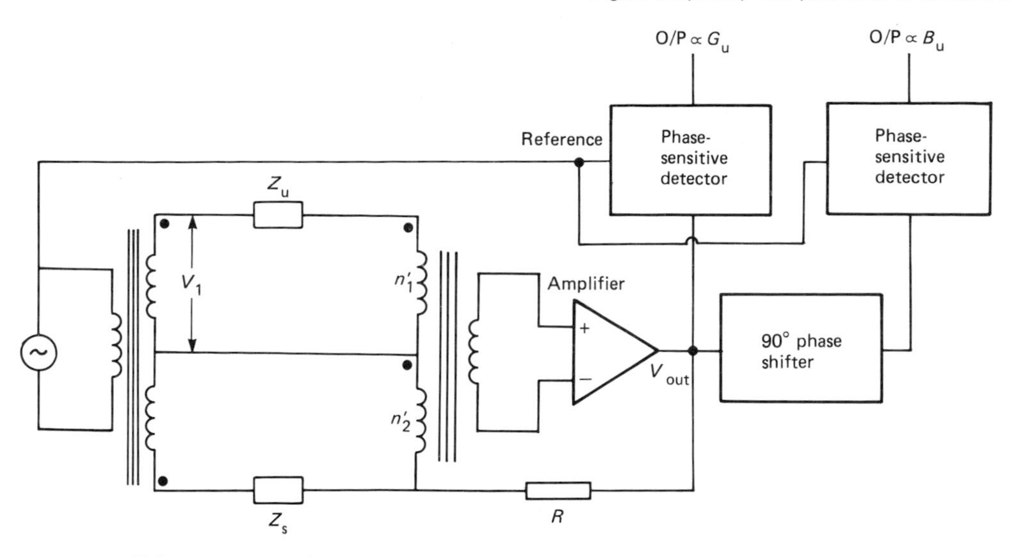

If $Z_s = \infty$ output gives conductance and susceptance of unknown impedance Z_u

If $Z_s \neq \infty$ output gives deviation of conductance and susceptance from the values of the standard impedance Z_s

Figure 1.72 Autobalancing ratio bridge.

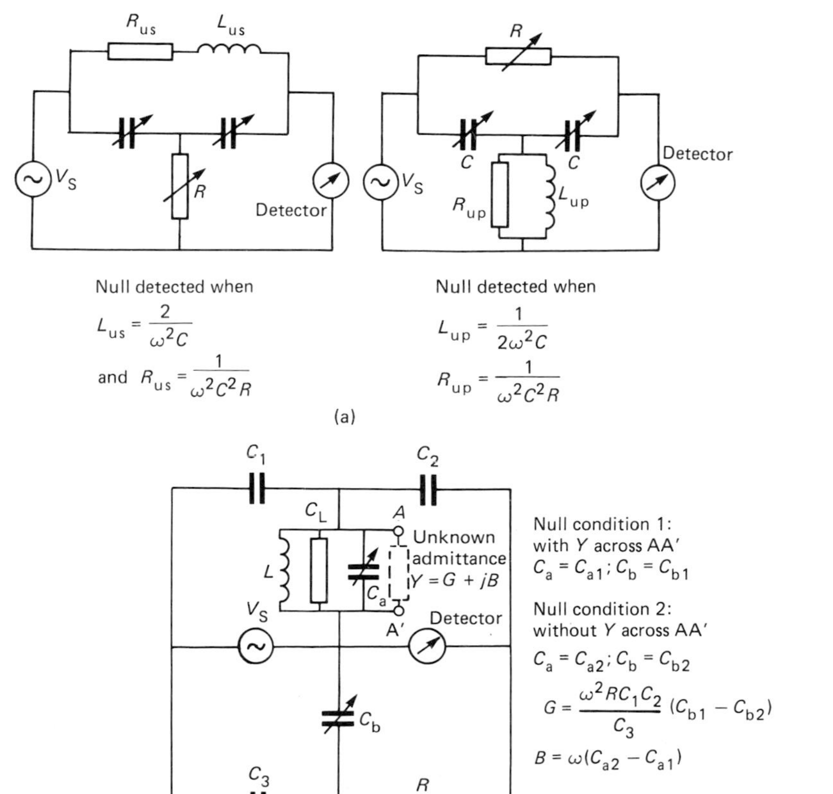

Figure 1.73 Bridged T (a) and parallel T (b) circuits for the measurement of impedance at high frequencies.

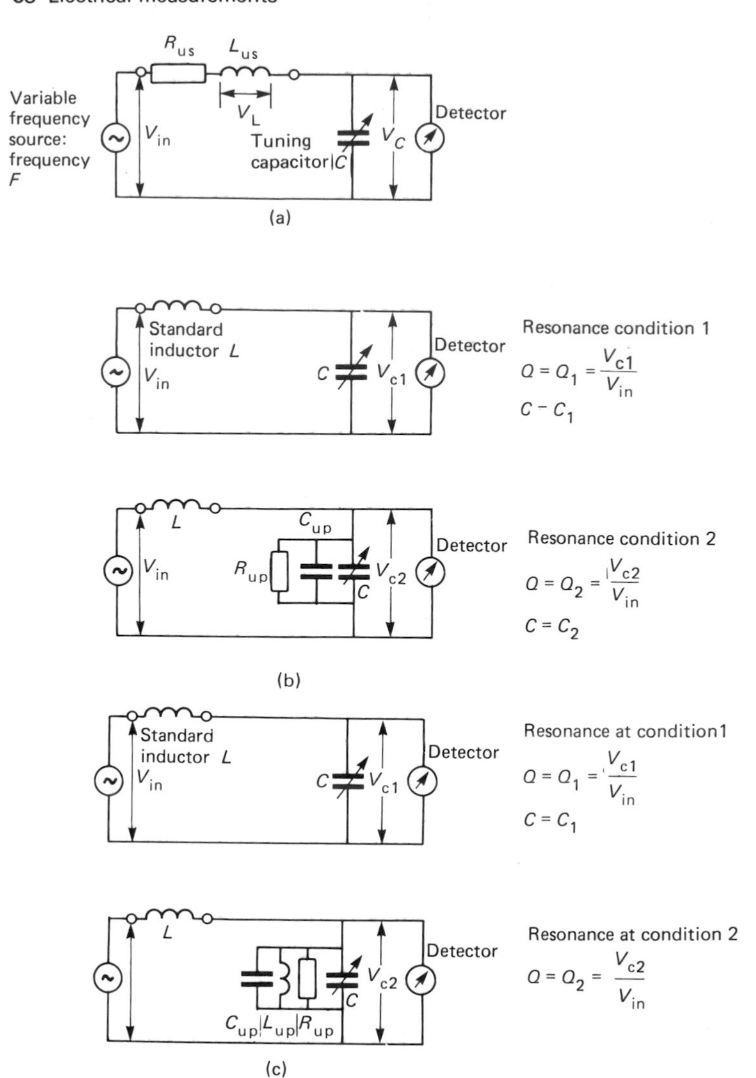

Figure 1.74 Q meter. (a) Inductance measurement; (b) capacitance measurement; (c) resistance measurement.

of the carrier frequencies of the standard frequency broadcasts and the characteristics of national time-signal bradcasts. Table 1.9 provides details of time broadcast facilities in the UK.

1.8.1 Frequency counters and universal timer/counters

Frequency measurements are undertaken by frequency counters whose functions (in addition to frequency measurement) may also include frequency ratio, period measurement and totalization. Universal timer/counters provide the functions of frequency counters with the addition of time-interval measurement. Figure 1.75 shows the elements of a microprocessor-controlled frequency counter. The input signal conditioning unit accepts a wide range of input signal levels typically with a maximum sensitivity corresponding to a sinusoid having an rms value of 20 mV and a dynamic range from 20 mV rms to 20 V rms. The trigger circuit has a trigger level which is either set automatically with respect to the input wave or can be continuously adjusted over some range. The trigger circuit generally employs hysteresis to reduce the effect of noise on the waveform as shown in Figure 1.76(a), although this can cause errors in

Table 1.9 UK time broadcasts

GBR 16 kHz radiated from Rugby (52°22′13″ N 01°10′25″ W)
Power: ERP 65 kW

Transmission modes: A1, FSK (16.00 and 15.95 kHz) and MSK (future)

Time signals: Schedule (UTC)	Form of the time signals
: 0255 to 0300 0855 to 0900 1455 to 1500 2055 to 2100 There is an interruption for maintenance from 1000 to 1400 every Tuesday	A1 type second pulses lasting 100 ms, lengthened to 500 ms at the minute The reference point is the start of carrier rise Uninterrupted carrier is transmitted for 24 s from 54 m 30 s and from 0 m 6 s DUT1:CCIR code by double pulses

MSF 60 kHz radiated from Rugby
Power: ERP 27 kW

Schedule (UTC)	Form of the time signals
Continuous except for an interruption for maintenance from 1000 to 1400 on the first Tuesday in each month	Interruptions of the carrier of 100 ms for the second pulses and of 500 ms for the minute pulses. The epoch is given by the beginning of the interruption BCD NRZ code, 100 bits/s (month, day of month, hour, minute), during minute interruptions BCD PWM code, 1 bit/s (year, month, day of month, day of week, hour, minute) from seconds 17 to 59 in each minute DUT1:CCIR code by double pulses

MSF 2.5, 5 and 10 MHz radiated from Rugby (service ends 1988)

ERP 1 kW	Schedule (UTC)	Form of the time signals
	Between minutes 0 and 5, 10 and 15, 20 and 25, 30 and 35, 40 and 45, 50 and 55	Second pulses of 5 cycles of 1 kHz modulation Minute pulses are prolonged DUT1:CCIR code by double pulses

The MSF and GBR transmissions are controlled by a caesium beam frequency standard. Accuracy $\pm 2 \times 10^{-12}$

time measurement, as shown in Figure 1.76(b).

The quartz crystal oscillator in a frequency counter or universal counter timer can be uncompensated, temperature compensated or oven stabilized. The frequency stability of quartz oscillators is affected by ageing, temperature, variations in supply voltage and changes in power supply mode, i.e. changing from line-frequency supply to battery supply. Table 1.10 gives comparative figures for the three types of quartz oscillator. The uncompensated oscillator gives sufficient accuracy for five- or six-digit measurement in most room-temperature applications. The temperature-compensated oscillator has a temperature-dependent compensating network for frequency correction and can give sufficient accuracy for a six- or seven-digit instrument. Oven-stabilized oscillators maintain the temperature of the crystal typically at 70 ± 0.01°C. They generally employ higher mass crystals with lower resonant frequencies and operate at an overtone of their fundamental frequency. They have better ageing performance than the other two types of crystal. They are suitable for use in seven- to nine-digit instruments.

The microprocessor provides control of the counting operation and the display and post-measurement computation.

Conventional frequency counters count the number of cycles, n_i, of the input waveform of frequency, f_i, in a gating period, t_g, which corresponds to a number of counts, n_{osc}, of the 10-MHz crystal oscillator. They have an uncertainty corresponding to ± 1 count of the input waveform. The relative resolution is given by

$$\text{Relative resolution} = \frac{\text{Smallest measurable change in measurement value}}{\text{Measurement value}}$$

and for the measurement of frequency is thus

$$\pm \frac{1}{\text{Gating period} \times \text{input frequency}} = \pm \frac{1}{t_g \cdot f_i}$$

In order to achieve measurements with good relative resolution for low-frequency signals long gating times are required. Reciprocal frequency counters synchronize the gating time to the input

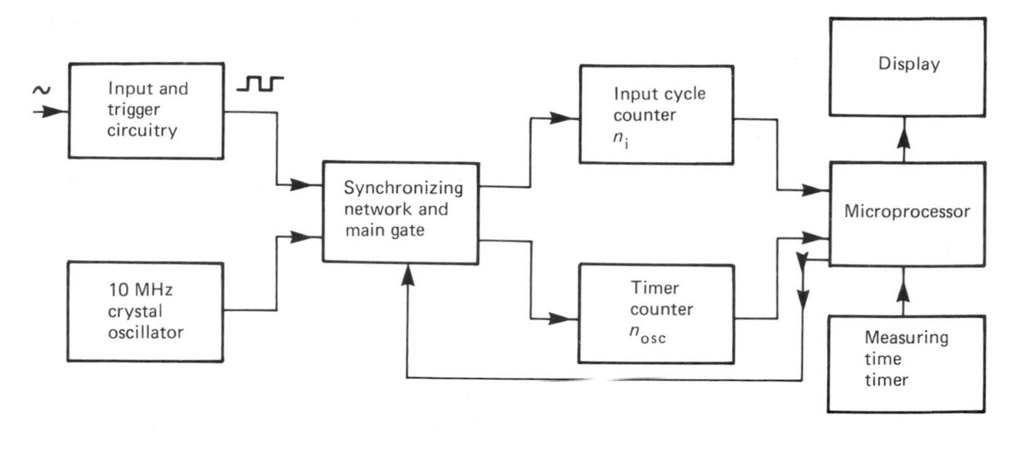

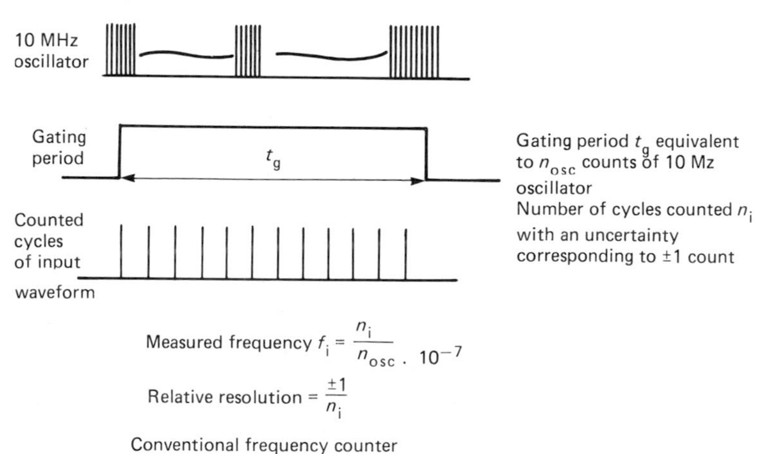

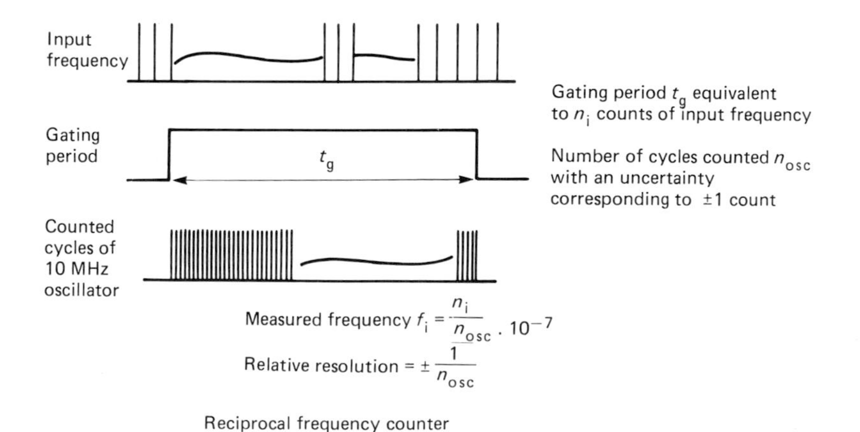

Figure 1.75 Digital frequency counter.

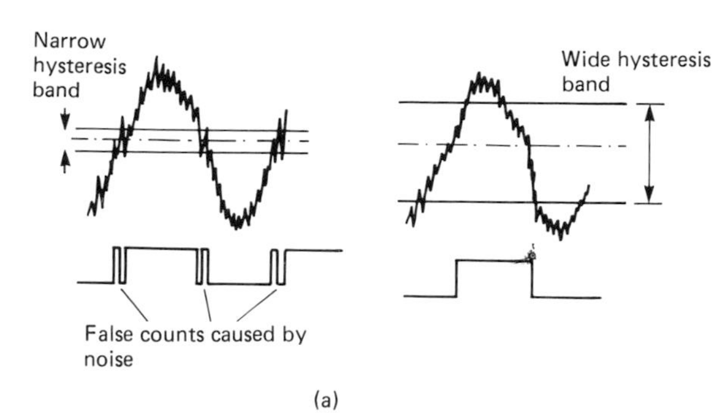

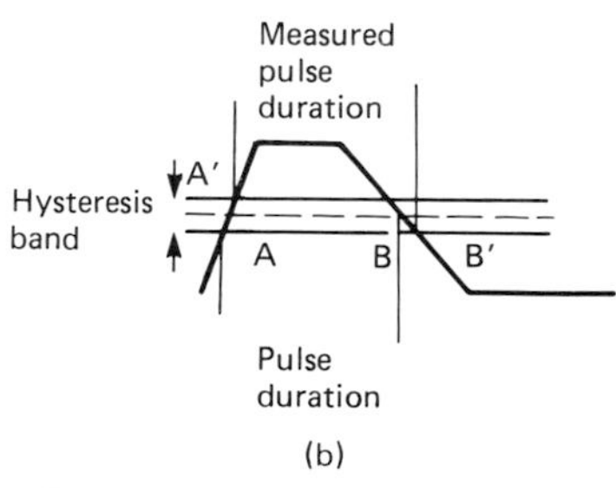

Figure 1.76 (a) The use of hysteresis to reduce the effects of noise; (b) timing errors caused by hysteresis.

waveform, which then becomes an exact number of cycles of the input waveform. The frequency of the input waveform is thus calculated as

$$f_i = \frac{\text{Number of cycles of input waveform}}{\text{Gating period}}$$

$$= \frac{n_i}{n_{osc}} \times 10^{-7} \quad \text{Hz}$$

The relative resolution of the reciprocal method is

$$\pm \frac{10^{-7}}{\text{Gating time}} = \pm \frac{10^{-7}}{t_g} = \pm \frac{1}{n_{osc}}$$

independent of the input frequency, and thus it is possible to provide high-resolution measurements for low-frequency signals. Modern frequency counters often employ both methods using the conventional method to obtain the high resolution at high frequencies.

The period, T_1, of the input wave is calculated from

$$T_i = \frac{1}{f_i} = \frac{\text{Gating period}}{\text{Number of cycles of input waveform}}$$

$$= \frac{n_{osc} \times 10^{-7}}{n_i}$$

with a relative resolution of ± 1 in n_{osc}.

The accuracy of frequency counters is limited by four factors. These are the system resolution, and trigger, systematic and time-base errors. Trigger error (TE) is the absolute measurement error due to input noise causing triggering which is too early or too late. For a sinusoidal input waveform it is given by

$$\text{TE} = \pm \frac{1}{\pi f_i} \quad \text{(input signal to noise ratio)}$$

and for a non-sinusoidal wave

$$\text{TE} = \pm \frac{\text{Peak-to-peak noise voltage}}{\text{Signal slew rate}}$$

Systematic error (SE) is caused by differential propagation delays in the start and stop sensors or amplifier channels of the counter, or by errors in the trigger level settings of the start and stop channels. These errors can be removed by calibration. The time-base error (TBE) is caused by deviation on the frequency of the crystal frequency from its calibrated value. The causes of the deviation have been considered above.

The relative accuracy of frequency measurement is given by

$$\pm \frac{\text{Resolution of } f_i}{f_i} \pm \frac{\text{TE}}{t_g} \pm \text{Relative TBE}$$

Table 1.10 Quartz oscillator characteristics

Stability against	*Uncompensated*	*Temperature compensated*	*Oven stabilized*
Ageing: /24 h	n.a.	n.a.	$<5\times10^{-10}$*
/month	$<5\times10^{-7}$	$<1\times10^{-7}$	$<1\times10^{-8}$
/year	$<5\times10^{-6}$	$<5\times10^{-7}$	$<7.5\times10^{-8}$
Temperature: 0–50°C ref. to +23°C	$<1\times10^{-5}$	$<1\times10^{-6}$	$<5\times10^{-9}$
Change in measuring and supply mode: line/int. battery/ext. D.C. 12–26 V	$<3\times10^{-7}$	$<5\times10^{-8}$	$<3\times10^{-9}$
Line voltage: ±10%	$<1\times10^{-8}$	$<1\times10^{-9}$	$<5\times10^{-10}$
Warm-up time to reach within 10^{-7} of final value	n.a.	n.a.	<15 min

* After 48 h of continuous operation.

Table 1.11 Universal timer/counter specifications

MEASURING FUNCTIONS	
Modes of operation	Frequency Single and multiple period Single and multiple ratio Single and double-line time interval Single and double-line time interval averaging Single and multiple totalizing
FREQUENCY MEASUREMENT	
Input	Channel A
Coupling	a.c. or d.c.
Frequency range	d.c. to 50 MHz (9902 and 9904) HF d.c. to 30 MHz VHF 10 MHz to 200 MHz pre-scaled by 4 (9906)
Accuracy	±1 count ± timebase accuracy
Gate times (9900 and 9902)	Manual: 1 ms to 100 s Automatic: gate times up to 1 s are selected automatically to avoid overspill Hysteresis avoids undesirable range changing for small frequency changes 1 ms to 100 s in decade steps (9904) HF: 1 ms to 100 s VHF: 4 ms to 400 s
SINGLE- AND MULTIPLE-PERIOD MEASUREMENT	
Input	Channel A
Range	1 μs to 1 s single period 100 ns to 1 s multiple period (9902 and 9904) 1 μs to 100 s single period 100 ns to 100 s multiple period (9906)
Clock unit	1 μs
Coupling	a.c. or d.c.
Periods averaged	1 to 10^5 in decade steps
Resolution	10 ps maximum
Accuracy	$\dfrac{\pm 0.3\%}{\text{Number of periods averaged}}$ ± count ± timebase accuracy (measured at 50 mV rms input with 40 dB S/N ratio)
Bandwidth	Automatically reduced to 10 MHz (3 dB) when period selected
TIME INTERVAL SINGLE AND DOUBLE INPUT	
Input	Single input: channel B Double input: start channel B stop channel A
Time range	100 ns to 10^4 s (2.8 h approx.) (9902) 100 ns to 10^5 s (28 h approx.) (9904) 100 ns to 10^6 s (280 h approx.) (9906)
Accuracy	±1 count ± trigger error ± timebase accuracy $\text{Trigger error} = \dfrac{5}{\text{Signal slope at the trigger point (V/}\mu\text{s)}}\ \text{ns}$
Clock units	100 ns to 10 ms in decade steps
Start/stop signals	Electrical or contact
Manual start/stop	By single push button on front panel
Trigger slope selection	Positive or negative slope can be selected on both start and stop
Manual start/stop (9900)	By single push button on front panel *N.B.* Input socket automatically biased for contact operation (1 mA current sink)
Trigger slope selection (9900)	Electrical-positive or negative slopes can be selected on both start and stop signals Contact-opening or closure can be selected on both start and stop signals
Bounce protection (9900)	A 10 ms dead time is automatically included when contact operation is selected

continued

Table 1.11—continued

TIME-INTERVAL AVERAGING SINGLE AND DOUBLE INPUT	
Input	Single input: channel B Double input: Start channel B Stop channel A
Time range	150 ns to 100 ms (9902) 150 ns to 1 s 9904) 150 ns to 10 s (9906)
Dead time between intervals	150 ns
Clock unit	100 ns
Time intervals averaged	1 to 10^5 in decade steps
Resolution	100 ns to 1 ps
Accuracy	±Timebase accuracy ± system error ± averaging error System error: 10 ns per input channel. This is the difference in delays between start and stop signals and can be minimized by matching externally $\text{Averaging error} = \frac{\text{Trigger error} \pm 100}{\sqrt{(\text{Intervals averaged})}}\ \text{ns}$ $\text{Trigger error} = \frac{5}{\text{Signal slope at the trigger point (V}/\mu\text{s)}}\ \text{ns}$
RATIO	
Higher-frequency input	Channel A
Higher-frequency range	10 Hz to 30 MHz (9900) d.c. to 50 MHz (9902, 9904)
Lower-frequency input	Channel B
Lower-frequency range reads	d.c. to 10 MHz $\frac{\text{Frequency } A}{\text{Frequency } B} \times n$
Multiplier n	1 to 10^5 in decade steps
Accuracy	$\frac{\pm 1 \text{ count} \pm \text{trigger error on Channel B}}{\text{No. of gated periods}}$ $\text{Trigger error} = \frac{5}{\text{Signal slope at the trigger point (V}/\mu\text{s)}}\ \text{ns}$
TOTALIZING	
Input	Channel A (10 MHz max.)
Max. rate	10^7 events per second
Pulse width	50 ns minimum at trigger points
Pre-scaling	Events can be pre-scaled in decade multiples (n) from 1 to 10^5
Reads	$\frac{\text{No. of input events} + 1 \text{ count} - 0}{n}$
Manual start/stop	By single push button on front panel
Electrical start/stop	By electrical signal applied to Channel B

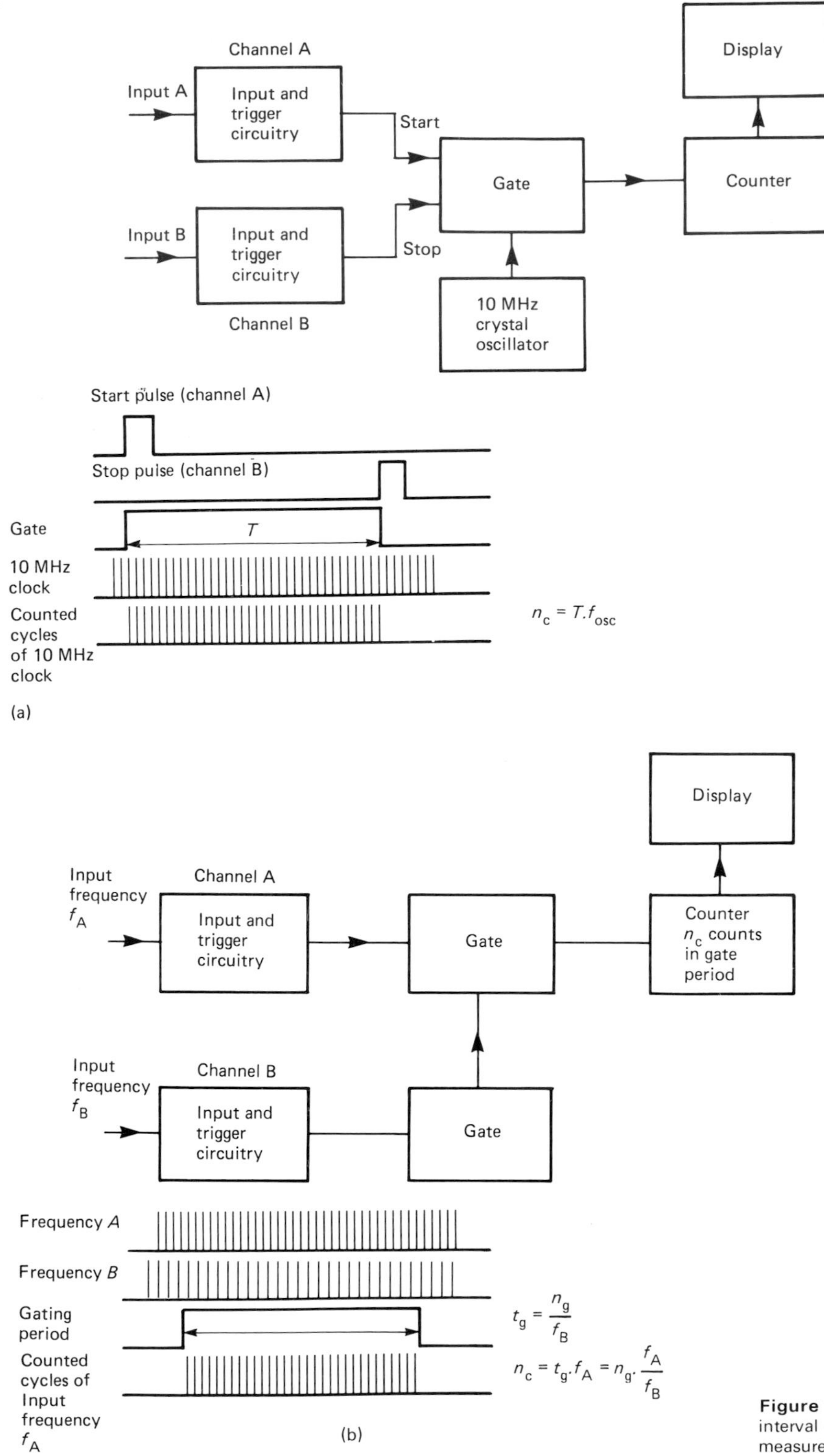

Figure 1.77 (a) Single-shot time-interval measurement; (b) frequency ratio measurement.

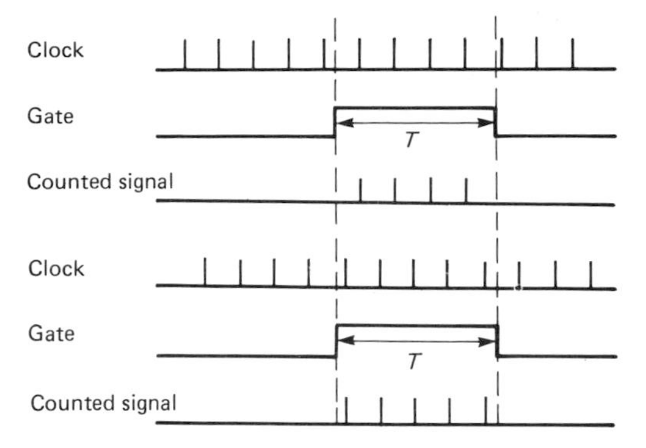

Figure 1.78 Resolution of one-shot time-interval measurement.

and the relative accuracy of period measurement is given by

$$\pm\frac{\text{Resolution of } T_i}{T_i}\pm\frac{\text{TE}}{t_g}\pm\text{Relative TBE}$$

Figure 1.77 shows the techniques employed in single-shot time interval measurement and frequency ratio measurement.

Table 1.11 gives the characteristics of a series of 200-MHz universal timer/counters (Racal-Dana 9902/9904/9906).

1.8.2 Time-interval averaging

Single-shot time interval measurements using a 10-MHz clock have a resolution of ±100 ns. However, by performing repeated measurements of the time interval it is possible to significantly improve the resolution of the measurement (Hewlett Packard, 1977b). As shown in Figure 1.78, the number of counts of the digital clock in the time interval, T, will be either n or $n+1$. It can be shown that if the measurement clock and the repetition rate are asynchronous the best estimate of the time interval, $\hat{T}$, is given by

$$\hat{T}=\bar{n}T_{osc}$$

where $\bar{n}$ is the average number of counts taken over N repetitions and T_{osc} is the period of the digital clock.

The standard deviation, σ_T, which is a measure of the resolution in time-interval averaging (TIA) for large N is given by

$$\sigma_T=\frac{T_{osc}}{\sqrt{N}}\sqrt{[F(F-1)]}$$

where F lies between 1 and 0 dependent on the time interval being measured. Thus the maximum standard deviation on the time estimate is $T_{osc}/(2\sqrt{N})$. By employing repeated measurements using a 10-MHz clock it is possible to obtain a resolution of 10 ps. Repeated measurements also reduce errors due to trigger errors caused by noise. The relative accuracy of TIA measurements is given by

$$\frac{\pm\text{Resolution of } T}{T}\pm\frac{\text{TE}}{\sqrt{(N)}\,.\,T}\pm\frac{\text{SE}}{T}\pm\text{Relative TBE}$$

With a high degree of confidence this can be expressed as

$$\pm\frac{1}{\sqrt{N}}\cdot\frac{1}{\bar{n}}\pm\frac{\text{TE}}{\sqrt{(N)}\,.\,\bar{n}\,.\,T_{osc}}\pm\frac{\text{SE}}{\bar{n}\,.\,T_{osc}}\pm\text{Relative TBE}$$

1.8.3 Microwave-frequency measurement

By the use of pre-scaling as shown in Figure 1.79, in which the input signal is frequency divided before it goes into the gate to be counted, it is possible to measure frequencies up to approximately 1.5 GHz. Higher frequencies typically up to 20 GHz can be measured using the heterodyne converter counter shown in Figure 1.80(a), in which the input signal is

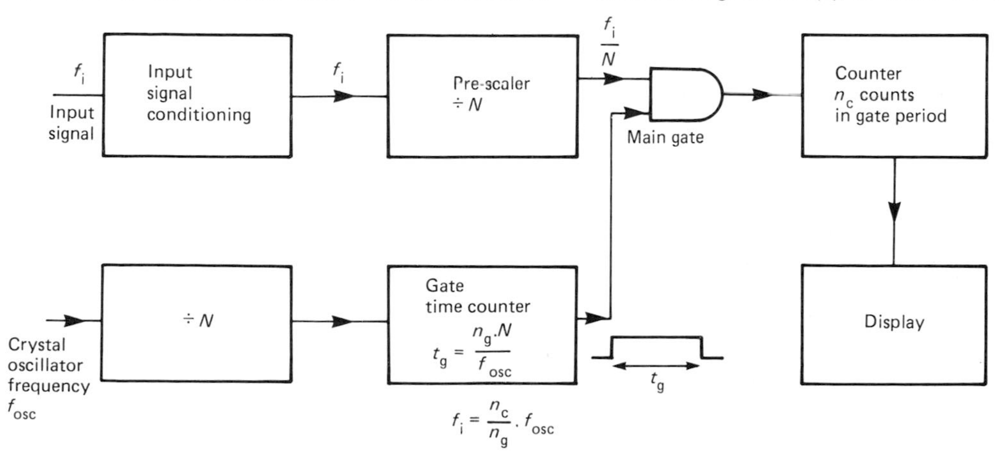

Figure 1.79 Frequency measurement range extension by input pre-scaling.

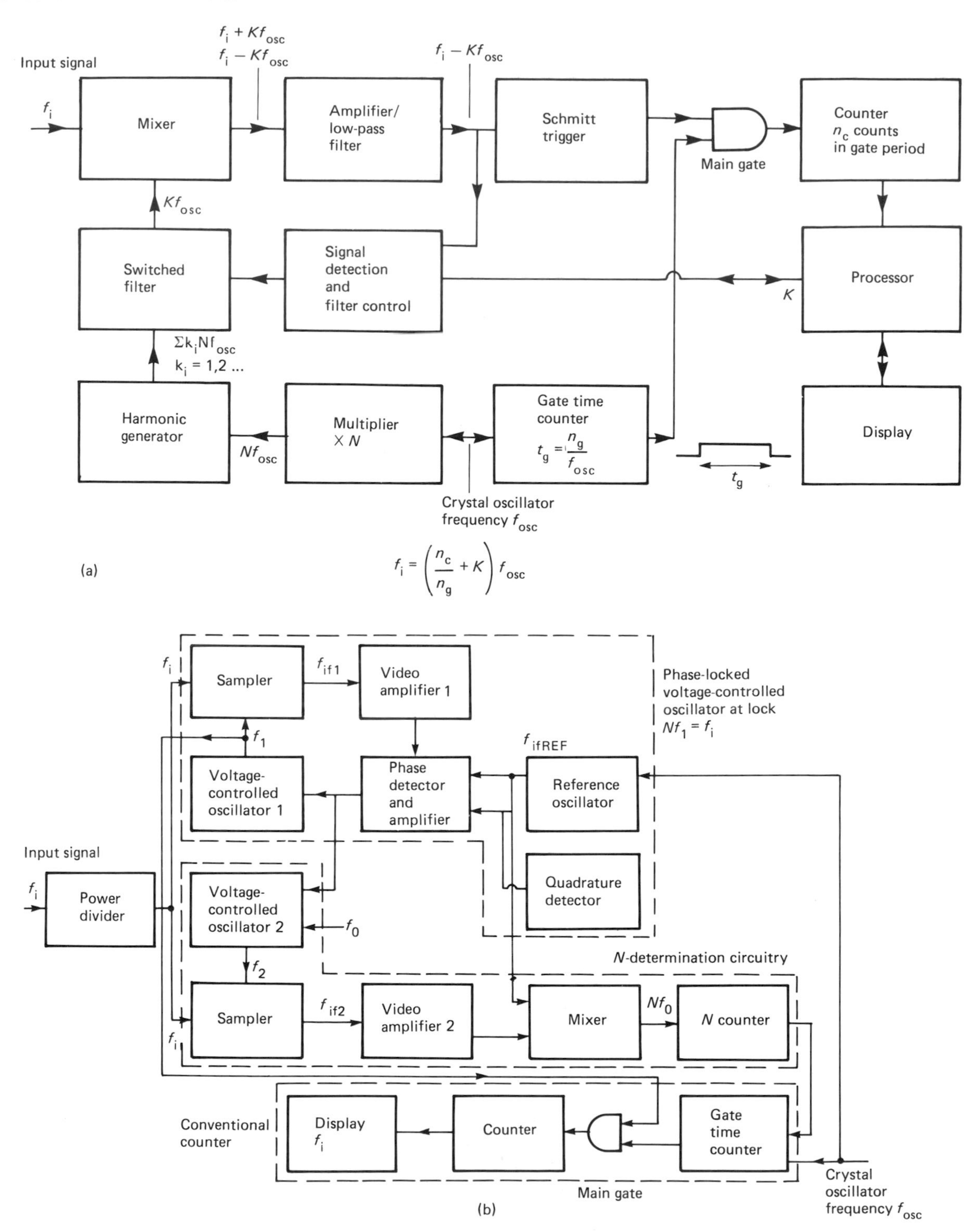

Figure 1.80 (a) Heterodyne converter counter; (b) transfer oscillator counter (from Hewlett Packard, 1977a).

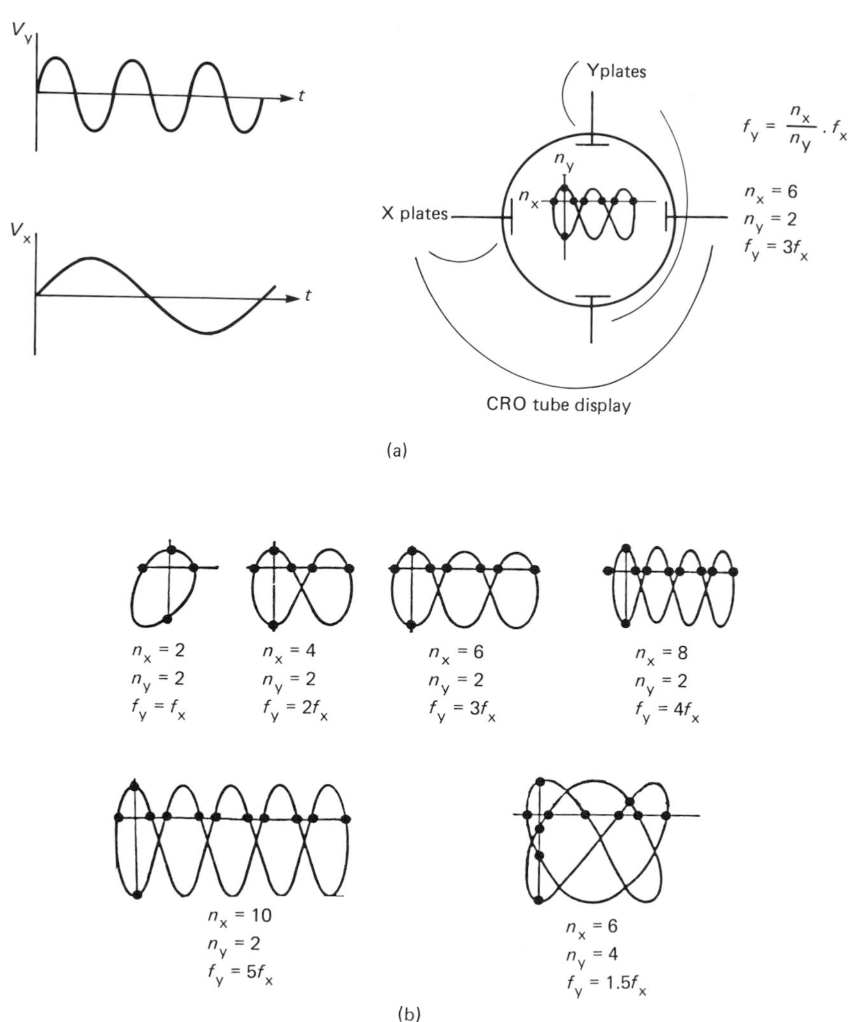

Figure 1.81 (a) Frequency measurement using Lissajous figures; (b) Lissajous figures for various ratios of f_X to f_Y;

down mixed by a frequency generated from a harmonic generator derived from a crystal-controlled oscillator. In the transfer oscillator technique shown in Figure 1.80(b) a low-frequency signal is phase locked to the microwave input signal. The frequency of the low-frequency signal is measured together with its harmonic relationship to the microwave signal. This technique typically provides measurements up to 23 GHz. Hybrid techniques using both heterodyne down conversion and transfer oscillator extend the measurement range typically to 40 GHz. Further details of these techniques can be found in Hewlett Packard (1977a).

1.9 Frequency and phase measurement using an oscilloscope

Lissajous figures can be used to measure the frequency or phase of a signal with respect to a reference source though with an accuracy much lower than for other methods described. Figure 1.81(a) shows the technique for frequency measurement. One signal is applied to the X plates of the oscilloscope and the other to the Y plates. Figure 1.81(b) shows the resulting patterns for various ratios of the frequency f_x applied to the X plates to the frequency f_y applied to

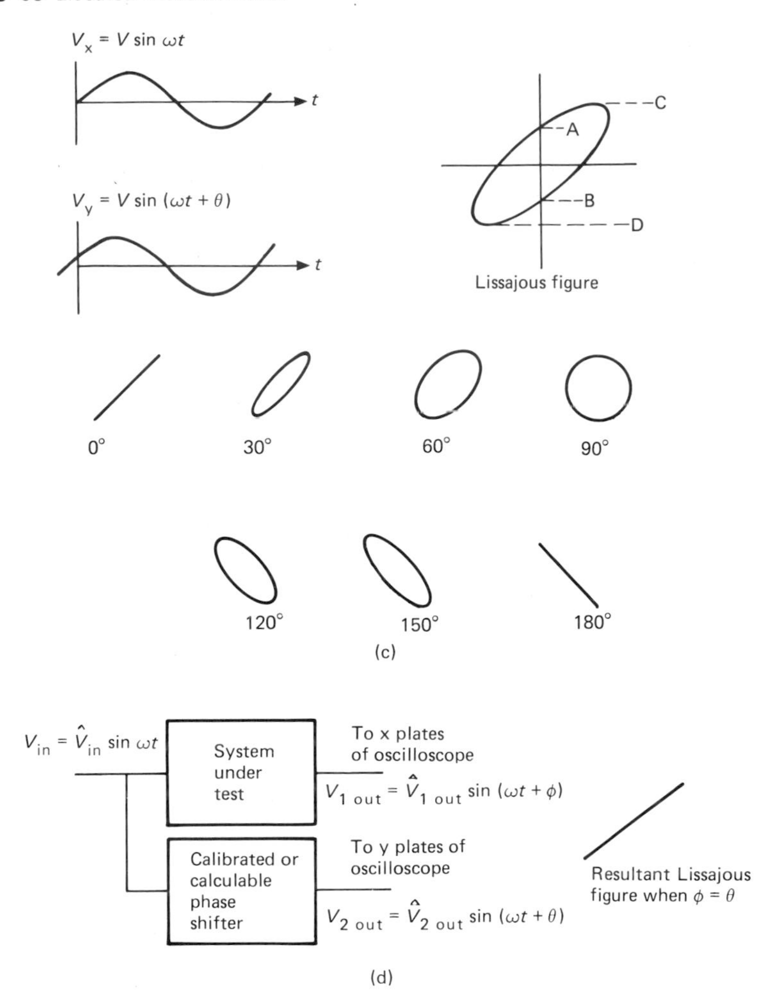

Figure 1.81 *Continued,* (c) phase measurement using Lissajous figures; (d) improved phase measurement using Lissajous figures.

the Y plates. If f_x is the known frequency then it is adjusted until a stationary pattern is obtained. f_y is then given by

$$f_y = \frac{f_x \cdot n_x}{n_y}$$

where n_x is the number of crossings of the horizontal line and n_y the number of crossings of the vertical line, as shown in Figure 1.81(b).

If the two signals are of the same frequency then their relative phases can be determined from the figures shown in Figure 1.81(c). The phase angle between the two signals is given by

$$\sin \theta = \frac{AB}{CD}$$

The accuracy of the method can be significantly increased by the use of a calibrated or calculable phase shift introduced to ensure zero phase shift between the two signals applied to the plates of the oscilloscope, as shown in Figure 1.81(d).

1.10 References

Analog Devices, *Data Acquisition Databook*, Analog Devices, Norwood, Mass. pp. 10, 123–125 (1984)

Arbel, A. F. *Analog Signal Processing and Instrumentation*, Cambridge University Press, Cambridge (1980)

Bailey, A. E. 'Units and standards of measurement', *J. Phys. E: Sci. Instrum.*, **15**, 849–856 (1982)

Bishop, J. and Cohen, E. 'Hall effect devices in power measurement', *Electronic Engineering (GB)*, **45**, No. 548, 57–61 (1973)

British Standards Institution, *BS 3938:1973: Specification for Current Transformers*, BSI, London (1973)
British Standards Institution, *BS 3941:1974: Specification for Voltage Transformers*, BSI, London (1974)
British Standards Institution, *BS 89:1977: Specification for Direct Acting Electrical Measuring Instruments and their Accessories*, BSI, London (1977)
Brodie, B. 'A 160 ppm digital voltmeter for use in ac calibration', *Electronic Engineering (GB)*, **56**, No. 693, 53–59 (1984)
Clarke, F. J. J. and Stockton, J. R. 'Principles and theory of wattmeters operating on the basis of regularly spaced sample pairs', *J. Phys. E: Sci. Instrum.*, **15**, 645–652 (1982)
Dix, C. H. 'Calculated performance of a digital sampling wattmeter using systematic sampling', *Proc. I.E.E.*, **129**, Part A, No. 3, 172–175 (1982)
Dix, C. H. and Bailey, A. E. 'Electrical Standards of Measurement—Part 1 D.C. and low frequency standards', *Proc. I.E.E.*, **122**, 1018–1036 (1975)
Fantom, A. E. *Microwave Power Measurement*, Peter Peregrinus for the IEE, Hitchin (1985)
Froelich, M. 'The influence of the pH value of the electrolyte on the e.m.f. stability of the international Weston cell', *Metrologia*, **10**, 35–39 (1974)
Goldman, D. T. and Bell, R. J. *SI: The International System of Units*, HMSO, London (1982)
Golding, E. W. and Widdis, F. C. *Electrical Measurements and Measuring Instruments*, 5th edn, Pitman, London (1963)
Gregory, B. A. *Electrical Instrumentation*, Macmillan, London (1973)
Gumbrecht, A. J. *Principles of Interference Rejection*, Solartron DVM Monograph No. 3, Solartron, Farnborough (1972)
Hague, B. and Foord, T. R. *Alternating Current Bridge Methods*, Pitman, London (1971)
Harris, F. K. *Electrical Measurements*, John Wiley, New York (1966)
Hewlett Packard. *Fundamentals of Time and Frequency Standards*, Application Note 52-1, Hewlett Packard, Palo Alto, CA (1974)
Hewlett Packard. *Fundamentals of Microwave Frequency Counters*, Application Note 200-1, Hewlett Packard, Palo Alto, CA (1977a)
Hewlett Packard. *Understanding Frequency Counter Specifications*, Application Note 200-4, Hewlett Packard, Palo Alto, CA (1977b)
Hewlett Packard. *Fundamentals of RF and Microwave Power Measurements*, Application Note 64-1, Hewlett Packard, Palo Alto, CA (1978)
Hugill, A. L. 'Displacement transducers based on reactive sensors in transformer ratio bridge circuits', in *Instrument Science and Technology*, Volume 2, ed. B. E. Jones, Adam Hilger, Bristol (1983)
Josephson, B. D. 'Supercurrents through barriers', *Phys. Letters*, **1**, 251 (1962)
Kibble, B. P., Smith, R. C. and Robinson, I. A. 'The NPL moving coil–ampere determination', *I.E.E.E. Trans.*, **IM-32**, 141–143 (1983)
Matouka, M. F. 'A wide-range digital power/energy meter for systems with non-sinusoidal waveforms', *I.E.E.E. Trans.*, **IE-29**, 18–31 (1982)
Miljanic, P. N., Stojanovic, B. and Petrovic, V. 'On the electronic three-phase active and reactive power measurement', *I.E.E.E. Trans.*, **IM-27**, 452–455 (1978)
NBS. NBS Monograph 84: *Standard Cells—Their Construction, Maintenance, and Characteristics*, NBS, Washington, DC (1965)
Neubert, H. P. K. *Instrument Transducers*, 2nd edn, Oxford University Press, London (1975)
Oliver, B. M. and Cage, J. M. *Electronic Measurements and Instrumentation*, McGraw-Hill, New York (1971)
Owens, A. R. 'Digital signal conditioning and conversion', in *Instrument Science and Technology*, Volume 2, ed. B. E. Jones, Adam Hilger, Bristol (1983)
Pearce, J. R. 'Scanning, A-to-D conversion and interference', *Solartron Technical Report Number 012/83*, Solartron Instruments, Farnborough (1983)
Pitman, J. C. 'Digital voltmeters: a new analogue to digital conversion technique', *Electronic Technology (GB)*, **12**, No. 6, 123–125 (1978)
Rathore, T. S. 'Theorems on power, mean and RMS values of uniformly sampled periodic signals', *Proc. I.E.E.*, **131**, Part A, No. 8, 598–600 (1984)
Rayner, G. H. 'An absolute determination of resistance by Campbell's method', *Metrologia*, **3**, 8–11 (1967)
Simeon, A. O. and McKay, C. D. 'Electronic wattmeter with differential inputs', *Electronic Engineering (GB)*, **53**, No. 648, 75–85 (1981)
Sheingold, D. H. *Analog/digital Conversion Notes*, Analog Devices, Norwood, Mass. (1977)
Somlo, P. I. and Hunter, J. D. *Microwave Impedance Measurement*, Peter Peregrinus for IEE, Hitchin (1985)
Spreadbury, P. J. 'Electronic voltage standards', *Electronics and Power*, **27**, 140–142 (1981)
Steele, J. A., Ditchfield, C. R. and Bailey, A. E. 'Electrical standards of measurement Part 2: r.f. and microwave standards', *Proc. I.E.E.*, **122**, 1037–1053 (1975)
Stone, N. W. B. *et al.* 'Electrical standards of measurement Part 3: submillimetre wave measurements and standards', *Proc. I.E.E.*, **122**, 1053–1070 (1975)
Tagg, G. F. *Electrical Indicating Instruments*, Butterworths, London (1974)
Thompson, A. M. and Lampard, D. G. 'A new theorem in electrostatics with applications to calculable standards of capacitance', *Nature (GB)*, **177**, 888 (1956)
Vigoureux, P. 'A determination of the ampere', *Metrologia*, **1**, 3–7 (1965)
Vigoureux, P. *Units and Standards for Electromagnetism*, Wykenham Publications, London (1971)

2 Optical measurements

A. W. S. TARRANT

2.1 Introduction

A beam of light can be characterized by its spectral composition, its intensity, its position and direction in space, its phase and its state of polarization. If something happens to it to alter any of those quantities, and the alterations can be quantified, then a good deal can usually be found out about the 'something' that caused the alteration. Consequently optical techniques can be used in a huge variety of ways and it would be quite impossible to describe them all here.

This chapter describes a selection of widely used instruments and techniques. Optical instruments can be conveniently thought of in two categories—those basically involving image formation (for example, microscopes and telescopes) and those which involve intensity measurement (for example, photometers). Many instruments involve both processes (for example, spectrophotometers) and it is convenient to regard these as falling in the second 'intensity measurement' category. For the purposes of this book we are almost entirely concerned with instruments in this second category. 'Image-formation' instruments are well described in familiar textbooks, such as R. S. Longhurst's *Geometrical and Physical Optics*.

The development of optical fibres and light guides has enormously broadened the scope of optical techniques. Rather than take wires to some remote instrument to obtain a meaningful signal from it, we can often now take an optical fibre, an obvious advantage where rapid response times are involved or in hazardous environments. This is discussed in Part 4.

In all branches of technology, it is quite easy to make a fool of oneself if one has no previous experience of the particular techniques involved. One purpose of this book is to help the non-specialist to find out what is and what is not possible. The author would like to pass on one tip to people new to optical techniques. When we consider what happens in an optical system we must consider what happens in the *whole optical system*; putting an optical system together is not quite like putting an electronic system together. Take, for example a spectrophotometer, which consists of a light source, a monochromator, a sample cell and a detector. An alteration in the position of the lamp will affect the distribution of light on the detector and upset its operation, despite the fact that the detector is several units 'down the line'. Optical systems must be thought of as a whole, not as a group of units acting in series.

It should be noted that the words 'optical' and 'light' are used in a very loose sense when applied to instruments. Strictly, these terms should only refer to radiation within the visible spectrum, i.e. the wavelength range 380–770 nm. The techniques used often serve equally well in the near ultra-violet and near infra-red regions of the spectrum, and naturally we use the same terms. It is quite usual to hear people talking about 'ultra-violet light' when they mean 'ultra-violet radiation', and many 'optical' fibres are used with infra-red radiation.

2.2 Light sources

Light sources for use in instruments may be grouped conveniently under two headings, (1) conventional or incoherent sources and (2) laser or coherent sources. Conventional sources are dealt with in sections 2.2.1 to 2.2.3 and laser sources in section 2.2.4.

The principal characteristics of a conventional light source for use in instruments are:

(1) Spectral power distribution;
(2) Luminance or radiance;
(3) Stability of light output;
(4) Ease of control of light output;
(5) Stability of position.

Other factors which may have to be taken into account in the design of an instrument system are heat dissipation, the nature of auxiliary equipment needed, source lifetime, cost and ease of replacement. By the 'radiance' of a light source we mean the amount of energy per unit solid angle radiated from a unit area of it. Very often this quantity is much more important than the actual power of the lamp. For example, if a xenon arc is to be used with a spectroscopic system, the quantity of interest is the amount of light which can be got through the slit of the spectrometer. A low-power lamp with a high radiance can be focused down to give a small, intense image at the slit and thus get a lot of light through; but if the lamp has a low radiance, it does not matter how powerful it is, it cannot be refocused to pass the same amount of radiation through the system. It can be easily shown in this case that the radiance is the only effective parameter of the source.

If light output in the visible region only is concerned, 'luminance' is sometimes used instead of 'radiance'. There is a strict parallel between units and definitions of 'radiant' quantities and 'luminous' quantities (BS Spec. 4727; IEC–CIE, *International Lighting Vocabulary*). The unit of light, the lumen, can be thought of as a unit of energy weighted with regard to wavelength according to its ability to produce a visible sensation of light (Walsh, 1958, p. 138).

2.2.1 Incandescent lamps

Incandescent sources are those in which light is generated by heating material electrically until it becomes white hot. Normally this material is tungsten but if only infra-red radiation is wanted it may be a ceramic material. In a tungsten lamp the heating is purely resistive, and the use of different filament diameters enables lamps to be made of similar power but different voltage ratings. The higher the voltage, the finer and more fragile is the filament. For instrument purposes small and compact filaments giving the highest radiance are usually needed, and so low-voltage lamps are often used. For lamps used as radiation standards it is customary to use a solid tungsten ribbon as a filament, but these require massive currents at low voltage (for example, 18 A, 6 V).

The spectral power distribution of a tungsten lamp corresponds closely to that of a Planckian radiator, as shown in Figure 2.1, and the enormous preponderance of red energy will be noted. Tungsten lamps have a high radiance, are very stable in light output provided the input power is stabilized and are perfectly stable in position. The light output can be precisely controlled by varying the input power, from zero to maximum, but as the filament has a large thermal mass there is no possibility of deliberately modulating the light output. If a lamp is run on an a.c. supply at mains frequency some modulation at twice that frequency invariably occurs and may cause trouble if other parts of the instrument system use mains — frequency modulation. The modulation is less marked with low-voltage lamps which have more massive filaments, but can only be overcome by using either a smoothed d.c. supply or a high-frequency power supply (10 kHz).

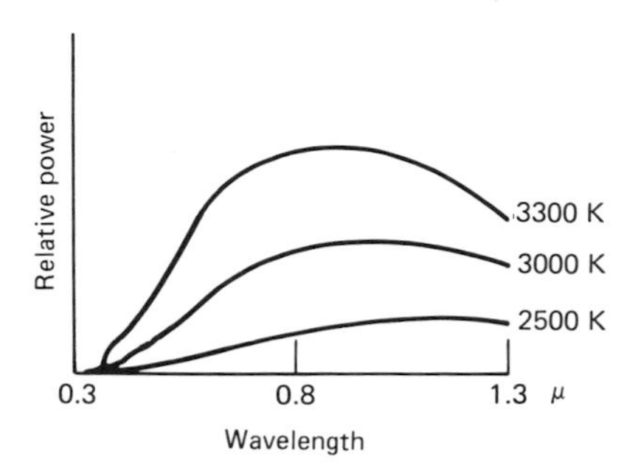

Figure 2.1 Spectral power distribution of tungsten lamp.

The main drawback to tungsten lamps is the limited life. The life depends on the voltage (Figure 2.2), and it is common practice in instrument work to under-run lamps to get a longer life.

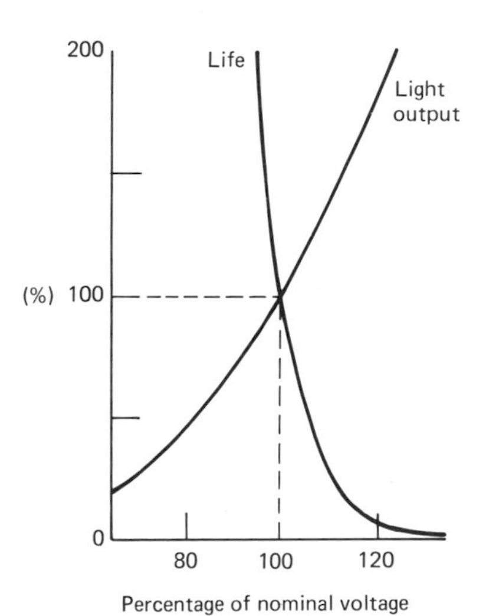

Figure 2.2 Variation with voltage of life and light output of a tungsten lamp (after Henderson and Marsden).

Longer lamp lives are obtained with 'tungsten halogen' lamps. These have a small amount of a halogen—usually bromine or iodine—in the envelope which retards the deterioration of the filament. It is necessary for the wall temperature of the bulb to be at least 300°C, and this entails the use of a small bulb made of quartz. However, this allows a small amount of ultra-violet radiation to escape, and with its small size and long life, the tungsten halogen lamp is a very attractive light source for instrument purposes.

2.2.1.1 Notes on handling and use

After lengthy use troubles can arise with lampholders,

usually with contact springs weakening. Screw caps are preferable to bayonet caps. The envelopes of tungsten halogen lamps should not be touched by hand—doing so will leave grease on them which will burn into the quartz when hot and ruin the surface.

2.2.2 Discharge lamps

Discharge lamps are those in which light is produced by the passage of a current through a gas or vapour, and hence produce mostly line spectra. Enclosed arcs of this kind have negative temperature/resistance characteristics and so current limiting devices are necessary. Inductors are often used if the lamp is to be run on an a.c. mains supply. Many types of lamp are available (Henderson and Marsden, 1972); some commonly met with in instrument work will be mentioned here.

2.2.2.1 Deuterium lamps

The radiation is generated by a low current density discharge in deuterium. Besides the visible line spectrum a continuous spectrum (Figure 2.3) is produced in the ultra-violet. The radiance is not high,

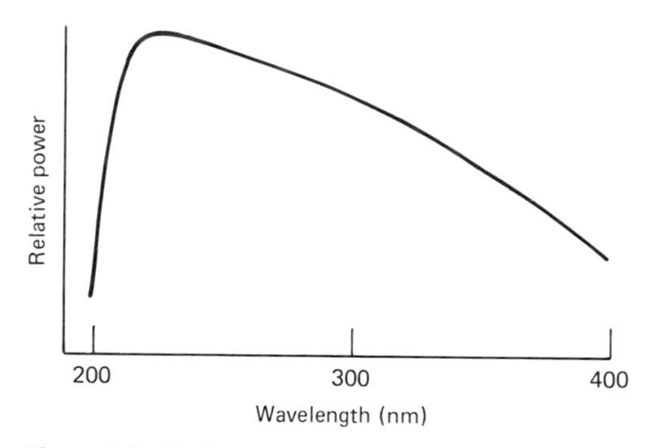

Figure 2.3 Typical spectral power distribution for a deuterium lamp.

but provided that the input power is stabilized these lamps are stable in both output and position. To obtain the necessary ultra-violet transmission either the whole envelope is made of silica or a silica window is used. These lamps are used as sources in ultra-violet spectrophotometers and are superior to tungsten lamps for that purpose at wavelengths below 330 nm.

2.2.2.2 Compact source lamps

Some types of discharge lamp offer light sources of extremely high luminance. These involve discharges of high current density in a gas at high pressure. The lamp-filling gases may be xenon, mercury, mercury plus iodine or a variety of other 'cocktails'. The light emitted is basically a line spectrum plus some continuous radiation, but in many cases the spectrum is, in effect, continuous. The spectrum of xenon, for example, has over 4000 lines and extends well down into the ultra-violet region. Xenon lamps are widely used in spectrofluorimeters and other instruments where a very intense source of ultra-violet radiation is required.

Many lamps of this kind require elaborate starting arrangements and are particularly difficult to re-start if switched off when hot. 'Igniter' circuits are available, but as these involve voltages up to 50 kV special attention should be given to the wiring involved. All these lamps contain gas under high pressure even when cold, and the maker's safety instructions should be rigidly followed.

Xenon arcs produce quite dangerous amounts of ultra-violet radiation—dangerous both in itself and in the ozone which is produced in the atmosphere. They must never be used unshielded, and to comply with the Health and Safety at Work Act, interlocks should be arranged so that the lamp is switched off if the instrument case is opened. Force-ducted ventilation should be used with the larger sizes, unless 'ozone-free' lamps are used. These are lamps with envelopes which do not transmit the shorter ultra-violet wavelengths.

2.2.3 Electronic sources: light-emitting diodes

By applying currents to suitably doped semiconductor junctions it is possible to produce a small amount of light. The luminance is very low but the light output can be modulated, by modulating the current, up to very high frequencies, and thus the light-emitting diode (LED) is a good source for a fibre optic communication system. LEDs are also commonly used in display systems. The spectral power distribution depends on the materials used in the junction. For electro-optical communication links there is no need to keep to the visible spectrum and wavelengths just longer than visible (for example, 850 nm) are often used. There are no serious operating problems and only low voltages are needed, but the light output is miniscule compared, say, with a tungsten lamp.

2.2.4 Lasers

Light from lasers differs from that from conventional sources by virtue of being 'coherent' whereas conventional sources produce 'incoherent' light. In an 'incoherent' beam there is no continuous phase relationship between light at one point of the beam and any other. The energy associated with any one quantum or wave packet can be shown to extend over a finite length—somewhere around 50 cm—as it travels through space. For that reason no interference effects can be observed if the beam is divided and subsequently superimposed if the path difference is longer than 50 cm or so. The same effect is responsible for the fact that monochromatic light from a conventional source in fact has a measurable band width.

However, in a laser the light is produced not from single events occurring randomly within single atoms but from synchronized events within a large number of atoms—hence the 'finite length of wave train' is not half a metre but can be an immense distance. Consequently, laser light is much more strictly 'monochromatic' than that from conventional sources; it is also very intense and is almost exactly unidirectional. Thus it is easy to focus a laser beam down to a very small spot at which an enormous density of energy can be achieved.

Lasers are valuable in applications where (1) the extended length of wave train is used (for example, holography, surveying), (2) a high energy density is needed (for example, cutting of sheet metal, ophthalmic surgery) and (3) the narrowness of the beam is used (for example, optical alignment techniques in engineering or building construction).

The operating principle of the laser is the stimulated emission of radiation. In any normal gas the number of electrons in atoms in each of the possible energy levels is determined by the temperature and other physical factors. In a laser this normal distribution is deliberately upset so as to overpopulate one of the higher levels. The excited atoms then not only release their excess energy as radiation but do so *in-phase*, so that the emission from vast numbers of atoms are combined in a single wave train. Lasing action can also be produced in solid and liquid systems; hundreds of atomic systems are now known which can be used in lasers, and so a wide range of types is available for either continuous or pulsed operation. A simple explanation of the principle is given in Heavens (1971) and numerous specialist textbooks (Dudley, 1976; Koechner, 1976; Mooradian *et al.*, 1976).

Although lasers are available in many types and powers, by far the most commonly used in laboratory work is the helium–neon laser operating at a wavelength of 632.8 nm. The power output is usually a few milliwatts. For applications where a high-energy density is needed (for example metal cutting) CO_2 lasers are often used. Their wavelength is in the infrared range (about 10.6 μm), and the output power may be up to 500 W. Their advantage in industrial work is their relatively high efficiency—about 10 per cent of the input power appears as output power.

For other wavelengths in the visible region krypton, argon or 'tuneable dye' lasers can be used. The krypton and argon types can be made to operate at a variety of fixed wavelengths. Tuneable dye lasers use a liquid system involving organic dyes. In such systems the operating frequency can be altered within a limited range by altering the optical geometry of the system.

2.2.4.1 Laser safety

Lasers are, by their nature, dangerous. The foremost risk is that of damage to the eyesight caused by burning of the retina. Even a moment's exposure may be catastrophic. Consequently safety precautions must be taken and strictly maintained.

It is a requirement of the Health and Safety at Work Act that all due precautions are taken. In practice this means that, amongst other things, all rooms in which lasers are used must be clearly marked with approved warning notices. The best precautions are to use the lowest laser powers that are possible, and to design equipment using lasers to be totally enclosed. A full description of safety requirements is given in *Standards for the Safe Use of Lasers*, published by the American National Standards Institute, which should be read and studied before any work with lasers is started. Useful guidance may also be obtained from BS 4803 and *Safety in Universities—Notes for Guidance*.

2.3 Detectors

The essential characteristics of a radiation detector are:

(1) The spectral sensitivity distribution;
(2) The response time;
(3) The sensitivity;
(4) The smallest amount of radiation that it can detect;
(5) The size and shape of its effective surface;
(6) Its stability over a period of time.

Other factors to be borne in mind in choosing a detector for an instrument are the precision and linearity of response, physical size, robustness and the extent of auxiliary equipment needed. Detectors can be used in three ways:

(1) Those in which the detector is used to effect an actual measurement of light intensity;
(2) Those where it is used to judge for equality of intensity between two beams; and
(3) Those in which it is required to establish the presence or absence of light.

In case (1) there needs to be an accurately linear relationship between the response and the intensity of radiation incident on the detector. Many detectors do not have this property and this fact may determine the

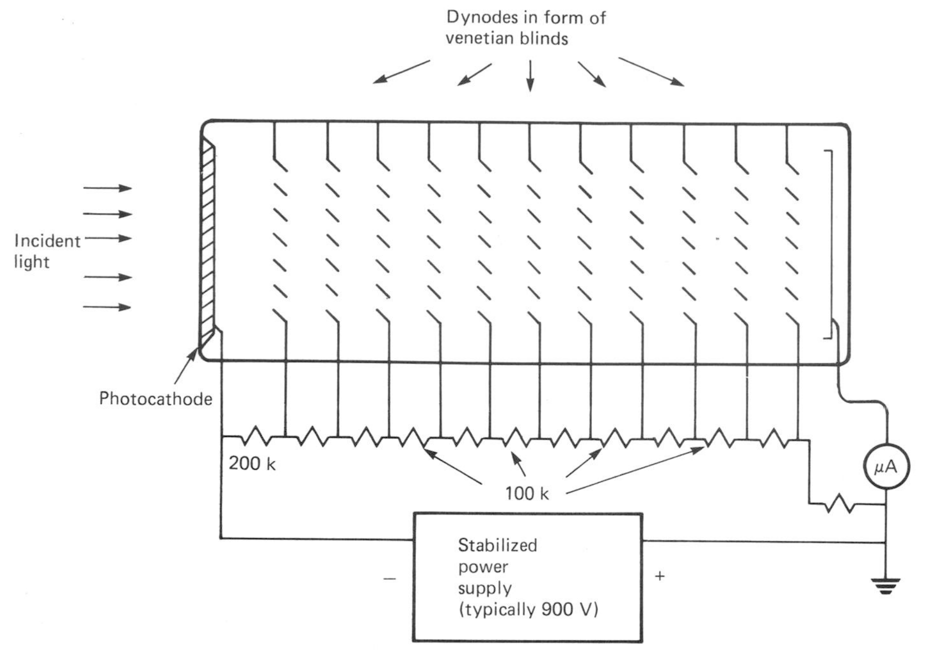

Figure 2.4 Construction of photomultiplier and typical circuit.

design of an instrument; for example many infra-red spectrophotometers have elaborate optical arrangements for matching the intensity in order that a non-linear detector can be used.

It should be noted that in almost all detectors the sensitivity varies markedly from point to point on the operative surface. Consequently if the light beam moves with respect to the detector the response will be altered. This effect is of critical importance in spectrophotometers and like instruments; if a solution cell whose faces are not perfectly flat and parallel is put into the beam it will act as a prism, move the beam on the detector and produce an erroneous result. Precautions should be taken against this effect by ensuring that imperfect cells are not used.

Some detectors are sensitive to the direction of polarization of incident light. Whilst in most optical instruments the light is randomly polarized, the distribution of intensities between the polarisation directions is by no means uniform, especially after passage through monochromators. This effect often causes no trouble, but it can be an extremely abstruse source of error.

2.3.1 Photomultipliers

Photomultipliers rely on the photoemissive effect. The light is made to fall on an emitting surface (the photocathode) (Figure 2.4) with a very low work function within a vacuum tube and causes the release of electrons. These electrons are attracted to a second electrode at a strongly positive voltage where each causes the emission of several secondary electrons, which are attracted to a third electrode, and so on. By repeating this process at a series of 'dynodes' the original electron stream is greatly multiplied, and results in a current of up to 100 μA or so at the final anode.

The spectral sensitivity is determined by the nature of the photocathode layer. The actual materials need not concern us here; different types of cathodes are referred to by a series of numbers from S1 upwards. The response is linear, response time rapid and the sensitivity is large. The sensitivity may be widely varied by varying the voltage applied to the dynode chain. Thus

$$S \propto V^{n/2}$$

where S = sensitivity, V = voltage applied and n = number of dynode stages. Conversely, where accurate measurements are needed, the dynode chain voltage must be held extremely stable, since between eight and fourteen stages are used. A variety of cathode shapes and multiplier configurations are available. There is no point in having a cathode of a much larger area than that actually to be used as this will add unnecessarily to the noise.

Emission from the photocathode also occurs as a result of thermionic emission, which produces a permanent 'dark current'. It is random variations in this dark current—noise—which limit the ultimate sensitivity.

Photomultipliers are also discussed in Chapter 3 and excellent information on their use is given in the makers' catalogues to which the reader is referred. It should be noted that they need very stable high-voltage power supplies, are fragile and are easily damaged by overloads. When used in instruments it is essential that interlocks are provided to remove the dynode voltage before any part of the case is opened. Moreover, photomultipliers must never be exposed to sunlight, even when disconnected.

2.3.2 Photovoltaic and photoconductive detectors (photodiodes)

When light falls on a semiconductor junction there is nearly always some effect upon the electrical behaviour of that junction, and such effects can be made use of in light detectors. There are two main categories, (1) those in which the action of the light is used to generate an emf and (2) those in which the action of light is used to effectively alter the resistance of the device. Type (1) are referred to as 'photovoltaic detectors'—sometimes called 'solar cells'—and type (2) are called 'photoconductive detectors'. There are some materials which show photoconductive effects but which are not strictly semiconductors (for example, lead sulphide) but devices using these are included in the category of photoconductive detectors.

In photovoltaic detectors the energy of the light is actually converted into electrical energy, often to such effect that no external energy source is needed to make a measurement; the solar cell uses this principle. The more sensitive photovoltaic detectors need an external power source, as do all photoconductive detectors.

2.3.2.1 Simple photovoltaic detectors

In many applications there is sufficient light available to use a photovoltaic cell as a self-powered device. This might be thought surprising, but is perfectly feasible. If we consider a detector of 15 mm diameter, illuminated to a level of 150 lux (average office lighting) then the radiant power received in the visible spectrum will be about 100 μW. If the conversion efficiency is only 5 per cent, that gives an output power of 5 μW and if that is fed to a galvo or moving coil meter of 50-Ω resistance, a current of around 300 μA will be obtained—more than enough for measurement purposes.

Detectors of this type were in use for many years before the advent of the semiconductor era in the 1950s and were then called 'rectifier cells' or 'barrier-layer' cells. The simplest type (see Figure 2.5) consists of a steel plate upon which a layer of selenium is deposited. A thin transparent film of gold is deposited on top of that to serve as an electrode. In earlier

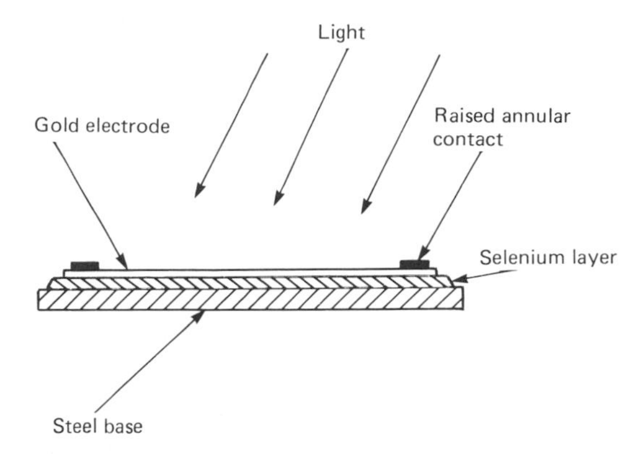

Figure 2.5 Simple photovoltaic detector.

models an annular contact electrode was sputtered on to facilitate contact with the gold film. Nowadays a totally encapsulated construction is used for cells of up to 30 mm diameter.

The semiconductor action takes place at the steel–selenium junction; under the action of light there is a build-up of electrons in the selenium, and if an external circuit is made via the gold electrode a current will flow. If there is zero resistance in that external circuit then the current will be proportional to the light intensity. Normally there will be some resistance, and the emf developed in it by the current will oppose the current-generation process. This means that the response will not be linear with light intensity (see Figure 2.6). If the external resistance is quite high (for example, 1000 Ω) then the response will approximate to a logarithmic one, and it is this feature which enables this type of detector to be used over a very

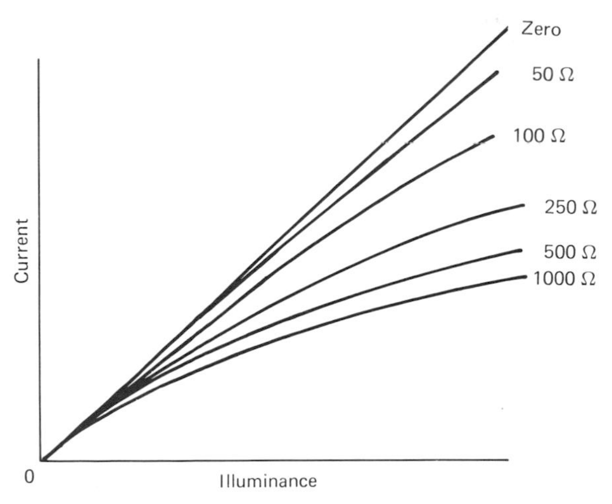

Figure 2.6 Non-linear response of photovoltaic detectors.

wide range of intensities in the photographic exposure meter and various daylight recording instruments.

A circuit known as the 'Campbell–Freeth Circuit' was devised to present zero resistance to the detector (Thewlis, 1961) so as to obtain a truly linear response. It is not widely used in current practice since if high accuracy is sought it is unlikely that a photovoltaic detector will be used.

Detectors of this variety, based on steel plates, are perhaps best described as 'cheap and cheerful'. The accuracy is not high—1 or 2 per cent at best—they are not particularly stable with temperature changes and they show marked fatigue effects. They are also extremely noisy and have a relatively long time response compared with more modern versions. However, they are cheap, robust, may be fabricated into a variety of shapes and have the outstanding advantage that no external power supply is required. Consequently they are widely used in exposure meters, street-lighting controls, photometers, 'abridged' spectrophotometers and colorimeters, flame photometers, simple densitometers and so on—all those cases where high accuracy is not needed but price is an important consideration.

The spectral response of this type of detector is broader than that of the human eye, but a filter can be used to produce a spectral response which is a good enough match for most photometric purposes. Such filters are usually encapsulated with the detectors into a single unit.

2.3.2.2 The silicon diode

The advent of doped semiconductor materials has enabled many other 'barrier-layer' systems to be used. One such is the diffused silicon photodiode; in this case the barrier layer is provided by a *p–n* junction near the surface of a silicon layer. This is very much less noisy than the detector described above and can thus be used for much lower light levels. External amplification is needed, but rapid response can be obtained (50 ns in some cases) so that the silicon diode is not far short of the performance obtained from the photomultiplier, with the advantages of a wider spectral range, less bulk and much less complexity.

2.3.2.3 Photoconductive detectors

Almost all semiconductor materials are light sensitive, as light falling on them produces an increased number of current carriers and hence an increase in conductivity; semiconductor devices have to be protected from light to prevent the ambient lighting upsetting their operation. Consequently it is possible to make a wide variety of light-sensitive devices using this principle.

These devices may be in the form of semiconductor diodes or triodes. In the former case the material is usually deposited as a thin film on a glass plate with electrodes attached; under the action of light the resistance between the electrodes drops markedly, usually in a non-linear fashion. Since there are dozens of semiconductor materials available, these devices can be made with many different spectral responses, covering the visible and near infra-red spectrum up to 5 μm or so. The response time is also dependent on the material, and whilst many are fast, some are very slow—notably cadmium sulphide, which has a response time of about 1 s.

Photoconductive detectors of the 'triode' type are in effect junction transistors that are exposed to light.* They offer in-built amplification but again usually produce a non-linear signal. Devices are now available in which a silicon diode is combined with an amplifier within a standard transistor housing.

All photoconductive devices are temperature sensitive and most drift quite badly with small changes of temperature. For that reason they are used with 'chopped' radiation (see p. 79 on detector techniques) in all critical applications. They are commonly used as detectors in spectrophotometers in the near infra-red range. A common technique is to use them as null-balance detectors comparing the sample and reference beams (see p. 83 on spectrophotometers) so that a non-linear response does not matter.

2.3.3 Pyroelectric detectors

Although they are strictly 'thermal' detectors, pyroelectric detectors are used very widely as light detectors. They rely on the use of materials which have the property of temperature-dependent spontaneous electric polarization. These may be in the form of crystals, ceramics or thin plastic films. When radiation falls on such a material thermal expansion takes place, minutely altering the lattice spacing in the crystal, which alters the electrical polarization and results in an emf and a charge being developed between two faces of the crystal. These faces are equipped with electrodes to provide connection to an external circuit and to an appropriate amplifier.

Pyroelectric detectors using this principle are extremely sensitive, and, being basically thermal detectors, they are sensitive to an enormous range of wavelengths. In practice the wavelength range is limited by the transmission of windows used, the absorption characteristics of the pyroelectric material and the reflection characteristics of its surfaces. The

* The joke is often made that the light-detecting properties of one famous type of phototransistor were only discovered when one batch of transistors was accidentally left unpainted. The author regrets that he cannot confirm this story.

last-mentioned can be adjusted by the deposition of thin films, so that relatively wide spectral responses may be obtained; alternatively the sensitivity to a particular wavelength can be considerably enhanced.

It is important to remember that pyroelectric detectors respond in effect only to *changes* in the radiation falling on them and not to steady-state radiation. Their response speed is extremely fast, but as they inherently have some capacity there is a compromise between sensitivity and response speed which can be determined by a appropriate choice of load resistor. The ability to respond only to changes in the radiation field enables them to be used for laser pulse measurements, and at a more mundane level they make excellent sensors for burglar alarms. They are usually made in quite small sizes (for example, 1 × 2 mm), but they can be made up to 1 cm in diameter with a lower response speed. They can also be made in the form of linear and two-dimensional arrays. If they are required to measure steady-state radiation, as in a spectrophotometer, then the usual technique of beam chopping (see section 2.4.3) can be used; the pyroelectric detector wll then respond only to the chopped beam and nothing else.

A variety of materials are used, notably lithium tantalate and doped lead zirconate titanate. The pyroelectric detector with its wide spectral range, fast response and relatively low cost is probably capable of further development and application than any other and will be seen in a very wide range of applications in the next few years.

2.3.4 Array detectors

The devices described above are suitable for making a single measurement of the intensity of a beam of light at any one instant. Often the need arises to measure the intensity of many points in an optical image, as in a television camera. In television camera tubes the image is formed on a photocathode resembling that of a photomultiplier which is scanned by an electron beam. Such tubes are outside the scope of this book, but they are necessarily expensive and require much supporting equipment.

In recent years 'array detectors' using semiconductor principles have been developed, enabling measurements to be made simultaneously at many points along a line or, in some cases, over a whole area comprising many thousands of image points. All these devices are based on integrated-circuit technology, and they fall into three main categories:

(1) Photo-diode arrays;
(2) Charge-coupled devices;
(3) Charge injection devices.

It is not possible to go into their operation in detail here, but more information can be found in the review article by Fry (1975) and in the book by Beynon (1979).

(1) A photodiode array consists of an array of photodiodes of microscopic dimensions, each capable of being coupled to a signal line in turn through an associated transistor circuit adjacent to it on the chip. The technique used is to charge all the photodiode elements equally; on exposure to the image, discharging takes place, those elements at the points of highest light intensity losing the most charge. The array is read by connecting each element in turn to the signal line and measuring the amount of charge needed to restore each element to the original charge potential. This can be carried out at the speeds normally associated with integrated circuits, the scan time and repetition rate depending on the number of elements involved—commonly one or two thousand. It should be noted that since all array detectors are charge-dependent devices, they are in effect time integrating over the interval between successive readouts.

(2) Charge-coupled devices (CCDs) consist of an array of electrodes deposited on a substrate, so that each electrode forms part of a metal-oxide-semiconductor device. By appropriate voltage biasing of electrodes to the substrate it is possible to generate a potential well under each electrode. Further, by manipulating the bias voltages on adjacent electrodes it is possible to transfer the charge in any one potential well to the next, and with appropriate circuitry at the end of the array the original charge in each well may be read in turn. These devices were originally developed as delay lines for computers, but since all semiconductor processes are affected by incident light, they make serviceable array detectors. The image is allowed to fall on the array, and charges will develop at the points of high intensity. These are held in the potential wells until the reading process is initiated at the desired interval.

(3) Charge-injection devices (CIDs) use a similar principle, but based on coupled pairs of potential wells rather than a continuous series. Each pair is addressed by using an X–Y coincident voltage technique. Light is measured by allowing photogenerated charges to build up on each pair and sensing the change as the potential well fills. Once reading has been completed this charge is injected into the substrate by removing the bias from both electrodes in each pair simultaneously.

At the time of writing all three types are under intensive development and significant improvements in performance may be expected. All may be made sensitive to visible or near infra-red radiation. The usual noise considerations apply and charge leakage between elements implies that the scan interval must be less than 1 s. CIDs offer the possibility of very rapid readout, so that they may be introduced for television purposes; they also permit sampling patterns other than straightforward roster scanning. There is also the possibility that, by using two variants of CCDs in a single device, a CCD may develop with dual red and blue sensitivities.

Outputs from array detectors may be displayed as a line scan on a CRO or as a raster display; alternatively, they may be fed into a microprocessor. Their applications are numerous; emission spectroscopy, size measurement, position measurement, robot guidance, facsimile recognition and document reading are amongst them. With improvement in manufacturing technique it is to be expected that their cost will be greatly reduced in the next few years and their use become widespread.

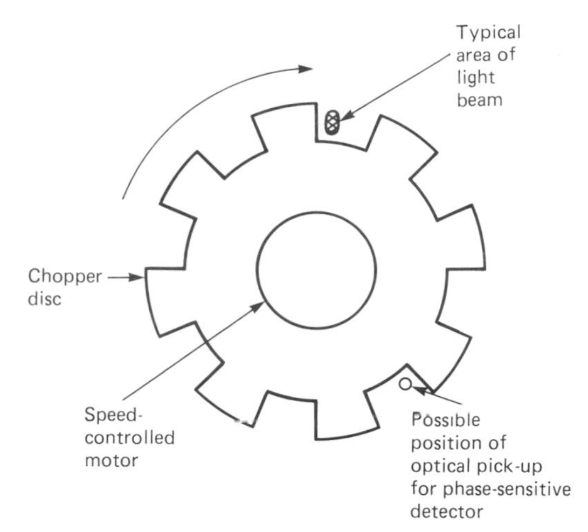

Figure 2.7 Chopper disc.

2.4 Detector techniques

In nearly all optical instruments which require a measurement of light intensity we rely upon the detector producing a signal (usually a current) which is accurately proportional to the light intensity. However, all detectors, by their nature, pass some current when in total darkness and we have to differentiate between the signal which is due to 'dark current' and that due to 'dark current plus light current'. A further problem arises if we try to measure very small light currents or larger light currents very accurately. In all detectors there are small random variations of the dark current—i.e. 'noise'—and it is this noise which limits the ultimate sensitivity of any detector.

The differentiation between 'dark' and 'dark plus light' signals is achieved by taking the difference of the signals when the light beam falls on the detector and when it is obscured. In some manually operated instruments two settings are made, one with and one without a shutter in the beam. In most instruments a chopper disc (Figure 2.7) is made to rotate in the light beam so that the beam is interrupted at a regular frequency and the signal is observed by a.c. circuits that do not respond to the continuous dark current. This technique is called 'beam chopping'.

The effects of noise can be reduced in three ways:

(1) Prolonging the time constant of the detector circuitry;
(2) Cooling the detector;
(3) Using synchronized techniques.

2.4.1 Detector circuit time constants

Dark current noise is due to random events on the atomic scale in the detector. Hence if the time constant of the detector circuit is made sufficiently long the variations in the output current will be smoothed out and it will be of a correct average value. The best choice of time constant will depend on the circumstances of any application; but clearly long time constants are not acceptable in many cases. Even in manually read instruments a time constant as long as 1 s will be irritating to the observer.

2.4.2 Detector cooling

It is not possible to go into a full discussion of detector noise here but in many detectors the largest contribution to the noise is that produced by thermionic emission within the detector. Thermionic emission from metals follows Langmuir's law:

$$i \propto e^{3T/2}$$

where T is absolute temperature and i represents the emission current. Room temperature is around 295 K in terms of absolute temperature, so a significant reduction in emission current and noise can be achieved by cooling the detector.

Detectors may be cooled by the use of liquid nitrogen (77 K, −196°C), solid CO_2 (195 K, −79°C) or by Peltier effect cooling devices. The use of liquid nitrogen or solid CO_2 is cumbersome; it usually greatly increases the complexity of the apparatus and requires recharging. Another problem is that of moisture from the air condensing or freezing on adjacent surfaces. Peltier-effect cooling usually prevents these problems but does not achieve such low temperatures. Unless special circumstances demand it, detector cooling is less attractive than other methods of noise reduction.

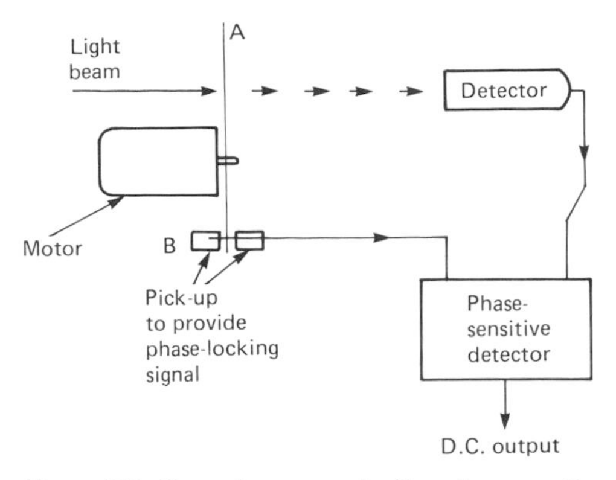

Figure 2.8 Beam chopper used with a phase-sensitive detector.

2.4.3 Beam chopping and phase-sensitive detection

Random noise may be thought of as a mixture of a large number of signals, all of different frequencies. If a beam chopper is used, running at a fixed frequency and the detector circuitry is made to respond preferentially to that frequency, a considerable reduction in noise can be achieved. Although this can be effected by the use of tuned circuits in the detector circuitry it is not very reliable, as the chopper speed usually cannot be held precisely constant. A much better technique is to use a phase sensitive detector, phase locked to the chopper disc by means of a separate pick-up.

Such a system is illustrated in Figure 2.8. The light beam to be measured is interrupted by a chopper disc, A, and the detector signal (Figure 2.9) is passed to a phase-sensitive detector. The gating of the detector is controlled by a separate pick-up, B, and the phase relationship between A and B is adjusted by moving B until the desired synchronization is achieved (Figure 2.10)—a double-beam oscilloscope is useful here.

The chopper speed should be kept as uniform as possible. Except when thermal detectors are used the chopping frequency is usually in the range 300–10 000 Hz. The phase-locking pick-up may be capacitative or optical, the latter being greatly preferable. Care must be taken that the chopper disc is not exposed to room light or this may be reflected from the back of the blades into the detector and falsely recorded as a 'dark' signal. It should be noted that because the beam to be measured has finite width it is not possible to 'chop square'.

By making the light pulses resemble the shape and phase of the gating pulses a very effective improvement in signal-to-noise ratio can be obtained—usually at least 1000:1. This technique is widely used, but stroboscopic trouble can occur when pulsating light sources are involved, such as fluorescent lamps or CRT screens, when the 'boxcar' system can be used.

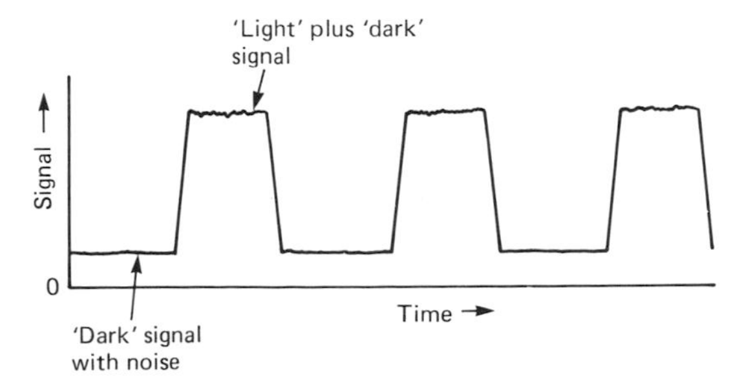

Figure 2.9 Typical output from a beam-chopped detector.

2.4.4 The boxcar detector

A further improvement in signal-to-noise ratio can be obtained with the 'boxcar' system. In some ways this is similar to the phase-sensitive detector system but instead of the chopping being done optically it is carried out electronically. When used with a pulsating source (for example, a fluorescent lamp) the detector signal is sampled at intervals phase locked with the source, the phase locking being provided from the power supply feeding the source (Figure 2.11). The sampling period is made very narrow, and the position

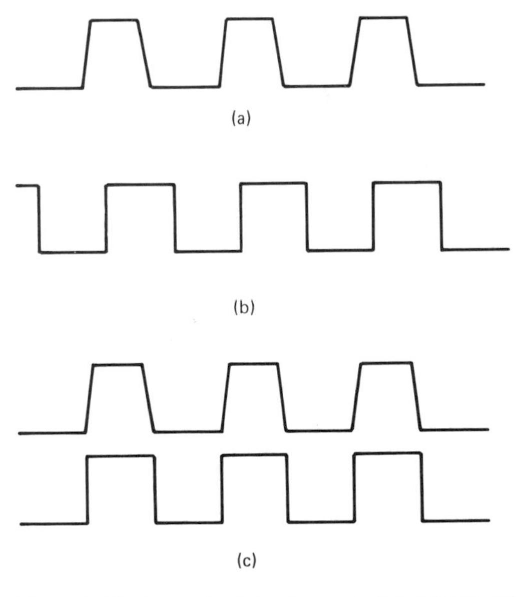

Figure 2.10 Input signals to phase-sensitive detector. (a) Signal from detector; (b) gating signal derived from pick-up, incorrectly phased; (c) detector and gating signals, correctly phased.

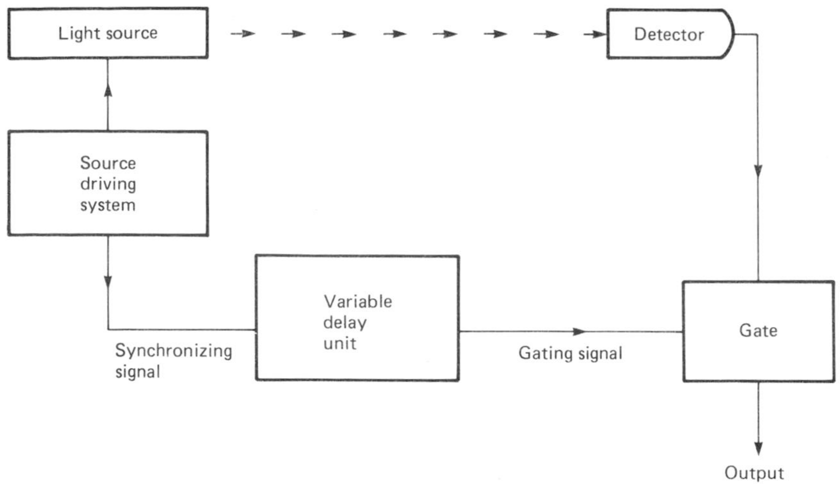

Figure 2.11 Boxcar detector system applied to a pulsating source.

of the sampling point within the phase is adjusted by a delay circuit (Figure 2.12). It is necessary to establish the dark current in a separate experiment. If a steady source is to be measured, an oscillator is used to provide the 'phase' signal.

This system both offers a considerable improvement in noise performance and enables the variation of light output of pulsating sources within a phase to be studied. Although more expensive than the beam chopping–PSD system it is very much more convenient to use, as it eliminates all the mechanical and optical problems associated with the chopper. These techniques are also discussed in Part 4. (The American term 'boxcar' arises from the fact that the gating signal, seen upside down on a CRT, resembles a train of boxcars.)

2.4.5 Photon counting

When measurement of a very weak light indeed with a photomultiplier is required the technique of 'photon counting' may be used. In a photomultiplier, as each electron leaves the photocathode it gives rise to an avalanche of electrons at the anode, of very short duration (see section 2.3.1 above). These brief bursts of output current can be seen with the help of a fast CRT. When very weak light falls on the cathode these bursts can be counted over a given period of time, giving a measure of the light intensity.

Although extremely sensitive, this system is by no means easy to use. It is necessary to discriminate between pulses due to photoelectrons leaving the cathode and those which have originated from spurious events in the dynode chain; this is done with a pulse-height discriminator, which has to be very

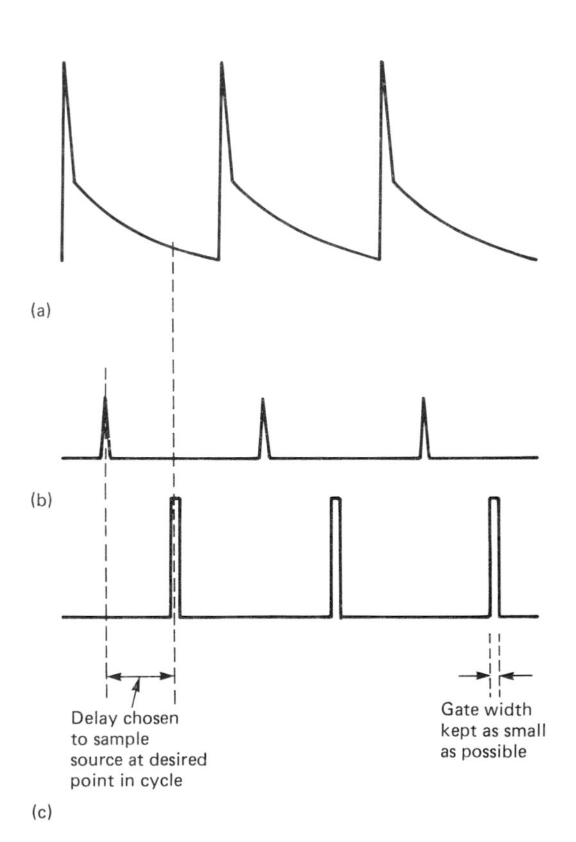

Figure 2.12 Signals used in boxcar detector system. (a) Signal from detector produced by pulsating source (e.g. CRT); (b) synchronizing signal derived from source; (c) gating signal.

carefully adjusted. Simultaneous double pulses are also a problem. Another problem in practice is the matching of 'photon counting' with 'electrometer' (i.e. normal) operation. It also has to be remembered that it is a counting and not a measuring operation, and since random statistics apply, if it is required to obtain an accuracy of ± 1 per cent on a single run then at least 10 000 counts must be made.

This technique is regularly used successfully in what might be termed 'research laboratory instrumentation'—Raman spectrographs and the like—but is difficult to set up and its use can really only be recommended for those cases where the requirement for extreme sensitivity demands it.

2.5 Intensity measurement

The term 'light-intensity measurement' can be used to refer to a large range of different styles of measurement. These can be categorized loosely into (1) those where the spectral sensitivity of the detector is used unmodified and (2) those where the spectral sensitivity of the detector is modified deliberately to match some defined response curve. Very often we are concerned with the comparison of light intensities where neither the spectral power distribution of the light falling on the detector nor its geometrical distribution with respect to the detector will change. Such comparative measurements are clearly in category (1), and any appropriate detector can be used. However, if, for example, we are concerned with the purposes of lighting engineering where we have to accurately measure 'illuminance' or 'luminance' with sources of any spectral or spatial distribution then we must use a photometer with a spectral response accurately matched to that of the human eye and a geometrical response accurately following a cosine law; that is a category (2) measurement.

Some 'measurements' in category (1) are only required to determine the presence or absence of light, as in a very large number of industrial photoelectric controls and counters, burglar alarms, street-lighting controls and so on. The only critical points here are that the detector should be arranged that it receives only light from the intended source and that it shall have sufficient response speed. Photodiodes and phototransistors are often used and are robust, reliable and cheap. Cadmium sulphide photoconductive cells are often used in lighting controls, but their long response time (about 1 s) restricts their use in many other applications.

2.5.1 Photometers

A typical photometer head for measurements of illuminance (the amount of visible light per square metre falling on a plane) is shown in Figure 2.13.

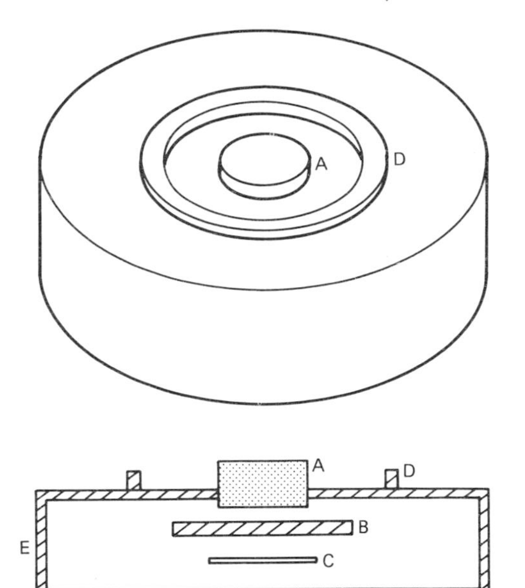

Figure 2.13 Cosine response photometer head. A: opal glass cylinder; B: filter; C: detector; D: light shield developed to produce cosine response; E: metal case.

Incident light falls on the opal glass cylinder A, and some reaches the detector surface C after passing through the filter layer B. The detector may be of either the photoconductive or photovoltaic type. The filter B is arranged to have a spectral transmission characteristic such that the detector–filter combination has a spectral sensitivity matching that of the human eye. In many instruments the filter layer is an integral part of the detector. The cosine response is achieved by careful design of the opal glass A in conjunction with the cylindrical protuberance D in its mounting. The reading is displayed on an appropriately calibrated meter connected to the head by a metre or two of a thin cable, so that the operator does not 'get in his own light' when making a reading.

Instruments of this kind are available at a wide range of prices, depending on the accuracy demanded; the best instruments of this type can achieve an accuracy of 1 or 2 per cent of the illuminance with normal light sources. Better accuracy can be achieved with the use of a 'Dresler filter' which is built up from a mosaic of different filters rather than a single layer, but this is considerably more expensive.

A widely used instrument is the 'Hagner' photometer, which is a combined instrument for measuring both illuminances and luminances. The 'illuminance' part is as described above, but the measuring head also incorporates a telescopic optical system focused on to a separate, internal detector. A beam divider enables the operator to look through the telescope and point the instrument at the surface whose luminance is required; the internal detector output is indicated on a meter also arranged to be within the field of view.

The use of silicon detectors in photometers offers the possibility of measuring radiation of wavelengths above those of the visible spectrum, which is usually regarded as extending from 380 to 770 nm. Many silicon detectors will operate at wavelengths up to 1170 nm. In view of the interest in the 700–1100 nm region for the purposes of fibre optic communications, a variety of dual-function 'photometer/radiometer' instruments are available. Basically these are photometers which are equipped with two interchangeable sets of filters, (1) to modify the spectral responsivity of the detector to match the spectral response of the human eye and (2) to produce a flat spectral response so that the instrument responds equally to radiation of all wavelengths and thus produces a reading of radiant power. In practice this 'flat' region cannot extend above 1170 nm and the use of the phrase 'radiometer' is misleading, because that implies an instrument capable of handling *all* wavelengths; traditional radiometers operate over very much wider wavelength ranges.

In construction, these instruments resemble photometers, as described above, except that the external detector head has to accommodate the interchangeable filters and the instrument has to have dual calibration. These instruments are usually restricted to the measurement of illuminance and irradiance, unlike the Hagner photometer, which can measure both luminance and illuminance. The effective wavelength operating range claimed in the 'radiometer' mode is usually 320–1100 nm.

2.5.2 Ultra-violet intensity measurements

Of recent years, a great deal of interest has developed in the non-visual effects of radiation on humans and animals, especially ultra-violet radiation. Several spectral response curves for photobiological effects (for example, erythema and photokeratitis) are now known (Steck, 1982). Ultra-violet photometers have been developed accordingly, using the same general principles as visible photometers but with appropriate detectors and filters. Photomultipliers are usually used as detectors, but the choice of filter materials is restricted; to date, nothing like the exact correlation of visible response to the human eye has been achieved.

An interesting development has been the introduction of ultra-violet film badges—on the lines of X-ray film badges—for monitoring exposure to ultra-violet radiation. These use photochemical reactions rather than conventional detectors (Young *et al.*, 1980).

2.5.3 Colour-temperature meters

The colour of incandescent light sources can be specified in terms of their 'colour temperature', i.e. the temperature at which the spectral power distribution of a Planckian blackbody radiator most closely resembles that of the source concerned. (*NB:* This is *not* the same as the actual temperature.) Since the spectral power distribution of a Planckian radiator follows the law

$$E_\lambda = c_1\{\lambda^s(e^{c_2/\lambda T} - 1)\}^{-1}$$

(see section 2.8) it is possible to determine T by determining the ratio of the E_λ values at two wavelengths. In practice, because the Planckian distribution is quite smooth, broad-band filters can be used.

Many photometers (see section 2.6.1) are also arranged to act as colour-temperature meters, usually by the use of a movable shade arranged so that different areas of the detectors can be covered by red and blue filters; thus the photometer becomes a 'red-to-blue ratio' measuring device.

Devices of this kind work reasonably well with incandescent sources (which include sunlight and daylight) but will give meaningless results if presented to a fluorescent or discharge lamp.

2.6 Wavelength and colour

2.6.1 Spectrophotometers

Instruments which are used to measure the optical transmission or reflection characteristics of a sample over a range of wavelengths are termed 'spectrophotometers'. This technique is widely used for analytical purposes in all branches of chemistry, and is the physical basis of all measurements of colour; thus it is of much interest in the consumer industries. Transmission measurements, usually on liquid samples, are much the most common.

All spectrophotometers contain four elements:

(1) A source of radiation;
(2) An optical system, or monochromator, to isolate a narrow band of wavelengths from the whole spectrum emitted by the source;
(3) The sample (and its cell if it is liquid or gaseous);
(4) A detector of radiation and its auxiliary equipment.

Note that theoretically it does not matter whether the light passes first through the monochromator and

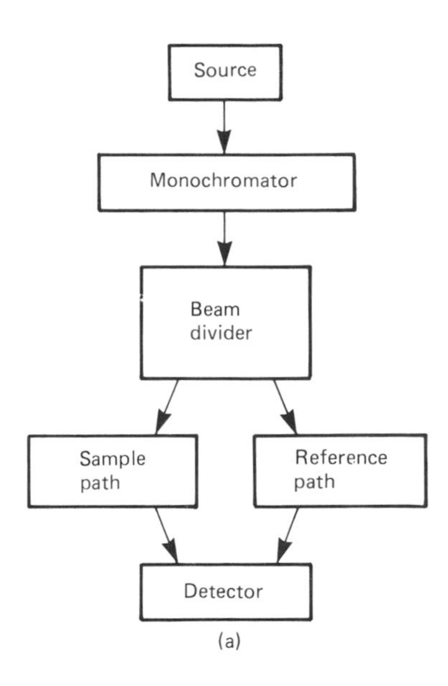

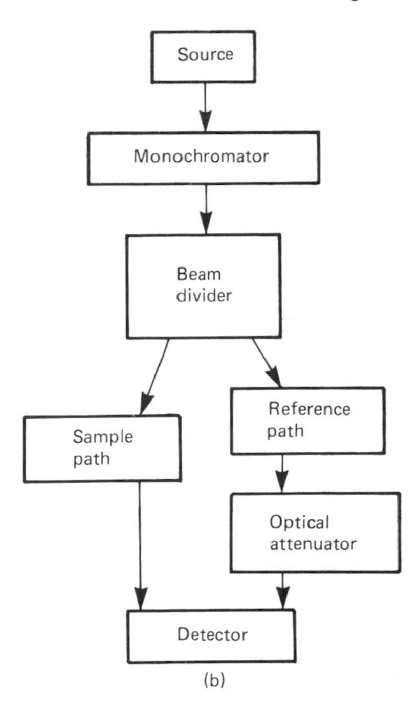

Figure 2.14 Double-beam techniques in spectrophotometry. (a) 'Linearity' method; (b) 'optical null' method.

then the sample or vice versa; the former is usual in visible or ultra-violet instruments, but for infra-red work the latter arrangement offers some advantages.

Spectrophotometers may be either 'single beam', in which the light beam takes a single fixed path and the measurements are effected by taking measurements with and without the sample present, or 'double beam', in which the light is made to pass through two paths, one containing the sample and the other a reference; the intensities are then compared. In work on chemical solutions the reference beam is usually passed through a cell identical to that of the sample containing the same solvent so that solvent and cell effects are cancelled out; in reflection work the reference sample is usually a standard white reflector. The single-beam technique is usual in manually operated instruments and the double-beam one in automatic instruments, nowadays the majority.

Two main varieties of 'double-beam' techniques are used (Figure 2.14). That shown in Figure 2.14(a) relies for accuracy on the linearity of response of the detector and is sometimes called 'the linearity method'. The light beam is made to follow alternate sample and reference paths and the detector is used to measure the intensity of each in turn; the ratio gives the transmission or reflection factor at the wavelength involved. The other method shown (Figure 2.14(b)) is called 'the optical-null method'. Here the intensity of the reference beam is reduced to equal that of the sample beam by some form of servocontrolled optical attenuator and the detector is called on only to judge for equality between the two beams. The accuracy thus depends on the optical attenuator, which may take the form of a variable aperture or a system of polarizing prisms.

Since spectrophotometric results are nearly always used in extensive calculations and much data can easily be collected, spectrophotometers are sometimes equipped with microprocessors. These microprocessors are also commonly used to control a variety of automatic functions, for example the wavelength-scanning mechanism, automatic sample changing and so on.

For chemical purposes it is nearly always the absorbance (i.e. optical density) of the sample rather than the transmission that is required:

$$A = \log_{10} \frac{1}{T}$$

where A is the absorbance or optical density and T is the transmission. The relation between Transmission and Absorbance is shown in Table 2.1.

Instruments for chemical work usually read in absorbance only. 'Wave number'—the number of waves in one centimetre—is also used by chemists in preference to wavelength. The relation between the two is shown in Table 2.2.

It is not economic to build all-purpose spectrophotometers in view of their use in widely

Table 2.1 Relation between transmission and absorbance

Transmission (%)	*Absorbance*
100	0
50	0.301
10	1.0
5	1.301
1	2.0
0.1	3.0

Table 2.2 Relation between wavelength and wavenumber

Wavelength	*Wavenumber (no. of waves/cm)*
200 nm	50 000
400 nm	25 000
500 nm	20 000
1 μ	10 000
5 μ	2 000
10 μ	1 000
50 μ	200

different fields. The most common varieties are:

(1) Transmission/absorbance, ultra-violet and visible range (200–600 nm);
(2) Transmission and reflection, near-ultra-violet and visible (300–800 nm);
(3) Transmission/absorbance, infra-red (2.5–25 μm).

The optical parts of a typical ultra-violet-visible instrument are shown in Figure 2.15. Light is taken either from a tungsten lamp A or deuterium lamp B by moving mirror C to the appropriate position. The beam is focused by a mirror on to the entrance slit E of the monochromator. One of a series of filters is inserted at F to exclude light of submultiples of the desired wavelength. The light is dispersed by the diffraction grating G, and a narrow band of wavelengths is selected by the exit slit K from the spectrum formed. The wavelength is changed by rotating the grating by a mechanism operated by a stepper motor.

The beam is divided into two by an array of divided mirror elements at L and images of the slit K are formed at R and S, the position of the reference and sample cells. A chopper disc driven by a synchronous motor M allows light to pass through only one beam at a time. Both beams are directed to T, a silica diffuser, so that the photomultiplier tube is presented alternately with sample and reference beams; the diffuser is necessary to overcome non-uniformities in the photomultiplier cathode (see section 2.4).

The signal from U is switched appropriately to sample or reference circuits by the signal driving the chopper M and the magnitudes compared. Their ratio gives the transmission which may be recorded directly, or an absorbance figure may be calculated from it and recorded. A microprocessor is used to control the functions of wavelength scanning, slit width, filter and lamp selection, and to effect any desired calculations on the basic transmission results.

2.6.2 Spectroradiometers

The technique of measuring the spectral power distribution (SPD) of a light source is termed 'spectroradiometry'. We may be concerned with the

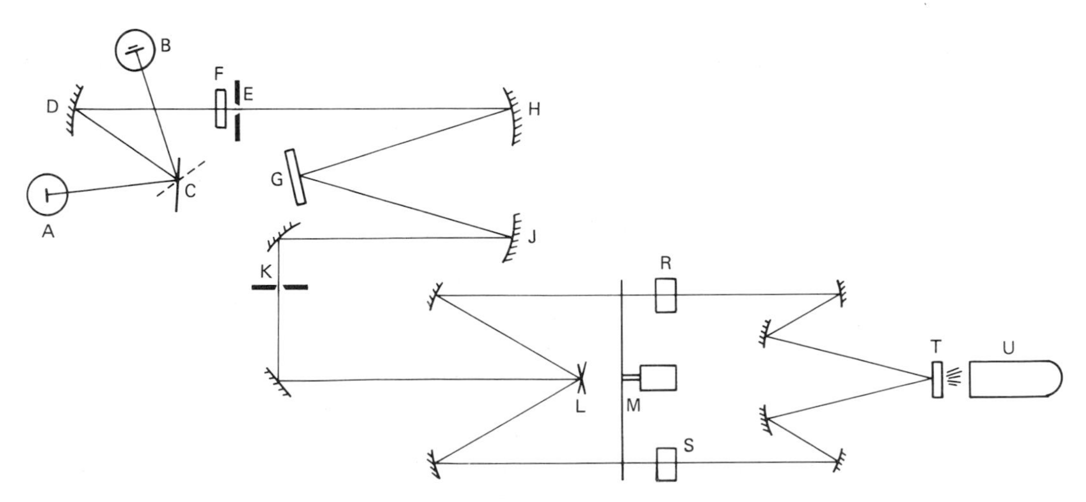

Figure 2.15 Typical ultra-violet–visible spectrophotometer.

SPD in relative or absolute terms. By 'relative' we refer to the power output per unit waveband at each wavelength of a range, expressed as a ratio of that at some specified wavelength. (For the visible spectrum this is often 560 nm). By 'absolute' we mean the actual power output per steradian per unit waveband at each wavelength over a range. 'Absolute' measurements are much more difficult than 'relative' ones and are not often carried out except in specialized laboratories.

Relative SPD measurements are effected by techniques similar to those of spectrophotometry (see section 2.6.1 above). The SPD of the unknown source is compared with that of a source whose SPD is known. In the single-beam method light from the source is passed through a monochromator to a detector whose output is recorded at each wavelength of the desired range. This is repeated with the source whose SPD is known, and the ratio of the readings of the two at each wavelength is then used to determine the unknown SPD in relative terms. If the SPD of the reference source is known in absolute terms then the SPD of the unknown can be determined in absolute terms.

In the double-beam method light from the two sources is passed alternately through a monochromator to a detector, enabling the ratio of the source outputs to be determined wavelength by wavelength. This method is sometimes offered as an 'alternative mode' of using double-beam spectrophotometers. The experience of the author is that with modern techniques the single-beam technique is simpler, more flexible, more accurate and as rapid as the double-beam one. It is not usually worthwhile to try to modify a spectrophotometer to act as a spectroradiometer if good accuracy is sought for.

A recent variation of the single-beam technique is found in the 'optical multichannel analyser'. Here light from the source is dispersed and made to fall not on a slit but on a multiple-array detector; each detector bit is read separately with the aid of a microprocessor. This technique is not as accurate as the conventional single-beam one but can be used with sources which vary rapidly with time (for example, pyrotechnic flares).

Both single-beam and double-beam methods require the unknown source to remain constant in intensity whilst its spectrum is scanned—which can take up to 3 min or so. Usually this is not difficult to arrange but if a source is inherently unstable (for example, a carbon arc lamp) the whole light output, or output at a single wavelength, can be monitored to provide a reference (Tarrant, 1967).

Spectroradiometry is not without its pitfalls, and the worker intending to embark in the field should consult suitable texts (Forsythe, 1941; Commission Internationale de l'Eclairage, 1984).

Particular problems arise (1) where line and continuous spectra are present together and (2) where the source concerned is time modulated (for example, fluorescent lamps, cathode ray tubes, etc.). When line and continuous spectra are present together they are transmitted through the monochromator in differing proportions and this must be taken into account or compensated for (Henderson, 1970; Moore, 1984). The modulation problem can be dealt with by giving the detector a long time constant (which implies a slow scan speed) or by the use of phase-locked detectors (Brown and Tarrant, 1981) (see section 2.4.3).

Few firms offer spectroradiometers as stock lines because they nearly always have to be custom built for particular applications. One recent instrument—the 'Surrey' spectroradiometer—is shown in Figure 2.16. This instrument was developed for work on cathode ray tubes but can be used on all steady sources (Brown and Tarrant, 1981). Light from the source is led by the front optics to a double monochromator which allows a narrow waveband to pass to a photomultiplier. The output from the photomultiplier tube is fed to a boxcar detector synchronized with the tube-driving signal so that the tube output is sampled only over a chosen period, enabling the initial glow or afterglow to be studied separately. The output from the boxcar detector is recorded by a desk-top computer which also controls the wavelength-scanning mechanism by means of a stepper motor. The known source is scanned first and the readings are also held in the computer, so that the SPD of the unknown source can be printed out as soon as the wavelength scan is completed. The colour can also be computed and printed out. The retro-illuminator is a retractable unit used when setting up. Light can be passed backwards through the monochromator to identify the precise area of the CRT face viewed by the monochromator system.

This instrument operates in the visible range of wavelengths 380–760 nm and can be used with screen luminances as low as 5 cd/m^2. When used for colour measurement an accuracy of ± 0.001 in x and y (CIE 1931 system) can be obtained.

2.6.3 The measurement of colour

2.6.3.1 Principles

The measurement of colour is not like the measurement of physical quantities such as pressure or viscosity because colour is not a physical object. It is a visual phenomenon, a part of the process of vision. It is also a *psychophysical phenomenon*, and if we attempt to measure it we must not lose sight of that fact.

The nature of colour is discussed briefly by the author elsewhere (Tarrant, 1981). The newcomer to the subject should consult the excellent textbooks on the subject by Wright (1969) and by Judd and Wysecki

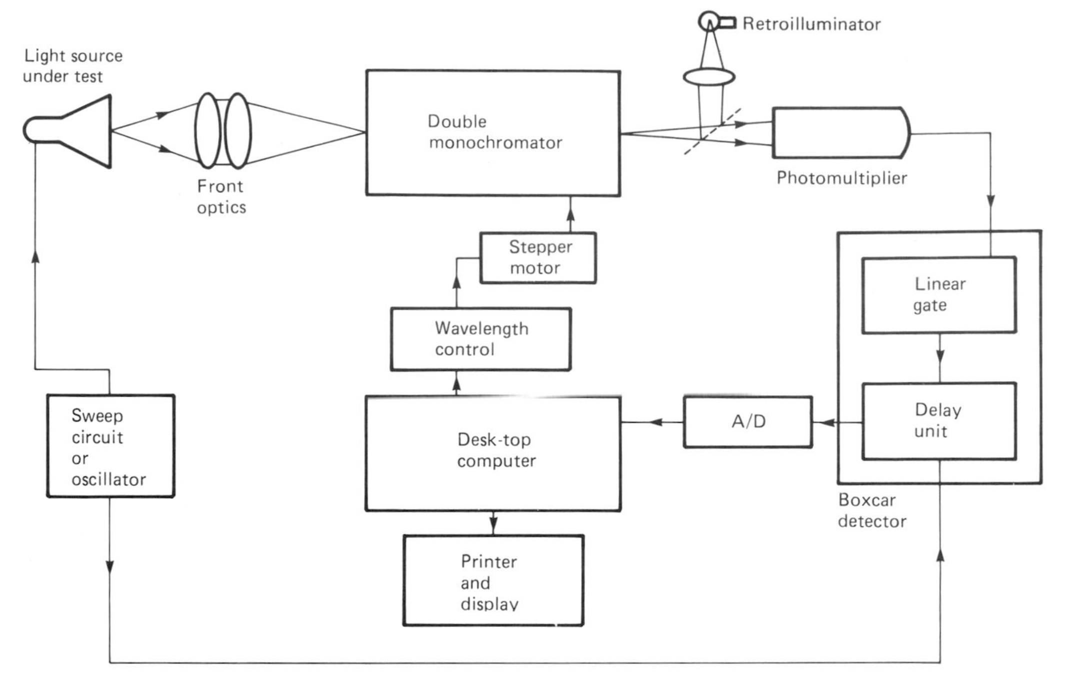

Figure 2.16 The 'Surrey' spectroradiometer.

(1975). Several systems of colour measurement are in use; for example, the CIE system and its derivatives, the Lovibond system (Chamberlin, 1979), and the Munsell system. It is not possible to go into details of these systems here and we shall confine ourselves to a few remarks on the CIE 1931 system. (The letters CIE stand for Commission Internationale de l'Eclairage and the 1931 is the original fundamental system. About six later systems based on it are in current use.)

There is strong evidence to suggest that in normal daytime vision our eyes operate with three sets of visual receptors corresponding to red, green and blue (a very wide range of colours can be produced by mixing these together) and that the responses to these add in a single arithmetical way. If we could then make some sort of triple photometer (each channel having spectral sensitivity curves corresponding to those of the receptor mechanisms) we should be able to make physical measurements to replicate the functioning of eyes. This cannot in fact be done, as the human visual mechanisms have *negative* responses to light of certain wavelengths. However, it is possible to produce photocell/filter combinations which correspond to red, green and blue, and which can be related to the human colour-matching functions (within a limited range of colours) by simple matrix equations.

This principle is used in photoelectric colorimeters, or 'tristimulus colorimeters' as they are sometimes called. One is illustrated in Figure 2.17. The sample is

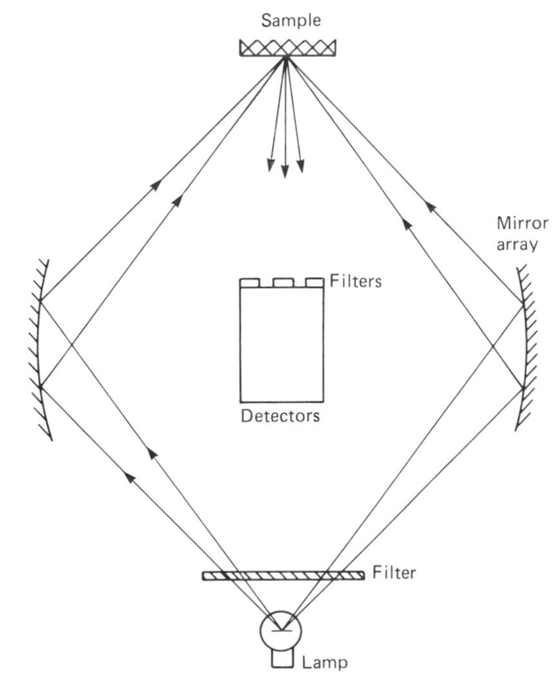

Figure 2.17 Tristimulus colorimeter.

illuminated by a lamp and f : combination which has the SPD of one of ιιιe defined 'Standard Illuminants'. Light diffusely reflected from the sample is passed to a photomultiplier through a set of filters, carefully designed so that the three filter/photomultiplier spectral sensitivity combinations can be related to the human colour-matching functions. By measuring each response in turn and carrying out a matrix calculation, the colour specification in the CIE 1931 system (or its derivatives) can be found. Most instruments nowadays incorporate microprocessors, which remove the labour from these calculations, so that the determination of a surface colour can be carried out rapidly and easily.

Consequently colorimetric measurements are used on a large scale in all consumer industries, and the colours of manufactured products can now be very tightly controlled. It is possible to achieve a high degree of precision, so that the minimum colour difference that can be measured is slightly smaller than that which the human eye can perceive. It should be noted that nearly all surfaces have markedly directional characteristics, and hence if a sample is measured in two instruments which have different viewing/illuminating geometry, different results must be expected.

Diagrams of these instruments (such as Figure 2.17), make them look very simple. In fact, the spectral sensitivity of the filter/photomultiplier combination has to be controlled and measured to a very high accuracy; it certainly is not economic to try and build a 'do-it-yourself' colorimeter. It is strongly emphasized that the foregoing remarks are no substitute for a proper discussion of the fascinating subject of colorimetery, and no-one should embark on colour measurement without reading at least one of the books mentioned above.

2.7 Measurement of optical properties

Transparent materials affect light beams passing through them, notably changing the speed of propagation; refractive index is, of course a measure of this. In this section we describe techniques for measuring the material properties which control such effects.

2.7.1 Refractometers

The precise measurement of the refractive index of transparent materials is vital to the design of optical instruments but is also of great value in chemical work. Knowledge of the refractive index of a substance is often useful in both identifying and establishing the concentration of organic substances, and by far the greatest use of refractometry is in chemical laboratories. Britton has used the refractive index of gases to determine concentrations of trilene in air, but this involves an interferometric technique which will not be discussed here.

When light passes from a less dense to a more dense optical medium, for example from air into glass, then the angle of the refracted ray depends upon the angle of incidence and the refractive indices of the two media (see Figure 2.18(a)) according to Snell's law:

$$\frac{n_2}{n_1}=\frac{\sin i_1}{\sin i_2}$$

In theory, then, we could determine the refractive index of an unknown substance in contact with air by measuring these angles and assuming that the refractive index of air is unity (in fact it is 1.000 27).

In practice for a solid sample we have to use a piece with two non-parallel flat surfaces and this involves measuring also the angle between them. This method can be used with the aid of a simple 'table spectrometer'. Liquid samples can be measured in this way with the use of a hollow prism, but it is a laborious method and requires a considerable volume of the liquid.

Most refractometers instead make use of the critical angle effect. When light passes from a more dense to a less dense medium it may be refracted (Figure 2.18(b)) but if the angle i_2 becomes so large that the ray cannot emerge from the dense medium then the ray is totally internally reflected (Figure 2.18(c)).

The transition from refraction to internal reflection occurs sharply and the value of the angle of i_2 at which this occurs is called the 'critical angle' (Figure 2.18(d)). If we call that angle i_c then

$$\frac{n_2}{n_1}=\frac{1}{\sin i_c}$$

Hence by determining i_c we can find n_1, if n_2 is known.

2.7.1.1 The Abbé refractometer

The main parts of the Abbé refractometer which uses this principle are shown in Figure 2.19. The liquid under test is placed in the narrow space between prisms A and B. Light from a diffuse monochromatic source (L), usually a sodium lamp, enters prism A, and thus the liquid layer, at a wide variety of angles. Consequently light will enter prism B at a variety of angles which is sharply limited by the critical angle. This light then enters the telescope (T), and on moving the telescope round, a sharp division is seen at the critical angle; one half of the field is bright and the other is almost totally dark. The telescope is moved to align the cross-wires on the light/dark boundary and

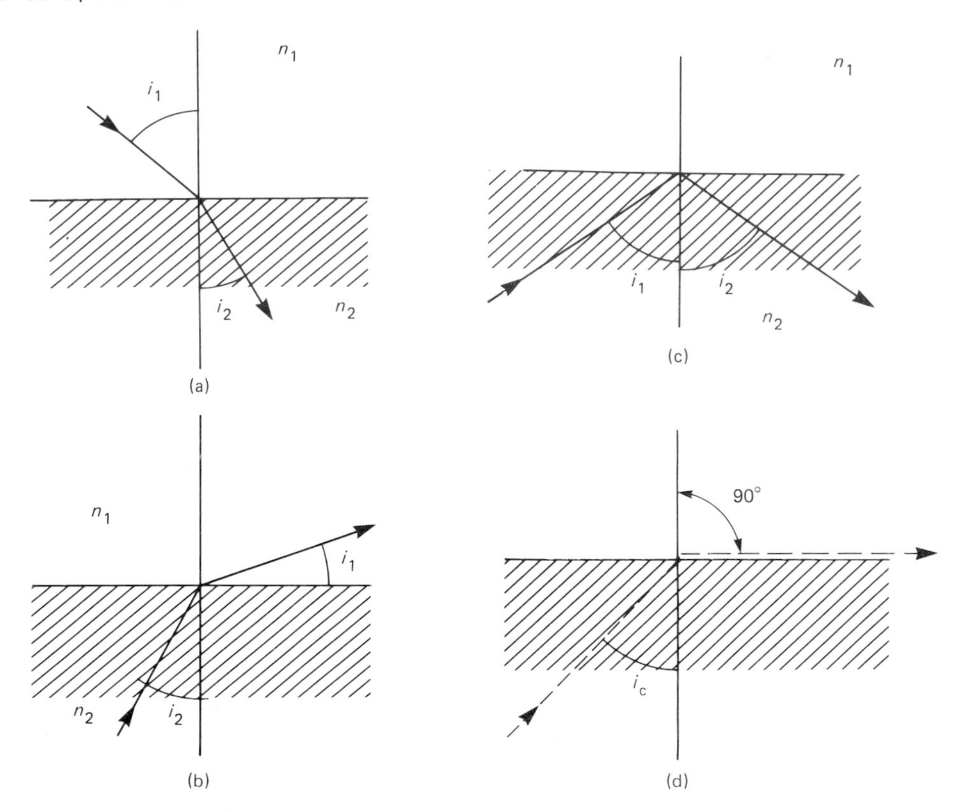

Figure 2.18 (a) Refraction of ray passing from less dense to more dense medium; (b) refraction of ray passing from more dense to less dense medium; (c) total internal reflection; (d) the critical angle case, where the refracted ray can just emerge.

the refractive index can be read off from a directly calibrated scale attached to it. This calibration also takes into account the glass/air refraction which occurs when the rays leave prism B.

Although simple to use, this instrument suffers from all the problems which complicate refractive index measurements. It should be noted that

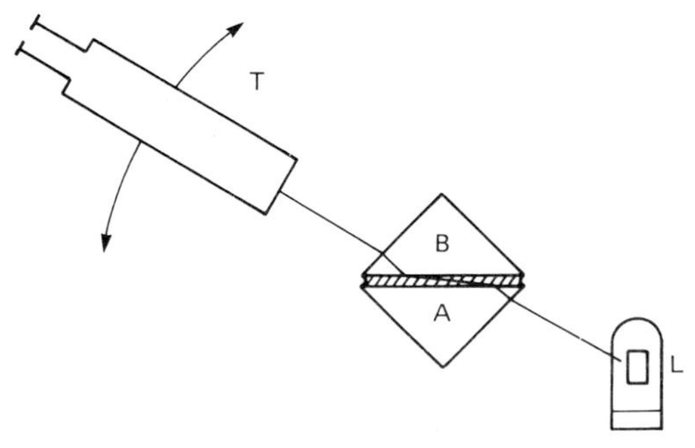

Figure 2.19 Abbé refractometer.

(1) In all optical materials the refractive index varies markedly with wavelength in a non-linear fashion. Hence either monochromatic sources or 'compensating' devices must be used. For high-accuracy work, monochromatic sources are invariably used.
(2) The refractive index of most liquids also varies markedly with temperature, and for accurate work temperature control is essential.
(3) Since the refractive index varies with concentration, difficulties may be encountered with concentrated solutions, especially of sugars, which tend to become inhomogeneous under the effects of surface tension and gravity.
(4) The range of refractive indices that can be measured in critical-angle instruments is limited by the refractive index of the prism A. Commercial instruments of this type are available for measuring refractive indices up to 1.74.
(5) In visual instruments the light/dark field boundary presents such severe visual contrast so that it is sometimes difficult to align the cross-wires on it.

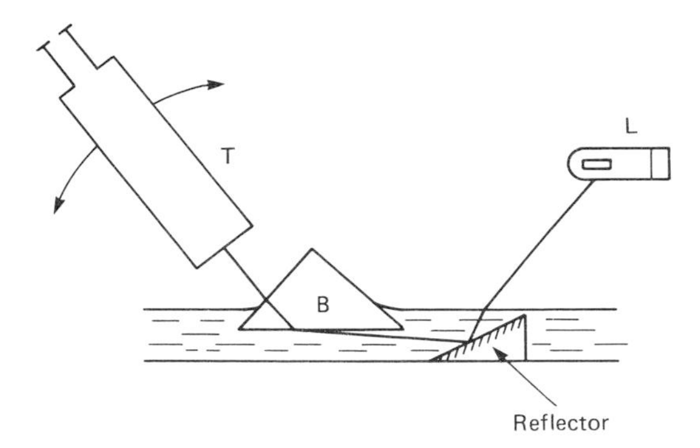

Figure 2.20 Dipping refractometer with reflector enabling light to enter prism B at near-grazing incidence.

2.7.1.2 Modified version of the Abbé refractometer

If plenty of liquid is available as a sample, the prism A of Figure 2.19 may be dispensed with and prism B simply dipped in the liquid. The illuminating arrangements are as shown in Figure 2.20. Instruments of this kind are usually called 'dipping refractometers'.

When readings are required in large numbers, or continuous monitoring of a process is called for, an automatic type of Abbé refractometer may be used. The optical system is essentially similar to that of the visual instrument, except that instead of moving the whole telescope to determine the position of the light/dark boundary the objective lens is kept fixed and a differentiating detector is scanned across its image plane under the control of a stepper motor. The detector's response to the sudden change of illuminance at the boundary enables its position, and thus the refractive index, to be determined. To ensure accuracy, the boundary is scanned in both directions and the mean position is taken.

2.7.1.3 Refractometry of solid samples

If a large piece of the sample is available it is possible to optically polish two faces at an angle on it and then to measure the deviation of a monochromatic beam that it produces with a table spectrometer. This is laborious and expensive. The Hilger–Chance refractometer has been designed for the determination of the refractive indices of optical glasses, and requires that only two roughly polished surfaces at right angles are available. It can also be used for liquids.

The optical parts are shown in Figure 2.21. Monochromatic light from the slit (A) is collected by the lens (B) and passes into the V-shaped prism block (C) which is made by fusing two prisms together to produce a very precise angle of 90 degrees between its surfaces. The light emerges and enters the telescope

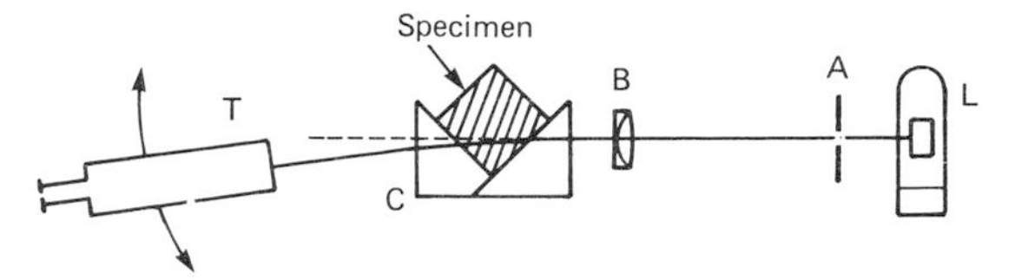

Figure 2.21 Hilger–Chance refractometer.

(T). When the specimen block is put in place the position of the emergent beam will depend on the refractive index of the sample. If it is greater than that of the V block the beam will be deflected upwards; if lower, downwards. The telescope is moved to determine the precise beam direction and the refractive index can be read off from a calibrated scale.

In the actual instrument the telescope is mounted on a rotating arm with the reflecting prism, so that the axis of the telescope remains horizontal. A wide slit with a central hairline is used at A, and the telescope eyepiece is equipped with two lines in its focal plane so that the central hairline may be set between them with great precision (Figure 2.22). Since it is the bulk of the sample which is responsible for the refraction and not only the surfaces, it is possible to place a few drops of liquid on the V-block so that perfect optical contact may be achieved on a roughly polished specimen. The v-block is equipped with side plates so that it forms a trough suitable for liquid samples.

When used for measuring optical glasses an accuracy of 0.0001 can be obtained. This is very high indeed, and the points raised about accuracy in connection with the Abbé refractometer should be borne in mind. It will be noticed that the instrument is arranged so that the rays pass the air/glass interfaces at normal or near-normal incidence, so as to reduce the effects of changes in the refractive index of air with temperature and humidity.

2.7.1.4 Solids of irregular shape

The refractive index of irregularly shaped pieces of solid materials may theoretically be found by immersing them in a liquid of identical refractive index. When this happens, rays traversing the liquid are not deviated when they encounter the solid but pass straight through, so that the liquid–solid

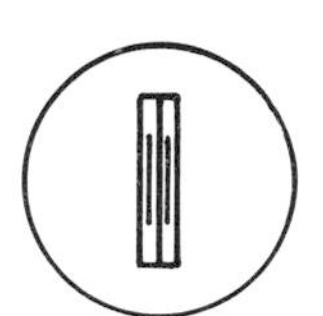
Figure 2.22 Appearance of field when the telescope is correctly aligned.

boundaries totally disappear. A suitable liquid may be made up by using liquids of different refractive index together. When a refractive index match has been found the refractive index of the liquid may be measured with an Abbé refractometer. Suitable liquids are given by Longhurst (1974):

	n
Benzene	1.504
Nitrobenzene	1.553
Carbon bisulphide	1.632
α-monobromonaphthalene	1.658

The author cannot recommend this process. Granted, it can be used for a magnificent lecture-room demonstration, but in practice these liquids are highly toxic, volatile and have an appalling smell. Moreover, the method depends upon both the solid and the liquid being absolutely colourless; if either has any trace of colour it is difficult to judge the precise refractive index match at which the boundaries disappear.

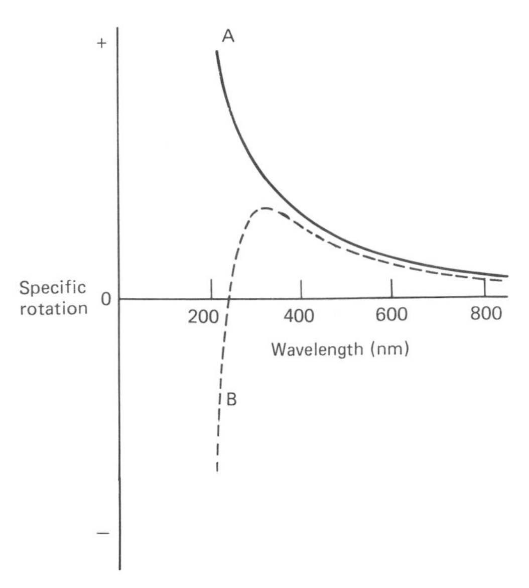

Figure 2.23 Variation of specific rotation with wavelength. A: typical sugar; B: steroid showing reversion.

2.7.2 Polarimeters

Some solutions and crystals have the property that when a beam of plane-polarized light passes through them the plane is rotated. This phenomenon is known as 'optical activity', and in liquids only occurs with those molecules that have no degree of symmetry. Consequently few compounds show this property, but one group of compounds of commercial importance does—sugars. Hence the measurement of optical activity offers an elegant way of determining the sugar content of solutions. This technique is referred to as 'polarimetry' or occasionally by its old-fashioned name of 'saccharimetry'.

Polarimetry is one of the oldest 'physical' methods applied to chemical analysis and has been in use for almost a hundred years. The original instruments were all visual, and though in recent years photoelectric instruments have appeared, visual instruments are still widely used because of their simplicity and low cost.

If the plane of polarization in a liquid is rotated by an angle θ on passage through a length of solution l, then

$$\theta = \alpha c l$$

where c is the concentration of the optically active substance and α is a coefficient for the particular substance called the 'specific rotation'. In all substances the specific rotation increases rapidly with decreasing wavelength (Figure 2.23) and for that reason monochromatic light sources are always used—very often a low-pressure sodium lamp. Some steroids show anomalous behaviour of the specific rotation with wavelength, but the 'spectro-polarimetry' involved is beyond the scope of this book.

2.7.2.1 The Laurent polarimeter

The main optical parts of this instrument are shown in Figure 2.24. The working is best described if the solution tube is at first imagined not to be present. Light from a monochromatic source (A) passes through a sheet of Polaroid (B) so that it emerges plane-polarized. It then encounters a half-wave plate (C) which covers only half of the area of the beam. The effect of the half-wave plate is to slightly alter the plane of polarization of the light that passes through it so that the situation is as shown in Figure 2.24(b). If the solution tube (D) is not present the light next encounters a second Polaroid sheet at E. This is mounted so that it can be rotated about the beam. On looking through the eyepiece (F) the two halves of the field will appear of unequal brilliance until E is rotated

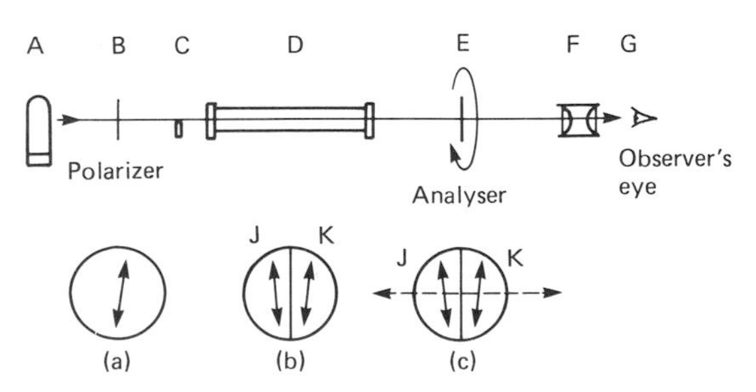

Figure 2.24 Laurent polarimeter. (a) Plane-polarized light after passage through B; (b) polarization directions in field after passage through C; (c) broken arrow shows plane of analyser at position of equal brilliance with no sample present.

until the plane it transmits is as shown in Figure 2.24(c). Since the planes in J and K differ only by a small angle, the position of equal brilliance can be judged very precisely. If the position of the analyser (E) is now read, the solution tube (D) can be put in position and the process repeated so that the rotation θ may be determined. Since the length of the solution tube is known the concentration of the solution may be determined if the specific rotation is known.

2.7.2.2 The Faraday-effect polarimeter

Amongst the many effects that Faraday discovered was the fact that glass becomes weakly optically active in a magnetic field. This discovery lay unused for over a hundred years until its employment in the Faraday-effect polarimeter. The main optical parts are shown schematically in Figure 2.25.

A tungsten lamp, filter and polaroid are used to provide plane-polarized monochromatic light. This is then passed through a 'Faraday cell' (a plain block of glass situated within a coil) which is energized from an oscillator at about 380 Hz, causing the plane of polarization to swing about 3 degrees either side of the mean position. If we assume for the time being that there is no solution in the cell and no current in the second Faraday cell this light will fall unaltered on the second polaroid. Since this is crossed on the mean position, the photomultiplier will produce a signal at *twice* the oscillator frequency, since there are two pulses of light transmission in each oscillator cycle (see Figure 2.25(a)). If an optically active sample is now put in the cell the rotation will produce the situation in Figure 2.25(b) and a component of the same frequency as the oscillator output. The photomultiplier signal is compared with the oscillator output in a phase-sensitive circuit so that any rotation produces a d.c. output. This is fed back to the second Faraday cell to oppose the rotation produced by the solution, and, by providing sufficient gain, the rotation produced by the sample will be completely restored. In this condition the current in the second cell will in effect give a measure of the rotation, which can be indicated directly with a suitably calibrated meter.

This arrangement can be made highly sensitive and a rotation of as little as 1/10 000th of a degree can be detected, enabling a short solution path length to be used—often 1 mm. Apart from polarimetry, this

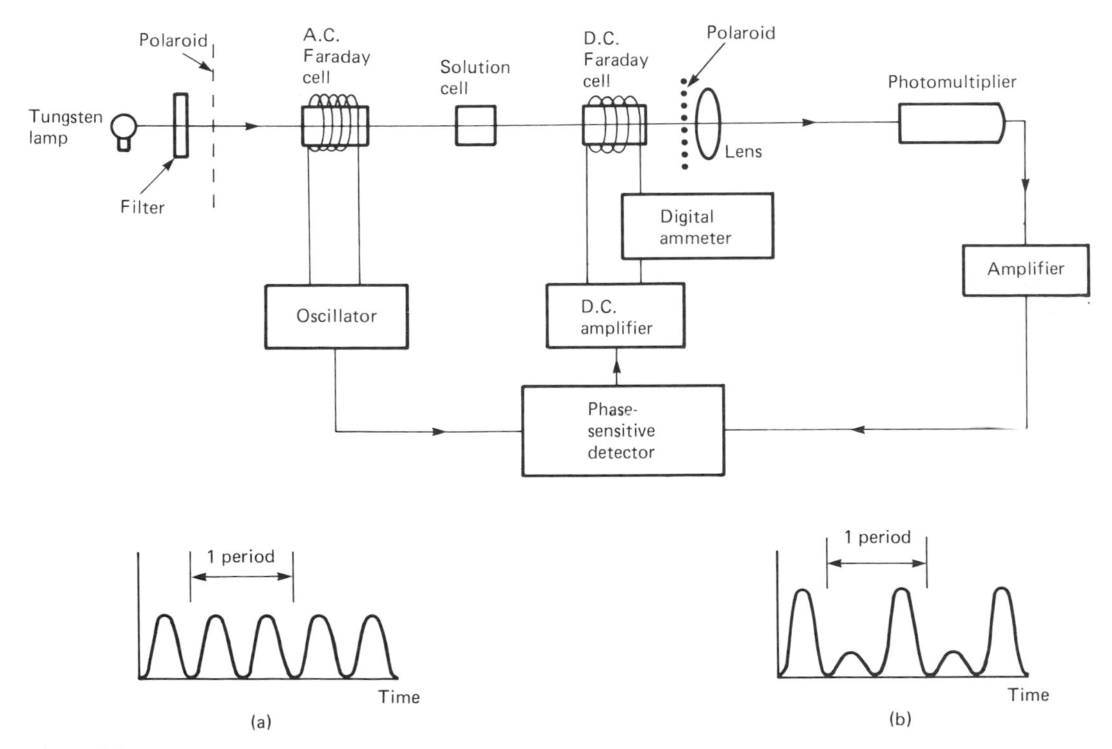

Figure 2.25 Faraday effect polarimeter. (a) Photomultiplier signal with no sample present; (b) photomultiplier signal with uncompensated rotation.

technique offers a very precise method of measuring angular displacements.

2.8 Thermal imaging techniques

Much useful information about sources and individual objects can be obtained by viewing them not with the visible light which they give off but by the infra-red radiation which they emit. We know from Planck's radiation law

$$P_\lambda = \frac{C_1}{\lambda^5[e^{C_2/\lambda T} - 1]}$$

(where P_λ represents the power radiated from a body at wavelength λ, T the absolute temperature and C_1 and C_2 are constants) that objects at the temperatures of our environment radiate significantly, but we are normally unaware of this because all that radiation is well out in the infra-red spectrum. The peak wavelength emitted by objects at around 20°C (293 K) is about 10 μm, whereas the human eye is only sensitive in the range 0.4–0.8 μm. By the use of an optical system sensitive to infra-red radiation this radiation can be studied, and since its intensity depends on the surface temperature of the objects concerned, the distribution of temperatures over an object or source can be made visible. It is possible to pick out variations in surface temperature of less than 1 K in favourable circumstances, and so this technique is of great value; for example, it enables the surface temperature of the walls of a building to be determined, and so reveals the areas of greatest heat loss. Figure 2.26 illustrates this: (a) is an ordinary photograph of a house, while (b) gives the same view using thermal imaging techniques. The higher temperatures associated with heat loss from the windows and door are immediately apparent. Thermal imaging can be used in medicine to reveal variations of surface temperature on a patient's body and thus reveal failures of circulation; it is of great military value since it will function in darkness, and has all manner of applications in the engineering field.

An excellent account of the technique is given by Lawson (1979) in *Electronic Imaging*; Figure 2.27 and Table 2.3 are based on this. Although the spectrum of a body around 300 K has its peak at 10 μm, the spectrum is quite broad. However, the atmosphere is effectively opaque from about 5–8 μm, and above 13 μm, which means that in practice the usable bands are 3–5 μm and 8–13 μm. Although there is much more energy available in the 8–13 μm band (see Table 2.3) and if used outdoors there is much less solar radiation, the sensitivity of available detectors is much better in the 3–5 μm band, and both are used in practice. Nicholas (1968) has developed a triple waveband system, using visible 2–5 μm and 8–13 μm ranges, but

(a)

(b)

Figure 2.26 Thermal imaging techniques applied to a house (courtesy, Agema Infrared Systems).

this technique has not so far been taken up in commercially available equipment.

In speaking of 'blackbody' radiations we must remember that, in practice, no surfaces have emissivities of 1.0, and often they have much less, so there is not a strict relationship between surface temperatures and radiant power for all the many surfaces in an exterior source. In daylight, reflected solar power is added to the emitted power and is a further complication. However, the main value of the technique is in recognizing differences of temperature rather than temperature in absolute terms.

The basic technique is to use some form of television camera sensitive to the appropriate waveband and to present its signal on a conventional television monitor. Often a colour presentation is used with a variety of colours to represent different intensities. The main problem lies in that phrase 'sensitive to the appropriate waveband'. The quantum energies involved are very low and elaborate detector systems

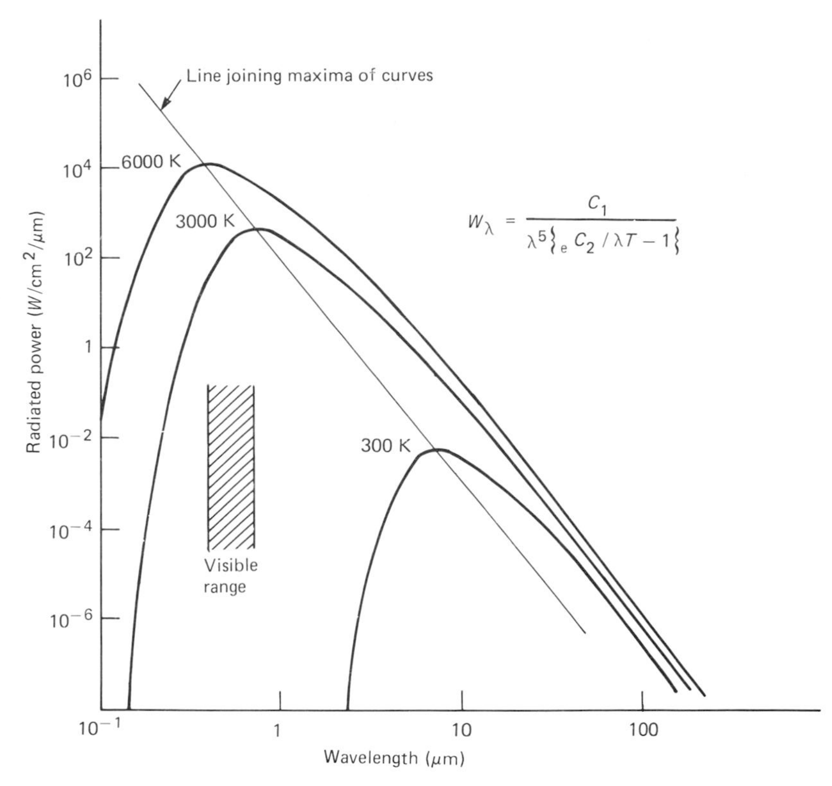

Figure 2.27 Log scale presentation of Planck's law.

Table 2.3 Solar and blackbody radiation at different wavelengths (after Lawson, 1979)

Wavelength band (μm)	*Typical value of solar radiation* (Wm^{-2})	*Emission from blackbody at 300 K* (Wm^{-2})
0.4–0.8	750	0
3–5	24	6
8–13	1.5	140

Note Complete tables of data relating to the spectral power distribution of black body radiators are given by M. Pivovonsky and M. Nagel.

have to be used to produce a response sufficiently good in both speed and resolution for a television-type presentation. The detector most widely used is the 'pyroelectric vidicon' tube, in which the infra-red image is formed on a piezoelectric target (producing a charge pattern thereon by the pyroelectric effect) which can be read with the conventional electron scanning technique. The camera resembles a conventional television camera except that the lenses have to transmit in the appropriate waveband; germanium is usually used.

A variety of line and matrix scanning techniques have been developed (see Lawson, 1979), and with the development of charge-coupled devices it is to be expected that effective detectors embodying this technique will become available in commercial equipment.

Systems of this type clearly have great possibilities beyond the fields of research (since the wavelengths used are so long they can, for example, be used by firemen to rescue people from smoke-filled buildings) and the greatest drawback to their use is their cost. Their resolution is not good by normal optical standards but in most applications this is of little importance; their great value lies in revealing small temperature differences between different items.

2.9 References

American National Standards Institute, Standards for the Safe Use of Lasers

Beynon, J. D. E. *Charge Coupled Devices and their Applications*, McGraw-Hill, New York (1979)

BS 4727, *Glossary of Terms Particular to Light and Colour* (1970)

BS 4803, *Guide on Protection of Personnel against Hazards from Laser Radiation* (1983)

Brown, S. and Tarrant, A. W. S. *A Sensitive Spectroradiometer using a Boxcar Detector*, Association International de la Couleur (1981)

Chamberlin, G. J. and Chamberlin, D. G. *Colour: its Measurement, Computation and Application*, Heyden, London (1979)

Commission Internationale de l'Eclairage, *The Spectroradiometric Measurement of Light Sources*, CIE Pub. No. 63 (1984)

Dudley, W. W. *Carbon Dioxide Lasers, Effects and Applications*, Academic Press, London (1976)

Dresler, A. and Frühling, H. S. 'Uber ein Photoelektrische Dreifarben Messgerät', *Das Licht*, **11**, 238 (1938)

Forsythe, W. E. *The Measurement of Radiant Energy*, McGraw-Hill, New York (1941)

Fry, P. W. 'Silicon photodiode arrays', *J. Sci. Inst.*, **8**, 337 (1975)

Geutler, G. *Die Farbe*, **23**, 191 (1974)

Heavens, O. S. *Lasers*, Duckworth, London (1971)

Henderson, S. T. *J. Phys. D.*, **3**, 255 (1970)

Henderson, S. T. and Marsden, A. M. *Lamps and Lighting*, Edward Arnold, London (1972)

IEC–CIE, *International Lighting Vocabulary*

Judd, D. B. and Wysecki, G. *Color in Business, Science and Industry*, Wiley, New York (1975)

Koechner, W. *Solid State Laser Engineering*, Springer, New York (1976)

Lawson, W. D. 'Thermal imaging', in *Electronic Imaging* (eds T. P. McLean and P. Schagen), Academic Press, London (1979)

Longhurst, R. S. *Geometrical and Physical Optics*, Longman, London (1974)

Mooradian, A., Jaeger, T. and Stokseth, P. *Tuneable Lasers and Applications*, Springer, New York (1976)

Moore, J. 'Sources of error in spectro-radiometry', *Lighting Research and Technology*, **12**, 213 (1984)

Nichols, L. W. and Laner, J. *Applied Optics*, **7**, 1757 (1968)

Pivovonsky, M. and Nagel, M. *Tables of Blackbody Radiation Functions*, Macmillan, London (1961)

Safety in Universities—Notes for Guidance, Part 2: 1, Lasers, Association of Commonwealth Universities (1978)

Steck, B. 'Effects of optical radiation on man', *Lighting Research and Technology*, **14**, 130 (1982)

Tarrant, A. W. S. *Some Work on the SPD of Daylight*, PhD thesis, University of Surrey (1967)

Tarrant, A. W. S. 'The nature of colour—the physicist's viewpoint', in *Natural Colours for Food and other Uses* (ed. Counsell, J. N.), Applied Science, London (1981)

Thewlis, J. (ed.) *Encyclopaedic Dictionary of Physics*, Vol. 1, p. 553, Pergamon, Oxford (1961)

Walsh, J. W. T. *Photometry*, Dover, London (1958)

Wright, W. D. *The Measurement of Colour*, Hilger, Bristol (1969)

Young, A. R., Magnus, I. A. and Gibbs, N. K. 'Ultraviolet radiation radiometry of solar simulation', *Proc. Conf. on Light Measurement*, SPIE, Vol. 262 (1980)

3 Nucleonic instrumentation technology

D. ALIAGA KELLY

3.1 Introduction

Nucleonic instruments can be classified as those which measure the various radiations or particles emitted by radioactive substances or nuclear accelerators, such as alpha particles, beta particles, electrons and positrons, gamma- and X-rays, neutrons and heavy particles such as protons and deuterons. A variety of other exotic particles also exist, such as neutrinos, mesons, muons, etc., but their study is limited to high-energy research laboratories and uses special detection systems; it will not be considered in this book.

An important factor in the measurements to be made is the energy of the particles or radiations. This is expressed in electron-volts (eV) and can range from below 1 eV to millions of eV (MeV). Neutrons of very low energies (0.025 eV) are called 'thermal neutrons' because their energies are comparable to those of gas particles at normal temperatures. However, other neutrons can have energies of 10 MeV or more. X-rays and gamma-rays can range from a few eV to MeV, and sometimes GeV (10^9 eV).

The selection of a particular detector and detection system depends on a large number of factors which have to be taken into account in choosing the optimum system for a particular project. One must first consider the particle or radiation to be detected, the number of events to be counted, whether the energies are to be measured and the interference with the measurement by background radiation of similar or dissimilar types. Then the selection of the detector can be made, bearing in mind cost and availability, as well as suitability for the particular problem. Choice of electronic units will again be governed by cost and availability, as well as the need to provide an output signal with the information required. It can be seen that the result will be a series of compromises, as no detector is perfect, even if unlimited finance is available.

The radioactive source to be used must also be considered, and a list of the more popular types is given in Tables 3.1 and 3.2. Some other sources, used particularly in X-ray fluorescence analysis, are given in Table 4.1 (p. 130).

3.1.1 Statistics of counting

The variability of any measurement is measured by the standard deviation (σ), which can be obtained from replicate determinations by well-known methods. There is an inherent variability in radioactivity measurements because the disintegrations occur in a random manner, described by the Poisson distribution. This distribution is characterized by the property that the standard deviation (σ) of a large number of events, N, is equal to its square root, i.e.

$$\sigma(N) = \sqrt{N} \tag{3.1}$$

For ease in mathematical application, the normal (Gaussian) approximation to the Poisson distribution is ordinarily used. This approximation, which is generally valid for numbers of events, N, equal to or greater than 20, is the particular normal distribution whose mean is N and whose standard deviation is $\sqrt{N}$.

Generally, the concern is not with the standard deviation of the number of counts but rather with the deviation in the rate (= number of counts per unit time):

$$R' = \frac{N}{t} \tag{3.2}$$

where t is the time of observation, which is assumed to be known with such high precision that its error may be neglected. The standard deviation in the counting rate, $\sigma(R')$, can be calculated by the usual methods for propagation of error:

$$\sigma(R') = \frac{\sqrt{N}}{t} = \left(\frac{R'}{t}\right)^{\frac{1}{2}} \tag{3.3}$$

In practice, all counting instruments have a background counting rate, B, when no radioactive source is present. When a source is present, the counting rate increases to R_0. The counting rate R due to the source is then

$$R = R_0 - B \tag{3.4}$$

By propagation-of-error methods, the standard

Table 3.1 Radiation sources in general use

Isotope	Half-life	Emissions (and energies—MeV) Beta E_{max}	E_{av}	Gamma	Alpha	
^{3}H (Tritium)	12.26 yr	0.018	0.006	Nil	Nil	
^{14}C	5730 yr	0.15	0.049	Nil	Nil	
^{22}Na	2.6 yr	0.55	0.21	1.28	Nil	Also 0.511 MeV annihilation
^{24}Na	15 h	1.39	1.37	2.75		
^{32}P	14.3 d	1.71	0.69	Nil	Nil	Pure beta emitter
^{36}Cl	3×10^5 yr	0.71	0.32	Nil	Nil	Betas emitted simulate fission products
^{60}Co	5.3 yr	0.31	0.095	1.17 (100) 1.33 (100)	Nil Nil	Used in radiography, etc.
^{90}Sr	28 yr	0.54	0.196	Nil	Nil	Pure beta emitter
^{90}Y	64.2 h	2.25	0.93	Nil	Nil	
^{131}I	8.04 d	0.61+	0.18+	0.36 (79)+	Nil	Used in medical applications
^{137}Cs	30 yr	0.5+	0.19+	0.66 (86)	Nil	Used as standard gamma calibration source
^{198}Au	2.7 d	0.99+	0.3+	0.41 (96)	Nil	Gammas adopted recently as universal standard
^{226}Ra	1600 yr	—	—	0.61 (22) 1.13 (13) 1.77 (25) + others		Earliest radioactive source isolated by Mme Curie—still used in medical applications
^{241}Am	457 yr	—	—	0.059 (35) + others	5.42 (12) 5.48 (85)	Alpha X-ray calibration source

Note: Figures in parentheses show the percentage of primary disintegrations that go into that particular emission (i.e. the abundance).
+ Indicates other radiations of lower abundance.

Table 3.2 Neutron sources

Source and type	Neutron emission n/s per unit activity or mass	Half-life	energies of neutrons emitted	
^{124}Sb–Be (γ, n)	5×10^{-5}/Bq	60 days	Low: 30 keV	For field assay of beryllium ores
^{226}Ra–Be (α, n)	3×10^{-4}/Bq	1622 yr	Max. 12 MeV Av. 4 MeV	Early n source, now replaced by Am/Be
^{210}Po–Be (α, n)	7×10^{-5}/Bq	138 days	Max. 10.8 MeV Av. 4.2 Mev	Short life is disadvantage
^{241}Am–Be (α, n)	6×10^{-5}/Bq	433 yr	Max. 11 MeV Av. 3–5 MeV	Most popular neutron source
^{252}Cf fission	$2.3 \times 10^{6}/\mu$g	2.65 yr	Simulates reactor neutron spectrum	Short life and high cost

deviation of R can be calculated as follows:

$$\sigma(R)=\left(\frac{R_0}{t_1}+\frac{B}{t_2}\right)^{\frac{1}{2}} \tag{3.5}$$

where t_1 and t_2 are the times over which source-plus-background and background counting rates were measured, respectively. Practical counting times depend on the activity of the source and of the background. For low-level counting one has to reduce the background by the use of massive shielding, careful material selection for the components of the counter, the use of such devices as anti-coincidence counters and as large a sample as possible.

The optimum division of a given time period for counting source and background is given by

$$\frac{t_2}{t_1}=\frac{1}{1+(R_0/B)^{\frac{1}{2}}} \tag{3.6}$$

Table 3.3 Limits of the Quantity χ^2 for sets of counts with random errors

Number of observations	*Lower limit for* χ^2	*Upper limit for* χ^2
3	0.103	5.99
4	0.352	7.81
5	0.711	9.49
6	1.14	11.07
7	1.63	12.59
8	2.17	14.07
9	2.73	15.51
10	3.33	16.92
15	6.57	23.68
20	10.12	30.14
25	13.85	36.42
30	17.71	42.56

3.1.1.1 Non-random errors

These may be due to faults in the counting equipment, personal errors in recording results or operating the equipment or errors in preparing the sample. The presence of such errors may be revealed by conducting a statistical analysis on a series of repeated measurements. For errors in the equipment or in the way it is operated the analysis uses the same source, and for source errors a number of sources is used.

While statistical errors will follow a Gaussian distribution, errors which are not random will not follow such a distribution, so that if a series of measurements cannot be fitted to a Gaussian distribution curve then non-random errors must be present. The chi-squared test allows one to test the goodness of fit of a series of observations to the Gaussian distribution. If non-random errors are not present, then the values of χ^2 as determined by the relation given in equation (3.7) should lie between the limits quoted in Table 3.3 for various groups of observations:

$$\chi^2 = \frac{\sum_{i=1}^{i=q} (\bar{n} - n_i)^2}{\bar{n}} \quad (3.7)$$

where $\bar{n}$ is the average count observed, n_i is the number counted in the ith observation and q is the number of observations. If a series of observations fits a Gaussian distribution then there is a 95 per cent probability that χ^2 will be greater than or equal to the lower limit quoted in Table 3.1, but only a 5 per cent probability that χ^2 will be greater than or equal to the upper limit quoted. Thus, for ten observations, if χ^2 lies outside the region of 3.33–16.92 it is very probable that errors of a non-random kind are present.

In applying the chi-squared test the number of counts recorded in each observation should be large enough to make the statistical error less than the accuracy required for the activity determination. Thus, if 10 000 counts are recorded for each observation, and for a series of observations χ^2 lies between the expected limits, it can be concluded that non-random errors of a magnitude greater than about ± 2 per cent are not present.

3.1.1.2 Radioactive decay

Radioactive sources have the property of disintegrating in a purely random manner, and the rate of decay is given by the law

$$\frac{dN}{dt} = -\lambda N \quad (3.8)$$

where λ is called the decay constant and N is the total number of radioactive atoms present at a time t. This may be expressed as

$$N = N_0 \exp(-\lambda t) \quad (3.9)$$

where N_0 is the number of atoms of the parent substance present at some arbitrary time zero. Combining these equations we have

$$\frac{dN}{dt} = -\lambda N_0 \exp(-\lambda t) \quad (3.10)$$

showing that the rate of decay falls off exponentially with time. It is usually more convenient to describe the decay in terms of the 'half-life' $T_{\frac{1}{2}}$ of the element. This is the time required for the activity, dN/dt, to fall to half its initial value and $\lambda = 0.693/T_{\frac{1}{2}}$.

When two or more radioactive substances are present in a source the calculation of the decay of each isotope becomes more complicated and will not be dealt with here.

The activity of a source is a measure of the frequency of disintegration occurring in it. Activity is measured in Becquerels (Bq), one Becquerel corresponding to one disintegration per second. The old unit, the Curie (Ci), is still often used and 1 Megabecquerel = 0.027 millicuries.

It is also often important to consider the radiation that has been absorbed—the *dose*. This is quoted in Grays, the gray being defined as that dose (of any ionizing radiation) which imparts 1 Joule of energy per kilogram of absorbing matter at the place of interest. So 1 Gy = 1 J kg^{-1}. The older unit, the rad, a hundred times smaller than the gray, is still often referred to.

3.1.2 Classification of detectors

Various features of detectors are important and have to be taken into account when deciding the choice of a particular system, notably

(1) Cost;
(2) Sizes available;
(3) Complexity in auxiliary electronics needed;
(4) Ability to measure energy and/or discriminate between various different types of radiations or particles;
(5) Efficiency—defined as the probability of an incident particle being recorded.

Detectors can be grouped generally into the classes listed below. Most of these are covered in more detail later, but Cherenkov detectors and cloud chambers are specialized research tools and are not discussed in this book.

3.1.2.1 Gas detectors

These include ionization chambers, gas proportional counters, Geiger counters, multiwire proportional chambers, spark counters, drift counters and cloud chambers.

3.1.2.2 Scintillation counters

Some substances have the property that, when bombarded with nuclear particles or ionizing radiation, they emit light which can be picked up by a suitable highly sensitive light detector which converts the light pulse into an electronic pulse that can be amplified and measured.

3.1.2.3 Cherenkov detectors

When a charged particle traverses a transparent medium at a speed greater than that of light within the medium, then Cherenkov light is produced. Only if the relative velocity $\beta = v/c$ and the refractive index n of the medium are such that $n\beta > 1$ will the radiation exist. When the condition is fulfilled, the Cherenkov light is emitted at the angle given by the relation

$$\cos\theta = \frac{1}{n\beta} \qquad (3.11)$$

where θ is the angle between the velocity vector for the particle and the propagation vector for any portion of the conical radiation wavefront.

3.1.2.4 Solid state detectors

Some semiconductor materials have the property that when a potential is applied across them, and an ionizing particle or ionizing radiation passes through the volume of material, ions are produced just as in the case of a gas-ionization chamber, producing electronic pulses in the external connections which can be amplified, measured or counted. A device can thus be made, acting like a solid ionization chamber. The materials which have found greatest use in this application are silicon and germanium.

3.1.2.5 Cloud chambers

These were used in early research work, and are still found in more sophisticated forms in high-energy research laboratories to demonstrate visually (or photographically) the actual paths of ionizing particles or radiation by means of trails of liquid droplets formed after the passage of such particles or radiation through supersaturated gas.

Photographic film can be used to detect the passage of ionizing particles or radiation, since they produce latent images along their paths in the sensitive emulsion. On development, the grains of silver appear along the tracks of the particles or ions.

3.1.2.6 Plastic film detectors

Thin (5 μm) plastic films of polycarbonate can be used as detectors of highly ionizing particles which can cause radiation damge to the molecules of the polycarbonate film. These tracks may be enlarged by etching with a suitable chemical and visually measured with a microscope. Alternatively, sparks can be generated between two electrodes, one of which is an aluminized mylar film, placed on either side of a thin, etched polycarbonate detector. The sparks that pass through the holes in the etched detector can be counted using a suitable electronic scaler.

3.1.2.7 Thermoluminescent detectors

For many years it was known that if one heated some substances, particularly fluorites and ceramics, they could be made to emit light photons and, in the case of ceramics, could be made incandescent. When ionizing radiation is absorbed in matter most of the absorbed energy goes into heat while a small fraction is used to break chemical bonds. In some materials a very minute fraction of the energy is stored in metastable energy states. Some of the energy thus stored can be recovered later as visible light photons if the material is heated, and this phenomenon is known as thermoluminescence (TL).

In 1950 Daniels proposed that this phenomenon could be used as a measurement of radiation dose, and in fact it was used to measure radiation after an atom-bomb test. Since then interest in TL as a radiation dosimeter has progressed to the stage that it could now well replace photographic film as the approved personnel radiation badge carried by people who may be involved with radioactive materials or radiation. It is being introduced by the National Radiological

Protection Board in the UK as an approved radiation monitor with the film badge.

3.1.2.8 Materials for TL dosimetry

The most popular phosphor for dosimetric purposes is lithium fluoride (LiF). This can be natural LiF, or with the lithium isotopes ^{6}Li and ^{7}Li enriched or depleted, as well as variations in which an added activator such as manganese (Mn) is added to the basic LiF. The advantages of LiF are:

(1) Its wide and linear energy response from 30 KeV up to and beyond 2 MeV;
(2) Its ability to measure doses from the mR to 10^5 R without being affected by the rate at which the dose is delivered—this is called 'dose-rate independence';
(3) Its ability to measure thermal neutrons as well as X-rays, gamma rays, beta rays and electrons;
(4) Its dose response is almost equivalent to the response of tissue, i.e. it has almost the same response as the human body;
(5) It is usable in quite small amounts, so can be used to measure doses to the fingers of an operator without impeding the operator's work;
(6) It can be re-used many times, so is cheap.

Another phosphor which has become quite popular in recent years is calcium fluoride with manganese (CaF_2:Mn), which has been found to be more sensitive than LiF for low-dose measurements (some ten times) and can measure a dose of 1 mR yet is linear in dose-rate response up to 10^5 R. However, it exhibits a large energy dependence and is not linear below 300 KeV.

Thermoluminescence has also been used to date ancient archaeological specimens such as potsherds, furnace floors, ceramic pots, etc. This technique depends on the fact that any object heated to a high temperature loses inherent thermoluminescent powers and, if left for a long period in a constant radioactive background, accumulates an amount of TL proportional to the time it has lain undisturbed in that environment.

3.1.3 Health and safety

Anyone who works with radioactive materials must understand clearly the kinds of hazards involved and their magnitude. Because radioactivity is not directly observable by the body's senses it requires suitable measuring equipment and handling techniques to ensure that any exposure is minimized, and, because of this, suitable legislation governs the handling and use of all radioactive material. In Chapter 4 an outline of the regulations is given as well as advice on contacting the local factory inspector before the use of any radioactive source is contemplated.

Because everyone in the world already receives steady radiation (from the natural radiopotassium in the human body and from the general background radiation to which all are subjected) the average human body acquires a dose of about 300 micro-grays (μGy) (equivalent to 30 millirads) per year. Hence, while it is almost impossible to reduce radiation exposure to zero it is important to ensure that using a radioactive source does not increase the dose to a level greater than many other hazards commonly met in daily life.

There are three main methods for minimizing the hazards due to the use of a radioactive source. These are:

(1) Shielding: a thickness of an appropriate material, such as lead, should be placed between the source and the worker;
(2) Distance: an increase in distance between source and worker reduces the radiation intensity;
(3) Time: the total dose to the body of the worker depends on the length of time spent in the radiation field. This time should be reduced to the minimum necessary to carry out the required operation.

The above notes are for sources which are contained in sealed capsules. Those which are in a form that might allow them to enter the body's tissues must be handled in ways which prevent such an occurrence (for example, by operating inside a 'glove box' which is a box allowing open radioactive sources to be dealt with while the operator stays outside the enclosure). Against internal exposure the best protection is good housekeeping, and against external radiation the best protection is good instrumentation, kept in operating condition and USED.

Instruments capable of monitoring radioactive hazards depend on the radiation or particles to be monitored. For gamma rays, emitted by the most usual radioactive sources to be handled, a variety of instruments is available. Possibly the cheapest yet most reliable monitor contains a Geiger counter, preferably surrounded by a suitable metal covering to modify the counter's response to the varied energies from gamma-emitting radioisotopes to make it similar to the response of the human body. There are many such instruments available from commercial suppliers. More elaborate ones are based on ionization chambers, which are capable of operating over a much wider range of intensities and are correspondingly more expensive.

For beta emitters, a Geiger counter with a thin window to allow the relatively easily absorbed beta particles to enter the counter is again the cheapest monitor. More expensive monitors are based on scintillation counters which can have large window areas, useful for monitoring extended sources or accidents where radioactive beta emitters have been spilt.

Alpha detection is particularly difficult, as most alphas are absorbed in extremely thin windows. Geiger counters with very thin windows can be used, or ionization chambers with an open front, used in air at normal atmospheric pressure. More expensive scintillation counters or semiconductor detectors can also be used.

Neutrons require much more elaborate and expensive monitors, ranging from ionization or proportional counters containing BF_3 or ^3He, to scintillation counters using ^6LiI, ^6Li-glass or plastic scintillators, depending on the energies of the neutrons.

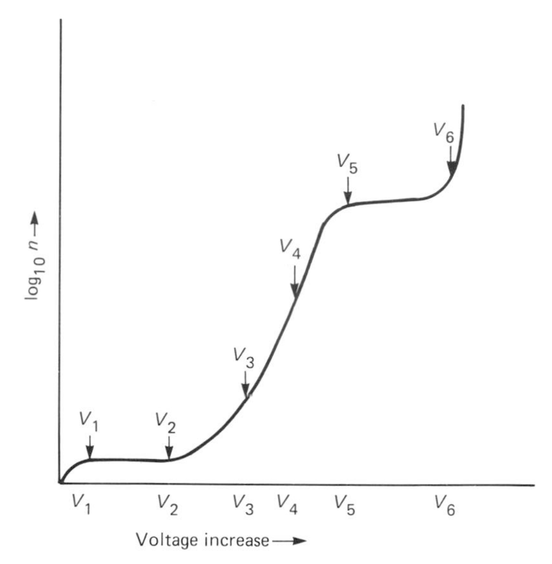

Figure 3.1 Response of gas counter to increase in voltage.

3.2 Detectors

3.2.1 Gas detectors

Gas-filled detectors may be subdivided into those giving a current reading and those indicating the arrival of single particles. The first class comprises the current ionization chambers, and the second the counting or pulse-ionization chambers, proportional counters and Geiger counters. The object of the ionization chamber is always the same—to measure the rate of formation of ion pairs within the gas. One must therefore be certain that the voltage applied to the electrodes is great enough to give saturation, i.e. to ensure that there will be no appreciable recombination of positive and negative ions.

To understand the relation between the three gas-filled detectors we can consider a counter of very typical geometry—two coaxial cylinders with gas between them. The inner cylinder, usually a fine wire (the anode) is at a positive potential relative to the outer cylinder (the cathode). Let us imagine ionization to take place in the gas, from a suitable radioactive source, producing, say, 10 electrons. The problem is to decide how many electrons (n) will arrive at the anode wire. Figure 3.1 shows the voltage applied across the counter V, plotted against the logarithm of n, i.e. $\log_{10} n$. When V is very small, of the order of volts or less, all 10 electrons do not arrive at the anode wire because of recombination. At V_1 the loss has become negligible as saturation has been achieved and the pulse contains 10 electrons. As V is increased, n remains at 10 until V_2 is reached, usually some tens or hundreds of volts. At this point the electrons begin to acquire sufficient energy between collisions at the end of their paths for ionization by collision to occur in the gas and this multiplication causes n to rise above 10, more or less exponentially with V as each initial electron gives rise to a small avalanche of secondary electrons by collision close to the wire anode. At any potential between V_2 and V_3 the multiplication is constant, but above this the final number of electrons reaching the wire is no longer proportional to the initial ionization. This is the region of limited proportionality, V_4. Above this, from V_5 to V_6 the region becomes that of the Geiger counter, where a single ion can produce a complete discharge of the counter. It is characterized by a spread of the discharge throughout the whole length of the counter, resulting in an output pulse size independent of the initial ionization. Above V_6 the counter goes into a continuous discharge. The ratio of n, the number of electrons in the output pulse, to the initial ionization at any voltage is called the gas-amplification factor A, and varies from unity in the ionization chamber region to 10^3–10^4 in the proportional region, reaching 10^5 just below the Geiger region.

Ionization-chamber detectors can be of a variety of shapes and types. Cylindrical geometry is the most usual one adopted but parallel plate chambers are often used in research. These chambers are much used in radiation-protection monitors, since they can be designed to be sufficiently sensitive for observing terrestrial radiation yet will not overload when placed in a very high field of radiation such as an isotopic irradiator. They can also be used, again in health physics, to integrate over a long period the amount of ionizing radiation passing through the chamber. An example is the small integrating chamber used in X-ray and accelerator establishments to observe the amount of radiation produced in personnel who carry the chambers during their working day.

Proportional counters are much more sensitive than ionization chambers and this allows weak sources of alpha and beta particles, and low energy X-rays to be counted. The end-window proportional counter is particularly useful for counting flat sources as it exhibits nearly 2π geometry—i.e. it counts particles entering the counter over a solid angle of nearly 2π. Cylindrical proportional counters are used in radiocarbon dating systems because of their sensitivity for the detection of low-energy ^{14}C beta particles ($E_{max} = 156$ keV) and even tritium ^{3}H beta particles ($E_{max} = 18$ keV).

3.2.1.1 Geiger–Muller detectors

The Geiger counter has been and is the most widely used detector of nuclear radiation. It exhibits several very attractive features, some of which are:

(1) Its cheapness—manufacturing techniques have so improved the design that Geiger–Muller tubes are a fraction of the cost of solid-state or scintillation detectors.
(2) The output signal from a Geiger–Muller tube can be of the order of 1 V, much higher than that from proportional, scintillation or solid-state detectors. This means that the cost of the electronic system required is a fraction of that of other counters. A Geiger–Muller tube with a simple high-voltage supply can drive most scaler units directly, with minimal or no amplification.
(3) The discharge mechanism is so sensitive that a single ionizing particle entering the sensitive volume of the counter can trigger the discharge.

With these advantages there are, however, some disadvantages which must be borne in mind. These include:

(1) The inability of the Geiger–Muller tube to discriminate between the energies of the ionizing particles triggering it.
(2) The tube has a finite life, though this has been greatly extended by the use of halogen fillings instead of organic gases. The latter gave lives of only about 10^{10} counts whereas the halogen tubes have lives of 10^{13} or more counts.
(3) There is a finite period between the initiation of a discharge in a Geiger–Muller counter and the time when it will accept a new discharge. This is called the dead time, and is of the order of 100 μs.

It is important to ensure that with Geiger counters the counting rate is such that the dead-time correction is only a small percentage of the counting rate observed. This correction can be calculated from the relation:

$$R' = \frac{R}{1 - R\tau}$$

where R is the observed counting rate per unit time, R' is the true counting rate per unit time and τ is the counter dead time.

Dead time for a particular counter may be evaluated by a series of measurements using two sources. Geiger tube manufacturers normally quote the dead time of a particular counter, and it is customary to increase this time electronically in a unit often used with Geiger counters, so that the total dead time is constant but greater than any variations in the tube's dead time, since between individual pulses the dead time can vary.

The counting rate characteristics can be understood by reference to Figure 3.2. The starting voltage V_s is

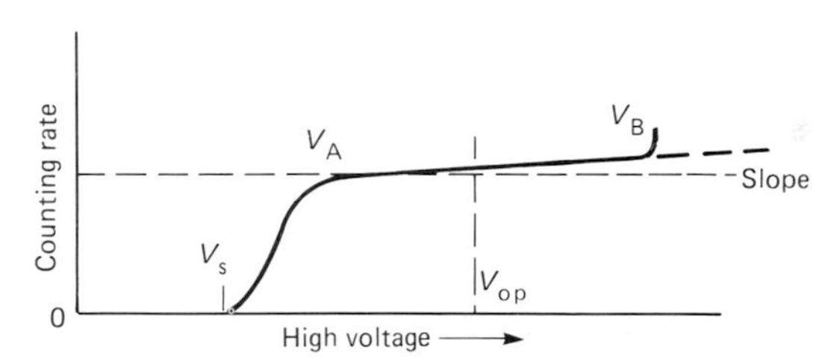

Figure 3.2 Geiger counter characteristic response.

the voltage which, when applied to the tube viewing a fixed radioactive source, makes it just start to count. As the high voltage is increased the counting rate rapidly increases until it reaches what is called the plateau. Here the increase in counting rate from V_A to V_B is small, of the order of 1–5 per cent per 100 V of high voltage. Above V_B the counting rate rises rapidly and the tube goes into a continuous discharge, which will damage the counter. An operating point is selected (V_{op}) on the plateau so that any slight variation of the high-voltage supply has a minimal effect on the counting rate.

In order to count low-energy beta or alpha particles with a Geiger counter a thin window must be provided to allow the particles to enter the sensitive volume and trigger the counter. Thin-walled glass counters can be produced, with wall thicknesses of the order of 30 mg/cm^2, suitable for counting high-energy beta particles (in this context, the mass per unit area is more important than the linear thickness: for glass, 25 mg/cm^2 corresponds to about 0.1 mm). For low-energy betas or alphas a very thin window is called for, and these have been made with thicknesses as low as 1.5 mg/cm^2. Figure 3.3 gives the transmission through windows of

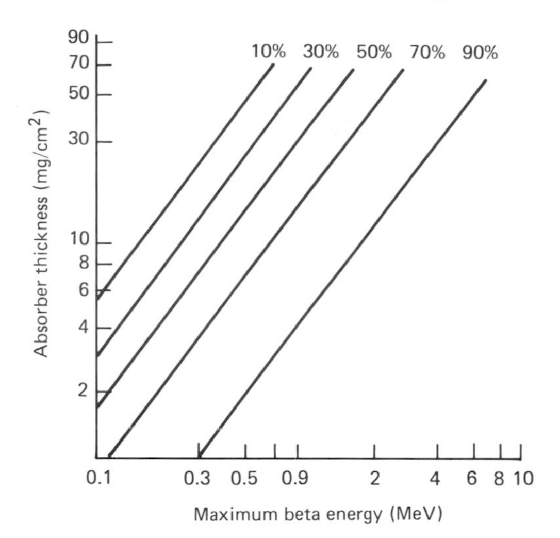

Figure 3.3 Transmission of thin windows.

different thickness, and Table 3.4 shows how it is applied to typical sources.

Alternatively, the source can be introduced directly into the counter by mixing as a gas with the counting gas, or if a solid source, by placing it directly inside the counter and allowing the flow of counting gas to pass through the counter continuously. This is the flow-counter method.

3.2.2 Scintillation detectors

Scintillation counters comprise three main items, the scintillator, the light-to-electrical pulse converter (generally a photomultiplier) and the electronic amplifier; we consider the wide variety of each of these items which are available today.

The scintillator can consist of a single crystal of organic or inorganic material, a plastic fluor, an activated glass or a liquid. Shapes and sizes can vary enormously, from the small NaI (Tl)—this is the way of writing 'sodium iodide with thallium additive'—crystal used in medical probes up to tanks of liquid scintillator of thousands of cubic metres used in cosmic ray research.

The selection of a suitable material—some of whose characteristics are given in Tables 3.5 and 3.6—depends on a number of competing factors. No material is perfect so compromises must be made in their selection for particular purposes.

3.2.2.1 Inorganic scintillators

NaI (Tl) is still after many years the best gamma and X-ray detector actually available, yet it is very hygroscopic and must be completely sealed from moisture. It is sensitive to shock, except in the new form of extruded NaI (Tl) called Polyscin, developed by Harshaw, and it is expensive, especially in large sizes. Its light output is, in general, proportional to the energy of the photon absorbed, so that the pulse height is a measure of the energy of the incident photon and, when calibrated with photons of known energy, the instrument can be used as a spectrometer. The decay lifetime of NaI (Tl) is relatively slow, being about 230 ns, although NaI without any thallium when operated at liquid-nitrogen temperatures (77 K) has a decay lifetime of only 65 ns.

The next most used inorganic scintillator is CsI. When activated with thallium as CsI(Tl) it can be used at ambient temperatures with a light output some 10 per cent lower than NaI (Tl) but with considerable resistance to shock. Its absorption coefficient is also greater than that of NaI (Tl) and these two characteristics have resulted in its use in space vehicles and satellites, since less mass is necessary and its resistance to the shock of launch is valuable. In thin layers it can be bent to match a circular light guide and has been used in this manner on probes to measure excited X-rays in soil. When activated with sodium instead of thallium the light output characteristics are changed. The light output is slightly higher than CsI (Tl) and the temperature/light output relation is different (see Figure 3.4). The maximum light output is seen to occur, in fact, at a temperature of about 80°C. This is of advantage in borehole logging, where increased temperatures and shock are likely to occur as the detector is lowered into the drill hole.

CsI (Tl) has been a popular detector for alpha particles, since it is not very much affected by moisture from the air and so can be used in windowless

Table 3.4 Transmission of thin windows

Nuclide	*Max. energy E_{max} (MeV)*	*Percentage transmission for window thickness of*			
		30 mg/cm²	*20 mg/cm²*	*7 mg/cm²*	*3 mg/cm²*
^{14}C	0.15	0.01	0.24	12	40
^{32}P	1.69	72	80.3	92	96

Table 3.5 Inorganic scintillator materials

Material[1]	*Density (g/cm^3)*	*Refractive index (n)*	*Light output*[2] *(% anthracene)*	*Decay constant (s)*	*Wavelength of maximum emission (nm)*	*Operating temperature (°C)*	*Hygroscopic*
NaI (Tl)	3.67	1.775	230	0.23×10^{-6}	413	Room	Yes
NaI (pure)	3.67	1.775	440	0.06×10^{-6}	303	—	Yes
CsI(Tl)	4.51	1.788	95	1.1×10^{-6}	580	Room	No
CsI (Na)	4.51	1.787	150–190	0.65×10^{-6}	420	Room	Yes
CsI (pure)	4.51	1.788	500	0.6×10^{-6}	400	—	No
CaF_2 (Eu)	3.17	1.443	110	1×10^{-6}	435	Room	No
LiI (Eu)	4.06	1.955	75	1.2×10^{-6}	475	Room	Yes
$CaWO_4$	6.1	1.92	36	6×10^{-6}	430	Room	No
ZnS (Ag)	4.09	2.356	300	0.2×10^{-6}	450	Room	No
ZnO(Ga)	5.61	2.02	90	0.4×10^{-9}	385	Room	No
$CdWO_4$	7.90			$0.9–20 \times 10^{-6}$	530	Room	No
$Bi_4Ge_3O_{12}$	7.13	2.15		0.3×10^{-6}	480	Room	No
CsF	4.64	1.48	40	5×10^{-12}	390	Room	No

1 The deliberately added impurity is given in parentheses.
2 Light output is expressed as a percentage of that of a standard crystal of Anthracene used in the same geometry.

Table 3.6 Properties of organic scintillators

Material	*Scintillator*	*Density (g/cm^3)*	*Refractive index (n)*	*Boiling melting or softening point (°C)*	*Light output (% anthracene)*	*Decay constant (ns)*	*Wavelength of max. emission (nm)*	*Loading content (% by weight)*	*H/C Number of H atoms/ C atoms*	*Attenuation length (l/e m)*	*Principal applications*
Plastic	NE102A	1.032	1.581	75	65	2.4	423	—	1.104	2.5	γ, α, β, fast *n*
	NE104	1.032	1.581	75	68	1.9	406	—	1.100	1.2	Ultra-fast counting
	NE104B	1.032	1.58	75	59	3.0	406	—	1.107	1.2	Ditto with BBQ[1] light guides
	NE105	1.037	1.58	75	46	—	423	—	1.098	—	Air-equivalent for dosimetry
	NE110	1.032	1.58	75	60	3.3	434	—	1.104	4.5	γ, α, β, fast *n*, etc.
	NE111A	1.032	1.58	75	55	1.6	370		1.103	—	Ultra-fast timing
	NE114	1.032	1.58	75	50	4.0	434		1.109	—	Cheaper for large arrays
	NE160	1.032	1.58	80	59	2.3	423		1.105	—	For use at higher temperatures—usable up to 150°C
	Pilot U	1.032	1.58	75	67	1.36	391		1.100	—	Ultra-fast timing
	Pilot 425	1.19	1.49	100	—		425		1.6		Cherenkov detector
Liquid	NE213	0.874	1.508	141	78	3.7	425		1.213		Fast *n* (PSD)[2]
	NE216	0.885	1.523	141	78	3.5	425		1.171		α, β (internal counting)
	NE220	1.306	1.442	104	65	3.8	425	29% O	1.669		Internal counting, dosimetry
	NE221	1.08	1.442	104	55	4	425	Gel	1.669		α, β (internal counting)
	NE224	0.887	1.505	169	80	2.6	425		1.330		γ, fast *n*
	NE226	1.61	1.38	80	20	3.3	430		0		γ, insensitive to *n*
	NE228	0.71	1.403	99	45	—	385		2.11		*n* (heptane-based)
	NE230	0.945	1.50	81	60	3.0	425	14.2% D	0.984		Deuterated
	NE232	0.89	1.43	81	60	4	430	24.5% D	1.96		Deuterated
	NE233	0.874	1.506	117	74	3.7	425		1.118		α, β (internal counting)
	NE235	0.858	1.47	350	40	4	420		2.0		For large tanks
	NE250	1.035	1.452	104	50	4	425	32% O	1.760		Internal counting, dosimetry

1 BBQ is a wavelength-shifter.
2 PSD means pulse shape discrimination.

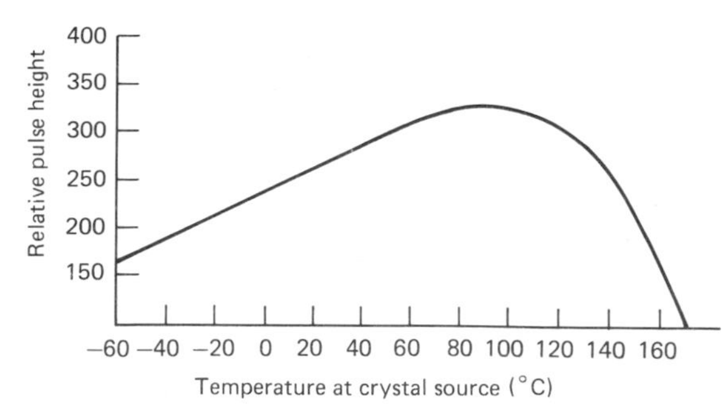

Figure 3.4 Relative output of CsI (Na) as a function of temperature.

counters. CsI (Na), on the other hand, quickly develops a layer impervious to alpha particles of 5–10 MeV energies when exposed to ambient air and thus is unsuitable for such use. CaF and CaF (Eu) are scintillators which have been developed to give twice the light output of NaI (Tl), but only very small specimens have been grown and production difficulties make their use impossible at present.

$Bi_4Ge_3O_{12}$ (bismuth germanate, or BGO) was developed to meet the requirements of the medical tomographic scanner which calls for large numbers of very small scintillators capable of high absorption of photons of about 170 keV energy yet able to respond to radiation changes quickly without exhibiting 'afterglow', especially when the detector has to integrate the current output of the photomultiplier. Its higher density than NaI (Tl) allows the use of smaller crystals but its low light output (8 per cent of NaI (Tl)) is a disadvantage. Against this it is non-hygroscopic, so it can be used with only light shielding.

CsF is a scintillator with a very fast decay time (about 5 ns) but a low light output (18 per cent of NaI (Tl)). It also has been used in tomographic scanners.

LiI (Eu) is a scintillator particularly useful for detecting neutrons, and as the lithium content can be changed by using enriched ^{6}Li to enhance the detection efficiency for slow neutrons, or by using almost pure ^{7}Li to make a detector insensitive to neutrons, it is a very versatile, if expensive, neutron detector. When cooled to liquid-nitrogen temperature the detection efficiency for fast neutrons is enhanced and the system can be used as a neutron spectrometer to determine the energies of the neutrons falling on the detector. LiI (Eu) has the disadvantage that it is extremely hygroscopic, even more so than NaI (Tl).

Cadmium tungstate ($CdWO_4$) and calcium tungstate ($CaWO_4$) single crystals have been grown, with some difficulty, and can be used as scintillators without being encapsulated as they are non-hygroscopic. However, their refractive index is high and this causes 60–70 per cent of the light emitted in scintillators to be entrapped in the crystal.

CaF_2 (Eu) is a non-hygroscopic scintillator which is inert towards almost all corrosives, and thus can be used for beta detection without a window or in contact with a liquid, such as the corrosive liquids used in fuel-element treatment. It can also be used in conjunction with a thick NaI (Tl) or CsI (Tl) or CsI (Na) crystal to detect beta particles in a background of gamma rays, using the 'Phoswich' concept, where events occurring only in the thin CaF_2 (Eu), unaccompanied by a simultaneous event in the thick crystal, are counted. That is, only particles which are totally absorbed in the CaF_2 (Eu) are of interest—when both crystals display coincident events, these are vetoed by coincidence and pulse-shape discrimination methods. Coincidence counting is discussed further in section 3.3.6.4.

3.2.2.2 Organic scintillators

The first organic scintillator was introduced by Kallman in 1947 when he used a naphthalene crystal to show that it would detect gamma rays. Later, anthracene was shown to exhibit improved detection efficiency and stilbene was also used. The latter has proved particularly useful for neutron detection. Mixtures of organic scintillators such as solutions of anthracene in naphthalene, liquid solutions and plastic solutions were also introduced, and now the range of organic scintillators available is very great. The plastic and liquid organic scintillators are generally cheaper to manufacture than inorganic scintillators, and can be made in relatively large sizes. They are generally much faster in response time than most inorganic scintillators, and being transparent to their own scintillation light can be used in very large sizes. Table 3.6 gives the essential details of a large range of plastic and liquid scintillators.

The widest use of organic scntillators is probably in the field of liquid-scintillation counting, where the internal counting of tritium, ^{14}C, ^{55}Fe and other emitters of low-energy beta particles at low activity levels is being carried out on an increasing scale. Biological samples of many types have to be incorporated into the scintillator and it is necessary to do this with the minimum quenching of the light emitted, minimum chemiluminescence as well as minimum effort—the last being an important factor where large numbers of samples are involved.

At low beta energies the counting equipment is just as important as the scintillator. Phosphorescence is reduced to a minimum by the use of special vials and reflectors. Chemiluminescence, another problem with biological and other samples, is not completely solved by the use of two photomultipliers in coincidence viewing the sample. This must be removed or reduced, for example, by removing alkalinity and/or peroxides by acidifying the solution before mixing with the scintillator.

3.2.2.3 Loaded organic scintillators

In order to improve the detection efficiency of scintillators for certain types of particles or ionizing or non-ionizing radiations small quantities of some substances can be added to scintillators without degrading greatly the light output. It must be borne in mind that in nearly all cases there is a loss of light output when foreign substances are added to an organic scintillator, but the gain in detection efficiency may be worth a slight drop in this output. Suitable loading materials are boron, both natural boron and boron enriched in ^{10}B and gadolinium—these are to increase the detection efficiency for neutrons. Tin and lead have been used to improve the detection efficiency for gamma rays and have been used in both liquid and plastic scintillators.

3.2.2.4 Plastic scintillators

Certain plastics such as polystyrene and polyvinyltoluene can be loaded with small quantities of certain substances such as p-terphenyl, etc., which cause them to scintillate when bombarded by ionizing particles or ionizing radiation. An acrylic such as methyl methacrylate can also be doped to produce a scintillating material but not with the same high light output as the polyvinyltoluene-based scintillators. It can be produced, however, much more cheaply and it can be used for many high-energy applications.

Plastic scintillators have the ability to be moulded into the most intricate shapes in order to suit a particular experiment, and their inertness to water, normal air and many chemical substances allows their use in direct contact with the activity to be measured. Being of low atomic number constituents, the organic scintillators are preferred to inorganics such as NaI (Tl) or CsI (Tl) for beta counting, since the number of beta particles which are scattered out of an organic scintillator without causing an interaction are about 8 per cent, whereas in a similar NaI (Tl) crystal the number scattered out would be 80–90 per cent.

When used to detect X- or gamma-rays organic scintillators differ in their response compared with inorganic scintillators. Where inorganic scintillators in general have basically three main types of response called photoelectric, Compton and pair production, because of the high Z (atomic weight) of the materials of the inorganic scintillators, the low-Z characteristics of the basic carbon and similar components in organic scintillators lead only to Compton reactions, except at very low energies, with the result that for a monoenergetic gamma emitter, the spectrum produced is a Compton distribution.

For study of the basic interactions between gamma- and X-rays, and scintillation materials, see Price (1964), as these reaction studies are beyond the scope of this chapter.

The ability to produce simple Compton distribution spectra has proved of considerable use in cases where one or two isotopes have to be measured at low intensities, and a large inorganic NaI (Tl) detector might be prohibitively expensive. Such is the case with whole-body counters used to measure the ^{40}K and ^{137}Cs present in the human body—the ^{40}K being the natural activity present in all potassium, the ^{137}Cs the result of fallout from the many atomic bomb tests. Similarly, when measuring the potassium content of fertilizer a plastic scintillator can carry this out more cheaply than an inorganic detector. The measurement of moisture in soil by gamma-ray transmission through a suitable sample is also performed more easily with a plastic scintillator, since the fast decay time of the organics compared with, say, NaI (Tl) allows higher counting rates to be used, with consequent reduction of statistical errors.

3.2.2.6 Scintillating ion-exchange resins

By treating the surfaces of plastic scintillating spheres in suitable ways the extraction and counting of very small amounts of beta-emitting isotopes may be carried out from large quantities of carrier liquid, such as rainwater, cooling water from reactors, effluents, rivers, etc., rather than having to evaporate large quantities of water in order to obtain a concentrated sample for analysis.

3.2.2.6 Flow cells

It is often necessary to continuously monitor tritium, ^{14}C and other beta-emitting isotopes in aqueous solution, and for this purpose the flow cell developed by Schram and Lombaert and containing crystalline anthracene has proved valuable. A number of improvements have been made to the design of this cell, resulting in the NE806 flow cell. The standard flow cell is designed for use on a single 2-in. diameter low-noise photomultiplier and can provide a tritium detection efficiency of 2 per cent at a background of 2 c/s and a ^{14}C detection efficiency of 30 per cent at a background of 1 c/s.

3.2.2.7 Photomultipliers

The photomultiplier is the device which converts the light flash produced in the scintillator into an amplified electrical pulse. It consists generally of an evacuated tube with a flat glass end onto the inner surface of which is deposited a semi-transparent layer of metal with a low 'work function', i.e. it has the property of releasing electrons when light falls on it. The most usual composition of this photocathode, as it is called, is caesium plus some other metal. A series of dynodes, also coated with similar materials, form an electrical optical system which draws secondary electrons away from one dynode and causes them to strike the next one with a minimum loss of electrons. The anode finally collects the multiplied shower of electrons and this forms the output pulse to the electronic system. Gains of 10^6–10^7 are obtained in this process.

Depending on the spectrum of the light emitted from the scintillator, the sensitivity of the light-sensitive photocathode can be optimized by choice of the surface material. One can detect single electrons emitted from the photocathode with gallium arsenide (GaAs) as the coating.

Silicon photodiodes can also be used to detect scintillation light, but as the spectral sensitivity of these devices is in the red region ($\sim$500–800 nm), as opposed to the usual scintillator light output of $\sim$400–500 nm, a scintillator such as CsI (Tl) must be used whose output can match the spectral range of a silicon photodiode. Light detectors are discussed further in Chapter 2.

3.2.3 Solid-state detectors

It was observed earlier that the operation of a solid state or semiconductor detector could be likened to that of an ionization chamber. A study of the various materials which were thought to be possible for use as radiation detectors has been carried out in many parts of the world, and the two materials which proved most suitable were silicon and germanium. Both these materials were under intense development for the transistor industry and so the detector researchers were able to make use of the work in progress. Other materials tested, which might later prove valuable, were cadmium telluride (CdTe), mercuric iodide (HgI_2), gallium arsenide (GaAs) and silicon carbide. CdTe and HgI_2 can be used at room temperature but, to date, have only been produced in relatively small sizes with great difficulty. They are suitable only for low-energy X-ray detection and spectrometry.

While in this book no attempt will be made to go deeply into the physics of semiconductor detectors it is useful to give a brief outline of their operation, especially compared with the gas-ionization chamber, of which they are often regarded as the solid-state equivalent.

First, a much smaller amount of energy is required to release electrons (and therefore holes) in all solids than in gases. An average energy of only 3 eV is required to produce an electron–hole pair in germanium (and 3.7 eV in silicon) while about 30 eV is required to produce the equivalent in gases. This relatively easy production of free holes and electrons in solids results from the close proximity of atoms which causes many electrons to exist at energy levels just below the conduction band. In gases the atoms are isolated and the electrons much more tightly bound. As a result of this, a given amount of energy absorbed from incident radiation produces more free charges in solids than in gas, and the statistical fluctuations become a much smaller fraction of the total charge released. This is the basic reason for semiconductor detectors producing better energy resolution than gas detectors, especially at high energies. At low energies, because the signal from the semiconductor is some ten times larger than that from the gas counter, the signal/noise ratio is enhanced.

In order to obtain an efficient detector from a semiconductor material we may consider what occurs should we take a slice of silicon 1 cm^2 in area and 1 mm thick and apply a potential across the faces. If the resistivity of this material is 2000 Ω cm, with ohmic contacts on each side, then the slice would behave like a resistor of 200 Ω, and if 100 V is applied across such a resistor then Ohm's Law states that a current of 0.5 A would pass. If radiation now falls on the silicon slice, a minute extra current will be produced, but this would be so small compared with the standing 0.5 A current that it would be undetectable. This is different from the gas-ionization chamber, where the standing current is extremely small. The solution to this problem is provided by semiconductor junctions. The operation of junctions depends on the fact that a mass action law compels the product of electron and hole concentrations to be constant for a given semiconductor at a fixed temperature. Therefore, heavy doping with a donor such as phosphorus not only increases the free electron concentration but also depresses the hole concentration to satisfy the relation that the product must have a value dependent only on the semiconductor. For example, silicon at room temperature has the relation $n \times p \approx 10^{20}$, where n is the number of holes and p is the number of electrons. Hence in a region where the number of donors is doped to a concentration of 10^{18} the number of holes will be reduced to about 10^2. McKay, of Bell Telephone Laboratories, first demonstrated in 1949 that if a reverse-biased p–n junction is formed on the semiconductor a strong electric field may be provided across the device which sweeps away free holes from the junction on the p-side (doped with boron) (Figure 3.5) and electrons away from it on the n side (doped with phosphorus). A region is produced which is free of holes or electrons, and is known as the depletion region. However, if an ionizing particle or quantum of gamma energy passes through the region, pairs of

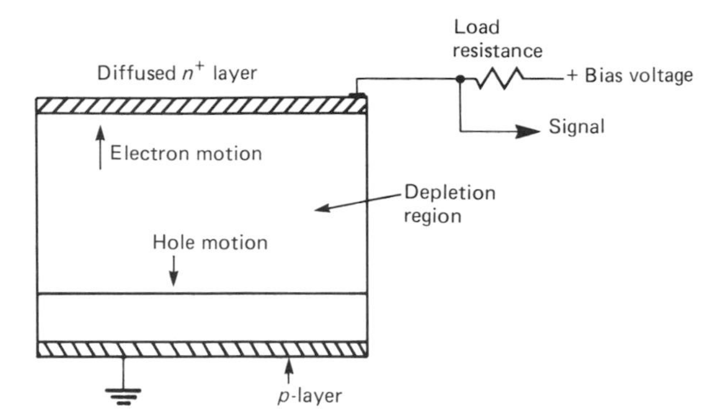

Figure 3.5 Schematic diagram of semiconductor detector.

holes and electrons are produced which are collected to produce a current in the external circuit. This is the basic operation of a semiconductor detector. The background signal is due to the collection of any pairs of holes plus electrons produced by thermal processes. By reducing the temperature of the detector by liquid nitrogen to about 77 K most of this background is removed, but in practical detectors the effect of surface contaminants on those surfaces not forming the diode can be acute. Various methods of avoiding these problems, such as the use of guard rings, have reduced much of this problem. However, the effects of very small amounts of oxygen, etc., can have devastating results on a detector, and most are enclosed in a high-vacuum chamber.

By doping a germanium or silicon crystal with lithium (an interstitial donor), which is carried out at moderate temperatures using an electric field across the crystal, the acceptors can be almost completely compensated in *p*-type silicon and germanium. This allows the preparation of relatively large detectors suitable for high-energy charged particle spectroscopy. By this means coaxial detectors with volumes up to about 100 cm^3 have been made, and these have revolutionized gamma-ray spectroscopy, as they can separate energy lines in a spectrum which earlier NaI (Tl) scintillation spectrometers could not resolve.

New work on purifying germanium and silicon has resulted in the manufacture of detectors of super-pure quality such that lithium drifting is not required. Detectors made from such material can be cycled from room temperature to liquid-nitrogen temperature and back when required without the permanent damage that would occur with lithium-drifted detectors. Surface-contamination problems, however, still require them to be kept *in vacuo*. Such material is the purest ever produced—about 1 part in 10^{13} of contaminants.

3.2.4 Detector applications

In all radiation-measuring systems there are several common factors which apply to measurements to be carried out. These are:

(1) Geometry;
(2) Scattering;
(3) Backscattering;
(4) Absorption;
(5) Self-absorption.

Geometry

Since any radioactive source emits its products in all directions (in 4π geometry) it is important to be able to calculate how many particles or quanta may be collected by the active volume of the counter. If we consider a point source of radiation as in Figure 3.6,

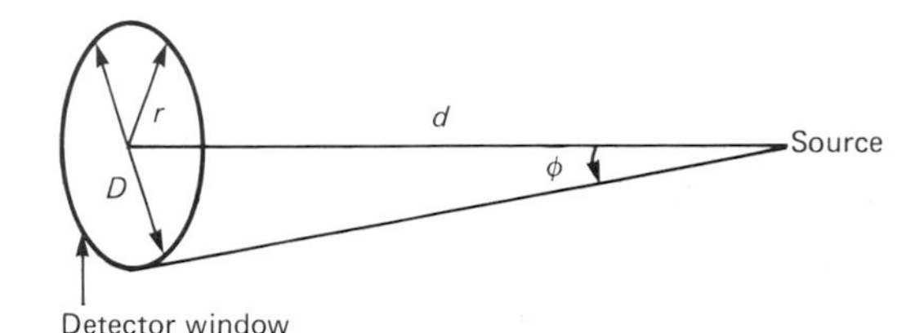

Figure 3.6 Geometry of radiation collection.

then all the emitted radiation from the source will pass through an imaginary sphere with the source at centre, providing there is no absorption. Also, for any given sphere size, the average radiation flux in radiations per unit time per unit sphere surface area is constant over the entire surface. The geometry factor, G, can be therefore written as the fraction of total 4π solid angle subtended by source and detector. For the case of the point source at a distance d from a circular window of radius r, we have the following relation:

$$G = 0.5(1 - \cos\phi) \tag{3.12}$$

$$= 0.5\left\{1 - \frac{1}{[1 + (r^2/d^2)]^{\frac{1}{2}}}\right\} \approx \frac{D^2}{16d^2} \tag{3.13}$$

Scattering

Particles and photons are scattered by material through which they pass, and this effect depends on the type of particle or photon, its energy, its mass, the type of material traversed, its mass and density. What we are concerned with here is not the loss of particle energy or particles themselves as they pass through a substance but the effects caused by particles deflected from the direct path between radioactive source and detector. It is found that some particles are absorbed into the material surrounding the source. Others are deflected away from it but are later re-scattered into the detector, so increasing the number of particles in the beam. Some of these deflected particles can be scattered at more than 90 degrees to the beam striking the detector—these are called 'backscattered' since they are scattered back in the direction from which they came. Scattering also occurs with photons, and this is particularly demonstrated by the increase in counting rate of a beam of gamma rays when a high-Z material such as lead is inserted into the beam.

Backscattering

Backscattering increases with increasing atomic number Z and with decreasing energy of the primary particle. For the most commonly used sample planchets (platinum) the backscattering factor has been determined as 1.04 for an ionization chamber inside which the sample is placed, known as a 50 per cent chamber.

Absorption

As the particles to be detected may be easily absorbed it is preferred to mount source and detector in an evacuated chamber, or insert the source directly into the gas of an ionization chamber or proportional counter. The most popular method at present uses a semiconductor detector, and this allows the energies to be determined very accurately when the detector and source are operated in a small evacuated cell. This is especially important for alphas.

Self-absorption

If visible amounts of solid are present in a source, losses in counting rate may be expected because of self-absorption of the particles emitted from the lower levels of the source which are unable to leave the surface of the source. Nader *et al.* give an expression for the self-absorption factor for alpha particles in a counter with 2π geometry (this is the gas counter with source inside the chamber):

$$f_s = 1 - \frac{s}{2pR} \quad \text{for } s < pR \tag{3.14}$$

and

$$f_s = \frac{0.5R}{s} \quad \text{for } s > pR \tag{3.15}$$

where s is the source thickness, R the maximum range of alpha particles in source material and p the maximum fraction of R which particles can spend in source and still be counted.

Radiation shield

The detector may be housed in a thick radiation shield to reduce the natural background which is found everywhere to a lower level where the required measurements can be made. The design of such natural radiation shields is a subject in itself, and can range from a few centimetres of lead to the massive battleship steel or purified lead used for whole-body monitors, where the object is to measure the natural ratiation from a human body.

3.2.4.1 Alpha-detector systems

The simplest alpha detector is the air-filled ionization chamber, used extensively in early work on radioactivity but now only used for alpha detection in health physics surveys of spilt activities on benches, etc. Even this application is seldom used, as more sensitive semiconductor or scintillation counters are usual for this purpose. Thin-window ionization or gas-proportional counters can also be used, or internal counters in which the sample is inserted into the active volume of a gas counter. Due to the intense ionization produced by alpha particles it is possible to count them in high backgrounds from other radiation such as betas and gamma rays by means of suitable discrimination circuits.

Ionization chambers Because alpha particles are very readily absorbed in a small thickness of gas or solid an early method used in counting them involved the radioactive source being placed inside a gas counter with no intermediate window. Figure 3.7 shows the schematic circuit of such a counter in which the gas, generally pure methane (CH_4), is allowed to flow through the counting volume. The counter can be operated in the ionization counter region, but more usually the high voltage supply is raised so that it operates in the proportional counter region, allowing discrimination to be made between alpha particles and beta particles which may be emitted from the same source and improving the ratio of signal to noise.

Ionization chambers are used in two different ways. Depending on whether the time constants of the associated electronic circuits are small or large, they are either 'counting', i.e. responding to each separate ionizing event, or 'integrating', i.e. collecting the ionization over a relatively long period of time. The counting mode is little used nowadays except for alpha particles.

Gas proportional counters Thin window proportional counters allow alphas to be counted in the presence of high beta and gamma backgrounds, since these detectors can discriminate between the high ionization density caused by alpha particles

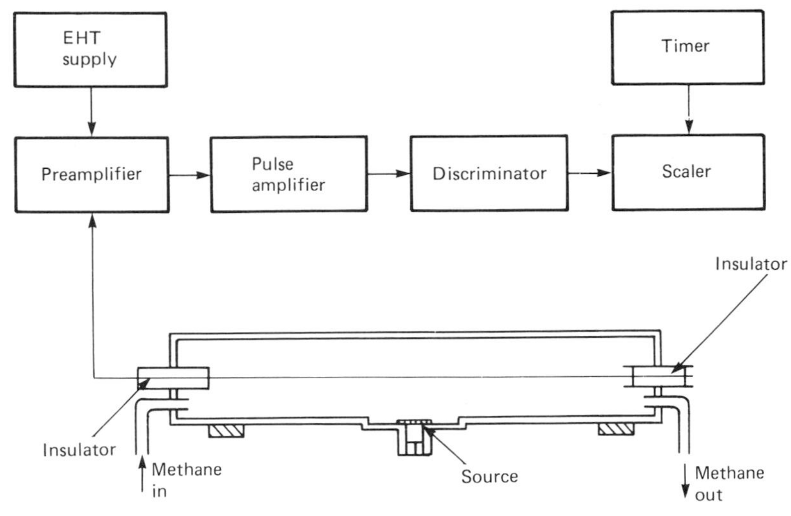

Figure 3.7 Gas flow-type proportional counter system.

passing through the gas of the counter and the relatively weak ionization produced by beta particles or gamma-ray photons.

With counters using flowing gas the source may be placed inside the chamber and the voltage across the detector raised to give an amplification of 5–50 times. In this case pure methane (CH_4) is used with the same arrangement as shown in Figure 3.7.

The detection efficiency of this system is considerably greater than that provided by the thin-window counter with source external to the counter. This is due (1) to the elimination of the particles lost in penetrating the window and (2) to the improved geometry whereby all particles which leave the source and enter the gas are counted.

Geiger counters The observations made about alpha-particle detection by gas proportional counters apply also to Geiger counters. That is, thin entrance windows or internally mounted sources are usable and the system of Figure 3.7 is also applicable to Geiger-counter operation. However, some differences have to be taken into account. Since operation in the Geiger region means that no differences in particle energy are measured, all particles are able to trigger the counter and no energy resolution is possible. On the other hand, the counter operating in the Geiger region is so sensitive that high electronic amplification is not required, and in many cases the counter will produce an output sufficient to trigger the scaler unit directly. In order to operate the system in its optimum condition, some amplification is desirable and should be variable. This is to set the operating high-voltage supply and the amplifier gain on the Geiger plateau (see p. 101), so that subsequent small variations in the high-voltage supply do not affect the counting characteristics of the system.

Scintillation counters Since alpha particles are very easily absorbed in a very short distance into any substance, problems are posed if alphas are to enter a scintillator through a window, however thin, which would also prevent light from entering and overloading the photomultiplier. The earliest scintillation detector using a photomultiplier tube had a very thin layer of zinc sulphide powder sprinkled on the glass envelope of a side-viewing photomultiplier of the RCA 931-A type, on which a small layer of adhesive held the zinc sulphide in place. Due to the low light transmission of powdered zinc sulphide a layer 5–10 mg/cm^2 is the optimum thickness. The whole system was enclosed in a light-tight box, source and detector together. Later experimenters used a layer of aluminium evaporated onto the zinc sulphide to allow the alphas in but to keep the light out. This was not successful, as an adequate thickness of aluminium to keep the light out also prevented the alphas from entering the zinc sulphide. When photomultipliers with flat entrance windows began to become available, the zinc sulphide was deposited on the face of a disc of transparent plastic such as Perspex/Lucite. This was easier to evaporate aluminium on than directly onto the photomultiplier, but the light problem was still present, and even aluminized mylar was not the answer. Operation in low or zero-light conditions was still the only satisfactory solution.

Thin scintillating plastics were also developed to detect alphas, and in thin layers, generally cemented by heat to a suitable plastic light guide, proved cheaper to use than zinc sulphide detectors. The light output from plastic scintillators was, however, less than that from zinc sulphide, but both scintillators, in their very thin layers, provide excellent discrimination against beta particles or gamma rays, often present as background when alpha-particle emitters are being measured. One major advantage of the plastic scintillator over the inorganic zinc sulphide detector is its very fast response, of the order of 4×10^{-9} s for the decay time of a pulse as compared with $4–10 \times 10^{-6}$ s for zinc sulphide.

Another scintillator which has been used for alpha detection is the inorganic crystal CsI (Tl). This can be used without any window (but in the dark) as it is non-hygroscopic, and, if suitably bevelled around the circumference, will produce an output proportional to the energy of the ionizing particle incident on it, thus acting as a spectrometer. Extremely thin CsI (Tl) detectors have been made by vacuum deposition and used where the beta and gamma background must be reduced to very small proportions yet the heavy particles of interest must be detected positively.

Inorganic single crystals are expensive, and when large areas of detector must be provided either the zinc sulphide powder screen or a large-area very thin plastic scintillator is used. The latter is much cheaper to produce than the former, as the thin layer of zinc sulphide needs a great deal of labour to produce and it is not always reproducible. In health physics applications, for monitors for hand and foot contamination, or bench-top contamination, various combinations of plastic scintillator, zinc sulphide or sometimes anthracene powder scintillators are used.

If the alpha emitter of interest is mixed in a liquid scintillator the geometrical effect is almost completely eliminated and, provided the radioactive isotope can be dissolved in the liquid scintillator, maximum detection efficiency is obtained. However, not all radioactive sources can be introduced in a chemical form suitable for solution, and in such cases the radioactive material can be introduced in a finely divided powder into a liquid scintillator with a gel matrix—such a matrix would be a very finely divided extremely pure grade of silica. McDowell (1980) has written a useful review of alpha liquid scintillation counting from which many references in the literature may be obtained.

3.2.4.2 Detection of beta particles

Ionization chambers Although ionization chambers were much used in the early days for the detection of beta particles, they are now used only for a few special purposes:

(1) The calibration of radioactive beta sources for surface dose rate. An extrapolation chamber varies the gap between the two electrodes of a parallel plate ionization chamber and the ionization current per unit gap versus air gap is plotted on a graph as the gap is reduced, and this extrapolates to zero gap with an uncertainty which is seldom as much as 1 per cent, giving a measure of the absolute dose from a beta source.
(2) Beta dosimetry. Most survey instruments used to measure dose rate incorporate some sort of device to allow beta rays to enter the ionization chamber. This can take the form of a thin window with a shutter to cut off the betas and allow gamma rays only to enter when mixed beta-gamma doses are being measured. However, the accuracy of such measurements leaves a great deal to be desired, and the accurate measurement of beta dose rates is a field where development is needed.

Proportional counters Beta particles may originate from three different positions in a proportional counter; first, from part of the gaseous content of the counter, called 'internal counting'; second, from a solid source inside the counter itself; and third, from an external source with the beta particles entering the counter by means of a thin window.

The first method, internal counting, involves mixing the radioactive source in the form of a gas with a gas suitable for counting; or the radioactive source can be transformed directly into a suitable gaseous form. This is the case when detection of ^{14}C involves changing solid carbon into gaseous carbon dioxide (CO_2), acetylene (CH) or methane (CH_4), any of which can be counted directly in the proportional counter. This method is used in measurement of radiocarbon to determine the age of the carbon: radiocarbon dating (see p. 140).

The second method involves the use of solid sources introduced by means of a gas-tight arm or drawer, so that the source is physically inside the gas volume of the counter. This method was much used in early days but now only in exceptional circumstances. The third method, with the source external to the counter, is now the most popular, since manufacturers have developed counters with extremely thin windows which reduce only slightly the energies of the particles crossing the boundary between the gas of the counter and the air outside. Thin mica or plastic windows can be of the order of a few mg/cm^2, and they are often supported by wire mesh to allow large areas to be accommodated.

A specialized form of proportional counter is the $2\pi/4\pi$ type used in the precise assay of radioactive sources. This is generally in the form of an open-ended pillbox, and the source is deposited on a thin plastic sheet either between two counters face to face (the 4π configuration) or with only a single counter looking at the same source (the 2π configuration). Such counters are generally made in the laboratory, using plastic as the body onto which a conducting layer of suitable

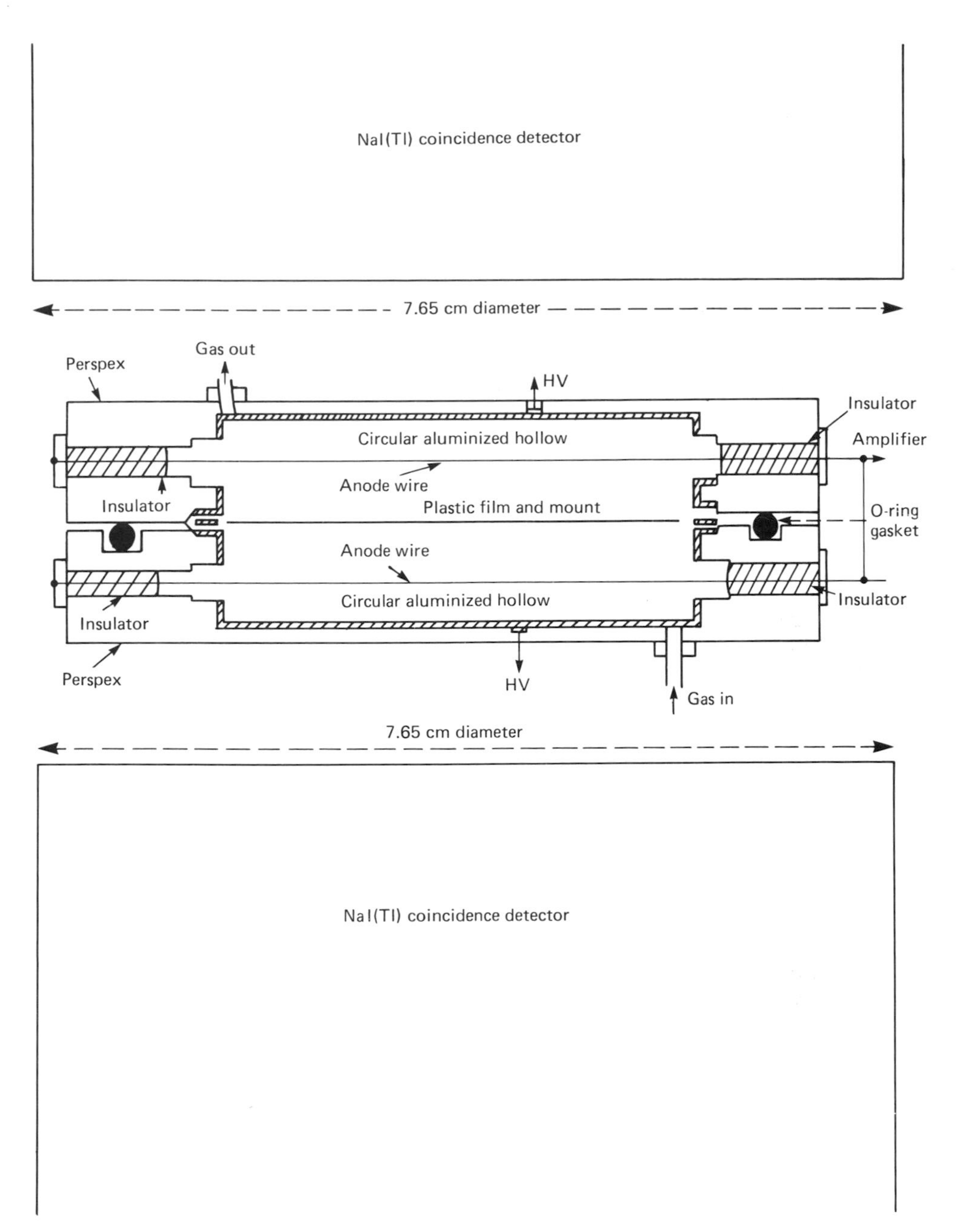

Figure 3.8 Typical design of 4π counter.

metal is deposited to form the cathode. One or more wires form the anode. A typical design is shown in Figure 3.8.

Geiger counters The Geiger counter has been the most popular detector for beta particles for a number of reasons. First, it is relatively cheap, either when manufactured in the laboratory or as a commercially available component. Second, it requires little in the way of special electronics to make it work, as opposed to scintillation or solid-state detectors. Third, the later halogen-filled versions have quite long working lives.

To detect beta particles some sort of window must be provided to allow the particles to enter the detector.

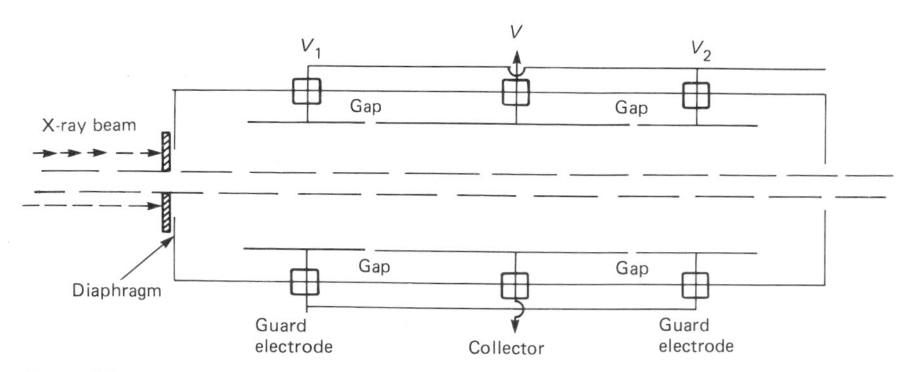

Figure 3.9 Free air ionization chamber.

The relations between the various parameters affecting the efficiency with which a particular radioactive source is detected by a particular Geiger counter have been studied extensively. Zumwalt (1950), in an early paper, provided the basic results which have been quoted in later books and papers, such as Price (1964) and Overman and Clark (1960). In general, the observed counting rate R and the corresponding disintegration rate A can be related by the equation

$$R = YA \tag{3.16}$$

where Y is a factor representing the counting efficiency. Some eleven factors are contained in Y and these are set out in those books mentioned above.

3.2.4.3 Detection of gamma rays (100 keV upwards)

Ionization chambers Dosimetry and dose-rate measurement of gamma rays are the main applications in which ionization chambers are used to detect gamma rays today. They are also used in some industrial measuring systems although scintillation and other counters are now generally replacing them.

A device called the free air ionization chamber (Figure 3.9) is used in X-ray work to calibrate the output of X-ray generators. This chamber measures directly the charge liberated in a known volume of air with the surrounding medium being air. In Figure 3.9 only ions produced in the volume $V_1 + V + V_2$ defined by the collecting electrodes are collected and measured, and any ionization outside this volume is not measured. However, due to the physical size of the chamber and the amount of auxiliary equipment required it is only used in standardizing laboratories, where small Bragg–Gray chambers may be calibrated by comparison with the free air ionization chamber.

The Bragg–Gray chamber depends on the principle that if a small cavity exists in a solid medium the energy spectrum and the angular distribution of the electrons in the cavity are not disturbed by the cavity, and the ionization occurring in the cavity is characteristic of the medium, provided:

(1) The cavity dimensions are small compared with the range of the secondary electrons;
(2) Absorption of the gamma radiation by the gas in the cavity is negligible;
(3) The cavity is surrounded by an equilibrium wall of the medium; and
(4) The dose rate over the volume of the medium at the location of the cavity is constant.

The best-known ionization chamber is that designed by Farmer, and this has been accepted worldwide as a sub-standard to the free air ionization chamber for measuring dose and dose rates for human beings.

Solid-state detectors The impact of solid-state detectors on the measurement of gamma-rays has been dramatic; the improvement in energy resolution compared with a NaI (Tl) detector can be seen in Figure 3.10, which shows how the solid-state detector can resolve the 1.17 and 1.33 MeV lines from ^{60}Co with the corresponding NaI (Tl) scintillation detector result superimposed. For energy resolution the solid-state detector provides a factor of 10 improvement on the scintillation counter. However, there are a number of disadvantages which prevent this detector superseding the scintillation detector. First, the present state of the art limits the size of solid-state detectors to some 100 cm^3 maximum, whereas scintillation crystals of NaI (Tl) may be grown to sizes of 76 cm diameter by 30 cm, while plastic and liquid scintillators can be even larger. Second, solid-state detectors of germanium have to be operated at liquid-nitrogen temperatures and *in vacuo*. Third, large solid-state detectors are very expensive. As a result, the present state of the art has tended towards the use of solid-state detectors when the problem is the determination of energy spectra, but scintillation counters are used in cases where extremely low levels of activity require to be detected. A very popular combined use of solid-state and scintillation detectors is embodied in the technique of

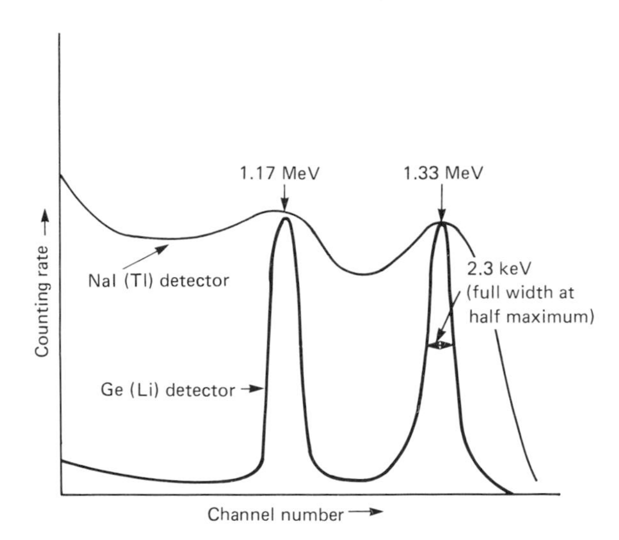

Figure 3.10 Comparison of energy resolution by different detectors.

surrounding the relatively small solid-state detector in its liquid nitrogen-cooled and evacuated cryostat with a suitable scintillation counter, which can be an inorganic crystal such as NaI (Tl) or a plastic or liquid scintillator. By operating the solid-state detector in anti-coincidence with the annular scintillation detector, Compton-scattered photons from the primary detector into the anti-coincidence shield (as it is often called) can be electronically subtracted from the solid-state detector's energy response spectrum.

3.2.4.4 The detection of neutrons

The detection of nuclear particles usually depends on the deposition of ionization energy and, since neutrons are uncharged, they cannot be detected directly. Neutron sensors therefore need to incorporate a conversion process by which the incoming particles are changed to ionizing species and nuclear reactions such as ^{235}U (fission) + ~200 MeV or $^{10}B\,(n, \alpha)Li^7$ + ~2 MeV are often used. Exothermic reactions are desirable because of the improved signal-to-noise ratio produced by the increased energy per event, but there are limits to the advantages which can be gained in this way. In particular, the above reactions are not used for neutron spectrometry because of uncertainty in the proportion of the energy carried by the reaction products, and for that purpose the detection of proton recoils in a hydrogenous material is often preferred.

There are also many ways of detecting the resultant ionization. It may be done in real time with, for example, solid-state semiconductor detectors, scintillators or ionization chambers, or it may be carried out at some later, more convenient, time by measuring the activation generated by the neutrons in a chosen medium (such as ^{56}Mn from ^{55}Mn). The choice of technique depends on the information required and the constraints of the environment. The latter are often imposed by the neutrons themselves. For example, boron-loaded scintillators can be made very sensitive and are convenient for detecting low fluxes, but scintillators and photomultipliers are vulnerable to damage by neutrons and are sensitive to the gamma fluxes which tend to accompany them. They are not suitable for the high-temperature, high-flux applications found in nuclear reactors.

Neutron populations in reactors are usually measured with gas-filled ionization chambers. Conversion is achieved by fission in ^{235}U oxide applied as a thin layer (~1 mg cm^{-2}) to the chamber electrode(s). The ^{10}B reaction is also used, natural or enriched boron being present either as a painted layer or as BF_3 gas. Ionization chambers can be operated as pulse devices, detecting individual events; as d.c. generators, in which events overlap to produce current; or in the so-called 'current fluctuation' or 'Campbell' mode in which neutron flux is inferred from the magnitude of the noise present on the output as a consequence of the way in which the neutrons arrive randomly in time. Once again the method used is chosen to suit operational requirements. For example, the individual events due to neutrons in a fission chamber are very much larger than those due to gammas, but the gamma photon arrival rate is usually much greater than that of the neutrons. Thus, pulse counters have good gamma rejection compared with d.c. chambers at low fluxes. On the other hand, the neutron-to-gamma flux ratio tends to improve at high reactor powers whilst counting losses increase and gamma pulse pile-up tends to simulate neutron events. D.C. operation therefore takes over from pulse measurement at these levels. The current fluctuation mode gives particularly good gamma discrimination because the signals depend on the mean square charge per event, i.e. the initial advantage of neutrons over gammas is accentuated. Such systems can work to the highest fluxes and it is now possible to make instruments which combine pulse counting and current fluctuation on a single chamber and which cover a dynamic range of more than 10 decades.

The sensitivity of a detector is proportional to the probability of occurrence of the expected nuclear reaction and can conveniently be described in terms of the cross-section of a single nucleus for that particular reaction. The unit of area is the barn, i.e. 10^{-24} cm^2. ^{10}B has a cross-section of order 4000 b to slow (thermal) neutrons whilst that of ^{235}U for fission is only ~550 b. In addition, the number of reacting atoms present in a given thickness of coating varies inversely with atomic weight so that, in principle, ^{10}B sensors are much more sensitive than those which depend on

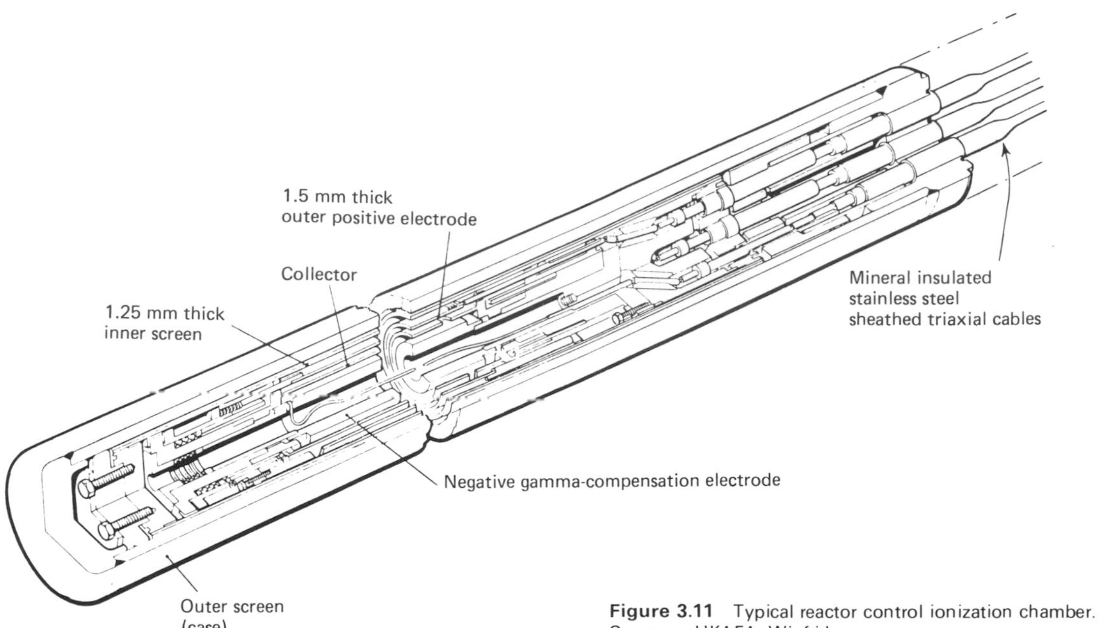

Figure 3.11 Typical reactor control ionization chamber. Courtesy, UKAEA, Winfrith.

fission. This advantage is offset by the lower energy per event and by the fact that boron is burnt up faster at a given neutron flux, i.e. that such detectors lose sensitivity with time.

Neutrons generate activation in most elements, and if detector constructional materials are not well chosen this activation can produce residual signals analogous to gamma signals and seriously shorten the dynamic range. The electrodes and envelopes of ion chambers are therefore made from high-purity materials which have small activation cross sections and short daughter half-lives. Aluminium is usually employed for low-temperature applications, but some chambers have to operate at $\sim 550°C$ (a dull red) and these use titanium and/or special low-manganese, low-cobalt stainless steels. Activity due to fission products from fissile coatings must also be considered and is a disadvantage to fission chambers. The choice of insulators is also influenced by radiation and temperature considerations. Polymers deteriorate at high fluences but adequate performance can be obtained from high-purity, polycrystalline alumina and from artificial sapphire, even at 550°C. Analogous problems are encountered with cables and special designs with multiple co-axial conductors insulated with high-purity compressed magnesia have been developed. Electrode/cable systems of this type can provide insulation resistances of order $10^9\ \Omega$ at 550°C and are configured to eliminate electrical interference even when measuring micro-amp signals in bandwidths of order 30 MHz under industrial plant conditions. Figure 3.11 shows the construction of a boron-coated gamma-compensated d.c. chamber designed to operate at 550°C and 45 Bar pressure in the AGR reactors. The gas filling is helium and β activity from the low-manganese steel outer case is screened by thick titanium electrodes. The diameter of this chamber is 9 cm; it is 75 cm long and weighs 25 kg. By contrast, Figure 3.12 shows a parallel plate design, some three hundred of which were used to replace fuel 'biscuits' to determine flux distributions in the ZEBRA experimental fast reactor at AEE Winfrith.

Boron trifluoride (BF_3) proportional counters are used for thermal neutron detection in many fields; they are convenient and sensitive, and are available commercially with sensitivities between 0.3 and $196\ s^{-1}$ (unit flux)$^{-1}$. They tend to be much more gamma sensitive than pulse-fission chambers because of the relatively low energy per event from the boron reaction, but this can be offset by the larger sensitivity. A substantial disadvantage for some applications is that they have a relatively short life in terms of total dose (gammas plus neutrons), and in reactor applications it may be necessary to provide withdrawal mechanisms to limit this dose at high power.

Proton-recoil counters are used to detect fast neutrons. These depend on the neutrons interacting with a material in the counter in a reaction of the (n, p) type, in which a proton is emitted which, being highly ionizing, can be detected. The material in the counter can be either a gas or a solid. It must have low-Z

Figure 3.12 Pulse-fission ionization chamber for flux-distribution measurement. Courtesy, UKAEA, Winfrith.

nuclei, preferably hydrogen, to allow the neutron to transfer energy. If a gas, the most favoured are hydrogen or helium, because they contain the greatest number of nuclei per unit volume. If a solid, then paraffin, polyethylene or a similar low-Z material can be used to line the inside of the counter.

This type of counter was used by Hurst *et al.* to measure the dose received by human tissue from neutrons in the range 0.2–10 MeV. It was a three-unit counter (the gas being methane) and two individual sections contained a thin (13.0 mg/cm^2) layer of polyethylene and a thick (100 mg/cm^2) layer of polyethylene. The energy responses of the three sources of protons combine in such a way as to give the desired overall response, which matches quite well the tissue–dose curve over the energy range 0.2–10 MeV. This counter also discriminates well against the gamma rays which nearly always accompany neutrons, especially in reactor environments.

Improvements in gas purification and counter design have led to the development of ^{3}He-filled proportional counters. ^{3}He pressures of 10–20 atm allow the use of these counters as direct neutron spectrometers to measure energy distributions, and in reactor neutron spectrum analysis they are found in most reactor centres all over the world.

As explained above, it is necessary for the measurement of neutrons that they shall interact with substances which will then emit charged particles. For thermal energy neutrons (around 0.025 eV in energy) a layer of fissionable material such as ^{235}U will produce reaction products in the form of alpha particles, fission products, helium ions, etc. However, more suitable materials for producing reaction products are ^{6}Li and ^{10}B. These have relatively high probability of a neutron producing a reaction corresponding to a cross-section of 945 barns for ^{6}Li and for ^{10}B of 3770 barns. By mixing ^{6}Li or ^{10}B with zinc sulphide powder and compressing thin rings of the mixture into circular slots in a methyl methacrylate disc, the reaction products from the ^{6}Li or ^{10}B atom disintegration when a neutron is absorbed strike adjacent ZnS particles and produce light flashes which can be detected by the photomultiplier to which the detector is optically coupled. Figure 3.13 shows such a neutron-detector system.

Neutron–proton reactions allow the detection of neutrons from thermal energies up to 200 MeV and higher. For this reaction to take place a hydrogen-type material is required with a high concentration of protons. Paraffin, polyethylene or gases such as hydrogen, ^{3}He, methane etc. provide good sources of

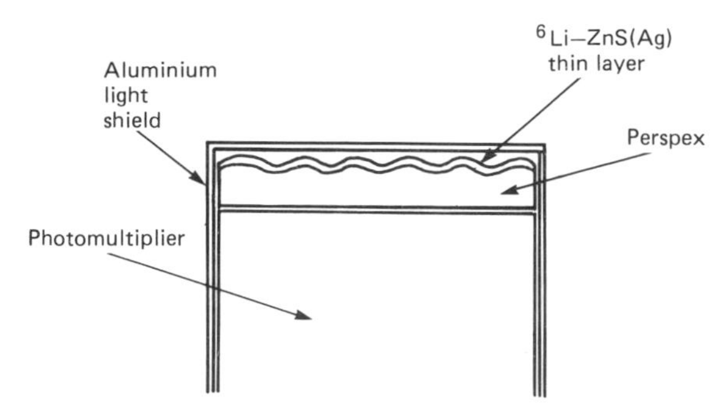

Figure 3.13 Thermal neutron scintillation counter.

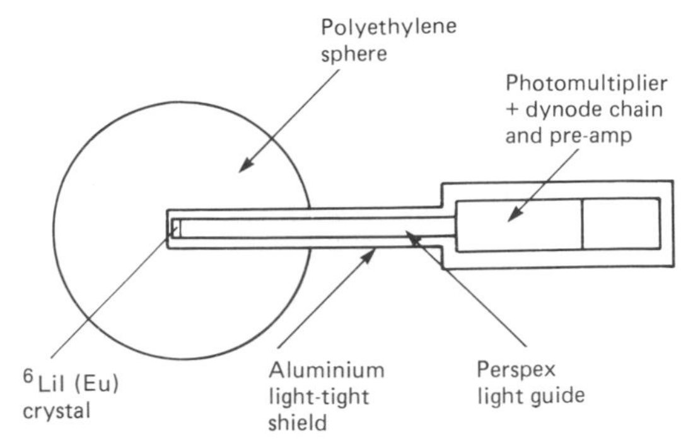

Figure 3.14 Sphere fast-neutron detector.

protons, and by mixing a scintillator sensitive to protons (such as ZnS) with such a hydrogenous material, the protons produced when a neutron beam interacts with the mixed materials can be counted by the flashes of light produced in the ZnS. Liquids can also be used, in which the liquid scintillator is mixed with boron, gadolinium, cadmium, etc. in chemical forms which dissolve in the scintillator. Large tanks of 500–1000-l sizes have been made for high-energy studies, including the study of cosmic ray neutrons. ^{6}Li can also be used dissolved in a cerium-activated glass, and this has proved a very useful neutron detector as the glass is inert to many substances. This ^{6}Li glass scintillator has proved very useful for studies in neutron radiography, where the neutrons are used in the manner of an X-radiograph to record on a photographic film the image produced in the glass scintillator by a beam of neutrons.

Another neutron detector is the single crystal of lithium iodide activated with europium ^{6}LiI (Eu) When this crystal is cooled to liquid-nitrogen temperature it can record the spectrum of a neutron source by pulse-height analysis. This is also possible with a special liquid scintillator NE213 (xylene-based), which has been adopted internationally as the standard scintillator for fast neutron spectrometry from 1 to 20 MeV.

One of the problems in neutron detection is the presence of a gamma-ray background in nearly every practical case. Most of the detectors described here do have the useful ability of being relatively insensitive to gamma rays. That is with the exception of LiI (Eu), which, because of its higher atomic number due to the iodine, is quite sensitive to gamma rays. By reducing the size of the scintillator to about 4 mm square by 1 mm thick and placing it on the end of a long thin light guide placed so that the detector lies at the centre of a polyethylene sphere, a detector is produced which can measure neutron dose rate with a response very close to the response of human tissue. Figure 3.14 shows a counter of this kind which is nearly isotropic in its response (being spherical) and with much-reduced sensitivity to gamma rays. This is known as the Bonner sphere, and the diameter of the sphere can be varied between 10 and 30 cm to cover the whole energy range.

For thermal neutrons ($E=0.025$ eV) an intermediate reaction such as ^{6}Li(n, α) or ^{10}B(n, α) or fission can be used, a solid state detector being employed to count the secondary particles emitted in the reaction. For fast neutrons a radiator in which the neutrons produce recoil protons can be mounted close to a solid-state detector and the detector counts the protons. By sandwiching a thin layer of ^{6}LiF between two solid-state detectors and summing the coincident alpha and tritium pulses to give an output signal proportional to the energy of the incident neutron plus 4.78 MeV, the response of the assembly with respect to incident neutron energy was found to be nearly linear.

3.3 Electronics

A more general treatment of the measurement of electrical quantities is given in Chapter 1 and in the chapter on Signal Processing, given in Part 4, where amplifiers and pulse-height analysers are especially considered. We concentrate here on aspects of electronics that are particularly relevant to nuclear instrumentation.

3.3.1 Electronics assemblies

Whilst it is perfectly feasible to design a set of electronics to perform a particular task—and indeed this is often done for dedicated systems in industry—the more usual system is to incorporate a series of interconnecting individual circuits into a common frame. This permits a variety of arrangements to be made by plugging in the required elements into this frame, which generally also provides the necessary power supplies. This 'building-block' system has become standardized worldwide under the title of

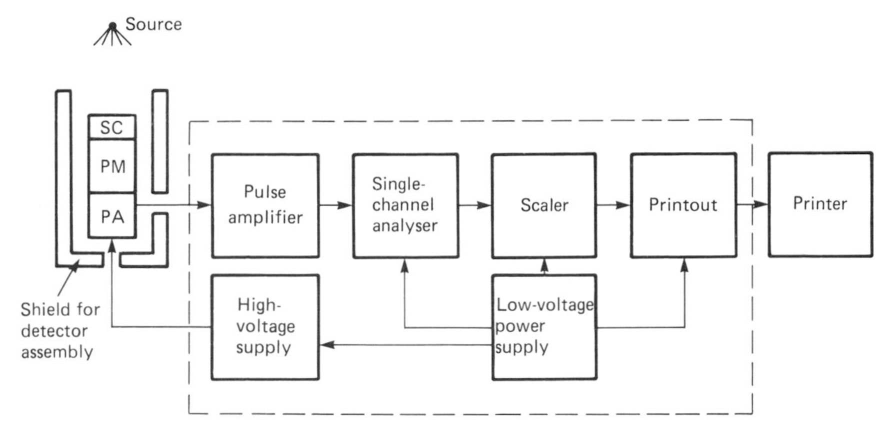

Figure 3.15 Typical arrangement of electronics. SC: scintillator; PM: photomultiplier; PA: preamplifier.

NIM and is based on the US Atomic Energy Commission (USAEC)–Committee on Nuclear Instrument Modules, presented in *USAEC Publication TID–20893*. The basic common frame is 483 mm (19 in.) wide and the plug-in units are of standard dimensions, a single module being 221 mm high × 34.4 mm wide × 250 mm deep (excluding the connector). Modules can be in widths of one, two or more multiples of 34.4 mm. Most standard units are single or double width. The rear connectors which plug into the standard NIM bin have a standardized arrangement for obtaining positive and negative stabilized supplies from the common power supply, which is mounted at the rear of the NIM bin. The use of a standardized module system allows the use of units from a number of different manufacturers in the same bin, since it may not be possible or economic to obtain all the units required from one supplier. Some seventy individual modules are available from one manufacturer, which gives some idea of the variety.

A typical arrangement is shown in Figure 3.15, where a scintillation counter is used to measure the gamma-ray energy spectrum from a small source of radioactivity. The detector could consist of a NaI (Tl) scintillator optically coupled to the photocathode of a photomultiplier, the whole contained in a light-tight shielded enclosure of metal, with the dynode resistor chain feeding each of the dynodes (see pp. 74, 106) located in the base of the photomultiplier, together with a suitable preamplifier to match the high-impedance output of the photomultiplier to the lower input impedance of the main pulse amplifier, generally of the order of 50 Ω. This also allows the use of a relatively long coaxial cable to couple the detector to the electronics if necessary.

The main amplifier raises the amplitudes of the input pulses to the range of about 0.5–10 V. The single-channel analyzer can then be set to cover a range of input voltages corresponding to the energies of the source to be measured. If the energy response of the scintillator–photomultiplier is linear—as it is for NaI (Tl)—then the system can be calibrated using sources of known energies, and an unknown energy source can be identified by interpolation.

3.3.2 Power supplies

The basic power supplies for nucleonic instruments are of two classes—the first supplies relatively low d.c. voltages at high currents (e.g. 5–30 V at 0.5–50 A) and the second high voltages at low currents (e.g. 200–5000 V at 200 μA–5 mA). Alternatively, batteries, both primary and secondary (i.e. rechargeable) can be used for portable instruments. In general, for laboratory use d.c. supplies are obtained by rectifying and smoothing the mains a.c. supply. In the UK and most European countries the mains a.c. power supply is 50 Hz while in the USA and South America it is 60 Hz, but generally a supply unit designed for one frequency can be used on the other. However, mains-supply voltages vary considerably, being 240 V in the UK and 220 V in most of the EEC countries, with some countries having supplies of 110, 115, 120, 125, 127, etc. The stability of some of these mains supplies can leave much to be desired, and fluctuations of plus and minus 50 V have been measured. As nucleonic instruments depend greatly on a stable mains supply, the use of special mains-stabilizing devices is almost a necessity for equipment which may have to be used on such varying mains supplies.

Two main types of voltage regulator are in use at present. The first uses a saturable inductor in the form of a transformer with a suitably designed air gap in the iron core. This is a useful device for cases where the voltage swing to be compensated is not large. The second type of stabilizer selects a portion of the output

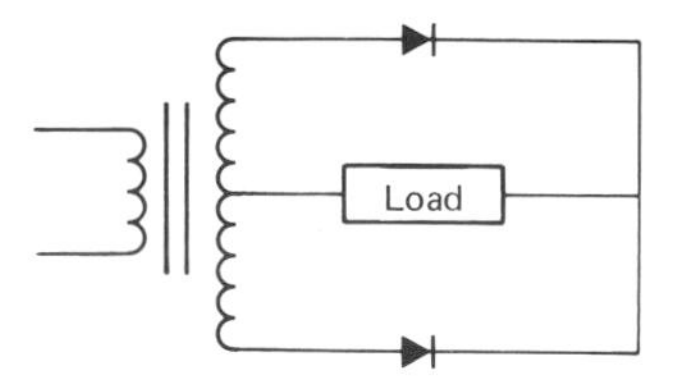

Figure 3.16 Full-wave rectifier.

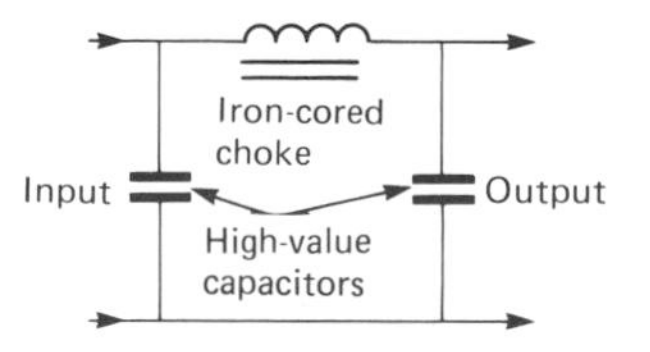

Figure 3.19 Smoothing filter.

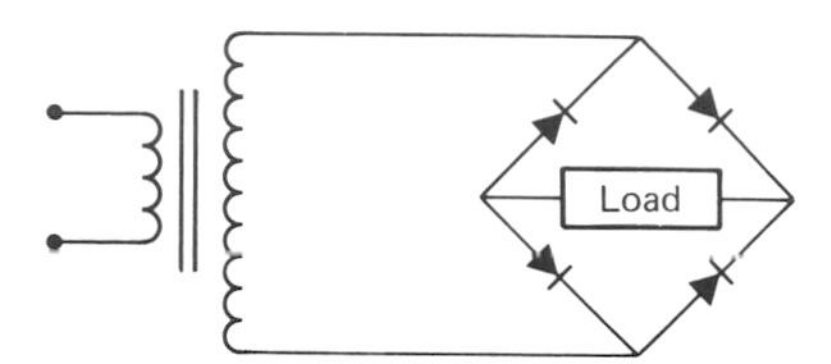

Figure 3.17 Bridge rectifier.

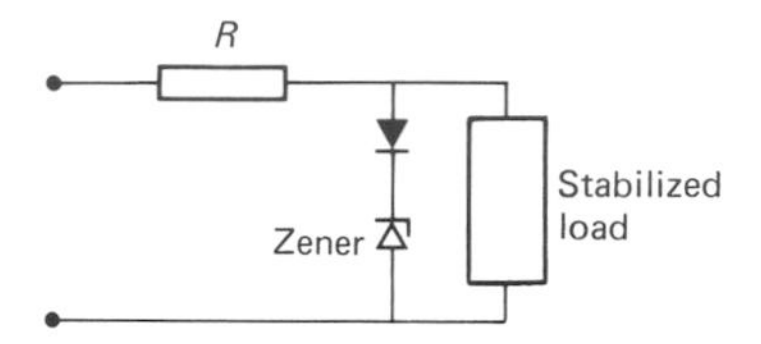

Figure 3.20 Simple stabilizer.

voltage, compares it with a standard and applies a suitable compensation voltage (plus or minus) to compensate. Some of these units use a motor-driven tapping switch to vary the input voltage to the system—this allows for slow voltage variations. A more sophisticated system uses a semi-conductor-controlled voltage supply to add to or subtract from the mains voltage.

The simplest power supply is obtained by a transformer and a rectifier. The best results are obtained with a full-wave rectifier (Figure 3.16) or a bridge rectifier (Figure 3.17). A voltage-doubling circuit is shown in Figure 3.18. The outputs from either system are then smoothed using a suitable filter as shown in Figure 3.19. A simple stabilizer may be fitted in the form of a Zener diode which has the characteristic that the voltage drop across it is almost independent of the current through it. A simple stabilizer is shown in Figure 3.20. Zeners may be used in series to allow quite high voltages to be stabilized.

An improved stabilizer uses a Zener diode as a reference element rather than an actual controller. Such a circuit is shown in Figure 3.21, where the sensing element, the transistor TR_3 compares a fraction of the output voltage which is applied to its base with the fixed Zener voltage. Through the series control transistor TR_4, the difference amplifier TR_1, TR_3 and TR_2, corrects the rise or fall in the output voltage which initiated the control signal.

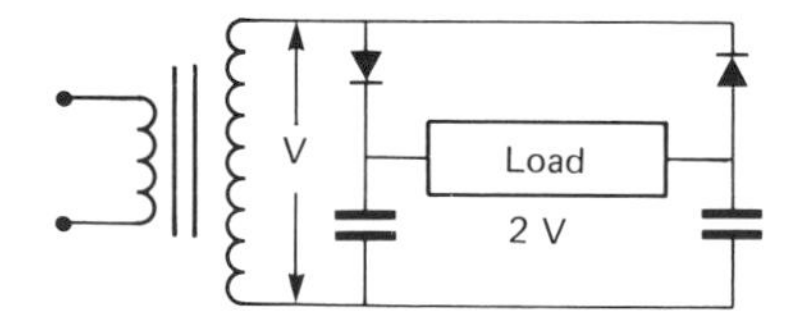

Figure 3.18 Voltage-doubling circuit.

3.3.2.1 High-voltage power supplies

High voltages are required to operate photomultipliers, semiconductor detectors, multi- and single-wire gas proportional counters, etc., and their output must be as stable and as free of pulse transients as possible. For photomultipliers the stability requirements are extremely important, since a variation in overall voltage across a photomultiplier of, say, 0.1 per cent, can make a 1 per cent change in output. For this reason stabilities of the order of 0.01 per cent over the range of current variation expected and 0.001 per cent over mains a.c. supply limits are typically required in such power supplies.

3.3.3 Amplifiers

3.3.3.1 Preamplifiers

Detectors often have to be mounted in locations quite distant from the main electronics, and if the cable is of any appreciable length, considerable loss of signal could occur. The preamplifier therefore serves more as an impedance transformer, to convert the high impedance of most detector outputs to a sufficiently low impedance which would match a 50- or 70-Ω connecting cable. If the detector is a scintillating counter, the output impedance of the photomultiplier is of the order of several thousand ohms, and there would be almost complete loss of signal if one coupled the output of the counter directly to the 50-Ω impedance of the cable—hence the necessity of providing a suitable impedance matching device.

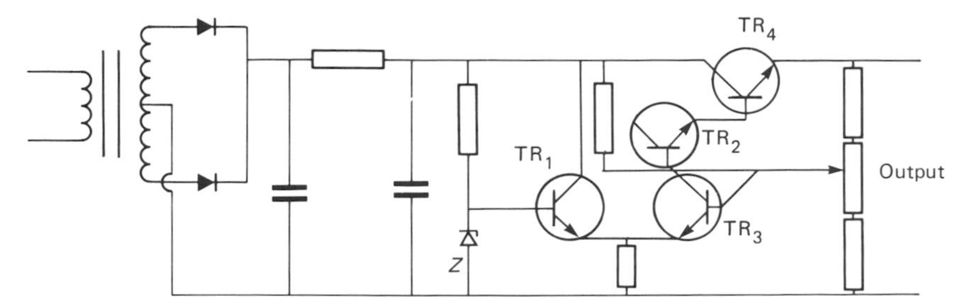

Figure 3.21 Improved stabilizer.

3.3.4 Scalers

From the earliest days it has been the counting of nuclear events which has been the means of demonstrating the decay of radioactive nuclei. Early counters used thermionic valves in scale-of-two circuits, which counted in twos. By using a series of these, scale-of-ten counters may be derived. However, solid-state circuits have now reduced such scale-of-ten counters to a single semiconductor chip, which is far more reliable than the thermionic valve systems and a fraction of the size and current consumption.

As scalers are all based on the scale-of-two, Figure 3.22 shows the arrangement for obtaining a scale-of-ten, and with present technology, many scales-of-ten may be incorporated on a single chip. The basic unit is called a J-K binary, because of the lettering on the original large-scale version of the binary unit.

Rates of 150–200 MHz may be obtained with modern decade counters which can be incorporated on a single chip together with the auxiliary units, such as standard oscillator, input and output circuits and means of driving light displays to indicate the count achieved.

3.3.5 Pulse-height analysers

If the detector of nuclear radiation has a response governed by the energy of the radiation to be measured then the amplitude of the pulses from the detector is a measure of the energy. To determine the energy, therefore, the pulses from the detector must be sorted into channels of increasing pulse amplitude.

Trigger circuits have the property that they can be set to trigger for all pulses above a preset level. This is acting as a discriminator. By using two trigger circuits, one set at a slightly higher triggering level than the other, and by connecting the outputs to an anti-coincidence circuit (*see* p. 122), the output of the anti-coincidence circuit will be only pulses which have amplitudes falling within the voltage difference between the triggering levels of the two discriminators. Figure 3.23 shows a typical arrangement and Figure 3.24 shows how an input pulse 1, below the triggering level V, produces no output, nor does pulse 3, above the triggering level $V+\Delta V$. However, pulse 2, falling between the two triggering levels, produces an output pulse. ΔV is called the channel width.

A multichannel analyser (MCA) allows the separation of pulses from a detector into channels determined by their amplitudes. Early analysers used up to twenty or more single-channel analysers set to successively increasing channels. These, however, proved difficult to stabilize, and the introduction of the Hutchinson-Scarrott system of an analogue-to-digital converter (ADC) combined with a computer memory enabled more than 8000 channels to be provided with good stability and adequate linearity. The advantages of the MCA are offset by the fact that the dead time (that is, the time during which the MCA is unable to accept another pulse for analysis) is longer than that of a single-channel analyser and so it has a lower

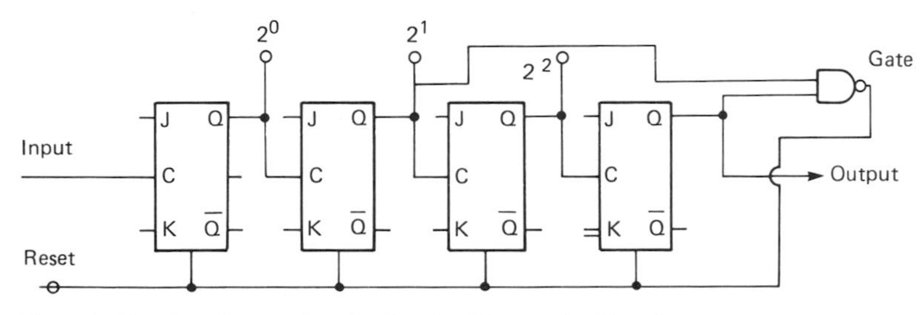

Figure 3.22 Decade-counting circuit using binary units (flip–flops).

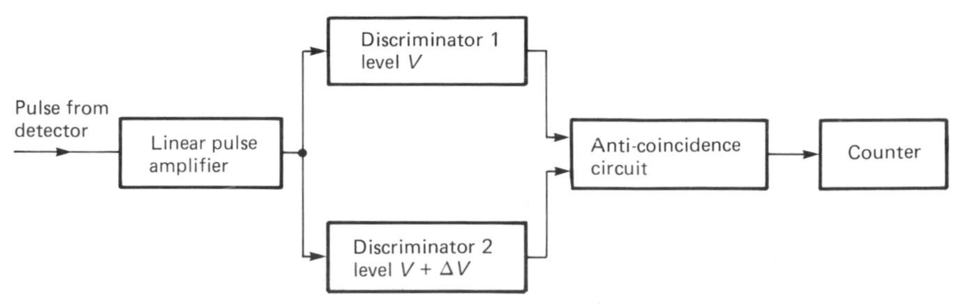

Figure 3.23 Single-channel pulse-height analyser.

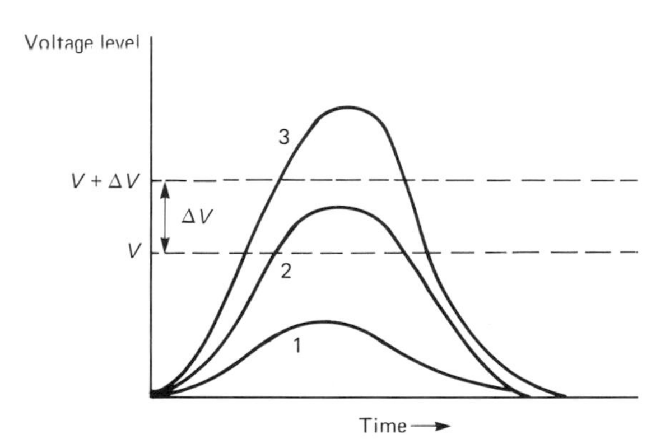

Figure 3.24 Waveforms and operation of Figure 3.23.

maximum counting rate. A block diagram of a typical multichannel analyser is shown in Figure 3.25. The original ADC was that of Wilkinson, in which a storage capacitor is first charged up so that it has a voltage equal to the peak height of the input pulse. The capacitor is then linearly discharged by a constant current so producing a ramp waveform, and during this period a high-frequency clock oscillator is switched on (Figure 3.26). Thus the period of the discharge, and the number of cycles of the clock are proportional to the magnitude of the input pulse. The number of clock pulses recorded during the ramp gives the channel number, and after counting these in a register the classification can be recorded, usually in a ferrite-core memory.

A later development is the use of the successive approximation analogue-to-digital converter, due to Gatti, Kandiah, etc., which provides improved channel stability and resolution. ADCs are further discussed in Chapter 1.

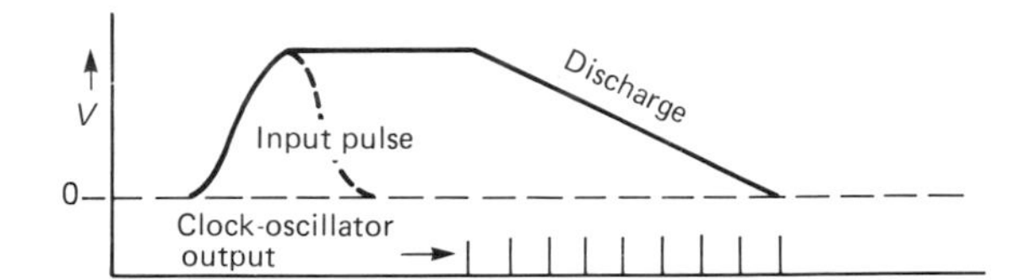

Figure 3.26 Principle of Wilkinson ADC using linear discharge of a capacitor.

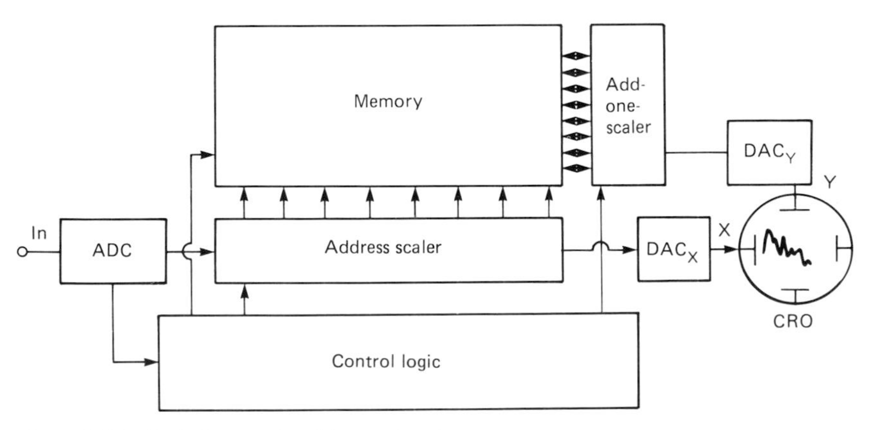

Figure 3.25 Block diagram of multichannel analyser.

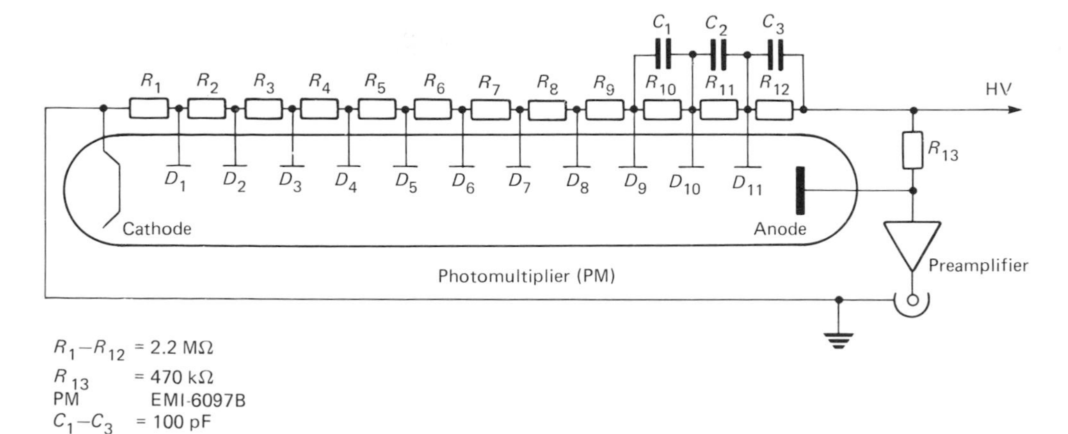

Figure 3.27 Dynode resistor chain for counting rates up to about 15 000 c/s.

3.3.6 Special electronic units

3.3.6.1 Dynode resistor chains

Each photomultiplier requires a resistor chain to feed each dynode an appropriate voltage to allow the electrons ejected by the scintillator light flash to be accelerated and multiplied at each dynode. For counting rates up to about 10^5 per second, the resistors are usually equal in value and high resistance, so that the total current taken by the chain of resistors is of the order of a few hundred microamperes. Figure 3.27 shows a typical dynode resistor chain. As has already been pointed out, the high voltage supplying the dynode chain must be extremely stable and free from voltage variations, spurious pulses, etc. A 0.1 per cent change in voltage may give nearly 1 per cent change of output. The mean current taken by the chain is small, and the fitting of bypass capacitors allows pulses of current higher in value than the standing current to be supplied to the dynodes, particularly those close to the anode where the multiplied electron cascade, which started from the photocathode, has now become a large number due to the multiplication effect. However, as the number of pulses to be counted becomes more than about 15 000–50 000 per second, it is necessary to increase the standing current through the dynode chain. Otherwise space charge effects cause the voltages on the dynodes to drop, so reducing the gain in the photomultiplier. When the counting rate to be measured is high, or very fast rise times have to be counted, then the dynode current may have to be increased from a few hundred microamperes to some 5–10 MA, and a circuit as shown in Figure 3.28 is used. Photomultipliers are also discussed in Chapter 2.

3.3.6.2 Adders/mixers

When a number of detector signals have to be combined into a single output, a mixer (or fan-in) unit is used. This sums the signals from up to eight

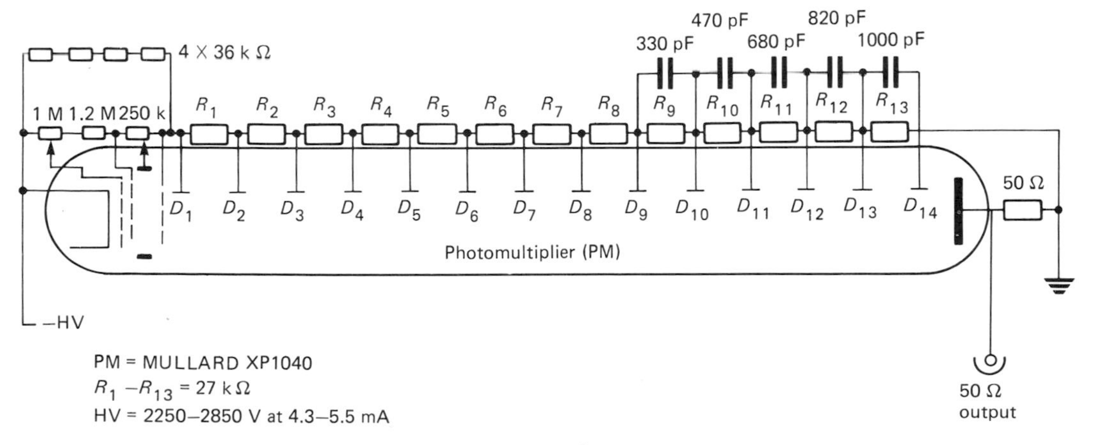

Figure 3.28 Dynode resistor chain for high counting rates (~10^6 c/s).

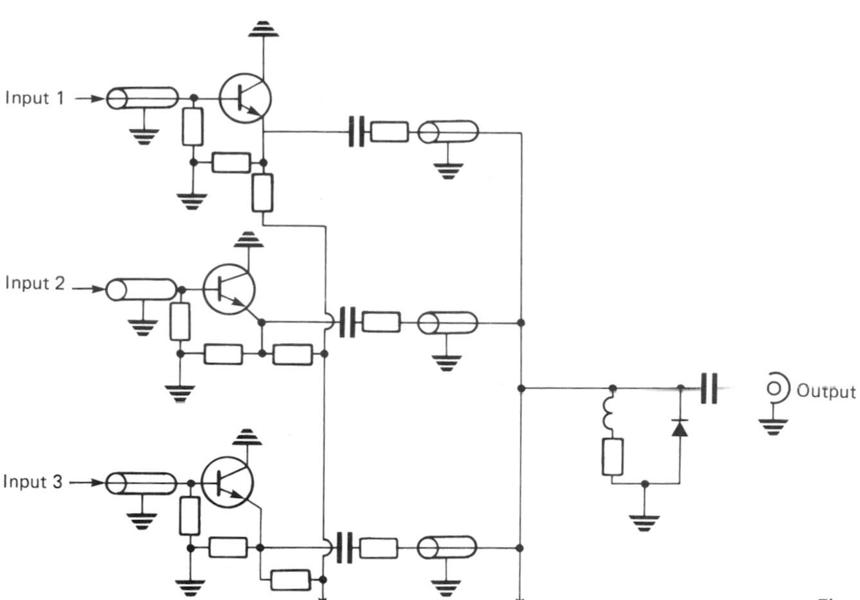

Figure 3.29 Fast-signal mixer/adder.

preamplifiers or amplifiers to give a common output. This is used, for example, when the outputs from several photomultipliers mounted on one large scintillator must be combined. Such a unit is also used in whole-body counters, where a number of separate detectors are distributed around the subject being monitored. Figure 3.29 shows the circuit of such a unit.

3.3.6.3 Balancing units

These units are, in effect, potentiometer units which allow the outputs from a number of separate detectors, often scintillation counters with multiple-photomultiplier assemblies on each large crystal, to be balanced before entering the main amplifier of a system. This is especially necessary when pulse-height analysis is to be performed, as otherwise the variations in output would prevent equal pulse heights being obtained for given energy of a spectrum line.

3.3.6.4 Coincidence and anti-coincidence circuits

A coincidence circuit is a device with two or more inputs which gives an output signal when all the inputs occur at the same time. These circuits have a finite *resolving time*—that is, the greatest interval of time τ which may elapse between signals for the circuit still to consider them coincident. Figure 3.30 shows a simple coincidence circuit where diodes are used as switches. If either or both diodes are held at zero potential then the relevant diode or diodes conduct and the output of the circuit is close to ground. However, if both inputs are caused to rise to the supply voltage level V_c by simultaneous application of pulses of height V_c, then both diodes cease to conduct and the output rises to the supply level for as long as the input pulses are present. An improved circuit is shown in Figure 3.31.

The use of coincidence circuits arose as a result of studies of cosmic rays, since it allowed a series of counters to be used as a telescope to determine the direction of path of such high-energy particles. By the use of highly absorbing slabs between counters, the nature and energies of these cosmic particles and the existence of showers of simultaneous particles were established. The anti-coincidence circuit was used in these measurements to determine the energies of particles which were absorbed in dense material such as lead, having triggered a telescope of counters before entering the lead, but did not trigger counters below the lead slab.

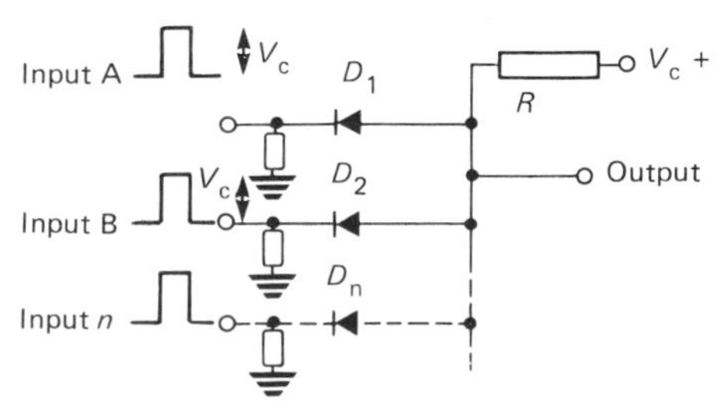

Figure 3.30 Simple coincidence circuit.

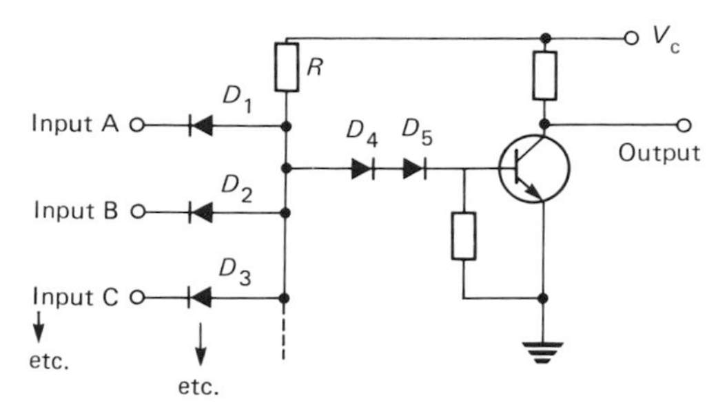

Figure 3.31 Improved coincidence circuit.

Nowadays coincidence circuits are used with detectors for the products of a nuclear reaction or particles emitted rapidly in cascade during radioactive decay or the two photons emitted in the annihilation of a positron. The latter phenomenon has come into use in the medical scanning and analysis of human living tissue by computer-activated tomography (CAT-scanning).

Anti-coincidence circuits are used in low-level counting by surrounding a central main counter, such as would be used in radiocarbon-dating measurements, with a guard counter such that any signal occurring simultaneously in the main and guard counters would NOT be counted, but only signals originating solely in the main central counter. An anti-coincidence circuit is shown in Figure 3.32, while Figure 3.33 gives a block diagram of the whole system.

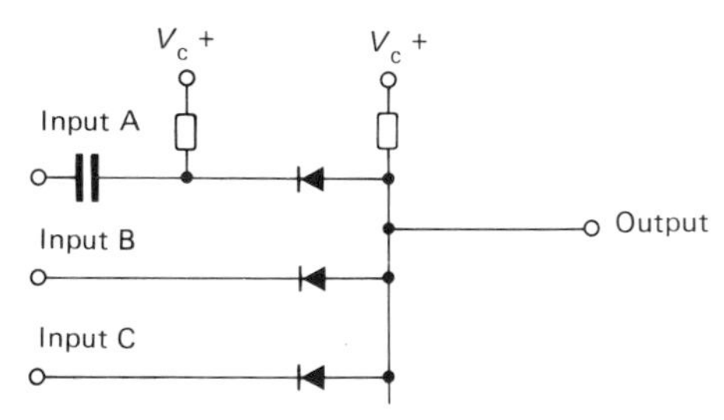

Figure 3.32 Anti-coincidence circuit.

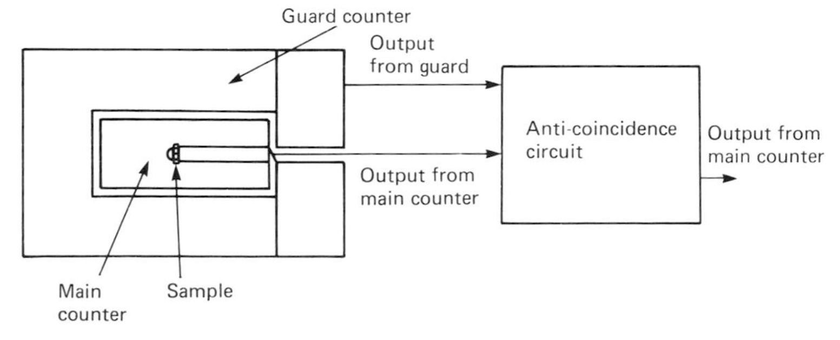

Figure 3.33 Use of anti-coincidence circuit.

3.4 References

Birks, J. B. *The Theory and Practice of Scintillation Counting*, Pergamon Press, Oxford (1964)

Dearnaley, G. and Northrup, D. C. *Semiconductor Counters for Nuclear Radiations*, Spon, London (1966)

Eichholz, G. G. and Poston, J. W. *Principles of Nuclear Radiation Detection*, Wiley, Chichester (1979)

Fremlin, J. H. *Applications of Nuclear Physics*, English Universities Press, London (1964)

Heath, R. L. *Scintillation Spectrometry Gamma-Ray Spectrum Catalogue*, Vols I and II: USAEC Report IDO-16880, Washington (1964)

Hoffer, P. B., Beck, R. N. and Gottschalk, A (eds) *Semiconductor Detectors in the Future of Nuclear Medicine*, Society of Nuclear Medicine, New York (1971)

Knoll, G. F., *Radiation Detection and Measurement*, Wiley, Chichester (1979)

McDowell, W. J., in *Liquid Scintillation Counting, Recent Applications and Developments* (ed. Peng, C. T.), Academic Press, London (1980)

Overman, R. T. and Clark, H. M. *Radioisotope Techniques*, McGraw-Hill, New York (1960)

Price, W. J. *Nuclear Radiation Detection*, McGraw-Hill, New York (1964)

Segrè, E., *Experimental Nuclear Physics*, Vol. III, Wiley, Chichester (1959)

Sharpe, J. *Nuclear Radiation Detection*, Methuen, London (1955)

Snell, A. H. (ed) *Nuclear Instruments and their Uses*, Vol. I, *Ionization Detectors*, Wiley, Chichester (1962) (Vol. II was not published)

Taylor, D. *The Measurement of Radio Isotopes*, Methuen, London (1951)

Turner, J. C. *Sample Preparation for Liquid Scintillation Counting*, The Radiochemical Centre, Amersham (1971, revised)

Watt, D. E. and Ramsden, D., *High Sensitivity Counting Techniques*, Pergamon Press, Oxford (1964)

Wilkinson, D. H. *Ionization Chambers and Counters*, Cambridge University Press, Cambridge (1950)

Zumwalt, L. R. *Absolute Beta Counting using End-Window Geiger–Muller Counters and Experimental Data on Beta-Particle Scattering Effects*, USAEC Report AECU-567 (1950)

4 Measurements employing nuclear techniques

D. ALIAGA KELLY

4.1. Introduction

There are two important aspects of using nuclear techniques in industry which must be provided for before any work takes place. These are:

(1) Compliance with the many legal requirements when using or intending to use radioactive sources; and
(2) Adequate health physics procedures and instruments to ensure that the user is meeting the legal requirements.

The legal requirements cover the proposed use of the radioactive source, the way in which it is delivered to the industrial site, the manner in which it is used, where and how it is stored when not in use and the way it is disposed of, either through waste disposal or return to the original manufacturer of the equipment or the source manufacturer. The local factory inspector must be advised when it is proposed to use a radioactive source on any industrial premises, and early contact with him will invariably result in the provision of useful advice on correctly complying with the legal requirements. The Legal Department of the UK Atomic Energy Authority produces a very useful list of current legal documents covering all aspects of radiation sources (Sim, 1986).

In order to comply with the legal requirements it will be necessary to establish a system of health physics measurements, first, to protect all personnel who may be in the vicinity of the radiation source, and second, to show proof that the radiation limits laid down by law are complied with. This generally calls for all persons concerned to carry suitable Film/thermoluminescent dosimeter badges which are obtainable from the National Radiological Protection Board; the films are generally processed monthly and the TLDs at longer intervals. In addition, depending on the type of radiation source used, a suitable monitoring instrument will be required to measure dose rate to allow the safe limits of operation to be defined. (See Chapter 3 for details.)

It is impossible to cover exhaustively all the applications of radioisotopes in this book; here we deal with the following:

(1) Density;
(2) Thickness;
(3) Level;
(4) Flow;
(5) Tracer applications;
(6) Material analysis.

Before considering these applications in detail we discuss some general points that are relevant to all or most of them.

One of the outstanding advantages of radioisotopes is the way in which their radiation can pass through the walls of a container to a suitable detector without requiring any contact with the substance being measured. This means that dangerous liquids or gases may be effectively controlled or measured without risk of leakage, either of the substance itself out from the container or of external contamination to the substance inside. Thus in the chemical industry many highly toxic materials may be measured and controlled completely without risk of leakage from outside the pipes, etc., conveying it from one part of the factory to another. (See also section 4.1.4.)

Another important advantage in the use of radioisotopes is that the measurement does not affect, for example, the flow of liquid or gas through a pipe, and such flow can be completely unimpeded. Thus the quantity of tobacco in a cigarette-making machine can be accurately determined as the continuous tube of paper containing the tobacco moves along, so that each cigarette contains a fixed amount of tobacco.

Speed is another important advantage of the use of radioisotopes. Measurements of density, level, etc. may be carried out continuously so that processes may be readily controlled by the measuring signal derived from the radioisotope system. Thus, a density gauge can control the mixing of liquids or solids, so that a constant density material is delivered from the machine. Speed in determining flow allows, for example, measurement of the flow of cooling gas to a nuclear reactor and observation of small local changes in flow due to obstructions which would be imperceptible using normal flow-measuring instruments.

The penetrating power of radiations from

radioisotopes is particularly well known in its use for gamma radiography. A tiny capsule of highly radioactive cobalt, for example, can be used to radiograph complex but massive metal castings where a conventional X-ray machine would be too bulky to fit. Gamma radiography is discussed further in Chapter 5. Also, leaks in pipes buried deep in the ground can be found by passing radioactive liquid along them and afterwards monitoring the line of the pipe for radiation from the active liquid which has soaked through the leak into the surrounding earth. If one uses a radioisotope with a very short half-life (half-life is the time taken for a particular radioisotope to decay to half its initial activity) the pipeline will be free of radioactive contamination in a short time, allowing repairs to be carried out without hazard to the workmen repairing it or to the domestic consumer when the liquid is, for example, the local water supply.

4.1.1 Radioactive measurement relations

When radiations from radioactive isotopes pass through any material, they are absorbed according to (1) their energy and (2) the density and type of material. This absorption follows in general the relationship

$$I = I_0 B \exp - (\mu_L x) \qquad (4.1)$$

where I_0 is the intensity of the incident radiation, I the intensity of the radiation after passing through the material, x the thickness of material (cm), μ_L the linear absorption coefficient (cm^{-1}) and B the build-up factor.

The *absorption coefficient* is a factor which relates the energy of the radiation and the density and type of material, and suitable tables are available (Hubbell) from which this factor may be obtained for the particular conditions of source and material under consideration. As the tables are usually given in terms of the *mass absorption coefficient* (μ_m) (generally in cm^2/g) it is useful to know that the *linear absorption coefficient* (μ_L) (in cm^{-1}) may be derived by multiplying the *mass absorption coefficient* (μ_m) (in cm^2/g) by the density (ρ) of the material (in g/cm^3). It must be borne in mind that in a mixture of materials each will have a different *absorption coefficient* for the same radiation passing through the mixture.

The *build-up factor*, B, is necessary when dealing with gamma- or X-radiation, where scattering of the incident radiation can make the intensity of the radiation which actually falls on the detector different from what it would be if no scattering were to take place. For electrons or beta particles this factor can be taken equal to 1. The complication of gamma ray absorption is illustrated by the non-linearity of the curves in Figure 4.1, which gives the thickness of different materials needed to effect a ten-fold attenuation in a narrow beam.

From equation (4.1) we can obtain some very useful information in deciding on the optimum conditions for making a measurement on a particular material, or, conversely, if we have a particular radioactive source, we can determine what are its limits for measuring different materials.

First, it can be shown that the maximum sensitivity for a density measurement is obtained when

$$x = 1/\mu_L \qquad (4.2)$$

that is, when the thickness of material is equal to the reciprocal of the *linear absorption coefficient*. This reciprocal is also called the *mean-free path*, and it can be shown that for any thickness of a particular substance there is an optimum type and intensity of radioactive source. For very dense materials, and for thick specimens, a source emitting high-energy radiation will be required. Therefore ^{60}Co, which emits gamma rays of 1.33 and 1.17 MeV, is frequently used. At the other end of the scale, for the measurement of very thin films such as, for example, *Melinex*, a soft beta or alpha particle emitter would be chosen. For measurement of thickness it is generally arranged to have two detectors and one source. The radiation from the source is allowed to fall on one detector through a standard piece of the material to be measured, while the other detector measures the radiation from the source which has passed through the sample under test. The signal from each of the pair of detectors is generally combined as two d.c. levels, the difference between which drives a potentiometer-type pen recorder.

4.1.2 Optimum time of measurement

The basic statistics of counting were outlined in the previous chapter (p. 95). We now consider how that is applied to particular measurements.

Suppose that the number of photons or particles detected per second is n and that the measurement is required to be made in a time t. The number actually recorded in t s will be $nt \pm \surd(nt)$, where $\surd(nt)$ is the standard deviation of the measurement according to Poisson statistics and is a measure of the uncertainty of the true value of nt. The relative uncertainty (coefficient of variation) is given by

$$\surd(nt)/(nt) = 1/\surd(nt)$$

A radioisotope instrument is used to measure some quality X of a material in terms of the output I of a radiation detector. The instrument sensitivity, or relative sensitivity, S, is defined as the ratio of the fractional change $\delta I/I$ in detector output which results from a given fractional change $\delta X/X$ in the quality

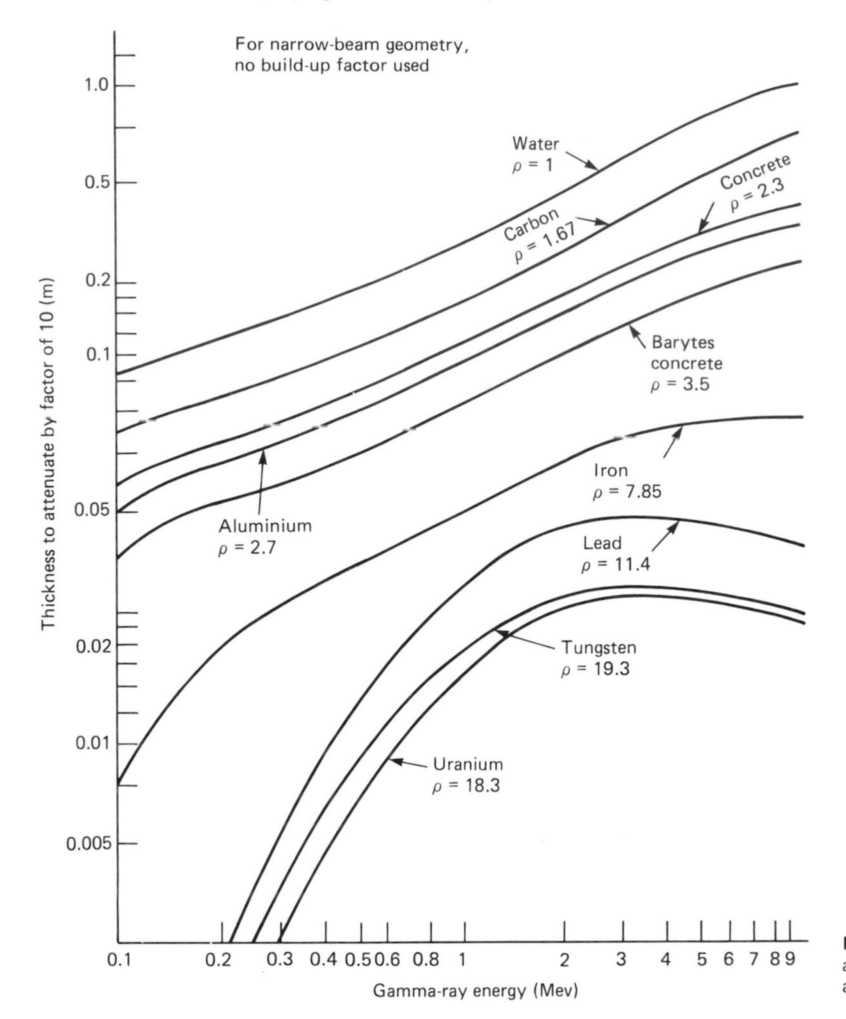

Figure 4.1 Thickness needed to attenuate a narrow gamma-ray beam by a factor of 10.

being measured, i.e.

$$S = \frac{\delta I}{I} \Big/ \frac{\delta X}{X} \tag{4.3}$$

If, in a measurement, the only source of error is the statistical variation in the number of recorded events, the coefficient of variation in the value of the quality measures

$$\frac{\delta X}{X} = \frac{1}{S} \cdot \frac{\sqrt{(nt)}}{nt} = \frac{1}{S\sqrt{(nt)}} \tag{4.4}$$

To reduce this to as small a value as possible then S, n or t or all three of these variables should be increased to as high a value as possible. In many cases, however, the time available for measurement is short. This is particularly true on high-speed production lines of sheet material, where only a few milliseconds may be available for the measurement.

It can now be seen how measurement time, collimation, detector size and absorber thickness may affect the error in the measurement. The shorter the measurement time, the greater the degree of collimation, the thicker the absorber and the smaller the detector, the greater will be the source activity required to maintain a constant error. A larger source will be more expensive, and in addition its physical size may impose a limit on the activity usable. Bearing in mind that a source radiates equally in all directions, only a very small fraction can be directed by collimation for useful measurement—the rest is merely absorbed in the shielding necessary to protect the source.

4.1.3 Accuracy/precision of measurements

The precision or reproducibility of a measurement is

defined in terms of the ability to repeat measurements of the same quantity. Precision is expressed quantitatively in terms of the standard deviation, σ, from the average value obtained by repeated measurements. In practice it is determined by statistical variations in the rate of emission from the radioactive source, instrumental instabilities and variations in measuring conditions.

The accuracy of a measurement is an expression of the degree of correctness with which an actual measurement yields the true value of the quantity being measured. It is expressed quantitatively in terms of the deviation from the true value of the mean of repeated measurements. The accuracy of a measurement depends on the precision and also on the accuracy of calibration. If the calibration is exact, then in the limit, accuracy and precision are equal. When measuring a quantity such as thickness it is relatively easy to obtain a good calibration. In analysing many types of samples, on the other hand, the true value is often difficult to obtain by conventional methods and care may have to be taken in quoting the results.

In general, therefore, a result is quoted along with the calculated error in the result and the confidence limits to which the error is known. Confidence limits of both one standard deviation, 1 σ (68 per cent of results lying within the quoted error), and two standard deviations, 2σ (95 per cent of results lying within the quoted error), are used.

In analytical instruments, when commenting on the smallest quantity or concentration which can be measured, the term 'limit of detection' is preferred. This is defined as the concentration at which the measured value is equal to some multiple of the standard deviation of the measurement.

In practice, the accuracy of radioisotope instruments used to measure the thickness of materials is generally within ± 1 per cent, except for very light-weight materials, when it is about ± 2 per cent. Coating thickness can usually be measured to about the same accuracy. Level gauges can be made sensitive to a movement of the interface of ± 1 mm. Gauges used to measure the density of homogeneous liquids in closed containers generally can operate to an accuracy of about ± 0.1 per cent, though some special instruments can reduce the error to ± 0.01 per cent. The accuracy of bulk density gauges is in the range ± 0.5 to ± 5 per cent, depending on the application and on the measuring conditions.

4.1.4 Measurements on fluids in containers

Nuclear methods may be used to make certain measurements on fluids flowing in pipes from 12.7 mm to 1 m in diameter. For plastics or thin-walled pipes up to 76 mm in diameter the combined unit of source and detector shown in Figure 4.2(a) is used, while for larger pipes the system consisting of a separate holder and detector shown in Figure 4.2(b) is used.

The gamma-ray source is housed in a shielded container with a safety shutter so that the maximum dose rate is less than 7.5 μGy/h, and is mounted on one side of the pipe or tank. A measuring chamber containing argon at 20 atm is fitted on the other side of the pipe or tank. It is fitted with a standardizing holder and has a detection sensitivity of ± 0.1 per cent.

The system may be used to measure density over a range of 500–4000 kg/m^3 with a sensitivity of 0.5 kg/m^3, specific gravity with an accuracy of ± 0.0005, percentage solids down to ± 0.05 per cent, and moisture content of slurries of a constant specific gravity to within ± 0.25 per cent.

The principal of the measurement is that the degree of absorption of the gamma rays in the flowing fluid is

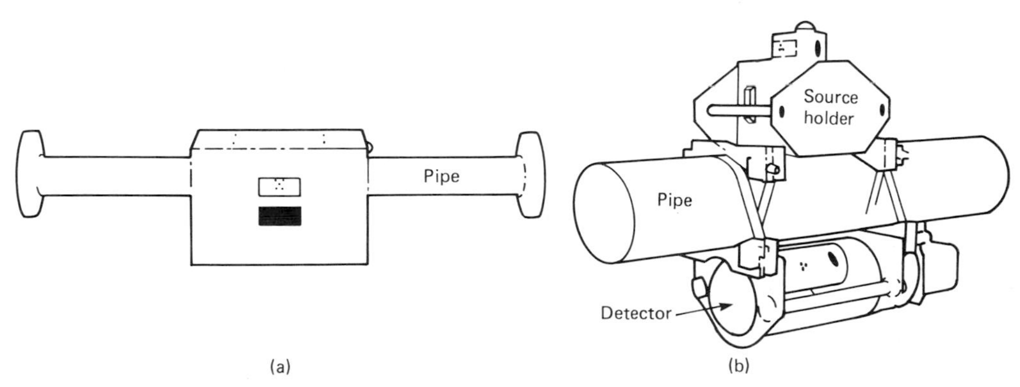

Figure 4.2 Fluid-density measuring systems. Courtesy, Nuclear Enterprises Ltd. (a) Combined detector/source holder; (b) separate units for larger pipes.

measured by the ionization chamber, where output is balanced against an adjustable source which is set by means of the calibrated control to the desired value of the material being measured. Deviations from this standard value are then shown on the calibrated meter mounted on the indicator front panel. Standardization of the system for the larger pipes is performed manually and a subsidiary source is provided for this purpose. The selection of the type of source depends on (1) the application, (2) the wall thickness and diameter of the pipe and (3) the sensitivity required. Sources in normal use are ^{137}Cs (source life 30 yr), ^{241}Am (460 yr) and ^{60}Co (5 yr).

The measuring head has a temperature range of -10 to $+55$°C and the indicator 5–40°C, and the response time minimum is 0.1 s, adjustable to 25 s.

4.2 Materials analysis

Nuclear methods of analysis, particularly neutron activation analysis, offer the most sensitive methods available for most elements in nature. However, no one method is suitable for all elements, and it is necessary to select the techniques to be used with due regard to the various factors involved, some of which may be listed as follows:

(1) Element to be detected;
(2) Quantities involved;
(3) Accuracy of the quantitative analysis required;
(4) Costs of various methods available;
(5) Availability of equipment to carry out the analysis to the statistical limits required;
(6) Time required;
(7) Matrix in which the element to be measured is located;
(8) Feasibility of changing the matrix mentioned in (7) to a more amenable substance.

For example, the environmental material sample may be analysed by many of the methods described, but the choice of method must be a compromise, depending on all the factors involved.

4.2.1 Activation analysis

When a material is irradiated by neutrons, photons, alpha or beta particles, protons, etc. a reaction may take place in the material depending on a number of factors. The most important of these are:

(1) The type of particle or photon used for the irradiation;
(2) The energy of the irradiation;
(3) The flux in the material; and
(4) The time of irradiation.

The most useful type of particle has been found to be neutrons, since their neutral charge allows them to penetrate the high field barriers surrounding most atoms and relatively low energies are required. In fact, one of the most useful means of irradiation is the extremely low energy neutrons of thermal energy (0.025 eV) which are produced abundantly in nuclear reactors. The interactions which occur cause some of the various elements present to become radioactive, and in their subsequent decay into neutral atoms again, to emit particles and radiation which are indicative of the elements present. Neutron activation analysis, as this is called, has become one of the most useful and sensitive methods of identifying certain elements in minute amounts without permanently damaging the specimen, as would be the case in most chemical methods of analysis. A detector system can be selected which responds uniquely to the radiation emitted as the excited atoms of the element of interest decay with emission of gamma rays or beta particles, while not responding to other types of radiation or to different energies from other elements which may be present in the sample. The decay half-life of the radiation is also used to identify the element of interest, while the actual magnitude of the response at the particular energy involved is a direct measure of the amount of the element of interest in the sample. Quantitatively the basic relation between A, the induced activity present at the end of the irradiation in Becquerels, i.e. disintegrations per second, is given by

$$A = N\sigma\phi\left[1-\exp\left(-\frac{0.693t_i}{T_{\frac{1}{2}}}\right)\right] \tag{4.5}$$

where N is the number of target atoms present, σ the cross section (cm^2), ϕ the irradiation flux (neutrons cm^{-2} s^{-1}), t_i the irradiation time and $T_{\frac{1}{2}}$ the half-life of product nuclide.

From this we may calculate N, the number of atoms of the element of interest present, after appropriate correction factors have been evaluated.

Beside the activation of the element and measurement of its subsequent decay products one may count directly the excitation products produced whilst the sample is being bombarded. This is called 'prompt gamma-ray analysis', and it has been used, for example, to analyse the surface of the moon.

Most elements do not produce radioactivity when bombarded with electrons, beta particles or gamma rays. However, most will emit characteristic X-rays when bombarded and the emitted X-rays are characteristic of each element present. This is the basis of X-ray fluorescence analysis, discussed below.

Electrons and protons have also been used to excite elements to emit characteristic X-rays or particles, but this technique requires operation in a very high vacuum chamber.

High-energy gamma rays have the property of exciting a few elements to emit neutrons—this occurs with beryllium when irradiated with gamma rays of energy greater than 1.67 MeV, and deuterium with gamma rays of energy greater than 2.23 MeV. This forms the basis of a portable monitor for beryllium prospecting in the field, using an ^{124}Sb source to excite the beryllium. Higher-energy gamma rays from accelerators etc. have the property of exciting many elements but require extremely expensive equipment.

4.2.2 X-ray fluorescence analysis

4.2.2.1 Dispersive X-ray fluorescence analysis

In dispersive X-ray fluorescence analysis the energy spectrum of the characteristic X-rays emitted by the substance when irradiated with X-rays is determined by means of a dispersive X-ray spectrometer, which uses as its analysing element the regular structure of a crystal through which the characteristic X-rays are passed. This property was discovered by Bragg and Bragg in 1913, who produced the first X-ray spectrum by crystal diffraction through a crystal of rock salt.

Figure 4.3 shows what happens when an X-ray is diffracted through a crystal of rock salt. The Braggs

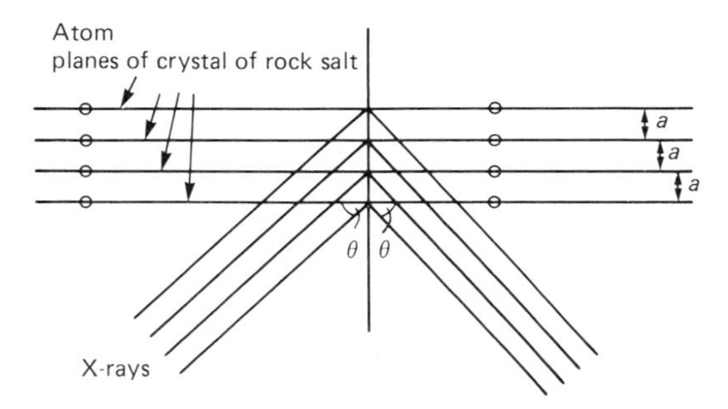

Figure 4.3 Diffraction of X-rays through crystal of rock salt.

showed that the X-rays were reflected from the crystal, meeting the Bragg relationship

$$n\lambda = 2a \sin \theta \qquad (4.6)$$

where λ is the wavelength of the incident radiation, a the distance between lattice planes and n the order of the reflection. θ is the angle of incidence and of reflection, which are equal.

To measure the intensity distribution of an X-ray spectrum by this method the incident beam has to be collimated and the detector placed at the corresponding position on the opposite side of the normal (see Figure 4.4). The system effectively selects all rays which are incident at the appropriate angle for

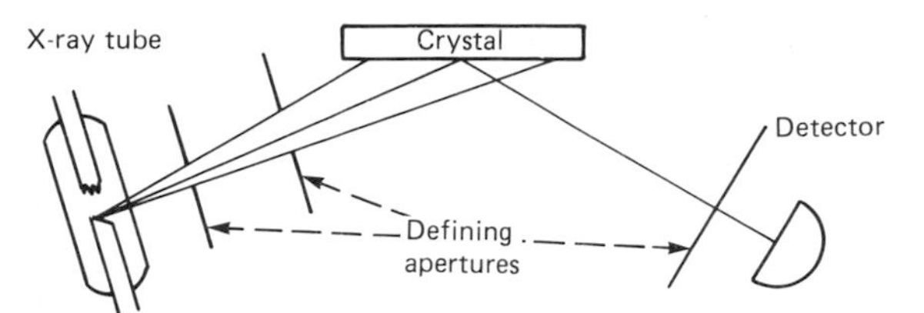

Figure 4.4 Dispersive X-ray spectrometer.

Bragg reflection. If the angle of incidence is varied (this will normally involve rotating the crystal through a controllable angle and the detector by twice this amount) then the detector will receive radiation at a different wavelength and a spectrum will be obtained. If the system is such that the angular range $d\theta$ is constant over the whole range of wavelength investigated, as in the geometry illustrated in Figure 4.4, we can write

$$n\, d\lambda = 2a\, d(\sin \theta) = 2a \cos \theta\, d\theta \qquad (4.7)$$

The intensity thus received will be proportional to $\cos \theta$ and to $d\theta$. After dividing by $\cos \theta$ and correcting for the variation with angle of the reflection coefficient of the crystal and the variation with wavelength of the detector efficiency, the recorded signal will be proportional to the intensity per unit wavelength interval, and it is in this form that continuous X-ray spectra are traditionally plotted.

4.2.2.2 X-ray fluorescence analysis (non-dispersive)

When a substance is irradiated by a beam of X-rays or γ-rays it is found that the elements present fluoresce, giving out X-rays of energies peculiar to each element present. By selecting the energy of the incident X-rays selection may be made of particular elements required. As an example, if a silver coin is irradiated with X-rays, the silver present emits X-rays of energies about 25.5 keV, and if other elements are also present, such as copper or zinc, 'characteristic' X-rays, as they are called, will be emitted with energies of 8.0 and 8.6 keV, respectively.

The Si (Li) or Ge (Li) detector, cooled to the temperature of liquid nitrogen, will separate the various spectral lines and a multichannel analyser will allow their intensity to be evaluated electronically. However, it must be pointed out that as the incident X-rays and the excited, emergent, characteristic X-rays have a very short path in metals the technique essentially measures the elemental content of the metal surface. The exciting X-rays are selected for their energy to exceed what is called the 'K-absorption edge' of the element of interest, so that elements of higher atomic weight will not be stimulated to emit their characteristic X-rays. Unfortunately, the range of

Table 4.1 Exciting sources for X-ray fluorescence analysis

Isotope	*Half-life*	*Principal photon energies (keV)*	*Emission (%)*
^{241}Am	433 yr	11.9–22.3	~40
		59.5	35.3
^{109}Cd	453 d	22.1, 25.0	102.3
		2.63–3.80	~10
		88.0	3.6
^{57}Co	270.5 d	6.40, 7.06	~55
		14.4	9.4
		122.0	85.2
		136.5	11.1
^{55}Fe	2.7 yr	5.89, 6.49	~28
^{153}Gd	241.5 d	41.3, 47.3	~110
		69.7	2.6
		97.4	30
		103.2	20
^{125}I	60.0 d	27.4, 31.1	138
		35.5	7
^{210}Pb	22.3 yr	9.42–16.4	~21
		46.5	~4
		+ Bremsstrahlung to 1.16 MeV	
^{147}Pm	2.623 yr	Characteristic X-rays of target	~0.4
+ target		+ Bremsstrahlung to 225 keV	
^{238}Pu	87.75 yr	11.6–21.7	~13
^{125m}Te	119.7 d	27.4, 31.1	~50
		159.0	83.5
^{170}Tm	128 d	52.0, 59.7	~5
		84.3	3.4
		+ Bremsstrahlung to 968 keV	
Tritium (^{3}H)	12.35 yr		
+ Ti target		4.51, 4.93	~10^{-2}
		+ Bremsstrahlung to 18.6 keV	
+ Zr target		1.79–2.5	~10^{-2}
		+ Bremsstrahlung to 18.6 keV	

exciting sources is limited and Table 4.1 lists those currently used. Alternatively, special X-ray tubes with anodes of suitable elements have been used to provide X-rays for specific analyses as well as intensities greater than are generally available from radioisotope sources.

Gas proportional counters and NaI (Tl) scintillation counters have also been used in non-dispersive X-ray fluorescence analysis, but the high resolution semiconductor detector has been the most important detector used in this work. While most systems are used in a fixed laboratory environment, due to the necessity of operating at liquid-nitrogen temperatures, several portable units are available commercially in which small insulated vessels containing liquid nitrogen, etc. give a period of up to 8 h use before requiring to be refilled.

The introduction of super-pure Ge detectors has permitted the introduction of portable systems which can operate for limited periods at liquid-nitrogen temperatures, but as long as they remain in a vacuum enclosure they can be allowed to rise to room temperature without damage to the detector, as would occur with the earlier Ge (Li) detector, where the lithium would diffuse out of the crystal at ambient temperature. As Si (Li) detectors are really only useful for X-rays up to about 30 keV, the introduction of the Ge (HP) detector allows X-ray non-dispersive fluorescence analysis to be used for higher energies in non-laboratory conditions.

4.2.3 Moisture measurement: by neutrons

If a beam of fast neutrons is passed through a substance any hydrogen in the sample will cause the fast neutrons to be slowed down to thermal energies, and these slow neutrons can be detected by means of a BF_3 or ^{3}He filled gas proportional counter system. As the major amount of hydrogen present in most material is due to water content, and the slowing down is directly proportional to the hydrogen density, this offers a way of measuring the moisture content of a great number of materials. Some elements such as cadmium, boron and the rare-earth elements chlorine and iron, however can have an effect on the measurement of water content since they have high thermal-neutron capture probabilities. When these elements are present this method of moisture measurement has to be used with caution. On the other hand, it provides a means of analysing substances for these elements, provided the actual content of hydrogen and the other elements is kept constant.

Since the BF_3 or ^3He counter is insensitive to fast neutrons it is possible to mount the radioactive fast neutron source close to the thermal neutron detector. Then any scattered thermal neutron from the slowing down of the fast neutrons, which enters the counter, will be recorded.

This type of equipment can be used to measure continuously the moisture content of granular materials in hoppers, bins and similar vessels. The measuring head is mounted external to the vessel and the radiation enters by a replaceable ceramic window (Figure 4.5).

The transducer comprises a radio-isotope source of fast neutrons and a slow neutron detector assembly, mounted within a shielding measuring head. The source is mounted on a rotatable disc. It may thus be positioned by an electropneumatic actuator either in the centre of the shield or adjacent to the slow neutron detector and ceramic radiation 'window' at one end of the measuring head, which is mounted externally on the vessel wall. Fast neutrons emitted from the source are slowed down, mainly by collisions with atoms of hydrogen in the moisture of material situated near the measuring head. The count rate in the detector increases with increasing hydrogen content and is used as an indication of moisture content.

The slow neutron detector is a standard, commercially available detector and is accessible without dismantling the measuring head. The electronics, housed in a sealed case with the measuring head, consist of an amplifier, pulse-height analyser and a line-driver module which feed pre-shaped 5-V pulses to the operator unit.

The pulses from the head electronic unit are processed digitally in the operator's unit to give an analogue indication of moisture content on a 100-mm meter mounted on the front panel, or an analogue or digital signal to other equipment. The instrument has a range of 0–20 per cent moisture and a precision of ± 0.5 per cent moisture, with a response time of 5–25 s. The sample volume is between 100 and 300 l and the operating temperature of the detector is 5–70°C and of the electronics 5–30°C. Figure 4.6 shows such a moisture gauge in use on a coke hopper.

A similar arrangement of detector and source is used

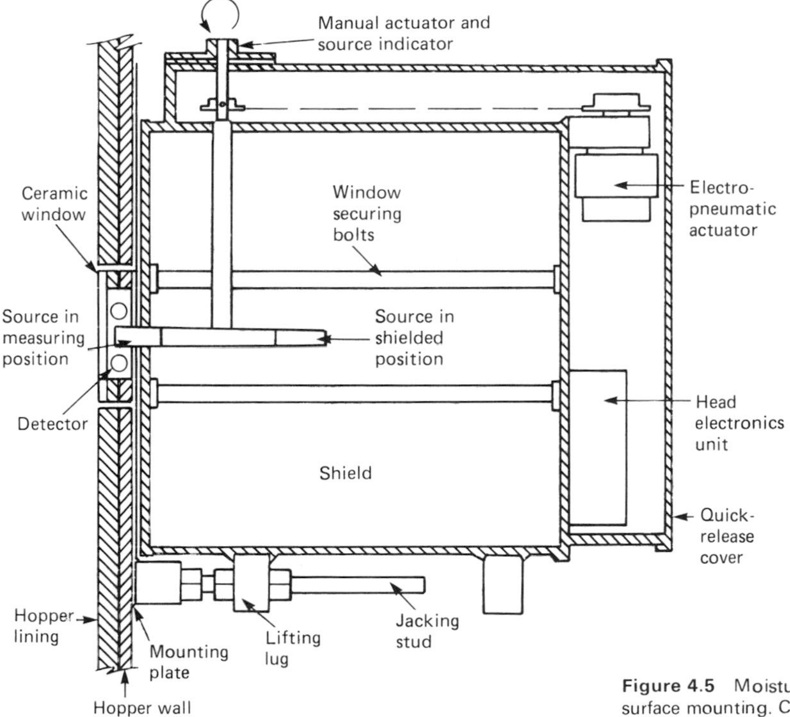

Figure 4.5 Moisture meter for granular material in hoppers, surface mounting. Courtesy, Nuclear Enterprises Ltd.

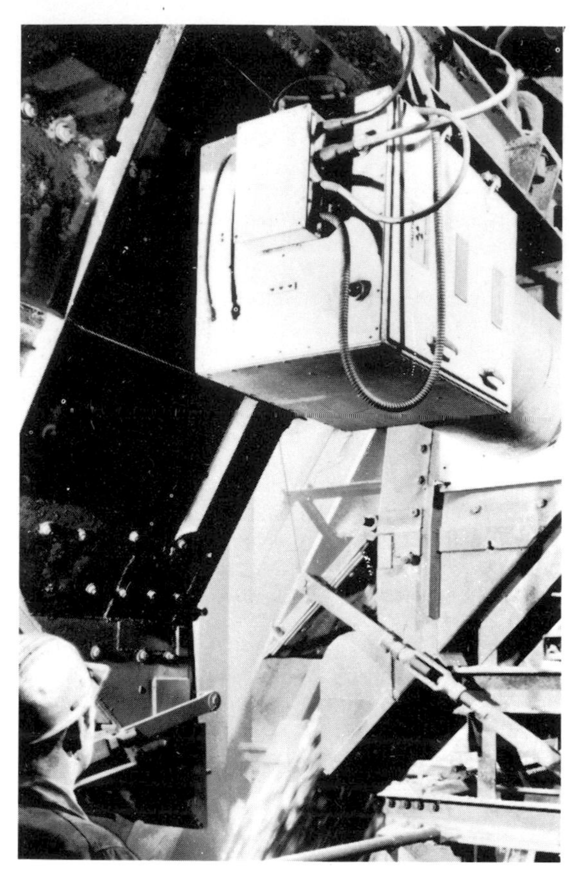

Figure 4.6 Coke moisture gauge in use on hoppers. Courtesy, Nuclear Enterprises Ltd.

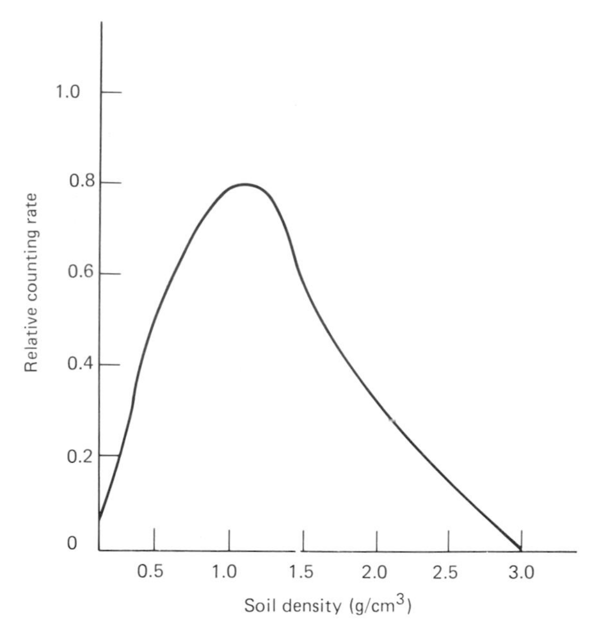

Figure 4.7 Response of a scattered-gamma-ray gauge to soil density.

in borehole neutron moisture meters. The fast neutron source is formed into an annulus around the usually cylindrical thermal neutron detector and the two components are mounted in a strong cylinder of steel, which can be let down a suitable borehole and will measure the distribution of water around the borehole as the instrument descends. This system is only slightly sensitive to the total soil density (provided no elements of large cross-section for absorbing neutrons are present), so only a rough estimate is needed of the total soil density to compensate for its effect on the measured value of moisture content. The water content in soil is usually quoted as percentage by weight, since the normal gravimetric method of determining water content measures the weight of water and the total weight of the soil sample. To convert the water content as measured by the neutron gauge measurement into percentage by weight one must also known the total density of the soil by some independent measurement. This is usually performed, in the borehole case, by a physically similar borehole instrument, which uses the scattering of gamma-rays from a source in the nose of the probe, around a lead plug to a suitable gas-filled detector in the rear of the probe—the lead plug shields the detector from any direct radiation from the source. Figure 4.7 shows the response of the instrument.

At zero density of material around the probe the response of the detector is zero, since there is no material near the detector to cause scattering of the gamma-rays and subsequent detection. In practice, even in free air, the probe will show a small background due to scattering by air molecules. As the surrounding density increases, scattering into the detector begins to occur, and the response of the instrument increases linearly with density but reaches a maximum for the particular source and soil material. The response then decreases until theoretically at maximum density the response will be zero. Since the response goes through a maximum with varying density, the probe parameters should be adjusted so that the density range of interest is entirely on one side of the maximum. Soil-density gauges are generally designed to operate on the negative-slope portion of the response.

4.2.3.1 Calibration of neutron-moisture gauges

Early models of neutron gauges used to be calibrated by inserting them into concrete blocks of known densities, but this could lead to serious error, since the response in concrete is quite different from that in soil. It has been suggested by Ballard and Gardner that in order to eliminate the sensitivity of the gauge to the composition of the material to be tested, one should use non-compactive homogeneous materials of known composition to obtain experimental gauge responses. This data is then fitted by a 'least-squares' technique to an equation which enables a correction to be obtained for a particular probe used in soil of a particular composition.

Dual gauges

Improved gauges have been developed in which the detector is a scintillation counter with the scintillator of cerium-activated lithium glass. As such a detector is sensitive to both neutrons and gamma rays, the same probe may be used to detect both slowed-down neutrons and gamma rays from the source—the two can be distinguished because they give pulses of different shapes. This avoids the necessity of having two probes, one to measure neutrons and the other to measure scattered gammas, allowing the measurement of both moisture and density by pulse-shape analysis.

Surface-neutron gauges are also available, in which both radioactive source and detector are mounted in a rectangular box which is simply placed on a flat soil surface. It is important to provide a smooth flat surface for such measurements, as gaps cause appreciable errors in the response.

4.2.4 Measurement of sulphur contents of liquid hydrocarbons

Sulphur occurs in many crude oils at a concentration of up to 5 per cent by weight and persists to a lesser extent in the refined product. As legislation in many countries prohibits the burning of fuels with a high sulphur content to minimize pollution, and sulphur compounds corrode engines and boilers and inhibit catalysts, it is essential to reduce the concentration to tolerable levels. Thus rapid measurement of sulphur content is essential and the measurement of the absorption of appropriate X-rays provides a suitable on-line method. In general, the mass absorption coefficient of an element increases with increase of atomic number (Figure 4.8) and decreases with shortening the wavelength of the X-rays. In order to make accurate measurement the wavelength chosen should be such that the absorption will be independent of changes in the carbon–hydrogen ratio of the hydrocarbon. When the X-rays used have an energy of 22 keV the mass attenuation for carbon and hydrogen are equal. Thus by using X-rays produced by allowing the radiation from the radio-element ^{241}Am to produce fluorescent excitation in a silver target which gives X-rays having an energy of 23 keV, the absorption is made independent of the carbon–hydrogen ratio. As this source has a half-life of 450 yr, no drift occurs owing to decay of the source. The X-rays are passed through a measuring cell through which the hydrocarbon flows and, as the absorption per unit weight of sulphur is many times greater than the absorption of carbon and hydrogen, the fraction of the X-rays absorbed is a measure of the concentration of the sulphur present. Unfortunately the degree of absorption of X-rays is also affected by the density of the sample and by the concentration of trace elements of high atomic weight and of water.

The concentration of X-rays is measured by a high resolution proportional counter so the accuracy will be a function of the statistical variation in the count

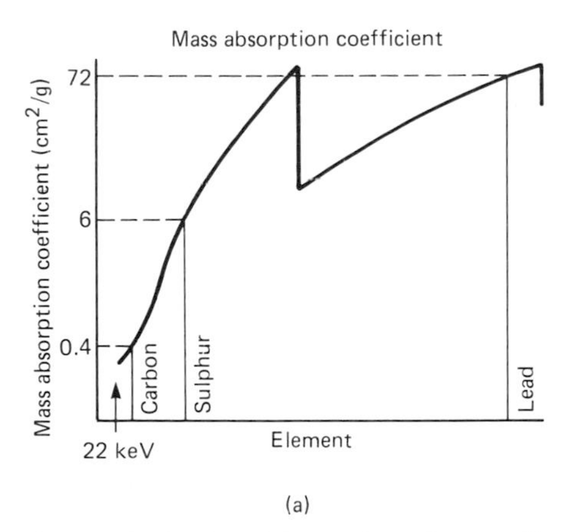

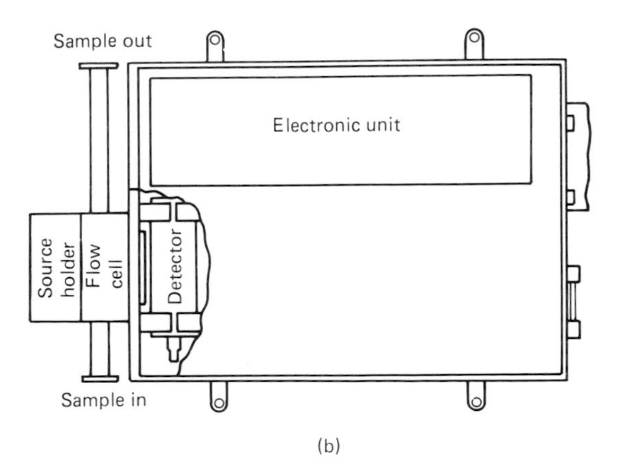

Figure 4.8 On-line sulphur analyser. Courtesy, Nuclear Enterprises Ltd. (a) Mass absorption coefficient; (b) arrangement of instrument.

rate and the stability of the detector and associated electronics, which can introduce an error of ±0.01 per cent sulphur. A water content of 650 ppm will also introduce an error of 0.01 per cent sulphur. Compensation for density variations may be achieved by measuring the density with a non-nucleonic meter and electronically correcting the sulphur signal.

Errors caused by impurities are not serious, since the water content can be reduced to below 500 ppm and the only serious contaminant, vanadium, seldom exceeds 50 ppm.

The stainless steel flow cell has standard flanges and is provided with high-pressure radiation windows and designed so that there are no stagnant volumes. The flow cell may be removed for cleaning without disturbing the source. Steam tracing or electrical heating can be arranged for samples likely to freeze. The output of the high-resolution proportional counter, capable of high count rates for statistical accuracy, is amplified and applied to a counter and digital-to-analogue convertor when required. Thus both digital and analogue outputs are available for display and control purposes.

The meter has a range of up to 0–6 per cent sulphur by weight and indicates with a precision of ±0.01 per cent sulphur by weight or ±1 per cent of the indicated weight, whichever is the larger. It is independent of carbon–hydrogen ratio from 6:1 to 10:1 and the integrating times are from 10 to 200 s. The flow cell is suitable for pressures up to 15 bar and temperatures up to 150°C, the temperature range for the electronics being −10 to +45°C.

The arrangement of the instrument is as shown in Figure 4.8.

4.2.5 The radio-isotope calcium monitor

The calcium content of raw material used in cement manufacture may be measured on-line in either dry powder or slurry form. The basis of the method is to measure the intensity of the characteristic K X-rays emitted by the flowing sample using a small ^{55}Fe radio-isotope source as the means of excitation. This source is chosen because of its efficient excitation of Ca X-rays in the region which is free from interference by iron.

In the form of instrument shown in Figure 4.9(a), used for dry solids, a sample of material in powder form is extracted from the main stream by a screw conveyor and fed into a hopper. In the powder-presentation unit the powder is extracted from the hopper and fed on to a continuously weighed sample presenter at a rate which is controlled so as to maintain a constant mass of sample per unit area on the latter within very close limits. After measurement, the sample is returned to the process. A system is fitted to provide an alarm if the mass per unit area of sample wanders outside preset limits, and aspiration is provided to eliminate airborne dust in the vicinity of the sample presenter and measuring head.

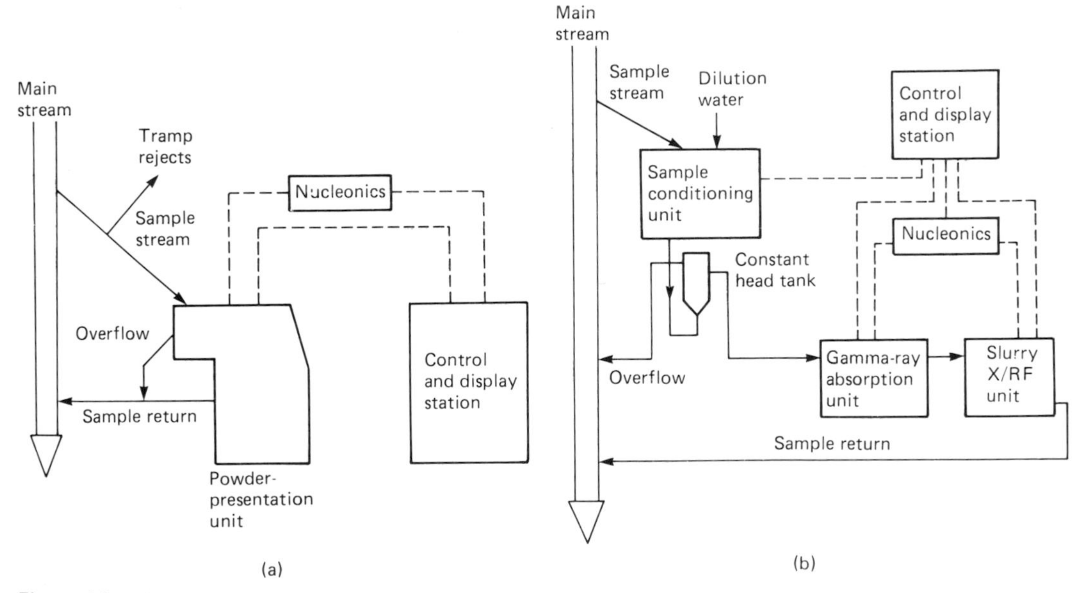

Figure 4.9 Block diagrams of calcium in cement raw material measuring instrument. Courtesy, Nuclear Enterprises Ltd. (a) Dry powder form of instrument; (b) slurry form.

Under these conditions it is possible to make precise reproducible X-ray fluorescence measurements of elements from atomic number 19 upwards without pelletizing.

The signal from the X-ray detector in the powder-presentation unit is transmitted via a head amplifier and a standard nucleonic counting chain to a remote display and control station. Analogue outputs can be provided for control purposes.

In the form used for slurries shown in Figure 4.9(b) an additional measurement is made of the density and hence the solids content of the slurry. The density of the slurry is measured by measuring the absorption of a highly collimated 660 keV gamma-ray beam. At this energy the measurement is independent of changes in solids composition.

The dry powder instrument is calibrated by comparing the instrument readings with chemical analysis carried out under closely controlled conditions with the maximum care taken to reduce sampling errors. This is best achieved by calibrating while the instrument is operating in closed loop with a series of homogeneous samples recirculated in turn. This gives a straight line relating percentage of calcium carbonate to the total X-ray count.

With slurry a line relating percentage calcium carbonate to X-ray count at each dilution is obtained producing a nomogram which enables a simple special purpose computer to be used to obtain the measured value indication or signal.

In normal operation the sample flows continuously through the instrument and the integrated reading obtained at the end of 2–5 min, representing about 2 kg of dry sample or 30 litres of slurry, is a measure of the composition of the sample.

An indication of CaO to within ± 0.15 per cent for the dry method and ± 0.20 per cent for slurries should be attainable by this method.

4.2.6 Wear and abrasion

The measurement of the wear experienced by mechanical bearings, pistons in the cylinder block or valves in internal combustion engines is extremely tedious when performed by normal mechanical methods. However, wear in a particular component may be easily measured by having the component irradiated in a neutron flux in a nuclear reactor to produce a small amount of induced radioactivity. Thus the iron in for example, piston rings, which have been activated and fitted to the pistons of a standard engine, will perform in an exactly similar way to normal piston rings, but when wear takes place, the active particles will be carried around by the lubrication system, and a suitable radiation detector will allow the wear to be measured, as well as the distribution of the particles in the lubrication system and the efficiency of the oil filter for removing such particles.

To measure the wear in bearings, one or other of the bearing surfaces is made slightly radioactive, and the amount of activity passed to the lubricating system is a measure of the wear experienced.

4.2.7 Leak detection

Leakage from pipes buried in the ground is a constant problem with municipal authorities, who may have to find leaks in water supplies, gas supplies or sewage pipes very rapidly and with the minimum of inconvenience to people living in the area, as essential supplies may have to be cut off until the leak is found and the pipe made safe again.

To find the position of large leaks in water distribution pipes two methods have been developed. The first uses an inflatable rubber ball with a diameter nearly equal to that of the pipe and containing 100 or so MBq of ^{24}Na which is inserted into the pipe after the leaking section has been isolated from the rest of the system. The only flow is then towards the leak and the ball is carried as far as the leak where it stops. As ^{24}Na emits a high-energy gamma ray its radiation can be observed on the surface through a considerable thickness of soil etc., by means of a sensitive portable detector. Alternatively, radioactive tracer is introduced directly into the fluid in the pipe.

After a suitable period the line of the pipe can be monitored with a sensitive portable detector, and the build-up of activity at the point of the leak can be determined. ^{24}Na is a favoured radioactive source for leak testing, especially of domestic water supply or sewage leaks, since it has a short half-life (15 h), emits a 2.7 MeV gamma ray and is soluble in water as ^{24}Na Cl. Thus, leak tests can be rapidly carried out and the activity will have decayed to safe limits in a very short time.

4.3 Mechanical measurements

4.3.1 Level measurement

4.3.1.1 Using X- or gamma-rays

Level measurements are usually made with the source and detector fixed in position on opposite sides of the outer wall of the container (Figure 4.10). Because many containers in the chemical engineering and oil-

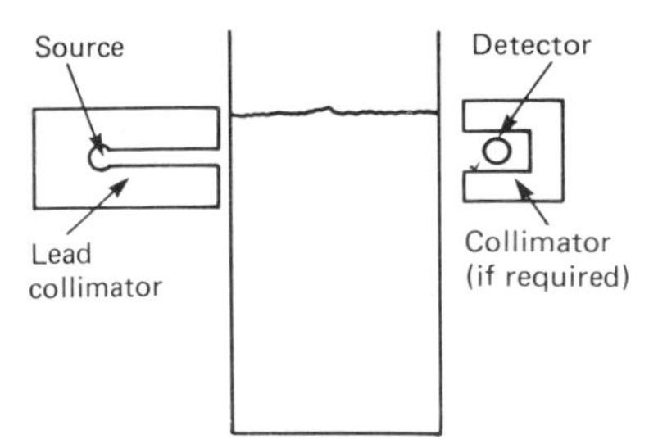

Figure 4.10 Level gauge (fixed).

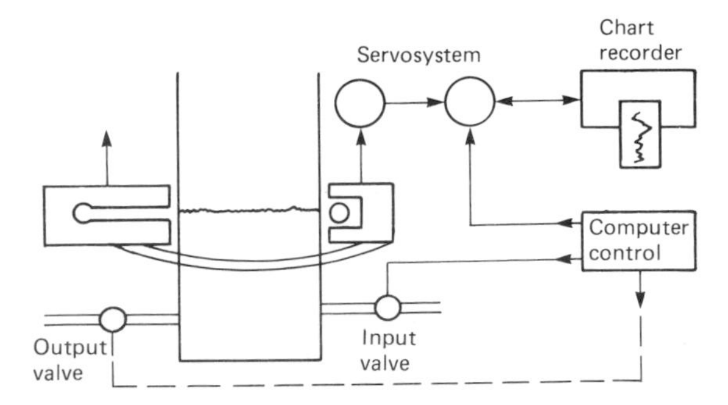

Figure 4.12 Level gauge (continuous) with automatic control.

refining industries, where most level gauges are installed, have large dimensions, high-activity sources are required and these have to be enclosed in thick lead shields with narrow collimators to reduce scattered radiation which could be a hazard to people working in the vicinity of such gauges. Because of cost, Geiger counters are the most usual detectors used, though they are not as efficient as scintillation counters. The important criterion in the design of a level gauge is to select a radioactive source to give the optimum path difference signal when the material or liquid to be monitored just obscures the beam from source to detector. The absorption of the beam by the wall of the container must be taken into account, as well as the absorption of the material being measured. A single source and single detector will provide a single response, but by using two detectors with a single source (Figure 4.11), it may be able to provide three readings—low (both detectors operating), normal (one detector operating) and high (neither detector operating). This system is used in papermaking systems to measure the level of the hot pulp.

Another level gauge to give a continuous indication of material or liquid level has the detector and the source mounted on a servocontrolled platform which follows the level of the liquid or material in the container (Figure 4.12). This provides a continuous readout of level, and it is possible to use this signal to control the amount of material or liquid entering or leaving the container in accordance with some pre-programmed schedule.

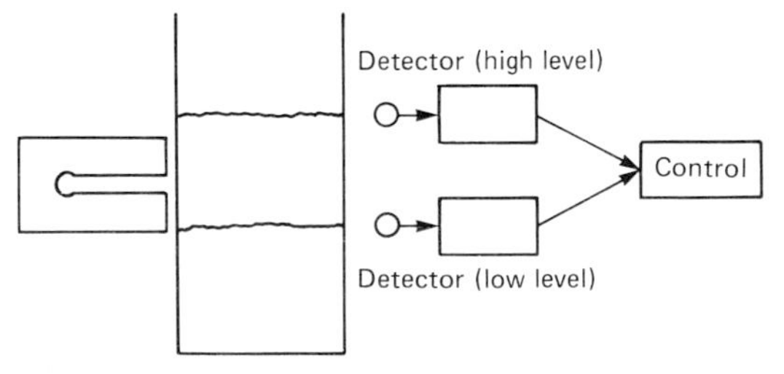

Figure 4.11 Dual detector level gauge.

Another level gauge uses a radioactive source inside the container but enclosed in a float which rises and falls with the level of the liquid inside the container. An external detector can then observe the source inside the container, indicate the level and initiate refilling procedure to keep the level at a predetermined point.

Portable level gauges consisting of radioactive source and Geiger detector in a hand-held probe, with the electronics and counting display in a portable box, have been made to detect the level of liquid CO_2 in high-pressure cylinders. This is a much simpler method than that of weighing the cylinder and subtracting this value from the weight when the cylinder was originally received.

4.3.1.2 Using neutrons

Some industrial materials have a very low atomic number (Z), such as oils, water, plastics, etc., and by using a beam of fast neutrons from a source such as ^{241}Am/Be or ^{252}Cf, and a suitable thermal-neutron detector such as cerium-activated lithium glass in a scintillation counter or $^{10}BF_3$ or ^{3}He gas-filled counters it is possible to measure the level of such material (Figure 4.13). Fast neutrons from the source

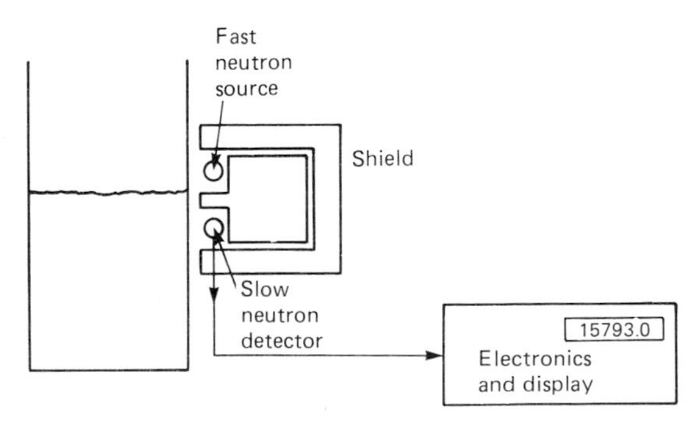

Figure 4.13 Level measurement by moderation of fast neutrons.

are moderated or slowed down by the low-Z material and some are scattered back into the detector. By mounting both the source of fast neutrons and the slow neutron detector at one side of the vessel the combination may be used to follow a varying level using a servocontrolled mechanism, or in a fixed position to control the level in the container to a preset position. This device is also usable as a portable detector system to find blockages in pipes, valves, etc., which often occur in plastics-manufacturing plants.

4.3.2 Measurement of flow

There are several methods of measuring flow using radioactive sources, as follows.

4.3.2.1 Dilution method

This involves injection of a liquid containing radioactivity into the flow line at a known constant rate; samples are taken further down the line where it is known that lateral mixing has been completed. The ratio of the concentrations of the radioactive liquid injected into the line and that of the samples allows the flow rate to be computed.

4.3.2.2 The 'Plug' method

This involves injecting a radioactive liquid into the flow line in a single pulse. By measuring the time this 'plug' of radioactive liquid takes to pass two positions a known distance apart, the flow can be calculated.

A variation of the 'plug' method again uses a single pulse of radioactive liquid injected into the stream, but the measurement consists in taking a sample at a constant rate from the line at a position beyond which full mixing has been completed. Here the flow rate can be calculated by measuring the average concentration of the continuous sample over a known time.

4.3.3 Mass and thickness

Since the quantitative reaction of particles and photons depends essentially on the concentration and mass of the particles with which they are reacting it is to be expected that nuclear techniques can provide means for measuring such things as mass. We have already (p. 132) referred to the measurement of density of the material near a borehole. We now describe some other techniques having industrial uses.

4.3.3.1 Measurement of mass, mass per unit area and thickness

The techniques employed in these measurements are basically the same. The radiation from a gamma ray source falls on the material and the transmitted radiation is measured by a suitable detector. In the nucleonic belt weigher shown in Figure 4.14, designed to measure the mass flow rate of granular material such as iron ore, limestone, coke, cement, fertilizers, etc., the absorption across the total width of the belt is

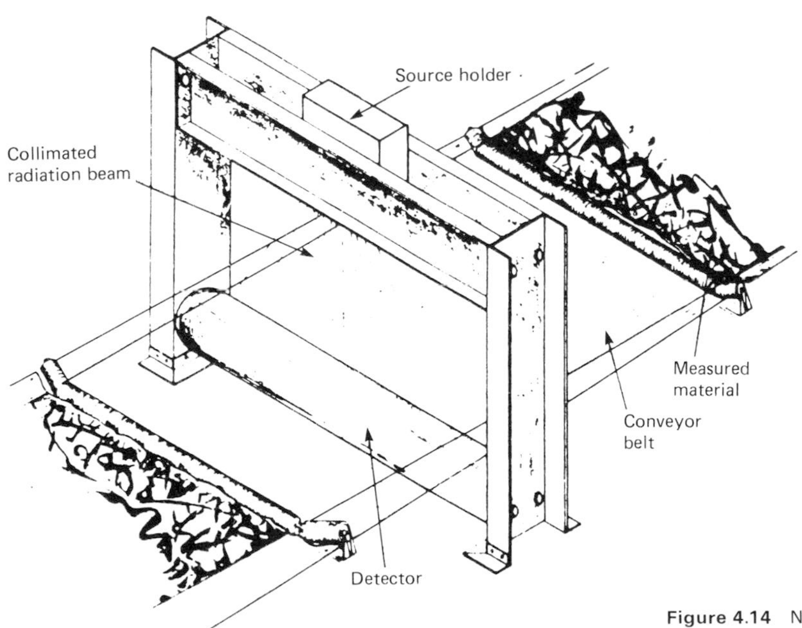

Figure 4.14 Nucleonic belt weigher. Courtesy, Nuclear Enterprises Ltd.

measured. The signal representing the total radiation falling on the detector is processed with a signal representing the belt speed by a solid-state electronic module and displayed as a mass flow rate and a total mass. The complete equipment comprises a C-frame assembly housing the source, consisting of ^{137}Cs enclosed in a welded steel capsule mounted in a shielding container with a radiation shutter, and the detector, a scintillation counter whose sensitive length matches the belt width, housed in a cylindrical flameproof enclosure suitable for Groups 11 A and B gases, with the preamplifier. A calibration plate is incorporated with the source to permit a spot check at a suitable point within the span. In addition, there is a dust- and moisture-proof housing for the electronics which may be mounted locally or up to 300 m from the detector.

The precision of the measurement is better than ± 1 per cent, and the operating temperature of the detector and electronics is -10 to $+40$°C. The detector and preamplifier may be serviced by unclassified staff, as the maximum dose rate is less than 7.5 μGy/h.

Similar equipment may be used to measure mass per unit area by restricting the area over which the radiation falls to a finite area, and if the thickness is constant and known the reading will be a measure of the density.

4.3.3.2 Measurement of coating thickness

In industry a wide variety of processes occur where it is necessary to measure and sometimes automatically control the thickness of a coating applied to a base material produced in strip form. Examples of such processes are the deposition of tin, zinc or lacquers on steel, or adhesives, wax, clay bitumen or plastics to paper, and many other processes.

By nucleonic methods measurement to an accuracy of ± 1 per cent of coating thickness can be made in a wide variety of circumstances by rugged equipment capable of a high reliability. Nucleonic coating-thickness gauges are based on the interaction of the radiation emitted from a radio-isotope source with the material to be measured. They consist basically of the radio-isotope source in a radiation shield and a radiation detector contained in a measuring head, and an electric console.

When the radiation emitted from the source is incident on the subject material, part of this radiation is scattered, part is absorbed and the rest passes through the material. A part of the absorbed radiation excites characteristic fluorescent X rays in the coating and/or backing.

Depending on the measurement required, a system is used in which the detector measures the intensity of scattered, transmitted or fluorescent radiation. The intensity of radiation monitored by the detector is a measure of the thickness (mass per unit area) of the coating. The electric console contains units which process the detector signal and indicate total coating thickness and/or deviation from the target thickness. The measuring head may be stationary or programmed to scan across the material. Depending on the type and thickness of coating and base materials, and machine details, one of four gauge types is selected: differential beta transmission, beta backscatter, X-ray fluorescence and preferential absorption.

Differential beta-transmission gauge (Figure 4.15)

The differential beta-transmission gauge is used to measure coatings applied to base materials in sheet form when the coating has a total weight of not less than about one-tenth of the weight of the base material, when both sides of the base and coated material are accessible and when the composition of coating and base is fairly similar. Here the thickness (mass per unit area) of the coating is monitored by measuring first the thickness of the base material before the coating is applied, followed by the total thickness of the material with its coating and then subtracting the former from the latter. The difference provides the coating thickness. The readings are obtained by passing the uncoated material through one measuring head and the coated material through the other, the coating being applied between the two positions. The intensity of radiation transmitted through the material is a measure of total thickness. Separate meters record the measurement determined by each head, and a third meter displays the difference between the two readings, which corresponds to the coating thickness.

Typical applications of this gauge are the measurements of wax and plastics coatings applied to paper and aluminium sheet or foil, or abrasives to paper or cloth.

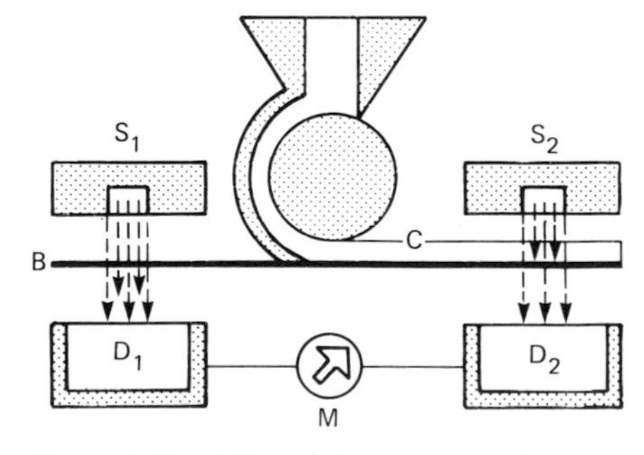

Figure 4.15 Differential beta-transmission gauge. Courtesy, Nuclear Enterprises Ltd. S_1: first source; D_1: first detector; S_2: second source; D_2: second detector; B: base material; C: coating; M: differential measurement indicator.

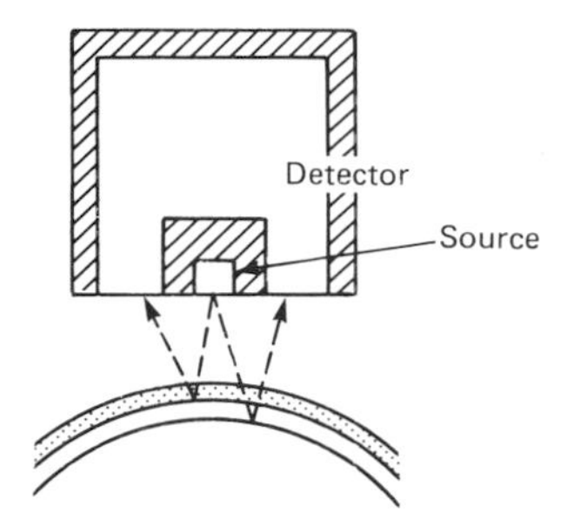

Figure 4.16 Beta-backscatter gauge. Courtesy, Nuclear Enterprises Ltd.

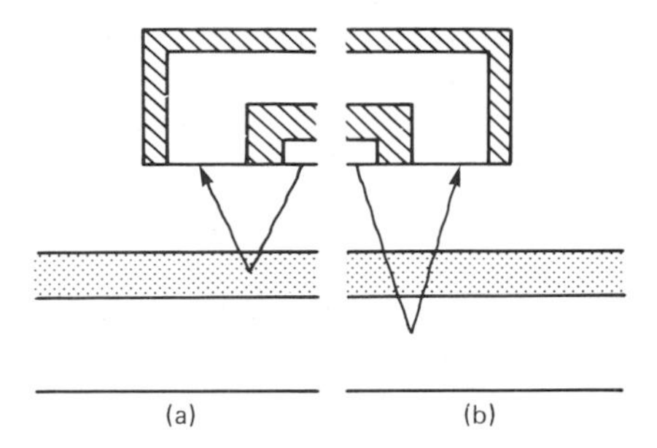

Figure 4.17 X-ray fluorescence gauge. Courtesy, Nuclear Enterprises Ltd. (a) Coating-fluorescence gauge which monitors the increase in intensity of X-rays excited in coating as its thickness increases; (b) backing-fluorescence gauge monitors the decrease in intensity of radiation excited in the backing material as the coating thickness increases.

Beta-backscatter gauge (Figure 4.16)

The beta-backscatter gauge is used to measure coating thickness when the process is such that the material is only accessible from one side and when the coating and backing material are of substantially different atomic number. The radio-isotope source and the detector are housed in the same enclosure. Where radiation is directed, for example, on to an uncoated calender roll it will be backscattered and measurable by the detector. Once a coating has been applied to the roll the intensity of the backscattered radiation returning to the detector will change. This change is a measure of the thickness of the coating. Typical applications of this gauge are the measurement of rubber and adhesives on calenders, paper on rollers, or lacquer, paint or plastics coatings applied to sheet steel.

Measurement of coating and backing by X-ray fluorescence

X-ray fluorescent techniques, employing radio-isotope sources to excite the characteristic fluorescent radiation, are normally used to measure exceptionally thin coatings. The coating-fluorescence gauge monitors the increase in intensity of an X-ray excited in the coating as the coating thickness is increased. The backing-fluorescence gauge excites an X-ray in the backing or base material and measures the decrease in intensity due to attenuation in the coating as the coating thickness is increased. The intensity of fluorescent radiation is normally measured with an ionization chamber, but a proportional or scintillation counter may sometimes be used.

By the use of compact geometry, high efficiency detectors and a fail-safe radiation shutter, the dose rates in the vicinity of the measuring head are kept well below the maximum permitted levels, ensuring absolute safety for operators and maintenance staff.

Figure 4.17(a) illustrates the principle of the coating-fluorescence gauge, which monitors the increase in intensity of an X-ray excited in the coating as the coating thickness is increased. Figure 4.17(b) illustrates the principle of the backing-fluorescence gauge which monitors the decrease in intensity of an X-ray excited in the backing material as the coating thickness increases.

The instrument is used to measure tin, zinc, aluminium and chromium coatings applied to sheet steel, or titanium coatings to paper or plastics sheet.

The preferential absorption gauge (Figure 4.18)

This gauge is used when the coating material has a higher mean atomic number than the base material. The gauge employs low-energy X-rays from a sealed radio-isotope source which are absorbed to a much greater extent by materials with a high atomic number, such as chlorine, than by materials such as paper or textiles, which have a low atomic number. It is thus possible to monitor variations in coating thickness by measuring the degree of preferential X-

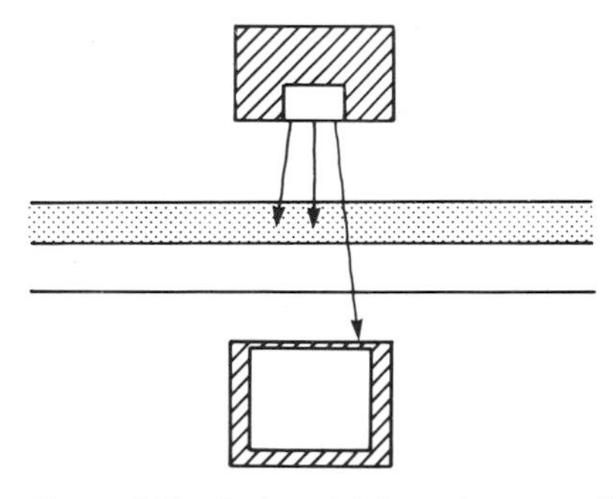

Figure 4.18 Preferential absorption gauge. Courtesy, Nuclear Enterprises Ltd.

ray absorption by the coating, using a single measuring head. The instrument is used to measure coatings which contain clay, titanium, halogens, iron or other substance with a high atomic number which have been applied to plastics, paper or textiles.

4.4 Miscellaneous measurements

4.4.1 Field-survey instruments

In prospecting for uranium, portable instruments are used (1) in aircraft, (2) in trucks or vans, (3) hand held and (4) for undersea surveys. Uranium is frequently found in the same geological formations as oil, and uranium surveys have been used to supplement other more conventional methods of surveying, such as seismic analyses. The special case of surveying for beryllium-bearing rocks was discussed in a previous section (p. 129).

As aircraft can lift large loads, the detectors used in such surveys have tended to be relatively large NaI (Tl) groups of detectors. For example one aircraft carried four NaI (T!) assemblies each 29.2 cm diameter and 10 cm thick, feeding into a four- or five-channel spectrometer which separately monitored the potassium, uranium, thorium and the background. Simultaneously a suitable position-finding system such as Loran-C is in operation, so that the airborne plot of the radioactivities, as the aircraft flies over a prescribed course, is printed with the position of the aircraft onto the chart recorder. In this way large areas of land or sea, which may contain ground-based survey teams with suitable instruments can survey the actual areas pinpointed in the aerial survey as possible sources of uranium.

4.4.1.1 Land-based radiometrical surveys

Just as the aircraft can carry suitable detector systems and computing equipment, so can land-based vehicles, which can be taken to the areas giving high-uranium indications. While similar electronics can usually be operated from a motor vehicle, the detectors will have to be smaller, especially if the terrain is very rugged, when manually portable survey monitors will be called for. These also can now incorporate a small computer, which can perform the necessary analyses of the signals received from potassium, uranium, thorium and background.

4.4.1.2 Undersea surveys

Measurement of the natural gamma radiation from rocks and sediments can be carried out using a towed seabed gamma-ray spectrometer. The spectrometer, developed by UKAEA Harwell in collaboration with the Institute of Geological Sciences, has been used over the last five years and has traversed more than 10 000 km in surveys of the United Kingdom continental shelf. It consists of a NaI (Tl) crystal-photomultiplier detector assembly, containing a crystal 76 mm diameter × 76 mm or 127 mm long, with EMI type 9758 photomultiplier together with a preamplifier and high-voltage generator which are potted in silicone rubber. The unit is mounted in a stainless steel cylinder, which in turn is mounted in a 30-m long flexible PVC hose 173 mm diameter, which is towed by a cable from the ship, and also contains suitable ballast in the form of steel chain to allow the probe to be dragged over the surface of the seabed without becoming entangled in wrecks or rock outcrops.

The electronics on board the ship provide four channels to allow potassium, uranium and thorium as well as the total gamma radioactivity to be measured and recorded on suitable chart recorders and teletypes, and provision is also made to feed the output to a computer-based multichannel analyser.

4.4.2 Dating of archaeological or geological specimens

4.4.2.1 Radiocarbon dating by gas proportional or liquid scintillation counting

The technique of radiocarbon dating was discovered by W. F. Libby and his associates in 1947, when they were investigating the radioactivity produced by the interaction of cosmic rays from outer space with air molecules. They discovered that interactions with nitrogen in the atmosphere produced radioactive ^{14}C which quickly transformed into $^{14}CO_2$, forming about 1 per cent of the total CO_2 in the world. As cosmic rays have been bombarding the earth at a steady rate over millions of years, forming some two atoms of ^{14}C per square centimetre of the earth's surface per second, then an equilibrium stage should have been reached, as the half-life (time for a radioactive substance to decay to half its original value) of ^{14}C is some 5000 yr. As CO_2 enters all living matter the distribution should be uniform. However, when the human being, animal, tree or plant dies, CO_2 transfer ceases, and the carbon already present in the now-dead object is fixed. The ^{14}C in this carbon will therefore start to decay with a half-life of some 5000 yr, so that measurement of the ^{14}C present in the sample allows one to determine, by the amount of ^{14}C still present, the time elapsed since the death of the person, animal, tree or plant. We expect to find two disintegrations per second for every 8 g of carbon in living beings or dissolved in sea water or in the atmosphere CO_2 for the total carbon in these three categories adds to 8 (7.5 in the oceans, 0.125 in the air, 0.25 in life forms and perhaps 0.125 in humus).

There are several problems associated with radiocarbon dating which must be overcome before one can arrive at an estimated age for a particular sample. First, the sample must be treated so as to release the ^{14}C in a suitable form for counting. Methods used are various, depending on the final form in which the sample is required and whether it is to be counted in a gas counter (Geiger or proportional) or in a liquid-scintillation counter.

One method is to transform the carbon in the sample by combustion into a gas suitable for counting in a gas proportional counter. This can be carbon dioxide (CO_2), methane (CH_4) or acetylene (C_2H_2). The original method used by Libby, in which the carbon sample was deposited in a thin layer inside the gas counter, has been superseded by the gas-combustion method. In this the sample is consumed by heating in a tube furnace or, in an improved way, in an oxygen bomb. The gas can be counted directly in a gas proportional counter, after suitable purification, as CO_2 or CH_4, or it can be transformed into a liquid form such as benzene, when it can be mixed with a liquid scintillator and measured in a liquid-scintillation counter.

Counting systems

When one considers that there are, at most, only two ^{14}C disintegrations per second from each 8 g of carbon, producing two soft beta particles ($E_{max} = 0.156$ MeV), one can appreciate that the counting is, indeed, very 'low-level'. The natural counting rate (unshielded) of a typical gas proportional counter 15 cm diameter × 60 cm long would be some 75.6 counts per second, whereas the signal due to 1 atm of live modern CO_2 or CH_4 filling the counter would be only 0.75 counts per second. In order to achieve a standard deviation of 1 per cent in the measurement, 10 000 counts would have to be measured, and at the rate of 0.75 counts per second the measurement would take 3.7 days.

It is immediately apparent that the natural background must be drastically reduced and, if possible, the sample size increased to improve the counting characteristics (Watt and Ramsden, 1964).

Background is due to many causes, some of the most important being:

(1) Environmental radioactivity from walls, air, rocks, etc.;
(2) Radioactivities present in the shield itself;
(3) Radioactivities in the materials used in the manufacture of the counters and associated devices inside the shield;
(4) Cosmic rays;
(5) Radioactive contamination of the gas or liquid scintillator itself;
(6) Spurious electronic pulses, spikes due to improper operation or pick-up of electromagnetic noise from the electricity mains supply, etc.

Calculation of a radiocarbon date

Since the measurement is to calculate the decay of ^{14}C, we have the relation

$$I = I_0 \exp - (\lambda t) \qquad (4.8)$$

where I is the activity of the sample when measured, I_0 the original activity of the sample (as reflected by a modern standard), λ the decay constant $= 0.693/T_{\frac{1}{2}}$ (where $T_{\frac{1}{2}} =$ half-life) and t the time elapsed. If $T_{\frac{1}{2}} =$ 5568 yr (the latest best value found for the half-life of ^{14}C is 5630 yr, but internationally it has been agreed that all dates are still referred to 5568 yr to avoid the confusion which would arise if the volumes of published dates required revision) then equation (4.8) may be re-written as

$$t = 8033 \log_e \frac{S_s - B}{S_0 - B} \qquad (4.9)$$

where S_s is the count rate of sample, S_0 the count rate of modern sample and B the count rate of dead carbon. A modern carbon standard of oxalic acid, 95 per cent of which is equivalent to 1890 wood, is used universally, and is available from the National Bureau of Standards in Washington, DC.

Anthracite coal, with an estimated age of 2×10^9 yr, can be used to provide the dead carbon background. Corrections must also be made for isotopic fractionation which can occur both in nature and during the various chemical procedures used in preparing the sample.

Statistics of carbon dating

The standard deviation σ of the source count rate when corrected for background is given by

$$\sigma = \left[\frac{S}{T - t_b} + \frac{B}{t_b}\right]^{\frac{1}{2}} \qquad (4.10)$$

where S is the gross count of sample plus background, B the background counted for a time t_b, T the total time available.

In carbon dating $S \approx 2B$ and the counting periods for sample and background are made equal, usually of the order of 48 h. Thus if $t = t_b = T/2$ and $S = D + B$ we have

$$\sigma_D = \left(\frac{D + 2B}{t}\right)^{\frac{1}{2}} \qquad (4.11)$$

The maximum age which can be determined by any

specific system depends on the minimum sample activity which can be detected. If the '2σ criterion' is used, the minimum sample counting rate detectable is equal to twice the standard deviation and the probability that the true value of D lies within the region $D \pm 2\sigma_D$ is 95.5 per cent. Some laboratories prefer to use the '4σ criterion', which gives a 99.99 per cent probability that the true value is within the interval $\pm 4\sigma$. If $D_{min} = 2\sigma_D$, then

$$D_{min} = 2\sqrt{\left(\frac{D_{min} + 2B}{t}\right)} \qquad (4.12)$$

then the maximum dating age T_{max} which can be achieved with the system can be estimated as follows. From equation (4.9)

$$T_{max} = \frac{T_{\frac{1}{2}}}{\log_e 2} \log_e \frac{D_0}{D_{min}}$$

$$= \frac{T_{\frac{1}{2}}}{\log_e 2} \log_e \left[\frac{D_0\sqrt{t}}{2\sqrt{(D_{min} + 2B)}}\right] \qquad (4.13)$$

As $D_{min} \ll 2B$ the equation can be simplified to

$$T_{max} = \frac{T_{\frac{1}{2}}}{\log_e 2} \log_e \left[\frac{D_0}{\sqrt{B}}\sqrt{t/8}\right] \qquad (4.14)$$

Where D_0 is the activity of the modern carbon sample, corrected for background. The ratio $D_0/\sqrt{B}$ is considered the factor of merit for a system.

For a typical system such as that of Libby (1985),

$t = 48$ h and $T_{\frac{1}{2}} = 5568$ yr

$D_0 = 6.7$ cpm and $B = 5$ cpm, so

$D_0/\sqrt{B} \approx 3$

Hence

$$T_{max} = \frac{5568}{\log_e 2} \log_e \left[3\sqrt{\left(\frac{48 \times 60}{8}\right)}\right]$$

$$= 8034 \times 4.038$$

$$= 32442 \text{ yr}$$

Calibration of the radiocarbon time scale

A number of corrections have to be applied to dates computed from the radiocarbon decay measurements to arrive at a 'true' age. First, a comparison of age by radiocarbon measurement with the age of wood known historically initially showed good agreement in 1949. When the radiocarbon-dating system was improved by the use of CO_2 proportional counters, in which the sample was introduced directly into the counter, closer inspection using many more samples showed discrepancies in the dates calculated by various methods. Some of these discrepancies can be accounted for, such as the effect on the atmosphere of all the burning of wood and coal since the nineteenth century—Suess called this the 'fossil-fuel effect'. Alternative methods of dating, such as dendrochronology (the counting of tree rings), thermoluminescent dating, historical dating, etc., have demonstrated that variations do occur in the curve relating radiocarbon dating and other methods. Ottaway (1983) describes in more detail the problems and the present state of the art.

4.4.3 Static elimination

Although not strictly instrumentation, an interesting application of radioactive sources is in the elimination of static electricity.

In a number of manufacturing processes static eelectricity is produced, generally, by friction between, for example, a sheet of paper and the rollers used to pass it through a printing process. This can cause tearing at the high speeds that are found in printing presses. In the weaving industry, when a loom is left standing overnight, the woven materials and the warp threads left in the loom remain charged for a long period. It is found that dust particles become attracted to the cloth, producing the so-called 'fog-marking', reducing thereby the value of the cloth. In rubber manufacture for cable coverings, motor-car tyres, etc. the rubber has to pass through rollers, where the static electricity so generated causes the material to remain stuck to the rollers instead of moving to the next processing stage.

All these static problems can be overcome, or at least reduced by mounting a suitable radioactive source close to the place where static is produced, so that ions of the appropriate sign are produced in quantities sufficient to neutralize the charges built up by friction, etc.

A wide variety of sources are now available in many shapes to allow them to be attached to the machines to give optimum ionization of the air at the critical locations. Long-strip sources of tritium, $^{90}Sr/^{90}Y$ and ^{241}Am are the most popular. The importance of preventing static discharges has been highlighted recently in two fields. The first is in the oil industry, where gas-filled oil tankers have exploded due to static discharges in the empty tanks. The second is in the microchip manufacturing industry, where static discharges can destroy a complete integrated circuit.

4.5 References

Clayton, C. G. and Cameron, J. F. *Radioisotope Instruments*, Vol. 1, Pergamon Press, Oxford (1971) (Vol. 2 was not published). This has a very extensive bibliography for further study

Gardner, R. P. and Ely, R. L. Jr. *Radioisotope Measurement Applications in Engineering*, Van Nostrand, New York (1967)

Hubbell, J. H. *Photon Cross-sections, Attenuation Co-efficients, and Energy Absorption Co-efficients from 10 keV to 100 GeV*, NS RDS NBS 29 (1969)

Libby, W. F. *Radiocarbon Dating*, University of Chicago Press (1955)

Ottaway, B. S. (ed.), 'Archaeology, Dendrochronology and the Radiocarbon Calibration Curve', *Edinburgh University Occasional Paper No. 9* (1983)

Shumilovskii, N. N. and Mel'ttser, L. V. *Radioactive Isotopes in Instrumentation and Control*, Pergamon Press, Oxford (1964)

Sim, D. F. *Summary of the Law Relating to Atomic Energy and Radioactive Substances*, UK Atomic Energy Authority (yearly). Includes (a) *Radioactive Substances Act 1960*; (b) *Factories Act—The Ionizing Radiations (Sealed Sources) Regulations 1969*. (HMSO Stat. Inst. 1969, No. 808); (c) *Factories Act—The Ionizing Radiations Regulations*, HMSO (1985)

Watt, D. E. and Ramsden, D. *High Sensitivity Counting Techniques*, Pergamon Press, Oxford (1964)

5 Non-destructive testing

SCOTTISH SCHOOL OF NON-DESTRUCTIVE TESTING

5.1 Introduction

The driving force for improvements and developments in non-destructive testing instrumentation is the continually increasing need to demonstrate the integrity and reliability of engineering materials, products and plant. Efficient materials manufacture, the assurance of product quality and re-assurance of plant at regular intervals during use represent the main need for non-destructive testing (NDT). This 'state-of-health' knowledge is necessary for both economic and safety reasons. Indeed, in the UK the latter reasons have been strengthened by legislation such as the Health and Safety at Work Act 1974.

Failures in engineering components generally result from a combination of conditions, the main three being inadequate design, incorrect use or the presence of defects in materials. The use of non-destructive testing seeks to eliminate the failures caused predominantly by defects. During manufacture these defects may, for example, be shrinkage and porosity in castings, laps and folds in forgings, laminations in plate material, lack of penetration and cracks in weldments. Alternatively, with increasing complexity of materials and conditions of service, less obvious factors may require to be controlled through NDT. For example, these features may be composition, microstructure and homogeneity.

Non-destructive testing is not confined to manufacture. The designer and user may find application to on-site testing of bridges, pipelines in the oil and gas industries, pressure vessels in the power-generation industry and in-service testing of nuclear plant, aircraft and refinery installations. Defects at this stage may be deterioration in plant due to fatigue and corrosion.

The purpose of non-destructive testing during service is to look for deterioration in plant to ensure that adequate warning is given of the need to repair or replace. Periodic checks also give confidence that 'all is well'.

In these ways, therefore, non-destructive testing plays an important role in the manufacture and use of materials. Moreover, as designs become more adventurous and as new materials are used, there is less justification for relying on past experience. Accordingly, non-destructive testing has an increasingly important role.

Methods of non-destructive testing are normally categorized in terms of whether their suitability is primarily for the examination of the surface features of materials (Figure 5.1) or the internal features of materials (Figure 5.2). Closely allied to this is the sensitivity of each method, since different situations invariably create different levels of quality required. Consequently the range of applications is diverse. In the following account the most widely used methods of non-destructive testing are reviewed, together with several current developments.

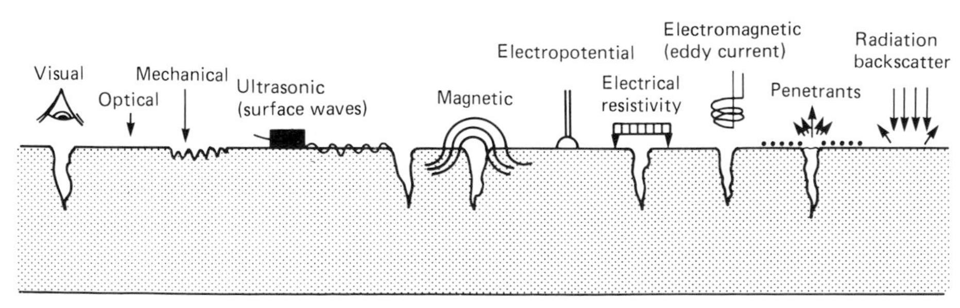

Figure 5.1 NDT methods for surface inspection.

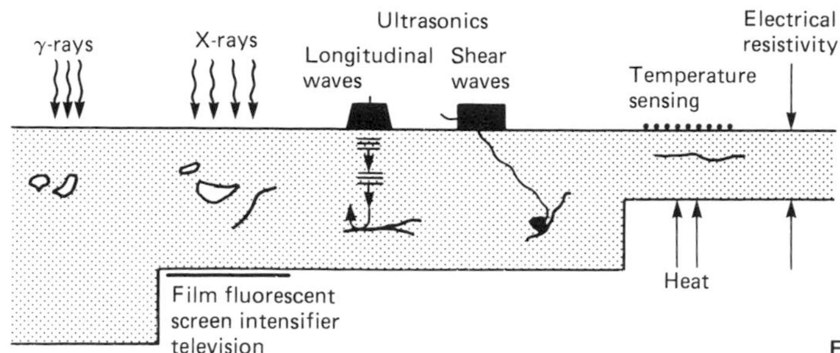

Figure 5.2 NDT methods for sub-surface inspection.

5.2 Visual examination

For many types of components, integrity is verified principally through visual inspection. Indeed, even for components that require further inspection using ultrasonics or radiography visual inspection still constitutes an important aspect of practical quality control.

Visual inspection is the most extensively used of any method. It is relatively easy to apply and can have one or more of the following advantages:

(1) Low cost;
(2) Can be applied while work is in progress;
(3) Allows early correction of faults;
(4) Gives indication of incorrect procedures;
(5) Gives early warning of faults developing when item is in use.

Equipment may range from that suitable for determining dimensional non-conformity such as the Welding Institute Gauges (Figure 5.3) to illuminated magnifiers (Figure 5.4) and the more sophisticated fibrescope (Figure 5.5). The instrument shown in Figure 5.5 is a high-resolution flexible fibrescope with end tip and focus control. Flexible lengths from 1 m to 5 m are available for viewing inaccessible areas in boilers, heat exchangers, castings, turbines, interior welds and other equipment where periodic or troubleshooting inspection is essential.

5.3 Surface-inspection methods

The inspection of surfaces for defects at or close to the surface presents great scope for a variety of inspection

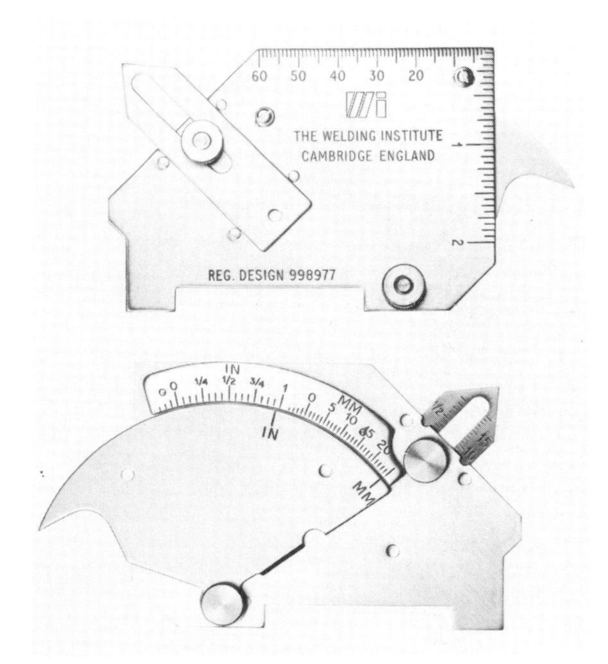

Figure 5.3 Gauges for visual inspection. Courtesy, The Welding Institute.

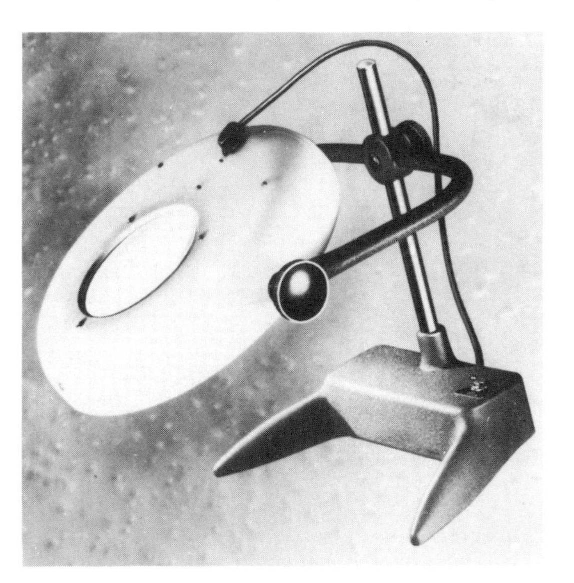

Figure 5.4 Illuminated magnifiers for visual inspection. Courtesy, P. W. Allen & Co.

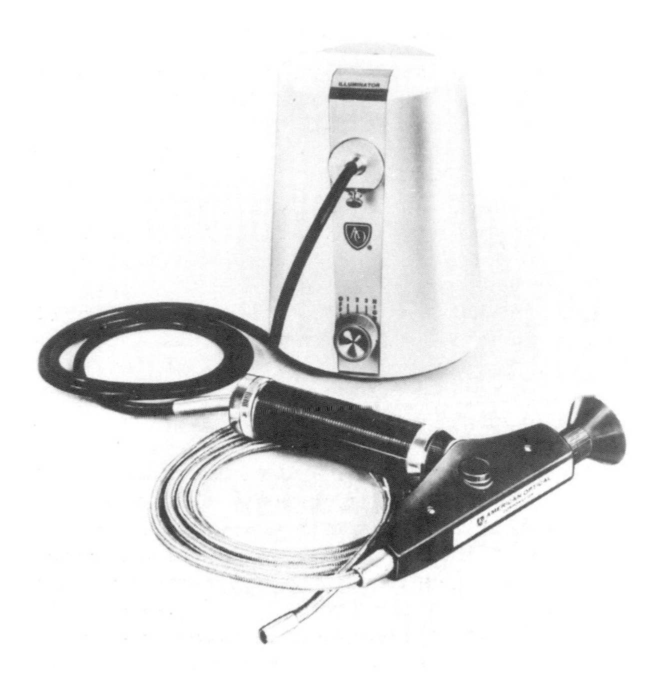

Figure 5.5 High-resolution flexible fibrescope. Courtesy, P. W. Allen & Co.

techniques. With internal-flaw detection one is often limited to radiographic and ultrasonic techniques, whereas with surface-flaw detection visual and other electromagnetic methods such as magnetic particle, potential drop and eddy current become available.

5.3.1 Visual techniques

In many instances defects are visible to the eye on the surface of components. However, for the purposes of recording or gaining access to difficult locations, photographic and photomicrographic methods can be very useful. In hazardous environments, as encountered in the nuclear and offshore fields, remote television cameras coupled to video recorders allow inspection results to be assessed after test. When coupled to remote transport systems these cameras can be used for pipeline inspection, the cameras themselves being miniaturized for very narrow pipe sections.

When surface-breaking defects are not immediately apparent their presence may be enhanced by the use of dye penetrants. A penetrating dye-loaded liquid is applied to a material surface where, due to its surface tension and wetting properties, a strong capillary effect exists, which causes the liquid to penetrate into fine openings on the surface. After a short time (about 10 min), the surface is cleaned and an absorbing powder applied which blots the dye penetrant liquid causing a stain around the defects. Since the dye is either a bright red or fluorescent under ultra-violet light, small defects become readily visible. The penetrant process itself can be made highly automated for large-scale production, but still requires trained inspectors for the final assessment. To achieve fully automated inspection, scanned ultra-violet lasers which excite the fluorescent dye and are coupled to photodetectors to receive the visible light from defect indications are under development.

5.3.2 Magnetic flux methods

When the material under test is ferromagnetic the magnetic properties may be exploited to provide testing methods based on the localized escape of flux around defects in magnetized material. For example, when a magnetic flux is present in a material such as iron, below magnetic saturation, the flux will tend to confine itself within the material surface. This is due to the continuity of the tangential component of the magnetic field strength, H, across the magnetic boundary. Since the permeability of iron is high, the external flux density, B_{ext}, is small (Figure 5.6(a)). Around a defect, the presence of a normal component of B incident on the defect will provide continuity of the flux to the air, and a localized flux escape will be apparent (Figure 5.6(b)). If only a tangential component is present, no flux leak occurs, maximum leakage conditions being obtained when B is normal to the defect.

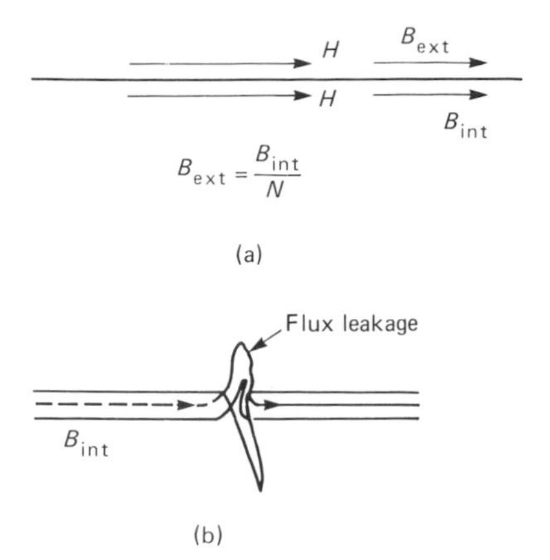

Figure 5.6 Principle of magnetic flux test.

5.3.2.1 Magnetization methods

To detect flux leakages, material magnetization levels must be substantial (values in excess 0.72 Tesla for the magnetic flux density). This, in turn, demands high current levels. Applying Ampere's current law to a 25-mm diameter circular bar of steel having a relative permeability of 240 the current required to achieve a magnetic field strength of 2400 A/m at the surface giving the required flux value is 188 A peak current.

Such current levels are applied either as a.c. current from a step-down transformer whose output is shorted by the specimen or as halfwave rectified current produced by diode rectification. Differences in the type of current used become apparent when assessing the 'skin depth' of the magnetic field. For a.c. current in a magnetic conductor, magnetic field penetration, even at 50 Hz, is limited; the d.c. component present in the half-wave rectified current produces a greater depth of penetration.

Several methods of achieving the desired flux density levels at the material surface are available. They include the use of threading bars, coils and electromagnets (Figure 5.7).

5.3.2.2 Flux-leakage detection

One of the most effective detection systems is the application of finely divided ferric-oxide particles to form an indication by its accumulation around the flux leakage. The addition of a fluorescent dye to the particle enables the indication to be easily seen under ultra-violet light. Such a system, however, does not sustain recording of defect information except through photographic and replication techniques such as strippable magnetic paint and rubber. Alternative flux-detection techniques are becoming available, such as the application of modified magnetic recording tape which is wrapped around the material before magnetization. After test, the tape can be unwound and played through a tape unit, which detects the presence of the recorded flux leakages.

For more dynamic situations, such as in the on-line testing of tube and bar materials, a faster detection technique for the recording of indications is required. A small detector head, comprising a highly permeable yoke on which a number of turns of wire are wrapped, can be used to detect small flux leakages by magnetic induction as the flux leak passes under the detector (Figure 5.8). The material motion can be linked to a chart recorder showing impulses as they occur.

5.3.3 Potential drop techniques

The measurement of material resistance can be related to measurements of the depth of surface-breaking cracks. A four-point probe head (Figure 5.9) is applied to a surface and current passed between the outer probes. The potential drop across the crack is measured by the two inner probes and, as the crack depth increases, the greater current path causes an increasing potential drop. By varying probe spacing, maximum sensitivity to changes in crack depth can be obtained. In addition, the application of a.c. current of varying frequency permits the depth of current penetration beneath the surface to be varied due to the 'skin effect' (see also p. 148).

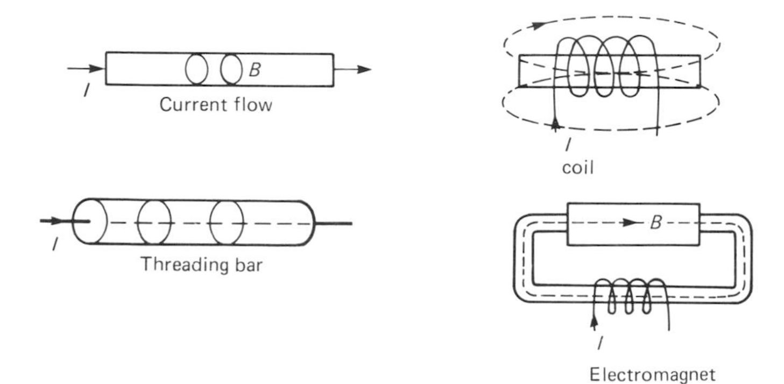

Figure 5.7 Ways of inducing flux.

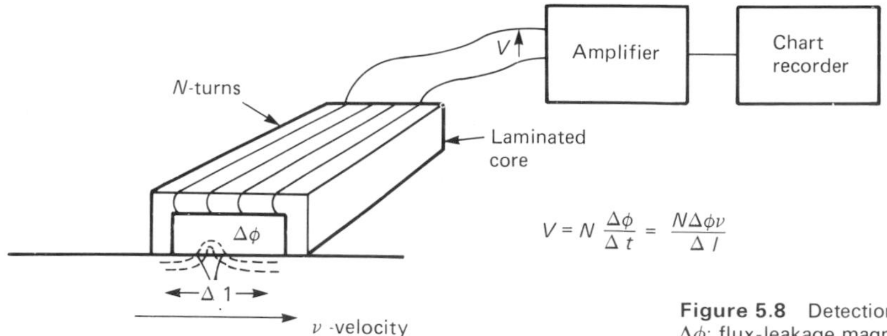

Figure 5.8 Detection of flux leakage. Δl: flux-leakage width; $\Delta\phi$: flux-leakage magnitude; V: induced voltage.

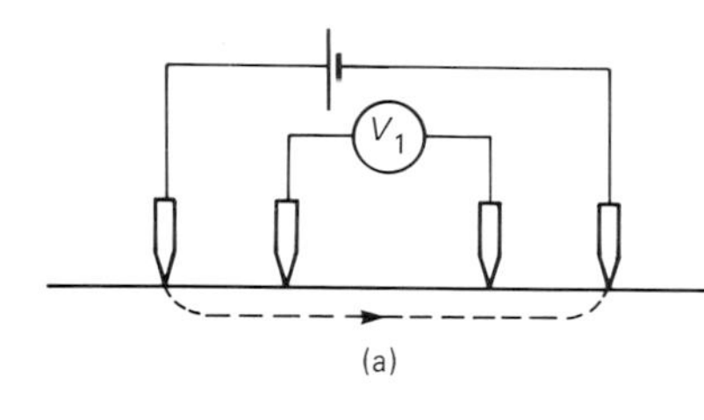

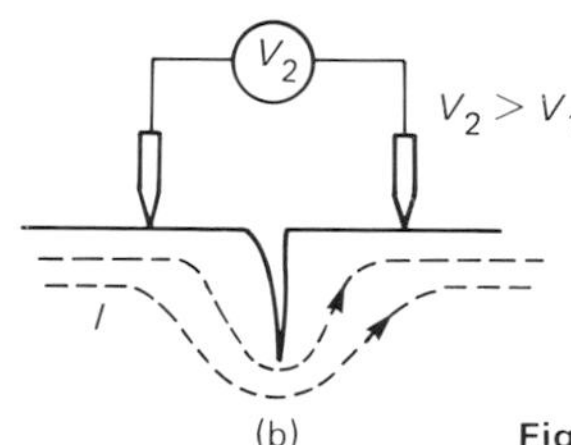

Figure 5.9 Probe for potential drop technique.

5.3.4 Eddy-current testing

A powerful method of assessing both the material properties and the presence of defects is the eddy-current technique. A time-changing magnetic field is used to induce weak electrical currents in the test material, these currents being sensitive to changes in both material conductivity and permeability. In turn, the intrinsic value of the conductivity depends mainly on the material composition but is influenced by changes in structure due to crystal imperfections (voids or interstitial atoms); stress conditions; or work hardening dependent upon the state of dislocations in the material. Additionally, the presence of discontinuities will disturb the eddy-current flow patterns giving detectable changes.

The usual eddy-current testing system comprises a coil which due to the applied current produces an a.c. magnetic field within the material. This, in turn, excites the eddy currents which produce their own field, thus altering that of the current (Figure 5.10). This reflects also in the impedance of the coil, whose resistive component is related to eddy-current losses and whose inductance depends on the magnetic circuit conditions. Thus, conductivity changes will be reflected in changes in coil resistance, whilst changes in permeability or in the presentation of the coil to the surface will affect the coil inductance.

The frequency of excitation represents an important test parameter, due to the 'skin effect' varying the depth of current penetration beneath the surface. From the skin-depth formula

$$\delta = \frac{1}{\sqrt{(\pi f \mu \sigma)}}$$

where f is the frequency, μ the permeability and σ the conductivity, it can be seen that in ferromagnetic

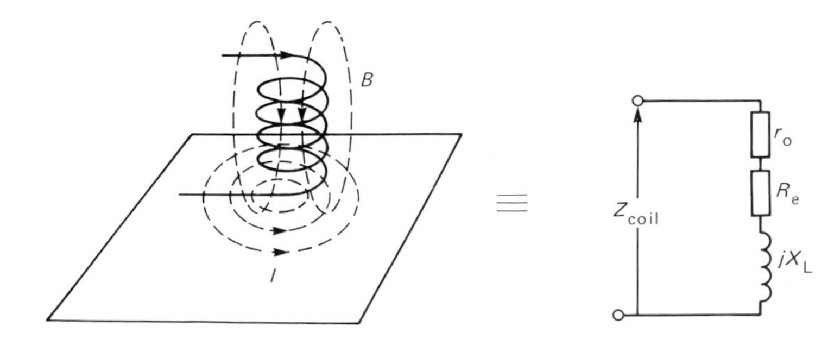

Figure 5.10 Principle of eddy-current testing. $Z_{coil} = (r_0 + R_e) + jX_L$; R_e are the additional losses due to eddy-current-flow.

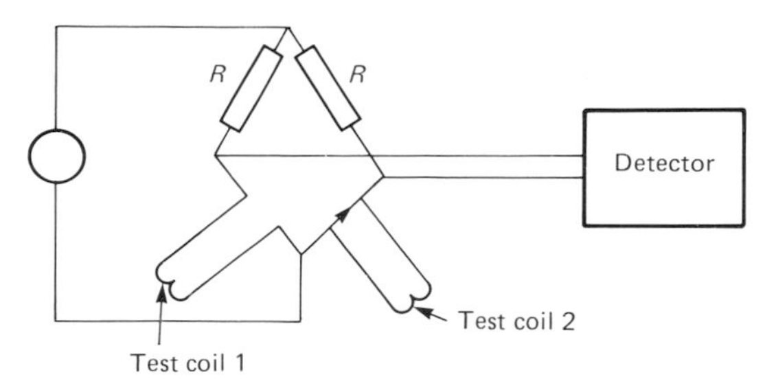

Figure 5.11 Simple type of eddy-current detector.

material the skin depth, δ, is less than in non-magnetic materials by the large factor of the square root of the permeability.

By selection of the appropriate frequency, usually in the range 100 kHz–10 MHz, the detection of discontinuities and other subsurface features can be varied. The higher the frequency, the less the depth of penetration. In addition, in ferromagnetic material the ability of an a.c. magnetic field to bias the material into saturation results in an incremental permeability close to that of the non-magnetic material. The eddy-current method therefore represents a very general testing technique for conducting materials in which changes in conductivity, permeability and surface geometry can be measured.

5.3.4.1 Eddy-current instrumentation

In eddy-current testing the coils are incorporated into a balanced-bridge configuration to provide maximum detection of small changes in coil impedance reflecting the material changes. The simplest type of detector is that which measures the magnitude of bridge imbalance. Such units are used for material comparison of known against unknown and for simple crack detection (Figure 5.11).

A more versatile type of unit is one in which the magnitude and phase of the coil-impedance change is measured (Figure 5.12), since changes in inductance will be 90 degrees out of phase with those from changes in conductivity. Such units as the vector display take the bridge imbalance voltage $V_0 e^{j(\omega t + \phi)}$, and pass it through two quadrature phase detectors. The 0° reference detector produces a voltage proportional to $V_0 \cos \phi$, whilst the 90° detector gives a voltage of $V_0 \sin \phi$. The vector point displayed on an X–Y storage oscilloscope represents the magnitude and phase of the voltage imbalance and hence the impedance change.

To allow positioning of the vector anywhere around the screen a vector rotator is incorporated using sine and cosine potentiometers. These implement the equation

$$V_x = V_x \cos \phi - V_y \sin \phi$$

$$V_y' = V_x \sin \phi + V_y \cos \phi$$

where V_x' and V_y' are the rotated X and Y oscilloscope voltages and V_x and V_y are the phase-detector outputs.

In setting up such a unit the movement of the spot during the lift-off of the probe from the surface identifies the magnetic circuit or permeability axis, whereas defect deflection and conductivity changes will introduce a component primarily along an axis at right angles to the permeability axis (Figure 5.13).

Additional vector processing can take place to

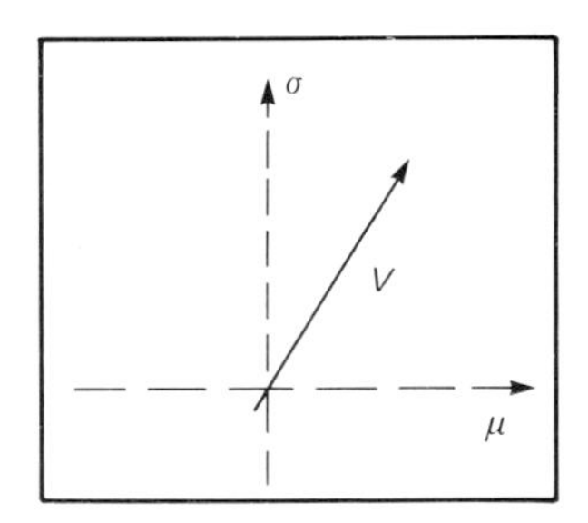

Figure 5.13 CRT display for eddy-current detection.

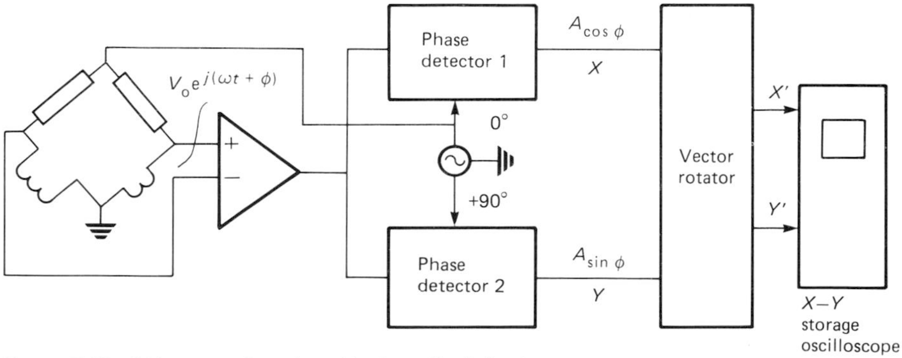

Figure 5.12 Eddy-current detection with phase discrimination.

remove the response from a known geometrical feature. Compensation probes can produce a signal from the feature, such as support members, this signal being subtracted from that of the defect plus feature. Such cancellation can record small defect responses which are being masked by the larger geometrical response. Also a number of different frequencies can be applied simultaneously (multi-frequency testing) to give depth information resulting from the response at each frequency.

The third type of testing situation for large-scale inspection of continuous material comprises a testing head through which the material passes. A detector system utilizing the phase detection units discussed previously is set up to respond to known defect orientations. When these defect signals exceed a predetermined threshold it is recorded along with the tube position on a strip chart or the tube itself is marked with the defect position.

5.4 Ultrasonics

5.4.1 General principles of ultrasonics

Ultrasonics, when applied to the non-destructive testing of an engineering component, relies on a probing beam of energy directed into the component interacting in an interpretable way with the component's structural features. If a flaw is present within the metal, the progression of the beam of energy is locally modified and the modification is detected and conveniently displayed to enable the flaw to be diagnosed. The diagnosis largely depends on a knowledge of the nature of the probing energy beam, its interaction with the structural features of the component under test and the manufacturing history of the component.

The ultrasonic energy is directed into the material under test in the form of mechanical waves or vibrations of very high frequency. Although its frequency may be anything in excess of the upper limit of audibility of the human ear, or 20 kHz, ultrasonic non-destructive testing frequencies normally lie in the range 0.5–10 MHz. The equation

$$\lambda = \frac{V}{f}$$

where λ is the wavelength, V the velocity and f the frequency, highlights this by relating wavelength and frequency to the velocity in the material.

The wavelength determines the defect sensitivity in that any defect dimensionally less than half the wavelength will not be detected. Consequently the ability to detect small defects increases with decreasing wavelength of vibration and, since the velocity of sound is characteristic of a particular material, increasing the frequency of vibration will provide the possibility of increased sensitivity. Frequency selection is thus a significant variable in the ability to detect small flaws.

The nature of ultrasonic waves is such that propagation involves particle motion in the medium through which they travel. The propagation may be by way of a volume change, the compression wave form or by a distortion process, the shear wave form. The speed of propagation thus depends on the elastic properties and the density of the particular medium.

Compression wave velocity

$$V_c - \sqrt{\left(\frac{E}{\rho} \cdot \frac{1-\mu}{(1+\mu)(1-2\mu)}\right)}$$

Shear wave velocity

$$V_s = \sqrt{\left[\frac{E}{\rho} \cdot \frac{1}{2(1+\mu)}\right]} = \sqrt{\frac{G}{\rho}}$$

where E is the modulus of elasticity, ρ the density, μ the Poisson's ratio and G the modulus of shear.

Other properties of ultrasonic waves relate to the results of ultrasound meeting an interface, i.e. a boundary wall between different media. When this occurs, some of the wave is reflected, the amount depending on the acoustic properties of the two media and the direction governed by the same laws as for light waves. If the ultrasound meets a boundary at an angle, the part of the wave that is not reflected is refracted, suffering a change of direction for its progression through the second medium. Energy may be lost or attenuated during the propagation of the ultrasound due to energy absorption within the medium and to scatter which results from interaction of the waves with microstructural features of size comparable with the wavelength. This is an important factor, as it counteracts the sensitivity to flaw location on the basis of frequency selection. Hence high frequency gives sensitivity to small flaws but may be limited by scatter and absorption to short-range detection.

The compression or longitudinal wave is the most common mode of propagation in ultrasonics. In this form, particle displacement at each point in a material is parallel to the direction of propagation. The propagating wavefront progresses by a series of alternate compressions and rarefactions, the total distance occupied by one compression and one rarefaction being the wavelength. Also commonly used are shear or transverse waves, which are characterized by the particle displacement at each point in a material being at right angles to the direction of propagation. In comparing these wave motions it should be appreciated that for a given material the shear waves have a velocity approximately five-ninths of that of compressional waves. It follows that for any frequency, the lower velocity of shear waves corresponds to a shorter wavelength. Hence, for a given frequency, the minimum size of defect detectable will be less in the case of shear waves.

Other forms of shear motion may be produced. Where there is a free surface a Rayleigh or surface wave may be generated. This type of shear wave propagates on the surface of a body with effective penetration of less than a wavelength. In thin sections bounded by two free surfaces a Lamb wave may be produced. This is a form of compressional wave which propagates in sheet material, its velocity depending not only on the elastic constant of the material but also on plate thickness and frequency. Such waveforms can be used in ultrasonic testing. A wave of a given mode of propagation may generate or transform to waves of other modes of propagation at refraction or reflection, and this may give rise to practical difficulties in the correct interpretation of test signals from the material.

Ultrasonic waves are generated in a transducer mounted on a probe. The transducer material has the property of expanding and contracting under an alternating electrical field due to the piezoelectric effect. It can thus transform electrical oscillations into mechanical vibrations and vice-versa. Since the probe is outside the specimen to be tested, it is necessary to provide a coupling agent between probe and specimen. The couplant, a liquid or pliable solid, is interposed between probe surface and specimen surface, and assists in the passage of ultrasonic energy. The probe may be used to transmit energy as a transmitter, receive energy as a receiver or transmit and receive as a transceiver. A characteristic of the transceiver or single-crystal probe is the dead zone, where defects cannot be resolved with any accuracy due to the transmission-echo width. Information on the passage of ultrasonic energy in the specimen under test is provided by way of the transducer, in the form of electrical impulses which are displayed on a cathode ray tube screen. The most commonly used presentation of the information is A-scan, where the horizontal base line represents distance or time intervals and the vertical axis gives signal amplitude or intensities of transmitted or reflected signals.

The basic methods of examination are transmission, pulse-echo and resonance. In the transmission method an ultrasonic beam of energy is passed through a specimen and investigated by placing an ultrasonic transmitter on one face and a receiver on the other. The presence of internal flaws is indicated by a reduction in the amplitude of the received signal or a loss of signal. No indication of defect depth is provided.

Although it is possible with the pulse echo method to use separate probes it is more common to have the transmitter and receiver housed within the one probe, a transceiver. Hence access to one surface only is necessary. This method relies on energy reflected from within the material, finding its way back to the probe. Information is provided on the time taken by the pulse to travel from the transmitter to an obstacle, backwall or flaw and return to the receiver. Time is proportional to the distance of the ultrasonic beam path, hence the cathode ray tube may be calibrated to enable accurate thickness measurement or defect location to be obtained (Figure 5.14(a)). For a defect suitably orientated to the beam direction an assessment of defect size can be made from the amplitude of the reflected signal. Compression probes are used in this test method to transmit compressional waves into the material normal to the entry surface. On the other hand, shear probes are used to transmit shear waves where it is desirable to introduce the energy into the material at an angle, and a reference called the skip distance may be used. This is the distance measured over the surface of the body between the probe index or beam exit point for the probe and the point where the beam axis impinges on the surface after following a double traverse path. Accurate defect location is possible using the skip distance and the beam path length (Figures 5.14(b) and (c)).

A condition of resonance exists when the thickness of a component is exactly equal to half the wavelength of the incident ultrasonic energy, i.e. the component will vibrate with its natural frequency. The condition causes an increase in the amplitude of the received pulse which can readily be identified. The condition of resonance can also be obtained if the thickness is a multiple of the half-wavelength. The resonance method consequently involves varying the frequency of ultrasonic waves to excite a maximum amplitude of vibration in a body or part of a body, generally for the purpose of determining thickness from one side only.

5.4.2 The ultrasonic test equipment controls and visual presentation

In most ultrasonic sets, the signal is displayed on a cathode ray tube (CRT). Operation of the instrument is similar in both the through transmission and pulse-echo techniques and block diagrams of the test equipment are shown in Figures 5.15(a) and (b).

The master timer controls the rate of generation of the pulses or pulse repetition frequency (PRF) and supplies the timebase circuit giving the base line whilst the pulse generator controls the output of the pulses or pulse energy which is transmitted to the probe. At the probe, electrical pulses are converted via the transducer into mechanical vibrations at the chosen frequency and directed into the test material. The amount of energy passing into the specimen is very small. On the sound beam returning to the probe, the mechanical vibrations are reconverted at the

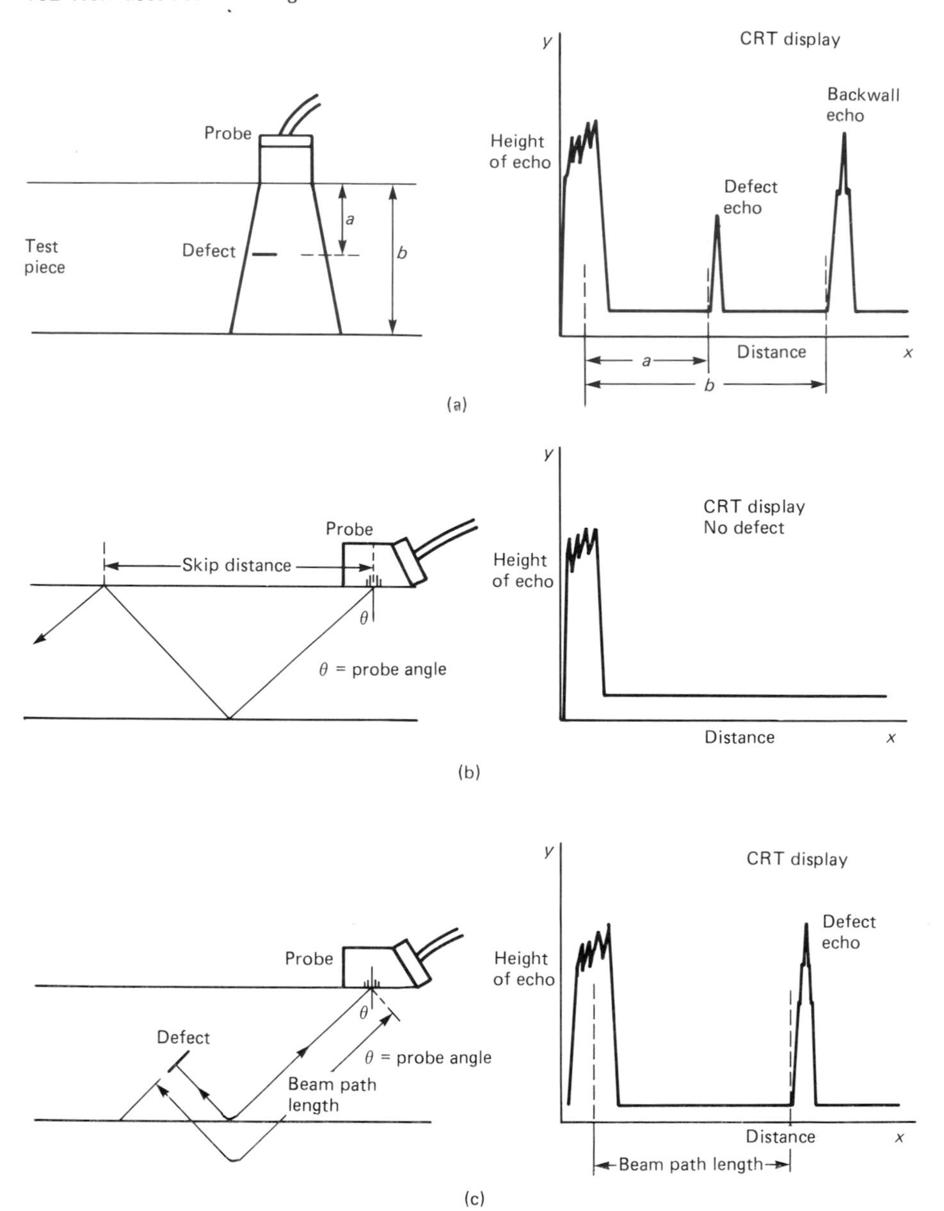

Figure 5.14 Displays presented by different ultrasonic probes and defects. (a) Distance (time) of travel – compression-wave examination; (b) skip distance – shear-wave examination; (c) beam path length (distance of travel) – shear-wave examination.

transducer into electrical oscillations. This is known as the piezoelectric effect. In the main, transmitter and receiver probes are combined. At the CRT the timebase amplifier controls the rate of sweep of the beam across the fact of the tube which, by virtue of the relationship between the distance travelled by ultrasonic waves in unit time, i.e. velocity, can be used as a distance depth scale when locating defects or measuring thickness. Signals coming from the receiver probe to the signal amplifier are magnified, this working in conjunction with an incorporated attenuator control. The signal is produced in the vertical axis. Visual display is by CRT with the transmitted and received signals in their proper time sequence with indications of relative amplitude.

A-, B- and C-scan presentations can be explained in

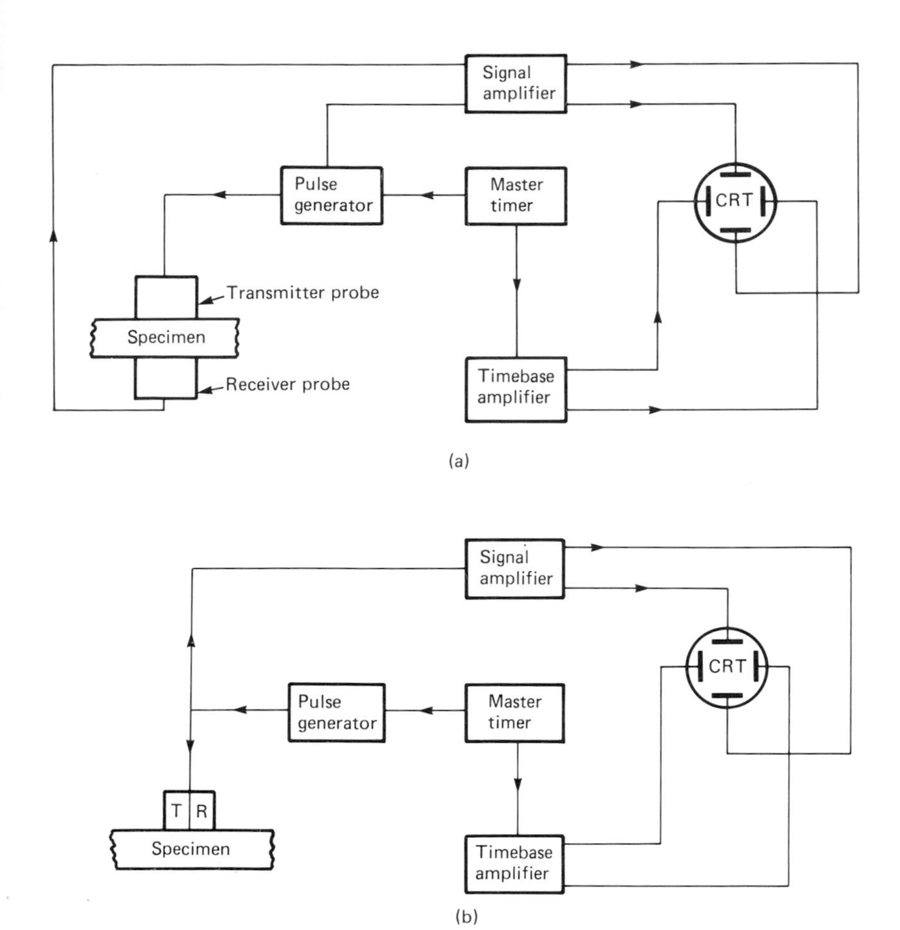

Figure 5.15 Block diagram of ultrasonic flaw detectors using (a) through transmission and (b) pulse-echo techniques.

terms of CRT display (Figure 5.16). Interposed between the cathode and anode in the CRT is a grid which is used to limit the electron flow and hence control ultimate screen brightness as it is made more positive or negative. There is also a deflector system immediately following the electron gun which can deflect the beam horizontally and vertically over the screen of the tube. In A-scan presentation, deflector plate X and Y coordinates represent timebase and amplified response. However, in B-scan the received and amplified echoes from defects and from front and rear surfaces of the test plate are applied, not to the Y deflector plate as in normal A-scan, but to the grid of the CRT in order to increase the brightness of the trace. If a signal proportional to the movement of the probe along the specimen is applied to the X deflector plates, the whole CRT display can be made to represent a cross section of the material over a given length of the sample (Figure 5.16(b)). A further extension to this type of presentation is used in C-scan, where both X and Y deflections on the tube follow the corresponding coordinates of probe traverse. Flaw echoes occurring within the material are gated and used to modulate the grid of the CRT and produce brightened areas in the regions occupied by the defects. A picture is obtained very like a radiograph but with greater sensitivity (Figure 5.16(c)). Figure 5.17 is a block diagram for equipment that can give an A- or B-scan presentation and Figure 5.18 does the same for C-scan.

5.4.3 Probe construction

In ultrasonic probe construction the piezoelectric crystal is attached to a non-piezoelectric front piece or shoe into which an acoustic compression wave is emitted (Figure 5.19 (a)). The other side of the crystal is attached to a material which absorbs energy emitted in the backward direction. These emitted waves

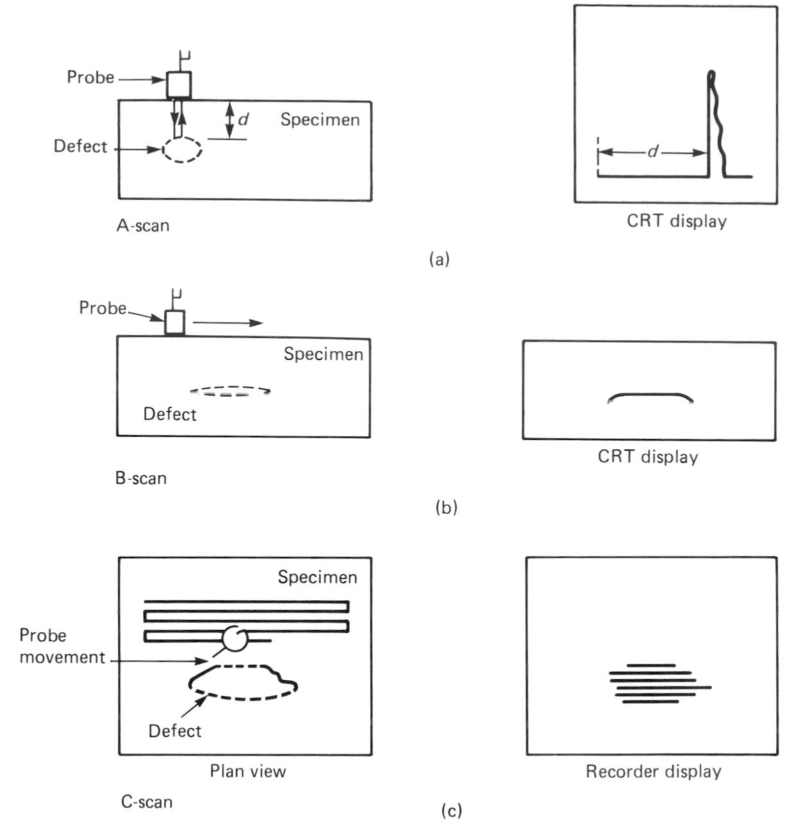

Figure 5.16 A-, B- and C-scan presentations. Courtesy, The Welding Institute.

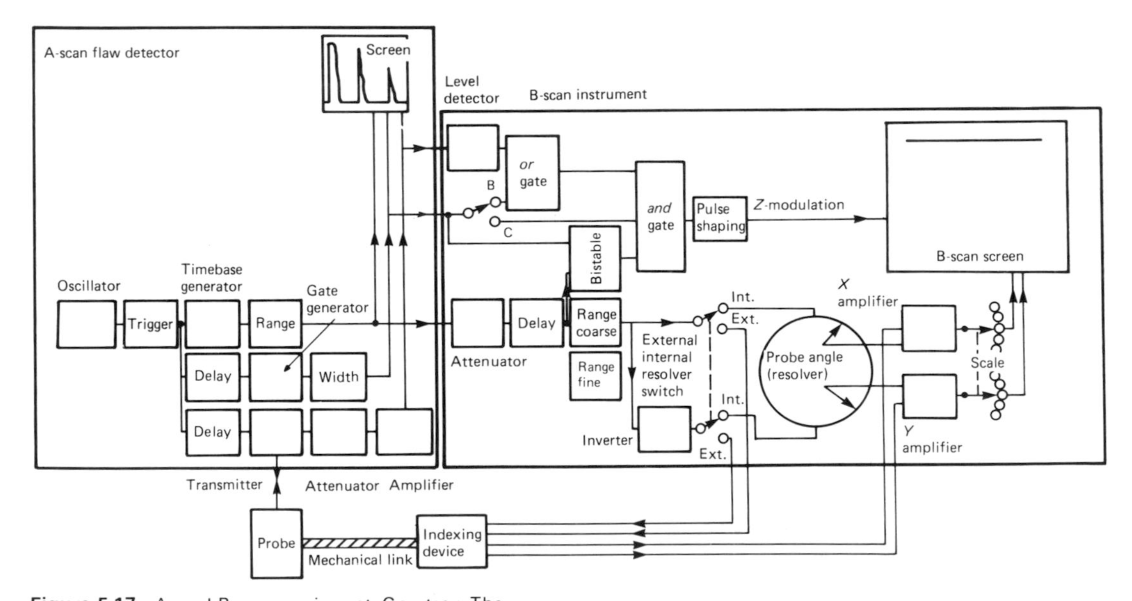

Figure 5.17 A- and B-scan equipment. Courtesy, The Welding Institute.

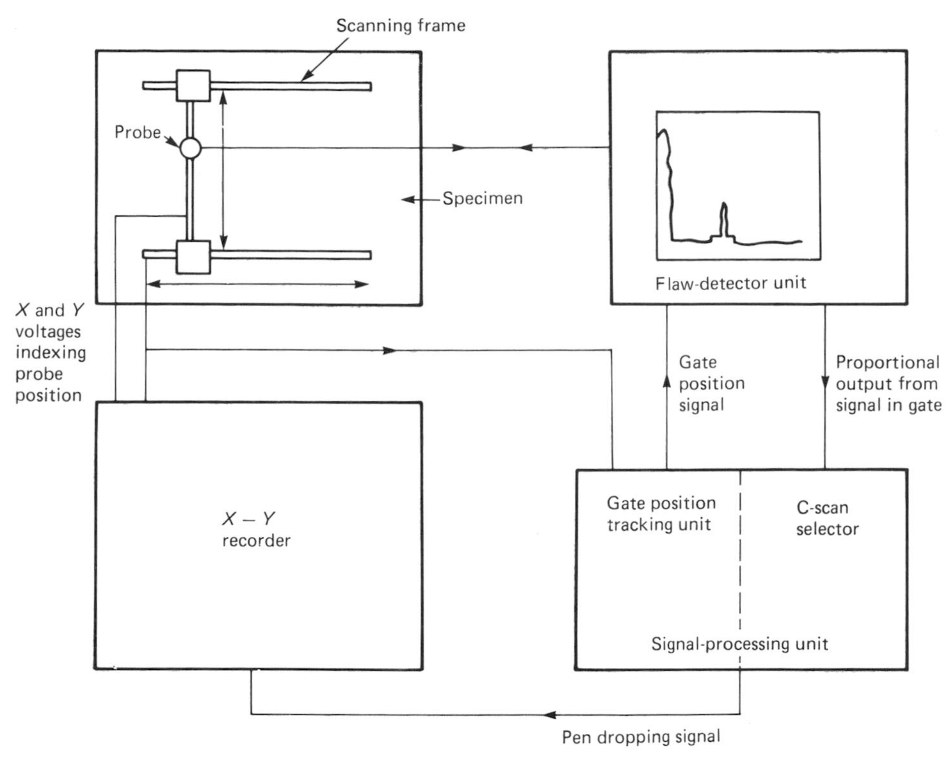

Figure 5.18 C-scan equipment. Courtesy, The Welding Institute.

correspond to an energy loss from the crystal and hence increase the crystal damping. Ideally, one would hope that the damping obtained from the emitted acoustic wave in the forward direction would be sufficient to give the correct probe performance. Probes may generate either compression waves or angled shear waves, using either single or twin piezoelectric crystals. In the single-crystal compression probe the zone of intensity variation is not confined within the perspex wear shoe. However, twin-crystal probes (Figure 5.19(b)) are mainly used since the dead zone or zone of non-resolution of the single form may be extremely long. In comparison, angle probes work on the principle of mode conversion at the boundary of the probe and workpiece. An angled beam of longitudinal waves is generated by the transducer and strikes the surface of the specimen under test at an angle. If the angle is chosen correctly, only an angled shear wave is transmitted into the workpiece. Again a twin crystal form is available as for compression probes, as shown in Figure 5.19(b).

Special probes of the focused, variable-angle, crystal mosaic or angled compression wave type are also available (Figure 5.20). By placing an acoustic lens in front of a transducer crystal it is possible to focus the emitted beam in a similar manner to an optical lens. If the lens is given a cylindrical curvature, it is possible to arrange that, when testing objects in immersion testing, the sound beam enters normal to a cylindrical surface. Variable-angle probes can be adjusted to give a varying angle of emitted sound wave, these being of value when testing at various angles on the same workpiece. In the mosaic type of probe a number of crystals are laid side by side. The crystals are excited in phase such that the mosaic acts as a probe of large crystal size equal to that of the mosaic. Finally, probes of the angled compression type have been used in the testing of austenitic materials where, due to large grain structure, shear waves are rapidly attenuated. Such probes can be made by angling the incident compressional beam until it is smaller than the initial angle for emission of shear waves only. Although a shear wave also occurs in the test piece it is rapidly lost due to attenuation.

5.4.4 Ultrasonic spectroscopy techniques

Ultrasonic spectroscopy is a technique used to analyse the frequency components of ultrasonic signals. The origins of spectroscopy date back to Newton, who

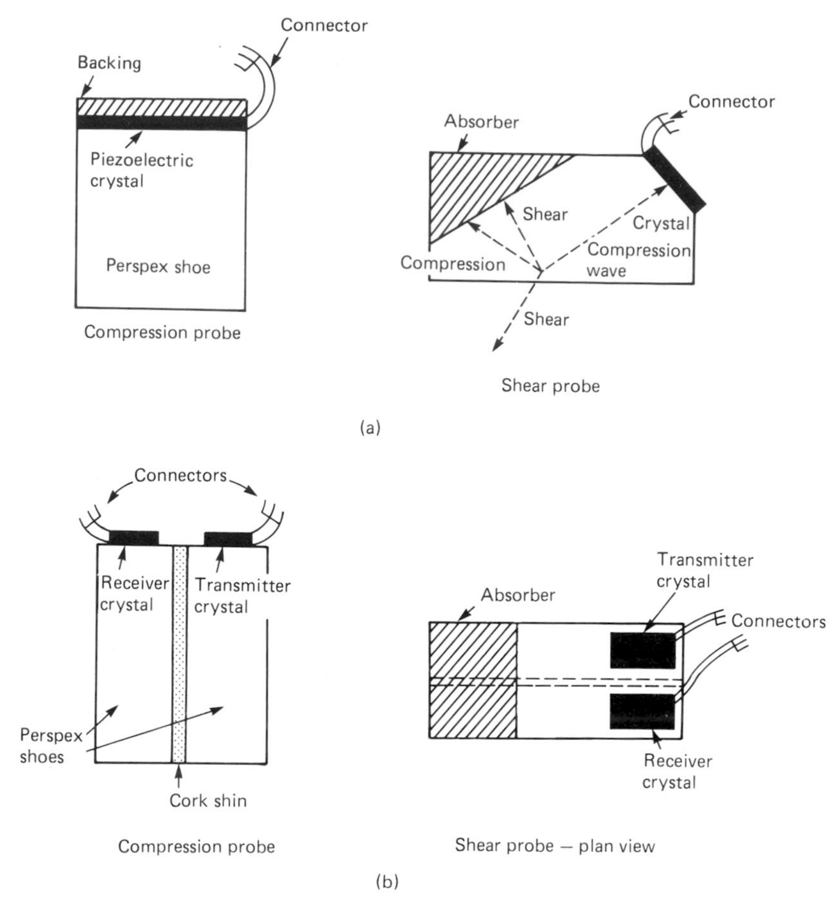

Figure 5.19 Single (a) and twin-crystal (b) probe construction.

showed that white light contained a number of different colours. Each colour corresponds to a different frequency, hence white light may be regarded as the sum of a number of different radiation frequencies, i.e. it contains a spectrum of frequencies. Short pulses of ultrasonic energy have characteristics similar to white light, i.e. they carry a spectrum of frequencies. The passage of white light through a material and subsequent examination of its spectrum can yield useful information regarding the atomic composition of the material. Likewise the passage of ultrasound through a material and subsequent examination of the spectrum can yield information about defect geometry, thickness, transducer frequency response and differences in microstructure. The difference in the type of information is related to the fact that ultrasonic signals are elastic instead of electromagnetic waves.

Interpretation of ultrasonic spectra requires a knowledge of how the ultrasonic energy is introduced into the specimen. Two main factors determine this:

(1) The frequency response characteristic of the transducer; and
(2) The output spectrum of the generator that excites the transducer.

The transducer response is important. Depending on the technique, one or two transducers may be used. The response depends on the mechanical behaviour and elastic constants and thickness normal to the disc surface are important. In addition, damping caused by a wear plate attached to the front of the specimen and transducer–specimen coupling conditions can be significant factors. Crystals which have a low Q value give a broad, flat response satisfactory for ultrasonic spectroscopy but suffer from a sharp fall-off in amplitude response, i.e. sensitivity.

Results indicate that crystal response is flat at frequencies well below the fundamental resonant frequency, f_0, suggesting that higher f_0 values would

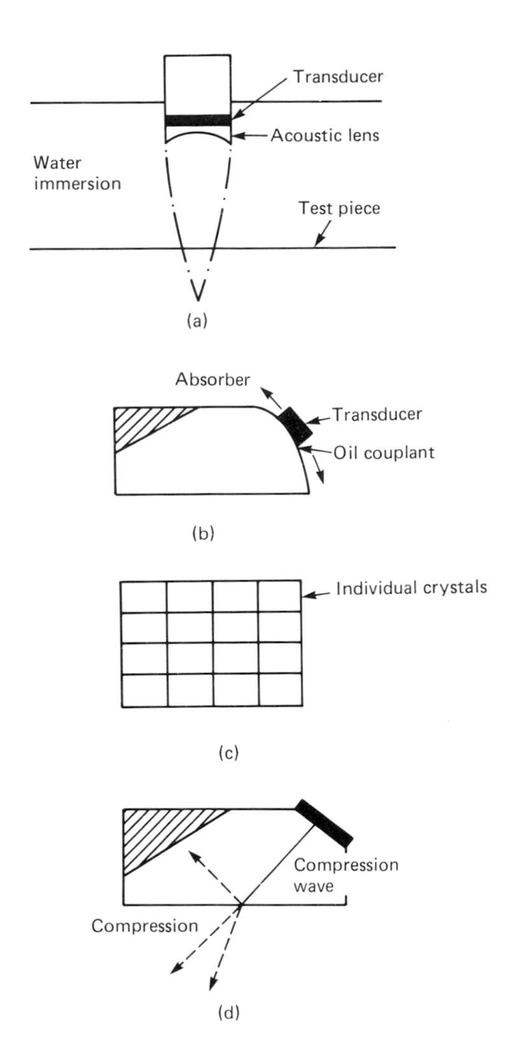

Figure 5.20 Special probe construction. (a) Focused probe; (b) variable-angle probe; (c) mosaic probe; (d) angled compression probe.

be an advantage. However. there is a limit to the thickness of crystal material which remains robust enough to be practical. In general, some attempt is made to equalize the transducer response's effect, for instance by changing the amplitude during frequency modulation.

A straightforward procedure can be adopted of varying frequency continuously and noting the response. Using such frequency-modulation techniques (Figure 5.21) requires one probe for transmission and one for reception. If a reflection test procedure is used, then both transducers may be mounted in one probe. Figure 5.21 shows a block outline of the system. The display on the CRT is then an amplitude versus frequency plot. Some modifications to the system include:

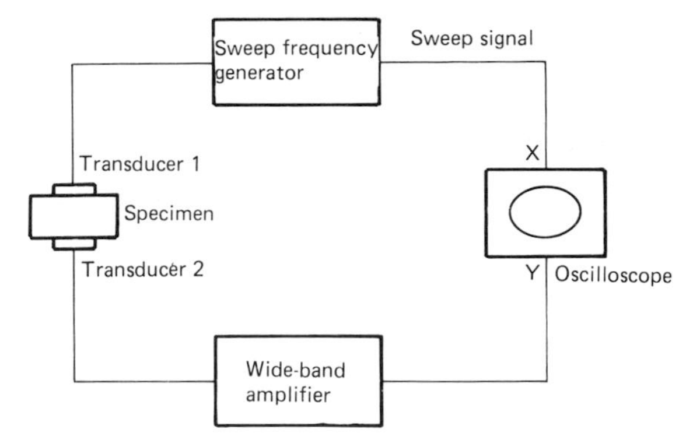

Figure 5.21 Frequency-modulation spectroscope.

(1) A detector and suitable filter between the wide-band amplifier and oscilloscope to allow the envelope function to be displayed; and
(2) Substitution of an electronically tuned rf amplifier to suppress spurious signals.

Alternatively, a range of frequencies may be introduced, using the spectra implicit in pulses. The output spectra for four types of output signal are shown in Figure 5.22. The types of output are:

(1) Single d.c. pulse with reactangular shape;
(2) Oscillating pulse with rectangular envelope and carrier frequency f_0;
(3) D.C. pulse with an exponential rise and decay;
(4) Oscillating pulse with exponential rise and decay and carrier frequency f_0.

From these results it can be seen that the main lobe of the spectrum contains most of the spectral energy and its width is inversely proportional to the pulse

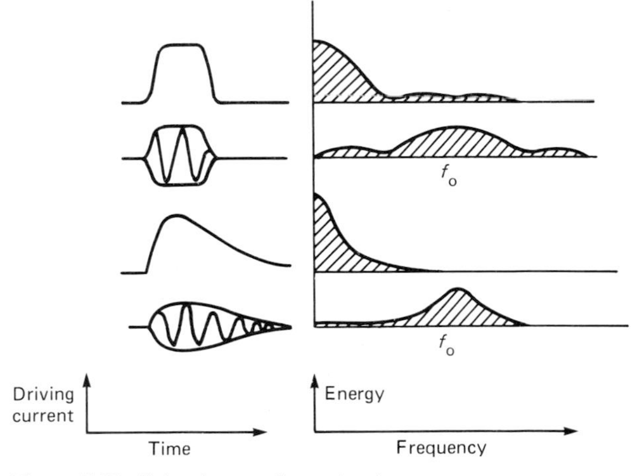

Figure 5.22 Pulse shape and associated spectra.

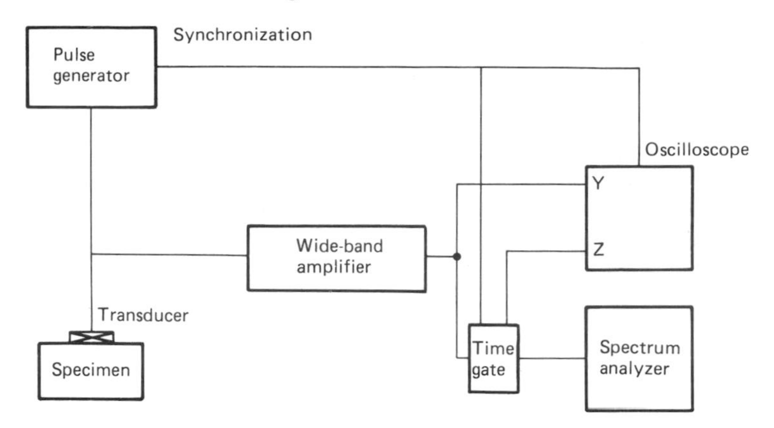

Figure 5.23 Pulse-echo spectroscope.

duration. In order to obtain a broad ultrasonic spectrum of large amplitude the excitation pulse must have as large an amplitude and as short a duration as possible. In practice, a compromise is required, since there is a limit to the breakdown strength of the transducer and the output voltage of the pulse generator. Electronic equipment used in pulse spectroscopy (Figure 5.23) differs considerably from that for frequency modulation, including a time gate for selecting the ultrasonic signals to be analysed, and so allowing observations in the time and frequency domains.

A pulse frequency-modulation technique can also be used which combines both the previous procedures (Figure 5.24). The time gate opens and closes periodically at a rate considerably higher than the frequency sweep rate, hence breaking the frequency sweep signals into a series of pulses. This technique has been used principally for determining transducer response characteristics.

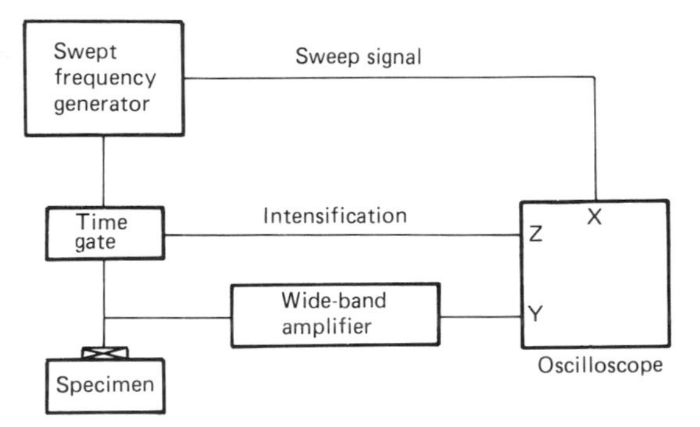

Figure 5.24 Pulsed frequency-modulation spectroscope.

5.4.5 Applications of ultrasonic spectroscopy

5.4.5.1 Transducer response

It is known that the results of conventional ultrasonic inspection can vary considerably from transducer to transducer even if element size and resonant frequency remain the same. To avoid this dificulty it is necessary to control both the response characteristics and the beam profile during fabrication. To determine the frequency response both the pulse and pulsed frequency-modulation techniques are used. The test is carried out by analysing the first backwall echo from a specimen whose ultrasonic attenuation has a negligible dependence on frequency and which is relatively thin, to avoid errors due to divergence of the ultrasonic beam. Analysis of the results yields response characteristics, typical examples of which are shown in Figure 5.25.

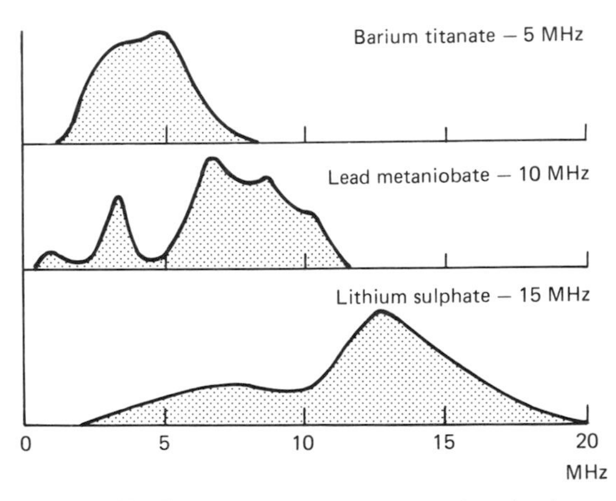

Figure 5.25 Frequency response of various piezoelectric transducers.

5.4.5.2 Microstructure

The attenuation of an ultrasonic signal as it passes through an amorphous or polycrystalline material will depend on the microstructure of the material and the frequency content of the ultrasonic signal. Differences in microstructure can therefore be detected by examining the ultrasonic attenuation spectra. The attenuation of ultrasound in polycrystalline materials with randomly oriented crystallites is caused mainly by scattering. The elastic properties of the single crystal and the average grain size determine the frequency dependence of the ultrasonic attenuation. Attenuation in glass, plastics and other amorphous materials is characteristic of the material, although caused by mechanisms other than grain-boundary scattering.

5.4.5.3 Analysing defect geometry

The assessment of the shape of a defect is often made by measuring changes in the ultrasonic echo height. Intepretation on this basis alone may give results that are less than accurate, since factors such as orientation, geometry and acoustic impedance could affect the echo size at least as much as the defect size. One technique that allows further investigation is to vary the wavelength of the ultrasonic beam by changing the frequency, hence leading to ultrasonic spectroscopy.

The pulse echo spectroscope is ideally suited for this technique. Examination of test specimens has shown that the use of defect echo heights for the purpose of size assessment is not advisable if the spectral 'signatures' of the defect echoes show significant differences.

Other ways of presenting information from ultrasonics

Many other techniques of processing ultrasonic information are available, and in the following the general principles of two main groups (1) ultrasonic visualization or (2) ultrasonic imaging will be outlined. More detailed information may be found in the references given at the end of this chapter.

Ultrasonic visualization makes use of two main techniques: (1) Schlieren methods and (2) photoelastic methods. A third method combining both of these has been described by D. M. Marsh in *Research Techniques in Non-destructive Testing* (1973). Schlieren methods depend on detecting the deviation of light caused by refractive index gradients accompanying ultrasonic waves. In most cases this technique has been used in liquids. Photoelastic visualization reveals the stresses in an ultrasonic wave using crossed polaroids to detect stress birefringence of the medium. Main uses have been for particularly intense fields, such as those in solids in physical contact with the transducer. Many other methods have been tried.

The principle of the Schlieren system is shown in Figure 5.26. In the absence of ultrasonics, all light is intercepted by E. Light diffracted by ultrasound and hence passing outside E forms an image in the dark field. Considerable care is required to make the system effective. Lenses or mirrors may be used in the system as focusing devices; mirrors are free from chromatic aberration and can be accurately produced even for large apertures. Problems may exist with layout and off-axis errors. Lens systems avoid these problems but can be expensive for large apertures.

The basic principles of photoelastic visualization (Figure 5.27) are the same as those used in photoelastic stress analysis (see Photoelastic Stress Analysis, Part 1). Visualization works well in situations where continuous ultrasound is being used. If a pulsed system is used then collimation of the light beam crossing the ultrasound beam and a pulsed light source are required.

The principal advantage of a photoelastic system is its compactness and lack of protective covering in contrast to the Schlieren system.

Photoelastic systems are also cheaper. However, the

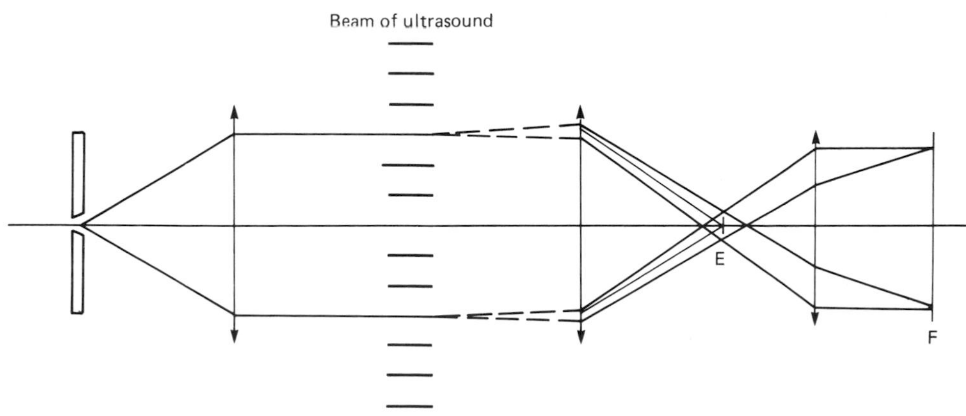

Figure 5.26 Schlieren visualization.

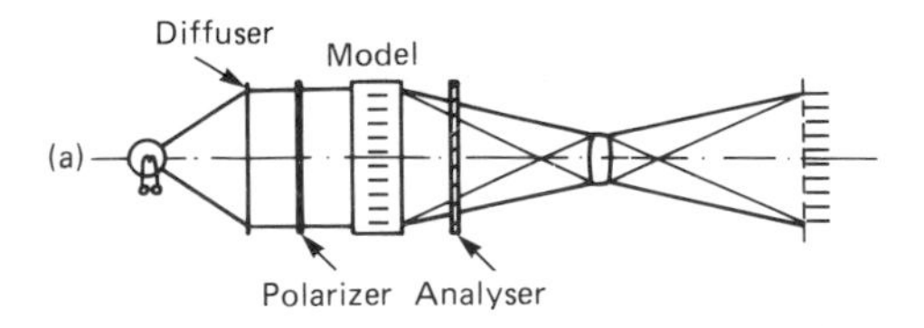

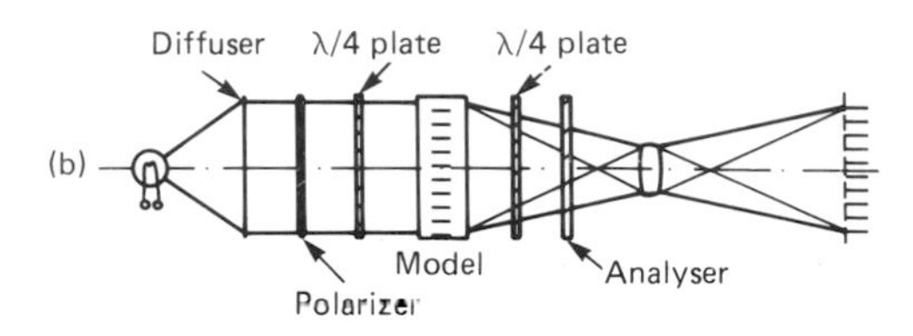

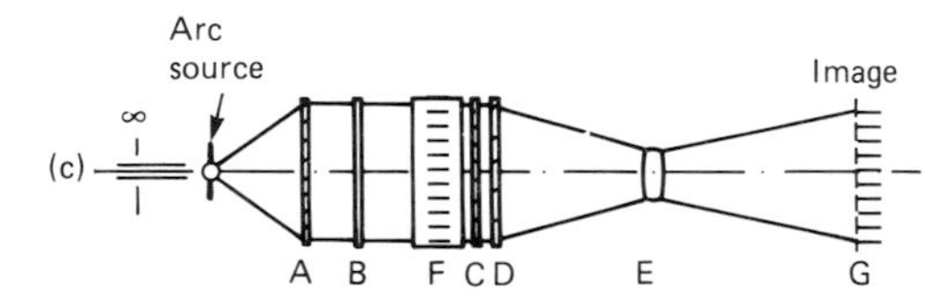

Figure 5.27 Diagrams of photoelastic visualization methods. (a) Basic anisotropic system; (b) circular polarized system for isotropic response; (c) practical large-aperture system using Fresnel lenses.

Schlieren system's major advantage lies in its ability to visualize in fluids as well as in solids. The sensitivity of photoelastic techniques to longitudinal waves is half that for shear waves. Measurements with Schlieren systems have shown that the sensitivity of ultrasound visualization systems is particularly affected by misalignment of the beam. For example, misalignment of the beam in water is magnified by refraction in the solid. It is therefore essential to ensure firm securing of transducers by clamps capable of fine adjustment.

An alternative approach to visualization of ultrasonic waves is the formation of an optical image from ultrasonic radiation. Many systems exist in this field and, once again, a few selected systems will be outlined.

Ultrasonic cameras of various types have been in existence for many years. In general, their use has been limited to the laboratory due to expense and limitations imposed by design. Most successful cameras have used quartz plates forming the end of an evacuated tube for electron-beam scattering, and the size of these tubes is limited by the need to withstand atmospheric pressure. This can be partly overcome by using thick plates and a harmonic, although there is a loss in sensitivity. A new approach to ultrasonic imaging is the use of liquid crystals to form colour images.

All these systems suffer from the need to have the object immersed in water, where the critical angle is such that radiation can only be received over a small angle.

Increasingly, acoustic holography is being introduced into ultrasonic flaw detection and characterization. A hologram is constructed from the ultrasonic radiation and is used to provide an optical picture of any defects. This type of system is theoretically complex and has practical problems associated with stability, as does any holographic system, but there is an increasing number of examples of its use (for example, medical and seismic analysis). Pasakomy has shown that real-time holographic systems may be used to examine welds in pressure vessels. Acoustic holography with scanned hologram systems has been used to examine double-'V' notch welds. Reactor pressure vessels were examined using the same technique and flaw detection was within the limits of error.

5.4.7 Automated ultrasonic testing

The use of microcomputer technology to provide semi-automatic and automatic testing systems has increased, and several pipe testers are available which are controlled by microprocessors. A typical system would have a microprocessor programmed to record defect signals which were considered to represent flaws. Depending on the installation, it could then mark these flaws, record them or transmit them via a link to a main computer. Testing of components on-line has been developed further by British Steel and by staff at SSNDT.

Allied to the development of automatic testing is the development of data-handling systems to allow flaw detection to be automatic. Since the volume of data to be analysed is large, the use of computers is essential. One danger associated with this development is that of over-inspection. The number of data values collected should be no more than is satisfactory to meet the specification for flaw detection.

5.4.8 Acoustic emission

Acoustic emission is the release of energy as a solid material undergoes fracture or deformation. In non-destructive testing two features of acoustic emission are important, its ability to allow remote detection of crack formation or movement at the time it occurs and to do this continuously.

The energy released when a component undergoes stress leads to two types of acoustic spectra; continuous or burst type. Burst-type spectra are usually associated with the leading edge of a crack being extended, i.e. crack growth. Analysis of burst spectra is carried out to identify material discontinuities. Continuous spectra are usually associated with yield rather than fracture mechanisms. Typical mechanisms which would release acoustic energy are (1) crack growth, (2) dislocation avalanching at the leading edge of discontinuities and (3) discontinuity surfaces rubbing.

Acoustic spectra are generated when a test specimen or component is stressed. Initial stressing will produce acoustic emissions from all discontinuities, including minor yield in areas of high local stress. The most comprehensive assessment of a structure is achieved in a single stress application; however, cyclic stressing may be used for structures that may not have their operating stresses exceeded. In large structures which have undergone a period of continuous stress a short period of stress relaxation will allow partial recovery of acoustic activity.

Acoustic emission inspection systems have three main functions; (1) signal detection, (2) signal conditioning and (3) analysis of the processed signals. It is known that the acoustic emission signal in steels is a very short pulse. Other studies have shown that a detecting system sensitive mainly to one mode gives a better detected signal. Transducer patterns may require some thought, depending on the geometry of the item under test. Typical transducer materials are PZT-5A, sensitive to Rayleigh waves and shear waves, and lithium niobate in an extended temperature range. Signal conditioning basically takes the transducer output and produces a signal which the analysis system can process. Typically a signal conditioning system will contain a low noise preamplifier and a filter amplification system.

The analysis of acoustic emission spectra depends on being able to eliminate non-recurring emissions and to process further statistically significant and predominant signals. A typical system is shown in Figure 5.28. The main operating functions are (1) a real-time display and (2) a source analysis computer. The real-time display gives the operator an indication of all emissions while significant emissions are processed by the analysis system. A permanent record of significant defects can also be produced.

In plant where turbulent fluid flow and cavitation are present, acoustic emission detection is best carried out in the low-megahertz frequency range to avoid noise interference.

Acoustic emission detection has been successfully applied to pipe rupture studies, monitoring of known flaws in pressure vessels, flaw formation in welds, stress corrosion cracking and fatigue failure.

5.5 Radiography

Radiography has long been an essential tool for inspection and development work in foundries. In

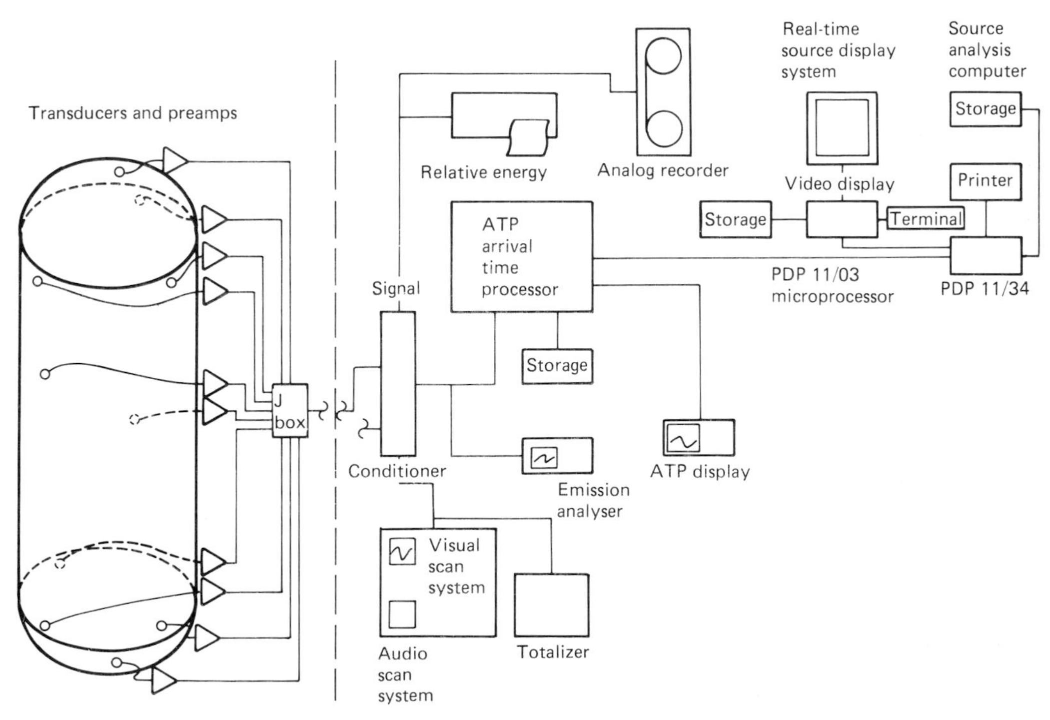

Figure 5.28 System for acoustic emission detection.

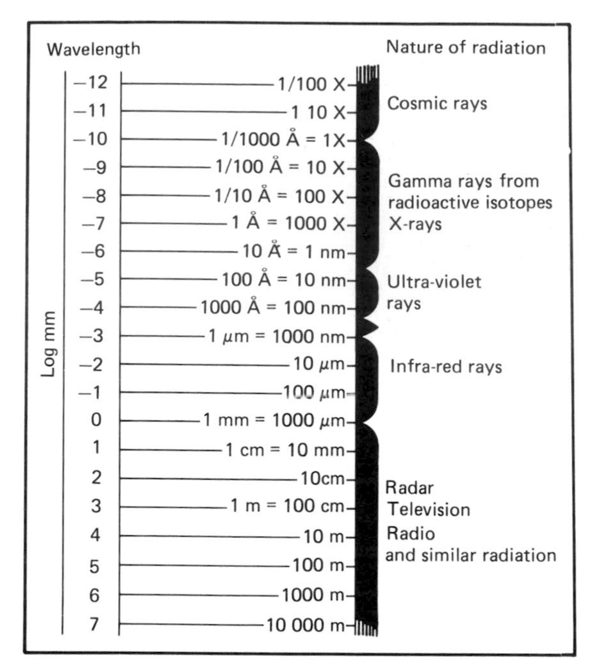

Figure 5.29 Electromagnetic spectrum.

addition, it is widely used in the pressure vessel, pipeline, offshore drilling platform and many other industries for checking the integrity of welded components at the in-process, completed or in-service stages. The technique also finds application in the aerospace industry.

The method relies on the ability of high-energy, short-wavelength sources of electromagnetic radiation such as X-rays, gamma rays (and neutron sources) to penetrate solid materials. By placing a suitable recording medium, usually photographic film, on the side of the specimen remote from the radiation source and with suitable adjustment of technique, a shadowgraph or two-dimensional image of the surface and internal features of the specimen can be obtained. Thus radiography is one of the few non-destructive testing methods suitable for the detection of internal flaws, and has the added advantage that a permanent record of these features is directly producted.

X- and gamma rays are parts of the electromagnetic spectrum (Figure 5.29). Members of this family are connected by the relationship velocity equals frequency times wavelength, and have a common velocity of 3×10^8 m/s. The shorter the wavelength, the higher the degree of penetration of matter. Equipment associated with them is also discussed in Chapter 3.

5.5.1 Gamma rays

The gamma-ray sources used in industrial radiography are artificially produced radioactive isotopes. Though many radioactive isotopes are available, only a few are suitable for radiography. Some of the properties of the commonly used isotopes are shown in Table 5.1. Since these sources emit radiation continuously they must be housed in a protective container which is made of a heavy metal such as lead or tungsten. When the container is opened in order to take the radiograph it is preferable that this be done by remote control in order to reduce the risk of exposing the operator to the harmful rays. Such a gamma-ray source container is shown in Figure 5.30.

5.5.2 X-rays

X-rays are produced when high-speed electrons are brought to rest by hitting a solid object. In radiography the X-rays are produced in an evacuated heavy-walled glass vessel or 'tube'. The typical construction of a tube is shown in Figure 5.31. In operation a d.c. voltage in the range 100 kV to 2 MV is applied between a heated, spirally wound filament (the cathode) and the positively charged fairly massive copper anode. The anode has embedded in it a tungsten insert or target of effective area 2–4 mm^2 and it is on to this target that the electrons are focused. The anode and cathode are placed about 50 mm apart and the tube current is kept low (5–10 mA) in order to prevent overheating of the anode. Typical voltages for a range of steel thicknesses are shown in Table 5.2.

The high voltage needed to produce the X-rays is obtained by relatively simply circuitry consisting of suitable combinations of transformers, rectifiers and capacitors. Two of the more widely used circuits are the Villard, which produces a pulsating output and the

Table 5.1 Isotopes commonly used in gamma radiography

Source	*Effective equiv. energy (MeV)*	*Half-life*	*Specific emission (R/h/Ci at 1 m)*	*Specific activity (Ci/g)*	*Used for steel thickness (mm) up to*
Thulium^{170}	0.08	117 days	0	0.0025	9
Iridium^{192}	0.4	75 days	0.48	25	75
Caesium^{137}	0.6	30 yr	0.35	25	80
Cobalt^{60}	1.1; 1.3	5.3 yr	1.3	120	140

Figure 5.30 Gamma-ray source container. Saddle is attached to the work piece. The operator slips the source container into the saddle and moves away. After a short delay the source moves into the exposed position. When the preset exposure time expires the source returns automatically into its safe shielding. Courtesy, Pantatron Radiation Engineering Ltd.

Greinacher, producing an almost constant potential output. These are shown in Figure 5.32 along with the form of the resulting voltage. A voltage stabilizer at the 240 V input stage is desirable.

Exposure (the product of current and time) varies from specimen to specimen, but with a tube current of 5 mA exposure times in the range 2–30 min are typical.

5.5.3 Sensitivity and IQI

Performance can be optimized by right choice of type of film and screens, voltage, exposure and film focal (target) distance. The better the technique, the higher the sensitivity of the radiograph. Sensitivity is a measure of the smallness of flaw which may be revealed on a radiograph. Unfortunately, this important feature cannot be measured directly, and so the relative quality of the radiograph is assessed using a device called an image indicator (IQI).

A number of different designs of IQI are in use, none of which is ideal. After extensive experimentation, the two types adopted by the British Standards Institution and accepted by the ISO (International Standards Organization) are the wire type and the step-hole type.

In use, the IQI should, wherever possible, be placed in close contact with the source side of the specimen, and in such a position that it will appear near the edge of whichever size of film is being used.

5.5.3.1 Wire type of IQI

This consists of a series of straight wires 30 mm long and diameters as shown in Table 5.3. Five models are available, the most popular containing seven consecutive wires. The wires are laid parallel and equidistant from each other and mounted between two thin transparent plastic sheets of low absorption for X- or gamma-rays. This means that although the wires are fixed in relation to each other, the IQI is flexible and thus useful for curved specimens. The wires should be of the same material as that being radiographed. IQIs for steel, copper and aluminium are commercially available. For other materials it may not be possible to obtain an IQI of matching composition. In this case, a material of similar absorptive properties may be substituted.

Also enclosed in the plastic sheet are letters and numbers to indicate the material type and the series of

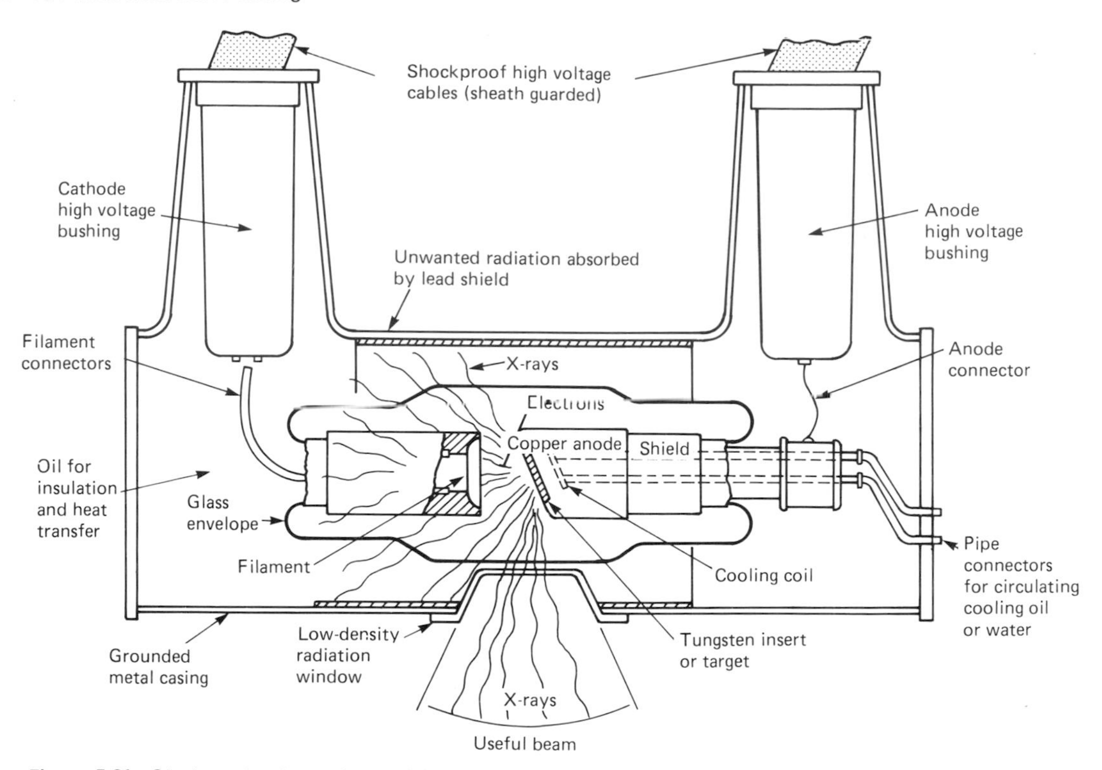

Figure 5.31 Single section, hot-cathode, high vacuum, oil-cooled, radiographic X-ray tube.

Table 5.2 Maximum thickness of steel which can be radiographed with different X-ray energies

X-ray energy (keV)	*High-sensitivity technique thickness (mm)*	*Low-sensitivity technique thickness (mm)*
100	10	25
150	15	50
200	25	75
250	40	90
400	75	110
1 000	125	160
2 000	200	250
5 000	300	350
30 000	300	380

wires (Figure 5.33). For example, 4Fe10 indicates that the IQI is suitable for iron and steel and contains wire diameters from 0.063 to 0.25 mm.

For weld radiography, the IQI is placed with the wires placed centrally on the weld and lying at right angles to the length of the weld.

The percentage sensitivity, S, of the radiograph is calculated from

$$S = \frac{\text{Diameter of thinnest wire visible on the radiograph} \times 100}{\text{Specimen thickness (mm)}}$$

Thus the better the technique, the more wires will be imaged on the radiograph. It should be noted that the smaller the value of S, the better the quality of the radiograph.

5.5.3.2 Step/hole type of IQI

This consists of a series of uniform thickness metal plaques each containing one or two drilled holes. The hole diameter is equal to step thickness. The step thicknesses and hole diameters are shown in Table 5.4.

For steps 1 to 8, two holes are drilled in each step, each hole being located 3 mm from the edge of the step. The remaining steps, 9 to 18 only have a single hole located in the centre of the step.

For convenience, the IQI may be machined as a single step wedge (Figure 5.34). For extra flexibility each plaque is mounted in line but separately between two thin plastic or rubber sheets. Three modules are available, the step series in each being as shown in Table 5.5.

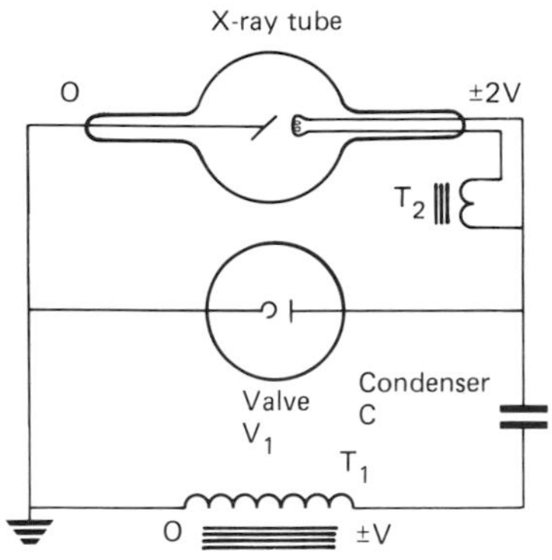

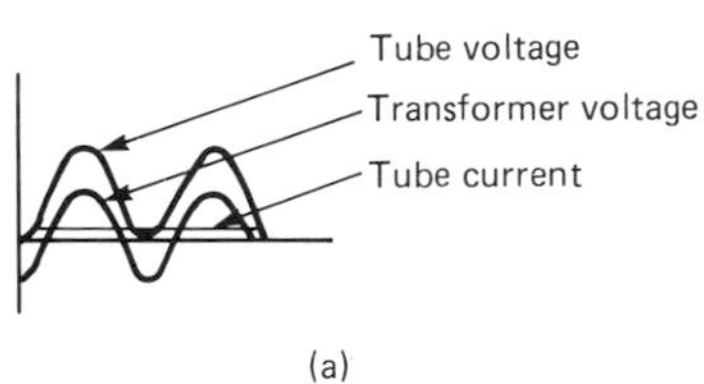

(a)

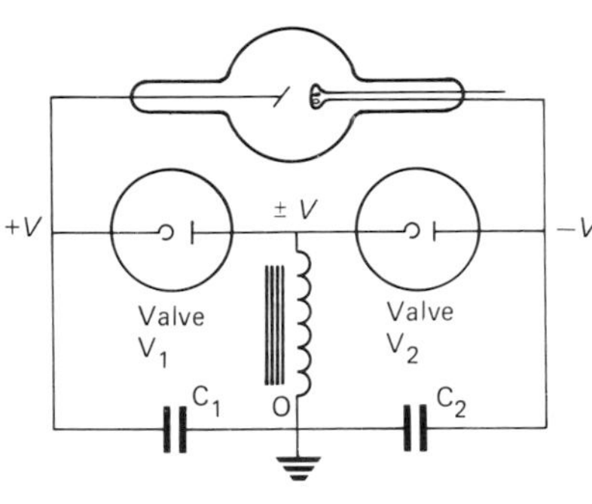

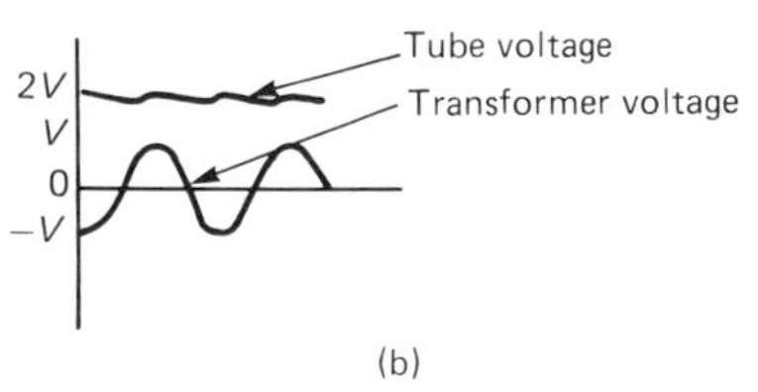

(b)

Figure 5.32 Voltage-doubling circuits. (a) Villard; (b) Greinacher constant-potential.

Table 5.3 Wire diameters

Wire no.	*Diameter (mm)*	*Wire no.*	*Diameter (mm)*	*Wire no.*	*Diameter (mm)*
1	0.032	8	0.160	15	0.80
2	0.010	9	0.200	16	1.00
3	0.050	10	0.250	17	1.25
4	0.063	11	0.320	18	1.00
5	0.080	12	0.400	19	2.00
6	0.100	13	0.500	20	2.50
7	0.125	14	0.63	21	3.20

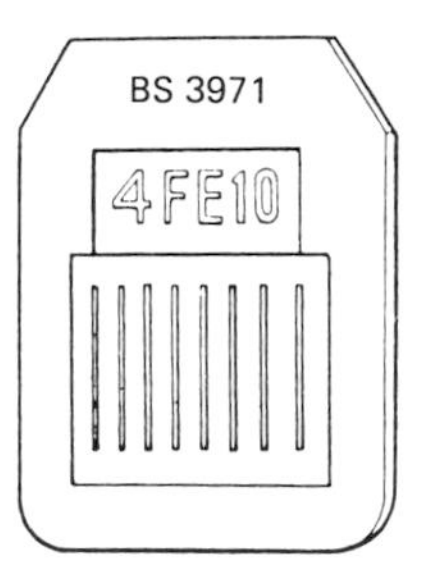

Figure 5.33 Wire-type IQI.

Table 5.4 Hole and step dimensions for IQI

Step no.	*Diameter and step thickness (mm)*	*Step no.*	*Diameter and step thickness (mm)*
1	0.125	10	1.00
2	0.160	11	1.25
3	0.200	12	1.60
4	0.250	13	2.00
5	0.320	14	2.50
6	0.400	15	3.20
7	0.500	16	4.00
8	0.630	17	5.00
9	0.800	18	6.30

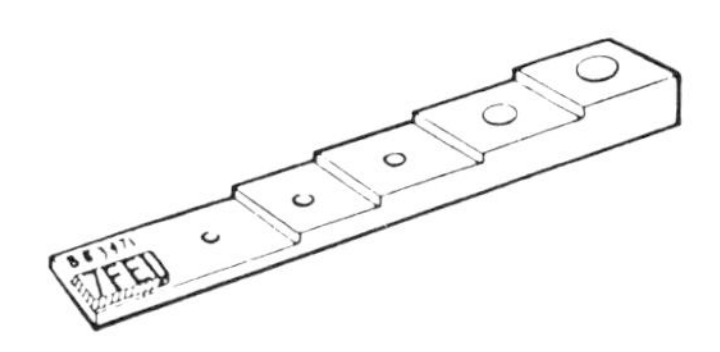

Figure 5.34 Step/hole type IQI.

Table 5.5 Step and holes included in different models of IQI

Model	*Step and hole sizes*
A	1–6 inclusive
B	7–12 inclusive
C	13–18 inclusive

In this type also, a series of letters and numbers identifies the material and thickness range for which the IQI is suitable. The percentage sensitivity, S, is calculated from

$$S = \frac{\text{Diameter of smallest hole visible on the radiograph} \times 100}{\text{Specimen thickness (mm)}}$$

For the thinner steps with two holes it is important that both holes are visible before that diameter can be included in the calculation. In use, this type is placed close to, but not on, a weld.

5.5.4 Xerography

Mainly because of the high price of silver, attempts have been made to replace photographic film as the recording medium. Fluoroscopy is one such method but is of limited application due to its lack of sensitivity and inability to cope with thick sections in dense materials. An alternative technique which has been developed and which can produce results of comparable sensitivity to film radiography in the medium voltage range (100–250 kV) is xerography. In this process, the X-ray film is replaced by an aluminium plate coated on one side with a layer of selenium 30—100 μm thick. A uniform positive electric charge is induced in the selenium layer. Since selenium has a high resistivity, the charge may be retained for long periods provided the plate is not exposed to light, ionizing radiations or unduly humid atmospheres.

Exposure to X- or gamma rays causes a leakage of the charge from the selenium to the earthed aluminium backing plate—the leakage at any point being proportional to the radiation dose falling on it. The process of forming the image is shown in Figure 5.35. This latent image is developed by blowing a fine white powder over the exposed plate. The powder becomes charged (by friction) and the negatively charged particles are attracted and adhere to the positively charged selenium, the amount attracted being proportional to the amount of charge at each point on the 'latent' selenium image.

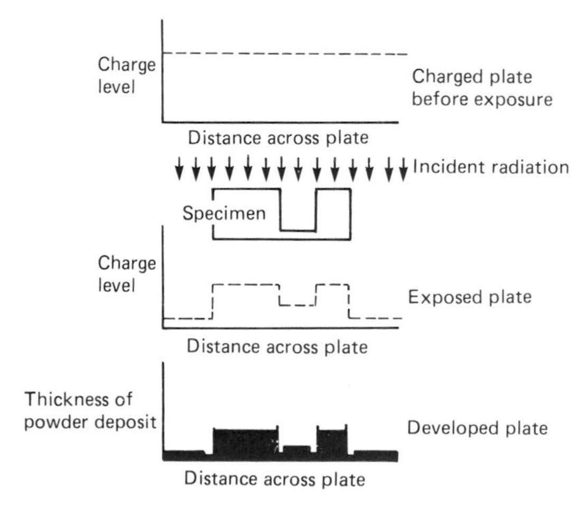

Figure 5.35 Diagrammatic representation of the process of xerography.

The image is now ready for viewing, which is best done in an illuminator using low-angle illumination, and the image will withstand vibration but it must not be touched.

The process is capable of a high degree of image sharpness since the selenium is virtually free of graininess and sharpness is not affected by the grain size of the powder. An apparent drawback is the low inherent contrast of the image, but there is an outlining effect on image detail which improves the rendering of most types of weld and casting flaws thus giving an apparent high contrast. The overall result is that sensitivity is similar to that obtained with film.

5.5.5 Fluoroscopic and image-intensification methods

In fluoroscopy the set-up of source, specimen and recording medium is similar to that for radiography. However, instead of film a specially constructed transparent screen is used which fluoresces, i.e. emits light when X-rays fall on it. This enables a positive image to be obtained since greater amounts of radiation, for example that passing through thinner parts of the specimen, will result in greater brightness.

Fluoroscopy has the following advantages over radiography:

(1) The need for expensive film is eliminated; and
(2) The fluorescent screen can be viewed while the specimen is moving, resulting in:
   (a) Easier image interpretation and
   (b) Faster throughput.

Unfortunately, the sensitivity possible with fluoroscopy is considerably inferior to that obtained with film radiography. It is difficult to obtain a sensitivity better than 5 per cent whereas for critical work a sensitivity of 2 per cent or better is required. Therefore, although the method is widely used in the medical field, its main use in industry is for applications where resolution of fine detail is not required. There are three reasons for the lack of sensitivity:

(1) Fluoroscopic images are usually very dim. The characteristics of the human eye are such that, even when fully dark adapted, it cannot perceive at low levels of brightness the small contrasts or fine detail which it can at higher levels.
(2) In an attempt to increase image brightness the fluorescent screens are usually constructed using a zinc sulphide–cadmium sulphide mixture which, although increasing brightness, gives a coarser-grained and hence a more blurred image.
(3) The image produced on a fluoroscopic screen is much less contrasty than that on a radiograph.

5.5.5.1 Image-intensification systems

In fluoroscopy the main problem of low screen brightness is due mainly to:

(1) The low efficiency—only a fraction of the incident X-rays are converted into light.
(2) The light which is produced at the screen is scattered in all directions, so that only a small proportion of the total produced is collected by the eye of the viewer.

In order to overcome these limitations, a number of image-intensification and image-enhancement systems have been developed.

The electron tube intensifier is the commonest type. Such instruments are commonly marketed by Philips and Westinghouse. In this system use is made of the phenomenon of photoelectricity, i.e. the property possessed by some materials of emitting electrons when irradiated by light.

The layout of the Philips system is shown in Figure 5.36. It consists of a heavy-walled glass tube with an

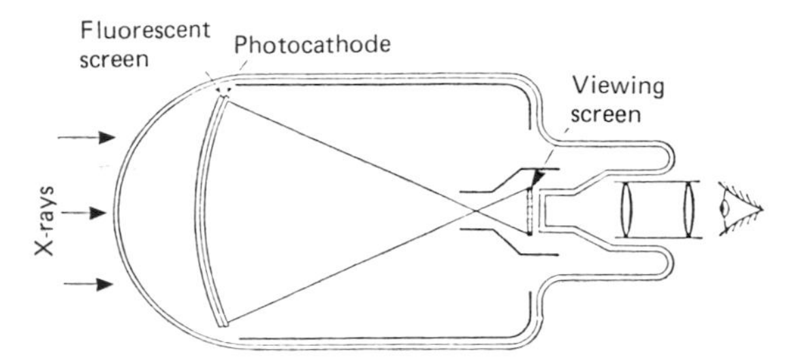

Figure 5.36 Diagram of 5-in Philips image-intensifier tube.

inner conducting layer over part of its surface which forms part of the electron-focusing system. At one end of the tube there is a two-component screen comprising a fluorescent screen and, in close contact with it, a photoelectric layer supported on a thin curved sheet of aluminium. At the other end of the tube is the viewing screen and an optical system.

The instrument operates as follows. When X-rays fall on the primary fluorescent screen they are converted into light which, in turn, excites the photoelectric layer in contact with it and causes it to emit electrons: i.e. a light image is converted into an electron image. The electrons are accelerated across the tube by a d.c. potential of 20–30 kV and focused on the viewing screen. Focusing of the electron image occurs largely because of the spherical curvature of the photocathode. However, fine focusing is achieved by variation of a small positive voltage applied to the inner conducting layer of the glass envelope. As can be seen from Figure 5.36, the electron image reproduced on the viewing screen is much smaller and hence much brighter than that formed at the photocathode. Further increase in brightness is obtained because the energy imparted to the electrons by the accelerating electric field is given up on impact.

Although brighter, the image formed on the final screen is small and it is necessary to increase its size with an optical magnifier. In the Philips instrument an in-line system provides a linear magnification of nine for either monocular or binocular viewing.

As can be seen from Figure 5.37, the Westinghouse instrument is somewhat similar to that of Philips but there are two important differences:

(1) Westinghouse use a subsidiary electron lens followed by a final main electron lens, fine focusing being achieved by varying the potential of the weak lens.
(2) Westinghouse use a system of mirrors and lens to prepare the image for final viewing.

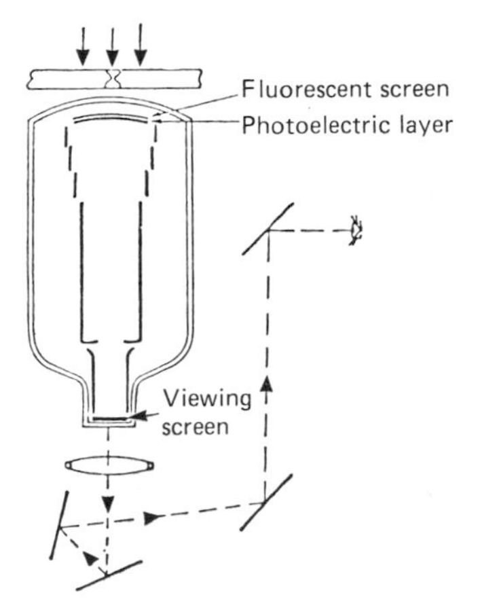

Figure 5.37 Diagram of Westinghouse image-intensifier tube and optical system.

The advantages of this system are that viewing is done out of line of the main X-ray beam and the image can be viewed simultaneously by two observers.

5.6 Underwater non-destructive testing

The exploration and recovery of gas and oil offshore based on large fabricated structures has created a demand for non-destructive testing capable of operation at and below the surface of the sea. Because of high capital costs, large operating costs and public interest the structures are expected to operate all the year round and to be re-certificated by relevant authorities and insurance companies typically on a five-year renewal basis.

The annual inspection programme must be planned around limited opportunities and normally covers:

(1) Areas of previous repair;
(2) Areas of high stress;
(3) Routine inspection on a five-year cycle.

The accumulated inspection records are of great importance. The inspection is designed to include checks on:

(1) Structural damage caused by operational incidents and by changes in sea-bed conditions;
(2) Marine growth accumulation both masking faults and adding extra mass to the structure;
(3) Fatigue cracking caused by the repeated cyclic loading caused by wind and wave action;
(4) Corrosion originating in the action of salt water.

The major area of concern is within the splash zone and just below, where incident light levels, oxygen levels, repeated wetting/drying and temperature differentials give rise to the highest corrosion rates. The environment is hostile to both equipment and operators in conditions where safety considerations make demands on inspection techniques requiring delicate handling and high interpretative skills.

5.6.1 Diver operations and communication

Generally, experienced divers are trained as non-destructive testing operators rather than inspection personnel being converted into divers. The work is fatiguing and inspection is complicated by poor communications between diver and the supervising surface inspection engineer. These constraints lead to the use of equipment which is either robust and provides indications which are simple for first-line interpretation by the diver or uses sophisticated data transmission to the surface for interpretation, and relegates the diver's role to one of positioning the sensor-head. Equipment requiring a high degree of interaction between diver and surface demands extensive training for optimum performance.

The communication to the diver is speech based. The ambient noise levels are high both at the surface, where generators, compressors, etc. are operating, and below it, where the diver's microphone interacts with his breathing equipment. The microphone picks up the voice within the helmet. Further clarity is lost by the effect of air pressure altering the characteristic resonance of the voice, effectively increasing the frequency band so that, even with frequency shifters, intelligible information is lost. Hence any communication with the diver is monosyllabic and repetitive.

For initial surveys and in areas of high ambient danger the mounting of sensor arrangements on remote-controlled vehicles (RCV) is on the increase.

Additional requirements for robust techniques and equipments are imposed by 24-h operation and continual changes in shift personnel. Equipment tends to become common property and not operated by one individual who can undertake maintenance and calibration.

Surface preparation prior to examination is undertaken to remove debris, marine growth, corrosion products and, where required, the existing surface protection. The preparation is commonly 15 cm on either side of a weld and down to bare metal. This may be carried out by hand scraping, water jetting, pneumatic/hydraulic needle descaler guns, or hydraulic wire brushing. Individual company requirements vary in their estimation of the effects of surface-cleaning methods which may peen over surface-breaking cracks.

5.6.2 Visual examination

The initial and most important examination is visual in that it provides an overall and general appreciation of the condition of the structure, the accumulation of marine growth and scouring. Whilst only the most obvious cracks will be detected, areas requiring supplementary non-destructive testing will be highlighted. The examination can be assisted by closed-circuit television (CCTV) and close-up photography. CCTV normally uses low-light level silicon diode cameras, capable of focusing from 4 in to infinity.

In order to assist visual examination, templates, mimics and pit gauges are used to relay information on the size of defects to the surface. Where extensive damage has occured a template of the area may be constructed above water for evaluation by the inspection engineer who can then design a repair technique.

5.6.3 Photography

Light is attenuated by scatter and selective absorption in seawater and the debris held in suspension so as to shift the colour balance of white light towards green. Correction filters are not normally used as they increase the attenuation. Correct balance and illumination levels are achieved with flood- or flashlights. The photography is normally on 35 or 70 mm format using colour film with stereoscopic recording where later analysis and templating will assist repair techniques.

Camera types are normally waterproofed versions

of land-based equipment. When specifically designed for underwater use, the infrequent loading of film and power packs is desirable with built-in flash and single-hand operation or remote firing from a submersible on preset or automatic control.

CCTV provides immediate and recordable data to the surface. It is of comparatively poor resolution and, in black and white, lacks the extra picture contrast given by colour still photography.

5.6.4 Magnetic particle inspection (MPI)

MPI is the most widely accepted technique for underwater non-destructive testing. As a robust system with wide operator latitude and immediate confirmation of a successful application by both diver and surface the technique is well suited to cope in the hostile environment.

Where large equipment cannot gain access, magnetization is by permanent magnets with a standard pull-off strength, although use will be limited away from flat material, and repeated application at 6-in intervals is required.

When working near to the air–sea boundary, the magnetization is derived from flexible coils driven from surface transformers and the leakage flux from cracks disclosed by fluorescent ink supplied from the surface. A.C. is used to facilitate surface crack detection. The flexible cables carrying the magnetization current are wrapped around a member or laid in a parallel conductor arrangement along the weld. At lower levels the primary energy is taken to a subsea transformer to minimize power losses. The transformer houses an ink reservoir which dilutes the concentrate 10:1 with seawater. Ink illumination is from hand-held ultra-violet lamps which also support the ink dispenser (Figure 5.38). At depth the low ambient light suits fluorescent inspection whilst in shallow conditions inspection during the night is preferred. Photographic recording of indications is normal along with written notes and any CCTV recording available.

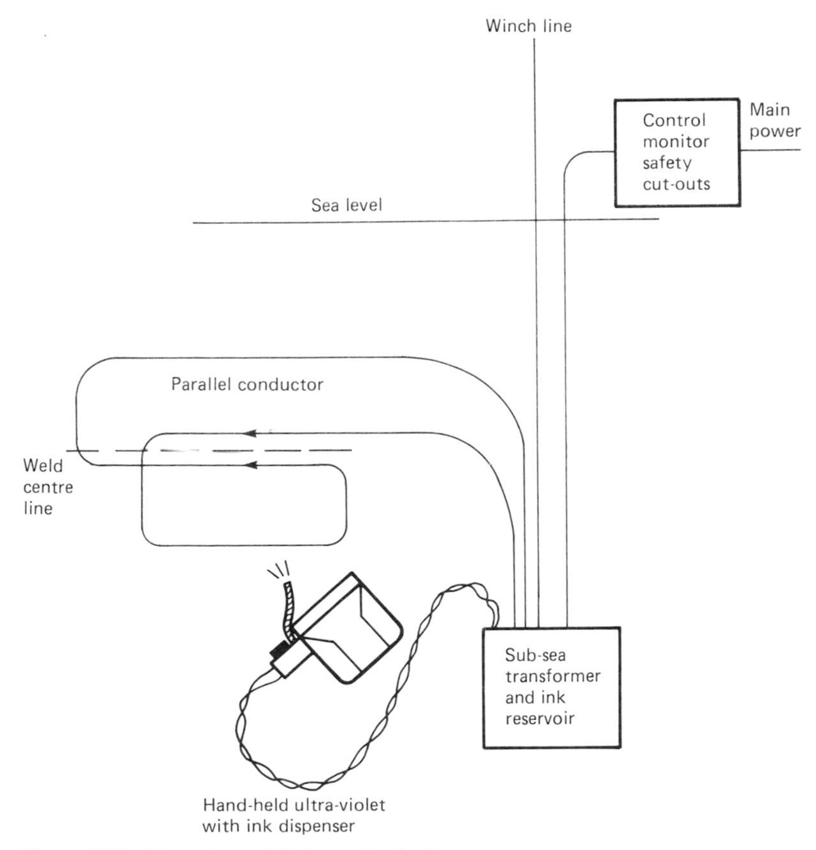

Figure 5.38 Magnetic particle inspection in deep water.

5.6.5 Ultrasonics

Ultrasonic non-destructive testing is mainly concerned with simple thickness checking to detect erosion and corrosion. Purpose-built probes detect the back-wall ultrasonic echo and will be hand held with rechargeable power supplies providing full portability. The units may be self-activating, switching on only when held on the test material, and are calibrated for steel although alternative calibration is possible. Ideally the unit therefore has no diver controls and can display a range of steel thickness from 1 to 300 mm on a digital readout. Where detailed examination of a weld is required then either a conventional surface A-scan set housed in a waterproof container with external extensions to the controls is used, or the diver operates the probes of a surface-located set under the direction of the surface non-destructive testing engineer. In either case, there is a facility for monitoring of the master set by a slave set. The diver, for instance, may be assisted by an audio feedback from the surface set which triggers on a threshold gate open to defect signals. The diver's hand movements will be monitored by a helmet video to allow monitoring and adjustment by the surface operator.

5.6.6 Corrosion protection

Because seawater behaves as an electrolytic fluid, corrosion of the steel structures may be inhibited by providing either sacrificial zinc anodes or impressing on the structure a constant electrical potential to reverse the electrolytic action. In order to check the operation of the impressed potential or the corrosion liability of the submerged structure, surface voltage is measured with reference to silver/silver chloride cell (Figure 5.39). Hand-held potential readers containing a reference cell, contact tip and digital reading voltmeter along with internal power supply will allow a survey to be completed by a diver which may be remotely monitored from the surface.

5.6.7 Other non-destructive testing techniques

There are a variety of other non-destructive techniques which have not yet gained common acceptance within the oil industry but are subject to varying degrees of investigation and experimental use. Some of these are described below.

5.6.7.1 Eddy current

Eddy-current techniques are described on pp. 148–150. The method can, with suitable head amplification, be used for a search-and-follow technique. Whilst it will not detect sub-surface cracks, the degree of surface cleaning both on the weld and to each side is not critical. Thus substantial savings in preparation and reprotection can be made.

5.6.7.2 A.C. potential difference (AC/PD)

As mentioned on p. 147, changes in potential difference can be used to detect surface defects. This is particularly valuable under water.

An a.c. will tend to travel just under the surface of a conductor because of the skin effect. The current flow between two contacts made on a steel specimen will approximately occupy a square with the contact points on one diagonal. In a uniform material there will be a steady ohmic voltage drop from contact to contact which will map out the current flow. In the presence of a surface crack orientated at right angles to the current flow there will be a step change in the potential which can be detected by two closely spaced voltage probes connected to a sensitive voltmeter (Figure 5.40). Crack penetration, regardless of attitude, will also influence the step voltage across the surface crack and allow depth estimation. The method

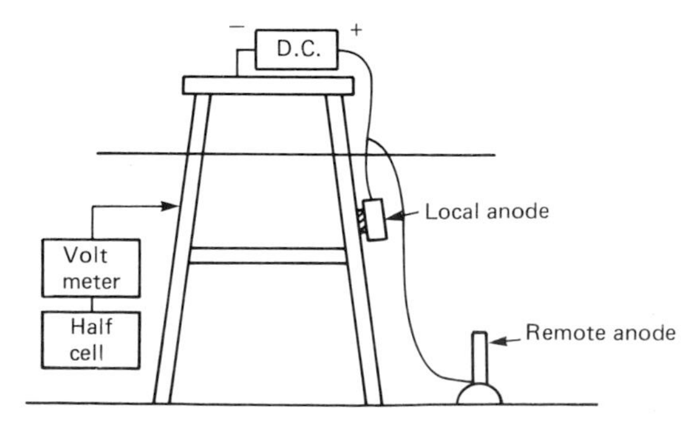

Figure 5.39 Measuring cathodic protection.

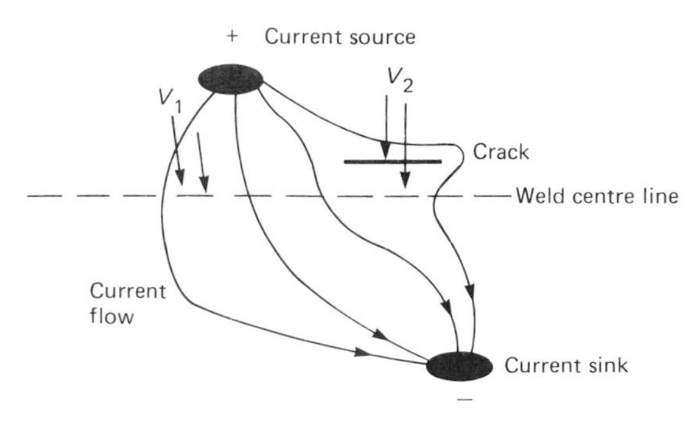

Figure 5.40 A.C. potential difference (AC/PD).

relies upon the efficiency of the contact made by the current driver and the voltage probe tips, and limitations occur because of the voltage safety limitations imposed on electrical sources capable of producing the constant current required. The voltages are limited to those below the optimum required to break down the surface barriers. The technique, however, is an obvious choice to first evaluate MPI indications in order to differentiate purely surface features (for example, grinding marks) from cracks.

5.6.7.3 Bulk ultrasonic scanning

The alternative to adapting surface non-destructive testing methods is to use the surrounding water as a couplant to transfer a large amount of ultrasonic energy into the structure and then monitor the returning energy by scanning either a single detector or the response from a tube from which acoustic energy can be used to construct a visual image (Figure 5.41). Such techniques are experimental but initial results indicate that very rapid inspection rates can be obtained with the diver entirely relegated to positioning the sensor. Analysis of the returning information is made initially by microcomputers, which attempt to remove the background variation and highlight the signals which alter the sensor position or scan. The devices do not attempt to characterize defects in detail, and for this other techniques are required.

5.6.7.4 Acoustic emission

Those developments which remove the need for a diver at every inspection are attractive but, as yet, are not fully proven. Acoustic emission is detected using probes fixed to the structure, 'listening' to the internal noise. As described on p. 160, the system relies upon the stress concentrations and fatigue failures to radiate an increasing amount of energy as failure approaches and this increased emission is detected by the probes.

5.7 Developments

Developments in non-destructive testing instrumentation are occurring rapidly not only to make existing inspection more reliable but also to prepare for future requirements. Acoustic emission which is one growth area has been considered earlier (p. 160). Another growth area is material monitoring techniques for automated production or in-service inspections. Space does not allow many of the existing developments currently underway to be considered. It can be said, however, that benefits to non-destructive testing will accrue when equipment has some degree of machine intelligence to enable storage of test conditions and so assist in maintaining reproducibility and similar functions. Similarly, equipment requires the recording of data and the aid of some form of 'electronic notebook'. Microsize (referred to below) allows a technician to concentrate on conducting the test while the equipment records the data for future analysis as

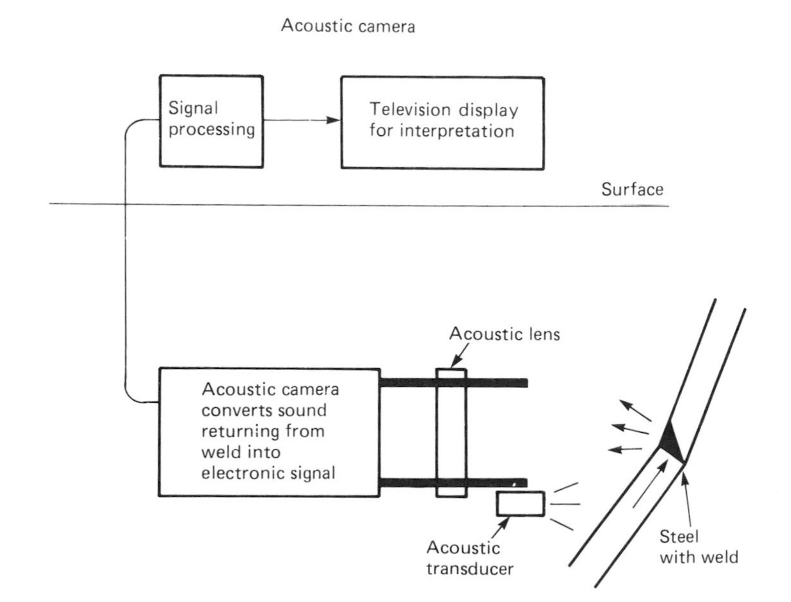

Figure 5.41 Bulk ultrasonic scanning.

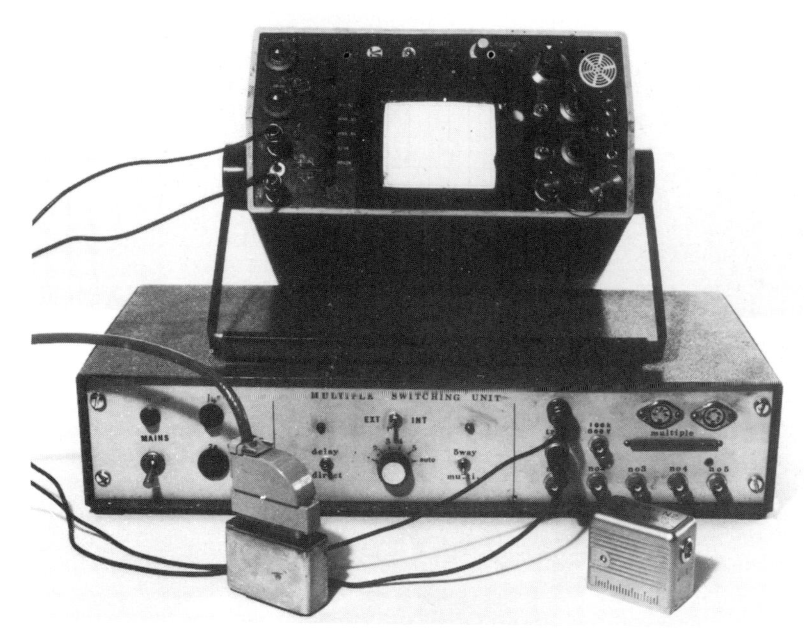

Figure 5.42 Variable-angle probe system. Courtesy, Babcock Power.

required. A permanent record is the objective for all methods of non-destructive testing.

Work at Babcock Power as described by Farley *et al.* (1982) gives details of the following:

(1) An electronically controlled, curved-array, variable-angle probe which has the potential for allowing more beam angles than normal to be applied in a shorter overall time (Figure 5.42).
(2) A microprocessor aid to flaw plotting and sizing (Microsize) which has been designed as a portable tool to complement the experience and skills of present ultrasonic technicians in applying conventional British Standard methods of flaw assessment (Figure 5.43); and
(3) An automated ultrasonic examination system utilizing a programmable ultrasonic flaw detector (Micropulse) under computer control via a highly flexible software package which allows rapid reprogramming for new geometries or ultrasonic techniques.

In parallel with this, improved methods of sizing defects are exemplified by the Harwell developed time-of-flight diffraction technique (TOFD) (Carter, 1984) while specific criteria for ultrasonic equipment and inspection procedures are well documented in the electrical supply industry (ESI) standards.

A new 'family' of eddy-current testing equipment is gaining acceptance. Its benefits lie in the incorporation of features to eliminate, as far as possible, the distracting number of variations in eddy-current testing. Equipment may have a vector display to show material changes due to defects along the vertical axis as opposed to changes associated with lift-off along the horizontal one.

In the case of ferritic steels with more conventional instrumentation it is often difficult to obtain a high degree of separation between the defect and lift-off channels; a small amount of lift-off often appears in the vertical displacement. Thus as the probe scans across a rough surface the continual lift-off changes will give rise to spurious defect indications. Examples of the equipment produced by Thornburn Technics are shown in Figure 5.44, in which the problem is overcome. An alternative to the Thorburn equipment is the Hocking Instruments AV10 unit.

5.8 Certification of personnel

No overview of non-destructive testing would be complete without some reference to the various operator-certification schemes currently in use worldwide. The range of products, processes and materials to which such methods have been applied has placed more and more demands on the skills and abilities of its practitioners. Quality assurance requires not only the products to have fitness-for-purpose but also the relevant personnel.

Developments in various countries in operator

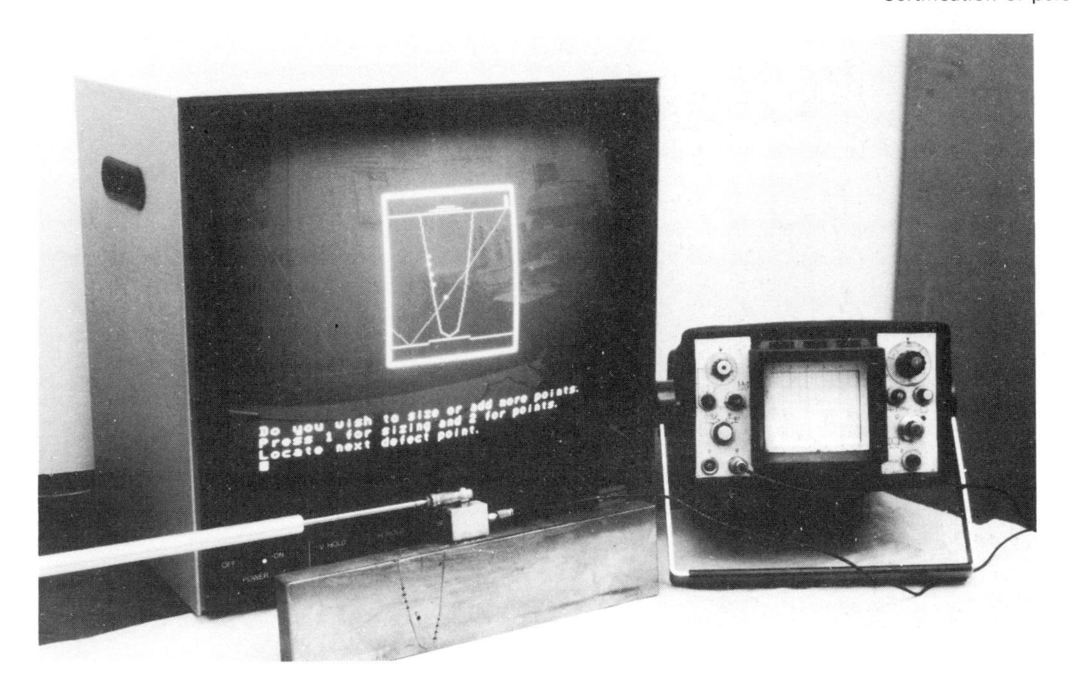

Figure 5.43 Microsize—display of simulated defects in a weld. Courtesy, Babcock Power.

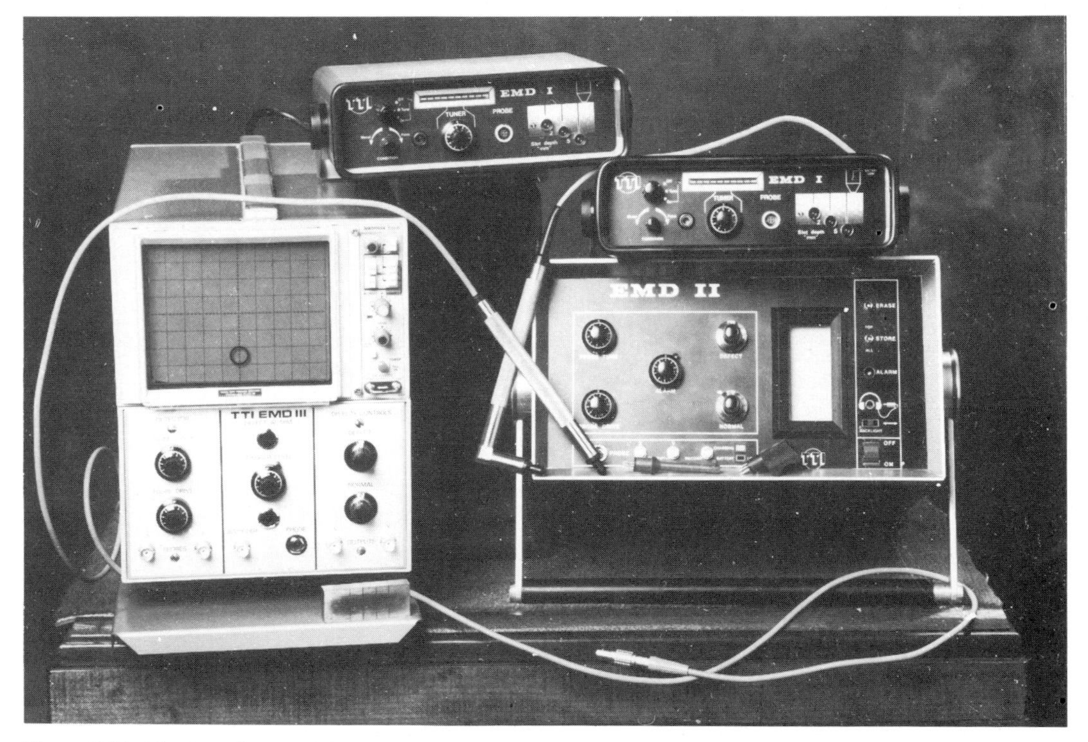

Figure 5.44 Thorburn Technic EMD equipment. Courtesy, Thorburn Technics (International) Ltd.

certification for non-destructive testing are given by Drury (1979). In the UK at the moment the most widely recognized certifying organization is the Certification Scheme for Weldment Inspection Personnel (CSWIP). This scheme, developed under the aegis of The Welding Institute and the British Institute of Non-destructive Testing, is operated for an independent management board representative of trade associations, insurance companies, classification societies, inspection organizations, nationalized industries, government departments and other bodies. Other UK approval authorities include:

British Rail
Central Electricity Generating Board
Gas Council
Ministry of Defence
Steel Castings Research Association
West Bromwich College

A new British Scheme introduced in 1987 and to which many of the above organizations have contributed, should unify UK interests in certification. The scheme is known as Personnel Certification in NDT (PCN) for the qualification and certification of personnel engaged in NDT and related inspection. The scheme, developed under the aegis of The British National Committee for Non-Destructive Testing and The British Institute of Non Destructive Testing is administered by an independent Central Certification Board (CCB) representative of trade associations, insurance companies, classification societies, inspection organizations, nationalized industries, government departments and other bodies. The CCB coordinates the activities of four specialist boards concerned with aerospace, castings, wrought products and welding. Each board is required to work to a common syllabus for the general theory part of written examinations, and to draft requirements for its own applications sector to satisfy the needs of the industrial activities it serves. Further information on PCN may be obtained from the Secretary, The British Institute of Non-Destructive Testing, 1 Spencer Parade, Northampton, NN1 5AA.

5.9 References

BS 3683, *Glossary of Terms used in Non-destructive Testing—Part 3 Radiological Flaw Detection* (1964)

BS 3971, *Image Quality Indications* (1966)

BS 4094, *Recommendation for Data on Shielding from Ionising Radiation* (1966)

Blitz, J. *Ultrasonic Methods and Applications*, Butterworths, London (1971)

Carter, P. 'Experience with the time-of-flight diffraction technique and accompanying portable and versatile ultrasonic digital recording system', *Brit. J. NDT*, **26**, 354 (1984)

Drury, J. 'Developments in various countries in operator certification for non-destructive testing', in *Developments in Pressure Vessel Technology*, Vol. 2, Applied Science, London (1979)

Electricity Supply Industry, Standards published by Central Electricity Generating Board

Ensminger, D. *Ultrasonics*, Dekker, New York (1973)

Erf, R. K. *Holographic Non-destructive Testing*, Academic Press, London (1974)

Farley, J. M. *et al.* 'Developments in the ultrasonic instrumentation to improve the reliability of NDT of pressurised components', *Conf. on In-service Inspection*, Institution of Mechanical Engineers (1982)

Filipczynski, L. *et al. Ultrasonic Methods of Testing Materials*, Butterworths, London (1966)

Greguss, P. *Ultrasonic Imaging*, Focal Press, London (1980)

Halmshaw, R. *Industrial Radiology: Theory and Practice*, Applied Science, London (1982)

HMSO, *The Ionising Radiations (Sealed Sources) Regulations* (1969)

Institute of Welding, *Handbook on Radiographic Apparatus and Techniques*

Krautkramer, J. and Krautkramer, H. *Ultrasonic Testing of Materials*, 3rd edn, Springer-Verlag, New York (1983)

Marsh, D. M. Means of visualizing ultrasonics, Chapter 10 in *Research Techniques in Non-destructive Testing*, Vol. 2 (ed. R. S. Sharpe). Academic Press, London (1973)

Martin, A. and Harbison, S. *An Introduction to Radiation Protection*, Chapman and Hall, London (1972)

McGonnigle, W. J. *Non-destructive Testing*, Gordon and Breach, London (1961)

Sharpe, R. S. (ed.). *Research Techniques in Non-destructive Testing*, Vols 1–5, Academic Press, London (1970–1982)

Sharpe, R. S. *Quality Technology Handbook*, Butterworths, London (1984)

Appendix 5.1 Fundamental standards used in ultrasonic testing

British Standards	(BS)	
BS 2704	(78)	Calibration Blocks
BS 3683 (Part 4)	(77)	Glossary of Terms
BS 3889 (Part 1A)	(69)	Ultrasonic Testing of Ferrous Pipes
BS 3923		Methods of Ultrasonic Examination of Welds
BS 3923 (Part 1)	(78)	Manual Examination of Fusion Welds in Ferritic Steel
BS 3923 (Part 2)	(72)	Automatic Examination of Fusion Welded Butt Joints in Ferritic Steels
BS 3923 (Part 3)	(72)	Manual Examination of Nozzle Welds
BS 4080	(66)	Methods of Non-Destructive Testing of Steel Castings

BS 4124	(67)	Non-Destructive Testing of Steel Forgings—Ultrasonic Flaw Detection
BS 4331		Methods for Assessing the Performance Characteristics of Ultrasonic Flaw Detection Equipment
BS 4331 (Part 1)	(78)	Overall Performance—Site Methods
BS 4331 (Part 2)	(72)	Electrical Performance
BS 4331 (Part 3)	(74)	Guidance on the In-service Monitoring of Probes (excluding Immersion Probes)
BS 5996	(81)	Methods of Testing and Quality Grading of Ferritic Steel Plate by Ultrasonic Methods
M 36	(78)	Ultrasonic Testing of Special Forgings by an Immersion Technique
M 42	(78)	Non-Destructive Testing of Fusion and Resistance Welds in Thin Gauge Material

American Standards		(ASTM)
E 114	(75)	Testing by the Reflection Method using Pulsed Longitudinal Waves Induced by Direct Contact
E 127	(75)	Aluminium Alloy Ultrasonic Standard Reference Blocks
E 164	(74)	Ultrasonic Contact Inspection of Weldments
E 213	(68)	Ultrasonic Inspection of Metal Pipe and Tubing for Longitudinal Discontinuities
E 214	(68)	Immersed Ultrasonic Testing
E 273	(68)	Ultrasonic Inspection of Longitudinal and Spiral Welds of Welded Pipes and Tubings
E 317	(79)	Performance Characteristics of Pulse Echo Ultrasonic Testing Systems
E 376	(69)	Seamless Austenitic Steel Pipe for High Temperature Central Station Service
E 388	(71)	Ultrasonic Testing of Heavy Steel Forgings
E 418	(64)	Ultrasonic Testing of Turbine and Generator Steel Rotor Forgings
E 435	(74)	Ultrasonic Inspection of Steel Plates for Pressure Vessels
E 50	(64)	Ultrasonic Examination of Large Forged Crank Shafts
E 531	(65)	Ultrasonic Inspection of Turbine-Generator Steel Retaining Rings
E 557	(73)	Ultrasonic Shear Wave Inspection of Steel Plates
E 578	(71)	Longitudinal Wave Ultrasonic Testing of Plain and Clad Steel Plates for Special Applications
E 609	(78)	Longitudinal Beam Ultrasonic Inspection of Carbon and Low Alloy Steel Castings

West German Standards	(DIN)—Translations available
17175	Seamless Tubes of Heat Resistant Steels: Technical Conditions of Delivery
17245	Ferritic Steel Castings Creep Resistant at Elevated Temperatures: Technical Conditions of Delivery
54120	Non-Destructive Testing: Calibration Block 1 and its Uses for the Ajustment and Control of Ultrasonic Echo Equipment
54122	Non-Destructive Testing: Calibration Block 2 and its Uses for the Adjustment and Control of Ultrasonic Echo Equipment

6 Noise measurement

J. KUEHN

6.1 Sound and sound fields

6.1.1 The nature of sound

If any elastic medium, whether it be gaseous, liquid or solid, is disturbed then this disturbance will travel away from the point of origin and be propagated through the medium. The way in which the disturbance is propagated and its speed will depend upon the nature and extent of the medium, its elasticity and density.

In the case of air, these disturbances are characterized by very small fluctuations in the density (and hence atmospheric pressure) of the air and by movements of the air molecules. Provided these fluctuations in pressure take place at a rate (frequency) between about 20 Hz and 16 kHz, then they can give rise to the sensation of audible sound in the human ear. The great sensitivity of the ear to pressure fluctuations is well illustrated by the fact that a peak-to-peak fluctuation of less than 1 part in 10^9 of the atmospheric pressure will, at frequencies around 3 kHz, be heard as audible sound. At this frequency and pressure the oscillating molecular movement of the air is less than 10^{-7} of a millimetre.

The magnitude of the sound pressure at any point in a sound field is usually expressed in terms of the rms (root-mean-square) value of the pressure fluctuations in the atmospheric pressure. This value is given by

$$P_{\text{ms}} = \sqrt{\left(\frac{1}{T}\int_0^T P^2(k)\,\mathrm{d}t\right)} \qquad (6.1)$$

where $P(t)$ is the instantaneous sound pressure at time t and T is a time period long compared with the periodic time of the lowest frequency present in the sound.

The SI unit for sound pressure is Newton/m^2 (N/m^2), which is now termed pascal, that is, 1 Newton/m^2 = 1 pascal. Atmospheric pressure is, of course, normally expressed in bars (1 bar = 10^5 pascal).

The great sensitivity of the human hearing mechanism, already mentioned, is such that at frequencies where it is most sensitive it can detect sound pressures as small as 2×10^{-5} pascals. It can also pick up sound pressures as high as 20 or even 100 pascals. When dealing with such a wide dynamic range (pressure range of 10 million to 1) it is inconvenient to express sound pressures in terms of pascals and so a logarithmic scale is used. Such a scale is the decibel scale. This is defined as ten times the logarithm to the base 10 of the ratio of two powers. When applying this scale to the measurement of sound pressure it is assumed that sound power is related to the square of the sound pressure. Thus

$$\text{Sound pressure level} = 10\log_{10}\left(\frac{W_1}{W_0}\right)$$

$$= 10\log_{10}\left(\frac{P^2}{P_0^2}\right)$$

$$= 20\log_{10}\left(\frac{P}{P_0}\right) \quad \text{dB} \qquad (6.2)$$

where P is the sound pressure being measured and P_0 is a 'reference' sound pressure (standardized at 2×10^{-5} pascals).

It should be noted that the use of the expression 'sound-pressure *level*' always denotes that the value is expressed in decibels. The reference pressure will, of course, be the 0 dB value on the decibel scale. The use of a reference pressure close to that representing the threshold of hearing at the frequencies where the ear is most sensitive means that most levels of interest will be positive. (A different reference pressure is used for underwater acoustics and for the 0 dB level on audiograms.) A good example of the use of the decibel scale of sound-pressure level and of the complicated response of the human ear to pure tones, is given in Figure 6.1, showing equal loudness contours for pure tones presented under binaural, free-field listening conditions.

The equal loudness level contours of Figure 6.1 (labelled in Phons) are assigned numerical values equal to that of the sound-pressure level at 1 kHz through which they pass. This use of 1 kHz as a 'reference' frequency is standardized in acoustics. It is the frequency from which the audible frequency range is 'divided-up' when choosing octave-band widths or one-third octave-band widths. Thus the octave band centred on 1 kHz and the third octave-band width

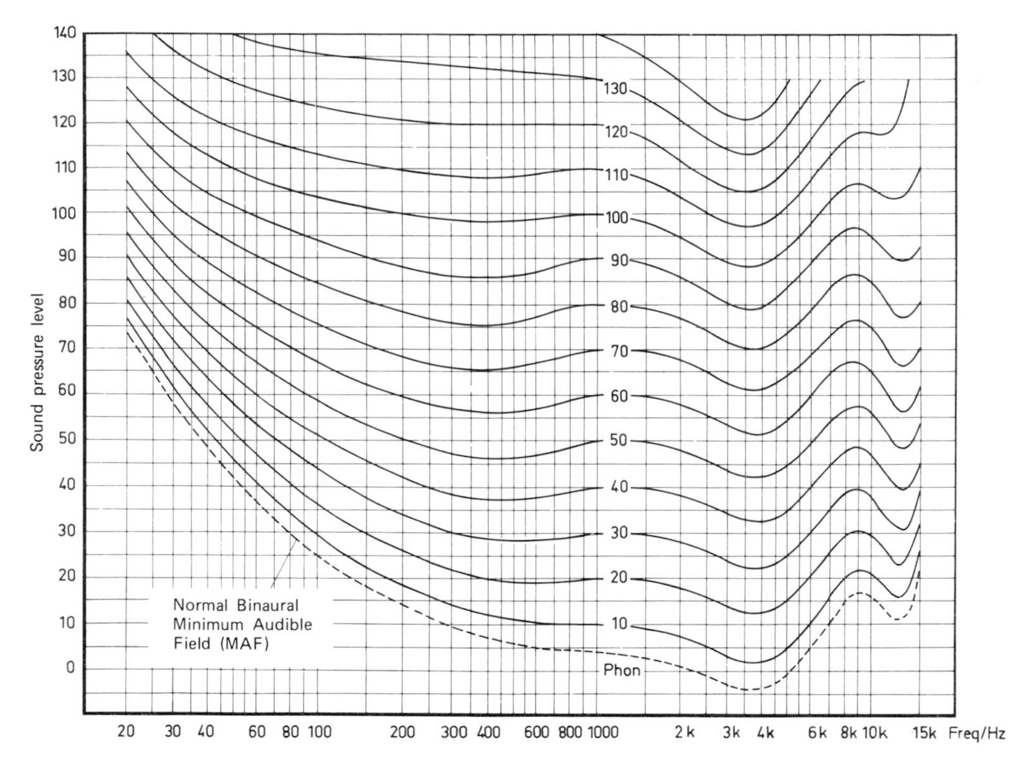

Figure 6.1 Normal equal loudness contours for pure tones. (Most of the figures in this chapter are reproduced by courtesy of Bruel & Kjaer (UK) Ltd.)

centred on 1 kHz are included in the 'standardized' frequency bands for acoustic measurements.

In daily life, most sounds encountered (particularly those designated as 'noise') are not the result of simple sinusoidal fluctuations in atmospheric pressure but are associated with pressure waveforms which vary with time both in frequency and magnitude. In addition, the spatial variation of sound pressure (i.e. the sound field) associated with a sound source is often complicated by the presence of sound-reflecting obstacles or walls.

6.1.2 Quantities characterizing a sound source or sound field

The definition of noise as 'unwanted sound' highlights the fact that the ultimate measure of a sound, as heard, involves physiological and often psychological measurements and assessments. Objective measurements can be used to evaluate the noise against predetermined and generally acceptable criteria.

When planning the installation of machinery and equipment it is often necessary to be able to predict the magnitude and nature of the sound fields at a distance from the noise source, which means using noise data associated only with the noise source and not with the environment in which the measurements were made.

Although the sound-pressure level, at a point, is the commonest measure of a sound field it is far from being a comprehensive measure. As a scalar quantity it provides no information on the direction of propagation of the sound and, except in the case of very simple sound fields (such as that from a physically small 'point' source or a plane wave—see BS 4727), it is not directly related to the sound power being radiated by the source. The importance, in some cases, of deducing or measuring the sound power output of a source and its directivity has received much attention in recent years, and equipment is now available (as research equipment and also commercially available for routine measurements) for measuring particle velocity or the sound intensity (W/m^2) at a point in a sound field.

In addition to some measure of the overall sound-pressure level at a point it is also usually necessary to carry out some frequency analysis of the signal to find out how the sound pressure levels are distributed throughout the audible range of frequencies.

6.1.3 Velocity of propagation of sound waves

As mentioned earlier, sound-pressure fluctuations (small compared with the atmospheric pressure) are propagated through a gas at a speed which is dependent on the elasticity and density of the gas. An appropriate expression for the velocity of propagation is

$$c=\sqrt{\left(\frac{\gamma P_0}{\rho_0}\right)} \qquad (6.3)$$

where γ is the ratio of the specific heat of the gas at constant pressure to that at constant volume (1.402 for air), P_0 is the gas pressure (Newtons/m^2) and ρ_0 is the density of the gas (kg/m^3). This expression leads to a value of 331.6 m/s for air at 0°C and a standard barometric pressure of 1.013×10^5 Newtons/m^2 (1.013 bar).

The use of the general gas law also leads to an expression showing that the velocity is directly proportional to the square root of the absolute temperature in K:

$$c=c_0\sqrt{\left(\frac{t+273}{273}\right)} \text{ m/s} \qquad (6.4)$$

where c_0 is the relocity of sound in air at 0°C (331.6 m/s) and t is the temperature (°C).

The speed of propagation of sound in a medium is an extremely important physical property of that medium, and its value figures prominently in many acoustic expressions. If sound propagates in one direction only then it is said to propagate as a 'plane free progressive wave' and the ratio of the sound pressure to the particle velocity, at any point, is always given by $\rho_0 c$.

Knowledge of the velocity of sound is important in assessing the effect of wind and of wind and temperature gradients upon the bending of sound waves. The velocity of sound also determines the wavelength (commonly abbreviated to λ) at any given frequency and it is the size of a source *in relation to the wavelength of the sound radiation* that greatly influences the radiating characteristics of the source. It also determines the extent of the near-field in the vicinity of the source. The screening effect of walls, buildings and obstacles is largely dependent on their size in relation to the wavelength of the sound. Thus a vibrating panel 0.3 m square or an open pipe of 0.3 m diameter would be a poor radiator of sound at 50 Hz [$\lambda=c/f\simeq6.6$ m] but could be an excellent, efficient radiator of sound at 1 kHz [$\lambda\simeq0.33$ m] and higher frequencies. The relation between frequency and wavelength is shown in Figure 6.2.

6.1.4 Selecting the quantities of interest

It cannot be emphasized too strongly that before carrying out any noise measurements the ultimate objective of those measurements must be clearly established so that adequate, complete and appropriate measurements are made.

The following examples indicate how the purpose of a noise measurement influences the acoustic quantities to be measured, the techniques to be employed and the supporting environmental or operational measurements/observations which may be needed.

6.1.4.1 Measurements to evaluate the effect of the noise on human beings

The ultimate aim might be to assess the risk of permanent or temporary hearing loss as a result of exposure to the noise or perhaps the ability of the noise to mask communication or to assess the likely acceptability of the noise in a residential area.

In these cases frequency analysis of the noise may be required, i.e. some measure of its fluctuation in level with time and a measure of the likely duration of the noise. Other factors of importance could be the level of other (background) noises at various times of the day

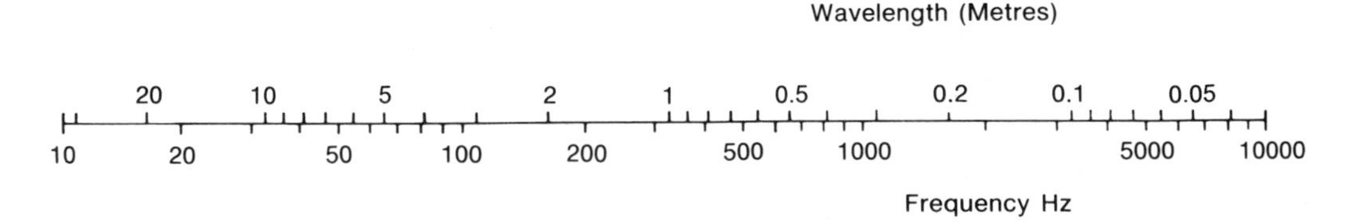

Figure 6.2 Relation between wavelength and frequency.

and night, weather conditions and the nature of the communication or other activity being undertaken. A knowledge of the criteria against which the acceptability or otherwise of the noise is to be judged will also usually indicate the appropriate bandwidths to be used in the noise measurements. In some cases, octave band or one-third octave band measurements could suffice whilst in other cases narrow-band analyses might be required with attendant longer sampling times.

6.1.4.2 Measurements for engineering design or noise-control decisions

Most plant installations and many individual items of machinery contain more than one source of noise and when the ultimate objective is plant-noise control the acoustic measurements have to:

(1) Establish the operational and machine installation conditions which give rise to unacceptable noisy operation;
(2) Identify and quantify the noise emission from the various sources;
(3) Establish an order of priority for noise control of the various sources;
(4) Provide adequate acoustic data to assist the design and measure the performance of noise control work.

These requirements imply that detailed noise analyses are always required and that these acoustic measurements must be fully supported by other engineering measurements to establish the plant operating and installation conditions and, where possible, assist in the design of the noise-control work that will follow. Such measurements could include measurements of temperatures, pressures, gas flows and vibration levels. All too often the value of acoustic measurements are much reduced because of the absence of these supporting measurements.

6.1.4.3 Noise labelling

In some cases 'noise labelling' of machines may be required by government regulations and by purchasers of production machinery, vehicles, plant used on construction sites, etc. Regulations, where they exist, normally specify the measurement procedure and there are British or international standards which apply (see Appendix).

6.1.4.4 Measurements for diagnostic purposes

Investigation of the often complex sources of noise of machines and of ways and means of reducing them may require very detailed frequency analysis of both the noise and vibrations and the advanced features offered by modern analysing and computing instrumentation.

In recent years, the use of acoustic measurements for 'fingerprinting' machinery to detect, in the early stages, changes in the mechanical condition of machinery has been further developed in conjunction with measurements of vibration for the same purpose. These measurements are also based on detailed frequency analyses.

6.2 Instrumentation for the measurement of sound-pressure level

The basic instrumentation chain consists of a microphone, signal-conditioning electronics, some form of filtering or weighting and a quantity indicator, analogue or digital. This is shown schematically in Figure 6.3.

6.2.1 Microphones

The microphone is the most important item in the

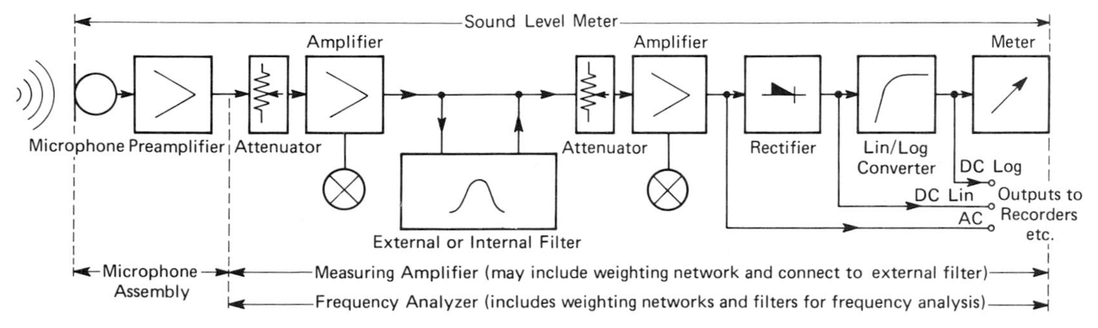

Figure 6.3 Block diagram of a noise-measuring system.

chain. It must be physically small so as not to disturb the sound field and hence the sound pressure which it is trying to measure, it must have a high acoustic impedance compared with that of the sound field at the position of measurement, it must be stable in its calibration, have a wide frequency response and be capable of handling without distortion (amplitude or phase distortion) the very wide dynamic range of sound-pressure levels to which it will be subjected. When used in instruments for the measurement of low sound-pressure levels it must also have an inherent low self-generated noise level. To permit of the use of closed cavity methods of calibration it should also have a well-defined diaphragm plane.

All these requirements are best met by the use of condenser or electret microphones, and with such high electrical impedance devices it is essential that the input stage for the microphone—which might take the form of a 0 dB gain impedance transforming unit—be attached directly to the microphone so that long extension cables may be used between the microphone and the rest of the measuring instrument.

6.2.1.1 Condenser microphone

The essential components of a condenser microphone are a thin metallic diaphragm mounted close to a rigid metal back plate from which it is electrically isolated. An exploded view of a typical instrument is given in Figure 6.4. A stabilized d.c. polarization voltage E_0 (usually around 200 V) is applied between the diaphragm and the backplate, fed from a high-resistance source to give a time constant $R(C_t+C_s)$ (see Figure 6.5) much longer than the period of the lowest frequency sound-pressure variation to be measured.

If the sound pressure acting on the diaphragm produces a change in the value of C_t—due to movement of the diaphragm—of $\Delta C(t)$ then the output voltage V_0 fed from the microphone to the preamplifier will be

$$V_0(t)=\frac{\Delta C(t)\,.\,E_0}{C_t+C_s}$$

since $C_t \gg \Delta C(t)$. It should be noted that the microphone sensitivity is proportional to the polarization voltage but inversely proportional to the total capacitance C_t+C_s. Moreover, if we wish the microphone to have a pressure sensitivity independent of frequency then the value of $\Delta C(t)$ (and hence the deflection of the diaphragm), for a given sound pressure must be independent of frequency, that is, the diaphragm must be 'stiffness' controlled. This requires a natural frequency above that of the frequency of the sounds to be measured.

Condenser microphones, which are precision instruments, can be manufactured having outstanding stability of calibration and sensitivities as high as 50 mV per pascal. The selection of a condenser microphone for a specific application is determined by the frequency range to be covered, the dynamic range of sound-pressure levels and the likely incidence of the sound.

A condenser microphone is a pressure microphone, which means that the electrical output is directly

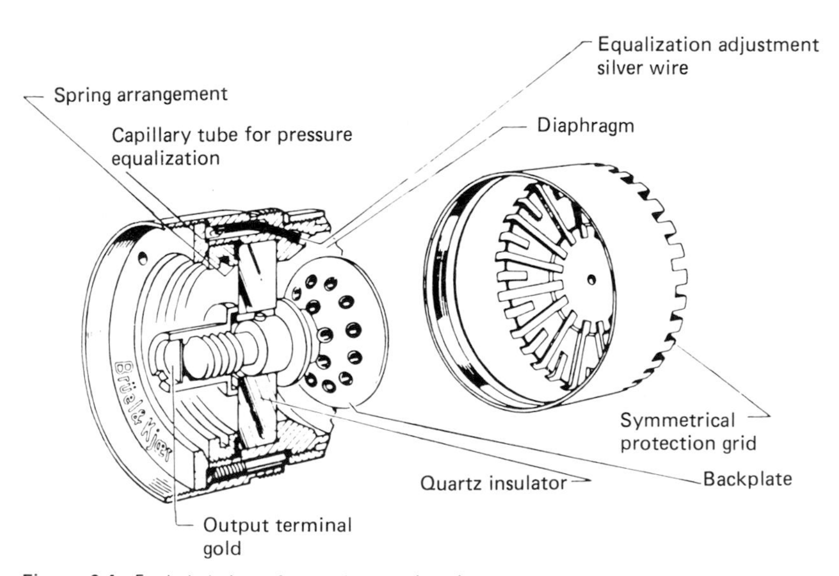

Figure 6.4 Exploded view of a condenser microphone.

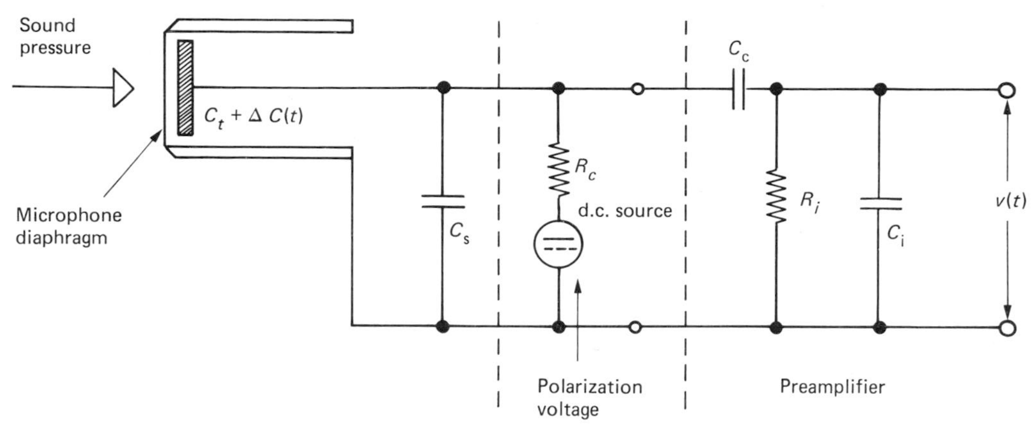

Figure 6.5 Equivalent circuit of a condenser microphone and microphone preamplifier.

proportional to the sound pressure acting on the diaphragm. At higher frequencies, where the dimensions of the diaphragm become a significant fraction of the wavelength, the presence of the diaphragm in the path of a sound wave creates a high impedance obstacle, from which the wave will be reflected. When this happens the presence of the microphone modifies the sound field, causing a higher pressure to be sensed by the diaphragm. A microphone with flat pressure response would therefore give an incorrect reading. For this reason, special condenser microphones known as free field microphones are made. They are for use in free field conditions, with perpendicular wave incidence on the diaphragm. Their pressure frequency response is so tailored as to give a flat response to the sound waves which would exist if they were not affected by the presence of the microphone. Free field microphones are used with sound level meters.

Effective pressure increases due to the presence of the microphone in the sound field are shown as free-field corrections in Figure 6.6 for the microphone whose outline is given. These corrections are added to the pressure response of the microphone. Figure 6.7 shows the directional response of a typical $\frac{1}{2}$-in condenser microphone. Microphones with flat pressure responses are mainly intended for use in couplers and should be used with grazing incidence in free-field conditions.

A complete condenser microphone consists of two parts, a microphone cartridge and a preamplifier, which is normally tubular in shape, with the same diameter as the cartridge. The preamplifier may be built into the body of a sound-level meter.

Generally, a larger diameter of microphone means higher sensitivity, but frequency-range coverage is inversely proportional to diaphragm dimensions. Most commonly used microphones are standardized 1 in and $\frac{1}{2}$ in, but $\frac{1}{4}$ in and even $\frac{1}{8}$ in are commercially available. The small-diaphragm microphones are suitable for the measurement of very high sound-pressure levels.

6.2.1.2 Electret microphone

Although requirements for precision measurement are best met by condenser microphones, the need to have a source of d.c. voltage for the polarization has led to the search for microphones which have an inherent polarized element. Such an element is called an electret. In the last ten years electrets have been produced which have excellent long-term stability and have been used to produce prepolarized condenser microphones. Many of the design problems have been overcome by retaining the stiff diaphragm of the conventional condenser microphone and applying the electret material as a thin coating on the surface of the backplate. The long-term stability of the charge-carrying element (after artificially ageing and stabilizing) is now better than 0.2 dB over a year.

The principle of operation of the electret condenser microphone is, of course, precisely the same as that of the conventional condenser microphone. At present, its cost is slightly in excess of that of a conventional condenser microphone of similar performance but it has application where lower power consumption and simplified associated electronics are at a premium.

6.2.1.3 Microphones for low-frequency work

The performance of condenser microphones at low frequencies is dependent on static pressure equalization and on the ratio of input impedance of the microphone preamplifier to the capacitive impedance of the microphone cartridge. For most measurements, good response to about 10 Hz is

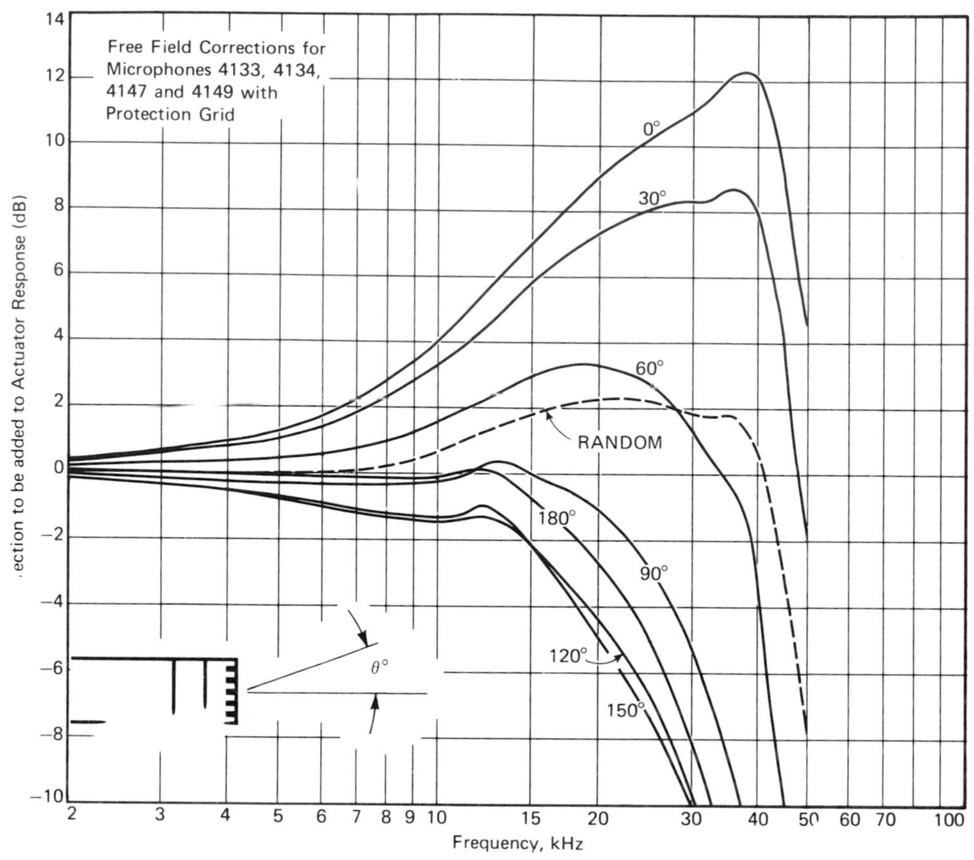

Figure 6.6 Directional characteristics of a half-inch microphone mounted on a sound level meter.

required and is provided by most microphones. Extension of performance to 1 or 2 Hz is avoided in order to reduce the sensitivity of a microphone to air movement (wind) and pressure changes due to opening and closing of doors and windows.

Microphones for low-frequency work exist, including special systems designed for sonic boom measurement, which are capable of excellent response down to a small fraction of 1 Hz.

6.2.2 Frequency weighting networks and filters

The perception of loudness of pure tones is known and shown in Figure 6.1. It can be seen that it is a function of both the frequency of the tone and of the sound pressure. Perception of loudness of complete sounds has attracted much research and methods have been devised for the computation of loudness or loudness level of such sounds.

Many years ago it was thought that an instrument with frequency-weighting characteristics corresponding to the sensitivity of the ear at low, medium and high sound intensities would give a simple instrument capable of measuring loudness of a wide range of sounds, simple and complex. Weightings now known as A, B and C were introduced, but it was found that, overall, the A weighting was most frequently judged to give reasonable results and the measurement of A-weighted sound pressure became accepted and standardized throughout the world. Their response curves are shown in Figure 6.8. Environmental noise and noise at work are measured with A weighting, and many sound-level measuring instruments no longer provide weightings other than A.

It cannot be claimed that A weighting is perfect, that the results of A-weighted measurements are a good measure of loudness or nuisance value of all types of noise, but at least they provide a standardized, commonly accepted method.

Where it is acknowledged that A weighting is a wrong or grossly inadequate description, and this is particularly likely when measuring low-frequency noise, sounds are analysed by means of octave or

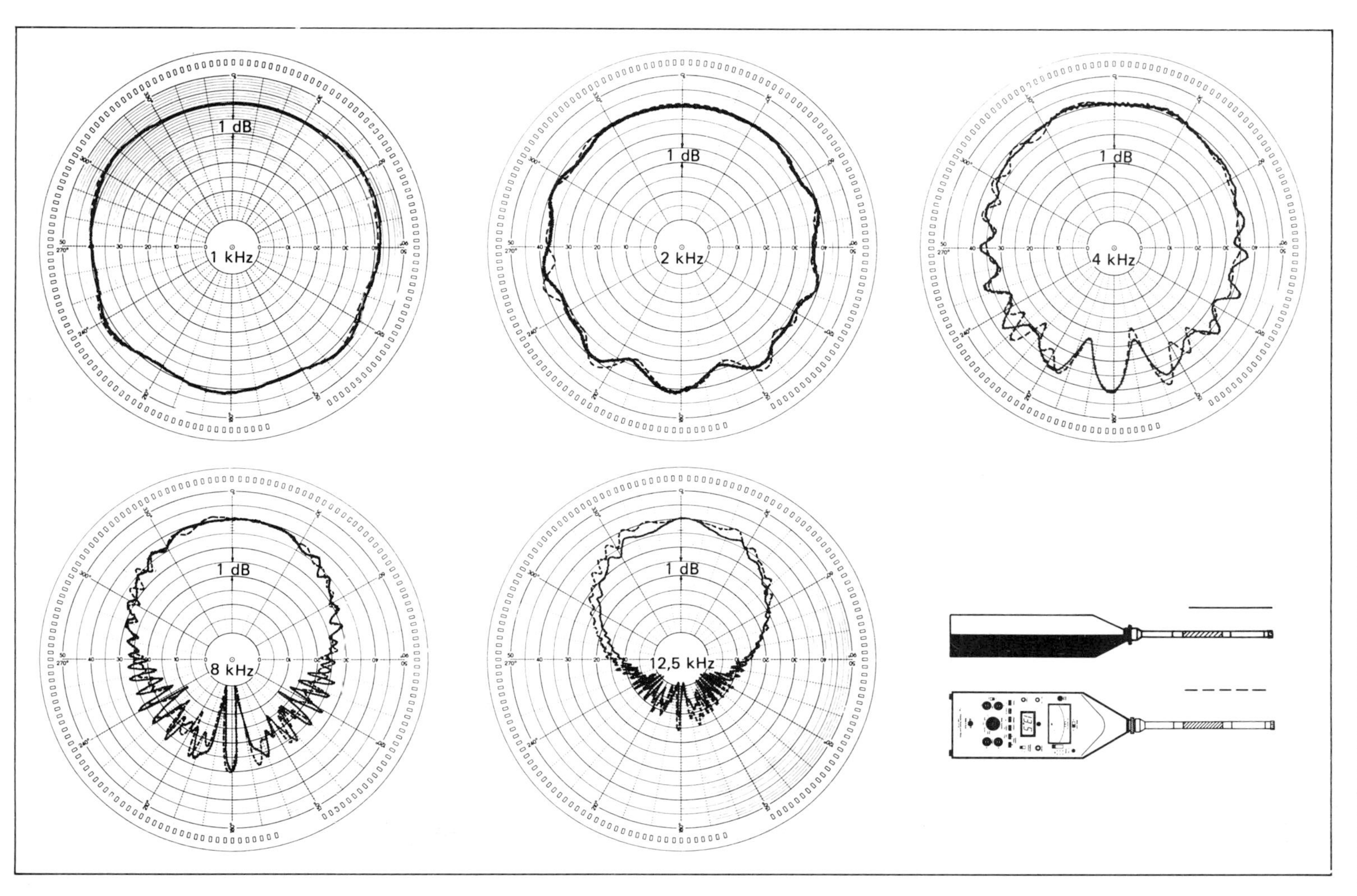

Figure 6.7 Free-field corrections to microphone readings for a series of B and K half-inch microphones.

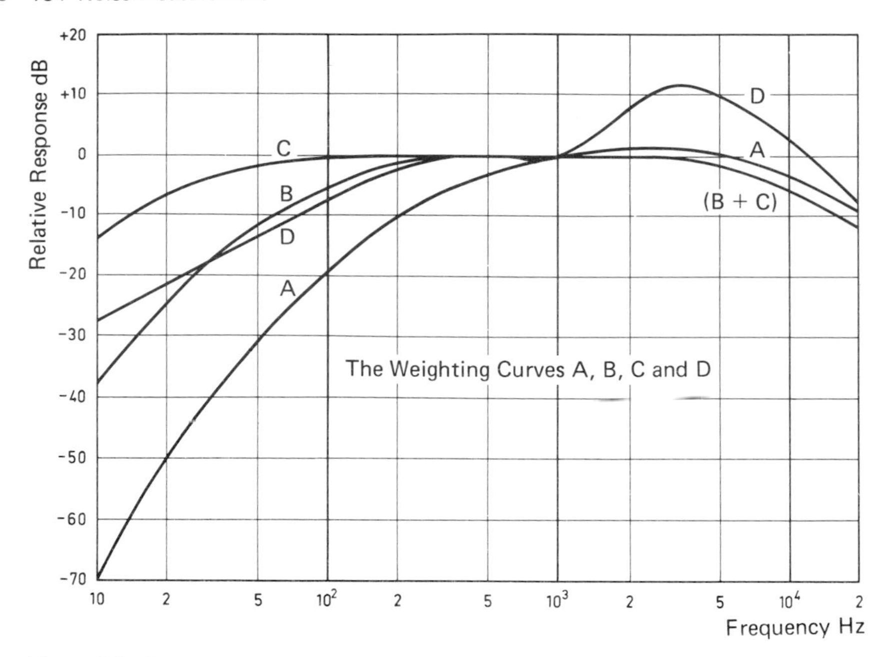

Figure 6.8 Frequency response curves of the A, B, C and D weighting networks.

third-octave filters and the results manipulated to give a desired value.

Aircraft noise is a special case, where the noise is analysed simultaneously in third-octave bands and individual band levels are sampled every 0.5 s, each one weighted according to a set of tables and summed, to give the answer in forms of a perceived noise level in decibels. An approximation to this figure may be obtained by the use of a simpler system, a sound-level meter with standardized 'D' weighting, also shown in Figure 6.8.

6.2.3 Sound-level meters

These are most widely used for the measurement of sound and can be purchased in forms ranging from simple instruments fitted with one frequency weighting network only and a fixed microphone to versions capable of handling a whole range of microphone probes and sophisticated signal processing with digital storage of data. The instrument therefore deserves special mention.

As with most other instruments, its cost and complication is related to the degree of precision of which it is capable. IEC Publication 651 and BS 5969 classify sound-level meters according to their degree of precision. The various classes (or types) and the intended fields of application are given in Table 6.1.

The performance specifications for all these types have the same 'centre value' requirements and they differ only in the tolerances allowed. These tolerances

Table 6.1 Classes of sound-level meter

Type	*Intended field of application*	*Absolute accuracy (at reference frequency) in reference direction at the reference sound-pressure level (dB)*
0	Laboratory reference standards	±0.4
1	Laboratory use and field use where the acoustical environment can be closely specified or controlled	±0.7
2	General field work	±1.0
3	Field noise surveys	±1.5

take into account such aspects as (1) variation of microphone response with the angle of incidence of the sound, (2) departures from design objectives in the frequency-weighting networks (tolerances allowed) and (3) level linearity of the amplifiers.

While it is not possible to state the precision of measurement for any type in any given application—since the overall accuracy depends on care with the measurement technique as well as the calibration of the measurement equipment—it is possible to state the instrument's absolute accuracy at the reference frequency of 1 kHz when sound is incident from the

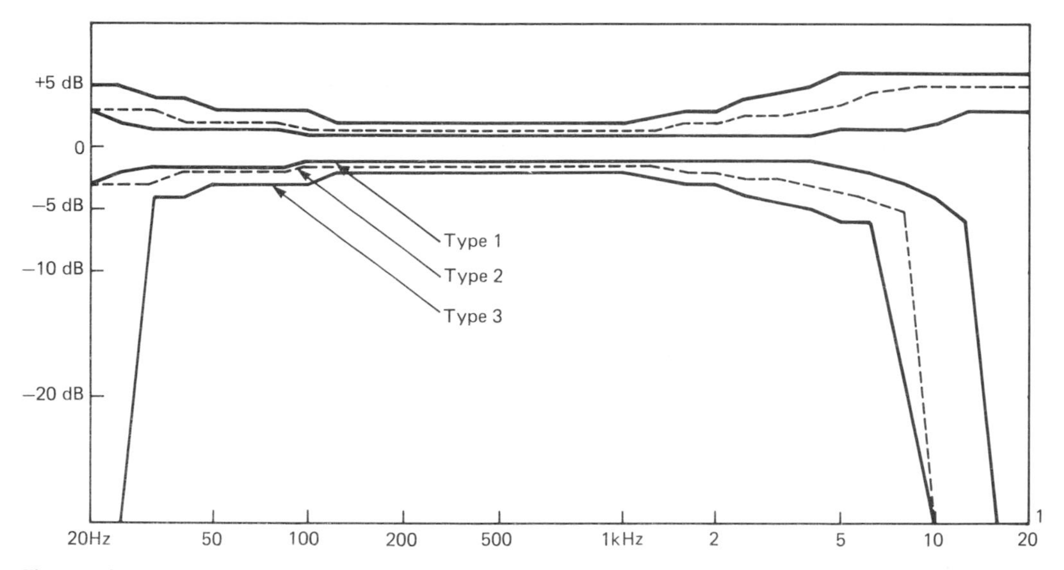

Figure 6.9 Tolerances in the response of sound-level meters at different frequencies.

reference direction at the reference sound-pressure level (usually 94 dB) and these values are given in Table 6.1.

Apart from the absolute accuracy at the reference frequency the important differences in performance are:

(1) Frequency range over which reliable measurements may be taken—with Type 2 and 3 instruments having much wider tolerances at both the low and the high ends of the audiofrequency range (see Figure 6.9).
(2) Validity of rms value when dealing with high crest factor (high ratio of peak to rms) signals, i.e. signals of an impulsive nature such as the exhaust of motorcycles and diesel engines.
(3) Validity of rms value when dealing with short-duration noises.
(4) Linearity of readout over a wide dynamic range and when switching ranges.

Prior to 1981 sound-level meters were classified either as 'Precision grade' (see IEC Publication 179 and BS 4197) or 'industrial grade' (see IEC 123 or BS 3489) with very wide tolerances. The current Type 1 sound-level meter is equivalent, broadly, to the old precision grade whilst Type 3 meters are the equivalent of the old industrial grade with minor modifications. A typical example of a simple sound-level meter which nevertheless conforms to Type 1 class of accuracy is shown in Figure 6.10. It has integrating facilities as discussed below. Figure 6.11 shows a more sophisticated instrument which includes filters but is still quite compact.

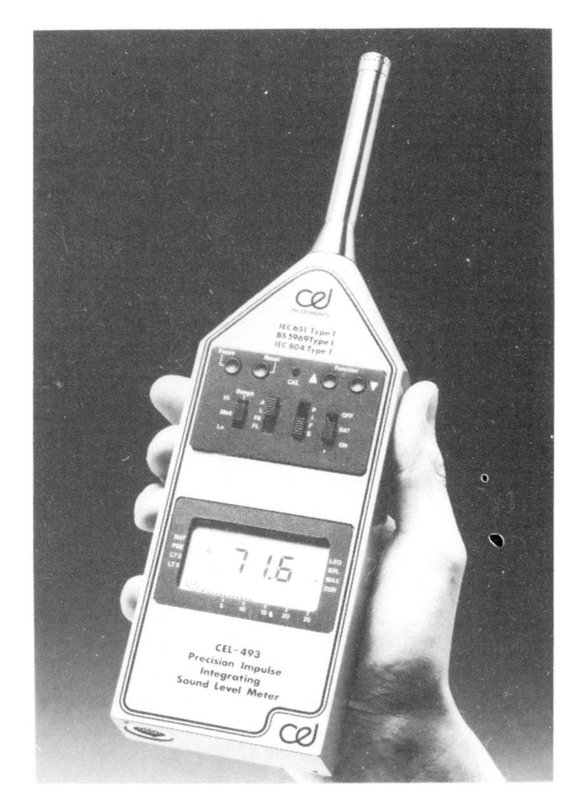

Figure 6.10 Simple sound-level meter. Courtesy, Lucas CEL Instruments Ltd.

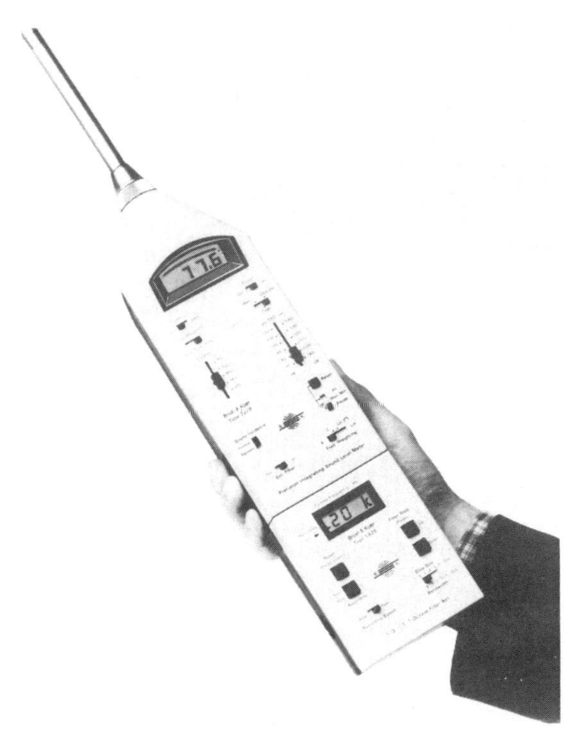

Figure 6.11 More sophisticated sound-level meter.

6.2.3.1 Integrating sound-level meters

Standard sound-level meters are designed to give the reading of sound level with fast or slow exponential time averaging. Neither is suitable for the determination of the 'average' value if the character of the sound is other than steady or is composed of separate periods of steady character. A special category of 'averaging' or 'integrating' meters is standardized and available which provide the 'energy average', measured over a period of interest.

The value measured is known as L_{eq} (the equivalent level), which is the level of steady noise which, if it persisted over the whole period of measurement, would give the same total energy as the fluctuating noise measured over the period.

As in the case of sound-level meters, Types 0 and 1 instruments represent the highest capability of accurately measuring noises which vary in level over a very wide dynamic range, or which contain very short bursts or impulse, for example gunshots, hammer-blows, etc. Type 3 instruments are not suitable for the accurate measurement of such events but may be used in cases where level variation is not wide or sudden.

The precise definition of the L_{eq} value of a noise waveform over a measurement period T is given by

$$L_{eq} = 10 \log \frac{1}{T} \int_0^T \frac{[p(t)]^2}{p_0^2} \, dt$$

where L_{eq} is the equivalent continuous sound level, T the measurement duration, $p(t)$ the instantaneous value of the (usually A-weighted) sound pressure and p_0 the reference rms sound pressure of 20 μPa.

A measure of the A-weighted L_{eq} is of great significance when considering the possibility of long-term hearing loss arising from habitual exposure to noise (see Barns and Robinson, 1970).

Most integrating sound-level meters also provide a value known as sound energy level or SEL, sometimes also described as single-event level. This represents the sound energy of a short-duration event such as an aircraft flying overhead or an explosion, but expressed as if the event lasted only 1 s. Thus

$$\text{SEL} = 10 \log \int \frac{[p(t)]^2}{p_0^2} \, dt$$

Many integrating sound-level meters allow the interruption of the integrating process or the summation of energy in separated periods.

6.2.3.2 Statistical sound-level meters

In the assessment of environmental noise—for example, that produced by city or motorway traffic—statistical information is often required. Special sound-level meters which analyse the sound levels are available. They provide probability and cumulative level distributions and give percentile levels, for example L_{10}, the level exceeded for 10 per cent of the analysis time as illustrated in Figures 6.12 and 6.13. Some of these instruments are equipped with built-in printers and can be programmed to print out a variety of values at preset time intervals.

6.2.4 Noise-exposure meters/noise-dose meters

The measurement of noise exposure to which a worker is subjected may be carried out successfully by an integrating sound-level meter if the worker remains in one location throughout the working day. In situations where he or she moves into and out of noisy areas or performs noisy operations and moves at the same time a body-worn instrument is required.

BS 6402 describes the requirements for such an instrument. It may be worn in a pocket but the microphone must be capable of being placed on the lapel or attached to a safety helmet. Figure 6.14 shows one in use with the microphone mounted on an earmuff and the rest of the equipment in a breast pocket. The instrument measures A-weighted

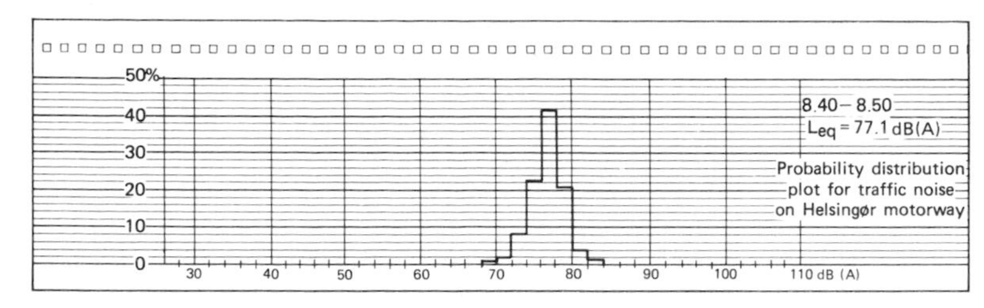

Figure 6.12 Probability distribution plot for motorway noise.

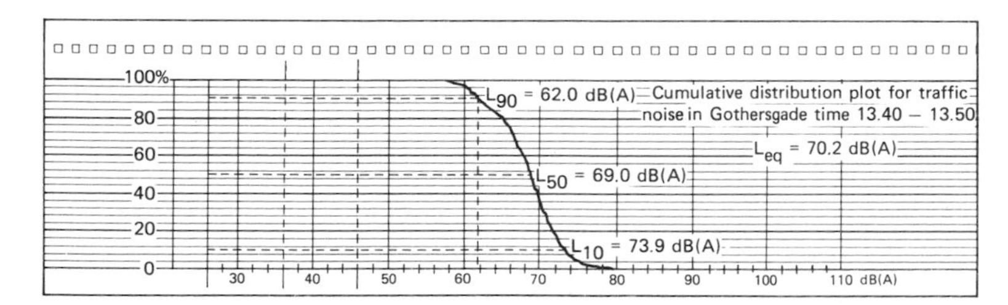

Figure 6.13 Cumulative distribution plot for traffic noise in a busy city street.

exposure according to

$$E = \int_0^T \mathrm{Pa}^2(t)\,\mathrm{d}t$$

where E is the exposure (Pa^2 . h) and T is the time (h).

Such an instrument gives the exposure in Pascal-squared hours, where 1 Pa^2 . h is the equivalent of 8 h exposure to a steady level of 84.9 dB. It may give the answers in terms of percentage of exposure as allowed by government regulations. For example, 100 per cent may mean 90 dB for 8 h, which is the maximum allowed with unprotected ears by the current regulations.

The actual performance requirements are similar to those of Type 2 sound-level meters and integrating sound-level meters, as accuracy of measurement is much more dependent on the clothing worn by the worker, the helmet or hat and the direction of microphone location in relation to noise source than on the precision of the measuring instrument.

6.2.5 Acoustic calibrators

It is desirable to check the calibration of equipment from time to time, and this can be done with acoustic calibrators, devices which provide a stable, accurately known sound pressure to a microphone which is part of an instrument system. Standards covering the requirements for such calibrators are in existence and they recommend operation at a level of 94 dB (1 Pa) or higher and a frequency of operation between 250 and 1000 Hz. The latter is often used, as it is a frequency at which the A-weighting network provides 0 dB insertion loss, so the calibration is valid for both A-weighted and unweighted measurement systems.

One type of calibrator known as a pistonphone offers extremely high order of stability, both short and long term, with known and trusted calibration capability to a tolerance within 0.1 dB. In pistonphones the level is normally 124 dB and the frequency 250 Hz.

The reason for the relatively high sound-pressure levels offered by calibrators is the frequent need to calibrate in the field in the presence of high ambient noise levels. Most calibrators are manufactured for operation with one specific microphone or a range of microphones. Some calibrators are so designed that they can operate satisfactorily with microphones of specified (usually standardized) microphone diameters but with different diaphragm constructions and protective grids (see Figures 6.15 and 6.16).

In a pistonphone the sound pressure generated is a function of total cavity volume, which means dependence on the diaphragm construction. In order to obtain the accuracy of calibration offered by a pistonphone the manufacturer's manual must be consulted regarding corrections required for specific types of microphones. The pressure generated by a pistonphone is also a function of static pressure in the cavity, which is linked with the atmosphere via a capillary tube. For the highest accuracy, atmospheric pressure must be measured and corrections applied.

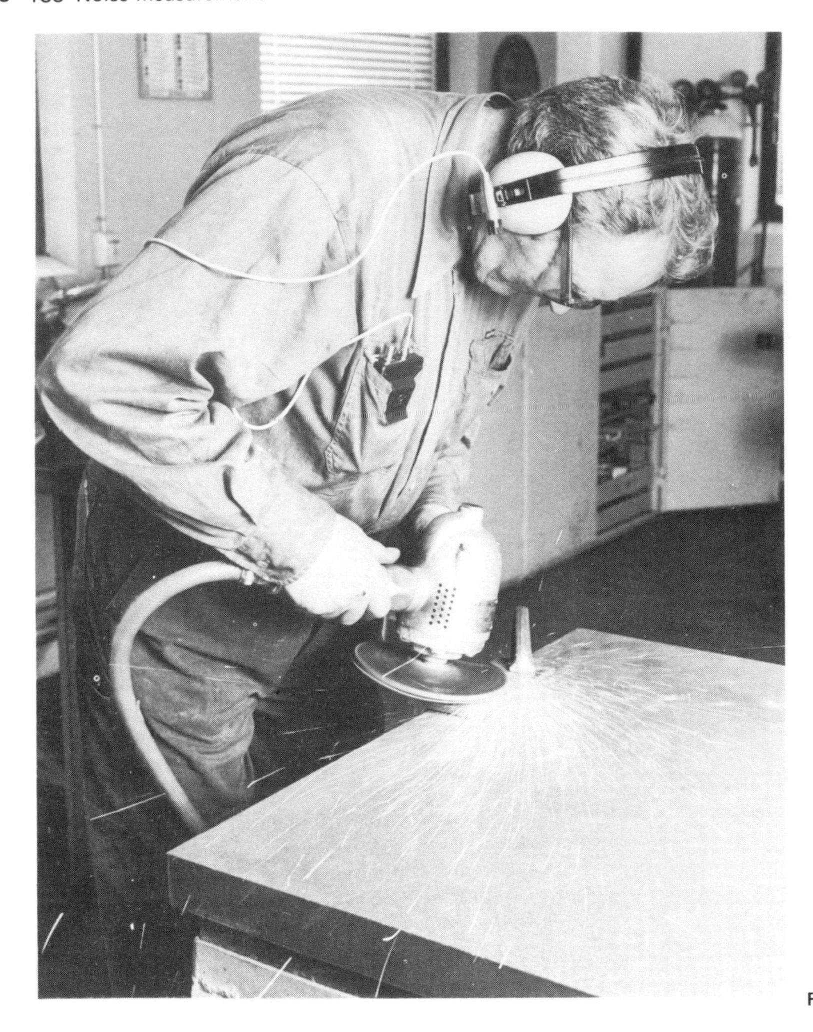

Figure 6.14 Noise-exposure meter in use.

Some calibrators offer more than one level and frequency. Calibrators should be subjected to periodic calibrations in approved laboratories. Calibration is discussed further in Section 6.6.

6.3 Frequency analysers

Frequently the overall A-weighted level, either instantaneous, time-weighted or integrated over a period of time, or even statistical information on its variations are insufficient descriptions of sound. Information on frequency content of sounds may be required. This may be in terms of content in standardized frequency bands, of octave or third-octave bandwidth. There are many standards which call for presentation of noise data in these bands.

Where information is required for diagnostic purposes very detailed, high-resolution analysis is performed. Some analogue instruments using tuned filter techniques are still in use—the modern instrument performs its analysing function by digital means.

6.3.1 Octave band analysers

These instruments are normally precision sound-level meters which include sets of filters or allow sets of filters to be connected, so that direct measurements in octave bands may be made. The filters are bandpass filters, whose passband encompasses an octave, i.e. the upper band edge frequency equals twice the lower band edge one. The filters should have smooth

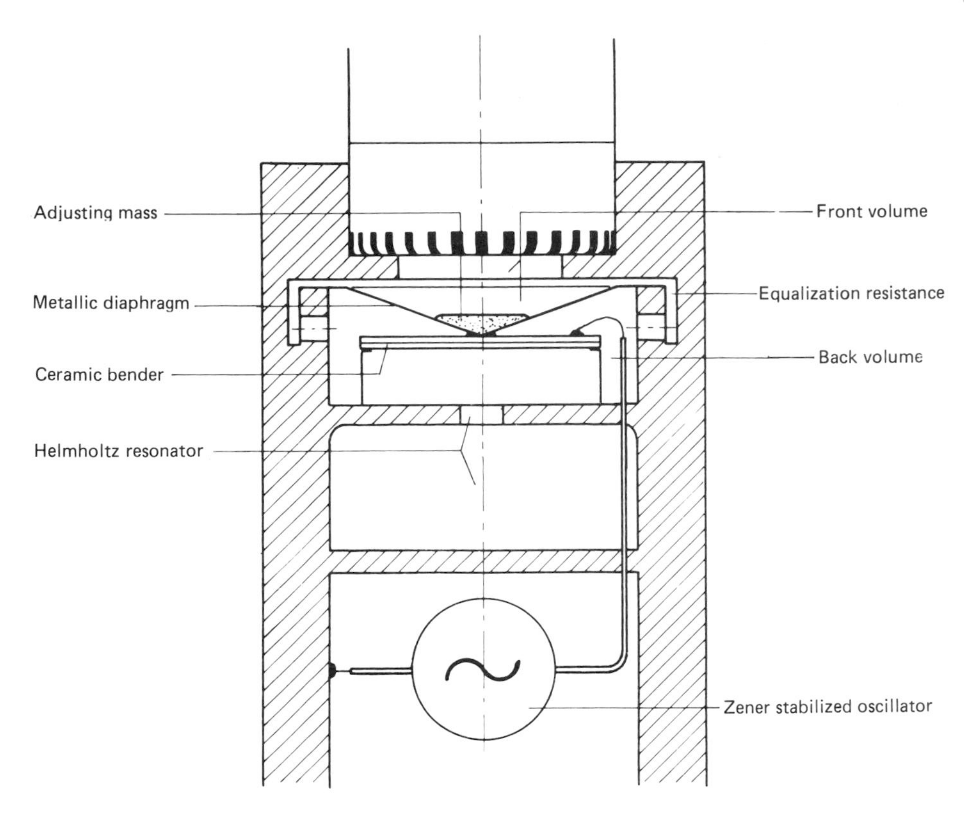

Figure 6.15 Principle of operation of a portable sound-level calibrator.

response in the passband, preferably offering 0 dB insertion loss. The centre frequency *fm* is the geometric mean

$$fm = \sqrt{(f_2 \cdot f_1)}$$

where f_2 is the upper band edge frequency and f_2 the lower band edge frequency.

Filters are known by their nominal centre frequencies. Attenuation outside the passband should be very high. ISO Standard 225 and BS 3593 list the 'preferred' frequencies while IEC 266 and BS 2475 give exact performance requirements for octave filters. A typical set of characteristics is given in Figure 6.17.

If a filter set is used with a sound-level meter then such a meter should possess a 'linear' response function, so that the only filtering is performed by the filters. Answers obtained in such analyses are known as 'octave band levels' in decibels, re 20 μPa.

From time to time, octave analysis is required for the purpose of assessing the importance of specific band levels in terms of their contribution to the overall A-weighted level. Octave band levels may then be corrected by introducing A-weighting attenuation at centre frequencies. Some sound-level meters allow the use of filters with A weighting included, thus giving answers in A-weighted octave band levels.

A simple portable instrument is likely to contain filters spanning the entire audiofrequency range. The analyser performs by switching filters into the system in turn and indicating results for each individual band. If noises tend to fluctuate, 'slow' response of the sound-level meter is used but this may still be too fast to obtain a meaningful average reading.

Special recorders, known as level recorders, may be used with octave band filters/sound-level meters which allow a permanent record to be obtained in the field. Parallel operation of all the filters in a set is also possible with special instruments, where the answers may be recorded in graphical form or in digital form,

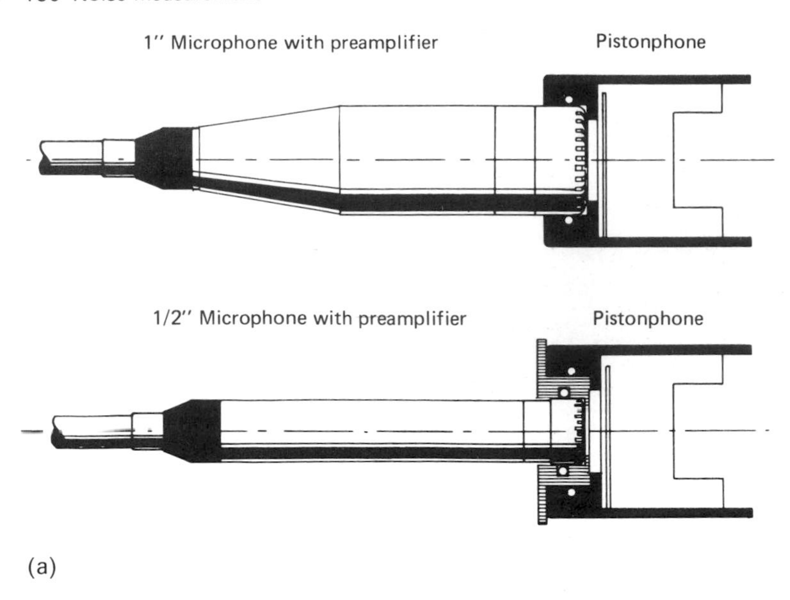

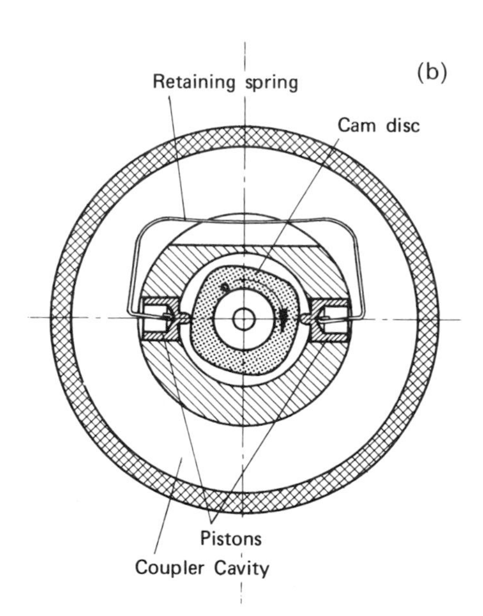

Figure 6.16 Pistonphone. (a) Mounting microphones on a pistonphone; (b) cross-sectional view showing the principle of operation.

and where the octave spectrum may be presented on a CRT.

6.3.2 Third-octave analysers

Octave band analysis offers insufficient resolution for many purposes. The next step is third-octave analysis by means of filters meeting ISO 266, IEC 225 and BS 2475. These filters may again be used directly with sound-level meters. As there are three times as many filters covering the same total frequency span of interest and the bandwidths are narrower, requiring longer averaging to obtain correct answers in fluctuating noise situations, the use of recording devices is almost mandatory. Typical characteristics are shown in Figure 6.18.

Apart from the field instruments described above, high-quality filter sets with automatic switching and recording facilities exist, but the modern trend is for parallel sets of filters used simultaneously with graphical presentation of third-octave spectra on CRT.

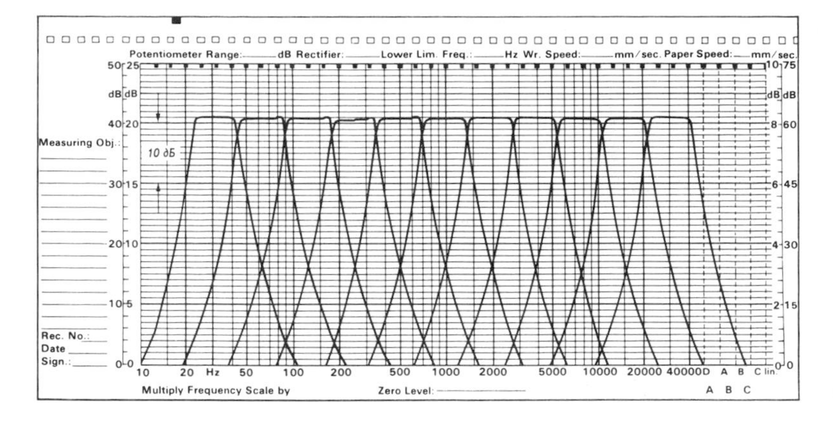

Figure 6.17 Frequency characteristics of octave filters in a typical analyser.

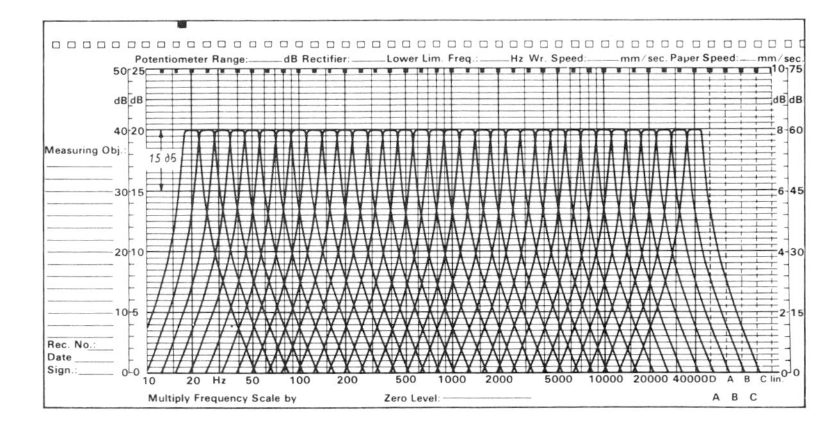

Figure 6.18 Frequency characteristics of third-octave filters in a typical analyser.

Digital filter instruments are also available, in which the exact equivalent of conventional, analogue filtering techniques is performed. Such instruments provide time coincident analysis, exponential averaging in each band or correct integration with time, presentation of spectra on a CRT and usually XY recorder output plus interfacing to digital computing devices (see Figure 6.19).

6.3.3 Narrow-band analysers

Conventional analogue instruments perform the analysis by a tunable filter, which may be of constant bandwidth type (Figure 6.20) or constant percentage (Figure 6.21). Constant percentage analysers may have a logarithmic frequency sweep capability, while constant bandwidth would usually offer a linear scan. Some constant bandwidth instruments have a synchronized link-up with generators, so that very detailed analysis of harmonics may be performed. The tuning may be manual, mechanical and linked with a recording device, or electronic via a voltage ramp. In the last case computer control via digital-to-analogue conversion may be achieved. With narrow band filtering of random signals, very long averaging times may be required, making analysis very time consuming.

Some narrow band analysers attempt to compute octave and third-octave bands from narrow-band analysis, but at present such techniques are valid only for the analyses of stationary signals.

6.3.4 Fast Fourier transform analysers

These are instruments in which input signals are digitized and fast Fourier transform (FFT) computations performed. Results are shown on a CRT and the output is available in analogue form for XY plotters or level recorders or in digital form for a variety of digital computing and memory devices.

Single-channel analysers are used for signal analysis and two-channel instruments for system analysis, where it is possible to directly analyse signals at two points in space and then to perform a variety of

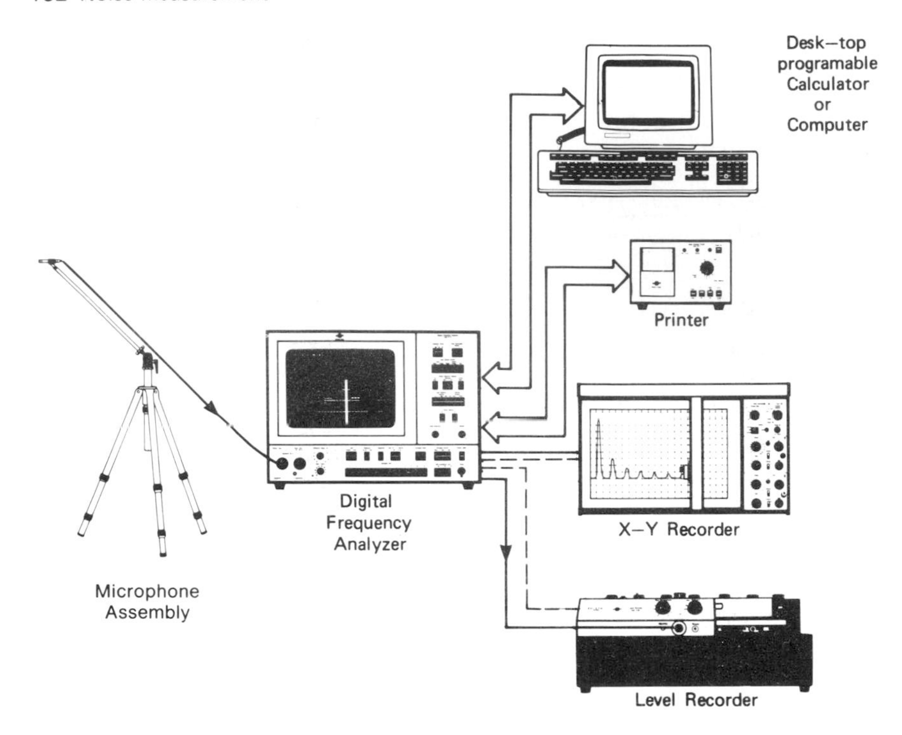

Figure 6.19 Output options with a typical digital frequency analyser.

computations in the instruments. In these, 'real-time' analysis means a form of analysis in which all the incoming signal is sampled, digitized and processed. Some instruments perform real-time analysis over a limited frequency range (for example, up to 2 kHz). If analysis above these frequencies is performed some data are lost between the data blocks analysed. In such a case, stationary signal analyses may be performed but transients may be missed.

Special transient analysis features are often incorporated (for example, sonic boom) which work with an internal or external trigger and allow the capture of a data block and its analysis. For normal operation, linear integration or exponential averaging are provided. The analysis results are given in 250–800 spectral lines, depending on type.

In a 400-line analyser, analysis would be performed in the range of, say, 0–20 Hz, 0–200 Hz, 0–2000 Hz and 0–20 000 Hz. The spectral resolution would be the upper frequency divided by number of spectral lines, for example, 2000/400, giving 5 Hz resolution.

Some instruments provide a 'zoom' mode, allowing the analyser to show results of analysis in part of the frequency range, focusing on selected frequency in the range and providing much higher resolution (see Figure 6.22).

Different forms of cursor are provided by different analysers, which allow the movement of the cursor to a part of the display to obtain a digital display of level and frequency at the point.

The frequency range of analysis and spectral resolution are governed by the internal sampling rate generator, and sampling rate may also be governed by an external pulse source. Such a source may be pulses proportional to a rotating speed of a machine, so that analysis of the fundamental and some harmonic frequencies may be observed, with the spectral pattern stationary in the frequency domain in spite of rotating speed changes during run-up tests.

Most analysers of the FFT type will have a digital interface, either parallel or serial, allowing direct use with desk-top computers in common use. In two-channel analysers, apart from the features mentioned above, further processing can provide such functions as transfer function (frequency response) correlation functions, coherence function, cepstrum, phase responses, probability distribution and sound intensity.

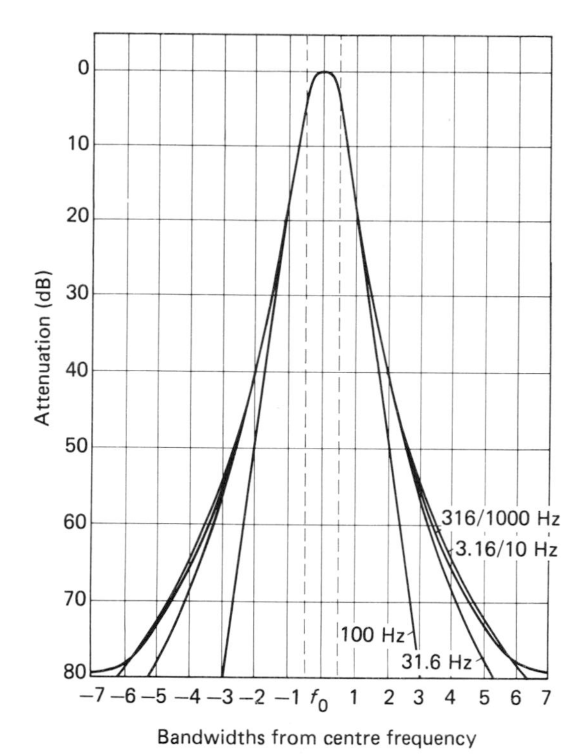

Figure 6.20 Characteristics of a typical constant bandwidth filter.

6.4 Recorders

6.4.1 Level recorders

Level recorders are recording voltmeters, which provide a record of d.c. or the rms value of a.c. signal on linear or logarithmic scale; some provide synchronization for octave and third-octave analysers, so that filter switching is timed for frequency-calibrated recording paper or synchronization with generators for frequency-response plotting. Such recorders are also useful for recording noise level against time.

6.4.2 XY plotters

These are recorders in which the Y plot is of d.c. input and X of a d.c. signal proportional to frequency. They are frequently used with FFT and other analysers for the recording of memorized CRT display.

6.4.3 Digital transient recorders

Digital transient recorders are dedicated instruments, specifically designed for the capture of transients such as gunshot, sonic boom, mains transient, etc. The incoming signals are digitized at preset rates and results captured in a memory when commanded by an internal or external trigger system, which may allow pre- and post-trigger recording. The memorized data may be replayed in digital form or via a D/A converter, usually at a rate hundreds or thousands of times slower than the recording speed, so that analogue recordings of fast transients may be made.

6.4.4 Tape recorders

These are used mainly for the purpose of gathering data in the field. Instrumentation tape recorders tend to be much more expensive than domestic types, as there are stringent requirements for stability of characteristics. Direct tape recorders are suitable for recording in the AF range, but where infrasonic signals are of interest FM tape recorders are used. The difference in range is shown in Figure 6.23, which also shows the effect of changing tape speed.

A direct tape recorder is likely to have a better signal-to-noise ratio, but a flatter frequency response and phase response will be provided by FM type.

IRIG standards are commonly followed, allowing recording and replaying on different recorders. The important standardization parameters are head geometry and track configuration, tape speeds, centre frequencies and percentage of frequency modulation. Tape reels are used as well as tape cassettes and the former offer better performance.

Recent developments in digital tape recorders suitable for field-instrumentation use enhance performance significantly in the dynamic range. Recorders are also discussed at some length in Part 4.

6.5 Sound-intensity analysers

Sound-intensity analysis may be performed by two-channel analysers offering this option. Dedicated sound-intensity analysers, based on octave or third-octave bands are available, mainly for the purpose of sound-power measurement and for the investigation of sound-energy flow from sources.

The sound-intensity vector is the net flow of sound energy per unit area at a given position, and is the product of sound pressure and particle velocity at the point. The intensity vector component in a given direction r is

$$I_r = \overline{p \times u_r}$$

where the horizontal bar denotes time average.

To measure particle velocity, current techniques rely on the finite difference approximation by means of integrating over time the difference in sound pressure

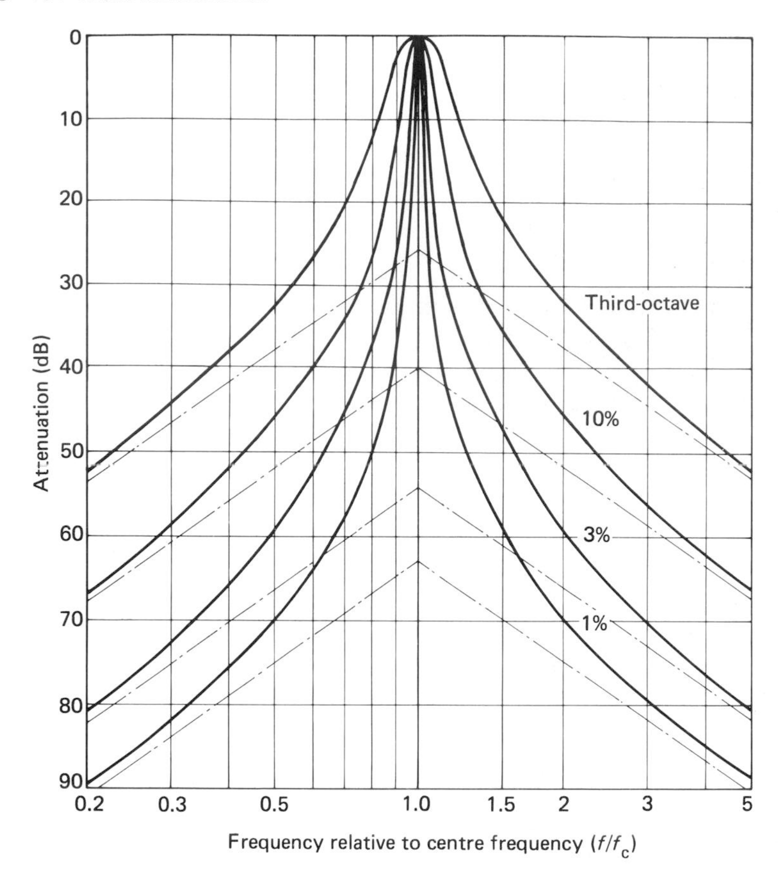

Figure 6.21 Bandpass filter selectivity curves for four constant percentage bandwidths.

at two points, A and B, separated by ΔR, giving

$$u_r = -\rho_0 \int \frac{(P_B - P_A)}{\Delta R} \, dt$$

where P_A and P_B are pressures at A and B and ρ_0 is the density of air. An outline of the system is given in Figure 6.24. This two-microphone technique requires the two transducers and associated signal paths to have very closely matched phase and amplitude characteristics. Only a small fraction of a degree in phase difference may be tolerated. Low frequencies require relatively large microphone separation (say, 50 mm) while at high frequencies only a few millimetres must be used. Rapid developments are taking place in this area.

6.6 Calibration of measuring instruments

6.6.1 Formal calibration

This is normally performed by approved laboratories, where instrumentation used is periodically calibrated and traceable to primary standards; the personnel are trained in calibration techniques, temperature is controlled and atmospheric pressure and humidity accurately measured.

Instruments used for official measurements, for example by consultants, environmental inspectors, factory inspectors or test laboratories, should be calibrated and certified at perhaps one- or two-year intervals. There are two separate areas of calibration, the transducers and the measuring, analysing and computing instruments.

Calibration of transducers, i.e. microphones, requires full facilities for such work. Condenser microphones of standard dimensions can be pressure calibrated by the absolute method of reciprocity. Where accurate free-field corrections are known for a specific type of condenser microphone, these can be added to the pressure calibration. Free-field reciprocity calibration of condenser microphones (IEC 486 and BS 5679) is very difficult to perform and requires a first-class anechoic chamber.

Microphones other than condenser (or non-standard size condenser) can be calibrated by comparison with absolutely calibrated standard

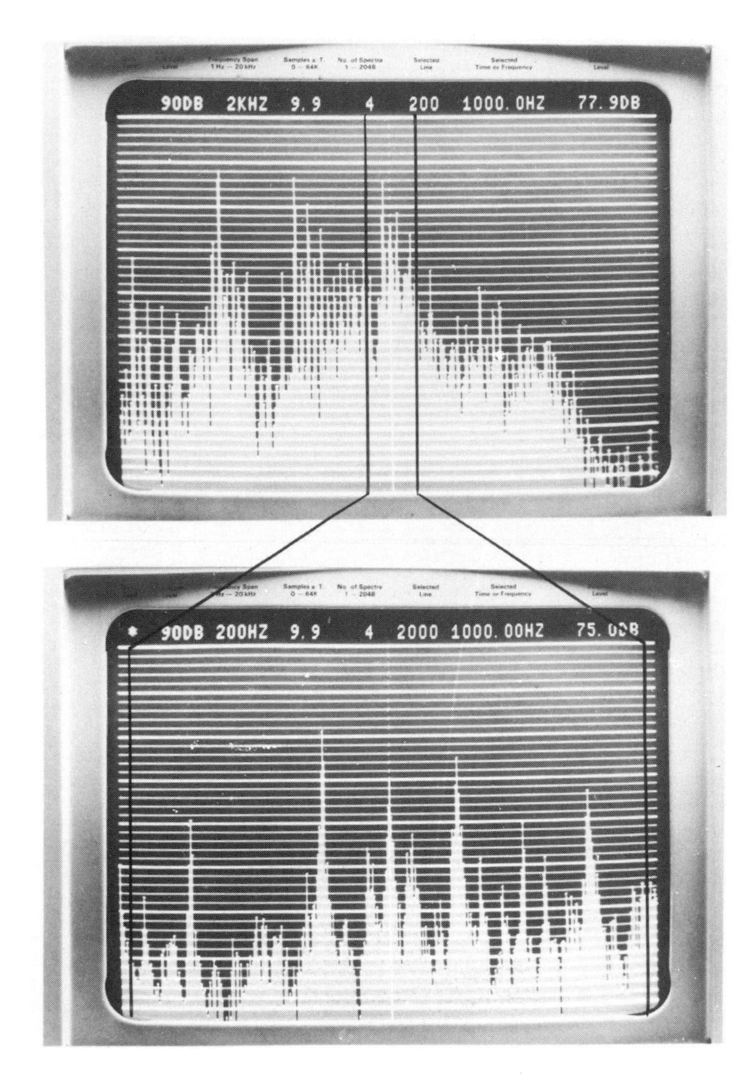

Figure 6.22 Scale expansion in presentation of data on spectra.

condenser microphones. This again requires an anechoic chamber.

Pressure calibration of condenser microphones may be performed by means of electrostatic actuators. Although this is not a standardized method, it offers very good results when compared with the reciprocity method, is much cheaper to perform and gives excellent repeatability. It is not a suitable method for frequencies below about 50 Hz. Calibration methods for two-microphone intensity probes are in the stage of development and standardization.

Sound-level meter calibration is standardized with a recommendation that acoustic tests on the complete unit are performed. This is a valid requirement if the microphone is attached to the body of the sound-level meter. Where the microphone is normally used on an extension or with a long extension cable it is possible to calibrate the microphone separately from the electronics. Filters and analysers require a purely electronic calibration, which is often time consuming if not computerized.

Microphone calibrators (see section 6.2.5) are devices frequently calibrated officially. This work is normally done with traceable calibrated microphones or by means of direct comparison with a traceable calibrator of the same kind. Sophisticated analysers, especially those with digital interfaces, lend themselves to computer-controlled checks and calibrations of all the complicated functions. The cost of this work by manual methods would be prohibitive.

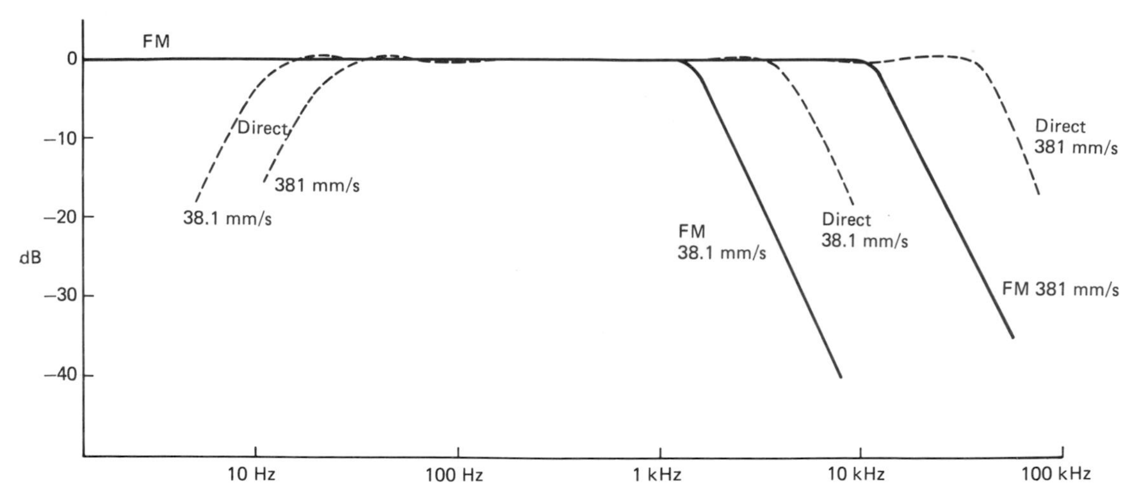

Figure 6.23 Typical frequency response characteristics of AM and FM recording.

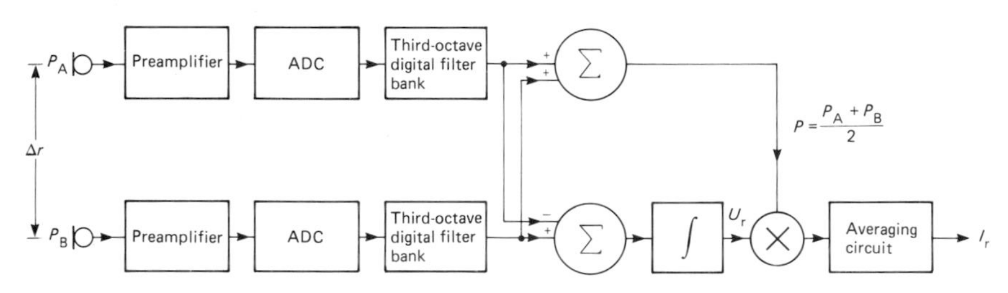

Figure 6.24 Sound-intensity analysis using two channels.

6.6.2 Field calibration

This normally implies the use of a calibrated, traceable microphone calibrator (for example, a pistonphone) which will provide a stable, accurately known sound pressure, at a known frequency, to a microphone used in the field instrument set-up (for example, a sound-level meter.

Although such 'calibration' is only a calibration at one point in frequency domain and at one level, it is an excellent check, required by most measurement standards. Its virtue is in showing departures from normal operation, signifying the need for maintenance and recalibration of instrumentation.

6.6.3 System 'calibration'

Complete instrument systems—for example, a microphone with an analyser or a microphone with a sound-level meter and a tape recorder—may be calibrated by a good-quality field calibrator. Again, the calibration would be at one frequency and one level, but if all instrument results line up within a tight tolerance a high degree of confidence in the entire system is achieved.

6.6.4 Field-system calibration

Some condenser microphone preamplifiers are equipped with a facility for 'insert voltage calibration'. This method is used mainly with multi-microphone arrays, where the use of a calibrator would be difficult and time consuming to perform. Permanently installed outdoor microphones may offer a built-in electrostatic actuator which allows regular (for example, daily) microphone and system checks.

6.7 The measurement of sound-pressure level and sound level

The measurement of sound-pressure level (spl) implies measurement without any frequency weighting, whereas sound level is frequency weighted spl. All measurement systems have limitations in the

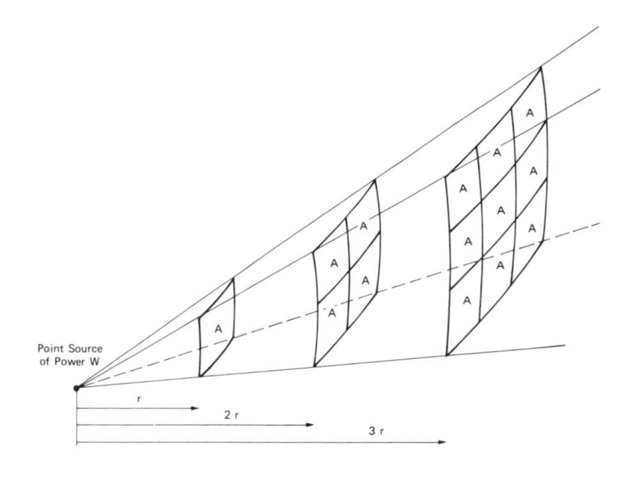

Figure 6.25 Flow of sound from a source.

frequency domain, and, even though many instruments have facilities designated as 'linear', these limitations must be borne in mind. Microphones have low-frequency limitations, as well as complex responses at the high-frequency end, and the total response of the measurement system is the combined response of the transducer and the following electronics.

When a broad-band signal is measured and spl quoted, this should always be accompanied by a statement defining the limits of the 'linear' range, and the response outside it. This is especially true if high levels of infrasound and/or ultrasound are present.

In the measurement of sound level, the low-frequency performance is governed by the shape of the weighting filter, but anomalies may be found at the high-frequency end. The tolerances of weighting networks are very wide in this region, the microphones may show a high degree of directionality (see section 6.2.1.1) and there may also be deviations from linearity in the response of the microphone.

Sound-pressure level or sound level is a value applicable to a point in space, the point at which the microphone diaphragm is located. It also applies to the level of sound pressure at the particular point in space when the location of sound source or sources is fixed and so is the location of absorbing or reflective surfaces in the area as well as of people.

In free space, without reflective surfaces, sound radiated from a source will flow through space, as shown in Figure 6.25. However, many practical sources of noise have complex frequency content and also complex radiation patterns, so that location of the microphone and distance from the source become important as well as its position relating to the radiation pattern. In free space sound waves approximate to plane waves.

6.7.1 Time averaging

The measurement of sound-pressure level or sound level (weighted sound-pressure level), unless otherwise specified, means the measurement of the rms value and the presentation of results in decibels relative to the standardized reference pressure of 20 μPa.

The rms value is measured with exponential averaging, that is, temporal weightings or time constants standardized for sound-level meters. These are known as F and S (previously known as 'Fast' and 'Slow').

In modern instruments, the rms value is provided by wide-range log . mean square detectors. The results are presented on an analogue meter calibrated in decibels, by a digital display or by a logarithmic d.c. output which may easily be recorded on simple recorders.

Some instruments provide the value known as 'instantaneous' or 'peak' value. Special detectors are used for this purpose, capable of giving the correct value of signals lasting for microseconds. Peak readings are usually taken with 'linear' frequency weighting.

In addition to the above, 'impulse'-measuring facilities are included in some sound-level measuring instruments. These are special detectors and use averaging circuits which respond very quickly to very short-duration transient sounds but which give very slow decay from the maximum value reached. At the time of writing, no British measurement standards or regulations require or allow the use of 'impulse' value other than for determination as to whether the sound is of an impulsive character or not.

One specific requirement for maximum 'fast' level is in vehicle drive-by test, according to BS 3425 and ISO 362. The 'slow' response gives slightly less fluctuating readout than the 'fast' mode—its main use is in 'smoothing' out the fluctuations.

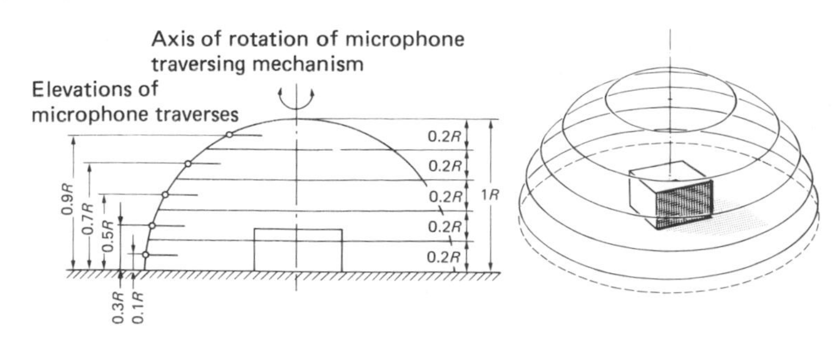

Figure 6.26 Coaxial circular paths for microphone traverses.

'Impulse' time weighting may be found in some sound-level meters, according to IEC 651 and BS 5969. This form of time weighting is mainly used in West Germany, where it originated. The value of meters incorporating this feature is in their superior rms detection capability when used in fast or slow mode.

Occasionally, there is a requirement for the measurement of maximum instantaneous pressure, unweighted. This can be performed with a sound-level meter with a linear response or a microphone and amplifier. The maximum instantaneous value, also known as peak value, may be obtained by displaying a.c. output on a storage oscilloscope or by using an instrument which faithfully responds to very short rise times and which has a 'hold' facility for the peak value. These hold circuits have limitations, and their specifications should be studied carefully. The peak value may be affected by the total frequency and phase response of the system.

6.7.2 Long time averaging

In the measurement of total sound energy at a point over a period of time (perhaps over a working day) as required by environmental regulations such as those relating to noise on construction and demolition sites or hearing-conservation regulations, special rules apply. The result required is in the form of L_{eq}, or equivalent sound level, which represents the same sound energy as that of the noise in question, however fluctuating with time.

The only satisfactory way of obtaining the correct value is to use an integrating/averaging sound-level meter to IEC 804 of the correct type. Noises with fast and wide-ranging fluctuations demand the use of Type 0 or Type 1 instruments. Where fluctuations are not so severe, Type 2 will be adequate.

The value known as SEL (sound-exposure level), normally available in integrating sound-level meters, is sometimes used for the purpose of describing short-duration events, for example an aircraft flying overhead or an explosion. Here the total energy measured is presented as though it occurred within 1 s. At present, there are no standards for the specific use of SEL in the UK.

6.7.3 Statistical distribution and percentiles

Traffic noise and environmental noise may be presented in statistical terms. Requirements for the location of microphones in relation to ground or facades of buildings are to be found in the relevant regulations. Percentile values are the required parameter. L_{10} means a level present or exceeded for 10 per cent of the measurement time and L_{50} the level exceeded for 50 per cent of the time. The value is derived from a cumulative distribution plot of the noise, as discussed in Section 6.2.3.2.

6.7.4 Space averaging

In the evaluation of sound insulation between rooms a sound is generated and measured in one room and also measured in the room into which it travels. In order to obtain information of the quality of insulation relating to frequency and compile an overall figure in accordance with relevant ISO and BS Standards, the measurements are carried out in octave and third-octave bands.

The sound is generated normally as a broad-band (white) noise or bands of noise. As the required value must represent the transfer of acoustic energy from one room to another, and the distribution of sound in each room may be complex, 'space averaging' must be carried out in both rooms. Band-pressure levels (correctly time averaged) must be noted for each microphone location and the average value found.

The space average for each frequency band is found by

$$L_p = 10 \log \left\{ \frac{1}{N} \left[\sum_{i=1}^{N} 10^{0.1 L_{pi}} \right] \right\}$$

L_p = space average level

where L_{pi} is the sound-pressure level at the ith measurement point and N the total number of measurement points. If large differences are found over the microphone location points in each band, a larger number of measurement points should be used. If very small variations occur, it is quite legitimate to use an arithmetic average value.

This kind of measurement may be carried out with one microphone located at specified points and levels noted and recorded, making sure that suitable time averaging was used at each point. It is possible to use an array of microphones and use a scanning (multiplexing) arrangement, feeding the outputs into an averaging analyser. The space average value may also be obtained by using one microphone on a rotating boom, feeding its output into an analyser with averaging facility. The averaging should cover a number of complete revolutions of the boom.

In integrating/averaging a sound-level meter may be used to obtain the space average value without the need for any calculations. It requires the activation of the averaging function for identical period of time at each predetermined location and allowing the instrument to average the values as total L_{eq}. With the same time allocated to each measurement point, $L_{eq} =$ space average.

6.7.5 Determination of sound power

As sound radiation from sources may have a very complex pattern, and sound in the vicinity of a source is also a function of the surroundings, it is not possible to describe or quantify the sound-emission characteristics in a simple way by making one or two measurements some distance away, nor even by making a large number of such measurements. The only method is to measure the total sound power radiated by the source. This requires the measurement to be carried out in specific acoustic conditions. ISO 3740–3746 and BS 4196 specify the acoustic conditions suitable for such tests and assign uncertainties which apply in the different conditions.

These standards describe the measurements of pressure, over a theoretical sphere or hemisphere or a 'box' which enclose the sound source, the computation of space average and finally the computation of sound-power level in decibels relative to 1 μW or sound power in watts. The results are often given in octave or third-octave bands or are A weighted. Figures 6.27 and 6.28 show the appropriate positions for a microphone.

The great disadvantage of obtaining sound power from the measurement of sound pressure is that special, well-defined acoustic conditions are required, and the achievement of such conditions for low uncertainties in measurements are expensive and often impossible. Large, heavy machinery would require a very large free space or large anechoic or reverberant chambers and the ability to run the machinery in them, i.e. provision of fuel or other source of energy, disposal of exhausts, coupling to loads, connection to gearboxes, etc.

With smaller sound sources it may be possible to perform the measurement by substitution or juxtaposition method *in situ*. These methods are based on measurements of noise of a source in an enclosed space, at one point, noting the results and then removing the noise source and substituting it by a calibrated sound source and comparing the sound-pressure readings.

If the unknown sound source gives a reading of X dB at the point and the calibrated one gives $X+N$ dB, then the sound-power level of the unknown source is N dB lower than that of the calibrated source.

The calibrated source may have variable but calibrated output. In situations where it is not possible to stop the unknown noise source its output may be adjusted so that the two sources together, working from the same or nearly the same location, give a reading 3 dB higher than the unknown source. When this happens, the power output of both sources is the same.

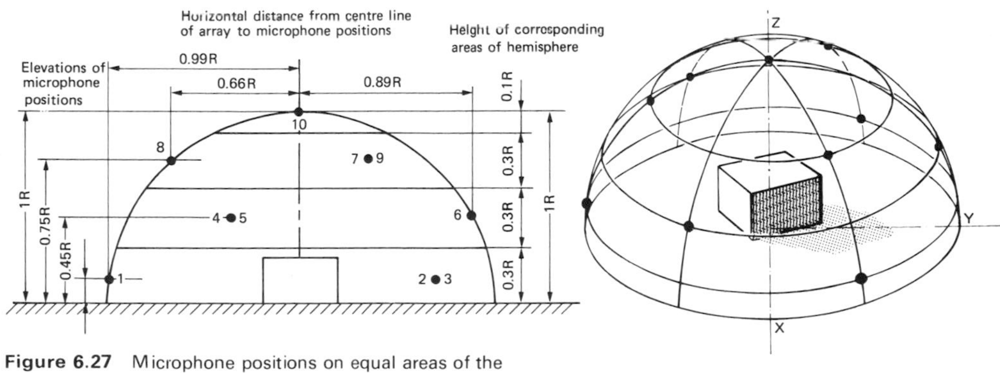

Figure 6.27 Microphone positions on equal areas of the surface of a hemisphere.

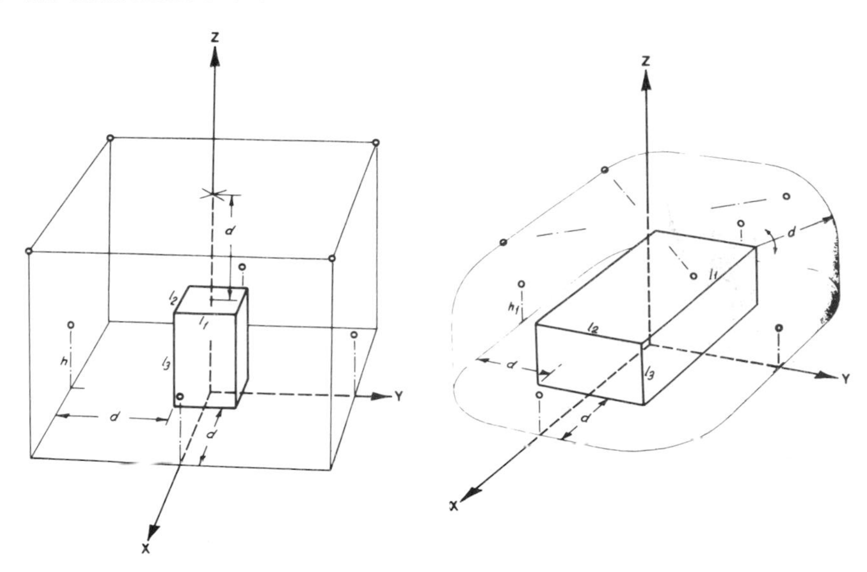

Figure 6.28 Microphone positions on a parallelepiped and a conformal surface.

The two methods described above are simple to use but must be considered as approximations only.

6.7.6 Measurement of sound power by means of sound intensity

Accurate and reliable sound-power measurement by means of pressure has the great disadvantage of requiring either free-field conditions in an open space or anechoic chamber or a well-diffused field in a reverberation chamber. Open spaces are difficult to find as are rain-free and wind-free conditions.

Anechoic chambers are expensive to acquire, and it is difficult to achieve good echo-free conditions at low frequencies; neither is it often convenient or practical to take machinery and install it in a chamber. Reverberation rooms are also expensive to build, and diffuse field conditions in them are not very easy to achieve.

In all above conditions machines require transporting, provision of power, gears, loads, etc., and this may mean that other sources or noise will interfere with the measurement.

The technique of measuring sound intensity, as discussed in Section 6.5, allows the measurement of sound power radiated by a source under conditions which would be considered adverse for pressure-measurement techniques. It allows the source to be used in almost any convenient location, for example on the factory floor where it is built or assembled or normally operating, in large or small enclosures, near refelcting surfaces and, what is extremely important, in the presence of other noises which would preclude the pressure method. It allows the estimation of sound power output of one item in a chain of several.

At present the standards relating to the use of this technique for sound-power measurement are being prepared but, due to its many advantages, the technique is finding wide acceptance. In addition, it has great value in giving the direction of energy flow, allowing accurate location of sources within a complex machine.

6.8 Effect of environmental conditions on measurements

6.8.1 Temperature

The first environmental condition most frequently considered is temperature. Careful study of instrument specifications will show the range of temperatures in which the equipment will work satisfactorily, and in some cases the effect (normally quite small) on the performance. Storage temperatures may also be given in the specifications. It is not always sufficient to look at the air temperature; metal instrument cases, calibrators, microphones, etc. may be heated to quite high temperatures by direct sunlight. Instruments stored in very low temperatures—for example, those carried in a boot of a car in the middle of winter—when brought into warm and humid surroundings

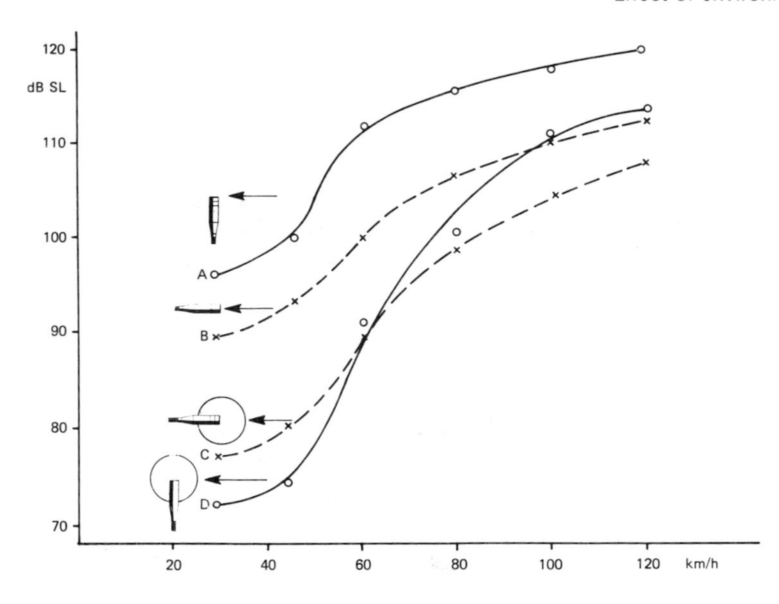

Figure 6.29 Noise levels induced in a half-inch free-field microphone fitted with a nosecone. A: With standard protection grid, wind parallel to diaphragm; B: as A but with wind at right angle to diaphragm; C: as B with windscreen; D: as A with windscreen.

may become covered by condensation and their performance may suffer. In addition, battery life is significantly shortened at low temperatures.

6.8.2 Humidity and rain

The range of relative humidity for which the instruments are designed is normally given in the specifications. Direct rain on instruments not specifically housed in showerproof cases must be avoided. Microphones require protective measures recommended by the manufacturers.

6.8.3 Wind

This is probably the worst enemy. Wind impinging on a microphone diaphragm will cause an output, mainly at the lower end of the spectrum (see Figure 6.29). Serious measurements in very windy conditions are virtually impossible, as are measurements of low-level noises even in slight wind conditions.

In permanent noise-monitoring systems it is now common practice to include an anemometer so that high-wind conditions may be identified and measurements discarded or treated with suspicion. Windscreens on microphones offer some help but do not eliminate the problems. These also have the beneficial effect of protecting the diaphragm from dust and chemical pollution, perhaps even showers. All windscreens have some effect on the frequency response of microphones.

6.8.4 Other noises

The measurement of one noise in the presence of another does cause problems if the noise to be

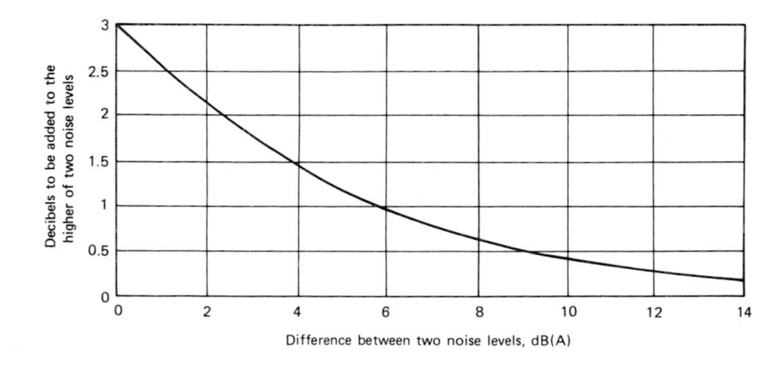

Figure 6.30 Noise-level addition chart.

measured is not at least 10 dB higher than other noise or noises. Providing the two noise sources do not have identical or very similar spectra, and both are stable with time, corrections may be made as shown in Figure 6.30. Sound-intensity measuring techniques may be used successfully in cases where other noises are as high as the noise to be measured, or even higher.

6.9 References

Burns, W. and Robinson, D. W. *Hearing and Noise in Industry*, HMSO, London (1970)

Hassall, J. R. and Zaveri, Z. *Acoustic Noise Measurements*, Bruel and Kjaer, Copenhagen (1979)

ISO Standards Handbook 4. *Acoustics, Vibration and Shock* (1985)

Randall, R. B. *Frequency Analysis*, Bruel and Kjaer, Copenhagen (1977)

Appendix 6.1

Standards play a large part in noise instrumentation. We therefore give below a list of international and British documents that are relevant.

BS 4727, part 3, Group 08 (1985), *British Standards Glossary of Acoustics and Electroacoustics Terminology particular to Telecommunications and Electronics*, is particularly helpful. This standard covers a very wide range of general terms, defines a wide variety of levels and deals with transmission and propagation, oscillations, transducers and apparatus, as well as psychological and architectural acoustics.

International standards	*British standard*	*Title*
IEC 225	BS 2475	Octave and third-octave filters intended for the analysis of sound and vibration
IEC 327	BS 5677	Precision method of pressure calibration of one inch standard condenser microphones by the reciprocity method
IEC 402	BS 5678	Simplified method for pressure calibration of one inch condenser microphone by the reciprocity method
IEC 486	BS 5679	Precision method for free field calibration of one inch standard condenser microphones
IEC 537	BS 5721	Frequency weighting for the measurement of aircraft noise
IEC 651	BS 5969	Sound level meters
IEC 655	BS 5941	Values for the difference between free field and pressure sensitivity levels for one inch standard condenser microphones
IEC 704		Test code for determination of airborne acoustical noise emitted by household and similar electrical appliances
IEC 804	BS 6698	Integrating/averaging sound level meters
ISO 140	BS 2750	Field and laboratory measurement of airborne and impact sound transmission in buildings
ISO 226	BS 3383	Normal equal loudness contours for pure tones and normal threshold of hearing under free field listening conditions
ISO 266	BS 3593	Preferred frequencies for acoustic measurements
ISO 362	BS 3425	Method for the measurement of noise emitted by motor vehicles
ISO 717	BS 5821	Rating of sound insulation in buildings and of building elements
ISO 532	BS 4198	Method of calculating loudness
ISO R.1996	BS 4142	Method of rating industrial noise affecting mixed residential and industrial areas
ISO 1999	BS 5330	Acoustics—assessment of occupational noise exposure for hearing conservation purposes
ISO 2204		Acoustics—guide to the measurement of airborne noise and evaluation of its effects on man
ISO 3352		Acoustics—assesment of noise with respect to its effects on the intelligibility of speech
ISO 3740–3746	BS 4196 (Parts 1–6)	Guide to the selection of methods of measuring noise emitted by machinery
ISO 3891	BS 5727	Acoustics—procedure for describing aircraft noise heard on ground
ISO 4871		Acoustics—noise labelling of machines
ISO 4872		Acoustics—measurement of airborne noise emitted by construction equipment intended for outdoor use—method for determining compliance with noise limits

Appendix 6.1 *continued*

International standards	*British standard*	*Title*
ISO 5130		Acoustics—measurement of noise emitted by stationary vehicles—survey method
ISO 6393		Acoustics—measurement of airborne noise emitted by earth moving machinery, method for determining compliance, with limits for exterior noise—stationary test conditions
	BS 5228	Noise control on construction and demolition sites
	BS 6402	Specification of personal sound exposure meters

Part 4

Instrumentation Systems

1 Design and construction of instruments

C. I. Daykin

1.1 Introduction

The purpose of this chapter is to give an insight into the types of components and construction used in commercial instrumentation.

To the designer, the *technology* in Instrument Technology depends on the availability of components and processes appropriate to his task. Being aware of what is possible is an important function of the designer, especially with the rapidity with which techniques now change. New materials such as ceramics and polymers will become increasingly important, as will semi-custom and large-scale integrated circuits. The need for low-cost automatic manufacture is having the greatest impact on design techniques, demanding fewer components and suitable geometries. Low-volume instruments and one-offs will be constructed in software.

The distinction between computer and instrument is already blurred, with many instruments offering a wide range of facilities and great flexibility. Smart sensors, which interface directly to a computer, will also shift the emphasis to software and mechanical aspects.

Historical practice, convention and the emergence of standards also contribute significantly to the subject. Standards, especially, benefit the designer and the user, and have made the task of the author and the reader somewhat simpler.

Commercial instruments exist because there is a market, and so details of their design and construction can only be understood in terms of a combination of commercial as well as technical reasons. A short section describes these trade-offs as a backdrop to the more technical information.

1.2 Instrument design

1.2.1 The designer's viewpoint

Many of the design features found in instruments are not obviously of direct benefit to the user. These can best be understood by also considering the designer's viewpoint.

The instrument designer's remit is to find the best compromise between cost and benefit to the users, especially when competition is fierce. For a typical, medium-volume instrument, its cost as a percentage of selling price, is distributed as follows:

Purchase cost	Design cost	20%
	Manufacturing cost	30%
	Selling costs	20%
	Other overheads	20%
	Profit	10%
		100%

Operating/maintenance cost may amount to 10% per annum. Benefits to the user can come from many features; for example:

Accuracy
Speed
Multi-purpose
Flexibility
Reliability
Integrity
Maintainability
Convenience

Fashion, as well as function, is very important since a smart, pleasing and professional appearance is often essential when selling instruments on a commercial basis.

For a particular product the unit cost can be reduced with higher volume production, and greater sales can be achieved with a lower selling price. The latter is called its 'market elasticity'. Since the manufacturer's objective is to maximize his return on investment, the combination of selling price and volume which yields the greatest profit is chosen.

1.2.2 Marketing

The designer is generally subordinate to marketing considerations, and consequently these play a major role in determining the design of an instrument and the method of its manufacture. A project will only go ahead if the anticipated return on investment is sufficiently high, and commensurate with the perceived level of risk. It is interesting to note that design accounts for a significant proportion of the total costs and, by its nature, involves a high degree of risk.

With the rapid developments in technology, product lifetimes are being reduced to as little as three years, and market elasticity is difficult to judge. In this way design is becoming more of a process of evolution, continuously responding to changes in market conditions. Initially, therefore, the designer will tend to err on the cautious side, and make cost reductions as volumes increase.

The anticipated volume is a major consideration, and calls for widely differing techniques in manufacture, depending on whether it is a low-, medium- or high-volume product.

1.2.3 Special instruments

Most instrumentation users configure systems from low-cost standard components with simple interfaces. Occasionally the need arises for a special component or system. It is preferable to modify a standard component wherever possible, since complete redesigns often take longer than anticipated, with an uncertain outcome.

Special systems also have to be tested and understood in order to achieve the necessary level of confidence. Maintenance adds to the extra cost, with the need for documentation and test equipment, making specials very expensive. This principle extends to software as well as hardware.

Table 1.1 Electronic components

Component and symbol	*Main types*	*Appearance*	*Range*	*Character*
Resistors	Metal oxide		0.1Ω–100 MΩ	1% general purpose
	Metal film		0.1Ω–100 MΩ	0.001% low drift
	Wirewound		0.01Ω–1 MΩ	0.01% high power
Capacitors	Air dielectric		0.01 pF–100 pF	0.01% high stability
	Ceramics		1 pF–10 μF	5% small size
	Polymer		1 pF–10 μF	1% general purpose
	Electrolytic		0.1 μF–1 F	10% small size
Inductors	Cylindrical core		0.1 μH–10 mH	10% general purpose
	Pot core		1 μH–1 H	0.1% high stability
	Toroidal core		100 μH–100 H	20% high values
				10^{-7} high accuracy ratios
Transformers	Cylindrical core		RF, IF types	
	Pot core		0.1 μH–1 mH	0.1% mutual inductor
	Toroidal core		0.1 H–10 H	20% high inductance
Diodes	PN junction		1 pA (on)–10^3 A (off)	Wide range
Transistors	Bipolar		10^{-17}W(input)–10^3W output	high freq, low noise
				high power etc.
	FETs		10^{-23}W(input)–10^3W output	as above.
Integrated circuits 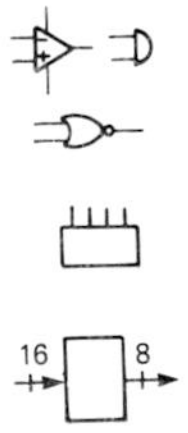	Analogue		operational amplifiers	wide range
			function blocks	multiply, divide
			amplifiers	high frequency
			switches	high accuracy
			semi custom	
	Digital		small scale integ. (SSI)	logic elements
			medium scale integ. (MSI)	function blocks
			large scale integ. (LSI)	major functions
			v. large scale integ. (VLSI)	complete systems
			semi custom	
16 8	Monolithic hybrid		A/D D/A conversion	
			special functions	
			semi custom	
Others	Thyristors		1A–10^3A	high power switch
	Triacs		1A–10^3A	high power switch
	Opto-couplers		singles, duals, quads	
	Thermistors		+ or − temp. coefficient	
	Relays		0.01Ω(on)–10^{12}Ω (off)	

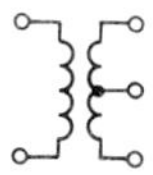

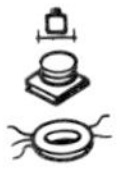

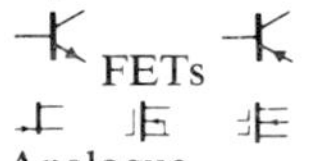

1.3 Elements of construction

1.3.1 Electronic components and printed circuits

Electronic circuitry now forms the basis of most modern measuring instruments. A wide range of electronic components are now available, from the simple resistor to complete data acquisition subsystems. Table 1.1 lists some of the more commonly used types.

Computer-aided design makes it possible to design complete systems on silicon using standard cells or by providing the interconnection pattern for a standard chip which has an array of basic components. This offers reductions in size and cost and improved design security.

The most common method of mounting and interconnecting electronic components is by double-sided through-hole plated, fibreglass printed circuit board (PCB) (Figure 1.1(a)). Component leads or pins are pushed through holes and soldered to tinned copper pads (Figure 1.1(b)). This secures the component and provides the connections. The solder joint is thus the most commonly used component and probably the most troublesome.

Tinning the copper pads with solder stops corrosion and makes soldering easier. Fluxes reduce oxidation and surface tension but a temperature-controlled soldering iron is indispensable. Large-

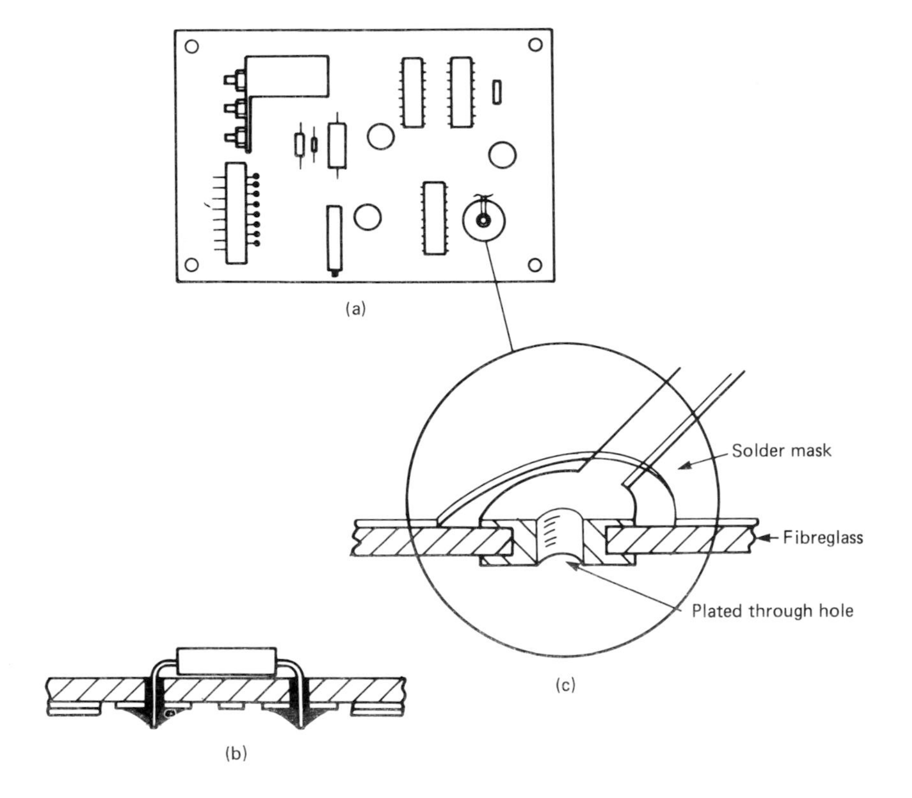

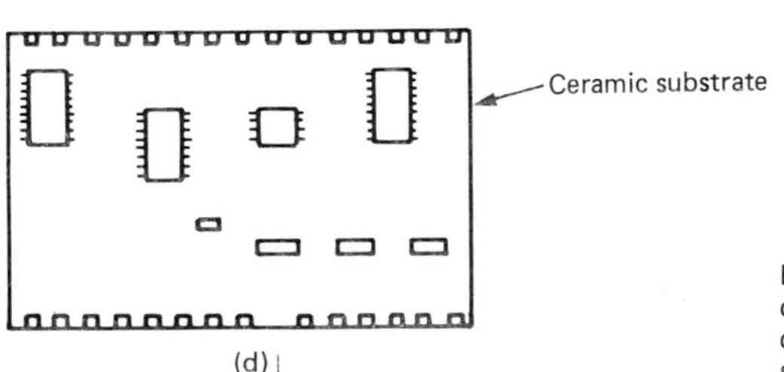

Figure 1.1 Printed electronic circuits. (a) Printed circuit board (PCB); (b) traditional axial component; (c) through-hole plating (close-up); (d) surface-mounted assemblies.

volume soldering can be done by a wave-soldering machine, where the circuit board is passed over a standing wave of molten solder. Components have to be on one side only, although this is also usually the case with manual soldering.

The often complicated routing of connections between components is made easier by having two layers of printed 'tracks', one on each surface, permitting cross-overs. Connections between the top and bottom conductor layers are provided by plated-through 'via' holes. It is generally considered bad practice to use the holes in which components are mounted as via holes, because of the possibility of damaging the connection when components are replaced. The through-hole plating (see Figure 1.1(c)) provides extra support for small pads, reducing the risk of them peeling off during soldering. They do, however, make component removal more difficult, often requiring the destruction of the component so that its leads can be removed individually. The more expensive components are therefore generally provided with sockets, which also makes testing and servicing much simpler.

The PCB is completed by the addition of a solder mask and a printed (silk-screened) component identification layer. The solder mask is to prevent solder bridges between adjacent tracks and pads, especially when using an automatic soldering technique such as wave soldering. The component identification layer helps assembly, test and servicing.

For very simple low-density circuits a single-sided PCB is often used, since manufacturing cost is lower. The pads and tracks must be larger, however, since without through-hole plating they tend to lift off more easily when soldering.

For very high-density circuits, especially digital, multilayer PCBs are used, as many as nine layers of printed circuits being laminated together with through-hole plating providing the interconnections.

Most electronic components have evolved to be suitable for this type of construction, along with machines for automatic handling and insertion. The humble resistor is an interesting example; this was originally designed for wiring between posts or tag-strips in valve circuits. Currently, they are supplied on long ribbons, and machines or hand tools are used for bending and cropping the leads, ready for insertion (see Figure 1.1(b)).

1.3.2 Surface-mounted assemblies (Figure 1.1(d))

The demand for greater complexity and higher density of circuits has resulted in important new developments which are predicted to overtake current methods by the year 1990. Semiconductors, chip resistors and chip capacitors are available in very small outline packages, and are easier to handle with automatic placement machines. Surface mounting eliminates the difficult problem of automatic insertion and, in most cases, the costly drilling process as well. Slightly higher densities can be achieved by using a ceramic substrate instead of fibreglass. Conductors of palladium silver, insulators and resistive inks are silk screened and baked onto the substrate to provide connections, cross-overs and some of the components. These techniques have been developed from the older 'chip and wire' hybrid thick film integrated circuit technique, used mainly in high-density military applications. In both cases, reflow soldering techniques are used due to the small size. Here, the solder is applied as a paste and silk screened onto the surface bonding pads. The component is then placed on its pads and the solder made to reflow by application of a short burst of heat which is not enough to damage the component. The heat can be applied by placing the substrate onto a hot vapour which then condenses at a precise temperature above the melting point of the solder. More simply, the substrate can be placed on a temperature-controlled hotplate or passed under a strip of hot air or radiant heat.

The technique is therefore very cost effective in high volumes, and with the increasing sophistication of silicon circuits results in 'smart sensors' where the circuitry may be printed onto any flat surface.

1.3.2.1 Circuit board replacement

When deciding servicing policy it should be realized that replacing a whole circuit board is often more cost effective than trying to trace a faulty component or connection. To this end, PCBs can be mounted for easy access and provided with a connector or connectors for rapid removal. The faulty circuit board can then be thrown away or returned to the supplier for repair.

1.3.3 Interconnections

There are many ways to provide the interconnection between circuit boards and the rest of the instrument, of which the most common are described below.

Connectors are used to facilitate rapid making and breaking of these connections, and simplify assembly test and servicing. Conventional wiring looms are still used because of their flexibility and because they can be designed for complicated routing and branching requirements. Termination of the wires can be by soldering, crimp or wire-wrap onto connector or circuit board pins. This, however, is a labour-intensive technique and is prone to wiring errors. Looms are given mechanical strength by lacing or sleeving wires as a tight bunch and anchoring to the chassis with cable ties.

Ribbon cable and insulation displacement connectors are now replacing conventional looms in many applications. As many as sixty connections can be made with one simple loom with very low labour costs. Wiring errors are eliminated since the routing is fixed at the design stage (see Figure 1.2).

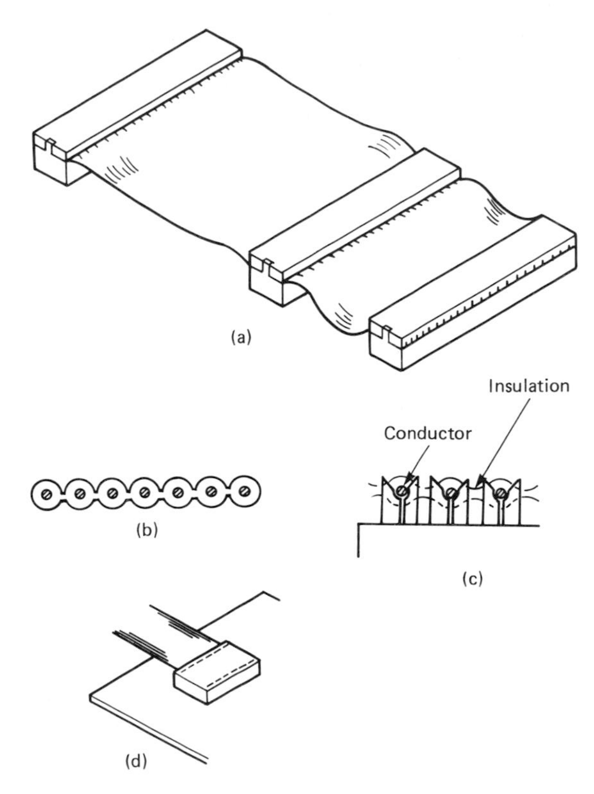

Figure 1.2 Ribbon cable interconnection. (a) Ribbon cable assembly; (b) ribbon cable cross-section; (c) insulation displacement terminator; (d) dual in-line header.

Connectors are very useful for isolating or removing a subassembly conveniently. They are, however, somewhat expensive and a common source of unreliability.

Another technique, which is used in demanding applications where space is at a premium, is the flexy circuit. Printed circuit boards are laminated with a thin, flexible sheet of Kapton which carries conductors. The connections are permanent, but the whole assembly can be folded up to fit into a limited space.

It is inappropriate to list here the many types of connectors. The connector manufacturers issue catalogues full of different types, and these are readily available.

1.3.4 Materials

A considerable variety of materials are available to the instrument designer, and new ones are being developed with special or improved characteristics, including polymers and superstrong ceramics. These materials can be bought in various forms, including sheet, block, rod and tube, and processed in a variety of ways.

1.3.4.1 Metals

Metals are usually used for strength and low cost as structural members. Aluminium for low weight and steel are the most common. Metals are also suitable for machining precise shapes to tight tolerances.

Stainless steels are used to resist corrosion and precious metal in thin layers helps to maintain clean electrical contacts. Metals are good conductors, and provide electrical screening as well as support. Mumetal and radiometal have high permeabilities and are used as very effective magnetic screens or in magnetic components. Some alloys—notably beryllium–copper—have very good spring characteristics, improved by annealing, and this is used to convert force into displacement in load cells and pressure transducers. Springs made of nimonic keep their properties at high temperatures, which is important in some transducer applications.

The precise thermal coefficient of the expansion of metals makes it possible to produce compensating designs, using different metals or alloys, and so maintaining critical distances independent of temperature. Invar has the lowest coefficient of expansion at less than 1 ppm per K over a useful range, but it is difficult to machine precisely.

Metals can be processed to change their characteristics as well as their shape; some can be hardened after machining and ground or honed to a very accurate and smooth finish, as found in bearings.

Metal components can be annealed, i.e. taken to a high temperature, in order to reduce internal stresses caused in the manufacture of the material and machining. Heat treatments can also improve stability, strength, spring quality, magnetic permeability or hardness.

1.3.4.2 Ceramics

For very high temperatures, ceramics are used as electrical and heat insulators or conductors (e.g. silicon carbide). The latest ceramics (e.g. zirconia, sialon, silicon nitride and silicon carbide) exhibit very high strength, hardness and stability even at temperatures over 1000°C. Processes for shaping them include slip casting, hot isostatic pressing

(HIP), green machining, flame spraying and grinding to finished size. Being hard, their grinding is best done by diamond or cubic boron nitride (CBN) tools. Alumina is widely used, despite being brittle, and many standard mechanical or electrical components are available.

Glass-loaded machinable ceramics are very convenient, having very similar properties to alumina, but are restricted to lower temperatures (less than 500°C). Special components can be made to accurate final size with conventional machining and tungsten tools.

Other compounds based on silicon include sapphires, quartz, glasses, artificial granite and the pure crystalline or amorphous substance. These have well-behaved and known properties (e.g. thermal expansion coefficient, conductivity and refractive index), which can be finely adjusted by adding different ingredients. The manufacture of electronic circuitry, with photolithography, chemical doping and milling, represents the ultimate in materials technology. Many of these techniques are applicable to instrument manufacture, and the gap between sensor and circuitry is narrowing—for example, in chemfets, in which a reversible chemical reaction produces a chemical potential that is coupled to one or more field-effect transistors. These transistors give amplification and possibly conversion to digital form before transmission to an indicator instrument with resulting higher integrity.

1.3.4.3 Plastics and polymers

Low-cost, lightweight and good insulating properties make plastics and polymers popular choices for standard mechanical components and enclosures. They can be moulded into elaborate shapes and given a pleasing appearance at very low cost in high volumes. PVC, PTFE, polyethylene, polypropylene, polycarbonates and nylon are widely used and available as a range of composites, strengthened with fibres or other ingredients to achieve the desired properties. More recently, carbon composites and Kevlar exhibit very high strength-to-weight ratio, useful for structural members. Carbon fibre is also very stable, making it suitable for dimensional calibration standards. Kapton and polyamides are used at higher temperatures and radiation levels.

A biodegradable plastic, (poly 3-hydroxybutyrate) or PHB is also available which can be controlled for operating life. Manufactured by cloned bacteria, this material represents one of many new materials emerging from advances in biotechnology.

More exotic materials are used for special applications, and a few examples are:

(1) Mumetal—very high magnetic permeability;
(2) PVDF—polyvinyledene fluoride, piezoelectric effect;
(3) Samarium/cobalt—very high magnetic remanence (fixed magnet);
(4) Sapphire—very high thermal conductivity;
(5) Ferrites—very stable magnetic permeability, wide range available.

1.3.4.4 Epoxy resins

Two-part epoxy resins can be used as adhesives, as potting material and paint. Parameters such as viscosity, setting time, set hardness and colour can be controlled. Most have good insulating properties, although conducting epoxies exist, and all are mechanically strong, some up to 300°C. The resin features in the important structure material: epoxy bonded fibreglass. Delicate assemblies can be ruggedized or passivated by a prophylactic layer of resin, which also improves design security.

Epoxy resin can be applied to a component and machined to size when cured. It can allow construction of an insulating joint with precisely controlled dimensions. Generally speaking, the thinner the glue layer, the stronger and more stable the join.

1.3.4.5 Paints and finishes

The appearance of an instrument is enhanced by the judicious use of texture and colour in combination with its controls and displays. A wide range of British Standard coordinated colours are available, allowing consistent results (BS 5252 and 4800).

Anodized or brushed aluminium panels have been popular for many years, although the trend is now back towards painted or plastic panels with more exotic colours. Nearly all materials, including plastic, can be spray painted by using suitable preparation and curing. Matt, gloss and a variety of textures are available.

Despite its age, silk-screen printing is used widely for lettering, diagrams and logos, especially on front panels.

Photosensitive plastic films, in one or a mixture of colours, are used for stick-on labels or as complete front panels with an LED display filter. The latter are often used in conjunction with laminated pressure pad-switches to provide a rugged, easy to clean splashproof control panel.

1.3.5 Mechanical manufacturing processes

Materials can be processed in many ways to produce the required component. The methods chosen depend on the type of material, the volume required and the type of shape and dimensional accuracy.

1.3.5.1 Bending and punching

Low-cost sheet metal or plastic can be bent or pressed into the required shape and holes punched with standard or special tools (Figure 1.3). Simple bending machines and a fly press cover most requirements, although hard tooling is more cost effective in large volumes. Most plastics are thermosetting and require heating, but metals are normally worked cold. Dimensional accuracy is typically not better than 0.5 mm.

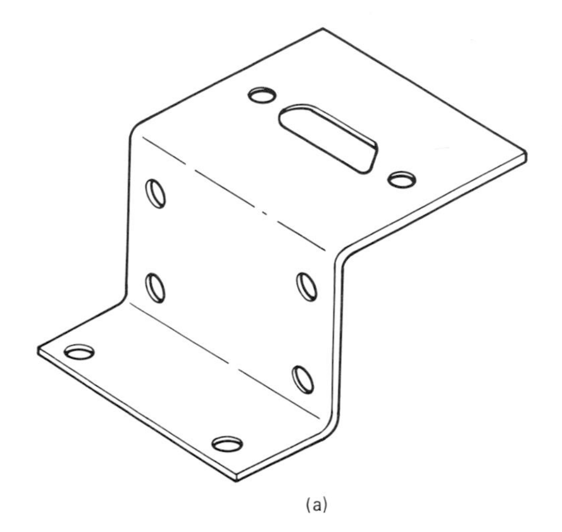
(a)

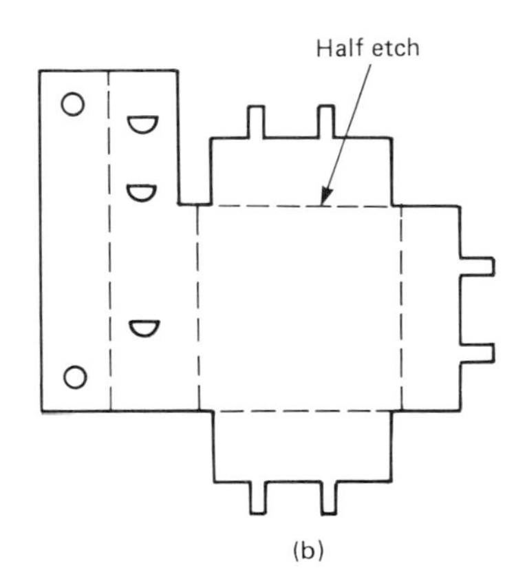

(b)

Figure 1.3 Sheet metal. (a) Bent and drilled or punched; (b) chemical milling.

1.3.5.2 Drilling and milling

Most materials can be machined, although glass (including fibreglass), ceramics and some metals require specially hardened tools. The hand or pillar drill is the simplest tool, and high accuracy can be achieved by using a jig to hold the workpiece and guide the rotating bit.

A milling machine is more complex, where the workpiece can be moved precisely relative to the rotating tool. Drills, reamers, cutters and slotting saws are used to create complex and accurate shapes. Tolerances of 1 μm can be achieved.

1.3.5.3 Turning

Rotating the workpiece against a tool is a method for turning long bars of material into large numbers of components at low unit cost. High accuracies can be achieved for internal and external diameters and length, typically 1 μm, making cylindrical components a popular choice in all branches of engineering.

A fully automatic machining centre and tool changer and component handler can produce vast numbers of precise components of many different types under computer control.

1.3.5.4 Grinding and honing

With grinding, a hard stone is used to remove a small but controlled amount of material. When grinding, both tool and component are moved to produce accurate geometries including relative concentricity and straightness (e.g. parallel) but with a poor surface finish. Precise flats, cylinders, cones and spherics are possible. The material must be fairly hard to get the best results and is usually metal or ceramic.

Honing requires a finer stone and produces a much better surface finish and potentially very high accuracy (0.1 μm). Relative accuracy (e.g. concentricity between outside and inside diameters) is not controllable, and so honing is usually preceded by grinding or precise turning.

1.3.5.5 Lapping

A fine sludge of abrasive is rubbed onto the workpiece surface to achieve ultra-high accuracy, better than 10 nm if the metal is very hard. In principle, any shape can be lapped, but optical surfaces such as flats and spherics are most common, since these can be checked by sensitive optical methods.

1.3.5.6 Chemical and electrochemical milling

Metal can be removed or deposited by chemical and electrochemical reactions. Surfaces can be selectively treated through masks. Complex shapes can be etched from sheet material of most kinds of metal using photolithographic techniques. Figure 1.3(b) shows an example where accuracies of 0.1 mm are achieved.

Gold, tin, copper and chromium can be deposited for printed circuit board manufacture or servicing of bearing components. Chemical etching of mechanical structures into silicon in combination with electronic circuitry is a process currently under development.

1.3.5.7 Extruding

In extruding, the material, in a plastic state and usually at a high temperature, is pushed through an orifice with the desired shape. Complex cross sections can be achieved and a wide range of standard items are available, cut to length. Extruded components are used for structural members, heat sinks and enclosures (Figure 1.4). Initial tooling is, however, expensive for non-standard sections.

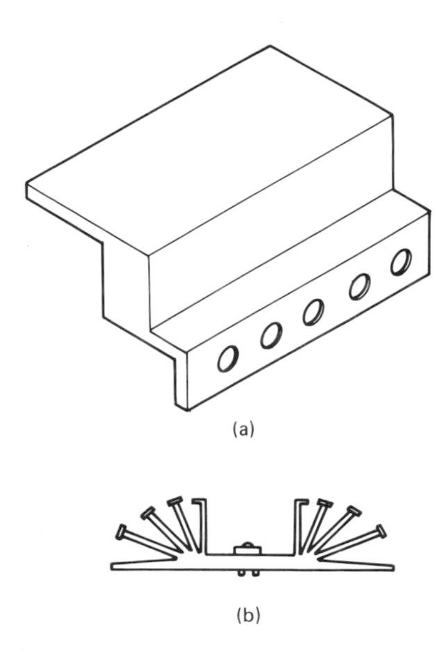

Figure 1.4 Extrusion. (a) Structural member; (b) heat sink.

1.3.5.8 Casting and moulding

Casting, like moulding, makes the component from the liquid or plastic phase but results in the destruction of the mould. It usually refers to components made of metals such as aluminium alloys and sand casts made from a pattern. Very elaborate forms can be made but further machining is required for accuracies better than 0.5 mm. Examples are shown in Figure 1.5. Plastics are moulded in a variety of ways and the mould can be used many times. Vacuum forming and injection moulding are used to achieve very low unit cost but tooling costs are high.

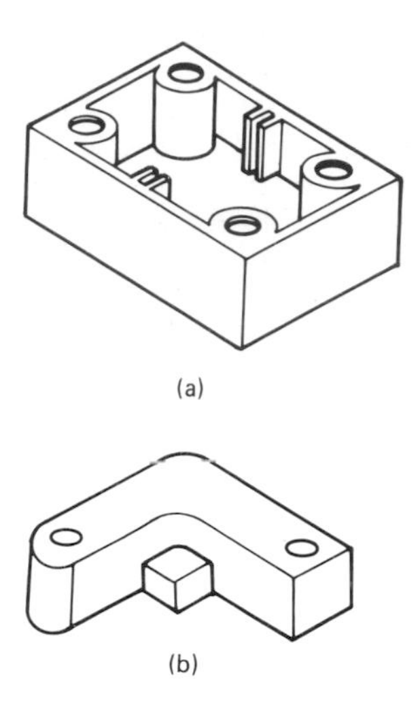

Figure 1.5 Examples of casting and moulding. (a) Moulded enclosure; (b) cast fixing.

1.3.5.9 Adhesives

Adhesive technology is advancing at a considerable rate, finding increasing use in instrument construction. Thin layers of adhesive can be stable and strong and provide electrical conduction or insulation. Almost any material can be glued, although high-temperature curing is still required for some applications. Metal components can be recovered by disintegration of the adhesive at high temperatures. Two-part adhesives are usually best for increased shelf life.

Jigs can be used for high dimensional accuracies, and automatic dispensing for high volume and low cost assembly.

1.3.6 Functional components

A wide range of devices are available, including bearings, couplings, gears and springs. Figure 1.6 shows the main types of components used and their principal characteristics.

1.3.6.1 Bearings

Bearings are used when a controlled movement either linear or rotary, is required. The simplest bearing consists of rubbing surfaces, prismatic for linear, cylindrical for rotation and spherical for universal movement. Soft materials such as copper and bronze and PTFE are used for reduced friction, and high precision can be achieved. Liquid or solid lubricants are sometimes used, including thin deposits of PTFE, graphite, organic and mineral oils. Friction can be further reduced by introducing a gap and rolling elements between the surfaces. The hardened steel balls or cylinders are held in cages or a recirculating mechanism. Roller bearings can be precise, low friction, relatively immune to contamination and capable of taking large loads.

The most precise bearing is the air bearing. A thin cushion of pressurized air is maintained between the bearing surfaces, considerably reducing the friction and giving a position governed by the average surface geometry. Accuracies of 0.01 μm are possible, but a source of clean, dry, pressurized air is required. Magnetic bearings maintain an air gap and have low friction but cannot tolerate side loads.

With bearings have evolved seals to eliminate contamination. For limited movement, elastic balloons of rubber or metal provide complete and possibly hermetic sealing. Seals made of low-friction polymer composites exclude larger particles and magnetic liquid lubricant can be trapped between magnets, providing an excellent low-friction seal for unlimited linear or rotary movement.

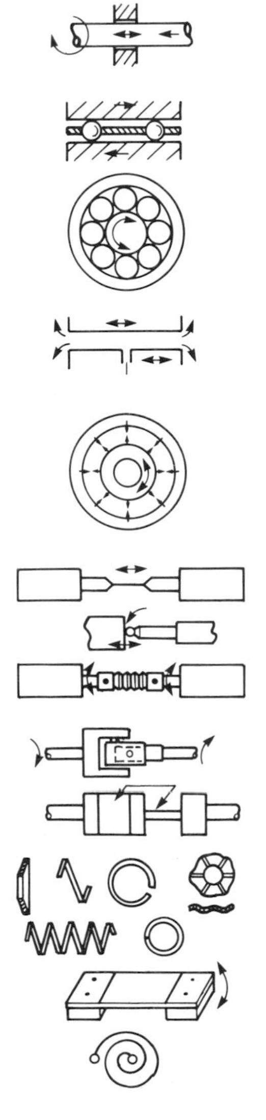

Figure 1.6 Mechanical components.

1.3.6.2 Couplings

It is occasionally necessary to couple the movement of two bearings, which creates problems of clashing. This can be overcome by using a flexible coupling which is stiff in the required direction and compliant to misalignment of the bearings. Couplings commonly used include:

(1) Spring wire or filaments;
(2) Bellows;
(3) Double hinges.

Each type is suitable for different combinations of side load, misalignment and torsional stiffness.

1.3.6.3 Springs

Springs are used to produce a controlled amount of force (e.g. for preloaded bearings, force/pressure transducers or fixings). They can take the form of a diaphragm, helix, crinkled washer or shaped flat sheet leaf spring. A thin circular disc with chemically milled Archimedes spinal slots is an increasingly used example of the latter. A pair of these can produce an accurate linear motion with good sideways stiffness and controllable spring constant.

1.4 Construction of electronic instruments

Electronic instruments can be categorized by the way they are intended to be used physically, resulting in certain types of construction:

(1) Site mounting;
(2) Panel mounting;
(3) Bench mounting;
(4) Rack mounting;
(5) Portable instruments.

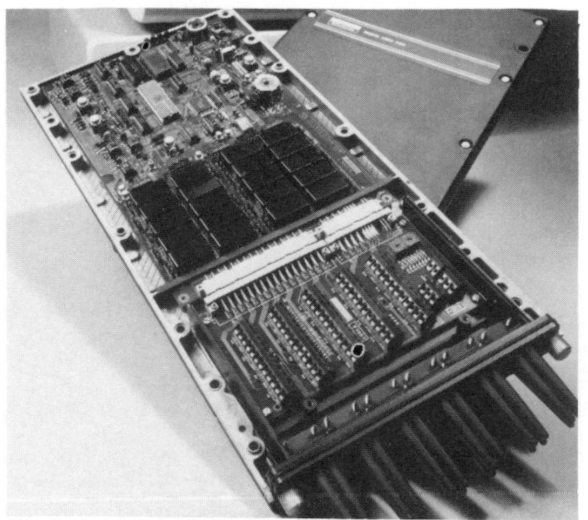

Figure 1.7 Instrument for site mounting (courtesy Solarton Instruments).

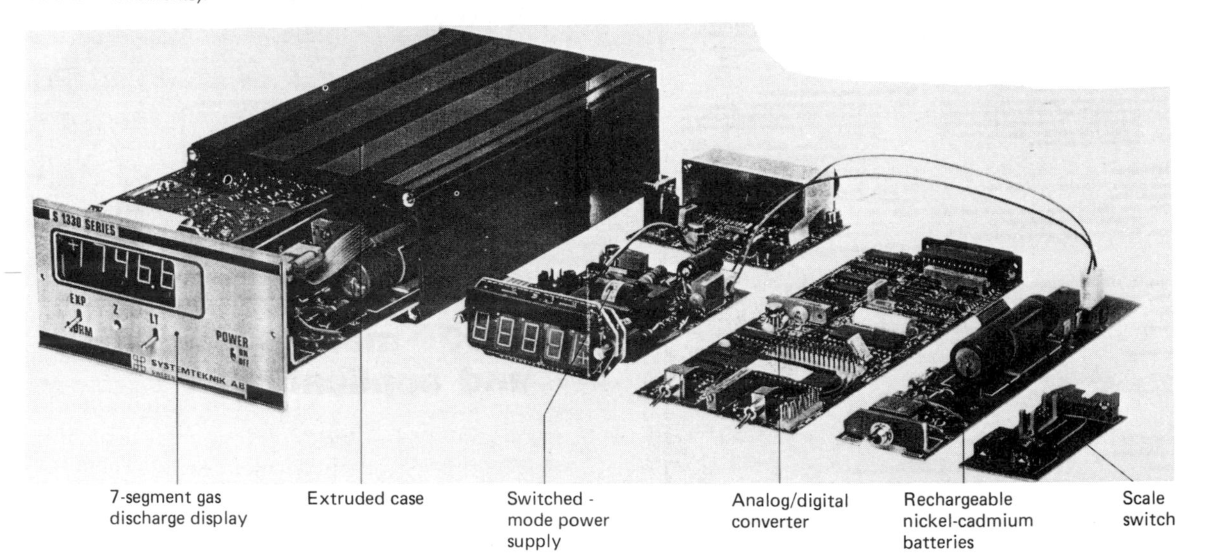

Figure 1.8 Panel-mounting instrument (courtesy Systemteknik AB).

1.4.1 Site mounting

The overriding consideration here is usually to get close to the physical process which is being measured or controlled. This usually results in the need to tolerate harsh environmental conditions such as extreme temperature, physical shock, muck and water. Signal conditioners and data-acquisition subsystems, which are connected to transducers and actuators, produce signals suitable for transmission over long distances, possibly to a central instrumentation and control system some miles away. Whole computerized systems are also available with ruggedized construction for use in less hostile environments.

The internal construction is usually very simple, since there are few, if any, controls or displays. Figure 1.7 shows an interesting example which tackles the common problem of wire terminations. The moulded plastic enclosure is sealed at the front with a rubber 'O' ring and is designed to pass the IPC 65 'hosepipe' test (see BS 5490). The main electronic circuit is on one printed circuit board mounted on pillars, connected to one of a variety of optional interface cards. The unit is easily bolted to a wall or the side of a machine.

1.4.2 Panel mounting

A convenient way for an instrument engineer to

Figure 1.9(a) Bench-mounting instrument (courtesy Automatic Systems Laboratories Ltd).

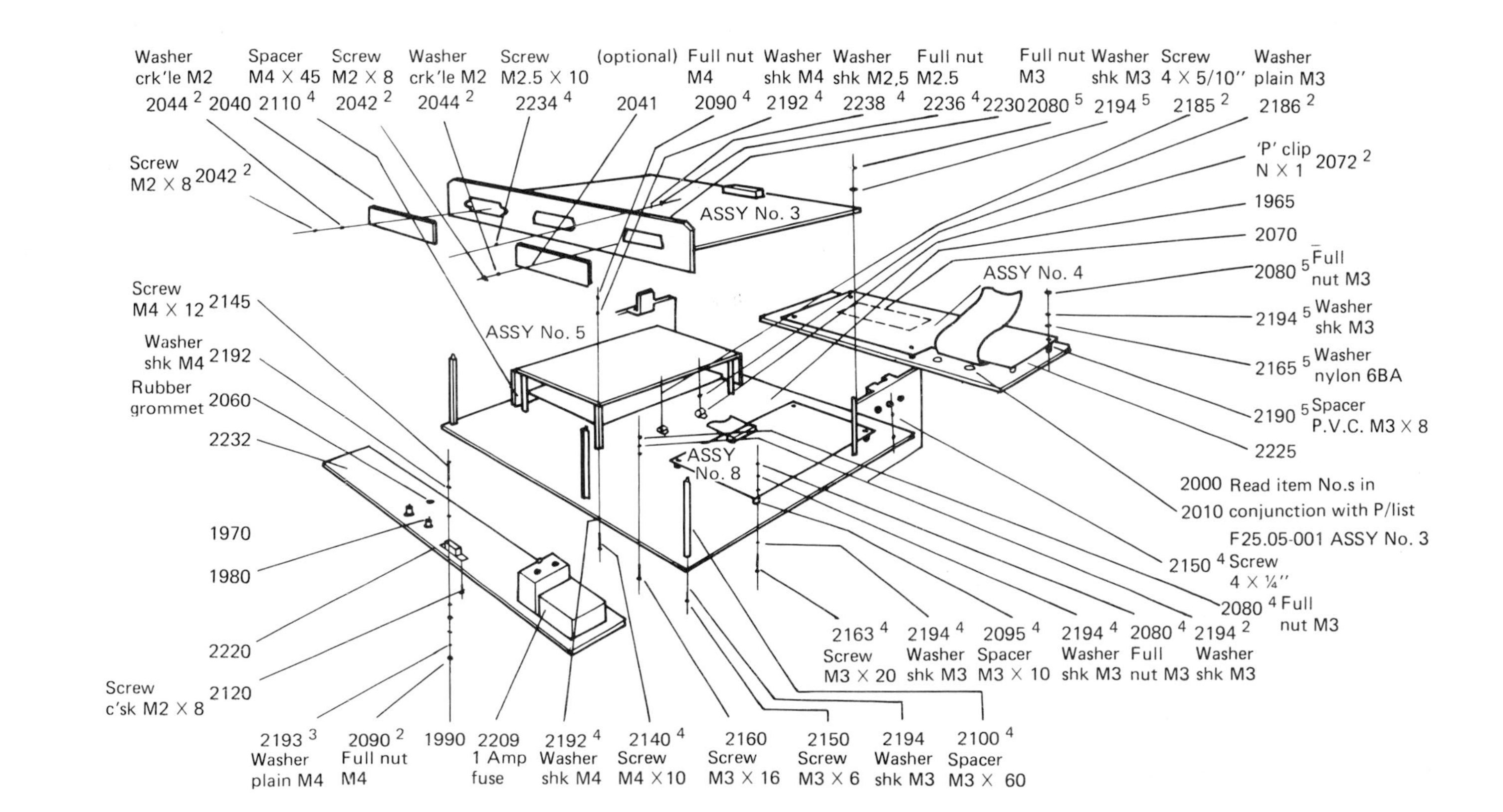

Figure 1.9(b) General assembly drawing of instrument shown in Figure 1.9(a).

construct a system is to mount the various instruments which require control or readout on a panel with the wiring and other system components protected inside a cabinet. Instruments designed for this purpose generally fit into one of a number of DIN standard cut-outs (see DIN 43 700). Figure 1.8 is an example illustrating the following features:

(1) The enclosure is an extruded aluminium tube;
(2) Internal construction is based around five printed circuit boards, onto which the electronic displays are soldered. The PCBs plug together and can be replaced easily for servicing;
(3) All user connections are at the rear, for permanent or semi-permanent installation.

1.4.3 Bench-mounting instruments

Instruments which require an external power source but a degree of portability are usually for benchtop operation. Size is important, since bench space is always in short supply.

Instruments in this category often have a wide range of controls and a display requiring careful attention to ergonomics. Figure 1.9(a) shows a typical instrument, where the following points are worth noting:

(1) The user inputs are at the front for easy access;
(2) There is a large clear display for comfortable viewing;
(3) The carrying handle doubles up as a tilt bar;
(4) It has modular internal construction with connectors for quick servicing.

The general assembly drawing for this instrument is included as Figure 1.9(b), to show how the parts fit together.

1.4.4 Rack-mounting instruments

Most large electronic instrumentation systems are constructed in 19-in wide metal cabinets of variable height (in units of 1.75 in = 1U). These can be for bench mounting, free standing or wall mounting. Large instruments are normally designed for bench operation or rack mounting for which optional brackets are supplied. Smaller modules plug into subracks which can then be bolted into a 19-in cabinet.

Figure 1.10 shows some of the elements of a modular instrumentation system with the following points:

(1) The modules are standard Eurocard sizes and widths (DIN 41914 or IEC 297);
(2) The connectors are to DIN standard (DIN 41612);
(3) The subrack uses standard mechanical components and can form part of a much larger instrumentation system.

The degree of modularity and standardization enables the user to mix a wide range of instruments and components from a large number of different suppliers worldwide.

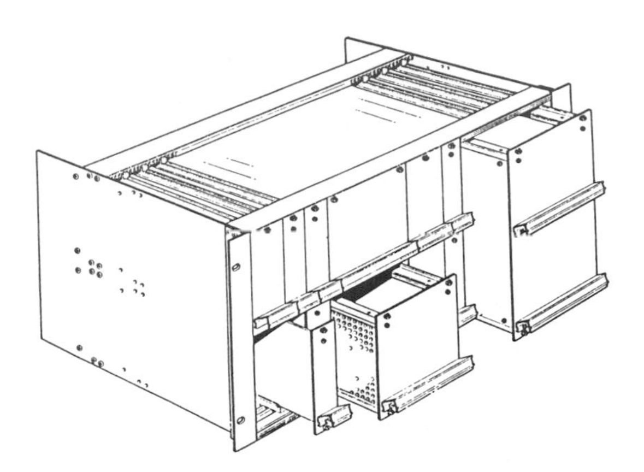

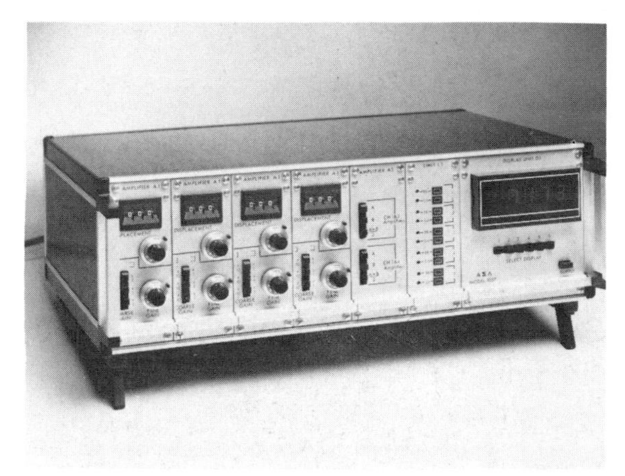

Figure 1.10 Rack-based modular instruments (courtesy Schroff (UK) Ltd and Automatic Systems Laboratories Ltd).

1.4.5 Portable instruments

Truly portable instruments are now common, due to the reduction in size and power consumption of electronic circuitry. Figure 1.11 shows good examples which incorporate the following features:

(1) Lightweight, low-cost moulded plastic case;
(2) Low-power CMOS circuitry and liquid crystal display (LCD);
(3) Battery power source gives long operating life.

Size reduction is mainly from circuit integration onto silicon and the use of small outline components.

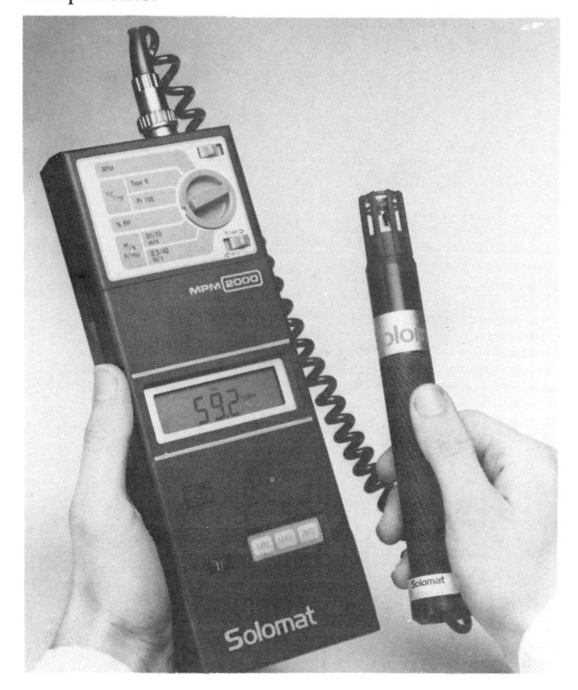

Figure 1.11 Portable instruments (courtesy Solomat SA).

1.4.6 Encapsulation

For particularly severe conditions, especially with regard to vibration, groups of electronic components are sometimes *encapsulated* (familiarly referred to as 'potting'). This involves casting them in a suitable material, commonly epoxy resin. This holds the components very securely in position and they are also protected from the atmosphere to which the instrument is exposed. To give further protection against stress (for instance, from differential thermal expansion) a complex procedure is occasionally used, with compliant silicone rubber introduced as well as the harder resin.

Some epoxies are strong up to 300°C. At higher temperatures (450°C) they are destroyed, allowing encapsulated components to be recovered if they are themselves heat resistant. Normally an encapsulated group would be thrown away if any fault developed inside it.

1.5 Mechanical instruments

Mechanical instruments are mainly used to interface between the physical world and electronic instrumentation. Examples are:

(1) Displacement transducers (linear and rotary);
(2) Force transducers (load cells);
(3) Accelerometers.

Such transducers often have to endure a wide temperature range, shock and vibration, requiring careful selection of materials and construction.

Many matters contribute to good mechanical design and construction, some of which are brought out in the devices described in other chapters of this book. We add to that here by showing details of one or two instruments where particular principles of design can be seen. Before that, however, we give a more general outline of kinematic design, a way of proceeding that can be of great value for designing instruments.

1.5.1 Kinematic design

A particular approach sometimes used for high-precision mechanical items is called kinematic design. When the relative location of two bodies must be constrained, so that there is either no movement or a closely controlled movement between them, it represents a way of overcoming the uncertainties that arise from the impossibility of achieving geometrical perfection in construction. A simple illustration is two flat surfaces in contact. If they can be regarded as ideal geometrical planes then the relative movement of the two bodies is accurately defined. However, it is expensive to approach geometrical perfection, and the imperfections of the surfaces mean that the relative position of the two parts will depend upon contact between high spots, and will vary slightly if the points of application of the forces holding them together are varied. The points of contact can be reduced, for instance, to four with a conventional chair, but it is notorious that a four-legged chair can rock unless the bottoms of the legs match the floor perfectly. *Kinematic design* calls for a three-legged chair, to avoid the *redundancy* of having its position decided by four points of contact.

More generally, a rigid solid body has 6 *degrees of freedom* which can be used to fix its position in space. These are often thought of as three Cartesian coordinates to give the position of one point of the body, and when that has been settled, rotation about three mutually perpendicular axes describes the body's attitude in space. The essence of kinematic design is that each degree of freedom should be constrained in an identifiable localized way.

Consider again the three-legged stool on a flat surface. The Z-coordinate of the tip of the leg has been constrained, as has rotation about two axes in the flat surface. There is still freedom of X and Y coordinates and for rotation about an axis perpendicular to the surface: 3 degrees of freedom removed by the three constraints between the leg-tips and the surface.

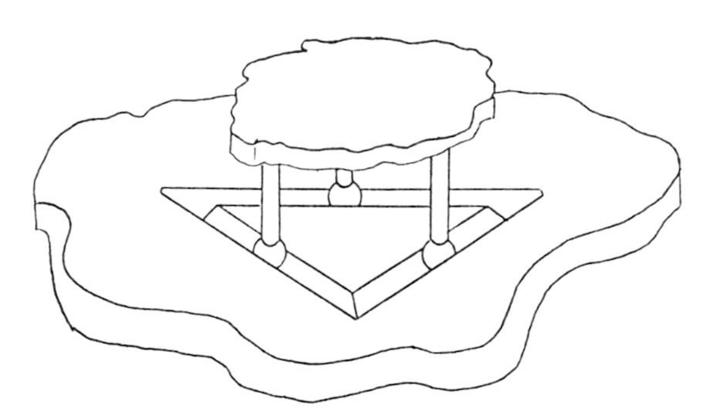

Figure 1.12 Kinematic design: six constraints locate a body.

A classical way of introducing six constraints and so locating one body relative to another is to have three V-grooves in one body and three hemispheres attached to the other body, as shown in Figure 1.12. When the hemispheres enter the grooves (which should be deep enough for contact to be made with their sides and not their edges) each has two constraints from touching two sides, making a total of six.

If one degree of freedom, say linear displacement, is required, five spheres can be used in a precise groove as in Figure 1.13. Each corresponds to a restricted movement.

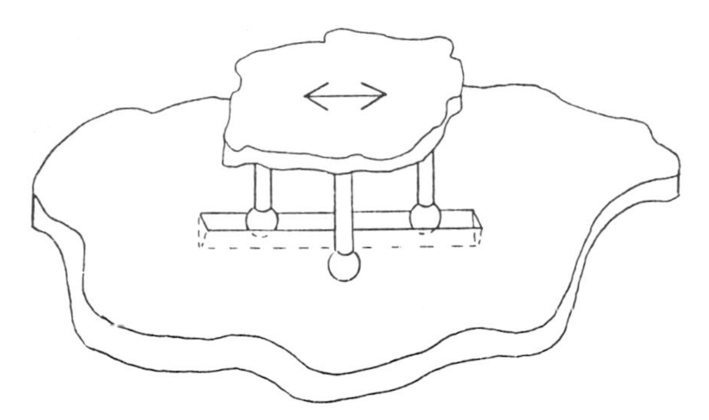

Figure 1.13 Kinematic design: five constraints allow linear movement.

For the required mating, it is important that contact should approximate to point contact and that the construction materials should be hard enough to allow very little deformation perpendicular to the surface under the loads normally encountered. The sphere-on-plane configuration described is one possible arrangement: crossed cylinders are similar in their behaviour and may be easier to construct.

Elastic hinges may be thought of as an extension of kinematic design. A conventional type of door hinge is expensive to produce if friction and play are to be greatly reduced, particularly for small devices. An alternative approach may be adopted when small, repeatable rotations must be catered for. Under this approach, some part is markedly weakened, as in Figure 1.14, so that the bending caused by a turning moment is concentrated in a small region. There is elastic resistance to deformation but very little friction and high repeatability.

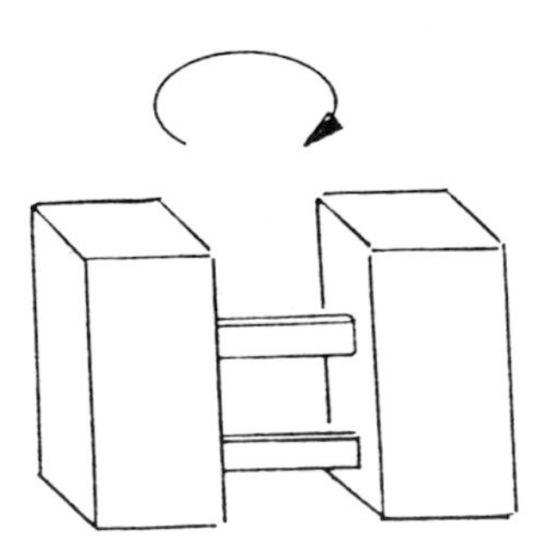

Figure 1.14 Principle of elastic hinge.

The advantages of kinematic design may be listed as:

(1) Commonly, only simple machining operations are needed at critical points.
(2) Wide tolerances on these operations should not affect repeatability, though they may downgrade absolute performance.
(3) Only small forces are needed. Often gravity is sufficient or a light spring if the direction relative to the vertical may change from time to time.
(4) Analysis and prediction of behaviour is simplified.

The main disadvantage arises if large forces have to be accommodated. Kinematically designed constraints normally work with small forces holding parts together and if these forces are overcome—even momentarily under the inertia forces of vibration—there can be serious malfunction. Indeed, the lack of symmetry in behaviour under kinematic design can prove a more general disadvantage (for instance, when considering the effects of wear).

Of course, the small additional complexity often means that it is not worth changing to kinematic design. Sometimes a compromise approach is adopted, such as localizing the points of contact between large structures without making them literal spheres on planes. In any case, when considering movements and locations in instruments it is helpful to bear the ideas of kinematic design in mind as a background to analysis.

1.5.2 Proximity transducer

This is a simple device which is used to detect the presence of an earthed surface which affects the capacitance between the two electrodes E1 and E2 in Figure 1.15. In a special application it is required to operate at a temperature cycling between 200°C and 400°C in a corrosive atmosphere and survive shocks of 1000 g. Design points to note are:

(1) The device is machined out of the solid to avoid a weak weld at position A.
(2) The temperature cycling causes thermal stresses which are taken up by the spring washer B (special nimonic spring material for high temperatures).
(3) The ceramic insulator blocks are under compression for increased strength.

1.5.3 Load cell

As discussed in Chapter 7 of Part 1, a load cell converts force into movement against the reaction of a spring. The movement is then measured by a displacement transducer and converted into electrical form.

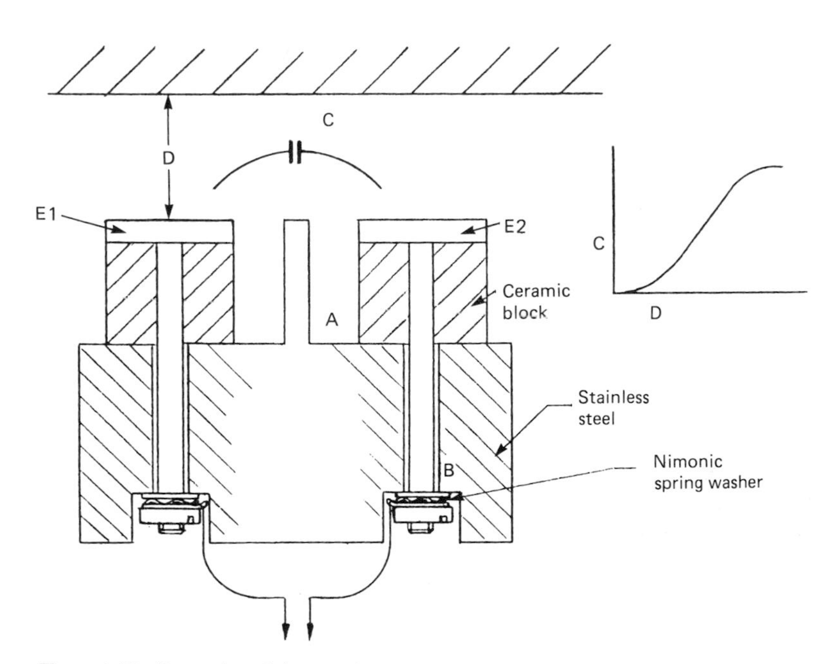

Figure 1.15 Rugged proximity transducer.

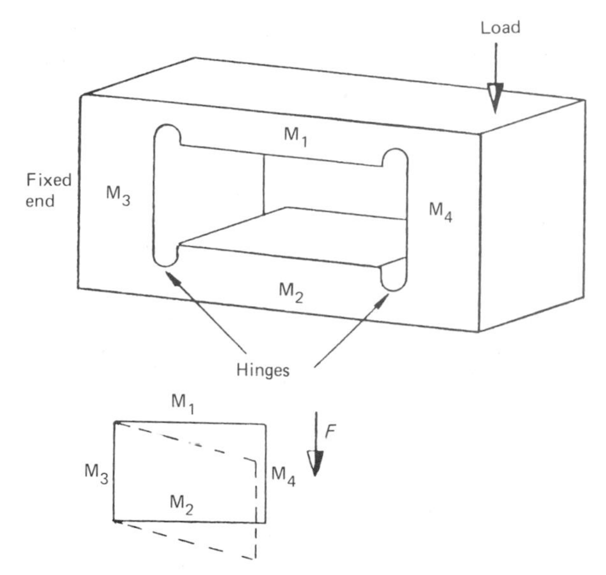

Figure 1.16 Load cell spring mechanism.

The load cell in Figure 1.16 consists of four stiff members and four flexures, machined out of a solid block of high-quality spring material in the shape of a parallelogram. The members M1, M2 and M3, M4 remain parallel as the force *F* bends the flexures at the thin sections (called hinges).

Any torque, caused by the load being offset from the vertical, will result in a small twisting of the load cell, but this is kept within the required limit by arranging the rotational stiffness to be much greater than the vertical stiffness. This is determined by the width.

The trapezoidal construction is far better in this respect than a normal cantilever, which would result in a non-linear response.

1.5.4 Combined actuator transducer

Figure 1.17 illustrates a more complex example, requiring a number of processing techniques to fabricate the complete item. The combined actuator

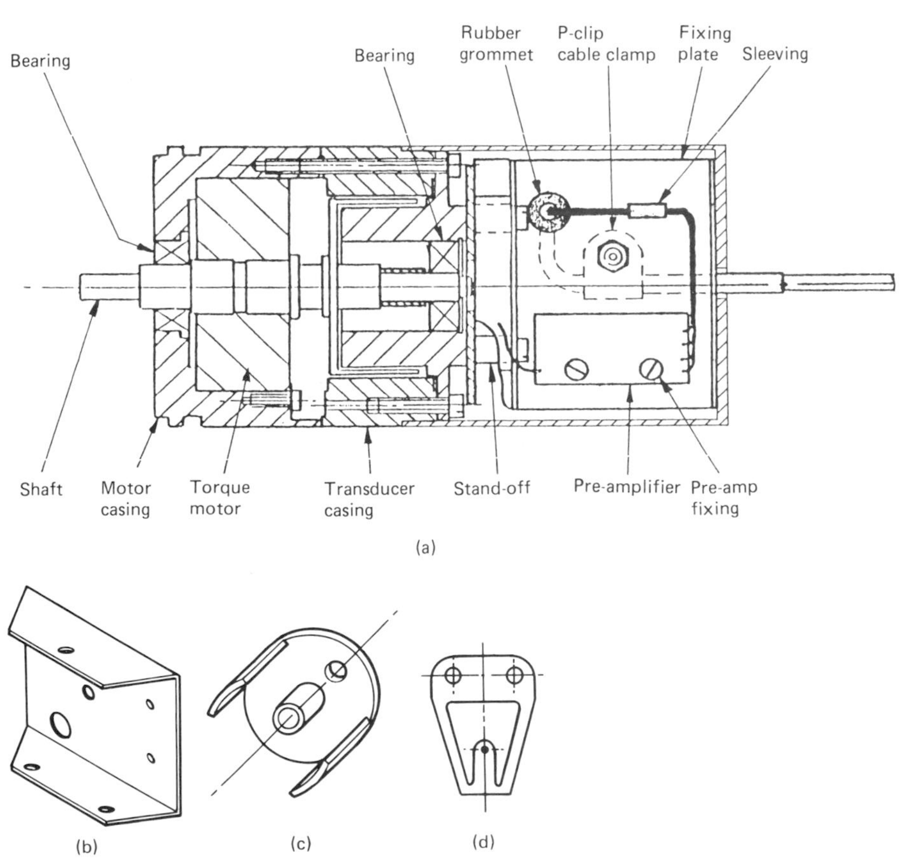

Figure 1.17 Combined actuator/transducer. (a) Whole assembly; (b) fixing plate; (c) transducer rotor; (d) spring contact.

transducer (CAT) is a low-volume product with applications in automatic optical instruments for mirror positioning. The major bought-in components are a torque motor and a miniature pre-amplifier produced by specialist suppliers. The motor incorporates the most advanced rare-earth magnetic materials for compactness and stability and the pre-amplifier uses small outline electronic components, surface mounted on a copper/fibreglass printed circuit board.

The assembled CAT is shown in Figure 1.17(a). It consists of three sections: the motor in its housing, a capacitive angular transducer in its housing and a rear-mounted plate (Figure 1.17(b)) for pre-amplifier fixing and cable clamping. The motor produces a torque which rotates the shaft in the bearings. Position information is provided by the transducer which produces an output via the pre-amplifier. The associated electronic servocontrol unit provides the power output stage, position feedback and loop-stabilization components to control the shaft angle to an arc second. This accuracy is attainable by the use of precise ballbearings which maintain radial and axial movement to within 10 μm.

The shaft, motor casing and transducer components are manufactured by precise turning of a non-magnetic stainless steel bar and finished by fine bead blasting. The motor and transducer electrodes (not shown) are glued in place with a thin layer of epoxy resin and in the latter case finished by turning to final size.

The two parts of the transducer stator are jigged concentric and held together by three screws in threaded holes A, leaving a precisely determined gap in which the transducer rotor (Figure 1.17(c)) rotates. The transducer rotor is also turned but the screens are precision ground to size, as this determines the transducer sensitivity and range.

The screens are earthed, via the shaft and a hardened gold rotating point contact held against vibration by a spring (Figure 1.17(d)). The spring is chemically milled from thin beryllium copper sheet.

The shaft with motor and transducer rotor glued in place, motor stator and casing, transducer stator and bearings are then assembled and held together with three screws as B. The fixing plate, assembled with cable, clamp and pre-amplifier separately, is added, mounted on standard standoffs, the wires made off, and then finally the cover put on.

In addition to the processes mentioned the manufacture of this unit requires drilling, reaming, bending, screen printing, soldering, heat treatment and anodizing. Materials include copper, PTFE, stainless steel, samarium cobalt, epoxy fibreglass, gold and aluminium, and machining tolerances are typically 25 μm for turning, 3 μm for grinding and 0.1 mm for bending.

The only external feature is the clamping ring at the shaft end for axial fixing (standard servo type size 20). This is provided because radial force could distort the thin wall section and cause transducer errors.

1.6 References

BIRKBECK, G. 'Mechanical Design', in *A Guide to Instrument Design*, SIMA and BSIRA, Taylor and Francis, London (1963)

CLAYTON, G. B. *Operational Amplifiers*, Butterworths, London (1979)

FURSE, J. E. 'Kinematic design of fine mechanisms in instruments', in *Instrument Science and Technology*, Volume 2, ed. E. B. Jones, Adam Hilger, Bristol (1983)

HOROWITZ, P. and HILL, W. *The Art of Electronics*, Cambridge University Press, Cambridge (1980)

KIBBLE, B. P. and RAYNER, G. H. *Co-axial AC Bridges*, Adam Hilger, Bristol (1984)

MORRELL, R. *Handbook of Properties of Technical and Engineering Ceramics*. Part 1, An introduction for the engineer and designer, HMSO, London (1985)

OBERG, E. and JONES, F. D. *Machinery's Handbook*, The Machinery Publishing Company, (1979)

SHIELDS, J. *Adhesives Handbook*, Butterworths, London (revised 3rd edn, 1985)

The standards referred to in the text are:

BS 5252 (1976) and 4800 (1981): Framework for colour co-ordination for building purposes
BS 5490 (1977 and 1985): Environmental protection provided by enclosures
DIN 43 700 (1982): Cutout dimensions for panel mounting instruments
DIN 41612: Standards for Eurocard connectors
DIN 41914 and IEC 297: Standards for Eurocards

2 Instrument installation and commissioning

A. Danielsson

2.1 Introduction

Plant safety and continuous effective plant operability are totally dependent upon correct installation and commissioning of the instrumentation systems. Process plants are increasingly becoming dependent upon automatic control systems owing to the advanced control functions and monitoring facilities that can be provided in order to improve plant efficiency, product throughput and product quality.

The instrumentation on a process plant represents a significant capital investment, and the importance of careful handling on site and the exactitude of the installation cannot be overstressed. Correct installation is also important in order to ensure long-term reliability and to obtain the best results from instruments which are capable of higher-order accuracies due to advances in technology. Quality control of the completed work is also an important function.

2.2 General requirements

Installation should be carried out using the best engineering practices by skilled personnel who are fully acquainted with the safety requirements and regulations governing a plant site. Prior to commencement of the work for a specific project, installation design details should be made available which define the scope of work and the extent of material supply and which give detailed installation information related to location, fixing, piping and wiring. Such design details should have already taken account of established installation recommendations and measuring technology requirements. The details contained in this chapter are intended to give general installation guidelines.

2.3 Storage and protection

When instruments are received on a job site it is of utmost importance that they are unpacked with care, examined for superficial damage and then placed in a secure store which should be free from dust and suitably heated. In order to minimize handling, large items of equipment such as control panels should be programmed to go directly into their intended location, but temporary anti-condensation heaters should be installed if the intended air-conditioning systems have not been commissioned.

Throughout construction, instruments and equipment installed in the field should be fitted with suitable coverings to protect them from mechanical abuse such as paint spraying, etc. Preferably, after an installation has been fabricated, the instrument should be removed from site and returned to the store for safe keeping until ready for precalibration and final loop checking. Again, when instruments are removed, care should be taken to seal the ends of piping, etc. to prevent ingress of foreign matter.

2.4 Mounting and accessibility

When instruments are mounted in their intended location, either on pipe stands, brackets or directly connected to vessels, etc., they should be vertically plumbed and firmly secured. Instrument mountings should be vibration free and should be located so that they do not obstruct access ways which may be required for maintenance to other items of equipment. They should also be clear of obvious hazards such as hot surfaces or drainage points from process equipment.

Locations should also be selected to ensure that the instruments are accessible for observation and maintenance. Where instruments are mounted at higher elevations, it must be ensured that they are accessible either by permanent or temporary means.

Instruments should be located as close as possible to their process tapping points in order to minimize the length of impulse lines, but consideration should be paid to the possibility of expansion of piping or vessels which could take place under operating conditions and which could result in damage if not properly catered for. All brackets and supports should be adequately protected against corrosion by priming and painting.

When installing final control elements such as control valves, again the requirement for maintenance access must be considered and clearance should be allowed above and below the valve to facilitate servicing of the valve actuator and the valve internals.

2.5 Piping systems

All instrument piping or tubing runs should be routed to meet the following requirements:

(1) They should be kept as short as possible;
(2) They should not cause any obstruction that would prohibit personnel or traffic access;
(3) They should not interfere with the accessibility for maintenance of other items of equipment;
(4) They should avoid hot environments or potential fire-risk areas;
(5) They should be located with sufficient clearance to permit lagging which may be required on adjacent pipework;
(6) The number of joints should be kept to a minimum consistent with good practice;
(7) All piping and tubing should be adequately supported along its entire length from supports attached to firm steelwork or structures (not handrails).

(*Note*: Tubing can be regarded as thin-walled seamless pipe that cannot be threaded and which is joined by compression fittings, as opposed to piping, which can be threaded or welded.)

2.5.1 Air supplies

Air supplies to instruments should be clean, dry and oil free. Air is normally distributed around a plant from a high-pressure header (e.g. 6–7 bar g), ideally forming a ring main. This header, usually of galvanized steel, should be sized to cope with the maximum demand of the instrument air users being serviced and an allowance should be made for possible future expansion or modifications to its duty.

Branch headers should be provided to supply individual instruments or groups of instruments. Again, adequate spare tappings should be allowed to cater for future expansion. Branch headers should be self draining and have adequate drainage/blow-off facilities. On small headers this may be achieved by the instrument air filter/regulators.

Each instrument air user should have an individual filter regulator. Piping and fittings installed after filter regulators should be non-ferrous.

2.5.2 Pneumatic signals

Pneumatic transmission signals are normally in the range of 0.2–1.0 bar g, and for these signals copper tubing is most commonly used, preferably with a PVC outer sheath. Other materials are sometimes used, depending on environmental considerations (e.g. alloy tubing or stainless steel). Although expensive, stainless steel tubing is the most durable and will withstand the most arduous service conditions.

Plastic tubing should preferably only be used within control panels. There are several problems to be considered when using plastic tubes on a plant site, as they are very vulnerable to damage unless adequately protected, they generally cannot be installed at subzero temperatures and they can be considerably weakened by exposure to hot surfaces. Also it should be remembered that they can be totally lost in the event of a fire.

Pneumatic tubing should be run on a cable tray or similar supporting steelwork for its entire length and securely clipped at regular intervals. Where a number of pneumatic signals are to be routed to a remote control room they should be marshalled in a remote junction box and the signals conveyed to the control room via multitube bundles. Such junction boxes should be carefully positioned in the plant in order to minimize the lengths of the individually run tubes. (See Figure 2.1 for typical termination of pneumatic multitubes.)

2.5.3 Impulse lines

These are the lines containing process fluid which run between the instrument impulse connection and the process tapping point, and are usually made up from piping and pipe fittings or tubing and compression fittings. Piping materials must be compatible with the process fluid.

Generally, tubing is easier to install and is capable of handling most service conditions provided that the correct fittings are used for terminating the tubing. Such fittings must be compatible with the tubing being run (i.e. of the same material).

Impulse lines should be designed to be as short as possible, and should be installed so that they are self draining for liquids and self venting for vapours or gases. If necessary, vent plugs or valves should be located at high points in liquid-filled lines and, similarly, drain plugs or valves should be fitted at low points in gas or vapour-filled lines. In any case, it should be ensured that there are provisions for isolation and depressurizing of instruments for maintenance purposes. Furthermore, filling plugs should be provided where lines are to be liquid sealed for chemical protection and, on services

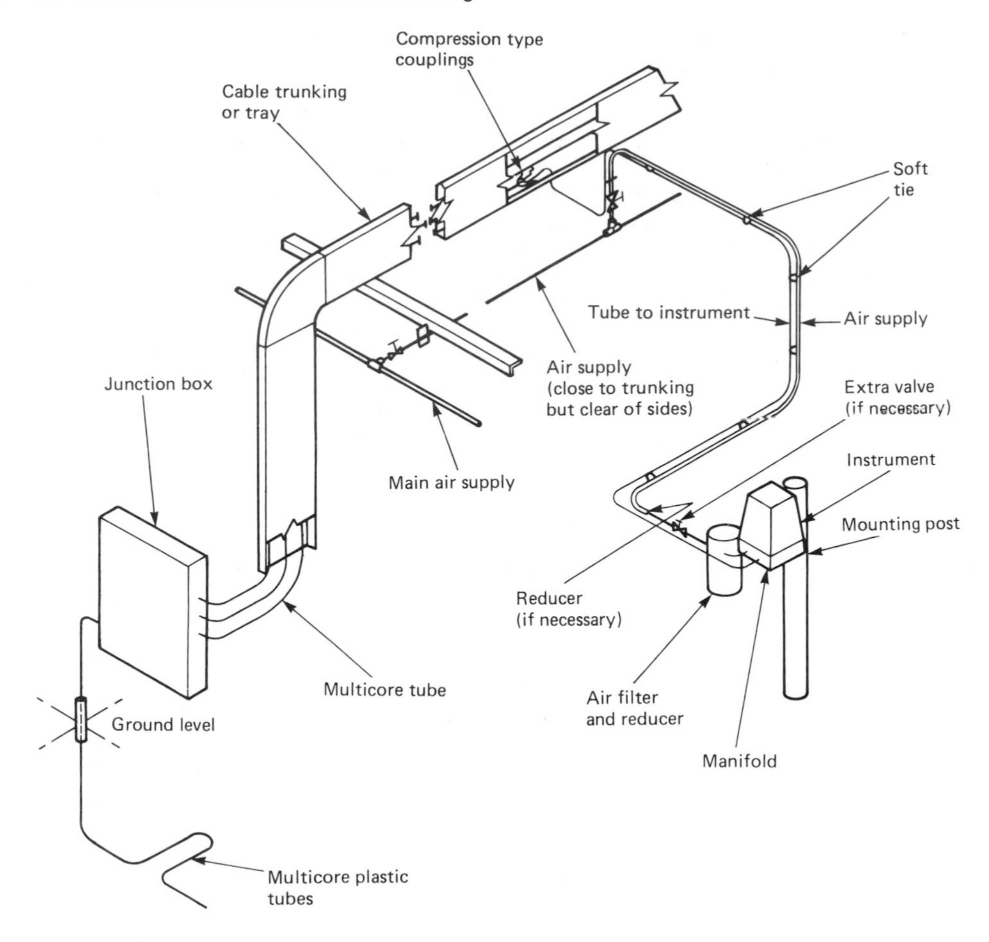

Figure 2.1 Typical field termination of pneumatic multitubes.

which are prone to plugging, rodding-out connections should be provided close to the tapping points.

2.6 Cabling

2.6.1 General requirements

Instrument cabling is generally run in multi-core cables from the control room to the plant area (either below or above ground) and then from field junction boxes in single pairs to the field measurement or actuating devices.

For distributed microprocessor systems the interconnection between the field and the control room is usually via duplicate data highways from remote located multiplexers or process interface units. Such duplicate highways would take totally independent routes from each other for plant security reasons.

Junction boxes must meet the hazardous area requirements applicable to their intended location and should be carefully positioned in order to minimize the lengths of individually run cables, always bearing in mind the potential hazards that could be created by fire.

Cable routes should be selected to meet the following requirements:

(1) They should be kept as short as possible.
(2) They should not cause any obstruction that would prohibit personnel or traffic access.
(3) They should not interfere with the accessibility for maintenance of other items of equipment.
(4) They should avoid hot environments or potential fire-risk areas.
(5) They should avoid areas where spillage is liable to occur or where escaping vapours or gases could present a hazard.

Cables should be supported for their whole run length by a cable tray or similar supporting steelwork. Cable trays should preferably be installed with their breadth in a vertical plane. The layout of cable trays on a plant should be carefully selected so that the minimum number of instruments in the immediate vicinity would be affected in the case of a local fire. Cable joints should be avoided other than in approved junction boxes or termination points. Cables entering junction boxes from below ground should be specially protected by fire-resistant ducting or something similar.

2.6.2 Cable types

There are three types of signal cabling generally under consideration, i.e.

(1) Instrument power supplies (above 50 V);
(2) High-level signals (between 6 and 50 V). This includes digital signals, alarm signals and high-level analogue signals (e.g. 4–20 mA).
(3) Low-level signals (below 5 V). This generally covers thermocouple compensating leads and resistance element leads.

Signal wiring should be made up in twisted pairs. Solid conductors are preferable so that there is no degradation of signal due to broken strands that may occur in stranded conductors. Where stranded conductors are used, crimped connectors should be fitted. Cable screens should be provided for instrument signals, particularly low-level analogue signals, unless the electronic system being used is deemed to have sufficient built-in 'noise' rejection. Further mechanical protection should be provided in the form of single-wire armour and PVC outer sheath, especially if the cables are installed in exposed areas, e.g. on open cable trays. Cables routed below ground in sand-filled trenches should also have an overall lead sheath if the area is prone to hydrocarbon or chemical spillage.

2.6.4 Cable segregation

Only signals of the same type should be contained within any one multicore cable. In addition, conductors forming part of intrinsically safe circuits should be contained in a multicore reserved solely for such circuits.

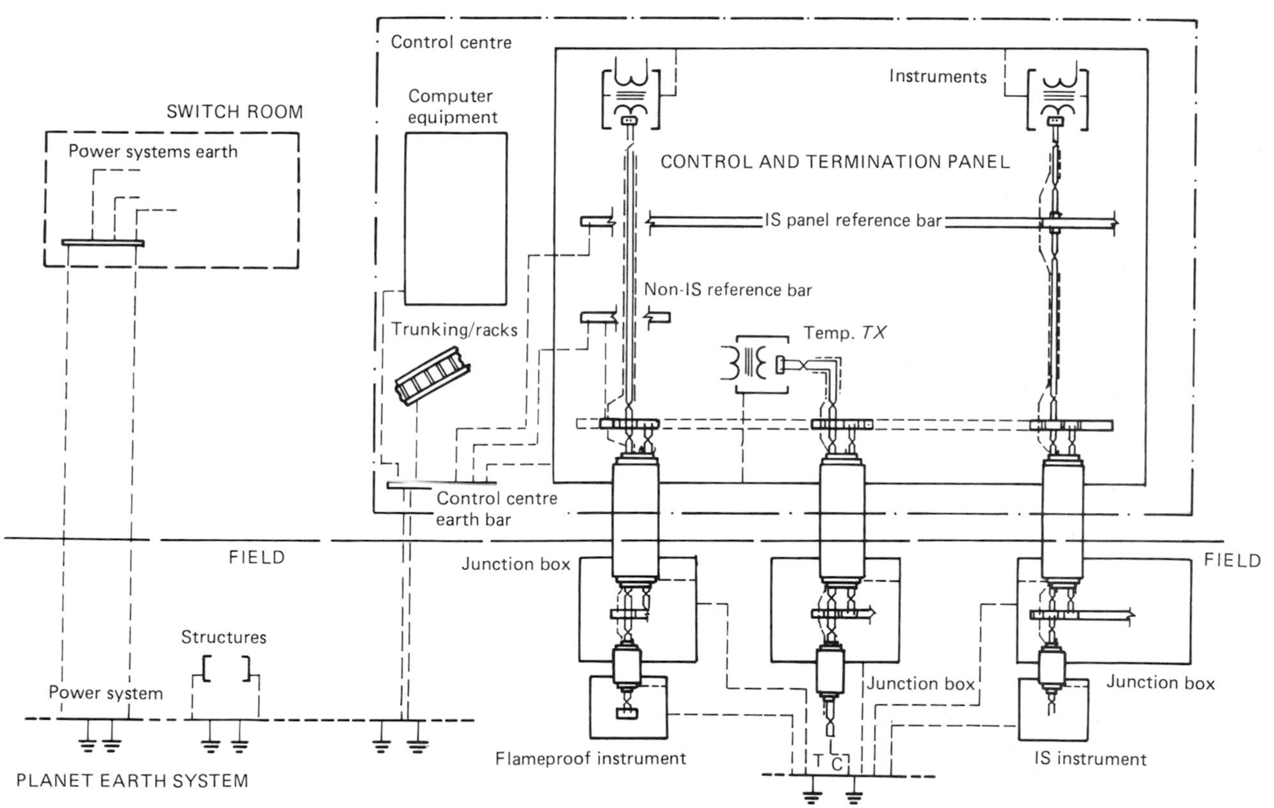

Figure 2.2 A typical control centre-earthing system.

When installing cables above or below ground they should be separated into groups according to the signal level and segregated with positive spacing between the cables. As a general rule, low-level signals should be installed furthest apart from instrument power supply cables with the high-level signal cables in between. Long parallel runs of dissimilar signals should be avoided as far as possible, as this is the situation where interference is most likely to occur.

Cables used for high-integrity systems such as emergency shutdown systems or data highways should take totally independent routes or should be positively segregated from other cables. Instrument cables should be run well clear of electrical power cables and should also, as far as possible, avoid noise-generating equipment such as motors. Cable crossings should always be made at right angles.

When cables are run in trenches, the routing of such trenches should be clearly marked with concrete cable markers on both sides of the trench, and the cables should be protected by earthenware or concrete covers.

2.7 Earthing

2.7.1 General requirements

Special attention must be paid to instrument earthing, particularly where field instruments are connected to a computer or microprocessor type control system. Where cable screens are used, earth continuity of screens must be maintained throughout the installation with the earthing at one point only, i.e. in the control room. At the field end the cable screen should be cut back and taped so that it is independent from earth. Intrinsically safe systems should be earthed through their own earth bar in the control room. Static earthing of instrument cases, panel frames, etc. should be connected to the electrical common plant earth. (See Figure 2.2 for a typical earthing system.)

Instrument earths should be wired to a common bus bar within the control centre and this should be connected to a remote earth electrode via an independent cable (preferably duplicated for security and test purposes). The resistance to earth, measured in the control room, should usually not exceed 1 Ω unless otherwise specified by a system manufacturer or by a certifying authority.

2.8 Testing and precommissioning

2.8.1 General

Before starting up a new installation the completed instrument installation must be fully tested to ensure that the equipment is in full working order. This testing normally falls into three phases, i.e. pre-installation testing; piping and cable testing; loop testing or precommissioning.

2.8.2 Pre-installation testing

This is the testing of each instrument for correct calibration and operation prior to its being installed in the field. Such testing is normally carried out in a workshop which is fully equipped for the purpose and should contain a means of generating the measured variable signals and also a method of accurately measuring the instrument input and output (where applicable). Test instruments should have a standard of accuracy better than the manufacturer's stated accuracy for the instruments being tested and should be regularly certified.

Instruments are normally calibration checked at five points (i.e. 0, 25, 50, 75 and 100 per cent) for both rising and falling signals, ensuring that the readings are within the manufacturer's stated tolerance.

After testing, instruments should be drained of any testing fluids that may have been used and, if necessary, blown through with dry air. Electronic instruments should be energized for a 24-h warm-up period prior to the calibration test being made. Control valves should be tested *in situ* after the pipework fabrication has been finished and flushing operations completed. Control valves should be checked for correct stroking at 0, 50 and 100 per cent open and at the same time the valves should be checked for correct closure action.

2.8.3 Piping and cable testing

This is an essential operation prior to loop testing.

2.8.3.1 Pneumatic lines

All air lines should be blown through with clean, dry air prior to final connection to instruments and they should also be pressure tested for a timed interval to ensure that they are leak free. This should be in the form of a continuity test from the field end to its destination (e.g. the control room).

2.8.3.2 Process piping

Impulse lines should also be flushed through and hydrostatically tested prior to connection of the instruments. All isolation valves or manifold valves should be checked for tight shutoff. On completion of hydrostatic tests, all piping should be drained and thoroughly dried out prior to reconnecting to any instruments.

2.8.3.3 Instrument cables

All instrument cables should be checked for continuity and insulation resistance before connection to any instrument or apparatus. The resistance should be checked core to core and core to earth. Cable screens must also be checked for continuity and insulation. Cable tests should comply with the requirements of Part 6 of the IEE Regulation for Electrical Installations (latest edition), or the rules and regulations with which the installation has to comply. Where cables are installed below ground, testing should be carried out before the trenches are back filled. Coaxial cables should be tested using sine-wave reflective testing techniques. As a prerequisite to cable testing it should be ensured that all cables and cable ends are properly identified.

2.8.4 Loop testing

The purpose of loop testing is to ensure that all instrumentation components in a loop are in full operational order when interconnected and are in a state ready for plant commissioning.

Prior to loop testing, inspection of the whole installation, including piping, wiring, mounting, etc., should be carried out to ensure that the installation is complete and that the work has been carried out in a professional manner. The control room panels or display stations must also be in a fully functional state.

Loop testing is generally a two-man operation, one in the field and one in the control room who should be equipped with some form of communication, e.g. field telephones or radio transceivers. Simulation signals should be injected at the field end equivalent to 0, 50 and 100 per cent of the instrument range and the loop function should be checked for correct operation in both rising and falling modes. All results should be properly documented on calibration or loop check sheets. All ancillary components in the loop should be checked at the same time.

Alarm and shutdown systems must also be systematically tested and all systems should be checked for 'fail-safe' operation including the checking of 'burn-out' features on thermocouple installations. At the loop-checking stage all ancillary work should be completed, such as setting zeros, filling liquid seals and fitting of accessories such as charts, ink, fuses, etc.

2.9 Plant commissioning

Commissioning is the bringing 'on-stream' of a process plant and the tuning of all instruments and controls to suit the process operational requirements. A plant, or section thereof is considered to be ready for commissioning when all instrument installations are mechanically complete and all testing, including loop testing, has been effected.

Before commissioning can be attempted it should be ensured that all air supplies are available and that all power supplies are fully functional, including any emergency standby supplies. It should also be ensured that all ancillary devices are operational, such as protective heating systems, air conditioning, etc. All control valve lubricators (when fitted) should be charged with the correct lubricant.

Commissioning is usually achieved by first commissioning the measuring system with any controller mode overridden. When a satisfactory measured variable is obtained, the responsiveness of a control system can be checked by varying the control valve position using the 'manual' control function. Once the system is seen to respond correctly and the required process variable reading is obtained it is then possible to switch to 'auto' in order to bring the controller function into action. The controller responses should then be adjusted to obtain optimum settings to suit the automatic operation of plant.

Alarm and shutdown systems should also be systematically brought into operation, but it is necessary to obtain the strict agreement of the plant operation supervisor before any overriding of trip systems is attempted or shutdown features are operated.

Finally, all instrumentation and control systems would need to be demonstrated to work satisfactorily before formal acceptance by the plant owner.

2.10 References

BS 6739, British Standard Code of Practice for Instrumentation in Process Control Systems: Installation Design and Practice (1986)

REGULATIONS FOR ELECTRICAL INSTILLATIONS 15TH ED. (1981) as issued by the Institution of Electrical Engineers

3 Sampling

J. G. Giles

3.1 Introduction

3.1.1 Importance of sampling

Any form of analysis instrument can only be as effective as its sampling system. Analysis instruments are out of commission more frequently due to troubles in the sampling system than to any other cause. Therefore time and care expended in designing and installing an efficient sampling system is well repaid in the saving of servicing time and dependability of instrument readings. The object of a sampling system is to obtain a truly representative sample of the solid, liquid or gas which is to be analysed, at an adequate and steady rate, and transport it without change to the analysis instrument, and all precautions necessary should be taken to ensure that this happens. Before the sample enters the instrument it may be necessary to process it to the required physical and chemical state, i.e. correct temperature, pressure, flow, purity, etc., without removing essential components. It is also essential to dispose of the sample and any reagent after analysis without introducing a toxic or explosive hazard. For this reason, the sample, after analysis, is continuously returned to the process at a suitable point or a sample-recovery and disposal system is provided.

3.1.2 Representative sample

It is essential that the sample taken should represent the mean composition of the process material. The methods used to overcome the problem of uneven sampling depend on the phase of the process sample, which may be in solid, liquid, gas or mixed-phase form.

3.1.2.1 Solids

When the process sample is solid in sheet form it is necessary to scan the whole sheet for a reliable measurement of the state of the sheet (e.g. thickness, density or moisture content). A measurement at one point is insufficient to give a representative value of the parameter being measured.

If the solid is in the form of granules or powder of uniform size a sample collected across a belt or chute and thoroughly mixed will give a reasonably representative sample. If measurement of density or moisture content of the solid can be made while it is in a vertical chute under a constant head, packing density problems may be avoided.

In some industries where the solids are transported as slurries it is possible to carry out the analysis directly on the slurry if a method is available to compensate for the carrier fluid and the velocities are high enough to ensure turbulent flow at the measurement point.

Variable-size solids are much more difficult to sample and specialist work on the subject should be consulted.

3.1.2.2 Liquids

When sampling liquid it is essential to ensure that either the liquid is turbulent in the process line or that there is at least 200 pipe diameters between the point of adding constituents and the sampling point. If neither is possible, a motorized or static mixer should be inserted into the process upstream of the sample point.

3.1.2.3 Gases

Gas samples must be thoroughly mixed and, as gas process lines are usually turbulent, the problem of finding a satisfactory sample point is reduced. The main exception is in large ducts such as furnace or boiler flues, where stratification can occur and the composition of the gas may vary from one point to another. In these cases special methods of sampling may be necessary such as multiple probes or long probes with multiple inlets in order to obtain a representative sample.

3.1.2.4 Mixed-phase sampling

Mixed phases such as liquid/gas mixtures or liquid/solids (i.e. slurries) are best avoided for any analytical method that involves taking a sample from the process. It is always preferable to use an in-line analysis method where this is possible.

3.1.3 Parts of analysis equipment

The analysis equipment consists of five main parts:

(1) Sample probe;
(2) Sample-transport system;
(3) Sample-conditioning equipment;
(4) The analysis instrument;
(5) Sample disposal.

3.1.3.1 Sample probe

This is the sampling tube that is used to withdraw the sample from the process.

3.1.3.2 Sample-transport system

This is the tube or pipe that transports the sample from the sample point to the sample-conditioning system.

3.1.3.3 Sample-conditioning system

This system ensures that the analyser receives the sample at the correct pressure and in the correct state to suit the analyser. This may require pressure increase (i.e. pumps) or reduction, filtration, cooling, drying and other equipment to protect the analyser from process upsets. Additionally, safety equipment and facilities for the introduction of calibration samples into the analyser may also be necessary.

3.1.3.4 The analysis instrument

This is the process analyser complete with the services such as power, air, steam, drain vents, carrier gases and signal conditioning that are required to make the instrument operational. (Analysis techniques are described in Part 2 of this book.)

3.1.3.5 Sample disposal

The sample flowing from the analyser and sample conditioning system must be disposed of safely. In many cases it is possible to vent gases to atmosphere or allow liquids to drain, but there are times when this is not satisfactory. Flammable or toxic gases must be vented in such a way that a hazard is not created. Liquids such as hydrocarbons can be collected in a suitable tank and pumped back into the process, whereas hazardous aqueous liquids may have to be treated before being allowed to flow into the drainage system.

3.1.4 Time lags

In any measuring instrument, particularly one which may be used with a controller, it is desirable that the time interval between occurrence of a change in the process fluid and its detection at the instrument should be as short as possible consistent with reliable measurement. In order to keep this time interval to a minimum the following points should be kept in mind.

3.1.4.1 Sample-transport line length

The distance between the sampling point and the analyser should be kept to the minimum. Where long sample transport lines are unavoidable a 'fast loop' may be used. The fast loop transports the sample at a flow rate higher than that required by the analyser, and the excess sample is either returned to the process, vented to atmosphere or allowed to flow to drain. The analyser is supplied with the required amount of sample from the fast loop through a short length of tubing.

3.1.4.2 Sampling components

Pipe, valves, filter and all sample-conditioning components should have the smallest volume consistent with a permissible pressure drop.

3.1.4.3 Pressure reduction

Gaseous samples should be filtered, and flow in the sample line at the lowest possible pressure, as the mass of gas in the system depends on the pressure of the gas as well as the volume in the system.

When sampling high-pressure gases the pressure-reducing valve must be situated at the sample point. This is necessary, because for a fixed mass flow rate of gas the response time will increase in proportion to the absolute pressure of the gas in the sample line (i.e. gas at 10 bar A will have a time lag five times that of gas at 2 bar A). This problem becomes more acute when the sample is a liquid which has to be vaporized for analysis (e.g. liquid butane or propane).

The ratio of volume of gas to volume of liquid can be in the region of 250:1, as is the case for propane. It is therefore essential to vaporize the liquid at the sample point and then treat it as a gas sample from then on.

3.1.4.4 Typical equations

(1) $t = \frac{L}{S}$

t = time lag

S = velocity (m/s)
L = line length (m)

(2) General gas law for ideal gases:

$$\frac{pv}{T} = \frac{8314 \times W}{10^5 \times M}$$

p = pressure
T = abs. temperature (K)
v = volume (l)
W = mass (g)
M = molecular weight

(3) Line volume:

$$\frac{\pi d^2}{4} = V_I$$

d = internal diameter of tube (mm)
V_I = volume (ml/m)

(4)

$$t = \frac{6L \times V_I}{100F}$$

L = line length (m)
V_I = internal volume of line (ml/m)
F = sample flow rate (l/min)
t = time lag (s)

(For an example of a fast loop calculation see Section 3.3.2.2 (Table 3.1.)

3.1.4.5 Useful data

Internal volume per metre (V_I) of typical sample lines:

⅛ in OD × 0.035 wall	= 1.5 ml/m
¼ in OD × 0.035 wall	= 16.4 ml/m
⅜ in OD × 0.035 wall	= 47.2 ml/m
½ in OD × 0.065 wall	= 69.4 ml/m
½ in nominal bore steel pipe (extra strong) (13.88 mm ID)	= 149.6 ml/m
3 mm OD × 1 mm wall	= 0.8 ml/m
6 mm OD × 1 mm wall	= 12.6 ml/m
8 mm OD × 1 mm wall	= 28.3 ml/m
10 mm OD × 1 m wall	= 50.3 ml/m
12 mm OD × 1.5 mm wall	= 63.6 ml/m

3.1.5 Construction materials

Stainless steel (Type 316 or 304) has become one of the most popular materials for the construction of sample systems due to its high resistance to corrosion, low surface adsorption (especially moisture), wide operating temperature range, high-pressure capability and the fact that it is easily obtainable. Care must be taken when there are materials in the sample which cause corrosion, such as chlorides and sulphides, in which case it is necessary to change to more expensive materials such as Monel.

When atmospheric sampling is carried out for trace constituents, teflon tubing is frequently used, as the surface adsorption of the compounds is less than stainless steel, but it is necessary to check that the compound to be measured does not diffuse through the wall of the tubing.

For water analysis (e.g. pH and conductivity) it is possible to use plastic (such as PVC or ABS) components, although materials such as Kunifer 10 (copper 90%, nickel 10%) are increasing in popularity when chlorides (e.g. salt water) are present, as they are totally immune to chloride corrosion.

3.2 Sample system components

3.2.1 Probes

The most important function of a probe is to obtain the sample from the most representative point (or points) in the process line.

3.2.1.1 Sample probe

A typical probe of this type for sampling all types of liquid and gases at low pressure is shown in Figure 3.1. It can be seen that the probe which is made of 21 mm OD (11.7 mm ID) stainless steel pipe extends to the centre of the line being sampled. However, if the latter is more than 500 mm OD the probe intrusion is kept at 250 mm to avoid vibration in use.

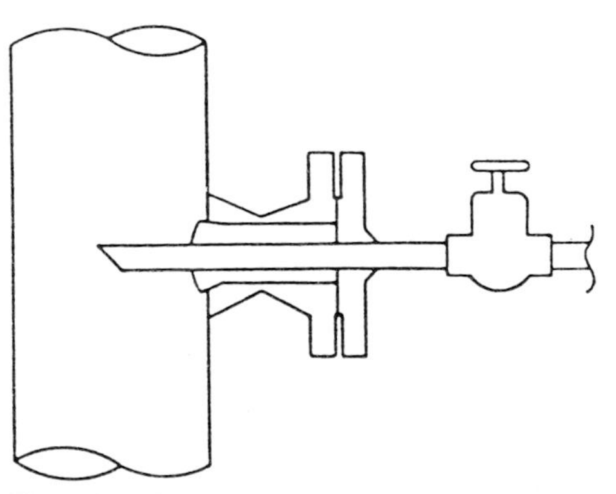

Figure 3.1 Sample probe (courtesy Ludlam Sysco).

3.2.1.2 Small-volume sample probe

This probe is used for sampling liquids which require

to be vaporized or for high-pressure gases (Figure 3.2). Typically, a 6 mm OD × 2 mm ID tube is inserted through the centre of a probe of the type described in section 3.2.1.1. The probe may be withdrawn through the valve for cleaning.

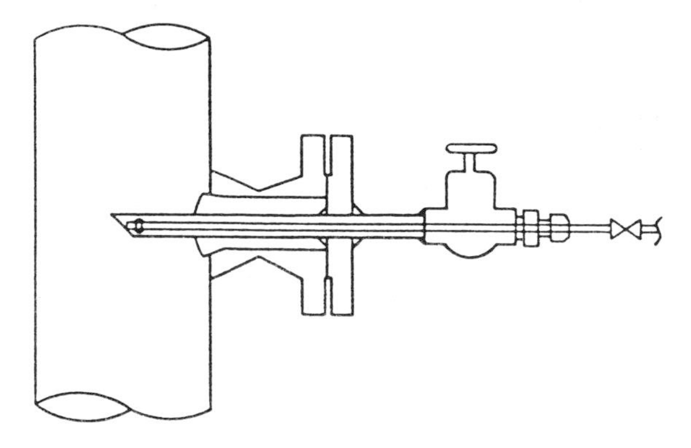

Figure 3.2 Small-volume sample probe (courtesy Ludlam Sysco).

3.2.1.3 Furnace gas probes

Low temperature probe Figure 3.3 shows a gas sampling probe with a ceramic outside filter for use in temperatures up to 400°C.

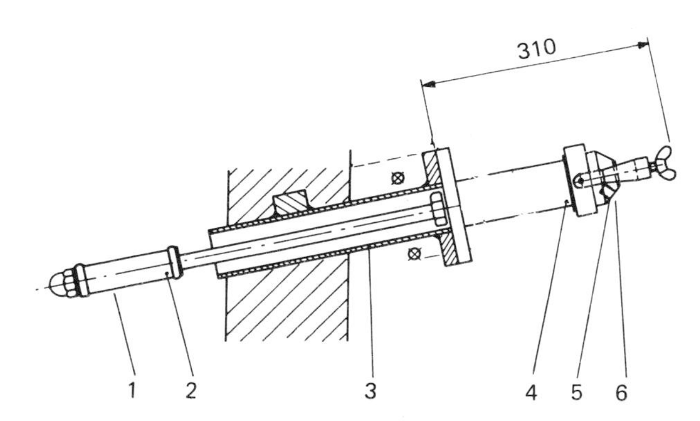

Figure 3.3 Gas-sampling probe (courtesy Hartmann and Braun). 1 Gas intake; 2 ceramic intake filter; 3 bushing tube with flange; 4 case with outlet filter; 5 internal screwthread; 6 gas outlet.

Water-wash probe This probe system is used for sampling furnace gases with high dust content at high temperature (up to 1600°C). (Figure 3.4). The wet gas sampling probe is water cooled and self priming. The water/gas mixture passes from the probe down to a water trap, where the gas and water are separated. The gas leaves the trap at a pressure of approximately 40 m bar, and precautions should be taken to avoid condensation in the analyser, either by ensuring that the analyser is at a higher temperature than the water trap or by passing the sample gas through a sample cooler at 5°C to reduce the humidity.

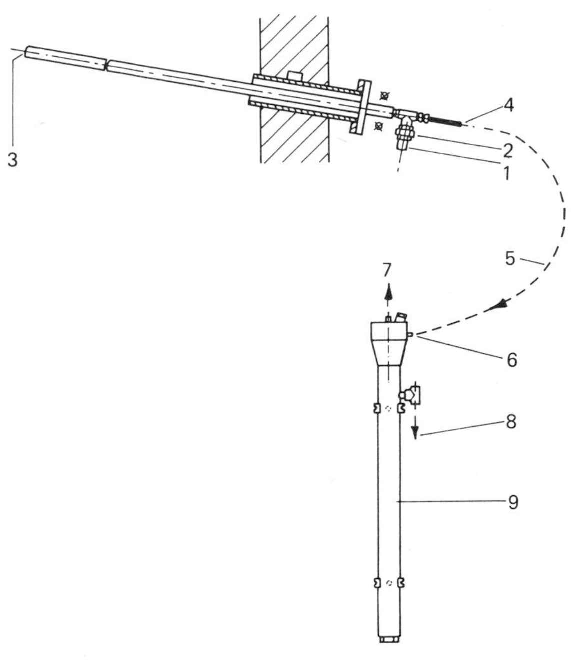

Figure 3.4 Water-wash probe (courtesy Hartmann and Braun). 1 Water intake; 2 water filter; 3 gas intake; 4 gas–water outlet; 5 connecting hose; 6 gas–water intake; 7 gas outlet; 8 water outlet; 9 water separator.

Note that this probe is not suitable for the measurement of water-soluble gases such as CO_2, SO_2 or H_2S.

Steam ejector The steam ejector illustrated in Figure 3.5 can be used for sample temperatures up to 180°C and, because the condensed steam dilutes the condensate present in the flue gas, the risk of corrosion of the sample lines when the steam/gas sample cools to the dew point is greatly reduced.

Dry steam is supplied to the probe and then ejected through a jet situated in the mouth of a Venturi. The flow of steam causes sample gas to be drawn into the probe. The steam and gas pass out of the probe and down the sample line to the analyser system at a positive pressure. The flow of steam through the sample line prevents the build-up of any corrosive condensate.

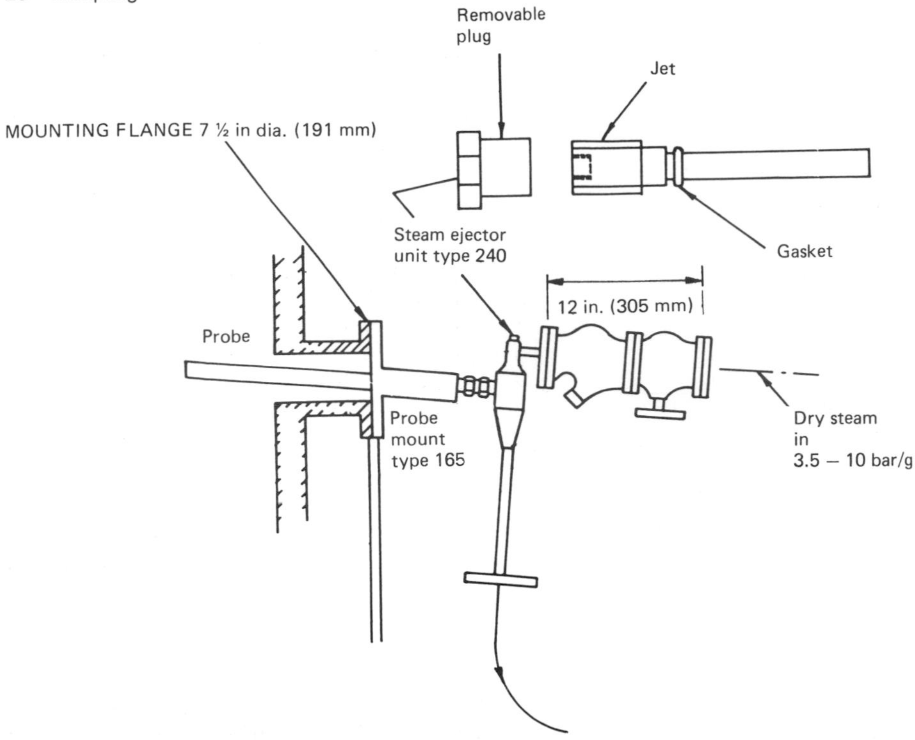

Figure 3.5 Steam ejection probe (courtesy Servomex).

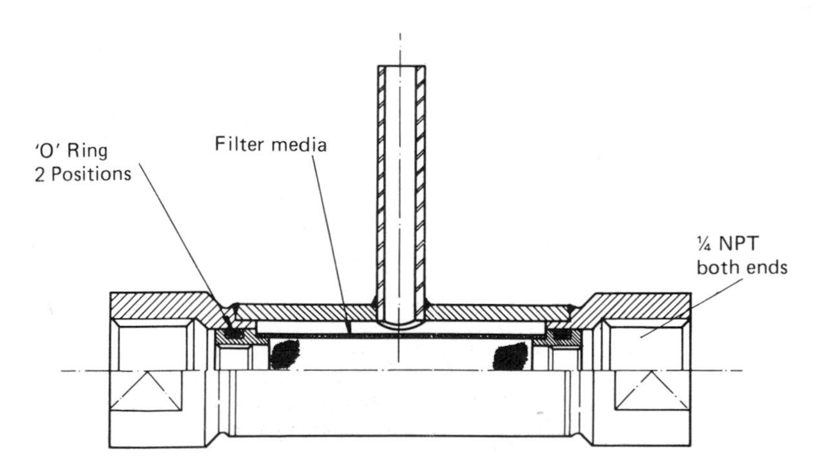

Figure 3.6 In-line filter (courtesy Microfiltrex).

3.2.2 Filters

3.2.2.1 'Y' strainers

'Y' strainers are available in stainless steel, carbon steel and bronze. They are ideal for preliminary filtering of samples before pumps or at sample points to prevent line scale entering sample lines. Filtration sizes are available from 75 to 400 μ (200 to 40 mesh). The main application for this type of filter is for liquids and steam.

3.2.2.2 In-line filters

This design of filter is normally used in a fast loop configuration and is self cleaning (Figure 3.6). Filtration is through a stainless steel or a ceramic element. Solid particles tend to be carried straight on in the sample stream so that maintenance time is very low. Filtration sizes are available from 150 μ (100 mesh) down to 5 μ. It is suitable for use with liquids or gases.

3.2.2.3 Filters with disposable glass microfibre element

These filters are available in a wide variety of sizes and porosities (Figure 3.7). Bodies are available in stainless steel, aluminium or plastic. Elements are made of glass microfibre and are bonded with either resin or fluorocarbon. The fluorocarbon-bonded filter is particularly useful for low-level moisture applications because of the low adsorption/desorption characteristic.

The smallest filter in this range has an internal volume of only 19 ml and is therefore suitable when a fast response time is required.

3.2.2.4 Miniature in-line filter

These are used for filtration of gases prior to pressure reduction and are frequently fitted as the last component in the sample system to protect the analyser (Figure 3.8).

Figure 3.7 Filter with disposable element (courtesy Balston).

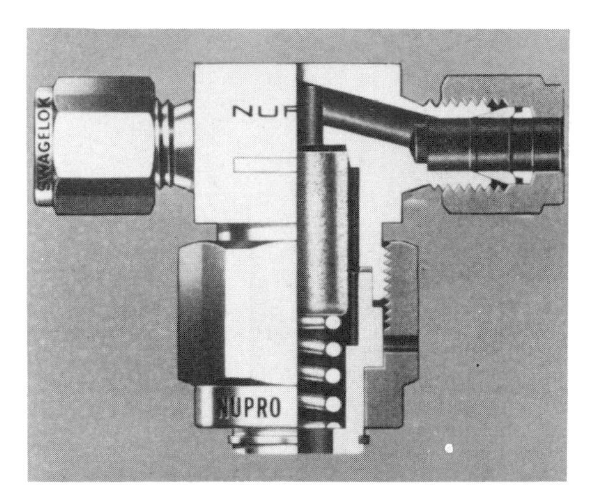

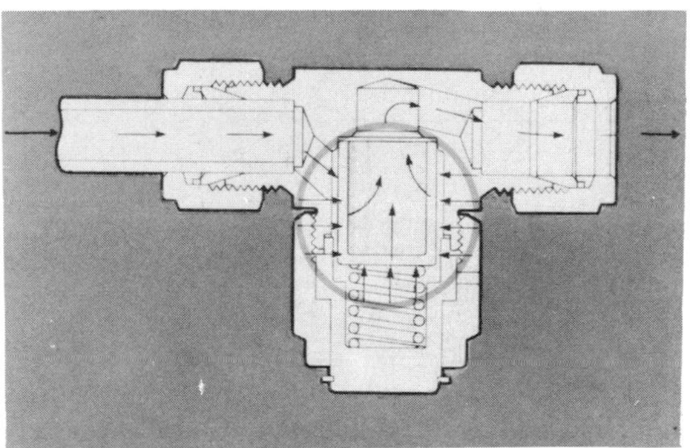

Figure 3.8 Miniature in-line filter (courtesy Nupro).

3.2.2.5 Manual self-cleaning filter

This type of filter works on the principle of edge filtration using discs usually made of stainless steel and fitted with a manual cleaning system (Figure 3.9). The filter is cleaned by rotating a handle which removes any deposits from the filter element while the sample is flowing. The main uses are for filtration of liquids where filter cleaning must be carried out regularly without the system being shut down; it is especially suitable for waxy material which can rapidly clog up normal filter media.

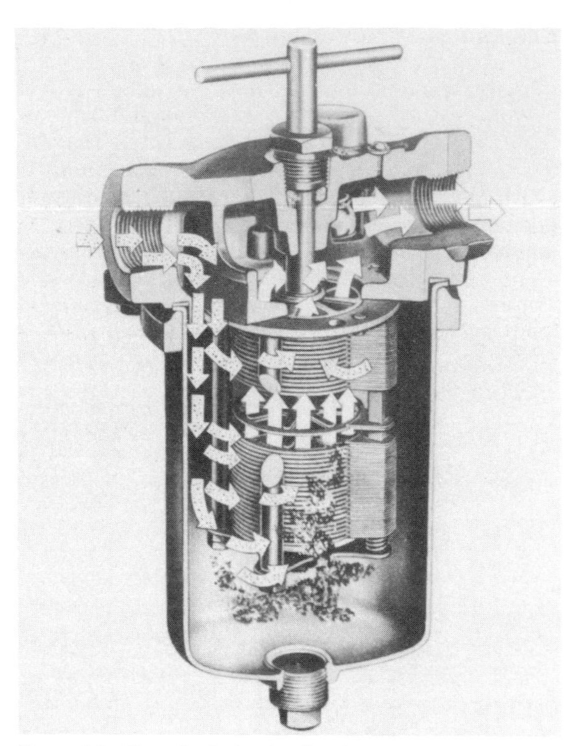

Figure 3.9 Manual self-cleaning filter (courtesy AMF CUNO).

3.2.3 Coalescers

Coalescers are a special type of filter for separating water from oil or oil from water (Figure 3.10). The incoming sample flows from the centre of a specially treated filter element through to the outside. In so doing, the diffused water is slowed down and coalesced, thus forming droplets which, when they reach the outer surface, drain downwards as the water is denser than the hydrocarbon. A bypass stream is taken from the bottom of the coalescer to remove the water. The dry hydrocarbon stream is taken from the top of the coalescer.

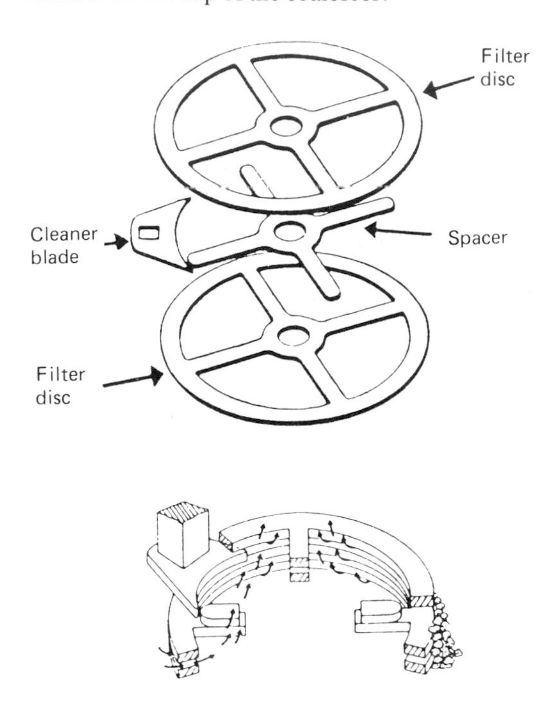

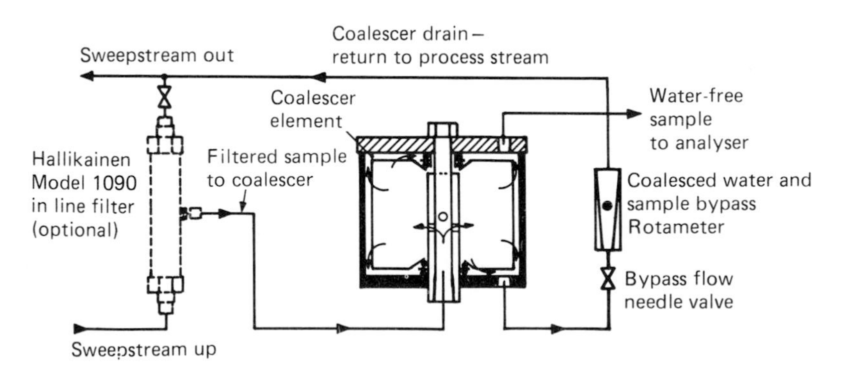

Figure 3.10 Coalescer (courtesy Fluid Data).

3.2.4 Coolers

3.2.4.1 Air coolers

These are usually used to bring the sample gas temperature close to ambient before feeding into the analyser.

3.2.4.2 Water-jacketed coolers

These are used to cool liquid and gas samples and are available in a wide range of sizes (Figure 3.11).

3.2.4.3 Refrigerated coolers

These are used to reduce the temperature of a gas to a fixed temperature (e.g. +5°C) in order to condense the water out of a sample prior to passing the gas into the analyser. Two types are available; one with an electrically driven compressor type refrigerator and another using a Peltier cooling element. The compressor type has a large cooling capacity whereas the Peltier type, being solid state, needs less maintenance.

3.2.5 Pumps, gas

Whenever gaseous samples have to be taken from sample points which are below the pressure required by the analyser a sample pump of some type is required. The pumps that are available can broadly be divided into two groups:

(1) The eductor or aspirator type; and
(2) The mechanical type.

3.2.5.1 Eductor or aspirator type

All of these types of pump operate on the principle of using the velocity of one fluid which may be liquid or gas to induce the flow in the sample gas. The pump may be fitted before or after the analyser, depending on the application. A typical application for a water-operated aspirator (similar to a laboratory vacuum pump) is for taking a sample of flue gas for oxygen measurement. In this case the suction port of the aspirator is connected directly to the probe via a sample line and the water/gas mixture from the outlet feeds into a separator arranged to supply the sample gas to the analyser at a positive pressure of about 300 mm water gauge.

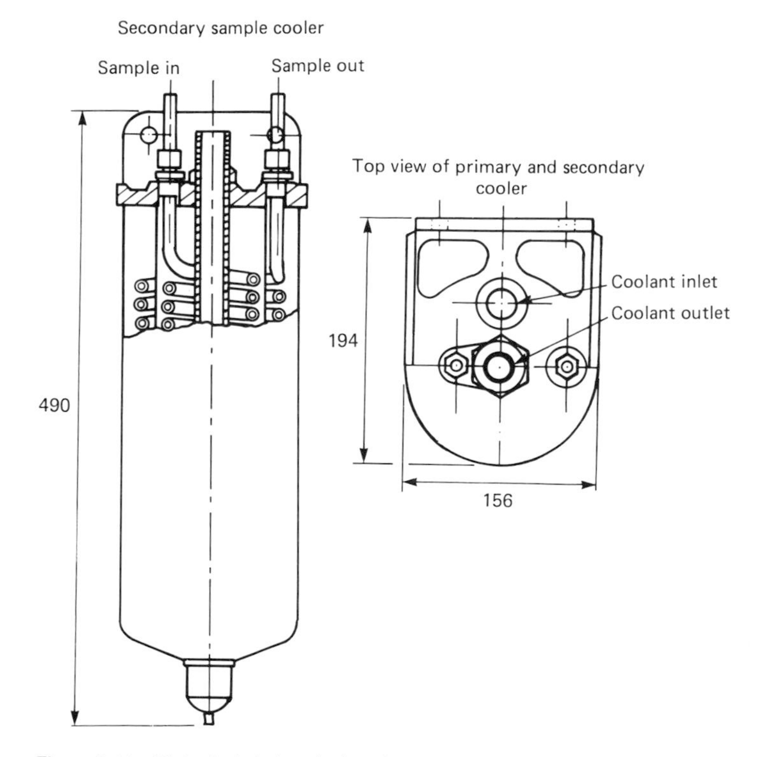

Figure 3.11 Water-jacketed cooler (courtesy George E. Lowe). Dimensions are shown in mm.

In cases where water will affect the analysis it is sometimes possible to place the eductor or aspirator after the analyser and draw the sample through the system. In these cases the eductor may be supplied with steam, air or water to provide the propulsive power.

3.2.5.2 Mechanical gas pumps

There are two main types of mechanical gas pump available:

(1) Rotary pumps; and
(2) Reciprocating piston or diaphragm pump.

Rotary pumps Rotary pumps can be divided into two categories, the rotary piston and the rotating fan types, but the latter is very rarely used as a sampling pump.

The rotary piston pump is manufactured in two configurations. The Rootes type has two pistons of equal size which rotate inside a housing with the synchronizing carried out by external gears. The rotary vane type is similar to those used extensively as vacuum pumps. The Rootes type is ideal where very large flow rates are required and, because there is a clearance between the pistons and the housing, it is possible to operate them on very dirty gases.

The main disadvantage of the rotary vane type is that, because there is contact between the vanes and the housing, lubrication is usually required, and this may interfere with the analysis.

Reciprocating piston and diaphragm pump Of these two types the diaphragm pump has become the most popular. The main reason for this is the improvement in the types of material available for the diaphragms and the fact that there are no piston seals to leak. The pumps are available in a wide variety of sizes from the miniature units for portable personnel protection analysers to large heavy duty industrial types.

A typical diaphragm pump (Figure 3.12) for boosting the pressure of the gas into the analyser could have an all stainless steel head with a Terylene reinforced viton diaphragm and viton valves. This gives the pump a very long service life on critical hydrocarbon applications.

Many variations are possible; for example, a Teflon-coated diaphragm can be fitted where viton

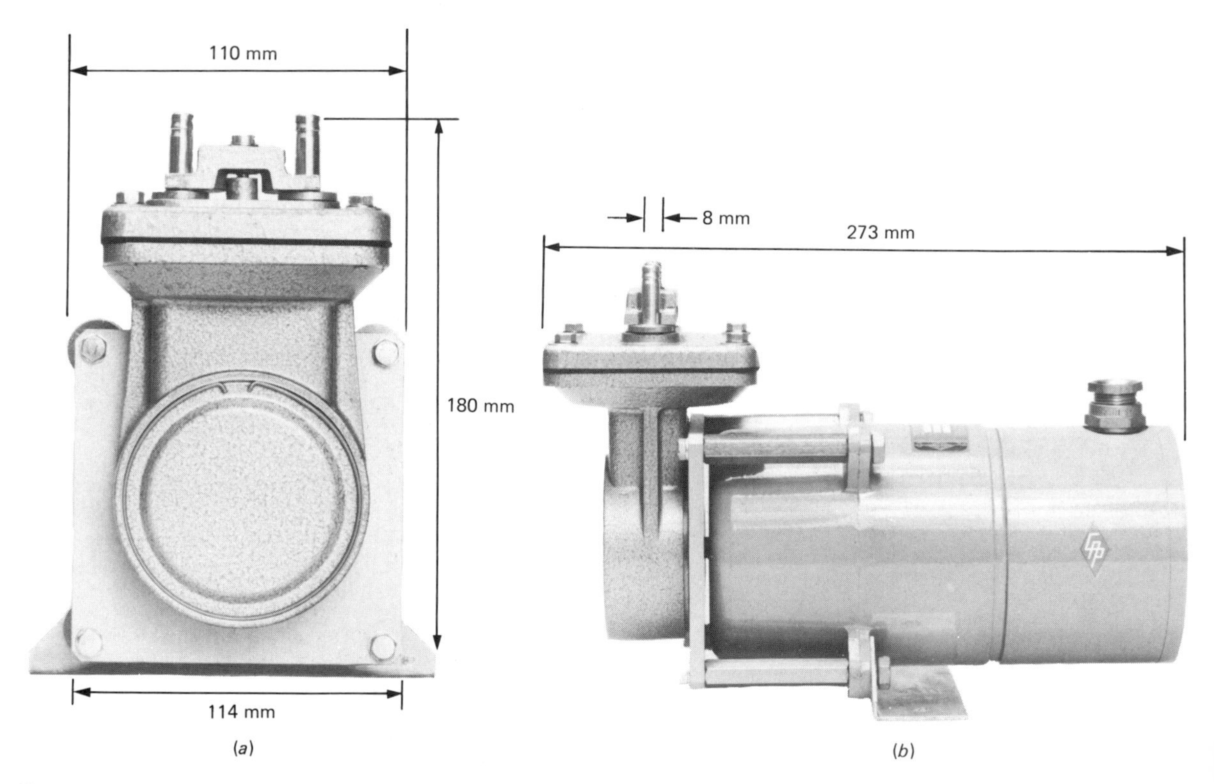

Figure 3.12 Diaphragm pump (courtesy Charles Austen Pumps).

may not be compatible with the sample and heaters may be fitted to the head to keep the sample above the dew point.

The piston pump is still used in certain cases where high accuracy is required in the flow rate (for example, gas blending) to produce specific gas mixtures. In these cases the pumps are usually operated immersed in oil so that the piston is well lubricated and there is no chance of gas leaks to and from the atmosphere.

3.2.6 Pumps: liquid

There are two situations where pumps are required in liquid sample systems:

(1) Where the pressure at the analyser is too low because either the process line pressure is too low, or the pressure drop in the sample line is too high or a combination of both.
(2) When the process sample has to be returned to the same process line after analysis.

The two most common types of pumps used for sample transfer are:

(1) Centrifugal (including turbine pump);
(2) Positive displacement (e.g. gear, peristaltic, etc.).

3.2.6.1 Centrifugal

The centrifugal and turbine pumps are mainly used when high flow rates of low-viscosity liquids are required. The turbine pumps are similar to centrifugal pumps but have a special impellor device which produces a considerably higher pressure than the same size centrifugal. In order to produce high pressures using a centrifugal pump there is a type available which has a gearbox to increase the rotor speed to above 20 000 rev/min.

3.2.6.2 Positive-displacement pumps

Positive-displacement pumps have the main characteristic of being constant flow devices. Some of these are specifically designed for metering purposes where an accurate flow rate must be maintained (e.g. process viscometers). They can take various forms:

(1) Gear pump;
(2) Rotary vane pump;
(3) Peristaltic pump.

Gear pumps Gear pumps are used mainly on high-viscosity products where the sample has some lubricating properties. They can generate high pressures for low flow rates and are used extensively for hydrocarbon samples ranging from diesel oil to the heaviest fuel oils.

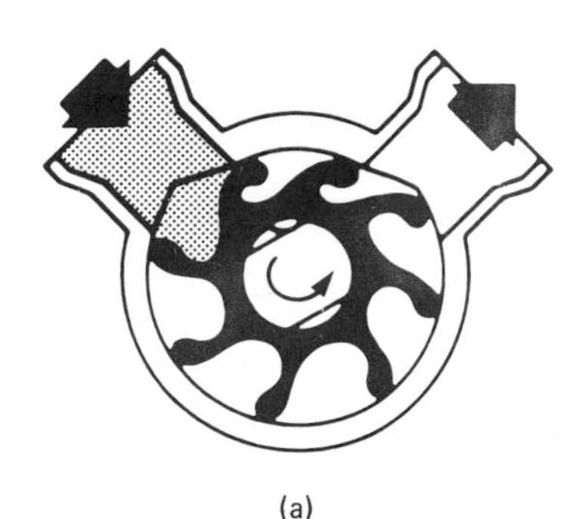

(a)

(b)

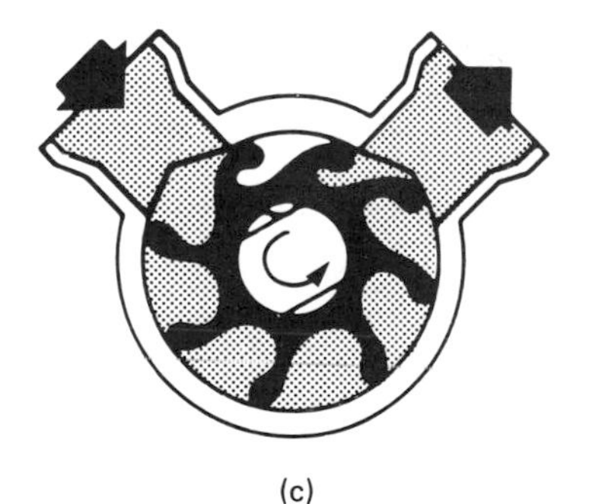

(c)

Figure 3.13 Flexible impeller pump (courtesy ITT Jabsco). (a) Upon leaving the offset plate the impeller blade straightens and creates a vacuum, drawing in liquid—instantly priming the pump. (b) As the impeller rotates it carries the liquid through the pump from the intake to outlet port, each successive blade drawing in liquid. (c) When the flexible blades again contact the offset plate they bend with a squeezing action which provides a continuous, uniform discharge of liquid.

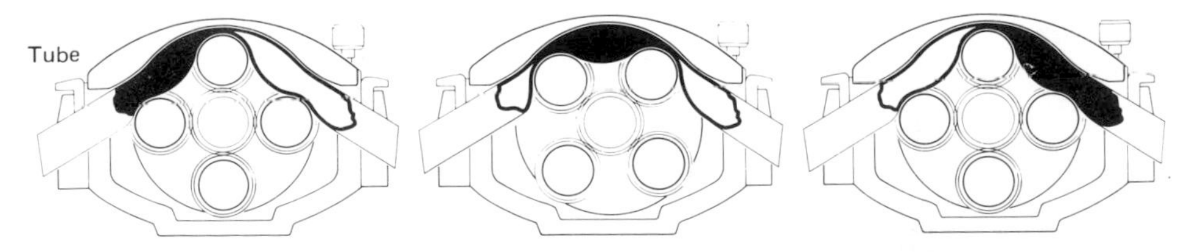

Figure 3.14 Peristaltic pump (courtesy Watson-Marlow). The advancing roller occludes the tube which, as it recovers to its normal size, draws in fluid which is trapped by the next roller (in the second part of the cycle) and expelled from the pump (in the third part of the cycle). This is the peristaltic flow-inducing action.

Rotary vane pumps These pumps are of two types, one having rigid vanes (usually metal) and the other fitted with a rotor made of an elastomer such as nitrile rubber or viton. The metal vane pumps have characteristics similar to the gear pumps described above, but can be supplied with a method of varying the flow rate externally while the pump is operating.

Pumps manufactured with the flexible vanes (Figure 3.13) are particularly suitable for pumping aqueous solutions and are available in a wide range of sizes but are only capable of producing differential pressures of up to 2 bar.

Peristaltic pumps Peristaltic pumps are used when either accurate metering is required or it is important that no contamination should enter the sample. As can be seen from Figure 3.14, the only material in contact with the sample is the special plastic tubing, which may be replaced very easily during routine servicing.

3.2.7 Flow measurement and indication

Flow measurement on analyser systems falls into three main categories:

(1) Measuring the flow precisely where the accuracy of the analyser depends on it;
(2) Measuring the flow where it is necessary to know the flow rate but it is not critical (e.g. fast loop flow);
(3) Checking that there is flow present but measurement is not required (e.g. cooling water for heat exchangers).

It is important to decide which category the flowmeter falls into when writing the specification, as the prices vary over a wide range, depending on the precision required.

The types of flowmeter available will be mentioned but not the construction or method of operation, as this is covered in *Instrument Technology*, Part 1, Chapter 1.

3.2.7.1 Variable-orifice meters

The variable-orifice meter is extensively used in analyser systems because of its simplicity, and there are two main types.

Glass tube This type is the most common as the position of the float is read directly on the scale attached to the tube and it is available calibrated for liquids or gases. The high-precision versions are available with an accuracy of ±1 per cent full-scale deflection (FSD), whereas the low-priced units have a typical accuracy of ±5 per cent FSD.

Metal tube The metal tube type is used mainly on liquids for high-pressure duty or where the liquid is flammable or hazardous. A good example is the fast loop of a hydrocarbon analyser. The float has a magnet embedded in it and the position is detected by an external follower system. The accuracy of metal tube flowmeters varies from ±10 per cent FSD to ±2 per cent FSD, depending on the type and whether individual calibration is required.

3.2.7.2 Differential-pressure devices

On sample systems these normally consist of an orifice plate or preset needle valve to produce the differential pressure, and are used to operate a gauge or liquid-filled manometer when indication is required or a differential pressure switch when used as a flow alarm.

3.2.7.3 Spinner or vane-type indicators

In this type the flow is indicated either by the rotation of a spinner or by the deflection of a vane by the fluid. It is ideal for duties such as cooling water flow, where it is essential to know that a flow is present but the actual flow rate is of secondary importance.

3.2.8 Pressure reduction and vaporization

The pressure-reduction stage in a sample system is often the most critical, because not only must the reduced pressure be kept constant but also provision must be made to ensure that under faulty conditions dangerously high pressures cannot be produced. Pressure reduction can be carried out in a variety of ways.

3.2.8.1 Simple needle valve

This is capable of giving good flow control if upstream and downstream pressures are constant.

Advantage: Simplicity and low cost.
Disadvantage: Any downstream blockage will allow pressure to rise.

They are only practical if downstream equipment can withstand upstream pressure safely.

3.2.8.2 Needle valve with liquid-filled lute

This combination is used to act as a pressure stabilizer and safety system combined. The maintained pressure will be equal to the liquid head when the needle valve flow is adjusted until bubbles are produced (Figure 3.15).

It is essential to choose a liquid which is not affected by sample gas and also does not evaporate in use and cause a drop in the controlled pressure.

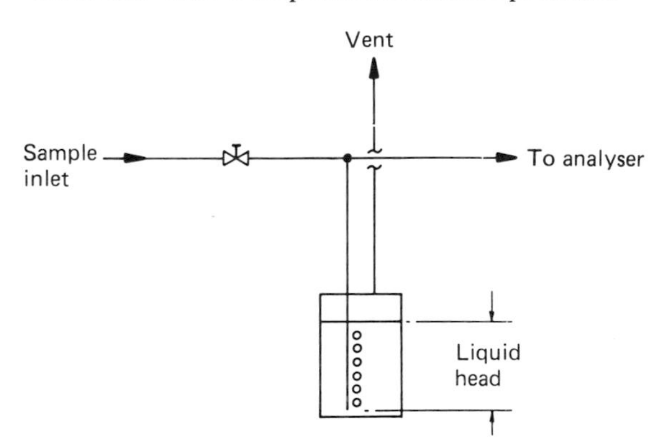

Figure 3.15 Lute-type pressure stabilizer (courtesy Ludlam Sysco).

3.2.8.3 Diaphragm-operated pressure controller

These regulators are used when there is either a very large reduction in pressure required or the downstream pressure must be accurately controlled (Figure 3.16). They are frequently used on gas cylinders to provide a controllable low-pressure gas supply.

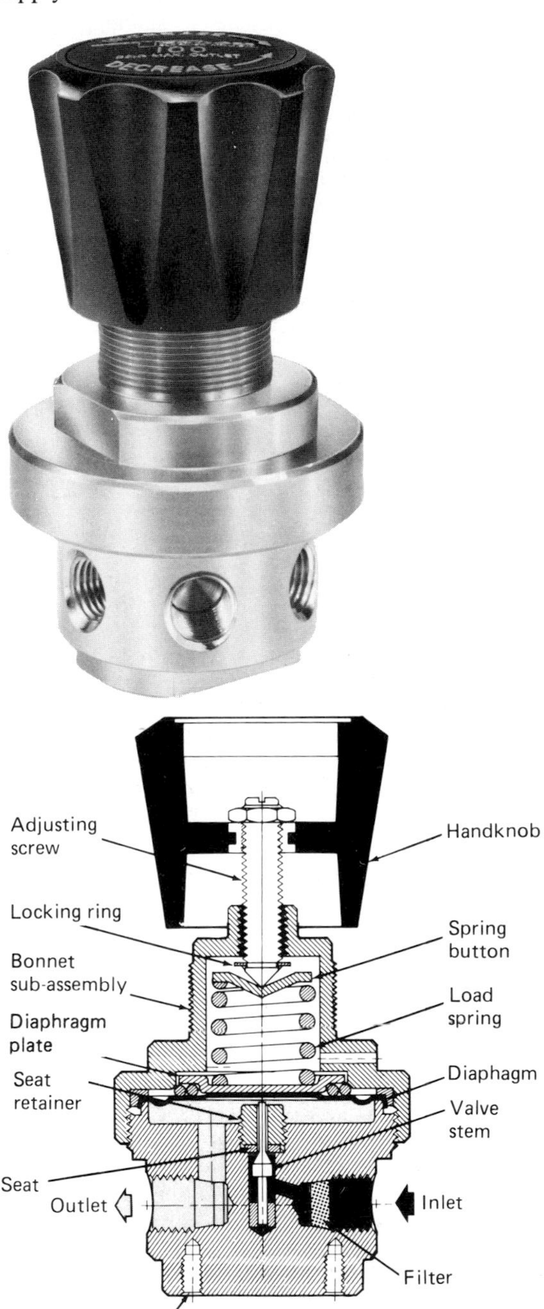

Figure 3.16 Diaphragm-operated pressure controller (courtesy Tescom).

3.2.8.4 Vaporization

There are cases when a sample in the liquid phase at high pressure has to be analysed in the gaseous phase. The pressure reduction and vaporization can be carried out in a specially adapted diaphragm-operated pressure controller as detailed above, where provision is made to heat the complete unit to replace the heat lost by the vaporization.

3.2.9 Sample lines, tube and pipe fitting

3.2.9.1 Sample lines

Sample lines can be looked at from two aspects: first, the materials of construction, which are covered in Section 3.1.5, and, second, the effect of the sample line on the process sample, which is detailed below.

The most important consideration is that the material chosen must not change the characteristics of the sample during its transportation to the analyser. There are two main ways in which the sample line material can affect the sample.

Adsorption and desorption Adsorption and desorption occur when molecules of gas or liquid are retained and discharged from the internal surface of the sample line material at varying rates. This has the effect of delaying the transport of the adsorbed material to the analyser and causing erroneous results.

Water and hydrogen sulphide at low levels are two common measurements where this problem is experienced. An example is when measuring water at a level of 10 ppm in a sample stream, where copper tubing has an adsorption/desorption which is twenty times greater than stainless steel tubing, and hence copper tubing would give a very sluggish response at the analyser.

Where this problem occurs it is possible to reduce the effects in the following ways:

(1) Careful choice of sample tube material;
(2) Raising the temperature of the sample line;
(3) Cleaning the sample line to ensure that it is absolutely free of impurities such as traces of oil;
(4) Increasing the sample flow rate to reduce the time the sample is in contact with the sample line material.

Permeability Permeability is the ability of gases to pass through the wall of the sample tubing. Two examples are:

(1) Polytetrafluoroethylene (PTFE) tubing is permeable to water and oxygen.
(2) Plasticized polyvinyl chloride (PVC) tubing is permeable to the smaller hydrocarbon molecules such as methane.

Permeability can have two effects on the analysis:

(1) External gases getting into the sample such as when measuring low-level oxygen using PTFE tubing. The results would always be high due to the ingress of oxygen from the air.
(2) Sample gases passing outwards through the tubing, such as when measuring a mixed hydrocarbon stream using plasticized PVC. The methane concentration would always be too low.

3.2.9.2 Tube, pipe and method of connection

Definition

(1) Pipe is normally rigid and the sizes are based on the nominal bore.

    Typical materials:
    Metallic: carbon steel, brass, etc.
    Plastic: UPVC, ABS, etc.

(2) Tubing is normally bendable or flexible, and the sizes are based on the outside diameter and wall thickness.

    Typical materials:
    Metallic: carbon steel, brass, etc.
    Plastic: UPVC, ABS, etc.

Methods of joining

Pipe (metallic): (1) Screwed
(2) Flanged
(3) Welded
(4) Brazed or soldered

Pipe (plastic): (1) Screwed
(2) Flanged
(3) Welded (by heat or use of solvents)

Tubing (metallic): (1) Welding
(2) Compression fitting
(3) Flanged

Tubing (plastic): (1) Compression
(2) Push-on fitting (especially for plastic tubing) with hose clip where required to withstand pressure.

General The most popular method of connecting metal tubing is the compression fitting, as it is capable of withstanding pressures up to the limit of the tubing itself and is easily dismantled for servicing as well as being obtainable manufactured in all the most common materials.

3.3 Typical sample systems

3.3.1 Gases

3.3.1.1 High-pressure sample to a process chromatograph

The example taken is for a chromatograph analysing the composition of a gas which is in the vapour phase at 35 bar (Figure 3.17). This is the case described in section 3.1.4.3, where it is necessary to reduce the pressure of the gas at the sample point in order to obtain a fast response time with a minimum wastage of process gas.

The sample is taken from the process line using a low-volume sample probe (section 3.2.1.2) and then flows immediately into a pressure reducing valve to drop the pressure to a constant 1.5 bar, which is measured on a local pressure gauge. A pressure relief valve set to relieve at 4 bar is connected at this point to protect downstream equipment if the pressure-reducing valve fails.

After pressure reduction the sample flows in small-bore tubing (6 mm OD) to the main sample system next to the analyser, where it flows through a filter (such as shown in section 3.2.2.3) to the sample selection system.

The fast loop flows out of the bottom of the filter body, bypassing the filter element, and then through a needle valve and flowmeter to an atmospheric vent on a low-pressure process line.

The stream selection system shown in Figure 3.17 is called a block and bleed system, and always has two or more three-way valves between each stream and the analyser inlet. The line between two of the valves on the stream which is not in operation is vented to atmosphere, so guaranteeing that the stream being analysed cannot be contaminated by any of the other streams. A simple system without block and bleed valve is described in section 3.3.2.1 below.

After the stream-selection system the sample flows through a needle valve flowmeter and a miniature in-line filter to the analyser sample inlet. The analyser sample outlet on this system flows to the atmospheric vent line.

3.3.1.2 Furnace gas using steam-injection probe inside the flue

This system utilizes a Venturi assembly located inside the flue (Figure 3.18). High-pressure steam enters the Venturi via a separate steam tube and a low-pressure region results inside the flue at the probe head. A mixture of steam and sample gas

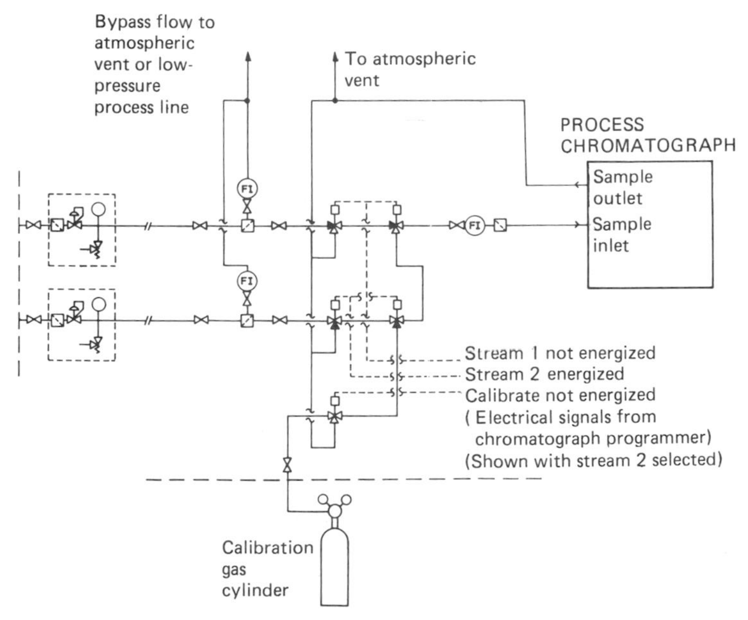

Figure 3.17 Schematic: high-pressure gas sample to chromatograph (courtesy Ludlam Sysco).

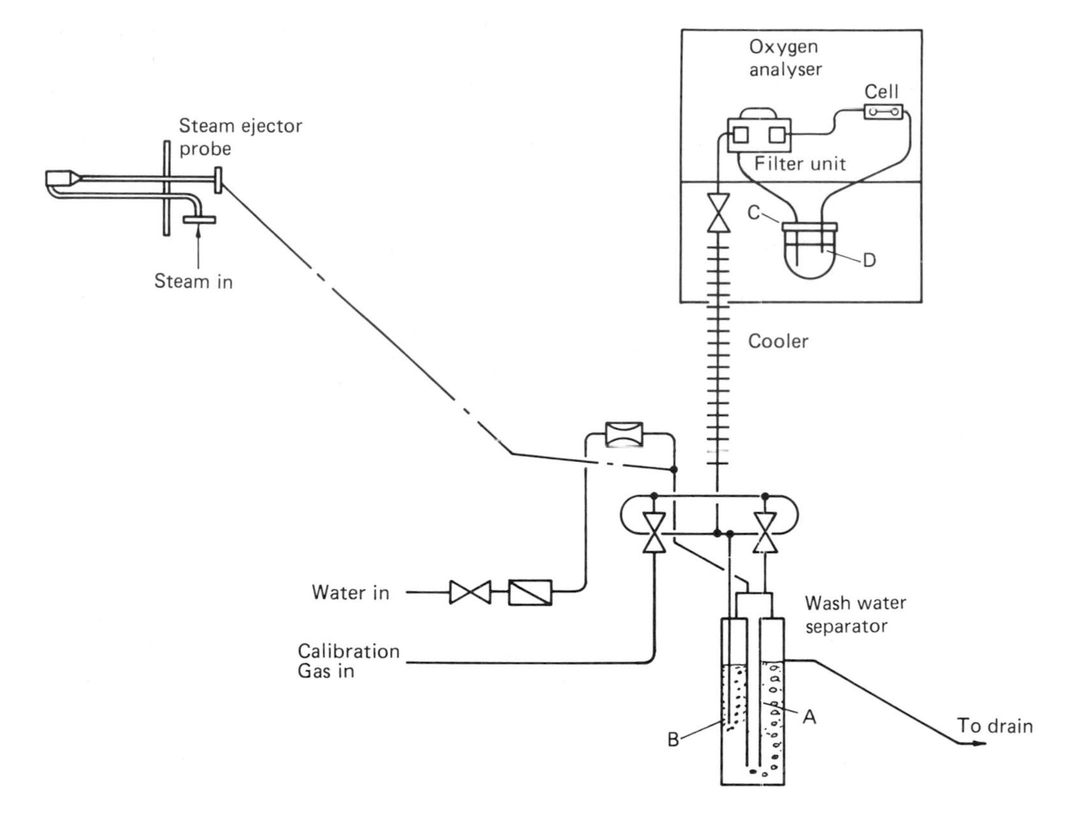

Figure 3.18 Schematic: furnace gas sampling (courtesy Servomex).

passes down the sample line. Butyl or EDPDM rubber-lined steam hose is recommended for sample lines, especially when high-sulphur fuels are used. This will minimize the effects of corrosion.

At the bottom end of the sample line the sample gas is mixed with a constant supply of water. The gas is separated from the water and taken either through a ball valve or a solenoid valve towards the sample loop, which minimizes the dead volume between each inlet valve and the analyser inlet.

The water, dust, etc. passes out of a dip leg (A) and to drain. It is assumed that the gas leaves the separator saturated with water at the temperature of the water. In the case of a flue gas system on a ship operating, for example, in the Red Sea this could be at 35°C. The system is designed to remove condensate that may be formed because of lower temperatures existing in downstream regions of the sample system.

At the end of the loop there is a second dip leg (B) passing into a separator. A 5 cm water differential pressure is produced by the difference in depth of the two dip legs, so there is always a continuous flow of gas round the loop and out to vent via dipleg (B).

The gas passes from the loop to a heat exchanger, which is designed so that the gas leaving the exchanger is within 1 K of the air temperature. This means that the gas leaving the heat exchanger can be at 36°C and saturated with water vapour. The gas now passes into the analyser which is maintained at 60°C.

The gas arrives in the analyser and enters the first chamber, which is a centrifugal separator in a stainless steel block at 60°C. The condensate droplets will be removed at this point and passed down through the bottom end of the separator into a bubbler unit (C). The bubbles in this tube represent the bypass flow. At the same time the gas is raised from 36°C to 60°C inside the analyser.

Gas now passes through a filter contained in the second chamber, the measuring cell and finally to a second dip leg (D) in the bubbler unit. The flow of gas through the analyser cell is determined by the difference in the length of the two legs inside the bubbler unit and cannot be altered by the operator.

This system has the following operating advantages:

(1) The system is under a positive pressure right from inside the flue, and so leaks in the sample line can only allow steam and sample out and not air in.
(2) The high speed steam jet scours the tube preventing build-up.
(3) The steam maintains the whole of the sample probe above the dew point and so prevents corrosive condensate forming on the outside of the probe.
(4) The steam keeps the entire probe below the temperature of the flue whenever the temperature of the flue is above the temperature of the steam.
(5) The actual sampling system is intrinsically safe as no electrical pumps are required.

3.3.1.3 Steam sampling for conductivity

The steam sample is taken from the process line by means of a special probe and then flows through thick-wall 316 stainless steel tubing to the sample system panel (Figure 3.19). The sample enters the sampling panel through a high-temperature, high-pressure isolating valve and then flows into the cooler, where the steam is condensed and the condensate temperature is reduced to a suitable temperature for the analyser (typically 30°C).

After the cooler, the condensate passes to a pressure-control valve to reduce the pressure to about 1 bar gauge. The temperature and pressure of the sample are then measured on suitable gauges and a pressure-relief valve (set at 2 bar g) is fitted to protect downstream equipment from excess pressure if a fault occurs in the pressure control valve. The constant-pressure, cooled sample passes through a needle valve, flowmeter and three-way valve into the conductivity cell and then to drain.

Facilities are provided for feeding water of known conductivity into the conductivity cell through the three-way valve for calibration purposes. The sample coolers are normally supplied with stainless steel cooling coils which are suitable where neither the sample nor the coolant contain appreciable chloride which can cause stress corrosion cracking.

When chlorides are known to be present in the sample or cooling water, cooling coils are available, made of alternative materials which are resistant to chloride-induced stress corrosion cracking.

3.3.2 Liquids

3.3.2.1 Liquid sample to a process chromatograph

The example taken is for a chromatograph measuring butane in gasoline (petrol) (Figure 3.20). The chromatograph in this case would be fitted with a liquid inject valve so that the sample will remain in the liquid phase at all times within the sample system.

In a liquid inject chromatograph the sample flow rate through the analyser is very low (typically 25 ml/min), so that a fast loop system is essential.

The sample flow from the process enters the sample system through an isolating valve, then through a pump (if required) and an in-line filter,

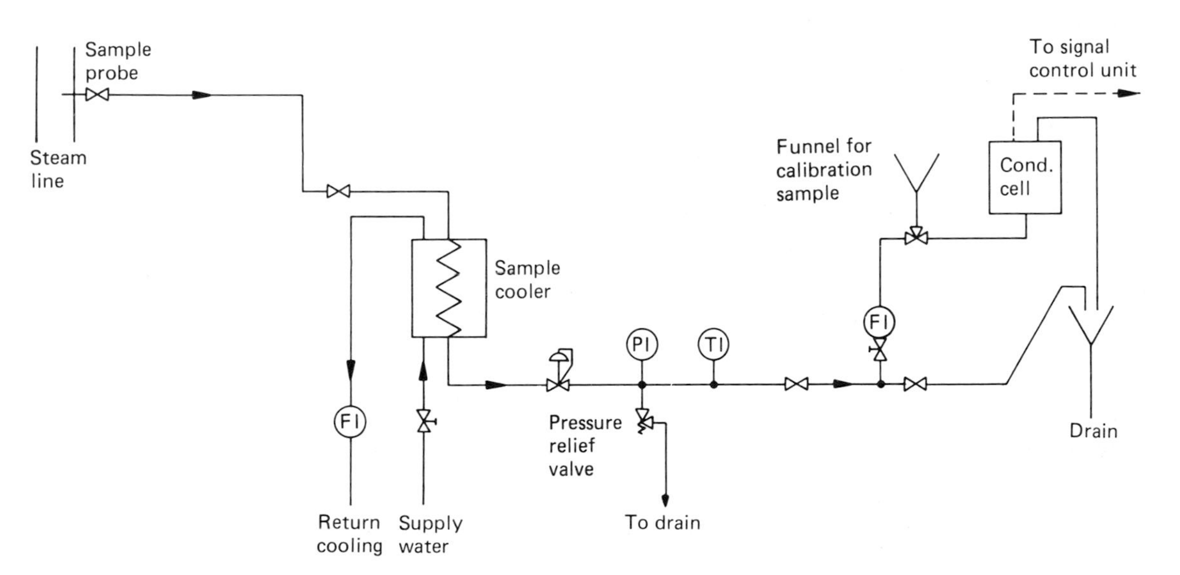

Figure 3.19 Schematic: steam sampling for conductivity (courtesy Ludlam Sysco).

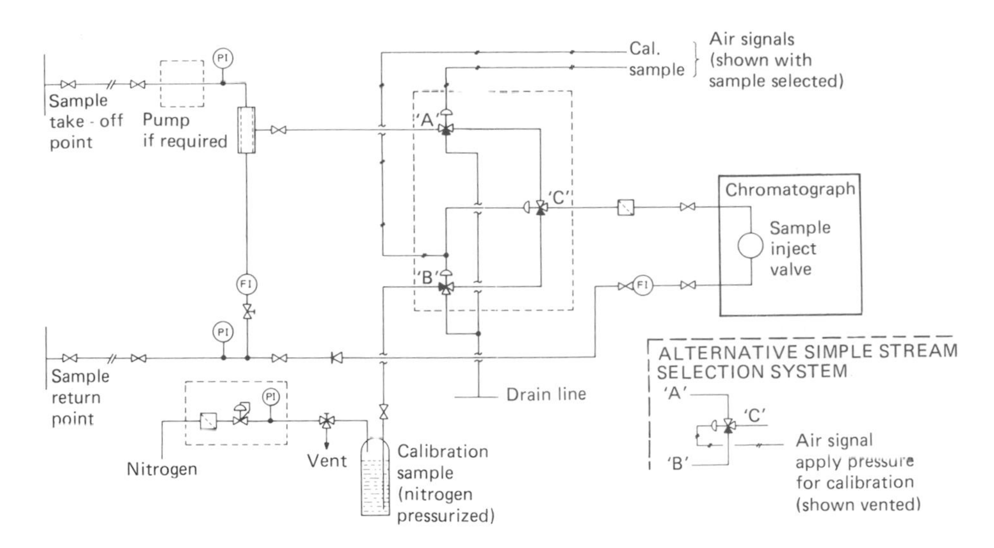

Figure 3.20 Schematic: liquid sample to process chromatograph (courtesy Ludlam Sysco).

from which the sample is taken to the analyser. After the filter the fast loop flows through a flowmeter followed by a needle valve, then through an isolating valve back to the process. Pressure gauges are fitted, one before the in-line filter and one after the needle valve, so that it is possible at any time to check that the pressure differential is sufficient (usually 1 bar minimum) to force the sample through the analyser.

The filtered sample flows through small-bore tubing (typically, 3 mm OD) to the sample/calibration selection valves. The system shown is the block and bleed configuration as described in Section 3.3.1.1. Where there is no risk of cross contamination the sample stream-selection system shown in the inset of Figure 3.20 may be used.

The selected sample flows through a miniature in-line filter (section 3.2.2.4) to the analyser, then through the flow control needle valve and non-return valve back to the fast loop return line.

When the sample is likely to vaporize at the sample return pressure it is essential to have the flow control needle valve after the flowmeter in both the fast loop and the sample through the analyser. This is done to avoid the possibility of any vapour flashing off in the needle valve and passing through the flowmeter, which would give erroneous readings.

The calibration sample is stored in a nitrogen-pressurized container and may be switched either manually or automatically from the chromatograph controller.

3.3.2.2 Gas oil sample to a distillation point analyser

In this case the process conditions are as follows (Figure 3.21):

Sample tap: Normal pressure: 5 bar g
Normal temperature: 70°C
Sample line length: 73 m

Sample return: Normal pressure: 5 bar g
Return line length: 73 m

This is a typical example of an oil-refinery application where, for safety reasons, the analyser has to be positioned at the edge of the process area and consequently the sample and return lines are relatively long. Data to illustrate the fast loop calculation, based on equations in Crane's Publication No. 410 M, are given in Table 3.1.

An electrically driven gear pump is positioned immediately outside the analyser house which pumps the sample round the fast loop and back to the return point. The sample from the pump enters the sample system cabinet and flows through an in-line filter from which the sample is taken to the analyser, through a needle valve and flowmeter back to the process. The filtered sample then passes through a water-jacketed cooler to reduce the temperature to that required for the coalescer and analyser. After the cooler the sample is pressure reduced to about 1 bar with a pressure-control valve.

The pressure is measured at this point and a relief valve is fitted so that, in the event of the pressure

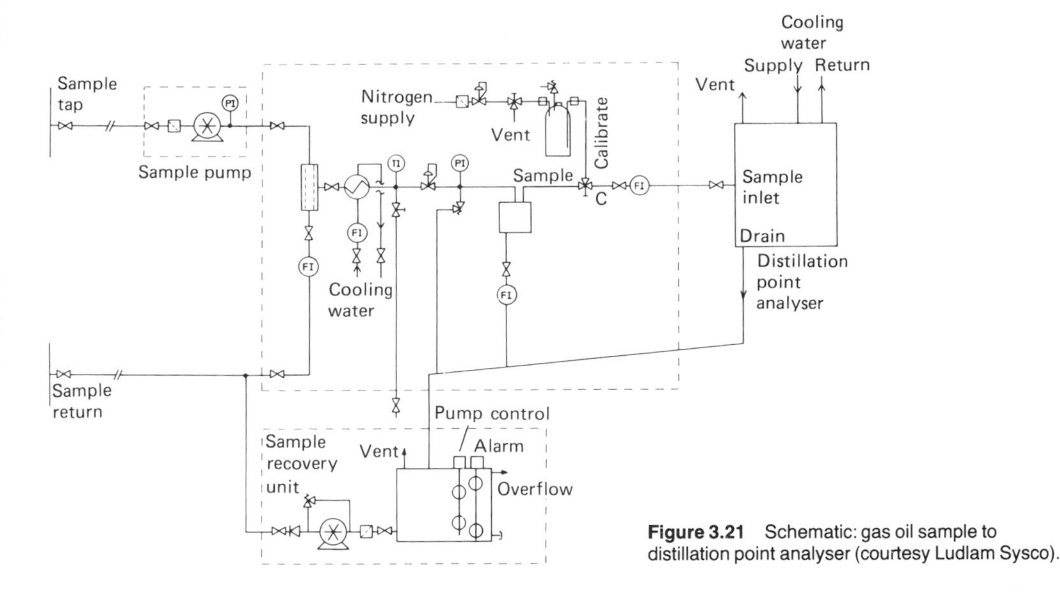

Figure 3.21 Schematic: gas oil sample to distillation point analyser (courtesy Ludlam Sysco).

Table 3.1 Fast-loop calculation for gas oil sample

Customer	Demo.
Date	Jan. 1986
Order No.	ABC 123
Tag number	Gas oil (sample line)
Density	825.00 kg/m^3
Viscosity	3.00 centipoise
Response time	39.92 s
Flow rate	16.60 l/min (1 m^3/h)
Length	73.00 m
Diameter (ID)	13.88 mm (½ in nominal bore (extra strong))
Velocity	1.83 m/s
RE=	6979
Flow	TURB
Friction factor	0.038
Delta P.	278.54 kPa (2.7854 bar)
Customer	Demo.
Date	Jan. 1986
Order No.	ABC 123
Tag number	Gas oil (return line)
Density	825.00 kg/m^3
Viscosity	3.00 centipoise
Response time	71.31 s
Flow rate	16.60 l/min (1 m^3/h)
Length	73.00 m
Diameter (ID)	18.55 mm (¾ in nominal bore (extra strong))
Velocity	1.02 m/s
RE=	5222
Flow	TURB
Friction factor	0.040
Delta P.	67.85 kPa (0.6785 bar)

control valve failing open, no downstream equipment will be damaged.

The gas oil sample may contain traces of free water and, as this will cause erroneous readings on the analyser, it is removed by the coalescer. The bypass sample from the bottom of the coalescer flows through a needle valve and flowmeter to the drain line. The dry sample from the coalescer flows through a three-way ball valve for calibration purposes and then a needle valve and flowmeter to control the sample flow into the analyser.

The calibration of this analyser is carried out by filling the calibration vessel with the sample which had previously been accurately analysed in the laboratory. The vessel is then pressurized with nitrogen to the same pressure as that set on the pressure-control valve and then the calibration sample is allowed to flow into the analyser by turning the three-way ball valve to the calibrate position.

The waste sample from the analyser has to flow to an atmospheric drain and, to prevent product wastage, it flows into a sample recovery unit along with the sample from the coalescer bypass and the pressure relief valve outlet.

The sample recovery unit consists of a steel tank from which the sample is returned to the process intermittently by means of a gear pump controlled by a level switch. An extra level switch is usually fitted to give an alarm if the level rises too high or falls too low.

A laboratory sample take-off point is fitted to enable a sample to be taken for separate analysis at any time without interfering with the operation of the analyser.

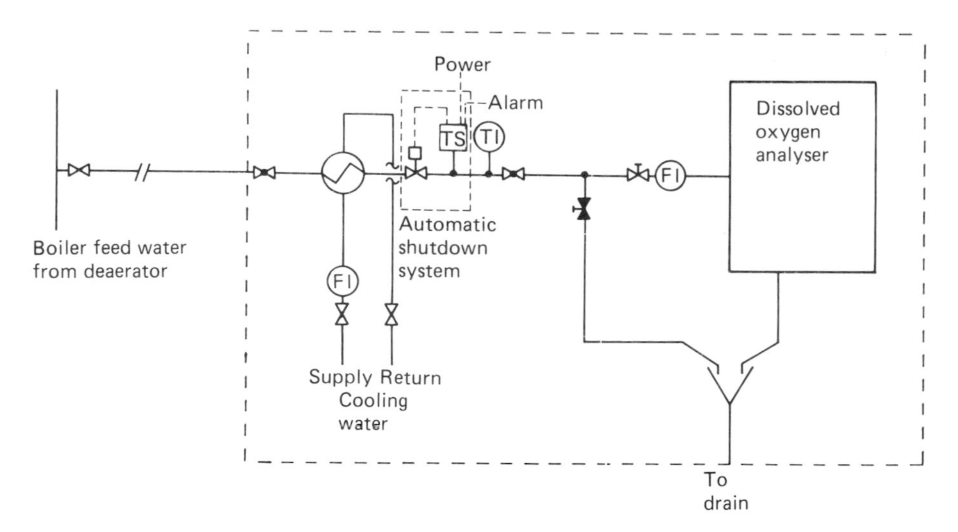

Figure 3.22 Schematic: water sample system for dissolved oxygen analyser (courtesy Ludlam Sysco).

3.3.2.3 Water-sampling system for dissolved oxygen analyser

Process conditions:

Sample tap: Normal pressure: 3.5 bar
Normal temperature: 140°C
Sample line length: 10 m

This system (Figure 3.22) illustrates a case where the sample system must be kept as simple as possible to prevent degradation of the sample. The analyser is measuring 0–20 μg/l oxygen and, because of the very low oxygen content, it is essential to avoid places in the system where oxygen can leak in or be retained in a pocket. Wherever possible ball or plug valves must be used, as they leave no dead volumes which are not purged with the sample.

The only needle valve in this system is the one controlling the flow into the analyser, and this must be mounted with the water flowing vertically up through it so that all air is displaced.

The sample line, which should be as short as possible, flows through an isolating ball valve into a water-jacketed cooler to drop the temperature from 140°C to approximately 30°C, and then the sample is flow controlled by a needle valve and flowmeter. In this case, it is essential to reduce the temperature of the water sample before pressure reduction otherwise the sample would flash off as steam. A bypass valve to drain is provided so that the sample can flow through the system while the analyser is being serviced.

When starting up an analyser such as this it may be necessary to loosen the compression fittings a little while the sample pressure is on to allow water to escape, and fill the small voids in the fitting and then finally tighten them up again to give an operational system.

An extra unit that is frequently added to a system such as this is automatic shutdown on cooling-water failure to protect the analyser from hot sample. This unit is shown dotted on Figure 3.22, and consists of an extra valve and a temperature detector, either pneumatic or electronically operated, so that the valve is held open normally but shuts and gives an alarm when the sample temperature exceeds a preset point. The valve would then be reset manually when the fault has been corrected.

3.4 References

CORNISH, D.C. *et al.* *Sampling Systems for Process Analysers*, Butterworths, London (1981)
Flow of Fluids, Publication No. 410M, Crane Limited (1982)
MARKS, J.W. *Sampling and Weighing of Bulk Solids*, Transtech Publications, Clausthal-Zellerfeld (1985)

4 Signal processing

M. L. Sanderson

4.1 Introduction

This chapter is concerned with the application of electronic signal-processing techniques to the signals generated by transducers and instruments. The techniques employed in analogue signal processing, digital-to-analogue conversion and analogue-to-digital conversion are considered together with time and frequency domain analysis of signals using specialized equipment such as frequency analysers, phase-sensitive detectors, correlators and multichannel analysers. The chapter concludes with a section on the use of computer-based systems in instrumentation.

Although the signals generated by most transducers are analogue, increasingly, with the advent of cheaper computing power in the form of microprocessors and microcomputers, the transmission and processing of much data and information is undertaken in a digital form. Digital systems show several advantages over analogue systems. They are not subject to drift; they suffer much less from the problems of data corruption during transmission; and that can be easily reduced to acceptable levels by the application of simple error-checking codes to the transmitted data. The level of signal degradation which can be tolerated in a digital system before the signal is no longer recoverable is considerably greater, and thus digital systems can operate with lower signal-to-noise ratios than comparable analogue systems. In terms of signal-processing, modification of the algorithm provided by the digital-processing scheme can usually be fairly easily effected by means of software changes or, at most, simple hardware changes. Thus the bandwidth and central frequency of a digital filter can be altered by simply adjusting the sampling frequency either by a software or a hardware modification. In an analogue filter the modification can usually only be undertaken by changing several components within the filter.

Analogue electronics are essential in the processing of either high-frequency or low-level signals. In general, they are capable of providing a larger dynamic range and greater resolution than their digital counterparts.

Hybrid systems combining both analogue and digital electronics can offer significant advantages. The performance of analogue signal processing can generally be improved by using digital electronics to supply control and computational facilities. These enable the overall system to be provided with such functions as auto-zeroing, online calibration and self diagnostics. Such combinations are often referred to as 'smart' or 'intelligent'.

4.2 Analogue signal processing

Transducers can be characterized in a number of ways, either by the fundamental physical principle they employ or by the quantity they measure (Jones, 1977; Arbel, 1980). From a signal-processing point of view, transducers can be classified as to whether they produce outputs which are voltage, current or charge; whether these quantities are unidirectional, bidirectional, alternating or transient in nature; whether the information is contained within the signal as a level, or as an amplitude, frequency or phase modulation of a carrier wave; and whether the relationship between the measured quantity and the output from the transducer is linear or non-linear. Associated with the signal there is an output impedance. The purpose of analogue signal conditioning is often to convert the output of the transducer into a voltage; to effect any impedance changes that are required; and to provide such operations as amplification, filtering, demodulation and linearization.

4.2.1 Operational amplifiers

One of the most common elements in any signal-conditioning scheme is the operational amplifier. By this is simply meant a high-gain d.c. coupled amplifier, which is used with negative feedback such that the closed-loop performance is defined by the feedback elements and not by the amplifier. It is usually, although not always, a differential input/single-ended output device as shown in Figure

4.1(a). The ideal operational amplifier characteristics are infinite gain and bandwidth, infinite common mode rejection ratio, infinite input impedance and zero output impedance and zero input offset voltage and bias currents, and that it provides noise-free gain. Figure 4.1(b) shows the internal structure of the general-purpose 741 integrated circuit operational amplifier. Table 4.1 gives the parameters of three operational amplifiers produced by different integrated circuit technologies. Figures 4.1(c) and 4.1(d) illustrate the two basic methods of applying feedback to the operational amplifier. Figure 4.1(c) shows the inverting mode which employs shunt feedback and Figure 4.1(d) the non-inverting mode which employs series feedback. Assuming that the amplifiers show ideal characteristics, then since they have infinite gain there can be no potential between the two inputs of the amplifier for any finite output, and since the amplifiers require no bias current and have infinite input impedance, then $i_1 = i_f$.

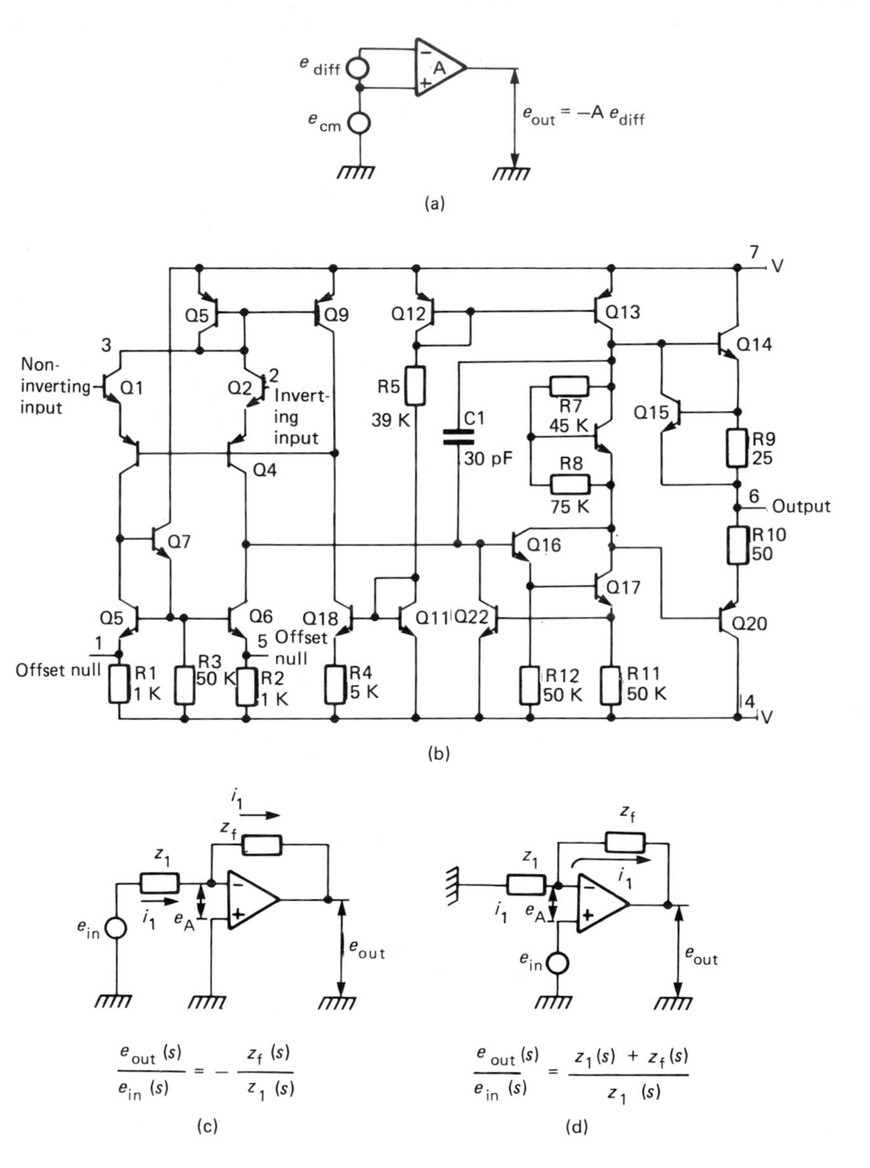

Figure 4.1 (a) Schematic diagram of an operational amplifier; (b) the internal structure of a 741 operational amplifier; (c) inverting mode feedback; (d) non-inverting mode feedback.

Table 4.1 Specifications of three operational amplifiers

Device	*741 Bipolar*	*μAF355 JFET*	*T071CP BIFET*
Supply voltage	±3 V to ±18 V	±4 V to ±18 V	±3 V to ±18 V
Open-loop gain	106 dB	106 dB	106 dB
Input resistance	2 MΩ	10^{12} Ω	10^{12} Ω
Offset voltage	1 mV	3 mV	3 mV
Offset voltage temperature coefficient	5 μV/K	5 μV/K	20 μV/K
Bias current	80 nA	30 pA	30 pA
Offset bias current	20 nA	10 pA	5 pA
Offset bias temperature coefficient	0.5 nA/K	doubles for every 20K	doubles for every 20K
CMRR	90 dB	100 dB	76 dB
Slew rate	0.5 V/μs	5 V/μs	13 V/μs
Full-power bandwidth	10 kHz	60 kHz	150 kHz
Supply voltage rejection ratio	30 μV/V	10 μV/V	158 μV/V

Thus for the inverting amplifier the transfer function between the input and output is given by:

$$\frac{e_{out}(s)}{e_{in}(s)} = \frac{-z_f(s)}{z_1(s)} = 1 - \frac{1}{\beta(s)};$$

$$\beta(s) = \frac{z_1(s)}{z_1(s) + z_f(s)} \tag{4.1}$$

where $\beta(s)$ is known as the feedback factor. Since the potential of the non-inverting input of the ideal amplifier does not change with input voltage it is called a 'virtual earth', and consequently the input impedance of the non-inverting amplifier is $z_1(s)$. Its output impedance is zero.

For the non-inverting amplifier:

$$\frac{e_{out}(s)}{e_{in}(s)} = \frac{z_1(s) + z_f(s)}{z_1(s)} = \frac{1}{\beta(s)} \tag{4.2}$$

The input impedance of the non-inverting amplifier is infinite since the input voltage is being applied to the non-inverting input of the operational amplifier. The closed-loop amplifier has zero output impedance. When $z_1(s) \to \infty$ or $z_f(s) \to 0$ then the closed-loop gain of the non-inverting amplifier $\to 1$, and the amplifier has the characteristics of a voltage follower, i.e. unity gain, infinite input impedance and zero output impedance.

It can be seen that by the application of feedback the closed-loop transfer function provided by the ideal operational amplifier is determined by the feedback elements and not the amplifier. By suitable choice of components, both linear and non-linear, for z_1 and z_f it is possible to produce a wide variety of signal-conditioning units. Tobey *et al.* (1971), Graeme (1973) and Clayton (1979) provide a wide range of examples of the use of operational amplifiers for signal processing. Figure 4.2 shows some commonly used operational amplifier circuits for linear signal processing and Figure 4.3 the application of operational amplifiers to non-linear signal processing providing half-wave and full-wave precision rectification, phase-sensitive detection and logarithmic and antilogarithmic amplification.

The following sections examine the effects of non-ideal operational amplifier characteristics on the closed-loop performance.

4.2.1.1 Gain and bandwidth

In practice, operational amplifiers do not have infinite gain or bandwidth, and consequently they produce phase shift which may interact with the phase shift within the feedback loop to produce overall positive feedback around the loop. If the amplifier is represented by a transfer function $A(s)$ then, re-analysing the inverting and non-inverting circuits in Figures 4.1(b) and 4.1(c) with:

$$e_A(s) = -\frac{e_{out}(s)}{A(s)} \tag{4.3}$$

then:

$$\frac{e_{out}(s)}{e_{in}(s)} = \text{ideal transfer function}\ \frac{1}{1 + 1/(A(s).\beta(s))} \tag{4.4}$$

where $A(s).\beta(s)$ is the loop gain.

For the closed-loop amplifier performance to be governed by the feedback components it is required that $|A(s).\beta(s)| >> 1$, i.e. that the loop gain should be significantly greater than unity. Stability of the closed-loop system requires that the roots of the characteristic equation, $1 + A(s).\beta(s)$, lie in the open left-half s plane. Closed-loop stability assessment for operational amplifiers and frequency compensation is generally undertaken using Bode diagrams (Clayton, 1979).

Many operational amplifiers can be approximated by a transfer function having a single low-frequency pole. Thus the transfer function $A(s)$ can be represented by:

$$A(s) = \frac{A_o}{(1 + s\tau_o)} \tag{4.5}$$

where A_o is the low-frequency gain of the amplifier and where the 3 dB break frequency is given by $f_o = 1/2\pi\tau_o$). This is shown in Figure 4.4. Typically, A_o is 100 dB and f_o is under 10 Hz.

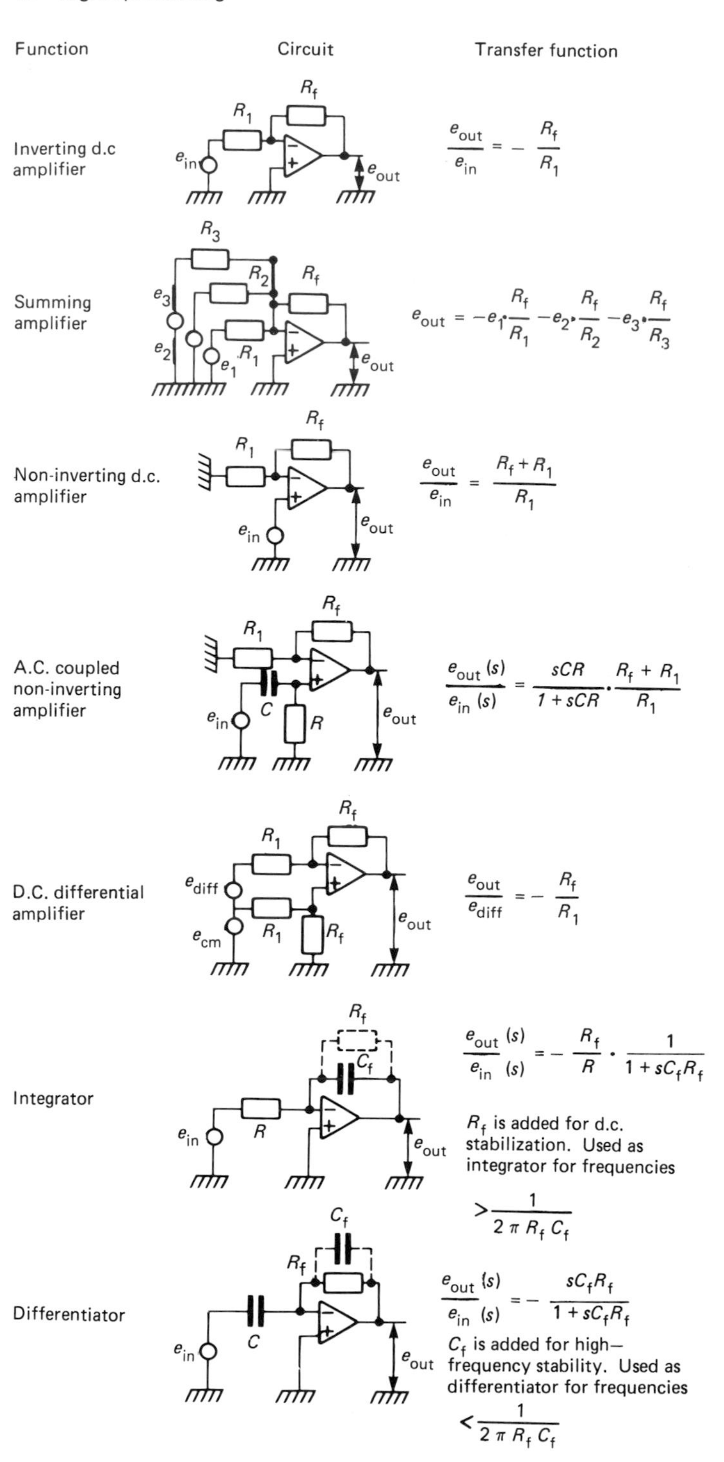

Figure 4.2 Linear signal processing using operational amplifiers.

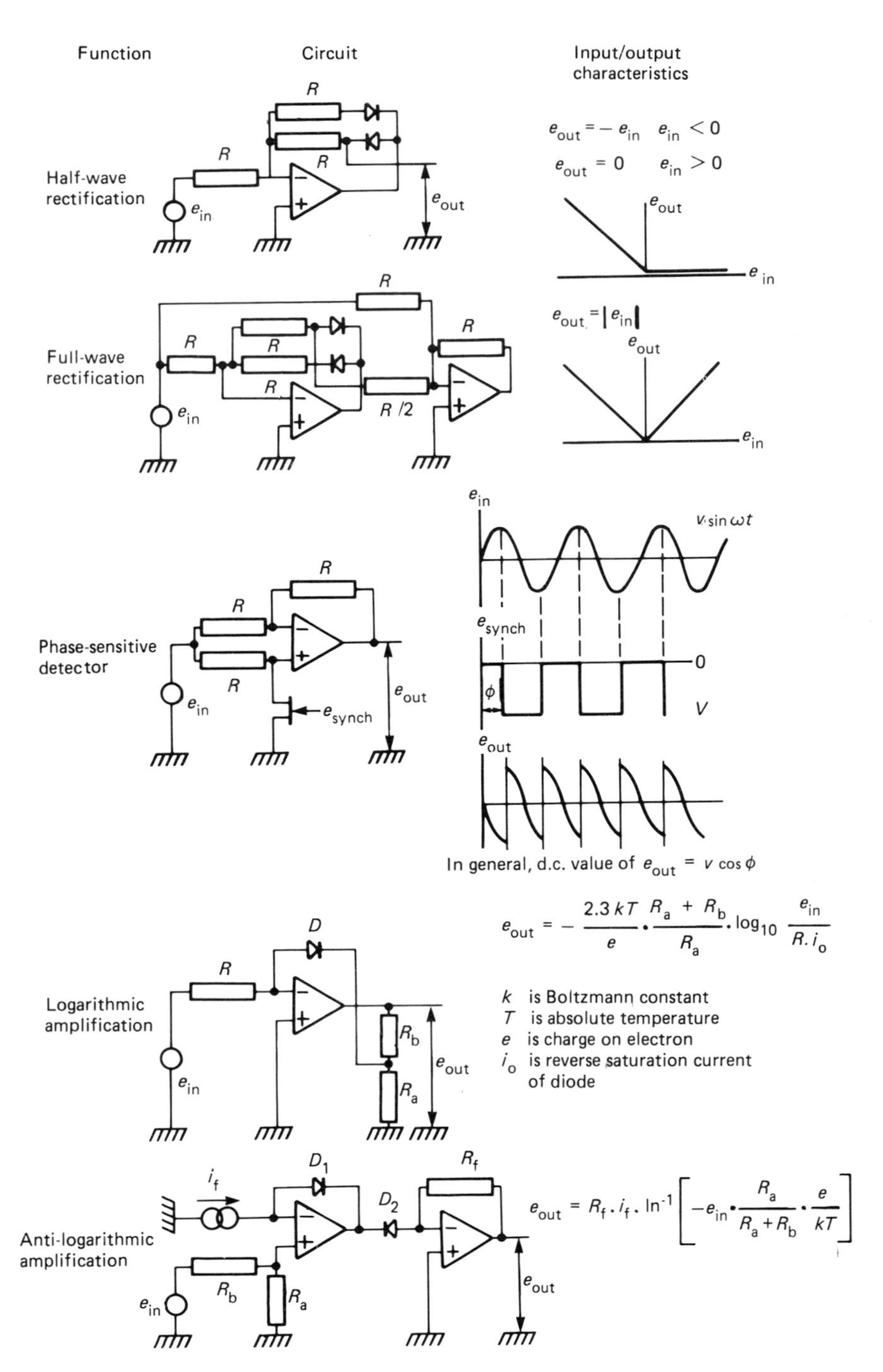

Figure 4.3 Non-linear signal processing using operational amplifiers.

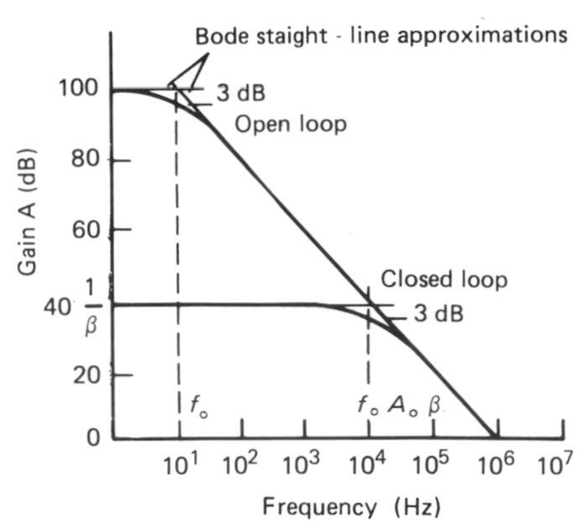

Figure 4.4 Gain–bandwidth product relationships in an operational amplifier.

If the amplifier is used as a non-inverting amplifier with resistive feedback, i.e. if $\beta(s)$ is real, then the closed-loop transfer function is given by:

$$A_{cl}(s) = \frac{e_{out}(s)}{e_{in}(s)} = \frac{1}{\beta} \cdot \frac{A_o\beta}{((A_o\beta + 1) + s\tau_o)} \tag{4.6}$$

and the 3 dB closed-loop break frequency, f_{cl}, is given by:

$$f_{cl} = \frac{(A_o\beta + 1)}{2\pi\tau_o} \simeq A_o\beta.f_o; \quad A_o\beta >> 1 \tag{4.7}$$

Thus the gain × bandwidth product for both the open-loop and closed-loop amplifiers is $A_o.f_o$, and the gain of the open-loop amplifier has been traded against the bandwidth of the closed-loop amplifier.

A two-pole approximation to the open-loop transfer function of an operational amplifier is given by:

$$A(s) = \frac{A_o}{(1 + \tau_o s)(1 + \tau_1 s)} \tag{4.8}$$

If such an amplifier, with $\tau_o >> \tau_1$, is employed in a non-inverting feedback amplifier configuration with a resistive feedback factor, then the closed-loop amplifier transfer function $A_{cl}(s)$ is given by:

$$A_{cl}(s) \simeq \frac{1}{\beta} \cdot \frac{A_o\beta}{\tau_o\tau_1[s^2 + (1/\tau_1)s + (A_o\beta/\tau_o\tau_1)]} \tag{4.9}$$

which can be compared with the standard second-order transfer function

$$H(s) = \frac{H_o\omega_n^2}{s^2 + 2\xi\omega_n s + \omega_n^2} \tag{4.10}$$

where H_o is the low-frequency gain, ω is the natural frequency and ξ is the damping factor, to give:

$$H_o = \frac{1}{\beta} \tag{4.11}$$

$$\omega_n = \sqrt{\left(\frac{A_o\beta}{\tau_o\tau_1}\right)} \tag{4.12}$$

$$\xi = \frac{1}{2}\sqrt{\left(\frac{\tau_o}{A_o\beta\tau_1}\right)} \tag{4.13}$$

If $\xi < 1$ the amplifier is said to be underdamped and the system has a response to a unit step input of the form:

$$e_{out}(t) = \frac{1}{\beta}\left[1 - \exp\left\{\frac{(-\xi\omega_n t)}{\sqrt{(1-\xi^2)}}\right\}\right] \times \sin(\omega_n\sqrt{(1-\xi^2)}.t + \cos^{-1}\xi)\Big\} \tag{4.14}$$

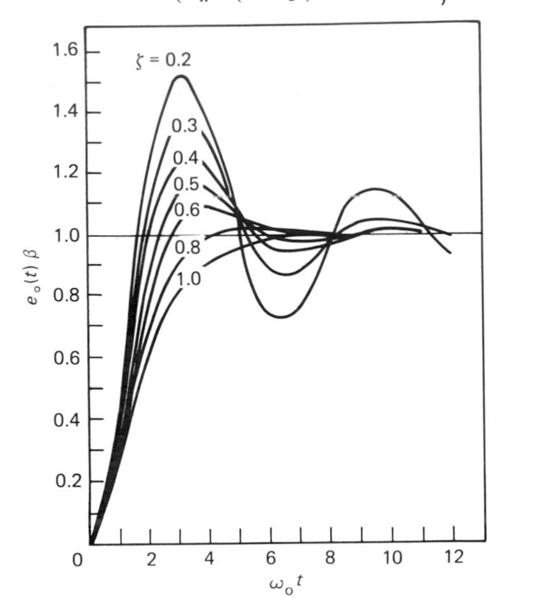

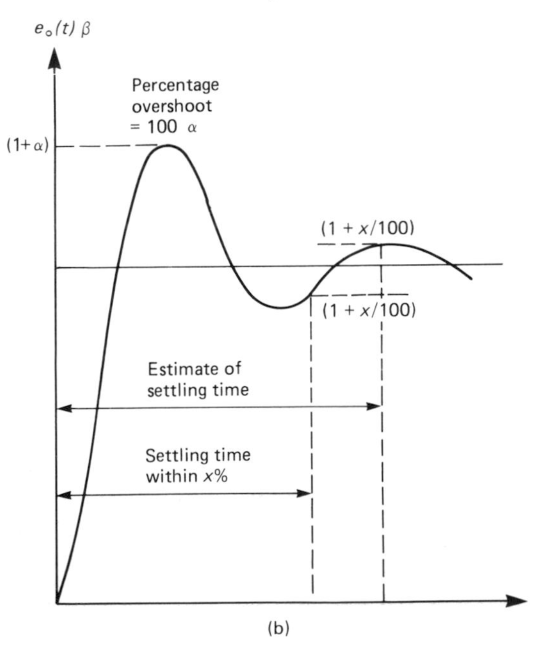

Figure 4.5 (a) Step responses for a second-order system with varying damping factor; (b) percentage overshoot and settling time for a second-order system.

The response of the system for various values of damping factor is shown in Figure 4.5. The maximum percentage overshoot (PO) is given by:

$$PO = 100 \exp\left[\frac{-\xi\pi}{\sqrt{(1-\xi^2)}}\right] ;\ \xi < 1 \qquad (4.15)$$

and an estimate of the time for the amplifier to settle to within $x\%$ of its final value is given by:

$$t_s = \frac{n\pi\sqrt{(1-\xi^2)}}{\omega_n} \qquad (4.16)$$

where the value of n is the smallest value which satisfies the equation:

$$100 \exp - \left[\frac{\xi n\pi}{\sqrt{(1-\xi^2)}}\right] \leq x \qquad (4.17)$$

As shown in Figure 4.5, this is generally an overestimate of the settling time.

For large-signal step changes the transient response of the closed-loop amplifier is governed by non-linear factors which arise as a consequence of the internal circuit design of the amplifier. Slew-rate limitation occurs, since there is a maximum rate at which the output of the amplifier can change. Figure 4.6 shows a typical large-signal step response. The

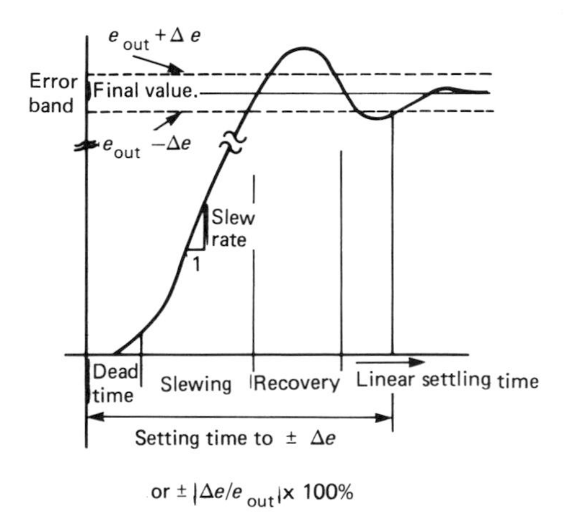

Figure 4.6 Large-signal step response.

settling time includes an initial propagation delay, together with the times required for the amplifier output to slew from its initial value to the vicinity of its final value, to recover from slew-rate limited overload and to settle to a given error in the linear range. The settling time is typically defined as the period required for an amplifier to swing from 0 V to full scale, typically 10 V, and settle within a specified percentage of the final output voltage. It is generally measured under conditions of unity gain, relatively low impedance levels and with no capacitive loading of the amplifier output.

4.2.1.2 Common mode rejection ratio (CMRR)

The operational amplifier in Figure 4.1(a) ideally should respond only to the differential signal e_{diff} applied between its input terminals and not to the common mode signal e_{cm}. CMRR is a measure of the ability of the amplifier to respond only to the differential signal and not to the common mode signal. It is defined as:

$$CMRR = 20 \log_{10} \frac{A_{diff}}{A_{cm}} \qquad (4.18)$$

where A_{diff} is the gain to the differential signal and A_{cm} is the gain to the common mode signal.

The common mode characteristic of the amplifier only causes error when the amplifier is being used in its non-inverting mode, since in the inverting mode there is no common mode signal applied to the amplifier. The CMRR is a function of several variables, including frequency, power supply, input signal level and temperature. Typically, an amplifier will have a common mode rejection of 100–120 dB at low frequencies.

4.2.1.3 Input and output impedances

Operational amplifiers, being differential devices, have both common mode and differential input impedances, as shown in Figure 4.7. The common

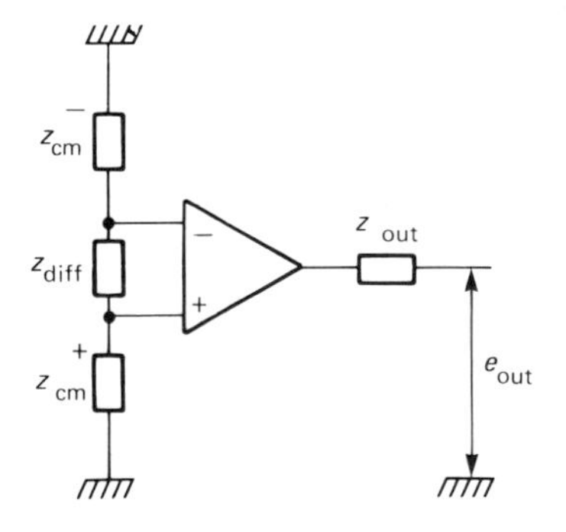

Figure 4.7 Input and output impedances for an operational amplifier.

mode input impedances z_{cm}^{+} and z_{cm}^{-} are the impedances between each input and ground or the power supply common. z_{diff} is the input impedance measured between the inputs. The dynamic impedances can be represented by resistance in parallel with capacitance. R_{cm}^{+} and R_{cm}^{-} are generally larger than R_{diff} by a factor of 100. Interaction of the input impedances with the feedback components to change the feedback can be minimized by employing feedback components with impedances significantly lower than the input impedances.

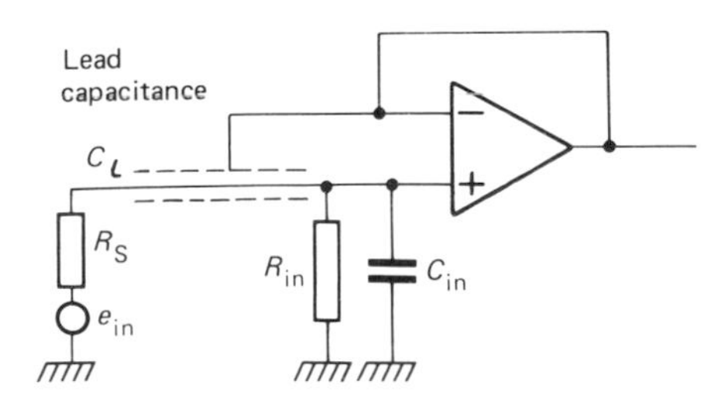

Figure 4.8 Driven shield for reducing the effect of lead capacitance.

For the non-inverting amplifier the input impedance is the common mode input impedance between the non-inverting input and ground, since the effect of the differential input impedance is increased by a factor $A(s).\beta(s)$. FET operational amplifiers can achieve common mode input impedances of up to $10^{15}\Omega$ in parallel with capacitances in the region of 1 or 2 pF, which makes them suitable for use as electrometer amplifiers, since, additionally, such amplifiers have low bias currents. In order to preserve the high input impedance such amplifiers provide it is often necessary to apply driven-shield techniques to the cable connecting the source to the input of the amplifier and to guard the input pins of the amplifier. Figure 4.8 shows a typical driven-shield arrangement. The effect of the lead capacitance has been eliminated by driving its outer sheath from the output of the voltage follower. If it is assumed that the amplifier is ideal then the overall amplifier has a transfer function given by:

$$\frac{e_{out}(s)}{e_{in}(s)} = \frac{1}{(1 + sR_sC_{in})} \tag{4.19}$$

Guld (1974) examines the use of such techniques and also the use of positive feedback techniques to eliminate the input capacitance of the amplifier.

The open-loop output impedance of the amplifier can usually be represented by a resistance. Typically, this has a value of between 100 and 500 Ω. Feedback reduces this output impedance by a factor $1/(A(s)\beta(s))$.

4.2.1.4 Input voltage offset and bias current

Because of mismatch in the components making up the differential input stage the amplifier gives a d.c. output for zero input. This is represented in the model of the real amplifier, as shown in Figure 4.9, by a voltage source, v_{os} in series with the non-inverting input of the ideal amplifier. v_{os} has a value equal to the voltage which, when applied to the input, would bring the output to zero. This offset voltage typically is 0.1–30 mV. It is a function of temperature, power supply and time.

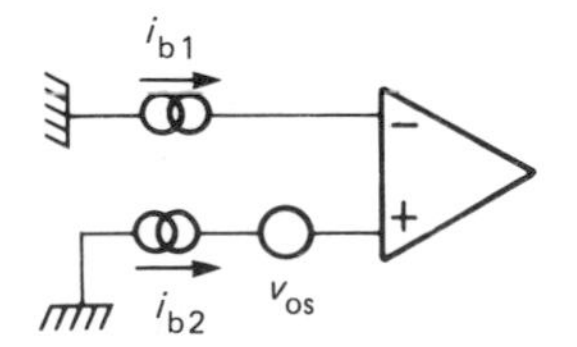

Figure 4.9 Input offset voltage and bias currents in an operational amplifier.

Most operational amplifiers have offset null adjustment pins which enable the initial offset to be nulled out. In precision d.c. measurements it is the temperature drift of the offset voltage which is of greater importance. The offset voltage temperature coefficient over a given range may be specified as:

$$\frac{\Delta v_{os}}{\Delta T} = \frac{v_{osH} - v_{osL}}{T_H - T_L} \tag{4.20}$$

where v_{osH} and v_{osL} are the measured values of the offset voltage measured at temperatures T_H and T_L. This is only an average measure of the temperature coefficient over the specified temperature range and, since the offset voltage is a non-linear function of temperature, it may be an over- or underestimate at any point, and it cannot be used to estimate the temperature drift from a given point in the temperature range. Some manufacturers (Analog Devices, 1982) use a butterfly specification of temperature coefficient, as shown in Figure 4.10, which allows bounds to be set on the offset voltage drift from some reference temperature T_{ref} at which it has been nulled. Using this technique, the offset voltage drift from the reference temperature T_{ref} to the working ambient temperature T_A is guaranteed to be less than $(T_{ref} - T_A)$ multiplied by the specified temperature coefficient. Typical values for the temperature coefficient of the offset voltage are in the range 0.5–20 μV/K.

It should be noted that offset nulling using the internal circuitry can considerably increase the offset voltage temperature coefficient, and it is recommended that in critical situations external nulling techniques should be employed. Generally, all offset voltage temperature specifications are for steady-state thermal conditions. Under transient thermal conditions such as warm-up or under conditions of thermal shock the value of offset voltage may be considerably increased.

The offset voltage is also a function of power supply voltage and time. Its variation with power supply voltage is specified as $\Delta v_{os}/\Delta V$ (in μV/V).

The drift of offset voltage with time occurs as a consequence of ageing. It is specified as μV/day, μV/month or μV/year, and is generally not linear with time. Extrapolation can often be made on the basis of a square-root estimate. The drift for any

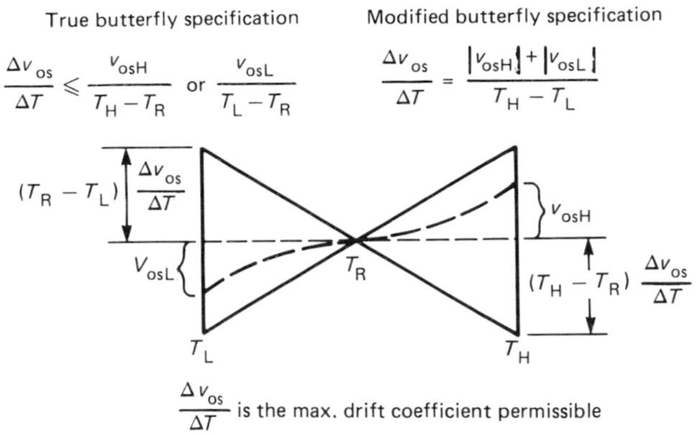

True butterfly specification

$$\frac{\Delta v_{os}}{\Delta T} \leqslant \frac{v_{osH}}{T_H - T_R} \text{ or } \frac{v_{osL}}{T_L - T_R}$$

Modified butterfly specification

$$\frac{\Delta v_{os}}{\Delta T} = \frac{|v_{osH}| + |v_{osL}|}{T_H - T_L}$$

$(T_R - T_L)\frac{\Delta v_{os}}{\Delta T}$ v_{osH} v_{osL} T_R $(T_H - T_R)\frac{\Delta v_{os}}{\Delta T}$ T_L T_H

$\frac{\Delta v_{os}}{\Delta T}$ is the max. drift coefficient permissible

Figure 4.10 Butterfly specification of the offset voltage temperature coefficient.

period is thus obtained by taking the product of the specified drift over a given period and multiplying it by the square root of the ratio of the time period for which the drift is required to the time period for which the drift is specified. The data for time variations tend to be somewhat more scanty than those for the temperature and power-supply variations.

The input stages of the operational amplifier have to be supplied with bias current. It is therefore necessary to provide d.c. bias paths through which these currents can flow. These bias currents are represented as shown in Figure 4.9 by two current generators, i_{b1} and i_{b2}, having values which are the average of the bias current into the positive and negative terminals. These current generators can have values from 10^{-7}A to 10^{-15}A, dependent on the technology employed in the fabrication of the operational amplifier. The bias currents are a function of temperature and power supply voltage.

The effect of offset voltage and bias currents on the output of an operational amplifier can be assessed in any circuit by shorting-out inductors, open-circuiting capacitors, short-circuiting voltage sources and open-circuiting current sources. Figure 4.11 shows the effect of the voltage offset and bias current on an inverting operational amplifier. The output offset due to bias currents can be minimized by making the source resistances at the two inputs equal, i.e.

$$R_s = \frac{R_1 \,.\, R_f}{R_1 + R_f}$$

Under such conditions the offset is now a function of the difference current $i_d = i_{b1} - i_{b2}$. This, in general, will be almost an order of magnitude smaller than either i_{b1} or i_{b2}, and since the two bias currents track each other with temperature the temperature coefficient for i_d will also be considerably smaller.

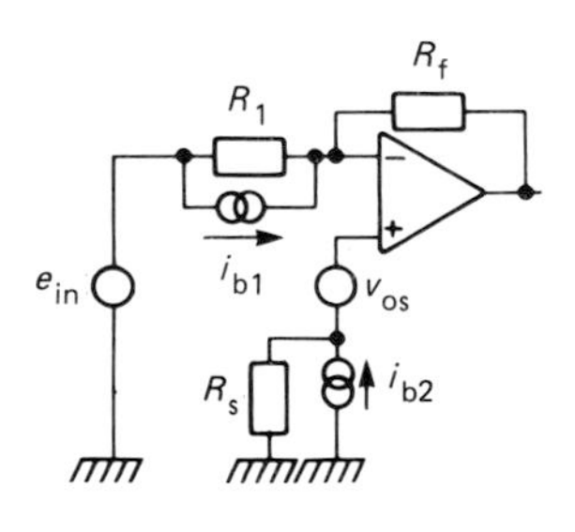

$$e_{out} = -e_{in} \cdot \frac{R_f}{R_1} + v_{os} \cdot \frac{R_f + R_1}{R_1} - (i_{b1} - i_{b2}) R_f$$

$$\text{if } R_s = \frac{R_1 \,.\, R_f}{R_1 + R_f}$$

Figure 4.11 Effects of offset voltage and bias current on the output of an operational amplifier.

4.2.1.5 Noise

In the detection of low-level signals the noise performance of the operational amplifier is extremely important, and it is therefore necessary to be able to calculate and minimize the noise contribution of the amplifier using data obtained from the manufacturers' specifications. Figure 4.12 shows the representation of the noise sources in an operational amplifier. The voltage source has an r.m.s. value given by e_n and the current generators have r.m.s. values given by i_n. The resistors in the circuit each have associated with them a thermal noise voltage e_R whose r.m.s. value is given by:

$$e_R = \sqrt{(4kTRf)} \qquad (4.21)$$

where k is the Boltzmann constant, T is the absolute

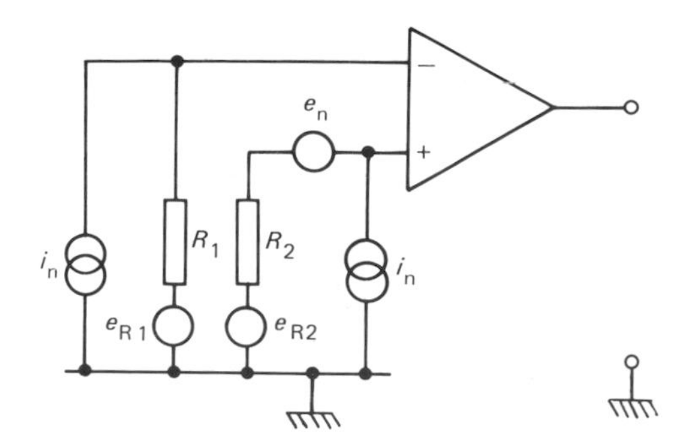

Figure 4.12 Noise sources in an operational amplifier.

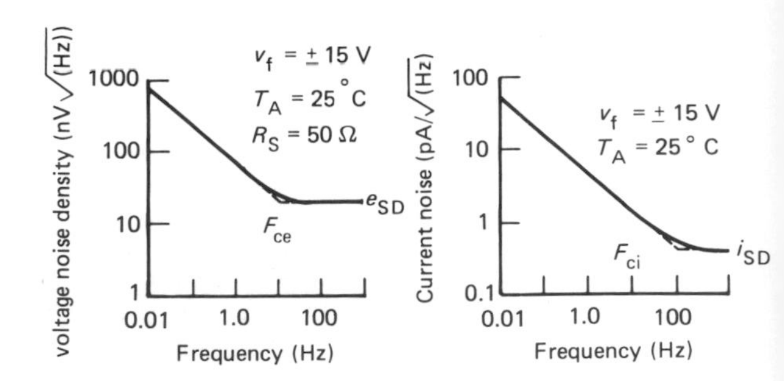

Figure 4.14 Voltage and current noise spectral densities for an OP-14 operational amplifier.

temperature, R is the resistance and f is the bandwidth over which the measurement is made.

It is therefore possible to calculate the total r.m.s. noise e_{ntot} referred to the input of the amplifier. It is given by:

$$e_{ntot} = \sqrt{(e_n^2 + R_1^2 i_n^2 + R_2^2 i_n^2 + 4kTR_1 f + 4kTR_2 f)} \quad (4.22)$$

The first term represents the voltage noise, the second and third terms the effect of the current noise interacting with the resistances associated with each input and the last two terms represent the thermal noise associated with the resistances. The addition is performed in a mean square sense, since it is assumed that the sources are statistically independent.

Manufacturers provide graphical representations of the r.m.s. level of the noise referred to the input as a function of the source resistance for a series of bandwidths, typically 10 Hz–100 kHz, 10 Hz–10 kHz or 10 Hz–1 kHz. Figure 4.13 shows such plots for a 741 operational amplifier. The peak-to-peak noise can be estimated to be $6.6e_{ntot}$, where e_{ntot} is the r.m.s. value of the noise which is assumed to be Gaussian.

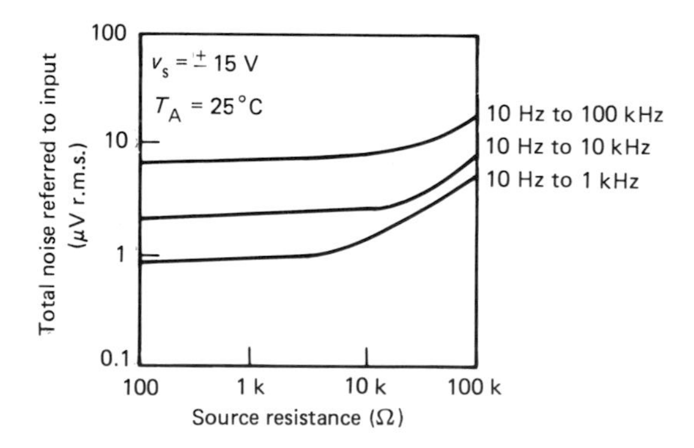

Figure 4.13 Noise characteristics of a 741 operational amplifier.

Manufacturers also provide spectral data on the voltage noise source and the current noise source. Figure 4.14 shows typical curves for the voltage spectral density in nV/√(Hz) and the current spectral density in pA/√(Hz) for a Precision Monolithics OP-14 operational amplifier. At high frequencies both the voltage and current noise exhibit white noise characteristics, i.e. have equal power in equal bandwidths. At low frequencies they exhibit a $1/f$ characteristic called flicker noise. This is characterized by having equal power in each decade of the bandwidth. By integrating these power spectra over a given frequency band it is possible to calculate the total r.m.s. noise in that band. Thus the r.m.s. value of the voltage noise e_n in a band as shown in Figure 4.14 can be calculated as:

$$e_n = e_{sd} \sqrt{\left[f_{ce}.\log_e \left(\frac{f_H}{f_L} \right) + (f_H - f_L) \right]} \quad (4.23)$$

where f_{ce} is the corner frequency of the voltage noise, f_H and f_L are the upper and lower frequencies of the bandwidth over which the noise is being measured and e_{sd} is the spectral density of the voltage noise at high frequencies.

A similar calculation can be performed for the r.m.s. value of the current i_n:

$$i_n = i_{sd} \sqrt{\left[f_{ci} \log_e \left(\frac{f_H}{f_L} \right) + (f_H - f_L) \right]} \quad (4.24)$$

where f_{ci} is now the corner frequency of the current noise and i_{sd} is the spectral density of the current noise at high frequencies.

It is possible to modify the spectral densities to take into account the frequency response of the amplifier in order to determine the noise at the output of the amplifier (Tobey *et al.*, 1971).

An alternative approach in noise assessment is to specify the amplifier in terms of a noise figure, NF (Arbel, 1980) and seek by noise matching of the source to the amplifier to minimize the noise figure. This approach is less useful in instrumentation applications, since it does not lead to the best signal-to-noise ratio at the output of the amplifier.

4.2.1.6 Operating parameters

In addition to the parameters specified above, there are also parameters specified for the operational amplifier which are designed to keep it within its linear regime. These include such parameters as input and output voltage swings and output drive current capability, which arise as a consequence of internal circuit design and which place limitations on the load and feedback impedances that the amplifier can drive. The full-power bandwidth of the amplifier is determined by the slew rate of the amplifier. It is specified as the maximum bandwidth at unity closed-loop gain for which a sinusoidal input signal will produce full rated output at rated load without exceeding a given level of distortion. Distortion in the amplifier is measured in terms of total harmonic distortion (THD). Overload recovery measures the ability of the amplifier to recover from a saturation condition caused by a specified amount of overdrive. For further details of these parameters the reader is referred to the manufacturers' amplifier specifications, which give the values of these parameters for particular amplifiers together with specifications of the conditions under which they are measured.

4.2.2 Instrumentation amplifiers

The single-ended inverting and non-inverting amplifiers shown in Figures 4.1(c) and 4.1(d) are not suitable for use as general-purpose instrumentation amplifiers, since in many transducers the low-level signal may be superimposed on a large-amplitude common mode signal, as in the case of the output signal from a strain gauge bridge or a line frequency interference signal, as shown in Figure 4.15(a). Measurement of the signal by the use of a single-ended amplifier leads to amplification of the

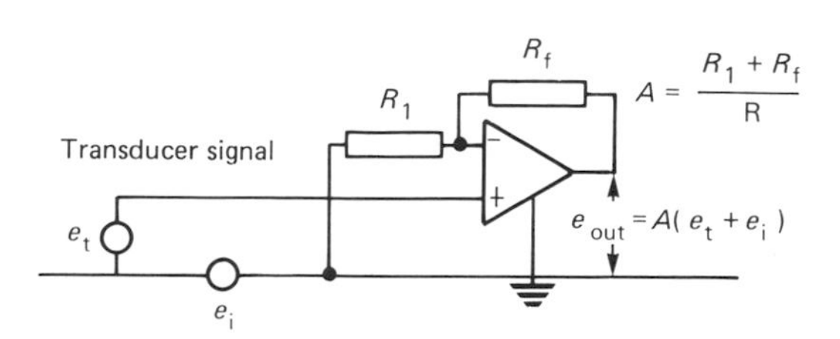

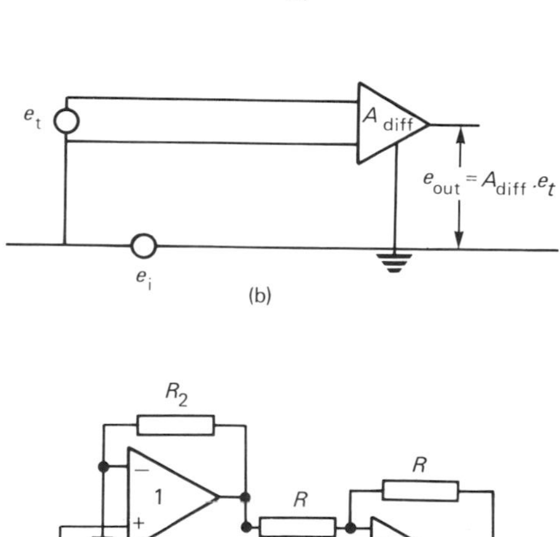

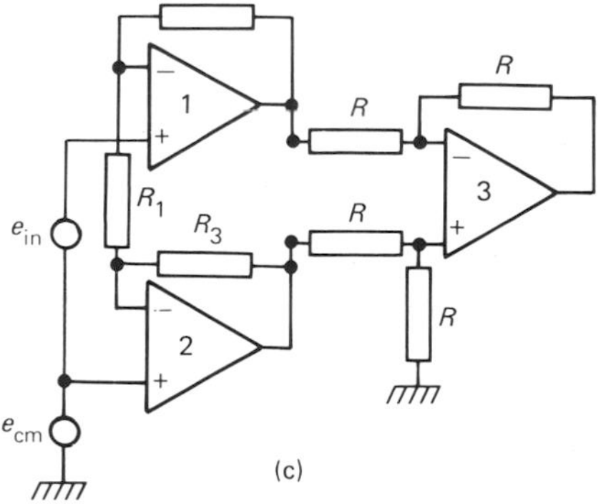

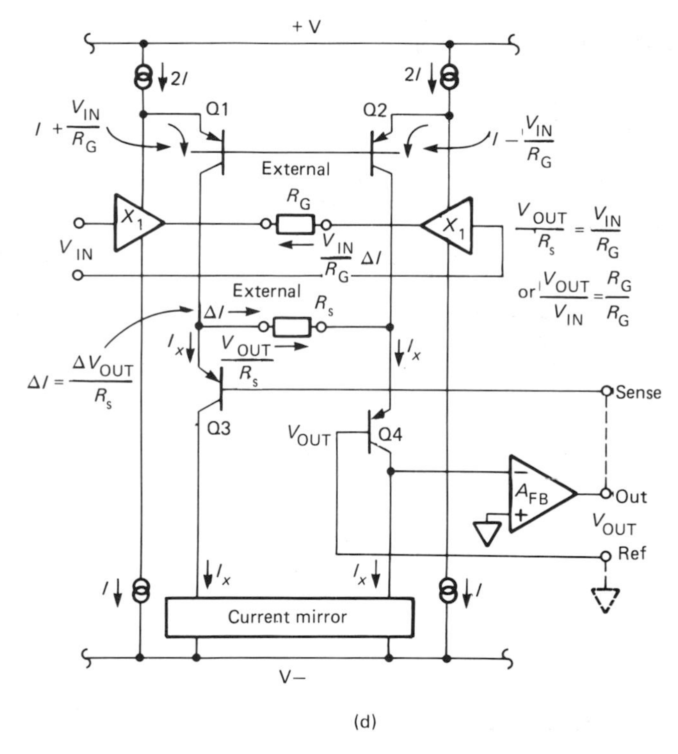

Figure 4.15 (a) Effect of common mode signals on a single-ended input amplifier; (b) the use of a differential amplifier to reduce the effects of common mode signals; (c) an instrumentation amplifier using three operational amplifiers; (d) the internal construction of a commercially available instrumentation amplifier AD.

common mode signal as well as the transducer signal. By using the differential amplifier shown in Figure 4.15(b), the CMRR of the amplifier reduces the effect of common mode signals. Instrumentation amplifiers are wideband, high input impedance differential amplifiers, having adjustable and stable differential gain, high common mode rejection ratio, low input drift characteristics and low output impedance. They are used for the amplification of signals from thermocouples, strain gauge bridges, current shunts and biological probes.

Figure 4.15(c) shows an instrumentation amplifier employing three operational amplifiers. Amplifiers A1 and A2 provide only unity gain to the common mode signal and amplifier A3 acts as a differential amplifier. The differential gain of the overall amplifier is given by:

$$A_{\text{diff}} = \frac{R_1 + R_2 + R_3}{R_1} \quad (4.25)$$

The common mode rejection ratio is governed by the common mode rejection ratio of the amplifiers making up the instrumentation amplifier and the matching of the resistors in the differential amplifier A3.

Figure 4.15(d) shows the internal construction of a commercially available instrumentation amplifier (Analog Devices AD521). The differential voltage e_{diff} appears across R_G and causes an imbalance in the currents flowing through the transistors Q_1 and Q_2. The current mirror consisting of Q_3 and Q_4 forces this imbalance current to flow through R_S. The output voltage e_{out} is given by R_S/R_G. Differential gains from 1 to 1000 can be provided by adjusting R_G. The characteristics of this amplifier are given in Table 4.2. Instrumentation amplifiers are available with software programmable gains.

Table 4.2 Instrumentation amplifier specification

Gain G	1–1000
Small-signal bandwidth	
$G = 1$	>2 MHz
$G = 1000$	40 kHz
Differential input voltage	±10 V
Common mode rejection ratio	
$G = 1$	70 dB min
$G = 1000$	100 dB min
Input impedances	
Differential	$3 \times 10^9\ \Omega \| 1.8$ pF
Common mode	$6 \times 10^{10}\ \Omega \| 3.0$ pF
Voltage offsets	
Input	3 mV max
vs temp.	15 μV/K max
Output	400 mV max
vs temp.	400 μV/K max
Input bias current	
Initial	80 nA max
vs temp.	1 nA/K
Noise	
Voltage noise wrt output	
0.1 Hz to 10 Hz	$\sqrt{[(0.5G)^2 + (225)^2]}$ μV p-p
10 Hz to 10 kHz	$\sqrt{[(1.2G)^2 + (50)^2]}$ μV r.m.s.
Input current noise	
10 Hz to 10 kHz	15 pA r.m.s.

4.2.3 Isolation amplifiers

The amplifier shown in Figure 4.15(d) can only allow the rejection of common mode signals which are within its power rails. By using an isolation technique as shown in Figure 4.16, which provides galvanic isolation of the front-end amplifier from the power supply and subsequent stages, it is possible to construct an amplifier which is capable of providing isolation capable of withstanding several kV. Isolation is generally provided by transformer coupling. In such amplifiers the primary source of leakage between input and output circuits is the capacitance between the windings of the transformer.

A power oscillator provides isolated power to the input amplifier, together with a carrier which is modulated by the amplified input signal and coupled across the isolation barrier. In the output section this signal is demodulated and filtered. The amplifier output is provided via a buffer amplifier. A second demodulator in the input stage is used to provide the input stage with feedback.

Applications for such amplifiers include the measurement of low-level d.c. and low-frequency voltage or current in the presence of high common mode voltages and the reception of signals transmitted at high impedance in a noisy environment. Isolation amplifiers are widely used in medical electronics where, for reasons of patient safety, it is necessary to limit d.c. and line frequency leakage currents. Table 4.3 gives the specification of a commercially available isolation amplifier (Analog Devices AD293).

4.2.4 Chopper-stabilized amplifiers and commutating auto-zero amplifiers

There is often a requirement to produce amplifiers which have a low offset voltage and also a low offset voltage temperature coefficient. Two amplifiers with such characteristics are chopper-stabilized amplifiers and commutating auto-zero (CAZ) amplifiers. These are shown in Figures 4.17(a) and 4.17(b), respectively. In the chopper-stabilized amplifier (Giacoletto, 1977) the signal is split into two paths on a frequency basis. The d.c. and low-frequency signals are routed via a low-pass filter to a modulator. This modulated signal (which is now an a.c. signal) is then amplified by means of an a.c.

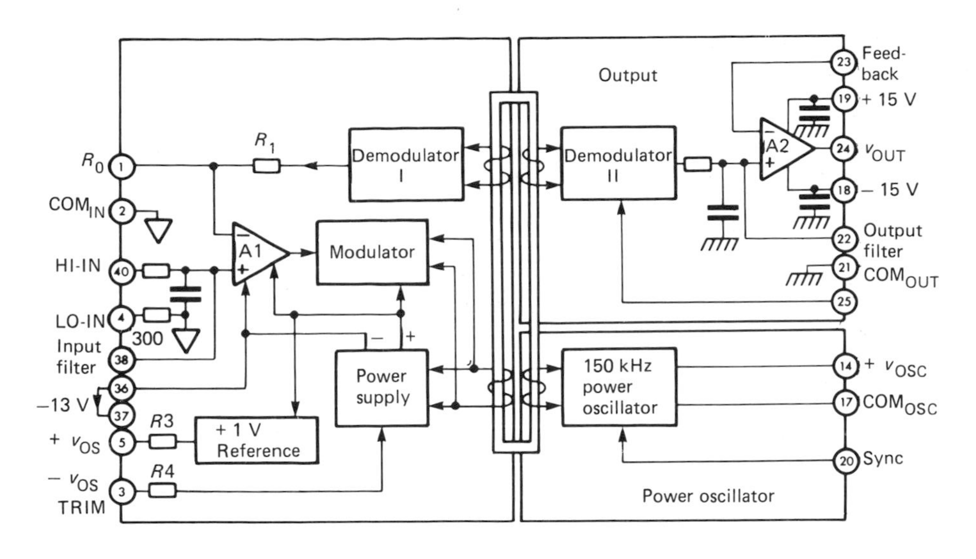

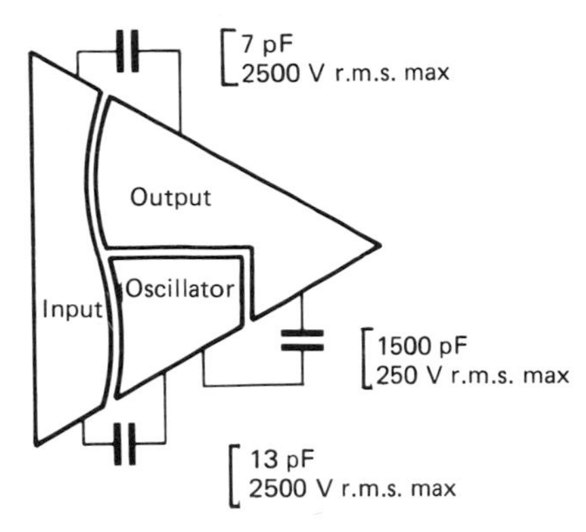

Figure 4.16 Isolation amplifier.

amplifier. After amplification it is demodulated, filtered and passed into the main amplifier. Thus the d.c. and low-frequency signals are amplified by an a.c. amplifier which ideally has no d.c. offset (the only offset being introduced by non-ideal modulation and demodulation of the signal). If the effective gain of the chopper stage is A_{chop} and the offset of the main amplifier is v_{osm}, then, referred to the overall amplifier input, the effective offset of the amplifier is v_{osm}/A_{chop}, since the d.c. signals have been given a d.c. offset-free gain of A_{chop} before they are applied to the main amplifier. Typically, such amplifiers have an offset voltage of between 15 and 50 μV and a temperature coefficient of between 0.1 and 1.0 μV/K, with a long-term stability of order 2 μV/month, 5 μV/year. They are normally used with heavy overall feedback to remove the effects of variations in the open-loop frequency response.

The commutating auto-zero amplifier consists of two operational amplifiers situated on the same substrate (Intersil, 1979). The amplifier is a two-state device and these two states are shown in Figure 4.17(b). In the state 1 operational amplifier 1 is being used as the active amplifier with the capacitor C_1 in series with its non-inverting input. Amplifier 2 is switched so that its offset is being stored on C_2. After a fixed period of time (typically 3 ms) the roles of the two amplifiers are reversed. The offset on amplifier 1 is now measured and the offset on amplifier 2 is removed by placing the capacitor C_2 in series with its non-inverting input. The commutation back to state 1 then occurs after a further 3 ms. CAZ amplifiers typically can have an initial offset voltage of 2 μV, a temperature coefficient of 0.05 μV/K and a time variation of the offset of 0.2 μV/year. CAZ amplifiers generate noise spikes at the commutation frequency, and thus it is necessary to limit the frequency of operation to one tenth of the commutation frequency.

Table 4.3 Isolation amplifier specification

Gain G	1–1000
Small-signal bandwidth	
($G = 1$ to 100)	2.5 kHz
Differential input voltage	±10 V
Max. common mode voltage	
(Inputs/Outputs)	
Continuous a.c. or d.c.	±2500 V peak
Common mode rejection	
(1 KΩ source impedance imbalance)	100 dB
Leakage current	
Input to output	
at 115 V a.c. 60 Hz	2 μA r.m.s. max
Input impedances $G = 1$	
Differential	$150\ \text{pG} \parallel 110^{8}\ \Omega$
Common mode	$30\ \text{pF} \parallel 5 \times 10^{10}\ \Omega$
Offset voltage referred to input	
initial	$(\pm 3 \pm 22/G_{in})$ mV
vs temp.	$(\pm 3 \pm 150/G_{in})$ μV/K
(G_{in} = gain of input stage)	
Input bias current	
Initial	2 nA
vs temp.	20 pA/K
Input noise	
Voltage	
0.05 Hz to 100 Hz	10 μV p-p
10 Hz to 1 kHz	5 μV r.m.s.
Current	
0.05 Hz to 100 Hz	50 pA p-p

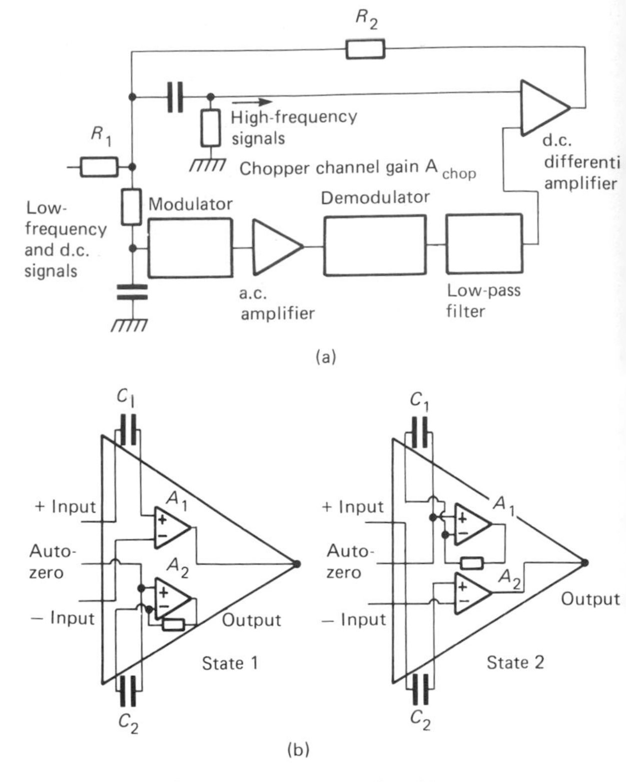

Figure 4.17 (a) Chopper-stabilized amplifier; (b) commutating auto-zero amplifier.

4.2.5 Charge and current amplifiers

Piezoelectric crystals, used in accelerometers, pressure transducers and load cells, produce charge outputs. A charge amplifier, as shown in Figure 4.18, can be used to convert the charge into a voltage. The non-inverting input of the amplifier acts as a virtual earth, and therefore the shunting effect of the capacitance of the cable connecting the crystal to the amplifier is removed, since there is no voltage drop across it. The large resistance R_f provides d.c. feedback for the amplifier and a path for its bias current. The charge amplifier has the advantage that its sensitivity and low-frequency cut-off are governed by the feedback components and not by the crystal. Typically, for a charge amplifier C_f and R_f will have values of 10 pF to 100 pF and 10^{11}–$10^{14}\,\Omega$, respectively. Using such feedback components it is possible to produce charge amplifiers which, for all practical purposes, have a response down to d.c.

Current-to-voltage conversion, for example, of the output of a photodiode can be provided by means of the current or transimpedance amplifier, as shown in Figure 4.19(a). The current from the

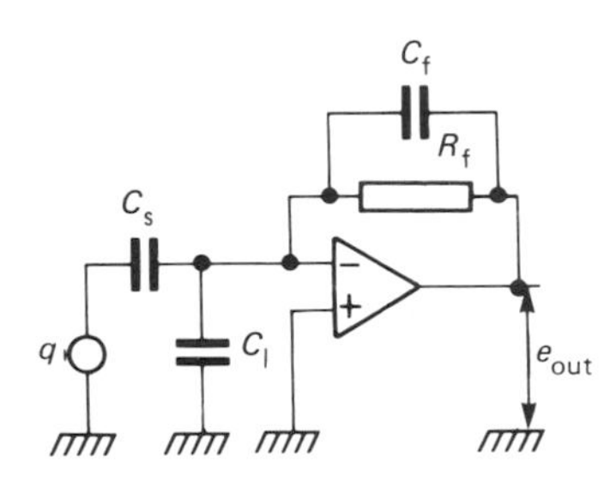

$$\frac{e_{out}(s)}{q\ (s)} = -\frac{sR_f}{1 + sR_f C_f}$$

if $sC_f R_f \gg 1$

$$\frac{e_{out}(s)}{q\ (s)} = -\frac{1}{C_f}$$

R_f is large resistance required for d.c. stabilization
C_s represents capacitance of source
C_l lead capacitance

Figure 4.18 Charge amplifier.

photodiode is detected under short-circuit conditions, since its cathode is connected to the virtual earth of the operational amplifier. The sensitivity of the amplifier is given by R_f. C_f is required for stability purposes and limits the bandwidth of the conversion. By the use of FET operational amplifiers with large values of R_f it is possible to measure currents of below 10^{-12}A. Since offset voltage and bias current errors can be nulled out, the measurement accuracy is then limited by the bias current temperature drift, which for FET amplifiers doubles for every 10K rise in temperature, and noise.

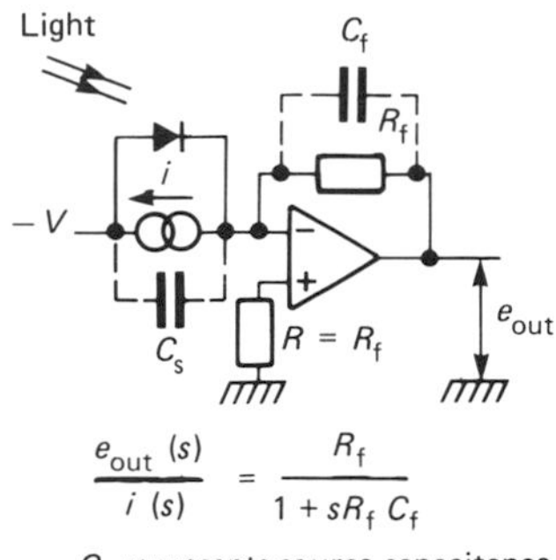

$$\frac{e_{out}(s)}{i(s)} = \frac{R_f}{1 + sR_f C_f}$$

C_s represents source capacitance

C_f is added for stability

(a)

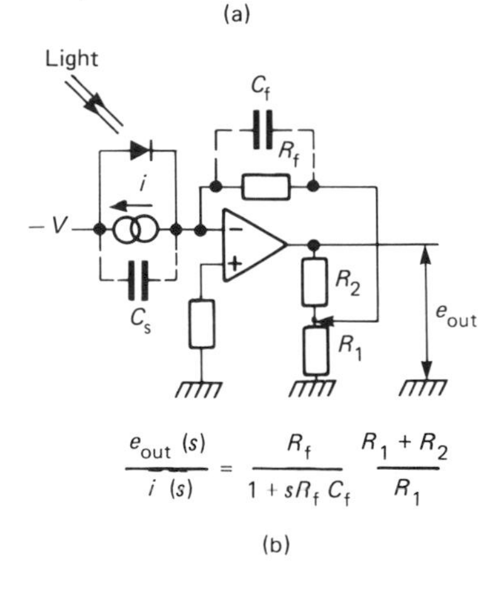

$$\frac{e_{out}(s)}{i(s)} = \frac{R_f}{1 + sR_f C_f} \quad \frac{R_1 + R_2}{R_1}$$

(b)

Figure 4.19 (a) Current (transimpedance) amplifier; (b) a current amplifier employing lower-value feedback resistors.

The need for large-feedback resistors can be eliminated by the use of the output divider circuit, as shown in Figure 4.19(b). This reduction in the value required for R_f is effected at the expense of a reduction in the amplifier loop gain and an increase in offset and noise gain.

4.2.6 Sample and hold amplifiers

Sample and hold (S and H) amplifiers, otherwise known as track and hold amplifiers, are used to maintain the input of an analogue-to-digital converter constant during the period in which the conversion is taking place. This is particularly required when the analogue signal is rapidly varying, when the signal is being quantized into a large number of levels or in systems where there is a high conversion throughput. Such an amplifier has two modes—the sample or track mode, in which the output follows the input, and the hold mode, in which the value is maintained at the value the input had at the instant at which the hold instruction was given. Logical signals, usually at TTL levels, determine which mode the amplifier is in at any one time.

A schematic diagram of such an amplifier is shown in Figure 4.20(a). In the sample or track mode the capacitor C is being charged or discharged by amplifier 1 at such a rate as to enable it to follow the input. The following action is ensured by means of the heavy overall feedback applied around the amplifier. In the hold mode the capacitor is isolated from amplifier 1 and it therefore retains its charge, being only discharged by the input bias current of amplifier 2, which acts as a follower. In the sampling or tracking mode ideally the capacitor should be small, since the maximum rate of change of voltage on the capacitor dV_c/dt is limited to i_{max}/C, where i_{max} is the maximum output current that amplifier 1 can supply. Thus for fast tracking the capacitor C should be small. In the hold mode the capacitor C should be large, since the droop rate is given by i_b/C, where i_b is the bias current drawn by amplifier 2. A compromise has therefore to be reached between these two conflicting requirements. The capacitor should be made of a material such as polystyrene or teflon in which dielectric absorption effects are small, thus minimizing errors due to 'creep'.

Figure 4.20(b) shows some of the specifications of the S and H amplifier which are of importance when the device is changing its state. The aperture time is the time delay between applying the hold command signal and the time at which the capacitor is isolated from amplifier 1. It is possible to allow for this time delay in precision timing circuits, but also associated with the aperture time there is a random uncertainty or jitter. Typically, the aperture time is 6–150 ns, dependent on the device. High-speed devices can have a jitter as low as 20 ps. After the settling time required for sample-to-hold transients to decay it is found that the signal in the hold mode has an offset with respect to the input signal. The major cause of this is charge transfer or an offset step, caused by capacitive coupling of the switching signal to the

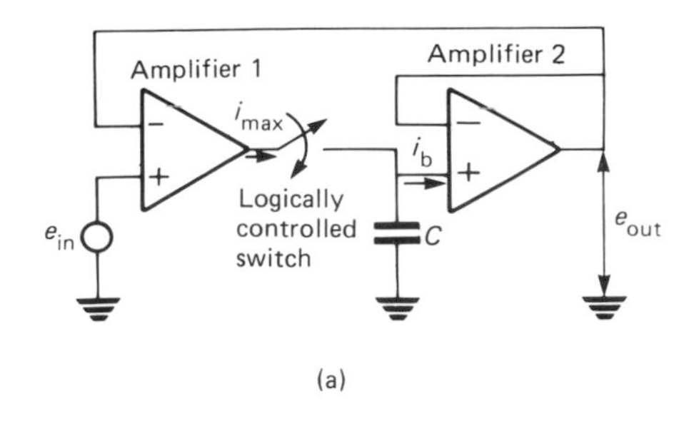

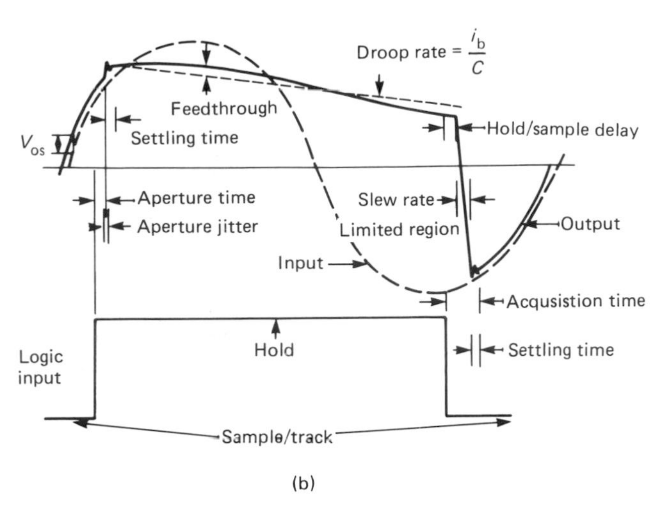

Figure 4.20 (a) Schematic diagram of a sample and hold amplifier; (b) specifications of a sample and hold amplifier.

hold capacitor. It can be reduced by using a large hold capacitor, at the expense of system response time, or cancelled by capacitively coupling a signal of the correct polarity to the hold capacitor. The signal in the hold mode will droop for the reasons stated above, and for a particular value of capacitor this is expressed in μV/μs or mV/μs. During the hold mode capacitive feedthrough of the input signal to the output will occur across the open switch, and this is usually measured as a rejection in dB for a signal at a given level and frequency. The acquisition time is the time for the output to reach its final value, within a specific error band, after the sample command has been given. For precision sample and hold amplifiers, typically this may be expressed as the time required to achieve an accuracy of ±0.01 per cent for a 10 V step. This time includes switching delay, slewing interval and settling time, and is dependent on the size of the hold capacitor employed. High speed S and H amplifiers can have acquisition times of order 250 ns.

4.2.7 Filtering

Filtering is used in signal conditioning in instrumentation systems to improve the signal-to-noise ratio by selectively reducing both random noise and also unwanted periodic signals such as line frequency interference. There are four types of filter—low pass, high pass, bandpass and bandstop. Figure 4.21 shows the ideal forms of these filters. In practice, it is not possible to have infinitely sharp transitions between the stop and pass bands, and the filter design problem thus becomes an approximation problem.

The simplest low-pass and high-pass filters are the single-pole filters whose transfer functions and amplitude and phase characteristics are shown in Figure 4.22(a). Such filters have a single real pole. Figure 4.22(b) shows the realizations of these filters using operational amplifiers.

Figure 4.23 shows the transfer functions and amplitude and phase characteristics of the four filter types using two-pole filters with complex conjugate pole pairs. Operational amplifiers are used extensively in the production of active filters having complex conjugate pole pairs. A wide variety of design techniques are available for such filters.

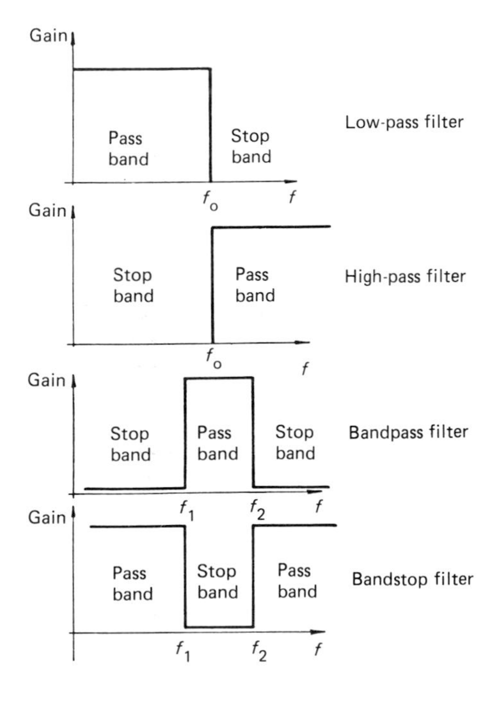

Figure 4.21 Ideal filters.

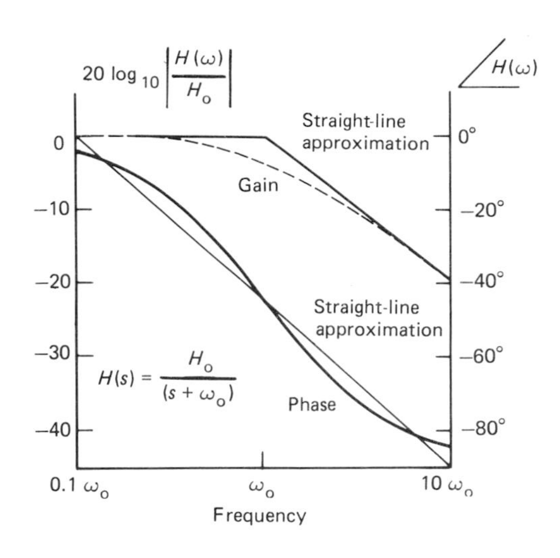

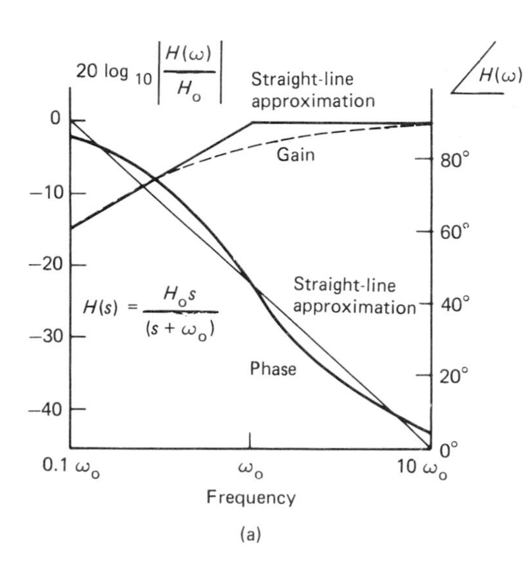

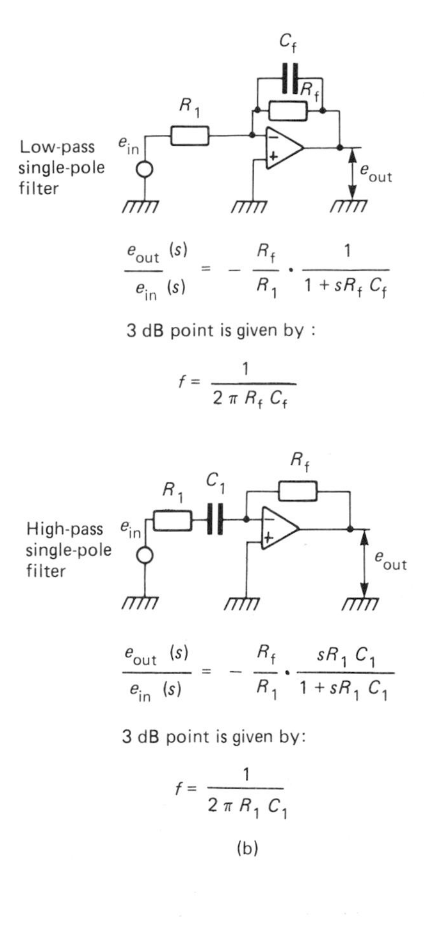

Figure 4.22 (a) Single-pole (a) low-pass and (b) high-pass filters; (b) realization of single-pole low-pass and high-pass filters using operational amplifiers.

Figure 4.24 shows two-pole complex conjugate filters realizations employing multiple feedback, voltage-controlled voltage source and state variable circuits. Tobey *et al.* (1971) and Bowron and Stephenson (1979) give detailed design procedures for the design of such filters.

In general, a filter can be represented by its transfer function $G(s)$, which can be represented as the ratio of two polynomials:

$$G(s) = \frac{K \prod_{i=1}^{m} (s - z_i)}{\prod_{j=1}^{n} (s - p_j)} \tag{4.26}$$

where in order for the filter to be realizable $n \geqslant m$.

The z_i are the zeros of the transfer function and the p_j are the poles of the transfer function. The number of poles of the transfer function is called the order of the filter, and, the larger the number of poles, the sharper the transition between the pass and stop bands. The problem in filter design is choosing the location of the poles according to some criteria. Figure 4.25 shows the amplitude characteristics of three approximations to the low-pass filter. These are Butterworth filters, which have a maximally flat response; Chebyshev filters, which have ripple in the pass band; and elliptic filters, which allow ripple in both the pass band and the stop band. In addition to such approximations, which are based on amplitude considerations, alternative filter realizations are available, such as the Thompson approximation, which provides a maximally flat

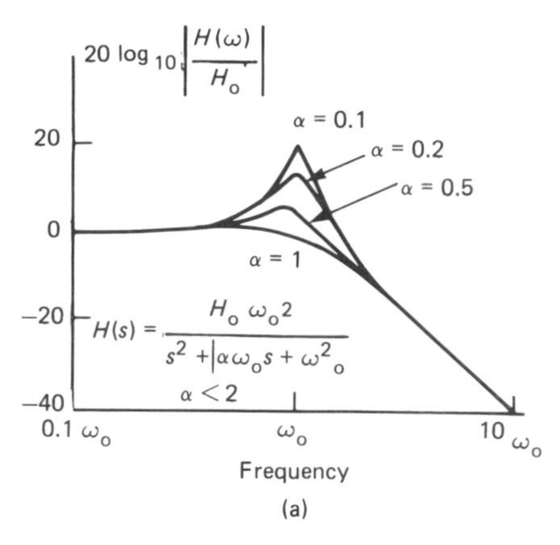

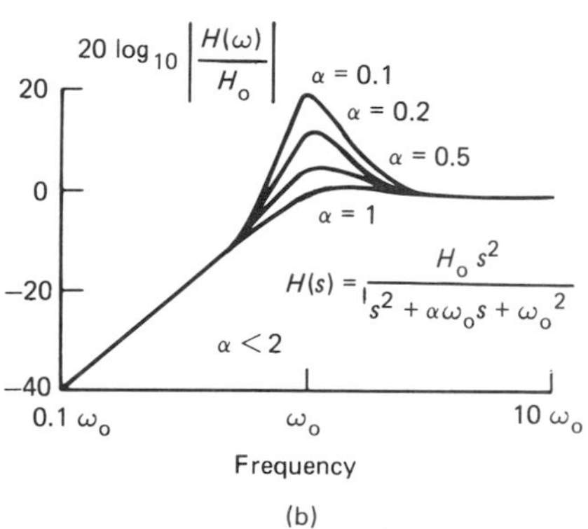

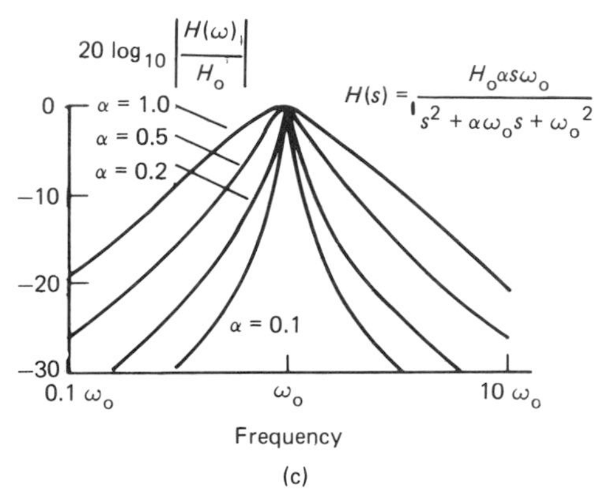

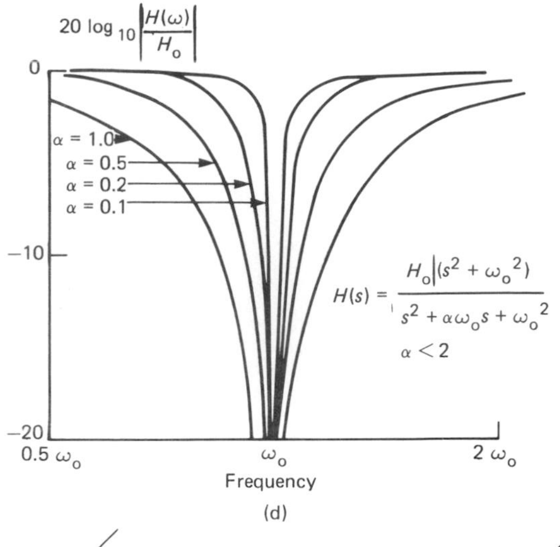

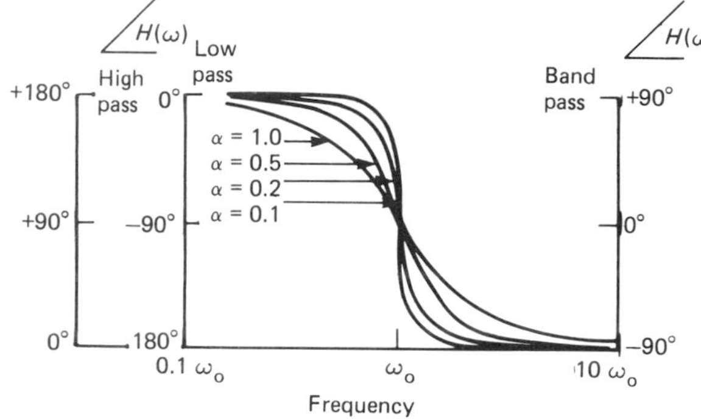

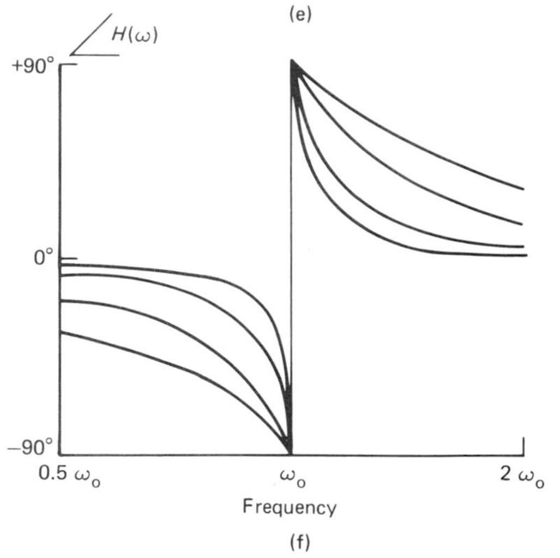

Figure 4.23 Amplitude and phase relations for filters. (a) low-pass, (b) high-pass, (c) bandpass and (d) bandstop filters employing two complex conjugate poles. (e) Phase relations corresponding to (a) (b) and (c); (f) phase relations corresponding to (d).

delay characteristic. Filters based on the Thompson approximation are of importance in digital transmission systems. Bowron and Stephenson (1979) give methods for locating the poles for low-pass filters using all the above approximations together with transformation techniques to enable the locations of the poles for the other filter types to be obtained. Tobey *et al.* (1971) give tables of the poles for low-pass Butterworth, Bessel and Chebyshev filters up to the tenth order.

Filters of order higher than two can be produced by cascading a series of active filters. Operational

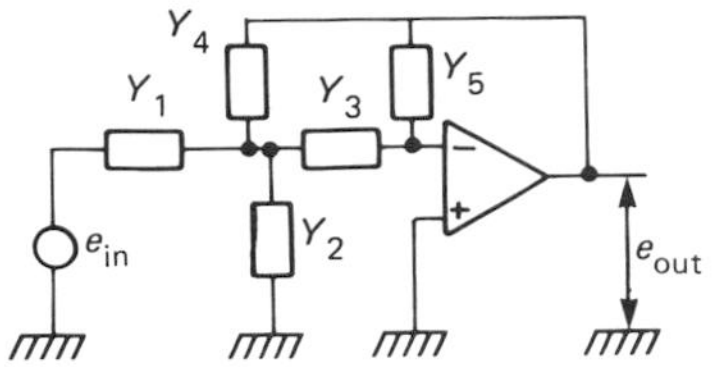

$$\frac{e_{out}(s)}{e_{in}(s)} = \frac{-Y_1 Y_3}{Y_5 (Y_1 + Y_2 + Y_3 + Y_4) + Y_3 Y_4}$$

Y_1, Y_2, Y_3, Y_4, Y_5 represent admittances of resistances and capacitances

	Element				
	1	2	3	4	5
Low pass	*R*	*C*	*R*	*R*	*C*
High pass	*C*	*R*	*C*	*C*	*R*
Bandpass	*R*	*R*	*C*	*C*	*R*

(a)

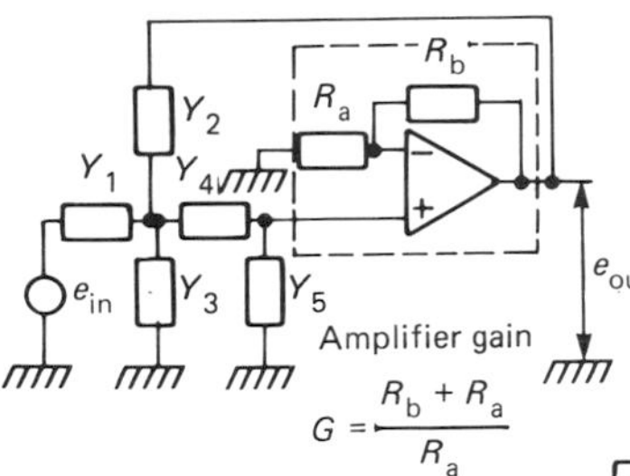

$$\frac{e_{out}(s)}{e_{in}(s)} = \frac{G Y_1 Y_4}{Y_5 (Y_1 + Y_2 + Y_3) + Y_4 (Y_1 + Y_2 [1-G] + Y_3)}$$

Y_1, Y_2, Y_3, Y_4, Y_5 represent admittances of resistances and capacitances

G is gain of non-inverting amplifier

	Element				
	1	2	3	4	5
Low pass	*R*	*C*	–	*R*	*C*
High pass	*C*	*R*	-	*C*	*R*
Bandpass	*R*	*R*	–	*C*	*R*/*C*

(b)

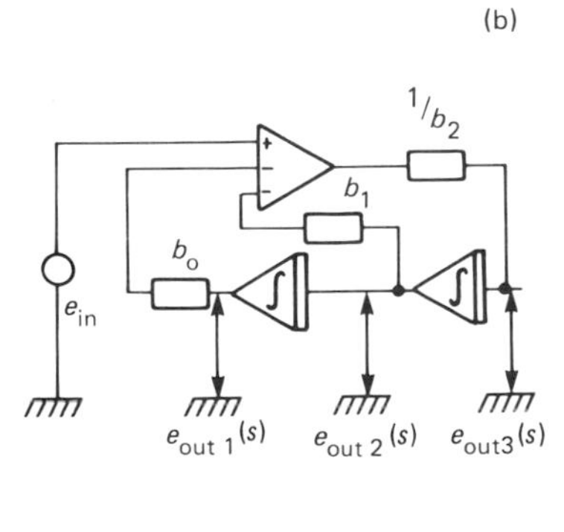

Low-pass filter:

$$\frac{e_{out\,1}(s)}{e_{in}(s)} = \frac{1}{b_o + b_1 s + b_2 s^2}$$

Bandpass filter:

$$\frac{e_{out\,2}(s)}{e_{in}(s)} = \frac{s}{b_o + b_1 s + b_2 s^2}$$

High-pass filter:

$$\frac{e_{out\,3}(s)}{e_{in}(s)} = \frac{s^2}{b_o + b_1 s + b_2 s^2}$$

(c)

Figure 4.24 Realization of two-pole complex conjugate filters using operational amplifiers. (a) Multiple feedback circuit; (b) voltage-controlled voltage source realization; (c) state-space realization.

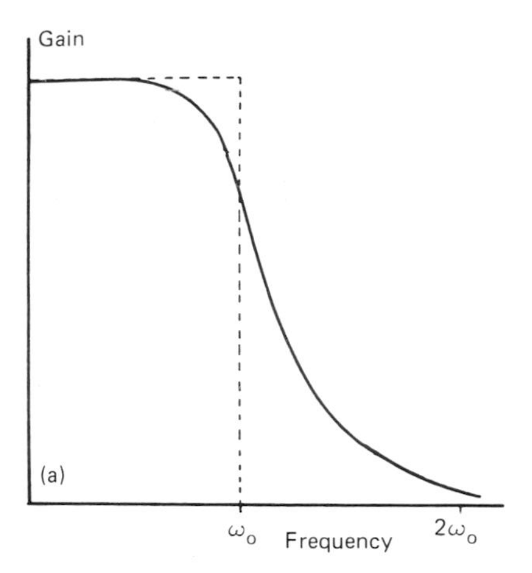

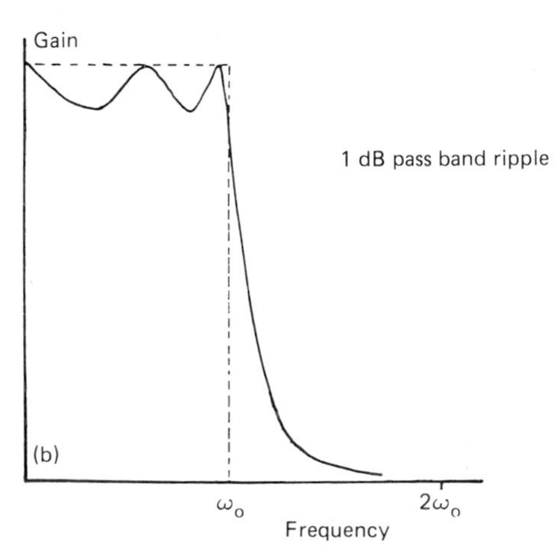

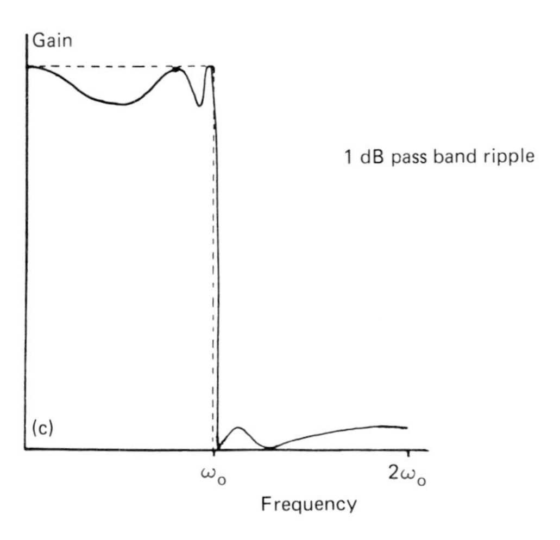

Figure 4.25 Fifth-order (a) Butterworth, (b) Chebyshev and (c) elliptic approximations to the low-pass filter.

amplifiers are ideally suited to such applications, since the low output impedance of the operational amplifier allows the filters to be cascaded with little interaction.

4.2.7.1 The Shannon sampling theorem and anti-aliasing filters

Data-transmission systems often operate by sampling the data from the transducer system, transmitting the sampled form of the data and reconstructing the original data at the other end of the data-transmission medium. The process of sampling the data modifies its spectrum as shown in Figure 4.26(a). If the original data has a maximum frequency content f_d, the question arises as to what is the slowest speed that the data can be sampled at for it to be still possible to reconstruct the data from the samples. The answer to this question is given by the Shannon sampling theorem, which states that for a band-limited signal, having frequencies up to f_d, in order to sample the data in such a manner that it is possible to reconstruct the data from the sampled form it is necessary to sample it with a frequency of at least $2f_d$. This is a theoretical limit and, as a

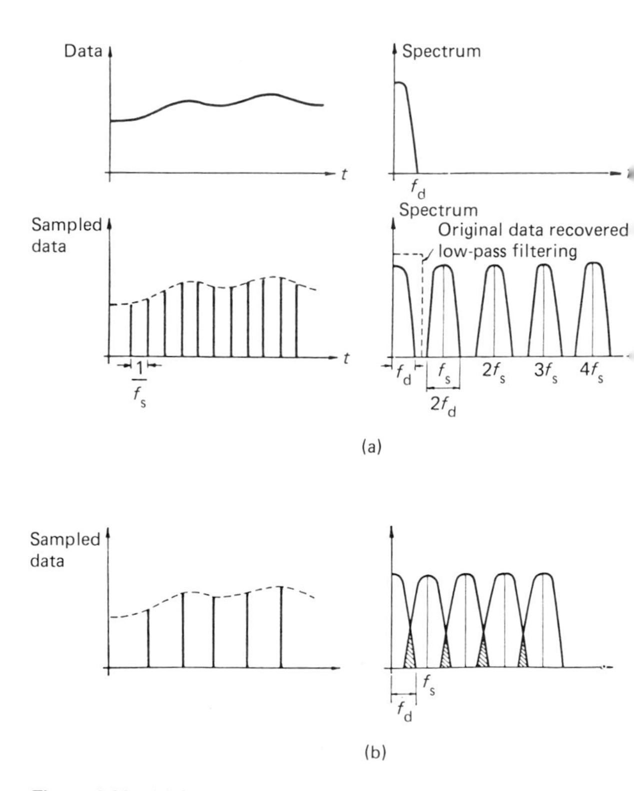

Figure 4.26 (a) Spectrum of sampled data; (b) aliasing errors in sampled data.

general rule of thumb, it is usual to sample signals with a frequency at least ten times the maximum frequency of interest. Figure 4.26(b) shows the effect of disobeying the Shannon sampling theorem. It can be seen that it is no longer possible to reconstruct the data by the application of a low-pass filter onto the sampled data. This effect caused by allowing the signal to have frequency components which have frequencies in excess of $f_s/2$ is called aliasing. It is usually prevented by the insertion of low-pass anti-aliasing filters prior to sampling the signal. The break frequencies of these filters are set according to the sampling rate.

4.2.8 Non-linear analogue signal processing

Although increasingly non-linear signal processing is provided by digital means, integrated circuits are available which enable the processing shown in Figures 4.27 and 4.28 to be applied to analogue signals.

Figure 4.27(a) shows the schematic diagram of a four-quadrant multiplier circuit. Four-quadrant devices allow both inputs to have either positive or negative polarity. The multiplier operates by converting the two input voltages to be multiplied into currents and generating an output current which is the ratio of the product of the two input currents to a reference current. Such multipliers provide typical multiplication errors of order 0.5 per cent of full scale and operate at frequencies in excess of 10 MHz. They can be used as modulators and demodulators, as gain control elements and in power measurement.

The multiplier can also be used to provide division and square rooting, as shown in Figures 4.27(b) and 4.27(c). The divider uses the multiplier in a feedback configuration. Division enables fixed or variable gain elements to be constructed and

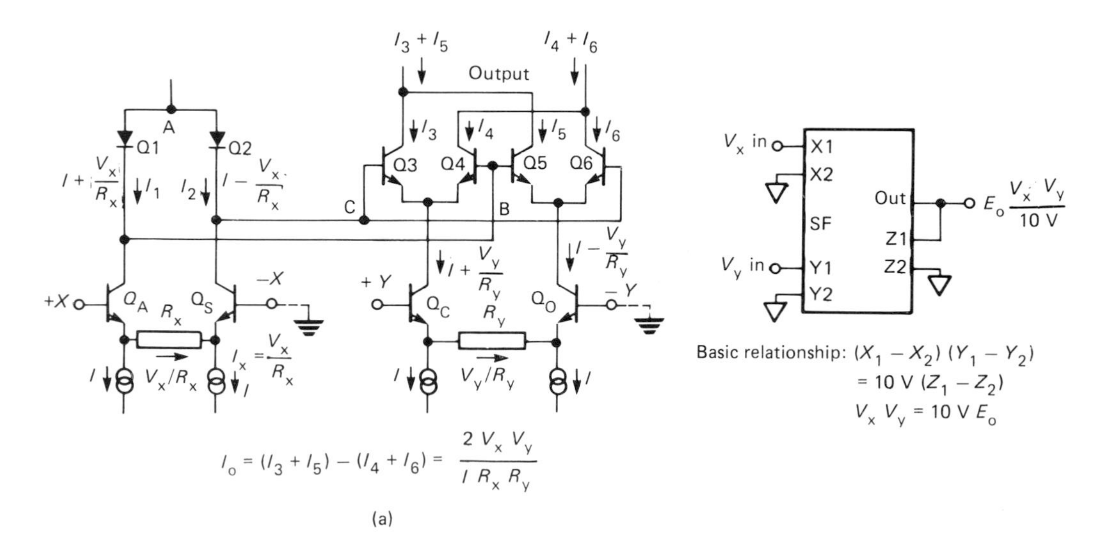

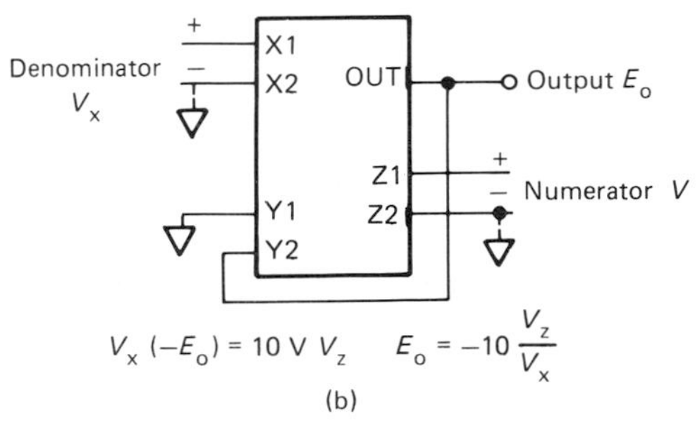

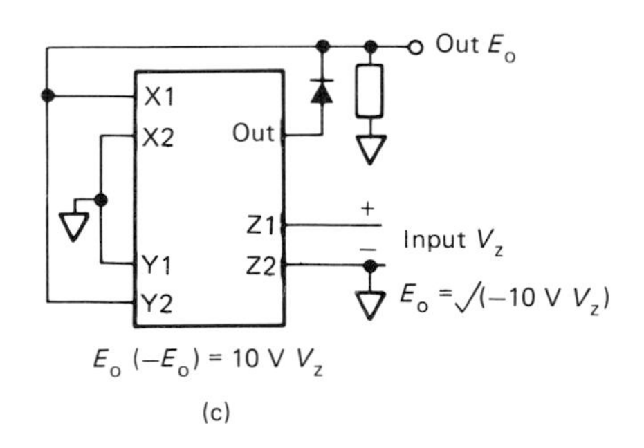

Figure 4.27 (a) Analogue multiplier; (b) analogue divider; (c) analogue square root extractor.

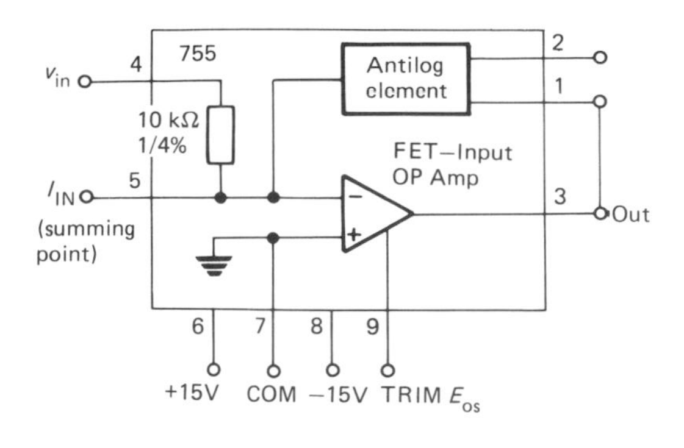

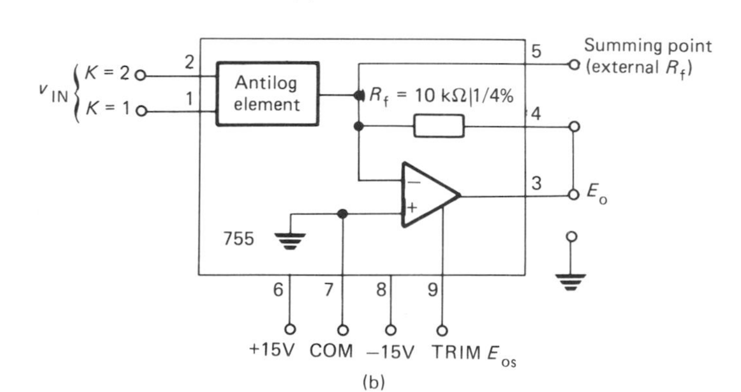

Figure 4.28 (a) Logarithmic amplifier; (b) antilogarithmic amplifier.

ratiometric measurements to be made. The square-root extractor, which is a single quadrant device allowing inputs of only one polarity, can be used in vector amplitude and r.m.s. computations and in linearizing the outputs from flowmeters based on differential pressure devices.

Integrated circuits are also available providing logarithmic and antilogarithmic amplification. They generally consist, as shown in Figures 4.28(a) and 4.28(b), of an operational amplifier and an exponential transconductance element. This element is usually provided by using the exponential current–voltage relationship of a transistor emitter–base junction. For logarithmic operation the element is applied in the feedback loop of the operational amplifier. In antilog operation the input voltage is applied directly to the input of the antilog element, which generates a current proportional to the exponential of the applied voltage. This current is then converted by the feedback resistance to a voltage at the output of the operational amplifier. Log-antilog amplifiers are available with a six-decade logarithmic conversion range and a typical log conformity error of 1.0 per cent. Logarithmic amplification is used for signal compression and for linearizing transducers with an exponential output characteristic. Antilogarithmic amplification is used to linearize transducers with a logarithmic output characteristic and in the expansion of compressed data.

4.3 Digital-to-analogue conversion

Decimal numbers can be represented in binary form as a series of bits which can take the value 1 or 0. Thus the number 191 is stored as an eight-bit binary number, as shown in Figure 4.29(a), and in general a decimal number, N, is represented in binary form by:

$$N = \sum_{n=0}^{M-1} a_n 2^n \tag{4.27}$$

where M is the number of bits used to represent the number and the constants a and n take the value 1 or 0.

To convert the digital number into an analogue voltage v_{out} requires the production of a voltage given by:

$$v_{out} = \frac{v_{ref}}{k} \sum_{n=0}^{M-1} a_n . 2^n \tag{4.28}$$

: significant MSB)	a_7	a_6	a_5	a_4	a_3	a_2	a_1	a_0	Least significant bit (LSB)
hting	2^7	2^6	2^5	2^4	2^3	2^2	2^1	2^0	
mal 191	1	0	1	1	1	1	1	1	

$$191 = 1.2^7 + 0.2^6 + 1.2^5 + 1.2^4 + 1.2^3 + 1.2^2 + 1.2^1 + 1.2^0$$

(a)

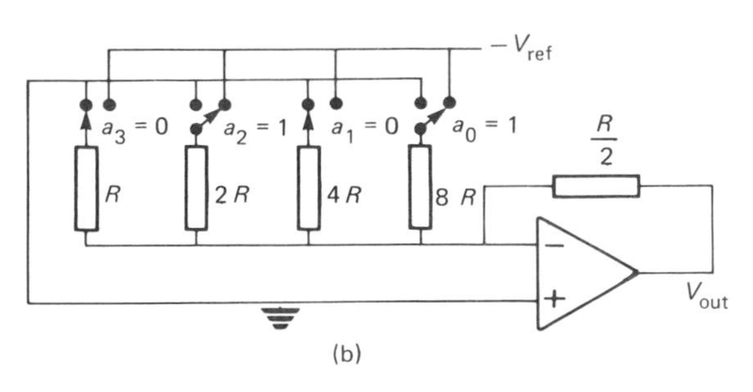

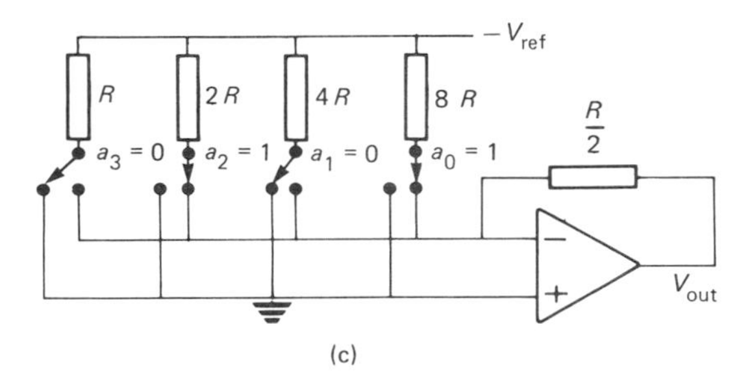

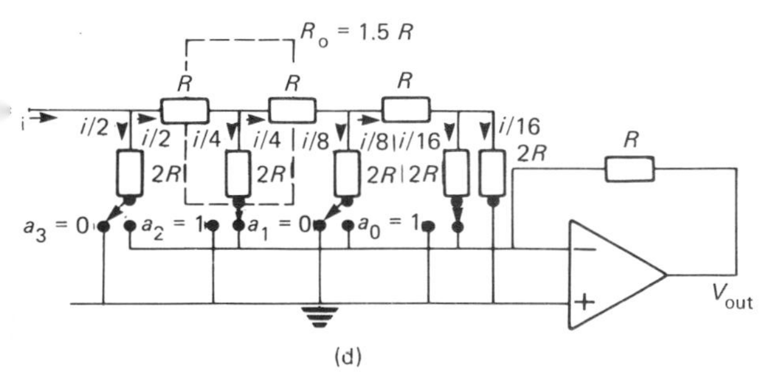

Figure 4.29 (a) Binary representation of a digital number; (b) voltage-switching DAC; (c) current-switching DAC; (d) *R-2R* ladder network.

where v_{ref} is the reference voltage of the digital-to-analogue converter (DAC) and k is a scaling constant. Thus an output voltage is produced by adding a series of weighted voltages together.

Figure 4.29(b) shows a voltage switching DAC using a weighted ladder network. The output from this device is given by:

$$v_{out} = v_{ref} \sum_{n=0}^{M-1} a_n.2^{(n-M)} \tag{4.29}$$

The transient response of such a device depends largely on charging and discharging parasitic capacitances, and can be improved by using current switching instead of voltage switching. This is shown in Figure 4.29(c). In current switching the potentials of all the switch points are maintained irrespective of whether a particular element is representing a 1 or a 0. One major disadvantage of the weighted ladder technique is the spread of resistance values which are required. Even for an 8-bit DAC there is a 256:1 spread in the resistance values, and the problem gets progressively worse as the number of bits in the DAC increases. It is difficult to produce sets of resistors which are accurately matched and with temperature coefficients which track for such a large spread in values. Figure 4.29(d) shows the *R*-2*R* ladder solution to this problem. Since the iterative impedance of the network shown within the dotted lines is 1.5*R* then there is an equal division of the current at the indicated nodes. Thus the *R*-2*R* ladder technique achieves the required weightings with a spread of resistance values of 2:1, irrespective of the number of bits in the converter. This technique forms the basis of most commercially available DACs. Digital-to-analogue conversion employing capacitive rather than resistive divider techniques suitable for fully integrated DACs employing MOS fabrication techniques are available.

DACs are available in which the reference voltage v_{ref} is no longer fixed but can be varied, and thus the output can be scaled by adjusting v_{ref}. This provides multiplication of the digital input by an analogue value. Such devices are called 'multiplying DACs'. DACs having a non-linear transfer characteristic are used in signal compression/expansion systems in order to allow efficient representation of analogue signals having a wide dynamic range. Devices such as the Precision Monolithic DAC-76 provide a dynamic range of 72 dB equivalent to a sign + 12-bit DAC in a sign + 7-bit format. When used in a feedback configuration in an analogue-to-digital converter such a DAC provides logarithmic conversion of the signal.

4.3.1 Accuracy of DACs

Since each digital code applied to the DAC should give rise to a unique analogue output it is possible to specify an accuracy of the DAC in terms of the error between the theoretically determined analogue output for that particular digital code and the actual analogue output. The error is caused by such factors as gain or calibration errors, zero offset errors and non-linearity errors. Two accuracies are usually specified by manufacturers—absolute accuracy and relative accuracy. Absolute accuracy is a measure of the deviation of the output of the DAC from the theoretical value for that digital code. Relative accuracy is the deviation from the theoretical value after the full-scale range has been calibrated. Both of these accuracies are specified as a percentage or parts per million (p.p.m.) of full-scale range or fractions of the least significant bit (LSB).

The ideal transfer characteristic of a DAC should be a series of analogue values, each one corresponding to a digital value, which, when plotted as shown in Figure 4.30(a), lie on a straight line. Gain or calibration error will mean that the slope of the transfer characteristic between the analogue output and the digital input will not have the correct slope. The zero error will have the effect that the digital code corresponding to zero will not give rise to zero output. These errors are shown in Figure 4.30(b).

The actual output from a DAC when plotted in the same manner as shown in Figure 4.30(a) will have a transfer characteristic which, after the gain and zero errors are removed, will not be a straight line. There are two measures of non-linearity—integral and differential. These are shown in Figures 4.30(c) and 4.30(d), respectively. Integral non-linearity is the deviation between the analogue output at a point and a straight-line transfer characteristic. The straight line can be chosen in several ways. It may simply be the straight line joining the end points of the transfer characteristic or it may be a line constructed in such a manner as to equalize the maximum positive and negative deviations from the line. As shown in Figure 4.30(c), by carefully choosing the line it is possible to improve

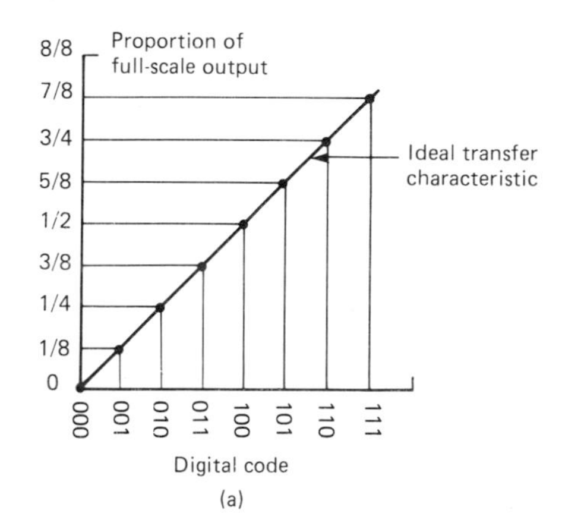

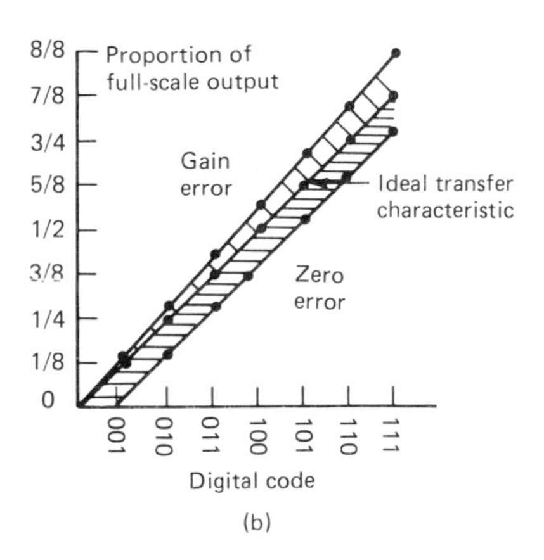

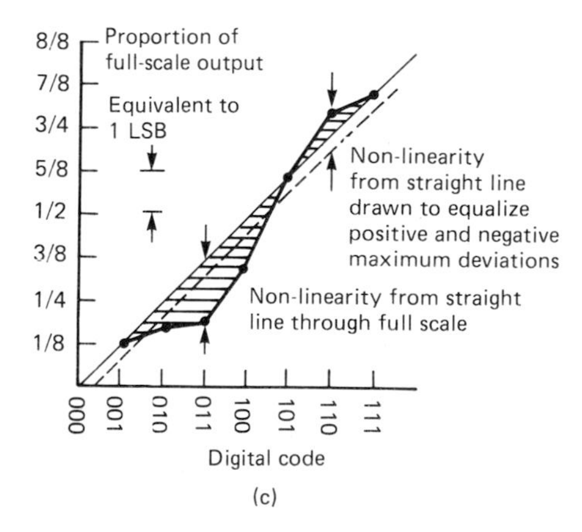

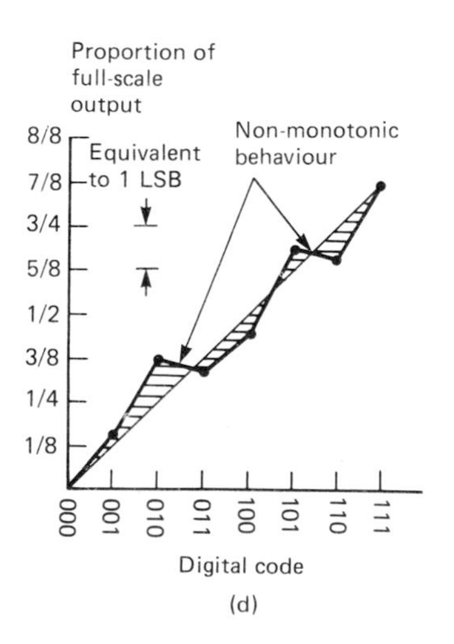

Figure 4.30 (a) Ideal transfer characteristic for DAC; (b) gain and zero error in a DAC; (c) integral non-linearity; (d) differential non-linearity.

the integral non-linearity specification. Integral non-linearity is expressed as a percentage or p.p.m. of full-scale range or as fractions of 1 LSB.

Ideally, as the digital input changes by 1 LSB the analogue output should change by a voltage equivalent to 1 LSB, no matter where on the digital input scale that change is effected. Differential non-linearity is the deviation of the change in the output, from the theoretical analogue output voltage equivalent to 1 LSB, as the digital input is changed by 1 LSB. It is usually expressed in fractions of 1 LSB. In order to ensure that the analogue output increases or remains constant as the digital input is increased, it is necessary that the differential non-linearity should be less than 1 LSB. If the differential non-linearity of a DAC is less than 1 LSB then the DAC is said to be monotonic. Monotonicity is particularly important when a DAC is used in an analogue-to-digital converter since non-monotonicity leads to missed codes in the analogue-to-digital conversion. It is also important in control applications where non-monotonicity can lead to 180 degrees of phase shifting of signals within the control loop, and can thus produce stability problems.

DACs are specified primarily by the number of bits they convert, the accuracy and linearity of conversion and the conversion time, which is typically measured by its settling time. This is the time required, following a prescribed data change, for the output of the device to reach and remain within a given fraction, usually ±½LSB, of its final value. The usual prescribed changes being full scale, 1 most significant bit (MSB) and 1 LSB at a major carry.

Table 4.4 DAC specification

Number of bits	8
Relative accuracy	±0.1% of FS
Monotonicity	Guaranteed
Settling time	250 ns
V_{ref}	−16.5 to +5.5 V
Output current	2 mA FS
Temperature coefficient of gain	20 p.p.m./K
CMOS or TTL compatible	

The output from a DAC may be either voltage or current and may be either unipolar or bipolar.

Manufacturers' specifications will also include the reference input, which can be internal or external, fixed or variable, single polarity or bipolar; the coding and data levels of the digital input of the DAC; the level, sense, loading and timing of control signals applied to the DAC; power supply requirements; output noise; offset; and the temperature coefficients of gain, linearity and offset. For further details of these the reader should consult Analog Devices (1984). Table 4.4 shows a typical specification for a medium-speed 8-bit DAC.

4.4 Analogue-to-digital conversion

Analogue-to-digital converters (ADCs) convert an analogue signal whose amplitude can vary continuously into a digital form which has a discrete number of levels. The number of levels is fixed by the number of bits employed in the conversion, and this sets the resolution of the conversion. For a binary code having n bits there are 2^n levels. Since the digital representation is discrete, there is a range of analogue values which all have the same digital representation. There is thus a quantization uncertainty in analogue-to-digital conversion of ±½LSB, and this is in addition to any other errors which may occur in the conversion itself. The methods which are available for analogue-to-digital conversion split into four general methods. These are voltage-to-time conversion, voltage-to-frequency conversion, feedback methods employing DACs and simultaneous comparison techniques (Owens, 1983; Arbel, 1980; Sheingold, 1977).

4.4.1 Voltage-to-time conversion

These techniques all employ some method of converting the voltage measurement into a corresponding time measurement. The available methods include voltage-to-pulsewidth conversion; single-slope conversion (which is sometimes called single-ramp conversion); and dual-slope conversion (which is also known as dual-ramp conversion). The voltage-to-pulsewidth converter in Figure 4.31(a) employs a voltage-controlled monostable to produce a pulse whose width is proportional to the input voltage. The width of this pulse is then measured by means of the reference clock. Thus the counter has within it at the end of the conversion a binary number which corresponds to the analogue input.

Two different single-ramp converters are shown in Figures 4.31(b) and 4.31(c). In Figure 4.31(b) the voltage to be measured is applied to the reference terminal of the comparator. The time measured is that for the ramp to climb from 0 V to the input voltage v_{in}. The start pulse enables the clock gate and the stop pulse, which is derived from the output of the comparator, disables it. This time is measured using the reference clock. In the converter in Figure 4.31(c) the input voltage to be measured is stored on a capacitor, and the time to be measured is that required to discharge the capacitor using a constant current source.

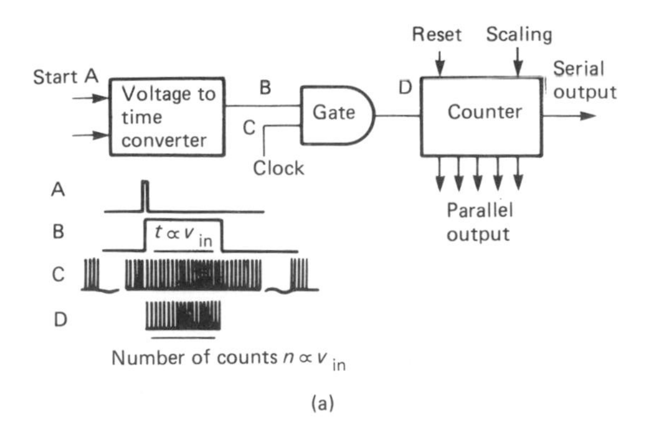

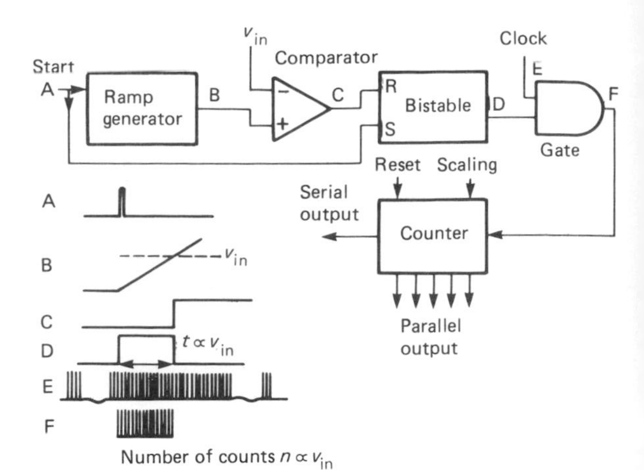

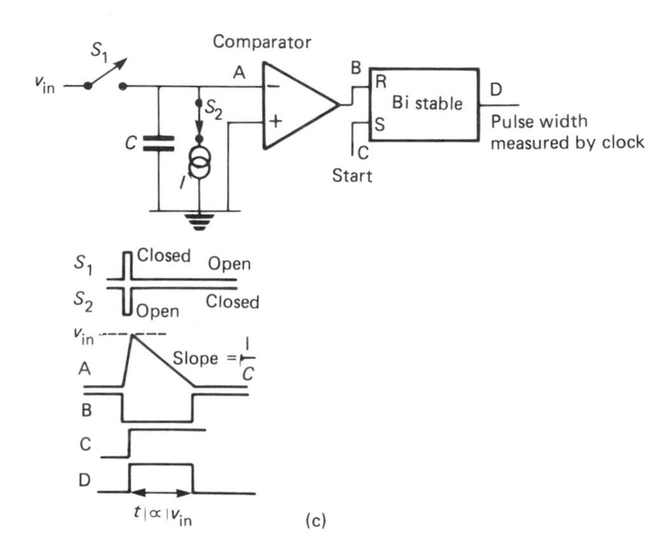

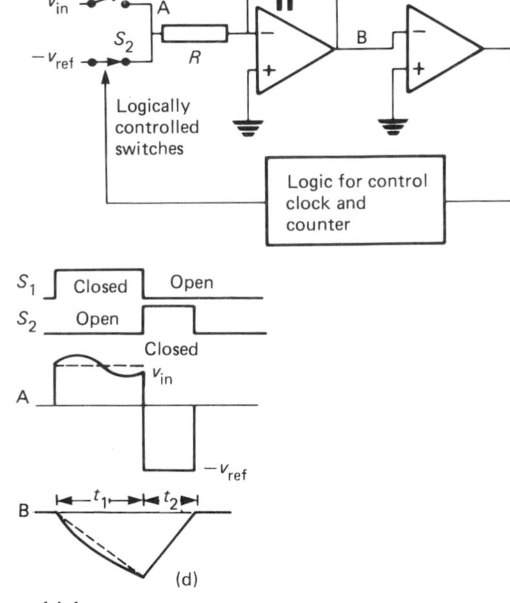

Figure 4.31 (a) Voltage-to-pulsewidth converter; (b) and (c) single-ramp converter; (d) dual-ramp converter.

The dual-ramp conversion technique shown in Figure 4.31(d) has the advantage of line frequency signal rejection. The operation of the converter is as follows.

The input voltage is electronically switched to the input of the integrator for a fixed period of time t_1, after which time the integrator output will be $-v_{in}t_1/RC$. The reference voltage $-v_{ref}$ is then applied to the integrator and the time is then measured for the output of the integrator to ramp back to zero. Thus:

$$\frac{v_{in}.t_1}{RC} = \frac{v_{ref}.t_2}{RC} \tag{4.30}$$

from which:

$$t_2 = \frac{v_{in}}{v_{ref}}.\, t_1 \tag{4.31}$$

If t_1 is a fixed number of counts, n_1, of a clock having a period t_c and t_2 is measured with the same clock, say, n_2 counts, then:

$$n_2 = \frac{v_{in}}{v_{ref}}.\, n_1 \tag{4.32}$$

The R and C components of the integrator do not appear in the defining equation of the ADC, neither does the frequency of the reference clock. The only variable which does appear in the defining equation is the reference voltage. The effect of offset voltage on the comparator will be minimized as long as its value remains constant over the cycle, and also providing it exhibits no hysteresis. Modifications of the technique employing quad slope integrators are available which reduce the effects of switch leakage current and offset voltage and bias current to second-order effects (Analog Devices, 1984). Errors caused by non-linearity of the integrator limit the conversion using dual-ramp techniques to five decades.

The rejection of line frequency is achieved as follows. If the input is a d.c. signal with an a.c. interference signal superimposed on it:

$$v_{in} = v_{d.c.} + v_{a.c.} \sin(\omega t + \phi) \tag{4.33}$$

where ϕ represents the phase of the interference signal at the start of the integration, then the value at the output of the integrator, v_{out}, at the end of the period t_1 is given by:

$$v_{out} = -\frac{v_{d.c.} t_1}{RC} - \frac{1}{RC}\int_0^{t_1} v_{a.c.} \sin(\omega t + \phi).dt \tag{4.34}$$

If the period, t_1 is made equal to the period of the line frequency then the integral of the line frequency signal or any harmonic of it over the period will be zero, as shown in Figure 4.31(d).

At any other frequency it is possible to find a value of ϕ such that the interference signal gives rise to no error. It is also possible to find a value ϕ_{max} such that the error is a maximum. It can be shown that the value of ϕ_{max} is given by:

$$\tan \phi_{max} = \frac{\sin \omega t_1}{(1 - \cos \omega t_1)} \tag{4.35}$$

The series mode rejection (SMR) of the ADC is given as the ratio of the maximum error produced by the sine wave to the peak magnitude of the sine wave. It is normally expressed in decibels as

series mode rejection (SMR) =

$$-20 \log_{10} \frac{\omega t_1}{\cos\phi_{max} - \cos(\omega t_1 + \phi_{max})} \tag{4.36}$$

A plot of the SMR of the ADC is shown in Figure 4.32. It can be seen that ideally the dual slope ADC provides infinite SMR for any frequency given by n/t_1, n = 1,2,3... . Practically, the amount of rejection such an ADC can produce is limited because of non-linear effects, due to the fact that the period t_1 can only be defined to a finite accuracy and that the frequency of the signal to be rejected may drift. However, such a technique can easily provide 40 dB of line-frequency rejection.

This technique is commonly used when low-speed conversion is required. Because of the integrating nature of the technique it is not suitable for rapidly changing or multiplexed signals. It is commercially available in integrated form, and forms the basis for most digital voltmeters.

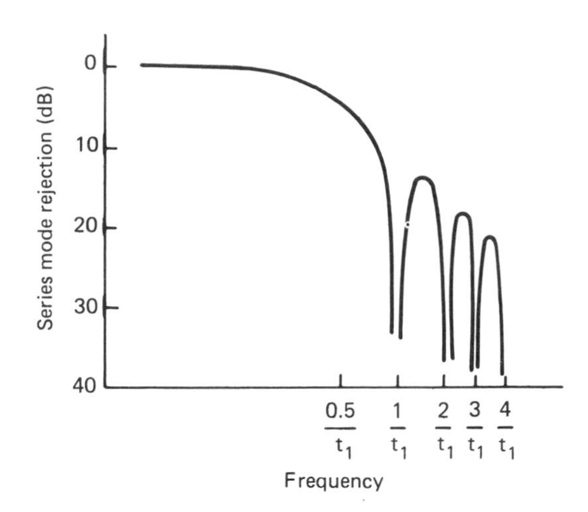

Figure 4.32 Series mode rejection for a dual-ramp converter.

4.4.2 Voltage-to-frequency conversion

Such techniques employ circuitry which converts the input voltage into a frequency which is linearly related to the input voltage. The analogue signal is converted to its digital equivalent by gating the frequency signal over a fixed period of time. Frequency transmission is often employed as a method of transmitting analogue data. A simple form of voltage-to-frequency conversion is shown in Figure 4.33(a). The input signal is integrated for the time that it takes the integrator output to go from 0 to $-v_{ref}$, i.e. for a time given by:

$$t = \frac{RCv_{ref}}{v_{in}} \tag{4.37}$$

If the discharge time for the capacitor is small, then the repetition frequency of the converter is given by:

$$f = \frac{v_{in}}{RCv_{ref}} \tag{4.38}$$

The accuracy of the conversion can be improved by using the frequency-to-voltage converter in a feedback configuration with a voltage-to-frequency converter (Owens, 1983).

Figure 4.33(b) shows a charge balance converter in which the output of the integrator is periodically reset, not by shorting it out but by injecting a charge onto it in the form of a current pulse containing a fixed charge. Thus over one cycle of operation the charge provided to the feedback capacitor by the current v_{in}/R exactly balances that supplied by the current pulse:

$$It_1 = \frac{v_{in}t_2}{R} \tag{4.39}$$

and

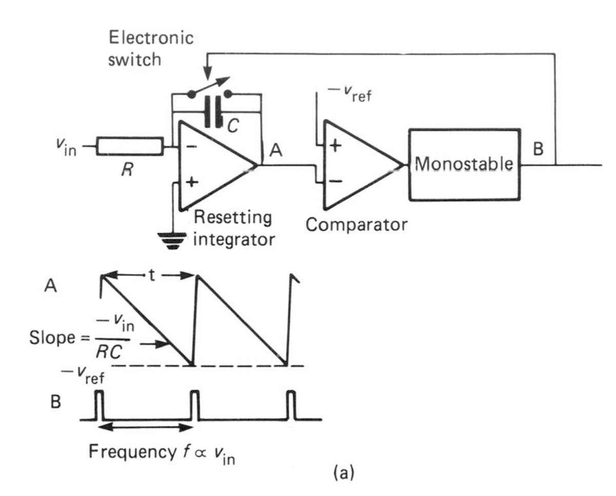

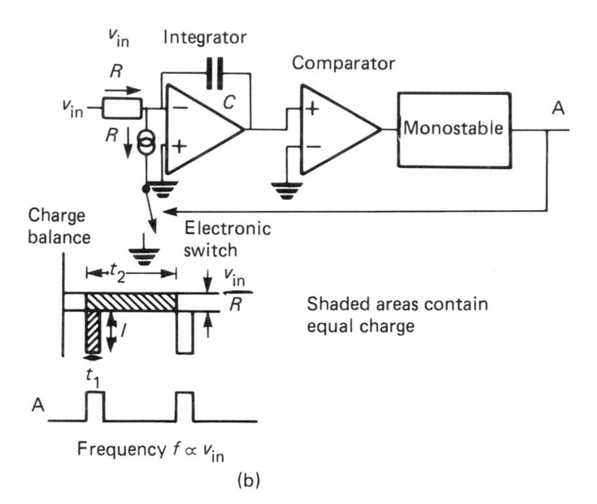

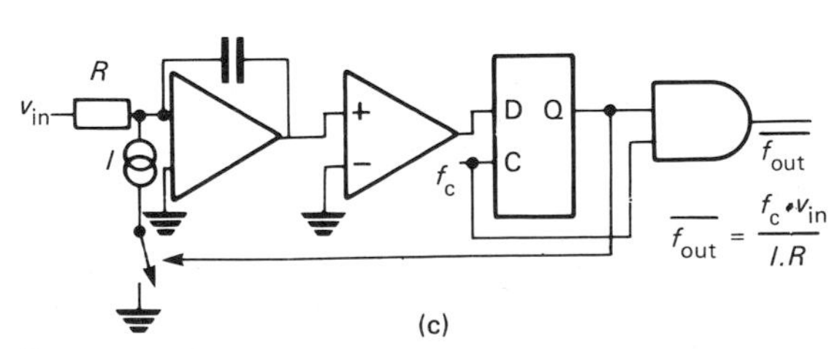

Figure 4.33 (a) Voltage-to-frequency converter; (b) charge balance converter; (c) delta–sigma modulation.

$$f = \frac{1}{t_2} = \frac{v_{in}}{R.I.t_1} \tag{4.40}$$

Delta–sigma modulation (Steele, 1975) is a form of voltage-to-frequency conversion which is often used in speech transmission. This is shown in Figure 4.33(c).

4.4.3 Feedback methods employing DACs

Feedback techniques adjust the digital input of a DAC in such a way as to find the digital input for the DAC whose analogue output most closely corresponds to the input voltage which is to be measured.

In the ramp or staircase generator shown in Figure 4.34(a) a counter is used to set the input of the DAC. Initially, the digital input is set to zero, and therefore the analogue output of the DAC is zero. The counter is then incremented by the clock. As the input to the DAC is incremented, the analogue output of the DAC also steps in value. Stepping continues until the output of the DAC exceeds the input voltage. The comparator output switches and inhibits further clock pulses from being fed into the counter. The counter now contains the digital equivalent of the analogue input. The conversion speed for this converter is dependent on the level of the input signal. The speed of conversion of slowly varying signals can be improved by the use of additional hardware and an up–down counter to produce a tracking ADC which will follow slowly changing signals.

The successive approximation technique shown in Figure 4.34(b) employs a decision-tree approach to the problem. The control circuitry on the first cycle of the conversion sets the MSB of the DAC to 1 and all the rest of the bits to 0. The output of the comparator is examined. If it is a 1, implying that the analogue input is greater than the output of the DAC, then the MSB of the DAC is maintained at a 1, otherwise it is changed to a 0. The next cycle determines whether the next most significant bit is a 1 or a 0. This process is then repeated for each bit of the DAC. The conversion period of the successive approximation analogue-to-digital conversion technique is fixed for a given ADC, irrespective of the signal level, and is equal to $n\tau$, where n is the number of bits and τ is the cycle time for determining a single bit. Integrated-circuit successive-approximation logic-generating chips are available to be used in conjunction with standard DACs and comparators to produce medium-speed DACs.

4.4.4 Simultaneous comparison

For high-speed conversion, the successive-approximation technique is often not fast enough,

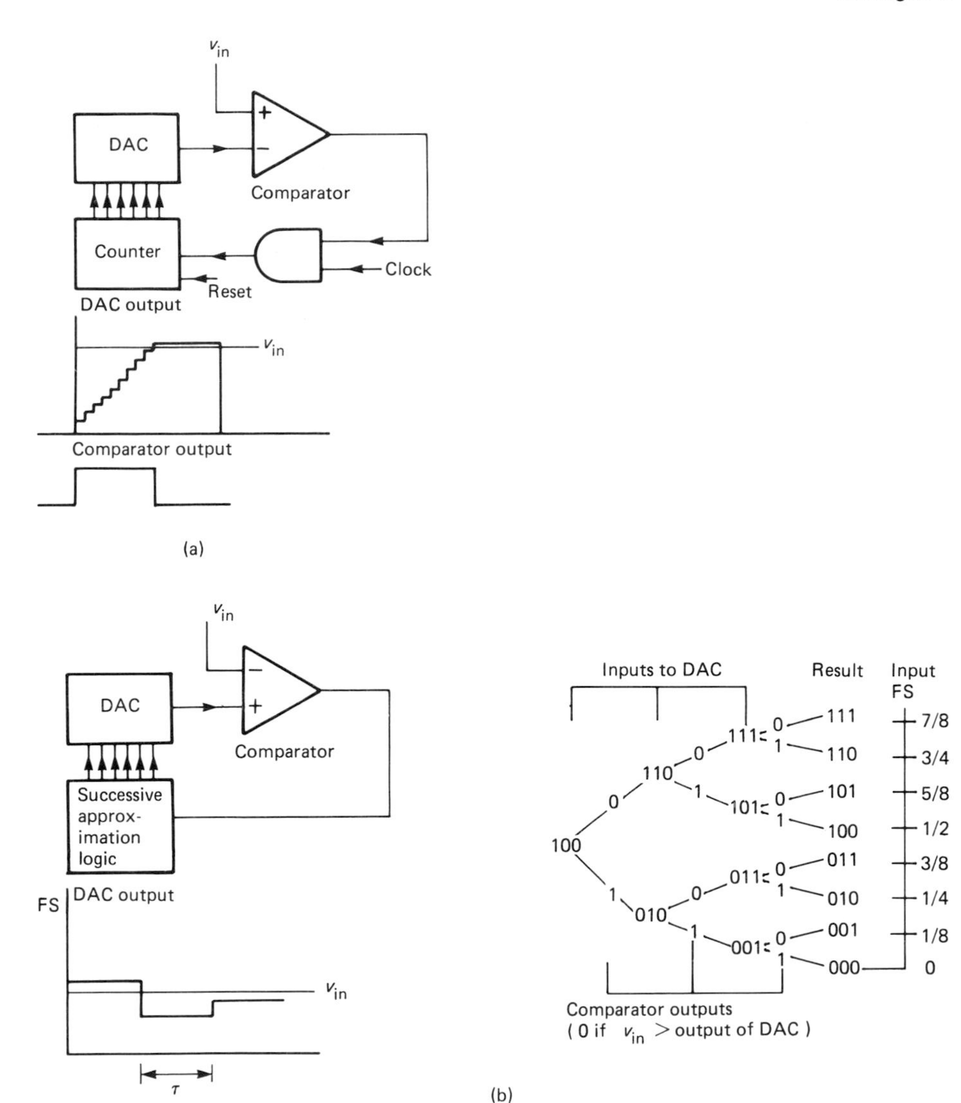

Figure 4.34 (a) Ramp or staircase generator; (b) successive approximation converter.

particularly in the digitization of video signals, and simultaneous comparison techniques are necessary. Such converters are called 'flash converters', since the conversion takes place in one cycle. Figure 4.35 shows the schematic diagram of a flash converter. The analogue values corresponding to each of the digital codes is simultaneously compared with the input. For an n-bit conversion, 2^{n-1} individual comparators are required. The logic then examines the output of the comparators and produces the digital code corresponding to the input. It is possible to reduce the number of comparators required in an n-bit conversion by means of subranging. In the first stage of the conversion the analogue signal is digitized using a set of comparators. This value is then stored in a latch and fed into a high-speed DAC. The output from this DAC is subtracted from the original analogue input and this difference or residue is amplified and redigitized using the same converters. By combining the two digitized values it is possible to produce an n-bit digital number corresponding to the analogue input.

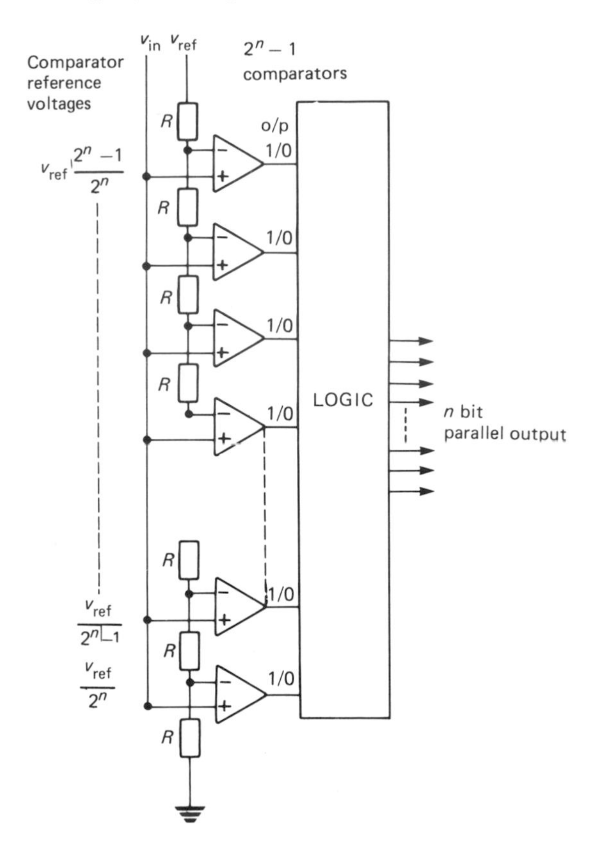

Figure 4.35 Flash converter.

4.4.5 ADC specifications

The most commonly encountered ADCs are dual-slope, successive-approximation and flash converters. These devices can be specified in terms of their analogue, digital and control sections, with the most important performance specifications being the resolution, accuracy and linearity of conversion, together with the conversion time or rate.

The analogue section of the ADC is usually specified in terms of the allowable analogue input signal range together with its associated source impedance. ADCs may accept either unipolar or bipolar inputs.

The digital section is specified by the technology employed and the output coding and format. The digital output may use TTL, CMOS or ECL technology. Unipolar inputs can be represented by binary coding. Bipolar inputs, which can have both positive and negative values, can be represented by a series of codings, including offset binary coding, two's complement coding (which is the technique used for representing negative numbers in most computers) or sign and magnitude coding. ADCs which are used for digital voltmeters often use BCD coding with sign and magnitude coding. Figure 4.36 shows the transfer characteristics of a unipolar ADC and a bipolar ADC using offset binary coding.

The formatting of the output is important from the point of view of interfacing the ADC to other equipment. The most common formats are parallel, serial and byte serial. The byte serial format is used particularly in microprocessor-compatible ADCs, where it may be necessary to communicate the output of a 12-bit ADC to an 8-bit microcomputer.

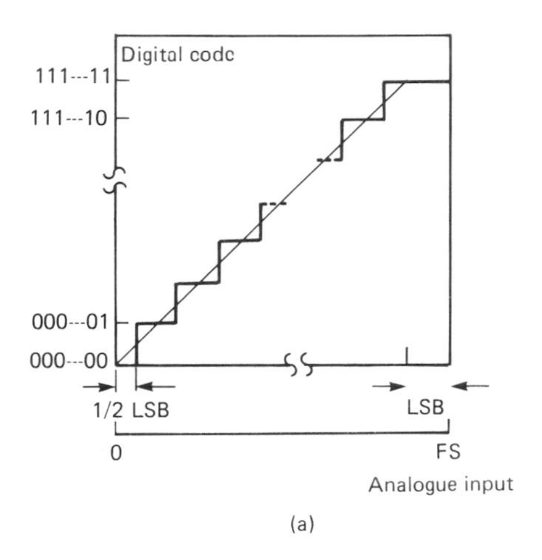

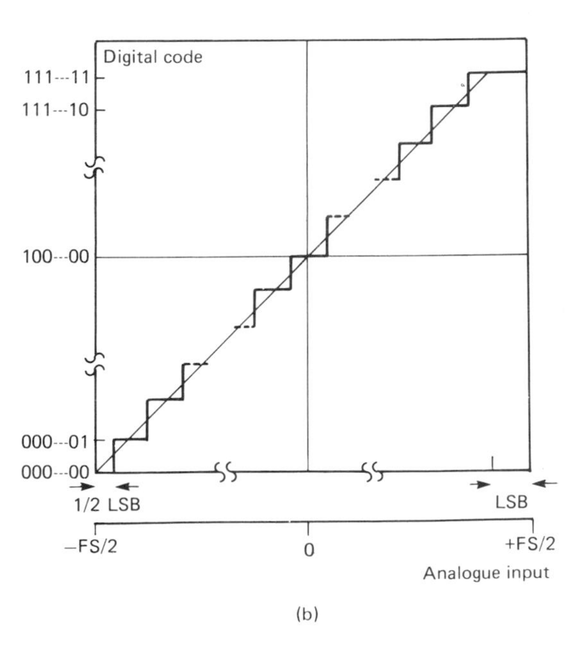

Figure 4.36 (a) Unipolar and (b) bipolar ADCs.

In such a case the data are transferred serially in two bytes, a high-order byte, which contains the more significant bits of the digital representation, and a low-order byte containing the remaining bits to complete the digital representation. For microprocessor applications it is necessary for the digital output of the converter to be tri-stated to provide the necessary isolation.

The control section is typically specified by the signals which are required by the ADC for such functions as start conversion, output the high-order byte, output the low-order byte and the control signals which the ADC gives to indicate the end of conversion.

The resolution provided by an ADC is fixed by the number of bits in the conversion.

The accuracy of the ADC is measured by its absolute and relative accuracies. The absolute error is the difference between the specified and actual values required to produce a given code. Since a range of analogue values will produce a given code, the analogue value is specified as the mid-range value. The error is caused by gain, zero and non-linearity errors together with noise. The relative accuracy is the error once the full-scale range has been calibrated.

The linearity of the converter is measured by its differential and integral non-linearity. Differential non-linearity, as shown in Figure 4.37, measures the difference in the range of values of the analogue input to produce a given code from the range equivalent to 1 LSB. A non-linearity of greater than 1 LSB in an ADC implies a missed code. This can occur particularly when a non-monotonic DAC is used in a feedback network to produce an ADC. If an ADC is specified as having no missed codes then its differential non-linearity is less than 1 LSB. Dual-slope ADCs do not exhibit missed codes. However, they are likely to exhibit integral non-linearity, which is concerned with the overall shape of the converter characteristic.

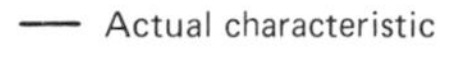

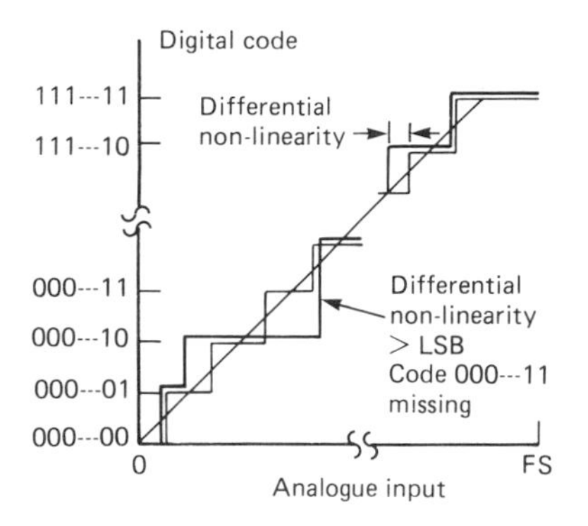

Figure 4.37 Differential non-linearity for ADC

The conversion time of an ADC is the time required for a complete measurement to be made. The conversion rate for most ADCs is the inverse of the conversion time. In high-speed converters pipelining may, however, allow one conversion to commence before the previous one is completed.

Table 4.5 Specification of three ADCs

Device	AD571J	ICL7109	AD5010KD
Technique employed	Successive approximation	Dual ramp	Flash conversion
Number of bits	10	12	6
Linearity	Differential non-linearity ≤1LSB	Non-linearity w.r.t. straight-line fit ≤1LSB	Linearity error ±¼LSB
Analogue input ranges	Unipolar 0 V to +10 V Bipolar −5 to +5 V	+6.2 V to −9 V	±2.5 V
Output coding	Unipolar: positive true binary Bipolar: positive true offset binary	12-bit binary + polarity and over range bits	6 binary + over range bit
Format	Parallel output	Two-byte parallel output UART compatible	Parallel output
Digital inputs/outputs	TTL/CMOS	TTL/CMOS	ECL
Conversion time/rate	25 μs	30 conversion/s	10 ns

Thus the conversion rate can exceed the rate determined from the conversion time.

Other specifications which are given for an ADC include gain, linearity and offset temperature coefficients. Table 4.5 compares typical characteristics of three commonly used converters.

4.5 Spectrum analysers and related equipment

Signal analysis in the frequency domain is used extensively in the testing of mechanical systems and in electronics and telecommunications. In mechanical systems it is used with suitable vibration transducers to analyse the modes of vibration of mechanical structures and to investigate vibrations in rotating machinery brought about by mechanical unbalance or worn bearings or gears.

In electronics and telecommunications, frequency domain analysis is applied to the analysis of both random and periodic signals. Signal analysers can provide measurements of the frequency stability and spectral purity of signal sources. When used in conjunction with a tracking frequency generator or a source of white or pseudo-random noise they can be used to measure the frequency response of amplifiers, filters and other networks. The operational characteristics of transceivers and communications systems are assessed by the measurement of such parameters as the spectral purity of the carrier wave, the spectral power distribution of the amplitude or frequency modulated wave, signal distortion and system signal-to-noise ratios.

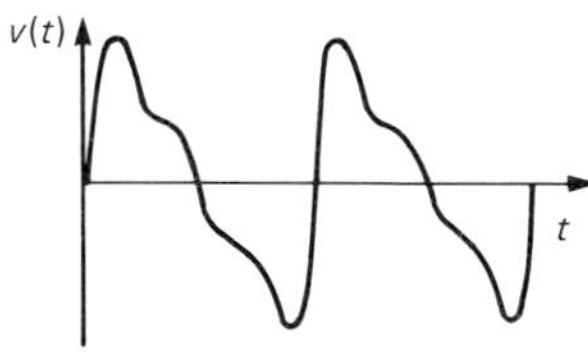

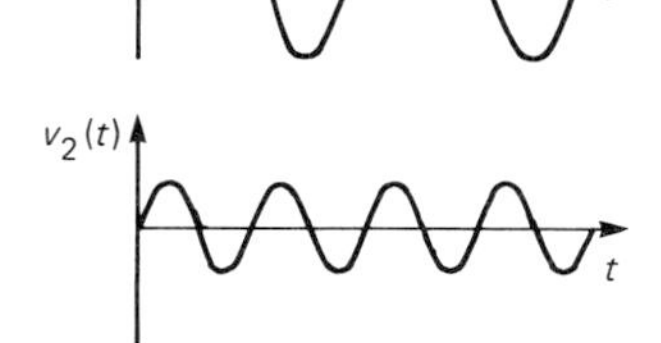

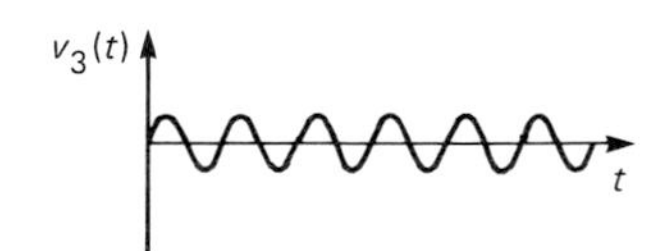

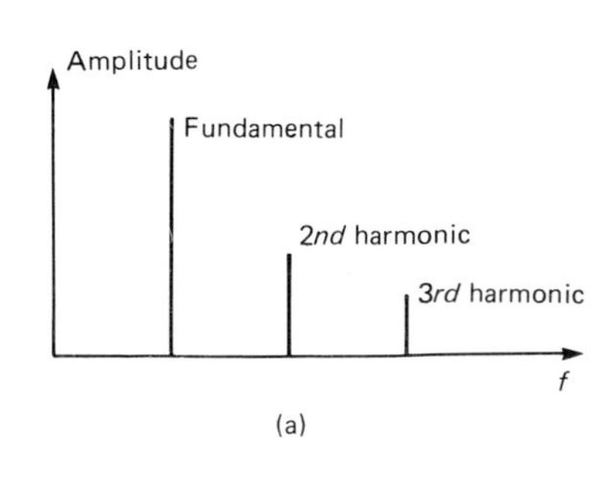

(a)

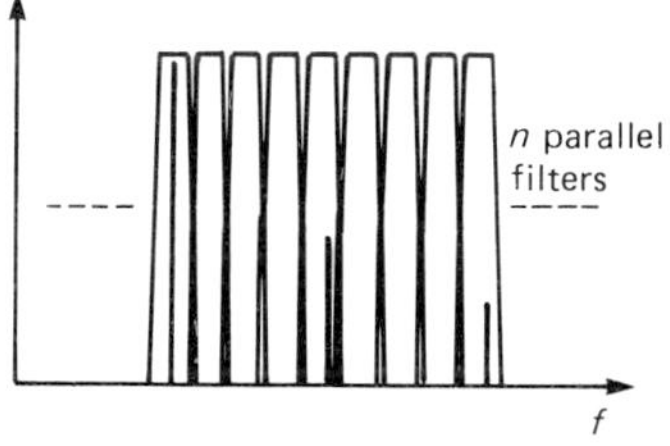

(i) Overlapping filter bank parallel filter bank analyser

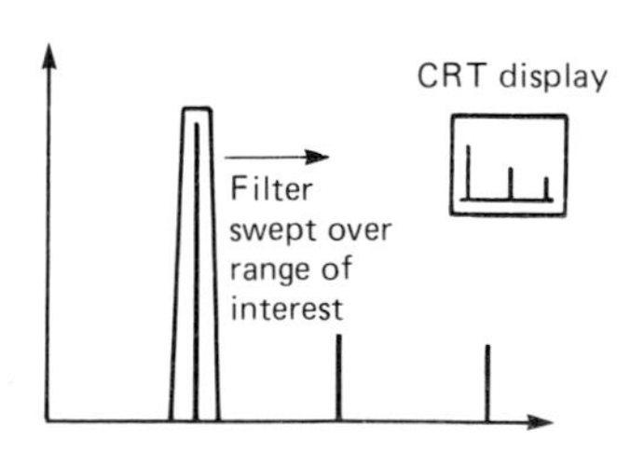

(ii) Swept-frequency analyser

(b)

Such analysis is provided by spectrum analysers, digital Fourier analysers, wave analysers and distortion analysers. Figure 4.38 shows how these different devices measure the various frequency parameters of a deterministic waveform whose time response $v(t)$ is composed of three frequency components $v_1(t)$, $v_2(t)$ and $v_3(t)$.

4.5.1 Spectrum analysers

These instruments provide displays of the frequency spectrum over a given frequency band. Spectrum analysers use either a parallel filter bank or a swept frequency technique. In the parallel filter bank analyser the frequency range is covered by a series of filters whose central frequencies and bandwidths are chosen so that they overlap. Typically, an audio analyser will have 32 of these filters, each covering one third of an octave. For wide band–narrow resolution analysis, particularly of r.f. or microwave signals, the preferred method of measurement is by means of the swept-frequency technique. In such analysers a narrow-band filter is centred around a frequency which is swept over the range for which the display is required. At a particular frequency the absolute value of the signal within that band is measured. The swept-frequency aspect of the spectrum analyser is usually achieved by heterodyning the input signal with the output of a local swept-frequency oscillator. A low-pass filter on the output of the mixer stage fixes the bandwidth of the narrow-band filter whilst the frequency of the local oscillator fixes its central frequency. By the use of a long-persistence CRT or a digital display it is

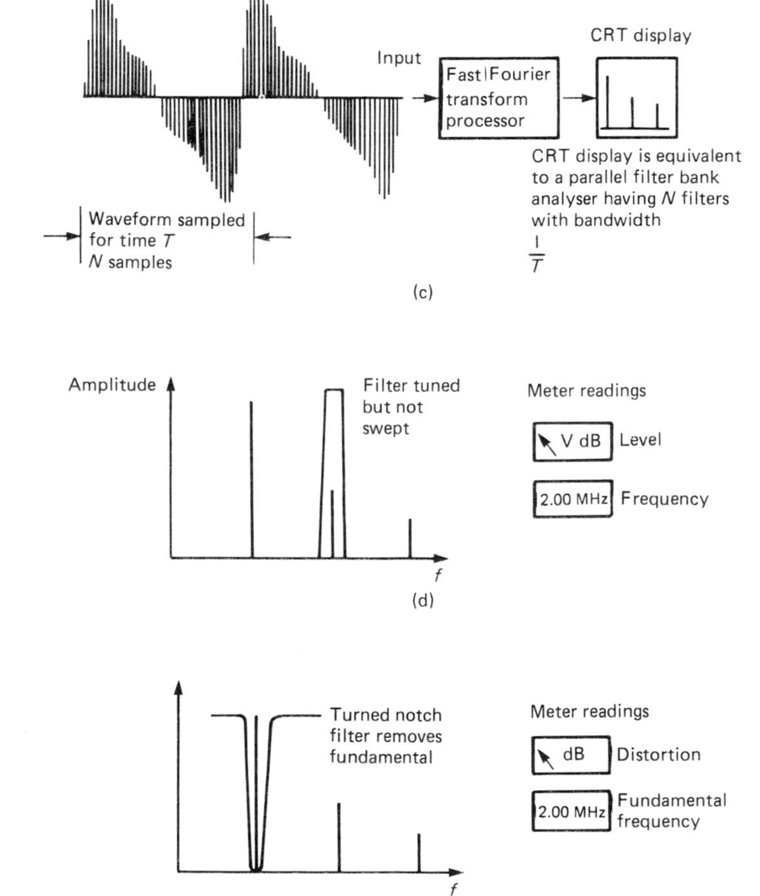

Figure 4.38 (a) Input waveform; (b) spectrum analyser; (c) digital Fourier analyser; (d) wave analyser; (e) distortion analyser.

possible to display the frequency spectrum as the swept-frequency oscillator sweeps over the required range. Figure 4.39 shows a schematic diagram of such an analyser suitable for analysing signals up to 1.25 GHz. The swept frequency is provided by sweeping either the 2–3.3 GHz local oscillator or the 500 MHz local oscillator, depending on the required range, which can be from 100 kHz to 1.25 GHz. The bandwidth of the filter is fixed by the band pass filters centred around 3 MHz at the output of the 3 MHz i.f. amplifier. This can be varied from 100 Hz to 300 kHz.

Typically, swept-frequency analysers are capable of operating with signals in the range from 5 Hz to 220 GHz with power levels in the range of −140 to +30 dBm. The bandwidth of the narrow-band filter varies with the frequency range over which the spectrum is being measured, and can vary from 1 Hz to 3 MHz.

4.5.2 Fourier analysers

Digital Fourier analysers convert the analogue waveform over a time period T into N samples. The discrete spectral response, $S_x(k\Delta f)$; $k = 1,2,\ldots,N$, which is equivalent to simultaneously obtaining the output from N filters having a bandwidth given by $\Delta f = 1/T$, is obtained by applying a discrete Fourier transform (DFT) to the sampled version of the signal. The spectral response is thus given by:

$$S_x(k.\Delta F) = \frac{T}{N}\sum_{n=1}^{N} x(n.\Delta t)\exp\left(\frac{-j.2\pi kn}{N}\right);\ k = 1,2,\ldots,N \quad (4.41)$$

$S_x(k\Delta f)$ is a complex quantity, which is obtained by operating on all the samples $x(n\Delta t)$; $n = 1,2,\ldots,N$ by the complex factor $\exp[-j(2\pi kn/N)]$. The discrete inverse transform is given by:

$$x(n\Delta t) = \frac{N}{T}\sum_{k=1}^{N} S_x(k.\Delta f)\exp\left(\frac{j.2\pi kn}{N}\right);\ n = 1,2,\ldots,N \quad (4.42)$$

Since $S_x(k\Delta f)$; $k = 1,2,\ldots,N$ is a complex quantity the DFT provides both amplitude and phase information at a particular point in the spectrum.

The discrete transforms are usually implemented by means of the fast Fourier transform (FFT) (Cooley and Tukey, 1965), which is particularly

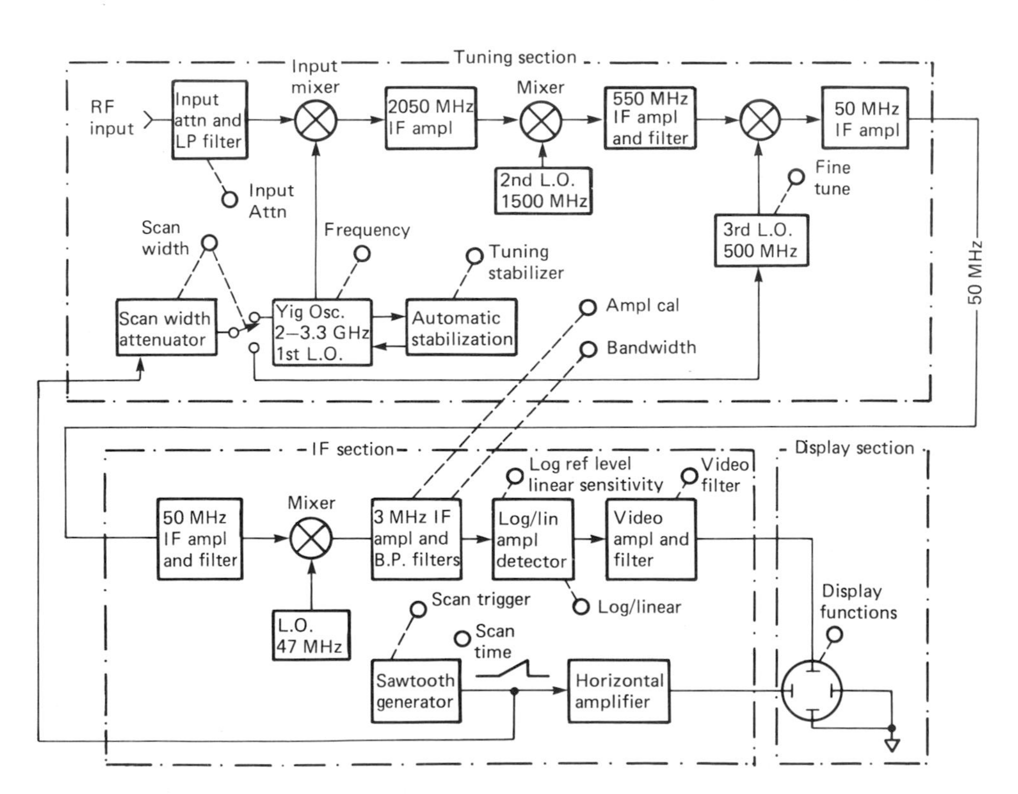

Figure 4.39 Swept-frequency analyser.

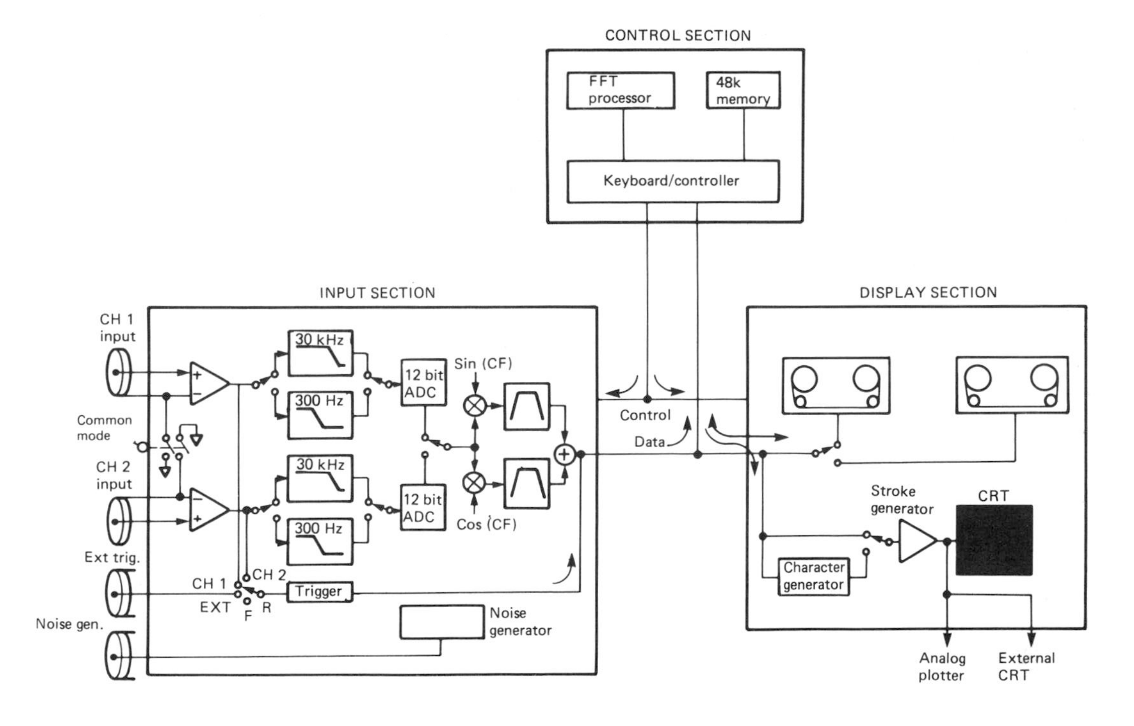

Figure 4.40 Digital signal analyser.

suitable for implementation in a digital computer since N is constrained to powers of 2 (e.g. 1024).

Figure 4.40 shows the block diagram representation of the HP5420 Digital Signal Analyser which employs a FFT algorithm. The input section consists of two identical channels. In each channel after the signal has been conditioned by the input amplifier the signal is passed through one of two anti-aliasing filters (see section 4.2.7.1). The cut-off frequency of these filters is chosen with respect to which of the sampling frequencies is being employed. The 30 kHz filter is used with a sampling rate of 102.4 kHz and the 300 kHz filter with a sampling rate of 1.024 MHz. The signal is then converted to a digital form using a 12-bit ADC. The multiplier and digital filter which follow allow the analyser to be used in either a baseband mode, in which the spectrum is displayed from d.c. to an upper frequency within the bandwidth of the analyser, or in a band-selectable mode, which allows the full resolution of the analyser to be focused in a narrow frequency band. The processing section of the analyser provides FFT processing on the input signal.

For one channel this can provide the real and imaginary or magnitude and phase components of the linear spectrum $S_x(f)$ of a time domain signal:

$$S_x(f) = F(x(t)) \tag{4.43}$$

where $F(x(t))$ is the Fourier transform of $x(t)$. The autospectrum, $G_{xx}(f)$, which contains no phase information is obtained from $S_x(f)$ by:

$$G_{xx}(f) = S_x(f).S_x(f)^* \tag{4.44}$$

where $S_x(f)^*$ indicates the complex conjugate of $S_x(f)$. $G_{xx}(f)$ normalized to a bandwidth of 1 Hz is the power spectral density (PSD), and represents the power in a bandwidth of 1 Hz centred around the frequency f.

In terms of the time domain characteristics of the signal $x(t)$ its autocorrelation function defined as:

$$R_{xx}(\tau) = \lim_{T\to\infty} \frac{1}{T} \int_0^T x(t)x(t+\tau).dt \tag{4.45}$$

is obtained by using the fact that the autocorrelation function is the inverse Fourier transform of $G_{xx}(f)$, i.e.:

$$R_{xx}(\tau) = F^{-1}(G_{xx}(f)) \tag{4.46}$$

and thus

$$R_{xx}(\tau) = F^{-1}(S_x(f).S_x(f)^*) \tag{4.47}$$

The inverse transform is obtained by use of equation (4.42).

Joint properties of two signals can be obtained by the use of the two channels. The cross power spectrum of two signals $x(t)$ and $y(t)$ can be computed as:

$$G_{yx}(f) = S_y(f).S_x(f)^* \tag{4.48}$$

where $S_y(f)$ is the linear spectrum of $y(t)$ and $S_x(f)^*$ is the complex conjugate linear spectrum of $x(t)$.

If $x(t)$ represents the input to a system and $y(t)$ the output of the system, then its transfer function, $H(f)$, which contains both amplitude and phase information, can be obtained by computing:

$$H(f) = \frac{\overline{G_{yx}(f)}}{\overline{G_{xx}(f)}} \tag{4.49}$$

where the bars indicate the time-averaged values. The input signal used for such measurements is often the internal random noise generator.

Coherence, given as:

$$\gamma^2 = \frac{\overline{G_{yx}(f)\,.\,G_{yx}(f)^*}}{\overline{G_{xx}(f)}\,.\,\overline{G_{yy}(f)}}\ ; 0 \leqslant \gamma^2 \leqslant 1 \tag{4.50}$$

measures the degree of causality between the input to a system and its output. If $\gamma < 1$ this indicates that the system output is caused by sources in addition to the system input or caused by system non-linearity.

The cross-correlation function between the two signals $x(t)$ and $y(t)$, $R_{xy}(\tau)$, defined as:

$$R_{xy}(\tau) = \lim_{T\to\infty} \frac{1}{T} \int_0^T y(t).x(t+\tau).dt \tag{4.51}$$

is also calculated within the analyser by means of the relationship which exists between the cross-correlation function in the time domain and the cross-spectral density in the frequency domain, namely:

$$R_{xy}(\tau) = F^{-1}(S_y(f).S_x(f)^*) \tag{4.52}$$

The HP 5420 Signal Analyser operates either in a base-band mode with a bandwidth from 0.8 Hz to 25.5 KHz, or in a passband mode, in which the central frequency of the frequency spectrum can be set between 0.016 Hz to 25.5 kHz, with bandwidths of between 0.008 Hz to 25.5 kHz. The input amplifiers have full-scale inputs variable between ±0.1 V and ±10 V.

For further details of the application of frequency analysis techniques in engineering the reader is directed to the work by Bendat and Piersol (1980). The series of applications notes on spectrum analysis produced by Hewlett Packard and referred to in the References provides an excellent introduction to the subject and to techniques for obtaining the best results from spectrum analysis equipment.

4.5.3 Wave and distortion analysers

Wave analysers (which are also known as frequency selective voltmeters, carrier frequency voltmeters or selective level meters) provide a finite bandwidth window filter which can be tuned over a frequency range. As the frequency range is swept, the analyser gives a visual indication of its central frequency and the amplitude at that frequency. Such analysers are used to provide amplitude measurements of a single component of a complex frequency signal, amplitude measurements in the presence of noise or interfering signals and measurement of energy in a well-defined bandwidth. Figure 4.38(d) shows the wave analyser measuring the amplitude of the $v_2(t)$ component in the waveform $v(t)$. Wave analysers typically operate from 15 Hz to above 32 MHz, with a selective band of between 3 Hz and 3.1 kHz, on signals having an absolute dynamic range from 0.1 μV to 300 V.

Distortion analysers provide band rejection around the central frequency to which it has been tuned and measure the energy in the rest of the spectrum. They are used to measure total harmonic distortion (THD), given by:

$$\text{THD} = \frac{\sqrt{[\Sigma(\text{harmonic amplitude})^2]}}{\text{fundamental amplitude}} \tag{4.53}$$

First, the energy in the total spectrum is measured and then a narrow-band filter is applied and the energy in the harmonics and noise is measured. THD is thus measured by:

$$\text{THD} = \frac{\sqrt{(\Sigma(\text{harmonics})^2 + \text{noise}^2)}}{\sqrt{((\text{fundamental})^2 + (\text{harmonics})^2 + (\text{noise})^2)}} \tag{4.54}$$

For harmonic content below 10 per cent this approximation will be within 0.5 per cent of the value given by the original definition.

Distortion analysers cover a frequency range from 5 Hz to 600 kHz and can measure distortion as low as 0.0018 per cent (−95 dB).

4.6 Lock-in amplifiers and phase-sensitive detection

Lock-in amplifiers provide a technique for the recovery of a coherent signal in the presence of noise. They can provide a very high degree of signal-to-noise ratio improvement without the drift associated with the production of high-Q bandpass filters. Their extensive range of applications includes signal processing from capacitive and inductive displacement transducers, radiometry, nuclear magnetic resonance and fringe position monitors (Blair and Sydenham, 1975).

The central element in the lock-in amplifier is the phase-sensitive detector (PSD), the operation of which is shown in Figure 4.41. The PSD consists of two channels having gain +1 and −1, the signal being synchronously switched between the two. The switching waveform can be represented by the Fourier series:

$$v_s(t) = \frac{4}{\pi}\left[\sin \omega t + \frac{1}{3}\sin 3\omega t + \frac{1}{5}\sin 5\omega t + \ldots\right] \quad (4.55)$$

If the input signal is:

$$v_{in}(t) + v \sin(\omega t + \phi) \quad (4.56)$$

then the output signal:

$$v_{out}(t) = v_{in}(t).v_s(t) \quad (4.57)$$

is given by:

$$v_{out}(t) = \frac{4v}{\pi}\sin(\omega t+\phi)\left[\sin \omega t + \frac{1}{3}\sin 3\omega t + \frac{1}{5}\sin 5\omega t \ldots\right] \quad (4.58)$$

and thus:

$$v_{out}(t) = 2v\left[\cos \phi - \frac{2}{3}\cos(2\omega t + \phi) + \ldots\right] \quad (4.59)$$

The low-pass filter on the output of the PSD provides a pass band for d.c. to f_{LPF} and attenuates the outputs at harmonics of the input frequency. Thus the d.c. output of the filter is given by:

$$v_{out}\ \text{d.c.} = \frac{2v}{\pi} \cdot \cos \phi \quad (4.60)$$

For input signals at frequencies other than the fundamental of the switching waveform the response of the PSD is as shown in Figure 4.42, where the effect of the bandwidth of the low-pass filter is

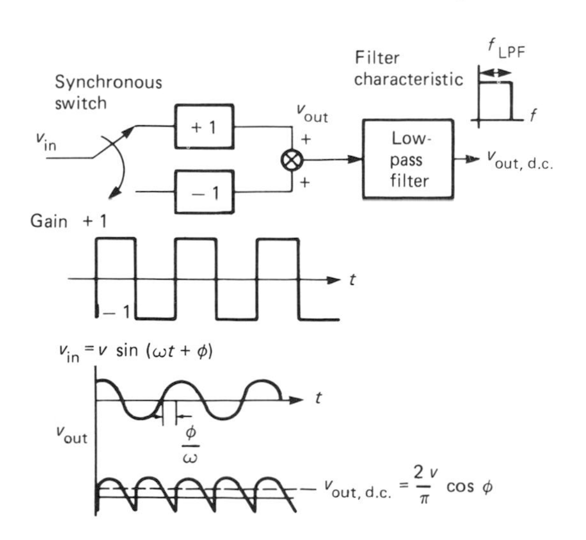

Figure 4.41 Phase-sensitive detector.

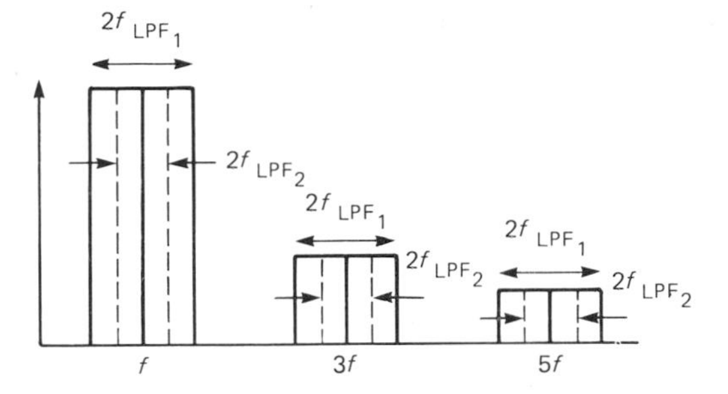

Figure 4.42 Frequency response of PSD.

shown. The PSD responds to signals in frequency bands centred around the fundamental and odd harmonics of the switching waveform. Signals whose frequencies correspond to the harmonic components of the switching waveform will give rise to d.c. signals. Those within the frequency bands centred around the harmonic frequencies will give rise to a.c. output signals whose frequency corresponds to a beat between the input signal and the harmonic of the switching waveform within that band. The amplitude response in each of these bands falls off as the number of the harmonic. By reducing the bandwidth of the low-pass filter it is thus possible to provide narrow-band filtering of the input signal. Figure 4.43 shows the block diagram representation of the Brookdeal Ortholoc SC 9505 two channel lock-in amplifier. The lock-in amplifier consists of a variable gain a.c. amplifier, a reference channel, two phase-sensitive detectors followed by d.c. amplification and filtering, and a vector computer. The input amplifier provides for either single-ended or differential voltage inputs or current input. These amplifiers are capable of providing an input sensitivity for full-scale output of between 10 nV and 500 mV or a current sensitivity as low as 10^{-14} A. The a.c. amplifier provides variable gain in order that the correct level of signal is presented to the PSDs together with high-pass and low-pass filtering. The reference channel generates two orthogonal reference waveforms for the two PSDs from a wide variety of input signal types and levels. Internal to the reference channel is a phase shifter which allows the phase of these two signals to be adjusted relative to the input signal. The lock-in amplifier has several modes of operation. These different modes adjust the distribution of gain between the a.c. gain prior to the PSD and the d.c. gain after the PSD.

In its HI-STAB mode the d.c. gain is minimized, and in such a mode the instrument has the best baseline stability. In its HI-RES mode the d.c. gain

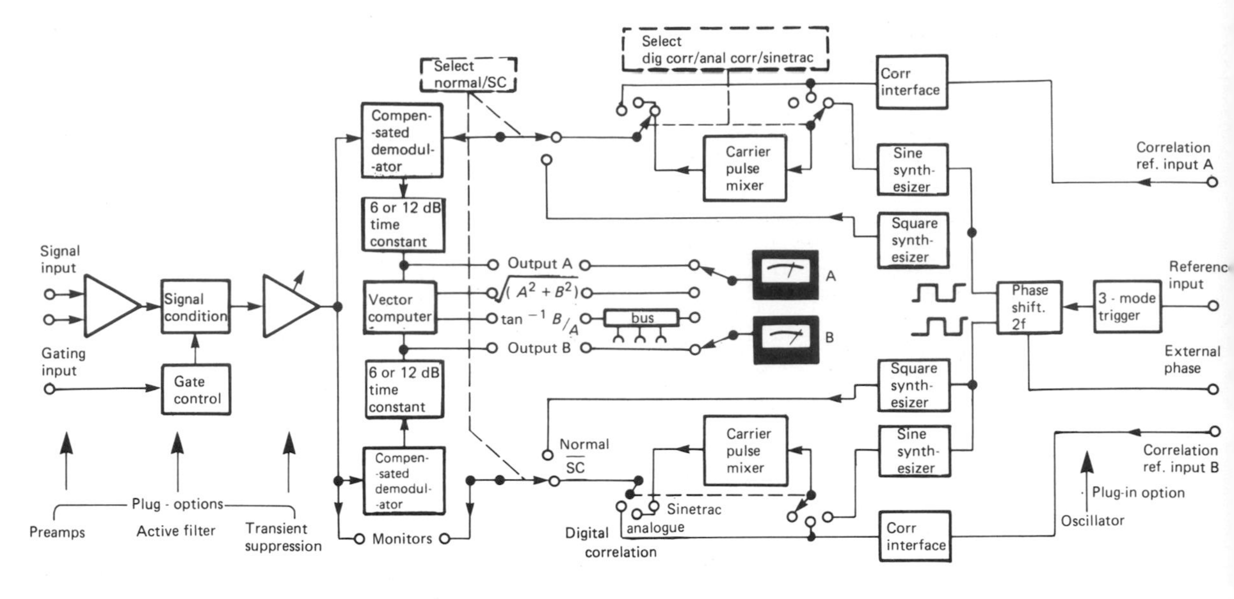

Figure 4.43 Two-channel lock-in amplifier.

is switched to its maximum value and consequently the PSD has its maximum overload capability and this allows the lock-in amplifier to extract the signal from a high level of noise. The lock-in amplifier output can be switched to indicate the magnitudes of the two orthogonal components A and B of the incoming signal or, by using the vector computer, the amplitude $\sqrt{(A^2 + B^2)}$ or the phase $\phi = \tan^{-1}(B/A)$ of the incoming signal.

By switching the reference channel with a variable mark to space ratio square wave whose mark-to-space ratio is determined by the input signal, the lock-in amplifier can be used in a SINETRAC mode, in which the response is to the fundamental only or as an analogue or digital correlator. The Brookdeal Ortholoc has a frequency range from 0.2 Hz to 200 kHz, a maximum broadband non-coherent input voltage for full-scale output of 10 000 × full scale and an input dynamic range of up to 160 dB.

For further details of the design of lock-in amplifiers and of their applications the reader is directed to the two works by Meade (1983a,b).

4.7 Box-car integrators

In the averaging of noisy repetitive signals which arise as a consequence of some external stimulus, as in the measurement of the signals within the neurological system, or which can be time referenced to some external triggering signal, the box-car

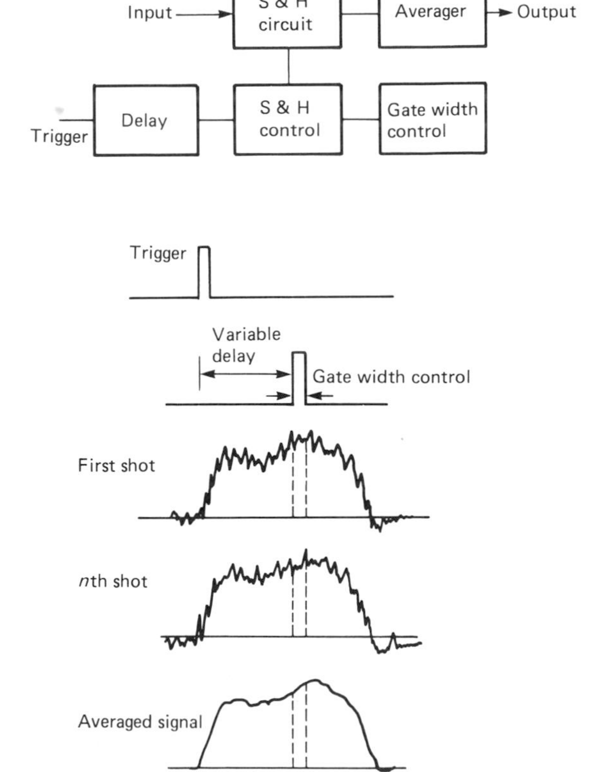

Figure 4.44 Box-car integrator.

integrator provides a method of averaging out the noise signals and of thus improving the signal-to-noise ratio (Arbel, 1980).

The operation of a box-car integrator is shown in Figure 4.44. The system employs a triggering signal which is time referenced to the signal being averaged. This, in neurological systems, may be the applied stimulus. The variable delay allows a particular section of the response signal to be selected. The sample and hold circuit is opened for a short aperture period and the signal within the gate width of the sample and hold circuit is stored. Repetition of the signal allows the portion of the signal at exactly the same time relative to the trigger to be selected. The system then averages the signal in this portion of the waveform. As the number of repetitions increases, the signal-to-noise ratio of the averaged signal improves. The improvement is proportional to the root of the number of repetitions. Automatic systems are available which scan the whole of the waveform, with an internally selected gate width to ensure that over a series of repetitions of the signal each section is averaged several times.

4.8 Correlators

Correlation is used in control and instrumentation for system identification, signal recovery and for the measurement of transport delays (for example, in the measurement of two-phase flows and steel strip and paper strip velocity measurement). Details of such applications can be found in Lange (1967) and Beck (1983).

The application of correlation techniques to flow velocity measurements is shown in Figure 4.45. These techniques depend upon measuring a random disturbance at one point in the process and then the length of time taken for the disturbance to pass the second measurement point. The transit time is estimated by cross correlating some signal generated by the disturbance which may be detected using optical, ultrasonic, thermal or impedance measurement techniques.

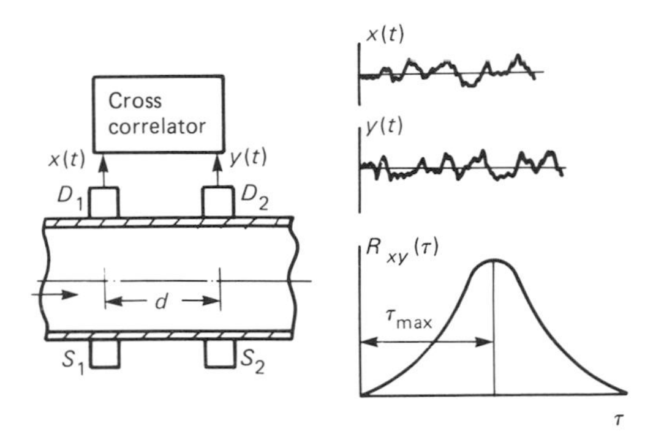

Figure 4.45 Flow velocity measurement using cross correlation.

The cross-correlation function of the two signals is given by:

$$R_{xy}(\tau) = \lim_{T\to\infty} \frac{1}{T} \int_0^T x(t-\tau) \,.\, y(t) \,.\, dt \tag{4.61}$$

$R_{xy}(\tau)$ has a maximum value at a time τ_{max} corresponding to the transit time. If the separation of the two measuring stations is d then the transport velocity is given by:

$$v = \frac{d}{\tau_{max}} \tag{4.62}$$

As shown in section 4.5, the cross-correlation function between two signals can be obtained by performing a Fourier transform on their cross-power spectrum, and thus a Fourier analyser can be used. Computation of the cross-correlation function can be provided by direct digital computation as:

$$R_{xy}(n.\Delta t) = \frac{1}{N} \sum_{k=1}^{N} x((k-n)\Delta t).y.(k.\Delta t) \tag{4.63}$$

However, quantizing the data into a large number of bits, providing the necessary delayed version of the x signal, and the digital computation of the cross-correlation function require systems which hitherto have been expensive. The problem can be somewhat simplified, and yields itself to solutions using VLSI technology or microprocessor-based systems by either quantizing the signal into only two levels (depending on whether the signal is positive or negative) or by correlating zero crossings of the two signals. Examples of such approaches have been given by Jordan (1979), Henry (1979) and Keech (1982).

Quantizing the signal into two levels is known as 'parity bit quantization', and the estimate of the correlation function in such a scheme is given by:

$$R_{xyp}(n.\Delta t) = \frac{1}{N} \sum_{k=1}^{N} \mathrm{sgn}[x(k-n)\Delta t]\mathrm{sgn}[y(k\Delta t)] \tag{4.64}$$

Thus multiplication is reduced to the logical exclusive OR between the bit representing sgn $x((k-n)\Delta t)$ and the bit representing sgn $y(k\Delta t)$.

Figure 4.46 shows the block diagram representation of a commercially available parity bit correlator designed for two-phase flow measurement. The correlator provides either a 256-, 512- or 1024-point correlation function with sampling intervals of 100,

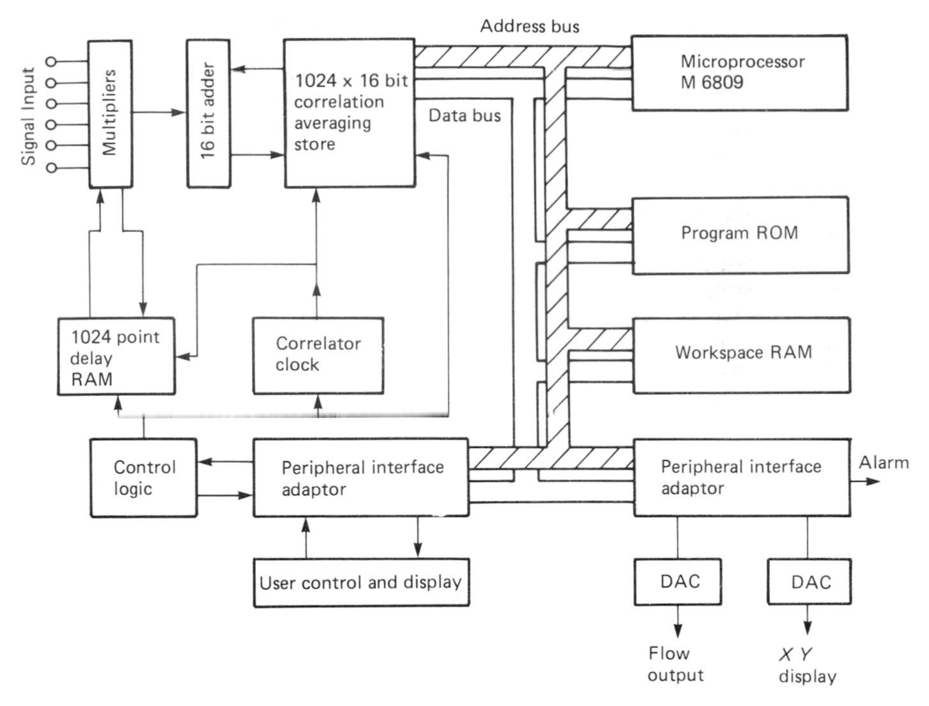

Figure 4.46 Parity-bit correlator (courtesy Kent Industrial Measurements).

200 or 400 μs, respectively. The correlation function, which can be represented by 1024 × 16 bits, is provided by means of a hard-wired logic. Computation on the correlation function is provided by the microprocessor, which provides adaptive filtering of the correlation function to improve the performance of the instrument under transient conditions; estimates the peak of the correlation function; and provides validation of the results by sensing such conditions as zero flow, shock waves, intense pressure waves and oscillating flow components.

4.9 Multichannel analysers

Multichannel analysers (MCAs) are used to measure the statistical properties of random signals. They can be used in the analysis of the energies of α, β and γ particles, or X-rays; for the measurement of the time intervals between events; or for obtaining count information of the number of occurrences of a particular event in sequential time intervals. They are used extensively in nuclear, chemical, medical and materials analysis.

The MCA operates in two modes, known as the pulse height analysis (PHA) mode and the multichannel scaling (MCS) mode. Both of these are shown in Figure 4.47.

In the PHA mode the MCA measures the amplitude frequency distribution of a series of input pulses. The MCA consists of a series of addressable memory locations which are referred to as channels. Initially, the contents of all the memory locations are zero. The ADC converts the amplitude of the input signal into a digital form. This in turn is used

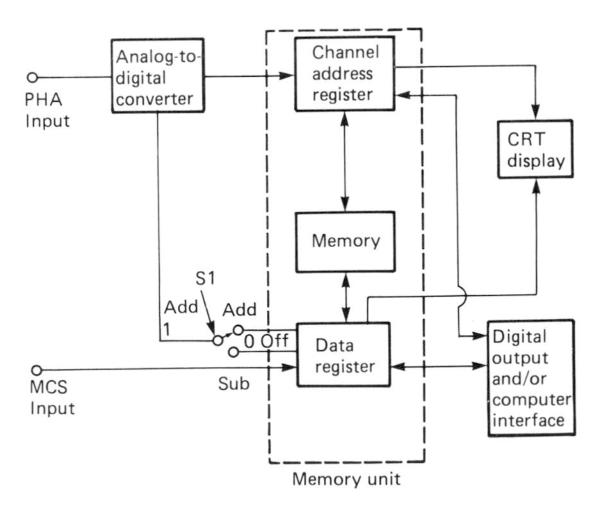

Figure 4.47 Multichannel analyser.

to address the memory. The value in the memory location corresponding to this address is loaded into the data register. A count of 1 is added to the contents of the data register and the data are then returned to the memory location from which it came. The count value in each channel or memory location at the end of the processing period is equal to the total number of pulses processed having amplitudes corresponding to that channel address.

In its MCS mode the channels act as a sequence of counters, with each channel counting events for a predetermined 'dwell time'. After this predetermined dwell time the counting is automatically passed to the next channel. The MCS thus provides a histogram of count-rate data, each channel representing a sequential time interval.

4.10 Computer systems

Much signal processing is now undertaken using a digital computer, with a wide variety of computing systems being used for the processing, ranging in size from embedded single-chip microprocessor-based systems up to large mainframe computers. It is not possible here to give more than an overview of the subject. There are a large number of specialist texts available dealing with digital computer systems and their applications, and for further details the reader should consult, for example, Bartree (1981), Lewin (1980) and Boyce (1977) for general system considerations; Paker (1980) for minicomputer systems and their application in instrumentation and control; and Aspinall (1980) and Barney (1985) for microcomputer systems and their applications.

There are two prime considerations in the specification of a digital computer system, the hardware and the software. The hardware of the system is the collection of electronic, electrical and mechanical devices making up the computer and its peripherals, which for a given system is fixed. The software is the set of stored programs which enable it to perform its prescribed tasks. The software is not fixed and within certain constraints (some of which are imposed by the particular language in which the programs are written and others which are imposed by the hardware of the computer) it is possible to modify the tasks performed by the computer. It is the stored-program aspect of the digital computer, giving the operator the ability to modify the tasks performed by the computer, which gives such a system its flexibility and power. The hardware aspects of the digital computer are dealt with in section 4.10.1 and the software is dealt with in section 4.10.2. Typical tasks undertaken by digital computers are considered in section 4.10.3. Section 4.10.4 gives examples of a single-chip microcomputer system suitable for use in an embedded microcomputer system, a control and data-logging subsystem based on a microcomputer and a measurement and control processor hosted by a minicomputer.

4.10.1 Hardware

Despite the wide variety of digital computer systems available, most systems use central processing units (CPUs) with the architecture shown in Figure 4.48(a) in computer systems with the overall structure as shown in Figure 4.48(b).

The digital computer system consists of a CPU, memory and input/output to peripheral devices. The CPU consists of three units. The arithmetic/logic unit (ALU), the control unit and the internal registers. The ALU performs both arithmetic operations such as addition, subtraction and possibly multiplication and division, together with logical operations on data brought from the memory. The control unit generates the timing and signals required for fetching, decoding and executing instructions stored in the memory and for input–output. As such, it provides the internal and external control signals required for correct operation of the complete system. In a microprocessor

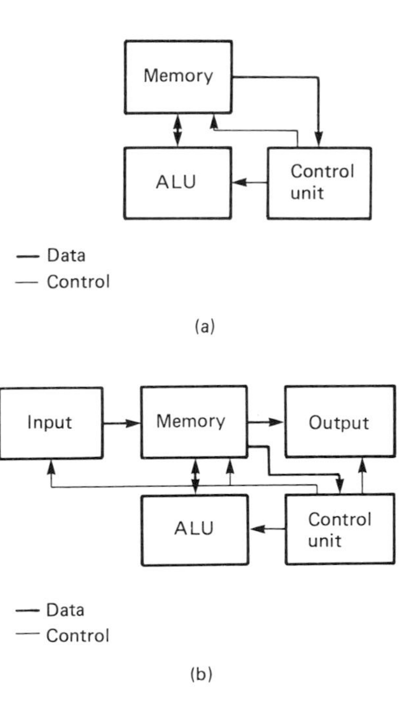

Figure 4.48 (a) Central processing unit; (b) schematic diagram of complete computer system.

such as the Rockwell 6502 (whose internal architecture is shown in Figure 4.49), the arithmetic/logic unit and the control unit consist of a single large-scale integrated (LSI) circuit. (The 6502 is used in the Apple, BBC Acorn and Commodore PET microcomputers.) Microcomputers in general use metal oxide semiconductor (MOS) technology for their CPUs, whereas minicomputers and main-frame machines have CPUs which are fabricated using the faster bipolar technologies such as TTL, or integrated injection logic (I^2L) or emitter coupled logic (ECL).

4.10.1.1 Memory

The memory or store consists of a series of storage locations which contain the program which the computer is executing and the data which are being operated on, together with the results of those operations. Each storage location is identified by having a unique address. The instructions and data in each of these addresses is stored as the bits of a binary number. The word length of the computer is the largest possible collection of binary bits the computer can manipulate, and in a microcomputer

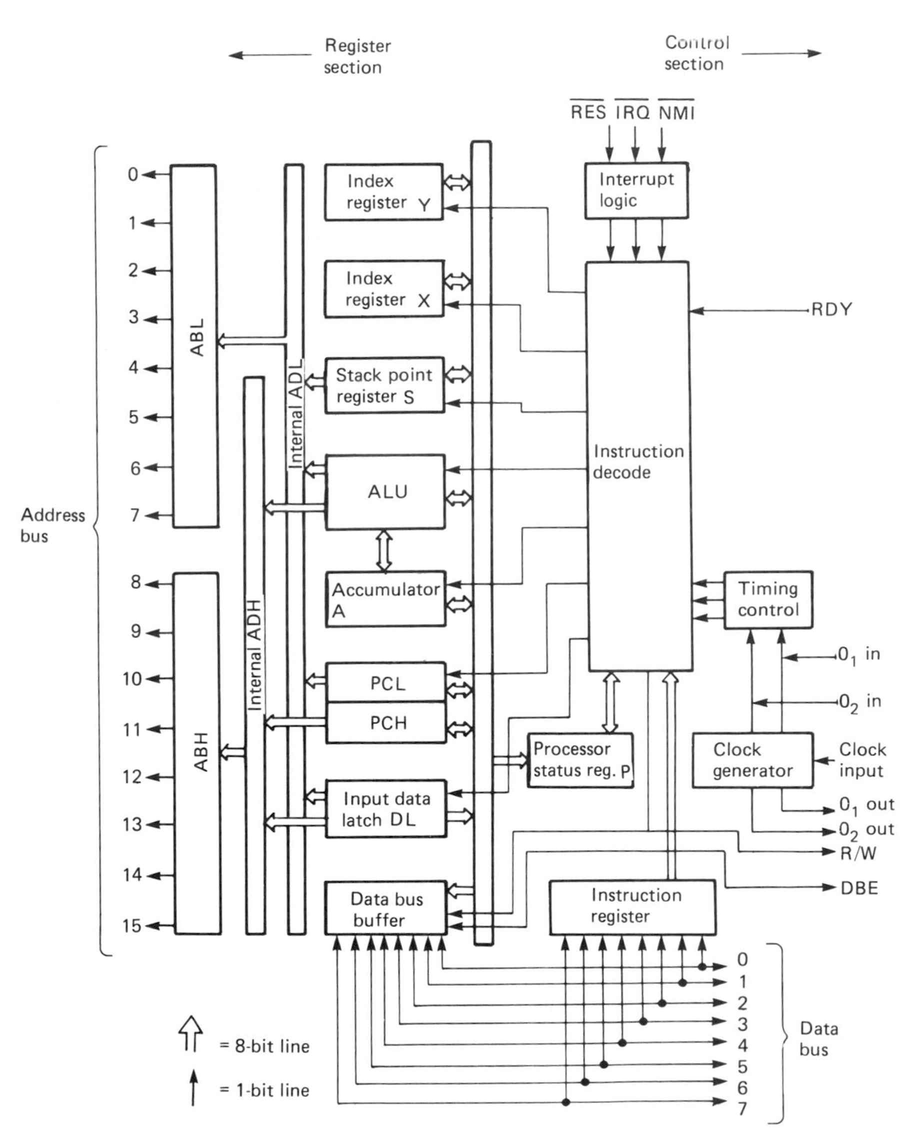

Figure 4.49 Internal architecture of 6502 microprocessor.

is typically either 8 or 16 bits. In larger machines the length is likely to be 16, 24, 32 or even 64 bits. The words in larger machines are often split down into smaller units called 'bytes'. Typically, a byte will consist of 8 bits.

There are a number of distinguishing features of different memory systems. The characteristic time associated with the memory is its access time, which is the time required to write into or read from a particular location in the store. The access time has two components: the time required for addressing and locating the required word in the store; and the switching or operation time for the storage element. Random access memory (RAM) is memory in which all the words are equally accessible, as opposed to a cyclic memory, in which the words are available only on a cyclic basis and for which the average access time is half the cycle period. The main memory of a computer system usually consists of RAM in the form of semiconductor LSI arrays.

Memory is either volatile, in which case its contents will be lost in the case of a power failure, or it is non-volatile or permanent. If reading of the memory destroys its contents then it is said to be destructive, otherwise it is said to be non-destructive. Static memories retain their information as long as the power is maintained, whereas dynamic memories are required to be constantly refreshed in order to retain their information. If it is only possible to read from the memory and not to write into it then the memory is said to be read-only memory (ROM). This is often used for permanent program storage. EPROM, Electrically Programmable ROM is a memory which can be electrically programmed but which under normal operating conditions behaves as ROM. The data are erased by the application of ultra-violet light. EAPROM is an alternative form of PROM, which is Electrically Alterable, hence its name. Both EPROM and EAPROM are used for program storage in development situations where it may be necessary to modify the program before committing it to ROM. Table 4.6 compares the characteristics of different storage media used in computer systems.

4.10.1.2 Peripheral devices

A complete computer system as shown in Figure 4.50 consists of a CPU together with a range of peripheral devices (Wilkinson and Horrocks, 1980).

Input devices The most common form of computer input terminal is the VDU, which consists of an alphanumeric keyboard together with a cathode ray tube (CRT) display which can provide either alphanumeric or graphical display. The alphanumeric input from the keyboard is usually converted to a 7-bit ASCII code and bi-directional data communication between the VDU and the computer is by means of two asynchronous serial lines at a

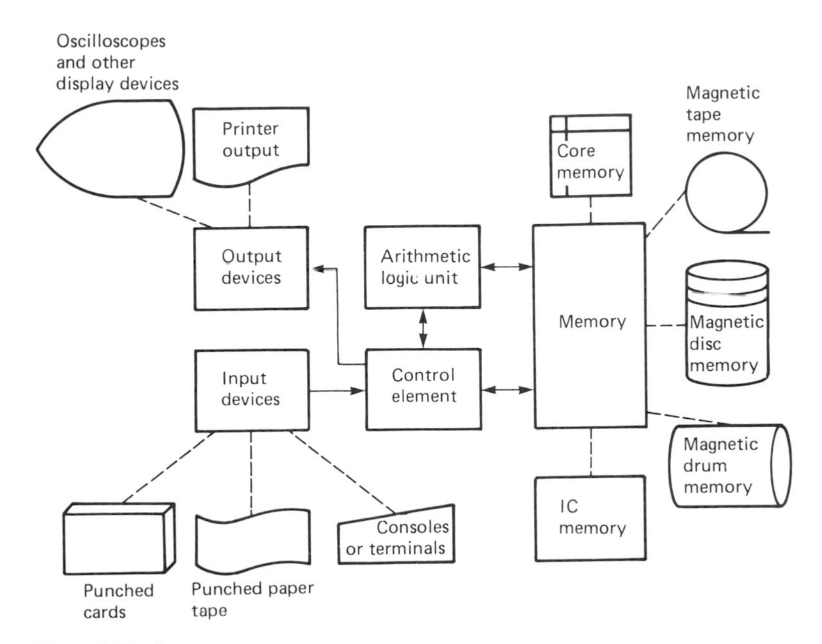

Figure 4.50 Computer system and its peripheral devices.

Table 4.6 Characteristics of memory systems

Storage device	*Access time and type*	*Capacity*	*Recording density*	*Data rate*	*Physical characteristics*	*Main application*
Magnetic drum	8.5–17 ms cyclic access	200 k–200 M bits	1600 BPI	2–8 M bits/s	Permanent storage	Backing storage
Magnetic disc	25–70 ms cyclic access	800 k–500 M bits	1000–6000 BPI	800 k–8 M bits/s	Permanent storage Rigid disc Fixed or moving heads	Backing and file storage
Floppy disc	100–400 ms cyclic access	2–5 M bits	3000 BPI	250 k bits/s	Permanent storage flexible disc	Cheap form of disc store; transportable packs
Magnetic tape	Several minutes serial access	30–800 M bits	1600–6250 BPI	10^7 bits/s	Permanent storage	Mass storage, on-line and archival
Cassette tape	50–500 s serial access	1–5 M bits	800 BPI	8–20 k bits/s	Permanent storage	Cheap high-capacity store; transportable packs
Magnetic cartridge	10–50 s serial access	20–22 M bits	1600 BPI	48–160 k bits/s	Permanent storage	Cheap high-capacity store; transportable packs
Magnetic core	250–750 ns (cycle times 650 ns – 1 μs) random access	4 k – 1 M bytes	—	—	Volatile unless protected; destructive read-out	Main store and backing store extension
Semiconductor						
Bipolar RAM	45–60 ns (cycle times 60–100 ns random access	256 – 1 k bits	—	—	Volatile	Registers, accumulators, buffers, push-down stores. Cache stores
CMOS RAM	50–100 ns (cycle times 500–600 ns random access	1 – 4 k bytes	—	—	Volatile	Cache stores Fast main memory
PMOS/NMOS RAM	150–400 ns (cycle times 350–800 ns) random access	1 – 16 k bytes	—	—	Volatile	Main memory Buffer stores
MOS/Bipolar ROM/PROM	55–500 ns random access	1–64 k bytes	—	—	Permanent/volatile	Read Only Memory Micro-program control stores, control programs, table-look-up
CCD	50–500 ns serial access	4–64 k bits	—	5–20 M bits/s	Volatile	Backing store
Magnetic bubbles	Several milliseconds serial access	16 k–64 k bits	—	100–300 k bits/s	Volatile	Backing store Mass memory

bit rate between 75 and 9600 bits/s. The VDU provides facilities for transmitting characters, lines or pages together with editing facilities for inserting or deleting characters or lines. Many VDUs are now intelligent terminals including a microprocessor and a local backing store in the form of a cassette or floppy disc. They can be programmed to perform specific tasks and may provide such facilities as a BASIC interpreter, utility programs for creating, deleting and editing files, and also communication packages.

Teletypes provide a keyboard and printer and can thus provide hard copy, but this is restricted to alphanumeric output. Additionally, teletypes provide a paper tape punch and reader. Paper tape input is usually by an 8-hole tape with 7-bit ASCII coding, the eighth bit being used as a parity check. Reading of the tape is by means of photoelectric sensing at rates of up to 500 characters/s (c/s). The punch is generally capable of operating at 300 c/s.

Larger mainframe machines may also provide means of input using card readers/punches. The data for these are usually formatted as 80 characters/card with each character using a 12-bit Hollerith code. The cards are read using optical techniques. Card readers are capable of reading at rates of between 100 and 2000 cards/min and punching cards at between 100 and 300 cards/min. Newer forms of data input to computer systems employ such techniques as light pens, graphics tablets and voice input.

Output devices In addition to output by means of VDUs or teletypes or other visual display systems employing light-emitting diodes (LEDs), liquid crystal displays (LCDs) and plasma displays, computer systems use both serial and line printers and digital plotters. Serial printers for alphanumeric output employ print head types such as a daisywheel printer or a dot matrix printer, which can also be programmed to provide graphical hard copy. Typically, such printers are capable of printing at rates of up to 200 c/s with up to 80 characters/line.

A line printer is used in applications where it is required to output a large amount of printed material. Line printers use such techniques as impact drum or dot printing, electrosensitive paper, electrostatic or electrophotographic printing, or laser printing. Such devices are capable of printing at up to 2000 lines/min.

Digital plotters provide hard copy of graphical and alphanumeric data from the computer. Such plotters, which are either flat-bed plotters or drum plotters, can have a step size as small as 0.05 mm and are capable of plotting at up to 10 000 steps/s. Display and recording devices are covered in Chapter 6.

Analogue and digital interfaces In measurement and control applications the input and output peripherals will also include analogue and digital interface units. The analogue input interface typically provides several analogue channels with signal conditioning and sample and hold amplifiers multiplexed to a single ADC. The analogue output interface may be either a single DAC whose output is connected to a series of sample and hold amplifiers or a bank of DACs. Input to a digital interface may be at TTL levels or higher voltages. The digital output may be required to drive TTL circuitry, lamps, relays or provide switching of a.c. power. Often both the analogue and digital interfaces provide electrical isolation between the computer and the external world by employing isolation amplifiers or optical couplers.

Mass data storage The peripheral back-up memory provides low-speed mass storage. Conventionally, this has been produced by either magnetic tape or disc systems, although newer forms of mass storage such as magnetic bubble memories and optical memories are now beginning to appear. The methods of digital information storage using magnetic surface storage are dealt with in Lewin, 1980.

Magnetic tape storage systems span a range from the large tape systems used on mainframe computers, which can provide up to 350 Mbytes of memory with recording densities of up to 12 500 bits per inch (bpi) and transfer rates of 1250 kbytes/s to audiocassette systems. Standard audiocassette recorders, used mainly with microprocessor systems, provide up to 1 Mbyte of memory storage on a 90-min cassette at a typical recording density of 800 bpi and transfer rates of 150 bytes/s. Access in magnetic tape storage systems is serial, with average access times in both large tape systems and cassette systems being of the order of minutes.

There are two varieties of magnetic discs, known as 'hard' and 'soft' discs. Hard discs are made of aluminium and coated with ferric oxide. Winchester disc drives, common on small and medium-size computer systems, are a form of hard disc having a high storage density and data-reading rate. The capacity of an 8 in Winchester drive lies between 10 and 50 Mbytes. Such systems typically have a density of 6000 bpi with 500 tracks/in (tpi) and data transfer rate of 800 kbytes/s. The average access time for such a system is between 40 and 50 ms.

Soft or 'floppy' discs derive their name from the fact that they are fabricated from Mylar with a ferric oxide coating, and hence are floppy. Floppy discs come in a range of sizes 8 in, 5.25 in and 3.5 in. 5.25 in discs, which are widely used with microcomputer systems and which are referred to as minidiscs, can provide storage of 800 kbytes of data with data transfer rates of up to 250 kbits/s. The average access time in a floppy disc system is between 100 and 400 ms.

4.10.2 Software

4.10.2.1 Machine code

Both data and instructions are stored in the memory of the CPU as binary numbers. Within the control unit of the CPU there is a decoding unit which takes the words in sequence from the memory, decodes them as to whether they are data or instructions and, if they are instructions, then provides the necessary control signals for the rest of the system for that operation to be executed. At this level in the memory the instructions are stored in a form that they can be executed by the CPU. This most fundamental level of language is called 'machine code'. In writing programs in machine code there can be difficulties in remembering the machine code instructions in binary or hexadecimal (to the base 16) form; in remembering the addresses to which branch, jump or loop instructions go; and in keeping track of these when altering the program. Mnemonic assemblers have therefore been devised in which each machine code instruction has a mnemonic such as ADD, SUB and INX and in which store locations are identified by labels. Table 4.7 lists the machine code instructions and mnemonic codes for the 6502 microprocessor system. Such assemblers ease the problems of programming in machine code since the assembler automatically takes care of the changes necessary in the addresses to which jump, loop and branch instructions refer. There is still, however, a one-to-one correspondence between assembly language instructions and the machine code instructions, and thus such a language is called a low-level language (LLL).

4.10.2.2 High-level languages

High-level languages (HLLs), which resemble English and employ mathematical notation, require no knowledge on the part of the programmer of the internal workings of the computer on which it is running but only of the syntax and semantics of the

Table 4.7 Machine code instructions and mnemonics for 6502 microprocessor

ADC	Add Memory to Accumulator with Carry	JSR	Jump to New Location Saving Return Address
AND	'AND' Memory with Accumulator		
ASL	Shift Left One Bit (Memory or Accumulator)	LDA	Load Accumulator with Memory
		LDX	Load index X with Memory
BCC	Branch on Carry Clear	LDY	Load Index Y with Memory
BCS	Branch on Carry Set	LSR	Shift Right One Bit (Memory or Accumulator)
BEQ	Branch on Result Zero		
BIT	Test Bits in Memory with Accumulator	NOP	No Operation
BMI	Branch on Result Minus	ORA	'OR' Memory with Accumulator
BNE	Branch on Result not Zero		
BPL	Branch on Result Plus	PHA	Push Accumulator on Stack
BRK	Force Break	PHP	Push Processor Status on Stack
BVC	Branch on Overflow Clear	PLA	Pull Accumulator from Stack
BVS	Branch on Overflow Set	PLP	Pull Processor Status from Stack
CLC	Clear Carry Flag	ROL	Rotate One Bit Left (Memory or Accumulator)
CLD	Clear Decimal Mode	ROR	Rotate One Bit Right (Memory or Accumulator)
CLI	Clear Interrupt Disable Bit	RTI	Return from Interrupt
CLV	Clear Overflow Flag	RTS	Return from Subroutine
CMP	Compare Memory and Accumulator		
CPX	Compare Memory and Index X	SBC	Subtract Memory from Accumulator with Borrow
CPY	Compare Memory and Index Y	SEC	Set Carry Flag
		SED	Set Decimal Mode
DEC	Decrement Memory by One	SEI	Set Interrupt Disable Status
DEX	Decrement Index X by One	STA	Store Accumulator in Memory
DEY	Decrement Index Y by One	STX	Store Index X in Memory
		STY	Store Index Y in Memory
EOR	'Exclusive-Or' Memory with Accumulator		
		TAX	Transfer Accumulator to Index X
INC	Increment Memory by One	TAY	Transfer Accumulator to Index Y
INX	Increment Index X by One	TXS	Transfer Stack Pointer to Index X
INY	Increment Index Y by One	TXA	Transfer Index X to Accumulator
		TXS	Transfer Index X to Stack Pointer
JMP	Jump to New Location	TYA	Transfer Index Y to Accumulator

HLL. HLLs are used for economy and convenience in program development, but this is achieved at the expense of compactness and efficiency of the programming. The compilers and interpreters of HLLs generate several machine code instructions for each statement of the HLL. Compiled languages differ from interpreted languages in that in a compiled language the source code is compiled into a binary code and at run time it is this code which is executed, whereas in an interpreted language at run time each instruction of the HLL is interpreted and executed. Consequently, interpreted languages are generally slower in execution than compiled languages.

In microcomputer systems HLLs are used either as a development tool (in which case the compiled code is installed on the microcomputer) or as the language of a system which has its own compiler for that particular language, either resident in ROM or available on disc.

There are at present a large number of HLLs suitable for running on micro- and minicomputers employed in real-time instrumentation and control applications. They are compared in some detail in the book by Bibbero and Stern (1982).

Although BASIC (Beginners All Purpose Symbolic Instruction Code) is to be found on a large number of micro- and minicomputers, it is a poor language for real-time applications. It is generally implemented as an interpreted language and therefore slow. As a language it is not internally structured and consequently concurrent task programming is difficult; also its input–output routines are awkward.

FORTRAN (FORmula TRANslation), which is a compiled language, has more highly developed input–output routines than BASIC but its large compiler requires disc storage. Real Time FORTRAN (RTF) is used extensively for control and military applications and cross compilers are available which will allow the compiled object code from a FORTRAN program to be installed in ROM or PROM on a microcomputer.

CORAL is based on ALGOL 60 (ALGOrithmic Language) and is commonly used industrially for online systems. CORAL provides a COMMON communicator facility which allows programs to be written as a number of compiled segments which each perform their own specified function when called by a control routine, and this therefore allows for flexible programming. The language offers advantages in that it allows for both fixed and floating point variables and also in its method of storing data in memory.

FORTH, which employs a technique known as 'indirect threaded coding', is significantly faster than BASIC and is suitable for multitask operations such as process control. A typical FORTH compiler requires 6k of memory. It has a lower program memory requirement than most other high-level languages.

C-CODE is a language which requires a large compiler but produces a compiled object code which requires typically only between 10 and 30 per cent more memory space than an efficient machine code program. As such, it is used extensively as a high-level development language for machine code programs for microcomputers. Suitable cross compilers allow the C-CODE programs to be targeted to a particular microcomputer system.

PASCAL, which is a derivative of ALGOL, is a structured language with a block structure capable of providing concurrent program support for the most complex real-time applications programs. P-CODE is an intermediate interpreted language between PASCAL and machine code, and thus PASCAL can be implemented by either compiling to P-CODE and interpreting the P-CODE with a small resident onboard interpreter, or by cross compiling to the machine code of the processor.

4.10.2.3 The operating system

The program which controls the overall operation of the computer is referred to as the operating system (OS), executive or monitor. In instrumentation and control systems the operating system has to be a real-time operating system which must keep in step with the external world. Events in the external world generate interrupt signals for the computer by causing contacts to close or signals to change levels. The events in the external world may be, for example, that one of the monitored variables has gone out of its permitted range or that there is a requirement for input to the system via the VDU or teletype. On receipt of the interrupt signal, after the present instruction has been executed, the sequence of instructions being executed is suspended in order to respond to the external event. This is usually done by means of an interrupt service routine. When the service routine has been executed control will return to the interrupted program. In large systems there may be several levels of interrupt priority with time-critical external events having the highest priority. The interrupt service routine has to detect the level of the interrupt which has been raised and in the presence of more than one interrupt has to respond to the interrupt having the highest priority level.

The operating system schedules and controls the handling of different input/output activities, including the processing of interrupts, and includes input–output organization, storage allocation, communication with the operator, interrupt control, queue control and display handling. The design of the operating system is fundamental to the efficient operation of the whole computer system.

4.10.3 Typical tasks performed by digital computers

In instruments which employ embedded computer systems the presence of the microcomputer may be transparent to the user, who will not have access to the programs resident in the microcomputer system and will therefore be unable to modify them. In such systems the program is generally written in machine code and stored in ROM or PROM.

The tasks which an embedded computer system provide depend largely on the individual application, but may include:

(1) Replacement of discrete digital circuitry for logic or sequence control of the overall system.
(2) Provision of dedicated computational facilities such as scaling, linearization and filtering.
(3) Improved system performance by the provision of auto-zeroing, auto-ranging and self-calibration facilities.
(4) Self diagnosis of its own operation and the operation of the total system.
(5) Control of the user interface, including keyboards, displays and printers.
(6) Communications with other systems and computers.

General-purpose computer systems for use in instrumentation and control are based on micro- or minicomputers and are programmed in a HLL. It is therefore open to the user to modify the tasks undertaken by the system.

The tasks performed by such systems are wide ranging (Sanderson, 1976; Paker, 1980) but may include:

(1) Data acquisition and logging of analogue and digital signals from instruments or processes together with the provision of alarm signals.
(2) Data reduction for recording or display.
(3) Computational facilities under user control to provide statistical analysis, digital filtering, spectral analysis or correlation measurements on the logged data.
(4) Sequence control of batch and continuous processes including start-up and emergency shutdown procedures.
(5) Direct digital control (DDC), in which analogue three-term controllers are replaced by discrete three-term controllers provided by software in the computer.
(6) Supervisory control, in which the set points of conventional analogue controllers are monitored and adjusted by the computer system.
(7) Optimization of system-operating parameters.
(8) Control of data communications, which may be either between the system and a supervisory main computer or intersystem communications in a distributed computer system.

For the theoretical basis of signal analysis the reader is referred to Papoulis (1977). Owens (1983) provides a large bibliography on the application of digital processing techniques. Digital control systems are covered in the book by Kuo (1980).

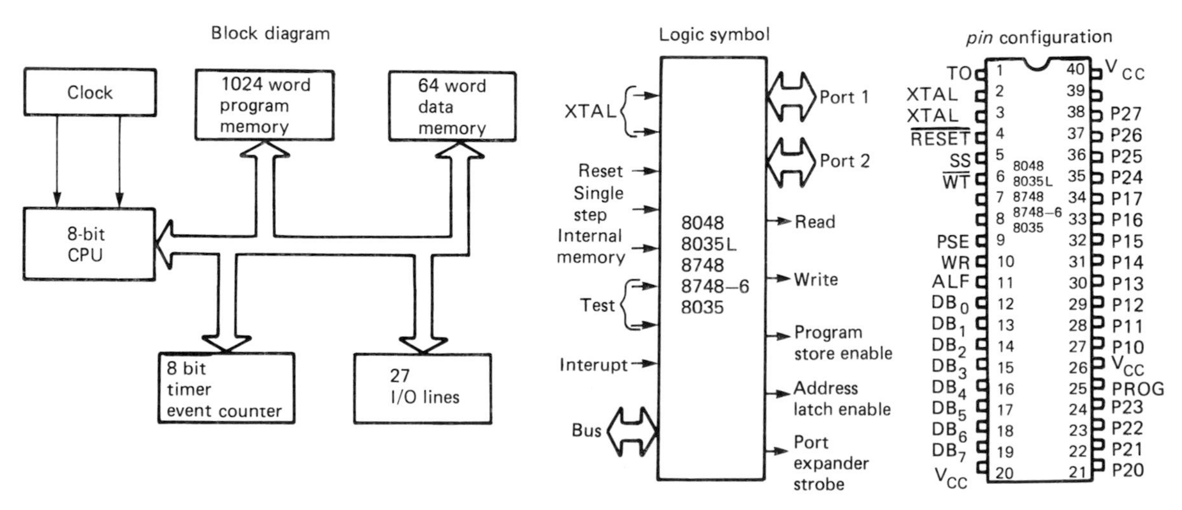

Figure 4.51 Block diagram, logic symbol and pin configuration of 8048 single-chip microcomputer.

4.10.4 A single-chip microcomputer for use in an imbedded computer system

A minimal computer system consists of an ALU, a control unit, memory for program and data storage, and input–output. A range of such systems using single-chip LSI circuits is available. Figure 4.51 shows the block diagram, logic symbol and pin configuration of the Intel 8048 single-chip microcomputer. Table 4.8 gives the key design parameters of this device which is packaged as a 40-pin dual in-line integrated circuit (Intel, 1980).

The 8048 is fabricated using an N channel MOS process and contains 1 K × 8 bit of ROM memory,

Table 4.8 Specification of 8085 single-chip microcomputer system

8-bit CPU, ROM, RAM, I/O in single package
Interchangeable ROM and EPROM versions
Single 5 V supply
2.5 μs and 5.0 μs cycle versions:
all instructions 1 or 2 cycles
Over 90 instructions: 70% single byte
1 k × 8 ROM/EPROM
64 × 8 RAM
27 I/O lines
Interval timer/event counter
Easily expandable memory and I/O
Compatible with 8080/8085 series peripherals
Single-level interrupt

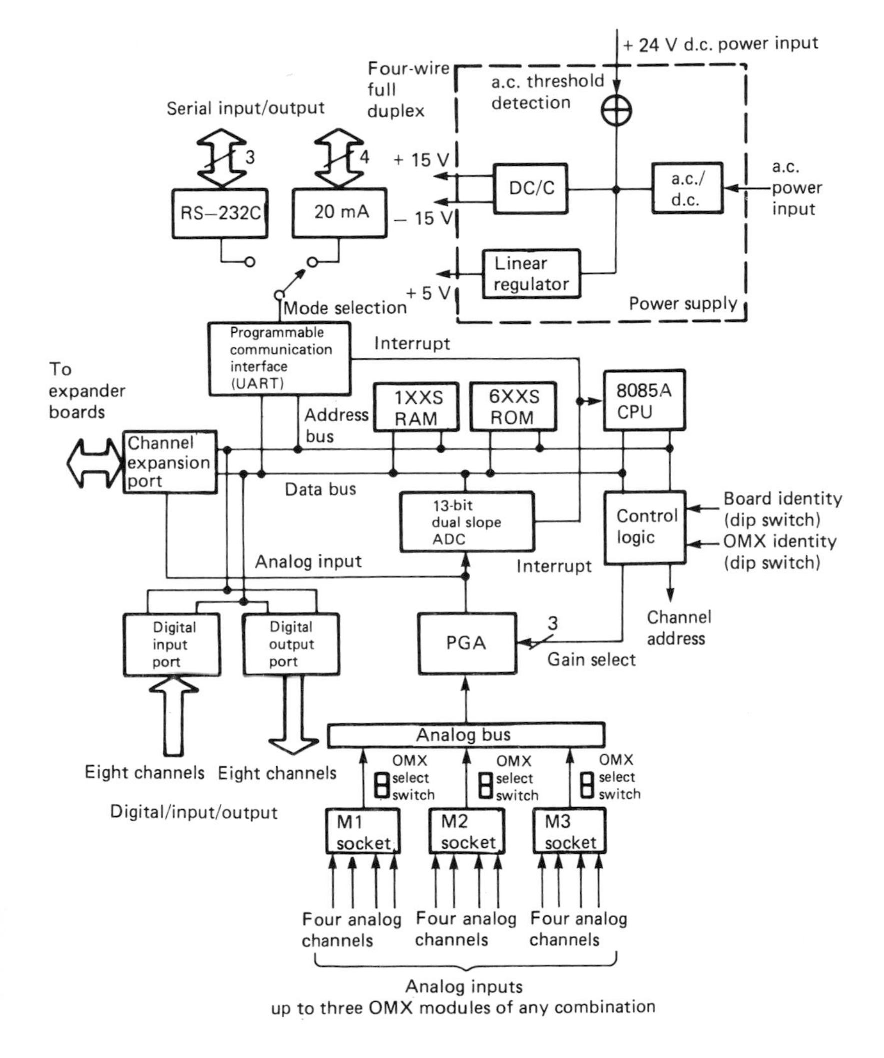

Figure 4.52 Master board for μMAC-4000 system.

Table 4.9 Input variables for μMAC-4000 system

Input	*Resolution*	*Accuracy*
Thermocouples Type J, K, T, S	0.1 K	±1.3 K
RTDs 100 Ω platinum	0.05 K	±0.3 K
Solid state temperature sensors		
AD590 or AC2626	0.1 K	±0.2 K
Strain gauge transducers		
±30 mV, ± 100 mV spans	0.025% of span	±0.01% of span
Low-level d.c. voltages		
±25 mV, ±50 mV, ±100 mV	6 μV	±0.005% of FS
High-level d.c. voltages		
±1 V, ±5 V, ±10 V	250 μV	±0.1% of FS
D.C.. Currents		
0 to ±1 mA, 0 to ±20 mA, 4–20 mA	0.031% of span	±0.025% of span

64 × 8 bit of RAM memory, 27 input–output lines and an 8-bit timer/counter. The 8048 is meant for use in high-volume production using factory programmed mask ROM. Development of such systems is undertaken using either the 8748, which uses EPROM instead of ROM, or the 8035, which can be used with external ROM or RAM. Both the 8748 and the 8035 are pin-for-pin compatible with the 8048. This family of microprocessors has been designed for efficient operation as either a controller or an arithmetic processor. Analogue signals can be processed by employing an ADC interfaced to one of the input ports of the 8048. Alternatively, single-chip microcomputers such as the 8022 provide an onboard 8-bit ADC.

4.10.4.1 A microprocessor-based measurement and control subsystem

The μMAC-4000, produced by Analog Devices (Analog Devices, 1984), is a modular measurement and control system based on an Intel 8085 microcomputer. It is designed to operate with any host computer having an RS-232-C or 20 mA serial communication port.

The basic building block of this modular system is the μMAC-4000 master board shown in Figure 4.52, which provides four, eight or 12 analogue input channels having a common mode voltage capability of ±1000 V and a common mode rejection ratio of 160 dB. Analogue-to-digital conversion is provided by means of a 13-bit ADC. A series of standard signal processing modules is available which allows the range of analogue input variables shown in Table 4.9. The eight channels of digital input provide isolation of up to ±300 V and are directly compatible with TTL signals or contact closures. The eight channels of digital output are compatible with TTL levels.

A cluster can be constructed consisting of a μMAC-4000 master board and up to six expander boards. The expansion boards can be the μMAC-4010, which has the same analogue and digital capability as the μMAC-4000 and acts as its slave; the μMAC-4030, which provides an eight-channel, 12-bit DAC capability with either voltage or current output; the μMAC-4020, which is a 16-channel digital input–output subsystem suitable for interfacing to high-level a.c. or d.c. signals and with 2500 V a.c. optical isolation; and the μMAC-4040, which provides 32 digital input channels and 32 digital output channels. A cluster can provide a maximum of 48 analogue input channels, 32 analogue output channels, 136 digital input channels and 136 digital output channels. It is possible to further expand the system by having a party line connection to up to eight clusters, as shown in Figure 4.53. In this configuration the system can have up to 384 analogue inputs, 256 analogue outputs, 1088 digital inputs and 1088 digital outputs.

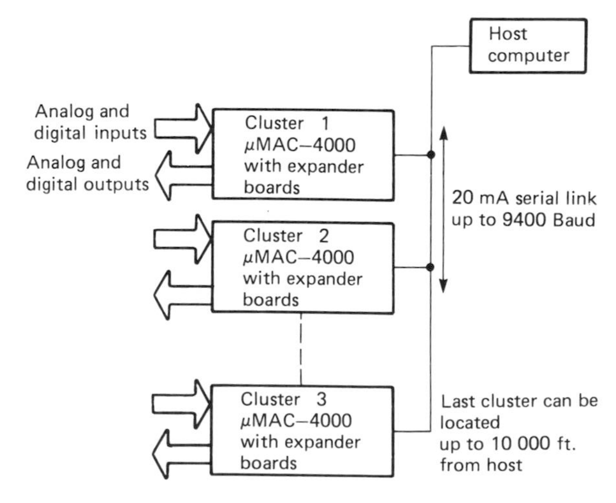

Figure 4.53 Cluster configuration for μMAC-4000 system.

The μMAC-4000 scales, linearizes and converts the input data to engineering units. It scans the analogue inputs fifteen or thirty times per second and stores the results in RAM. Communication between the subsystem and the host computer is by means of a full duplex UART which allows transmission rates of between 110 and 9600 Baud at distances of up to 10 000 ft. A powerful command set is resident in the firmware of the μMAC-4000 which allows simple communication between the subsystem and the host computer. Table 4.10 gives examples of these commands together with the T and C protocols which can be used with the system.

4.10.4.2 A measurement and control processor hosted by a minicomputer

A more powerful data acquisition and control system is provided by a system such as the HP 2250, the components of which are shown in Figure 4.54 (Hewlett Packard, 1983). The HP 2250 derives its intelligence from the onboard HP 1000 computer with its custom-designed operating system. The system consists of two sections: the processor unit and the measurement and control unit (MCU). The processor unit consists of five cards: Battery Back-up, ROM/RAM, CPU, Measurement and Control Interface (MCI), and HP–IB Interface. The

Table 4.10 μMAC-4000 commands and protocols

Command set

Command	Function
CHANNELn	Transmit channel n data
SCANn,m	Transmit channel n through channel m data
SETp.b.	Set digital output bit of port p
LIMITn,LL,HL	Sets 'HI' and 'LOW' limits of channel n
SCACn,v	Set channel n to analog value v

Protocols

Two serial protocols can be used with the μMAC-4000. The 'T' protocol is designed for use with CRT and TTY terminals where familiarization, debugging, system calibration and manual control is required. Simple 'English-like' commands are used with this protocol. The 'C' protocol is designed for use with computers and controllers where communication efficiency, reliability and adaptability to a wide variety of hosts is required

'T' protocol command

SCAN 0,2

This command requests the μMAC to transmit the latest data for channel 0 through 2

'T' protocol reply

CH0 = +0025.4 CH1 = +0653.5 CH2 = +0085.2

If thermocouples were connected to these channels, the data are in terms of °C or °F

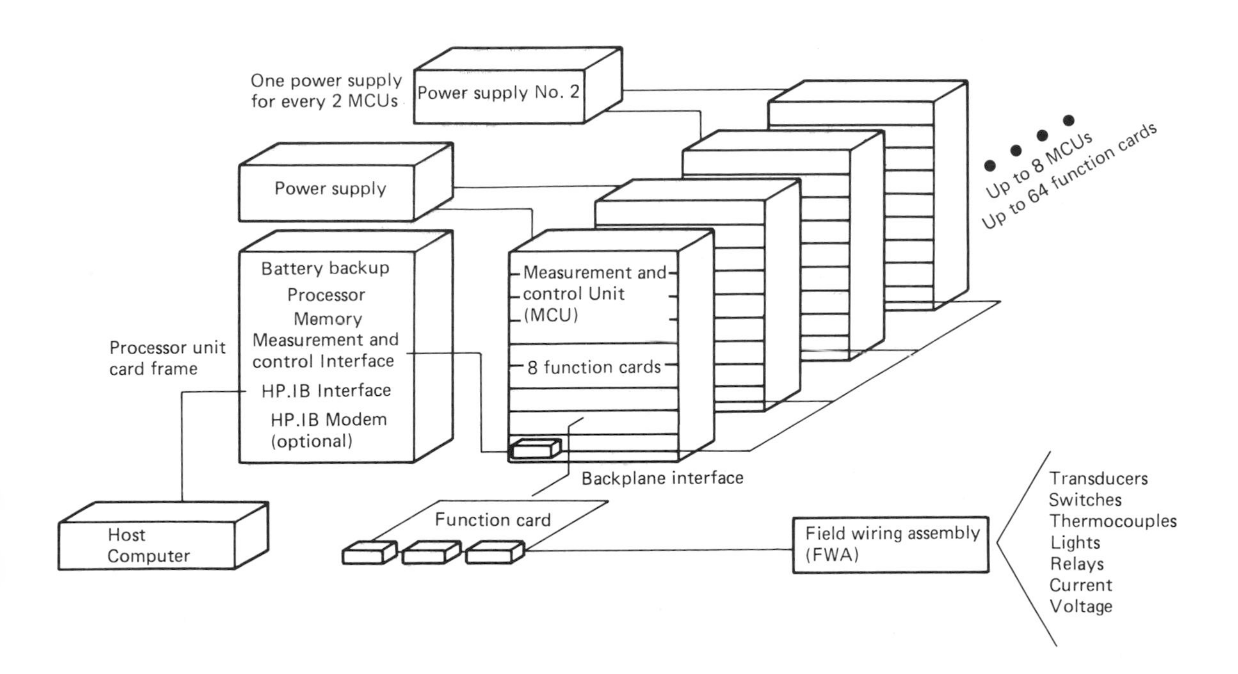

Figure 4.54 Components of HP 2250 measurement and control processor.

Battery Back-up card sustains the memory for up to 30 min in the case of a power failure. The memory card contains both the operating system for the HP 2250 in firmware and Read/Write memory for holding HP 2250 programs. The system has 28 kbytes of user memory. The CPU card employs an HP 1000 L series computer. The Measurement and Control Interface (MCI) provides control of the MCU. The MCI receives groups of control words from the CPU memory and causes signals to be output to the MCU. Data are written to the MCU or retrieved from the MCU through the MCI card.

The fifth card in the processor unit is the HP-IB Interface card. This provides the communication link between the HP 2250 and the host computer, which is typically either one of the HP 1000 or HP 200 series computers. This card allows communication between the HP 2250 and the host computer at distances up to 20 m. The separation can be increased to up to 1000 m using HP-IB extenders and fibre-optic cables.

The MCU is a cage into which eight function cards can be inserted. Each HP 2250 system may have up to eight MCUs. Table 4.11 shows the analogue and digital function cards provided for the system. Each function card is provided with slots to take signal-conditioning modules which filter, attenuate and isolate incoming and outgoing a.c. and d.c.

Table 4.11 Analogue and digital function cards for HP2250 system

Analog function cards

Function card	*No. of channels*	*Resolutions*	*Purpose of card*
High-speed analog input (HP 25501)	16	156 μV max. ±10 V	Convert analog input signal to digital representation
High-level FET multiplexer (HP 25502)	32	156 μV max. ±10 V	Extend HP 25501 ADC by 32 input channels
Low-level FET multiplexer (HP 25503)	32	1.56 μV max. ±10 V	Extend HP 25501 ADC by 32 input channels and provide gain
Relay multiplexer (HP25504)	16	1.56 μV max. ±100 V	Extend HP 25501 ADC by 16 input channels, provide gain and electrical isolation
Voltage/current output HP (25510)	4	5 mV or 5 μA max. ±10.24 V or 20 mA at 20 V	Provide analog voltage and current output

Digital function cards

Function card	*No. of channels*	*Purpose of card*
Digital input (HP25511)	32	Provide digital inputs for monitoring a.c. or d.c. signals (0–120 V d.c., 0–230 V a.c.) through required signal conditioning modules (SCMs). Inputs can be monitored individually or collectively and can generate interrupts upon transitions
Counter input (HP 25512)	4	Provide independently programmable counters which totalize, count up or down and measure periods, time intervals and frequencies. Interrupts can be generated upon overflow or completion of counting
Digital output (HP 25513)	32	Provides solid-state switching of a.c. and d.c. loads of up to 60 V (peak) at 300 mA and zero-voltage switching up to 120 V a.c. at 800 mA through required Signal Conditioning Modules (SCMs). Outputs can be programmed individually or as two 16-bit fields
Relay output (HP 25514)	16	Provides switching of high-current loads (up to 2 A at 30 V d.c. or 3 A at 125 V a.c.). SCMs suppress transients to protect the relays and prevent noise. The 16 channels can be programmed individually or collectively
Output pulse (HP 25515)	4	Provides digital pulses of programmable frequency, width and acceleration for controlling stepper motors. Limit switch inputs on this card can be programmed to abort pulse trains
Digital multifunction (HP 25516)	16-input, 16-output	Provides independently programmable digital inputs and outputs, counting and interrupt capabilities. SCMs provide a wide range of interfacing for monitoring and controlling transducers, instruments and switches

signals, enabling the cards to operate in the most electrically noisy environments. Forty-two different signal-conditioning modules are available. The system is capable of monitoring up to 1920 analogue points and 2048 digital points. The ADC provides a 14-bit conversion at a rate of 50 000 measurements/s.

The HP 2250 employs a measurement and control language called MCL/50, which provides the user with over 100 commands for high-level programming of all the analogue and digital function cards together with features found in FORTRAN and BASIC. These MCL/50 instructions can be embedded in applications programs written in FORTRAN or BASIC on the host computer. All MCL/50 commands sent to the HP 2250 pass through the HP-IB interface card and are placed in the ROM/RAM card. The CPU card then compiles and executes the commands and returns the results back to the host computer. The HP 2250 can also store the sampled data locally in its memory and operate on the data without reference back to the host computer.

Because the HP 2250 has on board a HP 1000 L series computer it is possible to download subroutines which have been written in HP 1000 FORTRAN or Assembly Language. This provides the system with increased flexibility and power and allows the manipulations provided by the HP 2250 to extend beyond the capabilities provided by the MCL/50 language.

An Automation Software Package is available for the HP 1000 host computer which contains the following programs: the External Subroutine Loader Routine, LINKR, which suitably formats files written in FORTRAN or Assembly Language for downloading from the host computer to the HP 2250; Continuous Data Acquisition (CDA) software, which allows for the collection of a large amount of data at high sampling rates; and the Measurement and Control Language Exerciser (MCX), which enables the programmer to have an interactive dialogue with the HP 2250.

The CDA software operates either in a History Mode or a Normal Mode. In the History Mode the HP 2250 samples the data and stores them locally in its memory. The storage available in this mode is limited to about 6500 readings. In the Normal Mode the software causes the data to be sent directly to the host computer where they are stored either in its memory or on high-speed, high-capacity disc drives. In such a mode it is possible to capture several minutes of data sampled at 50 kHz.

The MCX software enables the programmer to test the tasks performed by the MCL/50 software without having to write FORTRAN or BASIC programs. The software includes provision for the reporting of status information, downloading of user subroutines, providing access HP 2250 to buffers or variables, and the reporting of syntax, run time and communication errors.

References

ANALOG DEVICES INC. *Databook*, Volumes 1 and 2, Analog Devices Inc., Norwood, Massachusetts (1984)

ARBEL, A. F. *Analog Signal Processing and Instrumentation*, Cambridge University Press, Cambridge (1980)

ASPINALL, D. *The Microprocessor and its Application*, Cambridge University Press, Cambridge (1980)

BARNEY, G. C. *Intelligent Instrumentation: Microprocessor Applications in Measurement and Control*, Prentice-Hall, London (1985)

BARTREE, T. C. *Digital Computer Fundamentals* (5th edn), McGraw-Hill, London (1981)

BECK, M. S. 'Correlation instruments: cross correlation flowmeters', in *Instrument Science and Technology*, Volume 2, ed. B. E. Jones, Adam Hilger, Bristol (1983)

BENDAT, J. S. and PIERSOL, A. G. *Engineering Applications of Correlation and Spectral Analysis*, Wiley-Interscience, New York (1980)

BIBBERO, D. P. and STERN, D. M. *Microprocessor Systems: Interfacing and Applications*, Wiley-Interscience, New York (1982)

BLAIR, D. P. and SYDENHAM, P. H. 'Phase sensitive detection as a means to recover signals buried in noise, *Journal of Physics E, Scientific Instruments*, **5**, 621–627 (1975)

BOWRON, P. and STEPHENSON, F. W. *Active Filters for Communication and Instrumentation*, McGraw-Hill, Maidenhead (1979)

BOYCE, J. C. *Digital Computer Fundamentals*, Prentice-Hall, Englewood Cliffs, New Jersey (1977)

CLAYTON, G. B. *Operational Amplifiers*, 2nd edn, Newnes–Butterworths, London (1979)

COOLEY, J. W. and TUKEY, J. W. 'An algorithm for the machine calculation of complex Fourier series', *Mathematics of Computation*, **19**, No. 90, 297–313 (1965)

GIACOLLETTO, L. J. *Electronic Designers Handbook*, McGraw-Hill, New York (1977), pp. 13-140–13-146

GRAEME, J. G. *Applications of Operational Amplifiers: Third Generation Techniques*, McGraw-Hill, New York (1973)

GULD, C. 'Microelectrodes and input amplifiers', in *IEE Medical Electronics Monographs*, Nos 7–12, eds D. W. Hill and B. W. Watson, Peter Peregrinus, London,(1974), pp. 1–27

HENRY, R. M. 'An improved algorithm allowing fast on-line polarity correlation by microprocessor or minicomputer', *IEE Conference Digest, No. 1979/32*, 3/1–4 (1979)

HEWLETT PACKARD. 150 Series: *Spectrum Analyser Series Applications Notes*, Palo Alto, California (1971 onwards)

HEWLETT PACKARD. *An Introduction to Programming the HP 2250*, Product Note 2250-1, Palo Alto, California (1983)

INTEL CORPORATION. *Component Data Catalog*, Santa Clara (1980)

INTERSIL INC. *Data Book*, Intersil Inc., Cupertino, California (1979)

JONES, B. E. *Instrumentation, Measurement and Feedback*, McGraw-Hill, Maidenhead (1977)

JORDAN, J. 'Correlation circuits for measurement systems', *IEE Conference Digest No. 1979/32*, 1/1-4 (1979)

KEECH, R. P. 'The KPC multichannel correlation signal processor for velocity measurement', *Transactions of the Institute of Measurement and Control*, **4**, No. 1, 43–52 (1982)

KUO, B. C. *Digital Control Systems*, 2nd edn, Holt, Rinehart and Winton, New York (1980)

LANGE, F. H. *Correlation Techniques*, Iliffe Books, London (1967)

LEWIN, D. *Theory and Design of Digital Computer Systems*, Nelson, Walton-on-Thames (1980)

MEADE, M. L. 'Advances in lock-in amplifiers', in *Instrument Science and Technology*, Volume 2, ed. B. E. Jones, Adam Hilger, Bristol (1983a)

MEADE, M. L. *Lock-in Amplifiers: Principles and Applications*, Peter Peregrinus, London (1983b)

OWENS, A. R. 'Digital signal conditioning and conversion', in *Instrument Science and Technology*, ed. B. E. Jones, Adam Hilger, Bristol (1983), pp. 57–78

PAKER, Y. *Minicomputers: A Reference Book for Engineers and Managers*, Abacus Press, Tunbridge Wells (1981)

PAPOULIS, A. *Signal Analysis*, McGraw-Hill, London (1977)

SANDERSON, P. C. *Minicomputers*, Newnes–Butterworth, London (1976)

SHEINGOLD, D. H. *Analog/digital Conversion Notes*, Analog Devices, Norwood, Massachusetts (1977)

STEELE, R. *Delta Modulation Systems*, Pentech Press, London (1975)

TOBEY, G. E., GRAEME, J. G. and HUELSMAN, L. P. *Operational Amplifiers*, McGraw-Hill, New York (1971)

WILKINSON, B. and HORROCKS, D. *Computer Peripherals*, Hodder, London (1980)

5 Telemetry

M. L. Sanderson

5.1 Introduction

Within instrumentation there is often a need for telemetry in order to transmit data or information between two geographical locations. The transmission may be required to enable centralized supervisory data logging, signal processing or control to be exercised in large-scale systems which employ distributed data logging or control subsystems. In a chemical plant or power station these subsystems may be spread over a wide area. Telemetry may also be required for systems which are remote or inaccessible such as a spacecraft, a satellite, or an unmanned buoy in the middle of the ocean. It can be used to transmit information from the rotating sections of an electrical machine without the need for slip rings. By using telemetry-sensitive signal processing and recording, apparatus can be physically remote from hazardous and aggressive environments and can be operated in more closely monitored and controlled conditions.

Telemetry has traditionally been provided by either pneumatic or electrical transmission. Pneumatic transmission, as shown in Figure 5.1, has been used extensively in process instrumentation and control. The measured quantity (pressure, level, temperature, etc.) is converted to a pneumatic pressure, the standard signal ranges being 20–100 kPa gauge pressure (3–15 lb/in^2/g) and 20–180 kPa (3–27 lb/in^2/g). The lower limit of pressure provides a live zero for the instrument which enables line breaks to be detected, eases instrument calibration and checking, and provides for improved dynamic response since, when venting to atmospheric pressure, there is still sufficient driving pressure at 20 kPa. The pneumatic signals can be transmitted over distances up to 300 m in 6.35 mm or 9.5 mm OD plastic or metal tubing to a pneumatic indicator, recorder or controller. Return signals for control purposes are transmitted from the control element. The distance is limited by the speed of response, which quadruples with doubling the

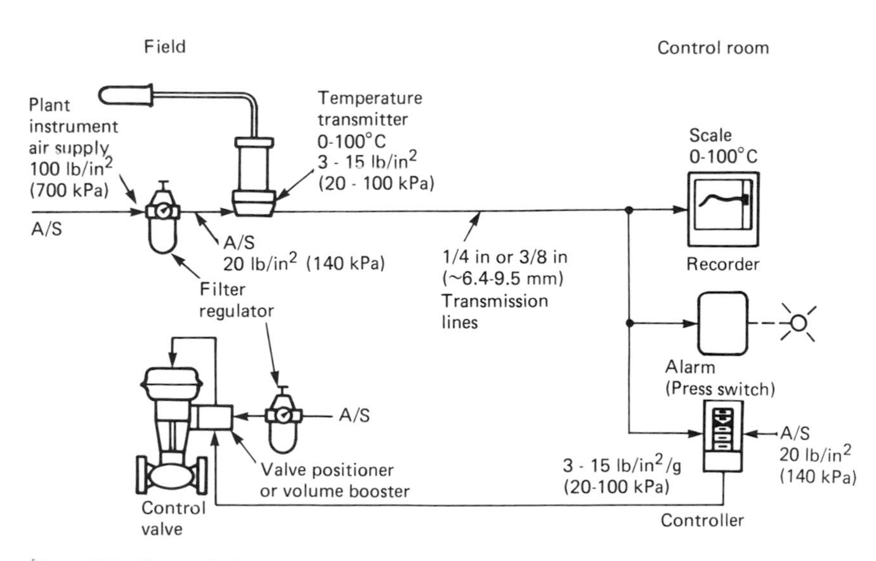

Figure 5.1 Pneumatic transmission.

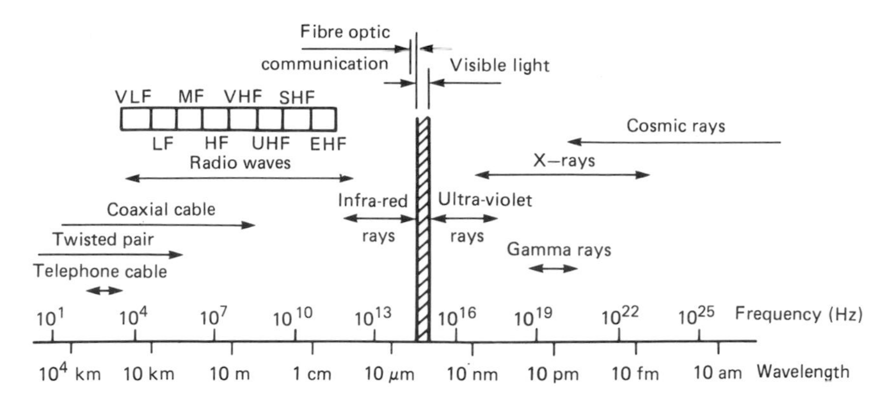

Figure 5.2 Electromagnetic spectrum.

distance. Pneumatic instrumentation generally is covered at greater length in Chapter 7.

Pneumatic instruments are intrinsically safe, and can therefore be used in hazardous areas. They provide protection against electrical power failure since systems employing air storage or turbine-driven compressors can continue to provide measurement and control during power failure. Pneumatic signals also directly interface with control valves which are pneumatically operated and thus do not require the electrical/pneumatic converters required by electrical telemetry systems, although they do suffer from the difficulty of being difficult to interface to data loggers. Pneumatic transmission systems require a dry, regulated air supply. Condensed moisture in the pipework at subzero temperatures or small solid contaminants can block the small passages within pneumatic instruments and cause loss of accuracy and failure. Further details of pneumatic transmission and instrumentation can be found in Bentley (1983) and Warnock (1985).

Increasingly telemetry in instrumentation is being undertaken using electrical, radio frequency, microwave or optical fibre techniques. The communication channels used include transmission lines employing two or more conductors which may be a twisted pair, a coaxial cable or a telephone line physically connecting the two sites; radio frequency (r.f.) or microwave links which allow the communication of data by modulation of an r.f. or microwave carrier; and optical links in which the data are transmitted as a modulation of light down a fibre-optic cable. All of these techniques employ some portion of the electromagnetic spectrum, as shown in Figure 5.2.

Figure 5.3 shows a complete telemetry system. Signal conditioning in the form of amplification and filtering normalizes the outputs from different transducers and restricts their bandwidths to those available on the communication channel. Transmission systems can employ voltage, current, position, pulse or frequency techniques in order to transmit analogue or digital data. Direct transmission of analogue signals as voltage, current or position

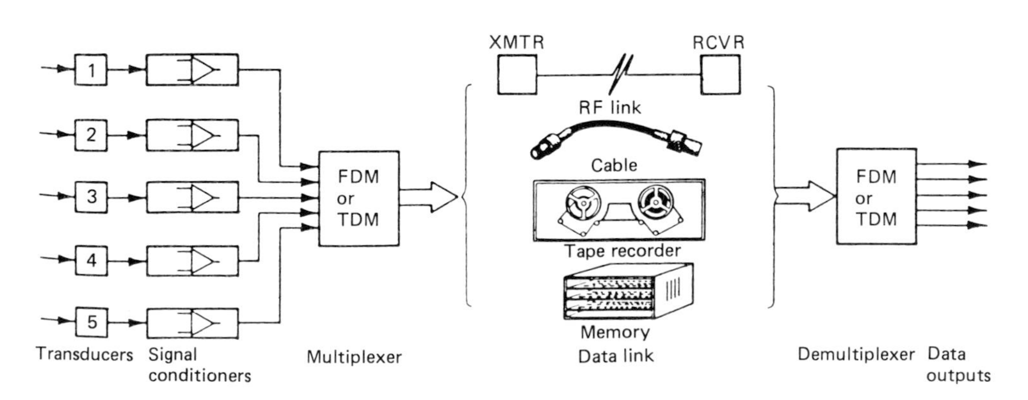

Figure 5.3 Telemetry system.

requires a physical connection between the two points in the form of two or more wires and cannot be used over the telephone network. Pulse and frequency telemetry can be used for transmission over both direct links and also for telephone, r.f., microwave and optical links. Multiplexing either on a time or frequency basis enables more than one signal to be transmitted over the same channel. In pulse operation the data are encoded as the amplitude, duration or position of the pulse or in a digital form. Transmission may be as a baseband signal or as an amplitude, frequency or phase modulation of a carrier wave.

In the transmission of digital signals the information capacity of the channel is limited by the available bandwidth, the power level and the noise present on the channel. The Shannon–Hartley theorem states that the information capacity, C, in bits/s (bps) for a channel having a bandwidth B Hz and additative Gaussian band-limited white noise is given by

$$C = B.\log_2\left(1 + \frac{S}{N}\right)$$

where S is the average signal power at the output of the channel and N is the noise power at the output of the channel.

This capacity represents the upper limit at which data can be reliably transmitted over a particular channel. In general, because the channel does not have the ideal gain and phase characteristics required by the theorem and also because it would not be practical to construct the elaborate coding and decoding arrangements necessary to come close to the ideal, the capacity of the channel is significantly below the theoretical limit.

Channel bandwidth limitations also give rise to bit rate limitations in digital data transmission because of intersymbol interference (ISI), in which the response of the channel to one digital signal interferes with the response to the next. The impulse response of a channel having a limited bandwidth of B Hz is shown in Figure 5.4(a). The response has zeros separated by $1/2B$ s. Thus for a second impulse transmitted across the channel at a time $1/2B$ s later there will be no ISI from the first impulse. This is shown in Figure 5.4(b). The maximum data rate for the channel such that no ISI occurs is thus $2B$ bps. This is known as the Nyquist rate. Figure 5.4(c) shows the effect of transmitting data at a rate in excess of the Nyquist rate.

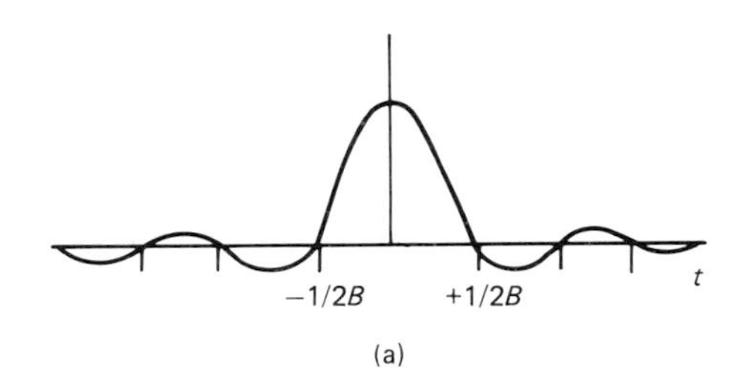

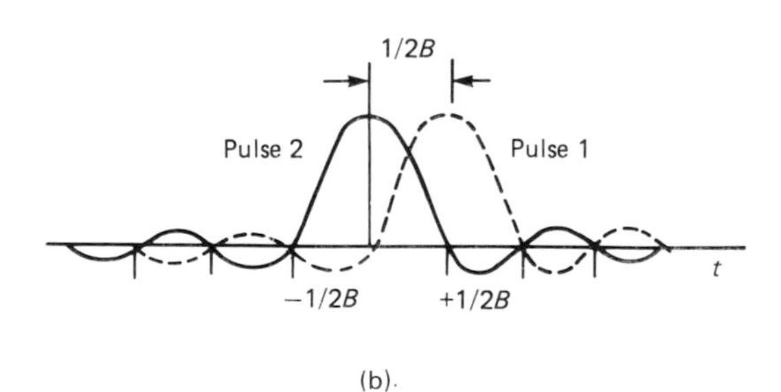

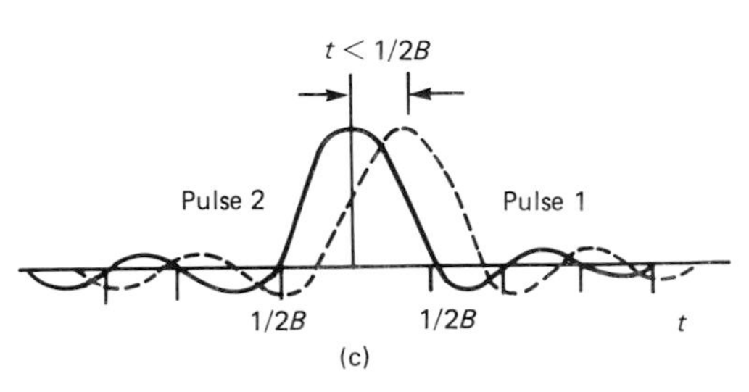

Figure 5.4 (a) Impulse response of a bandlimited channel; (b) impulse responses delayed by $1/2B$ s; (c) impulse responses delayed by less than 1/2B **s.**

5.2 Communication channels

5.2.1 Transmission lines

Transmission lines are used to guide electromagnetic waves, and in instrumentation these commonly take the form of a twisted pair, a coaxial cable or a telephone line. The primary constants of such lines in terms of their resistance, leakage conductance, inductance and capacitance are distributed as shown in Figure 5.5. At low frequencies, generally below 100 kHz, a medium-length line may be represented by the circuit shown in Figure 5.6, where R_L is the resistance of the wire and C_L is the lumped capacitance of the line. The line thus acts as a low-pass filter. The frequency response can be extended by loading the line with regularly placed lumped inductances.

Transmission lines are characterized by three secondary constants. These are the characteristic impedance, Z_0; the attenuation, α, per unit length of line which is usually expressed in dB/unit length; and the phase shift, β, which is measured in

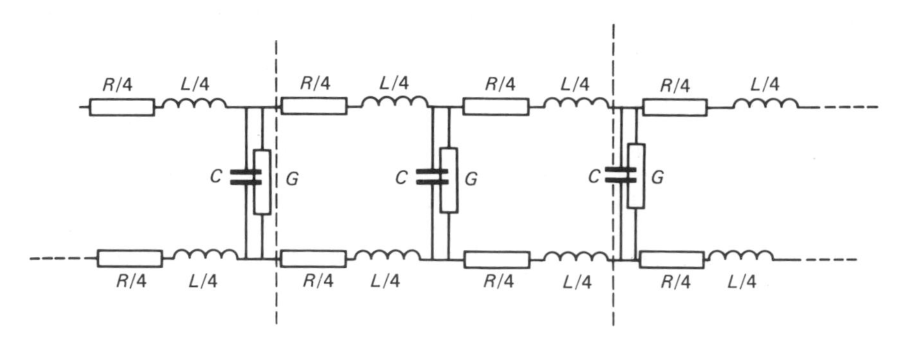

Figure 5.5 Distributed primary constants of a transmission line.

radians/unit length. The values of Z_0, α, β, are related to the primary line constants by:

$$Z_0 = \sqrt{\left(\frac{R + j\omega L}{G + j\omega C}\right)}\ \Omega$$

$$\alpha = 8.68.[0.5(\{(R^2 + \omega^2 L^2)(G^2 + \omega^2 C^2)\}^{1/2} + (RG - \omega^2 LC))]^{1/2} \text{ dB/unit length}$$

$$\beta = [0.5(\{(R^2 + \omega^2 L^2)(G^2 + \omega^2 C^2)\}^{1/2} - (RG - \omega^2 LC))]^{1/2} \text{ radians/unit length}$$

where R is the resistance per unit length, G is the leakage conductance per unit length, C is the capacitance per unit length and L is the inductance per unit length.

It is necessary to terminate transmission lines with their characteristic impedance if reflection or signal echo is to be avoided. The magnitude of the reflection for a line of characteristic impedance Z_0 terminated with an impedance Z_T is measured by the reflection coefficient, ρ, given by:

$$\rho = \frac{Z_T - Z_0}{Z_T + Z_0}$$

Twisted pairs are precisely what they say they are, namely two insulated conductors twisted together. The conductors are generally copper or aluminium and plastic is often used as the insulating material. The twisting reduces the effect of inductively coupled interference. Typical values of the primary constants for a 22 gauge copper twisted pair are R = 100 Ω/km, L = 1 mH/km, G = 10^{-5} S/km and C = 0.05 μF/km. At high frequencies the characteristic impedance of the line is approximately 140 Ω. Typical values for the attenuation of a twisted pair are 3.4 dB/km at 100 kHz, 14 dB/km at 1 MHz and 39 dB/km at 10 MHz. The high-frequency limitation for the use of twisted pairs at approximately 1 MHz occurs not so much as a consequence of attenuation but because of crosstalk caused by capacitive coupling between adjacent twisted pairs in a cable.

Coaxial cables which are used for data transmission at higher frequencies consist of a central core conductor surrounded by a dielectric material which may be either polythene or air. The construction of such cables is shown in Figure 5.7. The outer conductor consists of a solid or braided sheath around the dielectric. In the case of the air dielectric the central core is supported on polythene spacers placed uniformly along the line. The outer conductor is usually covered by an insulating coating. The loss at high frequencies in coaxial cable is due to the 'skin effect', which forces the current in the central core to flow near to its surface and thus increases the resistance of the conductor. Such cables have a characteristic impedance of between 50 and 75 Ω. The typical attenuation of a 0.61 cm diameter coaxial cable is 8 dB/100 m at 100 MHz and 25 dB/100 m at 1 GHz.

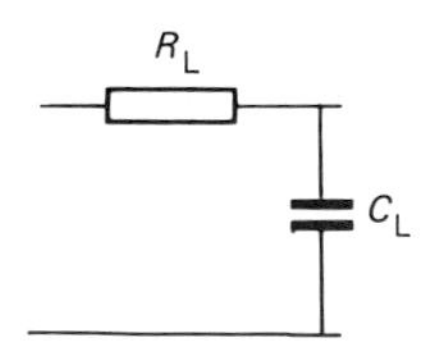

Figure 5.6 Low-frequency lumped approximation for a transmission line.

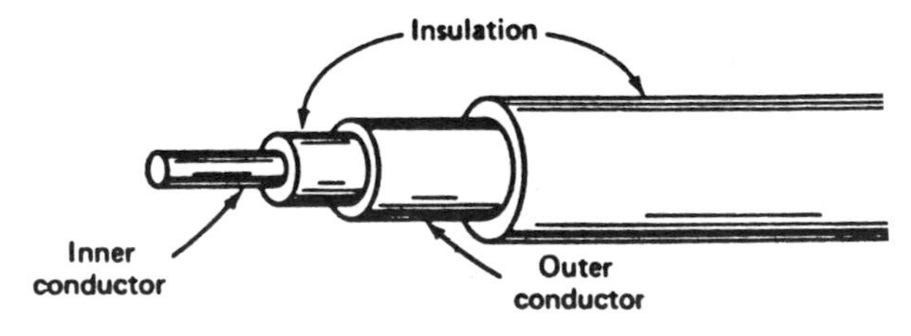

Figure 5.7 Coaxial cable.

Trunk telephone cables connecting exchanges consist of bunched twisted conductor pairs. The conductors are insulated with paper or polyethylene, the twisting being used to reduce the crosstalk between adjacent conductor pairs. A bunch of twisted cables is sheathed in plastic and the whole cable is given mechanical strength by binding with steel wire or tape which is itself sheathed in plastic. At audiofrequencies the impedance of the cable is dominated by its capacitance and resistance. This results in an attenuation which is frequency dependent and also phase delay distortion, since signals of different frequencies are not transmitted down the cable with the same velocity. Thus a pulse propagated down a cable results in a signal which is not only attenuated (of importance in voice and analogue communication) but which is also phase distorted (of importance in digital signal transmission). The degree of phase delay distortion is measured by the group delay $d\beta/d\omega$. The bandwidth of telephone cables is restricted at low frequencies by the use of a.c. amplification in the repeater stations used to boost the signal along the line. Loading is used to improve the high-frequency amplitude response of the line. This takes the form of lumped inductances which correct the attenuation characteristics of the line. These leave the line with a significant amount of phase delay distortion and also give the line attenuation at high frequencies. The useable frequency band of the telephone line is between 300 Hz and 3 kHz. Figure 5.8 shows typical amplitude and phase or group delay distortions relative to 800 Hz for a typical leased line and a line to which equalization or conditioning has been applied.

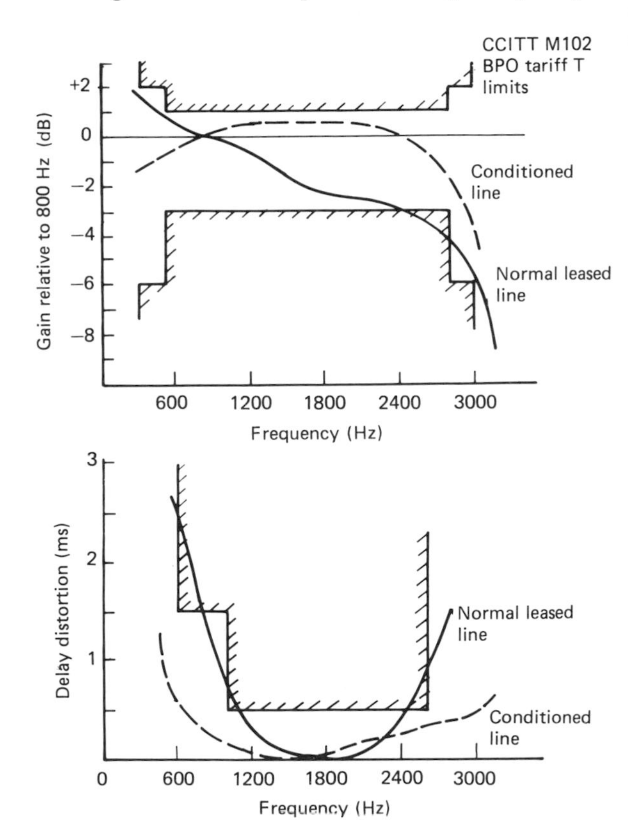

Figure 5.8 Gain and delay distortion on telephone lines.

In order to transmit digital information reliably the transmission equipment has to contend with a transmission loss which may be as high as 30 dB; a limited bandwidth caused by a transmission loss which varies with frequency; group delay variations with frequency; echoes caused by impedance mismatching and hybrid crosstalk; and noise which may be either Gaussian or impulsive noise caused by dial pulses, switching equipment or lightning strikes. Thus it can be seen that the nature of the telephone line causes particular problems in the transmission of digital data. Devices known as modems (MOdulators/DEModulators) are used to transmit digital data along telephone lines. These are considered in section 5.9.1.

5.2.2 Radiofrequency transmission

Radiofrequency (r.f.) transmission is widely used in both civilian and military telemetry and can occur from 3 Hz (which is referred to as very low frequency (VLF)) up as high as 300 GHz (which is referred to as extremely high frequency (EHF)). The transmission of the signal is by means of line-of-sight propagation, ground or surface wave diffraction, ionospheric reflection or forward scattering (Coates, 1982). The transmission of telemetry or data signals is usually undertaken as the amplitude, phase or frequency modulation of some r.f. carrier wave. These modulation techniques are described in section 5.5. The elements of an r.f. telemetry system are shown in Figure 5.9.

The allocation of frequency bands has been internationally agreed under the Radio Regulations of the International Telecommunication Union based in Geneva. These regulations were agreed in 1959 and revised in 1979 (HMSO, 1980). In the UK the Radio Regulatory Division of the Department of Trade approves equipment and issues licences for the users of radio telemetry links. For general-purpose low-power telemetry and telecontrol there are four bands which can be used. These are 0–185 kHz and 240–315 kHz, 173.2–173.35 MHz and 458.5–458.8 MHz. For high-power private point systems the allocated frequencies are in the UHF band 450–470 MHz.

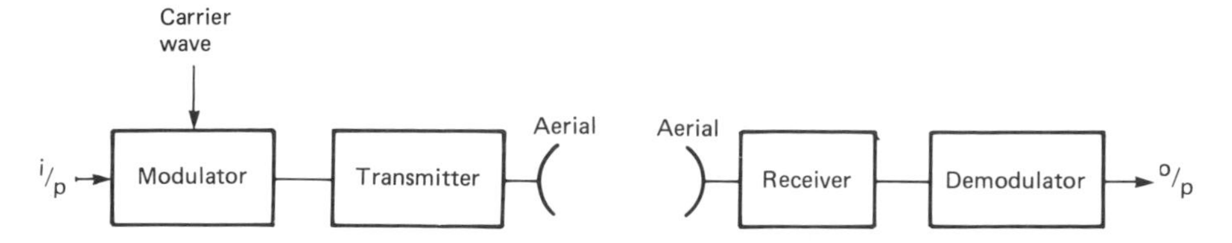

Figure 5.9 R.F. telemetry system.

For medical and biological telemetry there are three classes of equipment. Class I are low-power devices operating between 300 kHz and 30 MHz wholly contained within the body of an animal or man. Class II is broad-band equipment operating in the band 104.6–105 MHz. Class III equipment is narrow-band equipment operating in the same frequency band as the Class II equipment. Details of the requirements for r.f. equipment can be found in the relevant documents cited in the References (HMSO, 1963, 1978, 1979).

5.2.3 Fibre-optic communication

Increasingly, in data-communication systems there is a move towards the use of optical fibres for the transmission of data. Detailed design considerations for such systems can be found in Keiser (1983). Wilson and Hawkes (1983) and Senior (1985). As a transmission medium fibre-optic cables offer the following advantages:

(1) They are immune to electromagnetic interference.
(2) Data can be transmitted at much higher frequencies and with lower losses than twisted pairs or coaxial cables. Fibre optics can therefore be used for the multiplexing of a large number of signals along one cable with greater distances required between repeater stations.
(3) They can provide enhanced safety when operating in hazardous areas.
(4) Earth loop problems can be reduced.
(5) Since the signal is confined within the fibre by total internal reflection at the interface between the fibre and the cladding fibre-optic links provide a high degree of data security and little fibre-to-fibre crosstalk.
(6) The material of the fibre is very much less likely to be attacked chemically than copper-based systems and it can be provided with mechanical properties which will make such cables need less maintenance than the equivalent twisted pair or coaxial cable.
(7) Fibre-optic cables can offer both weight and size advantages over copper systems.

5.2.3.1 Optical fibres

The elements of an optical fibre as shown in Figure 5.10 are the core material, the cladding and the buffer coating. The core material is either plastic or glass. The cladding is a material whose refractive index is less than that of the core. Total internal reflection at the core/cladding interface confines the light to travel within the core. Fibres with plastic cores also have plastic cladding. Such fibres exhibit high losses but are widely used for short distance transmission. Multicomponent glasses containing a number of oxides are used for all but the lowest loss fibres which are usually made from pure silica. In low- and medium-loss fibres the glass core is surrounded by a glass or plastic cladding. The buffer coating is an elastic, abrasion-resistant plastic material which increases the mechanical strength of the fibre and provides it with mechanical isolation from geometrical irregularities, distortions or roughness of adjacent surfaces which could otherwise cause scattering losses when the fibre is incorporated into cables or supported by other structures.

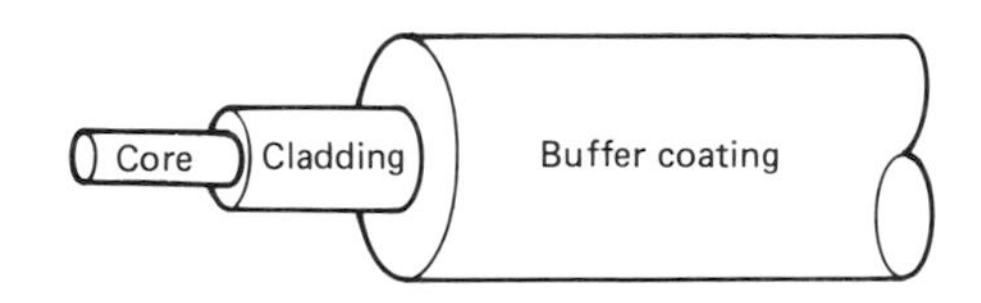

Figure 5.10 Elements of an optical fibre.

The numerical aperture (NA) of a fibre is a measure of the maximum core angle for light rays to be reflected down the fibre by total internal reflection.

By Snell's Law:

$$NA = \sin\theta = \sqrt{(\mu_1^2 - \mu_2^2)}$$

where μ_1 is the refractive index of the core material and μ_2 is the refractive index of the cladding material.

Fibres have NAs in the region of 0.15–0.4, corresponding to total acceptance angles of between 16 and 46 degrees. Fibres with higher NA values

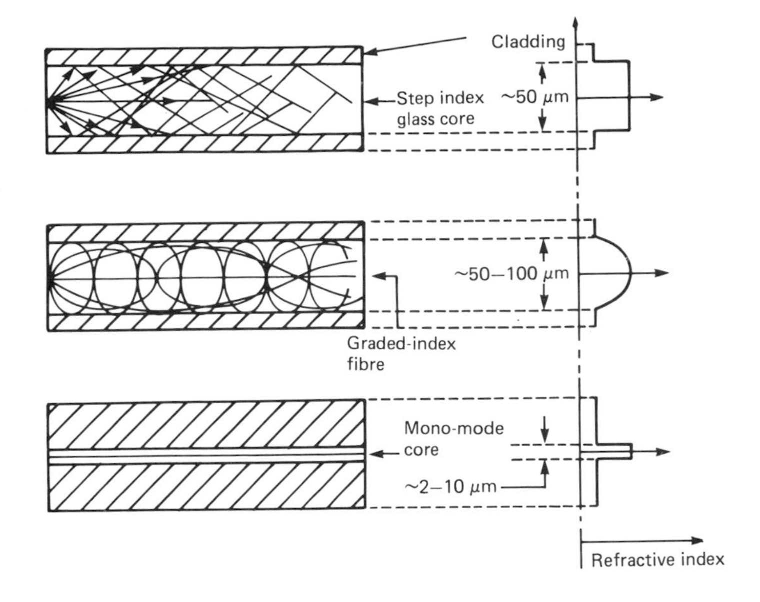

Figure 5.11 Propagation down fibres.

generally exhibit greater losses and low bandwidth capabilities.

The propagation of light down the fibres is described by Maxwell's equations, the solution of which gives rise to a set of bounded electromagnetic waves called the 'modes' of the fibre. Only a discrete number of modes can propagate down the fibre, determined by the particular solution of Maxwell's equation obtained when boundary conditions appropriate to the particular fibre are applied. Figure 5.11 shows the propagation down three types of fibre. The larger-core radius multimode fibres are either step index or graded index fibres. In the step index fibres there is a step change in the refractive index at the core/cladding interface. The refractive index of the graded index fibre varies across the core of the fibre. Monomode fibres have a small-core radius, which permits the light to travel along only one path in the fibre.

The larger-core radii of multimode fibres make it much easier to launch optical power into the fibre and facilitate the connecting of similar fibres. Power can be launched into such a fibre using light-emitting diodes (LEDs), whereas single-mode fibres must be excited with a laser diode.

Intermodal dispersion occurs in multimode fibres because each of the modes in the fibres travels at a slightly different velocity. An optical pulse launched into a fibre has its energy distributed amongst all its possible modes, and therefore as it travels down the fibre the dispersion has the effect of spreading the pulse out. Dispersion thus provides a bandwidth limitation on the fibre. This is specified in MHz.km. In graded index fibres the effect of intermodal dispersion is reduced over that in step index fibres because the grading bends the various possible light rays along paths of nominal equal delay. There is no intermodal dispersion in a single-mode fibre, and therefore these are used for the highest-capacity systems. The bandwidth limitation for a plastic clad step index fibre is typically 6–25 MHz.km. Employing graded index plastic-clad fibres this can be increased to the range of 200–400 MHz.km. For monomode fibres the bandwidth limitation is typically 500–1500 MHz.km.

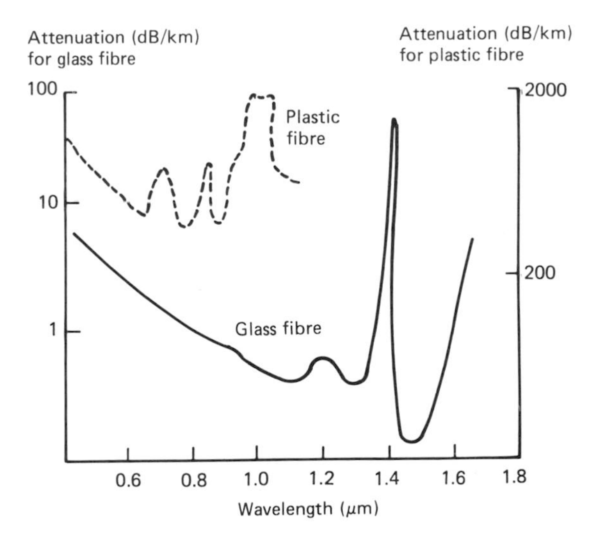

Figure 5.12 Attenuation characteristics of optical fibres.

Attenuation within a fibre, which is measured in dB/km, occurs as a consequence of absorption, scattering and radiative losses of optical energy. Absorption is caused by extrinsic absorption by impurity atoms in the core and intrinsic absorption by the basic constituents of the core material. One impurity which is of particular importance is the OH (water) ion, and for low-loss materials this is controlled to a concentration of less than 1 ppb. Scattering losses occur as a consequence of microscopic variations in material density or composition, and from structural irregularities or defects introduced during manufacture. Radiative losses occur whenever an optical fibre undergoes a bend having a finite radius of curvature.

Attenuation is a function of optical wavelength. Figure 5.12 shows the typical attenuation versus wavelength characteristics of a plastic and a monomode glass fibre. At 0.8 μm the attenuation of the plastic fibre is 350 dB/km and that of the glass fibre is approximately 1 dB/km. The minimum attenuation of the glass fibre is 0.2 dB/km at 1.55 μm. Figure 5.13 shows the construction of the light- and medium-duty optical cables.

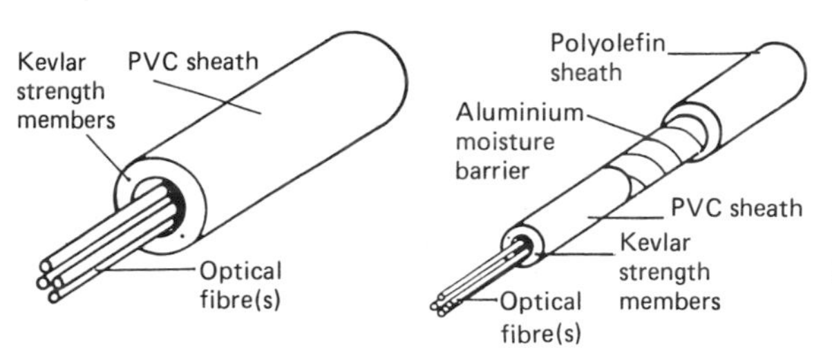

Figure 5.13 Light- and medium-duty optical cables.

5.2.3.2 Sources and detectors

The sources used in optical fibre transmission are LEDs and semiconductor laser diodes. LEDs are capable of launching a power of between 0.1 and 10 mW into the fibre. Such devices have a peak emission frequency in the near infra-red, typically between 0.8 and 1.0 μm. Figure 5.14 shows the typical spectral output from a LED. Limitations on the transmission rates using LEDs occur as a consequence of its rise time, typically between 2 and 10 ns, and chromatic dispersion. This occurs because the refractive index of the core material varies with optical wavelength, and therefore the various spectral components of a given mode will travel at different speeds.

Semiconductor laser diodes can provide significantly higher power, particularly with low-duty

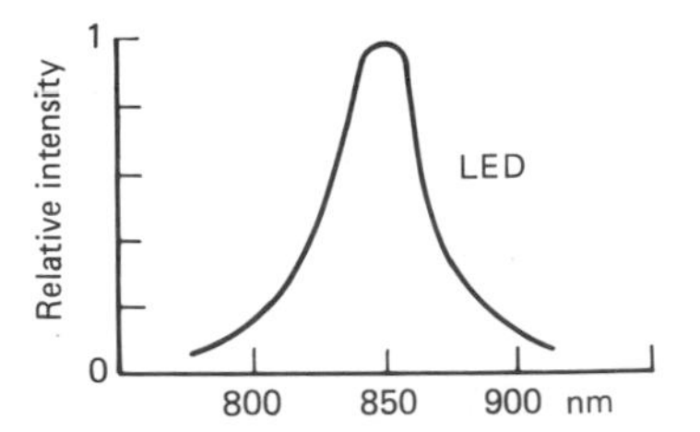

Figure 5.14 Spectral output from a LED.

cycles, with outputs typically in the region of 1 to 100 mW. Because they couple into the fibre more efficiently they offer a higher electrical to optical efficiency than do LEDs. The lasing action means that the device has a narrower spectral width compared with a LED, typically 2 nm or less, as shown in Figure 5.15. Chromatic dispersion is therefore less for laser diodes which also have a faster rise time, typically 1 ns.

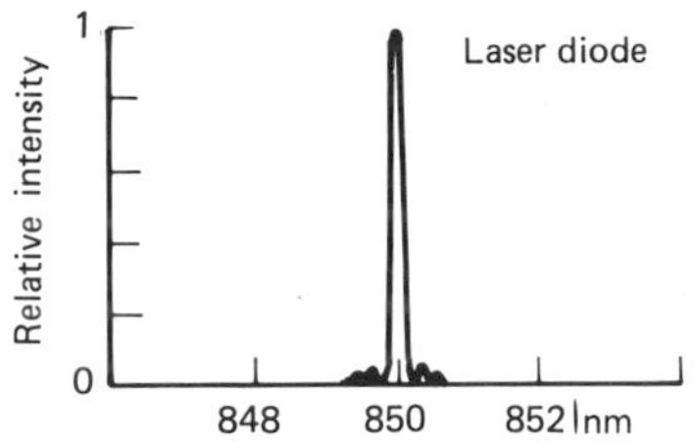

Figure 5.15 Spectral output from a laser diode.

For digital transmissions of below 50 Mbps LEDs require less complex drive circuitry than laser diodes and require no thermal or optical power stabilization.

Both p–i–n (p material–intrinsic–n material) diodes and avalanche photodiodes are used in the detection of the optical signal at the receiver. In the region 0.8–0.9 μm silicon is the main material used in the fabrication of these devices. The p–i–n diode has a typical responsivity of 0.65 A/W at 0.8 μm. The avalanche photodiode employs avalanche action to provide current gain and therefore higher detector responsivity. The avalanche gain can be 100, although the gain produces additional noise. The sensitivity of the photodetector and receiver system is determined by photodetector noise which occurs as a consequence of the statistical nature of the production of photoelectrons, and bulk and dark surface current, together with the thermal noise in the detector resistor and amplifier. For p–i–n diodes the thermal noise of the resistor and amplifier dominates, whereas with avalanche photodiodes the detector noise dominates.

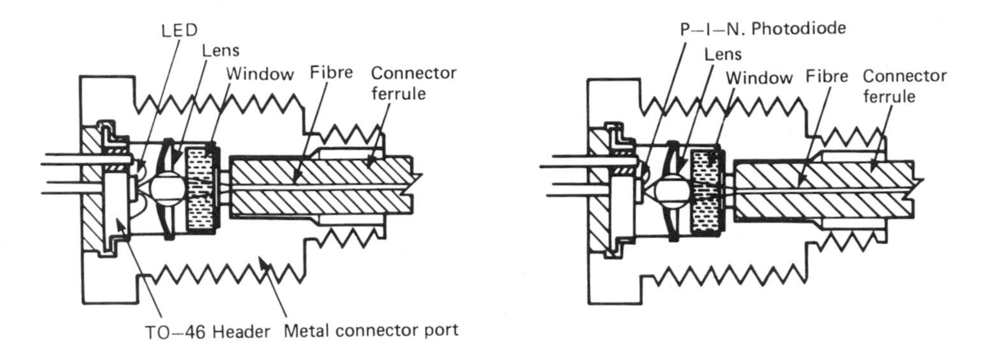

Figure 5.16 LED and p–i–n diode detector for use in a fibre-optic system.

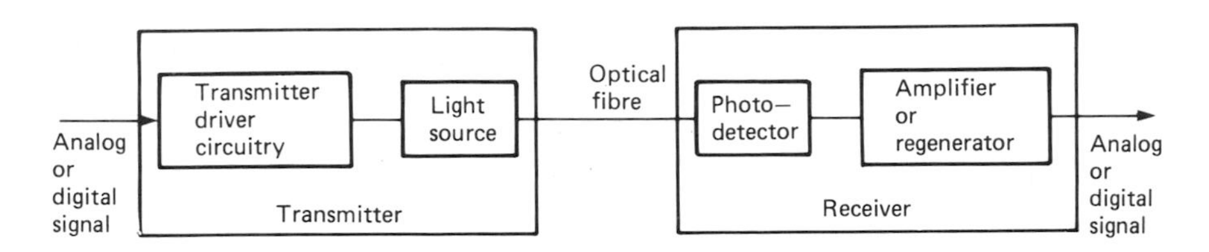

Figure 5.17 Fibre-optic communication system.

Figure 5.16 shows a LED and p–i–n diode detector for use in a fibre-optic system.

5.2.3.3 Fibre-optic communication systems

Figure 5.17 shows a complete fibre-optic communications system. In the design of such systems it is necessary to compute the system insertion loss in order that the system can be operated using the minimum transmitter output flux and minimum receiver input sensitivity. In addition to the loss in the cable itself, other sources of insertion loss occur at the connections between the transmitter and the cable and the cable and the receiver; at connectors joining cables; and at points where the cable has been spliced. The losses at these interfaces occur as a consequence of reflections, differences in fibre diameter, *NA* and fibre alignment. Directional couplers and star connectors also increase the insertion loss.

5.3 Signal multiplexing

In order to enable several signals to be transmitted over the same medium it is necessary to multiplex the signals. There are two forms of multiplexing—frequency-division multiplexing (FDM) and time-division multiplexing (TDM). FDM splits the available bandwidth of the transmission medium into a series of frequency bands and uses each of the frequency bands to transmit one of the signals. TDM splits the transmission into a series of time slots and allocates certain time slots, usually on a cyclical basis, for the transmission of one signal.

The basis of FDM is shown in Figure 5.18(a). The bandwidth of the transmission medium f_m is split into a series of frequency bands, having a bandwidth f_{ch}, each one of which is used to transmit one signal. Between these channels there are frequency bands, having bandwidth f_g, called 'guard bands' which are used to ensure that there is adequate separation and minimum crosstalk between any two adjacent channels. Figure 5.18(b) shows the transmission of three band-limited signals having spectral characteristics as shown, the low-pass filters at the input to the modulators being used to bandlimit the signals. Each of the signals then modulates a carrier. Any form of carrier modulation can be used, although it is desirable to use a modulation which requires minimum bandwidth. The modulation shown in Figure 5.18(b) is amplitude modulation (see section 5.5). The individually modulated signals are then summed and transmitted. Bandpass filters after reception are used to separate the channels, by providing attenuation which starts in the guard bands. The signals are then demodulated and smoothed.

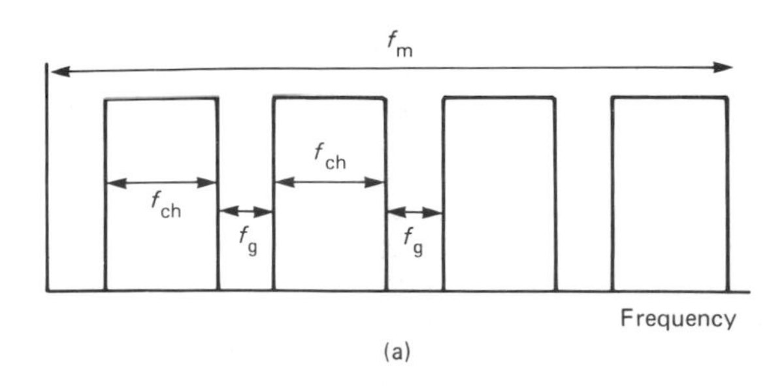

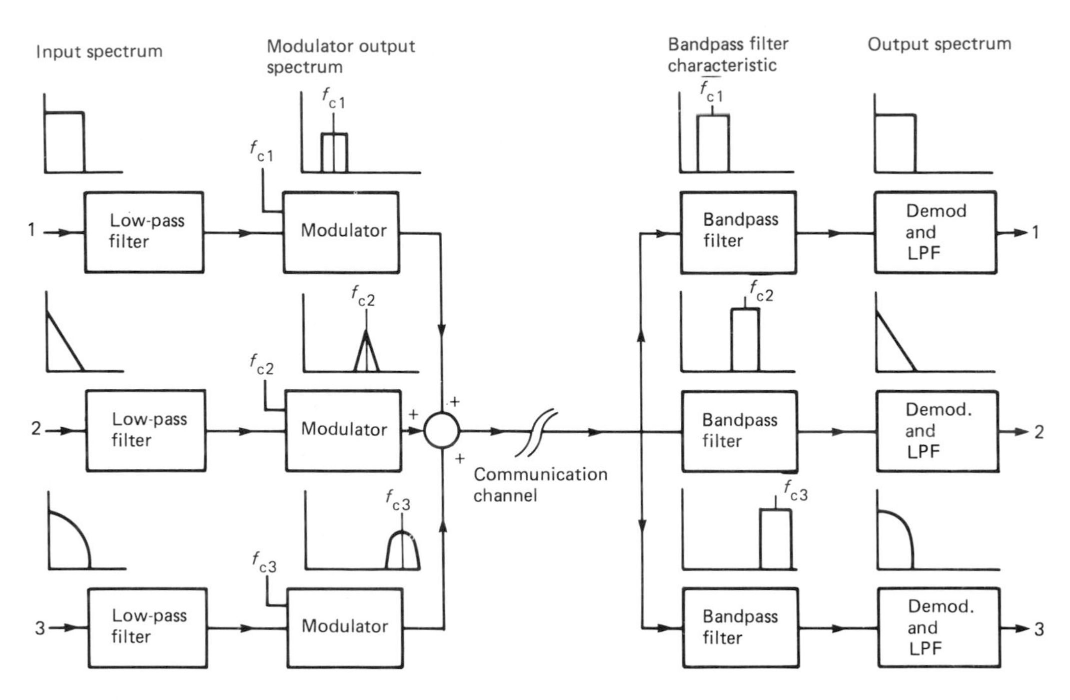

Figure 5.18 (a) Frequency-division multiplexing; (b) transmission of three signals using FDM.

TDM is shown schematically in Figure 5.19. The multiplexer acts as a switch connecting each of the signals in turn to the transmission channel for a given time. In order to recover the signals in the correct sequence it is necessary to employ a demultiplexer at the receiver or to have some means inherent within the transmitted signal to identify its source. If N signals are continuously multiplexed then each one of them is sampled at a rate of $1/N$ Hz. They must therefore be bandlimited to a frequency of $1/2N$ Hz if the Shannon sampling theorem is not to be violated.

The multiplexer acts as a multi-input–single output switch, and for electrical signals this can be done by mechanical or electronic switching. For high frequencies electronic multiplexing is employed, with integrated circuit multiplexers which use CMOS or BIFET technologies.

TDM circuitry is much simpler to implement than FDM circuitry, which requires modulators, bandpass filters and demodulators for each channel. In TDM only small errors occur as a consequence of circuit non-linearities, whereas phase and amplitude non-linearities have to be kept small in order to limit intermodulation and harmonic distortion in FDM systems. TDM achieves its low channel crosstalk by using a wideband system. At high transmission rates errors occur in TDM systems due to timing jitter, pulse accuracy and synchronization problems. Further details of FDM and TDM systems can be found in Johnson (1976) and Shanmugan (1979).

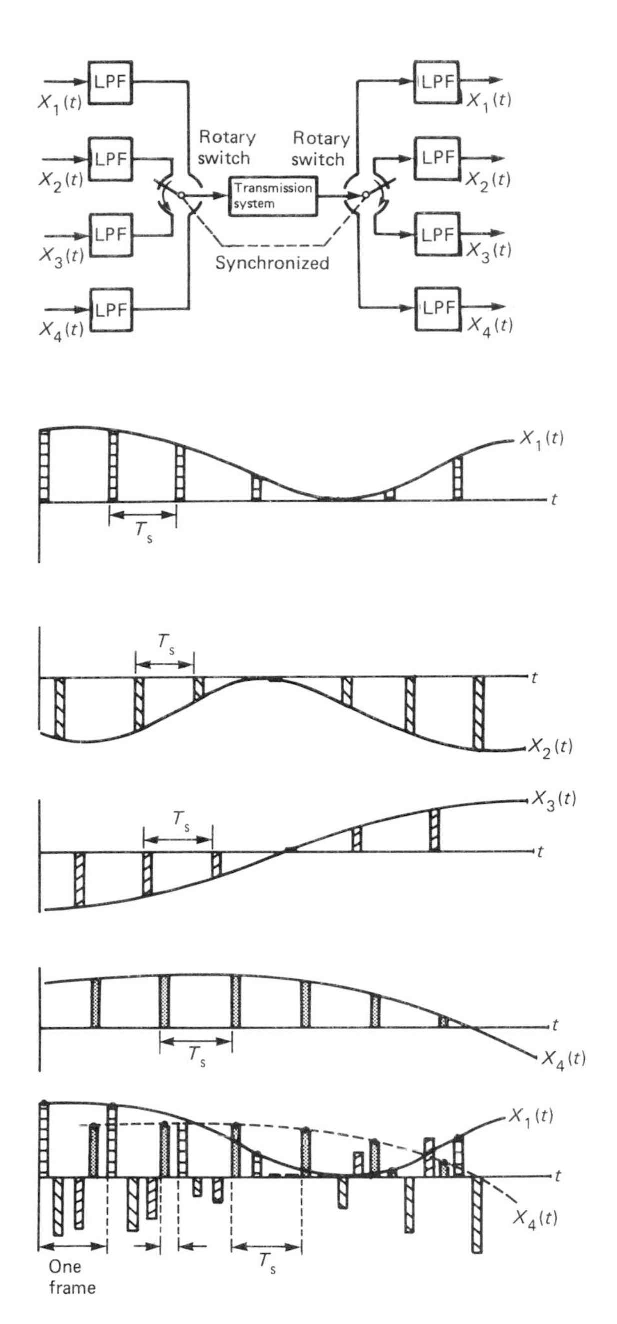

Figure 5.19 Time-division multiplexing.

5.4 Pulse encoding

Pulse code modulation (PCM) is one of the most commonly used methods of encoding analogue data for transmission in instrumentation systems. In PCM the analogue signal is sampled and converted into binary form by means of an ADC, and these data are then transmitted in serial form. This is shown in Figure 5.20. The bandwidth required for the transmission of a signal using PCM is considerably in excess of the bandwidth of the original signal. If a signal having a bandwidth f_d is encoded into an N-bit binary code the minimum bandwidth required to transmit the PCM encoded signal is $f_d.N$ Hz, i.e. N times the original signal bandwidth.

Several forms of PCM are shown in Figure 5.21. The non-return to zero (NRZ-L) is a common code that is easily interfaced to a computer. In the non-return to zero mark (NRZ-M) and the non-return to zero space (NRZ-S) codes level transitions represent bit changes. In bi-phase level (BIΦ-L) a bit transition occurs at the centre of every period. One is represented by a '1' level changing to a '0' level at the centre transition point, and zero is represented by a '0' level changing to a '1' level. In bi-phase mark and space code (BIΦ-M) and (BIΦ-S) a level change occurs at the beginning of each bit period. In BIΦ-M one is represented by a mid-bit transition; a zero has no transition. BIΦ-S is the converse of BIΦ-M. Delay modulation code DM-M and DM-S have transitions at mid-bit and at the end of the bit time. In DM-M a one is represented by a level change at mid-bit; a zero followed by a zero is represented by a level change after the first zero. No level change occurs if a zero precedes a one. DM-S is the converse of DM-M. Bi-phase codes have a transition at least every bit time which can be used for synchronization, but they require twice the bandwidth of the NRZ-L code. The delay modulation codes offer the greatest bandwidth saving but are more susceptible to error, and are used if bandwidth compression is needed or high signal-to-noise ratio is expected.

Alternative forms of encoding are shown in Figure 5.22. In pulse amplitude modulation (PAM) the amplitude of the signal transmitted is proportional to the magnitude of the signal being transmitted, and it can be used for the transmission of both analogue and digital signals. The channel bandwidth required for the transmission of PAM is less than that required for PCM, although the effects of ISI are more marked. PAM as a means of transmitting digital data requires more complex decoding schemes, in that it is necessary to discriminate between an increased number of levels.

Other forms of encoding which can be used include pulse-width modulation (PWM), otherwise referred to as pulse-duration modulation (PDM), which employs a constant height variable width pulse with the information being contained in the width of the pulse. In pulse-position modulation (PPM) the position of the pulses corresponds to the width of the pulse in PWM. Delta modulation and

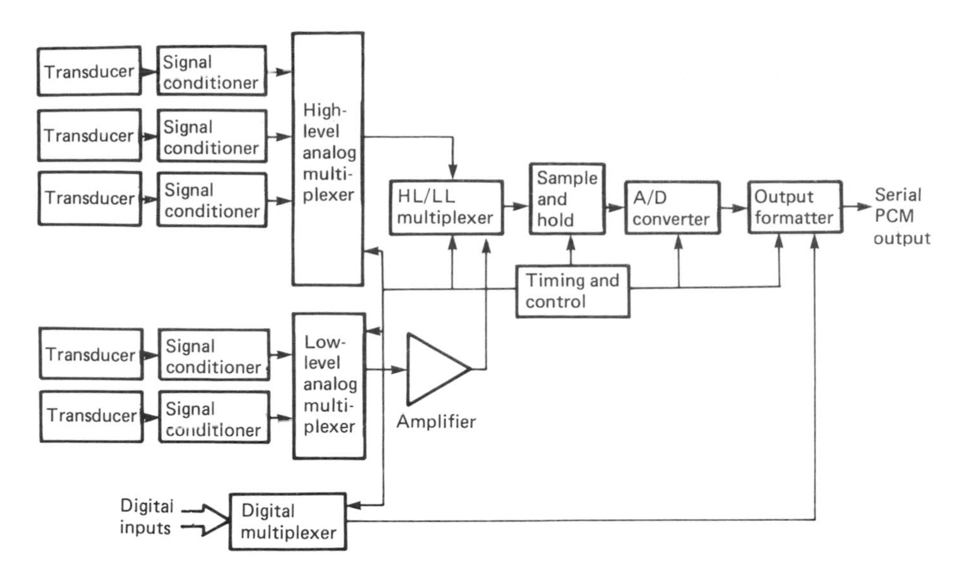

Figure 5.20 Pulse code modulation.

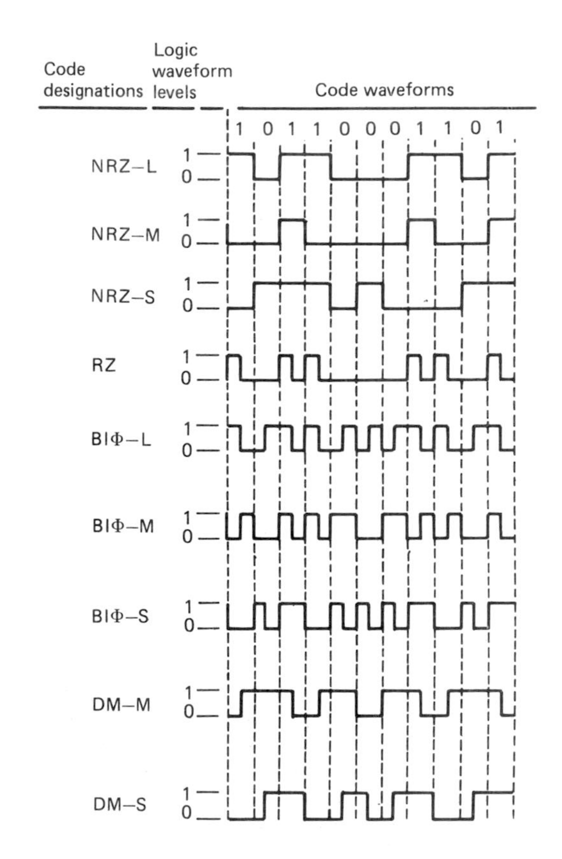

Figure 5.21 Types of pulse code modulation.

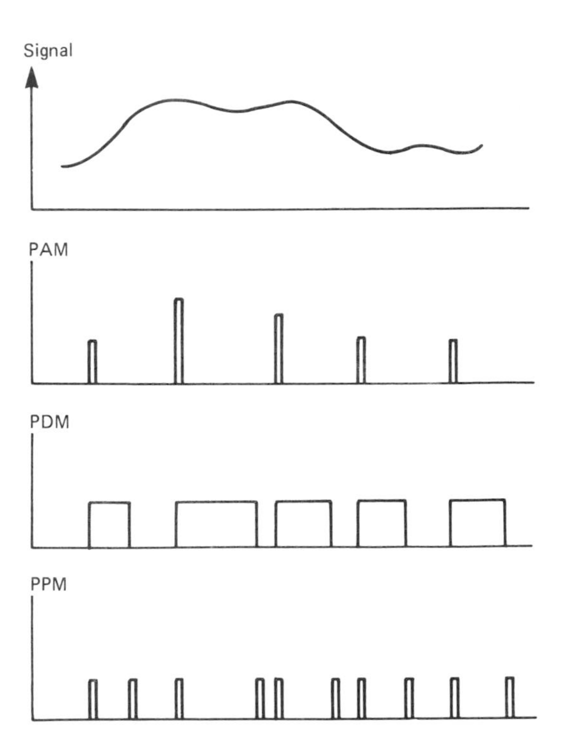

Figure 5.22 Other forms of pulse encoding.

sigma–delta modulation use pulse trains, the frequencies of which are proportional to either the rate of change of the signal or the amplitude of the signal itself. For analyses of the above systems in terms of the bandwidth required for transmission, signal-to-noise ratios and error rate, together with practical details the reader is directed to Hartley *et al.* (1967), Cattermole (1969), Steele (1975) and Shanmugan (1979).

5.5 Carrier wave modulation

Modulation is used to match the frequency characteristics of the data to be transmitted to those of the transmission channel, to reduce the effects of unwanted noise and interference and to facilitate the efficient radiation of the signal. These are all effected by shifting the frequency of the data into some frequency band centred around a carrier frequency. Modulation also allows the allocation of specific frequency bands for specific purposes such as in a FDM system or in r.f. transmission systems, where certain frequency bands are assigned for broadcasting, telemetry, etc. Modulation can also be used to overcome the limitations of signal-processing equipment in that the frequency of the signal can be shifted into frequency bands where the design of filters or amplifiers is somewhat easier, or into a frequency band that the processing equipment will accept. Modulation can be used to provide the bandwidth against signal-to-noise trade-offs which are indicated by the Hartley–Shannon theorem.

Carrier-wave modulation uses the modulation of one of its three parameters, namely amplitude, frequency or phase, and these are all shown in Figure 5.23. The techniques can be used for the transmission of both analogue and digital signals.

In amplitude modulation the amplitude of the carrier varies linearly with the amplitude of the signal to be transmitted. If the data signal $d(t)$ is represented by a sinusoid $d(t) = \cos 2\pi f_d t$ then in amplitude modulation the carrier wave $c(t)$ is given by:

$$c(t) = C(1 + m.\cos 2\pi f_d t)\cos 2\pi f_c t$$

where C is the amplitude of the unmodulated wave, f_c its frequency and m is the depth of modulation which has a value lying between 0 and 1. If $m = 1$ then the carrier is said to have 100 per cent modulation. The above expression for $c(t)$ can be rearranged as:

$$c(t) = C.\cos 2\pi f_c t + \frac{Cm}{2}[\cos 2\pi(f_c + f_d)t + \cos 2\pi(f_c - f_d)t]$$

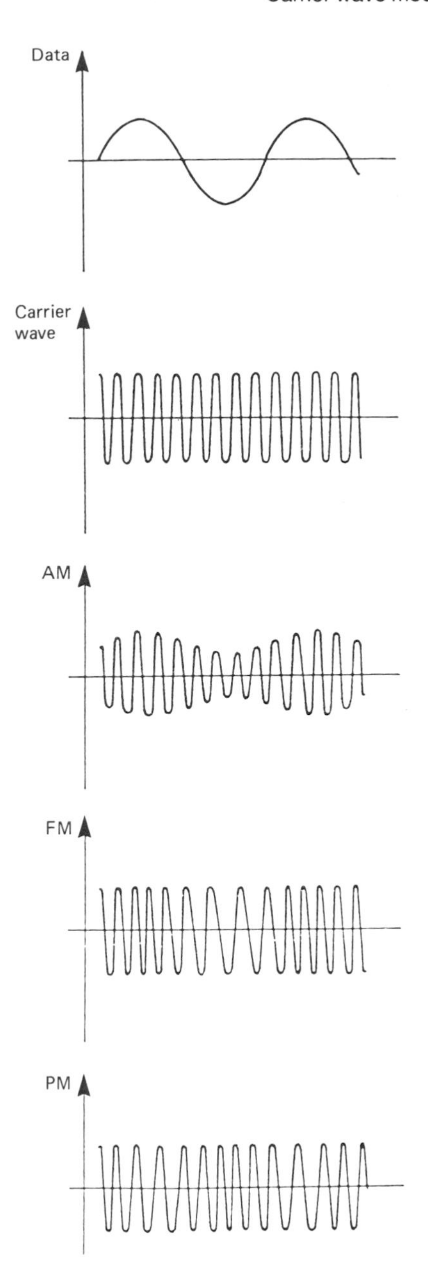

Figure 5.23 Amplitude, frequency and phase modulation of a carrier wave.

showing that the spectrum of the transmitted signal has three frequency components at the carrier frequency f_c and at the sum and difference frequencies $(f_c + f_d)$ and $(f_c - f_d)$. If the signal is represented by a spectrum having frequencies up to f_d then the transmitted spectrum has a bandwidth of $2f_d$ centred around f_c. Thus in order to transmit data using AM a bandwidth equal to twice that of the data is required. As can be seen, the envelope of the AM signal contains the information, and thus demodulation can be effected simply by rectifying and smoothing the signal.

Both the upper and lower sidebands of AM contain sufficient amplitude and phase information to reconstruct the data, and thus it is possible to reduce the bandwidth requirements of the system. Single side-band modulation (SSB) and vestigial side-band modulation (VSM) both transmit the data using amplitude modulation with smaller bandwidths than straight AM. SSB has half the bandwidth of a simple AM system; the low-frequency response is generally poor. VSM transmits one side band almost completely and only a trace of the other. It is very often used in high-speed data transmission, since it offers the best compromise between bandwidth requirements, low-frequency response and improved power efficiency.

In frequency modulation consider the carrier signal $c(t)$ given by:

$$c(t) = C\cos(2\pi f_c t + \phi(t))$$

Then the instantaneous frequency of this signal is given by:

$$f_i(t) = f_c + \frac{1}{2\pi} \cdot \frac{d\phi}{dt}$$

The frequency deviation

$$\frac{1}{2\pi} \cdot \frac{d\phi}{dt}$$

of the signal from the carrier frequency is made to be proportional to the data signal. If the data signal is represented by a single sinusoid of the form:

$$d(t) = \cos 2\pi f_d t$$

then:

$$\frac{d\phi(t)}{dt} = 2\pi k_f . \cos 2\pi f_d t$$

where k_f is the frequency deviation constant which has units of Hz/V. Thus:

$$c(t) = C\cos\left(2\pi f_c t + 2\pi k_f \int_{-\infty}^{t} d(\tau).d\tau\right)$$

and assuming zero initial phase deviation, then the carrier wave can be represented by:

$$c(t) = C\cos(2\pi f_c t + \beta \sin 2\pi f_d t)$$

where β is the modulation index and represents the maximum phase deviation produced by the data. It is possible to show that $c(t)$ can be represented by an infinite series of frequency components $f \pm nf_d$, $n = 1,2,3,\ldots$, given by:

$$c(t) = C \sum_{n=-\infty}^{\infty} J_n(\beta)\cos(2\pi f_c + n2\pi f_d)t$$

where $J_n(\beta)$ is a Bessel function of the first kind of order n and argument β. Since the signal consists of an infinite number of frequency components, limiting the transmission bandwidth distorts the signal, and the question arises as to what is a reasonable bandwidth for the system to have in order to transmit the data with an acceptable degree of distortion. For $\beta << 1$ only J_0 and J_1 are important, and large β implies large bandwidth. It has been found in practice that if 98 per cent or more of the FM signal power is transmitted then the signal distortion is negligible. Carson's rule indicates that the bandwidth required for FM transmission of a signal having a spectrum with components up to a frequency of f_d is given by $2(f_\Delta + f_d)$, where f_Δ is the maximum frequency deviation. For narrow-band FM systems having small frequency deviations the bandwidth required is the same as that for AM. Wide-band FM systems require a bandwidth of $2f_\Delta$. Frequency modulation is used extensively in r.f. telemetry and in FDM.

In phase modulation the instantaneous phase deviation ϕ is made proportional to the data signal. Thus:

$$\phi = k_p . \cos 2\pi f_d t$$

and it can be shown that the carrier wave $c(t)$ can be represented by:

$$c(t) = C\cos(2\pi f_c + \beta \cos 2\pi f_d t)$$

where β is now given by k_p. For further details of the various modulation schemes and their realizations the reader should consult Shanmugan (1979) and Coates (1982).

5.6 Error detection and correction codes

Errors occur in digital data communications systems as a consequence of the corruption of the data by noise. Figure 5.24 shows the bit error probability as a function of signal-to-noise ratio for a PCM transmission system using NRZ-L coding.

In order to reduce the probability of an error occurring in the transmission of the data, bits are added to the transmitted message. These bits add redundancy to the transmitted data, and since only part of the transmitted message is now the actual data, the efficiency of the transmission is reduced. There are two forms of error coding, known as forward error detection and correction coding (FEC), in which the transmitted message is coded in such a way that errors can be both detected and corrected continuously, and automatic repeat request coding (ARQ), in which if an error is detected then a request is sent to repeat the transmission. In terms of data-throughput rates FEC codes are more efficient than AQR codes because of the need to retransmit the data in the case of error in an ARQ code, although the equipment required to detect errors is somewhat simpler than that required to correct the errors from the corrupted message. ARQ codes are commonly used in instrumentation systems.

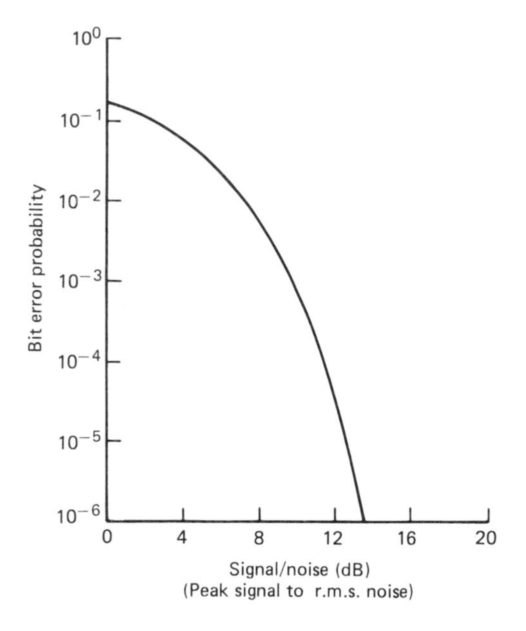

Figure 5.24 Bit-error probability for PCM transmission using a NRZ-L code.

Parity-checking coding is a form of coding used in ARQ coding in which $(n-k)$ bits are added to the k bits of the data to make an n-bit data stream. The simplest form of coding is parity-bit coding in which the number of added bits is one, and this additional bit is added to the data stream in order to make the total number of ones in the data stream either odd or even. The received data are checked for parity. This form of coding will only detect an odd number of bit errors. More complex forms of coding include linear block codes such as Hamming codes, cyclic codes such as Bose–Chanhuri–Hocquenghen codes and geometric codes. Such codes can be designed to detect multiple-burst errors, in which two or more successive bits are in error. In general the larger the number of parity bits, the less efficient the coding is, but the larger are both the maximum number of random errors and the maximum burst length that can be detected. Figure 5.25 shows examples of some of these coding techniques. Further details of coding techniques can be found in Shanmugan (1979), Bowdell (1981) and Coates (1982).

5.7 Direct analogue signal transmission

Analogue signals are rarely transmitted over transmission lines as a voltage since the method suffers from errors due to series and common mode inductively and capacitively coupled interference signals and those due to line resistance. The most common form of analogue signal transmission is as current.

Current transmission as shown in Figure 5.26 typically uses 0–20 or 4–20 mA. The analogue signal is converted to a current at the transmitter and is detected at the receiver either by measuring the potential difference developed across a fixed resistor or using the current to drive an indicating instrument or chart recorder. The length of line over which signals can be transmitted at low frequencies is primarily limited by the voltage available at the transmitter to overcome voltage drop along the line and across the receiver. With a typical voltage of 24 V the system is capable of transmitting the current over several kilometres. The percentage error in a current transmission system can be calculated as 50 × the ratio of the loop resistance in ohms to the total line insulation resistance, also expressed in ohms. The accuracy of current transmission system systems is typically ±0.5 per cent.

The advantage of using 4–20 mA instead of 0–20 mA is that the use of a live zero enables instrument or line faults to be detected. In the 4–20 mA system zero value is represented by 4 mA and failure is indicated by 0 mA. It is possible to use a 4–20 mA system as a two-wire transmission system in which both the power and the signal are transmitted along the same wire, as shown in Figure 5.27. The 4 mA standing current is used to power the remote instrumentation and the transmitter. With 24 V drive the maximum power available to the remote station is 96 mW. Integrated-circuit devices such as the Burr–Brown XTR 100 are available for providing two-wire transmission. This is capable of providing a 4–20 mA output span for an input voltage as small as 10 mV, and is capable of transmitting at frequencies up to 2 kHz over a distance of 600 m. Current transmission cannot be used over the public telephone system because it requires a d.c. transmission path, and telephone systems use a.c. amplifiers in the repeater stations.

Position telemetry transmits an analogue variable by reproducing at the receiver the positional information available at the transmitter. Such devices employ null techniques with either resistive or inductive elements to achieve the position telemetry. Figure 5.28 shows an inductive 'synchro'.

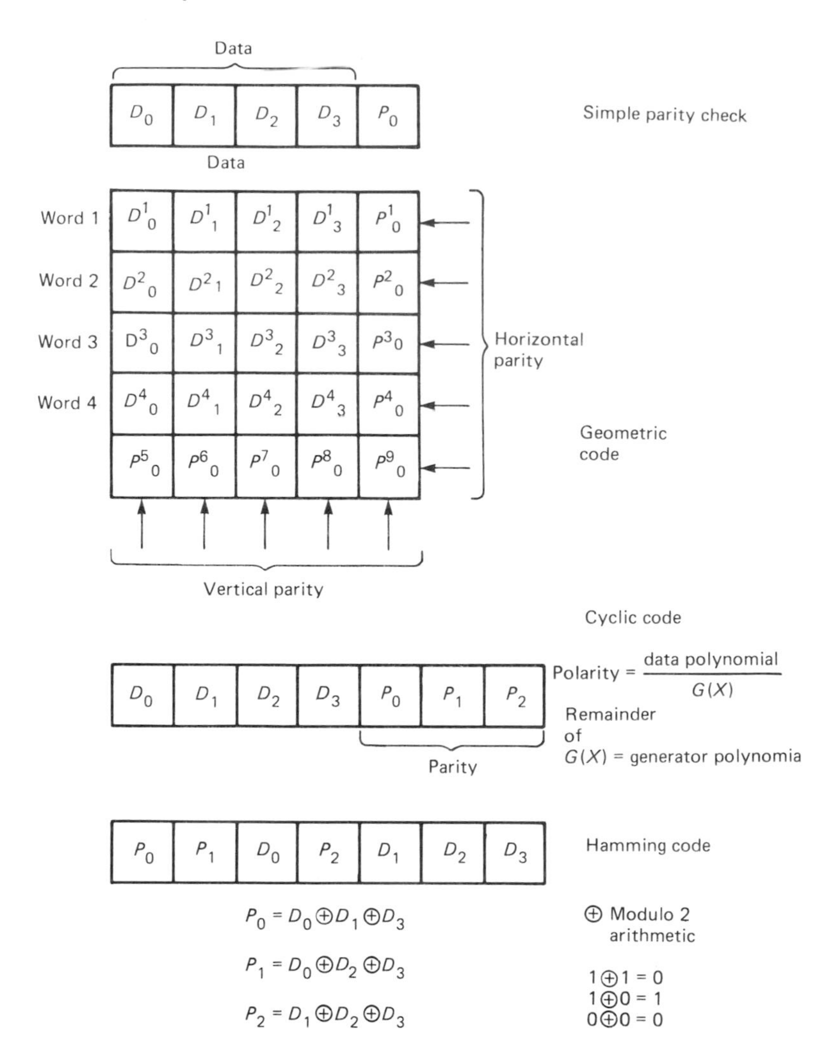

Figure 5.25 Error-detection coding.

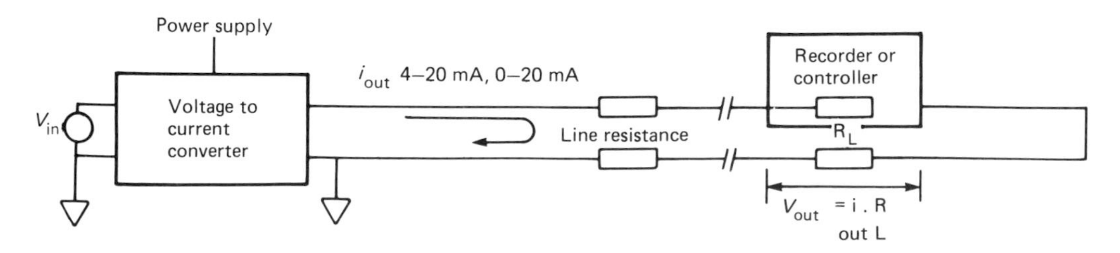

Figure 5.26 4-20 mA current transmission system.

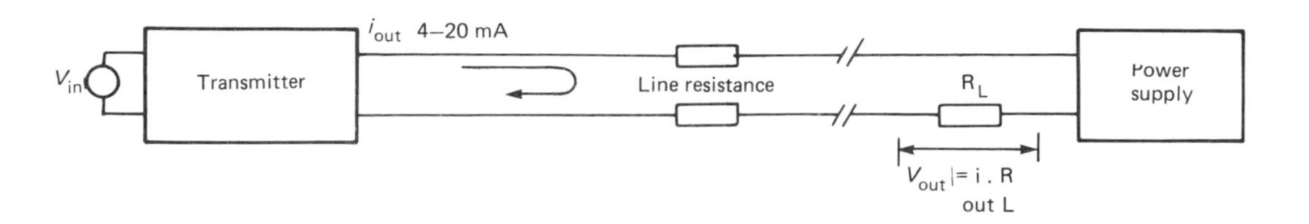

Figure 5.27 Two-wire transmission system.

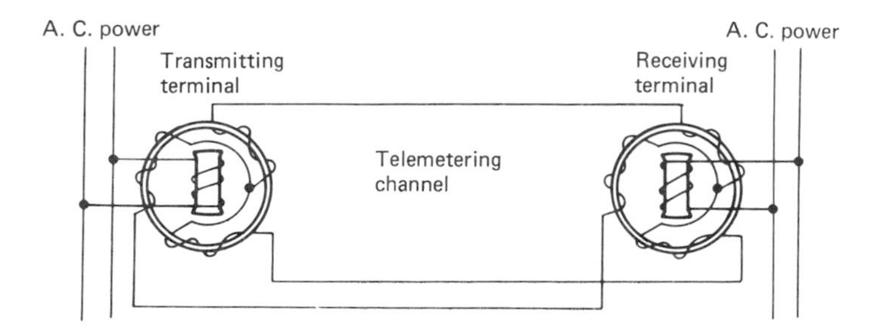

Figure 5.28 Position telemetry using an inductive 'synchro'.

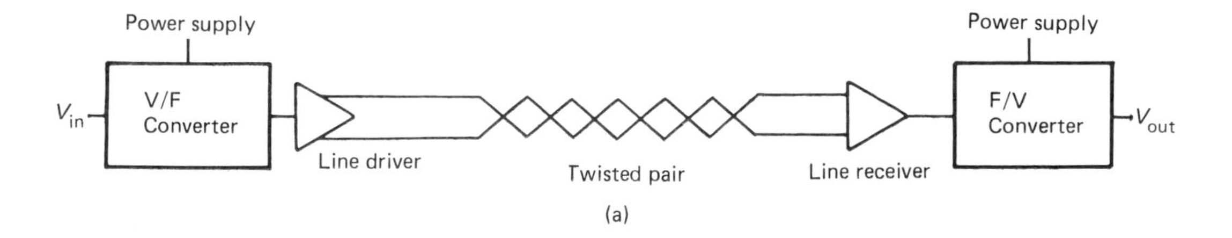

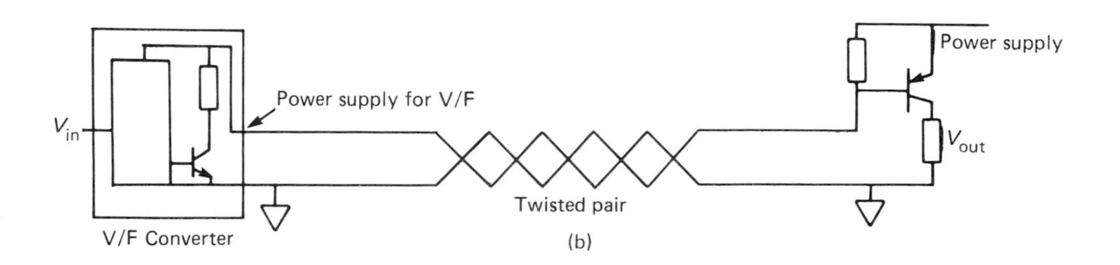

Figure 5.29 (a) Frequency-transmission system; (b) two-wire frequency-transmission system.

The a.c. power applied to the transmitter induces in the three stator windings, e.m.f.s, the magnitude of which are dependent upon the position of the transmitter rotor. If the receiver rotor is aligned in the same direction as the transmitter rotor then the e.m.f.s. induced in the stator windings of the receiver will be identical to those on the stator windings of the transmitter. There will therefore be no resultant circulating currents. If the receiver rotor is not aligned to the direction of the transmitter rotor then the circulating currents in the stator windings will be such as to generate a torque which will move the receiver rotor in such a direction as to align itself with the transmitter rotor.

5.8 Frequency transmission

By transmitting signals as frequency the characteristics of the transmission line in terms of amplitude and phase characteristics are less important. On reception the signal can be counted over a fixed period of time to provide a digital measurement. The resolution of such systems will be 1 count in the total number received. Thus for high resolution it is necessary to count the signal over a long time period, and this method of transmission is therefore unsuitable for rapidly changing or multiplexed signals but is useful for such applications as batch control, where, for example, a totalized value of a variable over a given period is required. Figure 5.29(a) shows a frequency-transmission system.

Frequency-transmission systems can also be used in two-wire transmission systems, as shown in Figure 5.29(b), where the twisted pair carries both the power to the remote device and the frequency signal in the form of current modulation. The frequency range of such systems is governed by the bandwidth of the channel over which the signal is to be transmitted, but commercially available integrated circuit V to f converters such as the Analog Devices AD458 convert a 0–10 V d.c. signal to a frequency in the range 0–10 kHz or 0–100 kHz with a maximum non-linearity of ±0.01 per cent of FS output, a maximum temperature coefficient of ±5 ppm/K, a maximum input offset voltage of ±10 mV and a maximum input offset voltage temperature coefficient of 30 μV/K. The response time is two output pulses plus 2 μs. A low-cost f to V converter such as the Analog Devices AD453 has an input frequency range of 0–100 kHz with a variable threshold voltage of between 0 and ±12 V, and can be used with low-level signals as well as high-level inputs from TTL and CMOS. The converter has a full-scale output of 10 V and a non-linearity of less than ±0.008 per cent of FS with a maximum temperature coefficient of ±50 ppm/K. The maximum response time is 4 ms.

5.9 Digital signal transmission

Digital signals are transmitted over transmission lines using either serial or parallel communication.

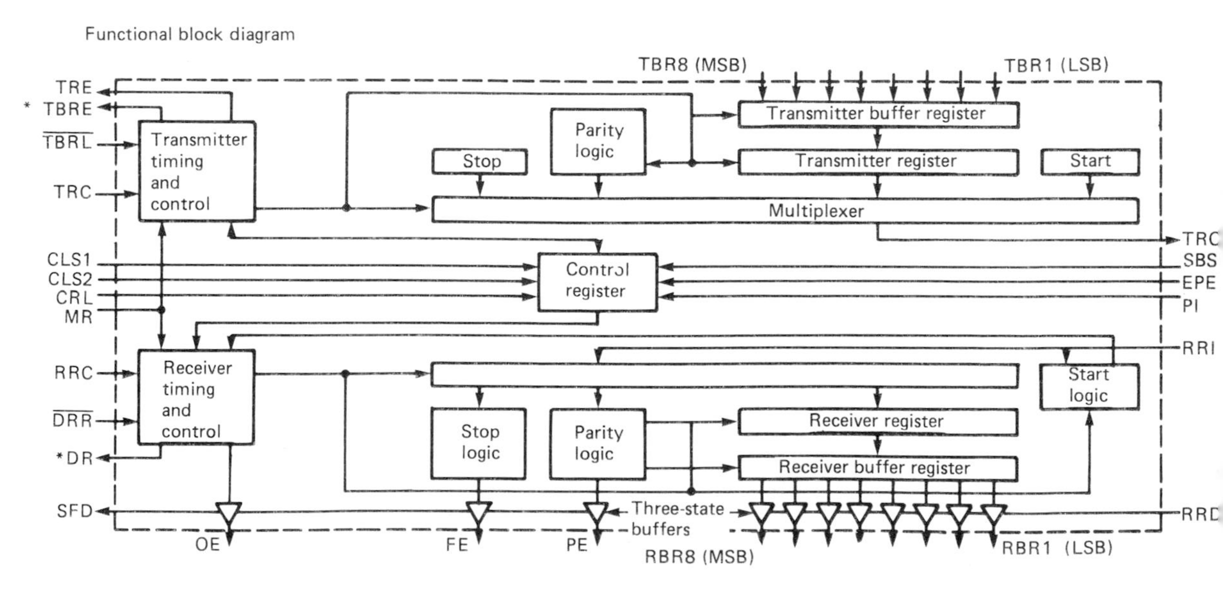

Figure 5.30(a) Universal asynchronous receiver transmitter (UART).

For long-distance communication serial communication is the preferred method. The serial communication may be either synchronous or asynchronous. In synchronous communication the data are sent in a continuous stream without stop or start information. Asynchronous communication refers to a mode of communication in which data are transmitted as individual blocks framed by start and stop bits. Bits are also added to the data stream for error detection. Integrated circuit devices known as universal asynchronous receiver transmitters (UARTS) are available for converting parallel data into a serial format suitable for transmission over a twisted pair or coaxial line and for reception of the data in serial format and reconversion to parallel format with parity-bit checking. The schematic diagram for such a device is shown in Figure 5.30.

Because of the high capacitance of twisted-pair and coaxial cables the length of line over which standard 74 series TTL can transmit digital signals is

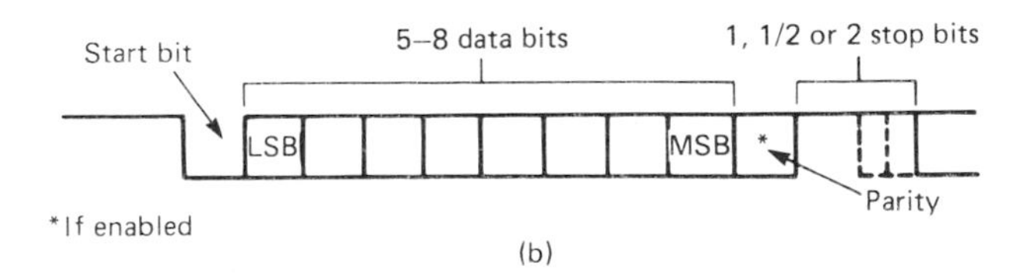

Figure 5.30(b) Serial data format.

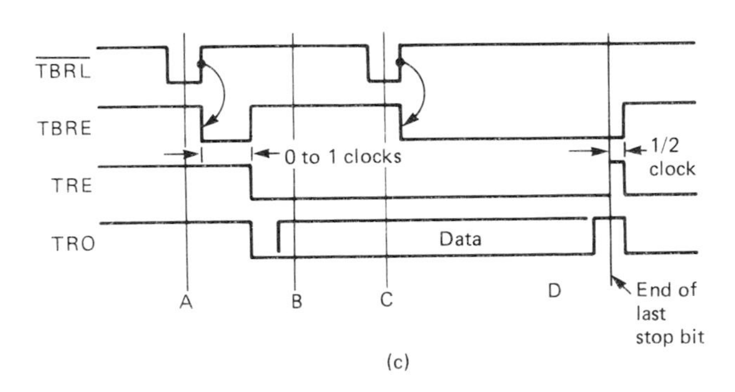

Figure 5.30(c) Transmitter timing (not to scale).

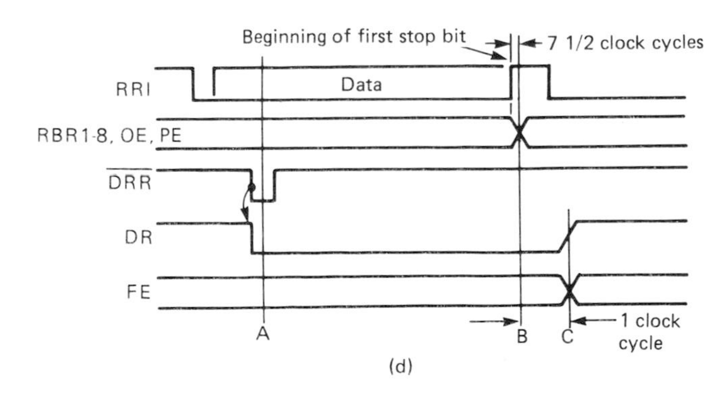

Figure 5.30(d) Receiver timing (not to scale).

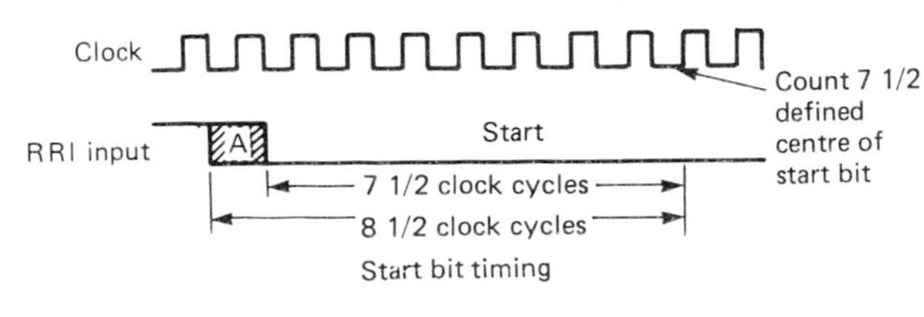

Figure 5.30(e) Start bit timing.

limited typically to a length of 3 m at 2 Mbit/s. This can be increased to 15 m by the use of open-collector TTL driving a low-impedance terminated line.

In order to drive digital signals over long lines special-purpose line driver and receiver circuits are available. Integrated circuit Driver/Receiver combinations such as the Texas Instruments SN75150/SN75152, SN75158/SN75157 and SN75156/SN75157 devices meet the internationally agreed EIA Standards RS-232C, RS-422A and RS-423A, respectively (see section 5.9.2).

5.9.1 Modems

In order to overcome the limitations of the public telephone lines digital data are transmitted down these lines by means of a modem. The two methods of modulation used by modems are frequency-shift keying (FSK) and phase-shift keying (PSK). Amplitude-modulation techniques are not used because of the unsuitable response of the line to step changes in amplitude. Modems can be used to transmit information in two directions along a telephone line.

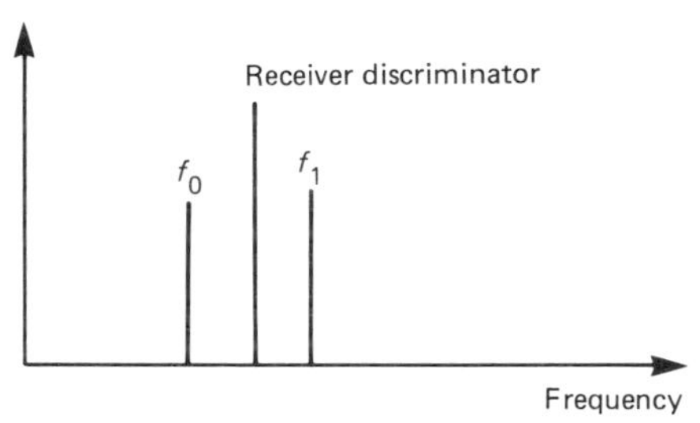

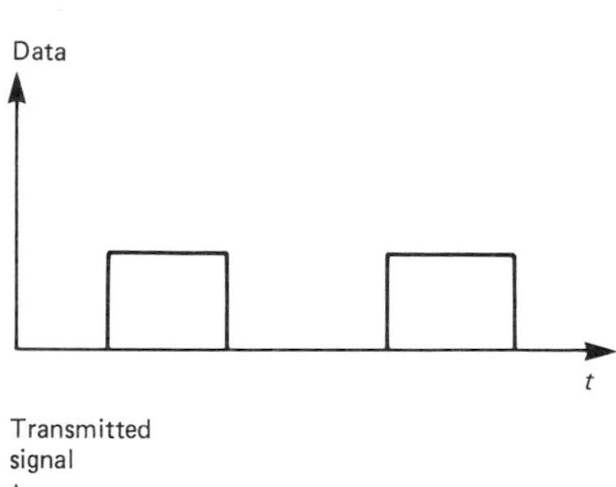

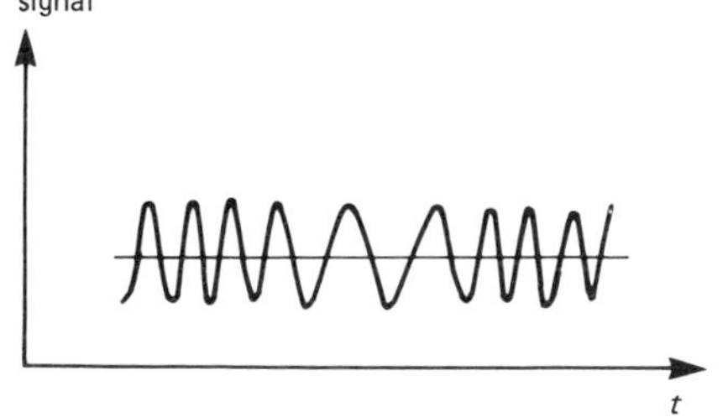

Figure 5.31 Frequency-shift keying.

Full-duplex operation is transmission of information in both directions simultaneously; half-duplex is the transmission of information in both directions but only in one direction at any one time; and simplex is the transmission of data in one direction only.

The principle of FSK is shown in Figure 5.31. FSK uses two different frequencies to represent a 1 and a 0, and this can be used for data transmission rates up to 1200 bits/s. The receiver uses a frequency discriminator whose threshold is set mid-way between the two frequencies. The recommended frequency shift is not less than 0.66 of the modulating frequency. Thus a modem operating at 1200 bits/s has a recommended central frequency of

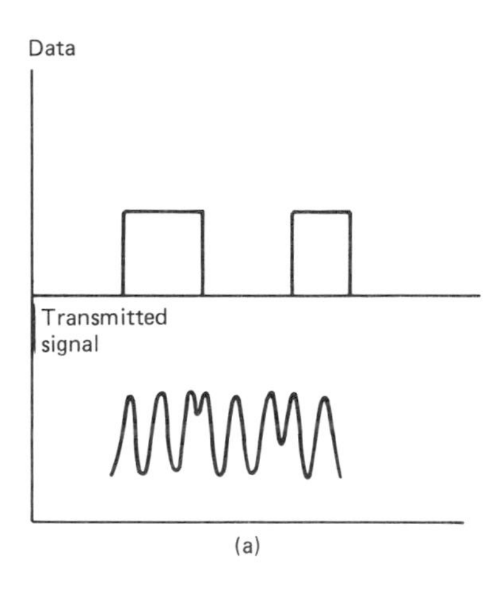

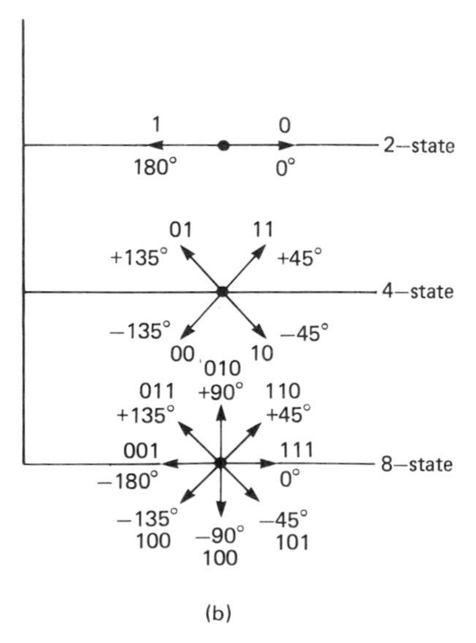

Figure 5.32 (a) Principle of phase-shift keying; (b) two-, four- and eight-state-shift keying.

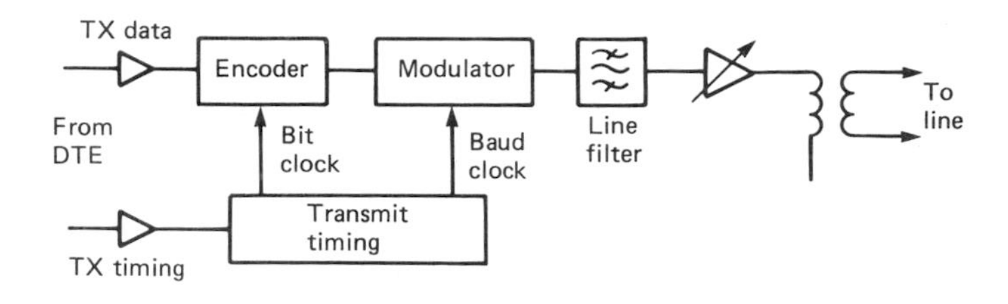

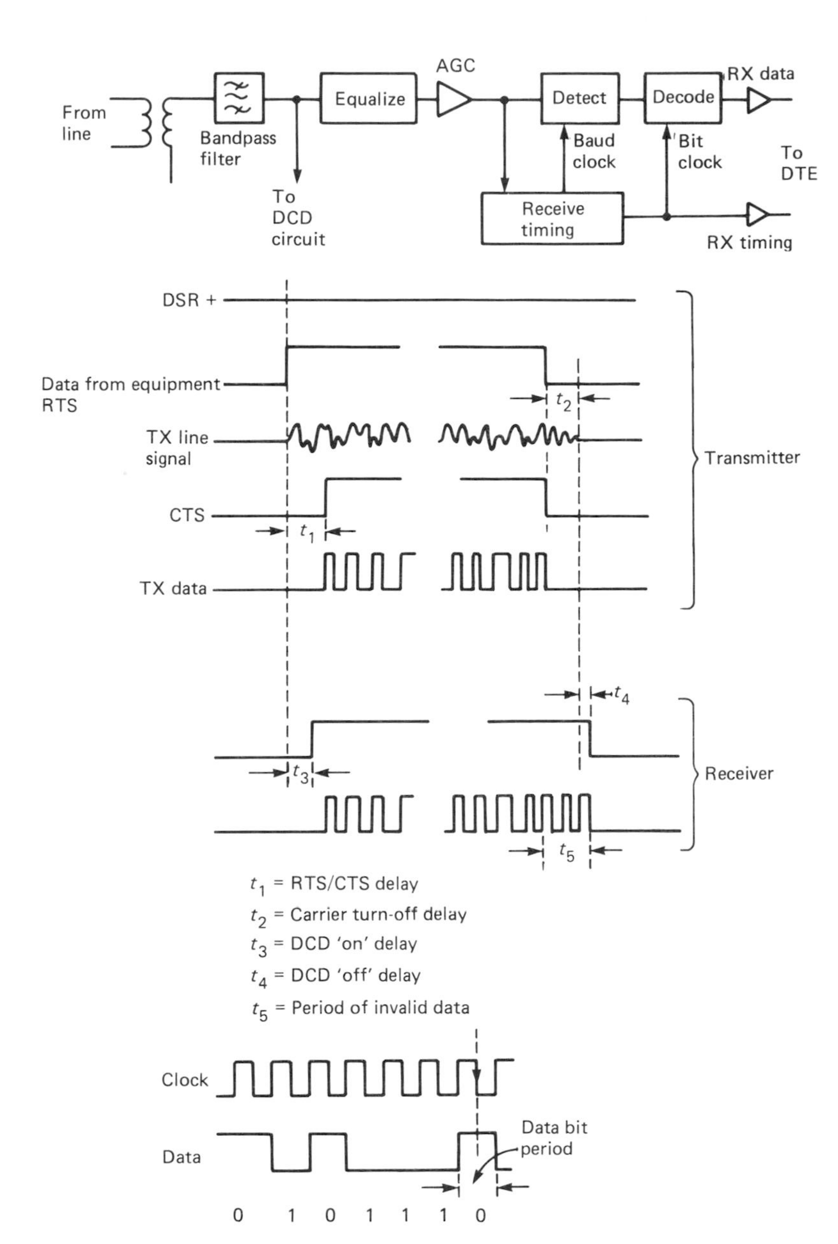

Figure 5.33 Modem operation.

1700 Hz and a frequency deviation of 800 Hz, with a 0 represented by a frequency of 1300 Hz and a 1 by a frequency of 2100 Hz. At a transmission rate of 200 bits/s it is possible to operate a full-duplex system. At 600 and 1200 bits/s half-duplex operation is used incorporating a slow-speed backward channel for supervisory control or low-speed return data.

At bit rates above 2400 bits/s the bandwidth and group delay characteristics of telephone lines make it impossible to transmit the data using FSK. It is necessary for each signal to contain more than one bit of information. This is achieved by a process known as phase shift keying (PSK), in which the phase of a constant-amplitude carrier is changed. Figure 5.32(a) shows the principle of PSK and Figure 5.32(b) shows how the information content of PSK can be increased by employing two-, four- and eight-state systems. It should now be seen that the number of signal elements/s (which is referred to as the Baud rate) has to be multiplied by the number of states to obtain the data transmission rate in bits/s. Thus an eight-state PSK operating at a Baud rate of 1200 bauds can transmit 9600 bits/s. This is the fastest transmission over telephone cables and is intended for operation over leased lines, i.e. lines which are permanently allocated to the user as opposed to switched lines. At the higher data transmission rates it is necessary to apply adaptive equalization of the line to ensure correct operation and also to have in-built error-correcting coding. Details of various schemes for modem operation are to be found in Coates (1982) and Blackwell (1981). The International Telephone and Telegraph Consultative Committee (CCITT) have made recommendations for the mode of operation of modems operating at different rates over telephone lines. These are set out in recommendations V21, V23, V26, V27 and V29. These are listed in the References.

Figure 5.33 shows the elements and operation of a typical modem. The data set ready (DSR) signal indicates to the equipment attached to the modem that it is ready to transmit data. When the equipment is ready to send the data it sends a request to send (RTS) signal. The modem then starts transmitting down the line. The first part of the transmission is to synchronize the receiving modem. Having given sufficient time for the receiver to synchronize, the transmitting modem sends a clear to send (CTS) signal to the equipment

Table 5.1 Pin assignments for RS-232

Pin number	*Signal nomenclature*	*Signal abbreviation*	*Signal description*	*Category*
1	AA	—	Protective ground	Ground
2	BA	TXD	Transmitted data	Data
3	BB	RXD	Received data	Data
4	CA	RTS	Request to send	Control
5	CB	CTS	Clear to send	Control
6	CC	DSR	Data set ready	Control
7	AB	—	Signal ground	Ground
8	CF	DCD	Received line signal detector	Control
9	—	—	—	Reserved for test
10	—	—	—	Reserved for test
11	—	—	—	Unassigned
12	SCF	—	Secondary received line signal detector	Control
13	SCB	—	Secondary clear to send	Control
14	SBA	—	Secondary transmitted data	Data
15	DB	—	Transmission signal element timing	Timing
16	SBB	—	Secondary received data	Data
17	DD	—	Received signal element timing	Timing
18	—	—	—	Unassigned
19	SCA	—	Secondary request to send	Control
20	CD	DTR	Data terminal ready	Control
21	CG	—	Signal quality detector	Control
22	CE	—	Ring indicator	Control
23	CH/CI	—	Data signal rate selector	Control
24	DA	—	Transmit signal element timing	Timing
25	—	—	—	Unassigned

and the data are then sent. At the receiver the detection of the transmitted signal sends the data carrier detected (DCD) line high and the signal transmitted is demodulated.

5.9.2 Data transmission and interfacing standards

To ease the problem of equipment interconnection various standards have been introduced for serial and parallel data transmission. For serial data transmission between data terminal equipment (DTE), such as a computer or a piece of peripheral equipment, and data communication equipment (DCE), such as a modem, the standards which are currently being used are the RS-232C standard produced in the USA by the Electronic Industries Association (EIA) in 1969 and their more recent RS-449 standard with its associated RS-422 and RS-423 standards.

The RS-232C standard defines an electromechanical interface by the designation of the pins of a 25-pin plug and socket which are used for providing electrical ground, data interchange, control and clock or timing signals between the two pieces of equipment. The standard also defines the signal levels, conditions and polarity at each interface connection. Table 5.1 gives the pin assignments for the interface and it can be seen that only pins 2 and 3 are used for data transmission. Logical 1 for the driver is an output voltage between -5 and -15 V with logical zero being between $+5$ and $+15$ V. The receiver detects logical 1 for input voltages <-3 V and logical 0 for input voltages >3 V, thus giving the system a minimum 2 V noise margin. The maximum transmission rate of data is 20 000 bits/s and the maximum length of the interconnecting cable is limited by the requirement that the receiver should not have more than 2500 pF across it. The length of cable permitted thus depends on its capacitance/unit length.

Table 5.2 Pin assignments for RS-449

Circuit mnemonic	*Circuit name*	*Circuit direction*	*Circuit type*
SG	Signal ground	—	Common
SC	Send common	To DCE	
RC	Receive common	From DCE	
IS	Terminal in service	To DCE	Control
IC	Incoming call	From DCE	
TR	Terminal ready	To DCE	
DM	Data mode	From DCE	
SD	Send data	To DCE	Primary channel data
RD	Receive data	From DCE	
TT	Terminal timing	To DCE	Primary channel timing
ST	Send timing	From DCE	
RT	Receive timing	From DCE	
RS	Request to send	To DCE	Primary channel control
CS	Clear to send	From DCE	
RR	Receiver ready	From DCE	
SQ	Signal quality	From DCE	
NS	New signal	To DCE	
SF	Select frequency	To DCE	
SR	Signal rate selector	To DCE	
SI	Signal rate indicator	From DCE	
SSD	Secondary send data	To DCE	Secondary channel data
SRD	Secondary receive data	From DCE	
SRS	Secondary request to send	To DCE	Secondary channel control
SCS	Secondary clear to send	From DCE	
SRR	Secondary receiver ready	From DCE	
LL	Local loopback	To DCE	Control
RL	Remote loopback	To DCE	
TM	Test mode	From DCE	
SS	Select standby	To DCE	Control
SB	Standby indicator	From DCE	

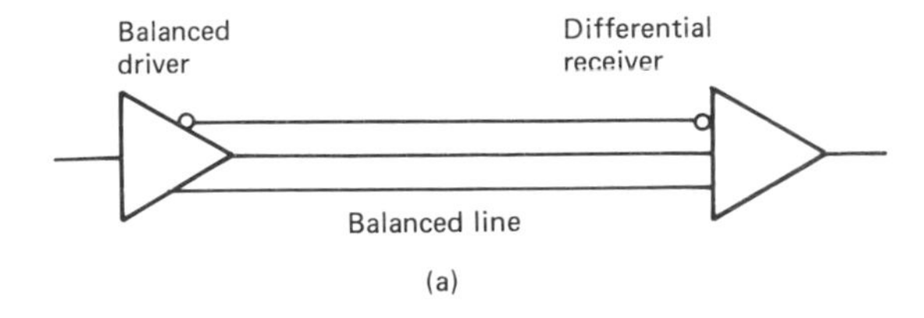

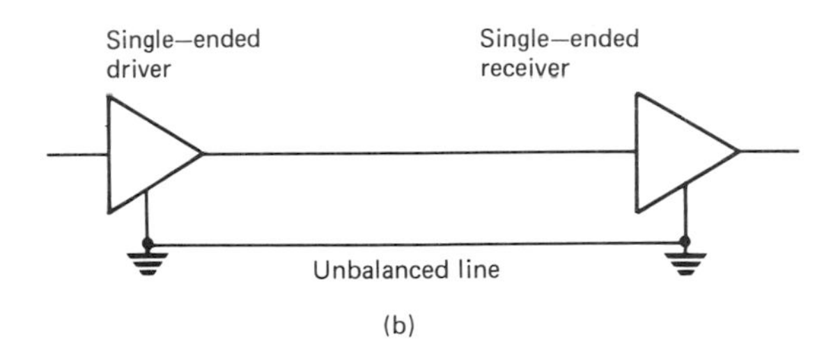

Figure 5.34 RS422 and RS423 driver/receiver systems.

The newer RS-449 interface standard which is used for higher data-transmission rates defines the mechanical characteristics in terms of the pin designations of a 37-pin interface. These are listed in Table 5.2. The electrical characteristics of the interface are specified by the two other associated standards, RS-422, which refers to communication by means of a balanced differential driver along a balanced interconnecting cable with detection being by means of a differential receiver, and RS423, which refers to communication by means of a single-ended driver on an unbalanced cable with detection by means of a differential receiver. These two systems are shown in Figure 5.34. The maximum recommended cable lengths for the balanced RS-422 standard are 4000 ft at 90 kbits/s, 380 ft at 1 Mbits/s and 40 ft at 10 Mbits/s. For the unbalanced RS-423 standard the limits are 4000 ft at 900 bits/s, 380 ft at 10 kbits/s and 40 ft at 100 kbits/s.

For further details of these interface standards the reader is directed to the standards produced by the EIA. These are listed in the References. Interfaces are also discussed in Volume 5.

The IEEE-488 Bus (IEEE, 1978), often referred to as the HPIB Bus (Hewlett Packard Interface Bus), is a standard which specifies a communications protocol between a controller and instruments connected onto the bus. The instruments typically connected on to the bus include digital voltmeters, signal generators, frequency meters and spectrum and impedance analysers. The bus allows up to 15 such instruments to be connected onto the bus. Devices talk, listen or do both, and at least one

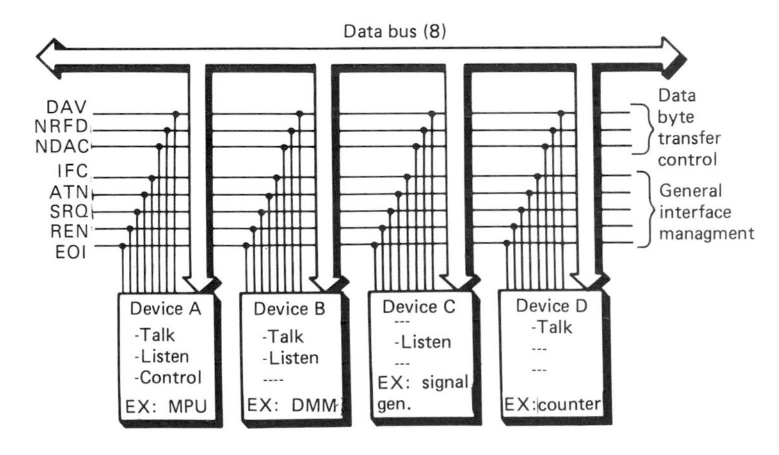

Figure 5.35 IEEE-488 bus system.

device on the bus must provide control, this usually being a computer. The bus uses 15 lines, the pin connections for which are shown in Table 5.3. The signal levels are TTL and the cable length between the controller and the device is limited to 2 m. The bus can be operated at a frequency of up to 1 MH. The connection diagram for a typical system are shown in Figure 5.35. Eight lines are used for addresses, program data and measurement data transfers, three lines are used for the control of data transfers by means of a handshake technique and five lines are used for general interface management.

CAMAC (which is an acronym for Computer Automated Measurement and Control) is a multiplexed interface system which not only specifies the connections and the communications protocol between the modules of the system which act as interfaces between the computer system and peripheral devices but also stipulates the physical dimensions of the plug-in modules. These modules are typically ADCs, DACs, digital buffers, serial to parallel converters, parallel to serial converters and level changers. The CAMAC system offers a 24-bit parallel data highway via an 86-way socket at the rear of each module. Twenty-three of the modules are housed in a single unit known as a 'crate', which additionally houses a controller. The CAMAC system was originally specified for the nuclear industry, and is particularly suited for systems where a large multiplexing ratio is required. Since each module addressed can have up to 16 subaddresses a crate can have up to 368 multiplexed inputs/outputs. For further details of the CAMAC system the reader is directed to Barnes (1981) and to the CAMAC standards issued by the Commission of the European Communities, given in the References.

The S100 Bus (also referred to as the IEEE-696 interface (IEEE, 1981)) is an interface standard devised for bus-oriented systems and was originally designed for interfacing microcomputer systems. Details of this bus can be found in the References. Questions of interfaces are dealt with at greater length in Chapter 10.

5.10 References

BARNES, R. C. M. 'A standard interface: CAMAC', in *Minicomputers: A Handbook for Engineers, Scientists, and Managers*, ed. Y. Paker, Abacus, London (1981), pp. 167–187

BENTLEY, J., *Principles of Measurement Systems*, Longman, London (1983)

BOWDELL, K. 'Interface data transmission', in *Minicomputers: A Handbook for Engineers, Scientists, and Managers*, ed. Y. Paker, Abacus, London (1981), pp. 148–166

Table 5.3 Pin assignments for IEEE-488 interface

Pin no.	*Function*	*Pin no.*	*Function*
1	DIO 1	13	DIO 5
2	DIO 2	14	DIO 6
3	DIO 3	15	DIO 7
4	DIO 4	16	DIO 8
5	EOI	17	REN
6	DAV	18	GND twisted pair with 6
7	NRFD	19	GND twisted pair with 7
8	NDAC	20	GND twisted pair with 8
9	IFC	21	GND twisted pair with 9
10	SRQ	22	GND twisted pair with 10
11	ATN	23	GND twisted pair with 11
12	Shield (to earth)	24	Signal ground

DIO= Data Input–Output
EOI= End Or Identify
REN= Remote Enable
DAV= Data Valid
NRFD= Not Ready For Data
NDAC= Not Data Accepted
IFC= Interface Clear
SRQ= Service Request
ATN= Attention
GND= Ground

BLACKWELL, J. 'Long distance communication', in *Minicomputers: A Handbook for Engineers, Scientists, and Managers*, ed. Y. Paker, Abacus, London (1981), pp. 301–316

CATTERMOLE, K. W. *Principles of Pulse Code Modulation*, Iliffe, London (1969)

CCITT. Recommendation V 24. List of definitions for interchange circuits between data-terminal equipment and data circuit-terminating equipment, in CCITT, Vol. 8.1, *Data Transmission over the Telephone Network*, International Telecommunication Union, Geneva (1977)

COATES, R. F. W. *Modern Communication Systems*, 2nd edn, Macmillan, London (1982)

EEC Commission: CAMAC. *A Modular System for Data Handling. Revised Description and Specification*, EUR 4100e, HMSO, London (1972)

EEC Commission: CAMAC. *Organisation of Multi-Crate Systems. Specification of the Branch Highway and CAMAC Crate Controller Type A*, EUR 4600e, HMSO, London (1972)

EEC Commission: CAMAC. *A Modular Instrumentation System for Data Handling. Specification of Amplitude Analogue Signals*, EUR 5100e, HMSO, London (1972)

EIA. *Standard RS-232C Interface between Data Terminal Equipment and Data Communications Equipment Employing Serial Binary Data Interchange*, EIA, Washington, DC (1969)

EIA. *Standard RS-449 General-purpose 37-position and 9-position Interface for Data Terminal Equipment and Data Circuit-terminating Equipment Employing Serial Binary Data Interchange*, EIA, Washington, DC (1977)

HARTLEY, G., MORNET, P., RALPH, F. and TARRON, D. J. *Techniques of Pulse Code Modulation in Communications Networks*, Cambridge University Press, Cambridge (1967)

HMSO. *Private Point-to-Point Systems Performance Specifications (Nos W.6457 and W.6458) for Angle-Modulated UHF Transmitters and Receivers and Systems in the 450–470 Mc/s Band*, HMSO, London (1963)

HMSO. *Performance Specification: Medical and Biological Telemetry Devices*, HMSO, London (1978)

HMSO. *Performance Specification: Transmitters and Receivers for Use in the Bands Allocated to Low Power Telemetry in the PMR Service*, HMSO, London (1979)

HMSO. International Telecommunication Union World Administrative Radio Conference, 1979, *Radio Regulations. Revised International Table of Frequency Allocations and Associated Terms and Definitions*, HMSO, London (1980)

IEEE. *IEEE-488-1978 Standard Interface for Programmable Instruments*, IEEE, New York (1978)

IEEE. *IEEE-696-1981 Standard Specification for S-100 Bus Interfacing Devices*, IEEE, New York (1981)

JOHNSON, C. S. 'Telemetry data systems', *Instrument Technology*, Aug. (1976), 39–53: Oct. (1976), 47–53

KEISER, G. *Optical Fibre Communication*, McGraw-Hill International, London (1983)

SENIOR, J. *Optical Fiber Communications, Principles and Practice*, Prentice-Hall, London (1985)

SHANMUGAN, S. *Digital and Analog Communications Systems*, John Wiley, New York (1979)

STEELE, R. *Delta Modulation Systems*, Pentech Press, London (1975)

WARNOCK, J.D. Section 16.27 in *The Process Instruments and Controls Handbook* 3rd edition, edited by D.M. Considine, McGraw-Hill, London (1985)

WILSON, J. and HAWKES, J. F. B. *Optoelectronics: An Introduction*, Prentice-Hall, London (1983)

6 Display and recording

M. L. Sanderson

6.1 Introduction

Display devices are used in instrumentation systems to provide instantaneous but non-permanent communication of information between a process or system and a human observer. The data can be presented to the observer in either an analogue or digital form. Analogue indicators require the observer to interpolate the reading when it occurs between two scale values, which requires some skill on the part of the observer. They are, however, particularly useful for providing an overview of the process or system and an indication of trends when an assessment of data from a large number of sources has to be quickly assimilated. Data displayed in a digital form require little skill from the observer in reading the value of the measured quantity, though any misreading can introduce a large observational error as easily as a small one. Using digital displays it is much more difficult to observe trends within a process or system and to quickly assess, for example, the deviation of the process or system from its normal operating conditions. Hybrid displays incorporating both analogue and digital displays combine the advantages of both.

The simplest indicating devices employ a pointer moving over a fixed scale; a moving scale passing a fixed pointer; or a bar graph in which a semi-transparent ribbon moves over a scale. These devices use mechanical or electromechanical means to effect the motion of the moving element. Displays can also be provided using illuminative devices such as light-emitting diodes (LEDs), liquid crystal displays (LCDs), plasma displays, and cathode ray tubes (CRTs). The mechanisms and configurations of these various display techniques are identified in Table 6.1.

Table 6.1 Commonly used display techniques

Display technique	*Mechanism*	*Configurations*
Indicating devices		
Moving pointer	Mechanical/electromechanical movement of pointer over a fixed scale	Horizontal/vertical, straight, arc, circular, or segment scales with edgewise strip, hairline, or arrow-shaped pointers
Moving scale	Mechanical/electromechanical movement of scale. Indication given by position of scale with respect to fixed pointer	Moving dial or moving drum analogue indicators, digital drum indicators.
Bar graph	Indication given by height or length of vertical or horizontal column	Moving column provided by mechanically driven ribbon or LED or LCD elements
Illuminative displays		
Light emitting diodes	Light output provided by recombination electroluminescence in a forward-biased semiconductor diode	Red, yellow, green, displays configured as lamps, bar graphs, 7- and 16-segment alphanumeric displays, dot matrix displays
Liquid crystal displays	The modulation of intensity of transmitted-reflected light by the application of an electric field to a liquid crystal cell	Reflective or transmissive displays, bar graph, 7-segment, dot matrix displays, alphanumeric panels
Plasma displays	Cathode glow of a neon gas discharge	Nixie tubes, 7-segment displays, plasma panels
CRT displays	Conversion into light of the energy of scanning electron beam by phosphor	Monochrome and colour tubes, storage tubes, configured as analogue, storage, sampling, or digitizing oscilloscopes, VDUs, graphic displays

Table 6.2 Commonly used recording techniques

Recording system	*Technique*	*Configurations*
Graphical recorders	Provide hard copy of data in graphical form using a variety of writing techniques, including pen-ink, impact printing, thermal, optical, and electric writing	Single/multichannel x–t strip chart and circular chart recorders, galvanometer recorders, analogue and digital x–y recorders, digital plotters
Printers	Provide hard copy of data in alphanumeric form using impact and non-impact printing techniques	Serial impact printers using cylinder, golf ball, daisywheel or dot matrix heads. Line printers using drum, chain-belt, oscillating bar, comb, and needle printing heads. Non-impact printers using thermal, electrical, electrostatic, magnetic, ink-jet, electrophotographic, and laser printing techniques
Magnetic recording	Use the magnetization of magnetic particles on a substrate to store information	Magnetic tape recorders using direct, frequency modulation or pulse code modulation technique for the storage of analogue or digital data. Spool to spool or cassette recorders. Floppy or hard discs for the storage of digital data
Transient recorders	Use semiconductor memory to store high-speed transient waveforms	Single/multichannel devices using analogue-to-digital conversion techniques. High-speed transient recorders using optical scanning techniques to capture data before transfer to semiconductor memory
Data loggers	Data-acquisition system having functions programmed from the front panel	Configured for a range of analogue or digital inputs with limited logical or mathematical functions. Internal display using LED, LCD, CRT. Hard copy provided by dot matrix ink or thermal or electrical writing technique. Data storage provided by semiconductor or magnetic storage using tape or disc

Recording enables hard copy of the information to be obtained in graphical or alphanumeric form or the information to be stored in a format which enables it to be retrieved at a later stage for subsequent analysis, display, or conversion into hard copy. Hard copy of graphical information is made by graphical recorders. x–t recorders enable the relationships between one or more variables and time to be obtained whilst x–y recorders enable the relationship between two variables to be obtained. These recorders employ analogue or digital drive mechanisms for the writing heads, generally with some form of feedback. The hard copy is provided using a variety of techniques, including ink pens or impact printing on normal paper, or thermal, optical, or electrical writing techniques on specially prepared paper. Alphanumeric recording of data is provided by a range of printers, including impact printing with cylinder, golf ball, daisywheel, or dot matrix heads; or non-impact printing techniques including ink-jet, thermal, electrical, electrostatic, electromagnetic, or laser printers.

Recording for later retrieval can use either magnetic or semiconductor data storage. Magnetic tape recorders are used for the storage of both analogue and digital data. Transient/waveform recorders (also called waveform digitizers) generally store their information in semiconductor memory. Data-logger systems may employ both semiconductor memory and magnetic storage on either disc or tape. Table 6.2 shows the techniques and configurations of the commonly used recording systems.

The display and recording of information in an instrumentation system provides the man/machine interface (MMI) between the observer and the system or process being monitored. It is of fundamental importance that the information should be presented to the observer in as clear and unambiguous a way as possible and in a form that is easily assimilated. In addition to the standard criteria for instrument performance such as accuracy, sensitivity, and speed of response, ergonomic factors involving the visibility, legibility, and organization and presentation of the information are also of importance in displays and recorders.

6.2 Indicating devices

In moving-pointer indicator devices (Figure 6.1) the pointer moves over a fixed scale which is mounted vertically or horizontally. The scale may be either straight or an arc. The motion is created by mechanical means as in pressure gauges or by electromechanical means using a moving coil movement. (For details of such movements the reader is directed to Electrical Measurements in Part 3 of this reference book.) The pointer consists of a knife edge, a line scribed on each side of a transparent member, or a sharp, arrow-shaped tip. The pointers are designed to minimize the reading error when the instrument is read from different angles. For precision work such 'parallax errors' are reduced by mounting a mirror behind the pointer. Additional pointers may be provided on the scale. These can be used to indicate the value of a set point or alarm limits.

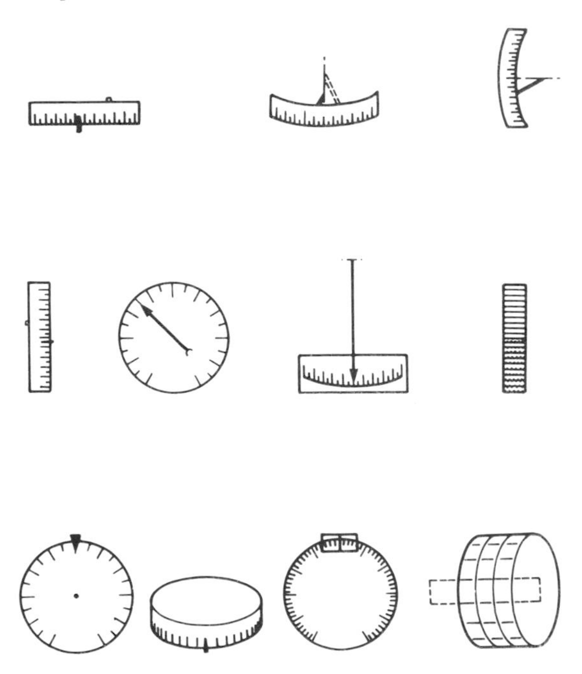

Figure 6.1 Moving-pointer and moving-scale indicators.

The pointer scales on such instruments should be designed for maximum clarity. BS 3693 sets out recommended scale formats. The standard recommends that the scale base length of an analogue indicating device should be at least $0.07D$, where D is the viewing distance. At a distance of 0.7 m, which is the distance at which the eye is in its resting state of accommodation, the minimum scale length should be 49 mm. The reading of analogue indicating devices requires the observer to interpolate between readings. It has been demonstrated that observers can subdivide the distance between two scale markings into five, and therefore a scale which is to be read to within 1 per cent of full-scale deflection (FSD) should be provided with twenty principal divisions. For electromechanical indicating instruments accuracy is classified by BS 89 into nine ranges, from ±0.05 per cent to ±5 per cent of FSD. A fast response is not required for visual displays since the human eye cannot follow changes much in excess of 20 Hz. Indicating devices typically provide frequency responses up to 1–2 Hz.

Moving-scale indicators in which the scale moves past a fixed pointer can provide indicators with long scale lengths. Examples of these are also shown in Figure 6.1.

In the bar graph indicator (Figure 6.2) a semitransparent ribbon moves over a scale. The top of the ribbon indicates the value of the measured quantity and the ribbon is generally driven by a mechanical lead. Arrays of LEDs or LCDs can be used to provide the solid state equivalent of the bar graph display.

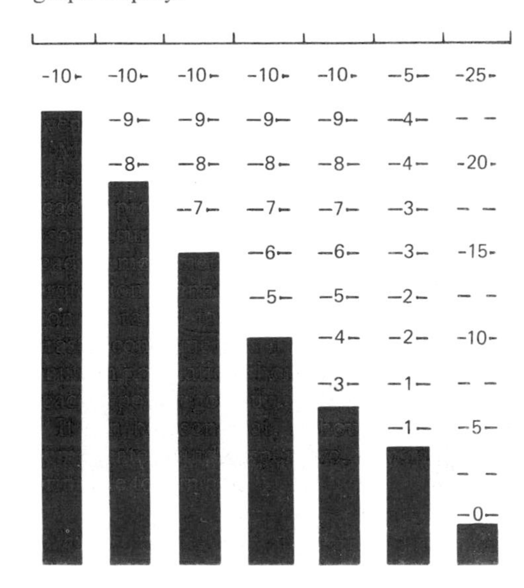

Figure 6.2 Bar-graph indicator.

6.3 Light-emitting diodes (LEDs)

These devices, as shown in Figures 6.3(a) and 6.3(b), use recombination (injection) electroluminescence and consist of a forward-biased p–n junction in which the majority carriers from both sides of the junction cross the internal potential barrier and enter the material on the other side, where they become minority carriers, thus disturbing the local minority carrier population. As the excess minority carriers diffuse away from the junction, recombination occurs. Electroluminescence takes place if the recombination results in radiation. The wavelength of the emitted radiation is inversely proportional to the band gap of the material, and therefore for the radiation to be in the visible region the band gap must be greater than 1.8 eV.

$GaAs_{0.6}P_{0.4}$ (gallium arsenide phosphide) emits red light at 650 nm. By increasing the proportion of phosphorus and doping with nitrogen the wavelength of the emitted light is reduced. $GaAs_{0.35}P_{0.65}N$ provides a source of orange light, whilst $GaAs_{0.15}P_{0.85}N$ emits yellow light at 589 nm. Gallium phosphide doped with nitrogen radiates green light at 570 nm. Although the internal quantum efficiencies of LED materials can be high, the external quantum efficiencies may be much lower. This is because the materials have a high refractive index and therefore a significant proportion of the emitted radiation strikes the material/air interface beyond the critical angle and is totally internally reflected. This is usually overcome by encapsulating the diode in a hemispherical dome made of epoxy resin, as shown in Figure 6.3(c). The construction of an LED element in a seven-segment display is shown in Figure 6.3(d).

The external quantum efficiencies of green diodes tend to be somewhat lower than those of red diodes, but, for the same output power, because of the

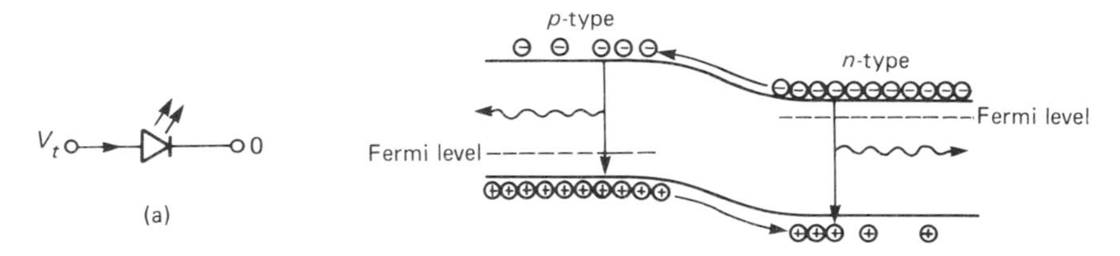

Injection of minority carriers and subsequent radiative recombination with the majority carriers in a forward biased p–n junction.

(b)

Plastic encapsulation
p–n Junction
Electrical contacts
(c)

2.5 cm (typically)
Chip
Reflector
Diffuser
(d)

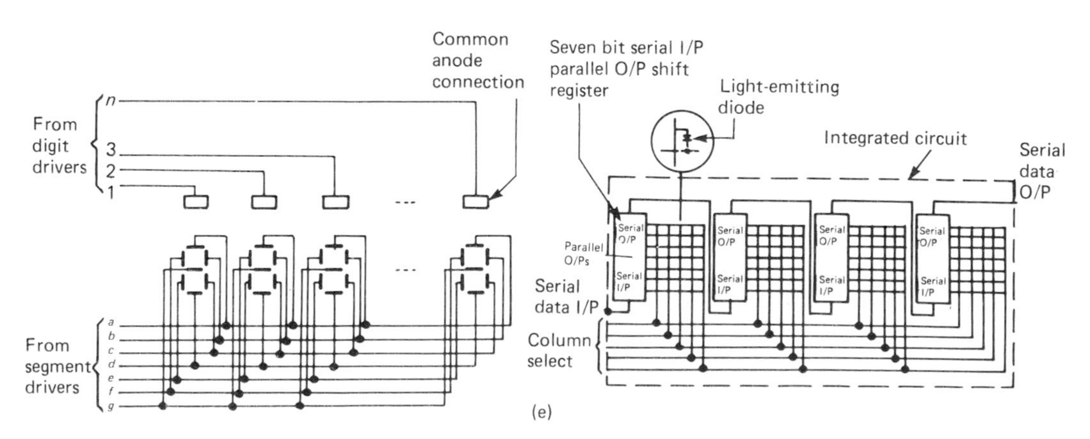

(e)

Figure 6.3 Light-emitting diodes.

sensitivity of the human eye the green diode has a higher luminous intensity.

Typical currents required for LED elements are in the range 10–100 mA. The forward diode drop is in the range 1.6–2.2 V, dependent on the particular device. The output luminous intensities of LEDs range from a few to over a hundred millicandela, and their viewing angle can be up to ±60°. The life expectancy of a LED display is twenty years, over which time it is expected that there will be a 50 per cent reduction in output power.

LEDs are available as lamps, seven- and sixteen-segment alphanumeric displays, and in dot matrix format (Figure 6.3(e)). Alphanumeric displays are provided with on-board decoding logic which enables the data to be entered in the form of ASCII or hexadecimal code.

6.4 Liquid crystal displays (LCDs)

These displays are passive and therefore emit no radiation of their own but depend upon the modulation of reflected or transmitted light. They are based upon the optical properties of a large class of organic materials known as liquid crystals. Liquid crystals have molecules which are rod shaped and which, even in their liquid state, can take up certain defined orientations relative to each other and also with respect to a solid interface. LCDs commonly use nematic liquid crystals such as *p*-azoxyanisole, in which the molecules are arranged with their long axes approximately parallel, as shown in Figure 6.4(a). They are highly anisotropic—that is, they have different optical or other properties in different directions. At a solid–liquid interface the ordering of the crystal can be either homogeneous (in which the molecules are parallel to the interface) or homeotropic (in which the molecules are aligned normal to the interface), as shown in Figure 6.4(b). If a liquid crystal is confined between two plates which without the application of an electric field is a homogeneous state, then when an electric field is applied the molecules will align themselves with this field in order to minimize their energy. As shown in Figure 6.4(c), if the E field is less than some critical value E_c then the ordering is not affected. If $E > E_c$ then the molecules furthest away from the interfaces are realigned. For values of the electric field $E >> E_c$ most of the molecules are realigned.

A typical reflective LCD consists of a twisted nematic cell in which the walls of the cell are such as to produce a homogeneous ordering of the molecules but rotated by 90 degrees (Figure 6.4(d)). Because of the birefringent nature of the crystal, light polarized on entry in the direction of alignment of the molecules at the first interface will leave the cell with its direction of polarization rotated by 90 degrees. On the application of an electric field in excess of the critical field (for a cell of thickness 10 μm a typical critical voltage is 3 V) the molecules align themselves in the direction of the applied field. The polarized light does not then undergo any rotation. If the cell is sandwiched between two pieces of polaroid with no applied field, both polarizers allow light to pass through them, and therefore incident light is reflected by the mirror and

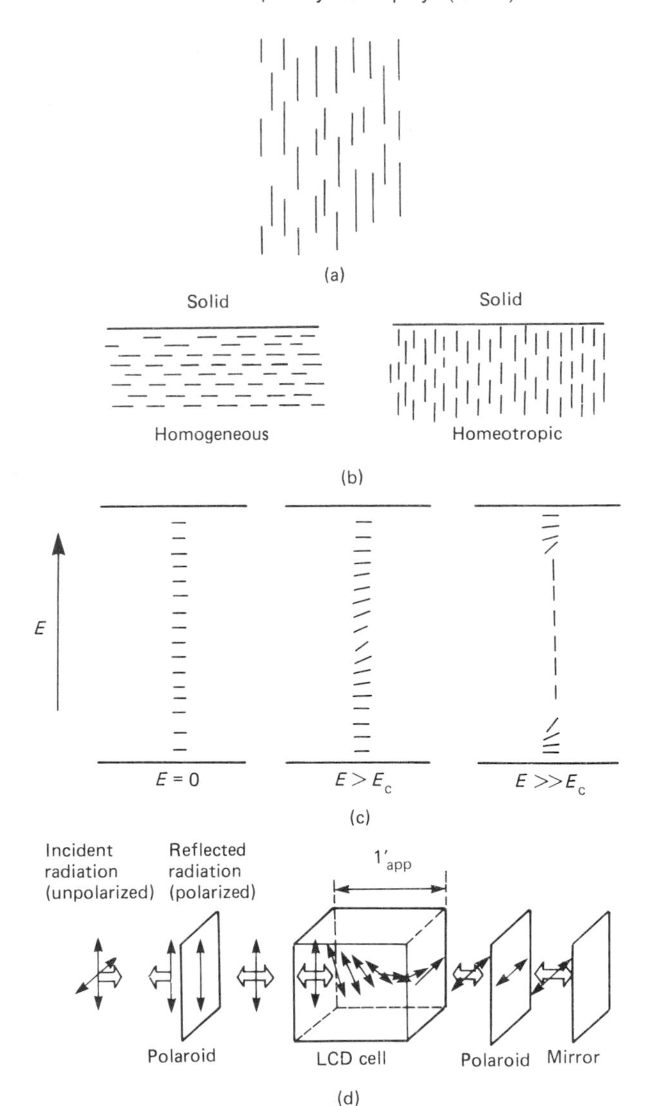

Figure 6.4 Liquid crystal displays. (a) Ordering of nematic crystals; (b) ordering at liquid crystal/solid interface; (c) application of an electric field to a liquid crystal cell; (d) reflective LCD using a twisted nematic cell.

the display appears bright. With the application of the voltage the polarizers are now crossed, and therefore no light is reflected and the display appears dark. Contrast ratios of 150:1 can be obtained.

D.C. voltages are generally not used in LCDs because of the electromechanical reactions. A.C. waveforms are employed with the cell having a response corresponding to the rms value of the applied voltage. The frequency of the a.c. is in the range 25 Hz to 1 kHz. Power consumption of LCD displays is low, with a typical current of 0.3–0.5 μA at 5 V. The optical switching time is typically 100–150 ms. They can operate over a temperature range of −10°C to +70°C.

Polarizing devices limit the maximum light which can be reflected from the cell. The viewing angle is also limited to approximately ±45 degrees. This can be improved by the use of cholesteric liquid crystals used in conjunction with dichroic dyes which absorb light whose direction of polarization is parallel to its long axis. Such displays do not require polarizing filters and are claimed to provide a viewing angle of ±90 degrees.

LCDs are available in bar graph, seven-segment, dot matrix, and alphanumeric display panel configurations.

6.5 Plasma displays

In a plasma display as shown in Figure 6.5(a) a glass envelope filled with neon gas, with added argon, krypton, and mercury to improve the discharge properties of the display, is provided with an anode and a cathode. The gas is ionized by applying a voltage of approximately 180 V between the anode and cathode. When ionization occurs there is an orange/yellow glow in the region of the cathode. Once ionization has taken place the voltage required to sustain the ionization can be reduced to about 100 V.

The original plasma display was the Nixie tube, which had a separate cathode for each of the figures 0–9. Plasma displays are now available as seven-segment displays or plasma panels as shown in Figures 6.5(a) and 6.5(b). The seven-segment display uses a separate cathode for each of the segments. The plasma panel consists, for example, of 512 × 512 vertical and horizontal x and y electrodes. By applying voltages to specific x and y electrode pairs (such that the sum of the voltages at the spot specified by the (x,y) address exceeds the ionization potential) then a glow will be produced at that spot. The display is operated by applying continuous a.c. voltages equivalent to the sustaining potential to all the electrodes. Ionization is achieved by pulsing selected pairs. The glow at a particular location is quenched by applying antiphase voltages to the electrodes.

The display produced by plasma discharge has a wide viewing angle capability, does not need back lighting, and is flicker free. A typical 50 mm seven-segment display has a power consumption of approximately 2 W.

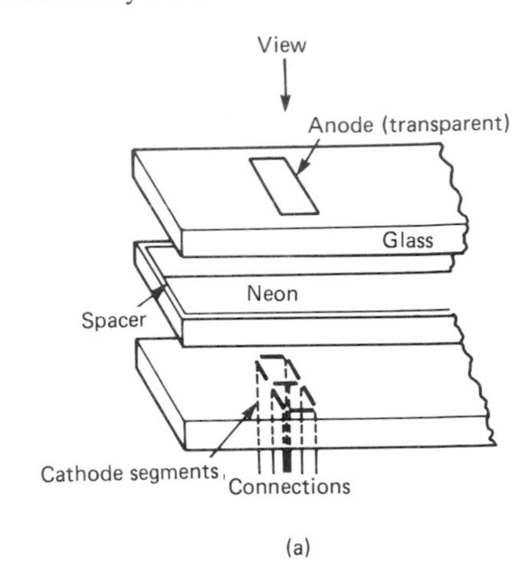

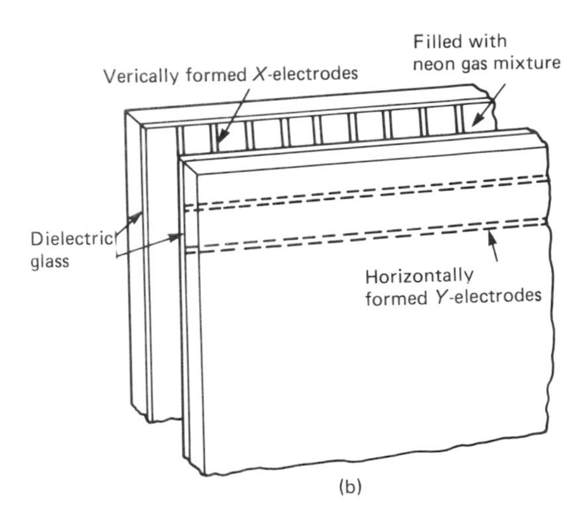

Figure 6.5 Plasma display. (a) Seven segment display; (b) plasma panel.

6.6 Cathode ray tubes (CRTs)

CRTs are used in oscilloscopes which are commonly employed for the display of repetitive or transient waveforms. They also form the basis of visual display units (VDUs) and graphic display units. The display is provided by using a phosphor which converts the energy from an electron beam into light at the point at which the beam impacts on the phosphor.

A CRT consists of an evacuated glass envelope in which the air pressure is less than 10^{-4} pascal (Figure 6.6). The thermionic cathode of nickel coated with oxides of barium, strontium, and calcium is indirectly heated to approximately 1100°K and thus gives off electrons. The number of electrons which strike the screen and hence control the brightness of the display is adjusted by means of the potential applied to a control grid surrounding the cathode. The control grid which has a pin hole through which the electrons can pass is held at a potential of between 0 and 100 V negative with respect to the cathode. The beam of electrons pass through the first anode A_1 which is typically held at a potential of 300 V positive with respect to the cathode before being focused, accelerated, and deflected.

Focusing and deflection of the beam can be by either electrostatic or magnetic means. In the electrostatic system shown in Figure 6.6(a) the cylindrical focusing anode A_2, which consists of disc baffles having an aperture in the centre of them, is between the first anode A_1 and the accelerating anode A_3, which is typically at a potential of 3–4 kV with respect to the cathode. Adjusting the potential on A_2 with respect to the potentials on A_1 and A_3 focuses the beam such that the electrons then travel along the axis of the tube. In a magnetic focusing system magnetic field coils around the tube create a force on the electrons, causing them to spiral about the axis and also inwardly. By employing magnetic focusing it is possible to achieve a smaller spot size than with electrostatic focusing. Deflection of the electron beam in a horizontal and vertical direction moves the position of the illuminated spot on the screen. Magnetic deflection provides greater deflection capability and is therefore used in CRTs for television, alphanumeric, and graphical displays. It is slower than electrostatic deflection, which is the deflection system commonly used in oscilloscopes.

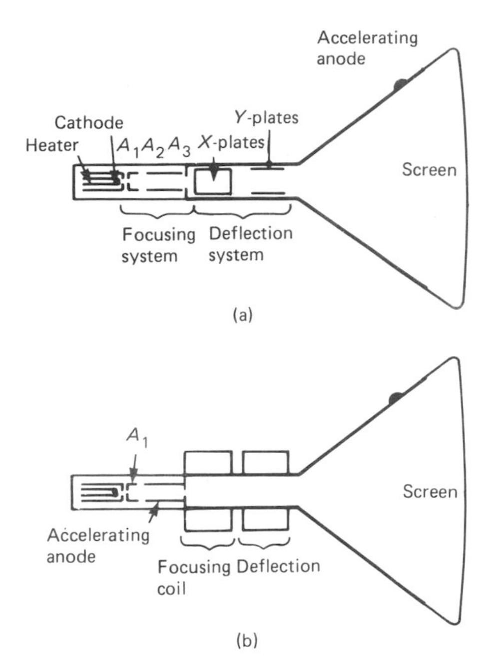

Figure 6.6 Cathode ray tube. (a) Electrostatic focusing and deflection; (b) electromagnetic focusing and deflection.

Acceleration of the beam is by either the use of the accelerating electrode A_3 (such tubes are referred to as monoaccelerator tubes) or by applying a high potential (10–14 kV) on to a post-deflection anode situated close to the CRT screen. This

Table 6.3 Characteristics of commonly used phosphors (courtesy of Tektronix)

Phosphor	*Fluorescence*	*Relative*[a] *luminance (%)*	*Relative*[b] *photographic writing speed (%)*	*Decay*	*Relative burn resistance*	*Comments*
P1	Yellow-green	50	20	Medium	Medium	In most applications replaced by P31
P4	White	50	40	Medium /short	Medium /high	Television displays
P7	Blue	35	75	Long	Medium	Long-decay, double-layer screen
P11	Blue	15	100	Medium /short	Medium	For photographic applications
P31	Green	100	50	Medium /short	High	General purposes, brightest available phosphor

[a]Measured with a photometer and luminance probe incorporating a standard eye filter. Representative of 10 kV aluminized screens with P31 phosphor as reference.
[b]P11 as reference with Polaroid 612 or 106 film. Representative of 10 kV aluminized screens.

technique, which is known as post-deflection acceleration (PDA), gives rise to tubes with higher light output and increased deflection sensitivity.

The phosphor coats the front screen of the CRT. A range of phosphors are available, the choice of which for a particular situation depends on the colour and efficiency of the luminescence required and its persistence time, that is, the time for which the afterglow continues after the electron beam has been removed. Table 6.3 provides the characteristics of some commonly used phosphors.

6.6.1 Colour displays

Colour displays are provided using a screen which employs groups of three phosphor dots. The material of the phosphor for each of the three dots is chosen such that it emits one of the primary colours (red, green, blue). The tube is provided with a shadow mask consisting of a metal screen with holes in it placed near to the screen and three electron guns, as shown in Figure 6.7(a). The guns are inclined to each other so that the beams coincide at

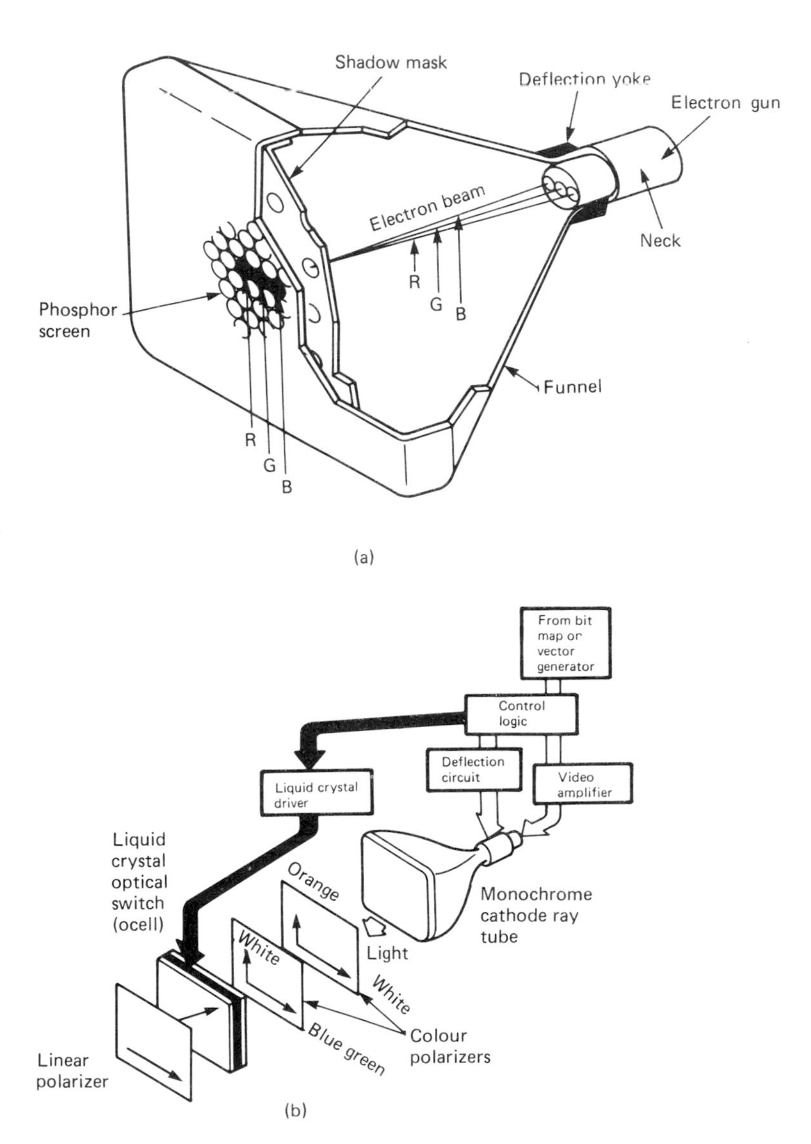

Figure 6.7 Colour displays. (a) Shadow mask colour tube; (b) liquid crystal colour display.

the plane of the shadow mask. After passing through the shadow mask the beams diverge and, on hitting the screen, energize only one of the phosphors at that particular location. The effect of a range of colours is achieved by adjusting the relative intensities of the three primary colours by adjusting the electron beam currents. The resolution of colour displays is generally lower than that of monochrome displays because the technique requires three phosphors to produce the effect of colour. Alignment of the shadow mask with the phosphor screen is also critical and the tube is sensitive to interfering magnetic fields.

An alternative method of providing a colour display is to use penetration phosphors which make use of the effect that the depth of penetration of the electron beam is dependent on beam energy. By using two phosphors, one of which has a non-luminescent coating, it is possible by adjusting the energy of the electron beam to change the colour of the display. For static displays having fixed colours this technique provides good resolution capability and is reasonably insensitive to interfering magnetic fields.

Liquid crystal colour displays employ a single phosphor having two separate emission peaks (Figure 6.7(b)). One is orange and the other blue-green. The colour polarizers orthogonally polarize the orange and blue-green components of the CRT's emission, and the liquid crystal cell rotates the polarized orange and blue-green information into the transmission axis of the linear polarizer and thus selects the colour of the display. Rotation of the orange and blue-green information is performed in synchronism with the information displayed by the sequentially addressed CRT. Alternate displays of information viewed through different coloured polarizing filters are integrated by the human eye to give colour images. This sequential technique can be used to provide all mixtures of the two primary colours contained in the phosphor.

6.6.2 Oscilloscopes

Oscilloscopes can be broadly classified as analogue, storage, sampling, or digitizing devices. Figure 6.8 shows the elements of a typical analogue oscilloscope. The signals to be observed as a function of time are applied to the vertical (Y) plates of the oscilloscope. The input stage of the vertical system matches the voltage levels of these signals to the drive requirements of the deflection plate of the oscilloscope which will have a typical deflection sensitivity of 20 V/cm. The coupling of the input stage can be either d.c. or a.c.

The important specifications for the vertical system include bandwidth and sensitivity. The bandwidth is generally specified as the highest frequency which can be displayed with less than 3 dB loss in amplitude compared with its value at low frequencies. The rise time, T_r, of an oscilloscope to a step input is related to its bandwidth, B, by $T_r = 0.35/B$. In order to measure the rise time of a

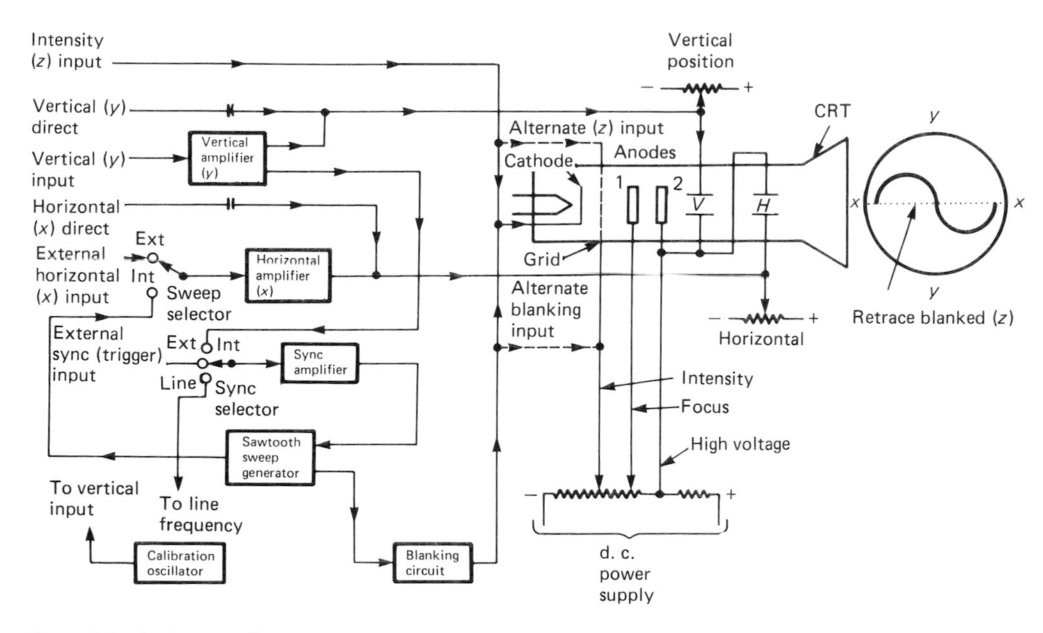

Figure 6.8 Analogue oscilloscope.

waveform with an accuracy of better than 2 per cent it is necessary that the rise time of the oscilloscope should be less than 0.2 of that of the waveform. Analogue oscilloscopes are available having bandwidths of up to 1 GHz.

The deflection sensitivity of an oscilloscope, commonly quoted as mV/cm, mV/div, or μV/cm, μV/div, gives a measure of the smallest signal the oscilloscope can measure accurately. Typically, the highest sensitivity for the vertical system corresponds to 10 μV/cm. There is a trade-off between bandwidth and sensitivity since the noise levels generated either by the amplifier itself or by pickup by the amplifier are greater in wideband measurements. High-sensitivity oscilloscopes may provide bandwidth-limiting controls to improve the display of low-level signals at moderate frequencies.

For comparison purposes, simultaneous viewing of multiple inputs is often required. This can be provided by the use of dual-trace or dual-beam methods. In the dual-trace method the beam is switched between the two input signals. Alternate sweeps of the display can be used for one of the two signals, or, in a single sweep, the display can be chopped between the two signals. Chopping can occur at frequencies up to 1 MHz. Both these methods have limitations for measuring fast transients since in the alternate method the transient may occur on one channel whilst the other is being displayed. The chopping rate of the chopped display limits the frequency range of the signals that can be observed.

Dual-beam oscilloscopes use two independent electron beams and vertical deflection systems. These can be provided with either a common horizontal system or two independent horizontal systems to enable the two signals to be displayed at different sweep speeds. By combining a dual-beam system with chopping it is possible to provide systems with up to eight inputs.

Other functions which are commonly provided on the Y inputs include facilities to invert one channel or to take the difference of two input signals and display the result. This enables unwanted common mode signals present on both inputs to be rejected.

For the display of signals as a function of time the horizontal system provides a sawtooth voltage to the X plates of the oscilloscope together with the blanking waveform necessary to suppress the flyback. The sweep speed required is determined by the waveform being observed. Sweep rates corresponding to as little as 200 ps/div can be obtained. In time measurements the sweep can either be continuous, providing a repetitive display, or single shot, in which the horizontal system is triggered to provide a single sweep. To provide a stable display in the repetitive mode the display is synchronized either internally from the vertical amplifier or externally using the signal triggering or initiating the signal being measured. Most oscilloscopes provide facilities for driving the X plates from an external source to enable, for example, a Lissajous figure to be displayed.

Delayed sweep enables the sweep to be initiated some time after the trigger. This delayed sweep facility can be used in conjunction with the timebase to provide expansion of one part of the waveform.

The trigger system allows the user to specify a position on one or more of the input signals in the case of internal triggering or on a trigger signal in the case of external triggering where the sweep is to initiate. Typical facilities provided by trigger level controls are auto (which triggers at the mean level of the waveform) and trigger level and direction control, i.e. triggering occurs at a particular signal level for positive-going signals.

6.6.3 Storage oscilloscopes

Storage oscilloscopes are used for the display of signals which are either too slow or too fast and infrequent for a conventional oscilloscope. They can also be used for comparing events occurring at different times.

The techniques which are used in storage oscilloscopes include bistable, variable persistence, fast transfer storage, or a combination of fast transfer storage with bistable or variable persistence storage. The storage capability is specified primarily by the writing speed. This is usually expressed in cm/μs or div/μs.

The phosphor in a bistable CRT has two stable states—written and unwritten. As shown in Figure 6.9(a), when writing the phosphor is charged positive in those areas where it is written on. The flood gun electrons hit the unwritten area but are too slow to illuminate it. However, in the written areas the positive charge of the phosphor attracts the electrons and provides them with sufficient velocity to keep the phosphor lit and also to knock out sufficient secondaries to keep the area positive. A bistable tube displays stored data at one level of intensity. It provides a bright, long-lasting display (up to 10 h), although with less contrast than other techniques. It has a slow writing speed with a typical value of 0.5 cm/μs. Split-screen operation, in which the information on one part of the screen is stored whilst the remainder has new information written onto it, can be provided using a bistable CRT.

The screen of the variable persistence CRT, shown in Figure 6.9(b), is similar to that of a conventional CRT. The storage screen consists of a fine metal mesh coated with a dielectric. A collector screen and ion-repeller screen are located behind

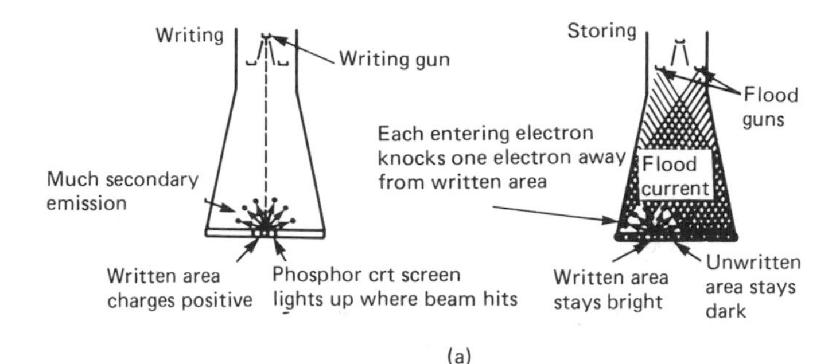

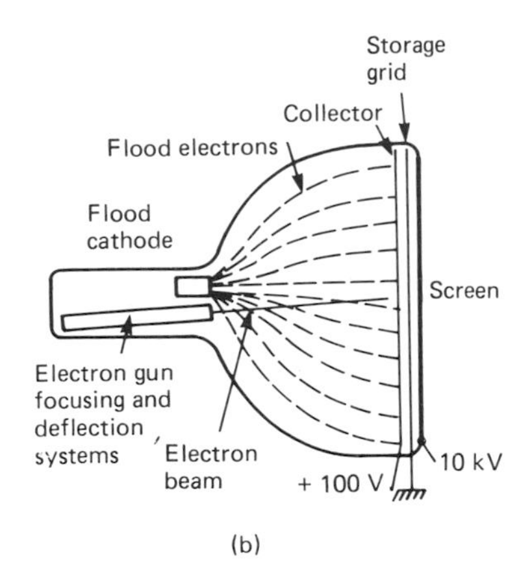

Figure 6.9 Storage CRTs. (a) Bistable; (b) variable persistence.

the storage screen. When the writing gun is employed a positive trace is written out on the storage screen by removing electrons from its dielectric. These electrons are collected by the collector screen. The positively charged areas of the dielectric are transparent to low-velocity electrons. The flood gun sprays the entire screen with low-velocity electrons. These penetrate the transparent areas but not other areas of the storage screen. The storage screen thus becomes a stencil for the flood gun. The charge on the mesh can be controlled, altering the contrast between the trace and the background and also modifying how long the trace is stored. Variable persistence storage provides high contrast between the waveform and the background. It enables waveforms to be stored for only as long as the time between repetitions, and therefore a continuously updated display can be obtained. Time variations in system responses can thus be observed. Integration of repetitive signals can also be provided since noise or jitter not common to all traces will not be stored or displayed. Signals with low repetition rates and fast rise times can be displayed by allowing successive repetitions to build up the trace brightness. The typical writing rate for variable persistence storage is 5 cm/μs. A typical storage time would be 30 s.

Fast transfer storage uses an intermediate mesh target which has been optimized for speed. This target captures the waveform and then transfers it to another mesh optimized for long-term storage. The technique provides increased writing speeds (typically up to 5500 cm/μs). From the transfer target storage can be by either bistable or variable persistence. Oscilloscopes are available in which storage by fast variable persistence, fast bistable, variable persistence, or bistable operation is user selectable. Such devices can provide a range of writing speeds, with storage time combinations ranging from 5500 cm/μs and 30 s using fast variable persistence storage to 0.2 cm/μs and 30 min using bistable storage.

6.6.4 Sampling oscilloscopes

The upper frequency limit for analogue oscilloscopes is typically 1 GHz. For the display of repetitive signals in excess of this frequency,

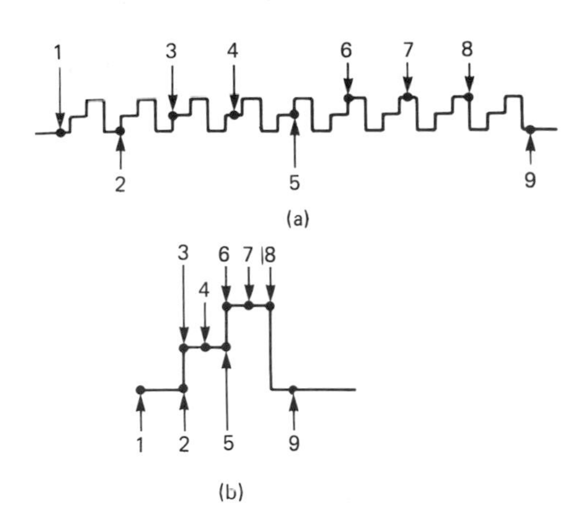

Figure 6.10 Sampling oscilloscope.

sampling techniques are employed. These extend the range to approximately 14 GHz. As shown in Figure 6.10, samples of different portions of successive waveforms are taken. Sampling of the waveform can be either sequential or random. The samples are stretched in time, amplified by relatively low bandwidth amplifiers, and then displayed. The display is identical to the sampled waveform. Sampling oscilloscopes are typically capable of resolving events of less than 5 mV in peak amplitude that occur in less than 30 ps on an equivalent timebase of less than 20 ps/cm.

6.6.5 Digitizing oscilloscopes

Digitizing oscilloscopes are useful for measurements on single-shot or low-repetition signals. The digital storage techniques they employ provide clear, crisp displays. No fading or blooming of the display occurs and since the data are stored in digital memory the storage time is unlimited.

Figure 6.11 shows the elements of a digitizing oscilloscope. The Y channel is sampled, converted to a digital form, and stored. A typical digitizing oscilloscope may provide dual-channel storage with a 100 MHz, 8-bit ADC on each channel feeding a 1 K × 8 bit store, with simultaneous sampling of the two channels. The sample rate depends on the timebase range but typically may go from 20 samples/s at 5 s/div to 100 M samples/s at 1 μs/div. Additional stores are often provided for comparative data.

Digitizing oscilloscopes can provide a variety of display modes, including a refreshed mode in which the stored data and display are updated by a triggered sweep; a roll mode in which the data and display are continually updated, producing the effect of new data rolling in from the right of the screen; a fast roll mode in which the data are continually updated but the display is updated after the trigger; an arm and release mode which allows a single trigger capture; a pre-trigger mode which in roll and fast roll allocates 0, 25, 75, and 100 per cent of the store to the pre-trigger signal; and a hold display mode which freezes the display immediately.

Communication to a computer system is generally provided by a IEEE-488 interface. This enables programming of the device to be undertaken, for stored data to be sent to an external controller, and, if required, to enable the device to receive new data for display. Digitizing oscilloscopes may also provide outputs for obtaining hard copy of the stored data on an analogue x–y plotter.

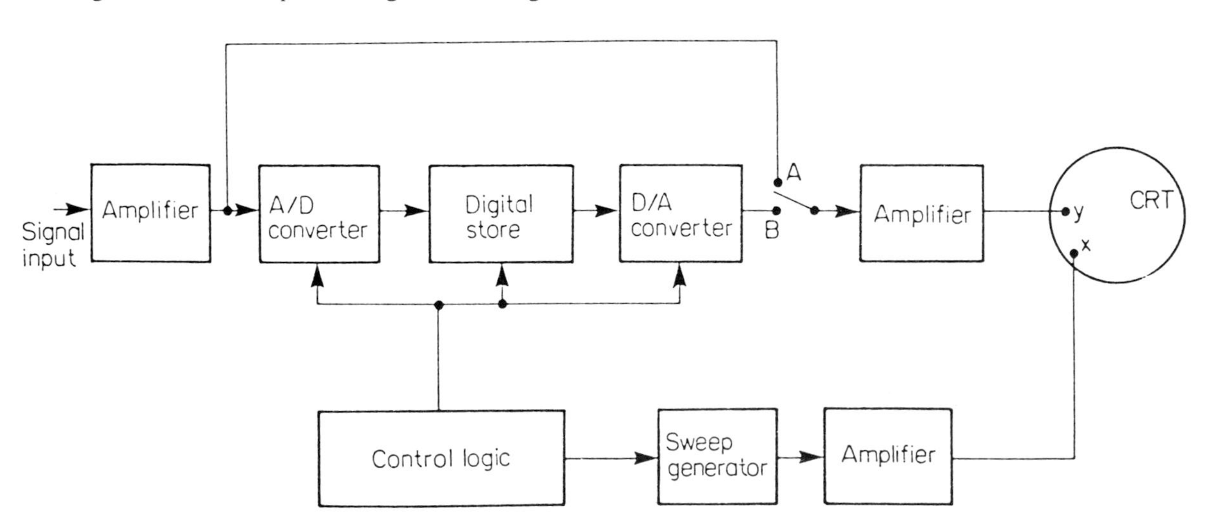

Figure 6.11 Digitizing oscilloscope.

6.6.6 Visual display units (VDUs)

A VDU comprising a keyboard and CRT display is widely used as a MMI in computer-based instrumentation systems. Alphanumeric VDU displays use the raster scan technique, as shown in Figure 6.12. A typical VDU will have 24 lines of 80 characters. The characters are made up of a 7 × 5 dot matrix. Thus seven raster lines are required for a single line of text and five dots for each character position. The space between characters is equivalent to one dot and that between lines is equivalent to one or two raster scans.

As the electron beam scans the first raster line the top row of the dot matrix representation for each character is produced in turn. This is used to

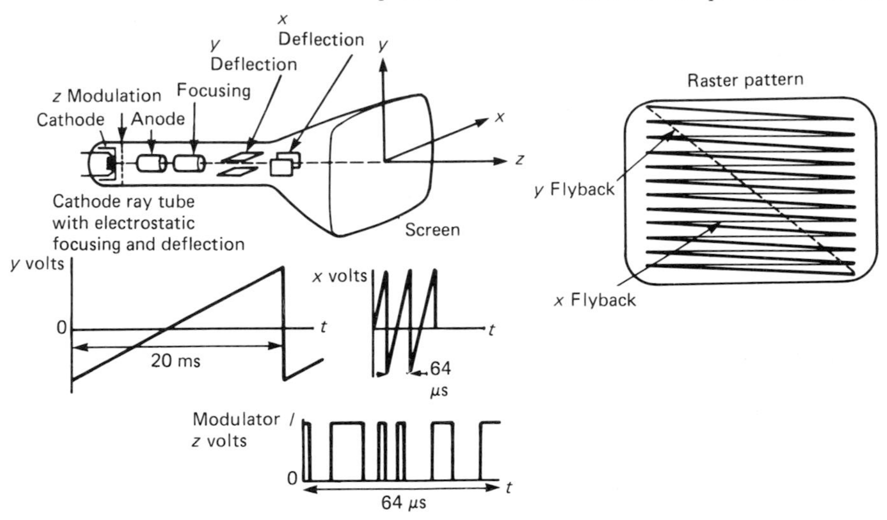

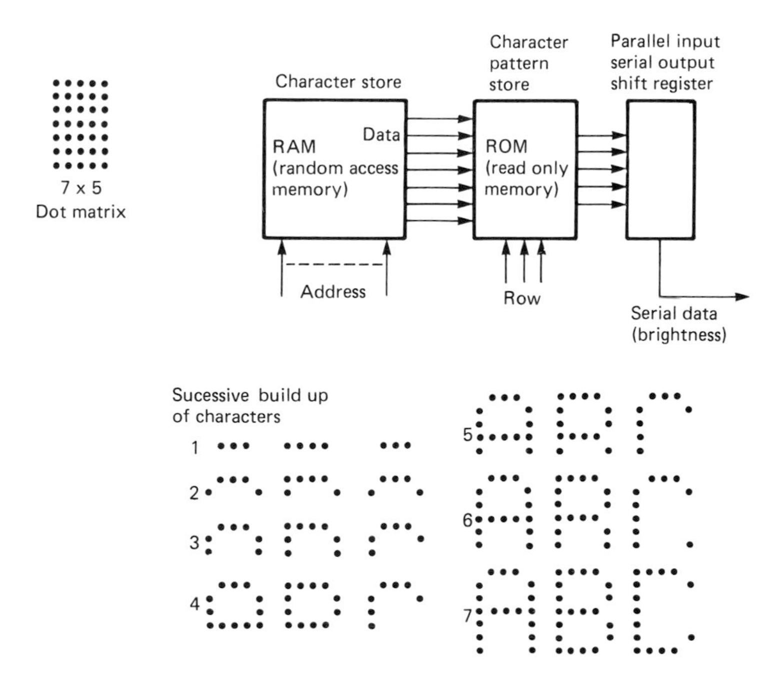

Figure 6.12 Visual display unit.

modulate the beam intensity. For the second raster scan the second row of the dot matrix representation is used. Seven raster scans generate one row of text. The process is repeated for each of the 24 lines and the total cycle repeated at a typical rate of 50 times per second.

The characters to be displayed on the screen are stored in the character store as 7-bit ASCII code, with the store organized as 24 × 80 words of 7 bits. The 7-bit word output of the character store addresses a character pattern ROM. A second input to the ROM selects which particular row of the dot matrix pattern is required. This pattern is provided as a parallel output which is then converted to a serial form to be applied to the brightness control of the CRT. For a single scan the row-selection inputs remain constant whilst the character pattern ROM is successively addressed by the ASCII codes corresponding to the 80 characters on the line. To build up a row of character the sequence of ASCII codes remains the same but on successive raster scans the row address of the character pattern ROM is changed.

6.6.7 Graphical displays

A graphical raster-scan display is one in which the picture or frame is composed of a large number of raster lines, each one of which is subdivided into a large number of picture elements (pixels). Standard television pictures have 625 lines, consisting of two interlaced frames of 313 lines. With each 313-line frame scanned every 20 ms the line scan rate is approximately 60 μs/line.

A graphic raster-scan display stores every pixel in random access or serial store (frame), as shown in Figure 6.13. The storage requirements in such systems can often be quite large. A system using 512 of the 625 lines for display with 1024 pixels on each line having a dual intensity (on/off) display requires a store of 512 Kbits. For a display having the same resolution but with either eight levels of brightness or eight colours the storage requirement then increases to 1.5 Mbits.

Displays are available which are less demanding on storage. Typically, these provide an on/off display having a resolution of 360 lines each having 720 pixels. This requires a 16K × 16 bit frame store. Limited graphics can be provided by alphanumeric raster scan displays supplying additional symbols to the ASCII set. These can then be used to produce graphs or pictures.

6.7 Graphical recorders

Hard copy of data from a process or system can be displayed in a graphical form either as the relationships between one or more variables and time using an x–t recorder or as the relationship between two variables using an x–y recorder. x–t recorders can be classified as either strip chart recorders, in which the data are recorded on a continuous roll of chart paper, or circular chart recorders, which, as their name implies, record the data on a circular chart.

6.7.1 Strip chart recorders

Most strip chart recorders employ a servo-feedback system (as shown in Figure 6.14) to ensure that the displacement of the writing head across the paper tracks the input voltage over the required frequency range. The position of the writing head is generally measured by a potentiometer system. The error signal between the demanded position and the actual position of the writing head is amplified using an a.c. or d.c. amplifier, and the output drives either an a.c. or d.c. motor. Movement of the writing head is effected by a mechanical linkage between the output of the motor and the writing head. The chart paper movement is generally controlled by a stepping motor.

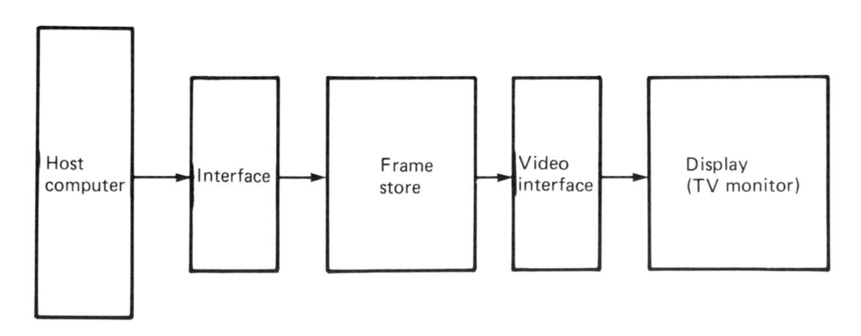

Figure 6.13 Elements of a graphical display.

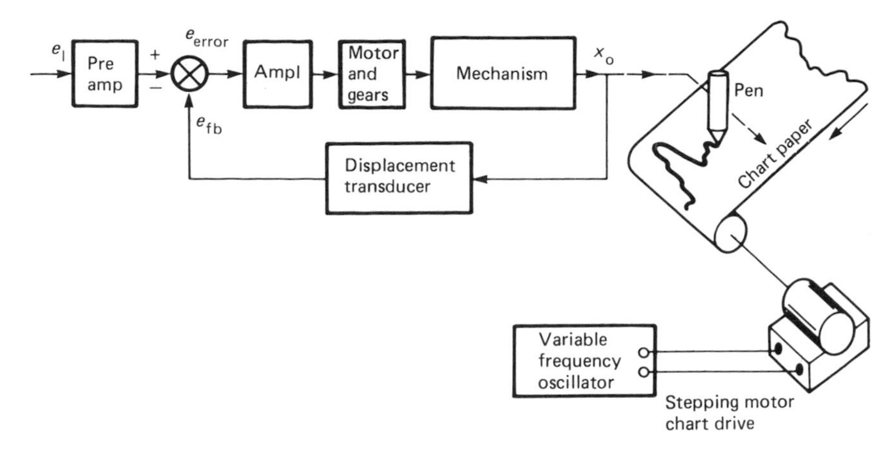

Figure 6.14 Strip chart recorder.

The methods used for recording the data onto the paper include:

(1) *Pen and ink*. In the past these have used pens having ink supplied from a refillable reservoir. Increasingly, such systems use disposable fibre-tipped pens. Multichannel operation can be achieved using up to six pens. For full-width recording using multiple pens, staggering of the pens is necessary to avoid mechanical interference.
(2) *Impact printing*. The 'ink' for the original impact systems was provided by a carbon ribbon placed between the pointer mechanism and the paper. A mark was made on the paper by pressing the pointer mechanism onto the paper. Newer methods can simultaneously record the data from up to twenty variables. This is achieved by having a wheel with an associated ink pad which provides the ink for the symbols on the wheel. By rotating the wheel different symbols can be printed on the paper for each of the variables. The wheel is moved across the paper in response to the variable being recorded.
(3) *Thermal writing*. These systems employ thermally sensitive paper which changes colour on the application of heat. They can use a moving writing head which is heated by an electric current or a fixed printing head having a large number of printing elements (up to several hundred for 100 mm wide paper). The particular printing element is selected according to the magnitude of the input signal. Multichannel operation is possible using such systems and time and date information can be provided in alphanumeric form.
(4) *Optical writing*. This technique is commonly used in galvanometer systems (see section 6.7.3). The source of light is generally ultra-violet to reduce unwanted effects from ambient light. The photographic paper used is sensitive to ultra-violet light. This paper develops in daylight or artificial light without the need for special chemicals. Fixing of the image is necessary to prevent long-term degradation.
(5) *Electric writing*. The chart paper used in this technique consists of a paper base coated with a layer of a coloured dye—black, blue, or red—which in turn is coated with a very thin surface coating of aluminium. The recording is effected by a tungsten wire stylus moving over the aluminium surface. When a potential of approximately 35 V is applied to this stylus an electric discharge occurs which removes the aluminium, revealing the dye. In multichannel recording the different channels are distinguished by the use of different line configurations (for example, solid, dashed, dotted). Alphanumeric information can also be provided in these systems.

The major specifications for a strip chart recorder include:

(1) *Number of channels*. Using a numbered print wheel, up to thirty channels can be monitored by a single recorder. In multichannel recorders, recording of the channels can be simultaneous or on a time-shared basis. The channels in a multichannel recording can be distinguished either by colour or by the nature of line marking.
(2) *Chart width*. This can be up to 250 mm.
(3) *Recording technique*. The recorder may employ any of the techniques described above.
(4) *Input specifications*. These are given in terms of the range of the input variable which may be

voltage, current, or from a thermocouple, RTD, pH electrode, or conductivity probe. Typical d.c. voltage spans are from 0.1 mV to 100 V. Those for d.c. current are typically from 0.1 mA to 1 A. The zero suppression provided by the recorder is specified by its suppression ratio, which gives the number of spans by which the zero can be suppressed. The rejection of line frequency signals at the input of the recorder is measured by its common and normal mode rejection ratios.

(5) *Performance specifications*. These include accuracy, deadband, resolution, response time, and chart speed. A typical voltage accuracy specification may be $+(0.3 + 0.1 \times$ suppression ratio) per cent of span, or 0.20 mV, whichever is greater. Deadband and resolution are usually expressed as a percentage of span, with 0.1 and ± 0.15 per cent, respectively, being typical figures. Chart recorders are designed for use with slowly varying inputs (< 1 Hz for full-scale travel). The response time is usually specified as a step response time and is often adjustable by the user to between 1 and 10 s. The chart speed may vary between mm/h and m/h, depending on the application.

6.7.2 Circular chart recorders

These recorders, as shown in Figure 6.15, generally use a 12-in diameter chart. They are particularly well suited for direct actuation by a number of mechanical sensors without the need for a transducer to convert the measured quantity into an electrical one. Thus they can be configured for a temperature recorder using a filled bulb thermometer system or as an absolute, differential, or gauge pressure recorder employing a bellows or a bourdon tube. Up to four variables can be recorded on a single chart. The rotation of the circular scale can be provided by a clockwork motor, which makes the recorder ideally suited in remote locations having no source of power.

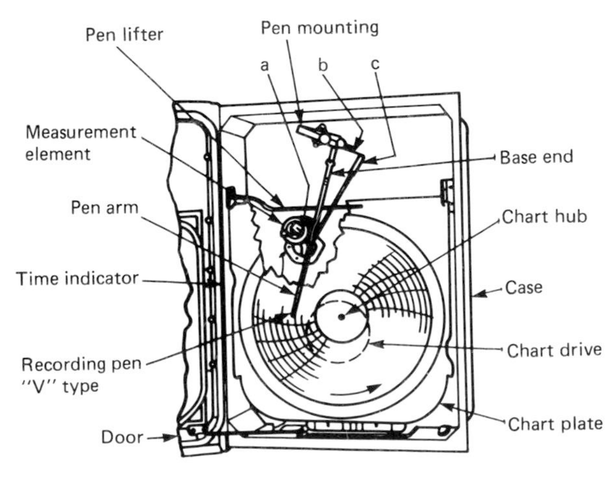

Figure 6.15 Circular chart recorder.

6.7.3 Galvanometer recorders

The D'Arsonoval movement used in the moving-coil indicating instruments can also provide the movement in an optical galvanometer recorder, as shown in Figure 6.16. These devices can have bandwidths in excess of 20 kHz. The light source in such devices is provided by either an ultra-violet or tungsten lamp. Movement of the light beam is effected by the rotation of a small mirror connected to the galvanometer movement. The light beam is focused into a spot on the light-sensitive paper. Positioning of the trace may also be achieved by mechanical means. A recorder may have several galvanometers and, since it is light that is being deflected, overlapping traces can be provided without the need for time staggering. Identification of individual traces is by sequential trace interruption in which the light from each trace in turn is interrupted by a series of pins passing in front of the galvanometers. Amplitude and time reference grids may also be provided.

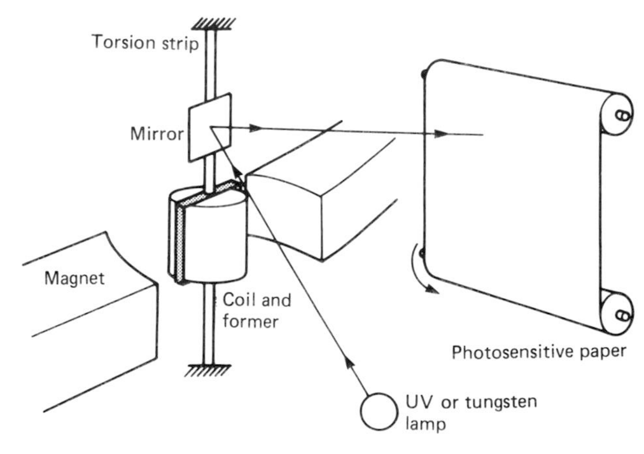

Figure 6.16 Galvanometer recorder.

Trade-offs are available between the sensitivity and frequency response of the galvanometers with high-sensitivity devices having low bandwidths. Manufacturers generally provide a range of plug-in galvanometers, having a range of sensitivities and bandwidths for different applications, which enables them to be quickly removed from the magnet block and replaced. Galvanometer systems are available

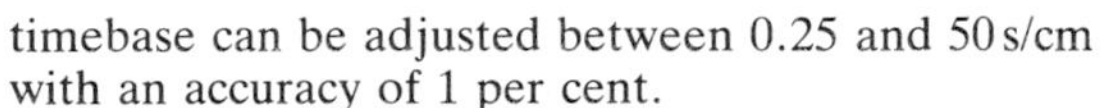

which enable up to 42 channels to be recorded on a 300 mm wide chart drive. The drive speed of the paper can be up to 5000 mm/s.

6.7.4 *x–y* recorders

In an analogue *x–y* recorder (Figure 6.17) both the *x* and *y* deflections of the writing head are controlled by servo feedback systems. The paper, which is usually either A3 or A4 size, is held in position by electrostatic attraction or vacuum. These recorders can be provided with either one or two pens. Plug-in timebases are also generally available to enable them to function as *x–t* recorders. A typical device may provide *x* and *y* input ranges which are continuously variable between 0.25 mV/cm and 10 V/cm with an accuracy of ±0.1 per cent of full scale and a deadband of 0.1 per cent of full scale. Zero offset adjustments are also provided. A typical timebase can be adjusted between 0.25 and 50 s/cm with an accuracy of 1 per cent.

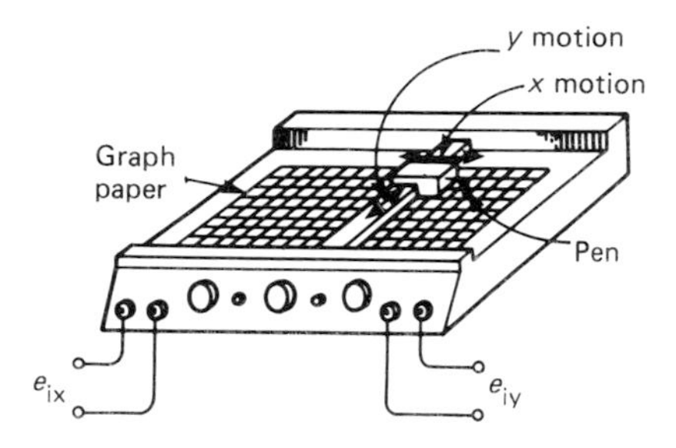

Figure 6.17 *x–y* recorder.

The dynamic performance of *x–y* recorders is specified by their slewing rate and acceleration. A very high-speed *x–y* recorder capable of recording a signal up to 10 Hz at an amplitude of 2 cm peak to peak will have a slewing rate of 97 cm/s and a peak acceleration of 7620 cm/s^2. Remote control of such functions as sweep start and reset, pen lift, and chart hold can be provided.

Digital *x–y* plotters replace the servo feedback with an open-loop stepping motor drive. These can replace traditional analogue recorders and provide increased measurement and graphics capabilities. A digital measurement plotting system can provide, for example, simultaneous sampling and storage of a number of input channels; a variety of trigger modes, including the ability to display pre-trigger data; multi-pen plotting of the data; annotation of the record with date, time, and set-up conditions; and an ability to draw grids and axes. Communication with such devices can be by means of the IEEE-488 or RS-232 interfaces.

For obtaining hard copy from digital data input, graphics plotters are available. With appropriate hardware and software these devices can draw grids, annotate charts, and differentiate data by the use of different colours and line types. They are specified by their line quality, plotting speed, and paper size. Intelligence built into the plotter will free the system's CPU for other tasks. The availability of a graphics language to control the plotter functions and graphics software packages also simplifies the user's programming tasks.

6.8 Magnetic recording

Using magnetic tape recording for data which are to be subsequently analysed has the advantage that, once recorded, the data can be replayed almost indefinitely. The recording period may vary from a

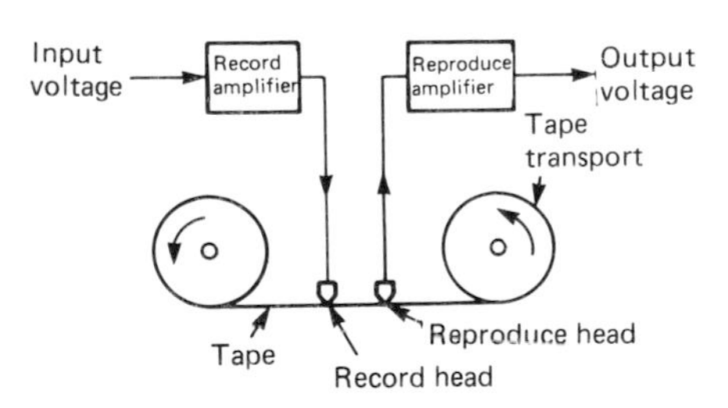

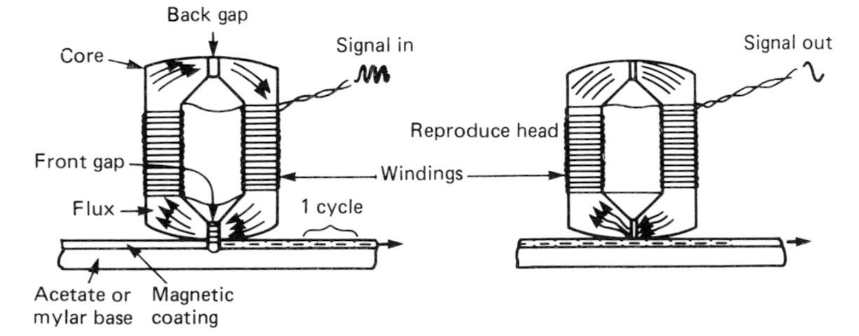

Figure 6.18 Elements of magnetic tape recorder system.

few minutes to several days. Speed translation of the captured data can be provided in that fast data can be slowed down and slow data speeded up by using different record and reproduce speeds. Multichannel recording enables correlations between one or more variables to be identified.

The methods employed in recording data onto magnetic tape include direct recording, frequency modulation (FM), and pulse code modulation (PCM). Figure 6.18 shows the elements of a magnetic tape recorder system. Modulation of the current in the recording head by the signal to be recorded linearly modulates the magnetic flux in the recording gap. The magnetic tape, consisting of magnetic particles on an acetate or mylar base, passes over the recording head. As they leave from under the recording head the particles retain a state of permanent magnetization proportional to the flux in the gap. The input signal is thus converted to a spatial variation of the magnetization of the particles on the tape. The reproduce head detects these changes as changes in the reluctance of its magnetic circuit, which induces a voltage in its winding. This voltage is proportional to the rate of change of flux. The reproduce head amplifier integrates the signal to provide a flat frequency characteristic.

Since the reproduce head generates a signal which is proportional to the rate of change of flux the direct recording method cannot be used down to d.c. The lower limit is typically 100 Hz. The upper frequency limit occurs when the induced variation in magnetization varies over distances smaller than the gap in the reproduce head. This sets an upper limit for direct recording of approximately 2 MHz, using gaps and tape speeds commonly available.

In FM recording systems the carrier signal is frequently modulated by the input signal (frequency modulation is discussed in Chapter 5). The central frequency is chosen with respect to the tape speed and frequency deviations of up to ±40 per cent of the carrier frequency are used. FM systems provide a d.c. response but at the expense of the high-frequency recording limit.

PCM techniques, which are also described in Chapter 5, are used in systems for the recording of digital data. The coding used is typically Non-Return to Zero Level or Delay Modulation.

A typical portable instrumentation tape recorder can provide up to 14 data channels using ½-in tape. Using direct record/reproduce, such a system can record signals in a frequency band from 100 Hz to 300 kHz with an input sensitivity of between 0.1 to 2.5 V rms and a signal-to-noise ratio of up to 40 dB. FM recording with such a system provides bandwidths extending from d.c. to as high as 40 kHz. A typical input sensitivity for FM recording is 0.1 to 10 V rms with a signal-to-noise ratio of up to 50 dB. TTL data recording using PCM can be achieved at data transfer rates of up to 480 kbits/s using such a system.

6.9 Transient/waveform recorders

These devices are used for the high-speed capture of relatively small amounts of data. The specifications provided by general-purpose transient recorders in terms of sampling rate, resolution, number of channels, and memory size are similar to those of the digitizing oscilloscope. Pre-trigger data capture facilities are also often provided. High-speed transient recorders, using scan conversion techniques in which the signal is written onto a silicon-diode target array, can provide sampling rates of 10^{11} samples/s (10 ps/point) with record lengths of 512 points and are capable of recording single-shot signals having bandwidths of up to 500 MHz.

6.10 Data loggers

Data loggers can be broadly defined as data-acquisition systems whose functions are programmed from the front panel. They take inputs from a mixture of analogue and digital signals, perform limited mathematical and logical operations on the inputs, and provide storage either in the form of semiconductor memory or magnetic tape or disc systems. Many data loggers are provided with an

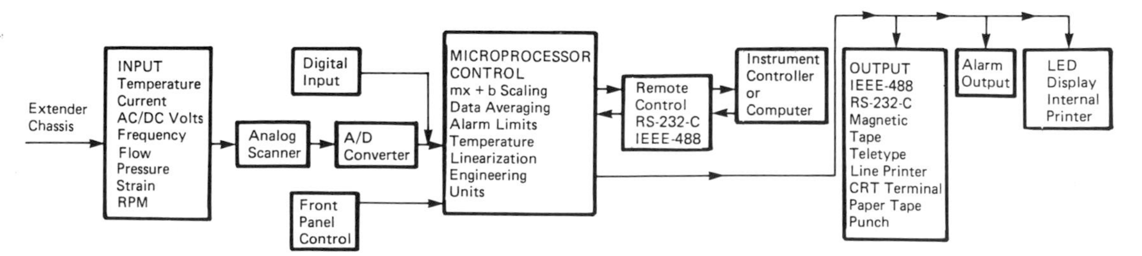

Figure 6.19 Data logger.

integral alphanumeric display and printer. Figure 6.19 shows the elements of a typical data-logging system. These systems can be characterized by:

(1) *The number and type of analogue or digital inputs which can be accepted.* The analogue inputs may be a.c. or d.c. current, d.c. current, or from thermocouples or RTDs.
(2) *The scanning rate and modes of scanning available.* These can include single-scan, monitor, interval scan, and continuous scan.
(3) *The method of programming.* This can be by one button per function on the front panel; by the use of a menu-driven approach by presenting the information to the user on a built-in CRT screen; or by using a high-level language such as BASIC.
(4) *The mathematical functions provided by the logger.* These may include linearization for thermocouple inputs, scaling to provide the user with an output in engineering units, and data averaging over a selectable number of scans.
(5) *The nature and capacity of the internal memory.* If the memory is semiconductor memory, is it volatile? If so, is it provided with battery back-up in case of power failure?
(6) *The printer technique and the width of the print output.* Data loggers typically use dot matrix ink printers or thermal or electric writing technique with continuous strip paper.
(7) *The display.* Data loggers generally provide display using LEDs or LCDs, although some incorporate CRTs.
(8) *Communication with other systems.* This is generally provided by RS 232/422/423 interfaces. Some data loggers are capable of being interrogated remotely using the public telephone network.

6.11 References

AGARD, P. J. *et al. Information and Display Systems in Process Instruments and Control Handbook*, eds D. M. Considine and G. Considine, McGraw-Hill, New York (1985)

BENTLEY, J. *Principles of Measurement Systems*, Longman, London (1983)

BOSMAN, D. 'Human factors in display design', in *Handbook of Measurement Science*, Volume 2, Practical Fundamentals, ed. P. Sydenham, John Wiley, Chichester (1983)

British Standards Institution, BS 89: 1977, Direct Acting Indicating Electrical Instruments and their Accessories (1977)

British Standards Institution, BS 3693 (Part 1: 1964 and Part 2: 1969). The Design of Scales and Indexes (1964 and 1969)

DOEBELIN, E. O. *Measurement Systems Application and Design*, 3rd edn, McGraw-Hill, London (1983)

LENK, J. D. *Handbook of Oscilloscopes, Theory and Application*, Prentice-Hall, Englewood Cliffs, New Jersey (1982)

WILKINSON, B. and HORROCKS, D. *Computer Peripherals*. Hodder and Stoughton, London (1980)

WILSON, J. and HAWKES, J.F.B. *Optoelectronics: An Introduction*. Prentice-Hall, Englewood Cliffs, New Jersey (1983)

7 Pneumatic instrumentation

E. G. Higham

7.1 Basic characteristics

The early evolution of process control was centred on the petroleum and chemical industries, where the process materials gave rise to hazardous environments and therefore the measuring and control systems had to be intrinsically safe. Pneumatic systems were particularly suitable in this connection, and, once the flapper/nozzle system and its associated pneumatic amplifier (customarily called relay) had been developed for the detection of small movements, sensors to measure temperature, flow, pressure, level and density were devised. These were followed by pneumatic mechanisms to implement the control functions whilst air cylinders or diaphragm motors were developed to actuate the final control elements.

The pneumatic systems had further advantages in that the installation did not involve special skills, equipment or materials. The reservoir tanks, normally provided to smooth out the pulsations of the air compressor, also stored a substantial volume of compressed air which, in the event of failure of the compressor or its motive power, stored sufficient energy to enable the control system to remain in operation until an orderly shutdown had been implemented. The provision of comparable features with electronic or other types of control systems involves a great deal of additional equipment.

Although pneumatic control systems have many attractive features they compare unfavourably with electronic systems in two particular respects, namely signal transmission and signal conditioning or signal processing.

For the majority of process applications a distance/velocity lag of about 1 s is quite acceptable. This corresponds typically to a pneumatic transmission distance of about 100 m and is not a limitation for a great many installations, but in large plants, where measurement signals may have to be transmitted over distances of a kilometre or more, electrical methods for transmission have to be adopted. When the successful operation of a process depends on conditioning the signal (e.g. taking the square root, multiplying, dividing, summing or averaging, differentiating or integrating the signals) the pneumatic devices are undoubtedly more complex, cumbersome and very often less accurate than the corresponding electronic devices.

Pneumatic systems are virtually incompatible with digital ones, and cannot compete at all with their advantages for signal transmission, multiplexing, computation, etc. On the other hand, when all these opportunities have been exploited and implemented to the full, the signal to the final control element is nearly always converted back to pneumatics so that the necessary combination of power, speed and stroke/movement can be achieved with a pneumatic actuator.

7.2 Pneumatic measurement and control systems

Pneumatic systems are characterized by the simplicity of the technology on which they are based and the relative ease with which they can be installed, operated and maintained. They are based on the use of a flapper/nozzle in conjunction with a pneumatic relay to detect a very small relative movement (typically, <0.002 mm) and to control a supply of compressed air so that a considerable force can be generated under precise control.

A typical flapper/nozzle system is shown in Figure 7.1. It comprises a flat strip of metal attached to the

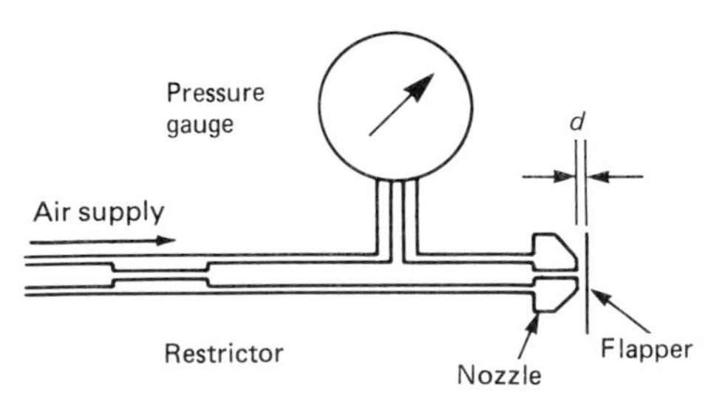

Figure 7.1 Typical flapper/nozzle system. (Most of the figures in this chapter are reproduced by courtesy of The Foxboro Company.)

device or member of which the relative motion is to be detected. This flapper is positioned so that when moved it covers or uncovers a 0.25 mm diameter hole located centrally in the 3 mm diameter flat surface of a truncated 90-degree cone. A supply of clean air, typically at 120 kPa, is connected via a restrictor and a 'T' junction to the nozzle. A pressure gauge connected at this 'T' junction would show that, with the nozzle covered by the flapper, the pressure approaches the supply pressure, but when the flapper moves away from the nozzle the pressure falls rapidly to a value determined by the relative values of the discharge characteristics of the nozzle and the restrictor, as shown in Figure 7.2.

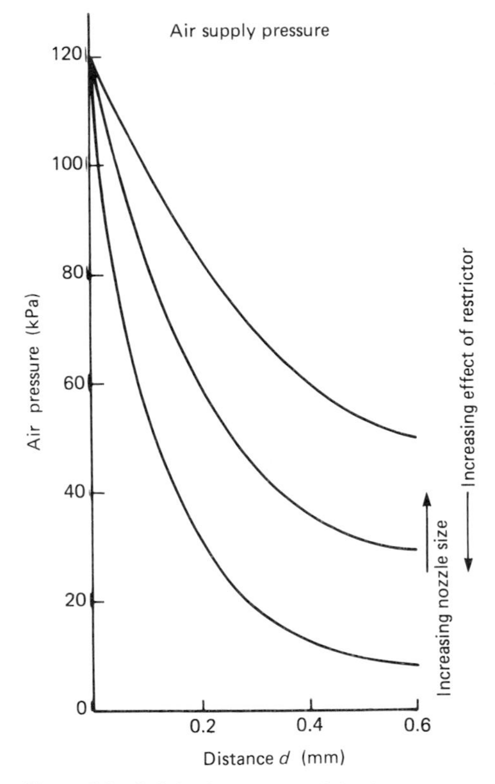

Figure 7.2 Relation between nozzle back pressure and flapper/nozzle distance.

For measurement purposes this change in pressure needs to be amplified, and this is effected by means of a pneumatic relay (which is equivalent to a pneumatic amplifier). In practice, it is convenient to incorporate the restrictor associated with the nozzle in the body of the relay, as shown in Figure 7.3. This figure also shows that the relay comprises two chambers isolated from each other by a flexible diaphragm which carries a cone and spigot that act as a valve to cover or uncover the exhaust port. The spigot acts against a small ball retained by the leaf spring so that it functions as a second valve which controls the flow of air from the supply to the output port.

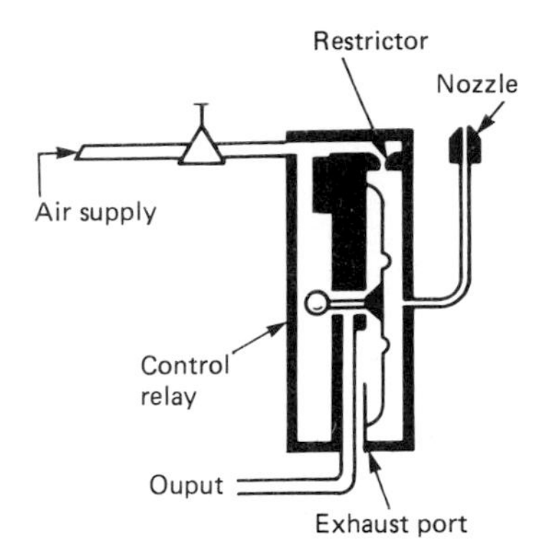

Figure 7.3 Pneumatic relay.

In operation, when the nozzle is covered the pressure in the associated chamber builds up, causing the conical valve to close the exhaust port and the ball valve to allow air to flow from the supply to the output port so that the output pressure rises.

When the nozzle is uncovered by movement of the flapper the flexible diaphragm moves so that the ball valve restricts the flow of air from the supply. At the same time the conical valve moves off its seat, opening the exhaust so that the output pressure falls.

With such a system the output pressure is driven from 20 to 100 kPa as a result of the relative movement between the flapper and nozzle of about 0.02 mm. Although the device has a non-linear character, this is relatively unimportant when it is included in the feedback loop of a measuring system.

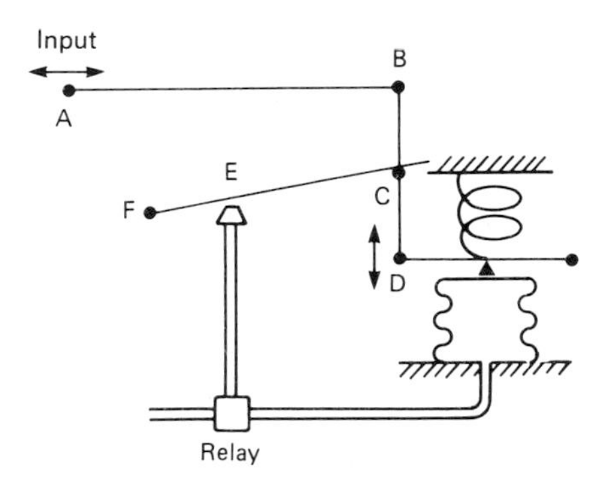

Figure 7.4 Basic pneumatic motion-balance system.

There are two basic schemes for utilizing the flapper/nozzle/relay system, namely the motion-balance and the force-balance systems. These are illustrated in Figures 7.4 and 7.5. Figure 7.4 shows a motion balance system in which the input motion is applied to point A on the lever AB. The opposite end (B) of this lever is pivoted to a second lever BCD which in turn has point D pivoted in a lever positioned by movement of the feedback bellows. At the centre (C) of the lever BD there is a spigot on which one end of the lever CEF is supported whilst it is pivoted at point F and has a flapper nozzle located at E. A horizontal displacement which causes A to move to the left is transmitted via B to C, and as a result the flapper at E moves off the nozzle so that the back pressure falls. This change is amplified by the relay so that the pressure in the bellows falls and the lever carrying the pivot D moves down until equilibrium is re-established. The output pressure is then proportional to the original displacement. By changing the inclination of the lever CEF the sensitivity or gain of the system may be changed.

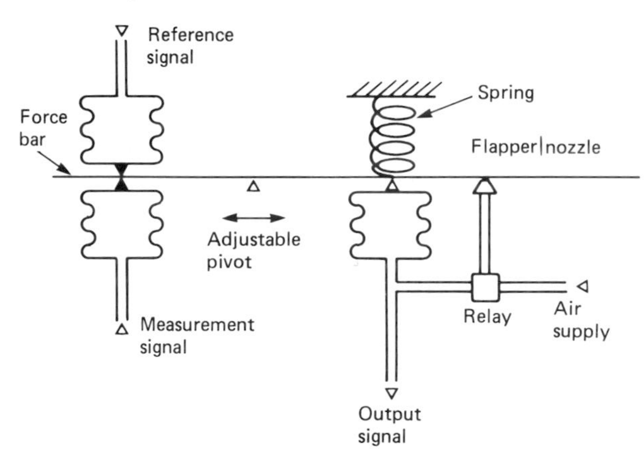

Figure 7.5 Basic pneumatic force-balance system.

Figure 7.5 illustrates a force-balance system. The measurement signal, in the form of a pressure, is applied to a bellows which is opposed by a similar bellows for the reference signal. The force difference applied to the lever supported on an adjustable pivot is opposed by a spring/bellows combination. Adjacent to the bellows is a flapper/nozzle sensor. In operation, if the measurement signal exceeds the reference signal the resultant force causes the force bar to rotate clockwise about the adjustable pivot so that the flapper moves closer to the nozzle, with the result that the pressure in the output bellows increases until equilibrium is re-established. The change in output pressure is then proportional to the change in the measurement signal.

7.3 Principal measurements

7.3.1 Introduction

Virtually all pneumatic measuring systems depend on a primary element such as an orifice plate, Bourdon tube, etc. to convert the physical parameter to be measured into either a force or a displacement which, in turn, can be sensed by some form of flapper/nozzle system or used directly to operate a mechanism such as an indicator, a recorder pen or a switch.

The measurements most widely used in the process industries are temperature, pressure, flow, level and density. In the following sections a description is given of the methods for implementing these measurements pneumatically, as opposed to describing the characteristics of the primary elements themselves. Also described are the pneumatic controllers which were evolved for the process industries and which are still very widely used.

7.3.2 Temperature

Filled thermal systems are used almost exclusively in pneumatic systems for temperature measurement and control. The sensing portion of the system comprises a bulb connected via a fine capillary tube to a spiral or helical Bourdon element or a bellows. When the bulb temperature rises, the increased volume of the enclosed fluid or its pressure is transmitted to the Bourdon element, which responds to the change by movement of its free end. This movement can be used directly to position a recording pen or an indicating pointer, either of which could also be coupled to a flapper/nozzle mechanism by a system of links and levers to actuate a pneumatic controller.

When a bellows is used to terminate the capillary the change in bulb temperature is converted into a force which, in turn, can be used to actuate a force-balance mechanism. Details of the materials used for filling the bulbs and their characteristics are given in Chapter 1 of Part 2 of this reference book.

In the motion balance systems the free end of the Bourdon tube is connected via adjustable links to a lever pivoted about the axis A in Figure 7.6. The free end of this lever bears on a second lever system pivoted about the axis B and driven by the feedback bellows. The free end of this second lever system is held in contact with a stem on a further lever system by a light spring. This latter lever is pivoted about the axis C and its free end is shaped to form the flapper of a flapper/nozzle system. The control relay associated with the flapper/nozzle system generates the output signal which is also applied to the feedback bellows.

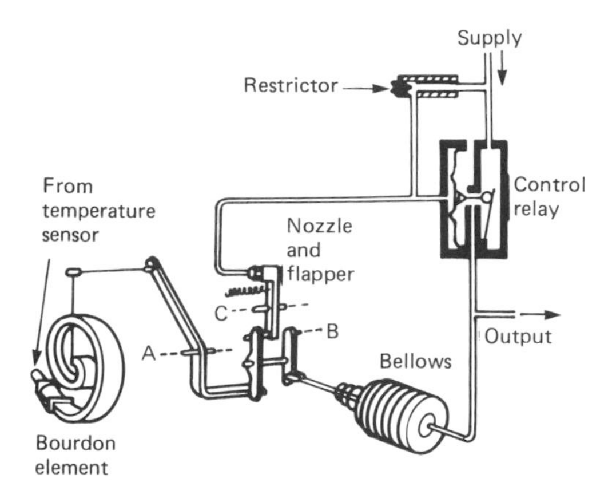

Figure 7.6 Motion-balance system for temperature measurement.

In operation, the links and levers are adjusted so that with the bulb held at the temperature corresponding to the lower range value the output signal is 20 kPa. If the measured temperature then rises, the lever pivoted at A moves in a clockwise direction (viewed from the left). In the absence of any movement of the bellows, this causes that associated lever pivoted about the axis C to rotate clockwise so that the flapper moves towards the nozzle. This, in turn, causes the nozzle back pressure to rise, a change which is amplified by the control relay, and this is fed back to the bellows so that the lever pivoted about the axis B moves until balance is restored. In this way, the change in the sensed temperature is converted into a change in the pneumatic output signal.

An example of the force-balance system is shown in Figure 7.7. There are two principal assemblies, namely the force-balance mechanism and the thermal system, which comprises the completely

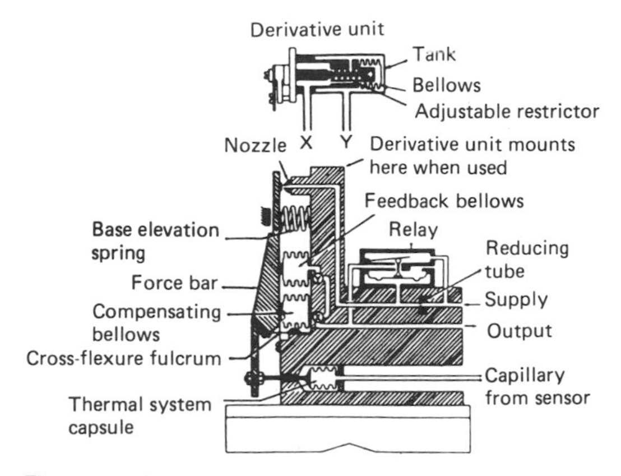

Figure 7.7 Force-balance system for temperature measurement.

sealed sensor and capillary assembly filled with gas under pressure. A change in temperature at the sensor causes a change in gas pressure. This change is converted by the thermal system capsule into a change in the force applied at the lower end of the force bar which, being pivoted on cross flexures, enables the force due to the thermal system to be balanced by combined forces developed by the compensating bellows, the feedback bellows and the base elevation spring. If the moment exerted by the thermal system bellows for the force exceeds the combined moments of the forces developed by the compensating bellows (which compensates the effect of ambient temperature at the transmitter), the feedback bellows (which develops a force proportional to the output signal) and the elevation spring (which provides means for adjusting the lower range value), the gap between the top of the force bar and the nozzle is reduced, causing the nozzle back pressure to rise. This change is amplified by the relay and applied to the feedback bellows so that the force on the force bar increases until balance is re-established. In this way, the forces applied to the force bar are held in balance so that the output signal of the transmitter is proportional to the measured temperature.

This type of temperature transmitter is suitable for measuring temperatures between 200 and 800 K, with spans from 25 to 300 K.

7.3.3 Pressure measurements

As explained in Part 1, the majority of pressure measurements utilize a Bourdon tube, a diaphragm or a bellows (alone or operated in conjunction with a stable spring) as the primary elements which convert the pressure into a motion or a force. Both methods are used in pneumatic systems; those depending on motion use a balancing technique similar to that described in the previous section for temperature measurements, whilst those which operate on the force-balance principle utilize a mechanism that has wider application. It was initially developed for differential pressure measurements which, because they provide a common basis for measuring flow, level and density, have very wide application in industry. By selecting alternative primary elements the same force-balance mechanism can be used for measurement of gauge or absolute pressures. The basic arrangement of the mechanism is shown in Figure 7.8.

The force developed by the primary element is applied via a flexure to one end of the force bar which comprises two sections threaded so that they can be joined together with a thin circular diaphragm of cobalt–nickel alloy clamped in the joint.

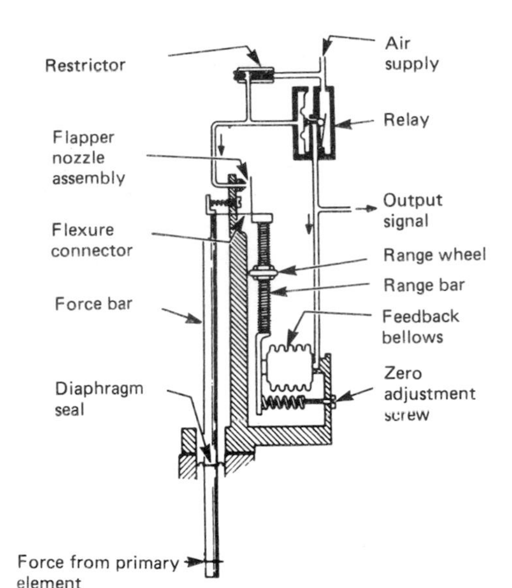

Figure 7.8 Basic arrangement of pneumatic force-balance mechanism.

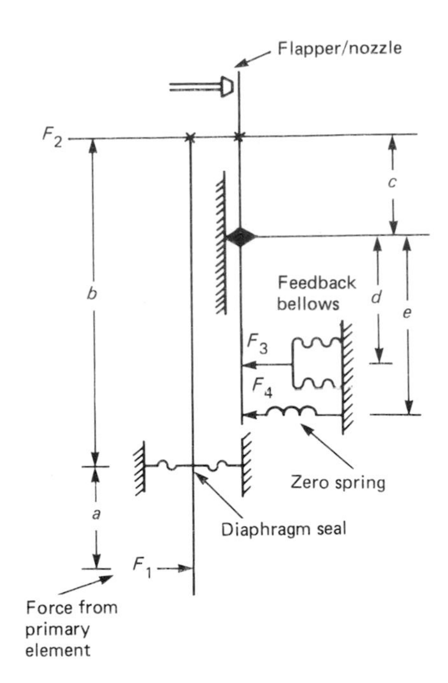

Figure 7.9 Diagram of forces involved in pneumatic force-balance mechanism.

This diaphragm serves as the pivot or flexure for the force bar as well as the seal between the process fluid and the force-balance mechanism, particularly in the differential pressure-measuring instruments.

The outer edge of the diaphragm seal is clamped between the body enclosing the primary element and the framework that supports the force-balance mechanism. This framework carries the zero adjustment spring, the feedback bellows, the pneumatic relay and the nozzle of the flapper/nozzle system.

At the lower end of the range bar is mounted the feedback bellows and zero spring, whilst at its upper end it carries the flapper and a flexure that connects it to the upper end of the force bar. The range bar is threaded so that the position of the range wheel (and hence the sensitivity of the system) can be varied. Figure 7.9 shows the diagram of forces for the mechanism.

In operation, force (F_1) from the primary element applied to the lower end of the force bar produces a moment (F_1a) about the diaphragm seal pivot which is transmitted to the upper end of the force bar where it becomes a force (F_2) applied via the flexural pivots and a transverse flexure connector to the upper end of the range bar. This force produces a moment (F_2c) about the range wheel, acting as a pivot, which is balanced by the combined moments produced by the forces F_3 and F_4 of the feedback bellows and zero spring, respectively.

Thus, at balance

$$F_1a = F_2b$$

and

$$F_2c = F_3d + F_4e$$

from which it follows that:

$$F_1 = \frac{bd}{ac} \cdot F_3 + \frac{be}{ac} \cdot F_4$$

By varying the position of the pivot on the range bar the ratio of F_1 to F_3 can be adjusted through a range of about 10 to 1. This provides the means for adjusting the span of an instrument.

The feedback loop is arranged so that when an increased force is generated by the primary element the resultant slight movement of the force bar causes the flapper to move closer to the nozzle so that its back pressure increases. This change is amplified by the relay and is applied to the bellows. As a result, the force that it applies to the range bar increases until a new equilibrium position is established and the various forces are balanced. Figure 7.10 illustrates the use of this force-balance mechanism in a gauge pressure transmitter if the low-pressure connection is open to the atmosphere or a high-range differential pressure transmitter if both high- and low-pressure signals are connected.

In this instrument the primary element is a bellows. Figure 7.11 shows essentially the same instrument but the bellows is evacuated and sealed on what was the low pressure side in the previous

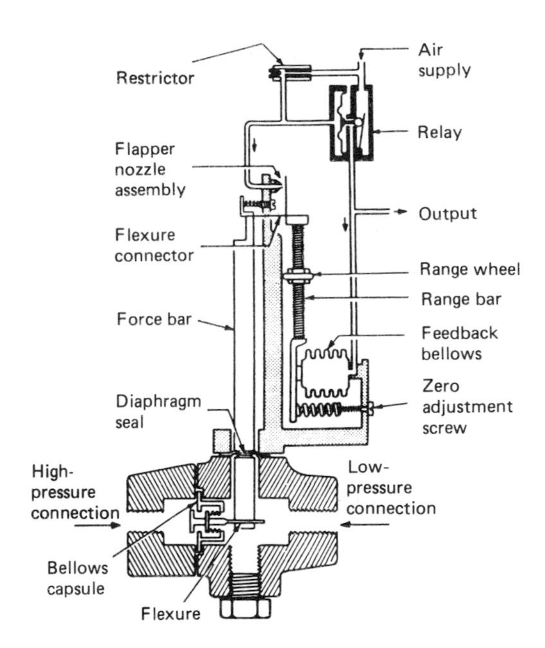

Figure 7.10 Force-balance mechanism incorporated into gauge pressure transmitter.

configuration, so that the instrument measures absolute pressure.

Figure 7.12 shows how the mechanism is used in a high-pressure transmitter in which the primary element is a Bourdon tube.

Figure 7.13 shows the use of the same mechanism attached to the differential pressure sensor which has become the most widely used form of pneumatic sensor.

The principal feature of the sensor is the diaphragm capsule shown in more detail in Figure 7.14. It comprises two identical corrugated diaphragms, welded at their perimeters to a back-up plate on which matching contours have been machined. At the centre, the two diaphragms are connected to a stem carrying the C-flexure by which the diaphragm is connected to the force bar. The small cavity between the diaphragms and the back-up plate is filled with silicone oil to provide damping.

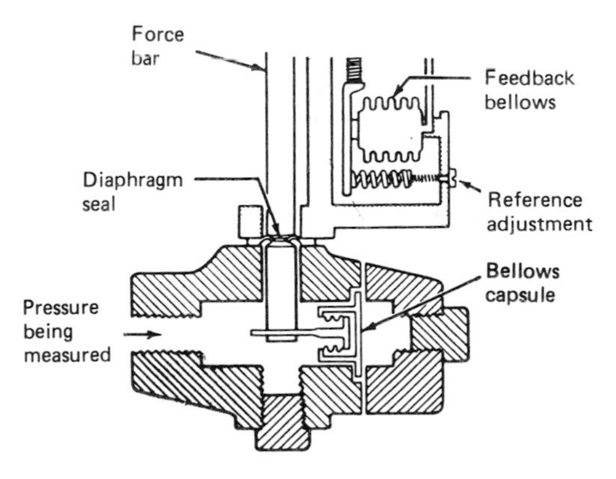

Figure 7.11 Force-balance mechanism incorporated into absolute pressure transmitter.

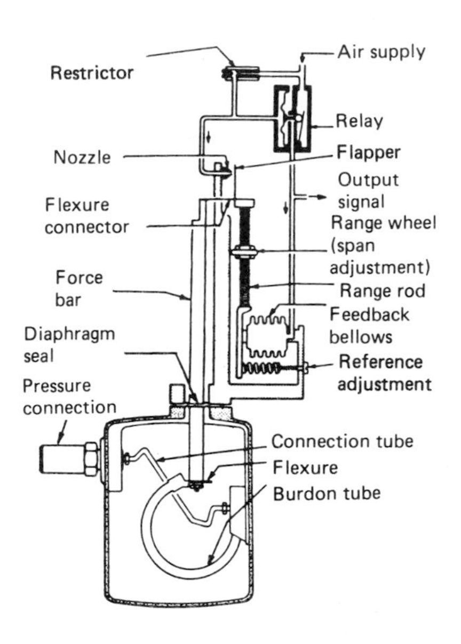

Figure 7.12 Force-balance mechanism incorporated into high-pressure transmitter incorporating a Bourdon tube.

If equal pressures are applied on either side of the capsule there is no resultant movement of the central stem. If, however, the force on one side is greater than that on the other, the resultant force is transmitted via the central stem to the C-flexure. If the force on one side greatly exceeds that on the other, then the diaphragm exposed to the higher pressure moves onto the matching contours of the back-up plate and is thereby protected from damage. This important feature has been a major factor in the successful widespread application of the differential pressure transmitter for measurement of many different process parameters, the most important being the measurement of flow in conjunction with orifice plates and the measurement of level in terms of hydrostatic pressures.

Figure 7.15 shows how the diaphragm capsule of the differential pressure transmitter is modified to convert it into an absolute pressure transmitter.

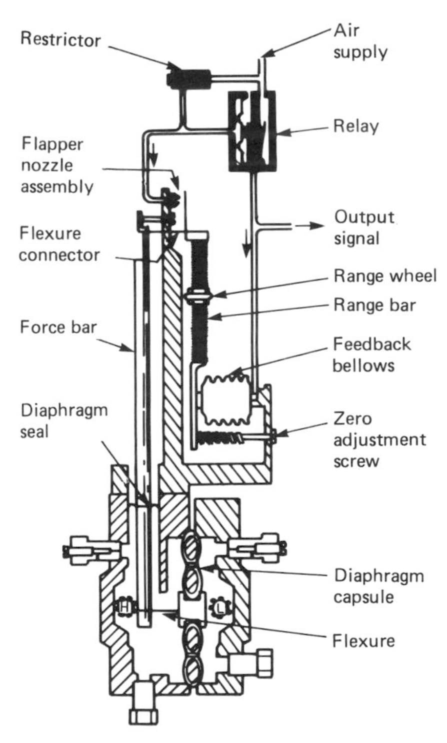

Figure 7.13 Force-balance mechanism incorporated into differential pressure transmitter.

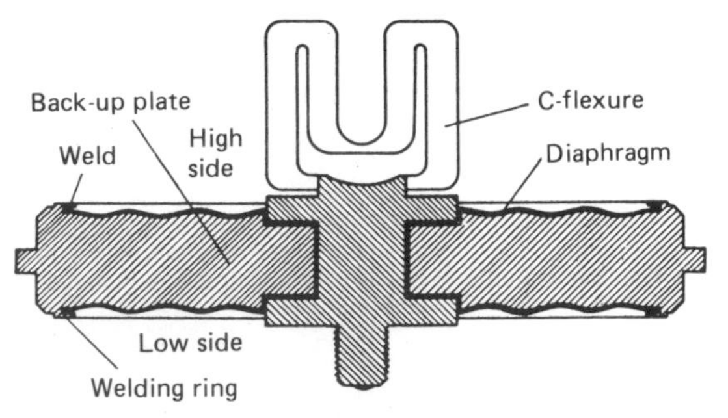

Figure 7.14 Construction of differential pressure sensor capsule.

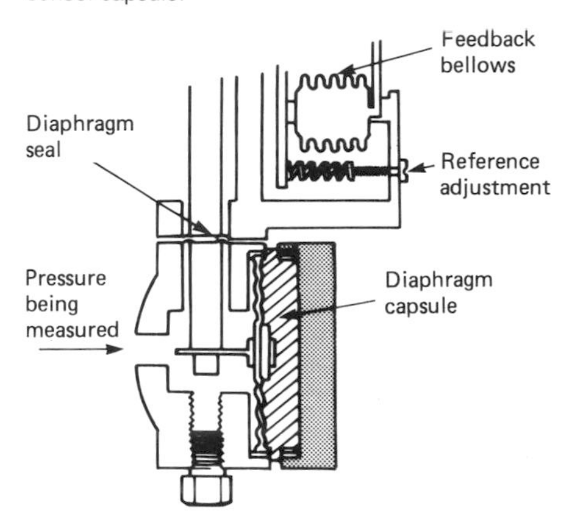

Figure 7.15 Modification of sensor capsule for absolute pressure measurements.

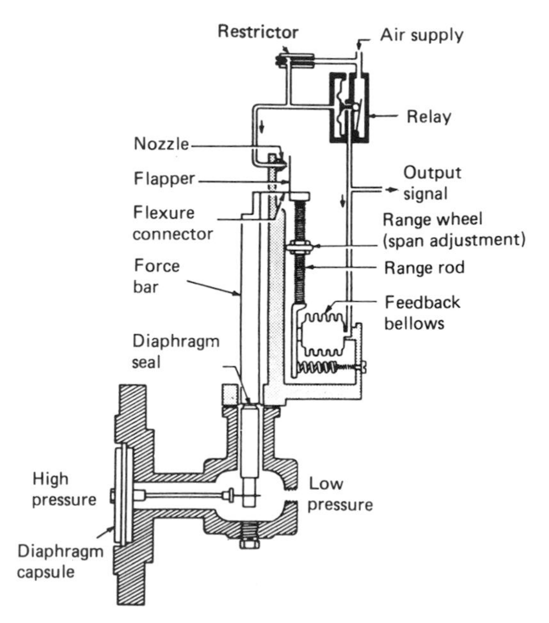

Figure 7.16 Force-balance mechanism incorporated into a flange-mounted level transmitter.

7.3.4 Level measurements

The diaphragm capsule and force balance mechanism can be adapted for measurement of liquid level. Figure 7.16 shows how it is mounted in a flange which, in turn, is fixed to the side of the vessel containing the liquid whose level is to be measured. If the tank is open then the low-pressure side of the capsule is also left open to the atmosphere. If, on the other hand, the measurement is to be made on a liquid in an enclosed tank, then the pressure of the atmosphere above the liquid must be applied to the rear of the capsule via the low-pressure connection. In operation, the hydrostatic pressure applied to the diaphragm being proportional to the head of liquid produces a force which is applied to the lower end of the force bar.

The same type of pneumatic force balance mechanism is used to measure this force and so generate an output signal which is directly proportional to the liquid level.

7.3.5 Buoyancy measurements

The same mechanism can be adapted to function as a buoyancy transmitter to measure either liquid density or liquid level, according to the configuration of the float which is adopted. This is shown in Figure 7.17 and further details of this method are given in Chapter 8 of Part 1 of this reference book.

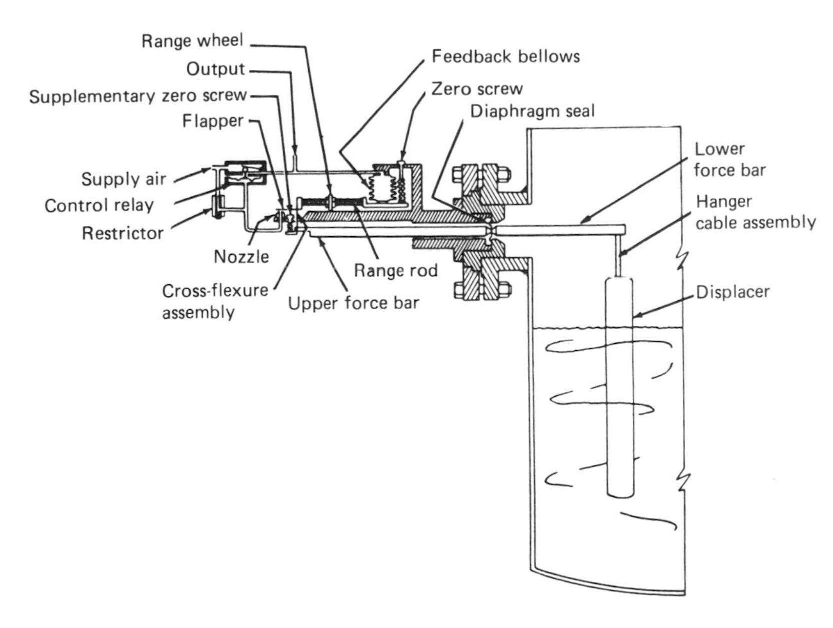

Figure 7.17 Force-balance mechanism incorporated into a buoyancy transmitter.

7.3.6 Target flow transmitter

This transmitter combines in a single unit both a primary element and a force-balance transmitter mechanism. As shown in Figure 7.18, the latter is essentially the same force-balance mechanism that is

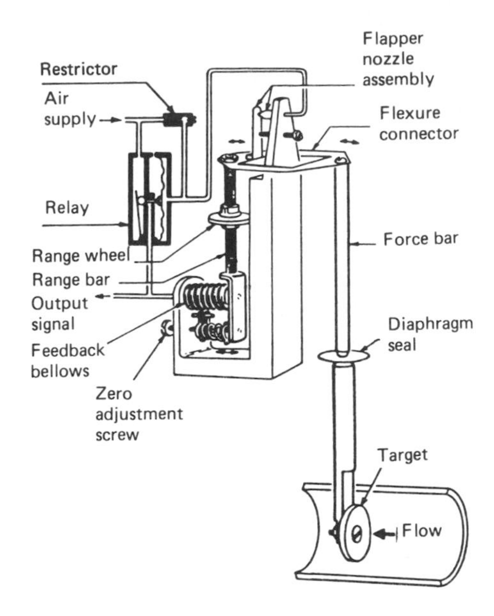

Figure 7.18 Force-balance mechanism incorporated into a target flow transmitter.

used in the other pneumatic transmitters described previously. The primary element is a disc-shaped target fixed to the end of the force bar and located centrally in the conduit carrying the process liquid.

As explained in Chapter 1 of Part 1, the force on the target is the difference between the upstream and downstream surface pressure integrated over the area of the target. The square root of this force is proportional to the flow rate.

7.3.7 Speed

A similar force-balance technique can be used to generate a pneumatic signal proportional to speed, as shown in Figure 7.19. As the transmitter input shaft rotates, the permanent magnet attached to it generates a magnetomotive force. This force tends to cause the non-magnetic alloy disc to rotate in the same direction. Since the disc is connected to the flapper by means of flexures, any rotation of the disc causes a corresponding movement of the flapper, thus changing the clearance between the flapper and the nozzle.

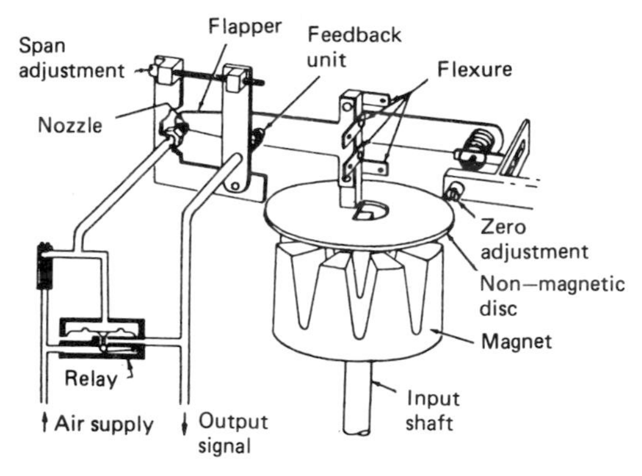

Figure 7.19 Modified force-balance mechanism incorporated into a speed transmitter.

As the flapper/nozzle relationship changes, the output pressure from the relay to the feedback unit changes until the force at the feedback unit balances the rotational force.

The output pressure which establishes the force balance is the transmitted pneumatic signal and is proportional to the speed of rotation. It may be used to actuate the pneumatic receiver in an indicator, recorder or controller.

7.4 Pneumatic transmission

In relatively simple control systems it is usually possible to link the primary sensing element directly to the controller mechanism and to locate this reasonably close to the final control element. However, when the process instrumentation is centralized, the primary sensing elements are arranged to operate in conjunction with a mechanism that develops a pneumatic signal for transmission between the point of measurement and the controller, which is usually mounted in a control room or sheltered area.

A standard for these transmission signals has been developed and is applied almost universally, with only small variations that arise as a result of applying different units of measure. If the lower range value is represented by a pressure P then the upper range value is represented by a pressure $5P$ and the span by $4P$. Furthermore, it is customary to arrange the nominal pressure of the air supplied to the instrument to be between $6.5P$ and $7.0P$.

In SI units, the zero or lower range value of the measurement is represented by a pressure of 20 kPa the upper range value by a pressure of 100 kPa and the span therefore by a change in pressure of 80 kPa. The corresponding Imperial units are 3 psi, 15 psi and 12 psi, whilst the corresponding metric units are 0.2 kg/cm^2 (0.2 bar), 1.0 kg/cm^2 (1.0 bar) and 0.8 kg/cm^2 (0.8 bar). (1 bar = 100 kPa = 14.7 psi.)

If a change in measured value occurs, this is sensed by the transmitter and its output pressure increases accordingly. When the transmitter is remote from the controller and connected to it by an appreciable length of tubing, there is a finite delay before the change of pressure generated at the transmitter reaches the controller. In the case of an increased signal, this delay is governed by the

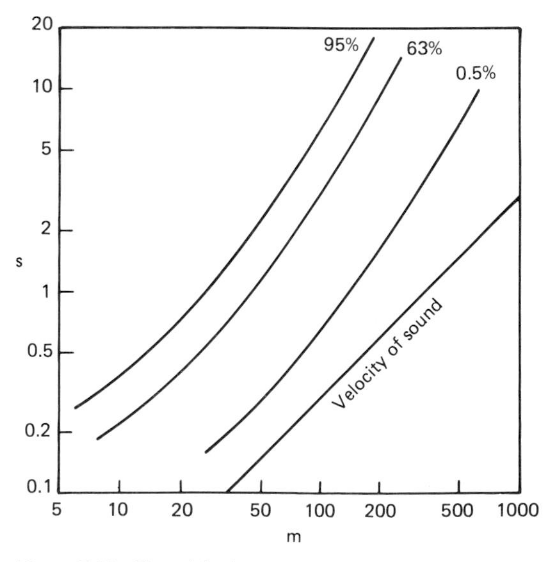

Figure 7.20 Time delay in response to step change of pneumatic pressure applied to various lengths of tube.

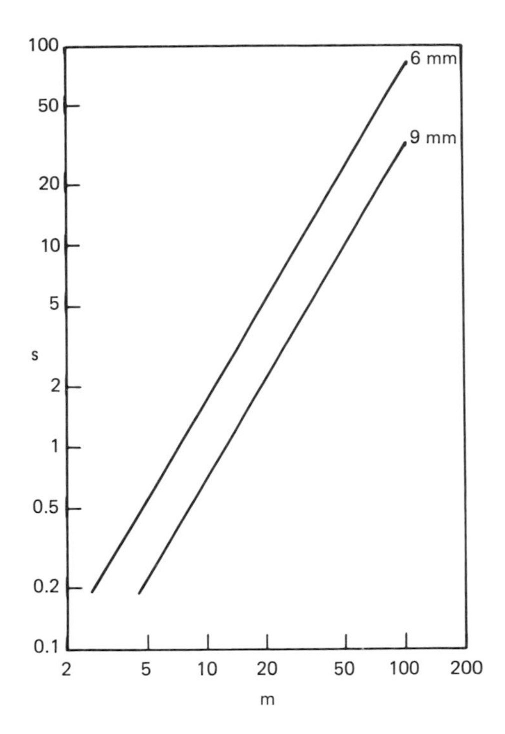

Figure 7.21 Time delay in response to a constant rate of change of pneumatic pressure applied to various lengths of tube.

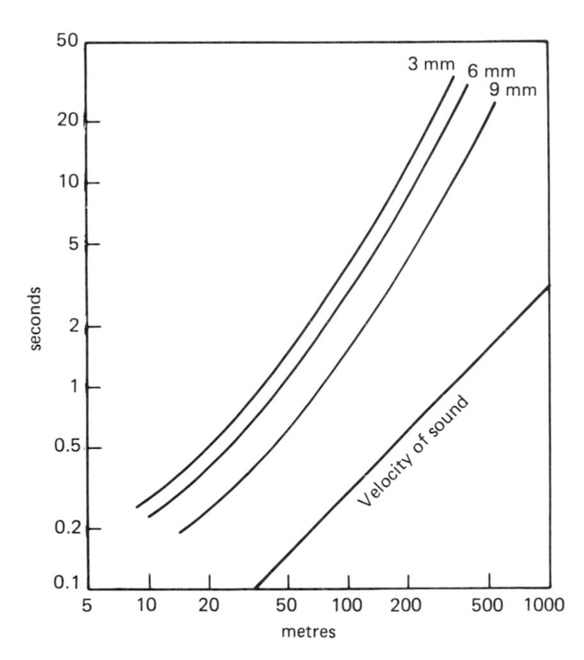

Figure 7.22 Time delay for 63.2 per cent response to step change of applied pressure for various lengths of tube.

magnitude of the change, the capability of the transmitter to supply the necessary compressed air to raise the pressure throughout the transmission line, the rate at which the change can be propagated along the line and finally the capacity of the receiving element.

A rigorous description of the overall characteristics of a pneumatic transmission system is beyond the scope of this book, but Figures 7.20–7.22 show the magnitude of the effects for reasonably representative conditions.

Figure 7.20 shows the time taken for the output to reach 95, 63 and 0.5 per cent of its ultimate value following a step change from 20 kPa to 100 kPa at the input according to the length of tube. Figure 7.21 shows the time lag versus tubing length when the input is changing at a constant rate. Figure 7.22 shows the time delay for a 63 per cent response to a step change of 20 to 100 kPa versus tubing length for 9 mm, 6 mm and 3 mm tubing.

7.5 Pneumatic controllers

7.5.1 Motion-balance controllers

Many of the early pneumatic controllers were based on the principle of motion balance because the sensor associated with the primary element produced a mechanical movement, rather than a force, to provide the signal proportional to the measured quantity. The technique is still widely used, and the Foxboro Model 43A, an example of a modern implementation of the concept embodying several variations of the control actions, is described in the following sections.

In most of these controllers a spring-loaded bellows or a Bourdon element converts the incoming signal into a motion. This motion is used directly to drive a pointer so that the measured value is displayed on a scale. On the same scale there is a second adjustable pointer the position of which identifies the set point or desired value for the controller action. For the majority of applications the required action is proportional plus integral, and is derived as follows.

As shown in Figure 7.23, a mechanical linkage between the two pointers is arranged so that movement of its midpoint is proportional to the difference between the measured value and the set point. This difference is applied horizontally at one end of the proportioning lever which is pivoted at its centre to a flat spring. A bellows connected to the pneumatic output of the controller and opposed by the integral action bellows applies a force to the flat spring, with the result that the proportioning lever is positioned virtually according to the magnitude of the output signal combined with the integral action.

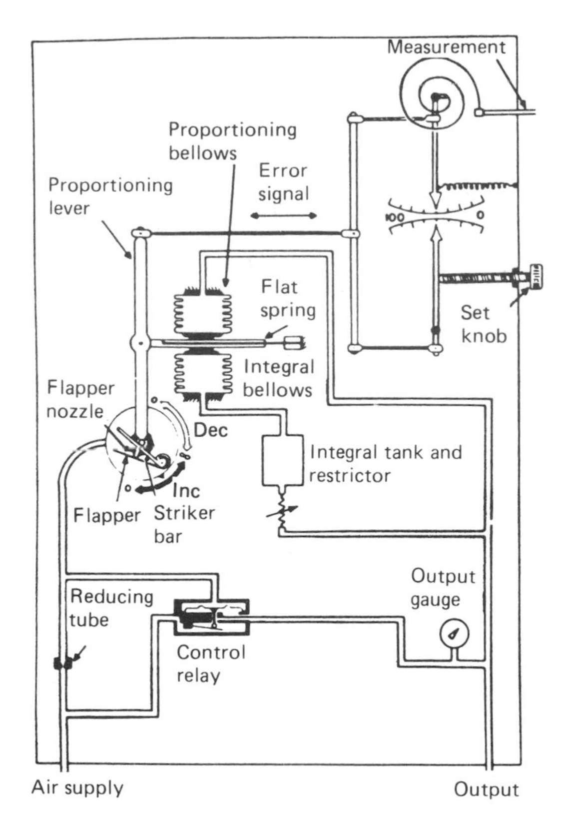

Figure 7.23 Basic configuration of pneumatic motion-balance two-term controller.

The lower free end of the proportioning lever bears on the striker bar of the flapper/nozzle system, and in so doing causes the flapper to cover or uncover the nozzle. Thus if the measurement exceeds the set point, the proportioning lever moves to the left (in the figure) and, being pivoted centrally, its lower free end moves to the right, off the striker bar, with the result that the flapper covers the nozzle. This causes the nozzle back pressure to increase, a change that is amplified by the control relay to a value that becomes the instrument output signal and is also applied to the proportional bellows, causing the proportioning lever to move downwards until it impinges again on the striker bar which, in turn, moves the flapper off the nozzle. Consequently the nozzle back pressure falls and the change, after being amplified by the control relay, is applied to the proportioning bellows as well as to the instrument output and via the integral tank and restrictor to the integral bellows. In this way any residual error between the measured value and the desired value is integrated and the output gradually adjusted until the error is eliminated.

This type of control action meets the majority of operational requirements. However, in a few instances proportional action alone is sufficient, in which case the integral bellows are replaced by a spring and the integral tank and restrictor are omitted.

On the other hand, there are a few instances where it is desirable to enhance the performance further by including derivative action, i.e. adding to the output signal a component that is proportional to the rate of change of the error signal. This is implemented by modifying the proportional bellows system as shown in Figure 7.24. In operation, a sudden change in the measurement signal actuates the flapper/nozzle mechanism so that the output signal changes. This change is fed back to the proportional bellows directly by the derivative mechanism, but the derivative restrictor causes the transient signal to die away so that the pressure in the proportioning bellows becomes equal to the

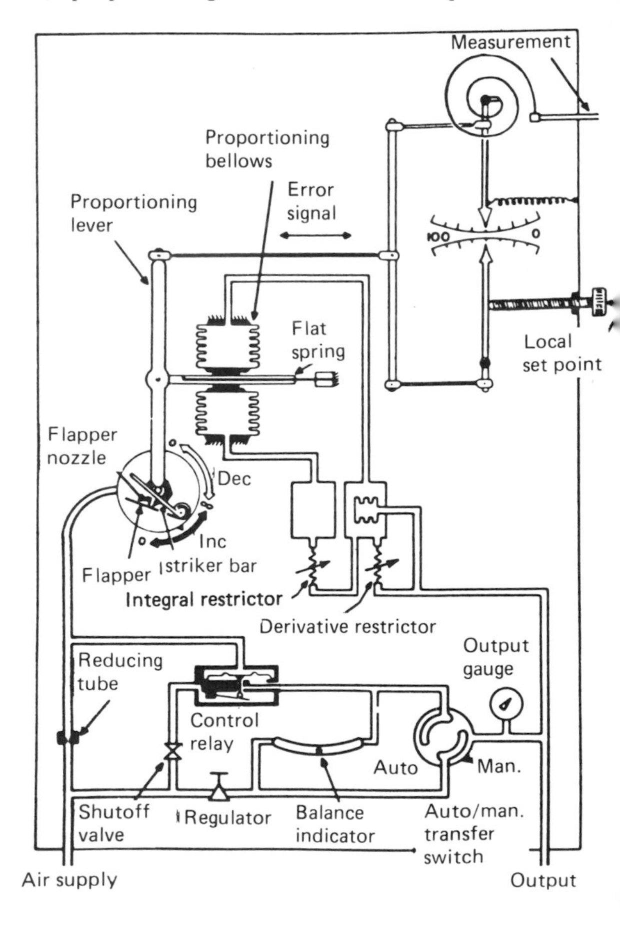

Figure 7.24 Basic configuration of pneumatic motion-balance three-term controller with auto/manual balance feature.

output pressure. In the meantime, the integral action builds up to reduce any difference between the measured value and the set point.

A further feature available in the instrument is the switch for transferring the controller from manual to automatic operation and vice versa. The switch itself is a simple two-way changeover device, but associated with it are a regulator and a sensitive balance indicator, comprising a metal ball mounted in a curved glass tube.

When it is desired to transfer the control loop from automatic to manual operation the regulator is adjusted until the ball in the balance indicator is positioned centrally to show that the regulator pressure has been set equal to the controller output. The process can then be transferred from automatic to manual control without imposing any transient disturbance. The reverse procedure is applied when the process has to be transferred from manual to automatic control, except that in this instance the set point control is adjusted to balance the controller output signal against the manually set value.

Other modes of operation which are available include differential gap action (also known as on/off control with a neutral zone), automatic shutdown (which is used to shut down a process when the measured value reaches a predetermined limit), on/off control (which is the simplest configuration of the instrument), remote pneumatic set (in which the manual set point control is replaced by an equivalent pneumatically driven mechanism) and batch operation (in which provision is made for avoiding saturation of the integral action circuit during the interval between successive batch process sequences). However, these configurations are rare compared with the simple proportional and proportional plus integral controllers.

7.5.2 Force-balance controllers

The Foxboro Model 130 Series of controllers serves to illustrate the method of operation of modern pneumatic force-balance controllers. The basic mechanism of the control unit is shown in Figure 7.25. There are four bellows, one each for the set point, measurement, proportional feedback and integral feedback, which bear on a floating disc with two bellows on each side of a fulcrum whose angular position can be varied. In operation, the forces applied via each bellows multiplied by the respective distances from the fulcrum are held in balance by operation of a conventional flapper/nozzle system which generates the pneumatic output signal.

If the angular position of the fulcrum is set so that it is directly above the proportional feedback bellows and the reset bellows (as shown in Figure 7.26(a)), then the unit functions as an on/off

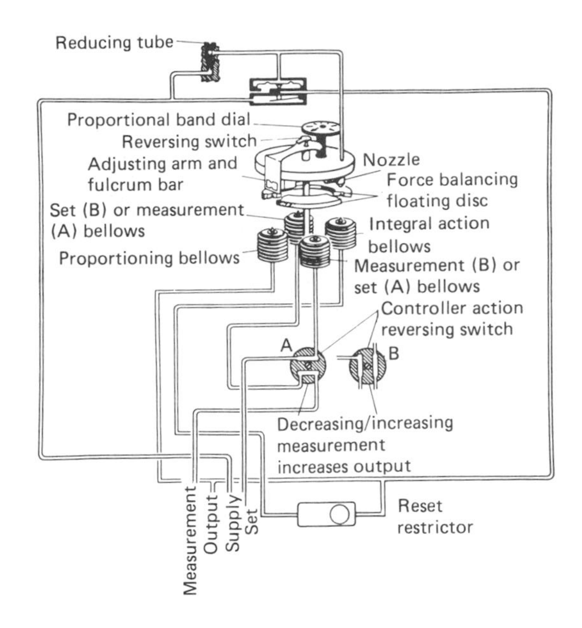

Figure 7.25 Basic mechanism of force-balance pneumatic controller.

controller. The slightest increase in measurement signal above the set point signal causes the nozzle to be covered so that output signal rises to the supply pressure. Any decrease in the measurement signal below the set point signal uncovers the nozzle so that the output signal falls to the zero level.

If the adjustable fulcrum is moved to the position shown in Figure 7.26(b), the proportional band is 25 per cent (or the gain is 4) (a/b = 1/4). If the measurement signal increases, the flapper/nozzle will be covered so that the output pressure rises until balance is restored. This will occur when the output pressure has increased by a factor of four (if the integral action is disregarded).

If the fulcrum is adjusted to the position shown in Figure 7.26(c), the proportional band is 400 per cent (or the gain is 0.25) (a/b = 4), then an increase in the measurement signal causes the flapper/nozzle to be covered so that the output pressure also rises until balance is restored. This will occur when the change in output pressure is one quarter that of the change in measurement signal.

Referring again to Figure 7.26 (b), if the measurement signal increases, so does the pressure in the output bellows by a factor of four. However, this change of output signal is applied via a restrictor to the reset bellows, which acts on the same side of the fulcrum. Thus as the pressure in the reset bellows rises, the output signal must rise proportionally more to restore balance. A difference in

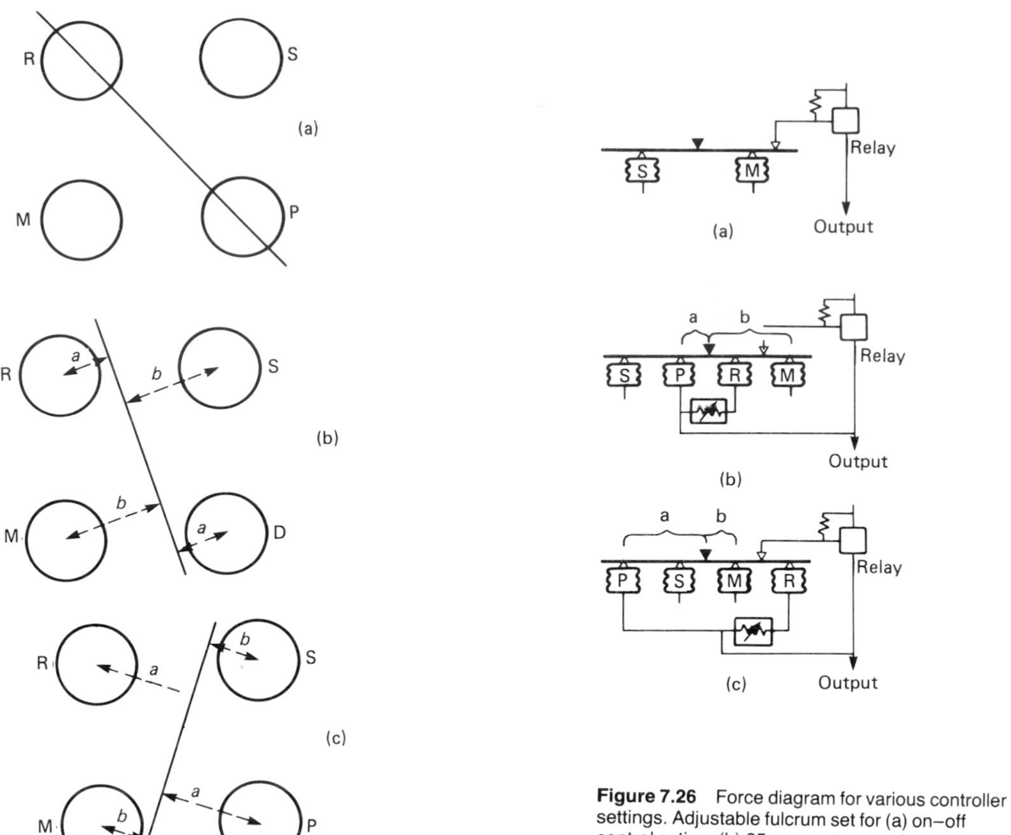

Figure 7.26 Force diagram for various controller settings. Adjustable fulcrum set for (a) on–off control action, (b) 25 per cent proportional band or (c) 400 per cent proportional band.

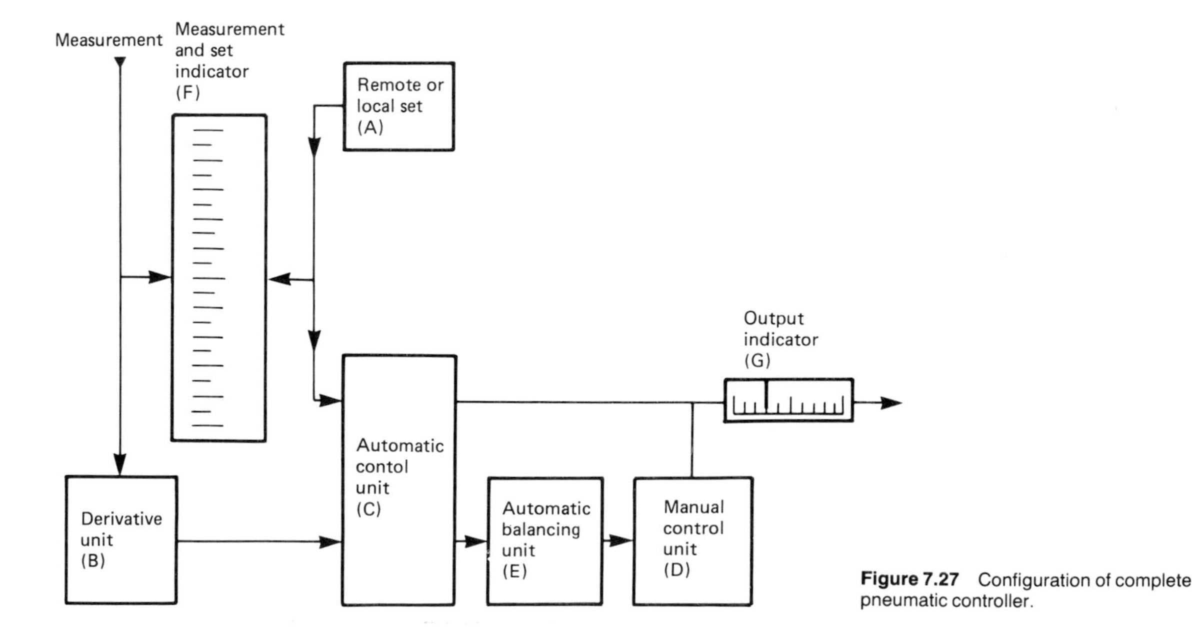

Figure 7.27 Configuration of complete pneumatic controller.

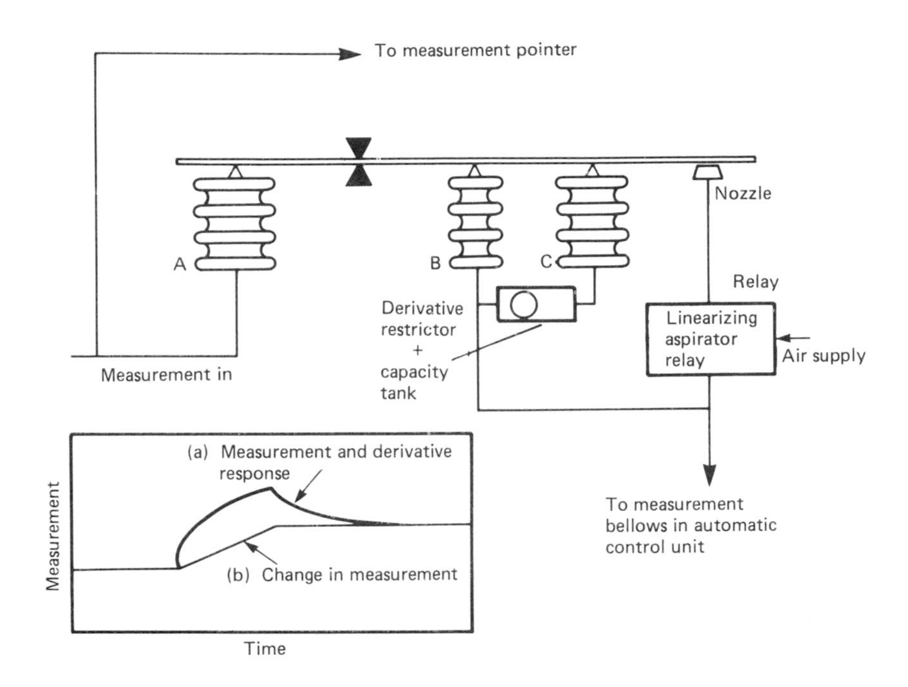

Figure 7.28 Basic mechanism of derivative function generation.

pressure then exists between the proportional bellows and the reset bellows until the process reacts to the controller output and reduces the measurement signal. This, in turn, reduces the pressure in the proportional bellows until the difference is eliminated. The reset action continues so long as there is a difference between the measurement and set point, and hence a difference between the reset and proportional bellows.

The configuration of the complete controller shown in Figure 7.27 is centred around the automatic control unit. The manual set point is adjusted by a knob in the front panel of the instrument. It drives a pointer on the scale and operates a mechanism that generates a proportional pressure to be applied to the automatic control unit.

The local/remote set point mechanism includes a receiver bellows which drives a second pointer to indicate the received value of the set point. It includes a switch which allows either the local or remote signal to be selected (Figure 7.27). The derivative unit is actuated from the measurement signal only and is not affected by changes of set point. As shown in Figure 7.28, it comprises a further force balance mechanism in which the moment generated by the signal to the measurement bellows A is balanced by the combined moments generated by two further bellows, one (C) arranged to generate a substantially larger moment than the other (B).

In steady conditions, the moment due to bellows A is equal to the combined moments of bellows B and C. Any increase in the measurement signal disturbs the balance, with the result that the relative position of the flapper/nozzle is reduced and the output of the linear aspirator relay rises. This signal is applied not only to the measurement bellows in the automatic control unit but also to bellows B and C.

However, the supply to bellows C passes through a restrictor so that, initially, balance is restored by the force generated by the bellows B alone. Subsequently, the force generated by bellows C rises and the effect is compensated by a fall in the output signal. After a time determined by the value of the restrictor, the pressures in bellows B and C become equal and they are then also equal to the pressure in bellows A. Thus, as a result of the transient, the effective measurement signal to the controller is modified by an amount proportional to the rate of change of the incoming signal, thereby providing the derivative controller action.

At the same time the full air supply pressure is applied via restrictors to chambers C and B so that the diaphragm takes up a position where the air vented to the atmosphere through the nozzle is reduced to a very low value.

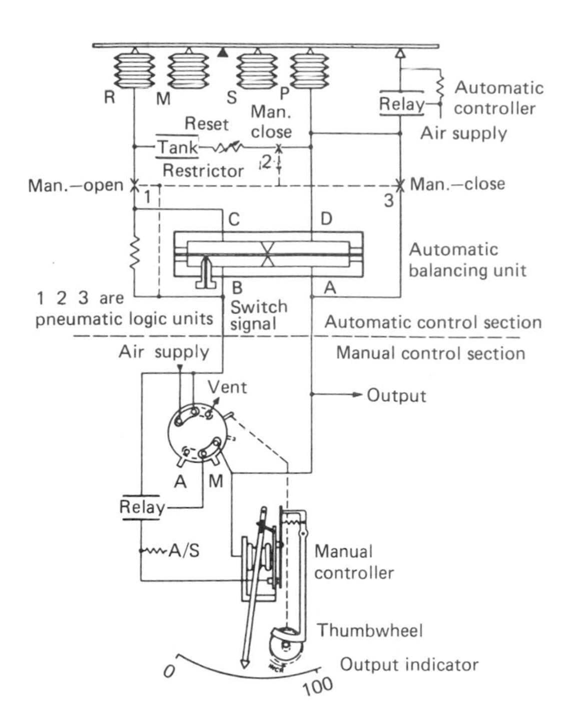

Figure 7.29 Manual-to-automatic transfer.

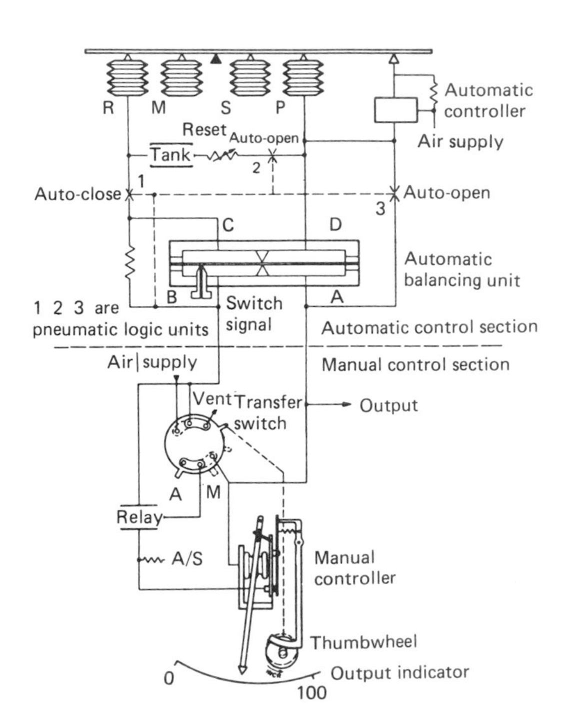

Figure 7.30 Automatic-to-manual transfer.

7.5.2.1 Automatic manual transfer switch

The design of previous pneumatic controllers has been such that, before transferring the control function from automatic to manual operation or vice versa, it is necessary to balance the two output signals manually, otherwise the process would be subjected to a 'bump', which may well cause unacceptable transient conditions.

To avoid this manual balancing operation, a unit has been devised and incorporated as a basic feature of the instrument. It is known as the automatic balancing unit, and, as shown in Figures 7.29 and 7.30, consists essentially of a chamber divided into four separate compartments isolated from each other by a floppy diaphragm pivoted at the centre. One of the chambers includes a nozzle through which air can escape slowly in normal operation as a result of the relative proximity of the diaphragm to the nozzle. When the position of the diaphragm moves away from the nozzle the flow of air increases and vice versa.

The interconnection of the unit when the controller is set in the manual operation mode is shown in Figure 7.29. In this mode, the manual controller generates the output signal which is also applied to the 'A' chamber in the automatic balancing unit. If the controller were in the automatic mode this same signal would be generated by the force balance unit operating in conjunction with the relay, in which case it would pass via the reset restrictor and tank to the reset bellows, but in the manual mode both these connections are closed by pneumatic logic switches.

Referring now to chamber D in the automatic balancing unit, this is connected directly to the proportional bellows (P) in the force-balance unit and so is held at that pressure. Any difference between the pressure in chambers A and D causes the floppy diaphragm to move with respect to the nozzle and so modify the pressure in chamber B which (in manual mode) is supplied via a restrictor. Thus as the output signal is changed by adjustment of the manual controller, the pressure in chamber A varies accordingly, causing the position of the floppy diaphragm to change. This, in turn, alters the rate at which air escapes via the nozzle from chamber B. This change is transferred via the restrictor to both chamber C and the reset bellows R in the force-balance unit so that the latter causes the pressure in its output bellows P to change until balance is restored through its normal *modus operandi*.

However, this same signal is also applied to chamber D, so that, in effect, the pressure in the proportional bellows P continuously follows that set by the manual controller. Consequently, when the

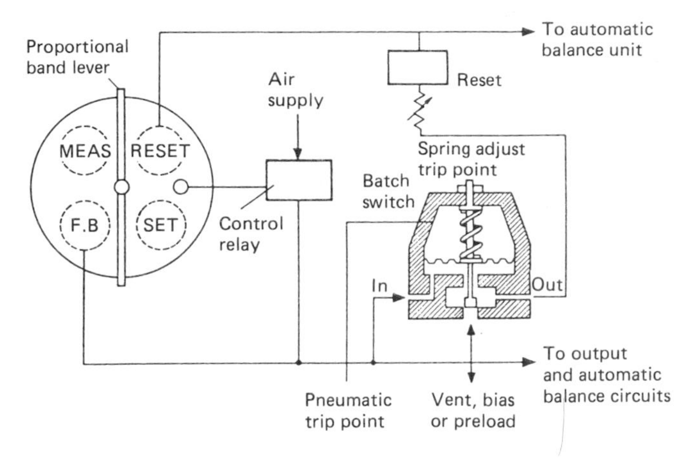

Figure 7.31 Functional diagram of batch-switch system. (*Notes*: (1) Bellows are beneath circular plate, nozzle is above; (2) an increase in measurement causes a decrease in output.

overall operation is transferred from manual to automatic, the force-balance control unit will have previously driven its own output signal to the value set by the Manual Controller so that the process is not subjected to a transient disturbance.

Before the controller action is transferred from automatic to manual operation the mechanism in the manual controller will have been receiving the output signal. Therefore the manual control lever is continually driven to a position corresponding to the output signal so that it is always ready for the controller to be transferred from automatic to manual control.

When the transfer switch is operated, the thumbwheel becomes engaged so that the output signal is now generated by the manual controller instead of the force-balance automatic controller. Prior to this, the automatic controller output signal is applied via the pneumatic logic switch to both chambers A and D in the automatic balancing unit so that half of the unit does not apply any moment.

7.5.2.2 Batch operation

When the controller is used to control a batch process or one in which the controller is held in the quiescent condition for an appreciable time it is necessary to include an additional feature which prevents saturation of the integral action that would otherwise occur because the measured value is held below the set point. (This is sometimes known as reset or integral wind-up.) When the process is restarted after being held in the quiescent state, the measured value overshoots the set point unless the normal *modus operandi* of the reset action is modified.

The modification involves inclusion of a batch switch, which, as shown in Figure 7.31, is essentially a pressure switch actuated by the output signal from the controller but with the trip point set by a spring or by an external pneumatic signal. While the controller is functioning as a controller, the batch switch has no effect, but when the process is shut down and the measurement signal moves outside the proportional band, the batch switch is tripped, whereupon it isolates the controller output from the reset bellows which is vented to atmosphere.

When the measurement returns within the proportional band the batch switch resets and the integral action immediately becomes operative in the normal manner.

7.6 Signal conditioning

7.6.1 Integrators

The most frequent requirement for integrators in pneumatic systems is to convert the signal from head-type flowmeters into direct reading or count of flow units. Since head-type flowmeters generate signals that are proportional to the square of flow rate, the integrator must extract the square root before totalizing the flow signal. In operation, the Foxboro Model 14A Integrator accepts a standard pneumatic signal, proportional to 0–100 per cent of differential pressure from a flow transmitter, which is applied to the integrator receiver bellows A in Figure 7.32. The force exerted by the bellows positions a force bar B in relation to a nozzle C. With an increase in differential pressure, the force bar approaches the nozzle and the resulting back pressure at the relay regulates the flow of air to drive the turbine rotor E. As the rotor revolves, the weights F, which are mounted on a cross-flexure assembly G on top of the rotor, develop a centrifugal force. This force feeds back through the thrust pin H to balance the force exerted on the force bar by the bellows.

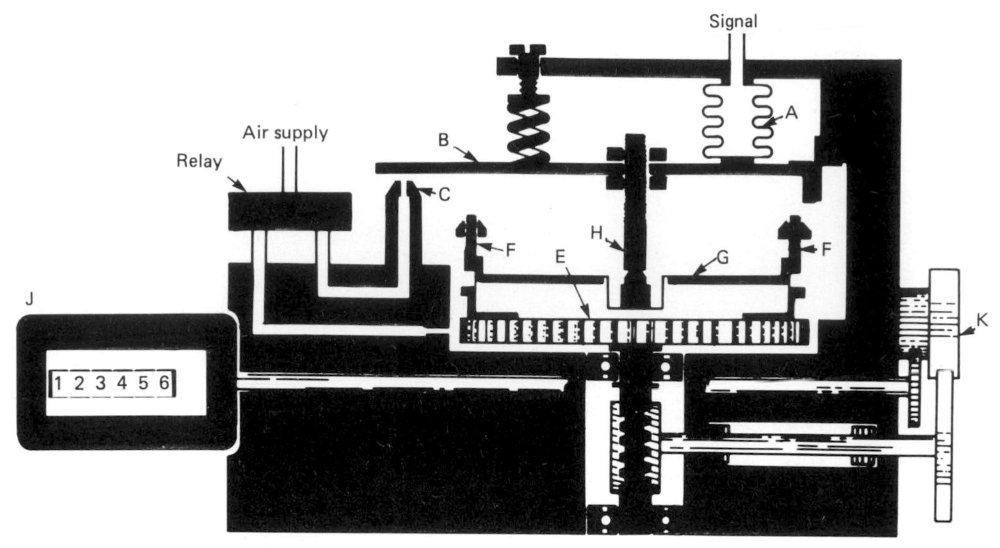

Figure 7.32 Functional diagram of pneumatic integrator.

The centrifugal force is proportional to the square of the turbine speed. This force balances the signal pressure which, for head-type flowmeters, is proportional to the square root of the flow rate. Therefore, turbine speed is directly proportional to flow and the integrator count, which is a totalization of the number of revolutions of the turbine rotor, is directly proportional to the total flow.

The turbine rotor is geared directly to counter J through gearing K. Changes in flow continuously produce changes in turbine speed to maintain a continuous balance of forces.

7.6.2 Analog square root extractor

The essential components of the square root extractor are shown in Figure 7.33, whilst Figure 7.34 shows the functional diagram. The input signal is applied to the A bellows and creates a force which disturbs the position of the nozzle with respect to the force arm. If the signal increases, the force arm is driven closer to the nozzle so that the back pressure increases. This change is amplified by the relay and its output is applied to both the C bellows and the B diaphragm.

Because the flexure arm is restrained so that the end remote from the flexure can only move along an arc, the combined effect of forces B and C is to drive the force arm away from the nozzle. This causes the nozzle back pressure to fall, which, in turn, reduces forces B and C until balance is restored.

Consideration of the vector force diagram shows that

$$\tan\theta = \frac{A}{B} \text{ or } B\tan\theta = A$$

For small changes, $\tan\theta$ is approximately equal to θ. Hence $B\theta = A$. Because (1) force A is proportional to the input signal, (2) force B is proportional to the

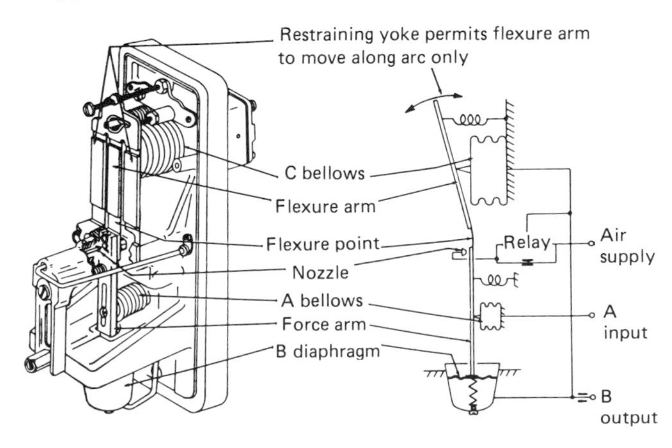

Figure 7.33 Functional diagram of pneumatic analog square root extractor.

output signal and (3) the position of the flexure arm to establish the angle θ is proportional to the force C, it follows that $B \times C \propto A$. As the bellows C is internally connected to B it follows that $B^2 \propto A$ or $B \propto \sqrt{A}$.

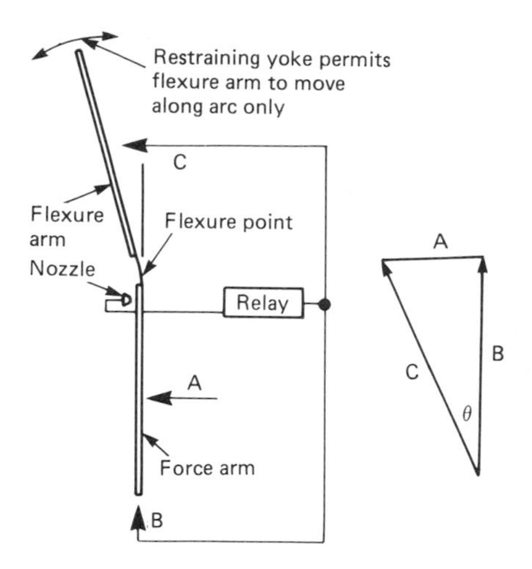

Figure 7.34 Force diagram for analog square root extractor.

7.6.3 Pneumatic summing unit and dynamic compensator

There are numerous process control systems which require two or more analog signals to be summed algebraically, and feed-forward control systems require lead/lag and impulse functions to be generated. The force-balance mechanism of the controller can be adapted to serve these functions. Figure 7.35 represents the arrangement of the bellows with respect to the fulcrum. It can be seen that with the gain set at unity (i.e. $a = b$) the two signals to be summed applied to the A and B bellows, respectively, the P and C bellows connected in parallel and supplied with air from the relay (which also supplies the output signal), then the output signal will be the average of the two input signals. If summing of the input signals is required, then only either the P or C bellows should be used. Similarly, signals may be subtracted by applying them to the A or B and C bellows, in which case the output is taken from the P bellows. In all these arrangements, the positioning of the fulcrum allows a constant to be introduced into the equation.

To generate a lead/lag, the mechanism is arranged as shown in Figure 7.36. The input signal is applied to the A bellows and the B and C bellows are connected together and supplied via a restriction and capacity tank from the P bellows, which, in common with the output, is supplied by the relay. As shown in Figure 7.37, the response of the system is determined by the setting of the gain and the restrictor.

Figure 7.38 shows the arrangement of the mechanism which generates an impulse response. The input signal is applied to bellows B, bellows P is

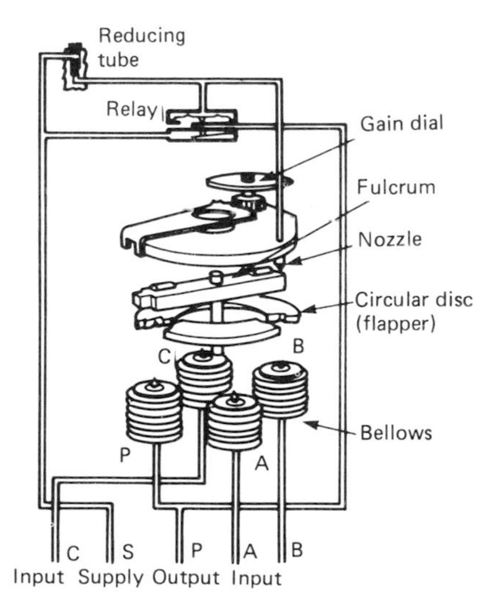

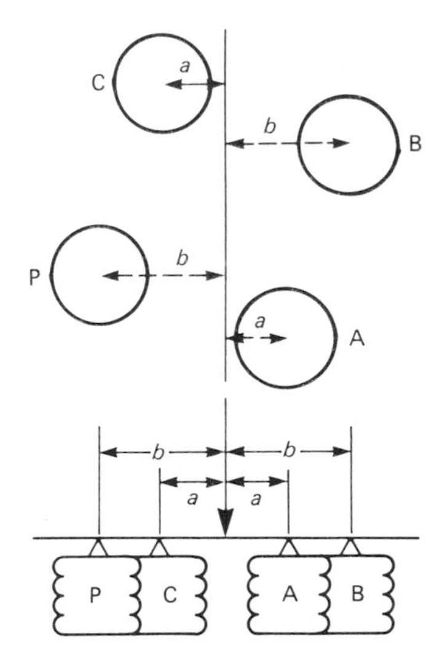

Figure 7.35 Basic configuration of pneumatic summing unit and dynamic compensator and force diagram.

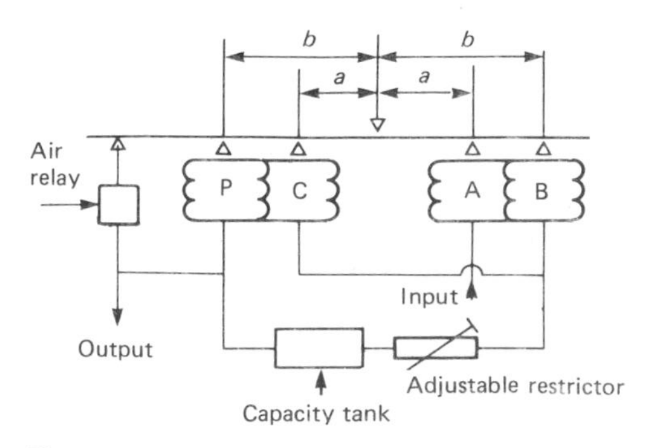

Figure 7.36 Force diagram for lead/lag unit.

connected to bellows B via a restrictor and a capacity tank, a plug valve is included so that one or the other of these two bellows can be vented, whilst the relay generates the signal for bellows C and the output from the relay. For reverse action, the roles of B and P are interchanged.

As shown in Figure 7.39, the output signal includes a positive- or negative-going impulse, according to the setting of the fulcrum and whether bellows P or B is vented. The recovery time is determined by the setting of the restrictor.

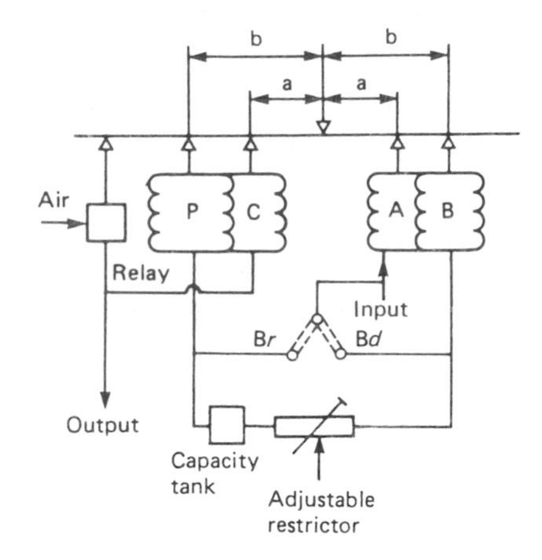

Figure 7.38 Force diagram for impulse generator.

7.6.4 Pneumatic-to-current converters

It is sometimes necessary to provide an interface between pneumatic and electronic control systems. This is particularly true when an existing pneumatic measurement and control system is being extended in a manner that involves the transmission of signals

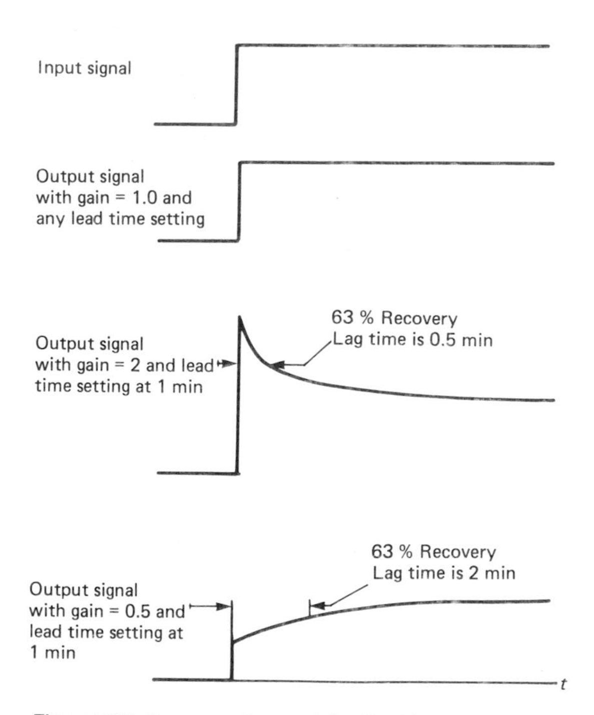

Figure 7.37 Response characteristic of lead/lag unit.

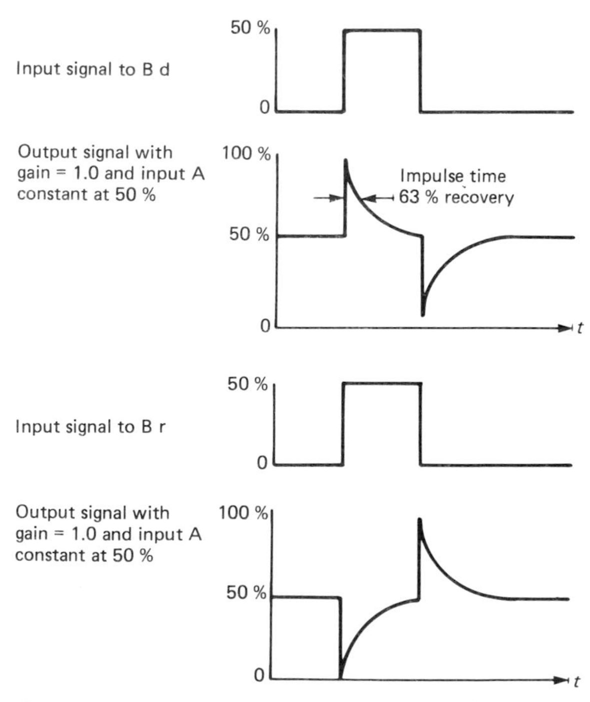

Figure 7.39 Response characteristic of impulse generator.

over long distances. In general, pneumatic transmission systems are more costly to install than the equivalent electronic current transmission systems, and they suffer from a transmission lag which detracts from the system performance.

Various methods have been used in the past to convert the pneumatic signals into a proportional electric current; the majority of these units are now based on the piezoresistive sensors described in Part 1. These are fabricated on a wafer of silicon where a circular portion at the centre is etched away or machined to form a thin diaphragm. One side of the diaphragm is exposed to the pressure to be measured while, on the reverse side, the strain and temperature sensors are formed by ion implantation or similar techniques which have been developed for semiconductor manufacture. The strain sensors are connected in the arms of a Wheatstone bridge from which the out-of-balance signal provides the measurement signal, but, because the gauge factor is affected by any change in the temperature of the silicon wafer, the temperature sensors are used to compensate for this by varying the excitation applied to the bridge network. There are several proprietary ways of implementing this compensation, but most require selection or adjustment of some components to optimize the compensation over a useful temperature range. Compared with most other pressure transmitters, these units are only intended for converting the output signal from a pneumatic transmitter into a proportional current, and, although they may be subjected to equally hostile environmental conditions, the sensor itself is only exposed to 'instrument type' air pressures. In the majority of cases, this is a pressure between 20 and 100 kPa or the equivalent in metric or imperial units of measure. Occasionally there is a requirement to measure signals in the range 20–180 kPa, or the equivalent in other units, when power cylinders or positioners are involved.

A typical converter is shown in Figure 7.40. Its accuracy, including linearity, hysteresis and repeatability, is ±0.25 per cent of span and the ambient temperature effect is less than ±0.75 per cent per 30 K change within an overall temperature range from 0 to 50°C and ±1 per cent per 30 K change within the range −30 to +80°C. In some instances the converters can be located in a clean or protected environment such as a room in which other instrumentation is accommodated. For these applications, the robust enclosure is not required and the sensors, together with the associated electronic circuits, can be assembled into smaller units so that up to 20 of them can be stacked together and mounted in a standard 19-in rack, as shown in Figure 7.40(b).

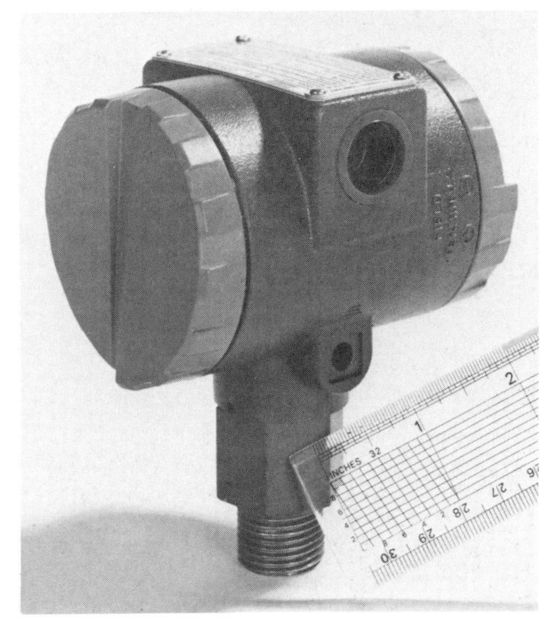

(a)

(b)

Figure 7.40 (a) Pneumatic-to-current converter; (b) pneumatic-to-current converters stacked in rack.

7.7 Electropneumatic interface

7.7.1 Diaphragm motor actuators

The pneumatic diaphragm actuator remains unchallenged as the most effective method of converting the signal from a controller into a force that can be used to adjust the setting of the final operator. In the

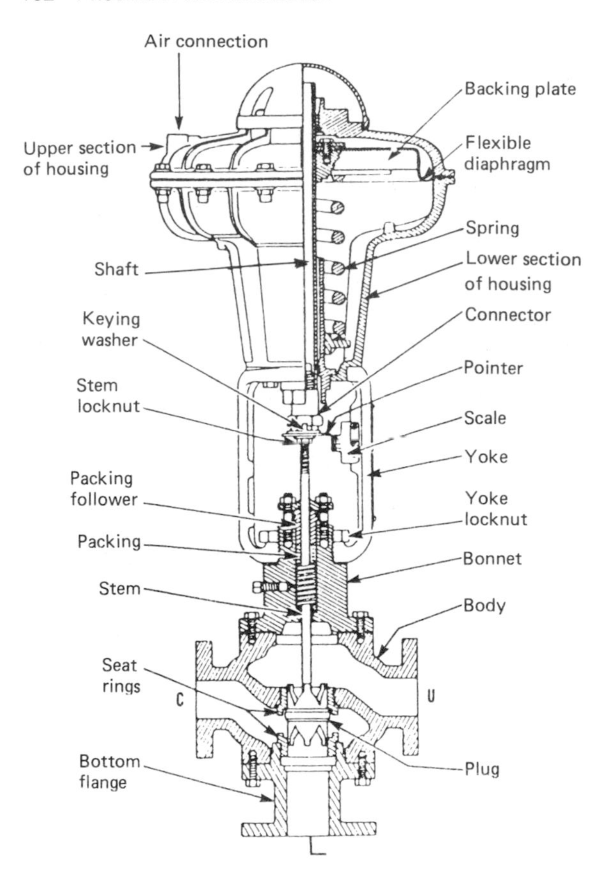

Figure 7.41 Pneumatic diaphragm motor actuator (mounted on two-way valve).

majority of process plants, the final control element is a control valve and the pneumatic diaphragm actuator is particularly well suited to provide the necessary force and stroke.

A typical actuator is shown in Figure 7.41. A flexible diaphragm that separates the upper and lower sections of the airtight housing is supported against a backing plate mounted on the shaft. A powerful spring, selected according to the required stroke and the air-operating pressure, opposes motion of the stem when compressed air is admitted to the upper section of the actuator housing. The stroke is limited by stops operating against the diaphragm backing plate.

The arrangement of the actuator and valve shown in Figure 7.41 is the customary configuration in which the spring forces the valve stem to its upper limit if the air supply or the air signal fails. A two-way valve is shown in the figure and in this case a spring would force the plug to direct the flow from the input port C to the exit port L in the event of an air failure. If the actuator were to be attached to a straightthrough valve, as would be the case if the exit port were blanked off, then the valve would close in the event of an air failure. In many instances this would represent a 'fail-safe' mode of operation (see Chapter 8), but in some plants the safe mode following failure of the air signal would be for the valve to be driven fully open. To deal with this eventuality, the actuator body can be inverted so that instead of the usual position shown in Figure 7.42(a) it is assembled as shown in Figure 7.42(b), in which case the spring would drive the valve stem downwards in the event of an air failure.

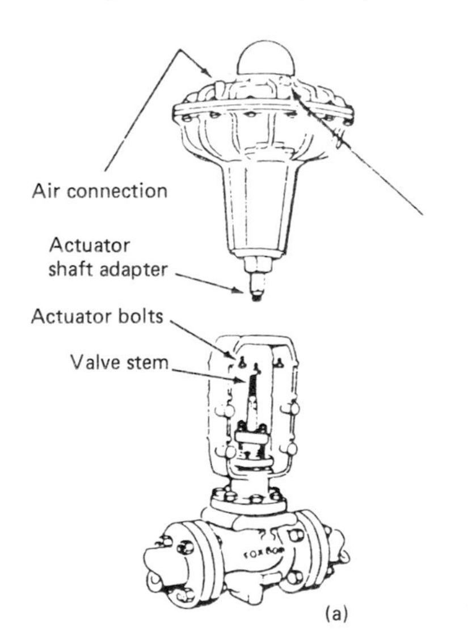

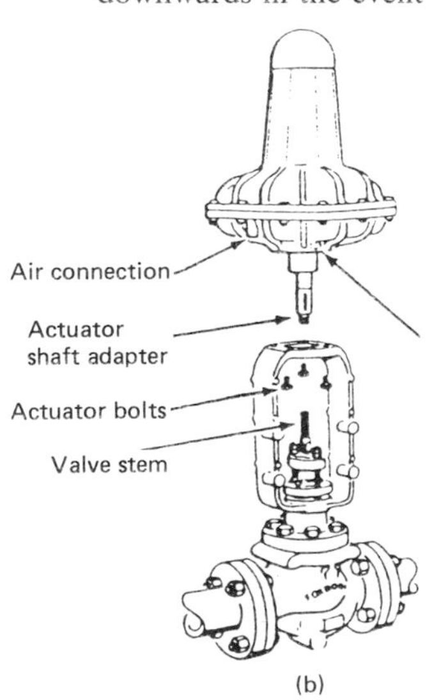

Figure 7.42 (a) Mounting for direct action; (b) mounting for reverse action.

Pneumatic valve stem positioners, electropneumatic converters or electropneumatic positioners, mounted on the valve yoke, can be used to enhance the precision or speed of response of the system.

7.7.2 Pneumatic valve positioner

A valve positioner is used to overcome stem friction and to position a valve accurately in spite of unbalanced forces in the valve body. One such unit is shown in Figure 7.43. It is usually mounted on the yoke of the valve and, in its normal mode of operation, a peg mounted on the valve stem and located in a slot in the feedback arm converts the stem movement into a shaft rotation.

Within the instrument, a flexure in the form of a 'U' with one leg extended is mounted on the shaft. The incoming pneumatic signal is fed to a bellows which applies a proportional force to the flexure at a point in line with the shaft. The free end of the flexure carries a steel ball which bears on a flapper which, together with the associated nozzle, is mounted on a disc that can be rotated about an axis perpendicular to that of the shaft.

In the normal mode of operation, an increase in the input signal causes the bellows to apply a force to the flexure so that the ball at its free end allows the flapper to move closer to the nozzle. This causes the back pressure to rise, and the change after amplification by the relay becomes the output signal, which is applied to the pneumatic actuator and so causes the valve stem to move until balance is re-established. The rotatable disc provides the means for adjusting the sensitivity as well as the direct or reverse action when used with either 'air-to-lift' or 'air-to-lower' actuators.

7.7.3 Electropneumatic converters

Although individual control loops may utilize electronic devices for sensing the primary measurement and deriving the control function, in the great majority of instances the final operator is required to provide a force or motion or both which can most readily be implemented pneumatically. Hence there

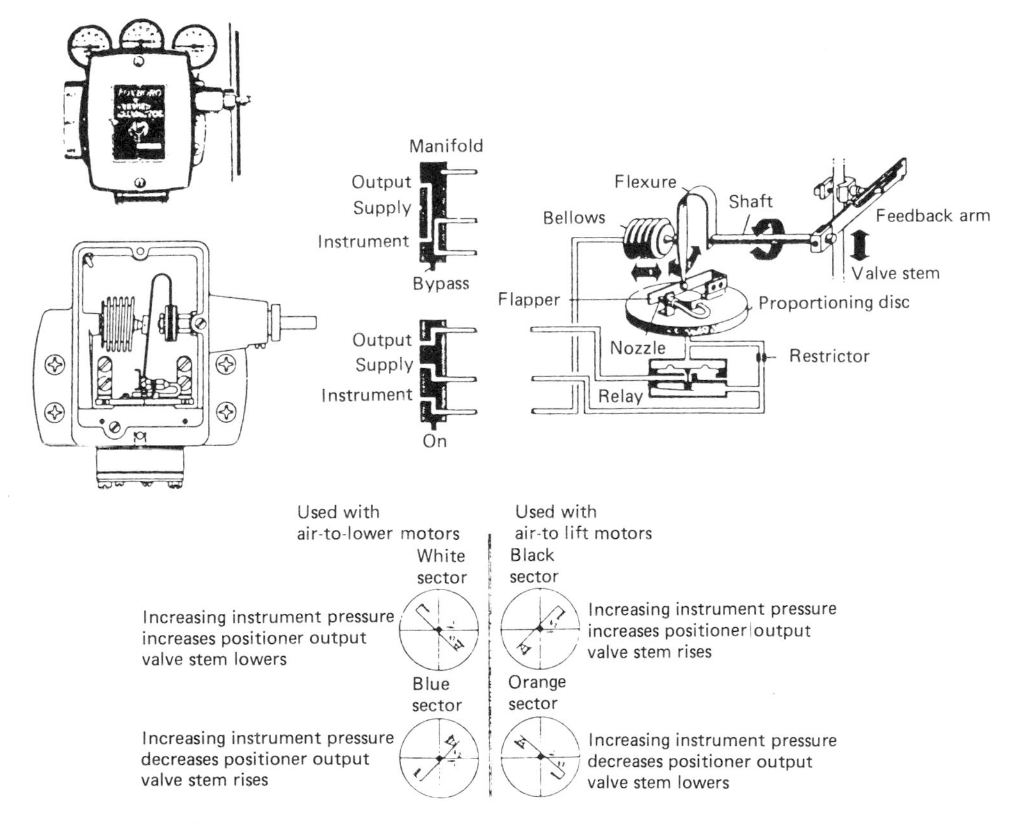

Figure 7.43 Pneumatic valve positioner.

is a requirement for devices which provide an interface between the electronic and pneumatic systems either to convert a current into a proportional air pressure or to convert a current into a valve setting by controlling the air supply to a diaphragm actuator. In most instances the devices have to be mounted on the process plants, where they are likely to be subjected to severe environmental conditions. Therefore it is important for them to be insensitive to vibration and well protected against mechanical damage and extremes of temperatures as well as corrosive atmospheres.

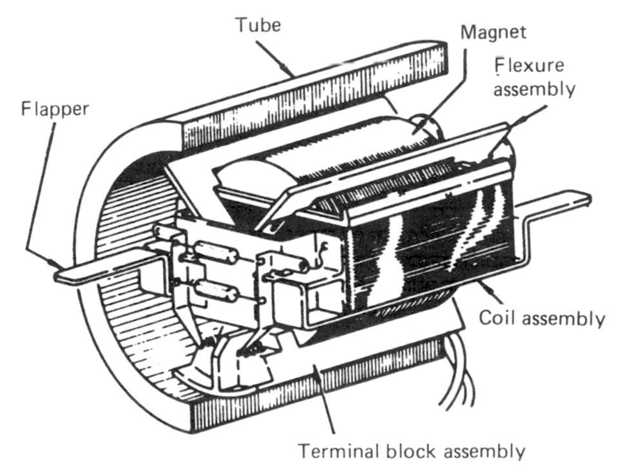

Figure 7.44 Galvanometer mechanism from electropneumatic converter.

The Foxboro E69F illustrates one method of achieving these requirements. It involves a motion-balance system based on a galvanometer movement, as shown in Figure 7.44. The system comprises a coil assembly mounted on cross flexures which allow it relative freedom to rotate about its axis but prevent any axial movement. The coil is suspended in a powerful magnetic field so that a current passing through the coil causes it to rotate. In the case of the electropneumatic converter, this rotation is sensed by a flapper/nozzle system, as shown in Figure 7.45.

The pneumatic section of the instrument comprises a feedback bellows and a bias spring which apply a force to a lever pivoted at one end by a flexure and carrying a nozzle at the free end. The nozzle is supplied via a restrictor and its back pressure is applied to a pneumatic relay whose output is applied to the feedback bellows and is also used as the pneumatic output from the unit.

With the current corresponding to the lower range value (e.g. 4 mA for a 4 to 20 mA converter) the relative position of the flapper and nozzle are adjusted so that the pneumatic output pressure is

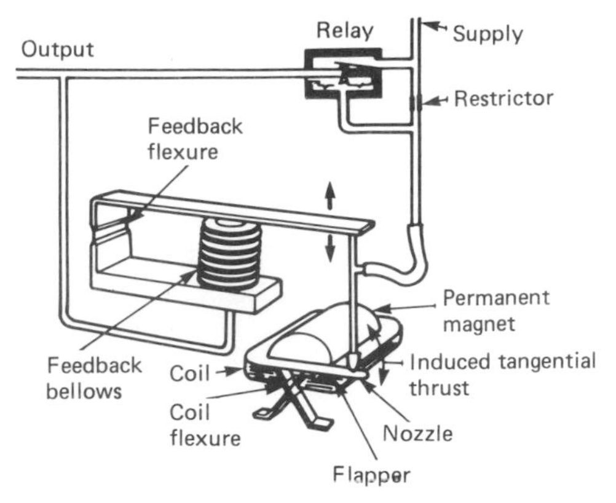

Figure 7.45 Basic configuration of electropneumatic converter.

equal to the required value (e.g. 20 kPa for a 20 to 100 kPa system).

An increase in the input current causes the coil to rotate, and in so doing moves the flapper towards the nozzle so that the back pressure is increased. The change is amplified by the relay and applied to the feedback bellows, so that the lever moves the nozzle away from the flapper until a new balance position is established.

The system is arranged so that when the coil current reaches the upper-range value (e.g. 20 mA) the pneumatic output signal reaches its corresponding upper-range value (e.g. 100 kPa). A limited range of adjustment is available by radial movement of the nozzle with respect to the axis of the coil.

7.7.4 Electropneumatic positioners

The Foxboro E69P shown in Figure 7.46 uses the same sensing mechanism in a positioner that drives the stem of a valve to a setting which is proportional to the incoming current signal. The positioner is

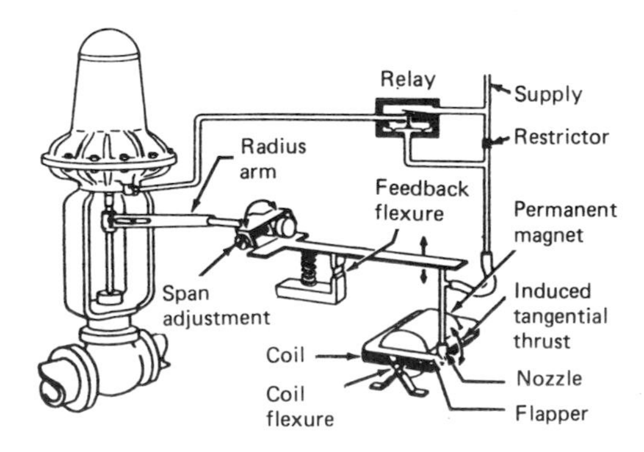

Figure 7.46 Basic configuration of electropneumatic valve-stem positioner.

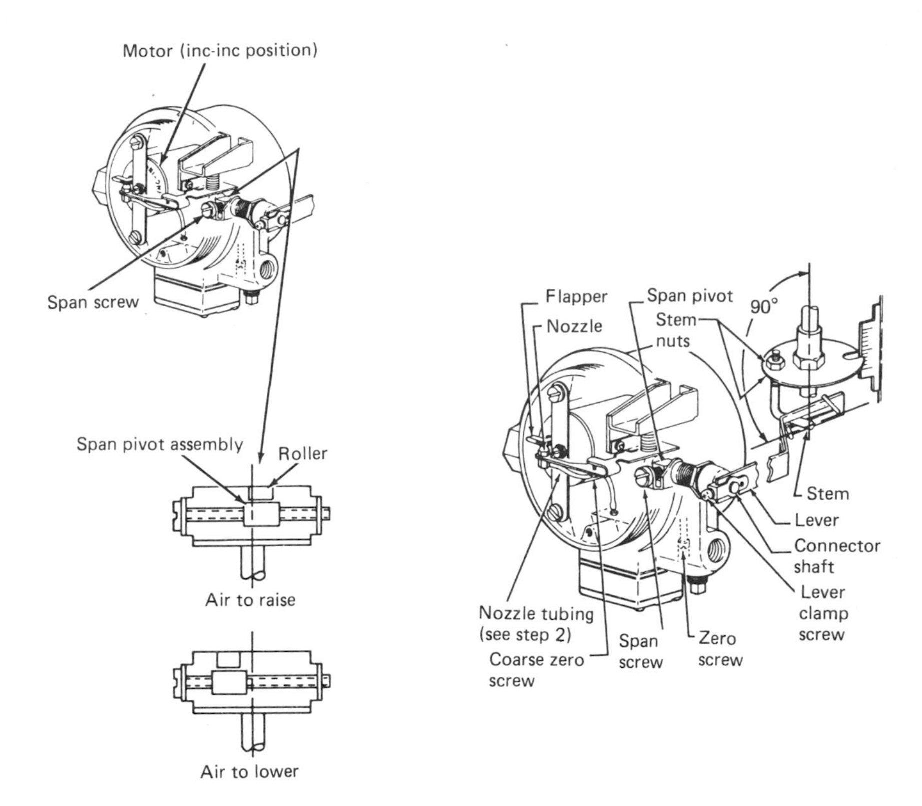

Figure 7.47 Mounting and adjustment of electro-pneumatic valve stem positioner.

mounted on the valve yoke and the valve stem is mechanically linked to it via a peg that is held by a spring against one side of a slot in the radius arm. As shown in Figure 7.47, movement of the valve stem causes the radius arm to rotate a shaft which passes through the instrument housing and the span pivot assembly. This comprises a roller whose axis is parallel to that of the shaft but the offset can be adjusted (to vary the span). The roller bears on the spring-loaded lever which carries the nozzle at its free end.

In operation, an increase in the current signal causes the coil to rotate so that the flapper moves closer to the nozzle. This increases the nozzle back pressure and the change, after being amplified by the relay, is applied to the valve actuator, causing the valve stem to move. This movement is transmitted via the radius arm to the shaft carrying the roller, with the result that the spring-loaded lever is repositioned until the valve stem has moved to a position corresponding to the new input current.

Compared with straightforward air-operated actuators, positioners have the advantage of virtually eliminating the effects of friction and stiffness as well as the reaction of the process pressure on the valve stem.

7.8 References

ANDERSON, N. A. *Instrumentation for Process Measurement and Control*, Chilton, London (1980)

CONSIDINE, D. M. (ed.). *Process Instruments and Controls Handbook*, 3rd edn, McGraw-Hill, New York (1985)

FOXBORO *Introduction to Process Control* (PUB 105B), Foxboro (1986)

MILLER, J. T. (ed.). *The Instrument Manual*, United Trades Press (1971)

8 Reliability

B. E. Noltingk

8.1 Introduction

The words 'reliable' and 'reliability' have been used in a general sense for a very long time. The value of the property they refer to has been increasingly recognized. In the military field, equipment with a superlative performance specification was worthless if it would not work when called upon to do so; for industrial processes involving some risk, there was a demand to quantify that risk instead of just calling it 'very low'. Consequently, since the 1950s the subject of reliability has been put on a sounder scientific and mathematical basis. In this brief account, attempting to relate it to instrumentation, we try to steer a middle course between broad qualitative generalities and a recapitulation of the sometimes abstruse mathematics that has been developed. As well as the whole military field, civil transport—particularly aeroplane systems—and process control plant, including that for nuclear power generation, have been the objects of extensive reliability analysis.

Much attention has been paid to the prediction of the performance of equipment from a reliability point of view. When considering different design approaches and when devising a strategy of management it is obviously of value to have figures for the rates at which different items will fail. There will probably be large statistical—and other—uncertainties, but the main purpose is often achieved if a failure rate is known to within an order of magnitude or even worse.

It might be thought that it is more important to increase reliability than to predict it, but the analysis inherent in prediction will often suggest economical ways to effect improvements. An atmosphere in which attention is focused on reliability will tend to produce reliable systems. Thought will also have to be given to the question of what is an acceptable risk. It will be realized that there are different modes of failure, some of which affect safety while others—through unnecessary shutdowns—influence the economical operation of plant; these two outcomes must both receive appropriate consideration.

It is convenient to think of equipment—particularly for instrumentation—at three different levels: components, modules and systems. They may be treated rather differently when their reliability is considered. There may be some uncertainty which category an item should be allotted to, but in general a component is an item that is replaced and never repaired, a module is a collection of components into something like an instrument while a system is the totality we are concerned with in a particular exercise, often including many modules with complex interconnections.

8.2 Components

The word 'component' most commonly brings to mind an electronic device—resistor, capacitor, etc. In integrated circuits many such are combined together, but with the definition that it is something that is replaced, not repaired, the whole device must be thought of as a component. The subject of reliability was first developed for equipment that was mainly electronic, partly because of the need and partly because the large population of nominally identical items used in electronics lent itself more obviously to a statistical approach. Note that soldered joints would be thought of as components, although their replacement would often be called a repair. In a similar way, small mechanical items are components, although there may be more uncertainty about classification: for example, an electric motor may be treated as a whole or it may be emphasized that the brushes within it can be replaced.

8.2.1 Physics of failure

How does a component fail? Three headings can be used: wear-out, misuse and inherent weakness. Some items are known from the outset to have limited lives; hot filaments must be expected to go open circuit eventually and gears to wear out. Components are designed to withstand only certain stresses of temperature, voltage, etc., and if these are exceeded they are likely to fail. During manufacture there may have been some mistake such as a foreign particle introduced or a locked-in mechanical stress. All three kinds of happening can cause failure in service—the overstressing may occur at some unpredictable time. All three interact with each other—manufacturing faults may shorten the life to wear-out and increase the sensitivity to overstressing.

The physical description of failure on a microscopic scale is commonly in chemical or mechanical terms. Corrosion changes composition and properties; a conducting foreign particle can move to bridge two conductors and short them out or stress can rupture and open circuit a conductor. A two-stage process is not uncommon, such as a broken seal leading subsequently to corrosion. In several such ways a delay can be introduced so that there is a time lapse between initiation and the failure becoming apparent, but it can be seen as plausible that the time scale is much less than the hoped-for life of the component.

8.2.2 Mathematical analysis

The instantaneous failure rate of an item at time t_1 can be thought of as the conditional probability that, having survived until t_1, it will fail during the short interval $t_2 - t_1$, divided by the duration of this interval. As in most probability concepts, probability for a single item can be replaced by the fraction of the total when considering a large number of identical items.

For many components, the instantaneous failure rate is not constant. At the beginning of life, there tend to be excessive failures, as discussed above, while items where there is a mechanism for wear-out will show a higher rate also as the end of life on that score is approached. This is illustrated in Figure 8.1.

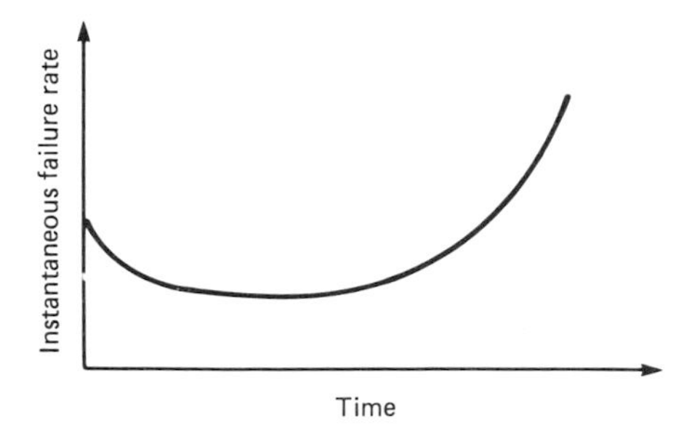

Figure 8.1 Variation of failure rate with time (the 'bath tub' curve).

The shape of the curve relating failure rate to time is often likened to a bath tub. However, it is much simpler to think in terms of a constant failure rate λ (dimension = 1/time), and we do that here. The justification is that the useful life occurs mainly between the periods when, on the one hand, early mortality and, on the other, wear-out are effective (i.e. the bottom of the 'bath tub' in Figure 8.1); λ is anyhow so imprecisely known that refinements are scarcely worth while. Then

$$\text{Mean time to failure} = \frac{1}{\lambda} \quad \text{(dimension = time)}$$

and reliability R, which must always be associated with some time, being defined as the probability of survival through that time, is

$$R(t) = \exp(-\lambda t) \qquad \text{(dimensionless)}$$

The probability density function (PDF) $f(t)$ giving the probability of failure during Δt at t, divided by Δt is

$$f(t) = \lambda \exp(-\lambda t)$$

(distinguished from the instantaneous failure rate in that it is not conditional on survival to t). If PDF is plotted against time, we get the exponential curve shown in Figure 8.2. The wide uncertainty in the statistics of small numbers is illustrated here, for if the mean time to failure is τ, there is still, for a single item, a 5 per cent probability that it will fail before 0.05τ and another 5 per cent probability that it will survive until after 3τ, corresponding to the areas shown under the curve.

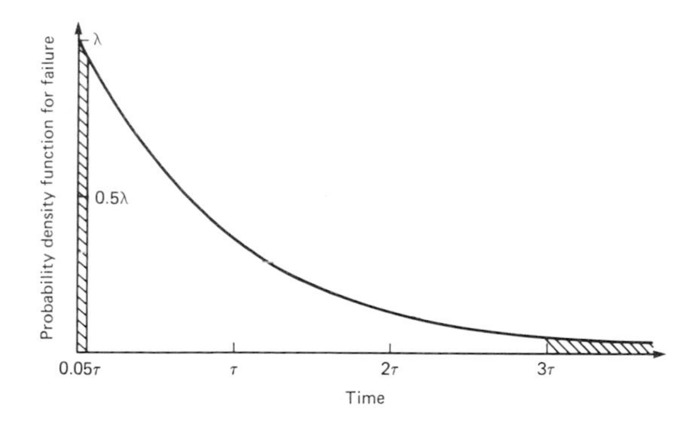

Figure 8.2 Probability density function for failure at different times.

8.2.3 Modes of failure

Failure of a component means that some characteristic changes in such a way as to prevent the component performing its required function, but there are several characteristics and several new values to which they can change. So we have to consider the *mode* of failure. Often, in electronic devices, it is to open circuit or to short circuit, but sometimes the shift is to a value not very different from the correct one. This commonly occurs relatively slowly, when it is described as 'degradation', whereas a sudden and complete failure is called 'catastrophic'. Degradation is observed more often in mechanical components than in electronic ones.

Different consequences of different modes of component failure are found when we consider the higher levels of modules and systems. There is a possibility of anticipating degradation errors by monitoring performance, whereas catastrophic failures are unpredictable (in individual cases, that is, as distinct from statistically).

8.2.4 Failure rates

In order to make practical use of theoretical concepts numerical values must be adopted for the failure rates of different components. It may be questioned whether there is such a thing as a failure rate that should be associated with any particular component. The concept of human mortality rate has been used by actuaries over many years and has been validated by the success of the predictions it has led to, but instrument components have an even wider variety than human beings. Probably, only the pragmatic answer can be given that, by adopting quantitative component reliability concepts, useful conclusions can be drawn. Much work has been devoted to establishing practical figures for failure rates, notably (in Britain) by the National Centre of Systems Reliability, part of the Safety and Reliability Directorate at Warrington; they maintain a continuously updated databank using contributions from many sources. Some observed failure rates reflect operational experience and some come from deliberate tests of isolated components.

Anyone seriously attempting to predict reliability should make use of comprehensive data from such publications as *Electronic Reliability Data* (1981), *Reliability and Maintainability* (1985) or MIL-HDBK-217D (1982) (see the References). The data should only be applied to items similar to those used to give the data.

Figure 8.3 gives an idea of the failure rates of different components and shows the enormous range over which they are spread. Some of the range is accounted for by operational conditions: temperature stress may introduce a factor of 3:1 for an increase from 0°C to 140°C, electrical stress a further

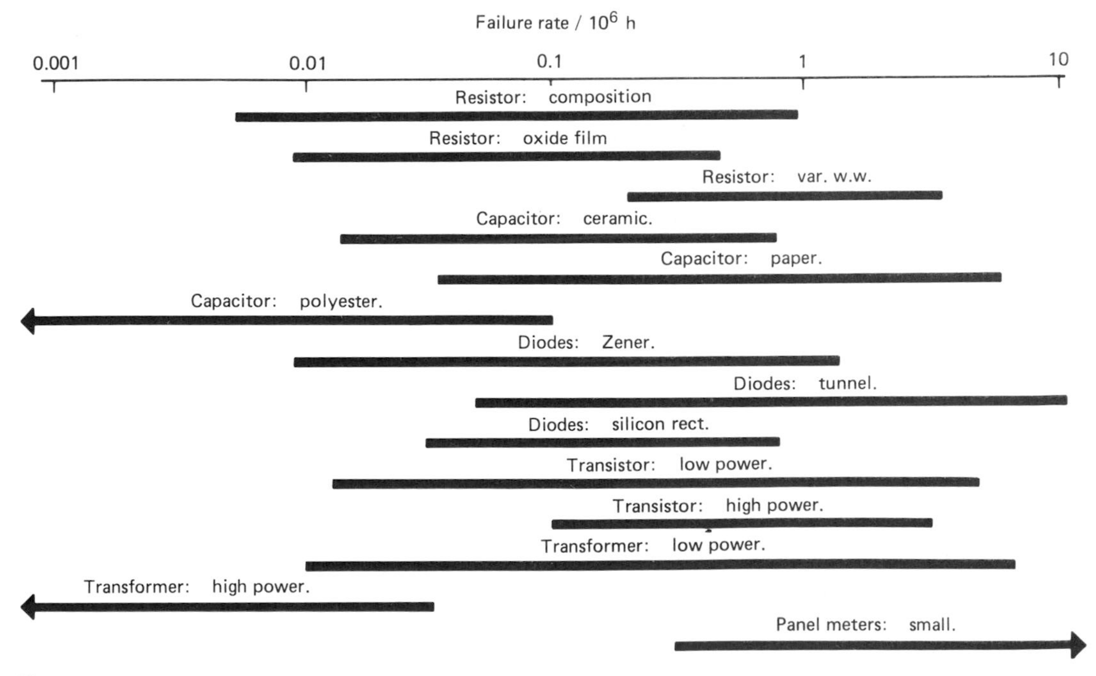

Figure 8.3 Failure rates of some components. (From Item 3 in the References, courtesy of the publishers.)

factor for many electronic components, while general severity of environment may alter the failure rate to be taken by as much as 50:1.

8.3 Modules

Ultimately, it is the behaviour of the whole system that is of interest, but it is helpful to consider also what happens at the module level, intermediate between component and system.

8.3.1 Failure rates

If all its components must be functioning correctly for the module to work, then its failure rate is determined by summing the failure rates of each component:

$$\lambda_{mod} = {}_1\lambda_{comp} + {}_2\lambda_{comp} + {}_3\lambda_{comp} + \cdots$$

and other parameters can be expressed in terms of λ_{mod}.

As we noted different modes of component failure, so we find different consequences ensue according to how a module fails. The commonest distinction is between fail-safe and fail-to-danger, the latter tending to damage either directly or if associated with some other eventuality, while the former tends only to cause an unnecessary shutdown. A simple example is a thermometer failing safe by reading too high when monitoring a process that becomes dangerous at excessive temperatures, but it is not always obvious which mode is fail-safe. I find this conveniently illustrated from a human situation. Thus, my wife worries about my safety if I have not arrived home by when I said I would, so a fail-safe feature of the information channel would lead to the expectation of a later-than-true return time. However, I can understand that with other wives and other worries the embarrassment would be associated with an unexpectedly early return, in which case the fail-safe failure would be in the opposite direction.

To determine module failure rates corresponding to different modes it is, of course, necessary to analyse the roles of different components—and the consequences of their failing in different modes—and sum the appropriate component rates.

8.3.2 Partial failures

Failure is defined as the termination of an item's ability to perform its required function. The precise function required, however, may vary from time to time. If an item can perform some of the functions required of it, but not meet its full specification, it is said to have undergone partial failure. The question arises sometimes with components, where it would commonly be associated with degradation, and more often with modules. Examples would be a capacitor with a high leakage current that still does not amount to a short circuit, a multimeter with one range out of action, a thermometer element with a slow response.

The possibility of partial failures complicates reliability analysis, which indeed becomes almost impossible if their number is likely to be comparable with that of total failures. A conservative value for failure rates will be derived if partial failures are counted in with total ones. The major problem they produce is in the field of monitoring the integrity of equipment, which we touch on under Systems (section 8.4) because, if some item has partially failed, it cannot be presumed that tests giving a satisfactory outcome under one set of conditions imply that the equipment will work equally well under a different set.

8.3.3 Design

A module that has been well designed will be more reliable because of lower failure rates in its components. This arises in several ways. Wear-out can be postponed, for instance, by underrunning incandescent lamps or by reducing the harshness of mechanical regimes. Understressed components were seen to have lower failure rates; stresses can generally be reduced by minimizing temperature rises and, for electronic components, reducing operating voltages. Design skills come into play in seeing that all this is done in a cost-effective way.

Good design also reduces the number of degradation failures. This can be appreciated from Figure 8.4. We plot there a typical probability density function showing the expected value of some component characteristic—say, the resistance of a

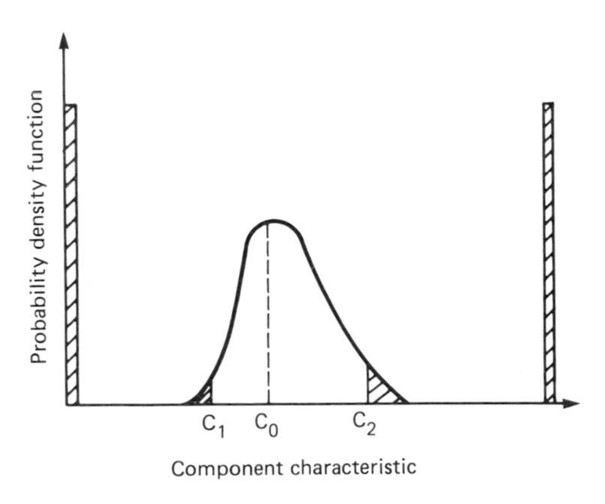

Figure 8.4 Spread of component characteristics.

resistor—after a certain time of operation. When installed, the characteristic had the value C_0 and for its module to work properly, the value has to lie between the limits C_1 and C_2. The probability that failure has occurred is thus given by the hatched area under the curve. This comprises the extremes, which correspond to catastrophic failure, and the flanks of the central curve, to which the characteristic may have drifted. The skill of the designer should ensure that the range of limits C_1–C_2 is wide enough for all the distribution curve for planned operating times to lie inside it.

In fact, weakness of design may account for some failures early in the life of some equipment, and some of the generally assumed high early failure rates. It is not unusual for development of a module to be completed at the same time as early trials, and if these suggest anomalously frequent failures the situation of a suspected component will be re-examined to see whether stresses are exceptionally high or whether there is an unusual sensitivity to drift. Component failure rates normally quoted are on the assumption of a good design from which teething troubles have been eliminated.

8.4 Systems

When we come to deal with complete systems there is scope for increasing reliability by introducing complex configurations which include 'redundancy'. The alternatives of fail-safe and fail-to-danger which we mentioned under modules now become important. The strategy adopted for repairs also plays a considerable part.

8.4.1 Redundancy

Redundancy in an item is formally defined just as the existence of more than one means of performing a given function. We can understand it best by illustrations. In the smoothing circuit of an electrical power supply unit, we need, say, at least 100 μF capacitance. If, then, we install two capacitors in parallel, each of this capacitance, we have a *redundant* component and the unit will continue to function should either of them go open circuit. However, should either develop a short circuit there is a total failure of the module. We have reduced the sensitivity to one mode of failure but increased the sensitivity to another mode. Incidentally, this shows how redundancy principles can be applied also at the module level.

Again, consider the protection of some plant against excessive temperature rise. We install a thermometer, arranging for the plant to be shut down if the thermometer gives a reading greater than some critical value. There is a certain small probability q_d that the thermometer will fail to show a high temperature if one arises. That would be

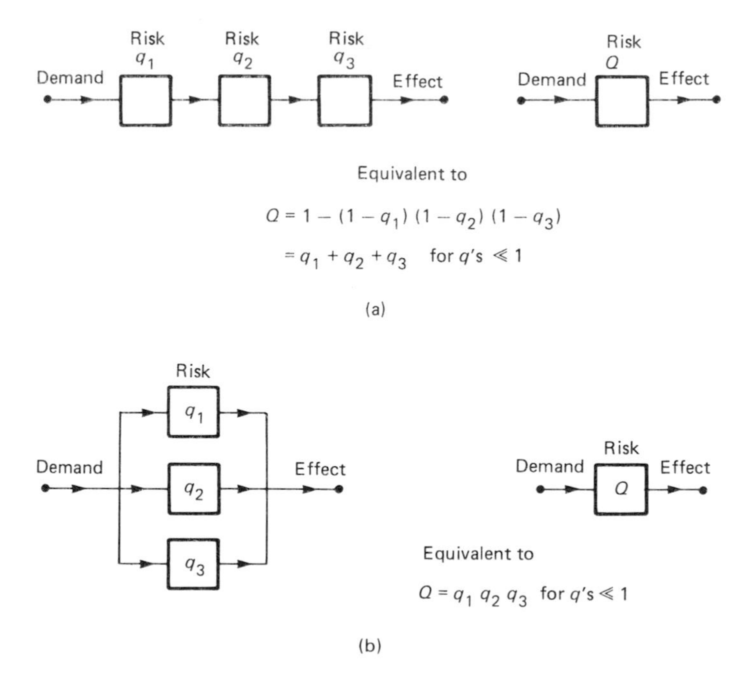

Figure 8.5 Reliability diagrams.

disastrous for the plant so we install another, identical thermometer and arrange for shutdown if either reads high; this reduces the probability of a dangerous failure to q_d^2 (q_d is assumed to be very small). However, there is also a finite probability q_s of a safe failure, corresponding to an incorrectly high reading of temperature and leading to an unnecessary shutdown. Introduction of the second thermometer has doubled this probability to $2q_s$, and we have a safer system at the cost of less economical working. All this is on the assumption that failures of two separate items—components or modules—are *independent*. It can be seen that the assumption is not always justified from examples such as a single overvoltage surge destroying two components or a fire in a cable duct damaging the connections between several modules. Much thought is given to ensuring independence at the stage of system design.

Arrangements in systems can be depicted in reliability diagrams, as shown in Figure 8.5. We generalize by talking of whether a demand gets through to cause an effect, the different blocks having probabilities q of not performing (hence $1 - q$ that they will perform) correctly. In the multi-thermometer case we have discussed, safe failures must be regarded as using the series configuration in Figure 8.5(a) because all elements must be performing for the signal 'Cool' to get through, but for failures to danger, the parallel arrangement in Figure 8.5(b) is appropriate, since any working element allows the signal 'Too hot' to pass.

'Partial redundancy' configurations are used to reduce the incidence of both safe and dangerous failures. A 'two out of three' example is depicted in Figure 8.6; it is arranged that when any two out of the three channels installed indicate danger, the danger signal is transmitted. For channels with equal risks q of failure, the risk of the total arrangement $Q = 3q^2$, appropriate values q_s or q_d being taken for the probability of elements having failed safe or to danger, respectively. In this simple case, $Q = 3q^2$ for both failure modes because a failure of two out of the three channels produces an effect in either mode; for the more general r out of n case, formulae will be different for the two modes. The importance of the logic element making the two out of three judgement should be noted: if that cannot be assumed to be 100 per cent reliable it must be given careful attention.

The total reliability of a complex control system can be examined by setting out the appropriate reliability diagram and inserting values for the risks associated with each element.

When some but not all redundant channels have failed, the system can be thought of as in a state of partial failure because it is operational but with a lower than specified reliability.

8.4.2 Repairs and availability

When considering the effect of redundancy on the performance of a whole system, we included values q for the probability that a unit was not working. If we start with a perfect system and simply let it deteriorate indefinitely, q will increase progressively from 0 (to 1 after infinite time) and can be calculated from failure rate data for each item at the time we are concerned with.

That unidirectional change is a correct picture for some situations, but with most large installations of instrumentation some maintenance and repair is undertaken. This plays an important part in the reliability achieved over long periods. To analyse the situation quantitatively, we think of the mean time to repair (MTTR) an assembly; this includes both the time elapsing before it is realized there has been a failure—which will be very short for 'self-revealing' faults—and the time for maintenance personnel to be deployed and achieve their repair mission.

In these continuing situations, we have to distinguish between a system's mean time to fail

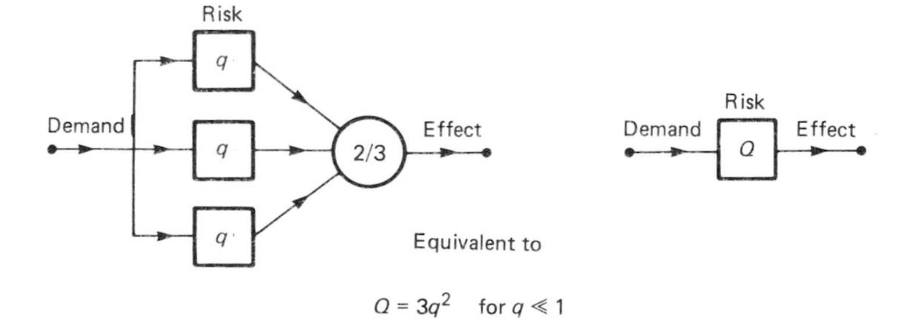

Figure 8.6 Partial redundancy of a 'two out of three' system.

(MTTF) and mean time between failures (MTBF). In fact, it can be seen that MTBF = MTTF + MTTR. We can also appreciate the concept of Availability, the fraction of time for which something can perform its design function:

$$\text{Availability} = \frac{\text{MTTF}}{\text{MTTF} + \text{MTTR}}$$

To avoid ambiguity of definition, it must be made clear whether shelf-life (i.e. periods when equipment is not switched on) is to be included.

The availability of modules or subsystems should be taken into consideration when calculating system reliability: it can be seen how this complicates calculations which have to take into account partial failures (and their modes) in redundant systems.

Maintainability is defined as 'the probability that a device that has failed will be restored to operational effectiveness within a given period of time when the maintenance action is performed in accordance with prescribed procedures'. With the (rather unrealistic) assumption of a *constant repair rate*, the situation can be analysed as were reliability and failure (section 8.2.2), giving

$$\text{Maintainability (in time } t) = 1 - \exp\left(\frac{-t}{\text{MTTR}}\right)$$

Good design can reduce MTTR as it can increase MTTF; the former is also influenced by maintenance procedures.

8.4.3 Software reliability

The expression *software reliability* is used because of the realization that when a complex computer system goes online it may contain errors that will only show up in certain infrequently occurring situations. The errors may be simple programmers' mistakes or they may arise from a misunderstanding of the physical system to which the computer is linked. It is, however, almost invariably the case for a large system that some software errors will only be discovered after the system has been brought into use.

There is no 'physics of failure' for software; perhaps there is a psychology of failure! Neither is there an instant of failure: software does not change with time except by human intervention. If we are contrasting the behaviour of hardware and software, a further difference is that in operational situations we are only concerned with partial failures of software because a system with it totally failed could never be launched.

It is difficult to describe the situation with a plausible mathematical model. The timescale for a failure is not that concerned with physical damage to a component, as in hardware failures. Instead, it is that for a set of circumstances to arise which will show up—and cause errors from—a partial failure that has been there all along.

Realization of the problem has led to attempts to devise techniques resulting in more reliable software. In practical instrumentation, the main conclusion is that it is wrong to presume on the perfect integrity of computer software.

8.5 Practical implementation

The foregoing sections give a brief outline of how the reliability of instruments and systems can be analysed and—very approximately—predicted. When setting up instrumentation, some practical steps should be taken which increase reliability and indeed must be presumed to have been effected if semi-quantitative predictions are to be valid.

8.5.1 Design and operation

We cannot do more than mention the vast question of good design. Some aspects are dealt with in Chapter 1 on Construction. Sound tolerancing of both electronic and mechanical components is important.

When availability was discussed, the need for a good maintenance procedure was apparent. If possible, components subject to wear-out (e.g. incandescent lamps and rubbing mechanical parts) should be replaced before they approach the ends of their lives—implying that the mean lives must be known and that the standard deviations about the means are not impossibly large. Adequate records, of course, should be kept; higher-than-expected failure rates for some components may imply poor design, which can be corrected.

In the reliability structure of a complete system, human action in response to a warning signal may play a part. This opens the door to human errors. It must certainly not be assumed that such errors are impossible, although it is very difficult to assign a probability in a particular situation. Smith suggests it ranges from 10^{-5} for overfilling the bath to 0.9 after 1 min in a situation of real emergency. Where the highest integrity is called for, the human element should be eliminated from the direct control chain and replaced by a fully automatic system. Human intervention should only be allowed when a time scale of hours gives an adequate period for leisurely thought.

8.5.2 Environment

A large part of providing reliable equipment consists

of ensuring that the equipment is suitable for the environment in which it will be operated. Many aspects of an environment are relevant.

High temperature is commonly recognized as an adverse environment. This is reflected in the factors by which failure rates are increased in quoted data. The useful properties of most materials degenerate as they get hotter. Not only temperature must be thought of; steep temperature gradients within a component or assembly can also give trouble, as can rapid and repeated temperature changes.

The atmosphere surrounding equipment may be the reason for early failure. Traces of corrosive gas or moisture can cause corrosion: remember that variations in temperature have an influence on whether vapour condenses out as liquid. Dust deposited from dust-laden atmospheres is often responsible for malfunctioning of moving parts or for changed electrical characteristics. Dust penetrates to the interior of imperfectly sealed containers through what is called 'breathing': when internal air cools and so contracts, air from outside is drawn in and this is repeated as cycles of temperature change are experienced.

Vibration and, indeed, exposure to any mechanical hazard should be avoided. The latter may originate from human tampering by those who do not realize the damage they cause—or by those who do, and welcome it. Appropriate protection is always important. On the other hand, inaccessibility may in practice reduce the maintenance given to an instrument, whether that inaccessibility arises from a remote location or from overprotective housing.

In some situations, radiation must be taken into account as an adverse environment. More likely to be overlooked is electrical interference. There is a danger, notably in cable runs, of electrical pick-up from one conductor to another. This can introduce false signals and so corrupt communication channels. It can also, at a higher level, cause physical damage; it might be thought difficult to transfer enough energy for this, but some active semiconductors are very sensitive, while if a protective system is engaged in protecting against a major disturbance, that disturbance itself can be the source of very energetic pulses.

Constancy of conditions is generally desirable, in particular a steady temperature. It is thus good practice to leave equipment switched on unless the consequent wear-out of components having a limited life is unacceptable.

Items will normally have specifications as to what environments they can withstand. From a reliability point of view, the limits indicated should not be thought of as precise go/no-go figures within which performance is perfect and outside which the item is useless. Rather is reliability increased progressively as the environment is made more favourable. Designer and user have to apply their skills so that this principle is used in the most cost-effective way.

8.5.3 Diversity

It was mentioned in section 8.4.1 that the technique of increasing reliability by introducing redundancy depended on failures being independent. Failures that have a common cause are a serious danger, because they can greatly reduce the protection given by redundancy.

To counter this, it is desirable to aim for *diversity*. The word is used to cover basic differences in doing things. If, for instance, two separate instruments provide redundant protection in a system, it could be specified that their components should come from different manufacturers or at least different batches; a potential source of failure that had occurred inadvertently would not then degrade the reliability of both instruments. In other situations, it is desirable that links between separate items of equipment should not follow the same path and so be susceptible to disruption by the same external event.

More fundamentally, diversity can provide some safeguard against faulty human analysis. In nuclear reactor protection systems, for instance, it may have been assumed that increased activity would always lead to more neutrons being found. If a redundant protection channel is based on the primary observation of raised temperature, this will prevent such drastic consequences if for any reason an increased neutron population should not be detected when the reactor power goes up.

It can be seen that, although it may tend to be complicated and expensive, diversity is a sound principle to aim at.

8.6 References

CLULEY, J. C. *Electronic Equipment Reliability*, Macmillan, London (1981)

GREEN, A.E. and BOURNE, A. J. *Reliability Technology*, John Wiley, Chichester (1972)

National Centre of Systems Reliability and Inspec. *Electronic Reliability Data*, Institution of Electrical Engineers (1981)

SMITH, D. J. *Reliability and Maintainability*, Macmillan, London (1985)

US Department of Defense. *Military Handbook: Reliability Prediction of Electronic Equipment* (MIL-HDBK-217D) (1982)

Reliability and Maintainability Data for Industrial Plants, A. P. Harris and Associates (1984)

Some British Standards relate to reliability, notably:
BS 4200, Part 2, Guide on the reliability of electronic equipment and parts used therein: terminology (1974)
BS 5760, Part 2, Guide on the reliability of engineering equipment and parts

9 Safety

L. C. Towle

9.1 Introduction

The interactions between the design and application of instrumentation and safety are many and diverse. The correct utilization of instrumentation for monitoring and control reduces risk. An obvious example is a fire detection and control system, but even a simple cistern control which prevents a water tank from overflowing affects overall safety. Any instrumentation which contributes to maintaining the designed status of an installation can arguably affect safety. However, instrumentation can increase the danger in an installation, usually by being incorrectly designed or used. The principal direct risks from electrical instrumentation are electrocution and the possibility of causing a fire or explosion by interaction between the electricity and flammable materials, which range from various insulating materials used on cables to the more sensitive oxygen-enriched hydrogen atmosphere of a badly ventilated battery charging room. Some aspects of the safety of lasers and the risks from radiation are dealt with elsewhere in this reference book, Part 3, Chapters 2, 3 and 5. Toxic materials should also be considered (see *Substances Hazardous to Health* in the References). These risks pale into insignificance when compared with the full range of possibilities of misapplying instrumentation to a process plant, but nevertheless in an overall safety analysis all risks must be minimized.

It is important to recognize that nowhere is absolute safety achievable, and that the aim is to achieve a socially acceptable level of safety. Quite what level has to be achieved is not well defined; it is perhaps sufficient to say that people are even more reluctant to be killed at work than elsewhere, and hence the level of safety must be higher than is generally accepted. For example, the risk level accepted by a young man riding a motorcycle for pleasure would not be acceptable to a process operator in a petrochemical plant. There are similar problems in determining how much financial expenditure is justified in achieving safety.

As well as the moral responsibilities implicit in not wishing to harm fellow mortals there are, in the majority of countries, strong legal sanctions, both civil and criminal, which can be used to encourage all designers to be careful. In the United Kingdom, the Health and Safety at Work Act 1974 together with the Electricity Regulations provides a framework for prosecuting anyone who carelessly puts at risk any human being, including himself. The act places responsibilities on manufacturers, users and individuals in some considerable detail, and the requirements are applied in almost all circumstances which can conceivably be regarded as work. For example, manufacturers are required to sell only equipment which is safe for its intended use, test it to check that it is safe, provide adequate installation instructions and be aware of the 'state of the art'. The Act was derived from the Robens Report, which is a very readable, well-argued discussion document which sets a reasonable background to the whole subject of industrial safety. The Act lays great stress on the need to recognize, record and evaluate levels of danger and the methods of reducing the risk to an acceptable level, and consequently there is a need for adequate documentation on the safety aspects of any installation. In the majority of installations the enforcing organization is the Factory Inspectorate, who have awesome powers to enter, inspect and issue various levels of injunction to prevent hazards. Fortunately, the majority of factory inspectors recognize that they do not have quite the infinite wisdom required to do their job, and proceed by a series of negotiated compromises to achieve a reasonable level of safety without having to resort to extreme measures. It is important to realize that the legal requirement in most installations is to take 'adequate precautions'. However, in the real world the use of certified equipment applied to the relevant British Standard Code of Practice is readily understood, easy to document and defensible, and is consequently the solution most frequently adopted.

9.2 Electrocution risk

In designing any electrical equipment it is necessary to reduce the risk of electrocution as far as possible. Many sectors of industry have special standards of construction and inspection combined with certification schemes to take into account their particular risks. For example, electro-medical equipment has to meet stringent standards, particularly in cases where sensors are inserted in the body.

It is useful to try to assess the equivalent circuit of the human body, and there are a large number of references on the subject which show quite wide discrepancies between experimental results. A few facts appear to be common. Figure 9.1 shows the generally accepted figures for the ability to detect the presence of current, and the level of current which causes muscular contraction, although it must again be stressed that individuals vary considerably. Muscular contraction is a fascinating process, involving an electrical impulse signal releasing a chemical which causes the mechanical movement. The currents required are about 15 mA, and to maintain a muscle contracted it requires about 10 pulses/s. When a direct current is applied it causes the muscle to contract once and then relax; consequently direct current tends to be safer. However, at higher levels direct current does cause paralysis, since variation in body resistance due to burns, etc. causes the current to fluctuate and hence contract the muscles. The 50–60 Hz normally used for domestic supplies is ideally chosen to make certain that paralysis occurs.

Body resistance is quite a complex picture, since much of the initial resistance is in the skin. A dry outer layer of skin, particularly in the areas which are calloused, gives quite high resistance at low voltage, typically 10–100 kΩ, but this falls to 1 kΩ at 500 V. Other, more sensitive areas of the body such as elbows have a much lower resistance (2 kΩ). Once the outer layer of skin is broken, the layer immediately below it has many capillaries filled with body fluid and has very low resistance. The bulk resistance of humans is mostly concentrated in the limbs and is taken to be 500 Ω. Figure 9.2 shows one curve of body resistance and a possible equivalent circuit of a human being at low voltage when the skin resistance is converted to a threshold voltage.

The process of killing someone directly by electricity is also quite complex. Generally, it is agreed that a current of 20–30 mA applied to the

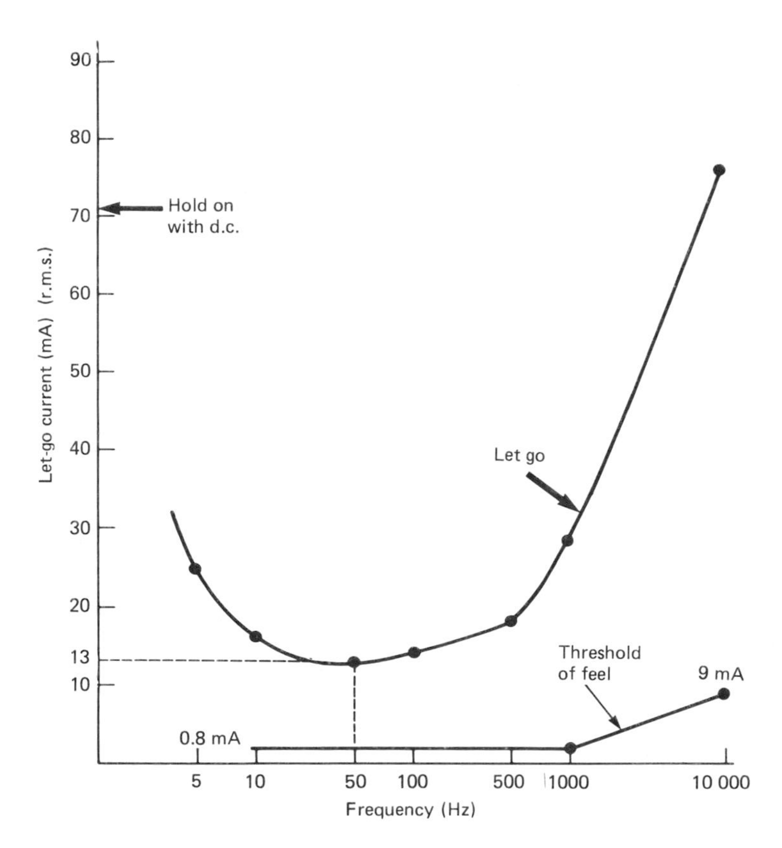

Figure 9.1 Variation with frequency of let-go current and threshold of feel.

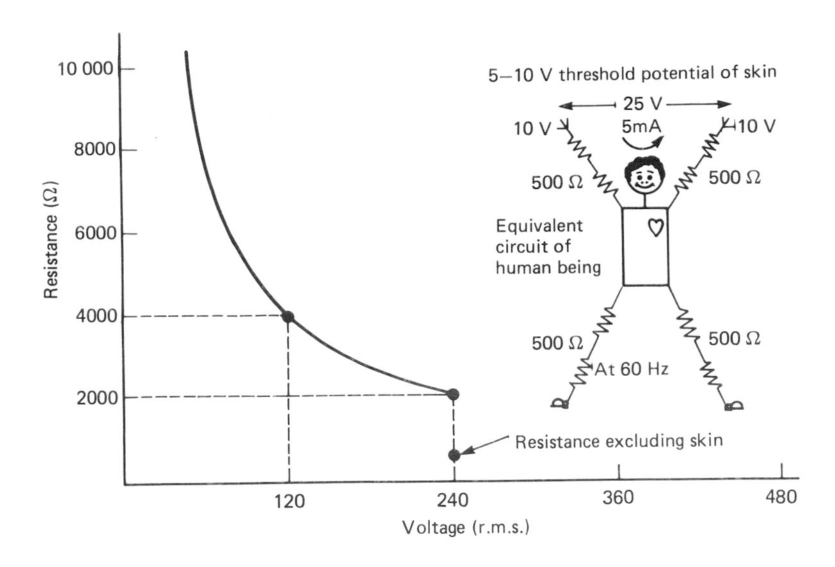

Figure 9.2 (a) Apparent increase of body resistance (hand to hand—dry) with reduction of voltage; (b) equivalent circuit of human being.

right muscles of the heart would stop it functioning. Just how to persuade this current to flow in the practical problem of hand-to-hand electrocution is widely discussed. Some sources suggest currents of the order of 10 A are necessary and others suggest there is a possibility of 40 mA being enough. The level of current is further complicated because there is a time factor involved in stopping the heart, and some protection techniques rely at least partially on this effect to achieve safety. The change is quite dramatic. For example, one reference suggests that heart fibrillation is possible at 50 mA if applied for 5 s and 1 A if applied for 10 ms. There seems little doubt, however, that the conventional 250 V 50 Hz supply used in the United Kingdom is potentially lethal, and that standing chest deep in a swimming pool with a defective under-water low-voltage lighting system is one very effective way of shortening a human being's life span.

The majority of modern instrumentation systems operate at 30 V or below, which to most people is not even detectable and is generally well below the accepted level of paralysis. There are, however, circumstances where even this voltage might be dangerous. Undersea divers are obviously at risk, but people working in confined hot spaces where sweat and moisture are high also need special care. Once the skin is broken, the danger is increased, and the possibilities of damage caused by electrodes fastened to the skull are so horrendous that only the highest level of expertise in the design of this type of equipment is acceptable. However, for the majority of conventional apparatus a level of 30 V is useable and is generally regarded as adequately safe. The design problem is usually to prevent the mains supply from becoming accessible, either by breaking through to the low-voltage circuitry, making the chassis live, or some other defect developing.

9.2.1 Earthing and bonding

It follows from the previous discussion that if all objects which can conduct electricity are bonded together so that an individual cannot become connected between two points with a potential difference greater than 30 V, then the installation is probably safe. The pattern of earthing and bonding varies slightly with the type of electrical supply available. Figure 9.3 illustrates the situation which arises if UK practice is followed. The supply to the instrument system is derived from the conventional 440 V three-phase neutral earthed distribution system, the live side being fused. A chassis connection to the neutral bond provides an adequate fault path to clear the fuse without undue elevation of the instrument chassis. All the adjacent metalwork, including the handrail, is bonded to the instrument chassis and returned separately (usually by several routes) to the neutral star point. Any personnel involved in the loop as illustrated are safe, because they are in parallel with the low-resistance bond XX′ which has no significant resistance. If the bond XX′ were broken then the potential of the handrail

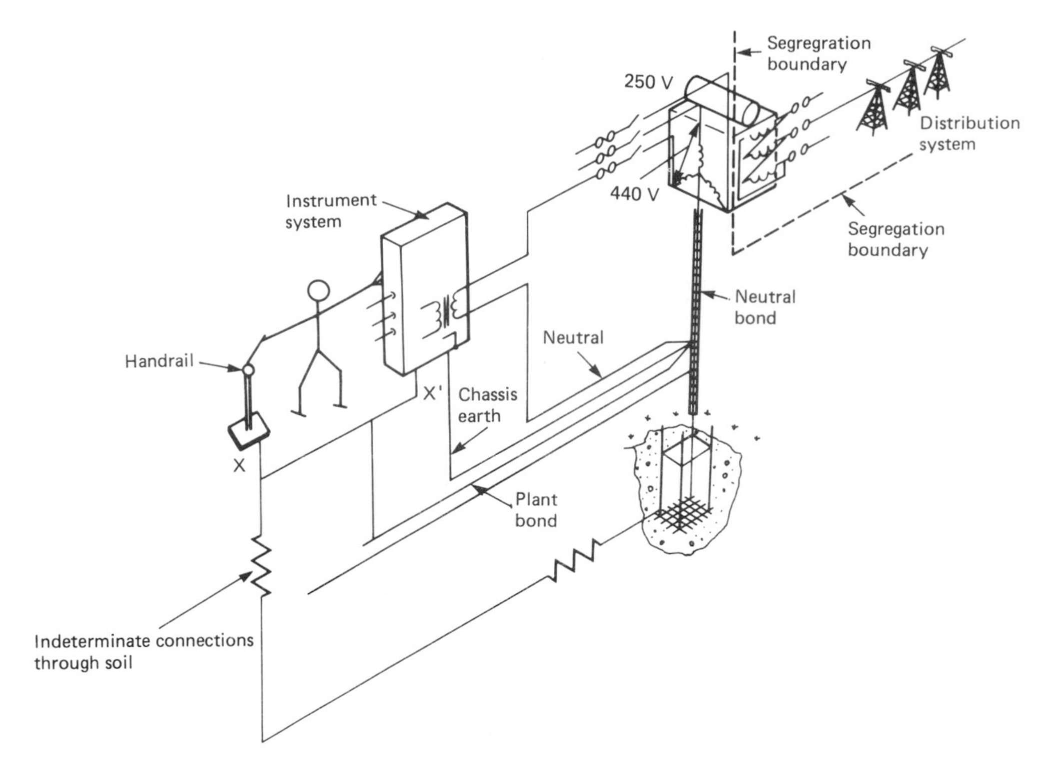

Figure 9.3 Normal UK installation with bonded neutral.

would be determined by the ill-defined resistance of the earth path. The instrument system would be elevated by the effect of the transient fault current in the chassis earth return, and the resultant potential difference across the human being might be uncomfortably high.

The fundamental earthing requirements of a safe system are therefore that there should be an adequate fault return path to operate any protective device which is incorporated, and that all parts of the plant should be bonded together to minimize potential differences.

There are, however, a number of circumstances where earthing is not used as a means of ensuring protection. Large quantities of domestic portable equipment are protected by 'double insulation', in which the primary insulation is reinforced by secondary insulation and there would need to be a coincident breakdown of two separate layers of insulation for danger to arise. Similarly, some areas for work on open equipment are made safe by being constructed entirely of insulating material, and the supplies derived from isolating transformers so as to reduce the risk of electrocution.

Where the environment is harsh or cables are exposed to rough treatment there is always the need to reduce working voltage, and there are many variants on the method of electrical protection, all of which have their particular advantages. Figure 9.4 shows the type of installation which is widely used in wet situations and, provided that the tools and cables are subject to frequent inspection, offers a reasonable level of protection.

The transformer is well designed to reduce the available voltage to 110 V which is then centre tapped to earth, which further reduces the fault voltage to earth to 55 V. Both phases of the supply are fused but a more sensitive detection of fault current is achieved by using an earth leakage circuit breaker (ELCB) which monitors the balance of the phase currents and if they differ by more than 20 mA triggers the circuit breaker. This sensitive fast detection combined with the lower voltage produces a reasonably safe system for most circumstances.

There are therefore many different techniques for reducing electrical shock risk. They all require consideration to be given to the nature of the supply, the design of the equipment, the environment, use, the method of installation and the frequency and effectiveness of inspection. These factors all interact so strongly that any safe installation must consider all these aspects.

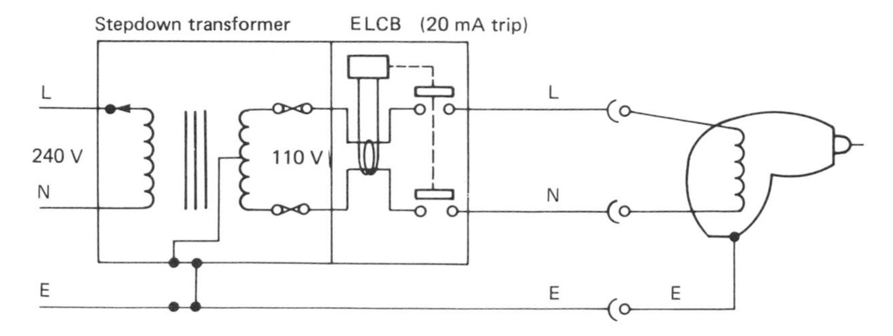

Figure 9.4 Isolating transformer supplying 110 V centre tapped to earth with earth leakage circuit breaker.

9.3 Flammable atmospheres

A large proportion of process control instrumentation is used in the petrochemical industry, where there is a possible risk of explosion if the equipment comes into contact with a flammable atmosphere. In practice, similar risks occur in all petrochemical and gas distribution sites, printing works, paint-spray booths and the numerous small stores of varnish, paint and encapsulating compounds which exist on most manufacturing sites.

The other related risk is that of dust explosions, which tend to attract less interest but are possibly more important. Almost any finely divided material is capable of being burned (most people are familiar with the burning steelwool demonstration) and, in particular, finely divided organic substances such as flour, sugar and animal feedstuffs all readily ignite. Dust explosions tend to be dramatic, since a small explosion normally raises a further dust cloud and the explosion rolls on to consume the available fuel. However, in general dusts need considerably more energy than gas to ignite them (millijoules rather than microjoules) and are usually ignited by temperatures in the region of 200°C. Frequently the instrumentation problem is solved by using T4 (135°C) temperature classified intrinsically safe equipment in a dust-tight enclosure.

The basic mechanism of a gas explosion requires three constituents: the flammable gas, oxygen (usually in the form of air) and a source of ignition (in this context an electrical spark or hot surface). A gas–air mixture must be mixed in certain proportions to be flammable. The boundary conditions are known as the lower and upper flammable limits, or in some documents the lower and upper explosive limits. The subject of explosion prevention concentrates on keeping these three constituents from coming together. The usual approach is to attempt to decide on the probability of the gas–air mixture being present and then to choose equipment which is protected adequately for its environment.

The study of the probability of gas–air mixture being present within the flammable limits is called 'area classification', and is without doubt the most difficult aspect of this subject. Expertise on all aspects of the plant and the behaviour of the gases present is required to carry out area classification well, and hence it is usually done by a committee on which the instrument engineer is only one member. Present practice is to divide the hazardous area according to the IEC Standard 79-10, as follows:

Zone 0: in which an explosive gas–air mixture is continuously present or present for long periods.
(*Note*: The vapour space of a closed process vessel or storage tank is an example of this zone.)

Zone 1: in which an explosive gas–air mixture is likely to occur in normal operation.

Zone 2: in which an explosive gas–air mixture is not likely to occur, and if it occurs it will only exist for a short term.

By inference, any location which is not a hazardous area is a safe area. Many authorities prefer the use of 'non-hazardous area', for semantic and legalistic reasons. The use of 'safe' is preferred

Table 9.1 Temperature classification

Class	*Maximum surface temperature (°C)*
T1	450
T2	300
T3	200
T4	135
T5	100
T6	85

in this document since it is a shorter, more distinctive, word than 'non-hazardous'.

American practice is still to divide hazardous areas into two divisions. Division 1 is the more hazardous of the two divisions and embraces both Zone 0 and Zone 1. Zone 2 and Division 2 are roughly synonymous.

The toxicity of many industrial gases means that an analysis of a plant from this aspect must be carried out. The two problems are frequently considered at the same time.

Having decided the risk of the gas being present then the nature of the gas from a spark ignition or flame propagation viewpoint is considered.

One of the better things that has happened in recent years is the almost universal use of the IEC system of grouping apparatus in a way which indicates that it can safely be used with certain gases. Pedantically, it is the apparatus that is grouped, but the distinction between grouping gases or equipment is an academic point which does not affect safety. The international gas grouping allocates the Roman numeral I to the underground mining activity where the predominant risk is methane, usually called firedamp, and coal dust. Historically, the mining industry was the initial reason for all the work on equipment for flammable atmospheres, and it retains a position of considerable influence. All surface industry equipment is marked with Roman numeral II and the gas groups are subdivided into IIA (propane), IIB (ethylene) and IIC (hydrogen). The IIC group requires the smallest amount of energy to ignite it, the relative sensitivities being approximately 1:3:8. The representative gas which is shown in parentheses is frequently used to describe the gas group.

This gas classification has the merit of using the same classification for all the methods of protection used. The boundaries of the gas groupings have been slightly modified to make this possible.

Unfortunately, the USA and Canada has opted to maintain their present gas and dust classification. The classifications and subdivisions are:

CLASS I:	Gases and vapours
	Group A (acetylene)
	Group B (hydrogen)
	Group C (ethylene)
	Group D (methane)
CLASS II:	Dusts
	Group E (metal dust)
	Group F (coal dust)
	Group G (grain dust)
CLASS III:	Fibres
	(No subgroups)

Gas–air mixtures can be ignited by contact with hot surfaces, and consequently all electrical equipment used in hazardous atmospheres must be classified according to its maximum surface temperature. BS 4683: Part 1 is the relevant standard in the United Kingdom, and this is almost identical to IEC 79-8. The use of temperature classification was introduced in the United Kingdom quite recently (late 1960s), and one of the problems of using equipment which was certified prior to this (e.g. equipment certified to BS 1259) is that somehow a temperature classification has to be derived.

For intrinsically safe circuits the maximum surface temperature is calculated or measured, including the possibility of faults occurring, in just the same way as the electrical spark energy requirements are derived. The possibility that flameproof equipment could become white hot under similar fault conditions is guarded against by generalizations about the adequate protective devices. All temperature classifications, unless otherwise specified, are assessed with reference to a maximum ambient temperature of 40°C. If equipment is used in a temperature higher than this, then its temperature classification should be reassessed. In the majority of circumstances regarding the temperature classification as a temperature-rise assessment will give adequate results. Particular care should be exercised when the 'ambient' temperature of a piece of apparatus can be raised by the process temperature (e.g. a pilot solenoid valve thermally connected to a hot process pipe). Frequently, equipment has a specified maximum working temperature at which it can safely be used, determined by insulating material, rating of components, etc. This should not be confused with the temperature classification and both requirements must be met.

When the probability of gas being present and the nature of gas has been established then the next step is to match the risk to the equipment used. Table 9.2 shows the alternative methods of protection which are described in the CENELEC standards and the areas of use permitted in the United Kingdom.

In light current engineering the predominant technique is intrinsic safety, but flameproof and increased safety are also used. The flameproof technique permits the explosion to occur within the enclosure but makes the box strong enough and controls any apertures well enough to prevent the explosion propagating to the outside atmosphere. Increased safety uses superior construction techniques and large derating factors to reduce the probability of sparking or hot spots occurring to an acceptable level. The other technique which is used to solve particular problems is pressurization and purging. This achieves safety by interposing a layer of air or inert gas between the source of ignition and the hazardous gas.

Table 9.2 Status of standards for methods of protection (as at January 84)

Technique	*IEC symbol Ex*	*Standard*			*UK code of practice part of BS5345*	*Permitted zone of use in UK*
		IEC 79–	*CENELEC EN 50*	*BRITISH BS 5501 Part*		
General requirement		Draft	014	1	1	
Oil immersion	o	6	015	2	None	2
Pressurization	p	2	016	3	5	1 or 2
Powder filling	q	5	017	4	None	2
Flameproof enclosure	d	1	018	5	3	1
Increased safety	e	7	019	6	6	1 or 2
Intrinsic safety	ia	3 Test apparatus	020 apparatus	7	4	0 ia
	or ib	11 Construction	039 system	9		1 ib
Non-incendive	n(N)	Voting draft	021 (Awaits IEC)	BS 4683 Pt3	7	2
Encapsulation	m	None	028 (Voting draft)	None	None	1
Special	s	None	None	SFA 3009	8	1

Where it can be used, intrinsic safety is normally regarded as the technique which is relevant to instrumentation. Intrinsic safety is a technique for ensuring that the electrical energy available in a circuit is too low to ignite the most easily ignitable mixture of gas and air. The design of the circuit and equipment is intended to ensure safety both in normal use and in all probable fault conditions.

There is no official definition of intrinsic safety. EN 50 020, the relevant CENELEC apparatus standard, defines an intrinsically safe circuit as:

> A circuit in which no spark or any thermal effect produced in the test conditions prescribed in this standard (which include normal operation and specified fault conditions) is capable of causing ignition of a given explosive atmosphere.

There are now two levels of intrinsic safety: 'ia' being the higher standard where safety is maintained with up to two fault and 'ib', where safety is maintained with up to one fault. Equipment certified to 'ib' standards is generally acceptable in all zones except Zone 0, and 'ia' equipment is suitable for use in all zones.

Intrinsic safety is, for all practical purposes, the only acceptable safety technique in Zone 0 (continuously hazardous) and the preferred technique in Zone 1 (hazardous in normal operation).

This technique is frequently used in Zone 2 (rarely hazardous) locations to ease the problems of live maintenance, documentation and personnel training. Intrinsic safety is essentially a low-power technique, and hence is particularly suited to industrial instrumentation. Its principal advantages are low cost, more flexible installations and the possibility of live maintenance and adjustment. Its disadvantages are low available power and its undeserved reputation of being difficult to understand. In general, if the electrical requirement is less than 30 V and 50 mA, then intrinsic safety is the preferred technique. If the power required is in excess of 3 W or the voltage greater than 50 V, or the current greater than 250 mA, the probability is that some other technique would be required. The upper limit is a rash generalization, because, with ingenuity, intrinsically safe systems can safely exceed these limits. Between these two sets of values intrinsically safe systems can frequently be devised.

When there is interconnection between more than one intrinsically safe apparatus, an analysis of the interactions and their combined effect on safety reveals that intrinsic safety is essentially a system concept. It can be argued that the other techniques rely on correct interconnection and the choice of the method of electrical protection. For example, a flameproof motor depends for its safety on having correctly rated switchgear for starting overload and fault protection, adequate provision for earthing and a satisfactory means of isolation, all of which constitute a system. However, the danger resulting from the failure of unsatisfactory safe-area equipment in an intrinsically safe system is more immediate and obvious, and hence there is a requirement for a more detailed consideration of all safety aspects which results in a system certificate and documentation. Where a system comprises intrinsically safe apparatus in the hazardous area and a certified source of power and receiving apparatus in the safe area, then the combination can be assessed against the CENELEC system standard EN 50 039. The agreed term for equipment intended for mounting in the safe area which is certified as having terminals which may be connected to the hazardous area is 'associated electrical apparatus'. This inelegant and quite forgettable expression is

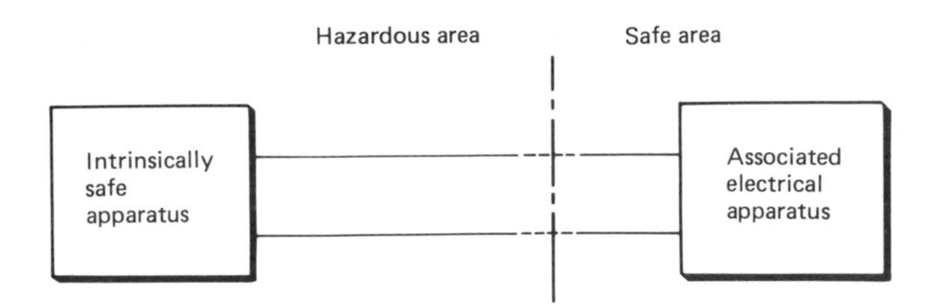

Figure 9.5 System with certified safe area equipment (associated apparatus).

very rarely used by anyone other than writers of standards, but it does distinguish certified safe-area equipment from equipment which can be mounted in the hazardous area.

Where an instrument loop is relatively simple, self contained and comprises the same equipment in the majority of applications, then it is usual for both the hazardous-area and safe-area equipment to be certified, and a system certificate for the specific combination to exist as illustrated in Figure 9.5.

In practice, there are only a few completely self-contained circuits, since the signal to or from the hazardous area is usually fed into or supplied from complex equipment. In these circumstances there is no real possibility of certifying the safe-area apparatus since it is complex, and there is a need to maintain flexibility in its choice and use. The solution in these circumstances is to introduce into the circuit an intrinsically safe interface which cannot transmit a dangerous level of energy to the hazardous area (see Figure 9.6). The majority of interfaces are designed to be safe with 250 V with respect to earth applied to them (i.e. the 440 three-phase neutral earth system commonly used in the United Kingdom).

Whatever the cause of the possible danger and the technique used to minimize it, the need to assess the risk, and to document the risk analysis and the precautions taken, is very important. There is a legal requirement to produce the documentation. There is little doubt that if the risks are recognized and documentary proof that they have been minimized is established, then the discipline involved in producing that proof will result in an installation which is unlikely to be dangerous and is infinitely easier to maintain in a safe condition.

9.4 Other safety aspects

The level of integrity of any interlock or instrument system depends upon the importance of the measurement and the consequences of a failure. It is not surprising that some of the most careful work in this area has been related to the control of atomic piles and similar sources of potential catastrophic failure. The majority of systems are less dramatic, and in the United Kingdom an excellent Code of Practice, BS 5304: 1975 discusses the techniques generally used for safeguarding machinery in non-hazardous circumstances. The general principles to be applied can be summarized as:

(1) The failure of any single component (including power supplies) of the system should not create a dangerous situation.

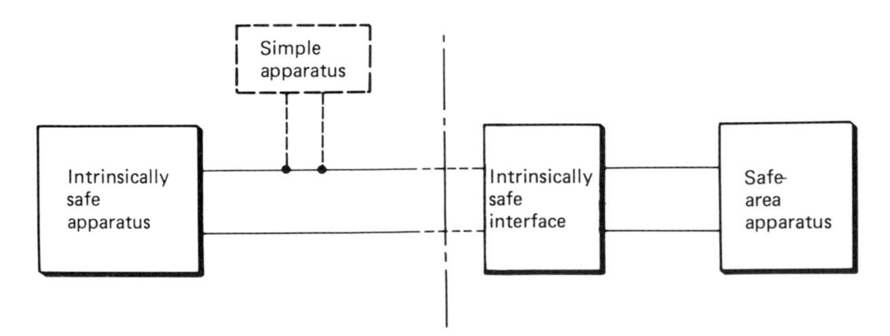

Figure 9.6 System with certified intrinsically safe interface.

(2) The failure of cabling to open or short circuit or short circuiting to ground of wiring should not create a dangerous situation. Pneumatic or electro-optic systems have different modes of failure but may have particular advantages in some circumstances.
(3) The system should be easily checked and readily understood. The virtue of simplicity in enhancing the reliability and serviceability of a system cannot be overstressed.
(4) The operational reliability of the system must be as high as possible. Foreseeable modes of failure can usually be arranged to produce a 'fail-safe' situation, but if the system fails and produces spurious shutdowns too frequently the temptation to override interlocks can become overwhelming. An interlock system to remain credible must therefore be operationally reliable and, if possible, some indication as to whether the alarm is real or a system fault may also be desirable.

These basic requirements, following up a fundamental analysis of the level of integrity to be achieved, form a framework upon which to build an adequate system.

9.5 Conclusion

It is difficult to adequately summarize the design requirements of a safe system. The desire to avoid accidents and in particular avoiding injuring and killing people is instinctive in the majority of engineers and hence does not need to be emphasized. Accident avoidance is a discipline to be cultivated, careful documentation tends to be a valuable aid and common sense is the aspect which is most frequently missing.

The majority of engineers cannot experience or have detailed knowledge of all aspects of engineering, and safety is not different from any other factor in this respect. The secret of success must therefore be the need to recognize the danger so as to know when to seek advice. This chapter has attempted to provide the background for recognizing the need to seek expert advice; it is not comprehensive enough to ensure a safe design.

9.6 References

BASS, H.G. *Intrinsic Safety*, Quartermaine House, Gravesend, Kent (1984)

COOPER, W. F. *Electrical Safety Engineering*, Butterworths, London (1978)

Electrical Safety in Hazardous Environments, Conferences, Institution of Electrical Engineers (1971), (1975) and (1982)

GARSIDE, R. H. *Intrinsically Safe Instrumentation: A Guide*, Safety Technology (1982). Predominantly applications, strong on UK and US technology and standards

HALL, J. *Intrinsic Safety*, Institution of Mining Electrical and Mining Mechanical Engineers (1985). A comprehensive treatise on mining applications of the art

ICI Engineering Codes and Regulations, ROSPA PUBLICATION No. IS 91. Now unfortunately out of print. Slightly dated but the most useful publication in this area. Beg, borrow or steal the first copy you find. Essential

MAGISON, E. C. *Electrical Instruments in Hazardous Locations*, 3rd edn. Instrument Society of America (1978). Comprehensive book portraying American viewpoint.

OLENIK, H. *et al. Explosion Protection Manual*, 2nd edn, Brown Boveri & Cie (1984). An excellent book on West German practice

REDDING, R. J. *Intrinsic Safety*, McGraw-Hill, New York (1971). Slightly dated but still relevant

ROBENS, LORD (chairman). *Safety and Health at Work*, Report of the Committee HMSO Cmnd. 5034 (1972)

Safety in Universities—Notes for Guidance, Association of Commonwealth Universities (1978)

Substances Hazardous to Health, Croner Publications, New Malden, Surrey (1986 with updates)

TOWLE, L. C. *Intrinsically Safe Installations on Ships and Offshore Structures*, Institute of Marine Engineers TP 1074 (1985)

Many British Standards have to do with safety. The following are the most relevant.

Electrical Apparatus for Potentially Explosive Atmospheres, Part 1: General Requirements BS 5501: Part 1: 1977 EN 50 014, British Standards Institution. Largely excluded by EN 50 020 but essential for completeness.

Part 3: Pressurized Apparatus 'p', BS 5501: Part 3: 1977 EN 50 016, British Standards Institution. Interlocks, etc. sets intrinsic safety problems. Useful.

Part 6: Increased Safety 'e', BS 5501: Part 6: 1977 EN 50 019, British Standards Institution. Useful information on terminals creepage and clearance

Part 7: Intrinsic safety 'i', BS 5501: Part 7: 1977 EN 50 020, British Standards Institution. Sets requirements for apparatus. Interpretations should also be read. Essential

Part 9: Specification for Intrinsically Safe Electrical Systems 'i', BS 5501: Part 9: 1982 EN 50 039, British Standards Institution. Sets requirements for combinations of apparatus. Second edition will follow soon to meet Group I political requirements. Essential

Selection, installation and maintenance of electrical apparatus for use in potentially explosive atmospheres (other than mining applications or explosive processing and manufacture).

Part 1: Basic Requirements for all Parts of the Code, BS 5345: Part 1: 1976, British Standards Institution. Includes essential definitions choice of equipment, etc. Table of gas characteristics. Essential

Part 2: Classification of Hazardous Areas, BS 5345: Part 2:

1983, British Standards Institution. Asks all the relevant questions but only gives a very few answers. Not much use but the best available. Useful
Part 4: Installation and Maintenance Requirements for Electrical Apparatus with Type of Protection 'i'. Intrinsically Safe Electrical Apparatus and Systems. BS 5345: Part 4: 1977, British Standards Institution. A bit dated, but the best current advice. Essential
Part 5: Installation and Maintenance Requirements for Electrical Apparatus Protected by Pressurization 'p', Continuous Dilution and Pressurized Rooms. BS 5345: Part 5: 1983, British Standards Institution. A useful technique for solving difficult problems; offers some suggestions
Part 6: Installation and Maintenance Requirements for Electrical Apparatus with Type of Protection 'e'. Increased Safety, BS 5345: Part 6: 1978, British Standards Institution. Not frequently used for instrumentation. The code of practice is very thin. Complementary
Part 11: Specific Industry Applications, BS 5345: Part 11, British Standards Institution. Draft contains many useful references
Code of Practice for Control of Undesirable Static Electricity, Part 1: General Considerations, BS 5958: Part 1: 1980, British Standards Institution. An interesting standard with fascinating sections on dusts, plastics, and human frailties. Essential
Code of Practice for Fire Precautions in Chemical Plant (formerly CP 3013), BS 5908: 1980, British Standards Institution. Standard which ranges widely through all aspects of subject. Appendices give a useful set of references. Controversial link between personnel risk and area classification. Useful
Safeguarding of Machinery (formerly CP 3004), BS 5304: 1975, British Standards Institution. A well-written and beautifully illustrated standard. Illustrates sound basic principles. Useful

10 Interface and backplane bus standards for instrumentation systems

E. G. Kingham

10.1 Introduction

Over the last decade the trend in instrument systems has been towards their assembly from a range of specific-purpose modules or building blocks. In a few cases the overall control of the assembly is vested in a purpose-designed controller, but more commonly a mini- or microcomputer is used for this purpose, with software written to suit the particular application. This modular approach has the advantages of enabling:

(1) Standard modules from a wide variety of sources to be used for a diverse range of applications;
(2) A system to be designed incorporating only the particular features required for the application;
(3) The system to be easily modified for a different performance or application; and
(4) A more speedy system design and assembly.

Such systems are based upon interfaces or backplane buses, of which a considerable number are now commercially available. They range from a number of self-contained instruments interconnected by a standard interface such as the RS 232 Serial connection or the IEC 625 parallel interface through modular systems contained in standard crates, like the CAMAC interface to card-based modules mounted in a backplane bus under the control of a processor.

In the various types there are often only minor differences in the electrical signals, timing and functional requirements of the different buses which, in the early designs, were often based upon similar processors. Nevertheless, these differences can be sufficient to prevent the use of modules from different suppliers and thus entrap a user with a particular manufacturer. Such a situation can have far-reaching consequences, depending on the technical and financial resources of the supplier and his ability to provide adequate levels of hardware and software support after purchase.

It is desirable to always consider the use of an interface or backplane bus standard supported by the international standards organizations. However, there are several in-house standards from companies of international repute which are, therefore, effectively *de facto* standards and for which modules or software can often be purchased from other sources. A knowledge of bus and interface specifications is necessary so as to be able to assess the suitability of a proposed system to carry out the required purpose. This chapter is not intended to be an extensive study of the topic, but sufficient information is given on the more commonly available interfaces and buses to enable an intelligent judgement to be made.

10.2 Principles of interfaces

The act of connecting two modules together creates the necessity for communicating paths and therefore establishes an interface. Within each module the components will communicate via the internal wiring in whatever form this exists, but for the modules to interact in a systematic manner some common intercommunication technique must be used.

These techniques fall into one of two broad classes: serial or parallel interfaces. In computer applications the latter are commonly called 'backplane buses' when applied to self-contained computer systems.

10.2.1 Serial interfaces

Serial interfaces require only one or two signal lines and data are transmitted as a stream of bits over one line. This has an advantage over parallel interfaces in that the number and cost of drivers and receivers to buffer the signals and process them is significantly reduced, as is the complexity of the interface connections. Converters from parallel to serial and back again are required in those applications where the data are presented or required in parallel form. Bi-directional lines for the transmission of both data and control signals are also necessary.

Serial interfaces connect commonly used, byte-orientated peripherals such as VDUs, printers and similar devices to the system, allowing the bit-by-bit transmission of data, addresses and control signals. They are also frequently used to transmit bus functions in networks. The relevant standards define features such as the data transmission rate, data format and electrical characteristics.

Interfaces often use an interlock handshake over the data flow control lines to confirm that data are only sent when the receiving device is ready and to confirm its correct receipt. Data handshaking is a sequence of interlocked signals, ensuring that each device waits for an acknowledgement of its last transmission before proceeding to the next stage. This protocol is normally a software function, as is parity generation and checking. The technique, which is illustrated in Figure 10.1 for a point-to-point application such as an interface line in a radial configuration (see Figure 10.3), can also be applied to parallel buses.

In all system applications care must be taken to cater for the condition where no response to a transmission command is received. To ensure that the entire system does not become locked up (hung up), and to indicate the occurrence of an error, provision, such as a timer mechanism, must be made.

The principal disadvantages of serial interface systems are a slow rate of data transfer, especially when compared with parallel transmission, the need to physically change wiring should the system require alterations and its inherent inability to incorporate features such as DMA.

Serial interfaces may be either synchronous or asynchronous (including isochronous). In the former case the data bit rate is determined by clocks in the transmitter and receiver and the data are therefore transmitted at a fixed rate. The clocks must be kept in synchronism with each other by transmitting a clock-synchronizing signal with the data (i.e. self-checking coding), or over a separate line.

Asynchronous systems are commonly used for low-speed terminals (< 1200 bits/s). They operate only when data are to be transmitted and the data characters are preceded and followed by start and stop framing bits. Such systems tend to be less efficient in their use of bus space than synchronous transmission because of the extra start and stop bits and the minimum time necessary between characters, even when using isochronous transmission.

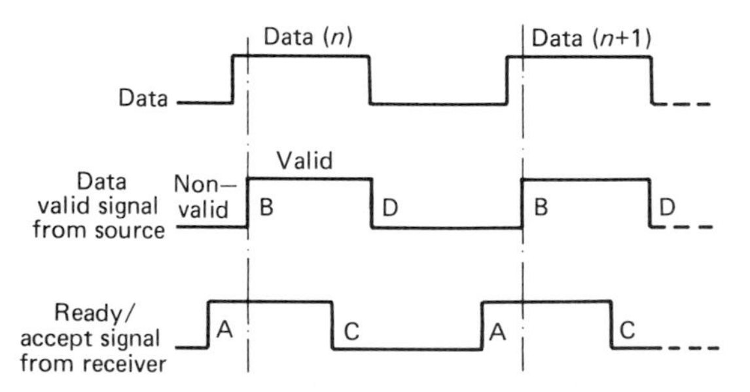

Figure 10.1 Handshake signals and timing sequence. A: ready to accept DATA; C: DATA accepted; A, B, C, D: full handshake sequence

10.2.2 Serial interface standards

The most common serial standards relevant to our interests are the ASCII and EBDIC standards and the CCITT V24 (EIA RS-232-C) synchronous standards. ASCII or EBDIC asynchronous transmissions are commonly used by most VDUs and teletypewriters. Eight serial data bits, preceded by the start bit and concluded by the stop bit(s), form a character and are sent over a cable. By the use of inexpensive parallel-to-serial and serial-to-parallel converters, i.e. universal asynchronous or synchronous receivers/transmitters (UART and USART), a simple twisted pair or a coaxial cable can be used for links up to 1 or 2 km in length at data rates of a few kbits/s. For longer distances modems must be used.

RS-232-C has been described in an earlier chapter. However, a brief description is also included here for the sake of completeness.

10.2.3 The RS-232-C interface

The Electric Industry Association (EIA) Standard RS-232-C is the most commonly used interface for bit-serial data transmission and is the most frequently available port provided in small computer systems. It is virtually identical to the CCITT Rec. V24 interface standard when used for these applications, and both were originally established to define the electrical physical and functional interface requirements of telecommunication authorities for connecting Data Terminal Equipment (DTE), Data Control Equipment (DCE) and modulators and demodulators (MODEMS). It is used as an asynchronous serial interface where data are transmitted over a pair of wires using nominal voltages of −12 V and +12 V driven, respectively, to represent 'mark' and 'space' (between −5 to −12 V and +5 to +12 V driven, respectively). Additional lines are used to carry data flow, control handshaking and other control signals and a 25-pin 'D' type connector is defined.

In practice, for the simplest bi-directional links only two lines, Transmitted Data and Received Data, are needed in addition to Signal Ground. Handshaking lines, Request to Send, Data Flow Control and Clear to Send are desirable but not always supported in all microcomputers. Data transfer rates up to 20 kbits/s are possible and standard data rates are:

50, 75, 110, 150, 350, 600, 1200, 2400, 4800, 9600, 19200 bits/s

Data transmitted over RS-232-C interfaces may use any convenient code, but ASCII is commonly used for computer systems.

EIA RS-232 has been supplemented by EIA RS-449,

RS-422 and RS-423. These later standards are designed to permit higher data rates (< 2 Mbits/s) and the use of longer lines. RS-449, which corresponds to a subset of CCITT Rec. V24, defines the mechanical and functional characteristics for synchronous and asynchronous serial binary data communication systems. Electrical characteristics for balanced and unbalanced circuits are specified in RS-422 (CCITT Rec. V11) and RS-423 (CCITT Rec. V10), respectively, and a 37-pin connector is used. A fuller description is given in Chapter 4 in Part 5.

Despite these more recent standards, RS-232-C has such extensive use that it seems likely that some considerable time will elapse before its eventual demise.

10.2.4 Current loop

In cases where transmission is required over longer distances or in particularly noisy environments it is also possible to use a current loop system. This consists basically of a total of two pairs of wires, each pair carrying serial data in Full Duplex Traffic mode (FDX). A loop possesses a transmitter at one end and a receiver at the other. The arrangement is reversed for the other loop, thus permitting the system to exchange information. In addition, a current generator is necessary in each loop and this may be incorporated into the transmitter, which controls the levels of the current. Additional receivers can be added into the system without complexity.

There is no clearly defined and internationally recognized standard available for current loop systems; however, the currents are normally 20 mA and the system is therefore often known as the '20 mA current loop'. Optocouplers are available from several manufacturers and are invariably used, allowing ground loops to be eliminated and common mode rejection to be greatly enhanced. Data rates of 20 kbits/s and transmission distances of 10 km are possible using these devices over suitable cables. Teleprinters normally operate with a 20 mA drive current, and the 20 mA current loop sees frequent usage for this purpose.

10.2.5 Hewlett–Packard interface loop (HP-1L)

To conclude our review of serial interfaces we should mention the Hewlett–Packard Interface Loop (HP-IL). This is a bit-serial interface particularly designed for low-cost systems, which may be composed of battery-supplied devices. It is designed to complement the IEC 625 (HP-IB) interface system, which is described shortly.

The HP-IL system makes use of a computer- or

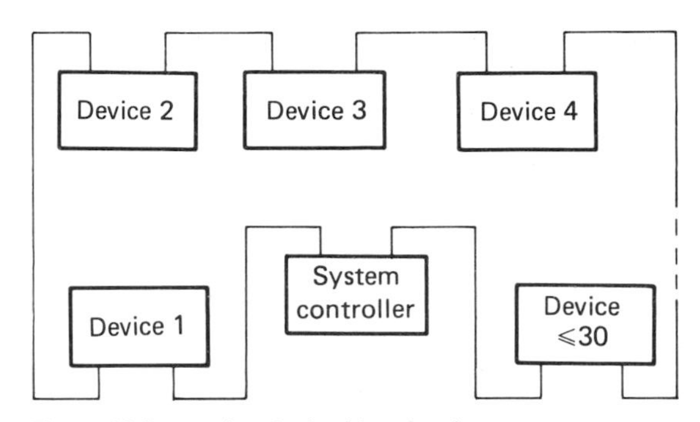

Figure 10.2 Hewlett–Packard interface loop

microprocessor-based device to act as a system controller and to manage the loop, controlling the devices connected to it. Up to 30 devices can be connected in the loop, as shown in Figure 10.2. Connections are effected by two-wire cables, running from the output port of one device to the input port of the next, the whole assembly comprising a closed loop as shown. The arrangement permits addresses, device capability identification, power on/off and error checking to be performed.

Of these, perhaps only the device capability identification needs clarifying. Each device possesses a unique capability number that identifies to the system controller the features that the device can undertake, e.g. 'print', etc. In executing a 'print' command, the controller polls each device to find the one that will respond with the appropriate identity. This procedure avoids the need for the user to know the address of each device and also simplifies the production and operation of system software and management, enabling devices of different speeds to operate in a unified manner.

The system is not the subject of an international standard. An increasing number of devices, varying from a hand-held computer to a digital multimeter, are manufactured to operate with the system and interfaces are available to connect to other systems such as IEC 625 (HP-IB) and RS-232-C.

10.3 Data-link control protocol

Protocol is a set of rules to be followed in operating a communication link or system. It covers aspects such as the detection and correction of errors, the sequence of message transmissions, the framing of characters, the control of the transmission line, etc. A number of protocols are in common usage including the ISO High Level Data Link Control (HLDC) and the more commonly met IBM Synchronous Data Link Control (SDLC).

10.3.1 Synchronous data-link control

Synchronous serial transmission has a number of sophisticated protocol variations under the general heading of Synchronous Data-Link Control (SDLC) when used in distributed systems requiring faster data links. Data are sent without start or stop bits being added to each character, and means are required to avoid the risk of the link receiver losing its timing when data are sent intermittently in this way. This is achieved by adding synchronizing characters every hundred bits; as these characters are decoded by the receiver, they can be used to keep its clock in synchronism with that of the transmitter. In SDLC practice, a number of transmitters and receivers may be connected to a link and data transmitted in large blocks of characters, called 'frames', each of which has a number of fields of one or more bytes of data. The start character indicates to all the receivers that a frame transmission has begun and an address follows. The appropriate receiver accepts the frame, checks it and returns a frame to confirm its accuracy or to highlight an error that has been detected.

A standard defines the protocol and covers such aspects as check characters, types of frame, system start/stop and error characters. Integrated circuits providing all the necessary protocol and other features are available from several manufacturers.

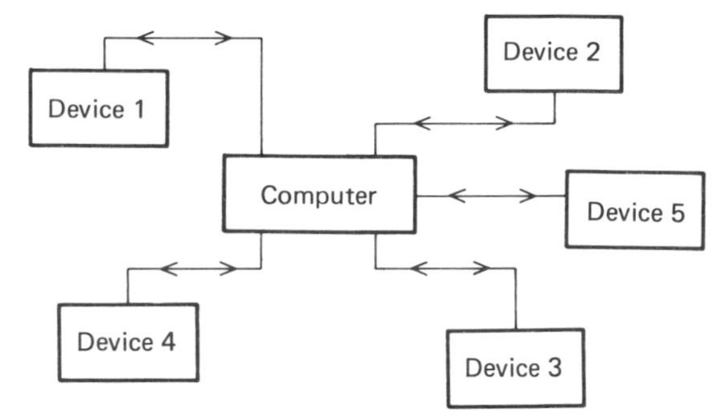

Figure 10.3 Radial or star connection

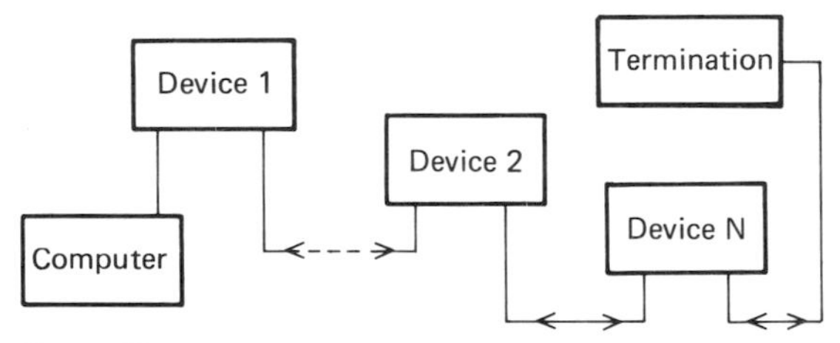

Figure 10.4 Daisy-chain or highway connection

10.4 Parallel interfaces and buses

In parallel interfaces the bits comprising data words are transmitted simultaneously over the interface on parallel lines. One line is used for each bit and an interface may be one or more words in width. Because the transfer is a total transfer of the word in a single cycle, the parallel interface operates in a fast manner, particularly in comparison with serial interfaces. However, in contrast to the serial interface, transceivers are required for each line.

Computers may use a radial or star form of interconnection, as illustrated in Figure 10.3, where each device connection is achieved by a dedicated, parallel interface. Such an approach is expensive for the reasons mentioned above, and is now used only rarely. A more common method is the 'daisy chain' or 'highway', illustrated in Figure 10.4. With this technique, only one driver and receiver are necessary in the computer.

During the 1960s several interface recommendations and standards began to emerge, using the above configurations. The objective of these proposals was to enable the interfacing of devices or instruments – supplied by various manufacturers for diverse applications – to each other and to the control unit, thus allowing easy expansion or modification of the resulting measuring system as the measurement requirements changed. The control unit could be a purpose-designed controller, a calculator or some form of computer. Several standard interfaces have emerged and are described below.

10.4.1 The BSI interface (BS 4421: 1969)

The British Standard Interface (BSI) system was amongst the first to be adopted as a standard (BS 4421: 1969). This uses a one-way, serial byte, asynchronous data traffic operational system. The BSI interface is a 'point-to-point' connecting system, i.e. only two devices, the SOURCE and the ACCEPTOR, can be connected to the interface. The roles of the two devices cannot be interchanged, owing to the one-way data traffic; if two-way data traffic is required, then two BSI interfaces, each complete with their interconnecting cables, must be provided between the two devices to function in the two opposite directions.

There are 18 signal lines in the BSI interface; of these, eight are data lines, the remainder providing certain simple control functions. As a consequence of its structure described above, instrument addressing or interrupt signals are not present in the BSI specification. The BSI system has not become widespread in practical applications, primarily because the electronic parameters of the system are not compatible with modern integrated circuit technology.

10.4.2 The SIAK interface

The SIAK interface was developed within the COMECON countries and was initiated by East Germany around the same time as was the BSI interface. Similar to that of the BSI system, it was a simple structure. The organization and timing of data transfer are less restricted in this system, owing to an arbitrary word length and asynchronous operation. Even so, this system has not gained widespread application.

10.5 Parallel bus

The parallel bus is commonly used in systems and is illustrated in Figure 10.5. This provides a common means of communication between a wide range of system modules, such as single-board computers, memory, digital and analogue I/O and controllers for peripheral devices. The bus structure must therefore accommodate all the necessary signals to allow the various system components to interact with each other.

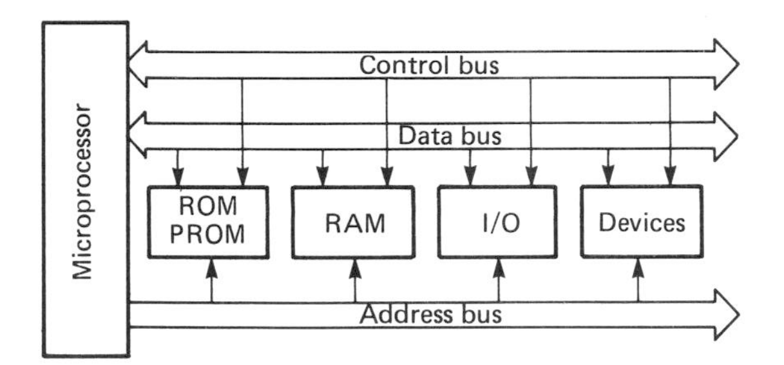

Figure 10.5 Typical parallel-bus system

The parallel bus transfers all the bits in a data word simultaneously across individual lines. Separate groups of these lines are allocated for the data bus, address bus and control bus and a typical, simple microcomputer will comprise eight data, 16 address and five (or more) control lines.

Two forms of the parallel bus exist, (1) the flexible, multi core cable connecting individual devices, and (2) the self-contained and integrated system centred around a backplane bus. The most popular form of the first type is the IEC interface system.

10.5.1 The IEC interface system (IEC 625-1)

The IEC interface is otherwise known as IEEE 488, HP-IB, GP-IB, IEC Bus or the ASCII bus. It is unique in that it was designed to ensure acceptance by both instrument users and manufacturers alike. In particular, it allows a number of programmable but free-standing instruments or devices to be interfaced so as to form a flexible and controllable data system. The instruments, which may come from different manufacturers, are able to be used alone for their intended purpose, e.g. as a voltmeter, frequency generator, etc. Alternatively, the inclusion of the standardized interface into their design allows them to be interconnected and controlled to a programmed requirement by a master device or controller.

The system uses byte-serial, bit-parallel means to transfer data in an asynchronous mode among the group of instruments. The individual devices can have three different functions, with respect to the interface system:

(1) *Talker:* These devices can be selectively addressed via an interface message and in their addressed state they can generate and transmit device-dependent messages to the bus. In a system only one device can be in the active talker state at any one time, and the message on the bus can come only from that device.
(2) *Listener:* These devices can be selectively addressed via an interface message and in their addressed (active) state can receive device-dependent messages from the bus. In a system several devices can be in the listener state simultaneously.
(3) *Controller:* These devices are capable of selectively addressing the other devices in the system: in other words, they can allocate the talker and listener roles and can transmit other interface messages to all or only some addressed devices of the system. Controllers can function in a controller mode only or also talk and listen. Those capable of control only do not transmit or receive device-dependent messages.

The roles of the devices in the operation of the interface system can be permanent or variable. Every device has at least one of the roles mentioned above. However, very often a single device can function in a different way at different times.

Up to 15 devices can be connected to the IEC interface system, which makes use of a cable bus consisting of 16 signal lines, grouped as follows:

- *data bus* (eight signal lines) DB
- *data transfer control bus* (three signal lines) DTCB
- *general control bus* (five signal lines) GCB

The signal line allocations are illustrated in Figure 10.6. The DTCB regulates the transfer of data bytes on the data bus from an addressed talker or from the controller to all listeners. The different devices each operate at various data transfer rates and the effective operational speed of the system must be regulated to

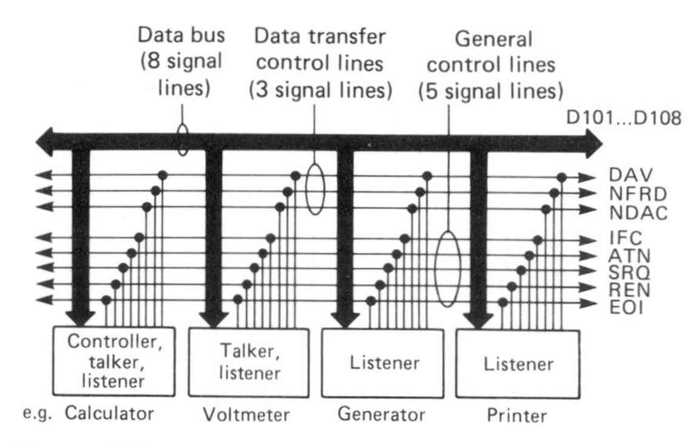

Figure 10.6 Structure of the IEC interface

that of the slowest listener. This is achieved by the asynchronous handshake cycle which takes place on the DTCB.

Signals on the GCB monitor the state of the system by resetting all devices, issuing 'attention' signals, responding to 'service requests', and enabling a device to assume remote control, etc. The system uses common screened connecting cables consisting of 16 signal lines and eight earth lines, with a combined maximum length of 20 m. Distances greater than this require the use of a proprietary bus extender which converts the data into serial form. Up to 15 instruments may be connected and a maximum data rate is 1 Mbytes/s with an effective rate of 200–500 kbytes/s.

Electrical signals are those of TTL for the driver and receiver parts of the interface system. All other parts can be built in any preferred technology. Signal levels are:

$$\text{High} = 0 \text{ (false)} > +2\text{V}$$

$$\text{Low} = 1 \text{ (true)} < 0.8\text{ V}$$

Two types of connector are specified in the American or European versions of the standard. The former, IEEE 488, uses the 24-pin Amphenol or Cinch series 57 Connector whilst the European IEC 625-1 uses the 25-pin trapezoidal-shaped cylindrical contact from the IEC 488 (Secretariat) 80A recommended list. Figure 10.7 shows the pin allocation of this IEC connector. Both specifications require a combined plug-and-socket connector to be used, thus permitting devices to be connected in piggyback fashion for either serial or radial systems.

The IEC 625-1 standard specifies only the interface dependent messages; a limited range of control and data transfer codes and formats are separately defined in a substantially device-free manner in IEC 625-2 and IEEE 728.

IEC interface system abridged specification

Type	Asynchronous, non-multiplexed
Addressability	31 talker and 31 listener addresses using primary (1st-byte) addressing; 961 and 961 using secondary (2nd-byte) addressing
Data transfer	Byte-serial, bit parallel, asynchronous
Connecting cable	Common screened, 16 signal and eight earth lines, <20 m total length or 2 m per device
Connector	Trapezoidal from IEC 48B, Secretariat 80A (MIL-C-24308)
Connection capacity	25 cylindrical pins
Signal level	TTL
Typical message length	10–20 characters
Data rate	Maximum speed 1 Mbyte/s Typical 200–500 kbytes/s
Maximum number of devices connected	15

10.6 Backplane buses

Backplane buses offer the same features as those already described for parallel buses but, because their physical properties are more precisely defined, it is possible to extend their performance in several ways. In some cases, and in particular in high-performance

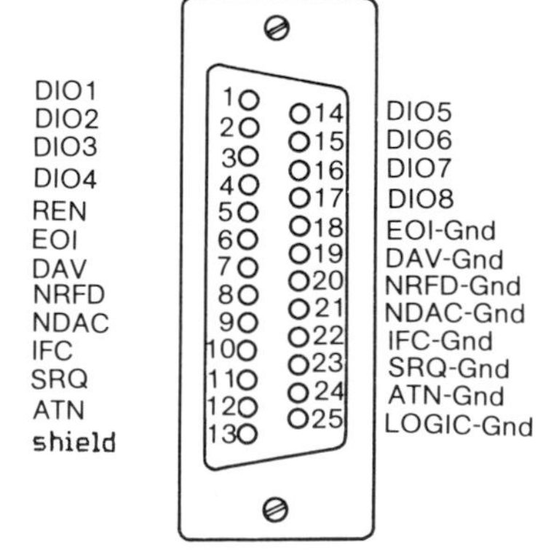

Figure 10.7 Wiring of the standard IEC interface connection

systems, the capacity of lines to accommodate more signals will be effectively increased by the use of multiplexing. Power lines will also be required. All this is achieved by the use of a motherboard technique, which forms the bus and comprises the bus lines, together with a means of mechanically accommodating and supporting the individual modules or

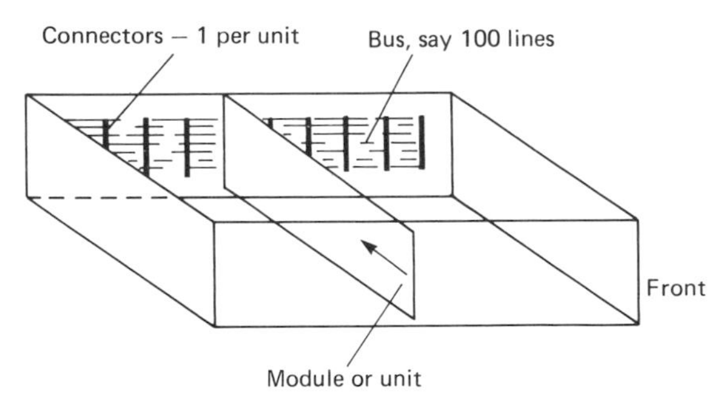

Figure 10.8 Schematic physical assembly of a backplane bus

device boards. This system is termed a 'backplane bus', and a basic example is shown in Figure 10.8.

The *data lines* are used for the transfer of all data in and out of the processor(s). The *address lines* identify the area of memory or I/O for which the transfer is intended. The *control lines* typically carry read or write instructions, address ready, wait or response signals, an interrupt signal or a DMA request.

Depending upon the particular system, the *power lines* may be common, with regulation provided at each device or individually regulated to each location.

10.6.1 Multiplexing

Buses using separate lines for addresses and data can transmit the next address whilst data are still being transmitted, thus allowing faster data transfers. However, the number of bus lines required for data and address can be reduced by the use of multiplexing when data and address information share the same bus lines in some prearranged sequence. For instance, address bus lines A_8 to A_{15} can carry the most significant 8 bits of the memory of I/O address, whilst the multiplexed data/address bus lines A_0 to A_7 carry the least significant 8 bits during the first clock cycle. Following this, during the subsequent two or three clock cycles, lines A_0 to A_7 become the data bus.

A multiplexed system thus requires two distinct transmissions together with synchronization facilities. Multiplexing has some cost and reliability benefits by virtue of the fewer components required, but the multiple transmissions slow down the system response.

10.6.2 Word length

An important feature of backplane buses is the size of word that they accommodate. The range of small computers currently available commonly use either 8- or 16-bit words, and the usefulness of a backplane bus is very dependent upon its ability to support either or both of these. However, 32-bit processors are now also becoming readily available and these can be supported by several of the buses currently available.

10.6.3 Interrupt priorities, multiprocessors and arbitration

The priority aspect of backplane buses is important because it determines which of the various devices on the bus can interrupt to gain control of the bus and at what time. Simple priority schemes depend upon the location of the cards in the bus relative to the microprocessor, those nearest receiving the highest priority.

More elaborate schemes are necessary for the more sophisticated buses which permit several 8- to 32-bit processors to share the bus. Such schemes generally use one microprocessor to act as a bus supervisor, monitoring requests from the various processors. These requests are made on separate bus-request lines and the grants made on corresponding separate bus-grant lines.

A priority system such as this requires a mechanism to arbitrate between simultaneous claims for control of the bus. Multiprocessor buses undertake this by queueing the requests – generally on the basis of sequential servicing. However, devices can share cycles on the bus or a particular device can take priority under stated conditions. Overall, no one device can keep permanent control of the bus and, if necessary, requesting devices can make use of a 'protest' line to initiate a software control response.

10.6.4 Errors and their detection

Information on the bus may become degraded by spurious signals from external circuits, crosstalk on the bus lines, mismatch in the bus, etc., leading to deterioration in an apparently satisfactory system.

Error detection and correction is necessary to prevent the complete disruption of the system and may be implemented by careful hardware design or by software. This may take the form of well-defined data test patterns, use of lateral or longitudinal parity checking on each data word or block, or by cyclic redundancy checks. The latter consists of carrying out a repetitive calculation on the data sequence being transmitted at each end (transmitter and receiver) of the transmission path and comparing these. It is a particularly powerful method, permitting a very low undetected error rate.

Table 10.1 Abridged Specifications for Backplane Buses

Bus	*Type*		*Address width or capacity*	*Data width (bits)*	*Card type & size (mm)*	*Connector*	*Connection capacity (No. of pins)*	*Power supplies (volts)*	*Interrupt levels*
	Synchronous	*Multiplexed*							
S100 (IEEE 696)	Yes	No	16 Mbytes 24 address lines	8 and 16	254 × 152.4 (max)	Edge	100	±8 ±16 (local regulation)	8 vectored
STD (IEEE P961)	Yes	No	64 kbytes	8	165 × 114	Edge	56	+5 +12	2
STE (IEEE P1000)	No	No	1 Mbyte (extended address memory)	8	Eurocard 160 × 100 (option 160 × 233)	DIN 41612	64	±5 ±12	4
G64	8-bit Yes 16-bit No	No No	16 (>256 kbytes)	8 and 16	Eurocard 160 × 100	DIN 41612	64	±5 ±12	3
Q (or Q22)	No	Yes	basic 64 kbytes extended 4 Mbytes	16	dual 132 × 226 quad 267 × 226	Double finger, edge	72	+5 +12	4
VAX BI	Yes	Yes	1 Gbytes for 30 bits	32	Double Eurocard 203 × 233	ZIF	120 (+180)	+5	4
Eurobus (ISO-DP 6951, BS 6475: (1984))	No	Yes	18 (10, 26, 34)	16 (8, 24, 32)	Eurocard 233 × 160	DIN 41612	64	+5	1 bus-line to arbiter
Multibus I (IEEE 796)	No	No	20 standard (1 Mbyte) 24 expanded (16 Mbytes)	8, 16	304.8 × 171.5	Edge	86 (+60)	+5 ±12	8
Multibus II (IEEE P1296)	Yes	Yes	32	32 (8, 16, 24)	Eurocard 100 × 220 233 × 220	DIN 41612	96	±5 ±12	
VME (IEEE P1014)	No	No	16 Mbytes (4 Gbytes with extended address)	16, 32	Eurocard 160 × 100 160 × 233	DIN 41612	96 (+96)	+5 ±12	8 assignable
Fastbus (IEEE 960)	No	Yes	32	32	367 × 400	AMP 2-532956 multi-sourced	130	+5 −5.2 −2	1
Futurebus (IEEE P896)	No	Yes	32 (4 Gbytes)	8, 16, 32	Eurocard preferred 280 × 367	DIN 41612	96 (+96)	+5	Virtual (by serial bus)

10.6.5 Bus length and termination

The physical length of the backplane bus is related to its performance and operating speed. In general, most reasonable performance direct buses are considerably less than 1 m in length, and the propagation time of the signal along the bus lines is highly relevant. In all cases the bus may be regarded as a transmission line, and proper terminations are necessary to avoid malfunctions due to mismatch, reflections, ringing, etc.

Bus handover	*Arbitration means*	*Error signals*	*Special cycles*	*Bandwidth (transfers)*	*Maximum board capacity*	*Some firms supplied by (see Part 6, 10.1)*
8 DMA	4 lines			6 MHz	22	205
1 DMA			Interrupt acknowledge	2 MHz		420
Multiple master	1 bus acknowledge, daisy-chain	Bus error	Read/Modify/Write, interrupt acknowledge, block transfer	2–3 MHz	21	17 120
Mono-processor with multiple DMA	1 bus acknowledge, daisy-chain	Parity, bus error	Interrupt acknowledge, Read/Modify/Write	8 bit: <1 MHz 16 bit: >1 MHz	20	173 353
	Master/slave	Parity, power failure		3.3 Mbytes/s	Typically 30	126 197
	Distributed	Parity		13.3 Mbytes/s	36	173
Multiple masters	2 (arbiter)	Cycle abort, reset, power fail	Read/Write, Read/Write/Hold, Vector, Read/Write/Vector, Retain	3–4 Mbytes/s	20	158
	1 bus acknoledge, daisy-chain		Lock	5 Mwords/s	16	42 116 231
	6 lines 2 levels	DC low, Timout, Protect, Bus error	Block transfer, Lock cycle, Message transfer	Burst: 10 MHz for 32-bit transfer Single: 29 MHz for 32-bit transfer	21	42 93
Multiple masters	Centralized 4 bus acknowledge, daisy-chain	AC fail, System fail	Read/Modify/Write, Block transfer, Interrupt acknowledge, Address privilege levels	10 MHz	20	197 205 294 337
	Distributed central timing		Read/Modify/Write Block transfers, Arbitration locked, Broadcast – sync	40 MHz – 32 bits	26	398
	32	Error reports	Read/Modify/Write, Block transfer, Split cycle, Event cycle	10 MHz – 32 bits	21	17 120

10.6.6 Bandwidth

Most backplane buses have an effective bandwidth, depending on the physical properties of the bus, of between 5 and 10 MHz. In practice, the transfer rate of data is more important, since a word may consist of 8 bits or multiples of 8 bits for each operating cycle, and this is expressed as Mwords/s.

10.7 Backplane bus standards

In the reviews that follow, a number of the more common and important backplane bus standards are described and their principal features summarized. The treatment is not exhaustive and the reader is referred to the specification documents for the full and usually comprehensive information. A number of the descriptions and abridged specifications contained in the following subsections are based on information provided by *Digital Equipment* and *Intel* and from the journals *Microprocessors and Microsystems* and *Electronics Industry*, and the permission of the publishers to use this material is gratefully acknowledged. The omission of some buses is not intended to imply any criticism, but merely indicates the magnitude of the problem facing the prospective system designer. The backplane buses surveyed are:

CAMAC (IEEE 583, etc.)
S100 (IEEE 696)
STD (IEEE P961)
STE (IEEE P1000)
G64
Q Bus
VAXBI Bus
Eurobus (ISO-DP 6951)
Multibus I (IEEE 796)
Multibus II (IEEE P1296)
VME (IEEE P1014)
Fastbus (IEEE 960)
Futurebus (IEEE P896)

CAMAC is not so immediately comparable. Abridged specifications for the other buses are given in Table 10.1.

The information given is necessarily brief – full details are given in the specifications available from the appropriate standards organizations or design authorities, whose addresses are given in the Bibliography.

10.7.1 The CAMAC interface system (IEEE 583)

CAMAC (*C*omputer *A*utomated *M*easurement and *C*ontrol) is a long-established (1969) modular interface system using a specified backplane and defined crate and module dimensions. Much development has taken place, resulting in standards covering systems ranging from single-crate (IEEE 583, IEC 482/516) through to multi-crate, fast parallel highway systems (IEEE 596, IEC 552), and serial highway systems (IEEE 595, IEC 640) and other relevant software and protocol documents. The basic features of CAMAC are:

(1) It is a modular system, with functional plug-in units (modules) that mount in a standard crate. Module and crate dimensions are specified.
(2) The plug-in units connect to a common internal backplane bus (Dataway) that is part of the crate and carries data, control systems and power.
(3) The system can connect to an online computer, or a controller or computer can be mounted in a module.
(4) Assemblies of crates may be interconnected by means of parallel or serial highways.

The Dataway has a minimum cycle time of 1 μs and is a parallel highway carrying the 24 + 24 read and write lines from the modules to and from a controller, so that words up to 24 bits long can be handled in a single data transfer.

The plug-in CAMAC unit (module) is a functional unit that converts the external functions to conform to the Dataway requirements. Each module therefore terminates in a double-sided printed circuit plug of 86 contacts connecting to a corresponding Dataway socket. A module may be any multiple of the single-unit width of 17 mm, and the standard defines only dimensions, electrical power supply requirements, signal standards and data transfer protocol of the interface to the Dataway. The standard leaves complete freedom regarding the inputs to modules and the choice of internal functions and the internal circuit technology of each module is left to the designer.

The CAMAC crate serves as the common housing for the plug-in units and its dimensions, power requirements and the Dataway are specified in the standard. It provides accommodation for 25 module widths and the right-hand two locations of each crate are allocated to the crate controller (or an interface to an external computer).

Due to these features, a wide variety of CAMAC modules with differing input features, each compatible with the system, have been available from more than 70 companies who manufactured some 600 CAMAC units. These can be used in a wide variety of industrial process control applications, such as electricity generation and steel and aluminium manufacturing, as well as in research and development.

As a result of its farsighted concept and despite its long life, CAMAC continues to be supported and is used particularly in high-energy research laboratories around the world, where its international acceptance has resulted in very significant financial savings.

CAMAC abridged specifications

Type	Synchronous, non-multiplexed
Write lines	24
Read lines	24
Module address lines	24
Sub-addresses	16
Demand (interrupt) lines	24
Data width	24

Cycle time	1 μs
Module size	222 × 305 × 17 mm
Connector	2 part, edge
Connection capacity	86 pins
Crate size	19 in
Power supplies	±6, ±12 V
Special cycles	Programmed transfer, DMA
Bandwidth	1 MHz-24 bits
Maximum board capacity	24
Supplied by firms	216 379

10.7.2 The S100 backplane bus (IEEE 696)

The S100 bus, otherwise known as the IEEE 696 bus, is an example of how an informal, loosely specified bus proposal can become widely accepted. Despite much initial incompatibility from a wide range of suppliers, the specification was refined and eventually, when supported by a respected standards organization, has become a usable tool.

The S100 bus comprises 100 bus lines which include 15 bi-direction data, 16 or 24 address, eight state, 11 bus control, six DMA control and eight vectored interrupt lines. It supports 8- and 16-bit processors, permitting up to 16 Mbytes of memory and 64 k I/O ports to be addressed. Its specification is such that almost any 8- or 16-bit processor from an 8080 onwards can be accepted and a data transfer rate of 6 Mwords/s is achieved. The devices are mounted on standard cards (254 × 135 mm) and use 100-way printed circuit edge connectors. Up to 22 slots are available to accommodate from four to 22 device cards.

Every system must have a permanent bus master, which is usually the CPU together with memory and I/O slave boards. The system organization is such that up to 16 temporary masters may also be accommodated. Memory slaves and I/O cards have 24 and 16 address lines, respectively.

Eight-bit transfers are made on two sets of uni-directional data buses, one for master-to-slave and the other for slave-to-master; 16-bit transfers on both buses are used together in a bi-directional mode and two extra bus lines are used for request and acknowledge. Both 8- and 16-bit slaves can therefore be mixed in a system subject to the controller possessing an enhanced byte serializer facility which serves four arbitration lines.

10.7.3 The STD backplane bus (IEEE P961)

The STD bus emerged around 1978 and has been improved over the years. It is an 8-bit, synchronous, non-multiplexed bus and is specified so as to allow it to support any 8-bit processor family and simple multiplexed systems. It accommodates a standard printed circuit card (165 × 114 mm) with a keyed integral printed circuit edge connector.

The bus is composed of a 56-line bused motherboard that permits any card to operate in any slot. Memory up to 64 k is supported on the bus and this can be expanded to 128 k by the use of the expansion line.

STD systems are configured to accommodate a diverse range of devices. Peripheral and I/O devices are connected using their own unique front-edge connectors and cable to the external device. No limit is placed on the number of slots on the bus but the backplane is sensitive to length and loading and each card represents one load per bus signal.

The extensive potential flexibility of the system requires caution in respect of compatibility. For instance, cards using peripheral chips usually depend upon specific timing signals for their satisfactory operation, and this sometimes prevents their use with other processor families. STD practice is to label cards that are processor dependent with a reference to the particular CPU, e.g. STD 6800, etc., and to specify any relevant waveform and timing requirements.

Interrupt servicing is available and is directly implemented by signal lines in systems with only a single interrupt device. In systems with a number of interrupt devices both serial and parallel schemes are defined.

10.7.4 The STE backplane bus (IEEE P1000)

The STE bus is a refined version of the STD bus. It makes use of Eurocards and is different from the STD bus in several fundamental ways. The bus is intended for implementing high-performance 8-bit micro-computer systems either in a stand-alone mode or in a multibus architecture. It may also serve as a high-speed I/O channel. In contrast to the STD bus, it functions asynchronously and is based upon the master–slave concept, where any master, having gained control of the bus, can address slaves via an asynchronous interlocked handshake. It is therefore possible to construct systems that incorporate devices of widely different speeds. Up to three masters may also be implemented within a single system.

Two independent system address spaces are supported, allowing memory of 1 Mbyte physical address space and 4 kbytes of I/O. Two daisy chains are defined; one for bus arbitration and the other for priority assignment to interrupting slaves. The system integrity has been improved with respect to the STD bus by the provision of a bus error line for monitoring data transfers and arbitration.

The bus makes use of single Eurocards with 64-pin connectors to DIN 41612 and up to 21 cards can be accommodated in it. There is a wide range of commercial boards available from different suppliers and these can be mixed as required.

10.7.5 The G64 backplane bus

The G64 bus was devised by GESPAC SA of Geneva in 1979, and a wide range of modules is available from a number of suppliers. It is a bus capable of supporting 8- or 16-bit microprocessors and the bus design is unrelated to any special CPU family. It is able to provide either synchronous or asynchronous transfers and comprises a non-multiplexed 64-line bus. Standard single Eurocards with a DIN 41612 type connector are used and between four and 20 slots are available.

Eight- and 16-bit memory modules are available, each being compatible with corresponding 8- and 16-bit processor modules, and memory addressing capability of 256 kbytes is possible. Up to 1 k × 8 or 1 k × 16 precoded field I/O addresses are also available and I/O modules can be used with either size of processor.

The bus is available as a two-layer device without proper terminations and capable of operating up to 1 MHz transfer rates, or as a four-layer bus terminated with 80 Ω, when more than 1 MHz is possible. Three interrupt lines are available together with one interrupt acknowledge line and one daisy-chain, thus allowing non-vectored or prioritized vectored interrupts to be serviced.

It operates as a monoprocessor system with the ability to service multiple DMA sources. These are serviced by Request Grant Acknowledge, using the interrupt daisy chain to establish priorities in multiple DMA configurations. It usually operates in a synchronous mode for 8-bit processors and in an asynchronous mode for 16 bits. Either mode is possible with a mixed system.

A 32-bit version of G64 (G128) is under development and, although making use of double Eurocards, it is otherwise planned to be compatible with existing G64 boards.

10.7.6 The Q backplane bus

The Q (or Q22) bus is a proprietary bus designed by Digital Equipment Corporation for use with their range of 16- and 32-bit MicroVax I and II and Micro PDP-11 Computers, where it functions as the 'low-end' bus. Another bus, the VAXBI bus, is designed to support VAX architecture and multiprocessor systems only as the 'high-end' or high-performance bus. The Q bus is a modified version of the Unibus concept used with their earlier PDP-11 computers and has been enhanced to provide a faster and more potent system capability. In each case the Q bus implements the architecture applicable to the particular computer series, and thus allows them to be compatible with larger and earlier systems and software.

DEC configure a range of system enclosures, each of which includes a Q bus as part of the backplane assembly. It receives, in common with much other DEC equipment, considerable commercial support in the provision of bus-compatible devices, thus making the equipment commercially attractive.

The Q bus includes 42 bidirectional and two unidirectional signal lines allocated as 16 multiplexed data/address lines, two multiplexed address/parity lines, four non-multiplexed extended address lines, six data transmission control lines, 10 interrupt control and DMA control lines, six system control lines and power, ground and spare lines. Operation is asynchronous on a master–slave handshake basis with a 10 μs timeout to avoid system lockup and bus arbitration is by the processor.

Interrupt is by vector and priority may be allocated by either distributed or position-defined arbitration. Cards are either dual or quad and use double finger-edge connectors with 72 or 144 connections, respectively. The maximum number of cards that can be accommodated depends on their electrical loading, but for a single backplane system up to 18 may be possible. Systems can be more complex and may comprise several backplanes with their enclosures.

The Q bus makes use of block-mode DMA to obtain a fast transfer rate to memory by reducing the amount of handshaking necessary in a transfer. Block-mode can be effected in either single-cycle mode or burst mode, when a DMA device can hold on to bus mastership for as long as necessary, albeit at the expense of other devices requiring access to the bus. Block mode offers a maximum rate of 3.3 Mbytes/s.

When used in MicroVAX systems the servicing of interrupts is possible by an automatic two-dimensional four-level system by distributed arbitration (i.e. priority levels allocated to the hardware). Micro PDP-11 systems have this possibility as well as by priority to the device electrically closest to the processor.

Comprehensive details of the Q bus and its applications are given in the relevant DEC publications.

10.7.7 VAXBI backplane bus

The VAXBI bus is Digital Equipment Company's high-performance 32-bit bus for use with their VAX computers and systems. It is a fully specified bus which uses a ZMOS chip as a synchronous transceiver interface between the bus and each module or card logic. The chip undertakes all the necessary bus protocol, ranging from system test, multiprocessing commands, interrupt services, parity and error checking to arbitration on distributed system facilities.

The VAXBI bus is a 32-bit, synchronous bus, with cycles occurring every 200 μs. Arbitration, address and data are all multiplexed over 32 signal lines and fully burdened transfer rates of 13.3 Mbytes/s are possible. No processor is dedicated to controlling the bus, arbitration logic occurring in each module. Single- or multi-responder module transactions can be initiated and interrupts from responder devices are treated as transactions and do not require interrupt request lines.

Distributed arbitration avoids the necessity of a dedicated arbiter and devices can seek bus control whenever the bus is idle or during arbitration intervals. Block-mode transfer is limited to 16 bytes and a round-robin arbitration scheme ensures fair bus access. Up to 16 modules (or VAXBI nodes) can be accommodated with arbitration access and up to 32 modules that interface to other hardware or power can otherwise be accommodated.

The standard module is a 203 × 233 mm Eurocard with all electrical and mechanical parameters fully specified. Only one corner of a card is used for the ZMOS chip interface, the remainder being available to the user. Sixty-pin zero insertion force (ZIF) connectors are used providing a total of 120 pins to the backplane, with a further 180 joining the user area of the board to associated I/O connections in the system housing or logic cage. Each logic cage can house up to six VAXBI modules and multi-cage systems can be configured.

Full details of the VAXBI bus and its system applications are available in the relevant DEC literature.

10.7.8 The Eurobus backplane bus (ISO-DP 6951, BS 6475: 1984)

The Eurobus is a general-purpose back-plane bus, originally designed jointly by the UK Ministry of Defence and the Ferranti Company. A chip set has also been designed to support the standard and is commercially available. The bus was originally designed for a 16-bit data width and was intended to join together processors, memory and I/O devices in a variety of single and multi-processor systems. The specification has now been extended to allow data widths of 8, 16, 24 and 32 bits to be supported; corresponding address widths of 10, 18, 26 and 34 bits are available with titles of Eurobus 10, etc.

The bus comprises a 64-pin connector covering the Eurobus 10 and 18 systems and two such connectors for the Eurobus 26 and 34 systems. Data transfer is asynchronous and fully handshaken and is multiplexed onto a single highway. Sixteen or 18 data/address lines are used together with nine control lines, one interrupt line and 11 power lines. The system uses double Eurocards (233 × 160 mm) with DIN 41612, 64-pin connectors, up to 20 devices in a maximum bus length of 500 mm and the bus must be terminated at each end.

Eurobus is based on the concept of central bus arbitration. The bus control is removed from the processor(s) and allocated to a dedicated bus arbiter serving all requirements. Devices are allocated a master–slave relationship according to their need and their potential. The arbiter has complete control of the bus on the basis of the master being the device initiating a transfer request and a slave responding. Slave status need not be a permanent one.

Eurobus caters for two fundamental and separate operations – interrupt and data transfer. The interrupt operation is very basic and fast, thanks to the control arbiter, which then performs any further system operations necessary. The protocol is designed to provide maximum speed, whilst still ensuring a handshake at every stage of the allocation and data transfer processes.

A particularly useful feature of Eurobus is its ability to allow two Eurobus systems to be linked together and operated as one. The system protocol caters for the inevitable conflicts where there are potential masters on both buses.

Eurobus does not support block transfers for reasons of reliability in its intended environment.

10.7.9 The Multibus I backplane bus (IEEE 796)

The Multibus backplane was originally devised by the Intel Corporation in 1974. After further development it is now widely used for microprocessor industrial system applications and receives wide commercial support. The bus makes use of an 86-pin connector and includes 16 bi-directional data lines, 20 standard (or 24 expanded) address lines, 18 control lines and eight lines of interrupts. The expanded addressing, when required, is accommodated by a 60-pin connector, which also carries other features such as power fail, memory initializing control signals and bus exchange lines. Power failure lines are optional.

The bus operates in an asynchronous, non-multiplexed manner and can support up to 1 Mbyte of direct addressing with 20 bits and 16 Mbytes using the expanded 24 bits. It can also address up to 64 k via I/O ports using 16-bit addressing. Both memory and I/O cycles can support 8- or 16-bit data transfers. The boards used are the original Intel cards of 304.8 × 171.5 mm and make use of printed circuit edge connections.

Multibus I is particularly used to interface the Intel 80/86 single-board computers and an extensive range of memory expansions, digital and analogue I/O boards and peripheral controllers.

The bus structure is based on the master–slave concept, where the master takes control, placing the slave address on the bus. The slave decodes the address before following its commands and a handshake between the two devices allows modules of widely different speeds to use the bus at data rates up to 5 Mwords/s.

The bus also allows multiple masters for multiprocessing and provides control signals on bus exchange lines for connecting them either in serial daisy chain priority or in parallel. In the latter case up to 16 masters may share the bus resources.

10.7.10 The Multibus II backplane bus (IEEE P1296)

In late 1983 Intel announced a new 32-bit bus architecture named Multibus II. This is based on principles similar to those of the original Multibus I and extends the multiple-bus approach to now use five buses, which are processor independent. The extra buses are:

Parallel System Bus (*iPSB*)
Local Bus Extension (*iLBX II*)
Serial System Bus (*iSSB*)

The original *I/O Expansion Bus* (*iSBX*) and the *Multichannel DMA I/O Bus* from Multibus I continue to be used.

The bus uses standard single- or double-height Eurocards and 96-pin, two-piece DIN (IEC 603-2) connectors. Mechanical dimensions are fully covered in the specification. The buses are interconnected by a common system interface that defines intermodule communication and data transfer protocol, and thus allows system designers to choose the most appropriate combination of the five buses to satisfy a particular requirement. In consequence, Multibus II uses a processor-independent, open-system architecture, which is suitable for a wide range of system designs. It provides a 32-bit parallel bus with a capability of 40 Mbytes/s transfer rate and high-speed access to large-capacity remote (off-board) memory. In addition, the serial system bus permits serial access to be achieved at low cost.

The advantages of the multiple-bus approach over a general purpose bus are claimed to be:

(1) A specialized bus carries out its functions more efficiently.
(2) Different functions can be carried out simultaneously on different buses.
(3) The fast bandwidth of the general-purpose bus is retained, thus providing an efficient bandwidth for communication and data transfer between processors.
(4) Only the buses necessary for particular system requirements need be incorporated, thus reducing costs.

The bus properties are summarized as follows. The *Parallel System Bus* is a general-purpose, multiple-processor-independent, synchronous bus that supports data movement and interprocessor communication using 8-, 16- or 32-bit processors. Arbitration and execution are also supported. Space is provided for 32-bit memory addresses and 16-bit I/O addresses. Sequential arbitration is provided for devices on a priority basis. Data transfers are clocked at 10 MHz and can be 8, 16, 24 or 32 bits wide. A burst data transfer rate permits a sustained bandwidth of 40 Mbytes/s. Up to 21 devices can be accommodated.

The *Local Extension Bus* functions as a fast processor execution bus which extends the local bus to remote memory. It possesses a 48 Mbytes/s bandwidth and a 12 MHz clock and provides local memory expansion, without arbitration, to 64 Mbytes, whilst supporting 8-, 16- and 32-bit processors. Up to six devices can be supported on this bus.

The *Serial System Bus* is a simplified version of the Parallel System Bus, being only 1 bit wide and running at 2 MHz. Accordingly, it is much cheaper and also allows up to 32 devices to be extended up to 10 m distant from the backplane.

The *Multichannel DMA I/O Bus* permits high-speed block data transfer between peripheral devices and single-board computers. It is an asynchronous bus, capable of supporting up to 16 devices of 8 or 16 bits with 16 Mbytes of memory to each device. It can support data transfers at 8 Mbytes/s at a distance of up to 15 m.

The *iSBX I/O Expansion Bus* allows the use of multimodule daughterboards to provide local (on-board) system expansion as and when necessary.

The Intel specification includes details of the different bus considerations and how separate Multibus II systems may be interconnected. Full mechanical details are also included.

10.7.11 The VME (IEEE P1014) and Versabus (IEEE P970) backplane buses

The VME bus is the Eurocard implementation of Versabus and is therefore smaller but still very suitable for industrial applications. The two systems are architecturally very similar, and this description is restricted to the VME bus, which was introduced and supported by Motorola, Mostek, Signetic/Philips and Thompson/EFCIS. The proposal is now out for public comment. An important attraction of the VME bus is the fact that it was fully specified before any cards were made available.

Whilst the VME bus harmonizes with signals of the 68000 microprocessor, other processors can be accepted. The specification allows for data transfers of words, 8, 16, or 32 bits wide and the use of 16-, 24- or 32-bit byte addresses.

The VME bus is a multi-master, asynchronous, non-multiplexed bus that can be visualized as four separate buses, namely:

(1) *Data Transfer Bus* (*DTB*);
(2) *Priority Interrupt Bus*;
(3) *DT Arbitration Bus*; and
(4) *Utility Bus*.

The specification also defines the module card size, connector type and pin disposition, power supplies, etc.

The *Data Transfer Bus* is a 16-bit data path capable of expansion to 32 bits by the use of a second, optional connector. A 24-bit address path, also expandable to 32 bits by the second connector, three control lines and six address modifiers together with other control lines, are also specified. The widths of data words are also identified by an address modifier code in the bus control signal. Accordingly, a variety of processors with differing data widths or address widths can be mixed in a single VME system, the bus control allowing efficient cooperation through selected common memory areas.

The *Priority Interrupt Bus* has seven interrupt lines and one daisy chain interrupt acknowledge line. The current master will acknowledge interrupt priorities via the low-order address lines and vectors associated with interrupts are passed via the low-order data lines.

The *Data Transfer Arbitration Bus* possesses four bus request lines and four grant daisy chain lines. Any master can hold one of four priority levels during normal bus operations; each of these levels has a request and grant line dedicated to it.

The *Utility Bus* is dedicated to timing signals, initialize and bus diagnostics, together with a 16 MHz system clock.

The *Interintelligence Bus*. In order to avoid the customary problem of slowing the system down by use of the Data Transfer Bus merely to allow communication between the processors (e.g. global memory) an interintelligence bus has been specified. This consists of a two-line synchronous serial bus capable of operating at 4 MHz between intelligent devices, thus permitting intercommunication to bypass the Data Transfer Bus.

As with most of the buses described, there are many companies offering support and cards, etc., for VME systems.

A multiprocessor system needs to use the main bus to access the I/O devices of the various processors simultaneously, and VME is no exception to this. In doing so it can overload the bus and the system can become bandwidth limited. To overcome this problem the processors can make use of a subsystem bus to access their own private resources or peripherals. Such a bus has been devised for VME under the term VSB. It retains its compatibility with the VME bus by using the two outer rows of pins in the J2 connector which are specified as 'User Defined'. It is necessary to multiplex 32 address/data lines as well as defining block transfers, arbitration, etc. Approval is currently being considered by the IEC.

10.7.12 The Fastbus backplane bus (IEEE 960)

The Fastbus has been designed to supersede much of the CAMAC specification. Its design commenced in 1977 and it became an accepted standard in 1984 (also IEC 45). Like CAMAC, it is a modular system based on a data bus and is capable of accommodating up to 26 modules in a single crate. Its performance is significantly better than CAMAC and it functions as a 32-bit multiprocessor system with multiple segments which operate independently but can be linked together for fast data transfers.

Fastbus is a high-speed system based on Emitter Coupled Logic (ECL) and is capable of >10 Mwords/s. It uses a multiplexed 32-bit parallel bus for both addressing and data transfer. Maximum speed of transfer is achieved by the use of block transfers when >25 Mwords/s in a non-handshake block transfer mode are possible.

Primarily designed for high-energy physics applications the physical design of Fastbus facilitates its application to quite arduous environments.

It allows systems to be configured as single-crate or multicrate with cable segments joining the segments via an interface (segment interconnect). Each segment can function as an autonomous bus, interconnecting one or several master devices with a number of slaves. Excessive demands by master devices is either avoided by logistical processes or by a time-out mechanism to avoid system lock-up.

Within each crate, modules can occupy any position and possess a unique address. Multiple processors may be plugged into a crate (i.e. a single segment) and share the backplane bus but must take turns in using it by means of an arbitration mechanism.

The essential feature of Fastbus is the bus protocol which is based on asynchronous operation with the ability to operate synchronously when maximum data transfer rates are necessary. Protocols are consistently uniform and allow the system to function in the different configurations possible.

A crate can accommodate up to 26 modules, each module being a printed circuit board of 367 × 400 mm with an optional, 16 mm wide front panel. A 130 pin (2 × 65) connector is used to connect to the backplane bus and a similar (3 × 65) connector is used to link to an auxiliary bus connecting to the outside world. A dissipation of 2 kW per crate (75 W per module) is designed into the specification and is distributed in the backplane, providing several hundred amperes at various d.c. voltages.

A considerable number of commercially made units are now available and the specification should be referred to for full details.

10.7.13 The Futurebus backplane bus (IEEE P896)

The Futurebus has been designed by the IEEE microprocessor standards committee after their work on S100 and Multibus I. It is intended to be a forward-looking, long-life manufacturer-independent standard. The draft of Futurebus is now (1987) out for public comment and some of the details quoted here may therefore be amended.

Futurebus comprises a high-performance, 32-bit highway operating with an asynchronous, multiplexed protocol; 8- and 16-bit data transmission can be supported and an independent serial highway is also included. It is designed to accommodate multiple processors and makes use of a distributed arbitration control. Daisy-chain arbitration and interrupt acknowledge systems are not possible, since the system does not rely on any particular cards being available in the system configuration. System security and comprehensive fault detection is possible by proper use of the extensive protocol, checking and fault reporting facilities provided in the specification.

The bus may be visualized as four buses, thus:

(1) *Parallel Data Bus*, providing 32 multiplexed address and data lines together with timing and state lines.
(2) *Parallel Arbitration Bus*, providing control and priority lines which allow a single master to take control of the Parallel Data Bus simultaneously with data transfer.
(3) *Serial Bus*, consisting of a serial line and a check line to permit separate transfers in addition to those of the Parallel Data Bus. The Serial Bus is an optional feature and not essential for Futurebus operation.
(4) *Power Bus*, which distributes the 5 V supplies and provides ground returns.

The family of Eurocards is recommended with a preferred choice of the maximum size card (280 × 367 mm) together with a single 96-pin DIN 41612 connector. Up to 32 cards can be accommodated on the bus and these can be inserted with the system live.

Any suitable task-scheduling software should be satisfactory for use with Futurebus and permit the operation of multiple software systems. Synchronization of parallel processors is provided and under these conditions a burst data transfer rate of 10 Mwords/s (32 bits) is possible. For such transfer rates the bus must be properly terminated to avoid mismatches, etc, and must use the protocol efficiently. This is achieved by ensuring that addresses are already known to the slave, thus avoiding address transmissions.

The philosophy of the bus is that of a communication bus between processors. Fetch instructions, etc, are minimized and complete messages are passed. In consequence, the bus will support virtual memory systems for disks and cache memories. Use of asynchronous protocol allows the speed of the current bus master to be achieved rather than slowing the system down to that of the slowest possible master. A distributed arbitration system offers sufficient flexibility in the allocation of priorities for at least 32 prospective bus masters.

10.8 Bibliography

1 International Telecommunications Union, CCITT: *Proposed New and Revised Series V Recommendations (Document AP VII No. 44-E)*, General Secretariat, Sales Section, ITU, CCITT, Place des Nations, CH-1211, Geneva 20, Switzerland
2 *The V-Series Report – Standards for Data Transmissions by Telephone*, Bootstrap Ltd, 1981
3 EIA Standards, RS Series, Electronic Industrial Association, Engineering Dept, Standards Sales Office, 2601 Eye Street NW, Washington, DC 20006, USA
4 Hewlett-Packard, Netherland BV, Central Mailing Dept, PO Box 529, 1180 AM, Amstelveen, The Netherlands (various publications)
5 MCNAMARA, J.E., *Technical Aspects of Data Communication*, Digital Equipment Corp., 1978
6 KORN, G.A., *Microprocessors and Small Digital Computer Systems for Engineers and Scientists*, McGraw-Hill (UK), 1978
7 WOOLARD, B.G., *Microprocessors and Microcomputers for Engineering Students and Technicians*, McGraw-Hill (UK), 1981

8 ZAKS, R. and LESEA, A., *Microprocessor Interfacing Techniques*, Sybex Inc., Berkeley, Calif., USA, 1979

9 CLULEY, J.C., *Minicomputer and Microprocessor Interfacing*, Edward Arnold, 1982

10 *Microprocessors and Microsystems*, Vol. 6, No. 9, Nov. 1982; Vol. 7, No. 6, July/Aug. 1983; Vol. 10, No. 2, March, 1986; Vol. 10, No. 6, July/Aug. 1986, Butterworths

11 'Board level design – VME allows for future expansion', *Electronics Industry*, April, 1983, E.S. Publications

12 IEEE Specifications – Service Center, 445 Hoes Lane, Piscataway, New Jersey, 08854, USA, or BSI Sales Dept, Linford Wood, Milton Keynes, MK14 6LE, UK (various publications)

13 IEEE Proposals, The IEEE Computer Society, Microprocessor Standards Committee, or IEE Working Party on Backplane Buses, The IEE, Savoy Place, London UK (various publications)

14 Eurobus, (ISO DP 6951 BS 6475: 1984), International Standards Organization, Case Postale 56, CH-1211, Geneva 20, Switzerland, or BSI Sales Dept, Linford Wood, Milton Keynes, MK14 6LE, UK

15 G64 – GESPA, SA, 3 Ch. des Ulx, CH-1228, Geneva, Plan les Ouates, Switzerland

16 *IBM Synchronous Data Link Control – General Information (GA 27 – 3093 File GEN-09)*

17 *Supermicrosystems Handbook*, Digital Equipment Corp., 1986

18 *VAXBI Options Handbook*, Digital Equipment Corp., 1986

19 *Multibus II Bus – Architecture Specifications Handbook*, Intel Corp. (UK) Ltd, 1984

20 RADNAI, R. and KINGHAM, E.G., *Jones' Instrument Technology*, Vol. 5, Automatic Instruments and Measuring Systems, Butterworths, 1986

Part 5

Further Scientific and Technical Information

In the first four parts of the *Instrumentation Reference Book*, each chapter has dealt with its own field without too much introduction or diversion, though simple scientific principles have been introduced where thought necessary or helpful. Sometimes, however, a reader will want to supplement the essential instrumentation ideas presented either by going more fully into the basic mathematics and physics or by considering the technology in rather more detail. We give assistance to those ends here in Part 5, in some cases simply by drawing together information which would be available to many readers if, but only if, they scoured through a number of other books to which they had access.

This has been facilitated by the ability to draw on material already published in *Electronics Engineer's Reference Book*, published by Butterworth under the editorship of F Mazda. There are always advantages in having a library of reference books built up under the auspices of one publishing house, and the interchangeability of material is particularly useful when subjects as closely related as electronic engineering and instrumentation are covered. The co-operation of editor and contributors is greatly appreciated.

1 Trigonometric functions and general formulae

J. Barron

Contents

1.1 Mathematical signs and symbols

Sign, symbol	*Quantity*
$=$	equal to
$\neq$ $\neq$	not equal to
$\equiv$	identically equal to
Δ	corresponds to
$\approx$	approximately equal to
$\rightarrow$	approaches
$\simeq$	asymptotically equal to
$\sim$	proportional to
∞	infinity
$<$	smaller than
$>$	larger than
$\leq$ $\leqslant$ $\leqq$	smaller than or equal to
$\geq$ $\geqslant$ $\geqq$	larger than or equal to
$\ll$	much smaller than
$\gg$	much larger than
$+$	plus
$-$	minus
$\cdot$ $\times$	multiplied by
$\frac{a}{b}$ a/b	a divided by b
$\lvert a\rvert$	magnitude of a
a^n	a raised to the power n
$a^{1/2}$ $\sqrt{a}$	square root of a
$a^{1/n}$ $\sqrt[n]{a}$	nth root of a
$\bar{a}$ $\langle a\rangle$	mean value of a
$p!$	factorial p, $1\times2\times3\times\ldots\times p$
$\binom{n}{p}$	binomial coefficient, $\dfrac{n(n-1)\ldots(n-p+1)}{1\times2\times3\times\ldots\times p}$
Σ	sum
Π	product
$f(x)$	function f of the variable x
$[+(x)]_a^b$	$f(b)-f(a)$
$\lim\limits_{x\to a} f(x)$; $\lim_{x\to a} f(x)$	the limit to which $f(x)$ tends as x approaches a
Δx	delta x = finite increment of x
δx	delta x = variation of x
$\frac{df}{dx}$; df/dx; $f'(x)$	differential coefficient of $f(x)$ with respect to x
$\frac{d^nf}{dx^n}$; $f^{(n)}(x)$	differential coefficient of order n of $f(x)$
$\frac{\partial f(x,y,\ldots)}{\partial x}$; $\left(\frac{\partial f}{\partial x}\right)_{y\ldots}$	partial differential coefficient of $f(x,y,\ldots)$ with respect to x, when $y,\ldots$ are held constant
df	the total differential of f
$\int f(x)\,dx$	indefinite integral of $f(x)$ with respect to x
$\int_a^b f(x)\,dx$	definite integral of $f(x)$ from $x=a$ to $x=b$
e	base of natural logarithms
e^x; exp x	e raised to the power x
$\log_a x$	logarithm to the base a of x
lg x; log x; $\log_{10} x$	common (Briggsian) logarithm of x
lb x; $\log_2 x$	binary logarithm of x
sin x	sine of x
cos x	cosine of x
tan x; tg x	tangent of x
cot x; ctg x	cotangent of x
sec x	secant of x
cosec x	cosecant of x
arcsin x	arc sine of x
arcos x	arc cosine of x
arctan x, arctg x	arc tangent of x
arccot x, arcctg x	arc cotangent of x
arcsec x	arc secant of x
arccosec x	arc cosecant of x
sinh x	hyperbolic sine of x
cosh x	hyperbolic cosine of x
tanh x	hyperbolic tangent of x
coth x	hyperbolic cotangent of x
sech x	hyperbolic secant of x
cosech x	hyperbolic cosecant of x
arcsinh x	inverse hyperbolic sine of x
arcosh x	inverse hyperbolic cosine of x
arctanh x	inverse hyperbolic tangent of x
arcoth x	inverse hyperbolic cotangent of x
arsech x	inverse hyperbolic secant of x
arcosech x	inverse hyperbolic cosecant of x
i, j	imaginary unity, $i^2=-1$
Re z	real part of z
Im z	imaginary part of z
$\lvert z\rvert$	modulus of z
arg z	argument of z
z^*	conjugate of z, complex conjugate of z
$\tilde{A}$, A′, A^t	transpose of matrix A
A*	complex conjugate matrix of matrix A
A^+	Hermitian conjugate matrix of matrix A
A, **a**	vector
$\lvert\mathbf{A}\rvert$, A	magnitude of vector
$\mathbf{A}\cdot\mathbf{B}$	scalar product
$\mathbf{A}\times\mathbf{B}$, $\mathbf{A}\wedge\mathbf{B}$	vector product
∇	differential vector operator
$\nabla\varphi$, grad φ	gradient of φ
$\nabla\cdot A$, div **A**	divergence of **A**
$\nabla\times\mathbf{A}$, $\nabla\wedge\mathbf{A}$, curl **A**, rot **A**	curl of **A**
$\nabla^2\varphi$, $\Delta\varphi$	Laplacian of φ

1.2 Trigonometric formulae

$$\sin^2 A+\cos^2 A=\sin A\,\text{cosec}\,A=1$$

$$\sin A=\frac{\cos A}{\cot A}=\frac{1}{\text{cosec}\,A}=(1-\cos^2 A)^{1/2}$$

$$\cos A=\frac{\sin A}{\tan A}=\frac{1}{\sec A}=(1-\sin^2 A)^{1/2}$$

$$\tan A=\frac{\sin A}{\cos A}=\frac{1}{\cot A}$$

$$1+\tan^2 A=\sec^2 A$$

$$1+\cot^2 A=\text{cosec}^2 A$$

$$1-\sin A=\text{coversin}\,A$$

$$1-\cos A=\text{versin}\,A$$

$$\tan\tfrac{1}{2}\theta=t;\quad \sin\theta=2t/(1+t^2);\quad \cos\theta=(1-t^2)/(1+t^2)$$

$\cot A = 1/\tan A$

$\sec A = 1/\cos A$

$\operatorname{cosec} A = 1/\sin A$

$\cos(A \pm B) = \cos A \cos B \mp \sin A \sin B$

$\sin(A \pm B) = \sin A \cos B \pm \cos A \sin B$

$$\tan(A \pm B) = \frac{\tan A \pm \tan B}{1 \mp \tan A \tan B}$$

$$\cot(A \pm B) = \frac{\cot A \cot B \mp 1}{\cot B \pm \cot A}$$

$\sin A \pm \sin B = 2 \sin \tfrac{1}{2}(A \pm B) \cos \tfrac{1}{2}(A \mp B)$

$\cos A + \cos B = 2 \cos \tfrac{1}{2}(A + B) \cos \tfrac{1}{2}(A - B)$

$\cos A - \cos B = 2 \sin \tfrac{1}{2}(A + B) \sin \tfrac{1}{2}(B - A)$

$$\tan A \pm \tan B = \frac{\sin(A \pm B)}{\cos A \cos B}$$

$$\cot A \pm \cot B = \frac{\sin(B \pm A)}{\sin A \sin B}$$

$\sin 2A = 2 \sin A \cos A$

$\cos 2A = \cos^2 A - \sin^2 A = 2\cos^2 A - 1 = 1 - 2\sin^2 A$

$\cos^2 A - \sin^2 B = \cos(A + B)\cos(A - B)$

$\tan 2A = 2 \tan A/(1 - \tan^2 A)$

$$\sin \tfrac{1}{2}A = \left(\frac{1 - \cos A}{2}\right)^{1/2}$$

$$\cos \tfrac{1}{2}A = \pm\left(\frac{1 + \cos A}{2}\right)^{1/2}$$

$$\tan \tfrac{1}{2}A = \frac{\sin A}{1 + \cos A}$$

$\sin^2 A = \tfrac{1}{2}(1 - \cos 2A)$

$\cos^2 A = \tfrac{1}{2}(1 + \cos 2A)$

$$\tan^2 A = \frac{1 - \cos 2A}{1 + \cos 2A}$$

$$\tan \tfrac{1}{2}(A \pm B) = \frac{\sin A \pm \sin B}{\cos A + \cos B}$$

$$\cot \tfrac{1}{2}(A \pm B) = \frac{\sin A \pm \sin B}{\cos B - \cos A}$$

1.3 Trigonometric values

Angle	*0°*	*30°*	*45°*	*60°*	*90°*	*180°*	*270°*	*360°*
Radians	0	$\pi/6$	$\pi/4$	$\pi/3$	$\pi/2$	π	$3\pi/2$	2π
Sine	0	$\frac{1}{2}$	$\frac{1}{2}\sqrt{2}$	$\frac{1}{2}\sqrt{3}$	1	0	-1	0
Cosine	1	$\frac{1}{2}\sqrt{3}$	$\frac{1}{2}\sqrt{2}$	$\frac{1}{2}$	0	-1	0	1
Tangent	0	$\frac{1}{2}\sqrt{3}$	1	$\sqrt{3}$	∞	0	∞	0

1.4 Approximations for small angles

$\sin\theta = \theta - \theta^3/6$; $\cos\theta = 1 - \theta^2/2$; $\tan\theta = \theta + \theta^3/3$;
(θ in radians)

1.5 Solution of triangles

$$\frac{\sin A}{a} = \frac{\sin B}{b} = \frac{\sin C}{c} \qquad \cos A = \frac{b^2 + c^2 - a^2}{2bc}$$

$$\cos B = \frac{c^2 + a^2 - b^2}{2ca} \qquad \cos C = \frac{a^2 + b^2 - c^2}{2ab}$$

where A, B, C and a, b, c are shown in *Figure 1.1*. If $s = \tfrac{1}{2}(a + b + c)$,

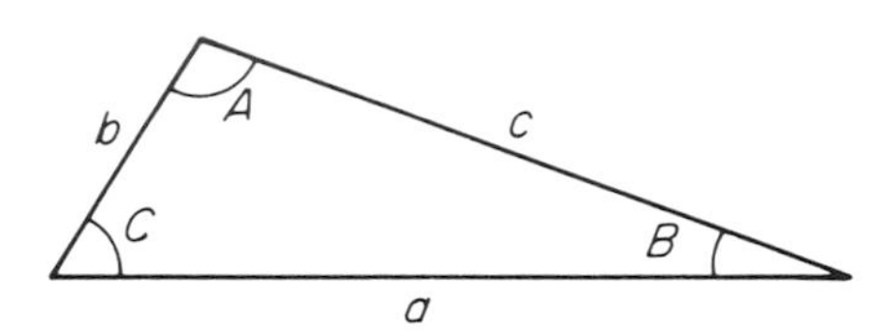

Figure 1.1 Triangle

$$\sin\frac{A}{2} = \sqrt{\frac{(s-b)(s-c)}{bc}} \qquad \sin\frac{B}{2} = \sqrt{\frac{(s-c)(s-a)}{ca}}$$

$$\sin\frac{C}{2} = \sqrt{\frac{(s-a)(s-b)}{ab}}$$

$$\cos\frac{A}{2} = \sqrt{\frac{s(s-a)}{bc}} \qquad \cos\frac{B}{2} = \sqrt{\frac{s(s-b)}{ca}}$$

$$\cos\frac{C}{2} = \sqrt{\frac{s(s-c)}{ab}}$$

$$\tan\frac{A}{2} = \sqrt{\frac{(s-b)(s-c)}{s(s-a)}} \qquad \tan\frac{B}{2} = \sqrt{\frac{(s-c)(s-a)}{s(s-b)}}$$

$$\tan\frac{C}{2} = \sqrt{\frac{(s-a)(s-b)}{s(s-c)}}$$

1.6 Spherical triangle

$$\frac{\sin A}{\sin a} = \frac{\sin B}{\sin b} = \frac{\sin C}{\sin c}$$

$\cos a = \cos b \cos c + \sin b \sin c \cos A$

$\cos b = \cos c \cos a + \sin c \sin a \cos B$

$\cos c = \cos a \cos b + \sin a \sin b \cos C$

where A, B, C and a, b, c are now as in *Figure 1.2.*

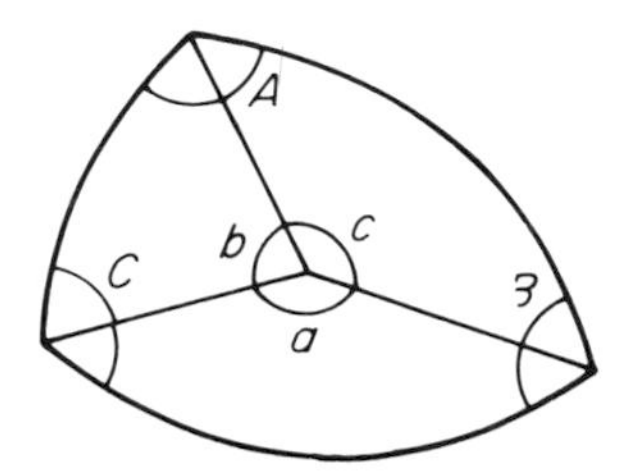

Figure 1.2 Spherical triangle

1.7 Exponential form

$\sin\theta = \frac{e^{i\theta} - e^{-i\theta}}{2i}$ $\cos\theta = \frac{e^{i\theta} + e^{-i\theta}}{2}$

$e^{i\theta} = \cos\theta + i\sin\theta$ $e^{-i\theta} = \cos\theta - i\sin\theta$

1.8 De Moivre's theorem

$(\cos A + i\sin A)(\cos B + i\sin B)$
$= \cos(A+B) + i\sin(A+B)$

1.9 Euler's relation

$(\cos\theta + i\sin\theta)^n = \cos n\theta + i\sin n\theta = e^{in\theta}$

1.10 Hyperbolic functions

$\sinh x = (e^x - e^{-x})/2$ $\cosh x = (e^x + e^{-x})/2$

$\tanh x = \sinh x / \cosh x$

Relations between hyperbolic functions can be obtained from the corresponding relations between trigonometric functions by reversing the sign of any term containing the product or implied product of two sines, e.g.:

$\cosh^2 A - \sinh^2 A = 1$

$\cosh 2A = 2\cosh^2 A - 1 = 1 + 2\sinh^2 A$
$= \cosh^2 A + \sinh^2 A$

$\cosh(A \pm B) = \cosh A \cosh B \pm \sinh A \sinh B$

$\sinh(A \pm B) = \sinh A \cosh B \pm \cosh A \sinh B$

$e^x = \cosh x + \sinh x$ $e^{-x} = \cosh x - \sinh x$

1.11 Complex variable

If $z = x + iy$, where x and y are real variables, z is a complex variable and is a function of x and y. z may be represented graphically in an Argand diagram (*Figure 1.3*).

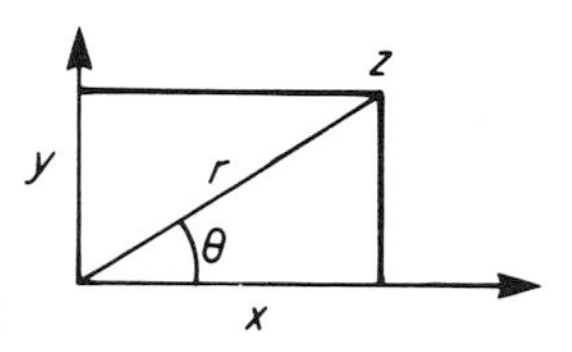

Figure 1.3 Argand diagram

Polar form:

$z = x + iy = |z| e^{i\theta} = |z|(\cos\theta + i\sin\theta)$

$x = r\cos\theta$ $y = r\sin\theta$

where $r = |z|$.

Complex arithmetic:

$z_1 = x_1 + iy_1$ $z_2 = x_2 + iy_2$

$z_1 \pm z_2 = (x_1 \pm x_2) + i(y_1 \pm y_2)$

$z_1 . z_2 = (x_1 x_2 - y_1 y_2) + i(x_1 y_2 + x_2 y_1)$

Conjugate:

$z^* = x - iy$ $z . z^* = x^2 + y^2 = |z|^2$

Function: another complex variable $w = u + iv$ may be related functionally to z by

$w = u + iv = f(x + iy) = f(z)$

which implies

$u = u(x, y)$ $v = v(x, y)$

e.g.,

$\cosh z = \cosh(x + iy) = \cosh x \cosh iy + \sinh x \sinh iy$
$= \cosh x \cos y + i \sinh x \sin y$

$u = \cosh x \cos y$ $v = \sinh x \sin y$

1.12 Cauchy–Riemann equations

If $u(x, y)$ and $v(x, y)$ are continuously differentiable with respect to x and y,

$\frac{\partial u}{\partial x} = \frac{\partial v}{\partial y}$ $\frac{\partial u}{\partial y} = -\frac{\partial v}{\partial x}$

$w = f(z)$ is continuously differentiable with respect to z and its derivative is

$$f'(z) = \frac{\partial u}{\partial x} + i\frac{\partial v}{\partial x} = \frac{\partial v}{\partial y} - i\frac{\partial u}{\partial y} = \frac{1}{i}\left(\frac{\partial u}{\partial y} + i\frac{\partial v}{\partial y}\right)$$

It is also easy to show that $\nabla^2 u = \nabla^2 v = 0$. Since the transformation from z to w is conformal, the curves u = constant and v = constant intersect each other at right angles, so that one set may be used as equipotentials and the other as field lines in a vector field.

1.13 Cauchy's theorem

If $f(z)$ is analytic everywhere inside a region bounded by C and a is a point within C

$$f(a) = \frac{1}{2\pi i} \int_C \frac{f(z)}{z - a} dz$$

This formula gives the value of a function at a point in the interior of a closed curve in terms of the values on that curve.

1.14 Zeros, poles and residues

If $f(z)$ vanishes at the point z_0 the Taylor series for z in the region of z_0 has its first two terms zero, and perhaps others also: $f(z)$ may then be written

$f(z) = (z - z_0)^n g(z)$

where $g(z_0) \neq 0$. Then $f(z)$ has a *zero* of order n at z_0. The reciprocal

$q(z) = 1/f(z) = h(z)/(z - z_0)^n$

where $h(z) = 1/g(z) \neq 0$ at z_0. $q(z)$ becomes infinite at $z = z_0$ and is said to have a *pole* of order n at z_0. $q(z)$ may be expanded in the form

$q(z)=c_{-n}(z-z_0)^n+\ldots+c_{-1}(z-z_0)^{-1}+c_0+\ldots$

where c_{-1} is the *residue* of $q(z)$ at $z=z_0$. From Cauchy's theorem, it may be shown that if a function $f(z)$ is analytic throughout a region enclosed by a curve C except at a finite number of poles, the integral of the function around C has a value of $2\pi i$ times the sum of the residues of the function at its poles within C. This fact can be used to evaluate many definite integrals whose indefinite form cannot be found.

1.15 Some standard forms

$$\int_0^{2\pi} e^{\cos\theta}\cos(n\theta-\sin\theta)\,d\theta=2\pi/n!$$

$$\int_0^{\infty}\frac{x^{a-1}}{1+x}\,dx=\pi\,\mathrm{cosec}\,a\pi$$

$$\int_0^{\infty}\frac{\sin\theta}{\theta}\,d\theta=\frac{\pi}{2}$$

$$\int_0^{\infty}x\exp(-h^2x^2)dx=\frac{1}{2h^2}$$

$$\int_0^{\infty}\frac{x^{a-1}}{1-x}\,dx=\pi\cot a\pi$$

$$\int_0^{\infty}\exp(-h^2x^2)\,dx=\frac{\sqrt{\pi}}{2h}$$

$$\int_0^{\infty}x^2\exp(-h^2x^2)\,dx=\frac{\sqrt{\pi}}{4h^3}$$

1.16 Coordinate systems

The basic system is the rectangular Cartesian system (x, y, z) to which all other systems are referred. Two other commonly used systems are as follows.

1.16.1 Cylindrical coordinates

Coordinates of point P are (x,y,z) or (r,θ,z) (see *Figure 1.4*), where

$x=r\cos\theta \qquad y=r\sin\theta \qquad z=z$

In these coordinates the volume element is $r\,dr\,d\theta\,dz$.

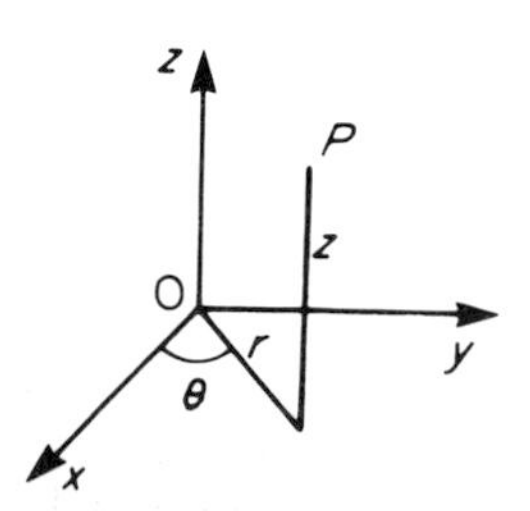

Figure 1.4 Cylindrical coordinates

1.16.2 Spherical polar coordinates

Coordinates of point P are (x,y,z) or (r,θ,φ) (see *Figure 1.5*), where

$x=r\sin\theta\cos\phi \qquad y=r\sin\theta\sin\phi \qquad z=r\cos\theta$

In these coordinates the volume element is $r^2\sin\theta\,dr\,d\theta\,d\phi$.

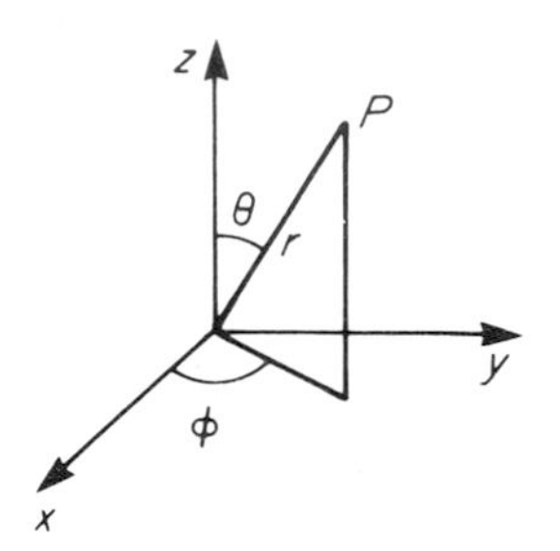

Figure 1.5 Spherical polar coordinates

1.17 Transformation of integrals

$$\iiint f(x,y,z)\,dx\,dy\,dz=\iiint \varphi(u,v,w)|J|\,du\,dv\,dw$$

where

$$J=\begin{vmatrix}\frac{\partial x}{\partial u} & \frac{\partial y}{\partial u} & \frac{\partial z}{\partial u}\\ \frac{\partial x}{\partial v} & \frac{\partial y}{\partial v} & \frac{\partial z}{\partial v}\\ \frac{\partial x}{\partial w} & \frac{\partial y}{\partial w} & \frac{\partial z}{\partial w}\end{vmatrix}=\frac{\partial(x,y,z)}{\partial(u,v,w)}$$

is the Jacobian of the transformation of coordinates. For Cartesian to cylindrical coordinates, $J=r$, and for Cartesian to spherical polars, it is $r^2\sin\theta$.

1.18 Laplace's equation

The equation satisfied by the scalar potential from which a vector field may be derived by taking the gradient is Laplace's equation, written as:

$$\nabla^2\phi=\frac{\partial^2\phi}{\partial x^2}+\frac{\partial^2\phi}{\partial y^2}+\frac{\partial^2\phi}{\partial z^2}=0$$

In cylindrical coordinates:

$$\nabla^2\phi=\frac{1}{r}\frac{\partial}{\partial r}\left(r\frac{\partial\phi}{\partial r}\right)+\frac{1}{r^2}\frac{\partial^2\phi}{\partial\theta^2}+\frac{\partial^2\phi}{\partial z^2}$$

In spherical polars:

$$\nabla^2\phi=\frac{1}{r^2}\frac{\partial}{\partial r}\left(r^2\frac{\partial\phi}{\partial r}\right)+\frac{1}{r^2\sin\theta}\frac{\partial}{\partial\theta}\left(\sin\theta\frac{\partial\phi}{\partial\theta}\right)+\frac{1}{r^2\sin^2\theta}\frac{\partial^2\phi}{\partial\phi^2}$$

The equation is solved by setting

$\phi=U(u)V(v)W(w)$

in the appropriate form of the equation, separating the variables and solving separately for the three functions, where (u, v, w) is the coordinate system in use.

In Cartesian coordinates, typically the functions are trigonometric, hyperbolic and exponential; in cylindrical coordinates the function of z is exponential, that of θ trigonometric and that of r is a Bessel function. In spherical polars, typically the function of r is a power of r, that of φ is trigonometric, and that of θ is a Legendre function of $\cos\theta$.

1.19 Solution of equations

1.19.1 Quadratic equation

$$ax^2+bx+c=0$$

$$x=-\frac{b}{2a}\pm\frac{\sqrt{b^2-4ac}}{2a}$$

In practical calculations if $b^2>4ac$, so that the roots are real and unequal, calculate the root of larger modulus first, using the same sign for both terms in the formula, then use the fact that $x_1x_2=c/a$ where x_1 and x_2 are the roots. This avoids the severe cancellation of significant digits which may otherwise occur in calculating the smaller root.

For polynomials other than quadratics, and for other functions, several methods of successive approximation are available.

1.19.2 Bisection method

By trial find x_0 and x_1 such that $f(x_0)$ and $f(x_1)$ have opposite signs (see *Figure 1.6*). Set $x_2=(x_0+x_1)/2$ and calculate $f(x_2)$. If

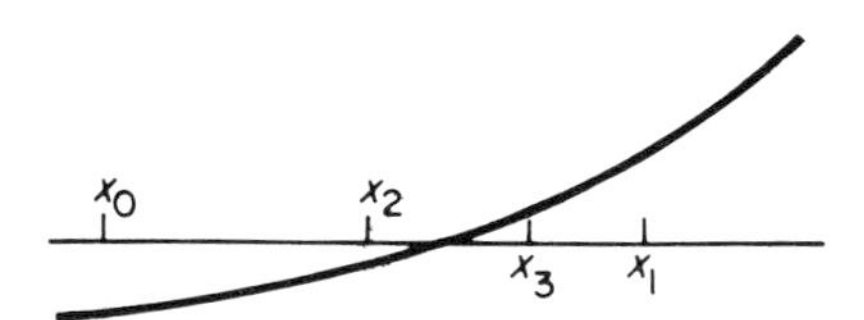

Figure 1.6 Bisection method

$f(x_0)f(x_2)$ is positive, the root lies in the interval (x_1, x_2); if negative in the interval (x_0, x_2); and if zero, x_2 is the root. Continue if necessary using the new interval.

1.19.3 Regula Falsi

By trial, find x_0 and x_1 as for the bisection method; these two values define two points $(x_0, f(x_0))$ and $(x_1, f(x_1))$. The straight line joining these two points cuts the x-axis at the point (see *Figure 1.7*)

$$x_2=\frac{x_0f(x_1)-x_1f(x_0)}{f(x_1)-f(x_0)}$$

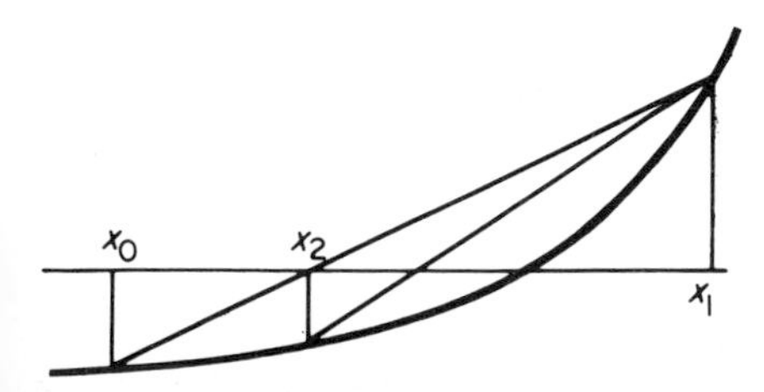

Figure 1.7 Regula Falsi

Evaluate $f(x_2)$ and repeat the process for whichever of the intervals (x_0, x_2) or (x_1, x_2) contains the root. This method can be accelerated by halving at each step the function value at the retained end of the interval, as shown in *Figure 1.8*.

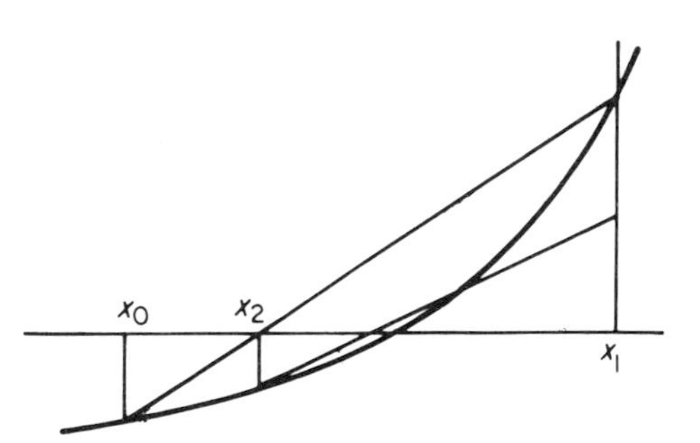

Figure 1.8 Accelerated method

1.19.4 Fixed-point iteration

Arrange the equation in the form

$$x=f(x)$$

Choose an initial value of x by trial, and calculate repetitively

$$x_{k+1}=f(x_k)$$

This process will not always converge.

1.19.5 Newton's method

Calculate repetitively (*Figure 1.9*)

$$x_{k+1}=x_k-f(x_k)/f'(x_k)$$

This method will converge unless: (a) x_k is near a point of inflexion of the function; or (b) x_k is near a local minimum; or (c) the root is

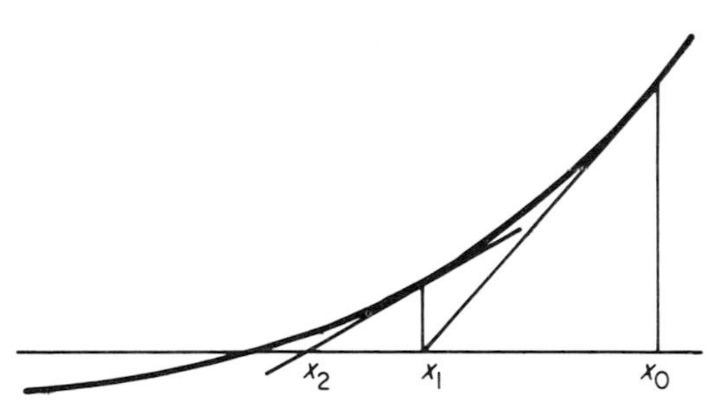

Figure 1.9 Newton's method

multiple. If one of these cases arises, most of the trouble can be overcome by checking at each stage that

$$f(x_{k+1})<f(x_k)$$

and if not, halving the preceding value of $|x_{k+1}-x_k|$.

1.20 Method of least squares

To obtain the best fit between a straight line $ax+by=1$ and several points (x_1, y_1), (x_2, y_2), ..., (x_n, y_n) found by observation, the coefficients a and b are to be chosen so that the sum of the squares of the errors

$e_i = ax_i + by_i - 1$

is a minimum. To do this, first write the set of inconsistent equations

$$ax_1 + by_1 - 1 = 0$$

$$ax_2 + by_2 - 1 = 0$$

$$\vdots$$

$$ax_n + by_n - 1 = 0$$

Multiply each equation by the value of x it contains, and add, obtaining

$$a \sum_{i=1}^{n} x_i^2 + b \sum_{i=1}^{n} x_i y_i - \sum_{i=1}^{n} x_i = 0$$

Similarly multiply by y and add, obtaining

$$a \sum_{i=1}^{n} x_i y_i + b \sum_{i=1}^{n} y_i^2 - \sum_{i=1}^{n} y_i = 0$$

Lastly, solve these two equations for a and b, which will be the required values giving the least squares fit.

1.21 Relation between decibels, current and voltage ratio, and power ratio

$$\mathrm{dB} = 10 \log \frac{P_1}{P_2} = 20 \log \frac{V_1}{V_2} = 20 \log \frac{I_1}{I_2}$$

dB	I_1/I_2 or V_1/V_2	I_2/I_1 or V_2/V_1	P_1/P_2	P_2/P_1
0.1	1.012	0.989	1.023	0.977
0.2	1.023	0.977	1.047	0.955
0.3	1.035	0.966	1.072	0.933
0.4	1.047	0.955	1.096	0.912
0.5	1.059	0.944	1.122	0.891
0.6	1.072	0.933	1.148	0.871
0.7	1.084	0.923	1.175	0.851
0.8	1.096	0.912	1.202	0.832
0.9	1.109	0.902	1.230	0.813
1.0	1.122	0.891	1.259	0.794
1.1	1.135	0.881	1.288	0.776
1.2	1.148	0.871	1.318	0.759
1.3	1.162	0.861	1.349	0.741
1.4	1.175	0.851	1.380	0.724
1.5	1.188	0.841	1.413	0.708
1.6	1.202	0.832	1.445	0.692
1.7	1.216	0.822	1.479	0.676
1.8	1.230	0.813	1.514	0.661
1.9	1.245	0.804	1.549	0.645
2.0	1.259	0.794	1.585	0.631
2.5	1.334	0.750	1.778	0.562
3.0	1.413	0.708	1.995	0.501
3.5	1.496	0.668	2.24	0.447
4.0	1.585	0.631	2.51	0.398
4.5	1.679	0.596	2.82	0.355
5.0	1.778	0.562	3.16	0.316
5.5	1.884	0.531	3.55	0.282
6.0	1.995	0.501	3.98	0.251
6.5	2.11	0.473	4.47	0.224
7.0	2.24	0.447	5.01	0.200
7.5	2.37	0.422	5.62	0.178
8.0	2.51	0.398	6.31	0.158
8.5	2.66	0.376	7.08	0.141
9.0	2.82	0.355	7.94	0.126
9.5	2.98	0.335	8.91	0.112
10.0	3.16	0.316	10.00	0.100
10.5	3.35	0.298	11.2	0.089 1
11.0	3.55	0.282	12.6	0.079 4
15.0	5.62	0.178	31.6	0.031 6
15.5	5.96	0.168	35.5	0.028 2
16.0	6.31	0.158	39.8	0.025 1
16.5	6.68	0.150	44.7	0.022 4
17.0	7.08	0.141	50.1	0.020 0
17.5	7.50	0.133	56.2	0.017 8
18.0	7.94	0.126	63.1	0.015 8
18.5	8.41	0.119	70.8	0.014 1
19.0	8.91	0.112	79.4	0.012 6
19.5	9.44	0.106	89.1	0.011 2
20.0	10.00	0.100 0	100	0.010 0
20.5	10.59	0.094 4	112	0.008 91
21.0	11.22	0.089 1	126	0.007 94
21.5	11.88	0.084 1	141	0.007 08
22.0	12.59	0.079 4	158	0.006 31
22.5	13.34	0.075 0	178	0.005 62
23.0	14.13	0.070 8	200	0.005 01
23.5	14.96	0.066 8	224	0.004 47
24.0	15.85	0.063 1	251	0.003 98
24.5	16.79	0.059 6	282	0.003 55
25.0	17.78	0.056 2	316	0.003 16
25.5	18.84	0.053 1	355	0.002 82
26.0	19.95	0.050 1	398	0.002 51
26.5	21.1	0.047 3	447	0.002 24
27.0	22.4	0.044 7	501	0.002 00
27.5	23.7	0.042 2	562	0.001 78
28.0	25.1	0.039 8	631	0.001 58
28.5	26.6	0.037 6	708	0.001 41
29.0	28.2	0.035 5	794	0.001 26
29.5	29.8	0.033 5	891	0.001 12
30.0	31.6	0.031 6	1 000	0.001 00
31.0	35.5	0.028 2	1 260	7.94×10^{-4}
32.0	39.8	0.025 1	1 580	6.31×10^{-4}
33.0	44.7	0.022 4	2 000	5.01×10^{-4}
34.0	50.1	0.020 0	2 510	3.98×10^{-4}
35.0	56.2	0.017 8		3.16×10^{-4}
36.0	63.1	0.015 8	3 980	2.51×10^{-4}
37.0	70.8	0.014 1	5 010	2.00×10^{-4}

2 Statistics

F. F. Mazda

Contents

2.1 Introduction

Data are available in vast quantities in all branches of electronic engineering. This chapter presents the more commonly used techniques for presenting and manipulating data to obtain meaningful results.

2.2 Data presentation

Probably the most common method used to present engineering data is by tables and graphs. For impact, or to convey information quickly, pictograms and bar charts may be used. Pie charts are useful in showing the different proportions of a unit.

A strata graph shows how the total is split amongst its constituents. For example, if a voltage is applied across four parallel circuits, then the total current curve may be as in *Figure 2.1*. This shows that the total current is made up of currents in the four parallel circuits, which vary in different ways with the applied voltage.

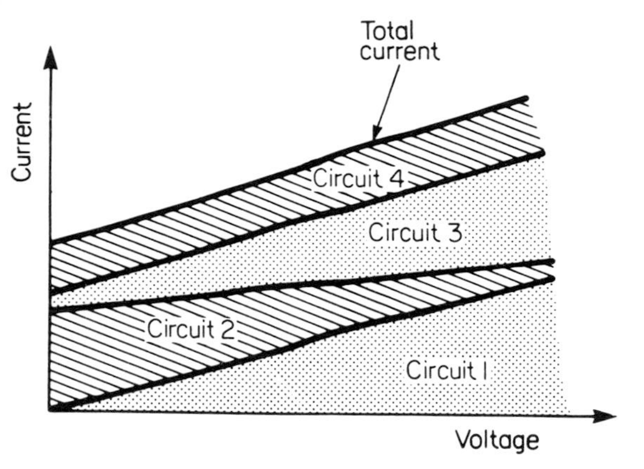

Figure 2.1 Illustration of a strata graph

Logarithmic or ratio graphs are used when one is more interested in the change in the ratios of numbers rather than their absolute value. In the logarithmic graph, equal ratios represent equal distances.

Frequency distributions are conveniently represented by a histogram as in *Figure 2.2*. This shows the voltage across a batch

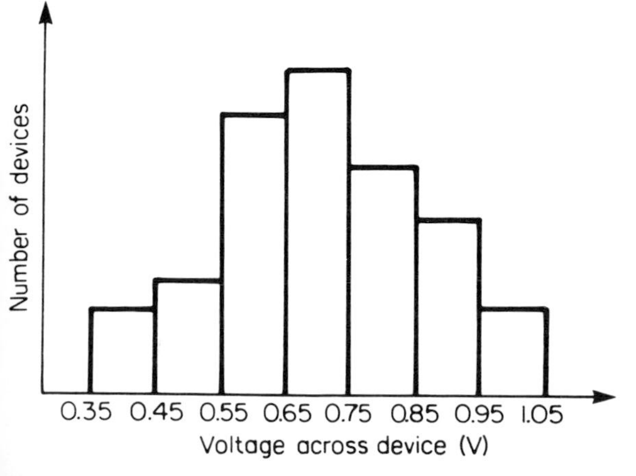

Figure 2.2 An histogram

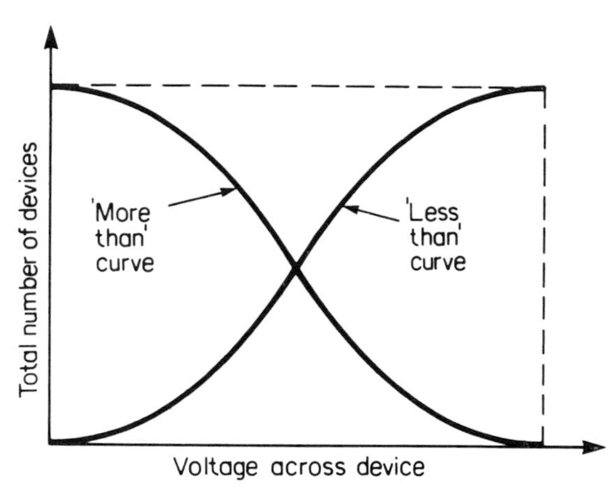

Figure 2.3 Illustration of ogives

of diodes. Most of the batch had voltage drops in the region 0.65 to 0.75 V, the next largest group being 0.55 to 0.65 volts. In a histogram, the areas of the rectangles represent the frequencies in the different groups. Ogives, illustrated in *Figure 2.3*, show the cumulative frequency occurrences above or below a given value. From this curve it is possible to read off the total number of devices having a voltage greater than or less than a specific value.

2.3 Averages

2.3.1 Arithmetic mean

The arithmetic mean of n numbers $x_1, x_2, x_3, \ldots, x_n$ is given by

$$\bar{x} = \frac{x_1 + x_2 + x_3 + \cdots + x_n}{n}$$

or

$$\bar{x} = \frac{\sum_{r=1}^{n} x_r}{n} \quad (2.1)$$

The arithmetic mean is easy to calculate and it takes into account all the figures. Its disadvantages are that it is influenced unduly by extreme values and the final result may not be a whole number, which can be absurd at times, e.g. a mean of $2\frac{1}{2}$ men.

2.3.2 Median and mode

Median or 'middle one' is found by placing all the figures in order and choosing the one in the middle, or if there are an even number of items, the mean of the two central numbers. It is a useful technique for finding the average of items which cannot be expressed in figures, e.g. shades of a colour. It is also not influenced by extreme values. However, the median is not representative of all the figures.

The mode is the most 'fashionable' item, that is, the one which appears the most frequently.

2.3.3 Geometric mean

The geometric mean of n numbers $x_1, x_2, x_3, \ldots, x_n$ is given by

$$x_g = \sqrt[n]{(x_1 \times x_2 \times x_3 \times \ldots \times x_n)} \quad (2.2)$$

This technique is used to find the average of quantities which follow a geometric progression or exponential law, such as rates of changes. Its advantage is that it takes into account all the numbers, but is not unduly influenced by extreme values.

2.3.4 Harmonic mean

The harmonic mean of n numbers $x_1, x_2, x_3, \ldots, x_n$ is given by

$$x_h = \frac{n}{\sum_{r=1}^{n} (1/x_r)} \tag{2.3}$$

This averaging method is used when dealing with rates or speeds or prices. As a rule when dealing with items such as A per B, if the figures are for equal As then use the harmonic mean but if they are for equal Bs use the arithmetic mean. So if a plane flies over three equal distances at speeds of 5 m/s, 10 m/s and 15 m/s the mean speed is given by the harmonic mean as

$$\frac{3}{\frac{1}{5}+\frac{1}{10}+\frac{1}{15}} = 8.18 \text{ m/s}$$

If, however, the plane were to fly for three equal times, of say, 20 seconds at speeds of 5 m/s, 10 m/s and 15 m/s, then the mean speed would be given by the arithmetic mean as $(5+10+15)/3 = 10$ m/s.

2.4 Dispersion from the average

2.4.1 Range and quartiles

The average represents the central figure of a series of numbers or items. It does not give any indication of the spread of the figures, in the series, from the average. Therefore, in *Figure 2.4*, both curves, A and B, have the same average but B has a wider deviation from the average than curve A.

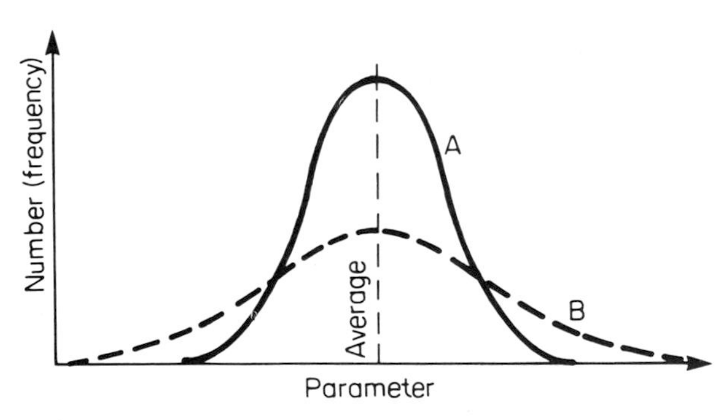

Figure 2.4 Illustration of deviation from the average

There are several ways of stating by how much the individual numbers, in the series, differ from the average. The range is the difference between the smallest and largest values. The series can also be divided into four quartiles and the dispersion stated as the interquartile range, which is the difference between the first and third quartile numbers, or the quartile deviation which is half this value.

The quartile deviation is easy to use and is not influenced by extreme values. However, it gives no indication of distribution between quartiles and covers only half the values in a series.

2.4.2 Mean deviation

This is found by taking the mean of the differences between each individual number in the series and the arithmetic mean, or median, of the series. Negative signs are ignored.

For a series of n numbers $x_1, x_2, x_3, \ldots, x_n$ having an arithmetic mean of $\bar{x}$ the mean deviation of the series is given by

$$\frac{\sum_{r=1}^{n} |x_r - \bar{x}|}{n} \tag{2.4}$$

The mean deviation takes into account all the items in the series. But it is not very suitable since it ignores signs.

2.4.3 Standard deviation

This is the most common measure of dispersion. For this the arithmetic mean must be used and not the median. It is calculated by squaring deviations from the mean, so eliminating their sign, adding the numbers together and then taking their mean and then the square root of the mean. Therefore, for the series in Section 6.4.2 the standard deviation is given by

$$\sigma = \left(\frac{\sum_{r=1}^{n} (x_r - \bar{x})^2}{n}\right)^{1/2} \tag{2.5}$$

The unit of the standard deviation is that of the original series. So if the series consists of the heights of a group of children in metres, then the mean and standard deviation are in metres. To compare two series having different units, such as the height of children and their weights, the coefficient of variation is used, which is unitless:

$$\text{coefficient of variation} = \frac{\sigma}{\bar{x}} \times 100 \tag{2.6}$$

2.5 Skewness

The distribution shown in *Figure 2.4* is symmetrical since the mean, median and mode all coincide. *Figure 2.5* shows a skewed distribution. It has positive skewness although if it bulges the other way, the skewness is said to be negative.

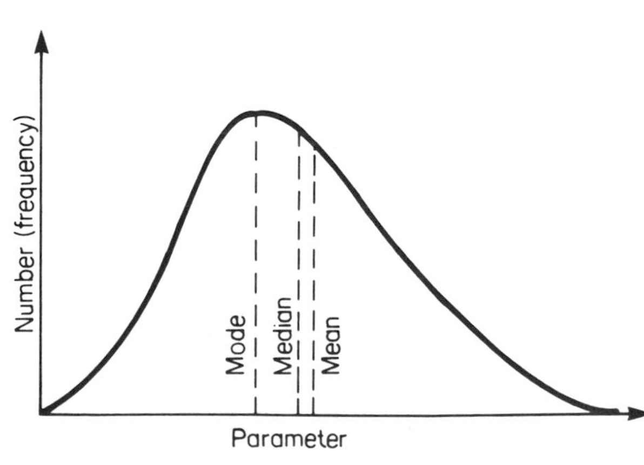

Figure 2.5 Illustration of skewness

There are several mathematical ways for expressing skewness. They all give a measure of the deviation between the mean, median and mode and they are usually stated in relative terms, for ease of comparison between series of different units. The Pearson coefficient of skewness is given by

$$P_k = \frac{\text{mean} - \text{mode}}{\text{standard deviation}} \tag{2.7}$$

Since the mode is sometimes difficult to measure this can also be stated as

$$P_k = \frac{3\,(\text{mean} - \text{median})}{\text{standard deviation}} \tag{2.8}$$

2.6 Combinations and permutations

2.6.1 Combinations

Combinations are the number of ways in which a proportion can be chosen from a group. Therefore the number of ways in which two letters can be chosen from a group of four letters A, B, C, D is equal to 6, i.e. AB, AC, AD, BC, BD, CD. This is written as

$$^4C_2 = 6$$

The factorial expansion is frequently used in combination calculations where

$$n! = n \times (n-1) \times (n-2) \times \cdots \times 3 \times 2 \times 1$$

Using this the number of combinations of n items from a group of n is given by

$$^nC_r = \frac{n!}{r!\,(n-r)!} \tag{2.9}$$

2.6.2 Permutations

Combinations do not indicate any sequencing. When sequencing within each combination is involved the result is known as a permutation. Therefore the number of permutations of two letters out of four letters A, B, C, D is 12, i.e. AB, BA, AC, CA, AD, DA, BC, CB, BD, DB, CD, DC. The number of permutations of r items from a group of n is given by

$$^nP_r = \frac{n!}{(n-r)!} \tag{2.10}$$

2.7 Regression and correlation

2.7.1 Regression

Regression is a method for establishing a mathematical relationship between two variables. Several equations may be used to establish this relationship, the most common being that of a straight line. *Figure 2.6* shows the plot of seven readings. This is called a scatter diagram. The points can be seen to lie approximately on the straight line AB.

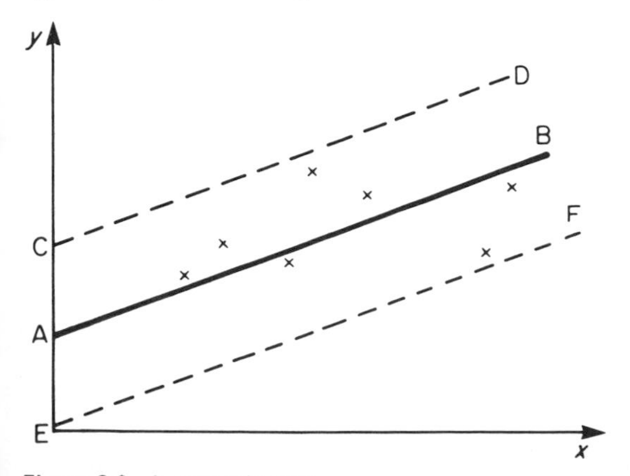

Figure 2.6 A scatter diagram

The equation of a straight line is given by

$$y = mx + c \tag{2.11}$$

where x is the independent variable, y the dependent variable, m is the slope of the line and c its interception on the y-axis. c is negative if the line intercepts the y-axis on its negative part and m is negative if the line slopes the other way to that shown in *Figure 2.6*.

The best straight line to fit a set of points is found by the method of least squares as

$$m = \frac{\sum xy - (\sum x \sum y)/n}{\sum x^2 - (\sum x)^2/n} \tag{2.12}$$

and

$$c = \frac{\sum x \sum xy - \sum y \sum x^2}{(\sum x)^2 - n \sum x^2} \tag{2.13}$$

where n is the number of points. The line passes through the mean values of x and y, i.e. $\bar{x}$ and $\bar{y}$.

2.7.2 Correlation

Correlation is a technique for establishing the strength of the relationship between variables. In *Figure 2.6* the individual figures are scattered on either side of a straight line and although one can approximate them by a straight line it may be required to establish if there is correlation between the x- and y-readings.

Several correlation coefficients exist. The product moment correlation coefficient (r) is given by

$$r = \frac{\sum (x-\bar{x})(y-\bar{y})}{n\sigma_x \sigma_y} \tag{2.14}$$

or

$$r = \frac{\sum (x-\bar{x})(y-\bar{y})}{[\sum (x-\bar{x})^2 \sum (y-\bar{y})^2]^{1/2}} \tag{2.15}$$

The value of r varies from $+1$, when all the points lie on a straight line and y increases with x, to -1, when all the points lie on a straight line but y decreases with x. When $r=0$ the points are widely scattered and there is said to be no correlation between x and y.

The standard error of estimation in r is given by

$$S_y = \sigma_y (1 - r^2)^{1/2} \tag{2.16}$$

In about 95% of cases, the actual values will lie between plus or minus twice the standard error of estimated values given by the regression equation. This is shown by lines CD and EF in *Figure 2.6*. Almost all the values will be within plus or minus three times the standard error of estimated values.

It should be noted that σ_y is the variability of the y-values, whereas S_y is a measure of the variability of the y-values as they differ from the regression which exists between x and y. If there is no regression then $r=0$ and $\sigma_y = S_y$.

It is often necessary to draw conclusions from the order in which items are ranked. For example, two judges may rank contestants in a beauty contest and we need to know if there is any correlation between their rankings. This may be done by using the Rank correlation coefficient (R) given by

$$R = 1 - \frac{6 \sum d^2}{n^3 - n} \tag{2.17}$$

where d is the difference between the two ranks for each item and n is the number of items. The value of R will vary from $+1$ when the two ranks are identical to -1 when they are exactly reversed.

2.8 Probability

If an event A occurs n times out of a total of m cases then the probability of occurrence is stated to be

$$P(\mathrm{A}) = n/m \tag{2.18}$$

Probability varies between 0 and 1. If $P(\mathrm{A})$ is the probability of occurrence then $1 - P(\mathrm{A})$ is the probability that event A will not occur and it can be written as $P(\bar{\mathrm{A}})$.

If A and B are two events then the probability that either may occur is given by

$$P(\mathrm{A\ or\ B}) = P(\mathrm{A}) + P(\mathrm{B}) - P(\mathrm{A\ and\ B}) \tag{2.19}$$

A special case of this probability law is when events are mutually exclusive, i.e. the occurrence of one event prevents the other from happening. Then

$$P(\mathrm{A\ or\ B}) = P(\mathrm{A}) + P(\mathrm{B}) \tag{2.20}$$

If A and B are two events then the probability that they may occur together is given by

$$P(\mathrm{A\ and\ B}) = P(\mathrm{A}) \times P(\mathrm{B}|\mathrm{A}) \tag{2.21}$$

or

$$P(\mathrm{A\ and\ B}) = P(\mathrm{B}) \times P(\mathrm{A}|\mathrm{B}) \tag{2.22}$$

$P(\mathrm{B}|\mathrm{A})$ is the probability that event B will occur assuming that event A has already occurred and $P(\mathrm{A}|\mathrm{B})$ is the probability that event A will occur assuming that event B has already occurred. A special case of this probability law is when A and B are independent events, i.e. the occurrence of one event has no influence on the probability of the other event occurring. Then

$$P(\mathrm{A\ and\ B}) = P(\mathrm{A}) \times P(\mathrm{B}) \tag{2.23}$$

Bayes' theorem on probability may be stated as

$$P(\mathrm{A}|\mathrm{B}) = \frac{P(\mathrm{A})P(\mathrm{B}|\mathrm{A})}{P(\mathrm{A})P(\mathrm{B}|\mathrm{A}) + P(\bar{\mathrm{A}})P(\mathrm{B}|\bar{\mathrm{A}})} \tag{2.24}$$

As an example of the use of Bayes' theorem suppose that a company discovers that 80% of those who bought its product in a year had been on the company's training course. 30% of those who bought a competitir's product had also been on the same training course. During that year the company had 20% of the market. The company wishes to know what percentage of buyers actually went on its training course, in order to discover the effectiveness of this course.

If B denotes that a person bought the company's product and T that he went on the training course then the problem is to find $P(\mathrm{B}|\mathrm{T})$. From the data $P(\mathrm{B}) = 0.2$, $P(\bar{\mathrm{B}}) = 0.8$, $P(\mathrm{T}|\mathrm{B}) = 0.8$, $P(\mathrm{T}|\bar{\mathrm{B}}) = 0.3$. Then from Equation (2.24)

$$P(\mathrm{B}|\mathrm{T}) = \frac{0.2 \times 0.8}{0.2 \times 0.8 + 0.8 \times 0.3} = 0.4$$

2.9 Probability distributions

There are several mathematical formulae with well-defined characteristics and these are known as probability distributions. If a problem can be made to fit one of these distributions then its solution is simplified. Distributions can be discrete when the characteristic can only take certain specific values, such as 0, 1, 2, etc., or they can be continuous where the characteristic can take any value.

2.9.1 Binomial distribution

The binomial probability distribution is given by

$$(p+q)^n = q^n + {}^nC_1 pq^{n-1} + {}^nC_2 p^2 q^{n-2} + \cdots + {}^nC_x p^x q^{n-x} + \cdots + p^n \tag{2.25}$$

where p is the probability of an event occurring, q $(=1-p)$ is the probability of an event not occurring and n is the number of selections.

The probability of an event occurring m successive times is given by the binomial distribution as

$$p(m) = {}^nC_m p^m q^{n-m} \tag{2.26}$$

The binomial distribution is used for discrete events and is applicable if the probability of occurrence p of an event is constant on each trial. The mean of the distribution $B(M)$ and the standard deviation $B(S)$ are given by

$$B(M) = np \tag{2.27}$$

$$B(S) = (npq)^{1/2} \tag{2.28}$$

2.9.2 Poisson distribution

The Poisson distribution is used for discrete events and, like the binomial distribution, it applies to mutually independent events. It is used in cases where p and q cannot both be defined. For example, one can state the number of goals which were scored in a football match, but not the goals which were not scored.

The Poisson distribution may be considered to be the limiting case of the binomial when n is large and p is small. The probability of an event occurring m successive times is given by the Poisson distribution as

$$p(m) = (np)^m \frac{\mathrm{e}^{-np}}{m!} \tag{2.29}$$

The mean $P(M)$ and standard deviation $P(S)$ of the Poisson distribution are given by

$$P(M) = np \tag{2.30}$$

$$P(S) = (np)^{1/2} \tag{2.31}$$

Poisson probability calculations can be done by the use of probability charts as shown in *Figure 2.7*. This shows the probability that an event will occur at least m times when the mean (or expected) value np is known.

2.9.3 Normal distribution

The normal distribution represents continuous events and is shown plotted in *Figure 2.8*. The x-axis gives the event and the y-

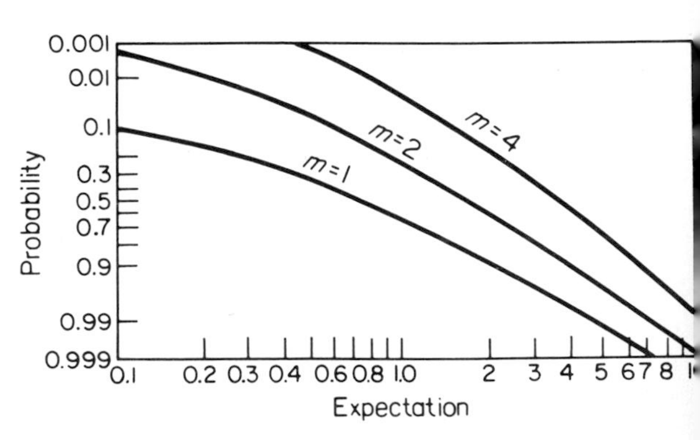

Figure 2.7 Poisson probability paper

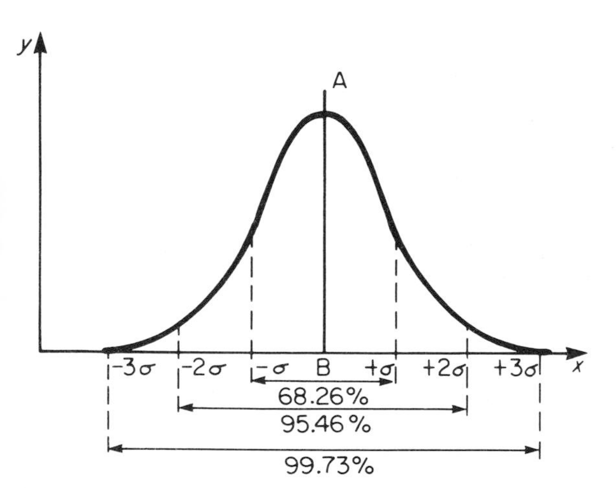

Figure 2.8 The normal curve

axis the probability of the event occurring. The curve shows that most of the events occur close to the mean value and this is usually the case in nature. The equation of the normal curve is given by

$$y=\frac{1}{\sigma(2\pi)^{1/2}}e^{-(x-\bar{x})^2/2\sigma^2} \quad (2.32)$$

where $\bar{x}$ is the mean of the values making up the curve and σ is their standard deviation.

Different distributions will have varying mean and standard deviations but if they are distributed normally then their curves will all follow Equation (2.32). These distributions can all be normalised to a standard form by moving the origin of their normal curve to their mean value, shown as B in *Figure 2.8*. The deviation from the mean is now represented on a new scale of units given by

$$\omega=\frac{x-\bar{x}}{\sigma} \quad (2.33)$$

The equation for the standardised normal curve now becomes

$$y=\frac{1}{(2\pi)^{1/2}}e^{-\omega^2/2} \quad (2.34)$$

The total area under the standardised normal curve is unity and the area between any two values of ω is the probability of an item from the distribution falling between these values. The normal curve extends infinitely in either direction but 68.26% of its values (area) fall between $\pm\sigma$, 95.46% between $\pm 2\sigma$, 99.73% between $\pm 3\sigma$ and 99.994% between $\pm 4\sigma$.

Table 2.1 gives the area under the normal curve for different values of ω. Since the normal curve is symmetrical the area from $+\omega$ to $+\infty$ is the same as from $-\omega$ to $-\infty$. As an example of the use of this table, suppose that 5000 street lamps have been installed in a city and that the lamps have a mean life of 1000 hours with a standard deviation of 100 hours. How many lamps will fail in the first 800 hours? From Equation (2.33)

$$\omega=(800-1000)/100=-2$$

Ignoring the negative sign, *Table 2.1* gives the probability of lamps not failing as 0.977 so that the probability of failure is $1-0.977$ or 0.023. Therefore 5000×0.023 or 115 lamps are expected to fail after 800 hours.

Table 2.1 Area under the normal curve from $-\infty$ to ω

ω	0.00	0.02	0.04	0.06	0.08
0.0	0.500	0.508	0.516	0.524	0.532
0.1	0.540	0.548	0.556	0.564	0.571
0.2	0.579	0.587	0.595	0.603	0.610
0.3	0.618	0.626	0.633	0.640	0.648
0.4	0.655	0.663	0.670	0.677	0.684
0.5	0.692	0.700	0.705	0.712	0.719
0.6	0.726	0.732	0.739	0.745	0.752
0.7	0.758	0.764	0.770	0.776	0.782
0.8	0.788	0.794	0.800	0.805	0.811
0.9	0.816	0.821	0.826	0.832	0.837
1.0	0.841	0.846	0.851	0.855	0.860
1.1	0.864	0.869	0.873	0.877	0.881
1.2	0.885	0.889	0.893	0.896	0.900
1.3	0.903	0.907	0.910	0.913	0.916
1.4	0.919	0.922	0.925	0.928	0.931
1.5	0.933	0.936	0.938	0.941	0.943
1.6	0.945	0.947	0.950	0.952	0.954
1.7	0.955	0.957	0.959	0.961	0.963
1.8	0.964	0.966	0.967	0.969	0.970
1.9	0.971	0.973	0.974	0.975	0.976
2.0	0.977	0.978	0.979	0.980	0.981
2.1	0.982	0.983	0.984	0.985	0.985
2.2	0.986	0.987	0.988	0.988	0.989
2.3	0.989	0.990	0.990	0.991	0.991
2.4	0.992	0.992	0.993	0.993	0.993
2.5	0.994	0.994	0.995	0.995	0.995
2.6	0.995	0.996	0.996	0.996	0.996
2.7	0.997	0.997	0.997	0.997	0.997
2.8	0.997	0.998	0.998	0.998	0.998
2.9	0.998	0.998	0.998	0.998	0.999
3.0	0.999	0.999	0.999	0.999	0.999

2.9.4 Exponential distribution

The exponential probability distribution is a continuous distribution and is shown in *Figure 6.9*. It has the equation

$$y=\frac{1}{x}e^{-x/\bar{x}} \quad (2.35)$$

where $\bar{x}$ is the mean of the distribution. Whereas in the normal distribution the mean value divides the population in half, for the exponential distribution 36.8% of the population is above the average and 63.2% below the average. *Table 2.2* shows the area under the exponential curve for different values of the ratio $K=x/\bar{x}$, this area being shown shaded in *Figure 2.9*.

As an example suppose that the time between failures of a piece

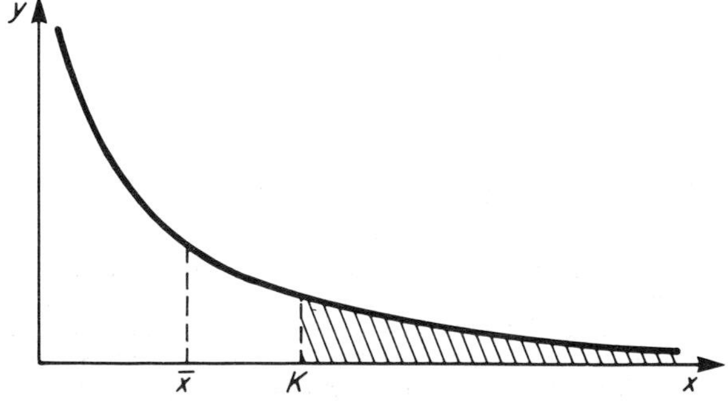

Figure 2.9 The exponential curve

Table 2.2 Area under the exponential curve from K to $+\infty$

K	0.00	0.02	0.04	0.06	0.08
0.0	1.000	0.980	0.961	0.942	0.923
0.1	0.905	0.886	0.869	0.852	0.835
0.2	0.819	0.803	0.787	0.771	0.776
0.3	0.741	0.726	0.712	0.698	0.684
0.4	0.670	0.657	0.644	0.631	0.619
0.5	0.607	0.595	0.583	0.571	0.560
0.6	0.549	0.538	0.527	0.517	0.507
0.7	0.497	0.487	0.477	0.468	0.458
0.8	0.449	0.440	0.432	0.423	0.415
0.9	0.407	0.399	0.391	0.383	0.375

of equipment is found to vary exponentially. If results indicate that the mean time between failures is 1000 hours, then what is the probability that the equipment will work for 700 hours or more without a failure? Calculating K as $700/1000 = 0.7$ then from *Table 2.2* the area beyond 0.7 is 0.497 which is the probability that the equipment will still be working after 700 hours.

2.9.5 Weibull distribution

This is a continuous probability distribution and its equation is given by

$$y = \alpha\beta(x-\gamma)^{\beta-1}\, e^{-\alpha(x-\gamma)^{\beta}} \qquad (2.36)$$

where α is called the scale factor, β the shape factor and γ the location factor.

The shape of the Weibull curve varies depending on the value of its factors. β is the most important, as shown in *Figure 2.10*, and

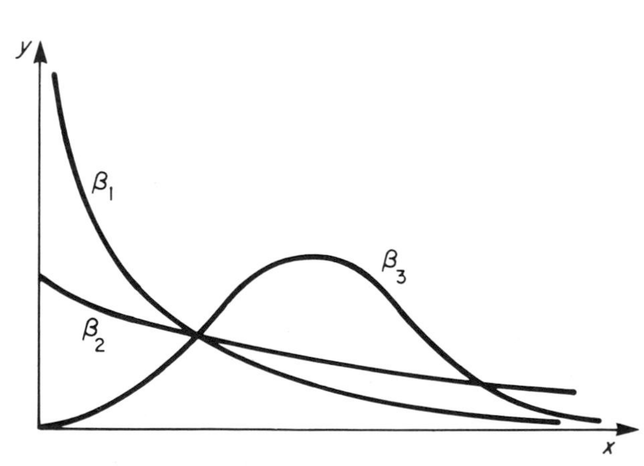

Figure 2.10 Weibull curves ($\alpha = 1$)

the Weibull curve varies from an exponential ($\beta = 1.0$) to a normal distribution ($\beta = 3.5$). In practice β varies from about $\frac{1}{3}$ to 5. Because the Weibull distribution can be made to fit a variety of different sets of data, it is popularly used for probability distributions.

Analytical calculations using the Weibull distribution are cumbersome. Usually predictions are made using Weibull probability paper. The data are plotted on this paper and the probability predictions read from the graph.

2.10 Sampling

A sample consists of a relatively small number of items drawn from a much larger population. This sample is analysed for certain attributes and it is then assumed that these attributes apply to the total population, within a certain tolerance of error.

Sampling is usually associated with the normal probability distribution and, based on this distribution, the errors which arise due to sampling can be estimated. Suppose a sample of n_s items is taken from a population of n_p items which are distributed normally. If the sample is found to have a mean of μ_s with a standard deviation of σ_s then the mean μ_p of the population can be estimated to be within a certain tolerance of μ_s. It is given by

$$\mu_p = \mu_s \pm \frac{\gamma\sigma_s}{n_s^{1/2}} \qquad (2.37)$$

γ is found from the normal curve depending on the level of confidence we need in specifying μ_p. For $\gamma = 1$ this level is 68.26%; for $\gamma = 2$ it is 95.46% and for $\gamma = 3$ it is 99.73%.

The standard error of mean σ_e is often defined as

$$\sigma_e = \frac{\sigma_s}{n_s^{1/2}} \qquad (2.38)$$

so Equation (6.37) can be rewritten as

$$\mu_p = \mu_s \pm \gamma\sigma_e \qquad (2.39)$$

As an example suppose that a sample of 100 items, selected at random from a much larger population, gives their mean weight as 20 kg with a standard deviation of 100 g. The standard error of the mean is therefore $100/(100)^{1/2} = 10$ g and one can say with 99.73% confidence that the mean value of the population lies between $20 \pm 3 \times 0.01$ or 20.03 kg and 19.97 kg.

If in a sample of n_s items the probability of occurrence of a particular attribute is p_s, then the standard error of probability p_e is defined as

$$p_e = \left(\frac{p_s q_s}{n_s}\right)^{1/2} \qquad (2.40)$$

where $q_s = 1 - p_s$.

The probability of occurrence of the attribute in the population is then given by

$$p_p = p_s \pm \gamma p_e \qquad (2.41)$$

where γ is again chosen to cover a certain confidence level.

As an example suppose a sample of 500 items shows that 50 are defective. Then the probability of occurrence of the defect in the sample is $50/500 = 0.1$. The standard error of probability is $(0.1 \times 0.9/500)^{1/2}$ or 0.0134. Therefore we can state with 95.46% confidence that the population from which the sample was drawn has a defect probability of $0.1 \pm 2 \times 0.0134$, i.e. 0.0732 to 0.1268; or we can state with 99.73% confidence that this value will lie between $0.1 \pm 3 \times 0.0134$, i.e. 0.0598 to 0.1402.

If two samples have been taken from the same population and these give standard deviations of σ_{s1} and σ_{s2} for sample sizes of n_{s1} and n_{s2} then Equation (2.38) can be modified to give the standard error of the difference between means as

$$\sigma_{de} = \left(\frac{\sigma_{s1}^2}{n_{s1}} + \frac{\sigma_{s2}^2}{n_{s2}}\right)^{1/2} \qquad (2.42)$$

Similarly Equation (2.40) can be modified to give the standard error of the difference between probabilities of two samples from the same population as

$$p_{de} = \left(\frac{p_{s1}q_{s1}}{n_{s1}} + \frac{p_{s2}q_{s2}}{n_{s2}}\right)^{1/2} \qquad (2.43)$$

2.11 Tests of significance

In taking samples we often obtain results which deviate from the expected. Tests of significance are then used to determine if this deviation is real or if it could have arisen due to sampling error.

2.11.1 Hypothesis testing

In this system a hypothesis is set up and is then tested at a given confidence level. For example, suppose a coin is tossed 100 times and it comes up heads 60 times. Is the coin biased or is it likely that this falls within a reasonable sampling error? The hypothesis is set up that the coin is not biased. Therefore one would expect that the probability of heads is 0.5, i.e. $p_s = 0.5$. The probability of tails, q_s, is also 0.5. Using Equation (6.40) the standard error of probability is given by $p_e = (0.5 \times 0.5/100)^{1/2}$ or 0.05. Therefore from Equation (6.41) the population probability at the 95.45% confidence level of getting heads is $0.5 + 2 \times 0.05 = 0.6$. Therefore it is highly likely that the coin is not biased and the results are due to sampling error.

The results of any significance test are not conclusive. For example, is 95.45% too high a confidence level to require? The higher the confidence level the greater the risk of rejecting a true hypothesis, and the lower the level the greater the risk of accepting a false hypothesis.

Suppose now that a sample of 100 items of production shows that five are defective. A second sample of 100 items is taken from the same production a few months later and gives two defectives. Does this show that the production quality is improving? Using Equation (2.43) the standard error of the difference between probabilities is given by $(0.05 \times 0.95/100 + 0.02 \times 0.98/100)^{1/2} = 0.0259$. This is less than twice the difference between the two probabilities, i.e. $0.05 - 0.02 = 0.03$, therefore the difference is very likely to have arisen due to sampling error and it does not necessarily indicate an improvement in quality.

2.11.2 Chi-square test

This is written as χ^2. If O is an observed result and E is the expected result then

$$\chi^2 = \sum \frac{(O - E)^2}{E} \tag{2.44}$$

The χ^2 distribution is given by tables such as *Table 2.3*, from which the probability can be determined. The number of degrees of freedom is the number of classes whose frequency can be assigned independently. If the data are presented in the form of a table having V vertical columns and H horizontal rows then the degrees of freedom are usually found as $(V-1)(H-1)$.

Returning to the earlier example, suppose a coin is tossed 100 times and it comes up heads 60 times and tails 40 times. Is the coin biased? The expected values for heads and tails are 50 each so that

$$\chi^2 = \frac{(60-50)^2}{50} + \frac{(40-50)^2}{50} = 4$$

The number of degrees of freedom is one since once we have fixed the frequency for heads that for tails is defined. Therefore entering *Table 2.3* with one degree of freedom the probability level for $\chi^2 = 4$ is seen to be above 2.5%, i.e. there is a strong probability that the difference in the two results arose by chance and the coin is not biased.

As a further example suppose that over a 24-hour period the average number of accidents which occur in a factory is seen to be as in *Table 6.4*. Does this indicate that most of the accidents occur during the late night and early morning periods? Applying the χ^2 tests the expected value, if there was no difference between the time periods, would be the mean of the number of accidents, i.e. 5.

Table 2.3 The chi-square distribution

Degrees of freedom	*Probability level*				
	0.100	0.050	0.025	0.010	0.005
1	2.71	3.84	5.02	6.63	7.88
2	4.61	5.99	7.38	9.21	10.60
3	6.25	7.81	9.35	11.34	12.84
4	7.78	9.49	11.14	13.28	14.86
5	9.24	11.07	12.83	15.09	16.75
6	10.64	12.59	14.45	16.81	18.55
7	12.02	14.07	16.01	18.48	20.28
8	13.36	15.51	17.53	20.09	21.96
9	14.68	16.92	19.02	21.67	23.59
10	15.99	18.31	20.48	23.21	25.19
12	18.55	21.03	23.34	26.22	28.30
14	21.06	23.68	26.12	29.14	31.32
16	23.54	26.30	28.85	32.00	34.27
18	25.99	28.87	31.53	34.81	37.16
20	28.41	31.41	34.17	37.57	40.00
30	40.26	43.77	46.98	50.89	53.67
40	51.81	55.76	59.34	63.69	66.77

Table 2.4 Frequency distribution of accidents in a factory during 24 hours

Time (24 hour clock)	*Number of accidents*
0–6	9
6–12	3
12–18	2
18–24	6

Therefore from Equation (2.44)

$$\chi^2 = \frac{(9-5)^2}{5} + \frac{(3-5)^2}{5} + \frac{(2-5)^2}{5} + \frac{(6-5)^2}{5}$$
$$= 6$$

There are three degrees of freedom, therefore from *Table 2.3* the probability of occurence of the result shown in *Table 2.4* is seen to be greater than 10%. The conclusion would be that although there is a trend, as yet there are not enough data to show if this trend is significant or not. For example, if the number of accidents were each three times as large, i.e. 27, 9, 6, 18 respectively, then χ^2 would be calculated as 20.67 and from *Table 2.3* it is seen that the results are highly significant since there is a very low probability, less than $\frac{1}{2}$%, that it can arise by chance.

2.11.3 Significance of correlation

The significance of the product moment correlation coefficient of Equations (2.14) or (2.15) can be tested at any confidence level by means of the standard error of estimation given by Equation (2.16). An alternative method is to use the Student t test of significance. This is given by

$$t = \frac{r(n-2)^{1/2}}{(1-r^2)^{1/2}} \tag{2.45}$$

where r is the correlation coefficient and n the number of items. Tables are then used, similar to *Table 2.3*, which give the probability level for $(n-2)$ degrees of freedom.

The Student t for the rank correlation coefficient is given by

$$t = R[(n-2)/(1-R^2)]^{1/2} \tag{2.46}$$

and the same Student t tables are used to check the significance of R.

Further reading

BESTERFIELD, D. H., *Quality Control*, Prentice Hall (1979)
CAPLEN, R. H., *A Practical Approach to Quality Control*, Business Books (1982)
CHALK, G. O. and STICK, A. W., *Statistics for the Engineer*, Butterworths (1975)
DAVID, H. A., *Order Statistics*, Wiley (1981)
DUNN, R. A. and RAMSING, K. D., *Management Science, a Practical Approach to Decision Making*, Macmillan (1981)
FITZSIMMONS, J. A., *Service Operations Management*, McGraw-Hill (1982)
GRANT, E. L. and LEAVENWORTH, R. S., *Statistical Quality Control*, McGraw-Hill (1980)
HAHN, W. C., *Modern Statistical Methods*, Butterworths (1979)
LYONS, S., *Handbook of Industrial Mathematics*, Cambridge University Press (1978)

3 Quantities and units

L. W. Turner

Contents

3.1 International unit system

The International System of Units (SI) is the modern form of the metric system agreed at an international conference in 1960. It has been adopted by the International Standards Organisation (ISO) and the International Electrotechnical Commission (IEC) and its use is recommended wherever the metric system is applied. It is now being adopted throughout most of the world and is likely to remain the primary world system of units of measurement for a very long time. The indications are that SI units will supersede the units of existing metric systems and all systems based on Imperial units.

SI units and the rules for their application are contained in *ISO Resolution* R1000 (1969, updated 1973) and an informatory document *SI-Le Systeme International d'Unités*, published by the Bureau International des Poids et Mesures (BIPM). An abridged version of the former is given in British Standards Institution (BSI) publication PD 5686 *The use of SI Units* (1969, updated 1973) and BS 3763 *International System (SI) Units*; BSI (1964) incorporates information from the BIPM document.

The adoption of SI presents less of a problem to the electronics engineer and the electrical engineer than to those concerned with other engineering disciplines as all the practical electrical units were long ago incorporated in the metre-kilogram-second (MKS) unit system and these remain unaffected in SI.

The SI was developed from the metric system as a fully coherent set of units for science, technology and engineering. A coherent system has the property that corresponding equations between quantities and between numerical values have exactly the same form, because the relations between units do not involve numerical conversion factors. In constructing a coherent unit system, the starting point is the selection and definition of a minimum set of independent 'base' units. From these, 'derived' units are obtained by forming products or quotients in various combinations, again without numerical factors. Thus the base units of length (metre), time (second) and mass (kilogram) yield the SI units of velocity (metre/second), force (kilogram-metre/second-squared) and so on. As a result there is, for any given physical quantity, only one SI unit with no alternatives and with no numerical conversion factors. A single SI unit (joule = kilogram metre-squared/second-squared) serves for energy of any kind, whether it be kinetic, potential, thermal, electrical, chemical..., thus unifying the usage in all branches of science and technology.

The SI has seven base units, and two supplementary units of angle. Certain important derived units have special names and can themselves be employed in combination to form alternative names for further derivations.

Each physical quantity has a quantity-symbol (e.g., *m* for mass) that represents it in equations, and a unit-symbol (e.g., kg for kilogram) to indicate its SI unit of measure.

3.1.1 Base units

Definitions of the seven base units have been laid down in the following terms. The quantity-symbol is given in italics, the unit-symbol (and its abbreviation) in roman type.

Length: l; metre (m). The length equal to 1 650 763.73 wavelengths in vacuum of the radiation corresponding to the transition between the levels $2p_{10}$ and $5d_5$ of the krypton-86 atom.

Mass: m; kilogram (kg). The mass of the international prototype kilogram (a block of platinum preserved at the International Bureau of Weights and Measures at Sèvres).

Time: t; second (s). The duration of 9 192 631 770 periods of the radiation corresponding to the transition between the two hyperfine levels of the ground state of the caesium-133 atom.

Electric current: i; ampere (A). The current which, maintained in two straight parallel conductors of infinite length, of negligible circular cross-section and 1 m apart in vacuum, produces a force equal to 2×10^{-7} newton per metre of length.

Thermodynamic temperature: T; kelvin (K). The fraction 1/273.16 of the thermodynamic (absolute) temperature of the triple point of water.

Luminous intensity: I; candela (cd). The luminous intensity in the perpendicular direction of a surface of 1/600 000 m^2 of a black body at the temperature of freezing platinum under a pressure of 101 325 newtons per square metre.

Amount of substance: Q; mole (mol). The amount of substance of a system which contains as many elementary entities as there are atoms in 0.012 kg of carbon-12. The elementary entity must be specified and may be an atom, a molecule, an ion, an electron, etc., or a specified group of such entities.

3.1.2 Supplementary angular units

Plane angle: α, β...; radian (rad). The plane angle between two radii of a circle which cut off on the circumference an arc of length equal to the radius.

Solid angle: Ω; steradian (sr). The solid angle which, having its vertex at the centre of a sphere, cuts off an area of the surface of the sphere equal to a square having sides equal to the radius.

Force: The base SI unit of electric current is in terms of force in newtons (N). A force of 1 N is that which endows unit mass (1 kg) with unit acceleration (1 m/s^2). The newton is thus not only a coherent unit; it is also devoid of any association with gravitational effects.

3.1.3 Temperature

The base SI unit of thermodynamic temperature is referred to a point of 'absolute zero' at which bodies possess zero thermal energy. For practical convenience two points on the Kelvin temperature scale, namely 273.15 K and 373.15 K, are used to define the Celsius (or Centigrade) scale (0°C and 100°C). Thus in terms of temperature *intervals*, 1 K = 1°C; but in terms of temperature *levels*, a Celsius temperature θ corresponds to a Kelvin temperature $(\theta + 273.15)$ K.

3.1.4 Derived units

Nine of the more important SI derived units with their definitions are given

Quantity	*Unit name*	*Unit symbol*
Force	newton	N
Energy	joule	J
Power	watt	W
Electric charge	coulomb	C
Electrical potential difference and EMF	volt	V
Electric resistance	ohm	Ω
Electric capacitance	farad	F
Electric inductance	henry	H
Magnetic flux	weber	Wb

Newton That force which gives to a mass of 1 kilogram an acceleration of 1 metre per second squared.

Joule The work done when the point of application of 1 newton is displaced a distance of 1 metre in the direction of the force.

Watt The power which gives rise to the production of energy at the rate of 1 joule per second.

Coulomb The quantity of electricity transported in 1 second by a current of 1 ampere.
Volt The difference of electric potential between two points of a conducting wire carrying a constant current of 1 ampere, when the power dissipated between these points is equal to 1 watt.
Ohm The electric resistance between two points of a conductor when a constant difference of potential of 1 volt, applied between these two points, produces in this conductor a current of 1 ampere, this conductor not being the source of any electromotive force.
Farad The capacitance of a capacitor between the plates of which there appears a difference of potential of 1 volt when it is charged by a quantity of electricity equal to 1 coulomb.
Henry The inductance of a closed circuit in which an electromotive force of 1 volt is produced when the electric current in the circuit varies uniformly at a rate of 1 ampere per second.
Weber The magnet flux which, linking a circuit of one turn, produces in it an electromotive force of 1 volt as it is reduced to zero at a uniform rate in 1 second.

Some of the simpler derived units are expressed in terms of the seven basic and two supplementary units directly. Examples are listed in *Table 3.1*.

Table 3.1 Directly derived units

Quantity	*Unit name*	*Unit symbol*
Area	square metre	m^2
Volume	cubic metre	m^3
Mass density	kilogram per cubic metre	kg/m^3
Linear velocity	metre per second	m/s
Linear acceleration	metre per second squared	m/s^2
Angular velocity	radian per second	rad/s
Angular acceleration	radian per second squared	rad/s^2
Force	kilogram metre per second squared	$kg\ m/s^2$
Magnetic field strength	ampere per metre	A/m
Concentration	mole per cubic metre	mol/m^3
Luminance	candela per square metre	cd/m^2

Units in common use, particularly those for which a statement in base units would be lengthy or complicated, have been given special shortened names (see *Table 3.2*).

Table 3.2 Named derived units

Quantity	*Unit name*	*Unit symbol*	*Derivation*
Force	newton	N	$kg\ m/s^2$
Pressure	pascal	Pa	N/m^2
Power	watt	W	J/s
Energy	joule	J	N m, W s
Electric charge	coulomb	C	A s
Electric flux	coulomb	C	A s
Magnetic flux	weber	Wb	V s
Magnetic flux density	tesla	T	Wb/m^2
Electric potential	volt	V	J/C, W/A
Resistance	ohm	Ω	V/A
Conductance	siemens	S	A/V
Capacitance	farad	F	A s/V, C/V
Inductance	henry	H	V s/A, Wb/A
Luminous flux	lumen	lm	cd sr
Illuminance	lux	lx	lm/m^2
Frequency	hertz	Hz	l/s

The named derived units are used to form further derivations. Examples are given in *Table 3.3*.

Table 3.3 Further derived units

Quantity	*Unit name*	*Unit symbol*
Torque	newton metre	N m
Dynamic viscosity	pascal second	Pa s
Surface tension	newton per metre	N/m
Power density	watt per square metre	W/m^2
Energy density	joule per cubic metre	J/m^3
Heat capacity	joule per kelvin	J/K
Specific heat capacity	joule per kilogram kelvin	J/(kg K)
Thermal conductivity	watt per metre kelvin	W/(m K)
Electric field strength	volt per metre	V/m
Magnetic field strength	ampere per metre	A/m
Electric flux density	coulomb per square metre	C/m^2
Current density	ampere per square metre	A/m^2
Resistivity	ohm metre	Ω m
Permittivity	farad per metre	F/m
Permeability	henry per metre	H/m

Names of SI units and the corresponding EMU and ESU CGS units are given in *Table 3.4*.

Table 3.4 Unit names

Quantity	*Symbol*	*SI*	*EMU & ESU*
Length	l	metre (m)	centimetre (cm)
Time	t	second (s)	second
Mass	m	kilogram (kg)	gram (g)
Force	F	newton (N)	dyne (dyn)
Frequency	f, v	hertz (Hz)	hertz
Energy	E, W	joule (J)	erg (erg)
Power	P	watt (W)	erg/second (erg/s)
Pressure	p	newton/metre^2 (N/m^2)	dyne/centimetre^2 (dyn/cm^2)
Electric charge	Q	coulomb (C)	coulomb
Electric potential	V	volt (V)	volt
Electric current	I	ampere (A)	ampere
Magnetic flux	Φ	weber (Wb)	maxwell (Mx)
Magnetic induction	B	tesla (T)	gauss (G)
Magnetic field strength	H	ampere turn/ metre (At/m)	oersted (Oe)
Magneto-motive force	F_m	ampere turn (At)	gilbert (Gb)
Resistance	R	ohm (Ω)	ohm
Inductance	L	henry (H)	henry
Conductance	G	mho (Ω^{-1}) (siemens)	mho
Capacitance	C	farad (F)	farad

3.1.5 Gravitational and absolute systems

There may be some difficulty in understanding the difference between SI and the Metric Technical System of units which has been used principally in Europe. The main difference is that while mass is expressed in kg in both systems, weight (representing a force) is expressed as kgf, a gravitational unit, in the MKSA system and as N in SI. An absolute unit of force differs from a gravitational unit of force because it induces unit acceleration in a unit mass whereas a gravitational unit imparts gravitational acceleration to a unit mass.

A comparison of the more commonly known systems and SI is shown in *Table 3.5*.

Table 3.5 Commonly used units of measurement

	SI (absolute)	*FPS (gravitational)*	*FPS (absolute)*	*cgs (absolute)*	*Metric technical units (gravitational)*
Length	metre (m)	ft	ft	cm	metre
Force	newton (N)	lbf	poundal (pdl)	dyne	kgf
Mass	kg	lb or slug	lb	gram	kg
Time	s	s	s	s	s
Temperature	°C K	°F	°F °R	°C K	°C K
Energy mech.	joule*	ft lbf	ft pdl	dyn cm = erg	kgf m
Energy heat		Btu	Btu	calorie	kcal
Power mech.	watt	hp	hp	erg/s	metric hp
Power elec.		watt	watt		watt
Electric current	amp	amp	amp	amp	amp
Pressure	N/m^2	lbf/ft^2	pdl/ft^2	dyn/cm^2	kgf/cm^2

* 1 joule = 1 newton metre or 1 watt second.

3.1.6 Expressing magnitudes of SI units

To express magnitudes of a unit, decimal multiples and submultiples are formed using the prefixes shown in *Table 3.6*. This method of expressing magnitudes ensures complete adherence to a decimal system.

Table 3.6 The internationally agreed multiples and submultiples

Factor by which the unit is multiplied		*Prefix*	*Symbol*	*Common everyday examples*
One million million (billion)	10^{12}	tera	T	
One thousand million	10^9	giga	G	gigahertz (GHz)
One million	10^6	mega	M	megawatt (MW)
One thousand	10^3	kilo	k	kilometre (km)
One hundred	10^2	hecto*	h	
Ten	10^1	deca*	da	decagram (dag)
UNITY	1			
One tenth	10^{-1}	deci*	d	decimetre (dm)
One hundredth	10^{-2}	centi*	c	centimetre (cm)
One thousandth	10^{-3}	milli	m	milligram (mg)
One millionth	10^{-6}	micro	μ	microsecond (μs)
One thousand millionth	10^{-9}	nano	n	nanosecond (ns)
One million millionth	10^{-12}	pico	p	picofarad (pF)
One thousand million millionth	10^{-15}	femto	f	
One million million millionth	10^{-18}	atto	a	

* To be avoided wherever possible.

3.1.7 Auxiliary units

Certain auxiliary units may be adopted where they have application in special fields. Some are acceptable on a temporary basis, pending a more widespread adoption of the SI system. *Table 3.7* lists some of these.

Table 3.7 Auxiliary units

Quantity	*Unit symbol*	*SI equivalent*
Day	d	86 400 s
Hour	h	3600 s
Minute (time)	min	60 s
Degree (angle)	°	$\pi/180$ rad
Minute (angle)	′	$\pi/10\,800$ rad
Second (angle)	″	$\pi/648\,000$ rad
Are	a	$1\ dam^2 = 10^2\ m^2$
Hectare	ha	$1\ hm^2 = 10^4\ m^2$
Barn	b	$100\ fm^2 = 10^{-28}\ m^2$
Standard atmosphere	atm	101 325 Pa
Bar	bar	$0.1\ MPa = 10^5\ Pa$
Litre	l	$1\ dm^3 = 10^{-3}\ m^3$
Tonne	t	$10^3\ kg = 1\ Mg$
Atomic mass unit	u	$1.660\,53 \times 10^{-27}$ kg
Angström	Å	$0.1\ nm = 10^{-10}$ m
Electron-volt	eV	$1.602\,19 \times 10^{-19}$ J
Curie	Ci	$3.7 \times 10^{10}\ s^{-1}$
Röntgen	R	2.58×10^{-4} C/kg

3.1.8 Nuclear engineering

It has been the practice to use special units with their individual names for evaluating and comparing results. These units are usually formed by multiplying a unit from the cgs or SI system by a number which matches a value derived from the result of some natural phenomenon. The adoption of SI both nationally and internationally has created the opportunity to examine the practice of using special units in the nuclear industry, with the object of eliminating as many as possible and using the pure system instead.

As an aid to this, ISO draft Recommendations 838 and 839 have been published, giving a list of quantities with special names, the SI unit and the alternative cgs unit. It is expected that as SI is increasingly adopted and absorbed, those units based on cgs will go out of use. The values of these special units illustrate the fact that a change from them to SI would not be as revolutionary as might be supposed. Examples of these values together with the SI units which replace them are shown in *Table 3.8*.

Table 3.8 Nuclear engineering

Special unit			*SI replacement*
Name		*Value*	
Angström	(Å)	10^{-10} m	m
Barn	(b)	$10^{-28}\ m^2$	m^2
Curie	(Ci)	$3.7 \times 10^{10}\ s^{-1}$	s^{-1}
Electron-volt	(eV)	$(1.602\,189\,2 \pm .000\,004\,6) \times 10^{-19}$ J	J
Röntgen	(R)	2.58×10^{-4} C/kg	C/kg

3.2 Universal constants in SI units

Table 3.9 Universal constants

The digits in parentheses following each quoted value represent the standard deviation error in the final digits of the quoted value as computed on the criterion of internal consistency. The unified scale of atomic weights is used throughout ($^{12}C=12$). C = coulomb; G = gauss; Hz = hertz; J = joule; N = newton; T = tesla; u = unified nuclidic mass unit; W = watt; Wb = weber. For result multiply the numerical value by the SI unit.

Constant	*Symbol*	*Numerical value*	*SI unit*
Speed of light in vacuum	c	2.997 925(1)	10^8 m s^1
Gravitational constant	G	6.670(5)*	10^{-11} N m^2 kg^2
Elementary charge	e	1.602 10(2)	10^{-19} C
Avogadro constant	N_A	6.022 52(9)	10^{26} $kmol^{-1}$
Mass unit	u	1.660 43(2)	10^{-27} kg
Electron rest mass	m_e	9.109 08(13)	10^{-31} kg
		5.485 97 (3)	10^{-4} u
Proton rest mass	m_p	1.672 52(3)	10^{-27} kg
		1.007 276 63(8)	u
Neutron rest mass	m_n	1.674 82(3)	10^{-27} kg
		1.008 665 4(4)	u
Faraday constant	F	9.648 70(5)	10^4 C mol^{-1}
Planck constant	h	6.625 59(16)	10^{-34} J s
	$h/2\pi$	1.054 494(25)	10^{-34} J s
Fine-structure constant	α	7.297 20(3)	10^{-3}
	$1/\alpha$	137.038 8(6)	
Charge-to-mass ratio for electron	e/m_e	1.758 796(6)	10^{11} C kg^{-1}
Quantum of magnetic flux	hc/e	4.135 56(4)	10^{-11} Wb
Rydberg constant	R_∞	1.097 373 1(1)	10^7 m^{-1}
Bohr radius	a_0	5.291 67(2)	10^{-11} m
Compton wavelength of	h/m_ec	2.426 21(2)	10^{-12} m
electron	$\lambda C/2\pi$	3.861 44(3)	10^{-13} m
Electron radius	$e^2/m_ec^2=r_e$	2.817 77(4)	10^{-15} m
Thomson cross section	$8\pi r_e^2/3$	6.651 6(2)	10^{-29} m^2
Compton wavelength of	$\lambda_{C,p}$	1.321 398(13)	10^{-15} m
proton	$\lambda_{C,p}/2\pi$	2.103 07(2)	10^{-16} m
Gyromagnetic ratio of	γ	2.675 192(7)	10^8 rad s^{-1} T^{-1}
proton	$\gamma/2\pi$	4.257 70(1)	10^7 Hz T^{-1}
(uncorrected for	γ'	2.675 123(7)	10^8 rad s^{-1} T^{-1}
diamagnetism of H_2O)	$\gamma'/2\pi$	4.257 59(1)	10^7 Hz T^{-1}
Bohr magneton	μ_B	9.273 2(2)	10^{-24} J T^{-1}
Nuclear magneton	μ_N	5.050 50(13)	10^{-27} J T^{-1}
Proton magnetic moment	μ_p	1.410 49(4)	10^{-26} J T^{-1}
	μ_p/μ_N	2.792 76(2)	
(uncorrected for diamagnetism in H_2O sample)	μ_p'/μ_N	2.792 68(2)	
Gas constant	R_0	8.314 34(35)	J K^{-1} mol^{-1}
Boltzmann constant	k	1.380 54(6)	10^{-23} J K^{-1}
First radiation constant ($2\pi hc^2$)	c_1	3.741 50(9)	10^{-16} W m^{-2}
Second radiation constant (hc/k)	c_2	1.438 79(6)	10^{-2} m K
Stefan–Boltzmann constant	σ	5.669 7(10)	10^{-8} W m^{-2} K^{-4}

* The universal gravitational constant is not, and cannot in our present state of knowledge, be expressed in terms of other fundamental constants. The value given here is a direct determination by P. R. Heyl and P. Chrzanowski, *J. Res. Natl. Bur. Std.* (*U.S.*) 29, 1 (1942).

The above values are extracts from *Review of Modern Physics* Vol. 37 No. 4 October 1965 published by the American Institute of Physics.

3.3 Metric to Imperial conversion factors

Table 3.10 Conversion factors

SI units	*British units*
SPACE AND TIME	
Length:	
1 μm (micron)	$=39.37\times10^{-6}$ in
1 mm	=0.039 370 1 in
1 cm	=0.393 701 in
1 m	=3.280 84 ft

Table 3.10 *continued*

SI units	*British units*
SPACE AND TIME	
Length:	
1 m	=1.093 61 yd
1 km	=0.621 371 mile
Area:	
1 mm^2	$=1.550\times10^{-3}$ in^2
1 cm^2	=0.155 0 in^2
1 m^2	=10.763 9 ft^2
1 m^2	=1.195 99 yd^2
1 ha	=2.471 05 acre
Volume:	
1 mm^3	$=61.023 7\times10^{-6}$ in^3
1 cm^3	$=61.023 7\times10^{-3}$ in^3
1 m^3	=35.314 7 ft^3
1 m^3	=1.307 95 yd^3
Capacity:	
10^6 m^3	$=219.969\times10^6$ gal
1 m^3	=219.969 gal
1 litre (l)	=0.219 969 gal
	=1.759 80 pint
Capacity flow:	
10^3 m^3/s	$=791.9\times10^6$ gal/h
1 m^3/s	$=13.20\times10^3$ gal/min
1 litre/s	=13.20 gal/min
1 m^3/kW h	=219.969 gal/kW h
1 m^3/s	=35.314 7 ft^3/s (cusecs)
1 litre/s	$=0.588 58\times10^{-3}$ ft^3/min (cfm)
Velocity:	
1 m/s	=3.280 84 ft/s=2.236 94 mile/h
1 km/h	=0.621 371 mile/h
Acceleration:	
1 m/s^2	=3.280 84 ft/s^2
MECHANICS	
Mass:	
1 g	=0.035 274 oz
1 kg	=2.204 62 lb
1 t	=0.984 207 ton=19.684 1 cwt
Mass flow:	
1 kg/s	=2.204 62 lb/s=7.936 64 klb/h
Mass density:	
1 kg/m^3	=0.062 428 lb/ft^3
1 kg/litre	=10.022 119 lb/gal
Mass per unit length:	
1 kg/m	=0.671 969 lb/ft=2.015 91 lb/yd
Mass per unit area:	
1 kg/m^2	=0.204 816 lb/ft^2
Specific volume:	
1 m^3/kg	=16.018 5 ft^3/lb
1 litre/tonne	=0.223 495 gal/ton
Momentum:	
1 kg m/s	=7.233 01 lb ft/s
Angular momentum:	
1 kg m^2/s	=23.730 4 lb ft^2/s
Moment of inertia:	
1 kg m^2	=23.730 4 lb ft^2
Force:	
1 N	=0.224 809 lbf
Weight (force) per unit length:	
1 N/m	=0.068 521 lbf/ft
	=0.205 566 lbf/yd
Moment of force (or torque):	
1 N m	=0.737 562 lbf ft

Table 3.10 *continued*

SI units	*British units*
MECHANICS	
Weight (force) per unit area:	
1 N/m^2	=0.020 885 lbf/ft^2
Pressure:	
1 N/m^2	=1.450 38 × 10^{-4} lbf/in^2
1 bar	=14.503 8 lbf/in^2
1 bar	=0.986 923 atmosphere
1 mbar	=0.401 463 in H_2O
	=0.029 53 in Hg
Stress:	
1 N/mm^2	=6.474 90 × 10^{-2} tonf/in^2
1 MN/m^2	=6.474 90 × 10^{-2} tonf/in^2
1 hbar	=0.647 490 tonf/in^2
Second moment of area:	
1 cm^4	=0.024 025 in^4
Section modulus:	
1 m^3	=61 023.7 in^3
1 cm^3	=0.061 023 7 in^3
Kinematic viscosity:	
1 m^2/s	=10.762 75 ft^2/s=10^6 cSt
1 cSt	=0.038 75 ft^2/h
Energy, work:	
1 J	=0.737 562 ft lbf
1 MJ	=0.372 5 hph
1 MJ	=0.277 78 kW h
Power:	
1 W	=0.737 562 ft lbf/s
1 kW	=1.341 hp=737.562 ft lbf/s
Fluid mass:	
(Ordinary) 1 kg/s	=2.204 62 lb/s=793 6.64 lb/h
(Velocity) 1 kg/m^2 s	=0.204 815 lb/ft^{2}s
HEAT	
Temperature:	
(Interval) 1 K	=9/5 deg R (Rankine)
1°C	=9/5 deg F
(Coefficient) 1°R^{-1}	=1 deg F^{-1}=5/9 deg C
1°C^{-1}	=5/9 deg F^{-1}
Quantity of heat:	
1 J	=9.478 17 × 10^{-4} Btu
1 J	=0.238 846 cal
1 kJ	=947.817 Btu
1 GJ	=947.817 × 10^3 Btu
1 kJ	=526.565 CHU
1 GJ	=526.565 × 10^3 CHU
1 GJ	=9.478 17 therm
Heat flow rate:	
1 W(J/s)	=3.412 14 Btu/h
1 W/m^2	=0.316 998 Btu/ft^2 h
Thermal conductivity:	
1 W/m °C	=6.933 47 Btu in/ft^2 h °F
Coefficient and heat transfer:	
1 W/m^2 °C	=0.176 110 Btu/ft^2 h °F
Heat capacity:	
1 J/°C	=0.526 57 × 10^{-3} Btu/°R
Specific heat capacity:	
1 J/g °C	=0.238 846 Btu/lb °F
1 kJ/kg °C	=0.238 846 Btu/lb °F
Entropy:	
1 J/K	=0.526 57 × 10^{-3} Btu/°R
Specific entropy:	
1 J/kg °C	=0.238 846 × 10^{-3} Btu/lb °F
1 J/kg K	=0.238 846 × 10^{-3} Btu/lb °R

SI units	*British units*
HEAT	
Specific energy/specific latent heat:	
1 J/g	=0.429 923 Btu/lb
1 J/kg	=0.429 923 × 10^{-3} Btu/lb
Calorific value:	
1 kJ/kg	=0.429 923 Btu/lb
1 kJ/kg	=0.773 861 4 CHU/lb
1 J/m^3	=0.026 839 2 × 10^{-3} Btu/ft^3
1 kJ/m^3	=0.026 839 2 Btu/ft^3
1 kJ/litre	=4.308 86 Btu/gal
1 kJ/kg	=0.009 630 2 therm/ton
ELECTRICITY	
Permeability:	
1 H/m	=10^7/4π μ_0
Magnetic flux density:	
1 tesla	=10^4 gauss=1 Wb/m^2
Conductivity:	
1 mho	=1 reciprocal ohm
1 siemens	=1 reciprocal ohm
Electric stress:	
1 kV/mm	=25.4 kV/in
1 kV/m	=0.025 4 kV/in

3.4 Symbols and abbreviations

Table 3.11 Quantities and units of periodic and related phenomena (based on ISO Recommendation R31)

Symbol	*Quantity*
T	periodic time
τ, (T)	time constant of an exponentially varying quantity
f, v	frequency
η	rotational frequency
ω	angular frequency
λ	wavelength
σ ($\tilde{v}$)	wavenumber
k	circular wavenumber
$\log_e (A_1/A_2)$	natural logarithm of the ratio of two amplitudes
$10 \log_{10} (P_1/P_2)$	ten times the common logarithm of the ratio of two powers
δ	damping coefficient
Λ	logarithmic decrement
α	attenuation coefficient
β	phase coefficient
γ	propagation coefficient

Table 3.12 Symbols for quantities and units of electricity and magnetism (based on ISO Recommendation R31)

Symbol	*Quantity*
I	electric current
Q	electric charge, quantity of electricity
ρ	volume density of charge, charge density (Q/V)
σ	surface density of charge (Q/A)
E, (K)	electric field strength

Table 3.12 *continued*

Symbol	*Quantity*
V, (φ)	electric potential
U, (V)	potential difference, tension
E	electromotive force
D	displacement (rationalised displacement)
D'	non-rationalised displacement
ψ	electric flux, flux of displacement (flux of rationalised displacement)
ψ'	flux of non-rationalised displacement
C	capacitance
ε	permittivity
ε_0	permittivity of vacuum
ε'	non-rationalised permittivity
ε'_0	non-rationalised permittivity of vacuum
ε_r	relative permittivity
χ_e	electric susceptibility
χ'_e	non-rationalised electric susceptibility
P	electric polarisation
p, (p_e)	electric dipole moment
J, (S)	current density
A, (α)	linear current density
H	magnetic field strength
H'	non-rationalised magnetic field strength
U_m	magnetic potential difference
F, (F_m)	magnetomotive force
B	magnetic flux density, magnetic induction
Φ	magnetic flux
A	magnetic vector potential
L	self-inductance
M, (L)	mutual inductance
k, (x)	coupling coefficient
σ	leakage coefficient
μ	permeability
μ_0	permeability of vacuum
μ'	non-rationalised permeability
μ'_0	non-rationalised permeability of vacuum
μ_r	relative permeability
k, (χ_m)	magnetic susceptibility
k', (χ'_m)	non-rationalised magnetic susceptibility
m	electromagnetic moment (magnetic moment)
H_i, (M)	magnetisation
J, (B_i)	magnetic polarisation
J'	non-rationalised magnetic polarisation
w	electromagnetic energy density
S	Poynting vector
c	velocity of propagation of electromagnetic waves *in vacuo*
R	resistance (to direct current)
G	conductance (to direct current)
ρ	resistivity
γ, σ	conductivity
R, R_m	reluctance
Λ, (P)	permeance
N	number of turns in winding
m	number of phases
p	number of pairs of poles
φ	phase displacement
Z	impedance (complex impedance)
[Z]	modulus of impedance (impedance)
X	reactance
R	resistance
Q	quality factor
Y	admittance (complex admittance)
[Y]	modulus of admittance (admittance)
B	susceptance
G	conductance
P	active power
S, (P_s)	apparent power
Q, (P_q)	reactive power

Table 3.13 Symbols for quantities and units of acoustics (based on ISO Recommendation R31)

Symbol	*Quantity*
T	period, periodic time
f, v	frequency, frequency interval
ω	angular frequency, circular frequency
λ	wavelength
k	circular wavenumber
ρ	density (mass density)
P_s	static pressure
p	(instantaneous) sound pressure
ε, (x)	(instantaneous) sound particle displacement
u, v	(instantaneous) sound particle velocity
a	(instantaneous) sound particle acceleration
q, U	(instantaneous) volume velocity
c	velocity of sound
E	sound energy density
P, (N, W)	sound energy flux, sound power
I, J	sound intensity
Z_s, (W)	specific acoustic impedance
Z_a, (Z)	acoustic impedance
Z_m, (w)	mechanical impedance
L_p, (L_N, L_w)	sound power level
L_p, (L)	sound pressure level
δ	damping coefficient
Λ	logarithmic decrement
α	attenuation coefficient
β	phase coefficient
γ	propagation coefficient
δ	dissipation coefficient
r, τ	reflection coefficient
γ	transmission coefficient
α, (α_a)	acoustic absorption coefficient
R	sound reduction index sound transmission loss
A	equivalent absorption area of a surface or object
T	reverberation time
L_N, (Λ)	loudness level
N	loudness

Table 3.14 Some technical abbreviations and symbols

Quantity	*Abbreviation*	*Symbol*
Alternating current	a.c.	
Ampere	A or amp	
Amplification factor		μ
Amplitude modulation	a.m.	
Angular velocity		ω
Audio frequency	a.f.	

Table 3.14 *continued*

Quantity	*Abbreviation*	*Symbol*
Automatic frequency control	a.f.c.	
Automatic gain control	a.g.c.	
Bandwidth		Δf
Beat frequency oscillator	b.f.o.	
British thermal unit	Btu	
Cathode-ray oscilloscope	c.r.o.	
Cathode-ray tube	c.r.t.	
Centigrade	c	
Centi-	C	
Centimetre	cm	
Square centimetre	cm^2 or sq cm	
Cubic centimetre	cm^3 or cu cm or c.c.	
Centimetre-gram-second	c.g.s.	
Continuous wave	c.w.	
Coulomb	C	
Deci-	d	
Decibel	dB	
Direct current	d.c.	
Direction finding	d.f.	
Double sideband	d.s.b.	
Efficiency		η
Equivalent isotropic radiated power	e.i.r.p.	
Electromagnetic unit	e.m.u.	
Electromotive force instantaneous value	e.m.f.	E or V, e or v
Electron-volt	eV	
Electrostatic unit	e.s.u.	
Fahrenheit	F	
Farad	F	
Frequency	freq.	f
Frequency modulation	f.m.	
Gauss	G	
Giga-	G	
Gram	g	
Henry	H	
Hertz	Hz	
High frequency	h.f.	
Independent sideband	i.s.b.	
Inductance-capacitance		L-C
Intermediate frequency	i.f.	
Kelvin	K	
Kilo-	k	
Knot	kn	
Length		l
Local oscillator	l.o.	
Logarithm, common		log or $\log_{10}$
Logarithm, natural		ln or $\log_e$
Low frequency	l.f.	
Low tension	l.t.	
Magnetomotive force	m.m.f.	F or M
Mass		m
Medium frequency	m.f.	
Mega-	M	
Metre	m	
Metre-kilogram-second	m.k.s.	

Table 3.14 *continued*

Quantity	*Abbreviation*	*Symbol*
Micro-	μ	
Micromicro-	p	
Micron		μ
Milli-	m	
Modulated continuous wave	m.c.w.	
Nano-	n	
Neper	N	
Noise factor		N
Ohm		Ω
Peak to peak	p–p	
Phase modulation	p.m.	
Pico-	p	
Plan-position indication	PPI	
Potential difference	p.d.	V
Power factor	p.f.	
Pulse repetition frequency	p.r.f.	
Radian	rad	
Radio frequency	r.f.	
Radio telephony	R/T	
Root mean square	r.m.s.	
Short-wave	s.w.	
Single sideband	s.s.b.	
Signal frequency	s.f.	
Standing wave ratio	s.w.r.	
Super-high frequency	s.h.f.	
Susceptance		B
Travelling-wave tube	t.w.t.	
Ultra-high frequency	u.h.f.	
Very high frequency	v.h.f.	
Very low frequency	v.l.f.	
Volt	V	
Voltage standing wave ratio	v.s.w.r.	
Watt	W	
Weber	Wb	
Wireless telegraphy	W/T	

Table 3.15 Greek alphabet and symbols

Name	*Symbol*		*Quantities used for*
alpha	A	α	angles, coefficients, area
beta	B	β	angles, coefficients
gamma	Γ	γ	specific gravity
delta	Δ	δ	density, increment, finite difference operator
epsilon	E	ε	Napierian logarithm, linear strain, permittivity, error, small quantity
zeta	Z	ζ	coordinates, coefficients, impedance (capital)
eta	H	η	magnetic field strength, efficiency
theta	Θ	θ	angular displacement, time
iota	I	ι	inertia
kappa	K	κ	bulk modulus, magnetic susceptibility
lambda	Λ	λ	permeance, conductivity, wavelength
mu	M	μ	bending moment, coefficient of friction, permeability

Table 3.15 *continued*

Name	*Symbol*		*Quantities used for*
nu	N	ν	kinematic viscosity, frequency, reluctivity
xi	Ξ	ξ	output coefficient
omicron	O	o	
pi	Π	π	circumference ÷ diameter
rho	P	ρ	specific resistance
sigma	Σ	σ	summation (capital), radar cross section, standard deviation
tau	T	τ	time constant, pulse length
upsilon	Y	u	
phi	Φ	φ	flux, phase
chi	X	χ	reactance (capital)
psi	Ψ	ψ	angles
omega	Ω	ω	angular velocity, ohms

References

1 COHEN, E. R. and TAYLOR, B. N., *Journal of Physical and Chemical Reference Data*, vol. 2, 663, (1973).
2 'Recommended values of physical constants', CODATA (1973).
3 McGLASHAN, M. L. *Physiochemical quantities and units*, London: The Royal Institute of Chemistry, (1971).

4 Electricity

M. G. Say

Contents

4.1 Introduction

Most of the observed electrical phenomena are explicable in terms of electric *charge* at rest, in motion and in acceleration. Static charges give rise to an *electric field* of force; charges in motion carry an electric field accompanied by a *magnetic field* of force; charges in acceleration develop a further field of *radiation*.

Modern physics has established the existence of elemental charges and their responsibility for observed phenomena. Modern physics is complex: it is customary to explain phenomena of engineering interest at a level adequate for a clear and reliable concept, based on the electrical nature of matter.

4.2 Molecules, atoms and electrons

Material substances, whether solid, liquid or gaseous, are conceived as composed of very large numbers of *molecules*. A molecule is the smallest portion of any substance which cannot be further subdivided without losing its characteristic material properties. In all states of matter molecules are in a state of rapid continuous motion. In a *solid* the molecules are relatively closely 'packed' and the molecules, although rapidly moving, maintain a fixed mean position. Attractive forces between molecules account for the tendency of the solid to retain its shape. In a *liquid* the molecules are less closely packed and there is a weaker cohesion between them, so that they can wander about with some freedom within the liquid, which consequently takes up the shape of the vessel in which it is contained. The molecules in a *gas* are still more mobile, and are relatively far apart. The cohesive force is very small, and the gas is enabled freely to contract and expand. The usual effect of heat is to increase the intensity and speed of molecular activity so that 'collisions' between molecules occur more often; the average spaces between the molecules increase, so that the substance attempts to expand, producing internal pressure if the expansion is resisted.

Molecules are capable of further subdivision, but the resulting particles, called *atoms*, no longer have the same properties as the molecules from which they came. An atom is the smallest portion of matter that can enter into chemical combination or be chemically separated, but it cannot generally maintain a separate existence except in the few special cases where a single atom forms a molecule. A molecule may consist of one, two or more (sometimes many more) atoms of various kinds. A substance whose molecules are composed entirely of atoms of the same kind is called an *element*. Where atoms of two or more kinds are present, the molecule is that of a chemical *compound*. At present 102 atoms are recognised, from combinations of which every conceivable substance is made. As the simplest example, the atom of hydrogen has a mass of 1.63×10^{-27} kg and a molecule (H_2), containing two atoms, has twice this mass. In one gram of hydrogen there are about 3×10^{23} molecules with an order of size between 1 and 0.1 nm.

Electrons, as small particles of negative electricity having apparently almost negligible mass, were discovered by J. J. Thomson, on a basis of much previous work by many investigators, notably Crookes. The discovery brought to light two important facts: (1) that atoms, the units of which all matter is made, are themselves complex structures, and (2) that electricity is atomic in nature. The atoms of all substances are constructed from particles. Those of engineering interest are: *electrons*, *protons* and *neutrinos*. Modern physics concerns itself also with *positrons*, *mesons*, *neutrons* and many more. An *electron* is a minute particle of negative electricity which, when dissociated from the atom (as it can be) indicates a purely electrical, nearly mass-less nature. From whatever atom they are derived, all electrons are similar. The electron charge is $e = 1.6 \times 10^{-19}$ C, so that 1 C $= 6.3 \times 10^{18}$ electron charges. The apparent rest mass of an electron is 1/1850 of that of a hydrogen atom, amounting to $m = 9 \times 10^{-28}$ g. The meaning to be attached to the 'size' of an electron (a figure of the order of 10^{-13} cm) is vague. A *proton* is electrically the opposite of an electron, having an equal charge, but positive. Further, protons are associated with a mass the same as that of the hydrogen nucleus. A *neutron* is a chargeless mass, the same as that of the proton.

4.3 Atomic structure

The mass of an atom is almost entirely concentrated in a nucleus of protons and neutrons. The simplest atom, of hydrogen, comprises a nucleus with a single proton, together with one associated electron occupying a region formerly called the K-shell. Helium has a nucleus of two protons and two neutrons, with two electrons in the K-shell. In these cases, as in all normal atoms, the sum of the electron charges is numerically equal to the sum of the proton charges, and the atom is electrically balanced. The neon atom has a nucleus with 10 protons and 10 neutrons, with its 10 electrons in the K- and L-shells.

The *atomic weight* A is the total number of protons and neutrons in the nucleus. If there are Z protons there will be $A - Z$ neutrons: Z is the *atomic number*. The nuclear structure is not known, and the forces that keep the protons together against their mutual repulsion are conjectural.

A nucleus of atomic weight A and atomic number Z has a charge of $+Ze$ and is normally surrounded by Z electrons each of charge $-e$. Thus copper has 29 protons and 35 neutrons ($A = 64$, $Z = 29$) in its nucleus, electrically neutralised by 29 electrons in an enveloping cloud. The atomic numbers of the known elements range from 1 for hydrogen to 102 for nobelium, and from time to time the list is extended. This multiplicity can be simplified: within the natural sequence of elements there can be distinguished groups with similar chemical and physical properties (see *Table 4.1*). These are the ***halogens*** (F 9, Cl 17, Br 35, I 53); the *alkali metals* (Li 3, Na 11, K 19, Rb 37, Cs 55); the *copper* group (Cu 29, Ag 47, Au 79); the *alkaline earths* (Be 4, Mg 12, Ca 20, Sr 38, Ba 56, Ra 88); the *chromium* group (Cr 24, Mo 42, W 74, U 92); and the *rare gases* (He 2, Ne 10, A 18, Kr 36, Xe 54, Rn 86). In the foregoing the brackets contain the chemical symbols of the elements concerned followed by their atomic numbers. The difference between the atomic numbers of two adjacent elements within a group is always 8, 18 or 32. Now these three bear to one another a simple arithmetical relation: $8 = 2 \times 2 \times 2$, $18 = 2 \times 3 \times 3$ and $32 = 2 \times 4 \times 4$. Arrangement of the elements in order in a periodic table beginning with an alkali metal and ending with a rare gas shows a remarkable repetition of basic similarities. The periods are I, 1–2; II, 3–10; III, 11–18; IV, 19–36; V, 37–54; VI, 55–86; VII, 87–?.

An element is often found to be a mixture of atoms with the same chemical property but different atomic weights (*isotopes*). Again, because of the convertibility of mass and energy, the mass of an atom depends on the energy locked up in its compacted nucleus. Thus small divergences are found in the atomic weights which, on simple grounds, would be expected to form integral multiples of the atomic weight of hydrogen. The atomic weight of oxygen is arbitrarily taken as 16.0, so that the mass of the proton is 1.007 6 and that of the hydrogen atom is 1.008 1.

Atoms may be in various energy states. Thus the atoms in the filament of an incandescent lamp may emit light when excited, e.g., by the passage of an electric heating current, but will not do so when the heater current is switched off. Now heat energy is the kinetic energy of the atoms of the heated body. The more vigorous impact of atoms may not always shift the atom as a whole, but may shift an electron from one orbit to another of higher energy level within the atom. This position is not normally stable, and the electron gives up its momentarily-acquired

Table 4.1 Elements

Period	*Atomic number*	*Name*	*Symbol*	*Atomic weight*
I	1	Hydrogen	H	1.008
	2	Helium	He	4.002
II	3	Lithium	Li	6.94
	4	Beryllium	Be	9.02
	5	Boron	B	10.82
	6	Carbon	C	12.00
	7	Nitrogen	N	14.008
	8	Oxygen	O	16.00
	9	Fluorine	F	19.00
	10	Neon	Ne	20.18
III	11	Sodium	Na	22.99
	12	Magnesium	Mg	24.32
	13	Aluminium	Al	26.97
	14	Silicon	Si	28.06
	15	Phosphorus	P	31.02
	16	Sulphur	S	32.06
	17	Chlorine	Cl	35.46
	18	Argon	A	39.94
IV	19	Potassium	K	39.09
	20	Calcium	Ca	40.08
	21	Scandium	Sc	45.10
	22	Titanium	Ti	47.90
	23	Vanadium	V	50.95
	24	Chromium	Cr	52.01
	25	Manganese	Mn	54.93
	26	Iron	Fe	55.84
	27	Cobalt	Co	58.94
	28	Nickel	Ni	58.69
	29	Copper	Cu	63.57
	30	Zinc	Zn	65.38
	31	Gallium	Ga	69.72
	32	Germanium	Ge	72.60
	33	Arsenic	As	74.91
	34	Selenium	Se	78.96
	35	Bromine	Br	79.91
	36	Krypton	Kr	83.70
V	37	Rubidium	Rb	85.44
	38	Strontium	Sr	87.63
	39	Yttrium	Y	88.92
	40	Zirconium	Zr	91.22
	41	Niobium	Nb	92.91
	42	Molybdenum	Mo	96.00
	43	Technetium	Tc	99.00
	44	Ruthenium	Ru	101.7
	45	Rhodium	Rh	102.9
	46	Palladium	Pd	106.7
	47	Silver	Ag	107.9
	48	Cadmium	Cd	112.4
	49	Indium	In	114.8
	50	Tin	Sn	118.0
	51	Antimony	Sb	121.8
	52	Tellurium	Te	127.6
	53	Iodine	I	126.9
	54	Xenon	Xe	131.3
VI	55	Caesium	Cs	132.9
	56	Barium	Ba	137.4
	57	Lanthanum	La	138.9
	58	Cerium	Ce	140.1
	59	Praseodymium	Pr	140.9
	60	Neodymium	Nd	144.3

Table 4.1 *continued*

Period	*Atomic number*	*Name*	*Symbol*	*Atomic weight*
VI	61	Promethium	Pm	147
	62	Samarium	Sm	150.4
	63	Europium	Eu	152.0
	64	Gadolinium	Gd	157.3
	65	Terbium	Tb	159.2
	66	Dysprosium	Dy	162.5
	67	Holmium	Ho	163.5
	68	Erbium	Er	167.6
	69	Thulium	Tm	169.4
	70	Ytterbium	Yb	173.0
	71	Lutecium	Lu	175.0
	72	Hafnium	Hf	178.6
	73	Tantalum	Ta	181.4
	74	Tungsten	W	184.0
	75	Rhenium	Re	186.3
	76	Osmium	Os	191.5
	77	Iridium	Ir	193.1
	78	Platinum	Pt	195.2
	79	Gold	Au	197.2
	80	Mercury	Hg	200.6
	81	Thallium	Tl	204.4
	82	Lead	Pb	207.2
	83	Bismuth	Bi	209.0
	84	Polonium	Po	210
	85	Astatine	At	211
	86	Radon	Rn	222
VII	87	Francium	Fr	223
	88	Radium	Ra	226.0
	89	Actinium	Ac	227
	90	Thorium	Th	232.1
	91	Protoactinium	Pa	234
	92	Uranium	U	238.1
	93	Neptunium	Np	239
	94	Plutonium	Pu	242
	95	Americium	Am	243
	96	Curium	Cm	243
	97	Berkelium	Bk	245
	98	Californium	Cf	246
	99	Einsteinium	Es	247
	100	Fermium	Fm	256
	101	Mendelevium	Md	256
	102	Nobelium	No	—

potential energy by falling back into its original level, releasing the energy as a definite amount of light, the *light-quantum* or *photon*.

Among the electrons of an atom those of the outside peripheral shell are unique in that, on account of all the electron charges on the shells between them and the nucleus, they are the most loosely bound and most easily *removable*. In a variety of ways it is possible so to excite an atom that one of the outer electrons is torn away, leaving the atom *ionised* or converted for the time into an *ion* with an effective positive charge due to the unbalanced electrical state it has acquired. Ionisation may occur due to impact by other fast-moving particles, by irradiation with rays of suitable wavelength and by the application of intense electric fields.

The three 'structures' of *Figure 4.1* are based on the former 'planetary' concept, now modified in favour of a more complex

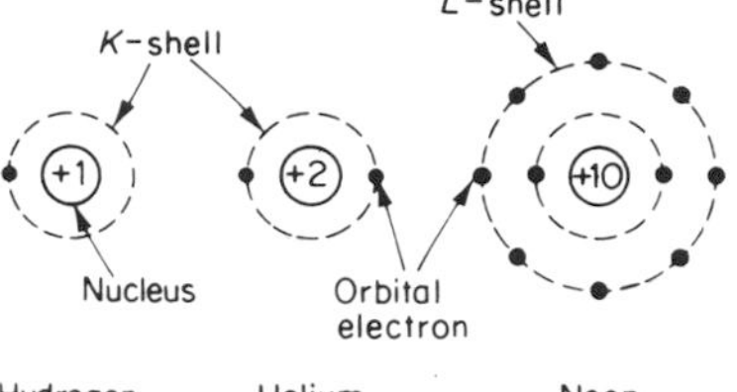

Figure 4.1 Atomic structure. The nuclei are marked with their positive charges in terms of total electron charge. The term 'orbital' is becoming obsolete. Electron: mass $m=9\times10^{28}$ g, charge $e=-1.6\times10^{-19}$ C. Proton: mass $=1.63\times10^{-24}$ g, charge $=+1.6\times10^{-19}$ C. Neutron = mass as for proton; no charge

idea derived from consideration of wave mechanics. It is still true that, apart from its mass, the chemical and physical properties of an atom are given to it by the arrangement of the electron 'cloud' surrounding the nucleus.

4.4 Wave mechanics

The fundamental laws of optics can be explained without regard to the nature of light as an electromagnetic wave phenomenon, and photo-electricity emphasises its nature as a stream or ray of corpuscles. The phenomena of diffraction or interference can only be explained on the wave concept. *Wave mechanics* correlates the two apparently conflicting ideas into a wider concept of 'waves of matter'. Electrons, atoms and even molecules participate in this duality, in that their effects appear sometimes as corpuscular, sometimes as of a wave nature. Streams of electrons behave in a corpuscular fashion in photo-emission, but in certain circumstances show the diffraction effects familiar in wave action. Considerations of particle mechanics led de Broglie to write several theoretic papers (1922–6) on the parallelism between the dynamics of a particle and geometrical optics, and suggested that it was necessary to admit that classical dynamics could not interpret phenomena involving energy quanta. Wave mechanics was established by Schrödinger in 1926 on de Broglie's conceptions.

When electrons interact with matter they exhibit wave properties: in the free state they act like particles. Light has a similar duality, as already noted. The hypothesis of de Broglie is that a particle of mass m and velocity u has wave properties with a wavelength $\lambda=h/mu$, where h is the Planck constant, $h=6.626\times10^{-34}$ J s. The mass m is relativistically affected by the velocity.

When electron waves are associated with an atom, only certain fixed-energy states are possible. The electron can be raised from one state to another if it is provided, by some external stimulus such as a photon, with the necessary energy-difference Δw in the form of an electromagnetic wave of wavelength $\lambda=hc/\Delta w$, where c is the velocity of free-space radiation (3×10^8 m/s). Similarly, if an electron falls from a state of higher to one of lower energy, it emits energy Δw as radiation. When electrons are raised in energy level, the atom is *excited*, but not ionised.

4.5 Electrons in atoms

Consider the hydrogen atom. Its single electron is not located at a fixed point, but can be anywhere in a region near the nucleus with some probability. The particular region is a kind of shell or cloud, of radius depending on the electron's energy state.

With a nucleus of atomic number Z, the Z electrons can have several possible configurations. There is a certain radial pattern of electron probability cloud distribution (or shell pattern). Each electron state gives rise to a cloud pattern, characterised by a definite energy level, and described by the series of quantum numbers n, l, m_l and m_s. The number n $(=1, 2, 3\ldots)$ is a measure of the energy level; l $(=0, 1, 2\ldots)$ is concerned with angular momentum; m_l is a measure of the component of angular momentum in the direction of an applied magnetic field; and m_s arises from the electron spin. It is customary to condense the nomenclature so that electron states corresponding to $l=0, 1, 2$ and 3 are described by the letters s, p, d and f and a numerical prefix gives the value of n. Thus boron has 2 electrons at level 1 with $l=0$, two at level 2 with $l=0$, and one at level 3 with $l=1$: this information is conveyed by the description $(1s)^2(2s)^2(2p)^1$.

The energy of an atom as a whole can vary according to the electron arrangement. The most stable state is that of minimum energy, and states of higher energy content are *excited*. By Pauli's *exclusion principle* the maximum possible number of electrons in states 1, 2, 3, 4…n are 2, 8, 18, 32, …, $2n^2$ respectively. Thus only 2 electrons can occupy the 1s state (or K-shell) and the remainder must, even for the normal minimum-energy condition, occupy other states. Hydrogen and helium, the first two elements, have respectively 1 and 2 electrons in the 1-quantum (K) shell; the next, lithium, has its third electron in the 2-quantum (L) shell. The passage from lithium to neon (*Figure 4.1*) results in the filling up of this shell to its full complement of 8 electrons. During the process, the electrons first enter the 2s subgroup, then fill the 2p subgroup until it has 6 electrons, the maximum allowable by the exclusion principle (see *Table 4.2*).

Table 4.2 Typical atomic structures

Element and atomic number		*Principal and secondary quantum numbers*									
		1s	*2s*	*2p*	*3s*	*3p*	*3d*	*4s*	*4p*	*4d*	*4f*
H	1	1									
He	2	2									
Li	3	2	1								
C	6	2	2	2							
N	7	2	2	3							
Ne	10	2	2	6							
Na	11	2	2	6	1						
Al	13	2	2	6	2	1					
Si	14	2	2	6	2	2					
Cl	17	2	2	6	2	5					
A	18	2	2	6	2	6					
K	19	2	2	6	2	6		1			
Mn	25	2	2	6	2	6	5	2			
Fe	26	2	2	6	2	6	6	2			
Co	27	2	2	6	2	6	7	2			
Ni	28	2	2	6	2	6	8	2			
Cu	29	2	2	6	2	6	10	1			
Ge	32	2	2	6	2	6	10	2	2		
Se	34	2	2	6		6	10	2	4		
Kr	36	2	2	6		6	10	2	6		
		1	2	3	*4s*	*4p*	*4d*	*4f*	*5s*	*5p*	
Rb	37	2	8	18	2	6			1		
Xe	54	2	8	18	2	6	10		2	6	

Very briefly, the effect of the electron-shell filling is as follows. Elements in the same chemical family have the same number of electrons in the subshell that is incompletely filled. The rare gases (He, Ne, A, Kr, Xe) have no uncompleted shells. Alkali metals (e.g., Na) have shells containing a single electron. The alkaline earths have two electrons in uncompleted shells. The good conductors (Ag, Cu, Au) have a single electron in the uppermost quantum state. An irregularity in the ordered sequence of filling (which holds consistently from H to A) begins at potassium (K) and continues to Ni, becoming again regular with Cu, and beginning a new irregularity with Rb.

4.6 Energy levels

The electron of a hydrogen atom, normally at level 1, can be raised to level 2 by endowing it with a particular quantity of energy most readily expressed as 10.2 eV. (1 eV = 1 electronvolt = 1.6×10^{-19} J is the energy acquired by a free electron falling through a potential difference of 1 V, which accelerates it and gives it kinetic energy). 10.2 V is the *first excitation potential* for the hydrogen atom. If the electron is given an energy of 13.6 eV it is freed from the atom, and 13.6 V is the *ionisation potential.* Other atoms have different potentials in accordance with their atomic arrangement.

4.7 Electrons in metals

An approximation to the behaviour of metals assumes that the atoms lose their valency electrons, which are free to wander in the ionic lattice of the material to form what is called an electron gas. The sharp energy-levels of the free atom are broadened into wide bands by the proximity of others. The potential within the metal is assumed to be smoothed out, and there is a sharp rise of potential at the surface that prevents the electrons from escaping: there is a potential-energy step at the surface that the electrons cannot normally overcome: it is of the order of 10 eV. If this is called W, then the energy of an electron wandering within the metal is $-W+\frac{1}{2}mu^2$.

The electrons are regarded as undergoing continual collisions on account of the thermal vibrations of the lattice, and on Fermi–Dirac statistical theory it is justifiable to treat the energy states (which are in accordance with Pauli's principle) as forming an energy-continuum. At very low temperatures the ordinary classical theory would suggest that electron energies spread over an almost zero range, but the exclusion principle makes this impossible and even at absolute zero of temperature the energies form a continuum, and physical properties will depend on how the electrons are distributed over the upper levels of this energy range.

4.8 Conductivity

The interaction of free electrons with the thermal vibrations of the ionic lattice (called 'collisions' for brevity) causes them to 'rebound' with a velocity of random direction but small compared with their average velocities as particles of an electron gas. Just as a difference of electric potential causes a drift in the general motion, so a difference of temperature between two parts of a metal carries energy from the hot region to the cold, accounting for thermal conduction and for its association with electrical conductivity. The free-electron theory, however, is inadequate to explain the dependence of conductivity on crystal axes in the metal.

At absolute zero of temperature (0 K = −273°C) the atoms cease to vibrate, and free electrons can pass through the lattice with little hindrance. At temperatures over the range 0.3–10 K (and usually round about 5 K) the resistance of certain metals, e.g., Zn, Al, Sn, Hg and Cu, becomes substantially zero. This phenomenon, known as *superconductivity*, has not been satisfactorily explained.

Superconductivity is destroyed by moderate magnetic fields. It can also be destroyed if the current is large enough to produce at the surface the same critical value of magnetic field. It follows that during the superconductivity phase the current must be almost purely superficial, with a depth of penetration of the order of 10 μm.

4.9 Electron emission

A metal may be regarded as a potential 'well' of depth $-V$ relative to its surface, so that an electron in the lowest energy state has (at absolute zero temperature) the energy $W=Ve$ (of the order 10 eV): other electrons occupy levels up to a height ε^* (5–8 eV) from the bottom of the 'well'. Before an electron can escape from the surface it must be endowed with an energy not less than $\varphi = W-\varepsilon^*$, called the *work function.*

Emission occurs by *surface irradiation* (e.g., with light) of frequency v if the energy quantum hv of the radiation is at least equal to φ. The threshold of photo-electric emission is therefore with radiation at a frequency not less than $v=\varphi/h$.

Emission takes place at *high temperatures* if, put simply, the kinetic energy of an electron normal to the surface is great enough to jump the potential step W. This leads to an expression for the emission current i in terms of temperature T, a constant A and the thermionic work-function φ:

$$i = AT^2 \exp(-\varphi/kT)$$

Electron emission is also the result of the application of a *high electric-field intensity* (of the order 1–10 GV/m) to a metal surface; also when the surface is bombarded with electrons or ions of sufficient kinetic energy, giving the effect of *secondary* emission.

4.10 Electrons in crystals

When atoms are brought together to form a crystal, their individual sharp and well-defined energy levels merge into energy *bands.* These bands may overlap, or there may be gaps in the energy levels available, depending on the lattice spacing and interatomic bonding. Conduction can take place only by electron migration into an empty or partly filled band: filled bands are not available. If an electron acquires a small amount of energy from the externally applied electric field, and can move into an available empty level, it can then contribute to the conduction process.

4.11 Insulators

In this case the 'distance' (or energy increase Δw in electronvolts) is too large for moderate electric applied fields to endow electrons with sufficient energy, so the material remains an insulator. High temperatures, however, may result in sufficient thermal agitation to permit electrons to 'jump the gap'.

4.12 Semiconductors

Intrinsic semiconductors (i.e., materials between the good conductors and the good insulators) have a small spacing of about 1 eV between their permitted bands, which affords a low conductivity, strongly dependent on temperature and of the order of one-millionth that of a conductor.

Impurity semiconductors have their low conductivity provided by the presence of minute quantities of foreign atoms (e.g., 1 in 10^8) or by deformations in the crystal structure. The impurities 'donate' electrons of energy-level that can be raised into a conduction band (n-type); or they can attract an electron from a filled band to leave a 'hole', or electron deficiency, the movement of which corresponds to the movement of a positive charge (p-type).

4.13 Magnetism

Modern magnetic theory is very complex, with ramifications in several branches of physics. Magnetic phenomena are associated with moving charges. Electrons, considered as particles, are assumed to possess an axial spin, which gives them the effect of a minute current-turn or of a small permanent magnet, called a Bohr *magneton*. The gyroscopic effect of electron spin develops a precession when a magnetic field is applied. If the precession effect exceeds the spin effect, the external applied magnetic field produces less magnetisation than it would in free space, and the material of which the electron is a constituent part is *diamagnetic*. If the spin effect exceeds that due to precession, the material is *paramagnetic*. The spin effect may, in certain cases, be very large, and high magnetisations are produced by an external field: such materials are *ferromagnetic*.

An iron atom has, in the $n=4$ shell (N), electrons that give it conductive properties. The K, L and N shells have equal numbers of electrons possessing opposite spin-directions, so cancelling. But shell M contains 9 electrons spinning in one direction and 5 in the other, leaving 4 net magnetons. Cobalt has 3, and nickel 2. In a solid metal, further cancellation occurs and the average number of unbalanced magnetons is: Fe, 2.2; Co, 1.7; Ni, 0.6.

In an iron crystal the magnetic axes of the atoms are aligned, unless upset by excessive thermal agitation. (At 770°C for Fe, the Curie point, the directions become random and ferromagnetism is lost.) A single Fe crystal magnetises most easily along a cube edge of the structure. It does not exhibit spontaneous magnetisation like a permanent magnet, however, because a crystal is divided into a large number of *domains* in which the various magnetic directions of the atoms form closed paths. But if a crystal is exposed to an external applied magnetic field, (i) the electron spin axes remain initially unchanged, but those domains having axes in the favourable direction grow at the expense of the others (domain-wall displacement); and (ii) for higher field intensities the spin axes orientate into the direction of the applied field.

If wall movement makes a domain acquire more internal energy, then the movement will relax again when the external field is removed. But if wall-movement results in loss of energy, the movement is non-reversible—i.e., it needs external force to reverse it. This accounts for hysteresis and remanence phenomena.

The closed-circuit self-magnetisation of a domain gives it a mechanical strain. When the magnetisation directions of individual domains are changed by an external field, the strain directions alter too, so that an assembly of domains will tend to lengthen or shorten. Thus readjustments in the crystal lattice occur, with deformations (e.g. 20 parts in 10^6) in one direction. This is the phenomenon of *magnetostriction*.

The practical art of magnetics consists in control of magnetic properties by alloying, heat-treatment and mechanical working to produce variants of crystal structure and consequent magnetic characteristics.

4.14 Simplified electrical theories

In the following paragraphs, a discussion of electrical phenomena is given in terms adequate for the purpose of simple explanation.

Consider two charged bodies separated in air (*Figure 4.2*).

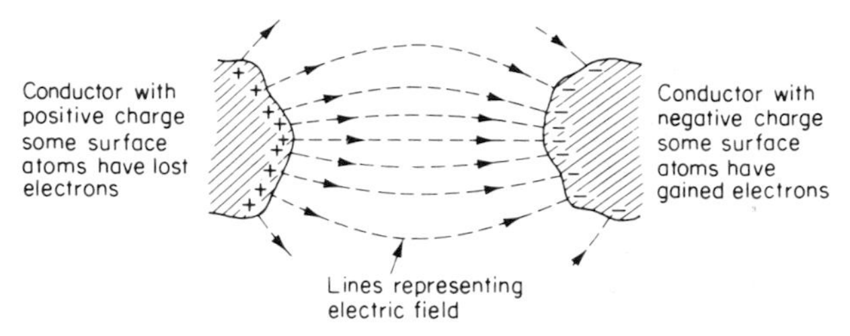

Figure 4.2 Charged conductors and their electric field

Work must have been done in a physical sense to produce on one an excess and on the other a deficiency of electrons, so that the system is a repository of potential energy. (The work done in separating charges is measured by the product of the charges separated and the difference of electrical potential that results.) Observation of the system shows certain effects of interest: (1) there is a difference of electric potential between the bodies depending on the amount of charge and the geometry of the system; (2) there is a mechanical force of attraction between the bodies. These effects are deemed to be manifestations of the *electric field* between the bodies, described as a special state of space and depicted by *lines of force* which express in a pictorial way the strength and direction of the force effects. The lines stretch between positive and negative elements of charge through the medium (in this case, air) which separates the two charged bodies. The electric field is only a concept—for the lines have no real existence—used to calculate various effects produced when charges are separated by any method which results in excess and deficiency states of atoms by electron transfer. Electrons and protons, or electrons and positively ionised atoms, attract each other, and the stability of the atom may be considered due to the balance of these attractions and dynamic forces such as electron spin. Electrons are repelled by electrons and protons by protons, these forces being summarised in the rules, formulated experimentally long before our present knowledge of atomic structure, that 'like charges repel and unlike charges attract one another'.

4.14.1 Conductors and insulators

In substances called *conductors*, the outer-shell electrons can be more or less freely interchanged between atoms. In copper, for example, the molecules are held together comparatively rigidly in the form of a 'lattice'—which gives the piece of copper its permanent shape—through the interstices of which outer electrons from the atoms can be interchanged within the confines of the surface of the piece, producing a random movement of free electrons called an 'electron atmosphere'. Such electrons are responsible for the phenomenon of electrical conductivity.

In other substances called *insulators* all the electrons are more or less firmly bound to their parent atoms so that little or no relative interchange of electron charges is possible. There is no marked line of demarcation between conductors and insulators, but the copper-group metals in the order silver, copper, gold, are outstanding in the series of conductors.

4.14.2 Conduction

Conduction is the name given to the movement of electrons, or ions, or both, giving rise to the phenomena described by the term *electric current*. The effects of a current include a redistribution of charges, heating of conductors, chemical changes in liquid solutions, magnetic effects, and many subsidiary phenomena.

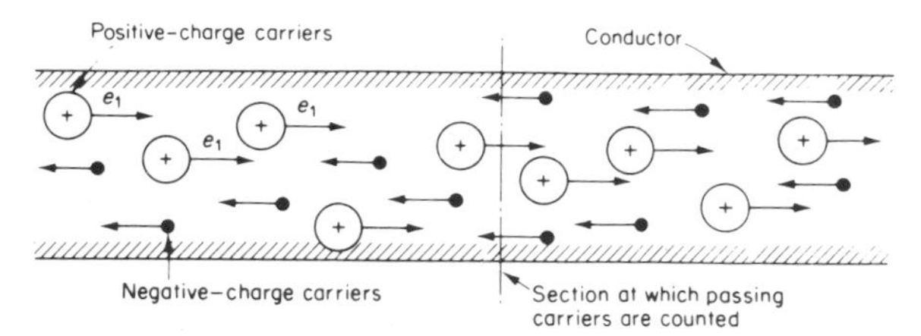

Figure 4.3 Electric current as the result of moving charges

If at some point on a conductor (*Figure 4.3*), n_1 carriers of electric charge (they can be water-drops, ions, dust particles, etc.) each with a positive charge e_1 arrive per second, and n_2 carriers (such as electrons) each with a negative charge e_2 arrive in the opposite direction per second, the total rate of passing of charge is $n_1e_1 + n_2e_2$, which is the charge per second or *current*. A study of conduction concerns the kind of carriers and their behaviour under given conditions. Since an electric field exerts mechanical forces on charges, the application of an electric field (i.e. a potential difference) between two points on a conductor will cause the movement of charges to occur, i.e., a current to flow, so long as the electric field is maintained.

The discontinuous particle nature of current flow is an observable factor. The current carried by a number of electricity carriers will vary slightly from instant to instant with the number of carriers passing a given point in a conductor. Since the electron charge is 1.6×10^{-19} C, and the passage of one coulomb per second (a rate of flow of one *ampere*) corresponds to $10^{19}/1.6 = 6.3 \times 10^{18}$ electron charges per second, it follows that the discontinuity will be observed only when the flow comprises the very rapid movement of a few electrons. This may happen in gaseous conductors, but in metallic conductors the flow is the very slow drift (measurable in mm/s) of an immense number of electrons.

A current may be the result of a two-way movement of positive and negative particles. Conventionally the direction of current flow is taken as the same as that of the positive charges and against that of the negative ones.

4.14.3 Conduction in metallic conductors

Reference has been made above to the 'electron atmosphere' of electrons in random motion within a lattice of comparatively rigid molecular structure in the case of copper, which is typical of the class of good metallic conductors. The random electronic motion, which intensifies with rise in temperature, merges into an average shift of charge of almost (but not quite) zero continuously (*Figure 4.4*). When an electric field is applied along the length of a conductor (as by maintaining a potential difference across its ends), the electrons have a *drift* towards the positive end superimposed upon their random digressions. The drift is slow, but such great numbers of electrons may be involved that very large currents, entirely due to electron drift, can be produced by this means. In their passage the electrons are impeded by the molecular lattice, the collisions producing heat and the opposition called *resistance*. The conventional direction of current flow is actually opposite to that of the drift of charge, which is exclusively electronic.

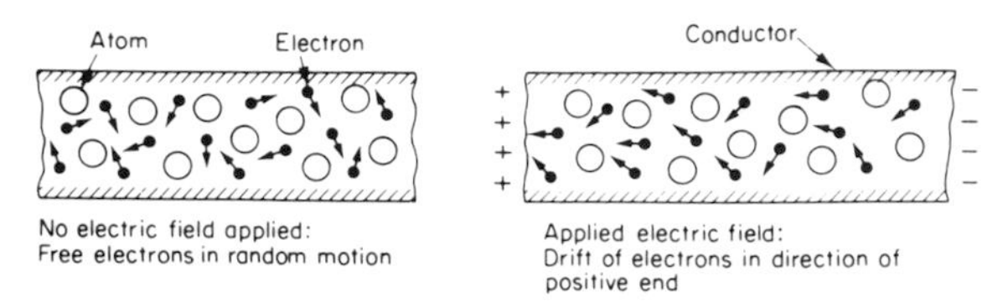

Figure 4.4 Electronic conduction in metals

4.14.4 Conduction in liquids

Liquids are classified according to whether they are *non-electrolytes* (non-conducting) or *electrolytes* (conducting). In the former the substances in solution break up into electrically balanced groups, whereas in the latter the substances form ions, each a part of a single molecule with either a positive or a negative charge. Thus common salt, NaCl, in a weak aqueous solution breaks up into sodium and chlorine ions. The sodium ion Na^+ is a sodium atom less one electron, the chlorine ion Cl^- is a chlorine atom with one electron more than normal. The ions attach themselves to groups of water molecules. When an electric field is applied the sets of ions move in opposite directions, and since they are much more massive than electrons the conductivity produced is markedly inferior to that in metals. Chemical actions take place in the liquid and at the electrodes when current passes. Faraday's electrolysis law states that the mass of an ion deposited at an electrode by electrolyte action is proportional to the quantity of electricity which passes and to the *chemical equivalent* of the ion.

4.14.5 Conduction in gases

Gaseous conduction is strongly affected by the pressure of the gas. At pressures corresponding to a few centimetres of mercury gauge, conduction takes place by the movement of positive and negative ions. Some degree of ionisation is always present due to stray radiations (light, etc.). The electrons produced attach themselves to gas atoms and the sets of positive and negative ions drift in opposite directions. At very low gas pressures the electrons produced by ionisation have a much longer free path before they collide with a molecule, and so have scope to attain high velocities. Their motional energy may be enough to *shock-ionise* neutral atoms, resulting in a great enrichment of the electron stream and an increased current flow. The current may build up to high values if the effect becomes cumulative, and eventually conduction may be effected through a *spark* or *arc*.

4.14.6 Conduction in vacuum

This may be considered as purely electronic, in that any electrons present (there can be no molecular *matter* present if the vacuum is perfect) are moved in accordance with the forces exerted on them by an applied electric field. The number of electrons is always small, and although high speeds may be reached the currents conducted in vacuum tubes are generally measurable only in milli- or micro-amperes.

4.14.7 Vacuum and gas-filled tubes

Some of the effects described above are illustrated in *Figure 4.5*. At the bottom is an electrode, the *cathode*, from the surface of which electrons are emitted, generally by heating the cathode material. At the top is a second electrode, the *anode*, and an electric field is established between anode and cathode, which are enclosed in a vessel which contains a low-pressure inert gas. The electric field causes electrons emitted from the cathode to move upwards. In their passage to the anode these electrons will encounter gas molecules. If conditions are suitable, the gas atoms are ionised, becoming in effect positive charges associated with the nuclear mass. Thereafter the current is increased by the detached electrons moving upwards and by the positive ions

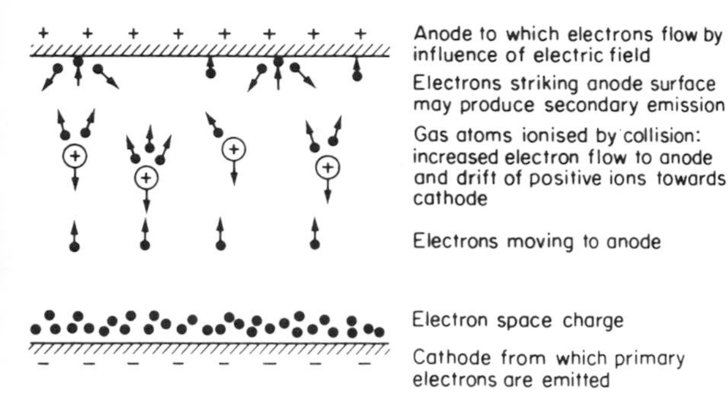

Figure 4.5 Electrical conduction in gases at low pressure

moving more slowly downwards. In certain devices (such as the mercury-arc rectifier) the impact of ions on the cathode surface maintains its emission. The impact of electrons on the anode may be energetic enough to cause the *secondary emission* of electrons from the anode surface. If the gas molecules are excluded and a vacuum established, the conduction becomes purely electronic.

4.14.8 Convection currents

Charges may be moved by mechanical means, on discs, endless belts, water-drops, dust or mist particles. A common example is the electron beam between anode and screen in the cathode-ray oscilloscope. Such a motion of charges, independent of an electric field, is termed a *convection* current.

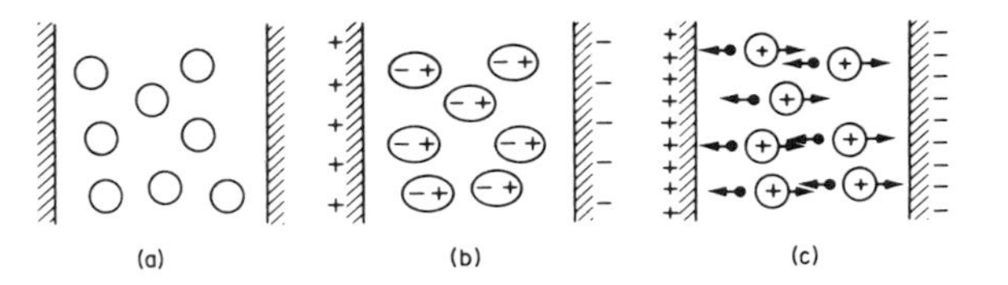

Figure 4.6 Polarisation, displacement and breakdown in a dielectric material: (a) no electric field; atoms unstrained; (b) electric field applied; polarisation; (c) intensified electric field; atoms ionised

4.14.9 Displacement and polarisation currents

If an electric field is applied to a perfect insulator, whether solid, liquid or gaseous, the electric field affects the atoms by producing a kind of 'stretching' or 'rotation' which displaces the electrical centres of negative and positive in opposite directions. This polarisation of the dielectric insulating material may be considered as taking place in the manner indicated in *Figure 4.6*. Before the electric field is applied, in (*a*), the atoms of the insulator are neutral and unstrained; (*b*) as the potential difference is raised the electric field exerts opposite mechanical forces on the negative and positive charges and the atoms become more and more highly strained. On the left face the atoms will all present their negative charges at the surface: on the right face, their positive charges. These surface polarisations are such as to account for the effect known as *permittivity*. The small displacement of the electric charges is an electron shift, i.e., a *displacement current* flows while the polarisation is being established. *Figure 4.6*(*c*) shows that under conditions of excessive electric field atomic disruption or ionisation may occur, converting the insulator material into a conductor, resulting in *breakdown*.

5 Light

D. R. Heath

Contents

5.1 Introduction

In recent years the growth of the field of opto-electronics has required the engineer to furnish himself with a knowledge of the nature of optical radiation and its interaction with matter. The increase in the importance of measurements of optical energy has also necessitated an introduction into the somewhat bewildering array of terminologies used in the hitherto specialist fields of radiometry and photometry.

5.2 The optical spectrum

Light is electromagnetic radiant energy and makes up part of the electromagnetic spectrum. The term *optical spectrum* is used to described the *light* portion of the electromagnetic spectrum and embraces not only the visible spectrum (that detectable by the eye) but also the important regions in optoelectronics of the ultraviolet and infrared.

The electromagnetic spectrum, classified into broad categories according to wavelength and frequency, is given in *Figure 6.1*, Chapter 6. It is observed that on this scale the optical spectrum forms only a very narrow region of the complete electromegnatic spectrum. *Figure 5.1* is an expanded diagram showing more detail of the ultraviolet, visible and infrared regions. By convention, optical radiation is generally specified according to its wavelength. The wavelength can be determined from a specific electromagnetic frequency from the equation:

$$\lambda = c/f \qquad (5.1)$$

where λ is the wavelength (m), f is the frequency (Hz) and c is the speed of light in a vacuum ($\sim 2.99 \times 10^8$ m s^{-1}). The preferred unit of length for specifying a particular wavelength in the visible spectrum is the nanometre (nm). Other units are also in common use, namely the angström (Å) and the micrometre or micron. The relation of these units is as follows:

1 nanometre (nm) $= 10^{-9}$ metre
1 angström (Å) $= 10^{-10}$ metre
1 micron (μm) $= 10^{-6}$ metre

The micron tends to be used for describing wavelengths in the infrared region and the nanometre for the ultraviolet and visible regions.

The wavenumber (cm^{-1}) is the reciprocal of the wavelength measured in centimetres, i.e. $1/\lambda$ (cm) = wavenumber (cm^{-1}).

5.3 Basic concepts of optical radiation

In describing the measurement of light and its interaction with matter, three complementary properties of electromagnetic radiation need to be invoked: ray, wave and quantum. At microwave and longer wavelengths it is generally true that radiant energy exhibits primarily wave properties while at the shorter wavelengths, X-ray and shorter, radiant energy primarily exhibits ray and quantum properties. In the region of the optical spectrum, ray, wave, and quantum properties will have their importance to varying degrees.

5.4 Radiometry and photometry

Radiometry is the science and technology of the measurement of radiation from all wavelengths within the optical spectrum. The basic unit of power in radiometry is the watt (W).

Photometry is concerned only with the measurement of light detected by the eye, i.e. that radiation which falls between the wavelengths 380 nm and 750 nm. The basic unit of power in photometry is the lumen (lm).

In radiometric measurements the ideal detector is one which has a flat response with wavelength whereas in photometry the ideal detector has a spectral response which approximates to that of the average human eye. To obtain consistent measurement techniques the response of the average human eye was established by the Commission Internationale de l'Eclairage (CIE) in 1924. The response known as the photopic eye response is shown in *Figure 5.2* and is observed to peak in the green/yellow part of the visible spectrum at 555 nm. The curve indicates that it takes approximately ten times as many units of blue light as green light to produce the same visibility effect on the average human eye.

The broken curve in *Figure 5.2* with a peak at 507 nm is termed the scotopic eye response. The existence of the two responses arises out of the fact that the eye's spectral response shifts at very low light levels. The retina of the human eye has two types of optical receptors, cones and rods. Cones are mainly responsible for colour vision and are highly concentrated in a 0.3 mm diameter spot, called the fovea, at the centre of the field of vision. Rods are not present in the fovea but have a very high density in the peripheral regions of the retina. They do not give rise to colour response but at low light levels are significantly more sensitive than cones. At normal levels of illumination (photopic response) the eye's response is determined by the cones in the retina whilst at very low light levels the retina's rod receptors take over and cause a shift in the response curve to the scotopic response.

In normal circumstances photometric measurements are based on the CIE photopic response and all photometric instruments must have sensors which match this response. At the peak wavelength of 555 nm of the photopic response one watt of radiant power is defined as the equivalent of 680 lumens of luminous power.

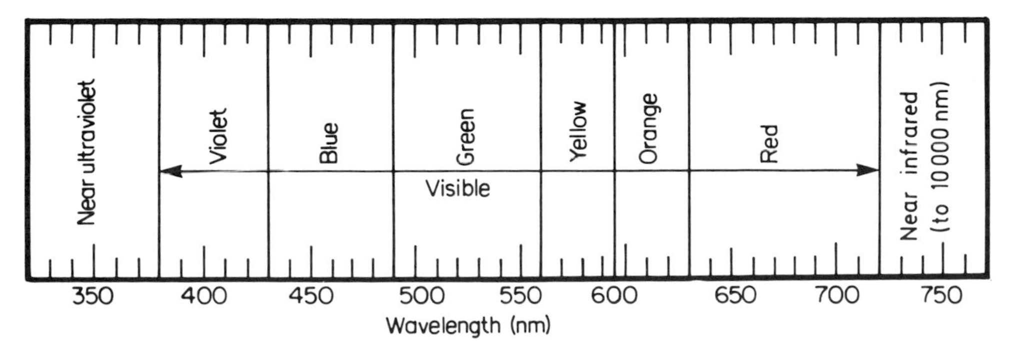

Figure 5.1 The visible spectrum

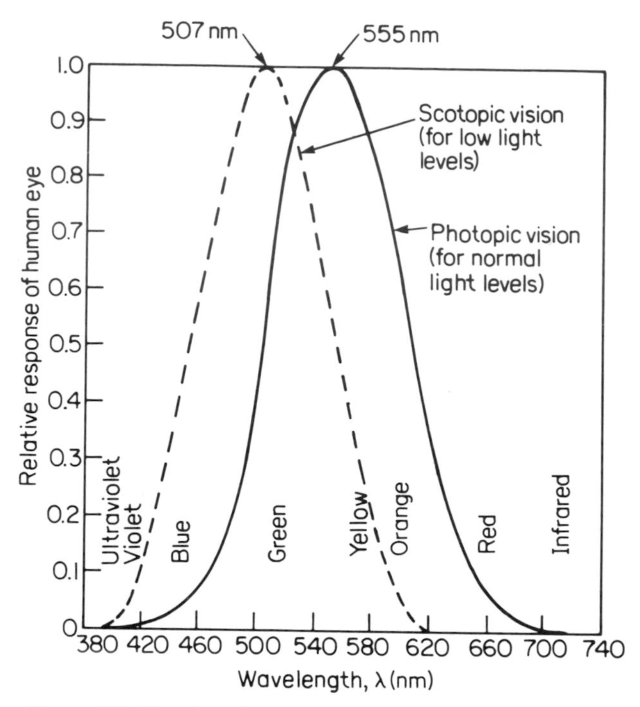

Figure 5.2 The photopic and scotopic eye responses

In order to convert a radiometric power measurement into photometric units both the spectral response of the eye and the spectral output of the light source must be taken into account. The conversion is then achieved by multiplying the energy radiated at each wavelength by the relative lumen/watt factor at that wavelength and summing the results. Note that in the ultraviolet and infrared portions of the optical spectrum although one may have high output in terms of watts the photometric value in lumens is zero due to lack of eye response in those ranges. However, it should be said that many observers can see the 900 nm radiation from a GaAs laser or the 1.06 μm radiation from a Nd:YAG laser since in this instance the intensity can be sufficiently high to elicit a visual response. Viewing of these sources in practice is not to be recommended for safety reasons and the moderately high energy densities at the eye which are involved.

5.5 Units of measurement

There are many possible measurements for characterising the output of a light source. The principles employed in defining radiometric and photopic measurement terms are very similar. The terms employed have the adjective radiant for a radiometric measurement and luminous for a photometric measurement. The subscript e is used to indicate a radiometric symbol and the subscript v for a photometric symbol. A physical visualisation of the terms to be defined is given in *Figure 5.3*. *Figure 5.4* illustrates the concept of solid angle required in the visualisation of *Figure 5.3*.

5.5.1 Radiometric terms and units

Radiant flux or **radiant power,** Φ_e The time rate of flow of radiant energy emitted from a light source. Expressed in $J\,s^{-1}$ or W.

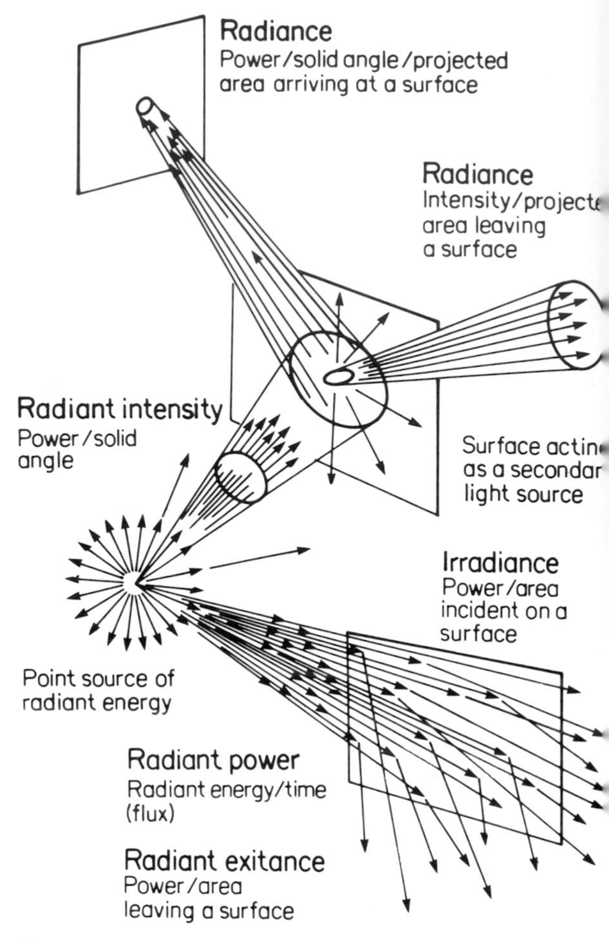

Figure 5.3 A visualisation of radiometric terms (from reference 1)

Irradiance, E_e The radiant flux density incident on a surface. Usually expressed in $W\,cm^{-2}$.

Radiant intensity, I_e The radiant flux per unit solid angle travelling in a given direction. Expressed in $W\,sr^{-1}$.

Radiant exitance, M_e The total radiant flux divided by the surface area of the source. Expressed in $W\,cm^{-2}$.

Radiance, L_e The radiant intensity per unit area, leaving, passing through, or arriving at a surface in a given direction. The surface area is the projected area as seen from the specified direction. Expressed in $W\,cm^{-2}\,sr^{-1}$.

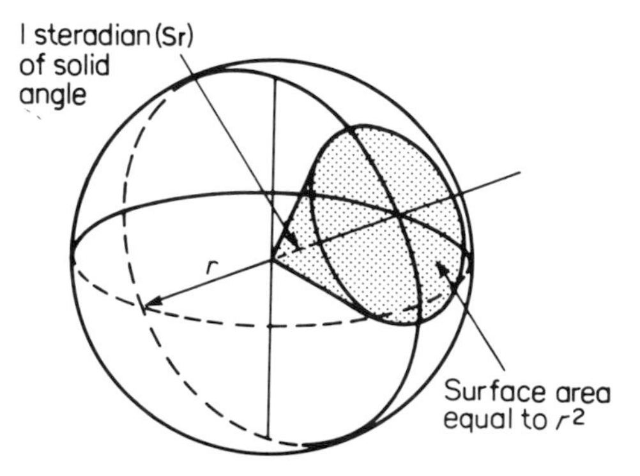

Figure 5.4 Diagram illustrating the steradian (from reference 1)

5.5.2 Photometric terms and units

The equivalent photometric terminologies to the radiometric ones defined above are as follows:
Luminous flux or **power**, Φ_v The time rate of flow of luminous energy emitted from a light source. Expressed in lm.
Illuminance or **illumination**, E_v The density of luminous power incident on a surface. Expressed in lm cm^{-2}. Note the following:

1 lm cm^{-2} = 1 phot
1 lm m^{-2} = 1 lux
1 lm ft^{-2} = 1 footcandle

Luminous intensity, I_v The luminous flux per unit solid angle, travelling in a given direction. Expressed in lm sr^{-1}. Note that 1 lm sr^{-1} = 1 cd.
Luminous exitance, M_v The total luminous flux divided by the surface area of the source. Expressed in lm cm^{-2}.
Luminance, L_v The luminous intensity per unit area, leaving, passing through or arriving at a surface in a given direction. The surface area is the projected area as seen from the specified direction. Expressed in lm cm^{-2} sr^{-1} or cd cm^{-2}.

Mathematically if the area of an emitter has a diameter or diagonal dimension greater than 0.1 of the distance of the detector it can be considered as an area source. Luminance is also called the photometric **brightness**, and is a widely used quantity. In *Figure 9.5* the projected area of the source, A_p varies directly as the cosine of θ, i.e. is a maximum at 0° or normal to the surface and minimum at 90°. Thus

$$A_p = A_s \cos\theta \tag{5.2}$$

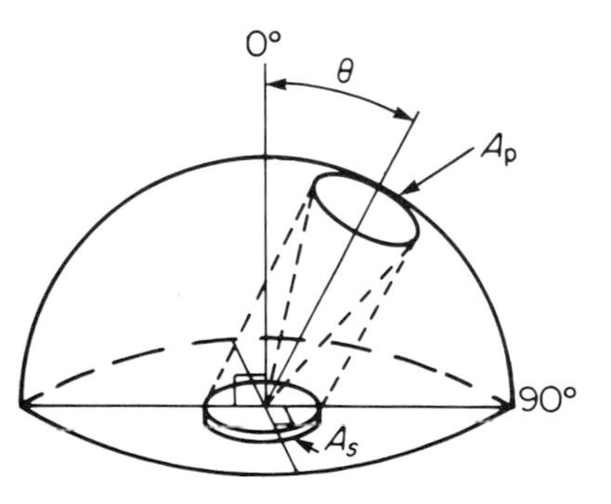

Figure 5.5 Diagram illustrating the projected area

Luminance is then the ratio of the luminous intensity (I_v) to the projected area of the source (A_p):

$$\text{luminance} = \frac{\text{luminous intensity}}{\text{projected area}} = \frac{I_v}{A_p}$$

$$= \frac{I_v}{A_s \cos\theta} \text{ lm sr}^{-1} \text{ per unit area}$$

since one lm sr^{-1} = one cd, depending on the units used for the area we have

1 cd cm^{-2} = 1 stilb
$1/\pi$ cd cm^{-2} = 1 lambert
$1/\pi$ cd ft^{-2} = 1 footlambert

Table 5.1 provides a summary of the radiometric and photometric terms with their symbols and units.

Table 5.1 Radiometric and photometric terms

Quantity	*Symbol*	*Unit(s)*
Radiant flux	Φ_e	W
Luminous flux	Φ_v	lm
Irradiance	E_e	W cm^{-2}
Illuminance	E_v	lm cm^{-2} = phot
		lm m^{-2} = lux
		lm ft^{-2} = footcandle
Radiant intensity	I_e	W sr^{-1}
Luminous intensity	I_v	lm sr^{-1} = cd
Radiant exitance	M_e	W cm^{-2}
Luminous exitance	M_v	lm cm^{-2}
Radiance	L_e	W cm^{-2} sr^{-1}
Luminance	L_v	lm cm^{-2} sr^{-1}
(Photometric brightness)		cd cm^{-2} = stilb
		$1/\pi$ cd cm^{-2} = lambert
		$1/\pi$ cd ft^{-2} = footlambert

Some typical values of natural scene illumination expressed in units of lm m^{-2} and footcandles are given in *Table 5.2*. *Table 5.3* gives some approximate values of luminance for various sources.

Table 5.2 Approximate levels of natural scene illumination (from reference 1)

	Footcandles	*lm* m^{-2}
Direct sunlight	$1.0–1.3 \times 10^4$	$1.0–1.3 \times 10^5$
Full daylight	$1–2 \times 10^3$	$1–2 \times 10^4$
Overcast day	10^2	10^3
Very dark day	10	10^2
Twilight	1	10
Deep twilight	10^{-1}	1
Full moon	10^{-2}	10^{-1}
Quarter moon	10^{-3}	10^{-2}
Starlight	10^{-4}	10^{-3}
Overcast starlight	10^{-5}	10^{-4}

Table 5.3 Approximate levels of luminance for various sources (from reference 1)

	Footlamberts	*cd* m^{-2}
Atomic fission bomb (0.1 ms after firing, 90 ft diameter ball)	6×10^{11}	2×10^{12}
Lightning flash	2×10^{10}	6.8×10^{10}
Carbon arc (positive crater)	4.7×10^6	1.6×10^7
Tungsten filament lamp (gas-filled, 16 lm W^{-1})	2.6×10^5	8.9×10^5
Sun (as observed from the earth's surface at meridian)	4.7×10^3	1.6×10^4
Clear blue sky	2300	7900
Fluorescent lamp (T-12 bulb, cool white, 430 mA medium loading)	2000	6850
Moon (as observed from earth's surface)	730	2500

5.6 Practical measurements

A wide variety of commercial instruments is available for carrying out optical radiation measurements.

The radiometer is an instrument which will normally employ a photodiode, phototube, photomultiplier or photoconductive cell as its detector. Each of these detectors has a sensitivity which varies with wavelength. It is therefore necessary for the instrument to be calibrated over the full range of wavelengths for which it is to be used. For measurement of monochromatic radiation the instrument reading is simply taken and multiplied by the appropriate factor in the detector sensitivity at the given wavelength. A result in units of power or energy is thereby obtained.

For the characterisation of broadband light sources, where the output is varying with wavelength, it is necessary to measure the source in narrow band increments of wavelength. This can be achieved by using a set of calibrated interference filters.

The spectroradiometer is specifically designed for broadband measurements and has a monochromator in front of the detector which performs the function of isolating all the wavelengths of interest. These can be scanned over the detector on a continuous basis as opposed to the discrete intervals afforded by filters.

The photometer is designed to make photometric measurements of sources. It usually consists of a photoconductive cell, silicon photodiode or photomultiplier with a filter incorporated to correct the total system response to that of the standard photopic eye response curve.

Thermopiles, bolometers and pyrometers generate signals which can be related to the incident power as a result of a change in temperature which is caused by absorption of the radiant energy. They have an advantage that their response as a function of wavelength is almost flat (constant with wavelength), but are limited to measurement of relatively high intensity sources and normally at wavelengths greater than 1 μm.

Calibration of most optical measuring instruments is carried out using tungsten lamp standards and calibrated thermopiles. The calibration accuracy of these lamp standards varies from approximately $\pm 8\%$ of absolute in the ultraviolet to $\pm 5\%$ of absolute in the visible and near infrared. Measurement systems calibrated with these standards will generally have accuracies of 8 to 10% of absolute. It is important to realise that the accuracy of optical measurements is rather poor compared to other spheres of physics. To obtain an accuracy of 5% in a measurement is very difficult; a good practitioner will be doing well to keep his errors to between 10 and 20%.

5.7 Interaction of light with matter

Light may interact with matter by being reflected, refracted, absorbed or transmitted. Two or more of these are usually involved.

5.7.1 Reflection

Some of the light impinging on any surface is reflected away from the surface. The reflectance varies according to the properties of the surface and the wavelength of the impinging radiation.

Regular or *specular reflection* is reflection in accordance with the laws of reflection with no diffusion (surface is smooth compared to the wavelength of the impinging radiation).

Diffuse reflection is diffusion by reflection in which on the microscopic scale there is no regular reflection (surface is rough when compared to the wavelength of the impinging radiation).

Reflectance (ρ) is the ratio of the reflected radiant or luminous flux to the incident flux.

Reflection (optical) density (D) is the logarithm to the base ten of the reciprocal of the reflectance.

$$D(\lambda) = \log_{10}\left(\frac{1}{\rho(\lambda)}\right) \tag{5.3}$$

where $\rho(\lambda)$ is the spectral reflectance.

5.7.2 Absorption

When a beam of light is propagated in a material medium its speed is less than its speed in a vacuum and its intensity gradually decreases as it progresses through the medium. The speed of light in a material medium varies with the wavelength and this variation is known as *dispersion*. When a beam traverses a medium some of the light is scattered and some is absorbed. If the absorption is true absorption the light energy is converted into heat. All media show some absorption—some absorb all wavelengths more or less equally, others show selective absorption in that they absorb some wavelengths very much more strongly than others. The phenomena of scattering, dispersion and absorption are intimately connected.

Absorption coefficient

Lambert's law of absorption states that equal paths in the same absorbing medium absorb equal fractions of the light that enters them. If in traversing a path of length dx the intensity is reduced from I to $I - dI$ then Lambert's law states that dI/I is the same for all elementary paths of length dx. Thus

$$\frac{dI}{I} = -K\,dx$$

where K is a constant known as the absorption coefficient. Therefore $\log I = -Kx + C$ where C is a constant. If $I = I_0$ at $x = 0$, $C = \log I_0$ and so

$$I = I_0 e^{-Kx} \tag{5.4}$$

Note that in considering a medium of thickness x, I_0 is not the intensity of incident light due to there being some reflection at the first surface. Similarly I is not the emergent intensity owing to reflection at the second surface. By measuring the emergent intensity for two different thicknesses the losses due to reflection may be eliminated.

5.7.3 Polarisation

For an explanation of polarisation of light we need to invoke the wave concept and the fact that light waves are of a transverse nature possessing transverse vibrations which have both an electric and magnetic character. *Figures 5.6* and *5.7* set out to illustrate the meaning of unpolarised and linearly polarised light.

In *Figure 5.6* a wave is propagating in the x direction with the vibrations in a single plane. Any light which by some cause possesses this property is said to be linearly polarised. Ordinary light, such as that received from the sun or incandescent lamps, is

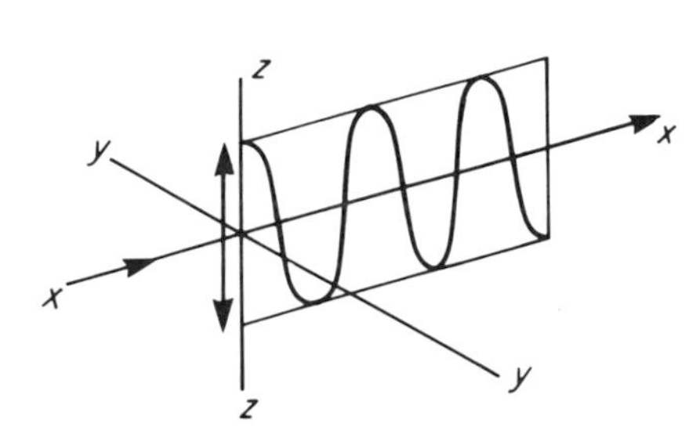

Figure 5.6 Linearly polarised light

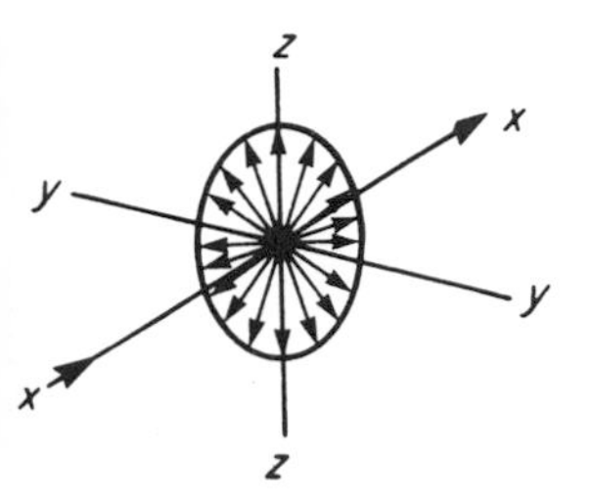

Figure 5.7 Unpolarised light

unpolarised and in this case the arrangement of vibrations is in all possible directions perpendicular to the direction of travel, as in *Figure 5.7*.

There are numerous methods for producing linearly polarised light, those most widely known being birefringence or double refraction, reflection, scattering and dichroism. Double reflection occurs in certain types of natural crystal such as calcite and quartz and will divide a beam of unpolarised light into two separate polarised beams of equal intensity. By eliminating one of the polarised beams a very efficient linear polariser can be made.

Dichroic polarisers make up the great majority of commercially produced synthetic polarisers. They exhibit dichroism, the property of absorbing light to different extents depending on the polarisation form of the incident beam.

The light emerging from a linear polariser can be given a 'twist' so that the vibrations are no longer confined to a single plane but instead form a helix. This is achieved by inserting a sheet of double-refracting material into the polarised beam which divides the beam into two beams of equal intensity but with slightly different speeds, one beam being slightly retarded. The light is said to be circularly polarised.

The application and uses of polarised light are very considerable—liquid crystal displays, control of light intensity, blocking and prevention of specular glare light, measuring optical rotation, measuring propagation of stress and strain are some notable ones.

References

1 ZAHA, M. A., 'Shedding some needed light on optical measurements', *Electronics*, 6 Nov., 91–6 (1972)

Further reading

CLAYTON, R. K., *Light and Living Matter*, Vol. 2, *The Biological Part*, McGraw-Hill, New York (1971)
GRUM, F. and BECHENER, R. J., *Optical Radiation Measurements*, Vol. 1, *Radiometry*, Academic Press, London (1979).
JENKINS, F. A. and WHITE, F. E., *Fundamentals of Optics*, 3rd edn, McGraw-Hill, New York (1957).
KEYS, R. J., *Optical and infrared detectors*, Springer Verlag (1980)
LAND, E. H., 'Some aspects on the development of sheet polarisers', *J. Opt. Soc. Am.*, **41**, 957 (1951)
LERMAN, S., *Radiant energy and the eye*, Macmillan (1980)
LONGHURST, R. S., *Geometrical and Physical Optics*, Longman, Green & Co., London (1957)
MAYER-ARENDT, J. R., *Introduction to Classical and Modern Optics*, Prentice Hall, Englewood Cliffs, NJ (1972)
RCA Electro-Optics Handbook, RCA Commercial Engineering, Harrison, NJ (1974)
WALSH, J. W. T., *Photometry*, Dover, New York (1965)

6 Radiation

L. W. Turner

Contents

6.1 Electromagnetic radiation

Light and heat were for centuries the only known kinds of radiation. Today it is known that light and heat radiation form only a very small part of an enormous range of radiations extending from the longest radio waves to the shortest gamma-rays and known as the electromagnetic spectrum, *Figure 6.1*. The wavelength of the radiations extends from about 100 kilometres to fractions of micrometres. The visible light radiations are near the centre of the spectrum. All other radiations are invisible to the human eye.

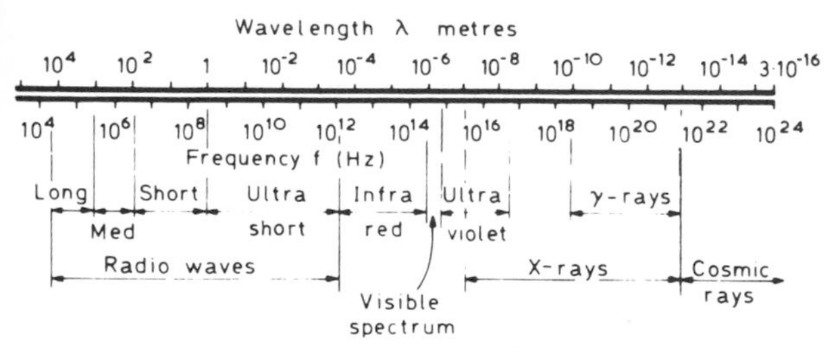

Figure 6.1 Electromagnetic wave spectrum

The researches into electromagnetic radiation can be traced back to 1680, to Newton's theory of the composition of white light. Newton showed that white light is made up from rays of different colours. A prism refracts these rays to varying degrees according to their wavelengths and spreads them out. The result is the visible spectrum of light.

In 1800, William Herschel, during research into the heating effects of the visible spectrum, discovered that the maximum heating was not within the visible spectrum but just beyond the red range. Herschel concluded that in addition to visible rays the sun emits certain invisible ones. These he called infrared rays.

The next year, the German physicist Ritter made a further discovery. He took a sheet of paper freshly coated with silver chloride and placed it on top of a visible spectrum produced from sunlight falling through a prism. After a while he examined the paper in bright light. It was blackened, and it was most blackened just beyond the violet range of the spectrum. These invisible rays Ritter called ultraviolet rays.

The next step was taken in 1805, when Thomas Young demonstrated that light consists of waves, a theory which the frenchman Fresnel soon proved conclusively. Fresnel showed that the waves vibrated transversely, either in many planes, or in one plane, when the waves were said to be plane-polarised. The plane containing both the direction of propagation and the direction of the electric vibrations is called the plane of polarisation.

In 1831 Faraday showed that when a beam of light was passed through a glass block to which a magnetic field was applied in the same direction as the direction of polarisation the plane of polarisation could be rotated. Moreover, when the magnetic field was increased, the angle of rotation also increased. The close relationship between light, magnetism and electricity was thus demonstrated for the first time.

In 1864 James Clerk Maxwell formulated his theory of electromagnetic waves and laid the foundation of the wave theory of the electromagnetic spectrum as it is known today.

The *fundamental Maxwell theory* includes two basic laws and the displacement-current hypothesis:

Ampere's law. The summation of the magnetic force H round a closed path is proportional to the total current flowing across the surface bounded by the path:

F = line-integral of $H.\mathrm{d}l = I$

Displacement current. The symbol I above includes polarisation and displacement currents as well as conduction currents.

Faraday's law. The summation of the electric force E round a closed path is proportional to the rate-of-change of the magnetic flux Φ across the surface bounded by the path:

e = line-integral of $E.\mathrm{d}l = -\mathrm{d}\Phi/\mathrm{d}t$

The magnetic flux is circuital, and representable by 'closed loops' in a 'magnetic circuit'. The electric flux may be circuital, or it may spring from charges. The total flux leaving or entering a charge Q is Q coulombs.

A metallic circuit is not essential for the development of an e.m.f. in accordance with Faraday's law. The voltage-gradient E exists in the space surrounding a changing magnetic flux. The conductor is needed when the e.m.f. is to produce conduction currents. Again, the existence of a magnetic field does not necessarily imply an associated conduction current: it may be the result of a displacement current.

Maxwell deduced from these laws (based on the work of Faraday) the existence of electromagnetic waves in free space and in material media. Waves in free space are classified in accordance with their frequency f and their wavelength λ, these being related to the free-space propagation velocity $c \simeq 3 \times 10^8$ m/s by the expression $c = f\lambda$. Radiant energy of wavelength between 0.4 and 0.8 μm (frequencies between 750 and 375 GHz) is appreciated by the eye as *light* of various colours over the visible spectrum between violet (the shorter wavelength) and red (the longer). Waves shorter than the visible are the *ultraviolet*, which may excite visible fluorescence in appropriate materials. *X-rays* are shorter still. At the longer-wave end of the visible spectrum is *infrared* radiation, felt as *heat*. The range of wavelengths of a few millimetres upward is utilised in *radio* communication.

In 1886 Heinrich Hertz verified Maxwell's theory. At that time Wimshurst machines were used to generate high voltages. A Leyden jar served to store the charge which could be discharged through a spark gap. Hertz connected a copper spiral in series with the Leyden jar; this spiral acted as a radiator of electromagnetic waves. A second spiral was placed a small distance from the first; this also was connected to a Leyden jar and a spark gap. When the wheel of the Wimshurst machine was turned sparks jumped across both gaps. The secondary sparks were caused by electromagnetic waves radiated from the first spiral and received by the second. These waves were what are today called *radio* waves. This experiment was the first of a series by which Hertz established the validity of Maxwell's theory.

In 1895, the German Roentgen found by chance that one of his discharge tubes had a strange effect on a chemical substance which happened to lie nearby: the substance emitted light. It even fluoresced when screened by a thick book. This meant that the tube emitted some kind of radiation. Roentgen called these unknown rays X-rays.

A year later the French physicist Henri Becquerel made a further discovery. He placed a photographic plate, wrapped in black paper, under a compound of uranium. He left it there over night. The plate, when developed, was blackened where the uranium had been. Becquerel had found that there exist minerals which give off invisible rays of some kind.

Later, research by Pierre and Marie Curie showed that many substances had this effect: radioactivity had been discovered. When this radiation was analysed it was found to consist of charged particles, later called alpha- and beta-rays by Rutherford. These particles were readily stopped by thin sheets of paper or metal.

In 1900, Villard discovered another radiation, much more

penetrating and able to pass even through a thick steel plate. This component proved to consist of electromagnetic waves which Rutherford called gamma-rays. They were the last additions to the electromagnetic spectrum as is known today.

Waves are classified according to their uses and the methods of their generation, as well as to their frequencies and wavelengths. Radio waves are divided into various bands: very low frequency (v.l.f.) below 30 kHz, low frequency (l.f.) 30–300 kHz, medium frequency (m.f.) 300–3000 kHz, high frequency (h.f.) 3–30 MHz, very high frequency (v.h.f.) 30–300 MHz, ultra high frequency (u.h.f.) 300–3000 MHz, super high frequency (s.h.f.) 3000–30 000 MHz, extra high frequency (e.h.f.) 30 000–300 000 MHz. Waves of frequencies higher than 100 MHz are generally called microwaves. The microwave band overlaps the infrared band. Actually, *all* wave bands merge imperceptibly into each other; there is never a clear-cut division. Next is the narrow band of visible light. These visible rays are followed by the ultraviolet rays and the X-rays. Again, all these bands merge into each other. Finally, the gamma-rays. They are actually part of the X-ray family and have similar characteristics excepting that of origin.

A convenient unit for the measurement of wavelengths shorter than radio waves is the micrometre (μm) which is 10^{-6} m. The micrometre is equal to the micron (μ), a term still used but now deprecated. Also deprecated is the term mμ, being 10^{-3} micron. For measurement of still shorter wavelengths, the nanometre (10^{-9} m) is used. The angström unit (Å), which is 10^{-10} m, is commonly used in optical physics. These units of wavelength are compared in *Table 6.1*.

Table 6.1 Comparison of units of length

		Å	*nm*	*μm*	*mm*	*cm*	*m*
Å	=	1	10^{-1}	10^{-4}	10^{-7}	10^{-8}	10^{-10}
nm	=	10	1	10^{-3}	10^{-6}	10^{-7}	10^{-9}
μm	=	10^4	10^3	1	10^{-3}	10^{-4}	10^{-6}
mm	=	10^7	10^6	10^3	1	10^{-1}	10^{-3}
cm	=	10^8	10^7	10^4	10	1	10^{-2}
m	=	10^{10}	10^9	10^6	10^3	10^2	1

Electromagnetic waves are generated by moving charges such as free electrons or oscillating atoms. Orbital electrons (see Chapter 8) radiate when they move from one orbit to another, and only certain orbits are permissible. Oscillating nuclei radiate gamma-rays.

The frequency of an electromagnetic radiation is given by the expression:

$$f = \frac{E}{h}$$

where E is the energy and h is Planck's constant ($h \simeq 6.6 \times 10^{-34}$ J s).

The identity of electromagnetic radiations has been established on the following grounds:

(1) The velocity of each *in vacuo* is constant.
(2) They all experience reflection, refraction, dispersion, diffraction, interference and polarisation.
(3) The mode of transmission is by transverse wave action.
(4) All electromagnetic radiation is emitted or absorbed in bursts or packets called quanta (or photons in the case of light).

In connection with (4), Planck established that the energy of each quantum varies directly with the frequency of the radiation (see the above expression).

Modern physics now accepts the concept of the dual nature of electromagnetic radiation, viz. that it has wave-like properties but at the same time it is emitted and absorbed in quanta.

Polarisation. An electromagnetic radiation possesses two fields at right angles to each other as viewed in the direction of the oncoming waves. These are the electric field and the magnetic field. The direction of either of these is known as the polarisation of the field, but the term is more usually related to the electric field, and this is at the same angle as the radiating source. For example, in the case of radio waves, a horizontally positioned receiving dipole will not respond efficiently to waves which are vertically polarised.

The same phenomenon occurs with light radiation which is normally unpolarised, i.e. it is vibrating in all transverse planes. A sheet of polaroid allows light to pass through in one plane, due to the molecular structure of the material, and the resulting plane-polarised light is absorbed if a second sheet of polaroid is set at right-angles to the first.

The 'optical window'. It is important to realise that our knowledge of the universe around us depends upon incoming electromagnetic radiation. However, from the entire spectrum of such radiation, only two bands effectively reach the earth's surface:

(1) the visible light spectrum, together with a relatively narrow band of the adjacent ultraviolet and infrared ranges;
(2) a narrow band of radio waves in the 1 cm to about 10 m band.

Thus the gamma-rays, X-rays, most of the ultraviolet and infrared rays, together with the longer ranges of radio waves fail to reach the earth's surface from outer space. This is mainly due to absorption in the ionosphere and atmosphere.

The applications of electromagnetic radiations range over an enormous field. Radio waves are used for telecommunication, sound and television broadcasting, navigation, radar, space exploration, industry, research, etc. Infrared rays have many applications including security systems, fire detection, dark photography, industry, medical therapy. Ultraviolet rays, of which the sun is the chief source, have wide industrial and medical applications. X-rays and gamma-rays have become the everyday tools of the doctor, the scientist and in industry. And the narrow band of visible rays not only enables the world around us to be seen but together with ultraviolet radiation makes possible the process of photosynthesis by which plants build up and store the compounds of all our food. Thus the laws governing electromagnetic radiation are relevant to life itself. Many of these applications are described in more detail in later sections.

6.2 Nuclear radiation

There are three main types of radiation that can originate in a nucleus: *alpha*, *beta*, and *gamma* radiation.

Alpha radiation. An alpha particle has a charge of two positive units and a mass of four units. It is thus equivalent to a helium nucleus, and is the heaviest of the particles emitted by radioactive isotopes. Alpha (α) particles are emitted mostly by heavy nuclei and can possess only discrete amounts of energy, i.e. they give a line energy spectrum. The probability of collision between particles increases with the size of the particles. Thus the rate of ionisation in a medium traversed by particles emitted from radioactive isotopes, and hence the rate of loss of energy of the particles, also increases with the size of the particles. Consequently the penetrating power of the large alpha particles is relatively poor.

Beta radiation. Beta particles can be considered as very fast electrons. They are thus much smaller than alpha particles and therefore have greater penetrating powers. Beta (β) radiation will be absorbed in about 100 inches of air or half an inch of Perspex.

Unlike α-particles, β-particles emitted in a nuclear process have a continuous energy spectrum, i.e. β-particles can possess any amount of energy up to a maximum determined by the energy equivalent to the change in mass involved in the nuclear reaction. This has been explained by postulating the existence of the *neutrino*, a particle having no charge and negligible mass. According to this theory, the energy is shared between the β-particle and the neutrino in proportions that may vary, thus giving rise to a continuous energy spectrum.

Gamma radiation. Gamma (γ) radiation is electromagnetic in nature and has, therefore, no charge or mass. Its wavelength is much shorter than that of light or radio waves, and is similar to that of *X-rays*. The distinction between γ-rays and X-rays is that γ-rays are produced within the nucleus while X-rays are produced by the transition of an electron from an outer to an inner orbit.

γ-radiation has well-defined amounts of energy—that is, it occupies very narrow bands of the energy spectrum—since it results from transitions between energy levels within the nucleus. Characteristic X-rays of all but the very lightest of elements also possess well-defined amounts of energy.

γ-radiation has very great penetrating powers. Significant amounts are able to pass through lead bricks 50 mm thick; γ-photons possessing 1 MeV of energy will lose less than 1% of their energy in traversing half a mile of air.

A rough comparison of the penetrating powers of α-, β-, and γ-radiation is given in *Figure 6.2*.

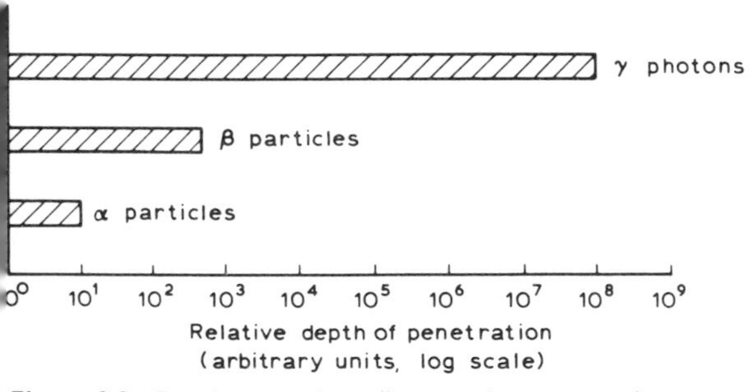

Figure 6.2 Rough comparison of penetrating powers of α-, β- and γ-radiation

6.2.1 Neutrons

Neutrons were discovered in 1932 as a result of bombarding light elements (for example, beryllium and boron) with α-particles. For laboratory purposes, this is still a convenient method of production, but the most useful and intense source is the nuclear reactor in which the neutrons are produced as a by-product of the fission of fissile materials such as uranium-235.

Free neutrons are unstable, and decay to give a proton and a low-energy β-particle. Neutrons when they are produced may have a wide range of energy, from the several millions of electron-volts of *fast neutrons* to the fractions of electron-volts of *thermal neutrons.*

Neutrons lose energy by elastic collision. An *elastic collision* is one in which the incident particles rebound—or are scattered—without the nucleus that is struck having been excited or broken up. An *inelastic collision* is one in which the struck nucleus is excited, or broken up, or captures the incoming particle. For neutrons, the loss of energy in an elastic collision is greater with light nuclei; for example, a 1 MeV neutron loses 28% of its energy in collision with a carbon atom, but only 2% in collision with lead.

By successive collisions, the energy of neutrons is reduced to that of the thermal agitation of the nucleus (that is, some 0.025 eV at 20°C) and the neutrons are then captured.

The consequence of the capture of a neutron may be a new nuclide, which may possibly be radioactive. That is, in fact, the main method of producing radioisotopes. Because they are uncharged, neutrons do not cause direct ionisation, and may travel large distances in materials having a high atomic number. The most efficient materials for shielding against neutron emission are those having light nuclei; as indicated above, these reduce the energy of neutrons much more rapidly than heavier materials. Examples of efficient shielding materials are water, the hydrocarbons and graphite.

6.2.2 Fission

Fission is the splitting of a heavy nucleus into two approximately equal fragments known as fission products. Fission is accompanied by the emission of several neutrons and the release of energy. It can be spontaneous or caused by the impact of a neutron, a fast particle or a photon:

$${}^{235}_{92}U + {}^{1}_{0}n \rightarrow {}^{93}_{38}Sr + {}^{140}_{54}Xe + {}^{1}_{0}n + {}^{1}_{0}n + {}^{1}_{0}n$$

The total number of sub-atomic particles is unchanged:

$$235 + 1 = 93 + 140 + 1 + 1 + 1$$

(other combinations of particles are possible)

The number of protons is unchanged.

Figure 6.3 is a diagrammatical representation of the foregoing.

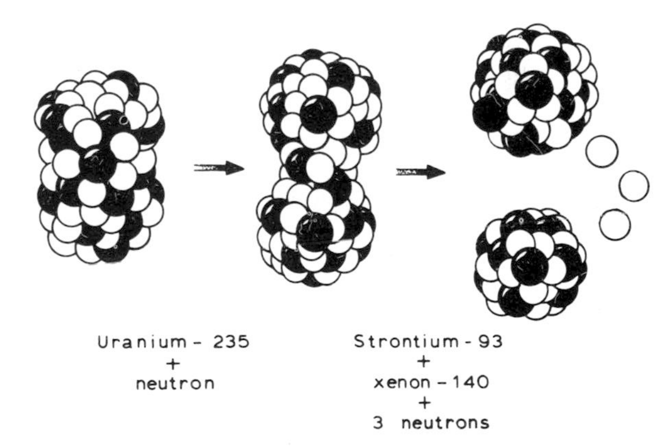

Figure 6.3 Diagrammatic representation of the fission of uranium-235

Other modes of nuclear disintegration

Mention must be made of two other methods of nuclear disintegration:

(1) emission of positively charged electrons or *positrons* (β^+);
(2) *electron capture.*

Positrons interact rapidly with electrons after ejection from the nucleus. The two electrical charges cancel each other, and the energy is released in a form of γ-radiation known as *annihilation radiation.*

In the process of electron capture, the energy of an unstable nucleus is dissipated by the capture into the nucleus of an inner orbital electron. The process is always accompanied by the emission of the characteristic X-rays of the atom produced by electron capture. For example, germanium-71, which decays in this manner, emits gallium X-rays.

6.2.3 Radioactive decay

Radioactive isotopes are giving off energy continuously and if the law of the conservation of energy is to be obeyed this radioactive decay cannot go on indefinitely. The nucleus of the radioactive atom undergoes a change when a particle is emitted and forms a new and often non-radioactive product. The rate at which this nuclear reaction takes place decreases with time in such a way that the time necessary to halve the reaction rate is constant for a given isotope and is known as its half-life. The half-life period can be as short as a fraction of a microsecond or as long as ten thousand million years.

Radioactive decay can be illustrated by considering a radioactive form of bismuth, $^{210}_{83}Bi$, which has a half-life of five days. If the number of particles emitted by a sample in one minute is recorded, then after five days, two minutes would be required for the same number to be emitted. After ten days, four minutes would be required, and so on. The amount of the radioactive bismuth, $^{210}_{83}Bi$, in the sample will diminish as the emission proceeds. The bismuth nuclei lose electrons as β-particles, and the radioactive bismuth is converted to polonium, $^{210}_{84}Po$.

In this particular case the product is itself radioactive. It emits α-particles and has a half-life of 138 days. The product of its disintegration is lead, $^{206}_{82}Pb$, which is not radioactive.

The disintegration of the radioactive bismuth can be represented as follows:

$$^{210}_{83}Bi \xrightarrow[\text{5 days}]{\beta} {}^{210}_{84}Po \xrightarrow[\text{138 days}]{\alpha} {}^{206}_{82}Pb \text{ (stable)}$$

6.2.4 Units

It is necessary to have units to define the quantity of radioactivity and its physical nature. The unit of quantity is the curie (Ci). This was originally defined as the quantity of radioactive material producing the same disintegration rate as one gram of pure radium.

The definition of quantity must be couched in different terms in modern times to include the many artificially produced radioisotopes. The curie is now defined as the quantity of radioisotope required to produce 3.7×10^{10} disintegrations per second. Quantity measurements made in the laboratory with small sources are often expressed in terms of disintegrations per second (d.p.s.). What is actually recorded by the detector is expressed in counts per second (c.p.s.). The weights of material associated with this activity can vary greatly. For example, 1 curie of iodine-131 weighs 8 micrograms, whereas 1 curie of uranium-238 weighs 2.7 tons.

The unit of energy is the electron-volt (eV). This is the kinetic energy acquired by an electron when accelerated through a potential difference of one volt. The electron-volt is equivalent to 1.6×10^{-19} joule. With α-, β-, and γ-radiation, it is usual to use thousands of electron-volts (keV) or millions of electron-volts (MeV).

Further reading

FOSTER, K. and ANDERSON, R., *Electro-magnetic Theory*, Vols. 1, 2, Butterworths, Sevenoaks (1970)

YARWOOD, J., *Atomic and Nuclear Physics*, University Tutorial Press (1973)

7 Connectors

C. Kindell, T. Kingham and J. Riley

Contents

7.1 Connector housings

Connector housings are of different shapes, sizes and form, being able to satisfy requirements for a range of applications and industries—commercial, professional, domestic and military. In a commercial low cost connector the insulator material can be nylon which can be used in a temperature range of $-40°C$ to $+105°C$ and is also available with a flame retardant additive.

The connector can be wire to wire using crimp snap-in contacts, wire to printed circuit board and also printed circuit board to printed circuit board. In the automotive industry it is necessary to have waterproof connectors to prevent any ingress of water thrown up by moving wheels and the velocity of the vehicle through rain. Wire seals are placed at the wire entry point in the connector. These seals, made from neoprene, grip the insulation of the wire very tightly thus preventing any water ingress. When the two halves of the connector are mated it is necessary to have facial seal thus precluding any water ingress between these two parts and preventing any capillary action of the water.

In the professional and military fields the housing material needs to be very stable and to counteract any attack by fluids. One type of material in this form is diallyl phthalate. The primary advantages of diallyl phthalate are exceptional dimensional stability, excellent resistance to heat, acids, alkalies and solvents, low water absorption and good dielectric strength. This combination of outstanding properties makes diallyl phthalate the best choice of plastics for high quality connectors.

Most connectors which are available in diallyl phthalate are also available in phenolic. While phenolic does not have outstanding resistance to acids, alkalies and solvents it nevertheless has many characteristics which make it a good choice for connector housings. Among these characteristics are excellent dimensional stability, good dielectric strength and heat resistance. In addition there are a number of fillers which can be added to phenolic to obtain certain desired properties.

7.2 Connector contacts

The contacts that are used with the majority of connectors are made from brass or phosphor bronze with a variety of platings from tin through to gold. The most common type of brass used in the manufacture of contacts is cartridge brass which has a composition of 70% copper and 30% zinc. This brass possesses good spring properties and strength, has excellent forming qualities and is a reasonably good conductor. Phosphor bronze alloys are deoxidised with phosphorus and contain from about 1 to 10% tin. These alloys are primarily used when a metal is needed with mechanical properties superior to those of brass and where the slightly reduced conductivity is of little consequence.

One extremely important use of phosphor bronze is in locations where the terminal may be exposed to ammonia. Ammonia environments causes stress corrosion cracking in cartridge brass terminals. On the other hand, phosphor bronze terminals are approximately 250 times more resistant to this type of failure. Associated with the materials used in the manufacture of the contacts is a variety of platings. Plating is a thin layer of metal applied to the contact by electrodeposition.

Corrosion is perhaps the most serious problem encountered in contacts and the plating used is designed to eliminate or reduce corrosion. Corrosion can spread uniformly over the surface of the contact covering it with a low conductivity layer, with the thickness of this layer being dependent upon environmental conditions, length of exposure and the type of metal being used. Brass contacts that are unplated and have been in service for a period of time have a reddish brown appearance rather than the bright yellow colour of cartridge brass. This reddish colour is a tarnish film caused by oxidation of the metal. Although this film may not impair conductivity at higher voltages, it does at the very least, destroy the appearance of the contact. To eliminate the problem of this tarnish film the contact is usually tin plated. Although oxides form on tin they are the same colour as the tin and the appearance remains the same. In addition tin is relatively soft, and if it is to be used as a contact plating, most of the oxides will be removed during mating and unmating of the contact.

Tin is the least expensive of the platings and is used primarily for corrosion protection and appearance on contacts which operate at a fairly high voltage. Another important feature of tin is that it facilitates soldering. Gold plating is always used on contacts which operate in low voltage level circuitry and corrosive environments. The presence of films caused by the combination of sulphur or oxygen with most metals can cause open circuit conditions in low voltage equipment. Since gold will not combine with sulphur or oxygen there is no possibility of these tarnish films forming.

7.3 Connector shapes and sizes

A family of connectors would need to include a range of rectangular and circular connectors associated with a variety of size-16 contacts including signal and power contacts, fibre optic contacts and subminiature coaxial contacts. The connectors shown in *Figure 7.1* are rectangular for in-line, panel mounting or rack and panel applications, and circular either for environmentally sealed application or for the commercial portion of MIL-C-5015 connector areas. The housings are available in nylon, diallyl phthalate or phenolic with, in certain areas, a metal shell. The contacts are available in brass or phosphor bronze with tin or gold plating finishes. Associated are a variety of accessories such as strain reliefs, pin hoods, jackscrews, guide pins and pin headers to give greater versatility to the connector ranges.

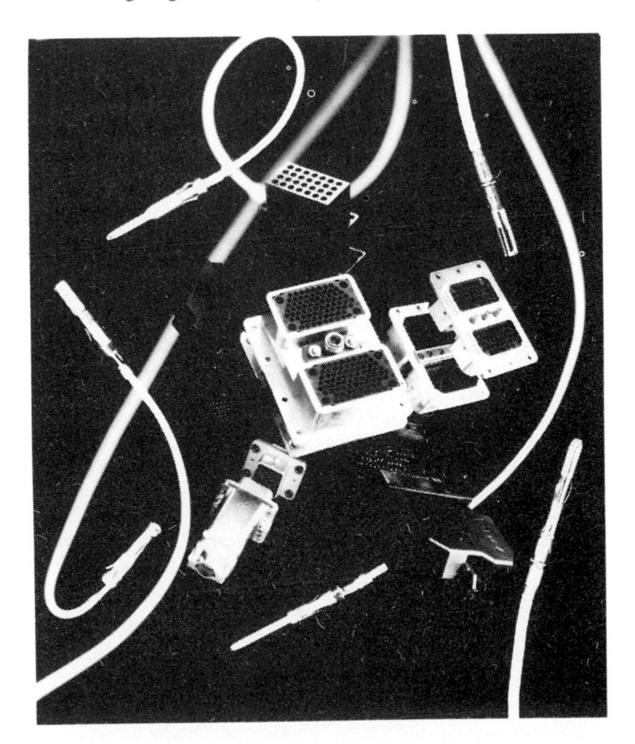

Figure 7.1 Multi-way pin and socket connectors (courtesy Amp of Great Britain Ltd)

7.4 Connector terminations

Crimping has long been recognised as an electrically and mechanically sound technique for terminating wires. Since crimping is a strictly mechanical process, it is relatively easy to automate. Because of this automation capability, crimping has become the accepted terminating technique in many industries.

Terminals or contacts designed for speed crimping in automatic or semi-automatic machines are often significantly different from those designed for handtool assembly, although most machine-crimpable terminals and contacts can also be applied with handtools.

Those designed for hand application cannot normally be used for automatic machinery. The selection of a crimping method is determined by a combination of five factors: (1) access of wire; (2) wire size; (3) production quantity; (4) power availability; and (5) terminal or contact design.

There are, of course, other factors which must be considered, e.g. if the finished leads are liable to be roughly handled, as in the appliance or automotive industries, the conductor insulation and the terminal or contact will have to be larger than electrically necessary in order to withstand misuse.

The user must remember the importance of maintaining the proper combination of wire terminal and tool, only then can the optimum crimp geometry and depth be obtained. In this respect it is best to follow the manufacturer's recommendations since most terminals have been designed for a specific crimp form. The effects of crimp depth are shown in *Figure 7.2*.

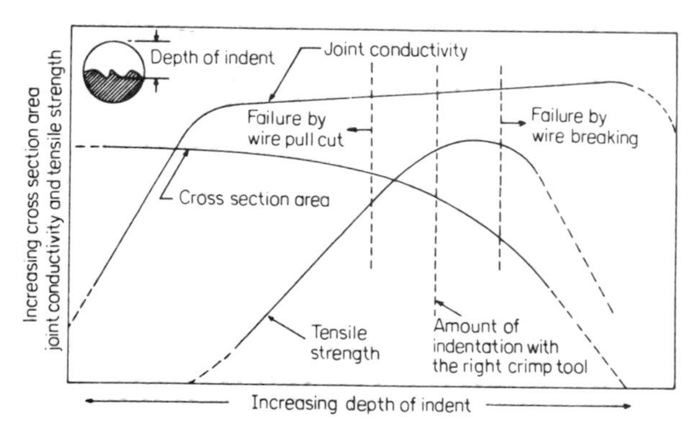

Figure 7.2 Effects of crimp depth

Tensile strength and electrical conductivity increase in proportion to the crimp depth. When the deformation is too great, tensile strength and conductivity suffer because of the reduced cross-sectional area. There is an optimum crimped depth for tensile strength and another for conductivity and, in general, these peaks do not coincide. Thus a design compromise is required to achieve the best combination of properties. However, the use of improper tooling can void the entire design.

Merely selecting the proper wire terminal and tool combination is not enough. The wire must be stripped to the recommended dimension, without nicking, and inserted into the terminal to the correct depth before crimping. A properly crimped terminal is shown in *Figure 7.3*.

It is possible to determine the relative quality of a crimp joint by measuring crimp depth in accordance with manufacturers' suggestions. Tensile strength also provides a relative indication of mechanical quality of crimped connections. This factor has been utilised as a 'user control test' in British and other international specifications.

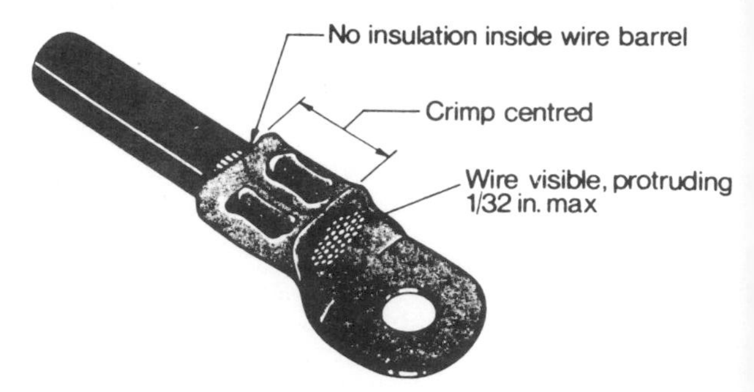

Figure 7.3 A correctly crimped un-insulated terminal

Crimped terminations do not require pre-soldering, and are designed to be used with untinned wires. Soldering may damage the crimp, burn the wire and produce a bad joint. Also the wire could stiffen from wicked solder, and break off later because of vibration. Soldering can effect the characteristics designed into the crimp and seriously affect the performance.

Many terminal designs have some means of supporting the wire insulation. These features are divided into two main categories: (a) insulation gripping; and (b) insulation supporting.

Insulation gripping terminals prevent wire flexing at the termination point and deter movement of the insulation. This feature improves the tensile strength of the crimp for applications where severe vibration is present.

Insulation supporting types of terminals have insulation that extends beyond the crimping barrel and over the wires own insulation. This only provides support and does not grip the insulation in a permanent manner. The latest feature for this type of termination is the funnel entry type. This, as its name implies, has a funnel form on the inside of the insulation sleeve, which aids in the correct placing of the stripped wire in the barrel of the terminal. It has long been possible to snag strands of wire on the edge of the wire barrel while putting the stripped wire into the terminal. This, depending on the number of snagged strands, could impair the electrical characteristics and cause 'hot spots' or under crimping of the terminal.

All these problems are resolved with the introduction of funnel entry, as the wire is able to go straight into the wire barrel without damage. Minimal operator skills, increased production rates and added benefits are possible.

Finally there is the introduction of insulation displacement or slotted beam termination, which has enjoyed wide use in the telecommunication and data systems industries. With the advent of connectors or terminal blocks designed for mass termination, this concept has spread and it is rapidly finding use in many other applications (see Section 7.6).

Although the appearance and materials in insulated displacement connectors vary the design of the slotted areas is basically the same in each.

Insulated wire fits loosely into the wider portion of the V-shaped slot and, as the wire is pushed deeper into the terminal, the narrowing slot displaces the insulation and restricts the conductor as in *Figure 7.4*. Additional downward movement of the insertion tool forces the conductor into the slot where electrical contact is made.

Insulation displacement has recently been applied to terminal blocks. One side has the insulation displacement contact whilst the other side would have the conventional screw terminal accepting ring tongues or bare wire. It would accept a wide range of wire sizes or stranded wire 0.3 mm^2 to 2 mm^2.

Wire termination is accomplished by two simple screwdriver type handtools with different insertion depths. The first is used to

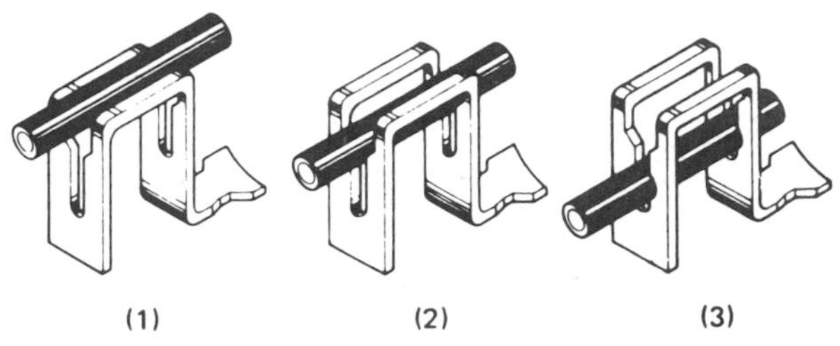

Figure 7.4 Insulation displacement contacts

insert a single wire into the terminal, the other for inserting the second wire.

7.5 Tooling

Crimping tools or machines should be selected after a thorough analysis, as with any other production system.

Generally, the following rates can be achieved with various types of tooling:

manual tools	100–175 per hour
power tools	150–300 per hour
semi-automatic machines	100–4000 per hour
lead making machine	up to 11 000 per hour

As these figures show, manual tools are intended for repair and maintenance, while powered tools and machines are designed for production applications. To the manufacturer whose output in wire terminations is relatively small, automated tooling is not necessary. While the economics of automated wire terminations may vary with different applications, an output of more than one million terminations a year can be taken as a guide for considering semi-automatic tooling.

The basics of a good crimp connection are the same whether the tool to be used is a simple plier type or a fully automatic lead making machine.

The basic type of crimping tool is the simple plier type. It is used for repair, or where very few crimps are to be made. These are similar in construction to ordinary pliers except that the jaws are specifically machined to form a crimp. Most of these tools are dependent upon the operator to complete the crimp properly by closing the pliers until the jaws bottom together. Many of the tools may be used for several functions such as wire stripping, cutting, and crimping a wide range of terminal sizes and types. Tools of this type are in wide use.

Other more sophisticated tooling is available, such as cycle controlled tools. This type normally contains a ratchet mechanism which prevents the tool opening before the crimp has been properly completed. This ratchet action produces a controlled uniform crimp every time regardless of operator skill. However, operator fatigue is normally a limiting factor in production with any manual tool.

Powered handtools, either pneumatic or electronic controlled, can be semi-portable or bench mounted. When larger production quantities of terminations are required, the need for this form of tooling is essential. They not only yield high rates of output at low installation cost but also give high standards of quality and are repeatable throughout the longest production run.

These tools offer the opportunity for the introduction of tape mounted products. A variety of tape mounted terminations are available in either reel or boxed form. Advanced tooling, with interchangeable die sets, gives a fast changeover with minimum downtime. During the crimping cycle the machine will automatically break the tape bonds and free the crimp product for easy extraction, at the same time indexing the termination into position for the next crimp operation.

7.6 Mass termination connectors

Mass termination is a method of manufacturing harnesses by taking wires directly to a connector and eliminating the steps of wire stripping, crimping and contact insertion into housings. It employs a connection technique known as insulation displacement, an idea developed many years ago for the telecommunications industry.

An unstripped insulated wire is forced into a slot which is narrower than the conductor diameter as in *Figure 7.3*. Insulation is displaced from the conductor. The sides of the slot deflect like a spring member and bear against the wire with a residual force that maintains high contact pressure during the life of the termination.

This is the basic principle, but it must be realised that each slot is carefully designed to accept dimensional changes without reducing contact forces. This is accomplished by designing enough deflection into the slot to compensate for creep, stress relaxation and differential thermal expansion.

The force required to push the wire into the slot is approximately 25 times less than for a conventional crimp and it is this factor, in conjunction with the facilitity of not having to strip the wire, that makes insulation displacement readily acceptable for mass termination, by taking several wires to the connector and terminating them simultaneously.

A typical system would employ a pre-loaded connector, with the receptacle having dual slots offering four regions of contact to the wire. The exit of the wire from the connector is at 90° to the mating pin, and can have a maximum current rating of up to 7.5 A.

The average tensile strength of the displacement connection when pulled along the axis of the wire is 70% of the tensile strength of the wire and 20% when pulled on axis parallel to the mating pin. Therefore, plastic strain ears are moulded into the connector to increase the wire removal force in this direction.

The different systems have been developed to accept a wide range of wire including 28 to 22 AWG (0.08 to 0.33 mm^2) wire and 26 to 18 AWG (0.13 to 0.82 mm^2) wire. The connectors are colour coded for each wire gauge since the dimensional difference of the slot width cannot be readily identified.

The pin headers for most systems are available for vertical and right angle applications in flat style for economical wire to post applications, polarised for correct mating and alighnment of housings and polarised with friction lock for applications in a vibration environment.

7.6.1 Types of tooling

To obtain all the benefits for harness manufacture, a full range of tooling is available, from simple 'T' handle tools to cable makers.

The 'T' handle tool would be used only for maintanance and repair. For discrete wires a self-indexing handtool, either manually or air operated, would be used for intermediate volumes.

For terminating ribbon cable, there are small bench presses for relatively low volumes of harnesses, and electric bench presses for higher production needs.

However, it is the innovation of the harness board tool and the cable maker that offers highest production savings. The harness board tool allows connectors to be mass terminated directly onto a harness board. The equipment consists of three parts: power tool, applicator and board mounted comb fixtures. The wires are routed on the harness board and placed through the appropriate comb fingers. The power tool and applicator assembly is placed on the combs to cut and insert the wires into the connectors. After binding with cable ties the harness can be removed from the board.

The cable maker, either double end or single end, will accept up to 20 wires which can be pulled from drums or reels on an appropriate rack. The individual wires can all be the same length, or variable, with a single connector on one end of the cable and multiple connectors on the other end.

In general, a complete cycle would take approximately 15 to 20 seconds according to how many connectors are being loaded. However, three double-ended cables, six-way at each end, able to be produced on the machine in one cycle, would be using the machine to its maximum capacity and the overall time would be expected to be longer.

A comparison can be made between an automatic cut, strip and terminating machine and the cable maker mentioned earlier. This comparison is on 100 000, six-way connectors:

Standard method

Cut, strip and terminate	3400/hour	176 hours
Manually insert contacts	900/hour	666 hours
	Total	842 hours

New method (mass termination)

Cable maker: assume conservative figure of two single-ended cables every 20 seconds		
	360 cables/hour	Total 278 hours

It can clearly be seen that labour savings of 67% are not unrealistic which must be the major benefit from using mass termination techniques. Other such benefits include no strip control, no crimp control, reduced wiring errors, no contact damage, reduced tooling wear.

7.6.2 Ribbon cable connectors

Connectors for 0.050″ pitch ribbon cable can also be considered as mass termination types. The basic four types of ribbon cable are: extruded, bonded, laminated, and woven, with extruded offering the best pitch tolerance and 'tearability'. The connectors, normally loaded with gold-plated contacts, are available in a standard number of ways up to 64, these being 10, 14, 16, 20, 26, 34, 40, 44, 50, 60 and 64.

There are various types of connectors used:

(1) Receptacle connectors, 0.100 grid for plugging to a header.
(2) Card edge connectors, 0.100 pitch to connect to the edge of a PCB.
(3) Pin connectors, to mate with receptacle connectors and offer ribbon to ribbon facility.
(4) Transition connectors, for soldering direct to PCB.
(5) DIL plug, 0.100 × 0.300 grid for either soldering to PCB or connecting to a DIP header.

The normal rating for these type of connectors is 1 ampere with an operating temperature range of −55°C to 105°C and a dielectric withstanding voltage of 500 V r.m.s.

7.7 Fibre optics connectors

When joining fibres light losses will occur in four ways:

(1) Surface finish. The ends of the fibres must be square and smooth and this is usually accomplished by polishing the cut ends.
(2) End separation. Ideally the fibre ends should touch but this could cause damage and so they are normally held between 0.001″ and 0.005″ apart.
(3) Axial misalignment. This causes the highest loss and must be controlled to within 50% of the smaller fibre diameter.
(4) Angular misalignment. The ends of the fibres should be parallel to within 2%.

Any connector system must therefore hold the fibre ends to within these limits, and several different variations have been developed:

(a) *Tube method.* This method uses a metal jack and plug which are usually held together by a threaded coupling. The fit of the plug into the jack provides the primary alignment and guides the fibre in the jack into a tapered alignment hole in the plug. The depth of engagement must be accurately controlled to ensure correct end separation. These connectors are normally made from turned metal parts and have to be produced to close tolerances (*Figure 7.5*).

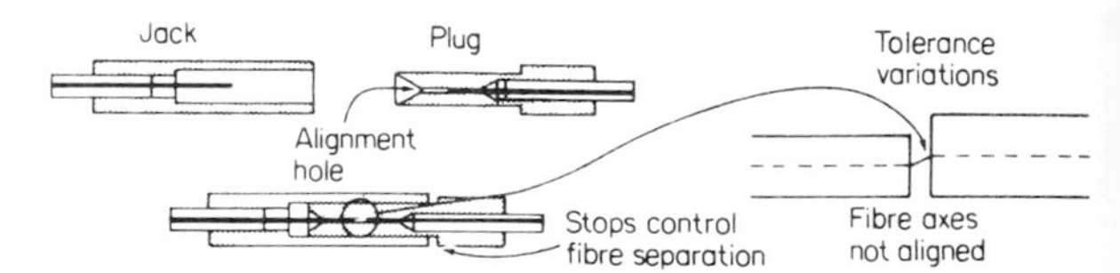

Figure 7.5 Fibre optic tube alignment connector

(b) *Straight sleeve method.* A precision sleeve is used to mate two plugs which are often designed similar to the SMA coaxial connectors and the sleeve aligns the fibres. These connectors are made from very tightly toleranced metal turned parts and due to the design, concentricity needs to be very good (*Figure 7.6*).

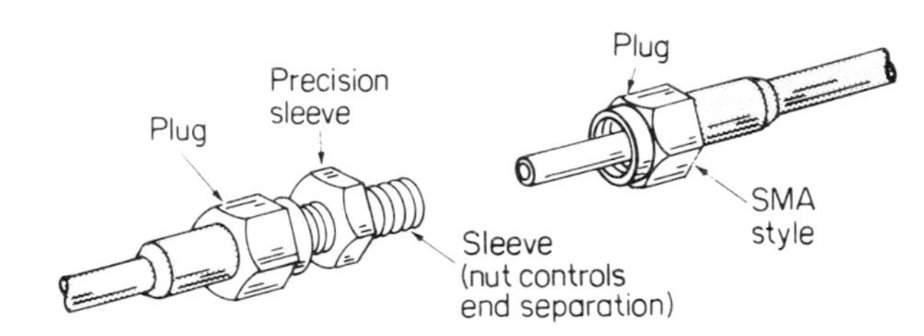

Figure 7.6 Fibre optic straight sleeve connector

(c) *Double eccentric method.* Here the fibres are mounted within two eccentrics which are then mated. The eccentrics are then rotated to bring the fibre axes into very close alignment and locked. This produces a very good coupling with much looser manufacturing tolerances but the adjustment can be cumbersome and must usually be done with some test equipment to measure maximum adjustment (*Figure 7.7*).

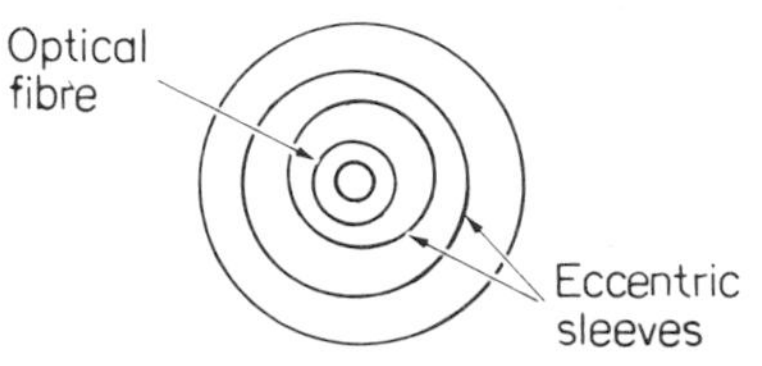

Figure 7.7 Fibre optic double eccentric connector

(d) *Three-rod method.* Three rods can be placed together such that their centre space is the size of the fibre to be joined. The rods, all of equal diameter, compress and centre the fibres radially and usually have some compliancy to absorb fibre variations. With this design it is important that the two mating parts overlap to allow both members to compress each fibre. The individual parts in this design can be moulded plastic but need to be well toleranced (*Figure 7.8*).

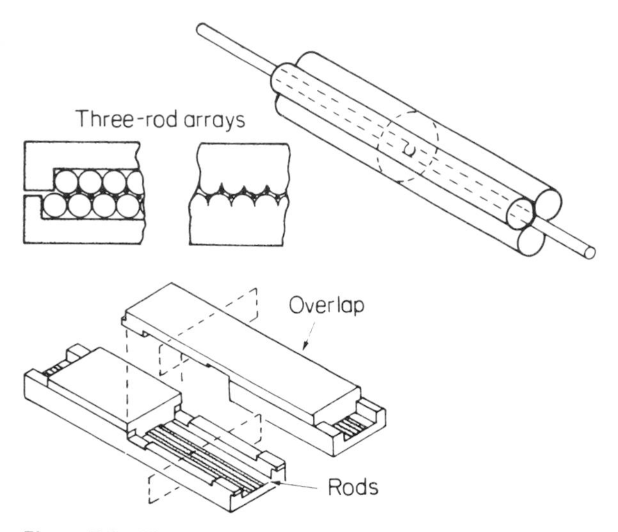

Figure 7.8 Fibre optic three-rod connector

(e) *Four-pin method.* Four pins can be used to centre a fibre and the pins are held in a ferrule. This method is sometimes used with the straight sleeve design when the pins are used to centre the fibre and the sleeve is used to align the mating halves. These parts are normally turned metal and held to tight tolerances (*Figure 7.9*).

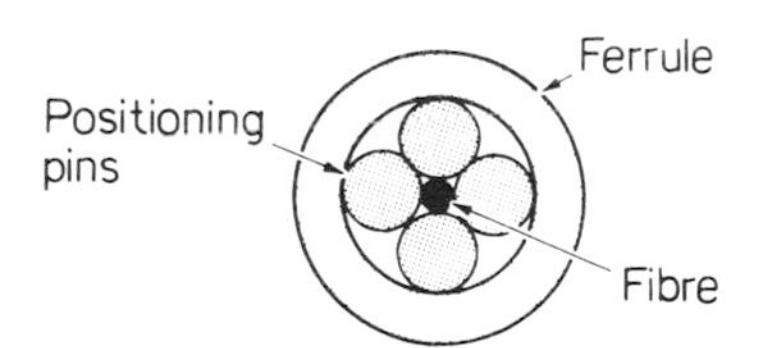

Figure 7.9 Fibre optic four-pin method of connection

(f) *Resilient ferrule.* This method utilises a ferrule and a splice bushing. The front of the ferrule is tapered to match a similar taper in the bush and the two parts are compressed together with a screw-on cap which forces the two tapers together. This moves the fibre in the ferrule on centre and provides a sealed interface between the two parts preventing foreign matter from entering the optical interface. The compression feature accommodates differences in fibre sizes and enables manufacturing tolerances to be considerably relaxed. The parts are plastic mouldings and typically produce a connector loss of less than 2 dB per through way at very low cost (*Figure 7.10*).

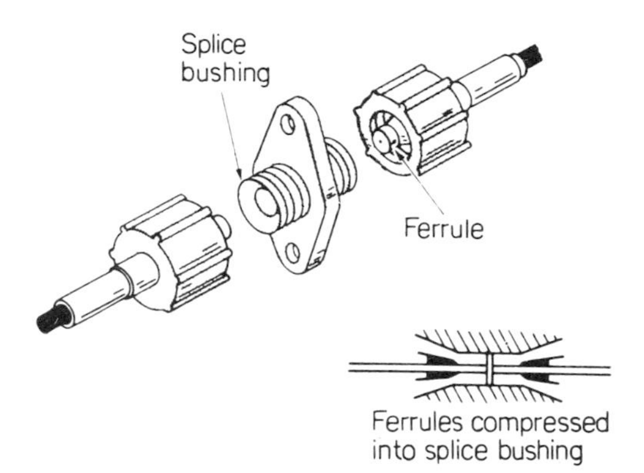

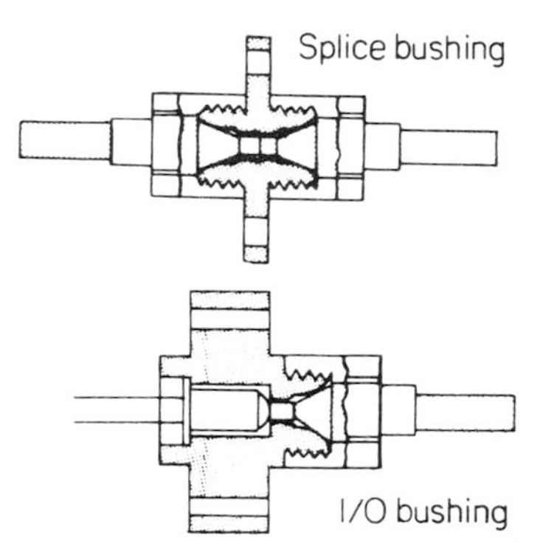

Figure 7.10 Fibre optic resilient ferrule alignment mechanism

7.8 Radio frequency connectors

Radio frequency connectors are used for terminating radio frequency transmission lines which have to be run in coaxial cables. These cables are available in sizes ranging from less than 3 mm diameter for low power applications of around 50 watts, to over 76.2 mm diameter for powers of 100 000 watts. In addition to power handling capabilities cables are also available for high frequency applications, high and low temperature applications, severe environmental applications and many other specialised uses which all require mating connectors.

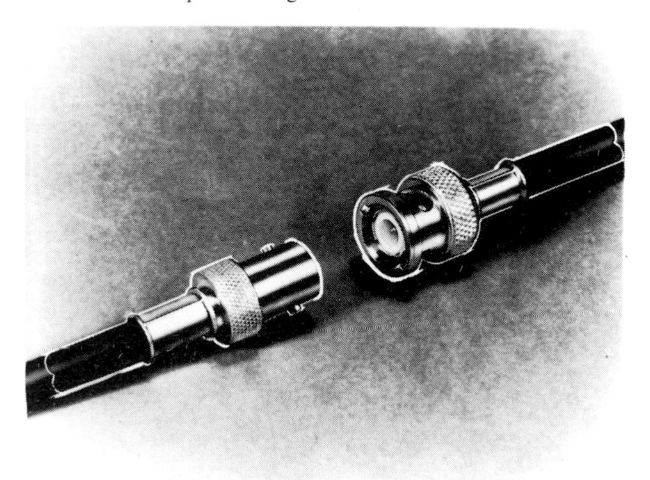

Figure 7.11 A typical r.f. connector (courtesy AMP of Great Britain Ltd)

Some of the more popular ranges are grouped in approximate cable diameter size with their operating frequency ranges as follows:

3 mm diameter	SMA 0–12.4 GHz, SMB 0–1 GHz, SMC 0–1 GHz
5 mm diameter	BNC 0–4 GHz, TNC 0–11 GHz, min. u.h.f. 0–2 GHz
7 mm diameter	N 0–11 GHz, C 0–11 GHz, u.h.f.

The design and construction of the range of connectors is very similar throughout as they all have to terminate a centre conductor and a woven copper braid screen. The variations are in size and materials. For example, to meet the MIL-C-39012 specification it is necessary to use high quality materials such as brass, silver plated for the shells, and teflon for the dielectric with gold-plated copper centre contacts. This is mainly due to the requirement for a temperature range of −65°C to +165°C. There are however, three distinct types of termination: (a) soldering and clamping; (b) crimping; and (c) soldering and crimping.

(a) With the soldering and clamping type of design the centre contact is soldered to the centre conductor and the flexible braid is then clamped to the shell of the connector by a series of tapered washers and nuts. The biggest advantage of this type of connector is that it is field repairable and replaceable without the use of special tools. The disadvantages are the possibility of a cold solder joint through underheating or melting the dielectric by overheating. Any solder which gets onto the outside of the centre contact must be removed otherwise the connector will not mate properly. It is easy to assemble the connector wrongly due to the large number of parts involved.

These connectors are used in large numbers by the military.

(b) In the crimping design the centre contact is crimped to the centre conductor and the flexible braid is then crimped between the connector shell and a ferrule. There are versions which require two separate crimps, normally to meet the MIL specification, and versions which can have both crimps made together. The advantages of this method of termination are speed and reliability together with improved electrical performance. Testing has shown that the SWR of a crimped connector is lower than the soldered and clamped version. The crimp is always repeatable and does not rely upon operator skill. The disadvantages of this design are that a special crimp tool is required and the connectors are not field repairable.

(c) With the soldering and crimping type of design the centre contact is soldered to the centre connector and the flexible braid is crimped. Obviously all the advantages and disadvantages of the previous two methods are involved with this design.

Further reading

EVANS, C. J., 'Connector finishes: tin in place of gold', *IEE Trans. Comps, Hybrid & Manf. Technol.*, **CHMT-3**, No. 2 (June 1980)

KINDELL, C. 'Ribon cable review', *Electronic Production* (Nov./Dec. 1980)

PEEL, M., 'Material for contact integrity', *New Electronics* (Jan. 1980)

McDERMOTT, J., 'Hardware and interconnect devices', *EDN* (July 1980)

TANAKA, T., 'Connectors for low-level electronic circuitry', *Electronic Engineering* (Feb. 1981)

McDERMOTT, J., 'Flat cable and its connector systems', *EDN* (Jan. 1981)

CLARK, R., 'The critical role of connectors in modern system design', *Electronics & Power* (Sept. 1981)

MILNER, J. 'LDCs for IDCs', *New Electronics* (Jan. 1982)

ROELOFS, J. A. M. and SVED, A., 'Insulation displacement connections', *Electronic Components and Applications*, **4**, No. 2 (Feb. 1982)

McDERMOTT, J., 'Flat cable and connectors', *EDN* (Aug. 1982)

SAVAGE, J. and WALTON, A., 'The UK connector scene—a review', *Electronic Production* (Sept. 1982)

8 Noise and communication

K. R. Sturley

Contents

8.1 Interference and noise in communication systems

Information transmission accuracy can be seriously impaired by interference from other transmission systems and by noise. Interference from other transmission channels can usually be reduced to negligible proportions by proper channel allocation, by operating transmitters in adjacent or overlapping channels geographically far apart, and by the use of directive transmitting and receiving aerials. Noise may be impulsive or random. Impulsive noise may be man-made from electrical machinery or natural from electrical storms; the former is controllable and can be reduced to a low level by special precautions taken at the noise source, but the latter has to be accepted when it occurs. Random (or white) noise arises from the random movement of electrons due to temperature and other effects in current-carrying components in, or associated with, the receiving system.

8.2 Man-made noise

Man-made electrical noise is caused by switching surges, electrical motor and thermostat operation, insulator flash-overs on power lines, etc. It is generally transmitted by the mains power lines and its effect can be reduced by:

(i) Suitable r.f. filtering at the noise source;
(ii) Siting the receiver aerial well away from mains lines and in a position giving maximum signal pick-up;
(iii) Connecting the aerial to the receiver by a shielded lead.

The noise causes a crackle in phones or loudspeaker, or white or black spots on a monochrome television picture screen, and its spectral components decrease with frequency so that its effect is greatest at the lowest received frequencies.

Car ignition is another source of impulsive noise but it gives maximum interference in the v.h.f. and u.h.f. bands; a high degree of suppression is achieved by resistances in distributor and spark plug leads.

8.3 Natural sources of noise

Impulsive noise can also be caused by lightning discharges, and like man-made noise its effect decreases with increase of received frequency. Over the v.h.f. band such noise is only evident when the storm is within a mile or two of the receiving aerial.

Cosmic noise from outer space is quite different in character and generally occurs over relatively narrow bands of the frequency spectrum from about 20 MHz upwards. It is a valuable asset to the radio astronomer and does not at present pose a serious problem for the communications engineer.

8.4 Random noise

This type of noise is caused by random movement of electrons in passive elements such as resistors, conductors and inductors, and in active elements such as electronic valves and transistors.

8.4.1 Thermal noise

Random noise in passive elements is referred to as thermal noise since it is entirely associated with temperature, being directly proportional to absolute temperature. Unlike impulsive noise its energy is distributed evenly through the r.f. spectrum and it must be taken into account when planning any communication system. Thermal noise ultimately limits the maximum amplification that can usefully be employed, and so determines the minimum acceptable value of received signal. It produces a steady hiss in a loudspeaker and a shimmering background to a television picture.

Nyquist has shown that thermal noise in a conductor is equivalent to a r.m.s. voltage Vn in series with the conductor resistance R, where

$$Vn=(4kTR\,\Delta f)^{1/2} \tag{8.1}$$

k = Boltzmann's constant, $1.372\times10^{-23}J/K$
T = absolute temperature of conductor
Δf = pass band (Hz) of the circuits after R

If the frequency response were rectangular the pass band would be the difference between the frequencies defining the sides of the rectangle. In practice the sides are sloping and bandwidth is

$$\Delta f=\frac{1}{E_o^2}\int_0^\infty [E(f)]^2\,df \tag{8.2}$$

where E_o = midband or maximum value of the voltage ordinate and $E(f)$ = the voltage expression for the frequency response.

A sufficient degree of accuracy is normally achieved by taking the standard definition of bandwidth, i.e. the frequency difference between points where the response has fallen by 3 dB.

Figure 8.1 allows the r.m.s. noise voltage for a given resistance and bandwidth to be determined. Thus for

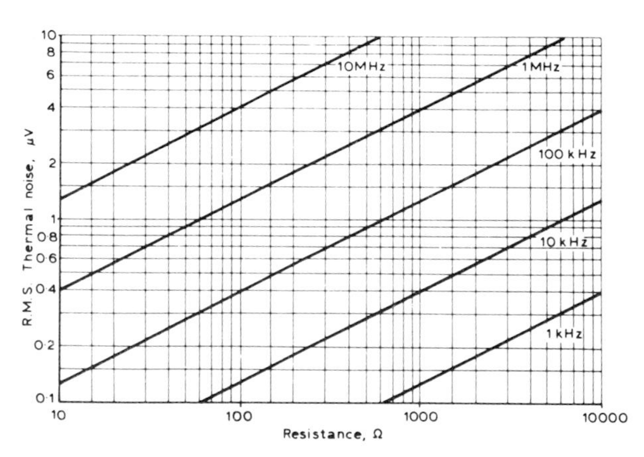

Figure 8.1 R.M.S. thermal noise (kV) plotted against resistance at different bandwidths. T=290, K=17°C

$R=10$ kΩ, $T=T_0=17$°C or 290 K

and

$\Delta f=10$ kHz, $V\simeq1.26\ \mu$V

When two resistances in series are at different temperatures

$$V_n=[4k\,\Delta f(R_1T_1+R_2T_2)]^{1/2} \tag{8.3}$$

Two resistances in parallel at the same temperature *Figure 8.2(a)* are equivalent to a noise voltage

$$V_n=[4kT\,\Delta f\,.\,R_1R_2/(R_1+R_2)]^{1/2} \tag{8.4}$$

in series with two resistances in parallel, *Figure 8.2(b)*.

The equivalent current generator concept is shown in *Figure 8.2(c)* where

$$I_n=[4kT\,\Delta f(G_1+G_2)]^{1/2} \tag{8.5}$$

If R is the series resistance of a coil in a tuned circuit of Q factor, Q_o, the noise voltage from the tuned circuit becomes

$$V_{no}=V_nQ_o=Q_o(4kTR\,\Delta f)^{1/2} \tag{8.6}$$

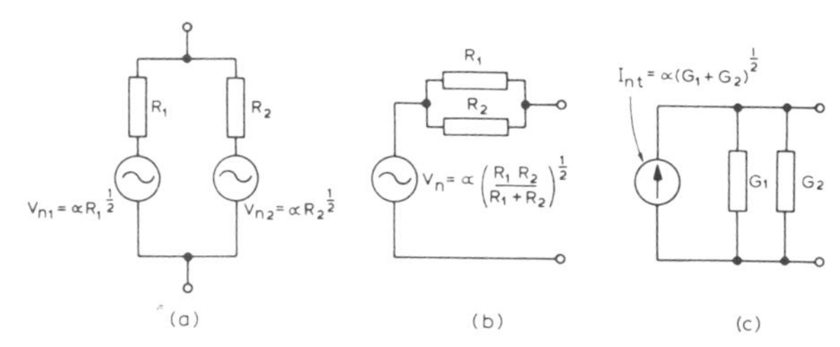

Figure 8.2 (a) Noise voltages of two resistances in parallel; (b) an equivalent circuit and (c) a current noise generator equivalent $\alpha=(4kT'f)^{1/2}$

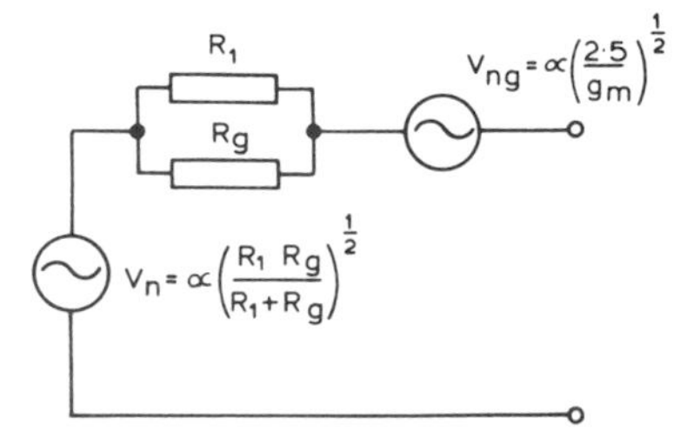

Figure 8.3 Noise voltage input circuit for a valve $\alpha=(4kT'f)^{1/2}$

The signal injected into the circuit is also multiplied by Q_o so that signal-to-noise ratio is unaffected.

8.5 Electronic valve noise

8.5.1 Shot noise

Noise in valves, termed shot noise, is caused by random variations in the flow of electrons from cathode to anode. It may be regarded as the same phenomenon as thermal (conductor) noise with the valve slope resistance, acting in place of the conductor resistance at a temperature between 0.5 to 0.7 of the cathode temperature.

Shot noise r.m.s. current from a diode is given by

$$I_n=(4k\alpha T_k g_d \Delta f)^{1/2} \tag{8.7}$$

where

T_k = absolute temperature of the cathode
α = temperature correction factor assumed to be about 0.66

$g_d=\dfrac{dI_d}{dV_a}$, slope conductance of the diode.

Experiment[1] has shown noise in a triode is obtained by replacing g_d in equation (8.7) by g_m/β where β has a value between 0.5 and 1 with a typical value of 0.85, thus

$$I_{na}=(4k\alpha T_k g_m \Delta f/\beta)^{1/2} \tag{8.8}$$

Since $I_a=g_m V_g$, the noise current can be converted to a noise voltage at the grid of the valve of

$$V_{ng}=I_{na}/g_m=(4k\alpha T_k \Delta f/\beta g_m)^{1/2}$$
$$=[4kT_o \Delta f \, . \, \alpha T_k/\beta g_m T_o]^{1/2} \tag{8.9}$$

where T_o is the normal ambient (room) temperature.

The part $\alpha T_k/\beta g_m T_o$ of expression (51.9) above is equivalent to a resistance, which approximates to

$$R_{ng}=2.5/g_m \tag{8.10}$$

and this is the equivalent noise resistance in the grid of the triode at room temperature. The factor 2.5 in R_{ng} may have a range from 2 to 3 in particular cases. The equivalent noise circuit for a triode having a grid leak R_g and fed from a generator of internal resistance R_1 is as in *Figure 8.3*.

8.5.2 Partition noise

A multielectrode valve such as a tetrode produces greater noise than a triode due to the division of electron current between screen and anode; for this reason the additional noise is known as partition noise. The equivalent noise resistance in the grid circuit becomes

$$R_{ng}(tet)=(I_a/I_k)(20I_s/g_m^2+2.5g_m) \tag{8.11}$$

Where I_a, I_k and I_s are the d.c. anode, cathode and screen currents respectively. I_s should be small and g_m large for low noise in tetrode or multielectrode valves. The factor $20I_s/g_m^2$ is normally between 3 to 6 times $2.5/g_m$ so that a tetrode valve is much noisier than a triode.

At frequencies greater than about 30 MHz, the transit time of the electron from cathode to anode becomes significant and this reduces gain and increases noise. Signal-to-noise ratio therefore deteriorates. Partition noise in multielectrode valves also increases and the neutralised triode, or triodes in cascode give much better signal-to-noise ratios at high frequencies.

At much higher frequencies (above 1 GHz) the velocity modulated electron tube, such as the klystron and travelling wave tube, replace the normal electron valve. In the klystron, shot noise is present but there is also chromatic noise due to random variations in the velocities of the individual electrons.

8.5.3 Flicker noise

At very low frequencies valve noise is greater than would be expected from thermal considerations. Schottky suggested that this is due to random variations in the state of the cathode surface and termed it *flicker*. Flicker noise tends to be inversely proportional to frequency below about 1 kHz so that the equivalent noise resistance at 10 Hz might be 100 times greater than the shot noise at 1 kHz. Ageing of the valve tends to increase flicker noise and this appears to be due to formation of a high resistance barium silicate layer between nickel cathode and oxide coating.

8.6 Transistor noise

Transistor noise exhibits characteristics very similar to those of valves, with noise increasing at both ends of the frequency scale. Resistance noise is also present due to the extrinsic resistance of the material and the major contributor is the base extrinsic resistance r_b'. Its value is given by expression (8.1), T being the absolute temperature of the transistor under working conditions.

Shot and partition noise arise from random fluctuations in the movement of minority and majority carriers, and there are four sources, viz.

(i) Majority carriers injected from emitter to base and thence to collector.
(ii) Majority carriers from emitter which recombine in the base.
(iii) Minority carriers injected from base into emitter.
(iv) Minority carriers injected from base into the collector.

Sources (i) and (ii) are the most important, sources (iii) and (iv) being significant only at low bias currents. Under the latter condition which gives least noise, silicon transistors are superior to germanium because of their much lower values of I_{co}.

A simplified equivalent circuit for the noise currents and voltages in a transistor is that of *Figure 8.4* where

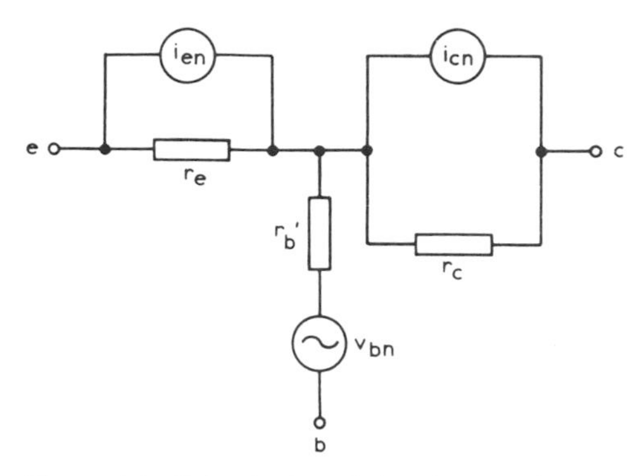

Figure 8.4 Noise circuit equivalent for a transistor

$i_{en}=(2eI_e\,\Delta f)^{1/2}$ the shot noise current in the emitter
$i_{cn}=[2e(I_{co}+I_c(1-\alpha_o))\Delta f]^{1/2}$, the shot and partition noise current in the collector

$v_{bn}=(4kT\,.\,r'_b\,\Delta f)^{1/2}$, the thermal noise due to the base extrinsic resistance

e = electronic charge = 1.602×10^{-19} coulomb

Since transistors are power amplifying devices the equivalent noise resistance concept is less useful and noise quality is defined in terms of noise figure.

Flicker noise, which is important at low frequencies (less than about 1 kHz), is believed to be due to carrier generation and recombination at the base–emitter surface. Above 1 kHz noise remains constant until a frequency of about $f_\alpha(1-\alpha_o)^{1/2}$ is reached, where f_α is the frequency at which the collector–emitter current gain has fallen to $0.7\alpha_o$. Above this frequency, which is about $0.15f_\alpha$, partition noise increases rapidly.

8.7 Noise figure

Noise figure (F) is defined as the ratio of the input signal-to-noise available power ratio to the output signal-to-noise available power ratio, where available power is the maximum power which can be developed from a power source of voltage V and internal resistance R_s. This occurs for matched conditions and is $V^2/4R_s$.

$$F=(P_{si}/P_{ni})/(P_{so}/P_{no})$$
$$=P_{no}/G_aP_{ni} \qquad (8.12)$$

where G_a = available power gain of the amplifier.

Since noise available output power is the sum of G_aP_{ni} and that contributed by the amplifier P_{na}

$$F=1+\frac{P_{na}}{G_aP_{ni}} \qquad (8.13)$$

The available thermal input power is $V^2/4R_s$ or $kT\,\Delta f$, which is independent of R_s, hence

$$F=1+P_{na}/G_akT\,\Delta f \qquad (8.14)$$

and

$$F(\text{dB})=10\log_{10}(1+P_{na}/G_akT\,\Delta f) \qquad (8.15)$$

The noise figure for an amplifier whose only source of noise is its input resistance R_1 is

$$F=1+R_s/R_1 \qquad (8.16)$$

because the available output noise is reduced by $R_1/(R_s+R_1)$ but the available signal gain is reduced by $[R_1/(R_s+R_1)]^2$. For matched conditions $F \mp 2$ or 3 dB and maximum signal-to-noise ratio occurs when $R_1=\infty$. Signal-to-noise ratio is unchanged if R_1 is noiseless because available noise power is then reduced by the same amount as the available gain.

If the above amplifier has a valve, whose equivalent input noise resistance is R_{ng},

$$F=1+\frac{R_s}{R_1}+\frac{R_{ng}}{R_s}\left[1+\frac{R_s}{R_1}\right]^2 \qquad (8.17)$$

Noise figure for a transistor over the range of frequencies for which it is constant is

$$F=1+\frac{r'_b+0.5r_e}{R_s}+\frac{(r'_b+r_e+R_s)^2(1-\alpha_o)}{2r_eR_s\alpha_o} \qquad (8.18)$$

At frequencies greater than $f_\alpha(1-\alpha_o)^{1/2}$, the last term is multiplied by $[1+(f/f_\alpha)^2/(1-\alpha_o)]$. The frequency f_T at which collector–base current gain is unity is generally given by the transistor manufacturer and it may be noted that $f_T\simeq f_\alpha$, the frequency at which collector–emitter current gain is $0.7\alpha_o$.

Expression (8.18) shows that transistor noise figure is dependent on R_s but it is also affected by I_c through r_e and α_o. As a general rule the lower the value of I_c the lower is noise figure and the greater is the optimum value of R_s. This is shown in *Figure 8.5* which is typical of a r.f. silicon transistor. Flicker noise causes

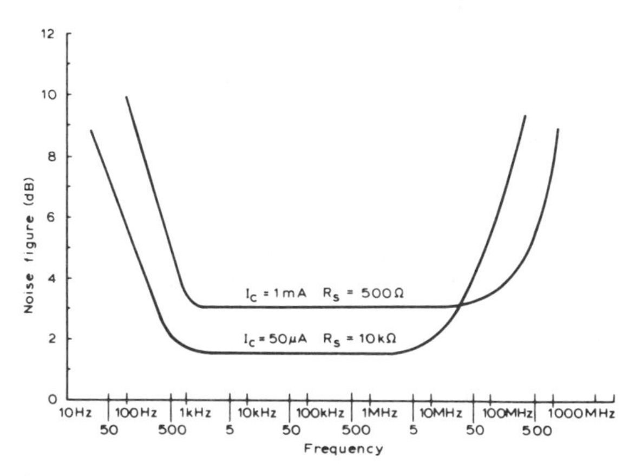

Figure 8.5 Typical noise figure—frequency curves for a r.f. transistor

the increase below 1 kHz, and decrease of gain and increase of partition noise causes increased noise factor at the high frequency end. The high frequency at which F begins to increase is about $0.15f_\alpha$; at low values of collector current f_α falls, being approximately proportional to I_c^{-1}. The type of configuration, common emitter, base or collector has little effect on noise figure.

Transistors do not provide satisfactory noise figures above about 1.5 GHz, but the travelling-wave tube and tunnel diode can achieve noise figures of 3 to 6 dB over the range 1 to 10 GHz.

Sometimes noise temperature is quoted in preference to noise figure and the relationship is

$$F=(1+T/T_o) \qquad (8.19)$$

T is the temperature to which the noise source resistance would have to be raised to produce the same available noise output power as the amplifier. Thus if $T=T_o=290$ K, $F=2$ or 3 dB.

The overall noise figure of cascaded amplifiers can easily be calculated and is

$$F_t=F_1+\frac{F_2-1}{G_1}+\frac{F_3-1}{G_1G_2}+\ldots\frac{F_n-1}{G_1\ldots G_{n-1}} \qquad (8.20)$$

where $F_1, F_2, \ldots, F_n$ and $G_1, G_2, \ldots, G_n$ are respectively the noise figures and available gains of the separate stages from input to output. From equation (8.20) it can be seen that the first stage of an amplifier system largely determines the overall signal-to-noise ratio, and that when a choice has to be made between two first-stage amplifiers having the same noise figure, the amplifier having the highest gain should be selected because increase of G_1 reduces the noise effect of subsequent stages.

8.8 Measurement of noise

Noise measurement requires a calibrated noise generator to provide a controllable noise input to an amplifier or receiver, and a r.m.s. meter to measure the noise output of the amplifier or receiver. The noise generator generally consists of a temperature-limited (tungsten filament) diode, terminated by a resistance R as shown in *Figure 8.6*. The diode has sufficient anode voltage to

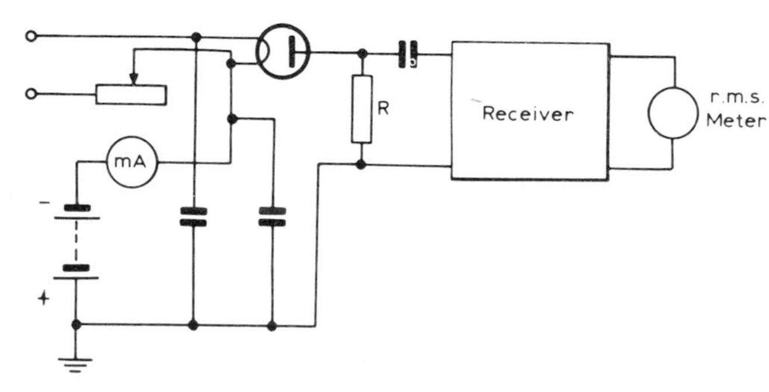

Figure 8.6 Noise figure measurements

ensure that it operates under saturation conditions and anode saturation current is varied by control of the diode filament current. A milliammeter reads the anode current I_d and the shot noise current component of this is given by $(2I_d e\,\Delta f)^{1/2}$ where e is electronic charge, 1.602×10^{-19} coulomb. The shot noise has the same flat spectrum as the thermal noise in R, and the meter is calibrated in dB with reference to noise power in R and so provides a direct reading of noise factor. R is generally selected to be 75 Ω, the normal input impedance of a receiver.

When measuring, the diode filament current is first switched off, and the reading of the r.m.s. meter in the receiver output noted. The diode filament is switched on and adjusted to increase the r.m.s. output reading 1.414 times (double noise power). The dB reading on the diode anode current meter is the noise figure, since

Noise output power diode off $= GP_{nR} + P_{na}$
Noise output power diode on $= G(P_{nR} + P_{nd}) + P_{na} = 2GP_{nd}$
Noise figure $= 10\log_{10}[(GP_{nR} + P_{na})/GP_{nR}] = 10\log_{10} P_{nd}/P_{nR}$

The diode is satisfactory up to about 600 MHz but above this value transit time of electrons begins to cause error. For measurements above 1 GHz a gas discharge tube has to be used as a noise source.

8.9 Methods of improving signal-to-noise ratio

There are five methods of improving signal-to-noise ratio, viz.,

(i) Increase the transmitted power of the signal.
(ii) Redistribute the transmitted power.
(iii) Modify the information content before transmission and return it to normal at the receiving point.
(iv) Reduce the effectiveness of the noise interference with signal.
(v) Reduce the noise power.

8.9.1 Increase of transmitted power

An overall increase in transmitted power is costly and could lead to greater interference for users of adjacent channels.

8.9.2 Redistribution of transmitted power

With amplitude modulation it is possible to redistribute the power among the transmitted components so as to increase the effective signal power. Suppression of the carrier in a double sideband amplitude modulation signal and a commensurate increase in sideband power increases the effective signal power, and therefore signal-to-noise ratio by 4.75 dB (3 times) for the same average power or by 12 dB for the same peak envelope power. Single sideband operation by removal of one sideband reduces signal-to-noise ratio by 3 dB because signal power is reduced to $\frac{1}{4}$ (6 dB) and the non-correlated random noise power is only halved (3 dB). If all the power is transferred to one sideband, single sideband operation increases signal-to-noise ratio by 3 dB.

8.9.3 Modification of information content before transmission and restoration at receiver

8.9.3.1 The compander

A serious problem with speech transmission is that signal-to-noise ratio varies with the amplitude of the speech, and during gaps between syllables and variations in level when speaking, the noise may become obtrusive. This can be overcome by using compression of the level variations before transmission, and expansion after detection at the receiver, a process known as companding. The compressor contains a variable loss circuit which reduces amplification as speech amplitude increases and the expander performs the reverse operation.

A typical block schematic for a compander circuit is shown in *Figure 8.7*. The input speech signal is passed to an amplifier

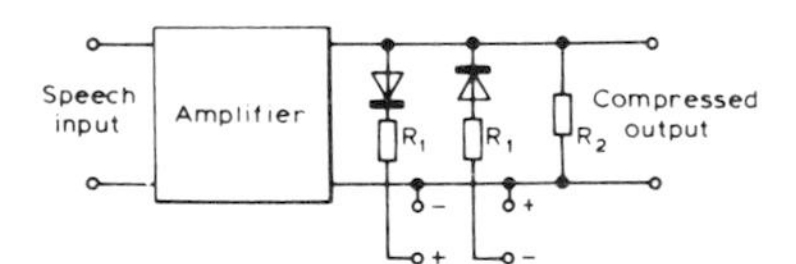

Figure 8.7 A compressor circuit

across whose output are shunted two reverse biased diodes; one becoming conductive and reducing the amplification for positive going signals and the other doing the same for negative-going signals. The input–output characteristic is S shaped as shown in *Figure 8.8*; the diodes should be selected for near identical

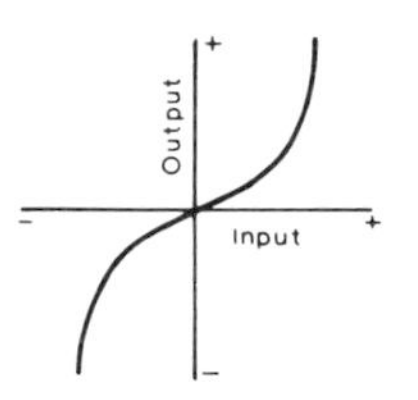

Figure 8.8 Compressor input–output characteristic

shunting characteristics. Series resistances R_1 are included to control the turn-over, and shunt resistance R_2 determines the maximum slope near zero.

A similar circuit is used in the expander after detection but as shown, *Figure 8.9*, the diodes form a series arm of a potential divider and the expanded output appears across R_3. The expander characteristic, *Figure 8.10*, has low amplification in the gaps between speech, and amplification increases with increase in speech amplitude.

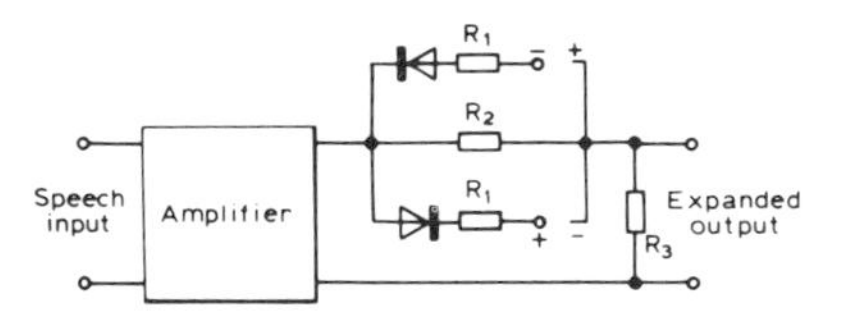

Figure 8.9 An expander circuit

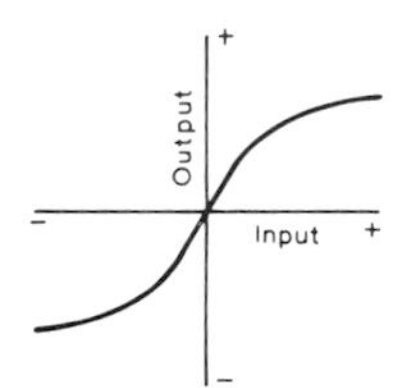

Figure 8.10 Expander input–output characteristics

The diodes have a logarithmic compression characteristic, and with large compression the dB input against dB output tends to a line of low slope, e.g., an input variation of 20 dB being compressed to an output variation of 5 dB. If greater compression is required two compressors are used in tandem.

The collector–emitter resistance of a transistor may be used in place of the diode resistance as the variable gain device. The collector–emitter resistance is varied by base-emitter bias current, which is derived by rectification of the speech signal from a separate auxiliary amplifier. A time delay is inserted in the main controlled channel so that high-amplitude speech transients can be anticipated.

8.9.3.2 Lincompex

The compander system described above proves quite satisfactory provided the propagation loss is constant as it is with a line or coaxial cable. It is quite unsuitable for a shortwave point-to-point communication system via the ionosphere. A method known as Lincompex[2] (linked compression expansion) has been successfully developed by the British Post Office. *Figure 8.11* is a block diagram of the transmit–receive paths. The simple form of diode compressor and expander cannot be used and must be replaced by the transistor type, controlled by a current derived from rectification of the speech signal. The current controls the compression directly at the transmitting end and this information must be sent to the receiver by a channel unaffected by any propagational variations. This is done by confining it in a narrow channel (approximately 180 Hz wide) and using it to frequency-modulate a sub-carrier at 2.9 kHz. A limiter at the receiver removes all amplitude variations introduced by the r.f. propagation path, and a frequency discriminator extracts the original control information.

The transmit chain has two paths for the speech signals, one (A) carries the compressed speech signal, which is limited to the range 250 to 2700 Hz by the low-pass output filter. A time delay of 4 ms is included before the two compressors in tandem, each of which has a 2 to 1 compression ratio, and the delay allows the compressors to anticipate high amplitude transients. The 2:1 compression ratio introduces a loss of $x/2$ dB for every x dB change in input, and the two in tandem introduce a loss of $2(x/2)=x$ dB for every x dB change of input. The result is an almost constant speech output level for a 60 dB variation of speech input. Another time delay (10 ms) is inserted between the compressors and output filter in order to compensate for the control signal delay due to its narrow bandwidth path.

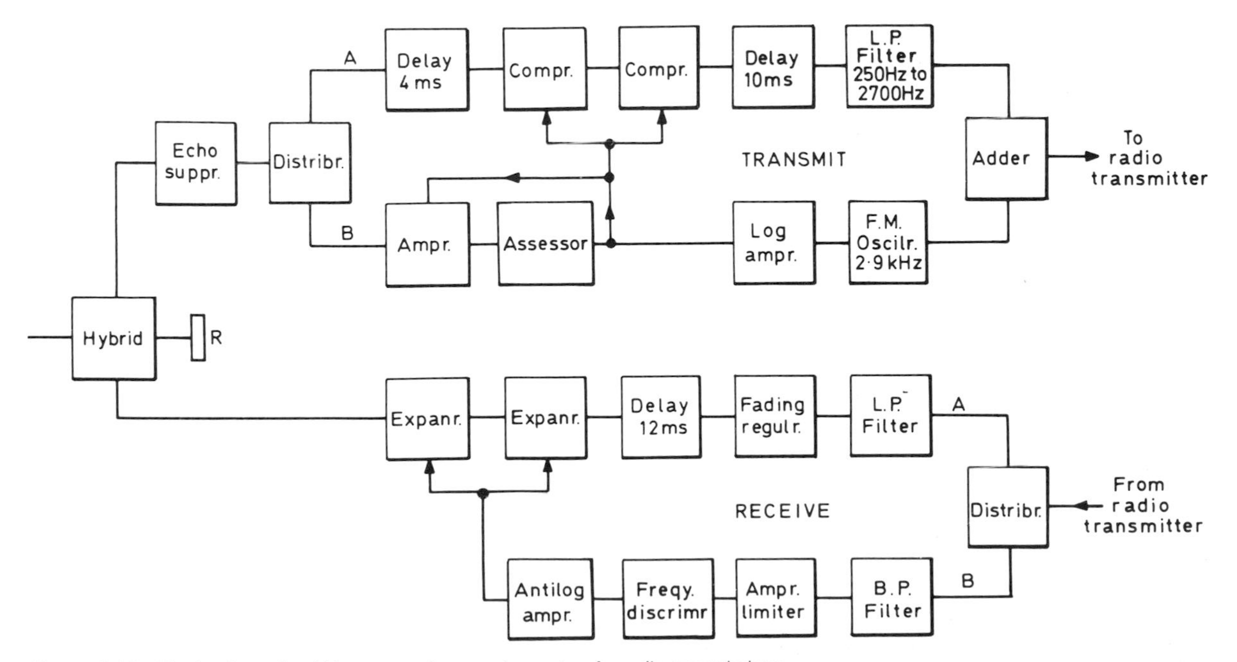

Figure 8.11 Block schematic of Lincompex Compander system for radio transmissions

The other transmit path (B) contains an amplitude-assessor circuit having a rectified d.c. output current proportional to the speech level. This d.c. current controls the compressors, and after passing through a logarithmic amplifier is used to frequency modulate the sub-carrier to produce the control signal having a frequency deviation of 2 Hz/dB speech level change. The time constant of the d.c. control voltage is 19 ms permitting compressor loss to be varied at almost syllabic rate, and the bandwidth of the frequency-modulated sub-carrier to be kept within ±90 Hz. The control signal is added to the compressed speech and the combined signal modulates the transmitter.

The receive chain also has two paths; path (A) filters the compressed speech from the control signal and passes the speech to the expanders via a fading regulator, which removes any speech fading not eliminated by the receiver a.g.c., and a time delay, which compensates for the increased delay due to the narrow-band control path (B). The latter has a band pass filter to remove the compressed speech from the control signal and an amplitude limiter to remove propagational amplitude variations. The control signal passes to a frequency discriminator and thence to an antilog amplifier, the output from which controls the gain of the expansion circuits. The time constant of the expansion control is between 18 ms and 20 ms.

8.9.3.3 Pre-emphasis and de-emphasis

Audio energy in speech and music broadcasting tends to be greatest at the low frequencies. A more level distribution of energy is achieved if the higher audio frequencies are given greater amplification than the lower before transmission. The receiver circuits must be given a reverse amplification-frequency response to restore the original energy distribution, and this can lead to an improved signal-to-noise ratio since the received noise content is reduced at the same time as the high audio frequencies are reduced. The degree of improvement is not amenable to measurement and a subjective assessment has to be made. The increased high-frequency amplification before transmission is known as pre-emphasis followed by de-emphasis in the receiver audio circuits. F.M. broadcasting (maximum frequency deviation ±75 kHz) shows a greater subjective improvement than a.m., and it is estimated to be 4.5 dB when the pre- and de-emphasis circuits have time-constants of 75 μs. A simple *RC* potential divider can be used for de-emphasis in the receiver audio circuits, and 75 μs time constant gives losses of 3 and 14 dB at 2.1 and 10 kHz respectively compared with 0 dB at low frequencies.

8.9.4 Reduction of noise effectiveness

Noise, like information, has amplitude and time characteristics, and it is noise amplitude that causes the interference with a.m. signals. If the information is made to control the time characteristics of the carrier so that carrier amplitude is transmitted at a constant value, an amplitude limiter in the receiver can remove all amplitude variations due to noise without impairing the information. The noise has some effect on the receiver carrier time variations, which are phase-modulated by noise, but the phase change is very much less than the amplitude change so that signal-to-noise ratio is increased.

8.9.4.1 Frequency modulation

If the information amplitude is used to modulate the carrier frequency, and an amplitude limiter is employed at the receiver, the detected message-to-noise ratio is greatly improved. F.M. produces many pairs of sidebands per modulating frequency especially at low frequencies, and this 'bass boost' is corrected at the receiver detector to cause a 'bass cut' of the low frequency noise components. This triangulation of noise leads to 4.75 dB signal-to-noise betterment. Phase modulation does not give this improvement because the pairs of sidebands are independent of modulating frequency. The standard deviation of 75 kHz raises signal-to-noise ratio by another 14 dB, and pre-emphasis and de-emphasis by 4.5 dB, bringing the total improvement to 23.25 dB over a.m.

The increased signal-to-noise performance of the f.m. receiver is dependent on having sufficient input signal to operate the amplitude limiter satisfactorily. Below a given input signal-to-noise ratio output information-to-noise ratio is worse than for a.m. The threshold value increases with increase of frequency deviation because the increased receiver bandwidth brings in more noise as indicated in *Figure 8.12*.

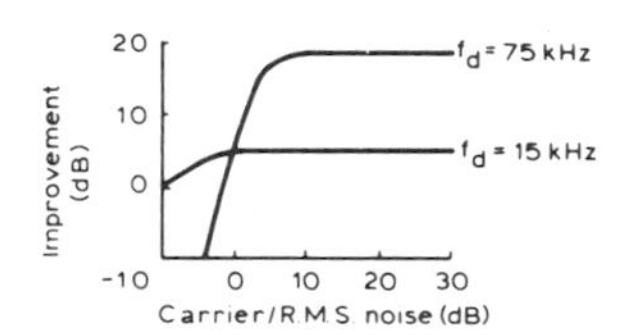

Figure 8.12 Threshold noise effect with f.m. compared with a.m.

8.9.4.2 Pulse modulation

Pulse modulated systems using change of pulse position (p.p.m.) and change of duration (p.d.m.) can also increase signal-to-noise ratio but pulse amplitude modulation (p.a.m.) is no better than normal a.m. because an amplitude limiter cannot be used.

8.9.4.3 Impulse noise and bandwidth

When an impulse noise occurs at the input of a narrow bandwidth receiver the result is a damped oscillation at the mid-frequency of the pass band as shown in *Figure 8.13(a)*. When a wide bandwidth is employed the result is a large initial amplitude with a very rapid decay, *Figure 8.13(b)*. An amplitude limiter is much more effective in suppressing the large amplitude near-single pulse than the long train of lower amplitude oscillations. Increasing reception bandwidth can therefore appreciably reduce interference due to impulsive noise provided that an amplitude limiter can be used.

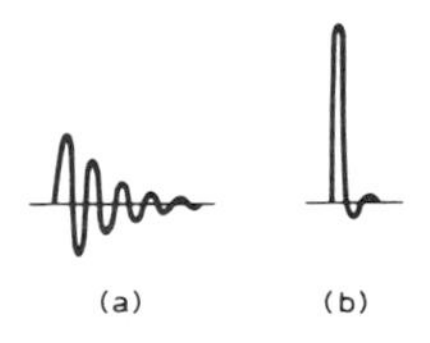

Figure 8.13 Output wave shape due to an impulse in a (a) narrow band amplifier; (b) a wideband amplifier

8.9.4.4 Pulse code modulation

A very considerable improvement in information-to-noise ratio can be achieved by employing pulse code modulation[3] (p.c.m.). P.C.M. converts the information amplitude into a digital form by sampling and employing constant amplitude pulses, whose presence or absence in a given time order represents the amplitude level as a binary number. Over long cable or microwave links it is possible to amplify the digital pulses when signal-to-noise ratio is very low, to regenerate and pass on a freshly constituted signal almost free of noise to the next link. With analogue or direct non-coded modulation such as a.m. and f.m., noise tends to be cumulative from link to link. The high signal-to-noise ratio of p.c.m. is obtained at the expense of much increased bandwidth, and Shannon has shown that with an ideal system of coding giving zero detection error there is a relationship between information capacity C (binary digits or bits/sec), bandwidth W (Hz) and average signal-to-noise thermal noise power ratio (S/N) as follows:

$$C = W \log_2 (1 + S/N) \qquad (8.21)$$

Two channels having the same C will transmit information equally well though W, and S/N may be different. Thus for a channel capacity of 10^6 bits/s, $W = 0.167$ MHz and $S/N = 63 \equiv 18$ dB, or $W = 0.334$ MHz and $S/N = 7 \equiv 8.5$ dB. Doubling of bandwidth very nearly permits the S/N dB value to be halved, and this is normally a much better exchange rate than for f.m. analogue modulation, for which doubling bandwidth improves S/N power ratio 4 times or by 6 dB.

In any practical system the probability of error is finite, and a probability of 10^{-6} (1 error in 10^6 bits) causes negligible impairment of information. Assuming that the detector registers a pulse when the incoming amplitude exceeds one half the normal pulse amplitude, an error will occur when the noise amplitude exceeds this value. The probability of an error occurring due to this is

$$Pe = \frac{1}{(2\pi)^{\frac{1}{2}}} \frac{2Vn}{V_p} \exp(-V_p^2/8Vn^2) \qquad (8.22)$$

where V_p = peak voltage of the pulse
and V_n = r.m.s. voltage of the noise.

The curve is plotted in *Figure 8.14*.

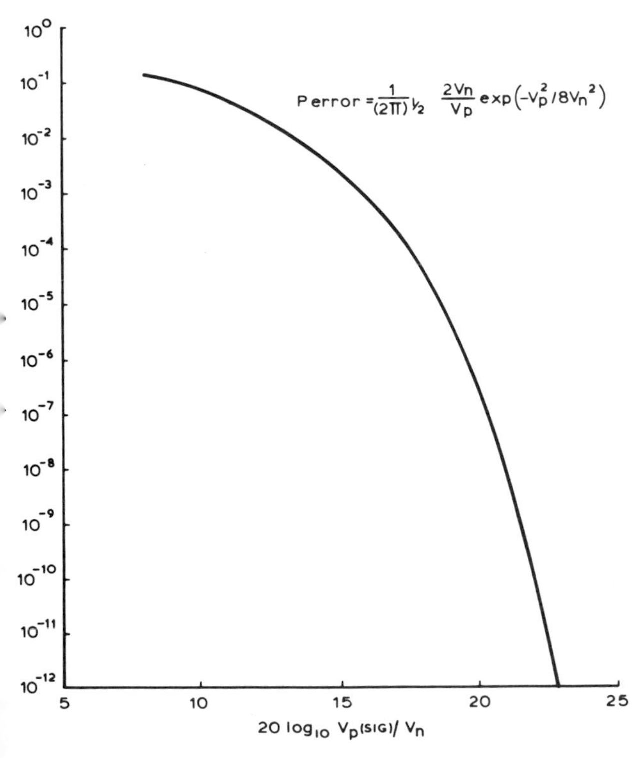

Figure 8.14 Probability of error at different V_p(sig)/r.m.s. noise ratios

An error probability of 10^{-6} requires a V_p/V_n of approximately 20 dB, or since $V_{av} = V_p/2$ a V_{av}/V_n of 17 dB. In a binary system 2 pulses can be transmitted per cycle of bandwidth so that by Shannon's ideal system

$$\frac{C}{W} = 2 = \log_2 (1 + S/N) \quad \text{and} \quad S/N = 3 \simeq 5 \text{ dB} \qquad (8.23)$$

Hence the practical system requires 12 dB greater S/N ratio than the ideal, but the output message-to-noise ratio is infinite, i.e. noise introduced in the transmission path and the receiver is completely removed. There will, however, be a form of noise present with the output message due to the necessary sampling process at the transmit end. Conversion of amplitude level to a digital number must be carried out at a constant level, and the reconstructed decoded signal at the receiver is not a smooth wave but a series of steps. These quantum level steps superimpose on the original signal a disturbance having a uniform frequency spectrum similar to thermal noise. It is this quantising noise which determines the output message-to-noise ratio and it is made small by decreasing the quantum level steps. The maximum error is half the quantum step, l, and the r.m.s. error introduced is $l/2(3)^{1/2}$. The number of levels present in the p.a.m. wave after sampling are 2^n where n is the number of binary digits. The message peak-to-peak amplitude is $2^n l$, so that the

$$\text{Message (pk-to-pk)/r.m.s. noise} = \frac{2^n l}{l/2(3)^{1/2}} = 2(3)^{1/2} 2^n \qquad (8.24)$$

$$\begin{aligned} M/N \text{ (dB)} &= 20 \log_{10} 2(3)^{1/2} 2^n \\ &= 20n \log_{10} 2 + 20 \log_{10} 2(3)^{1/2} \\ &= (6n + 10.8) \text{ dB} \end{aligned} \qquad (8.25)$$

Increase of digits (n) means an increased message-to-noise ratio but also increased bandwidth and therefore increased transmission path and receiver noise; care must be exercised to ensure that quantising noise remains the limiting factor.

Expression (8.25) represents the message-to-noise ratio for maximum information amplitude, and smaller amplitudes will give an inferior noise result. A companding system should therefore be provided before sampling of the information takes place.

8.9.5 Reduction of noise

Since thermal noise power is proportional to bandwidth, the latter should be restricted to that necessary for the objective in view. Thus the bandwidth of an a.m. receiver should not be greater than twice the maximum modulating frequency for d.s.b. signals, or half this value for s.s.b. operation.

In a f.m. system information power content is proportional to $(\text{bandwidth})^2$ so that increased bandwidth improves signal-to-noise ratio even though r.m.s. noise is increased. When, however, carrier and noise voltages approach in value, signal-to-noise ratio is worse with the wider band f.m. transmission (threshold effect).

Noise is reduced by appropriate coupling between signal source and receiver input and by adjusting the operating conditions of the first stage transistor for minimum noise figure.

Noise is also reduced by refrigerating the input stage of a receiver with liquid helium and this method is used for satellite communication in earth station receivers using masers. The maser amplifies by virtue of a negative resistance characteristic and its noise contribution is equivalent to the thermal noise generated in a resistance of equal value. The noise temperature of the maser itself may be as low as 2 K to 10 K and that of the other parts of the input equipment 15 K to 30 K.

Parametric amplification, by which gain is achieved by periodic variation of a tuning parameter (usually capacitance), can provide the relatively low noise figures 1.5 to 6 dB over the range 5 to 25 GHz. Energy at the 'pump' frequency (f_p) operating the variable reactance, usually a varactor diode, is transferred to the signal frequency (f_s) in the parametric amplifier or to an idler frequency ($f_p \pm f_s$) in the parametric converter. It is the resistance component of the varactor diode that mainly determines the noise figure of the system. Refrigeration is also of value with parametric amplification.

References

1 NORTH, D. O., 'Fluctuations in Space-Charge-Limited Currents at Moderately High Frequencies', *RCA. Rev.*, **4**, 441 (1940), **5**, 106, 244 (1940)

2 WATT-CARTER, D. E. and WHEELER, L. K., 'The Lincompex System for the Protection of HF Radio Telephone Circuits', *P.O. Elect. Engrs. J.*, **59**, 163 (1966)

3 BELL SYSTEM LABORATORIES, *Transmission Systems for Communications* (1964)

Part 6

General and Commercial Information

General and Commercial Information

1 Classification

The 'directory' part of the Reference Book is to help find organizations relevant to instrumentation. A broad classification has been adopted for the fields of instrumentation, based initially on the chapters in Parts 1–4. By not breaking subjects down into fine detail, we avoid the directory becoming excessively large and reduce the risk of information becoming out of date. The categories used are:

1 Flow
2 Viscosity
3 Length, position and speed
4 Strain
5 Level and volume
6 Vibration
7 Weight and force
8 Density
9 Pressure
10 Vacuum
11 Particle sizing
12 Temperature
13 Chemical analysis, general
14 Spectroscopy
15 Electro-chemical techniques for analysis
16 Gas analysis
17 Moisture, humidity
18 Electrical quantities
19 Optical quantities
20 Nucleonics
21 Measurements using nuclear techniques
22 Non-destructive testing
23 Noise
24 Optical instruments primarily producing images
25 Optical fibres
26 Surfaces
27 Medical and other biological instruments
28 Meteorological instrumentation
29 Civil engineering
30 Geology
31 Instrument design and construction
32 Sampling
33 Data processing
34 Telemetry
35 Display and recording
36 Instrument interfaces
37 Pneumatics
38 Reliability
39 Safety
40 Installation
41 Rental
42 General

The 'General' category is used for those organizations whose diversity of interests makes it inappropriate for them to be included elsewhere.

2 Addresses

A long address list has been compiled. It is divided into four parts:

1 Private firms
2 Academic institutions, i.e. university and college departments and other organizations in the public sector where relevant research is done.
3 Contacts at all British universities for enquiries about industrial application of their expertise.
4 Learned societies and engineering institutions.

To save extensive duplication, each of the bodies in parts 1 and 2 has been given a number and that number is used elsewhere to refer to the body. Under the private firms are shown their concerns in terms of manufacture, calibration, repair, consultancy and research, and for both private firms and academic institutions the numbers of the instrumentation categories in which they are interested are given.

3 Categories of instrumentation

To facilitate following up any instrumentation category of interest, we give here the reference numbers of the firms concerned with that particular category in their capacity as Manufacturer etc.

1 Flow

Manufacturers: 2 8 11 19 21 22 25 33 34 36 37 39 41 43 48 55 56 57 60 61 62 67 70 74 75 79 85 86 100 104 107 112 113 114 121 124 136 142 146 148 149 152 154 162 163 165 167 169 174 180 186 187 194 198 201 202 210 216 221 232 239 242 247 248 253 256 258 265 273 275 276 280 281 283 284 285 300 302 303 304 306 308 311 312 324 325 326 329 332 339 345 346 351 352 367 373 382 391 396 397 400 402 404 405 406 412 415 416 418 427 429 438 439 440 443 445 446 453 454

Calibration: 21 22 55 62 67 70 79 85 107 162 163 167 174 180 201 210 248 253 265 281 283 284 290 303 308 312 324 326 329 352 382 400 418 440 443 445 454

Repairs: 22 37 61 62 67 79 85 107 152 162 167 174 180 201 210 248 253 265 281 283 284 285 290 303 312 324 328 352 382 400 405 418 440 443 445

Consultants: 2 11 22 36 39 41 56 60 79 83 85 107 149 152 162 167 180 194 202 208 210 245 246 248 258 265 281 283 284 285 290 293 303 311 312 324 326 373 382 391 402 405 415 418 427 438 440 443 445

Contract Research: 11 85 90 169 180 194 201 202 210 248 258 281 283 285 290 324 346 391 396 405 415 427 438 440 443 445

Academic Institutions: 464 465 467 470 472 478 480 481 490 495 500 505 506 507 510 516 517 520 532 534

2 Viscosity

Manufacturers: 7 19 43 87 97 104 108 149 164 172 190 248 256 284 300 322 332 373 380

Calibration: 108 190 248 284

Repairs: 164 190 248 284 322

Consultants: 149 164 248 284 322 373

Contract Research: 248

Academic Institutions: 471 492 495 510 516 527

3 Length, position and speed

Manufacturers: 2 7 8 20 38 49 75 89 94 98 103 104 108 144 151 156 177 190 212 219 222 234 237 251 283 295 332 335 351 357 359 368 375 376 434 440 444 447

Calibration: 108 177 190 212 234 251 283 359 440

Repairs: 177 190 212 251 283 359 440

Consultants: 2 20 94 144 177 212 234 251 283 357 359 440

Contract Research: 177 234 283 295 357 440

Academic Institutions: 472 473 477 478 483 490 491 493 501 506 507 511 516 518 524 534

4 Strain

Manufacturers: 4 11 25 37 40 41 56 57 68 86 104 119 127 145 169 177 188 189 190 212 216 224 251 258 272 273 282 283 296 300 304 311 317 326 332 335 365 368 376 380 386 399 458

Calibration: 40 68 177 188 189 190 212 224 251 282 283 326

Repairs: 37 40 68 177 188 189 190 212 251 283

Consultants: 11 41 56 119 177 212 251 258 282 283 296 311 326 386

Contract Research: 11 169 177 188 189 258 272 282 283

Academic Institutions: 470 477 481 484 493 507 510 514 522 524

5 Level and volume

Manufacturers: 2 8 20 25 33 34 37 39 48 49 51 55 57 58 60 61 65 66 67 70 75 78 86 91 104 112 121 124 127 129 134 139 142 146 150 154 156 163 165 181 186 187 191 194 198 202 209 210 211 212 217 240 248 256 258 265 274 280 283 285 300 306 307 308 309 313 320 324 326 331 332 342 345 351 365 366 373 376 382 385 397 402 415 416 422 431 438 440 443 453 459

Calibration: 55 66 67 70 129 134 139 163 210 211 212 240 248 265 283 290 308 309 320 324 326 366 382 422 431 440 443

Repairs: 37 61 66 67 91 134 210 211 212 240 248 265 283 285 290 309 320 324 331 366 382 422 431 440 443 459

Consultants: 2 20 39 51 60 83 91 134 194 202 208 210 211 212 246 248 258 265 274 283 285 290 293 309 320 324 326 366 373 382 385 402 415 431 438 440 443

Contract Research: 134 194 202 210 248 258 283 285 290 324 385 415 438 440 443

Academic Institutions: 464 472 478 490 501

6 Vibration

Manufacturers: 4 11 25 36 40 66 68 69 80 87 94 104 106 108 114 145 169 177 191 199 212 220 254 268 273 283 295 310 317 326 332 335 365 368 375 376 397 399 443 456

Calibration: 40 66 68 106 108 177 199 212 220 283 290 326 443

Repairs: 40 66 68 177 199 212 220 254 283 290 443

Consultants: 11 36 94 106 177 199 212 220 254 268 283 290 326 443 456

Contract Research: 11 169 177 199 268 283 290 295 443

Academic Institutions: 463 464 469 470 472 477 481 490 492 493 495 503 507 514 516 524 525

7 Weight and force

Manufacturers: 4 7 11 37 40 57 68 78 98 103 104 109 119 132 145 151 169 212 224 251 254 271 272 282 283 300 304 313 316 332 335 371 375 376 380 458 460

Calibration: 40 68 212 224 251 271 282 283 290 316

Repairs: 37 40 68 212 251 254 271 283 290 316

Consultants: 11 119 212 251 254 271 282 283 290 316 460

Contract Research: 11 169 271 272 282 283 290

Academic Institutions: 470 478 481 505 506 510 511 524

8 Density

Manufacturers: 19 26 37 67 74 104 149 163 164 165 172 187 190 209 210 215 217 256 274 280 285 298 300 316 322 326 332 333 334 365 366 400 405 408 415 419 459

Calibration: 26 67 163 190 210 316 326 366 400

Repairs: 26 37 67 164 190 210 285 316 322 334 366 400 405 459

Consultants: 149 164 210 215 246 274 285 316 322 326 366 405 408 415

Contract Research: 210 215 285 334 405 408 415

Academic Institutions: 470 495

9 Pressure

Manufacturers: 3 4 8 11 25 29 33 34 36 37 39 41 44 47 48 56 57 59 60 61 62 65 74 75 78 79 80 84 85 86 96 98 104 106 107 108 113 115 121 127 129 131 134 136 139 142 145 150 152 154 164 165 167 169 172 176 180 181 184 185 186 187 191 194 195 196 209 210 211 212 232 239 240 243 244 247 248 251 251 254 256 258 265 268 271 273 279 282 283 284 285 296 300 304 306 307 311 312 320 321 324 329 332 335 338 339 342 346 351 358 366 372 373 375 376 385 391 395 397 400 402 405 412 413 415 416 418 422 423 431 433 438 440 441 443 445 446 449 454 455 456 458 459 461

Calibration: 59 62 79 84 85 96 106 107 108 115 129 134 139 167 180 196 210 211 212 225 240 243 248 251 265 271 282 283 284 290 312 320 321 324 329 366 400 413 418 422 431 440 443 445 449 454

Repairs: 37 61 62 79 85 96 107 115 134 140 152 164 167 180 184 196 210 211 212 225 240 243 248 251 254 265 271 283 284 285 290 312 320 321 324 328 366 400 405 418 422 431 440 443 445 449 459

Consultants: 11 29 36 39 41 56 60 79 84 85 96 106 107 115 134 152 164 167 180 184 194 208 210 211 212 246 248 251 254 258 265 268 271 282 283 284 285 290 293 296 311 312 320 321 324 366 373 385 391 402 405 415 418 431 438 440 443 445 449 456 461

Contract Research: 11 29 59 85 96 115 131 134 169 180 194 210 248 258 268 271 282 283 285 290 324 346 358 385 391 405 413 415 438 440 443 445

Academic Institutions: 470 478 481 486 490 493 511 514 524

10 Vacuum

Manufacturers: 11 25 34 36 37 47 59 60 61 65 75 81 84 85 86 98 104 108 121 129 131 134 136 139 142 154 165 172 176 185 187 194 195 196 210 248 267 268 282 284 300 304 332 338 339 342 351 361 366 373 375 384 385 391 395 412 413 416 422 431 438 440 443 454 461

Calibration: 59 84 85 108 129 134 139 196 210 248 282 284 366 413 422 431 440 443 454

Repairs: 37 61 85 134 196 210 248 284 328 366 422 431 440 443

Consultants: 11 36 60 84 85 134 194 210 246 248 268 282 284 293 366 373 385 391 431 438 440 443 461

Contract Research: 11 59 85 131 134 194 210 248 268 282 385 391 413 438 440 443

Academic Institutions: 493 511 513 517

11 Particle sizing

Manufacturers: 20 37 62 85 89 114 251 258 265 272 360 418

Calibration: 62 85 265 418

Repairs: 37 62 85 265 418

Consultants: 20 85 258 265 418

Contract Research: 85 258 272

Academic Institutions: 464 470 472 475 481 487 490 495 502 507 516 526

12 Temperature

Manufacturers: 5 8 12 13 24 25 29 36 37 38 39 41 44 46 48 49 51 56 60 61 65 66 68 72 74 75 78 79 80 84 85 86 87 95 96 99 102 104 106 107 113 114 118 121 125 127 134 135 137 141 142 145 152 154 160 164 165 166 168 169 172 176 179 180 181 184 185 186 187 188 194 195 196 207 210 213 216 221 222 232 235 238 240 241 243 244 247 251 258 259 260 261 262 263 264 265 266 268 275 279 283 285 289 296 300 304 306 311 315 318 320 321 322 324 331 332 336 347 348 351 358 362 365 366 368 372 373 378 384 385 391 395 397 399 405 408 409 410 413 415 416 419 422 423 425 427 429 431 432 440 443 445 446 449 452 456 458 459

Calibration: 66 68 72 79 84 85 95 96 106 107 134 166 179 180 188 196 210 225 235 240 241 243 260 261 262 264 265 283 290 320 321 324 348 366 378 413 422 431 432 440 443 445 449

Repairs: 37 61 66 68 72 79 85 95 96 107 130 134 137 140 152 164 179 180 184 188 196 207 210 213 225 238 240 241 243 260 262 264 265 283 285 290 320 321 322 324 331 348 366 378 405 422 431 440 443 445 449 459

Consultants: 12 29 36 39 41 51 56 60 72 79 84 85 95 96 102 106 107 134 152 164 179 180 184 194 208 210 241 246 258 260 262 264 265 268 283 285 290 293 296 311 315 320 321 322 324 366 373 385 391 405 408 415 427 431 432 440 443 445 449 456

Contract Research: 12 29 85 96 102 134 169 179 180 188 194 210 258 262 264 268 283 285 290 318 324 358 385 391 405 408 413 415 427 440 443 445

Academic Institutions: 472 477 478 481 482 489 490 493 506 507 511 514 524

13 Chemical analysis, general

Manufacturers: 21 26 30 37 61 80 86 87 88 89 142 159 163 164 165 183 194 215 241 252 257 258 259 265 274 285 305 316 319 330 332 345 366 367 382 389 391 408 418 440 446

Calibration: 21 26 163 241 265 316 366 382 418 440

Repairs: 26 37 61 164 241 265 285 316 366 382 418 440

Consultants: 159 164 183 194 215 241 258 265 274 285 316 366 382 391 408 418 440

Contract Research: 194 215 258 285 305 391 408 440

Academic Institutions: 464 468 470 471 485 489 494 498 502 506 513 516 531

14 Spectroscopy

Manufacturers: 31 32 51 89 133 206 216 222 253 257 271 319 330 389 391 408 418

Calibration: 253 271 290 418

Repairs: 133 253 271 290 418

Consultants: 51 271 290 391 408 418

Contract Research: 31 271 290 391 408

Academic Institutions: 463 464 471 485 489 493 494 496 497 498 506 513 516 526 527 531

15 Electro-chemical techniques for analysis

Manufacturers: 26 27 32 37 80 142 164 165 183 192 194 236 241 251 253 257 258 259 260 269 285 331 332 345 355 366 382 390 399 429 454

Calibration: 26 241 253 260 366 382 390 454

Repairs: 26 27 37 164 241 253 260 285 331 355 366 382 390

Consultants: 27 164 183 192 194 241 246 258 260 285 355 366 382

Contract Research: 27 192 194 236 258 285

Academic Institutions: 465 477 483 489 494 498 500 506 516 524 527 531

16 Gas analysis

Manufacturers: 21 26 27 28 30 36 37 52 68 74 80 86 89 95 105 113 118 128 136 143 159 164 165 175 179 182 194 215 248 251 253 258 260 265 267 269 274 277 285 300 311 325 330 345 366 383 388 389 390 391 392 393 397 403 404 408 412 414 418 425 429 443 448

Calibration: 21 26 68 95 105 128 175 179 248 253 260 265 290 366 390 418 443 448

Repairs: 26 27 37 68 95 105 128 164 175 179 248 253 260 265 285 290 366 390 418 443 448

Consultants: 27 36 95 128 159 164 175 179 194 215 248 258 260 265 274 285 290 311 366 391 403 408 418 443 448

Contract Research: 27 90 179 194 215 248 258 285 290 391 403 408 414 443

Academic Institutions: 464 470 471 472 477 481 486 489 513 516 524 530 531

17 Moisture, humidity

Manufacturers: 12 19 26 27 29 30 34 36 37 51 65 66 68 80 86 87 104 107 142 151 152 164 172 174 215 222 244 251 263 264 283 286 291 292 311 316 321 325 332 348 366 372 387 389 393 404 414 423 429 440 449

Calibration: 26 66 68 107 174 264 283 286 290 291 316 321 348 366 387 440 449

Repairs: 26 27 37 66 68 107 140 152 164 174 264 283 286 290 291 316 321 348 366 440 449

Consultants: 12 27 29 36 51 107 152 164 215 246 264 283 290 311 316 321 366 440 449

Contract Research: 12 27 29 215 264 283 286 290 414 440

Academic Institutions: 470 472 477 493 505 506 509 516 526

18 Electrical quantities

Manufacturers: 29 35 36 46 49 63 66 75 80 86 91 104 107 110 117 125 131 134 141 147 155 157 171 185 188 194 196 201 210 215 233 236 238 239 243 258 266 268 289 297 300 301 304 332 346 349 350 356 358 362 364 366 399 436 440 445 452 460

Calibration: 66 107 134 161 171 188 196 201 210 225 243 290 349 364 366 440 445

Repairs: 66 91 107 130 134 161 188 196 201 210 225 238 243 290 349 356 364 366 440 445

Consultants: 29 36 91 107 134 194 210 215 258 268 290 349 366 440 445 460

Contract Research: 29 131 134 147 188 194 201 210 215 236 258 268 290 346 358 364 440 445

Academic Institutions: 463 470 477 478 482 484 490 493 498 499 500 505 506 511 514 518 528 530

19 Optical quantities

Manufacturers: 20 26 35 68 80 89 136 149 151 164 181 213 215 234 258 267 334 361 382 399 440 451

Calibration: 26 68 234 290 382 440

Repairs: 26 68 164 213 290 334 382 440

Consultants: 20 149 164 215 234 258 290 382 440

Contract Research: 215 234 258 290 334 440

Academic Institutions: 463 466 471 478 482 484 493 497 498 506 511 513 514 522 534

20 Nucleonics

Manufacturers: 6 50 76 133 142 151 174 194 217 218 237 314 326 415

Calibration: 174 326

Repairs: 76 133 174

Consultants: 194 218 326 415

Contract Research: 6 194 218 415

Academic Institutions: 464 470 474 477 507 511 512 523 534

21 Measurements using nuclear techniques

Manufacturers: 37 76 133 142 149 151 164 174 206 215 217 218 237 256 300 314 319 326 365 366 415 439

Calibration: 174 326 366

Repairs: 37 76 133 164 174 366

Consultants: 149 164 215 218 326 366 415

Contract Research: 215 218 415

Academic Institutions: 464 470 472 474 498 507 516 523 526 532

22 Non-destructive testing

Manufacturers: 6 40 71 87 89 94 109 119 133 134 137 144 147 169 177 190 201 206 210 213 216 224 255 283 311 315 317 326 355 361 368 369 380 386 391 418 439 457 459

Calibration: 40 134 177 190 201 210 224 283 290 326 418 457

Repairs: 40 133 134 137 177 190 201 210 213 283 290 355 418 457 459

Consultants: 94 119 134 144 177 210 283 290 311 315 326 355 369 386 391 418

Contract Research: 6 134 147 169 177 201 210 283 290 391

Academic Institutions: 463 464 468 470 471 474 475 477 478 483 484 486 488 490 499 514 515 516 519 522 526 532

23 Noise

Manufacturers: 36 46 65 66 68 80 86 106 107 145 169 273 283 310 356 364 399 405 440

Calibration: 66 68 106 107 225 283 290 364 440

Repairs: 66 68 107 225 283 290 356 364 405 440

Consultants: 36 106 107 283 290 405 440

Contract Research: 169 283 290 364 405 440

Academic Institutions: 470 481 492 507 514 516 524 525

24 Optical instruments primarily producing images

Manufacturers: 2 20 31 89 151 213 234 311 336 343 355 361

Calibration: 234

Repairs: 213 355

Consultants: 2 20 234 311 343 355

Contract Research: 31 234 343

Academic Institutions: 463 471 474 475 478 481 495 497 500 514 521 522 530

25 Optical fibres

Manufacturers: 35 45 258

Calibration: 45

Repairs: 45

Consultants: 45 258

Contract Research: 258

Academic Institutions: 464 465 470 473 478 481 484 490 493 496 499 500 504 507 511 522 524

26 Surfaces

Manufacturers: 1 2 66 137 144 148 156 219 234 267 283 295 321 322 331 336 359 368 418 439 444

Calibration: 66 234 283 290 321 359 418

Repairs: 66 137 283 290 321 322 331 359 418

Consultants: 1 2 144 234 283 290 321 322 359 418

Contract Research: 1 234 283 290 295

Academic Institutions: 463 469 471 475 491 495 510 511 513 534 535

27 Medical and other biological instruments

Manufacturers: 14 28 31 39 52 56 65 68 80 86 87 89 98 175 213 241 258 269 272 274 278 280 283 295 305 310 326 330 336 338 361 364 404 419 448

Calibration: 68 175 241 283 290 326 364 448

Repairs: 68 175 213 241 278 283 290 364 448

Consultants: 14 39 56 175 241 258 274 283 290 326 448

Contract Research: 14 31 258 272 283 290 295 305 364

Academic Institutions: 465 466 477 483 486 496 498 500 506 516 523 524 526 529 530 533

28 Meteorological instrumentation

Manufacturers: 34 62 65 66 79 80 85 86 89 104 124 129 148 176 186 188 191 268 332 342 364 368 449 453

Calibration: 62 66 79 85 129 188 290 364 449

Repairs: 62 66 79 85 188 290 364 449

Consultants: 79 85 268 290 449

Contract Research: 85 188 268 290 364

Academic Institutions: 500 516 526

29 Civil engineering

Manufacturers: 40 65 71 91 255 271 284 296 316 326 350 411 436

Calibration: 40 271 284 316 326

Repairs: 40 91 271 284 316

Consultants: 91 271 284 296 316 326 411

Contract Research: 271 411

Academic Institutions: 468 483 507 516 519 534

30 Geology

Manufacturers: 71 91 108 192 284 342 343 418 436 459

Calibration: 108 284 418

Repairs: 91 284 418 459

Consultants: 91 192 284 343 418

Contract Research: 192 343

Academic Institutions: 468 476 519

31 Instrument design and construction

Manufacturers: 1 4 8 12 14 27 29 46 50 66 74 77 79 85 91 100 102 104 106 107 117 118 119 123 136 138 144 164 180 184 187 194 207 210 212 216 218 223 224 226 228 229 230 236 242 248 251 258 263 264 265 267 268 272 274 283 285 303 305 309 311 322 330 334 342 343 346 347 355 356 357 360 368 383 396 397 403 405 412 414 415 427 432 436 437 443 445 452 460

Calibration: 66 79 85 106 107 180 210 212 224 248 251 264 265 283 290 303 309 432 443 445

Repairs: 27 66 79 85 91 107 164 180 184 193 207 210 212 228 248 251 264 265 283 285 290 303 309 322 328 334 355 356 405 443 445

Consultants: 1 12 14 27 29 77 79 85 91 101 102 106 107 119 123 144 164 180 184 193 194 200 208 210 212 218 226 229 230 248 251 258 264 265 268 274 283 285 290 303 309 311 322 340 343 354 355 357 403 405 415 424 427 432 443 445 460

Contract Research: 1 12 14 27 29 77 85 90 102 138 180 194 210 218 226 230 236 248 258 264 268 272 283 285 290 305 334 340 343 346 357 396 403 405 414 415 427 437 443 445

Academic Institutions: 463 464 465 471 472 474 477 481 486 495 497 498 501 504 505 507 514 516 519 521 522 524 534

32 Sampling

Manufacturers: 21 27 28 30 37 79 80 81 88 93 95 104 142 146 163 164 183 196 199 215 226 230 242 252 258 262 265 274 285 287 300 305 316 332 356 383 385 395 404 412 418 427 448 459

Calibration: 21 79 95 163 196 199 262 265 290 316 418 448

Repairs: 27 37 79 95 164 196 199 262 265 285 290 316 356 418 448 459

Consultants: 27 79 95 164 183 199 215 226 230 258 262 265 274 285 290 316 354 385 417 418 427 448

Contract Research: 27 199 215 226 230 258 262 285 290 305 385 417 427

Academic Institutions: 470 474 477 493 506 510 523 524 530

33 Data processing

Manufacturers: 1 13 15 29 35 37 49 56 58 61 75 80 93 98 102 104 113 125 126 131 133 139 144 145 170 177 180 185 196 202 210 216 218 222 230 257 258 259 265 268 283 285 296 300 316 318 322 324 330 332 344 368 377 385 388 391 402 405 408 427 429 430 434 437 440 443 456 458 459

Calibration: 139 177 180 196 210 265 283 290 316 324 440 443

Repairs: 37 61 133 140 177 180 196 210 265 283 285 290 316 322 324 405 440 443 459

Consultants: 1 29 56 102 144 177 180 200 202 210 218 230 258 265 268 283 285 290 293 296 316 322 324 340 385 391 402 405 408 427 440 443 456

Contract Research: 1 29 102 131 177 180 202 210 218 230 258 268 283 285 290 318 324 340 385 391 405 408 427 437 440 443

Academic Institutions: 463 464 468 470 471 472 474 475 477 478 479 481 491 492 493 494 495 496 501 516 517 523 524 530 531 534

34 Telemetry

Manufacturers: 13 37 56 58 61 93 102 104 125 131 134 160 165 170 172 185 202 210 212 216 218 230 248 258 263 268 272 283 285 296 312 318 323 324 332 333 356 370 381 385 402 405 410 427 429 430 437 442 443 445 449 451 459

Calibration: 134 210 212 248 283 290 312 324 443 445 449

Repairs: 37 61 134 210 212 248 283 285 290 312 323 324 356 405 443 445 449 459

Consultants: 56 102 134 200 202 210 212 218 230 248 258 268 283 285 290 296 312 324 370 385 402 405 427 443 445 449

Contract Research: 102 131 134 202 210 218 230 248 258 268 272 283 285 290 318 324 385 405 427 437 443 445

Academic Institutions: 464 465 470 472 483 495 504 516 524

35 Display and recording

Manufacturers: 8 12 13 15 16 24 25 29 34 36 37 43 46 48 49 53 58 60 61 63 65 66 68 70 75 80 85 86 89 93 98 102 104 107 118 125 127 129 131 133 134 142 144 145 155 160 165 171 180 185 186 194 199 202 210 212 213 216 218 233 238 251 253 258 265 268 271 279 283 285 287 296 300 303 304 313 316 318 321 324 326 332 335 344 356 365 366 368 373 374 381 385 388 391 397 407 410 415 427 428 430 431 432 437 439 440 441 443 445 449 456 458 459

Calibration: 66 68 70 85 107 129 134 171 180 199 210 212 225 251 253 265 271 283 290 303 316 321 324 326 366 431 432 440 443 445 449

4 Manufacturers

Manufacturers of instruments of a certain type can be found by looking up the appropriate category to find numbers which are translated into names and addresses in the address list. As remarked above, this is only a coarse breakdown of subject. If it is known exactly what is required, the search may be reduced by having recourse to one of the directories listed at the end of Section 8. These split up topics into finer detail, so reducing the number of potential suppliers quoted, at the cost of having a much larger publication that needs more frequent updating. The aim here has been, not to include every firm that might have some claim to an entry, but to provide a list of manageable proportions. Thus it is restricted to manufacturers. Stockists, distributors and agents are excluded. Also, it is confined to firms based in Britain.

Instruments are difficult to define. In general we concentrate on the hardware side of complete units for making measurements. Components, as well as software and systems, are omitted. There is a deliberate attempt to concentrate on instruments for making practical, industrial, scientific measurements while leaving out, on the one hand, very advanced equipment only used in academic research and, on the other hand, very simple devices for domestic and educational use.

5 Services

5.1 Hiring

Many firms will, of course, rent out their own equipment. Those that we list under Category 41 specialize in this field and cover a wide range of equipment.

5.2 Installation

Installation is an aspect of the use of instrumentation that needs attention to many points of detail, so much so that we have devoted a whole chapter to it (Chapter 2 of Part 4). Large users will often have accumulated their own experience and do their own installation, but under Category 40 we list those firms that specialize in this field and whose services are generally available.

5.3 Calibration

The calibration of instruments is a broad and deep subject. Ultimately it concerns relating the reading of a particular item to fundamental standards. This is called *traceability*. The definitions of the standards are given in several of our technical chapters.

The National Measurement Accreditation Service (NAMAS) plays the major part in organizing the calibration of instruments in Britain. It has two branches, the British Calibration Service (BCS) and the National Testing Laboratory Accreditation Scheme (NATLAS). It is operated from the National Physical Laboratory.

Laboratories apply to BCS for recognition, and if the Service is satisfied as to their competence it gives them accreditation. A periodically updated directory, currently including some 150 laboratories, is maintained and this indicates the field of measurement in which each laboratory can give authenticated certification of calibration. Because testing is related to calibration, the similar NATLAS directory is complementary.

The soundest way to gain information about calibration services and by whom they are offered is thus to consult NAMAS directories. Many of the firms listed there are users rather than manufacturers or service organizations, but the following firms in our directory have BCS accreditation for themselves or their affiliates:

40 166 171 189 224 262 269 359 366 399 400 435 464 472 477 510 522

Some manufacturers and other firms offer facilities independent of NAMAS. The numbers of those appearing in this Reference Book are:

18 19 21 22 26 45 55 59 62 66 67 68 70 72 79 84 85 95 96 105 106 107 108 115 128 129 134 139 161 162 163 167 174 175 177 179 180 188 190 196 199 201 210 211 212 220 225 234 235 240 241 243 248 251 253 260 261 264 265 271 281 282 283 284 286 290 291 303 308 309 312 316 320 321 324 326 329 348 349 352 364 378 382 387 390 413 418 422 431 432 440 443 445 448 449 454 457 462

5.4 Repair

Most manufacturers are prepared to repair their own products. Some firms undertake repairs on a broader basis, and there are advantages in employing them, particularly for users who have to maintain a large pool of equipment. They include:

22 26 27 37 40 45 61 62 66 67 68 72 76 79 85 95 96 105 107 115 128 130 133 134 137 140 152 162 164 167 174 175 177 179 180 184 188 189 190 193 196 199 201 207 210 211 212 213 220 225 228 238 240 241 243 248 251 253 260 262 264 265 271 278 281 283 284 285 286 290 291 303 309 312 316 320 321 322 323 324 328 331 334 341 348 349 352 355 356 359 364 366 378 382 390 400 405 418 422 431 440 443 445 448 449 457 459

6 Research in instrumentation

It is useful to give interested parties the opportunity to set up links with other organizations that are doing relevant research. That is the aim of this part of the directory. It is difficult to decide what should be included. There are only a few university departments, for instance, whose name immediately indicates work on instrumentation. However, almost the whole of experimental science has some involvement in this field, because science is so closely concerned with making measurements. The aim here has been to exclude research where the measurement technique is much less important than the results observed and also, in accordance with our emphasis on industrial applications, those instrument developments that appear at present to be usable only in pure, academic research.

The chief source of information for compiling this part of the directory has been the book *Current Research in Britain*. We also acknowledge valuable help from the Science and Engineering Research Council (Mr D. J. Tallantire) and the Institution of Chemical Engineers (Mr E. H. Higham).

As in the case of identifying relevant manufacturers, a university or polytechnic department working in a particular field of instrumentation research may be found by looking up that category and then consulting the master address list to find which departments the numbers given refer to. It should be pointed out that this directory is concerned with research, not educational courses. Some information on the latter is collected in the Yearbook published by the Institute of Measurement and Control (see page 15).

It is not only at universities and polytechnics that research is undertaken, so we also give here – under the heading of Institutes – some other organizations. These include nationalized industries and laboratories and some large private concerns aimed primarily at research. Any of these are potential consultants and sources for contract research, as are a number of manufacturers and other small firms whose numbers may again be found under the appropriate category.

All British universities are increasingly conscious of the value of collaborative research (see Section 10.3). The University Directors of Industrial Liaison (UDIL) play a big part in co-ordinating such activities. British Expertise in Science and Technology (BEST) is another useful source of information, as is the book *Register of Consulting Scientists* Adam Hilger 6th ed. (1984).

7 Learned societies and engineering institutions

The Institute of Measurement and Control is the professional body most directly concerned with instrumentation. It is also the British point of contact for *IMEKO* (the International Measurement Confederation).

A number of larger bodies cover different aspects of instrumentation within their wider interests, notably the following:

The Institute of Physics has an Instrument Science and Technology group with a strong concern for instrumentation. The Materials and Testing and the Stress Analysis groups also have some relevance.

The Royal Society of Chemistry fulfils the responsibilities and activities formerly undertaken by The Chemical Society, The Society for Analytical Chemistry, the Royal Institute of Chemistry and The Faraday Society. It is divided into six divisions, of which the Analytical Division has a wide concern for instrumentation. Mössbauer Spectroscopy and High Resolution Spectroscopy are covered in other divisions.

The Institution of Electrical Engineers is divided into professional groups, including: C 11 Instrumentation Systems and Concepts; E 1 Measurements and Instruments in Electronics; S 4 Fundamental Aspects of Measurement; and S 6 Non-destructive Testing. It is currently merging with the Institution of Electronic and Radio Engineers.

The Institution of Mechanical Engineers is divided into divisions, of which the following have instrumentation interests: Power Industries, Process Industries, Engineering Manufacture, Aerospace, Automobile.

Two smaller bodies with specialized instrumentation concerns, as indicated by their names, are:

The British Institute of Non-Destructive Testing
The British Society for Strain Measurement

The address of all these bodies are given at the end of the master address list.

8 General instrumentation books

For the most part, individual technical chapters of the *Instrumentation Reference Book* have given references for further reading to books with a particular relevance to the topic of that chapter. Here we list some more general books, which each give an overview of a wide range of instrumentation subjects. To help the reader decide whether any particular book will help him with the problem he faces, we include a table of contents for each of these books and a brief description of the background and style.

Potvin, J. *Applied Process Control Instrumentation*, Reston Publishing Co. (Prentice-Hall), 1985, 262 pp.

1. Boilers
2. Food Industry
3. Water treatment
4. Petroleum
5. Pulp and paper
6. Textiles
7. Miscellaneous processes

Gives easily understood descriptions of some selected operations. The content is largely provided through the courtesy of Taylor Instrument Company.

Holman, J. P. *Experimental Methods for Engineers*, McGraw-Hill, 4th edition, 1984, 490 pp.

1. Introduction
2. Basic concepts
3. Analysis of experimental data
4. Basic electrical measurements and sensing devices
5. Displacement and area measurements
6. Pressure measurement
7. Flow measurement
8. The measurement of temperature
9. Thermal- and transport-property measurements
10. Force, torque and strain measurements
11. Motion and vibration measurement
12. Thermal- and nuclear-radiation measurements
13. Air-pollution sampling and measurement
14. Data acquisition and processing

Appendices include Conversion factors, Properties of metals, Thermal properties of some non-metals, Properties of some liquids and gases (in SI and other units), Diffusion co-efficients, Emissivities and Moments of inertia. The book is suitable for use in an engineering educational course. Problems are set for students at the ends of chapters. A good deal of material is covered, without appearing to suffer from condensation.

Bentley, J. P. *Principles of Measurement Systems*, Construction Press (Longmans), 1983, 424 pp.

A General principles

1. The general measurement system
2. Static characteristics of measurement system elements
3. The accuracy of measurement systems in the steady state
4. Dynamic characteristics of measurement systems
5. Loading effects in measurement systems
6. Signals and noise in measurement systems
7. Reliability, choice and economics of measurement systems

B Typical measurement system elements

8. Sensing elements
9. Signal conditioning elements
10. Signal processing elements
11. Data presentation elements

C Specialized measurement systems

12. Flow measurement systems
13. Pneumatic measurement systems
14. Heat transfer effects in measurement systems
15. Thermal radiation measurement systems
16. Ultrasonic measurement systems
17. Gas chromatography
18. Data acquisition and telemetry systems

There is a systematic educational emphasis, with problems to solve for exercises and appropriately mathematical. Repeated references to systems aims to unify the approach.

Jones, B. E. (ed) *Instrument Science and Technology*, Adam Hilger, Vol. 1, 1982, 144 pp; Vol 2, 1983, 147 pp; Vol. 3, 1985, 183 pp.

Volume 1

Finkelstein, L., Introduction

Finkelstein, L. and Watts, R. D., Mathematical models of instruments – fundamental principles

van den Bos, A., Applications of statistical parameter estimation methods to physical measurements

Woschni, E.-G., Dynamics of measurement – relations to system and information theory

Bosman, D., Systematic design of instrumentation systems

Jones, B. E., Feedback in instruments and its applications

Wright, R. I., Instrument reliability

Pearce, B. G. and Schackel, B., The ergonomics of scientific instrument design

Institute of Physics, Thermal noise

Usher, M. J., Noise and bandwidth

Faulkner, E. A., The principles of impedance optimization and noise matching

Fellgett, P. B. and Usher, M. J., Fluctuation phenomena in instrument science

Blair, D. P. and Sydenham, P. H., Phase sensitive detection as a means to recover signals buried in noise

Volume 2

Cunningham, M. J., Measurement errors and instrument inaccuracies
Furse, J. E., Kinematic design of fine mechanisms in instruments
Hugill, A. L., Displacement transducers based on reactive sensors in transformer ratio bridge circuits
Middelhoek, S. and Noorlag, D. J. W., Silicon microtransducers
Woolvet, G. A., Digital transducers
Owens, A. R., Digital signal conditioning and conversion
Meade, M. L., Advances in lock-in amplifiers
Beck, M. S., Correlation in instruments: cross correlation flowmeters
Palmer, R. B. J., Nucleonic instrumentation applied to the measurement of physical parameters by means of ionizing radiation
Verhagen, C. J. D. M., Measurement for and by pattern recognition
Noltingk, B. E., Creative instrument design
Shaw, R., Future instrumentation: its effect on production and research
Sydenham, P. H., The literature of instrument science and technology

Volume 3

Bailey, A. E., Units and standards of measurement
Weichert, L., The avoidance of electrical interference in instruments
Richards, J. C. S., Some aspects of transducer immitance measurement
Medlock, R. S., Sensors for mechanical properties
Garratt, J. D., Survey of displacement transducers below 50 mm
Erdem, U., Force and weight measurement
Brain, T. J. S. and Scott, R. W. W., Survey of pipeline flowmeters
Bentley, J. P., Temperature sensor characteristics and measurement system design
Hahn, C. E. W., Techniques for measuring the partial pressures of gases in the blood: Part 1 *in vitro* measurements; Part 2 *in vivo* measurements
Crecraft, D. E., Ultrasonic instrumentation: principles, methods and applications
Asher, R. C., Ultrasonic sensors in the chemical and process industries
Blitz, J., Nondestructive testing using electromagnetic instrumentation
Roughton, J. E., Non-invasive measurements
Duncombe, E., Some instrumental techniques for hostile environments

These three volumes are an anthology of invited articles which were previously published in the *Journal of Scientific Instruments*.

Sydenham, P. H. (ed) *Handbook of Measurement Science*, John Wiley, Vol. 1, 1982, 654 pp; Vol. 2, 1983, 758 pp.

Volume 1 Theoretical Fundamentals

Finkelstein, L., Theory and philosophy of measurement
Sydenham, P. H., Measurements, models and systems
Sydenham, P. H., Standardization of measurement fundamentals and practices
Woschni, E.-G., Signals and systems in the time and frequency domains
Miller, M. J., Discrete signals and frequency spectra
Hofmann, D., Measurement errors, probability and information theory
Verhagen, C. J. D. M. *et al.*, Pattern recognition
van den Bos, A., Parameter estimation
Kerwin, W. J., Analog signal filtering and processing
Bolton, A. G., Filtering and processing of digital signals
Munroe, D. M., Signal-to-noise ratio improvement
Zuch, E. L., Signal data conversion
Grimes, R. W., Transmission of data
Atkinson, P., Closed-loop systems

Volume 2 Practical Fundamentals

Sydenham, P. H., Physiology of measurement systems
Sydenham, P. H., Static and steady-state considerations
Sydenham, P. H., Measurement systems dynamics
Finkelstein, L. and Watts, R. D., Fundamentals of transducers: description by mathematical models
Bosman, D., Human factors in display design
Schnell, L., Measurement of electrical signals and quantities
Trylinski, W., Mechanical regime of measuring instruments
Sydenham, P. H., Electrical and electronic regime of measuring instruments
Prasad, J. and Mitra, G., Optical regime of measuring instruments
Sydenham, P. H., Transducer practice: displacement
Brain, T. J. S. and Blake, K. A., Transducer practice: flow
Sydenham, P. H., Transducer practice: thermal
Watts, R. S., Transducer practice: chemical analysis
Peuscher, F. G., Design and manufacture of measurement systems
Bonollo, E., Quality control and inspection of products
Hobson, J. W., Management of existing measurement systems
Sydenham, P. H. *et al.*, Calibration, evaluation and accreditation

Sydenham, P. H., Sources of information on measurement

These two extensive volumes aim to get away from the 'summary or catalogue' that might be expected in a handbook and to deal with matters pertinent to a science of instrumentation.

Huskins, D. J., *Quality Measuring Instruments in On-line Process Analysis*, Ellis Horwood, 1982, 455 pp.

1. Fluid density monitors
2. Consistency and viscosity measurement
3. Surface tension measurement
4. Cloud-point and pour-point analysers
5. Thermal conductivity gas analysers
6. Paramagnetic oxygen analysers
7. Dissolved oxygen analysers
8. Vapour pressure analysers
9. Vapour–liquid ratio analysers
10. Boiling point analysers
11. Freezing point analysers
12. Calorimeters, flame-speed monitors and thermal analysers
13. Flash-point monitors
14. Octane rating measurement
15. Sorptiometry and porosimetry
16. Chemical type analysers

It gives a good account of how the particular properties it covers can be measured.

Beckwith, T. G., Buck, N. L. and Mavangoni, R. D., *Mechanical Measurements*, Addison-Wesley, 3rd edition, 1982, 730 pp.

1. Fundamentals of technical measurement

1. The process of measurement: an overview
2. The analog measurand: its time-dependent characteristics
3. Measuring system response
4. Sensors
5. Signal conditioning
6. Application of digital techniques to mechanical measurements
7. Readout and data processing
8. Standards of measurement
9. Treatment of uncertainties

2. Applied mechanical measurements

10. Determination of count, events per unit time, and time interval
11. Displacement and dimensional measurement
12. Strain and stress: measurement and analysis
13. Measurement of force and torque
14. Measurement of pressure
15. Measurement of fluid flow
16. Temperature measurements
17. Measurement of motion
18. Acoustical measurements

Appendices: Standards and conversion equations; Theoretical basis for Fourier analysis; Number systems; Some useful data; Stress and strain relationships; Further consideration of class intervals and goodness of fit.

Concentrating on the measurement of mechanical quantities, though extending into subjects on the fringe and into the electronic side, the book is able to deal with its subject thoroughly and effectively. Problems are included for exercises.

Sydenham, P. H., *Transducers in Measurement and Control*, Adam Hilger, 2nd edition, 1980, 99 pp.

General Principles, Microdisplacement Sensors
Industrial and Surveying Range Length Transducers
Acoustic Angle Measurements
Tilt and Alignment
Using Transducers to Measure and Control Position
History and Technique of Temperature Measurement
Non-Contact and Lesser Known Methods of Determining Temperature
Measuring Moisture
Flow
Force, Weight and Torque
Pollution Monitors – 1
Pollution Monitors – 2
Measurement Difficulties, Instrument Information

The chapters in this popular book were originally published as a series of articles in *Electronics To-day International* in 1972 to 1973.

Johnson, C. D., *Process Control Instrumentation Technology*, Wiley, 1977, 428 pp.

1. Introduction to process control
2. Analog signal conditioning
3. Digital signal conditioning
4. Thermal transducers
5. Mechanical transducers
6. Optical transducers
7. Final control element
8. Controller principles
9. Analog controllers
10. Digital control principles
11. Control loop characteristics

Appendices: Units; Digital review; Thermocouple tables; Mechanical review.

The book is written with a systematic educational approach, including a summary and problems at the end of each chapter. It gives a sound foundation but does not aspire to advanced material

Considine, D. M., *Encyclopaedia of Instrumentation and Control*, McGraw-Hill, 1971, 788 pp.

A Table of Contents is inappropriate as a description, because this encyclopedia contains nearly 700 entries arranged alphabetically. The material under one entry ranges from a few lines to a few pages. The five principal information categories are: Measurands; Measurement systems; Instrument data processing; Control systems; and Applications. More than 100 contributors are named.

The following directories contain detailed information relevant to instrumentation, notably where different products can be obtained. They are updated periodically.

Electronic Engineering Index, Technical Indexes Ltd, Bracknell, 300 pp.

Contents
Products and Trade Names
Distributors
Selector Charts
Information on Services from Technical Indexes
Foreign Trade Offices
Professional and Trade Organizations
Addresses

Instrument Engineer's Yearbook, The Institute of Measurement and Control, London, 200 pp.

Contents
The Institute of Measurement and Control
The Engineering Profession
Education and Training
Products and Services
Names and Addresses
Trade Names
Agents and Distributors
Standards
Acronyms

Physics Bulletin Buyer's Guide, Institute of Physics, Bristol, 170 pp.

Contents
Product Index
Products
Suppliers
Appointed Agents

9 British Standards

Some British Standards relevant to instrumentation are shown below. A number of them have more than one part. When publication is spread over several years, a dash is shown after the first date. The list includes some general standards, as well as repeating those quoted in earlier chapters.

Number	*Date*	*Title*
89	1977	Specification for direct acting indicating electrical measuring instruments and their accessories
90	1975	Direct acting electrical recording instruments and their accessories
188	1977	Methods for the determination of the viscosity of liquids
229	1957	Flameproof enclosure of electrical apparatus
410	1976	Specification for test sieves
907	1965	Dial gauges for linear measurement
1041	1943	Code for temperature measurement
1042	1965–	Methods of measurement of fluid flow in closed conduits
1259	1958	Intrinsically safe electrical apparatus and circuits for use in explosive atmospheres
1523	1954–	Glossary of terms used in automatic controlling and regulating systems
1646	1979	Symbolic representation for process measurement control functions and instrumentation
1780	1960	Specification for Bourdon tube pressure and vacuum gauges
1794	1952	Chart ranges for temperature recording instruments
1843	1952	Colour code for twin compensating cables for thermocouples
1900	1976	Specification for secondary reference thermometers
1904	1984	Specification for industrial platinum resistance thermometer sensors
1986	1964	Design and dimensional features of measuring and control instruments for industrial processes
2082	1954	Code for disappearing-filament optical pyrometers
2475	1964	Octave and one third octave band-pass filters
2643	1955	Glossary of terms relating to the performance of measuring instruments
2740	1969	Simple smoke alarms and alarm metering devices

Number	*Date*	*Title*
2765	1969	Dimensions of temperature detecting elements and corresponding pockets
2811	1969	Smoke density indicators and recorders
3048	1958	Code for the continuous sampling and automatic analysis of flue gases. Indicators and recorders
3145	1978	Specification for laboratory pH meters
3166	1959	Thermographs (liquid-filled and vapour pressure types) for use within the temperature range −20 °F to 220 °F (−30 °C to 105 °C)
3231	1960	Thermographs (bimetallic type) for air temperatures within the range 0 °F to 140 °F (−20 °C to 60 °C)
3282	1969	Glossary of terms relating to gas chromatography
3383	1961	Normal equal-loudness contours for pure tones and normal threshold of hearing under free-field listening conditions
3403	1972	Indicating tachometers and speedometer systems for industrial, railway and marine use
3406	1961–	Methods for the determination of size distribution
3425	1966	Method for the measurement of noise emitted by motor vehicles
3512	1962	Method of evaluating the performance of pneumatic transmitters with 3 to 15 lb/in^2 (gauge) output
3527	1976–	Glossary of terms used in data processing
3593	1963	Recommendation on preferred frequencies for acoustical measurements
3680	1964–	Methods of measurement of liquid flow in open channels
3683	1977–	Glossary of terms used in non-destructive testing
3693	1964	Recommendations for the design of scales and indexes
3792	1964	Recommendations for the installation of automatic liquid level and temperature measuring instruments on storage tanks
3938	1973	Current transformers
3941	1975	Voltage transformers
3971	1980	Specification for image quality indicators for industrial radiography (including guidance on their use)
4094	1966	Recommendation for data on shielding from ionizing radiation
4137	1967	Guide to the selection of electrical equipment for use in Division 2 areas
4142	1967	Method of rating industrial noise affecting mixed residential and industrial areas
4161	1967–	Gas meters
4196	1981	Sound power levels of noise sources
4198	1967	Method for calculating loudness
4200	1967–	Guide on the reliability of electronic equipment and parts used therein
4308	1979	Specification for documentation to be supplied with electronic measuring apparatus
4359	1969–	Methods for determination of specific surface of powders
4509	1980	Method for evaluating transmitters for use in process control systems
4587	1970	Recommendations for the selection of apparatus and techniques for the analysis of gases by gas chromatography
4671	1971	Method of evaluating analogue chart recorders and indicators for use in process control systems
4683	1971	Electrical apparatus for explosive atmospheres
4727	1971–	Glossary of electrotechnical, power, telecommunication, electronics, lighting and colour terms
4743	1979	Specification for safety requirements for electronic measuring apparatus
4803	1983	Radiation safety of laser products and systems
4877	1972	Recommendations for general principles of nuclear reactor instrumentation
4889	1973	Method for specifying the performance of electronic measuring equipment
4892	1973	Guide to the measurement of thermal radiation by means of the thermopile radiometer
4937	1981	International thermocouple reference tables
5164	1975	Indirect acting electrical indicating and recording instruments and their accessories

Number	*Date*	*Title*
5233	1975	Glossary of terms used in metrology
5235	1975	Dial-type expansion thermometers
5308	1975	Instrumentation cables
5309	1976	Methods for sampling chemical products
5345	1976–	Code of practice for the selection, installation and maintenance of electrical apparatus for use in potentially explosive atmospheres (other than mining applications or explosive processing and manufacture)
5490	1977	Specification for degrees of protection provided by enclosures
5501	1977	Electrical apparatus for potentially explosive atmospheres
5677	1979	Method (precision) for pressure calibration of one-inch standard condenser microphones by the reciprocity technique
5678	1979	Method (simplified) for pressure calibration of one-inch condenser microphones by the reciprocity technique
5679	1979	Method (precision) for free-field calibration of one-inch standard condenser microphones by the reciprocity technique
5760	1979–	Reliability of systems, equipments and components
5781	1979	Measurement and calibration systems
5792	1980	Specification for electromagnetic flow-meters
5844	1980	Methods of measurement of fluid flow: estimation of uncertainty of a flow-rate measurement
5857	1980	Methods for measurement of fluid flow in closed conduits using tracers
5875	1980	Glossary of terms and symbols for measurement of fluid flow in closed conduits
5967	1980–	Operating conditions for industrial-process measurement and control equipment
5969	1981	Specification for sound level meters
6020	1981	Instruments for the detection of combustible gases
6169	1981	Methods for volumetric measurement of liquid hydrocarbons
6174	1982	Specification for differential pressure transmitters with electrical outputs
6175	1982	Specification for temperature transmitters with electrical outputs
6199	1981	Methods of measurement of liquid flow in closed conduits using weighing and volumetric methods
6739	1986	Code of Practice for Instrumentation in Process Control Systems: Installation design and practice

10 Master Address List

10.1 Private firms

1 **3D Digital Design & Development Ltd**
Duchess House
18-19 Warren Street
LONDON
W1P 3DB
Tel 01 387 7388 Tlx 8953742
MAN CON RES
26 31 33 36

2 **AB Controls & Technology Ltd**
Sanderson Street
SHEFFIELD
S9 2UA
Tel 0742 442424 Tlx 547700
MAN CON
1 3 5 24 26

3 **A G C O Unicell**
Lowick Close
Newby Road Industrial Estate
Hazel Grove
STOCKPORT
Cheshire SK7 5DA
Tel 061 456 5666 Tlx 668379
MAN
9

4 **A J B Associates (Electronics) Ltd**
54 High Street
WELLS
Somerset BA5 2SN
Tel 0749 74932 Tlx 444516
MAN
4 6 7 9 31

5 **A J Thermosensors Ltd**
Martlets Way
Goring-by-Sea
WORTHING
West Sussex BN12 4EG
Tel 0903 502471 Tlx 877510
MAN
12

6 **A M Lock & Co Ltd**
PO Box OL82
Neville Street
OLDHAM
Lancs OL9 6LF
Tel 061 624 0333 Tlx 669971
MAN RES
20 22

7 **A Macklow-Smith Ltd**
Watchmoor Road
off Moorlands Road
CAMBERLEY
Surrey GU15 3AH
Tel 0276 24459
MAN
2 3 7

8 **A T Controls**
Moss Farm
Nipe Lane
UPHOLLAND
Lancs WN8 9PY
Tel 0942 223570
MAN
1 3 5 9 12 31 35 36 40

9 **A V T Group**
Avtech House
Birdmall Lane
Cheadle Heath
STOCKPORT
Cheshire SK3 0XU
Tel 061 491 2222 Tlx 669028
MAN
41

10 **A W R Instruments**
Aughton Group
Interlock Business Centre
Knight Road
STROOD
Kent ME2 2AT
Tel 0634 74751 Tlx 966644
MAN
41

11 **Acam Instrumentation Ltd**
23 Thomas Street
NORTHAMPTON
Northants NN1 3EW
MAN CON RES
1 4 6 7 9 10

12 **Accuron Ltd**
34 Newnham Road
CAMBRIDGE
CB3 9EY
Tel 0223 461462
MAN CON RES
12 17 31 35

13 **Action Instruments Europe Inc**
St James Works
St Pancras
CHICHESTER
West Sussex PO19 4NN
Tel 0243 774022 Tlx 869331
MAN
12 33 34 35

14 **Actorset Instruments Ltd**
3 Ashcroft Court
Northfield Avenue
CAMBRIDGE
Cambs CB4 25N
Tel 0223 64346
MAN CON RES
27 31

15 **Adac Corp (Europe) Ltd**
PO Box 42
EPSOM
Surrey KT18 7SP
Tel 03727 42577 Tlx 918179
MAN
33 35 36

16 **Advance Bryans Instruments Ltd**
14/16 Wates Way
MITCHAM
Surrey
Tel 01 640 5624 Tlx 946097
MAN
35

17 **Advantec Systems Ltd**
21 Buryfield Road
SOLIHULL
W Midlands B91 2DS
Tel 021 743 4647
MAN
36

18 **Alan Cobham Engineering Ltd**
Holland Way
BLANDFORD FORUM
Dorset DT11 7BJ
Tel 0258 51441 Tlx 417254
CAL

19 **Alexander Cardew Ltd**
2 Studio Place
Kinnerton Street
Knightsbridge
LONDON
SW7
Tel 01 235 3785 Tlx 916078
MAN
1 2 8 17

20 **Allen Bennett Controls Ltd**
Sanderson Street
SHEFFIELD
South Yorks S9 2UA
Tel 0742 442424 Tlx 547700
MAN CON
3 5 11 19 24 39

21 **Allison Engineering Ltd**
10 High Street
BILLERICAY
Essex CM12 9BE
Tel 02774 59519 Tlx 995965

MAN CAL
1 13 16 32 39

22 **Alpeco Ltd**
Chinnor Road
Bledlow
PRINCES RISBOROUGH
Bucks HP17 9PH
Tel 08444 2666 Tlx 837192

MAN CAL REP CON
1

23 **Alpha Electronics Ltd**
Unit 5, Linstock Trading Estate
Linstock Way
Wigan Road
ATHERTON
Lancs M29 0QA
Tel 0942 873434

MAN
41

24 **Alpha Instrumentation Ltd**
29 Brunswick Street East
HOVE
East Sussex BN3 1AU
Tel 0273 775701

MAN
12 35 37

25 **Amot Controls Ltd**
Elton Works
Western Way
BURY ST EDMUNDS
Suffolk IP33 3SZ
Tel 0284 62222 Tlx 81283

MAN
1 4 5 6 9 10 12 35 36 39 40

26 **Anacon (Instruments) Ltd**
St Peters Road
MAIDENHEAD
Berkshire SL6 7QA
Tel 0628 39711 Tlx 847283

MAN CAL REP
8 13 15 16 17 19

27 **Analysis Automation Ltd**
Southfield House
Southfield Road
Eynsham
OXFORD
OX8 1JD
Tel 0865 881888 Tlx 837509

MAN REP CON RES
15 16 17 31 32 40

28 **Analytical Development Co Ltd**
Pindar Road
HODDESDON
Herts EN11 0AQ
Tel 0992 469638 Tlx 266952

MAN
16 27 32

29 **Ancom Ltd**
Devonshire Street
CHELTENHAM
Glos GL50 3LT
Tel 0242 513861

MAN CON RES
9 12 17 18 31 33 35 36

30 **Applied Automation UK Ltd**
Ashlyn House
Terrace Road North
Binfield
BRACKNELL
Berks RG12 5JA
Tel 0344 489575 Tlx 849732

MAN
13 16 17 32

31 **Applied Photophysics Ltd**
18-21 Corsham Street
LONDON
N1 6DR
Tel 01 251 4068 Tlx 263641

MAN RES
14 24 27

32 **Applied Research Laboratories Ltd**
Wingate House
Wingate Road
LUTON
Beds LU4 8PU
Tel 0582 573474 Tlx 82298

MAN
14 15

33 **Ashdown Process Control Ltd**
Unit 3
Huntsfield North Close
Shorncliffe Industrial Estate
FOLKESTONE
Kent CT20 3UH
Tel 0303 40691 Tlx 966355

MAN
1 5 9

34 **Ashridge Engineering Ltd**
47 Gates Green Road
WEST WICKHAM
Kent BR4 9DE
Tel 01 462 4481

MAN
1 5 9 10 17 28 35

35 **Aspen Electronics Ltd**
1/3 Kildare Close
Eastcote
RUISLIP
Middx HA4 9DE
Tel 01 868 1311 Tlx 8812727

MAN
18 19 25 33

36 **Aughton Hire**
Woodward Road
29 Kirkby Industrial Estate
Kirkby
LIVERPOOL
L33 7UZ
Tel 051 548 0000 Tlx 628681

MAN CON
1 6 9 10 12 16 17 18 23 35 37

37 **Auriema Ltd**
442 Bath Road
SLOUGH
Bucks SL1 6BB
Tel 06286 4353 Tlx 847155

MAN REP
1 4 5 7 8 9 10 11 12 13 15 16 17 21 32 33 34 35 39 41

38 **Automatic Systems Laboratories Ltd**
Breckland
Saxon Street
Linford Wood
MILTON KEYNES
Bucks MK14 6LD
Tel 0908 320666 Tlx 826349

MAN
3 12

39 **Auxitrol Ltd**
Auxitrol House
8 Chapel Grove
ADDLESTONE
Surrey KT15 1UG
Tel 0932 58111 Tlx 929822

MAN CON
1 5 9 12 27 37

40 **Avery Denison Ltd**
Moor Road
LEEDS
West Yorkshire LS10 2DE
Tel 0532 708011 Tlx 557526

MAN CAL REP
4 6 7 22 29

41 **B I C C Pyrotenax Ltd**
Hedgeley Road
HEBBURN
Tyne & Wear NE31 1XR
Tel 091 483 2244 Tlx 53573

MAN CON
1 4 9 12

42 **B I C C Vero Electronics Ltd**
Flanders Road
SOUTHAMPTON
Hants SO3 3LG

MAN
36

43 **B Rhodes & Son Ltd**
Danes Road
Crow Lane
ROMFORD
Essex EM7 0HR
Tel 0708 62333 Tlx 896084

MAN
1 2 35

44 **B S & B Safety Systems**
Gable House
Turnham Green Terrace
LONDON
W4 1QP
Tel 01 994 1083/4 Tlx 892924

MAN
9 12

45 **B W Instrument Services Ltd**
2 Batemans Lane
Wythall
BIRMINGHAM
B47 6NG
Tel 0564 822808

MAN CAL REP CON
25

46 **Bach-Simpson (UK) Ltd**
Trenant Estate
WADEBRIDGE
Cornwall PL27 6HD
Tel 020881 2031 Tlx 45451

MAN
12 18 23 31 35

47 **Bailey & Mackey Ltd**
Baltimore Road
BIRMINGHAM
West Midlands B42 1DE
Tel 021 357 5351 Tlx 339931

MAN
9 10

48 **Bailey Controls Co**
43-51 Wembley Hill Road
Middx HA9 8JD
Tel 01 900 1633 Tlx 8811318

MAN
1 5 9 12 35

49 **Barbie Engineering Ltd**
10 Beresford Avenue
TWICKENHAM
Middx TW1 2PZ
Tel 01 892 3431 Tlx 934999

MAN
3 5 12 18 33 35

50 **Bayham Ltd**
Daneshill West
BASINGSTOKE
Hants RG24 0PG
Tel 0256 464911 Tlx 858318

MAN
20 31 39

51 **Beaconsfield Instrument Co**
Council Hall
BEACONSFIELD
Bucks HP9 2PP
Tel 04946 77833 Tlx 83573

MAN CON
5 12 14 17

52 **Bedfont Technical Instruments Ltd**
Bedfont House
Holywell Lane
Upchurch
SITTINGBOURNE
Kent ME9 7HN
Tel 0634 375614 Tlx 261507

MAN
16 27 39

53 **Beka Associates Ltd**
PO Box 39
HITCHIN
Herts SG5 2DD
Tel 0462 38301

MAN
35

54 **Beresford Hartwell & Associates**
78 Manor Road
WALLINGTON
Surrey SM6 8RZ
Tel 01 647 1686 Tlx 946729

CON RES
42

55 **Bestobell Mobrey**
190-196 Bath Road
SLOUGH
Berks SL1 4DN
Tel 0753 34646 Tlx 878875-7

MAN CAL
1 5

56 **Biodata Ltd**
10 Stocks Street
MANCHESTER
M8 8QG
Tel 061 834 6688 Tlx 665608

MAN CON
1 4 9 12 27 33 34 36

57 **Bofors Electronics Ltd**
Murdock Road
BEDFORD
Beds MK41 7PQ
Tel 0234 49241 Tlx 826524

MAN
1 4 5 7 9

58 **Bowthorpe Controls**
124 Electric Avenue
Witton
BIRMINGHAM
B6 7DZ
Tel 021 328 0940 Tlx 338346

MAN
5 33 34 35

59 **Bradbury Controls Ltd**
1 Worcester Street
MONMOUTH
Gwent NP5 3DF
Tel 0600 4511 Tlx 497737

MAN CAL RES
9 10 39

60 **Brimac Automation Ltd**
Fountain Square
DISLEY
Cheshire SK12 2AB
Tel 0663 64595 Tlx 668165

MAN CON
1 5 9 10 12 35 36 37

61 **Bristol Babcock Ltd**
218 Purley Way
Brownfields
CROYDON
Surrey CR9 4HE
Tel 01 686 0400 Tlx 262335

MAN REP
1 5 9 10 12 13 33 34 35 36 37 40

62 **Bristol Industrial & Research Assoc Ltd**
PO Box 2
Portishead
BRISTOL
BS20 9JB
Tel 0272 847787 Tlx 444214

MAN CAL REP
1 9 11 28

63 **British Brown-Boveri Ltd**
Darby House
Lawn Central
TELFORD
Shropshire TF3 4JB
Tel 0952 502000 Tlx 35284

MAN
18 35

64 **British Fluidics and Controls Ltd**
Forest Road
Hainault
ILFORD
Essex 1G6 3HJ
Tel 01 500 3300

MAN
37

65 **British Rototherm Co Ltd, The**
Kenfig Industrial Estate
Margam
PORT TALBOT
West Glamorgan SA13 2PW
Tel 0656 740551 Tlx 497341

MAN
5 9 10 12 17 23 27 28 29 35 37 40

66 **Brooks and Sons (Instruments) Ltd**
1 Back West Avenue
Gosforth
NEWCASTLE UPON TYNE
Tyne and Wear NE3 4ES
Tel 091 285 2052

MAN CAL REP
5 6 12 17 18 23 26 28 31 35 36

67 **Brooks Instruments Division**
Emerson Electric UK Ltd
Brooksmeter House
Stuart Road Bredbury
STOCKPORT
Cheshire SK6 2SR
Tel 061 430 7100 Tlx 667393

MAN CAL REP
1 5 8

68 **Bruel & Kjaer (UK) Ltd**
92 Uxbridge Road
HARROW
Middx HA3 6BZ
Tel 01 954 2366 Tlx 934150

MAN CAL REP
4 6 7 12 16 17 19 23 27 35

69 **C F R Giesler Ltd**
Industrial Estate
Empson Street
Bromley-by-Bow
LONDON
E3 3LT
Tel 01 987 2161 Tlx 262636

MAN
6

70 **C G F Automation Ltd**
62 Valley Way
MARKET HARBOROUGH
Leics LE16 7PS
Tel 0858 62253 Tlx 342217

MAN CAL
1 5 35

71 **C N S Electronics Ltd**
61-63 Holmes Road
LONDON
NW5 3AL
Tel 01 485 1003 Tlx 299110

MAN
22 29 30

72 **Calex Instrumentation Ltd**
PO Box 2
Bassett Road
LEIGHTON BUZZARD
Beds LU7 7AG
Tel 0525 373178 Tlx 82454

MAN CAL REP CON
12

73 **Cambridge Consultants Ltd**
Science Park
Milton Road
CAMBRIDGE
Cambs CB4 4DW
Tel 0223 358855 Tlx 81481

CON RES
42

74 **Cambridge Instruments Ltd**
Industrial Divison
Oakfield Road
LONDON
SE20 8EW
Tel 01 659 2424 Tlx 262716

MAN
1 8 9 12 16 31 37

75 **Camille Bauer Controls Ltd**
Priest House
Priest Street
Cradley Heath
WARLEY
West Midlands B64 6JN
Tel 0384 638822 Tlx 334063

MAN
1 3 5 9 10 12 18 33 35 36 39

76 **Canberra Packard**
Brook House
14 Station Road
PANGBOURNE
Berkshire RG8 7DT

MAN REP
20 21

77 **Capelrig Ltd**
Norton Centre
Poynernook Road
ABERDEEN
AB1 2RW
Tel 0224 574384 Tlx 739169

MAN CON RES
31

78 **Carel Components Ltd**
24 Endeavour Way
Wimbledon Park
LONDON
SW19 8UH
Tel 01 946 9882 Tlx 928717

MAN
5 7 9 12 36

79 **Casella London Ltd**
Regent House
Britannia Walk
LONDON
N1 7ND
Tel 01 253 8581 Tlx 261641

MAN CAL REP CON
1 9 12 28 31 32 39 40

80 **Channel Electronics (Sussex) Ltd**
PO Box 58
SEAFORD
Sussex BN25 3JB
Tel 0323 894961 Tlx 877825

MAN
6 9 12 13 15 16 17 18 19 23 27 28 32 33 35

81 **Charles Austen Pumps Ltd**
100 Royston Road
Byfleet
WEYBRIDGE
Surrey KT14 7PB
Tel 09323 43224/5 Tlx 928368

MAN
10 32 37

82 **Charles M Christie**
Industrial Instrument Consultant
2 Winnington Road
Marple
STOCKPORT
Cheshire SK6 6PD
Tel 061 941 5571

CON
42

83 Charles Wells & Associates
The Old Coach House
Hopes Yard
UPPINGHAM
Leics LE15 9QQ
Tel 0572 821377

CON
1 5 37 39

84 Charnwood Instrumentation Services
Unit 11
The Ivanhoe Industrial Estate
Smisby Road
ASHBY-DE-LA-ZOUCH
Leics LE6 5UU
Tel 0530 415809 Tlx 341194

MAN CAL CON
9 10 12

85 Chell Instruments Ltd
Tudor House
Grammar School Road
NORTH WALSHAM
Norfolk NR28 9JH
Tel 0692 402488 Tlx 97198

MAN CAL REP CON RES
1 9 10 11 12 28 31 35 36 40

86 Chessell Ltd
Dominion Way
WORTHING
West Sussex BN 14 8QL
Tel 0903 205222 Tlx 877296

MAN
1 4 5 9 10 12 13 16 17 18 23 27 28 35 36

87 Clandon Scientific Ltd
Lysons Avenue
Ash Vale
ALDERSHOT
Hants GU12 5QR
Tel 0252 514711 Tlx 858210

MAN
2 6 12 13 17 22 27

88 Clifford & Snell Ltd
512 Purley Way
CROYDON
Surrey CRO 4NZ
Tel 01 681 3331 Tlx 946507

MAN
13 32 39

89 Coherent (UK) Ltd
Cambridge Science Park
Milton Road
CAMBRIDGE
Cambs CB4 4FR
Tel 0223 68501 Tlx 817466

MAN
3 11 13 14 16 19 22 24 27 28 35

90 Colston James Consultants
3 The Butts
WARWICK
Warks CV34 4SS
Tel 0926 493616

RES
1 16 31 36

91 Computing Techniques (MFG) Ltd
Brookers Road
BILLINGSHURST
West Sussex RH14 9RZ
Tel 040381 3171 Tlx 877479

MAN REP CON
5 18 29 30 31 40

92 Comtalk Communications Ltd
1 Carlyle Road
Manor Park
LONDON E12
Tel 01 478 8938

MAN
41

93 Concurrent Technologies Ltd
5th Floor
Fairfax house
Causton Road
COLCHESTER
Essex CO1 1RJ
Tel 0206 42996 Tlx 94012560

MAN
32 33 34 35 36

94 Condition Monitoring Ltd
Unit 2
Tavistock Estate
Ruscombe Lane
TWYFORD
Berks RG10 9NJ
Tel 0734 342636 Tlx 847151

MAN CON
3 6 22 36 40

95 Consolidated Ceramic Products UK
PO Box 145
SHEFFIELD
Yorks S12 2NT
Tel 0742 644471 Tlx 547201

MAN CAL REP CON
12 16 32 40

96 Contral Instrument Services
131F Church Walk
Denny
FALKIRK
Stirlingshire FK6 6HS
Tel 0324 826496 Tlx 265871

MAN CAL REP CON RES
9 12

97 Contraves Industrial Products Ltd
Times House
Station Approach
RUISLIP
Middx HA4 8LH
Tel 0895 630196 Tlx 935129

MAN
2

98 Control Transducers
North Lodge
25 Kimbolton Road
BEDFORD
Beds MK40 2NY
Tel 0234 217704 Tlx 825619

MAN
3 7 9 10 27 33 35

99 Controls & Automation Ltd
Bury Mead Road
HITCHIN
Herts SG5 1RT
Tel 0462 36161 Tlx 826495

MAN
12

100 Cook Variometers Ltd
PO Box 36
HIGH WYCOMBE
Bucks HP13 6BB
Tel 0494 33171

MAN
1 31

101 Cordforth Control Engineering
159/160 High Street
STOCKTON ON TEES
Cleveland TS18 1PL
Tel 0642 670510 Tlx 58448

CON
31

102 Coutech, Products Division
Parametric Ltd
Jensen Court
Astmoor Industrial Estate
RUNCORN
Cheshire WA7 1SQ
Tel 09285 67321 Tlx 628372

MAN CON RES
12 31 33 34 35 36

103 Crane Electronics Ltd
Station Road
Stoke Golding
NUNEATON
Warks CV13 6HA
Tel 0455 212157 Tlx 312242

MAN
3 7

104 Crompton Instruments
Freebournes Road
WITHAM
Essex CM8 3AH
Tel 0376 512601 Tlx 987961

MAN
1 2 3 4 5 6 7 8 9 10 12 17 18 28 31 32 33 34 35 36

105 Crowcon Instruments Ltd
Temple Road
Cowley
OXFORD
Oxon OX4 2EL
Tel 0865 776707 Tlx 837688

MAN CAL REP
16

106 D J Birchall Ltd
102 Bath Road
CHELTENHAM
Glos GL53 7JX
Tel 0242 518588 Tlx 43531

MAN CAL CON
6 9 12 23 31 39 40

107 D M G Control Systems
19 Bridgewater Road
Hertburn Industrial Estate
WASHINGTON
Tyne & Wear NE37 2SG
Tel 091 417 9888 Tlx 537286

MAN CAL REP CON
1 9 12 17 18 23 31 35 37 40

108 Daco Scientific Ltd
Vulcan House
Calleva Industrial Park
ALDERMASTON
Berks RG7 4QW
Tel 07356 77311 Tlx 847208

MAN CAL
2 3 6 9 10 30

109 Dage (GB) Ltd
Rabans Lane
AYLESBURY
Bucks HP19 3RG
Tel 0296 33200 Tlx 847208

MAN
7 22

110 Dale Electronics Ltd
Dale House
Wharf Road
Frimley Green
CAMBERLEY
Surrey GU16 6LF
Tel 0252 835094 Tlx 858663

MAN
18

111 Danesbury Instruments
22 Parkway
WELWYN GARDEN CITY
Herts AL8 6HG
Tel 07073 38623 Tlx 825633

MAN
41

112 Danfoss Flowmetering Ltd
Magflo Works
Bowbridge
STROUD
Glos GL5 2NN
Tel 04536 71631 Tlx 43692

MAN
1 5

113 Daniel Industries Ltd
Library House
The Green
Datchet
SLOUGH
Berks SL3 9AV
Tel 0753 48587 Tlx 847223

MAN
1 9 12 16 33

114 Dantec Electronics Ltd
Techno House
Redcliffe Way
BRISTOL
BS1 6NU
Tel 0272 291436 Tlx 449695

MAN
1 6 11 12

115 Dart Instrumentation Ltd
Unit 8
Milland Road Industrial Estate
NEATH
W Glam SA11 1NJ
Tel 0639 54281 Tlx 48690

MAN CAL REP CON RES
9 40

116 Data Translation Ltd
Unit 13
The Business Centre
Molly Millars Lane
WOKINGHAM
Berks RG11 2QZ
Tel 0734 793838

MAN
36

117 Datron Instruments
Hurricane Way
Norwich Airport
NORWICH
Norfolk NR6 6JB
Tel 0603 404824 Tlx 975173

MAN
18 31

118 David Bishop Instruments Ltd
Station Road
HEATHFIELD
East Sussex TN21 8DR
Tel 04352 65773889 Tlx 95137

MAN
12 16 31 35

119 Davis (Decade) Ltd
30 Spring Lane
Erdington
BIRMINGHAM
B24 9BX
Tel 021 384 8208

MAN CON
4 7 22 31

120 Dean Microsystems
7 Horseshoe Park
Pangbourne
READING
Berks RG8 7JW
Tel 07357 5155

MAN
36

121 Delta Controls Ltd
Island Farm Avenue
EAST MOLESEY
Surrey KT8 0UZ
Tel 01 941 5166 Tlx 27109

MAN
1 5 9 10 12 39

122 Derrick A Patient
114 Lower Ham Road
KINGSTON UPON THAMES
Surrey KT2 5BD
Tel 01 546 9875

CON
42

123 Design Automation Ltd
September House
Cox Green Lane
MAIDENHEAD
Berks SL6 3EL
Tel 0628 24929 Tlx

MAN CON
31 36 39

124 Detectronic Ltd
Unit 2c Limefield
Leamington Road
BLACKBURN
Lancs BB2 6ER
Tel 0254 63416 Tlx 635091

MAN
1 5 28

125 DI-AN Micro Systems Ltd
Mersey House
Heaton Mersey
STOCKPORT
Cheshire SK4 3EA
Tel 061 442 9768 Tlx 669592

MAN
12 18 33 34 35 36

126 Digital Equipment Co Ltd
PO Box 110
Worton Grange
READING
Berks RG2 0TR
Tel 0734 868711 Tlx 848327/8

MAN
33 36

127 Digitron Instrumentation Ltd
Mead Lane
HERTFORD
Herts SG13 7AW
Tel 0992 57441 Tlx 81517

MAN
4 5 9 12 35

128 Draeger Ltd
The Willows
Mark Road
HEMEL HEMPSTEAD
Herts HP2 7BW
Tel 0442 3542 Tlx 826093

MAN CAL REP CON
16 39

129 Druck Ltd
Fir Tree Lane
GROBY
Leics LE6 0FH
Tel 0533 878551 Tlx 341743

MAN CAL
5 9 10 28 35

130 Duval Electronics Ltd
112 Walton Street
OXFORD
Oxon OX2 6AJ
Tel 0865 512464

REP
12 18

131 E A O UK Ltd
Hook Rise Business & Industrial Estate
Unit 2
225 Hook Rise South
SURBITON
Surrey KT6 7LD
Tel 01 397 7041/4 Tlx 928574

MAN RES
9 10 18 33 34 35

132 E E L Ltd
Kings Building
Castle Street
EAST COWES
Isle of Wight PO32 6RH
Tel 0983 291515 Tlx 869180

MAN
7

133 E G & G Instruments Ltd
Doncastle House
Doncastle Road
BRACKNELL
Berks RG12 4PG
Tel 0344 423931 Tlx 847164

MAN REP
14 20 21 22 33 35 36 38 39

134 E V O Measurement & Control Ltd
Unit 111E Hartlebury Trading Estate
Hartlebury
KIDDERMINSTER
Worcs DY10 4JB
Tel 0299 250018 Tlx 339153

MAN CAL REP CON RES
5 9 10 12 18 22 34 35 36 37

135 Eagle Controls International Ltd
PO Box 42
Knap Close
LETCHWORTH
Herts SG6 1HQ
Tel 04626 70566 Tlx 825845

MAN
12

136 Edwards High Vacuum
Manor Royal
CRAWLEY
West Sussex RH10 2LW
Tel 0293 28844 Tlx 87123

MAN
1 9 10 16 19 31 36

137 Elcometer Instruments Ltd
Edge Lane
Droylsden
MANCHESTER
M35 6BU
Tel 061 370 7611 Tlx 668960

MAN REP
12 22 26

138 Electronic Design & Development
54 Burnage Lane
Burnage
MANCHESTER
M19 2NL
Tel 061 225 6783

MAN RES
31

139 Electronic Sensors International Ltd
Guildford Road
FARNHAM
Surrey GU9 9PZ
Tel 0252 714666 Tlx 858623

MAN CAL
5 9 10 33

140 Electroplan Limited
PO Box 19
Orchard Road
ROYSTON
Herts SG8 5HH
Tel 0763 41171 Tlx 81337

REP
9 12 17 33 35 36

141 Electroserv (Inst) Ltd
Athey Street Works
Athey Street
MACCLESFIELD
Cheshire SK11 8EE
Tel 0625 618526

MAN
12 18 41

142 Endress & Hauser Ltd
Ledson Road
Wythenshawe
MANCHESTER
M23 9PH
Tel 061 998 0321 Tlx 668501

MAN
1 5 9 10 12 13 15 17 20 21 32 35 39

143 Energy Technology & Control Ltd
25 North Street
LEWES
East Sussex BN7 2PE
Tel 0273 47601 Tlx 87110

MAN
16

144 Engineering & Scientific Equipment Ltd
22-26 Mount Pleasant
Alperton
WEMBLEY
Middx HA0 1TU
Tel 01 903 4721 Tlx 261577

MAN CON
3 22 26 31 33 35 36

145 Entran Ltd
Sales and Technical Centre
5 Albert Road
CROWTHORNE
Berks RG11 7LT
Tel 0344 778848 Tlx 847422

MAN
4 6 7 9 12 23 33 35

146 Epic Products Ltd
Unit F2
Broadway Industrial Area
SALFORD
M5 2UQ
Tel 061 872 1487 Tlx 665781

MAN
1 5 32

147 Erlebach Technology Ltd
Dixies
High Street
ASHWELL
Herts SG7 5NT
Tel 046274 2881

MAN RES
18 22 36

148 Erwin Sick Optic-Electronics Ltd
39 Hedley Road
ST ALBANS
Herts AL1 5BN
Tel 0727 31121 Tlx 263631

MAN
1 26 28 39

149 Eur-Control GB Ltd
26 Breakfield
COULSDON
Surrey CR3 2XW
Tel 01 668 5287 Tlx 24904

MAN CON
1 2 8 19 21

150 Eurogauge Co Ltd
Imberhorne Lane
EAST GRINSTEAD
West Sussex RH19 1RF
Tel 0342 23641 Tlx 95606

MAN
5 9

151 Eurostem Systems Ltd
Northern Sales/Service
51A East Parade
HARROGATE
North Yorkshire HG1 5LQ
Tel 0423 524488 Tlx 57701

MAN
3 7 17 19 20 21 24

152 Eurotherm Ltd
Faraday Close
Durrington
WORTHING
West Sussex BN13 3PL
Tel 0903 68500 Tlx 87114

MAN REP CON
1 9 12 17 36

153 Expo Safety Systems
Summer Road
THAMES DITTON
Surrey KT7 0RH
Tel 01 398 8011 Tlx 8954824

MAN
36 39

154 F Bamford (Instruments) Ltd
Ajax Works
Whitehill
STOCKPORT
Cheshire SK4 1NT
Tel 061 480 6507 Tlx 668518

MAN
1 5 9 10 12

155 Farnell Instruments Ltd
Sandbeck Way
WETHERBY
West Yorkshire LS22 4DH
Tel 0937 61961 Tlx 557294

MAN
18 35 36

156 Federal Gauges Ltd
Brick Knoll Park
Ashley Road
ST ALBANS
Herts AL1 5PL
Tel 0727 65622 Tlx 21391

MAN
3 5 26

157 Feedback Instruments Ltd
Park Road
CROWBOROUGH
Sussex TN6 2QR
Tel 95255

MAN
18

158 Ferranti PLC
Microelectronics Group
Ferry Road
EDINBURGH
Lothian EH5 2XS
Tel 031 3322411 Tlx 72141

MAN
36

159 Field Analytical Co Ltd
PO Box 113
WEYBRIDGE
Surrey KT13 9UZJ
Tel 0932 43311 Tlx 934755

MAN CON
13 16

160 Field Electronics Ltd
Gill House
Conway Street
HOVE
East Sussex BN3 3LW
Tel 0273 729361 Tlx 877159

MAN
12 34 35 36

161 Fieldtech Heathrow Ltd
Huntavia House
420 Bath Road
Longford
WEST DRAYTON
Middx UB7 0LL
Tel 23734

CAL REP
18 42

162 Flow Measurement & Automation (F M A) Ltd
Premier House
Victoria Way
WOKING
Surrey GU21 1DG
Tel 04862 69011 Tlx 859440

MAN CAL REP CON
1

163 Flowline Systems Ltd
183 Ballards Lane
Finchley
LONDON
N3 1LL
Tel 01 349 1742 Tlx 24103

MAN CAL
1 5 8 13 32

164 Fluid Data (UK) Ltd
20 Bourne Industrial Park
Bourne Road
CRAYFORD
Kent DA1 4BZ
Tel 0322 528125 Tlx 8951085

MAN REP CON
2 8 9 12 13 15 16 17 19 21 31 32

165 Foxboro Great Britain Ltd
REDHILL
Surrey RH1 2HL
Tel 0737 65000 Tlx 892852

MAN
1 5 8 9 10 12 13 15 16 34 35 37 40

166 Furnace Instruments Ltd
Pyro Works
Harwood Street
SHEFFIELD
S2 4SE
Tel 0742 731608 Tlx 547676

MAN CAL
12

167 Furness Controls Ltd
Beeching Road
BEXHILL
East Sussex TN39 3LJ
Tel 0424 210316 Tlx 957012

MAN CAL REP CON
1 9

168 Furse Engineering
Wilford Road
NOTTINGHAM
Notts NG2 1EB
Tel 0602 863471 Tlx 377065

MAN
12

169 Fylde Electronic Laboratories Ltd
49/51 Fylde Road
PRESTON
Lancs PR1 2XQ
Tel 0772 57560

MAN RES
1 4 6 7 9 12 22 23 36

170 G E C Measurements (Leicester)
Scudmore Road
New Parks
LEICESTER
Leics LE3 1UF
Tel 0538 871491 Tlx 34552

MAN
33 34

171 G E C Measurements (Stafford)
St Leonards Works
STAFFORD
Staffs ST17 4LX
Tel 0785 223251 Tlx 36240

MAN CAL
18 35

172 G H Zeal Ltd
Lombard Road
Merton
LONDON
SW19 3UU
Tel 01 542 2283 Tlx 929519

MAN
2 8 9 10 12 17 34

173 G S M Syntel Ltd
Queens Mill Road
HUDDERSFIELD
HD1 5PG
Tel 0484 535101

MAN
36

174 G W Thornton & Sons Ltd
Grether House
Crown Royal Industrial Park
Shawcross Street
STOCKPORT
Cheshire SK1 3HB
Tel 061 477 1010 Tlx 669937

MAN CAL REP
1 17 20 21

175 Gas Measurement Instruments Ltd
Inchinnan Estate
RENFREW
PA4 9RG
Tel 041 812 3211 Tlx 779748

MAN CAL REP CON
16 27

176 Gauges Bourdon (GB) Ltd
Bourdon House
Middleton Industrial Estate
GUILDFORD
Surrey GU2 5XW
Tel 0483 67776 Tlx 859410

MAN
9 10 12 28

177 Gaydon Technology
Gaydon Proving Ground
Banbury Road
Lighthorne
WARWICK
Warks CV35 0BL
Tel 0926 641111 Tlx 3131

MAN CAL REP CON RES
3 4 6 22 33

178 GEC Reliance Ltd
Turnells Mill Lane
WELLINGBOROUGH
Northants NN8 2RB
Tel 0933 225000 Tlx 311492

MAN
41

179 General Monitors Ireland Ltd
Queens Avenue
Hurdsfield Industrial Estate
MACCLESFIELD
Cheshire SK10 2BN
Tel 0625 619583 Tlx 669443

MAN CAL REP CON RES
12 16

180 Gervase Instruments Ltd
Britannia Works
Littlemead Industrial Estate
CRANLEIGH
Surrey GU6 8ND
Tel 0483 275566 Tlx 859473

MAN CAL REP CON RES
1 9 12 31 33 35 40

181 Gestra (UK) Ltd
Bancroft Court
HITCHIN
Herts SG5 1PH
Tel 0462 31681 Tlx 825741

MAN
5 9 12 19

182 Gow-Mac Instrument Co (UK) Ltd
PO Box G13
GILLINGHAM
Kent ME7 4MA
Tel 0634 57661 Tlx 965181

MAN
16

183 Gradwell Associates
84 Clifton Road
RUNCORN
Cheshire WA7 4TD
Tel 09285 73675

MAN CON
13 15 32

184 Gulton Ltd - West Division
The Hyde
BRIGHTON
East Sussex BN2 4JU
Tel 0273 606271 Tlx 87172

MAN REP CON
9 12 31 36 40

185 H A Wainwright & Co Ltd
Hawco House
Cathedral Hill Industrial Estate
GUILDFORD
Surrey GU2 5YB
Tel 0483 60606 Tlx 859364

MAN
9 10 12 18 33 34 35 36

186 H E S Group Ltd
Bentley Way
DAVENTRY
Northants NN11 5PH
Tel 0327 77511 Tlx 311903

MAN
1 5 9 12 28 35

187 H N L Instruments & Controls Ltd
Dukesway Teesside Industrial Estate
Thornaby
STOCKTON ON TEES
Cleveland TS17 9LT
Tel 0642 765553 Tlx 587148

MAN
1 5 8 9 10 12 31 36 37 39

188 H Tinsley and Co Ltd
61 Imperial Way
CROYDON
Surrey CR0 4RR
Tel 01 681 8431 Tlx 8952453

MAN CAL REP RES
4 12 18 28

189 H Tinsley Strain Measurements Ltd
61 Imperial Way
CROYDON
Surrey CR0 4RR
Tel 01 651 8431 Tlx 8952453

MAN CAL REP RES
4

190 H W Wallace & Co Ltd
172 St James's Road
CROYDON
Surrey CR9 2HR
Tel 01-686 4954 Tlx 946300

MAN CAL REP
2 3 4 8 22

191 Haenni Limited
Invincible Road
FARNBOROUGH
Hants GU14 7QU
Tel 0252 515151 Tlx 858093

MAN
5 6 9 28

192 Hanson Research Limited
6-10 Swan Street
EYNSHAM
Oxon OX8 1HU
Tel 0865 880692 Tlx 83138

MAN CON RES
15 30

193 Hartley Measurements Ltd
Unit 4 Bear Court
Daneshill East
BASINGSTOKE
Hants RG24 0QT
Tel 0256 56695 Tlx 858733

REP CON
31 42

194 Hartmann & Braun (UK) Ltd
Moulton Park
NORTHAMPTON
Northants NN3 1TF
Tel 0604 46311 Tlx 311056

MAN CON RES
1 5 9 10 12 1315 16 18 20 31 35 36 40

195 Haven Automation Ltd
Cwmdu Industrial Estate
Gendros
SWANSEA
W Glam SA5 5LQ
Tel 0792 588722 Tlx 48479

MAN
9 10 12

196 Hawco Ltd
Hawco House
Cathedral Hill Industrial Estate
GUILDFORD
Surrey GU2 5YB
Tel 0483 60606 Tlx 859364

MAN CAL REP
9 10 12 18 32 33 36

197 Hawke Systems
Newlands Drive
Poyle
SLOUGH
Berks SL3 0DX

MAN
36

198 Hawker Electronics Ltd
250 Melchett Road
Kings Norton
BIRMINGHAM
B30 3HP
Tel 021 459 8911 Tlx 336579

MAN
1 5

199 Hawker Siddley Dynamics Engineering Ltd
Bridge Road East
WELWYN GARDEN CITY
Herts AL7 1LR
Tel 0707 331299 Tlx 262079

MAN CAL REP CON RES
6 32 35

200 Heap Laverack and Partners
44 Wallington Square
WALLINGTON
Surrey SM6 8RY
Tel 01 647 1025

CON
31 33 34 38 39

201 Heme International Ltd
Pimbo Road
West Pimbo
SKELMERSDALE
Lancs WN8 9PD
Tel 0695 20535 Tlx 629792

MAN CAL REP RES
1 18 22

202 Henry Williams Ltd
Dosworth Street
DARLINGTON
Durham DL1 2NJ
Tel 0325 462722 Tlx 58421

MAN CON RES
1 5 33 34 35 36 40

203 Hewlett Packard Ltd
Eskdale Road
Winnersh Triangle
WOKINGHAM
Berks RG11 5DZ
Tel 0734 696622 Tlx 848884

MAN
36

204 Hexagon Technology Ltd
2 West End
Weston Turville
AYLESBURY
Bucks HP22 5TT
Tel 029 661 2310 Tlx 826715

CON
36 39

205 High Technology Electronics Ltd
303-305 Portswood Road
SOUTHAMPTON
Hants SO2 1LD
Tel 0703 581555

MAN
36

206 Hilger Analytical Ltd
Westwood
MARGATE
Kent CT9 4JL
Tel 0843 225131 Tlx 96252

MAN
14 21 22

207 Hillcrest Engineering Instrumentation Ltd
Upperhulme
LEEK
Staffs ST13 8TY
Tel 053834 693688 Tlx 87616

MAN REP
12 31

208 Hinton, James R B
3 Copyhold Lane
Gt Bedwyn
MARLBOROUGH
Wilts SN8 3LR

CON
1 5 9 12 31 37 38 39 40

209 Holledge (Instruments) Ltd
Sandy Lane
CRAWLEY DOWN
Sussex RH10 4HS
Tel 0342 716411 Tlx 95737

MAN
5 8 9 36 37 40

210 Honeywell Control Systems Ltd
Charles Square
BRACKNELL
Berks RG12 1EB
Tel 0344 424555 Tlx 847064

MAN CAL REP CON RES
1 5 8 9 10 12 18 22 31 33 34 35 37 40 41

211 Hopkinsons Ltd
Britannia Works
PO Box B27
HUDDERSFIELD
West Yorks HD2 2UR
Tel 0484 22171 Tlx 51682

MAN CAL REP CON
5 9

212 Hottinger Baldwin Messtechnik
c/o Schenk Ltd
Station Approach
BICESTER
Oxon OX6 7B7
Tel 0869 243351 Tlx 837669

MAN CAL REP CON
3 4 5 6 7 9 31 34 35 36

213 Hughes Aircraft Systems
Clive House
12-18 Queens Road
WEYBRIDGE
Surrey KT13 9XD
Tel 0932 57855 Tlx 929727

MAN REP
12 19 22 24 27 35

214 Hurnsound Ltd
12 Mons Avenue
BILLERICAY
Essex CM11 2HG
Tel 02774 52296

CON
42

215 Huskins Analyser Engineering Ltd
64b St Swithin Street
ABERDEEN
Scotland AB1 6XJ
Tel 0224 324650

MAN CON RES
8 13 16 17 18 19 21 32 38 39

216 Hytec Electronics Ltd
Sales and Service
94 Regent Road
LEICESTER
Leics LE1 7DJ
Tel 0533 542428

MAN
1 4 12 14 22 31 33 34 35 36 40

217 I C I Physics & Radioisotope Services
PO Box 1
Billingham
STOCKTON ON TEES
Cleveland TS23 1LD
Tel 0642 522555/523260 Tlx 587443

MAN
5 8 20 21 39

218 I D E C Systems Ltd
Yarm Road
STOCKTON ON TEES
Cleveland TS18 3RP
Tel 0642 677333 Tlx 58161

MAN CON RES
20 21 31 33 34 35 36

219 I P L
The Grove Trading Estate
DORCHESTER
Dorset DT1 1SY
Tel 0305 63673 Tlx 41166

MAN
3 26 36

220 I R D Mechanalysis (UK) Ltd
Bumpers Lane
Sealand Industrial Estate
CHESTER
CH1 4LT

MAN CAL REP CON
6

221 I S A Controls Ltd
Dale Road Industrial Estate
SHILDON
Co Durham DL4 2QZ
Tel 0388 773065 Tlx 587188

MAN
1 12

222 Infrared Engineering Ltd
Galliford Road
The Causeway
MALDON
Essex CM9 7XD
Tel 0621 52244 Tlx 995266

MAN
3 12 14 17 33

223 Instamec
21 Susan Grove
MORETON
Merseyside L46 OTZ
Tel 051 678 3965

MAN
31 36 40

224 Instron Ltd
Coronation Road
HIGH WYCOMBE
Bucks HP12 3SY
Tel 0494 33333 Tlx 83222

MAN CAL
4 7 22 31

225 Instrument Centre Ltd, The
53 Fairfax Road
NEWPORT
Gwent NP9 0HR
Tel 0633 280566 Tlx 498352

CAL REP
9 12 18 23 35 39

226 Instrument Construction & Maintenance (ICAM) Ltd
Dock Road Industrial Estate
Deeside
CONNAH'S QUAY
Clwyd
Tel 0244 817143/819400

MAN CON RES
31 32 36 37 39 40

227 Instrument Rentals (UK) Ltd
Dorcan House
Meadfield Road
LANGLEY
Slough SL3 8AL
Tel 0753 44878 Tlx 935371

MAN
41

228 Instrument Services (Northern)
14-16 Railway Avenue
Woodside Lairage
BIRKENHEAD
Merseyside L41 1EH
Tel 051 647 4814/5

MAN REP
31 37 40 41

229 Integrated Electronic Systems
Colchester Factory Estate
CARDIFF
S Glam CF3 7AP
Tel 0222 493475

MAN CON
31 36

230 Integrated Measurement Systems Ltd
Unit 306 Solent Business Centre
Millbroook Road West
SOUTHAMPTON
Hants SO1 0HW
Tel 0703 771143 Tlx 477575

MAN CON RES
31 32 33 34 36

231 Intel Corporation (UK) Ltd
Pipers Way
SWINDON
Wilts SN3 1RJ
Tel 0793 488388 Tlx 444447

MAN
36

232 International Controls Corp Ltd
PO Box 20
Burrell Way
THETFORD
Norfolk IP24 3QZ
Tel 0842 64151 Tlx 817710

MAN
1 9 12

233 International Recorders Ltd
92 High Street
BERKHAMSTED
Herts HP4 2BL
Tel 04427 5959/74070 Tlx 825946

MAN
18 35

234 International Robomation/ Intelligence Ltd
Stowe House
1688 High Street
KNOWLE
West Midlands B93 0ND
Tel 05645 2054 Tlx 333234

MAN CAL CON RES
3 19 24 26

235 Ircon Ltd
27 Essex Road
DARTFORD
Kent DA1 2AU
Tel 0322 77531 Tlx 896197

MAN CAL
12

236 Irwin Desman Ltd
294 Purley Way
CROYDON
Surrey CR9 4QL
Tel 01 686 6441 Tlx 946422

MAN RES
15 18 31 36 39

237 Isotope Measuring Systems Ltd
22 Quarry Park Close
Moulton Park Industrial Estate
NORTHAMPTON
Northants NN3 1QB
Tel 0604 42107/8 Tlx 312653

MAN
3 20 21

238 J W Instruments
38 High Green Road
NORMANTON
West Yorkshire WF6 2LG
Tel 0924 891049

MAN REP
12 18 35

239 Jackson Instruments & Controls Co Ltd
101 Knighton Fields Road West
LEICESTER
Leics LE2 6LH

MAN
1 9 18 39

240 James Hugh Group Ltd
150-152 West End Lane
LONDON
NW6 1SD
Tel 01 328 3121 Tlx 299170

MAN CAL REP
5 9 12 39

241 Jenway Ltd
Gransmore Green
Felsted
DUNMOW
Essex CM6 3LB
Tel 0371 820122 Tlx 817766

MAN CAL REP CON
12 13 15 27

242 Jiskoot Autocontrol Ltd
85 Goods Station Road
TUNBRIDGE WELLS
Kent TN1 2DJ
Tel 0892 2291 Tlx 95223

MAN
1 31 32

243 John Firth Instruments Ltd
Beech House
Hob Hey Lane
Culcheth
WARRINGTON
Cheshire WA3 4NJ
Tel 635091

MAN CAL REP
9 12 18 39

244 Johnson Control Systems Ltd
PO Box 79 Stonehill Green
Westlea Down
SWINDON
Wilts SN5 7DD
Tel 0793 26141 Tlx 444459

MAN
9 12 17 37

245 Jordan Kent Metering Systems Ltd
Arnolds Field Industrial Estate
Wickwar
WOTTON UNDER EDGE
Glos GL12 8ND
Tel 0454 24556 Tlx 43290

CON
1

246 K C Controls Ltd
83 Bell Street
REIGATE
Surrey RH2 7YU
Tel 07372 22931 Tlx 935212

CON
1 5 8 9 10 12 15 17 35 36 37

247 K Controls Ltd
Stone Close
Horton Road
WEST DRAYTON
Middx UB7 8JU
Tel 0895 449601

MAN
1 9 12 36 37

248 K D G Instruments Ltd
Crompton Way
CRAWLEY
West Sussex RH10 2Y4
Tel 0293 25151 Tlx 87235

MAN CAL REP CON RES
1 2 5 9 10 16 31 34 40

249 K D G Maxseal
Wood Road
Kingswood
BRISTOL
BS15 2DU
Tel 0272 606464 Tlx 449462

MAN
37

250 Kane-May Ltd
Swallowfield
WELWYN GARDEN CITY
Herts AL7 1JP
Tel 0707 331051 Tlx 25724

MAN
9 11 12 15 16 17

251 Kelsey Instruments
University of Warwick Science Park
COVENTRY
Warks CV4 7EZ
Tel 0203 471321

MAN CAL REP CON
3 4 7 9 31 35 36

252 Kent Industrial Measurements Ltd (Chertsey)
Analytical Instruments
Hanworth Lane
CHERTSEY
Surrey KT16 9LF
Tel 09328 62671 Tlx 264022

MAN
13 32

253 Kent Industrial Measurements Ltd (Stonehouse)
Oldends Lane
STONEHOUSE
Glos GL10 3TA
Tel 045 382 6661 Tlx 43127

MAN CAL REP
1 14 15 16 35

254 Kistler Instruments Ltd
Whiteoaks
The Grove
HARTLEY WINTNEY
Hants RG27 8RN
Tel 025 1263555 Tlx 858545

MAN REP CON
6 7 9

255 Kolectric Ltd
Brent Road
SOUTHALL
Middx UB2 5NH
Tel 01 574 6002 Tlx 935459

MAN
22 29

256 Krohne Measurement and Control Ltd
Osyth Close
Brackmills
NORTHAMPTON
Northants NN4 0ES
Tel 0604 66144 Tlx 311544

MAN
1 2 5 8 9 21

257 L D C/Milton Roy
Milton Roy House
Diamond Way
Stone Business Park
STONE
Staffs ST15 0HH

MAN
13 14 15 33

258 Labcon Ltd
24 Northfield Way
Aycliffe Industrial Estate
NEWTON AYCLIFFE
Co Durham DL5 6EJ

MAN CON RES
1 4 5 9 11 12 13 15 16 18 19 25 27 31 32 33 34 35 36 37 39 40

259 Labfacility Ltd
26 Tudor Road
HAMPTON
Middx TW12 2NQ
Tel 01 941 4849 Tlx 8952151

MAN
12 13 15 33 36

260 Land Combustion Ltd
Stubley Lane
Dronfield
SHEFFIELD
S18 6NQ
Tel 0246 417691 Tlx 547360

MAN CAL REP CON
12 15 16

261 Land Infrared Ltd
Dronfield
SHEFFIELD
S18 6DJ
Tel 0246 417691 Tlx 54457

MAN CAL
12

262 Land Pyrometers Ltd
Dronfield
SHEFFIELD
South Yorks S18 6DJ
Tel 0246 417691 Tlx 54102

MAN CAL REP CON RES
12 32

263 Lee Dickens Ltd
Rushton Road
Desborough
KETTERING
Northants NN14 2QW
Tel 0536 760156 Tlx 341227

MAN
12 17 31 34 36

264 Lee Integer Ltd
31 Commercial Road
KETTERING
Northants NN16 8DQ
Tel 0536 511010 Tlx 295141

MAN CAL REP CON RES
12 17 31

265 Leeds & Northrup Ltd
Whardale Road
Tyseley
BIRMINGHAM
B11 2DJ
Tel 021 706 6171 Tlx 336577

MAN CAL REP CON
1 5 9 11 12 13 16 31 32 33 35 40

266 Levell Electronics Ltd
Moxon Street
BARNET
Herts EN5 5SD
Tel 01 449 5028

MAN
12 18

267 Leybold-Heraeus Limited
Waterside Way
Plough Lane
LONDON
SW17 7AB
Tel 01 947 9744 Tlx 896430

MAN
10 16 19 26 31 40

268 Lintronic Systems Ltd
54/58 Bartholomew Close
LONDON
EC1A 7HB
Tel 01 606 0791/2 Tlx 888657

MAN CON RES
6 9 10 12 18 28 31 33 34 35 36 40

269 Lion Laboratories Ltd
Ty Verlon Industrial Estate
BARRY
South Glam CF6 3BE
Tel 0446 744244 Tlx 498289

MAN
15 16 27

270 Livingston Hire Ltd
The Rental Centre
99 Waldegrave Road
TEDDINGTON
Middlesex TW11 8LL
Tel 01 977 8866 Tlx 23920

MAN
41

271 Lloyd Instruments
Brook Avenue
Warsash
SOUTHAMPTON
Hants SO3 6HP

MAN CAL REP CON RES
7 9 14 29 35

272 Loughborough Projects Limited
New Campus
Ashby Road
LOUGHBOROUGH
Leics LE11 3TF
Tel 0509 262042

MAN RES
4 7 11 27 31 34

273 Lucas Dawe Ultrasonics Ltd
Concord Road
Western Avenue
LONDON
W3 0SD
Tel 01 992 6751 Tlx 934848

MAN
1 4 6 9 23

274 Ludlam Sysco Ltd
Broadway
Market Lavington
DEVIZES
Wiltshire SN10 5RQ
Tel 0380 818411 Tlx 44649

MAN CON
5 8 13 16 27 31 32 40

275 Luke Martyn & Co
1 Sandyhill Park
Drumbeg
Dunmurry
BELFAST
BT17 9LS
Tel 0232 612141

MAN
1 12 40

276 Lurmark Ltd
Longstanton
CAMBRIDGE
Cambs CB4 5BS
Tel 0954 60097 Tlx 817714

MAN
1

277 M S A (Britain) Ltd
East Shawhead
COATBRIDGE
Lanarkshire ML5 4TD
Tel 0236 24966 Tlx 778396

MAN
16 39

278 M S E Scientific Instrument
Sussex Manor Park
Gatwick Road
CRAWLEY
RH10 2QQ
Tel 0293 31100 Tlx 87119

MAN REP
27

279 M S W Jucker Instruments Ltd
2 Baird Close
Maxwell Way
CRAWLEY
Sussex RH10 2SY
Tel 0293 513636 Tlx 87228

MAN
9 12 35 37

280 Magnetrol International UK
Idenden House
Medway Street
MAIDSTONE
Kent ME14 1JT
Tel 0622 671929 Tlx 965020

MAN
1 5 8 27 37

281 Maurer Instruments Ltd
Unit 4
Browells Lane
FELTHAM
Middx TW13 7EQ
Tel 01 890 7300 Tlx 887326

MAN CAL REP CON RES
1

282 Maywood Instruments Ltd
Rankine Road
Daneshill West
BASINGSTOKE
Hants RG24 0PP
Tel 0256 57572 Tlx 858165

MAN CAL CON RES
4 7 9 10

283 Meclec Company
5 & 6 Towerfield Close
SHOEBURYNESS
Essex SS3 9QP
Tel 03708 5047 Tlx 99273

MAN CAL REP CON RES
1 3 4 5 6 7 9 12 17 22 23 26 27 31 33 34 35 36 37 38 39 40

284 Meclubricon Ltd
Unit 14 Webner Industrial Estate
Ettingshall Road
WOLVERHAMPTON
West Midlands WV2 2LD
Tel 0902 404643/4 Tlx 337343

MAN CAL REP CON
1 2 9 10 29 30 37 38

285 Mescon I C S Ltd
24 Abbotsinch Road
GRANGEMOUTH
FK3 9UX
Tel 0324 471550 Tlx 778559

MAN REP CON RES
1 5 8 9 12 13 15 16 31 32 33 34 35 37 38 39 40

286 Michell Instruments Ltd
Unit 9
Nuffield Close
Nuffield Road
CAMBRIDGE
Cambs CB4 1SS
Tel 0223 312427 Tlx 817741

MAN CAL REP RES
17

287 Microdata Ltd
Unit 7
The Hertfordshire Business Centre
Alexander Road
LONDON COLNEY
Herts AL2 1JG
Tel 0727 26118 Tlx 924937

MAN
32 35

288 Microlease PLC
Forbes House
Whitefriars Estate
Tudor Road
HARROW
Middx HA3 5SS
Tel 01 427 8822 Tlx 8953165

MAN
41

289 Microtherm Ltd
Hawco House
Cathedral Hill Industrial Estate
GUILDFORD
Surrey GU2 5YB
Tel 0483 60606 Tlx 859364

MAN
12 18

290 Midland Instrument Centre Ltd
Unit 14 Webner Industrial Estate
Ettingshall Road
WOLVERHAMPTON
West Midlands WV2 2LD
Tel 0902 44341 Tlx 337343

CAL REP CON RES
1 5 6 7 9 12 14 16 17 18 19 22 23 26 27 28 31 32 33 34 35 36 37 39 40

291 Moisture Control & Measurement Ltd
Thorp Arch Trading Estate
WETHERBY
Yorks LS23 7BJ
Tel 0937 843927 Tlx 557654

MAN CAL REP
17

292 Moisture Systems Ltd
The Old School
Station Road
COGENHOE
Northants NN7 1LT
Tel 0604 890606 Tlx 312463

MAN
17

293 Moore Products Co (UK) Ltd
Copse Road
Lufton
YEOVIL
Somerset BA22 8R
Tel 0935 706262 Tlx 94011272

CON
1 5 9 10 12 33 35 36 37

294 Motorola Ltd
Fairfax House
69 Buckingham Street
AYLESBURY
Bucks HP20 2NF
Tel 0296 395252

MAN
36

295 Movement Techniques Ltd
12-14 The Technology Centre
Epinal Way
LOUGHBOROUGH
Leics LE11 0QE
Tel 0509 267637 Tlx 341995

MAN RES
3 6 26 27

296 Mowlem Microsystems Ltd
Eastman Way
HEMEL HEMPSTEAD
Herts HP2 7HB
Tel 0442 218444 Tlx 825239

MAN CON
4 9 12 29 33 34 35 36

297 N E I Electronics Ltd - Edgcumbe Instruments
Main Street
Bothwell
GLASGOW
G71 8E2
Tel 0698 852574 Tlx 778873

MAN
18

298 N E I-Thompson Valves
17 Balena Close
Creekmoor Industrial Estate
POOLE
Dorset BH17 7EF
Tel 0202 697521 Tlx 41256

MAN
8

299 N G Bailey & Co Ltd
Cutler Heights Lane
BRADFORD
West Yorkshire BD4 9JF
Tel 0274 682856 Tlx 517293

MAN CON
40

300 Negretti Automation Ltd
Stocklake
AYLESBURY
Bucks HP20 1DR
Tel 0296 395931 Tlx 83285

MAN
1 2 4 5 7 8 9 10 12 16 18 21 32 33 35

301 NEI Electronics Ltd - Edgcumbe Instruments
Main Street
Bothwell
GLASGOW
G71 8EZ
Tel 0698 852574 Tlx 778873

MAN
18

302 Neles International Ltd
Neles House
60 High Street
BURNHAM
Bucks SL1 7JT
Tel 06286 4499 Tlx 848273

MAN
1

303 Neptune Measurement Ltd
PO Box 2
Dobcross
OLDHAM
OL3 5BD
Tel 04577 4822 Tlx 668064

MAN CAL REP CON
1 31 35

304 Newport Electronics Ltd
Cavendish Road
STEVENAGE
Herts SG1 2ET
Tel 0438 365671 Tlx 826454

MAN
1 4 7 9 10 12 18 35

305 Newton Instrument Co Ltd
Carr Lane
HOYLAKE
Merseyside
Tel 051 632 4780

MAN RES
13 27 31 32

306 Newtronic Controls International Ltd
Stag Industrial Estate
Atlantic Street
Broadheath
ALTRINCHAM
Cheshire WA14 5NN
Tel 061 928 4275 Tlx 665781

MAN
1 5 9 12

307 Nivotek Ltd
7/8 New Street
STOURPORT ON SEVERN
Worcs DY13 8UJ
Tel 02993 5132 Tlx 334303

MAN
5 9

308 Nixon Instrumentation Ltd
Charlton Kings Industrial Estate
CHELTENHAM
Glos GL53 8DZ
Tel 0242 43006 Tlx 43379

MAN CAL
1 5

309 Normond Instruments Ltd
Hydrex house
Garden Road
RICHMOND
Surrey TW9 4NR
Tel 01 392 1355 Tlx 919193

MAN CAL REP CON
5 31 36 40

310 Northern Acoustic Equipment
9 Braeside
Kirklevington
YARM ON TEES
Cleveland TS15 9NB
Tel 0642 782391

MAN
6 23 27 39

311 Norwood Instruments Ltd
New Mill Road
Honley
HUDDERSFIELD
Yorks HD7 2QD
Tel 0484 661318

MAN CON
1 4 9 12 16 17 22 24 31

312 Nottingham Flow Controls Ltd
Charlotte Street
MELTON MOWBRAY
Leics LE13 1NA
Tel 0664 67797 Tlx 342337

MAN CAL REP CON
1 9 34

313 Novatech Measurements Ltd
83 Castleham Road
ST LEONARDS ON SEA
East Sussex TN38 9NT
Tel 0424 52744 Tlx 957400

MAN
5 7 35

314 Nuclear Enterprises Ltd
Bath Road
Beenham
READING
RG7 5PR

MAN
20 21 39

315 Nulectrohms Ltd
Meppershall
SHEFFORD
Beds SG17 5LX
Tel 0462 813000 Tlx 825471

MAN CON
12 22

316 Oertling
Smethwick
WARLEY
West Midlands B66 2LP
Tel 021 558 1112/2161 Tlx 336490

MAN CAL REP CON
7 8 13 17 29 32 33 35 36

317 Ometron Ltd
Worsley Bridge Road
LONDON
SE26 5BX
Tel 01 461 5555 Tlx 262461

MAN
4 6 22

318 Oxbridge Technology Ltd
Breckland
Saxon Street
Linford Wood
MILTON KEYNES
Bucks MK14 6LD
Tel 0908 314626 Tlx 825835

MAN RES
12 33 34 35 36

319 Oxford Analytical Instruments Ltd
20 Nuffield Way
ABINGDON
Oxon OX14 1TX
Tel 0235 32123 Tlx 83621

MAN
13 14 21

320 Oxford Pressure Systems Ltd
6-10 Swan Street
EYNSHAM
Oxon OX8 1HU
Tel 0865 880692 Tlx 83138

MAN CAL REP CON
5 9 12

321 P J D Instruments Ltd
Dominion Road Industrial Estate
SOUTHALL
Middx UB2 5DP
Tel 01 843 0442 Tlx 21879

MAN CAL REP CON
9 12 17 26 35 40 41

322 Paar Scientific Ltd
594 Kingston Road
Raynes Park
LONDON
SW20 8DN
Tel 01-540 8553 Tlx 938292

MAN REP CON
2 8 12 26 31 33 36 40

323 Packs Infotel Ltd
25 Invincible Road Industrial Estate
FARNBOROUGH
Hants GU14 7QE
Tel 0252 540678

MAN REP
34

324 PACS Ltd
Chilworth Research Centre
Chilworth
SOUTHAMPTON
Hants SO1 7NP
Tel 0703 766231 Tlx 477380

MAN CAL REP CON RES
1 5 9 12 33 34 35 36 39

325 Panametrics Ltd
Justin Manor
341 London Road
MITCHAM
Surrey CR4 4BE
Tel 01 640 2252 Tlx 8811645

MAN
1 16 17

326 Pantatron Radiation Engineering Ltd
H & M Engineering Works
Castlemaine Avenue
GILLINGHAM
Kent
Tel 0634 52359 Tlx 965061

MAN CAL CON
1 4 5 6 8 20 21 22 27 29 35 39

327 Parker Hannifin (UK) Ltd
Wetherby Road
DERBY
DE2 8JH
Tel 0332 365631 Tlx 37427

MAN
37 40

328 Penny & Giles Transducers Ltd
Airfield Road
CHRISTCHURCH
Dorset BH23 3TH
Tel 0202 476621 Tlx 41566

REP
1 9 10 31

329 Perflow Instruments Ltd
Unit 3 Chapmans Park High Road
Willesden
LONDON
NW10 2DY
Tel 01 451 4577 Tlx 923620

MAN CAL
1 9

330 Perkin-Elmer Ltd
Post Office Lane
BEACONSFIELD
Bucks HP9 1QA
Tel 04946 6161 Tlx 83257

MAN
13 14 16 27 31 33 36

331 Pfaudler-Balfour Ltd
Riverside Place
LEVEN
Fife KY8 4RW
Tel 0333 23020 Tlx 72304

MAN REP
5 12 15 26

332 Philips Industrial Automation
Pye Unicam
York Street
CAMBRIDGE
CB1 2PX
Tel 0223 358866 Tlx 817331

MAN
1 2 3 4 5 6 7 8 9 10 12 13 15 17 18 28 32 33 34 35 36 37 40

333 Photain Controls Ltd
Ford Aerodrome
ARUNDEL
West Sussex BN18 0BE
Tel 0903 721531 Tlx 87325

MAN
8 34 39

334 Photoelectronics (Arcall) Ltd
Titan Works
Limpsfield Road
WARLINGHAM
Surrey
Tel 08832 4183 Tlx 418203

MAN REP RES
8 19 31

335 Pioden Controls Ltd
Graham Bell House
Roper Close
Roper Road
CANTERBURY
Kent CT2 7EP
Tel 0227 463641 Tlx 965816

MAN
3 4 6 7 9 35

336 Planer Products Ltd
110 Windmill Road
SUNBURY ON THAMES
Middx TW16 7HD
Tel 0932 786262 Tlx 928185

MAN
12 24 26 27

337 Plessey Microsystems Ltd
Water Lane
TOWCESTER
Northants WN12 7JN
Tel 0327 50312 Tlx 31628

MAN
36

338 Pneupac Ltd
Crescent Road
LUTON
Beds LU2 0AH
Tel 0582 453303 Tlx 825512

MAN
9 10 27 37

339 Poddymeter Ltd
52 Mill Place
KINGSTON UPON THAMES
Surrey KT1 2RW

MAN
1 9 10

340 Porut Engineers Ltd
49 Norval Road
NORTH WEMBLEY
Middx HA0 3TD
Tel 01 904 7654

CON RES
31 33 35 36 37 39 40

341 Power Conversion Ltd
Fir Tree Lane
GROBY
Leics LE6 0FH
Tel 0533 878881 Tlx 341 401

REP
42

342 Pressure Sensor (Mktg) Ltd
Chequers Chambers
The Square
FOREST ROW
E Sussex RH18 5ES
Tel 0342 824466 Tlx 957547

MAN
5 9 10 28 30 31

343 Prior Scientific Instruments Ltd
London Road
BISHOPS STORTFORD
Herts CM23 5NB
Tel 0279 506414

MAN CON RES
24 30 31

344 Process Control & Instrumentation Ltd
7B Herald Industrial Estate
Botley Road
Hedge End
SOUTHAMPTON
Hants SO3 3JW
Tel 04892 4888 Tlx 477033

MAN
33 35

345 Process Measurement & Analysis Ltd
First Floor Office Block
Falcongate Industrial Centre
Old Gorsey Lane
WALLASEY
Merseyside L44 4HD
Tel 051 639 6021 Tlx 629393

MAN
1 5 13 15 16

346 Prosser Scientific Instruments Ltd
Lady Lane Industrial Estate
Hadleigh
IPSWICH
Suffolk IP7 6BQ
Tel 0473 823005 Tlx 987703

MAN RES
1 9 18 31 36

347 Protech Instruments and Systems Ltd
241 Selbourne Road
LUTON
Beds LU4 8NU
Tel 0582 596181 Tlx 825274

MAN
12 31 36

348 Protimeter PLC
Meter House
Fieldhouse Lane
MARLOW
Bucks SL7 1LX
Tel 06284 72722 Tlx 849305

MAN CAL REP
12 17

349 Pullman Instruments Ltd
Hornbeam Park
Hookstone Lane
HARROGATE
HG2 8QN
Tel 0423 873109/873114

MAN CAL REP CON
18

350 Pulsar Developments Ltd
Spracklen House
Dukes Place
MARLOW
Bucks SL7 2QH
Tel 06284 73555/74324 Tlx 849131

MAN
18 29 36 40

351 Pyropress Engineering Company Ltd, The
Bell Close
Plympton
PLYMOUTH
PL7 4JH
Tel 0752 339866 Tlx 45458

MAN
1 3 5 9 10 12

352 Quadrina Ltd
Flint Road
LETCHWORTH
Herts SG6 1HS
Tel 0462 673486 Tlx 826726

MAN CAL REP
1

353 Quin Systems Ltd
35 Broad Street
WOKINGHAM
Berks RG11 1AU
Tel 0734 783114

MAN
36

354 R C M Consulting Services
19 Barns Dene
HARPENDEN
Herts AL5 2HH
Tel 05827 5921

CON
31 32 36 39 40

355 R L Barton & Partners
Test House
Chandos Road
BRISTOL
Avon BS6 6PG
Tel 0272 732521

MAN REP CON
15 22 24 31

356 Racal Instrumentation Ltd
Hardley Industrial Estate
Hythe
SOUTHAMPTON
Hants SO4 6ZH
Tel 0703 843265 Tlx 47600

MAN REP
18 23 31 32 34 35 36 40

357 Radun Controls Ltd
Taffs Well
CARDIFF
CF4 7QA
Tel 0222 810046 Tlx 498514

MAN CON RES
3 31

358 Raleigh Instruments Ltd
Raytel House
19 Brook Road
RAYLEIGH
Essex SS6 7XH
Tel 0268 775656 Tlx 99440

MAN RES
9 12 18

359 Rank Taylor Hobson Ltd
PO Box 36
New Star Road
Thurmaston Lane
LEICESTER
Leics LE4 7JQ
Tel 0533 763771 Tlx 34411

MAN CAL REP CON
3 26

360 Research Engineers Ltd
Orsman Road
LONDON
N1 5RD
Tel 01 739 7811 Tlx 261973

MAN
11 31

361 Research Instrument Ltd
Kernick Road
PENRYN
Cornwall TR10 9DQ
Tel 0326 72753

MAN
10 19 22 24 27 38

362 Rochester Instrument Systems Ltd
Maxim Road
CRAYFORD
Kent DA1 4BG
Tel 0322 526211 Tlx 27831

MAN
12 18

363 Rockhall Ltd
128 Station Road
Glenfield
LEICESTER
Leics LE3 8BR
Tel 0533 313531

MAN
41

364 Rohde & Schwarz UK Ltd
Roebuck Road
CHESSINGTON
Surrey KT9 1LP
Tel 01 397 8771 Tlx 928479

MAN CAL REP RES
18 23 27 28 36

365 Ronan Engineering Ltd
1 Tilley Road
Crowther Industrial Estate
WASHINGTON
Tyne & Wear NE38 OEA
Tel 091 4161689 Tlx 537746

MAN
4 5 6 8 12 21 35

366 Rosemount Ltd
Heath Place
BOGNOR REGIS
West Sussex PO22 9SH
Tel 0243 863121 Tlx 86218

MAN CAL REP CON
5 8 9 10 12 13 15 16 17 18 21 35 36

367 Rossell Fluid Control Ltd
Mead Lane
LYDNEY
Glos GL15 5EU
Tel 0594 42219 Tlx 437150

MAN
1 13

368 Rostol Ltd
123 Lynchford Road
FARNBOROUGH
Hants GU14 6ET
Tel 0252 512469 Tlx 946240

MAN
3 4 6 12 22 26 28 31 33 35

369 Roxby Engineering International Ltd
Roxby House
Station Road
SIDCUP
Kent DA15 7EJ
Tel 01 300 3393 Tlx 896172

MAN CON
22 40

370 Safety Technology Ltd
Osborn Way
HOOK
Hants RG27 9HX
Tel 025672 4081 Tlx 858415

MAN CON
34 36 39

371 Salter Industrial Measurements Ltd
George Street
WEST BROMWICH
West Midlands B70 6AD
Tel 021 5531855 Tlx 337314

MAN
7

372 Satchwell Control Systems Ltd
PO Box 57
Farnham Road
SLOUGH
Berks SL1 4UH
Tel 0753 23961 Tlx 848186

MAN
9 12 17 37

373 Satt Control UK Ltd
The Brook Trading Estate
Deadbrook Lane
ALDERSHOT
Hants GU12 4XB
Tel 0252 27252 Tlx 858200

MAN CON
1 2 5 9 10 12 35

374 Scama Ltd
Eastern Way
BURY ST EDMUNDS
Suffolk IP32 7AQ
Tel 81357

MAN
35

375 Schaevitz EM Ltd
543 Ipswich Road
SLOUGH
Berks SU 4EG
Tel 0753 376222/9 Tlx 847818

MAN
3 6 7 9 10

376 Schenk Ltd
Station Approach
BICESTER
Oxon OX6 7BZ
Tel 837669

MAN
3 4 5 6 7 9

377 Sellars Datasystems Ltd
Edge Lane
DROYLSDEN
Manchester M35 6BU
Tel 061 301 2317 Tlx 668960

MAN
33 36

378 Sensing Devices Ltd
97 Tithebarn Road
SOUTHPORT
Merseyside PR8 6AG
Tel 0704 35739 Tlx 677250

MAN CAL REP
12

379 Sension Systems Ltd
Denton Drive Industrial Estate
NORTHWICH
Cheshire CW9 7LU
Tel 0606 44321

MAN
36

380 Sensor Technology Ltd
PO Box 36
BANBURY
Oxon OX15 6JB
Tel 0295 73746 Tlx 946240

MAN
2 4 7 22

381 Serck Controls
Rowley Drive
COVENTRY
West Midlands CV3 4FH
Tel 0203 305050 Tlx 311970

MAN
34 35

382 Serck Glocon
St Lukes Street
Southgate Street
GLOUCESTER
Glos GL1 5RE
Tel 0452 28631 Tlx 43139

MAN CAL REP CON
1 5 13 15 19

383 Severn Science (Instruments) Ltd
Short Way
Thornbury Industrial Estate
Thornbury
BRISTOL
BS12 2UT
Tel 0454 414723

MAN
16 31 32 39

384 Severn Science Ltd
Cooper Road
Thornbury Industrial Estate
Thornbury
BRISTOL
BS12 2UL
Tel 0454 414723

MAN
10 12

385 Shape Instruments Ltd
Transducer Works
Toutley Road
WOKINGHAM
Berks RG11 5SS
Tel 0734 791111 Tlx 847160

MAN CON RES
5 9 10 12 32 33 34 35 40

386 Sharples Stress Engineers Ltd
Unit 331
Walton Summit Centre
Bamber Bridge
PRESTON
Lancs PR5 8AR
Tel 0772 323359

MAN CON
4 22

387 Shaw Moisture Meters
Rawson Road
Westgate
BRADFORD
W Yorks BD1 3SQ
Tel 0274 733582 Tlx 51598

MAN CAL
17

388 Shawcity Ltd
Units 12/13 Pioneer Road
FARINGDON
Oxfordshire SN7 7BU
Tel 0367 21675 Tlx 444406

MAN
16 33 35 39

389 Sieger Analysers
31 Nuffield Estate
POOLE
Dorset BH17 7PZ
Tel 0202 676161 Tlx 41138

MAN
13 14 16 17

390 Sieger Ltd
31 Nuffield Estate
POOLE
Dorset BH17 7RZ
Tel 0202 676161 Tlx 41138

MAN CAL REP
15 16 39

391 Siemens Ltd
Siemens House
Eaton Bank
CONGLETON
Cheshire CW12 1PH
Tel 0260 278311 Tlx 8951091

MAN CON RES
1 9 10 12 13 14 16 22 33 35 36 39

392 Signal Instrument Co Ltd
Standards House
Doman Road
CAMBERLEY
Surrey GU15 3DW
Tel 0276 682841 Tlx 859035

MAN
16

393 Simac Instrumentation Ltd
Lyon Road
Hersham
WALTON-ON-THAMES
Surrey KT12 3PU
Tel 0932 243444 Tlx 928360

MAN
16 17

394 Simon-Drake Automation Ltd
10 Whitehouse Street
BRISTOL
Avon BS3 4AU
Tel 0272 637405 Tlx 449210

MAN CON RES
40

395 Sirco Controls Ltd
Sweynes Industrial Estate
Ashingdon Road
ROCHFORD
Essex SS4 1RQ
Tel 0702 545125 Tlx 99325

MAN
9 10 12 32

396 Skeltonhall Ltd
70 Carwood Road
SHEFFIELD
South Yorks S4 7SD
Tel 0742 431332 Tlx 547868

MAN RES
1 31

397 Skil Controls Ltd
Greenhey Place
East Gillibrands
SKELMERSDALE
Lancs WN8 9SB
Tel 0695 23671 Tlx 627825

MAN
1 5 6 9 12 16 31 35 37 40

398 Smith and Jones Systems
6 Ashford Court
Elmsleigh Avenue
LEICESTER
LEICS LE2 2DJ

MAN
36

399 Solartron Instruments
Victoria Road
FARNBOROUGH
Hants GU14 7PW
Tel 0252 544433 Tlx 858245

MAN
4 6 12 15 18 19 23 36 38

400 Solartron Transducers
Victoria Road
FARNBOROUGH
Hants GU14 7PW
Tel 0252 544433 Tlx 858245

MAN CAL REP
1 8 9

401 South Coast Electronics Ltd
6 Monks Avenue
LANCING
West Sussex BN15 9DJ
Tel 0903 753754

MAN
41

402 Space Age Electronics (Industrial) Ltd
Spalding Hall
Victoria Road
LONDON
NW4 2BE
Tlx 8811085

MAN CON
1 5 9 33 34 36 40

403 Spantech International Ltd
Africa House
64-78 Kingsway
LONDON
WC2B 6AH
Tel 01 405 2085/1931 Tlx 27445

MAN CON RES
16 31 40

404 Spantech Products Ltd
Spantech House
Lagham Road
SOUTH GODSTONE
Surrey RH9 8HP
Tel 0342 893239 Tlx 957576

MAN
1 16 17 27 32 39

405 Spectra-Tek
Swinton Grange
MALTON
N Yorks YO17 0QR
Tel 0653 695551 Tlx 57850

MAN REP CON RES
1 8 9 12 23 31 33 34 39

406 Spirax-Sarco Ltd
Charlton House
CHELTENHAM
Glos GL53 8ER
Tel 0242 521361 Tlx 43123

MAN
1

407 Stack Ltd
Unit 8
Wedgwood Road
BICESTER
Oxon OX6 7UL
Tel 0869 240404

MAN
35

408 Stanton Redcroft Ltd
Copper Mill Lane
LONDON
SW17 0BN
Tel 01 946 7731 Tlx 8811417

MAN CON RES
8 12 13 14 16 33 39

409 Starr Pyrometers Ltd
36 Whiteways Road
SHEFFIELD
S Yorks S4 8EX
Tel 0742 439565

MAN
12

410 Staveley Electrical Ltd
Gatefoot Mill
Staveley
KENDAL
Cumbria LA8 9PN
Tel 0539 821370

MAN
12 34 35 37 40

411 Structural Statics Ltd
21 Southgate Street
WINCHESTER
Hants SO23 9EB
Tel 0962 66091 Tlx 477104

MAN CON RES
29 40

412 Swagelock UK Ltd
3 Kelvin Close
Science Park North
Birchwood
WARRINGTON
Cheshire WA3 7PB
Tel 0925 818454 Tlx 628535

MAN
1 9 10 16 31 32

413 Sydney Smith Dennis Ltd
Crossgate Drive
Queens Drive Industrial Estate
NOTTINGHAM
Notts NG2 1LQ
Tel 0602 861341 Tlx 37563

MAN CAL RES
9 10 12

414 Systech Instruments Ltd
Goodsons Industrial Mews
Wellington Street
THAME
Oxon OX9 3BX
Tel 084421 6838 Tlx 937400

MAN RES
16 17 31

415 T M G Instrumentation
Unit 7
Rees House
Burn Hall Industrial Estate
FLEETWOOD
Lancs FY7 8RS
Tel 0253 864459

MAN CON RES
1 5 8 9 12 20 21 31 35 36

416 Taylor Instrument
A Division of Combustion Engineering Ltd
Gunnels Wood Road
STEVENAGE
Herts SG1 2EL
Tel 0438 312366 Tlx 82281

MAN
1 5 9 10 12 36 37 39 40

417 Tech-Writer Associates
6 James Cottages
Kew Road
KEW GARDENS
Surrey TW9 3DX
Tel 01 940 5301

CON RES
32

418 Techmation Ltd
58 Edgware Way
EDGWARE
Middlesex HA8 8JP
Tel 01 958 3111 Tlx 262245

MAN CAL REP CON
1 9 11 13 14 16 22 26 30 32

419 Techne (Cambridge) Ltd
Duxford
CAMBRIDGE
Cambs CB2 4PZ
Tel 0223 832401 Tlx 817257

MAN
8 12 27

420 Technitron Ltd
Church Crookham
FLEET
Surrey
Tel 0252 851085

MAN
36

421 Technivise Ltd
South Hill
CHISLEHURST
Kent BR7 5EH
Tel 01 467 5555

CON
42

422 Tecnomatic Controls Ltd
119/120 Western Road
HOVE
East Sussex BN3 1DB
Tel 0273 727373 Tlx 878176

MAN CAL REP
5 9 10 12

423 Teddington Controls Ltd
Holmbush
ST AUSTELL
Cornwall PL25 3HS
Tel 0726 74400 Tlx 45242

MAN
9 12 17 36 38 39

424 Teeside Automation Services Ltd
Tas House
37/39 Norton Road
STOCKTON ON TEES
Cleveland TS18 2BU
Tel 0642 613622 Tlx 587241

CON
31

425 Telegan Ltd
Legion House
Godstone Road
KENLEY
Surrey CR2 5YS
Tel 01 668 8251 Tlx 946587

MAN
12 16

426 Telematic Systems Ltd
Unit 17
PO Box 54
Alban Park
ST ALBANS
Herts AL4 0XY
Tel 0727 33147 Tlx 261083

MAN

427 Tensor Computers Ltd
Hail Weston House
Hail Weston
ST NEOTS
Cambs PE19 4J4
Tel 0480 215580 Tlx 32392

MAN CON RES
1 12 31 32 33 34 35 36 40

428 Terminal Display Systems Ltd
Lower Philips Road
Whitebirk Industrial Estate
BLACKBURN
Lancs BB1 5TH
Tel 0254 676921 Tlx 635693

MAN
35 36

429 Testoterm Ltd
Old Flour Mill
Queens Street
EMSWORTH
Hants PO10 7BT

MAN
1 12 15 16 17 33 34

430 Texas Instruments
EIC Division
Mail Station 45
Manton Lane
BEDFORD
Beds MK41 7PA
Tel 0234 270111 Tlx 82178

MAN
33 34 35

431 Thelpe Instruments Ltd
Unit 9 Maurice Road
WALLSEND
Tyne & Wear
Tel 091 262 0682

MAN CAL REP CON
5 9 10 12 35 37 40

432 Thermal Detection
1 Mary Street
STOCKTON ON TEES
Cleveland TS18 4AN
Tel 0642 602878

MAN CAL CON
12 31 35 36 40

433 Theta Pneumatics Ltd
6-10 Swan Street
EYNSHAM
Oxon OX8 1HU
Tel 0865 880692 Tlx 83138

MAN
9

434 Thomas Mercer Ltd
Eywood Road
ST ALBANS
Herts AL1 2ND
Tel 0727 55313 Tlx 27514

MAN
3 33

435 Thorn EMI Datatech Ltd
Spur Road
Feltham Trading Estate
FELTHAM
Middx TW14 0TD
Tel 01 890 1477 Tlx 23995

MAN
41

436 Thorn EMI Instruments Ltd
Archcliffe Road
DOVER
Kent CT17 9EN
Tel 0304 202620 Tlx 96283

MAN
18 29 30 31 36

437 Thurnall Engineering
Northbank Industrial Estate
Cadishead
MANCHESTER
Lancs M30 5BL
Tel 061 775 5416

MAN RES
31 33 34 35 40

438 Tierway Ltd
3 Washington Road
West Wilts Trading Estate
WESTBURY
Wilts BA13 4JP
Tel 0373 823599 Tlx 44294

MAN CON RES
1 5 9 10 36 40

439 Toshiba International Co Ltd
Audrey House
Ely Place
LONDON
EC1N 6SN
Tel 01 242 7295 Tlx 265062

MAN
1 21 22 26 35

440 Trace Automation
Capital House
89 Herne Road
Bushey
WATFORD
Herts WD2 3LP
Tel 01 950 9898 Tlx 8950511

MAN CAL REP CON RES
1 3 5 9 10 12 13 17 18 19 23 33 35 36

441 Transamerica Instruments Ltd
Lennox Road
BASINGSTOKE
Hants RG22 4AW
Tel 0256 20244 Tlx 858103

MAN
9 35

442 Transmitton Ltd
M/O BICC Group
Smisby Road
ASHBY DE LA ZOUCH
Leics LE6 5UG
Tel 0530 415941 Tlx 341628

MAN
34

443 Trolex Products Ltd
10a Newby Road
Hazel Grove
STOCKPORT
Cheshire SK7 5DY
Tel 061 483 1435 Tlx 666170

MAN CAL REP CON RES
1 5 6 9 10 12 16 31 33 34 35 39 40

444 Trumeter Co Ltd
Milltown Street
Radcliffe
MANCHESTER
M26 9NX
Tel 061 724 6311 Tlx 667088

MAN
3 26

445 Turnbull Control Systems Ltd
Broadwater Trading Estate
WORTHING
W Sussex BN14 8NW
Tel 0903 205277 Tlx 87437

MAN CAL REP CON RES
1 9 12 18 31 34 35 36 40

446 U C C International Ltd
PO Box 3
THETFORD
Norfolk IP24 3RT
Tel 0842 4251 Tlx 81258

MAN
1 9 12 13

447 Unimatic Engineers Ltd
122 Granville Road
Cricklewood
LONDON
NW2 2LN
Tel 01 455 0012 Tlx 922396

MAN
3

448 V G Gas Analysis Systems Ltd
Aston Way
Holms Chapel Road
MIDDLEWICH
Cheshire CW10 0HT
Tel 060 684 4731 Tlx 668061

MAN CAL REP CON
16 27 32 39

449 Vaisala (UK) Ltd
Cambridge Science Park
Milton Road
CAMBRIDGE
Cambs CB4 4BH
Tel 0223 862112 Tlx 817204

MAN CAL REP CON
9 12 17 28 34 35

450 Vickers Systems Ltd
Lang Pneumatic Division
68-70 Old Bedford Road
LUTON
Beds LU2 7PA
Tel 0582 413023/413024

MAN
37

451 Visolux Ltd
28 Clifton Industrial Estate
CAMBRIDGE
Cambs CB1 4ZG
Tel 0223 242020 Tlx 81434

MAN
19 34 39

452 W H Tew Group
Crocus Street
NOTTINGHAM
NG23 3DR
Tel 0602 865027 Tlx 377404

MAN
12 18 31

453 W S Ocean Systems
Unit 101 Blackdown Industrial Estate
Haste Hill
HASLEMERE
Surrey GU27 3AY
Tel 0428 54500 Tlx 859447

MAN
1 5 28

454 Wallace and Tiernan Ltd
Priory Works
TONBRIDGE
Kent TN11 0QL
Tel 0732 364481 Tlx 95278

MAN CAL
1 9 10 15

455 Watson Smith Control Instrumentation Division
Norgren Martonair Limited
Box 22 Eastern Avenue
LICHFIELD
Staffs WS13 6SB
Tel 0543 414333 Tlx 338555

MAN
1 9 37 39

456 Weir Pumps Ltd
Electronics Division
Newlands Road
Cathcart
GLASGOW
G44 4EX
Tel 041 637 7141 Tlx 77161/2

MAN CON
6 9 12 33 35 36

457 Wells-Krauterkramer Ltd
Blackhorse Road
LETCHWORTH
Herts SG6 1HF
Tel 04626 78151 Tlx 82329

MAN CAL REP
22

458 Welwyn Strain Measurement Ltd
Armstrong Road
BASINGSTOKE
Hants RG24 0QA
Tel 0256 62131 Tlx 858520

MAN
4 7 9 12 33 35

459 Whessoe Systems and Controls Ltd
Brinkburn Road
DARLINGTON
Co Durham DL3 6DS
Tel 0325 460188 Tlx 58110

MAN REP
5 8 9 12 22 30 32 33 34 35 39 40

460 White Electrical Instrument Co Ltd
Spring Lane North
MALVERN LINK
Worcs WR14 1BL
Tel 06845 3218 Tlx 335294

MAN CON
7 18 31 36

461 Wika Pressure Gauges (UK) Ltd
Ullswater Crescent
Marlpit Lane
COULSDON
Surrey CR3 2HR
Tel 01 668 0924 Tlx 946152

MAN CON
9 10

462 Wirral Instrumentation and Electrical Co Ltd
81 Canning Street
BIRKENHEAD
Merseyside L41 1AF
Tel 051 647 5005/6 Tlx 627110

CAL CON RES
40

10.2 Academic institutions, i.e. university and college departments and other organizations in the public sector where relevant research is done.

463 Aberdeen University
Physics Dept
Meston Walk
OLD ABERDEEN
AB9 2UE
Tel 0224 40241

6 14 18 19 22 24 26 31 33

464 Atomic Energy Research Establishment
(UKAEA)
Harwell
DIDCOT
Oxon OX11 0RA
Tel 0235 24141 Tlx 83135

1 5 6 11 13 14 16 20 21 22 25 31 33 34 35 39

465 Bangor University College
University of Wales
Electronic Engineering Dept
Dean Street
BANGOR
Gwynedd LL57 1UT
Tel 0248 351151 Tlx 61100

1 15 25 27 31 34

466 Birmingham University
Clinical Chemistry Dept
PO Box 363
BIRMINGHAM
B15 2TT

19 27

467 Bradford University
Control Engineering Dept
BRADFORD
West Yorks BD7 1DP
Tel 0274 733466

1

468 Brighton Polytechnic (a)
Civil Engineering Dept
Moulsecomb
BRIGHTON
BN2 4GJ
Tel 0273 693655

13 22 29 30 33

469 Brighton Polytechnic (b)
Mechanical and Production Engineering Dept
Moulsecomb
BRIGHTON
BN2 4GJ
Tel 0273 693655

6 26

470 British Coal
Headquarters Technical Department
Ashby Road
STANHOPE BRETBY
Burton-on-Trent DE15 0QD
Tel 0283 216161 Tlx 341741

1 4 6 7 8 9 11 13 16 17 18 20 21 22 23 25 32 33 34 38 39 40

471 British Rail Research
Railway Technical Centre
London Road
DERBY
DE2 8UP
Tel 0332 42442

2 13 14 16 19 22 24 26 31 33

472 British Steel Corporation
Teesside Laboratories
PO Box 11
Grangetown
MIDDLESBOROUGH
Cleveland TS6 6UB
Tel 0642 467144 Tlx 58347

1 3 5 6 11 12 16 17 21 31 33 34 40

473 Brunel University (a)
Manufacturing Metrology Dept
UXBRIDGE
Middx UB8 3PH
Tel 0895 74000 Tlx 261173

3 25 37

474 Brunel University (b)
Physics Dept
Kingston Lane
UXBRIDGE
Middx UB8 3PH
Tel 0895 74000 Tlx 261173

20 21 22 24 31 32 33 36

475 Cardiff University College (a)
University of Wales
Materials Institute
Newport Road
CARDIFF
CF2 1TA
Tel 0222 874823 Tlx 498635

11 22 24 26 33

476 Cardiff University College (b)
University of Wales
Mineral Exploitation Dept
Newport Road
CARDIFF
CF2 1TA
Tel 0222 874823

30

477 Central Electricity Generating Board
Technology Planning & Research Division
15 Newgate Street
LONDON
EC1
Tel 01 248 1202 Tlx 883141

3 4 6 12 15 16 17 18 20 22 27 31 32 33 36

478 City University
Measurement and Instrumentation Centre
School of Elec Eng & Applied Science
Northampton Square
LONDON
EC1V 0HB
Tel 01 253 4399 Tlx 263896

1 3 5 7 9 12 18 19 22 24 25 33

479 Cranfield Institute of Technology (a)
Aeronautics Dept
Cranfield
BEDFORD
Beds MK43 0AL
Tel 0234 750111 Tlx 825072

33 40

480 Cranfield Institute of Technology (b)
Fluid Engineering and Instrumentation Department
Cranfield
BEDFORD
Beds MK43 0AL
Tel 0234 750111 Tlx 825072

1

481 Cranfield Institute of Technology (c)
Mechanical Engineering Dept
Cranfield
BEDFORD
Beds MK43 0AL
Tel 0234 750111 Tlx 825072

1 4 6 7 9 11 12 16 23 24 25 31 33 40

482 Durham University
Engineering and Applied Science Dept
Electronics Science Laboratories
SOUTH ROAD
Durham DH1 3LE
Tel 0385 64971 Tlx

12 18 19 35

483 Edinburgh University
Electrical Engineering Dept
The King's Buildings
Mayfield Road
EDINBURGH
EH9 3JL

3 15 22 27 29 34 38

484 ERA Technology
Cleeve Road
LEATHERHEAD
Surrey KT22 7SA
Tel 264045

4 18 19 22 25 39

485 Essex University
Chemistry Dept
Wivenhoe Park
COLCHESTER
Essex CO4 3SQ
Tel 0206 862286 Tlx 98440

13 14

486 Fulmer Research Ltd
Stoke Poges
SLOUGH
Berks SL2 4QDS
Tel 02816 2181 Tlx 849374

9 16 22 27 31 35 36 38

487 Institute of Occupational Medicine
Physics Branch
8 Roxburgh Place
EDINBURGH
EH8 9SU
Tel 031 667 5131

11

488 Keele University
Physics Dept
KEELE
Staffs ST5 5BG
Tel 0782 621111 Tlx 36113

22

489 Kent University at Canterbury (a)
Dept of Chemistry
CANTERBURY
Kent CT2 7NH
Tel 0227 66822

12 13 14 15 16

490 Kent University at Canterbury (b)
Physics Laboratory
CANTERBURY
Kent CT2 7NR
Tel 0227 66822 Tlx 965449

1 3 5 6 9 11 12 18 22 25

491 Leeds University
Dept of Mechanical Engineering
LEEDS
LS2 9JT

3 26 33

492 Liverpool University (a)
Architecture & Building Engineering Dept
PO Box 147
LIVERPOOL
L69 3BX
Tel 051 709 6022

2 6 23 33

493 Liverpool University (b)
Electrical Eng and Electronics Dept
Brownlow Hill
PO Box 147
LIVERPOOL
L69 3BX
Tel 051 709 6022 Tlx 627095

3 4 6 9 10 12 14 17 18 19 25 32 33

494 London University (a)
Imperial College of Science and Technology
Chemistry Dept
South Kensington
LONDON
SW7 2AY

13 14 15 33

495 London University (b)
Imperial College of Science and Technology
Chemical Engineering and Chemical Technology Dept
LONDON
SW7 2BT
Tel 01 589 5111 Tlx 261503

1 2 6 8 11 24 26 31 33 34 36

496 London University (c)
Imperial College of Science and Technology
Electrical Engineering Dept
Exhibition Road
LONDON
SW7 2BT
Tel 01 589 5111 Tlx 261503

14 25 27 33

497 London University (d)
Imperial College of Science and Technology
Physics Dept, The Blackett Laboratory
Prince Consort Road
LONDON
SW7 2BZ
Tel 01 589 5111 Tlx 261503

14 19 24 31

498 London University (e)
King's College
Chemistry Dept
Strand
LONDON
WC2R 2LS
Tel 01 836 5454 Tlx

13 14 15 18 19 21 27 31 36

499 London University (f)
King's College
Electronic and Electrical Engineering Dept
Strand
LONDON
WC2R 2LS
Tel 01 836 5454 Tlx

18 22 25

500 London University (g)
University College
Electronic and Electrical Engineering Dept
Torrington Place
LONDON
WC1E 7JE
Tel 01 380 7302

1 15 18 24 25 27 28

501 London University (h)
University College
Photogrammetry and Surveying Dept
Gower Street
LONDON
WC1E 6BT
Tel 01-387 7050 Tlx 296273

3 5 31 33

502 Loughborough University of Technology (a)
Chemical Engineering Dept
LOUGHBOROUGH
Leics LE11 3TU
Tel 0509 263171 Tlx 34319

11 13

503 Loughborough University of Technology (b)
Electrical and Electronic Engineering Dept
LOUGHBOROUGH
Leics LE11 3TU

6

504 Manchester Polytechnic (a)
Chemistry Dept
Chester Street
MANCHESTER
M1 5GD
Tel 061 228 6171

25 31 34 36

505 Manchester Polytechnic (b)
Mathematics & Physics Dept
Mathematics & Computing
MANCHESTER
M1 5GD
Tel 061 228 6171 Tlx

1 7 17 18 31 36

506 Manchester University (a)
Institute of Science & Technology
Dept of Instrumentation and Analytical Science
PO Box 88
MANCHESTER
M60 1QD
Tel 061 236 3311

1 3 7 12 13 14 15 17 18 19 27 32 39

507 Manchester University (b)
Engineering Dept
Oxford Road
MANCHESTER
M13 9PL
Tel 061 273 7121

1 3 4 6 11 12 20 21 23 25 29 31 36 38

508 N A M A S
National Physical Laboratory
TEDDINGTON
Middx TW11 0LW

42

509 N E R C
Institute of Hydrology
Maclean Building
Crowmarsh Gifford
WALLINGFORD
Oxon OX10 8BB

17

510 National Engineering Laboratory
East Kilbride
GLASGOW
G75 0QU
Tel 03552 20222 Tlx 777888

1 2 4 7 26 32

511 National Physical Laboratory
Queens Road
TEDDINGTON
Middx TW11 0LW
Tel 01 977 3222 Tlx 262344

3 7 9 10 12 18 19 20 25 26 39

512 National Radiological Protection Board
Chilton
DIDCOT
Oxon OX11 0RQ

20

513 Nottingham University
Chemistry Dept
University Park
NOTTINGHAM
NG7 2RD

10 13 14 16 19 26

514 Open University
Dept of Electronics
Walton Hall
MILTON KEYNES
Beds MK7 6AA
Tel 0908 74066

4 6 9 12 18 19 22 23 24 31 35 36 39

515 Paisley College of Technology
Scottish School of Non-Destructive Testing
High Street
PAISLEY
Renfrewshire PA1 2BE
Tel 041 887 1241 Tlx 778951

22

516 Reading University
Research Unit for Instrument Physics
5 Earley Gate
Whiteknights
READING
Berks RG6 2AL
Tel 0734 875123

1 2 3 6 11 13 14 15 16 17 21 22 23 27 28 29 31 33 34 35 38 39

517 Salford University
Electrical and Electronic Engineering Dept
SALFORD
M5 4WT
Tel 061 736 5843 Tlx 668680

1 10 33

518 Sheffield University (a)
Electrical & Electronic Engineering Dept
Mappin Street
SHEFFIELD
S1 3JD

3 18

519 Sheffield University (b)
Civil & Structural Engineering Dept
Mappin Street
SHEFFIELD
S1 3JD
Tel 0742 768555 Tlx 547216

22 29 30 31 40

520 Sheffield University (c)
Mechanical Engineering Dept
SHEFFIELD
S10 2TN
Tel 0742 768555

1

521 Shirley Institute
Wilmslow Road
MANCHESTER
M20 8RX
Tel 061 4458141 Tlx 668417

24 31

522 Sira Ltd
South Hill
CHISLEHURST
Kent BR7 5EH
Tel 01 467 2636 Tlx 896649

4 19 22 24 25 31 39

523 South Bank Polytechnic
Physical Sciences & Scientific Computing Dept
Borough Road
LONDON
SE1 0AA
Tel 01 928 8989

20 21 27 32 33

524 Southampton University (a)
Electronics & Computer Science Dept
Highfield
SOUTHAMPTON
SO9 5NH
Tel 0703 559122 Tlx 47661

3 4 6 7 9 12 15 16 23 25 27 31 32 33 34 35 36 40

525 Southampton University (b)
Sound & Vibration Research Dept
SOUTHAMPTON
SO9 5NH
Tel 0703 559122 Tlx 47661

6 23

526 Strathclyde University (a)
Applied Physics Dept
John Anderson Building
107 Rottenrow
GLASGOW
G4 0NG
Tel 041 552 4400

11 14 17 21 22 27 28

527 Strathclyde University (b)
Pure & Applied Chemistry Dept
Thomas Graham Building
295 Cathedral Street
GLASGOW
G1 1XL
Tel 041 552 4400

2 14 15

528 Sunderland Polytechnic
Physics Dept
Langham Tower
Ryhope Road
SUNDERLAND

18

529 Sussex University
School of Engineering and Applied Sciences
Falmer
BRIGHTON
BN1 9QT
Tel 0273 606755 Tlx 878358

27

530 Swansea University College (a)
University of Wales
Electrical & Electronic Engineering Dept
Singleton Park
SWANSEA
SA2 8PP
Tel 0792 295568 Tlx 48358

16 18 24 27 32 33

531 Swansea University College (b)
University of Wales
Chemistry Dept
Singleton Park
SWANSEA
SA2 8PP
Tel 0792 205678

13 14 15 16 33

532 Teesside Polytechnic
Electrical, Instrumentation & Control Engineering Dept
MIDDLESBROUGH
Cleveland TS1 3BA
Tel 0642 218121

1 21 22

533 Wales Polytechnic
Electrical & Electronic Engineering Dept
PONTYPRIDD
Mid Glam CF37 1DL
Tel 0443 405133 Tlx

27

534 Warwick University (a)
Engineering Dept (Computing and Control Engineering)
COVENTRY
CV4 7AL
Tel 0203 523523 Tlx 311904

1 3 19 20 26 29 31 33

535 Warwick University (b)
Dept of Engineering (Mechanical Eng)
COVENTRY
CV4 7AL
Tel 0203 24011

26

10.3 Contacts at all British universities for enquiries about industrial application of their expertise. As well as those of UDIL and BEST, we give below the addresses at the different universities to which enquiries about consultancy etc. should be directed. Those marked with an asterisk particularly use the word *instrument* or *instrumentation* in descriptive literature. However, too much emphasis should not be placed on this because instrumentation of one kind or another features very widely in university activities, and we have not attempted to do any categorization here.

UDIL
University of Manchester Institute of Science and Technology, PO Box 88, Sackville Street, Manchester M60 1QD, Tel. 061 236 3311

BEST
Longman Cartermill Ltd. The Technology Centre, St Andrews, Fife KY16 9EA, Tel. 0334 77660

*Aberdeen University**
Industrial Liaison Office, 48 College Bounds, Aberdeen, AB9 2TT, Tel. 0224 492681, Telex 73458

Aberystwith University College
Industrial Liaison Office, The University College of Wales, Penglais, Aberystwith, Tel. 0970 3111

Aston University
Aston Technical Management and Planning Services Ltd., University of Aston, Aston Triangle, Birmingham B4 7ET, Tel. 021 359 4647, Telex 336997

*Bangor University College**
Technology Enterprise Centre, UCNW Bangor, Gwynedd, LL57 2DG, Tel. 0248 351151

Bath University
South Western Industrial Research Ltd., The University of Bath, Claverton Down, Bath, BA2 7AY, Tel. 0225 63637

Belfast University
QUBIS (Queen's University Business and Industrial Services), The Queen's University of Belfast, Belfast, BT7 1NN, Tel. 0232 661111, Telex 74487

Birmingham University
U and I, Chancellor's Court, University of Birmingham, Birmingham B15 2TT, Tel. 021 472 1301, Telex 338938

Bradford University
Industrial Liaison Office, University of Bradford, Bradford, West Yorkshire, BD7 1DP, Tel. 0274 733466, Telex 51309

Bristol University
Industrial Liaison Office, University of Bristol, Room 1.19, Senate House, Myndall Avenue, Bristol, BS8 1TH, Tel. 0272 303991

Brunel University
Industrial Services Bureau, Brunel University, Uxbridge, Middlesex, UB8 3PH, Tel. 0895 39234, Telex 261173

Cambridge University
Wolfson Cambridge Industrial Unit, 20 Trumpington Street, Cambridge CB2 1QA, Tel. 0223 334755

Cardiff University College
University Industry Centre, 57 Park Place, Cardiff, CF1 1XL, Tel. 9222 874838

*City University**
Bureau for Industrial Enterprise, Northampton Square, London EC1V 0HB, Tel. 01 253 4399, Telex 263896

Cranfield Institute of Technology
IT Institute, Cranfield, Bedford, MK43 0AL, Tel. 0234 750111

Dundee University
Industrial Liaison Office, University of Dundee, Dundee, DD1 4HN, Tel. 0382 23181

Durham University
Industrial Research Laboratories, University of Durham, South Road, Durham, DH1 3LE, Tel. 0385 64971

East Anglia University
Industrial Liaison, The Registry, University of East Anglia, Norwich, NR4 7TJ, Tel. 0603 58738

Edinburgh University
UnivEd Technologies Ltd., 16 Buccleugh Place, Edinburgh, EH8 9LN, Tel. 031 667 1011

Essex University
Industrial Liaison and Research Development, University of Essex, Wivenhoe Park, Colchester, Essex, CO4 3SQ, Tel. 0206 862286

Exeter University
Exeter Enterprises Ltd., Northcote House, The Queen's Drive, University of Exeter, EX4 4QJ, Tel. 0392 214085

Glasgow University
Industrial Liaison Office, University of Glasgow, 2 The Square, Glasgow, G12 8QQ, Tel. 041 330 5199

Heriot-Watt University
UNILINK, Heriot-Watt University, Riccarton, Edinburgh, EH14 4AS, Tel. 031 4495111

Hull University
Industrial and Commercial Development Agency, The University of Hull, Cottingham Road, Hull, HU6 7RX, Tel. 0482 465510

Keele University
Industrial Liaison Unit, Department of Physics, University of Keele, Staffs, ST5 5BG, Tel. 0782 621111

Kent University
KSIP Ltd. (Kent Scientific and Industrial Projects Ltd.), Physics Laboratory, The University, Canterbury, Kent, CT2 7NR, Tel. 0229 66822, Telex 965449

Lancaster University
Commercial and Industrial Development Bureau, University of Lancaster, Bailrigg, Lancaster, LA1 4YX, Tel. 0524 65201

Leeds University
University of Leeds Industrial Services Ltd., 175 Woodhouse Lane, Leeds, LS2 3AR, Tel. 0532 454084

Leicester University
University Administration, University of Leicester, University Road, Leicester, LE1 7RH, Tel. 0533 556662

Liverpool University
Research and Development Advisory Service, Senate House, Abercrombie Square, PO Box 147, Liverpool L69 3BX, Tel. 051 709 6002

London University Birbeck College
Industrial Liaison Office, Malet Street, London, WC1E 7HX, Tel. 01 631 6554, Telex 269400

London University Imperial College of Science and Technology
Industrial Liaison Office, Imperial College of Science and Technology, London, SW7 2AZ, Tel. 01 589 5111, Telex 261503

London University King's College (KQC)
KCL Research Enterprises, King's College London, Campden Hill Road, London, W8 7AH, Tel. 01 937 5411

*London University Queen Mary College**
QMC Industrial Research Ltd., 229 Mile End Road, London E1 4AA, Tel. 01 790 0066

London University Royal Holloway and Bedford New College
Information Office, Royal Holloway and Bedford New College, Egham Hill, Egham, Surrey, TW20 0EX, Tel. 0784 34455

London University University College
UCL Academic Services Unit, University College London, Gower Street, London, WC1E 6BT, Tel. 01 380 7456, Telex 269018

*Loughborough University of Technology**
Loughborough Consultants Ltd., University of Technology, Loughborough, Leics., LE11 3TF, Tel. 0509 222597, Telex 34319

Manchester University
Office of Research Exploitation, The University, Oxford Road, Manchester, M13 9PL, Tel. 061 272 3333

Newcastle upon Tyne University
Research and Industry Liaison, 6 Kensington Terrace, Newcastle upon Tyne, NE1 7RU, Tel. 091 232 8511

Nottingham University
Industrial and Business Liaison Office, University of Nottingham, University Park, Nottingham, NG7 2RD, Tel. 0602 506101

Open University
The Office of Industrial Services, The Open University, Walton Hall, Milton Keynes, MK7 6AA, Tel. 098 653945

Oxford University
Industrial Liaison Office, 56 Banbury Road, Oxford, OX2 6PA, Tel. 0865 59295

*Reading University**
Reading University Industry Link, University of Reading, Whiteknights, PO Box 217, Reading, RG6 2AH, Tel. 0734 875123, Telex 847813

St Andrews University
Industrial Liaison Service, University of St Andrews, College Gate, St Andrews, Scotland, KY16 9AJ, Tel. 0334 76161, Telex 76212

Salford University
CAMPUS, University of Salford, 43 The Crescent, Salford, M5 4WT, Tel. 061 743 1727

Sheffield University
The Commercial and Industrial Development Bureau (CIDB), The University of Sheffield, Western Bank, Sheffield, S10 2TN, Tel. 0742 768 555

Southampton University
Industrial Liaison Unit ISVR, University of Southampton, Highfield, Southampton, SO9 5NH, Tel. 0703 559122

Stirling University
Industrial Projects Service, University of Stirling, Stirling, Scotland, FK9 4LA, Tel. 0786 73171

*Strathclyde University**
Strathclyde Technology Transfer Ltd., 5.16 Livingstone Tower, 26 Richmond Street, Glasgow, G1 1XH, Tel. 041 552 8771

Surrey University
Bureau of Industrial and External Liaison, University of Surrey, Guildford, GU2 5XH, Tel. 0483 509110

*Sussex University**
Industrial Liaison Office, University of Sussex, Sussex House, Falmer, Brighton, East Sussex, BN1 9RH, Tel. 0273 606755

Swansea University College
Industrial Liaison Office, University College of Swansea, Singleton Park, Swansea, SA2 8PP, Tel. 0792 295276

Ulster University
Industrial Unit, University of Ulster, Coleraine, Co. Londonderry, Northern Ireland, BT52 1SA, Tel. 0265 4141

*University of Manchester Institute of Science and Technology**
UMIST Research and Consultancy Services, UMIST, PO Box 88, Manchester, M60 1QD, Tel. 061 236 3311

University of Wales Institute of Science and Technology
Industrial Liaison Office, UWIST, PO Box 25, Cardiff, CF1 3XE, Tel. 0222 42588

Warwick University
Technology Transfer Office, Senate House, University of Warwick, Coventry, CV4 7AL, Tel. 0203 523716

York University
External Relations Office, University of York, Heslington, York, North Yorkshire, YO1 5DD, Tel. 0904 430000

10.4 Learned Societies and Engineering Institutions

British Institute of Non-Destructive Testing
1 Spencer Parade, Northampton, NN1 5AA, Tel. 0604 30124

British Society for Strain Measurement
281 Heaton Road, Newcastle upon Tyne, NE6 5QB, Tel. 0632 655273

Institute of Measurement and Control
87 Gower Street, London, WC1E 6AA, Tel. 01 387 4949

Institute of Physics
47 Belgrave Square, London, SW1X 8QX, Tel. 01 235 6111, Telex 918453

Institution of Electrical Engineers
Savoy Place, London, WC2R 0BL, Tel. 01 240 1871, Telex 261176

Institution of Mechanical Engineers
1 Birdcage Walk, London, SW1H 9JJ, Tel. 01 222 7899, Telex 917944

Royal Society of Chemistry
Burlington House, Piccadilly, London, W1V 0BN, Tel. 01 437 8656, Telex 268001

Index

Note The initial number refers to the relevant part of the book.

Welwyn Strain Measurement
The widest range of quality products and services in the U.K.
MEM FOIL STRAIN GAUGES
HIGH PERFORMANCE ADHESIVES & ACCESSORIES
STRAIN GAUGE INSTRUMENTATION
PHOTOELASTIC POLARISCOPES AND MATERIALS
TENSLAC BRITTLE LACQUERS
EDUCATIONAL AIDS
WELBOND STRAIN GAUGE INSTALLATION SERVICES
TRAINING FACILITIES - Regular 2-day 'hands-on' training workshops for technicians and engineers. One-day or three day seminars.
1. Seminar in progress.
2. Strain gauge installation service
3. Portable strain indicator
4. Typical foil strain gauge
WSM
Welwyn Strain Measurement
A member of the Crystalate Group
ARMSTRONG ROAD BASINGSTOKE RG24 0QA ENGLAND
Telephone: Basingstoke (0256) 462131 Fax: 471441 Telex: 858520